Handbook of
Thin Film
Technology

Other McGraw-Hill Handbooks of Interest

AMERICAN INSTITUTE OF PHYSICS · American Institute of Physics Handbook

BAUMEISTER AND MARKS · Standard Handbook for Mechanical Engineers

BEEMAN · Industrial Power Systems Handbook

BLATZ · Radiation Hygiene Handbook

BRADY · Materials Handbook

BURINGTON AND MAY · Handbook of Probability and Statistics with Tables

COCKRELL · Industrial Electronics Handbook

CONDON AND ODISHAW · Handbook of Physics

COOMBS · Printed Circuits Handbook

CROFT, CARR, AND WATT · American Electricians' Handbook

ETHERINGTON · Nuclear Engineering Handbook

FINK AND CARROLL · Standard Handbook for Electrical Engineers

GRUENBERG · Handbook of Telemetry and Remote Control

HAMSHER · Communication System Engineering Handbook

HARPER · Handbook of Electronic Packaging

HENNEY · Radio Engineering Handbook

HENNEY AND WALSH · Electronic Components Handbook

HUNTER · Handbook of Semiconductor Electronics

HUSKEY AND KORN · Computer Handbook

IRESON · Reliability Handbook

JASIK · Antenna Engineering Handbook

JURAN · Quality Control Handbook

KLERER AND KORN · Digital Computer User's Handbook

KOELLE · Handbook of Astronautical Engineering

KORN AND KORN · Mathematical Handbook for Scientists and Engineers

KURTZ · The Lineman's and Cableman's Handbook

LANDEE, DAVIS, AND ALBRECHT · Electronic Designers' Handbook

MACHOL · System Engineering Handbook

MARKUS · Electronics and Nucleonics Dictionary

MARKUS · Handbook of Electronic Control Circuits

MARKUS AND ZELUFF · Handbook of Industrial and Electronic Circuits

PERRY · Engineering Manual

SHEA · Amplifier Handbook

SKOLNIK · Radar Handbook

SMEATON · Motor Application and Maintenance Handbook

STETKA · NFPA Handbook of the National Electrical Code

TERMAN · Radio Engineers' Handbook

TRUXAL · Control Engineers' Handbook

Handbook of Thin Film Technology

EDITED BY

LEON I. MAISSEL and REINHARD GLANG

International Business Machines Corporation
Components Division, East Fishkill Facility
Hopewell Junction, N.Y.

A McGRAW-HILL
CLASSIC
HANDBOOK
REISSUE

McGRAW-HILL BOOK COMPANY

New York St. Louis San Francisco Auckland
Bogotá Hamburg Johannesburg London Madrid Mexico
Montreal New Delhi Panama Paris São Paulo
Singapore Sydney Tokyo Toronto

Handbook of Thin Film Technology, *1983 Reissue*

Contributors

G. S. ANDERSON
3M Company
St. Paul, Minnesota

I. H. BLECH
Israel Institute of Technology
Haifa, Israel

R. BROWN
Pyrofilm Corporation
Whippany, New Jersey

D. S. CAMPBELL
The Plessey Company Limited
West Lothian, Scotland

M. COHEN
Micro-Bit Corporation
Burlington, Massachusetts

N. FOSTER
Bell Telephone Laboratories
Allentown, Pennsylvania

D. GERSTENBERG
Bell Telephone Laboratories
Allentown, Pennsylvania

R. GLANG
IBM Corporation
E. Fishkill, New York

L. V. GREGOR
IBM Corporation
E. Fishkill, New York

P. J. HARROP
The Plessey Company Limited
West Lothian, Scotland

R. HOLMWOOD
IBM Corporation
E. Fishkill, New York

E. JOYNSON
General Electric Corporation
Schenectady, New York

I. KHAN
NASA Research Center
Cambridge, Massachusetts

J. A. KURTZ
Cogar Corporation
Poughkeepsie, New York

S. MADER
IBM Corporation
Yorktown Heights, New York

L. I. MAISSEL
IBM Corporation
E. Fishkill, New York

v

C. A. NEUGEBAUER
General Electric Corporation
Schenectady, New York

W. A. PLISKIN
IBM Corporation
E. Fishkill, New York

J. RAFFEL
Lincoln Laboratories
Lexington, Massachusetts

H. SELLO
Fairchild Semiconductor Corporation
Palo Alto, California

J. G. SIMMONS
University of Toronto
Toronto, Canada

G. K. WEHNER
University of Minnesota
Minneapolis, Minnesota

P. K. WEIMER
David Sarnoff Research Center
Princeton, New Jersey

S. J. ZANIN
IBM Corporation
E. Fishkill, New York

Preface

This Handbook is intended to serve as a detailed introduction and guide to thin film technology. The subject has grown rapidly in the last decade and plays a key role in many segments of industry today. The book provides a comprehensive source of information which will be useful to scientific workers who are engaged or have a peripheral interest in this field. It could also serve as the basis for a university course at the graduate level, since the fundamentals and principles of vacuum and thin film technology are treated extensively.

The book is divided into four sections. The first two discuss the methods of preparation and the nature of thin films, respectively. Section Three is concerned with film properties and Section Four with practical applications. The latter pertain primarily to the electronics industry where thin films have recently reached the large-scale manufacturing stage. Familiarity with the calculus is assumed and, in certain chapters, some degree of knowledge of solid state-physics is desirable. In no place, however, have advanced concepts been used without suitable introduction.

While every attempt has been made to keep the material presented as current as possible, it should be remembered that thin film and vacuum techniques are still in a stage of rapid development and change. In addition, the wide scope of the subject makes it difficult to provide the amount of detail which some readers might desire. However, those wishing to delve further into particular areas will find appropriate references to help them locate the more specialized literature.

In writing and editing this Handbook, the continuing support and interest of many of our colleagues within as well as outside IBM have been most helpful and stimulating. We would also like to thank the IBM Corporation for providing a working environment which makes an undertaking such as this possible.

L. I. Maissel
R. Glang

Contents

Part One

Preparation of Thin Films

Chapter **1**

Vacuum Evaporation

REINHARD GLANG

IBM Components Division, East Fishkill, New York

LIST OF SYMBOLS

a first coefficient in series expressions for heat capacity
a_B activity of constituent B in an alloy
A constant in molecular-distribution functions
A_w wall area upon which gas is impinging
A_e surface area from which evaporation takes place
A_r surface area receiving deposit
A_m effective cross-sectional area of an evaporation-rate monitor

α	plane angle
α_v	evaporation coefficient
α_c	condensation coefficient
α_T	thermal-energy accommodation coefficient
b	second coefficient in series expressions for heat capacity
β	plane angle
c	(1) third coefficient in series expressions for heat capacity; (2) speed of a gas atom or molecule
$\overline{c^2}$	mean-square speed ⎫
c_m	most probable speed ⎬ in Maxwellian speed distributions
$\bar{c}$	arithmetic-average speed ⎭
c_t	propagation velocity of an elastic wave normal to the major surfaces of a thin quartz wafer
C_p	heat capacity at constant pressure
$\mathbf{C}_p$	molar heat capacity at constant pressure
C	conductance (gas-flow rate divided by pressure difference)
C_{visc}	conductance for viscous flow
C_{mol}	conductance for molecular flow
C_{trans}	conductance for transition flow
C_f	mass-determination sensitivity of resonating quartz-crystal wafer, $\text{Hz g}^{-1}\,\text{cm}^2$
d	film thickness
d_q	thickness of quartz wafer
d'	film-deposition rate, Å s^{-1}
D	diffusion coefficient
D'	ionization-rate monitor gauge constant
E_k	kinetic energy
E_v	evaporation energy
η	viscosity coefficient of gases
f	statistical factor in gas-viscosity equation
f_B	activity coefficient of constituent B in an alloy
f_0	resonance frequency of a quartz-crystal wafer
Δf	change of f_0 due to accumulated mass
F	force
φ	angle of vapor emission
g	gravitational constant
G	weight of deposit, g wt
G	Gibbs' free energy
$\Delta_e\mathbf{G}^\circ$	standard free energy of evaporation
$\Delta_d\mathbf{G}$	molar free energy of decomposition
Γ	mass-evaporation rate, $\text{g cm}^{-2}\,\text{s}^{-1}$
h	height, normal source-to-substrate distance
H	enthalpy or heat content at constant pressure
$\mathbf{H}$	molar enthalpy
$\Delta_e\mathbf{H}$	molar heat of evaporation
$\Delta_e\mathbf{H}^\circ$	standard heat of evaporation
$\Delta_e\mathbf{H}^*$	molar heat of evaporation at 0°K
I	electron current
I_e	electron-emission current
I_i	ion current
k	Boltzmann's constant
K	(1) material specific factor in the evaporation equation of binary alloys; (2) torsional constant for impingement-rate monitors; (3) deposit-distribution factor (≈ 1) in quartz-crystal oscillators
$K_p(T)$	temperature-dependent thermodynamic constant for equilibria involving gaseous constituents
l	lengths, distances
λ	(1) mean free path of gas particles; (2) wavelength of light
m	mass of an atom or molecule

M	molar mass ($\mathbf{N}_A$ molecules)
$\mathfrak{M}_e$	mass of material evaporated from an area A_e
$\mathfrak{M}_r$	mass of deposit received by an area A_r
μ	chemical potential
μ°	standard chemical potential (at 1 atm)
n	number of moles of material
n_f	refractive index of film material
N	(1) number of gas particles; (2) frequency constant of AT cut quartz wafers
$\mathbf{N}_A$	Avogadro's number
N_i	number of gas particles impinging on a surface
N_u	number of gas particles having the velocity component u in the x direction
N_c	number of gas particles having the speed c in any direction
N_e	number of gas particles evaporating from a surface
N_m	number of gas particles arriving at a rate-monitor device
ν	oscillation frequency of surface atoms
ω	(1) solid angle; (2) angular velocity
p	gas pressure
p^*	thermodynamic equilibrium pressure
p_B	partial pressure of constituent B in a gas mixture
p_r	evaporation pressure at the substrate surface (or at distance r from source)
P	probability of evaporation
π	3.14159
Q	macroscopic flow rate of gases, torr l s^{-1}
Q_v	kinetic-energy partition function for a vapor phase
Q_c	kinetic-energy partition function for a condensed phase
Q_R^*	rotational-energy partition function for activated molecules in the liquid state
Q_{R_v}	rotational-energy partition function for molecules in the vapor state
r	direct distance between a small vapor source and a receiving point (substrate)
$\mathbf{R}$	universal gas constant
R	resistance, ohms
R_s	sheet resistance, ohms/sq
ρ	(1) resistivity, ohm-cm; (2) density of film material
ρ_q	density of crystalline quartz
s	(1) radius of a thin-ring source; (2) radius of the receiving cylinder in impingement-rate monitors
S	entropy at constant pressure
$\mathbf{S}$	molar entropy at constant pressure
$\Delta_e \mathbf{S}^\circ$	standard entropy of evaporation
S_0	dynamic sensitivity of ionization-rate monitor gauges
σ	(1) diameter of gas atoms or molecules; (2) mass per unit wall area in impingement-rate monitors
t	time
T	temperature, °K
τ	time constant
θ	(1) angle of vapor or light incidence; (2) surface coverage of metal films with oxygen, whereby $\theta = 1$ corresponds to formation of stoichiometric oxide
u	velocity component of gas particle in the x direction
v	(1) velocity component of gas particle in the y direction; (2) rate, cm s^{-1}, at which an evaporating solid surface recedes from its original position
V	(1) volume; (2) voltage
$\mathbf{V}$	molar volume
w	(1) velocity component of gas particle in the z direction; (2) lateral dimension, width
x	axis in an orthogonal coordinate system
x_B	mole fraction of constituent B in an alloy
y	axis in an orthogonal coordinate system
z	(1) axis in an orthogonal coordinate system; (2) number of nearest surface sites in the adsorption of oxygen on metal films
Z	numerical factor characterizing transition flow of gases

1. INTRODUCTION

The first evaporated thin films were probably the deposits which Faraday[1] obtained in 1857 when he exploded metal wires in an inert atmosphere. Further experimentation in the nineteenth century was stimulated by interest in the optical phenomena associated with thin layers of materials and by investigations of the kinetics and diffusion of gases. The possibility of depositing thin metal films in a vacuum by Joule heating of platinum wires was discovered in 1887 by Nahrwold[2] and a year later adapted by Kundt[3] for the purpose of measuring refractive indices of metal films. In the following decades, evaporated thin films remained in the domain of academic interest until the development of vacuum equipment had progressed far enough to permit large-scale applications and control of film properties. During the last 25 years, evaporated films have found industrial usage for an increasing number of purposes. Examples are antireflection coatings, front-surface mirrors, interference filters, sunglasses, decorative coatings on plastics and textiles, in the manufacture of cathode-ray tubes, and most recently in electronic circuits, the topic of prime interest in this book.

Although commonly referred to as a single process, the deposition of thin films by vacuum evaporation consists of several distinguishable steps:

1. Transition of a condensed phase, which may be solid or liquid, into the gaseous state

2. Vapor traversing the space between the evaporation source and the substrate at reduced gas pressure

3. Condensation of the vapor upon arrival on the substrates

Accordingly, the theory of vacuum evaporation includes the thermodynamics of phase transitions from which the equilibrium vapor pressure of materials can be derived, as well as the kinetic theory of gases which provides models of the atomistic processes. Further investigations of the sometimes complex events occurring in the exchange of single molecules between a condensed phase and its vapor led to the theory of evaporation, a specialized extension of the kinetic theory. From this basis, the distribution of deposits on surfaces surrounding a vapor source can be derived. The kinetic aspects of condensation processes are a topic in their own right and are therefore treated separately in Chap. 8, which is devoted to nucleation and growth phenomena.

Experimentally, vacuum evaporation and its applications have benefited from various disciplines which have contributed toward solutions of practical problems. These pertain to the construction of suitable vapor sources, the development of special techniques for the evaporation of alloys, compounds, and mixtures, and to questions of process control and automation. Finally, thin film deposition requires a properly evacuated chamber; however, vacuum technology is such a complex topic that it is treated separately in Chap. 2.

2. THERMODYNAMIC AND KINETIC FOUNDATIONS

The transition of solids or liquids into the gaseous state may be treated as a macroscopic or as an atomistic phenomenon. The former approach is based on thermodynamics and yields a quantitative understanding of evaporation rates, interactions between evaporants and their containers, stability of compounds, and compositional changes encountered during the evaporation of alloys. The atomistic approach is derived from the kinetic theory of gases and provides models which describe evaporation processes in terms of properties of individual particles. The latter theory also applies to the evacuation of vessels and thus relates to Chap. 2. Although thermodynamic and kinetic theories are treated in various textbooks, aspects pertinent to vacuum evaporation will be reviewed, and some of the widely used equations will be introduced.

a. The Equilibrium Vapor Pressure of Materials

In thermodynamics, the condensed and the gaseous states of materials are characterized by functions which depend on the macroscopic variables of pressure, temperature, volume, and mass. Of particular significance is the thermodynamic equilibrium, a situation where two states, for instance, a condensed phase and its vapor, exist at the same temperature and in contact with each other without undergoing net changes. This means that the amounts of evaporating and condensing material are equal at all times as long as the equilibrium is maintained. Under these conditions, solids and liquids have characteristic vapor pressures which are unique functions of temperature.

At first sight the equilibrium vapor pressure may appear unrelated to vacuum evaporation since the latter is not an equilibrium process but involves transfer of material from one state to the other. However, atomic theory and extensive experiments have shown that evaporation rates cannot exceed an upper limit which is proportional to the equilibrium vapor pressure. Therefore, the saturation pressure of a vapor over its condensed phase is an important quantity to assess the amenability of a substance to evaporation and the temperatures required to achieve practical transfer rates.

The transformation of a condensed phase into vapor involves the conversion of thermal energy supplied to the evaporant into mechanical energy as represented by the expansion into vapor. According to the second law of thermodynamics, the conversion efficiency of thermal energy is limited because a fraction of it must serve to increase the entropy of the system and is not available for the production of mechanical energy. The conversion takes place with the greatest efficiency if the system changes reversibly.[4] Depending on the macroscopic variables under consideration, there are several ways to express the energy balance of such reversible phase transitions. The most commonly used form is

$$\Delta G = \Delta H - T \Delta S \tag{1}$$

where ΔG, ΔH, and ΔS are the changes in Gibbs' free energy G, in enthalpy H, and entropy S associated with the process. Equation (1) leads to simple thermodynamic relationships if the pressure p and the temperature T are treated as independent macroscopic variables; the volume V is thereby defined as a dependent quantity. The characteristic functions occurring in Eq. (1) are extensive properties, and hence the amount of substance involved is an additional variable.

For practical applications, it is necessary to know the dependence of the thermodynamic functions on the macroscopic variables which can be measured. These relationships are obtained by considering a closed system undergoing an infinitesimal change at equilibrium which leads to the following fundamental and often used partial derivatives:[4]

$$\frac{\partial G}{\partial T} = -S \qquad (p = \text{const}) \tag{2}$$

$$\frac{\partial G}{\partial p} = V \qquad (T = \text{const}) \tag{3}$$

V is the volume of the substance in the closed system. The total differential of the free energy is accordingly

$$dG = -S \, dT + V \, dp \tag{4}$$

The temperature dependences of the enthalpy and the entropy at constant pressure are defined by the partial differential equations

$$\frac{\partial H}{\partial T} = C_p \tag{5a}$$

and

$$\frac{\partial S}{\partial T} = \frac{C_p}{T} \tag{5b}$$

where C_p is the heat capacity of the system at constant pressure.

Per definition, the free-energy change ΔG is obtained only with reversible processes occurring at the thermodynamic equilibrium between the two states. An actual process conducted under such conditions would be infinitely slow and is therefore hypothetical. The importance of ΔG lies in the fact that it constitutes a quantitative measure for the driving force associated with a possible system change from one state to another. Knowledge of ΔG thus makes it possible to determine the stability of one state relative to another. If there is no driving force between two states, they can coexist without change and therefore are at equilibrium. Consequently, the particular values which the macroscopic variables assume under equilibrium conditions can be derived by stipulating $\Delta G = 0$. In the case of a pure condensed phase in contact with its own vapor, the resulting relationship defines the equilibrium pressure of a material as a function of temperature.

For systems containing more than one phase or several substances, it is practical to consider the free energy of each component separately. This is done by introducing the chemical potential $\mathbf{\mu}$, which has the same significance and variables as G but refers to 1 g mol of material in a particular state. In regard to the vapor-solid or vapor-liquid equilibrium containing only a single substance, the relations established for G are also valid for $\mathbf{\mu}$ without any modification other than designating the extensive properties as molar quantities. This will be done by using the notations $\mathbf{H}$, $\mathbf{S}$, $\mathbf{V}$, and $\mathbf{C}_p$.

(1) Clausius-Clapeyron's Equation The vapor-solid or vapor-liquid equilibrium can be defined as an equality of the chemical potentials

$$\mathbf{\mu}_c = \mathbf{\mu}_g \tag{6}$$

whereby the subscripts c and g identify quantities associated with the condensed and gaseous states, respectively. Equation (6) also holds in differential form,

$$d\mathbf{\mu}_c = d\mathbf{\mu}_g \tag{7}$$

because these increments refer to changes from one equilibrium point to another. Substitution of the total differentials according to Eq. (4) into Eq. (7) yields

$$\frac{dp^*}{dT} = \frac{\mathbf{S}_g - \mathbf{S}_c}{\mathbf{V}_g - \mathbf{V}_c} \tag{8}$$

where dp^* is the change in equilibrium pressure resulting from the small temperature change dT. The symbol p^* has been introduced to denote that the pressure defined by Eq. (8) is no longer a variable to be chosen at liberty.

Equation (6) can also be expressed in terms of the enthalpy and entropy functions,

$$\mathbf{H}_c - T\mathbf{S}_c = \mathbf{H}_g - T\mathbf{S}_g$$

which may be substituted into Eq. (8) to yield Clausius-Clapeyron's equation:

$$\frac{dp^*}{dT} = \frac{\mathbf{H}_g - \mathbf{H}_c}{T(\mathbf{V}_g - \mathbf{V}_c)} \tag{9}$$

To solve Eq. (9), two approximations are made. The molar volume of the condensed phase is neglected since it is very small compared with the molar volume of the vapor. Furthermore, the vapor is assumed to obey the ideal-gas law, thereby neglecting deviations due to van der Waals forces between gaseous particles:

$$\mathbf{V}_g - \mathbf{V}_c \simeq \mathbf{V}_g = \frac{\mathbf{R}T}{p}$$

The difference of the enthalpies in Eq. (9) is the molar heat of evaporation,

$$\mathbf{H}_g - \mathbf{H}_c = \Delta_e\mathbf{H}$$

With these modifications, Eq. (9) transforms into

$$\frac{dp^*}{p^*} = \frac{\Delta_e\mathbf{H}\,dT}{\mathbf{R}T^2} \tag{10}$$

As a first approximation, the heat of evaporation may be assumed to be constant so that Eq. (10) can be integrated:

$$\ln p^* \simeq - \frac{\Delta_e \mathbf{H}}{\mathbf{R} T} + \text{const} \tag{11}$$

Equation (11) holds only for small temperature intervals. A more accurate vapor-pressure function is obtained by considering the temperature dependence of the molar heat of evaporation. The latter is determined by changes which the molar specific heats in both states undergo with temperature:

$$\frac{d(\Delta_e \mathbf{H})}{dT} = \mathbf{C}_{p,g} - \mathbf{C}_{p,c} = \Delta \mathbf{C}_p \tag{12}$$

Near and above room temperature, the specific heats may be expressed in the general form $\mathbf{C}_p = a + bT + cT^{-2}$, where the coefficients a, b, and c are numerical factors specific for a particular substance. Insertion of the $\mathbf{C}_p$ functions into Eq. (12) and subsequent integration give the heat of vaporization as a function of temperature:

$$\Delta_e \mathbf{H} = \Delta_e \mathbf{H}^* + \Delta a \, T + \frac{\Delta b}{2} \, T^2 - \Delta c \, T^{-1} \tag{13}$$

where $\Delta_e \mathbf{H}^*$ is an integration constant which must be determined experimentally. Further substitution of Eq. (13) into Eq. (10) and integration of the latter yield

$$\ln p^* = - \frac{\Delta_e \mathbf{H}^*}{\mathbf{R} T} + \frac{\Delta a}{\mathbf{R}} \ln T + \frac{\Delta b}{2\mathbf{R}} \, T + \frac{\Delta c}{2\mathbf{R}} \, T^{-2} + I \tag{14}$$

I is another constant of integration which is specific for a given material.

Theoretically, Eq. (14) permits the calculation of vapor pressures from fundamental material constants. However, the constants $\Delta_e \mathbf{H}^*$ and I cannot be derived from first principles. Instead they are commonly calculated from vapor pressures which have been either measured by one of several experimental methods[5,6] or computed from spectroscopically determined energy levels.[7] The compilations of thermochemical data rarely list these constants. Therefore, an expression equivalent to Eq. (14) but containing only standard thermochemical data is desirable to calculate vapor pressures for different temperatures.

(2) Calculation of Vapor Pressures from Standard Thermochemical Data Numerical values of thermodynamic functions are commonly listed for the standard state, and the symbols referring to it are identified by adding the superscript $^\circ$. The standard state is represented by 1 g mol of material at 1 atm of pressure and 298°K.[8] Thus, $\mathbf{\mu}_c^\circ(T)$ and $\mathbf{\mu}_g^\circ(T)$ are the chemical potentials of a pure substance at the temperature T and 1 atm in the condensed and in the gaseous states. Equilibrium between the two states can be established hypothetically in van't Hoff's equilibrium box by expanding (or compressing) both phases isothermally and reversibly. In this operation, the two chemical potentials change until they become equal at the equilibrium pressure p^*:

$$\mathbf{\mu}_g^\circ(T) + \int_1^{p^*} \left(\frac{\partial \mathbf{\mu}_g}{\partial p} \right)_T dp = \mathbf{\mu}_c^\circ(T) + \int_1^{p^*} \left(\frac{\partial \mathbf{\mu}_c}{\partial p} \right)_T dp \tag{15}$$

The effect of pressure on the chemical potential of the condensed phase is again neglected, and partial differentiation of $\mathbf{\mu}_g^\circ$ yields the molar volume according to Eq. (3)

$$\left(\frac{\partial \mathbf{\mu}_g}{\partial p} \right)_T = \mathbf{V} = \frac{\mathbf{R} T}{p}$$

Integration of Eq. (15) leads to the equilibrium condition

$$\mathbf{\mu}_g^\circ(T) + \mathbf{R} T \ln p^* = \mathbf{\mu}_c^\circ(T)$$

Introducing $\mathbf{\mu}_g°(T) - \mathbf{\mu}_c°(T) = \Delta_e\mathbf{G}°(T)$, the standard free energy of evaporation, the vapor pressure becomes

$$\log p^* = -\frac{\Delta_e\mathbf{G}°(T)}{RT}\log e \tag{16}$$

Thus, the equilibrium pressure (in atmospheres) can be calculated for any temperature at which the free energy of evaporation is known, if the latter is divided by $4.575T$ cal mol^{-1}. Conversion into torr is easily possible according to

$$\log p^* \text{ (torr)} = \log p^* \text{ (atm)} + \log 760$$

Since $\Delta_e\mathbf{G}°$ is often listed only for the reference state at 298°K, it is useful to know the temperature dependence of Eq. (16). Starting with

$$\Delta_e\mathbf{G}°(T) = \Delta_e\mathbf{G}°(298) + \int_{298}^{T}\left(\frac{\partial\Delta_e\mathbf{G}°}{\partial T}\right)_p dT$$

and substituting the standard entropy of evaporation according to Eq. (2)

$$\left(\frac{\partial\Delta_e\mathbf{G}°}{\partial T}\right)_p = -\Delta_e\mathbf{S}°(T)$$

whereby the temperature dependence of the entropy follows from Eq. (5b)

$$\Delta_e\mathbf{S}°(T) = \Delta_e\mathbf{S}°(298) + \int_{298}^{T}\frac{\Delta C_p}{T}dT$$

one obtains

$$\Delta_e\mathbf{G}°(T) = \Delta_e\mathbf{G}°(298) - \int_{298}^{T}\left[\Delta_e\mathbf{S}°(298) + \int_{298}^{T}\frac{\Delta C_p}{T}dT\right]dT$$

Further substitution of the standard enthalpy and entropy for $\Delta_e\mathbf{G}°(298)$ according to Eq. (1) and solving the first integral term yield

$$\Delta_e\mathbf{G}°(T) = \Delta_e\mathbf{H}°(298) - T\,\Delta_e\mathbf{S}°(298) - \iint_{298}^{T}\frac{\Delta C_p}{T}dT^2$$

Insertion of this expression for the temperature dependence of the standard free energy into Eq. (16) leads to

$$\log p^* = -\frac{\Delta_e\mathbf{H}°(298)}{4.575T} + \frac{\Delta_e\mathbf{S}(298)}{4.575} + \frac{1}{4.575T}\iint_{298}^{T}\frac{\Delta C_p}{T}dT^2 \tag{17}$$

In contrast to the former Eq. (14), this expression relates the vapor pressure to the standard enthalpy and entropy, the quantities which are most commonly given in thermodynamic tables.[9-15] If these data are not available either, it is often possible to estimate their approximate values by methods as discussed, for instance, by Kubaschewski.[6]

The complete solution of Eq. (17) depends on the form of the specific-heat term. A first-order approximation is obtained by neglecting specific-heat changes and assuming $\Delta C_p = 0$. More accurate values result if a constant average of $\overline{\Delta C_p} = \Delta a$ is applied over the entire interval from 298 to T°K. The integral term is then given by

$$\Delta a \iint_{298}^{T}\frac{dT^2}{T} = \Delta a\,T\left(\ln\frac{T}{298} + \frac{298}{T} - 1\right) = \Delta a\,Tf\left(\frac{T}{298}\right)$$

and can be evaluated by means of Table 1.

TABLE 1   Numerical Values of the Function $Tf\left(\dfrac{T}{298}\right)$ [16]

T, °K	$Tf\left(\dfrac{T}{298}\right)$	T, °K	$Tf\left(\dfrac{T}{298}\right)$
400	15.7	1400	1063
500	56.6	1500	1222
600	118	1600	1386
700	195	1700	1558
800	288	1800	1734
900	392	1900	1917
1000	508	2000	2105
1100	634	2500	3115
1200	769	3000	4224
1300	912	3500	5418

If the specific-heat functions are given as series for both the gaseous and the condensed state, the temperature dependence of the heat-capacity change may be considered,

$$\Delta C_p = \Delta a + \Delta b\, T + \Delta c\, T^{-2}$$

and consequently the following terms occur:

$$\frac{1}{T}\int\!\!\int_{298}^{T} \frac{\Delta C_p}{T}\, dT^2 = \Delta a\, f\left(\frac{T}{298}\right) + \frac{\Delta b}{2}\, T\left(1 - \frac{298}{T}\right)^2 + \frac{\Delta c}{2}\left(\frac{1}{T} - \frac{1}{298}\right)^2 \quad (18)$$

Equation (18) can be solved and terms with equal powers of T combined with the standard enthalpy and entropy terms in Eq. (17). This operation leads to vapor-pressure expressions of the form

$$\log p^* \text{ (torr)} = AT^{-1} + B + C \log T + DT + ET^{-2} \quad (19)$$

The contributions of the first-, second-, and third-order specific-heat terms to the coefficients in Eq. (19) are as follows:

$$A = -\frac{\Delta_e H°(298) - 298(\Delta a + 149\,\Delta b) + 3.35 \times 10^{-3}\,\Delta c}{4.575}$$

$$B = \frac{\Delta_e S°(298) - 6.70\,\Delta a - 298\,\Delta b + 5.6 \times 10^{-6}\,\Delta c}{4.575} + 2.8808 \quad (20)$$

$$C = \frac{\Delta a}{1.987} \qquad D = \frac{0.5\,\Delta b}{4.575} \qquad E = \frac{0.5\,\Delta c}{4.575}$$

Vapor-pressure data are often given in the form of Eq. (19). The coefficients C and D are considered only if the specific heats are known with sufficient accuracy and confidence in the standard enthalpy and entropy values merit refined calculations. There are few cases where the latter conditions warrant the use of the coefficient E.

(3) Vapor-pressure Data   The vapor pressures of all common elements have been determined, usually by several authors independently. Temperatures producing vapor pressures sufficiently high for practical evaporation rates range from about 600°C on up. With increasing temperatures, vapor-pressure measurements are more and more subject to experimental problems such as errors in the evaporation temperature, excessive residual gas pressure, and reaction of the condensed evaporant phase with the container. Consequently, the accuracy of vapor-pressure data is limited, and numerical values vary from author to author. A judicious choice is therefore necessary, particularly if the data are relatively old. To illustrate the situation, the vapor pressure of liquid aluminum will be examined. Figure 1 shows experimental points which Nesmeyanov[5] selected from several investigations as being the most reliable. Based on these data, he derived the equation

$$\log p^* \text{ (torr)} = -15{,}993T^{-1} + 12.409 - 0.999 \log T - 3.52 \times 10^{-6}T$$

which is also shown in Fig. 1. Kelley[17] gives the free energy of evaporation for liquid aluminum as

$$\Delta_e G^\circ(T) = 65{,}680 - 43.72T + 4.61 \log T$$

from which the vapor pressure can be derived according to Eq. (16). Kubaschewski's[6] vapor-pressure equation

$$\log p^* \text{ (torr)} = -16{,}450T^{-1} + 12.36 - 1.023 \log T$$

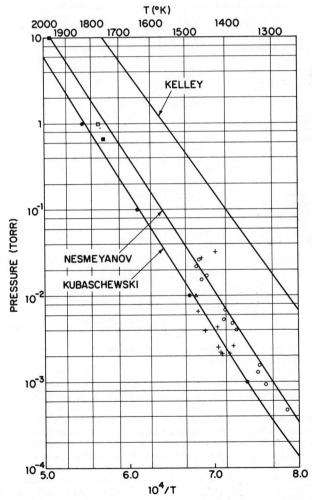

Fig. 1 Vapor pressure of liquid aluminum as measured by various authors (symbols), according to different vapor-pressure equations (lines), and tabulated by Honig (full circles).

is based on data[18] not included in Nesmeyanov's[5] review. As Fig. 1 shows, the discrepancies are substantial even if the older values of Kelley are disregarded. The values used by Honig[19] (full circles in Fig. 1) agree closely with Kubaschewski's curve.

The three vapor-pressure equations reflect different standard enthalpies and entropies, which have been evaluated according to Eqs. (20) and are listed below.

	$\Delta_e H°$ (298), kcal mol^{-1}	$\Delta_e S°$ (298), Cl mol^{-1}	ΔC_p, cal mol^{-1} deg^{-1}
Nesmeyanov[5]...........	72.58	30.28	$-1.99 - 3.2 \times 10^{-5}\ T$
Kelley[17]................	65.08	30.32	-2.0
Kubaschewski[6].........	74.65	29.77	-2.03

The specific heat for liquid aluminum is 7.00 cal mol^{-1} deg^{-1} while that of aluminum vapor is known more accurately: $C_p = 4.97 + 1.2 \times 10^4 T^{-2}$ cal mol^{-1} deg^{-1}.[11] However, it would be pointless to consider the third-order C_p term in view of the uncertainty in the enthalpy value. The latter problem is illustrated in Fig. 2, which shows

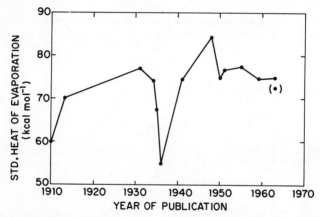

Fig. 2 Standard heat of evaporation of liquid aluminum as calculated by Nesmeyanov[5] from reported vapor-pressure data.

standard enthalpies calculated by Nesmeyanov[5] from vapor-pressure measurements of different authors in historical sequence. The more recent values converge at about 75 kcal mol^{-1}, and therefore, the vapor pressures given by Kubaschewski[6] and Honig[19] seem to be the most reliable ones. The enthalpy obtained from Nesmeyanov's vapor-pressure equation (parenthetical dot in Fig. 2) is at variance with this conclusion, although he, too, lists 75 kcal mol^{-1} as the most probable value (see Ref. 5, p. 433).

The vapor pressures of other metals are often affiliated with uncertainties similar to those demonstrated for aluminum. Data compilations in the form of tables or coefficients for Eq. (19) have been published by various authors.[5,6,17,20,21] The widely used vapor-pressure charts by Honig[19] allow a quick assessment of pressures to be expected at a given temperature; his numerical values are reproduced in Table 2 and generally preferred in this chapter.

b. The Kinetic Theory of Gases

The atomistic model of the evaporation process is based on the kinetic theory, which treats gases as an assembly of a large number of atoms or molecules of identical mass m

and size. For many applications, the shape and the structure of these particles may be ignored and the molecules can be considered as elastic spheres with diameters very small compared with their average distance of separation. Furthermore, the gas molecules are assumed to be in a constant state of random motion, colliding with one another and with the walls of the vessel containing them. In analogy to the ideal-gas equation describing the macroscopic behavior of gases, no allowance is made for forces between molecules except at the moment of collision. According to this model, the pressure exerted by a gas on its confining vessel results from the momentums which individual molecules impart when they strike the walls.

(1) Atomistic Concept of Gas Pressure and Temperature The relationships between the macroscopic gas pressure and the translational energy of individual molecules can be derived in several more or less rigorous ways.[22,23] The particle speed c is thereby thought to have three velocity components u, v, w perpendicular to each other. As the speeds vary among the different particles, mean-square velocities

$$\overline{u^2} = \frac{\Sigma u^2}{N} \tag{21}$$

and mean-square speeds

$$\overline{c^2} = \frac{\Sigma c^2}{N} = \overline{u^2} + \overline{v^2} + \overline{w^2} \tag{22}$$

are defined to represent the state of all molecules N in the enclosure. It can be shown by theoretical arguments[22,23] that the molecules having velocities u exert the pressure

$$p = \frac{N}{V} m\overline{u^2} \tag{23}$$

on a wall perpendicular to the direction of u.

The velocities in different directions are assumed to be uniformly distributed, $\overline{u^2} = \overline{v^2} = \overline{w^2}$, and therefore

$$\overline{u^2} = \tfrac{1}{3}\overline{c^2} \tag{24}$$

Substitution of Eq. (24) into Eq. (23) yields the gas pressure as the force which the impinging molecules exert upon the walls:

$$p = \frac{1}{3}\frac{N}{V} m\overline{c^2} \tag{25}$$

The temperature of the gas is introduced by comparing Eq. (25) with the universal-gas equation $p = nRT/V$, where n is the number of gram moles of gas. According to Avogadro's law, the number of molecules per gram mole N_A is a constant for all substances. Consequently, the total number of molecules N in a volume containing n moles of gas is nN_A, and therefore

$$p = \frac{N}{V}\frac{R}{N_A} T = \frac{N}{V} kT \tag{26}$$

where k is the Boltzmann constant.[4]

From Eq. (26), the number of molecules per cubic centimeter at various pressures and temperatures can be calculated. It is often referred to as Loschmidt's number, and for pressures given in torr the relationship is

$$\frac{N}{V} = 9.656 \times 10^{18} \frac{p}{T} \quad \text{cm}^{-3}$$

The two expressions for the pressure [Eqs. (25) and (26)] establish the identity

$$\tfrac{1}{3}m\overline{c^2} = kT$$

or, after multiplication with $\tfrac{3}{2}$,

$$\tfrac{1}{2}mc^2 = \tfrac{3}{2}kT \tag{27}$$

TABLE 2 Vapor-pressure Data for the Solid and Liquid Elements*

Symbol	Element	Data temp range, °K	Temperatures (°K) for vapor pressures, torr														
			10^3	10^2	10^1	1	10^{-1}	10^{-2}	10^{-3}	10^{-4}	10^{-5}	10^{-6}	10^{-7}	10^{-8}	10^{-9}	10^{-10}	10^{-11}
Ac	Actinium	1873, est.	3510	3030	2660	2350	2100	1905	1740	1605	1490	1390	1305	1230	1160	1100	1045
Ag	Silver	958–2200	2490	2100	1815	1605	1435	1300	1195	1105	1025	958	899	847	800	759	721
Al	Aluminium	1220–1468	2800	2370	2050	1830	1640	1490	1355	1245	1160	1085	1015	958	906	860	815
Am	Americium	1103–1453	2970	2400	2020	1745	1540	1375	1245	1140	1050	971	905	848	797	752	712
As_4	Arsenic(s)	Est.	900	795	712	645	590	550	510	477	447	423	400	377	358	340	323
At_2	Astatine	Est.	620	540	480	434	398	364	338	316	296	280	265	252	241	231	221
Au	Gold	1073–1847	3130	2680	2320	2040	1840	1670	1525	1405	1305	1220	1150	1080	1020	964	915
B	Boron	1781–2413	4000	3500	3100	2780	2520	2300	2140	1980	1855	1740	1640	1555	1480	1405	1335
Ba	Barium	1333–1419	1930	1570	1310	1125	984	883	800	735	675	627	583	545	510	480	450
Be	Beryllium	1103–1552	2810	2390	2080	1830	1650	1500	1370	1270	1180	1105	1035	980	925	878	832
ΣBi	Bismuth		1900	1570	1350	1170	1050	945	860	790	732	682	640	602	568	540	510
ΣC	Carbon(s)	1820–2700	4190	3780	3450	3170	2930	2730	2560	2410	2260	2140	2030	1930	1845	1765	1695
Ca	Calcium	730–1546	1800	1475	1250	1075	962	870	795	732	678	630	590	555	524	495	470
Cd	Cadmium	411–1040	1060	885	762	665	593	538	490	450	419	392	368	347	328	310	293
Ce	Cerium	1611–2038	3830	3220	2780	2440	2180	1970	1795	1650	1525	1420	1325	1245	1175	1110	1050
Co	Cobalt	1363–1522	3220	2790	2440	2180	1960	1790	1655	1530	1430	1340	1265	1195	1130	1070	1020
Cr	Chromium	1273–1557	3000	2550	2240	2010	1825	1670	1540	1430	1335	1250	1175	1110	1055	1010	960
ΣCs	Cesium	300–955	980	775	643	553	482	428	387	351	322	297	274	257	241	226	213
Cu	Copper	1143–1897	2920	2460	2140	1890	1690	1530	1405	1300	1210	1125	1060	995	945	895	855
Dy	Dysprosium	1258–1773	2780	2300	1965	1710	1535	1390	1270	1170	1090	1020	955	898	847	801	760
Er	Erbium	1773, est.	2920	2420	2060	1800	1605	1450	1325	1220	1125	1050	981	922	869	822	779
Eu	Europium	696–900	1800	1500	1260	1100	981	884	805	739	682	634	592	556	523	495	469
Fr	Francium	Est.	980	760	620	528	462	410	368	334	306	280	260	242	225	210	198
Fe	Iron	1356–1889	3200	2740	2390	2130	1920	1750	1615	1500	1400	1305	1230	1165	1105	1050	1000
Ga	Gallium(l)	1179–1383	2730	2300	1980	1745	1555	1405	1280	1180	1090	1015	950	892	841	796	755
Gd	Gadolinium	Est.	3100	2580	2220	1955	1760	1600	1465	1350	1250	1170	1100	1035	980	930	880
ΣGe	Germanium	1510–1885	3180	2680	2320	2050	1830	1670	1530	1410	1310	1220	1150	1085	1030	980	940
Hf	Hafnium	2035–2277	4780	4130	3630	3240	2930	2670	2450	2270	2120	1980	1865	1760	1665	1580	1505
Hg	Mercury	193–575	642	535	458	398	353	319	289	266	246	229	214	201	190	180	170
Ho	Holmium	923–2023	2910	2410	2060	1800	1605	1450	1325	1220	1125	1050	981	922	869	822	779
In	Indium(l)	646–1348	2430	2030	1740	1520	1355	1220	1110	1015	937	870	812	761	716	677	641
Ir	Iridium	1986–2600	4900	4250	3750	3360	3040	2770	2560	2380	2220	2080	1960	1850	1755	1665	1585
K	Potassium	373–1031	1070	858	720	618	540	481	434	396	364	338	315	294	276	260	247
La	Lanthanum	1655–2167	3680	3150	2760	2450	2200	2000	1835	1695	1570	1465	1375	1295	1220	1155	1100
Li	Lithium	735–1353	1620	1370	1170	1020	900	810	740	677	623	579	541	508	480	452	430
Lu	Lutetium	Est.	3370	2910	2550	2270	2030	1845	1685	1550	1440	1345	1260	1185	1120	1060	1000
Mg	Magnesium	626–1376	1400	1170	1000	878	782	712	650	600	555	519	487	458	432	410	388

Table — columns unlabeled in source; each element row gives its range followed by 15 tabulated values.

Symbol	Element	Range															
Mn	Manganese	1523–1823	660	695	734	778	827	884	948	1020	1110	1210	1335	1490	1695	1970	2370
Mo	Molybdenum	2070–2504	1610	1690	1770	1865	1975	2095	2230	2390	2580	2800	3060	3390	3790	4300	5020
Na	Sodium	496–1156	294	310	328	347	370	396	428	466	508	562	630	714	825	978	1175
Nb	Niobium	2304–2596	1765	1845	1935	2035	2140	2260	2400	2550	2720	2930	3170	3450	3790	4200	4710
Nd	Neodymium	1240–1600	846	895	945	1000	1070	1135	1220	1320	1440	1575	1770	2000	2300	2740	3430
Ni	Nickel	1307–1895	1040	1090	1145	1200	1270	1345	1430	1535	1655	1800	1970	2180	2430	2770	3230
Os	Osmium	2300–2300	1875	1965	2060	2170	2290	2430	2580	2760	2960	3190	3460	3800	4200	4710	5340
P_4	Phosphorus(s)		283	297	312	327	342	361	381	402	430	458	493	534	582	642	715
Pb	Lead	1200–2028	516	546	580	615	656	702	758	820	898	988	1105	1250	1435	1700	2070
Pd	Palladium	1294–1640	945	995	1050	1115	1185	1265	1355	1465	1590	1735	1920	2150	2450	2840	3380
∑Po	Polonium	711–1286	332	348	365	384	408	432	460	494	537	588	655	743	862	1040	1250
Pr	Praseodymium	1423–1693	900	950	1005	1070	1140	1220	1315	1420	1550	1700	1890	2120	2420	2820	3370
Pt	Platinum	1697–2042	1335	1405	1480	1565	1655	1765	1885	2020	2180	2370	2590	2860	3190	3610	4170
Pu	Plutonium(l)	1392–1793	931	983	1040	1105	1180	1265	1365	1480	1615	1780	1975	2230	2550	2980	3590
Ra	Radium	Est.	436	460	488	520	552	590	638	690	755	830	920	1060	1225	1490	1840
Rb	Rubidium		227	240	254	271	289	312	336	367	402	446	500	568	665	802	1000
Re	Rhenium	2494–2999	1900	1995	2100	2220	2350	2490	2660	2860	3080	3340	3680	4080	4600	5220	6050
Rh	Rhodium	1709–2205	1330	1395	1470	1550	1640	1745	1855	1980	2130	2310	2520	2780	3110	3520	4070
Ru	Ruthenium	2000–2500	1540	1610	1695	1780	1880	1990	2120	2260	2420	2620	2860	3130	3480	3900	4450
∑S	Sulfur		230	240	252	263	276	290	310	328	353	382	420	462	519	606	739
∑Sb	Antimony	693–1110	477	498	526	552	582	618	656	698	748	806	885	1030	1250	1560	1960
Sc	Scandium	1301–1780	881	929	983	1045	1110	1190	1280	1380	1505	1650	1835	2070	2370	2780	3360
∑Se	Selenium	550–950	286	301	317	336	356	380	406	437	472	516	570	636	719	826	972
∑Si	Silicon	1640–2054	1090	1145	1200	1265	1340	1420	1510	1610	1745	1905	2090	2330	2620	2990	3490
Sm	Samarium	789–833	542	573	608	644	688	738	790	853	926	1015	1120	1260	1450	1715	2120
Sn	Tin(l)	1424–1753	805	852	900	955	1020	1080	1170	1270	1380	1520	1685	1885	2140	2500	2960
Sr	Strontium		433	458	483	514	546	582	626	677	738	810	900	1005	1160	1370	1680
Ta	Tantalum	2624–2948	1930	2020	2120	2230	2370	2510	2680	2860	3080	3330	3630	3980	4400	4930	5580
Tb	Terbium	Est.	900	950	1005	1070	1140	1220	1315	1420	1550	1700	1890	2120	2420	2820	3370
Tc	Technetium		1580	1665	1750	1840	1950	2060	2200	2350	2530	2760	3030	3370	3790	4300	5000
Te	Tellurium	481–1128	366	385	405	428	454	482	515	553	596	647	706	791	905	1065	1300
Th	Thorium	1757–1956	1450	1525	1610	1705	1815	1935	2080	2250	2440	2680	2960	3310	3750	4340	5130
Ti	Titanium	1510–1822	1140	1200	1265	1335	1410	1500	1600	1715	1850	2010	2210	2450	2760	3130	3640
Tl	Thallium	519–924	473	499	527	556	592	632	680	736	803	882	979	1100	1255	1460	1750
Tm	Thulium	809–1219	624	655	691	731	776	825	882	953	1030	1120	1235	1370	1540	1760	2060
U	Uranium	1630–2071	1190	1255	1325	1405	1495	1600	1720	1855	2010	2200	2430	2720	3080	3540	4180
V	Vanadium	1666–1882	1235	1295	1365	1435	1510	1605	1705	1820	1960	2120	2320	2560	2850	3220	3720
W	Tungsten	2518–3300	2050	2150	2270	2390	2520	2680	2840	3030	3250	3500	3810	4180	4630	5200	5900
Y	Yttrium	1774–2103	1045	1100	1160	1230	1305	1390	1490	1605	1740	1905	2105	2355	2670	3085	3650
Yb	Ytterbium	Est.	436	460	488	520	552	590	638	690	755	830	920	1060	1225	1490	1840
Zn	Zinc	422–1089	336	354	374	396	421	450	482	520	565	617	681	760	870	1010	1210
Zr	Zirconium	1949–2054	1500	1580	1665	1755	1855	1975	2110	2260	2450	2670	2930	3250	3650	4170	4830

* With permission, from Ref. 19.
⊙ indicates melting point.

which shows that the temperature of a gas is proportional to the average kinetic energy of the molecules. Furthermore, since the mean-square speed $\overline{c^2}$ is composed of three equal mean-square velocities parallel to the orthogonal coordinates, Eq. (27) also implies that the total translational energy of gas molecules is equally distributed among three mutually perpendicular directions of motion with average increments of $\frac{1}{2}kT$ per degree of freedom. This is the equipartition principle of kinetic energy as it applies to the three degrees of translational freedom. Polyatomic molecules can store additional energy in the form of rotation and oscillation. These internal energies are also distributed in average increments of $\frac{1}{2}kT$ per degree of freedom, provided kT is not small compared with the energy of the excited state.

(2) Molecular-distribution Functions Instead of characterizing the motion of gas molecules by a mean-square value, the entire range of possible values and their relative frequencies will now be considered. The functions governing the distributions of molecular properties were first derived by J. C. Maxwell and L. Boltzmann. Maxwell started with the assumption that the number of molecules having a certain speed between c and $c + dc$ is determined only by the kinetic energy and is therefore a function of $c^2 = u^2 + v^2 + w^2$. The use of even exponents allows for the fact that positive and negative velocities occur with equal probability, an obvious requirement since gases never accumulate preferentially near any one particular wall. Furthermore, the distribution function for the speed c must be the product of three independent but identical distribution functions for the components u, v, w, which determine the direction and magnitude of c. These considerations lead to the following definition equations:

$$\frac{dN_u}{N} = \phi(u^2)\,du \qquad \frac{dN_v}{N} = \phi(v^2)\,dv \qquad \frac{dN_w}{N} = \phi(w^2)\,dw$$

and
$$\frac{dN_{u,v,w}}{N} = \psi(u^2 + v^2 + w^2)\,du\,dv\,dw = \phi(u^2)\phi(v^2)\phi(w^2)\,du\,dv\,dw \qquad (28)$$

dN_u/N is the fraction of all molecules having velocities between u and $u + du$, and $\phi(u^2)$ is their distribution function. Analogous relations hold for the velocities v and w. $dN_{u,v,w}/N$ is the fraction of all molecules having simultaneously velocities between u and $u + du$, v and $v + dv$, w and $w + dw$, and ψ is the corresponding distribution function.

The mathematical solution to satisfy these differential equations is an exponential function for ϕ with two constants A and c_m:

$$\phi(u^2) = A \exp\left(-\frac{u^2}{c_m^2}\right)$$

Analogous expressions hold for $\phi(v^2)$ and $\phi(w^2)$. From these follows

$$\psi(u^2 + v^2 + w^2) = A^3 \exp\left(-\frac{c^2}{c_m^2}\right)$$

The constant A can be determined by considering that the integrals of Eqs. (28) must represent the total number of molecules N:

$$\int_{-\infty}^{+\infty} dN_u = \int_{-\infty}^{+\infty} NA \exp\left(-\frac{u^2}{c_m^2}\right) du = N$$

Therefore, $A = (\pi c_m^2)^{-\frac{1}{2}}$, and the distribution functions become

$$\phi(u^2) = (\pi c_m^2)^{-\frac{1}{2}} \exp(-u^2/c_m^2)$$

and
$$\psi(u^2 + v^2 + w^2) = (\pi c_m^2)^{-\frac{3}{2}} \exp\left(-\frac{c^2}{c_m^2}\right)$$

The distribution function $\psi(u^2 + v^2 + w^2)$ gives the fraction of molecules with speeds $c = (u^2 + v^2 + w^2)^{\frac{1}{2}}$ in one particular direction. There are other molecules with speeds of the same magnitude c but composed of different velocities and moving in different directions. If molecular speeds are represented in a three-dimensional velocity space, the function $\psi(u^2 + v^2 + w^2)$ can be interpreted as indicating the frequency of speeds at a point with the coordinates u, v, w. Since the volume in the velocity space between the speeds c and $c + dc$ is $4\pi c^2\, dc$, a new distribution function $\Phi(c^2)$ can be derived which represents the fraction of molecules with a random speed c:

$$\frac{dN_c}{N} = \Phi(c^2)\, dc = 4\pi c^2\psi(u^2 + v^2 + w^2)\, dc \tag{29}$$

Explicitly, the random speed distribution is

$$\Phi(c^2) = \frac{4}{c_m{}^3\sqrt{\pi}}\, c^2 \exp\left(-\frac{c^2}{c_m{}^2}\right) \tag{30}$$

Differentiation of $\Phi(c^2)$ with respect to c shows that the distribution has a maximum at c_m. Therefore, c_m is the most probable speed. The relationship to the mean-square speed $\overline{c^2}$ can be derived by defining the latter through an integral rather than a summation:

$$\overline{c^2} = \frac{1}{N}\int_0^\infty c^2\, dN_c \equiv \frac{\Sigma c^2}{N} \tag{31}$$

Substituting dN_c/N from Eqs. (29) and (30) into Eq. (31) and solving the integral yields $c_m{}^2 = \frac{2}{3}\overline{c^2}$. Since $\overline{c^2} = 3kT/m$, the most probable speed is $c_m = (2kT/m)^{\frac{1}{2}}$. Restating the distribution functions in terms of gas temperature and mass rather than c_m, the fraction of molecules having velocities between u and $u + du$ in one particular direction is

$$\frac{dN_u}{N} = \phi(u^2)\, du = \left(\frac{m}{2\pi kT}\right)^{\frac{1}{2}}\exp\left(-\frac{mu^2}{2kT}\right) du \tag{32}$$

The fraction of molecules having random speeds between c and $c + dc$ is

$$\frac{dN_c}{N} = \Phi(c^2)\, dc = 4\pi\left(\frac{m}{2\pi kT}\right)^{\frac{3}{2}} c^2 \exp\left(-\frac{mc^2}{2kT}\right) dc \tag{33}$$

In addition to $\overline{c^2}$ and c_m, there is a third characteristic speed, the arithmetic average $\bar{c}$, which is defined as

$$\bar{c} = \frac{1}{N}\int_0^\infty c\, dN_c \tag{34}$$

Substituting dN_c/N from Eq. (33) into Eq. (34) and solving the definite integral yields

$$\bar{c} = \left(\frac{8kT}{\pi m}\right)^{\frac{1}{2}} = 14{,}551\left(\frac{T}{M}\right)^{\frac{1}{2}} \qquad \text{cm s}^{-1} \tag{35}$$

where M = molar mass in grams. The three characteristic speeds form the numerical ratios

$$\sqrt{\overline{c^2}} : \bar{c} : c_m = \sqrt{\tfrac{3}{2}} : \sqrt{4/\pi} : 1 = 1.225 : 1.128 : 1$$

Speed distributions $\Phi(c^2)$ for aluminum vapor at 1200°C ($p^* \approx 10^{-2}$ Torr) and for hydrogen at the same temperature as well as at 25°C are shown in Fig. 3. The molecular speeds are of the order of 10^5 cm s^{-1}. Because of their smaller mass, hydrogen molecules travel faster than aluminum atoms. The effect of temperature is to increase the dispersion of the speed distribution.

The distribution of kinetic energies E_k among molecules is similar to that of the random speeds and can be derived from Eq. (33) by substituting $E_k = \frac{1}{2}mc^2$ and $dE_k = mc\, dc$, which yields

$$\frac{dN_{E_k}}{N} = \Phi(E_k)\, dE_k = \frac{2}{kT\sqrt{\pi}}\left(\frac{E_k}{kT}\right)^{\frac{1}{2}}\exp\left(-\frac{E_k}{kT}\right) dE_k$$

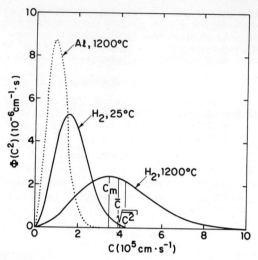

Fig. 3 Speed distributions for aluminum vapor and hydrogen gas.

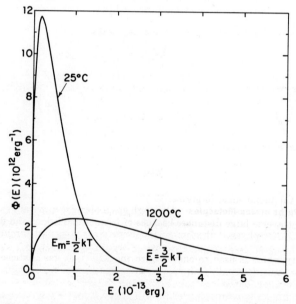

Fig. 4 Kinetic-energy distributions for two gas temperatures.

Figure 4 shows the kinetic-energy distributions corresponding to the speed distributions given in Fig. 3. The kinetic energies are distributed only according to the temperature of the gas but independent of the molecular mass. It is also of interest to note that the average energy $\overline{E_k} = \frac{3}{2}kT$ is three times greater than the most common energy $\frac{1}{2}kT$. This is due to the tail of the distribution at high multiples of kT. Particularly at higher temperatures, some molecules have energies far in excess of the average value.

(3) Impingement Rate of Molecules on a Surface The impingement rate is the number of molecules which strike 1 cm² of a surface in unit time in a gas at rest. It is often needed in calculations of atomistic events and may be derived by considering only velocities perpendicular to the surface. If the volume V contains N molecules, the number of those having a certain velocity u is given by Eq. (32):

$$dN_u = N\Phi(u^2)\,du$$

Only a fraction of these can reach the surface within a time interval dt, namely, those which are within the striking distance $u\,dt$. If A_w is the wall area under consideration, the fraction of the total volume which contributes to the impingement rate is $(u\,dt\,A_w)/V$. Therefore,

$$d^2N_u = \frac{N}{V}\,A_w u\Phi(u^2)\,du\,dt \tag{36}$$

is the number of particles impinging with velocities u. Since the molecular velocities range from zero to infinity, Eq. (36) must be integrated for all velocities. Proceeding from Eq. (32), this operation yields

$$\int_0^\infty u\Phi(u^2)\,du = \left(\frac{kT}{2\pi m}\right)^{\frac{1}{2}}$$

and therefore the impingement rate is

$$\frac{dN_i}{A_w\,dt} = \frac{N}{V}\left(\frac{kT}{2\pi m}\right)^{\frac{1}{2}}$$

or, after substituting the average molecular speed from Eq. (35),

$$\frac{dN_i}{A_w\,dt} = \frac{1}{4}\frac{N}{V}\,\bar{c}\qquad \mathrm{cm^{-2}\,s^{-1}}$$

The relation between impingement rate and gas pressure is obtained by inserting p/kT for N/V:

$$\frac{dN_i}{A_w\,dt} = (2\pi mkT)^{-\frac{1}{2}}p \tag{37}$$

For pressures given in torr, the impingement rate is

$$\frac{dN_i}{A_w\,dt} = 3.513 \times 10^{22}(MT)^{-\frac{1}{2}}p \qquad \mathrm{cm^{-2}\,s^{-1}}$$

where M is the molar mass in grams.

(4) Free Paths of Gas Molecules Although gas molecules move at very high speeds, they do not traverse large distances because of frequent collisions and deflection from their course. For physical phenomena involving mass transport through the gaseous state, it is of interest to know the distances which molecules travel between collisions. The quantity which is used to characterize the ability of gas particles to cover distances is the mean free path (mfp) λ. It is defined as the average distance of travel between subsequent collisions. An oversimplified but instructive derivation of the mfp follows from the assumption that all gas molecules are at rest except one. If the traveling molecule has the average speed $\bar{c}$, its path length after a time dt is $c\,dt$. Assuming further that all molecules have the same diameter σ, the traveling molecule collides with all particles whose centers are within a cross-sectional area $\pi\sigma^2$ along its path. Thus, the volume within which collisions are possible is $\pi\sigma^2\,dt$. By multiplying with the number of molecules per cubic centimeter, the collision frequency becomes $(N/V)\pi\sigma^2\bar{c}\,dt$. These collisions divide the molecule's traveling distance $\bar{c}\,dt$ into equal intervals which are considered as the mean free path:

$$\lambda \approx \left(\frac{N}{V}\pi\sigma^2\right)^{-1}$$

A rigorous derivation, which takes into account the relative motion of all gas molecules, yields the more accurate expression

$$\lambda = \left(\frac{N}{V} \pi \sigma^2 \sqrt{2} \right)^{-1}$$

or, if the gas pressure is substituted for N/V,

$$\lambda = \frac{kT}{p\pi\sigma^2 \sqrt{2}} \tag{38}$$

Thus, the mean free path is inversely proportional to the gas pressure.

Since intermolecular collisions are statistical events, the actual free paths vary. Their distribution is given by an exponential decay function

$$\frac{N}{N_0} = \exp\left(\frac{-l}{\lambda} \right) \tag{39}$$

where N/N_0 is the fraction of molecules which have not suffered a collision as yet after traveling a free path of length l. According to Eq. (39), only 37% of all molecules traverse a distance as long as the mfp without being deflected. On the other hand, 1% of the molecules reach distances of at least 4.5λ before they collide.

To obtain numerical values of the mfp, it is necessary to know the molecular diameter σ. For some materials, this quantity has been determined by directing a beam of metal vapor into a gas and analyzing the deposit obtained at various distances from the vapor source. The molecular diameter plays a role in several other physical processes which involve interaction between gas molecules. An important example is the flow of gases through narrow tubes where the observed property, the coefficient of viscosity, is related to σ through the mechanism of momentum transfer during intermolecular collisions. Another example is the deviation of real gases from the ideal-gas law.

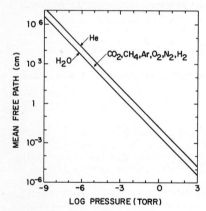

Fig. 5 Mean free paths of common gases as functions of pressure.

Since one of the coefficients in van der Waals' equation represents the actual volume of the gas particles, molecular diameters can also be derived from the empirical volume correction of van der Waals' equation of state. Additional methods for the determination of σ are based on the density of solids consisting of closely packed molecules, or on observations of the intensity decrease of an electron beam passing through a gas.

While the principal effects of molecular diameters in physical phenomena such as scattering of particle beams or viscous flow of gases are well established, the derivation of exact relationships poses theoretical difficulties. This is due to the fact that the hard-sphere model for gas molecules is an oversimplification. Gas molecules have complex structures and are not necessarily spherical. Furthermore, molecules exert attractive and repulsive forces upon each other which vary with distance. Thus, instead of visualizing molecules as hard spheres of well-defined size, it is more realistic to associate them with effective collision cross sections whose diameters may vary depending on the type of experiment made. A comparison of molecular diameters as obtained by different methods has been given by Dushman (Ref. 21, p. 39). For the common gases He, H_2, O_2, N_2, Ar, CH_4, CO_2, and H_2O, the effective cross-sectional diameters range from 2 to 5 Å. The mean free paths corresponding to these values at various pressures according to Eq. (38) are shown in Fig. 5. Since the particle diameters are not too different, the mean free paths of all common gases fall into a narrow band bounded by the largest and the smallest molecules, H_2O and He.

(5) Gas Viscosity and Flow The mean free paths and viscosities of gases are of interest in assessing the vapor distribution around evaporation sources, but their role in vacuum technology is more important. The evacuation of a vessel by means of pumps involves connecting passages through which the gas must flow in order to be removed. The behavior of streaming gases, however, is determined by the motion of individual molecules and their collisions. For a broad treatment of the subject, the reader is referred to Dushman (Ref. 21, p. 80). The following section merely introduces the concept of viscosity and characterizes the different types of gas flow as related to molecular events.

A gas streaming through a narrow tube encounters resistance at the tube walls. As a result, the gas layer closest to the wall is slowed down and does not participate in the collective motion. This layer in turn delays the propagation of the next layer, and so forth. Consequently, every layer of gas moves at a different velocity depending upon its distance from the wall. This situation is shown schematically in Fig. 6. Since there is a velocity gradient between adjacent layers, the latter exert forces in the direction of gas flow (tangential) upon each other. The tangential force per unit area F_{tg} is transmitted by collisions between molecules from adjacent layers and results from the ensuing exchange of momentum. It is proportional to the radial velocity gradient dv/dx:

$$F_{tg} = \eta \frac{dv}{dx}$$

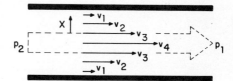

Fig. 6 Schematic representation of particle velocities for viscous flow through a narrow tube.

The proportionality factor η is the viscosity coefficient and expresses that property of a moving gas which results from internal friction caused by intermolecular collisions. It is specific for different gases and is measured in poise. A substance has a viscosity of 1 Poise if a tangential force of 1 dyn cm^{-2} arises from a radial velocity gradient of 1 cm s^{-1} cm^{-1}. The viscosities of the common gases are of the order of 10^{-4} Poise.

In order to relate η to properties of individual gas particles such as mass, velocity, and mean free path, the tangential force F_{tg} must be derived from kinetic theory. This has been attempted by different authors, and their results are of the general form

$$\eta = fm \frac{N}{V} \bar{c} \lambda \tag{40}$$

The factor f varies between 0.3 and 0.5 depending on the assumed model of molecular interaction. The derivation of Chapman and Cowling[24] yields $f = 0.499$, and this value is now preferred over previous ones which were obtained without consideration of attractive and repulsive forces between molecules. By substituting $\bar{c}$ from Eq. (35) and λ from Eq. (38), Eq. (40) can be transformed into

$$\eta = \frac{2f}{\pi \sigma^2} \left(\frac{mkT}{\pi} \right)^{\frac{1}{2}}$$

From this relation follow the conclusions that the viscosity of gases is independent of pressure and, in contrast to the viscosity of liquids, increases with temperature. Both these predictions have been confirmed experimentally for that region of pressures where intermolecular collisions are more frequent than wall collisions. Furthermore, since the viscosity η is amenable to experimental determination by macroscopic gas-flow measurements, numerical values of effective molecular diameters can be derived. These are shown in Table 3.

The macroscopic flow rate of gases Q is that quantity of gas which passes through the cross section of a duct in unit time. It may be expressed as a volume of gas at a certain pressure or as a molecular current

$$Q = p \frac{dV}{dt} = kT \frac{dN}{dt}$$

The most common units for Q are torr $l\ s^{-1}$. At relatively high pressures, the flow rate is dependent on the viscosity of the gas and is characterized by a distribution of particle velocities as pictured in Fig. 6. Accordingly, this type of gas flow is called *viscous laminar flow*. The rate of viscous flow can be derived from the laws of hydrodynamics, which lead to different expressions depending on the geometrical shape of

TABLE 3   Effective Molecular Diameters from Viscosity Measurements at 0°C*

Gas	σ, 10^{-8} cm	Gas	σ, 10^{-8} cm
He..........	2.18	Ar..........	3.67
H_2..........	2.75	Kr..........	4.15
Ne..........	2.60	Xe..........	4.91
O_2..........	3.64	CO_2........	4.65
Air..........	3.74	H_2O........	4.68

* From Ref. 22, p. 149.

the duct. The simplest ducts are straight cylindrical tubes of uniform cross section for which the rate of viscous flow was first given by Poiseuille,

$$Q_{\text{visc}} = \frac{\pi r^4}{8\eta l} \bar{p}(p_2 - p_1)$$

where r is the radius and l the length of the tube; $(p_2 - p_1)$ is the pressure difference across the length of the tube, and $\bar{p}$ the average of p_2 and p_1.

In analogy to Ohm's law, the quantity

$$C = \frac{Q}{p_2 - p_1}$$

is called the conductance and is commonly used to characterize the gas-carrying ability of ducts. Thus, the conductance of a cylindrical tube in the range of viscous flow is

$$C_{\text{visc}} = \frac{\pi r^4}{8\eta l} \bar{p}$$

i.e., it is inversely proportional to the viscosity and decreases toward lower pressures. For air, which has a viscosity of 1.845×10^{-4} Poise at 25°C, numerical values can be derived from

$$C_{\text{visc}} = 2{,}840 \frac{r^4}{l} \bar{p}_{\text{torr}} \qquad l\ s^{-1} \qquad (41)$$

(r and l in cm).

The laws of viscous flow follow from the assumption of frequent intermolecular collisions which are responsible for the behavior of the gas as a coherent medium. These laws cannot be extended to very low pressures where wall collisions are more frequent than intermolecular collisions. As shown schematically in Fig. 7, gas flow at very low particle densities is characterized by molecules traveling independently of each other with random motion superimposed upon the direction of gas transport.

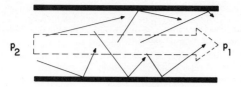

Fig. 7 Schematic representation of particle velocities for free molecular flow through a narrow tube.

This behavior is called *free molecular flow*. It was first analyzed by Knudsen,[25] who derived the molecular-flow equation for long cylindrical tubes:

$$Q_{\text{mol}} = \frac{2}{3}\pi\frac{r^3}{l}\bar{c}(p_2 - p_1)$$

Substitution of $\bar{c}$ from Eq. (35) and dividing by the pressure difference $p_2 - p_1$ across the tube length l yields the conductance in the molecular-flow region,

$$C_{\text{mol}} = 30.48\frac{r^3}{l}\left(\frac{T}{M}\right)^{\frac{1}{2}} \qquad l\ \text{s}^{-1}$$

It should be noted that the conductance in the region of molecular flow is independent of the gas pressure. Numerical values for air flow ($\sqrt{T/M} = 3.207$ at 25°C) can be derived from

$$C_{\text{mol}} = 97.75\frac{r^3}{l} \qquad l\ \text{s}^{-1} \tag{42}$$

(r and l in cm).

A comparison of Eqs. (41) and (42) shows that the conductance of a cylindrical tube is much greater in the region of viscous flow than for molecular flow. The transition from one type of flow to the other is not sharp but gradual. The movement of gases in this interval is referred to as *transition flow*. The gas flow in this intermediate range cannot be derived from first principles. It is customarily described by empirical equations which assume that the total gas flow consists of a viscous and a molecular contribution. For long cylindrical tubes, the conductance for transition flow has been formulated by Knudsen,

$$C_{\text{trans}} = C_{\text{visc}} + ZC_{\text{mol}} \tag{43}$$

where

$$Z = \frac{1 + 2.507r/\bar{\lambda}}{1 + 3.095r/\bar{\lambda}}$$

and $\bar{\lambda}$ is the mean free path associated with the average pressure $\bar{p}$.

To assess the fractions of flow attributable to viscous and molecular flow, Eq. (43) may be rewritten as

$$C_{\text{trans}} = C_{\text{mol}}\left(\frac{C_{\text{visc}}}{C_{\text{mol}}} + Z\right) = C_{\text{mol}}\left(\frac{0.1472r}{\bar{\lambda}} + Z\right) \tag{44}$$

The parameter Z is 0.810 for $r/\bar{\lambda} \gg 1$ (viscous flow) and approaches 1.000 for $r/\bar{\lambda} \ll 1$ (molecular flow). Since Z does not vary widely, the first term of Eq. (44) determines the predominant type of flow. Values of the mean free paths may be taken from Fig. 5 or calculated from Eq. (38). The latter yields for air at 25°C

$$\bar{\lambda} = \frac{5.10^{-3}}{\bar{p}_{\text{torr}}} \qquad \text{cm}$$

which transforms Eq. (44) into

$$C_{\text{trans}} = C_{\text{mol}}(29r\bar{p}_{\text{torr}} + Z) \tag{45}$$

Comparison of the two terms in Eq. (45) shows that the gas flow is at least 95% viscous if $r\bar{p}_{torr} > 0.5$ while for $r\bar{p}_{torr} < 0.005$ the flow may be treated as molecular. Thus, the region of transition flow encompasses two decades of pressure.

The flow of gas induced by high-vacuum pumps (below 10^{-3} Torr) is molecular, and the paths of molecules are frequently interrupted by wall collisions. The presence of pipes, elbows, and baffles makes it difficult to derive a mathematical expression for the conductance of that region which connects a vacuum chamber with its pump. Methods of estimating the transmission probability through a complex system of tubes, diaphragms, and baffles are of great interest in determining the effective speed of pumps and have been reviewed by Steckelmacher.[26] Most powerful are Monte Carlo techniques, which were first applied to the problem by Levenson, Milleron, and Davis.[27] In this analysis, the molecular conductance of irregularly shaped channels is computed from possible trajectories of individual molecules. If a sufficiently large number of molecular paths is taken into account, the conductance of channels as complex as diffusion-pump baffles can be derived fairly accurately.

3. EVAPORATION THEORY

The application of the kinetic-gas theory to interpret evaporation phenomena resulted in a specialized evaporation theory. Early attempts to express quantitatively the rates at which condensed materials enter the gaseous state are connected mainly with the names of Hertz, Knudsen, and Langmuir. The observation of deviations from the originally postulated ideal behavior led to refinements of the transition mechanisms, which became possible as the understanding of molecular and crystalline structures increased. As a result, the evaporation theory includes concepts of reaction kinetics, thermodynamics, and solid-state theory. The questions pertaining to the directionality of evaporating molecules are primarily answered by statistical considerations derived from gas kinetics and sorption theory.

a. Evaporation Rates

(1) The Hertz-Knudsen Equation The first systematic investigation of evaporation rates in a vacuum was conducted by Hertz in 1882.[28] He distilled mercury at reduced air pressure and observed the evaporation losses while simultaneously measuring the hydrostatic pressure exerted on the evaporating surface. By using an evaporant of good thermal conductivity such as mercury, he excluded the possibility of limiting the evaporation rates because of insufficient heat supply to the surface. For all conditions chosen, Hertz found evaporation rates proportional to the difference between the equilibrium pressure of mercury at the surface temperature of the reservoir p^* and the hydrostatic pressure p acting on that surface.

From these observations, he drew the important and fundamental conclusion that a liquid has a specific ability to evaporate and cannot exceed a certain maximum evaporation rate at a given temperature, even if the supply of heat is unlimited. Furthermore, the theoretical maximum evaporation rates are obtained only if as many evaporant molecules leave the surface as would be required to exert the equilibrium pressure p^* on the same surface while none of them must return. The latter condition means that a hydrostatic pressure of $p = 0$ must be maintained. Based on these considerations, the number of molecules dN_e evaporating from a surface area A_e during the time dt is equal to the impingement rate given by Eq. (37) with the equilibrium pressure p^* inserted, minus a return flux corresponding to the hydrostatic pressure p of the evaporant in the gas phase:

$$\frac{dN_e}{A_e\,dt} = (2\pi mkT)^{-\frac{1}{2}}(p^* - p) \qquad \text{cm}^{-2}\,\text{s}^{-1} \qquad (46)$$

The evaporation rates originally measured by Hertz were only about one-tenth as high as the theoretical maximum rates. The latter were actually obtained by Knudsen[29] in 1915. Knudsen argued that molecules impinging on the evaporating surface may be reflected back into the gas rather than incorporated into the liquid. Consequently, there is a certain fraction $(1 - \alpha_v)$ of vapor molecules which contribute to the evaporant pressure but not to the net molecular flux from the condensed into the vapor phase. To account for this situation, he introduced the evaporation coefficient α_v, defined as the ratio of the observed evaporation rate in vacuo to the value theoretically possible according to Eq. (46). The most general form of the evaporation rate equation is then

$$\frac{dN_e}{A_e\,dt} = \alpha_v(2\pi mkT)^{-\frac{1}{2}}(p^* - p)$$

which is commonly referred to as the Hertz-Knudsen equation.

Knudsen found the evaporation coefficient α_v to be strongly dependent on the condition of the mercury surface. In his earlier experiments, where evaporation took place from the surface of a small quantity of mercury, he obtained values of α_v as low as 5×10^{-4}. Concluding that the low rates were attributable to surface contamination as manifest in the discolored appearance of the metal, he allowed carefully purified mercury to evaporate from a series of droplets which were falling from a pipette and thus continually generated fresh, clean surfaces. This experiment yielded the maximum evaporation rate

$$\frac{dN_e}{A_e\,dt} = (2\pi mkT)^{-\frac{1}{2}}p^* \tag{47}$$

(2) Free Evaporation and Effusion It was first shown by Langmuir[30] in 1913 that the Hertz-Knudsen equation also applies to the evaporation from free solid surfaces. He investigated the evaporation of tungsten from filaments in evacuated glass bulbs and assumed that the evaporation rate of a material at pressures below 1 Torr is the same as if the surface were in equilibrium with its vapor. Since recondensation of evaporated species was thereby excluded, he derived the maximum rate as stated by Eq. (47). Knowing the weight loss which the filaments had suffered during a certain time of evaporation, Langmuir was able to calculate the vapor pressure of tungsten. In order to do so, the molecular evaporation rate [Eq. (47)] is multiplied with the mass of an individual molecule. This yields the mass evaporation rate

$$\Gamma = m\,\frac{dN_e}{A_e\,dt} = \left(\frac{m}{2\pi kT}\right)^{\frac{1}{2}}p^* \tag{48}$$

or
$$\Gamma = 5.834 \times 10^{-2}\left(\frac{M}{T}\right)^{\frac{1}{2}}p^* \qquad \text{g cm}^{-2}\text{ s}^{-1}$$

for pressures in torr.

The mass evaporation rate per unit area Γ is related to the total amount of evaporated material $\mathfrak{M}_e$ through the double integral

$$\mathfrak{M}_e = \iint\limits_{t\ A_e} \Gamma\,dA_e\,dt \tag{49}$$

Assuming a uniform evaporation rate across the entire evaporation area and no variations with time, Γ can be determined from experimental data and inserted into Eq. (48) to obtain the vapor pressure. Numerical values of mass evaporation rates for metals at various vapor pressures have been tabulated by Dushman.[21] At $p^* = 10^{-2}$ Torr, Γ is of the order of 10^{-4} g cm^{-2} s^{-1} for most elements.

Phase transitions of this type, which constitute evaporation from free surfaces, are commonly referred to as Langmuir or free evaporation. Since the assumption $\alpha_v = 1$ is not always valid, as will be discussed later, an evaporation coefficient $\alpha_v < 1$ is often needed in Eq. (48). An alternate evaporation technique which avoids the uncertainty introduced by α_v being possibly smaller than 1 was established by Knudsen[31] and is associated with his name. In Knudsen's technique, evaporation occurs as effusion from an isothermal enclosure with a small orifice (Knudsen cell). The evaporating surface within the enclosure is large compared with the orifice and maintains the equilibrium pressure p^* inside. The diameter of the orifice must be about one-tenth or less of the mfp of the gas molecules at the equilibrium pressure p^*, and the wall around the orifice must be vanishingly thin so that gas particles leaving the enclosure are not scattered or adsorbed and desorbed by the orifice wall. Under these conditions, the orifice constitutes an evaporating surface with the evaporant pressure p^* but without the ability to reflect vapor molecules; hence $\alpha_v = 1$. If A_e is the orifice area, the total effusion from the Knudsen cell into the vacuum is $A_e(2\pi mkT)^{-\frac{1}{2}}(p^* - p)$ molecules per second.

Langmuir's and Knudsen's techniques have been employed in many experimental versions to determine the vapor pressure of materials and heats of vaporization.[5] Knacke and Stranski[32] have reviewed these methods, and a critical examination of both techniques with their limitations has been published by Rutner.[33] Langmuir's method suffers from the uncertainty of whether or not an observed rate of weight loss truly reflects the equilibrium rate of evaporation. It is often used, however, to determine α_v by comparing its results with independently known vapor-pressure data or with evaporation-rate measurements from Knudsen cells. The principal problem with Knudsen's technique is that an ideal cell with an infinitely thin-walled orifice yielding free molecular flow can only be approximated. In practice, orifices of finite thickness must be used, which necessitates the application of corrective terms in the effusion equation. Expressions for the conductance of various orifices and tubes are derived and discussed in Dushman's book.[21] Freeman and Edwards[34] solved the transmission problem for nonideal orifices by machine computations.

b. Evaporation Mechanisms

The evaporation coefficient α_v was introduced into the Hertz-Knudsen equation to account for observed evaporation rates smaller than those permitted by the equilibrium pressure. Its theoretical justification, however, was obtained from the kinetics of a condensation process which included the possibility of elastic reflection of vapor molecules on the evaporant surface. Because of this derivation, the same coefficient is sometimes referred to as the condensation coefficient α_c and is defined as the ratio of molecules condensing to those colliding with the surface of the condensed phase.* In evaporations of the Langmuir type, however, recondensation of evaporated molecules is excluded by definition, and an understanding of the evaporation coefficient in the sense of α_c is therefore fictitious. The assumption $\alpha_v = \alpha_c$, which is implied when the terms evaporation and condensation coefficients are used synonymously, is contingent upon identical mechanisms governing both processes and therefore is not generally true.

The necessity of having to resort to a condensation model to interpret the evaporation coefficient α_v has led many investigators to attempt the derivation of an expression equivalent to the Hertz-Knudsen equation strictly from the point of evaporation. The aim of these approaches is to establish an atomistic model for the evaporation process capable of yielding $\alpha_v < 1$ without the assumption that evaporation and condensation are the results of oppositely directed but otherwise identical mechanisms. Hence it is necessary to consider conditions which may inhibit or delay the departure of atoms or molecules from the surface of a condensed phase. Various assumptions about such constraining factors and different approaches to treat them mathematically have been tried, and the interested reader is referred to the review article by Knacke and Stranski,[32] to the more recent and shorter treatment by Dettorre, Knorr, and Hall,[37] or to Hirth and Pound's book,[35] which offers the most comprehensive coverage and bibliography on the subject. The following sections about evaporation mechanisms and representative experimental results are largely based on Hirth and Pound's work.

---

* Neither α_v nor α_c is identical with the accommodation coefficient α_T, which is used to describe the degree of energy exchange toward equilibrium between incident gas molecules and a condensed phase (see Chap. 8). If energy exchange does play a retarding role in the process of evaporation, α_T enters the evaporation coefficient as one of several possible constraining factors. Definitions of the various coefficients and their relationships among each other are discussed by Hirth and Pound (Ref. 35, p. 1) and by Winslow.[36]

(1) Liquids In the simplest model for the evaporation process, the condensed phase is considered as a system of oscillators whose surface molecules are bound by a certain energy of evaporation E_v. Transition into the gas phase is assumed to occur if a surface molecule possesses an energy of oscillation equal to or greater than E_v. It is also implied that all surface molecules have the same binding energy and an equal chance to evaporate. Based on this model, the probability of evaporation P was first estimated by Polanyi and Wigner.[38] Taking into account interaction among molecules by interference of eigenfrequencies, they arrived at

$$P \approx \frac{2E_v}{kT} \, \nu \exp\left(-\frac{E_v}{kT}\right)$$

where ν is the oscillation frequency of the surface atoms.

A more refined statistical model evolved from the work of Herzfeld,[39] Pelzer,[40] and Neumann.[41] They characterized the state of the surface molecules by a Maxwellian energy distribution and a spatial distribution, the latter relating the displacement of a molecule from its equilibrium position to its potential energy. A molecule evaporates if it is displaced far enough for its potential energy to equal the energy of evaporation. This treatment yields an evaporation probability per unit area and unit time of

$$P = \left(\frac{kT}{2\pi m}\right)^{\frac{1}{2}} \frac{Q_v}{Q_c} \exp\left(-\frac{E_v}{kT}\right) \tag{50}$$

where Q_v and Q_c are the partition functions of the vapor and of the condensed phase, respectively. Equation (50) can be transformed by equating the evaporation probability P with the molecular evaporation rate $dN_e/A_e\,dt$ and substituting $\frac{1}{4}\bar{c}$ for $(kT/2\pi m)^{\frac{1}{2}}$ [see Eq. (35)]. Furthermore, it can be shown that

$$\frac{Q_v}{Q_c} \exp\left(-\frac{E_v}{kT}\right) = \frac{N}{V}$$

is the statistical expression for the law of mass action. Consequently, Eq. (50) is equivalent to the Hertz-Knudsen equation for maximum evaporation rates as stated previously in Eq. (47). The same solution has also been obtained by Penner,[42] who treated the evaporation process in terms of the reaction-rate theory.[43]

The significance of these results is that the simple oscillator model of an evaporating surface is capable of explaining only maximum rates with a coefficient $\alpha_v = 1$ in the Hertz-Knudsen equation. Experience has shown that the model applies to liquids which evaporate by exchange of single atoms with a monatomic vapor. Most molten metals fall into this category, and $\alpha_v = 1$ has been confirmed experimentally for liquid mercury, potassium, and beryllium. Certain organic liquids whose molecules have spherical symmetry and small entropies of evaporation follow the same behavior; an example is CCl_4.

To explain evaporation coefficients smaller than 1, the possibility of changes in the internal energy states of molecules has been examined. In substances where molecules in the condensed state have fewer degrees of freedom than in the gaseous state, evaporation requires the exchange of internal energy in addition to translational energy. Herzfeld[44] was the first to consider the effect of internal energy changes on evaporation rates, and several authors have treated the problem thereafter (see Refs. 41, 42, 44 to 48). It is statistically unlikely that a surface molecule receives the internal and translational energies necessary for evaporation under the thermal-equilibrium conditions at exactly the same instance. More probable is a sequence of events in which the molecule is first activated by obtaining sufficient translational energy and then must wait for an internal energy quantum before evaporation can occur. Among the different forms of internal energy, rotational terms are considered to have the strongest effect on α_v because the relaxation time required to activate a rotational degree of freedom in the translationally activated state is estimated to be longer than for any other process.[47]

The mathematical treatment of evaporation involving internal energy changes is again based on the theory of absolute reaction rates,[43] with some variations among the different authors in the characterization of the activated state and in the assignment of partition functions. An illustrative example is the case where a rotational degree of freedom is restricted in the liquid but operative in the gaseous state. Derivation based on the rate theory yields the Hertz-Knudsen equation in the form

$$\frac{dN_e}{A_e\,dt} = \left(\frac{Q_R{}^*}{Q_{R_v}}\right) (2\pi mkT)^{-\frac{1}{2}}(p^* - p)$$

where $Q_R{}^*$ and Q_{R_v} are the rotational partition functions for the activated liquid state and for the vapor, respectively. Thus, the evaporation coefficient α_v takes on the significance of the ratio of two partition functions. Here, restricted evaporation ($\alpha_v < 1$) is due to the lack of one degree of freedom, which reduces the number of possible states for molecules in the liquid. This form of hindered phase transition is referred to as an entropy constraint.[35]

Hirth and Pound (Ref. 35, p. 83) have compiled and critically evaluated experimental α_v data from several authors. They conclude that the mechanism of activation in two steps with $\alpha_v < 1$ holds for liquids with small polar molecules evaporating from static surfaces. Examples are benzene, chloroform, ethanol, methanol, glycerol, and water. Conflicting results of $\alpha_v \approx 1$ have also been published for glycerol and water; the discrepancy may arise from differences in the experimental conditions such as turbulence in the vapor stream which could continually disturb the vapor-liquid interface. Liquids consisting of polar molecules which are either of the long-chain type or large and planar show a different behavior insofar as their evaporation coefficients are approximately equal to 1. Examples are 2-ethyl-hexyl phthalate, lauric acid, hexadecanol, n-dibutyl phthalate, tridecyl methane, n-hexadecane, and similar compounds. A possible explanation is that in these molecules the rotational degrees of freedom are operative in the translationally activated state; hence $Q_R{}^* = Q_{R_v}$ and $\alpha_v = 1$.

The evaporation rate of liquids is also retarded by surface contamination, as was experienced by Knudsen[29] in his earlier experiments ($\alpha_v \approx 10^{-3}$). The mechanisms by which impurities act as interface constraints during evaporation are presently understood only qualitatively. The most plausible explanations are that impurity atoms affect surface molecules by increasing the activation energy of evaporation, or that slow diffusion of evaporant molecules through an impurity film is the rate-determining step. Understandably, little experimental evidence exists since great care is generally taken to avoid surface contamination in evaporation-rate studies.

(2) Crystalline Solids While an evaporation model based on equal binding energies for all surface molecules is a reasonable point of departure for liquids, it is not applicable to crystalline solids. The model for crystal surfaces as established by Kossel[49] and Stranski[50,51] distinguishes different sites with variable numbers of nearest and second-nearest neighbor atoms. Since the binding forces exerted by adjacent atoms are additive, the evaporation energies for atoms on different sites must also be different. As shown in Fig. 8, the atoms S in a completed surface plane and L' in a ledge are surrounded by more than the average number of neighbors and therefore can be removed only by supplying an energy greater than the lattice energy. Conversely, a free surface atom like A or one at a ledge like L has less than the average number of neighbors and therefore may be removed by relatively small energies. The atom K in the kink or half-crystal position is most significant because its binding energy is equal to the lattice energy. Consequently, a crystal can be built in repeatable steps of equal energy by adding kink atoms. The removal or addition of a kink atom is therefore the microscopically reversible step which controls the evaporation and growth equilibrium on a crystal surface.

The probability of direct transitions of atoms from kink sites into the vapor as calculated from the binding energies is still too small to account for actually observed evaporation rates. This was first shown by Volmer[40] and led to the theory of step-wise evaporation. According to this concept, evaporation is initiated by an atom dissociating from a kink position, followed by dissociation from the ledge position and diffusion across the crystal surface. This sequence is indicated by the arrows in Fig. 8. Before desorption can occur, the diffusing adatom must be activated by receiving the required translational energy. The theory of stepwise evaporation has been further developed by Hirth and Pound,[52] who predicted that crystal edges act as sources for monomolecular ledges which subsequently move across the crystal surface. It is therefore possible to evaporate perfect crystals bounded by singular surfaces without having to nucleate evaporation steps within the surfaces. This important function of crystal edges in the kinetics of evaporation is distinctly different from the minor role which edges play as nucleation sites in the kinetics of crystal growth.

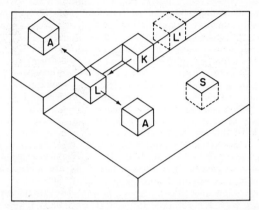

Fig. 8 Schematic of a crystal surface with atomic sites having different numbers of nearest neighbors. S: surface atom within a completed crystal plane. L': atom within a monatomic ledge. K: kink position. L: atom at a monatomic ledge. A: single atoms adsorbed on the surface.

A rigorous treatment of the metastable equilibrium population of adatoms on the crystal surface as determined by the supply (dissociation and diffusion) and annihilation (activation and desorption) processes has been given by Hirth and Pound (Ref. 35, p. 92). Simultaneous solution of the equations governing the individual process steps yields a general evaporation-flux equation from which numerical predictions for evaporation coefficients can be obtained in certain specific cases. An important parameter in the solutions is the average spacing between monatomic ledges on the crystal surface. On large low-index surfaces of single crystals, the ledge spacing is wider than the mean random-walk distance of the adatoms. Consequently, the evaporation rate is controlled by the interplay of ledge generation at crystal edges and surface diffusion of adatoms. Under these conditions, the evaporation coefficient for monatomic vapors should tend toward a minimum value of $\alpha_v = \frac{1}{3}$.

On high-index crystal planes, the surface is likely to form many ledges with spacings comparable with the mean diffusion distance which adatoms can cover before they are activated and desorbed. This situation leads to $\alpha_v \approx 1$. Other cases which have been considered include the possibilities of an entropy constraint for atoms dissociating from a ledge and of a high activation energy required for dissociation of atoms from a kink site. Both cases give values of $\alpha_v < \frac{1}{3}$. Evaporation coefficients smaller than 1 may also be due to adsorbed impurities. Their restraining influence on the evaporation process is attributed to the protective action of firmly adsorbed molecules on crystal edges so that the formation of new evaporation ledges is inhibited. Consequently, adatoms are supplied only from already existing ledges which grow to macroscopic size and thereby lower the rate of evaporation.

Experimental evidence pertaining to perfect single crystals to confirm these concepts is sparse. Sears[53] was able to show that evaporation from crystal edges occurs readily if the external vapor pressure is lowered as little as 2% below the equilibrium pressure. Conversely, no evaporation took place from a singular surface of the same crystal at external pressures as low as half the equilibrium value. This behavior supports the argument that crystal edges serve as sources for evaporation ledges.

The theory of stepwise evaporation can be extended to imperfect crystals and polycrystalline materials (Ref. 35, p. 107). Such solids follow the same evaporation kinetics as perfect crystals except for differences in the spacing of evaporation ledges. Spiral dislocations, for example, serve as additional sources for monatomic ledges, but the spacing of the latter should eventually assume the same values during evaporation as those between edge-generated ledges. Therefore, imperfect crystals should have evaporation coefficients similar to those of perfect crystals; i.e., α_v should approach one-third. In polycrystalline materials, ledge sources in the form of grain boundaries, cracks, crystal edges, and dislocations are numerous. The average ledge spacing will therefore be small, which leads one to expect evaporation coefficients approaching 1 unless the process is subject to entropy or impurity constraints.

Hirth and Pound have critically reviewed published evaporation coefficients for a number of polycrystalline solids and imperfect single crystals.[54] These data were obtained by comparing the results of Langmuir- and Knudsen-type vapor-pressure measurements. In agreement with the theory, polycrystalline metals with monatomic vapors such as Ag, Cu, Be, Cd, Hg, Fe, Cr, and Pt yielded evaporation coefficients of about 1. However, this result is contingent upon the availability of clean crystal surfaces, and coefficients smaller than 1 have been obtained in a number of instances where this condition was not satisfied. The effect of impurities on evaporation has been demonstrated for beryllium, whose coefficient α_v was reduced to 0.02 by allowing the metal surface to chemisorb oxygen.

Evaporation rates for low-index planes on imperfect single crystals have been determined for Ag, KCl, KI, rhombic sulfur, CsI, CsBr, NaCl, and LiF. Except for the latter two compounds, evaporation coefficients between 0.3 and 0.8 have been found. While these data are compatible with the theoretical value of $\frac{1}{3} < \alpha_v < 1$ for surface-diffusion-controlled evaporation, the possibility of entropy constraints has also been considered and cannot be excluded in several cases. Uncertainties in the interpretation of some data also arise from the use of vacuum pressures too high to rule out impurity adsorption. The evaporation coefficient of NaCl ($\alpha_v = 0.1$ to 0.2) and LiF ($\alpha_v = 0.1$ to 1) were found to be temperature-dependent, which is attributed to varying degrees of formation of polyatomic gaseous species.

The evaporation behavior of polycrystalline solids which evaporate as polymers (Se, Te, As, P, C) and of compounds which dissociate upon evaporation (NH_4Cl, AlN, BN, Al_2O_3) is governed by more complex kinetic processes than heretofore discussed. The observed values vary from $\alpha_v \approx 1$ for Te_2 vapor to $\alpha_v \approx 10^{-7}$ for P_2 and P_4 vapor. A detailed theoretical analysis of these cases has not been possible yet, one reason being the lack of sufficient information about the identity and concentration of the evaporating species as dependent on time, temperature, and crystal morphology.

c. Directionality of Evaporating Molecules

In the preceding discussion of evaporation mechanisms, only the total number of molecules leaving a surface was of interest. It is to be expected that the interaction between the surface of the condensed phase and an individual atom about to evaporate also determines the direction into which the particle will be emitted. The answer to this question must be statistical in nature and contingent upon assumptions about the energy states of molecules at the instant of evaporation. Since the distribution of kinetic energies among molecules is well known for the gaseous state, the spatial distribution of particles can be derived for the effusion of gases from an ideal Knudsen cell.

(1) The Cosine Law of Emission An isothermal enclosure with an infinitesimally small opening dA_e bounded by vanishingly thin walls is shown schematically in Fig. 9. It is assumed to contain N molecules which have a Maxwellian speed distribution.

Most of these molecules will impinge on the enclosure walls and be reflected without causing a net change in the total speed distribution. Those molecules moving toward the opening, however, will leave the enclosure in the same direction and at the same speed that they possessed immediately prior to their escape. The total population of gas molecules in the enclosure is assumed to remain constant because of the presence of a condensed phase. The distribution of molecules in the evaporant stream is then determined by the distribution of molecular speeds inside the effusion cell and may be described by an expression which gives the number of molecules within a small solid angle $d\omega$ for every direction of emission. The latter is defined by its angle φ with the normal to the surface element dA_e.

Considering first molecules of only one particular speed c, their total number in the enclosure is $dN_c = N\Phi(c^2)\,dc$. Within a time dt, only those molecules can reach dA_e and leave in the direction φ which are in the slanted prism indicated in Fig. 9. Thus, the fraction of molecules within striking distance of the opening is $c\,dt\,\cos\varphi\,dA_e/V$. The number of molecules actually crossing the opening is further reduced because their speeds c within the slanted prism are randomly distributed in all directions. Hence, only the fraction $d\omega/4\pi$ is actually moving toward dA_e. Multiplication of dN_c with the volume fraction and the angular fraction yields the number of molecules having a speed c and leaving in the direction φ:

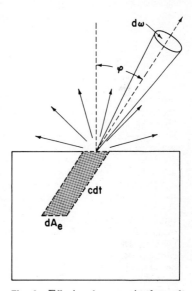

$$d^4N_{e,c}(\varphi) = \frac{N}{V}\,c\,\Phi(c^2)\,dc\,dA_e\,dt\,\cos\varphi\,\frac{d\omega}{4\pi}$$

Integration over all speeds c gives the total number of molecules per angle increment $d\omega$, whereby

$$\int_0^\infty c\,\Phi(c^2)\,dc = \bar{c}$$

and therefore

$$d^3N_e(\varphi) = \frac{1}{4}\frac{N}{V}\,\bar{c}\,dA_e\,dt\,\cos\varphi\,\frac{d\omega}{\pi}$$

Fig. 9 Effusion from an isothermal enclosure through a small orifice.

The small increment of mass carried by these molecules is $d^3\mathfrak{M}_e(\varphi) = m\,d^3N_e(\varphi)$, and since $m \times \frac{1}{4}(N/V)\bar{c} = \Gamma$, one obtains

$$d^3\mathfrak{M}_e(\varphi) = \Gamma\,dA_e\,dt\,\cos\varphi\,\frac{d\omega}{\pi} \qquad (51a)$$

or, substituting the total mass of evaporated material $\mathfrak{M}_e$ according to Eq. (49),

$$d\mathfrak{M}_e(\varphi) = \mathfrak{M}_e\,\cos\varphi\,\frac{d\omega}{\pi} \qquad (51b)$$

Equations (51a) and (51b) represent the cosine law of emission and are equivalent to Lambert's law in optics. According to the cosine law, emission of material from a small evaporating area does not occur uniformly in all directions but favors directions approximately normal to the emitting surface where $\cos\varphi$ has its maximum values.

The amount of material which condenses on an opposing surface also depends on the position of the receiving surface with regard to the emission source. As shown in Fig. 10, the material contained in an evaporant beam of solid angle $d\omega$ covers an area which increases with distance as well as with the angle of incidence θ. The element of the receiving surface which corresponds to $d\omega$ is $dA_r = r^2\,d\omega/\cos\theta$. Therefore, the mass deposited per unit area is

$$\frac{d\mathfrak{M}_r(\varphi,\theta)}{dA_r} = \frac{\mathfrak{M}_e}{\pi r^2} \cos\varphi \cos\theta \tag{52}$$

The validity of the cosine law for the effusion of gases was first tested by Knudsen,[55] who showed that the expected distribution is obtained at vapor pressures low enough to permit a mean free path of at least ten times the diameter of the effusion opening. Mayer[56] confirmed the cosine-law distribution for the effusion of air, hydrogen, and carbon dioxide by measuring the momentum of impinging molecules at various angles φ opposite the opening with a torsion balance.

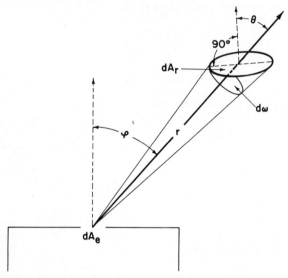

Fig. 10 Surface element dA_r receiving deposit from a small-area source dA_e.

The extension of the cosine law to emission from liquid or solid surfaces is generally taken to be permissible, thereby implying that free-evaporating molecules, too, have Maxwellian speed distributions. This assumption can be justified theoretically in those cases where evaporation takes place without surface constraints, i.e., if $\alpha_v = 1$. Then, the distribution of energies and directions of molecules emerging from a surface cannot differ from those which would impinge from an equilibrated vapor without violating the second law of thermodynamics. However, if the evaporation process is subject to entropy constraints, this argument cannot be applied, and deviations from the cosine law are possible (Ref. 35, p. 8). In practice, the condition $\alpha_v = 1$ holds for the free evaporation of simple solids and liquids with monatomic vapors, and therefore the cosine law may be extended to the greater portion of evaporant materials of current interest.

Experimental evidence to prove the validity of the cosine law for free evaporation was furnished by Knudsen,[57] who evaporated crystals of sulfur, zinc, silver, and antimony trisulfide from a point on the surface of a spherical glass flask. The experimental arrangement, which is shown schematically in Fig. 11, provides for $\cos\varphi = \cos\theta = r/2r_0$ at every point of the receiving surface. Thus, the amount of deposit according to Eq. (52) should be independent of the location on the receiving surface, and Knudsen observed that the glass sphere was indeed uniformly coated. More recently, Heald and Brown[58] analyzed the angular distribution of CO_2 molecules impinging on and being reemitted from a polished copper surface; their results, too, were in accord with the cosine law.

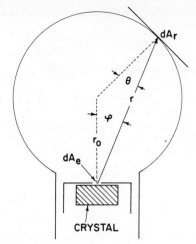

Fig. 11 Evaporation from a small-area source dA_e onto a spherical receiving surface.

(2) Emission from a Point Source As shown in the preceding section, molecules evaporating from a small surface element yield a directional emission pattern. At least hypothetically it is possible to change that condition and visualize molecules emerging from an infinitesimally small sphere ("point") of surface area dA_e. Provided the molecules have again a Maxwellian speed distribution at the moment of departure, they will move away from the point source in all directions with equal probability, as shown schematically in Fig. 12. Consequently, the mass evaporation

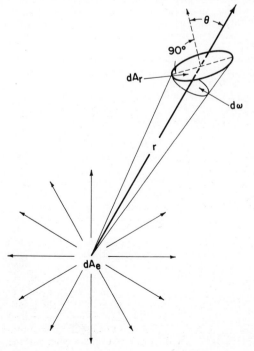

Fig. 12 Evaporation from a point source dA_e onto a receiving surface element dA_r.

rate Γ is uniform in all directions, and the mass contained in a narrow beam of solid angle $d\omega$ is

$$d^3\mathfrak{M}_e = \Gamma \, dA_e \, dt \, \frac{d\omega}{4\pi} \tag{53a}$$

or, since $\displaystyle\iint_{t, A_e} \Gamma \, dA_e \, dt$ is the total evaporated mass $\mathfrak{M}_e$,

$$d\mathfrak{M}_e = \mathfrak{M}_e \frac{d\omega}{4\pi} \tag{53b}$$

For a receiving element dA_r within the solid angle $d\omega$, the dependence on source distance and beam direction is the same as for a small surface source, $dA_r = r^2 \, d\omega / \cos \theta$, and therefore the amount of deposit received from a point source is

$$\frac{d\mathfrak{M}_r}{dA_r} = \frac{\mathfrak{M}_e}{4\pi r^2} \cos \theta \tag{54}$$

The receiving surface of uniform deposit is a sphere with the point source in the center so that $\cos \theta = 1$ and $r = \text{const.}$

The nondirectional emission pattern of an idealized point source is of limited interest insofar as it is a fair approximation for only a few practical sources. Equation (53a) is sometimes used to derive deposit distributions obtained by evaporating from linear or ring-shaped wire sources, which are then considered to form an array of point sources.

4. THE CONSTRUCTION AND USE OF VAPOR SOURCES

The evaporation of materials in a vacuum system requires a vapor source to support the evaporant and to supply the heat of vaporization while maintaining the charge at a temperature sufficiently high to produce the desired vapor pressure. The rates utilized for film deposition may vary from less than 1 to more than 1,000 Å s^{-1}, and the vaporization temperatures of any particular element differ accordingly. Rough estimates of source operating temperatures are commonly based on the assumption that vapor pressures of 10^{-2} Torr must be established to produce useful film condensation rates. For most materials of practical interest, these temperatures fall into the range from 1000 to 2000°C.

To avoid contamination of the deposit, the support material itself must have negligible vapor and dissociation pressures at the operating temperature. Suitable materials are refractory metals and oxides. Further selection within these categories is made by considering the possibilities of alloying and chemical reactions between the support and evaporant materials. Alloying is often accompanied by drastic lowering of the melting point and hence may lead to rapid destruction of the source. Chemical reactions involving compounds tend to produce volatile contaminants such as lower oxides which are incorporated into the film. Additional factors influencing the choice of support materials are their availability in the desired shape (wire, sheet, crucible) and their amenability to different modes of heating. The former determines the amount of evaporant which the source can hold (capacity), while the latter affects the complexity of the source structure and the external power supply.

Attempts to reconcile these demands have led to numerous source designs, which have been reviewed by Holland (Ref. 59, p. 108) and Behrndt.[60] Specialized sources of rather complex construction are nearly always required to evaporate quantities in excess of a few grams. Table 4 shows the temperatures of interest in this context for the more common elements, along with existing experience concerning suitable support materials. The melting points and approximate temperatures required to produce 10^{-2} Torr are taken from Honig.[19] For several elements, these data differ from the older values given in Holland's[59] widely used table. Pirani and Yarwood's[61] compilation of evaporation techniques for the elements has also been used. Most of the evaporation coefficients α_v are from Hirth and Pound[35] except those of Sb and and As.[62] Individual source materials and types of vapor sources are reviewed in more detail in the following sections.

TABLE 4   Temperatures and Support Materials Used in the Evaporation of the Elements

Element and predominant vapor species	Temp, °C		Support materials		Remarks
	mp	$p^* = 10^{-2}$ Torr	Wire, foil	Crucible	
Aluminum (Al)	659	1220	W	C, BN, TiB$_2$-BN	Wets all materials readily and tends to creep out of containers. Alloys with W and reacts with carbon. Nitride crucibles preferred
Antimony (Sb$_4$, Sb$_2$)	630	530	Mo, Ta, Ni	Oxides, BN, metals, C	Polyatomic vapor, $\alpha_v = 0.2$. Requires temperatures above mp. Toxic
Arsenic (As$_4$, As$_2$)	820	~300	....	Oxides, C	Polyatomic vapor, $\alpha_v = 5.10^{-5}$– 5.10^{-2}. Sublimates but requires temperatures above 300°C. Toxic
Barium (Ba)...	710	610	W, Mo, Ta, Ni, Fe	Metals	Wets refractory metals without alloying. Reacts with most oxides at elevated temperatures
Beryllium (Be)	1283	1230	W, Mo, Ta	C, refractory oxides	Wets refractory metals. Toxic, particularly BeO dust
Bismuth (Bi, Bi$_2$)	271	670	W, Mo, Ta, Ni	Oxides, C, metals	Vapors are toxic
Boron (B).....	2100 ± 100	2000	....	C	Deposits from carbon supports are probably not pure boron
Cadmium (Cd)	321	265	W, Mo, Ta, Fe, Ni	Oxides, metals	Film condensation requires high supersaturation. Sublimates. Wall deposits of Cd spoil vacuum system
Calcium (Ca)..	850	600	W	Al$_2$O$_3$	
Carbon (C$_3$, C, C$_2$)	~3700	~2600	....	....	Carbon-arc or electron-bombardment evaporation. $\alpha_v < 1$
Chromium (Cr)	~1900	1400	W, Ta	....	High evaporation rates without melting. Sublimation from radiation-heated Cr rods preferred. Cr electrodeposits are likely to release hydrogen
Cobalt (Co)...	1495	1520	W	Al$_2$O$_3$, BeO	Alloys with W, charge should not weigh more than 30 % of filament to limit destruction. Small sublimation rates possible
Copper (Cu)...	1084	1260	W, Mo, Ta	Mo, C, Al$_2$O$_3$	Practically no interaction with refractory materials. Mo preferred for crucibles because it can be machined and conducts heat well
Gallium (Ga)..	30	1130	....	BeO, Al$_2$O$_3$	Alloys with refractory metals. The oxides are attacked above 1000°C
Germanium (Ge)	940	1400	W, Mo, Ta	W, C, Al$_2$O$_3$	Wets refractory metals but low solubility in W. Purest films by electron-gun evaporation
Gold (Au).....	1063	1400	W, Mo	Mo, C	Reacts with Ta, wets W and Mo. Mo crucibles last for several evaporations
Indium (In)...	156	950	W, Mo	Mo, C	Mo boats preferred
Iron (Fe)......	1536	1480	W	BeO, Al$_2$O$_3$, ZrO$_2$	Alloys with all refractory metals. Charge should not weigh more than 30 % of W filament to limit destruction. Small sublimation rates possible
Lead (Pb).....	328	715	W, Mo, Ni, Fe	Metals	Does not wet refractory metals. Toxic
Magnesium (Mg)	650	440	W, Mo, Ta, Ni	Fe, C	Sublimates
Manganese (Mn)	1244	940	W, Mo, Ta	Al$_2$O$_3$	Wets refractory metals

TABLE 4 Temperatures and Support Materials Used in the Evaporation of the Elements (Continued)

Element and predominant vapor species	Temp, °C		Support materials		Remarks
	mp	$p^* = 10^{-2}$ Torr	Wire, foil	Crucible	
Molybdenum (Mo)	2620	2530	....	....	Small rates by sublimation from Mo foils. Electron-gun evaporation preferred
Nickel (Ni)....	1450	1530	W, W foil lined with Al_2O_3	Refractory oxides	Alloys with refractory metals; hence charge must be limited. Small rates by sublimation from Ni foil or wire. Electron-gun evaporation preferred
Palladium (Pd)	1550	1460	W, W foil lined with Al_2O_3	Al_2O_3	Alloys with refractory metals. Small sublimation rates possible
Platinum (Pt)	1770	2100	W	ThO_2, ZrO_2	Alloys with refractory metals. Multistrand W wire offers short evaporation times. Electron-gun evaporation preferred
Rhodium (Rh)	1966	2040	W	ThO_2, ZrO_2	Small rates by sublimation from Rh foils. Electron-gun evaporation preferred
Selenium (Se_2, Se_n: $n = 1-8$)[63]	217	240	Mo, Ta, stainless steel 304	Mo, Ta, C, Al_2O_3	Wets all support materials. Wall deposits spoil vacuum system. Toxic. $\alpha_v = 1$
Silicon (Si)....	1410	1350	....	BeO, ZrO_2, ThO_2, C	Refractory oxide crucibles are attacked by molten Si and films are contaminated by SiO. Small rates by sublimation from Si filaments. Electron-gun evaporation gives purest films
Silver (Ag)....	961	1030	Mo, Ta	Mo, C	Does not wet W. Mo crucibles are very durable sources
Strontium (Sr).	770	540	W, Mo, Ta	Mo, Ta, C	Wets all refractory metals without alloying
Tantalum (Ta)	3000	3060	....	....	Evaporation by resistance heating of touching Ta wires, or by drawing an arc between Ta rods. Electron-gun evaporation preferred
Tellurium (Te_2)	450	375	W, Mo, Ta	Mo, Ta, C, Al_2O_3	Wets all refractory metals without alloying. Contaminates vacuum system. Toxic. $\alpha_v = 0.4$
Tin (Sn)......	232	1250	W, Ta	C, Al_2O_3	Wets and attacks Mo
Titanium (Ti)	1700	1750	W, Ta	C, ThO_2	Reacts with refractory metals. Small sublimation rates from resistance-heated rods or wires. Electron-gun evaporation preferred
Tungsten (W)	3380	3230	....	....	Evaporation by resistance heating of touching W wires, or by drawing an arc between W rods. Electron-gun evaporation preferred
Vanadium (V).	1920	1850	Mo, W	Mo	Wets Mo without alloying. Alloys slightly with W. Small sublimation rates possible
Zinc (Zn).....	420	345	W, Ta, Ni	Fe, Al_2O_3, C, Mo	High sublimation rates. Wets refractory metals without alloying. Wall deposits spoil vacuum system
Zirconium (Zr)	1850	2400	W	....	Wets and slightly alloys with W. Electron-gun evaporation preferred

a. Wire and Metal-foil Sources

The simplest vapor sources are resistance-heated wires and metal foils of various types, as shown in Fig. 13. They are commercially available in a variety of sizes and shapes at sufficiently low prices to be discarded after one experiment if necessary. Materials of construction are the refractory metals, which have high melting points and low vapor pressures. Most commonly used are tungsten, molybdenum, and tantalum. Platinum, iron, or nickel are sometimes employed for materials which evaporate below 1000°C. As shown in Table 5, the volatility of the support materials will generally not be a problem except for molybdenum at temperatures above 1800°C. However, tungsten and molybdenum form volatile oxides by reaction with residual water vapor or dissolved oxygen and thus may contribute condensable impurities.

TABLE 5 Properties of Refractory Metals

Property	Tungsten	Molybdenum	Tantalum
Melting point, °C[19]...................	3380	2610	3000
T, °C, for $p^* = 10^{-6}$ Torr[19]..........	2410	1820	2240
Electrical resistivity,[64] 10^{-6} ohm-cm:			
At 20°C...........................	5.5	5.7	13.5
At 1000°C........................	33	32	54
At 2000°C........................	66	62	87
Thermal expansion, %[64]:			
From 0–1000°C...................	0.5	0.5	0.7
From 0–2000°C...................	1.1	1.2	1.5

The sources shown in Fig. 13a and b are commonly made from 0.02 to 0.06-in.-diameter tungsten wire. Their utility is limited to evaporants which can be affixed to the source, typically in the form of wire. Upon melting, the evaporant must wet the filament and be held by its surface tension. Spreading of the molten evaporant across the wire is desirable to increase the evaporation surface and thermal contact. This is encouraged by distributing the initial charge evenly over the entire length of the source. Suitable techniques are electroplating of the evaporant onto the filament, multistrand filaments where the evaporant is added as one of the fibers, or hanging short loops of evaporant wire onto each coil of a helix (Fig. 13b). Multistrand filaments are generally preferred because they offer a greater surface area than single-wire filaments. Because of their simplicity and low cost, wire sources are also used to evaporate metals such as Al, Ni, Fe, or Pt which are known to alloy with tungsten. Under these conditions, the evaporant charge should be small compared with the mass of the filament, and the latter is quickly destroyed if the molten evaporant collects in one spot.

Even if the evaporant does not alloy with the filament, the capacity of wire sources is small. Provided that uniform wetting and distribution of the melt over a few centimeters of filament length are obtained, such sources may hold amounts up to 1 g. Wire baskets as shown in Fig. 13c are used to evaporate pellets or chips of dielectrics or metals which either sublimate or do not wet the wire material upon melting. If wetting occurs, the turns of the basket are shorted and the temperature of the source drops.

Metal foils as shown in Fig. 13d, e, and f have capacities of a few grams and are the most universal types of sources for small evaporant quantities. They are fabricated from 0.005- to 0.015-in.-thick sheets of tungsten, molybdenum, or tantalum. The dimpled sources have reduced widths in the center to concentrate the heating in the area of the evaporant. The recessed circles or ovals are only ⅛ to ¼ in. deep since refractory metals are not easily drawn to greater depths. Canoe or boat sources may be made in the laboratory by bending metal sheets into the desired shape. This is readily done with tantalum and not too difficult with molybdenum. Tungsten, however, is very brittle and breaks if it is bent at room temperature. Deformation is possible at elevated temperatures in a nonoxidizing atmosphere (Ref. 59,

p. 109). All three metals embrittle after having been heated in vacuum, especially if alloying with the evaporant has occurred. Their mechanical properties are reviewed in detail by Kohl.[64]

Wetting of the metal-foil surface by the molten evaporant is desirable in the interest of good thermal contact. However, molten metals also lower the electrical resistance of the foil in the contact area, thereby causing the temperature to drop. This problem is avoided with oxide-coated foils as shown in Fig. 13e. Oxide-coated metal sources were first described by Olsen et al.,[65] who embedded tungsten-wire baskets in sintered Al_2O_3 or BeO. Nowadays oxide coatings are applied by plasma-spray processes. Typical commercial sources consist of 0.010-in. Mo foils (or sometimes Ta) covered with an equally thick layer of alumina. The oxide layer is claimed to be non-porous and withstands temperature cycling without chipping. Maximum operating temperatures of 1850 to 1900°C are possible, but the vapor pressure of molybdenum

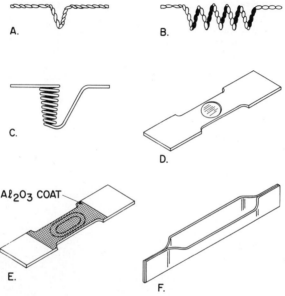

Fig. 13 Wire and metal-foil sources. (A) Hairpin source. (B) Wire helix. (C) Wire basket. (D) Dimpled foil. (E) Dimpled foil with alumina coating. (F) Canoe type.

exceeds 10^{-6} Torr in this range and that of alumina is even higher. The power requirements of such sources are 30 to 50% above those of comparable uncoated foils since the thermal contact between the resistance-heated metal and the evaporant is reduced by the oxide coat. Molten metals do not wet the alumina surface but coalesce into a sphere. Whereas alloying is inhibited, the possibility of forming volatile oxides from the Al_2O_3 and the evaporant metal should be considered.

Electrical connections to wire and foil sources are made by attaching their ends to heavy copper or stainless-steel clamps. The latter are usually part of massive metal bars connected rigidly to a pair of electrical feedthroughs. To avoid warping of the source due to thermal expansion (see Table 5), flexible cables such as braided copper wire or multilayered copper foil are often used. Since the electrical resistance of wire and foil sources is small, low-voltage power supplies rated at 1 to 3 kW are required. Typical arrangements consist of a step transformer (5 to 20 V) whose primary side is connected to a variable 110- or 220-V transformer. The latter is necessary to raise the operating voltage as the resistance of the refractory metals increases strongly with temperature (see Table 5). The secondary current may be as low as 20 A for some of the wire sources, or as high as 500 A for some foil sources.   If the source current exceeds 100 A, it is advisable to use water-cooled feedthroughs.

There are wire or foil sources to evaporate small charges of nearly all elements except the refractory metals themselves.   Films of W, Mo, and Ta have been prepared by spring-loading two wires (0.02 in. diameter) of the respective metals such that their ends touch and form a high-resistance junction.   Melting of the junction and evaporation are induced by passing an electric current through the arrangement.[66] An alternate technique is vacuum-arc evaporation.   Lucas et al.[67] established short arcs between zone-purified rods of Ta, Nb, and V by means of a dc welding generator. A similar arrangement has been described by Massey,[68] who sustained an arc between two carbon rods and produced carbon films up to 1 μ thick.   In general, however, heating by electron bombardment is the preferred technique to evaporate refractory metals.

b. Sublimation Sources

The problem of finding nonreactive supports for materials evaporating above 1000°C can be circumvented in those cases where the 10^{-2} Torr vaporization temperature is near the melting point of the evaporant.   The elements Cr, Mo, Pd, V, Fe, and Si reach vapor pressures of 10^{-2} Torr before they melt and hence can be sublimated. This allows evaporation from wires or foils of the film materials by direct resistance heating without contact with any foreign material.   High-purity nickel and iron films have been prepared in this manner by Behrndt.[69]   Kilgore and Roberts[70] produced silicon films by sublimation from resistance-heated ribbons at 1300 to 1350°C.   The technique is of particular interest for silicon because the molten element rapidly alloys with and destroys refractory-metal filaments.   Although silicon has also been evaporated from ThO_2, Al_2O_3, and SiO_2 boats,[71] it has a strong tendency to attack refractory oxides and to form volatile SiO.   Thus, without resorting to electron-gun evaporation, high-purity silicon films are best obtained by sublimation.

The sublimation technique can be extended to a few more metals with vapor pressures in the 10^{-3} Torr range at their respective melting points.   Examples are Ni,[72] Rh, and Ti.   Sublimation of titanium is widely practiced in vacuum technology to getter chemically reactive gases, an application which requires only small deposition rates (see Chap. 2).   For the purpose of depositing films with thicknesses of 1,000 Å or more, however, the sublimation rates are generally too small.   This disadvantage can be somewhat compensated by using filaments of large surface area and by reducing the source-to-substrate distance to a few centimeters.   But an enlarged sublimation surface increases the power requirement of the source, and the short evaporation distance reduces the area of uniform deposit thickness while it enhances radiation heating of the substrates.   For these reasons, sublimation of metals is not widely practiced.

An important exception from this rule is chromium because its vapor pressure reaches 10^{-2} Torr about 500°C below the melting point.   Consequently, high deposition rates can be obtained by sublimation from compact sources and with substrate distances of the order of 50 cm.   The simplest arrangement is to place chromium chips in a tungsten or tantalum boat.[73]   However, since the chips do not melt, their thermal contact with the filament is poor and variable, which makes it difficult to maintain constant evaporation rates.   Better heat contact is obtained if the chromium is electroplated onto a tungsten filament.[74]   Heavily chrome-plated tungsten rods are also commercially available.   However, electrodeposited metals tend to occlude significant amounts of hydrogen, which may contribute substantially to the residual gas load of the vacuum system.

A practical chromium sublimation source has been developed by Roberts and Via.[75] As shown in Fig. 14, a chromium rod is supported within a dual-wall cylinder.   The latter is fabricated from pieces of 0.002 in. (inner cylinder) and 0.005 in. (outer cylinder) thick tantalum sheet which are spot-welded together to form a continuous current path.   Since the source is compact and well shielded, the power requirements are less than 750 W.   Sublimation occurs from the entire surface area of the rod, which is uniformly heated by radiation.   Therefore, ejection of particles does not occur, and the evaporation rates are constant over extended periods of time.   By increasing the voltage, condensation rates of 100 Å s⁻¹ or more across normal bell-jar distances may be obtained.   The source capacity amounts to at least 50% of the initial rod charge. The latter may be prepared from chromium powder by pressing and sintering, but the purest films are obtained from zone-refined rods.

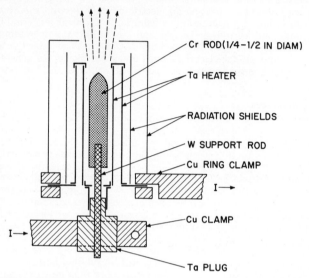

Cr ROD(1/4-1/2 IN DIAM)

Ta HEATER

RADIATION SHIELDS

W SUPPORT ROD

Cu RING CLAMP

I→

Cu CLAMP

I→

Ta PLUG

Fig. 14 Chromium sublimation source after Roberts and Via.[75]   The electric current flows through the tantalum cylinder (heavy lines).

Sublimation sources are frequently used for the vaporization of thermally stable compounds.   These are commonly obtained as powders or loosely sintered chunks containing large quantities of absorbed or occluded gases.   Upon heating, spontaneous release of gases often causes violent ejection of evaporant particles which are incorporated into the films.   To avoid this problem, baffled sources have been designed which inhibit direct line-of-sight transmission from the evaporant charge to the substrates. The simplest of these sources are made by spot welding two dimpled tantalum foils together to form an enclosure.   The upper dimple has an aperture to allow the vapor to escape.   Within the enclosure and opposite the aperture is a small sheet-metal baffle to restrain evaporant particles.   These sources are suitable for the evaporation of moderate quantities of II-VI compounds.[76]

Large source capacities and high evaporation rates are often an important requirement in the deposition of dielectric films.   The first source to satisfy these objectives was the well-known Drumheller[77] design shown in Fig. 15a.   It is charged with SiO chunks packed around a resistance-heated tantalum cylinder.   The latter is perforated and acts as a chimney through which the externally generated SiO vapors escape. A ring-shaped lid prevents the ejection of oxide particles.   In many ways, it is an example of good source design: compact, well shielded for minimum power requirements, of large capacity, demanding little maintenance, and capable of yielding high rates because of its large internal surface area.   Accordingly, it has become one of the most widely used sources in the laboratory.

Since 0.001- to 0.005-in. tantalum sheet can easily be cut, bent, and spot-welded, dielectric sublimation sources of the "optically dense" type exist in a great variety of designs.   Vergara et al.[78] introduced a new family of these sources, where the generation and emission of vapor take place in two separate parts of the source.   An example is shown in Fig. 15b.   A disadvantage compared with the Drumheller source is the need for a second power supply to heat the vapor-emitting box. But present vendor catalogs list many designs also based on functional separation by internal baffles, where both compartments form one continuous current path and are energized by only one power supply. Although primarily developed for SiO, these sources are also suitable to evaporate the sulfides, selenides, and tellurides of Zn and Cd as well as MgF_2.

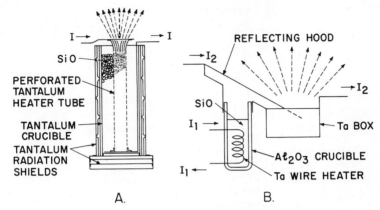

Fig. 15 Optically dense SiO sources. (A) The Drumheller[77] source. (B) Compartmentalized source. (After Vergara, Greenhouse, and Nicholas.[78])

c. Crucible Sources and Materials

(1) Refractory Metals Crucible sources are required to support molten metals in quantities of a few grams or more. Since the melt is in contact with the container for prolonged periods of time, the selection of thermally stable and noncontaminating support materials must be made very carefully. The compatibility of molten evaporants with refractory metals can be assessed by studying the phase diagram of the two materials.[79] The metal combination should have very small mutual solubilities and no low-melting eutectic. If a suitable refractory metal can be found, it is preferable over oxide crucibles since slight interaction of the latter with molten metals and vaporization of contaminant oxides are sometimes difficult to recognize. Moreover, metal crucibles can be obtained in a greater variety of shapes and sizes, and they are better heat conductors and also less sensitive to rapid temperature changes than oxide crucibles.

Referring back to Table 4, a number of metals can be evaporated from molybdenum crucibles. Of the three refractory metals, molybdenum is the least expensive and can be machined more easily than tungsten. A source design which is particularly useful for the evaporation of Cu, Ag, or Au is shown in Fig. 16. The solubilities of these three metals in molybdenum and vice versa are very small. Hence the crucible is not attacked, and the author has obtained up to 50 copper depositions from the same container. The stability toward silver appears to be equally good, while gold has been observed to penetrate the molybdenum walls after fewer than 10 evaporations. The resistance-heated filament consists of two halves cut from tantalum foil and spotwelded together as shown in the upper part of Fig. 16. Held by copper clamps at both ends, it maintains its cylindrical shape without contacting other parts of the source. An additional advantage of the meandering sheet-metal filament over a wire helix is its larger radiating surface. Because of the multiple heat shields, a power input of about 500 W is sufficient for Cu or Ag evaporations. The rates are very stable and respond well to feedback control from ionization or crystal-oscillator rate monitors.

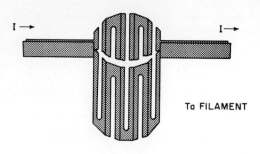

Ta FILAMENT

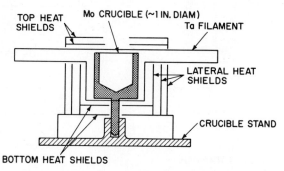

Fig. 16 Molybdenum crucible source with meandering tantalum sheet filament.

(2) Refractory Oxides Refractory-oxide crucibles are available in certain standard shapes and sizes from manufacturers of vacuum accessories.[80] A thorough review of refractory oxide properties has been given by Kohl,[64] and the data in Table 6 are selected from that source. The physical properties of ceramic oxides show consider-

TABLE 6 Properties of Refractory Oxides*

Oxide	Porosity, vol %	Fusion temp, °C	Max normal use temp, °C	Thermal conductivity, cal s⁻¹ deg⁻¹ cm⁻² cm	Thermal stress resistance
Sintered ThO_2............	3–7	3050	2500	0.007	Fair–poor
Sintered BeO.............	3–7	2570	1900	0.046	Excellent
Sintered ZrO_2 (stabilized[92]:					
ZrO_2, $4HfO_2$, $4CaO$)	3–10	2550	2200	0.005	Fair–good
Sintered Al_2O_3............	3–7	2030	1900	0.014	Good
Sintered MgO.............	3–7	2800	1900	0.016	Fair–poor
Vitreous SiO_2.............	0	1710	1100	0.012	Excellent
Sintered TiO_2.............	3–7	1840	1600	0.008	Fair–poor

* From Kohl.[64]

able variations depending on the manufacturing process. Generally, the purest products have superior thermal stability and mechanical strength. Oxide crucibles for vacuum evaporation are typically sintered from powder of at least 99.8% purity and retain a small pore volume. Wall thicknesses are of the order of 1 to 3 mm, and high thermal conductivities are desirable for reasons of heat transfer as well as thermal shock resistance. The fusion temperatures in Table 6 are somewhat lower than the melting points of the pure oxides but still far above those temperatures at which these materials can be safely used. Similarly, the maximum normal use temperatures in Table 6 do not necessarily hold for vacuum applications since there may be additional restrictions.

An important factor in assessing the suitability of an oxide support material is its thermodynamic stability. The first five oxides listed in Table 7 have sufficiently large free energies of dissociation to maintain oxygen partial pressures of less than 10^{-12} Torr at 2000°C. TiO_2 and NiO are not useful evaporation-support materials because they decompose too readily. The utility of SiO_2 is limited because it gives up its oxygen to several metals which are capable of forming oxides of greater thermodynamic stability. Thus, SiO_2 in contact with Al or Mg is rapidly attacked even below 1000°C. Other metals such as Ag, Au, or Pt do not react chemically but require evaporation temperatures which are too close to or even above the softening point of vitreous silica. Consequently, only the first five oxides in Table 7 merit serious

TABLE 7 Thermodynamic Stability of Refractory Oxides as Represented by Their Free Energies of Dissociation at 1800°C[81]

Dissociation Process	Std Free Energy at 1800°C, kcal/g-at of oxygen
$\frac{1}{2}ThO_2(s) = \frac{1}{2}Th(s) + \frac{1}{2}O_2(g)$	+111
$BeO(s) = Be(l) + \frac{1}{2}O_2(g)$	+100
$\frac{1}{2}ZrO_2(s) = \frac{1}{2}Zr(s) + \frac{1}{2}O_2(g)$	+ 84
$\frac{1}{3}Al_2O_3(s) = \frac{2}{3}Al(l) + \frac{1}{2}O_2(g)$	+ 82
$MgO(s) = Mg(g) + \frac{1}{2}O_2(g)$	+ 53
$\frac{1}{2}SiO_2(s) = \frac{1}{2}Si(l) + \frac{1}{2}O_2(g)$	+ 49
$3TiO_2(s) = Ti_3O_5(s) + \frac{1}{2}O_2(g)$	+ 18
$NiO(s) = Ni(l) + \frac{1}{2}O_2(g)$	+ 13

consideration. Within this group, the sublimation pressure of some of the oxides imposes an upper temperature limit on their utility in vacuum. MgO, for instance, sublimates noticeably at 1600 to 1900°C, Al_2O_3 at 1900°C, and BeO at 1900 to 2100°C.

In principle, the compatibility of an evaporant metal with an oxide support material can be judged by comparing the free energies of dissociation of both metal oxides at the anticipated use temperature. However, the kinetics of the reactants cannot be predicted, and in some cases the reaction may be slow enough to permit limited use of a combination whose thermodynamic stability is marginal. Unfortunately, not all the metal–metal oxide interface reactions of potential interest have been investigated. Johnson[82] studied the stability of 21 metal–metal oxide pairs in vacuum at temperatures from 1500 to 2300°C. Economos and Kingery,[81,83] in a similar study, classified the observed interface reactions according to their severity. The slightest forms of interaction are discoloration of the oxide surface and metal penetration along the oxide grain boundaries. More severe reaction may result in visible corrosion of the oxide surface or in the formation of a new phase at the interface. The significance of these observations for the evaporation of metals from oxide crucibles is not always obvious. Relatively heavy interaction with new-phase formation, for example, may be acceptable if none of the reaction products is volatile. Conversely, the absence of even discoloration at the interface may obscure the fact that a volatile suboxide has been emitted. A stability matrix which combines the observations of Johnson,[82] Economos and Kingery[81,83] with data from Kohl[64] is shown in Table 8.

The stability matrix demonstrates that the refractory metals W, Mo, and Ta cannot be evaporated from oxide supports. The data are also of interest in judging the stability of refractory-metal filaments in contact with oxide crucibles or boats. Among the

TABLE 8 Thermal Stability of Refractory Oxides in Contact with Metals*

Metal	Temp, °C for 10^{-2} Torr	Refractory oxides				
		ThO$_2$	BeO	ZrO$_2$	Al$_2$O$_3$	MgO
W....	3230	Slight reduction at 2200°C	Stable at 1700°C; reaction >1800°C	Stable <1600°C; little reaction up to 2000°C	Limited by the onset of Al$_2$O$_3$ sublimation at 1900°C	Limited by the onset of MgO sublimation at 1600–1900°C
Mo...	2530	Little reaction up to 2300°C	Stable at 1900°C but not above	Stable at 2000°C; ZrO$_2$ decomposes at 2300°C		
Ta....	3060	Stable up to 1900°C	Stable up to 1600°C	Stable up to 1600°C		
Zr....	2400	Interaction begins at 1800°C	Interaction begins at 1600°C	Slight interaction at 1800°C	Oxide attacked at 1600°C	
Be....	1230	Only slight interaction at **1600°C**			Oxide discolored at **1400°C**	Stable at **1400°C** but not at **1600°C**
Si.....	1350	Little or no attack at **1400°C**; noticeable reaction at 1600°C			Slight reaction at 1400°C	Slight reaction at 1400°C, strong at 1600°C
Ti....	1750	Slight interaction at **1800°C**	Little reaction at 1600°C but considerable at 1800°C		Little reaction at 1400°C but considerable at 1800°C	
Ni....	1530	Ni(l) is stable in contact with all oxides at **1800°C**				

Boldface type indicates metal-oxide pairs which can be used for thin film deposition.
* After Johnson,[82] Economos and Kingery,[81,83] and Kohl.[64]

other metals, only Ni and Be are either nonreactive or have sufficiently slow reaction kinetics to be safely evaporated from oxide boats.   The cases of Si and Ti in some of the most stable oxides are marginal.   Although the oxide supports are not strongly attacked, both elements are known to form volatile suboxides, which leads one to expect deposits with a certain oxide content.   There are, however, many other metals not listed in Table 8 which can be evaporated from refractory oxide enclosures. Examples are As, Sb, Bi, Te, Ca, Mn, and others with vaporization temperatures below 1000°C, as well as Co, Fe, Pd, Pt, Rh which require temperatures between 1500 and 2100°C.

Oxide crucibles are commonly heated by radiation from refractory-metal filaments. The simplest arrangement is a wire coil as shown in Fig. 17.   The filament touches the crucible in several places, thus enhancing the heat transfer by conduction.   However, this involves the risk of interaction between filament and oxide.   The source should be surrounded by radiation shields since the filament emits only a small fraction of its energy in the direction of the crucible.

Fig. 17 Oxide crucible with wire-coil heater.

A more efficient design is the DaSilva[84] source shown in Fig. 18. It is made from 0.0025-in. Ta foil rolled into a cylinder and spot-welded to circular top and bottom electrodes. The source requires cylindrical crucibles and yields rates which can be well controlled through the power input to the cylinder. With proper heat shielding, the power requirements are less than 500 W. Many other heater arrangements, for example, modifications of the crucible source shown in Fig. 16, are possible.

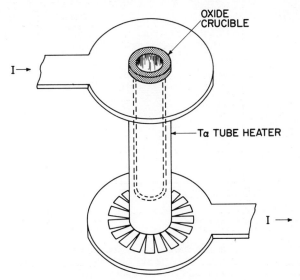

Fig. 18 DaSilva[84] crucible source.

(3) Boron Nitride A number of borides, carbides, silicides, and nitrides have excellent thermal stabilities. Reviews of their physical properties and availability have been given by Hauck[85] and Kohl.[64] One of these "space-age materials," boron nitride (BN), has found application for evaporation crucibles. It is a white insulating material with a structure similar to graphite. Like the latter, it is relatively soft and can be machined with ordinary cutting tools into a great variety of shapes. The thermal conductivity of BN is similar to that of alumina, and its heat-shock resistance is excellent. Its mechanical strength is about one-half that of alumina. It has a tendency to absorb water and therefore requires outgassing prior to metal evaporation. The maximum use temperature is determined by the nitrogen dissociation pressure, which reaches 10^{-2} Torr at 1600°C and exceeds 10^{-1} Torr at 1800°C.[86]

The density and also the amount of gas evolving when boron nitride is heated depend on the manufacturing process. Pyrolytic boron nitride* deposited from boron halide and ammonia has the highest purity and density and is impervious to gases. Manufacturing processes which sinter powders with organic binders yield products with a stronger tendency to contribute impurities to the evaporant. A composition consisting of 50% of BN and titanium diboride (TiB_2) each is also well established as a crucible material.† Titanium diboride has a thermal conductivity similar to BN but is harder and melts at 2940°C. The composite material is gray and machinable like the pure nitride. Its electrical resistivity at 1100 to 1200°C is of the order of a few milliohm-centimeters.[87]

The most important application of boron nitride crucibles is in the evaporation of aluminum. Contrary to nearly all ceramic oxides, neither the nitride nor the com-

* For example, Boralloy, Union Carbide Corp., New York, N.Y.
† Marketed as HDA Composite Ceramic by Union Carbide Corp., New York, N.Y.

posite is appreciably attacked by molten aluminum. Evaporation of the latter from a boron nitride crucible has been described by Thompson and Libsch,[88] while Ames et al.[87] used the composite ceramic. Both groups of investigators employed rf induction heating. The crucible developed by Ames et al.[87] is of particular interest and is shown in Fig. 19. It is designed to avoid the well-known problem of molten aluminum

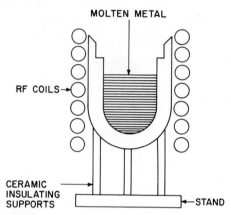

MOLTEN METAL

RF COILS→

CERAMIC
INSULATING
SUPPORTS

←STAND

Fig. 19 RF heated aluminum source with boron nitride–titanium diboride crucible. (*After Ames, Kaplan, and Roland.*[87])

migrating upward and spilling over the brim of the container. This is overcome by the reduced thickness of the upper portion of the crucible, which is approximately one-third of the skin depth of the rf field. Consequently, coupling in this region is sufficiently strong to affect complete vaporization of the migrating aluminum layer. The wall thickness of the lower portion is about equal to the skin depth. It permits adequate coupling to evaporate the metal but shields the melt from the rf field enough to minimize turbulence and droplet ejection. The source is capable of producing reliably spatter-free aluminum films of bulk resistivity.

Rf induction heating, whether in conjunction with boron nitride, refractory oxide, or metal crucibles, has some advantages over radiation heating. Since at least part of the energy is coupled directly into the evaporant metal, it is not necessary to produce filament or crucible temperatures in excess of the vaporization temperature for the purpose of maintaining a heat flow. Thus, interactions between evaporant and container walls are minimized. Also, the utilization of the energy supplied to the source is more efficient than in the case of filament heating since radiation and conduction heat losses are smaller. The need for radiation shields is obviated by the water-cooled coupling coils surrounding the crucible.

Disadvantages are the greater cost and space requirement of the rf power supply, the need for more thoroughly insulated vacuum feedthroughs, and the necessity of having to experiment with coil spacing, diameter, and external tuning for efficient coupling. For guidelines in the design of rf heated sources, their power requirements and depth of penetration as functions of frequency, the reader is referred to the book by Pirani and Yarwood.[61] The essential consideration is that most of the power is induced near the surface of the charge. The thickness of the layer, within which two-thirds of the energy is generated, is called the skin depth. The latter is inversely proportional to the square root of the radio frequency [see Chap. 2, Sec. 4a(3)]. Therefore, uniform heating of large metal charges is effectively accomplished with lower frequencies, while small or thin objects require very high frequencies. To give some examples, commercial equipment operating at about 10 kHz is suitable to heat metal charges of several pounds. Quantities of several grams require generators operating at a few hundred kilohertz, while satisfactory energy transfer into charges of a few cubic millimeter volume (milligram quantities) is obtained at frequencies greater than 50 MHz.[89]

(4) Carbon Another material frequently used for the evaporation of metals is carbon. It is available in several forms with properties varying widely as a result of different manufacturing processes. An extensive review of carbon products has been given by Kohl,[64] and the properties of possible interest in vacuum evaporation are listed in Table 9.

TABLE 9 Properties of Carbon[64,82]

Vapor pressure:
At 1900°C	$\sim 10^{-6}$ Torr
At 2500°C	$\sim 10^{-2}$ Torr

Density, commercial graphites:
	1.6–1.8
Vitreous carbon	1.3–1.5
Pyrolytic graphite	Up to 2.26
Theoretical	2.265

Electrical resistivity, 10^{-3} ohm-cm:
Commercial graphites	0.5–2.5
Vitreous carbon	1–5
Pyrolytic graphite	0.5–500

Coefficient of thermal expansion, 10^{-6} deg^{-1}:
Commercial graphites	2–6
Vitreous carbon	2–4
Pyrolytic graphite	<1–10

Thermal stability in contact with metals and oxides:
W	Carbide formation >1400°C
Mo	Carbide formation >1200°C
Ta	Carbide formation >1000°C
Pt, Pd, Rh	Reactions at temperatures closely below the eutectic melting points
Al	The molten metal forms carbides
Si, Ti	The molten metals react rapidly and form nonvolatile carbides
Cu, Be	Can be evaporated without reaction
Alkali metals	The molten metals penetrate graphite and form compounds
ThO_2	Reduction observed after 4 min at 2000°C
BeO	Reduction observed after 2 min at 2300°C
ZrO_2	Reduction observed after 4 min at 1600°C
MgO	Reduction observed after 8 min at 1800°C

Ordinary commercial graphites have a large pore volume and are relatively soft. Pyrolytic graphite is denser and harder. A relatively new material is vitreous carbon,[90] which is prepared by thermal decomposition of organic polymers. Contrary to graphite, which has a dull surface and tends to flake or abrade easily, vitreous carbon has a high luster, does not flake, and yields conchoidal fractures. Its density is relatively low because of a pore volume of about 30%.[91] But the pores are of spherical shape and not accessible to gases; hence the material has a very small permeability even for helium. The purity and ash content (<50 ppm) are superior to ordinary graphites (several hundred ppm). Vitreous carbon has been recommended for melting of III-V compounds, tellurides, and MgF_2. It is not wetted by molten aluminum and even resists the attack of molten Na_2O_2 for a limited time. Although it appears quite promising as an evaporant-support material, very little has been reported to date about its actual usage. Vitreous carbon is available in the form of crucibles or thin plates which can be machined.*

All carbon products are electrically conductive and may be resistance or induction heated. They are also good heat conductors and have low coefficients of thermal expansion; hence their thermal shock resistance is excellent. The various graphites can be machined into intricate shapes such as slotted bars or crucibles for direct resistance heating. However, objects of small cross sections are fragile.

* For example, from The Plessey Company (U.K.) Ltd.; or from Gallard-Schlesinger Chemical Mfg. Corp., Carle Place, L.I., N.Y.

While the vapor pressure of carbon permits vacuum applications up to 2000°C, the practical limits are often much lower.   As shown in Table 9, many metals including the refractories form carbides or eutectics at moderate temperatures.   Another problem is the large internal surface area of ordinary graphites.   The gases released upon heating are CO_2, CO, N_2, and H_2, and complete outgassing is obtained only at temperatures above 1700°C.   In evaporating copper from graphite crucibles, the author encountered very persistent gas evolution, which was sufficiently violent to eject micron-size droplets of copper onto substrate surfaces at 20 in. distance.   This difficulty should not occur with some of the denser "vacuum-tight" graphites.

A study of aluminum evaporation from graphite boats has been published by Moriya et al.[92]   The molten metal wets graphite and flows over the brim, thus continually changing the effective evaporation area.   The flow can be prevented by fitting the brim with an Al_2O_3 barrier since the contact angle between liquid aluminum and the oxide is nearly 90°.   Another problem is the formation of Al_4C_3, which floats on the melt and thereby reduces the free surface area.   The addition of 1 to 2% titanium to the melt was found to suppress this reaction.   Finally, to achieve a constant liquid level for uniform evaporation rates, the charge was continually replenished by feeding aluminum wire into the boat.   Wire-feed mechanisms are commercially available[93] and also useful to enhance the capacitance of small sources.[94,95]

The application of carbon heaters can be extended to materials which normally form carbides if ceramic inserts are used.   As shown at the bottom of Table 9, the reduction temperatures of refractory oxides approach or exceed 2000°C.   Boron nitride, too, is stable in contact with carbon.   Hemmer and Piedmont[96] built the evaporation source shown in Fig. 20, which consisted of a dual-wall resistance-heated graphite cylinder with a boron nitride liner.   The tantalum reflectors allowed temperatures of 2000°C to be obtained at 700 W power input.   The maximum temperature achieved was 2600°C, and compounds such as SiO_2, CeO_2, and Si_3N_4 were evaporated.   However, in view of the vapor pressure of carbon and the dissociation of BN, the resulting films were probably not very pure.

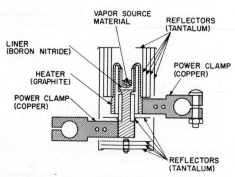

Fig. 20. Graphite evaporation source with boron nitride liner. (*After Hemmer and Piedmont.*[96])

d. Electron-bombardment Heated Sources

Instead of supplying energy by resistance or induction heating, vaporization of substances can also be accomplished by electron bombardment.   A stream of electrons is accelerated through fields of typically 5 to 10 kV and focused onto the evaporant surface.   Upon impingement, most of the kinetic particle energy is converted into heat, and temperatures exceeding 3000°C may be obtained.   Since the energy is imparted by charged particles, it can be concentrated on the evaporating surface while other portions of the evaporant are maintained at lower temperatures.   Hence, interactions between evaporant and support materials are greatly reduced.

Experimentally, the method may be implemented in a number of ways.   This is reflected in the great variety of source arrangements described in the literature and reviewed by Holland[97] and Behrndt.[60]   The sources may be classified according to the modes of electron acceleration or by the techniques used to support the evaporant. The choice of the latter is fairly independent of the former and allows considerable freedom of combination, as will be illustrated by examples in each category.

Devices operating on the principle of electron-bombardment heating are referred to as electron guns.   A hot cathode is universally employed as the electron source, with tungsten wire the preferred filament material because it retains its strength and shape at the high temperatures required for efficient electron emission.[98]   The filament life is limited by possible reactions with the evaporant vapors and by sputtering due to impingement of high-energy positive ions.   Hence, electron guns should be constructed so that the filament can be replaced easily.

A potential of several kilovolts is needed to accelerate the electrons emitted from the cathode.   If the electric field is maintained between the cathode and the evaporant (the "work"), the structure is called work-accelerated.   The alternative is self-accelerated guns.   These have a separate anode with an aperture through which the electron beam passes toward the work.   The electron energy is sufficient to ionize residual gas or evaporant molecules encountered along the way.   Since ionization causes loss of beam energy and focus, the pressure in the vacuum chamber must be below 10^{-4} Torr.   Additional problems due to ionization are specific for particular types of gun structures.

(1) Work-accelerated Electron Guns   Typically, work-accelerated structures have a hot cathode in form of a wire loop in close proximity to the evaporant.   The electrons converge radially upon the work.   The simplest arrangement is the pendant-drop configuration introduced by Holland[99] and shown in Fig. 21a.   The metal to be evaporated must be in rod or wire form and centered within the cathode loop.   Evaporation takes place from the molten tip, and the substrates are located below the source.   The drop of molten metal at the tip is held by its surface tension and requires careful control of the electric energy supplied.   If the drop temperature exceeds the melting point too far, the metal rod melts back and the drop falls off.   Therefore, applicability of the method is limited to metals with high surface tension and vapor

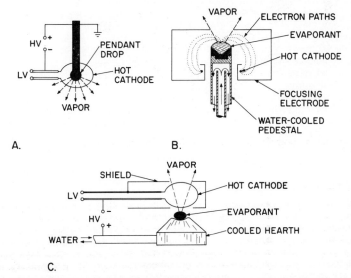

Fig. 21 Work-accelerated electron-bombardment sources. (*A*) Pendant-drop method. (*B*) Shielded filament (*Unvala*[103]).   (*C*) Shielded filament (*Chopra and Randlett*[104]).

pressures greater than 10^{-3} Torr at their melting points. The evaporation rates from molten drops of maximum supportable size have been measured by Heavens.[100] They decrease in the order Fe, W, Ni, Ta, Ag, Mo, Pt. Titanium[97] and zirconium[101] are also amenable to this technique.

In addition to being rate-limited, the pendant-drop method poses other difficulties. Some of these are apparent in a more complex design described by Milleron.[102] In this version, Mo or Ti wire is supplied from a spool and fed upward through a cooled copper tip which aids in maintaining a stable molten ball of metal. The wire-feed rate and the power input must be matched to the evaporation rate as determined by the vapor pressure and size of the molten drop. Heavy outgassing upon melting is suppressed by resistance heating of the wire on its way up from the spool to the guiding tip.

Another problem arises from exposure of the cathode to high vapor densities generated by the hot source. The resulting ionization leads to cathode erosion by sputtering with ensuing film contamination. To avoid these difficulties, it has become customary to offset the cathode filament away from the work and to surround it by metal shields. These are arranged so that straight line-of-sight transmission from the filament to the source as well as to the substrates is eliminated. Furthermore, the shields are at cathode potential to repel electrons and force them into curved paths ending on the evaporant.

Although cathode shielding and electrostatic focusing have been used in conjunction with the pendant-drop method,[101] it has found much wider application in work-accelerated guns with water-cooled supports. As shown in Fig. 21b and c, the cathode loop may be below (Unvala and Booker[103]) or above (Chopra and Randlett[104]) the work. Configurations similar to Fig. 21c are relatively easy to implement and have been used to evaporate such metals as Ta,[105] Si,[71] and Mo.[106] The high voltage is generally applied to the cathode, whose power supply must therefore be electrically insulated, while the evaporant support is grounded. If the cathode is grounded, the strongest electric field coincides with the region of highest vapor density surrounding the work, a condition which often results in a glow discharge.

Even with the potential applied to the cathode, work-accelerated guns are rate-limited because the vapor density decreases only moderately across the short distance between source and filament. The ionization of vapor within the electrostatic shield produces a space charge which reduces the accelerating voltage. At higher evaporation rates, a glow discharge between the cathode and the bell-jar walls may be established. This instability makes the control of power input and evaporation rates somewhat difficult. The close proximity of vapor source and filament shield is also responsible for heavy accumulation of deposits on parts of the shield. Hence, frequent cleaning or change of shields is necessary. To avoid these shortcomings, electron guns have been designed where the cathode and accelerating structure are remote from the evaporant source.

(2) Self-accelerated Electron Guns This class of guns is similar to and actually evolved from x-ray tubes. A small tungsten helix or hairpin filament constitutes the electron source. Rough focusing may be achieved electrostatically by means of a cylindrical shield and an anode disk as shown in Fig. 22a. The space between the anode and the work is field-free. Increased pressure due to outgassing in the anode-cathode region, however, may lead to ionization and defocusing. If the beam is to traverse longer distances, or focusing into a very small spot is desired, better control is obtained with a magnetic lens. An example is shown in Fig. 22b, where the electron beam is focused by a negatively biased filament shield, a conical anode, and a magnetic coil. For purposes of evaporation, focal spots of a few millimeters diameter are used. Telefocus guns of the type shown in Fig. 22b have been successfully employed to evaporate refractory metals such as Nb[109] or Ta[110] which require temperatures above 3000°C. Their history and design have been reviewed by Denton and Greene.[108] Thun and Ramsey[111] developed a different gun structure which utilized only electrostatic focusing between spherically shaped cathode and anode surfaces.

Contrary to work-accelerated guns, which operate with voltages of less than 10 kV, considerably higher voltages are used in telefocus guns. Consequently, the danger of x-rays emanating from the vacuum system must be considered. Aluminum or glass walls absorb x-rays generated by electrons of up to about 10 keV energy. Radiation generated by more energetic electrons is only partially retained by stainless-steel walls.[98] It is therefore advisable to monitor the radiation near high-energy electron-gun evaporation systems and to use lead shields if necessary.

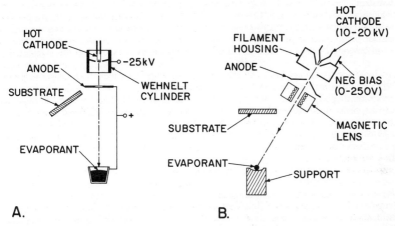

A. **B.**

Fig. 22 Self-accelerated electron guns. (A) Electrostatically focused. (*Reichelt and Mueller.*[107]) (B) Electrostatically and magnetically focused. (*Denton and Greene.*[108])

(3) Bent-beam Electron Guns In the electron guns shown in Fig. 22, the path of the electron beam is a straight line. Therefore, either the gun or the substrates must be mounted off to the side. This restriction in the arrangement of electron source and substrate can be removed by bending the electron beam through a transverse magnetic field.[108] Forcing the electrons into curved paths also allows effective separation of gun structure and vapor source without resorting to long distances. This possibility was first realized by Holland,[99] and compact electron guns which combine the advantages of other structures have been developed since.

An example of a bent-beam electron gun is shown in Fig. 23. The transverse field is provided by an electromagnet which permits focusing during operation. Other models have permanent magnets and variable operating voltage to adjust the beam. Relatively large area elongated cathodes are employed to increase the electron-emission current. This facilitates gun operation at voltages below 10 kV without sacrificing power. The cathode, although close to the evaporant, is in an offset posi-

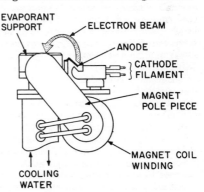

Fig. 23 Bent-beam electron gun with water-cooled evaporant support. (*With permission of Temescal Metallurgical Co., Berkeley, Calif.*)

tion and shielded, hence protected against deposits and erosion by ion bombardment.

Work-accelerated guns similar to those shown in Fig. 21b and c as well as telefocus structures of the types shown in Fig. 22 are commercially available. For laboratory- and bell-jar-type operations, however, the bent-beam guns have become most popular because they are compact, of universal applicability, and not rate-limited. Commercial models offer powers between 2 and 10 kW with accelerating voltages from 3 to 10 kV. They all use a water-cooled copper hearth to support the evaporant, and most of the guns are bakeable to facilitate outgassing. Depending upon the degree of thermal contact between evaporant and support, temperatures up to 3500°C may be achieved so that refractory metals as well as oxides can be evaporated.

A general problem encountered in electron-gun evaporation is rate monitoring. The uncontrolled generation of positive ions interferes with the performance of most ionization-rate monitors. Crystal-oscillator monitors, too, tend to give erroneous rate indications because of the accumulation of charges. However, it is possible to remove charged particles from the vapor entering the monitoring device by means of an electrostatic grid. Stray electrons may also be collected by the substrates and, if these are insulating, establish potentials of several hundred volts. Rairden and Neugebauer[112] observed catastrophic discharges associated with surface damage upon formation of a thin continuous metal film. This can be avoided if the substrates are either allowed to float electrically or kept at the filament potential. The latter situation results in a moderate substrate bombardment by positive ions, which may improve film purity but also affects nucleation and growth processes.[104]

(4) Evaporant Support and Materials The various methods of supporting materials during electron-bombardment heating have already been introduced in Figs. 21 to 23. Because of the limitations of the pendant-drop technique, the water-cooled copper hearth is the preferred type of support. Although the evaporant is in contact with the support surface, chemical interaction is negligible. This is due to the fact that a "skull" of solid evaporant is maintained at the interface and separates the melt from the support.[113] Overheating and spreading of such a molten button may, however, lead to excessive heat losses and reduced evaporation rates. As shown in Fig. 21b, Unvala and Booker[103] avoided this problem by evaporating from a short rod to separate the molten surface from the pedestal. Denton and Greene[108] used a massive tungsten heat sink without water cooling, but the duration of the experiment had to be limited to a few minutes. Siddall and Probyn[114] reduced heat losses by employing refractory compound rather than metal hearths.

The advantage of evaporation without support interaction is lost if the container is allowed to heat up or if the vaporization energy is supplied indirectly by bombarding the crucible. The choice of support materials for a given evaporant is then subject to the same rules as in resistance or induction heating. Yet, electron-beam heating sometimes offers other advantages such as greater simplicity of construction, more directional heat supply, or less stirring of the melt. Examples are the evaporation of Ni and Fe from molybdenum crucibles lined with Al_2O_3 or ZrO_2,[115] of Ge from SiO_2[107] or tungsten[116] containers, and of Be, Cu, Au from BeO, Al_2O_3, or ZrO_2 crucibles.[117] Electron-bombardment heating of dimpled molybdenum and tantalum boats[118] as well as the use of graphite crucibles[111] have also been described.

Electron-bombardment heating may be used to evaporate compounds, provided these do not decompose upon heating. This problem is discussed in more detail in Sec. 6 of this chapter. Examples of compounds which have been evaporated from electron guns are SiO, SiO_2, MgF_2, Al_2O_3, CeO_2, TiO_2, ZrO_2,[108] and ZnS.[119] When dielectric materials are evaporated, it is necessary to use an electrically conductive support to avoid accumulation of charges and deflection of oncoming electrons.

Although electron-bombardment heating is a very versatile and almost universal evaporation technique, it is generally not chosen if the more easily controlled alternative of resistance heating is available. The method is of practical importance in those cases where the greatest film purity is desired and no suitable support material exists. Examples are the evaporation of expitaxial silicon films[103] and of high-purity refractory metals such as Ta, Nb, W, and Mo[109,110,112,120] which require temperatures between 3000 and 3500°C. Of special interest are also the evaporation of carbon[121] and of boron[122] by means of electron bombardment. Finally, electron guns with water-cooled hearths are used with increasing frequency to evaporate platinum metals as well as some of the more reactive metals like Al, Ni, and Fe.

5. THICKNESS DISTRIBUTION OF EVAPORATED FILMS

In addition to materials and modes of construction, an important consideration in the design of vapor sources is their emission characteristics. The latter manifests itself in the distribution of condensate on the substrates. Since most practical applications require a certain film thickness, the substrate area which can be covered uniformly is a significant figure of merit for vapor sources. Assuming condensation coefficients of $\alpha_c = 1$, the distribution of deposits can be derived in principle from the emission laws, Eqs. (51) and (52) in Sec. 3c. In practice, however, it is found that the emission patterns of many useful sources deviate from the idealized behavior. The causes of such modified emission patterns are collisions of evaporant molecules among each other and with walls constituting part of the source. The mathematical formulation of modified emission laws has often been attempted but is difficult, and the resulting expressions apply only to effusion cells with well-defined orifices. No emission theory exists for sources of the open-crucible type, and their behavior can be discussed only on an empirical basis.

a. Thickness Profiles Resulting from the Basic Emission Laws

The distribution of evaporated material is given by Eqs. (52) and (54) (Sec. 3c) as a function of source-to-substrate distance and angle of incidence. Hence, thickness profiles can be derived for substrate areas of any given shape or position relative to the source. Normally, however, the substrates are planar and parallel to the plane of the emitting surface, and only this case is treated here.

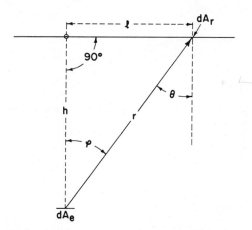

Fig. 24   Evaporation onto a plane-parallel receiver.

(1) Small Surface and Point Sources   The mass received at a point on the substrate defined in terms of r, φ, and θ is given directly by Eqs. (52) and (54). To convert mass into film thickness d, one may visualize the small mass $d\mathfrak{M}_r$ occupying a volume of $dA_r\, d$. If ρ is the density of the film material, then

$$d = \frac{1}{\rho} \frac{d\mathfrak{M}_r}{dA_r} \tag{55}$$

For a plane-parallel receiver at a distance h from the source, the angle of incidence θ equals the emission angle φ, and $\cos \theta = \cos \varphi = h/r$ as shown in Fig. 24. The evaporation distance r for a surface element dA_r varies with the distance l from the center of the substrate area according to $r^2 = h^2 + l^2$. Substituting these relations and Eq. (55) into Eqs. (52) and (54) yields

For the small-area source:

$$d = \frac{\mathfrak{M}_e}{\pi\rho h^2[1 + (l/h)^2]^2}$$

For the point source:

$$d = \frac{\mathfrak{M}_e}{4\pi\rho h^2[1 + (l/h)^2]^{\frac{3}{2}}}$$

Since both sources have infinitesimally small evaporation areas dA_e, the evaporated mass $\mathfrak{M}_e$ and hence d, too, are differential quantities. Without assuming that Eqs. (52) and (54) also apply for small but finite areas, the emission characteristics of the two source types may be defined in terms of the ratio d/d_0, where d_0 is the center thickness at $l = 0$ (see Fig. 24). This leads to the expressions

For the small-area source:

$$\frac{d}{d_0} = \left[1 + \left(\frac{l}{h}\right)^2\right]^{-2} \tag{56}$$

For the point source:

$$\frac{d}{d_0} = \left[1 + \left(\frac{l}{h}\right)^2\right]^{-\frac{3}{2}} \tag{57}$$

The thickness distributions corresponding to Eqs. (56) and (57) are shown in Fig. 25. Because of the smaller negative exponent, the thickness obtained from a point source decreases less rapidly than that of a small-surface source. Characteristic for the small-surface source is a thickness decrease of 2% at a center distance l equal to 10% of the evaporation distance h.

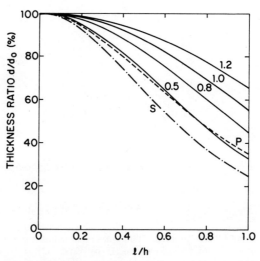

Fig. 25 Film-thickness distributions for a small-surface source (S) and a point source (P). The drawn lines are profiles for circular-disk sources, whereby the numbers indicate the ratios of source radius to substrate distance s/h.

(2) Ring and Circular-disk Sources In practice evaporations are performed with sources whose evaporating surfaces are not vanishingly small. The distributions obtained from sources of finite dimensions can be derived by integrating Eqs. (51a) and (53a) in Sec. 3c, i.e., by adding the thickness increments received at one point from all emitting elements dA_e. In doing so it is assumed that the entire source surface emits at a uniform rate Γ. Such integrations have been performed by Steckelmacher and Holland[123] for infinitely thin linear and ring-shaped sources consisting of many emitting points or surface elements. The case of a truly two-dimensional evaporation source was first solved by von Hippel[124] and is treated in the following part.

The source is assumed to be a circular disk of radius s emitting toward a plane-parallel receiver. Therefore, the distribution of material on the substrate plane is centrosymmetrical and can be described by one coordinate, the center distance l. The arrangement is shown schematically in Fig. 26. A particular emitting element representing a small fraction of a thin ring may be expressed as $dA_e = s\, d\alpha\, ds$, where α is the angle between l and the projection of s onto the substrate plane. Since the latter is parallel to the source plane, it is again $\cos \varphi = \cos \theta = h/r$. Substituting

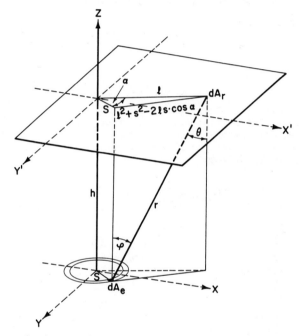

Fig. 26 Evaporation from an element dA_e of a ring to a point dA_r on the substrate plane x', y'.

this relationship and Eq. (55) into the small-surface source Eq. (52), the thickness received from a disk source is given by

$$d = \iiint_{t,s,\alpha} \frac{\Gamma s\, d\alpha\, ds}{\pi \rho} \frac{h^2}{r^4} dt$$

The triple integral arises from Eq. (49) [Sec. 3a(2)] since the total evaporated mass $\mathfrak{M}_e$ of all surface elements dA_e and their time dependence must be considered. After substituting the evaporation distance r by quantities which characterize the position of the receiving point relative to the source, $r^2 = h^2 + l^2 + s^2 - 2ls \cos \alpha$ (see Fig. 26), integration over α from 0 to 2π can be performed. From the resulting expression

$$d = \iint_{t,s} \frac{2\Gamma h^2}{\rho} \frac{h^2 + l^2 + s^2}{[(h^2 - l^2 + s^2)^2 + (2lh)^2]^{\frac{3}{2}}} s\, ds\, dt \tag{58}$$

the distribution obtained from an infinitely thin ring source is easily derived since $2\pi s\,ds \int_t \Gamma\,dt = \mathfrak{M}_e$ is the total mass evaporated from such a source. Therefore, Eq. (58) becomes

$$d = \frac{\mathfrak{M}_e}{\pi \rho h^2}\frac{1 + (l/h)^2 + (s/h)^2}{\{[1 - (l/h)^2 + (s/h)^2]^2 + 4(l/h)^2\}^{\frac{3}{2}}} \qquad \text{for thin-ring-surface sources} \quad (59)$$

The thickness uniformity resulting from thin-ring-surface sources is again best described by introducing the center thickness (at $l = 0$)

$$d_0 = \frac{\mathfrak{M}_e}{\pi \rho h^2}\frac{1}{[1 + (s/h)^2]^2} \qquad (60)$$

and forming the ratio d/d_0. In addition to the relative center distance l/h, Eqs. (59) and (60) contain a second parameter, the relative source radius s/h. Machine-calculated distributions for thin-ring-surface sources having large ratios of s/h are

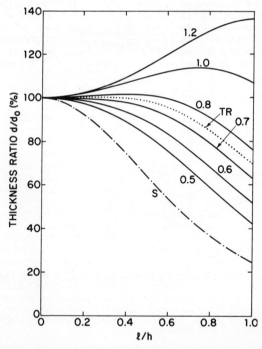

Fig. 27 Film-thickness distributions for a small-surface source (S) and for infinitesimally thin ring-surface sources (drawn lines). The numbers indicate the ratios of source radius s to shortest evaporation distance h. The dotted curve TR represents a ring source of finite dimensions with an inner radius $s_i = 0.7h$ and an outer radius $s_o = 0.8h$.

shown in Fig. 27. If the source radius is relatively small ($s \leqq 0.1h$), the ring-source distributions are nearly the same as obtained from a single-element-surface source. This conclusion can be derived from Eq. (59) directly by assuming $(s/h)^2 \ll 1$. Substantial improvements in thickness uniformity across large substrate areas result only if ring sources are employed whose diameter is comparable with the source-to-substrate spacing h. The most uniform thickness around the center of the substrate area is obtained with $s/h = 0.7$ to 0.8. At larger radii, the throw of ring sources toward the substrate center becomes smaller than the emission toward more peripheral areas.

To continue the derivation of material distributions for circular-disk sources, Eq. (58) must be integrated over the disk radius s. Integration by parts yields

$$d = \int_t \frac{\Gamma}{2\rho} \left[1 - \frac{h^2 + l^2 - s^2}{\sqrt{(h^2 - l^2 + s^2)^2 + (2lh)^2}} \right] dt \tag{61}$$

In this case, the total evaporated mass $\mathfrak{M}_e$ can be substituted for Γ according to $\mathfrak{M}_e = \pi s^2 \int_t \Gamma \, dt$, which gives the thickness expressions for circular-disk sources:

$$d = \frac{\mathfrak{M}_e}{2\pi\rho s^2} \left\{ 1 - \frac{1 + (l/h)^2 - (s/h)^2}{\sqrt{[1 - (l/h)^2 + (s/h)^2]^2 + 4(l/h)^2}} \right\} \tag{62}$$

and

$$d_0 = \frac{\mathfrak{M}_e}{2\pi\rho s^2} \frac{2(s/h)^2}{1 + (s/h)^2}$$

Machine-calculated thickness distributions for disk sources of comparatively large diameters are shown in Fig. 25. Surprisingly, a source of radius $s = 0.5h$ yields a distribution which is similar to that of a single-point source. Further increases of the source radius contribute toward better thickness uniformity around the center but not as effectively as ring sources. For small-diameter-disk sources, Eq. (62) can be approximated by a series development which leads to

$$\frac{d}{d_0} \approx \frac{1 + (s/h)^2}{[1 + (l/h)^2]^2 + [1 - (l/h)^2](s/h)^2} \quad \text{for } s \leqq 0.1h$$

Numerical evaluation of this expression gives distributions which are within a few tenths of 1% of the single-surface-element source.

Equation (62) may also be used to derive distributions for ring sources of finite width by subtracting the emission of an inner disk from that of an outer disk. The profile resulting from such a "true" ring source of optimum dimensions ($s = 0.7$ to $0.8h$) is indicated in Fig. 27 (dotted line, TR). Although the width of this source is relatively large, the distribution is very similar to that which an infinitely thin ring of $s = 0.75h$ would yield.

In conclusion, the thickness distributions obtained from disk and ring sources whose diameters or widths are finite but smaller than the source-to-substrate spacing are adequately represented by the formulas derived for single-element and thin-ring-surface sources. While large-diameter ring sources produce uniform film thickness over a much greater deposition area than a small source at the center, one cannot improve the deposit distribution significantly by enlarging the size of an ordinary flat filament or crucible source. The advantage of a source with a large evaporation surface is primarily the attainment of reasonably high deposition rates at moderate filament temperatures with correspondingly low vapor pressures. Thereby, the chances of chemical interaction between the evaporant and the container material are also reduced.

b. Emission Characteristics of Practical Vapor Sources

In applying the results of the previous section to practical source design, limitations are encountered which are best illustrated by an example. In order to obtain a deposition rate of 10 Å s^{-1} for a material of density $\rho = 10$ g cm^{-3} with an assumed mass evaporation rate of $\Gamma = 10^{-4}$ g cm^{-2} s^{-1}, which would be typical for an evaporant pressure $p^* = 10^{-2}$ Torr, one needs a disk source with a radius of $s = 0.1h$ according to Eq. (61). Since source-to-substrate distances in bell-jar systems are usually of the order of 10 in. or larger, a source with a diameter amounting to 20% of that distance poses several problems: It is difficult to maintain a uniform evaporation temperature across such a large area; the radiation heat losses are likely to overtax the capability of the power supply, and they will also cause outgassing of various surfaces inside the system. For these reasons, relatively small area sources are nearly always used, and one may conclude that evaporant pressures in excess of 10^{-2} Torr occur frequently in practical deposition processes. Such high vapor densities in the immediate vicinity of the source cause reduced evaporation rates because of back diffusion, and deviations from the basic emission law due to the scattering of colliding molecules. These problems are aggravated if the source is not flat but has sidewalls to enhance the evaporant capacity.

(1) Approximate Small-area and Point Sources There are several evaporation sources whose distribution patterns follow the basic emission laws. Flat metal strips or shallow dimpled boats of the type shown in Fig. 13d and e, for example, have repeatedly been found to yield cosine-law emission.[125,126] Even if the molten evaporant does not wet the filament but forms an approximate sphere as does silver on tungsten, the emission characteristics of the small-surface source are still maintained. This is attributable to the fact that vapor molecules impinging on the hot filament surface are adsorbed and reemitted diffusely within a time period corresponding roughly to the Debye frequency of the lattice of the filament material.[127] Consequently, hot surfaces which are not in contact with the condensed evaporant phase but are exposed to its vapor act effectively as extensions of the emitting surface. As long as these surfaces are within the source plane, no deviations from the cosine emission law result. This is still true if the plane of the source is not perfectly parallel to the substrate plane but is tilted. However, the latter condition affects the symmetry of the distribution around the center of the substrate area.[126]

Distributions approaching that of an ideal point source are obtained from two configurations. One is the hairpin filament shown in Fig. 13a; the other is the pendant-drop arrangement shown in Fig. 21a. Both sources are of limited practical value, either because the source capacity is too small or because the instability of the drop does not permit sufficiently high evaporation temperatures. The small surface of such sources prevents the attainment of high evaporation rates in principle.

(2) Effusion Cells Directional emission patterns which do not obey the cosine law but favor the substrate region opposite to the source ("beaming") were first investigated in conjunction with the effusion of gases from nonideal orifices. Clausing[128] derived a distribution equation for effusion from short tubes which is based on cosine-law emission but modified to account for the effect of tube walls preventing direct passage of some of the molecules. The resulting distributions have cusp-shaped maxima in the center and fall off more rapidly at higher emission angles than the straight cosine-law distribution. While fairly successful in describing the effusion of gases, Clausing's approach has often been modified for orifices of different shapes or to achieve a more adequate representation of observed distribution patterns. A review of this subject may be found in Dushman's book.[21]

As for evaporant distributions emanating from effusion cells, Clausing's formula is at best an approximation. This is due to the many forms of molecular interaction—within the orifice, in the beam, with residual gases, and on neighboring surfaces—which cannot always be prevented and whose effects on the emission pattern are difficult to assess quantitatively. An illustrative example is the complex evaporation mechanism suggested by Ruth and Hirth,[129] which takes into account adsorption, surface diffusion, and desorption of evaporant vapor in the orifice region. As a result of this interaction with the walls, the evaporant beam contains molecules of different history. There are those which come directly from the interior of the effusion cell. Others have been adsorbed on the orifice walls, diffused toward the opening, and reemitted. A third fraction arises from the fact that the concentration of adsorbed molecules does not decrease to zero at the upper orifice rim. Instead, the diffusion process continues and covers part of the outer lid surface, where further desorption occurs. Assuming diffuse reemission of adsorbed molecules according to the cosine law, the authors used machine calculations to determine emission patterns of SiO for various shapes of effusion openings. Their results for a tubular opening of equal length and diameter are shown in Fig. 28 and demonstrate that the contributions of adsorbed and reemitted molecules are far from negligible. Moreover, the computed curve agrees very well with SiO effusion measurements carried out independently by Guenther.[130] Figure 28 also shows Clausing's distribution, which was calculated for the same type of tubular opening but does not take into account surface diffusion and yields an exaggeratedly steep thickness decrease.

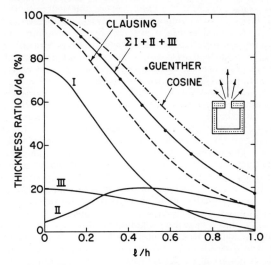

Fig. 28 Thickness distributions for a short tubular orifice (length = diameter). After Clausing,[128] experimental points from Guenther,[130] and calculated for Guenther's effusion cell by Ruth and Hirth[129]: I = direct effusion from the interior of the cell; II = reemitted from the orifice wall; III = diffused onto and reemitted from the lid of the effusion cell.

The curvature and steepness of the profiles in Fig. 28, however, cannot be generalized. Learn and Spriggs,[131] for example, found a distribution for lead and tin vapor effusing from a tubular opening, which falls off even more rapidly than Clausing's equation applied to their case would yield. The factors which, in addition to sorption and surface diffusion, cause deviations from the ideal cosine effusion are not fully understood as yet. Ehrler and Kraus[132] conducted effusion experiments with SiO at cell pressures in excess of 10^{-1} Torr. They observed not only beam dispersion but also a distribution of molecular speeds which was narrower than the Maxwellian. They argue that both effects are due to scattering within the relatively dense molecular beam where molecules collide as the faster ones try to pass slower particles and thus bring about deflection as well as a tendency to assume more uniform speeds. Another factor pertaining to the evaporation of compounds such as KCl, which enter the gas phase partly as monomer, partly as di- and trimer molecules, was recently reported by Grimley and LaRue.[133] For the same effusion cell, they observed different angular flux distributions for the different molecular species as well as a dependence on cell temperature.

In conclusion, the only generalization presently possible is that emission patterns from effusion cells tend to be more directional and have a more pointed maximum in the center than the cosine emission pattern. The severity of the effect is strongly influenced by the orifice geometry, but other factors such as molecular species, cell temperature, and vapor density, too, play a role.

(3) Surface Sources with Perpendicular Sidewalls This category includes most practical sources, particularly the various crucible types, since sidewalls in one form or another are required to hold evaporant quantities in excess of a few grams. As previously discussed, hot surfaces exposed to evaporant vapors act as extended emitting surfaces. If these protrude from the primary source plane, they are likely to affect the emission pattern through sorption and surface-diffusion processes like effusion orifice walls. Contrary to effusion cells, however, typical crucible sources have openings which are as large as the evaporating surface, and the evaporant pressure inside is below the equilibrium value corresponding to the source temperature. Therefore, one would expect only moderate beaming effects from relatively short crucible sources. This is indeed the case. A crucible-source distribution reported by Behrndt and Jones,[134] for example, shows only slight deviations from the cosine law near the substrate center and falls off somewhat more rapidly toward the periphery. The chromium sublimation source shown in Fig. 14, although considerably deeper, is also only slightly directional. The long and narrow inner tube of the Drumheller SiO source, however, has a strong beaming effect and yields an emission pattern which resembles a Clausing distribution.[126]

An interesting source and its distribution have been reported by Spriggs and Learn.[135] These authors evaporated CdS from an array of cylindrical bores in a relatively large ($s/h \approx 0.25$) molybdenum block. The thickness profiles obtained deviate noticeably from the cosine distribution at distances $l > 0.4h$. Beaming also manifests itself in deposition rates around the center which are about twice as high as the calculated values. The effect of the source walls is further enhanced because CdS is a poor heat conductor and therefore evaporates predominantly from the hot walls rather than from the entire cross section of the bores. The latter phenomenon produced a doughnut-shaped thickness profile if the evaporant beam from a single bore was made to pass through a small aperture some distance above the source.

The effects of obstacles in the vapor path, either alone or in conjunction with high residual gas pressure, have been investigated by Rohn.[136] When evaporating LiF from a small surface source at a residual gas pressure of 10^{-4} Torr, he obtained a nearly perfect cosine distribution. Under these conditions, a small obstacle in the form of a wire loop placed in the vapor stream as shown in Fig. 29 did not distort

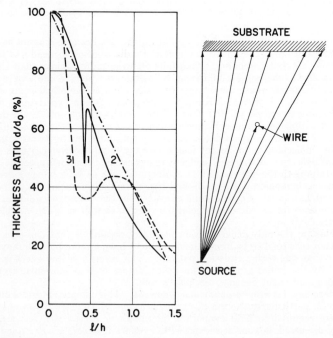

Fig. 29 Distribution profiles as affected by a heated wire and increased gas pressure. (*After Rohn.*[136]) 1. Residual gas pressure $< 10^{-4}$ Torr, wire hot or cold. 2. Residual gas pressure $= 10^{-1}$ Torr, wire cold. 3. Residual gas pressure $= 10^{-1}$ Torr, wire heated.

the cosine distribution except for casting a geometrical shadow. If the residual gas pressure was raised to 10^{-1} Torr without heating the wire, the evaporant molecules were scattered toward larger emission angles and gave an almost linear thickness profile as shown by curve 2 in Fig. 29. Very strong beaming as represented by curve 3 was obtained by heating the wire at 10^{-1} Torr gas pressure. The drastic distortion of this profile and the relatively flat (not cusp-shaped) thickness profile around the center spot are clearly not due to vapor adsorption and reemission from the surface of the wire loop. Instead, the effect must be attributed to rising currents of hot gas generated by the wire loop ("chimney action"). A hot crucible wall around a surface from which evaporation takes place at relatively high vapor pressures may cause similar thermal currents and thereby distort the basic emission pattern.

The combined effects of reemission from sidewalls, intermolecular collisions, thermal currents, and nonuniform temperature of different source parts are generally not predictable. Hence, an empirical approach to source design is often necessary and may, through minor variations of the source geometry, lead to more uniform distributions. An example is the two thickness profiles in Fig. 30 which were obtained by flash evaporation of Cr-SiO pellets.[137] The complete flash filament is shown in Sec. 6c(3), Fig. 43. Evaporation of the pellets occurs almost exclusively from the flat bottom (filament) of the source, but the sidewalls necessary to prevent ejection of the evaporant exert a significant influence on the shape of the profile. It is estimated that the pellet feed rates (about 1 g min^{-1}) generate momentary and local evaporant pressures in excess of 2.10^{-1} Torr. Consequently, interactions with the walls as well as in the released vapor itself can be substantial. Comparison of the two profiles in Fig. 30 shows a significant reduction of the beaming effect if the deflection shield is conical and thus offers a wider path for the vapor bursts. The absolute evaporation rates reflect the degree of beaming insofar as the cylindrical shield yielded center

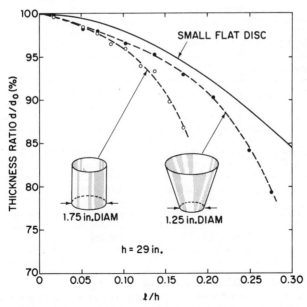

Fig. 30 Thickness profiles obtained by flash evaporation of sintered Cr-20 at. % SiO pellets from cylindrical and cone-shaped crucible sources. For comparison, the regular cosine-law distribution expected from a small flat disk is also shown.

thicknesses of 1.45 times the value calculated from Eq. (62), whereas the cone-shaped source gave 1.35 times the theoretical maximum.

In this context, an observation of Beavitt, Turnell, and Campbell[138] is of interest. When flash-evaporating gold from a filament at 2000°C, these authors found a particle-speed distribution in the vapor that was narrower than the Maxwellian. This result is similar to Ehrler and Kraus'[132] findings for the effusion of SiO at high pressures, and if interpreted in terms of their concept, one would conclude that flash evaporation is especially inducive to heavy molecular interaction in the vapor.

In conclusion, the most uniform thickness distributions which one can hope to attain with small-area crucible or similar sources are slightly inferior to or at best approach those of flat-surface sources. While moderate improvements in uniformity can be achieved by making the ratio of sidewall height to source diameter as small as possible, the most effective measure is to increase the evaporation distance and thereby enlarge the relatively uniform area circumscribed by incidence angles near 0°. This method is, of course, limited by the available system dimensions and by the absolute rates which the source can produce. Therefore, different source arrangements capable of yielding better uniformity are of considerable interest.

c. Sources for Uniform Coverage of Large Areas

Successful attempts to achieve uniform large-area coverage are almost all based on the emission patterns obtained from large-diameter ring sources and have been reviewed by Behrndt.[139] In its simplest form a ring source consists of a circular wire loop emitting material from its entire surface. The integration of the emission flux from such a wire-ring source has been performed by Oberg et al.,[140] while Bugenis and Preuss[141] measured actual thickness profiles of evaporated gold films containing a radioactive isotope. They found slightly greater thickness variations and center thicknesses than Oberg's formula predicts, both indications of possible beaming effects or, as suggested by the authors, due to deviations of the source geometry from the ideal circular shape and planar position. Behrndt and coworkers[134,142-145] experimented extensively with more complex configurations such as multiple rings and crisscrossing wires to obtain uniform large-area coverage with Ni-Fe films. Although film-thickness variations as low as ±1% over substrate areas of several inches diameter could be achieved, these types of wire sources are difficult to operate and limited in their application to those metals which can either be sublimated or loaded uniformly onto refractory-metal wires. Therefore, alternate possibilities to produce distributions similar to those of ring sources have been studied.

One approach to achieve uniform coverage utilizes a rotating shutter in the evaporant beam to alter the amount of vapor admitted to the substrate.[146] Evaporations may be performed from crucibles, which are the most universal sources, and the resulting emission patterns are modified by the contour of the shutter. It is essential that the axis of rotation coincides with the centers of the source and the substrate area. The design of the shutter is determined by the distribution pattern of the source and the size of the area to be coated. To minimize the inevitable reduction of evaporation rates, no material is intercepted at the most peripheral points of the deposition area. From there toward the geometrical center, the shutter blade must represent arcs of increasing length such that the amount of material intercepted at any center distance reduces the deposition rate to the rate at the most peripheral points. The shutter contours required for uniform distributions are spirals whose exact outlines for deposition areas of various diameters have been calculated by Behrndt.[146] He was able to coat substrate areas of 10×10 cm² uniformly within ±0.35%.

The most stringent demands concerning thickness uniformity arise in the fabrication of optical coatings,[147] and therefore, the techniques for accomplishing this objective have been developed in conjunction with the deposition of multilayer dielectric films. The methods originated from the idea that distributions identical to those of ring sources are also obtained if either an eccentrically located source is rotated opposite stationary substrates or if the substrates are rotated vis-à-vis a stationary eccentric source. Since rotating the source poses experimental problems in regard to power leads and temperature control, rotation of substrates is the preferred solution.

The latter arrangement also allows uniform coating of several substrates positioned at equal center distances.

The identity of thickness distributions resulting from an eccentric small-area source on rotating substrates and those from thin-ring sources was first recognized by Fisher and Platt,[148] and the principle has been utilized by several investigators.[139] The distribution equation (59) for ideal ring sources is immediately applicable to rotating substrates if the parameter s is taken to be the horizontal distance of the small-area source from the axis of rotation, and l as the radial distance of a point on the substrate plane from the center of rotation. Behrndt[149] has shown that minimum thickness variations at a radial substrate distance $l = 0.1h$ are obtained if the eccentricity s of the small-area source equals $0.71h$. He introduced a further refinement by placing several substrates equidistant from a common axis of rotation and making them revolve individually around their own axes.[150] The resulting planetary motion of individual substrates reduces even those thickness variations which would be caused by beaming of crucible sources. Another advantage is the greater homogeneity of those film properties which may be affected by different angles of incidence. The thickness of films produced under these conditions varied by as little as $\pm 0.16\%$ on substrates of 6 in. diameter. The construction of such a revolving substrate holder is, however, a major mechanical task and rarely warranted by the more modest uniformity requirements in electronic-film applications.

6. THE EVAPORATION OF COMPOUNDS, ALLOYS, AND MIXTURES

Relatively few inorganic compounds, alloys, or mixtures evaporate congruently because the constituents which are present in the solid or liquid state usually differ in their vapor pressures. Consequently, the composition of the vapor and hence of the condensate is not the same as that of the source material. In principle, compositional changes associated with the transition into the gaseous state can be predicted from thermodynamics. In practice, however, the available thermochemical data are rarely sufficient to describe quantitatively the complex processes which may occur. Empirical information is therefore the most reliable guide in determining the experimental conditions necessary to produce films of the desired composition. This objective cannot always be reached by direct evaporation, and special techniques such as reactive, two-source, and flash evaporation have been developed. These are attempts to control the vapor composition regardless of differences in volatility of the constituents, and sometimes they are the only way to prepare certain films of considerable practical interest.

a. Evaporation Phenomena of Compounds

In the evaporation of metals, the predominant species in the gas phase are single metal atoms, and only a small fraction—usually less than 0.1%—associate into diatomic molecules. As indicated in Table 4, there are a few elements—C, S, Se, Te, P, As, Sb—whose vapors consist of polyatomic species. In the evaporation of compounds, however, transition into the gas phase rarely occurs without changes of the molecular species. This has been established by mass-spectroscopic investigations of the vapors evolving from compounds heated in vacuo.[151] The evidence shows that vaporization of compounds is usually accompanied by dissociation or association or both processes. Whereas association does not affect the stoichiometry of the constituents, dissociation often does, namely, if one of the dissociation products is not volatile. The latter case represents thermal decomposition and makes direct evaporation impractical. Thus, deposition of compound films from a single vapor source requires that the material enters the gaseous state either in the form of complete molecules or—if dissociation occurs—that the constituents are equally volatile. If this condition is satisfied, one speaks of congruent evaporation. Table 10 lists pertinent information for a number of compounds which are of interest in thin film work and amenable to direct evaporation. The various mechanisms involved are discussed in the following.

TABLE 10   Direct Evaporation of Inorganic Compounds

Compound	Vapor species observed (in order of decreasing frequency)	mp, °C	T, °C, at which $p^* = 10^{-2}$ Torr	Comments on actual evaporation temperatures, support materials used, and related experience
		Oxides		
Al_2O_3.....	Al, O, AlO, Al_2O, O_2, $(AlO)_2$	2030^{154}	$\sim 1800^{17,152}$	From W and Mo supports at 1850–2250°C.[159] With telefocus gun at 2200°C, no decomposition[152] From W support: Al_2O_3 films have small oxygen deficits.[153] O_2-dissociation pressure at 1780°C: 1.5 × 10^{-18} Torr[153]
B_2O_3......	$B_2O_3^{151}$	450^{154}	$\sim 1700^{154}$	From Pt and Mo supports at 940–1370°C[151]
BaO......	Ba, BaO, Ba_2O, $(BaO)_2$, Ba_2O_3, O_2^{159}	1925^{154}	$1540^{17,154}$	From Al_2O_3 crucible at 1200–1500°C.[159] From Pt crucible with only slight decomposition, p_{O_2} (1540°C) = 3.5 × 10^{-18} Torr[153]
BeO......	Be, O, $(BeO)_n$, $n = 1–6$, Be_2O^{151}	2530^{154}	2230^{154}	From W support at 2070–2230°C.[151] With telefocus gun at 2400–2700°C, no decomposition[152]
Bi_2O_3.....	....	817^{154}	1840^{153}	From Pt support[153]
CaO......	Ca, CaO, O, O_2^{151}	$\sim 2600^{154}$	$\sim 2050^{17}$	Support materials: ZrO_2, Mo, W. The latter two form volatile oxides, molybdates, and wolframates at 1900–2150°C[151]
CeO_2......	CeO, CeO_2^{151}	1950^{153}	....	From W support without decomposition[153]
In_2O_3.....	In, In_2O, O_2^{151}	....	....	From Pt support with only little decomposition.[153] Vapor species observed at 1100–1450°C. At 1000–1450°C from Al_2O_3 crucible, more In_2O than In[151]
MgO......	Mg, MgO, O, O_2^{151}	2800^{154}	$\sim 1560^{152}$	Mo or W supports at 1840–2000° form volatile oxides, molybdates, and wolframates.[151] With telefocus gun at 1925°C, no decomposition.[152] From Al_2O_3 at 1670°C[159]
MoO_3.....	$(MoO_3)_3$, $(MoO_3)_n$, $n = 4,5^{155,159}$	795^{154}	610^{155}	From Mo oven at 500–700°C, the trimer is the main species. Above 1000°C, there is some decomposition into $MoO_2(s)$ + $O_2(g)$.[155] At 730°C, the oxygen-decomposition pressure is 1.1 × 10^{-14} Torr.[153] From Pt at 530–730°C[158]
NiO......	Ni, O_2, NiO, O^{151}	2090^{153}	1586^{153}	From Al_2O_3 crucible at 1300–1440°C.[151] Heavy decom-

TABLE 10 Direct Evaporation of Inorganic Compounds (Continued)

Compound	Vapor species observed (in order of decreasing frequency)	mp, °C	T, °C, at which $p^* = 10^{-2}$ Torr	Comments on actual evaporation temperatures, support materials used, and related experience
		Oxides		
Sb_2O_3.....	....	656[154]	~450[154]	position with $p_{O_2} = 4 \times 10^{-1}$ Torr at 1586°C[153] Lower oxides result if evaporated from W supports. Pt heaters do not produce decomposition[153]
SiO.......	SiO	....	1025[156,157]	Usually evaporated from Ta or Mo heaters at residual gas pressures below 10^{-5} Torr and at temperatures between 1150 and 1250°C. Dissociation into Si and O_2 begins above 1250°C and may lead to oxygen-deficient films[153]
SiO_2......	SiO, O_2[153,158]	1730[152]	~1250[152]	With telefocus gun at 1500–1600°C, no decomposition.[152] Ta, Mo, W supports are attacked by SiO_2 and contribute volatile oxides.[153] From Al_2O_3 at 1630°C, SiO_2 vapor species is present[159]
SnO_2......	SnO, O_2[151]	....	....	From SiO_2 crucible at 975–1250°C.[151] Films directly evaporated from W support are slightly oxygen-deficient[153]
SrO.......	Sr, O_2, SrO[159]	2460[154]	~1760[17]	From Al_2O_3 at 1830°C.[159] Evaporation from Mo or W at 1700–2000°C produces volatile Mo and W oxides, molybdates, and wolframates[151]
TiO_2......	TiO, Ti, TiO_2, O_2[153,159]	1840[154]	....	TiO_2 source material decomposes into lower oxides upon heating.[152,153] p_{O_2} at 2000°C is 10^{-10} Torr. Nearly stoichiometric films by pulsed electron-beam heating[169]
WO_3......	$(WO_3)_3$, WO_3[155]	1473[154]	1140[155]	From Pt oven at 1040–1300°C.[155] From Pt support at 1220°C.[159] From W heater with only slight decomposition; p_{O_2} at 1120°C is 3×10^{-10} Torr[153]
ZrO_2......	ZrO, O_2	2700[154]	....	From Ta support at 1730°C, volatile TaO.[159] From W support, oxygen-deficient films.[153] ZrO_2 source material loses oxygen when heated by electron beams[152]

TABLE 10   Direct Evaporation of Inorganic Compounds (Continued)

Compound	Vapor species observed (in order of decreasing frequency)	mp, °C	T, °C, at which $p^* = 10^{-2}$ Torr	Comments on actual evaporation temperatures, support materials used, and related experience
		Sulfides, Selenides, Tellurides		
ZnS......	....	1830[168] ($p \approx 150$ atm)	1000[154]	From Mo support. Minute deviations from stoichiometry if allowed to react with residual gases. From Ta at 1050°C[167]
ZnSe......	....	1520[168] ($p \approx 2$ atm)	820[160]	
CdS......	S$_2$, Cd, S, S$_3$, S$_4$[151]	1750[168] ($p \approx 100$ atm)	670[161]	From Pt oven at 740°C.[151] Films tend to deviate from stoichiometry.[153] Suitable support materials: graphite, Ta, Mo, W, SiO$_2$, Al$_2$O$_3$-coated W; evaporation at 600–700°C[167]
CdSe.....	Se$_2$, Cd	1250[168]	660[162,163]	From Al$_2$O$_3$ crucible[167]
CdTe.....	Te$_2$, Cd[164]	1100[168]	570[164]	From Ta boat at 750–850°C; film stoichiometry depends on condensation temperature[76]
PbS......	PbS, Pb, S$_2$, (PbS)$_2$[151]	1112[154]	675[154]	From quartz crucible at 625–925°C.[151] From Mo support.[153] Purest films from quartz furnace at 700°C; Fe or Mo boats react and form volatile sulfides[165]
Sb$_2$S$_3$.....	....	546[154]	550[153]	From Mo support[153]
Sb$_2$Se$_3$....	Sb$_4$, (SbSe)$_2$, Sb$_2$, SbSe[151]	611[166]	....	From graphite at 725°C.[151] From Ta oven at 500–600°C, fractionation and films of variable stoichiometry[166]
		Halides		
NaCl.....	NaCl, (NaCl)$_2$,[151] (NaCl)$_3$	801[154]	670[17,154]	From Ta, Mo, or Cu ovens at 550–800°C[151]
KCl......	KCl, (KCl)$_2$[151]	772[154]	635[17,154]	From Ni or Cu ovens at 500–740°C[151]
AgCl......	AgCl, (AgCl)$_3$[151]	455[154]	690[154]	At 710–770°C.[151] From Mo support, $p^* = 10^{-2}$ Torr at 790°C[153]
MgF$_2$.....	MgF$_2$, (MgF$_2$)$_2$, (MgF$_2$)$_3$[151]	1263[154]	1130[154]	From Pt oven at 950–1230°C.[151] From Mo support.[153] Very little dissociation into the elements[158]
CaF$_2$......	CaF$_2$, CaF[151]	1418[154]	~1300[154]	From Ta oven at 980–1400°C.[151] From Mo support[153]
PbCl$_2$.....	....	678[154]	~430[154]	Direct evaporation possible[153]

(1) Evaporation without Dissociation The simplest transition of a compound AB into the gas phase is described by the equation

$$AB \ (s \text{ or } l) = AB(g)$$

As in the evaporation of the elements, the free energy of evaporation associated with this process is only a function of the temperature and hence is given by the standard value $\Delta_e \mathbf{G}°(T)$. In analogy to Eq. (16), vapor pressures can be derived from

$$\log p_{AB}{}^* = - \frac{\Delta_e \mathbf{G}°(T)}{4.575T} \qquad p_{AB}{}^* \text{ in atm}$$

A glance at the observed vapor species in Table 10 shows that very few compounds evaporate in this simple mode. Examples are SiO and MgF_2, which are among the most widely used film materials. The evaporation behavior of B_2O_3, CaF_2, and most of the divalent group IV oxides (SiO homologs like GeO or SnO) is very similar. The degree of dissociation or association which some of these compounds show is negligible for most practical purposes and hardly affects the vapor pressure. Generally, the tendency to dissociate is greater the higher the evaporation temperature and the lower the pressure. An example is the onset of the dissociation of SiO into Si and O_2 at 1250°C.

Most metal halides are known to form polymolecular species according to

$$nAB(s \text{ or } l) = A_n B_n(g)$$

where typically $n = 2$ or 3. The fraction of di- and trimers in alkali chloride vapors is significant, particularly at lower temperatures. The tendency to associate is even stronger in the case of MoO_3 and WO_3 where the trimer represents more than 80% of the vapor species. This fact must be considered when vapor pressures and molecular weights are inserted into Eq. (48) to determine mass-evaporation rates. The evaporation coefficients α_v are seldom known except for some of the alkali halides which have values between 0.1 and 1.

(2) Evaporation with Dissociation: Chalcogenides Evaporation by dissociation into the elements according to

$$AB(s) = A(g) + \tfrac{1}{2}B_2(g) \tag{63}$$

is often found with binary compounds. If the heat of evaporation into A and B atoms and the heat of dissociation of AB molecules are both known, it is possible to assess the thermal stability of the compound molecules and to predict the resulting gas species.[170]

Well-known examples of complete dissociation are the II-VI compounds. Somorjai[171] investigated the free evaporation of CdS and CdSe at 600 to 900°C. He concluded that the kinetics involves diffusion of A and B surface atoms with recombination of the latter to form B_2 molecules prior to entering the gas phase. According to the law of mass action, equilibrium is obtained when the product of the two gas pressures equals a temperature-dependent constant $K_p(T)$,

$$p_A p_{B_2}{}^{\frac{1}{2}} = K_p(T)$$

whereby
$$\log K_p(T) = - \frac{\Delta_e \mathbf{G}(T)}{4.575T}$$

The total pressure over the compound AB, $p_t = p_A + p_{B_2}$, may vary according to the ratio of partial pressures and assumes its minimum value if the vapor composition is stoichiometric: $p_{B_2} = \tfrac{1}{2}p_A$. For this case, the total equilibrium pressure is given by

$$\log p_t{}^* = - \frac{2}{3} \frac{\Delta_e \mathbf{G}(T)}{4.575T} + \log 3 - \frac{2}{3} \log 2 \qquad p_t{}^* \text{ in atm}$$

Comparing the free evaporation rates with vapor pressures obtained under equilibrium conditions, Somorjai[171] found coefficients α_v between 0.01 and 0.1, which indicates that one of the processes preceding the evaporation of Cd and S_2 or Se_2 particles is constrained by an activation energy. This is the reason why source temperatures measured during the evaporation of II-VI compounds tend to be somewhat higher than those expected from equilibrium-vapor-pressure data (see Table 10).

Volatility of both constituents of a compound is only a necessary but not a sufficient requirement for the deposition of stoichiometric films. When arriving at the substrate, A and B_2 particles must also become absorbed in the proper ratio and recombine to form the AB compound. Differences in the sticking coefficients may lead to films which either are deficient in the nonmetallic constituent or contain excess metal as a second phase. Inhomogeneous films may also be the result of incomplete recombination. The presence of second phases in II-VI compounds can be detected by measuring the optical transmittance of the films. The pure compounds have a sharp absorption edge at that wavelength which corresponds to the forbidden energy gap of the semiconductor. In CdTe films condensed at room temperature, for example, the absorption is shifted toward greater wavelengths by the presence of free tellurium, whereas pure compound films were obtained at 150 to 250°C.[76] The importance of the substrate temperature in determining the film composition has also been observed with other II-VI compounds. CdS films deposited at 150°C are transparent and yellow whereas room-temperature condensates are black, probably because of the presence of excess Cd.[167]

The other II-VI compounds are less well investigated but should show similar behavior; i.e., they can be evaporated directly from single sources at temperatures below 1000°C, and the substrate temperature is an important control parameter to ensure proper film composition. The chalcogenides of divalent group IV elements are also volatile but differ in the degree of dissociation. Thus, compounds like SiS, GeS, SnS, and PbS evaporate primarily as molecular species.[151] As the thermodynamic stability of the sulfides decreases from Si toward Pb, one has to be increasingly concerned about possible reactions with the support material. In this respect, quartz containers are safer than metal boats.[165] The same considerations apply to the selenides and tellurides of group IV elements.[172] The chalcogenides of group V have been of little practical interest as thin films, and their evaporation behavior is exemplified by Sb_2Se_3 in Table 10. Fractionation due to differences in volatility and stratified films with variable composition are typical for this class of compounds.

(3) Evaporation with Dissociation: Oxides The evaporation of most oxides requires temperatures in excess of 1500°C. The binary oxides of Be, Mg, Ca, Sr, Ba, and Ni evaporate predominantly by dissociation according to Eq. (63). But as shown in Table 10, their vapors also contain molecular species, oxygen atoms, and in some cases lower oxides. The tendency to form suboxides is stronger among group III metals as exemplified by Al or In, and most pronounced in the fourth group whose elements are capable of forming relatively stable and volatile binary oxides:

$$AB_2(s) = AB(g) + \tfrac{1}{2}B_2(g)$$

Because of the high temperatures required, the evaporation kinetics of the oxides is essentially unexplored. Only in the case of Al_2O_3 has an evaporation coefficient of $\alpha_v = 2.10^{-4}$ been published (Ref. 35, p. 139). Generally, congruent evaporation is more likely to be attained with binary oxides than with sesqui- or dioxides. This is due to the fact that the higher oxides tend to lose oxygen at temperatures which are too low for the volatilization of the resulting suboxides. Hence, slightly oxygen-deficient and therefore discolored films are often obtained.

Another difficulty is the interaction of oxides with support materials. Tungsten and molybdenum are generally unstable because they recombine with the dissociated oxygen and thereby affect the stoichiometry of the evaporant. Moreover, the reaction products are volatile either in the form of MoO_3 and WO_3 or as molybdates and wolframates which are incorporated into the films. Tantalum, too, forms a volatile suboxide as, for example, in contact with ZrO_2:[159]

$$ZrO_2(s) + Ta = ZrO(g) + TaO(g)$$

Many metal oxides are also reduced if heated in graphite boats. Evaporation from platinum containers minimizes the risk of secondary reactions but is possible only with some of the more volatile oxides such as BaO. The less stable oxides such as In_2O_3 yield slightly oxygen-deficient films even if evaporated from platinum. Refractory oxide crucibles have been used, but they involve the risk of chemical attack by the highly reactive alkali-earth oxides, or of evaporating themselves at the high temperatures required.

The best method to evaporate refractory oxides directly is electron-beam heating. Reichelt and Mueller[152] avoided the container problem by focusing the beam on a small part of relatively large source crystals. Because of the poor heat conductivity of the oxides, the focal spot assumed temperatures which were 200 to 400°C higher than those corresponding to equilibrium pressures of 10^{-2} Torr. Under these conditions, MgO, BeO, Al_2O_3, SiO_2, and ThO_2 films could be directly evaporated. The oxides of Ti, Zr, Nb, Ta, and Cr, however, did not evaporate congruently but decomposed as indicated by the discoloration of the source crystals. The deposition of stoichiometric TiO_2 films is of particular interest because of the high index of refraction. However, TiO_2 loses oxygen very easily, and the nonstoichiometric products normally obtained are intensely colored as well as semiconducting. Reichelt[169] succeeded in depositing insulating TiO_2 films which were stoichiometric within a fraction of 10^{-10} by utilizing electron-beam pulses of 0.5 s duration. With this technique, decomposition of the source material adjacent to the focal spot is minimized since the latter is rapidly heated to temperatures where the vapor pressure of the suboxide is sufficiently high for congruent evaporation to take place. The same technique has also been used to prepare SiO_2 films.[169] Cariou et al.[173] obtained HfO_2, ZrO_2, and Ta_2O_5 films of good dielectric properties by direct electron-beam evaporation. An oil-free vacuum was used to avoid oxide reduction by residual hydrocarbons. The direct evaporation of SiO_2, Al_2O_3, and ThO_2 with an electron gun has also been reported by Cox et al.[174] It is noteworthy that most oxides are very effective getters for residual water vapor whose presence in the films may cause variations in density, stress, and optical properties.[175] Pliskin[176] found electron-gun evaporated SiO_2 films to be very reactive because of bond straining and porosity. They also contained some Si_2O_3 and tended to absorb water. Finally, it has been observed that electron-beam-evaporated Al_2O_3 films incorporate free electrical charges.[177]

In summary, a number of inorganic compounds, including most halides, the II-VI compound semiconductors, and several important oxides, are amenable to direct evaporation. The evaporation kinetics in most cases is complex and involves dissociation or association of molecules so that the vapor consists of different species. This must be taken into account when mass-evaporation rates are derived from vapor-pressure data. Figure 31 shows the vapor pressures for compounds as far as they are presently known. The curves represent total pressures including all species in stoichiometric proportions ($p_t{*}$). Because of the experimental difficulties in establishing true vapor-solid equilibria at high temperatures, the accuracy of the data is seldom very high. Furthermore, it should be kept in mind that the temperatures required for free evaporation will often be higher than the curves suggest because of evaporation constraints as reflected in α_v-values < 1. Additional vapor-pressure data covering the 1- to 760-Torr range of many inorganic compounds have been compiled by Stull.[178]

(4) Decomposition A large number of compounds evaporate incongruently; i.e., the volatilities of their constituents at elevated temperatures differ so strongly that the vapor and hence the condensate do not have the same chemical composition as the source. The simplest decomposition reactions are of the type

$$AB(s) = A(s) + \tfrac{1}{2}B_2(g)$$

where the metal A forms a separate phase. In this case, the free energy of the decomposition process $\Delta_d\mathbf{G}$ is fixed by the standard chemical potentials of both solid phases and dependent merely on temperature. The decomposition pressure is therefore given by

$$\tfrac{1}{2} \log p_{B_2}{}^* = -\frac{\Delta_d \mathbf{G}°(T)}{4.575T} \qquad p^* \text{ in atm}$$

The free energy of decomposition $\Delta_d \mathbf{G}°(T)$ is identical with the free energy of formation of AB from the elements except for its sign. Hence, the decomposition pressure can be calculated from the appropriate free-energy-of-formation data, which have been tabulated by several authors for most common compounds.[6,159,179,180]

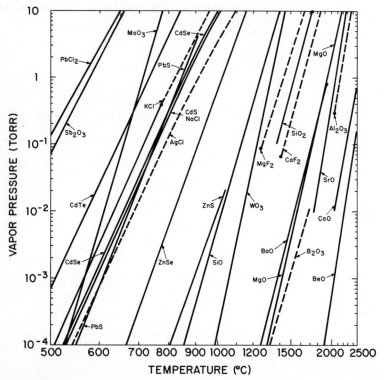

Fig. 31 Equilibrium vapor pressures of inorganic compounds. Drawn lines: over solids; broken lines: over liquids. The full circles indicate melting points. The sources of the data are referenced in the fourth column of Table 10.

The kinetics of decomposition reactions has been investigated by Rickert[181] using Ag_2S, Ag_2Se, and CuI as model substances. Evaporation is preceded by formation of B_2 molecules in the absorbed state, while the metal atoms diffuse and form nuclei and crystallites. Most of the metal borides,[182] carbides, and nitrides[151] follow similar mechanisms. Incongruent evaporation involving the formation of lower carbides $(WC \rightarrow W_2C)$ or nitrides $(CrN \rightarrow Cr_2N)$ has also been observed. In all these cases, either no films are obtained or their composition is different from the source material. Known exceptions of nearly congruent evaporation in this category are TiC, ZrC,[183] and AlN.[86]

Among the oxides, Cr_2O_3 and Fe_2O_3 are examples of decomposition into suboxides of low volatility. Attempts to deposit films of these compounds have been described by Holland.[153] If evaporated from tungsten boats, lower oxide films of undefined composition are obtained. In the case of Fe_2O_3, the source material decomposes into a mixture of Fe_3O_4 and metallic iron, whereas Cr_2O_3 yields oxygen-deficient films although volatile oxides do exist.[151] Generally, lower oxides or oxygen-deficient deposits can be converted into higher oxides by heating in air. However, the required temperatures are often too high for the film substrate, or the post-deposition reaction may yield highly stressed, porous, or nonadherent films. Therefore, methods other than direct evaporation such as reactive evaporation, reactive sputtering, or rf sputtering are preferable for these materials.

Another important family of compounds whose evaporation is preceded by decomposition are the III-V compounds. Their pressure-temperature-composition diagrams have been reviewed by Weiser,[184] and their predominant mode of decompositions is

$$AB(s) = A(l) + \frac{1}{n} B_n(g) \qquad (n = 2, 4) \qquad (64)$$

The relatively volatile group V constituents (P, As, and Sb) reach pressures of 10^{-2} Torr at temperatures of 700 to 900°C. In this interval, the group III metals Al, Ga, and In are molten, but their vapor pressures are orders of magnitude lower than those of the group V elements. Therefore, the vapor consists predominantly of B_n molecules and contains virtually no metal atoms or, in the case of indium compounds, a nonstoichiometric amount.[151]

The direct evaporation of III-V compounds is further complicated by the fact that the residual source materials $AB(s)$ and $A(l)$ form solid or liquid solutions. Consequently, the free energy of the decomposition reaction Eq. (64) is also dependent on the concentration of residual B in the source and therefore varies as decomposition proceeds. Accordingly, the vapor pressure of the constituent B can only be derived if the free energy is given as a function of both the temperature and the activity a_B in the remaining solution:

$$\frac{1}{n} \log p_{B_n} = - \frac{\Delta_d \mathbf{G}(T, a_B)}{4.575T} \qquad p_{B_n} \text{ in atm}$$

An experimental investigation leading to the determination of thermodynamic constants and of a vapor-pressure equation for GaP has been published by Lee and Schoonmaker.[185]

Attempts to deposit III-V compound films by direct evaporation in spite of these difficulties have been reported by Paparoditis.[186] The indium compounds are most promising because In is a little more volatile than Ga or Al. Evaporating InSb and InAs to completion, Paparoditis obtained fractionation and stratified films. The films could be homogenized and their stoichiometry restored by post-deposition annealing at temperatures around 300°C. Dale et al.[187] tried to counteract the fractionation of InSb by replenishing the source repeatedly with new material, thereby approaching an indium-rich melt from which InSb evaporates nearly in the proper stoichiometric ratio.[188] However, these techniques are not applicable to the other III-V compounds which have greater vapor-pressure differences and higher melting points. Because of the latter, restoration of the compound by interdiffusion of the condensed elements requires relatively high temperatures which involve the risk of losing the volatile constituent during the crystallization process.[186] Therefore, special techniques such as two-source and flash evaporation are usually considered when III-V compounds are to be deposited.

b. Evaporation Phenomena of Alloys

(1) Raoult's Law The constituents of alloys evaporate independently of each other and, like the pure metals, mostly as single atoms. Monatomic vapor may be observed over alloys even in those cases where the pure element is known to form molecules. An example is the occurrence of monatomic antimony vapor over Pt-Sb[189] and Au-Sb[190] alloys. The vapor pressure of an alloy constituent, however, is different from that of the pure metal at the same temperature. This is due to the change in chemical potential which a metal B experiences when it is dissolved in another metal A to form an alloy A-B. If the concentration is given by the mole fraction x_B, the chemical potential in the alloy, $\mu_{B,c}(T)$, differs from that of the pure metal, $\mu°_{B,c}(T)$, by the energy spent to disperse B[191]:

$$\mu_{B,c}(T) = \mu°_{B,c}(T) + RT \ln x_B \tag{65}$$

The use of the mole fraction in Eq. (65) implies an ideal solution; i.e., the energy of atomic interactions between two different constituents must be the same as between two equal constituents.

The chemical potential of an ideal gas at the pressure p is related to the potential of the standard state, $\mu°_{B,g}(T)$ at 1 atm, by the equation [see Sec. 2a(2)]:

$$\mu_{B,g}(T) = \mu°_{B,g}(T) + RT \ln p_B \tag{66}$$

Equilibrium between the alloy and its adjacent vapor phase $B(g)$ is obtained if the two potentials [Eqs. (65) and (66)] are equal. This yields the vapor-pressure relation

$$RT \ln p_B = -[\mu°_{B,g}(T) - \mu°_{B,c}(t)] + RT \ln x_B \tag{67}$$

The difference in standard chemical potentials is the standard free energy of evaporation for the pure metal, $\Delta_e G°(T)$, whch according to Eq. (16) must be divided by RT to yield the vapor pressure of the pure metal. Thus, Eq. (67) simplifies to

$$p_B = x_B p_B^*$$

This expression is one form of Raoult's law, which states that the pressure of an element over an ideal solution is reduced in proportion to its mole fraction. It can be verified experimentally by measuring the vapor pressure over alloys of known composition. Deviations from Raoult's law are common because most alloy systems are not ideal solutions. In order to describe the behavior of real solutions, the activity a_B has been introduced. It is defined as the actual ratio of vapor pressures,

$$a_B = \frac{p_B}{p_B^*}$$

and related to the mole fraction by the activity coefficient f_B:

$$a_B = f_B x_B$$

By substituting a_B for x_B, the usual thermodynamic relationships can be applied to systems which are not ideal solutions. Therefore, the activity and its coefficient are important empirical parameters which have been tabulated for a number of binary systems.[192] Since the activity of a material is temperature- as well as system-dependent, the existing data are far from being complete. The examples shown in Figs. 32 and 33 are of particular interest because Ni-Fe and Ni-Cr are the most widely used alloy systems in the thin film field. The deviations from Raoult's law in the nickel-iron system are relatively small and negative. In the nickel-chromium system, positive as well as negative deviations occur. They are especially large at small concentrations. The diagram also shows the activity behavior in a two-phase solid.

Since the vapor pressure of alloys is affected by the energy state of the atoms in the condensed phase, the activity is related to the phase diagram of the system. Activity plots and phase diagrams for a few binary systems have been compared by Koller.[194] Ideal behavior ($f_B \approx 1$) can be expected only but not always from simple systems which show complete miscibility. Compound formation in the solid state is likely to reduce the vapor pressure since stronger bonds have to be broken. This usually causes negative deviations ($f_B < 1$) even when the temperature is above the melting point of the compound.   Positive deviations from Raoult's law indicate that the attractive forces between unlike atoms in an alloy are weaker than the interaction in the pure phase (Ref. 6, p. 36).   If there is a miscibility gap in the liquid state, the activity of the lighter component floating on top of the melt will be close to 1.

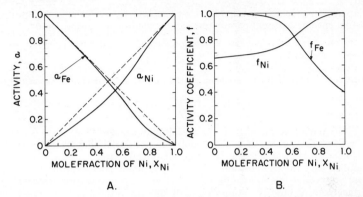

Fig. 32   (*A*) Activities of Ni and Fe in liquid nickel-iron alloys at 1600°C.   Ideal behavior according to Raoult's law: broken lines.   Experimentally determined: solid lines.   (*B*) Activity coefficients for the same conditions.   (*After data from Zellars, Payne, Morris, and Kipp.*[193])

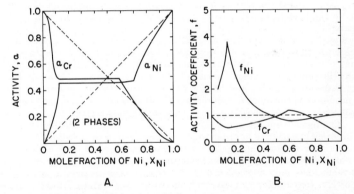

Fig. 33   (*A*) Activities of Ni and Cr in solid nickel-chromium alloys at 1100°C.   Raoult's law: broken lines.   Experimentally determined: solid lines.   (*B*) Activity coefficients for the same conditions.   (*After data from Hultgren et al.*[192])

(2) Compositional Changes during Evaporation   The application of Raoult's law to the evaporation of *liquid alloys* introduces the mole fraction and—for real solutions—the activity coefficient into the Hertz-Knudsen equation.   After dividing Eq. (48) by the molecular weight, the evaporation rates for the constituents of a binary alloy are

$$\frac{dn_A}{dt} = \frac{5.834 \times 10^{-2}}{\sqrt{M_A T}} f_A x_A p_A^* \qquad \text{mol cm}^{-2} \text{ s}^{-1} \qquad (68a)$$

and

$$\frac{dn_B}{dt} = \frac{5.834 \times 10^{-2}}{\sqrt{M_B T}} f_B x_B p_B^* \qquad \text{mol cm}^{-2} \text{ s}^{-1} \qquad (68b)$$

The ratio of A and B particles in the vapor stream at any moment is therefore

$$\frac{dn_A}{db_B} = \frac{f_A x_A p_A^*}{f_B x_B p_B^*} \left(\frac{M_B}{M_A}\right)^{\frac{1}{2}}$$ (69)

whereby $x_A + x_B = 1$. The ratio is dependent on time since x_A and x_B change as the evaporation proceeds. Assuming that the activity coefficients remain constant, the system specific material parameters can be combined into one factor

$$K = \frac{f_A p_A^*}{f_B p_B^*} \left(\frac{M_B}{M_A}\right)^{\frac{1}{2}}$$

The integration of Eq. (68a) has been performed by Zinsmeister[188] and leads to

$$\frac{x_A}{x_A^\circ} \left(\frac{1 - x_A^\circ}{1 - x_A}\right)^K = \left(\frac{n}{n^\circ}\right)^{K-1}$$ (70)

where x_A° = mole fraction of A in the initial alloy charge

n° = total number of moles A and B in the initial alloy charge

$n = n_A + n_B$ = number of moles left in the source

It is implied that the composition of the alloy remains homogeneous throughout the evaporation process, in other words, that there are no concentration gradients between the surface and the interior. From Eq. (70), one can compute x_A as a function of the fraction of material already evaporated, $(n^\circ - n)/n^\circ$. The mole fraction x_B is obtained from an expression analogous to Eq. (70). Substitution of x_A and x_B into Eq. (69) permits computation of the molar ratio dn_A/dn_B in the vapor and hence at the moment of condensation as a function of the evaporated fraction of the charge. Such calculations have been performed by Zinsmeister,[188] and his results are shown in Fig. 34. The change of the molar ratio during the evaporation is significantly affected by the system specific parameter K and by the mole fraction of the initial alloy charge x_A°.

From Fig. 34, several conclusions can be drawn. Unless K happens to be 1, in which case an alloy evaporates congruently, the composition of the vapor deviates from that of the melt. Initially, the more volatile constituent evaporates preferentially, but the proportions are reversed as the charge is used up. Directly evaporated films are therefore stratified and have a vertical concentration gradient. This effect is more pronounced the more K differs from 1. If K is close to 1, it may be attempted to deposit nearly homogeneous films by evaporating only a small fraction of the total charge. The initial concentration x_A° must then be carefully selected to compensate for the different volatilities in order to obtain the desired composition.

When the curves shown in Fig. 34 are used to assess the chances of direct evaporation, a number of complications must be considered. The assumption of constant activity coefficients may not be warranted. It is then necessary to perform the integration of Eq. (68) in steps for intervals of constant K values. Furthermore, mass transfer from the interior to the surface of the melt may not be instantaneous. In this case, the surface concentration and its relation to the composition in the interior enter the calculation.[188] Lastly, it should be remembered that surface contaminations such as oxide or slag films act as evaporation constraints and may introduce evaporation coefficients $\alpha_v \ll 1$ into the starting equations (68).

Compositional changes during the *sublimation of alloys* have been treated by Huijer et al.[195] In the absence of convection and stirring effects as encountered in melts, the source surface is depleted of the more volatile constituent B. The evaporation rate of B therefore decreases. At the same time, B atoms diffuse from the interior of the solid to the surface. Eventually, the two processes balance and reach a stationary state.

During the transient state, the diffusion process is governed by Fick's second differential equation

$$\frac{\partial x_B}{\partial t} = D \frac{\partial^2 x_B}{\partial y^2}$$ (71)

where y is the distance of a point in the interior of the alloy from the surface.   Since both constituents are continually evaporating, the alloy surface recedes from its original position.   Huijer et al.[195] treated this complication by letting the origin of the y axis move along a parallel z axis whose origin is fixed at the initial surface.

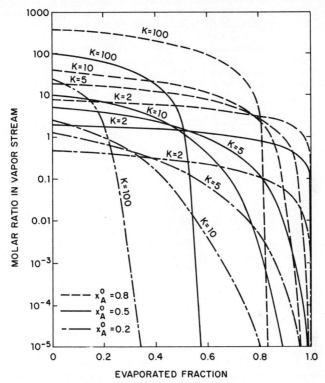

Fig. 34 Molar ratio in the vapor of binary alloys as function of the fraction already evaporated. *(From Zinsmeister.*[188]*)*

Assuming the surface recedes at a constant rate of v cm s^{-1}, the relationship between the coordinates is $z = y - vt$.   Equation (71) is thereby transformed into

$$\frac{\partial x_B}{\partial t} = D \frac{\partial^2 x_B}{\partial z^2} + v \frac{\partial x_B}{\partial z} \tag{72}$$

In the steady state, $\partial x_B/\partial t = 0$, and Eq. (72) has the solution

$$x_B = x_B{}^\circ - C \exp\left(-\frac{v}{D} z\right)$$

where $x_B{}^\circ$ is the mole fraction of B in the initial charge and C a constant which depends on the evaporation rates and densities of the constituents.

An analytic solution of Eq. (72) for the transient states is not available.   Huijer et al.[195] obtained numerical solutions with the aid of a computer.   The assumed boundary conditions are a constant initial concentration $x_B{}^\circ$ throughout the alloy, and a material-removal rate v determined by the evaporation equations (68a) and (68b) with varying values of x_A and x_B at the surface.   Figure 35 shows their results for one specific value of v/D.   During the transient state, the concentration profiles intersect the receding material surface at successively smaller values until the steady state is reached.   The shaded area represents the amount of B evaporated during a steady-state period.   Since the surface concentration is constant, the ratio of $A:B$ in the vapor does not change any more.   Furthermore, the profiles at t_6 and t_7 are parallel to each other and therefore include exactly the same amount of constituent B as would have been contained in the equally thick layer which was actually removed, assuming the latter had had a uniform concentration $x_B°$ (area between dotted lines in Fig. 35).   It follows that an alloy evaporates in its original composition once the steady state has been obtained.   The latter happens faster, and the surface layer affected by the outward-diffusion process is thinner, the larger the ratio v/D.   Thus, high evaporation rates and small diffusion coefficients favor the applicability of this method.

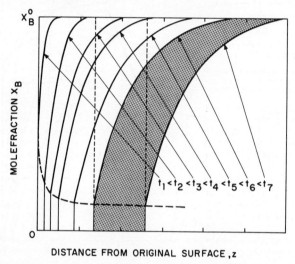

Fig. 35   Mole fraction of constituent B in a solid alloy as function of distance from the original source surface and time.   The broken line represents the intersects of the distributions with the receding surface and hence the momentary surface concentration.   The shaded area indicates losses of B during a period of steady-state evaporation.   (*After Huijer, Langendam, and Lely.*[195])

(3) Examples of Direct Alloy Evaporation   The direct evaporation of alloys for the purpose of thin film fabrication has been investigated primarily with the systems Ni-Fe and Ni-Cr.   Nickel-alloy films containing around 15% Fe (Permalloy) are widely utilized for magnetic memory elements, while 80% Ni–20% Cr films (Nichrome) often serve as microminiature resistors.   The two systems are eminently suited to illustrate the evaporation behavior of alloys because one has a low and the other a high ratio of vapor pressures.   Both are often evaporated from tungsten boats, although the melts react with the support material.   Alumina or zirconia crucibles are therefore preferred for Ni-Fe alloys.   This does not appear to be true for Nichrome, probably because of the affinity of chromium for oxygen and the noticeable volatility of its suboxides.   Resistance, electron bombardment, and induction heating have all been tried whereby the latter has the advantage of stirring the melt while the other methods provide only thermal convection.

A number of methods are available to determine the ratio of evaporated constituents.   The direct approach is to analyze the vapor with a mass spectrometer.[196] More common is the use of shuttering techniques to condense separate fractions and then analyze the deposits. Time resolution of deposit fractions has also been obtained by condensation on a moving Mylar tape.[197] The film composition may be determined by x-ray fluorescence[197] or chemical microanalysis[198] with techniques such as emission spectroscopy[199] or colorimetry.[200]

The preceding section about compositional changes allows an a priori assessment of the results to be expected from the direct evaporation of alloys, and Table 11 shows such data for Permalloy.

TABLE 11 Evaporation Data for Permalloy (85% Ni–15% Fe) at Two Temperatures*

	1600°C (liquid)		2000°C (liquid)	
Activity coefficients..............	$f_{Ni} = 1$ $f_{Fe} = 1$	$f_{Ni} = 0.97$ $f_{Fe} = 0.52$	$f_{Ni} = 1$ $f_{Fe} = 1$	$f_{Ni} = 0.97$ $f_{Fe} = 0.52$
Material parameter K.............	2	1.1	1.6	0.9
Initial evaporation rate $\Gamma(Ni + Fe)$, g cm^{-2} s^{-1}..........	3×10^{-4}	2.5×10^{-4}	2.5×10^{-2}	2.2×10^{-2}
Initial Ni:Fe ratio in the vapor.....	75:25	85:15	78:22	87:13

* Based on vapor pressures by Honig[19] and activity coefficients from Fig. 32.

Since the material parameter K introduced in Sec. 6b(2) is close to 1 and the evaporation rates are 10^{-4} g cm^{-2} s^{-1} or higher, one would conclude that Permalloy is amenable to direct evaporation. If ideal behavior is assumed, the films should be somewhat richer in iron than the source material. However, the negative deviation from Raoult's law of Fe in Ni should allow nearly 1:1 transfer of the source composition to the films. This has been confirmed experimentally. The fact that Ni-Fe films are less iron-rich than expected on the basis of $f_{Fe} = 1$ was first reported by Blois[201] for evaporation at 1600°C. Brice and Pick[115] evaporated at 2000°C and established that charges of 85% or more Ni produce initial deposits of nearly the same compositions. When using large melts which maintain their composition fairly well for longer periods of time, the iron content of the films is typically within 2 to 4% of the charge. Evaporation of a major portion of the melt (70 to 75%), however, yields stratified films with concentration gradients of about 5% per 1,000 Å.[202] Thus, theory and experiment agree very well in the case of Ni-Fe, and Permalloy films may indeed be fabricated by direct evaporation.

TABLE 12 Evaporation Data for Nichrome (80% Ni–20% Cr) at Two Temperatures*

	1100°C (solid)		1450°C (liquid)
Activity coefficients......................	$f_{Ni} = 1$ $f_{Cr} = 1$	$f_{Ni} = 0.9$ $f_{Cr} = 0.8$	$f_{Ni} = 1$ $f_{Cr} = 1$
Material parameter K....................	11	10	7.5
Initial evaporation rate $\Gamma(Ni + Cr)$, g cm^{-2} s^{-1}.................	6.5×10^{-8}	5.3×10^{-8}	1×10^{-4}
Initial Ni:Cr ratio in the vapor	30:70	31:69	50:50

* Based on vapor pressures by Honig[19] and activity coefficients from Fig. 33. Activities in the liquid state are not known.

Evaporation data for Nichrome are listed in Table 12. At the lower temperature, evaporation is impractical in view of the small rates. Lakshmanan[203] obtained deposition rates of 0.17 Å s^{-1} from a sublimation source operating at 1000°C. The most common source temperatures for Nichrome are between 1400 and 1500°C (mp = 1395°C). Although the spread in the vapor pressures of Ni and Cr is smaller at higher temperatures, the parameter K is still too high to expect constant film composition from a significant fraction of the melt. This has been confirmed by Degenhart and Pratt,[198] who evaporated 12% of a small Nichrome charge at 1400 to 1500°C. The initial fractions of the deposit had the approximate composition 15% Ni–85% Cr, whereas the last fractions consisted of about 80% Ni and 20% Cr. Similar results were obtained when Nichrome vapors were analyzed by a mass spectrometer.[196] Here, the initial vapor composition was 17:83, and after 12 min of evaporation 80:20. These results indicate that the material-transfer ratio is even worse than calculated in Table 12, which means the activity coefficients in this system tend to increase the K value. Deviations from Raoult's law have been reported for Ni-Cr alloys containing small amounts of Al and Cu.[199,204] The activity of chromium is greater than its mole fraction ($f_{Cr} > 1$), which agrees with Degenhart and Pratt's results. The evaporation of Ni appears to be somewhat retarded ($f_{Ni} < 1$), whereas the copper evaporation is enhanced ($f_{Cu} > 1$). Aluminum shows large negative deviations from the ideal behavior.

These studies demonstrate that the controlled deposition of homogeneous Nichrome films by direct evaporation from the melt is not easily accomplished. It would be necessary to use a very large charge of substantially lower Cr content than desired for the films, and to evaporate only a small fraction of it. This approach has been tried by Wied and Tierman,[205] but the results are not described in terms of composition.

An interesting alternative is the evaporation of the solid alloy. The analysis of compositional changes during the sublimation of alloys given by Huijer et al.[195] was applied to Nichrome. Evaporating from a wire at 1300°C, the steady-state distribution was reached after 3 h of evaporation. From the decrease in wire diameter and an analysis of the steady-state profile, Huijer et al. determined the diffusion coefficient of Cr in Ni to be 6.7×10^{-10} cm² s⁻¹. The steady-state surface concentration of Cr was found to be 3%, which upon substitution into the evaporation equation yields a vapor ratio Ni:Cr of 83:17 in close agreement with the 80:20 ratio in the wire. However, the vapor pressures of the two metals at 1300°C are 0.25×10^{-3} and 1.8×10^{-3} Torr, respectively, which amounts to a total evaporation rate as low as 3×10^{-6} g cm⁻² s⁻¹. Therefore, the sublimation technique, too, is of little practical value. More successful methods of depositing Nichrome films with controlled composition are flash evaporation[200] and dc bias sputtering.[206]

Compositional changes during the evaporation of other alloys are not as thoroughly investigated as those of the two nickel systems. An example of successful direct evaporation is Cr-Ge. Since the vapor pressures of these two elements are nearly identical, Riddle[207] obtained almost congruent evaporation at 1310°C from melts containing equal amounts of Cr and Ge. Sauer and Unger[208] studied the evaporation of Ag-Cu, Ag-Al, Ag-Sn, Cu-Sn, and Pb-Sn alloys. The compositions of the deposited fractions agreed quite well with predictions based on Raoult's law. Alloys of the brass type like Cu-Zn and Cu-Cd cannot be directly evaporated because Cu is nonvolatile at temperatures where Zn and Cd have vapor pressures exceeding 1 Torr. The behavior of several bronzes has been described by Holland (Ref. 59, p. 182); there are some compositions in the Cu-Sn and Cu-Al systems which can be evaporated with little fractionation.

c. Special Evaporation Techniques

The preceding discussions of the behavior of compounds and alloys show that several materials which are of interest as thin films are not amenable to direct evaporation. Heavy fractionation, compositional changes, and stratified films are also predictable if mixtures such as cermets (ceramics and metals) are heated jointly in one source. For these situations, special evaporation techniques have been developed which utilize different experimental principles to obtain films of controlled composition.

(1) Reactive Evaporation In the deposition of metal films, the background pressure is usually kept as low as possible because the interaction of residual gases with the evaporant has mostly detrimental effects on film properties.[209] In reactive evaporation, however, relatively high oxygen pressures from 10^{-5} to 10^{-2} Torr are deliberately maintained to produce fully oxidized metal films. Thus, the technique is applicable in those cases where metal oxides cannot be evaporated directly because of complete or partial decomposition. A controlled oxygen leak is generally used to provide the desired atmosphere, but the thermal decomposition of MnO_2 has also been employed to establish the oxygen pressure.[210] For the kinetics of reactive evaporation, it is important to remember that the mean free path of gas particles at 10^{-4} Torr is of the order of 50 cm (see Fig. 5). Hence, the probability of forming metal oxide molecules through collisions in the gas phase is very small. Instead, recombination takes place on the substrate surface, which is exposed to high incidence rates of metal and oxygen particles.

The formation of an oxide film by reactive evaporation begins with the impingement of metal atoms and O_2 molecules on the substrate surface. Some of these are adsorbed; others are reflected or desorbed after brief contact with the surface. The ratio of permanently adsorbed to total impinging particles is the condensation coefficient α_c. The adsorbed vapors must then react to form the metal oxide. This reaction is essentially an ordering process whereby the adsorbed atoms diffuse across the surface until they fall into the potential wells represented by the regular lattice sites. Consequently, the film-growth process is controlled by the impingement rates of metal atoms and O_2 molecules, the condensation coefficients of the two species, and the substrate temperature.

For practical purposes, the rate at which metal atoms arrive at the substrate is best expressed in terms of the deposition rate as obtained from the same source at the same temperature and distance but in the absence of oxygen:

$$\frac{dN_{Me}}{A_r\,dt} = \mathbf{N}_A \frac{\rho_{Me}}{M_{Me}} d' \qquad \text{atoms cm}^{-2}\,\text{s}^{-1}$$

where ρ_{Me} = density of the metal film, g cm^{-3}
M_{Me} = molar mass of the metal, g mol^{-1}
d' = pure metal condensation rate, cm s^{-1}
A_r = receiving (substrate) surface, cm^2

The impingement rate of oxygen molecules is given by the Eq. (37) of Sec. 2b(3):

$$\frac{dN_{O_2}}{A_r\,dt} = 3.513 \times 10^{22}(M_{O_2}T)^{-\frac{1}{2}}p_{O_2} \qquad \text{molecules cm}^{-2}\,\text{s}^{-1}$$

where M_{O_2} = molar mass of O_2, 32 g mol^{-1}
T = gas temperature, usually 300°K
p_{O_2} = O_2 partial pressure, torr

Impingement rates for a number of evaporants and common gases can be determined with the aid of the nomogram in Fig. 36. As indicated by the broken lines, a chromium condensation rate of 1 Å s^{-1} corresponds to 8×10^{14} atoms cm^{-2} s^{-1}. If the evaporation is carried out at an oxygen pressure of 10^{-6} Torr, the competing rate is about 4×10^{14} O_2 molecules cm^{-2} s^{-1}. Since oxygen is strongly chemisorbed by chromium, a substantial fraction will be incorporated into the film, which therefore consists partially of chromium oxide.[211] O_2:Cr impingement ratios of less than 10^{-2} are required to produce essentially metallic films.[212] If the impinging gas is not adsorbed, however, it may be present in much greater concentrations without being incorporated. An example is the evaporation of Cr at similar rates in 3×10^{-3} Torr of He, which resulted in metallic films of nearly bulk resistivity.[213]

The condensation coefficient α_c may reflect two mechanisms which are responsible for particles being rejected by the film surface. One is the immediate rebounding of molecules whose translational energy was not taken up by the solid upon impact. The other mechanism is rapid desorption as caused by either a small energy of adsorption or high substrate temperatures (see Chap. 2, Sec. 3a). Oxygen is strongly chemisorbed by all metals of interest for reactive evaporation (see Chap. 2, Fig. 37). Even if these energies are substantially reduced because the metal surface is partially oxidized, the corresponding residence times of adsorbed oxygen at room temperature are still long compared with the film-growth rate. Hence, condensation coefficients $\alpha_c < 1$ indicate mostly lack of energy accommodation (see Chap. 8, Sec. 2a).

The accommodation coefficients of vapors on solids are largely unknown. Available data pertain to the common gases on clean surfaces of a few metals such as Mo and W.[214-217] The values at room temperature are between 0.1 and 0.5, with oxygen at the lower end of the range. The condensation coefficients of metals are usually assumed to be 1 because the impingement rates used in vacuum evaporation cor-

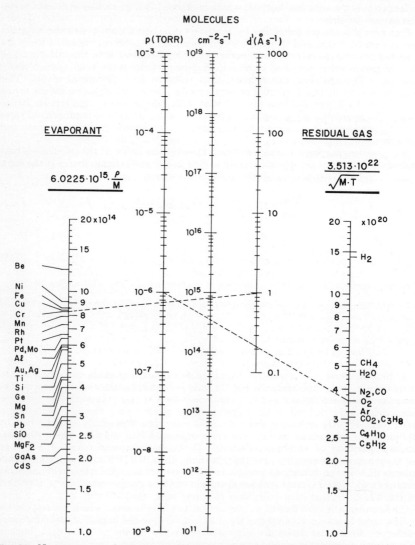

Fig. 36 Nomogram to determine impingement rates. The appropriate point on the evaporant axis is to be connected with the observed (pure) deposition rate on the d' axis. For residual gases, points on the far right and on the pressure axes should be connected. The intersects with the center axis give the impingement rates. (A gas temperature of $300°K$ has been assumed in locating the common gases on the far right axis.)

respond to pressures far greater than the equilibrium values at ordinary substrate temperatures. Experimentally, the condensation rates of metals evaporated from the same source under identical conditions are usually found to be independent of the substrate temperature from $-195°C$ up to several hundred degrees Celsius. For a few metals, condensation coefficients of 1 have actually been measured (Ref. 35, p. 117).

The reactive evaporation of metals or lower metal oxides onto substrates at moderate temperatures produces amorphous or poorly crystallized films whose stoichiometry is largely determined by the impingement ratio of the constituents. However, because of differences in the condensation of species, the film composition is not necessarily the same as the impingement ratio. This is best illustrated by the mechanism proposed by Ritter[218,219] for the reactive evaporation of Si and SiO. He assumed that the state of film oxidation is determined by the rate at which oxygen is chemisorbed:

$$\left(\frac{dN_{O_2}}{dt}\right)_{ads} = \left(\frac{dN_{O_2}}{dt}\right)_{imp} \frac{z(1-\theta)^2}{z-\theta} \alpha_c \exp\left(-\frac{E}{RT}\right)$$

where θ = surface coverage with oxygen
z = number of nearest surface sites
E = activation energy for the chemisorption of oxygen
α_c = condensation coefficient of O_2 molecules

The Si or SiO particles which arrive simultaneously are thought to condense with $\alpha_c = 1$. Accordingly, the ratio of O : Si atoms in the film is

$$\left(\frac{N_O}{N_{Si}}\right)_{film} = 2\frac{(dN_{O_2})_{ads}}{(dN_{Si})_{imp}} = 2\left(\frac{dN_{O_2}}{dN_{Si}}\right)_{imp} \frac{z(1-\theta)^2}{z-\theta} \alpha_c \exp\left(-\frac{E}{RT}\right) \quad (73a)$$

and

$$\left(\frac{N_O}{N_{Si}}\right)_{film} = 2\frac{(dN_{O_2})_{ads}}{(dN_{SiO})_{imp}} + 1 = 2\left(\frac{dN_{O_2}}{dN_{SiO}}\right)_{imp} \frac{z(1-\theta)^2}{z-\theta} \alpha_c \exp\left(-\frac{E}{RT}\right) + 1 \quad (73b)$$

The composition of film samples condensed at room temperature was determined by chemical microanalysis and plotted against the known impingement ratios as shown in Fig. 37. Equations (73a) and (73b) were evaluated by assuming $z = 4$ nearest neighbors and by relating the coverage to the state of film oxidation such that $\theta = 1$ for fully oxidized films with $(N_O/N_{Si})_{film} = 2$, and $\theta = 0.5$ for SiO films with $(N_O/N_{Si})_{film} = 1$. The product $\alpha_c \exp(-E/RT)$ was derived empirically by seeking the best fit for the experimental data. Values of 0.6 for Si and 0.2 for SiO were thus obtained. Further analysis led to the conclusion that the activation energy E was less than 1 kcal mol^{-1} for both reactions, which agrees with the known fact that the chemisorption of oxygen on metals is usually not activated (see Chap. 2, Sec. 3a).

As Fig. 37 demonstrates, Ritter's mechanism describes the empirical relation between film composition and impingement ratio quite well. To produce Si_2O_3 films, an impingement ratio of 20 as represented by a deposition rate of 5 Å s^{-1} in 10^{-4} Torr of oxygen is required. In a subsequent study, Anastasio[220] found a ratio slightly greater than 10 in substantial agreement with Ritter's data. Anastasio's investigation also showed that the same relationship between film composition and impingement ratio holds for a range of deposition rates from 10 to 110 Å s^{-1} and for oxygen pressures between 2×10^{-5} and 2×10^{-4} Torr.

The large excess of impinging oxygen molecules required might suggest that completely oxidized films are obtained with the greatest assurance if the oxygen pressure is increased beyond the necessary minimum. While this is true, it has detrimental consequences for the film properties. When evaporating TiO, Ritter[219] found a decrease in hardness and refractive index of the TiO films condensed at 300°C as the oxygen pressure was increased from 10^{-4} to 10^{-3} Torr. He attributes this to the greater frequency of collisions and associated energy losses which TiO molecules suffer at the higher oxygen pressure. The effect should be less significant at higher substrate temperatures, which automatically supply the thermal energy required for surface diffusion and ordering processes. This expectation is confirmed by the study of Feuersanger et al.,[221] who evaporated Ba and Ti at 10^{-2} Torr of oxygen. Here, the probability of collisions in the gas phase was high, yet crystalline $BaTiO_3$ films were obtained by maintaining condensation temperatures of 800 to 1000°C.

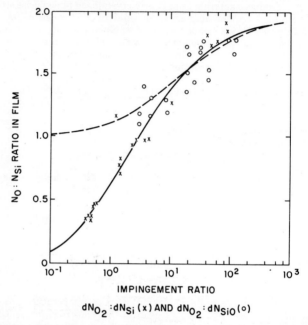

$$dN_{O_2} : dN_{Si} \, (\times) \text{ AND } dN_{O_2} : dN_{SiO} \, (\circ)$$

Fig. 37 Oxygen-to-silicon ratio in reactively evaporated films as determined experimentally (points) and calculated from Eqs. (73a) and (73b) (curves). (*After Ritter.*[219])

The reactive evaporation of dielectric films whose properties depend on the attainment of a crystalline structure requires elevated substrate temperatures even if the metal atoms do not lose their kinetic energy by collisions. In these cases, the surface reaction or atomic ordering process is the rate-controlling step. This process is thermally activated in contrast to Ritter's mechanism for Si_2O_3, where nonactivated chemisorption determines the degree of oxygen incorporation. Surface-reaction-controlled deposition has been observed by Krikorian[222] in her study of epitaxial oxide films formed by reactive evaporation onto sapphire substrates. Her results in Fig. 38 show that the growth rates of crystalline Ta_2O_5 films increase with temperature. Since the impingement rates of metal atoms and oxygen molecules are constant for each curve, a significant fraction must be rejected by the growing surface. At the substrate temperatures used, one can expect relatively poor energy accommodation of impinging particles. This holds even for Ta atoms, as the declining growth rates for the pure metal in Fig. 38 show. In addition, the residence times of adsorbed oxygen atoms at these temperatures are short enough to reduce α_c further.

Although the film-growth process is surface-reaction-controlled, the oxygen-to-metal impingement ratio must still be carefully chosen. A minimum O_2 impingement rate is required for complete oxidation. With Ta rates of the order of 2 Å s^{-1}, the minimum oxygen pressure at 700°C is about 10^{-3} Torr, whereas a somewhat lower pressure is sufficient at 900°C, where the reaction rate is higher.[222] The expectation that an excessive amount of oxygen is tolerable because the film surface rejects it is true only in regard to film composition. The film structure, however, is adversely affected. In the case of Ta_2O_5 growing on single crystalline substrates, epitaxy could not be achieved if the oxygen pressure approached 10^{-3} Torr. For Al_2O_3 films growing at 400 to 500°C, the pressure dividing single-crystal and polycrystalline growth conditions was about 10^{-5} Torr.[222]

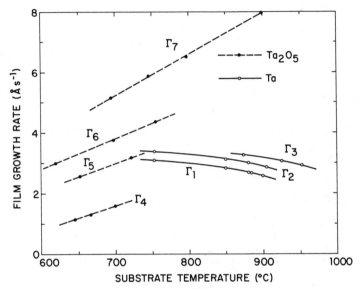

Fig. 38 Film growth rates vs. substrate temperature for constant evaporation rates Γ. The solid lines are for pure tantalum evaporation (Γ_1, Γ_2, Γ_3), the broken lines for reactive evaporation (Γ_4, Γ_5, Γ_6, Γ_7). (*After Krikorian.*[222])

Examples of metal oxides which have been reactively evaporated and the conditions employed are listed in Table 13. In nearly all cases, the deposition rates are small to ensure a high impingement ratio without having to increase the oxygen pressure beyond 10^{-4} Torr. Elevated substrate temperatures are used if film properties such as crystallinity, density, hardness, optical absorption, or dielectric constants are of interest rather than merely film composition. Although the existing studies pertain mostly to oxides, reactive evaporation is applicable to other classes of compounds. An example is CdS, which tends to yield slightly nonstoichiometric, cadmium-rich films of low resistivity if evaporated alone. Pizzarello[230] obtained stoichiometric films of high resistivity by evaporating CdS in the presence of sulfur vapor. Titanium and zirconium nitrides have been prepared by evaporating the respective metals in a nitrogen atmosphere.[231] The formation of carbide films by metal evaporation in the presence of hydrocarbons has apparently not been investigated as yet, possibly because the required substrate temperatures are likely to be impractically high.

(2) Two-source Evaporation The installation of two or more sources for the evaporation of different materials in the same vacuum system is widely practiced to produce multilayer films. By operating two sources simultaneously, it is possible to deposit multiconstituent films which are not amenable to direct evaporation. The types of sources employed are the same as in single-source evaporation. Resistance-heated effusion cells,[232-235] electron guns,[236,237] and wire-ring sources[238] have been successfully used. To avoid cross contamination, the sources must be separated by shields which shadow one source from the vapor of the other without blocking passage to the sub-

TABLE 13 Reactive Evaporation of Metal Oxides

Deposition process	Oxygen pressure, Torr	Deposition rate, Å s^{-1}	Substrate temp, °C	Comments on film properties	References
$2Al(g) + \tfrac{3}{2}O_2 = Al_2O_3(s)$	5×10^{-4}	......	......	50-Å-thick pinhole-free films of 10^8 ohms cm^{-1} dc resistance	223
$2Cr(g) + \tfrac{3}{2}O_2 = Cr_2O_3(s)$	10^{-5}–10^{-4}	4–5	......		224
	2×10^{-6}–5×10^{-5}	~1	400–500	Epitaxial Al_2O_3 films on sapphire at $p_{O_2} \leqq 10^{-5}$ Torr	222
$2Fe(g) + \tfrac{3}{2}O_2 = Fe_2O_3(s)$	2×10^{-5}	~2	300–400	Dielectric films of poor crystallinity	225
	~10^{-5}	~1	25	Film density approaches that of bulk Cr_2O_3	211
$2Cu(g) + \tfrac{1}{2}O_2 = Cu_2O(s)$	10^{-4}–10^{-3}	......	......	Nearly amorphous Fe_2O_3 films	226
	10^{-4}–10^{-2}	......	......	Cu_2O films converted into CuI films by exposure to I_2 vapors	227
$SiO(g) + \tfrac{1}{2}O_2 = SiO_2(s)$	10^{-5}–10^{-4}	......	−195 to 500	Absorption coefficient is a function of Si:O ratio in films	228, 229
	9×10^{-5}	4.5	Up to 350°C	Essentially SiO_2 films without absorption of visible light	174
	10^{-4}	5	25	O_2:Si impingement ratio must be about 20 to obtain SiO_2 films	218, 219
	2×10^{-5}–2×10^{-4}	10–110	......	Film composition depends only on the O_2:Si impingement ratio. This must be ~10 for SiO_2 films	220
$2Ta(g) + \tfrac{5}{2}O_2 = Ta_2O_5(s)$	10^{-4}–10^{-3}	~2	700–900	Epitaxial Ta_2O_5 films on sapphire and CaF_2 substrates	222
$TiO(g) + \tfrac{1}{2}O_2 = TiO_2(s)$	~10^{-4}	......	300	Hardness and refractive index of films suffer if $p_{O_2} > 10^{-4}$ Torr	219
$Ba(g) + Ti(g) + \tfrac{1}{2}O_2 = BaTiO_3(s)$	10^{-2}	2–8	770–1025	Films on polished Pt-Rh substrates had dielectric constants up to 1,330	221

TABLE 14 Two-source Evaporation, Experimental Conditions, and Types of Films Obtained

Evaporated constituents	Evaporation conditions and method of control	Substrate temp, °C	Films obtained	References
		Alloy and Multiphase Films		
Cu + Ni............	Sequential evaporation	Low	Stratified films. Annealing at 200°C yields two-phase alloy films	241
Cu, Ag, Au, Mg, Sn, Fe, Co.	Simultaneous evaporation from two sources. Ionization-rate monitor control, $\pm 1\%$	−193	Binary alloy films of metastable structures	240
Cu, Ag, Au, Al, Ni, and others	Simultaneous evaporation from two sources. Rates adjusted by varying source temperatures	25–600	Binary alloy films of varying composition and structure	242
Ni + Fe............	Two wire-ring sources, evaporation rates controlled by quartz-crystal oscillator	300	Permalloy films. $d' \approx 10$ Å s^{-1}	238
Nb + Sn...........	Two sources, rates monitored by particle impingement-rate monitor. Impingement ratio $N_{Nb}:N_{Sn} = 3$	25–700	Superconducting Nb$_3$Sn films. $d' \approx 2$ Å s^{-1}, $\alpha_c \approx 1$	243
V, Nb + Si, Sn.........	Two electron-gun sources, rates monitored by measuring ionization current	......	Superconducting films of approx composition Nb$_3$Sn and V$_3$Si	237
ZnS + LiF...........	Two sources, rates monitored by a microbalance. Variable impingement ratios	30–40	Mixed dielectric films of different composition. $d' = 10$–30 Å s^{-1}	244
Au, Cr + SiO, MgF$_2$.......	Two-source evaporation with ionization-rate monitor control (± 1–2%)	25–300	Au–SiO, Au–MgF$_2$, Cr–SiO, and Cr–MgF$_2$ resistor films of different compositions	245
Cr + SiO...........	SiO source at 1100°C, Cr source at 1500°C. Impingement ratio varied with location on substrate	400	High-resistivity Cr–SiO films of variable composition	246

TABLE 14 Two-source Evaporation, Experimental Conditions, and Types of Films Obtained (Continued)

Evaporated constituents	Evaporation conditions and method of control	Substrate temp, °C	Films obtained	References
		Compound Films		
Cd + S..........	Two effusion ovens, Cd at 400–450°C, S at 120–150°C. Cd excess	400–650	Stoichiometric CdS crystals	232
Cd + Se.........	Impingement fluxes controlled by source temperature. $N_{Cd} = 2 \times 10^{16}$, $N_{Se} = 10^{16}$–10^{17} cm^{-2} s^{-1}	200	Stoichiometric CdSe films	247, 248
PbSe + PbTe......	Source temperatures varied around 700°C	300	Epitaxial films of PbSe$_{1-x}$Te$_x$ on NaCl crystals	233
Bi + Te..........	Bi source at 750°C, Te source temperature variable. $N_{Te}:N_{Bi} = 10$–40	400–500	Stoichiometric films of Bi$_2$Te$_3$. n-type, 2×10^{19} electrons cm^{-3}	249
Bi + Se..........	Rate control by quartz-crystal oscillator. Se source at 250°C; Bi source temperature variable	52	Vitreous, semiconducting films of nonstoichiometric composition	250
Al + Sb..........	Source temperatures adjusted by quartz-crystal oscillator to yield $N_{Sb}:N_{Al}$ ratios of 1.6–16	550	Stoichiometric AlSb films. $d' \approx 10$ Å s^{-1}	234
Ga + As..........	Ga source at 940–970°C, As source at 300°C. $N_{As}:N_{Ga} \approx 10$; Ga impingement flux: 10^{15} cm^{-2} s^{-1}	550	Stoichiometric GaAs films. Epitaxial on (100) NaCl, polycrystalline on quartz	251
Ga + As..........	Ga source at 910°C, As source at 295°C. Deposition rate: <2 Å s^{-1}	375–450	Stoichiometric GaAs films on GaAs, Ge, and Al$_2$O$_3$ single-crystal substrates. Fiber texture to single-crystalline	252
In + As..........	Incident fluxes: $N_{In} = 5 \times 10^{16}$ cm^{-2} s^{-1}, $N_{As} = 5 \times 10^{16}$–$5 \times 10^{17}$ cm^{-2} s^{-1}	230–680	Stoichiometric InAs films; n-type	247, 248
In + Sb..........	Incident fluxes: $N_{In} = 5 \times 10^{16}$–$5 \times 10^{17}$ cm^{-2} s^{-1}	400–520	Stoichiometric InSb films; n-type	247, 248
	Source temperatures adjusted by microbalance to yield $N_{Sb}:N_{In} = 1.1$	250	Stoichiometric InSb films; α_e of Sb ≈ 0.6	253

strate. Thermal insulation of the sources is also important to facilitate independent temperature adjustments.[233]

The evaporation of two materials at different temperatures followed by joint condensation on the same substrate circumvents the problems of fractionation and decomposition encountered in the direct evaporation of most alloys and certain compounds. It is also possible to codeposit materials which form neither compounds nor solid solutions. Because of the random arrival of single atoms and their limited mobility in the absorbed state, these multiphase films consist of very small particles which are intimately mixed. Important examples are resistor films of ceramic oxides and metals (cermets), whose multiphase nature becomes apparent only after annealing and recrystallization.[239] Segregation of phases can also be suppressed in binary metal systems. This was demonstrated by Mader,[240] who obtained metastable alloy films by condensation onto liquid-nitrogen-cooled substrates. A list of coevaporated materials, their experimental conditions, and the resulting films is shown in Table 14.

The objective of two-source evaporation is the deposition of films having one particular composition. Therefore, the central problem in this technique is the control of condensation rates in the exact constituent ratio desired. These rates are primarily determined by the equilibrium pressure of the evaporant, $p^*(T_1)$, at the source temperature. As the evaporant spreads omnidirectionally, the particle density decreases with increasing distance from the source. Assuming a small-surface source with an emission characteristic according to Eq. (52), and expressing $\mathfrak{M}_r$ by the corresponding impingement pressure [Eq. (37)] and $\mathfrak{M}_e$ by the mass-evaporation rate [Eqs. (48) and (49)], the vapor pressure p_r at the receiving surface is given by

$$p_r = \frac{A_e \cos \varphi \cos \theta}{\pi r^2} \, p^*(T_1) \tag{74}$$

where A_e = area of vapor source
  φ, θ = angles of emission and incidence, respectively
  r = distance between source and point receiving deposit

Most metals and congruently evaporating compounds have very small equilibrium pressures at practical substrate temperatures T_3. Accordingly, p_r is much larger than the equilibrium pressure $p^*(T_3)$. If a vapor is highly supersaturated with respect to the surface temperature, condensation occurs generally with $\alpha_c = 1$. Consequently the composition of alloy and metal-dielectric films is determined by the constituent pressures p_r at the substrate. The control of the film composition may then be achieved by direct measurements of the particle density in the vapor stream, using ionization-rate monitors,[236,240,245] quartz-crystal oscillators,[238,250] electromagnetic microbalances,[253] or mechanisms which are actuated by the momentum received from impinging particles.[243,244] The two monitoring devices required must be located so that they are exposed to the vapor of only one but not the other constituent. The source temperatures are adjusted manually to yield predetermined condensation rates, or the electric signal of the monitor may be used to maintain the rates constant at the desired levels (see Sec. 7d). Feedback control in conjunction with ionization-rate monitors has allowed film-composition control within ± 1 to 2%.[240,245]

A general problem is the positioning of two sources such that the substrate area is exposed uniformly to both vapor streams. According to Eq. (74), the effective vapor pressure p_r depends on the angle of incidence θ. While a few percent thickness variation across the substrate area is often acceptable, differences in angles of incidence from two sources also cause compositional deviations, as illustrated in Fig. 39. This effect is minimized if concentric two-source arrangements are used.[238,247,248] An alternate solution is to tilt the effusion ovens such that both are aimed at the substrate surface under the same angle of inclination.[232] In general, angle-of-incidence effects can be reduced by increasing the source-to-substrate distance. If films of continuously varying composition are to be deposited, this may be accomplished by deliberately spacing the two sources far apart.[246]

The composition of compound films is less sensitive to source positioning and requires less stringent control of the impingement fluxes than alloy or metal dielectric films.   The reason is that the condensation rate of compounds is not solely a function of the ratio of constituent vapor pressures but is also determined by surface reaction. Thus, two-source evaporation of compounds is similar to reactive evaporation where the substrate temperature controls the film growth.   The preferential condensation of compounds from binary vapors has been analyzed by Guenther[247,248] and developed into an experimental technique known as the three-temperature method.   The latter exploits the fact that the free energies required for the dissociation of compounds according to

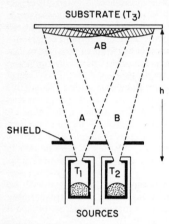

$$AB(s) = A(g) + \tfrac{1}{2}B_2(g)$$

are greater than the free energies of evaporation of the constituents

$$A(s) = A(g) \qquad \text{and} \qquad B(s) = \tfrac{1}{2}B_2(g)$$

This difference is reflected in constituent vapor pressures which are lower over the compound than over the pure elements.   Figure 40 illustrates the situation for CdSe.   At fixed constituent pressures p_r over the substrate surface, there exists a wide temperature interval within which the vapors are supersaturated only in regard to the compound but not for the elements.

Fig. 39 Two-source evaporation arrangement yielding variable film composition.

The degree of supersaturation required to deposit the compound but not the pure elements depends further on the condensation coefficients α_c, which are generally not known a priori.   Guenther[248] determined the condensation coefficients of Cd and

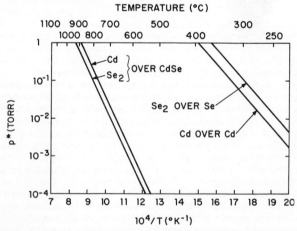

Fig. 40 Equilibrium vapor pressures of Cd and Se₂ over the compound (*Burmeister and Stevenson*[163]) and over the pure elements (*Honig*[19]).

Se₂ experimentally and derived temperatures and impingement rates which allow only compound formation.   As shown in Table 14, the stoichiometry of CdSe films condensed at 200°C with a Cd impingement rate of 2×10^{16} cm^{-2} s^{-1} is maintained if the selenium rates vary from 10^{16} to 10^{17} cm^{-2} s^{-1}.   This degree of freedom in the constituent vapor-pressure ratio has been termed the stoichiometric interval.[248]   Its extent is a function of the substrate temperature and the vapor pressures p_r.

The three-temperature method is also applicable if one of the constituents is not volatile at the substrate temperature.   This is the case with III-V compounds and is illustrated for InAs in Fig. 41.   In contrast to CdSe, indium vapor which is super-saturated with respect to the compound is also supersaturated in regard to the pure-metal phase.   Therefore, stoichiometric intervals for III-V compounds require an excess of the more volatile constituent to ensure that the condensing group III element reacts to completion.   The evaporation conditions listed in Table 14 show that the impingement ratios may vary from 1 to 10.   The absolute rates generally used are of the order 10^{15} to 10^{17} cm^{-2} s^{-1}, which corresponds to pressures of 10^{-5} to 10^{-3} Torr over the substrate.   Higher pressures are not recommended because of excessive scattering as the two vapor streams mix.

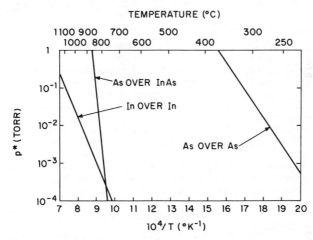

Fig. 41   Equilibrium vapor pressures of In and As over the elements (*Honig*[19]), and decomposition pressure of As over InAs (*Guenther*[248]).

Considerable freedom also exists in the choice of substrate temperatures.   However, the useful temperature interval is often narrowed by factors other than film stoichiometry.   To obtain highly ordered crystalline films, temperatures somewhat above the minimum values for single-phase condensation are required.   Upper limits are imposed by the necessity to achieve supersaturation below about 10^{-3} Torr, or by the melting point of the compound as in the case of InSb.   More subtle effects are slight deviations from stoichiometry which vary with temperature and determine the type and degree of semiconductivity.   Thus, InAs films deposited at lower temperatures contain an As excess, while condensation at the upper end of the temperature range leads to an As deficit.[248]

The risk of obtaining large deviations from the stoichiometric composition is potentially greater if one or both constituents have substantial solubilities in the compound. This is the case for Bi and Te in Bi$_2$Te$_3$.   However, Haenlein and Guenther[249] established that such compounds, too, may have stoichiometric intervals.   As shown in Table 14, a large excess of Te vapor can be rejected and the film stoichiometry maintained.   These considerations do not apply at low substrate temperatures, where impinging molecules are indiscriminately incorporated into the film as exemplified by the Bi-Se system in Table 14.   The pseudo-binary system Pb (Se,Te) is comparable with Bi$_2$Te$_3$ insofar as one constituent, PbTe, has a fairly broad range of existence.[79] Mixed compound films of the III-V family have also been prepared by simultaneous evaporation from three sources.[235]

(3) Flash Evaporation Flash evaporation is another technique for the deposition of films whose constituents have different vapor pressures. In contrast to two-source evaporation, it does not require provisions to monitor the vapor density, nor is the control of the source temperature particularly critical. The objective of film-composition control is accomplished by evaporating to completion small quantities of the constituents in the desired ratio. Only one filament is used at a temperature sufficiently high to evaporate the less volatile material. Although fractionation occurs during the evaporation of each particle, the latter are so small that stratification in the film is limited to a few atomic layers. These potential inhomogeneities are further reduced by dispensing the evaporant in a steady trickle. Thus, there are several particles in different stages of fractionation residing on the filament at all times. The net result of these simultaneous, discrete evaporations is a vapor stream whose composition is uniform and identical to that of the source material. Accordingly, excellent control of film composition can be achieved. Young and Heritage[254] reported that Ni-Fe-Cr films prepared by flash evaporation showed less compositional variation than sputtered films.

The technique is applicable for the evaporation of alloys, metal-dielectric mixtures, and compounds, as shown by the examples in Table 15. In most cases, the vapors impinging on the substrate are highly supersaturated so that the film composition is not affected by condensation coefficients. The cuprous chalcogenides[271] are an exception, but generally the control of film composition is determined by how well the objective of complete evaporation of the source material is accomplished. For this purpose, several experimental techniques are available. They can be characterized by three criteria, namely, the form in which the evaporant material is introduced, the mechanism used to dispense the evaporant, and the type of flash filament employed.

In the case of metals and alloys, continuous evaporation of small quantities is possible by feeding wire from a spool through a guide tube against the hot filament. Such mechanisms have been described by Tandeski et al.[257] for Permalloy and by Siddall and Probyn[114] for Nichrome. A modified arrangement, which allows alternate flash evaporation of metals and organic insulators, utilizes knife-edges to cut short lengths of wire which are dropped onto the filament.[259] In general, however, the evaporant is not available in wire form, and the greatest freedom in regard to materials is obtained by utilizing powders.

Various types of powder dispensers are shown in Fig. 42. They all have storage hoppers but differ in the powder release and transport mechanisms. The moving-belt feeder (Fig. 42a) was the first device employed in flash evaporation, but it has rarely been used after its introduction by Harris and Siegel in 1948,[255] since alternate approaches are mechanically simpler. The worm-drive feeder in Fig. 42b moves the powder from the hopper to a release opening by means of an Archimedean screw. Its application has been restricted to metal-dielectric mixtures because the worm gear tends to compress metal powders into lumps, while brittle compounds are ground into very small particles. These effects make uniform and complete evaporation impossible. The ratchet gear and spring mechanism is widely used in other types of feeders, too, to prevent the powder from sticking and accumulating in the chute.

Since particle grinding and jamming are common problems with worm-drive feeders, alternate solutions have been sought. The disk feeder shown in Fig. 42c exerts no mechanical force on the powder. Its feed rate is determined by the gap between hopper and disk and by the number of revolutions per minute. The device can be modified to dispense pellets rather than powder by providing an orifice instead of a gap at the lower end of the hopper.[137] This design offers a greater loading capacity than the disk magazine feeder in Fig. 42d, which is also used to dispense metal pellets or spheres.[258]

Powder dispensers which operate by either mechanically[266] or electromagnetically[276] induced vibrations (Fig. 42e and d) have been employed mostly for the evaporation of compounds. They do not allow the same latitude in feed-rate adjustment as disk devices because the trickle of powder tends to become discontinuous or nonuniform under various conditions. This is not too critical, since most compounds evaporate more easily and rapidly than refractory metals, and since each compound

TABLE 15 Flash Evaporation of Materials

Materials	Form of evaporant	Feeder mechanism	Filament temp., °C	Substrate temp., °C	Comments on films	Reference
Metals and Alloys						
Au(64)-Cd(36), Cu(52)-Zn(48)...	Powdered alloys, 80/100 mesh	Moving belt	....	....	Au-Cd and β-brass films have composition of source	255
Ni + Fe...	Mixed powders	Disk and wiper	~1930	....	Ni-Fe films with ±1% control of composition	256
Ni(86)-Fe(14)...	Alloy wire	Spool and guide tube	2000	300	Ni-Fe film composition equal to that of source ±0.2%	257
Ni; Fe; Cu; constantan; chromel; alumel	Pellets	Disk magazine	2000	200–250	Thin film thermocouples. Rates: 5–300 Å s^{-1} across 12 cm distance	258
Ni(80)-Cr(20)...	Alloy wire	Spool and guide tube	1620	....	Nichrome films, Cr content varies with filament temperature	114
Ni + Cr (20, 55, 70 % Cr)...	Mixed powders, 100/300 mesh	Vibrating chute	1800	300	Alloy films within 1 % of source composition. Rates: 1–10 Å s^{-1}	200
Sn; nylon...	Wire; strands	Two spools and cutting knives	200 (nylon)	~0	Alternate layers of metal and insulator	259
Metal-Dielectric Mixtures						
Cu + SiO (1:5)...	Mixed powders	Rotating tube	....	−269	Highly disordered films of high resistivity	260
Cr + 30 mol % SiO...	Mixed powders, 325/400 mesh	Worm drive	~2000	200	250 ohms/sq resistor films with +20 to 50 % deviations	261
Cr + SiO (62 and 74 mol % Cr)...	Mixed powders, 125/325 mesh	Worm drive	2000	400	Resistor films; SiO content less than source. Rates: 4 Å s^{-1} across 23 cm distance	262
Cr + SiO (50–100 mol % Cr)...	Sintered pellets, ~0.7 mm size	Disk and wiper	2050 ± 50	200	Resistor films, SiO content equals that of source ±1–3 %. Rates: 20–30 Å s^{-1} across 70 cm distance	137, 239
Cr(15)-Si(85)...	Powdered alloy	Vibrating chute	2000	200–500	Resistor films	263
Cr$_3$Si + TaSi$_2$ + Al$_2$O$_3$...	Mixed powders	Worm drive	2500	200–400	Resistor films with ± 10 % control	264

TABLE 15 Flash Evaporation of Materials (Continued)

Materials	Form of evaporant	Feeder mechanism	Filament temp, °C	Substrate temp, °C	Comments on films	Reference
		Compounds				
AlSb...........	Powder, 100/150 mesh	Vibrating trough	1400–1600	700	Imperfect epitaxial films on Ge	265, 266
GaP...........	Powder, 100/150 mesh	Vibrating trough	1500	540	Epitaxial films on Ge crystals	265, 266
GaAs...........	Powder, 100/200 mesh	Vibrating trough	1450	300–670	Epitaxial films on Ge crystals above 600°C. Rates: 2–30 Å s^{-1}	267
	Powder, 100/200 mesh	Worm drive	1400–1800	530 ± 10	Epitaxial films on GaAs crystals. Rates: 2–5 Å s^{-1} across 21 cm distance	268
	Powder, 100/150 mesh	Vibrating trough	1300–1800	475–525	Epitaxial films on Ge crystals	265, 266
	Powder, 40/60 mesh	Micrometer screw and piston	1325	525–575	Highly oriented films on Ge and GaAs crystals. Rates: 10–25 Å s^{-1} across 10 cm distance	269
GaSb...........	Powder, 100/150 mesh	Vibrating trough	1650	500	Epitaxial films on Ge crystals	265, 266
InP...........	Powder, 100/150 mesh	Vibrating trough	1400–1650	300	Epitaxial films on Ge crystals	265, 266
InAs...........	Powder, 100/150 mesh	Vibrating trough	1500	500	Epitaxial films on Ge crystals	265, 266
InSb...........	Granules	Vibrating trough	1600	450–460	Epitaxial films on InSb crystals, n-type, 10^{15}–10^{17} donors per cm³	270
	Powder, 100/150 mesh	Vibrating trough	1650	300–400	Epitaxial films on Ge crystals	265, 266
CuS; Cu₂Se...........	Powders of 250–300 μ	Vibrating chute	1400	25	Semitransparent, conductive films of Cu₁.₈S and Cu₁.₈Se	271
BaTiO₃...........	Sintered powder, 100/200 mesh	Vibrating trough	2300	500–700	Crystalline films, dielectric constants of 400–700. Rate: 3 Å s^{-1} across 8 cm distance	272, 272
	?	?	2100	900–1000	Epitaxial films on sapphire and Si crystals, $p(O_2) = 10^{-4}$ Torr. Rates: 0.1–0.3 Å s^{-1}	273
Various perovskites...........	Sintered powder, 100/200 mesh	Vibrating trough	2050–2300	500–700	Perovskite films, epitaxial on LiF crystals. Rates: 1–3 Å s^{-1} across 8 cm distance	274

particle contains the constituents in the proper ratio.   An ultrasonic particle feeder consisting of a stainless-steel tube attached to a barium titanate transducer has been described by Eckardt and Peacock.[277]   A vibratory feeder of large storage capacity is commercially available.*   It is a cylindrical container with a helical grove cut into its interior wall.   Vibrations induced by a solenoid convey a continuous stream of

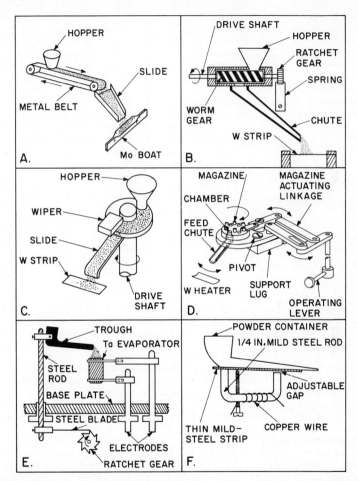

Fig. 42 Flash-evaporation mechanisms. (*A*) Belt feeder. (*Harris and Siegel*.[255]) (*B*) Worm-drive feeder with mechanical vibrator. (*Himes, Stout, and Thun*,[275] *Braun and Lood*.[262]) (*C*) Disk feeder. (*Beam and Takahashi*.[256]) (*D*) Disk magazine feeder. (*Marshall, Atlas, and Putner*.[258]) (*E*) Mechanically vibrated trough and cylindrical source. (*Richards*.[266]) (*F*) Electromagnetically vibrated powder dispenser. (*Campbell and Hendry*.[200])

evaporant upward in the helix and to an exit channel.   The material must be available in the form of granules or pellets of 0.1- to 0.8-mm size.

Of considerable importance is the proper choice of material and design for the flash-evaporation filament.   The latter must be capable of attaining temperatures of

* Bendix Corporation, Vacuum Division, Cincinnati, Ohio; in Europe: Balzers A.G., Liechtenstein.

typically 2000°C without volatilization or heavy reaction with the evaporant. Flat posts of refractory materials which are heated by electron bombardment have been used in conjunction with wire feeders.[114,257] Powder flash evaporation from an electron-beam-heated tungsten disk has been reported by Wilson and Terry.[264] By far the most common technique, however, is the evaporation from resistance-heated refractory-metal sources. The simplest filaments are flat strips of 0.005-in. tungsten sheet which are easily made and may be discarded after one experiment. The latter consideration is significant because tungsten alloys with most of the materials in Table 15 so that the useful life of the filament is limited. For the III-V compounds, which require somewhat lower evaporation temperatures, tantalum filaments are also suitable.[266,270] $BaTiO_3$ and the other perovskites, however, severely attack tungsten through oxidation. Therefore, these materials have been evaporated from iridium filaments.[272–274]

A universal problem encountered in the flash evaporation of powders from flat filaments is incomplete evaporation due to particle ejection and deflection. Since the

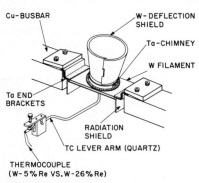

Cu-BUSBAR

W-DEFLECTION SHIELD

Ta-CHIMNEY

W FILAMENT

Ta END BRACKETS

RADIATION SHIELD

TC LEVER ARM (QUARTZ)

THERMOCOUPLE
(W-5% Re VS.W-26% Re)

Fig. 43 Flash-evaporation filament with spot-welded chimney and separate deflection cone. (*Glang, Holmwood, and Maissel.*[137])

evaporant has a large surface-to-volume ratio but can usually not be degassed prior to evaporation, the sudden release of gases upon impact on the filament is often sufficient to expel particles. Another loss mechanism is the deflection of falling powder away from the filament by the current of rising vapors. The latter effect is stronger the higher the evaporation rates, which are thereby limited. An example is the maximum feed rate of 0.15 g min⁻¹ for Nichrome alloys found by Campbell and Hendry.[200] Small particles are more strongly deflected than larger ones,[268] which is the reason for using graded powders. The losses of unevaporated source material may be as high as 50%,[256] particularly if long source-to-substrate distances necessitate high feed and evaporation rates.[137]

In the evaporation of homogeneous powders such as compounds or alloys, particle ejection and deflection affect only the economy of the process and the maintenance of the vacuum system. There is, of course, also the risk that particles may be thrown against the substrate surface. To avoid these difficulties, coarser powders and cylindrical (see Fig. 42e) or conical crucibles have been used in the evaporation of perovskites[274] and III-V compounds.[266,270] Ellis[271] folded a tungsten sheet to form a resistance-heated filament shaped like an open envelope. When alloy or metal-dielectric films are evaporated from mixed powders, evaporant losses are also associated with unpredictable changes in film composition. This has been observed with Ni-Fe films, which tend to contain less iron than the source material,[256] and with Cr-SiO films, which are deficient in the lighter component, SiO.[137,262] Since the properties of resistor films, in particular the degree of resistance change during post-deposition annealing and stabilization, are very sensitive to variations in composition, flash evaporation is practical only if evaporant losses are avoided.

In the case of Cr-SiO, this has been accomplished by sintering the mixed powders into pellets which contain the constituents in the proper ratio and are sufficiently large to fall through the rising vapors without deflection.[137] To avoid losses from rebounding or decrepitating pellets, the flash filament may be provided with a spot-welded tantalum chimney and a separate metal cone, as shown in Fig. 43. The filament can be changed without having to discard the cone. Temperature control is facilitated by a thermocouple whose tip is spring-loaded against the filament bottom. With this arrangement, pellets containing up to 30 mol % SiO yielded films of identical composition with deviations of only ±1%. At higher SiO concentrations, the degree of control is slightly inferior.   Furthermore, since deflection losses are eliminated, the flash source is only power-limited; i.e., the feed rate can be increased until the filament temperature drops because of an overload of evaporant.   Feed rates of about 1 g min^{-1} with deposition rates between 20 and 30 Å s^{-1} across 70 cm substrate distance have been obtained.

Flash evaporations have mostly been performed in poor vacua of 10^{-5} to 10^{-4} Torr. This is attributable to the high gas content of the evaporant powder and outgassing from the surfaces surrounding the relatively large-area flash filament.   The effect of the high background pressure on film properties may be inconsequential, especially in the case of oxide films.   However, this is not so if strongly electropositive metals are deposited.   Siddall and Probyn[114] found that the chromium content of Nichrome films evaporated at 10^{-4} Torr varied significantly as a function of the filament temperature.   At 1400°C, the films were deficient in chromium, while an excess was found at 1700°C.   It is believed that the evaporation of Cr at low temperatures is retarded by the formation of chromium oxide which becomes volatile at the higher temperatures.   In a few instances, pressures of 10^{-6} Torr or less have been maintained, for example, during the flash evaporation of InSb[270] and CrSiO.[137,262]

The temperatures of the substrates in Table 15 determine primarily the degree of order and crystallinity in the films.   The temperatures listed for compounds are those required for homo- or heteroepitaxial films.   In the case of III-V compounds, the upper limit is determined by reevaporation of the group V constituent, whereas lower temperatures yield polycrystalline films.[265]   An excess of the group V constituent in the source material does not affect the film stoichiometry adversely since the condensation process is reaction-controlled as in two-source evaporation.[266]   Flash evaporation has also been used to prepare single-phase pseudo-binary compounds of the III-V family.[278]

7. DEPOSITION MONITORING AND CONTROL

The objective in vacuum evaporation is nearly always to deposit films to certain specifications.   If the latter pertain to film properties which are primarily extensive such as thickness or sheet resistance, it is sufficient to determine when the accumulated deposit has reached the desired value so that the process can be terminated. However, intensive film properties such as density, resistivity, stress, or crystallinity depend on the rates at which evaporant and residual gas molecules arrive at the substrate.   It is therefore often necessary to maintain specified evaporation rates. The sensing devices which allow measurements during the evaporation process are referred to as either thickness or rate monitors.   They exploit different physical effects to determine the density of the evaporant stream, the mass of the deposit, or a thickness-dependent film property.

Traditionally, monitoring devices have been constructed by individual investigators to suit their particular purpose, and numerous design modifications have been reported in each class.   Several types of monitors have become commercially available during the last few years, after sufficient operational experience had been accumulated to judge their ruggedness and ranges of application and to select the most practical design features.   Reviews of thickness and rate monitors have been given by Steckelmacher[279] and Behrndt.[139]   The reader is referred to these articles for more detailed descriptions of individual devices, since the following treatment is limited to the operating principles and gives only a few representative examples.

a. Monitoring of the Vapor Stream

There are two methods of measuring the density of the evaporant vapor stream.   In one technique, the vapor molecules are ionized by collisions with electrons, and the ions are collected.   The other approach is based on measuring the dynamic force which impinging particles exert on a surface.   Both methods indicate the *evaporation rate* at a particular instance.   To derive the accumulated film thickness, the rates must be integrated.   Furthermore, the methods require empirical calibration; i.e., the film thicknesses obtained for known deposition times have to be determined independently to correlate rate-meter indication and actual deposition rate. The resulting calibration curves are specific for the individual monitoring device and the material evaporated. Their reproducibility is also contingent upon leaving the source, rate meter, and substrates in fixed positions.

(1) Ionization-gauge Rate Monitors Rate-monitoring devices based on the ionization of evaporant vapors are shown in Fig. 44. All have a hot-tungsten emission

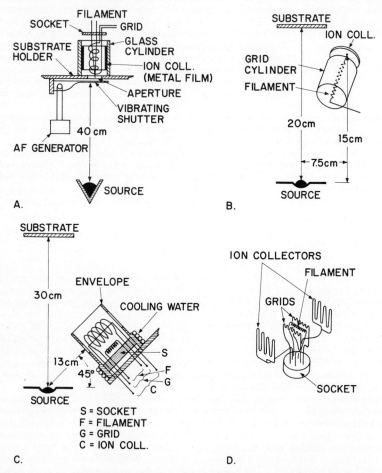

Fig. 44 Ionization-rate monitor designs and arrangements. (*A*) After Schwarz.[280] (*B*) After Giedd and Perkins.[284] (*C*) After Perkins.[285] (*D*) After Dufour and Zega.[282]

filament to provide electrons. The latter are accelerated with voltages of 150 to 200 V by means of grids (anodes) in the form of a helix, a cylinder, or a wire electrode. The ion collector may be a cylinder, a disk, or a simple wire to which negative voltages of the order of 20 to 50 V are applied. The monitors are mounted so that the ion collector and the entrance aperture point toward the vapor source. The distance to the source may be the same as the source-to-substrate separation or made shorter to enhance the sensitivity.

The ion current I_i which is produced by an electron-emission current I_e is given by[280]

$$I_i = D'I_e \left(\frac{N}{V}\right)_m$$

where $(N/V)_m$ is the number of vapor molecules per cubic centimeter in the monitor and D' is a gauge constant. The vapor density in the monitor depends on the rate of molecules arriving at the entrance aperture (area $= A_m$) and on their average velocity $\bar{c}$:

$$\frac{dN_m}{A_m \, dt} = \left(\frac{N}{V}\right)_m \bar{c}$$

The arrival rate can be derived from the emission law which is applicable to the vapor source. For a small-surface source it is given by Eq. (52) in Sec. 3c(1):

$$\frac{d\mathfrak{M}}{A_m \, dt} = m \frac{dN_m}{A_m \, dt} = \Gamma \frac{A_e \cos \varphi_m \cos \theta_m}{\pi r_m^2}$$

where A_e = effective vapor-source area
$\quad r_m$ = source-to-monitor distance
$\quad \varphi_m, \theta_m$ = angles of emission and incidence for the monitor
$\quad \Gamma$ = mass-evaporation rate per unit source area

By substituting $\bar{c}$ according to Eq. (35) and Γ from Eq. (48), the monitor ion current can be expressed as a function of the vapor pressure p^* of the evaporant at the source temperature T:

$$I_i = D'I_e \frac{A_e \cos \varphi_m \cos \theta_m}{\pi 4kTr_m^2} p^* \tag{75}$$

The significance of the gauge constant becomes more apparent in the relation[280]

$$D' = 7.5 \times 10^{-4} S_0 k T_0 \qquad \text{(for } p^* \text{ in torr, } T_0 = 300°\text{K)}$$

Here, S_0 is the dynamic gauge sensitivity, a parameter which measures the response of an ion gauge to the flow of vapor, as opposed to the static sensitivity, which is usually quoted for ion gauges but refers to stationary gases.[281] Dynamic gauge sensitivities as determined for rate monitors range from 0.05 to 0.5 Torr^{-1}.[280,282]

For practical purposes, it is of interest to correlate the measured ion current I_i with the deposition rate d' as observed on the thin film substrate. For emission from a small-area source, the deposition rate follows from Eq. (52) [Sec. 3c(1)] by differentiating with respect to time and also considering Eqs. (49) and (55):

$$d' = \left(\frac{m}{2\pi kT}\right)^{\frac{1}{2}} \frac{A_e \cos \varphi_s \cos \theta_s}{\rho \pi r_s^2} p^* \tag{76}$$

The subscript s refers to the substrate position, and ρ is the density of the deposit. Combination of Eqs. (75) and (76) yields the equation

$$\frac{I_i}{d'} = D'I_e \frac{r_s^2 \cos \varphi_m \cos \theta_m}{r_m^2 \cos \varphi_s \cos \theta_s} \rho \mathbf{N}_A \left(\frac{\pi}{8MRT}\right)^{\frac{1}{2}} \tag{77}$$

Thus, the response of the monitor gauge in terms of measured ion current depends on the gauge structure as represented by D', on the distances and angular relationships, and on the material constants ρ and M, and it is proportional to the emission current I_e. The $T^{-\frac{1}{2}}$ dependence is usually neglected and considered a material constant since the range of useful evaporation temperatures encompasses at the most a few hundred degrees for any specific evaporant.

Although the performance ratio I_i/d' of a monitor is explicitly stated by Eq. (77), the gauge constant D' is not known a priori. Therefore, rate monitors must be empirically calibrated. For several experimental devices and specific materials, calibration curves have been published, and numerical values are listed in Table 16. The I_i/d' ratios are constant throughout the investigated ranges of deposition rates,

TABLE 16 Operational Features and Performance Ratios of Ionization-rate Monitors

Reference	Gauge structure	Residual gas discrimination	Emission current I_e	I_i/d', μA/Å s⁻¹	Evaporant	Readout
Haase[283]	Commercial ionization gauge	Chopper wheel, 6 Hz	?	$\sim 10^{-4}$	Ag	Rates displayed on oscilloscope
Schwarz[280]	Cylindrical collector (Fig. 44a)	Vibrating shutter, 10–20 Hz	3 mA	1.5×10^{-4} 4.0×10^{-4} 4.5×10^{-4}	Al Cr Ni	Current indicator, and integrator to show thickness directly
Giedd and Perkins[284]	Cylindrical grid, disk collector (Fig. 44b)	$p < 10^{-6}$ Torr: neglect; $p > 10^{-6}$ Torr: sub. $I_i = I_{total} - I_{res.\ gas}$	40 mA	10^{-1}	Sn	Rates registered on strip-chart recorder
Perkins[285]	Straight-wire collector (Fig. 44c)	Compensation with second gauge	10 mA	1.6×10^{-2} 4.7×10^{-2} 8.0×10^{-2} 12.0×10^{-2}	SiO Sn Pb CdTe*	Rates registered on strip-chart recorder
Dufour and Zega[282]	Dual grid and collector structure (Fig. 44d)	Compensating gauge, chopper wheel, 20 Hz	$\leqq 4$ mA	8×10^{-5}	Al	Rates registered on strip-chart recorder
Brownell at al.[286]	Straight-wire collector	Vibrating shutter, 170 Hz	?	0.4 V/Å s⁻¹ 1.4 V/Å s⁻¹ 1.7 V/Å s⁻¹	Ge SiO Au	Strip-chart recorder, and integrator to show thickness directly

* From Ref. 76.

and they are higher the greater the electron-emission current. The latter must be stabilized to facilitate stable monitor operation. A potentially disturbing factor is stray electrons emitted from the hot-vapor source. To guard against these, the source may be positively biased with 150 to 200 V,[283,284] or the monitor aperture may have magnetic and electric deflection plates to strip the arriving vapor of charged particles.[287]

A common difficulty with ionization-rate monitors is residual gas contributions to the ion current. The effect is illustrated by the data of Perkins,[285] whose gauge had a linear characteristic for residual gases with an ion current of 0.04 μA at 10^{-6} Torr. The evaporation of SiO at a rate of 20 Å s^{-1} yielded a current of 0.32 μA. Thus, even under rather favorable conditions, the background contribution to the total ion current amounts to 11%. One solution to the problem is to modulate the vapor stream entering the monitor by a chopper wheel or a vibrating shutter, thereby producing an alternating current which can be discriminated from the dc component attributable to the background. The alternate approach is to install a second, identical gauge which is exposed only to the residual gases but shielded against the vapor stream. Its output signal can be used to compensate for the background effect of the monitor gauge. Examples of both methods are listed in Table 16. The gauge of Dufour and Zega[282] shown in Fig. 44d utilizes the dual grid-collector structure for this compensatory function in combination with the beam-modulation technique.

Several additional points are essential for the successful operation of ionization-gauge rate monitors. If dielectric materials are to be evaporated, accumulation of deposits on grid and collector must be avoided. In Perkins'[285] device, both these filaments are electrically heated so that the evaporant vapors do not condense. In dc monitors of this type, it is also necessary to select a socket material of high insulation resistance ($>10^{12}$ ohms) and thereby maintain leakage currents between collector and grid at levels which are negligibly small compared with the ion current. The Joule heating of all three filaments tends to increase the temperature and therefore lower the insulation resistance of the alumina socket. This is prevented by water cooling of the socket holder. Shielding of the monitor structure against unnecessary deposits, particularly against metal films on the socket surface, is a general requirement which applies to all types. Finally, provisions for baking the monitor at temperatures around 300°C are desirable to minimize outgassing and background effects.

Since rate-monitor signals are very weak—typically a few tenths of 1 μA or less—they must be amplified for display and control purposes. Examples of instruments used are given in the last column of Table 16; for circuit diagrams, the reader is referred to the original publications. Electronic integrators which indicate accumulated thickness directly have been employed by Schwarz[280] and Brownell et al.[286] The latter were able to control film thicknesses within ± 10 Å. The rate monitor described by Perkins[285] gave thickness control of ± 2 to 5%.

A new type of ionization-rate monitor has recently been introduced by Zega.[288] It is distinguished from other models by the utilization of a small electron beam ($I_e = 0.2$ mA) to effect vapor ionization. The beam is swept at a frequency of 750 Hz across an entrance aperture through which the vapor passes. Thus, an ac signal is generated whose amplitude is proportional to the evaporant density while the background pressure contributes a dc reference level. The device is capable of controlling evaporation rates greater than a few angstroms per second, and its electronic controls include an integrator to monitor film thickness. The functional parts are enclosed in a small metal box and do not accumulate any deposits. Other advantages are the attainment of background discrimination without any moving parts, and the fact that no water cooling is required.

(2) Particle-impingement-rate Monitors Devices of this type were introduced by Neugebauer in 1964[243] but have not found wide usage as yet. Two different designs are shown in Fig. 45. Both consist of a light, thin-walled aluminum cylinder which is mounted off to the side of the source and partially shielded against the vapor. The exposed area of the cylinder receives a momentum whenever an atom impinges. The resulting torque tends to turn the cylinder. In Beavitt's[289] model (Fig. 45b), the cylinder is pivoted and free to turn whereas the one in Fig. 45a is suspended from a wire whose torsion force opposes rotation. Thus, the pivot device revolves as long as molecules are impinging, whereas the torsion device turns only by a certain angle until torque and torsion forces balance. Both devices use magnetic damping to avoid oscillations.

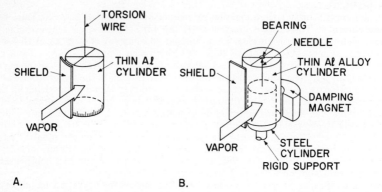

Fig. 45 Particle-impingement-rate monitors. (*A*) Torsion-wire device. (*After Neugebauer.*[243]) (*B*) Pivot-supported device. (*After Beavitt.*[289])

The equations relating deposition rate and meter response of impingement-rate monitors have been derived by Beavitt.[289] For wire-suspended cylinders, the deposition rate d' is proportional to the turning angle β:

$$d' = \frac{2K}{\rho \bar{c} s^2 h} \beta$$

where K = torsional constant of the suspension
ρ = density of deposit
$\bar{c}$ = average molecular velocity
s = cylinder radius
h = cylinder height

Neugebauer determined a sensitivity of 1.5 Å s^{-1} rad^{-1} for his device during the evaporation of tin.

The pivoted cylinder indicates deposition rates by its angular velocity ω. This quantity is difficult to measure unless one resorts to pulse-counting methods. It is, however, an integrating device which allows convenient monitoring of the accumulated deposit if the total number of revolutions ($2\pi\omega t$) is counted:

$$d = 2\pi\omega t \frac{s\sigma(\tfrac{1}{2}s + 2h)}{\rho \bar{c} \tau h}$$

Here, σ = mass per unit wall and top area
τ = time constant of the damped rotor

A device made by Beavitt[289] turned 1 rad if 5.3 mg of Al or Au was evaporated from a source at 10 cm distance.

In comparing the two monitors, the torsion-wire device has a relatively slow response time and therefore does not indicate sudden changes in evaporation rate. The pivoted device does not suffer from this shortcoming as much as from the fact that a minimum impingement rate is required to overcome the friction of the bearing. The threshold value for Beavitt's model was 2 Å s^{-1} of Al. The $T^{-\frac{1}{2}}$ dependence, which enters through the presence of $\bar{c}$ in the denominators, can be neglected for practical purposes since the impingement rate rises exponentially with the source temperature. Both gauges measure the true impingement rate regardless of α_c since reevaporation of molecules occurs randomly in all directions and therefore does not cause additional torque.

Common advantages of both models are the simplicity of their construction, the absence of residual gas effects, independence from previously accumulated deposits, and the fact that no adjustment to obtain a null position is required. The principle is not limited to cylindrical rotors but has also been applied to rotating disks which can be mounted in closer proximity to the substrates. Although the vapor impinges nearly perpendicular onto the surface of such a device, Beavitt[289] showed that disk devices are more sensitive than cylinders. He constructed a disk monitor from thin aluminum foil which had a sensitivity of 0.04 Å s^{-1} rad^{-1} for Al and a response time of 0.5 min. The response time may be shortened but only at the expense of losing sensitivity.

b. Monitoring of the Deposited Mass

Mass-sensing devices may be used for all evaporant materials. They operate either by determining the weight of the deposit, or by detecting the change in oscillating frequency of a small quartz crystal on which the evaporant condenses. Both types of devices can be made very sensitive to the point where extraneous effects interfere critically with their operation. Since the measured quantity is the accumulated mass of deposit, conversion into film thickness requires knowledge of the material's density. The latter is usually somewhat lower for films than for bulk materials. Furthermore, the locations of the sensing device and of the substrates with respect to the source are often different so that the geometric factors must be considered in the same manner as discussed for ionization-gauge-rate monitors. Lastly, the response of the device per unit mass of deposit is not always predictable with the required accuracy. For these reasons, mass-sensing devices, too, are best calibrated after their installation by comparing their readings with independently measured film thicknesses on the substrates.

(1) Microbalances Instruments suitable for the gravimetric determination of small quantities of mass are summarily referred to as microbalances. Their design may be based on different principles such as the elongation of a thin quartz-fiber helix, the torsion of a wire, or the deflection of a pivot-mounted beam.[290] Their applications encompass many areas other than thin film measurements, and the literature about microbalances and related topics is extensive. A detailed review about the construction of such instruments including a table of different models and their sensitivities was given by Behrndt in 1956.[291] For more recent developments, especially vacuum applications, the reader is referred to "Vacuum Microbalance Techniques," the proceedings of annual symposia on the same topic which have appeared regularly since 1961 (Plenum Press, New York). The models selected for this review illustrate some of the more commonly used design features and are of established utility for thin film work, but they are by no means the only ones for which this claim can be made.

Campbell and Blackburn[244] constructed a relatively simple microbalance by converting the movements of a microammeter. A later version of this instrument is due to Hayes and Roberts[292] and is shown in Fig. 46. The meter arm is balanced by an adjustable weight and the meter spring. In the null position, the two photodiodes receive equal amounts of light through the aperture in the centrally located shutter. Deviations from this position alter the resistance of the diodes and thereby unbalance the bridge circuit. The resulting current flows through the meter coil and produces a restoring force on the meter arm. The restoring current is proportional to the force acting on the mica vane at the other end and is amplified in the control system. A strip-chart recorder provides a permanent record of the evaporation.

In the position shown in Fig. 46a, the vane experiences two forces due to its exposure to the vapor stream. The weight of the deposit,

$$G = A_m \rho d \quad \text{g wt} \quad \text{or} \quad \text{Equivalent force } F_1 = g A_m \rho d \quad \text{dyn} \quad (78)$$

where A_m = vane area
  ρ = deposit density
  d = deposit thickness, cm
  g = gravitational constant

tends to pull the meter arm down. An upward-directed force arises from the momentum imparted by the impinging vapor molecules. Assuming a sufficiently large source-to-monitor distance so that all molecules arrive at the same angle of

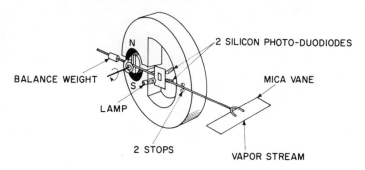

A.

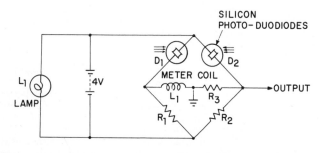

B.

Fig. 46 (A) Schematic drawing and (B) circuit diagram of a microbalance constructed from a microammeter movement. (*Hayes and Roberts.*[292])

incidence θ, the dynamic force generated by the vapor stream is

$$F_2 = \frac{d\mathfrak{M}}{dt} \bar{c} \cos \theta \qquad \text{dyn}$$

where $d\mathfrak{M}/dt$ = mass deposition rate on vane area, g s^{-1}

$$\bar{c} = 14{,}551 \sqrt{T/M}, \text{ cm s}^{-1} = \text{avg molecular velocity}$$

Since $d\mathfrak{M}/dt = A_m \rho d'$, the dynamic force is related to the deposition rate d', Å s^{-1} by

$$F_2 = 1.4551 \times 10^{-4} A_m \rho \sqrt{T/M} \, d' \cos \theta \qquad \text{dyn} \qquad (79)$$

The net force $F_1 - F_2$, which is observed during evaporation at constant rate, is traced in Fig. 47. The line of gradual weight increase, $F_1 = gA_m \rho t d'$, is shifted downward by the constant impingement force F_2. A true weight indication is obtained when the evaporation stops.

The response to both deposit weight and particle-impingement forces is also found in microbalances of different construction. To separate the two effects, the vane must be oriented in a vertical direction and the vapor impinge horizontally. If the beam axis is then left in the horizontal position as shown in Fig. 46a, the balance indicates only weight. Conversely, if the meter movement is turned 90° so that its axis points toward the center of gravity, the instrument functions as a particle-impingement device.[292] This latter option is usually not available on other microbalances.

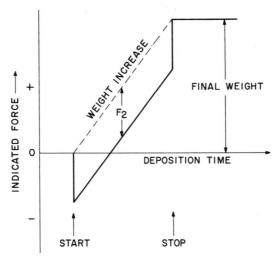

Fig. 47 Net force on a microbalance with horizontal axis and vane, for vertical vapor incidence.

To give an idea of the forces originating in film-deposition experiments, numerical values must be inserted into Eqs. (78) and (79). A metal of density 10 contributes 10^{-7} g wt cm^{-2} of vane area for every angstrom of film thickness, which is equivalent to a force of 10^{-4} dyn $cm^{-2}/\text{Å}$. The material constant $\rho \sqrt{T/M}$ in Eq. (79) may vary from 20 to 85 for different metals, whereby temperatures necessary to achieve 10^{-2} Torr evaporant pressure are assumed. Values between 40 and 50 are most common. Thus, for perpendicular impact $(\theta = 0)$, 1 cm^2 of vane surface experiences a force of about 7×10^{-3} dyn for every Å s^{-1} of deposition rate. Since practical vane areas are of the order of 1 cm^2, accuracies of ± 10 Å for thickness and ± 1 Å s^{-1} for rates require instruments which indicate accurately 10^{-6} g wt or about 10^{-2} dyn. Most instruments exceed these specifications. The meter-movement device of Roberts and Hayes had a maximum sensitivity of 2×10^{-6} g wt and could sense a dynamic force of 5×10^{-3} dyn.

The most common mode of microbalance construction utilizes a crossbeam which is balanced on a taut torsion fiber. A number of models require that the beam deflection is observed—for instance, with a cathetometer or on the scale micrometer of a microscope—to derive the weight change.[293] Sensitivities of 1.5×10^{-7} g wt have been obtained with such instruments.[294,295] For monitoring and controlling processes in vacuum systems, it is more practical if the beam deflection generates a compensating force which provides a proportional electric signal. Such balances are referred to as electromagnetic or electrostatic depending on the means employed to generate the restoring force. The zero position of the beam is usually sensed with a light beam, but capacitive or inductive effects may also be utilized.

A schematic drawing of an electromagnetic torsion balance is shown in Fig. 48. The model is entirely constructed from quartz rods with a 40-μ-diameter quartz torsion fiber.[296] A small permanent magnet is fused to one end of the crossbeam and attracted by the solenoid which supplies the restoring force. Oscillations are damped by the copper cylinder surrounding the magnet. The balance reportedly indicates 10^{-7} g wt with an accuracy of $\pm 10\%$. For comparison, a monolayer of iron weighs 1.8×10^{-7} g wt cm^{-2}.

The balance shown in Fig. 48 has also been modified into a model with electrostatic force compensation.[297] This version had a thin metallized quartz wafer instead of a magnet at the beam end. The wafer formed the center plate of a three-plate capacitor and could be charged electrostatically because a tungsten torsion wire was used instead of the quartz fiber. The restoring force was produced by applying voltage to the three plates. While mechanically quite rugged, the balance achieved high sensitivity—2.5×10^{-7} g wt gave an amplified output signal of 1 V—through the use of a very thin (10-μ-diameter) tungsten wire. If installed vibration-free, the zero position fluctuated by no more than the equivalent of 4×10^{-9} g wt. Oscillations were eliminated by a feedback circuit. Since it was constructed entirely from refractory mate-

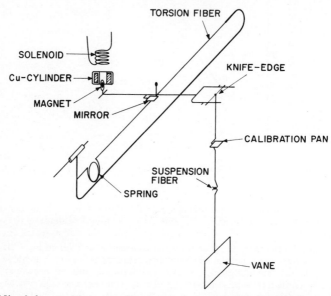

Fig. 48 Microbalance with torsion-fiber suspension and electromagnetic-force compensation at beam end. (*Mayer et al.*[296])

rials such as fused quartz and tungsten, the balance could be baked at temperatures above 400°C.

Figure 49 shows an electromagnetic balance with a moving coil to restore the zero position. This instrument, which was reported by Cahn and Schultz,[298] has a hollow aluminum beam suspended on a torsion ribbon. The latter is more rugged than a wire and self-centering so that no arresting mechanism is required. The deposit is received by a $\frac{3}{4}$-in.-diameter mica scale pan which may be suspended into the substrate area by means of a stirrup. Such an arrangement, which also includes a water-cooled shield to protect the balance against source radiation, has been described by Houde.[253] The zero position of the balance is detected by a phototube whose current varies according to the degree of illumination admitted by the shutter. The amplified current is applied to the coil and, in restoring the zero position, generates a voltage proportional to the weight change. This signal is then recorded. On the most sensitive range, the instrument gives a 5-μV output for 10^{-7} g wt of deposit (about 1.5 Å of SiO).

Microbalances of similar construction with sensitivities from 10^{-6} to 10^{-7} g wt and capacities of typically 0.2 to 2 g wt are commercially available from several manufacturers which are listed in Ref. 279.   Among the more recent developments, an instrument described by Gast[299,300] is of particular interest because the sample holder is suspended magnetically and physically separated from the balance mechanism. A remarkably sensitive torsion balance has been described by Pearson and Wadsworth.[301]   It can detect forces of 2×10^{-8} dyn, which is equal to the rms effects caused by the Brownian motion.   Other disturbing effects are often encountered with less sensitive balances as well.   Foremost among them is the shift of the zero position due to temperature inhomogeneities along the balance beam.   Wolsky et al.[302] discuss modes of beam construction which enhance the chances of maintaining the arm-lengths constant within the desired 0.01 ppm.   Mechanical vibrations, electrostatic effects, and gas adsorption, too, interfere with precision weighting.[291]

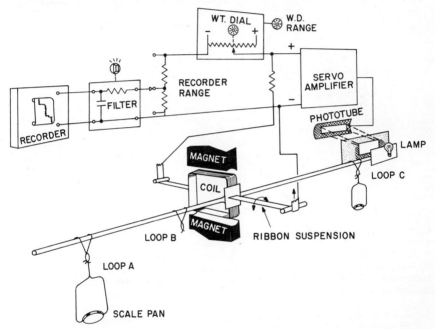

Fig. 49   Microbalance with torsion-ribbon suspension, coil movement, and signal recorder. (*Cahn and Schultz.*[298])

(2) Crystal Oscillators   The use of quartz-crystal oscillators to determine small quantities of deposited matter was first explored by Sauerbrey[303,304] and Lostis.[305] The transducers required to monitor film thickness are of relatively simple construction and about as sensitive as microbalances while practically unaffected by mechanical shocks and external vibration.   Therefore, crystal oscillators have received much interest during the last years and are currently the most widely used means for monitoring thin film depositions.   The crystal-oscillator monitor utilizes the piezoelectric properties of quartz.   A thin crystal wafer is contacted on its two surfaces and made part of an oscillator circuit.   The ac field induces thickness-shear oscillations in the crystal whose resonance frequency is inversely proportional to the wafer thickness d_q,

$$f = \frac{c_t}{2d_q}$$

where c_t is the propagation velocity of the elastic wave in the direction of thickness. The major surfaces of the wafer are antinodal.

An important consideration in preparing the quartz wafer is the temperature dependence of the resonance frequency. The temperature coefficient of frequency (TCF) of quartz is related to the elastic constants. It has positive and negative terms whose magnitudes depend on the direction of the vibration with respect to the natural crystal axes. Since frequency changes resulting from temperature fluctuations affect the accuracy of mass determinations, the quartz crystals are cut in an orientation where the TCF terms compensate each other. This is the case if the plane of the cut forms an angle of about 35° with the c, x plane as illustrated in Fig. 50. The orientation is designated as an AT cut and is used in all thickness monitors.

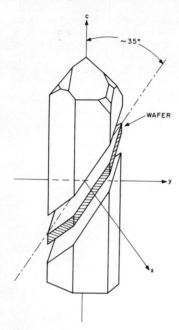

The temperature dependence of the TCF for a number of AT crystals of slightly different orientation has been measured by Phelps.[306] Figure 51 shows the function for a cut made at 35°20′. The TCF reaches zero at about 30°C and remains smaller than $\pm 5.10^{-6}$ deg^{-1} in a temperature interval of about ± 30°C. Changes in the cutting angle of only 10′ shift the temperature of zero TCF by as much as 50°C and also narrow the interval of small TCF values. The cutting angles generally preferred are between 35°10′ and 35°20′. The resonance frequencies for thickness-shear mode oscillations of AT crystals are given by

$$f_0 = \frac{N}{d_q} \qquad (80)$$

Fig. 50 Quartz crystal with AT cut wafer.

where $N = 1.67 \times 10^6$ Hz mm..

If a small $\Delta \mathfrak{M}$ is added to either one or both sides of the wafer, it may be assumed that the original crystal surfaces remain antinodes of vibration; i.e., the foreign matter does not store elastic-deformation energy during the vibration cycle. Hence, the deposit affects the resonance frequency only through its mass whereas material

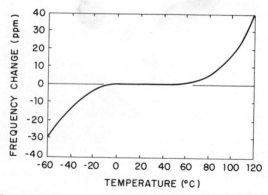

Fig. 51 Frequency change vs. temperature for an AT crystal cut at 35°20′.

specific properties such as density or elastic constants are inconsequential. The effect of mass loading on the frequency may be derived by differentiating Eq. (80) with respect to d_q and substituting the added increment of quartz Δd_q by an identical mass $\Delta\mathfrak{M}$ of foreign matter and different thickness d_f.[304] An alternate derivation was given by Stockbridge,[307] who applied perturbation analysis to a resonating plate and obtained essentially the same relation:

$$\Delta f = -\frac{Kf_0^2}{N\rho_q}\frac{\Delta\mathfrak{M}}{A_m} = -\frac{KN}{\rho_q d_q^2}\frac{\Delta\mathfrak{M}}{A_m} \qquad (81)$$

Here, ρ_q is the density of quartz (2.65 g cm^{-3}) and $K \approx 1$ is a constant which depends on the distribution of deposit over the monitor area A_m. The latter should be the total surface area of the wafer. However, if the electrodes applied to the wafer cover

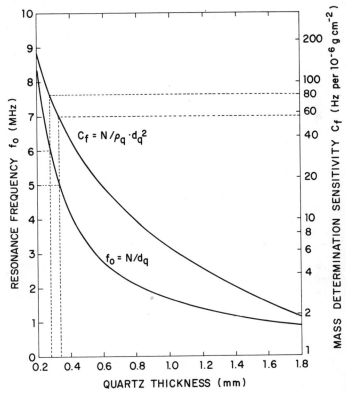

Fig. 52 Resonance frequency and mass-determination sensitivity of AT crystals as functions of wafer thickness.

only part of the crystal surface, as is often the case, then A_m is to be taken as the electrode area. It has been shown that oscillations outside the electrode area are negligible and deposits there contribute only 1% to the frequency change Δf.[304]

The proportionality factor in Eq. (81),

$$\frac{f_0^2}{N\rho_q} = \frac{N}{\rho_q d_q^2} = C_f \qquad \text{Hz g}^{-1}\text{ cm}^2 \qquad (82)$$

is called the mass-determination sensitivity of the crystal. As shown in Fig. 52, the thinner crystals with the higher resonance frequencies give larger frequency changes per deposited mass per unit area. Hence, in the interest of sensitivity, the crystal wafers should be made as thin as their fragility permits. There is, however, another limitation. The relationship between Δf and $\Delta\mathfrak{M}$ ceases to be linear if the thickness of the deposited mass is no longer small compared with the wafer thickness. Estimates as to how far linearity may be assumed without committing significant errors vary from $\Delta f_{max} = 0.5\%$[139] to 5%[279] of the initial frequency f_0. Although the maximum permissible frequency change is larger for the thinner crystals, the mass which causes this Δf_{max} is greater for the less sensitive (thicker) crystals; this is illustrated in Fig. 53. Crystals of about 0.3 mm thickness with initial frequencies of 5 to 6 MHz are generally considered to offer the best compromise between high sensitivity and high mass-loading capacity. This range is indicated (dashed lines) in Figs. 52 and 53.

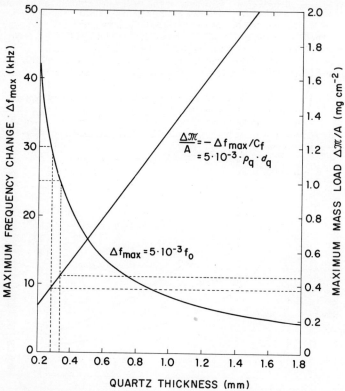

Fig. 53 Maximum frequency change and mass loading for AT crystals as functions of wafer thickness.

The mass-frequency equation (81) has repeatedly been verified by control experiments with microbalances. Eschbach and Kruidhof[308] obtained an experimental value of C_f which agreed within 0.4% with the value derived from the material parameters of quartz [Eq. (82)]. It is therefore not necessary to calibrate monitor wafers empirically, but Niedermayer et al.[309] point out that the temperature of the crystal should be taken into account. The interval of approximately linear frequency changes has also been investigated. Empirical values of Δf_{max} for 5-MHz crystals range from 50 to 100 kHz,[308,310,311] which is equivalent to deposits of 1 to 2 mg cm^{-2}. The corresponding maximum thickness

$$d = \frac{\Delta \mathfrak{M} \times 10^8}{\rho A_m} \quad \text{Å}$$

varies from 5,000 to 50,000 Å depending on the density ρ of the film material.

The onset of nonlinear frequency response may be due to second-order terms neglected in the perturbation analysis of Stockbridge,[307] or the assumption of no energy storage in the deposit may be invalid. In the former case, deviations from linearity should be independent of the deposit material, whereas the latter possibility leads one to expect differences according to the mechanical properties of the film material.[307] If the nonlinear portion of the frequency-mass characteristics is known for the particular crystal, the useful range may be extended to thicknesses of several microns. Calibration of a monitor wafer by independent thickness measurements is possible because accumulated deposits can be etched off and the crystal may be used again. However, when the film thickness is no longer small compared with the thickness of the crystal, the deposit begins to store elastic energy and introduces vibrations other than in the thickness-shear mode. At some point, the oscillations of the crystal in the direction of thickness change irregularly (the frequency "jumps"), or they may cease altogether. In view of this limitation and the modest price of the quartz wafers, it is usually not worthwhile to extend the thickness range by an elaborate calibration procedure.

The crystals used in deposition monitors are either circular or square platelets of typically 13- to 14-mm size. They are mounted in holders which prevent them from shifting position, yet allow easy removal and replacement. Two examples of crystal holders are shown in Fig. 54. The electric field is applied through thin gold or some-

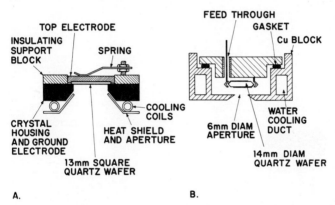

Fig. 54 Oscillator crystal holders for deposition monitoring. (*A*) After Behrndt and Love.[145] (*B*) After Pulker.[310]

times silver films evaporated onto both sides of the wafer. In Fig. 54a, the film electrodes extend over the entire surface whereas in Fig. 54b, only the center area is covered by 6- to 8-mm dots. In the latter case, evaporated leads permit contacts to be made at the wafer edge. In other arrangements, only the exposed surface of the wafer is metallized whereas the opposite side is contacted directly by a metal block. A holder whose materials of construction allow outgassing of the entire transducer at 450°C has been described by Langer and Patton.[312] The frequency characteristics of quartz crystals which have been heated to 400°C are fully restored after cooling.[310] Utilization of crystal oscillators in ultrahigh vacuum is therefore possible. The effect of pressure on the oscillation frequency has been found to be negligible throughout the entire high- and ultrahigh-vacuum range, provided there are no adsorption or desorption phenomena on the crystal surface.[313]

Although AT cut wafers have the smallest possible temperature coefficient of frequency, it is still necessary to protect the crystal against temperature changes due to radiation from the source and heat of condensation. Therefore, the crystal housing is usually water-cooled and forms a radiation shield which surrounds the entire crystal except for the deposition area. The heat received by this necessarily exposed area still causes temperature increases of several degrees Celsius with ensuing frequency changes of 10 to 100 Hz[145,310,311] which are equivalent to mass changes of 10^{-7} to 10^{-6} g cm^{-2}. The effect can be minimized by using a small-entrance aperture as in Fig. 54b, but it is not negligible when making precision measurements. Behrndt[139] recommends delaying the opening of the substrate shutter for some time while exposing the monitor to the source and vapors. With this procedure, the greater part of the crystal-temperature change takes place prior to the accumulation of deposit on the substrates.

The instrumentation required to operate a monitor consists basically of an oscillator and a suitable frequency meter. While it is possible to measure the frequency change

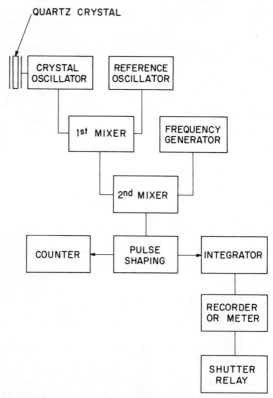

Fig. 55 Block diagram for quartz-crystal oscillator instrumentation. (*After Pulker.*[310])

at the level of the resonance frequency f_0, this requires expensive equipment capable of giving accurate six-digit readings. It is therefore common practice to use a reference oscillator, for instance, a second crystal, with a fixed frequency and to generate a difference frequency. The latter is mixed again with the signal from a variable-frequency generator, as indicated in the block diagram in Fig. 55. The advantage of this arrangement is that it permits operation in the most sensitive frequency range regardless of the film thickness accumulated on the crystal. The output signal is in the audio-frequency range and drives the counter circuit.

The various methods of frequency measurement have been discussed recently by Steckelmacher[279] and Langer and Patton.[312] Digital as well as analog-type instruments are capable of measuring frequencies accurately to within ±1 ppm. If digital equipment is used, a printer may serve to provide a permanent record, and deposition rates may be derived by comparing subsequent readings. The differentiation may also be performed electronically, and rates as well as total thickness may be displayed on separate indicators. For analog-type instruments, the frequency must be converted into a dc signal which may be displayed on a meter or recorded on a strip chart. Automatic termination of the deposition process by actuating a shutter relay at a preselected signal level is easily possible.

In practice, the accuracy of crystal-oscillator monitors is determined by the stability of the oscillator circuit. The effects of hydrostatic gas pressure and gas adsorption on oscillator frequency have been studied by Stockbridge[313] and are negligible for most practical purposes. If a frequency change of 1 Hz in a 5-MHz oscillator can be detected, the corresponding sensitivity of the monitor is about 2×10^{-8} g cm^{-2}. Ordinarily, however, the stability of the circuits is only of the order of 10 to 100 Hz h^{-1}. Thus, the practical mass-detection limit is 10^{-7} to 10^{-6} g cm^{-2}. The thickness and rate-control figures reported for these conditions are typically $\pm 2\%$.[279,314] The attainment of $\pm 1\%$ accuracy for masses in the 10^{-7} to 10^{-6} g cm^{-2} range has been reported by Hillecke and Niedermayer[315] but requires consideration of the temperature dependence of the frequency. Warner and Stockbridge[316] describe instrumentation capable of measuring frequencies with an accuracy of 1 part in 10^{10}. In addition, the temperature of the quartz crystal was maintained constant within ± 0.1 to 0.01°C, which allowed the detection of mass changes in the 10^{-10} to 10^{-12} g cm^{-2} range. While this sensitivity exceeds that of microbalances, one cannot expect to attain such a temperature constancy during vacuum evaporation. Good experimental practices concerning the use of quartz-crystal oscillators for film-deposition control have been discussed in detail by Riegert.[317]

c. Monitoring of Specific Film Properties

In preparing thin films, the investigator is often interested in only one particular film property to assume a certain final value, whereas the deposition rate is of secondary importance. Such properties may be optical or electrical ones, and their attainment can be accomplished more assuredly by direct observation during the deposition process than by monitoring the mass and correlating it to the desired property. While often simpler in regard to equipment, these methods are not universally applicable, and some of them require especially prepared monitoring substrates to facilitate in situ measurements. If the latter is the case, the monitoring substrate must be positioned so that it experiences the same incidence of vapor molecules as the regular substrates. Furthermore, its thermal conductivity and mode of heating should be identical to those of the other substrates to ensure equal temperatures.

(1) Optical Monitors Optical phenomena such as light absorption, transmittance, reflectance, and related interference effects can be utilized to monitor the growth of films during vacuum deposition. The necessary equipment is relatively simple and consists basically of a light source and a photocell. Both items are preferably located outside the vacuum system, with suitable windows and optical paths provided for communication with the substrates. The choice of the quantity to be measured depends on the type of substrate and the film to be monitored. Metal films, for example, may be observed by transmittance measurements, provided they are deposited onto transparent substrates. The amount of transmitted light T_r, however, decreases rapidly with thickness so that sensitive measurements are limited to rather thin films. Furthermore, the extinction law

$$T_r = T_0 \exp(-\alpha d)$$

where α, cm^{-1}, is the adsorption coefficient, is not obeyed during the nucleation and island stage of film growth.[139] Similar considerations hold for reflectance measurements on metal films. Therefore, optical monitoring techniques are primarily used for dielectric films.

The monitoring of transparent films exploits the periodic fluctuations in light intensity which arise from multiple reflections within the film and subsequent interference. The conditions for constructive and destructive interference vary depending on the refractive indices of film and substrate materials. Also, interference maxima in transmission coincide with minima in reflectance, and vice versa. This is discussed more extensively in the context of film thickness measurements in Chap. 11, Sec. 1. The difference in thickness monitoring is that the extreme values of intensity occur sequentially as the film thickness increases by increments which lengthen the optical path of the light beam by a quarter of the wavelength λ. Therefore, the order of a particular interference maximum or minimum can be observed directly.

Transmittance and reflectance monitoring techniques are most often employed when films are deposited for optical purposes such as beam splitters, mirrors, antireflection coatings, and interference filters. These applications usually require film thicknesses of λ/4 or multiples thereof, and the advantage of monitoring the desired film property directly is obvious. The various monitor designs developed for such purposes are discussed in Behrndt's review.[139] Steckelmacher et al.[318,319] describe a monitor arrangement with a modulated light beam so that the output of the photocell can be selectively amplified and the effect of light scattering eliminated. The design was successfully used in the fabrication of multilayer interference systems.

As for film applications in the microelectronics field, optical thickness monitoring has not been a widespread practice because the emphasis there is on electrical rather than optical properties. Transmittance measurements are usually not possible, either because the substrates are opaque or because they are mounted against a metal holder-heater plate. This leaves reflectance measurements as the most suitable approach. However, the more common types of optical monitors including commercial models[320] are intended for small (nearly perpendicular) angles of light incidence on the substrates. Consequently, the light source and photocell compete for installation space with such essential parts as pumping port, evaporation source, and heat shields. Although this problem can be resolved, the ensuing difficulties have generally discouraged the use of optical and favored other monitoring devices.

The advent of rf sputtering methods for the deposition of dielectric films has changed this situation because all the monitoring techniques based on electrical measurements are severely disturbed by the glow discharge. The renewed interest in optical monitors in conjunction with sputtering (see also Schaible and Standley[321]) led to the development of a technique which permits installation of the illuminator and photocell off to the sides of the vacuum system.[322] The arrangement is shown in Fig. 56 for an rf sputtering system as described by Davidse and Maissel.[323] The large quartz cathode in close proximity to the substrate plane necessitates a large angle of incidence θ of about 80°. Since θ approaches or exceeds Brewster's angle

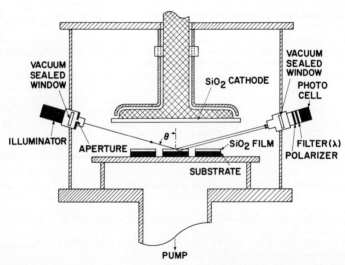

Fig. 56 Schematic of an RF sputtering system (after Davidse and Maissel[323]) with optical-thickness monitor.[322]

for the dielectric film material, the conditions for destructive and constructive interference are opposite for the parallel and perpendicular polarized components of the light beam.[324] A polarization filter in front of the photocell admits only the perpendicular component, which yields a greater difference between maximum and minimum intensity than the parallel component. An interference filter singles out one particular wavelength to be received by the photocell. If the substrate surface—in this case, silicon—is reflective, the intensity of the reflected beam as a function of thickness is given by the Eq. (16) in Chap. 11, Sec. 1c(2). The output voltage of the photocell varies in proportion and yields a recorder trace as shown in Fig. 57.

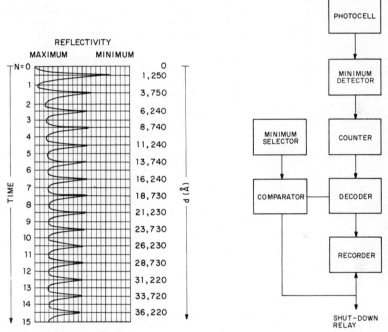

Fig. 57 Recorder trace of optical-thickness monitor shown in Fig. 56 for the deposition of RF sputtered SiO_2 on silicon ($\lambda =$ 5,500 Å; $\theta = 79°$).

Fig. 58 Block diagram for automatic thickness control with an optical monitor.[322]

The continuous recorder trace permits the thickness to be determined for every desired fraction of a wavelength. Automatic termination of the process, however, is most easily accomplished by stopping at a preselected minimum. A suitable circuit for this function is shown in the block diagram in Fig. 58. In that case, one can obtain only film thicknesses which are multiples of $\lambda/4$ according to Eq. (15) in Chap. 11, Sec. 1c(2),

$$d = \frac{N\lambda}{2n_f \cos \theta_f} + \Delta t_\varphi + \Delta t_r$$

where $N = \frac{1}{2}, \frac{3}{2}, \frac{5}{2}, \ldots$ for interference minima, or $N = 0, 1, 2, 3$, for maxima (see Fig. 57)

n_f = refractive index of film material (1.470 for sputtered SiO_2)

θ_f = angle of refraction in the film

Δt_φ = phase-shift correction

Δt_r = reflectivity correction

The reflectivity correction can be neglected in this application since neither the angle of incidence θ nor the wavelength λ is variable.[325,326] The phase-shift correction Δt_φ arises from the fact that the incident light beam suffers two phase shifts, one of 2π at the film surface and another of different magnitude at the film-substrate interface. For SiO$_2$ films on silicon, the two phase shifts are nearly identical and Δt_φ may be neglected.[326,327] The phase-shift corrections for SiO$_2$, Al$_2$O$_3$, and silicon nitride films on several common metal surfaces including germanium have been calculated by Pliskin.[328] Their values vary from -60 to -300 Å for the perpendicular component of the light beam and hence should be considered in most cases.

The thickness monitor described has been used in the rf sputter deposition of SiO$_2$ onto silicon wafers. Film thicknesses between 2,000 and 30,000 Å have been controlled to within ± 1 to 2%, an accuracy which is fairly typical for optical monitors. By virtue of its installation in rather uncontested locations of the vacuum system, the monitor is also adaptable for vacuum evaporation systems. There, the most widely used dielectric material is SiO, which absorbs part of the visible spectrum. Consequently, the intensity of subsequent reflectivity maxima will decrease. Experimental curves of light transmittance through growing SiO films show an exponential decrease superimposed on the intensity variations arising from interference. This tends to reduce the difference between maxima and minima so that their distinction becomes increasingly difficult. For example, Zerbst[329] shows a transmittance curve which was essentially flat from about 8,000 Å of SiO on up. With a more sophisticated electronic sensing circuit, it is claimed that 1-μ-thick SiO films could be controlled to within ± 20 Å.[330]

The utilization of interference techniques can be extended into the infrared part of the spectrum. This is of particular interest when monitoring the thickness of epitaxial silicon films, and several references are given by Steckelmacher.[279] Because of the elevated temperatures required for epitaxy, no external light source is necessary. The radiation emitted from the substrate—which may be silicon[324] or sapphire[331]— suffers interference due to partial reflection at the film surface. In the case of silicon growing on silicon, the effect is contingent upon differences in the impurity levels of film and substrate, since the refractive index of silicon varies with the free carrier concentration [see also Chap. 11, Sec. 1c(4)].

(2) Resistance Monitors The resistance R of a conductive film pattern of fixed length l and width w is given by

$$R = \rho \frac{l}{wd} \tag{83}$$

Hence, if the resistivity ρ of a film material remains constant throughout the deposition, the film thickness d can be continuously monitored by in situ resistance measurements. In its simplest form, the technique requires a glass slide or strip with two metal contacts at the ends to determine l, and an evaporation mask which confines the film deposition to a path of width w between the end contacts. The contacts must be applied prior to the experiment, either in a separate metal evaporation or by painting on a metal paste and burning it in. The monitor slide and its mask are mounted side by side with the substrates to ensure equal exposure to the evaporant vapor. The metal lands are contacted by a pair of probes to facilitate current-voltage measurements inside the vacuum system.

While easily implemented, resistance monitoring is not a reliable method to observe and control the thickness of metal films. During the early stages of film growth, the deposit consists of nuclei or partially connected islands (see Chap. 8) whose resistance does not obey Eq. (83). As the film becomes continuous, it goes through a thickness range in which the resistivity varies as a result of diminishing contributions from surface scattering (see Chap. 13). Furthermore, during the deposition of the first few hundred angstroms, the resistance vs. thickness function is also affected by the application of electric fields as shown in Fig. 18, Chap. 18. At greater film thicknesses, the inverse proportionality between R and d is satisfied, but the thickness can only be deduced if the resistivity ρ is known. Since the latter often varies substantially as the result of fluctuations in the residual gas pressure, the deposition rate, and the substrate temperature, film-thickness values derived from resistance measurements are rarely accurate.

In view of these uncertainties, in situ resistance measurements are mainly used to monitor and control the deposition of films intended for the fabrication of resistors. In this application, film thickness and constancy of resistivity are of secondary importance. The quantity of prime interest is the sheet resistance R_s, which is defined by

$$R = R_s \frac{l}{w}$$

and represents the ratio of resistivity over thickness ρ/d. For a fixed monitor aspect ratio l/w, current-voltage measurements give a direct indication of the sheet resistance, and the desired value is obtained regardless of resistivity variations within the film because these are compensated by adjusting the final film thickness.

The usual resistance monitor arrangement involves a Wheatstone-bridge circuit as shown in the block diagram in Fig. 59. In the simplest form, the film resistance as

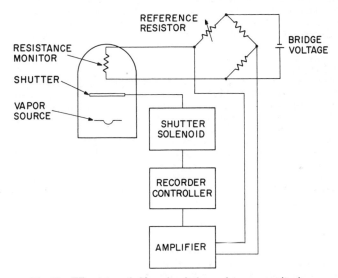

Fig. 59 Wheatstone-bridge circuit for resistance monitoring.

indicated by the unbalance of the bridge is continuously recorded on a strip chart and the shutter solenoid is actuated when the balance point is reached. During deposition, the resistance of the monitor decreases over many decades and the sensitivity of the bridge varies accordingly. Therefore, Bennett and Flanagan[332] utilized a variable reference resistor which decreased automatically in steps whenever the bridge went through the balance point. The resistance of the monitor was recorded only at these points. Steckelmacher et al.[333] employed a preset 10-turn potentiometer and altered the resistance in the ratio arm of the bridge. It has also been suggested that the reference potentiometer be driven by a motor and the evaporation rate be adjusted automatically so as to maintain the bridge balanced at all times.[279]

A more complex control system has been described by Turner et al.,[334] who used the monitor as the input resistor of a feedback amplifier. With a constant voltage applied to the resistance monitor, the output voltage of the amplifier was proportional to the film thickness (assuming constant film resistivity throughout the deposition). The same signal was given into an operational differentiator whose output voltage was proportional to the deposition rate. A trigger circuit allowed termination of the process at a preselected level of thickness, while constant evaporation rates could be maintained through a servo-control circuit governed by the rate signal.

The conventional monitoring techniques employing Wheatstone bridges in one form or other are two-terminal resistance measurements and therefore subject to errors due to lead and contact resistances. These errors become significant only if the resistance values to be monitored are smaller than a few hundred ohms. One method to avoid this problem is the use of more complicated monitor masks which yield a long and narrow, meandering resistance path.[334] However, as the resistance path is made narrower, the definition of its width by means of an evaporation mask becomes less accurate. Slight variations in the mask-to-monitor–slide spacing produce varying degrees of vapor scattering and corresponding fluctuations in the pattern width.

A monitor configuration which allows four-terminal resistance measurements and thereby eliminates contact and lead-resistance contributions has been described by Glang, Holmwood, and Maissel.[137] As shown in Fig. 60, it consists of a round

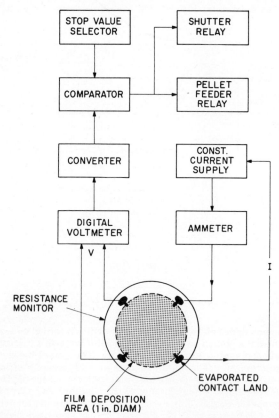

Fig. 60 Circular resistance monitor with four symmetrically located contacts, and block diagram of associated control circuit. (*After Glang, Holmwood, and Maissel.*[137])

substrate with four preevaporated metal lands. These are contacted by spring-loaded clips while the film-deposition area is defined by a mask with a circular opening. The contact lands extend narrow tips toward the center just far enough to touch the perimeter of the circular deposition area. Generally, peripheral four-point-probe resistance measurements on irregularly shaped objects require at least two different current-voltage readings, as discussed by van der Pauw.[335] However, because of the symmetry of this particular configuration, no cyclic switching of current and voltage terminals is necessary. If connected as shown in Fig. 60, the sheet resistance of the film is obtained from one measurement:

$$R_s = 4.53 \frac{V}{I}$$

Evaluation is further simplified by using a constant current of 4.53 mA which makes the voltage displayed by the DVM numerically identical to the sheet resistance.

In addition to being a four-terminal measurement, the technique has other advantages. It can be shown[335] that small variations in the diameter of the circular film area as caused by misregistration or variable spacing of the mask have only negligible effects on the indicated resistance value. The latter is also independent of the contact pressure applied to the lands. The technique is particularly suitable for the fabrication of integrated semiconductor circuits, an application where circular silicon wafers are the standard substrates and methods of surface insulation are readily available. Thus, the identity of regular and monitor substrates in regard to thermal conductivity, heat absorption, and surface temperature is assured. Automatic termination of the process is initiated by a comparator which receives its second input from the stop-value selector shown in Fig. 60. The latter is a four-digit binary-coded decimal switch into which the desired termination value can be dialed.

One problem common to all resistance monitors is the danger of load damage at the beginning of a deposition, when electrical continuity between the preevaporated contacts is first established by an extremely thin film. Crittenden and Hoffman[336] introduced a technique whereby the substrate shutter remains closed while the monitor receives a certain amount of deposit. Thus, when the films on the substrates go through the critically thin stage, the condensate on the monitor is already beyond that thickness. However, their method is really a thickness-control technique which relies on the film resistivity being constant, and it does not lend itself to direct monitoring of the sheet resistance. In the flash evaporation of cermet films, the film-condensation rate is controlled fairly well by the evaporant feed rate. It is therefore possible to estimate the time required to establish a "safe" thickness of about 100 to 200 Å and delay the application of the monitor voltage accordingly.[137] The circuit employed by Steckelmacher et al.[333] had provisions for varying the bridge voltage so that the power to be dissipated by the monitor never exceeded 50 mW.

The accuracy with which a preselected sheet-resistance value can be obtained is of the order of 1 to 2%. The deposition stop value as represented by a preset reference resistor or binary-coded-decimal switch, and the sensitivity of the detection circuits permit, of course, a greater accuracy. In practice, these values are always slightly exceeded. If the terminating signal cuts off the source current,[337] evaporation continues at diminishing rates until the source has cooled. The faster and more common technique is to trigger the substrate shutter, but because of its inertia the latter does not close at the same instant when the solenoid receives the actuating signal. With some experience, it is often possible to anticipate the degree of overshooting and compensate for it by slightly changing the setting of the stop-value selector. However, freshly deposited films are likely to incur resistance changes during subsequent cooling, exposure to air, and the customary annealing processes. Hence, empirical control of the final sheet-resistance value may very well be established without considering the relatively small increment due to overshooting.

Since film resistor materials such as tantalum or Nichrome are often deposited by sputtering techniques, it is important to be aware of the fact that resistance monitors exposed to a glow discharge are subject to unpredictable fluctuations. Maissel et al.[338] found that true resistance indications ($\pm 1\%$) could be obtained only if the aspect ratio of the monitor was chosen to produce readings of less than 20 ohms. In sputtering Nichrome, Stern[206] measured the ion current and relied essentially on constant sputtering yields to control the process. However, the system also had a built-in four-point probe which allowed in situ resistance measurements when the discharge was interrupted.

(3) Capacitance Monitors There have been few attempts to utilize capacitance measurements for evaporation monitoring. The main reason is that the most sensitive arrangement to measure the capacitance of a dielectric film would require the application of a top electrode to the growing film surface and therefore is unfeasible. The possibility of using a planar capacitor as a sensing element has been investigated by Keister and Scapple.[339] Their monitor consisted of a series of narrow parallel lines etched out of aluminum film and spaced 0.0075 in. apart on a quartz substrate. Alternate lines were connected in parallel so as to form the two electrodes of a capacitor. Upon application of an ac voltage, the displacement current spreads throughout the adjacent dielectric media. The quartz substrate alone yielded a capacitance of about 65 pF whereas the deposition of SiO added 1 to 1.5% capacitance per micron of film thickness. Thus, the technique is not very sensitive and is applicable only to relatively thick films. The error associated with thickness monitoring was between 2 and 4%. By relating the capacitance change to a change in voltage amplitude, deposition-rate measurements with an accuracy of about 4% were also made.

A monitor based on capacitance measurements in the evaporant vapor which indicates rates directly has been described by Riddle.[340] The sensing element was formed by two parallel plates, and the vacuum in between served as the dielectric. During evaporation, the vapor passing between the electrodes alters the dielectric constant and hence the capacitance. The device is reportedly capable of indicating rates from 0.1 to 100 Å s^{-1}. However, its merit would have to be judged against other and better-established vapor monitors such as ionization-rate meters, and it does not offer the advantage of monitoring a film property directly.

d. Evaporation-process Control

(1) Thickness Control Provisions for monitoring an evaporation process are usually combined with means to control film deposition. Frequently, the only requirement is to terminate the process when the film thickness or a thickness-related property has reached a certain value. The simplest way to do this is to evaporate a weighed amount of source material to completion. If the emission characteristic of the source is known, the film thickness to be expected from a given quantity of evaporant can be calculated by means of the equations and methods discussed in Sec. 5a. Nomographs for this purpose have also been published.[341] In practice, the technique requires empirical adjustment, and its utilization is limited to applications where thickness control is not critical. Greater accuracy and more positive control are offered by methods which employ a monitoring device.

As the discussion of monitors has shown, some of them give indications of the accumulated deposit thickness and are therefore directly applicable to determine when the process should be discontinued. Other devices are primarily rate indicators and require integration over the deposition time. For automatic thickness control, monitors which generate an electric signal are necessary. As shown in Table 17, most devices produce an electric output suitable for control purposes. With ionization-rate monitors, the signal must be integrated to obtain the film thickness, and circuits as well as instruments for electronic integration have been reported by Schwarz[280] and Brownell et al.[286] Process termination may be implemented by analog or digital techniques, depending on the nature of the signal and the preference of the investigator. In the former case, a recorder with an adjustable on-off control point is a convenient means for recognizing the stop value and triggering the closure of the substrate shutter. A digital control circuit is shown in Fig. 60 in conjunction with resistance monitoring.

The only devices which do not yield an electric signal are the particle-impingement monitors, and while the use of an electronic revolution counter has been suggested,[289] the idea has not been implemented as yet. Optical monitors require specialized circuitry to sense the occurrence and register the number of reflectivity maxima or minima.[318,319,330]

(2) Rate Control The control of evaporation rates is a more complex task than thickness control because it requires adjustment of the source temperature. The

TABLE 17 Evaporation Process Control

Control method	Electrical signal		Prerequisites for	
	Available	Related to	Thickness control	Rate control
Evaporation to completion........	No	....	Weighed evaporant charge	Not applicable
Control of source temperature, power, or current........	Yes	Evaporation rate	Rates must be integrated	Error-signal feedback
Ionization-rate monitor........	Yes	Evaporation rate	Rates must be integrated	Error-signal feedback
Particle impingement devices.....	No	....	Direct observation on pivoted models, manual stop	Direct observation on torsion-wire models, manual power adjustment
Microbalances..........	Electromagnetic or electrostatic models: yes	Deposit thickness	Strip-chart recorder, preset microswitch for automatic termination	Programmed weight increase, or second signal by electronic differentiation
Crystal oscillator.........	Yes	Deposit thickness	Counter, meter, or recorder; preset microswitch for automatic termination	Second signal by differentiation; servoloop
Optical monitors..........	With photocell: yes	Deposit thickness	Strip-chart recorder; fairly complex circuitry for automatic termination	Has not been implemented yet
Resistance monitors..........	Yes	Deposit thickness	Strip-chart recorder and preset microswitch, or digital techniques	Second signal by electronic differentiation; servoloop
Capacitance monitors..........	Yes	Deposit thickness[339] or rate[340]	Planar capacitor substrate; oscillator circuit with reference capacity and comparator[339]	Direct rate indication from Riddle's vapor capacitor[340]

simplest approach is to assume that the emission of a source remains constant throughout an evaporation if the power input does not vary and hence that the evaporant temperature does not change. This concept is applicable only if the effective source area does not vary as a result of slag formation, spreading, or contraction of the evaporant, and if there are no changes in the degree of heat transfer between the evaporant and its support. An example of rate control through a wattmeter is the radiation-heated source described by DaSilva[84] (see Fig. 18). The feasibility of this control principle has also been investigated by Bath and Steckelmacher,[342] who studied the constancy of evaporation rates produced by a resistance-heated molybdenum boat. If the source was maintained at constant voltage, fluctuations of the molten evaporant (bismuth) caused resistance variations sufficiently severe to affect power input and rates. Although the rms current variations observed were relatively small, the exponential temperature dependence of the vapor pressure makes evaporation rates very sensitive to fluctuations of the source temperature. Stable rates were obtained, however, if the rms variations were used to generate an error signal which controlled the power supply and stabilized the current. The method worked satisfactorily for Bi, Cu, and MgF_2, provided relatively large charges of evaporant were employed and slag formation avoided.

One may expect to achieve the same or even better rate constancy by sensing the source temperature directly with a thermocouple and utilizing the latter to control the power supply. However, the conditions of constant evaporant surface area and heat transfer are often difficult to meet. During the evaporation of CdTe from a small effusion cell, for example, the maintenance of constant evaporation rates was accompanied by a gradual rise of the source temperature.[76] Therefore, the control of evaporation rates is generally implemented by means of deposition-monitoring devices which, for this application, must yield an electric signal proportional to rates.

Referring back to Table 17, again, the only devices which provide rate signals directly are ionization-rate monitors and the capacitance vapor monitor described by Riddle.[340] With microbalances, crystal oscillators, and resistance monitors, it is necessary to differentiate the primary signal and derive a secondary signal proportional to the deposition rate. Control circuits incorporating electronic differentiation have been described for these three types of thickness monitors (see, for example, Cahn and Schultz[298] for microbalances; Behrndt and Love,[145] Bath,[343] and Bath, English, and Steckelmacher[344] for crystal oscillators; Turner, Birtwistle, and Hoffman[334] for resistance monitors). Rate control may also be achieved by programming the weight increase, a technique introduced by Hayes and Roberts.[292]

Whether the rate signal is primary or secondary with respect to the monitor, it must be compared with a stable reference signal to produce a differential output suitable for controlling the source power. The latter function requires signal amplification. Continuous adjustment of the source power supply in response to the error signal received from the sensing device may be accomplished by means of saturable reactors or silicon-controlled rectifiers (SCRs or thyristors). The former have been used extensively in conjunction with ionization-rate monitors, and circuit diagrams have been given by various authors.[284,286,345] A servoloop control system employing crystal oscillators and saturable reactors has been described by Behrndt and Love.[145]

The use of SCRs for automatic evaporation-rate control is of more recent origin. Their mode of operation is discussed in Steckelmacher's review,[279] and a system for rate as well as thickness control utilizing a crystal oscillator as the sensing element has been described by Bath et al.[343,344] The latter article is of general interest because the system components are designed as individual modules which offer considerable flexibility in performing various evaporation-control functions. For instance, the drive signal may be derived from monitoring devices other than crystal oscillators. Furthermore, although the system was primarily intended for the control of resistance-heated sources, the control loop is sufficiently flexible to permit stable operation of other source types with different rate vs. power characteristics. The evaporation rates of electron-gun sources, for example, are commonly varied by altering the electron-emission current. Induction-heated vapor sources, too, are amenable to feedback control, although the higher cost and the space and safety requirements of the generator and saturable reactor usually make other source types preferable. An induction-heated nickel-iron source which was servo-controlled from a resistance monitor has been described by Turner et al.[334]

Generally, the development of deposition-control equipment tends toward further automation to eliminate operator judgment and thereby increase the reproducibility of film properties. The objective is to perform the entire pump-down and deposition cycle by combining pressure and temperature sensors, rate monitors, and servomechanisms with adjustable electronic delay devices so that every step of the process is automatically initiated when certain conditions—pressures, temperatures, or evaporation rates—have been established. This trend is exemplified by the system described by English, Putner, and Holland.[346]

REFERENCES

1. Faraday, M., *Phil. Trans.*, **147**, 145 (1857).
2. Nahrwold, R., *Ann. Physik*, **31**, 467 (1887).
3. Kundt, A., *Ann. Physik*, **34**, 473 (1888).
4. See, for example, Guggenheim, E. A., "Thermodynamics. An Advanced Treatment for Chemists and Physicists," 4th ed., North Holland Publishing Company, Amsterdam, 1959.
5. Nesmeyanov, A. N., "Vapor Pressure of the Chemical Elements," Elsevier Publishing Company, New York, 1963.
6. Kubaschewski, O., and E. L. Evans, "Metallurgical Thermochemistry," Pergamon Press, New York, 1965.
7. See, for example, Lewis, G. N., and M. Randall, "Thermodynamics," 2d ed., p. 419, McGraw-Hill Book Company, New York, 1961.
8. See, for example, Darken, L. S., and R. W. Gurry, "Physical Chemistry of Metals," p. 233; McGraw-Hill Book Company, New York, 1953.
9. Rossini, F. D., "Selected Values of Thermodynamic Properties," *NBS Circular* 500, Government Printing Office, Washington, D.C., 1952.
10. Wagman, D. D., "Selected Values of Chemical Thermodynamic Properties," *NBS Technical Note* 270-1 (1965), 270-2 (1966), Government Printing Office, Washington, D.C.
11. Kelley, K. K., "High Temperature Heat Content, Heat Capacity, and Entropy Data for the Elements and Inorganic Compounds," *Bureau of Mines Bulletin* 584, Government Printing Office, Washington, D.C., 1960.
12. Kelley, K. K., and E. G. King, "Entropies of the Elements and Inorganic Compounds," *Bureau of Mines Bulletin* 592, Government Printing Office, Washington, D.C., 1961.
13. Stull, D. R., and G. C. Sinke, "Thermodynamic Properties of the Elements," Advances in Chemistry Series, no. 18, American Chemical Society, 1956.
14. Stull, D. R., "JANAF Thermochemical Data," The Dow Chemical Co., U.S. Clearinghouse, Springfield, Va., 1965/66.
15. Wicks, C. E., and F. E. Block, "Thermodynamic Properties of 65 Elements—Their Oxides, Halides, Carbides, and Nitrides," *Bureau of Mines Bulletin* 605, Government Printing Office, Washington, D.C., 1963.
16. Ulich, H., "Kurzes Lehrbuch der physikalischen Chemie," 5th ed., p. 100, Theodor Steinkopff Verlag, Leipzig, 1948.
17. Kelley, K. K., "The Free Energies of Vaporization and Vapor Pressures of Inorganic Substances," *Bureau of Mines Bulletin* 383, Government Printing Office, 1935.
18. Hohmann, E., and H. Bommer, *Z. Anorg. Allgem. Chem.*, **248**, 383 (1941).
19. Honig, R. E., *RCA Rev.*, **23**, 567 (1962).
20. Hultgren, R., R. L. Orr, P. D. Anderson, and K. K. Kelley, "Selected Values of Thermodynamic Properties of Metals and Alloys," John Wiley & Sons, Inc., New York, 1963.
21. Dushman, S., "Scientific Foundations of Vacuum Technique," John Wiley & Sons, Inc., New York, 1962.
22. See, for example, Kennard, E. H., "Kinetic Theory of Gases," McGraw-Hill Book Company, New York, 1938; or Ref. 23.
23. Parker, P., "Electronics," p. 935, Edward Arnold (Publishers) Ltd., London, 1955.
24. Chapman, S., and T. G. Cowling, "The Mathematical Theory of Non-uniform Gases," Cambridge University Press, New York, 1939.
25. Knudsen, M., *Ann. Physik*, **28**, 75, 999 (1909); **35**, 389 (1911).

26. Steckelmacher, W., *Vacuum*, **16**, 561 (1966).
27. Levenson, L. L., N. Milleron, and D. H. Davis, *Trans. 7th AVS Symp.*, 1960, p. 372, Pergamon Press, Oxford, England.
28. Hertz, H., *Ann. Physik*, **17**, 177 (1882).
29. Knudsen, M., *Ann. Physik*, **47**, 697 (1915).
30. Langmuir, I., *Physik. Z.*, **14**, 1273 (1913).
31. Knudsen, M., *Ann. Physik*, **29**, 179 (1909).
32. Knacke, O., and I. N. Stranski, *Progr. Metal Phys.*, **6**, 181 (1956).
33. Rutner, E., in E. Rutner, P. Goldfinger, and J. P. Hirth (eds.), "Condensation and Evaporation of Solids," p. 149, Gordon and Breach, Science Publishers, Inc., New York, 1964.
34. Freeman, R. D., and J. G. Edwards, Ref. 33, p. 127.
35. Hirth, J. P., and G. M. Pound, "Condensation and Evaporation, Nucleation and Growth Kinetics," The Macmillan Company, New York, 1963.
36. Winslow, G. H., Ref. 33, p. 29.
37. Dettorre, J. F., T. G. Knorr, and E. H. Hall, in C. F. Powell, J. H. Oxley, and J. M. Blocher, Jr. (eds.), "Vapor Deposition," p. 62, John Wiley & Sons, Inc., New York, 1966.
38. Polanyi, M., and E. Wigner, *Z. Physik. Chem.*, **139A**, 439 (1928).
39. Herzfeld, K. F., "Kinetische Theorie der Waerme," p. 229, vol. 3, pt. 2 of Mueller-Pouilletta, "Lehrbuch der Physik," F. Vieweg & Sons, Braunschweig, 1925.
40. Pelzer, H., in M. Volmer (ed.), "Kinetik der Phasenbildung," Theodor Steinkopff Verlag, Dresden (1939).
41. Neumann, K., *Z. Physik. Chem.*, **197A**, 16 (1950).
42. Penner, S. S., *J. Phys. Chem.*, **52**, 367, 949, 1262 (1948); **56**, 475 (1952); **65**, 702 (1961).
43. Glasstone, S., K. J. Laidler, and H. Eyring, "The Theory of Rate Processes," McGraw-Hill Book Company, New York, 1941.
44. Herzfeld, K. F., *J. Chem. Phys.*, **3**, 319 (1935).
45. Stearn, A. E., and H. Eyring, *J. Chem. Phys.*, **5**, 113 (1937).
46. Shultz, R. D., and A. O. Dekker, *J. Chem. Phys.*, **23**, 2133 (1955).
47. Mortensen, E. M., and H. Eyring, *J. Phys. Chem.*, **64**, 847 (1960).
48. Eyring, H., F. M. Wanlass, and E. M. Eyring, Ref. 33, p. 3.
49. Kossel, W., *Nachr. Ges. Wiss. Goettingen*, 1927, p. 135; *Ann. Physik*, **33**, 651 (1938).
50. Stranski, I. N., *Z. Physik. Chem.*, **136**, 259 (1928); **11B**, 421 (1931).
51. Knacke, O., and I. N. Stranski, *Ergeb. Exakt. Naturw.*, **26**, 383 (1952). For a review, see also Mayer, H., "Physik duenner Schichten," vol. II, p. 14, Wissenschaftliche Verlagsgesellschaft m.b.H., Stuttgart, 1955.
52. Hirth, J. P., and G. M. Pound, *J. Chem. Phys.*, **26**, 1216 (1957).
53. Sears, G. W., *J. Chem. Phys.*, **24**, 868 (1956); **27**, 1308 (1957).
54. Hirth, J. P., and G. M. Pound, Ref. 35, p. 138; see also Winterbottom, W. L., and J. P. Hirth, Ref. 33, p. 347.
55. Knudsen, M., *Ann. Physik*, **28**, 75, 999 (1909).
56. Mayer, H., *Z. Physik*, **52**, 235 (1929).
57. Knudsen, M., *Ann. Physik*, **52**, 105 (1917).
58. Heald, J. H., Jr., and R. F. Brown, *Ext. Abstr. 14th AVS Symp.*, 1967, p. 63, Herbick and Held Printing Co., Pittsburgh, Pa.
59. Holland, L., "Vacuum Deposition of Thin Films," John Wiley & Sons, Inc., New York, 1961.
60. Behrndt, K. H., in Bunshah, R. F. (ed.), "Techniques of Metals Research," vol. I, pt. 3, p. 1225, Interscience Publishers, Inc., New York, 1968.
61. Pirani, M., and J. Yarwood, "Principles of Vacuum Engineering," Reinhold Publishing Corporation, New York, 1961.
62. Rosenblatt, G. M., P. K. Lee, and M. B. Dowell, *J. Chem. Phys.*, **45**, 3454 (1966).
63. Yamdagni, R., and R. F. Porter, *J. Electrochem. Soc.*, **115**, 601 (1968).
64. Kohl, W. H., "Handbook of Materials and Techniques for Vacuum Devices," Reinhold Publishing Corporation, New York, 1967.
65. Olsen, L. O., C. S. Smith, and E. C. Crittenden, *J. Appl. Phys.*, **16**, 425 (1945).
66. Nicholson, J. L., *Rev. Sci. Instr.*, **34**, 118 (1963).
67. Lucas, M. S. P., C. R. Vail, W. C. Stewart, and H. A. Owen, *Trans. 8th AVS Symp.*, 1961, p. 988, The Macmillan Company, New York.
68. Massey, B. J., *Trans. 8th AVS Symp.*, 1961, p. 992; The Macmillan Company, New York.
69. Behrndt, K. H., *J. Appl. Phys.*, **33**, 193 (1962).
70. Kilgore, B. F., and R. W. Roberts, *Rev. Sci. Instr.*, **34**, 11 (1963).
71. Hale, A. P., *Vacuum*, **13**, 93 (1963).
72. Neugebauer, C. A., *J. Appl. Phys.*, **31**, 1525 (1960).

73. Layton, W. T., and H. E. Culver, *Proc. Electron. Components Conf.*, 1966, p. 225, Washington, D.C.
74. Hoffman, D. M., and J. Riseman, *Trans. 6th AVS Symp.*, 1959, p. 218, Pergamon Press, New York.
75. Roberts, G. C., and G. G. Via, U.S. Patent 3,313,914, 1967.
76. Glang, R., J. G. Kren, and W. J. Patrick, *J. Electrochem. Soc.*, **110**, 408 (1963).
77. Drumheller, C. E., *Trans. 7th AVS Symp.*, 1960, p. 306, Pergamon Press, New York.
78. Vergara, W. C., H. M. Greenhouse, and N. C. Nicholas, *Rev. Sci. Instr.*, **34**, 520 (1963); for instructions on how to make a baffled SiO source, see also De Tuerk, J. J., Jr., *J. Vacuum Sci. Technol.*, **5**, 88 (1968).
79. For binary phase diagrams, see Hansen, M., and K. Anderko, "Constitution of Binary Alloys," 2d ed., McGraw-Hill Book Company, New York, 1958; Elliott, R. P., "Constitution of Binary Alloys, First Supplement," McGraw-Hill Book Company, New York, 1965.
80. See, for example, "Guide to Scientific Instruments—1967/68," American Association for the Advancement of Science, New York.
81. Economos, G., and W. D. Kingery, *J. Am. Ceram. Soc.*, **36**, 403 (1953).
82. Johnson, P. D., *J. Am. Ceram. Soc.*, **33**, 168 (1950).
83. Kingery, W. D., *J. Am. Ceram. Soc.*, **36**, 362 (1953).
84. DaSilva, E. M., *Rev. Sci. Instr.*, **31**, 959 (1960).
85. Hauck, J. E., *Mater. Design Eng.*, no. 208, p. 85, July, 1963.
86. Hildenbrand, D. L., and W. F. Hall, Ref. 33, p. 399.
87. Ames, I., L. H. Kaplan, and P. A. Roland, *Rev. Sci. Instr.*, **37**, 1737 (1966).
88. Thompson, F. E., and J. F. Libsch, *SCP Solid State Technol.*, December, 1965, p. 50.
89. Picard, R. G., and J. E. Joy, *Electronics*, **24**, 126 (April, 1951).
90. Lewis, J. C., B. Redfern, and F. C. Cowlard, *Solid-State Electron.*, **6**, 251 (1963).
91. Rothwell, W. S., *J. Appl. Phys.*, **39**, 1840 (1968).
92. Moriya, Y., N. Okuma, and K. Sugiura, *Trans. 8th AVS Symp.*, 1961, vol. 2, p. 1055, The Macmillan Company, New York.
93. Wikel, V., *Solid State Technol.*, **11**, (2), 12 (1968).
94. Holland, L., *Vacuum*, **6**, 161 (1956).
95. Toombs, P. A. B., and A. J. Jeal, *J. Sci. Instr.*, **42**, 722 (1965).
96. Hemmer, F. J., and J. R. Piedmont, *Rev. Sci. Instr.*, **33**, 1355 (1962).
97. Holland, L., in "Thin Film Microelectronics," p. 143, John Wiley & Sons, Inc., New York, 1965.
98. Candidus, E. S., M. H. Hablanian, and H. A. Steinherz, *Trans. 6th AVS Symp.*, 1959, p. 185, Pergamon Press, New York.
99. Holland, L., British Patent 754,102, 1951.
100. Heavens, O. S., *J. Sci. Instr.*, **36**, 95 (1959).
101. Kelly, J. C., *J. Sci. Instr.*, **36**, 89 (1959).
102. Milleron, N., *Trans. 4th AVS Symp.*, 1957, p. 148, Pergamon Press, New York.
103. Unvala, B. A., and G. R. Booker, *Phil. Mag.*, **9**, 691 (1964).
104. Chopra, K. L., and M. R. Randlett, *Rev. Sci. Instr.*, **37**, 1421 (1966).
105. Berry, R. W., Proc. 3d Symp. on Electron Beam Technology, 1961, p. 359, Alloyd Electronics Corp., Cambridge, Mass.
106. Holmwood, R. A., and R. Glang, *J. Electrochem. Soc.*, **112**, 827 (1965).
107. Reichelt, W., and G. F. P. Mueller, *Trans. 8th AVS Symp.*, 1961, p. 956, Pergamon Press, New York.
108. Denton, R. A., and A. D. Greene, *Proc. 5th Electron Beam Symp.*, 1963, p. 180, Alloyd Electronics Corp., Cambridge, Mass.
109. Fowler, P., *J. Appl. Phys.*, **34**, 3538 (1963).
110. Gerstenberg, D., and P. M. Hall, *J. Electrochem. Soc.*, **111**, 936 (1964).
111. Thun, R. E., and J. B. Ramsey, *Trans. 6th AVS Symp.*, 1959, p. 192, Pergamon Press, New York.
112. Rairden, J. R., and C. A. Neugebauer, *Proc. IEEE*, **52**, 1234 (1964).
113. Brunner, W. F., and H. G. Patton, *Trans. 8th AVS Symp.*, 1961, p. 895, The Macmillan Company, New York.
114. Siddall, G., and B. A. Probyn, *Trans. 8th AVS Symp.*, 1961, p. 1017, Pergamon Press, New York.
115. Brice, J. C., and U. Pick, *J. Sci. Instr.*, **41**, 633 (1964).
116. Davey, J. E., R. J. Tiernan, T. Pankey, and M. D. Montgomery, *Solid-State Electron.*, **6**, 205 (1963).
117. Voigt, J. W., and K. W. Foster, *Rev. Sci. Instr.*, **35**, 1087 (1964).
118. Brownell, R. B., W. D. McLennan, R. L. Ramsey, and E. J. White, *Rev. Sci. Instr.*, **35**, 1147 (1964).

119. Reames, J. P., *Trans. 6th AVS Symp.*, 1959, p. 215, Pergamon Press, New York.
120. Maskalick, N. J., and C. W. Lewis, *Trans. 8th AVS Symp.*, 1961, vol. 2, p. 874, Pergamon Press, New York.
121. Blackburn, D. H., and W. Haller, *Rev. Sci. Instr.*, **36**, 901 (1965).
122. Erdman, K. L., D. Axen, J. R. McDonald, and L. P. Robertson, *Rev. Sci. Instr.*, **35**, 122 (1964).
123. Steckelmacher, W., and L. Holland, *Vacuum*, **2**, 346 (1952); see also Ref. 59.
124. von Hippel, A., *Ann. Physik*, **81**, 1043 (1926).
125. Holland, L., and W. Steckelmacher, *Vacuum*, **2**, 346 (1952).
126. Anastasio, T. A., and W. J. Slattery, *J. Vacuum Sci. Technol.*, **4**, 203 (1967).
127. Holland, L., and N. J. Newman, *Rev. Sci. Instr.*, **23**, 642 (1952).
128. Clausing, P., *Z. Physik*, **66**, 471 (1930).
129. Ruth, V., and J. P. Hirth, Ref. 33, p. 99.
130. Guenther, K. G., *Z. Angew. Phys.*, **9**, 550 (1957).
131. Learn, A. J., and R. S. Spriggs, *Rev. Sci. Instr.*, **34**, 179 (1963).
132. Ehrler, F., and Th. Kraus, *Trans. 3d Intern. Vacuum Congr.*, 1965, vol. 2, pp. 131, 135, Pergamon Press, New York.
133. Grimley, R. T., and J. LaRue, *Ext. Abstr. 14th AVS Symp.*, 1967, p. 59, Herbick and Held Printing Co., Pittsburgh, Pa.
134. Behrndt, K. H., and R. A. Jones, *Vacuum*, **11**, 129 (1961).
135. Spriggs, R. S., and A. J. Learn, *Rev. Sci. Instr.*, **37**, 1539 (1966).
136. Rohn, K., *Z. Physik*, **126**, 20 (1949).
137. Glang, R., R. A. Holmwood, and L. I. Maissel, *Thin Solid Films*, **1**, 151 (1967).
138. Beavitt, A. R., R. C. Turnell, and D. S. Campbell, *Thin Solid Films*, **1**, 3 (1967).
139. Behrndt, K. H., in G. Hass and R. E. Thun (eds.), "Physics of Thin Films," vol. 3, p. 1, Academic Press Inc., New York, 1966.
140. Oberg, P. E., R. M. Sander, and E. J. Torok, *Vacuum*, **13**, 53 (1963).
141. Bugenis, C., and L. E. Preuss, *Trans. 10th AVS Symp.*, 1963, p. 374, The Macmillan Company, New York.
142. Behrndt, K. H., and R. A. Jones, *Trans. 5th AVS Symp.*, 1958, p. 217, Pergamon Press, New York.
143. Behrndt, K. H., *Trans. 6th AVS Symp.*, 1959, p. 242, Pergamon Press, New York.
144. Behrndt, K. H., *Trans. 7th AVS Symp.*, 1960, p. 137, Pergamon Press, New York.
145. Behrndt, K. H., and R. W. Love, *Vacuum*, **12** (1962).
146. Behrndt, K. H., *Trans. 9th AVS Symp.*, 1962, p. 111, The Macmillan Company, New York.
147. Hass, G., and E. Ritter, *J. Vacuum Sci. Technol.*, **4**, 71 (1967).
148. Fisher, R. A., and J. R. Platt, *Rev. Sci. Instr.*, **8**, 505 (1937).
149. Behrndt, K. H., *Trans. 10th AVS Symp.*, 1963, p. 379, The Macmillan Company, New York.
150. Behrndt, K. H., and D. W. Doughty, *J. Vacuum Sci. Technol.*, **3**, 264 (1966).
151. Drowart, J., Ref. 33, p. 255.
152. Reichelt, W., and P. Mueller, *Vakuum-Tech.*, 1962, no. 8.
153. Holland, L., Ref. 59, p. 464.
154. Kubaschewski, O., and E. L. Evans, Ref. 6, p. 326.
155. Blackburn, P. E., M. Hoch, and H. L. Johnston, *J. Phys. Chem.*, **62**, 769 (1958).
156. Schaefer, H., and R. Hoernle, *Z. Anorg. Allgem. Chem.*, **263**, 26 (1950).
157. Tombs, N. C., and A. J. E. Welch, *J. Iron Steel Inst.*, **172**, 69 (1952).
158. Hacman, D., W. K. Huber, and G. Rettinghaus, *Ext. Abstr. 14th AVS Symp.*, 1967, p. 27, Herbick and Held Printing Co., Pittsburgh, Pa.
159. Simons, S. L., Ref. 21, p. 760.
160. Woesten, W. J., and M. G. Geers, *J. Phys. Chem.*, **66**, 1252 (1962).
161. Neuhaus, A., and W. Retting, *Z. Elektrochem.*, **62**, 33 (1958).
162. Woesten, W. J., *J. Phys. Chem.*, **65**, 1949 (1961).
163. Burmeister, R. A., Jr., and D. A. Stevenson, *J. Electrochem. Soc.*, **114**, 394 (1967).
164. de Nobel, D., *Philips Res. Rept.*, **14**, 361, 430 (1959).
165. Schoolar, R. B., and J. N. Zemel, *J. Appl. Phys.*, **35**, 1849 (1964).
166. Efstathion, A., D. M. Hoffman, and E. R. Levin, *Ext. Abstr. 13th AVS Symp.*, 1966, p. 143, Herbick and Held Printing Co., Pittsburgh, Pa.
167. Vecht, A., in G. Hass and R. E. Thun (eds.), "Physics of Thin Films," vol. 3, p. 165, Academic Press Inc., New York, 1966.
168. Hamilton, P. M., *SCP Solid State Technol.*, June, 1964, p. 15.
169. Reichelt, W., *Proc. 3d Intern. Vacuum Congr.*, 1965, vol. 2, p. 25, Pergamon Press, New York.
170. Colin, R., and P. Goldfinger, Ref. 33, p. 165.

171. Somorjai, G. A., Ref. 33, p. 417.
172. Goswami, A., and S. S. Koli, in R. Niedermayer and H. Mayer (eds.), "Basic Problems in Thin Film Physics," p. 646, Vandenhoeck and Ruprecht, Goettingen, 1966.
173. Cariou, F. E., V. A. Cajal, and M. M. Gajary, *Proc. 1967 Electron. Components Conf.*, p. 60.
174. Cox, J. T., G. Hass, and J. B. Ramsey, *J. Phys. (Paris)*, **25**, 250 (1964).
175. Reichelt, W., *Vide*, no. 106, p. 390, 1963.
176. Pliskin, W. A., and P. P. Castrucci, *Electrochem. Technol.*, **6**, 85 (1968).
177. Eisele, K. M., Ref. 171, p. 672.
178. Stull, D. R., *Ind. Eng. Chem.*, **39**, 540 (1947). The tables may also be found in the "Handbook of Chemistry and Physics," 47th ed., p. D-109, The Chemical Rubber Co., Cleveland, Ohio, 1966.
179. Darken, L. S., and R. W. Gurry, "Physical Chemistry of Metals," p. 342, McGraw-Hill Book Company, New York, 1953.
180. Quill, L. L. (ed.), "The Chemistry and Metallurgy of Miscellaneous Materials," McGraw-Hill Book Company, New York, 1950.
181. Rickert, H., Ref. 33, p. 201.
182. Knauff, K. G., Ref. 172, p. 207.
183. Lyon, T. F., Ref. 33, p. 435.
184. Weiser, K., in R. K. Willardson and H. L. Goering (eds.), "Compound Semiconductors," vol. 1, "Preparation of III-V Compounds," p. 471, Reinhold Publishing Corporation, New York, 1962.
185. Lee, P. K., and R. C. Schoonmaker, Ref. 33, p. 379.
186. Paparoditis, C., Ref. 183, p. 326.
187. Dale, E. B., G. Senecal, and D. Huebner, *Trans. 10th AVS Symp.*, 1963, p. 348, The Macmillan Company, New York.
188. Zinsmeister, G., *Vakuum-Tech.*, no. 8, p. 223, 1964.
189. Sommer, A. H., *J. Appl. Phys.*, **37**, 2789 (1966).
190. Lowe, R. M., *J. Appl. Phys.*, **39**, 2476 (1968).
191. See, for example, Kortuem, G., "Treatise on Electrochemistry," 2d ed., p. 54, Elsevier Publishing Company, New York, 1965.
192. Hultgren, R., R. L. Orr, P. D. Anderson, and K. K. Kelley, "Selected Values of Thermodynamic Properties of Metals and Alloys," John Wiley & Sons, Inc., New York, 1963.
193. Zellars, G. R., S. L. Payne, J. P. Morris, and R. L. Kipp, *Trans. Met. Soc. AIME*, **215**, 181 (1959).
194. Koller, L. R., Ref. 21, p. 691.
195. Huijer, P., W. T. Langendam, and J. A. Lely, *Philips Tech. Rev.*, **24**, 144 (1963).
196. An example of this technique is described in the leaflet BRL-No. 67-6, issued by the Bendix Research Laboratories, Southfield, Mich.
197. Finegan, J. J., and P. R. Gould, *Trans. 9th AVS Symp.*, 1962, p. 129, The Macmillan Company, New York.
198. Degenhart, H. J., and I. H. Pratt, *Trans. 10th AVS Symp.*, 1963, p. 480, The Macmillan Company, New York.
199. Swift, R. A., B. A. Noval, and K. M. Merz, *J. Vacuum Sci. Technol.*, **5**, 79 (1968).
200. Campbell, D. S., and B. Hendry, *Brit. J. Appl. Phys.*, **16**, 1719 (1965).
201. Blois, M. S., *J. Appl. Phys.*, **26**, 975 (1955).
202. Penn, T. C., and F. G. West, *J. Appl. Phys.*, **38**, 2060 (1967).
203. Lakshmanan, K., *Trans. 8th AVS Symp.*, 1961, vol. 2, p. 868, The Macmillan Company, New York.
204. Lewis, C. W., and M. Schick, Ref. 33, p. 699.
205. Wied, O., and M. Tierman, *Proc. 3d Ann. Microelectron. Symp.*, St. Louis, 1964, p. 1-C-1.
206. Stern, E., *Proc. Electron. Components Conf.*, 1966, p. 233.
207. Riddle, G. C., *Ext. Abstr. 13th AVS Symp.*, 1966, p. 89, Herbick and Held Printing Co., Pittsburgh, Pa.
208. Sauer, E., and E. Unger, *Z. Naturforsch.*, **13a**, 72 (1958).
209. For a review of residual gas effects on thin film properties, see Caswell, H. L., in G. Hass (ed.), "Physics of Thin Films," vol. 1, p. 1, Academic Press Inc., New York, 1963.
210. Clapham, P. B., *J. Sci. Instr.*, **39**, 596 (1962).
211. Wolter, A. R., *J. Appl. Phys.*, **36**, 2377 (1965).
212. Schwartz, H. J., *J. Appl. Phys.*, **34**, 2053 (1963).
213. Lu, Chih-Shun, and A. A. Milgram, *J. Vacuum Sci. Technol.*, **4**, 49 (1967).

214. Roberts, R. W., and T. A. Vanderslice, "Ultrahigh Vacuum and Its Applications," p. 167, Prentice-Hall, Inc., Englewood Cliffs, N.J., 1963.
215. Gibson, R., B. Bergsnov-Hansen, N. Endow, and R. A. Pasternak, *Trans. 10th AVS Symp.*, 1963, p. 88, The Macmillan Company, New York.
216. Hansen, N., and W. Littman, *Trans. 3d Intern. Vacuum Congr.*, 1965, vol. 2, p. 465, Pergamon Press, New York.
217. Singleton, J. H., *Trans. 3d Intern. Vacuum Congr.*, 1965, vol. 2, p. 441, Pergamon Press, New York.
218. Ritter, E., *Monatsh. Chem.*, **95**, 795 (1964).
219. Ritter, E., *J. Vacuum Sci. Technol.*, **3**, 225 (1966).
220. Anastasio, T. A., *J. Appl. Phys.*, **38**, 2606 (1967).
221. Feuersanger, A. E., A. K. Hagenlocher, and A. L. Solomon, *J. Electrochem. Soc.*, **111**, 1387 (1964).
222. Krikorian, E., *Ext. Abstr. 13th AVS Symp.*, 1966, p. 175, Herbick and Held Printing Co., Pittsburgh, Pa.
223. Schilling, R. B., *Proc. IEEE*, **52**, 1350 (1964).
224. Novice, M. A., J. A. Bennett, and K. B. Cross, *J. Vacuum Sci. Technol.*, **1**, 73 (1964).
225. Frank, R. I., and W. L. Moberg, *J. Vacuum Sci. Technol.*, **4**, 133 (1967).
226. Holland, L., Ref. 59, p. 476.
227. Herring, C. S., and A. D. Tevebaugh, *J. Electrochem. Soc.*, **110**, 119 (1963).
228. Cremer, E., and H. Pulker, *Monatsh. Chem.*, **93**, 491 (1962).
229. Ritter, E., *Opt. Acta*, **9**, 197 (1962).
230. Pizzarello, F. A., *J. Appl. Phys.*, **35**, 2730 (1964).
231. Itoh, A., *Proc. 4th Intern. Vacuum Congr.*, 1968, pt. 2, p. 536, The Institute of Physics and the Physical Society, London.
232. Miller, R. J., and C. H. Bachman, *J. Appl. Phys.*, **29**, 1277 (1958).
233. Bis, R. F., A. S. Rodolakis, and J. N. Zemel, *Rev. Sci. Instr.*, **36**, 1626 (1965).
234. Johnson, J. E., *J. Appl. Phys.*, **36**, 3193 (1965).
235. Potter, R. F., *Ext. Abstr. 13th AVS Symp.*, 1966, p. 81, Herbick and Held Printing Co., Pittsburgh, Pa.
236. Frankl, D. R., A. Hagenlocher, E. D. Haffner, P. H. Heck, A. Sandor, E. Both, and H. J. Degenhart, *Proc. Electron. Components Conf.*, 1962, p. 44.
237. Edgecumbe, J., L. G. Rosner, and D. E. Anderson, *J. Appl. Phys.*, **35**, 2198 (1964).
238. Behrndt, K. H., *J. Metals*, **14**, 208 (1962).
239. Glang, R., R. A. Holmwood, and S. R. Herd, *J. Vacuum Sci. Technol.*, **4**, 163 (1967).
240. Mader, S., *J. Vacuum Sci. Technol.*, **2**, 35 (1965).
241. Sachtler, W. M. H., G. J. H. Dorgelo, and R. Jongepier, Ref. 171, p. 218.
242. Belser, R. B., *J. Appl. Phys.*, **31**, 562 (1960).
243. Neugebauer, C. A., *J. Appl. Phys.*, **35**, 3599 (1964).
244. Campbell, D. S., and H. Blackburn, *Trans. 7th AVS Symp.*, 1960, p. 313, Pergamon Press, New York.
245. Beckerman, M., and R. E. Thun, *Trans. 8th AVS Symp.*, 1961, p. 905, Pergamon Press, New York.
246. Ostrander, W. J., and C. W. Lewis, *Trans. 8th AVS Symp.*, 1961, p. 881, Pergamon Press, New York.
247. Guenther, K. G., *Z. Naturforsch.*, **13a**, 1081 (1958).
248. Guenther, K. G., in J. C. Anderson (ed.), "The Use of Thin Films in Physical Investigations," p. 213, Academic Press Inc., New York, 1966.
249. Haenlein, E., and K. G. Guenther, *Naturwiss.*, **46**, 319 (1959).
250. Schottmiller, J. C., F. Ryan, and T. Taylor, *Ext. Abstr. 14th AVS Symp.*, 1967, p. 29, Herbick and Held Printing Co., Pittsburgh, Pa.
251. Steinberg, R. F., and D. M. Scruggs, *J. Appl. Phys.*, **37**, 4586 (1966); see also Steinberg, R. F., *Ext. Abstr. 13th AVS Symp.*, 1966, p. 171, Herbick and Held Printing Co., Pittsburgh, Pa.
252. Davey, J. E., and T. Pankey, *J. Appl. Phys.*, **39**, 1941 (1968); see also Ref. 253.
253. Houde, A. L., in K. H. Behrndt (ed.), "Vacuum Microbalance Techniques," vol. 3, p. 109, Plenum Press, New York, 1963.
254. Young, A. S., and R. J. Heritage, *Proc. 4th Intern. Vacuum Congr.*, 1968, pt. 2, p. 496, The Institute of Physics and the Physical Society, London.
255. Harris, L., and B. M. Siegel, *J. Appl. Phys.*, **19**, 739 (1948).
256. Beam, W. R., and T. Takahashi, *Rev. Sci. Instr.*, **35**, 1623 (1964).
257. Tandeski, D. A., M. M. Hanson, and P. E. Oberg, *Vacuum*, **14**, 3 (1964).
258. Marshall, R., L. Atlas, and T. Putner, *J. Sci. Instr.*, **43**, 144 (1966).
259. Androes, G. M., R. H. Hammond, and W. D. Knight, *Rev. Sci. Instr.*, **32**, 251 (1961).
260. Feldtkeller, E., *Z. Physik*, **157**, 65 (1959).

261. Beckerman, M., and R. L. Bullard, *Proc. Electron. Components Conf.*, 1962, p. 53.
262. Braun, L., and D. E. Lood, *Proc. IEEE*, **54**, 1521 (1966).
263. Layer, E. H., *Trans. 6th AVS Symp.*, 1959, p. 210, Pergamon Press, New York.
264. Wilson, R. W., and L. E. Terry, *Proc. Electron. Components Conf.*, 1967, p. 397.
265. Richards, J. L., P. B. Hart, and L. M. Gallone, *J. Appl. Phys.*, **34**, 3418 (1963).
266. Richards, J. L., Ref. 248, p. 71.
267. Mueller, E. K., *J. Appl. Phys.*, **35**, 580 (1964).
268. Zyetz, M. C., and A. M. Despres, *Ext. Abstr. 13th AVS Symp.*, 1966, p. 169, Herbick and Held Printing Co., Pittsburgh, Pa.
269. Light, T. B., E. M. Hull, and R. Gereth, *J. Electrochem. Soc.*, **115**, 857 (1968).
270. Holloway, H., J. L. Richards, L. C. Bobb, and J. Perry, Jr., *J. Appl. Phys.*, **37**, 4694 (1966).
271. Ellis, S. G., *J. Appl. Phys.*, **38**, 2906 (1967).
272. Mueller, E. K., B. J. Nicholson, and M. H. Francombe, *Electrochem. Technol.*, **1**, 158 (1963).
273. Brown, V. R., *Ext. Abstr. 13th AVS Symp.*, 1966, p. 139, Herbick and Held Printing Co., Pittsburgh, Pa.
274. Mueller, E. K., B. J. Nicholson, and G. L. E. Turner, *J. Electrochem. Soc.*, **110**, 969 (1963).
275. Himes, W., B. F. Stout, and R. E. Thun, *Trans. 9th AVS Symp.*, 1962, p. 144, The Macmillan Company, New York.
276. Carter, E. E., *Rev. Sci. Instr.*, **34**, 588 (1963).
277. Eckardt, J. R., and R. N. Peacock, *J. Vacuum Sci. Technol.*, **3**, 356 (1966).
278. Mueller, E. K., and J. L. Richards, *J. Appl. Phys.*, **35**, 1233 (1964).
279. Steckelmacher, W., in L. Holland (ed.), "Thin Film Microelectronics," p. 193, John Wiley & Sons, Inc., New York, 1965.
280. Schwarz, H., *Rev. Sci. Instr.*, **32**, 194 (1961).
281. Schwarz, H., *Arch. Tech. Messen*, **5**, 1341 (1960).
282. Dufour, C., and B. Zega, *Vide*, no. 104, p. 180, 1963.
283. Haase, O., *Z. Naturforsch.*, **12a**, 941 (1957).
284. Giedd, G. R., and M. H. Perkins, *Rev. Sci. Instr.*, **31**, 773 (1960).
285. Perkins, M. H., *Trans. 8th AVS Symp.*, 1961, p. 1025, Pergamon Press, New York.
286. Brownell, R. B., W. D. McLennan, R. L. Ramey, and E. J. White, *Rev. Sci. Instr.*, **35**, 1147 (1964).
287. Hammond, R. H., G. M. Kelly, C. H. Meyer, Jr., and J. H. Perene, Jr., "Vacuum Deposition of Refractory Thin Films and Thin Film Compounds," distributed by General Dynamics, San Diego, Calif.
288. Zega, B., *Proc. 4th Intern. Vacuum Congr.*, 1968, pt. 2, p. 523, The Institute of Physics and the Physical Society, London.
289. Beavitt, A. R., *J. Sci. Instr.*, **43**, 182 (1966).
290. Poulis, J. A., P. J. Meeusen, W. Dekker, and J. P. de Mey, in "Vacuum Microbalance Techniques," vol. 6, p. 27, Plenum Press, New York, 1967.
291. Behrndt, K. H., *Z. Angew. Phys.*, **8**, 453 (1956).
292. Hayes, R. E., and A. R. V. Roberts, *J. Sci. Instr.*, **39**, 428 (1962).
293. Gulbransen, E. A., *Rev. Sci. Instr.*, **15**, 201 (1944).
294. Wolsky, S. P., *Phys. Rev.*, **108**, 1131 (1957).
295. Wolsky, S. P., and E. J. Zdanuk, in "Vacuum Microbalance Techniques," vol. 1, p. 35, Plenum Press, New York, 1961.
296. Mayer, H., W. Schroen, and D. Stuenkel, *Trans. 7th AVS Symp.*, 1960, p. 279, Pergamon Press, New York.
297. Mayer, H., R. Niedermayer, W. Schroen, D. Stuenkel, and H. Goehre, in "Vacuum Microbalance Techniques," vol. 3, p. 75, Plenum Press, New York, 1963.
298. Cahn, L., and H. R. Schultz, in "Vacuum Microbalance Techniques," vol. 3, p. 29, Plenum Press, New York, 1963.
299. Gast, Th., *Vakuum-Tech.*, **14**, 41 (1965).
300. Gast, Th., in "Vacuum Microbalance Techniques," vol. 6, p. 59, Plenum Press, New York, 1967.
301. Pearson, S., and N. J. Wadsworth, *J. Sci. Instr.*, **42**, 150 (1965).
302. Wolsky, S. P., E. J. Zdanuk, C. H. Massen, and J. A. Poulis, in "Vacuum Microbalance Techniques," vol. 6, p. 37, Plenum Press, New York, 1967.
303. Sauerbrey, G., *Phys. Verhandl.*, **8**, 113 (1957).
304. Sauerbrey, G., *Z. Physik*, **155**, 206 (1959).
305. Lostis, M. P., *J. Phys. Radium*, **20**, 25 (1959).
306. Phelps, F. P., *Proc. 11th Ann. Symp. Frequency Control*, 1957, p. 256, Fort Monmouth, N.J.

307. Stockbridge, C. D., in "Vacuum Microbalance Techniques," vol. 5, p. 193, Plenum Press, New York, 1966.
308. Eschbach, H. L., and E. W. Kruidhof, in "Vacuum Microbalance Techniques," vol. 5, p. 207; Plenum Press, New York, 1966.
309. Niedermayer, R., N. Gladkich, and D. Hillecke, in "Vacuum Microbalance Techniques," vol. 5, p. 217, Plenum Press, New York, 1966.
310. Pulker, H. K., Z. Angew. Phys., 20, 537 (1966).
311. Pulker, H. K., and W. Schaedler, Ext. Abstr. 14th AVS Symp., 1967, p. 79, Herbick and Held Printing Co., Pittsburgh, Pa.
312. Langer, A., and J. T. Patton, in "Vacuum Microbalance Techniques," vol. 5, p. 231, Plenum Press, New York, 1966.
313. Stockbridge, C. D., in "Vacuum Microbalance Techniques," vol. 5, p. 147, Plenum Press, New York, 1966.
314. Bakos, J., G. Nagy, and J. Szigeti, J. Appl. Phys., 37, 4433 (1966).
315. Hillecke, D., and R. Niedermayer, Vakuum-Tech., 3, 69 (1965).
316. Warner, A. W., and C. D. Stockbridge, in "Vacuum Microbalance Techniques," vol. 3, p. 55, Plenum Press, New York, 1963.
317. Riegert, R. P., Proc. 4th Intern. Vacuum Congr., 1968, pt. 2, p. 527, The Institute of Physics and the Physical Society, London.
318. Steckelmacher, W., J. M. Parisot, L. Holland, and T. Putner, Vacuum, 9, 171 (1959).
319. Steckelmacher, W., and J. English, Trans. 8th AVS Symp., 1961, p. 852, Pergamon Press, New York.
320. Ross, A., Vakuum-Tech., 8, 1 (1959).
321. Schaible, P. M., and C. L. Standley, IBM Tech. Disclosure Bull. 6, no. 1, p. 112, 1963.
322. Description and recorder trace courtesy of J. S. Logan and E. S. Ward, IBM Corp., East Fishkill Facility, Hopewell Junction, N.Y.
323. Davidse, P. D., and L. I. Maissel, Trans. 3d Intern. Vacuum Congr., 1965, vol. 2, p. 651, Pergamon Press, New York; Davidse, P. D., and L. I. Maissel, J. Appl. Phys., 37, 574 (1966).
324. Boss, D. W., W. A. Pliskin, and M. Revitz, IBM Tech. Disclosure Bull. 9, no. 10, p. 1389, 1967.
325. Pliskin, W. A., and R. A. Wesson, IBM J. Res. Develop., 12, 192 (1968).
326. Wesson, R. A., H. W. Young, and W. A. Pliskin, Appl. Phys. Letters, 11, 105 (1967).
327. Wesson, R. A., R. P. Phillips, and W. A. Pliskin, J. Appl. Phys., 38, 2455 (1967).
328. Pliskin, W. A., Solid-State Electron., 11, 957 (1968).
329. Zerbst, H., Vakuum-Tech., 12, 173 (1963).
330. Fury, A. M., and C. L. Smith, U.S. Patent 3,059,611, 1962.
331. Dumin, D. J., Rev. Sci. Instr., 38, 1107 (1967).
332. Bennett, J. A., and T. P. Flanagan, J. Sci. Instr., 37, 143 (1960).
333. Steckelmacher, W., J. English, H. H. A. Bath, D. Haynes, J. T. Holden, and L. Holland, Trans. 10th AVS Symp., 1963, p. 415, The Macmillan Company, New York.
334. Turner, J. A., J. K. Birtwistle, and R. G. Hoffman, J. Sci. Instr., 40, 557 (1963).
335. van der Pauw, L. J., Philips Res. Rept., 13, 1 (1958).
336. Crittenden, E. C., and R. W. Hoffman, Rev. Mod. Phys., 25, 310 (1953).
337. Bishop, F. W., Rev. Sci. Instr., 20, 527 (1949).
338. Maissel, L. I., R. J. Hecht, and N. W. Silcox, Proc. Electron. Components Conf., 1963, p. 190.
339. Keister, F. Z., and R. Y. Scapple, Trans. 9th AVS Symp., 1962, p. 116, The Macmillan Company, New York.
340. Riddle, G. C., Proc. 4th Symp. Electron Beam Technol., 1962, p. 340, Alloyd Electronics Corp., Cambridge, Mass.
341. Bond, W. L., J. Opt. Soc. Am., 44, 429 (1954).
342. Bath, H. H. A., and W. Steckelmacher, J. Sci. Instr., 42, 144 (1965).
343. Bath, H. H. A., J. Sci. Instr., 43, 374 (1966).
344. Bath, H. H. A., J. English, and W. Steckelmacher, Electron. Components, 7, 239 (1966).
345. Yaffe, T. H., W. C. Vergara, and H. M. Greenhouse, U.S. Patent 3,316,386, 1967.
346. English, J., T. Putner, and L. Holland, Proc. 4th Intern. Vacuum Congr., 1968, pt. 2, p. 491, The Institute of Physics and the Physical Society, London.

Chapter **2**

High-vacuum Technology

REINHARD GLANG
RICHARD A. HOLMWOOD
and
JOHN A. KURTZ*

IBM Components Division, East Fishkill, New York

* Present address: Cogar Co., Poughkeepsie, N.Y.

LIST OF SYMBOLS

a	(1) constant in Henry's law; (2) heating rate, deg s^{-1}
A	area, cm^2
α_{solid}	linear coefficient of thermal expansion
α_s	sticking coefficient of gas particles impinging on surface
b	constant in Langmuir isotherm, $Torr^{-1}$
β	empirical exponent in pump-down equations
C_0	solubility of gases in solids, cm^3 (STP) per cm^3 of solid, or vol %
C	(1) gas concentration in a solid; (2) conductance of a vacuum chamber section, $l\ s^{-1}$
d	thickness (walls, parts, etc.)
d_s	skin depth in rf heating
D	diffusion coefficient of gases in solids, $cm^2\ s^{-1}$
E_a	energy of adsorption
E_d	energy of desorption (including activation)
E_D	activation energy of gas diffusion in solids
E_P	activation energy of gas permeation in solids
I	electric current, A
l	length, distance, cm
M	molar mass
μ	relative magnetic permeability
N	number of gas particles:
	N_a adsorbed on a surface
	N_0 adsorbed on a surface at time zero
	N_g in a volume of gas
	N_i impinging on a surface
	N_m adsorbed in a monolayer
	N_s leaving a solid by diffusion and desorption
ν	frequency of alternating current, Hz
p	gas pressure, Torr, or atm
P	permeability of solids for gases
Q	gas-flow rates, Torr $l\ s^{-1}$, or cm^3 (STP) s^{-1}:
	Q_a throughput (admittance) of an unbaffled pump
	Q_c conductance-limited throughput
	Q_D gas-evolution rate from outdiffusion
	Q_O outgassing rate (diffusion and desorption)
	Q_P permeation rate of gases through solids
R	universal gas constant
ρ	volume resistivity, ohm-cm
S	speed of vacuum pumps, $l\ s^{-1}$
	S_e effective pumping speed (conductance-limited)
	S_s pumping speed due to surface adsorption
σ	ionization probability of gas particles, $cm^{-1}\ Torr^{-1}$
t	time
T	temperature, °K
	T_m temperature of maximum desorption rate for a given E_d
τ_0	thermal-vibration period of an adatom
$\bar{\tau}$	average residence time of adatoms on a surface
θ	surface coverage (N_a/N_m)
V	gas volume
x	(1) length coordinate; (2) $x = E_d/RT$

1. INTRODUCTION

The evaporation of thin films with controlled properties requires an operating environment which interferes as little as possible with the process of film formation. Much research and development have been devoted to the attainment of high vacuum to minimize the interaction between residual gases and the surfaces of growing films. Because of these efforts, a wide variety of vacuum components, materials, and assembly techniques are available today. This chapter is intended as a review of the commoner types of vacuum hardware and established operating techniques. Emphasis is on the physical principles governing the behavior of components and materials and their practical limitations. Therefore, the extent of the treatment of specific materials or parts is not always proportional to their importance and utilization in practice. The proliferation of vacuum equipment during the last few years makes an exhaustive discussion impossible. Thus, only representative examples are given in each class. A complete listing of manufacturers of vacuum equipment and related instrumentation may be found in the "Guide to Scientific Instruments" published annually by the American Association for the Advancement of Science.[1]

Like other specialized fields of science, vacuum technology has adopted a number of technical terms which have assumed specific meaning. These are defined in the "Glossary of Terms Used in Vacuum Technology" issued by the American Vacuum Society Committee on Standards (Pergamon Press, New York, 1958) and will be introduced when needed.* At this point it is sufficient to clarify the terminology for the various degrees of vacuum, which are distinguished according to pressure ranges:

Low vacuum.................	760 –25 Torr
Medium vacuum.............	25 -10^{-3} Torr
High vacuum................	10^{-3}–10^{-6} Torr
Very high vacuum...........	10^{-6}–10^{-9} Torr
Ultrahigh vacuum...........	Below 10^{-9} Torr

The term "high vacuum" is also used in a broader sense to designate any environment with a pressure smaller than 10^{-3} Torr. The term "fore vacuum" is not associated with a specific pressure range but refers to the gas-filled space on the outlet side of a pump, especially between a high- and a low- or medium-vacuum pump.

2. VACUUM PUMPS

Two different principles may be employed to reduce the pressure in a vacuum enclosure. The first one involves physical removal of gases from the vessel and exhausting the gas load to the outside. Examples of this operating mode are the mechanical and the vapor-stream pumps. The other methods of evacuation rely on condensation or trapping of gas molecules on some part of the inner surface of the enclosure without discharging the gas. Cryogenic, cryosorption, sublimation, and getter-ion pumps belong in this category.

A quantitative measure of the gas-transport or trapping ability of a pump is its throughput Q_a expressed in torr l s^{-1} at 20°C It is that quantity of gas which flows through the intake cross section of the operating pump in unit time, and it is pressure-dependent. Another quantity characterizing the performance of a pump is its speed S. This is the ratio of the throughput to the partial pressure of a specific gas at a point near the intake port of the pump: $S = Q_a/p$. The units of the pumping speed are liters per second at 20°C. Most vacuum pumps maintain nearly constant speed over several decades of pressure. Above and below this region, their speed drops drastically and the pumps become inefficient.

While the throughput is a quantity which can be clearly defined, the concept of pumping speed in the molecular-flow region is problematical. This is due to the fact that the definition of pressure as a scalar force acting on a surface becomes ambiguous

---

* A comprehensive multilingual glossary and dictionary is "Elsevier's Dictionary of High Vacuum Science and Technology" by F. Weber, Elsevier Publishing Company, New York, 1968.

in a vacuum chamber with different net flow of gas at different points. Using Monte Carlo computations, several investigators have shown that the speeds of diffusion[2] and getter-ion[3] pumps vary as a result of different angular distributions of molecules approaching and leaving the intake opening. Differences in angular distribution arise from the different shapes of vacuum chambers and pumps, which may be responsible for speeds up to 25% higher than those obtained with the regular cosine input distribution.[2] To avoid these inconsistencies, it has been suggested that pumping speeds be defined not based on gas pressure but in terms of the rate of molecular incidence at the mouth, multiplied by the probability that an arriving molecule will be captured by the pump.[4] Another approach to derive true pumping speeds for directional rather than random molecular motion in vacuum systems utilizes network analysis in analogy to radiant-heat-transfer problems.[5]

In addition to throughput and speed, an important criterion for selecting a pump is its tendency to emit gases into the vacuum chamber. In principle, all pumps may act as sources of residual gases under certain conditions, but they differ greatly in the amount and nature of the gases released. For some applications, the traces of organic vapors stemming from the operating fluids of pumps are detrimental, and the choice is then limited to oil-free models. In the following sections, the various types of pumps are discussed in regard to their operating principles, performance ranges, and limitations.

a. Mechanical Pumps

Mechanical pumps move gases by the cyclic motion of a system of mechanical parts. The objective of gas transport by means of rotating bodies can be accomplished in different ways, and various operating principles are reviewed by Dushman.[6] A survey of commercially available models and their performance has been published by Lucas.[7] For thin film evaporators, three types of mechanical pumps are of interest: oil-sealed rotary pumps, Roots pumps, and molecular-drag pumps.

(1) Oil-sealed Rotary Pumps These are most widely used to establish the necessary fore vacuum for high-vacuum pumps. Current models affect gas transport through rotating cylinders and differ mainly in the slide mechanisms which trap and carry out a volume of gas. An example is the sliding-vane pump shown in Fig. 1. An eccen-

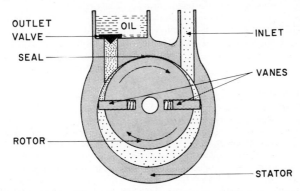

Fig. 1 Schematic of a vane-type rotary oil pump.

trically positioned rotor fits tightly against a cylindrical seat machined into the stator. Two spring-loaded vanes sliding in diametrically opposed slots in the rotor press against the inner surface of the stator. Friction and wear are minimized by a thin oil film which lubricates all parts of the pump and also seals the minute gap at the seat. The exhaust end is normally closed by a pressure valve leading into an oil reservoir. When turning, the rotor draws air from the inlet side into the pump. The crescent-shaped air volume is then compressed, thus forcing the outlet valve open and permitting the gas to be discharged. Technical refinements to improve pump performance are operation of two units in series (two-stage pumps), and a vented outlet valve admitting a small air flow into the compression section. This practice is called gas ballasting and reduces the condensation of water vapor by lowering the compression ratio.

The pumping speed of single-stage rotary oil pumps* declines rapidly at pressures below 1 Torr. Since lower fore pressures are needed at the outlet side of high-vacuum pumps, it is customary to build rotary oil pumps with two stages in series. Typical speed vs. pressure curves of such compound pumps are shown in Fig. 2. Full pump-

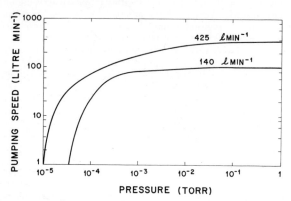

Fig. 2 Two-stage rotary oil pump speeds as functions of pressure (for models from Welch Scientific Co., Skokie, Ill.).

ing efficiency is maintained down to 10^{-3} Torr, and ultimate pressures of about 10^{-4} Torr can be obtained. Commercially available models have speeds ranging from about 20 up to several thousand liters per minute. Relatively small pumps such as the 140 l min^{-1} pump of Fig. 2 are sufficient to maintain medium vacuum on the discharge side of a 6-in.-diameter diffusion pump. The evacuation of an 18-in.-diameter bell jar of 100 l volume within a reasonably short time requires a larger pump. An estimate of the pump-down time needed to reduce the pressure from p_0 to p in a volume V (in liters) can be made by means of the formula†

$$t = \frac{4.6V}{S_0 + S} \log \frac{p_0}{p} \quad \text{min}$$

where S_0 and S are the pumping speeds (in liters per minute) at the initial and final pressures. Starting at atmospheric pressure, the 425 l min^{-1} pump of Fig. 2 establishes 10^{-2} Torr in a 100-l system within about 5 min. At lower pressures, this pump-down equation becomes inapplicable, as it neglects outgassing from the system walls. A more refined expression for the exhaust time of vacuum systems with rotary pumps has been derived by Bunn and Ward.[8]

The fluids in rotary oil pumps are either mineral oils or diphenyl ethers with relatively high vapor pressures of 10^{-3} to 10^{-5} Torr at 50°C. Vapor emission from mechanical pumps is enhanced by the gradual decomposition of the fluids during operation.[9] These vapors can enter the vacuum chamber through the fore line and contaminate the interior.[10] Therefore, it has been suggested that low-vapor-pressure diffusion-pump oils be used in mechanical pumps.[11] However, experience has shown that the lubricating action of diffusion-pump fluids is not sufficient to prevent binding

* The method for measuring pumping speeds of mechanical vacuum pumps has been defined by an American Vacuum Society Standard published in *J. Vacuum Sci. Technol.*, **5**, 93 (1968).

† This relationship can be derived from Eq. (21) in Sec. 5a if all outgassing and other gas-contributing processes are neglected.

of the moving parts.[12] Extended operation is possible if lubricants such as MoS_2 or Teflon powder are dispersed in the fluid, or if the surfaces of the moving parts are coated with Teflon.[12] However, these methods have not been sufficiently proved as yet, and the problem of back-streaming oil vapors is commonly solved by installing fore-line traps. These may operate either by condensation on liquid-nitrogen-cooled surfaces, or by adsorption on surface-active materials.

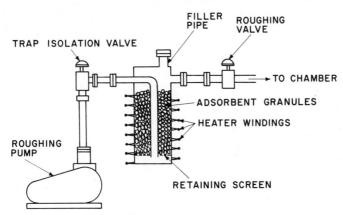

Fig. 3 Bakeable fore-line trap with pump and isolation valves.

The installation of an adsorption fore-line trap is shown in Fig. 3. The trap must be bakeable to restore the adsorption capacity of the trapping material periodically. Comparative pumping tests without and with different traps have been made by Holland and coworkers,[13,14] who found back-streaming rates of the order of 10^{-5} g cm^{-2} min^{-1} if no traps were used. A liquid-nitrogen trap reduced rates to less than 0.1% of the untrapped value. Adsorption traps containing activated alumina proved superior to those charged with zeolite or activated-charcoal granules. They reduced back-streaming rates by 99% with only 10 to 20% loss of the untrapped pumping speed. When rotary oil pumps are used to back a diffusion pump, vapors of the former have been observed to reach the vacuum chamber by streaming through the diffusion pump.[11] However, if a trap is placed in the diffusion-pump line, a separate fore-line trap is often unnecessary.

(2) Roots Pumps Pumps without a discharge valve which move gases by the propelling action of rapidly rotating members are called rotary blowers. A fairly common representative of this type is the Roots pump, which has a pair of two-lobe interengaging impellers. Their characteristic contour and operating mode are illustrated in Fig. 4. In the first position, air enters on the inlet side. Part of this air is trapped by the lower impeller, as indicated in the second position. This volume is discharged in the third and fourth positions. The latter also shows another quantity of gas trapped by the upper impeller which will be discharged during the next quarter revolution. The two counterrotating impellers are ma-

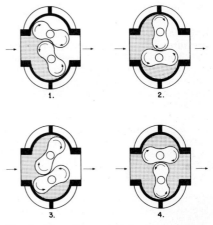

Fig. 4 Operating principle of the Roots pump.

chined to close tolerances and operate without liquid seal or internal lubrication. Backflow of gases limits the compression ratio, and therefore the ultimate pressure of Roots pumps depends strongly on the gas pressure at the discharge side.

Pumping against atmospheric pressure is possible, but the resulting friction generates heat and requires provisions for cooling. It is customary to back Roots pumps with rotary oil pumps, whereby the recommended ratio of pumping speeds is 10:1.[15] Thus, Roots pumps must have high pumping speeds, and commercial models rated at 10^4 to 10^6 l min^{-1} are available. Their operating ranges are 1 to 10^{-3} Torr in conjunction with single-stage and 10^{-3} to 10^{-5} Torr with two-stage rotary pumps. Although supposedly oil-free, Roots pumps backed by a rotary oil pump show some back streaming. The residual gases at 10^{-4} Torr are mostly CO and hydrocarbons plus lesser amounts of H_2O and H_2.[16] The main advantage of Roots pumps is their ability to handle large gas loads in a pressure region where neither rotary oil nor diffusion pumps are fully efficient. They are rarely used in high-vacuum systems but offer economic advantages in industrial applications requiring 10 to 10^{-2} Torr.[17] A more detailed performance analysis and pumping-speed curves have been published by Winzenburger.[18]

(3) Molecular-drag Pumps   Molecular-drag pumps create a gas flow toward a suitable fore pump by means of a rapidly moving surface which imparts a tangential momentum to impinging gas molecules. At or near atmospheric pressure, molecules which have struck the rotating surface and received a velocity component in the direction of rotation transmit their momentum to more remote particles by collisions. Hence, the resulting gas flow is proportional to the viscosity of the gas. At low pressures, the gas flow is molecular and nearly all molecules impinge on the rotating surface. The result is greater pumping efficiency and a compression ratio independent of the pressure on the discharge side. The ratio is higher the narrower the channel between rotating surface and wall through which the gas must pass. For an infinitely thin gap, the theoretical pressure ratio is $1:10^4$.[19] Early molecular-drag pumps had gap widths as small as 30 to 50 μ, but the tight fit of rotor and stator parts involved the risk of binding due to thermal expansion or large dust particles.

Pumps operating on the molecular-drag principle began to enter the high-vacuum field after Becker[19] introduced a new design which is now known as a turbomolecular pump and is shown schematically in Fig. 5. Stator and rotor are segmented into a series of slotted disks similar to turbine blades. The gap between opposing surfaces is 1 mm wide and hence does not pose a tolerance or binding problem. Consequently, the pressure gradient between adjacent pairs of disks is small, but since there are many disks in series, the pump is capable of maintaining high total pressure ratios. Operating at 16,000 rpm, the compression ratios are 1:250 for hydrogen, $1:10^6$ for air, and even higher for vapors of greater molecular weight. The pumps are available in dif-

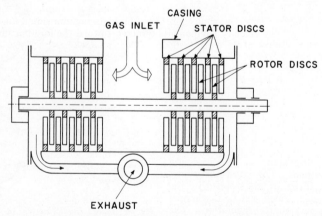

Fig. 5   Turbomolecular pump after Becker.[19]

ferent sizes with air speeds from 260 to 4,000 l s⁻¹.[20] At pressures greater than 10^{-2} Torr the pumping speed declines drastically, and it amounts to only a few percent of the rated value from 1 Torr on up. Therefore, the recommended fore pressure for turbomolecular pumps is 10^{-2} Torr. However, the pumps are not damaged if continuously operated at 0.3 Torr, and they can even withstand brief exposure to atmosphere.

The ultimate vacuum capability of turbomolecular pumps has probably not been fully realized as yet. Based on the compression ratio for air, it should be about 10^{-10} Torr, but the reported pressures are 10^{-8} to 10^{-9} Torr.[21,22] The residual gases are those with the lowest molecular weights, i.e., hydrogen and possibly helium. Although the stator and rotor are not oil-sealed, a lubricant is used in the bearings of the turbine shaft. Therefore, the possibility of organic vapors entering the vacuum chamber exists. Milleron[23] found oil molecules issuing from the mouth of a turbomolecular pump at rates of about 10^{-10} per second.

A molecular-drag pump of different design has been described by Williams and Beams.[24] This pump had a magnetically suspended rotor and could be baked at 400°C. The lowest ultimate pressure obtained was 4×10^{-10} Torr against a fore pressure of 6×10^{-8} Torr.

b. Diffusion Pumps

The idea of evacuating a vessel by momentum transfer from streaming to diffusing molecules was first patented and described by Gaede.[25] The ensuing technical evolution of the diffusion pump is the subject of an interesting review article by Jaeckel,[26] and the operating principles are analyzed by Dushman.[27] The basic features of modern diffusion pumps appeared in a design described by Langmuir in 1916,[28] but development to improve their performance is still continuing. In particular, various traps and baffles have been introduced to eliminate the problem of back streaming of pump-fluid vapors. The performance of present-day diffusion pumps not only depends on the pump design but is significantly influenced by the type of baffle used in conjunction with the pump.

(1) Operating Principle The basic elements of a diffusion pump are shown schematically in Fig. 6. The work fluid is heated by a boiler, and its hot vapor rises in a chimney. The direction of flow is reversed at the jet cap so that the vapor issues through a nozzle pointing away from the high-vacuum side. In passing from a region of comparatively higher into one of lower pressure, the vapor expands. The expansion of the gas alters the normal distribution of molecular velocities by creating a component in the direction of expansion which is larger than the velocities associated with a static gas at thermal equilibrium. Hence the vapor jet moves at a velocity which is supersonic with respect to its temperature. This fact is significant because molecules emerging with a normal velocity distribution would spread omnidirectionally and yield no pumping action. Gas molecules from the high-vacuum side diffuse through the throat and receive velocity components directed toward the fore-vacuum side by colliding with molecules of the working fluid. Thus, a zone of reduced gas pressure is generated in the vicinity of the nozzle, and more gas from the high-vacuum side diffuses toward this region.

As the vapor jet expands further below the nozzle, it becomes less dense and, through collisions, loses part of its directionality. Consequently, the carried gas, too, loses streaming velocity and creates a region of increased pressure in the lower part of the pump from which the accumulated gas load must be removed by a backing pump. This pressure distribution has been confirmed experimentally, most recently by Hablanian and Landfors.[29] To prevent back diffusion of gas from the densified into the rarefied zone, the vapor jet should retain as much of its density as possible. To reconcile this requirement with a wide throat area for maximum gas intake, the cross section of the lower zone is narrowed through aerodynamically shaped tapering stacks. Furthermore, the outer walls are water-cooled to recover the work fluid and to produce a denser boundary layer by removing vapor molecules which travel laterally without contributing to the jet action. To enhance the directionality and speed of the vapor, most pumps employ multistage stacks, typically with three jets working in series.

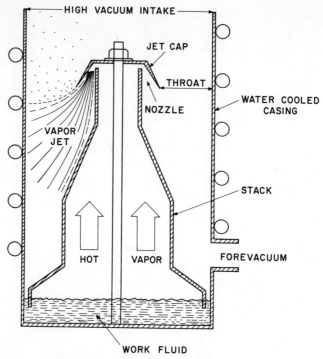

Fig. 6 Basic elements of a diffusion pump.

To minimize the emission of contaminating vapors into the high vacuum, the fluids used in diffusion pumps must satisfy two requirements. One is stability in regard to thermal decomposition and oxidation at operating temperatures; the other is a low vapor pressure near room temperature. Originally, the only suitable fluid known was mercury, but since the successful introduction of high-boiling petroleum fractions,[30] a large number of organic working fluids (oils) have been developed. Compounds of

TABLE 1   Properties of Diffusion-pump Fluids

Trade name	Type of compound	Avg molecular wt	Vapor pressure, Torr, at 25°C	Boiling point, °C at ~0.5 Torr	Density
Octoil*......	Diethyl hexyl phthalate	391	2×10^{-7}	183	0.983
Octoil S*....	Diethyl hexyl sebacate	427	5×10^{-8}	199	0.912
Convoil 20*..	Hydrocarbons	400	8×10^{-6}	190	0.86
Convalex 10*.	5-ring polyphenyl ether	454	2×10^{-9}	~275	1.2
Santovac 5†..	5-ring polyphenyl ether	446	~10^{-9}	260	
Santovac 6†..	6-ring polyphenyl ether	538	~10^{-9}	300	
D.C. 702‡....	Polymethyl siloxanes	530	5×10^{-7}	180	1.071
D.C. 704‡....	Tetraphenyl tetramethyl siloxane	484	2×10^{-8}	215	1.066
D.C. 705‡....	Pentaphenyl trimethyl siloxane	546	3×10^{-10}	245	1.095

* Consolidated Vacuum Corp., Rochester, N.Y.
† Monsanto Co., St. Louis, Mo.
‡ Dow Corning Corp., Midland, Mich.

high molecular weight possess the desired low vapor pressure, and the necessary thermal stability has been found in certain hydrocarbons, polyesters, polyethers,[31] and siloxanes. Commercially available pump oils representing each class and their physical properties as specified by the manufacturers are listed in Table 1. The operating temperatures of diffusion-pump oils are either at or somewhat below the 0.5-Torr boiling points.

(2) Back Streaming Contrary to expectations, the ultimate vacuum attainable by diffusion pumps is not as low as the vapor pressure of the oil at the temperature of the upper part of the pump. Alpert[32] found that diffusion pumps emit as much contaminant gas as they remove in the pressure region of 10^{-7} to 10^{-8} Torr. Even if the contaminant vapors are not supersaturated with respect to the temperatures inside the vacuum chamber, monolayer formation through adsorption occurs within a few seconds at a pressure of 10^{-7} Torr. Back streaming of pump-oil vapors is a serious problem in high-vacuum experiments because it adds impurities to the deposit and thereby degrades the film properties. Moreover, an oil film accumulated on the substrate surface prior to film deposition is often the reason for subsequent lack of film adhesion. Therefore, considerable effort has been devoted to reducing back streaming of vapors by identifying and eliminating its sources.

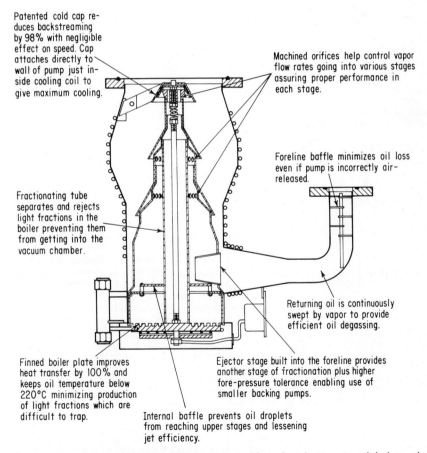

Patented cold cap reduces backstreaming by 98% with negligible effect on speed. Cap attaches directly to wall of pump just inside cooling coil to give maximum cooling.

Machined orifices help control vapor flow rates going into various stages assuring proper performance in each stage.

Fractionating tube separates and rejects light fractions in the boiler preventing them from getting into the vacuum chamber.

Foreline baffle minimizes oil loss even if pump is incorrectly air-released.

Returning oil is continuously swept by vapor to provide efficient oil degassing.

Finned boiler plate improves heat transfer by 100% and keeps oil temperature below 220°C minimizing production of light fractions which are difficult to trap.

Ejector stage built into the foreline provides another stage of fractionation plus higher fore-pressure tolerance enabling use of smaller backing pumps.

Internal baffle prevents oil droplets from reaching upper stages and lessening jet efficiency.

Fig. 7 Schematic of a high-speed diffusion pump with various features to minimize emission of contaminant vapors into the high vacuum. (*With permission of Edwards High Vacuum International, Ltd.*)

The origin of back streaming oil vapors has been investigated by Power and Crawley,[33] who collected the condensate on the walls of a vacuum system. While some back streaming was found to arise from hot condensate on the surfaces of the lower part of the diffusion pump, the major contributions originated at the top jet cap. Foremost among the causes were minor leaks at the tightening nut and oil condensation on the top jet cap. While the former can be avoided by careful assembly, oil condensation may be eliminated by heating the jet cap. If "wet running" is inhibited, there is still considerable residual back streaming from the outer rim of the nozzle. This is attributable to molecule-wall and intramolecular collisions in the boundary layer of the jet. As indicated by the broken lines and dots in Fig. 6, the vapor jet loses its coherence in the boundary layer and thereby creates molecules with undesirable velocities. Effective measures against this mechanism are a cooled guard ring around the nozzle rim[33] or a water-cooled shield above the top jet cap.[34] With these prime sources of vapor emission eliminated, irregular pressure fluctuations above the jet cap of diffusion pumps became apparent. These vapor bursts are attributed to eruptive boiling of the liquid,[35-37] a phenomenon which is minimized in modern diffusion pumps by improved boiler design[38] and by internal splash baffles. Okamoto and Murakami[39] increased the heat input to the boiler to raise the compression ratio of the pump and thereby reduced the pressure fluctuations to negligible proportions.

Another source of back streaming is the gradual decomposition of pump oils into more volatile fractions and noncondensable gases. When measuring the vapor pressure of Octoil, Reich[40] found an anomalous pressure contribution of about 5×10^{-7} Torr. Analysis of the residual gases with a mass spectrometer revealed the presence of H_2O, CO, and significant amounts of higher hydrocarbons.[16] Silicone oils, too, are subject to deterioration. For example, D.C. 702 fluid at 200°C liberated decomposition products at rates of 10^{-6} Torr l s^{-1}, half of which were noncondensable at liquid-nitrogen temperature.[41] Condensation products from the back streaming of polyphenyl ethers have been analyzed by gas chromatography.[42] In addition to isomers of the fluid itself, various fragments of the parent molecule were identified.

The occurrence of volatile fractions in diffusion-pump vapors has led to further refinements in the design. A modern diffusion pump embodying those features which are essential for optimum performance is shown schematically in Fig. 7. The cylindrical tube in the center of the three-stage stack is characteristic for fractioning pumps. It serves to reject volatile decomposition products in the lower stages and allows only the highest-boiling fractions to reach the top jet. Such pumps emit considerably less contaminating vapor into the vacuum chamber than older models. Currently quoted unbaffled back-streaming rates are of the order of 0.1 mg per hour of operation per square centimeter of cross section at the intake. This is equivalent to about 0.25 cm^3 of fluid loss per day in a 6-in.-diameter diffusion pump. These levels of contamination are still harmful in high-vacuum work, and it is universal practice to interpose traps or baffles between the diffusion pump and the vacuum chamber to intercept back-streaming molecules.

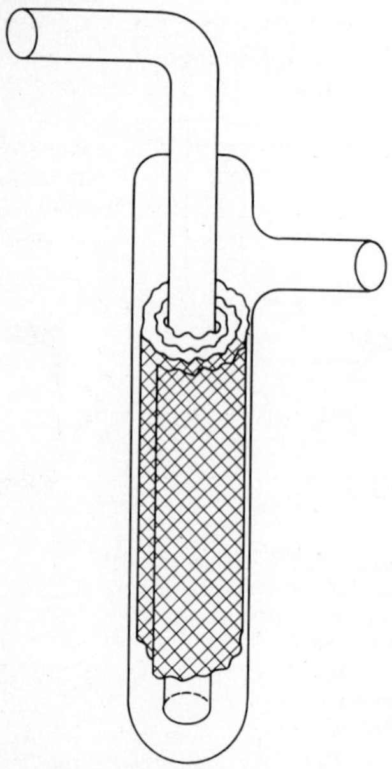

(3) Traps and Baffles The objective of restraining vapor molecules from passing into the vacuum chamber must be accomplished without overly impeding the flow of pumped gas in the opposite direction. A lowering of the conductance in the intake area by about 50% is tolerable in view of the available pumping speeds. To enhance the condensation of streaming molecules, the obstructing surfaces are often cooled. Figure 8 shows a simple reentrant glass trap

Fig. 8 Reentrant copper-foil trap after Alpert.[43]

with a spiral of corrugated copper foil. The trap combines relatively high conductance with a large internal surface area and forces frequent molecule-wall collisions. It proved effective in restraining oil vapors even without refrigeration and allowed pressures of 10^{-9} Torr to be maintained in a small glass system.[43]

A liquid-nitrogen container trap for metal systems is shown schematically in Fig. 9. Its shortcoming is the possibility of fluid condensation on the uncooled wall where an oil film can spread toward the vacuum system and reevaporate. This problem is

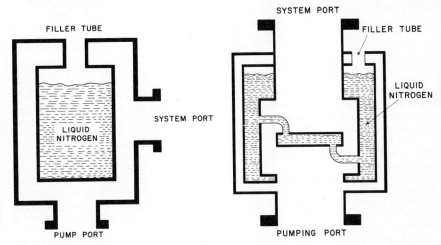

Fig. 9 Liquid-nitrogen container trap for oil diffusion pumps.

Fig. 10 Dewar trap with cooled baffle.

reduced in the baffled Dewar trap shown in Fig. 10 where all the passage walls are cooled. The rate of surface migration and reevaporation at liquid-nitrogen temperature is negligible. Traps of similar design in glass or stainless steel must be baked periodically to drive out accumulated condensate.

Baffled traps which block completely any line-of-sight transmission of molecules are called optically dense. The effectiveness of this concept depends on the sticking probability of molecules impinging on the baffle surface. An analysis of back-streaming rates through an optically dense elbow trap has been performed by Jones and Tsonis,[44] who showed that back streaming at pressures below 10^{-6} Torr is primarily due to molecules rebounding from walls even if these have a sticking probability as high as 0.99. Therefore, optically dense traps should require at least two molecule-wall collisions (double bounce) before passage is possible. This multiple-collision feature is used in a variety of stainless-steel baffles which are sometimes an integral part of the diffusion pump. Most common are chevron baffles as shown schematically in Fig. 11. They may be merely air-cooled or have a cooling jacket for

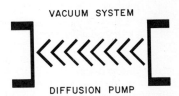

Fig. 11 Optically dense baffle of the chevron type.

water, Freon, or liquid-nitrogen circulation. In comparative tests, Tolmie[45] found refrigerated baffles superior to water-cooled ones. He was able to maintain 10^{-9} Torr using a single refrigerated chevron baffle over a 6-in.-diameter diffusion pump with D.C. 705 fluid. The residual gases transmitted were water vapor and hydrogen with lesser amounts of carbon monoxide and light hydrocarbons. The same ultimate pressure has also been reached by thermoelectric cooling of an ordinarily air-cooled baffle.[46]

Nonrefrigerated adsorbent traps as shown in Fig. 12 were introduced by Biondi.[47]

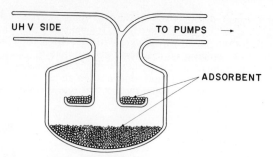

UHV SIDE   TO PUMPS →

ADSORBENT

Fig. 12   Adsorbent trap after Biondi.[47]

Filled with zeolite or activated alumina, these traps maintained pressures of 10^{-10} to 10^{-9} Torr for 75 days of continuous operation without having to be regenerated. Levenson and Milleron[48] constructed a metal adsorption trap as shown in Fig. 13 and achieved ultimate pressures of 2×10^{-9} Torr with Convoil 20 pump fluid. External cooling with liquid nitrogen brought a further pressure reduction by nearly a factor of 10. Deliberate admission of gases showed that CO_2 and *n*-butane were readily adsorbed but not H_2, He, Ar, N_2, CO, and CH_4. Convoil 20 vapors were adsorbed, too, but the zeolite gradually saturated, with a noticeable increase in pressure. Gosselin and Bryant[49] trapped D.C. 705 pump-oil vapor with zeolite and obtained an ultimate pressure of 2×10^{-9} Torr. The refrigerated zeolite trap reduced the emission of decomposition products such as benzene derivatives by three to four orders of magnitude.

The effectiveness of traps and baffles has been evaluated by determining the con-

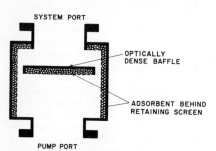

SYSTEM PORT

OPTICALLY
DENSE BAFFLE

ADSORBENT BEHIND
RETAINING SCREEN

PUMP PORT

Fig. 13 Metal version of adsorbent trap after Levenson and Milleron.[48]

densation rates of back-streaming fluid vapors. Conventional methods such as collecting the condensate from large internal wall areas in burettes,[50] or weighing of the oil film condensed on a cooled surface[51] have sensitivities of at best a few angstroms per hour. However, there are several techniques with lower limits of detection. Examples of back-streaming rates obtained with different types of traps and baffles are listed in Table 2. The data represent optimum operating conditions and a variety of experimental methods. Holland et al.[52] converted the condensate into a solid polymer film by electron bombardment during its deposition and measured the thickness interferometrically. The rates for D.C. 705 oil were about half of those observed with D.C. 704. Considerably higher rates up to several hundred angstroms per hour occurred when oil vapor trapped in previous experiments was released during the bakeout cycle. Langdon and Fochtman[53] measured the ultraviolet adsorption of the condensed oil films to determine their thickness. They found lower rates if the traps were cooled but little difference whether the coolant was water of 5°C or liquid nitrogen. Their observations suggest that back streaming through well-designed traps may be more attributable to minute droplets than to vapor. Carter et al.[54] used irradiated pump oil and measured the radioactivity of the condensate arising from the presence of Si^{31}. This technique allowed observations in short time intervals and revealed high back-streaming rates when the diffusion pump was turned on with substantial decreases during the next 2 hours of operation. Also, the differences in rates for water and liquid-nitrogen cooling were significant. Another tracer technique, where the fluid was labeled with tritium, has been described by Huber et al.[55] The

TABLE 2 Condensation Rates Due to Back Streaming for Various Pump Oil/Baffle Combinations

Reference	Baffles, traps, and cooling	Pump oils	Condensation rates	Operating pressure, Torr
Holland et al.[52]	Chevron and Dewar traps in series, both with liquid N_2 cooling	D.C. 704 D.C. 705	<5 Å h^{-1} ...	2×10^{-9} 1×10^{-9}
Langdon and Fochtman[53]	Chevron baffle with 5°C water or liquid N_2 cooling	D.C. 704 D.C. 705 Convalex 10	~5 Å h^{-1} ~2.5 Å h^{-1} ~0.6 Å h^{-1}	$\sim10^{-8}$
	Optically tight elbow trap with 5°C water or liquid N_2 cooling	D.C. 705 Convalex 10	~0.6 Å h^{-1} ~0.2 Å h^{-1}	
Carter et al.[54]	No baffle Chevron baffle with air or water cooling Chevron baffle with liquid N_2 cooling	D.C. 704	400 Å h^{-1} $100 \to 1$ Å h^{-1} $4 \to 0.1$ Å h^{-1}	2×10^{-5} 2×10^{-5} 2×10^{-6}
Huber et al.[55]	Optically dense water and Freon-cooled baffles in series	D.C. 704	$<10^{12}$ molecules cm^{-2} h^{-1}	

condensation rate quoted corresponds to about 1 % of a monolayer or a fraction of an angstrom per hour.   The quartz-crystal oscillator[56] and combined chromatographic–mass spectrometric[57] techniques have also been used to determine the amount and nature of back-streaming pump-oil vapors.

(4) Performance Ranges The ultimate pressures obtained with oil diffusion pumps are generally limited not by the pump's compression ratio but by gas emission from various parts of the vacuum system.   If all interior surfaces have been thoroughly degassed and suitable adsorbent traps or cooled baffles are used, it is possible to achieve pressures as low as 10^{-11} Torr by means of diffusion pumps.   Examples of such systems have been described by Hablanian and Vitkus[58] (D.C. 705 pump oil and room-temperature baffle), by Singleton[59] (Octoil or D.C. 704 fluid with liquid-nitrogen trapping), and by Steinrisser[60] (polyphenyl ether fluid and refrigerated zeolite trap).   Silicone oil D.C. 705 and polyphenyl ethers are most often chosen as their vapor pressures are the lowest.   If effectively trapped or baffled, the condensation rates due to back streaming are smaller than 1 Å h^{-1} and thus negligible for many applications.

Because of their ability to pump over a wide pressure range and their essentially maintenance-free operation, oil diffusion pumps are the most widely used means of attaining high vacuum.   The majority of commercial pumps employ three-stage jets with steel casing and aluminum nozzles.   Since pumping speeds are related to the size of the intake area, the approximate pump diameter indicates the capability of a pump.   A familiar example is the 6-in. diffusion pump whose speed characteristics are shown in Fig. 14.   Pumps of other sizes have similar curves except that their plateau speeds are different.   Full pumping speeds are reached at 10^{-3} Torr and remain constant well into the ultrahigh-vacuum region.   Fore pressures of 10^{-3} to 10^{-2} Torr are recommended, but up to 0.5 Torr are tolerable.   The power requirements of a nominal 6-in. pump are 1.5 to 2 kW and proportionately greater for larger sizes.

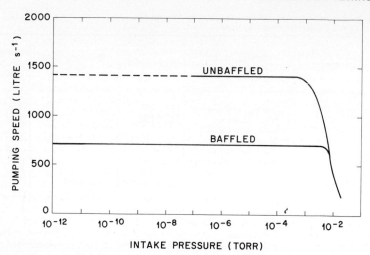

Fig. 14 Speed of a 6-in.-diameter diffusion pump as a function of pressure. A speed reduction of 50 % has been assumed for the baffled curve.

Metal-encased diffusion pumps are available from 2 to about 50 in. diameter with corresponding unbaffled speeds of 100 to 100,000 l s⁻¹. They dominate the field because of their compatibility with medium- to large-sized vacuum systems of metal construction. Pyrex-glass diffusion pumps for special applications such as microchemistry, electron-tube production, or small-scale ultrahigh-vacuum studies are offered commercially in smaller sizes. Typical glass pumps have intake diameters of about 1 in. and pumping speeds of 10 to 30 l s⁻¹.

The increasing usage of organic pump fluids has not replaced mercury diffusion pumps completely. The latter are still used for applications where even minute traces of organic vapor are thought to be harmful, or simply because of the preference of some investigators. Commercially available metal-encased (iron) mercury pumps have speeds similar to oil pumps up to about 50,000 l s⁻¹. Mercury glass pumps are smaller and offer speeds below 20 l s⁻¹. Most mercury pumps operate with two- or three-stage jets and reach full pumping speed at 10^{-2} to 10^{-3} Torr. As for ultimate vacua, the inherent limitation of the high vapor pressure of mercury (2×10^{-3} Torr at 25°C) can be overcome with refrigerated traps since the vapor pressure is immeasurably small at liquid-nitrogen temperature. Recommended are two or three traps in series with the first one held at a temperature above the freezing point of the mercury to avoid depleting the pump of its fluid. Simple reentrant glass traps[59,61] or more elaborate designs with cooled baffle and wall surfaces[62,63] have allowed the attainment of 10^{-11} to 10^{-12} Torr.

Closely related to diffusion pumps are two other types of vapor-stream pumps, the ejector pump and the vapor booster pump. The principle of *ejector pumps* is similar to that of diffusion pumps insofar as gas is removed by entrainment in a high-velocity stream of vapor emerging from a nozzle. However, ejector pumps operate at higher pressures up to 1 atm, and the gas is sucked into the vapor stream as a turbulent viscous flow created primarily by aerodynamic forces. Condensation of vapor plays no part in the pumping action but merely serves to recover the working fluid. Ejector pumps are most efficient in the 10^{-1} to 10^{-2} Torr region, and a single ejector stage as shown in Fig. 7 is commonly used to discharge the gas load of a diffusion pump into the fore vacuum.

Vapor booster pumps are large diffusion pumps specialized for heavy duty in the medium-vacuum range. Typically, they consist of two or three jet stages and a final ejector stage. The jet stages operate on the diffusion-condensation principle; i.e., they entrain gas by molecular diffusion. Characteristic of their construction is the pronounced reduction in cross section from the intake area to the ejector stage. By discharging the working fluid at high rates into a small-diameter outlet, vapor booster pumps can operate against considerably higher backing pressures (several torr) and handle larger gas loads than diffusion pumps. The working fluids may be of the same families as diffusion-pump oils, but more volatile compounds such as water, lower alcohols, or chlorinated diphenyl derivatives are also suitable.[64] Typical pumping speeds are 1,000 to 10,000 l s^{-1}, and two-stage pumps produce vacua down to 10^{-4} Torr. Vapor booster pumps are recommended for large-scale industrial applications at pressures of 10^{-1} to 10^{-3} Torr.[65] Their capability to remove large volumes of gas as well as their construction from corrosion-resistant materials are of particular value in chemical distillation processes and vacuum smelting. Compared with Roots pumps, which are competing for similar applications, the vapor booster pumps offer operating pressures about an order of magnitude lower.

c. Cryogenic Pumps

The operating parts of cryogenic pumps are cooled metal surfaces in the form of disks, tubing, or cylinders. They are installed inside the vacuum chamber, or they may form a separate container communicating with the chamber through a high-conductance opening. The pumps operate at low pressures where the mean free path is of the order of the chamber dimensions, and most residual gas molecules strike the cryosurface without encountering prior collisions. Hence, cryogenic pumps reduce the system pressure quite rapidly. The events following impingement are the same as in the evaporation of crystalline solids but in reverse sequence. Some molecules are immediately reflected back into the vacuum, whereas others become adsorbed and exist in this state for some time, diffusing across the surface until they are permanently trapped at an energetically favorable site. While in the adsorbed mobile state, some molecules may escape back into the vacuum by desorption. The ratio of molecules which are ultimately condensed to the initial impingement flux is the condensation or capture coefficient α_c.

The pumping speed of a cryogenic pump is given by the product of the kinetic impingement rate, the cryosurface area, and the condensation coefficient. After an initial nucleation period during which the original metal surface becomes covered with a few layers of condensed molecules, the pumping process is determined by the thermal accommodation of vapor molecules on their own crystal lattice. Thus, cryogenic pumping is a continuous nonsaturating process, limited at worst by poor heat conductivity of the accumulated condensate. An experimental technique to determine the pumping speed of cryosurfaces has been described by Mullen and Hiza.[66]

Since the condensation coefficient is one of the factors determining cryopumping speeds, it is of considerable practical interest. Yu and Soo[67] derived capture coefficients from Polanyi's[68] model of the surface potential of crystalline solids. Their theoretical value of α_c for CO_2 at 77°K agrees well with experimental data. Methods for measuring condensation coefficients have been reviewed by Chubb.[69,70] Levenson[71] used a quartz-crystal microbalance to determine the sticking probabilities of Ar and CO_2 at surface temperatures from 4.2 to 77°K and found values between 0.8 and 1.0. Heald and Brown[72] employed a mass spectrometer to observe the reemission from a polished copper surface upon which a molecular beam of CO_2 was impinging. They found the condensation coefficient to be dependent on the impingement rate, on the surface temperature, and on the coverage of the surface with condensed molecules. When the impingement flux approached the thermodynamic vapor pressure of CO_2 at the temperature of the condensing surface, α_c became 1. The condensation coefficient is also dependent on the temperature of the impinging gas. According to experimental data collected by Hengevoss,[73a] the condensation coefficients of atmospheric gases at 10°K fall into the range from 0.6 to 0.8 if the gas temperature is 300°K. Values of $\alpha_c = 1$ are approached at gas temperatures below 200°K. Condensation coefficients of atmospheric gases at surface temperatures of 77°K or below have recently been compiled by Hobson and Redhead.[74] The numerical values of α_c range from 0.5 to 1.0 depending on the experimental conditions.

Although cryogenic pumping is contingent upon supersaturation with respect to the cryosurface temperature, unsaturated gases, too, may be removed along with condensable vapors. This effect is called cryotrapping and may be considered as an adsorption process which does not saturate because of continuous renewal of the condensate surface. The amount of cryotrapped gas is usually smaller than the quantity of condensed vapor. For example, Wang et al.[75] trapped nitrogen while condensing water vapor at 77°K in molecular ratios of $1:10^4$ to $1:10^2$. Cryotrapping of hydrogen by CO_2 in ratios between $1:10^3$ and $1:10$ has been reported by Dawbarn.[76] Hydrogen has also been cryotrapped by condensates of H_2O,[76] Ar, and N_2.[73a,73b] Thus, cooperative condensation is not an uncommon phenomenon. It is, however, a selective process which does not apply to all gases alike. Helium, for instance, is not trapped by H_2O, and the presence of nitrogen reduces the cryotrapping of H_2 by CO_2.[76]

In assessing the applicability of cryogenic pumping, the question of available coolants is important. The only fluids whose boiling points are sufficiently low for effective vapor condensation are liquefied gases. The ability of the three commonly used coolants to condense various gases can be determined from Fig. 15. Cooling

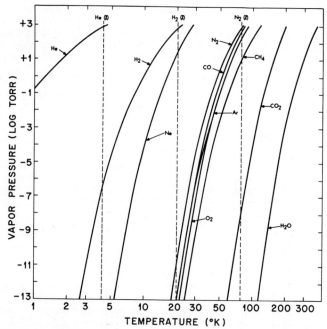

Fig. 15 Vapor pressures of common gases at low temperatures. (*After Honig and Hook.*[77]) The dashed lines indicate the boiling temperatures of N_2, H_2, and He at 760 Torr.

with liquid nitrogen is inexpensive and efficient because of its high heat of vaporization (38 cal cm^{-3}). A widely used embodiment are Meissner traps,[78] coils of copper tubing through which liquid nitrogen is circulated. They are often installed in vacuum chambers evacuated by rotary oil and diffusion pumps, a combination which normally yields pressures of 10^{-5} to 10^{-6} Torr. A Meissner trap reduces this pressure by a factor of about 10. The effect of the trap on the residual gases is illustrated by Caswell's[79] data in Fig. 16. As expected from Fig. 15, the two condensable gases H_2O and CO_2 constitute the greater part of the gases removed. But the partial pressures of the noncondensable gases have also been lowered, which indicates that cryotrapping plays a significant role.

The final pressure of 10^{-6} Torr obtained by Caswell[79] does not represent the lower limit of liquid-nitrogen pumps since the predominant residual gas is still water vapor.

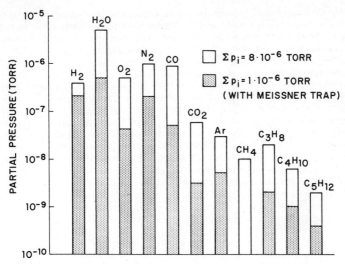

Fig. 16 Partial gas pressures in two similar demountable vacuum systems, one without and one with a Meissner trap. (*After Caswell.*[79])

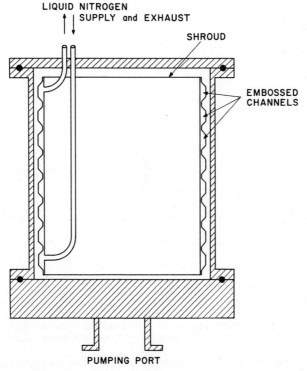

Fig. 17 Vacuum chamber with liquid-nitrogen shroud.

The latter condenses more efficiently if a liquid-nitrogen-cooled shroud as shown in Fig. 17 is used instead of copper coils. These shrouds are made from two sheets of stainless steel welded together and carrying a meandering pattern of embossed channels. Pressures of 10^{-7} Torr are readily obtained with such cryopanels.[80] A performance evaluation of a liquid-nitrogen trap under conditions of heavy outgassing in a plastic metallizing system has been given by Holland and Barker.[81]

The attainment of pressures below 10^{-8} Torr by cryogenic pumping requires either liquid hydrogen or helium cooling. According to Fig. 15, hydrogen at its boiling point condenses effectively all gases except He, Ne, and its own vapor. Although cryogenic disks operating at 20°K have achieved pumping speeds of the order of 10,000 l of N_2 per second, liquid-hydrogen cooling is not widely used. Liquid-helium-cooled pumps, however, are of considerable practical interest. The use of helium cooling is somewhat restricted by the high cost of the gas and its low heat of vaporization (0.9 cal cm^{-3}). To minimize evaporation losses, helium traps are designed such that the condensation surface is surrounded by liquid-nitrogen-cooled radiation shields. Examples of such designs are given by Caswell,[82] who obtained a reduction of liquid-helium consumption by a factor of 200 through liquid-nitrogen-cooled shields. Since the flow of gas molecules to be pumped is obstructed by radiation shields, their shape and placement are critical.

Greater operating economy is achieved if the evaporated helium is condensed and recirculated. A cryostat based on the closed-cycle principle has been described by Forth[83] and is shown in Fig. 18. The outer helix of the coil carrying the return flow of helium serves as a radiation shield. These cryostats are available in different sizes,

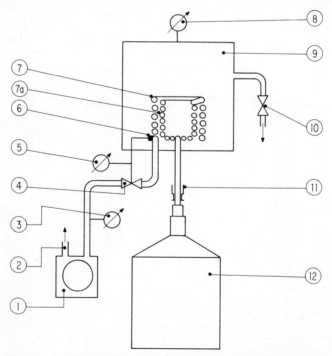

Fig. 18 Schematic of a liquid-helium condensation pump operating on the closed-cycle principle. (1) Helium-pump; (2) connection for helium-recovery plant; (3) pressure gauge; (4) control valve; (5) temperature gauge; (6) temperature-sensing element; (7) radiation shield; (7a) cold surface; (8) vacuum gauge; (9) vacuum container; (10) valve with connection for preevacuation pump set; (11) vacuum-jacketed feed pipe; (12) helium-storage tank. (*With permission of H. J. Forth.*[83])

and a pumping speed for nitrogen of 5,000 l s⁻¹ is quoted for the smallest unit. The cryostats are also capable of operating at reduced helium pressure, thereby achieving condensation temperatures down to 2.5°K. The importance of this feature is obvious from Fig. 15, which shows a saturation pressure of about 10^{-7} Torr for hydrogen at 4.2°K. A reduction of the cryosurface temperature to 2.5°K should lower the H_2-saturation pressure to about 10^{-13} Torr. However, ultrahigh vacua of this order are not easily obtained by cryogenic pumping alone. Experimental studies of hydrogen condensation on helium-cooled surfaces by Chubb[69,70] yielded ultimate pressures of only 10^{-9} Torr, although temperatures down to 2.2°K were obtained and the condensation coefficient of H_2 was as high as 0.9. In an earlier investigation, Farkass and Vanderschmidt[84] were able to lower the pressure of a large preevacuated chamber through liquid-helium cooling by a factor of 10. But they, too, did not reach pressures significantly below 10^{-9} Torr. In practice cryogenic pumping is rarely used alone but rather in combination with other techniques of evacuation. An example of this kind was reported by Sheldon and Hablanian,[85] who reached ultimate pressures near 10^{-13} Torr in an all-metal system exhausted by a baffled diffusion pump and a liquid-helium trap.

d. Cryosorption Pumps

Cryosorption pumps are trapping devices which remove gases by adsorption on cooled surfaces. The process is primarily intended for the removal of unsaturated vapors and permanent gases. Saturation of the trapping surface occurs as available sites are occupied. Thus, cryosorption is a batch operation and requires periodic regeneration of the sorbing surface by thermally induced desorption. For maximum trapping capacity, cryosorption devices employ porous materials with large internal surface areas. Furthermore, refrigeration of the adsorbent is essential since the heats of adsorption of noncondensable gases are so small that desorption may occur even below room temperature. This is illustrated in Table 3, which shows the boiling

TABLE 3 Boiling Points and Desorption Temperatures of Common Gases for Molecular Sieve 13X*

Gas	760 Torr boiling point, °K	Temp range of desorption, °K
H_2O	373	450–480
CO_2	195	280–300
CH_4	112	150–180
O_2	90	120–150
Ar	87	110–140
CO	82	190–210
N_2	77	140–190
Ne	27	Very little
H_2	20	adsorption
He	4.2	at 78°K

* After Murakami and Okamoto.[86]

points of common gases in decreasing order together with the desorption ranges as determined by Murakami and Okamoto.[86] These authors adsorbed gases at 77°K on artificial zeolite and observed their desorption while raising the temperature at a uniform rate. The data demonstrate that adsorption requires temperatures not too far above the boiling points of the gases. Ne, H_2, and He can be adsorbed only at temperatures below 78°K.

In addition to surface area and temperature, the amount of adsorbed gas depends on its pressure. The relationship between the concentration of adsorbed species and its equilibrium vapor pressure at constant temperature is called adsorption isotherm (see also Sec. 3a) and is specific for a given solid-gas system. While adsorption isotherms at relatively high pressures have been known for a long time,[87] low-pressure isotherms are of more recent interest stimulated primarily by high-vacuum applications.[73a]   The derivation of low-pressure adsorption isotherms from theory has been attempted by Hobson and Armstrong,[88] who applied Polanyi's[68] surface-potential model of solids and were able to interpret the isotherms of N_2 and Ar at 77°K.   Ehlers[89] described an experimental technique to determine simultaneously the sticking probability of molecules on the surface of the sorbent, the diffusion coefficient of molecules through the sorbent, and the adsorption isotherm.   From these parameters, cryosorption processes and the resulting pressures for different vacuum conditions can be estimated.   However, these data are seldom completely known.   In practice one has to rely on empirical isotherms or pumping curves for specific conditions.

(1) **Properties of Adsorbent Materials**   Various adsorbent materials and their properties are listed in Table 4.   Activated charcoal has widely varying properties depending on the starting material and fabrication process, and its pore size is not uniform. Coconut charcoal adsorbs about ten times its own volume of gas (STP) at 77°K.[90] The reforming catalyst adsorbs rather effectively in the very-high-vacuum range,

TABLE 4   Properties of Adsorbent Materials

Material	Form	Internal surface area, $m^2 \ g^{-1}$	Limiting diam for sorbed molecules	Reference
Activated charcoal:				
From coconuts.......	Granules	Up to 2,500	~50 Å	90
From anthracite.......	$\frac{1}{32}$- to $\frac{1}{16}$-in. granules	500–580	~20 Å	91
Re-forming catalyst (0.35 % Pt on Al_2O_3)...	$\frac{1}{16}$-in. pellets	460	~32 Å	91
Activated alumina.......		230–380		93
Silica gel..............		500–600		93
Molecular sieves:				
Type 3A (potassium)..	$\frac{1}{16}$- and $\frac{1}{8}$-in. pellets		3 Å	Linde Company, Division of Union Carbide Corp., Moorestown, N.J.
Type 4A (sodium).....			4 Å	
Type 5A (sodium and calcium)........		700–800	5 Å	
Type 13X (sodium)...			13 Å	

reducing the pressure from 10^{-8} to 3×10^{-10} Torr.[91]   Since one of the major residual gases at this pressure is hydrogen, which is not readily adsorbed by oxides, the presence of finely dispersed platinum seems to be of particular value for sorption in ultrahigh vacuum.   Palladium, too, has been suggested as a selective sorbent for hydrogen and residual hydrocarbons.[92]

Most widely used and best investigated are the molecular sieves or synthetic zeolites. These are alkali alumino silicates whose structure consists of a network of polyhedra with oxygen bridging atoms.   The polyhedra have voids in their centers which are connected through a system of intracrystalline pores arising from stacking of the polyhedra.   The alkali metal can be replaced by ion-exchange techniques which allow variation of the effective pore size.   During the synthesis of zeolites, the pores are occupied by water of crystallization, which can be driven out by heating without disintegration of the structure.   Hence, molecular sieves are adsorbent materials of highly reproducible structure with uniform pores of molecular dimensions.

Since adsorption on the large internal surface of the sieves requires diffusion of the gas through the pores, molecules beyond a certain size are not admitted.   The common gases have molecular diameters between 2 and 4 Å; hence most of them can enter the sieves.   The adsorption characteristics of the different types of molecular sieves are very similar except that the 13X type has a slightly higher capacity and can adsorb aromatic hydrocarbons.   Calculations of the adsorptive capacity based on the specified internal surface area and molecular diameters must consider the possibility of preferential adsorption of one species over another. Furthermore, the entire theoretical surface area is not always available. Experimental studies of Bailey[94] yielded effective areas of only 400 m^2 g^{-1} for 5A sieve at 77°K for the gases O_2, N_2, Ar, and CH_4. Figure 19 shows a decrease of the adsorption capacity for Ar and N_2 below certain temperatures while oxygen is increasingly adsorbed. It has been suggested that the zeolite pores contract upon cooling and therefore exclude the larger N_2 and Ar species but admit the smaller oxygen.[95]

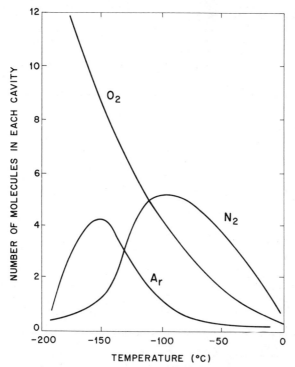

Fig. 19 Adsorptive capacities of molecular sieve 5A for O_2, N_2, and Ar as functions of temperature. (*After Barrer.*[95])

In a comparative study of molecular sieves, Bannock[90] found the type 5A to be superior in regard to pumping speed and adsorption capacity for air. The adsorption isotherms of the common gases on 5A sieve are shown in Fig. 20. The more easily condensed gases reach saturation when about 100 Torr l g^{-1} have been adsorbed. This is interpreted as monolayer coverage of the adsorbent surface. The steep rise of the methane curve at about 10 Torr represents the onset of multilayer adsorption (see Sec. 3a). The adsorption capacities for H_2, Ne, and He at 77°K are significantly smaller because of their lower condensation temperatures. The adsorption of all gases decreases rapidly below 10^{-3} Torr in continuation of the trend indicated in Fig. 20.[96] Stern and DiPaolo[97] found that the capacity for N_2 in this pressure range increased significantly after repeated adsorption-desorption cycles.

Utilization of the maximum adsorption capacity is contingent upon the absence of significant amounts of water vapor. Even at room temperature, the 5A sieve adsorbs as much as 18% of its own weight, or nearly 200 Torr l of water vapor per gr wt of sieve.[94] Whereas all the other common gases are readily desorbed by returning the

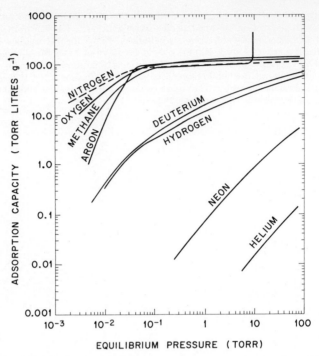

Fig. 20   Adsorption isotherms for the common gases on molecular sieve 5A at 77°K. (*After Bailey.*[94])

cryosorption pump to room temperature (see Table 3), regeneration of a sieve containing water vapor requires several hours of baking at 350°C.   It is generally not recommended to exceed this temperature because of the onset of zeolite disintegration, but some investigators have employed outgassing temperatures as high as 450°C.[98]

Another factor to be considered in the application of cryosorption pumps is the poor thermal conductivity of molecular sieves.   Since their adsorption efficiency depends on refrigeration, the sieve charge in the trap container is applied either as a thin liner retained by a metal screen, or distributed in narrow channels.   Bannock[90] used tube elements of 2 cm diameter and 60 cm length.   Better heat contact was obtained by Sands and Dick,[93] who produced well-adhering adsorbent layers by plasma spraying of molecular sieve onto metal tubes.   The method requires a second and cold stream of sieve pellets since material from the plasma stream loses its adsorptive properties and acts merely as a binder.   This technique should also eliminate the problem of dust which occurs in loosely packed sieve beds and leads to contamination of the vacuum chamber.   Bailey[94] identified molecular-sieve dust particles of 3 to 8 μ diameter which he was able to re-strain by use of glass frits, but he does not comment on the resulting loss of conductance.

(2) Vacuum Applications   Cryosorption pumps containing molecular sieves are utilized in three major applications: as traps against back diffusion of oil vapor [see Sec. 2b(3)], as fore pumps in oil-free vacuum systems, and to achieve ultrahigh vacuum in degassed systems which have been preevacuated by another technique.   Because of the pressure dependence of the adsorption isotherms as shown in Fig. 20, molecular-sieve pumps are more efficient as fore pumps than in the ultrahigh-vacuum region. Accordingly, the former application is far better investigated than the latter, and commercially available cryosorption traps have become quite popular to establish oil-free fore vacua in ion-pumped systems.

The performance of molecular sieves in reducing atmospheric pressure has been investigated by Turner and Feinleib,[96] who introduced a significant parameter, the ratio of volume to be pumped to the amount of sieve required. From the adsorption isotherms of the common gases at 77°K, they deduced the final equilibrium pressure to be expected from a given charge of adsorbent. Their results are shown in Fig. 21. The graph demonstrates that the attainment of pressures below 10^{-3}

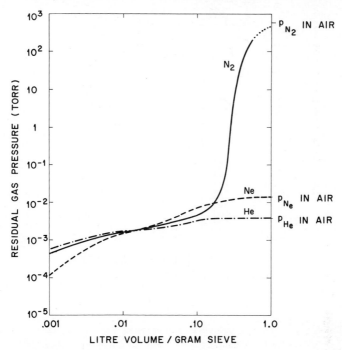

Fig. 21 Residual partial pressures to be expected from the adsorption of a finite volume of air onto fixed quantities of molecular sieve 5A at 77°K. The sieve had been regenerated at 300°C. (*After Turner and Feinleib.*[96])

Torr requires a disproportionately large amount of adsorbent. A fore vacuum of 10^{-2} Torr, however, can be obtained efficiently with ratios of 0.1 to 0.01 l per gr wt, and these are indeed the ratios used in commercial cryosorption pumps. A consequence of the poor adsorption of He and Ne is that the small concentrations of these gases in the atmosphere are sufficient to make them major constituents of the residual gas load. Mass-spectrometric analysis of the residual gases after cryosorption pumping confirms this expectation.[96] As shown in Fig. 22, the nitrogen and helium pressures are reduced to approximately the predicted levels. Neon, however, is sorbed even less than expected, probably because it is displaced by some of the more readily sorbed gases.

The concept of pumping speed as encountered in other types of pumps is of limited value in assessing the performance of cryosorption pumps. The main reason is the variability of adsorption rates with adsorbate concentration. Varadi and Ettre[99] measured the speed of air adsorption at 10^{-1} Torr starting pressure on molecular sieve 13X at 77°K. On freshly baked material, they found relatively slow adsorption rates which increased and reached a maximum when about 15 Torr l of air per gr wt of sieve had been taken up. Thereafter, the rates decreased again and fell toward very small values after adsorption of about 75 Torr l per gr wt. The pumping speeds varied from about 1 to 5 Torr l s^{-1} for a 100-g zeolite charge. Since regeneration was

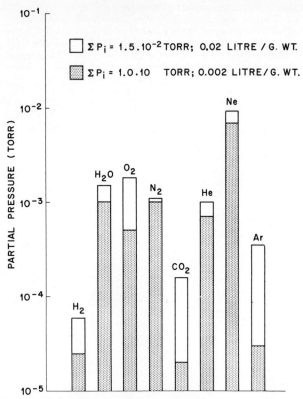

Fig. 22 Residual-gas composition after 10 min of cryosorption pumping of air with molecular sieve 5A at 77°K, for two volume-to-weight ratios. (*After Turner and Feinleib.*[96])

performed by desorption at room temperature, the adsorption capacity as well as the pumping speed deteriorated from cycle to cycle because of the accumulation of water in the sieve. Decreasing pumping speeds as a function of increasing saturation have also been found on molecular sieve 5A.[100]

The performance of commercial pumps from different manufacturers is relatively uniform. The pumps vary primarily in the distribution of sieve material within the refrigerated enclosure and in their provisions for efficient cooling. The predominant sorbent is molecular sieve 5A, and charges of 3 to 5 lb per pump are employed. Larger pumping capacities are obtained by installing several units as shown in Fig. 23. Safety valves are essential since saturated traps develop dangerous overpressure as the liquid nitrogen evaporates. A 40- to 60-l system can be evacuated with a single pump, 100- to 150-l chambers require two stages, and three stages are sufficient for 180- to 350-l volumes. All manufacturers recommend prechilling the sieve with nitrogen for 10 to 15 min before connecting the pump to the system. The most efficient mode of pumping is to operate multistage traps in sequence rather than simultaneously[96] so that the second or third stages are not presaturated with the more readily adsorbed gases.

Typical pumping curves for one- and two-stage operation are shown in Fig. 24. While pressures of 10^{-2} to 10^{-3} Torr are readily obtained, multistage pumping, too, is limited by the accumulation of neon. Therefore, a single-stage trap on a chamber prepumped with a rotary oil pump is a more efficient combination and capable of yielding 10^{-4} to 10^{-5} Torr[96] or even lower pressures.[98] In this case, pump-oil vapors

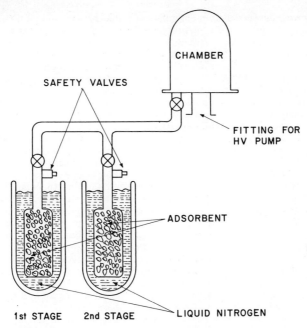

Fig. 23 Schematic of a two-stage cryosorption-pump system.

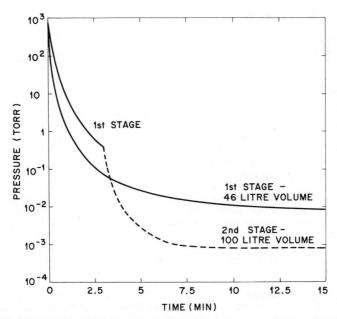

Fig. 24 Pumping curves after 15 min of prechilling for single- and two-stage evacuation with cryosorption traps. The volume-to-weight ratio is about 0.02 l per gr wt in both cases. (*With permission of Ultek, Palo Alto, Calif.*)

are unlikely to reach the vacuum chamber since the cryosorption pump also acts as a fore-line trap.   In spite of the neon problem, Bannock[90] obtained ultimate pressures of 10^{-6} Torr with cryosorption trapping alone, but it appears that he used very small volume-to-weight ratios.

The application of cryosorption traps for obtaining ultrahigh vacuum is less well established than their utilization as fore pumps.   This is due to the rapidly decreasing adsorption capacity of molecular sieves for all gases at very low pressures and their inability to trap significant amounts of Ne, He, and H_2 at 77°K.   The extension of the operating range of cryosorption traps requires preevacuation with another type of pump to lower the total gas load, in particular the initial pressures of the lighter gases. This has been investigated by Read,[101] who prepumped a small bakeable vacuum system with an oil diffusion pump to 10^{-9} Torr.   Subsequent sorption onto molecular sieve 13X in a liquid-nitrogen-cooled stainless-steel trap resulted in pressures as low as 5×10^{-11} Torr.   However, it is questionable how much of this reduction can be attributed to the presence of the sieve material.   Inkley and Coleman[91] investigated cryosorption pumping in a small Pyrex-glass system preevacuated to 10^{-8} Torr with a mercury diffusion pump.   Their trap consisted of a stainless-steel container with a monel metal cage to hold different adsorbents.   Activation of the trap by cooling with liquid nitrogen reduced the pressure to 5×10^{-10} Torr, but the effect was obtained regardless of the type of absorbent used and even when the cage was empty.   Thus, adsorption on the metal surface of the trap aided substantially in the pumping action of noncondensable gases.   The limited adsorption capacity of molecular sieves in ultrahigh vacuum is also apparent in Inkley and Coleman's experiment insofar as they observed a gradual pressure increase to 2×10^{-9} Torr during 20 h, a rate consistent with permeation of the glass envelope by atmospheric helium.

An attempt to overcome the restrictions of both cryosorption and cryogenic pumping is the cryogenic array described by Gareis and Hagenback.[102]   This unit combines liquid-nitrogen- and liquid-helium-cooled chevron baffles, the former shielding the latter, with a helium-cooled molecular-sieve panel.   The function of the cooled chevron baffles is to condense the heavier gases and leave the innermost surface of the cryosorption panel free to adsorb He, Ne, and H_2 at 4.2°K.   It is claimed that pressures as low as 10^{-13} Torr can be attained.

e. Getter Pumps

Getter pumps are trapping devices which employ reactive materials to bind residual gases on the inner surfaces of the system.   The trapping agents are freshly prepared metal films which capture gases by several mechanisms, a phenomenon referred to as cleanup.   Since getter pumps do not require lubricants and do not employ work fluids other than condensable metal vapor, they are free of organic contaminants and thus have an advantage over diffusion pumps.   Like the latter, getter pumps are applicable over the entire range from high to ultrahigh vacuum.   They require starting pressures between 10^{-2} and 10^{-4} Torr but no continuous backing.   Fore vacuum may be established either by cryosorption or rotary oil pumps which are valved off as soon as the starting pressure has been reached.   The pumping action of getter pumps is based on two types of interaction between residual gases and metal surfaces, one being chemical and the other electrostatic in nature.   Devices which utilize both mechanisms concurrently are called getter-ion-pumps.

(1) Chemical Cleanup and Sublimation Pumps   Chemical cleanup occurs if residual gas molecules impinging on an atomically clean metal surface become permanently attached.   This requires energies of interaction which are large compared with thermal energies.   Hence, chemical cleanup is effective between strongly electropositive metals and gases which are chemisorbed or form a chemical compound.   Since the rare gases are merely physisorbed with energies of less than 10 kcal mol^{-1}, they cannot be trapped at room temperature or above.   However, O_2, N_2, H_2, H_2O, CO, CO_2, and light hydrocarbons interact strongly with a large number of metals.[103]   A wealth of knowledge about these interactions has been accumulated in the manufacturing of electron tubes, where flash evaporation of reactive metals is used to establish very high vacua prior to seal-off.   Summaries of known gettering properties of metals have been published by Wagener[104] and more recently by Stout and Vanderslice.[105]

Chemisorption as the initial step in the gettering process may be followed by diffusion of the gas into the interior of the metal and by formation of stoichiometric compounds. The widely used getter metals Ba, Ti, Mo, Ta, for example, are capable of forming stable oxides, carbides, hydrides, and nitrides. However, the full utilization of the gettering potential of bulk metals demands elevated temperatures and considerable diffusion times to complete the reaction.[106] These requirements are incompatible with the simplicity of operation and rapid gas removal desired for getter pumping. Therefore, the preferred mode of operation is to deposit a thin film of the getter metal onto the inner walls of the vacuum system, thereby achieving a high surface-to-volume ratio. The condensing surfaces are at room temperature or refrigerated to enhance the sticking probability of the impinging gas atoms. Under these conditions the gettering capacity of metals for most gases is limited to monolayer formation while diffusion into the interior, solid solubility, and compound formation are less important. As continued adsorption leads to saturation of the metal surface, the gettering action ceases. Therefore, atomically clean metal surfaces must be restored by continuous or intermittent metal deposition at rates which are determined by the amount of gas to be pumped. Thus, getter pumping is a complex nonequilibrium process in which residual gas molecules and metal atoms impinge either alternately or simultaneously. Accordingly, gettering speeds and capacities vary widely depending on the experimental conditions and modes of operation.

To qualify as a getter material for high-vacuum applications, a metal must satisfy several requirements. First it must be possible to evaporate the metal for extended periods of time without encountering source problems such as chemical attack of the container or emission of impurities. These difficulties are greatly simplified if the metal can be sublimated from resistance-heated wires, rods, or sheets. Furthermore, the metal should adsorb all common gases except the rare gases, and the adsorption products should be stable. Against these criteria, the gettering properties of several metals have been evaluated by Jackson and Haas,[107] who evaporated flashes of metal in a gettering bulb and observed the cleanup rates for H_2, CO, CH_4, and H_2O at initial pressures of 10^{-9} Torr. The most effective getter metal was tantalum, which takes up H_2 and CO twice as fast as the next best materials, titanium and molybdenum. The three other metals investigated, zirconium, vanadium, and niobium, are only moderately effective because they do not getter H_2 and CH_4 very well. As for the first criterion, titanium and vanadium are readily evaporated, as shown by the vapor pressures in Fig. 25. Tantalum, although the most effective getter, is rarely used because it requires very high evaporation temperatures. Molybdenum film gettering

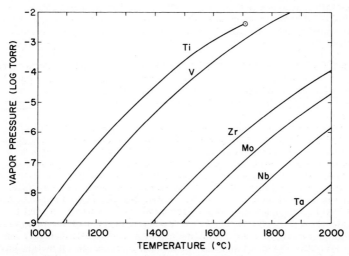

Fig. 25 Vapor pressures of gettering metals. (*After Nesmeyanov; see Ref. 5, Chap. 1.*)

has been tried by Milleron and Popp[108] and gave respectable pumping speeds of about 10^4 l s⁻¹ for hydrogen. There have also been getter pumps operating with barium films[109,110] which are particularly effective in trapping oxygen. However, the majority of practical gettering devices utilize titanium films.

Titanium begins to sublimate noticeably at 1100°C[111] and can maintain vapor pressures of the order of 10^{-3} Torr without melting. Consequently, most titanium getter pumps operate by sublimation as does the device shown in Fig. 26. The walls of the pump module may be designed for circulating-water or for liquid-nitrogen cooling. A resistance-heated titanium wire serves as the vapor source. To obtain sufficiently high sublimation rates from the relatively small wire surface, the source is operated just under the melting point of titanium. This involves the risk of destruction if the wire develops a hot spot. To avoid this problem, McCracken[112] introduced filaments of Ti 15% Mo alloy. Sublimation from these alloy sources is self-regulating: If a section of the wire is overheated, titanium evaporates preferentially and its vapor pressure decreases proportional to the residual concentration, while the melting points of the more molybdenum-rich alloys increase. The resulting uniformity of sublimation extends the useful filament life. Operating limits are imposed by gradual recrystallization which leads to uncontrollable evaporation rates.[113]

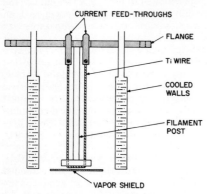

CURRENT FEED-THROUGHS

FLANGE

Ti WIRE

COOLED WALLS

FILAMENT POST

VAPOR SHIELD

Fig. 26 Titanium-wire sublimation pump.

Other titanium-vapor sources employ continuous feeding of wire onto a heated pedestal,[114] electron-bombardment heating of replaceable cartridges,[115,116] or resistance-heated rods.[117] The latter sublimation sources, because of their larger surface areas, operate effectively at temperatures safely below the melting point. Typical metal-evaporation rates are 0.1 to 1 g h⁻¹ with related pumping speeds from 10^{-3} to 10^{-1} Torr l s⁻¹ for nitrogen. Depending upon the required evaporation rates, the useful operating life of a 100-g titanium charge may vary from several hundred to several thousand hours.

As stated earlier, the gettering capacities of metals are very dependent on the experimental conditions and mode of operation. Therefore, specific figures hold only for the individual case investigated. An example of relative gettering capacities of titanium films for different gases gathered under identical conditions is the data of Klopfer and Ermrich:[118]

Gas	H_2	O_2	N_2	CO_2	CO	H_2O
Getter capacity, 10^{-3} Torr l per mg Ti	600	45	5.1	7.5	10	5

The high capacity for H_2 reflects the large solubility of hydrogen in titanium. The latter is of dubious value insofar as hydrogen dissolved in the film is also easily released, thus turning the getter pump into a source of contaminant gas. Hydrogen is indeed often found in the residual gas spectrum of titanium-pumped systems because it is normally present as an impurity of the order of 100 ppm in the initial metal charge.[119] Another common contaminant in high-purity titanium is traces of carbon. If sufficient energy is supplied as during the operation of getter-ion pumps, the carbon reacts with residual hydrogen gas and forms CH_4.[119,120]

The adsorption of O_2, N_2, CO, and CO_2 onto titanium films at room temperature is irreversible and therefore does not involve the risk of future gas release. However, if adsorption takes place on liquid-nitrogen-cooled surfaces, the adsorption process is reversible.   Elsworth et al.[121] showed that the greater part of nitrogen adsorbed by a thin titanium film at 77°K could be recovered upon subsequent heating to 20°C.   In spite of this reversibility, titanium sublimation onto liquid-nitrogen-cooled surfaces (cryogettering) is often used because the sticking coefficients of impinging gas molecules are greater at the lower temperature and therefore yield higher sorption rates, as demonstrated by Clausing.[122]   The same author also found that titanium films evaporated in the presence of 10^{-3} Torr of helium are more effective in sorbing hydrogen than films evaporated in high vacuum, a phenomenon attributed to the greater porosity and surface roughness of the films deposited in helium.

If the gettering film is kept at room temperature or below, continued pumping action depends on renewed deposition of titanium to restore a clean surface.   Since the adsorption of some gases, notably hydrogen, is reversible, a subsequent temperature increase can cause thermal release of previously trapped gases.   Therefore, deliberate outgassing of titanium getter pumps during the pump-down cycle is widely practiced. Denison[123] recommends baking at 300°C to release reversibly sorbed gases while firmly chemisorbed species like nitrogen diffuse into the interior and form titanium compounds.   The accumulation of titanium films containing significant amounts of compounds may result in a gradual loss of adhesion with subsequent peeling or flaking.[124]   It is therefore recommended to occasionally clean the inner walls of sublimation pumps.   Titanium films may be dissolved in 10 % HNO_3 2 % HF solution.[123]

An interesting version of a titanium getter pump has been described by Hirsch.[125] Here, the fresh metal surface is generated by a wire brush rotating inside a titanium cylinder.   The device reduces the pressure from 5×10^{-3} to 5×10^{-5} Torr in about 30 s and attains speeds of 1 to 10 l s^{-1} in this range.   The advantage of this operating principle is the absence of thermal outgassing effects.

Since the rare gases are not amenable to chemical cleanup, titanium gettering is merely an auxiliary method of vacuum pumping.   It is employed regularly in ion pumps to enhance their pumping speeds.   Values up to 10^4 Torr l s^{-1} [112] and ultimate pressures down to 10^{-12} Torr[126] have been achieved by auxiliary cryogettering.   The effectiveness of combined chemical gettering and cryogenic pumping at extremely low pressures is not limited to ion-pumped systems.   Ultimate pressures of 10^{-13} to 10^{-14} Torr have been reported for stainless-steel chambers evacuated by oil diffusion pumps if back streaming of oil vapors was inhibited by cryosublimation traps.[127]   These devices are installed between the pump and the vacuum chamber.   They represent an optically dense baffle in which titanium is evaporated against liquid-nitrogen-cooled surfaces.*

(2) Electrical Cleanup and Ion Pumps   If gas ions are accelerated by an electric field toward a solid surface, there is a high probability of their becoming embedded in the lattice.   This type of electrical cleanup was first observed to occur in ionization gauges by Penning in 1937.[128]   Penning gauges as well as ion beams of uniform energy have been used extensively to investigate the mechanisms of ion trapping under various experimental conditions.   Reviews of the numerous results have been given by Redhead et al.[129] and by Grant and Carter.[130]   The latter article summarizes the available data on trapping efficiencies for rare gases as dependent on target material, temperature, and ion energy, information which is of great practical interest since getter pumping of the rare gases is entirely dependent upon electrical cleanup.

The most important parameter in electrical cleanup processes is the trapping efficiency, i.e., the ratio of ions being trapped by the solid to the total number of ions impinging.   Since ion trapping is contingent upon penetration of the surface, the ions must have a certain threshold energy which Teloy[131] determined to be 30 eV for Ar and 10 eV for He ions.   Generally, ion energies above 100 eV are required to achieve trapping efficiencies greater than 0.1.   Values close to unity are obtained if ions are accelerated through fields of 1 kV and larger.[132]   Energies of this magnitude are sufficient to cause sputtering of the target-film material and may liberate previously trapped gases.   The depths of ion penetration into the solid are of the order of a few atomic layers.   The number of ions which can be taken up by a surface is limited and

---

* Aero Vac Corp., Troy, N.Y., "Ultra High Vacuum Cryosublimation Traps."

depends on the density of the bombarding-particle stream. Typically, saturation is reached after trapping 10^{14} to 10^{15} argon ions per square centimeter of surface, which is equivalent to one monolayer or less.

Considerable effort has been devoted to deriving pumping equations based on various models of ion trapping.[130] If the trapping efficiency of a surface were to remain constant throughout the process, the pressure in a closed system being ion-pumped should decrease exponentially with time, an expectation which is not borne out by experiments. Attempts to account for experimental pump-down curves by assuming a decreasing number of available trapping sites have not met with success. This is due to the fact that ion trapping is always accompanied by gas-release processes which increase in frequency as the target surface approaches saturation. Two gas-release mechanisms have been identified. Spontaneous desorption of trapped gases occurs because of thermal activation and may be observed after cessation of ion pumping.[133] During the pumping process, outdiffusion and desorption are also induced by impinging ions. Direct experimental evidence for gas release due to ion bombardment is the so-called "memory effects." These are observed if an ion pump operates against different gases in subsequent experiments. During the second pump-down, the initial concentration of previously trapped species on the target surface does not correspond to the new partial pressure. Hence, primary trapped species are released into the vacuum until a new equilibrium has been reached. Accordingly, the ultimate pressure obtained with an ion pump does not represent a condition where all available trapping sites are occupied but constitutes a dynamic equilibrium where ion-trapping and gas-release processes balance.

Although ions can be driven into glasses, metallic-target materials such as Ni, W, Mo, Ta, Ti, Zr, and Al are more efficient trapping surfaces.[130] Regardless of the material used, surface saturation and release processes limit the amount of gas which can be trapped. Therefore, the gettering capacity of the target surface must be restored by metal deposition. Previously trapped gas atoms are thereby covered and can no longer leave the surface without diffusing through new layers of metal. When buried at some depth below the surface, the release of trapped atoms is greatly reduced. For example, outdiffusion of Ar forced into molybdenum with a concentration of 1 % is insignificant at temperatures up to 1500°C.[134] Ambridge et al.[135] found that the thermal release of rare gases trapped in titanium took place only at temperatures above 500°C.

In getter-ion pumps, chemical cleanup by freshly deposited metal films and ion pumping occur simultaneously. The greatest pumping efficiency is obtained if the ion-target film is also a good chemical getter. Although electrical gettering is primarily aimed at the removal of the rare gases, the trapping of chemically reactive species is also promoted. Teloy[131] observed that N_2 and CO molecules dissociated after colliding with energetic electrons and were subsequently chemisorbed by the normally nonreactive walls of the system. The excitation of metastable states in N_2 molecules bombarded with low-energy electrons may also enhance the chemical cleanup.[136]

The interest in getter-ion devices as ultrahigh-vacuum pumps was greatly stimulated by the work of Alpert[32] in 1953, when he evacuated a small diffusion-pumped glass system to 10^{-10} Torr by means of a Bayard-Alpert ionization gauge. The ensuing developments have brought forth pumps which combine chemical with electrical gettering at high efficiency. Commercial getter-ion pumps are now available under a variety of names.[137] Generically, they may be divided into evapor-ion* and sputter-ion pumps according to the methods of metal-vapor generation.

(3) Evapor-ion Pumps The principal feature which distinguishes evapor-ion from sputter-ion pumps is that in the former the functions of metal-vapor generation and gas ionization are separated while in the latter they are inseparable. In evapor-ion pumps, ions are produced by collision of gas atoms with electrons extracted from a hot

---

* Evapor-ion pump is not universally accepted as a generic designation. Alternate names such as radial-electric-field pump[138] or electrostatic getter-ion pump[139] have also been suggested.

tungsten filament and accelerated by an electric field. The ions are driven into the negatively charged walls of the pump, which also act as the receiving surface for the gettering metal film. Pumps of this type were first described by Herb and coworkers,[114,140] and one of their early versions is shown schematically in Fig. 27. The outer grid carries a potential of −500 V which accelerates the inert gas ions toward the wall. Titanium is evaporated by continuously feeding wire onto a heated source.

Pumps of this design have been built in sizes from 1 to 24 in. diameter with speeds up to 10,000 l s⁻¹.[141] The speed ratings refer to those gases which are chemically gettered. The pumping speeds for the inert gases are much lower, as illustrated in Table 5. Water vapor is removed at relatively low speeds since it is not very well gettered by titanium. The 4-in. pump in Table 5 is relatively more efficient in regard to inert ion pumping than the older

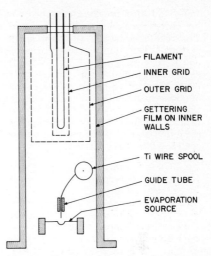

FILAMENT
INNER GRID
OUTER GRID
GETTERING FILM ON INNER WALLS
Ti WIRE SPOOL
GUIDE TUBE
EVAPORATION SOURCE

Fig. 27 Evapor-ion pump after Herb, Davis, Divatia, and Saxon.[114]

18-in. pump because of improvements in the geometrical arrangement of the internal elements. By injecting electrons with appropriate energies such that they are forced to orbit in the electrical field, their mean free paths are increased to several hundred centimeters and their ionization efficiency is enhanced. The orbiting electrons finally strike a titanium cartridge and thereby also serve as the heat source for the sublimation

TABLE 5 Pumping Speeds of Evapor-ion Pumps, l s⁻¹

	H_2	N_2	O_2	CO_2	CO	Ar	He
18-in.-diam pump (Swartz[142])	3,300	2,000	1,000	. . . . .	. . . . .	5	
4-in.-diam pump (Douglas et al.[143]) . .	900	500	. . . . .	. . . . .	. . . . .	7	
6-in.-diam Electro-Ion* pump	4,800	1,600	1,500	1,600	1,300	24	2.5

* Trademark of Granville-Phillips Co., Boulder, Colo.

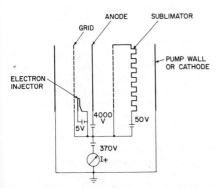

ANODE SUBLIMATOR
GRID
PUMP WALL OR CATHODE
ELECTRON INJECTOR
4000 V
50V
5V
370V
I+

Fig. 28 Schematic of an electrostatic getter-ion pump cell after Bills[139] and Denison.[123]

process.[143] This design does not require a grid since the electric field maintained between a central metal rod and the pump walls accelerates both electrons and ions. A modified version utilizes a glass tube as the surrounding surface with a wire-mesh screen.[144] The latter prevents the accumulated titanium film from peeling off in large flakes which lower the inert-gas pumping speed and may ground the anode.

The principle of orbiting electrons is also used in the 6-in. pump of Table 5. This is an enlarged model of a pump developed by Bills[139] and Denison.[123] Figure 28 shows one of four identical cells contained in the pump. The positive grid potential of 370 V drives ions toward the walls and also prevents electrons from leaving the grid volume. Therefore, the ion current may be used as a pressure indicator down to about 10^{-8} Torr pressure. The resistance-heated titanium sublimator is independent of the electron-emission current. The pump walls are cooled during operation but can be baked at 300°C to outgas previously accumulated deposit. Pumping speeds for two gases as functions of pressure are shown in Fig. 29. Argon is removed at constant rates down to the lowest pressures, while the predominantly chemical gettering rate of nitrogen shows two discontinuous decreases. These reflect deliberate reductions of the sublimator power, an economy measure intended to extend the lifetime of the titanium charge by avoiding unnecessarily high evaporation rates. The power reduction at 10^{-8} Torr does not affect the speed because the latter is limited by the conductance of the intake.

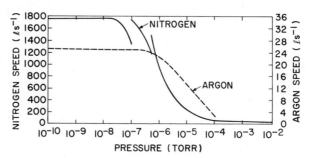

Fig. 29 Electrostatic getter-ion pump speeds for N₂ and Ar as functions of pressure. The titanium sublimation rates are lowered at 5.5×10^{-7}, 1×10^{-7}, and 1×10^{-8} Torr. (*After Denison.*[123])

The fore pressures which have been reported as necessary to start evapor-ion pumps vary considerably. While some authors have found 10^{-2} Torr sufficiently low, others prefer 10^{-4} Torr. In this transition interval, the pumping speeds are still small, as Fig. 29 shows. Therefore, pump-down times are considerably shorter the lower the starting pressure. Furthermore, the ion currents associated with electrical pumping at high fore pressures are relatively large and may cause excessive heating of the pump with ensuing release of adsorbed gases. This risk is even greater if the combination of pressure, electric field, and electrode spacing is such that a glow discharge is generated. The pumping procedure outlined by Adam and Baechler[145] avoids this difficulty by operating the titanium sublimator alone until the fore pressure is reduced to 10^{-4} Torr.

The residual gases found in evapor-ion pumped systems are mostly attributable to outgassing of the chamber walls. In unbaked vacuum systems, the ultimate pressures obtained with evapor-ion pumps are 10^{-7} to 10^{-8} Torr.[115] Except for the absence of higher hydrocarbons, the gas composition is similar to that of diffusion-pumped systems with water vapor the predominant constituent and variable amounts of N₂, CO, Ar, CH₄, H₂.[16,145] In an ion-pumped bakeable vacuum system capable of achieving 3×10^{-11} Torr, the residual gases consisted primarily of hydrogen with lesser amounts of CO, H₂O, and CH₄.[123] Gases previously trapped in the pump are not released in significant quantities since the ion energies in evapor-ion pumps are not high enough to cause severe memory effects. The presence of lower hydrocarbons such as CH₄ or C₂H₆ is attributed to surface reactions between hydrogen and carbon impurities on the pump walls.[146]

(4) Sputter-ion Pumps Sputter-ion pumps evolved from the Penning ionization gauge. Their functional elements are cells consisting of a cylindrical anode between two cathodes in a magnetic field as shown in Fig. 30. The cathodes carry a negative dc potential of several kilovolts with respect to the anode. Electrons emitted from the cathode surfaces are accelerated toward the anode by the electric field. The magnetic field imparts radial-velocity components and forces the electrons into spiraling trajectories. Because of the long free paths of the orbiting electrons, the ionization efficiency is high and permits the gaseous discharge to be sustained down to pressures in the ultrahigh-vacuum range. The positively charged gas ions are attracted toward the cathodes where some of them become trapped at the surface. Since the ions impinge with energies up to several keV, they also cause sputtering of the cathode material. The metal vapor thus generated spreads throughout the cell and condenses on all surfaces including the cathodes. Thus, chemical as well as electrical cleanup of residual gases are initiated simultaneously. While trapping by chemisorption occurs predominantly on the inner surface of the cylindrical anode, electrical cleanup is primarily confined to the cathodes. By pumping radioactive krypton, Lafferty and Vanderslice[147] demonstrated that ion gettering occurred mostly on the periphery of the cathode opposite to the anode walls, whereas the cathode center served as the source of sputtered metal. This nonuniformity is essential to the functioning of sputter-ion pumps since no perma-

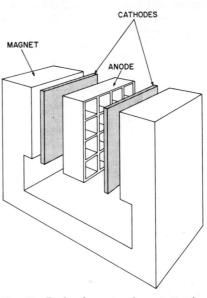

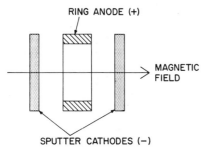

Fig. 30 Schematic of a Penning discharge cell.

Fig. 31 Basic elements of a sputter-ion pump with multicell anode.

nent burial of inert-gas particles could occur if the ion current was uniformly distributed.

The pumping capability of a single Penning discharge cell is too small to evacuate systems of practical size. The sputter-ion pump introduced by Hall[148] represents a significant technological advance insofar as it had a much greater pumping speed than a Penning cell. This was achieved by utilizing an anode consisting of many cells which are opposed by two cathode plates as shown in Fig. 31. The effectiveness of the multicellular structure is due to the fact that the maximum charge containable in a hollow anode is proportional to the voltage across the cell and to its length but independent of the diameter.

Hall's pump was able to remove air at 10 l s^{-1} and achieved ultimate pressures of 10^{-11} Torr. These performance data stimulated rapid further development so that a variety of commercial sputter-ion pumps became available within a few years. The honeycomb structure of the anode is a common characteristic of all pumps, whereas cell size and geometry are variables whose influence on the performance has been discussed by Baechler.[149] Another common feature is the necessity of having to maintain magnetic fields of 1 to 2 kG in a large volume which requires bulky external magnets. Since sputter-ion pumps are primarily intended for the ultrahigh-vacuum range, they must be bakeable. Commercial pumps can be heated up to 400 or 500°C, whereby the higher temperatures usually require that the external magnets are temporarily removed. As for cathode materials, titanium with its excellent chemical gettering properties is the commonest metal. Molybdenum cathodes have been investigated by Jackson and Haas[150] and found to be similarly effective.

Sputter-ion pumps of the type introduced by Hall[148] suffer from a peculiar problem arising from the fact that sputtering of metal and burial of trapped gases occur in adjacent areas of the cathode. The discharges within individual anode cells are somewhat unstable and tend to oscillate, thereby striking cathode areas which have previously served as trapping regions. Besides, gases driven into the central cathode area are subsequently released because of continued sputtering of metal. Consequently, these pumps produce periodic pressure bursts of up to two orders of magnitude, which decay within about 1 min as the released gases are again cleaned up. The gas species primarily responsible for this phenomenon is argon since it is the commonest of the rare gases in air. Accordingly, this particular memory effect has come to be known as "argon instability."

The further development of sputter-ion pumps has been guided by the objective of avoiding argon instability and other memory effects. This has been accomplished by several design modifications which distinguish currently available commercial pumps from each other. The most obvious departure from the original design is the introduction of a third pair of electrodes. Accordingly, sputter-ion pumps are designated as either *diode* or *triode* pumps.[151] The two modes of construction are shown schematically in Fig. 32.

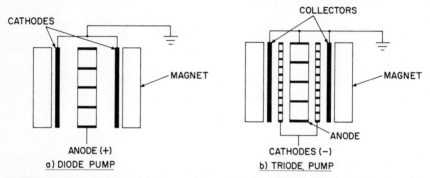

Fig. 32 Schematic representations of sputter-ion pumps with diode and triode structure.

Jepsen et al.[152] constructed a diode pump where the argon instability was suppressed by employing slotted or grooved cathodes. The surface of such a cathode presents itself under different angles to the electric field. Since the sputtering yield of ions with identical energies is greater for glancing incidence than for perpendicular impact, grooving or slotting stabilizes the subdivision of the cathode surface into sputtering and trapping regions. Maximum sputtering yield is obtained if the angles of ion incidence on the donor areas of the cathode are 10 to 25°.[153]

Another stabilized diode pump is the differential sputter-yield ion pump described by Tom and James.[154] Here the cathodes consist of two different metals such as titanium and tantalum. According to Carter,[153] tantalum sputters at higher yields than titanium. By utilizing these two metals in the same pump, better pumping speed and stability than with ordinary diode pumps are obtained.

Sputter-ion pumps of the triode type were first described by Brubaker.[155] Their performance characteristics have been reviewed by Bance and Craig.[151] The distinguishing features of the triode structure are titanium cathodes in the form of an open honeycomb. These are positioned between the multicell anode and the collector plates which are kept at ground potential as shown in Fig. 32b. Since anode and collector materials are not sputtered, these electrodes may be made from stainless steel. The negative cathode potential is several kilovolts. Positive gas ions generated in the space between cathode and anode are accelerated toward the cathode. Because of the open structure, most of the ions pass through the cathode but lose energy as they move toward the collector. Hence, the ions arrive at the collector with relatively low energies which are inducive to trapping but insufficient to sputter metal. Hence, gases already buried are not released. Those ions which do collide with the cathode strike with high energies and at small angles of incidence. Therefore, they are highly effective in sputtering titanium. The glancing impact favors forward scattering of the sputtered material onto the collector electrode. Accordingly, electrical cleanup takes place on the cathode surfaces facing away from the anode and also on the collector.[156]

In spite of technological advances in sputter-ion pumps, the mechanisms by which inert gases are trapped and released are not yet so well understood as to account quantitatively for the observed behavior of these pumps. An attempt in this direction has recently been made by Jepsen,[164] who proposed a new trapping mechanism involving "energetic neutrals." According to his hypothesis, a fraction of the rare-gas ions striking the cathode is neutralized but does not stick. These atoms are backscattered toward the anode and sidewalls with relatively high energies and sticking probabilities. In contrast to the ions driven into the cathode, the energetic neutrals are permanently buried under sputtered material. For a particular type of diode pump, Jepsen estimated that about 10% of the argon ions were permanently pumped, which was in good agreement with the observed argon pumping speed.

Commercial sputter-ion pumps embodying different modes of construction are nevertheless similar in regard to their performance. Pump-down curves for several models have been published by Milleron.[23] An objective comparison is difficult because the pumping speeds depend on the gas species, the operating pressure, and the history of the pump. The effects of these variables have been investigated by Andrew.[157] The pumping curve for air in Fig. 33 is typical insofar as it shows a speed

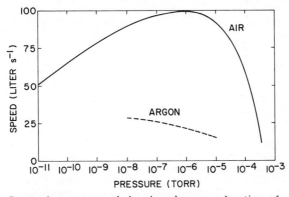

Fig. 33 Sputter-ion pump speeds for air and argon as functions of pressure.

increase from intermediate to lower pressures with a maximum around 10^{-6} Torr. The gradual decrease of pumping speeds toward very low pressures can be compensated by operating a titanium sublimation pump simultaneously. Rare gases such as argon are removed at considerably lower rates than the chemically gettered species. Commercial pumps are usually rated in terms of airspeed at the 10^{-6} Torr maximum. Standard pump sizes range from 1 to 1,200 l s^{-1}, but smaller as well as larger pumps have also been reported. The pumping speeds for gases other than air are given in percent of the airspeed. Table 6 shows the ranges most widely quoted.

TABLE 6 Pumping Speeds of Sputter-ion Pumps for Various Gases

Gas	Relative speed, % of airspeed
H_2	150–250
Air, CO_2, CO, H_2O	100
Light hydrocarbons	90–180
N_2	85–95
O_2	60–100
He, Ar	5–30

It is often claimed that sputter-ion pumps may be turned on at 10^{-2} Torr, but some diode pumps are notoriously difficult to start at such high fore pressures. Triode pumps do not have this problem and may even be started at 10^{-1} Torr. However, the pumping speed in this region is small, and it takes an equally long time to reduce the pressure from 10^{-1} to 10^{-2} Torr as is required for the 10^{-2} to 10^{-7} Torr interval.[151] In addition, the titanium consumption at high starting pressures is high since the ion current increases with the gas pressure, thus shortening the useful life of the cathode. The required fore vacuum is commonly established by means of cryosorption traps. Holland et al.[158] found well-trapped rotary oil pumps to be superior since they do not discriminate against the rare gases and facilitate easier starts of the ion pump.

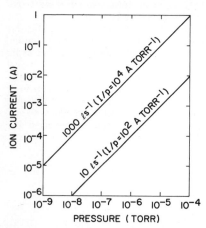

Fig. 34 Ion current vs. pressure for sputter-ion pumps rated at 10 and 1,000 l s^{-1}, respectively. A pumping efficiency of 0.1 Torr l C^{-1} has been assumed.

The relationship between ion current and pressure is linear over a wide range, as shown in Fig. 34. Therefore, sputter-ion pumps may be calibrated to indicate the pressure. Deviations from strict proportionality as well as declining pumping speeds are encountered at very low pressures, as was discussed by Lamont.[159] By optimizing the pump design in regard to electrode geometry, magnetic and electric fields, the useful operating range can be extended down to 10^{-11} or 10^{-12} Torr.[160] In diode pumps, it has been observed that the current in the center of the cell becomes relatively larger as the pressure decreases from 10^{-4} to 10^{-10} Torr.[161] Hence, at declining pressure sputtering is focused into a small area in the center of the cathode cells. From about 10^{-10} Torr on down, the Penning discharge tends to become unstable and may extinguish. To assure the starting and sustaining of a discharge at these low pressures, it has been suggested that a β-ray source such as Ni^{63} be inserted into the pump.[162] Another potential problem is the formation of titanium whiskers or protruding flakes on the cathode surface. These give rise to field-emission currents which destroy the established relationship between ion current and pressure. Methods to remove such field emitters have been discussed by Hamilton.[163]

The ion current I which a pump sustains at a given pressure p is a measure of the pump's speed.[160] As shown in Fig. 34, higher pumping speeds require proportionately greater discharge intensities I/p. However, not every ion contributing to the measured current becomes trapped and removed from the gas load. A meaningful figure of merit can be derived by dividing the actual pumping speed S by the I/p ratio. A pumping efficiency of $Sp/I \approx 0.2$ Torr l C^{-1}, for example, indicates the removal of one gas particle for every ion striking the cathode. The pumping efficiency of sputter-ion pumps depends on design factors such as electrode geometry, cathode voltage, and magnetic field. Practical values for nitrogen pumping are 0.05 to 0.1 Torr l C^{-1}; i.e., 2 to 4 ions are required to remove one N$_2$ molecule.[164] Projected lifetimes of commercial diode pump cathodes are of the order of 30,000 h for continuous operation at 10^{-6} Torr, or proportionately shorter at higher pressures.

Since the greater part of the gas load normally pumped is amenable to chemical gettering, operating life and pumping speeds may be increased by combined sputter-ion and sublimation pumping. Speeds up to 50,000 l s^{-1} have been achieved with this mode of operation.[165] The effects of combination pumping on residual gases and ultimate pressures have been investigated by Newton and O'Neill.[166] Holland et al.[138] demonstrated that simultaneous operation of evapor-ion and sputter-ion pumps, too, yielded superior pumping speeds.

Sputter-ion pumps are not damaged by accidental exposure to air. However, they are sensitive to certain impurities. Kelly and Vanderslice[167] observed that the pumping action ceased after prolonged operation in benzene vapors of 10^{-3} Torr. The cause of this degradation is decomposition of the organic compound and formation of carbonaceous residues which cover the titanium surface and reduce the sputtering yield to zero. The same failure occurred during prolonged exposure to oil vapors from a fore pump,[168] but it was possible to restore the pumping efficiency by baking. Mercury vapors, which may come in contact with sputter-ion pumps when using a mercury diffusion pump on the same system, were found to be harmless at low pressures. However, mercury pressures in excess of 10^{-5} Torr lead to early deterioration of sputter-ion pumps by causing the titanium deposit to flake off.[168] Since sputter-ion pumps are expensive, the availability of replacement electrodes from the manufacturer may be an important economic consideration in certain cases.

The ultimate pressures and residual-gas compositions obtained with sputter-ion pumps are similar to those of evapor-ion pumps. Hydrocarbons are normally absent except for methane, which is synthesized from hydrogen and carbon impurities in the titanium.[169] Since methane is cleaned up at reasonable rates, its dynamic-equilibrium concentration does not dominate the residual-gas atmosphere. In bakeable systems operating in the ultrahigh-vacuum range, hydrogen is the prevailing constituent.[170] Carbon monoxide, nitrogen, or water vapor may be present in smaller amounts depending on the degree of outgassing from the inner surfaces of the system. Ultimate pressures below 10^{-12} Torr have been reported by Davis[171] for a small bakeable metal system.

3. VACUUM MATERIALS

The construction of vacuum systems requires a variety of materials with different mechanical properties to serve as chamber walls, feedthroughs, and internal components. The walls of a vacuum system must withstand atmospheric pressure without collapsing. Therefore, rigid materials like metals and glasses are used to enclose the evacuated operating space. Access to the latter must be provided by means of seals, gaskets, feedthroughs, and valves. Repeated demounting, opening, or closing of the system is most conveniently accomplished if elastomers are utilized. Within the three classes of metals, glasses, and elastomers, the suitability of specific materials is determined by the amount of volatile contaminants which they contribute to the residual-gas load. Materials having a noticeable vapor pressure at the highest intended operating temperature must obviously be avoided. An example is brass, which has a zinc partial pressure of about 10^{-3} Torr at 300°C.

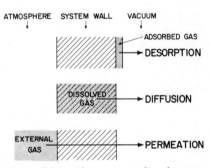

Fig. 35 Schematic representation of sources of residual gases associated with vacuum materials.

But even materials with negligible vapor pressures still act as sources of residual gases. The origin of these gases and the processes by which they enter the vacuum system are summarized in Fig. 35.

The processes of desorption, diffusion, and permeation associated with materials of construction are of fundamental importance in vacuum technology because they counteract the gas-removal action of the pumps. In the absence of leaks, these processes together with gas emission from the pumps determine the ultimate pressure and the composition of the residual atmosphere in the system. Consequently, the material properties related to these contributing processes are of great interest in the design of vacuum systems and in assessing the possibility of interactions between deposited films and residual gases.

a. Adsorption of Gases

A gas atom or molecule approaching an atomically clean solid surface experiences a potential field as shown schematically in Fig. 36. The first shallow potential well arises from weak interaction due to van der Waals and polarization forces. The transition into this state is called physisorption and is associated with small adsorption energies of less than 10 kcal mol^{-1}. The adsorption process stops in this state if the

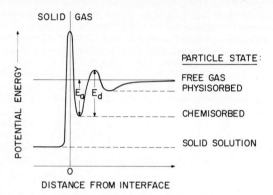

Fig. 36 Potential energy of a gas particle at a solid-gas interface. (*After Jaeckel.*[172])

physisorbed atom (adatom N_a) and the solid are chemically inert. Examples are H_2, N_2, O_2 on glass and the rare gases on all surfaces.[129] Some gas-solid systems are capable of further interaction whereby the gas enters a deeper potential well. This process involves electronic interaction between gas and solid in various forms as reviewed, for instance, by Nasini and Ricca.[173] The resulting energy loss suffered by the adatom, the heat of adsorption E_a, is greater than 10 kcal mol^{-1} and proportional to the heat of formation of the respective chemical compound. This is illustrated in Fig. 37, where the heats of adsorption for various gases on metals are plotted against

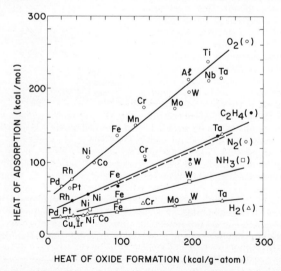

Fig. 37 Comparison of heats of adsorption on clean metal surfaces and heats of metal-oxide formation. (*After Sachtler.*[174])

the heat of metal oxide formation.[174] Gas-solid interaction of this more intense form is referred to as chemisorption. For a more extensive treatment of adsorption phenomena, the reader is referred to the literature.[175]

The distinction of sorption processes according to their heat effects is based on different bonding mechanisms. The dividing value of 10 kcal mol^{-1} is arbitrary since it falls into a region of mixed binding forces. Furthermore, heats of adsorption are sensitive to the state of the surface in regard to crystallinity, impurities, and concentration of adatoms already present. Hence, their measured values vary considerably under different experimental conditions. Reviews of experimental techniques and data have been given by various authors.[176-178]

Whereas physisorption is not an activated process, chemisorbed particles often require a desorption energy E_d greater than the adsorption energy E_a in order to return into the gas phase. This situation is shown in Fig. 36. The common gases H_2, CO, O_2, and N_2, however, are usually adsorbed by transition metals either without activation, or $E_d - E_a$ is negligibly small.[175]

The desorption energy determines the average residence time $\bar{\tau}$ which an adatom spends on the surface according to

$$\bar{\tau} = \tau_0 \exp\left(\frac{E_d}{RT}\right) \tag{1}$$

where τ_0 is the adatom's period of thermal vibration normal to the surface. In numerical estimates, values for τ_0 of the order of the Debye frequency (10^{-14} to 10^{-12} s) are commonly assumed. At room temperature, Eq. (1) yields average residence times of 2 μs for $E_d = 10$ kcal mol^{-1}, whereas about 30 years are obtained for $E_d = 30$ kcal mol^{-1}.[129] Under equilibrium conditions, the density of adatoms is the product of the impingement rate dN_i/dt and the average residence time:

$$\frac{N_a}{A} = \alpha_s \frac{dN_i}{A\,dt} \bar{\tau} \qquad \text{atoms cm}^{-2} \tag{2}$$

The factor α_s is the sticking coefficient, which has been introduced because a fraction $(1 - \alpha_s)$ of the impinging atoms rebounds upon impact. For most gas-solid combinations, $0.1 < \alpha_s < 1$; it varies with gas species, temperature, and coverage of the surface.[129,179] Substitution of the impingement rate according to Eq. (37) of Chap. 1 and of $\bar{\tau}$ from Eq. (1) leads to

$$\frac{N_a}{A} = \alpha_s p \frac{3.5 \times 10^{22} \tau_0}{\sqrt{MT}} \exp\left(\frac{E_d}{RT}\right) \tag{3}$$

or

$$\frac{N_a}{A} = ap \qquad \text{atoms cm}^{-2}$$

where $a = $ const if $T = $ const. Hence, at constant temperature the adsorbate density should be proportional to the gas pressure, a relationship referred to as Henry's law.

Henry's law is the simplest form of an adsorption isotherm. It is obeyed at low gas pressures or at high temperatures, i.e., under conditions where only a small fraction of the surface is covered by adatoms. Theoretical adsorption models based on the statistical theory of imperfect gases and on the Kirkwood-Muller potential for the gas-solid interaction have been developed by Huang[180] and Hobson.[181] The former assumes a homogeneous surface, while Hobson considers the possibility that the adsorption energies of individual surface sites vary. Both approaches lead to expressions which are equivalent to Henry's law at pressures in the ultrahigh-vacuum range.

At higher pressures, the greater density of adatoms causes deviations from the simple linear relationship. A considerable number of empirical expressions have been derived to account for the various observed adsorption isotherms (see, for example, Dushman[182] or Trapnell[175]). Among these, Langmuir's isotherm is of fundamental importance because it introduces the concept of a monomolecular layer of adsorbed particles as the result of the short-range nature of the surface forces.[183] The physical assumptions underlying Langmuir's isotherm are that all surface sites have equal adsorption energies, that adatoms do not interact with each other, and that only one atom or molecule can be accommodated per site. Therefore, there is a maximum number of adatoms N_m which constitute a close-packed single layer. The number of particles in such a complete monolayer is determined by the diameter of the adsorbed species. Approximate values are 5×10^{14} cm^{-2} for H_2O, CH_4, and CO_2; 8×10^{14} cm^{-2} for CO, O_2, N_2, and Ar; and 2×10^{15} cm^{-2} for H_2 and He.

Langmuir's isotherm is obtained if the adsorption and desorption rates are assumed to be equal. Referring to 1 cm^2 of surface area, the adsorption rate is

$$\frac{dN_a}{dt} = \alpha_s \frac{dN_i}{dt} (1 - \theta)$$

where $\theta = N_a/N_m$ is the surface coverage. The factor $(1 - \theta)$ accounts for the reduced probability of adsorption, assuming that gas particles impinging on occupied sites are reflected. The impingement rate at constant temperature is proportional to the pressure, $dN_i/dt = c_1 p$, and hence the adsorption rate becomes

$$\frac{dN_a}{dt} = \alpha_s c_1 p (1 - \theta) \tag{4}$$

If the equilibrium concentration of adatoms is N_a and their average residence time is $\bar{\tau}$, the number of particles leaving the surface per unit time is $N_a/\bar{\tau}$. Using Eq. (1), the desorption rate can be expressed as

$$-\frac{dN_a}{dt} = \frac{N_a}{\tau_0} \exp \left(\frac{-E_d}{\mathbf{R}T} \right) \tag{5}$$

Substituting θN_m for N_a and consolidating all other factors into a constant c_2, Eq. (5) becomes

$$-\frac{dN_a}{dt} = c_2 N_m \theta \tag{6}$$

The equality of Eqs. (4) and (6) leads to Langmuir's isotherm,

$$\theta = \frac{bp}{1 + bp} \tag{7}$$

where $b = \alpha_s c_1 / c_2 N_m$ is a temperature and material dependent constant. The pressure dependence of the surface coverage θ for an assumed value of b is shown in Fig. 38.

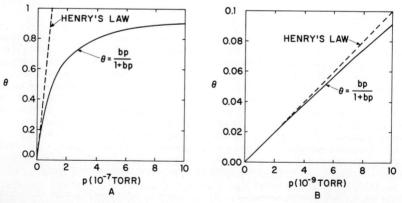

Fig. 38 Langmuir's adsorption isotherm for $b = 10^7$ Torr^{-1}. (*a*) Approaching monolayer coverage at $p > 10^{-7}$ Torr. (*b*) Transforming into Henry's law at $p < 10^{-8}$ Torr.

The hyperbolic approach of monolayer coverage has often but not always been found to give an adequate representation of experimental adsorption data. As shown in Fig. 38b, Langmuir's isotherm transforms into Henry's law at very low pressures.

The majority of adsorption measurements have been made in the course of investigations pertaining to heterogeneous catalysis or to determine actual rather than apparent (geometrical) surface areas of solids. These experiments were carried out mostly at pressures above 10^{-3} Torr, and the resulting isotherms are often complex. Many of them have been successfully interpreted by the theory of multilayer adsorption. This is an extension of Langmuir's concept introduced by Brunauer, Emmett, and Teller,[184] which assumes that adsorption on top of adatoms leading to two or more layers of adsorbed species is possible. Since the second and subsequent layers interact with layers of adatoms and not with the free surface of the solid, the adsorption energies vary, and the isotherms may show two or more saturation regions.[87]

Langmuir's assumption of equal desorption energies for all surface sites must also be questioned. In many cases the binding forces acting upon adatoms depend on crystal orientation or on the number of nearest neighbors within the crystal surface. Consequently, sites of different adsorption energies exist on the surface. If a surface of this kind is covered with a monolayer and then the pressure is reduced, the sites with the smallest desorption energies will release their adatoms first. Accordingly, E_d is sometimes found to increase with decreasing coverage.

The fact that solid surfaces exposed to atmospheric pressure acquire one or more layers of adsorbed gas is of great importance in high-vacuum work. This is readily demonstrated by comparing the amounts of adsorbed and free gas in a vacuum system. Assuming a cylindrical chamber of 100 l volume and 18 in. diameter, the surface-to-volume ratio A/V is 0.12. Considering the presence of fittings, connections, machined grooves, and fixtures inside the system, this value is obviously too small. The roughness of the internal surfaces also increases the actual surface area. Thus, a value of $A/V = 1$ is more realistic. If the entire surface area A is covered with one monolayer of gas, the number of adsorbed atoms is $N_a = 5 \times 10^{14} \times A$. According to the universal gas equation, the number of free gas molecules in the volume V is $N_g = 3.24 \times 10^{16} p V$ at room temperature. Thus, the ratio of adsorbed to free-gas particles in a system of surface-to-volume ratio 1 is

$$\frac{N_a}{N_g} = \frac{1.5 \times 10^{-2}}{p} \qquad [p \text{ in Torr}]$$

The ratio is even larger for smaller systems. Consequently, at pressures where the mean free path approaches the dimensions of the apparatus (typically 10^{-3} Torr), the surfaces of an evacuated system hold much more gas than is present in the gas phase.

b. Outgassing of Materials

Gas release by desorption and by diffusion are referred to jointly as outgassing since they are difficult to distinguish experimentally. Both processes draw from a limited reservoir of gas stored either in the absorbed layer or in the interior of the wall. When the system pressure is reduced, the rates of desorption and diffusion decrease with time as the reservoirs of stored gas are depleted. Eventually, equilibrium concentrations corresponding to the new pressure are reached such that gas release is balanced by gas adsorption. This consideration shows that the largest outgassing effects are to be expected during the transient state of the pump-down. The discrimination between desorptive and diffusive gas release as well as their quantitative assessment require a closer analysis of the possible individual processes.

(1) Isothermal Desorption Although Langmuir's adsorption isotherm has been derived from kinetic equations, it is a thermodynamic relation pertaining to equilibrium states with no information about the atomic mechanisms leading from one state to another. The current knowledge of desorption kinetics is limited to a few simple model reactions which do not adequately represent the complex desorption processes during an actual pump-down. However, some of the simpler kinetic relationships are useful to demonstrate several principal effects.

The desorption rate for a first-order reaction of the type $N_a \rightarrow N_g$ with a uniform desorption energy E_d is given by Eq. (5). The use of Eq. (5) for desorption rates during a pump-down implies that every desorbed gas particle is removed from the system; i.e., there is no readsorption ($\alpha_s = 0$). For such an idealized situation the desorption rate can be written in the form

$$- \frac{dN_a}{dt} = N_a K_1$$

where $K_1 = \tau_0 \exp(-E_d/\mathbf{R}T)$ and $T = \text{const}$. Integration of this expression yields N_a as a function of time and of the initial adatom density N_0 at $t = 0$:

$$N_a = N_0 \exp(-K_1 t)$$

Substitution of N_a into the rate equation leads to an exponential law:

$$- \frac{dN_a}{dt} = N_0 K_1 \exp(-K_1 t) \tag{8}$$

To show the significance of this dependency, several functions $K_1 \exp(-K_1 t)$ have been plotted in Fig. 39. The sensitivity of the rates to the values of E_d is striking. At 25°C, the desorption of adatoms with $E_d = 20$ kcal mol^{-1} becomes negligible within 2 hours. For $E_d = 10$ kcal mol^{-1}, the surface layer is released in fractions of a second; hence physisorbed gases do not contribute to prolonged pump-down cycles but are desorbed in a quick initial burst. The other extreme is firmly chemisorbed gases with desorption energies greater than 35 kcal mol^{-1}. These are released at very small rates which are persistent but contribute negligible amounts to the residual-gas load. The troublesome gases which are responsible for prolonged and noticeable outgassing are those with desorption energies around 25 kcal mol^{-1}. Although derived from an oversimplified model, these conclusions have been verified experimentally. An example is water vapor, which has desorption energies between 22 and 24 kcal mol^{-1} on metal surfaces and is notoriously difficult to remove at room temperature.[185]

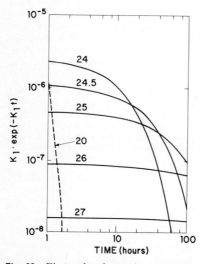

Fig. 39 First-order desorption-rate functions vs. time for desorption energies E_d between 20 and 27 kcal mol^{-1}. A desorption temperature of 25°C and a thermal vibration period of 10^{-12} s have been assumed.

The first-order desorption equation is not applicable if gases dissociate upon chemisorption. Examples are diatomic gases on metals,[129] which are desorbed according to $2N_a \rightarrow N_g$ and therefore obey a second-order rate equation:

$$- \frac{dN_a}{dt} = K N_a^2 \exp\left(\frac{-E_d}{\mathbf{R}T}\right) = K_2 N_a^2 \tag{9}$$

Integration of Eq. (9) yields

$$N_a = N_0 (1 + N_0 K_2 t)^{-1}$$

so that the time function of the desorption rate becomes

$$- \frac{dN_a}{dt} = K_2 N_0^2 (1 + N_0 K_2 t)^{-2}$$

Although the time dependence of second-order rates is hyperbolic rather than exponential, E_d enters the factor K_2 exponentially and therefore affects the desorption rates strongly.

(2) Desorption at Increasing Temperature Since desorption rates depend exponentially on temperature, they increase drastically upon heating. Characteristic desorption spectra are obtained when surfaces are steadily heated. Assuming that the idealized desorption equation (5) is valid (i.e., E_d = const, α_s = 0, and first-order kinetics), a linear temperature schedule ($T = at$) starting at 0°K yields the following temperature function:

$$-\frac{dN_a}{dt} = -a\frac{dN_a}{dT} = \frac{N_a}{\tau_0}\exp\left(\frac{-E_d}{RT}\right) \tag{10}$$

The temperature dependence of N_a follows by integration of Eq. (10),

$$N_a = N_0 \exp\left\{-\frac{E_d}{a\tau_0 R}\left[\frac{e^{-x}}{x} + Ei(-x)\right]\right\} \tag{11}$$

where N_0 is the number of adatoms at $t = 0$ and $x = E_d/RT$. Significant desorption occurs if $x \geq 30$, and for that case the exponential integral[186] can be approximated by a series,

$$Ei(-x) = e^{-x}\left(-\frac{1}{x} + \frac{1}{x^2} - \frac{2}{x^3} + \frac{6}{x^4} - \cdots\right)$$

Substitution of the series transforms Eq. (11) into

$$N_a \approx N_0 \exp\left[-\frac{E_d}{a\tau_0 R}\frac{e^{-x}}{x^2}\left(1 - \frac{2}{x} + \frac{6}{x^2}\right)\right]$$

The desorption rate as a function of increasing temperature is obtained by substituting N_a into Eq. (10). The numerical example in Fig. 40 shows that desorption occurs in a narrow temperature interval with a pronounced maximum.

In practice, desorption rates resulting from a linear temperature schedule do not peak as sharply as Fig. 40 suggests. Considerable broadening of the desorption

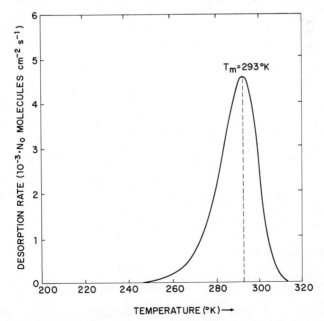

Fig. 40 First-order kinetics desorption rate as a function of uniformly rising temperature ($a =$ 0.1 deg s^{-1}) for $\tau_0 = 10^{-13}$ s and $E_d = 20$ kcal mol^{-1}.

maximum is usually taken as an indication that the surface trapping sites are associated with a spectrum of binding energies. This situation has been treated mathematically by Grant et al.[187,188] Somewhat broader desorption maxima are also obtained for second-order kinetics, i.e., if Eq. (9) is integrated for a linear time schedule. An example of second-order desorption for linearly increasing temperature is nitrogen on rhenium.[189]

Differentiation of the first-order desorption-rate equation (10) leads to a relationship between the desorption energy E_d and the temperature of maximum release rates T_m:

$$\frac{E_d}{RT_m{}^2} = \frac{1}{a\tau_0} \exp\left(\frac{-E_d}{RT_m}\right)$$

The function E_d vs. T_m has been evaluated numerically in Fig. 41. It is nearly linear and little affected by the preexponential factor. This correlation between desorption

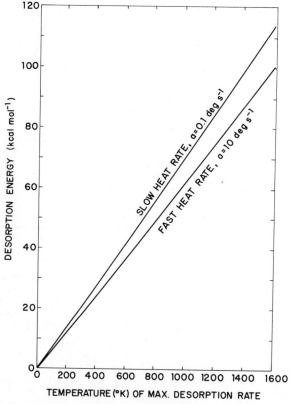

Fig. 41 Desorption energy vs. temperature of maximum desorption rate for two linear temperature schedules (first-order kinetics, $\tau_0 = 10^{-13}$ s).

energies and their characteristic release temperatures is often used to investigate the interaction of gases with metal surfaces. Redhead[190] observed the release of H_2, N_2, CO, and O_2 from gradually heated tungsten filaments and found that chemisorption of the first three gases occurred with two different energies. Other gas-metal systems have been investigated by Richardson and Strehlow[191] and by Pagano.[185]

(3) Gas Release by Diffusion Typically metals and glasses contain 1 to 100 volume % of gas in solid solution, and part of this amount may be released when the materials are exposed to very low pressures. It is assumed that the equilibrium concentrations of dissolved gases in solids obey Henry's law since they are generally smaller than 0.1 atomic %. In glasses and elastomers, Henry's law takes the form $C = C_0 p$, where C is the gas concentration and C_0 the solid solubility. The latter is dimensionless and represents the amount of gas in cm³ (STP) which dissolves under equilibrium conditions in 1 cm³ of solid at an external gas pressure of 1 atm. C_0 may also be expressed in volume percent. In metals, the solubility of the monatomic rare gases is zero, whereas diatomic gases dissociate upon entering. It follows from the law of mass action that the equilibrium concentration of diatomic gases in metals is given by $C = C_0 \sqrt{p}$.

To assess the gas-evolution rates due to diffusion from the interior of a solid slab, the hypothetical case illustrated in Fig. 42 is considered. The solid slab of thickness

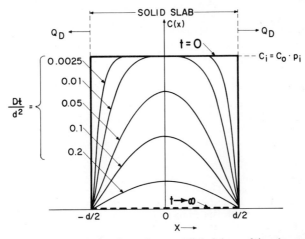

Fig. 42 Residual-gas concentration in a plane-parallel slab resulting from diffusion into adjacent vacuum on both sides. $t = 0$ and $t \to \infty$ are equilibrium states; the other distributions are for transient states.

d contains an initially uniform concentration of gas C_i which represents an equilibrium distribution established, for instance, during the manufacture of the material. If the partial pressure of the dissolved gas is lowered on both sides of the slab, the solid is no longer in equilibrium with its environment, and the gas atoms diffuse out on both sides. The quantity of gas which evolves if dN_s particles leave the solid is

$$Q_D = -\frac{RT}{N_A} \frac{dN_s}{dt} \quad \text{cm}^3 \text{ (STP) s}^{-1}$$

According to Fick's first diffusion law, the gas-evolution rate Q_D depends on the concentration gradient in the solid at the phase boundary,

$$Q_D = -AD \left(\frac{dC}{dx}\right)_{x=d/2} \tag{12}$$

where A is the surface area and D the diffusion coefficient. To obtain the gas-evolution rate as a function of time, the time dependence of dC/dx must be known. The solute distribution $C(x,t)$ during transient states may in principle be derived from Fick's second diffusion law:

$$\frac{\partial C}{\partial t} = D \frac{\partial^2 C}{\partial x^2} \tag{13}$$

Solutions of this differential equation for various boundary conditions are given by Jost[192] or may be taken from analogous heat-flow problems.[193] For the case of uniform initial concentration, Eq. (13) can be satisfied by the distribution function.[194]

$$C(x,t) = C_i \frac{4}{\pi} \sum_{n=0}^{\infty} \frac{(-1)^n}{2n+1} \cos \frac{\pi x (2n+1)}{d} \exp \left\{ -[\pi(2n+1)]^2 \frac{Dt}{d^2} \right\}$$

Solute distributions $C(x)$ at several times during the transient period are also shown in Fig. 42. To obtain outgassing rates according to Eq. (12), the function $C(x,t)$ must be differentiated at $x = d/2$, which yields

$$Q_D = \frac{4ADC_0 p_i}{d} \sum_{n=0}^{\infty} \exp \left\{ -[\pi(2n+1)]^2 \frac{Dt}{d^2} \right\}$$

Since diffusion coefficients in solids are very small, the condition $Dt/d^2 < 0.05$ is often satisfied, and then the gas-evolution rate may be approximated by[195]

$$Q_D \approx AC_0 p_i \sqrt{\frac{D}{\pi t}}$$

Although the $1/\sqrt{t}$ relationship has been derived for idealized boundary conditions, it has frequently been verified by experiments and is considered to be characteristic for outgassing processes which are diffusion-controlled. In practice, initially uniform gas concentrations may be encountered with new materials which have not been degassed as yet. When these materials are heated in vacuum, the dissolved gases stemming from the manufacturing process are released irreversibly. The diffusion of atmospheric gases into the solid at room temperature is too slow to reestablish a uniform equilibrium concentration. Therefore, diffusive gas release is most significant with new materials and tends to decrease with time as the gas distribution approaches the low-pressure equilibrium corresponding to $t \rightarrow \infty$ in Fig. 42.

The diffusion coefficients of atmospheric gases at room temperature are too small to attain uniform concentrations throughout the bulk during exposure to air, but gas penetration into a limited depth may occur. The resulting gas concentrations decrease toward the interior of the solid and represent reversible transient states. An exact solution of Eq. (13) for such complicated initial distributions has not been attempted as yet. Rogers[196] approximated transient-state distributions by a step function and derived gas-evolution rates which were again proportional to $1/\sqrt{t}$. Nonequilibrium distributions of limited depth are also obtained if gas ions are driven into solids. An example has been reported by Low and Argano,[197] who forced hydrogen into Vycor glass by activating the sorption process with an rf discharge. The subsequent thermal-release rates were also proportional to $1/\sqrt{t}$. The same result was obtained by Maddix and Allen[198] after activated sorption of rare gases on quartz surfaces.

Diffusive gas release at linearly increasing temperature has been investigated by Erents et al.,[199] who implanted rare-gas ions into metals and observed their subsequent thermal release. The resulting gas evolution vs. temperature curves had maxima which are indicative of the energy E_D governing the temperature dependence of the diffusion coefficient:

$$D(T) = D_0 \exp \left(\frac{-E_D}{RT} \right)$$

A mathematical analysis of diffusive gas release at increasing temperature has been performed by Farrell et al.,[200,201] who derived a solution of Eq. (13) for the simplified initial condition that all gas atoms are located at the same depth below the surface. This model predicts a rapid increase of gas-evolution rates near some characteristic temperature, followed by an approximately exponential decrease. If gas distribution over different depths is taken into account, the outgassing peaks broaden but not enough to agree with experimentally observed curves of argon in glass.[202] It has therefore been concluded that gas diffusion out of solids may involve atomistic processes with distributed energies E_D. Similar results have been reported for the thermal release of ion-implanted argon from nickel whereby the broad thermal-release spectrum required the assumption of six discrete activation energies.[203]

As Fig. 42 demonstrates, the magnitude of diffusive gas release from specific materials depends on the solid solubility C_0. For gases in elastomers, this quantity falls typically into the range of 3 to 10 volume %.[204] In glasses, the solubility of helium has been extensively studied, and it amounts to 0.1 to 1 volume %.[205] The solubilities of nearly all other gases in glasses are smaller than that of helium.[129] Thus, dissolved gases in glasses are of little practical importance. An exception is water, which is a minor constituent of some glasses.

A wealth of data exist about the solubilities of gases in metals. Detailed reviews have been given by Smithells,[206] Barrer,[207] and Norton.[208] The gas of greatest concern in regard to the vacuum behavior of metals is hydrogen, which is also the most widely investigated one. It forms true solid solutions with a large number of metals including Fe, steel, Ni, Co, Cu, Ag, Cr, Mo, and W. Between room temperature and 400°C, the solubility of H_2 in most of these metals is less than 1 volume %, and it increases with increasing temperature. Nickel has a somewhat larger solubility of nearly 10 volume %, whereas hydrogen is insoluble in solid aluminum. Well known for its unusually high solubility is palladium, which takes up as much as 900 times its own volume of hydrogen. Similarly large amounts of the order of 1,000 cm³ (STP) of H_2 per cm³ of solid are taken up by Ti, Zr, V, and Ta. The solubility in these metals is due to their tendency to form hydrides, and it decreases with increasing temperature.

The solubilities of other gases in metals are considerably less than 1 volume % at or near room temperature. Examples are O_2 in Cu and Co, N_2 in Al, Cu, and Fe. Nitrogen is known to be soluble only in metals with which it forms nitrides, and the solubility of CO is similarly restricted to carbonyl-forming metals. Oxygen in silver has a relatively high solubility of about 100 volume % at room temperature.

In assessing the potential outgassing effects of gases dissolved in metals, palladium and silver are obviously poor choices for the construction of vacuum systems. Al, Fe, steel, and Cu have the lowest solubilities for the most critical gas, hydrogen. The diffusion coefficient of hydrogen, which is also the most mobile gas species in metals, is of the order of 10^{-8} cm² s⁻¹ at room temperature.[129] Consequently, evolution of hydrogen is significant during the bakeout cycle but negligible after the system has been brought back to room temperature for film deposition. The complete release of gases initially present in metals is not possible by baking alone but requires vacuum melting. The amounts obtained are typically smaller than the volume of the metal sample[208] but often greater than 1 volume % since the solubilities at the temperatures of the manufacturing process are greater than those at room temperature.

c. Outgassing Rates of Vacuum Materials

Several experimental techniques are available to determine the outgassing rates of materials. To ensure consistency of data reported by different investigators, these methods are defined in a tentative standard of the American Vacuum Society.[209] It specifies that standard outgassing rates should be given in Torr l s⁻¹ per square centimeter of geometrical surface area and for room temperature. [1 cm³ (STP) = 0.760 Torr l.] A sample of the material to be tested is placed into a vacuum system, and outgassing is initiated by evacuation. The gas-evolution rates may be measured by the so-called throughput method. This technique utilizes low pumping speeds which are obtained by constricting the gas flow to the pump with an orifice or narrow pipe of known conductance. The dynamic balance between the outgassing rate Q_0 and the conductance-limited throughput Q_c determines the pressure change in the system according to

$$Q_0 - Q_c = V \frac{dp}{dt} \qquad \text{Torr l s}^{-1} \qquad (14)$$

The gas-evolution rate as a function of time is deduced from the observed pressures by differentiation.

An alternate technique is the rate-of-rise method, where the evacuated chamber containing the sample is isolated from the pump, and the pressure rise due to outgassing is observed. This procedure is repeated at successively lower pressures since only the rate of rise immediately after closing the pumping port is indicative of the outgassing rate at that pressure. In both techniques the contributions from the system walls to the measured outgassing rates must be subtracted to obtain the gas-evolution rate of the sample in the chamber.

Comprehensive reviews of outgassing rate vs. time curves have been given by Dayton[204] and Jaeckel,[172] and examples are shown in Fig. 43. Since the measurements do not distinguish between gas release by diffusion as opposed to desorption, it is often attempted to infer the responsible process from the shape of the curve.[172] For comparison, the exponential dependence resulting from first-order desorption and the $1/\sqrt{t}$ relationship for diffusion-controlled processes are also shown in Fig. 43. In

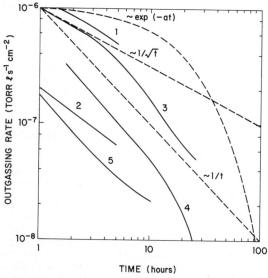

TIME (hours)

Fig. 43 Measured outgassing rates: (1) Plexiglass (methyl methacrylate).[172] (2) Teflon.[172] (3) Viton A, unbaked.[210] (4) Mild steel, sandblasted.[204] (5) Stainless steel.[204] The broken lines are calculated curves for simple rate laws.

most instances neither relationship holds. Instead it has often been found that outgassing rates of metal surfaces are approximately proportional to $1/t$ during the first 10 to 20 h.[204] Since metal surfaces exposed to moist air are known to adsorb several monolayers of water vapor, Dayton[204] tried to explain the empirical relationship by assuming desorption of H_2O with distributed energies E_d. The summation of many first-order desorption-rate curves of the type shown in Fig. 39 yields indeed a function proportional to $1/t$.

The validity of outgassing data from the throughput method is not undisputed. In a series of investigations on outgassing rates in large ultrahigh-vacuum systems, Farkass and coworkers[211,212] found values which were two to four decades lower than those obtained from the same materials in conventional systems of smaller size. To explain this discrepancy, Farkass[213] pointed out that no allowance is made in Eq. (14) for readsorption of desorbed particles, although sticking coefficients of gases at very low pressures are known to be greater than zero.[179]

Experimental evidence furnished by Hobson and Earnshaw[214] indicates that the differential equation (14) describes satisfactorily the dynamic pressure response of some gases such as He, CH_4, N_2, and Ar. However, it does not explain the pressure changes observed when either H_2 or CO are trapped by the sudden addition of a cold trap. The authors assume the presence of a surface phase which may consist of condensed, adsorbed, or dissolved gas, and which emits as well as adsorbs free molecules. This model leads to predictions which agree semiquantitatively with Hobson's experiments and which demonstrate that varying degrees of error may have been introduced into previously reported data because of the simple transient equation (14).

(1) Metals Outgassing rates of metals for the purpose of estimating the residual-gas contributions from vacuum chamber walls are shown in Table 7. Mild steel tends to form a corrosion layer with a high sorption capacity and therefore exhibits relatively large outgassing rates. The data also demonstrate the effects of prior treatment on

TABLE 7   Outgassing Rates of Metals at Room Temperature

Material	Treatment	Pumping time, h	Outgassing rate after pumping, $Torr\ l\ s^{-1}\ cm^{-2}$	Reference
Mild steel......	Unpolished, slightly rusty	8	2×10^{-8}	215
	Polished	8	2×10^{-9}	215
Stainless steel..	Machined casting	5	2×10^{-9}	215
	Alcohol-vapor degreased	5	9×10^{-10}	215
	Alcohol-vapor degreased	24	2×10^{-10}	215
	Baked at 400°C	16	3×10^{-14}	215
	Chemically cleaned	12	$<10^{-12}$	23
	Baked at 300°C	24	10^{-12}	216
	Baked at 1000°C	3	10^{-17} (extrapolated)	216
Aluminum.....	Polished	5	7×10^{-10}	215
	Alcohol degreased	24	1.5×10^{-10}	215
	Baked at 400°C	16	2×10^{-14}	215

the resulting gas-release rates. Polishing of vacuum materials reduces the effective surface area and hence the adsorption capacity. Outgassing rates of steel, copper, and aluminum surfaces after various cleaning treatments have been investigated by Flecken and Noeller[217] and by Varadi[218] for nickel. Without prior heat treatment, these metals desorb mostly water vapor, typically at rates between 10^{-8} and 10^{-10} Torr $l\ s^{-1}\ cm^{-2}$.[219] The outgassing rates can be reduced by solvent or chemical cleaning procedures but usually by less than an order of magnitude. An exception is the chemical treatment reported by Milleron,[23] who immersed stainless-steel parts in a commercial cleaning solution* and obtained rates below 10^{-12} Torr $l\ s^{-1}\ cm^{-2}$. The procedure involves chemical etching and leaves a bright, smooth surface. Contamination of cleaned parts by dust, dirt, exposure to humid air, or handling without nylon gloves causes increased outgassing rates.

The most effective treatment to reduce outgassing of metals is heating under vacuum. Desorption of water vapor is essentially completed after baking at 300 to 400°C. Thereafter, outgassing of stainless steel is predominantly due to the diffusive release of hydrogen[216] at rates which are very small at room temperature. During the heat treatment, CO and CO_2 are often emitted in addition to hydrogen.[217,218] The carbon oxide release rates are diffusion-limited,[220] and the process has been explained as diffusion of carbon in iron (or nickel) followed by surface reaction with gaseous oxygen.[221] The degassing of 304 stainless steel has also been investigated by Strausser,[222] and his results confirm that hydrogen is the predominant gas released. After baking, H_2O, CO_2, and O_2 are minor constituents. Typical outgassing rates of such metals as steel, copper, or aluminum after several hours of baking in vacuum are 10^{-12} to 10^{-14} Torr $l\ s^{-1}\ cm^{-2}$.[219]

* DS-9; Diversey Co., Chicago, Ill.

(2) Glasses The principal gases which evolve from glasses upon heating are water vapor and carbon dioxide. The sorption of water vapor on clean glass surfaces exposed to humid air does not stop with the formation of a monolayer but continues for very long times. Accordingly, the gas quantities released into vacuum are significantly greater for aged than for freshly drawn glass.[223] The outgassing rates at room temperature are slow, and most studies are therefore carried out at steadily increasing temperature. Early examples are the outgassing curves of Sherwood,[224] who found total amounts equivalent to 10 to 50 monolayers of H_2O and up to 5 monolayers of CO_2.

More recent outgassing studies are mostly concerned with borosilicate glasses. A typical desorption curve for Pyrex glass is shown in Fig. 44. The peak structure

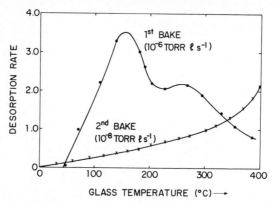

Fig. 44 Desorption of water vapor from a Pyrex-glass surface of 180 cm² area at increasing temperature. (*After Erents and Carter.*[225])

obtained during the first baking after exposure to air indicates desorption of at least two monolayers with different energies E_d.[225] After heating to 400°C, the desorption process has gone to completion. This is evident from the much lower outgassing rates observed during repeated baking. Characteristic for glasses is the gradual onset of another outgassing mechanism which increases sharply at about 400°C as shown by the second bake curve in Fig. 44. Todd[223] measured the evolution of water vapor at 480°C on soda-lime glass and found a $\sqrt{t}$ law indicative of a diffusion-controlled mechanism. The same process is responsible for the release of water vapor from Pyrex glass at temperatures above 350°C.[226] Since this form of outgassing arises from water incorporated into the structure of the glass, it is considered as gradual decomposition. The diffusion release of structural water continues for very long periods of time, and it is therefore recommended not to exceed 350°C when baking Pyrex-glass systems.[82]

The amounts of CO_2 released from Pyrex glass are smaller than the quantities of water vapor. However, if the CO_2 evolved during the bakeout cycle is condensed in a liquid-nitrogen-cooled trap, the latter may become a persistent source of residual gas. This is due to the fact that CO_2 has a vapor pressure of about 10^{-8} Torr at 77°K, which is sufficiently high to prevent the attainment of a stable vacuum of 10^{-9} Torr.[227,228] The CO_2 can be desorbed by briefly returning the trap to room temperature after the bakeout.

Another source of glass-surface contamination is oil-diffusion-pump vapors which are firmly chemisorbed by dehydrated (baked) glass surfaces.[229] The resulting surface film may weaken the adhesion of subsequently deposited films on glass substrates or may become a source of gas evolution from glass walls during the next bakeout. Cleaning of glass surfaces with hydrofluoric acid solutions has also been reported to be detrimental, since it resulted in very persistent outgassing of fluorine.[230]

(3) Elastomers Outgassing rates of elastomers suitable for O rings and of other organic materials have been measured by a number of investigators. A selection of data for those materials which are of prime interest in vacuum technology is given in Table 8. Additional outgassing curves for elastomers and for epoxy resins may be found in Refs. 234 and 235.

TABLE 8 Outgassing Rates of Elastomers and Vacuum Grease at Room Temperature

Material	Treatment	Pumping time, h	Outgassing rate after pumping 10^{-8} Torr l s^{-1} cm^{-2}	Reference
O rings:				
Neoprene.............	Not baked	24	5.4	231
Silicone rubber........	Not baked	24	0.44	231
Teflon................	Not baked	24	2.5	231
Butyl rubber..........	Not baked	24	1.1	231
Natural rubber........	Not baked	24	2.0	231
Viton A..............	Not baked	24	2.0	231
P.T.F.E (Teflon)........	Not baked	48	0.035	232
Silastomer 80............	Not baked	51	2.2	232
	Baked 200°C, 24 h	101	0.11	232
	After exposure to air, 24 h	51	1.0	232
Viton A................	Not baked	51	10.0	232
	Baked 200°C, 24 h	122	0.2	232
	After exposure to air, 24 h	48	3	232
Viton A (Edwards high-vacuum O ring)	As manufactured	15	20	215
	Second pumping	15	10	215
	Baked 150°C, 4 h	15	0.01	215
	Baked 200°C, 16 h	15	0.01–0.02	215
	As above after 64 h exposure to air	15	10	215
Viton A................	Baked 200°C, 12 h	>12	0.02	233
Vespel................	Baked 300°C, 12 h	>12	0.004	233
Buna N (G. Angus, P 60 O ring)	As manufactured	15	100	215
	Second pumping	15	30	215
	Baked 100°C, 4 h	15	3	215
	As above after 64 h exposure to air	15	5	215
Apiezon M grease........	....	8	3	215
Silicone H.V. grease.......	....	8	0.8	215

Generally, elastomers which have not been baked have outgassing rates as high as 10^{-8} to 10^{-6} Torr l s^{-1} cm^{-2}. The rate of polytetrafluorethylene (P.T.F.E.) is significantly lower, but unfortunately this polymer is not cross-linked and flows on compression [see Sec. 4b(2) and Table 17]. An important consideration in choosing O-ring materials from the available elastomers is their thermal stability. Ideally, an elastomer should be capable of withstanding without decomposition temperatures up to 400°C as are required to degas metals and glasses. Such a material does not exist. Viton A,* a copolymer of vinylidene fluoride and hexachloropropylene, has been the most widely accepted compromise. It may be heated up to 200°C, which is more than most other synthetic rubbers can be subjected to.

As Table 8 shows, mild baking at 100 to 200°C reduces the outgassing rates of elastomers by one to two orders of magnitude. The major gas constituent released is water vapor in amounts which are equivalent to 100 or more monolayers. The water is sorbed in the interior of the elastomers and released by diffusion. In the case of Viton, this process is largely reversed during exposure to air of normal humidity.[232]

* Trademark of E. I. du Pont de Nemours & Company.

Silastomer 80 is as temperature-resistant as Viton and also has a low outgassing rate after baking. However, it has an unusually high permeability for gases and is therefore unsuitable for gasketing. Teflon differs from the other elastomers insofar as its major gas contribution is CO rather than H_2O.[232]

Because of its superior thermal stability, Viton A and its outgassing characteristics have been examined in great detail. De Csernatony[236,237] reported decomposition pressures of 10^{-10} Torr at room temperature and 10^{-8} Torr at 180°C. Some hydrocarbon fragments are released during the first bakeout, but these do not appear in later heat cycles. Therefore, they do not indicate decomposition but the initial presence of incompletely polymerized material.[232] Significant decomposition was observed by Addiss et al.[238] during heating at 300°C. Most prominent among the evolving gases was HF. Although the outgassing rates may vary because of slight differences in the manufacturing process, the lowest values of about 10^{-10} Torr l s^{-1} cm^{-2} have been repeatedly confirmed. To achieve these rates, degassing at optimum temperatures in the range 150 to 200°C[210] is necessary since Viton A has a significant solubility for the atmospheric gases H_2O and CO_2.[237]

A more recent entry into the elastomer field are the polyimids, which are available under the trade name Vespel.* The main advantage of Vespel is its reported ability to withstand baking at 300°C.[233] Vespel absorbs water vapor more eagerly than Viton and therefore requires baking. The released gases consist mainly of H_2O accompanied by lesser amounts of CO, CO_2, and H_2. After returning to room temperature, only CO and CO_2 are evolved, with rates as low as 4×10^{-11} Torr l s^{-1} cm^{-2}

d. The Permeation Process

The atomistic events leading to penetration of solids by gases have been formulated by Norton.[239] The process is initiated by the adsorption of gas particles impinging on the outer surface of an evacuated enclosure. Provided the solid has a finite solubility, adsorbed atoms or molecules enter the interior of the material. The resulting concentration gradient drives the dissolving gas particles deeper into the solid until they reach the inner surface of the vacuum enclosure. The emerging particles go through an adsorbed state from which they are desorbed into the interior because their concentration exceeds the quantity in equilibrium with the low internal gas pressure.

It is generally assumed that adsorption and desorption occur rapidly compared with gas transport by solid-state diffusion. Therefore, permeation rates are determined by the diffusion rates of the gas through the solid. Their derivation can be reduced to a one-dimensional diffusion problem by considering permeation through a plane parallel wall of thickness d as illustrated

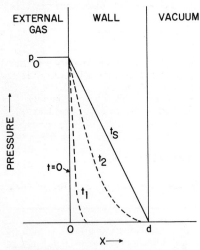

Fig. 45 Gas concentration (equivalent pressure) in a solid wall, before the onset of permeation ($t = 0$), in two transient states (t_1 and t_2) and in the steady state (t_s).

in Fig. 45. According to Henry's law, the concentration of dissolved gas at any depth in the wall may be represented by the equivalent pressure which the solid of that concentration would show if it were in equilibrium with the free gas. With this understanding, Fick's second diffusion law [Eq. (13)] takes the form

$$\frac{\partial p}{\partial t} = D \frac{\partial^2 p}{\partial x^2}$$ (15

* Trademark of E. I. du Pont de Nemours & Company.

Concentration-profile functions $p(x,t)$ for the interior of the wall must satisfy the following boundary conditions: At $t = 0$, gas of the pressure p_0 is admitted to the external wall while the interior of the solid is free of gas $[p(x,0) = 0$ for $0 < x < d]$. The free gas pressure on the vacuum side is assumed to be negligibly small compared with p_0. As diffusion proceeds at $t > 0$, the initial pressures are maintained constant on both sides of the wall.

The solution of Eq. (15) for these boundary conditions is a sum of complementary error functions.[193] Concentration profiles for two transient states are shown in Fig. 45 and demonstrate the gradual buildup of gas in the solid. After sufficiently long times t_s, the wall is saturated with gas and the concentration reaches its steady-state distribution. The gas concentration at the inner wall surface is of the greatest interest because the gradient dC/dx at $x = d$ determines the rate of gas release into the vacuum:

$$Q_P = AD \left(\frac{dC}{dx} \right)_{x=d} \quad \text{cm}^3 \text{ (STP) s}^{-1} \quad (16a)$$

Q_P is the permeation rate, A the surface area, cm^2, D the diffusion coefficient, cm^2 s^{-1}, and $dC = C_0 \, dp$, atm, according to Henry's law. Equation (16a) may therefore be rewritten as

$$Q_P = ADC_0 \left(\frac{dp}{dx} \right)_{x=d} \quad (16b)$$

A general solution of Eq. (16b) in the form of a Fourier series has been given by Rogers, Buritz, and Alpert.[240] For moderately short times covering the greater part of the transient state, these authors introduced the so-called "early approximation":

$$Q_P \approx 2AC_0 p_0 \sqrt{\frac{D}{\pi t}} \exp \left(\frac{-d^2}{4Dt} \right) \quad \text{cm}^3 \text{ (STP) s}^{-1} \quad (17)$$

The permeation rate for the steady state follows from Eq. (16b) directly by a simple argument: In the steady state, the concentration profile in the wall does not change with time. Therefore, the amounts of gas entering and leaving a thin layer dx of the wall must be the same for all depths x: $Q_P(x) = $ const. This is possible only if the gradient dp/dx is constant at all depths. Therefore, the steady-state distribution is linear as shown in Fig. 45 with $dp/dx = p_0/d$. Substitution into Eq. (16b) yields the steady-state permeation equation:

$$Q_P = ADC_0 \frac{p_0}{d} \quad \text{cm}^3 \text{ (STP) s}^{-1} \quad (18)$$

Permeation rates for the transient and steady states are shown in Fig. 46. Contrary to outgassing processes, the gas contribution from permeation increases during the transient state and reaches its greatest value after the pump-down cycle.

e. Permeability of Vacuum Materials

Dynamic measurements yielding the permeation rate Q_P directly have been described by Altemose.[205] The method of rise was used by Rogers, Buritz, and Alpert.[240] The pressure increase dp due to a small quantity of gas Q_P entering a vacuum enclosure of volume V is given by

$$Q_P = V \frac{dp}{dt}$$

This relation can be substituted into Eq. (17) or (18) for the purpose of analyzing

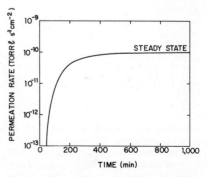

Fig. 46 Permeation rate during transient state (early approximation) and steady state. Assumed values: $D = 10^{-7}$ cm^2 s^{-1}, $d = 1$ mm, $C_0 = 0.01$, $p_0 = 10$ Torr.

the measured pressure vs. time functions. Since the times required to obtain steady-state rates at room temperature may be as long as several weeks, observation of transient-state rates is more practical. The "early approximation" [Eq. (17)] has been verified experimentally by both the dynamic method and the method of rise. Transient-state rates also allow a separate determination of the diffusion coefficient D and the solid solubility C_0,[240] whereas steady-state rates yield only the product $P = DC_0$. The quantity P is called the permeability and is used as a measure of gas penetration through solids. It is defined according to Eq. (18) as

$$P = Q_P \frac{d}{A p_0} \quad \text{for glasses and elastomers} \qquad (19a)$$

and should be given in cm^3 (STP) s^{-1} cm^{-2} (cross section) for a wall of 1 mm thickness and for a pressure drop of 10 Torr (American Vacuum Society Standard). However, the use of these dimensions is not uniform, and data representing P in Torr l for a wall thickness of 1 cm or for a pressure differential of 1 Torr are often encountered. The permeability of metals differs from Eq. (19a) because of molecular dissociation,

$$P = Q_P \frac{d}{A \sqrt{p_0}} \quad \text{for metals} \qquad (19b)$$

and here the most commonly used reference pressure is 1 atm.

Since the permeabilities at 25°C are small, measurements are often made at elevated temperatures. The experimental permeability values obey the relationship

$$P(T) = P_0 \exp\left(\frac{-E_P}{RT}\right)$$

indicating that permeation is a thermally activated process. The temperature dependence is analogous to that of the diffusion coefficient D, and since $P = DC_0$, one may anticipate $E_P = E_D$ if the permeation is truly diffusion-controlled and if C_0 does not vary with temperature. This expectation has been confirmed in the case of helium permeation through several glasses.[240] However, there are known instances of strongly temperature-dependent solubilities of gases in metals and also in some glasses.[129] In these cases the two energies E_P and E_D are different. Numerical values of E_P for He, Ne, and H_2 through glasses are 5 to 12 kcal mol^{-1}.[205] Similar values hold for elastomers. The permeation of larger gas molecules such as N_2 and O_2 through glass is associated with energies of 20 to 30 kcal mol^{-1}, while E_P values in metals are 20 to 60 kcal mol^{-1}.[172,241] A review of gas-solid systems whose permeabilities have been investigated appears in Redhead's article.[129] Two general rules about the permeation of gases through solids have been formulated by Norton,[242] who stated that all polymers are penetrated by all gases and that no metals are penetrated by any of the rare gases. So far, these generalizations have not been contradicted by experimental results.

(1) **Metals** Since the permeability is proportional to the solubility of gases in solids, hydrogen is the gas with the highest permeation rate in metals. As illustrated in Fig. 47, hydrogen permeation in iron and Ni is relatively high. Most other metals including steel have lower permeabilities with the exception of palladium. The two metals Pd and Pt have similarly large solubilities for hydrogen but vastly different permeabilities. This is due to the diffusion energy E_D, which is nearly twice as large in Pt as in Pd.[243] The systems H_2-W and H_2-Mo have been investigated at temperatures above 800°C.[241] The permeabilities of other gases are considerably smaller than those of hydrogen in accordance with their lower solubilities and diffusion coefficients. Measurable permeabilities have been found for N_2 through Mo, Fe, Cr; for CO through Fe, Ni; and for O_2 through Ni and Cu.[82] Collins and Turnbull[244] observed that type 304 stainless steel was not permeated by N_2, CO, and CO_2 at temperatures up to 800°C. An unusually high permeability is that of oxygen in silver.[245]

(2) **Glasses** The gas with the highest permeability through glasses is helium. Permeation rates for twenty glasses at different temperatures have been measured

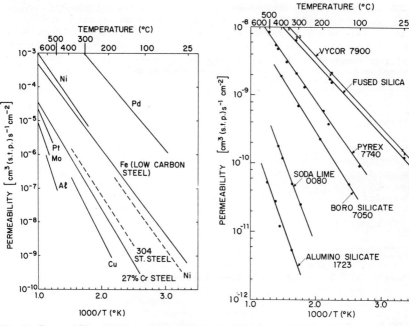

Fig. 47 Permeability of hydrogen in metals. The data are permeation rates through 1-mm-thick walls at a pressure drop of 1 atm. (*Drawn lines: Norton;*[239] *broken lines: Collins and Turnbull.*[244])

Fig. 48 Permeability of helium through glasses. The data represent permeation rates through 1-mm-thick walls at a pressure drop of 10 Torr. (*After Altemose.*[205])

by Altemose,[205] and some of his data are reproduced in Fig. 48. His results demonstrate that the permeability of glasses is proportional to the concentration of network formers such as SiO_2, B_2O_3, and P_2O_5 in the glass while the addition of basic metal oxides reduces the permeability. The dependence of permeation rates on glass composition was first recognized by Norton[242] and is one reason why Pyrex glass is preferred over fused quartz or Vycor as a material for vacuum enclosures. Fused silica and Vycor are also the most pervious glasses for gases other than helium. However, as shown in Table 9, the permeability for gas atoms or molecules of larger diameters is much smaller than for helium.[205]

TABLE 9 Permeability of Gases through Fused Silica and Vycor (Wall Thickness: 1 mm; Pressure Drop: 10 Torr)

Gas	Permeability, cm³ (STP) s⁻¹ cm⁻², through	
	Fused silica at 25°C[239]	Vycor at 400°C[246]
He	5×10^{-11}	6×10^{-10}
Ne	2×10^{-15}	7×10^{-12}
H₂	3×10^{-14}	2×10^{-11}
N₂	2×10^{-29}	8×10^{-14}
O₂	1×10^{-28} } estimated	
Ar	2×10^{-29} }	

(3) Elastomers   Representative data for the permeabilities of elastomers are shown in Table 10.   Compared with glasses and metals, elastomers are not as selective and allow all gases to pass at significant rates.   Most striking is the high permeability of silicone rubber which disqualifies this material from high-vacuum applications.

TABLE 10   Permeability of Gases through Elastomers.   The Data Represent Permeation Rates in 10^{-9} cm³ (STP) s⁻¹ cm⁻² for 1-mm-thick Walls and for a Pressure Drop of 10 Torr*

Elastomer	At 25°C						At 80°C
	H_2	He	N_2	O_2	Ar	CO_2	Air
Silicone rubber......	700	400	300	600	700	600–3000	600
Buna N...........	10	10	0.1–1	1–4	2	20	10
Neoprene.........	10	10	1	3–4	3	....	>10
Natural rubber.....	50	30	7–9	20	20	100	60
Viton A...........	3	10–20	0.1–0.4	0.1–2	0.1	5–8	10
Butyl rubber.......	7	8	0.3	1–5	.....	5–10	4

* From Holland.[215]

Viton A and butyl rubber have the lowest permeabilities.   Although they are penetrated by water vapor, the permeation rates are smaller than the outgassing rates. Generally, permeation data may vary from case to case because of slight differences in the chemical composition of different products.[237]

(4) Permeation of Atmospheric Gases   The severity of permeation effects in vacuum systems also depends on the abundance of gases in the atmosphere.   The partial pressures of the constituents of air are listed in Table 11.   These pressures must be

TABLE 11   Partial Pressures of Gases in Air[239]

Gas	Pressure, Torr	Gas	Pressure, Torr
Nitrogen............	595	Helium...........	4.0×10^{-3}
Oxygen.............	159	Krypton..........	8.4×10^{-4}
Argon..............	7.05	Hydrogen.........	3.8×10^{-4}
Carbon dioxide......	2.5×10^{-1}	Water............	11.9*
Neon...............	1.4×10^{-2}	Methane..........	1.5×10^{-3}

* At 25°C and 50% relative humidity.

divided by the reference pressure used in the numerical permeability value P for the purpose of calculating actual permeation rates:

$$\frac{Q_P}{A} = P \frac{p_0}{d}$$

Several significant examples are listed in Table 12.   These estimates demonstrate that elastomer O rings including those made of Viton A are a limiting factor in attaining ultrahigh vacuum.   The permeability of the main atmospheric gases is greater than the outgassing rate of Viton after baking (see Table 8).   Accordingly, the use of elastomers in ultrahigh-vacuum systems is generally avoided.   There is, however, an exception to this rule.   Farkass and Barry[231] observed that cooling of butyl rubber, Neoprene, Buna, and Viton A O rings to $-25°C$ lowered their gas contributions significantly.   The effect of decreasing the temperature on the structure of elastomers is to increase the fraction of crystalline material at the expense of amorphous regions.[236] Since gases diffuse faster through amorphous than through crystalline solids, permea-

TABLE 12   Permeation Rates at 25°C for Those Atmospheric Gases Which Penetrate Vacuum Materials Most Severely

Wall material	Permeating gas	Permeation rate, cm^3 (STP) s^{-1} cm^{-2}
2 mm of 304 stainless steel...	H_2	3×10^{-15}
1 mm of Pyrex 7740........	He	4×10^{-15}
1 mm of fused silica........	He	2×10^{-14}
	Ne	3×10^{-18}
	H_2	1×10^{-18}
5 mm of Viton A...........	N_2	$1-5 \times 10^{-9}$
	O_2	$0.3-6 \times 10^{-9}$
	$N_2 + O_2$	$10^{-9}-10^{-8}$

tion as well as release of internally sorbed gases are reduced by cooling.   The effect is further enhanced by the exponential temperature dependence of P and D.   A practical limit to cooling is the fact that elastomers loose their elasticity and become brittle at very low temperatures.   A suitable compromise is Freon cooling, and at that temperature the permeability of butyl rubber is significantly lower than that of Viton A. Using Freon-cooled butyl rubber gaskets, Farkass[211] was able to achieve ultimate pressures of 10^{-11} Torr.

The permeation rates of atmospheric hydrogen through steel and of helium through glasses as shown in Table 12 are smaller than the outgassing rates of the same materials. Only at extremely low pressures in thoroughly degassed systems may permeation of atmospheric gases become the limiting factor.   An example has been reported by Alpert and Buritz,[247] who found the ultimate pressure in a small sealed-off Pyrex-glass system to be determined by the influx of helium through the walls.

4. VACUUM-ASSEMBLY TECHNIQUES

The assembly of vacuum systems involves parts constructed from materials of different physical properties and joined together by various sealing techniques.   The primary objective in making vacuum connections is the prevention of atmospheric-gas penetration through the joint.   Vacuum seals are classified according to the materials joined (i.e., glass-to-metal), the fabrication techniques used, their permanence, and the functions performed.   Functional seals are subassemblies designed to transmit electric current, radiation, or mechanical motion from the outside to the interior of the vacuum chamber.   They are commonly referred to as feedthroughs.   Consideration of such temperature-dependent material properties as vapor pressure and permeability becomes increasingly important as the seal requirements go up from high- to ultrahigh-vacuum applications.

The selection of a seal is determined mainly by the functional requirement and the intended vacuum range.   Within each category of seals, the choice may be narrowed down by factors such as cost and availability.   It is important to realize that inadequate sealing techniques are a major cause of problems in vacuum equipment.   The greater expense for a strong, reliable seal is justified against the cost of leak detection, repair work, loss of experiments or process product.   The following section should serve as an introduction to commonly encountered vacuum sealing and assembly techniques.   An extensive treatment of the subject may be found in the book of Roth.[248]

a. Permanent Vacuum Joints

Permanent joints are primarily of interest in the design and construction of vacuum systems.   However, since deposition equipment is often built for particular applications or investigations, the system operator, too, should have a working knowledge of the principal joining techniques and the properties of the resulting joints.

Joining techniques are material-dependent. Permanent metal-to-metal seals may be either welded or brazed. Soldered joints are semipermanent since the members can be separated and remade without destroying the original sealing surfaces. This advantage is usually outweighed by the poor quality of soldered vacuum seals. Fused-glass seals are losing importance as all-metal vacuum systems are replacing the formerly popular glass systems. For special vacuum components, however, glass-to-metal or ceramic-to-metal joints are indispensable. The third group of vacuum materials, the elastomers, play a role only in demountable seals.

(1) **Welding Processes** In the welding process, the parts to be joined are placed in intimate contact and locally heated until the adjoining edges are molten. Because of interdiffusion and mixing, the original surface boundaries are destroyed. Subsequent solidification produces a fused or welded joint. The members to be joined are designed and machined to fit closely so that joining by fusion is possible. Vacuum chambers or other large components constructed from heavy-gauge rolled plate are not formed to such close tolerances and require the addition of a filler metal to obtain a tight seal. Some welding processes are also contingent upon the application of pressure to promote fusion, but these are less common in the construction of vacuum joints. Welding processes are further differentiated according to the methods by which heat is supplied to the joint area. Gas (flame), electric-arc, and electron-beam welding are all utilized. These processes are treated in great detail in text and handbooks.[249-251]

Gas welding utilizes the combustion of a fuel gas mixed with oxygen in a torch as the heat source. The highest flame temperatures—up to 3100°C—are developed by the oxyacetylene torch. Adjustment of the oxygen-to-fuel gas ratio permits the ambient in which the work is heated to be varied from oxidizing to neutral or reducing. However, the shielding effect of a reducing flame is not sufficient to protect materials like stainless steel, copper, or aluminum completely against oxidation. The inclusion of oxides in the welded seam leads to porous joints. While this problem may be alleviated with fluxes, these, too, can become incorporated in the weld. So-called "flux joints" appear vacuum-tight after fabrication but often develop leaks during service. Organic or mineral fluxes having high vapor pressures pose the additional risk of system contamination. Another shortcoming of gas welding arises from the fact that flames are not very concentrated sources of energy. Hence, gas-welding speeds are slower, heat-affected zones are larger, and the distortion of the work is greater than in arc or electron-beam welding. For these reasons, gas-welding processes are not recommended for vacuum joining.

Arc welding processes generate heat by an electric arc maintained either between an electrode and the workpiece or between two electrodes. Various techniques exist employing dc or ac arcs, consumable or nonconsumable electrodes. The arc temperatures range from 4000 to 7000°C depending on the conditions employed. At these temperatures, the consumable electrodes melt and are incorporated into the joint.

In those processes where the workpiece serves as one electrode, the polarity of the latter plays an important role. This is because the mobility of free electrons in the arc plasma is higher than that of the positive ions. Thus, if the electrode is negative (direct-current straight polarity, or DCSP welding), about twice as much energy is transmitted to the workpiece as in the reverse case (direct-current reverse polarity, or DCRP welding). Consequently, the fusion zones in DCSP welding are deeper and narrower than in the DCRP mode. The latter, however, has the advantage of superior cleaning action since the ions bombarding the workpiece remove surface films and oxides. As the electrode is subject to intense electron bombardment, overheating must be avoided by using lower power levels, larger electrodes, or water cooling.

A common problem in dc welding is the deflection of the arc away from the intended direction ("arc blow"). This is due to the arc current's generating a magnetic field which is distorted in the discharge region because the magnetic flux in the joint members is unbalanced. The occurrence of arc blow is furthered by certain weld geometries such as corners, and it can be reduced by shorter arc distances, lower current levels, or better electrical grounding of the workpiece. It is eliminated if the arc is sustained with an ac power supply. The penetration depths in ac arc welding are in between those of the two dc techniques. The cleaning action during the negative half cycle of the workpiece is sufficient for the welding of aluminum and magnesium alloys.

The nonconsumable electrodes for arc-welding processes consist of materials having high melting points and high electron emission. Examples are carbon, tungsten, and thoriated tungsten. A further distinction into shielded and unshielded processes is based on the presence or absence of a protective gas. Table 13 lists the characteristic features of several common arc-welding processes.

TABLE 13 Common Arc-welding Processes

Welding process	Electrodes		Type	Protective gas
	No.	Material		
Carbon arc	1	Carbon	Nonconsumable	Unshielded
Twin carbon arc	2	Carbon	Nonconsumable	Unshielded
Atomic-hydrogen arc	2	Tungsten	Nonconsumable	H_2
Tungsten inert gas (Tig)	1	Tungsten	Nonconsumable	Ar, He
Metal inert gas (Mig)	1	Filler metal	Consumable	Mixtures
Plasma arc	1	Tungsten	Nonconsumable	Inert

The *carbon-arc processes* are unshielded and hence applicable only to metals which do not readily oxidize. The *atomic-hydrogen* arc offers some protection against oxidation. It is a very concentrated heat source since the hydrogen dissociates in the plasma and imparts the energy of recombination to the workpiece. However, many metals have high solubilities for hydrogen near their melting points. Upon cooling, the fusion zone becomes supersaturated and often porous or brittle. Iron, nickel, and copper are likely to show this behavior. The carbon- and atomic-hydrogen-arc techniques offered early alternatives to the oxyacetylene torch. Because of their shortcomings, they are now largely replaced by the other processes in Table 13, which are all gas-shielded.

The *shielded-arc-welding* processes protect the workpiece against oxidation by a flow of chemically inactive gases which are insoluble in the molten metal. Argon is the most economic choice, but helium is preferred for the welding of thick sections or for materials of high thermal conductivity such as copper. The reason is the higher ionization potential of helium, which yields higher arc voltages and greater penetrating power. Gas purities in excess of 99.75% are sufficient for copper and stainless steel, whereas aluminum, titanium, and zirconium require 99.95% or even higher purity. The Mig processes utilize mixtures of inert gases with oxygen and carbon monoxide in various proportions. In some cases, the protection by the gas flow around the workpiece is not sufficient, and the welding operation must then be performed in a glove-box-type chamber. This additional precaution is necessary for the most reactive metals like Be, Ti, Zr, and Ta. It is also advisable for light-gauge materials and metals of high thermal conductivity since the quality of a weld is very sensitive to accidental oxidation.

The *Tig process* (also argon-arc or heli-arc) is most commonly used for the joining of vacuum components. It permits welding of many industrially important metals without fluxes and yields smooth, tight joints. A Tig welding torch is shown schematically in Fig. 49. The inert gas flows through the collet body, typically at rates of 10 to 15 ft³ h⁻¹, and is confined by a ceramic cup for directionality and to shield the electrode. A separately held filler rod may be used to supply additional metal. The arc is started by short circuiting the electrode to the workpiece and drawing it away. As the operation is manual, experience and skill are required to control the weld zone by adjusting distance, arc current, voltage, and welding speed. In the DCSP mode,

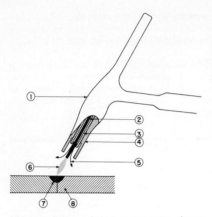

Fig. 49 Schematic of a tungsten inert gas (Tig) welding torch. (1) Torch body; (2) tungsten electrode; (3) collet body; (4) ceramic cup; (5) inert-gas flow; (6) arc discharge; (7) weld pool; (8) workpiece.

an open-circuit voltage of about 80 V is sufficient, whereas the DCRP mode requires higher voltages (150 V). In ac welding, the arc must be restarted every half cycle. A common practice is to superimpose a high-frequency voltage on the electrode to ensure reignition of the arc. This is known as ACHF welding, but high-frequency stabilizers are also used in dc welding to avoid accidental transfer of tungsten during the touch starting.

Metal-inert-gas welding (Mig) is similar to the Tig process except that a consumable electrode is used. The electrode material is transferred across the arc either in large globules (short-circuit method) or in a fine spray of small droplets. The type of transfer depends on the arc current and the electrode diameter. The latter is typically 0.030 in., and the consumption rates exceed 100 in. min^{-1}. The electrode wire is supplied from a spool and fed through the torch semiautomatically. The quality of the welded joints is as good as in the Tig process, but the Mig welding equipment is more expensive. Its use is indicated where higher welding speeds or long joints are required.

Plasma arc welding is a more recent development of the tungsten-inert-gas process.[252] Its distinguishing feature is a narrower nozzle or cup, which constricts the arc and provides a more concentrated heat source. A high-frequency voltage between the electrode and the nozzle serves to ionize the gas. The ions are blown out through the orifice and establish the arc. To avoid turbulence in the molten weld zone, relatively low flow rates (1 to 15 ft^3 h^{-1}) are used. These are not sufficient to shield the workpiece, and additional gas flow at rates up to 35 ft^3 h^{-1} must be provided from an outer nozzle. The plasma torch can be operated either in the DCSP or in the DCRP mode with tungsten or water-cooled copper electrodes, respectively. It is suitable for all metals which can be Tig welded with the exception of aluminum and magnesium. Compared with Tig welding, the plasma process is less sensitive to torch-workpiece separation, it produces a more controlled arc, and the weld penetration is deeper (up to 0.250 in. in stainless steel). Preferred vacuum applications are the joining of thin-walled sections such as bellows assemblies, or the fabrication of narrow welds as required for tube butt joints. For a more detailed description of the process and related equipment, the reader is referred to the literature.[253]

Electron-beam welding utilizes electron bombardment to heat the workpiece in a chamber evacuated to less than 10^{-5} Torr. The electrons are emitted from a thermionic cathode and accelerated through a potential of roughly 100 kV. Upon impingement on the workpiece, their kinetic energy—corresponding to about half the velocity of light—is transformed into heat. Since x-rays are also generated, shielding to protect the operator is mandatory. The beam is focused by electron optics into a small area of 0.010 to 0.060 in. diameter, yielding power densities of the order of 10^7 W in.$^{-2}$.

The penetration of the electron beam depends on various experimental parameters such as welding speed and beam power as well as on the physical properties of the material to be welded. The mechanisms of electron penetration in solids and weld formation have been discussed by Schwartz[254] and Meier.[255] Typical are very narrow, deeply penetrating joints as shown in Fig. 50. The fusion zone in electron-beam welding may have a depth-to-width ratio of 20:1 or greater. Because of the sharp focusing of the electrons, the heat-affected zones are also smaller, and hence distortion of the workpiece as well as changes of the material characteristics are kept

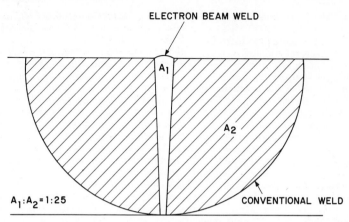

Fig. 50 Comparison of fusion zones: A_1 = produced by electron-beam welding; A_2 = produced by conventional arc welding. (*After Meier.*[255])

at a minimum. In some cases it has even been possible to join dissimilar metals with considerable differences in their thermal-expansion coefficients. Another advantage is the virtually contamination-free environment in which the operation is performed. This makes the method suitable for the welding of thin metal joints or small-diameter wires which would burn out if subjected to the more conventional processes. Various steels, aluminum alloys, and also the highly reactive refractory metals have been successfully welded.[255-258]

Commercial electron-beam-welding equipment with 3- to 25-kW beam power and operating voltages from 30 to 150 kV is available.[259] The development has proceeded in the direction of higher voltages, which yield smaller beam diameters and thus greater welding depths, narrower zones, and higher speeds. The welding of 4-in.-thick pieces of stainless steel and of 5-in.-thick sections of aluminum alloy at 150 kV/25 kW with speeds of 7 and 10 in. min^{-1}, respectively, illustrates the capability of present high-voltage units.[259] Generally, the electron beam remains fixed and the workpiece is either rotated or moved on an X-Y table underneath by means of suitable feedthroughs.

Electron-beam welding is still a relatively new method, and the equipment considerably more expensive than arc-welding apparatus. Thus, most vacuum joining is performed with the less expensive Tig and chamber-welding techniques. However, electron-beam welding can be expected to increase in importance as the standards for weld quality are raised. The disadvantage of having to work in a vacuum chamber is about to be overcome by the development of equipment operating in air. Because of the scattering of electrons, the fusion-zone depth-to-width ratios are smaller and the working distances are limited to $\frac{1}{4}$ to $\frac{1}{2}$ in. Units capable of delivering 6.5 kW power to the workpiece as well as high-quality welds in steel, aluminum alloys, Inconel, molybdenum, and copper have already been reported.[259,260]

(2) Welding Materials and Joint Design In the construction of vacuum equipment, the material most widely used is the 300 series austenitic stainless steel. These are low-carbon ($<0.08\%$) high-alloy steels with approximately 18% Cr and 8% Ni. As the chromium forms a continuous well-adhering oxide film, it imparts excellent corrosion resistance to the steel. Examples are the types 304 and 316 stainless steels, which are available in a variety of shapes suitable for vacuum assemblies. A potential problem in welding these materials is the "sensitizing" of the heat-affected zone. The term refers to the formation of chromium carbide, $Cr_{23}C_6$, at the grain boundaries in the temperature range from 425 to 800°C. Although the presence of the carbide does not reduce the mechanical strength of the weld, the corrosion resistance in the region around the fusion zone is lowered because of the ensuing chromium deficit.

The corrosion resistance can be restored by postweld annealing. Heating the assembly to 1000 to 1100°C with subsequent quenching through the sensitive temperature range results in redistribution of the chromium.

There are several stainless steels in which the chromium depletion associated with carbide formation is suppressed by slight changes in composition. The types 304L and 316L, for example, contain less than 0.03% carbon. Other steels contain additional metals with a higher affinity for carbon than chromium. Type 321 with about 0.4% Ti and type 347 with 0.8% Nb are examples of such stabilized steels. While the regular 304 and 316 steels can be welded without a filler, this is not always the case with stabilized steels. Titanium, for instance, is readily oxidized and thus lost from the fusion zone. Hence, filler material must be added in welding type 321 stainless steel.

The ability to weld dissimilar metals or alloys depends on the metallurgical reactions in the fusion zone, thermal conductivity, and differences in thermal expansion of the two members. The stainless steels 304, 304L, 316, and 316L are sufficiently similar to be compatible. They can also be Tig-welded to Kovar, nickel, and nickel-base alloys such as Invar and Inconel. Stainless steel-to-monel welds tend to crack, and the two metals are better brazed. If two metals differ significantly in their thermal-expansion coefficients, the joint members are highly stressed after cooling, and this may lead to distortion or cracking of the weld.

When selecting parts for a welded joint, the materials should be clearly identified to avoid compatibility problems and to ascertain whether a filler is needed or not. The similarity in appearance of many metals may lead to undesirable choices. The stainless-steel types 303 and 303 Se, for example, are often encountered in machine shops but should not be used for vacuum components because of their sulfur and selenium contents. Component parts consisting of chromium or nickel-plated brass are also similar in appearance to stainless steel. While brass cannot be Tig-welded, the introduction of such parts into a vacuum system is also problematical because of the high vapor pressure of zinc.

Proper design and preparation of welded joints are fundamental to the achievement of mechanically sound, vacuum-tight seals. The following rules are of general validity for all types of welding processes. Special considerations apply in the case of electron-beam welding.[261] Several common types of welded joints are illustrated in Fig. 51. When placed together for welding as shown in the first column, the parts should be carefully cleaned and free of all traces of machining oil. Ultrasonic cleaning in a detergent solution followed by water rinsing and vapor degreasing is a suitable technique.

The second and third columns in Fig. 51 illustrate wrong and right practices in placing the welded seam. The latter should always be on the vacuum side of the assembly and fully penetrating. The incorrect practices produce crevices or pockets of trapped gas between the welds. When such assemblies are evacuated later on, the trapped gas may escape very slowly and thus constitute a virtual leak.

The surface of the welded seam on the vacuum side should be smooth and uniform to facilitate thorough cleaning. Mechanical treatments such as brushing, grinding, or polishing are to be discouraged as they tend to seal off pinholes and cracks temporarily. For similar reasons, the welding should be done in long uninterrupted passes. A final bakeout of the cleaned welded assembly at 1100°C in dry hydrogen is beneficial to reduce outgassing in vacuum [see Sec. 3c(1)]. Furthermore, the discovery of leaks is more easily accomplished if the welded assembly can be tested separately rather than after having been incorporated into the vacuum system. In general, the probability of repairing leaky welds by simply rewelding the joint is low, particularly if the part has already been in service. The best action to take in this situation is to grind off the old weld and to remachine the joint member surfaces.

One of the tasks often encountered in the assembly of vacuum components is the joining of a metal tube to a flange. The joint designs shown in Fig. 52A and B require little machining and can be used to attach tubing of less than 2.5 in. diameter to flanges intended for elastomer gaskets. Larger-diameter tubing or flanges intended for metal gaskets are particularly sensitive to warpage. Therefore, such joints must be designed to minimize the extent of heating required. In Fig. 52C and D, this is

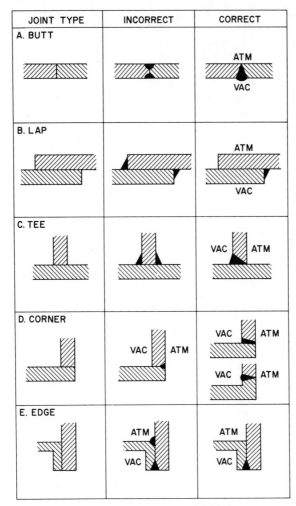

Fig. 51 Common types of welded joints.

accomplished by machining the heavy flange to match the wall thickness of the thinner tube.   Figures 52*A* and *D* illustrate another common practice, the use of tack welds to hold the two members of the joints safely in place during the welding process. Tack welds are applied on the nonvacuum surface of the joint.

An example of a more difficult vacuum weld is the joining of thin-walled bellows to heavier tubing. A suitable assembly is shown in Fig. 53. The heavy-walled section is notched to minimize heating and warping. The bellows part is inserted into a heavy copper sleeve which acts as a heat sink. The edge of the heavy tube has a small-angle chamfer so that pressure exerted through the copper sleeve ensures contact between the members at the point of the weld. In this as well as in other special cases, the thin film worker will usually have to rely on the experience and skill available in the design and shop areas.

(3) Soldering and Brazing   Contrary to welding, the temperatures employed in soldering and brazing are below the melting point of the parts to be joined.   It is only

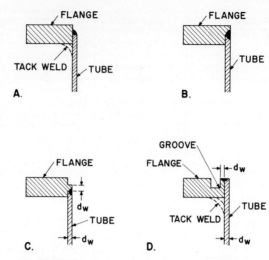

Fig. 52 Four common tube-to-flange weld designs.

necessary to exceed the melting temperature of a third material, the filler metal. The latter is added to the matched joint members and, when liquefied, spreads over the adjacent surfaces and fills the narrow gap between them. In this stage, diffusion and alloying between filler and joint materials occur. Subsequent solidification yields the soldered or brazed joint.

The difference between the two techniques is in the degree of interaction obtained. Soldering is performed with low-melting-point materials, usually below 500°C, and alloying and diffusion affect only a thin surface layer of the adjacent parts. After solidification, most of the solder is found in the joint interstices. Brazing utilizes filler materials of higher melting points, the depth of the reaction zone in the joint members is greater, and only a small fraction of the brazing metal remains unaltered in the interstices. The latter factor explains why members for a brazed joint must be machined to closer tolerances than is required for soldering. For the same reasons, brazed joints are mechanically strong, whereas soldered joints are sensitive to shock and tend to develop leaks. To avoid ambiguities, the terms "silver solder" and "hard solder" should not be used when referring to a brazing operation.[262]

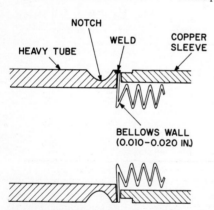

Fig. 53 Welding assembly for joining bellows to heavier tubing.

Soldering is usually performed in air, and the parts are heated by a gas torch. Fluxes are required to remove oxide films and to ensure wetting of the surfaces. The fluxes consist of inorganic acids and salts or organic compounds such as acids, bases, and rosins. They are a potential source of contamination and must be removed before incorporating the part into a vacuum system. Common solder materials are alloys containing two or more of such metals as Sn, Pb, P, Sb, Bi, Cd, Zn, Ag, In, and Ga. Most of these metals have comparatively high vapor pressures and low melting points (see Chap. 1, Table 2), which makes them unsuitable for the majority of high-vacuum applications. However, ince soldered joints are easily dissembled, they are quite practical for semipermanent joints, particularly outside vacuum systems. Examples are water-cooling lines and thermocouple-wire connections. Inside vacuum systems, low-vapor-pressure solders like 60-40 Sn-Pb, 50-50 In-Sn, or pure In may be used with discrimination, but it is generally advisable to resort to brazing. For additional information on soldering processes and practices, the reader is referred to the literature.[263-265]

The *brazing* process[248,263,266,267] relies on capillary action to spread the filler material throughout the joint interstices. The concomitant metallurgical reactions are often complex and may involve grain-boundary diffusion or formation of intermetallic compounds in addition to diffusion and alloying between filler and joint materials. The rates at which these processes occur vary with the temperature and duration of the brazing cycle. Identification of the resulting metallurgical phases becomes increasingly difficult as the number of constituents in the system increases. In addition to the metallurgical reactions, factors to be considered in the selection of brazing materials are their melting points, vapor pressures, and ability to wet and flow on the parent metal surface.

If brazing is to be used to form vacuum-tight joints, fillers containing metals with high vapor pressures such as Cd, Zn, P, Bi, or Pb are obviously poor choices. Other metals like Pt, Rh, Ta, Nb, and Re with melting points above 1700°C are also unsuitable for the commoner vacuum construction materials. As a rule, the melting point of the filler should be at least 100°C below the solidus line of the joint material. The

TABLE 14

Common Brazing Materials

No.	Brazing-material constituents and composition, wt %	Temp, °C		
		mp	Liquidus	Solidus
1	Copper (100)	1083		
2	Gold (100)	1063		
3	Pd(10)-Ag (90)	....	1065	1000
4	Au(35)-Cu(62)-Ni(3)	....	1030	975
5	B(2.9)-Cr(7)-Si(4.5)-Fe(3)-Ni(82.6)	....	1000	971
6	Au(40)-Cu(60)	....	980	950
7	Au(82)-Ni(18)	950		
8	Pd(10)-Ag(58)-Cu(32)	....	852	825
9	Ag(72)-Cu(28)	780		
10	In(10)-Cu(27)-Ag(63)*	....	730	685

Metal Couples Which Can Be Brazed with the Fillers Above

Metal 2 \ Metal 1	W	Mo	Ni	Monel	Kovar	Stainless steel	Cu
Copper.............	4, 7	6, 7, 8, 9	6, 7, 9	4, 9	6, 7, 9	7, 8	8, 9, 10
Stainless steel........	5, 7	3, 5	5, 7	1	2, 7, 9	1, 8	
Kovar.............	4, 7	7	2, 6, 7	4, 9	2, 9		
Monel.............	4, 7	4	4, 9	4, 9			
Nickel.............	4, 7	4, 7	4, 6, 9				
Molybdenum........	4, 7, 8	4, 7, 8					
Tungsten............	7, 8						

* Used as step braze after No. 9.

elements or alloys of the group Cu, Ag, Au, and Ni satisfy this requirement and are most often utilized for the joining of vacuum parts.   Representative brazing materials which are commercially available in the form of wire, foil, rings, powder, or paste are listed in Table 14.   Additional braze filler and parent-metal combinations are given in the literature.[248,262,266]   If a choice between different brazing materials exists, the one with the lowest melting temperature is preferable to minimize metallurgical interactions, grain growth, and warpage.   However, there are situations where higher temperatures are advantageous.   Possible reasons are the desire to achieve stress relief, reduction of oxide films, or vaporization of surface contaminants.

The ability of a molten brazing material to wet the joint members is characterized by the contact angle between a sessile drop and the solid part surface.   Contact angles smaller than 90° are required for wetting to occur.   Perfect wetting and spreading are attained if the contact angle is 0°.   The latter is usually the case if there is metallurgical interaction between the braze and part materials.   Surface cleanliness is, of course, a prerequisite to facilitate uniform wetting.

To further assess the flow characteristics of a filler material, the phase diagram of the respective system must be taken into account.   These are found in metallurgical texts and handbooks.[268]   An example of practical importance is the phase diagram of Ag-Cu shown in Fig. 54.   Brazes which are either pure metals (α phases) or eutectics

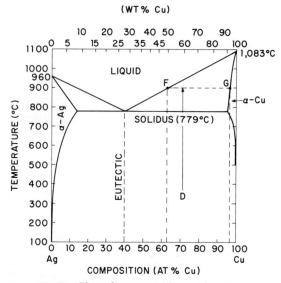

Fig. 54   Phase diagram of silver and copper.

have discrete melting points and flow rapidly into the joint as soon as their melting temperatures are exceeded.   In these cases, relatively narrow joint clearances of 0.001 to 0.003 in. are commonly used.

A different situation arises if the filler is of a noneutectic composition as indicated by the line D in Fig. 54.   As the temperature rises above the solidus line, a portion of the filler alloy liquefies into a melt of composition F and flows into the joint, leaving behind a spongy solid of composition G (Fig. 54).   Consequently, the remaining filler becomes copper-richer and requires a higher temperature to melt.   Upon further heating, the composition and temperature of the molten filler continue to change according to the liquidus line until the melting point of pure copper is reached.   The flow of noneutectic filler alloys is therefore more sluggish than that of discrete-melting-point materials and requires somewhat larger joint clearances such as 0.002

to 0.005 in. The flow of eutectic filler materials may be impeded in a similar manner if the composition of the melt changes because of alloying with the joint material.

Several problems may arise from these metallurgical processes. If the fraction of molten material withdrawn from the initial brazing charge is high early in the brazing cycle, liquefaction of the remainder may be difficult. Excessive alloying and dissolution of the part material occur because of either too high temperatures or prolonged heating. The results are joints with an eroded appearance and inferior mechanical strength. In joining thin sections, overalloying is particularly dangerous, as it can lead to voids in the joint members. On the other hand, alloying with the joint material is sometimes utilized to form successive or step brazes. This technique exploits the fact that the liquidus temperature of the joint after brazing is higher than the melting point of the initial alloy. Finally, filler materials which are known to yield intermetallic compounds when alloying with the part material should not be used because the resulting joints are likely to be brittle.

As in welding, the means of heating parts to be joined by brazing vary. The simplest brazing tool is the gas torch, but since the operation is performed in air, fluxes are needed to maintain a wettable surface. Furthermore, torch brazing is most easily done with low-melting (high-vapor-pressure!) brazes since the gas flame is not a very concentrated heat source. Consequently, the technique is not particularly suitable for vacuum joints. Alternate and preferred methods are induction and

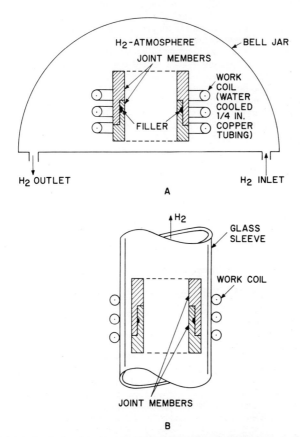

Fig. 55 Brazing by induction heating in hydrogen. A = bell-jar arrangement. B = in a glass tube.

furnace heating. These may be performed either in a protective atmosphere or in vacuum. The latter technique is of little practical importance in this context.

Brazing by induction heating[250] utilizes high-frequency currents of 100 to 1,000 kHz which are coupled to the joint members by a work coil. It is performed in a reducing atmosphere, usually in hydrogen. Two possible arrangements are shown in Fig. 55. In both cases the work coil is wrapped closely around the joint members. Joule heating occurs predominantly near the surface of the joint parts as the induced current decreases rapidly toward the interior. The distance at which the current density has fallen to 37% of its value at the surface is called the skin depth d_s and is given by

$$d_s = 1,980 \sqrt{\frac{\rho}{\mu\nu}} \quad \text{in.}$$

where ρ = resistivity of the metal, ohm-cm
μ = relative magnetic permeability ($= 1$ for nonmagnetic metals)
ν = current frequency, Hz
The skin depth in stainless steel ($\rho \approx 80$ μohm-cm), for example, is about 0.025 in. for the widely used frequency of 450 kHz.

In practice, where a variety of differently shaped parts are encountered, it is generally necessary to form coils to fit the particular workpiece. Also, lower frequencies are required if heavier sections are to be joined and a greater skin depth is desired to achieve more uniform heating. The joining of magnetic to nonmagnetic materials is complicated by the widely differing magnetic susceptibilities. The magnetic part heats up and expands faster and thus tends to draw away the molten brazing material. In general, however, the method permits rapid, localized heating and is especially suitable for the joining of thin-walled sections requiring close control of the brazing cycle. Since the equipment is rather expensive and the experimental arrangements are somewhat inflexible, induction-brazing facilities are less often used in the laboratory than in production plants.

The more flexible and therefore more common technique for the fabrication of laboratory vacuum equipment is *furnace brazing*. This is performed mostly in dry hydrogen, although other gases or mixtures including nitrogen, the rare gases, or

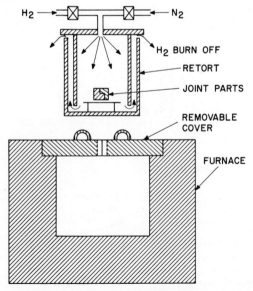

Fig. 56 Schematic of a retort furnace.

dissociated ammonia are occasionally used. The parts to be brazed are assembled with the filler added and securely supported to prevent misalignment in the subsequent handling. Two types of ovens are widely used, namely, pusher and retort furnaces. The pusher furnace has three zones of different temperatures for preheating, brazing, and cooling, through which the joint assembly passes. Fairly high hydrogen-purge rates are required, particularly during loading and unloading when the furnace doors are open.

A retort furnace is shown schematically in Fig. 56. The retort containing the parts is purged successively with N_2 and H_2 before being lowered into the oven. Low hydrogen-purge rates equivalent to two to three volume changes during the heating cycle are sufficient. The hydrogen exits under the retort lid and is burned off. Cooling is initiated by opening the furnace cover and withdrawing the retort. This furnace is well suited for the brazing of objects of different shapes and sizes as encountered in the laboratory.

An important feature of brazing in a hydrogen atmosphere is that oxide films on the metal-part surfaces are reduced prior to the melting of the filler. Whether this will occur or not depends on the purity of the gas, the thermodynamic stability of the oxide, the temperature, and the reaction rate.[269] The oxides of Be, Mg, Al, Si, and Ti, for example, are not reduced as they require higher temperatures than those which are practical in brazing. The protective chromium oxide film on stainless steel is reduced only if the dew point of the hydrogen is below $-60°C$ and the temperature exceeds $950°C$. Accordingly, high-temperature fillers such as pure Cu or Ni 18–Au 82 (weight %) are used to braze steel. If lower brazing temperatures are desired, the steel may be nickel-plated and a filler material like Ag-Cu eutectic may be applied. Of the different types of stainless steel available, the low-carbon-content varieties 304L and 316L, or the stabilized steels 321 and 347 are recommended, as they are less subject to sensitizing [see Sec. $4a(2)$ on Welding Materials].

The metals and alloys of Ti, Zr, W, Nb, and Ta are embrittled when brazed in hydrogen. As a result, their tensile properties are adversely affected. Hydrogen embrittlement also occurs in electrolytic tough-pitch copper, as the latter contains small amounts of oxygen. To be suitable for hydrogen brazing, copper must be designated as oxygen-free high-conductivity grade (OFHC copper).

Several joint designs commonly used in brazing are shown in Fig. 57. The lap,

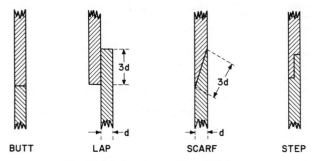

BUTT   LAP   SCARF   STEP

Fig. 57   Various types of brazed joints.

scarf, and step types are preferred over butt joints since they offer a larger interface area. As indicated in the drawings, the overlapping portion should be at least three times as long as the material is thick. For very thin sheet metal, 0.100 in. is considered to be the minimum length. If corners exist between the joint parts, they should be square and tight-fitting as in Fig. 58A. Rounded corners as in Fig. 58B disrupt the capillary flow of the filler material. In some cases, this may be desirable to maintain certain parts of the joint unbrazed. An alternate method to confine the molten filler is the application of carbon coatings or graphite rings.

Whereas in torch brazing the filler may be applied during the operation by means of a rod, induction and furnace brazing necessitate placement of the charge before loading. Application of a filler ring from the outside as shown in Fig. 59A requires a shield to prevent direct heating and premature melting. Insertion of filler wire into a groove within the joint as in Fig. 59B is the more reliable technique. If the joint members are of unequal mass, the heavier section should be grooved. The filler quantities are determined by the diameter of the wire and should be chosen according to surface area and clearance of the joint.[270]

In joining dissimilar metals, differences in the coefficients of thermal expansion must be considered. Two pieces of tubing, for example, require placement of the part with the higher expansion rate on the outside so that the filler is compressed upon cooling.

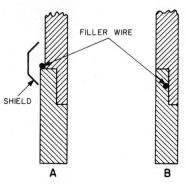

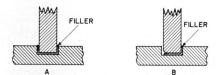

Fig. 58 Flow of brazing metal. A = at square corners. B = at a rounded corner.

Fig. 59 Placement of filler charge for induction or furnace brazing. A = external with heat shield. B = in internal groove.

The general rules pertaining to brazing are similar to those in welding. For instance, the formation of crevices between the joint surfaces must be avoided as these may trap a small volume of air and thus yield virtual leaks. For this reason, only one filler ring per joint should be applied. Cleaning of parts prior to brazing is essential for vacuum-tight joints. The use of sulfur-base cutting oils in the machining of parts is not recommended because nickel, especially Ni-Cu alloys, tends to form sulfides in the grain boundaries. The result is weak, brittle joints which are susceptible to cracks. Sulfur embrittlement is effecively suppressed, however, if chromium is present as in stainless steel.

(4) Glass-to-Metal Seals Glass-to-metal seals were widely used in the past, when most vacuum systems were made from glass. Their utilization has decreased considerably as metal became the preferred material of construction. In other applications, glass-to-metal seals have been replaced by ceramic-to-metal seals because the latter offer greater mechanical strength and better electrical, thermal, and chemical stability. Presently, the use of glass-to-metal seals is largely confined to ionization gauges and optical windows in bakeable metal systems.

The joining of two materials as different as glasses and metals requires careful consideration of their physical and chemical properties. The latter are reviewed extensively along with common sealing techniques by Roth[248] and Kohl.[263] Since permanent seals are formed at high temperatures and the two materials contract by different amounts upon subsequent cooling, considerable stress develops at the interface. To accommodate this stress and ensure adhesion, it is usually necessary to produce an intermediate metal oxide layer. The latter reacts partially with the molten glass to form a strong bond. It may also relieve some of the stress by yielding. The gas atmosphere and temperature at which the oxide is formed are critical because they affect the layer thickness and composition. Thick oxide films tend to be porous and are likely to fail mechanically. Similar problems occur if the oxidation process leads to the formation of higher metal oxides which do not bond as well as the lower ones. This applies to multivalent metals such as Cu, Fe, Cr, Mo, and W, and the color imparted by the oxide to the glass is a criterion of successful seal formation. Examples are shown in Table 15; colors darker than those listed generally indicate overoxidation and weak seals.

TABLE 15   Glass-to-Metal Seal Colors Indicative of Good Joints

Metal	Color observed through the glass
Pt...............	Metallic (no oxide formed, limited strength)
Cu...............	Gold to purple (color of Cu_2O; CuO is black)
Ni...............	Green to gray
FeNiCo.........	Gray to brown
Cr...............	Green
FeNiCr..........	Brown to green
Mo..............	Brown
W...............	Yellow to brown

The most important factor determining the mechanical strength of glass-to-metal seals is the interfacial stress arising from differences in thermal expansion of the two materials joined.   At the temperature where the seal is made, the glass is liquid and flows without setting up stress.   This condition continues upon cooling until the viscosity of the glass becomes too high to permit further flow.   This occurs typically about 20°C below the annealing temperature of the glass (defined as resulting in complete stress relief in 15 min) and is referred to as the set point.   Consequently, the resulting stress at room temperature is due to the difference in contraction when cooling below the set point.

It is generally not possible to find glass and metal pairs whose expansion characteristics are identical throughout the entire temperature range from 25°C to the set point. As shown in Fig. 60, a typical metal expands almost linearly, whereas the coefficients of expansion of glasses increase at temperatures approaching the annealing point. From such expansion curves, the stress in a seal made from metal-glass pairs can be derived.   Graphically, this is done by translating the glass-expansion curves in Fig. 60 vertically until the set points a or b fall on the expansion curve of the metal at points c or d, respectively.   The resulting displacements ΔA and ΔB of the glass-expansion curves at 25°C represent the thermal mismatch or strain of the seal.   If the total expansion of the glass is smaller than that of the metal (glass A in Fig. 60), the glass becomes compressed parallel to the interface.   Conversely, glass B, expanding more than the metal, will be subject to tensile stress.

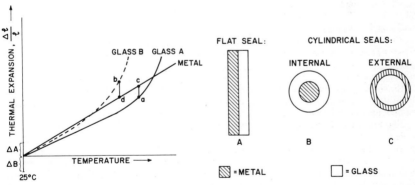

Fig.   60   Thermal-expansion   curves   of glasses and metals.

Fig. 61   Cross sections of basic glass-to-metal seal geometries.

The condition represented by glass A is generally preferred because glasses are far less likely to break under compression than in tension.   Scratches and imperfections in the surface are known to reduce the intrinsic tensile strength of glasses, which is of the order of 1,000 kg mm^{-2}, by two to three orders of magnitude.   If seals are fabricated from materials selected for their relative thermal compatibility, tensile stresses of less than 0.5 kg mm^{-2} are considered favorable.   This corresponds to an expansion mismatch of 0.01% or less in the temperature range from the set point to 25°C.

Stresses up to 1.5 kg mm^{-2} (up to about 0.05% difference in thermal expansion) are tolerable, but the seals must be carefully annealed and are more likely to break.

So-called "matched seals," which meet the criteria discussed in the previous paragraph, can be made from several material combinations. The "soft" glasses such as soda lime or lead silicates are compatible with Pt, FeNi, FeCr, and FeNiCr alloys. The "hard" glasses, as represented by the borosilicates, require higher sealing temperatures and metals with low thermal-expansion coefficients. Examples are Mo, W, and FeNiCo alloys such as Kovar. However, some of these alloys undergo phase transformations and concurrent changes of their expansion coefficients at certain temperatures. Kovar has such an inflection point at 435°C. In these cases, glasses with set points below the inflection point of the alloy must be used. No matched seals are possible with fused silica because of its high melting temperature and small coefficient of thermal expansion (5 $\times$ 10^{-7} deg^{-1}).

The stresses developing in glass-to-metal seals are also dependent on the geometry of the seal. Most configurations involving beads, wires, ribbons, tubes, or disks can be reduced to one of the three basic geometries shown in Fig. 61. In flat seals, stress forces occur normal as well as parallel to the interface. Figure 61B may represent a sealed-in metal rod or tube, while Fig. 61C in disk form would be a window seal. Both cylindrical configurations are subject to stresses in three mutually perpendicular directions, namely, normal and tangential to the interface, and axial, i.e., parallel to the rotational axis of the cylinder. Depending on the relative magnitudes of the expansion coefficients, these stress components may be tensile or compressive. The various possible stress distributions are shown in Table 16.

TABLE 16 Possible Stress Distributions in the Glass Part of Glass-to-Metal Seals for the Three Types Shown in Fig. 61

Seal geometry	$\alpha_{glass} < \alpha_{metal}$			$\alpha_{glass} > \alpha_{metal}$		
	Normal	Parallel or tangential	Axial	Normal	Parallel or tangential	Axial
Type A.....	Small	Compressive	....	Small	Tensile	
Type B.....	Tensile	Compressive	Compressive	Compressive	Tensile	Tensile
Type C......	Compressive	Compressive	Compressive	Tensile	Tensile	Tensile

Since compressive stresses are preferred because of the greater strength of the glass, type A or C seals with $\alpha_{glass} < \alpha_{metal}$ are most favorable for matched seals. Internal cylindrical seals (type B) can be satisfactory, provided the thermal mismatch is small ($<0.05\%$) and the glass-to-metal bond strong.

If the conditions for a matched seal cannot be met, an alternate type of seal may be possible which relies on the excess stress being relieved by plastic deformation of the metal. Such unmatched glass-to-metal pairs are known as *Housekeeper seals*.[271] They can be fabricated in the form of wire, ribbon, disk, or feather-edge tubular seals from soft as well as hard glasses. Copper is the preferred metal since it forms strong bonds with glasses and has a low yield strength. Pt, Fe, and Mo have also been used in seals of this type.

A common form of the Housekeeper seal is the feather-edge tubular seal shown in Fig. 62. The sealed end of the metal tube is tapered to minimize the resistance to yielding when the glass contracts below the set point. Typically, the thickness of the feather edge is 0.002 to 0.003 in., and the length of the tapered section is about twice the length of the sealed zone. Depending on the application of the glass, internal (glass only on the outside of the metal tube), external (glass only on the inside of the metal tube), and double-sided seals (Fig. 62) are distinguished. In the latter case, it is recommended to make the internal overlap of the glass about twice as long as the external overlap. Seals of this type have been made with copper tubing of 0.5 to 5 in. diameter.

Housekeeper seals are subject to failure from repeated temperature cycling because the yield strength of the metal is exceeded during each cycle. Copper feather-edge seals survive only a few hundred excursions from room temperature to 400°C. They also have low mechanical shock resistance and tend to corrode. For these reasons, Housekeeper seals have been replaced in most applications by matched seals of various iron alloys, which are easier to fabricate, mechanically stronger, and less sensitive to temperature cycling.

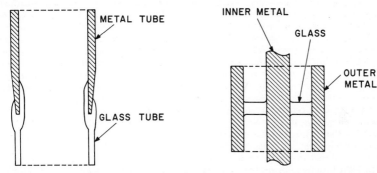

Fig. 62 Housekeeper seal of the double-sided feather-edge tubular type.

Fig. 63 Coaxial compressive glass-to-metal seal.

Another type of unmatched seals are *compression seals*. They are designed so that the glass is subject to compressive forces, thus exploiting the greater strength of the material in this mode of stress. An example is the coaxial conductor shown in Fig. 63. Compression of the glass is attained if the outer metal expands more than the glass and the latter more than the inner metal. Seals of this type are often made such that one interface is a matched seal and $\alpha_{outer\ metal} > \alpha_{inner\ metal}$. They can also be fabricated with the same metal on both sides of the glass, but then the outer ring must be thicker than the core to keep the glass under compression.

For those glasses and metals which differ strongly in their expansion characteristics, the fabrication of seals requires distribution of the stress forces over a greater length. This is accomplished with *graded seals*, which consist of several glass segments varying slightly in thermal expansion. They are fused together so that the expansion on one end matches that of the glass, whereas the other end is compatible with the metal. Graded seals of this type are expensive and tend to be fairly long as well as highly stressed.

More recent and shorter versions are *multiform* and *impregnated seals*. Multiform seals are fabricated by pressing glass powders of different thermal expansion into rings which are then fused together. Impregnated seals achieve a graded composition by dipping a porous body sintered from quartz powder into an alkali borate solution. The latter rises in the sintered-quartz part by capillary action. Subsequent firing produces a fused body of graded expansion characteristics. Graded seals are used for joining quartz to metals like Kovar and tungsten, or to join hard and soft glasses to each other.

(5) Ceramic-to-Metal Seals Ceramics are available as flat, cylindrical, or tubular parts of good dimensional tolerances which can be joined to various metals. The resulting seals combine the excellent insulation properties, chemical inertness, thermal and mechanical shock resistance of ceramics with the conductivity and processability of a metal. Accordingly, ceramic-to-metal seals are widely used in the assembly of vacuum systems for electrical leads such as thermocouples, substrate heaters, and—most important—to bring in high voltages as required for sputtering or electron-beam heating. Within these applications, high-alumina ceramics ($>95\%$ Al_2O_3) play the dominant role. These have mechanical strengths around 20 kg mm^{-2} in tension and about ten times as high in compression. Among the metals which are commercially joined to ceramics are OFHC copper, molybdenum, nickel, NiFe alloys, monel, Kovar, and occasionally some types of stainless steel.

The processes by which ceramic oxides and metals can be joined were mostly developed in the vacuum-tube industry. Reviews of such sealing techniques have been given by Kohl,[263,272] Roth,[248] and Espe.[273] In principle, the ceramic part of the seal is first metallized and then brazed to the metal part, or metallization of the surface and joining by brazing are performed in one heating operation. Mechanisms such as mechanical interlocking on the rough surface, chemical reaction, solid-state diffusion, and formation of glassy interface layers are responsible for establishing good adhesion and a vacuum-tight seal between the ceramic and the metal. As in glass-to-metal seals, matching of the expansion characteristics of the two seal materials is essential to avoid excessive stress. The temperature interval to be considered here extends from 25°C up to the melting point of the braze material. Furthermore, as ceramics are mechanically stronger in compression than in tension, the relative magnitudes of the expansion coefficients and the seal geometry should be chosen to give predominantly compressive stresses.

The commonest ceramic-to-metal sealing technique is the sintered metal powder or *molybdenum-manganese process*. Here, the ceramic component is first coated with a paint consisting of 4 parts (by weight) of molybdenum and 1 part manganese powder mixed with a binder and solvents. The 0.001- to 0.002-in.-thick layer is dried and sintered in wet hydrogen for about 30 min at temperatures ranging from 1300 to 1600°C depending on the materials involved. Subsequent plating with 0.0002 to 0.0004 in. of nickel or copper and renewed firing in hydrogen at 1000°C furnish a layer on which the braze metal can wet and flow. The tensile strength of such fired-on metal coatings is of the order of 10 kg mm^{-2}. Commonly used brazing alloys are the eutectics of Ag-Cu and Au-Ni (see Table 14).

Alternate joining techniques are the *active metal* and the *hydride processes*. The former technique relies on establishing a firm bond to the ceramic oxide by means of reactive metals such as titanium or zirconium which are present in the brazing alloy. Hence, premetallization of the ceramic is not necessary. In the hydride process, the seal is also formed in one heating cycle, but the reactive metal is interposed between the ceramic and the brazing alloy in the form of hydride powder.

Geometrically, ceramic-to-metal joints may be classified as either cylindrical or butt seals. Examples are shown in Fig. 64. If the expansion coefficient of the ceramic is greater than that of the metal, internal cylindrical designs as in Fig. 64A to C are required to exert compressive stress on the ceramic. Opposite expansion behavior calls for external cylindrical seals as shown in Fig. 64D. Feedthroughs of the type illustrated in Fig. 64B are possible only if the diameter of the metal rod or wire does not exceed 0.040 in. Heavier rods require an adapter tube as in Fig. 64C. In making butt seals, it is recommended practice to join ceramic parts on both sides of the metal so that the shear stresses at the two interfaces compensate each other. This principle is observed in Fig. 64E and F. The latter seal consists of a flexible hollow metal ring with ceramic rings joined to both surfaces of the bottom wall. This assembly may then be brazed to a heavier metal part. The advantage of such seals is their greater reliability during temperature cycling as differences in thermal-expansion rates are taken up by the thin, flexible metal ring.

If additional joining operations on the metal part of the seal are required, the latter should be designed sufficiently long to allow heating as far away from the seal as possible. While brazing by induction heating may be used in such situations, welding of assembled ceramic-to-metal seals is the preferred method. Another design possibility is to reduce the metal wall thickness in the seal area, which permits partial stress relief by yielding as in Housekeeper seals. Plastic yielding of the braze alloy, too, may aid in relieving interfacial stress. This has been reported to occur in the widely used Kovar-to-alumina seals which are bonded with Ag-Cu eutectic braze.[273]

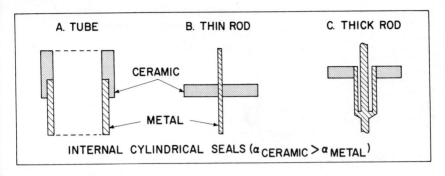

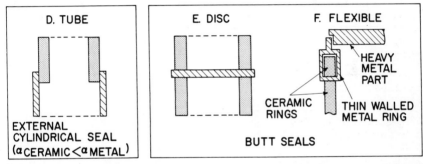

Fig. 64 Common types of ceramic-to-metal seals.

b. Demountable Seals

Demountable seals are used in vacuum systems where access to the chamber or anticipated changes in component parts are the primary consideration. The materials required for these types of seals include various organic substances as well as metals. The former have relatively high vapor pressures, permeabilities, and outgassing rates, as discussed in Secs. $3c(3)$ and $3e(3)$. Hence, their utilization is generally restricted to nonbakeable high-vacuum systems intended to operate at pressures above 10^{-8} Torr. Furthermore, seals involving organic materials should be designed so that the area exposed to the vacuum is as small as possible. In very-high- and ultrahigh-vacuum systems, metal gasket seals are used almost exclusively.

(1) Waxes, Resins, and Adhesives Waxes which melt at relatively low temperatures without significant decomposition are sometimes used to join small metal, glass, or ceramic parts temporarily. A possible reason for resorting to this technique is the risk of damaging delicate parts by the higher temperatures required to make a permanent seal. However, such applications must be restricted to locations where the joined parts are not heated significantly above room temperature. More common than low-melting waxes is the use of epoxy resins since some of these are cold-setting and do not require heating at all. They also allow service over a wider temperature range, with usage reported from liquid-helium temperature up to 250°C.[274] Their vapor pressures are comparatively low, and outgassing rates tend to decrease with continued usage. The tensile strengths of epoxy joints are in the range of 1 to 5 kg mm^{-2}, the higher values being obtained if the resins are cured at 150 to 200°C. Above or below room temperature, the strength of epoxy joints decreases to an extent which depends on the type of resin and the curing cycle. Dismantling of epoxy joints is not very easily done but requires heating to at least 150°C or prolonged soaking in solvents such as trichloroethylene.

Small pores or pinholes in vacuum enclosures may be sealed with adhesives or lacquers which are available as liquids and aerosols. However, this is only a temporary measure and involves certain risks. The sealants are soluble in organic solvents such as often used to clean vacuum parts; hence the leaks may be exposed accidentally. Furthermore, the chances of subsequently repairing such leaks by brazing or welding are reduced by incomplete removal of sealant residues. Consequently, the use of sealants in spots or over large external surface areas is not a recommended practice.

Another application of sealants is the tightening of threaded pipe joints as used in cooling ducts or in connecting thermocouple vacuum gauges to the chamber. Typically, one layer of 0.004-in.-thick Teflon tape is wrapped around $\frac{1}{8}$-in.-diameter male pipe threads to produce a vacuum-tight joint. For larger thread sizes, several layers of tape are required. Additional information about sealing practices and properties of waxes, resins, and adhesives may be found in Ref. 248 (p. 236).

(2) Elastomer Gasket Seals   Gasket seals are obtained by deforming a soft material, the gasket, between two hard surfaces, the flanges. The soft material should ideally be plastic to fill small irregularities in the flange surface, as well as elastic to maintain the required sealing pressure. There is no material which possesses both properties in the desired degree. Therefore, gasket seals are designed to rely primarily on one or the other type of deformation. The two basic alternatives are shown schematically in Fig. 65. Seal A arises from compression of the gasket while seal B results from

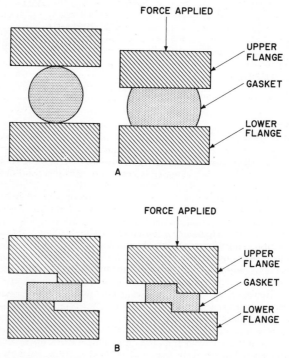

Fig. 65   Gasket sealing. A = by compression. B = by shearing.

plastic deformation. For compression seals, elastomers are used as gasket materials, whereas certain metals are sufficiently plastic to serve in seals made by shearing. In the former case, the sealing surfaces are perpendicular to the applied force, whereas they are parallel in the latter case. Generally, the compression seals require a greater force and a higher degree of deformation than do the sheared gasket seals.

Materials suitable for compression-type gaskets are listed in Table 17. Outgassing

TABLE 17 Properties of Elastomer Gasket Materials*

Material		Service temp, °C		Compression set
Trade name	Generic name	Lowest	Highest	
Natural rubber......	Isoprene	−30 to −65	60–75	Fair
Nitrile, Buna N, Hycar, Perbunan...	Acrylonitrile/butadiene	−25 to −50	85–150	Good
Neoprene..........	Chloroprene	−50	120	Poor
Silicone, Silastic.....	Polysiloxane	−120	300	Excellent
Viton A and B......	Vinylidene fluoride/ hexafluoropropylene	....	150–260	Good
Teflon, Fluon.......	Polytetrafluorethylene	−190	280–400	Poor
Kel-F, Hostaflon....	Polytrifluorchlorethylene	....	....	Poor

* Data from Roth.[248]

and permeation rates were previously shown in Tables 8 and 10. The service temperatures are limited at the lower end by embrittlement and loss of elastic properties. The upper limit is imposed by outgassing and permeation rates, which increase with temperature (see Secs. 3c and 3e). These limits are not exactly defined but depend on the duration of the exposure and the specific vacuum requirements.

Another important criterion of elastomer materials is their ability to return to their original shape after having been compressed. The degree of compression required for a tight seal depends primarily on the hardness of the elastomer. The latter varies not only among the different compounds but also within each class as a result of different types and amounts of fillers used. Typically, the hardness of elastomers ranges from 30 to 100 durometer (Shore). As a rule, the harder materials need less compression than the soft ones to form a satisfactory seal. A gasket of a durometer hardness of 50, for example, should be compressed by at least 15%, and 10% compression is considered to be the absolute minimum under any conditions (see Ref. 248, p. 313). When the compressive force is released, the gasket should return to its original dimension. In practice, this is not fully the case, and the residual deformation is known as compression set. Figures of merit are determined at compression ratios slightly higher than those applied in actual service. Test ratios of 20% are typical for materials of 85 to 95 durometer hardness, whereas softer elastomers (30 to 45 durometer) are subjected to compression-test ratios as high as 40%. Under these conditions, most elastomers experience a permanent reduction in gasket height of less than 10%. Exceptions are Neoprene and the fluorinated hydrocarbons, which suffer considerably larger plastic deformations, as indicated by the "poor" rating in the last column of Table 17. Silicone rubbers have the lowest compression set, but their utility is impeded by the high permeability of this material [see Sec. 5e(3)]. Consequently, Viton, Nitrile, Buna N, and natural rubber are the most commonly used gasket materials. However, they, too, show higher compression set if used at temperatures significantly below 20°C or above 75°C.

Since typical elastomer gaskets have only a limited ability to flow plastically, the achievement of a vacuum-tight seal is contingent upon smooth flange surfaces. For machined parts, surface finishes of 30 to 40 μin. rms are typical, and these are more than adequate to provide an elastomer gasket seal. However, accidentally made indentations or radial scratches across the flange surface are unlikely to be sealed and thus constitute potential leaks. In smoothing out such defects with an abrasive, the direction of the motion should not be radial across the seal surface to avoid creation of leakage channels.

The commonest type of elastomer gasket used in vacuum systems is the circular-cross-section O ring. As shown in Fig. 65A, a round gasket yields to the applied force by expanding laterally. In the absence of external constraints, excessive force leads

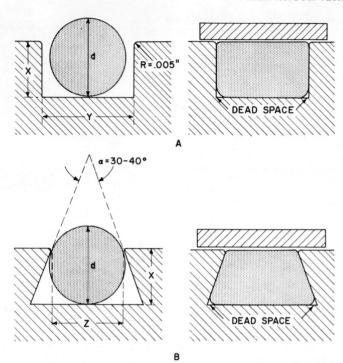

Fig. 66 Limited-compression flange seals. A = rectangular groove. B = trapezoidal groove.

to unlimited deformation of the gasket and hence to excessive compression set. This risk is avoided if grooved flanges are utilized. Two examples of such limited-compression seals are shown in Fig. 66. The grooves must be carefully designed to allow for lateral expansion of the gasket. Typical for elastomer gaskets of 40 to 60 durometer hardness is a compression ratio of $x/d = 0.7$. Since the cross-sectional area of the gasket remains constant during compression, the necessary dimensions of rectangular or trapezoidal grooves can be calculated. In doing this, one has to consider

TABLE 18   Common O-ring Sizes and Groove Dimensions (See Fig. 66)

O-ring dimensions			Groove dimensions, in.			
Cross-section thickness d, in.		Range of available ID, in.	Rectangular		Trapezoidal (dovetail), $\alpha = 40°$	
Nominal	Actual		x	y	x	z
$\frac{1}{16}$	0.070 ± 0.003	$\frac{1}{16}$–$5\frac{1}{4}$	0.050	0.080	0.049	0.059
$\frac{3}{32}$	0.103 ± 0.003	$\frac{3}{8}$–$9\frac{3}{4}$	0.074	0.118	0.074	0.088
$\frac{1}{8}$	0.139 ± 0.004	$\frac{3}{4}$–18	0.100	0.160	0.100	0.120
$\frac{3}{16}$	0.210 ± 0.005	$1\frac{1}{2}$–26	0.147	0.242	0.149	0.185
$\frac{1}{4}$	0.275 ± 0.006	$4\frac{1}{2}$–26	0.198	0.317	0.202	0.252

that the elastomer does not flow into sharp corners. Therefore, the cross-sectional area of the groove is typically designed to be about 5% larger than the O-ring cross section. The resulting unfilled corners are referred to as dead space. The advantage of the trapezoidal over the rectangular groove is that it better retains the O ring when the upper member of the flange is raised.

O rings are standardized commercial items specified by their cross-sectional thickness d and inside diameter (ID). Common sizes and the corresponding groove dimensions for 5% dead space are listed in Table 18. For more detailed design information, the reader is referred to Ref. 248 (pp. 348 and 357).

O rings are also available with metal retainer rings of matched sizes. These permit the assembly of limited-compression seals without grooved flanges. As shown in Fig. 67, the retainer ring serves as a spacer and thus prevents excessive gasket deformation. Bolt-clearance holes are provided for aligning the parts.

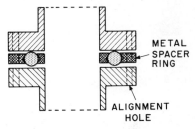

Fig. 67 Limited-compression flange seal with gasket confined by metal retainer ring.

The sizes of circular flanges in the range from 4 to 24 in. diameter on vacuum equipment and major components are standardized. Generally, they conform to specifications issued by the American Vacuum Society.[275] Unfortunately, this uniformity has not yet been attained with smaller flanges as used for porthole and feedthrough assemblies. These vary in size among different manufacturers, and compatibility problems with parts from different sources are not uncommon.

To be distinguished from flange seals, where the compressive force acts perpendicular to the plane of the O ring, are shaft or piston seals, which are compressed radially. As shown in Fig. 68, one speaks of internal or external shaft seals depending

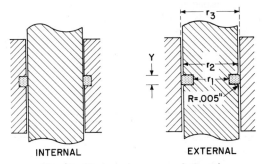

Fig. 68 Rectangular-groove shaft seals.

on the position of the groove. The latter is commonly rectangular, and the design dimensions in Table 18 apply if $(r_2 - r_1)/2$ is taken to be x. Shaft seals may serve for a variety of purposes. Common are push-pull rotary seals of this type, which require a clearance $r_3 - r_2$ of typically 0.005 in. Less clearance is needed if the seals are used in the static mode, whereas heavy demands in regard to sliding motion require somewhat increased groove dimensions to reduce the compression and wear of the gasket. The moving surfaces should have finishes of better than 16 μin. rms and be without visible scratches. Lubrication with vacuum grease is generally necessary.

An alternate way of sealing an O ring against the outer surface of a round shaft or tube is illustrated in Fig. 69. Here, the gasket is compressed by tightening a nut which pushes a metal washer against the O ring. The inner surface of the washer is

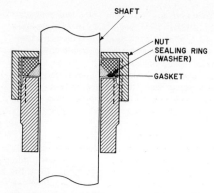

Fig. 69 Demountable compression coupling with conical seal.

conical and hence presses the gasket into a triangular cross section. The design rules for proper confinement of such conical seals are discussed in Ref. 248 (p. 369).

In general, properly designed O-ring gasket seals provide reliable and readily demountable vacuum seals. In assembling these seals, it is recommended practice to clean the flange surfaces carefully and inspect them for damage, particularly radial scratches. Likewise, the O ring itself must be clean and should be examined. Surface imperfections are best recognized if the gasket is slightly stretched. Greasing of O rings is fairly common; it permits the gasket to slip into place properly when the sealing pressure is applied and may also seal some minute leakage channels. However, the effect is temporary, and the

practice is at best a compromise. Another important factor is the proper O ring size. If either the cross section or the diameter is too large for the confining groove, the excess material is extruded when compressed and the O ring is permanently damaged. Conversely, an O ring whose cross section or diameter is too small may be made to fit the groove but will not achieve the proper compression ratio.

A very important seal which cannot always be made with an O ring is that between the chamber and the baseplate of a vacuum system. In the case of metal bell jars, grooving of the flange surface is possible, and limited-compression seals of rectangular or trapezoidal profiles are therefore common. Glass bell jars, however, require a different approach. One possibility is a relatively thick, flat gasket which is laterally confined but free to expand upward as shown in Fig. 70A. The function of the groove here is to prevent slippage of the gasket when the chamber is opened or closed. The same purpose is also accomplished without a retaining groove by using an L-shaped gasket which fits the bell jar tightly, as shown in Fig. 70B. In either case, the sealing surface of the glass bell jar must be ground and polished to provide a leaktight enclosure. The sealing force is atmospheric pressure and may thus vary depending on the diameter and wall thickness of the chamber. Slight greasing of the gasket to improve the tightness of only moderately compressed seals is

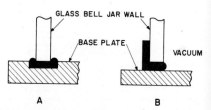

Fig. 70 Glass bell jar-to-base plate seals. A = flat gasket of rectangular cross section. B = gasket with L-shaped cross section.

often practiced. In every case, however, the relatively large exposed gasket surface constitutes a significant source of outgassing and permeation.

The problem of leakage between gasket and flange surfaces has led to some special developments. Since leak rates are proportional to the pressure difference across pores or channels, reduction of the external pressure yields an improvement in the ultimate internal pressure. Implementation of this approach requires a double gasket seal with two concentric O rings. The space between these two seals is referred to as the guard vacuum and may be evacuated by a separate rotary oil pump. Establishing a guard vacuum also reduces the rates of permeation through the gasket. If outgassing has been reduced to very low levels, permeation may become the predominant factor contributing to the residual-gas load. This may be the case at 10^{-9} Torr or below, and under these conditions the guard vacuum can result in reduced ultimate pressures. Further improvement can be obtained by moderate cooling ($-20°C$) of elastomer gaskets, as was shown by Farkass and Barry[276] [see also Sec. 3e(4)].

The non-cross-linked fluorinated hydrocarbons listed in Table 17 require different seal designs if they are to be used as gasket materials. The relatively low permeability and wide range of service temperatures of Teflon, for example, have stimulated attempts to employ it in high-vacuum seals. As the material flows away from the flange surface when compressed, the seal must be of the unlimited type, as shown in Fig. 65A. However, some means of maintaining the seal pressure such as sprig-loaded tightening bolts is required. Round O rings consisting of a rubber core with a Teflon coating are also available. These take advantage of the elastic properties of the core and the excellent lubricating properties of Teflon. They are often used in rotary-shaft seals.

Other types of Teflon seals circumvent the flow problem by fully confining the gasket. A groove of rectangular cross section with the gasket compressed by a tight-fitting flange as shown in Fig. 71A makes a vacuum-tight seal. Compression of less than 10% of the original gasket height is sufficient but requires considerably greater force than rubber gaskets. This sealing force is reduced if the flange surfaces have mating concentric ridges and grooves as shown in Fig. 71B. Because of their high compression-set values, Teflon gaskets do not revert to their original shapes, and the remaining deformation is especially strong after temperature cycling. However, the gaskets can be reused if indentations and other irregularities are removed by annealing at 330°C (see Ref. 248, p. 323).

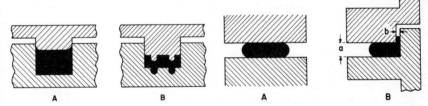

Fig. 71 Teflon gasket seals. A = groove seal. B = ridged-groove seal.

Fig. 72 Metal-wire seals. A = planar seal. B = corner seal.

(3) Metal Gasket Seals Metal gaskets are utilized to seal vacuum systems intended for pressures below 10^{-8} Torr which require outgassing at temperatures around 400°C. The commonest gasket material is OFHC copper, followed by aluminum and gold. Indium gaskets are used occasionally, but the low melting point of 156°C permits only mild bakeout cycles. In general, metal gaskets have lower permeabilities than elastomers but require higher sealing pressures and greater care in seal preparation, and they are rarely reusable. Hence, metal gasket seals are relatively expensive.

The equivalent of elastomer O rings are metal gaskets of circular cross section which are used in so-called wire seals. The simplest configuration is a wire gasket between two flat flanges as shown in Fig. 72A. Indium wire is particularly suitable because the necessary deformation is obtained with relatively small sealing forces. Aluminum-wire seals of this type require greater force but may be heated up to 400°C. When heat-cycled, the aluminum flows and forms a firm bond to stainless-steel flange surfaces because of frictional breakdown of the surface oxide films. These so-called cemented joints are strong enough to withstand the differential contraction upon subsequent cooling.[215] Satisfactory seals can be made with flanges of up to 25 μin. surface finish while the wire gasket is compressed by 80 to 85%. A maximum temperature of 400°C should not be exceeded to avoid the formation of Al-Fe intermetallic compounds.

Gold-wire gaskets must be annealed to a ductile state prior to their use in a seal. If placed between flat flanges, the latter require fairly good surface finishes such as 8 μin. More common is the use of gold-wire gaskets in corner seals as shown in Fig. 72B. Typical wire diameters range from 0.020 to 0.040 in., and these are compressed by about 50% (dimension a in Fig. 72B). The recommended radial clearance (dimension b in Fig. 72B) is 0.001 to 0.003 in. (Ref. 248, p. 380). Flange surface finishes of 16 μin. are adequate. The seals may be cycled repeatedly up to 450°C without forming leaks. An alternative to corner seals are gold-wire spacer seals as described by Hawrylak.[277]

Copper-wire gaskets are rarely used because the difference in thermal expansion between the gasket and stainless-steel flanges produces leaks during heat cycling. Seals can be made, however, if the wire gasket is confined in a groove which limits the flow of the wire metal. An example is the copper-wire seal described by Wheeler,[278] which remained leaktight after heat cycling up to 450°C.

To obtain high sealing pressure between flat flanges without resorting to excessive force, metal gaskets of different cross sections are sometimes used. They are normally made of copper. An example is the coined gasket seal shown in Fig. 73. When compressed, the raised center portions are pushed into the gasket body. It is necessary to apply the sealing pressure uniformly and controllably, for example, by tightening the bolts with an adjustable-torque wrench. A gasket may then be used repeatedly, whereby the torque applied must be successively increased. With a copper gasket whose cross section was similar to the one in Fig. 73, Wheeler and Carlson[279] obtained 35 seals, but they consider only 4 seals to be practical. The sealing force was of the order of 2,000 lb per inch of seal circumference. Flange surface finishes of 32 μin. are satisfactory, but surface flatness is very critical (0.001 in.). The results obtained in sealing and temperature cycling with metal gaskets of various cross-sectional shapes are reviewed in Ref. 248 (p. 410).

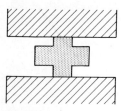

Fig. 73 Coined metal gasket seal.

The most ubiquitous types of metal seals are those made by shearing a flat copper ring of rectangular cross section. Various seal geometries based on this principle are shown in Fig. 74. The step seals A and B are designed to shear the gasket to about one-half its original thickness. One distinguishes clearance and overlapped-step

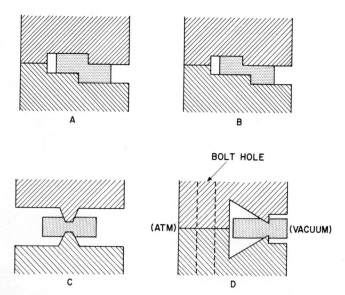

Fig. 74 Seals with flat metal gaskets of rectangular cross section. A = clearance step seal. B = overlapped step seal. C = knife-edge seal. D = ConFlat* seal.

* Trademark of Varian Associates, Palo Alto, Calif.

seals depending on whether the vertical flange edges are offset (Fig. 74A) or over-lapping (Fig. 74B). Comparative tests with 0.040-in.-thick copper gaskets have been described by Wheeler and Carlson.[279] Although a gasket may be used several times if the tightening force is controlled and successively increased, this is not recommended practice.

An alternate seal geometry for metal gaskets of rectangular cross section is shown in Fig. 74C. The seal is formed at the sides of the knife-edges, whereby the torque must be controlled to limit the depth of the bite. The copper gaskets are 0.010 to 0.040 in. thick, and the knife-edges form angles of 30 to 45° with the normal to the surface (Ref. 248, p. 427). The blunt knife-edge is necessary to ensure good contact between flange and gasket. Relatively narrow edge widths of 0.003 to 0.010 in. are recommended to keep the sealing force small. Surface finishes of 32 μin. are satisfactory.[279] The ConFlat* seal in Fig. 74D is also made between conical flange surfaces, but the depth of the bite is limited to 0.012 to 0.015 in. on both sides of the copper gasket.[279] The latter is 0.080 in. thick and 0.250 in. wide. If the gasket is to be reused, the flanges must not be tightened until their surfaces meet as shown in Fig. 74D. Although as many as 22 successive seals have been reported, 3 sealing cycles are a more realistic figure.[279]

c. Vacuum Feedthroughs

The transfer of electric current and mechanical motion into the vacuum chamber is a prerequisite for nearly all types of vacuum work including thin film deposition. Transmission of energy through the vacuum enclosure is facilitated by subassemblies known as feedthroughs. These are flange-mounted to the bell jar or baseplate, or to a vacuum collar of matching diameter which provides additional portholes. Elastomer gaskets are commonly used to seal the flange, and thus demounting for service or repair work is easily accomplished, but for the same reason feedthroughs are often the cause of leaks. It is therefore advisable to select and install feedthroughs carefully for the specific purpose on hand. A variety of electrical and mechanical motion feedthroughs are offered as stock items by vacuum-equipment manufacturers. Most of them include at least one permanent seal of the glass-to-metal or ceramic-to-metal type. Some motion feedthroughs also incorporate metal bellows assemblies which are difficult to join in nonspecialized machine shops. Thus, selection of a suitable feedthrough requires some familiarity with the different varieties available and their specific capabilities.

(1) Electrical Feedthroughs The design and construction of electrical feedthroughs vary according to the currents and voltages which they are intended to transmit. *High-current low-voltage feedthroughs* are needed to energize resistance-heated evaporation sources. For this purpose, voltages of less than 50 V are generally sufficient. The current requirements depend on the size and type of the evaporation source and range from 50 to 200 A for small boat or crucible sources, up to 1,000 A for some large-area low-resistance sources. OFHC copper rods with diameters between $\frac{1}{4}$ and $\frac{3}{4}$ in. are used to transfer currents of 150 to 600 A into vacuum systems. The demands made on electrical insulation against the chamber walls are slight because of the relatively low voltages.

Different types of high-current feedthroughs are shown in Fig. 75. The copper rod in Fig. 75A is insulated from the baseplate by the O ring in its flange. An external water-cooled jacket prevents overheating of the elastomer gasket and permits continuous operation at 400 to 500 A. The cooling water must be supplied without grounding the feedthrough; rubber or plastic tubing is suitable for this purpose. Cooling by conduction through the length of the copper rod is not always adequate. Examples are situations where the evaporation-filament support clamp should remain at a low temperature, or where passage of currents greater than 500 A is required. In these cases, the internally cooled feedthrough shown in Fig. 75B is preferable. It consists of a $\frac{1}{2}$-in. closed copper tube with a smaller internal tube for water circula-

---

* Trademark of Varian Associates, Palo Alto, Calif.

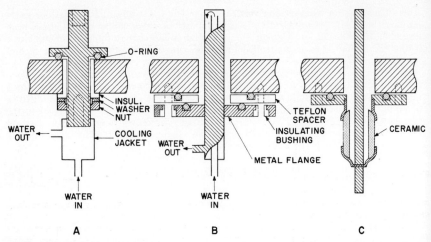

Fig. 75 High-current feedthroughs. A = massive copper rod with external water cooling. B = copper tubing with internal water cooling. C = copper rod with ceramic insulation.

tion. The feedthrough tube is brazed to a flange, and the latter is electrically insulated by a Teflon spacer and phenolic-resin bushings. Feedthroughs of this type may be used up to 1,000 A.

Current feedthroughs of the type shown in Fig. 75C involve a ceramic-to-metal seal. They are available in different sizes for passage of currents from 100 to 600 A. While water cooling is not required, there are similar versions with hollow rather than solid rods to permit water cooling of internal parts. An advantage of the solid-rod types is their ability to withstand baking at 450°C. However, ceramic-to-metal seals are sensitive to rapid temperature cycles and should not be exposed to changes faster than 25 deg min⁻¹.

In the design of *high-voltage feedthroughs*, the breakdown voltage of the insulator is the primary consideration. In addition, the electrode material and its shape and spacing from surrounding surfaces are of importance. High-voltage feedthroughs are normally rated for breakdown in air. It should be borne in mind, however, that breakdown may occur at lower voltages because of reduced gas pressure. This applies in particular to the operation of sputtering systems in the 1- to 100-mTorr range.

In fabricating high-voltage feedthroughs, ceramic-to-metal seals in one form or another are invariably used. Current feedthroughs of the type shown in Fig. 75C, for example, may also serve as voltage feedthroughs. Some designs are capable of transmitting up to 10 kV. However, many high-voltage applications do not require high currents and heavy conductor rods. In sputtering, for example, the currents rarely exceed a few amperes, and conductors of less than 0.060 in. diameter are sufficient. Shielding of the high-voltage-carrying internal lead is necessary to prevent it from being sputtered. The feedthroughs for electron-gun sources, on the other hand, often carry 10- to 50-A currents and require larger diameters such as $\frac{1}{8}$ to $\frac{1}{4}$ in.

For voltages from 15 to 25 kV, feedthroughs of the type shown in Fig. 76 are commercially available. Their current-carrying capability ranges from 1 to 25 A depending on the design and the diameter of the metal conductor. The shape of the ceramic body is determined by the objectives of reducing surface leakage currents

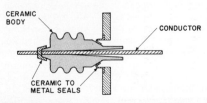

Fig. 76 High-voltage (>15 kV) feedthrough for low-current (< 25 A) requirements.

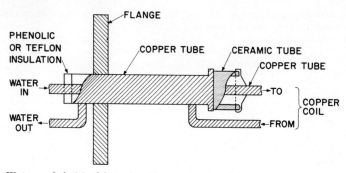

Fig. 77 Water-cooled rf feedthrough with ceramic-to-metal seal on the vacuum side only.

and of increasing flashover and corona-start voltages. Such feedthroughs are used for electron-gun sources and for glow-discharge cleaning arrangements.

The feedthrough shown in Fig. 77 is designed for transmitting radio-frequency (400 kHz to 10 MHz) power into a vacuum chamber. Two concentric copper tubes serve as power leads and also to circulate cooling water. Internally they are connected to an rf induction coil which may be used to heat an evaporation source or to braze vacuum components. Whereas a ceramic-to-metal seal is utilized on the vacuum side, a less expensive organic insulator is often adequate to separate the two leads on the external section and to seal the water-circulation system. For power levels in excess of 10 kW and frequencies higher than 1 MHz, however, ceramic-to-metal seals on both sides are recommended.

Electrical feedthroughs for *low current and low voltage* allow close spacing of leads without much concern for insulation resistance. Therefore, it is practical to group them closely together in header assemblies which provide for many leads with only one flange seal. Their current-carrying ability is limited by the temperature rise due to Joule heating of the wire and the resulting stress in the metal-to-insulator seal. To assess the current capacity of the copper leads, the current levels which produce 100°C wire temperatures under continuous duty are listed in Table 19. The values for single wire in air are considered permissible for copper sealed to glass. The values for cables in conduits are more representative for vacuum conditions and often

TABLE 19 Current Levels Which Produce a Temperature of 100°C in Copper Wire as Dependent on Wire Size*

Copper-wire size		Current level, A, for	
AWG (B&S)	Diam, 10^{-3} in.	Single wire in air	Cables in conduits
8	128.5	73	46
10	101.9	55	33
12	80.8	41	23
14	64.0	32	17
16	50.8	22	13
18	40.3	16	10
20	32.0	11	7.5
22	25.3	...	5.0

* From B. Goodman (ed.), "Radio Amateur's Handbook," 42d ed., p. 509, The American Radio Relay League, Inc., 1965.

identical with the manufacturer's suggested safe operating limits.　Maximum current levels for other metals can be found in Ref. 248 (p. 454).

Multiple feedthroughs are widely used to bring thermocouple, rate monitor, ionization gauge, and substrate heater leads into vacuum systems.　Several commercially available assemblies are shown in Fig. 78.　Type A represents the so-called hermetic

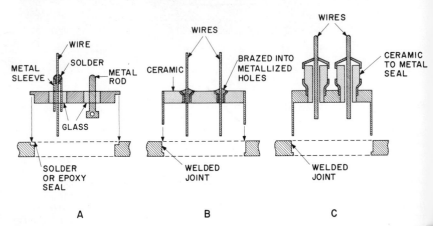

Fig. 78　Multiple feedthroughs for low current and low voltage.　A = hollow and solid pins with glass-to-metal seal.　B and C = bakeable versions utilizing ceramic-to-metal seals.

or compression seals which contain lead wires of nickel-iron alloys in either tubular or solid-rod form.　The tubular type is particularly suitable for thermocouple leads since it allows insertion of the wires without discontinuity.　However, it is sometimes difficult to produce leaktight seals because of uneven wetting of the solder.　The solid-wire feedthroughs are available in diameters up to 0.090 in. with solder lugs on the external side.　They are rated for currents up to 20 A and for 2 kV.　Pin configurations corresponding to the well-known octal-tube socket are most popular since the latter allows for quick external cable connections.　The feedthrough header may be sealed to a suitable flange either by soldering or with an epoxy resin.

Figure 78B and C shows multiple feedthroughs involving ceramic-to-metal seals. These can be welded to demountable flanges and therefore withstand baking temperatures of 300 to 450°C.　They may be ordered with specific lead wires for thermocouples or with current-carrying wires (nickel, Kovar) already brazed into the metal sleeves.　Feedthrough assemblies of this type are rated for 5 to 7 A current and voltages up to 6 kV.

(2) Motion Feedthroughs　The transmission of linear, rotary, and tilting motion into vacuum systems is essential for the operation of certain internal parts such as shutters, beam choppers, mask and substrate changers, and other functions.　A great variety of motion feedthroughs has been developed and tried (see Ref. 248, p. 491). The commoner designs utilize metal shafts sealed with elastomer gaskets, metal bellows, or magnetic coupling to move internal vacuum parts.

A number of *shaft seals with elastomer gaskets* are shown in Fig. 79.　Type A is an internal double O-ring shaft seal [see Sec. 4b(2)] which allows push-pull as well as rotary operation.　The two O rings center the shaft, but external ball bearings are sometimes provided in addition to ensure alignment.　The shaft may be lubricated with a silicone grease of low vapor pressure.　This is particularly important if the feedthrough is used extensively for linear motion.　For rotary motion, self-lubricating Teflon-coated gaskets may be utilized.　The space between the shaft and the body may either be evacuated to provide a guard vacuum, or it may be filled with oil or grease for lubricating purposes.　The latter mode of construction is typical for high

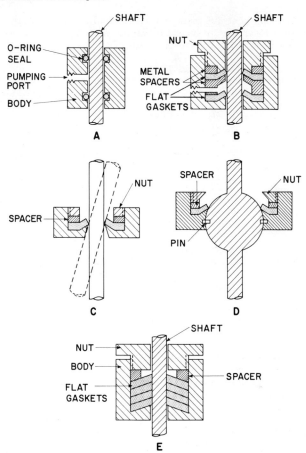

Fig. 79 Motion feedthroughs with elastomer-sealed shafts. A = internal shaft seal. B, C, and D = Wilson or lip seals. E = chevron seal.

vacuum valves actuated by linear motion. Feedthroughs of this type are commercially available with shaft diameters from 0.125 to 2.00 in. Linear travel up to 4 in. and rotational speeds of 500 rpm are typical specifications. Some oil-lubricated shaft seals permit speeds greater than 1,000 rpm at leak rates of less than 10^{-5} cm³ (STP) s⁻¹ but no linear motion. Utilization of shaft seals on vacuum systems maintaining pressures below 10^{-7} Torr is problematical, in particular if push-pull operation is required. The latter often causes pressure bursts of up to two orders of magnitude depending on the length of travel, speed of movement, and type of lubricant.

The feedthrough in Fig. 79B represents a Wilson or lip seal.[280] Characteristic for this category are flat elastomer gaskets with a center hole whose diameter is 20 to 35 % smaller than the shaft. The gaskets are supported at an angle of 30° and forced against the shaft by a compression nut. A good surface finish and lubrication of the shaft are essential to minimize leakage. The space between the two parallel seals is usually filled with a low-vapor-pressure oil, but silicon-grease lubrication and guard vacua are also employed. Shaft diameters range from $\frac{1}{16}$ to $2\frac{1}{2}$ in. Static leak rates through Wilson seals are of the order of 10^{-6} cm³ (STP) s⁻¹. They increase only by a factor of 2 during slow (about 60 rpm) rotation (Ref. 248, p. 520). Linear motion causes higher leak rates, particularly when the shaft is moved into the chamber.

Wilson seals of somewhat different design may also be utilized to provide rocking or tilting motion. Examples are shown in Fig. 79C and D. The former type can be rotated and pushed as well, whereas the latter permits only rocking motion around the pin axis.

The push-pull or rotary feedthrough in Fig. 79E is known as a chevron seal. The stacked gaskets provide a large seal area and good alignment of the shaft. The sealing force is supplied by the tightening nut. Gasket lubrication is essential. Chevron seals are mainly used to seal large-diameter shafts in situations where only small linear or rotational speeds are required. Modifications of the basic type shown in Fig. 79E include separation of gaskets to establish a guard vacuum, and spring-loaded washers between the gaskets to increase the sealing pressure.[281]

In general, the use of elastomer-gasket shaft seals is limited to systems operating in the high-vacuum range. They are particularly satisfactory where only intermittent rotary motion is required. Increasing leak rates are encountered at high speeds of rotation and even more so during sliding motion. Continuous use aggravates the problem because of wear of the gasket and loss of the lubricant. To avoid these difficulties, another type of motion feedthrough which is based on metal bellows may be used. Some of these are bakeable and can therefore be applied in ultrahigh-vacuum systems.

Bellows are thin-walled metal tubes consisting of many tubular segments which render them flexible as well as stretchable. They are fabricated either by welding rings (diaphragms) alternately on their inner and outer circumferences together (Fig. 80A), or by forming a thin tube under hydraulic pressure on a die (Fig. 80B). For the first type, weldable materials like stainless steel and nickel alloys are commonly used, whereas the formed bellows require ductile metals such as phosphor bronze, brass, or beryllium copper. The flexibility of the bellows depends on the elasticity and thickness of the metal and the number and depth of the convolutions. The latter can be made deeper in welded bellows, which therefore allow greater flexibility. Ratios of extended to compressed lengths up to 6:1 are possible.

With few exceptions, bellows-type feedthroughs are designed either for push-pull action or for rotation. The former type is of rather limited utility insofar as com-

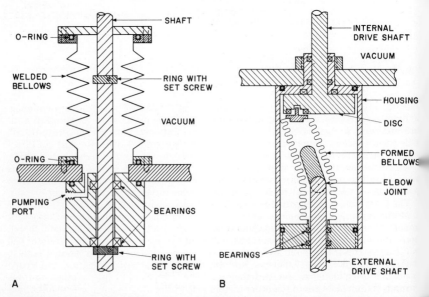

Fig. 80 Motion feedthroughs with metal bellows. *A* = welded-bellows design for linear motion. *B* = formed-bellows design for rotary motion.

mercial versions permit linear motion only up to 1 in. In addition, they are expensive and tend to be bulky because the shaft has to be long enough to accommodate the compressed bellows, and the outer diameter of the bellows must be much wider than the shaft to provide maximum extension.

One type of linear-motion bellows seal is shown in Fig. 80A. On the vacuum side, the bellows are welded to a demountable flange seal to facilitate leak testing and repairs. The shaft is supported externally by two bearings. The more flexible bellows often require evacuation of the internal space to lower the pressure difference across their walls. This reduces the effort required for actuation, but more importantly, failure to do so causes twisting of the bellows, which may overstress the material. To guard against accidental overextension or compression, the upward and downward motions are limited by pins or rings with setscrews on both ends of the shaft. The maximum allowable pressure gradients and extended lengths are specified by the manufacturer. When ordering a bellows seal for a linear-motion feedthrough, it is advisable to buy the bellows already attached to the desired flanges since assembly from individual parts involves welding of the thin-walled tube, which is a delicate operation [see Sec. 4a(2)].

A rotary-motion feedthrough sealed with bellows is shown in Fig. 80B. The externally produced torque is transmitted by the bellows section, which is eccentrically coupled to a disk at the end of the internal drive shaft. In some designs, additional mechanical coupling through the external shaft is possible by means of two elbow joints. Since rotation causes the bellows to flex continually, the eccentricity of the disk coupling should be small in order to increase the permissible rotational speed and the life expectancy of the bellows. However, the transmitted torque is thereby reduced. The clearance space between the internal drive shaft and the housing is often not sealed but contains pumping portholes so that the housing can be evacuated. Thus, the bearings supporting the upper drive shaft are exposed to the vacuum in the system. Rotary-motion feedthroughs of this design are typically rated for speeds up to 500 rpm and torques of 10 to 25 in.-lb. Completely welded models and others having metal gaskets are also available and withstand baking at temperatures of 250 to 300°C and higher.

An alternate method of inducing motion in a vacuum is by magnetic coupling through the wall of the vacuum enclosure. This requires permanent or electromagnets external to the system which can be manipulated in the desired mode of motion, and a magnetic object such as a soft iron rod or ring within the chamber. Close spacing and a nonmagnetic wall are essential for effective energy transmission. Arrangements for various special applications are described in Ref. 248 (p. 540). Induction of rotary motion is the most frequent form of utilization, and a simple way of implementation is shown schematically in Fig. 81. Rotation of the external magnets is transmitted to the interval drive shaft. In commercial versions of this type, the magnets can be removed, and the remaining assembly may be heated up to 400°C. Rotational speeds of 750 rpm at 6 ft-lb torque are attainable. Higher speeds up to 10,000 rpm are also possible with some designs, but they are rarely required in the deposition of thin films. At speeds higher than 1,000 rpm, the available torque is generally small and

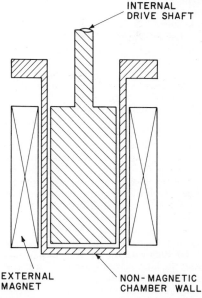

INTERNAL DRIVE SHAFT

EXTERNAL MAGNET

NON-MAGNETIC CHAMBER WALL

Fig. 81 Magnetically coupled rotary feedthrough.

of the order of a few inch-ounces. By virtue of their construction, magnetically coupled rotary feedthroughs ensure leak-free operation and are therefore preferred in ultrahigh-vacuum systems.

d. Vacuum Valves

Vacuum valves are needed to separate different parts of a vacuum system from each other and from the surrounding atmosphere. Depending on the specific function required, different mechanisms are utilized to control the gas flow through the valve. Motion feedthroughs to actuate the opening and closing mechanism are a necessary part of all valves. General valve requirements are maximum conductance in the open state and minimum outgassing and leakage. Adequate conductance is ensured if the cross-sectional area of the open valve is as large as that of the entrance port. Outgassing is kept at low levels by using mostly metals for valve construction and by avoiding exposure of the internal surfaces to air as much as possible. While vacuum grease is often required for lubrication, it should be applied sparingly. Another source of outgassing is elastomer gaskets, which are used in all valves except those specifically designed to be bakeable.

Leakage through valves is often associated with the seal around the shaft of the motion feedthrough. Therefore, the type of shaft seal utilized in a valve is an important distinguishing feature. One speaks of packed valves if the shaft is sealed with elastomer gaskets of the types shown in Fig. 79. In contrast, packless valves are characterized by motion transmitted through bellows seals or by magnetic coupling. Gas leakage through a well-constructed packed valve stem should be less than 10^{-5} cm^3 (STP) s^{-1}, whereas packless construction yields leak rates which are at least two orders of magnitude lower (see Ref. 248, p. 616). For this reason, packless valves are essential to the assembly of ultrahigh-vacuum systems, but they are also used often in conventional systems to minimize leakage. While the type of shaft seal is a general characteristic, more specific designations refer to the functions of valves in vacuum systems.

(1) Roughing and Fore-line Valves Roughing and fore-line valves serve to connect or disconnect fore-vacuum pumps with the vacuum chamber or with a high-vacuum pump. Stainless steel or brass are commonly used in their construction, and connections to the vacuum lines are made by welding or brazing. Two valve types which are suitable for roughing as well as fore-line applications are shown in Fig. 82.

Disk valves are sealed at the valve seat by means of an elastomer gasket. This may be a flat disk held in place by a washer and screw as shown in Fig. 82A or an O ring contained in a groove in the metal disk. The valve stem may also be sealed by elastomer gaskets (packed), but this mode of construction tends to develop leakage when the shaft is moved. The packless design shown in Fig. 82A is less prone to leak and therefore more widely used. Commercial versions with motor-driven or manually or pneumatically actuated shafts and port sizes ranging from $\frac{1}{2}$ to 2 in. diameter are available. Depending on the presence or absence of an elbow section (dashed outline in Fig. 82A), the valves are referred to as either in-line or right-angle types.

In *ball valves*, the main seal is achieved by two elastomer rings which are pressed against the surface of a metal sphere as shown in Fig. 82B. The sphere has a large-diameter bore which can be rotated by the stem to open or close the valve. The shaft and body are sealed by O rings. Ball valves are susceptible to leakage, particularly during opening and closing. Periodic lubrication of the gaskets is necessary to reduce wear and deterioration of vacuum-tightness. For these reasons, they are less often used in modern vacuum systems than disk valves. Commercial ball valves have portholes from $\frac{1}{2}$ to 2 in. diameter and are usually for manual operation, but pneumatically actuated versions, too, are available.

(2) High-vacuum Valves High-vacuum valves are located between the chamber and the high-vacuum pump. Their prime requirement is high conductance to ensure maximum pumping speed. Furthermore, the internal surfaces of high-vacuum valves are a part of the high-vacuum enclosure in the open position; hence minimum leakage and outgassing are essential. To reduce adsorption of air as much as possible, the

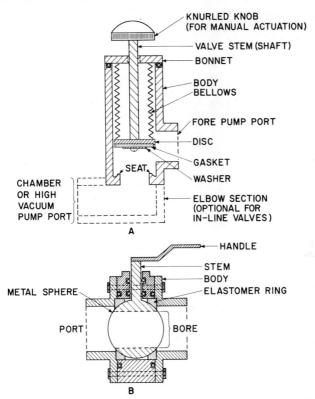

Fig. 82 Roughing and fore-line valves. *A* = packless disk valve. *B* = packed ball valve.

valves are generally installed so that their interior remains connected to the high-vacuum pump when the valve is closed and the chamber vented.

The most widely encountered high-vacuum valves are of the *gate or sliding type.* Although their internal mechanisms vary among different manufacturers, their operating principle is basically that illustrated in Fig. 83. Valve closure is facilitated by a disk carrying an O ring in a retaining groove. This assembly is pressed against the chamber port surface when the stem is moved down all the way, transmitting force through the linkage between carriage and disk. Ball bearings as shown in Fig. 83 are often utilized to permit sliding against the valve body wall. The disk falls back on the carriage upon retraction of the shaft. If installed as indicated in Fig. 83, the interior of the valve with the exception of the disk surface remains under vacuum while the chamber is opened to air. This is the preferred mode of utilization, but it requires a sealing force normal to the disk which is greater than the atmospheric pressure on the chamber side.

Packed-shaft seals of either the double-O-ring or the Wilson type (see Fig. 79) are common because leakage through the seal is negligible as long as the shaft is not moved. Although the leak rates increase during opening or closing of the valve, this is usually tolerable as it occurs either at the beginning of the pump-down or shortly before venting the system. The added expense of packless construction is warranted only in special situations. An example is sputtering systems, which require actuation of the valve at a more critical time of the cycle. In this case, the high-vacuum valve must be partially closed (throttled) when the filling gas is admitted in order to

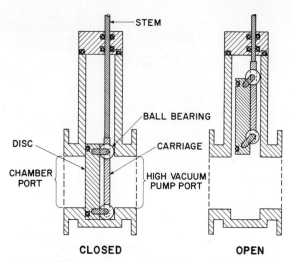

CLOSED   **OPEN**

Fig. 83   High-vacuum gate valve with packed-shaft seal.

maintain the gas throughput within the rated limits of the high-vacuum pump.   As this occurs after high vacuum has been established, bursts of air due to movement of the shaft adversely affect the purity of the sputtering gas.

Gate-valve bodies and internal parts are fabricated from mild-steel, stainless-steel, or cast-aluminum alloys.   They are usually designed for flange mounting.   Available sizes encompass porthole diameters from 2 to 60 in.; the larger valves are mostly for pneumatic rather than manual operation.   The principal advantage of gate valves is their high conductance, since retraction of the disk yields a wide, unobstructed through-hole.   Furthermore, the flange-to-flange distance of gate valves is short, typical valve thicknesses being less than 3 in. for porthole diameters up to 8 in.   In addition to serving as high-vacuum valves, small gate valves are sometimes used in roughing or fore-vacuum lines.

Other types of high-vacuum valves are shown in Fig. 84.   The *disk valve* in Fig. 84*A* is often employed if a right angle between port holes is desired.   This is the case, for instance, with large (>10 in.) diffusion pumps where chamber installation in a cantilevered position gives a more convenient working height above the floor.   Generally, right-angle disk valves have packed-shaft seals of the double-O-ring type.   An additional gasket in the bonnet and a back seal plate on the stem as shown in Fig. 84*A* offer additional protection against leakage along the shaft while the valve is in the open position (see Ref. 281, p. 354).   High-vacuum disk valves with metal bellows seals are also available and preferable if the valve is to be used for throttling of the gas throughput.   In the interest of keeping gas adsorption at a minimum, installation with the disk sealing against the air-filled chamber as in Fig. 84*A* is recommended.   Porthole sizes from 2 to 35 in. diameter are commercially available.   Typically, the smaller valves (<6 in.) are made from cast aluminum, whereas the larger ones are fabricated from mild steel.   As in the case of gate valves, large disk valves are usually for pneumatic operation; the smaller versions may be for either pneumatic or manual actuation.

The principle of *flap valves* is illustrated in Fig. 84*B*.   The valve closes by means of an O-ring gasket in a grooved disk.   The latter is attached to a lever- or cam-type mechanism which permits tilting and lifting of the disk from the valve seat.   The actuating stem may be a rotary or a push-pull feedthrough.   Right-angle as well as in-line flap valves have been made.   The best designs are those which allow complete retraction of the disk from the gas passage, and a comparison of the conductances of otherwise similar models will reveal to what extent this has been achieved.   Another

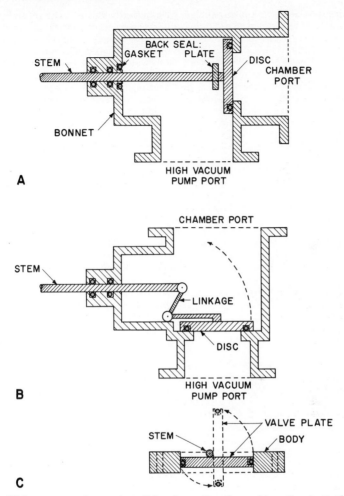

Fig. 84 High-vacuum valves. A = disk valve. B = flap valve. C = butterfly valve.

design problem with flap valves is the transmission of a sufficient sealing force to the disk.   This is the reason why flap valves are generally installed as shown in Fig. 84B, where atmospheric pressure on the chamber side aids in keeping the valve seal tight. However, the internal surfaces of the valve and of the actuator are thereby allowed to adsorb air, which is desorbed when the chamber is evacuated.   Therefore, flap valves are not as extensively used as gate and disk types.   They have recently come into more widespread usage in vacuum locks designed for horizontal loading of substrates into evacuated work chambers.

The *butterfly valve* shown in Fig. 84C is of relatively simple construction.   Its body consists of a ring whose inner surface makes a seal against an O ring recessed in the perimeter of a circular valve plate.   Opening or closing is facilitated by making a 90° turn of the stem (perpendicular to the plane of the drawing).   External shaft seals (see Fig. 68) are often used in this type of valve.   Their main advantage is the small size, since the ring-shaped body is usually less than $1\frac{1}{2}$ in. thick.   However, clearance on both sides of the valve is necessary to allow the plate to rotate.   Furthermore, the conductance of the through hole is somewhat reduced by the presence of the plate in the open position. A known problem with butterfly valves is the reliability of the O-ring seal. A well-designed retaining groove and gasket lubrication are essential for this purpose. However, repeated opening and closing of the valve wipes off the lubricant, and the resulting friction tends to abrade the gasket or even force it out of the groove. Hence, butterfly valves are not as widely used as the other types. They also serve occasionally as roughing, fore-line, or throttle valves. Common sizes are of 1 to 6 in. inner diameter and may be either pneumatically or manually actuated.

All the valves described in the preceding paragraphs employ elastomer gaskets and are therefore not suitable for ultrahigh-vacuum systems. The latter application requires *bakeable valves* constructed completely of metal and capable of withstanding temperatures up to at least 400°C. The use of elastomer gaskets in the motion feed-through seal can be circumvented by resorting to metal bellows. The substitution of metal gaskets for the elastomer at the valve seat is possible in only some of the designs discussed previously. A common type of bakeable valve is similar to the disk valve illustrated in Fig. 82A. The valve seat is machined into a knife-edge, and closure is made by forcing a copper gasket or a soft metal piston against the seat. The torque necessary to produce a vacuum-tight seal is quite high and increases further with repeated closures. Therefore, bakeable valves are mostly small with portholes of less than 2 in. diameter. The useful life of the seal is limited and may permit as few as 10 or in some cases as many as 1,000 closures, depending on the design, the torque required, and the severity of the bakeout cycle. For other types of bakeable high-vacuum valves, the reader is referred to the literature (see, for example, Ref. 248, p. 638).

(3) Gas-admittance Valves To vent a vacuum chamber or to admit a controlled flow of gas, small valves of relatively simple construction as shown in Fig. 85 are

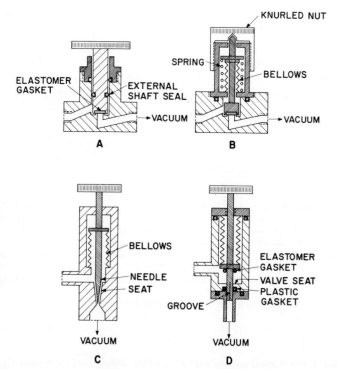

Fig. 85 Gas-admittance valves. A = packed disk valve. B = packless disk valve. C = packless needle valve. D = packless variable-leak valve.

utilized. The types A and B are on-off valves which seal by means of elastomer disks. They are connected to the vacuum chamber by short lines of $\frac{1}{4}$ to $\frac{3}{8}$ in. diameter. Threaded pipe joints wrapped with Teflon tape [see Sec. $4b(1)$] provide a reliable seal. The packed valve in Fig. 85A is used exclusively for venting of vacuum chambers since the shaft seal may leak when the valve is open. The packless type shown in Fig. 85B does not leak in the open position and is therefore preferable as an admittance valve for high-purity gases.

Figure 85C and D are two common designs of variable-leak valves (see Ref. 248, p. 659). Their purpose is to adjust and maintain a certain gas-flow rate into the vacuum chamber, a requirement which occurs in sputtering, reactive evaporation, calibration of mass spectrometers, and pumping-speed tests. With *needle valves*, flow control is achieved by varying the space between a hard stainless-steel pointed shaft (needle) and a copper or brass seat of matching shape. Very small flow rates can be obtained by forcing the needle into the soft seat, but excessive pressure leads to wear and permanent damage of the seat. It is not recommended to use needle valves to cut off the gas flow completely. In applications such as sputtering, needle valves in conjunction with a packless disk valve as in Fig. 85B are most practical. This combination permits regulation of the gas flow with little adjustment of the needle while the gas can be turned on and off independently. Since needle valves are often used for small leak rates, gasket-sealed shafts are generally undesirable because of the high permeability of elastomers.

Another type of *adjustable-leak valve* is shown in Fig. 85D. It combines both on-off and flow-rate control. Complete closure is made by pressing an elastomer O ring against the valve seat. In the open position, a channel is formed between a narrow groove in the shaft and a hard plastic gasket below the seat. The depth of the groove increases toward the end of the shaft so that the diameter of the leakage channel increases the more the stem is retracted. These valves are suitable for moderately high throughputs of 1 to 100 cm^3 (STP) s^{-1} and serve very well as argon leaks in sputtering systems.

5. DESIGN AND PERFORMANCE OF VACUUM SYSTEMS

In this section, the various types of vacuum systems resulting from combinations of vacuum pumps, parts, materials, and their performance are discussed. The factors determining system performance can be identified in principle, but numerical data for the various gas-contributing processes usually refer to equilibrium situations. Hence, fairly reliable estimates can be made for the ultimate pressures attained. A priori assessments of pump-down times, however, suffer from the inability to formulate exactly the initial conditions of the system and the transient processes involved.

a. The Gas Balance in High-vacuum Systems

The evacuation of a vessel initially at atmospheric pressure begins with turbulent and viscous gas flow. Its duration, however, is only of the order of a few minutes, and little attention is given to this phase of the pump-down. High vacuum is established by various molecular-flow processes which counteract each other. At any moment during the pump-down, the pressure in the system is the result of a dynamic balance between gas removal and gas contributions as shown schematically in Fig. 86. Connecting tubes, traps, or baffles between the chamber and the pump reduce the intrinsic speed S of the latter. If the conductance of the interconnecting section is C, the effective speed S_e at which gas is removed from the chamber is

$$\frac{1}{S_e} = \frac{1}{S} + \frac{1}{C}$$

The resulting loss of gas $S_e p$ in the vessel is partly compensated by outgassing from the walls Q_O, permeation Q_P, air leakage through small cracks or pores Q_L, and vapor emission (back streaming, thermal or ion impact release) from the pump Q_B. The

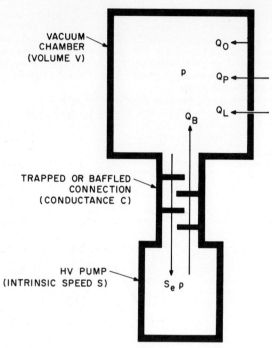

Fig. 86 Schematic representation of the gas balance in high-vacuum systems.

net change in the amount of gas contained in the vessel is

$$V \frac{dp}{dt} = -S_e p + Q_O + Q_P + Q_L + Q_B \qquad (20)$$

To be exact, Eq. (20) has to be formulated for the partial pressures of all gases present since the pumping speeds and contributing processes vary for different molecular species. However, in practice there is usually one process which dominates in a certain pressure range.

To find the pump-down function $p(t)$, Eq. (20) is simplified by assuming that S_e is pressure-independent and by considering only two gas-contributing processes:

$$V \frac{dp}{dt} + S_e p = Q_{\text{const}} + Q(t) \qquad (21)$$

Here, Q_{const} represents all time-independent processes such as air leakage, steady-state permeation, or constant vapor emission (for example, from pump fluids or other high-vapor-pressure materials). Processes whose rates change with time are included in $Q(t)$. This term is usually dominated by outgassing phenomena and may decay rapidly or slowly depending on whether the desorption energy is small or large[282] [see also Sec. 3b(1)]. Eventually, every vacuum system reaches a constant pressure and the time-dependent gas contributions become negligibly small against Q_{const}. Hence the ultimate pressure in a vacuum system is determined by constant-rate (steady-state) processes:

$$p_{\text{ult}} = \frac{Q_{\text{const}}}{S_e} \qquad \left(\text{for } \frac{dp}{dt} = 0, \, t \to \infty \right)$$

The general solution of Eq. (21) for initial pressures of $p_0 < 10^{-3}$ Torr at $t = 0$ is[276]

$$p(t) = \frac{Q_{const}}{S_e} + \left(p_0 - \frac{Q_{const}}{S_e} \right) \exp \left(\frac{-S_e}{V} t \right)$$
$$+ \frac{1}{V} \exp \left(\frac{-S_e}{V} t \right) \int_0^t Q(t) \exp \left(\frac{S_e}{V} t \right) dt \quad (22)$$

The exponential time constant S_e/V is usually of the order of 1 s^{-1}. Considering only the second term of Eq. (22), the system pressure should drop by a factor of 10 about every 2 s. In reality it takes several hours of pumping to approach the ultimate pressure because the gas load during the pump-down is dominated by the time-dependent processes $Q(t)$ represented by the third term in Eq. (22).

Since pressure vs. time observations are widely used to determine outgassing rates of vacuum materials, considerable effort has been devoted toward finding analytical expressions for $Q(t)$. Empirical relations of the general form $Q = Kt^{-\beta}$ have been discussed by Kraus,[283] who assumed $\beta = 1$ for desorption and $\beta = 0.5$ for diffusion-controlled processes, and by Dayton,[284] who admitted variations of β from 0.7 to 2 for desorption. In many cases, these expressions agreed with the actual pump-down curves only for the first 10 h. Attempts to derive the same relationship from theory have been made by Dayton[284] and Kraus.[285] The former assumed processes with distributed desorption energies while the latter developed a first-order desorption-rate equation based on a statistical relation between the surface coverage and the vapor pressure of the adsorbed layer.

As mentioned earlier in Sec. 3c, the failure to consider readsorption of desorbed species in the gas balance led to serious discrepancies with experimental observations.[213] A differential equation with an adsorption term was formulated by Hobson and Earnshaw:[214,286]

$$V \frac{dp}{dt} + (S_e + S_s)p = Q_{const} + \frac{N_a}{\bar{\tau}} \quad (23)$$

Here the desorption rate $N_a/\bar{\tau}$ [Eq. (5)] has been substituted for the time-dependent outgassing rate $Q(t)$, and the product $S_s p$ represents the pumping action of the sorbing surface. The latter includes the kinetic impingement rate and the sticking coefficient and hence is a function of pumping time. Hobson and Earnshaw measured the time constants of both the adsorption and the desorption processes and thus were able to substitute numerical values into Eq. (23). The same two processes also determine the mass balance of the adsorbed layer or "surface phase":

$$\frac{dN_a}{dt} = S_s p - \frac{N_a}{\bar{\tau}} \quad (24)$$

The simultaneous solution of Eqs. (23) and (24) yields several exponential terms whose contributions vary depending on the relative magnitudes of S_e and S_s. Experimental pressure vs. time curves for CO and H_2 pumped in a glass vessel are in better agreement with the solutions derived from Eqs. (23) and (24) than with those based on the simpler gas balance equation (21).[214,286] Similar results for the pumping of Ar at 77°K support the assumption that this surface-phase theory is of general validity.[287]

The discussion of the mass balance in evacuated vessels shows that the performance of vacuum systems not only depends on the capability of the pumps but is also a function of the gas-contributing processes. The latter are determined by the materials utilized in the construction of the system and by their conditioning. Because of the many choices among pumps, materials, and parts, vacuum systems are often unique, which makes it difficult to establish common criteria for their classification. The usual division according to performance into high- and ultrahigh-vacuum systems leads to distinctions based essentially on methods of joining and gasketing, as discussed in Sec. 4. There is, however, considerable freedom in the choice of pumps and chamber materials in each performance category.

b. Demountable High-vacuum Systems

The Committee on Standards of the American Vacuum Society defines this category as "systems with demountable joints and seals, usually containing materials which do not have a very low outgassing rate and in which limiting pressures of 5×10^{-5} to 5×10^{-7} Torr may be obtained by the use of extremely fast pumps in spite of the outgassing load or even the presence of small leaks."[288] They are also referred to as *dynamic vacuum systems, conventional,*[82] or *general-purpose evaporators.*[215] Their mode of construction allows great flexibility in the design of internal fixtures, and the pump-down cycle is relatively short. Therefore, these systems are commonly used to deposit thin films for laboratory or industrial purposes.

A typical demountable system is illustrated in Fig. 87. The use of elastomer

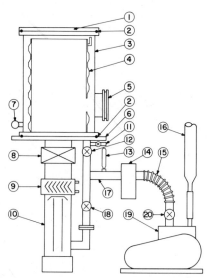

Fig. 87 Demountable high-vacuum system. (1) Top plate; (2) elastomer seal; (3) bell jar; (4) liquid-nitrogen shroud; (5) viewing port; (6) baseplate; (7) pressure gauge; (8) high-vacuum isolation valve; (9) diffusion-pump baffle; (10) diffusion pump; (11) vacuum-release valve; (12) roughing valve; (13) fore-vacuum gauge; (14) fore-line trap; (15) bellows (or rubber hose); (16) exhaust duct; (17) fore-line; (18) fore-vacuum valve; (19) rotary oil pump; (20) fore-line trap isolation valve.

gaskets to seal the chamber and also in valves or feedthroughs is characteristic. The vacuum chamber is usually an 18-in.-diameter bell jar of either glass or stainless steel with a volume of the order of 100 l. Rotary oil and oil diffusion pumps are the preferred means of evacuation for reasons of economy and pump-down speed. An empirical rule for the selection of the diffusion pump is that its speed should be about 5 l s^{-1} for every liter of bell-jar volume.[204] This ratio is reduced by typically a factor of 2 because of the presence of baffles and traps. These requirements are generally met by 4- or 6-in. diffusion pumps with water-cooled baffles. A more detailed discussion of the design and assembly of demountable systems including the factors determining the choice of fore and backing pumps may be found in Ref. 289.

It is common practice to enhance the performance of demountable systems by various means such as refrigerated or multiple baffles, traps, and cryogenic pumps. An example has been described by Caswell,[79] who lowered the ultimate pressure from 5×10^{-6} to 1×10^{-6} Torr by incorporating a Meissner trap. The effectiveness of a liquid-nitrogen-cooled shroud as shown in Fig. 87 is illustrated by the pump-down curve in Fig. 88. The speed of the 4-in. diffusion pump is somewhat marginal for a

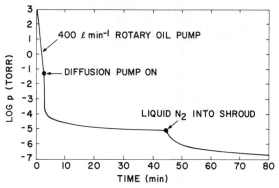

Fig. 88 Pump-down curve of a demountable high-vacuum system.   18-in.-diameter bell jar, 75-l volume, 4-in. diffusion pump with water-cooled baffle, and liquid-nitrogen shroud.[80]

chamber of this size, but the large cryogenic surface area of the shroud compensates this weakness.   A pump-down time of about 1 h is typical for demountable systems. Several hours are required to reach the ultimate pressure, which was about 1×10^{-7} Torr in the example shown.[80]

The residual gases in demountable systems are predominantly due to outgassing and consist mostly of water vapor.   Lesser constituents are CO_2, CO, N_2, and H_2 plus hydrocarbons in quantities which depend on the type of fluid and baffle used with the diffusion pumps (see Fig. 16 and Table 2).   Hydrogen is sometimes found to be the leading constituent, but in unbaked systems this is probably the result of water vapor dissociating on hot surfaces, for example, at mass-spectrometer or ion-gauge filaments, and gettering of oxygen by reactive metals.   The role of elastomers as the foremost contributors to the outgassing process is best illustrated by data reported by Holland (Ref. 215, p. 132).   In his system, an elastomer gasket was used only to seal the high-vacuum valve whereas the baseplate-to-chamber seal consisted of indium.   After outgassing from the chamber walls had been reduced by mild internal heating to about 150°C, Holland obtained pressures of 4×10^{-7} Torr corresponding to an overall outgassing rate of 2.5×10^{-8} Torr l s^{-1} cm^{-2}.   Comparison with known degassing rates for stainless steel showed that about 95% of the total outgassing load Q_{const} originated from the Buna N gasket in the valve, although it formed only 1% of the internal surface area.   Substitution of Viton A seals reduced the average outgassing rate to about 1×10^{-8} Torr l s^{-1} cm^{-2} and allowed the attainment of 5×10^{-8} Torr if a Meissner trap was used.

An attempt to reduce the residual-gas load in a Pyrex-glass bell-jar system by means other than baking has been reported by Gregg and Reid.[290]   These authors established a high-frequency glow discharge during the pump-down to clean the substrate and chamber-wall surfaces by ion bombardment.   The glow discharge could be maintained down to about 10^{-5} Torr, and its extinction was followed by a rapid drop in pressure. The latter is due to the onset of adsorption onto the freshly cleaned walls, a process which was enhanced by precoating the bell jar with a surface-active film such as zinc sulfide.   However, the technique was only moderately successful, yielding ultimate pressures of 10^{-6} Torr and a residual-gas load of mostly CO_2, H_2O, and O_2.

The use of getter-ion instead of diffusion pumps does not change the performance of demountable systems significantly.   Caswell[79] achieved pressures of 2×10^{-7} Torr with a sputter-ion pump and a Meissner trap.   The resulting gas composition was similar to that in diffusion-pumped systems.   Maynard[291] employed an evapor-ion pump on a demountable system which could be baked at 100°C by internal heaters. He achieved an ultimate pressure of 3×10^{-7} Torr.   The principal gases were CO, N_2, Ar, H_2, CH_4, and H_2O in proportions varying with the operating conditions.   The absence of hydrocarbons other than methane is an advantage of usually little significance in this type of system.   However, in the presence of a glow discharge or an electron beam, even small back-streaming rates from oil diffusion pumps may lead to the accumulation of solid polymer or carbon films.

As stated in the definition, the performance of demountable vacuum systems is contingent upon high pumping speeds to counteract heavy outgassing. Because of the use of different construction materials, the gas emission within such a system varies locally, and therefore the pressure in the vessel is not constant. Deceivingly low readings may be obtained if an ionization gauge is placed in a location favoring removal of gases, for instance, in the vicinity of a cryogenic surface. If this is the case, the indicated pressure does not reflect the intensity of the residual-gas bombardment to which other parts of the system including the substrates are exposed. Thus, the approach of achieving very low pressures by high pumping speeds has practical limitations. To obtain ultrahigh vacuum, the gas-contributing processes must be reduced to levels significantly lower than those in dynamic systems.

c. Ultrahigh-vacuum Systems

An ultrahigh-vacuum system is defined as "a system constructed with materials having very low outgassing rates and which can be thoroughly degassed so that pressures less than 10^{-9} Torr can be achieved with pumps of moderate speed."[288] The procedure which yields the greatest and most rapid reduction of outgassing rates is baking of the entire system or at least of its chamber. An additional advantage of baking is that upon subsequent cooling, residual gases are adsorbed until the surfaces are saturated; thus, the pumping speed is temporarily enhanced.

The effect of the baking temperature on the limiting pressure was demonstrated by Hickmott,[292] who subjected a Pyrex-glass system to different heat treatments and found decreasing ultimate pressures, as shown in Table 20. The necessity of baking

TABLE 20   Effect of Baking on the Ultimate
Pressure in a Pyrex-glass System[292]

Baking temp	Baking time	Ultimate pressure, Torr
Room temp	2 weeks	8×10^{-7}
125°C	4 h	4×10^{-7}
225°C	12 h	5×10^{-9}
425°C	Repeated cycling	2×10^{-10}

at relatively high temperatures to minimize outgassing rates imposes restrictions on the use of vacuum materials. Furthermore, baking and subsequent cooling extend the duration of the pump-down cycle. The time required to reach the ultimate pressure is of the order of 10 to 50 h. Therefore, ultrahigh-vacuum systems have not found industrial usage as yet, but they are common in research applications such as the study of surface phenomena, space simulation, and plasma physics.

(1) Partially Bakeable Systems   A number of investigators have tried to reconcile the necessity of baking with the convenient access afforded by elastomer base plate-to-chamber gaskets. Typically, these systems are evacuated by rotary oil and thoroughly baffled oil diffusion pumps. The vacuum enclosures are stainless-steel tanks or bell jars which can be baked while the gaskets are maintained at temperatures below their stability limit.

The first example of this kind was reported by Turner et al.,[293] who used an indium gasket which was water-cooled while the chamber was degassed at 425°C. Ultimate pressures of 10^{-8} Torr were obtained. In an analysis of this experiment, Holland[294] pointed out that the outgassing rate of the unbaked surface area was several orders of magnitude greater than that of the baked walls, and that a relatively small unbaked area is sufficient to prevent the attainment of ultrahigh vacuum.

Farkass and coworkers[212] constructed large (30- by 50-in.) stainless-steel tank systems with butyl-rubber gaskets. Outgassing and permeation rates of the elastomer were kept at low levels by Freon cooling [see Sec. 3e(4)]. The gas-evolution rates of the tank walls were sufficiently low to achieve pressures of 10^{-11} Torr with only occasional baking and with a normal ratio of effective pumping speed to chamber volume. The addition of a cryogenic helium pump further reduced the ultimate pressure to 5×10^{-12} Torr.

A stainless-steel tank bakeable at 400°C with Freon-cooled (-25°C) elastomer gaskets has also been utilized by Adler,[295] who achieved pressures of 2×10^{-10} Torr. The same ultimate pressure was reported by de Csernatony[236] for a metal chamber with Viton A double-O-ring seals. While the chamber was baked at 300°C, the temperature of the elastomer seals was controlled at 150°C by means of external water-cooling coils.

The pump-down times of cooled-gasket systems are similar to those of fully bakeable ones; hence the utilization of this principle in thin film evaporators does not permit rapid cycling. However, bakeout and cooling may be performed overnight. Systems which either can be used with water-cooled Viton gaskets (10^{-10} Torr) or made fully bakeable with metal gaskets (10^{-12} Torr) have been described by Reisinger.[296]

(2) Fully Bakeable Systems This category encompasses systems designed such that not only the vacuum chamber but also the baseplate with its seal and the parts attached below the plate can be baked. The problems created by heating concern primarily the seals. One approach is the use of glass vessels with permanent fused-glass or metal-to-glass seals. However, the utility of such systems is restricted by their small size and by the tedious opening and closing procedure. The demountable feature is retained if metal gaskets are employed [see Sec. 4b(3)]. The type of vacuum enclosure is largely determined by the choice of the joining technique. Fused-glass joints necessitate the use of small glass tubes or bulbs, whereas metal gaskets permit large-diameter metal chambers of greater experimental utility. Baking is accomplished by surrounding the chamber, baseplate, and adjoining components with a removable resistance-heated enclosure.

Getter-ion and diffusion pumps are about equally common in fully bakeable systems. The design of systems with either type of pump has been discussed by Zaphiropoulos and de Taddeo.[297] If diffusion pumps are chosen, they must be baffled and trapped more carefully than is necessary for conventional evaporators. Auxiliary pumping by means of cryogenic surfaces or chemical gettering to enhance the pumping speed and to lower the ultimate pressure is widely practiced. The installation of a titanium getter pump in line with a molecular-sieve trap on the high-vacuum side of the system has been reported to inhibit back streaming of oil vapors very effectively.[298]

The task of designing an ultrahigh-vacuum system whose materials and components are compatible and which can be fully baked without sacrificing the required operational freedom tends to produce highly individualized solutions. Consequently, ultrahigh-vacuum systems are less standardized than conventional evaporators, and their performance depends on several factors pertaining to their mode of construction. Operating data for small *Pyrex-glass systems* as reported by four different authors are listed in Table 21. These systems had permanent glass or metal-to-glass[299] joints.

TABLE 21 Small Pyrex-glass Systems

Reference	Baking temp, °C	Ultimate pressure, Torr	Predominant residual gases
Venema[63].....................	450°C	$<10^{-12}$	?
Young and Hession[299]..........	250°C	10^{-10}–10^{-11}	H_2, CO, CO_2
Steinrisser[60]..................	350°C	10^{-11}	CO, (CO_2, H_2)
Singleton[59]...................	410°C	5×10^{-11}	H_2, CO_2, CO

They were evacuated by combinations of rotary oil and mercury or oil diffusion pumps. If properly trapped and baffled [see Secs. 2b(3) and 2b(4)], the pump fluid has little effect on the ultimate pressure obtained.[59] No auxiliary pumps were used in the examples shown in Table 21. The absence of demountable joints and the small volume of the vessels (a few liters) are decisive in attaining low outgassing rates and ultrahigh vacuum. Opinions about the optimum baking temperature for Pyrex glass are divided. Young and Hession[299] do not consider it necessary to exceed 250°C (baking time: 15 h), a temperature sufficiently low to dismiss the possibility of glass decomposition [see Sec. 3c(2)], while the other authors prefer higher temperatures.

The next group of systems, whose operating data are shown in Table 22, was built with *demountable metal gaskets* and evacuated by thoroughly baffled, refrigerated *oil diffusion pumps*. The bakeable seals included aluminum wire,[215] aluminum foil,[300] and copper gaskets.[58] The system of Power et al.[301] required only one aluminum gasket since the pump and baffle section was fabricated as one permanently joined unit. The vacuum chambers are mostly of stainless steel, and degassing is accomplished by maintaining temperatures around 400°C for several hours. Batzer and Ryan's[300] system consisted of an aluminum alloy whose outgassing rate after baking at 200°C was as low as that of stainless steel baked at 400°C. The systems of Holland and Hablanian listed in Table 22 established fore vacuum by trapped rotary oil pumps; the others had cryosorption traps. Evidence of back streaming was reported by Hablanian and Vitkus,[58] who observed the cracking pattern of the D.C. 705 pump oil, particularly if the system was not baked. A mode of construction which minimizes back streaming has been described by Reisinger,[296] who used a small diffusion pump to back the main diffusion pump. This lowers the fore pressure of the main diffusion pump and thereby reduces its tendency to emit gases back into the system.

TABLE 22 Diffusion-pumped Systems with Metal Gaskets

Reference	Baking temp, °C	Ultimate pressure, Torr	Predominant residual gases	Comments
Holland (Ref. 215, p. 135).	425°	10^{-9}		
Holland[302]....	370°	10^{-10}	....	With Ti sublimation pump
Hablanian and Vitkus[58]....	Not baked	8×10^{-8}	Phenyl compounds,	Chamber with liquid-nitrogen cold finger
	400°	5×10^{-11}	$M < 50$	
Power et al.[301]	400°	6×10^{-11}	H_2, H_2O, CO	Hg diffusion pump with baffle and trap in one unit
Batzer and Ryan[300]	200°	10^{-10}	H_2, H_2O, CO/N_2	Aluminum chamber
	200°	3×10^{-11}	H_2, H_2O, CO/N_2	with Mo sublimation pump

Ultimate pressures below 10^{-9} Torr are difficult to obtain by diffusion pumping and system degassing alone. Hence, auxiliary pumping with cryosurfaces or getter metals is widely practiced. These techniques extend the attainable pressure range down to about 10^{-11} Torr. Most effective is the combination of both auxiliary pumping methods in the form of cryosublimation traps.[303] These units consist of a stainless-steel shell enclosing a liquid-nitrogen-cooled copper baffle. Titanium filaments within the structure permit metal sublimation onto the cooled baffle surfaces. The device is installed between the vacuum chamber and the oil-vapor trap of the diffusion pump. Within a few minutes of operation, the cryosublimation trap lowers the pressure by a factor of 200 to 1,000, and ultimate pressures of 10^{-14} Torr can be attained.

Although similar in performance, *ultrahigh-vacuum systems with getter-ion pumps* are sufficiently different in their design to be considered separately. One of the commoner versions is shown in Fig. 89. Typical features are stainless-steel chambers of 20 to 120 l volume, demountable chamber joints with aluminum[215] or gold-wire[304] seals, and a removable baking enclosure. The sputter-ion pump, which may be attached to a side port or to the baseplate, has its own heating element for degassing. Although the pressure is indicated by the ion current of the pump, an independent gauge is usually provided in the upper part of the chamber. The high-vacuum valve

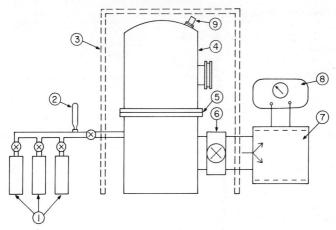

Fig. 89 Fully bakeable ultrahigh-vacuum system with sputter-ion pump. (1) Cryosorption pumps; (2) fore-vacuum gauge; (3) removable oven; (4) bell jar; (5) demountable metal seal; (6) bakeable high-vacuum valve; (7) sputter-ion pump; (8) high-voltage power supply with ion-current meter indicating pressure; (9) pressure gauge.

must be all-metal and bakeable. Its presence is not absolutely necessary since gases absorbed by the sputter-ion pump during exposure to air are removed in the bakeout cycle. Cryosorption pumps to establish fore vacuum are generally preferred over rotary oil pumps. An alternate possibility is the use of a water aspirator.[304]

The pump-down and bakeout cycle of a sputter-ion pumped system is shown in Fig. 90. The time required to reach the ultimate pressure is of the order of 20 h.

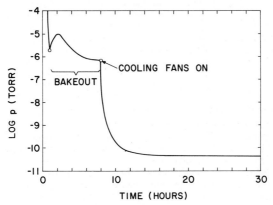

Fig. 90 Pump-down curve of a bakeable ultrahigh vacuum system. 18-in.-diameter bell jar, 120-l volume, 500 l s^{-1} sputter-ion pump, and titanium sublimation pump. (*After Zaphiropoulos and de Taddeo.*[297])

Forced-air cooling of the system after baking is often used to shorten the procedure. Auxiliary pumping by sublimation of titanium is also typical. Examples illustrating the performance of such systems are listed in Table 23. The design and characteristic of Holland's system are discussed in detail in Ref. 215. The stainless-steel chamber evacuated by Boebel and Babjak[305] had quartz and lithium fluoride windows which did not prevent the attainment of pressures as low as 10^{-12} Torr. Denison's[123] system demonstrates that evapor-ion pumps, too, are capable of establishing ultrahigh

TABLE 23 Getter-ion Pumped Systems with Metal Gaskets

Reference	Baking temp, °C	Ultimate pressure, Torr	Predominant residual gases	Comments
Holland (Ref. 215, p. 140).	400°C	10^{-9}	. . . .	Sputter-ion and Ti sublimation pumps
Boebel and Babjak[305]. . . .	Not given	10^{-12}	. . . .	Sputter-ion and Ti sublimation pumps plus Meissner trap
Denison[123].	300°C	3×10^{-11}	$H_2(CO,H_2O,CH_4)$	Electrostatic getter-ion pump with Ti sublimator
Caswell[79].	430°C	3×10^{-9}	H_2, CO, N_2	2-l stainless-steel chamber with sputter-ion and cryogenic helium pump

vacuum. Caswell[79] used a liquid-helium trap to augment the effectiveness of the sputter-ion pump. Commercial equipment operating in the 10^{-11} to 10^{-12} Torr range is available.

An unusual solution to the problem of bakeable joints has been reported by Blahnik.[306] He utilized a vacuum enclosure consisting of two stainless-steel chambers forming a guard vacuum. The inner chamber was sealed by a moat of molten boron oxide. After baking at 900°C, the seal was solidified by cooling to 50°C. The vacuum in the inner chamber was maintained by sputter-ion pumping and intermittent molybdenum evaporation. Pressures below 10^{-10} Torr were thus obtained.

Ultrahigh vacuum may also be established by means of turbomolecular pumps. Outlaw[22] reported a system consisting of a stainless-steel bell jar with rotary oil and turbomolecular pumps. After baking the vacuum enclosure at 400°C, an ultimate pressure of 10^{-9} Torr was obtained. Auxiliary pumping with a titanium sublimator, a Meissner trap, and a cold-cathode ionization gauge produced pressures as low as 5×10^{-12} Torr.

The composition of the residual gas in bakeable ultrahigh-vacuum systems has been investigated in several cases. A comparison of Tables 21 and 23 shows that the predominant gases are nearly the same in Pyrex-glass and metal chambers. They are also quite independent of the types of pumps used. Hydrogen and carbon oxides stemming from slow outdiffusion processes or from secondary reactions at ionization-gauge filaments constitute the greater part of the load. Their presence as the prevailing constituents is not limited to systems with titanium pumps, although the latter are prime sources of hydrogen and carbon compounds. In getter-ion pumped systems, argon is commonly encountered, whereas hydrocarbons are additional constituents in oil-diffusion-pumped chambers.

d. Gas Evolution and Gettering Due to Evaporation Sources

The residual-gas load in vacuum evaporators is also affected by the evaporant source. Pressure increases of about an order of magnitude are often observed when vapor sources are heated. After an initial outgassing burst, the pressure decreases again and may even drop below the original level because of gettering (see Ref. 215, p. 112). In the case of metal evaporants, the pressure increase is primarily due to thermal release of dissolved gases such as H_2, CO, CO_2, or N_2. Methane stemming from secondary reactions of carbon impurities is also often detected. Oxygen is rarely identified because it reacts with carbon and forms CO. Wagener and Marth[307] analyzed the gases evolved from several refractory metals at 1000 to 2000°C. They found that titanium released about equal amounts of H_2 and H_2O, tantalum mostly CO and H_2, while the gases evolved from nickel, iron, and molybdenum consisted of over 80% carbon monoxide.

The outgassing behavior of metal charges has been thoroughly investigated in the case of copper. Guenther and Lamatsch[21] observed several gas-evolution maxima while raising the temperature of small test samples. Gas release due to adsorbed layers and surface oxide occurred between 300 and 700°C, whereas diffusive release required temperatures above 900°C. The cleanest samples evolved about 10^{-4} Torr l cm^{-2} of carbon monoxide, whereas the quantities of H_2 and CO_2 were an order of magnitude lower. The samples had relatively small dissolved-gas concentrations of less than 1 ppm so that the gas quantities originating from the adsorbed surface layer were comparable. In a similar study, Kirkendall et al.[308] examined the effects of various prior treatments on the degassing rates of OFHC copper. Samples which had been vapor-degreased and chemically etched evolved a total of 10^{-2} Torr l cm^{-2} of gas. Firing in hydrogen was quite effective in reducing all gas contributions except H_2, which was subsequently released in an amount of 10^{-3} Torr l cm^{-2}. A total gas quantity of 10^{-4} Torr l cm^{-2} was obtained from samples previously degassed by vacuum heating.

After an evaporation source has reached its operating temperature, the amount of available gas is gradually exhausted and the system pressure decreases. This decay process is accelerated if the evaporant is chemically reactive. Many substances are capable of gettering action, although to a lesser degree than titanium. An example is chromium, which reacts so strongly with oxygen that at pressures above 10^{-6} Torr the deposits contain increasing amounts of Cr_2O_3.[309] Caswell[79] observed changes in the residual-gas composition during the evaporation of tin. While hydrogen was released by the source, the oxygen pressure in the chamber decreased because of gettering by the fresh tin deposit. An example of gettering by compounds is the evaporation of SiO.[310] Priest et al.[311] showed that SiO films getter H_2O and O_2 rapidly and hence may significantly reduce the pressure in conventional evaporators.

When thin films of high purity are to be deposited, it is common practice to degas the source material for some time at a temperature below the vaporization range. Quite often this treatment is insufficient, and additional gas is released at the evaporation temperature. A further precaution against film contamination is to start the evaporation but shield the substrates by a shutter until the pressure has decreased. In conventional evaporators, the expected pressure reduction may be partially offset by outgassing from adjacent surfaces which receive radiation from the source and desorb gases at increasing rates as their temperature rises. This problem is alleviated by surrounding the vapor source with refractory-metal heat shields and by water cooling of the electric-power feedthroughs. A liquid-nitrogen shroud is also beneficial because it prevents exposure of the chamber walls to the source radiation.

In ultrahigh-vacuum systems, the effects of source outgassing exceed all other residual-gas contributions. The advantage of having established a very low starting pressure is lost unless the evaporant and source construction materials are of the highest purity and have been thoroughly degassed prior to deposition. An increasing number of ultrapure metals including refractories are now prepared by multiple-zone melting.[312] The purity of such materials is further enhanced if the refining process is conducted in an ultrahigh vacuum.[313]

e. Evaporation Systems with Automated Features

Thin film deposition is usually performed as a batch operation in demountable high-vacuum systems as described in Sec. 5b. There are, however, situations where continuous operation is desirable for either economic or technical reasons. For example, if large numbers of substrates—particularly those having large surface areas—are to be coated, the pump-down cycles associated with batch operation limit the volume that can be processed. Another motive may be the necessity to apply several films on top of each other without exposing the interfaces to air. Consequently, several attempts have been made to develop systems for large-volume coating or multilayer deposition. Such systems tend to be mechanically complex and offer special design problems.

(1) General Considerations The decision to utilize an evaporator with automated features should be preceded by a thorough economic and technical evaluation. Because of the greater complexity and the necessity for some custom-designed components, the cost for equipment and engineering time is much higher than for a conventional batch-type evaporator. Since applicable prior experience is generally not available, the risk of encountering unpredictable expenditures due to redesigning, excessive debugging, or maintenance is high. These disadvantages must be weighed against the expected economic or technical benefits. Such a study may demonstrate that several batch-type evaporators offering an equivalent production capacity are more economical than one automated system.

This alternative may not always be available, namely, if exposure to air between subsequent depositions has been shown to be detrimental to the quality or reliability of the product. The reduction in materials handling, uniformity of automatically controlled process steps, and elimination of operator error have indeed been strong incentives for the development of continuous evaporators. Examples include the fabrication of interconnections of superconducting films in cryogenic circuits and the application of protective coatings on films sensitive to exposure to air. In the case of sputtered thin film tantalum resistors, Cook[314] obtained the most reproducible properties after sputtering had been in process for 4 days in a continuous in-line system.

The design of continuous systems for sputtering is somewhat less critical than for evaporation because the chamber is continually flushed with high-purity gas. Consequently, impurities from sources outside the deposition station must diffuse upstream before they can enter the workspace. Continuous sputtering systems are therefore often based on the differential-pumping principle, which has the advantage of allowing access to the chamber at all times. However, special design restrictions arise from the bombardment of internal system parts and walls with ions and from the electrical conductivity of the plasma. Continuous evaporators, on the other hand, permit random diffusion of impurity gases with comparatively long mean free paths. Consequently, the film properties may be adversely affected by impurities stemming from system parts other than the deposition station. Therefore, continuous or multiple evaporation systems are mostly designed with vacuum locks or substrate magazines and thus allow only intermittent operation. In either case, the assembly techniques for continuous systems are the same as for demountable systems; i.e., elastomer seals are used for feedthroughs, valves, connection of modules, and to provide access to the chamber. Consequently, the systems are not bakeable and can attain only moderate high vacuum.

(2) Continuous and Multiple Deposition Systems Vacuum systems for continuous or multiple deposition are generally unique because they are designed for one specialized function. The following examples represent various solutions which evolved from different practical requirements.

Figure 91 shows the operating principle of a commercial evaporator intended for the deposition of aluminum onto a ribbon of plastic foil.[315] As the ribbon is unwound, it passes over the evaporation source and is coated. Other treatments such as pre-

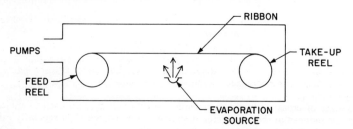

Fig. 91 Continuous evaporation onto a plastic ribbon.

heating can be incorporated into the design. The use of a flexible ribbon avoids the problems which are otherwise associated with substrate loading and transport.

A system which has been used for the coating of steel flat stock is shown schematically in Fig. 92. The strip enters and leaves a series of differentially pumped chambers through roller seals. The latter allow the occasional passage of sections

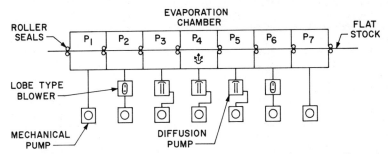

Fig. 92 Differentially pumped multichamber system for continuous evaporation onto flat stock.

with double thickness, a necessity arising from the welding of strips to ensure continuous operation. Thus, atmospheric gases leak into the system continually, but their effect on the pressure in the central chamber can be minimized by proper design and choice of pumps.[316,317]

For the fabrication of multilayer microelectronic circuits, batch-type demountable high-vacuum systems have been most widely used. Their distinguishing features are rotary mask and substrate changers and multiple evaporation sources.[318-320] In the example shown in Fig. 93, the sources are stationary while the substrate carrier can be

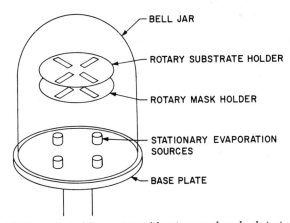

Fig. 93 Multilayer deposition system with rotary mask and substrate changers.

rotated. The film patterns are defined by masks which are rotated under and indexed onto the substrates in the appropriate sequence. Movable shutters over the evaporation sources, provisions for substrate heating, rate monitoring, and glow-discharge cleaning are often incorporated. Vacuum equipment for these functions is commercially available.

The danger of cross contamination from the different sources used in multilayer deposition is reduced in the arrangement shown in Fig. 94, which is generally referred to as a "four-chamber in-line" system.[321] Each chamber is individually pumped and

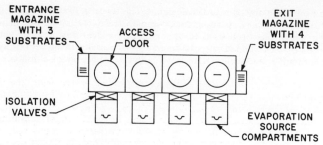

Fig. 94 Four-chamber in-line system for multilayer deposition (isolation valves between magazines and chambers not shown).

capable of performing one evaporation step. The chambers are interconnected to allow passage of the substrates, but they can also be valved off and isolated from each other. The evaporation sources are located in the bottom compartments, which can be valved off for recharging or other services. The vacuum-locked entrance and exit magazines were designed to hold 24 flat glass substrates. The latter are moved and positioned by a conveyor mechanism. Provisions for radiation heating of substrates, mask changing, and automatic mask indexing are contained in each chamber. The system has been used to fabricate thin film RC networks in substantial quantities for computer applications. It has the advantage of great flexibility since masks and evaporants may be changed, or additional modules may be added.

An 11-chamber differentially pumped sputter system which is open to air at both ends has been used in the production of tantalum-film resistors.[314,322,323] The pumping system is similar to that shown in Fig. 92. The substrates are mounted on carriers which are pushed through the system from the outside so that no internal mechanism is needed. The clearance between carrier and opening must be a compromise between leak rate and reliability of carrier movement. At the entrance and exit seals, the leakage is reduced by lengthening the passage through which the carriers travel. A process rate of two substrates per minute has been reported.

Another approach uses vacuum locks for entering and withdrawing of substrates as shown in Fig. 95. While differential pumping is not required, the locks must be

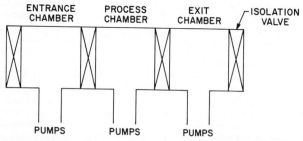

Fig. 95 In-line system with vacuum locks for intermittent loading of substrates.

alternately evacuated and vented for the insertion or removal of new workpieces. This design is considerably less expensive than open-end systems, but sealing and operating difficulties with the moving parts of the locks may be anticipated. If the central process chamber is sufficiently large to accommodate many substrates, only one air lock to slide the substrates in and out is required. Systems of this type are commercially available.*

* For example, from Airco Temescal, Berkeley, Calif., Model FC-1100 fast-cycle vacuum coater.

Other automated systems are basically variations of the types described. Whether differentially pumped or operating with vacuum locks, their practicality is determined by problems of design and technology. A detailed comparison of the performance of both system designs and a discussion of the difficulties encountered have been published by Hanfmann.[324] Automatic control of evaporation processes is further discussed in Chap. 1, Sec. 7d.

(3) Special Design Problems A common problem arises from the fact that all continuous deposition systems require moving parts, either for substrate transport or for mask and substrate changers. Friction in vacuum, however, is more severe than in air. Thus, mechanical malfunctions in the form of jamming, imperfect mask-to-substrate contact, or misregistration are often encountered. These difficulties are aggravated if the substrates are heated, which is almost always necessary. The use of lubricants such as grease is to be avoided because of outgassing, particularly if the parts in question are heated. Dry lubricants such as Teflon films or molybdenum disulfide have been used in vacuum, but their ability to withstand heating without outgassing is also limited.

Further difficulties are associated with the monitoring of the process. The measurement of important control parameters such as resistivity or substrate temperature is usually predicated upon intimate contact between probes or thermocouples and the substrate. This is not easily achieved since the contact must be broken while the substrates move. Consequently, wide variations of substrate temperatures are fairly common. Radiation-type measurements, which do not rely on physical contact to the substrates, are therefore preferable for temperature control.

Another important design consideration is the fact that continuous deposition systems require special evaporation sources of large capacity to avoid frequent interruption of the operation. Upscaled versions of baffled sublimation and large-crucible sources have been developed for this purpose. Since large sources dissipate much power, radiation shields—possibly even water-cooled—are necessary to prevent excessive outgassing from the surrounding chamber walls. An alternative to increasing the source size is a small source with a replenishing mechanism. Various wire, powder, and pellet feeders of the types described in Chap. 1, Sec. 6c(3) are commercially available. Preformed metal charges which are placed near the source, picked up by a means of a vacuum manipulator (e.g., a tilting-motion feedthrough), and dropped into the source crucible have been used successfully in the in-line system shown in Fig. 94.

The chances of utilizing continuous deposition systems are best if the films are to cover the entire surface such as in the metallization of ribbons, sheet stock, plastic articles, or in the optical coating of lenses. If film patterns are required, it is necessary to apply evaporation masks or post-deposition etching techniques. In conjunction with multilayer evaporation, which has been one of the prime objectives in the development of highly mechanized systems, differential etching techniques are generally inapplicable. Therefore, multiple-deposition systems for microelectronic circuitry require masks and mask changers. Precision indexing of consecutive masks within a vacuum system is a formidable problem. Intimate contact between substrate and mask as well as mask registration are adversely affected by thermal expansion and warping. Most restrictive, however, is the fact that the demand for smaller and smaller thin film components has led to dimensions which are no longer attainable with evaporation masks but require precision photoetching techniques (see Chap. 7). Consequently, the use of multilayer evaporation systems for microelectronics applications has been declining over the last years. However, the growing volume of monolithic circuits manufactured by the semiconductor industry makes single-film continuous deposition systems economically attractive. While the initial cost of such a system is several times higher than that of a batch-type coater, the gain in wafer throughput is even greater yet. As more operating experience with presently available components such as vacuum entry locks, large-capacity evaporation sources, and deposition-control equipment is accumulated, continuous wafer processing under computer control appears more and more feasible.

6. VACUUM MEASUREMENTS

Having assembled a vacuum system from various components and materials, it becomes necessary to test and continually monitor its performance. Three types of measurements are required to characterize the quality of the vacuum attained. They are aimed at establishing the integrity of the vacuum enclosure, determining the total gas pressure within the chamber, and knowing the residual gas species present. The experimental methods commonly used to gain this information are discussed in this section.

a. Leak Detection

A common problem encountered in the operation of high-vacuum systems is the inability to reach the ultimate pressure for which the system was designed, coupled with unusually long pump-down times. These symptoms may be an indication of a leak somewhere in the vacuum enclosure. The gas flow into the vacuum through narrow pores or channels is of the molecular or transition type [see Chap. 1, Sec. 2b(5)], and although the amounts of gas transmitted are small, they are sufficient to affect the performance of the system. For example, the transition flow through a channel of 10μ diameter and 1 mm length is 5×10^{-3} Torr l s^{-1} (see Ref. 6, p. 108). Assuming that the system is pumped out at a speed of 500 l s^{-1}, the leak rate becomes equal to the throughput of the pump when a pressure of 10^{-5} Torr has been reached. Actually, the ultimate pressure obtained will be even higher yet because of additional gas contributions from desorption. To obtain or restore maximum performance of a vacuum system, the presence and location of vacuum leaks must be established before repair can be attempted. A number of leak-detection methods have been developed for this purpose and are reviewed extensively in the literature.[325,326]

(1) Identification of Vacuum Leaks The task of identifying small leaks in a vacuum enclosure is complicated by the fact that high residual gas pressures may also be caused by so-called *virtual leaks*. These are small volumes of gas trapped within the system and released slowly when the internal pressure is reduced. Possible sources of virtual leaks are blind threaded holes with nonvented screws, improperly made permanent or double-O-ring seals, or other component parts which may occlude gas and communicate with the high vacuum through small openings. Materials with a high capacity to adsorb gases may also be responsible for anomalously high residual gas pressures. Examples are vacuum greases which adsorb gases heavily, and porous materials such as certain ceramics or wood which may have been left in the system unintentionally. Cold traps, too, may act as virtual leaks because the vapor pressures of condensates like water or CO_2 at common refrigeration temperatures are not negligible under high-vacuum conditions.[227,228]

Leak testing may become a very time-consuming and frustrating experience if the source of the observed pressure rise is a virtual leak. Hence, the presence or absence of the latter should first be established. Unfortunately, discriminating between real and virtual leaks is not an easy matter. Close observation of the pressure response to selective cooling or heating of parts of the vacuum enclosure is sometimes useful to distinguish between condensable and permanent gases. However, this does not necessarily lead to positive identification of their origin. If the deterioration of the ultimate pressure occurs gradually over a period of time, excessive desorption from accumulated contamination is the more likely cause. Thorough cleaning of all internal surfaces and changing the pump oil should correct this condition. On the other hand, sudden changes in the pump-down characteristics, particularly after the installation of new vacuum components, are more likely to be caused by real leaks. The best policy is to prevent the formation of virtual leaks by following recommended assembly procedures. Specifically, this calls for venting of blind threaded holes and proper design of permanent and double-O-ring seals (see Sec. 4). Furthermore, the use of porous materials in vacuum systems should be avoided and grease applied only sparingly. Cold traps should be refrigerated only after the main gas load has been pumped out, or if this is not possible, they should be baked out periodically to release accumulated condensable gases.

If the presence of a real leak has been made reasonably certain, a first indication of its location and magnitude can be obtained by isolating the chamber from the high-vacuum pump and observing the *pressure rise* Δp for a period of time Δt. If the volume of the chamber is V, the leak rate is $\Delta p\ V/\Delta t$. This gives an idea of the size of the leak and also shows whether it is above the high-vacuum valve or in the pumping section of the system. Recording the rate of pressure rise for a leaktight system is also useful because it provides a basis for comparison with future leak tests of the same kind.

A convenient method to leak-test individual vacuum components is to *pressurize* them prior to assembly. The part may then be submersed in water or painted with a soap solution. The presence of a leak is indicated by bubbles originating from the defective spot. By allowing several minutes for close observation, leaks as small as 10^{-4} Torr l s^{-1} can be detected.

Another simple technique, which is, however, restricted to glass systems or components, employs a *Tesla coil* for leak testing. It requires internal pressures of less than 2 Torr. The coil is moved over the glass surface at a distance of about 1 cm. When passing a leak, ionized molecules are drawn into the system and produce a glow discharge. Fairly rapid movement of the coil is advisable to avoid damage to the glass. Leaks down to about 10^{-3} Torr l s^{-1} can be identified.

The most universal methods of leak testing are based on local and external application of a *tracer gas or liquid* onto the vacuum enclosure and observing the change of the internal gas pressure. Thermocouple or Pirani gauges may serve to indicate changes of the internal gas composition which result when the tracer substance hits the area of the leak. Gases such as hydrogen, helium, carbon dioxide, and butane have been found to penetrate small holes rapidly and cause differences in the thermal conductivity of the residual gas which are sufficient to produce a noticeable gauge response.[327] Liquid tracers such as acetone or ether are also suitable. However, they should be used with caution as they attack elastomer gaskets and may create additional leaks. They may also temporarily obscure a leak by swelling the gasket, or a gasket soaked with a solvent may become a virtual leak. The sensitivity of the tracer method in conjunction with thermocouple or Pirani gauges is about 10^{-3} Torr l s^{-1}. At pressures below 10^{-4} Torr, the gauge response becomes so small as to render the technique useless.

The range of the tracer method can be extended by orders of magnitude if an ionization gauge is utilized to sense the pressure change. Vapors of the tracer substance entering the system through the leak change the ionization rate in the gauge and thus indicate when a defective wall area is exposed to the spray. The sensitivity of this method is greater the lower the pressure and may be as high as 10^{-10} Torr l s^{-1} at 10^{-10} Torr. An upper limit is imposed by the operating range of the ionization gauge.

(2) The Helium Leak Detector   The helium leak detector (HLD) is the most sensitive (up to 10^{-14} Torr l s^{-1}) and versatile instrument available for leak testing. It, too, uses an externally applied tracer gas, but a mass spectrometer designed to respond only to helium serves as an indicator. Commercial HLDs are self-contained vacuum systems constructed as shown in Fig. 96. A cold trap protects the mass spectrometer from contamination with condensable gases. The pressure indicator is generally a cold-cathode Penning ionization gauge whose circuitry shuts off the spectrometer filament if the pressure rises excessively. The gas to be sampled is drawn into the system from the inlet by means of the pumps, and its flow can be adjusted with the throttle valve. The pumps are designed only for evacuating

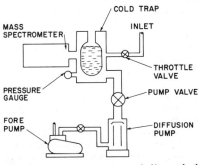

Fig. 96 Schematic of a helium leak detector.

the small volume of the leak detector and a short connecting line. Thus, if the inlet is connected to a larger volume, the latter must be pumped out independently. An HLD with a titanium sublimation pump instead of the cold trap has also been described.[328] The advantage of this model is that it traps hydrogen as well as condensable gases. Therefore, the slit in front of the helium-ion collector can be widened without interference from hydrogen ions, and the sensitivity of the instrument is increased.

On demountable high-vacuum systems of the type shown in Fig. 87, two locations are commonly used for connecting to the inlet of an HLD. One is the roughing line between the chamber and the fore pump, the other is the fore line connecting the diffusion pump with the fore pump. Connection of the HLD requires a vacuum valve with about 2 in. length of tubulation and $\frac{3}{4}$ in. OD in one or both of these lines. A heavy-wall rubber hose of $\frac{5}{8}$ in. ID facilitates convenient coupling of the valve to the inlet of the leak detector. If the leak is suspected to be fairly large, sampling through the roughing line is preferable to avoid flooding of the HLD with helium. For the same reason, it is recommended to open the HLD throttle valve only gradually when starting to probe with helium. For small leaks, the HLD should be attached to the fore line since sampling at the discharge end of the diffusion pump causes a greater amount of helium to be taken in by the leak detector. Maximum sensitivity is obtained if the fore-line valve is closed and the HLD serves as a fore vacuum for the diffusion pump.

It is important to avoid getting an excess of helium into the vacuum system and particularly into the HLD. Many materials adsorb helium and then release it over prolonged periods of time, during which the sensitivity of the tester is reduced by a continuous background signal. Vacuum grease is one of these strongly adsorbent materials, and it should therefore be used only sparingly and as a thin layer. Rubber can cause the same problem, and metal tube rather than hose connections are sometimes recommended for this reason. However, rubber-hose connections have the advantage of being flexible and convenient to install, and they can easily be replaced if accidental saturation with helium has occurred.

Once the leak tester has been connected and both systems are evacuated, probing with helium may start. A fine nozzle or glass capillary attached to a helium tank makes a suitable spray. The gas flow should be small and adjusted such that discrete bubbles can be distinguished when the spray nozzle is held under water. Since helium is lighter than air, probing should begin at the uppermost portion of the vacuum system and then move downward. Also, the nozzle should initially be held a few inches away from the vacuum enclosure and moved in closer only if no signal is obtained. Since it takes a few seconds for the helium to flow through the leak and into the detector, rapid movement of the spray makes it difficult to recognize the defective area. A probing speed of 1 to 2 ft min^{-1} is about right. Somewhat faster movement is indicated for narrowing down the exact position of a leak once the general area has been identified and the point of maximum response is sought. When a leak has been found, it may be sealed temporarily [see Sec. 4b(1)] so that other parts of the system can be checked out for additional leaks.

In leak testing of vacuum systems, the possibility of pores or channels in the internal water or gas lines should not be overlooked. A general test is readily performed by draining these lines and filling them with helium. If the test is positive, the vacuum system must be opened to provide access to the lines. The inlet of the HLD is then connected through a flexible hose with a fine capillary tube often referred to as a sniffer. The latter is passed over the suspected line filled with helium under pressure. When the capillary approaches a leak, some helium is drawn into the HLD together with the continuous stream of air. However, the sensitivity of this method is three to four orders of magnitude lower than in the commoner mode of testing under vacuum.

Helium leak detectors should be calibrated and adjusted for maximum sensitivity in frequent intervals, particularly if the instrument is suspected of not working properly. A standard helium leak of the type shown in Fig. 97 lends itself readily for this purpose. The device consists of a container from which helium escapes at a constant rate by permeation of a glass membrane. Since permeation is a thermally

activated process, the temperature dependence
of the leak rate, which is usually given by the
manufacturer, should be considered.   The leak
rates may vary from 10^{-10} to 10^{-6} Torr l s^{-1}.
This is sufficiently small compared with the
amount of gas in the container to ensure nearly
constant leak rates for long periods of time,
sometimes for years.   The standard leak is connected to the intake of the HLD and its flow
regulated by the needle valve.   Methods of leak-
rate calibration have been described by Work[329]
and Bicknell.[330]

(3) Leak Repair   If a vacuum system has been
built properly and its components have been
leak-tested individually prior to assembly whenever possible, the most likely parts to be leaking
are the gasket seals.   The leak-testing procedure can often be shortened considerably by
starting to probe around the gaskets.   Leaking
in these areas is sometimes corrected by tightening the flange bolts.   Where this method fails,

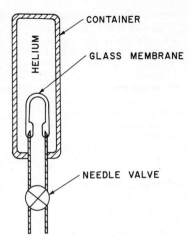

Fig. 97   Standard helium leak.

the seal must be dismantled and inspected.   Removal of foreign particles, polishing
of the flange surfaces, replacement of a damaged gasket, or better alignment of the
sealing parts are corrective steps to be considered.   If a service ring ("dutchman")
is used between the baseplate and the bell jar, it is generally not necessary to disturb
both seals when the chamber is opened.   To maintain pressure continually on the
lower seal, the service ring may be mechanically clamped to the baseplate.

Leaks in permanent joints call for thorough repair.   Welded or brazed parts must
be taken out, remachined, and rewelded or brazed as discussed in Sec. 4a.   Leaky
glass-to-metal or ceramic-to-metal seals will generally have to be discarded and
replaced.   The use of adhesive paint, putty, waxes, etc., is to be discouraged, as they
may contaminate the system.   Moreover, since they temporarily seal the leak, the
temptation is great to defer permanent repair until the sealing substance becomes
dry and starts cracking, thus exposing the old leak again.

b. Pressure Measurement

The total gas pressure is the quantity most often measured to characterize the
degree of vacuum attained in a system.   It may vary over many orders of magnitude,
and different methods of determination are required to cover the entire range.   In
principle, two types of measuring instruments may be distinguished.   The first
group measures the pressure directly as the force exerted by a gas per unit area and
exploits such phenomena as the elastic deformation of a membrane or the level
difference of a liquid in a pair of interconnected tubes.   These instruments indicate
pressure independently of the gas composition, but they cannot be used in the high-
vacuum range where the forces to be measured are very small.   The second group
of instruments measures physical properties which are well-defined functions of the gas
density.   Examples are the thermal conductivity or the degree of ionization of a
gas.   Since these properties vary among different gas species, the measurements
depend on the gas composition.   Furthermore, measurements of this type must be
related to direct pressure measurements by calibration experiments.   Results of the
latter can then be extrapolated to very low pressures and thereby permit the determination of gas densities down to the ultrahigh-vacuum range.

The different gas properties which are pressure-dependent and can be utilized to
make measurements have led to a great variety of instruments and measuring techniques.   Some of these are listed in Table 24.   For an exhaustive treatment, the
reader is referred to the book by Leck.[331]   A brief characterization of the more widely
used instruments has been given by Diels and Jaeckel.[332]   Chapter 5 in Dushman's
book (Ref. 6, p. 220) offers a comprehensive treatment with emphasis on the physical

TABLE 24 Instruments for Measuring Reduced Gas Pressures, and Their Ranges

Instrument	Measured quantity	Reading depends on gas species?	Approx pressure range, Torr	Comments
U-tube manometer.	Height of a Hg column	No	760–1	Some instruments down to 10^{-2} Torr
Mechanical manometers, Bourdon gauges	Deflection of a thin wall or diaphragm	No	760–1	Can be extended to about 10^{-2} Torr
Capacitance manometers	Capacitance as function of diaphragm position	No	760–10^{-4}	Requires reference vacuum. Suitable transfer standard.
Spark coil or discharge tube	Appearance and color of a glow discharge	Yes	10–10^{-3}	Not quantitative. Requires experience
McLeod manometer	Volume of known gas quantity after compression	No	10–10^{-5}	Condensable vapors must be absent. Important standard for gauge calibration
Thermocouple gauge	Filament temperature as function of gas pressure	Yes	10–10^{-3}	Also dependent on ambient temperature. Used mostly as fore-vacuum gauges
Pirani gauge		Yes	1–10^{-3}	
Hot-cathode ionization gauges:	Ion current produced by constant electron emission current	Yes (all gauges)		Hot-emission filament (all gauges)
Triode tube	....	....	10^{-3}–10^{-7}	Cylindrical collector
Schulz-Phelps gauge	....	....	1–10^{-5}	Short electron paths
Bayard-Alpert gauge (BAG)	....	....	10^{-3}–10^{-10}	Central collector wire, filament outside grid
Redhead gauge (modulated BAG)	....	....	10^{-3}–10^{-11}	Auxiliary electrode to modulate ion current
Klopfer gauge	....	....	10^{-2}–10^{-11}	Ion collector remote and shielded from electron collector. Magnetic field
Helmer gauge*	....	....	10^{-4}–10^{-12}	Remote ion collector; ion paths bent 90° by an electric field
Lafferty gauge	....	....	10^{-6}–10^{-13}	Magnetic field, very low electron-emission current
Cold-cathode ionization gauges:	Glow-discharge current	Yes (all gauges)	....	Magnetic field (all gauges)
Penning gauge	....	....	10^{-2}–10^{-5}	Simple, rugged, but not as accurate as BAG
Redhead gauge	....	....	10^{-3}–10^{-12} and lower	Auxiliary cathode separates emission current, $I_i = \text{const } p^n$

* Trademark of Varian Associates.

principles involved.   The following review is more limited in scope and not intended to be all-inclusive.

(1) Pressure Gauges for Low and Medium Vacuum   Pressures above 1 Torr can be measured easily and accurately with the common *U-tube manometer,* which involves reading the height of a barometric column of mercury.[333]   There are also manometers of more complex design capable of measuring pressures down to about 10^{-2} Torr,[334] but these require more refined experimental techniques and are not as convenient to operate as the simple U-tube.

Alternate instruments for the low-vacuum range are *mechanical manometers* and

Bourdon gauges. These measure the mechanical displacement which thin walls or diaphragms suffer when different pressures exist on both sides. Their lower limit is about 1 Torr but may be extended to about 10^{-2} Torr if special measuring techniques are employed. For a discussion of individual instruments, see Ref. 6 (p. 235).

The displacement of a diaphragm can also be sensed electrically if it is made part of a capacitor.[335] Such a *capacitance manometer* is shown schematically in Fig. 98.

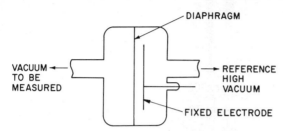

Fig. 98 Schematic of a capacitance manometer.

The diaphragm divides the instrument housing into two halves, one of which is kept at a reference pressure at least two orders of magnitude lower than the minimum pressure to be measured. The distance of the diaphragm from the fixed electrode is a function of the unknown pressure and can be determined by capacitance measurements. Instruments of this type with sensitivities down to 10^{-4} Torr are commercially available.* Their advantage is the ability to measure pressure directly and independently of gas composition. Although they require initial calibration against another gauge, their characteristics are so constant that they are useful as transfer standards.[336]

Mechanical and U-tube manometers operating above 1 Torr are of little importance in vacuum systems because this pressure range is quickly traversed during the pump-down cycle and does not require continuous monitoring. The medium-vacuum range (25 to 10 ^{3} Torr), however, is of considerable interest because it includes the fore pressure of high-vacuum pumps as well as the chamber pressure in sputtering systems. A simple qualitative test may be performed by inducing an *electric discharge* in the vacuum system. This is accomplished either by holding a spark coil to the glass wall or by connecting a small discharge tube to the vacuum system. Various discharge phenomena such as appearance of the positive column, striations, and fluorescence of the glass walls are characteristic of different air pressures and may serve to divide the interval from 10 to 10^{-3} Torr into decades (see, for example, Ref. 32, p. 102). An experienced observer can also draw some conclusions about the presence of gases other than air from the color of the plasma. In most situations, however, more quantitative information about the gas pressure is desired, and measurements in the medium-vacuum range are performed mostly by McLeod manometers or by observing the thermal conductivity of the gas.

(2) The McLeod Manometer McLeod manometers rely on Boyle's law to measure the gas pressure directly. They consist of a system of glass bulbs and tubes which are interconnected so that a large volume of gas of the pressure to be measured p is compressed into a smaller volume of higher pressure p_c by means of a mercury column. In the model shown in Fig. 99A, a gas quantity of volume V is trapped in the glass bulb and capillary I by adjusting the mercury level to the bottom of the bulb. More mercury from an additional reservoir is then forced into the system until the height of the column in capillary II is level with the upper end of capillary I as shown in Fig. 99B. Both capillaries must have identical inner diameters to eliminate capillarity effects. The volume of the compressed gas V_c can be expressed as

$$V_c = ah \qquad \text{cm}^3$$

* One such instrument has been evaluated by J. P. Bromberg, *J. Vacuum Sci. Technol.*, 801 (1969).

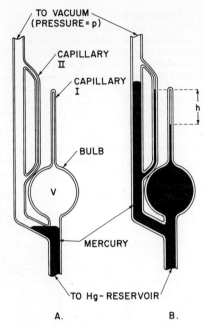

TO VACUUM
(PRESSURE = p)

CAPILLARY
II

CAPILLARY
I

h

BULB

V

MERCURY

TO Hg - RESERVOIR

A.   B.

Fig. 99 McLeod manometer before (A) and after (B) compression of trapped gas volume V.

where a is the unit volume of the capillary in terms of cm³ per mm length. The corresponding pressure is given by the height difference of the two capillary columns h (in mm), and the system pressure p. The latter is usually small compared with h, and hence

$$p = h + p \approx h$$

According to Boyle's law, $p_c V_c = pV$, and therefore

$$p \approx \frac{a}{V} h^2 \qquad \text{Torr}$$

The factor a/V is the gauge constant, which determines the sensitivity of the instrument. Typical manometers used in the laboratory have capillaries of 1 to 3 mm diameter and bulb volumes of the order of 100 cm³ which yield gauge constants of about 10^{-5} mm⁻¹. Assuming 10 mm as the smallest height which can be read with good accuracy, the corresponding low-pressure limit is about 10^{-3} Torr. Instruments capable of measuring down to 10^{-5} Torr have been constructed by using a larger bulb volume and smaller capillaries. In these cases, the purity of the mercury and the condition of the inner capillary walls are extremely critical. The bulb volume V is limited by the impracticality of raising and supporting a large quantity of mercury. This is accomplished either by lifting a mercury reservoir which is connected to the lower end of the bulb by means of a flexible hose, or by pushing the mercury up with a tube-and-plunger type of arrangement.

While Boyle's law holds accurately for permanent gases such as N_2, O_2, and H_2 others like H_2O, CO_2, and hydrocarbons may condense on the capillary walls when compressed. Their presence must be excluded to obtain accurate readings. This is done by inserting a liquid-nitrogen cold trap in the line connecting the McLeod gauge with the system. Unfortunately, the vapor pressure of mercury at room temperature is not negligible (2×10^{-3} Torr) so that a noticeable amount of vapor streams from the gauge to the cold trap. This phenomenon gives rise to a pressure gradient along the connecting line. The vapor stream effect may be taken into account either by using a corrected formula for the compression ratio of the gauge[337] or by putting a constriction of 1 to 2 mm diameter into the connecting line.[338]

McLeod manometers are of considerable practical importance because they yield accurate and direct pressure measurements over a range which extends into high vacuum. Therefore, they offer one of the few experimental ways to calibrate ionization gauges. Although the pressure range common to both types of instruments is small, McLeod gauges are widely used as primary standards because of the relative simplicity of the procedure. Alternate calibration techniques, particularly those suitable for lower pressures, require special vacuum and gas-flow systems of considerably greater complexity.[339]

(3) Thermal-conductivity Gauges The kinetic theory predicts that the thermal conductivity of gases is independent of pressure in the region of viscous flow just like the gas viscosity [see Chap. 1, Sec. 2b(5)]. Experimentally, only small decreases in thermal conductivity are found when the pressure is lowered from 760 to about 5 Torr. At lower pressures, however, where the mean free path of molecules is larger than the spacing between opposite surfaces, the conduction of heat from a hot to a cold object becomes linearly proportional to the gas pressure. This behavior is used in several types of medium-vacuum gauges.   In general, they employ one or more electrically heated filaments whose temperature is determined by the balance of the energy supplied and the heat losses to the surrounding walls.   The power dissipation of the filament is only partly due to conduction through the gas; another part is represented by radiation losses which are proportional to the emissivity and the temperature of the filament surface.   Since the latter mechanism is pressure-independent, it is desirable to minimize radiation losses to make the filament temperature as pressure-sensitive as possible.   Hence, filament materials of relatively low emissivity such as tungsten or platinum are used.   The filament temperature as an indicator of gas pressure can be measured either directly by a thermocouple or indirectly by monitoring the resistance of the filament (Pirani gauges).

Thermocouple gauges are available in metal or glass envelopes and with different internal wire arrangements.   Figure 100A shows a typical metal-enclosed gauge.

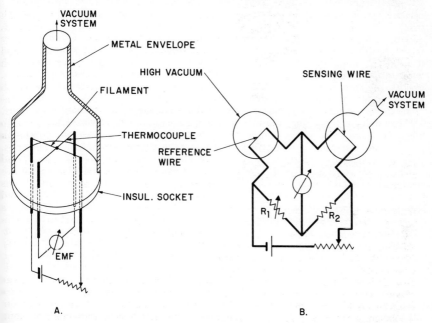

Fig. 100   Pressure gauges based on thermal conductivity of gases. A = thermocouple gauge. B = Pirani gauge.

The thermocouple is spot-welded to the center of the filament wire, which receives a constant energy from an external power supply.   The filament assumes maximum temperature (less than 200°C) at pressures below 10^{-3} Torr where the conductivity of the gas is very small.   The latter increases with pressure, and the emf produced by the thermocouple decreases accordingly.   The characteristic is nonlinear, but the pressure can be read directly from an appropriately calibrated moving-coil ammeter. Since the emf of the thermocouple also depends on the cold-junction temperature, the gauges are sensitive to changes in ambient temperature, but these can be compensated.[340,341]   The pressure range depends on the construction of the gauge and covers typically 10 to 10^{-3} Torr.

Pirani gauges are constructed as shown schematically in Fig. 100B.   A resistance wire enclosed in a glass or metal envelope is exposed to the pressure in the vacuum system while being part of a Wheatstone-bridge circuit.   An identical wire in a similar but thoroughly evacuated and sealed enclosure serves as the reference.   Both wires are heated by the current from a constant-voltage supply.   The other resistors in the bridge are adjusted for zero current through the central ammeter while the envelope of the sensing wire is evacuated to 10^{-4} Torr or less. At higher pressures, the temperature of the sensing wire decreases as the thermal conductivity of the gas increases. Consequently the resistance of the sensing wire decreases, and the current through the unbalanced bridge indicates the change in pressure. This method is known as the constant-voltage mode and is often used in commercial gauges. The applicable range is about 10^{-3} to 10^{-1} Torr. Other types of Pirani gauges are designed to maintain the temperature of the sensing wire constant and utilize the energy input as a measure of the pressure.

In general, the current-pressure characteristics of Pirani gauges are nonlinear and also sensitive to differences in ambient temperature. To reduce the effects of temperature variations, the reference wire is frequently sealed in a glass tube at 10^{-5} Torr and placed into the same envelope with the sensing wire. The characteristics of such instruments are likely to change if the system is often filled with helium because the latter permeates glass and slowly fills the vacuum around the reference wire.

Some gauges utilize a bead of semiconducting material in contact with a heated wire. Such instruments are referred to as *thermistor gauges* (see, for example, Ref. 6, p. 297). They offer greater pressure sensitivity because of the strong negative temperature dependence of the semiconductor's resistance. Thermistor gauges indicate pressures from 10^{-3} to 50 Torr, but their response is slower than that of Pirani gauges because of the greater thermal inertia of the bead.

All thermal-conductivity gauges require calibration against a standard such as a McLeod manometer. Since the thermal conductivity in the molecular-flow region is inversely proportional to the square root of the molar mass (see, for example, Ref. 6, p. 44), the instrument readings depend on the gas species present. Commercial gauges are generally calibrated for dry air or nitrogen whose conductivities are very similar. The readings of such instruments obtained in other gases such as argon can be converted into true pressures by means of calibration curves. Once calibrated, the gauges do not necessarily retain their characteristics because the filament surfaces and hence their emissivities tend to change. However, high accuracy is rarely required since the main application of these gauges is to monitor the fore vacuum of pumping systems. For this purpose, they are very well suited since an electric output signal is available which may be utilized for automatic control.

(4) Pressure Gauges for High to Ultrahigh Vacuum Practically all pressure measurements below 10^{-3} Torr are based on the phenomenon of residual-gas ionization. Alternate methods, which utilize special versions of McLeod and Pirani manometers or other types of gauges (see, for example, Ref. 6, Chap. 5), are far less convenient and cannot be extended down to the lowest pressures of interest. To induce ionization, electrons with energies greater than the ionizing potential of the gas—usually above 50 eV—are injected. If they collide with gas particles, the latter may become ionized. The probability of such an event depends on the energy of the electrons and on the type of gas. Ionization probabilities are expressed as the number of ionizing collisions per centimeter of electron path at 1 Torr gas pressure and 0°C. Their energy dependence is shown in Fig. 101 for a number of com-

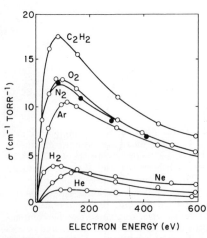

Fig. 101 Ionization probabilities of common gases as functions of the electron energy. (*After data from Smith and Tate,*[342] *figure from Ref.* 331 *with permission of The Institute of Physics and The Physical Society, London.*)

mon gases. Electron energies of about 150 eV, which yield the highest ionization rate, are employed in most ionization gauges.

The positive gas ions generated by the electron collisions are accelerated toward a collector, and the resulting current is taken as an indication of the particle density in the gauge. The relationship between the ion current I_i, the electron-emission current I_e, and the gas pressure is (Ref. 129, p. 405)

$$I_i = I_e l\sigma p \quad A \tag{25}$$

where l = length of electron paths in the gauge, cm

σ = ionization probability, cm^{-1} Torr^{-1}

p = gas pressure, Torr

Equation (25) defines an important figure of merit, the gauge constant K:

$$K = l\sigma = \frac{I_i}{I_e p} \quad \text{Torr}^{-1} \tag{26}$$

The quantity

$$l\sigma I_e = \frac{I_i}{p} \quad A \text{ Torr}^{-1} \tag{27}$$

is also often used to express the sensitivity of a gauge since most instruments operate at a constant emission current I_e. As the equations show, both gauge parameters include the ionization probability σ and are therefore gas specific.

Experimentally, gas ionization may be implemented by two different methods. The first one relies on the acceleration of electrons from a hot-emission filament, and the corresponding instruments are referred to as hot-cathode ionization gauges. In instruments of the second type, the cold-cathode ionization gauges, ionization is obtained by a glow discharge whose positive ion current is monitored. Both types of gauges have been constructed in various designs, and many of them are commercially available. The objective of these modifications has usually been the extension of the instrument range down to very low pressures. A performance comparison of several gauges has been given by Kornelsen.[343] Recent developments of ionization gauges have been reviewed by Groszkowski.[344]

(5) Hot-cathode Ionization Gauges
The earliest and most straightforward embodiment of the ionization principle is the *triode ionization gauge*. As shown in the schematic in Fig. 102, it has a central filament which is heated to emit electrons. Its potential to ground and also to the cylindrical ion collector is about +30 V. The surrounding, loosely wound grid is positively biased at about 180 V. The electrons from the filament are attracted toward the grid, where most of them pass through. Between grid and collector, the direction of the electrons is reversed because of the oppositely directed electric field. Thus, the electrons cycle back and forth until they strike the grid. Along their paths, some of them collide with residual gas atoms and produce an ion-electron pair.

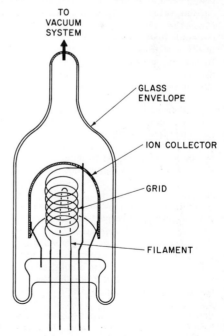

Fig. 102 Schematic of a triode ionization gauge.

The positively charged gas ions are attracted to the collector and thus generate a small current. If the emission current is kept constant, the ion current is proportional to the gas pressure as shown by Eq. (25). The gauge constant $l\sigma$ is of the order of 10 Torr^{-1}, and typical emission currents are between 1 and 10 mA. Hence, ion currents of 10^{-8} to 10^{-7} A are obtained at pressures of 10^{-6} Torr.

The linear relationship between ion current and pressure in these types of gauges fails above approximately 10^{-3} Torr. This is due to the high frequency of collisions which produce many low-energy electrons incapable of ionizing gas molecules. These electrons are collected by the grid and thus measured as part of the emission current. Since the latter is regulated at a constant level, the ionization efficiency and hence the sensitivity of the gauge are reduced. If the pressure increases further, a space charge of positive ions forms around the filament and arclike discharges may be observed.

The range of ionization gauges can be widened toward higher pressures if the electron paths are kept short so that the ionization frequency is reduced. A gauge capable of measuring up to 0.6 Torr has been developed by *Schulz and Phelps*.[345] Its characteristic features are an emission filament between two closely spaced parallel molybdenum plates. These are biased such that one of them collects the positive ions whereas the electrons go directly to the other plate without oscillatory motion. More recently, another ionization gauge for medium vacuum, the MilliTorr,* has found increasing utilization. It functions on the same principle, attaining short electron paths by closely spaced wire electrodes. Operating at low emission currents of 0.1 mA or less, the gauge has a constant of 0.5 Torr^{-1} and indicates pressures from 10^{-5} to 1 Torr. A thoria-coated iridium filament permits operation at air pressures above 1 Torr without burnout. The gauge also allows measurements in a glow-discharge environment and thus makes it suitable for sputtering systems. The electric signal may be used for automatic pressure control through a servo valve.[346]

At very low pressures, the range of ionization gauges is limited by the so-called *x-ray effect*. Referring back to Fig. 102, the grid electrode of the triode tube is continually bombarded with electrons of about 150 eV at an intensity equal to the emission current. These electrons generate soft x-rays, which may strike the ion collector and cause the emission of photoelectrons. The meter measuring the collector current does not distinguish between negative charges leaving the collector electrode and positive charges received. Hence, the collector current ceases to be proportional to the gas pressure when the secondary electron emission is no longer negligible against the ion current. In triode tubes, the magnitude of the secondary-emission current is equivalent to an ion current obtained at about 10^{-8} Torr. Therefore, reasonably accurate pressure readings below 10^{-7} Torr cannot be expected. The development of modern ionization gauges has been dominated by the desire to lower the x-ray limit through modifications of the electrode structure. Three approaches have been successful in this respect: a drastic reduction of the ion-collector area, physical separation and shielding of electron- and ion-collecting electrodes, and the use of a magnetic field to lengthen the electron paths, which permits a reduction of the electron-emission current without loss of sensitivity [see Eq. (27)].

The first approach was taken by Bayard and Alpert,[347] whose design is shown schematically in Fig. 103. The *Bayard-Alpert gauge (BAG)* differs from the triode tube described previously in the reversed positions of the emission filament and the ion collector with respect to the grid. The ion collector has been reduced to a thin (0.005 in. diameter) wire in the center of the grid helix. The electrode potentials and the emission current are very similar to those in the triode tube with the cylindrical collector. While the central wire collects most of the positive ions formed within the grid helix, it presents a solid angle to the x-rays from the grid which is two to three orders of magnitude smaller than that of the cylinder in the triode tube. Hence, the practical x-ray limit of the BAG lies at about 10^{-10} Torr.[344] The gauge constants vary from 10 to 25 Torr^{-1} for nitrogen, depending on the electrode structures. The linearity of the ion current–pressure relationship [Eq. (25)] has been demonstrated by Alpert and Buritz[247] for the interval from 10^{-3} to 10^{-10} Torr.

* Trademark of Varian Associates, Vacuum Division, Palo Alto, Calif.; see *Varian Vacuum Views*, May/June, 1969.

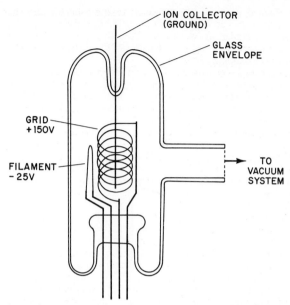

ION COLLECTOR
(GROUND)

GLASS
ENVELOPE

GRID
+150V

FILAMENT
-25V

TO
VACUUM
SYSTEM

Fig. 103 Schematic of a Bayard-Alpert ionization gauge.

By extending the range of measurable pressures down into the ultrahigh-vacuum region, the Bayard-Alpert gauge has found immediate and widespread usage. Gauges from various manufacturers differ mostly in their electrode design. The version shown in Fig. 103 has the collector wire and the other electrode seals at opposite ends of the glass tube to minimize leakage currents. Since the collector current is very small, it is essential that the tube socket and the connecting cable have high insulation resistance. A common feature is the availability of a second built-in filament to be used if the first one burns out. The effects of changes in the dimensions of the gauge electrodes have been investigated by Kinsella[348] and Groszkowski.[349] Redhead[350] extended the measurable pressure range downward by approximately one decade. This is accomplished by placing an additional electrode close to the grid helix. The potential of this auxiliary electrode varies periodically so as to modulate the positive-ion current. The photoemission current due to soft x-rays remains unaffected, and the two current contributions can therefore be separated.*

The second method of reducing the x-ray limit in ionization gauges, namely, by removing the ion collector from the vicinity of the electron-collecting grid, has been implemented in several ways. One version, the *Klopfer gauge*,[351] is shown schematically in Fig. 104. The electrons emitted by the filament are collimated by the electrostatic potentials of three slits and an axial magnetic field. While passing through the gauge chamber, they have energies of 100 to 200 eV, enough to ionize some gas molecules before they reach the electron collector. The gas ions are attracted toward the ion-collector plate at the bottom. The latter is shielded against x-rays from the electron collector by the housing. The positively biased metal grid opposite the collector prevents the escape of positive ions into the vacuum system but allows the exchange of neutral gas particles. The gauge has a linear ion current–pressure characteristic from 10^{-2} to 10^{-11} Torr.

* Note added in proof: Efforts to improve the performance of BAG's through changes in electrode design and positioning continue to be successful. P. A. Redhead recently described a new gauge with a sensitivity for nitrogen of about 50 $Torr^{-1}$ [*J. Vacuum Sci. Technol.*, **6**, 848 (1969)].

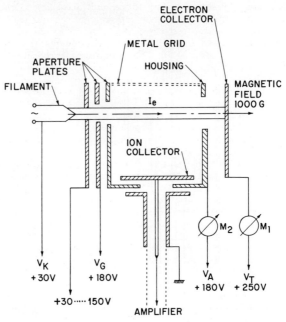

Fig. 104 Principle of the Klopfer ionization gauge.[351]

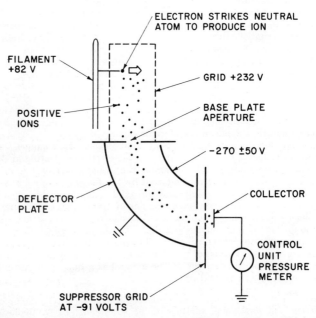

Fig. 105 Principle of the Helmer ionization gauge. (*With permission of Varian Associates Vacuum Division, Palo Alto, Calif.*)

The Helmer gauge[*352] shown in Fig. 105 is built according to the same principle. Its filament and grid structure resemble the BAG, but the ions are extracted from the grid space through the base-aperture plate, which is at ground potential. Their further path is determined by electrostatic deflection plates which achieve a 90° curvature without a magnetic field. The suppressor grid in front of the collector repels secondary electrons which may be emitted when the ions strike the collector plate. The position of the latter is such that it cannot be reached by x-rays from the grid. The collector does receive x-rays generated by energetic electrons or ions striking the deflector plates. However, their intensity is about 100 times smaller than in the BAG, and the resulting emission current is equal to the ion current at about 10^{-13} Torr. The sensitivity of the gauge is 0.1 A Torr^{-1} for $I_e = 6$ mA.

The third possibility to lower the x-ray limit of ionization gauges is to lengthen the electron paths so much that a sufficiently large number of ionizing collisions can be supported with a drastically reduced emission current. This condition can be implemented by applying a magnetic field perpendicular to the electric field which accelerates the electrons. The principle has been utilized by *Lafferty*[353] in the gauge shown in Fig. 106. Electrons emitted from the central filament are forced into spiraling trajectories whose radii are functions of the anode potential and the magnetic field. If the latter is sufficiently high, most of the electrons do not reach the anode until they have made many orbits. Thus, the electron density and ionizing probability are greatly increased. Both quantities are further enhanced by secondary-electron emission so that the gauge operates effectively at filament emission currents as low as 10^{-7} to 10^{-9} A. The positive ions generated by collisions travel in essentially straight paths to the collector plate at the top of the gauge. The sensitivity is of the order of 0.1 A Torr^{-1},[343] and the x-ray limit is very low, about 10^{-14} Torr, by virtue of the small electron-impingement rates. Ion-current measurements near the lower pressure limit require the use of an electron multiplier.[354] Toward higher pressures, the gauge operation becomes space-charge-limited at about 10^{-6} Torr.

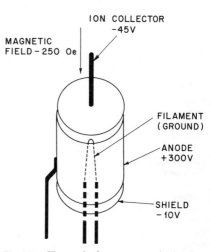

Fig. 106   Hot-cathode magnetron ionization gauge after Lafferty[353] (glass envelope not shown).

(6) Cold-cathode Ionization Gauges   The idea of measuring the current of a glow discharge as an indicator of the gas pressure was first implemented by *Penning*[128] in 1937. The principal arrangement is shown in Fig. 107. A dc potential of about 2 kV is maintained between a ring-shaped anode and two cathode plates. Because of the ubiquitous presence of cosmic rays and stray radioactivity, a few secondary electrons are liberated from the cathodes. These initiate some ionization, whereupon positive ions strike the cathodes with enough energy to cause substantial secondary emission with subsequent further ionization of the gas. As a result, a self-sustaining glow discharge is obtained. The charged particles are confined within the electrode space by a magnetic field of approximately 400 Oe. This forces the electrons to travel in long helical paths before reaching the anode and to make many ionizing collisions. Therefore, the discharge can be sustained down to about 5×10^{-6} Torr pressure. The positive ions are little affected by the magnetic field and go to the cathodes directly. The total discharge current, which consists of positive-ion and electron-emission currents, is used as a measure of the gas pressure in the gauge.

* Trademark of Varian Associates, Vacuum Division, Palo Alto, Calif.

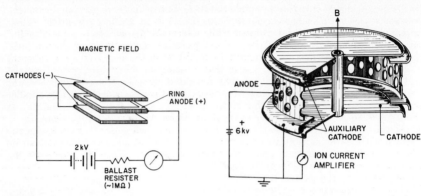

Fig. 107   Schematic of a Penning ionization gauge.

Fig. 108   Cold-cathode magnetron ionization gauge after Redhead.[355]   (*Courtesy of P. A. Redhead, with permission.*)

The principal advantage of Penning gauges is the absence of a hot filament.   Their simple and rugged construction makes them insensitive to exposure to air.   But it is often difficult to start the discharge at low pressures, and the relationship between discharge current and pressure is not linear.   Moreover, temporary irregularities in the discharge current occur frequently as a result of oscillations in the plasma.   Therefore, Penning gauges are not as accurate as hot-cathode gauges.   However, they are commercially used in applications where accuracy is not essential.   An example is their utilization as relay actuators in the high-pressure protection circuits of leak detectors.

An improved type of cold-cathode ionization gauge is the design shown in Fig. 108, which was developed by *Redhead*.[356]   Its cathode has the form of a spool consisting of a small central cylinder with two end disks.   The anode is a cylinder with many holes to admit gas into the gauge.   The electrode potentials and the magnetic field (about 2,000 Oe) are chosen so that the electrons are trapped in the magnetron-like cavity.   When ionization of a gas atom occurs, the ion goes to the cathode while the electron may either become trapped and orbit, or spiral outward until it impinges on the anode.   In doing so, it may cause x-ray emission, but this sequence of events is proportional to the particle density and comparatively rare.   Hence the secondary-emission current remains small against the ion current at all pressures, and the gauge has no predictable x-ray limit.

Another potential operating limit of the Redhead magnetron gauge is avoided by the auxiliary cathodes which shield the cathode disks against the high electric field at their periphery.   Thus, field-emission currents do not involve the ion-collecting cathode but merely the auxiliary electrodes which are independently grounded.   The gauge has been shown to be useful in the range from 10^{-3} to 10^{-12} Torr where it has a slightly nonlinear characteristic of the form

$$I_i = \text{const } p^n$$

with n being about 1.1.   Application at lower pressures is largely a question of instrumentation to measure very small currents since the sensitivity is of the order of 1 A Torr^{-1}.   While such measurements can be performed with electron multipliers, the cost of such equipment becomes comparable with residual-gas analyzers, which are usually preferred in this pressure region.

(7) Operation of High-vacuum Gauges   The purpose of low-pressure measurements is generally to determine the number of particles per unit volume in order to derive the rate at which molecules impinge on some specified surface within the vacuum system.   The area of interest may be a thin film, a substrate, or an instrument.   It is usually assumed that the gas pressure indicated by the gauge also represents the conditions existing at other points in the vacuum system. This assumption, however, is at best an approximation because the behavior of gases at very low pressures is dominated by molecule-wall interactions while intermolecular collisions are relatively rare. Consequently, the distributions of gas particles and their velocities are not uniform and Maxwellian. Further limitations in the measurement of gas pressures are due to problems associated with ionization gauges. Most of these potential sources of error cannot be eliminated. An understanding of the mechanisms involved is necessary to minimize their effects and to appreciate the uncertainty of low-pressure readings.

The problems of nonuniform gas distributions in vacuum systems have been reviewed by Moore,[357] who enumerated the causes which may lead to variable gas densities. These are pumps acting as traps and as sources of directional gas particles, nonrandom reflections of impinging molecules from walls, surface migration of adsorbed species, varying adsorption and desorption rates at certain parts of the internal walls, and gas-density variations arising from temperature differences. Although attempts have been made to describe real gas distributions analytically, these are restricted to systems of simple geometries. An experimental study has been conducted by Holland,[358] who considered the total residual gas pressure as due to a Maxwellian and a directional component. By mounting an ionization gauge such that its inlet tubulation could be rotated, Holland was able to demonstrate significant pressure differences for different gauge orientations.

While it is not possible to eliminate all sources of directional pressure contributions, the major ones can be taken into account when installing an ionization gauge in a vacuum system. If the gauge is positioned so that it faces a pump, a cryogenic surface, or a heavily outgassing area such as a heated part, it will predictably indicate pressures which are either lower or higher than the particle density to which the substrate is subjected. For more truly representative pressure readings, the gauge tubulation should generally be pointed upward or sideways in a location where beaming effects are similar to those in the substrate location. The danger of deceptively low pressure readings is greater in systems which use high pumping speeds against large desorption rates. Under these conditions, one can expect a large directional pressure component which is unlikely to be reflected in the gauge reading.

Other causes of misinterpreting ionization-gauge readings are associated with the gauge itself. First it should be remembered that the gauges are usually calibrated for dry nitrogen, which is rarely the predominant gas in vacuum systems. Thus, the gauge indicates the total pressure only in terms of its "dry-nitrogen equivalent." Conversion into true pressures may be attempted by considering the sensitivities of other gases relative to nitrogen.[359] However, this requires some information about the composition of the residual gas in the system.

Ionization gauges also act as pumps and sources of residual gases. The pumping action is due to the electrical cleanup mechanism discussed in Sec. 2e(2). It is considerably stronger in cold- than in hot-cathode gauges. Gas release by desorption from the electrodes is usually induced deliberately when the gauge is turned on. In addition to heating the emission filament, the grid electrode, too, is baked out by passing a current through it for a few minutes (degassing). The energy released in this procedure is sufficient to heat and partially outgas the gauge envelope. However, it does not reduce the outgassing rates to levels sufficiently low for ultrahigh-vacuum operation. Gas contributions from a clean but unbaked gauge tubulation may prevent the attainment of readings below 10^{-8} Torr.[360] This difficulty may also be encountered if oil vapors from the pump are allowed to become adsorbed by the tubulation.[361] To minimize pressure differences between the gauge interior and the vacuum system, the tubulation should have a high conductance.[362] A diameter of about $\frac{3}{4}$ in. is considered to be adequate for pressures down to 10^{-6} Torr, whereas 1 in. is recommended for lower pressures.

Another important phenomenon which determines the lowest pressure measurable with ionization gauges is the desorption of gas particles from the electron collector due to its bombardment with electrons. It has been observed by Redhead[363] that a small fraction of the oxygen desorbed from a molybdenum grid struck by 90-eV electrons becomes ionized. The positive oxygen ions may cause a background ion current equivalent to 10^{-8} Torr in a typical BAG. Consequently, the low-pressure limit posed by electronic desorption may often be more problematical than the x-ray effect.[364] According to Redhead, the adsorbed oxygen can be removed if the electrode is initially bombarded with 1-kV electrons at a power density of 12 W cm^{-2} for a few minutes.[363] This procedure is claimed to be more effective than the common Joule heating of the grid wire.

c. Residual-gas Analysis

To understand and control the interaction of residual gases with thin film growth processes, it is often necessary to determine not only the total pressure in a system but also the gas composition. Knowledge of partial pressures of individual gas species is of interest in recognizing gas contributions from pumps, in identifying the sources of outgassing processes, in relating film properties to the presence of specific gases, and in evaluating the effects of experimental procedures such as baking or glow-discharge cleaning.

The measurement of partial pressures in high-vacuum systems is performed by means of mass spectrometers, instruments which are capable of separating ionized molecules of different mass-to-charge ratios and measuring the respective ion currents. Mass spectrometers specifically designed for high-vacuum investigations are referred to as residual-gas analyzers (RGA). Their distinguishing features are high sensitivity and the ability to withstand baking so that gases of low partial pressure can be identified without being obscured by contributions from the analyzer itself. A great variety of such instruments is commercially available. They employ different combinations of magnetic, electrostatic, or electrodynamic fields to achieve separation of ions. Accordingly, the characteristics of RGAs differ significantly, and several performance parameters must be taken into account to select an instrument suitable for a particular application.

(1) Performance Parameters of Residual-gas Analyzers All RGAs consist of three functional parts: an ionizer, an analyzer, and a detection system. The principal differences between various instruments are in the analyzers which employ different methods of sorting ions. These are discussed in the next section together with their performance ranges. The ionization and detection systems are fairly independent of the type of analyzer used. The most important performance parameters needed to assess the suitability of an instrument for a particular experiment are the sensitivity, the resolution, the mass range, and the scan rate.

The conversion of residual-gas particles into positive ions is accomplished through collisions with an electron beam. The latter is extracted from an emission filament by means of an electric field and suitable apertures. Since the emission filament is rapidly destroyed if exposed to reactive gases such as oxygen, all RGAs require operating pressures of less than 10^{-4} Torr, sometimes less than 10^{-5} Torr.

The ion current I_i originating from an electron current I_e passing a distance l through a gas of pressure p is given by Eq. (25) of Sec. 6b(4):

$$I_i = I_e l \sigma p$$

Values of the ionization probability σ for most of the common gases lie between 1 and 10 cm^{-1} Torr^{-1} (see Fig. 101). Practical values for the emission current and the electron-path lengths in RGAs are 0.1 to 1 mA and 1 to 2 cm, respectively. Thus, the ion currents generated in RGAs are of the order of $10^{-4}p$ to $10^{-2}p$ ampere. Since most instruments operate with emission currents of less than 1 mA, and also because of imperfect transmission through the analyzer, the ion currents arriving at the collector are typically around $10^{-4}p$ ampere. Hence, to detect a partial pressure of 10^{-11} Torr, the collector and output system must respond to an ion current of 10^{-15} A, which is equivalent to about 6,000 ions per second. Sensitive electrometers and amplifiers are adaptable for this range. The minimum detectable pressure can be lowered if the detection system utilizes an electron multiplier. It is then possible to register currents as low as 10^{-17} A corresponding to pressures of 10^{-13} Torr. Further lowering of the minimum detectable pressure is contingent upon increasing the ion-source yield I_i/p. This has been done in some of the more recent instruments like the quadrupole, where emission currents up to 10 mA are used to register pressures as low as 10^{-15} Torr.

The ratio of ion current to pressure is often called the *sensitivity* of an RGA. Although it is a figure of merit in comparing different analyzers, it is not a directly observable quantity. More meaningful to the experimentalist are sensitivities quoted in torr per instrument scale division. These take into account the capability of the detection system and relate directly to the minimum pressure which the output device can display. The sensitivity varies for different gas species since the ion current depends on the ionization probability σ. RGA manufacturers usually quote the sensitivity for nitrogen ($M = 28$).

Most RGAs have an output meter which indicates the magnitude of the ion current due to the mass in focus. However, it is generally desired to have a permanent record of the results. Thus, RGAs are commonly equipped with strip-chart recorders or X-Y plotters to trace the complete spectrum of the mass range scanned by the instrument. If rapid changes in the composition of the residual-gas atmosphere are to be observed, the response of mechanical-type recorders is too slow. For such applications, recording oscillographs are available. These are mirror galvanometers which record by means of a light beam on a strip of rapidly moving photosensitive paper. The spectrum may also be displayed on an oscilloscope.[365] The speed with which a gas analyzer can sweep the ion beams of all masses within a given mass range across the collector and record the resulting spectrum is the *scan rate*. It is expressed in either seconds per mass peak or seconds per mass range. The fast-scanning instruments allow downward adjustment of the scan rate to match readout equipment of slower response time.

The *resolution* of RGAs defines their ability to identify separately particles differing by one atomic mass unit. The use of this term is not uniform. One definition, which emphasizes the qualitative identification of species, considers the relative height of the valley between two adjacent, equally strong and overlapping peaks. This is illustrated in Fig. 109A, where a valley resolution of 10% at mass M is shown. Another definition of resolution is more meaningful for the quantitative evaluation of RGA spectra. It considers the contribution of the tail of one mass peak to an equally strong adjacent peak. Figure 109B illustrates unit resolution (less than 1% "crosstalk") for the mass M. In most analyzers, the separation of adjacent peaks decreases with

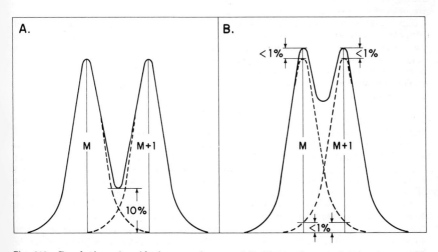

Fig. 109 Resolution of residual-gas analyzers. (A) 10% valley resolution at mass M. (B) Unit ("crosstalk") resolution at mass M.

increasing mass. Hence, the resolving power of an instrument is better the greater the mass for which a particular value of resolution is quoted.

Finally, the *mass range* of an RGA is that range of atomic or molecular masses which an instrument can cover regardless of resolution. The mass range is usually larger than the mass interval for which unit resolution is claimed. It is of interest to note that most gases commonly encountered in high-vacuum systems have masses of less than 50 atomic mass units. Therefore, the acquisition of an RGA with a larger mass range may be an unnecessary expense for many applications.

The operating parts of RGAs, the ionizer, analyzer, and collector, are mounted in a vacuum-tight enclosure. The latter consists of glass or, in more recent instruments, stainless steel and is bakeable, usually up to 450°C. The envelope of most analyzers is tubular, and the lengths of the units vary from 15 to 50 cm. The size is an important consideration in assessing the space requirements and the outgassing contributions of an RGA.

(2) Common Types of Residual-gas Analyzers A variety of techniques to separate ions has been explored, and the literature describing operating principles and characteristic features of RGAs is extensive. Reviews have been given by Caswell[82] and more recently by Blauth.[366] McDowell's[367] book on mass spectrometry includes an article by Dibeler on instruments available from various manufacturers. However, commercial models are continually improving with ensuing changes in performance ratings and model designations. The current interest in developing more compact units of greater stability and simplicity in operation is reflected in numerous publications.[368] Therefore, the following summary considers only the different principles involved and is confined to general rather than specific performance data.

Generically, two groups of analyzers may be distinguished. The first category employs *magnetic fields* in conjunction with *electric fields* which may be either *static* (Dempster, sector, cycloidal analyzers) or *radio-frequency* (Omegatron). Ions traversing the magnetic field are forced into paths whose radii of curvature are proportional to their momentum. If all ions have the same energy, the radii become mass-dependent. By varying the accelerating voltage, ions of different mass are made to arrive successively at the same point, the collector. Although the analyzer tube of these instruments can be made quite compact, considerable bulk is added by the external magnets. Instruments based on this principle have long been known and are still more numerous than those of the second group, the *dynamic analyzers*.

Dynamic analyzers or mass filters were developed after World War II. They utilize electric fields which are varied such that ions of different mass-to-charge ratios arrive at the collector in predictable succession. The foremost representatives of this group are the quadrupole, the monopole, and the time-of-flight analyzers. Their advantages are excellent resolution at high masses and fast scan rates. However, they are generally larger than static-magnetic-field analyzers and also more expensive. Instruments of this category have been reviewed by Harrington.[369]

The methods of ion separation used in the various types of instruments are shown schematically in Fig. 110. In the *Dempster*[370] analyzer, the ions are first accelerated by an electric field toward a slit and then enter the magnetic field. After 180° deflection, they pass through a second slit and strike the collector. The same principle is utilized in the 60° *sector* analyzer shown in Fig. 110B. The resolving power of these instruments is related to their size. Small models of about 15 cm length with unit resolution up to mass 20 as well as larger ones of about 40 cm length with resolution up to mass 150 are available. The sensitivities are of the order of 10^{-10} to 10^{-11} Torr per division if electrometers are used; they may be extended to 10^{-13} or 10^{-14} Torr per division with electron multipliers. The scan rates are generally slow and of the order of minutes per mass range.

Cycloidal analyzers were introduced by Hipple and Bleakney[371] in 1936. As shown in Fig. 110C, the ions travel in crossed magnetic and electric fields. They acquire a drift velocity which is proportional to the ratio of the fields and independent of the ion mass. The combination of this drift velocity and the circular motion induced by the magnetic field focuses ions of equal mass into the collector slit after a deflection of 360°. Cycloidal analyzers have twice the theoretical resolving power of Dempster

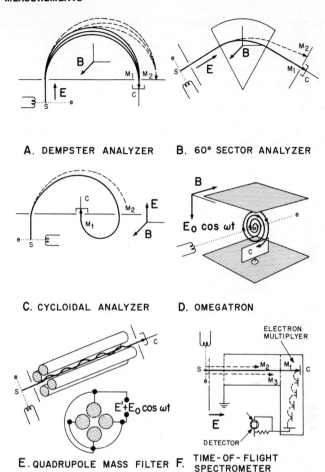

A. DEMPSTER ANALYZER   B. 60° SECTOR ANALYZER

C. CYCLOIDAL ANALYZER   D. OMEGATRON

E. QUADRUPOLE MASS FILTER   F. TIME-OF-FLIGHT SPECTROMETER

Fig. 110 Principles of ion separation in residual-gas analyzers: ..... e = ionizing electron beam; S = source of ion beam; E = electrostatic field; $E_0 \cos \omega t$ = rf field; B = magnetic field; C = ion collector.

or sector analyzers of comparable size, and some achieve unit resolution up to mass 150. The sensitivities are of the order of 10^{-11} to 10^{-12} Torr per division. Adjustable scan rates from 10 s to several minutes per mass range are customary.

The *omegatron* was developed by Sommer, Thomas, and Hipple.[372] It is an inexpensive compact instrument of relatively simple construction. As sketched in Fig. 110D, ions are generated along the direction of the magnetic field. As long as they have only thermal energy, they spiral in very tight orbits. By applying a small rf voltage to the top and bottom electrodes, most ions alternately gain and lose energy. However, if the radio frequency equals the cyclotron frequency for one particular mass, then these ions continue to gain energy and spiral outward until they impinge upon the collector plate. Thus, ion separation is accomplished by varying the frequency of the rf voltage.

Specific instruments have been described by several authors.[373,374] The resolving power of omegatrons is greater the more revolutions the ions can make. However, long trajectories produce high ion densities, which in turn lead to space-charge effects and operational instability. Therefore, the instrument works best with ions of low mass whose angular velocities are high. Typically, omegatrons have unit resolution up to mass 30 or 50. Provisions to extend the mass range have been discussed by Barz.[375] The sensitivities approach 10^{-11} Torr per division, and the scan rates are normally of the order of seconds per mass peak.

Mass filters of the *quadrupole* type were introduced by Paul and coworkers in 1953,[376] and several versions have been described since then.[377-380] Their theory of operation assumes hyperbolic field-forming surfaces, but experience has shown that excellent results can be obtained with round rods. As illustrated in Fig. 110E, pairs of opposing rods are connected to dc and rf voltage supplies, and the ions are accelerated along the axis between the rods. For a given spacing between rod centers, the dc and rf voltages are chosen to allow ions of only one mass to oscillate in stable orbits and reach the collector at the end of the tube. The trajectories of all other ions spiral outward and terminate at the rod surfaces. Ion separation and mass scanning are attained by varying the frequency of the rf voltage. The ratio of dc to rf voltages may be altered to gain sensitivity at the expense of resolving power or vice versa. A more recent development based on the same principle is the *monopole*,[381,382] which employs a single rod mounted in a V-shaped block.

Commercial quadrupoles or monopoles are about 30 cm long and may be used directly in vacuum systems since they are open at both ends and do not require magnets. They are usually equipped with electron multipliers and achieve sensitivities of 10^{-14} to 10^{-15} Torr per division. Ten percent valley resolution is obtained from mass 1 to 250, in some cases even up to 600.[383] Rapid scanning of the order of milliseconds per mass peak as well as slow scanning (minutes per mass peak) is possible.

The *time-of-flight* spectrometer was proposed by Stephens[384] in 1946. Its principle is illustrated in Fig. 110F. The ions are accelerated into a field-free drift tube with equal energies. Hence, the lighter particles travel at greater speeds and reach the collector in shorter times than ions of greater mass. Furthermore, the electron beam is pulse-modulated to produce short bursts of ionized particles. Each burst is separated according to mass by virtue of the different flight times. Consequently, the instrument scans very rapidly and requires only about 100 μs per spectrum. Useful resolution has been demonstrated up to mass 400. However, this is contingent upon rather long drift tubes of the order of 1 m, a feature which is not always compatible with the desired application. Also, the sensitivities are not particularly high (order of 10^{-12} Torr per division) because of the pulsed mode of operation.

(3) Interpretation of RGA Spectra Since residual-gas analyzers separate ions according to their mass-to-charge ratio, the resulting mass spectra have to be identified with the originating gas species. Ambiguity arises in cases where different molecules have the same mass. This difficulty can be overcome by taking into account dissociation and double ionization of molecules, processes which are specific for different species. The fractions of particles which dissociate or are doubly ionized under the conditions of residual-gas spectroscopy are fairly constant and therefore cause peaks of predictable intensity relative to the base peak. The resulting family of peaks is called the cracking pattern.

The commonest examples of gases with equal masses are N_2 and CO. Distinction between the two species is possible by analyzing their cracking patterns. The parent molecule N_2 is mostly collected as N_2^+ at mass 28, but 6 to 7% are converted into N_2^{++} and N^+, which appear at a mass-to-charge ratio of 14. The isotope N^{15} may also be recognizable as a very small peak. The molecule CO, on the other hand, produces several dissociation products such as C^+, O^+, and CO^{++} which give rise to comparatively weak peaks at the masses 12, 16, and 14. The relative intensities of fragment and isotope ions have been published by several authors[385-387] or may be obtained from the instrument manufacturers. Composite mass peaks can be analyzed by apportioning their total intensity to the different contributing species according to the intensities of the subsidiary peaks. It is also possible to identify species of high molecular weight by observing merely the cracking pattern in the lower-mass range.[388] Hydrocarbon fragment ions characteristic of common pump oils have been listed to allow identification of backstreaming and to discriminate against other sources of system contamination.[389] The quantitative analysis of a mass spectrum requires knowledge of the numerical relationship between the collector current due to a particular ion species and the corresponding partial pressure of the residual gas. These calibration factors are instrument-specific and may be determined by operating the system with controlled leaks to establish known partial pressures. As far as present investigations show, the relationships between partial pressure and ion current are linear at least in the range below 10^{-5} Torr. They are, however, instrument-specific, and it requires a very substantial amount of work to calibrate for a variety of different gases. This task is greatly reduced by utilization of relative sensitivities. These are based on the known ionization efficiencies of gases, i.e., on the constant ratio of ion currents obtained for equal amounts of different gases. If the relative sensitivities for a particular analyzer design and for the experimental arrangement are known, calibration for one gas species allows derivation of the absolute sensitivities for the other gases. Most manufacturers supply relative-sensitivity data as well as cracking patterns with their instruments. Published data are also available, particularly for the older instrument types operating with static electromagnetic fields.

REFERENCES

1. "Guide to Scientific Instruments 1967/68," American Association for the Advancement of Science, New York.
2. Chubb, J. N., *Vacuum*, **16**, 591 (1966).
3. Fischer, E., and H. Mommsen, *Vacuum*, **17**, 309 (1967).
4. Buhl, R., *Vacuum*, **16**, 585 (1966).
5. Barnes, C. B., Jr., *Vacuum*, **14**, 429 (1964).
6. Dushman, S., "Scientific Foundations of Vacuum Technique," 2d ed., p. 118, John Wiley & Sons, Inc., New York, 1962.
7. Lucas, T. E., *Vacuum*, **15**, 227 (1965).
8. Bunn, J. P., and L. Ward, *Vacuum*, **15**, 407 (1965).
9. Baker, M., L. Holland, and L. Laurenson, "Evaluation of Organic Vapor and Decomposition Products from Oil Seal Rotary Pumps," Technical Report issued by Edwards High Vacuum, Inc., Grand Island, N.Y., 1968.
10. Serwatzky, G., *Trans. 3d Intern. Vacuum Congr.*, 1965, vol. 2.1, p. 201, Pergamon Press, New York.
11. Haas, T. W., and A. G. Jackson, *J. Vacuum Sci. Technol.*, **3**, 168 (1966).
12. Hamilton, N. E., E. H. Stroberg, R. St. Pierre, and G. Shinopulos, *Ext. Abstr. 14th AVS Symp.*, 1967, p. 101, Herbick and Held Printing Co., Pittsburgh, Pa.
13. Holland, L., L. Laurenson, and M. A. Baker, *Ext. Abstr. 13th AVS Symp.*, 1966, p. 29, Herbick and Held Printing Co., Pittsburgh, Pa.
14. Fulker, M. J., "Backstreaming from Rotary Pumps," Technical Report issued by Edwards High Vacuum, Inc., Grand Island, N.Y., 1968.
15. Lorenz, A., *Proc. 1st Intern. Vacuum Congr.*, 1958, vol. 1, p. 177, Pergamon Press, New York.
16. Reich, G., and H. G. Noeller, *Proc. 1st Intern. Vacuum Congr.*, 1958, vol. 1, p. 93, Pergamon Press, New York.
17. Thees, R., *Proc. 1st Intern. Vacuum Congr.*, 1958, vol. 1, p. 164, Pergamon Press, New York.
18. Winzenburger, E. A., *Trans. 4th AVS Symp.*, 1957, p. 1, Pergamon Press, New York.
19. Becker, W., *Proc. 1st Intern. Vacuum Congr.*, 1958, vol. 1, p. 173, Pergamon Press, New York.
20. Becker, W., *Vacuum*, **16**, 625 (1966).
21. Guenther, K. G., and H. Lamatsch, *Trans. 3d Intern. Vacuum Congr.*, 1965, vol. 2.1, p. 161, Pergamon Press, New York.
22. Outlaw, R. A., *J. Vacuum Sci. Technol.*, **3**, 352 (1966).
23. Milleron, N., *IEEE Trans. Nucl. Sci.*, **NS-14**(3), 794 (1967).
24. Williams, C. E., and J. W. Beams, *Trans. 8th AVS Symp.*, 1961, p. 295, The Macmillan Company, New York.
25. Gaede, W., Deutsches Reichspatent 286,404, 1913; and *Ann. Physik*, **46**, 357 (1915).
26. Jaeckel, R., *Proc. 1st Intern. Vacuum Congr.*, 1958, vol. 1, p. 21, Pergamon Press, New York.

27. Dushman, S., in Ref. 6, p. 137.
28. Langmuir, I., *Gen. Elec. Rev.*, **19**, 1060 (1916).
29. Hablanian, M. H., and A. A. Landfors, *Ext. Abstr. 14th AVS Symp.*, 1967, p. 65, Herbick and Held Printing Co., Pittsburgh, Pa.
30. Burch, C. R., *Proc. Roy. Soc. London*, **123**, 27 (1929).
31. Hickman, K. C. D., *Trans. 8th AVS Symp.*, 1961, vol. 1, p. 307, The Macmillan Company, New York.
32. Alpert, D. J., *J. Appl. Phys.*, **24**, 860 (1953).
33. Power, B. D., and D. J. Crawley, *Vacuum*, **4**, 415 (1954).
34. Milleron, N., and L. L. Levenson, *Trans. 7th AVS Symp.*, 1960, p. 213, Pergamon Press, New York.
35. Kennedy, P. B., *Trans. 8th AVS Symp.*, 1961, vol. 1, p. 320, The Macmillan Company, New York.
36. Bance, U. R., and E. H. Harden, *Vacuum*, **15**, 437 (1965).
37. Tolmie, E. D., and D. J. Crawley, *Ext. Abstr. 13th AVS Symp.*, 1966, p. 97, Herbick and Held Printing Co., Pittsburgh, Pa.
38. Milleron, N., and L. L. Levenson, *Trans. 8th AVS Symp.*, 1961, vol. 1, p. 342, The Macmillan Company, New York.
39 Okamoto, H., and Y. Murakami, *Ext. Abstr. 14th AVS Symp.*, 1967, p. 67, Herbick and Held Printing Co., Pittsburgh, Pa.; see also *Vacuum*, **17**, 79 (1967).
40. Reich, G., *Z. Angew. Phys.*, **9**, 23 (1957).
41. Pagano, F., *Trans. 8th AVS Symp.*, 1961, p. 86, The Macmillan Company, New York.
42. O'Neill, H. J., W. M. Langdon, J. Frerichs, and A. Somora, *J. Vacuum Sci. Technol.*, **3**, 273 (1966).
43. Alpert, D., *Rev. Sci. Instr.*, **24**, 1004 (1953).
44. Jones, D. W., and C. A. Tsonis, *J. Vacuum Sci. Technol.*, **1**, 19 (1964).
45. Tolmie, E. D., *Vacuum*, **15**, 497 (1965).
46. Crawley, D. J., and J. M. Miller. *Vacuum*, **15**, 183 (1965).
47. Biondi, M. A., *Rev. Sci. Instr.*, **30**, 831 (1959).
48. Levenson, L. L., and N. Milleron, *Trans. 8th AVS Symp.*, 1961, p. 91, The Macmillan Company, New York.
49. Gosselin, C. M., and P. J. Bryant, *J. Vacuum Sci. Technol.*, **2**, 293 (1965).
50. Buhl, R., and O. Winkler, *Vacuum*, **16**, 697 (1966).
51. Oswald, R. D., and D. J. Crawley, *Vacuum*, **16**, 623 (1966).
52. Holland, L., L. Laurenson, and C. Priestland, *Rev. Sci. Instr.*, **34**, 377 (1963).
53. Langdon, W. M., and E. G. Fochtman, *Trans. 10th AVS Symp.*, 1963, p. 128, The Macmillan Company, New York.
54. Carter, J. G., J. A. Elder, R. D. Birkhoff, and A. K. Roecklein, *J. Vacuum Sci. Technol.*, **2**, 59 (1965).
55. Huber, W. K., R. Dobrozemsky, and F. Viehboeck, *Ext. Abstr. 13th AVS Symp.*, 1966, p. 119, Herbick and Held Printing Co., Pittsburgh, Pa.
56. Baker, M. A., "A Cooled Quartz Crystal Microbalance Method for Measuring Diffusion Pump Back Streaming," Technical Report issued by Edwards High Vacuum, Inc., Grand Island, N.Y., 1968.
57. Olejniczak, J. S., "Identification of Some Organic Vapors from Rotary Pump Oil by the Combined Chromatographic–Mass Spectrometric Technique," Technical Report issued by Edwards High Vacuum, Inc., Grand Island, N.Y., 1968.
58. Hablanian, M. H., and P. L. Vitkus, *Trans. 10th AVS Symp.*, 1963, p. 140, The Macmillan Company, New York.
59. Singleton, J. H., *J. Vacuum Sci. Technol.*, **3**, 354 (1966).
60. Steinrisser, F., *J. Vacuum Sci. Technol.*, **4**, 44 (1967).
61. Trickett, B., *Vacuum*, **14**, 397 (1964).
62. Venema, A., and M. Bandringa, *Philips Tech. Rev.*, **20**, 145 (1958).
63. Venema, A., *Vacuum*, **9**, 54 (1959).
64. Malpas, E., *Vacnique*, **5**(18), 2 (1965), Edwards High Vacuum Intern. Ltd., Crawley, Sussex, England.
65. Barrett, A. S. D., and N. T. M. Dennis, *Proc. 1st Intern. Vacuum Congr.* (1958), p. 212, Pergamon Press, New York.
66. Mullen, L. O., and M. J. Hiza, *J. Vacuum Sci. Technol.*, **4**, 219 (1967).
67. Yu, J. S., and S. L. Soo, *J. Vacuum Sci. Technol.*, **3**, 11 (1966).
68. Polanyi, M., *Trans. Faraday Soc.*, **28**, 316 (1932).
69. Chubb, J. N., and I. E. Pollard, *Vacuum*, **15**, 491 (1965).
70. Chubb, J. N., *Vacuum*, **16**, 681 (1966).
71. Levenson, L. L., *Ext. Abstr. 14th AVS Symp.*, 1967, p. 95, Herbick and Held Printing Co., Pittsburgh, Pa.

72. Heald, J. H., Jr., and R. F. Brown, *Ext. Abstr. 14th AVS Symp.*, 1967, p. 63, Herbick and Held Printing Co., Pittsburgh, Pa.
73a. Hengevoss, J., *Trans. 3d Intern. Vacuum Congr.*, 1965, vol. 1, p. 51, Pergamon Press, New York.
73b. Hengevoss, J., and E. A. Trendelenburg, *Vacuum*, **17**, 495 (1967).
74. Hobson, J. P., and P. A. Redhead, *Proc. 4th Intern. Vacuum Congr.*, 1968, vol. 1, p. 3, The Institute of Physics and the Physical Society, London.
75. Wang, E. S. J., J. A. Collins, and J. D. Haygood, *Advan. Cryogenic Eng.*, **7**, 44 (1966).
76. Dawbarn, R., *Ext. Abstr. 14th AVS Symp.*, 1967, p. 97, Herbick and Held Printing Co., Pittsburgh, Pa.
77. Honig, R. E., and H. O. Hook, *RCA Rev.*, **21**, 360 (1960).
78. Meissner, C. R., *Trans. 7th AVS Symp.*, 1960, p. 196, Pergamon Press, New York.
79. Caswell, H. L., *IBM J. Res. Develop.*, **4**, 130 (1960).
80. Holmwood, R. A., and R. Glang, *J. Electrochem. Soc.*, **112**, 827 (1965).
81. Holland, L., and D. W. Barker, *Vacuum*, **15**, 289 (1965).
82. Caswell, H. C., in G. Hass (ed.), "Physics of Thin Films," vol. 1, p. 1, Academic Press Inc., New York, 1963.
83. Forth, H. J., *Vacuum*, **17**, 21 (1967).
84. Farkass, I., and G. F. Vanderschmidt, *Trans. 6th AVS Symp.*, 1959, p. 45, Pergamon Press, New York.
85. Sheldon, D. P., and M. H. Hablanian, *Ext. Abstr. 14th AVS Symp.*, 1967, p. 159, Herbick and Held Printing Co., Pittsburgh, Pa.
86. Murakami, Y., and H. Okamoto, *Trans. 10th AVS Symp.*, 1963, p. 93, The Macmillan Company, New York.
87. See, for example, Brunauer, S., "The Adsorption of Gases and Vapors," Princeton University Press, Princeton, N.J., 1958.
88. Hobson, J. P., and R. A. Armstrong, *J. Phys. Chem.*, **67**, 2000 (1963).
89. Ehlers, H., *Trans. 3d Intern. Vacuum Congr.*, 1965, vol. 2, p. 457, Pergamon Press, New York.
90. Bannock, R. R., *Vacuum*, **12**, 101 (1962).
91. Inkley, F. A., and J. W. Coleman, *Vacuum*, **15**, 401 (1965).
92. Noe, P., M. Constant, and G. Troadec, *Proc. 4th Intern. Vacuum Congr.*, 1968, vol. 1, p. 397, The Institute of Physics and The Physical Society, London.
93. Sands, A., and S. M. Dick, *Vacuum*, **16**, 691 (1966).
94. Bailey, J. R., *Trans. 3d Intern. Vacuum Congr.*, 1965, vol. 2, p. 447, Pergamon Press, New York.
95. Barrer, R. M., *Brit. Chem. Eng.*, **4**(5), 267 (1959).
96. Turner, F. T., and M. Feinleib, *Trans. 8th AVS Symp.*, 1961, vol. 1, p. 300, The Macmillan Company, New York.
97. Stern, S. A., and F. S. DiPaolo, *J. Vacuum Sci. Technol.*, **4**, 347 (1967).
98. Lukač, P., *Proc. 4th Intern. Vacuum Congr.*, 1968, vol. 1, p. 389, The Institute of Physics and The Physical Society, London.
99. Varadi, P. F., and K. Ettre, *Trans. 7th AVS Symp.*, 1960, p. 248, Pergamon Press, New York.
100. Boers, A. L., *Proc. 4th Intern. Vacuum Congr.*, 1968, vol. 1, p. 393, The Institute of Physics and The Physical Society, London.
101. Read, P. L., *Vacuum*, **13**, 271 (1963).
102. Gareis, P. J., and G. F. Hagenback, *Ind. Eng. Chem.*, **57**(5), 27 (1965).
103. Bond, G. C., "Catalysis by Metals," p. 66, Academic Press Inc., New York, 1962.
104. Wagener, S., *Proc. Inst. Elec. Engrs.*, pt. 3, **99**, 135 (1952).
105. Ref. 6, p. 622.
106. Stout, V. G., and M. D. Gibbons, *J. Appl. Phys.*, **26**, 1488 (1955).
107. Jackson, A. G., and T. W. Haas, *J. Vacuum Sci. Technol.*, **4**, 42 (1967).
108. Milleron, N., and E. C. Popp, *Trans. 5th AVS Symp.*, 1958, p. 153, Pergamon Press, New York.
109. Sibata, S., C. Hayashi, and H. Kumagai, *Proc. 1st Intern. Vacuum Congr.*, 1958, p. 430, Pergamon Press, New York.
110. Cloud, R. W., H. Milde, and S. F. Philip, *Trans. 8th AVS Symp.*, 1961, p. 357, The Macmillan Company, New York.
111. Morrison, J., *Trans. 6th AVS Symp.*, 1959, p. 291, Pergamon Press, New York.
112. McCracken, G. M., *Vacuum*, **15**, 433 (1965).
113. Lawson, R. W., and J. W. Woodward, *Vacuum*, **17**, 205 (1967).
114. Herb, R. G., R. H. Davis, A. S. Divatia, and D. Saxon, *Phys. Rev.*, **89**, 897 (1953).
115. Pauly, T., R. D. Welton, and R. G. Herb, *Trans. 7th AVS Symp.*, 1960, p. 51, Pergamon Press, New York.

116. Gould, C. L., and P. Mandel, *Trans. 9th AVS Symp.*, 1962, p. 360, The Macmillan Company, New York.
117. Warren, K. A., D. R. Denison, and D. G. Bills, *Ext. Abstr. 13th AVS Symp.*, 1966, p. 17, Herbick and Held Printing Co., Pittsburgh, Pa.
118. Klopfer, A., and W. Ermrich, *Vakuum-Tech.*, **8**, 162 (1959).
119. Klopfer, A., and W. Ermrich, *Proc. 1st Intern. Vacuum Congr.*, 1958, p. 427, Pergamon Press, New York.
120. Holland, L., L. Laurenson, and P. G. W. Allen, *Trans. 8th AVS Symp.*, 1961, p. 208, The Macmillan Company, New York.
121. Elsworth, L., L. Holland, and L. Laurenson, *Vacuum*, **15**, 337 (1965).
122. Clausing, R. E., *Trans. 8th AVS Symp.*, 1961, p. 345, The Macmillan Company, New York.
123. Denison, D. R., *J. Vacuum Sci. Technol.*, **4**, 156 (1967).
124. Reich, G., and H. G. Noeller, *Proc. 1st Intern. Vacuum Congr.*, 1958, p. 443, Pergamon Press, New York.
125. Hirsch, R. L., *J. Vacuum Sci. Technol.*, **5**, 61 (1968).
126. Rivière, J. C., J. B. Thompson, J. E. Read, and I. Wilson, *Vacuum*, **15**, 353 (1965).
127. Das, K. B., *Ext. Abstr. 14th AVS Symp.*, 1967, p. 161, Herbick and Held Printing Co., Pittsburgh, Pa.
128. Penning, F. M., *Physica*, **4**, 71 (1937).
129. Redhead, P. A., J. P. Hobson, and E. V. Kornelsen, *Advan. Electron. Electron. Phys.*, **17**, 323 (1962).
130. Grant, W. A., and G. Carter, *Vacuum*, **15**, 477 (1965).
131. Teloy, E., *Trans. 3d Intern. Vacuum Congr.*, 1965, vol. 2.3, p. 613, Pergamon Press, New York.
132. Varnerin, L. J., and J. H. Carmichael, *J. Appl. Phys.*, **28**, 913 (1957).
133. Smeaton, G. P., and G. Carter, *J. Vacuum Sci. Technol.*, **3**, 208 (1966).
134. Blodgett, K. B., and T. A. Vanderslice, *J. Appl. Phys.*, **31**, 1017 (1960).
135. Ambridge, T., A. R. Bayly, K. Erents, and G. Carter, *Vacuum*, **17**, 329 (1967).
136. Jaeckel, R., and E. Teloy, *Trans. 8th AVS Symp.*, 1961, p. 406, The Macmillan Company, New York.
137. For a listing of a number of commercial getter-ion pumps and their manufacturers, see, for example, Ref. 7.
138. Holland, L., L. Laurenson, and M. J. Fulker, *Vacuum*, **16**, 663 (1966).
139. Bills, D. G., *J. Vacuum Sci. Technol.*, **4**, 149 (1967).
140. Davis, R. H., and A. S. Divatia, *Rev. Sci. Instr.*, **25**, 193 (1954).
141. Herb, R. G., *Proc. 1st Intern. Vacuum Congr.*, 1958, vol. 1, p. 45, Pergamon Press, New York.
142. Swartz, J. C., *Trans. 2d AVS Symp.*, 1955, p. 83, Pergamon Press, New York.
143. Douglas, R. A., J. Zabritski, and R. G. Herb, *Rev. Sci. Instr.*, **36**, 1 (1965).
144. Naik, P. K., and R. G. Herb, *J. Vacuum Sci. Technol.*, **5**, 42 (1968).
145. Adam, H., and W. Baechler, *Trans. 8th AVS Symp.*, 1961, p. 374, The Macmillan Company, New York.
146. Maliakal, J. C., *Ext. Abstr. 14th AVS Symp.*, 1967, p. 89, Herbick and Held Printing Co., Pittsburgh, Pa.
147. Lafferty, J. M., and T. A. Vanderslice, *Proc. IRE*, **49**, 1136 (1961).
148. Hall, L. D., *Rev. Sci. Instr.*, **29**, 367 (1958).
149. Baechler, W., *Trans. 3d Intern. Vacuum Congr.*, 1965, vol. 2.3, p. 609, Pergamon Press, New York.
150. Jackson, A. G., and T. W. Haas, *J. Vacuum Sci. Technol.*, **3**, 80 (1966).
151. Bance, U. R., and R. D. Craig, *Vacuum*, **16**, 647 (1966).
152. Jepsen, R. L., A. B. Francis, S. L. Rutherford, and B. E. Kietzmann, *Trans. 7th AVS Symp.*, 1960, p. 45, Pergamon Press, New York.
153. Carter, G., *Trans. 9th AVS Symp.*, 1962, p. 351, The Macmillan Company, New York.
154. Tom, T., and B. D. James, *J. Vacuum Sci. Technol.*, **6**, 304 (1969).
155. Brubaker, W. M., *Trans. 6th AVS Symp.*, 1959, p. 302, Pergamon Press, New York.
156. Hamilton, A. B., *Trans. 8th AVS Symp.*, 1961, p. 388, The Macmillan Company, New York.
157. Andrew, D., *Vacuum*, **16**, 653 (1966).
158. Holland, L., M. J. Fulker, and L. Laurenson, *Suppl. Nuovo Cimento*, **5**, 242 (1967).
159. Lamont, L. T., Jr., *Ext. Abstr. 14th AVS Symp.*, 1967, p. 149, Herbick and Held Printing Co., Pittsburgh, Pa.
160. Rutherford, S. L., *Trans. 10th AVS Symp.*, 1963, p. 185, The Macmillan Company, New York.
161. Young, J. R., *J. Vacuum Sci. Technol.*, **5**, 102 (1968).

162. Komiya, S., H. Sato, and C. Hayashi, *Ext. Abstr. 13th AVS Symp.*, 1966, p. 19, Herbick and Held Printing Co., Pittsburgh, Pa.
163. Hamilton, A. R., *Ext. Abstr. 14th AVS Symp.*, 1967, p. 91, Herbick and Held Printing Co., Pittsburgh, Pa.
164. Jepsen, R. L., *Proc. 4th Intern. Vacuum Congr.*, 1968, vol. 1, p. 317, The Institute of Physics and The Physical Society, London.
165. Brothers, C. F., T. Tom, and D. F. Munro, *Trans. 10th AVS Symp.*, 1963, p. 202, The Macmillan Company, New York.
166. Newton, S. W., and J. O'Neill, *Vacuum*, **16**, 677 (1966).
167. Kelly, J. E., and T. A. Vanderslice, *Vacuum*, **11**, 205 (1961).
168. Ward, B. J., and J. C. Bill, *Vacuum*, **16**, 659 (1966).
169. Lichtman, D., *J. Vacuum Sci. Technol.*, **1**, 23 (1964).
170. Rivière, J. C., and J. D. Allinson, *Vacuum*, **14**, 97 (1964).
171. Davis, W. D., *Trans. 9th AVS Symp.*, 1962, p. 363, The Macmillan Company, New York.
172. Jaeckel, R., *Trans. 8th AVS Symp.*, 1961, vol. 1, p. 17, The Macmillan Company, New York.
173. Nasini, A. G., and F. Ricca, *Trans. 8th AVS Symp.*, 1961, vol. 1, p. 160, The Macmillan Company, New York.
174. Sachtler, W. M. H., *Trans. 3d Intern. Vacuum Congr.*, 1965, vol. 1, p. 41, Pergamon Press, New York.
175. See, for example, Trapnell, B. M. W., "Chemisorption," Butterworth & Co. (Publishers), Inc., London, 1955. A more recent review of gas-solid interactions in vacuum systems has been given by J. P. Hobson and P. A. Redhead in *Proc. 4th Intern. Vacuum Congr.*, 1968, vol. 1, p. 3, The Institute of Physics and The Physical Society, London.
176. Dillon, J. A., *Trans. 8th AVS Symp.*, 1961, vol. 1, p. 113, The Macmillan Company, New York.
177. Halsey, G. D., Jr., *Trans. 8th AVS Symp.*, 1961, vol. 1, p. 119, The Macmillan Company, New York.
178. Ehrlich, G., *Trans. 8th AVS Symp.*, 1961, vol. 1, p. 126, The Macmillan Company, New York.
179. Becker, J. A., *Solid State Phys.*, **7**, 379 (1958).
180. Huang, A. B., *J. Vacuum Sci. Technol.*, **2**, 6 (1965).
181. Hobson, J. P., *J. Vacuum Sci. Technol.*, **3**, 281 (1966).
182. Dushman, S., Ref. 6, p. 376.
183. Langmuir, I., *Chem. Rev.*, **13**, 147 (1933).
184. Brunauer, S., P. H. Emmett, and E. Teller, *J. Am. Chem. Soc.*, **60**, 309 (1938).
185. Pagano, F., *Ext. Abstr. 13th AVS Symp.*, 1966, p. 103, Herbick and Held Printing Co., Pittsburgh, Pa.
186. Jahncke, E., and F. Emde, "Tables of Functions," 4th ed., p. 1, Dover Publications, Inc., New York, 1945.
187. Grant, W. A., and G. Carter, *Vacuum*, **15**, 13 (1965).
188. Erents, K., W. A. Grant, and J. Carter, *Vacuum*, **15**, 529 (1965).
189. Kammer, D. W., J. H. Leck, and E. E. Donaldson, *Ext. Abstr. 14th AVS Symp.*, 1967, p. 121, Herbick and Held Printing Co., Pittsburgh, Pa.
190. Redhead, P. A., *Trans. 6th AVS Symp.*, 1959, p. 12, Pergamon Press, New York.
191. Richardson, D. M., and R. A. Strehlow, *Trans. 10th AVS Symp.*, 1963, p. 97, The Macmillan Company, New York.
192. Jost, W., "Diffusion in Solids, Liquids, Gases," Academic Press Inc., New York, 1952.
193. See, for example, Carslaw, H. S., and J. C. Jaeger, "Conduction of Heat in Solids," p. 310, Oxford University Press, Fair Lawn, N.J., 1959.
194. Crank, J., "The Mathematics of Diffusion," p. 45, Oxford University Press, Fair Lawn, N.J., 1956.
195. Lewin, G., *Ext. Abstr. 14th AVS Symp.*, 1967, p. 117, Herbick and Held Printing Co., Pittsburgh, Pa.
196. Rogers, K. W., *Trans. 10th AVS Symp.*, 1963, p. 84, The Macmillan Company, New York.
197. Low, M. J. D., and E. S. Argano, *J. Vacuum Sci. Technol.*, **3**, 324 (1966).
198. Maddix, H. S., and M. A. Allen, *Trans. 10th AVS Symp.*, 1963, p. 197, The Macmillan Company, New York.
199. Erents, K., R. P. W. Layson, and G. Carter, *J. Vacuum Sci. Technol.*, **4**, 252 (1967).
200. Farrell, G., and G. Carter, *Vacuum*, **17**, 15 (1967).
201. Farrell, G., W. A. Grant, K. Erents, and G. Carter, *J. Vacuum Sci. Technol.*, **16**, 295 (1966).

202. Konjevic, R., W. A. Grant, and G. Carter, *Vacuum*, **17**, 501 (1967).
203. Truhlar, J. F., and E. E. Donaldson, *Ext. Abstr. 14th AVS Symp.*, 1967, p. 119, Herbick and Held Printing Co., Pittsburgh, Pa.
204. Dayton, B. B., *Trans. 6th AVS Symp.*, 1959, p. 101, Pergamon Press, New York.
205. Altemose, V. O., *J. Appl. Phys.*, **32**, 1309 (1961).
206. Smithells, C. J., "Gases and Metals," Chapman & Hall, Ltd., London, 1937.
207. Barrer, R. M., "Diffusion in and through Solids," p. 146, Cambridge University Press, London, 1951.
208. Norton, F. J., Ref. 6, p. 516.
209. American Vacuum Society, Tentative Standard for Reporting of Outgassing Data, *J. Vacuum Sci. Technol.*, **2**, 314 (1965).
210. Crawley, D. J., and L. de Csernatony, *Vacuum*, **14**, 7 (1964).
211. Farkass, I., "Problems of Producing a Clean Surface in Ultra-high Vacuum," presented at the Second Annual Symposium on Fundamental Materials in Sciences, Boston, 1964.
212. Farkass, I., L. J. Bonis, and G. W. Horn, presented at the Eleventh AVS Symposium, Chicago, Ill., 1964; also Ilikon Paper AVS 64.
213. Farkass, I., Problems of Producing a Clean Surface by Out-gassing in Ultra-high Vacuum, in L. J. Bonis and H. H. Hausner (eds.), "Fundamental Phenomena in the Materials Sciences," vol. 2, "Surface Phenomena," Plenum Press, New York, 1965; see also Farkass, I., L. J. Bonis, and G. W. Horn, Ilikon Paper IV 65.
214. Hobson, J. P., and J. W. Earnshaw, *J. Vacuum Sci. Technol.*, **4**, 257 (1967).
215. Holland, L., "Thin Film Microelectronics," p. 116, John Wiley & Sons, Inc., New York, 1965.
216. Calder, R., and G. Lewin, *Ext. Abstr. 13th AVS Symp.*, 1966, p. 107, Herbick and Held Printing Co., Pittsburgh, Pa.
217. Flecken, F. A., and H. G. Noeller, *Trans. 8th AVS Symp.*, 1961, vol. 1, p. 58, The Macmillan Company, New York.
218. Varadi, P. F., *Trans. 8th AVS Symp.*, 1961, vol. 1, p. 73, The Macmillan Company, New York.
219. Zhilnin, V. S., L. P. Zhilnina, and A. A. Kuzmin, *Proc. 4th Intern. Vacuum Congr.*, 1968, vol. 2, p. 801, The Institute of Physics and The Physical Society, London.
220. Degras, D. A., A. Schram, B. Lux, and L. A. Petermann, *Trans. 8th AVS Symp.*, vol. 1, p. 177, The Macmillan Company, New York.
221. Collins, R. H., and J. C. Turnbull, *Vacuum*, **11**, 119 (1961).
222. Strausser, Y. E., *Ext. Abstr. 14th AVS Symp.*, 1967, p. 131, Herbick and Held Printing Co., Pittsburgh, Pa.
223. Todd, B. J., *J. Appl. Phys.*, **26**, 1238 (1955).
224. Sherwood, G., *J. Am. Chem. Soc.*, **40**, 1645 (1968).
225. Erents, K., and G. Carter, *Vacuum*, **15**, 573 (1965).
226. Bills, D. G., and A. A. Evett, *J. Appl. Phys.*, **30**, 564 (1959).
227. Singleton, J. H., and W. J. Lange, *J. Vacuum Sci. Technol.*, **2**, 93 (1965).
228. Dawson, P. H., and N. R. Whetten, *J. Vacuum Sci. Technol.*, **4**, 382 (1967).
229. Holland, L., *Brit. J. Appl. Phys.*, **9**, 410 (1958).
230. Schreiner, D. G., *Ext. Abstr. 14th AVS Symp.*, 1967, p. 127, Herbick and Held Printing Co., Pittsburgh, Pa.
231. Farkass, I., and E. J. Barry, *Trans. 7th AVS Symp.*, 1960, p. 35, Pergamon Press, New York.
232. Barton, R. S., and R. P. Govier, *J. Vacuum Sci. Technol.*, **2**, 113 (1965).
233. Hait, P. W., *Ext. Abstr. 13th AVS Symp.*, 1966, p. 125, Herbick and Held Printing Co., Pittsburgh, Pa.
234. Markley, F., R. Roman, and R. Vosecek, *Trans. 8th AVS Symp.*, 1961, vol. 1, p. 78, The Macmillan Company, New York.
235. Brown, R. D., *Vacuum*, **17**, 505 (1967).
236. de Csernatony, L., *Ext. Abstr. 13th AVS Symp.*, 1966, p. 123, Herbick and Held Printing Co., Pittsburgh, Pa.
237. de Csernatony, L., *Vacuum*, **16**, 13, 129, 247 (1966).
238. Addiss, R. R., L. Pensak, and N. J. Scott, *Trans. 7th AVS Symp.*, 1960, p. 39, Pergamon Press, New York.
239. Norton, F. J., *Trans. 8th AVS Symp.*, 1961, vol. 1, p. 8, The Macmillan Company, New York.
240. Rogers, W. A., R. S. Buritz, and D. Alpert, *J. Appl. Phys.*, **25**, 868 (1954).
241. Frauenfelder, R., *Ext. Abstr. 13th AVS Symp.*, 1966, p. 101, Herbick and Held Printing Co., Pittsburgh, Pa.
242. Norton, F. J., *J. Appl. Phys.*, **28**, 34 (1957).
243. Ref. 6, p. 573.

244. Collins, R. H., and J. C. Turnbull, *Vacuum*, **11**, 114 (1961).
245. Whetten, N. R., and J. R. Young, *Rev. Sci. Instr.*, **30**, 472 (1959).
246. Leiby, C. C., and C. L. Chen, *J. Appl. Phys.*, **31**, 268 (1960).
247. Alpert, D., and R. S. Buritz, *J. Appl. Phys.*, **25**, 202 (1954).
248. Roth, A., "Vacuum Sealing Techniques," Pergamon Press, New York, 1966.
249. Lancaster, J. F., "The Metallurgy of Welding, Brazing, and Soldering," George Allen & Unwin, Ltd., London, 1965.
250. Patton, W. J., "The Science and Practice of Welding," Prentice-Hall, Inc., Englewood Cliffs, N.J., 1967.
251. Phillips, A. L. (ed.), "Welding Handbook," 5th ed., American Welding Society, New York, 1962.
252. Filipski, S. P., *Welding J.*, **43**(1), 937 (1964).
253. *Welding Metals Fabrication*, **34**(5), 162 (1966).
254. Schwartz, H., *Trans. 8th AVS Symp.*, 1961, p. 699, Pergamon Press, New York.
255. Meier, J. W., *Trans. 8th AVS Symp.*, 1961, p. 670, Pergamon Press, New York.
256. Lander, H. J., W. T. Hess, R. Bakish, and S. S. White, *Trans. 8th AVS Symp.*, 1961, p. 685, Pergamon Press, New York.
257. Hess, W. T., H. J. Lander, S. S. White, and L. McD. Schetky, *Trans. 8th AVS Symp.*, 1961, p. 679, Pergamon Press, New York.
258. Ito, G., H. Suzuki, T. Hashimoto, and F. Matsuda, *Trans. 8th AVS Symp.*, 1961, p. 693, Pergamon Press, New York.
259. Meier, S. W., *Welding J.*, **43**(11), 925 (1964).
260. Duhamel, R. F., *Welding J.*, **44**(6), 465 (1965).
261. Birnie, J. V., *Welding Metals Fabrication*, **31**(1), 27 (1966).
262. Kohl, W. H., *Vacuum*, **14**, 175 (1964).
263. Kohl, W. H., "Handbook of Materials and Techniques for Vacuum Devices," Reinhold Publishing Corporation, New York, 1967.
264. Fenton, E. A., "Soldering Manual," American Welding Society, New York, 1959.
265. Barber, C. L., "Solder, Its Fundamentals and Usage," 2d ed., Kester Solder Co., 1961.
266. Espe, W., "Materials of High Vacuum Technology," vol. 1, Pergamon Press, New York, 1966.
267. Fenton, E. A., "Brazing Manual," American Welding Society, New York, 1963.
268. See, for example, R. E. Reed-Hill, "Physical Metallurgy Principles," D. Van Nostrand Company, Inc., Princeton, N.J., 1964; or Brick, R. M., R. B. Gordon, and A. Phillips, "Structure and Properties of Alloys," McGraw-Hill Book Company, New York, 1965.
269. Chang, W. H., *Welding J.*, vol. 35, res. suppl. 622-S, 1956.
270. *Mater. Methods*, **43**(1), 129 (1956).
271. Housekeeper, W. G., *J. Am. Inst. Elec. Engrs.*, **42**, 870 (1923).
272. Kohl, W., *Vacuum*, **14**, 333 (1964).
273. Espe, W., *Vacuum*, **16**, 1 (1966).
274. Balain, K. S., *Trans. 8th AVS Symp.*, 1961, p. 1300, Pergamon Press, New York.
275. American Vacuum Society, Committee on Standards, *J. Vacuum Sci. Technol.*, **3**, 170 (1966).
276. Farkass, I., and E. J. Barry, *Trans. 8th AVS Symp.*, 1961, vol. 1, p. 66, Pergamon Press, New York.
277. Hawrylak, R. A., *J. Vacuum Sci. Technol.*, **4**, 364 (1967).
278. Wheeler, H. R., *Trans. 10th AVS Symp.*, 1963, p. 159, The Macmillan Company, New York.
279. Wheeler, H. R., and M. Carlson, *Trans. 8th AVS Symp.*, 1961, p. 1309, Pergamon Press, New York.
280. Wilson, R. R., *Rev. Sci. Instr.*, **12**, 91 (1941).
281. Guthrie, A., "Vacuum Technology," p. 380, John Wiley & Sons, Inc., New York, 1963.
282. Hobson, J. P., *Trans. 8th AVS Symp.*, 1961, vol. 1, p. 26, The Macmillan Company, New York.
283. Kraus, T., *Trans. 6th AVS Symp.*, 1959, p. 204, Pergamon Press, New York.
284. Dayton, B. B., *Trans. 8th AVS Symp.*, 1961, vol. 1, p. 42, The Macmillan Company, New York.
285. Kraus, T., *Trans. 10th AVS Symp.*, 1963, p. 77, The Macmillan Company, New York.
286. Hobson, J. P., *Ext. Abstr. 14th AVS Symp.*, 1967, p. 51, Herbick and Held Printing Co., Pittsburgh, Pa.
287. Earnshaw, J. W., and J. P. Hobson, *J. Vacuum Sci. Technol.*, **5**, 19 (1968).
288. American Vacuum Society, Committee on Standards, "Glossary of Terms Used in Vacuum Technology," Pergamon Press, New York, 1958.
289. Brunner, W. F., Jr., and T. H. Batzer, "Practical Vacuum Techniques," p. 77, Reinhold Publishing Corporation, New York, 1965.
290. Gregg, A. H., and D. J. Reid, *Vide*, **122**, 171 (1966).

291. Maynard, C. P., *Vacuum*, **15**, 239 (1965).
292. Hickmott, T. W., *J. Appl. Phys.*, **31**, 128 (1960).
293. Turner, J. A., R. M. Pickard, and G. R. Hoffman, *J. Sci. Instr.*, **39**, 26 (1962).
294. Holland, L., *J. Sci. Instr.*, **39**, 247 (1962).
295. Adler, J. E. M., *J. Vacuum Sci. Technol.*, **2**, 209 (1965).
296. Reisinger, H., *Balzers High Vacuum Rept.* 11, August, 1967.
297. Zaphiropoulos, R., and D. de Taddeo, *Trans. 3d Intern. Vacuum Congr.*, 1965, vol. 2.2, p. 373, Pergamon Press, New York.
298. Gretz, R. D., *J. Vacuum Sci. Technol.*, **5**, 49 (1968).
299. Young, J. R., and F. P. Hession, *J. Vacuum Sci. Technol.*, **1**, 65 (1964).
300. Batzer, T. H., and J. F. Ryan, *Trans. 10th AVS Symp.*, 1963, p. 166, The Macmillan Company, New York.
301. Power, B. D., N. T. M. Dennis, and L. de Csernatony, *Trans. 10th AVS Symp.*, 1963, p. 147, The Macmillan Company, New York.
302. Holland, L., *Trans. 7th AVS Symp.*, 1960, p. 168, Pergamon Press, New York.
303. Holkeboer, D. H., *Proc. 4th Intern. Vacuum Congr.*, 1968, vol. 1, p. 405, The Institute of Physics and The Physical Society, London.
304. Caswell, H. L., *Trans. 6th AVS Symp.*, 1959, p. 66, Pergamon Press, New York.
305. Boebel, C. P., and S. J. Babjak, *Ext. Abstr. 13th AVS Symp.*, 1966, p. 129, Herbick and Held Printing Co., Pittsburgh, Pa.
306. Blahnik, C. E., *J. Vacuum Sci. Technol.*, **4**, 378 (1967).
307. Wagener, J. W., and P. T. Marth, *J. Appl. Phys.*, **28**, 1027 (1957).
308. Kirkendall, T. D., P. F. Varadi, and H. D. Doolittle, *J. Vacuum Sci. Technol.*, **3**, 214 (1966).
309. Wolter, A. R., *J. Appl. Phys.*, **36**, 2377 (1965).
310. Kirchner, F., *Naturwiss.*, **48**, 548 (1961).
311. Priest, J., H. L. Caswell, and Y. Budo, *Vacuum*, **12**, 301 (1962).
312. Heil, R. H., Jr., *Solid State Technol.*, **11**(1), 21 (1968).
313. *Varian Vacuum Views*, September, 1967, Varian Vacuum Division, Palo Alto, Calif.
314. Cook, H. C., *J. Vacuum Sci. Technol.*, **4**, 80 (1967).
315. Clough, P. J., *Trans. 4th AVS Symp.*, 1957, p. 136, Pergamon Press, New York.
316. Dayton, B. B., "Handbook of Vacuum Physics," vol. 1, pt. 1, The Macmillan Company, New York, 1964.
317. Williams, B. J., B. Fletcher, and J. A. A. Emery, *Proc. 4th Intern. Vacuum Congr.*, 1968, vol. 2, p. 753, The Institute of Physics and The Physical Society, London.
318. Ames, I., M. F. Gendron, and H. Seki, *Trans. 9th AVS Symp.*, 1962, p. 133, The Macmillan Company, New York.
319. Caswell, H. L., and J. R. Priest, *Trans. 9th AVS Symp.*, 1962, p. 138, The Macmillan Company, New York.
320. Gleed, W. L., J. S. Hill, K. H. Hursey, and P. R. Stuart, *Vacuum*, **18**, 213 (1968).
321. Himes, W., B. F. Stout, and R. E. Thun, *Trans. 9th AVS Symp.*, 1962, p. 144, The Macmillan Company, New York.
322. Charschan, S. S., R. W. Glenn, and H. Westgaard, *Western Elec. Engr.*, **7**, 9 (1963).
323. Bolde, J. W., S. S. Charschan, and J. J. Dineen, *Bell System Tech. J.*, **43**, 127 (1964).
324. Hanfman, A. M., *Solid State Technol.*, **11**(12), 37 (1968); see also *Proc. 4th Intern. Vacuum Congr.*, 1968, vol. 2, p. 549, The Institute of Physics and The Physical Society, London.
325. Guthrie, A., and R. K. Wakerling, "Vacuum Equipment and Techniques," chap. 5, McGraw-Hill Book Company, New York, 1949.
326. Turnbull, A. H., "Leak Detection and Detectors," in A. H. Beck (ed.), "Handbook of Vacuum Physics," vol. 1, pt. 3, The Macmillan Company, New York, 1964.
327. Blears, J., and J. H. Leck, *J. Sci. Instr. Suppl.*, **1**, 20 (1951).
328. Nemeth, R., U.S. Patent 3,227,872, 1966.
329. Work, R. H., *Trans. 5th AVS Symp.*, 1958, p. 126, Pergamon Press, New York.
330. Bicknell, C. B., *Trans. 6th AVS Symp.*, 1959, p. 98, Pergamon Press, New York.
331. Leck, J. H., "Pressure Measurement in Vacuum Systems," 2d ed., Chapman & Hall, Ltd., London, 1964; paperback edition by Science Paperbacks, London, 1967, distributed by Barnes & Noble, Inc., New York.
332. Diels, K., and R. Jaeckel, "Leybold Vakuum-Taschenbuch," 2d ed., pp. 96, 120, Springer-Verlag OHG, Berlin, 1962.
333. Brombacher, W. G., D. P. Johnson, and J. L. Cross, National Bureau of Standards Monograph 8, U.S. Department of Commerce, Washington D.C., 1960.
334. See, for example, Thomas, A. M., and J. L. Cross, *J. Vacuum Sci. Technol.*, **4**, 1 (1967).
335. Alpert, D., C. G. Matland, and A. O. McConbrey, *Rev. Sci. Instr.*, **22**, 370 (1951).
336. Utterbach, N. G., and T. Griffith, Jr., *Rev. Sci. Instr.*, **37**, 866 (1966).

337. Dadson, R. S., K. W. T. Elliott, and D. M. Woodman, *Proc. 4th Intern. Vacuum Congr.*, 1968, vol. 2, p. 679, The Institute of Physics and The Physical Society, London.
338. Ishii, H., and K. Nakayama, *Trans. 8th AVS Symp.*, 1961, p. 519, Pergamon Press, New York.
339. See, for example, Session III B, *Proc. 4th Intern. Vacuum Congr.*, 1968, p. 591, The Institute of Physics and The Physical Society, London.
340. Benson, J. M., *Trans. 3d AVS Symp.*, 1956, p. 87, Pergamon Press, New York.
341. Harvey, P. C., *J. Vacuum Sci. Technol.*, **4**, 339 (1967).
342. Smith, P. T., *Phys. Rev.*, **36**, 1293 (1930); **37**, 808 (1931); also J. T. Tate and P. T. Smith, *Phys. Rev.*, **39**, 270 (1932).
343. Kornelsen, E. V., *Proc. 3d Intern. Congr. Vacuum Techniques*, 1965, vol. 1, p. 65, Pergamon Press, New York.
344. Groszkowski, J., *Proc. 4th Intern. Vacuum Congr.*, 1968, vol. 2, p. 631, The Institute of Physics and The Physical Society, London.
345. Schulz, G. J., and A. V. Phelps, *Rev. Sci. Instr.*, **28**, 1051 (1957).
346. See, for example, *Product Rept.* 213-5, Granville-Phillips Company, Boulder, Colo.
347. Bayard, R. T., and D. Alpert, *Rev. Sci. Instr.*, **21**, 571 (1950).
348. Kinsella, J. J., *Trans. 1st AVS Symp.*, 1954, p. 65, Pergamon Press, New York.
349. Groszkowski, J., *Proc. 3d Intern. Congr. Vacuum Techniques*, 1965, vol. 2, p. 241, Pergamon Press, New York.
350. Redhead, P. A., *Rev. Sci. Instr.*, **31**, 343 (1960).
351. Klopfer, A., *Trans. 8th AVS Symp.*, 1961, p. 439, Pergamon Press, New York.
352. Helmer, J. C., and W. H. Hayward, *Rev. Sci. Instr.*, **37**, 1652 (1966).
353. Lafferty, J. M., *Trans. 9th AVS Symp.*, 1962, p. 438, The Macmillan Company, New York.
354. Lafferty, J. M., *Rev. Sci. Instr.*, **34**, 467 (1963).
355. Redhead, P. A., *Can. J. Phys.*, **37**, 1260 (1959).
356. Redhead, P. A., *Advan. Vacuum Sci. Technol.*, p. 410, Pergamon Press, New York, 1961.
357. Moore, B. C., *J. Vacuum Sci. Technol.*, **6**, 246 (1969).
358. Holland, L., *Vacnique*, vol. 5, no. 17, October, 1965, Edwards High Vacuum International, Ltd., Manor Royal, Crawley, Sussex, England.
359. See Ref. 331, p. 82.
360. Santeler, D. J., *Trans. 8th AVS Symp.*, 1961, p. 549, Pergamon Press, New York.
361. Blears, J., *Proc. Roy. Soc. London*, **A188**, 62 (1947).
362. Redhead, P. A., *Trans. 7th AVS Symp.*, 1960, p. 108, Pergamon Press, New York.
363. Redhead, P. A., *Vacuum*, **13**, 253 (1963).
364. Redhead, P. A., *J. Vacuum Sci. Technol.*, **3**, 173 (1966).
365. Kendall, B. R. F., *Trans. 10th AVS Symp.*, 1963, p. 278, The Macmillan Company, New York.
366. Blauth, E. W., *Proc. 4th Intern. Vacuum Congr.*, 1968, p. 21, The Institute of Physics and The Physical Society, London.
367. McDowell, C. A., "Mass Spectrometry," McGraw-Hill Book Company, New York, 1963.
368. See, for example, Session III C, *Proc. 4th Intern. Vacuum Congr.*, 1968, p. 685, The Institute of Physics and The Physics Society, London.
369. Harrington, D. B., in J. D. Waldron (ed.), "Advances in Mass Spectroscopy," p. 249, Pergamon Press, New York, 1959.
370. Dempster, A. J., *Phys. Rev.*, **11**, 316 (1918).
371. Hipple, J. A., and W. Bleakney, *Phys. Rev.*, **49**, 884 (1936).
372. Sommer, H., M. A. Thomas, and J. A. Hipple, *Phys. Rev.*, **76**, 1877 (1949).
373. Reich, G., and M. J. Noeller, *Z. Angew. Phys.*, **9**, 617 (1957).
374. Lafferty, J. M., and T. A. Vanderslice, *Proc. IRE*, **49**, 1144 (1961).
375. Barz, E., *Trans. 3d Intern. Vacuum Congr.*, 1965, vol. 2.2, p. 473, Pergamon Press, New York.
376. Paul, W., and H. Steinwedel, *Z. Naturforsch.*, **8a**, 448 (1953); Paul, W., H. P. Reinhard, and U. von Zahn, *Z. Physik*, **152**(2), 143 (1958).
377. Guenther, K. G., and W. Haenlein, *Trans. 8th AVS Symp.*, 1961, p. 573, The Macmillan Company, New York.
378. Brunnée, C., L. Delgmann, and K. Kronenberger, *Vakuum-Tech.*, **13**, 35 (1964).
379. Bueltemann, H. J., and L. Delgmann, *Vacuum*, **15**, 301 (1965).
380. Brubaker, W. M., *Ext. Abstr. 14th AVS Symp.*, 1967, p. 23, Herbick and Held Printing Co., Pittsburgh, Pa.
381. von Zahn, U., *Rev. Sci. Instr.*, **34**, 1 (1963).

382. Hudson, J. B., "R. F. Mass Spectrometer Partial Pressure Analyzer," Technical Report distributed by General Electric Co., Schenectady, N.Y.
383. Grande, R. E., R. L. Watters, and J. B. Hudson, *J. Vacuum Sci. Technol.*, **3**, 329 (1966).
384. Stephens, W. E., *Phys. Rev.*, **69**, 691 (1946).
385. Klopfer, A., and W. Schmidt, *Vacuum*, **10**, 363 (1960).
386. See Ref. 82, p. 46.
387. Craig, R. D., and E. H. Harden, *Vacuum*, **16**, 67 (1966).
388. Coulson, J. A., R. D. Craig, and R. G. Johnson, *Ext. Abstr. 14th AVS Symp.*, 1967, p. 73, Herbick and Held Printing Co., Pittsburgh, Pa.
389. Wood, G. M., Jr., and R. J. Roenigk, Jr., *J. Vacuum Sci. Technol.*, **6**, 871 (1969).

Chapter 3

The Nature of Physical Sputtering

GOTTFRIED K. WEHNER

University of Minnesota, Minneapolis

and

GERALD S. ANDERSON

3M Company, St. Paul, Minnesota

1. INTRODUCTION

a. Playing Billiards with Atoms

When a solid (or liquid) is bombarded with single atoms, ions, or molecules, many phenomena can arise. Which ones are possible or predominate depends largely on the kinetic energy of the bombarding particles.

At very low kinetic energies (<5 eV) the interaction is essentially confined to the outermost surface layer of the target material. When a noble-gas atom of such low kinetic energy bombards an atomically clean metal surface, hardly anything is likely to happen. The noble-gas atom may be reflected or come into thermal equilibrium with the surface and subsequently be evaporated. This is the area in which accommodation, sticking, and momentum-transfer coefficients describe the situation. The potential energy of the bombarding species (excited atoms or ions) plays an important part because it is responsible for electronic transitions which can give rise to the ejection of γ electrons (secondary electrons) or, in the case of compound materials or adsorbed surface impurities, to the breaking or rearranging of chemical bonds. This leads to desorption, chemical reactions, polymerization, etc., similar effects being induced by bombarding electrons or photons.

At kinetic energies which exceed the binding energy of the atoms (related to the heat of sublimation of the target material) a new phenomenon arises. Atoms of the lattice are pushed into new positions, giving rise to surface migration of atoms and to surface damage. At energies exceeding roughly $4H$ (where H = heat of sublimation of target material) the dislodging of atoms and their ejection into the gas phase begin to play a decisive role. This process is called physical sputtering. Electrons on a clean metal surface would need much higher kinetic energies (about 500 keV for Cu) to accomplish physical sputtering because the energy transfer between the light electron and a heavy target atom is very inefficient. One can readily drill holes in alkali-halide targets with low-energy electrons (100 eV), but this process is based on dissociation and evaporation rather than on physical sputtering.

Ions rather than neutral atoms are used for bombardment because one can accelerate these to any desired kinetic energy with electric fields. Although claims to the contrary have been made, one would not expect physical sputtering effects (well above threshold) to differ between ions and neutral atoms. In fact an ion, at least on a clean metal surface, becomes neutralized by a field-emitted electron just before impact, as is known from field ion microscopy. The neutralization energy is then transferred in a radiationless (Auger-type) transition to the lattice electrons and may cause the ejection of a γ electron. Thus in general one can state that the potential energy of the ion goes into electron transitions while its kinetic energy goes mostly into lattice-atom vibrations or displacements. Sputtering is always coupled with surface migration of atoms and permanent or temporary damage to the lattice.

Still somewhat puzzling is the fact that the sputtering threshold energies are essentially independent of the masses of the collision partners or of the energy-transfer coefficient. This seems to indicate that at very low kinetic energies one can no longer treat the collisions as independent binary collisions. Other neighboring atoms may already become involved while a collision is still in progress, and this would require the introduction of modified masses.

At kinetic energies well above threshold there is very strong evidence that sputtering is the result of a sequence of independent binary collisions just as if the ion (or neutralized ion) had collided with the atoms of a gas cloud. Here, of course, the individual masses of the collision partners play a decisive role. This, then, is the region in which sputtering resembles most closely a three-dimensional billiard game with atoms. For bombardment of a surface under normal incidence more than one collision is necessary for the ejection of a sputtered atom because the momentum vector has to be changed by more than 90°. Only under oblique bombardment can one detect forward-sputtered atoms or ions which result from a single collision between an ion and a surface atom.

Billiard-game considerations also apply to the fate of the bombarding ion. If its mass is lower than that of a target atom, it may be reflected or scattered backward in a single collision event. If its mass is higher, it can only be reflected backward as the result of more than one collision. At kinetic energies in excess of ~ 100 eV, ions begin to become embedded in the lattice. For Ar^+ in Cu the penetration depth is roughly 10 Å keV^{-1}. Crystal structure and orientation are important factors in determining penetration depth. As sputtering proceeds, an equilibrium situation is reached where the embedded ions are sputtered just as the target atoms are. The sputtering effects caused by molecular ion bombardment can be readily interpreted in terms of single-atom bombardment if one assumes that the atoms of a molecule arrive with the same velocity as the molecule.

When the bombarding ions reach velocities of 10^6 to 10^7 cm s^{-1}, kinetic emission of γ electrons becomes superimposed on the Auger-type potential emission of γ electrons. The γ coefficients can now amount to many electrons per impinging ion. If one performs sputtering experiments at energies above about 8 keV, one should not overlook the x-ray hazard which can arise from these γ electrons. For still higher kinetic energy, surface and sputtering effects give way to increasing bulk lattice damage and deeper ion penetration. In the billiard-ball analogy one would have to space the balls that represent the lattice atoms farther apart from each other. As a consequence, the sputtering yields (atoms per ion) cease to increase in proportion to ion energy and reach flat maxima which are located at much lower energy for lighter ions (H^+ at 2 keV) than for heavy ions (Hg^+ at 50 to 100 keV). If the target is a thin foil, one can actually observe more sputtering from the far side than from the front side at sufficiently high energies (MeV protons). At very high ion or particle energies the subject falls more into the realm of radiation damage than of sputtering. The energy region of primary interest for the sputter deposition of thin films stretches from threshold (as, for example, during bias sputtering) to about 5 keV.

Measurements of sputtering yield and of the average velocities of sputtered atoms (which are much higher than those of evaporated atoms) show that sputtering is a rather inefficient process. Usually, more than 95% of the ion energy appears as heat in the target. Sputtering yields rarely differ by more than a factor of 10. For instance, the sputtering rate for tungsten in Ar is only a factor of 2 lower than that of aluminum, whereas (at 2000°C) the evaporation rates for these two metals differ by nine orders of magnitude.

The billiard-game analogy also explains (to a certain degree at least) the preferential ejection of sputtered atoms in the close-packed direction of a crystal lattice as shown by computer calculations or by billiard-game-like experiments. One should not, however, overlook some important differences between sputtering and a conventional billiard game. Not only have the sizes of the billiard balls to be decreased at higher bombardment velocities (if one retains their position) but the balls are not hard spheres and may not remain "undamaged." Some of the atoms may be ejected as excited atoms or as negative or positive ions, expecially in the case of metal surfaces that are contaminated with adsorbed impurities or in the case of insulator targets. As a consequence of this, there is a frequently overlooked difference between ion-beam sputtering in a field-free space and plasma sputtering in which the target is negative with respect to the surrounding plasma. In the latter case, the ejected positive ions are pulled back to the target while the negative ions (as well as the γ electrons) are accelerated in a direction away from the target surface. These negative ions (often oxygen or hydrocarbons) can cause resputtering of material at the substrate or anywhere else in the apparatus that they may reach. Mass-spectrometric analysis of the positive ions ejected through ion-beam sputtering has revealed a surprising and not yet understood phenomenon: often, many of the sputtered ions consist of whole clusters of atoms.

b. History and Milestones

The first observation of metal deposits sputtered from the cathode of a glow discharge was reported in 1852 by Grove.[1]   This original article is worth reading if one wants to appreciate fully the progress which has been made in this field since then. The apparatus shown in that article is reproduced in Fig. 1.

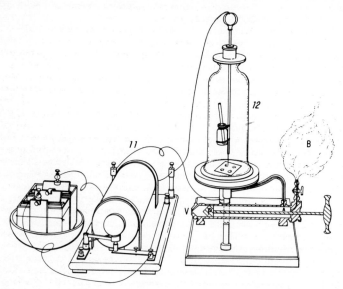

Fig. 1   Grove's sputtering apparatus.

The history of sputtering can probably be traced most rapidly by reading a selection of survey articles which show this field in its proper perspective in time and corresponding state of physics knowledge.[2–6]

Regular articles on the subject of sputtering are now published in so many journals that one has to keep track of new contributions by regularly monitoring those abstracting journals which list "Sputtering" separately in their subject index.   Examples are *Chemical Abstracts* or *Physics Abstracts*.   For those more interested in thin film applications of sputtering the abstract section of *Vacuum* gives a fairly complete survey of current publications.   A thorough bibliography of papers in atomic and molecular processes with sputtering listed as a separate subject index is published twice a year by the Atomic and Molecular Processes Information Center of the Oak Ridge National Laboratory.   Every second year the same agency publishes a very useful International Directory of Workers in the field of atomic and molecular collisions.

Sputtering literature has mushroomed in recent years.   In 1956, *Physics Abstracts* listed five abstracts; this had swelled to more than 100 by 1967.   However, as the basic phenomenon gradually becomes better understood, emphasis has begun to shift more and more from the target to the substrate or from the basic to the applications side.   As a consequence of this, papers on sputtering are shifting in ever-increasing numbers from the *Physical Review* to the *Journal of Applied Physics* or to such specialized journals as *Surface Science, Thin Solid Films, Thin Films,* and the *Journal of Vacuum Science and Technology.*   For many years the biannual International Conference on Ionization Phenomena in Gases has been the favorite meeting for scientists interested in the basic physics of sputtering.

Looking back over the past century one can list a number of papers which stand out as highlights in the progress of our knowledge of sputtering. Keeping such a list concise is a difficult task, and we are well aware that many papers not explicitly mentioned were essential for providing data and observations on which others could build their progress. The simple glow discharge with an anode and a cathode at a gas pressure ranging from 0.1 to several torr was the workhorse for cathode-sputtering studies up to 1923. A series of papers by von Hippel and Blechschmidt[7-9] contains a very thorough study of glow-discharge sputtering with many interesting details such as spectroscopic observations of the sputtered neutral atoms and measurements of the velocity distribution of the bombarding ions.

The glow discharge, although useful and popular in many thin film applications, is not well suited for basic studies of sputtering for several reasons: (1) The mean free path of ions and sputtered atoms is so short that ions arrive at the target with a large energy spread and an undetermined angle of incidence. (2) Some sputtered material diffuses back to the cathode. (3) Bombarding-ion energy, ion current density, and gas pressure cannot be controlled independently of each other. (4) It is not possible to work with very low bombarding-ion energies such as those near threshold because a self-sustained glow discharge cannot be maintained below some characteristic voltage. (5) Charge exchange between ions and gas atoms in the cathode region complicates the interpretation of results. These shortcomings began to become recognized when Hg pool-type rectifiers and thyratrons came into being. These operate at much lower gas pressures than the glow discharge. The observation that thyratron cathodes become damaged when the cathode is not heated to full power so that the discharge voltage (and hence the bombarding-ion energy) rises above a critical value led to the first reasonable data on sputtering threshold energies.[10]

The first attempt to collect meaningful data on sputtering yields as a function of bombarding-ion energy was made by Guentherschulze and Meyer.[11,12] They established a discharge of several amperes between a thermionic oxide cathode and an anode, thereby supporting a dense plasma at gas pressures low enough (less than 10 mTorr) for the mean free path of ions or sputtered atoms to become comparable with or larger than the tube dimensions. By immersing the target in the plasma as a third independent electrode, they laid the groundwork for what is now known as triode sputtering.

The beneficial use of magnetic fields for suppressing γ electrons or for increasing plasma density was taught by Penning and Moubis.[13]

In retrospect, it is somewhat surprising how long the hot-spot, or evaporation, theory of sputtering was favored, even though Stark[14] discussed the sputtering process in terms of impact and atomic-collision laws as long ago as 1908 and many experimental observations furnished strong arguments against an evaporation process. Such evidence includes the high ejection energies of sputtered atoms, the lack of any thermionic electron emission from the impact points, and the insensitivity of sputtering yields to the thermal conductivity of the target material or to the target temperature.

In 1942, Fetz[15] found that sputtering yields increase for more oblique incidence of the ions. In addition, Wehner showed that atoms tend to be sputtered in the forward direction when the target surface is under oblique low-energy (<1 keV) ion bombardment.[16] Both these observations are incompatible with any evaporation process.

The interesting etch effects which result from ion bombardment of polycrystalline metal targets were observed as early as 1912 by Stark and Wendt,[17] and results strongly suggested that the sputtering yield must be sensitive to crystal orientation. It is therefore somewhat surprising that the first single-crystal sputtering studies were performed only 30 years later. In 1955, a rather unexpected phenomenon relating to single-crystal sputtering was discovered by Wehner,[18] namely, that the sputtered atoms are preferentially ejected in the close-packed-crystal directions. Nothing similar had ever been observed in vacuum sublimation of single crystals. Preferential atom ejection during single-crystal sputtering and the resulting deposit patterns have been the subject of many detailed experimental as well as theoretical studies since then.

Bombarding a single crystal with an ion beam and varying the angle of incidence led Fluit[19] to another important discovery. The sputtering yield exhibits maxima and minima as a function of the angle of incidence. The minima in yield were correlated with those lattice directions in which the crystal model shows maximum transparency and in which the ions can penetrate most deeply. Almen and Bruce,[20,21] studying the sputtering yield of many ion-metal combinations as well as the number of ions which are collected in a target, found a direct correlation between these two quantities. High-sputtering-yield materials or ions which give high sputtering yields always have low collection efficiencies, and vice versa.

Panin[22] and, in much more detail, Datz and Snoek[23] analyzed the energy and angular distribution of secondary ions, reflected or sputtered from metal surfaces under oblique ion bombardment, and found that their results could be exactly described on the basis of biparticle collisions between ions and single isolated surface atoms, just as in gas-scattering experiments.

With the momentum-transfer model fairly well established and accepted, it was only natural to look for atoms sputtered from the back side of a thin foil when the front side is under ion bombardment. Experiments along this line at low bombarding energies remained inconclusive, however, because the sputter attack is so much more rapid at the front side that difficulties arise with sputter etching and pinhole formation. The decisive experiment was performed by Thompson,[24] who bombarded single-crystal Au foils with protons in the MeV region. At these high energies and with the proper foil thickness he was able to achieve more sputtering at the back than at the front side.

A powerful technique for observing the lattice damage or the removal of atoms by individual collision events is provided by Mueller's field ion microscope. This was first shown by Sinha and Mueller,[25] who bombarded liquid-hydrogen-cooled W tips with 20-keV He and Hg atoms (charge exchanged). The resulting damage consisted of vacancies (possibly sputtered atoms), interstitials, and clusters of such defects.

It is clear from what has been said above that the sputtering mechanism is not simple. It is difficult to formulate a theory which would enable one to predict the experimental data without the use of adjustable parameters. This is because so many parameters are involved: kinetic energy of the ions, electronic structure of the collision partners, lattice structure and orientation, binding energy of lattice atoms, etc. A primary difficulty seems to center around the potential functions which one should apply for the bombarding ion–target atom and target atom–target atom interactions.

The first attempts to try to understand sputtering in terms of radiation-damage theory were made by Keywell.[26] Gibson et al.[27] introduced the first computer simulation of the collision events. The ejection patterns which are observed in single-crystal sputtering have been interpreted to be direct evidence of focusing-collision chains. Well-pronounced ejection patterns are obtained, however, even in lattices where focusing collisions are impossible (Ge, Si, or in the zigzag directions of a cph lattice) and ejection patterns are obtained even at energies where no more than two or three atoms could be involved in a collision chain. All this has recently forced some rethinking. Both Harrison et al.[28] and Lehmann and Sigmund[29] came to the conclusion that long focusing chains contribute very little to sputtering. Sputtering is mainly the result of collisions near the surface, and ejection patterns can be explained by considering only collision events in the first three surface layers. Harrison has stated that present computers have neither enough memory capacity nor enough computation speed to simulate the sputtering process completely. The most comprehensive theoretical studies of the sputtering process are presently performed by Sigmund.[30]

c. Implications—Applications

Interest in sputtering has been nourished over the years from many different sides. In many gas-discharge applications sputtering is considered a nuisance because it causes erosion of the electrodes, damage to active surface layers in thermionic low-work-function cathodes, undesired deposits which blacken walls and observation windows, or gas-pressure changes through gettering—in short, in many cases it limits the useful life of such devices (including gas lasers). Thus the goal was and often is to find ways and means for reducing sputtering. One interesting solution to this problem, applicable to glow-discharge display tubes, was to add a small amount of Hg to the noble gas.   The Hg film which condenses at the cathode is subject to sputtering but is continuously replenished, thereby protecting the underlying cathode material from attack.   Electrode-erosion problems are common in all kinds of ion devices, such as mass analyzers or separators, ion engines, and plasmatrons.   If charge exchange plays a role, one has no control over the trajectories of the fast neutrals, and sputtering effects may be observed at surfaces where they are least expected.   An insulating wall in contact with a plasma always becomes negatively charged.   If this charging causes floating potentials which exceed the sputtering threshold, sputtering occurs.   This may happen, for instance, at glass-container walls.

In thermonuclear fusion one is concerned about possible sputtering of the container walls by fast hydrogen or deuterium ions even though the sputtering yield for such light ions is about two orders of magnitude lower than that of Ar, since they penetrate deeply and have a poor energy transfer to the metal atoms of wall materials like stainless steel.   The requirements for purity of such plasmas are extremely high since even very small numbers of the heavier atoms can cause undesired cooling of the plasma.

At one time there was serious concern about sputtering of windows, thermal-control coatings, solar-cell surfaces, etc., in space vehicles which sweep up the constituents of the outer atmosphere and which impinge with energies in the 5-eV range.   Fortunately it has turned out that such energies are below the sputtering threshold and the effects become negligibly small.

A most useful application of sputtering has been found in sputter-getter pumps. These are multiple-cell Penning discharges with a strong magnetic field such that a discharge can be maintained at pressures as low as 10^{-12} Torr.   The pumped gases accumulate in the sputter-deposited Ti films, especially at those places where the deposit is under some ion bombardment.

Sputtering provides a method of obtaining atomic vapor for absorption or emission spectroscopy, and hollow-cathode discharge lamps of various materials have become popular for this application.   Mass analysis of sputtered atoms has been perfected to the extent that routine parts per billion trace analysis of conducting or nonconducting solids can be performed.[31]

At the lunar surface or on other bodies in the solar system which have no protective atmosphere, sputtering by the solar wind (mostly protons and α particles in the low keV range) must have modified the surface.   Although the rates are very small, one can estimate that the moon must have lost about 5 to 10 cm of thickness during the 4×10^9 years of its existence.   Simulation, using greatly accelerated rates, shows that many powdered minerals are darkened by the solar-wind bombardment, and sputtered or sputter-deposited atoms are responsible for crusting of the surface.

In low-energy electron diffraction (LEED), Farnsworth introduced sputtering as one method for preparing atomically clean surfaces.   Ar ions with energies of less than 500 eV are usually used, and it is necessary to drive out the embedded noble-gas ions as well as to anneal the bombardment damage before a well-ordered surface is obtained. An interesting metallurgical electron microscope is on the market in which the kinetic γ electrons released under ion bombardment are used for image formation.   This allows the direct observation of sputter etching for variously oriented crystallites. In transmission electron microscopy sputtering is often used for thinning the specimen.

Sputter etching of materials which are difficult to etch by chemical methods has recently become popular in microcircuit fabrication.   In combination with photo-resist methods one can prevent undercutting and obtain much better pattern definition. This subject is treated in detail in other chapters of this book.

By far the most active area at present concerns the sputter deposition of films and coatings.   As early as 1877 sputtering was used to coat mirrors.[32]   Sputter coating of flimsy fabrics with Au enjoyed a certain popularity in the 1920s, and the Ediphone phonograph wax masters were coated by Western Electric (1930) with sputter-deposited Au which served as a base for subsequent electrolytic thickening.   With improvements in vacuum technology, however, sputtering was later largely replaced by the much faster vacuum evaporation.   With the advent of electron-beam evaporation this became true even for those metals or materials which are difficult to evaporate such as Ta, W, and Al_2O_3. However, increased sophistication combined with a better understanding of the sputtering process and the invention of rf sputtering of insulators has brought about the realization that sputtering offers a number of unique advantages which make it particularly attractive in certain applications. Of course, each individual case needs to be evaluated in comparison with other possible deposition methods which may be faster or cheaper.

The main points to consider in evaluating the merits of sputter deposition for films are as follows:

1. Sputtering yields or corresponding deposition rates are not much different for different metals, alloys, or even insulators. The compatibility and applicability to various materials make sputtering attractive in multilayer deposition.

2. One can sputter-deposit films of complicated materials such as stainless steel, Evanohm, Permalloy, or even Pyrex glass without composition changes provided one keeps the target temperature sufficiently low and prevents ion bombardment of the substrate and provided the sticking coefficients of the various species at the substrate and the angular distribution of the ejected species are the same.

3. Film-thickness control becomes relatively simple. One determines the deposition rate by laying down a film of a thickness that will be easy to measure, and thereafter with the same geometric arrangement and under the same operating conditions (gas pressure, target current, and voltage) it becomes largely a matter of adjusting the deposition time to achieve the desired film thickness.

4. Sputtering can be accomplished from large-area targets. This often simplifies the problem of film-thickness uniformity and is beneficial with respect to pinholes because shadowing effects from dust particles become less pronounced. From the inside of cylindrical targets one can uniformly sputter-coat shafts mounted at the axis or one can sputter films uniformly onto the inside walls of tubing or cylinders from target rods in the center.

5. In sputtering there are no difficulties with "spitting" (the ejection of larger agglomerates), which often occurs in vacuum evaporation, and there are no restrictions with respect to gravitational forces in electrode or substrate arrangement. In many systems downward sputtering is often preferred because it simplifies substrate mounting.

6. Cleaning of substrates is much simplified because one can sputter-clean the surface of the substrates by ion bombardment before sputter deposition. Shutters between target and substrate are often very useful or necessary for presputtering in order to achieve thermal or background pressure-equilibrium conditions.

7. Most targets are such that there is sufficient material to last for many deposition runs.

8. The plasma can be manipulated with magnetic fields to achieve greater film-thickness uniformity, and fast electrons can be kept away from the substrate so as to reduce substrate heating.

9. The presence of the plasma offers other unique possibilities for achieving films with desired properties. Negative biasing of the substrate before film deposition can be used to remove oxide films and to improve film adherence in the case of metal films on metal substrates. With negative biasing during film deposition one can achieve preferential oxygen removal and obtain metal films of high purity.[33]

10. Sputtering is an atom-by-atom process. As a consequence one can obtain nonporous films with surfaces which closely reproduce the surface finish of the substrate, provided one avoids major surface migration of atoms by keeping the substrate sufficiently cold and sputters at sufficiently low gas pressure in order to avoid clustering or agglomeration of atoms already in the gas phase. Thus a highly polished surface retains its finish after film deposition, at least for thin films. For thicker films interesting columnar-growth features which are sensitive to biasing and substrate temperature have recently been reported.[34]

11. If the discharge is operated at gas pressures of less than 5 mTorr, the sputtered atoms arrive with their high kinetic ejection energy at the substrate, and this may be beneficial with respect to film structure or adherence of the film. For instance, much

lower epitaxial temperatures have been observed for sputter deposition than for vacuum evaporation, as discussed in Chap. 4.

12. Through dc or rf sputtering of metals or semiconductors in reactive gases such as oxygen or nitrogen one can form well-oxidized or nitrided films of such materials as SiO_2, TiO_2, Si_3N_4, and SnO_2.

The main disadvantage of sputtering is the fact that the deposition rates are relatively low. Typical deposition rates are in the range of 50 to 3,000 Å min^{-1}. For higher sputtering rates it is necessary to cool the target. If one tries to achieve higher sputtering rates by increasing the plasma density, one also has to find means for efficient substrate cooling. This is sometimes difficult or at least cumbersome.

Scanning the literature of recent years, one finds an enormous variety of thin film problems which investigators have attempted to solve through sputter deposition. This ranges from superconducting films to cermets (sputtering simultaneously from a metal and an insulator target); from ferro- and piezoelectric to ferromagnetic films; from resistive, conductive, and insulator films for passive microcircuit elements to protective and passivating films in active devices; from coatings for better corrosion, abrasion, and wear resistance to solid lubricating films; from coatings of plastics for flexible circuitry or connectors to coatings of razor blades; from photoemissive films to optical coatings; from attempts to form new metastable alloys to preparing pinhole-free Cr masks for photoetching. (These aspects are discussed in many other chapters of this book.)

2. METHODS OF SPUTTERING

a. Plasmas

Historically, sputtering was discovered at the cathodes of cold-cathode glow discharges, and these have been the main workhorses for sputter deposition of films since then. Although it is very simple to set up a dc glow discharge, the detailed explanation of its various glow-zoned and dark spaces and of the interrelationship between discharge voltage, discharge current, and gas pressure continues to be a challenging task.[172]

The primary progress in the understanding of the basic sputtering process has come from studies in triode systems where the plasma is formed independently as the positive column of a discharge maintained between a thermionic cathode and an anode. Sputtering is accomplished by inserting the target in this plasma as a separate negative electrode. The main advantage of incorporating a thermionic cathode is that even without a magnetic field a plasma can be maintained at much lower gas pressure (low millitorr region) than in a dc glow discharge which requires gas pressures above about 30 mTorr.

By operating at gas pressures where the mean free path of ions and of sputtered atoms becomes comparable with or larger than the ion-accelerating region or the tube dimensions, one reduces or eliminates many of the complications inherent in glow discharges such as diffusion of sputtered material back to the target, poorly defined bombarding-ion energies and angles of incidence, and charge-exchange effects in the ion-accelerating region. The discharge is fed and maintained by the electrons released from the thermionic cathode and not by the γ electrons from the glow-discharge cold cathode. Thus whatever happens at the target in a triode system is unimportant for maintaining the plasma. One is therefore free to select the bombarding-ion energy even down to very low energy independent of other parameters like discharge current or gas pressure and one can regulate the bombarding-ion current density independently by the main discharge current.

Langmuir's probe theory[35] furnishes all the information needed for understanding the simple and straightforward processes which take place at a target in contact with a low-gas-pressure plasma. The situation in a plasma tube is quite different from that in a high-vacuum tube. The plasma has such high electrical conductivity that a voltage applied to a probe or target does not result in field changes throughout the whole tube but only in the immediate vicinity of the probe. When, for instance, a negative voltage is applied with respect to the plasma (or with respect to the anode because anode potential is usually close to plasma potential), the plasma electrons in the electrode vicinity are repelled, a positive-ion sheath is formed through which the ions stream from the plasma toward the electrode, and the major part of this voltage drop is localized in this sheath. The sheath thickness d after reaching steady-state conditions is controlled by Langmuir's space-charge equation

$$j^+ \sim \frac{U^{\frac{3}{2}}}{d^2}$$

where j^+ is the ion-current density furnished by the plasma at the sheath edge (and controlled by the main discharge current) and U the voltage difference between target and plasma. The sheath thicknesses for various noble gases can be directly read from graphs.[36]

The sheath is essentially dark and therefore clearly visible since there are no plasma electrons present to cause excitation of gas atoms. Only the γ electrons released from the target under bombardment could cause excitation in the sheath while in transit from target to plasma, but their numbers are few and they soon acquire velocities beyond the excitation maximum. The cross section for excitation by ions is so small that this can be neglected. The ion current to a flat probe (with guard ring) is independent of the applied voltage. The only result of increasing the applied voltage is to push the sheath edge farther away from the electrode such that more positive ions in the sheath can compensate for the increased negative charge at the electrode. The ion-current density furnished by the plasma is

$$j^+ = \tfrac{1}{4}n^+v^+e$$

where n^+ is the density of ions in the plasma, v^+ their average random velocity, and e the electronic charge. The current measured at a negative probe contains, of course, not only the ion current but also the superimposed γ electron current. At bombarding-ion energies of less than 1 keV, only the potential emission of γ electrons is important. For Ar^+ on a clean W surface, the emission amounts to roughly 10%. Note that the ion-sheath thickness is independent of gas pressure as long as the random ion-current density in the plasma (or the main discharge current) remains constant. The ions bombard a flat electrode under essentially normal incidence. The bombarding-ion energy (with very little energy spread because the ion temperature is not much higher than the gas temperature) corresponds to the voltage applied between electrode and plasma. At higher gas pressures, charge exchange with corresponding formation of fast neutrals in the ion sheath may become important.

In a very first approximation, the conditions at the cathode of a dc glow discharge are not much different from those of such a negative electrode in plasma, except that the γ electrons released from the cathode are now most important for maintaining the glow-discharge plasma.

In order to understand what happens in the anode vicinity or later in the case of rf sputtering, it is necessary to discuss the electron-collecting part of Langmuir's probe characteristic. If the probe voltage becomes less and less negative with respect to the plasma, with the ion sheath consequently shrinking in thickness, fast electrons from the plasma begin to reach the probe. Electrons in a low-pressure plasma are not in thermal equilibrium with the gas (poor energy exchange due to the large difference in masses) and have therefore a wide velocity spread which corresponds to temperatures of tens of thousands of degrees. As the probe's potential approaches that of the plasma, the probe current, consisting of collected electrons (as well as positive ions), now rises exponentially. The electron current becomes finally more than two orders of magnitude higher than the ion current because the random electron-current density

$$j^- = \tfrac{1}{4}n^-v^-e$$

and with $n^+ = n^-$, the ratio

$$\frac{j^-}{j^+} = \frac{v^-}{v^+} = \left(\frac{M}{m}\right)^{\frac{1}{2}}$$

where M = ion mass and m = electron mass.

At plasma potential one reaches the well-known knee in the probe characteristic because from here on the roles of the charge carriers are reversed. At positive potentials, positive ions are repelled, an electron sheath forms around the probe, and one enters the region of a saturated electron current to the probe. At more positive potentials of the probe, however, an important difference from the negative probe branch is seen: As soon as the electrons are accelerated to energies above the ionization energy of the gas, the probe becomes covered with a region of increased plasma density and the probe current rises still further very steeply. The whole probe characteristic resembles that of a rectifier. One can build up large voltage drops in front of a negative electrode but never in front of a positive electrode. With the electron current being so much larger than the ion current and having such a large velocity spread, it becomes clear that zero current to the probe (i.e., equal ion and electron current) occurs at a somewhat negative voltage with respect to plasma. Consequently every insulator or floating electrode (or wall) charges up somewhat negatively in order to repel some of the (more abundant) arriving electrons. It should be clear from the probe characteristic that one can bias an electrode as negatively as one wants to (unless the ion sheath becomes so thick that it fills the tube and extinguishes the plasma), but one can never bias much more positively than the ionization potential of the gas; otherwise the electrode becomes a new anode, draws large electron currents, and becomes very hot.

The stage is now set for a discussion of rf triode sputtering of insulators. Let us insert a metal electrode in a dc plasma (between a thermionic cathode and an anode) and apply a rf voltage (in the megahertz range) between this electrode and anode. From the foregoing we saw that the electrode-plasma region behaves like a rectifying junction. We can therefore insert somewhere in the rf circuit a blocking capacitor which will charge up so that the electrode becomes mostly negative with respect to the plasma. We can now replace the external capacitor with an internal one by enclosing the metal electrode with an insulator. Now the insulator surface at the plasma side becomes negatively charged with respect to the plasma (and with respect to the metal electrode) and will be sputtered. Only short bursts of electrons are pulled to the insulator surface while for the major part of one rf cycle the ions are accelerated toward the surface.

The negative charge at the insulator surface is such that the overall ion current in one cycle just equals the electron current so that the net current becomes zero—as required for an insulator. An Ar^+ ion on its way from the plasma to target is not able to traverse the full ion sheath in one cycle at frequencies of 10^7 Hz or higher and becomes immersed in a neutral plasma for a short period in each cycle. It is therefore repeatedly accelerated during successive cycles but is never substantially decelerated, because an electron sheath with its high electrical conductivity can never sustain high field gradients. With respect to the frequency range most suitable for rf sputtering in a triode system the following considerations apply: If the capacitor or the internal capacitor has a very small capacitance, it loses its charge very fast under ion bombardment. Thus at 10 pF cm^{-2} with an ion current of 10 mA cm^{-2} the capacitor loses voltage at a rate of 10^9 V s^{-1}. If one does not want to lose more than 100 V cycle^{-1}, one has to operate with frequencies above 10 MHz. With lower ion-current density and higher capacitance one can operate in a lower frequency range. An upper limit for the radio frequency arises when transit times of electrons become important.

In recent years the simple rf diode system has become popular for sputtering applications, and several manufacturers have equipment of this type now on the market. Here plasma excitation is accomplished by the rf fields which accelerate the plasma electrons to ionizing energies. This kind of plasma excitation becomes very efficient at higher frequencies, and one can maintain an rf-excited plasma at lower gas pressures than a dc glow discharge. It is advantageous to work in the higher-frequency range such as 10 to 50 MHz for more efficient plasma excitation. With the γ electrons being no longer essential for maintaining the rf diode discharge, the latter becomes in some ways simpler to understand than the dc glow discharge. Let us consider two equal-sized electrodes in a noble gas at a pressure in the millitorr range. When a symmetrical rf voltage in the megahertz range is applied between these two electrodes and a plasma is established, one observes that each electrode becomes covered with a dark sheath and both are therefore being sputtered. To a first approximation one can assume the ions to be stationary (later we must modify this assumption; otherwise there would be no sputtering) and the situation is as if in one half cycle a column of plasma electrons moves part way into the temporarily positive electrode, leaving behind the electron-depleted positive-ion sheath in front of the other electrode, which takes up the whole voltage difference. In the other half cycle the functions of the two electrodes become just reversed. There is not sufficient time for the plasma to become deionized. It remains now positive with respect to the electrodes most of the time, and one measures a dc voltage close to the rf amplitude between the plasma and either one of the two electrodes. The ions in the sheaths are, however, not stationary if one considers more than one cycle. They are jerked in successive cycles[37] toward the electrodes and cause sputtering upon impingement. At lower frequencies the main differences are that the plasma needs the γ electrons for excitation and each electrode becomes an anode in one half cycle and a cathode in the other, just as if one had two alternating dc glow discharges.

So far we have discussed two equal-sized electrodes. A blocking capacitor in the rf circuit would, in such a symmetrical situation, have no influence. One can therefore cover one or both electrodes with insulators which are then sputtered under ion bombardment. Various such symmetrical systems are commercially available. Let us now increase the area of one electrode, creating an unsymmetrical situation. Since no dc drop can occur in the external circuit, one finds that the maximal ion-sheath thickness remains still equal at both electrodes when they are at their negative peaks provided that the plasma densities are equal at both locations. But when the electron block shifts toward the smaller electrode and the ion sheath builds up in front of the other larger electrode, more electrons enter the small electrode than enter the large electrode in the other half cycle. A blocking capacitor inserted somewhere in the rf circuit charges up now with the unsymmetrical current such that the smaller electrode becomes more negative than the larger one. As a consequence of this, one observes that the ion-sheath thickness and bombarding-ion energy at the larger electrode are less than those of the smaller electrode. With large electrode-area ratios one can readily reduce the bombarding-ion energy at the larger electrode to below the sputtering threshold. This will be discussed more fully in Chap. 4. In cylindrical geometry (small wire in center of large-area cylindrical electrode) the arrangement and operation are identical to that of the early glow-discharge rectifier tubes. As discussed above, the capacitor can be made internal by covering the smaller electrode with the insulator to be sputtered. This then is the basis of the asymmetrical parallel-plate rf glow-discharge devices which were first developed by Davidse and Maissel.[38] The "large" electrode in these systems consists of the grounded shield around the target, the grounded substrate support plate, the baseplate, and all the other plasma-exposed grounded parts in the system.[39] (See Chap. 4 for additional discussion.)

The picture presented so far is, of course, a very oversimplified one and is intended only to give the reader a grasp of the basic phenomena, especially those involved in rf sputtering. When it comes to details much room remains for arguments, measurements, and refinements.

We do not agree, for instance, with the explanations of Toombs,[40] who concluded in disagreement with Holland et al.[41] that no difference exists between single-ended and symmetrical rf sputtering systems with respect to the voltage which the substrate acquires with respect to the rf plasma. In single-ended systems the grounded substrate holder is an integral part of the rf circuit and essential for increasing the area of the "anode." In symmetrical systems, one may let the substrate float, and provided it does not get immersed in the ion sheaths of the targets, the floating substrate should assume a potential close to and only slightly negative with respect to the plasma just like a floating electrode in a dc plasma.   Although the electron temperature in a rf low-pressure plasma is fairly high, one would not expect this negative voltage to exceed by much the sputtering threshold energy, unless the γ electrons from the target bombard the substrate and change this situation.

So far, we have completely ignored the γ electrons in rf sputtering because they are not essential for plasma generation.   They are, however, there, and may reach substantial numbers when insulators are sputtered with ion energies of several keV under which conditions the kinetic emission coefficients can exceed unity.   These γ electrons are accelerated across the ion sheaths in directions normal to the target surface and traverse the plasma with high velocities.   They can become a major source of substrate heating and charging unless the substrate is arranged to be outside their paths or their trajectories are bent away from the substrate with magnetic fields.

If the γ electrons bombard an electrode with energies that are large enough for the secondary-electron-emission coefficient to exceed unity (at most insulators this happens above about 40 eV), this electrode will become less negative and its potential will become closer to that of the plasma.

It may be desirable to ion-bombard a film as it deposits (see Chap. 4); however, there are several reasons why one may want to minimize substrate bombardment during film deposition: As one knows, for instance, from getter triode pumps, more noble-gas atoms are likely to accumulate in a film under bias deposition, especially at higher biasing voltages.   The sputter-deposited films may become contaminated with substrate or substrate-holder material because sputtering of these materials can take place from areas which see the plasma but not the target and more is sputtered from there than is deposited from the target.   And finally one knows in the case of insulator films that ion bombardment of the film during deposition often leads to oxygen-deficient brownish-looking films of poor optical quality, unless one provides additional oxygen with a noble gas–oxygen mixture as the sputtering gas.

Another source for resputtering of material from the substrate is negative ions released from the target under ion bombardment and accelerated to high velocities in its ion sheath.   With clean metal targets the number of negative ions is negligibly small, but this is not so with many insulating materials.   Resputtering from the substrate by fast negative ions can readily be demonstrated by placing a small obstacle halfway between target and substrate.   The negative ions are well aligned in a direction normal to the flat target surface and therefore cast a well-defined shadow of the obstacle at the substrate, causing film-thickness differences.   With a moderate magnetic field parallel to the target surface one can separate the effects caused by γ electrons and negative ions bombarding the substrate.

The determination of the plasma potential with a Langmuir probe is a straightforward matter in a dc discharge but can be full of pitfalls in the case of a rf plasma. Fetz and Oechsner[42] describe a method for determining the plasma potential in a rf-excited plasma by means of a Langmuir probe with variable probe area.   This should prove to be most useful for accurate measurements on substrate charging, sputtering from parts other than the target, and measuring potential differences between various parts of the rf plasma.

Since sputtering is a very inefficient process, it is necessary at higher deposition rates or power levels to find ways and means for efficient target cooling.   This may pose problems, especially in the case of insulator targets.   By far the major part of the substrate heating arises from the γ electrons.   An obvious solution is to prevent the γ electrons from bombarding the substrate.   This can be more readily accomplished with the proper magnetic fields in a system with cylindrical geometry (as in Ref. 13).

Sputtering of various materials often leads to surprises.   A typical case is TiO_2. While a smooth Al_2O_3 target surface essentially retains its original whitish color during sputtering, it is observed that a TiO_2 target assumes a dark metallic appearance (Ti enrichment) and develops a high electrical surface conductivity.   As a result, sputtering soon takes place from all over the target, even from those areas which are far away from the metal backing plate.   Another surprise arises when a high-yield material (such as Cu, Ag, Au) and a low-yield one (such as Mo, Ta) are sputtered simultaneously. Even if only very small amounts of low-sputtering-yield materials arrive at the high-yield target during sputtering, one observes there the development of closely spaced cones which give the surface a velvetlike appearance. The sputtering yield from such a cone-covered surface becomes very small because most material is sputtered back and forth between the cones and cannot escape from the surface.[43]

Difficulties often arise with materials such as many of the rare-earth metals or their oxides which have high vapor pressure and low electronic work function. These have a tendency for spark—or cathode-arc spot—formation. Materials with high bulk electrical conductivity provide breakdown-prone "capacitors," and one should never overlook the fact that the full sputtering voltage appears across the insulator to be sputtered. In materials with high dielectric constant it is important that no gap should exist at any place between target and target metal backing; otherwise the local decrease in capacitance will cause a lower sputtering yield there. This is particularly important in lower-frequency rf triode sputtering where a low-capacitance region loses its negative charge much faster under ion bombardment.

If a complicated material such as Pyrex glass is sputtered it is, at least at first glance, most surprising that the deposited material appears to be Pyrex again. Although it is known that material is transferred from target to vapor in an unchanged composition after reaching equilibrium conditions, it must be assumed that most molecules are dissociated in the sputtering process and that the material arrives at the substrate predominantly in atomic form. It remains to be determined whether the atomic oxygen, the ultraviolet from the plasma, the low-energy noble-gas ion bombardment, or the high kinetic energies of the sputtered atoms are responsible for providing the activation energies needed for reassembly into the proper molecules.

b. Ion Beams

Sputtering with ion beams, particularly with mass-separated ion beams in high vacuum, has proved to be the ideal method for collecting reliable data at higher bombarding-ion energies, but it has seldom been used for thin film deposition. We will therefore devote here only a few short film-deposition-related remarks to this subject.

The attractive features of ion-beam sputtering are that one can (with differential pumping) sputter under high-vacuum conditions, can maintain a field-free region between target and substrate, and can bombard the target under oblique incidence. Very little is known so far on the effects of noble-gas embedment on nucleation and growth of sputter-deposited films.[44] It would be very surprising if the noble-gas content of the films in plasma sputtering, which is unavoidable, although small, did not have a major influence on certain film properties of certain materials. Ion-beam sputtering offers the possibility of controlling the gas content of films and studying this parameter in detail (although we are not aware of any systematic studies performed so far).

In plasma sputtering the target is always negative with respect to the plasma, and as a consequence positively charged ejected ions will be forced to return to the target while negative ions are accelerated away from the target. These separation effects can, of course, be completely eliminated in the field-free working volume achievable in ion-beam sputtering. Other advantages of a field-free target surface are that one can sputter materials which are available only in powder form without the disturbing forces which otherwise act on the powder particles. One can furthermore avoid difficulties with sparking or arcing at the target surface, which are very annoying with certain materials like Pb, Zn, and Ca when sputtering is performed in a plasma. At lower beam energies one uses an auxiliary thermionic cathode as electron source for complete charge neutralization at the target surface and for reducing the space-charge effects (which limit the current density) in the beam.

Target cooling becomes a severe problem if one wants to achieve high deposition rates because the efficiency of sputtering decreases with increasing bombarding-ion energy. Substrate cooling, on the other hand, should hardly pose major problems because the γ electrons are not accelerated toward the substrate.

With oblique incidence one achieves substantial increases in sputtering yield in many materials. This can at least partly compensate for the lower ion-current densities which one has to live with in ion-beam sputtering. Of particular interest should be a study of the consequences of the higher ejection energies of those atoms which are sputtered under oblique bombardment in the forward direction. Problems can arise with contamination of the films by atoms sputtered from the beam-accelerating electrodes (possibly by charge-exchanged neutrals).

As ion sources, modifications of von Ardenne's Duoplasmatron have been used mostly.[45] Von Ardenne et al.[46] reported deposition rates for Ta or SiO_2 of 300 Å min^{-1} obtained with a 10-keV 1-mA Ar ion beam at target-substrate distances of 2 cm. Chopra and Randlett[47] worked with a beam of 0.8 cm diameter, 50 mA, 2 keV and obtained an Ag deposition rate of 400 Å min^{-1} at 8 cm distance. Nablo and King[48] sputtered various metals, semiconductors, and insulators with a 30-keV, 10-mA, Ar ion beam onto 5- by 5-cm substrate areas and obtained deposition rates of 30 to 150 Å min^{-1}. A major drawback of beam schemes is the fact that the bombarded target area is small and the method is not well suited for high rate deposition of films with uniform thickness over larger substrate areas. However, large-diameter ion beams (Hg) composed of many smaller beams have received much attention in recent years as thrusters in space-propulsion applications, and have been developed to a high degree of sophistication. Such beams could conceivably partly overcome these limitations, if applied for sputter deposition of films, and efforts in this direction are in progress.[49]

3. SPUTTERING YIELDS AND THRESHOLDS

a. Yields

A detailed presentation of sputtering-yield data, especially those of interest in thin film deposition, is given in Chap. 4. We will therefore confine our discussion here to only a few general remarks.

The sputtering yield measured in atoms per incident ion is the most important parameter for characterizing the sputtering process, and the foremost goal of every sputtering theory is to be able to calculate yields from atomic or lattice data without adjustable parameters.

Experimentally the sputtering yields have been determined with a wide variety of techniques such as radioactive tracers, spectroscopic methods, observation of the changes in the resonant frequency of a quartz crystal which is used either as a target or as a substrate, and measuring the changes of the electrical resistance of a target or of its electronic work function (thermionic electron emission). All methods are usually calibrated against direct weight-loss experiments.

In the next section we will present yield data for very low ion energies (sputtering thresholds). Above the threshold the sputtering yield rises first exponentially, then linearly, then less than linearly, approaches a flat maximum, and decreases again with increasing ion energy. Some typical curves for polycrystalline Cu bombarded by Ar ions are shown in Fig. 2. One notices rather wide discrepancies in the data of different investigators. These were traced back to differences in the preferred crystal orientation of the polycrystalline samples. Especially at higher bombarding energies sputtering studies with single crystals show wide yield differences which depend on the orientation of the crystal with respect to the beam.[19] Whenever the beam sees a low density of projected lattice points, the ions penetrate more deeply, and as a consequence the collection efficiency for ions becomes higher and the sputtering yields decrease.[20,21]

With respect to the target material one finds under otherwise identical conditions that the sputtering yields show a periodicity which is closely linked to their position in the periodic system of elements and as a result of periodicities in heat of sublimation, atomic shell, and crystal structure. The masses of target atoms enter through the energy-transfer coefficient $4mM/(m + M)^2$ (m = mass of target atom, M = mass of ion). Sputtering yields of metals in solid or liquid form are not much different,[50,51] and the target temperature has only a minor influence on sputtering yields (unless one raises the temperature to values where evaporation becomes superimposed on sputtering, in which case Thompson and Nelson[52] found some anomalies).

On the variation of sputtering yields with ion species the most complete data were collected by Almen and Bruce[20,21] in the higher-energy region (45 keV). Figure 3,

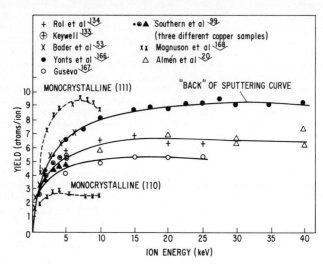

Fig. 2 Variation of sputtering yield with energy for argon ions bombarding copper (high-energy region).

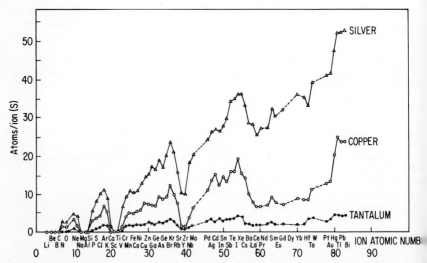

Fig. 3 Variation of sputtering yield with atomic number of the bombarding ion for 45-keV bombardment of copper, silver, and tantalum targets.[18,19]

which is taken from this work, shows clearly that the noble-gas ions give the highes sputtering yields while elements from the center columns of the periodic chart (Al, T Zr, Hf, and including the rare-earth series) have the lowest yields. Of interest an unexplained so far is the fact that the yields vary with ion species much more (facto

100 or more) than with target-atom species (factor 10). Sputtering yields can reach rather high values, as is shown in the case of Ag-Hg with 50 atoms per ion.

Molecular species behave with respect to sputtering yield as if the atoms of the molecule arrived separately with the same velocity as the molecule and initiated their own sputtering events. The sputtering yield under neutral-atom bombardment should be the same as that of the corresponding ion. This was confirmed in a number of studies,[53,54] although in at least one paper it was claimed that differences were found.[55] Reliable experimental data on this point are difficult to gather.

Especially at higher bombarding energies one might expect that the noble-gas ions which become embedded in the target (average depth of penetration for a 1-keV Ar ion in Cu is about 10 Å) have an effect on the sputtering yield. After equilibrium conditions are reached, the sputtered species must contain sputtered atoms of the bombarding species unless these leave the near-surface layers by bombardment-enhanced diffusion and evaporation. It was observed with the electron microscope that noble-gas ions which were embedded at high energy congregate into microscopic noble-gas bubbles.[56] In general it is believed that these effects have no large influence on sputtering yields, but this remains to a certain degree conjecture until more concrete experimental data have been collected.

An important parameter in sputtering yields is the angle of incidence of the ions. As expected in a billiard-game model, the sputtering yields increase with more oblique incidence of the ions because less directional change of the momentum is required for ejecting atoms in a forward direction. Oblique incidence, especially at higher bombarding energy, furthermore confines the action closer to the target surface, and thus sputtering is enhanced. Some typical examples of yield variation vs. impinging angle are shown in Fig. 4. The wide differences of different target materials with respect

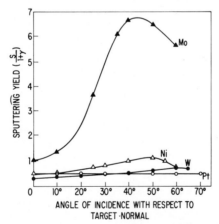

Fig. 4 Variation of sputtering yield with angle of incidence for 200-eV mercury-ion bombardment of nickel, molybdenum, tungsten, and platinum.[169]

to yield variation with ion-impingement angle remain unexplained so far. The yield maximum in these curves is generally attributed to the influence of the roughness of the target surface, which tends to decrease the yield at very oblique angles. The roughness of the target surface affects the yield appreciably even under normal ion incidence, because sputtered atoms may not be able to clear the surface and may become reattached on neighboring elevations.

We now have a wealth of yield data for metals, but data for insulators, especially in the energy range which is of interest in thin film sputter deposition, are badly lacking. In rf sputtering, of course, one has to live with a wide spectrum of bombarding-ion energies, and this makes it impossible to collect meaningful data on the yield-energy dependence. With ion beams, however, one should have no basic difficulties provided one employs a sensitive method for measuring the sputtered atoms and knows the potential which the insulator surface assumes under the bombardment. Insulator yields (see Chap. 4) are usually lower than those of metals.

b. Thresholds

The subject of sputtering thresholds, or the minimum ion energies which are able to dislodge target atoms, has been a very controversial subject for many years. Up to 1912, a value of 495 eV is repeatedly mentioned as valid for all gas-metal combinations.[2]

Indications of the existence of a sputtering-threshold energy were obtained in gas rectifier tubes, in which it was observed that thermionic Th-W cathodes were damaged when the voltage drop or the bombarding-ion energy exceeded a critical value of the order of 20 to 30 V.[10] In some later studies much higher sputtering thresholds were again reported,[57] but one can readily explain such high values as being either due to the formation of impurity layers at the target surface from background gases or due to an insufficient sensitivity for measuring very low sputtering yields. One can further be misled by the fact that background gases may convert the sputter-deposited metal films at a glass wall of the tube to transparent oxide films. One just plainly does not observe any metal deposits even in very long sputtering runs if the deposition rate becomes very low. With improving vacuum techniques the measured and reported thresholds reverted to the original values of Hull et al.[10] Morgulis and Tischenko[58] found unusually low threshold energies with a radioactive-tracer method which allowed yield measurements down to 10^{-4} atom/ion. Without more detailed data on the discharge used in their experiments, the suspicion remains that the low values (Ni-Ar 8 eV, Ag-Ar 3 eV) may have been due to multiply charged ions or a neglected correction in ion energy due to plasma-potential differences.

The methods which have been employed for achieving a very high sensitivity in yield measurements range from radioactive-tracer methods to the use of sensitive built-in microbalances or oscillating-quartz detectors or spectroscopic methods.

The widest range of data covering many materials and gases has been collected with the spectroscopic method by Stuart and Wehner.[59] This method and the results of the measurements are described here in a little more detail.

The target is immersed like a large negative Langmuir probe in a low-pressure noble gas or Hg plasma of very high density. The high density is essential to the success of this method and is achieved by magnetic fields, with which one can increase the bombarding-ion current density to 100 mA cm^{-2}. The sputtered atoms, which are ejected mostly in the neutral unexcited state, find favorable excitation conditions in the plasma, and one observes the emission spectrum of the target atoms superimposed on the emission spectrum of the discharge gas. With a monochromator one singles out a strong emission line of the target material and measures its intensity with a photomultiplier. It is a reasonable assumption that the intensity of the spectral line is approximately proportional to the sputtering yield. The relative spectral yield vs. energy curve can be converted to an absolute-yield curve by matching to values at higher bombarding energies where absolute data are available. The overlapping parts of the curves show that the proportionality assumption is justified. Absolute curves do not extend much below yields of 0.1 atom/ion, whereas the spectral in situ method allows instantaneous measurements down to 10^{-5} atom/ion.

Figure 5 shows the sputtering yields of a number of metals vs. Ar ion energy, and Fig. 6 shows yields for W under bombardment by various ion species. One can extrapolate these semilog curves to an infinite slope and rather loosely define a threshold energy for sputtering. A surprising result is the fact that the thresholds differ very little for different ions but seem to be more characteristic for different target materials. In Table 1 are listed the thresholds for 23 metals under bombardment by Ne, Ar, Kr, Xe, and Hg. One notices that the masses of the collision partners play hardly any role and that the threshold values are roughly $4H$ (H = heat of sublimation). This surprising result probably indicates that at such low bombarding energy the collision events cannot be treated independently of each other because the col-

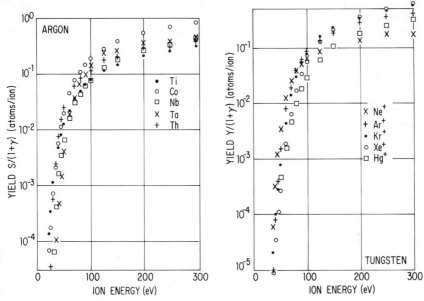

Fig. 5 Yield curves in argon.   **Fig. 6** Tungsten yield curves.

TABLE 1 Threshold Energies

	Ne	Ar	Kr	Xe	Hg		Ne	Ar	Kr	Xe	Hg
Be.....	**12**	15	15	15		Mo...	24	24	**28**	**27**	32
Al.....	**13**	**13**	15	18	18	Rh....	25	24	**25**	**25**	
Ti.....	**22**	**20**	**17**	18	25	Pd....	20	20	**20**	15	20
V......	21	**23**	**25**	28	25	Ag....	12	15	**15**	17	
Cr.....	22	**22**	**18**	20	23	Ta....	25	26	30	**30**	30
Fe.....	22	**20**	**25**	23	25	W....	35	33	30	**30**	30
Co.....	20	**25**	**22**	22		Re....	35	35	25	**30**	35
Ni.....	23	**21**	**25**	20		Pt....	27	25	22	**22**	25
Cu.....	17	**17**	**16**	15	20	Au....	20	20	20	**18**	
Ge.....	23	**25**	**22**	18	25	Th....	20	24	25	**25**	
Zr.....	23	22	**18**	**25**	30	U.....	20	23	25	**22**	27
Nb.....	27	25	**26**	**32**							

Boldface values are those for which the energy-transfer factor $4m_1m_2/(m_1 + m_2)^2$ is 0.9 or higher.

lision times proper are probably so long that the neighbor atoms come into play before the primary collision is over. In such a case one may have to introduce modified masses.

Some sputtering thresholds for insulators were measured by Jorgenson and Wehner[60] with a Langmuir-probe method. The probe characteristic is very sensitive to deposits of insulating films because this changes the electronic work function and the electron acceptance of the probe. A thin insulator film is deposited on the probe by rf sputtering. Such a thin film has sufficient electrical conductivity that it can be removed by applying a dc bias voltage on the probe. Different bias voltages require different times to remove the same thickness of insulator film (as monitored in the probe characteristic), and one can use this as a very sensitive method for collecting dc sputtering yield data. Figure 7 shows the result of such measurements for SiO_2 sputtered with

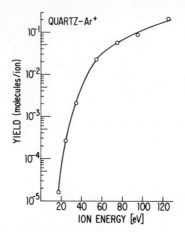

Fig. 7 Sputtering yields in molecules per ion vs. bombarding energy.

Ar ions. The threshold value for this case was found to be about 16 eV. The value for Pyrex 7740 in Ar was found to be approximately the same.

4. VELOCITIES OF SPUTTERED ATOMS

The subject of energies and energy distributions of sputtered atoms has gained much attention in recent years for various reasons. For a better understanding of the basic concepts in sputtering it would be of much interest if energy-distribution data would show any evidence of focusing-collision chains and indicate the maximal energy which can be focused in various crystal directions. In sputtering theory Sigmund[61] concluded that the combined kinetic energy of sputtered atoms per incident ion represents a more meaningful parameter to calculate than the sputtering yield. For practical purposes a knowledge of the average energies of sputtered atoms is of interest in film nucleation and growth, in particular in epitaxy, or with respect to the question of resputtering of material by sputtered atoms, or of solar-wind sputtering of material from the lunar surface (sputtered atoms may or may not escape from the moon depending on velocity),[62] or in energy considerations in electric propulsion[63] or forces on ion-bombarded electrodes.[64]

a. Various Experimental Methods

The various schemes which have been used for measuring velocities of neutral sputtered atoms fall into two groups: those which allow only the determination of average energies (forces, calorimetric) and those which measure the velocity distribution (mechanical velocity selectors, time-of-flight methods, post-ionization with subsequent retarding-field analysis, etc.). The first group does not allow the identification of the species which cause the forces on the target or the heating of a collector of the sputtered particles. The methods in this group are suspect of large potential errors because neutralized ions reflected from the target or negative sputtered ions accelerated in the ion sheath in front of the target can contribute substantial amounts to the energies to be measured. This is especially true for nonnoble metals where oxide or other films, which may form on the target surface from background impurities in the gas, are known to be a notorious source of negative ions. Methods of the second group not only give data on the much more informative velocity distribution rather than only average velocities but they usually can be set to detect and measure only the atom species in which one is interested so that the above-mentioned possible errors are avoided.

Another distinction between the various methods arises from the range of bombarding-ion energies at which the measurements are made. At high ion energies one can bombard the target with ion beams under any desired angle of incidence. With differential pumping one can keep the target chamber at a low gas pressure and use a mass spectrometer for analysis of the sputtered species and for velocity selection. This scheme has been most successful in the detection and measurement of reflected or of sputtered ions, especially those which result from biparticle collisions.[22,23] In the case of neutral sputtered atoms one encounters difficulties with post-ionization especially of the fast neutrals.[65] Other schemes such as a time-of-flight technique whereby the ion beam is pulsed to cause short bursts of sputtered atoms to be deposited on a fast-spinning magnetically suspended rotor (3,000 rps) have been more successful in the case of neutrals.[66-69]

At lower ion energies (<1 keV) the achievable ion-beam current density becomes very small because of space-charge limitations and schemes where one immerses the target like a large negative Langmuir probe in a low-pressure plasma (millitorr range) have been more successful. The angle of incidence is in this case confined to the direction normal to the target surface.

Mayer[70] and later Sporn[71] got the first indication of the high ejection energies of sputtered atoms by studying the thin luminous layer which covers an ion-bombarded target when it is coated with a thin oxide layer or a layer of an alkaline metal. This layer arises from the de-excitation of atoms which are ejected in an excited state. From the layer thickness and the known lifetime of an excited state ($\sim 10^{-8}$ s) they were able to obtain a rough estimate of the ejection velocity. Sporn arrived at energies of 5 to 7 eV, and Mayer presented his results in terms of a surface temperature of about 30,000°K.

The various methods based on measurements of forces either at the target[64] or at a collector,[72,73] or the calorimetric methods where the temperature rise of the substrate due to the impact of sputtered atoms is measured[74] are likely to contain the above-mentioned systematic errors and will not be discussed here in further detail.

An interesting method which allows the determination of the velocity distribution of sputtered atoms in a plasma was developed by Oechsner and Reichert.[75] In a high-frequency-excited low-pressure plasma (27 MHz, 10^{-3} to 10^{-4} Torr) one can achieve very high electron temperatures (7 to 12×10^{5}°K). It is therefore possible to ionize a substantial part of the atoms which are sputtered from the target and enter into this plasma. By incorporating a retarding-field detector, consisting simply of a collector screened against the discharge by a highly transparent metallic grid, to which a retarding potential can be applied with respect to the collector, one can analyze the collector current with and without sputtered atoms in the plasma. In the retardation curves one can then readily separate the current which arises from the slow plasma gas ions from the one of the superimposed faster sputtered ions. In this experimental method one determines only the energy distribution of sputtered neutral particles, because positive ions ejected from the target are stopped in the Langmuir sheath between target and plasma while negative sputtered ions are accelerated in the sheath to such high velocities that they are outside the measured range. Results obtained with this method are in excellent agreement with results obtained in the spectroscopic time-of-flight method discussed below.

The method used by Stuart and Wehner[76,77] is a time-of-flight technique based on emission spectroscopy. The target immersed in a low-pressure high-density plasma is pulsed to a fixed negative voltage for 1 μs so that atoms are sputtered from the target as a group. These atoms are ejected mostly as neutral unexcited atoms, but they undergo excitational collisions with the plasma electrons and emit their characteristic spectra. Those atoms traveling in a particular direction become spatially dispersed as a consequence of the distribution of velocities. This dispersion is observed as a time distribution of photons emitted by the sputtered atoms as they pass through a small observation volume a known distance away (6 cm) from the target. The time distribution can be readily converted to a velocity or energy distribution of sputtered atoms.   A confirmation of data obtained with this time-of-flight method was possible with another spectroscopic technique, namely, the observation of the Doppler shift of spectral lines from excited sputtered atoms when they travel toward the spectrograph. The distribution of velocities ranging from zero to about 10^6 cm s^{-1} results in both a shift and a broadening of the spectral line extending from 0 to 0.1 Å toward shorter wavelength.

b. Results

A typical set of energy-distribution curves for various bombarding-ion energies is shown for an Ag(110) surface under Hg-ion bombardment in Fig. 8.   Figure 9 shows a comparison of energy distributions for four materials from the same period of the

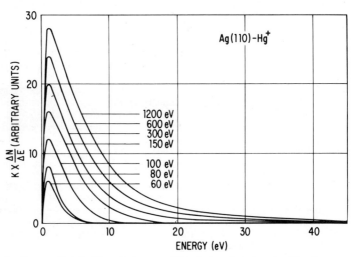

Fig. 8   Energy distribution of atoms sputtered at various bombarding-ion energies.

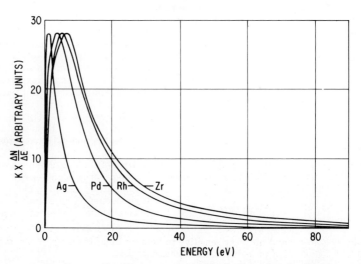

Fig. 9   Energy distributions under 1,200-eV Kr-ion bombardment.

periodic system (Kr⁺ bombardment a⁺ ¹ 200 eV). It is of interest to note that Rh, Pd, and Ag, which are adjacent elements and have nearly identical atomic weights and the same crystal structure, have markedly different energy distributions just as the sputtering yields do. The average ejection energy of various polycrystalline materials bombarded by 1,200-eV Kr⁺ ions is plotted in Fig. 10 as a function of the atomic number of the target material. One notices clearly two features: higher

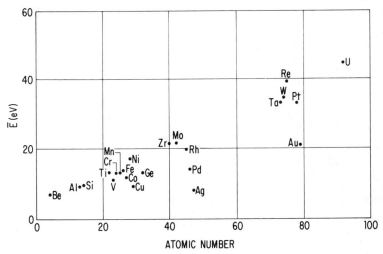

Fig. 10   Average ejection energies under 1,200-eV Kr-ion bombardment.

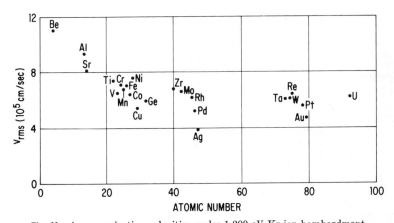

Fig. 11   Average ejection velocities under 1,200-eV Kr-ion bombardment.

ejection energies for heavier target atoms and within each period lower ejection energies for higher-sputtering-yield materials. Figure 11 shows a plot of the average (rms) ejection velocity for the same polycrystalline materials as in Fig. 10. A striking feature is the fact that the ejected atoms from most metals ($Z > 20$) have average ejection velocities which are not much different and lie between 4 and 8×10^5 cm s⁻¹. For Ar⁺, the most commonly used sputtering gas, these values are about 20% lower and not very sensitive to bombarding-ion energy in the range from 600 to 1,200 eV (and probably much higher energies).

We can summarize the experimental results as follows:

1. More oblique ejection results in higher ejection energies, as predictable from momentum and energy-conservation laws.

2. With increasing bombarding-ion energy the average ejection energies approach a constant value at about 1 keV, while sputtering yields at these ion energies still increase nearly proportionately with ion energy. Only the high-energy tail which must result from the most favorable near-surface collisions becomes more pronounced.

3. Ejection energies decrease with lighter bombarding ions just as the sputtering yields do. This is probably related to the deeper ion penetration and consequently higher losses.

4. Neither the crystal orientation nor the crystal structure plays an important role in ejection energies. This result seems to indicate that focusing collisions are not important in sputtering.

5. Higher-sputtering-yield materials tend to have lower average ejection energies.

6. Heavy target atoms have the highest ejection energies and light target atoms the highest ejection velocities.

Results for sputtered ions are very similar to those of sputtered neutral atoms, except that the most probable ejection energy is shifted toward higher energies. Veksler[78] suggested that ions may have a higher chance to escape without being neutralized by field-emitted electrons when they are ejected with higher velocities.

At much higher bombarding energies the average ejection energies of sputtered atoms do not seem to be much different. The unusually high ejection energies found by Kopitzki and Stier[73] could not be confirmed by others such as Ben'yaminovich and Veksler.[79] Thompson found indications that a small fraction of the sputtered atoms seem to have velocities of the order of those anticipated for thermal evaporation. They explained this effect as due to the formation of thermal spikes which begin to contribute to sputtering at higher ion energies (40 keV).

Benninghoven[65] found that the velocities of Al ions sputtered from an Al alloy are considerably lower than those of ions sputtered from pure Al.

The high ejection energy of sputtered atoms has been used for creating atomic beams with much higher than thermal energy.[80]

5. ANGULAR DISTRIBUTION OF SPUTTERED MATERIAL

a. Polycrystalline Targets

The angular distribution of sputtered particles from targets is of theoretical importance since it helps establish the mechanism of sputtering; it is also of practical importance since its knowledge helps one to predict the expected film-thickness distribution from a given target.

The first detailed study on the angular distribution of sputtered material was reported in 1935 by Seeliger and Sommermeyer.[81] They bombarded Ag with 10-keV Ar ions and found that Knudsen's cosine law satisfactorily described the ejected atom distribution. Furthermore, this distribution was observed to be independent of the angle of incidence. As we shall note in a later section, this work had for a long time been erroneously considered to constitute a proof that sputtering is actually an evaporation process.

It was, in fact, some 25 years later that the ejected particle distributions from polycrystalline material were again investigated seriously. Wehner and Rosenberg[82] studied the angular distribution of material sputtered from polycrystalline targets under 100 to 1,000 eV normally incident Hg-ion bombardment. At higher energies the distribution approached the cosine law, while at lower energies less material was ejected normal to the surface than would be expected for the cosine law. This deviation from the cosine law appeared to be more pronounced for Mo and Fe than for Ni or Pt. Furthermore, under oblique incidence, material was noted to be ejected much more strongly in the forward direction (away from the surface normal). Koedam,[83] who studied the sputtering of polycrystalline Ag with 100- to 250-eV Ne⁺,

Ar$^+$, and Kr$^+$ ions, found that the angular distribution followed a cosine law.   On the other hand, Rol et al.[84] found that the distribution of material sputtered from polycrystalline Cu by 20-keV Ar$^+$ ions followed a Gaussian distribution more closely than a cosine distribution.   They also noted that this distribution was essentially independent of the angle of incidence.   Grønlund et al.[85] bombarded Ag with 4-keV Ne$^+$ ions at an angle incidence of 60° and observed a cosine distribution of ejected particles. For 9-keV D$^+$ ion bombardment at 60° they also observed essentially a cosine distribution except that more material seemed to be ejected in the forward direction. For their studies they exercised special care to obtain highly polished target surfaces. They suggested this was necessary since they expected that scattering from a rough surface would always yield a cosine distribution.   The effects of temperature, ion dose, and ion mass on ejected atom distributions were studied by Cobic and Perovic,[86] who bombarded Cu and Pb with 17-keV Ar$^+$, Kr$^+$, and Xe$^+$ ions.   They reported that they had observed a temperature dependence of the angular distribution of ejected material.   At higher temperatures ($\sim$100°C) they observed a cosine distribution. They then cooled targets with water and liquid nitrogen and found that at lower temperatures more material was ejected nearer the surface normal than would be expected for the cosine law.   However, even at their lowest temperature (~ -160°C) they noted that with greater bombardment doses the distribution would again approach the cosine law.   This could be related to changes in surface topography.

From a theoretical point of view, these studies on polycrystalline material are of much less interest than studies on single-crystal material, which are discussed in the next section.   From a practical point of view, one will generally not go too far wrong if one assumes a cosine distribution for ejected material.   Variations in resulting distributions will generally depend strongly on the targets selected.   For instance, sheet or foil material is usually rolled and has a preferred orientation.   This can be a principal factor in determining the angular distribution of ejected material.   Further, the surface topography, which may vary during a target lifetime, can be an important factor affecting observed results.

b. Single Crystals

The spatial distribution of atoms ejected by the sputtering of single crystals has been very important for developing a theoretical understanding of the mechanism of sputtering.

Early theories of sputtering could be divided into two general types: a momentum-transfer mechanism and a hot-spot evaporation mechanism.   For about 20 years the cosine distribution of sputtered material observed by Seeliger and Sommermeyer[81] was commonly accepted as evidence for the evaporation theory of sputtering.   Then in 1954, Wehner[16] studied the angular distribution of ejected material for low-energy bombardment.   For obliquely incident ions he noted that material was ejected preferentially in a forward direction, away from the surface normal.   This result is not compatible with the evaporation process.   Shortly after this work, Wehner[18] made another discovery which firmly established that low-energy sputtering was in fact the result of momentum transfer.   He sputtered single-crystal W balls with 150-eV Hg$^+$ ions and observed that material was ejected preferentially in certain crystallographic directions.   Such preferred directions have not been seen for sublimation from a single crystal.[87]

A large number of papers have been published subsequently describing atom-ejection patterns observed for various crystal structures while varying such parameters as bombarding-ion mass and energy, angle of incidence, and temperature. Since this book's primary concern is thin films and not sputtering, we will just summarize the important results of the studies and will not discuss them in detail.   In general the most preferred ejection directions are the nearest-neighbor directions and next the second-nearest-neighbor directions.   This point will be discussed in more detail in Sec. 8.

Studies made of atom-ejection patterns for fcc metals demonstrate that for this structure the most preferred atom-ejection direction is the $\langle 110 \rangle$ direction.[83,88-92] Preferred ejection is also observed in the $\langle 100 \rangle$ direction, although generally to a lesser extent.   Under very favorable conditions, preferred ejection has also been observed in the ⟨111⟩ direction.

For bcc metals[93-95] the closest-packed ⟨111⟩ direction is the predominant preferred ejection direction.   At higher energies preferred ejection is also observed in the ⟨100⟩ direction.   At 50-keV Ar-ion bombardment of W and Mo, Nelson also observed preferred ejection in the ⟨110⟩ direction.

A few studies have also been made for the hcp metals Zn,[96] Ti, and Re[97] for low-energy ion bombardment.   In this case the most pronounced preferred atom-ejection direction is the ⟨11$\bar{2}$0⟩ direction, which is the nearest-neighbor direction lying in the basal plane.   Preferred atom ejection is also observed in the ⟨20$\bar{2}$3⟩ direction, which is the direction of the line of centers of nearest-neighbor atoms on adjacent basal planes.

Figure 12 shows an example of some spot patterns obtained for the sputtering of the (111), (100), and (110) planes of fcc crystals.   All spots correspond to ⟨110⟩ directions, as indicated by the rods on the crystal models.

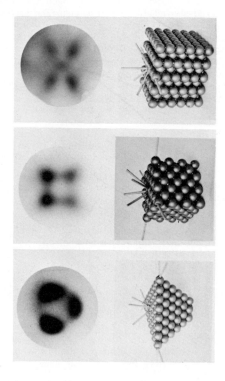

Fig. 12   Some spot patterns obtained for the sputtering of the (111), (100), and (110) planes of fcc crystals.

Atom-ejection deposited patterns have also been observed for semiconductors, although the anisotropy is less pronounced than it is for metals.   The preferred ejection directions are first the ⟨111⟩ and second the ⟨100⟩.[88,89]   The amount of material sputtered in a preferred ejection direction and the orientation of this direction depend sensitively on the angle of incidence.   This suggests that the observed preferential ejection directions result from near-surface collision sequences rather than focusing chains.   Preferred ejection in the ⟨100⟩ direction does not seem possible for an unaltered structure. Anderson and Wehner[88] suggested this might be expected if any of the natural interstitial sites were occupied.

A most interesting aspect of the atom-ejection patterns for semiconductors is the pronounced temperature dependence. It was discovered by Anderson et al.[98] that preferred ejection directions appeared only if the target temperature was above a characteristic temperature. This is evidently the reason other investigators had difficulty in observing preferred ejection directions for semiconductors.[83,99] The explanation for this result is as follows: Anisotropy of ejected material in preferred ejection directions requires a single-crystal target. At low target temperatures the lattice damage resulting from the ion bombardment remains frozen in and results in an amorphous surface. Therefore, no preferred ejection directions are observed. For sufficiently high target temperatures, the damage produced by an impinging ion anneals out sufficiently rapidly so a subsequent impinging ion also strikes an ordered surface. The characteristic annealing temperature depends on such parameters as kind of target, mass of impinging ion, energy of impinging ion, and rate of bombardment.[100,101]

The anisotropy of atom ejection from single crystals has not found much practical application. Cunningham used the deposit patterns for orienting single crystals.[102]

Another application which could be made would be to orient a single-crystal target so that a highly preferred atom-ejection direction pointed toward the substrate. This might allow one to increase the deposition rate to some extent.

At present the real importance of atom-ejection patterns from single crystals is in their contribution to a theoretical understanding of sputtering.

6. ANALYSIS OF EJECTED MATERIALS

a. State of Ejected Particles

One of the early mass analyses of material ejected from the targets by sputtering was reported by Woodyard and Cooper.[103] They bombarded polycrystalline Cu with low-energy (0 to 100 eV) Ar ions and studied the neutral particles which were ejected. They detected Cu atoms and Cu_2 molecules. The ratio of Cu_2 to Cu increased with bombarding-ion energies to about 5% for ion energies of 100 eV. No Cu_3 molecules were detected. Woodyard later reported on the ejection of Cu from a Cu(100) target bombarded by 0.5- to 8-keV Ar ions.[104] He studied both the sputtered neutral particles and the sputtered ions. In the case of the ejected neutrals for 2-keV Ar-ion bombardment, he detected Cu and Cu_2 molecules. No Cu_3 was detected. In the case of ejected ions, which constitute only about 1% of the ejected species, the situation is more complicated. Here he observed Cu_N^+ ions with N in the range of 1 to 11 and noted that the peak heights of the ions alternated. For N odd, the Cu_N^+ peak height was always greater than the Cu_{N-1}. The ratios of the peak heights for 3.5-keV Ar-ion bombardment were given for Cu up to Cu_7. The ratios were constant down to 2.5-keV bombarding-ion energy. Below 2.5 keV, the ratios for N greater than 3 decreased until at 500 eV only Cu^+, Cu_2^+, and Cu_3^+ were detected. These were in the same ratios as observed at 3.5 keV.

Herzog et al.[105] reported on the ejection of ions under 12-keV ion bombardment. In the case of Ar-ion bombardment of Al, molecules up to Al_7 were detected. Al_8 was barely detected. The situation is surprisingly different in the case of Xe-ion bombardment where clusters of Al atoms up to Al_{18} were observed. On the basis of the energy distribution of the clusters, these authors felt quite confident that the material was sputtered away as clusters and that these clusters were not formed in the vapor phase. Herzog did not notice any change in the relative intensities of the Al molecules under Xe bombardment for ion energies from 12 to 4 keV.

Somewhat similar results were reported by Jurela and Perovic.[106] They studied the ejection of positive ions emitted from polycrystalline Al and Co targets bombarded by 40-keV Ar ions. In the case of Co, for instance, they observed Co^+, Co_2^+, Co_3^+, and Co^{2+}. At present it is not clear what mechanism is responsible for the ejection of these clusters of atoms.

Comas and Cooper[107] studied material ejected from GaAs targets under low-energy 0- to 140-eV Ar-ion bombardment. Approximately 99% of the collected particles were neutral Ga and As atomic species. The balance were neutral GaAs molecules. No neutral Ga_2, As_2, or $(GaAs)_2$ molecules or negative Ga^-, As^-, or $(GaAs)^-$ ions were detected.

Hardly any work seems to have been reported for compounds such as oxides. We have looked at the emission spectra in a glow discharge of particles sputtered from oxides such as $BaTiO_3$ and have observed pronounced line-spectra characteristics of atomic-metal atoms. It seems that the major part of the material is sputtered in atomic rather than molecular form.

b. Composition of Sputtered Particles

The sputtering of alloys is one of the more suitable techniques for the preparation of alloy films because it can provide good composition control. The composition of the deposited film can be almost identical with the composition of the target. This can be intuitively seen by considering the mechanism of sputtering wherein target atoms are dislodged from the target by momentum transfer from the impinging ion. Since this process is generally not strongly dependent on temperature, the target temperature can be maintained sufficiently low to prevent diffusion within the target with essentially no sacrifice in deposition rate. Since there is no mass transport within the target, one can easily see that the composition of the sputtered constituents must be identical with the composition of the target. Of course, the composition of the deposit can still vary if the sticking coefficient or the angular distribution is different for each ejected species.

Flur and Riseman[108] found that films sputtered from a Permalloy (81 Ni-19 Fe) target had the same composition as the target. Francombe and Noreika[109] reported 79.7% Ni in films sputtered from an 81 Ni-19 Fe target. It was therefore necessary to use a target slightly deficient in Fe to produce a film of the desired composition. Khan and Francombe[110] found that the sputtering of an 85 Au-15 Ni target gave films with a lattice constant essentially the same as that of the bulk, which indicated that the composition of the film was the same as the composition of the target. Other workers have found rather substantial compositional differences between the deposited films and the target. In the sputtering of 79.88 Ni-20.19 Cr, for example, Pratt[111] reported Ni film content to be in the range 82 to 90%. Michalak[112] gave some results on the sputtering of CrCo and NiCr alloys. A 75 Cr-25 Co target gave a film containing 69% Cr while an 85 Ni-15 Cr film was deposited from an 80 Ni-20 Cr target. More recently, Patterson and Shirn[113] reported some work on the sputtering of NiCr alloys. They sputtered alloys containing 22, 42, and 80% Cr. Ion energies varied from 200 to 1,000 eV. Results showed that the target composition was preserved in a sputtered film to ±1% for each component. Near the surface of the sputtered 78 Ni-22 Cr target, a relative Cr enrichment was found.

Recently G. S. Anderson has performed some studies on the sputtering of alloys where the composition of the sputtered constituents in flight was determined by the spectroscopic-emission technique. This has an important advantage in that the species being sputtered from the target are analyzed directly and factors such as the sticking coefficients do not enter. Furthermore, one can determine the instantaneous sputtering-constituent ratio. This is also, of course, a very convenient method for studying the effect of temperature on the composition of the sputtered material. This method has been applied for the study of the temperature dependence of the Ag/Cu sputtering ratio for the Ag/Cu eutectic.[114] The targets were bombarded by 100-eV Ar ions for target temperatures in the range of 80 to 285°C. At low target temperatures, the Ag and Cu sputtering yields varied according to a simple model. At higher temperatures, Ag/Cu sputtering ratios differing by a factor greater than 10 were noted. Steady-state conditions were not obtained for sputter removal of several microns from the targets. These experiments demonstrated that good composition control can be obtained by sputtering provided that target temperatures are maintained sufficiently low.

Another technique for obtaining alloy films is the simultaneous sputtering of multiple targets. This has an added advantage in that a wide range of compositions can be studied with the same target arrangement by varying the sputtering rates of the targets independently. One should not overlook, however, that very unexpected and detrimental effects can occur.[115] For example, if a low- and a high-sputtering-yield target material are sputtered simultaneously, the low-sputtering-yield material will stay clean whereas the high-sputtering-yield target will tend to become coated with the low-sputtering-yield material.

7. MODIFICATIONS OF THE SPUTTERED SURFACE

a. Metals

The etching of metal surfaces by ion bombardment is a process which has been quite extensively studied.[116] As an etching process, sputtering has certain advantages such as being applicable to all materials as a universal etchant and being well controllable, since the important parameters such as ion energy, bombardment rate, and target temperature can be easily controlled.

A common effect of the sputter etching of polycrystalline material is the development of grain boundaries, as noted by Haymann.[117,118] Another feature which is usually observed on the surface is the etch hillock.[119] These results plus the fact that different crystallographic planes sputter at different rates have been utilized by Haymann for metallurgical studies of metal surfaces.[120]

Ion bombardment can also affect the orientation of surfaces. For instance, Ogilvie[121] observed that after ion bombardment Ag crystals develop disoriented crystallites in their surface which tend to have preferred orientations.

b. Semiconductors

A more dramatic structural change is observed for semiconductors, where ion bombardment results in an amorphous surface layer.[122] This structure change is readily observable optically.[123] The crystalline-to-amorphous transition under ion bombardment occurs only if the target temperature is so low that the damage does not anneal out as rapidly as it is being produced. Therefore, this critical temperature will depend on the target material, the kind of bombarding ion, the bombarding-ion energy, and the bombardment flux.[100,124] The bombardment of Ge crystals with Hg ions of energy 100 eV results in preferentially etched pits which reveal the crystal orientation.[125] These pits were more abundant at lower temperatures. At higher temperatures and energies, these pits developed a truncated appearance. The location of the pits did not seem to correspond to defects which can be revealed by chemical etching, which suggests they may be connected with dislocations created by the bombardment.

c. Alloys and Multitargets

Very few studies have been performed where the target itself was analyzed after sputtering. Asada and Quasebarth[126] performed some early work on sputtered alloys of Cu containing a small amount of Au. They found that the Au was preferentially removed, leaving the surface Cu-enriched. Later Gillam[127] sputtered a Cu_3Au and found that initially more Cu was removed than would correspond to the 3:1 Cu:Au ratio, leaving behind a layer richer in Au than in Cu_3Au. This altered layer then progressed into the target with further sputtering, as to be expected at sufficiently low target temperatures.

In the case of multitargets, unusual and unexpected results are often observed, especially on the high-sputtering-yield target. The nature of the surface of the high-sputtering-yield material will often vary with the rate of arrival of low-sputtering-yield material—from a wartlike appearance for high arrival rates to a surface containing widely separated sharp cones for low arrival rates. These target effects can result in large sputtering yield changes and pronounced alterations in the spatial distribution of ejected material. It should be emphasized that the arrival rates which can cause such effects are so low that it is not even necessary for the target surfaces to "see" each other in a typical sputtering system.   Gas scattering of ejected atoms, even at pressures of a few microns, can cause these effects.

d. Compounds

The sputtering of compounds such as oxides is certainly more complex than the sputtering of simple elemental targets.   The transfer of momenta via several collisions from the impinging ion to the ejected atom is a violent process.   The energies involved can certainly exceed chemical bonding energies, which are typically a few eV, and so would be expected to rupture many bonds.   In the case of an oxide, for instance, one might expect that the breaking of a bond would result in the oxygen preferentially leaving the target.   This leaves the target surface deficient in oxygen; that is, the oxide surface will have been reduced.   Studies of Hg-ion bombardment of oxide powders have demonstrated, for instance, that CuO is converted to Cu_2O and finally to pure Cu while Fe_2O_3 gets reduced to Fe_3O_4, then to FeO, and finally to pure Fe.[128]   This reduction is even more pronounced, as to be expected, if a reducing gas such as hydrogen is used for the bombardment.[129]   The resulting amount of reduction might be expected to depend on the bond strength, or dissociation energy. For lower energies there appears to be indeed greater reduction.[39]   The target darkening is greater for a rougher surface because the oxygen loss is enhanced when atoms are sputtered back and forth in crevices.   Furthermore, if the sputtering is performed in an oxidizing atmosphere, e.g., one containing a partial pressure of oxygen, less reduction (hence less darkening) is observed.

8. THEORETICAL MODELS

An early interpretation of sputtering was that localized heating in the vicinity of an ion impact was sufficient to cause target material to be released through evaporation.[8,130]   Such a theory would predict that the ejection of material could be described by Knudsen's cosine law.   This appeared to be substantiated by Seeliger and Sommermeyer, who observed this distribution for the ejected material and also found it to be independent of the angle of incidence.[81]   Later, however, Wehner[16] noted that at lower bombarding-ion energies the distribution of ejected material did depend on the angle of incidence.   This indicated that sputtering was actually a momentum-transfer process, as was first suggested by Stark[14] and later reemphasized by Lamar and Compton.[131]   Later work on atom-ejection patterns from single crystals and the energy distributions of ejected atoms, as described in earlier sections of this chapter, clearly demonstrated that sputtering is indeed a momentum-transfer process.   However, the formulation of this concept in an adequate theory which would explain these experimental results as well as many others such as the dependence of the sputtering yield on the kind of target, mass and energy of the bombarding ion, and direction of ion impact relative to crystallographic directions is a highly complicated problem. The earlier theories in particular employed many concepts taken from radiation-damage theory to explain the dependence of sputtering yield on the masses of particles involved and on the energy of the impinging ion.

The energy of the incident particle determines the type of interaction with a lattice atom.   At very high energies the effects of electron screening can be ignored and the interaction can be treated as a repulsion between nuclear charges.   These interactions are Rutherford-collision events.   At lower energies the nuclear charges are partially screened by the electron clouds and the collisions can be considered as partially screened Coulomb collisions.   At low energies there is very little penetration of the electron clouds; so the collisions were considered as hard-sphere collisions.   It should be realized that while such a model can be useful for visualizing certain events, this is a drastic assumption which is very far from physical reality.   Essentially no collisions are very near hard-sphere.   Furthermore, when the velocity of an impinging ion is greater than the approximate velocity of electrons at the Fermi level, the energy loss due to electron excitations becomes dominant.[132]

One of the earliest theories of sputtering employing radiation-damage theory was described by Keywell,[133] who proposed a modified neutron-cooling theory.   He assumed hard-sphere collisions and estimated that the energy of the incident ion decays exponentially with the number of collisions. The number of displacements produced at each collision is assumed to be related to the energy transferred; the probability of a displaced atom's escaping as a sputtered particle is related to the depth below the surface at which it originated. The result was an expression indicating that the sputtering yield is proportional to $E^{\frac{1}{2}}$, where E is the kinetic energy of the incident particle. It can be seen from the section on sputtering yields that this relation is approximately correct in some cases and certain energy ranges. This theory ignores energy losses due to electronic excitations and assumes a random solid, so that crystallographic effects obviously could not be explained.

An even simpler model was postulated by Rol et al.[134] They assumed that only the first collision made by an impinging ion contributes to sputtering. They postulated that subsequent collisions occur too deep within the target to lead to sputtering events. The sputtering yield was assumed to be proportional to the energy transferred in the first collision and to vary inversely with a quantity which they called the mean free path. This was actually a parameter introduced so as to include the energy dependence of the average distance that an ion would travel before making its first collision and resulted in a sputtering yield which varied slightly less than linearly with impinging ion energy—which generally agrees with yield data in this range. This theory also gave a qualitative interpretation of the dependence of the sputtering yield of single crystals on the direction of impingement relative to crystallographic axes.[135-138] The sputtering yield for keV ion bombardment exhibits minima for those crystallographic directions where the lattice is quite open, that is, for those directions in which the impinging ion may become channeled and thereby experience extraordinarily long ranges. The yield exhibits maxima for those directions where the crystal appears more opaque. Thus when the range of the impinging ion is great, the sputtering yield will be less since there will be fewer collisions near the surface which can lead to sputtering events. This range dependence of impinging ions on crystallographic direction has been directly demonstrated.[139-141] The inverse relationship between ion range and sputtering yield was pointed out quite early by Almen and Bruce.[21] Although this model has been quite useful for estimating relative yields vs. crystallographic orientation, it has been criticized by some as being insufficient.[142]

These theories were attempts to explain how the sputtering yield varied with the energy of the impinging ion and with the direction of impingement, but could not explain the atom-ejection patterns resulting from the sputtering of single crystals, as described in an earlier section of this chapter.

In considering correlated collision events in single crystals, Silsbee pointed out that momentum would be focused along a row of equally spaced hard spheres if the distance between centers is less than twice the atomic diameter.[143] Now consider a fcc crystal. The closest-packed row of atoms is the ⟨110⟩ direction. At sufficiently low interaction energies (which means that the effective atomic diameter is large), Silsbee focusing is possible in this direction. As an example of the maximum focusing energies involved, calculations for Cu by Thompson[144] indicated 60 eV and by Gibson et al.[27] 30 eV. Since the ⟨110⟩ direction is the most favorable ejection direction for fcc crystals (see earlier section), it would appear logical to interpret the observed preferential ejection in terms of focused-collision sequences. This viewpoint appears to be strengthened by the fact that momentum focusing is also possible in the fcc ⟨100⟩ and ⟨111⟩ directions, which are also directions in which preferred atom ejection is observed. In these directions, however, the atom spacing is too great for Silsbee focusing. The computer calculations of Gibson et al.[27] demonstrated that an atom moving in a direction near the ⟨100⟩ direction experiences a deflection in the ⟨100⟩ direction when it passes through a ring of four atoms in the (100) plane. This can best be seen by examining a crystal model. Thus in this direction there is a lens-type momentum focusing. A similar focusing is possible in the ⟨111⟩ direction. The lens-type focusing is generally less efficient than Silsbee focusing, and so the observation that higher bombarding-ion energies are required for appearance of sputtered-atom ejection in these directions seems very reasonable. Similar arguments can be made for the bcc structure. Because of such results, many physicists have concluded that focusing must be included in any sputtering theory. Such a theory was proposed by Sanders and Thompson,[145] for example. The evidence for the importance of these focusing events seemed so strong that Southern et al.[99] remarked: "The ejection patterns leave little doubt that focusing collision chains are primarily responsible for the transport of momentum to the surfaces of close-packed metals." Papers were published which were concerned with some of the details of this mechanism, such as defocusing resulting from an increased atomic-vibration amplitude at higher temperatures,[146] the effect of defects on the collision events,[147] and attempts at measurements of the defocusing energy.[148]

At the same time several experiments have indicated that focusing chains may *not* be the only mechanism resulting in preferred atom ejection in certain directions. For instance, detailed atom-ejection studies on Ge led Anderson to suggest that focusing chains do not seem to play as important a role for semiconductors as for metals.[89] Studies on the temperature dependence of the sputtering yields of the low-index planes of Ge showed that the yield for a (111) surface is greater than that of the (110) surface, less than that of the (100) surface, and similar to the yield for an amorphous surface. This demonstrated that near-surface collision sequences dominate in this case.[149,150] Furthermore, studies on hcp metals showed that preferred atom ejection occurs in $\langle 20\bar{2}3 \rangle$ directions (the directions of the line of centers of nearest neighbors on adjacent basal planes in which linear chains can be only two atoms long) as well as in the $\langle 11\bar{2}0 \rangle$ directions (nearest-neighbor directions lying in the basal plane which are continuous rows of close-packed atoms).[96,97,151]

Harrison et al.[28,152] set up a computer program to simulate sputtering and concluded that spot patterns can be explained without the use of focusing chains. According to their model, preferred ejection in certain directions occurs as the result of glancing collisions initiated by the incident ion in the first few atomic layers of an ordered crystal. They found that if an ion does not make a hard collision within a first lattice-repeat distance of the surface, it will become channeled and penetrate too deeply into the target to cause sputtering. Spot patterns were found to be insensitive to the interatomic potential, whereas the calculated yields were sensitive. However, this conclusion is probably not surprising here since bombardment was considered only in channeling directions. Furthermore, the correlation between their calculated sputtering yields and experimental yield values is not particularly good. Their calculations have recently been seriously criticized by Robinson[153] as being unrealistic because of the choice of a much too hard potential function in the Ar-Cu collision.

Lehmann and Sigmund[154] also showed theoretically that the spot patterns in single-crystal sputtering are not necessarily the result of focusing chains. They conclude that spots arise as a consequence of a regular surface structure. The sputtered intensity has maxima in those crystal directions where ejection requires a minimum of energy. They give several experimental observations which seem to be in agreement with their conclusions that most focusing events are caused by "focusons" of range of only a couple of lattice distances. Among these are the following: For hcp substrates, the sputtering yield is similar for the (0001) and (10$\bar{1}$0) surfaces while focusons can contribute only for the (10$\bar{1}$0) surface; the sputtering yield of bcc and fcc metals depends much less on temperature than does the range of focusons; the fact that the sputtering yield seems to vary inversely as the heat of sublimation (which is closely related to the surface binding energy) suggests that near-surface events play a stronger role in sputtering than long-range focusons. Although this model appears to be the most satisfactory approach at present, it should be pointed out that agreement is by no means unanimous. For instance, von Jan and Nelson have criticized the Lehmann and Sigmund approach and claim to show that focusing collisions do in fact play an important role in sputtering, especially for higher energies.[155]

The concept of sputtering efficiency γ, defined as the fraction of bombarding-ion energy leaving the target via sputtering and backscattering, was first introduced by Sigmund.[156] His calculation surprisingly indicated that $\gamma = 0.024$ for self-sputtering, independent of ion energy and target material, for the elastic-collision range above 1 keV. Andersen[157] verified this for lead self-sputtering.

These theoretical studies have recently been expanded by Sigmund.[158] He uses methods of transport theory and considers a random target with a plane surface. As we have already noted, the evidence is mounting that focused-collision sequences are important only for secondary effects and can be ignored in the first approximation.[155] The lack of a significant temperature dependence of the sputtering yield and the relatively small contribution of "spot patterns" to the sputtering yield of single crystals are cited as reasons.[159-161] Gurmin et al.[162] added more evidence for the unimportance of focusons for sputtering when they found that the sputtering yield is not much different for Zn from that for ZrC, even at energies up to 17 keV. A Boltzmann-type integrodifferential equation is employed using the power approximation of the Thomas-Fermi cross section and a planar-potential barrier. He obtains the following expression for the sputtering yield from a planar target:

$$S = \frac{\alpha S_n(E)}{16\pi^3 a^2 U_0}$$

where α is a function of (M_2/M_1), $S_n(E)$ is the nuclear stopping cross section of the ion, $a = 0.219$ Å = the screening radius proposed by Andersen and Sigmund[163] for the Born-Mayer interaction between two atoms, and U_0 is the surface-barrier energy, which is taken as the heat of sublimation. The form of $S_n(E)$ depends on the ion energy. For lower ion energies, where the Born-Mayer potential is more appropriate than the Thomas-Fermi potential, Sigmund approximates the B-M potential to obtain

$$S_n(E) = 12\pi a^2 \lambda E$$

where $\lambda = 4M_1 M_2/(M_1 + M_2)^2$. Thus he obtains for the sputtering yield

$$S = \frac{3\alpha \lambda E}{4\pi^2 U_0}$$

Thus the sputtering yield for low energies is proportional to the bombarding-ion energy and inversely proportional to the heat of sublimation. At higher energies he uses the stopping cross section for the Thomas-Fermi potential as derived by Lindhard et al.[164] The equation he obtains for the sputtering yield is

$$S(E) = 0.042 \frac{\alpha S_n(E)}{U_0}$$

for perpendicular incidence. Values for α are given in graphical form and $S_n(E)$ in tabular form. The theoretical sputtering yields for this range (keV heavy and medium mass ions) can thus be rather quickly determined.

The battle over theoretical interpretation of sputtering will probably continue for some time since the problem is unquestionably highly complex. Furthermore, there is a considerable quantity of data covering various aspects of sputtering, which must be explained by a complete model. The tendency has been to select only those data which support a theory and ignore other information. In defense of the theorists it should be emphasized, however, that a certain amount of data selectivity is justified and necessary since many incorrect experimental data have been reported. In the case of sputtering, as with other complex problems, it may well be that computers will play an increasing role as an aid for certain calculations, although the best current theory, that of P. Sigmund,[158] does not require a computer, and sputtering yields over a wide range of energies of bombarding ions can be calculated with the aid of a graph and a table.

9. MISCELLANEOUS

Sometimes the question is raised as to who introduced the word "sputtering." A literature search shows that Sir J. J. Thomson[170] used the word "splittering" but that I. Langmuir and K. H. Kingdon eliminated the "l" in their publications in the years

1920 to 1923. In 1923, the Research Staff of the General Electric Co., London,[171] still used only the term "cathode disintegration." Kay,[172] following a suggestion by Guentherschulze,[173] tried without much success to introduce the term "impact evaporation."

Keller[174] searched for atoms sputtered by fast neutrons, a topic which is of practical interest in thermonuclear plasmas. For Mo bombarded with 14-MeV neutrons the sputtering yield was found to be less than 10^{-4} atom/neutron, which is not in serious disagreement with the theoretical estimate of 10^{-6} atom/neutron.

Transmission sputtering of thin films by low-energy hydrogen ions was studied by Robinson.[175] The motivation for this work was to investigate the possibility of injecting charge-exchanged neutral hydrogen or deuterium atoms through a thin foil into a thermonuclear plasma and to find out how detrimental sputtering would be for the life of such a foil. Results showed that a 500-Å-thick Au foil loses about 10^{-2} atom per H^+ ion or 2×10^{-2} atom per D^+ ion at the far side. These values are of the same order as those for front-side sputtering, where it was found that Cu sputters at a rate of 2×10^{-2} atom per 15-keV H^+ ion.[54]

A 340-page book by Pleshivtsev on "Cathode Sputtering," with 742 references, appeared in Russian.[176]

Although it is not a physical sputtering process, two recent papers on the disintegration of alkali halides under electron bombardment[177,178] are of interest. One may encounter surprisingly high yields in this process, such as one ejected ion pair (NaCl, KCl at 280°C) per each 300-eV electron!

Sputtering due to negative oxygen ions has been studied at Au-covered anodes in oxygen discharges.[179] The results from this work are not of much basic value because the authors were not able to determine the number of bombarding ions since the main current to the anode is carried by electrons.

The sputtering of porous metals (W, Mo) was studied by Martynenko.[180] This subject is of practical interest when the target material is available only in powder form or with respect to the influence of surface roughness on sputtering yields. With Cd^+ ions of 200 to 500 eV the sputtering yields fell rapidly when the porosity increased from 16 to 40% and remained thereafter constant at roughly half the value for the bulk material.

Benninghoven[181] studied the sputter emission of negative ions from compounds like Ag_2CO_3, Ag_2SO_4, and $AgNO_3$ under 3-keV Ar^+-ion bombardment. He found that the anion parent peak is often the most pronounced line in the mass spectrum (SO_4^-, CO_3^-, etc.), accompanied by singly negatively charged fragments of the anion. The cation appears as a singly positively charged ion. His studies give no information on the ratio of ejected neutrals to ions.

Mass spectroscopy was used by Benninghoven for monitoring the appearance of TaO^+ ions when predeposited thin films of Au, Ag, In, Zn are sputter-removed from a Ta substrate.[182] This method was used for collecting relative sputtering yields for 1-keV Ar^+ ions as a function of various parameters. A most interesting phenomenon is the fact that the film-metal signal, which naturally decays as the film is sputtered away, begins to rise again steeply when the TaO^+ signal appears. This is believed to be the result of the increased ionization probability of the film-metal atoms when they are sputtered from a Ta surface instead of from a surface of their own kind. Another recent paper by Benninghoven[183] is an excellent and detailed discussion of ion formation and emission during sputtering.

Some very interesting work on high-rate sputtering of stainless steel was performed by Dahlgren and McClanahan.[184] Sputtering was accomplished in a krypton triode discharge with 22 A discharge current. The target current density of the 1,500-eV ions is 5 mA cm^{-2}. With a sputtering yield of 2.2 atoms/ion, sputtering rates of 5,000 Å min^{-1} were achieved, and in a 160-h run a deposit of 1.6-mm thickness was built up on a temperature-controlled Cu substrate. The authors studied composition, microstructure, crystal structure, density, hardness, and surface appearance of the deposit as a function of substrate biasing and substrate temperature. Aside from the wealth of interesting deposit and growth data, which are not the subject of this chapter, they found that the deposit contained no detectable krypton (detection sensitivity was 20 ppm) and the composition of a film deposited without bias was very close to the composition of the target material but became noticeably deficient in Ni (which decreased by 23%) under 100-V negative substrate bias.

The last decade has certainly been a most fruitful one for our better understanding of the sputtering process. But still much remains to be done and is not yet fully understood, as pointed out in various sections of this chapter. On the experimental side we expect that the data will continue to become more and more reliable with cleaner and better-controlled vacuum and discharge conditions which became possible with ion-getter or turbomolecular pumps and the progress in vacuum technology in general. The collection of reliable basic data in insulator sputtering is probably one of the most urgent tasks. Much rewarding information on sputtering of multi-component materials and the enrichment of the low-sputtering species at the target surface is bound to come from a fairly new technique: Auger electron spectroscopy,[185] which allows in situ composition analysis in the outermost 10 atom layers with a sensitivity down to one-hundredth of a monolayer. On the theoretical side the interaction will continue to be most fertile between radiation-damage and particle-solid-collision-interaction physicists, and much stimulus for sputtering research in general will be provided by the numerous practical applications, of which many can barely be foreseen at present.

REFERENCES

1. Grove, W. R., *Phil. Trans. Roy. Soc. London*, **142**, 87 (1852).
2. Kohlschuetter, V., *Jahr. Radioaktivitaet*, **9**, 355 (1912).
3. Wehner, G. K., "Advances in Electronics and Electron Physics," vol. VII, p. 239, Academic Press Inc., New York, 1955.
4. Behrisch, R., *Ergeb. Exact. Naturw.*, **35**, 295 (1964).
5. Kaminsky, M., "Atomic and Ionic Impact Phenomena on Metal Surfaces," p. 24, Springer-Verlag, Berlin, 1964.
6. Carter, G., and J. S. Colligon, "Ion Bombardment of Solids," p. 310, Heinemann Educational Books, London, 1968.
7. Blechschmidt, E., *Ann. Physik*, **81**, 999 (1926).
8. v. Hippel, A., *Ann. Physik*, **81**, 1043 (1926).
9. Blechschmidt, E., and A. v. Hippel, *Ann. Physik*, **86**, 1006 (1928).
10. Hull, A. W., *Trans. Am. Inst. Elec. Engrs.*, **47**, 753 (1928).
11. Guentherschulze, A., and K. Meyer, *Z. Physik*, **62**, 607 (1930).
12. Meyer, K., and A. Guentherschulze, *Z. Physik*, **71**, 279 (1931).
13. Penning, F. M., and J. H. A. Moubis, *Koninkl. Ned. Akad. Wetenschap. Froc.*, **43**, 41 (1940).
14. Stark, J., *Z. Elektrochem.*, **14**, 752 (1908); **15**, 509 (1909).
15. Fetz, H., *Z. Physik*, **119**, 590 (1942).
16. Wehner, G. K., *J. Appl. Phys.*, **25**, 270 (1954).
17. Stark, J., and G. Wendt, *Ann. Physik*, **38**, 921 (1912).
18. Wehner, G. K., *J. Appl. Phys.*, **26**, 1056 (1955).
19. Fluit, J. M., *Colloq. Intern. Centre Natl. Rech. Sci.*, *Bellevue*, 1961.
20. Almen, O., and G. Bruce, *Nucl. Instr. Methods*, **11**, 279 (1961).
21. Almen, O., and G. Bruce, *Nucl. Instr. Methods*, **11**, 257 (1961).
22. Panin, B., *Soviet Phys. JETP English Transl.*, **15**, 215 (1962).
23. Datz, S., and C. Snoek, *Phys. Rev.*, **134**, A347 (1964).
24. Thompson, M. W., *Phil. Mag.*, **4**, 139 (1959).
25. Sinha, M. K., and E. W. Mueller, *J. Appl. Phys.*, **35**, 1256 (1964).
26. Keywell, F., *Phys. Rev.*, **87**, 160 (1952).
27. Gibson, J. B., A. N. Goland, M. Milgram, and G. H. Vineyard, *Phys. Rev.*, **120**, 1229 (1960).
28. Harrison, D. E., N. S. Levy, J. P. Johnson, and H. M. Efron, *J. Appl. Phys.*, **39**, 3742 (1968).
29. Lehmann, C., and P. Sigmund, *Phys. Stat. Solidi*, **16**, 507 (1966).
30. Sigmund, P., submitted to *Phys. Rev.* for publication.
31. Barrington, A. E., R. F. K. Herzog, and W. P. Poschenrieder, *J. Vacuum Sci. Technol.*, **3**, 239 (1966).
32. Wright, *Am. J. Sci.*, **13**, 49 (1877).
33. Nordman, J. E., *J. Appl. Phys.*, **40**, 2111 (1969).

34. Johnston, J. D., and A. G. Graybeal, American Vacuum Society Rocky Mountain Section Symposium, May 15, 1969.
35. Loeb, L. B., "Basic Processes of Gaseous Electronics," p. 329, University of California Press, Berkeley, 1955.
36. v. Ardenne, M., "Tabellen zur angewandten Physik," vol. 1, p. 316, VEB Verlag der Wissenschaften, Berlin, 1962.
37. Tsui, R. T. C., *Phys. Rev.*, **168**, 107 (1968).
38. Davidse, P. D., and L. I. Maissel, *J. Appl. Phys.*, **37**, 574 (1966).
39. Vossen, J. L., and F. J. O'Neill, *RCA Rev.*, **29**, 149 (1968).
40. Toombs, P. A. B., *Brit. J. Appl. Phys. (J. Phys. D)*, ser. 2, **1**, 662 (1968).
41. Holland, L., T. Putner, and G. N. Jackson, *J. Sci. Instr. (J. Phys. E)*, ser. 2, **1**, 32 (1968).
42. Fetz, H., and H. Oechsner, *Z. Angew. Phys.*, **12**, 250 (1960).
43. Guentherschulze, A., and W. Tollmien, *Z. Physik*, **119**, 685 (1942).
44. Winters, H. F., and E. Kay, *J. Appl. Phys.*, **38**, 3928 (1967).
45. v. Ardenne, M., "Tabellen zur angewandten Physik," VEB Verlag der Wissenschaften, Berlin, 1962.
46. v. Ardenne, M., S. Schiller, and U. Heisig, *Nachrichtentech.*, **15**, 306 (1965).
47. Chopra, K. L., and M. R. Randlett, *Rev. Sci. Instr.*, **38**, 1147 (1967).
48. Nablo, S. V., and W. J. King, IEEE Conference Paper 64, p. 484, 1964.
49. Fiedler, O., G. Reisse, B. Schoeneich, and C. Weissmantel, *Proc. 4th Intern. Vacuum Congr.*, 1968, p. 569.
50. Wehner, G. K., Annual Report on Sputtering Yields, ONR Contract Nonr 1589 (15), 1959.
51. Garvin, H. L., Final Report, NASA Contract NAS 3-6273, 1968.
52. Thompson, M. W., and R. S. Nelson, *Phil. Mag.*, **7**, 84, 2015 (1962).
53. Bader, M., F. Witteborn, and T. W. Snouse, *NASA Rept.* TR-R-105, 1961.
54. KenKnight, C. E., and G. K. Wehner, *J. Appl. Phys.*, **35**, 322 (1964).
55. Mahadevan, P., J. L. Layton, A. R. Comeaux, and D. B. Medved, *J. Appl. Phys.*, **34**, 2810 (1963).
56. Nelson, R. S., and C. J. Beevers, *Phil. Mag.*, **9**, 343 (1964).
57. Guentherschulze, A., *Z. Physik*, **118**, 145 (1941).
58. Morgulis, N. D., and V. D. Tischenko, *Izv. Akad. Nauk SSSR Ser. Fiz.*, **20**, 1190 (1956).
59. Stuart, R. V., and G. K. Wehner, *J. Appl. Phys.*, **33**, 2345 (1962).
60. Jorgenson, G. V., and G. K. Wehner, *J. Appl. Phys.*, **36**, 2672 (1965).
61. Sigmund, P., *Can. J. Phys.*, **46**, 731 (1968).
62. Wehner, G. K., C. KenKnight, and D. L. Rosenberg, *Planet. Space Sci.*, **11**, 885 (1963).
63. Wehner, G. K., *Astronaut. Acta*, **14**, 65 (1968).
64. Wehner, G. K., *J. Appl. Phys.*, **31**, 1392 (1960).
65. Benninghoven, A., *Ann. Physik*, **15**, 113 (1965).
66. Thompson, M. W., B. W. Farmery, and P. A. Newson, *Phil. Mag.*, **18**, 361 (1968).
67. Thompson, M. W.; *Phil. Mag.*, **18**, 377 (1968).
68. Farmery, B. W., and M. W. Thompson, *Phil. Mag.*, **18**, 361 (1968).
69. Beuscher, H., and K. Kopitzki, *Z. Physik*, **184**, 382 (1965).
70. Mayer, H., *Phil. Mag.*, **16**, 594 (1933).
71. Sporn, H., *Z. Physik*, **112**, 278 (1939).
72. Wehner, G. K., *Phys. Rev.*, **114**, 1270 (1959).
73. Kopitzki, K., and H. E. Stier, *Z. Naturforsch.*, **17a**, 346 (1962).
74. Weijsenfeld, C. H., *Phys. Letters*, **2**, 295 (1962).
75. Oechsner, H., and L. Reichert, *Phys. Letters*, **23**, 90 (1966).
76. Stuart, R. V., and G. K. Wehner, *J. Appl. Phys.*, **35**, 1819 (1964).
77. Stuart, R. V., G. K. Wehner, and G. S. Anderson, *J. Appl. Phys.*, **40**, 803 (1969).
78. Veksler, V. I., *Soviet Phys. JETP English Transl.*, 1960, p. 235.
79. Ben'yaminovich, M. B., and V. I. Veksler, *Izv. Akad. Nauk USSR Ser. Fiz.-Mat. Nauk*, **3**, 29 (1963).
80. Politiek, J., P. K. Rol, J. Los, and P. G. Ikelaar, *Rev. Sci. Instr.*, **39**, 1147 (1968).
81. Seeliger, R., and K. Sommermeyer, *Z. Physik*, **93**, 692 (1935).
82. Wehner, G. K., and D. Rosenberg, *J. Appl. Phys.*, **31**, 177 (1960).
83. Koedam, M., Cathode Sputtering by Rare Gas Ions of Low Energy Bombardment of Polycrystalline and Monocrystalline Material, Thesis, State University of Utrecht, March, 1961.
84. Rol, P. K., J. M. Fluit, and J. Kistemaker, *Physics*, **26**, 1000 (1960).
85. Grønlund, F., and W. J. Moore, *J. Chem. Phys.*, **32**, 1540 (1960).

86. Cobic, B., and B. Perovic, *Proc. 4th Intern. Symp. Ionization Phenomena in Gases*, 1960, vol. 1, p. 260, North Holland Publishing Company, Amsterdam.
87. Cooper, C. B., and J. Comas, *J. Appl. Phys.*, **36**, 2891 (1965).
88. Anderson, G. S., and G. K. Wehner, *J. Appl. Phys.*, **31**, 2305 (1960).
89. Anderson, G. S., *J. Appl. Phys.*, **33**, 2017 (1962).
90. Koedam, M., and A. Hoogendoorn, *Physica*, **26**, 351 (1960).
91. Yurasova, V. E., N. V. Pleshivtsev, and I. V. Orfanov, *Soviet Phys. JETP English Transl.*, **37**, 689 (1960).
92. Nelson, R. S., and M. W. Thompson, *Proc. Roy. Soc. London*, A259, 458 (1961).
93. Anderson, G. S., *J. Appl. Phys.*, **34**, 659 (1963).
94. Nelson, R. S., *Phil. Mag.*, **8**, 693 (1963).
95. Ashmyanskii, R. A., M. B. Ben'yaminovich, and V. I. Veksler, *Soviet Phys. Solid State English Transl.*, **7**, 1314 (1965).
96. Hasiguti, P. P., R. Hanada, and S. Yamaguchi, *J. Phys. Soc. Japan*, **18** (suppl. III), 164 (1963).
97. Wehner, G. K., G. S. Anderson, and C. E. KenKnight, Surface Bombardment Studies, Annual Report 2483, General Mills Electronics Division, Minneapolis, Minn., November, 1963, AEC Contract AT-11-1-722.
98. Anderson, G. S., G. K. Wehner, and H. J. Olin, *J. Appl. Phys.*, **34**, 3492 (1963).
99. Southern, A. L., W. R. Willis, and M. T. Robinson, *J. Appl. Phys.*, **34**, 153 (1963).
100. Anderson, G. S., and G. K. Wehner, *Surface Sci.*, **2**, 367 (1964).
101. Anderson, G. S., *J. Appl. Phys.*, **37**, 3455 (1966).
102. Cunningham, R. L., and J. Ng-Yelim, *J. Appl. Phys.*, **35**, 54 (1965).
103. Woodyard, J. R., and C. B. Cooper, *J. Appl. Phys.*, **35**, 1107 (1964).
104. Woodyard, J. R., paper given at 15th Annual Conference on Mass Spectrometry and Allied Topics, Denver, Colo., 1967.
105. Herzog, R. F. K., W. P. Poschenrieder, F. G. Ruedenauer, F. G. Satkiewicz, 15th Annual Conference on Mass Spectrometry, May, 1967, Denver, Colo.
106. Jurela, Z., and B. Perovic, paper given at 8th International Conference on Phenomena in Ionized Gases, Vienna, Austria, 1967.
107. Comas, J., and C. B. Cooper, *J. Appl. Phys.*, **38**, 2956 (1957).
108. Flur, B. L., and J. Riseman, *J. Appl. Phys.*, **35**, 344 (1964).
109. Francombe, N. H., and A. J. Noreika, *J. Appl. Phys.*, **32**, 97S (1961).
110. Khan, I. H., and M. H. Francombe, *J. Appl. Phys.*, **36**, 1699 (1965).
111. Pratt, R. H., National Electronics Conference, 1964.
112. Michalak, E. M., Symposium on the Deposition of Thin Films by Sputtering, University of Rochester, June, 1966.
113. Patterson, W. L., and G. A. Shirn, *J. Vacuum Sci. Technol.*, **4**, 343 (1967).
114. Anderson, G. S., *J. Appl. Phys.*, **40**, 2884, 1969.
115. Anderson, G. S., G. V. Jorgenson, and G. K. Wehner, Applied Science Division, Litton Annual Report on Sputtering, ONR Contract Nonr 1589(15), November, 1966.
116. Kaminsky, M., "Atomic and Ionic Impact Phenomena on Metal Surfaces," pp. 211–218, Academic Press Inc., New York, 1965.
117. Haymann, P., *Compt. Rend.*, **248**, 2472 (1958).
118. Haymann, P., and J. J. Trillat, *Compt. Rend.*, **251**, 85 (1960).
119. Wehner, G. K., *Phys. Rev.*, **102**, 690 (1956).
120. Haymann, P., and H. Gervais, "Le Bombardment ionique," International Conference of Centre National de la Recherche Scientifique, no. 113, p. 91, Bellevue, 1962.
121. Ogilvie, G. J., and A. A. Thomson, *J. Phys. Chem. Solids*, **17**, 203 (1961).
122. Jacobson, R. L., and G. K. Wehner, *J. Appl. Phys.*, **36**, 2674 (1965).
123. Mazey, D. J., R. S. Nelson, and R. S. Barnes, *Phil. Mag.*, **17**, 1145 (1968).
124. Anderson, G. S., *J. Appl. Phys.*, **38**, 1607 (1967).
125. Wehner, G. K., *J. Appl. Phys.*, **29**, 217 (1958).
126. Asada, T., and K. Quasebarth, *Z. Phys. Chem.*, A143, 435 (1929).
127. Gillam, E., *J. Phys. Chem. Solids*, **11**, 55 (1959).
128. Wehner, G. K., C. E. KenKnight, and D. Rosenberg, *Planet. Space Sci.*, **11**, 1257 (1963).
129. Rosenberg, D. L., and G. K. Wehner, *J. Geophys. Res.*, **69**, 3307 (1964).
130. Townes, C. H., *Phys. Rev.*, **65**, 310 (1944).
131. Lamar, E. S., and K. T. Compton, *Science*, **80**, 541 (1934).
132. Seitz, F., and J. S. Koehler, *Solid State Phys.*, **2**, 305 (1956).
133. Keywell, F., *Phys. Rev.*, **97**, 1611 (1955).
134. Rol, P. K., J. M. Fluit, and J. Kistemaker, *Physica*, **26**, 1009 (1960).
135. Fluit, J. M., and P. K. Rol, *Physica*, **30**, 857 (1964).
136. Fluit, J. M., P. K. Rol, and J. Kistemaker, *J. Appl. Phys.*, **34**, 690 (1963).

137. Onderdelinden, D., F. W. Saris, and P. K. Rol, *Nucl. Instr. Methods*, **38**, 269 (1965).
138. Onderdelinden, D., F. W. Saris, and P. K. Rol, *Proc. 7th Intern. Conf. Phenomena in Ionized Gases*, vol. 1, Belgrade, 1966.
139. Piercy, G. R., et al., *Can. J. Phys.*, **42**, 1116 (1964).
140. Kornelsen, E. V., et al., *Phys. Rev.*, **135**, A849 (1964).
141. Nelson, R. S., and M. W. Thompson, *Phil. Mag.*, **8**, 1677 (1963).
142. Snouse, T. W., and L. C. Houghney, *J. Appl. Phys.*, **37**, 700 (1966).
143. Silsbee, R. H., *J. Appl. Phys.*, **28**, 1246 (1957).
144. Thompson, M. W., *Proc. 5th Intern. Conf. Ionization Phenomena in Gases*, Munich, 1961, p. 85, North Holland Publishing Company, Amsterdam.
145. Sanders, J. B., and M. W. Thompson, *Phil. Mag.*, **17**, 211 (1968).
146. Thompson, M. W., B. W. Farmery, and M. J. Hall, The Influence of Thermal Vibration on Focused Collision Sequences, Report AERG-R-4044, March, 1962, Harwell, Berkshire, England.
147. Bulgakov, Yu. V., *Soviet Phys. Solid State English Transl.*, **6**, 912 (1964).
148. Thompson, M. W., *Phys. Letters*, **6**, 24 (1963).
149. Anderson, G. S., *J. Appl. Phys.*, **37**, 2838 (1966).
150. Anderson, G. S., *J. Appl. Phys.*, **38**, 1607 (1967).
151. Robinson, M. T., and A. L. Southern, *J. Appl. Phys.*, **39**, 3463 (1968).
152. Harrison, D. E., Jr., J. P. Johnson III, and N. S. Levy, *Appl. Phys. Letters*, **8**, 33 (1966).
153. Robinson, M. T., *J. Appl. Phys.*, **40**, 2670 (1969).
154. Lehmann, C., and P. Sigmund, *Phys. Stat. Solidi*, **16**, 507 (1966).
155. von Jan, R., and R. S. Nelson, *Phil. Mag.*, **17**, 1017 (1968).
156. Sigmund, P., *Can. J. Phys.*, **46**, 731 (1968).
157. Andersen, H. H., *Appl. Phys. Letters*, **13**, 85 (1968).
158. Sigmund, P., to be published in *Phys. Rev.*
159. Cuderman, J. F., and J. J. Brady, *Surface Sci.*, **10**, 410 (1968).
160. Musket, R. G., and H. P. Smith, *J. Appl. Phys.*, **39**, 3579 (1968).
161. Higgins, T. B., N. T. Olson, and H. P. Smith, *J. Appl. Phys.*, **39**, 4849 (1968).
162. Gurmin, B. M., T. P. Martynenko, and Yu. A. Ryzhov, *Fiz. Tverd. Tela*, **10**, 411 (1968).
163. Andersen, H. H., and P. Sigmund, Danish Atomic Energy Commission, Riso Report 103, 1965.
164. Lindhard, J., V. Nielsen, and M. Scharff, *Mat. Fys. Medd. Dan. Vid. Selsk*, vol. 36, no. 10, 1968.
165. Rol, P. K., J. M. Fluit, and J. Kistemaker, "Electromagnetic Isotope Separation of Radioactive Isotopes," p. 207, Springer-Verlag OHG Vienna, 1960.
166. Yonts, O. C., C. E. Norman, and D. E. Harrison, *J. Appl. Phys.*, **31**, 447 (1960).
167. Guseva, M. I., *Soviet Phys. Solid State English Transl.*, **1**, 1410 (1959).
168. Magnuson, G. D., and C. E. Carlson, *J. Appl. Phys.*, **34**, 3267 (1963).
169. Wehner, G. K., *J. Appl. Phys.*, **30**, 1762 (1959).
170. Thomson, J. J., "Rays of Positive Electricity," Longmans, Green & Co., Ltd., London 1921.
171. Research Staff of the General Electric Co., London, *Phil. Mag.*, **45**, 98 (1923).
172. Kay, Eric, Impact Evaporation and Thin Film Growth in a Glow Discharge, *Advan. Electron.*, **17**, 245 (1962).
173. Guentherschulze, A., *Vacuum*, **3**, 360 (1953).
174. Keller, K., *Plasma Phys.*, **10**, 159 (1968).
175. Robinson, H., *J. Appl. Phys.*, **39**, 3441 (1968).
176. Pleshivtsev, N. V., "Cathode Sputtering," Atomizdat, Moscow, 1968.
177. Steffen, H., R. Niedermayer, and H. Mayer, *Thin Films*, **1**, 223 (1968).
178. Townsend, P. D., and J. C. Kelly, *Phys. Letters*, **26A**, 138 (1968).
179. Jennings, T. A., and W. McNeill, *Appl. Phys. Letters*, **12**, 25 (1968).
180. Martynenko, T. P., *Zh. Tekhn. Fiz.*, **38**, 759 (1968).
181. Benninghoven, A., *Z. Naturforsch.*, **24a**, 859 (1969).
182. Benninghoven, A., *Z. Angew. Phys.*, **27**, 51 (1969).
183. Benninghoven, A., *Z. Physik*, **220**, 159 (1969).
184. Dahlgren, S. D., and E. D. McClanahan, *3rd Symposium on the Deposition of Thin Films by Sputtering*, University of Rochester, N.Y., Sept., 1969, p. 20.
185. Riviere, J. C., *Bull. Phys. Soc.*, 1969, p. 85.

Chapter **4**

Application of Sputtering to the Deposition of Films

LEON MAISSEL

IBM Components Division, East Fishkill, New York

1. INTRODUCTION

Although application of the sputtering process to the deposition of thin films has been known and practiced for well over a century, it is only relatively recently that this method has become a serious competitor to vacuum evaporation. This long delay for sputtering to mature as a method for depositing films has been mainly the result of the greater tendency of sputtered films to be contaminated as well as the large number of process parameters involved, the latter having made good process control difficult to achieve. In recent years, the demands of the electronics industry for films with qualities difficult or even impossible to attain by any other means have stimulated a good deal of fruitful effort directed toward solving these problems, and today sputtering as a deposition technique is receiving an increasing level of acceptance, not only for electronic applications but for all areas where films may be needed

The general nature of the sputtering process is discussed in Chap. 3, and we shall be concerned here only with those aspects which are important for the deposition of films. The reader who is interested in additional information on the broad subject of sputtered films is referred to previous reviews.[1-5]

Sputtering offers advantages over most other deposition techniques if any of the following characteristics or film types are important:
1. Multicomponent films (alloys, compounds, etc.)
2. Refractory materials
3. Insulating films
4. Good adhesion
5. Low-temperature epitaxy
6. Thickness uniformity over large planar areas

Sputtering may not be a suitable method to use if any of the following limitations are unacceptable:
1. Source material must be available in sheet form.
2. Deposition rates are usually less than 2,000 Å min^{-1}.
3. Substrate must be cooled (unless low deposition rates are used).

2. THE SELF-SUSTAINED GLOW DISCHARGE

An understanding of the processes that occur within the self-sustained glow discharge is vital for the design and interpretation of sputtering experiments. Accordingly, the following section contains a description of the principal phenomena involved with special emphasis on those relating to sputtering. The interested reader who wishes to pursue the subject in greater detail is referred to the literature.[6-10]

If a dc voltage is applied between two electrodes spaced some distance apart in gas at low pressure, the current flow will be negligibly small unless some minimum value, called the breakdown voltage, is used. If, however, the cathode is caused to emit electrons (for example, by irradiating it with ultraviolet light), a small current will flow between the electrodes. This current, which is immediately extinguished when the source of electron emission at the cathode is removed, may be observed to increase with increasing electrode spacing provided the applied voltage exceeds the ionization potential of the gas concerned. It is assumed that the electrodes are large enough so that neither ions nor electrons are lost by diffusion out of the interelectrode space.

The increase in current with spacing occurs because, at a given pressure, electrons on their way from the cathode to the anode make a fixed number of ionizing collisions per unit length so that the greater the distance they travel the more ionizing collisions are made. With each collision a new electron is generated so that the total number of electrons reaching the anode is greater than the number originally produced at the cathode by the postulated external source. The ions that result from these collisions are also accelerated by the applied field and move toward the cathode. If the total applied voltage is high enough, some of these ions will, on striking the cathode, eject secondary electrons from its surface.

The two processes of ionization by electron impact and secondary emission of electrons by ions thus control the current I in the system, as described by Eq. (1):

$$I = \frac{I_0 \exp (\alpha d)}{1 - \gamma[\exp (\alpha d) - 1]} \tag{1}$$

where I_0 is the primary electron current generated at the cathode by the external source, α is the number of ions per unit length produced by the electrons, d is the spacing between the electrodes, and γ is the number of secondary electrons emitted per incident ion.

This equation, which is due to Townsend, gives a good description of the current through a gas at voltages well below breakdown, provided α and γ are known. In Fig. 1 are shown some curves of α as a function of electron energy for a gas temperature of 0°C and 1 Torr pressure, whereas Fig. 2 shows some values of γ for molybdenum for argon and helium ions as a function of their energy.[12]

If the voltage between the electrodes is raised, both α and γ are increased and the denominator of (1) rapidly approaches zero. The current I then becomes infinite, (1) no longer holds, and gas breakdown is said to have occurred. In practice, breakdown oc-

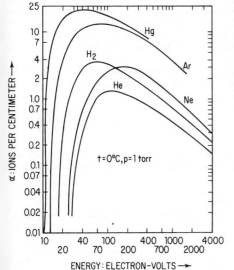

Fig. 1 Variation of ionization per unit length α as a function of electron energy.

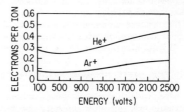

Fig. 2 Coefficient of secondary emission of electrons γ by argon and helium ions for molybdenum.

curs at voltages lower than would be predicted by (1) assuming a uniform field between the electrodes. This is a consequence of the fact that when the ion current reaches significant proportions the ions tend to accumulate in front of the cathode and produce a localized space charge there. This has the net effect of greatly increasing the electric field immediately in front of the cathode so that both α and γ change very much more rapidly with applied voltage than (1) might suggest. Once breakdown has occurred, the number of secondary electrons produced at the cathode is sufficient to maintain the discharge and the glow is said to be self-sustained.

Visual examination of a self-sustained glow discharge shows it to have a fair degree of structure, as illustrated in Fig. 3. Of primary interest is the region marked as the Crookes dark space. This is the region in which the positive ions have accumulated to form the space charge discussed above. Its thickness is approximately the mean distance traveled by an electron from the cathode before it makes an ionizing collision. This is not to be confused with the electronic mean free path, which includes elastic collisions and is five to ten times smaller.

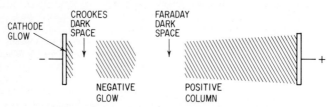

Fig. 3   Cross-sectional view of glow discharge.

Because of the large electric field in the Crookes space, the electrons traverse it very rapidly. However, once they reach the edge of the negative glow and begin to produce significant numbers of ion-electron pairs, the preponderance of positive charge falls off very rapidly and a neutral region consisting of approximately equal numbers of ions and electrons begins. Such a region, where the number of ions and electrons are very nearly the same, is called a plasma. Since the electrons screen the ions, the latter are not "aware" of the existence of the electrodes and move through the negative glow by diffusion, rather than by drift in a field. However, once an ion reaches the edge of the Crookes dark space, it "sees" the potential at the cathode and is rapidly accelerated toward it.

At relatively low voltages the cross-sectional area of the glow will be less than the available cathode area. This arises because the glow requires a minimum current density if it is to produce sufficient secondary electrons to maintain itself. When only a small amount of power is available, the cross-sectional area of the glow is reduced until this minimum current density is attained. If additional power is applied to the discharge tube, the glow adjusts to this by increasing its cross-sectional area, thus raising the total current but keeping the current density at the cathode constant. As long as this current density does not increase, neither does the voltage across the dark space. This minimum voltage drop needed to maintain the glow is called the normal cathode fall, and the corresponding glow is referred to as the normal glow. This type of discharge is therefore a constant-voltage device, and many practical applications of this feature have been made.

When the power to the discharge tube is increased past the point where the glow covers the entire available cathode area, the current density at the cathode must increase. This the glow can do only by increasing the number of secondary electrons coming out of the cathode. This, in turn, implies an increase in the coefficient of secondary electron emission γ, which, as seen in Fig. 2, requires an increase in the cathode fall. A glow discharge operating in this mode is referred to as "abnormal." In practice, this is the only mode of interest for the deposition of films by sputtering.

As we saw above, when electrons enter the negative glow, they possess essentially the full cathode fall of potential. This energy is then lost through a series of either ionizing or excitation collisions (electrons losing virtually no energy in elastic collisions with gas atoms). Eventually, the electrons' energy is reduced to the point that they are no longer able to produce additional ions. The region in the discharge where this happens defines the far edge of the negative glow, and since no more ions are being produced, the electrons begin to accumulate there, forming a region of slightly negative space charge. In this region the electrons have insufficient energy to cause either ionization or excitation. Consequently, it is a dark region. It is known as the Faraday dark space.

The remainder of the discharge, from the edge of the Faraday dark space to the anode, represents a situation which is quite similar to the Townsend discharge discussed above. That is, there is a steady supply of electrons and a small electric field. Thus, after passing through the Faraday dark space by diffusion, electrons are accelerated toward the anode in a manner similar to that described by Eq. (1) except that $\gamma = 0$ and (unless the anode is very close to the Faraday dark space) edge losses due to diffusion cannot be neglected. This region is called the positive column.

Since the self-sustaining feature of the discharge depends only on the emission of sufficient electrons at the cathode by positive ions from the negative glow, the exact location of the anode normally makes very little difference to the electrical characteristics of the glow. Thus, if the anode is moved closer and closer in toward the cathode, the positive column will be extinguished, the Faraday dark space will disappear, and finally, a large fraction of the negative glow may be extinguished before any appreciable effect on the electrical characteristics is seen. As the anode approaches the edge of the Crookes dark space, however, the number of ions being produced becomes insufficient and the voltage required to sustain the glow begins to rise rapidly since a higher secondary-emission coefficient is needed to compensate for the reduced number of ions. Such a glow is referred to as an obstructed glow. If the anode is pushed right to the edge of the dark space (and therefore inside the mean distance for ionization), no ions can be formed and no glow can be maintained, even for very large applied voltages. As was stated above, sputtering experiments are invariably performed with glows that are strongly abnormal. This is primarily because the current density in the normal glow is too low for atoms to be sputtered out of the cathode at useful rates and also because the voltage drop in the normal glow is quite low so that sputtering yields are similarly low.

The situation is further complicated by the phenomenon of charge transfer. An ion moving in its own gas has a relatively high probability of transferring its charge to a neutral atom. The neutral, which was formerly the ion, continues with the momentum which it possessed prior to charge transfer while the newly formed ion has only thermal energy. The influence of the charge-transfer process on glow discharges has been investigated in some detail by Davis and Vanderslice.[13] Their results for argon ions moving in the dark space of a glow discharge having an applied voltage of 600 V at 60 mTorr pressure are shown in Fig. 4. As can be seen, most of the ions reach the cathode with essentially zero energy and very few of them reach the cathode with an energy of even 60% of the applied voltage. Thus (at least for these conditions) very few of the ions

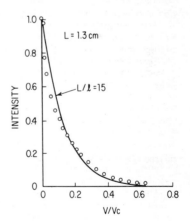

Fig. 4 Number of ions reaching cathode as a function of their energy.

that cross the dark space are able to assume the full cathode fall of potential and a great many energetic neutrals bombard the cathode.

These results indicate that, for Ar^+ in Ar, the mean free path for charge transfer is one-fifteenth the thickness of the dark space. On the other hand, for doubly ionized argon the probability for charge transfer is much less and the energy distribution of Ar^{++} ions striking the cathode is such that the majority of the Ar^{++} ions reach the cathode having acquired the full applied voltage. Since the mean free path for charge transfer increases with increasing ion energy for most systems and since the fraction of doubly ionized ions increases as the applied voltage is increased, the number of energetic neutrals decreases with applied voltage.

In addition to causing sputtering, neutrals that strike the cathode may have a surprisingly high probability of being reflected with little or no loss in energy. Very few quantitative data are available, but for argon on molybdenum, the fraction of reflected neutrals at an energy of 750 eV has been estimated by Medved[14] as ∼20%. This number is expected to be even greater at lower particle energies. Energetic neutrals moving away from the cathode can also be generated by an entirely different mechanism. Hagstrum[15] has shown that energetic noble-gas ions have a finite probability of being reflected as neutrals excited into metastable states. For argon ions striking a tungsten surface at 1,000 eV the probability that this will happen is 0.5%. This number increases with the square root of the energy. The significance of these energetic neutrals for the sputtering process will be discussed later.

Since, for the most effective collection of sputtered material, the substrate (often also the anode) should be as close to the cathode as possible (see later), in most sputtering systems it is placed just far enough away from the cathode not to obstruct the negative glow. As a consequence of this, many of the electrons that arrive at the substrate have substantial energies. The influence of high-energy electrons on the properties of depositing films has been studied to a limited extent and will be discussed in a later section. Also, if some of the material leaving the cathode is in the form of negative ions, these will be accelerated across the dark space and will also arrive at the substrate with appreciable energies.

If an additional electrode is inserted into the glow and is biased with respect to the anode, the effect of negative bias is that a second cathode is created with its own ion sheath around it. However, since the glow is already being sustained by the flow of secondary electrons from the primary cathode, the second cathode can operate at as low a voltage as desired since the glow is not dependent on it for a supply of secondary electrons. The dark space that forms around such an additional cathode (sometimes referred to as a probe) is completely analogous to the Crookes dark space and is referred to as the Langmuir dark space.

The thickness of the sheath is a balance between the attractive forces due to the negative charge on the probe and the repulsive forces introduced by the presence of a higher-than-normal density of positive ions. The sheath therefore grows thicker for increasing probe voltages but thins for increasing ion densities in the external plasma. A quantitative relation between sheath thickness d and the other quantities is given by

$$d^2 = \frac{4\epsilon_0}{9j} \left(\frac{2q}{m}\right)^{\frac{1}{2}} V^{\frac{3}{2}}$$

where V is the probe potential (relative to the plasma), j is the current density due to ions, m is the mass of an ion, q is the electronic charge, and ϵ_0 is the permittivity of free space.

If the bias given to an additional electrode is positive with respect to the anode, it becomes a new anode for the discharge, the original anode (and everything electrically connected to it) now becoming a second cathode. Whatever the area of the anode, it must collect an electron current equal to the total current at the various cathodes. If it should have an area significantly less than that of the cathodes, sufficient electrons may not reach it by diffusion through the plasma. Under these circumstances a negative space charge will accumulate in its vicinity and an anode fall of potential will appear which is just sufficient to create additional electrons through ionization of the gas atoms surrounding the anode. The fall of potential at a small anode is thus in the order of the ionization potential of the gas, and no more.

3. THERMIONICALLY AND MAGNETICALLY SUPPORTED GLOW DISCHARGES

As the pressure in a self-sustained glow is reduced, the number of ionizing collisions per unit length made by electrons leaving the cathode diminishes and consequently the ion density in the glow is reduced. This reduced ion density leads, in turn, to a lower number of ions striking the cathode, and the glow compensates for this deficiency by increasing the voltage drop across the Crookes dark space (as this leads to an increase in the secondary-emission coefficient of the ions).

As the pressure is still further reduced, the length of the dark space as well as the voltage required to maintain the glow will increase. Eventually the dark space will extend all the way to the anode and the available voltage will not be sufficient to start or maintain the glow. For most sputtering systems this occurs at pressures in the order of 10 to 20 mTorr. In order for a glow to operate at lower pressures than this, it is necessary to provide some source of electrons other than those emitted through secondary emission by ions. Alternatively, some way may be found to increase the ionizing efficiency of the available electrons.

a. Thermionically Supported Glow

To supply additional electrons it is possible to use a hot cathode which emits electrons through thermionic rather than secondary emission. Such a system, which is sometimes called a low-voltage arc, will provide finite current even in high vacuum. In the latter case, however, the available current is space-charge-limited because of the electron cloud present in the immediate vicinity of the cathode, and the full emission current can reach the anode only if rather substantial voltages are applied. In the presence of a gas at low pressure, however, ions will be produced by collision provided the applied voltage exceeds the ionization potential of the gas. As in the case of the self-sustained glow discharge, the slow-moving ions tend to accumulate in front of the cathode and a dark space is produced.

In detail, this dark space is quite different from that in front of the cathode of a self-sustained discharge. It consists of an electron sheath immediately in front of the cathode followed by a positive-ion sheath adjacent to the plasma. It is often referred to as a double sheath. This sheath serves two functions. To a lesser extent it neutralizes the electronic space charge, but more important, it acts as though it were a virtual anode very close to the cathode so that the electric field there becomes comparable with what would be obtained in high vacuum if a much higher voltage were applied across the entire tube.

Thus, by the application of a relatively small voltage in the order of 20 V, the full emission current of the emitter can be drawn. If the total applied voltage is further increased, additional current may be drawn from the cathode both because of additional electrons ejected by secondary emission and also because of enhanced heating of the cathode by the high-energy ions. In most sputtering systems, the thermionic cathode is a heated tungsten filament which can tolerate ion bombardment for long periods. However, if an oxide-coated or thoriated cathode is used, exceeding the minimum voltage required to maintain the glow will greatly shorten the cathode's life. In sputtering systems employing a thermionically supported discharge, 50 to 100 V is commonly used, and under these conditions, a current of several amperes can readily be drawn.

If the pressure in a thermionically supported discharge is decreased, the ion density decreases in the same way as it does in the self-sustained glow. The ionic space charge at the thermionic cathode therefore decreases and, consequently, so does the electric field there. To compensate for this, additional external voltage must be applied to the system, and at pressures below about 1 mTorr the total current that is drawn begins to approach the vacuum value. A large fraction of the emission current can be collected at the anode even at pressures substantially below 1 mTorr if the electric field at the cathode is artificially maintained instead of relying on the ionic space charge. This is done most simply by inserting a grid bearing the anode potential a very short distance in front of the cathode. The system is now nothing more than a low-voltage electron gun with the main anode serving to collect the electrons.[16]

When a low-pressure arc is used as an aid for sputtering, the material to be sputtered is inserted as a target, or probe, into the plasma, as shown in Fig. 5. Note that the anode must be biased positively with respect to the substrate. If it were at the same potential as the substrate, some electrons from the thermionic cathode would be deflected toward it and would be collected there rather than at the anode. The result would be gross inhomogeneities in the plasma density at the target. The role of the magnetic field will be discussed presently.

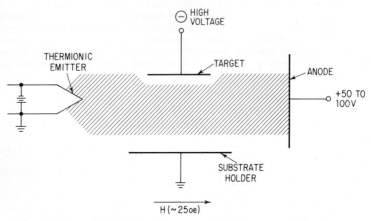

Fig. 5 Schematic view of thermionically supported glow discharge adapted for sputtering (triode sputtering).

b. Magnetically Sputtered Glow *Supported*

If, instead of increasing the number of electrons emitted at the cathode, the efficiency of the available electrons is increased, it will be possible to sputter at lower pressure or, if the pressure is not reduced, to obtain greater current for a given applied voltage.

In general, the effect of a magnetic field on a glow discharge is via the electrons, the influence on the much heavier ions being negligible. One effect of the magnetic field is to cause electrons that are not moving parallel to it to describe helical paths around the lines of magnetic force. The radius of the helix decreases with increasing magnetic field. Thus, the electrons must travel over a longer path in order to advance a given distance from the cathode. In this way the magnetic field acts as though the gas pressure had been increased.

Another important mechanism is the so-called magnetic-bottle effect whereby radial diffusion of the electrons out of the glow is greatly hindered so that the loss of electrons which would otherwise be available for ionization is reduced. This will be discussed further in the section on rf discharges. It is the reason for the relatively weak field used in the thermionically supported sputtering system shown in Fig. 5. A stronger field distorts the glow, as will be discussed presently.

(1) Longitudinal Fields (Parallel to Electric Field in Dark Space) By definition, a longitudinal field has no effect on the electrons moving through the Crookes dark space. The magnetic-bottle effect just discussed is relatively unimportant in a self-sustained dc glow. There is, however, an additional effect which allows the longitudinal field to play a role in dc discharges. It will be recalled that electrons enter the negative glow with close to the full cathode fall of potential and then proceed to lose this energy through a series of ionizing collisions until they reach the edge of the Faraday dark space. Since there is no significant electric field in the negative glow, electrons move in random directions after they have made a few collisions there. These electrons will therefore spiral around the magnetic-field lines, effectively enhancing their ionization probability for a given linear distance traveled. As a result, the number of ions produced by these electrons at the near edge of the negative glow is increased so that for the same electrical conditions a greater number of ions are available for bombarding the cathode.

The longitudinal magnetic field is the preferred mode for sputtering purposes because there is no distortion of the glow and the uniformity of films deposited through sputtering is not affected, except for certain edge effects which will be discussed later.

(2) Transverse Field (Normal to Electric Field in Dark Space) In this case, electrons travel through the dark space along curved paths instead of in straight lines, and a reduction in the thickness of the dark space results.   This is particularly important when the curvature of the electron path is short enough so that electrons originating at the cathode do not reach the edge of the negative glow.   Unfortunately, although the increase in the ionizing efficiency of the electrons is considerable, in the parallel-plate geometry most frequently used for sputtering, a transverse field will cause the discharge to be swept to one side.   Consequently, it is of little interest for sputtering applications.

If a cylindrical, rather than a planar, cathode is used, a transverse field is now one that is applied along the axis of the cylinder.   With this configuration, the discharge cannot be swept to one side since the cathode surface is semi-infinite.   Electrons from the cathode circle it with a cycloidal motion, and the uniformity of the discharge over the surface of the cylinder is not affected.   The effect of a magnetic field on the I-V characteristics of this configuration has been studied in detail by Penning and Moubis.[17]   With a magnetic field of about 300 G, a current of several amperes could be obtained at 500 V as compared with a current of only a few tenths of an ampere at 1,500 V without a magnetic field.

In the work of Penning and Moubis the sputtered surface was the outside of the cathode.   However, it is also possible to sputter off the inside surface of a cylindrical cathode (with or without an axial magnetic field).   If the anode is external to the cathode, the discharge can be made to reside inside the cylinder over only a rather narrow range of pressure (hollow-cathode effect).   If, however, the anode of the system is a wire lying along the axis of the cylinder, a discharge that will operate at very low pressure is obtained.   This is sometimes called the inverted magnetron.   At relatively high pressures the discharge operates in a similar manner to the cylindrical geometry in which the discharge is on the outside of the cylinder.   However, at low pressures, enhanced ionization is obtained because electrons leaving the cathode are caused to spiral around the anode.   Because of the small relative area of the anode, the electrons may make many revolutions around it before "finding" it.   Because of the resulting very long paths traversed by the electrons, sufficient ionization to maintain discharges is produced even at very low pressures.   Discharges have been reported down to pressures as low as 10^{-12} Torr.[18]   However, when operating in this mode the main voltage drop across this system is at a negative space charge around the anode rather than at a positive space charge at the cathode, so that in this mode of the discharge much less sputtering of the cylinder material will occur.   Attempts to insert additional negative probes into this system do not appear to have been made as yet, although some preliminary investigations of sputtering in an inverted magnetron have been reported.[19]

A magnetic field which is radially symmetric but which has strong transverse components may be obtained if the polarities of the solenoids or permanent magnets are arranged so that they oppose rather than supplement one another, as shown in Fig. 6.   The influence of such a field on a glow discharge was first described by Rokhlin in 1949[20] and was later studied for its influence on sputter-

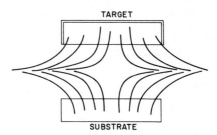

TARGET

SUBSTRATE

Fig. 6 Lines of force in a quadrupole magnetic field.

ing under dc conditions by Kay.[21]   While an increase in deposition rate by a factor of about 30 was reported, the thickness uniformity of the deposit over a $4\frac{1}{2}$-in.-diameter area was only $\pm 25\%$.

4. GROWTH OF FILMS IN THE SPUTTERING ENVIRONMENT

Although films deposited through sputtering are generally indistinguishable from films deposited by other methods, there is evidence that the way in which sputtered films grow may be quite different from the way in which films grow during, for example, vacuum evaporation. The first evidence for this appears to have been provided by Ditchburn,[22] who compared sputtered and evaporated films of cadmium in a study of whether or not the critical-nucleation-density phenomenon found in evaporation was also present during sputtering. Within the limitations of his experiments, he could find no minimum accumulation rate below which cadmium films could not be deposited by sputtering. This is quite different from what is observed for evaporation.

As discussed below, there are a number of reasons why we expect the growth conditions during sputtering to be different from those that prevail during evaporation and other deposition methods.

a. Influence of Inert-gas Background

This is very difficult to separate out from the other variables which we will discuss below since high pressures will reduce the energy of arrival of the sputtered atoms and at the same time increase the charged-particle density at the substrate. In addition, unless special precautions are taken, an increase in the inert-gas pressure generally means an increase in the background-impurity concentration (see later). The few experimental data that do exist on the effect of pressure, per se, appear to be contradictory. Thus, for example, Francombe and Schlachter,[23] when attempting to grow epitaxial gold on rock salt, found that at low pressures only amorphous films could be obtained but that as the sputtering pressure was increased preferred orientation became more perfect until, at pressures of 300 mTorr and deposition rates of 0.2 Å min^{-1}, epitaxial growth could be obtained at substrate temperatures as low as 30°C. On the other hand Krikorian and Sneed[24] found the epitaxial temperature for growth of germanium on germanium by sputtering to be lower the lower the sputtering pressure. The lowest epitaxial temperatures observed were for evaporated films, suggesting that evaporation represents the best situation. It was, however, pointed out that the influence of sputtering pressure might be associated with a higher concentration of background impurities. This would correlate with the results of Evans and Noreika,[25] who studied the epitaxial growth of thin films of gallium arsenide on sodium chloride substrates and found the role of background impurities to be very significant in controlling the lowest temperature at which epitaxy occurred as well as the degree of epitaxy at the optimum substrate temperature. A partial pressure of water vapor in the order of 3×10^{-5} Torr was sufficient to cause a significant deterioration in epitaxial behavior.

The trapping of inert gases in sputtered films will be discussed in a later section, while further consideration to the growth of single-crystal films by sputtering is given in Chap. 10.

b. Influence of Energy of Sputtered Particles

Whether or not the high energy of sputtered atoms has any influence on the nucleation and growth of films has yet to be clearly established. Such evidence as exists[26] suggests that there are only minor differences in nucleation associated with the particle energy per se. It is particularly difficult to separate the influence of energetic sputtered atoms from that of high-energy electrons since both will vary with cathode voltage in about the same way (see Sec. 4c).

There is some evidence that many of the sputtered atoms have sufficient energy to penetrate one to two atomic layers into surfaces on which they land.[27] One of the consequences of this is widely believed to be the generally superior adhesion of sputtered films compared with films deposited by other methods. Indirect confirmation of this has been given in an experiment of Mattox and McDonald,[28] who found that they could obtain adherent films of cadmium on iron provided the sputtering voltage exceeded 1,500 V. Cadmium and iron are mutually insoluble, and evaporated films of cadmium on iron have very poor adhesion.

The influence of cathode voltage on the structure of sputtered films has been pointed out by Schuetze et al.,[29] who found that tantalum films deposited at low cathode voltages had high resistivity and an open, networklike structure as well as a very low density, whereas films deposited at high cathode voltages had properties close to that of bulk tantalum. It is significant that the only important parameter was found to be cathode voltage and that large variations in sputtering pressure and/or sputtering current had little influence on the structure. This might imply that the energy of the atoms arriving at the substrate is not much affected by their passage through the gas (at least for tantalum).

It has been speculated[30] that the high energy of the arriving sputtered atoms can be accommodated by the substrate surface. This would facilitate the growth of epitaxial films at low temperatures. It has also been suggested that some of the sputtered atoms may generate point defects in the substrate as a result of their penetration into it and that point defects then act as nucleation sites for the formation of stable nuclei. Finally, it has been suggested that the high energy of sputtered atoms plays a significant role in the lowering of the epitaxial temperature through enhanced substrate cleaning.[31]

c. Role of Charged-particle Bombardment

As part of the investigation already discussed above, Ditchburn[22] performed an experiment in which cadmium vapor was generated inside an evacuated tube whose walls were at too high a temperature for nucleation to occur, even though the partial pressure of cadmium was in the order of 10^{-3} Torr. When, however, an electrodeless discharge was initiated inside a part of the apparatus, a cadmium film was seen to condense in that part only. This suggested that bombardment by charged particles allowed the cadmium to nucleate, although the possibility that cadmium ions were being produced and accelerated into the walls could not be entirely eliminated.

Recent experiments with evaporated films have confirmed that a significant role is played by charged particles in the nucleation and growth of thin films. Chopra, for example, has shown[32] that increased continuity in thin evaporated-metal films is obtained if they are grown in the presence of an electric field applied in the plane of the substrate, while Stirland[33] has shown that electron bombardment of a rocksalt substrate prior to the deposition of gold films has a pronounced influence on the continuity of epitaxial films. In the latter work, electron energies in the range 9 to 300 V were used. The energy threshold for influence by electron bombardment was found to lie between 12 and 75 V.

Related effects have been seen for bombardment by positive ions during growth. For example, Wehner[34] was able to grow single-crystal films of germanium on germanium (100) by subjecting the substrate to positive-ion bombardment (100 V). The deposition rate was around 0.5 μ h^{-1}, at temperatures between 250 and 300°C, which is significantly lower than the "evaporation limit" of Krikorian and Sneed[24] of approximately 350°C. In the case of positive-ion bombardment, however, it is not clear whether it is the positive charge that is responsible for the effects seen or whether partial resputtering is responsible for improving the quality of the films. This will be covered more fully below.

Effects of the type just discussed could explain why the nucleation and growth of thin films during sputtering are often dependent on the position of the substrate in the sputtering chamber. For example, Molnar et al.[35] compared the growth of gallium arsenide films sputtered onto vitreous silica and polished calcium fluoride substrates as a function of both temperature and their position in the sputtering system. In particular, substrates located just inside the Crookes dark space were compared with those situated well inside the negative glow. It was found that when the substrate was inside the negative glow a predominantly (111) orientation was present over the whole temperature range (up to 600°C), whereas when the film was deposited onto a substrate placed in the dark space the films were amorphous up to 400°C and showed a (110) texture up to 510°C, the latter changing abruptly to a

(111) texture for further temperature increases. Unfortunately, the present level of our understanding of these effects is not sufficient for us to make general predictions as to what type of film will grow under any given set of discharge conditions.

d. Bombardment by Energetic Particles

We have already seen that there may be present in a glow discharge a substantial number of energetic neutral-gas atoms moving in a direction away from the cathode. Inevitably, these will bombard the depositing film and may resputter some fraction of it. Jones et al.[36] have observed that silicon films deposited through both rf and dc sputtering have a reemission coefficient of about 0.1 (i.e., 10% of the film being deposited was being reemitted). Under their conditions there were no energetic ions drawn from the plasma into the film. It was also shown that insulating films of SiO_2 deposited through rf sputtering could develop substantial floating potentials at the surface during deposition (approximately 100 V). Ions accelerated out of the plasma as a result of these large floating potentials could then cause considerable resputtering, and reemission coefficients as high as 0.7 were observed.

Another example of the influence of the discharge conditions on the structure of films is seen in the case of β-tantalum. First reported by Read and Altman,[37] this is a polymorph of tantalum which does not exist in bulk form. The exact conditions for obtaining β-tantalum do not appear to have been completely specified as yet. However, β-tantalum (as opposed to the normal bcc structure) tends to be obtained when sputtering is performed at high pressures and at relatively low substrate temperatures (less than about 600°C). If there is any appreciable contamination of the argon sputtering ambient, the β form will not be obtained.

As might be expected from the above, if sputtered films are allowed to deposit outside the glow discharge, they will have properties which are quite different from those of films sputtered in the conventional manner. Thus, for example, Chopra et al.[38] used an ion beam which they generated in a modified duoplasmatron and sputtered a small target. The sputtered material was collected on substrates some 8 cm from the target. Since a relatively narrow beam was used for sputtering, the substrate was not exposed to the plasma during film deposition. A number of materials normally having the bcc crystal structure were studied, and it was found that, provided the substrate temperature was not too high (typically less than 450°C), the films that were deposited had a structure which approximated very closely an fcc structure. In addition, the films had densities that were considerably lower than normal. These results are summarized in Table 1. There was considerable evidence to show that these new structures were true polymorphs and not impurity-supported phases.

TABLE 1   Experimental and Calculated Structural Parameters
(For sputtered metal films deposited outside the plasma)

	Ta	Mo	W	Re	Hf	Zr
Experimental fcc lattice constants, Å	4.39	4.19	4.13	4.04	5.02	4.61
Normal lattice constants, Å:						
a_0	3.30	3.15	3.16	2.76	3.19	3.23
c_0	....	....	....	4.46	5.06	5.15
Experimental density, g/cm⁻³	14.2	8.66	17.3	18.8	9.37	6.18
Normal density, g cm⁻³	16.6	10.2	19.2	21.0	13.2	6.51
Density change, %	14.5	15.1	9.8	10.5	29.1	5.2

Some mention should also be made of the influence of substrate temperature. There is ample evidence that temperature plays at least as important a role in determining the structure of sputtered films as it does for evaporated films. However, the mechanisms through which temperature exercises its influence may be quite different for the two types of deposition process. As we have seen, films being deposited through sputtering are commonly subject to bombardment by ions, energetic neutrals, or both. Such bombardment is likely to cause damage to the films, mainly through the introduction of point defects. For a growing film, low temperatures will suffice to anneal out these defects as quickly as they are formed. Ogilvie and Thompson[39] studied the disorientation produced in the surface of single-crystal silver as a result of bombardment by argon ions, as a function of temperature. Their results showed that the disorientation that was produced was highly dependent on the temperature at which the bombardment was carried out. They concluded that point defects were being introduced by the impact of the ions at the surface and suggested the possibility that annealing was accelerated by the presence of such defects. If annealing was performed subsequent to bombardment, the disorientation produced by the bombardment was not removed.

5. SPUTTERING EQUIPMENT AND THE INFLUENCE OF DEPOSITION PARAMETERS

a. General Parameters

(1) Cathode Shields and Uniformity of the Deposit It is common practice in sputtering equipment to arrange for sputtering to take place from only one side of the target. This is due to the fact that the back side of the cathode often contains cooling coils, attachment fixtures, etc., which it would be highly undesirable to have sputter. It also serves to economize on the total amount of current that must be supplied to the cathode. Suppression of sputtering is most commonly accomplished by use of a shield of metal at anode potential placed at a distance less than the Crookes dark space away from the cathode.[1] As we have already seen, no discharge will take place between two surfaces spaced a distance less than the Crookes dark space. Clearly, the cathode shield must follow all the contours of the cathode so that at no point is the Crookes-dark-space distance exceeded. Even though a discharge somewhere inside the shield-cathode combination might not produce any undesired sputtered material, it could easily develop into an arc. Dielectric material may be used to suppress sputtering instead of a cathode shield. However, there is usually a danger of outgassing, and the shaping of a dielectric suppressor is invariably more difficult. In addition, there is frequently a problem with sputtered material which coats the dielectric.

The emission of sputtered material from polycrystalline or amorphous targets bombarded at normal incidence closely approximates a cosine (see Chap. 3). Consequently, some material will always be ejected at a small enough angle to the cathode surface to be able to emerge from the space between the cathode and substrate planes. In addition, collision of sputtered atoms with the gas atoms may cause some deflection of the former. The chance that sputtered material will be lost thus increases with cathode-substrate spacing as well as with distance away from the center of the cathode. In Fig. 7 are shown some theoretical distributions of material in the substrate plane as a function of the ratio of cathode diameter to electrode spacing. The curves were calculated assuming that emission of material from the cathode follows a cosine distribution, that all material remained where it landed, and that there were no collisions with gas atoms. The cathode was disk-shaped.

The distribution at the substrate of material sputtered from a metallic target has been studied by Schwartz et al.[40] Using a "bull's-eye" target consisting of four concentric rings of different metals and subsequently analyzing the distribution of each metal in the deposit by means of x-ray fluorescence, they were able to identify, on the substrate, material from the various zones in the cathode. The resulting curve for a cathode diameter-to-spacing ratio of 6.8 is shown in Fig. 8. The slight increase at the center was believed to be due to a slight increase in current density toward the axis of the cathode which had been reported by others.[41] Of most interest, however,

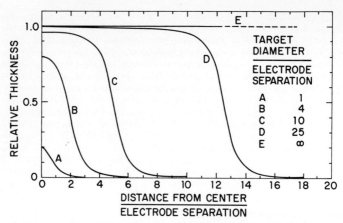

Fig. 7   Theoretical distribution of sputtered material in substrate plane.

is the sharp rise in the amount of material at the edge of the deposit—quite different from the theoretical distribution of Fig. 7.

This excessive sputtering from the outer edge of the cathode, as though there were a ring source superimposed on the main disk, has been observed by many workers and is most clearly manifested in greater corrosion of the cathode at the rim.[42]   The effect results from changes in the ion trajectories at the rim, as illustrated in Fig. 9. This shows the potential distribution in the vicinity of the cathode-shield combination.   As can be seen, the dark space has a focusing effect on ions near the edge so that there is an increase in current density at the rim.   Also of importance is the fact

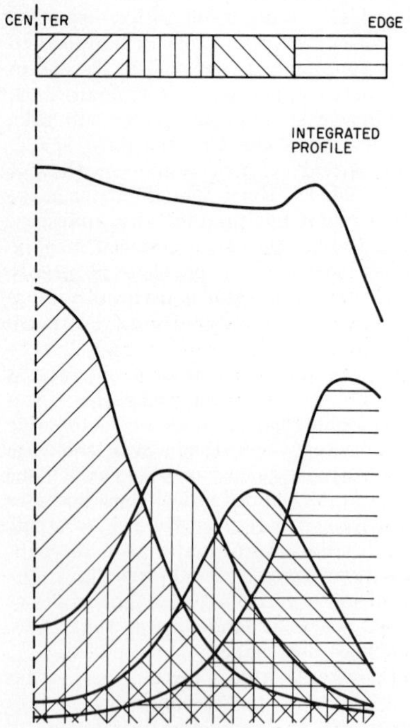

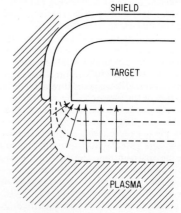

Fig. 8   Distribution of bull's-eye cathode material in substrate plane.

Fig. 9   Potential distribution in vicinity of cathode shield.

that, at the edge, ions arrive at oblique angles of incidence and as a result their sputtering yields are considerably greater than those of ions originating away from the edge which arrive at normal incidence. This enhanced sputtering at the rim can be minimized by increasing the length of the shield,[43] as shown in Fig. 10a, which mechanically blocks some of the ions that would otherwise be available to bombard at the rim or by bending the cathode shield around to partly cover the cathode,[40] as shown in Fig. 10b. As a limiting case of the latter, the shield may completely cover the face of the cathode and have one or more apertures in it which allow selected areas of the cathode to be sputtered where desired.[44]

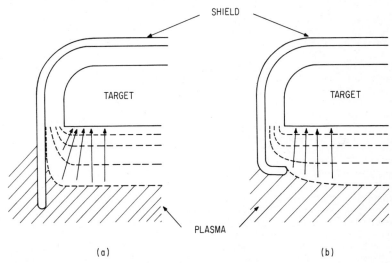

Fig. 10 (a)   Reducing rim effect by extending cathode shield.   (b)   Reducing rim effect by wrapping shield around the cathode.

If a longitudinal magnetic field is used with the sputtering system, it is usually found that the edge effect is further enhanced. This is mainly the result of an increase in current density at the edge, since electrons leaving the cathode in the vicinity of the shield are not parallel to the magnetic field and therefore spiral around the lines of force, leading to enhanced ionization, and hence further increased current density, in the vicinity of the cathode edge. In practice, the edge-enhancement effect is not necessarily disadvantageous as it helps to smooth out the distribution of material at the substrate which would otherwise be as shown in Fig. 7.

The effect of increasing pressure is to decrease the edge effect; however, the effect of pressure is not large until relatively high pressures are reached (see below). Thus, in any particular sputtering system, a certain amount of empirical experimentation with the design of the shield is advisable to optimize the uniformity of the deposit for that particular system along with its particular operating conditions.

(2) Energy Dissipation and Substrate-temperature Control Although sputtering is basically a low-temperature process, considerable amounts of energy are dissipated as a by-product, and unless precautions are taken for the effective removal of the resulting heat, substantial and usually undesired temperatures will be generated inside the sputtering system. Less than 1% of the total applied power goes into the ejection of sputtered material and secondary electrons. About 75% of the remaining, nonproductive power is dissipated as heat at the cathode by the ions and energetic neutrals that strike it. The remaining power is absorbed by the secondary electrons as they are accelerated across the Crookes dark space and then is dissipated by them through collision with the substrate. If the substrate is given sufficient negative bias to repel the energetic electrons, it will still heat up because it attracts an equivalent current of positive ions which then dissipate *their* energy there. Thus, in sputtering systems that operate at low pressure (such as rf or triode systems), particularly when the substrate is very close to the cathode, the energy per unit area dissipated at the substrate may be comparable with that being dissipated at the cathode.

It follows from the above that cooling of the substrate during sputtering is usually necessary. While it is easy to cool the substrate holder, it is often very difficult to effect good heat transfer between it and the substrate itself. This is because, at the pressures used for sputtering, thermal conduction between two surfaces lightly in contact is effectively zero and heat is transmitted only by radiation. For example, under glow-discharge conditions of 3 W cm^{-2} at the cathode at a pressure of 60 mTorr, a glass substrate 1 mm thick resting on a water-cooled holder reaches a surface temperature of about 300°C. Of this 300°C, only 50°C is dropped across the substrate, the remaining 250°C appearing as a discontinuity between the underside of the glass and the substrate holder.[3]

In practice, the only really satisfactory means for achieving the required heat transfer is to use a thin layer of a liquid between the substrate and the substrate holder.[45] Clearly, the number of suitable liquids is limited, and most success has been achieved with gallium or gallium-indium alloys. Molybdenum and, to a lesser extent, stainless steel are suitable materials for the surface of the substrate holder; the former is not corroded by gallium to temperatures as high as 500°C. The gallium is applied at a temperature slightly above its melting point by wiping it onto both the substrate holder and the underside of the substrate, after which the two are pressed together. For situations where the substrate temperature is not expected to rise above about 150°C, very satisfactory results can be obtained by the use of silicone-based vacuum grease as the heat-transfer medium. When an intermediate liquid medium cannot be used, less satisfactory, but often acceptable, results can be obtained by placing a soft metal foil such as indium, tin, or lead between the substrate and the substrate holder and applying uniform pressure to the substrate, thus pressing it up against the foil and squeezing the foil between it and the holder.[46]

In some cases the exact substrate temperature is not important but uniformity of temperature across the substrate is nevertheless required. This can be achieved by complete thermal isolation of the substrate holder, which allows equilibrium to be established at the substrate between the energy dissipated by the discharge and the heat lost by radiation. Under reproducible glow-discharge conditions a uniform substrate temperature will then be obtained which is a function of the discharge conditions.

Although the sputtering yield for most materials increases with temperature, it is nevertheless not usually advisable to allow the cathode temperature to rise to any extent during sputtering because of possible problems with outgassing. Cooling of high-voltage cathodes does introduce problems, however, since the coolant that is used will also be at high voltage. If a liquid coolant is used, care must be taken to see that electrolysis effects do not lead to internal corrosion of the cathode support.[44] This particular problem can be avoided if a gas, such as forming gas, is used for the coolant.

Since most of the heat produced at the substrate is the result of bombardment by high-energy electrons, any measure taken to divert these electrons away from the substrate will be effective in reducing substrate temperature. One method for doing this is to apply a magnetic field parallel to the substrate.[47] Note that, in the cylindrical configuration with axial magnetic field, this is achieved automatically.

(3) Pumping Systems Because of the large amount of heat dissipation and because of the possibility of ionic bombardment of any exposed surfaces, there is usually considerably more outgassing in sputtering systems than in comparable evaporation systems. The problem is further aggravated by the fact that any small increases in the background pressure resulting from outgassing (or any other cause) will normally go undetected, being masked by the large background pressure of the sputtering gas. It is therefore advisable to arrange for all surfaces that are exposed to the plasma to be cooled and grounded, if at all possible. It is also advantageous for most systems to arrange for fresh gas to be constantly admitted and removed during sputtering rather than to seal the system off and merely maintain the sputtering pressure.   The steady passage of relatively large amounts of fresh, high-purity gas through the system serves mainly to dilute the concentration of outgassed impurities present in the gas, although it also helps cooling.   Since the amount of outgassing generally diminishes with time, it is common practice to insert a shutter between the substrate and the cathode, opening it after a time period which experience has shown allows the system to be reasonably well cleaned up (see Sec. 6b).

The flow of large quantities of inert gas at relatively high pressure through a system does, however, impose considerable strain on most pumps.   Thus, for example, Vacion pumps are not suitable for sputtering applications, except in low-pressure systems.   Diffusion pumps are probably the most widely used type for sputtering, and provided care is taken to see that no significant backstreaming of the pump fluid occurs, they are quite satisfactory.   The two main criteria which determine the extent of the backstreaming of diffusion pumps under given pressure and flow conditions are the fore-pressure tolerance and the throughput.   For further discussion, see Chap. 2.

If there is reason to believe that the diffusion pump is introducing impurities, these can be minimized by partially throttling the system until the dynamic pressure at the diffusion pump itself drops to an acceptable value.   It has also been shown that sputtering can be accomplished with the main valve completely shut off if a titanium sublimation pump is attached to the sputtering chamber.   The pressure is then maintained through a small leak of the sputtering gas.[48]   The advantage of this arrangement is that the titanium sublimation pump does not pump the inert gases but has substantial pumping rates for oxygen, hydrogen, nitrogen, etc.   An auxiliary glow discharge using a reactive-metal cathode in an adjoining chamber can also be employed to perform much the same function (see Sec. 6b). 4-21

b. Factors Influencing the Deposition Rate

(1) Current and Voltage   In the energy range employed for the deposition of most films, the sputtering yield increases relatively slowly with increasing ion energy. On the other hand the number of particles striking the cathode is proportional to the current density.   Consequently, current is a significantly more important parameter for determining the deposition rate than is voltage.   This is illustrated in Fig. 11,

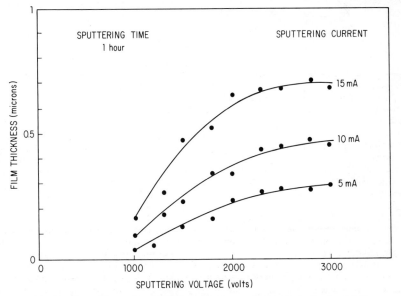

Fig. 11 Deposition rate of tantalum films vs. voltage at constant current.

which shows the deposition rate of tantalum films[29] as a function of voltage, for conditions of constant current. As can be seen, beyond about 2,500 V the increase in deposition rate is negligible even though the total power to the system is being steadily increased. With a limited amount of power available, it is therefore best to sputter at high current and low voltage. This is achieved either by use of the thermionically or magnetically supported glow or by going to higher pressure (see below).

(2) Pressure and Source-to-Substrate Distance As the pressure in a sputtering system is raised, the ion density and, therefore, the sputtering-current density increases. Therefore, in agreement with what was just said above, it is not surprising that (for constant power input) the deposition rate increases linearly with pressure provided the latter is not too high. This is illustrated[49] in Fig. 12. The figure shows that the current also increases linearly with pressure but at a slightly greater rate. This is because the fraction of the sputtered deposit that returns to the cathode by back diffusion also increases with pressure. Below pressures of about 20 mTorr, however, back diffusion is negligible, and it is only at pressures in excess of approximately 130 mTorr that more than half the material that leaves the cathode returns to it through diffusion.[50] Thus, until pressures of this order are reached, the deposition rate will continue to increase with increasing pressure. As expected, when still higher pressures (in the torr range) are used, the deposition rate falls off rapidly with increasing pressure.[51]

In practice, the optimum pressure to use for sputtering may be lower than the break-even point of around 130 mTorr just mentioned. This is because of the problem of backstreaming already discussed above and also because the reduced size of the Crookes dark space at the higher pressures makes effective shielding of the cathode increasingly more difficult. For the self-sustained dc glow discharge in argon many workers have found the best compromise for sputtering to be at 50 to 60 mTorr.

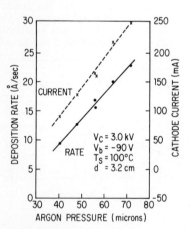

Fig. 12 Deposition rate of molybdenum vs. pressure.

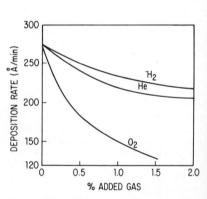

Fig. 13 Effect of H_2, He, and O_2 on deposition rate of SiO_2.

(3) Gaseous Impurities Any of several impurity gases, when present in the argon during sputtering, may cause an appreciable reduction in deposition rate. Figure 13 shows some data on the relative deposition rates of SiO_2 films (rf sputtering) for various percentages of hydrogen, helium, and oxygen. Gases such as carbon dioxide and water vapor which are decomposed in the glow discharge to produce oxygen had an effect similar to pure oxygen, while carbon monoxide showed a very slight tendency to increase the deposition rate, possibly because it removed traces of residual oxygen from the gas. Nitrogen had essentially no effect.

Stern and Caswell,[52] in a study involving Nichrome films, have explained the influence of hydrogen on the deposition rate as due to a current-robbing effect. Because of its high mobility, the hydrogen would be expected to carry a fraction of the current which is substantially larger than its concentration in the gas phase. On the other hand, for practical purposes the sputtering yield of hydrogen may be considered to be zero.   Consequently, a large fraction of the measured current derives from carriers which produce no sputtered material.   Similar reasoning would explain the effect of helium.

The mechanism by which oxygen affects the deposition rate is quite different and has been explained for the case of rf-sputtered SiO_2 by Jones et al.[53]   If it is supposed that the SiO_2 target is sputtered away layer by layer, then alternating layers of oxygen and silicon atoms would be successively removed from the target.   If, however, an oxygen layer that is removed by sputtering is immediately replaced by oxygen out of the sputtering ambient, sputtering could, in principle, never proceed beyond the first layer.   In practice, the effect saturates at an oxygen partial pressure of about 5×10^{-4} Torr at which point the deposition rate is approximately half that obtained with pure argon.   This suggests that at pressures greater than this the oxygen sputtered away is replaced as fast as it is removed but that roughly 50% of the sputtered material comes from below the immediate surface.

Similar effects to those for SiO_2 are seen if small percentages of oxygen are present during the sputtering of metals,[127] whether by dc or rf methods.   The mechanism described above will be operative in these cases, but the effect will be complicated by the formation on the surface of the target of an oxide layer which will probably have a lower sputtering yield.

(4) Substrate Temperature   A strong dependence of the deposition rate on the substrate temperature has been reported for a number of sputtered materials.   These include SiO_2,[54] gallium arsenide,[25] and germanium,[24] for all of which there was a marked decrease in deposition rate with increasing substrate temperature.   Krikorian and Sneed[24] have made a direct comparison of the temperature dependence of both sputtered and evaporated germanium films and conclude that the mechanism is the same in both cases.   For example, for polycrystalline films, an activation energy of 0.12 eV was found for both the sputtering and evaporation cases, independent of the deposition conditions.   They concluded that a common growth-rate mechanism applied to both cases so that the phenomenon was not peculiar to sputtering.   These authors have also shown that the temperature dependence for material deposited by reactive sputtering is reversed.   That is, deposition rate increases with increasing substrate temperature.   This will be discussed again in Sec. 7.

Glang et al.[49] found that the deposition rate of molybdenum by sputtering was also dependent on temperature, particularly if the film was subjected to ionic bombardment (bias sputtering) during deposition. Their results are shown in Fig. 14.   The apparent increase in the temperature dependence under condition of ionic bombardment is probably due to the accompanying increase in deposition rate.

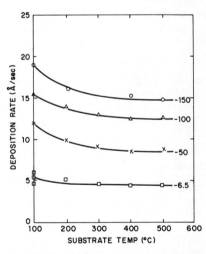

Fig. 14   Deposition rate vs. temperature for molybdenum films at several substrate biases.

c. Other Types of Sputtering Equipment

Although systems employing two plane-parallel electrodes are the most widely used for sputtering, a number of variations exist.   The motivation behind these is usually to enable sputtering to proceed at lower pressures on the mistaken premise that films sputtered at lower pressures necessarily have higher purity (for example, see Fig. 20).

The thermionically supported glow discharge for sputtering has already been discussed, and most equipment[55] is a variation on Fig. 5. By using a large filament and allowing the plasma to diffuse out of a suitably positioned filament shield, a uniform plasma may be obtained over a relatively large area. For example, Toombs[56] was able to deposit copper at 400 Å min^{-1} at 2 mTorr pressure onto a 400-cm^2 substrate with $\pm 20\%$ thickness variation. A somewhat similar arrangement has also been described by Huss.[11] The cylindrical geometry (with or without magnetic field) has already been discussed above, as well as elsewhere.[57]

In recent years some workers have begun to experiment with the use of high-intensity ion beams for sputtering[58,59] as well as with beamlike constructions in which a very intense plasma is generated in an auxiliary chamber (50 to 60 A) and allowed to diffuse into the main sputtering chamber. Because of a strong auxiliary magnetic field, the plasma emerges from the primary ionization chamber as a relatively narrow beam. Additional voltage is then applied at the target where the ions receive the energies they require in order to perform sputtering.[60,61]

Finally, we may make brief mention of several production-oriented sputtering systems which have been described in the literature in recent years. These vary from single-chamber systems with internal substrate-feeder arrangements[62] to small moving-belt systems[63] to large systems capable of handling a steady stream of substrates entering directly from atmospheric pressure. After passing through a series of chambers at successively lower pressures, they enter the sputtering chamber and then emerge, following the same sequence in reverse, all this being on a continuously moving belt.[64,65]

6. TRAPPING OF IMPURITIES IN FILMS DURING SPUTTERING

a. General Considerations

The fraction f_i of impurity of species i trapped in a film is given by

$$f_i = \frac{\alpha_i N_i}{\cdot \alpha_i N_i + R} \qquad (2)$$

where N_i is the number of atoms of species i bombarding unit area of film in unit time during deposition, α_i is the effective sticking coefficient of the species i during deposition, and R is the deposition rate of the film. It is clear from (2) that there are three possible ways to reduce f_i: for example, the deposition rate R can be increased. While such an approach is practical for certain materials deposited through evaporation or pyrolysis, it is only of limited use for sputtering. A significant improvement in the purity level of films deposited by sputtering will therefore depend on reducing either α_i or N_i (or both).

During sputtering, both α_i and N_i are likely to be higher than their counterparts for other deposition methods. For α_i, this is due to the fact that many of the impurities will be present as atomic rather than molecular species and also because many of them will be ionized.

Fortunately, as will be discussed below, methods are available for reducing both N_i (getting sputtering) and α_i (bias and ac sputtering). Note, however, that no methods are currently available to prevent nongaseous impurities present in the cathode from finding their way into the film.

Usually, in thin film investigations, the worker is concerned with some specific physical property of the film such as resistivity or coercive force rather than with the impurity level per se, and it is often more convenient, as well as more meaningful, to assess any particular film-preparation method in terms of that property. Much work on the deposition of high-purity films through sputtering has involved resistivity as a measurement criterion. We should therefore point out that at relatively low impurity levels the resistivity of films is a function of structure as well as of the impurity concentration. This will become evident in the section on bias sputtering.

b. Getter Sputtering

This technique turns to advantage the fact that a depositing film is an active getter for impurities. It has been widely observed that the purity of thin films improves as they grow in thickness. In a getter-sputtering system, the sputtering gas is made to pass over an area of freshly deposited film material before reaching the region where the film which is to be retained is being deposited. In its simplest form, therefore, a getter-sputtering system would consist of a very large cathode and an equally large substrate holder with the proviso that the only film material that is used is taken from what is deposited right near the center of this assembly. Any impurity atoms in the sputtering gas would have made a large number of collisions with depositing film material before reaching the central region and would therefore have a very high probability of being removed. Such an arrangement thus cleanses the sputtering gas of its impurities, whether they were present in the original gas or whether they were generated through outgassing from the walls. However, the technique can do nothing about gaseous impurities that originate in the cathode itself.

In more practical getter-sputtering arrangements, the cathode is of finite size but the walls of the sputtering chamber are arranged to be as close as possible to both the substrate and cathode, subject only to not interfering with the glow discharge. The compact sputtering chamber is constructed so as to be moderately gastight and is placed inside the main vacuum chamber. This arrangement has two advantages: Since a positive pressure of sputtering gas is maintained inside the sputtering chamber, any tendency for oil vapors or other impurities from the main chamber to find their way into the sputtering environment is reduced.[66] In addition, if the inner chamber is reasonably tight, an appreciable pressure differential can be maintained between it and the primary vacuum chamber so that any tendency for the backstreaming of oil from the diffusion pumps is significantly reduced. Another feature that is sometimes incorporated in getter-sputtering systems is the elimination of the cathode shield so that material is sputtered from all sides of the cathode. While the cooling of such a cathode is considerably more complicated than in conventional designs, very efficient gettering results if the sputtering gas is admitted at the top of the chamber close to the axis of this unshielded cathode. A getter-sputtering system showing these various features[67] is illustrated in Fig. 15.

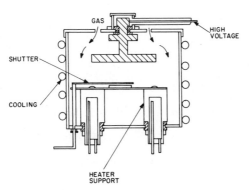

GAS

HIGH VOLTAGE

SHUTTER

COOLING

HEATER SUPPORT

Fig. 15 Getter sputtering system, after Cooke et al.

One of the most important features of the apparatus shown in Fig. 15 is the shutter. Without this it is usually impossible to operate a getter-sputtering system satisfactorily. Although the arrangement of Fig. 15 is very effective for removing the impurities that are brought in by the sputtering gas itself (including those coming from the pumping system), it is relatively ineffective in coping with impurities knocked off the walls. The emission of impurities from the walls of the chamber is inevitable when the glow is first switched on and a period of time which varies from minutes to hours (depending on the overall state of the system) must be expected during which these impurities will be present. After a time period has elapsed which experience has shown to be sufficient for these internally generated impurities to be down to insignificant levels, the shutter can be opened. A similar function is served by the shutter with respect to impurity gases initially present at or near the surface of the cathode.

The first description of a getter-sputtering apparatus appears to have been published in 1943 by Berghaus and Burkhardt,[68] whose equipment was remarkably similar to that shown in Fig. 15. More recently, getter sputtering has been developed as a film-deposition technique by Theuerer and Hauser, who have also described a number of different versions of getter-sputtering equipment.[69,70] They have been primarily interested in superconducting materials and have been able to deposit films of materials such as niobium and tantalum with critical temperatures which agree well with the accepted bulk values.

c. Bias Sputtering and Asymmetric AC Sputtering

It was discussed in an earlier section that films being deposited by sputtering may be subject to a certain amount of resputtering. This occurs either through the action of energetic neutrals, negative ions that originate at, or near, the cathode, or in the case of insulating films, because positive ions may have been accelerated toward the film surface as a result of a significant negative floating potential there. If, instead of leaving its potential to chance, the film is deliberately given a negative potential with respect to the plasma, the resulting technique is referred to as bias sputtering.[71,72]

The basis for expecting bias sputtering to lead to films of higher purity is that, during resputtering, most impurities should be preferentially removed relative to the atoms of the main film. Whether or not this will occur depends on the relative strengths of the metal-to-impurity and the metal-to-metal bonds. Thus, for example, as a means for removing oxygen from sputtered films, bias sputtering is very effective for materials such as tantalum, molybdenum, and niobium but has little effect for materials such as aluminum and magnesium in which the oxide bond is stronger than the metallic bond.

The first description of an apparatus in which the substrate is given a negative bias with respect to the anode in order to improve film purity appears to have been given by Berghaus and Burkhardt.[73] However, they mistakenly thought that the improved properties exhibited by their films were a consequence of the additional heating of the film by the ion bombardment. A description of a bias-sputtering apparatus for improving the epitaxial growth of germanium films was given by Wehner[34] in 1962, while a detailed analysis of the processes occurring during bias sputtering (with particular reference to tantalum) was provided by Maissel and Schaible[74] in 1965. In the latter publication it was shown that (2) is modified in the presence of bias to

$$f_i = \frac{\alpha_i N_i - (j/q)(AS - \beta)}{\alpha_i N_i - (j/q)(AS - \beta) + R}$$

where

$$A = \frac{\alpha_i N_i + \beta j/q}{\alpha_i N_i + j(S + \beta)/q}$$

and β is the fraction of the bias current due to impurity ions, j is the bias current density, q is the electronic charge, and S is the sputtering yield for the impurities.

The results of this theory are compared with experiment in Fig. 16, while in Fig. 17 are shown the effects of bias against different background levels of impurity. The increase in impurity concentration at very low bias is not quantitatively understood at this time; however, the very rapid falloff in impurity concentration to the right of the peak at approximately 20 V is believed to represent the onset for sputtering of the impurities out of the tantalum. The increase in resistivity at high bias voltages is believed to be due to an increase in the concentration of trapped argon and will be discussed in the next section.

While bias has a pronounced effect on the impurity content of sputtered films, it can also exert an influence on their structure. The influence of bias on the structure of

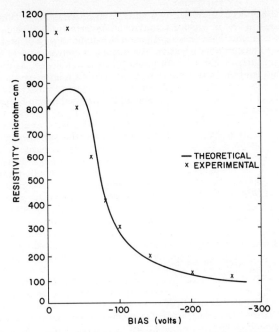

Fig. 16 Resistivity vs. bias voltage for tantalum films comparing theory with experiment

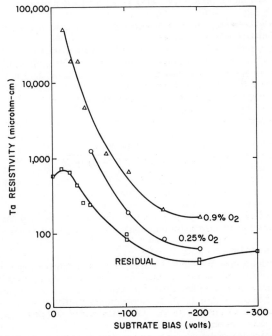

Fig. 17 Resistivity of tantalum films vs. bias for several levels of oxygen contamination during deposition.

tantalum films has been described by Vratny et al.,[75] whose data (shown in Fig. 18) indicate the range of bias in which β-tantalum is converted to the normal bcc structure. It is of interest to note that these authors studied the effects of positive as well as negative bias. The curve for negative bias is qualitatively similar to that of Fig. 16, but the exact meaning of the curve for positive bias is not clear. As discussed earlier, it is not possible to obtain any appreciable positive bias relative to the plasma. Any surface so biased simply becomes a new anode, the original anode becoming a secondary cathode. It is possible that the sharp decrease in resistivity at about +10 V is due to heating of the substrate associated with the relatively large electron flux pulled in under these conditions. It should be noted here that electron bombardment can cause desorption of surface impurities. However, the process requires a minimum electron energy and is quite inefficient relative to sputtering. Thus, for example, in the case of oxygen and molybdenum, a threshold of about 19 V is required and the maximum desorption probability (at 90 eV) is 5×10^{-4} atoms/electron.[77]

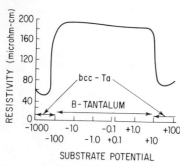

Fig. 18 Resistivity vs. bias for tantalum showing change in phase (β to bcc).

In a study of the bias sputtering of molybdenum films, Glang et al.[49] observed that their films, when deposited with little or no bias, tended to show a preferred orientation in the (110) direction. As the bias during deposition was increased, this preferred orientation changed to the (211) direction for films deposited at 100°C, and into equal intensities of the (211) and (110) directions for films deposited at 500°C. They also observed that bias had a pronounced effect on the stress level of their films. Films deposited at or slightly below the optimum voltage of 100 V tended to be under tensile stress, whereas films deposited with no bias, or with biases of around 150 V, tended to be under compressive stress.

The resistivity and structure of sputtered films were further studied by d'Heurle,[76] who compared molybdenum films deposited by getter sputtering and/or bias sputtering. Films deposited at 350°C using bias sputtering alone had a resistivity of 10 μohm-cm, whereas films deposited using getter sputtering alone had a resistivity of 30 μohm-cm. Films deposited using both bias and getter sputtering had resistivities of about 7 μohm-cm. Bias sputtering has also been shown to have a significant influence on the magnetic properties of thin films.[78,79]

Closely related to bias sputtering is asymmetric ac sputtering, a process which was described by Frerichs[80] in 1962. In this method, the film is also partially resputtered during deposition, but this is accomplished by applying alternating current between substrate and "cathode." By employing a suitable resistance network, it is arranged that a significantly larger current flows during the half cycle when the "cathode" is negative so that a net transfer of material to the substrate results. Asymmetric ac sputtering appears to be in all ways equivalent to bias sputtering except that, since the bias is applied for only half the total time, the concentration of trapped impurities will always be less in films deposited by bias sputtering than in ac sputtered films.[74]

d. Trapping of Inert Gases

The concentration of argon that is trapped in a film during deposition should, in principle, be given by (2) above. It has, however, been shown that for inert gases having thermal energies, α is effectively equal to zero.[81] Despite this fact, it is more the rule than the exception to find appreciable quantities of the sputtering gas trapped in films deposited by sputtering. Winters and Kay[82] have studied this aspect of the system nickel-argon in some detail. They found that argon ions were not trapped in nickel unless they arrived at the surface with an energy of approximately 100 eV, or more. This energy was presumed to be sufficient to drive the argon ions into the surface, where they would then be held. A similar situation would be expected to hold for argon atoms. They then deposited nickel films in a conventional sputtering apparatus and studied the concentration of trapped argon as a function of bias during film deposition.

The results are shown in Fig. 19. The curve can be seen to have the same general form as Fig. 16 above. The increase in argon concentration for voltages greater than 100 V follows from what has just been said. The form of the curve for voltages between 0 and 100 is also of considerable interest. The relatively high concentration of argon in films deposited at zero bias was explained as due to energetic neutrals reflected from the cathode and arriving at the substrates with energies of 100 V or more. These have already been discussed in earlier sections. The decrease in argon concentration between 0 and 100 V was explained as being due to the resputtering of some of the trapped argon.

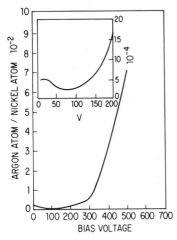

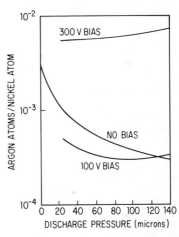

Fig. 19 Concentration of argon trapped in nickel films as a function of bias during deposition.

Fig. 20 Concentration of argon trapped in nickel films as a function of sputtering pressure.

Also of considerable interest is the argon concentration as a function of the pressure used during sputtering. This is shown in Fig. 20. As can be seen, for the no-bias case the concentration of trapped argon continues to decrease with increasing pressure. This confirms the role of energetic neutrals originating at the cathode since at higher pressure progressively fewer of these energetic argon atoms would arrive at the substrate with energies greater than 100 eV. It is thus clear that, provided the use of high pressures for sputtering does not of itself lead to the presence of excess impurities, the purest films are obtained by sputtering at the highest pressure. The concentration of argon as a function of substrate temperature was also studied and found to be relatively independent of temperature to 200°C, above which a falloff with temperature began which became very steep at about 400°C.

The trapping of argon in insulating films, specifically SiO$_2$, has been studied by Jones et al.[83] Results in general agreement with Winters and Kay were obtained with the added complication that, when high floating potentials were obtained, energetic ions were pulled in and trapped in the film as well as the energetic neutrals. The concentration of trapped argon was seen to have a strong temperature dependence, which was quantitatively explained in terms of a model in which the argon was bound to the surface at sites having a continuous distribution of binding energies in the range 0 to 1.8 eV.

7. REACTIVE SPUTTERING

a. Principles and Techniques

In the preceding section we discussed the trapping of impurities during sputtering and the ways by which this can be avoided. In certain instances, however, a compound of the sputtered material and the sputtering gas may actually be the desired end product. Reactive gas may therefore be deliberately added to the sputtering system to produce this compound in film form. When this is done, the process is termed "reactive sputtering." Most work on reactive sputtering has involved oxides, although work has also been done on nitrides, sulfides, etc. Reviews of reactive sputtering include those by Schwartz[84] and Holland,[1] while Holland and Siddall[85] have described a small plant for the reactive sputtering of various materials.

A question of continuing interest to people studying reactive sputtering is where in the sputtering system is the compound formed. Does the reactive gas form a layer on the cathode surface, following which molecules of the compound are sputtered out, or does the reaction occur at the substrate itself? The bulk of the available evidence suggests that the latter mechanism is the most important.[132] Reaction in the gas phase is unlikely because of problems of conserving momentum while also dissipating the heat of reaction.

The ejection of molecular species from a cathode is, however, by no means unknown. Thus, for example, McHugh and Sheffield[86] performed a mass-spectrometric analysis of the ions ejected from a tantalum surface when it was sputtered in mercury vapor under conditions such that contamination of the cathode could not be eliminated. They found the three most prominent species to be TaO^+, Ta^+, and TaO_2^+ which were ejected in an abundance ratio of 10:9:5.6. Note, however, that only 0.1% of the sputtered material was in the form of ions and that the molecular nature of the neutrals could have been quite different.

Where only one compound between the cathode material and the reactive gas exists, reactive sputtering is a straightforward process provided sufficient reactive gas is present to form the compound. In many cases, however, a series of compounds may be formed depending on the conditions prevailing during sputtering. This problem has been addressed in a series of papers by Perny and his coworkers.[87,88,89] Their work shows that the particular phase that is formed depends both on the percentage of reactive gas in the sputtering environment and on a quantity E^*, called the reduced field, which is given by

$$E^* = \frac{\text{cathode voltage}}{\text{cathode-anode spacing} \times \text{pressure}}$$

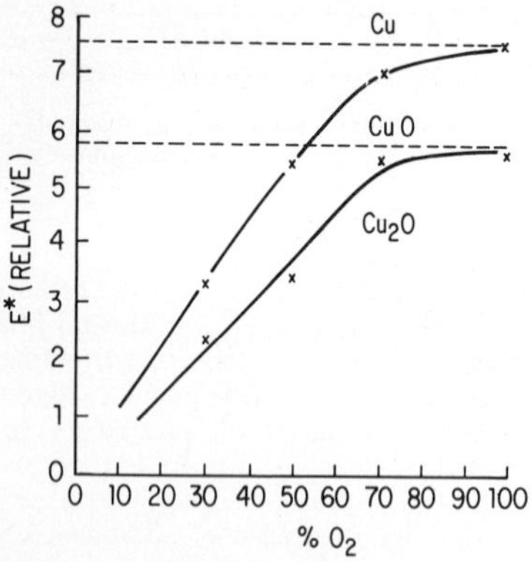

Fig. 21 "Phase diagram" showing formation of various copper oxide phases as a function of percent O_2 and the reduced field.

They have studied a number of oxide systems, including tantalum and silicon, but perhaps the most interesting is the copper-oxygen system. Figure 21 shows how the phase which forms is a function of both the reduced field and the percentage of oxygen in the sputtering gas (remaining gas was argon). Figure 21 reveals several unexpected features. For example, the phase that is formed at the highest oxygen concentration is Cu_2O rather than the CuO phase, which has the larger fraction of oxygen. In addition, it is seen that for any given oxygen concentration, including 100% oxygen, any of the three phases (even metallic copper) can be obtained by choosing the right reduced field.

Perny has put forward a phenomenological theory to explain the role of the reduced field[89] in which it is suggested that formation of the reactive film material takes place in the gas phase but at a very short distance from the substrate in a layer which he calls the "virtual thin film." Although the exact significance of the reduced field is not clear at this time, it may be noted that the energies of either neutrals or negative ions originating at the cathode would be expected to be very nearly proportional to the reduced field. The latter would therefore strongly influence the reemission coefficient. It would be expected that a copper oxide film being subjected to some resputtering during its growth would preferentially reemit the oxygen atoms since the copper-copper bond is considerably stronger than the copper-oxygen bond. By measuring the deposition rate at various reduced fields and oxygen concentrations, Perny was able to plot three-dimensional "phase diagrams" for copper-oxygen and other systems.

A detailed comparison between growth of films during reactive sputtering and growth directly through rf sputtering (see next section) as well as through direct and reactive evaporation has been made by Krikorian and Sneed.[90] Two systems studied in detail were tantalum oxide and aluminum oxide. A special feature of their study was to determine the conditions under which single-crystal films of insulating material could be grown by reactive sputtering. It was found that epitaxy occurred over only a relatively narrow range of partial pressure of the reactive gas. At partial pressures below the minimum of this range, the films that were obtained were neither single-crystal nor were they insulators (but rather two-phase systems). At pressures exceeding the top of this range, the films were insulating but were no longer epitaxial. This maximum partial pressure of reactive gas was termed the epitaxial pressure. Similar epitaxial pressures were found for both reactive evaporation and reactive sputtering.

Through consideration of the activation energies required for the various types of growth process, it was concluded that formation of the oxide phase during reactive sputtering was occurring almost exclusively at the substrate. Thus, for example, the increase in deposition rate with substrate temperature for the reactive case (see above) was considered to be strong proof that the growth of an insulating film at the substrate was being limited by the arrival rate of the reactive-gas atoms whose rate of reaction with the metallic atoms was increasing with substrate temperature. Thus, for example, for aluminum oxide films being deposited on sapphire by direct (rf) sputtering, the deposition rate changed from 75 Å min⁻¹ at a 375°C substrate temperature to 38 Å min⁻¹ at a 520°C substrate temperature, whereas for aluminum oxide deposited on sapphire by reactive sputtering in a dc supported glow discharge, the deposition rate changed from 95 Å min⁻¹ at a substrate temperature of 375°C to 145 Å min⁻¹ at a substrate temperature of 500°C.

In an earlier section it was shown that the deposition rate for sputtered films becomes relatively insensitive to cathode voltage at high voltages but, over a wide range, is directly proportional to the current density at the cathode. See, for example, Fig. 12. Peters and Mantell[91] have shown that during reactive sputtering (at least for the case of lead-tellurium oxide films) the deposition rate actually decreases at high cathode voltages. This is shown in Fig. 22, which displays the deposition rate per unit cathode current vs. the ion energy at the target. In separate experiments it was established that the deposition rate increased linearly with current at the cathode and was not greatly dependent on the sputtering pressure. It was also shown that the sputtering yield of material at the cathode increased almost linearly with ion

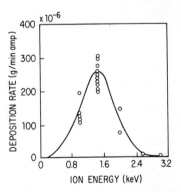

Fig. 22 Deposition rate of lead–tellurium oxide films vs. cathode voltage.

energy to at least 3 keV; so the shape of the curve in Fig. 22 could not be ascribed either to an anomalous sputtering yield or to back diffusion.

Peters and Mantell also found that the relative amounts of heat removed at the cathode and at the substrate were dependent on the oxygen concentration. At the higher oxygen concentrations, more heat was removed at the anode relative to the cathode. Depending on conditions, 35 to 60% of the total applied energy was not dissipated at either electrode and presumably went into heating up the sputtering gas. In the presence of oxygen the relative amount of heat dissipated at the anode also increased slightly with total applied power. All these observations are consistent with the generation of negative ions at the cathode which are accelerated across the Crookes dark space and then proceed through the negative glow and strike anode and substrate with substantial energies.

In general, the procedures followed during reactive sputtering are no different from those of conventional sputtering except that in many cases reduced life for the pressure gauges must be expected. In addition, for discharges that are supported thermionically by hot filaments, the life of the filaments will be severely impaired by the presence of reactive components unless the partial pressure of these is kept to very low values.

It is important to note that if a mixture of reactive and inert gases is used (see below), it is advisable that the gases be premixed before admission to the sputtering chamber. If the gases are admitted separately and their partial pressures are deduced from pressure values read in the main chamber, their actual ratio will be a function of the order in which they were admitted into the chamber. This is because in the pressure range used for sputtering the pumping rate in most vacuum systems is very sensitive to pressure.

b. Oxide Systems

By far the largest fraction of the reactive systems that have been studied have been oxides. This is both because of the useful properties of many metal oxides and because of the ease with which reactive sputtering in oxygen can be made to take place. The addition of oxygen to a glow discharge in argon has a noticeable effect on its appearance. In particular, the discharge appears as though its pressure had been raised appreciably even though only a small percentage of oxygen was added. That is, the cathode dark space contracts and the impedance of the glow decreases. This effect is a consequence of the fact that the drift velocity of electrons moving through the dark space is reduced by the presence of the oxygen atoms which temporarily attach electrons and form negative ions. The net result is that the electrons form positive ions more efficiently (see Fig. 1).

A second consequence of the presence of negative ions in the glow is that an accumulation region of ions may form around the anode in an analogous manner to the positive-ion sheath at the cathode. The exact role of such an accumulation region of negative ions with respect to reactive sputtering is not known, but some workers believe that it represents an important zone in a reactive-sputtering system.[92] For additional discussion on the influence of electronegative gases on glow discharges, see Ref. 7.

It has been found for many oxide systems that, for complete reactive sputtering to occur, it is not necessary to perform the sputtering in pure oxygen, and a sputtering ambient consisting of argon containing only 1 or 2% oxygen can produce the same product as pure oxygen. See, for example, Fig. 21. By keeping the percentage of oxygen down to the minimum required for the complete oxidation of the depositing film, a substantial increase in the deposition rate can be obtained. A typical example of this is provided in the data of Valletta et al.,[93] as shown in Fig. 23. Similar results have been reported by Vratny[94] for reactive sputtering of tantalum oxide. The latter found that the deposition rate dropped from 100 to 1 Å min^{-1} when the partial pressure of O_2 in 20 mTorr of Ar changed from 1×10^{-4} to 1×10^{-2} Torr.

This is not universally true, however, as shown, for example, in observations by Jackson et al.[95] They found that the deposition rate of lead–titanium oxide films was essentially the same whether pure oxygen or a mixed gas was used, provided the gas pressure and current density were the same.   Since there appeared to be other advantages associated with the use of pure oxygen, this was the preferred gas.   For example, there were indications that the highest oxidation state was not achieved unless sputtering was performed in pure oxygen.   By contrast, Valletta and Pliskin[96] studied the manganese oxygen system and found that (above 1 % O_2) the composition of the sputtering gas made very little difference to either the deposition rate or the stoichiometry of the film.   The latter was found to depend almost exclusively on the substrate temperature.   For substrate temperatures below approximately 350°C the more conductive MnO_2 phase was obtained, whereas at temperatures above this the high-resistivity Mn_2O_3 phase was the product.   In general, poor film quality was seen for high deposition rates and/or low substrate temperatures.

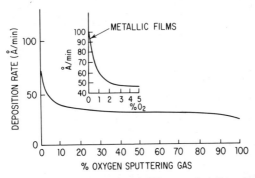

Fig. 23 Deposition rate of reactively sputtered SiO_2 as a function of percent oxygen in sputtering gas.

All in all, a large variety of metal oxide films have been deposited by reactive sputtering.   Among the earliest studies was that of Hiesinger and Koenig,[97] who examined oxides of aluminum, iron, cadmium, titanium, silicon, and thorium.   Some other systems that have been investigated are tin oxide,[98] lead–tellurium oxide,[99] vanadium dioxide,[100] and a variety of glass[101] and ceramic-like films.[102]   The latter paper contains some very interesting data relating to the influence of the type of substrate on the deposition rate.   For example, the deposition rate of an aluminum-silicate-type glass was 0.07 Å mA^{-1} min^{-1} for deposition onto a molybdenum sheet substrate and 0.64 Å mA^{-1} min^{-1} for deposition onto a substrate of single-crystal sapphire.   At this time there appears to be no explanation for this unusual behavior. Holland and Siddall[103] studied a number of oxides including indium oxide, tantalum oxide, and tin oxide, while Smith and Kennedy[104] studied a variety of reactively sputtered oxide films as possible dielectrics for capacitor applications.   Oxides investigated include those of silicon, tantalum, niobium, zirconium, and titanium. In a similar study, Goldstein and Leonhard[105] extended this inventory to include hafnium, lanthanum, and yttrium.

c. Other Systems

After oxygen, nitrogen appears to have received the most attention as an active gas for reactive sputtering.   Thus, for example, the tantalum-nitrogen system has been extensively investigated[106,107] because of the useful properties of nitrogen-bearing tantalum films.   Both dissolved nitrogen and tantalum nitrides are of interest.   The usual method for controlling the percentage of nitrogen in the tantalum-nitrogen films is to adjust the percentage of nitrogen in the sputtering atmosphere.   The titanium system has also been similarly studied, though to a lesser extent.[108]

In recent years considerable interest has been generated in the silicon-nitrogen system because of the very useful properties of silicon nitride in integrated circuits. Unlike the tantalum or titanium systems mentioned above, it has been found that low sputtering pressures are necessary if good-quality silicon nitride films are to be obtained. It is therefore necessary to use either a thermionically supported glow discharge[109] or rf sputtering.[110] Sputtering pressures below 5 to 10 mTorr appear to be satisfactory. In contrast to the silicon-oxygen system, it is necessary to sputter in pure nitrogen if good results are to be obtained. The most likely explanation for this appears to be the need for nitrogen atoms to be present for the formation of silicon nitride since there is no chemisorption of N_2 on silicon. The main species in the plasma in the vicinity of the substrate is N_2^+, N^+ being formed primarily in the dark space en route to the cathode[111] or through the dissociation of N_2^+ on striking the cathode surface. Consequently, nitrogen atoms can only reach the substrate either through being sputtered out of the target, possibly from a silicon nitride layer on the target surface,[109] or through reflection at the target after traversing the dark space. In both cases, nitrogen atoms must undergo a relatively small number of collisions in traveling from cathode to substrate if they are to reach the substrate surface. It is also possible that some energetic N_2 is driven into the growing film and remains there in the form of trapped gas.[112]

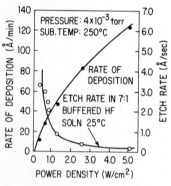

Fig. 24 Influence of sputtering power on quality of silicon nitride films.

In addition to the pressure effect, it has been observed that the best-quality silicon nitride films are obtained at maximum power, that is, at the highest deposition rates. This is illustrated in, for example, Fig. 24, where low etch rate implies high quality, or dense, films. The much greater reactivity of oxygen over nitrogen with respect to silicon is demonstrated by studies in which varying amounts of oxygen were added to a nitrogen sputtering atmosphere. It was found that for only 2% oxygen in the nitrogen a mixture of nitride and oxide was observed, while for an oxygen content greater than 5% the resulting films showed no evidence of containing any nitrogen at all.[113]

Ammonia instead of nitrogen as the active gas for the production of silicon nitride films has also been investigated.[114] Unlike nitrogen, it was found that heavy dilution with argon was possible without causing changes in the stoichiometry of the films. Thus, for example, an atmosphere containing as little as 2% ammonia in argon gave films which were largely silicon nitride, while only 5% ammonia was necessary to get films which were entirely silicon nitride. By contrast films sputtered in 5% nitrogen, 95% argon were found to be essentially pure silicon. This increased reactivity of the ammonia is ascribed to the much larger percentage of N^+ ions present in an ammonia discharge. The main disadvantage of silicon nitride films sputtered in ammonia is that they contain trapped hydrogen which can affect some of the electrical properties of the film.

If care is taken to see that no toxic gases escape into the laboratory atmosphere, many elements may be introduced as hydrides which react with the film. For example, Lakshmanan and Mitchell[115] have deposited a variety of sulfide films including cadmium, lead, copper, tin, and molybdenum by reactive sputtering in H_2S. Similarly carbon can be used for reactive sputtering by introducing it in the form of methane. This has been done in the case of, for example, the tantalum-carbon system.[106] Hydrogen will almost certainly be trapped in films prepared in this manner unless high substrate temperature and/or some form of bias sputtering is used during deposition.

Pompei[92] has described an unusual approach by which reactive sputtering of a wide variety of materials can be performed. These include compounds of sulfur, selenium, tellurium, phosphorus, arsenic, antimony, and bismuth. To understand his method, it is necessary first to say a few words about a technique called "chemical sputtering." This name is an unfortunate one since the process has, in reality, nothing at all to do with sputtering. To perform chemical sputtering, the material concerned is placed inside a low-pressure discharge in a hydrogen atmosphere. The hydrogen ions then react with the surface of the material to produce the hydride. The latter has a relatively high vapor pressure; so its molecules diffuse away from the material, eventually to decompose back into hydrogen and the element in other portions of the discharge tube where the ion density is sufficiently low. The process exhibits none of the special characteristics associated with film deposition by "true" sputtering. For example, alternating current may be used (in fact, the glow may be excited through an externally wound coil), no relationship between deposition rate and cathode voltage is seen, and the surface that is "sputtered" need not be the cathode or even electrically a part of the discharge.

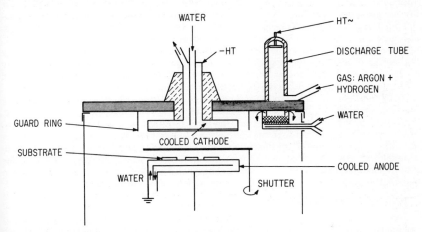

Fig. 25 System for deposition of sulfide films by combining reactive sputtering with chemical sputtering.

In Pompei's system, which is illustrated in Fig. 25, chemical sputtering is made to occur in an auxiliary discharge tube, but instead of decomposing to form a metal film, the hydride is allowed to pass from the auxiliary discharge tube into the main discharge chamber where reactive sputtering occurs between it and the metal of the main target. The main advantages of this method compared with the direct admission of the hydride into the chamber from a storage vessel are that high purity of the hydride can be guaranteed and also that the hydride is prepared in situ as needed. It is particularly important to note that the hydrides of the materials mentioned above are all extremely poisonous; so the advantages of preparing only small amounts, most of which are completely reacted in the main chamber, are obvious.

8. RF SPUTTERING

a. Direct Sputtering of Insulators

Although it has been known for some time that insulators as well as metals are subject to sputtering,[116] the preparation of thin film insulators by this technique has, until recently, been limited to reactive sputtering. Simple substitution of an insulator for the metal target in a conventional dc sputtering system is doomed to failure because of the immediate buildup of a surface charge of positive ions on the front side of the insulator. This then prevents any further ion bombardment. In principle, positive-charge buildup can be overcome by several methods such as simultaneously bombarding the insulator with both an ion beam and an electron beam,[117] leakage of the charge across the surface of a target of small area, leakage directly through the insulator target if it is hot enough,[118] or placing a metal grid at or near the surface of the target. In the latter approach, the metal grid supports the field needed to attract the ions as well as providing secondary electrons which serve to neutralize those ions which land in the spacings of the grid, that is, on the insulator surface.[119] While these methods may be suitable in varying degrees for the etching of insulator surfaces by ion bombardment, they are unsuitable for film deposition either because of thickness-uniformity problems or because of the high level of film contamination that is introduced as a by-product of the process.

b. History

The first recorded observation that material is removed from the walls of a glass tube if it is subjected to a high-frequency discharge excited through external electrodes appears to have been made by Robertson and Clapp in 1933.[120] In a follow-up investigation of their work, Hay[121] recognized that the removal of material was due to sputtering and that this occurred only if the frequency used was sufficiently high, although the reason for this was not understood. Some 10 years later, Lodge and Stewart[122] provided further evidence that the material was removed through sputtering and associated the latter with the appearance of a negative charge on the surface of the insulator underneath the high-frequency electrode. In 1957, Levitskii[123] made probe measurements and sputtering studies in a high-frequency discharge containing internal metal electrodes. In 1962, Anderson et al.,[124] pursuing an earlier suggestion by Wehner,[125] showed how the application of radio frequency to the outside of a triode sputtering discharge tube could be used to effect cleaning of the inside walls of the tube and suggested that the method could be used for the deposition of insulating films. This general approach was subsequently developed by Davidse and Maissel[54,126] into a practical method for the deposition of insulating films at reasonable rates over substantial areas. They also showed that a triode system was unnecessary and that apparatus resembling a dc sputtering system could be used.

c. Principles of RF Sputtering

Consider a glass discharge tube having two metal electrodes of equal area facing one another a few inches apart. We have already seen that, following the application of sufficient dc voltage to this configuration, a dark space develops at the cathode, being in fact, the controlling region of the discharge. If, instead of direct current, a low-frequency alternating voltage is applied to this tube, it is observed that the system behaves as though it had two cathodes since a dark space is seen at both electrodes. In actual fact this system is really a succession of short-lived dc discharges with the polarity alternating since, at low frequencies, there is ample time for a dc discharge to become fully established within each cycle. Such a discharge is thus no different from the dc case in that it is dependent on secondary electrons from either or both electrodes to sustain itself and will be extinguished at the same pressure as the equivalent dc discharge.

If the frequency of the applied voltage is raised sufficiently, it is observed that the minimum pressure at which the discharge will operate is gradually reduced, the effect being detectable above about 50 kHz and leveling off for frequencies in excess of a few megahertz. It is thus apparent that the discharge is receiving electrons from some source other than secondaries out of the electrodes. The additional electrons that are generated by the high-frequency discharge are a consequence of the fact that electrons oscillating in an rf field can pick up sufficient energy from the field to cause ionization in the body of the gas.

It is well known that a free electron in vacuum will oscillate in an alternating field with its velocity 90° out of phase with the applied field. Under these circumstances it will absorb on the average no power from the applied field. The electron can, however, gain energy from the field if it undergoes collisions with gas atoms while it is oscillating so that its ordered simple harmonic motion is changed to a random motion. The electron can increase the random component of its velocity with each collision until it builds up sufficient energy to make an ionizing collision with a gas atom. The high voltage at the cathode which is essential in a dc glow for the generation of secondary electrons is no longer needed by the rf glow in order to maintain itself. The fact that an electron can continue to gain energy in the field even though it may be moving either with or against it can be understood by observing that the energy absorbed is proportional to the square of the electric field and hence is independent of its sign. For further details on the quantitative aspects of high-frequency breakdown the interested reader is referred to the literature.[9]

Since the applied rf field appears mainly between the two electrodes, an electron which escapes from the interelectrode space as a result of a random collision will no longer oscillate in the rf field and will therefore not acquire sufficient energy to cause ionization. Hence it will be lost to the glow. Because of this, a magnetic field parallel to the rf field can perform a valuable function in that it constrains the electrons and reduces their chance of being lost, particularly at lower pressures. Thus the magnetic field is considerably more important for enhancing the rf discharge in terms of improving its efficiency than it is for the dc case.

As pointed out by Levitskii,[123] the mechanism by which electrons absorb energy from the rf field does not apply to the ions, which would be expected to pick up at most a few volts of energy from the field. In addition, since the rf discharge is not dependent on secondary electrons from the electrodes, large fields are not necessary for maintaining it. At first glance we would not, therefore, expect an rf discharge to contain ions with sufficient energy to cause sputtering.

Experimentally, it is found that the plasma of an rf discharge of the type under discussion develops an appreciable potential which is positive with respect to both electrodes.[123] This high plasma potential is a consequence of the very much higher mobility of the electrons as compared with the ions. Because of its very brief duration, relatively few ions are able to reach either electrode within the span of one cycle of the applied field. Many electrons, however, do succeed in reaching an electrode during each half cycle. As a result, the rf current that is measured in the external circuit is due almost entirely to electrons from the glow which reach the electrodes on alternate cycles. The electrons that are extracted from the glow during each half cycle are those that happen to be within a distance A from the electrode at the beginning of a cycle. A is the amplitude of oscillation of an electron in the rf field. If the latter is given by

$$E = E_m \cos \omega t$$

then A is given by

$$A = \mu \frac{E_m}{\omega} \tag{3}$$

where μ is the electron mobility at the pressure concerned.

It follows from (3) that in this configuration the electrode separation should be of the general order of A, or larger, if the discharge is to operate efficiently. Otherwise, any electrons generated as a result of gaseous ionization will be swept out of the discharge and collected by the electrodes within one cycle.

Thus, as explained by Levitskii,[123] the plasma of the rf discharge becomes concentrated in a zone symmetrically located between the electrodes and having a width equal to $d - 2A$ (where d is the spacing between the electrodes). This plasma region oscillates between the electrodes at the applied angular frequency ω with an amplitude A and touches each electrode alternately at the instant when $\omega t = \pi/2$ and $\frac{3}{2}\pi$. Within the central region the electron concentration is reasonably constant, but it falls off rapidly outside of this as we approach within a distance A of either electrode. Levitskii determined the magnitude of the plasma potential by using a static probe and showed, as expected, that it varied with the applied rf voltage as well as increasing with decreasing pressure, as shown in Fig. 26. (V applied was about 350 V in Fig. 26.)

In addition to using the static probe, Levitskii confirmed that some of the ions acquired the potential difference existing between the plasma and either electrode. He did this by directly measuring the energy of the ions striking the electrodes and obtained the interesting result that the maximum energy of ions arriving at the electrodes was greater than the space potential measured by his probe. He blamed this discrepancy on inaccuracies in his probe measurements. Recent calculations by Tsui,[128] however, have shown that this is a consequence of the fact that some of the ions may enter the cathode fall region at just the right time so as to "see" the full peak-to-peak voltage for most of the time that they spend in transit to the target. Similarly, ions that begin their journey across the sheath when the rf field is approaching zero arrive at the target with less than the self-bias voltage. The more rf cycles required by the ion to traverse the dark space, the greater the chance that its energy will average out to the self-bias voltage. As a result, the spread in energy decreases with decreasing cathode potential.

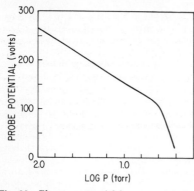

Fig. 26 Plasma potential in symmetrical rf system with metal electrodes, as a function of pressure.

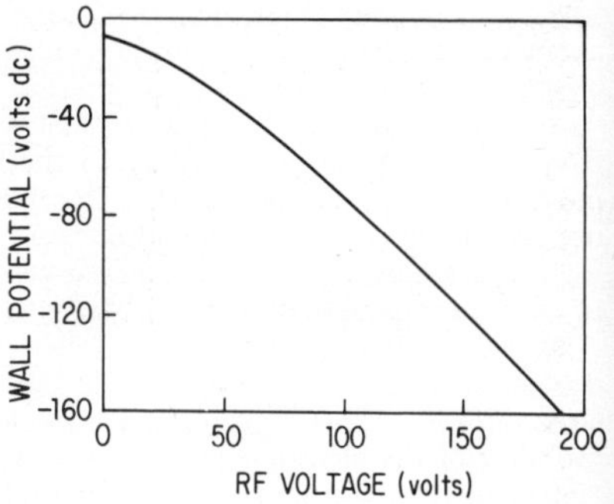

Fig. 27 DC wall potential (self-bias voltage) as a function of rf voltage as measured by Butler and Kino.

The above considerations were for electrodes of equal area. Let us now consider a configuration in which one electrode is appreciably larger than the other. Since the total current through each electrode must be the same, it follows that the current density at the larger electrode will be less. Thus, at the larger electrode the number of electrons lost per unit area from the plasma will be less than at the smaller one. Since, however, no dc voltage can be maintained in the external rf circuit, the potential must be the same at both electrodes, and hence such a configuration will, to a first approximation, still behave as though it had two electrodes whose areas are equal. If, however, a capacitor is placed somewhere in the external circuit,[129] the dc voltage at the smaller electrode is no longer required to equal that at the larger electrode, and because of the higher current density there, it may be significantly higher than that at the larger electrode. The ratio of the voltages at the two electrodes changes more rapidly than in simple inverse proportion to the ratio of the areas. This is because a reduction in voltage at a given electrode causes a corresponding decrease in the thickness of the ion sheath at the electrode. This, in turn, reduces the capacitive impedance of the larger electrode by a factor larger than that due to area alone. If we place a layer of insulation over one of the electrodes, instead of using a capacitor in the external circuit, the situation is electrically unchanged and an appreciable voltage can be developed between the plasma and the smaller electrode. If the latter is the one with the sheet of insulation placed over it, then ion bombardment, and hence sputtering, of the insulator can take place.

The dc voltage that develops between two electrodes of different area in an rf glow discharge, when one is an insulator, has been studied by Butler and Kino.[130] By placing a metal layer opposite the rf electrode on the inside of the glass wall, they were able to measure the potential difference between the surface of the insulator and the plasma. The latter was generated through direct current with the aid of a thermionic cathode. RF voltage was applied between the insulator electrode and the thermionic cathode. Their experimental results showing the dc wall potential as a function of the amplitude of the applied rf signal are seen in Fig. 27. As expected, they confirmed that the dc potential of the insulator surface was very nearly equal to the full value of the rf voltage. The finite value of the dc potential at zero rf voltage represents the floating potential of the wall in the dc glow. Butler and Kino reasoned that the magnitude of the dc voltage on the insulator was such that most of the electrons from the plasma were repelled, only the most energetic ones being collected. This allowed the ion and electron currents to the insulator to be equal, a condition which must prevail since the presence of the insulator means that no direct current can flow in the external circuit.

d. Techniques of RF Sputtering

As we have already discussed, the physical appearance of an rf sputtering system is much like that of a dc system. It has also been pointed out that the low-operating-pressure feature associated with triode systems for dc conditions largely disappears under rf conditions. We will therefore consider only self-sustained (or diode) systems.

Perhaps the most important difference between a dc and an rf system is that the latter requires an impedance-matching network between power supply and discharge chamber. Two examples of the most commonly used types are shown in Fig. 28, including formulas to calculate the component values.[131] R_g is the impedance of the generator (which is almost always 50 ohms and is given by the manufacturer) while R is the impedance of the rf glow discharge. The latter must be measured for each

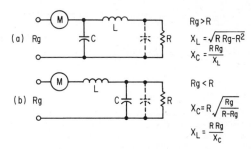

$$R_g > R$$
$$X_L = \sqrt{R\,R_g - R^2}$$
$$X_C = \frac{R\,R_g}{X_L}$$

$$R_g < R$$
$$X_C = R\sqrt{\frac{R_g}{R - R_g}}$$
$$X_L = \frac{R\,R_g}{X_C}$$

Fig. 28 Typical networks for impedance matching between an rf generator and its load.

particular system. In most cases, the impedance of the generator will be higher than that of the glow and the L network shown in Fig. 28a will be used.[94] The dotted capacitance shown represents that which exists between the rf electrode and its shield. In principle, this may be considered to be a part of the matching network and so tuned out. In practice, however, this capacitor has rather low Q so that, as pointed out by Davidse and Maissel,[54] if the shield is located very close to the rf electrode (as is common practice for dc sputtering), a significant loss of power can occur.

It is also common practice to include a blocking capacitor[129] in the tuning network. Although an insulating target will serve this purpose, an external capacitor removes any uncertainties resulting from leakage through or around the edges of the target in the presence of the plasma.

Another problem peculiar to rf systems is that adequate grounding of the substrate assembly cannot be taken for granted. What is a short circuit under dc conditions may exhibit considerable rf impedance. Thus, long leads will exhibit inductance while overhanging flanges and projections will exhibit capacitance. If the substrate assembly is not adequately grounded, undesirable rf voltages can develop on its surface. These may have a drastic effect on the life of, for example, heating elements placed inside such an assembly. Finally, it may be remarked that in rf sputtering systems (particularly when the target is an insulator) cooling of the cathode assembly is strongly recommended. A cross section of a typical cooled-cathode assembly is shown in Fig. 29.

Although, as we have seen, cathode current is the best quantity to use for control of deposition rate when using direct current, current is not suitable for this purpose under conditions of radio frequency. In addition to being a difficult quantity to measure, the rf current is not related in a simple manner to the number of ions striking the target. Good control of deposition rate can, however, be obtained by monitoring either rf electrode potential or total rf power. The electrode potential in an asymmetric system is, as we have seen, very nearly equal to twice the self-bias voltage. Alternatively, the latter quantity may be measured directly and used as the controlling parameter.

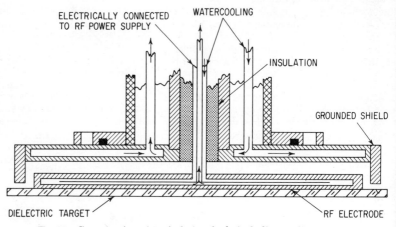

Fig. 29 Cross section of typical rf cathode including cooling system.

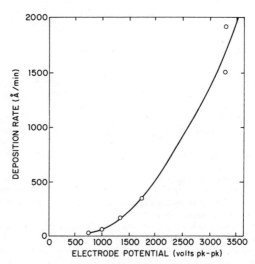

Fig. 30 Influence of rf potential on deposition rate of SiO_2.

The influence of the rf electrode potential on the deposition rate of SiO_2 is shown in Fig. 30. Note that this curve, unlike Fig. 11 or Fig. 22, does not represent conditions of constant current where voltage would be closely proportional to power. The deposition rate is seen to rise approximately as the square of the electrode potential in this case. Since we would expect the number of ions striking the target to be proportional to the electrode potential, the quadratic dependence suggests that the number of atoms ejected per ion is increasing approximately linearly with voltage. This is in fact the case for SiO_2 in this energy range.[133] On the other hand, for a material such as aluminum, whose sputtering yield changes relatively little in this voltage range, the deposition rate is approximately linearly dependent on power.[129]

In contrast to the asymmetric systems which we have described above, several symmetrical systems have been developed in which the two sputtering electrodes have equal area. As we have already discussed above, with such an arrangement a capacitor is not needed in the external circuit for sputtering to occur. The chief motivation behind these designs, however, appears to be to eliminate any possible sputtering of the grounded electrode. In addition, since two target electrodes are used, they may be positioned in any manner desired with respect to one another so that improved thickness uniformity of the deposited film is in principle possible.

Two embodiments of these ideas are shown in Fig. 31. In Fig. 31a a disk electrode is placed concentrically inside an annular electrode of equal area and a single dielec-

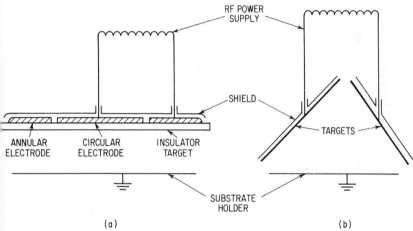

Fig. 31 Symmetrical rf sputtering system. (a) Using concentric electrodes. (b) Using adjacent electrodes at right angles.

tric disk is shared by both.[134] In Fig. 31b, the electrodes may be sputtered directly or they may be used as backing plates for insulator targets.[135] Note that it is common practice to ground the electrical center of the rf power supply in such a system for safety reasons. If this is done, the system is no longer symmetrical but rather must be considered as consisting of two electrodes 180° out of phase, both sputtering with respect to the grounded baseplate (which now constitutes the large electrode).

Another variation on rf-sputtering-system design is to place a one- or two-turn rf coil either around the outside of the discharge chamber or inside it around the interelectrode space.[135] This coil has a separate rf power supply while the target electrode is energized by its own power supply, which may be rf or dc. With such a coil a glow discharge can be maintained down to very low pressures—in the order of 0.5 mTorr. This is a consequence of the fact that the high-frequency magnetic field generated by the coil produces a circulating current of electrons which are able to travel substantial distances during each half cycle and therefore are able to cause ionization, even at very low pressures. Closely related to this is a system described by Gawehn[136] in which the rf plasma is excited a short distance away from the interelectrode space which it then reaches through diffusion.

The influence of the deposition conditions on the properties of insulating films deposited through rf sputtering is similar to what might be expected in terms of the general considerations already discussed. Typical results are shown in Figs. 32 and 33, which are for SiO_2. P etch is a chemical etch which has been found to give a good measure of film quality. In general, the effect of deposition rate is understandable in terms of the time required by an arriving atom to find an optimum

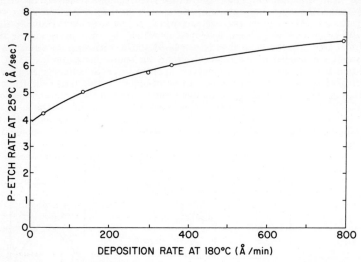

Fig. 32   Influence of deposition rate on quality of rf-sputtered SiO_2 films.

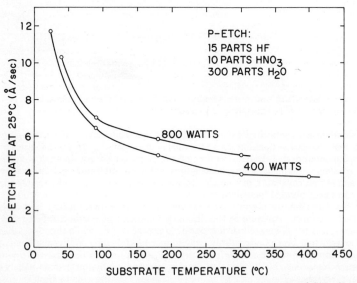

Fig. 33   Influence of substrate temperature on quality of rf-sputtered SiO_2 films.

position, this time being less at higher temperature. It is also possible that atoms not in optimum positions are more likely to be reemitted at higher temperatures. The power effect seen in Fig. 33 is considered to be a reflection of the small influence of deposition rate seen in Fig. 32. For further details, at least for SiO_2, the reader is referred to the literature.[137]

It is possible to perform sputtering in a system where both direct current and a superimposed rf voltage are used. This type of system is functionally similar to the rf coil system discussed above in that the presence of radio frequency at the target allows generation of plasma at lower pressures than could be obtained by direct current alone, whereas the bulk of the sputtering power still comes from direct current. This allows for economy in power supplies and reduces the power-handling capacity required for the components of the impedance-matching network. This arrangement has been shown to offer advantages over sputtering in direct current alone for the reactive sputtering of a number of metals in oxygen.[94] However, no comparison was made with sputtering using radio frequency alone at a power equivalent to the direct current.

9. SPUTTERING OF MULTICOMPONENT MATERIALS

An important feature of the sputtering process is that the chemical composition of a sputtered film will often be the same as that of the cathode from which it was sputtered. This will be true even though the components of the system may differ considerably in their relative sputtering rates (see Chap. 3). It will not be true if a significant amount of resputtering occurs (see below). The very first time that sputtering is performed from a multicomponent cathode, the component with the highest sputtering rate will come off faster, but a so-called "altered region" soon forms at the surface of the cathode. This region becomes sufficiently deficient in the higher-sputtering yield component to just compensate for its greater removal rate; so subsequent deposits have the composition of the parent material.

Evidence for this correspondence between a sputtered film and the cathode composition has been reported by many workers. Typical of these observations are those of Fisher and Weber[138] on stainless steel and brass, and of Hanau[139] on a number of aluminum alloys. Gillam[140] has made a detailed study of the formation and structure of the above-mentioned altered surface layer in the system AuCu₃. He suggested that the thickness of the layer is a measure of the depth to which the bombarding ions penetrate. In his experiments, an estimated thickness of about 40 Å was seen for bombardment by 400-eV ions (He, Ar, and Xe were used, and all gave similar results).

In cases where compositional differences between sputtered films and their parent cathodes are found, the cause may usually be traced to one of three mechanisms:

1. The cathode temperature is too high. Not only can this cause diffusion into the altered layer, but one or more of the components of the system may, under this condition, have a significant vapor pressure which might lead to its evaporation from the cathode.

2. Oxidation effects prevail. In most real systems some oxide is present on the surface of the cathode, particularly at the beginning of a sputtering run. The exact amount is a balance between the partial pressure of oxidants in the system and the rate of material removal by sputtering, but as long as the oxide covers an appreciable fraction of the total cathode area, the relative sputtering rates will be those not of the pure materials but of their oxides.[127,132]

3. Resputtering of the film is occurring as, for example, during bias sputtering. If this is the case, material with the highest sputtering yield will be preferentially removed from the film (principle of bias sputtering!). Unlike the cathode, the surface of the growing film is constantly being replenished with material of the cathode composition so an "altered layer" cannot form. The back-and-forth movement of an excess of high-yield material is not itself important except that some will inevitably be lost through diffusion out of the interelectrode space. Note that in computing the relative sputtering rates of various atoms out of the film surface, their relative atomic radii must be taken into account since a larger atom will have a higher probability of being "struck" by a bombarding ion.[141]

Wolsky et al.[142] have conducted experiments which suggest that in some cases multicomponent sputtering may occur through a diatomic or multiatomic sputtering mechanism which maintains stoichiometry of the deposited film even without the formation of an altered layer at the cathode surface.

It is also possible to "synthesize" multicomponent films by sputtering. This has been described by, for example, Sinclair and Peters,[143] who used a separate cathode for each component. In their paper, a number of designs for interleaving the cathodes are shown. The method depends on the fact that not all the sputtered material deposits directly beneath its point of origin on the cathode (see Fig. 8) but may end up some distance away from this. This effect leads to a fair degree of intermingling of the components sputtered from the interleaving cathodes. To obtain any particular alloy composition, the relative voltages to the cathodes are suitably

TABLE 2 Sputtering Yields of Various Materials in Argon
(Bombarding energy of Ar⁺ in volts)

Target	200	600	1,000	2,000	5,000	10,000	Reference
			Sputtering Yields, Atoms/Ion				
Ag....................	1.6	3.4	....	....	....	8.8	145, 146
Al....................	0.35	1.2	....	....	2.0	....	145, 147
Au....................	1.1	2.8	3.6	5.6	7.9	....	145, 148
C.....................	0.05*	0.2*	....	....	....	....	149
Co....................	0.6	1.4	....	....	....	....	145
Cr....................	0.7	1.3	....	....	....	....	145
Cu....................	1.1	2.3	3.2	4.3	5.5	6.6	145, 150, 152, 153
Fe....................	0.5	1.3	1.4	2.0†	2.5†	....	145, 151
Ge....................	0.5	1.2	1.5	2.0	3.0	....	145, 150
Mo....................	0.4	0.9	1.1	....	1.5	2.2	145, 151, 154, 157
Nb....................	0.25	0.65	....	....	....	....	145
Ni....................	0.7	1.5	2.1	....	....	....	145, 151
Os....................	0.4	0.95	....	....	....	....	145
Pd....................	1.0	2.4	....	....	....	....	145
Pt....................	0.6	1.6	....	....	....	....	145
Re....................	0.4	0.9	....	....	....	....	145
Rh....................	0.55	1.5	....	....	....	....	145
Si....................	0.2	0.5	0.6	0.9	1.4	....	145, 150
Ta....................	0.3	0.6	....	....	1.05	....	145, 147
Th....................	0.3	0.7	....	....	....	....	145
Ti....................	0.2	0.6	....	1.1	1.7	2.1	145, 157
U.....................	0.35	1.0	....	....	....	....	145
W.....................	0.3	0.6	....	....	1.1	....	145, 147
Zr....................	0.3	0.75	....	....	....	....	145
KCl(100).............	....	....	....	0.9	1.6	1.95	155
KBr(100).............	....	....	....	0.3	0.55	0.6	155
LiF(100).............	....	....	....	1.3	1.9	2.2	155
NaCl(100)............	....	....	....	0.35	0.75	1.0	155
			Sputtering Yields, Molecules/Ion				
Cds(1010)............	0.5	1.2	....	....	....	....	156
GaAs(110)............	0.4	0.9	....	....	....	....	156
GaP(111).............	0.4	1.0	....	....	....	....	156
GaSb(111)............	0.4	0.9	1.2	....	....	....	142
InSb(110)............	0.25	0.55	....	....	....	....	156
PbTe(110)............	0.6	1.40	....	....	....	....	156
SiC(0001)............	....	0.45	....	....	....	....	156
SiO₂.................	....	....	0.13	0.4	....	....	133
Al₂O₃...............	....	....	0.04	0.11	....	....	133

* Kr⁺ ions.
† Type 304 stainless steel.

adjusted. A functionally similar arrangement was used by Bickley and Campbell[144] to prepare lead titanate films by reactive sputtering. Their cathode consisted of a titanium sheet on whose surface they photoprinted a pattern of lead dots. It was found necessary to keep the center-to-center spacing between the latter to less than 4 mm, and for best results, it was necessary that the substrate holder be rotated during film deposition.

10. DATA ON SPUTTERING

As would be expected, the deposition rates for different materials sputtered under identical conditions of current density, system geometry, etc., correlate closely with their sputtering yields. The computation of deposition rates directly from the yield values is, however, extremely difficult because of uncertainties in secondary-electron yield, reemission coefficient, cathode-current-density profile, etc. The values given in Table 2 should therefore be regarded only as a general guide. Many of the published data are not corrected for secondary-electron emission, and in addition, the actual numbers given in the table were usually obtained by reading off a curve. They may therefore be in error by as much as 20%.

Results are given only for argon-ion bombardment, as this gas is by far the most widely used when film deposition is performed through sputtering. Sputtering-yield data for other bombarding species have been summarized in Ref. 3. Targets are polycrystalline or amorphous unless the target orientation is quoted.

REFERENCES

1. Holland, L., "The Vacuum Deposition of Thin Films," John Wiley & Sons, Inc., New York, 1956.
2. Kay, E., *Advan. Electron. Electron Phys.*, **17**, 245 (1962).
3. Maissel, L. I., "Physics of Thin Films," vol. 3, p. 61, Academic Press Inc., New York, 1966.
4. Francombe, M. H., "Basic Problems in Thin Film Physics," p. 52, Vandenhoeck and Ruprecht, Goettingen, 1966.
5. Chopra, K. L., "Thin Film Phenomena," McGraw-Hill Book Company, New York, 1969.
6. Cobine, J. D., "Gaseous Conductors: Theory and Engineering Applications," p. 205, Dover Publications, Inc., New York, 1958.
7. Francis, A., in S. Flügge (ed.), "Handbuch der Physik," vol. 22, p. 53, Springer-Verlag OHG, Berlin, 1956.
8. Parker, P., "Electronics," chap. 15, Edward Arnold (Publishers) Ltd., London, 1950.
9. Brown, S. C., "Introduction to Electrical Discharges in Gases," John Wiley & Sons, Inc., New York, 1966.
10. Acton, J. R., and J. D. Swift, "Cold Cathode Discharge Tubes," Academic Press Inc., New York, 1963.
11. Huss, W. N., *SCP Solid State Technol.*, December, 1966, p. 50.
12. Mahadevan, P., J. K. Layton, and D. B. Medved, *Phys. Rev.*, **129**, 79 (1963).
13. Davis, W. D., and T. A. Vanderslice, *Phys. Rev.*, **131**, 219 (1963).
14. Medved, D. B., *J. Appl. Phys.*, **34**, 3142 (1963).
15. Hagstrum, H. D., *Phys. Rev.*, **123**, 758 (1961).
16. Muly, E. C., and A. J. Aronson, *Extended Abstract, 14th Natl. Symp.*, 1967, p. 145.
17. Penning, S. M., and J. H. A. Moubis, *Proc. Koninkl. Ned. Akad. Wetenschap.*, **43**, 41 (1940).
18. Kervalishvili, M. A., and A. V. Zharinov, *Soviet Phys. Tech. Phys. English Transl.*, **10**, 1682 (1966).
19. Gill, W. D., and E. Kay, *Rev. Sci. Instr.*, **36**, 277 (1965).
20. Rokhlin, G. N., *USSR*, **1**, 347 (1949).
21. Kay, E., *J. Appl. Phys.*, **34**, 760 (1963).
22. Ditchburn, R. W., *Proc. Roy. Soc. London*, **A141**, 169 (1933).
23. Francombe, M. H., in J. C. Anderson (ed.), "The Use of Thin Films in Physical Investigations," p. 55, Academic Press Inc., New York, 1966.
24. Krikorian, E., and R. J. Sneed, *J. Appl. Phys.*, **37**, 3665 (1966).
25. Evans, P., and A. J. Noreika, *Phil. Mag.*, **13**, 717 (1966).
26. Campbell, D. S., and B. N. Chapman, *J. Phys. C. (Solid St. Phys.)*, **2**, 200, 1969.

27. MacDonald, R. J., and E. Haneman, *J. Appl. Phys.*, **37**, 3049 (1966).
28. Mattox, B. M., and J. E. McDonald, *J. Appl. Phys.*, **34**, 2493 (1963).
29. Schuetze, H. J., H. W. Ehlbeck, and G. G. Doerbeck, *Trans. 10th Natl. Vacuum Symp.*, 1963, p. 434.
30. Chopra, K. L., *J. Appl. Phys.*, **37**, 3405 (1966).
31. Layton, C. K., and K. B. Cross, *Thin Solid Films*, **1**, 169 (1967).
32. Chopra, K. L., *Appl. Phys. Letters*, **7**, 140 (1965).
33. Stirland, D. J., *Appl. Phys. Letters*, **8**, 326 (1966).
34. Wehner, G. K., U.S. Patent 3,021,271, February, 1962.
35. Molnar, B., J. J. Flood, and M. H. Francombe, *J. Appl. Phys.*, **35**, 3554 (1964).
36. Jones, R. E., C. L. Standley, and L. I. Maissel, *J. Appl. Phys.*, **38**, 4656 (1967).
37. Read, M. H., and C. Altman, *Appl. Phys. Letters*, **7**, 51 (1965).
38. Chopra, K. L., A. R. Randlett, and R. H. Duff, *Phil. Mag.*, **16**, 2601 (1967).
39. Ogilvie, G. J., and A. H. Thompson, *J. Phys. Chem. Solids*, **17**, 203 (1961).
40. Schwartz, G. C., R. E. Jones, and L. I. Maissel, *J. Vacuum Sci. Technol.*, **6**, 351, 1969.
41. Chiplonkar, V. T., and S. J. Joshi, *Physica*, **30**, 1746 (1964).
42. Altman, C., *Trans. 9th Natl. Vacuum Symp.*, 1962, p. 174.
43. Rodit, R. R., and R. E. Dreikorn, *Proc. 2d Ann. Sputtering Symp.*, Rochester, N.Y., p. 81.
44. Humphries, R. S., *Rev. Sci. Instr.*, **37**, 1734 (1966).
45. Maissel, L. I., U.S. Patent 3,294,661, December, 1966.
46. Maissel, L. I., and J. H. Vaughn, *Vacuum*, **13**, 421 (1963).
47. Schwartz, E., British Patent 685,288, 1952.
48. Stern, E., to be published.
49. Glang, R., R. A. Holmwood, and P. C. Furois, *Trans. 3d Intern. Vacuum Congr. Stuttgart 1965*, Pergamon Press, vol. 2, p. 643, New York, 1967.
50. Laegried, N., and G. K. Wehner, *J. Appl. Phys.*, **32**, 365 (1961).
51. Stocker, B. J., *J. Appl. Phys.*, **32**, 465 (1961).
52. Stern, E., and H. L. Caswell, *J. Vacuum Sci. Technol.*, **4**, 128 (1967).
53. Jones, R. E., H. F. Winters, and L. I. Maissel, *J. Vacuum Sci. Technol.*, **5**, 84 (1968).
54. Davidse, P. D., and L. I. Maissel, *J. Appl. Phys.*, **37**, 574 (1966).
55. Nickerson, J., *SCP Solid State Technol.*, **8**, 30 (1965).
56. Toombs, P. A., *Extended Abstract, 14th Natl. Vacuum Symp.*, 1967, p. 137.
57. Sinclair, W. R., and F. G. Peters, *J. Vacuum Sci. Technol.*, **2**, 1 (1965).
58. Gaydou, F., *Vide*, **126**, 454 (1966).
59. Dugdale, R. A., and S. D. Ford, *Brit. Ceram. Soc. Trans.*, **65**, 165 (1966).
60. Chopra, K. L., and M. R. Randlett, *Rev. Sci. Instr.*, **38**, 1147 (1967).
61. Gaydou, F., *Vakuum-Tech.*, **15**, 161 (1966).
62. Needham, J. G., *Trans. 10th Natl. Vacuum Symp.*, 1963, p. 402.
63. Shockley, W. L., and E. L. Geissinger, *IRE Trans. Components Pts.*, **11**, 34 (1964).
64. Charsan, S. S., and H. Westgaard, U.S. Patent 3,294,670, December, 1966.
65. Hanfmann, A. M., *Electron Packag. Prod.*, **7**, Special Thin Film Supplement, p TF 3, 1967.
66. Sosniak, J., *J. Vacuum Sci. Technol.*, **4**, 87 (1967).
67. Cooke, H. C., C. W. Covington, and J. F. Libsch, *Trans. Met. Soc. AIME*, **236**, 31⁴ (1966).
68. Berghaus, B., and W. Burkhardt, U.S. Patent, Alien Property Custodian 283,312 March, 1943.
69. Theuerer, H. C., and J. J. Hauser, *J. Appl. Phys.*, **35**, 554 (1964).
70. Theuerer, H. C., and J. J. Hauser, *Trans. Met. Soc. AIME*, **233**, 588 (1965).
71. Maissel, L. I., and P. M. Schaible, *J. Vacuum Sci. Technol.*, **1**, 79 (1964).
72. Vratny, F., and N. Schwartz, *J. Vacuum Sci. Technol.*, **1**, 79 (1964).
73. Berghaus, B., and W. Burkhardt, U.S. Patent 2,305,758, December, 1942.
74. Maissel, L. I., and P. M. Schaible, *J. Appl. Phys.*, **35**, 237 (1965).
75. Vratny, F., B. H. Vromen, and A. J. Harendza-Harinxma, *Electrochem. Technol.*, **5** 283 (1967).
76. d'Heurle, F. M., *Trans. Met. Soc. AIME*, **236**, 312 (1966).
77. Redhead, P. A., *Can. J. Phys.*, **42**, 886 (1964).
78. Flur, B. L., *Proc. Intern. Mag. Conf.*, 2.4-1 Washington, D.C., 1965.
79. Griest, A. J., and B. L. Flur, *J. Appl. Phys.*, **38**, 1431 (1967).
80. Frerichs, R., *J. Appl. Phys.*, **33**, 1898 (1962).
81. Winters, H. F., E. E. Horne, and E. E. Donaldson, *J. Chem. Phys.*, **41**, 2766 (1964)
82. Winters, H. F., and E. Kay, *J. Appl. Phys.*, **38**, 3928 (1967).
83. Jones, R. E., G. C. Schwartz, and M. Zuegel, *Trans. 4th Intern. Vacuum Congr. Manchester, 1968.*

84. Schwartz, N., *Trans. 10th Natl. Vacuum Symp.*, 1963, p. 235.
85. Holland, L., and G. Siddall, *Vacuum*, **3**, 245 (1953).
86. McHugh, J. A., and J. C. Sheffield, *J. Appl. Phys.*, **35**, 512 (1964).
87. Perny, G., and B. Laville Saint Martin, *Proc. Intern. Symp.*, Clausthal, 1965, Vanden-hoeck and Ruprecht, Goettingen, 1966.
88. Perny, G., B. Laville Saint Martin, and M. Samirant, *Compt. Rend.*, **263**, 265 (1966).
89. Perny, G., *Vide*, October, 1966, p. 106.
90. Krikorian, E., and R. J. Sneed, *Tech. Rept.* AFAL-TR-67-139, August, 1967, Air Force Avionics Lab., Wright Patterson Base, Ohio.
91. Peters, F. G., and C. L. Mantell, *Trans. 9th Natl. Vacuum Symp.*, 1962, p. 184.
92. Pompei, J., *Proc. 2d Ann. Sputtering Symp.*, 1967, p. 127, Rochester, N.Y.
93. Valletta, R. M., J. A. Perri, and J. Riseman, *Electrochem. Technol.*, **4**, 402 (1966).
94. Vratny, F., *J. Electrochem. Soc.*, **114**, 505 (1967).
95. Jackson, N. F., E. J. Hollands, and D. S. Campbell, *Proc. Joint IERE/IEE Conf. Applications of Thin Films in Electronic Engineering*, Imperial College, July, 1966, p. 13-1.
96. Valletta, R. M., and W. A. Pliskin, *J. Electrochem. Soc.*, **114**, 944 (1967).
97. Hiesinger, L., and H. Koenig, "Festschrift 100 Jahre Heraeus Platinschmelze," p. 376, Hanau, 1951.
98. Sinclair, W. R., F. G. Peters, D. W. Stillinger, and S. E. Koonce, *J. Electrochem. Soc.*, **112**, 1 (1965).
99. Peters, F. G., *Am. Ceram. Soc. Bull.*, **45**, 1017 (1966).
100. Fuls, E. M., E. H. Hensler, and A. R. Ross, *Appl. Phys. Letters*, **10**, 199 (1967).
101. Sinclair, W. R., and F. G. Peters, *J. Am. Ceram. Soc.*, **46**, 1 (1963).
102. Williams, J. C., W. R. Sinclair, and S. F. Coons, *J. Am. Ceram. Soc.*, **46**, 161 (1963).
103. Holland, L., and G. Siddall, *Vacuum*, **3**, 375 (1953).
104. Smith, E. E., and D. R. Kennedy, *Proc. Inst. Elec. Engrs. London*, **109**, 504 (1962).
105. Goldstein, R. M., and F. W. Leonhard, *Proc. Electron. Components Conf.*, 1967, p. 312.
106. Gerstenberg, D., and C. J. Calbick, *J. Appl. Phys.*, **35**, 402 (1964).
107. Krikorian, E., and R. J. Sneed, *J. Appl. Phys.*, **37**, 3674 (1966).
108. Gerstenberg, D., *Ann. Physik*, **11**, 354 (1963).
109. Janus, A. R., and G. S. Shirn, *J. Vacuum Sci. Technol.*, **4**, 57 (1967).
110. Hu, S. M., and L. V. Gregor, *J. Electrochem. Soc.*, **114**, 826 (1967).
111. Shahin, M. M., *J. Chem. Phys.*, **43**, 1798 (1965).
112. Cordes, L. F., *Appl. Phys. Letters*, **11**, 383 (1967).
113. Burkhardt, P. J., and L. V. Gregor, *Extended Abstract, 14th Natl. Vacuum Symp.*, 1967, p. 31.
114. Hu, S. M., L. V. Gregor, and L. I. Maissel, *IBM Tech. Discl. Bull.*, **10**, 100 (1967).
115. Lakshmanan, T. K., and J. M. Mitchell, *Trans. 10th Natl. Vacuum Symp.*, 1963, p. 335.
116. Stark, J., and G. Wendt, *Ann. Physik*, **3**, 921 (1912).
117. Hines, R. L., and R. Wallor, *J. Appl. Phys.*, **32**, 202 (1961).
118. Dugdale, R. A., and S. D. Ford, *Brit. Ceram. Soc. Trans.*, **65**, 165 (1966).
119. Spivak, G. V., A. I. Krokhin, and L. V. Lazarev, *Dokl. Akad. Nauk SSSR*, **104**, 579 (1955).
120. Robertson, J. K., and C. W. Clapp, *Nature*, **132**, 479 (1933).
121. Hay, R. H., *Can. J. Res.*, **A16**, 191 (1938).
122. Lodge, J. I., and R. W. Stewart, *Can. J. Res.*, **A26**, 205 (1948).
123. Levitskii, S. N., *Soviet Phys. Tech. Phys. English Transl.*, **27**, 913 (1957).
124. Anderson, G. S., W. N. Mayer, and G. K. Wehner, *J. Appl. Phys.*, **33**, 2991 (1962).
125. Wehner, G. K., *Advan. Electron. Electron Phys.*, **7**, 239 (1955).
126. Davidse, P. D., and L. I. Maissel, *Trans. 3d Intern. Vacuum Congr.*, Stuttgart, 1965.
127. Hasseltine, E. H., F. C. Hurlbut, N. T. Olson, and H. P. Smith, *J. Appl. Phys.*, **38**, 4313 (1967).
128. Tsui, R. T. C., *Phys. Rev.*, vol. **168**, April, 1968.
129. Davidse, P. D., *SCP Solid State Technol.*, December, 1966, p. 36.
130. Butler, H. S., and G. S. Kino, *Phys. Fluids*, **6**, 1346 (1963).
131. "Radio Amateurs Handbook," p. 49, American Radio Relay League, 1966.
132. Hollands, E., and D. S. Campbell, *J. Mater. Sci.*, **3**, 544 (1968).
133. Davidse, P. D., and L. I. Maissel, *J. Vacuum Sci. Technol.*, **4**, 33 (1967).
134. Putner, T., *Thin Solid Films*, **1**, 165 (1967).
135. Kloss, F., and L. Herte, *SCP Solid State Technol.*, December, 1967, p. 45.
136. Gawehn, H., *Z. Angew. Phys.*, **14**, 458 (1962).
137. Pliskin, W. A., P. D. Davidse, H. S. Lehman, and L. I. Maissel, *IBM J. Res. Develop.*, **11**, 461 (1967).

138. Fisher, T. F., and C. E. Weber, *J. Appl. Phys.*, **23**, 181 (1952).
139. Hanau, R., *Phys. Rev.*, **76**, 153 (1949).
140. Gillam, E., *Phys. Chem. Solids*, **11**, 55 (1959).
141. Winters, H. F., D. L. Raimondi, and D. E. Horne, *J. Appl. Phys.*, **40**, 2996, 1969.
142. Wolsky, S. P., D. Shooter, and E. J. Zdanuk, *Trans. 9th Natl. Vacuum Symp.*, 1962, p. 164.
143. Sinclair, W. R., and F. G. Peters, *Rev. Sci. Instr.*, **33**, 744 (1962).
144. Bickley, W. P., and D. S. Campbell, *Vide*, **99**, 214 (1962).
145. Laegreid, N., and G. K. Wehner, *J. Appl. Phys.*, **32**, 365 (1961).
146. Almén, O., and G. Bruce, *Nucl. Instr. Methods*, **2**, 257 (1961).
147. Carlston, C. E., G. D. Magnuson, A. Comeaux, and P. Mahadevan, *Phys. Rev.*, **138**, A759 (1965).
148. Robinson, M. T., and A. L. Southern, *J. Appl. Phys.*, **38**, 2969 (1967).
149. Rosenberg, D., and G. K. Wehner, *J. Appl. Phys.*, **33**, 1842 (1962).
150. Southern, A. L., W. R. Willis, and M. T. Robinson, *J. Appl. Phys.*, **34**, 153 (1963).
151. Weijsenfeld, C. H., A. Hoogendoorn, and M. Koedam, *Physica*, **27**, 963 (1961).
152. Yonts, O. C., C. E. Normand, and D. E. Harrison, *J. Appl. Phys.*, **31**, 447 (1960).
153. Rol, P. K., J. M. Fluit, and J. Kistemaker, in D. B. Langmuir, E. Stuhlinger, and J. M. Sellen (eds.), "Electrostatic Propulsion," p. 203, Academic Press Inc., New York, 1961.
154. Pitkin, E. T., in D. B. Langmuir, E. Stuhlinger, and J. M. Sellen (eds.), "Electrostatic Propulsion," p. 195, Academic Press Inc., New York, 1961.
155. Navinsek, B., *J. Appl. Phys.*, **36**, 1678 (1965).
156. Comas, J., and C. B. Cooper, *J. Appl. Phys.*, **37**, 2820 (1966).
157. Kurbatov, O. K., *Soviet Phys. Tech. Phys. English Transl.*, **12**, 1328 (1968).

Chapter **5**

The Deposition of Thin Films by Chemical Methods

DAVID S. CAMPBELL

The Plessey Company Limited, Whiteside Works, Bathgate, West Lothian, Scotland
also: Visiting Senior Lecturer in Materials Science, Electrical Engineering
Department, Imperial College, London University, London, England

1. INTRODUCTION

A wide choice of preparation techniques is open to the scientist who is interested in thin films. Broadly, these methods may be divided into two classes. One class depends on the physical evaporation or ejection of material from a source, i.e., evaporation or sputtering, while the other class depends on a chemical reaction. The latter may be dependent on electrical separation of ions as in electroplating and anodization or may depend on thermal effects as in vapor-phase deposition and thermal growth, but in all cases a definite chemical reaction is required to obtain the final film.

The chemical methods of film deposition have been summarized in previous papers; some of these have also examined and compared the complete range of both chemical and physical techniques.[1-3] However, physical methods of preparing films are considered in detail in other chapters in this book. This chapter is therefore concerned with giving a broad outline of the various chemical methods that are available.

TABLE 1 Chemical Methods of Thin Film Preparation

Basic class	Method	Author
Formation from the medium....	Electroplating	Milazzo,[4] Lowenheim,[5] Brenner,[6] West[7]
	Ion plating	Mattox[8]
	Chemical reduction	Lowenheim,[5] Gorbunova and Nikiforova[9]
	Vapor phase	Joyce,[10] Schaefer,[11] Gregor[12]
	Plasma reaction*	Gregor[12]
	Hydrophilic	Holt[14,15]
Formation from the substrate...	Anodization	Young[16]
	Gaseous anodization	Miles and Smith,[17] Jackson[18]
	Thermal	Evitts et al. (for Si),[19] Evans[20]
	Plasma reaction*	Ligenza[13]

* Plasma reactions are difficult to classify. They are sometimes considered under sputtering, but they are essentially vapor-phase or thermal-growth reactions, with the discharge supplying the energy necessary to effect the chemical change.

Table 1 summarizes the methods; in certain cases there is considerable overlap with physical methods, but they are included here for completeness (e.g., ion plating and evaporation, plasma reaction and sputtering, gaseous anodization and reactive sputtering). Also included in Table 1 are major references to texts on the methods, where they are available, as a help in further reading.

Some of the methods summarized in Table 1 are capable of producing both thin (<10,000 Å) and thick (>10,000 Å) films. However, there are various techniques that are capable of producing only thick films, and these are summarized in Table 2. Only glazing is basically a chemical technique; these methods do not come within the scope of this chapter.

TABLE 2   Methods of Thick Film Preparation

Method	*Author*
Glazing....................	Pliskin and Conrad,[21] Davis et al.,[22] Singer and German[23]
Electrophoretic.............	Audubert and de Mende[24]
Flame spraying............	Huffadine and Thomas[25]
Painting	

Finally it should be noted that the nature of the resultant films, i.e., nucleation, growth, etc., and the properties of the films are all considered in subsequent sections of this book and therefore will not be examined here.

2. ELECTROPLATING[4-7]

a. Aqueous Media

(1) General Electroplating has been known for a considerable time, and many standard textbooks now exist on the subject.[4-6] The apparatus involved is basically simple, consisting of an anode and cathode immersed in a suitable electrolyte. Metal is deposited on the cathode, and the relationship between the weight of material deposited and the various parameters can be expressed by the first and second laws of electrolysis. These state:

1. The weight of the deposit is proportional to the amount of electricity passed.
2. The weight of material deposited by the same quantity of electricity is proportional to the electrochemical equivalent E.

Expressed as an equation, the weight deposited per unit area G/A is given by

$$\frac{G}{A} = JtE\alpha \qquad \text{g cm}^{-2} \qquad (1)$$

where J is the current density and t the time. This equation introduces another term, the current efficiency α, which is the ratio of the experimental to theoretical weight deposited; it can generally be expected to be between unity and 0.5.

Equation (1) can be written in a slightly different form to give the rate of deposition. If a thickness l is deposited in time t, then the rate of deposition l/t is given by

$$\frac{l}{t} = \frac{JE\alpha}{\rho} \qquad \text{cm s}^{-1} \qquad (2)$$

where ρ is the film density.

The rate of deposition values can be very high at high current densities. For example, silver will deposit at 10 Å s^{-1} at a current density of 1 mA cm^{-2}, and this will rise to a 1 μ s^{-1} at 1 A cm^{-2}. Such a proportionality to current density holds only if α remains unchanged. This can be expressed in another way by stating that no secondary reactions must occur.

It is to be noted that other types of units are often used by those practiced in the art to express rate of deposition. For example, quite often ampere-hours (Ah) per thousandth of an inch per square foot are used. (In the case of silver, 1 mA cm^{-2}, which is very nearly the same as 1 A ft^{-2}, gives 1 thousandth of an inch of deposit in 6.1 h.)

Of the 70 metallic elements, it is found possible to plate only 33 successfully, and of this latter number only 14 are deposited commercially. A large variety of baths can be used for the possible elements to improve the adhesion, crystalline structure, current efficiency, etc. However, it is not possible to plate elements outside the group of 33, as other reactions (e.g., formation of hydrogen) can more readily occur. This can be illustrated by considering the I-V characteristics of a plating solution. For a simple system the curve will be as shown in Fig. 1.

Such a curve is obtained with a probe placed near the cathode. The equilibrium potential of the cathode in the solution is indicated by the intercept value on the voltage axis. A negative intercept implies that the cathode will dissolve in the electrolyte at zero voltage (i.e., it will corrode). Equilibrium potentials for the different metals vary from $+1.7$ to -1.66 V (see Table 3). Saturation is seen in the curve—at high cathode voltages ions cannot get to the cathode fast enough.

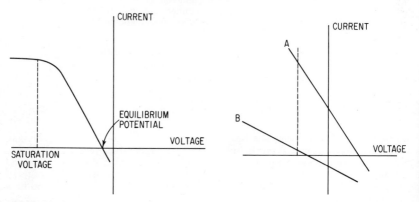

Fig. 1 Ideal I-V characteristic for a single reaction deposition.

Fig. 2 Ideal I-V characteristics for two reactions.

If two reactions are possible, each will affect the other. Two I-V curves can now be drawn for the two reactions, and Fig. 2 shows this case. In the case of alloy plating this means that the composition of the alloy will depend on the voltage used, as indicated by the dashed line in Fig. 2. If one of the reactions is the formation of hydrogen, the curves will now be as in Fig. 3.

As mentioned earlier, the curves of Figs. 1, 2, and 3 must be obtained by a probe just by the cathode. Since the actual distribution of voltage in a bath will be as

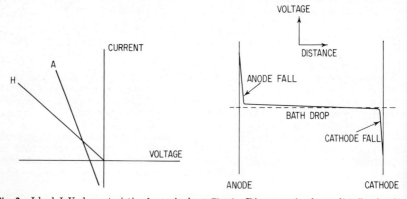

Fig. 3 Ideal I-V characteristics for a single reaction plus hydrogen evolution.

Fig. 4 Diagram of voltage distribution in a plating bath.

shown in Fig. 4, an incorrect relationship between I and V will be obtained if the voltage drop was measured across the whole bath. The drop in the bath itself must be reduced to as low a value as possible to reduce waste of power in heating, and this is usually achieved by the addition of conducting salts.

The cathode fall distance usually extends some 3,000 Å into the bath, and fields in this region will therefore be very high. Three separate reactions may be distinguished in this region; these are shown diagrammatically in Fig. 5.

The structure that is obtained can vary from single-crystal[26] or crystalline aggregates right through fiber-growth deposits to unoriented deposits of very fine grain

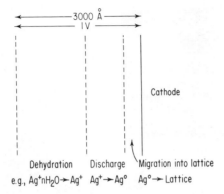

Dehydration   Discharge   Migration into lattice

e.g., $Ag^+nH_2O \rightarrow Ag^+$ $Ag^+ \rightarrow Ag^\circ$ $Ag^\circ \rightarrow Lattice$

Fig. 5   Diagram of reaction regions in the cathode fall.   (Ag° is an adsorbed silver atom.)

size and disordered structure.   Figure 6 shows a typical columnar growth obtained. The columns grow normal to the surface of the cathode.   It is to be noted that the periodicity of the surface roughness of the film is given by the column sizes (Fig. 6b). That this effect is due to the columnar growth can be seen by comparing the substrate and film surface in Fig. 6a.   Furthermore it is possible to find film specimens that

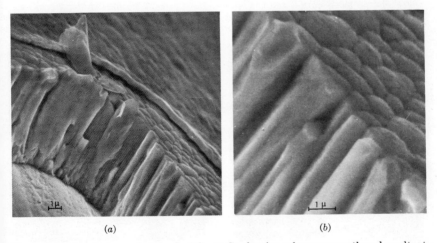

(a)   (b)

Fig. 6   Scanning electron micrograph of Cu on Cu showing columnar growth and resultant surface roughness.[29]   (a) General view of substrate and film showing both a cross section through the film and the top surface.   (b) Top surface of film showing surface roughness produced by columnar growth.

have become detached from the substrate during preparation so that the underside of the film is visible.   A scanning electron micrograph of such a situation is shown in Fig. 7, and it can be seen that the underside of the film has in fact, replicated the substrate surface.

Figures 8 and 9 show further examples of electroplated films. Figure 8 is of the top surface of a 2-μ film of gold on nickel-iron; the deposition conditions did not produce columnar growth, and so flaws and scratches in the original nickel-iron surface are reproduced in the surface of the gold. The micrograph of Fig. 9a shows a similar

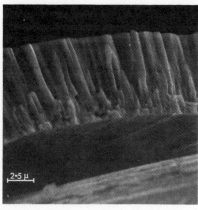

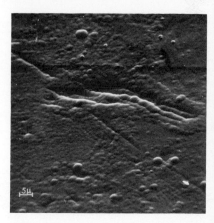

Fig. 7 Scanning electron micrograph of a film of Cu on Cu. The film has become partially detached and the underside of the Cu film is visible.[29]

Fig. 8 Scanning electron micrograph of the top surface of Au on Ni-Fe showing the reproduction of substrate flaws in the film (film 2 μ thick).[43]

layer but deposited at a much higher current density. Under these circumstances gas evolution has occurred and the bubbles have become trapped in the deposit, with many blowholes resulting. Figure 9b shows an enlargement of one of the areas from Fig. 9a.

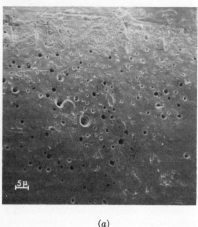

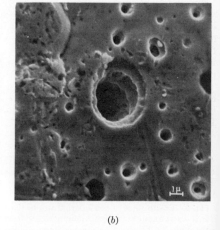

(a)

(b)

Fig. 9 Scanning electron micrographs of the top surface of Au on Ni-Fe showing the effect of plating at high current densities (film 5 μ thick).[43] (a) 700 ×. (b) 3,500 ×.

(2) Deposition of Elements Table 3 gives a list of the elements that can be easily deposited by electroplating. Included in the table are the density and electrochemical equivalent of each element. The valency state in the most easily used bath is quoted,

TABLE 3 Table of Metal Elements That Can Be Electroplated*

Metal	Density	Electrochemical equivalent, mg C⁻¹	Valency state	Equilibrium potential, V‡
Al.......	2.7	0.093	3	−1.66
Ag.......	10.5	1.118	1	+0.80
Au.......	19.3	2.043	1	+1.70
Cd.......	8.6	0.582	2	−0.40
Co.......	8.7	0.306	2	−0.28
Cr†......	7.1	0.180	3	−0.70
Cu.......	8.9	0.329	2	+0.34
Fe.......	7.9	0.289	2	−0.44
Ni.......	8.9	0.304	2	−0.25
Pb.......	11.4	1.074	2	−0.13
Pt.......	21.5	1.011	2	+1.20
Rh.......	12.4	0.356	3	+0.80
Sn.......	7.3	0.615	2	−0.14
Zn.......	7.1	0.339	2	−0.76

* For the appropriate baths, see Ref. 15.
† The current efficiency is normally only about 0.13.
‡ The quantity, equilibrium potential, is often referred to as the standard electrode potential. A standard temperature of 25°C and a standard solution concentration of 1 g ion/1,000 g solvent are taken. The arbitrary zero of potential is taken as that of a reversible hydrogen electrode with gas at 1 atm pressure, in a solution of hydrogen ions of 1 g/1,000 g solvent.

but change of bath can change the state of the depositing ions and there will be a proportional change in the value of E. The growth rate can then be calculated from Eq. (2) provided the current efficiency is known, and as stated previously it is usually between unity and 0.5 for a viable plating system. The effect of solution temperature will be noticed only if α changes with temperature, with α generally increasing with temperature rise. As the deposition rate can be high, it is possible to use electrochemical deposition for forming thick layers—a process known as electroforming—and for refining. An example of electroforming is the preparation of master disks for gramophone records. In this case the initial deposit to give a suitable cathode is obtained either by using a colloidal suspension of metal or by chemical-reduction plating.

Single-crystal films have been grown successfully by Lawless[26] using Cu on single-crystal Ni, Ni on single-crystal Cu, and Au on single-crystal Cu. The films are highly perfect up to 1,000 Å. All the films examined were continuous at 50 Å. Island-structure films were found at smaller thicknesses [<15 Å for Au on (111) Cu, <50 Å for Au on (100) Ni].*

(3) Anodic Deposition It is possible to form oxides of elements successfully on the anode of the electrode system. The oxides are deposited from the solution (cf. anodization, where it is the oxide of the anode material that is formed). Films of Pb and Mn oxides have been grown in this way, but the method has little importance in thin film technology.

(4) Deposition of Alloys The many alloys that have been successfully deposited (100 or so) are discussed in detail in Brenner's two volumes on the electrodeposition of alloys.[6] Nickel-iron-alloy films can be noted as an example, as these films are particularly important for magnetic-storage devices (see Chap. 21). It is not possible to deposit every combination because of the characteristics of the separate elements, although it is often possible to slow one of the reactions down by suitable chemical complexing. The effect of complexing is to lower the equilibrium potential

* A new review of the nucleation, growth, and structure of electrodeposited films has been given by Lawless with regard to both single-crystal and polycrystalline deposits.[44]

to a more negative value, and this results in a crowding together of the *I-V* character-istics for the separate elements (Fig. 2). It is not necessarily the case that the potentials of the separate metals come sufficiently close to permit codeposition, and it may also be necessary to vary the individual concentrations or the concentration of the complexing agent if it affects both elements. Cyanide is a typical complexing ion for Ag, Cd, Zn, and Cu.

Ternary alloys can also be deposited by electroplating; Brenner lists 15 that can be easily formed.[6]

(5) Displacement Reactions It has already been noted that an equilibrium potential exists for a metal and an electrolyte such that the rate of ion deposition is the same as the rate of ions leaving the cathode. Thus at zero volts on the cathode, metal will, if the equilibrium potential is negative, go into solution in the electrolyte. If a metal with a more negative equilibrium potential is immersed in an electrolyte which would deposit a metal with a less negative equilibrium potential, the ions of the electrolyte will deposit on the original metal and the original metal will go into solution. Equilibrium-potential values are included in Table 3 against $H/H^+ = 0$ V. Such a reaction is known as a displacement reaction. Although the metal in contact with the electrolyte is at all times overall electrically neutral, it is found that different areas of the metal become anodic and cathodic. Figure 10 shows a typical distribution of reactions for the deposition of a copper film on iron. Because of the anodic-cathodic nature of the reaction, an uneven deposit is obtained with the thicker film at the cathodic areas. Moreover, the reaction will stop if a nonporous continuous film of metal is deposited over all the surface.

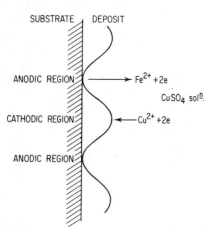

Fig. 10 Diagram of the displacement reaction for deposition of Cu on Fe.

The importance of such reactions does not lie in the production of metal layers, however, but in the preparation of oxide layers and corrosion in general (see Evans[20] and Sec. 9 on Thermal Growth).

b. Nonaqueous Media

As has already been discussed, some metals cannot be deposited because other reactions are more probable. This can sometimes be avoided by the use of nonaqueous media. Al, Cr, Ti, and the platinum metals, for example, can be deposited from baths of the fused salts. As an illustration, Al can be prepared from a molten bath at 150 to 175°C of $AlCl_3 + NaCl + LiCl$. In this case it is most important to use dry, pure starting materials and to agitate during deposition. If the conditions are not optimum, porous films result. In certain cases it is possible to use organic solvents with either the metal salts or organometallic compounds dissolved in them. Al is again a suitable example; $AlCl_3$–ethyl pyridinium bromide in toluene can be used. There are many other examples that can be quoted—the reader is referred to suitable texts for details.[5]

3. ION PLATING[8]

Ion plating is a method of deposition that is difficult to classify as either chemical or physical. It has been developed by Mattox[8] and can be described as discharge-assisted evaporation. On the other hand it has a certain similarity to electroplating inasmuch as the metal is in the form of positive ions that are attracted to the cathode substrate. Figure 11 shows a diagram of the apparatus. The material to be deposited is evaporated from a suitable filament, as in ordinary evaporation techniques. A glow discharge is maintained at a pressure between 10^{-1} and 10^{-2} Torr between the filament as anode and the substrate as cathode, so that the evaporated atoms are ionized in the plasma. The discharge potential is maintained at as high a value as possible so that the ionized atoms are accelerated to the substrate. Adhesion of the deposit is found to be very good because of the high energy of arrival of the deposit ions. It is normal to use an inert-gas atmosphere in the chamber. One of the advantages of the system is that the discharge will keep the substrate clean, but this also means that the deposit itself will be sputtered away. The conditions therefore have to be chosen correctly so that the deposition rate is higher than the sputtering rate. In certain cases such as Al, which does not sputter easily, this problem is not important (rates of deposition of between 10 and 100 Å s^{-1} are possible with Al), but with other materials that do sputter easily, such as Au, the problem matters. Au can

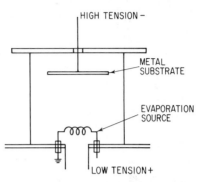

Fig. 11 Diagram of ion-plating apparatus.

be successfully deposited at around 20 Å s^{-1} using an accelerating voltage of 5 kV and a cathode current density of 0.3 mA cm^{-2}. Various metals have been deposited by the technique, including Ag, Au, Al, Cu, Cr, and Ni.

4. CHEMICAL-REDUCTION PLATING[5,9]

Films of metals may be deposited directly without any electrode potentials being involved, by the chemical reduction of a suitable compound in solution. Such deposition is known as chemical-reduction plating or electroless deposition. Four different types of reaction may be distinguished.

a. Noncatalytic Reactions

These take place at any surface submersed in the bath. Silver mirrors are usually formed in this way, by the use of a mild reducing agent such as formaldehyde in a solution of silver nitrate. Very thick layers may be built up.

b. Catalytic Reactions

The ability of the metal to deposit on anything can sometimes be a considerable nuisance, and more controlled reactions are often more useful. In these the metal will deposit only on certain surfaces of other metals and nowhere else. The deposition of nickel, for example, can be achieved by such techniques by the reduction of $NiCl_2$ by sodium hypophosphite, when the metal will grow on a surface of nickel itself, cobalt, iron, and aluminum—the metal acts as the catalyst. (NOTE: The use of sodium hypophosphite as the reducing agent means that between 5 and 10% of phosphorus will become incorporated in the film.) This type of reaction has become so important that a complete book is now available on chemical-reduction plating of just nickel.[9] Other metals, particularly the Pt group, can be deposited in this manner.

c. Catalytic Reactions Using Activators

The number of metal surfaces that will catalyze deposition is limited. It is found, however, that it is possible to activate the surfaces of noncatalytic metals so that deposition will take place on these surfaces. The role of the activator is to lower the activation energy for the reduction reaction at particular points on the surface so that deposition will occur at these points. Islands of metal will thus grow and spread and eventually give a continuous film. The best activators to be used for particular metals are listed in standard texts on the subject;[5,9] PdCl₂ is often used for Cu and Ni. Very little of the activator is required—in the case of PdCl₂ a dip in a 0.01% solution followed by a rinse in water is all that is required.

d. Catalytic Reactions Using Activators and Sensitizers

For nonmetallic surfaces, a sensitization before activation is required. For Ni this takes the form of a dip in a 0.1% solution of SnCl₂ followed by a rinse. The activation is then carried out in the normal way. The advantage of such reactions is that it is possible to plate onto glass and other nonconducting surfaces. Also, and this applies in general, it is possible to plate surfaces that are difficult of access such as the inside of tubes.

e. Summary

Chemical-reduction plating uses very simple apparatus, provided a suitable reaction and, if necessary, catalysts are available. The rates of deposition depend on solution

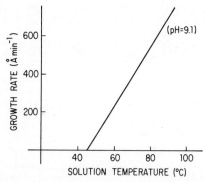

Fig. 12 Rate of deposition vs. bath temperature for Ni-Co on glass (see Refs. 27 and 28). (pH = 9.1.)

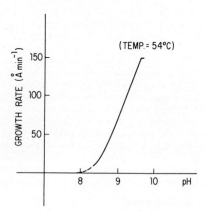

Fig. 13 Rate of deposition vs. pH for Ni-Co on glass (see Refs. 27 and 28). (Bath temperature = 54°C.)

pH and temperature, and typical characteristics are shown in Figs. 12 and 13 for the chemical-reduction plating of Ni-Co.[27,28] Structurally the films grow as nucleated

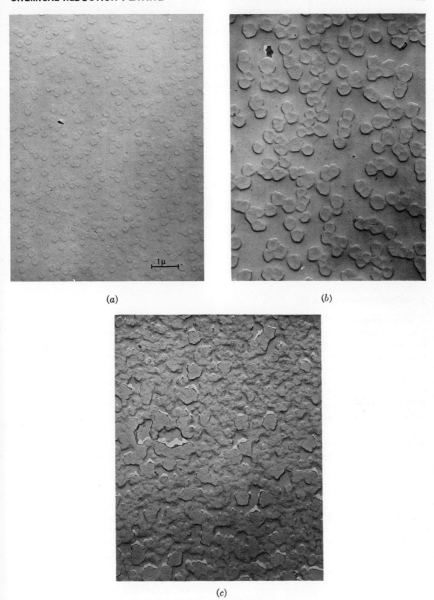

Fig. 14 Growth sequence of Ni-Co on mica.[29] (a) 5 Å thick. (b) 10 Å thick. (c) 100 Å thick.

islands, and Fig. 14 shows a typical growth sequence.[29] The degree of crystallinity depends on the material being deposited and the bath temperature; on deposition at room temperature the films can be amorphous (e.g., Ni, if phosphorus is included in the structure) but will crystallize on heating. Figure 15 shows such a change in a Ni-Co film between 60 and 200°C.

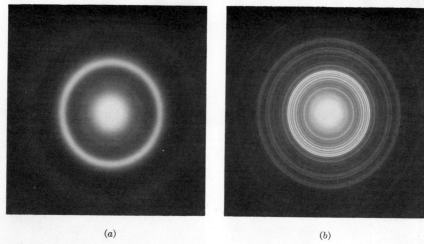

(a) (b)

Fig. 15   Electron-diffraction patterns of Ni-Co films deposited on glass.[29] (a) As deposited at 60°C. (b) Annealed at 200°C.

5. VAPOR-PHASE GROWTH[10-12,30]

The deposition of a film on a surface composed of the same or of a different substrate by means of a chemical reaction occurring from a gaseous phase at the surface is known as vapor-phase growth or vapor plating.   Usually the surface is hotter than the surroundings so that a heterogeneous reaction occurs at the surface; otherwise the reaction may occur in a homogeneous manner in the gas phase.   However, other means of activating the chemical reaction may be used such as a glow discharge or ultraviolet radiation.   For specific details the texts already cited should be consulted, together with the surveys by Powell for metals and oxides.[31]

Four different types of reaction may be distinguished.   These will now be examined.

a. Disproportionation

The reaction is typified by the equation

$$A + AB_2 \rightleftharpoons 2AB$$

where A and B are two elements.   The higher-valency state is more stable at lower temperatures, so that if a hot gas of AB is passed into a colder region, deposition of A can occur.

Such a method is often employed for the preparation of silicon and germanium.   A halide is formed and the reaction is usually used to transport silicon and germanium from a high-temperature source zone to a lower-temperature substrate zone.   For silicon, the iodide is best, and suitable quantities of the iodide can be generated at a hot-zone temperature of 1000°C.   The substrate temperature may be as low as one wishes for the straight deposition of silicon, but to obtain an epitaxial deposit on a single-crystal silicon substrate, the temperature must be 950°C or more.   Figure 16 shows a diagram of a typical closed-tube system that may be employed.   Alternatively a continuous-flow,

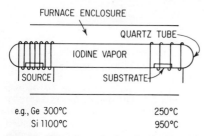

Fig. 16   Diagram of closed-tube vapor-transport system using disproportionation.

i.e., open, system may be used in which iodine vapor, usually diluted with hydrogen, is continuously passed over the source and then the substrate. Growth rates can be high—Ge can be deposited at rates up to 400 Å s^{-1} by the open system using the GeI$_2$ → GeI$_4$ + Ge reaction.

b. Polymerization[12]

Both organic and inorganic polymers may be prepared from monomer vapor by the use of electron beam, ultraviolet irradiation, or glow discharge. Insulating films prepared in this manner can have very desirable properties (see Chap. 16 on Dielectric Films and Gregor[12]).

Electron-beam irradiation has been applied to a large number of materials including styrene, butadiene, divinylbenzene, etc. Apart from the production of insulating films, electron-beam polymerization has been studied in terms of the buildup of contamination from pump-oil vapor in the electron microscope. Recent workers[12,32] have described an apparatus for the production of polymer films from evaporated epoxy resin.

Ultraviolet irradiation techniques are widely known in photoresist etching. In the etching, a relatively thick layer of photosensitive material is spread evenly over the surface and then irradiated through a mask. The irradiated areas polymerize to give a material that is insoluble in the solvent used for the unpolymerized film. A similar technique has been used to prepare insulating films. White,[33] for example, exposed metal layers in butadiene vapor to irradiation and thus built up thin dielectric layers. Various other vapors have also been used—methyl methacrylate, acrolein, and divinylbenzene, for example. The rate of growth has been plotted as a function of substrate temperature (Fig. 17) for polyacrolein by Gregor,[12] who has also described suitable apparatus.

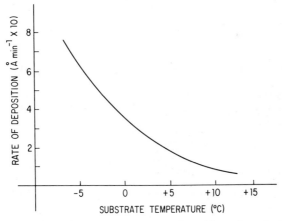

Fig. 17 Growth rate vs. substrate temperature for ultraviolet polymerization of acrolein. (*After Gregor.*[12])

Glow-discharge techniques have been used by various workers, with the monomer introduced into the discharge chamber. Bradley and Hammes[34] have listed over 40 monomers that have been examined in this way. RF excitation has also been used, and an apparatus has been described by Connell and Gregor[12,35] that can produce insulating films from styrene, etc., at relatively high rates of deposition (1,200 Å min^{-1}).

c. Reduction, Oxidation, Nitriding

This is usually undertaken using a halide of the required metal or metal oxide because of the high vapor pressure of the halide and the ease of removal of the by-prod-

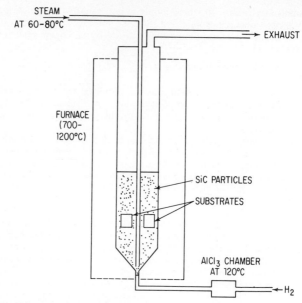

Fig. 18 Diagram of fluidized-bed system for oxidation of $AlCl_3$ to give Al_2O_3.

ucts. The two possible reactions are

$$2AX \xrightarrow{H_2O} A_2O + 2HX$$

$$2AX \xrightarrow{H_2} 2A + 2HX$$

Metals, semiconductors, and oxides may be deposited using this technique, by

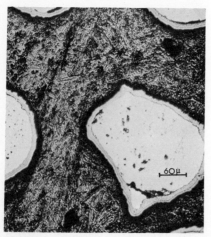

Fig. 19 Particle coating obtained in fluidized-bed column. (SiC particles with Al_2O_3 coating; dark matrix is potting compound used for holding the particles during sectioning and polishing.)

the use of either hydrogen or steam (often diluted in argon) onto hot substrates. To ensure thorough mixing and an even temperature in the reactor, a fluidized bed is often employed. An example of such a system for the case of Al_2O_3 growth on metal sheet is shown in Fig. 18, using silicon carbide particles in the column. Naturally the particles will become coated as well, which is illustrated in Fig. 19.

This type of reaction is widely used to prepare Si or Ge. In the Si case, either trichlorsilane ($SiHCl_3$) or silicon tetrachloride ($SiCl_4$) may be used. High growth rates are possible (200 Å s^{-1} at 1100°C for Si from either $SiHCl_3$ or $SiCl_4$), and the growth kinetics has been studied in detail by various workers.[10]

Nitrides can also be prepared in this manner using an atmosphere of ammonia (e.g., $SiH_4 + NH_3 + H_2 \rightarrow Si_3N_4$)[12]. Glow-discharge conditions can also be used to effect nitriding in a similar gas mixture.[36]

d. Decomposition

Decomposition as represented by

$$AB \rightarrow A + B$$

can be effected by heat (pyrolysis) and glow discharge.

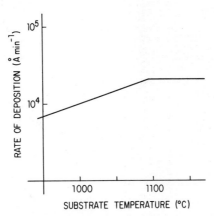

Fig. 20 Growth rate vs. substrate temperature for the growth of Si by the pyrolysis of SiH₄. (*After Theuerer.*[39])

Pyrolytic reactions have been widely applied to the preparation of Si (from silane, SiH₄), Ni [from nickel carbonyl, Ni(CO)₄], and SiO₂ (from the decomposition of silicone esters[12,37]). Growth-rate curves for the first and last of these reactions are shown in Figs. 20 and 21. The rates for Ni can be very high at moderate substrate temperatures (1,000 Å s⁻¹ at 200°C).

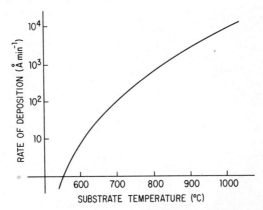

Fig. 21 Growth rate vs. substrate temperature for the pyrolytic decomposition of ethyl silicate (see Ref. 12).

Glow discharges have also been used to prepare insulating films by decomposition; silicone esters can be decomposed in an oxygen plasma. The discharge may be excited by radio frequency.[38]

e. Summary

The various possible reactions for vapor plating have been briefly examined. The apparatus required is sometimes more complex than for electroplating or chemical-reduction plating, with gas-pumping systems or vacuum systems being required, and for pyrolytic work furnaces that may be required to run at fairly high temperatures (2000°C for carbon by the decomposition of toluene and benzene). This high temperature often limits the type of substrate that may be used. Very high growth rates are possible [i.e., Ni by the decomposition of $Ni(CO)_4$ at 200°C]. The growth at slow rates of deposition is generally of a nucleated type as illustrated in Fig. 22, whereas at fast rates the form of growth is still under debate. Thick (10 μ or more) single-crystal films may be obtained if a single-crystal substrate is used. The high temperatures involved are often a help in growing single-crystal films as they enable the arriving atoms to move around on the surface easily and so assume an ordered structure position.

Fig. 22 Electron micrograph showing growth of Si on (100) Si (pyrolysis of SiH_4). (Growth rate 0.1 Å s^{-1}; average deposit thickness 300 Å.)[40]

Table 4 summarizes some of the films that have been grown by vapor-phase methods. This list is, however, not exhaustive, and the reader is referred to the general texts for more detailed information.

TABLE 4   Summary of Materials That Can Be Deposited by Vapor-phase Reactions (Parent Material in Parentheses)*

Method	Material
Disproportionation $(A + AB_2 \rightleftharpoons 2AB)$....	$Al(AlI_3)$; $Ge(GeI_2)$; $Si(SiI_2)$; III-V compounds (iodides). $C(CO)$
Polymerization $(AB \rightarrow nAB)$............	Polymers of methyl methacrylate, styrene, divinylbenzene, butadiene, acrolein, epoxy resins, allylglycidyl ether, etc. (by electron-beam, photolysis, or glow-discharge techniques)
Reduction $(AX \xrightarrow{H_2} A + HX)$............	$Al(AlCl_3)$; $Ti(TiBr_4)$; $Sn(SnCl_4)$; $Ta(TaCl_5)$; $Nb(NbCl_5)$; $Cr(CrCl_2)$; $Si(SiHCl_3$ or $SiCl_4)$; $Ge(GeCl_4)$
Oxidation $AX \xrightarrow{H_2O} AO + HX$............	$Al_2O_3(AlCl_3)$; $TiO_2(TiCl_4)$; $Ta_2O_5(TaCl_5)$; $SnO_2(SnCl_4)$
Nitriding, etc. $(AX \xrightarrow{NH_3} AN + HX)$......	Si_3N_4 (from SiH_4 by pyrolysis or glow discharge); $TiN(TiCl_4)$; $TaN(TaCl_5)$; SiC $(SiCl_4 + CH_4)$
Decomposition $(AB \rightarrow A + B)$...........	SiO_2 (from silicon esters by pyrolysis or glow discharge); $Ti(TiI_4)$; Pb (Pb-organics); $Mo(MoCl_5)$; $Fe[Fe(CO)_5]$; $Ni[Ni(CO)_4]$; C(toluene); $Si(SiH_4)$; $MnO_2[Mn(NO_3)_2]$; BN (B trichloroborazole)

*For details see Refs. 10, 11, 12, 30.

6. HYDROPHILIC METHOD[14]

It has recently been reported by Holt[14,15] that a surface chemical process known as the Langmuir-Blodgett technique can be used to produce multimonolayers of long-chain fatty acids. One end of long-chain molecules of certain fatty acids is hydrophilic and the other end is hydrophobic so that they normally float upright on the surface of water. If such a film is compressed by a moving raft, an immersed substrate pulled through the surface will be coated by a monolayer of the acid. One end, the hydrophilic one, will be chemically bonded to the surface. Such a layer will be ~25 Å thick. Recompression of the film on the water and reimmersion of the substrate deposits another monolayer with the hydrophobic ends of the two layers joined. By successive withdrawal and immersion of the substrate, a film can be built up in monolayer steps.

Films have been prepared over the thickness range 200 to 20,000 Å, and they have good electrical properties ($\epsilon = 3$). Breakdown strengths of 2.6 MV cm^{-1} have been observed with films 5,000 Å thick.

Such films are extremely stable. Any base electrode can be used as the substrate, and the top electrode may be applied by conventional evaporation techniques.

7. ANODIZATION[16]

a. Introduction

The methods that have been examined so far have used various chemical methods of depositing films on foreign substrates. The choice of substrates has been limited by electrical and thermal considerations, but in spite of this the choice is quite wide. However, as can be seen from Table 1, a second class of film-preparation techniques is available in which oxides, nitrides, etc., are prepared on substrates of the parent metal. The electrochemical method of doing this is by anodization. As the name implies, the film grows on the anode in an electrolytic cell. The basic equations that govern the process can be written as

$$M + nH_2O \rightarrow MO_n + 2nH^+ + 2ne \qquad \text{at anode}$$
$$2ne + 2nH_2O \rightarrow nH_2 \uparrow + 2nOH^- \qquad \text{at cathode}$$

Thus an oxide grows on the metal anode surface and hydrogen is evolved at the cathode. The equations imply the presence of water; anodization usually is undertaken in an aqueous electrolyte, but it is possible to use other media such as pure alcohols or fused salts (e.g., $NaNO_3$).

At the anode the reaction with the larger electronegative potential will occur; with some materials this may not be the anodization reaction—the metal of the anode may go into solution or oxygen may be evolved. The pH of the solution determines

TABLE 5   Summary of Metals That Can Be Anodized to Give Non-porous Adhesive Oxides[12]

Metal	Thickness-voltage ratio, Å V^{-1}	Max thickness attainable, μ
Al*.......	3.5	1.5
Ta........	16	1.1
Nb.......	43	
Ti........	15 (using aqueous electrolyte) 50 (using fused NaCl electrolyte)	
Zr........	12–30	>1.0
Si........	3.5	0.12

* The pH of the electrolyte must be correct or parts of the oxide will redissolve. This is an advantage in making porous layers for decorative purposes as there is then no thickness limit, but such films are electrically useless.

which reaction will occur: for example, Cu can be anodized in a highly alkaline solution, although the films are not electrically useful. There is thus a limit to the number of metals that can be anodized. Even those which will anodize may not form electrically useful layers because the oxide is nonadherent or too porous. The metals that can be anodized to give useful films are listed in Table 5.

b. Method of Formation

Oxide films can be prepared under one of two different conditions. For most metals, a constant current is passed through the cell and the film thickness is proportional to the time for which the current is passed. Figure 23 shows typical curves for Al and Ta. The voltage developed across the films is a measure of the film thickness from which are derived the thickness-voltage ratios quoted in Table 5. As can be seen from Fig. 23, the growth rate under these conditions is found to be comparable with those found in electroplating (e.g., Ta at 2 mA cm^{-2} will grow oxide at the rate of 10 Å s^{-1}; the rate for Al under the same conditions is 11 Å s^{-1}).

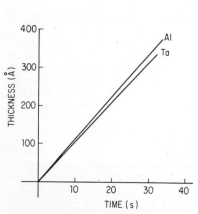

Fig. 23 Thickness vs. time for anodic oxide growing on Al and Ta. Constant-current growth (2 mA cm^{-2}).

Fig. 24 Field recrystallization in Ta$_2$O$_5$ anodic films[42] due to application of high field.

There is, however, a limit to the thickness that can be obtained. Either the film will break down by arcing, or field recrystallization will occur. The first of these is the limitation for Al and Ta, and the maximum voltage and hence thickness attainable is a function of purity of substrate, composition of electrolyte, and various other features. Field recrystallization, too, occurs because of the applied voltage. Although all the anodic oxides show this to some extent (Al$_2$O$_3$ formed at 500 V will contain 10% by volume of crystalline material), it is of major importance in Ta and Nb. Figure 24 shows the effect in Ta$_2$O$_5$; the recrystallized areas will be electrical shorts through the film.

A second method of growing films is under constant-voltage conditions, and the formation characteristics are shown in Fig. 25. This shows the thickness-time curve with the final thickness given by the applied voltage (see Table 5). The current will fall to zero as the voltage across the film rises to the applied total voltage. The curves of Fig. 25 may be compared with thermal-growth curves. This method of formation is often used with Al substrates, as breakdown at the impurities in the Al which would occur at the high fields that are possible in constant-current growth can be controlled.

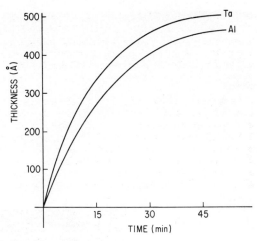

Fig. 25 Formation characteristics of anodic oxide films for constant-voltage growth (30 V).

c. Structure and Growth Behavior

The main feature that is common to anodic oxide films is that they grow in a non-nucleated manner as continuous layers of amorphous material. Generally the oxide surface is smooth and featureless, although if the metal surface is insufficiently clean or smooth, craterlike features are possible at the oxide-metal interface, and domelike features are found on the oxide-electrolyte interface. Figures 26 and 27 show such features. The grain boundaries of the substrate can also give irregularities on the oxide-electrolyte surface, and an example is shown in Fig. 28.

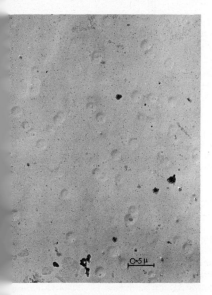

Fig. 26 Al-Al$_2$O$_3$ interface for an anodic film grown on an imperfect surface[42] showing craterlike features.

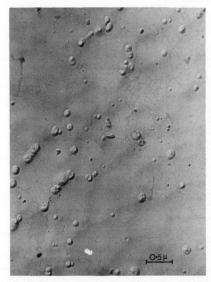

Fig. 27 Al$_2$O$_3$-electrolyte interface for an anodic film grown on an imperfect surface[42] showing domelike features.

Fig. 28   Oxide-electrolyte interface for anodic film showing effect of grain boundaries in the zirconium substrate.[42]

The theory of the growth process is still under discussion.   An analysis of the situation was given by Cabrera and Mott,[41] who considered that a potential barrier exists between the metal and the oxide and that the growth occurred by the movement of ions over the barrier.   Which ion moves, whether the metal or oxygen, appears to depend on the system; as is noted in Chap. 12 on Mechanical Properties of Thin Films, the stresses in the film and the resultant electrical characteristics are also governed by the growth kinetics.

8. GASEOUS ANODIZATION[17,18]

In gaseous anodization the liquid electrolyte of the conventional wet process is replaced by a low-pressure (10^{-2} Torr) glow discharge.   The experimental arrangement is shown schematically in Fig. 29.   As originally demonstrated by Miles and

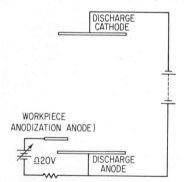

Fig. 29   Schematic circuit for gaseous anodization.

Smith,[17] the metal specimen to be anodized is positioned in the most conductive region of the discharge and is positively polarized with respect to the anode.   Various metals have been anodized in this way[17] (e.g., Al, Ta, and La-Ti).   The thickness voltage ratio is higher than for conventional anodization (e.g., 26 Å V^{-1} for Ta) and this has been explained in terms of the higher temperature of the anode in the gaseous discharge.   In general gaseous anodization may be applied to smooth anodiz able metals. The system has a very low current efficiency and the throwing power is small (i.e., it is not possible to anodize the inside of porous structures). It is very important to use a nonreactive anode in the discharge circuit; otherwise all the voltage in the anodizing circuit will be dropped across the oxide formed on the discharge anode.

. THERMAL GROWTH[20]

a. Nonassisted Growth

A large number of films can be formed on metal substrates by heating them in gases of the required type (O_2 for oxides, CO for carbides, N_2 for nitrides). As with anodized films, the best films have amorphous structures and are coherent. However, they are limited in thickness because the reaction will become very slow as the film thickness increases. If the thickness is plotted against time, a parabolic relationship similar

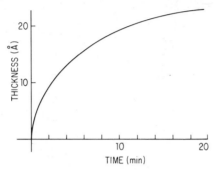

Fig. 30 Thickness vs. time curve for thermal growth of Al_2O_3 on Al in air at 20°C.

to that of anodization under constant voltage (see Fig. 30) is typically obtained. However, if a noncoherent oxide is formed, oxide will continue to be formed but the thickness will reach a limit, with material continuously flaking off from the surface above this limit. In certain cases (e.g., Al at 500°C) recrystallization can occur exposing the underlying substrate, which is then reoxidized.

Since the mobilities of ions through the oxide are dependent on temperature, and since the stress in the film will depend on the thickness of the oxide, it follows that at low temperatures coherent continuous films may be formed. At higher temperatures, the thickness can reach the limit at which the stress causes the film to break up and expose the original substrate. This is illustrated in Fig. 31, where the total thickness plotted against time for a general oxidation case.

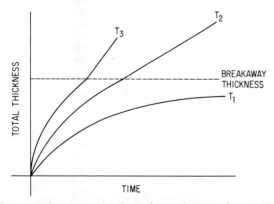

Fig. 31 Thickness vs. time curves for thermal growth (general case) ($T_1 < T_2 < T_3$).

TABLE 6 Summary of Ways of Preparing Thin Films

Method	Order of rate of deposition, Å s^{-1}	Rate control	Type*	Advantages	Disadvantages
Electroplating.........	Up to 10^4, normally 10^2–10^3	Current density	M	Simple apparatus	Metallic substrate
Chemical reduction...	10	Solution temp., pH	M	Simple apparatus. Can put metals on insulators	Limited number of materials with suitable reactions
Vapor phase..........	1–10^3	Pressure, temp.	M, S, I	Single crystal, clean films are possible	Substrate temp may have to be high (>1000°C). Low-pressure gas systems may be required
Anodization..........	10	Current density	I	Simple apparatus. Amorphous films, which are continuous at low thicknesses, can be obtained	Metallic substrate. Only limited number of metals can be anodized. Total thickness limited
Thermal.............	1	Pressure, temp.	I	Simple apparatus	Metallic substrate. Total thickness limited. Limited number of metals give coherent films
Evaporation..........	10–10^3	Source temp.	M, S, I	Large range of materials and substrates possible	Vacuum apparatus required. Some materials decompose on heating
Sputtering...........	10	Current density. Target potential	M, S, I	High adhesion. Using rf techniques, very wide range of materials possible	Suitable target required. Vacuum apparatus required

* M = metals; S = semiconductors; I = insulators.

TABLE 7 Applicability of Preparation Methods to Microelectronics

	Electroplating	Chemical reduction	Vapor phase	Anodization	Thermal	Evaporation	Sputtering
Conductors, resistors	/	/	/			X	/
Insulators, capacitors				X	/	X	/
Active devices			X			/	
Magnetic materials	X	/				/	/
Superconductors			/			X	

Single hatching indicates that the component can be prepared by the method; crosshatching indicates that the method is widely used.

5-23

There are only a limited number of metals that give coherent oxides which are continuous and useful in electrical terms; of these the most important are Al, Ta, and Si. The reaction involved can often be described in electrochemical terms, but it is not the intention of this chapter to consider this further. For information on the whole subject of thermal growth and oxide layers, the reader is referred to Evans' work on corrosion.[20]

b. Plasma Oxidation

One of the disadvantages of thermal growth is the high temperatures that may have to be reached before appreciably thick films are produced. For example, with silicon, Evitts[19] has shown that a temperature of 1100°C is needed to grow a film 3,500 Å thick in about 1 h. This can be avoided by oxidizing in a plasma of oxygen. Ligenza[13] has shown that using an rf-excited discharge and an oxygen pressure of between 0.1 and 1 Torr, silicon may be oxidized to a thickness of around 3,500 Å at 300° C. If the silicon is made the anode of a 50-V system, then negatively charged oxygen ions will bombard the surface and thicknesses above 3,500 Å can be observed. In this latter case material will also be sputtered from the cathode, but the atoms can be prevented from landing on the anode by placing a suitable bend in the discharge tube so that the cathode cannot "see" the anode.

The exact mechanism of growth involved is still under discussion, and it may well be that the whole process is best described as gaseous anodization. The process has found some interest in integrated-circuitry technology.

10. GENERAL SUMMARY OF THIN FILM PREPARATION TECHNIQUES

Having discussed the various chemical ways of preparing thin films, we can now draw up a table reviewing the various techniques. This has been done in Table 6, where the advantages and disadvantages of the main methods have been briefly summarized. Also included for comparison are evaporation and sputtering.

From such a table it is not possible to point to the best way of preparing a thin film. The method used must depend on the type of film required, the limitations present on choice of substrate, and quite often in the case of multiple deposition, the general compatibility of the various processes to be used.

Table 7 summarizes the applicability of preparation methods to microelectronics. The shaded areas indicate applicability, and the crosshatching indicates that the method is widely used. Evaporation and sputtering processes are currently preferred because of their universality. However, the potential usefulness of chemical methods must not be overlooked, as they may well offer cheaper ways of doing the same thing.

ACKNOWLEDGMENTS

The author is grateful for the help given by numerous colleagues and to The Plessey Company for permission to publish.

REFERENCES

1. Behrndt, K. H., in H. G. F. Wilsdorf (ed.), "Thin Films," p. 1, American Society for Metals, Metals Park, Ohio, 1964.
2. Campbell, D. S., in J. C. Anderson (ed.), "The Use of Thin Films in Physical Investigations," p. 11, Academic Press Inc., New York, 1966.
3. Campbell, D. S., *Proc. Conf. Electrical Uses of Thin Films, IERE Conf. Proc.*, no. 7 p. 35/1, 1966.
4. Milazzo, G., "Electrochemistry," Elsevier Publishing Company, Amsterdam, 1963.
5. Lowenheim, F. A., "Modern Electroplating," John Wiley & Sons, Inc., New York, 1963
6. Brenner, A., "Electrodeposition of Alloys," vols. 1, 2, Academic Press Inc., New York 1963.

7. West, J. M., "Electrodeposition and Corrosion Processes," D. Van Nostrand Company, Inc., Princton, N.J., 1965.
8. Mattox, G. M., *Electrochem Tech.*, **2**, 295 (1964).
9. Gorbunova, K. M., and A. A. Nikiforova, "Physichemical Principles of Nickel Plating," Israel Programme for Scientific Translations, Jerusalem, 1963.
10. Joyce, B. A., in J. C. Anderson (ed.), "The Use of Thin Films in Physical Investigations," p. 87, Academic Press Inc., New York, 1966.
11. Schaefer, H., "Chemical Transport Reactions," Academic Press Inc., New York, 1964.
12. Gregor, L. V., in G. Hass and R. Thun (eds.), "Physics of Thin Films," vol. 3, p. 131, Academic Press Inc., New York, 1966.
13. Ligenza, J. R., *J. Appl. Phys.*, **36**, 2703 (1965).
14. Holt, L., *Nature*, **214**, 1105 (1967).
15. Lewis, B., *Bull. IPPS*, **18**, 226 (1967).
16. Young, L., "Anodic Oxide Films," Academic Press Inc., New York, 1961.
17. Miles, J. L., and P. H. Smith, *J. Electrochem. Soc.*, **110**, 12 (1963).
18. Jackson, N. F., *J. Mater. Sci.*, **2**, 12 (1967).
19. Evitts, H. C., H. W. Copper, and S. S. Flaschen, *J. Electrochem. Soc.*, **111**, 688 (1964).
20. Evans, U. R., "An Introduction to Metallic Corrosion," Edward Arnold (Publishers), Ltd., London, 1958.
21. Pliskin, W. A., and E. E. Conrad, *Electrochem. Technol.*, **2**, 196 (1964).
22. Davis, E. M., W. E. Harding, R. S. Schwartz, and J. J. Coming, *IBM J. Res. Develop.*, **8**, 102 (1964).
23. Singer, F., and W. L. German, "Ceramic Glazes," Borax Consolidated, London, 1964.
24. Audubert, R., and S. de Mende, "The Principles of Electrophoresis" (translated by A. J. Pomerans), Hutchinson & Co. (Publishers), Ltd., London, 1959.
25. Huffadine, J. B., and A. G. Thomas, *Powder Met.*, **7**, 290 (1964).
26. Lawless, K. R., *J. Vacuum Sci. Technol.*, **2**, 1 (1965).
27. Heritage, R. J., and M. T. Walker, *J. Electron. Control*, **7**, 542 (1960).
28. Hendy, J. C., H. D. Richards, and A. W. Simpson, *J. Mater. Sci.*, **1**, 127 (1966).
29. Richards, H. D., private communication, 1966.
30. Powell, C. F., J. H. Oxley, J. M. Blocher, Jr., "Vapour Deposition," John Wiley & Sons, Inc., New York, 1966.
31. Powell, C. F., Ref. 30, pp. 277 and 343.
32. Brennemann, A. E., and L. V. Gregor, *J. Electrochem. Soc.*, **112**, 1194 (1965).
33. White, P., *Electronics Reliability and Micromin.*, **2**, 161 (1963).
34. Bradley, A., and J. P. Hammes, *J. Electrochem. Soc.*, **110**, 15 (1963).
35. Connell, R. A., and L. V. Gregor, *J. Electrochem. Soc.*, **112**, 1198 (1965).
36. Sterling, H. F., and R. C. G. Swann, *Solid-State Electron.*, **8**, 653 (1965).
37. Pensak, L., *Phys. Rev.*, **75**, 472 (1949).
38. Alt, L. L., S. W. Ing, Jr., and K. W. Laendle, *J. Electrochem. Soc.*, **110**, 465 (1963).
39. Theuerer, H. C., *J. Electrochem. Soc.*, **108**, 649 (1961).
40. Joyce, B. A., private communication, 1967.
41. Cabrera, N., and N. F. Mott, *Rept. Progr. Phys.*, **12**, 163 (1948–1949).
42. Jackson, N. F., private communication, 1967.
43. Davies, T. A., private communication, 1967.
44. Lawless, K. L., in, G. Hass and R. E. Thun (eds.), "Physics of Thin Films," vol. 4, p. 191, Academic Press Inc., New York, 1967.

Chapter 6

Thin Film Substrates

RICHARD BROWN*

Bell Telephone Laboratories, Murray Hill, New Jersey

* Present address: Pyrofilm Corporation, Whippany, N.J.

1. INTRODUCTION

Although it accompanies the thin film from "cradle to grave," the substrate is often ignored or only obliquely referred to. This chapter focuses on substrates to provide the thin film worker with criteria for selecting the proper material to fill his particular need. Unfortunately, an ideal substrate does not exist. Specific applications require different substrate materials which offer an acceptable compromise for the purpose on hand. Ideally, the substrate should provide only mechanical support but not interact with the film except for sufficient adhesion. In practice, however, the substrate exerts considerable influence on the thin film characteristics

Various substrate requirements and their implications are summarized in Table 1 The major properties considered in this chapter are substrate surfaces, chemical composition and stability, thermal conductivity, and cost. Additional factors which are discussed more briefly include thermal stability, porosity, mechanical strength, and ease of manufacture. Thermal stability requirements virtually eliminate the use of organic substrate materials since these decompose below 250°C. Therefore the following discussion deals exclusively with glasses, poly- and single-crystalline ceramics, semiconductors, and metals.

2. COMMON SUBSTRATE MATERIALS

a. Substrate Forming Processes

Substrates are either manufactured as shaped pieces with controlled tolerances o separated into appropriately dimensioned modules from larger ware. The method of forming a substrate determine, among other things, its tolerances, surface character istics, and cost. For the thin film worker, forming processes are of interest only inso far as they determine some of the substrate properties. The material presented here may be found in considerable detail in a number of books.[1-3]

(1) Glasses Glasses are generally produced by mixing the raw materials at on end of a continuous tank-type furnace and withdrawing the molten glass from the other end. Optimum melting and fabricating conditions depend largely on the viscosity of the glass. The American Society for Testing and Materials (ASTM has established methods for determining temperatures at which glasses have certai reference viscosities. These are shown in Fig. 1 for a number of common glasses

Most glass-forming processes start at the "flow point," which is defined as yielding a viscosity of 10^5 Poise. Although not yet recognized as a standard, the flow point is also indicated in Fig. 1. The softening point is the temperature at which a glass deforms under its own weight. "Soft" glasses have a large temperature interval between the flow point and the softening point and are said to have "long working ranges." They are easier to fabricate than glasses such as the aluminosilicates. Fused silica has a short working range in practice, although the temperature interval between the characteristic viscosities is long. The reason is its high working temperature; as the glass cools, its viscosity increases rapidly, which makes ordinary fabricating operations difficult.

TABLE 1 Substrate Requirements and Related Considerations

Atomically smooth surfaces...	Important for precision capacitors and resistors. Ensures low coercive forces and angular dispersion in magnetic films. Undesirable substrate features are often retained in epitaxial films and grown oxides
Inertness to chemicals used in processing	Hydrofluoric acid is frequently used as an etchant. It attacks most silicate-based glasses
High volume and surface resistivity	To minimize component degradation due to ion migration, for high stability against weathering, and for inertness during processing involving high field strengths
High thermal conductivity....	To provide surface-temperature control during processing and to permit high component density through the ability to extract heat
Coefficient of thermal expansion similar to film	To reduce film stress due to differences in expansion coefficients
High mechanical strength.....	To withstand substrate strain in processing and packaging, and to allow complex shapes including holes
High thermal shock resistance.	Sudden large changes in substrate temperature during operations such as bonding may cause substrate fracture or induce strain
Zero porosity...............	To minimize outgassing during vacuum deposition, and to reduce contamination of the film from occluded residues
High recrystallization temperature	To prevent changes in the surface structure during the polishing operations in metallic substrates, and to inhibit interdiffusion at the substrate-metal interface if the polishing temperature approaches the recrystalization temperature
Good dimensional tolerances: Low warp Surface waviness not exceeding 1 mil Good planarity	To allow for proper alignment and fit into automatic processing equipment and substrate assemblies. To provide surface conditions necessary for accurate pattern registration with photographic and mechanical masking procedures
Excellent scribing, cutting, and module-separation properties	Larger substrates are frequently reduced in size after processing
Low cost..................	Especially important where discrete film components are substituted for nonfilm types

Glasses containing alkali-metal oxides such as soda-lime glass can readily be drawn at moderate pressures and annealed at conveniently low temperatures between 700 and 1000°C. Thickness control for thin glass sheets (up to 0.060 in.) can be achieved by adjusting the rate of draw. Thicker glasses (up to 0.250 in.) require grinding and polishing, processes which perturb the surface. An alternate method is to pull the molten glass across a pool of molten tin.[4] While traversing the tin bath, the glass gradually loses heat and hardens, thereby assuming the flatness of the molten tin. However, the "tin-float" technique is restricted to certain glass compositions of 0.25 in. thickness.

The viscosities of the borosilicate glasses (Nos. 5 and 6 in Fig. 1) do not change rapidly enough in the working range to permit control of sheet thickness by rate of draw. It is therefore necessary to pass the molten ribbon of glass between preset rollers. However, because of variations in glass composition and cooling rate, the rollers frequently produce drawing lines or an "orange-peel" effect. This condition can be corrected by additional processing steps such as grinding and polishing. The modified borosilicate glasses (Nos. 13 and 14 in Fig. 1) have somewhat steeper viscosity curves and are more amenable to drawing but offer other problems. Their high resistivities and viscosities make it difficult to control their temperatures by electrical-resistance heating. Proper annealing is also difficult, and therefore these glasses frequently suffer from waviness, strain, gas inclusions, and devitrification. However, the borosilicates have so many other attractive properties that they are presently the most widely used material for glass substrates.

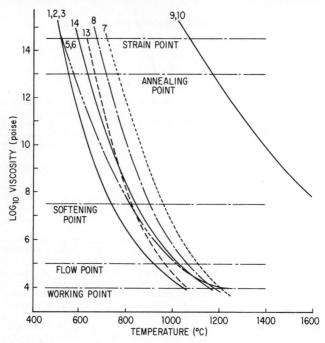

Fig. 1 Viscosities of commercial glasses as functions of temperature. The numbers refer to the glasses in Table 3.

Although flat glass plates are the more common type of thin film substrates, certain applications require cylindrical rods. Glass rod is manufactured by drawing through a nozzle at a constant rate. The nozzle is small compared with the area through which sheet glass is drawn, and certain glasses which cannot be manufactured as sheets are therefore available in rod form. These include the aluminosilicates (No. 8 in Fig. 1), which are used for tin oxide resistors, and the borosilicates (Nos. 5 and 6 in Fig. 1), which are common materials for laboratory ware.

(2) Polycrystalline Ceramics In contrast to glasses, ceramic materials do not permit extensive reworking like rolling and drawing. Each ceramic piece must be initially formed and fired to the desired tolerances. The latter depend on the forming methods selected, which in turn possess varying abilities to form substrates of different shapes and with different ranges of physical properties. Usually the raw materials for ceramic substrates are available as purified oxide powders. The oxides are mixed, reground, and then combined with organic compounds functioning as plasticizers, binders, or lubricants. A flow chart indicating three alternate methods to form ceramic substrates is shown in Fig. 2. These techniques, their advantages and limitations, are briefly reviewed in the following paragraphs.

Powder Pressing. In powder pressing, dry or slightly dampened powder is packed into an abrasion-resistant die under a sufficiently high pressure (8,000 to 20,000 psi) to form a dense body. This process allows rapid or automatic production of parts with reasonably controlled tolerances since the shrinkage during the sintering process is slight. There are, however, limitations. Holes cannot be located too close to an outside edge. Pressure variations from uneven filling of long or complex dies lead to defects like inhomogeneous properties and excessive warp. As a rule, powder pressing is not recommended for parts greater than 6 in. square.

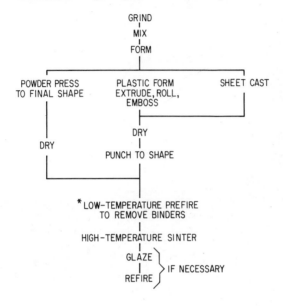

Fig. 2 Ceramic-substrate forming processes.

Isostatic Pressing. In contrast to dry pressing, this method applies uniform pressure to the powders. Here, dry powders are enclosed in an elastic container and inserted into a cavity which is first evacuated and then filled with a liquid like water or glycerin. Pressures ranging from 5,000 to 10,000 psi are applied through the liquid and yield a uniformly compacted piece. Production of 1,000 to 1,500 pieces per hour is feasible. An important advantage of isostatic pressing is that it permits fabrication of pieces with relatively large length-to-width ratios. For example, cylinders of 14 in. diameter and 24 in. length can be made. However, the surfaces which contact the elastic container must be machined since they are not smooth enough for substrate use. This extra step adds to the cost of isostatic pressing and makes it uneconomical for the production of most substrates.

Extrusion. If oxide powders are mixed with certain organic materials, the mixture becomes plastic enough to be forced through a die. Extrusion is particularly useful in forming long pieces of uniform cross section such as cylindrical resistor cores. Extrusion is a fast process, it can reproduce fine detail, and it is economical. However, considerable shrinkage occurs during drying and firing so that close final tolerances are hard to obtain. Thin sheets up to 0.1 in. thick can be prepared by passing the extrusion through successive preset rollers until the proper prefiring thickness is obtained.

Sheet Casting. Sheets up to 6 in. wide and 0.1 in. thick after firing may be formed by sheet casting.[5] The oxides are prepared as a slurry by adding liquid or organic binders, plasticizers, and solvents. The slurry is spread onto a carrier film of Mylar or cellulose acetate which moves at a constant speed under a metal knife blade positioned a short distance above the film. Such a mechanism is shown in Fig. 3.

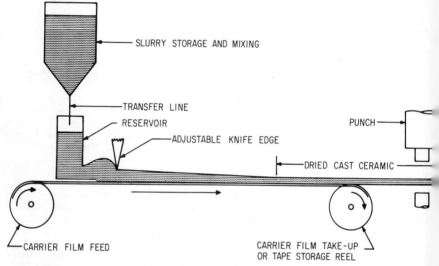

SLURRY STORAGE AND MIXING

TRANSFER LINE
RESERVOIR
ADJUSTABLE KNIFE EDGE
PUNCH
DRIED CAST CERAMIC

CARRIER FILM FEED

CARRIER FILM TAKE-UP
OR TAPE STORAGE REEL

Fig. 3 Schematic of sheet-casting method for forming thin ceramic substrates.

As the film and slurry move under the blade, a thin sheet of wet ceramic forms. The thickness of this sheet is controlled by adjusting the height of the blade over the carrier film. The resultant ceramic sheet—usually referred to as "green sheet" or in the "green state"—is air-dried to remove the solvents. Typically, a 30 to 50% reduction in green-sheet thickness occurs during solvent evaporation. Holes, alignment edges, and orientation marks are subsequently punched out of the flexible green sheet. The sheets may then be used either individually or as laminates.[6]

Sintering of Oxides. After the substrate has been formed by one of the above processes, it is often necessary to prefire at 300 to 600°C to remove organic binders, lubricants, and plasticizers used as forming aids. Further heating at higher temperatures to densify the aggregate of small particles is called sintering. This results in a ceramic body of considerable mechanical strength. Sintering of solids consolidates the fine particles by recrystallization, a process whereby the larger grains grow at the expense of the smaller ones. Excessive grain growth is responsible for the undesirable roughness exhibited by many "as-fired" ceramics. Recently, it has been possible to inhibit grain growth during firing and to produce high-purity dense alumina with markedly improved surfaces and improved mechanical and electrical properties.

During sintering, ceramics shrink as much as 18 to 25% in their linear dimensions or up to 50% in volume because of loss of binder and particle coalescence. This rather startling change in dimensions is illustrated in Fig. 4. Numerical examples of the nominal shrinkage accompanying various sintering processes are given in Table 2 along with the tolerances that can be commercially obtained.

Glazing. Undesirable "as-fired" surfaces may be improved by coating one or both sides of the substrate with thin glassy layers called glazes. These consist of low-

TABLE 2 Substrate Shrinkage and Tolerances for Various Ceramic Forming Processes

Forming process	Material	% firing shrinkage	Commercial tolerances
Powder pressing...............	94–99 % alumina	16	±1 % but not less than ±0.005 in. in any dimension
	steatite	10	
	94–98 beryllia	16	
	barium titanate	14	
Extrusion....................	Same as for powder pressing	13–14	±½ % but not less than ±0.003 in. in any dimension
Isostatic pressing..............	94–99 % alumina	16–18	Same as for powder pressing
	steatite	10–12	
	94–98 % beryllia	16–18	
Sheet casting.................	94–99 % alumina	} 18–22	Length and width: ±½ %. Thicknesses up to 0.040 in.: ±10 %
	steatite		
	94–98 % beryllia		
	barium titanate		

melting silicate glasses which are squeegeed, dipped, or sprayed on. Another firing cycle melts and fuses the glaze to the body. Glazes may also be applied by the so-called "transfer-tape" process.[7] This consists of mixing the glaze in the form of a fine powder with suitable organic binders and plasticizers. The resulting slurry is cast onto a carrier sheet such as Mylar or cellulose acetate. The glaze layer may be coated with an organic adhesive if a sticky surface is desired. It is then pressed over the substrate surface to be covered, the carrier layer is stripped, and the glaze is fired as described above. This method permits selective application of the glaze at thicknesses between 15 and 250 μ (0.0006 to 0.010 in.) to virtually any material. While adherence of glazes to substrates is not a problem, the coefficients of expansion of the body and glaze materials should be close to prevent excessive strain [see Sec. 5b(1)].

Hot Pressing. In this method, powders are densified by the simultaneous application of high temperatures and pressures. It is used to form materials which normally

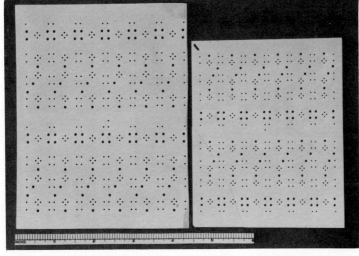

Fig. 4 Photograph of 96 % alumina substrate in the green state (left), and after sintering (right).

TABLE 3 Approximate Compositions (in Weight %) of Substrate Glasses and Glazes

No.	Supplier	Code No.	Designation	SiO₂	Na₂O	K₂O	CaO	BaO	MgO	Al₂O₃	B₂O₃	PbO	As₂O₃	BiO	TiO₂	ZnO
1	American Saint Gobain		Soda-lime	67.7	15.6	0.6	5.6	2.0	4.0	2.8	1.5					
2	Corning	0080														
3	Kimble	R6														
4	Corning	0211	Alkali-zinc borosilicate	64.4	6.2	6.9				4.1	10.3	...	...	...	3.1	5.4
5	Corning	7740	Borosilicate	80.5	4.1	0.5				2.5	12.8					
6	Kimble	KG23														
7	Kimble	EE2	Aluminosilicate	61.5	0.1		11.4		8.2	18.7						
8	Corning	1723		57.0	1.0		5.5		12.0	20.5	4.0					
9	Corning	7940	Fused silica	99.5+												
10	General Electric															
11	Typical commercial glaze		Lead-alkali borosilicate	40–50	1–2		8–15			10	6–10	20				
12	Bell Telephone Laboratories	4B19T		40		2.0	5.0	5.0		8.0	10	18	0.5	10	2	
13	Corning	7059	Modified aluminosilicate	50.2				25.1		10.7	13	...	0.4			
14	Intellux			49.9			4.3	25.1		10.3	10.5					

6-8

TABLE 4 Properties of Vitreous Substrate Materials

Glass property	Condition	Units	Type of glass							
			Soda-lime	Alkali-zinc boro-silicate	Alkali-boro-silicate	Lime-alumino silicate	Fused silica	Alkali-lead boro-silicate	Alumino-boro-silicate	Photo-sensitive glass
Composition (as in Table 3)			1,2,3	4	5,6	7,8	9,10	11,12	13,14	......
Strain point		°C	472	506	520	670	990	......	613	422
Anneal point		°C	512	539	565	710	1050	725	650	452
Softening point		°C	696	696	720	910	1580	......	820	350*
Thermal expansion	25–300°C	ppm deg^{-1}	9.2	7.2	3.25	4.6	0.56	5.5–6.1	4.5	8.3–8.6
Density	25°C	g cm^{-3}	2.47	2.51	2.23	2.63	2.2	1.58	2.76	2.36
Refractive index	Na$_D$ line		1.51	1.53	1.47	......	1.458	1.58	1.53	......
Volume resistivity	300°C	ohm-cm	$10^{5.6}$	$10^{7.9}$	$10^{7.2}$	$10^{13.8}$	$10^{11.2}$	10^{12}	$10^{12.4}$	......
Dielectric constant	25°C, 1 MHz		6.5	6.7	4.6	6.4	3.9	8.9	5.8	6.5
Power factor (tan δ)	25°C, 1 MHz	70	0.01	0.0046	0.0062	0.0013	0.00002	0.004	0.0011	0.033
Young's modulus		10^6 psi	10	10.8	9.1	12.4	10.5	......	9.8	
Poisson's ratio			0.24	......	0.20	0.26	0.17	......	0.28	
Chemical durability										
Losses at: in 5% HCl	100°C, 24h	mg/cm^2	0.02	0.03	0.005	0.4	0.001	......	5.5	
5% NaOH	99°C, 6h		0.5	2.0	1.1	0.3	0.7	......	3.7	
0.02N Na$_2$CO$_3$	100°C, 6h		0.1	0.1	0.1	0.1	0.03	......	0.3	
Relative cost			Low	Low	Low	Average	High	Low	Average	High
Availability	Rod		✓	✓	✓	✓	✓	✓		
	Sheet				✓		✓		✓	

* Maximum safe operating temperature.

TABLE 5 Properties of Crystalline Substrate Materials

Property	Test method	Conditions	Units	Alumina	Alumina	Alumina
Composition.............	Chemical analysis	....	....	80–90 % Al_2O_3	90–96 % Al_2O_3	>96 % Al_2O_3
Density..................	ASTM* C 20–46	25°C	g cm⁻³	3.3–3.6	3.5–3.8	3.7–3.9
Flexural strength (modulus of rupture).....	ASTM* D 667–44	5-in. span	10^3 psi	35–45	40–50	45–65
Compressive strength.......	ACMA† Test no. 1	....	10^3 psi	140–275	250–400	300–425
Tensile strength............	ASTM* D 116–44	....	10^3 psi	8–22	10–28	21–28
Impact resistance..........	ASTM* D 256–56	(Charpy)	in.-lb	6–7	6–8	7–8
Young's modulus...........	Resonant frequency	....	10^6 psi	31–40	40–46	45–52
Hardness..................	Scratch	....	mohs	>9	>9	>9
Max. safe temperature for continuous use..........	....	In air	°C	1400–1500	1500–1700	1600–1725
Thermal shock resistance...	(Relative values)	....	....	Good	Good	Good
Thermal conductivity......	ASTM* C 408–68	....	cal s⁻¹ cm⁻¹ deg⁻¹	0.03–0.04	0.03–0.06	0.04–0.07
Specific heat..............	ASTM* C 351–54T	25°C	cal g⁻¹ deg⁻¹	0.2	0.2	0.2
Thermal expansion........	ASTM* C 327–53T	25–300°C	ppm deg⁻¹	6.5–8	6.5–8	7–10
Dielectric strength........	ASTM* D 116–44	....	V mil⁻¹	180–260	180–340	220–340
Volume resistivity.........	ASTM* D 257–54T	300°C	ohm-cm	10^{13}–10^{12}	10^{10}–10^{13}	10^{13}–10^{15}
Dielectric constant........	ASTM* D 150–54T	25°C, 10^6 Hz	....	7.5–9.0	8.5–9.5	9.0–10.0
Power factor (loss tangent)	ASTM* D 150–54T	25°C, 10^6 Hz	....	0.0002–0.001	0.0001–0.0005	0.0003–0.0015
Loss factor................	ASTM* D 150–54T	25°C, 10^6 Hz	....	0.0015–0.009	0.00085–0.0048	0.0027–0.015
Relative cost.............	....	....	....	Average	Average	Average

sinter with difficulty or which require very high sintering temperatures. In contrast to normal sintering, densification occurs without grain growth, but the process is slow. It requires long heating and cooling times for the sample as well as for the dies and associated equipment. Therefore, hot pressing is a relatively expensive method of substrate production.

(3) Single-crystal Substrates Substrates of single-crystalline materials like sapphire (α-alumina), silicon, and germanium are obtained by cutting larger crystals (boules) into wafers. Most crystals are grown by the Czochralski technique, where a small seed crystal is dipped into the surface of the molten substrate material and slowly withdrawn. Since the seed is slightly cooler than the melt, the latter solidifies and forms a larger single crystal. Presently, silicon crystals of 2.0 in. diameter and 12 in. length are routinely prepared, but larger crystals may also be grown.

Another method of crystal growth is flame fusion (Verneuil technique), which was originally developed for the preparation of artificial gems. An example is sapphire boules up to $1\frac{3}{8}$ in. diameter and 6 in. length, which are made by dropping fine alumina powder through an oxyhydrogen flame directed at a refractory pillar. The alumina powder melts and forms a bead on the pillar. As more powder is added and the pillar is lowered, the sapphire crystal grows. Alumina melts at 2030 to 2050°C and boils only 100°C above this. Therefore, maintaining proper growth conditions is critical.

Large crystals of high purity and perfection have been grown by the Czochralski method, which is more promising with regard to volume production and cost. Crystals grown by the Verneuil method are limited in size and characteristically display more defects per unit volume. Both methods require slicing into wafers with subsequent grinding and polishing of the resulting surfaces. Specific crystal orientations are attainable but at additional cost.

(4) Metal Substrates Metal substrates are primarily used for magnetic films. To obtain uniform film properties, it is essential to provide smooth surfaces by

Beryllia	Beryllia	Forsterite	Steatite	Glass mica	Alkaline-earth porcelain	Glass ceramics	Titanate ceramics
98% BeO	99.5% BeO	2MgO–SiO₂	MgO–SiO₂				
2.80	2.85–2.88	2.8–3.1	2.4–2.8	3.1–3.4	2.6	2.5–2.6	3.5–5.5
25–28	30	12–30	17–22	13–21	18	13–50	10–22
>225	>225	80–85	68–90	30–36	~100	....	40–120
	~20	9–10	8.5–13	8–9	~10	~3	4–10
		4–7.5	3.8–4.3	1.7–1.95‡	4–5	21	3
45	51	21	15–16	9.5	....	13–14	10–15
9	9	7.5	7.5	4–6	8	9	
1500	1500	1000	1000	1200	1050	500	1200
Good	Very good	Poor	Poor–good	Very good	Good	Excellent	Poor
0.25	0.35	0.012	0.009–0.010	0.0012	0.010	0.004–0.009	0.008–0.01
0.34	0.31	0.2	....	....	....	0.2	~0.1
6.5	8.0′	6.8–10	6.0–11.1	10.8–14.4	4–6	§	7–10
350	350	240	210–230	320–350	....	....	50–300
10^{14}	10^{14}	10^{12}	10^8–10^{11}	10^{15}	10^8	10^8–10^{10}	10^6–10^{11}
6.5	6.7	6.2	5.9–6.3	6.5–7.5	4.5–6.3	5.5–9.1	15–10,000
0.0001	0.0003	0.0004	0.0008–0.0035	0.0002–0.001	....	0.006	0.0002–0.05
0.001	0.002	0.002	0.005–0.021	0.010–0.016	0.026	0.008–0.02	
High	High	Low	Low	Low	Low	High	Low

* American Society for Testing and Materials, Philadelphia, Pa.
† Alumina Ceramic Manufacturing Association, New York, N.Y.
‡ Izod.
§ Depends on heat treatment.

mechanical or electropolishing. The former is often unsuitable since it leaves polishing compound trapped in the surface of the metal. Electrolytic, chemical, or vibratory methods to remove surface material are therefore preferred. Polished sheets of aluminum,[8] copper, and silver[9] have been used successfully as ground planes for magnetic-film arrays. For high-power, low-frequency applications, porcelain-enamel finishes on metal plates are also practical.

b. Substrate Properties

Now that methods of substrate manufacturing have been discussed, the chemical, physical, and mechanical properties of the various material categories will be reviewed. As stated before, the properties of glasses vary widely depending on their chemical composition. Typical compositions of commonly used glasses are listed in Table 3. The resulting properties are shown in Table 4. The major advantage of glasses is that smooth surfaces can be achieved directly by drawing, which is reflected in generally low cost. Individually, glasses vary significantly in regard to volume resistivity, loss tangent, and softening points. Their poor thermal conductivity and the difficulty of obtaining intricate shapes including holes preclude the use of glasses for many electronic applications.

The properties of several polycrystalline ceramic materials are shown in Table 5. Compared with the glasses, they offer higher softening temperatures, greater mechanical strength, better thermal conductivity, and superior chemical stability. Disadvantages are the rougher surfaces or, if that condition is somewhat improved, the greater cost. Special cases are glass mica and glass ceramics. The latter are recrystallized glasses made by melting a glass batch with a nucleating agent such as lithia

TABLE 6 Properties of Single-crystal Substrates

Material	α-aluminum oxide	Magnesium oxide	Titanium dioxide	Silicon dioxide	Magnesium fluoride	Silicon	Germanium	Mica (muscovite)	Gallium arsenide
Chemical formula	Al_2O_3	MgO	TiO_2	SiO_2	MgF_2	Si	Ge	$KH_2Al_3(SiO_4)_3$	GaAs
Crystal system	Hexagonal	Cubic	Tetragonal	Hexagonal	Tetragonal	Cubic	Cubic	Monoclinic	Cubic
Density, g cm^{-3}	3.98	3.58	4.25	2.65	3.18	2.33	5.36	2.76–3.0	5.3
Hardness, mohs	9	5.5	5.5–6	7	6	7	6.5	2.5–3.0	......
Melting point, °C	2040	2800	1825	~1425§	1255	1420	936	......	1238
Thermal conductivity at 25°C, cal cm^{-1}s^{-1}deg^{-1}	0.09	0.06	0.03* 0.02†	0.03* 0.02†	......	0.20	0.13	0.0016†	0.10
Thermal-expansion coefficient at 25–300°C, ppm deg^{-1}	8* 7.5†	13.8	9.2* 7.1†	12* 22†	18.8* 13.1†	2.5–3.5	5.5–6.4	~11* ~20†	5.93
Specific heat, cal g^{-1} deg^{-1}	0.20	0.276	0.17	......	0.25	0.18	0.073	0.21	0.086
Young's modulus, 10^6 psi	50–56	36–52	36–41	......	15.5–16.3	15–23	15–23‡	......	8.6–17‡
Flexural strength, 10^3 psi	65–100	20	20	......	~9	......	......	......	......
Dielectric constant at 25°C, 10^4 Hz	11.5* 9.4†	9.65	165* 86†	4.5	4.87* 5.45†	12	15.7	6.5–9	12.5
Dielectric strength at 25°C, V mil^{-1}	1,200	......	......	......	......	......	......	3,250–6,250	......
Power factor (loss tangent) at 25°C, 10^4 Hz	0.002	<3 × 10^{-4}	0.008* 2 × 10^{-4}†	......	......	......	......	1–2 × 10^{-4}	......
Poisson's ratio	0.2	0.2–0.36	......	......	......	......	0.28	......	......

* Parallel to c axis.
† Perpendicular to c axis.
‡ Depends on orientation.
§ Transition from quartz to crystobalite.

or titania. The batch is formed with transparent glass by conventional techniques. As the glass cools, it reaches a temperature where precipitation of the nucleating agent occurs. The nucleated glass is reheated to a temperature range where growth of the nucleated crystals continues. The glass composition and the nature of the heat treatment determine the type of crystallization and the final properties. Glass ceramic materials may be made to close dimensional tolerances because of the very small changes in density which occur during the crystallization process. Also, the expansion coefficient may be tailored to some degree. However, their cost is relatively high, and their thermal conductivities and electrical properties are not as good as those of true ceramic materials.

Glass mica is made by combining either natural (muscovite) or synthetic (fluorophlogopite) mica powder with high-resistivity glass powder. The powders are blended, pressed, and dried like ordinary ceramics. They are then fired and cooled under pressures of about 25,000 psi. The parts shrink less than $\frac{1}{2}\%$, in contrast to ceramic parts, which may shrink as much as 22%. This low shrinkage permits accurate dimensioning of complex shapes as well as the inclusion of metal parts in the molded form. The latter advantage may obviate the need for plated through-holes in applications requiring wiring on both surfaces of the substrate.

Finally, Table 6 lists the properties of some materials which are available in single-crystalline form. Here, silicon and, to a lesser degree, germanium and sapphire are by far the most widely applied materials. They combine desirable properties such as great chemical and thermal stability with excellent thermal conductivity. In addition, the electrical resistivity of silicon can be modified controllably over a wide range. Common disadvantages are the necessity of finishing operations to obtain smooth surfaces and the relatively small sizes available. Individual properties of substrate materials are discussed in more detail in the following sections.

3. SUBSTRATE SURFACES

The quality of the surface is the most important property of a substrate since it is here that the film-substrate interaction occurs. Various types of irregularities make up the overall surface texture. The following categories of surface defects may be encountered:

1. On the atomic scale: point defects, dislocation lines, monatomic ledges on cleavage planes
2. Submicron features: polishing scratches, glass-drawing asperities, pores due to less than theoretical density of the body
3. Micron scale: grinding scratches, crystallite boundaries in polycrystalline materials, pores, glass-drawing lines
4. Macrodefects: surface warp, glaze menisci, fused particles

The different nature of these defects requires a variety of methods to determine or to characterize quantitatively the condition of substrate surfaces.

a. Surface Characterization Techniques

Techniques to resolve surface defects on the atomic scale such as LEED, or decoration methods to reveal cleavage steps are mostly of interest in nucleation and growth experiments. They are treated in Chaps. 8 and 9. Electron microscopy and reflection electron diffraction are discussed in Chap. 9. The latter technique, although primarily used to yield structural information, is of interest in this context because an incident electron beam forming an angle of 1° with the surface is capable of detecting asperities down to 5 Å height.[10] Various optical methods such as light-section microscopy, multiple-beam interference, and phase-contrast microscopy (interference using polarized light and a birefringent prism) are also useful to examine substrate surfaces. The principles involved are the same as described for thickness measurements in Chap. 11. Instead of limiting the observation of fringes to a step in the film, the entire substrate is viewed whereby the interference fringes form a contour map of the surface.

By far the most widely used technique to characterize the smoothness of a substrate is the stylus method described in Sec. 1e(4), Chap. 11. Its sensitivity is adequate for those defects which are of greatest concern in thin film work, and it is rapid and simple to use. A particular advantage over other methods of surface examination is the fact that the stylus technique yields a quantitative measure of surface roughness. The statistical significance of the various figures of merit derived from stylus instruments warrants a brief review.

(1) The Root-Mean-Square (RMS) Roughness A typical stylus trace of a substrate surface is shown in Fig. 5a. The irregularities of the surface may be thought to

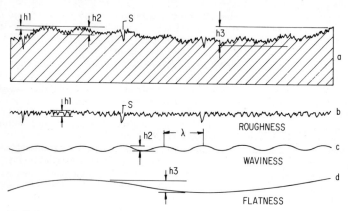

Fig. 5 (*a*) Stylus trace of a surface with several features. (*b*) Roughness, h 1. (*c*) Waviness, h 2. (*d*) Flatness deviation, h 3.

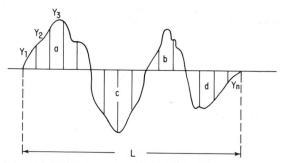

Fig. 6 Schematic of a surface trace.

consist of three components of different periodicities. These are referred to as roughness, waviness, and flatness in Fig. 5b, c, and d. Roughness is the property which stylus instruments measure; it may be characterized numerically by the average deviation of the trace from an arbitrary mean. Referring to the trace in Fig. 6, the rms value is derived by dividing the peaks and valleys into narrow segments of height y, summation over all y^2, and subsequent averaging:

$$\text{RMS} = \sqrt{\frac{y_1{}^2 + y_2{}^2 + y_3{}^2 + \cdots + y_n{}^2}{n}}$$

As this illustration shows, the rms value is an integral quantity which is most useful to assess the total of a quantity fluctuating over a period of time. An example is the power conveyed by an alternating current. However, the rms value does not constitute a direct measure of the surface roughness, although it is occasionally used for that purpose.

(2) The Arithmetic Average (AA) Roughness   The present standard of expressing surface roughness is called the arithmetic average (AA) in America (ASA B 46.1–1962) or the centerline average (CLA) in England.   To derive this value, the abscissa of a trace as shown in Fig. 6 must be drawn over an assigned length of surface and so that the areas under the curve above and below the centerline are equal.   The mathematical definition of this quantity is

$$AA = \frac{a + b + c + d + \cdots}{ML}$$

where $a, b, c, d, \ldots$ = areas under the peaks or above the valleys of the trace (see Fig. 6)
$\quad\quad\quad\quad L$ = assigned length of the stylus travel
$\quad\quad\quad\quad M$ = vertical magnification used

Typically, the AA value is 10 to 30% lower than the rms value for the same trace.   It has the advantage of an unambiguous mathematical definition, it can readily be measured from a stylus trace by means of a planimeter, or it may be derived electronically if an integrating instrument is available.   Most stylus instruments are equipped for this option and provide the AA value directly.

In using statistical averages to characterize substrate surfaces, it should be clear that a single number can give only a very incomplete description of the surface texture. Low values may mask the presence of deep scratches (feature s in Fig. 5) which may later cause discontinuities in deposited thin films.   Similarly, surface waviness and flatness are not reflected in rms or AA values, although both properties are important in photoresist work.   Finally, there is some ambiguity in both methods because surface profiles having the same amplitudes but different periodicities yield identical average values.   Other factors to consider when stylus traces are interpreted are the tip radius—typically 0.05 to 0.1 mil—which limits the fine resolution in the direction of travel, and the different magnifications utilized in the horizontal and vertical axes of the profile.

b. Surface Texture of Substrate Materials

The thin film worker is primarily interested in surface features in the micron and submicron ranges.   Therefore, stylus traces are widely used to evaluate the adequacy of substrates for a particular purpose.   Representative surface profiles are shown in Fig. 7.   Polished single-crystal wafers such as silicon or sapphire and polished vitreous materials such as fused silica possess very uniform surfaces and yield smooth traces as indicated in Fig. 7a.   Drawn glasses and glazes have surfaces which are smooth, as shown in Fig. 7b, except for an occasional surface irregularity.   The latter may be as high as 1,000 Å and stem from the forming operation.   Modified borosilicate glasses tend to be wavy because of their high drawing temperatures and short working ranges; however, the deviations from a flat plane are less than 0.0005 in. and do not create a flatness problem.   Sintered ceramic materials, particularly alumina, are of great practical interest because of their mechanical strength and high thermal conductivity.   A trace of a typical as-fired 96% alumina is shown in Fig. 7c.   This type of surface roughness is generally unsatisfactory as it exceeds the thickness of most films.   Polishing such materials does not significantly improve the surface, presumably because there is little cohesion between the grain boundaries.   Hence, chipping occurs during the finishing operation.   As shown in Fig. 7d, flat smooth plateaus alternate with depressions several thousand angstroms deep in the regions of grain boundaries.

To overcome these limitations, very fine-grained ceramics have been developed. In addition to less textured surfaces, such bodies have higher mechanical strength and volume resistivity than large-grained materials of comparable compositions.   In Fig. 8a and b are shown an electron micrograph and a stylus trace of a 7-μin. alumina surface.   Even finer grains have been obtained more recently by Stetson and Gyurk.[11]

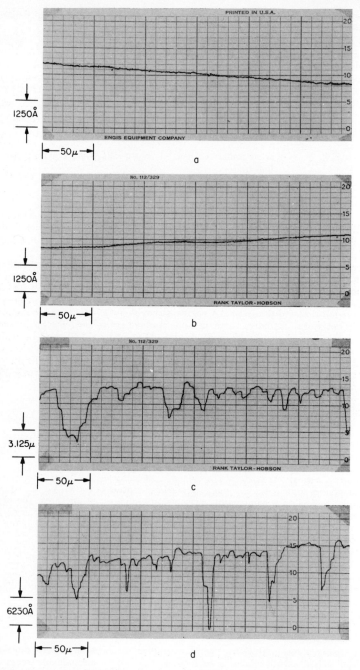

Fig. 7 Stylus traces of substrates. (a) Polished glass or single crystal. (b) Drawn glass or glaze. (c) As-fired large-grain ceramic (note change in vertical scale). (d) Same as (c) but polished (note change in vertical scale).

The electron micrograph and stylus trace of this material are shown in Fig. 8c and d. The improvement in surface uniformity is significant. The roughness of 2.5 μin. approaches the threshold value below which thin film capacitors may be fabricated. Polished surfaces of these materials yield traces similar to those shown in Fig. 7a except for a few small pull-outs or exposed pores.

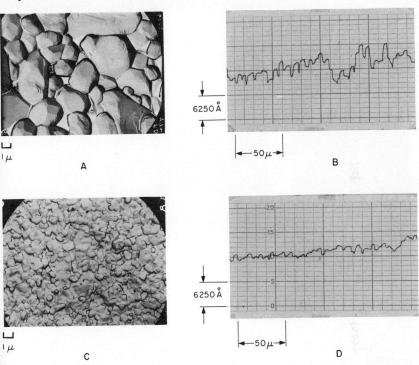

Fig. 8 Surfaces of small-grain alumina. (A) and (B) are 7-μin. finishes. (C) and (D) are 2.5-μin. finishes.

For many applications, characterization of the surface texture by the average roughness value is sufficient. If a more detailed description of the surface is required, for example, to facilitate comparison with electron-microscope pictures, the stylus trace must be more fully evaluated. This can be done by separating the surface profile into its component parts, as previously shown in Fig. 5. Another method has been described by Schwartz and Brown,[12] who derived four variables from the stylus trace to define the surface. The parameters selected were the average peak-to-valley height, the maximum peak-to-valley height, the number of peaks or valleys per unit length of stylus travel, and the average peak half-width. With these four parameters, the authors were able to differentiate between surfaces whose traces were very similar, to obtain a measure of the crystallite size, and to note single damaging defects that would go unnoticed if only a single number is used. To determine the effects of surface texture on thin film components, it is necessary to separate the different factors which contribute to the overall surface profile.

c. Effects of Surface Texture on Thin Film Components

In the preparation of thin film components for microelectronic applications, surface roughness and flatness are crucial properties. Flatness is required to facilitate close contact with photomasks and can be obtained satisfactorily with most substrate materials. The surface microfinish requirements, however, vary with the film thickness to be deposited and are difficult to meet in cases where extremely thin films such as 100 Å or less are desired.

As stated in the introduction, atomically flat surfaces over large substrate areas are unattainable. Even the best single-crystalline substrates acquire defects in the fabrication and manufacturing processes. These defects are replicated by the films or propagate into the deposit as, for example, dislocations continue into epitaxial layers.[13] Defects on the submicron scale such as fine scratches on silicon wafer surfaces also generate irregularities in thin films. Thus, Balk et al.[14] reported that mechanical damage of the original surface affects the rate of oxidation, and that the thermal oxide retains the gross surface features. This is illustrated in Fig. 9A

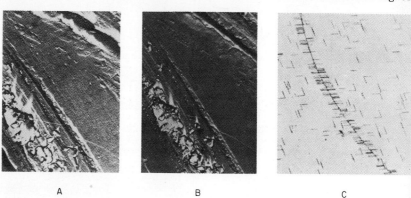

A   B   C

Fig. 9 Electron micrographs of scratched silicon surfaces. (A) Etched. (B) Same area oxidized in dry oxygen. (*From Balk, Aliotta, and Gregor,*[14] *courtesy of the Metallurgical Society, AIME.*) (C) Stacking faults on (110)-oriented surface. (*From Mendelson,*[15] *courtesy of the Macmillan Company, New York.*)

and B. While the "sandiness" of the original surface disappeared after oxidation, the coarser features including the scratch remained. Figure 9C shows another type of defect which occurs in epitaxial films. These are stacking faults generated in the grown layer at sites where the underlying substrate surface was mechanically damaged.[15] Characteristic is the high stacking-fault density along the polishing scratch.

In the fabrication of thin film electronic components, glass and ceramic substrates are more common than single-crystal wafers. Hence, there is more concern about the coarser surface imperfections than about those of merely atomic dimensions. The various effects of surface irregularities on thin films used for microelectronic applications are reviewed in the following sections.

(1) Thin Film Resistors Thin film resistors are relatively insensitive to surface roughness as long as the latter does not exceed the film thickness. The substrates used for this application include glasses, polished fused silica, ceramics, and single-crystal wafers. A comparison of Nichrome films deposited on sintered ceramics and glass[16-18] revealed that the rougher surfaces yielded films of higher sheet resistances, lower temperature coefficients, and poorer stability during thermal aging. Similar behavior was found for silicon films deposited on as-fired alumina.[19] A compilation of data illustrating the influence of surface roughness on the sheet resistance of tantalum nitride films is shown in Table 7.

The data of Brown[20] and Coffman and Turnauer[21] in Table 7 are in good agreement and demonstrate the increasing effect which surface roughness has on the sheet resistance. A more detailed examination of the latter authors on substrates with high glass-to-crystalline ratios led to further conclusions. It was found that the crystal shape, the ratio of the glassy matrix to crystalline matter, and the density of

TABLE 7 Effect of Surface Roughness on Sheet Resistance of Tantalum Nitride Films

Substrate	Polished sapphire	Glazed 96% alumina	Polished 99% alumina	Polished 96% alumina	As-fired 96% alumina	Polished steatite	As-fired alkali-earth porcelain	As-fired 99% alumina	As-fired barium titanate	As-fired steatite	Reference
AA value, μin..........	1	1	2	6	18	30	32	40	65	120	21
R_s relative to smooth glass	1	1	1.07	1.4	1.7	2.0	2.2	1.91	4.0	1.3	20
			3.0	1.23	1.59						22

crystallites affected the sheet resistance more strongly than the roughness measured by the stylus instrument.

(2) Thin Film Capacitors Dielectric films are particularly sensitive to isolated defects protruding from an otherwise perfect surface. The reason is the nonuniform electric field of such points, which is likely to cause dielectric breakdown. Schwartz and Brown[12] investigated the effects of surface roughness on Ta-Ta_2O_5-Au capacitors. They concluded that high fields caused by protrusions and thin spots on the electrode surfaces were in many cases responsible for short component life and low breakdown voltages. Silicon monoxide[23-25] and Al-Al_2O_3-Al[26] film capacitors are similarly affected by surface imperfections. McLean and Rosztoczy[27] were able to produce thin film capacitors on comparatively rough substrates by utilizing counterelectrodes of manganese oxide and metal. Premature breakdown of the dielectric film was suppressed by the semiconducting manganese oxide.

(3) Magnetic Films The properties of magnetic films are also affected by the treatment of the substrate surface.[28-32] Scratches in the substrate surface, for example, reduce the switching energy.[33] Local stresses affect the direction of magnetization so that the area surrounding the scratch resembles a Néel wall, the only difference being that the magnetization direction is parallel instead of antiparallel.

Prosen et al.[34] were the first to quantitatively relate magnetic properties to surface roughness. They measured the surface area by adsorbing a monolayer of a surfactant labeled with C^{14}.[35] Further investigations of Kobale[36] and Wiehl[37] demonstrated the dependence of the magnetic properties of evaporated Permalloy films on surface roughness. In the latter investigation, electron and interference contrast microscopy were used to examine the condition of the surface. Since polycrystalline substrates expose crystallites of different orientations, the evaporant vapor impinges at various angles of incidence and may thus lead to local stresses.[38] Uniaxial anisotropy in magnetic films may also be caused by shadowing effects associated with protruding crystallites. Similar considerations hold for electrodeposited films, and a dependence of magnetostriction on crystallite orientation has been reported.[39]

Summarizing the results of these investigations, the effects of surface roughness on magnetic films can be described as follows: Surface roughness characterized by AA values of the order of the Néel-wall width ($\sim$1,000 Å) affects the magnetic field H_c necessary to induce wall motion. Higher fields are required the rougher the surface. Surface-roughness values of about 1μ alter the angular dispersion α_{90}. The latter is particularly sensitive to large crystallites in the substrate surface. Finally, surface roughness as induced by unidirectional polishing influences the uniaxial anisotropy H_k.

Other magnetic properties such as magnetostriction and ferromagnetic resonance are also affected by surface roughness. Ondris[40] pointed out that the texture of the surface influences the validity of static measurements. An example is the saturation magnetism of very thin films. Many anomalous and conflicting results or measurements on various substrates may be traced to a lack of appreciation of substrate-surface effects.

(4) Interconnections Surface roughness also plays a role in techniques used to interconnect thin film components. Electroplating, for example, is employed in plated-wire memories or to increase the thickness of evaporated metal films. Garte[41] found that gold layers plated onto rough surfaces tend to occlude gases, which leads to porous films of variable thickness.

Of great practical importance are the effects of roughness on wire-bonding processes. Riben et al.[42] examined a variety of substrate materials and concluded that for split-tip welding of thin wires (less than 0.002 in. diameter), the surface roughness should not exceed 30 μin.; with 0.005-in. wire, roughnesses greater than 30 μin. can be tolerated but only at the price of reduced bonding efficiency. Ultrasonic bonding is far more sensitive to surface roughness than welding, and finishes of 3 μin. or better are recommended for high bond yields.

d. Substrate Flatness

The flatness or planarity of substrates is of importance mainly in two situations encountered in thin film processing. The first occurs when the substrate is placed in contact with a substrate holder or heater during film deposition; the second arises in subtractive etching operations, where intimate contact with a photomask is required.

(1) Substrate Heating Most substrates are heated during film deposition by mounting them against a flat metal heater plate. Although the objective is to obtain heat exchange by conduction, this is generally achieved in a very incomplete way. The reason is that the two surfaces, the heater plate and the substrate back, do not match perfectly but touch only in spots. As a result, the heat flow into the substrate is reduced, and the surface temperature is nonuniform.

Furthermore, control of the latter is difficult to achieve because of the temperature differential between the heater plate, where a thermocouple can easily be installed, and the substrate surface, whose temperature is hardly amenable to direct measurement. With materials of poor thermal conductivity such as glasses, gradients of 50 to 100°C across the substrate thickness are frequently observed even at moderate deposition temperatures around 400°C. At 800°C, a temperature difference as high as 250°C has been found.[43]

Experimental techniques to equalize the temperature of substrates have been developed. These rely on providing the substrate with a thermally conductive backing. An example has been reported by Hanson et al.,[44] who used conductive tin oxide on very thin glass plates. The authors were able to limit lateral temperature variations to ±5°C across an area of 2.5 cm diameter. A common practice in many laboratories is the application of gallium-indium mixtures to the back surface of the substrate. The eutectic of these two metals is liquid at room temperature and can easily be spread into a coherent layer thick enough to provide thermal contact across the entire substrate back. The useful temperature range of this technique is rarely limited by the vapor pressure of the two metals, as these are small enough to be neglected in most situations. However, loss of heat contact (solidification!) may occur as a result of alloying with the holder or with metal and semiconductor substrates. Robbie and Stoddart[45] reported a maximum use temperature of 250°C. Organic materials such as high-vacuum grease may be used in the same manner, but because of the onset of evaporation and decomposition, the temperature limits are lower. The utilization of a high-temperature acrylic cement has also been reported.[46]

(2) Image Widening Separation between the substrate and a negative photomask during the exposure of a photoresist film leads to widening of the positive regions in

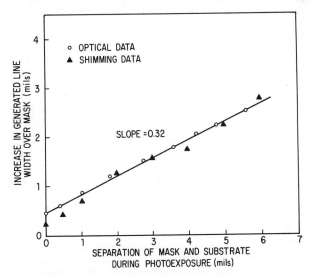

Fig. 10 Effect of mask-substrate separation on photoresist pattern dimensions. (*From Brown,*[20] *courtesy of the American Ceramic Society.*)

the pattern. This phenomenon introduces dimensional errors into subtractively etched film patterns which increase with increasing separation. In the absence of dust, lack of substrate flatness is the prime cause of insufficient mask contact. Aggravating the effect are light scattering and multiple reflections at the optical boundaries, and lack of collimation of the light source. On the other hand, image widening is somewhat reduced if thin flexible substrates are used, since these flatten out while held by the vacuum frame of the exposure table.

In view of the many variables involved, the degree of image widening cannot be predicted in general terms. However, several investigators have reported data pertaining to specific situations. Culver and Layton[47] calculated that a 0.25-in. light source at 25 in. distance would increase the line width as determined by the mask by 0.001 in. for 0.001 in. of separation. Glang and Schaible[48] separated mask and silicon wafer surfaces during the exposure of a 0.001-in.-wide line and obtained 0.00025 in. widening for 0.001 in. separation. Brown[20] found a similar increase of 0.00032 in. for the same separation but with 0.010-in.-wide lines. His data are shown in Fig. 10 and demonstrate the proportionality of the widening effect and the separation distance.

e. Porosity

Although porosity is a volume property, it also affects substrate surfaces. Where pores intersect the surface, they retain dirt or contamination from cleaning solutions. During subsequent vacuum processing, the release of occluded gases or decomposition of organic residues is often a problem. Whereas glasses are generally nonporous and their outgassing is limited to desorption and outdiffusion of water vapor [see Chap. 2, Sec. 3c(2)], polycrystalline ceramics invariably have pores. Their sizes and distribution vary depending on the conditions of preparation and are an important criterion of the material quality.

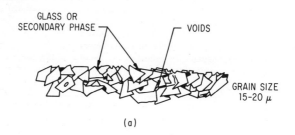

(a)

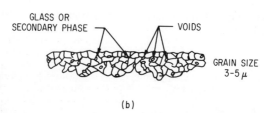

(b)

Fig. 11 Schematic cross sections of high-density alumina prepared with glass. (a) 96% Al_2O_3 + 4% glass. (b) 99% Al_2O_3 + 1% glass.

As stated in Sec. 2, polycrystalline ceramics are sintered rather than melted. This requires densification by diffusion of atoms into, or of vacancies out of initially present voids. Usually, the densification process is incomplete with some pores remaining. Filling of the residual pores can be accomplished by adding glass as a flux, and the resulting structures are shown in Fig. 11. The comparatively large, recrystallized grains are interspersed with glassy regions. Inhibited grain growth was employed in preparing the finer-grained body of Fig. 11b, which has a lower glass content. However, there is also a higher probability for pores to remain unfilled in spite of the greater density of the material. This does not mean that all fine-grain ceramics are porous. It shows, however, that density and porosity cannot be mutually inferred but must be distinguished from each other.

4. SUBSTRATE PROPERTIES RELATED TO CHEMICAL COMPOSITION

In electronic applications the substrate serves not only as a mechanical support for the film but also as an insulator between different electric components. Thus, attention should be given to the compatibility of the substrate constituents and the deposited film. The processing chemicals must also be carefully selected to avoid undesirable reactions which adversely affect the deposited film. These considerations require a closer examination of the composition of substrate materials and their possible chemical reactions. This is particularly important in the case of glass substrates, whose composition varies more widely than those of ceramics and single-crystalline materials.

The constituent elements of glasses and ceramics are shown in the periodic table in Fig. 12. It demonstrates that glasses are composed of elements in two separate

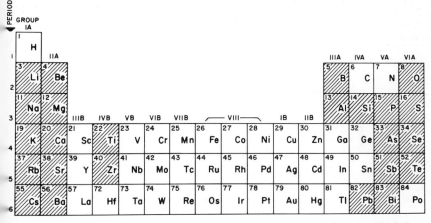

Fig. 12   Periodic chart showing common glass- and ceramic-forming elements (shaded).

regions. The principal glass-forming elements are in group IVA. Sulfides, selenides, and tellurides, too, form low-melting glasses, but these are unsuitable for most substrate applications, mainly because of their insufficient chemical stability. Many other elements from adjacent groups can be substituted into the chain or network structure of silicate glasses and thereby provide a great variety in the types of glasses obtainable. The elements in groups IA and IIA serve primarily as fluxing materials which help control the glass viscosity as well as the temperatures of melting, firing, and forming.

a. Volume Resistivity

Many glasses contain alkali metals to make them more amenable to drawing. A common constituent is sodium oxide, for which potassium is sometimes substituted. These metals form ions of relatively small size and high mobility which impart undesirable properties to the glass such as low chemical durability and high dielectric loss. If one compares the properties of the glasses in Table 4 with their chemical compositions in Table 3, one finds that resistivity and composition are related.

In Fig. 13 the volume resistivities of several glasses are plotted as functions of temperature. The slopes of these curves indicate thermally activated conduction which is thought to be due to ions moving through the glass. Accordingly, the glasses with the least alkali content have the highest resistivities. Owen and Douglis[49] showed that trace impurities of alkali metals contribute to the conductivity of fused silica, too. All other oxides except alumina tend to increase the resistivity of glasses. A comparison between the effects of boron oxide and alumina is of particular interest. In general, the resistivity of glasses is proportional to their viscosity. However, an increasing boron oxide concentration decreases the viscosity while raising the resistivity,

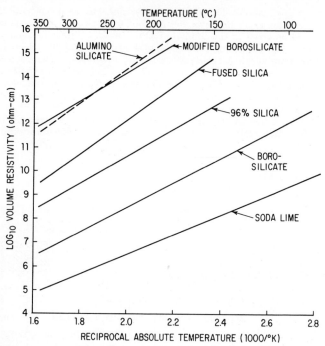

Fig. 13 Volume resistivities for various glasses as functions of temperature.

whereas the converse is true for glasses containing alumina. A careful balance must be maintained in alumino-borosilicate glasses to permit ease of drawing and yet retain high resistivity.

Kraus and Darby[50] investigated the effects of temperature on the motion of sodium ions in soda glass. They found the activation energy to be of the order of 30 to 60 kcal mole^{-1}. Hence a temperature increase from 25 to 300°C may enhance the ion mobility by a factor of 10^7. A tantalum nitride resistor dissipating 40 W in.$^{-2}$ on a glass substrate may raise the temperature by about 200°C. This temperature in conjunction with high electric fields (800 to 1,000 V cm^{-1} for some meandering patterns) is sufficient to produce considerable ion migration. The mobile ions on the substrate surface travel to the oppositely charged electrode and may cause secondary reactions. Erosion of metal films near the negative terminals of resistors has been observed in Nichrome[18] as well as in tantalum nitride resistors.[20] Orr and Masessa[51] studied the effects of ion migration on tantalum nitride and Cox[52] on tin oxide resistors. Snow et al.[53] investigated the motion of alkali ions within a protective oxide layer on silicon. They found an exponential temperature dependence with activation energies of 32 kcal mole^{-1} for sodium and 22 kcal mole^{-1} for lithium

Martin[26] attributed early failures of thin film capacitors on soda-lime glass to alkali migration from the substrate into the dielectric.

The temperature dependence of the resistivities of ceramic materials is shown in Fig. 14. In general, ceramic oxides have large band gaps and are insulators at room temperature. Conduction by mobile ions is usually small except at high temperatures. However, certain impurities or the presence of cations of variable valency may impart electronic conductivity. Even a small fraction of such cations is sufficient to cause a drastically lowered resistivity since the mobility of electrons is much greater than that of ions.

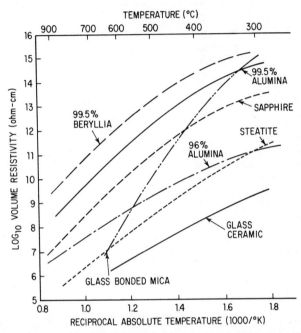

Fig. 14   Volume resistivities of ceramic materials as functions of temperature.

Another factor to be considered is the porosity of the substrate. In polycrystalline materials, porosity always increases the resistivity, but if the pores are filled with a secondary phase, the latter may increase or decrease the resistivity depending on composition and impurity levels.

b. Dielectric Properties

Dielectric constants from 4 to 10 are typical for glasses and ceramic materials, as can be seen in Tables 4 and 5. An important parameter in the selection of an insulating substrate material is its dielectric loss. At frequencies below about 10^{10} Hz, vibrational and deformation losses at room temperature are small; thus, ceramics are usually good dielectrics. The two important sources of dielectric losses in ceramics are ion migration and space-charge polarization. In glasses, the field-induced migration of alkali ions is the main contributor to dielectric loss up to 10^6 Hz. A similar type of loss is observed in crystalline materials. In addition, polarization and dielectric loss may occur as a result of space-charge buildup at interfaces in two-phase systems, for example, those consisting of glassy and polycrystalline regions.

In the fabrication of functional thin film assemblies, substrates with a high dielectric constant are frequently desired to achieve a high capacitance per unit area.[54] Titanates, for example, with ϵ values of 1,700 have been used as substrates for thin film amplifiers.   On such substrates, interelectrode capacitance becomes significant as the distance between conducting paths decreases.   This effect has been treated by Kaiser and Castro[55] and by Happ.[56]   The latter derived a series of curves to assess the stray capacitance between adjacent conductors.   One set of these curves, which permit the calculation of interelectrode capacitances, is shown in Fig. 15.   One can derive from Fig. 15 that stray capacitances between adjacent conductors may present a serious constraint in the design of thin film functional assemblies, particularly of those with high-frequency characteristics.

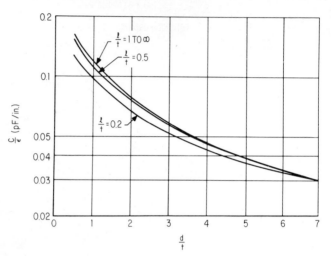

Fig. 15 Effect of conductor geometry and substrate dielectric constant on interelectrode capacitance. C = stray capacitance between adjacent conductors; ϵ = dielectric constant; t = substrate thickness; d = interelectrode spacing; l = electrode length.   (*From Happ,*[56] *courtesy of Chilton Company, Publishers.*)

c. Durability

The effects of chemicals on substrate materials are of importance in regard to changes induced by exposure to the atmosphere and etching and cleaning solutions. Most often encountered and well known for its detrimental effects in many cases is water vapor.   The latter is adsorbed by nearly all surfaces in proportion to its vapor pressure in the ambient.   Upon application of an electric field, ions stemming from the adsorbed water molecules become mobile and cause surface leakage.   This is shown in Fig. 16, where the (hypothetical) sheet resistance of a monolayer of pure water is indicated as a reference point.   The figure demonstrates that the insulation resistance of substrates may fall off by orders of magnitude if the relative humidity exceeds about 30%.   The fact that the observed surface resistance may be lower than the value of pure water indicates an enhanced mobility of ions due to interaction between the solid and the adsorbed species.   Surfaces which adsorb water readily are called hydrophilic and are of no value in humid environments unless they can be encapsulated.   The protection achieved by coating a surface with a water repellent or hydrophobic film is demonstrated by curve G in Fig. 16.

The mobilization of adsorbed water by an electric field may lead to corrosion of the substrate.   Possible mechanisms may involve reactive materials giving rise to formation of new compounds on the surface, preferential dissolution of one substrate constituent, or total dissolution with continuous exposure of fresh material.[58]   Atmospheric gases other than water vapor may also be detrimental, but their effects are essentially limited to glasses.   For example, alkali ions formed under humid conditions may recombine with carbon dioxide or sulfur dioxide.   The resultant crystalline carbonates and sulfates adversely affect film adhesion and stability and should be removed prior to deposition. In view of the eminent practical importance of substrate stability under atmospheric conditions, it is often attempted to predict the durability of glasses by means of accelerated laboratory tests. However, the results tend to be specific for the conditions chosen and do not necessarily permit extrapolation into the range of conditions anticipated for actual use.

During cleaning in preparation of thin film deposition, substrates are often subjected to inorganic chemicals. Some of these are highly reactive and must not be

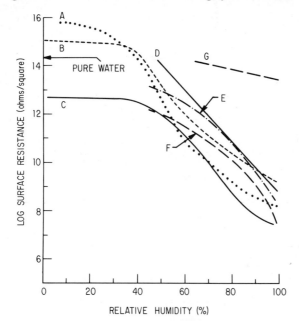

ig. 16 Surface resistance of various substrate materials as affected by relative humidity. ▲ = fused silica; B and E = borosilicate glasses; C = soft glass; D = aluminosilicate lass; F = 99.5% alumina; G = silicone-coated borosilicate glass. (Curves A, B, C, and G rom Guyer,[57] courtesy of IEEE.)

sed in conjunction with certain types of substrate materials. For example, most f the commercial glasses are attacked by fluorides and strong alkali solutions. The ttack of silicate glasses by mineral acids proceeds by neutralizing and preferentially issolving alkali ions, thus leaving behind a more resistant, silica-rich surface. Boro-licate glasses in particular form thin, highly protective layers which slow the rate f acid attack. In alkaline solutions, silicate glasses are more rapidly attacked than ▲ acidic or neutral media because the hydroxyl ions react with the silicate network. ▲ence, a protective layer is not formed, and dissolution proceeds at a constant rate. f interest in this context are also specific cases where particular elements are prefer-ntially dissolved. An example is the leaching of barium oxide from a glass substrate uring the anodization of a tantalum film in an aqueous solution of oxalic acid and ▲hylene glycol at 105°C.[20] This behavior contrasts with observations of other ▲vestigators,[59] who found ordinary soda-lime glass to be stable under identical ▲nditions. The attack was attributed to chelation of the barium ions by the oxalic ▲id and could be avoided if other electrolytes, for example, citric acid, were used.

Lead is another constituent frequently added to glazes in order to lower the melting ▲int, permit maturing at relatively low temperatures, or improve the electrical ▲sistance.[60] Its cations, however, form complexes with dibasic acids and may therefore disqualify lead-bearing glasses from certain usages. An example is shown in Fig. 17, where negative deviations from the proportionality between film thickness and anodization time occur if lead oxide glass substrates are used in conjunction with citric acid solutions. The shaded area marked 2 reflects the range of values obtained for samples fired under various conditions. The oxidation state and concentration of the lead are probably the parameters in this case. The reduced anodization rates are attributed to the incorporation of the extracted lead into the anodic oxide. The reaction rate of phosphoric acid with lead (curve 3) is much slower, and hence no loss in efficiency is observed within the anodization time of 50 min. Recently, a lead glaze has been developed which is stable in citric acid,[61] as demonstrated by curve 4 in Fig. 17.

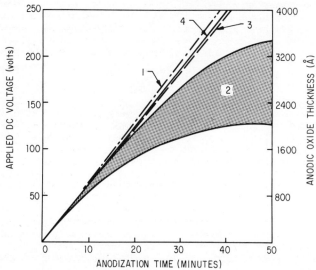

Fig. 17 Effect of electrolyte contamination on anodic-film growth. (1) Film deposited on lead-free glass and anodized in 0.01 % citric acid. (2) Films deposited on high (20 %) lead-bearing glasses and anodized in 0.01 % citric acid. (3) Same as 2 but anodized in 0.01 % phosphoric acid. (4) Films deposited on a substrate coated with a lead-bearing glaze (glass no. 12 in Table 2) and anodized in 0.01 % citric acid.

The preceding discussion was primarily concerned with glasses because they are chemically less stable than ceramic bodies. However, the latter are not entirely inert to chemical attack, particularly if they contain silicates as secondary phases. The crystalline phases of the magnesium silicates (forsterite and steatite) are also attacked by glass etchants, although at a considerably slower rate than glasses. Crystalline alumina and beryllia are highly inert. Barium and lead oxides are susceptible to leaching from their respective titanates.

5. THERMAL AND MECHANICAL PROPERTIES

In the course of thin film processing and packaging, and during actual use, substrates are subjected to thermal and mechanical stresses. Since heating is coupled with expansion, it also introduces mechanical strain, and the two groups of properties should therefore be considered together to assess the possibilities of substrate damage.

a. Thermal Properties

The basic properties in this category, the coefficient of expansion and the thermal conductivity, are of importance in selecting substrates for thin film components and circuits because they determine dimensional changes and heat flow during thermal cycling. There are two types of situations where a substrate encounters temperature changes. One occurs during processing when the entire substrate is heated—for example, to deposit a film—and subsequently cooled. The problem of heat transfer from the holder to the substrate has already been discussed in Sec. $3d(1)$, and the desirability of high thermal conductivity to minimize temperature gradients is obvious. The other type of thermal stress arises when thin film components are under an electrical load and the substrate must dissipate or permit extraction of the Joule heat. Here, the stresses exerted are greater than in heating the entire substrate because the energy is released locally and thus causes some regions to expand while others maintain their dimensions.

(1) Thermal Expansion If solids are heated, the additional thermal energy causes an increase in the vibrational amplitudes of the individual atoms. The average separation between atoms is thereby increased and the solid expands. Quantitatively, this phenomenon is described by the coefficient of linear expansion α, which is defined as the ratio of change in length Δl per degree to the length of the body l_0 at $0°C$:

$$\alpha = \frac{\Delta l}{l_0(T - 273)}$$

This definition is an approximation because α is slightly temperature-dependent. However, this is of little consequence as long as only modest temperature intervals are considered. It is common practice to characterize the expansion of solids by an average coefficient for the range from 0 to 300°C. The expansion of several substrate materials within this temperature interval is shown in Fig. 18.

Whereas amorphous materials like glasses and crystalline solids with a cubic lattice structure expand uniformly in all directions, such is not the case for crystalline bodies of lower lattice symmetry. An example is alumina, which has a trigonal structure and different expansion coefficients in the directions parallel and perpendicular to the c axis. This behavior is called anisotropy and manifests itself in nonisotropic stress. The coefficients of expansion of two-phase solids like glass ceramics can be varied deliberately by heat treatment. The available range encompasses large negative to large positive values and thus allows better matching of substrate and film expansion.

One can infer from Fig. 18 that most combinations of substrates and films will be more or less mismatched in regard to thermal expansion. This may result in loss of adhesion during thermal cycling. Differences in the adhesion of 1-μ-thick Nichrome films on different substrates, for example, have been attributed to expansion mismatch.[62] Another consequence of the same phenomenon is changes in the electrical resistivity of thin films. Settzo[63] and Greenough[64] investigated tin oxide resistors on a number of substrates and concluded that the predominant factor in the aging of their resistors was the difference in expansion coefficients between films and substrates. In contrast with these results, Belser[65] reported that the temperature coefficients of resistance (TCR) of titanium and zirconium films on fused silica were not greatly affected by thermal-expansion differences.

An attempt to calculate the effect of thermal-expansion mismatch between substrate and film on the TCR of thin films has been made by Hall.[66] He derived the relationship

$$\alpha_{R,f} = \alpha_R - 4.2(\alpha_f - \alpha_s)$$

where the TCR of a thin metal film $\alpha_{R,f}$ is obtained from the TCR of the bulk metal α_R by introducing a correction term involving the coefficient of expansion of the film α_f and the substrate α_s. Since thin metal films tend to have small temperature coefficients of resistance, the correction term is often significant.

Magnetic films which exhibit zero magnetostriction are not affected by thermal-expansion differences. Likewise, Takahashi[67] found no difference in the temperature dependence of the anisotropy of films deposited on different substrates.

Matching of thermal-expansion coefficients is most important if the thin film substrate also carries device chips made from a different material. Well-known examples are silicon transistor or circuit chips mounted on ceramic or glass modules. Here differences in thermal expansion result in considerable shear stresses at the interface between chip and substrate when the structure is undergoing temperature changes. Consequently, separation may occur if the shear stress exceeds the strength of the bonding material or the adhesion at any one of the interfaces. In these cases, the

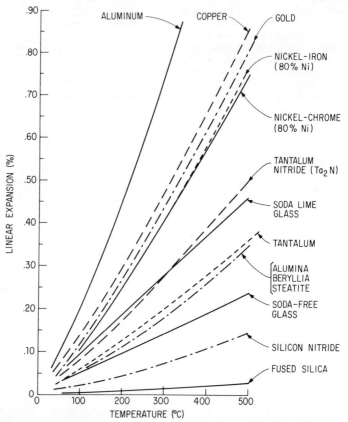

Fig. 18 Linear expansion as a function of temperature for various substrate and film materials.

necessity to match expansion coefficients as closely as possible narrows the choice of substrate materials much more than in situations where only passive thin film components are present.

(2) Thermal Conductivity The temperature of thin film circuits or components increases in proportion to the electric power applied. At elevated temperatures, the films become susceptible to recrystallization and oxidation which may lead to resistance changes or even destruction of the film. The rates of both processes are exponential functions of temperature, and the processes are irreversible. Therefore, temperature control of thin film circuits is an important design consideration. The substrate itself, too, may be affected by excessive temperature rises because substrate electrolysis or reaction with the film material are also accelerated by heating. The factor which is most often relied upon to dissipate the heat of thin film circuit elements and thereby to limit the temperature increase is the thermal conductivity κ of the substrate.

Thermal conductivities of substrate materials vary over three orders of magnitude, as shown in Fig. 19. They are also somewhat temperature-dependent, but this is often neglected in making approximate calculations of thermal effects. The lowest conductivities are those of the glasses, which range from 0.002 to 0.004 cal cm^{-1} s^{-1} deg^{-1} at room temperature and increase by about 10% per 100°C temperature rise.

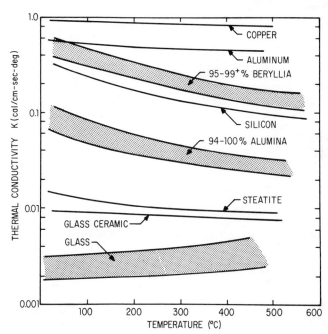

g. 19 Thermal conductivities of various substrate materials as functions of temperature.

he thermal conductivities of ceramics vary more widely depending on the material. hile some of them are as low as 0.02 cal cm^{-1} s^{-1} deg^{-1} (steatite), others are much gher. Beryllia, for example, is as good a heat conductor as aluminum, with a value 0.6 cal cm^{-1} s^{-1} deg^{-1}. In contrast to glasses, the thermal conductivities of ceramics crease with temperature; they are reduced by about 50% if the temperature rises 0°C. The conductivities of metals are practically independent of temperature.

Glazed ceramic substrates are of particular interest in this context because they mbine the smooth surface of glasses with the superior thermal conductivity of ramic bodies. Depending on the thickness of the glaze, the thermal conductivity the composite may be an order of magnitude above that of glasses of comparable ickness.[68] The effects of substrate conductivity on the temperature distribution ound specific components have been studied experimentally by several authors. cLean et al.[69] used temperature-indicating crayons to examine the thermal profiles tantalum nitride resistors. Thermocouples on Nichrome[17] and carbon film resis- rs[70] have been employed for the same purpose. Presently, most investigators use ermal microscopy. In this method, the infrared emission from the sample surface focused onto a sensitive bolometer whose electrical output is proportional to the ident radiation. Indium antimonide or gold-doped germanium photodetectors are mmonly used in this application. Their signal is amplified and displayed on a ter or recorder. The method can be as accurate as ±1°C and is capable of resolving as down to 1 mil².[71]

(3) Resistance to Thermal Shock and Stress The ability of substrates to withstand rapid temperature changes without damage is generally referred to as their thermal shock resistance. This complex quantity is related to the thermal conductivity, the specific heat, and the density of the material since these properties together determine the temperature of the material during the change. It is also dependent on the coefficient of thermal expansion and the elastic modulus of the substrate, because these are the factors which govern the resulting stress. Thermal shock resistance is often used empirically by referring it to the behavior of different materials in a particular thermal cycle. To obtain a more quantitative and universal figure of merit, a coefficient of thermal endurance F has been defined by Winklemann and Schott:[72]

$$F = \frac{P}{\alpha E} \sqrt{\frac{\kappa}{\rho c}}$$

where P = tensile strength
  α = linear coefficient of thermal expansion
  E = Young's modulus
  κ = thermal conductivity
  ρ = density
  c = specific heat

Although the equation yields only qualitative agreement with experimental observations, it is of value in comparing different substrate materials with each other. Table 8 shows the relative shock resistances of four materials together with their expansion coefficients, since the latter exert the greatest influence on the substrate's ability to withstand shock. It is of interest to note that fused silica has the highest shock resistance, while other glasses are at the bottom of the list.

TABLE 8   Relative Shock Resistance of Substrate Materials

Material	Thermal-endurance factor, F	Thermal coefficient of linear expansion, ppm/°C, for 0–300°C
Silica...........	13.0	0.56
Alumina.........	3.7	6.0
Beryllia.........	3.0	6.1
Glass...........	0.9	9

The resistance to thermal stress differs from the shock resistance insofar as it refers to the ability of a body to withstand nonuniform stress. Thermal stress is generated if the expansion of one part of a substrate is impeded by adjacent material which does not expand. This condition may arise from a number of causes.[73] Examples are anisotropic polycrystalline bodies and two-phase materials such as glazed ceramics, glazed ceramics, or glazed metals. Thermal stresses are also produced in homogeneous bodies if temperature gradients are present. During heating and cooling, for instance, the substrate surface responds more rapidly to the induced change than the interior; hence, a temperature gradient exists perpendicular to the surface. Transverse gradients are caused by electrically loading thin film components which cover only a part of the substrate.

It is difficult to assign quantitative values to the stress resistance of substrates. Qualitatively it is probably related to the thermal shock resistance, and the same order of merit prevails as in Table 8. Failures due to excessive stress alone are probably very rare. Substrate cracking or chipping are more likely to be due to a combination of factors including mechanical and thermal shock as well as thermal stress.

b. Thermal Design Considerations

After discussing thermal properties of substrates as essentially independent material parameters, it is of interest to examine to what extent these factors can be taken into account in the design of thin film circuits. Two problems have been singled out for treatment because of their practical importance. One is the warping of glazed substrates; the other deals with the question of power dissipation in thin film networks.

(1) Stresses in Glazed Substrates As stated in Sec. 5a(2), glazed ceramics are attractive as substrate materials for thin film circuits. However, they are also thermally stressed since they are a composite body of two different materials. If the

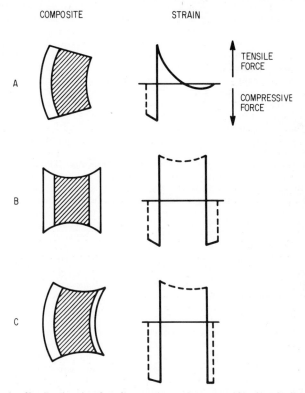

Fig. 20 Strain distribution in glazed ceramics. A = one side glazed; B = both sides glazed uniformly; C = both sides glazed asymmetrically. *(From Hoens and Kreuter,*[74] *courtesy of Academic Press Inc.)*

strain generated during the cooling of the glaze is excessive, chipping or warping occurs and renders the substrate useless. The expected warpage and strain distribution in three types of composite substrates are shown in Fig. 20. It is assumed that the glaze is under compression. This is desirable because the resistance of glasses to compressive forces is higher than to tensile forces. Experimentally, this can be achieved by selecting a glaze which has a smaller coefficient of thermal expansion than the substrate. Assuming that the two materials are matched at the temperature of glaze formation, the stronger contraction of the substrate during cooling to room temperature will tend to compress the glaze.

Figure 20 demonstrates that substrate warpage can be supressed by symmetrically glazing both sides. However, the steep stress gradients at the interfaces cannot be avoided. This is tolerable to some degree if the adhesion of the two materials is firm enough to withstand the shear forces involved. Examples are combinations of ceramics and matched glazes. Hoens and Kreuter[74] pointed out that the strain due to the relative expansion of the glaze and the substrate body varies along the section because of the presence of two dissimilar materials and the formation of an interfacial layer.

The more desirable combination of metallic substrates coated with an insulating layer poses problems which have not yet been overcome in a reliable form. One reason is the relatively large coefficients of expansion associated with metals, which lead to large stress gradients at the interface. Another factor is the rather weak adhesion which is generally obtained between metal surfaces and deposited oxide or glass films. An alternative which is satisfactory for many applications is silicon-crystal wafers with an insulating layer of thermally grown or sputtered SiO_2. Their stress situation corresponds to Fig. 20a, and they provide excellent substrates for thin film resistors because of the high thermal conductivity of the silicon.[75]

(2) Power Dissipation in Thin Film Circuit Networks The electric energy supplied to a circuit is partly stored in the capacitive and inductive elements and partly dissipated in the form of heat. In the absence of heat sinks, the temperature of thin film components would rise without limit as long as power is added. This is obviously not the case. Instead, the component reaches a steady-state temperature which is indicative of the balance between power input and energy losses from various dissipative mechanisms. To ensure stable circuit operation and long component life, it is necessary to design a circuit such that a modest maximum temperature is not exceeded.

This task is easily defined but difficult to accomplish. The complexity of the problem stems from a variety of factors. To begin with, power dissipation by conduction is difficult to compute because the heat flows laterally as well as vertically into the substrate. It may involve different materials whose thermal properties are not always accurately known, as well as variations in cross section of the conductive paths. Furthermore, there are additional dissipation mechanisms such as radiation and convection, the latter being particularly complex because of its dependence on type and pressure of the ambient gas as well as on the speed of gas movement. The attempts of different authors to predict temperature distributions and maximum values on substrates vary in the degree to which these mechanisms are taken into account. The most advanced approaches utilize extensive computer iterations to arrive at temperature maps of the substrate–film component assembly. However, this treatment also requires simplifying assumptions to reduce the problem to manageable proportions.

Although the presence of more than one component on the substrate further complicates the issue, attempts have been made to predict at least approximately the temperature distribution on substrates carrying film-circuit networks. Peek[76] approached the problem by lumping all resistors into one equivalent device. While his model does not yield the temperatures of the individual resistors, it identifies the location and temperature of the hottest point ("hot spot"). One result of this study is the conclusion that more effective power dissipation is obtained the smaller the aspect ratio of the resistor.

Abraham and Poehler[77] treated the heat-flow problem for thin film transistors and hot-electron devices on glass substrates. Simplifying assumptions made were that heat losses occurred only by conduction through the substrate, and that rectangular film devices were uniform heat sources. They concluded that Mead devices will suffer thermal destruction before they achieve a figure of merit M of 10 MHz. M is defined as

$$M = \frac{g_m}{2\pi C_I}$$

where $g_m = \partial I_O / \partial V_I$ = transconductance
I_O = output current
V_I = signal input voltage
C_I = input capacitance

In contrast, thin film transistors are not limited by thermal considerations until a figure of merit of several hundred megahertz is exceeded. While power dissipation is not the only consideration in evaluating the upper limit of the transconductance g_m, the authors used this criterion to screen experimental devices.

An electrical-analog treatment of the heat-flow problem has been given by Burks.[78] He considered the circuit to be isothermal and added substrate effects as series resistances. Martin et al.[79] used digital-computer analysis to determine the temperature distribution along leads over a single encapsulated resistor on a ceramic substrate. The same technique was also applied to four resistor networks on both encapsulated and bare substrates. This method accurately predicts temperature distributions over complex parts once the program is established. A disadvantge is that the technique requires the accurate determination of material constants and empirical coefficients by actual measurements in order to substantiate the program results initially.

In view of the simplifications necessary to predict temperatures a priori, attempts to supplement theoretical treatments with experimental measurements are more promising. An example is the study of Burks et al.[80] on hybrid integrated circuits. These authors used an infrared probe for direct measurements and electrical analogs for simulation. Furthermore, they derived heat-balance equations for which numerical control solutions could be obtained. Finally, semiempirical correction factors for such parameters as substrate area, thermal conductivity, and uniformity of resistor power dissipation were derived by graphically evaluating a large number of experimental data. The accuracy claimed for this method is $\pm 10\%$. In general, it has been shown by Young[81] that changing the size and shape of a resistor pattern also changes the temperature distribution of the resistor and hence affects the resistor stability. He was able to flatten the thermal profile of resistors by such geometrical changes, thereby reducing the hot-spot temperature.

A number of substrate-selection and circuit-design criteria can be derived from the studies made so far. These may be summarized as follows: Since high thermal conductivity is of prime importance, high-density alumina and beryllia are desirable substrate materials. Metal plates insulated with oxide, glaze, or porcelain layers are preferable if available. Very thin glass wafers mounted on effective heat sinks are also a possibility. Power-dissipating elements should be as close to heat sinks as possible and evenly distributed over the substrate. In the case of thin film resistors with aspect ratios <1, large-area metal-film contacts aid in dissipating the power. In general, leads should have high thermal conductivity and bonds low thermal impedance. For thin film conductors, this means they should be as wide and thick as possible. Finally, the formation of interfacial oxides should be avoided.

Although these considerations were mostly derived with steady-state power dissipation in mind, they also hold for pulsed operation. However, transient voltages superimposed on the normal operating voltage are an additional factor to guard against. Thin film devices often have small signal rise times and cannot dissipate sudden power surges quickly enough. To avoid circuit destruction from such causes, it is advisable to base the thermal design on the anticipated peak rather than average power.

c. Mechanical Properties

The mechanical strength of substrates is of interest in regard to the handling which they experience throughout processing and in the mounting of thin film modules into assemblies. Related to the strength is the ability to separate large processed substrates mechanically into individual modules.

It is characteristic for glasses and ceramics that they are quite brittle at ordinary temperatures, a property which distinguishes them from most other materials. They follow Hooke's law accurately up to the point of fracture, which, in contrast to the behavior of metals, occurs without plastic deformation. The glasses are also elastically homogeneous and isotropic. The use of glasses is limited to temperatures below the strain point above which viscous flow becomes perceptible. This viscous flow differs from the plastic flow of metals because it continues for the duration of the applied force.

(1) Strength Tests For practical reasons, the strength of brittle materials should not be measured by tensile tests. Instead, the transverse-bending or modulus-of-rupture test is recommended.[82] In this test the specimen is laid across two supports, and increasing loads are applied at the center until fracture occurs. The modulus of rupture is computed from the fracture load. The latter exerts maximum tensile stress at the lower (opposite) surface of the specimen. The compressive and tensile elastic moduli are assumed to be equal.

The results of mechanical-strength tests are sensitive to several factors such as method of testing, shape, and surface condition of the sample. For instance, microfissures or flaws in the surface of glass samples reduce the measured strength considerably. For these reasons, many of the published data do not apply to thin film substrates although the materials may be identical. However, the data do give an indication of relative or approximate strengths.

Glasses and ceramics are many times stronger under compressive than under tensile loads. Compressive tests on glasses give strength of 150,000 to 200,000 psi while ceramics yield values as high as 400,000 psi. These should be considered as minimum values, since it is not possible to obtain absolute data of compressive strength by the usual methods.

Other mechanical properties such as hardness, creep, fatigue, and elastic constants are of secondary importance in substrate applications. The interested reader is referred to a more extensive review article by Parker[83] on the mechanical properties of glasses and ceramics.

(2) Separation of Substrates One advantage of thin film circuitry is the ability to fabricate many components or circuits simultaneously on a common substrate. This practice often necessitates separation of the fully processed substrate into smaller parts or modules. Suitable mechanical-separation techniques include sawing, scribing and breaking, and prescoring. Well known are the dicing processes for silicon wafers by means of multiple-diamond-wheel or wire-and-slurry saws.

Glass substrates, which are typically 1 mm (0.040 in.) thick, can be conveniently separated by scribing. The yield depends on numerous factors such as flaw distribution on the surface, thermal history of the glass, the pressure of the scribing tool stresses in the deposited film, and operator skill. Surface flaws or minute scratches for example, are a recognized cause of irregular glass fracture. While such flaws can often be removed by etching the glass, this procedure also changes the dimension of the substrate, or it may impart other undesirable features to the surface. Glass substrates may, of course, also be separated by sawing with diamond blades. However, this technique is more expensive to set up for than scribing.

Cutting with diamond wheels is frequently used to separate glazed and unglazed ceramics. It requires a number of preparatory and finishing operations such as mounting of the substrate on a flat holder plate, aligning, rotation of the mounted assembly for the second cut, and removal of the mounting cement. Chipping of the glaze along the cut edges and contamination of the entire surface by sawdust are common occurrences in this technique. To avoid these problems, prescored substrates have been introduced.[84] These are subdivided into modules by grooves deep enough to permit breaking under applied pressure, yet not so deep as to cause accidental separation during processing.

The parameters affecting the scoring procedure have been studied by Kleiner.[8] He prepared samples by different heat treatments and with score marks of varying depths and angles. The modulus of rupture (MOR) was used as a measure of the forces required for separation. His observations led to the conclusion that the optimum range of MOR values is 10,000 to 30,000 psi. Values in this range were consistently obtained on 0.030-in.-thick substrates if these were scored 0.003 to 0.010 in deep with blades made from 0.002-in. shim stock having angles of 30 to 45°. However, the scoring with shim stock also produced a tendency of the ceramic to tear apart nonuniformly. Annealing had no significant effect on the MOR values.

The use of lasers to separate brittle substrate materials may soon become a practical and economical alternative. Recently, linear cutting speeds of 1 in. s^{-1} on 0.025-in.

thick alumina substrates have been reported.[86] This rate was limited by the power output of the laser and not imposed by processing problems.

6. SUBSTRATE CLEANING

The cleanliness of the substrate surface exerts a decisive influence on film growth and adhesion. A thoroughly cleaned substrate is a prerequisite for the preparation of films with reproducible properties. The choice of cleaning techniques depends on the nature of the substrate, the type of contaminants, and the degree of cleanliness required. Residues from manufacturing and packaging, lint, fingerprints, oil, and airborne particulate matter are examples of frequently encountered contaminants. Accordingly, the thin film worker has to address the questions of contaminant identification and effective removal. Suitable answers can seldom be derived from general principles but usually require an empirical approach.

a. Surface Cleanliness Tests

Several tests which are useful to assess the cleanliness of substrate surfaces are listed in Table 9. The first three methods are based on the wettability of surfaces and may help in recognizing the presence of hydrophobic contaminants on normally hydrophilic materials or vice versa. Impurities having the same affinity to water as the substrate material are not detected. Although quite sensitive, the interpretation of the observed surface condition is somewhat subjective. Furthermore, the characteristic wetting patterns are more difficult to recognize the rougher the surface. The contact-angle method, too, relies on surface wettability but yields somewhat more quantitative information. It is also affected by surface roughness, as has been shown by Wenzel.[90] He found that the average contact angle $\bar{\theta}$, which is representative of the true surface wettability, was related to the measured contact angle θ by

$$\cos \bar{\theta} = x \cos \theta$$

Here, x is the factor by which the true surface area exceeds the geometrical area.

The coefficient-of-friction method is one of the few techniques which may be applied within vacuum systems as well as outside. Although it yields numerical values, a correlation between the measured coefficient and the concentration of contaminants on the surface is not easily established. Besides, the test may be destructive.

Adhesion tests in one form or another are invariably destructive insofar as one cannot assume the original surface condition to be retained throughout the test. Therefore, this method requires test samples representing the same cleaning technique as applied to the substrates. Also, adhesion tests are strictly functional and merely indicate whether a particular cleaning method yields the desired result or not. This consideration applies more or less to all other tests, too, insofar as quantitative values of cleanliness—for example, in terms of contaminant molecules per unit area—are extremely difficult to obtain and rarely needed. Consequently, the usual practice is to establish one test empirically as being indicative of a surface condition which will yield the desired film property.

In addition to the optical methods which are listed at the bottom of Table 9 and which are self-explanatory, other tests have been used by various authors. A review article of Linford and Saubestre[91] describes methods such as electrolytic copper deposition, the use of potassium ferricyanide paper, and gravimetric determinations. Measurements of electrode potentials[92] and radioactive-tracer methods[93,94] have also been tried. The most stringent cleanliness requirements occur in the epitaxial growth of films. In some of these cases, low-energy electron diffraction has been applied to identify the desired atomically clean state of the surface.[95] Although LEED techniques yield reproducible results, it was recently pointed out by Bauer[96] that its sensitivity is limited to one monolayer. Using Auger spectroscopy, Bauer suggested that nickel and iron contamination was responsible for the structure obtained by low-energy electron diffraction on "clean silicon" surfaces. Four other physical

TABLE 9   Methods to Test Surface Cleanliness

Method	Description of test	Sensitivity	References
Black-breath figures	Surfaces which are free of hydrophobic contamination allow condensation of a smooth specular water film of low reflectance. The smooth water film has a lower refractive index than the glass and forms an antireflection coating. As it evaporates, the film thickness decreases, and interference occurs. Contact angles of droplets in the black-breath figure approach zero, while on less clean surfaces, the contact angles are appreciably larger. Such water films are called "gray-breath figures." Clean glass showing black-breath figures has an abnormally high coefficient of friction ($\mu_s = 1$, see below)	. . . .	Holland[58]
Atomizer test	The dry substrate surface is exposed to a fine water spray. On a contaminated surface, the sprayed droplets coalesce into larger ones, whereas on a clean surface, the fine mist remains	$\frac{1}{10}$ monolayer	ASTM F21–62T
Water-break test	If a clean substrate is slowly withdrawn from a container filled with pure water, a continuous film of water remains on the surface	1 monolayer	ASTM F22–62T
Contact angle	The contact angle between a water droplet and a surface is a quantitative measure of the surface wettability. On completely wettable surfaces, the contact angle is 0°. On nonwettable surfaces, the droplets assume spherical shapes with contact angles approaching 180°	. . . .	Longman and Palmer[87]
Coefficient of friction μ_s	The resistance to sliding of a metal or glass object on a substrate surface is dependent on the cleanliness of the surface. The closer μ_s approaches 1, the cleaner is the surface	$\frac{1}{2}$ monolayer	Holland[58]
Coefficient of indium adhesion σ	The surface is tested by measuring the coefficient of adhesion σ between it and a clean piece of indium. σ is defined as the ratio of tensile force required to produce adhesion failure to the joining force. It ranges from 0 for dirty surfaces to as high as 2 for clean ones. The method works equally well for hydrophobic and hydrophilic surface contaminants. Freshly fractured silicon of high purity and rigorously cleaned glass yield coefficients of 2	$\frac{1}{2}$ monolayer	Krieger and Wilson[88]
Film adhesion	The degree of adhesion of metal films on substrate surfaces also indicates surface cleanliness. However, lack of adhesion is not always caused by surface contamination		

TABLE 9   Methods to Test Surface Cleanliness (Continued)

Method	Description of test	Sensitivity	References
Fluorescent dyes	Certain surface contaminants such as oils absorb fluorescent dyes, or dyes may be added to such materials as resists. Residues of this type are readily detected when illuminated with ultraviolet light	1 monolayer	Missel et al.[89]
Edge lighting	If glass is illuminated at one edge, most of the light is transmitted by internal reflections to the opposite side. The remainder is reflected from the surface. Surface contamination which interrupts reflection from the surface is frequently detectable as light areas on an otherwise dark surface	Several monolayers	

phenomena which are sensitive to surface contamination in the monolayer range have been utilized by Allen et al.[108] to identify atomically clean silicon-crystal surfaces.

b. Cleaning Procedures

The process of substrate cleaning requires that bonds are broken between contaminant molecules as well as between the contaminant and the substrate. This may be accomplished by chemical means as in solvent cleaning, or by supplying sufficient energy to vaporize the impurity, for example, by heating or particle bombardment. It is generally desirable to limit the removal process to the contaminant layer, but a mild attack of the substrate material itself is often tolerable and ensures completeness of the cleaning operation. Some cleaning methods require substrate handling or the use of solvents and must therefore be applied outside the vacuum system. The physical cleaning methods are generally conducted in situ by installing provisions for substrate heating or particle bombardment in the deposition system.

(1) Solvent Cleaning Suitable reagents for substrate cleaning include aqueous solutions of acids and alkalies as well as organic solvents such as alcohols, ketones, and chlorinated hydrocarbons. The cleaning effect of acids is due to the conversion of some oxides and greases into water-soluble compounds. Alkaline agents dissolve fatty materials by saponification, which renders them wettable. The use of acids and alkalies, however, is subject to limitations. Their ability to attack glasses has been discussed in Sec. 4c. In cases where the substrate is chemically inert or where a slight attack can be tolerated, one has to guard against the formation of residues and adsorption of solvent molecules. The inorganic compounds retained are often nonvolatile and cannot be removed by subsequent heating in vacuum. An example is the retention of adsorbed chromium on glass surfaces which have been cleaned in hot hydrosulfuric–chromic acid mixtures. Hydrofluoric acid solutions, which are often used to remove insoluble deposits by dissolving a thin layer of the underlying glass,[97] have also been found to impart contamination in the form of firmly adsorbed fluorine. Indicative of the effect is the observation of fluorine by mass spectrometry even after the treated glass had been baked in vacuum at 325°C for 36 h.[98]

The residue problem may also be encountered with organic solvents. Putner[99] observed poor film adhesion on glass substrates cleaned with carbon tetrachloride and trichloroethylene. After cleaning, the surface was covered with a whitish deposit which could not be removed by heating. It is therefore inferred that a chloride film was formed by reaction of the glass with the solvents.

To increase the rate of contaminant removal and to achieve completeness, heating or ultrasonic agitation of the solvent are commonly employed. Hot-solvent cleaning is mostly conducted in the form of vapor degreasing; i.e., the substrate is suspended above the boiling liquid in a closed container. The rising solvent vapor condenses on the object to be cleaned, thereby heating it and increasing the rate of dissolution of surface contaminants. As the spent solution drips back into the bath, fresh and clean distillate repeats the process.

In ultrasonic cleaning, dissolution of residues is enhanced by the intense local stirring action of the shock waves created in the solvent. Thus, solvent saturated with impurities is continually carried away from the substrate surface and fresh, less saturated liquid is admitted. Mechanical vibrations induced in the substrate further aid in loosening gross contaminants such as particulate-matter flakes. The parameters which affect the efficiency of ultrasonic cleaning are numerous. The frequency of vibration, applied power, type and temperature of the solvent, its surface tension and viscosity, and the presence of nucleating particles and dissolved gases are factors which play a role. The lowering of the gas pressure above the agitated liquid may be detrimental in some cases but helpful in others.[99] An example of the former is the removal of dissolved gases which provide nuclei for cavitation centers so that the cavitation effects are reduced.[100] On the other hand, the removal of air trapped in recesses of intricately shaped or slightly porous substrates aids in the cleaning action by permitting more complete wetting.

Comparative tests of vapor degreasing with a variety of solvents and ultrasonic cleaning in isopropyl alcohol have been conducted by Putner.[99] While he obtained the cleanest surfaces by vapor degreasing in pure isopropyl alcohol, low-frequency ultrasonic agitation was most effective in removing gross surface contaminants such as particles and fingerprints. Cleaning in detergent solution was less effective than either of the other techniques. Discussing these results, Holland[58] suggested that the decisive factor was the higher temperature of solvent and substrate attained in vapor degreasing.

In view of the variability of contaminants, the wide choice of cleaning agents, and the different purposes to be accomplished, solvent cleaning is more an empirical art rather than a science. Accordingly, there are innumerable cleaning procedures which vary from one laboratory to another. Some of the published treatments may serve to illustrate the variety of procedures and to provide guidelines in selecting suitable chemicals.

To clean borosilicate glasses, Tichane[101] recommended a prepolish with precipitated calcium carbonate and a subsequent two-step etch in dilute sodium hydroxide and hydrochloric acid. Bateson[102] utilized a chalk and detergent treatment followed by water rinses and brush buffing. Other authors cleaned a variety of glass, ceramic, and single-crystal substrates by first washing them in concentrated nitric acid, then in concentrated sodium hydroxide.[103] The substrates were subsequently rinsed in water and alcohol and stored in vacuum desiccators. Koontz and Sullivan[104] employed a continuous flow of deionized water to clean semiconductor wafers. Filby and Nielsen[105] were able to grow single-crystalline silicon films on fused-silica substrates which had been previously washed sequentially in trichloroethylene, chromic acid, deionized water, and acetone to remove surface contamination introduced during the cutting process. Behrndt and Maddocks[106] found that the properties of evaporated nickel-iron films deposited on glass substrates cleaned with detergent alone were not reproducible. This condition could be remedied if an additional ultrasonic cleaning step in ammonium bifluoride was introduced. About 5 μ of glass was thereby removed.

The drying of wet-cleaned substrates is also critical because recontamination can occur unless stringent precautions are taken. Drying may be accomplished in a vapor degreaser, a clean oven, or with hot filtered air or nitrogen. Cleaning tanks and baskets should be immaculate and the surrounding atmosphere free from airborne contamination. Pressurized hoods with filtered air are recommended. Typically, the last step in substrate cleaning is rinsing in deionized water. As the latter may contain traces of salts or organic matter, withdrawal of the substrate should be conducted such that a minimum of liquid adheres to the surface. Visible stains occur if the remaining drops are left to evaporate. Therefore, residual solvent drops should either be blown off or removed by centrifuging. It is furthermore good practice to clean substrates immediately prior to insertion into the vacuum system. If storage cannot be avoided, dust-free containers with a lid or desiccators may be utilized.

(2) Substrate Cleaning by Heating An effective cleaning procedure for unglazed ceramics is firing at high temperatures (1000°C). Glasses and glazed ceramics may also be heated but to lower temperatures. If their geometry permits, they can be fired with a gas-air torch. Here, the removal of surface contaminants is brought about by the flame's imparting sufficient energy for the desorption of surface molecules. Organic matter is oxidized and thereby transformed into volatile constituents. Since the flame contains ionized particles which impinge on the surface, the energy of their recombination further aids in the removal of adsorbed molecules, a mechanism which is also encountered in glow-discharge cleaning.[58] It is essential, however, to adjust the gas mixture properly and thus avoid incomplete combustion; the latter may lead to deposition of soot on the substrate surface. If the flame is too hot, it may cause melting or warping of the substrate. Nonuniform heating is also detrimental because it can induce stress and subsequent cracking of the substrate.

Surface cleanliness as obtained by different treatments was studied by Nielsen[107] in conjunction with evaporated Permalloy films on glass. A scratch test served to measure the cleanliness obtained prior to film deposition. The weight applied to a titanium point and required to scratch the surface was the quantitative indicator of cleanliness. It was found that glasses melted in a platinum tray under vacuum had the cleanest surfaces. Heating in a high vacuum may also be employed in cleaning silicon surfaces. The removal mechanism involves the formation of the volatile monoxide according to $SiO_2 + Si = SiO$. Temperatures of at least 1280°C are required to produce silicon surfaces which are atomically clean.[108]

(3) Glow-discharge Cleaning This is the most widely used technique to clean substrates in situ and immediately prior to film deposition. It is effected by exposing the substrates to the plasma of a glow discharge. Experimental arrangements to implement the technique have been described by Holland.[109] Typically, the discharge is established between two electrodes positioned in the vicinity of the substrates such that the surface of the latter is immersed in the plasma. The discharge voltages may vary from 500 to 5,000 V. The electrodes are traditionally made of aluminum since this metal sputters very slowly in the presence of oxidizing gases and therefore does not significantly deposit on the substrate surfaces. If the two aluminum electrodes are spaced closely together, ac high-voltage supplies can be used, but dc discharges are more common.

The substrate itself is not part of the glow-discharge circuit as it is in sputter cleaning. Although the latter is an effective cleaning method, it involves bombardment of the substrate with high-energy particles, sputtering, and possibly roughening of the substrate surface, as well as deposition of foreign material from the counterelectrode. If it is felt that these effects are not detrimental and sputter cleaning is preferred over glow-discharge cleaning, it will generally be necessary to employ rf voltage because of the insulating nature of most substrates.

In glow-discharge cleaning, removal of impurities and other beneficial changes of the substrate surface are brought about by one or more of the following mechanisms:

1. Straightforward heating due to impingement of charged particles and their recombination

2. Impurity desorption through electron bombardment

3. Impurity desorption resulting from low-energy ion or neutral-particle bombardment

4. Volatilization of organic residues by chemical reaction with dissociated oxygen

5. Modification of glass surfaces through the addition of oxygen

6. Enhanced nucleation during subsequent film deposition

The mechanisms (4) and (5) are most important, a fact from which early attempts at glow-discharge cleaning benefited insofar as they were conducted at rather high residual gas pressures. It has been shown in a number of investigations that glow-discharge cleaning is effective only if oxygen is present. A good although not widely known example is the study of Hirai et al.[110] Mechanism (5) is particularly important for glasses with high SiO_2 contents, where it appears to aid in the formation of "oxide bridges" between the glass and reactive metals like Al or Cr. However, improved adhesion due to glow-discharge cleaning in oxygen has also been reported for a non-reactive metal like gold.[111] This may be related to observations of enhanced gold diffusion into bulk SiO_2 in the presence of oxygen, but the present understanding of adhesion mechanisms between films and substrates is as yet incomplete (see Sec. 2c, Chap. 12). Also of interest in this context is the work of Florescu,[112] who measured the adhesion of aluminum films on glass. He found that the effects of glow-discharge cleaning persisted if he removed the substrate from the vacuum system and evaporated after subsequent reinsertion. Florescu concluded that this behavior was due to a negative surface charge formed during glow-discharge cleaning; if the charge was removed between the cleaning and deposition steps, the film adhesion was as weak as if no discharge cleaning had taken place.

In the absence of oxygen, if, for example, argon is used for glow-discharge cleaning, mechanism (1) is generally the reason for improvements of the substrate surface. Mechanism (2) is effective in certain configurations. However, bombardment of the substrate surface with electrons has been shown to produce polymer films whose constituents stem from pump-oil vapors in the vacuum chamber. In this respect, silicone oils are worse than hydrocarbons. This topic has been discussed by Holland and coworkers.[109,113] Mechanism (3) applies only in situations where the substrate assumes a fairly high floating potential. It involves the risk of substrate sputtering and, if neutral particles bombard the surface, of coating the latter with material from the high-voltage cathode. Therefore, (3) is generally an undesirable mechanism.

(4) Other Cleaning Methods In addition to the more general and widely used techniques discussed in the previous sections, there are less common cleaning methods which are applicable in certain situations. An example is the cleaving of single crystals to produce an intrinsically clean surface. This procedure is, of course, limited to materials available in single-crystalline form which have a suitable cleavage plane such as rock salt. The technique is important only in nucleation and growth studies, where accurate definition of the surface state is the paramount consideration. To prevent adsorption of atmospheric gases once the atomically clean surface is obtained, cleavage is often performed in the vacuum system at pressures of 10^{-8} Torr or less. However, cases have been reported where little difference was found in thin gold films grown on rock-salt surfaces cleaved in air and vacuum.[114]

A special dust-removal technique involves coating of the substrate surface with an adhesive or lacquer which is subsequently stripped, hopefully taking the dust with it. Results of this method have been published by Jorgenson and Wehner,[115] who considered dust to be the prime cause of pinholes in their films. They found nitrocellulose in amylacetate to be effective in stripping dust without leaving a residue. Attempts to strip the lacquer in vacuo were only partly successful insofar as the decomposition products from the heated lacquer were responsible for poor adhesion of subsequently deposited films.

7. COST AND AVAILABILITY OF SUBSTRATES

In selecting substrates for practical applications such as thin film circuits, chemical and physical properties are not the only consideration, but economy, too, plays a role. The cost of the substrate may represent a significant fraction of the total cost of the finished module. In addition to the type of material, factors which influence substrate cost are the required dimensions and tolerances, the number of substrates ordered, and the expense of such additional processing as inspection, cleaning, and separation.

Because of excessive warping, the size of ceramic substrates is presently limited to areas of 16 to 25 in.[2] There are indications that warping may be reduced by improved processing, and 60-in.[2] substrates of acceptable flatness may become available in the foreseeable future. However, there are many applications where the present sizes are entirely sufficient. Another limitation of ceramic substrates arises from their uneven shrinkage during firing. Excessive warp occurs when length-to-width ratios exceed a factor of 4. For drawn hard glass, the size-limiting factor is a width of 9 to 10 in., which is again more than adequate for most purposes.

Requirements arising from processing or packaging of substrates sometimes call for dimensional tolerances which cannot be met by the substrate fabrication process. Additional finishing operations such as grinding and polishing tend to increase the substrate price quite drastically by factors of $2\frac{1}{2}$ to 3. However, uniformity or reliability advantages may outweigh the extra cost.[116-118] For glass substrates, the usual supplier specifications are ± 0.015 in. for planarity and ± 0.005 in. for thickness tolerances.

Like with many other industrial articles, the prices of substrates are subject to volume discounts. For most ceramics, the minimum price is typically obtained with

TABLE 10 Cost of Selected Flat Substrates Relative to Soda-lime Glass = 1

Material	Cost
Soda-lime glass (drawn)	1
Alkali-free glass (drawn)	3
Fused silica (polished)	6
Glass-mica (polished)	5
Glass-ceramic (polished)	15
94–96% alumina (as-fired)	5
94–96% alumina (glazed)	7–8
94–96% alumina (polished)	15
99.5% alumina (as-fired)	7
99.5% alumina (polished)	21
99.9% alumina (as-fired)	200
99.9% alumina (polished)	300
0.040-in. sapphire (polished one side)	400
0.040-in. sapphire (polished both sides)	450
0.020-in. sapphire (polished one side)	350
0.020-in. sapphire (polished both sides)	380
99% beryllia (as-fired)	25
99% beryllia (glazed)	30
99% beryllia (polished)	125
Steatite (as fired)	5
Forsterite (as fired)	5
Barium titanate (as-fired)	12
Barium titanate (polished)	35

TABLE 11 Substrate-upgrading Processes

Substrate	Problems	Remedy	References
As-fired poly-crystalline ceramic	Surface roughness	Application of 25–100 μ of glass (glazing)	See Sec. 2a(2)
		Application of silicon monoxide film	Degenhart and Pratt[25]
		Rheotaxial deposition of silicon onto molten glaze	Rasmanis[121]
Glass	Surface scratches and craters	Deposition of several microns of silicon monoxide	Maddocks and Thun[122]
	Low volume resistivity	Deposition of silicon monoxide	Intellux, Inc.[123]
	Solubility in hydrofluoric acid solutions	Application of tantalum oxide layer	McLean[124]
Polished metal	Surface roughness and conductivity	Deposition of up to 1 μ of silicon monoxide	Bertelson[125]
fcc polished metal	Diffusion of sputtered fcc metals into the substrate	Deposition of barrier layer of bcc metal such as tantalum	Kay[126]

TABLE 12 Substrate Manufacturers and Processors

	*	Soda-lime	Alkali boro-silicate	Alkali-free	Fused silica	Special	Glass ceramic	Glass mica	Glazed ceramic	$\leq 96\%$ alumina	$> 96\%$ alumina	Beryllia	Steatite	Forsterite	Alkaline-earth porcelain	Titanate	Sapphire	Silicon
		Glasses							Polycrystalline materials								Single-crystal materials	
American companies:																		
American Feldmuehle, Glenbrook, Conn.	M										✓							
American Lava, Chattanooga, Tenn.	M,P								✓	✓	✓	✓	✓	✓	✓	✓		
Amersil, Hillside, N.J.	M				✓													
Basic Ceramics, Hawthorne, N.J.	M,P								✓	✓	✓	✓	✓	✓				
Brush Beryllium, Elmore, Ohio	M,P										✓	✓						
Carborundum, Latrobe, Pa.	M,P								✓	✓	✓							
Centralab, Milwaukee, Wis.	M,P								✓	✓	✓		✓			✓		
Cermetron, San Diego, Calif.	M,P									✓	✓		✓			✓		
Coors Porcelain, Golden, Colo.	M,P								✓	✓	✓	✓	✓					
Corning Glass, Corning, N.Y.	M,P		✓	✓	✓	✓	✓		✓	✓								
Crownover Manufacturing, La Jolla, Calif.	M								✓	✓								
Degussa, New York, N.Y.	M,P									✓	✓							
Diamonite, Schreve, Ohio	M,P									✓	✓							
Dow Chemical, Midland, Mich.	M								✓	✓								✓
Electro-Ceramics, Salt Lake City, Utah	M,P								✓	✓								
Electra Manufacturing, Independence, Kans.	M,P									✓					✓	✓		
Erie Technological, State College, Pa.	M,P															✓		

Company	Type
Frenchtown Porcelain (CFI), Frenchtown, N.J.	M,P
General Electric, Schenectady, N.Y.	M
Haveg Industries, Taunton, Mass.	M,P
INSACO, Quakertown, Pa.	P
Intellux, Goleta, Calif.	M
J. M. Freed, Perkasie, Pa.	P
Lapp Industries, Leroy, N.Y.	M,P
Linde, Union, N.J.	M
Meller, Providence, R.I.	P
Molecular Dielectrics, Clifton, N.J.	M,P
Monsanto Chemicals, St. Louis, Mo.	M
Mycalex, Clifton, N.J.	M,P
National Beryllia, Haskell, N.J.	M,P
NGK Spark Plugs, Los Angeles, Calif.	M,P
Owens-Illinois, Toledo, Ohio	M,P
Rosenthal China, New York, N.Y.	M,P
Royal Worcester, Charlotte, N.C.	M,P
Semi-Metals, Westbury, L.I., N.Y.	M
Texas Instruments, Dallas, Tex.	M
Trans-Tech, Gaithersburg, Md.	M,P
United Mineral and Chemical, New York, N.Y.	M,P
Ventron, Bradford, Pa.	M
Vitta, Wilton, Conn.	M
Western Gold and Platinum, Belmont, Calif.	M,P

TABLE 12 Substrate Manufacturers and Processors (Continued)

	*	Glasses					Polycrystalline materials										Single-crystal materials	
		Soda-lime	Alkali boro-silicate	Alkali-free	Fused silica	Special	Glass ceramic	Glass mica	Glazed ceramic	≤96% alumina	>96% alumina	Beryllia	Steatite	Forsterite	Alkaline-earth porcelain	Titanate	Sapphire	Silicon
European companies:																		
Andermann and Ryder, London, England	M,P	✓																
Englass, Leicester, England	M,P		✓							✓	✓							
Hoboken Chemical, Hoboken, Belgium	M,P								✓		✓							✓
Plessey, Chessington, Surrey, England	M,P								✓	✓			✓			✓		
Rosenthal Technische Werke, Frankfurt, Germany	M,P									✓	✓	✓	✓					
Royal Worcester, Tonyrefail, Glamorgan, England	M,P									✓	✓	✓	✓	✓	✓			
Smiths Industries, Rugby, Warwickshire, England	M,P								✓	✓	✓							
Steatite and Porcelain, Stourport-on-Severn, Worcestershire, England	M,P									✓	✓	✓	✓		✓			
Wacker, Munich, Germany	M,P														✓	✓		✓
Japanese companies:																		
Chisso, Chiyoga-ku, Tokyo	M,P																	
Komatsu Electronic, Hiratsuka-shi, Kanagawa-ken	M,P																	✓
Kyocera, Nakagyo-ku, Kyoto	M,P									✓	✓	✓	✓	✓	✓			✓
NGK Spark Plugs, Mizuho-ku, Nagoya	M,P									✓	✓		✓			✓		

orders of 25,000 pieces or more per year. In the case of hard glass substrates, the maximum discount may require purchase of 1.2×10^6 in.^{2}.[119]

The cost for substrate separation varies according to material and cutting method employed. Diamond sawing is expensive because of the additional handling operations and yield losses due to chipping. Prescored substrates, although more expensive than ordinary ones, are often more economical than diamond sawing, even if the yield suffers somewhat in the breaking process.

To facilitate a comparison of substrate prices, the costs of various materials with different finishes are listed in Table 10. The prices are relative to soda-lime glass and apply to sizes of 1 in.^{2}. The higher cost of as-fired titanates and of beryllia compared with as-fired 94 to 99.5% alumina is due to the price of the raw materials. Glazed substrates are usually about 1.5 times more expensive than the as-fired bodies. Polishing typically increases the cost by a factor of 3. Beryllia is an exception on the high side because stringent precautions must be taken owing to the toxicity of its dust. This is a case where prescored substrates which obviate the need for safety enclosures during substrate separation offer a decisive advantage. Polishing of the second side adds about 12% to the finished cost of most ceramics. The 99.9% alumina and sapphire substrates must be sliced, ground, and polished to thickness. If sapphire substrates are required in a specific crystal orientation as for the deposition of heteroepitaxial films, the price of the thin (0.020-in.) wafers approaches that of the thicker (0.040-in.) substrates.

Very often a substrate material has desirable bulk properties but its surface is unsatisfactory because of roughness or lack of chemical stability. In these cases, it may be possible to upgrade the surface by an extra finishing step. The principle of these procedures is to coat the substrate with a layer of another material which adheres well and adds the desired property without losing the attractive features of the original substrate. Glazing of ceramics is the best-known example and has been discussed before. Other types of upgrading processes are listed in Table 11 along with the undesirable surface features to be masked. Needless to say, these methods must be used with discretion. Films of silicon monoxide, for instance, are heavily stressed and tend to chip off. This danger increases with film thickness[120] and is particularly acute on metal substrates. In rheotaxial deposition of silicon for active devices, it should be ensured that the glaze does not contain elements which are electronically active in the semiconductor.

Finally, the reader who has considered his substrate requirements and decided which material he wants to use may find it helpful to consult Table 12. Here are listed manufacturers of common substrates and companies which "customize" substrates to the user's specifications. The list may serve as a guide in finding sources of supply but is by no means exhaustive or intended as an endorsement.

ACKNOWLEDGMENTS

In the compilation of material used in this chapter, the author has enjoyed the cooperation of many colleagues. In particular, the technical staffs of the Bell Telephone Laboratories and the Western Electric Engineering Center contributed data and advice, as did many companies whose commercial literature provided information unavailable elsewhere. The author is further indebted to D. A. McLean for his recommendations and interest, and especially to J. C. Williams for many stimulating discussions. Thanks are also due to his wife Judy for preparation and proofreading of the manuscript.

REFERENCES

1. Shand, E. B., "Glass Engineering Handbook," 2d ed., McGraw-Hill Book Company, New York, 1958.
2. Kingery, W. D., "Introduction to Ceramics," John Wiley & Sons, Inc., New York, 1960.

3. Hove, J. E., and W. C. Riley (eds.), "Ceramics for Advanced Technologies," John Wiley & Sons, Inc., New York, 1965.
4. *Glass Ind.*, **48**, 309 (June, 1967).
5. Park, J. L., Jr., assignor to American Lava Corporation, U.S. Patent 2,966,719, Jan. 3, 1961.
6. Stetson, H., and B. Schwartz, presented at Electronics Division Meeting, American Ceramic Society, San Francisco, October, 1961, *Abstr. Am. Ceram. Soc. Bull.*, **40**, 584 (1961).
7. Ettre, K., H. D. Doolittle, P. F. Varadi, and R. F. Spurr, *Proc. Electron. Components Conf.*, Washington, D.C., May 7–9, 1963.
8. Bradley, F. M., *J. Brit. Inst. Elec. Engrs.*, **20**, 765 (1960).
9. Proebster, W., *Proc. Intern. Solid Circuits Conf.*, p. 28, AIEE, 1962.
10. Patel, J. R., R. S. Wagner, and S. Moss, *Acta Met.*, **10**, 759 (1962).
11. Stetson, A. W., and W. J. Gyurk, presented at the 69th Annual Meeting American Ceramic Society, Apr. 29, 1967, New York, *Abstr.* 29-E-67, *Am. Ceram. Soc. Bull.*, **46**, 387 (1967).
12. Schwartz, N., and R. Brown, *Trans. 8th AVS Symp.*, 1961, p. 836, Pergamon Press, London, 1962.
13. Manasevit, H., A. Miller, F. L. Morritz, and R. Nolder, *Trans. Met. Soc. AIME*, p. 233, 1965.
14. Balk, P., C. F. Aliotta, and L. V. Gregor, *Trans. Met. Soc. AIME*, p. 563, 1965.
15. Mendelson, S., in M. H. Francombe and H. Sato (eds.), "Single Crystal Films," The Macmillan Company, New York, 1964.
16. Lasko, W. R., and H. A. Roth, *IRE Trans. Component Pts.*, **CP-8**, 160 (December, 1961).
17. Degenhart, H. J., and I. H. Pratt, *Proc. IEEE Intern. Conv.*, pt. 6, p. 59, 1963.
18. Siddall, G., and B. A. Probyn, *Brit. J. Appl. Phys.*, **12**, 668 (1961).
19. Collins, F. M., ASTIA 239282, 1960.
20. Brown, R., *Am. Ceram. Soc. Bull.*, **45**, 720 (1966).
21. Coffman, B., and H. Turnauer, *Trans. 9th AVS Symp.*, 1962, p. 89, The Macmillan Company, New York, 1963.
22. Degenhart, H. J., and I. H. Pratt, *Trans. 10th AVS Symp.*, 1963, p. 480, The Macmillan Company, New York, 1964.
23. Clark, R. S., *Trans. Met. Soc. AIME*, **233**(3), 592 (1965).
24. Chaikin, S. W., and G. A. St. John, *Electrochem. Technol.*, **1**, 291[9–10] (1963).
25. Degenhart, H. J., and I. H. Pratt, *Trans. 8th AVS Symp.*, 1961, p. 859, Pergamon Press, New York, 1962.
26. Martin, J. H., *IEEE Trans. Parts, Mater. Packag.*, **PMP-1**, 267 (1965).
27. McLean, D. A., and F. E. Rosztoczy, *Electrochem. Technol.*, **4**, 523 (1966).
28. Moore, A. C., and A. S. Young, *J. Appl. Phys.*, **31**, suppl. 287–8S (1960).
29. Stillwell, G. R., and D. B. Dove, *J. Appl. Phys.*, **34**, 1941 (1963).
30. Ahn, K. Y., and J. Freedman, *IEEE Trans. Mag.*, **MAG-3**, 157 (1967).
31. Lloyd, J. C., and R. S. Smith, *J. Appl. Phys.*, **30**, suppl. 274–5S (1959).
32. Lemke, J. S., IEEE International Conference, Non-linear Magnetics, Washington, D.C., Apr. 17, 1963.
33. Methfessel, S., S. Middelhoek, and H. Thomas, *J. Appl. Phys.*, **31**, suppl. 302–4S (1960).
34. Prosen, R. J., B. E. Gran, J. Kivel, C. W. Searle, and A. H. Morrish, *J. Appl. Phys.*, **34**(2), 1147 (1963).
35. Kivel, J., F. C. Albers, D. A. Olsen, and R. E. Johnson, *J. Phys. Chem.*, **67**, 1235 (1963).
36. Kobale, M., and W. Kuny, *Z. Angew. Phys.*, **15**, 39 (1963).
37. Wiehl, H. E., *Z. Angew. Phys.*, **18**, 541, 5/6 (1965).
38. Smith, D. O., *J. Appl. Phys.*, **30**, suppl. 264–5S (1959).
39. Girard, R., *J. Appl. Phys.*, **38**, 1423 (1967).
40. Ondris, M., in R. Niedermayer and H. Mayer (eds.), "Proceedings of the International Symposium on Basic Problems in Thin Film Physics," Vandenhoeck and Ruprecht, Goettingen, 1966.
41. Garte, S. M., *Plating*, **53**, 1335 (1966).
42. Riben, A. R., S. L. Sherman, W. V. Land, and R. Geisler, in T. S. Schilliday and J. Vaccaro (eds.), "Physics of Failure in Electronics," vol. 5, p. 534, 1967.
43. Williams, J. D., and L. E. Terry, *J. Electrochem. Soc.*, **114**, 158 (1967).
44. Hanson, M. M., P. E. Oberg, and C. H. Tolman, *J. Vacuum Sci. Technol.*, **3**, 277 (1966).
45. Robbie, C. J., and C. T. Stoddart, *J. Sci. Inst.*, **1**, ser. 2, 56 (1968).

46. Seitchik, J. A., and B. F. Stein, *Rev. Sci. Instr.*, **39**, 1062 (1968).
47. Culver, H. E., and W. T. Layton, presented at the 69th Annual Meeting, American Ceramic Society, Apr. 29, 1967, New York, *Abstr.* 33-E-67, *Am. Ceram. Soc. Bull.*, **46**, 387 (1967).
48. Glang, R., and P. M. Schaible, *Thin Solid Films*, **1**, 309 (1967/1968).
49. Owen, A. E., and R. W. Douglis, *J. Soc. Glass Technol.*, **43**, 159T (1959).
50. Kraus, H., and B. Darby, "The Properties of Electrically Conducting Systems," American Chemical Society Monograph Series, 1922.
51. Orr, W. H., and A. J. Masessa, presented at the Buffalo Meeting of the Electrochemical Society, Oct. 13, 1965, *Abst.* 95, *J. Electrochem. Soc.*, **112**(8), 183C (1965).
52. Cox, S. M., *Phys. Chem. Glasses*, **5**(6), 161 (1964).
53. Snow, E. H., A. S. Grove, B. E. Deal, and C. T. Sah, *J. Appl. Phys.*, **36**, 1664 (1965).
54. Fuller, W. D., *NEC Proc.*, **17**, 32 (1961).
55. Kaiser, H. R., and P. S. Castro, *Proc. IRE*, **50**, 2142 (1962).
56. Happ, W. W., *Electron. Ind.*, vol. E4-7, June, 1962.
57. Guyer, E. M., *Proc. IRE*, **32**, 743 (1944).
58. Holland, L., "The Properties of Glass Surfaces," Chapman & Hall, Ltd., London, 1964.
59. Berry, R. W., and D. J. Sloan, *Proc. IRE*, **47**, 1070 (1959).
60. Leiser, C. F., *Glass Ind.*, **44**, 509 (1963).
61. DiMarcello, F. V., A. W. Treptow, and L. A. Baker, *Am. Ceram. Soc. Bull.*, **47**, 511 (1968).
62. Alderson, R. H., and F. Ashworth, *Brit. J. Appl. Phys.*, **8**, 205 (1957).
63. Settzo, R. J., *Electron. Reliability Microminiaturization*, **1**, 347 (1962), Pergamon Press, New York.
64. Greenough, K. F., in M. F. Goldberg and J. Vaccaro (eds.), "Physics of Failure in Electronics," vol. 3, p. 20, Spartan Books, Baltimore, 1963.
65. Belser, R. B., and W. H. Hicklin, *J. Appl. Phys.*, **30**, 313 (1959).
66. Hall, P. M., *Appl. Phys. Letters*, **12**(6), 212 (1968).
67. Takahashi, M., *J. Appl. Phys.*, **30**, suppl. 1101 (1962).
68. Rigterink, M. D., and J. C. Williams, *Am. Ceram. Soc. Bull.*, **43**, 894 (1964).
69. McLean, D. A., N. Schwartz, and E. D. Tidd, *Proc. IEEE*, **52**, 1450 (1964).
70. McLean, W. E., J. A. Thornton, and H. R. Aschan, *Proc. 1963 Electron. Components Conf.*, Washington, D.C., 1963.
71. Walker, M., J. Roschen, and E. Schlegel, *IEEE Trans. Electron Devices*, **ED-10**(4), 263 (1963).
72. Winklemann, A., and O. Schott, *Ann. Physik. Chem.*, **51**, 730 (1894).
73. Kingery, W. D., *J. Am. Ceram. Soc.*, **38**, 3 (1955).
74. Hoens, M. F. A., and J. C. Kreuter, in G. H. Stewart (ed.), "Science of Ceramics," vol. 2, p. 383, Academic Press, Inc., New York, 1965.
75. Schaible, P. M., J. Overmeyer, and R. Glang, "Physics of Failure in Electronics," vol. 5, p. 143, Air Force Cato Show Printing Co., 1966.
76. Peek, J. R., *Electron. Design*, **14**(24) (October, 1966).
77. Abraham, D., and T. O. Poehler, *J. Appl. Phys.*, **36**, 2013 (1965).
78. Burks, D. P., *Semicond. Prod. Solid State Technol.*, **9**, 33 (1966).
79. Martin, J. H., V. T. Guntlow, and D. P. Burks, *Proc. 1967 Electron. Components Conf.*, p. 354, Washington, D.C., 1967.
80. Burks, D. P., V. T. Guntlow, and J. H. Martin, *Proc. Natl. Electron. Conf.*, **23**, 413 (Oct. 23, 1967), Chicago, Ill.
81. Young, P. R., *Trans. Infrared Sessions*, Spring Convention of SNT, Feb. 22, Los Angeles, Society for Nondestructive Testing, Evanston, Ill., October, 1965.
82. Duckworth, W. H., *J. Am. Ceram. Soc.*, **34**, 1 (1951).
83. Parker, C. J., *Glass Ind.*, **44**, 489 (1963).
84. Paulley, D. S., and D. L. Lockwood, assignors to Coors Porcelain Co., U.S. Patent 3,324,212, June 6, 1967.
85. Kleiner, R. N., presented at 70th Annual Meeting, American Ceramic Society, Apr. 23, 1968, Chicago, Ill., *Am. Ceram. Soc. Bull.*, **48**, 1139 (1969).
86. Lumley, R. M., presented at 70th Annual Meeting, American Ceramic Society, Apr. 23, 1968, Chicago, Ill., *Abstr.* 13-E-68, *Am. Ceram. Soc. Bull.*, **47**, 385 (1968).
87. Longman, G. W., and R. P. Palmer, *J. Colloid Interface Sci.*, **24**, 185 (1967).
88. Krieger, G. L., and G. J. Wilson, *Mater. Res. Std.*, July, 1965, p. 341.
89. Missel, L., D. R. Torgeson, and H. M. Wagner, *Electron. Packag. Prod.*, 1966, p. 70.
90. Wenzel, R. N., *Ind. Eng. Chem.*, **28**, 988 (1936).
91. Linford, H. B., and E. B. Saubestre, *Electrochem. Soc. Res. Rept.* 26, 1954.
92. Feder, D. O., and E. S. Jacob, *ASTM Spec. Tech. Publ.* 300, p. 53, 1961.
93. Bulat, T. J., *Ref.* 92, p. 77.

94. Stern, H. A., Ref. 92, p. 82.
95. Thomas, R. N., and M. H. Francombe, *Appl. Phys. Letters*, **11**, 108, 134 (1967).
96. Bauer, E., *Phys. Letters*, **26A**(11), 530 (1968).
97. Cawley, R. H. A., *Chem. Ind.*, **45**, 1205 (1953).
98. Schreiner, D. G., *Extended Abstracts, 14th AVS Symp.*, 1967, p. 127, Herbick and Held Printing Co., Pittsburgh, Pa.
99. Putner, T., *Brit. J. Appl. Phys.*, **10**, 332 (1959).
100. Neppiras, E. A., *Research*, **6**, 276 (1953).
101. Tichane, R. M., *Bull. Am. Ceram. Soc.*, **42**, 441 (1963).
102. Bateson, S., *Vacuum*, **2**, 365 (1952).
103. DeKlerk, J., and E. F. Kelly, *Rev. Sci. Instr.*, **36**, 506 (1965).
104. Koontz, D. E., and M. J. Sullivan, *ASTM Spec. Tech. Publ.* 246, p. 183, 1958.
105. Filby, J. D., and S. Nielsen, *J. Electrochem. Soc.*, **112**, 957 (1965).
106. Behrndt, K. H., and F. S. Maddocks, *J. Appl. Phys.*, **30**, suppl. 276 S (1959).
107. Nielsen, S., *Trans. 7th AVS Symp.*, 1960, p. 293, Pergamon Press, London, 1961.
108. Allen, F. G., J. Eisinger, H. D. Hagstrom, and J. T. Low, *J. Appl. Phys.*, **30**, 1563 (1959).
109. Holland, L., "Vacuum Deposition of Thin Films," Chapman & Hall, Ltd., London, 1956; *Brit. J. Appl. Phys.*, **9**, 410 (1958).
110. Hirai, H., K. Ando, and Y. Maekawa, *Mem. Fac. Eng., Osaka City Univ.*, **8**, 103 (1966), in English.
111. Mattox, D. M., *J. Appl. Phys.*, **37**, 3613 (1966).
112. Florescu, N. A., *Vacuum*, **7-8**, 46 (1955).
113. Holland, L., L. Laurenson, and C. Priestland, *Rev. Sci. Instr.*, **34**, 377 (1963).
114. Wasserman, E. F., and R. L. Hines, *J. Appl. Phys.*, **38**, 196 (1967).
115. Jorgenson, G. J., and G. K. Wehner, *Trans. 10th AVS Symp.*, 1963, p. 388, The Macmillan Company, New York, 1964.
116. Svec, J. J., *Ceram. Ind.*, **87**, 54 (September, 1966).
117. Berry, R. W., W. H. Jackson, G. I. Parisi, and A. H. Schafer, *Proc. Electron. Components Conf.*, Washington, D.C., May 5, 1964.
118. Robinson, P. H., and C. W. Mueller, *Trans. Met. Soc. AIME*, **236**, 268 (1966).
119. Paulsen, J. F., and E. F. Steiner, *Ceram. Age*, **82**, 36 (July, 1965).
120. Budo, Y., and J. Priest, *Solid-State Electron.*, **6**, 159 (1963).
121. Rasmanis, E., *J. Electrochem. Soc.*, **110**, 57C (1963).
122. Maddocks, F. S., and R. E. Thun, *J. Electrochem. Soc.*, **109**, 99 (1962).
123. Intellux, Inc., *Rept.* AD 401-397.
124. McLean, D. A., U.S. Patent 3,220,938.
125. Bertelson, B. J., *J. Appl. Phys.*, **33**, 2026 (1962).
126. Kay, E., *Nature*, **202**, 788 (1964).

Chapter **7**

Generation of Patterns
in Thin Films

REINHARD GLANG
and
LAWRENCE V. GREGOR

IBM Components Division, East Fishkill, New York

1. INTRODUCTION

In addition to deposition techniques and control of physical properties, the utilization of thin films in microelectronics is predicated on the ability to distribute the deposited material in well-defined patterns. The trend toward more complex and smaller integrated circuits has stimulated considerable development, with resulting improvements in the techniques for generating thin film patterns. Accuracy and high resolution have been the primary objectives.

Most of the methods commonly-used for generating thin film patterns fall into one of two categories. The first one relies on a physical barrier of appropriate shape being present in close contact with the substrate during film deposition. Portions of the evaporant beam are thus intercepted and prevented from condensation on the corresponding areas of the substrate. This technique is referred to as evaporation masking. The second category encompasses techniques of forming the desired film pattern after the deposition by selective removal of portions of the film. This is accomplished by the use of organic lacquers or resists which protect the film in the desired configuration while the superfluous and unprotected regions are etched away.

These methods are referred to as subtractive etching or, since they employ photographic techniques to produce the pattern, as photolithography. In addition to these well-established techniques, there are several more recent methods of film-pattern formation. Some of these are becoming very useful in a variety of applications. They are reviewed in Sec. 4 of this chapter.

2. FILM DEPOSITION THROUGH MASKS

The direct deposition of thin film patterns requires a suitably shaped aperture commonly referred to as a mask. The latter may be a metal, graphite, or glass plate with the desired pattern cut or etched into it. The mask is placed in close proximity to the substrate, thereby allowing condensation of the evaporant vapor only in the exposed substrate areas. A typical deposition arrangement is shown in Fig. 1. The film thickness can be controlled by a rate monitor and a shutter.

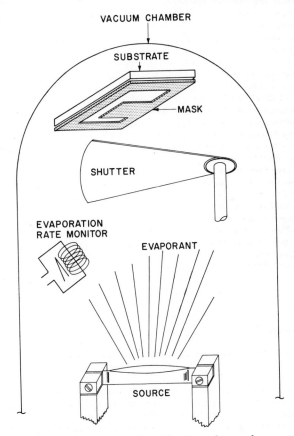

Fig. 1 Film-pattern deposition through a mask.

Accurate replication of the mask pattern by the deposited film is contingent upon two conditions. First, the mean free path of the evaporant particles must be long compared with the mask-to-substrate spacing to avoid random condensation caused by intermolecular collisions. Second, the sticking coefficient of the impinging vapor must be close to unity to prevent reevaporation and lateral spreading under the mask. When evaporating in a high vacuum at moderate substrate temperatures and with the mask in physical contact with the substrate as shown in Fig. 1, these prerequisites are usually satisfied. In sputtering, however, the higher gas pressure in the vacuum system and, in some cases, resputtering of condensed species prevent the attainment of sharply delineated film patterns.

The intricacy and precision of film patterns made by evaporation through masks are limited by the mechanical properties of the mask material and the accuracy of mask-fabrication procedures. Since thin film components and circuits are the product of several evaporations, dimensional agreement between different masks—generally known as registration—is also a critical requirement. Finally, evaporation masks are subject to topological constraints insofar as completely connected patterns cannot be cut out of a mask without leaving the area circumscribed by such a pattern unsupported. These limitations are primarily responsible for the inapplicability of evaporation masks in many of the present microcircuit manufacturing processes.

a. Fabrication of Evaporation Masks

To fabricate an evaporation mask, material must be removed from a blank selectively and accurately in the desired pattern. This may be done either mechanically or chemically depending on the type and thickness of the material chosen. The choice of material is governed by the desire to produce a mask which maintains its planarity and dimensions as much as possible even when heated during film deposition, and which has sufficient mechanical strength to be handled. Metals and alloys having low thermal-expansion coefficients are most often used because of their strength and amenability to mechanical forming processes. Graphite is also easily machined and maintains its dimensions very well; however, only rather coarse masks are possible since thin blanks with intricate patterns are too fragile. Other materials which qualify from the point of planarity, strength, and low thermal expansion are glasses. But glasses cannot be readily machined, and only certain compositions which lend themselves to photochemical processing have been used as masks.

(1) Machining Techniques Several techniques are available for making openings of various shapes in metal or alloy blanks. The simplest approach is to punch holes, but the resulting patterns tend to be irregular and the masks lack planarity.[1] Better accuracy is obtained if milling machines with suitable cutting tools are used. The time required to machine complex masks manually makes this a very tedious and expensive method. This problem can be overcome by automation of the machining process. An automated milling system was described by DeLano[1] in 1961. It employed a motor-driven X-Y table moving in increments of 0.005 in. with a positioning accuracy of 0.0002 in. The table held an aluminum blank into which the pattern was cut. The cutting tool was stationary and lowered into the blank on command from a punch-card reader. Patterns thus produced consist of holes and lines along rectangular coordinates. A similar automated system has been used to machine masks out of high-density graphite blanks.[2] This technique is claimed to be quite economical and to have yielded line-width tolerances of ±0.0002 in. The main limitation is that only relatively crude patterns of simple geometry can be produced.

Another precise method of cutting patterns into metal plates is arc erosion.[3] Here, the mask blank is submerged in a dielectric oil and an eroding arc is established between it and a suitably shaped toolhead. The process has been used to fabricate masks from Invar,[4] but it is fairly expensive because the eroding electrode must be a positive relief of the pattern to be cut. Hence, a special toolhead has to be machined for each configuration, and although it may cut many identical patterns, the edges of the relief wear off with usage.

Electron-beam machining has also been investigated as a mask-fabrication process.[5] Although very fine lines have been obtained, the edges were somewhat irregular, and the technique was not developed far enough to yield complete masks.

(2) Etching Techniques Masks of higher resolution than obtainable by mechanical processes can be fabricated by selective etching of thin metal sheets or foils. These methods usually require a photographic mask of the pattern and application of a photosensitive resist, as described later in Sec. 3. The resolution of etched masks is higher the thinner the foil since undercutting during the etch is the determining factor [see Sec. 3f(2)].　Foil thicknesses from 0.004 in. down to 0.0002 in. have been used. Suitable metals are copper,[6,7] stainless steel,[8] nickel, and molybdenum.　Foils of the latter material are most widely used.　They can be etched in dilute nitric acid or other solutions [see Sec. 3d(7)] and permit lines as narrow as 0.003 in. to be fabricated. Since the etched foils are very thin, they lack stiffness and planarity.　Hence, they must be supported by a heavier metal frame or a thick backing plate with large windows in the fine pattern areas.　Molybdenum-foil masks are often bonded to Kovar frames.

Other mask-making processes combine chemical etching with electroplating to produce thin laminated sheets.[9]　This so-called electroforming process is illustrated in Fig. 2.　The film pattern to be evaporated is first applied as a photoresist image on

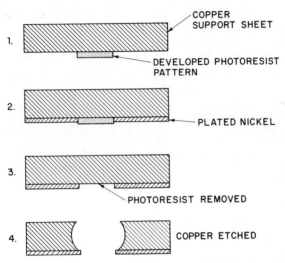

Fig. 2　Fabrication of an evaporation mask by electroforming.

a flat copper sheet which is at least 0.002 in. thick.　A nickel film is then electroplated to a thickness slightly less than that of the resist film to avoid lateral overgrowth. After resist removal, the copper is selectively etched.　By deliberately overetching, an aperture defined by the thin nickel film alone is obtained, whereas the copper foil serves as a support.　Pattern tolerances of ±0.0001 in. are possible.　The mask shown in Fig. 3 was made by a similar process.[10]　In this case, the photoresist and nickel films were applied to a 0.031 in.-thick copper-beryllium sheet.　The narrow lines in the nickel film (Fig. 3A) are 0.002 in. wide with ±0.0002 in. tolerances. Larger windows were cut into the support sheet (Fig. 3B) by differential etching so that the deposited film pattern was defined only by the thin nickel mask.

(3) Glass Masks　Evaporation through masks has often been used in conjunction with glass substrates of 1- to 10-in. linear dimensions.　In such cases, the greater expansion of metal masks during substrate heating leads to noticeable misregistration of the deposited film pattern.　Consequently, attempts have been made to utilize glass masks.　The possibilities of shaping glass by mechanical processes such as abrasive powder driven by compressed gas or ultrasonic cutting are limited.　Chemical etching, however, offers acceptable resolution and edge definition.　For actual mask fabrication, a particular type of photosensitive glass known as Fotoform B* has been employed.[11]　This material is fabricated in thin sheets which are photosensitive prior to firing.　The "green" sheets are exposed photographically with the desired mask pattern, which is then etched out chemically.　Thereafter, the etched sheets are

* Manufactured by Corning Glass Works, Corning, N.Y.

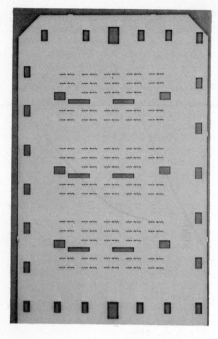   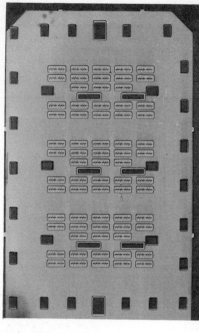

                    (A)   (B)

Fig. 3 Electroformed nickel mask. A = high-resolution side facing substrate; B = reverse side facing the evaporation source and showing the large windows in the support sheet. (*Courtesy of H. M. Greenhouse and R. T. Galla,*[10] *The Bendix Corp., with permission.*)

fired.   The composition of the material and the entire processing are proprietary, and it is necessary to order the masks fully specified from the manufacturer.

Because of the graininess of the glass-ceramic sheet and its thickness, which is comparable with that of microscope slides, the smallest openings attainable are 0.005 in. wide with etched tolerances of ±0.001 in.   Dimensional control also suffers because considerable shrinkage—up to 10%—of the green sheets occurs during the firing.   The masks should be examined for warpage since lack of physical contact with the substrate may contribute further to dimensional deviations of the film pattern.   Extreme care in handling and cleaning of glass masks is necessary to avoid breakage.

(4) Mask Cleaning Regardless of mask material and fabrication process, all masks should be thoroughly cleaned and inspected prior to use.   Dust particles, fibers, or debris from the machining process may cause discontinuities in the deposited film pattern.   Surface contaminants, particularly oil, grease, or other organic matter, may become volatile when the mask is heated and then be adsorbed by the substrate, thus being a likely cause of weak film adhesion.   Common mask-cleaning procedures involve vapor degreasing or ultrasonic agitation in organic solvents or aqueous detergent solutions.

If masks accumulate deposits whose thicknesses are not negligible compared with the smallest opening dimension, the accuracy of subsequently deposited film patterns is affected.   Therefore, periodic removal of deposits is necessary.   This is best done by chemical means to avoid mechanical damage of the mask.   For most metallic deposits, selective etchants which do not attack the mask material are available. However, some inert materials such as SiO or polymer films become soluble only in very corrosive and nonspecific chemicals.   In these cases, cleaning can be implemented by precoating the mask with an easily soluble film, for example, a metal.   Subsequent deposits of inert material may then be removed by attacking the underlying coat until the inert layer is left unsupported. Deposit removal by chemical-vapor reactions has also been suggested.[12]

b. Mask Holders

Mechanical provisions are required to hold mask and substrate securely and in close contact. Since most masks have insufficient mechanical rigidity, they are mounted in heavier frames, so-called mask holders, which provide planarity and facilitate attachment to the substrate holder by means of bolts or clamps. If friction between mask and substrate surfaces is anticipated—caused, for example, by movement during registration—or if the film thickness is comparable with the thickness of the mask, it is advisable to allow for some clearance by adding shims. A spacing of 0.005 to 0.020 in. between mask and substrate is adequate for this purpose, yet small enough to keep lateral spreading of the evaporant under the mask at tolerable levels. Since most evaporations are performed on substrates which already have a deposited film pattern, proper registration of the mask is essential. Therefore, mask and substrate holder combinations must have provisions for small lateral movements of mask and substrate relative to each other.

Mask holders vary widely in their design depending on the type of substrate to be processed. Figure 4 shows an assembly for the deposition of small metal dots on

Fig. 4 Molybdenum-foil mask for the deposition of small metal dots. (*International Business Machines Corp.*)

silicon wafers. The mask is a molybdenum foil with an array of circular holes. The mask frame and the silicon wafer can be moved with respect to each other to line up the holes with the circuit pattern on the wafer. This is done under a microscope prior to insertion into the vacuum chamber.

A mask-substrate holder assembly for rectangular glass substrates is shown in Fig. 5. Here, a metal or glass mask is mounted within the mask area of the frame. The substrate is attached to the underside of the holder plate and held against the mask by one stationary and one spring-loaded pin. Registration is achieved by viewing the substrate through the mask openings under a microscope and translating either the mask or the substrate by means of lateral setscrews. Both the assemblies shown in Figs. 4 and 5 can be heated by radiation from filaments mounted behind the substrate holder to facilitate film deposition at elevated temperatures.

Mask-to-substrate contact and registration are much more difficult to achieve if the masks are to be changed in the vacuum system as is required for multiple evaporations.

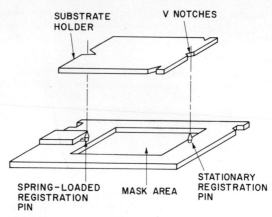

Fig. 5 Mask and substrate holder assembly for rectangular glass substrates. (*From Lessor, Thun, and Maissel,*[4] *with permission of IEEE Spectrum.*)

The contacting and aligning motions must then be carried out automatically and without the aid of a microscope. Provisions for automatic alignment include guide pins, cam surfaces, or other mechanical devices. Furthermore, automatic mask or substrate changers or both are needed, in the form of either rotating disks or stacking magazines. All these mechanisms involve moving parts and therefore the risk of excessive friction and binding. Lubrication of internal vacuum parts is generally discouraged, particularly if the parts are to be heated, since vaporization or decomposition of lubricants severely contaminates the vacuum system. Low-vapor-pressure lubricants of good chemical stability such as MoS_2 mixed with vacuum-pump oil are occasionally used.[13] The design of vacuum systems for multiple evaporations and their special problems are further discussed in Chap. 2, Sec. 5e.

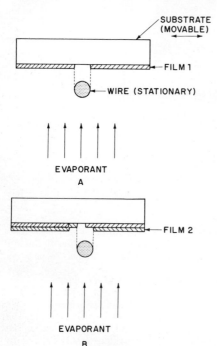

Fig. 6 Evaporation with a wire mask. A = single wire producing a gap in the film. B = reduced gap width obtained by translating the substrate to the left, followed by a second evaporation.

c. Wire-grill Masks

Some of the limitations pertaining to resolution and fabrication of metal or glass masks are overcome by the wire-grill masks developed by Weimer and coworkers.[14] These are based on the utilization of thin (about 0.0015 in. diameter) Nichrome wire mounted in close proximity to the substrate. As shown in Fig. 6A the wire shields an area as wide as its diameter from receiving deposit. The wire must be within less than 0.001 in. from the substrate surface, and the directionality of the vapor stream must not be affected by poor vacuum or close source-to-substrate spacing. Scattering due to sticking coefficients of less than unity or increased temperature of the mask must also be absent. If one or more of these conditions are not satisfied, the edges of the gap in the film are not sharply defined ut taper off gradually. In extreme cases, the entire gap may be bridged by a thin veil of deposit. Figure 6B illustrates how gaps narrower than the wire diameter can be produced. This requires a small displacement of the substrate perpendicular to he wire and a second film deposition. The resolution limit is about 1 μ.

To deposit thin film patterns, a grill of many parallel wires is necessary. Parallelism and even spacing are attained by drawing the wire through a die and winding it upon a frame which consists of a split screw threaded with 480 turns/in. Deposition hrough such a grill yields a pattern of parallel lines whose separation and width are determined by the diameter and spacing of the wire, or modified through lateral substrate movements as shown in Fig. 6B. To terminate the lines and form useful rectangular patterns, interchangeable metal masks with relatively simple, easily machined cutouts are utilized. The metal masks are located below and in close proximity to the wire grill. They can be moved parallel to the stationary wires. Thus, the problem of changing and registering masks has been reduced to two linear motions, namely, hose of the metal mask and those of the substrate. Both displacements can be closely controlled by means of micrometer screws. Although the freedom of thin film pattern ormation is limited to parallel lines and rectangles, many useful circuit designs can be implemented. Evaporated thin film field-effect devices are particularly amenable to abrication with wire-grill masks and have been deposited in several different configurations (see Chap. 20).

4. Applications of Evaporation Masks

The utilization of evaporation masks is primarily determined by their cost, pattern etail, and tolerances attainable. For a review, these factors are listed in Table 1. A

ABLE 1 Characteristic Dimensions of Evaporation Masks

Type of mask	Smallest aperture, 10^{-3} in.	Tolerances, 10^{-3} in.	Comments
tched glass (Fotoform B)......	5	± 1	Fragile, fairly expensive
unched metal................	3	± 1	Inexpensive, but lack planarity
ngraved (milled) metal or graphite............	2	± 0.2	Manual process: expensive; automatic process: patterns limited
rc-eroded metal..............	2	± 0.2	Special tool required; expensive
tched metal foil..............	2	± 0.2	Thickness-dependent; mostly Mo foil
lectroformed metal film........	0.5	± 0.1	Fragile
ire grill....................	0.04	*	Pattern limitations

* Depend on wire to substrate spacing, temperature, and type of evaporant.

omparison of the smallest apertures and the associated tolerances shows that the lative accuracies of dimensions at the process limits are generally poorer than $\pm 10\%$. nce accurate line widths of 0.001 in. and smaller are common requirements in the brication of integrated circuits, evaporation masks are not widely used in this major ea of thin film applications. An exception are etched molybdenum-foil masks for etal-contact dots of the type shown in Fig. 4. Here, the dimensional requirements 0.004 in. diameter ± 0.0005 in. tolerance are well within the capability of the fabrication process. Another example is the evaporation of metal-film electrodes on thin O$_2$ films for capacitance-voltage measurements to evaluate the quality of the electric. In this case, small and accurate dimensions are not as important as the sire to maintain the purity of the oxide, which is likely to suffer if the electrodes were rmed instead by photoresist and etching processes.

Evaporation masks have been most widely used in situations where relatively large substrate areas need to be covered and where the smallest dimensions are greater than 0.002 in. Thin film circuit networks consisting of resistors, capacitors, and interconnections, as well as cryogenic circuits on fairly large glass substrates, are examples of successful mask utilization.[4,15] For cryogenic devices, pattern definition by means of mask changers is a necessity since freshly deposited films of superconducting metal must not be exposed to air (see Chap. 22). The cryogenic circuit shown in Fig. 7 exemplifies the complexity and process control that have been achieved.

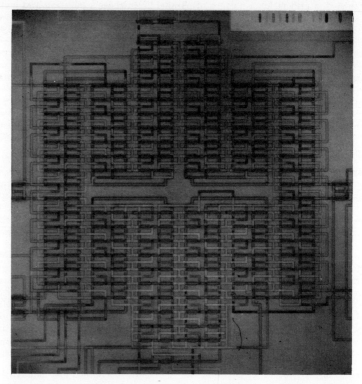

Fig. 7 Cryogenic thin film circuit (100-stage ring) made by evaporation through nine separate masks on a 2- by 2-in. substrate. The line widths are 0.010 in. (0.25 mm) and 0.004 in. (0.10 mm), respectively. (*Courtesy of I. Ames, IBM Corp., Yorktown Heights, N.Y.*)

In addition to these applications, evaporation masks have also been utilized in less conventional ways to form thin film patterns. One area is the deposition of polymer films from monomer vapors. Since condensation occurs only where the adsorbed monomer species is irradiated with ultraviolet light, patterns can be defined by masking the radiation.[16] Such polymer film patterns may be used as dielectrics in thin film capacitors,[17] or the method may serve to apply a photoresist mask in a vacuum system.[18] Evaporation masks have also been employed to deposit extremely thin films which affect the nucleation of subsequently deposited thicker films. After the very thin film pattern is formed, the masks are removed and the entire substrate surface is exposed to the vapors of the second film material. Depending on the choice of evaporants and the deposition conditions, nucleation of the second film can be confined to[19] or inhibited within[20] the pretreated pattern areas. The attainment of regions as narrow as 0.0001 in. by controlled nucleation along the edges of mask-evaporated patterns has been reported by Ames et al.[21]

3. PHOTOLITHOGRAPHY

Pattern formation by photolithographic techniques is based on the application of a polymer film in the desired configuration to metal or insulator films covering the entire substrate surface. The pattern of the polymer stencil mask is repeated in the metal or insulator film by etching away the unprotected regions. The stencil mask is generated by using a photosensitive polymer material referred to as photoresist, whose molecular structure and solubility are changed by irradiation with photons. A transparency with the desired film pattern, the photographic mask, is required to confine the exposure to those regions where the solubility of the photoresist is to be changed. The process is analogous to photographic contact printing except that the development of the photoresist pattern is followed by etching of the film material and subsequent removal of the polymer stencil mask.

Since photolithographic patterns are defined by light, configurations of much greater complexity and finer detail can be obtained than film deposition through metal masks can yield. Therefore, photolithography has become the dominant technique in the fabrication of microelectronic circuitry where precise definition of fine line patterns is a prime requirement. Examples are the opening of holes in SiO_2 films for diffusion or metallization of silicon wafers, and the formation of metal-film contacts and interconnection patterns.

The demands for extremely small pattern dimensions in integrated circuits have greatly stimulated the development of all the techniques which are involved in photolithography. These encompass microphotography, high-resolution photoresists, precision exposure, developing, and etching processes. To these must be added entirely new methods of image multiplication which were developed specifically for microelectronic applications. These multiplication techniques make it possible to process many small circuit chips or modules of the same kind together in one operation, an important factor in regard to economy and uniformity of the product.

Although the development of photolithographic techniques is still continuing and current practices vary from laboratory to laboratory, the principal steps and the capability of the processes described in this section are well established. A general review of photolithography as used in semiconductor-circuit manufacturing has been written by Castrucci et al.[22] More comprehensive sources of information on experimental techniques and suitable instrumentation are the ICE Technology Handbook[23] and a brochure on photofabrication published by the Eastman Kodak Co.[24] Since the resolution, accuracy, and definition of the etched patterns depend critically on special photographic techniques, the fabrication of photomasks is an important part of this technology.

a. Fabrication of Photographic Masks

Photographic masks for contact printing are precision images of the pattern to be etched. Because of the small dimensions required, the pattern must first be produced at a greatly enlarged size. It is then reduced to the final scale in one or more photographic-reduction steps. In nearly all cases, the size of the substrate is far greater than the final image. Hence, it is most economical to reproduce the original image many times and form an array of the pattern. Many circuit chips or modules of identical configuration can then be produced on the same substrate by one contact printing.

To provide accurate pattern dimensions, it is essential that the original image, the artwork, is prepared with the greatest precision. The same requirement applies to the reduction process, and the photographic emulsions used must match the resolving power of the camera. Furthermore, all steps in the mask-fabrication process must be conducted in a controlled environment since dust particles or mechanical damage introduced at any point propagate and lead to defects in the etched pattern. Constancy of temperature is equally important because thermal expansion of instruments or materials causes noticeable changes in pattern dimensions.

Another problem arises from the fact that circuit patterns require more than one etching step and hence several different masks. These are referred to as a set. When producing several patterns on the same substrate, it is of great importance that the successively etched film patterns are exactly in the right positions relative to each other.   This can be achieved only if the set of masks has been prepared at exactly the same reduction ratio and with perfect alignment of successive images in all mask-fabrication and contact-printing steps.   The alignment process and the degree to which correct superposition of images is accomplished are called registration.   Registration marks in the form of dots or line patterns are commonly placed in a few locations of the pattern array to facilitate precise alignment over the entire substrate area from one layer to the next.   Several registration techniques have been described by Ostapkovich.[25]

(1) Artwork Generation   The generation of the artwork begins with the translation of the engineer's circuit diagram into a layout.   First, a composite of all components and interconnections in their actual shapes and configurations is needed. This is normally done by making a drawing on graph paper.   Component sizes and line widths must be determined according to the electrical-performance requirements and the available layout area.   The same factors also impose restrictions on the placement and geometrical arrangement of all circuit elements.   This is a complex task which requires observation of many empirical ground rules pertaining to minimum pattern dimensions and line spacings attainable by masking and etching, component interaction, and parasitic effects (see, for example, Ref. 23, p. 291).   The problem can usually not be solved in a straightforward manner but requires repeated attempts before an electrically functional as well as technically feasible layout is obtained.   The composite circuit layout must then be broken down into individual mask layouts. Each of these possesses only those portions of the circuit pattern which are to be etched in one particular step and from the same film.

The mask layouts are transformed into large-scale images of the patterns to be etched.   To obtain a high degree of dimensional accuracy, these images are produced at typically 200 to 1,000 times the actual size.   Since ink-drawn images have relatively poor contrast, the patterns are prepared from laminated plastic sheets.   These are Mylar or polyester-base materials whose dimensions are little affected by fluctuations of temperature or humidity.   They consist of a clear transparent sheet laminated with a red, photographically opaque film.   The pattern is carefully cut into the red film,[26] and the circumscribed areas are peeled off.   The cutting may be performed with a knife guided by hand.   This yields an accuracy of 0.5 to 1 mm or, after a 200:1 reduction, dimensional errors of 2 to 5 μ.

Better results are obtained if the cuts are made by precision drafting-scribing machines (coordinatographs).   Commercial models and their suppliers may be found in a review article by Maple.[27]   Coordinatographs are capable of cutting with an accuracy of ± 25 μ within the X-Y plane.   Hence, statistical errors committed in the artwork preparation are reduced to about 0.1 μ after 200 times reduction.   Typically these machines cover a cutting area of 36 by 36 in., which is sufficient for most layouts requiring 200:1 reduction.   For higher reduction ratios or very large circuit chips coordinatographs with 50- by 50-in. or larger cutting areas are available.   Modern systems are also equipped with digital control and thus allow programmed cutting

An important point to be observed in the preparation of the original artwork is the type of image required.   The original artwork undergoes positive-negative as well as left-right (mirror-image) reversals in every reduction and contact-printing step Therefore, the number of mask-processing steps planned must be considered so that the cut artwork represents a positive or negative of the right polarity with respect to the final mask.[28]   The nature of the photoresist (positive or negative), too, has to be taken into account.

Layout design with pencil and paper is time-consuming and requires attention to many details.   Therefore, methods of computer-aided artwork design have been developed.[29,30]   The method described by Strickland et al.[29] utilizes a graphic data processing system consisting of a small digital computer and a large-screen buffered display with a light pen.   The latter is used to assemble the circuit diagram on the cathode-ray tube.   At the same time, the circuit schematic is stored in the computer memory.   Thereafter, the actual artwork is laid out on the screen also by means of the light pen.   Thus, different layout configurations can be tried without tedious erasures and redrawing. The final version prepared on the screen may be transformed into drawings on paper by means of a computer-controlled X-Y plotter, or into cut artwork via a numerically controlled coordinatograph. In addition to considerable time savings, computer-aided design has the advantages of avoiding errors by checking the layout against the stored circuit schematic, and the relative ease with which engineering changes can be implemented. The preparation of cut artwork and some of the subsequent photographic reductions can be eliminated if the stored pattern is transferred into an artwork generator, which produces a 10:1 photographic transparency directly. Such systems are more fully discussed in Sec. $3a(3)$.

(2) Optical Principles of Photoreduction The conversion of the artwork into photographic masks with the final pattern dimensions requires a drastic reduction in size. This is done by photographing reduced images of the original artwork. The requirements of pattern resolution, accuracy, contrast, and image field in mask fabrication are among the most stringent in microphotography and have initiated several new developments. To understand the problems and limitations encountered, it is necessary to examine the parameters which characterize and limit the optical performance of lenses in general. For a broader treatment of microphotographic principles, the reader is referred to Stevens' book.[31] The Kodak pamphlet on the same subject[32] and recent review articles by Rottmann[33] and by Altman[33a] provide suitable introductions.

A reduced image of an object is obtained if object and image plane are placed at

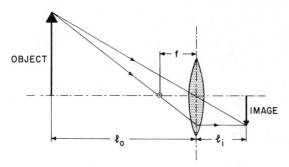

Fig. 8 Formation of a reduced image.

conjugate foci of a lens as shown in Fig. 8. The *reduction ratio* follows from geometrical optics:

$$\text{Reduction ratio } R = \frac{l_o}{l_i} = \frac{l_o - f}{f} \tag{1}$$

Solving Eq. (1) for the lens-to-image distance $l_i = f(1 + 1/R)$ shows that reduction ratios of 10:1 or greater place the image very close to the focal length f of the lens. The object distance l_o is then given approximately by the product of focal length and reduction ratio. Assuming the latter to be 100:1, a lens with a relatively long focus of $f = 100$ mm would require an object distance of 10 m. This presents a mechanical problem insofar as cameras providing such lengths with sufficient accuracy are extremely difficult to build and maintain. The reduction ratio 100:1 can be implemented if the focal length of the lens is reduced to about 20 mm, but then the useful image field becomes very small, as will be discussed later. Consequently, reduction ratios in microphotography are limited and related to image-field size. Values from 10:1 to 50:1 are common. As a result, the reduction of artwork has to be done in two or more steps.

The *optical quality* of an aerial image must be evaluated in terms of three properties: resolution, contrast, and size of the useful image field. These are all dependent on lens-design parameters which include, in addition to focal length, the numerical

aperture and the type and severity of aberrations.   The *numerical aperture* characterizes the ability of a lens to transmit light and is approximately proportional to the
size of the lens:

$$\text{N.A.} = \frac{\text{lens diameter}}{2 \times \text{focal length}}$$

It is related to the more familiar F number engraved on camera lenses by

$$\text{N.A.} = \frac{1}{2 \times F \text{ No.}}$$

Aberrations are small deviations of light rays from the ideal paths postulated by
geometrical optics.   They arise from different physical causes and require corrections.
One group of aberrations stems from the fact that light of different wavelengths is
focused at different distances from the lens.   Since color rendition is immaterial in the
fabrication of photographic masks, wavelength-dependent aberrations can be avoided
by using monochromatic light.   The green line in the emission spectrum of mercury
vapor at 5,460 Å provides a sufficiently high intensity and falls into the spectral region
where the photographic plates have their sensitivity maximum.

Aberrations also arise from rays which travel at some distance from the optical axis
of the lens and whose foci deviate from the ideal points in the image plane.   Lenses of
good quality are so designed that these aberrations are minimized for the particular
wavelength to be used.   However, even the best corrected lenses retain some aberrations which are apparent as distortions, astigmatism, and curvature of the image field.
Because of the latter, the aerial image curves away from the ideal focal plane.   The
displacement along the optical axis is small near the center but increases with center
distance.   The area around the optical axis within which a sharp image can be
obtained depends on the depth of focus of the lens and is called the *useful image field*.
Since the depth of focus is proportional to $\lambda/(\text{N.A.})^2$, it follows that the useful image
field of a lens is greater the smaller the numerical aperture, i.e., if the more peripheral
rays are excluded.   Furthermore, since the numerical aperture and the focal length
of a lens are inversely proportional, the useful image field is also related to the focal
length.   The latter relationship is of practical value for rough estimates.   It has been
found that the useful fields of good camera lenses are typically one-fifth of their focal
lengths.[27,31,33]   The useful fields of microscope objectives are even smaller and usually
less than one-tenth of the focal lengths.[27,31]   This explains the previously described
problem of reconciling high reduction ratios with large image fields.

Having obtained a well-focused image within the useful field of a properly corrected
lens, the question of the smallest resolvable dimension becomes of interest.   Under
these conditions, the attainable *resolution* of a lens is limited by the diffraction of light
waves.   The image of a small point produced by an aberration-free lens in the focal
plane is not a sharp point but is spread out into a diffraction pattern known as the
Airy disk.   It has a bright center surrounded by concentric rings of decreasing
intensity.   The images of two point sources can be resolved only if the corresponding
Airy disks are so far apart that they do not overlap.   The smallest resolvable dimension L_{min} can therefore be defined from the radius of the Airy disk.   The theory
yields[31,33]

$$L_{min} = \frac{0.61\lambda}{\text{N.A.}}$$

for resolution of point images (Airy formula) and

$$L_{min} = \frac{0.5\lambda}{\text{N.A.}}$$

for resolution of line images, where λ is the wavelength of the light.   Consequently
the theoretical resolving power of a lens is greater the larger the numerical aperture.
The effect of aperture size on the diameter of the Airy disk is illustrated in Ref. 32.
Experimentally, the resolution of lenses is commonly determined by photographing

test patterns on high-resolution plates. A widely used test pattern consists of arrays of alternate light and dark bars of equal width but different sizes (see, for example, Ref. 32 or 34). The array with the smallest bars still clearly resolved indicates the resolution attained. The latter is expressed in terms of lines per millimeter. Since the lines are separated by equally wide spaces, it is

$$\text{Resolution} = \frac{1}{2L_{min}} \quad \text{lines/mm}$$

where L_{min} must be given in millimeters.

The third property which characterizes the performance of a lens is its ability to project fine line patterns with sufficient contrast. Assuming a narrow illuminated slit in the object plane, some of the light is diffracted away from the optical axis and does not enter the aperture of the lens. The narrower the slit, the stronger becomes the angular dispersion of the light, and the intensity available for image formation decreases. Consequently, the contrast of an image is dependent on the pattern dimensions to be resolved. Plots of relative contrast vs. lines per millimeter resolved are made to indicate lens performance. An example is shown in Fig. 9. Since lens

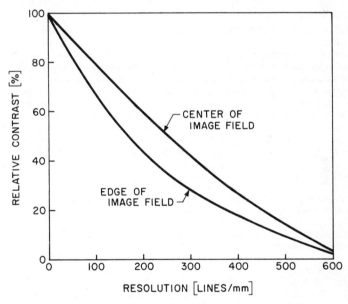

Fig. 9 Relative contrast of a lens as a function of resolution.

resolution is often quoted for 4% contrast at the center of the image field, it is important to realize that this condition does not produce sufficient black-and-white contrast on photographic-emulsion plates. For purposes of mask making, image contrasts of 40 to 50% are needed, a requirement which reduces the useful resolution considerably.[25] Similarly, the contrast associated with the theoretical resolution obtained from the Airy formula is far from sufficient for photographic-mask fabrication.

(3) Photoreduction Techniques As shown in the previous section, the reduction of the original artwork to the final pattern size has to be done in more than one step because of limitations in camera length and useful image-field diameter. This requires at least two camera systems, one capable of yielding fairly large images with moderate fine resolution and one yielding very high resolution but over a relatively small area. In addition, the small image of the final size must be repeatedly photographed to

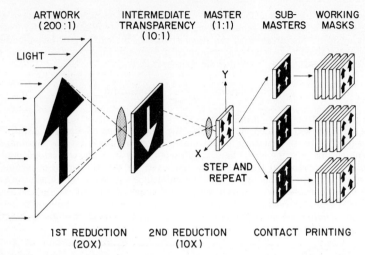

ARTWORK   INTERMEDIATE MASTER   SUB-   WORKING
(200:1)   TRANSPARENCY   (1:1)   MASTERS   MASKS
                    (10:1)

LIGHT

Y

X

STEP AND
REPEAT

1ST REDUCTION   2ND REDUCTION   CONTACT PRINTING
   (20X)   (10X)

Fig. 10 Two-step photoreduction process as used in the fabrication of photographic masks.

produce the desired array of identical patterns on the mask. A typical reduction sequence is shown schematically in Fig. 10. If the artwork is larger that 200:1, an additional reduction step may be needed prior to the final reduction. The intermediate transparency is usually a 10:1 image of the artwork on a fairly large photographic plate. It is repeatedly photographed on the same plate, which is translated into the proper positions. This is done by special optical systems which combine the features of a camera with a precision X-Y table (step-and-repeat cameras). The result of this process is the so-called master mask, which contains the pattern array on a 2- by 2-in. photographic plate. Since emulsion plates are easily damaged and withstand only a limited number of contact printings, several submasters are made, and from these an even greater number of actual working masks is derived, as indicated in Fig. 10.

TABLE 2 Examples of Resolution, Useful Field, and Artwork Requirements for Photographic Masks

	$1\ \mu$	$2.5\ \mu$	$5\ \mu$
Smallest line width desired in pattern:			
Lens resolution needed to resolve, lines/mm:			
1:1 mask	500	200	100
10:1 intermediate transparency	50	20	10
Smallest line width, mm, to be cut on artwork:			
200:1	(0.2)	0.5	1.0
500:1	0.5	1.25	2.5
1,000:1	1.0	2.5	5.0
Size of square pattern area to be processed:			
On mask (chip size), 10^{-3} in.	50	100	200
On 10:1 intermediate transparency, in.	0.5	1.0	2.0
On artwork 200:1, in.	10	20	40
On artwork 500:1, in.	25	50	(100)
On artwork 1,000:1, in.	50	(100)	(200)

To visualize the lens requirements in photomask fabrication, resolution and field sizes associated with several small line widths and chip sizes are listed in Table 2. For the initial reduction steps which produce the 10:1 intermediate transparency, the resolution requirements are fairly modest and rarely exceed 50 lines/mm. If the chip size is not greater than 0.2 in., a useful field of 2 by 2 in. is satisfactory. The demand in the final reduction step are far more stringent. First it should be noted that artwork of at least 500:1 is needed to reproduce 1-μ-wide lines with 5% cutting error or less. This is not possible in conjunction with larger chip sizes which exceed the capability of the coordinatograph. More stringent, however, is the necessity to combine resolution of several hundred lines per millimeter with useful field sizes of 0.200 in. As shown in the previous section, high resolution is contingent upon a *large* numerical aperture, whereas the useful field is greater for *small* apertures.

Optical systems with resolution of several hundred lines per millimeter have been available for some time in the form of microscope objectives. However, their useful fields are too small for mask fabrication.[27,35] An alternate possibility are 8- and 16-mm movie-camera lenses, but their fields, too, are only about 1 to 2 mm.[35] Camera objectives of 35 mm are more favorable, and certain types have been used in step-and-repeat systems.[27] However, all these lenses are not fully optimized for photoreduction purposes. Furthermore, the lens requirements are continually becoming more stringent. The development of large-scale integration tends to push line widths down and chip sizes up. In response to these needs, new types of optical systems have been developed which try to combine high resolution with large useful fields. They are often referred to as *process lenses*.

Contrary to camera lenses, process lenses are corrected for only one wavelength. If they are intended for the exposure of photographic-emulsion plates, the aberrations are minimized for the 5,460-Å mercury line. They have comparatively large apertures of typically $F/1.8$ to $F/3$. The trade-off between resolution and useful image field is demonstrated by Fig. 11. The curves of Tibbets and Wilczynski[36] show the loss of resolution with increasing contrast. The measured performance of the commercial process lenses[37] is similar to the values of the design curves from Ref. 36. Efforts to push the design limits upward by increasing the apertures and the focal lengths continue and promise to be successful.[35,38] Desired are 1- to 4-in.-diameter image fields with resolutions of at least 200 lines/mm for large-scale integration, and greater image

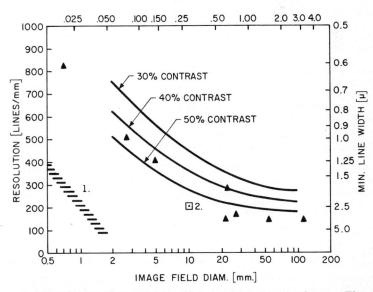

Fig. 11 Relationship between resolution and image field of process lenses. The curves are lenses designed by Tibbets and Wilczynski[36] for $\lambda = 5,460$ Å and 10:1 reduction. The triangles indicate the actual performance of eight commercial process lenses (after Dey and Harrell[37]). Region 1 marks the performance of low-power microscope objectives,[27,35] and point 2 is a 35-mm camera lens.[27]

fields than presently available for $1-\mu$ and smaller lines in projection printing [see Sec. 4c(1)].

The mechanical design of precision reduction cameras has been reviewed by Cooperman.[39] *First-reduction cameras* which are commercially available have been listed by Maple.[27] Their basic elements are the camera body with the lens, an illumination system, the artwork holder (copyboard), and a precision focusing mechanism. Typical lenses have focal lengths of 105 to 155 mm[40] with resolutions of 75 to 150 lines/mm.[39] Although the reduction ratio may be variable, the lenses are fully corrected for only one object distance. The size of the copyboard matches the size of the artwork and may range from 36 by 36 in. to 72 by 72 in. It must hold the artwork perfectly flat and is illuminated from the back, which yields higher contrast than photography in reflected light. Large-area light sources behind a diffusing screen are preferred over point sources because they do not require critical alignment. A disadvantage is the loss of 95% of the light emerging from the transparency.[33] A microscope and micrometer usually serve to adjust the focusing mechanism. All these parts are mounted on a massive frame whose rigidity and vibration-free mounting are most important to maintain the reduction ratio and focus. Some cameras offer object-to-lens distances up to 4.5 m (14 ft).[39]

Step-and-repeat systems are precision reduction cameras with an X-Y table mechanism to position the photographic plate with respect to the image. The first model was built by Helmers and Nall[41] in 1959 for the specific purpose of making microminiaturization masks. A number of commercial models have been developed since that time.[27] Usually, these cameras have reduction ratios of 10:1 or 4:1 and employ lenses with 28- to 75-mm focal length.[40,42,43] Their useful image fields are about 10 mm or smaller, and resolution is of the order of 200 to 400 lines/mm. The 10:1 (or 4:1) transparency is back-lighted from a point source with a reflector and condenser lens. This arrangement yields more uniform illumination and allows for shorter exposure times than diffuse lighting, but it requires precise alignment on the optical axis. The X-Y table holding the photographic plate moves on precision-bearing cross slides which are driven by lead screws and motors. The lead screws determine the accuracy with which the motion can be controlled.[25] On some of the larger machines with X-Y travel up to 30 in., the repeatability of the table position is of the order of 5 to 25 μ. For shorter travel distances such as 2 in., a repeatability of ± 0.13 μ (5 μ in.) has been claimed,[43] but this appears somewhat optimistic. The sequence of events is controlled by a punched tape. The program determines the length of X or Y travel— for example, in increments of 12.5 μ (0.0005 in.)—the stopping, and the exposure between motions.

A more recent development is the *photorepeater* of the D. W. Mann Company. In this machine, the photographic plate moves under a lens which resembles an inverted microscope objective and therefore yields high resolution. Its special features are very short exposures of about 5 μs duration by means of a xenon flash lamp. The distance of travel during such a brief interval is so short that the plate motion need not be stopped. Successive exposures either occur at predetermined intervals or are controlled by tape. The stepping precision is 1.25 μ (50 μin.). Instruments with multiple objectives allow simultaneous exposure of several masks in a set.

Since pattern alignment becomes increasingly critical as device dimensions are reduced, new techniques to sense and control X-Y-table positions have been developed. These include optical scanners and laser interferometers.[25] Particularly the latter method permits very accurate length measurements over long distances. Repeatabilities of ± 0.25 μ (10 μin.) have been reported for a digital-controlled interferometric positioning system.[44] Another photorepeater with a laser positioning system is claimed to provide a positioning accuracy better than 0.1 μ (3 μin.).[45]

A further step toward automatic mask fabrication is the so-called *artwork generator*.[46] These machines do not require artwork cut into laminated plastic sheets but accept computer instructions to generate the 10:1 intermediate transparency directly. The design and operating principles of one model have been described by Cook et al.[47] Basically, it is a projection camera with an X-Y table to hold the photographic plate. Light beams of different cross sections are produced by programmed selection of simple geometrical patterns on a photographic plate in the aperture plane. By combination of suitably shaped light beams with appropriate table motion, a complete circuit-chip pattern can be photographically synthesized at 10 × magnification. The resulting transparency is reduced and multiplied into an array mask via a 10 × step-and-repeat camera.

Artwork generation by this method is considerably faster than by the conventional cutting technique and quite accurate since the pattern is synthesized optically at 10 × magnification. It lends itself well to highly complex patterns as required for large-scale integration, where the cutting process is very tedious and likely to introduce errors. However, considerable effort has to be devoted to devising a machine language suitable for instructing the artwork generator. This problem is also discussed in the article of Cook et al.[47] Commercial models are presently available from two companies.* One of these has been described in some detail by Beeh.[44] It includes an automated step-and-repeat camera with lenses for 10:1 and 4:1 reduction, an X,Y table operating over a 4- by 4-in. area with an interferometric positioning system, and a small general-purpose computer. The latter receives instructions from punched or magnetic tape. Ultimate line widths of $1 \pm 0.25 \mu$ are said to be attainable.

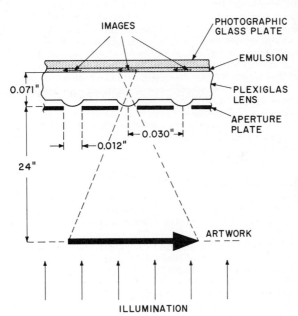

(4) Multilens Cameras Multilens cameras are an alternative to step-and-repeat cameras. They were developed in 1963 by Rudge, Harding, and Mutter[48] and are often referred to as "fly's eye" cameras. Their principle is illustrated in Fig. 12. The optics consists of an array of convex-plano lenses each of which forms a reduced image of the object in the back plane of the lens. The dimensions given in Fig. 12 are for 500 × reduction and image areas on 0.030-in. centers. Thus, the reduction of the original artwork can be done in one step, and no definition or contrast is lost by repeated transfer of images from one plate to the next. The aperture of the individual lenses corresponds to $F/4$, and image formation is nearly diffraction-limited. Focus-

* D. W. Mann Co., Burlington, Mass., and Optomechanics, Inc., Plainview, N.Y.

ing is assured automatically since the photographic plate contacts the back plane of the lens.

The lenses are relatively easy to manufacture and inexpensive. A polished copper block with an array of indentations made by a precision steel ball serves as the form for compression molding of the Plexiglas (methyl methacrylate) lens. A slight misplacement of an indentation is not critical because all patterns from subsequent masks will have an identical error. Thus, registration within a mask set is inherently assured. The apertures are punched into 0.001-in. steel shim stock. The lens and photographic plate are mounted on a rugged vertical shaft attached to a light box which illuminates the artwork.

The resolution of the lens shown in Fig. 12 is 400 lines/mm on the optical axis but drops to about 100 lines/mm toward the periphery. Barrel distortion becomes increasingly noticeable as objects are removed from the center. For fine line patterns, the useful image field is at best 0.3 mm (0.012 in.) in diameter. Within this circle, however, transistor contacts in the form of 3-μ-wide lines are possible.

Arrays with up to about 1,000 lens elements have been made. By covering some of the aperture holes and photographing different artwork in succession, mixed pattern arrays may be produced. This allows incorporation of registration marks or test sites into the masks. The time savings compared with the 3 to 4 h required for step-and-repeat processing are substantial. Limitations are the fixed center distances of the small lenses and increasing distortions when trying to use larger lenses and cell sizes. A more sophisticated multilens consisting of several lens elements has been designed by Dill.[49] It was calculated to give diffraction-limited performance at $F/4.5$ over an image area of 2 mm (0.080 in.). However, with the demand for larger chip sizes and smaller, more accurately controlled line dimensions, the utilization of multilens cameras has sharply decreased.

If the Plexiglas lens in Fig. 12 is removed, the artwork can be photographed through the aperture plate. In this case, one speaks of a pinhole camera. These have actually been used to produce masks,[50] but the resolution of the pinhole is limited and the reduction of the original pattern must be done in two steps (see Ref. 23, p. 46). Comparative tests with and without the Plexiglas lens yielded inferior results from the pinhole-camera technique.[48]

(5) Photographic-emulsion Plates   Photographic plates capable of high resolution and contrast are needed for the intermediate transparency and the masks. Silver-halide-emulsion plates with very fine grains are suitable for this purpose. They are often referred to as Lippmann-emulsion plates and are characterized by grains of 0.01 to 0.1 μ diameter embedded in gelatin.[51] The emulsion layers are 5 to 7 μ thick and absorb about 50% of the incident light. They are most sensitive in the 4,500- to 5,500-Å region. Since the depth of focus of the image is very small, the emulsion must be supported by an optically flat carrier. Glass plates are the preferred substrate material as they offer the highest dimensional stability and rigidity.

Lippmann-emulsion plates are commercially available in a variety of sizes and from different vendors.[51,52] High-resolution emulsions on flexible Estar-base films and on methyl methacrylate plates which can be drilled or cut are also available.[24] The first plates to be utilized for microminiature-device fabrication were Kodak spectroscopic plates,[26] but they have generally been replaced by special high-resolution (HR) plates. These are capable of resolving about 2,000 lines/mm, which is more than common camera systems yield. Procedures and chemicals for contact printing and developing of HR plates have been described by Holthaus.[26] For applications where only wide lines of 50 μ (0.002 in.) or greater are required, Strauss[53] recommends the Estar-base Kodalith Ortho film, type 3. This material has a maximum resolution of only 200 lines/mm,[26] but it has the advantage of conforming well to out-of-flat substrates.

Besides resolution, the most important property of the developed transparency or mask is the optical density in supposedly opaque and transparent areas. The light intensity transmitted by the photographic plate is proportional to the mass of reduced silver per unit area, which is determined by the amount of light received during the exposure. To evaluate the quality of the photographic image, the transmittance of different regions on the photographic plate can be measured with a microdensitometer[54] for image details down to about 150 to 200 lines/mm. A diffraction method for images of higher resolution has been described by De Belder et al.[52] The optical density D is expressed numerically as the logarithm of the ratio of incident (I_o) to transmitted (I_t) light:

$$D = \log_{10} \frac{I_o}{I_t}$$

HR plates generally contain enough silver to produce $D \geq 3$ in fully exposed and developed regions.[51] Achieving this density, however, requires local exposures which are often incompatible with those suitable for other areas on the same mask. Densities of 2.0 to 2.5 are therefore more typical for the opaque regions on a photomask.

The sharpest image with clearly defined edges would be obtained if the density changed abruptly to zero when going from an opaque to a transparent area. This is, unfortunately, not the case, because of the diffraction effects mentioned in Sec. 3a(2). Spreading of the opaque areas into adjacent regions is further enhanced by light scattering as the focused image is superimposed on the relatively thick silver emulsion. Consequently, the transition from opaque to transparent regions is gradual, and on masks with closely spaced lines there may be a veil of low-density silver covering all the interstices. This phenomenon is referred to as fog, and subsequent exposure of photoresist through such a mask is very critical if not impossible to adjust. The difference in optical densities between intentionally and inadvertently exposed regions is the contrast of the mask. It can be determined by microdensitometry and should be at least 1, preferably 2, to yield well-defined photoresist patterns.[55]

The proper exposure of photographic masks including coarse as well as fine lines with dimensions near the diffraction limit of the optical system is problematical. If the image of a line is thought to consist of dense arrays of Airy disks, it will be understood that wide lines require less exposure than narrow lines because more disks overlap in the former case than in the latter. Furthermore, narrow lines of the order of 10 μ or less are subject to relatively significant dimensional variations depending on the intensity and duration of the exposure. The reason is that the amount of light diffracted out of the dark and into the adjacent transparent regions is not negligible compared with what the dark areas receive. Thus, overexposure causes line broadening while underexposure yields too narrow lines. The effect of diffraction as determined by the aperture of the optical system on the width of 1- to 15-μ lines is very convincingly demonstrated in Ref. 32.

To minimize diffraction effects, the HR plates are usually exposed for optimum sharpness and definition of the finest lines in the pattern. If such lines are viewed microscopically at 1,200 times magnification, it can be recognized that they consist of unconnected silver particles of about 0.25 μ size.[55] Although thin photoresist coats are theoretically capable of resolving this structure, the slightly divergent light normally used as well as diffraction and scattering effects prevent the formation of discontinuous lines. However, at optimum conditions for fine lines, the wider ones are overexposed. Consequently, they lack definition and have a halo around them which becomes more noticeable when making resist exposures through such a mask.[55]

The problem of photographing fine lines near the diffraction limit has been discussed by Altman.[56] Proposed solutions are to either put a small neutral-density filter behind the wide lines prior to photographing (masking), or sequential and graded exposure of different mask regions. Neither approach is very practical, but the former one has been implemented in an experimental way. Hance[57] coated the entire artwork surface with a film of dyed photoresist, thereby providing a uniform low-density background. This film was then stripped in the areas of small openings or lines.

High-resolution plates are very delicate and easily damaged. Upon receipt, they may be contaminated with fibrous material and require careful cleaning.[58] Stringent inspection prior to use is recommended to reduce yield losses later on.[59] Clean processing techniques and suitable laboratory equipment have been described by Levine.[60] Special care is indicated in handling the master plates since one damaged mask may necessitate fabrication of an entire new set to ensure registration. Rules to be observed in contact printing and problems arising from lack of planarity have been discussed by Geikas and Ables.[61]

The working plates are subject to considerable wear and damage since they must contact the substrate. During contact printing, they are pressed against the photoresist film, which often leads to transfer of photoresist particles to the emulsion mask. This is an important mode of mask deterioration since the photoresist particles cannot be readily removed. Other causes are scratches in the emulsion from hard particles or structural irregularities in evaporated or sputtered films protruding through the photoresist layer. Consequently, the life of the working masks is limited and ranges from 20 to 50 contacts depending on the conditions of the substrate and the care exercised. Techniques to extend the useful life of emulsion masks have occasionally been reported. These consist of applying thin layers of commercial waxes[26] or a soft thin plastic foil[62] on the emulsion side of the mask to serve as a cushion between its surface and the rough spots on the substrate. However, these methods have not found wide acceptance.

Custom-made photomasks can be purchased from several companies. If the original artwork is supplied, the cost for a master plate may vary from $50 to $200 depending on the complexity and size of the pattern, the quantity ordered, and other factors. Copies of the master mask cost about $5 per plate. A complete set of integrated-circuit masks may consist of 4 to 20, more typically of 7 to 10, different plates. Such a complete set may cost as much as $20,000 to $50,000 with the fabrication of the artwork included.[45] Thus, masks contribute substantially to the cost of integrated circuits.

A new development in the fabrication of emulsion masks has been reported by Kerwin and Stanionis.[63] The technique is called image integration because it permits composition of photographic images in subsequent steps. The carrier of the image is a glass plate with a gelatin emulsion. Sensitizer and metal salts are incorporated by immersion of the plates in solutions. Exposure and development produce an image of silver amalgam. Suggested uses of this technique are registration tests of mask sets by making a composite image through subsequent exposures, the filling-in of pinholes in dark mask areas, and pattern alterations by adding new features to an existing mask.

(6) Metal-film Masks Given the optical systems developed for emulsion mask fabrication and a high-resolution photoresist, it is natural to think of masks where the pattern is etched into a thin metal film. Published accounts of this approach have so far been confined to chromium, although it is not the only metal which yields firmly adherent and hard films. The early development of such chromium masks has been reviewed by Stelter.[64] Process descriptions have also been published by Rogel[65] and by the Eastman Kodak Co.[66] The metal is evaporated in a high-vacuum system onto a thoroughly cleaned, optically flat glass plate. One recommendation is to deposit the chromium in two coats with vigorous cleaning (swabbing) after the first coat to uncover potential pinholes due to weakly adherent spots. The metal plate is then covered with photoresist and ready for exposure. Such metal-clad plates with or without photoresist coat are commercially available. The chromium film is about 700 Å thick and yields an optical density greater than 2.2.[66]

Photoresist-coated chromium plates may be exposed either by contact printing with silver-halide-emulsion masks[67] or by projecting images directly onto the plate.[64,66,6] In both cases, the optical systems designed for halide-emulsion-mask exposure are not directly applicable because they are designed for a wavelength outside the sensitivity range of most photoresists. While mercury lamps are suitable sources of illumination optical filters which transmit either the 4,358-Å or the 4,047-Å emission line are required. The camera lenses must be corrected accordingly.

If halide-emulsion masks are used to expose chromium masks, there is no gain in resolution. However, by proper exposure it is possible to reproduce the pattern without the variable optical density found at the line edges on emulsion masks. The chromium film is either completely etched off, or it retains the initial thickness in the protected regions. Thus, chromium masks offer good contrast and edge definition

While some of the optical defects in the emulsion mask may be transferred in the printing process, this is not always the case. Many emulsion-mask defects are either too small or do not have sufficient photographic contrast to be resolved by the photoresist. A major advantage of chromium masks is their greater durability. Factors contributing to their longevity are the hardness, the excellent adhesion, and the chemical inertness of the metal film. The former reduces the probability of introducing scratches into the pattern; the latter two factors permit the removal of transferred photoresist by solvents which would destroy emulsion images. Claims made for the useful life of chromium masks vary from 100 to 500 contact exposures, depending on the conditions of use and the skill of the operator.

The photoresist films which protect the chromium film during the pattern etch are only about one-tenth as thick as a silver-halide emulsion, and they are not inherently grainy. Therefore, chromium masks are potentially capable of higher resolution than Lippmann plates. However, to realize this advantage, the pattern image must be projected directly onto the resist-coated chromium plate by optical systems capable of the desired resolution. Inverted microscope objectives of very high theoretical resolution but with extremely limited fields have been successfully used. Reported pattern resolutions range from 900 to 1,500 lines/mm corresponding to 0.6- to 0.3-μ-wide lines.[64,66,68] Projection-exposure techniques in general and their specific problems are more fully discussed in Sec. 4c(1).

b. Photoresists

There are many organic compounds whose structure and solubility change when exposed to light, particularly in the ultraviolet region. The first materials found to have this property were natural products such as fish glue, asphaltum, sugars, or gelatin sensitized with dichromates.[69] Organic colloids of this type have been used for some time to produce scales and graticules, and for other photoengraving applications. In addition to being light-sensitive, a practical photoresist system must also have the ability to form well-adherent uniform coats which are not physically or chemically destroyed by the etching process. Lastly, selective solvents are required to develop and finally remove the resist pattern.

The development of modern photoresist systems which possess all these properties in a highly controlled and reproducible way originated at the Eastman Kodak Research Laboratory. The latter formulated a light-sensitive polymer for photoengraving purposes in 1953 which was well qualified by its excellent shelf life and broad latitude to variations in exposure and development, and which was similar in chemical nature to the photoresists used later in microminiaturization.[70] In the following decade, other photoresists were developed by several companies, so that there is a considerable choice of products. Their compositions and solvent systems are proprietary and generally not publicized. However, the nature of the constituents can often be inferred from the patent literature of the manufacturing companies. In addition, several users concerned with the consistency of the products and reproducibility of their properties have analyzed the most widely used photoresists and published their results. Besides these relatively well known materials, there are many other light-sensitive systems which have not yet found commercial applications. Examples are the photosensitive dyes whose reactions have been described by Oster.[71]

Generically, there are two types of photoresists. They differ in their responses to light and the solubility behavior induced toward the developing solvent system. This is illustrated in Fig. 13. Materials which are rendered less soluble by illumination yield a negative pattern of the mask and are called negative photoresists. Conversely, positive photoresists become more soluble when subject to light and therefore yield a positive image of the mask. Both types of resist are utilized in practice and have their peculiar advantages and limitations.

(1) Negative Photoresists Commercially available negative photoresists, recommended diluting agents (thinners), and developers are listed in Table 3. Since the resist manufacturers do not release chemical and compositional data, the users have to rely on the reagents and recipes furnished by the respective vendors. Practical information of this type is readily available. The Kodak products have been in

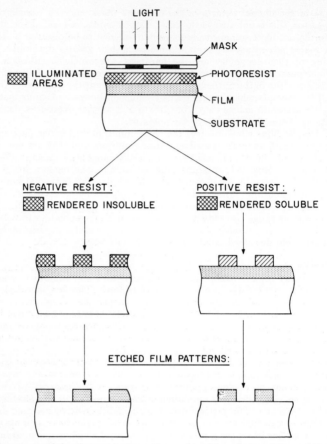

Fig. 13 Exposure and development of negative and positive photoresists, and the resulting etched film patterns.

widespread use, and a wealth of information and experience with Kodak resists has been published. This is not true for the photoresists of other manufacturers, and the discussion of negative resists, their nature and behavior, applies primarily to Kodak materials.

The principal constituents of a photoresist solution are a polymer, a sensitizer, and the solvent system. The polymers are characterized by unsaturated carbon bonds capable of reacting further and forming longer or cross-linked molecules. This reaction, however, must be stimulated by energy transferred by a sensitizer. The latter are chromophoric organic molecules which absorb and reemit light quanta. The degree to which reaction and insolubilization occur depends on the exposure of the resist film. With the exception of Emulsitone No. 150, which requires mixing of the resin solution and the sensitizer just prior to use, polymerization of the resists while still in solution is negligibly slow at the recommended storage temperatures. Hence, resists may be kept in brown bottles for long periods of time. Small amounts of antioxidant compounds to stabilize the solutions further, and surfactants to enhance wetting of the substrate surface are sometimes added.

The solvents used to keep the polymers in solution are mixtures of organic liquids. They include aliphatic esters such as butyl acetate and cellosolve acetate (2-ethoxy ethyl acetate), aromatic hydrocarbons like xylene and ethyl benzene, chlorinated

TABLE 3 Negative Photoresist Systems and Their Suppliers

Resist	Thinner	Developer	Supplier
KPR	KPR	KPR	Eastman Kodak Co., Rochester, N.Y.
KPR 2	KOR	KOR	
KPR 3	KOR	KOR	
KPL	KPR	KPR	
KOR	KOR	KOR	
KMER	KMER	KMER	
KTFR	KTFR	KTFR	
DCR 3140	3140	Dynachem	Dynachem Corp., Downey, Calif.
DCR 3154	3154	Dynachem	
DCR 3118		Dynachem	
DCR 3170		DCR 3170	
Waycoat No. 10	No. 10	Waycoat	Philip A. Hunt Chemical Co., Palisades
Waycoat No. 20	No. 20	Waycoat	Park, N.J.
Emulsitone No. 150		deionized water (50°)	Emulsitone Co., Livingston, N.J.

hydrocarbons like chlorobenzene and methylene chloride, and ketones such as cyclohexanone. The same solvents are also used for thinners and developers. The only water-base resist is Emulsitone No. 150, which is believed to be a mixture of polyvinyl alcohol and a dichromate. It is distinguished from other resins by its hydrophilic nature, which ensures good adhesion of the resist coat to SiO_2 surfaces and therefore is claimed to afford long etching times.

The composition and properties of negative photoresists vary according to the manufacturers. Little is known about the Waycoat and Dynachem products except that the latter is believed to be an aliphatic polyester of phthalic acid. The Kodak products, however, have been widely investigated, and their physical properties as determined by the manufacturer are shown in Table 4. General discussions pertaining to the evolution of these resists and their intended applications have been published by Bates[70] and Martinson.[55]

The KPR (Kodak photoresist) family contains polyvinyl cinnamates with sulfur compounds of the naphthothiazol group as sensitizers.[72-76] The different varieties are distinguished primarily by their solids contents and viscosities, as shown in Table 4. Since these properties determine the flow characteristics and thus the thickness of the coating, their applications vary accordingly. Only KPR and KPR 2 are intended for thin film etching. KPL (Kodak photosensitive lacquer) is of the same chemical nature except for a more volatile solvent system (methylene chloride). Its chemical similarity to KPR is reflected in the identical index of refraction and spectral sensitivity. Like KPR 2, KPL was formulated for the etching of relatively thick films such as copper-clad circuit boards.

KOR (Kodak Ortho resist) is thought to be a polyvinyl cinnamylidene acetate.[77] Contrary to the other photoresists, whose sensitivity is confined to the ultraviolet and adjacent blue region of the spectrum, the response of KOR extends farther into the green part of the spectrum. It was designed to yield faster exposures but is otherwise similar in performance to KPR.

The development of KMER (Kodak metal-etch resist) was motivated by the demand for thin high-resolution coatings with good adhesion to metal films. It contains a natural polyisoprene resin and an aromatic diazide as a sensitizer.[76,78-80] Because of its origin from a natural product, KMER solutions vary somewhat in purity and composition. This problem was overcome with the advent of KTFR (Kodak thin film resist), which is chemically almost identical but is derived from a synthetic polyisoprene product. The good adhesion to metal surfaces and the resis-

TABLE 4 Physical Properties of Kodak Photosensitive Resists

Property	KPR	KPR 2	KPR 3	KPL	KOR	KMER	KTFR
Viscosity at 25°C, centipoise	11.7–13.1	22.6–27.6	34.0–38.5	620–800	46–54	396–504	465–535
Solids content, weight %	6.4–7.6	8.4–9.2	9.1–9.9	23–27	9.2–9.8	24.7–27.3	27.2–28.8
Density at 20°C*	1.010	1.090	1.070	1.2406	1.040	0.901	0.893
Index of refraction of dried, unexposed films*	1.610	1.611	......	1.610	1.616	1.554	1.541
Surface tension,* dynes cm^{-1}	30.1	33.2	31.7	......	31.2	28.9	29.2
Spectral-sensitivity range, Å	2,600–4,600				2,500–5,500	3,100–4,800	

* These data are typical values and may vary from batch to batch.

tance to strongly oxidizing acidic and alkaline solutions exhibited by KMER have been retained in KTFR.

Since KPR/KPR 2 and KTFR/KMER are the most widely used negative photoresists in the formation of thin film patterns, their properties are of special interest. Data sheets pertaining to their electrical properties are distributed by the manufacturer, and several selected values are listed in Table 5. The breakdown fields are

TABLE 5 Electrical Properties of Kodak Photoresist Films

Property	KPR, KPR 2 unexposed	KTFR	
		Unexposed	Exposed and developed
Breakdown voltage, kV/0.001 in.	2.1 (2.7-mil film) 1.2 (6.8-mil film)	1.7 (1.8-mil film)	1.2 (1.8-mil film)
Dielectric constant (up to 10 MHz)................	2.9–3.3	2.4–2.6	2.4–2.9
Volume resistivity, ohm-cm......	....	1.8×10^{14}	2.3×10^{14}
Water absorption, %, at:			
20 % RH	0.2		
50 % RH	0.4	} Negligible	} Negligible
70 % RH	0.6		

about 3.10^5 V cm^{-1}, the dielectric constants are relatively small, and the insulation resistances are quite high. In contrast to KPR, which absorbs small amounts of water, KTFR is hydrophobic. Chemically, these photoresists have been characterized by a number of analytical techniques such as centrifuging,[78] spectroscopy,[79] chromatography, and classical chemical methods.[76] The results of a series of analyses performed by Levine et al.[76] are shown in Table 6. The prevalent mechanism of insolubilization in both resists is cross linking.

TABLE 6 Composition of KPR 2 and KTFR*

	KPR 2	KTFR
Solvent......................	86–87 % chlorobenzene 13–14 % cyclohexanone	~82 % o-, m-, p-xylene ~12 % ethyl benzene ~6 % methyl cellosolve
Polymer....................	Polyvinyl cinnamate	Partially cyclized poly-cis-isoprene with about 1 double bond among 10 carbon atoms
Avg mol. wt................	315,000–350,000	~120,000
Sensitizer....................	1.4 g per 100 g of polymer. Naphthothiazol derivative with 10 % sulfur	Aromatic diazide
Minority constituents.........	Polyvinyl alcohol, cinnamic acid, hydroquinone, traces of water	

* After Levine, Lesoine, and Offenbach.[76]

(2) Positive Photoresists Commercially available positive photoresists are listed in Table 7. The AZ formulations are based on a number of patents granted to the Azoplate Corp., Murray Hill, N.J.[81] From these it is surmised that AZ photoresists are Novolak resins with one of several possible naphthoquinone diazides functioning as sensitizers. The solvent systems include some of the same liquids as encountered among the negative resists such as cellosolve acetate, butylacetate, xylene, and toluene. Of the resists listed in Table 7, AZ-1350 has been shown to be suitable for fine-line etching of thin films. AZ-1350H is chemically identical but has a higher solids content and yields thicker coats. Other resists like AZ-340 are intended for deep etching and photoengraving. The Kodak autopositive resist KAR 3 has come out only recently, and actual working experience has not been published as yet. The exposed films are developed in an aqueous solution.

TABLE 7 Positive Photoresist Systems

Resist	Thinner	Developer	Supplier
AZ-1350	AZ	AZ	Shipley Company, Inc.,
AZ-1350H	AZ	AZ	Newton, Mass.
AZ-340	AZ	AZ-300	
AZ-111	AZ	AZ-330	
AZ-119	AZ-119	AZ-303	
AZ-345	AZ	AZ-303	
AZ-165	....	AZ-303	
KAR 3	KAR thinner	KAR developer conc.	Eastman Kodak Co., Rochester, N.Y.

Some of the physical properties of AZ-1350 and KAR 3 as given by their manufacturers are listed in Table 8. The low viscosity of the AZ resist is noteworthy and indicative of its relatively low molecular weight ($\sim$1,000). The solubility of the resin after evaporation of the solvent is strongly dependent on the functional groups present in the macromolecules. These groups are responsible for the initial insolubility of the resist film in the developer. The latter is a buffered aqueous solution of sodium hydroxide. When the resist film is illuminated, the sensitizer transfers energy to the functional groups of the polymer, which thereupon change, for example, into carboxyl groups, and render the resist film soluble.[82] The solubilization process is narrowly confined to the immediate vicinity of the absorbing chromophores. Therefore, the AZ-1350 resist yields very good resolution even in relatively thick coats. The manufacturer claims that 5-μ-wide lines can be developed in a 6-μ-thick coat.

TABLE 8 Physical Properties of Two Positive Photoresists

Property	AZ-1350	KAR 3
Viscosity, centipoise................	4.5–4.6	10.0–14.0
Solids content, wt. %	$\sim$19.5	25.5–27.5
Useful spectral sensitivity, Å........	3,400–4,500	3,200–4,600

c. Photoresist Pattern Formation

Application and processing of photoresist films are a highly empirical art. Although the same procedures are used everywhere, they vary in detail from laboratory to laboratory. Thus, when a photoresist processing facility is established, it is generally necessary to adjust the conditions and process parameters until satisfactory results for the purpose on hand are obtained. Regardless of the resist used, several precautions are universally observed. They pertain to the environment in which the operations are performed. Since dust and lint particles settling on substrate or resist surfaces are a potential cause for pinholes, their presence must be minimized. This is accomplished by performing all the critical operations such as substrate cleaning, coating, mask registration, and inspection in laminar-flow clean hoods. The chemical operations, particularly developing, rinsing, and resist thinning, involve organic solvents whose fumes are flammable and often toxic. Hence, they require an adequately ventilated work area.

Another critical factor is the humidity of the ambient. Coatings of the KPR group tend to leave thin residues in unexposed developed regions if they are subject to high relative humidity at any time in the application cycle. The adsorption of moisture on substrate surfaces prior to coating may weaken the adhesion of the resist film. On the other hand, an extremely dry atmosphere is inducive to electrostatic charging and furthers the accumulation of dust on the substrate surface. Consequently, most laboratories control the relative humidity at moderate levels between 30 and 50%.

Accidental exposure of resist coats to light containing wavelengths shorter than 4,800 Å produces "fogging," a thin veil of developed photoresist covering the entire substrate including the regions to be etched. Therefore, safelights must be used in all areas where resist coatings are handled. Suitable sources are gold fluorescent tubes or yellow incandescent lights. White fluorescent tubes with suitable yellow or orange filters are also widely used.

A well-defined etched pattern with controlled dimensions and free from defects is obtained only if the substrate surface is perfectly clean and dry prior to coating. Removal of particles and organic residues is commonly achieved by cleaning with solvents such as trichloroethylene. The substrates may be immersed in the solvent, sprayed, or put into a vapor degreaser. Solvent residues may be blown off with pure dry compressed air. If storage is necessary, the substrates should be protected against the accumulation of dust and adsorption of moisture. Heating at 200°C prior to photoresist application is often recommended to ensure a dry surface. If desorption of volatile adsorbates occurs later when the coated substrates are baked, the adhesion of the resist film is likely to be affected. As soon as a clean, dry surface has been obtained, the resist coat should be applied without delay.

The precautions outlined in the preceding paragraph are generally sufficient for freshly deposited thin film surfaces. However, if the resist application has been preceded by a chemical treatment, the film surface may be in a condition where good adhesion of the coating cannot be attained. In general, acid treatments and high-temperature annealing are inducive to good adhesion [see also Sec. 3f(2)]. Alkaline treatments often produce poor adhesion of the resist coat and should therefore be followed by a rinse in dilute acid whenever this is possible.

As stated earlier, the photoresists most widely used are the KMER/KTFR and KPR/KPR 2 systems. More recently, AZ-1350 has gained wider acceptance by virtue of its high-resolution capability. The choice among these resists depends on the materials to be etched, the chemicals involved, and the preference of the investigator. There are few absolute rules about resist selection. One is that the KMER/KTFR resists should not be applied to copper or copper-alloy surfaces because this metal affects the stability of the sensitizer.[83] KMER, although very similar to KTFR, is not as pure and has a higher content of gel and foreign particles. Hence, if it is to be used, it should be thoroughly purified by a process such as described in Ref. 78. KTFR is said to be more resistant to strong alkali and acid solutions than KPR; it can even withstand the attack of aqua regia long enough to permit etching of thin gold films.[62] The positive resists of the AZ family are not fully resistant to alkaline etches, although they are developed in a dilute sodium hydroxide solution. The resistance of AZ-1350 toward alkali etches can be enhanced by post-baking at temperatures approaching 200°C, but this makes the removal of the cured stencil mask more difficult. The Kodak autopositive resist KAR 3, too, should not be used with alkaline etch solutions.

(1) Application of Photoresist As received from the manufacturer, all photoresist solutions contain varying amounts of gel and foreign particles. If these are incorporated into the resist coat, they degrade the quality of the developed pattern considerably. Therefore, it is universally recommended to filter photoresist solutions prior to use.[84-86] This is commonly done by means of solvent-resistant millipore filters which are available in materials such as nylon, cellulose, and Teflon with pore sizes ranging from 14 down to 0.25 μ. To avoid clogging, two filtration steps are used.

First the large particles are removed by gravity or pressure filtration through a relatively coarse filter. This is followed by filtration under pressure through a fine filter having about 1-μ-wide pores. A widely practiced technique is to build the fine filter into the syringe with which the resist is dispensed onto the substrate. An alternate cleaning technique has been described by Taylor,[87] who removed globular particles from KMER by electrophoresis.

The most radical and effective cleaning procedure has been developed by one of the microminiature-circuit manufacturers.[78] It involves two chemical treatments in the form of liquid-liquid extractions, followed by centrifuging. Detailed analyses of the purified products have shown that this process removes particulate matter as well as trace metal impurities. The latter are present in KMER up to about 60 ppm, in KTFR below the 1 ppm level. The distribution of molecular weights and the concentration of sensitizer in the purified resists are not altered by this treatment. Such purified solutions yield photoresist coatings of greatly improved quality which has been demonstrated by dark-field and electron microscopy.[88] The technique is of particular value for KMER, whereas the conventional filtration methods are generally sufficient to purify KTFR.

Another property of commercial resist solutions which is often altered prior to use is their solids content. As shown in Fig. 14, the solids content determines the viscosity of photoresist solutions and hence their flow characteristics. To achieve coatings of controlled and uniform thickness, it is often desirable to lower the viscosity of the resist solution. The manufacturers supply suitable diluting agents (thinners), but some laboratories prefer to use pure solvents of their own choice such as toluene, xylene, and cellosolve. While diluting, it is essential to avoid introducing air bubbles as the escape of entrapped gases from the viscous solutions is slow. The practice is common with KMER/KTFR as these are the most viscous resists. A thinner for AZ-1350 photoresist is also available but rarely needed because the viscosity is low to

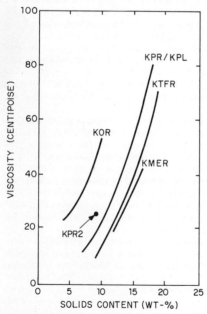

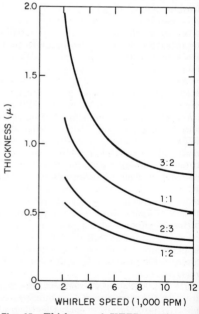

Fig. 14 Viscosity of Kodak photoresists as determined by solids content. (*After Damon,*[89] *copyright 1967, Eastman Kodak Co.*)

Fig. 15 Thickness of KTFR coatings as dependent on whirler speed. The curves are for different ratios of resist to thinner as indicated. (*After Damon,*[89] *copyright 1967, Eastman Kodak Co.*)

bégin with. Dilution is usually performed by mixing the resist and its thinner in certain empirically determined proportions. More positive control is obtained if the viscosity of the thinned product is measured as the control parameter. This requires accurately maintained temperatures since the viscosity of KTFR, for example, changes about 2% per degree.[86]

Methods of resist application to substrates range from spraying to dipping, flowing, and rolling.[24] In the manufacture of thin film circuitry, whirler coating has become the standard technique as it yields the most uniform and reproducible thicknesses. It is executed by placing the substrate on a motor-driven rotary vacuum chuck mounted in a bowl. The surface is then flooded with the liquid resist and the motor is turned on. The centrifugal forces distribute the viscous liquid evenly before much solvent evaporation takes place, and excess material is thrown off at the substrate edges. Application of the resist while the substrate is rotating is not recommended, as this yields uneven coverage. For square or rectangular substrates, low-speed whirlers of 50 to 1,000 rpm are recommended. These yield about 2.5-μ coatings at 75 rpm with full-strength KPR.[24] The coatings are thicker at the substrate edges and corners.

Most substrates processed for thin film circuitry are circular wafers, and these are coated in high-speed whirlers. Such instruments capable of coating one or more wafers simultaneously are available from several manufacturers. The speeds most often utilized are between 2,000 and 5,000 rpm. The resulting resist thickness is determined by the viscosity of the solution, the speed of rotation, and the whirler acceleration. In general, the thickness is a hyperbolic function of speed, as shown in Fig. 15. Similar results have also been found by other authors.[86] The thicknesses obtained from the more dilute formulations are the least sensitive to revolution speed, and spinning above 6,000 rpm has little effect on the coating.

Damon[89] conducted an extensive study comparing the thicknesses obtained on five different whirlers under seemingly identical conditions and found significant differences among these models. This led to the realization that the acceleration of the whirler, too, is an important control parameter. The time to reach the final speed of revolution varied from about 0.1 to 3.6 s among the whirlers tested. The fast-accelerating models gave somewhat thinner coatings, but more importantly, the uniformity of the thickness across the wafers was better. This observation has also been made by Schwartz.[86] Finally, the slightly thicker rim of photoresist along the wafer edge is reduced if fast acceleration is used.[89] Consequently, models with accelerations of less than 1 s are now generally preferred. The same considerations and similar whirler-speed curves apply to AZ-1350.

The photoresist thicknesses which are actually used to etch thin film circuits vary from 0.3 to about 2 μ. Thin coatings of 0.3 to 0.5 μ of KTFR are still effective etch masks in some but not all situations. A lower limit is encountered when diluting the resist too much, which leads to discontinuous films. Thick films of 1 to 2 μ give added protection against etch penetration and pinhole formation at the expense of slightly less resolution and some graininess. Such films are commonly prepared by applying two thinner coats on top of each other. AZ-1350 coatings suffer the smallest loss in resolution if made thicker; this is more fully discussed in Sec. 3f(1). A rule of thumb for Kodak photoresists is that the resist thickness should be about one-third of the width of the finest line.[55]

The last step in the application of resist coatings is the so-called pre-baking. The purpose of this operation is to drive out residual solvents which, if left in the film, may weaken the adhesion of the photoresist to the substrate. Incompletely baked samples may also have variable exposure requirements because residual solvents inhibit cross linking or conversion of functional groups. Excessive baking, on the other hand, has undesirable consequences such as "fogging" due to thermally activated cross linking in negative resists, and melting or decomposition in positive resists. The manufacturers of photoresists make definite recommendations as to the temperature and duration of the pre-baking cycles. The latter is, of course, somewhat dependent on the resist thickness. In practice, there is a tendency to use somewhat different conditions than specified. Table 9 shows recommended and extreme baking cycles.

TABLE 9 Pre-baking Cycles as Recommended by the Manufacturers and Extreme Cycles Mentioned in the Literature

Photoresist	Manufacturer's recommendations	Extreme cycles
KPR, KPR 2....	10 min at 120°C[24]	30 min at 110°C
KMER..........	10 min at 120°C[24]	
KTFR..........	At least 20 min at 82°C; baking above 104°C may affect adhesion[24]	30 min at 120°C
AZ-1350........	5–30 min at 60–65°C	15 min at 100°C

(2) Exposure of Photoresist Coatings The photoresist-coated substrates are exposed in a system which includes a light source, provisions for holding the substrate and mask in close contact, a table allowing X, Y, and rotational adjustments of the relative positions of substrate and mask, and a microscope to observe the latter. Binocular microscopes with split optics have been developed which allow simultaneous viewing of different portions of the surface. Proper mask registration is attained by checking pattern registration at high magnification in two fairly remote areas concurrently. The registration accuracy is, again, limited by the resolution of the optical system. It has been estimated that a minimum misregistration of $\pm 1.25\ \mu$ in the alignment of two subsequent patterns should be allowed for.[27] Commercial mask-alignment machines and their suppliers may be found in Maple's review article.[27] To avoid unintentional exposure during the alignment procedure, a second light source devoid of short-wavelength radiation or a suitable filter must be used. Particular registration problems are encountered if very dense positive-resist masks are utilized. These may make it impossible to view substrate areas large enough to discern previously etched patterns. To guard against this failure, suitable registration marks have to be designed into the entire mask set.

Before new masks are used to expose good substrates, it is advisable to check if they are free of defects and if the entire set registers well. Effective mask-inspection procedures have been discussed by Henriksen.[90] They include microscopical inspection at 600× magnification to verify critical dimensions as well as examination for pattern defects, contrast, edge acuity, fog, pinholes, and other flaws. The registration of a mask set and the absence of layout errors can be tested by superimposing all photomasks onto the same substrate and comparing the composite pattern with the original layout.

The photoresist may be exposed with any light source having a significant output in the near ultraviolet part of the spectrum. Large-area sources are rarely used and only in conjunction with large substrates. Since their light is diffuse, they do not resolve fine details and are acceptable only for rather coarse (50-μ or 0.002-in. and wider) lines. Fine-line patterns are exposed with point sources such as carbon arc, high-pressure mercury vapor, or xenon flash lamps which produce nearly collimated light if they are remote from the substrate. Collimation by lenses must be empirically adjusted because light which is too well collimated resolves minute defects in the mask or even silver grains, thus causing island or pinhole defects in the resist.[62] The uniformity of the light intensity across the substrate area is generally better with remote point sources than with collimating lenses unless the latter are larger than the image field and equipped with an aperture to block out the peripheral portions of the beam.[91]

Commercial ultraviolet lamps are not identical in light output, and their spectral distribution changes with age. Correct exposure is also dependent on the type of mask and the nature, thickness, and processing variables of the resist. According to Damon,[91] proper exposure of typical KTFR coats requires light energies of about 100 mW-s cm^{-2}, while KPR needs about twice as much. Htoo[92] found somewhat lower energies but the same order of sensitivities. The positive resist AZ-1350 requires a slightly higher light energy than KPR.[92] The investigation of Damon[91] also showed that the product of light intensity and exposure time is not a constant, but that satisfactory exposure at relatively low intensities requires disproportionately longer times. Exposure times of 1 to 10 s permit sufficiently accurate timing as well as reasonably fast processing and are commonly used. Considering the required exposure energies, the light source must be capable of yielding irradiation intensities of at least 10 mW cm^{-2} at the substrate surface.

The fine adjustment of an exposure system has to be done empirically by means of test patterns.[32] To ensure reproducibility, the power supply of the light source must be regulated. For good fine-line definition, the exposure energy has to be controlled to within ±5 to 10% of the optimum value.[55] Hence, the light intensity at the substrate surface should be either continually monitored or checked periodically. Although the resists have a wider exposure latitude, diffraction effects and mask quality necessitate this precision. Overexposure, for example, produces cross linking of negative resists in regions under the mask outside of the maximum density pattern. Light spreading under dark mask areas is also caused by diffuse scattering of light in the resist and by reflection from the film or substrate surface underneath. The net effect of these conditions is line broadening, which may amount to as much as 2.5 μ (100 μin.). On the other hand, underexposure is even more dangerous because only the resist near the surface cross-links. Hence, the entire pattern may be washed off when the image is developed. With positive resists, the result is the opposite, namely, an insufficiently exposed pattern leaving behind a veil of insoluble polymer. These considerations also show why the optimum exposure is thickness-dependent.

Other potential causes of line broadening are lack of substrate planarity or smoothness and insufficient contact between mask and resist surface. Since the exposing light sources are usually somewhat divergent, it follows from geometrical optics that any gap between the emulsion of the mask and the resist coat causes spreading of the exposed region. The extent of this effect is system-dependent and proportional to the mask-to-substrate spacing. Total broadening (both line edges) of about 6 μ or 250 μin. has been observed in one particular case, where mask and substrate were intentionally separated by 25 μ (0.001 in.).[93] With perfectly smooth surfaces, the effect is minimized by the use of vacuum to hold the mask against the substrate. The latter deforms elastically when pressed against the mask.

A peculiarity of KMER/KTFR resists is the so-called oxygen effect. It arises from the fact that the sensitizer in these materials decomposes without cross linking the polymer if exposure takes place in the presence of oxygen.[62] This reaction is essentially confined to a thin layer of resist near the free surface, which therefore remains soluble in the developer. Except for variable exposure requirements, the oxygen effect may not be noticed when working with resist coatings of 1 μ and thicker. However, coatings of 0.5 μ or less may be thinned down so much that they do not offer enough protection in the subsequent etch. The oxygen effect is best avoided if the vacuum frame holding mask and substrate together establishes reduced pressure at the resist-to-mask interface. This is not possible in projection exposures, and covering the resist surface with nitrogen gas is recommended[55] but not as effective as vacuum.

Even the most intimate contact and reduced pressure do not eliminate the optical interface between the mask and the resist coat. This can be accomplished by interposing a soft medium of suitable index of refraction. Exposure with oil immersion has been tried and is claimed to yield fewer defects in the resist pattern since scratches in the glass mask are thereby made invisible.[62] However, this has not become a widespread practice.

(3) Photoresist Development and Post-baking Although the development of photoresist images may be performed by immersion and soaking with mild agitation, spraying of the developer onto the coated substrate is widely preferred and recommended. The reasons are the continued renewal of the solvent and the gentle brushing action of the spray, both leading to shorter development times and hence less swelling and better dimensional control of the remaining pattern.[55] In general, the procedures involve a short exposure (10 to 60 s) to the spray followed by several nonaqueous rinses. Residual solvent is blown off the surface with pure compressed air or nitrogen. The operations are most conveniently and controllably performed in automatic spray-developing machines which are commercially available.* These are capable of processing several

---

* See, for example, R. Price, *SCP Solid State Technol.*, **9**(8), 47 (1966).

wafers simultaneously through a number of automatically timed developing and rinsing cycles. The solvents are dispensed through built-in spray nozzles while the wafers are held on vacuum chucks.

Photoresist development procedures are distributed by the manufacturers for their own chemicals (see Tables 3 and 7). Some alternate processes and published experience are of interest in this context. KPR, for example, may also be developed in a trichloroethylene vapor degreaser equipped with a spray nozzle. Timing is critical (about 15 s) as the resist coat swells and softens strongly in this solvent. The procedure is definitely not applicable to KMER/KTFR. The subsequent rinse involves mixtures of KPR developer with absolute ethyl or isopropyl alcohol.[24]

KMER may be developed in xylene,[55] a mixture of Stoddard solvent (a high-boiling-point petroleum fraction) and xylene,[85] or in KMER developer with 40% dipropyl carbonate added.[87] The latter solution is claimed to yield superior line resolution. The rinsing solutions are mixtures of KMER thinner with ethyl or isopropyl alcohol, or 90% Stoddard solvent with 10% isopropyl alcohol.[85]

KTFR, although similar to KMER, should not be subjected to KMER developer as the latter is too strong and rapid. The commercial KTFR developer contains some xylene, too, but it is thinned with petroleum fractions. An alternate developer is Stoddard solvent. Suitable rinses are Stoddard solvent–isopropyl alcohol mixtures or butyl acetate.[62]

Typical development times of AZ-1350 whirler coats in the full-strength aqueous AZ developer are of the order of 5 s if executed at 16 to 22°C as directed by the manufacturer. They are prolonged by a factor of about 10 if the developer is used at half strength to facilitate more accurate timing. The development times depend on the thickness of the coat. The surface is rinsed in deionized water.

In conjunction with negative resists, leaching effects may be encountered.[62] These arise from small quantities of resist in regions which have not become fully insoluble because of underexposure. The effect is more likely to be encountered with thick rather than thin coatings and with developing treatments in "active" solvents such as trichloroethylene. These leach out strings of loose polymer from the swelled image which may extend into areas intended to remain clear. Rinsing in nonsolvents such as alcohols shrinks the swelled images rapidly and confines leaching to the surface.

Post-baking of the developed resist images is generally performed to evaporate residual solvents, to enhance the chemical stability of the polymer coat, and to further its adhesion to the underlying film. As in pre-baking, the manufacturers recommend certain cycles,[24] but there is a tendency to exceed the specified temperatures. Experience with the particular system under investigation is the best guide in this question. Post-baking temperatures in excess of those used to dry the substrate surface prior to coating should not be applied to avoid gas desorption at the interface and loss of adhesion.[55]

The recommended minimum baking cycles for KPR, KPR-2, KMER, and KTFR are 10 min at 120°C. KPR coatings are thermally the most stable and can withstand up to 5 min baking at 260°C.[24] Heating cycles of half an hour and more at 180°C are fairly common, and Kelley[94] used even 190°C. With KMER or KTFR coats, the manufacturer cautions against exceeding 148°C, since oxidation begins above this temperature.[24] However, baking cycles of 10 to 20 min at 160°C seem to be safe or even yield superior performance.[62]

The literature pertaining to the AZ-1350 resist claims that no bake is required if mild chemical etchants are used. The advantage of this omission is ease of removal of the polymer after etching. Stronger etchants such as buffered hydrofluoric or nitric acid solutions, however, should be preceded by baking at 150°C for 30 min. The highest chemical resistance is obtained by baking at 200°C, but the removal of the polymer mask thereafter is difficult.[95]

d. Chemical Etchants for Thin Film Patterns

The reagents used to etch thin film patterns are principally the same chemicals which are known to dissolve the materials in bulk form. For thin film etching, these chemicals are applied in fairly dilute solutions to reduce the etch rate and avoid attack of material under the polymer stencil mask. While aqueous solutions are by far the most common, alcohols like methanol, ethylene glycol, or glycerin are sometimes added to moderate the reaction by reducing the dissociation of the active chemical or by increasing the viscosity of the solution. In general, the slower etches produce less undercutting than the rapid ones. If a choice between different reagents exists, metal ions changing their valency state like iron:

$$2Fe^{3+} + Cu \text{ (film)} = 2Fe^{++} + Cu^{++} \text{ (solution)}$$

are preferable to oxidizers which evolve gas as, for example, dilute nitric acid:

$$2HNO_3 + 6H^+ + 3Cu \text{ (film)} = 2NO \text{ (gas)} + 4H_2O + 3Cu^{++} \text{ (solution)}$$

Although NO is a colorless gas, the evolving fumes are usually brown because NO_2 forms upon contact with air. The reason for preferring reactions of the former type is that gas bubbles often cling to incompletely etched film surfaces and thus prevent clean pattern definition. Another criterion in selecting reagents and solvents is their effect on the photoresist mask.

Considerable experience has been accumulated with thin film etches for various materials. Only the smaller part of these empirical data has been published since etch compositions are often carefully kept laboratory secrets. The following review of etchants is therefore far from complete, and it should be realized that the solutions listed may vary in composition from laboratory to laboratory. It is good practice to experiment with different etch compositions and adjust them until the best results for the task on hand are obtained. As for etching techniques, immersion with or without agitation is most common. Spray etching has the advantages of continually contacting the film surface with fresh solution of constant concentration, good wetting of areas between closely spaced mask portions, and rapid removal of the reaction products. It is often considered to yield better definition and less undercutting than immersion etching. However, a well-ventilated and corrosion-resistant work area for spraying of reactive chemicals is not always available.

When dealing with freshly deposited and coated thin films, special cleaning techniques prior to etching are rarely required. The necessity may arise, however, if multilayer films are etched subsequently in different baths. An example is a copper film with a flash of chromium on top. The chromium etch may leave the copper surface stained, thus causing uneven attack in the following copper etch. A non-oxidizing acid treatment alleviates this problem. Commercial copper cleaners such as Neutra-Clean,* a mildly acidic solution containing chelating compounds, are also eminently suitable but usually require diluting.

Unusual difficulties are sometimes encountered when the metal film to be etched is in electrical contact with another metal which is also exposed to the electrolyte. Under these conditions, the more electropositive metal becomes the anode and goes into solution while the less electropositive metal acts as the cathode and merely discharges hydrogen ions. Thus, one may find well-proved etchants to be inert toward metals that are normally attacked rapidly. The situation is likely to be encountered with thin film resistors having terminals of a different metal. The fact that the terminal surface may be resist-coated is no insurance against electrolytic-cell formation since the exposed sides (film thickness) are sufficient to induce passivation. An example of this type encountered by one of the authors is the passivation of Cr-SiO cermet films in ferricyanide etch due to the presence of molybdenum. The inverse problem of corrosion of metal films in normally nonreactive solutions due to the presence of another (cathodic) metal may also be observed.

(1) Aluminum Aluminum films dissolve readily in alkaline as well as in acidic etches. KTFR baked at 120°C is said to give the least undercutting,[83] but excellent results are also obtained with AZ-1350 in acid etches.

An aqueous solution of 20% NaOH to be used at 60 to 90°C is given in Ref. 83. This is probably a rapid etch of little use for thin film etching. One of the authors found a mixture of equal parts of 26 weight % aqueous NaOH and methyl alcohol

* Shipley Company, Inc., Newton, Mass.

applied at 30°C to be still very rapid with a strong tendency to undercut when used as an immersion etch. In general, NaOH solutions are likely to affect the adhesion of the resist coat and yield poorly defined line edges, particularly on thicker (1 μ) films. Gas evolution and bubbles clinging to the surface are also a problem.

Hydrochloric acid etches are listed in Ref. 83: 1 part concentrated HCl + 4 parts H_2O, and in Ref. 82: 20% HCl at 80°C. The latter seems to have given satisfactory results in fine-line etching. Mixtures of concentrated HCl and commercial $FeCl_3$ solutions are also recommended[83] but are probably too fast for thin films.

Another widely used acid system consists of 80 to 95 ml H_3PO_4, 5 ml of HNO_3, and 0 to 20 ml of water. If used around 40°C, the etch rates vary from 1,500 to 2,500 Å min^{-1} depending on the water content. Since gas evolution occurs, some form of agitation to remove bubbles from the surface should be used.

(2) Chromium The behavior of chromium in acids is characterized by surface passivation. Depassivation can be induced by physical contact with electropositive metals (aluminum wire,[96] zinc rod or pellets) which leads to brief hydrogen evolution at the Cr surface and subsequent rapid etch. Cr^{++} ions in aqueous solution accomplish the same, and so does the application of a cathodic potential. Once depassivated, Cr dissolves rapidly in nearly all mineral acids. The dissolution rates have been studied by Mueller[97] and decrease in the order HCl > HBr > H_2SO_4 > $HClO_4$.

Practical acid etches which have been published are

Or: 50 ml glycerin
     50 ml concentrated HCl
     Etch rates are about 800 Å min^{-1} [98]

Concentrated HCl, depassivation with zinc rod, yielding an etch rate of 1,500 Å min^{-1}.[65] This probably includes the depassivation period, and the true rates may be 2 to 3 times higher.

Or: 9 parts saturated ceric sulfate solution
     1 part concentrated HNO_3
     Etch rates are about 800 Å min^{-1} [98]

A composition similar to the latter is recommended by the manufacturer to etch Kodak metal-clad chromium masks:[99]

     164.5 g ceric ammonium nitrate
     43 ml of concentrated (70%) perchloric acid, and water to make 1 liter

This etch may be used at 50°C where the etch rate is three times as high as at room temperature.

Another chromium etch which gives satisfactory results is the following solution

     454 g $AlCl_3·6H_2O$
     135 g $ZnCl_2$
     30 ml H_3PO_4
     400 ml H_2O

The Eastman Kodak pamphlet[83] also lists an alkaline ferricyanide etch which does not require depassivation for the etching of Cr masks. It is to be used at room temperature:

1 part of a solution consisting of 50 g NaOH and 100 ml of deionized water, plus 3 parts of a solution consisting of 100 g $K_3[Fe(CN)_6]$ and 300 ml of deionized water

The etch rates vary from 250 to 1,000 Å min^{-1}.

(3) Cr-Si and Cr-SiO Films The chemical behavior of mixed films is generally similar to that of the predominant constituent. Assuming that the film components form two or more finely dispersed phases, dissolution of the major phase is somewhat slower than with the pure material but nevertheless effective. The minor phase is left behind as a loose network which can be washed away with little mechanical action

Examples of this type are Cr-Si films which can be dissolved in an etchant effective on pure silicon:

$$60 \text{ parts concentrated } H_3PO_4$$
$$5 \text{ parts concentrated } HNO_3$$
$$1 \text{ part concentrated } HF$$

The etch rates are 600 to 900 Å min^{-1}.[100]

Cr-SiO films are increasingly difficult to attack chemically the higher their SiO content. A practical limit lies between 20 and 30 atomic % SiO. Acidic chromium etches such as 6N HCl or 25% H_2SO_4 at 50 to 70°C work with depassivation, but gas bubbles cling tenaciously to the surface. Therefore, removal of films with substantial SiO concentrations tends to remain incomplete and to leave spots. Line edges are irregular, not smooth and straight. Good edge definition is obtained in the following solution:

$$20 \text{ g } K_3 [Fe(CN)_6]$$
$$10 \text{ g NaOH}$$
$$100 \text{ ml } H_2O$$

On Cr-20 atomic % SiO films, this solution is used at 50 to 60°C and yields etch rates of about 1,000 Å min^{-1}.[93] Room temperature is sufficient for pure chromium films.

(4) Copper The sensitizer in KMER/KTFR is subject to decomposition in the presence of copper. Hence, these resists should not be utilized on copper or copper alloys.[83] Stained (oxidized) copper surfaces may be cleaned in dilute (5 to 8 volume %) HCl, dilute ($\sim$50%) H_2SO_4, 10% ammonium persulfate solution, or in dilute commercial copper cleaners such as Neutra-Clean.*

The most common reagents are ferric chloride solutions, which are well established as deep etchants in photoengraving and for heavy copper-clad circuit boards. Commercial $FeCl_3$ solutions are slightly acidic and typically 36 to 42° Baumé (density at 15°C = 1.33 to 1.41). This is usually too concentrated for thin film etching and yields high removal rates.[101] A suitable immersion etch is obtained with about 300 g $FeCl_3$ per liter. For spray etching, greater dilution (60 g $FeCl_3$ per liter) is advisable.[102]

Alternate copper etches such as chromic sulfuric acid or ammonium persulfate solutions have been investigated by Sayers and Smit.[103] They found, however, that ferric chloride solutions of suitable concentration produce the least undercutting on 0.001- to 0.003-in.-thick copper films. A very rapid (10 to 15 μ min^{-1}) chromic sulfuric acid spray etch is commercially available.†

The iodine etch described by Zyetz and Despres[104] for gold etching may be modified to etch copper as well. A suitable solution consists of 200 g KI, 100 g I_2, and 400 ml of water. It is dark and not transparent, which makes recognition of the end point difficult. The etch is very rapid, but undercutting is effectively limited by the formation of an insoluble copper compound which occurs predominantly at the line edges. The procedure is to dip the substrate for a few seconds into the etch, rinse, and remove the dark compound residue in Neutra-Clean. If subsequent inspection shows residual copper film, another dip cycle will remove it.

(5) Gold The Kodak data booklet[83] suggests the use of KTFR and aqua regia (1 part concentrated HCl plus 3 parts concentrated HNO_3) for gold etching. However, the technique is limited to very thin gold films since the photoresist does not withstand the highly reactive mixture very long.

An alternative is the iodine etch given by Zyetz and Despres.[104] The solution is made from 400 g KI, 100 g I_2, and 400 ml of water. Attack is rapid, and the opaque nature of the solution prevents recognition of the end point. Hence, a quick sequence of dipping, rinsing, and examination is recommended. If the full-strength solution is diluted with water 1:4, a slower etch (0.5 to 1 μ min^{-1}) is obtained which yields better control of line edges. Alkali cyanide solutions containing hydrogen peroxide are also suitable for gold etching.

---

* Shipley Company, Inc., Newton, Mass.
† CR-10, Shipley Company, Inc., Newton, Mass.

(6) Hafnium It has been reported that hafnium films masked with KPR can be etched in 1 to 2% HF solutions without difficulty.[105]

(7) Molybdenum The Kodak data booklet[83] recommends KTFR stencil masks and two etches without commenting on rates and results:

1 part concentrated HNO_3
1 part concentrated H_2SO_4
3 parts H_2O

Or:

200 g $K_3 [Fe(CN)_6]$
20 g NaOH
3 to 3.5 g sodium oxalate
Water to make 1 liter solution

Reagents of the alkali ferricyanide type are universally used in molybdenum film etching. A similar composition known to give very fast rates of nearly 1 μ min^{-1} and used with KPR has been published by Dalton:[106]

92 g $K_3 [Fe(CN)_6]$
20 g KOH
300 ml H_2O

(8) Nichrome Procedural details and particular problems encountered in Nichrome etching are rarely published for proprietary reasons. Nichrome is attacked only by relatively strong chemicals, and alternate methods such as reverse photolithography[107] (see Sec. 4a) or sputter etching (see Sec. 4b) are often preferred. For chemical etching, the following mixture has been suggested in conjunction with KTFR:[83]

1 part concentrated HNO_3
1 part concentrated HCl
3 parts water

A ferric chloride solution of 36 to 42° Baumé has also been reported as suitable.[83] Etching in equal parts of concentrated HCl and water at about 50°C is a fairly common practice.

(9) Platinum One of the very few chemicals which attack platinum is free chlorine. Accordingly, platinum films may be etched in aqua regia (1 part concentrated HCl plus 3 parts concentrated HNO_3). It is recommended to use KTFR and precede the etch by a 30-s immersion in 48% HF solution.[83] As in the case of gold films, etching times are limited because the resist mask is destroyed. Platinum can also be dissolved in aerated alkali cyanide solutions or better yet in mixtures of hydrogen peroxide and alkali cyanides.

(10) Silicon There are probably as many silicon etches as there are laboratories. Most of them are rapidly acting compositions for polishing of single-crystal surfaces and of little value for pattern etching. Others are slow and preferential-acting formulations to develop etch pits indicative of crystal defects. A composition aimed at yielding well-defined patterns is the mesa etch published by Lawson:[85]

5 parts HNO_3
3 parts acetic acid
3 parts HF

Another example is the Cr-Si etch given in Sec. 3d(3). The three common ingredients in etches of this type are an oxidizing agent such as HNO_3, Br_2, or H_2O_2, a small amount of HF to bring the oxidized silicon in solution, and various amounts of water with moderating agents such as phosphoric or acetic acids. Considerable latitude exists in formulating an etch of the desired rate and characteristics.

(11) Silicon Monoxide For some time, no satisfactory etchants for SiO films have have been known. Predictably, an etchant would have to contain some HF to ensure dissolution of silicon dioxide. Hydrofluoric acid alone, however, does not etch SiO films uniformly but leaves residues and produces irregular edges. A satisfactory etch was developed by Chance.[108]   It consists of a 10 to 12 molar aqueous solution of NH_4F with NH_4OH or alkali hydroxide added to yield a pH of about 9.   If used at 80 to 90°C, the etch rates are of the order of 5,000 Å min^{-1}.   With the common negative photoresists, clean film removal and sharp line edges showing no undercutting are obtained.

(12) Silicon Dioxide SiO_2 films grown by thermal oxidation of silicon or deposited by rf sputtering are generally etched in buffered hydrofluoric acid.   The etch is commonly prepared by mixing a 40% (by weight) NH_4F solution with concentrated (48%) HF.   The proportions vary typically from 9 to 12 volume % HF and 88 to 91 volume % buffer solution (see, for example, Ref. 82).   The etch rates of thermally grown and high-quality sputtered SiO_2 films are similar and increase with the amount of free acid.   For the concentration range given above, the rates are between 600 and 1,200 Å min^{-1}.   They are also dependent on impurities present in the oxide, the degree of agitation, and the temperature of the solvent.[109]

KTFR or KPR may be used as stencil masks.   KPR and even more so KPR-2 yield thicker coats which cover rough surfaces more reliably than the normally thin KTFR coats.   The advantage is better protection against etch penetration and pinholing [see Sec. 3f(3)].   The limiting factor is usually resist adhesion, which tends to weaken during the etching and thereby promotes undercutting.   Well-processed and baked KTFR or KPR films can withstand immersion in buffered HF with little deterioration up to about 10 to 12 min.[85,110]   If the SiO_2 film requires longer etching times because it is relatively thick, an often successful expedient is to interrupt the etching and rebake the resist.

The AZ-1350 positive resist is also well suited for SiO_2 etching.   The manufacturer recommends a buffered etch consisting of 389 g NH_4F, 140 ml HF, and deionized water to make 1 liter.   The etch rate is said to be 1,000 Å min^{-1}.

Another widely used etch for SiO_2 films is the so-called P-etch.[111,112]   Its composition is

$$15 \text{ ml } 49\% \text{ HF}$$
$$10 \text{ ml } 70\% \text{ HNO}_3$$
$$300 \text{ ml } H_2O$$

The etch has two applications: One is to remove controllably very thin layers of SiO_2 as the etch rate is very slow (120 Å min^{-1} on thermally grown SiO_2).   The other use is diagnostic insofar as the etch rate is very sensitive to nonstoichiometry, porosity, and impurities in SiO_2 films.   Thus, an approximate assessment of the quality of SiO_2 films can be made by determining the P-etch rate.

(13) Silicon Nitride Si_3N_4 films are chemically inert and difficult to etch.   Concentrated (48%) HF is a suitable solvent whose rates of attack vary from 150 Å min^{-1} [110] to 500 to 1,000 Å min^{-1} [98] depending on the method of film preparation and thickness.   Buffered HF attacks at an even slower rate of about 15 Å min^{-1}.[110]   Another possible etchant is a solution of 3 ml concentrated HF and 10 ml concentrated HNO_3 at 70°C.[98]   The common problem of all these reagents is that the existing photoresists cannot withstand their attack for sufficiently long times to etch more than a few hundred angstroms of Si_3N_4.   To circumvent this limitation, stencil masks of molybdenum[110] or chromium-silver[98] films prepared by photoresist patterning may be used.   These metals are not attacked or penetrated by concentrated HF so that relatively thick, defect-free Si_3N_4 patterns can be prepared.   Holes with diameters as small as 7.5 μ and straight, sharp line edges have been reported.[110] SiO_2 films, too, are widely used as masks, and the Si_3N_4 is then etched in hot concentrated phosphoric acid.[113]

(14) Silver   For the etching of silver films, KTFR masks and a solution of 55 weight % aqueous ferric nitrate to be applied at 43 to 49°C are recommended.[83]   However, one of the authors found this mixture to be too rapid for thin film etching.   A more workable solution is obtained by dissolving 55 g of $Fe(NO_3)_3$ in ethylene glycol to make 100 ml and then adding 25 ml of water.   Once thoroughly homogenized, the mixture may be stored up to one week in a brown bottle.   It is still a rapid etch, yielding rates of about 3,000 Å min^{-1} if used as a spray etch.   This is advisable, since observation of the film surface during immersion is difficult because of the color of the liquid.   The high etch rate requires quick rinsing with water to prevent over-etching.   The KI-I$_2$ etches listed for copper and gold also yield fair results.   A brief immersion of the film sample for 1 to 3 s followed immediately by a water rinse is usually sufficient to remove films of 1 μ thickness or less.

Other etches consisting of equal amounts of concentrated nitric acid and water, or nitric acid and glycerin* have been published by Woitsch.[98]   No etch rates are given, but one may infer that excellent pattern definition was obtained since the silver film was used as a stencil mask to etch silicon nitride.   Another possible etching system which has apparently not yet been exploited for thin films is ferricyanide solutions with sodium thiosulfate added to prevent precipitation of silver ferro- and ferricyanides.[69]

(15) Tantalum   Etching of tantalum films is generally done in mixtures of nitric and hydrofluoric acids.   A moderately fast etch to be used at room temperature has the composition:[114]

> 2 parts concentrated HNO$_3$
> 1 part 48% HF
> 1 part water

The water may be omitted to obtain a faster etch, which is useful if the tantalum film contains significant amounts of oxygen and tends to resist etching, or if the photoresist adhesion is weak and a faster etch is less likely to tax the resist than the milder but slower solution.   Timing of the etch process is critical to avoid overetching.   Because of the presence of hydrofluoric acid, the etch also attacks substrates such as glasses and SiO$_2$-passivated wafers.

Substrate attack can be avoided with an alkali etch reported by Grossman and Herman.[115]   Nine to ten parts of 30% NaOH or KOH solution are heated to 90°C, and 1 part of 30 to 35% hydrogen peroxide is added.   This mixture etches tantalum, tantalum nitride, and tantalum oxide at rates of 1,000 to 2,000 Å min^{-1}.   However, it also attacks photoresists and must be used in conjunction with a metal-film mask, for example, gold.

(16) Titanium   The Kodak booklet[83] recommends KTFR masks for titanium and lists two etch compositions, both of which are too rapid for thin film work.

> 90 ml H$_2$O
> 10 ml 48% HF
> ―――――――――――――
> Rate: 12 μ min^{-1} at 32°C

> 70 ml H$_2$O
> 10 ml 48% HF
> 20 ml concentrated HNO$_3$
> ―――――――――――――
> Rate: 18 μ min^{-1} at 32°C

e. Removal of Photoresist Masks

(1) Resist Strippers   The final removal of cross-linked polymer films is an onerous task because these compounds are not truly soluble.   The degree of difficulty encountered depends on the nature of the photoresist film, its thickness, and the underlying substrate.   In general, stripping of resist films becomes more difficult the higher the post-baking temperature.   The most widely used resist-removal techniques rely on hot chlorinated hydrocarbons to swell the polymer, in conjunction with acids to loosen the adhesion of the resist film to its substrate.   Oxidizing agents like hot H$_2$SO$_4$ may be used to decompose the organic material, but film corrosion often prohibits such drastic action.

In spite of considerable experimentation, no truly effective universal resist stripper has been found.   Immersion in hot "active" solvents, sometimes preceded by swelling in hot trichloroethylene, and subsequent swabbing or brushing are typical procedures.   The latter operations are nearly always necessary to remove tenaciously

---

* Mixtures of concentrated nitric acid and certain organic compounds such as alcohols are known to explode occasionally after prolonged standing.   They should not be stored

clinging fibers or patches of the resist coat at the risk of mechanically damaging the film pattern. The jet action of spray rinses is generally no satisfactory alternative. Solvent mixtures which have been found to be more or less effective include tri- and tetrachloroethylene, methylene chloride, dichlorobenzene, chlorotoluene, in varying proportions with formic, trifluoroacetic, and phosphoric acids or phenol. Ketones such as cyclohexanone, acetone, and butyral acetone are also common.

A number of resist-stripper compositions have been published by various authors. If film oxidation is no problem, Caro's acid consisting of equal parts of concentrated H_2SO_4 and a 30% hydrogen peroxide solution is an effective stripper.[84] Sullivan[116] removed highly baked KTFR films without apparent residue by 10 min immersion in nearly boiling triethylenetetramine, followed by a hot-water rinse. Corrosion of metal films such as aluminum did not occur, but a nitrogen atmosphere above the hot liquid is required to avoid oxidation of the solvent.

Widely used are commercial stripper solutions, and among these J-100* is probably the best known. It seems to have been developed primarily for KMER/KTFR but works also with KPR and AZ-1350 coats. Recommended solvent temperatures vary from 80°C[84] to 130°C[85] with immersion times of 2 to 3 min. Spraying with acetone, immersion in boiling trichloroethylene, and final rinses in xylene or isopropyl alcohol are used to remove the stripper.[85]

Corrosion is not an uncommon problem if commercial strippers are used on metal films. J-100, for example, contains a substantial amount of phenol in addition to tetrachloroethylene and dichlorobenzene. Therefore, it reacts acidic, and the hot solution may attack metal films involving chromium or aluminum.[116] As with chemical etches, the corrosion action is enhanced if two metals of different electrochemical potentials are exposed to the stripper simultaneously and in electrical contact with each other. An example experienced by one of the authors is the attack of Cr-SiO resistor films with silver film contacts. It was concluded that the corrosive action is due to the presence of phenol since mixtures of the chlorinated hydrocarbons alone did not produce the attack, but neither were these mixtures effective as strippers. The degree of attack in phenol-containing solutions was found to vary from negligible to severe depending on whether the solvent was dry or contaminated with water. Thus, the corrosion is probably galvanic in nature and due to the dissociation of the phenolic acid in the presence of water.

The completeness of resist removal is not an easy matter to judge. Film surfaces of perfectly clean appearance may still have a residual layer of molecular thickness left. The nature of such residues has never been determined; they may consist of surfactants such as sulfonic acids which are present in some commercial strippers, or of other firmly adsorbed organic material. The presence of residues may be determined by wettability tests since they render the film surfaces strongly hydrophobic. In some cases, the adhesion of subsequently deposited films is severely affected. Removal of these residues is, however, exceedingly difficult because they tend to yield only to strong oxidizing agents such as hot chromic-sulfuric acid.

The removal of positive resist AZ-1350 is generally easier than that of the common negative resists. If the resist has not been baked above 95°C, solubilization can be induced without mechanical action by immersion in alkaline solutions such as the commercial AZ stripper.[95] Exposure to ultraviolet light aids in the process. If the resist coat has been baked up to 120°C, heating of the solvent to 50 to 65°C and ultrasonic agitation are recommended. Other effective solvents are acetone or ethyl acetate. Resist coats baked above 120°C are difficult to remove. The manufacturer recommends hot (60°C) concentrated sulfuric acid or burning off in oxygen.[95] Immersion in hot J-100 followed by swabbing and rinsing in acetone is also effective.

The Kodak autopositive resist KAR 3 can be removed in its developer after ultraviolet exposure, or it may be stripped in such solvents as acetone, cyclohexanone, or cellosolve. No mechanical action is required.[117]

(2) Volatilization of Photoresists A universal alternative to solvent stripping is the conversion of the stencil mask into volatile oxides. Heating of resist coats to 300 to

---

* Indust-Ri-Chem Laboratory, Richardson, Tex.

500°C for about 20 min in oxygen is required to produce a film surface free of organic residue.[83] However, these conditions are often too severe and may oxidize the film surface or cause undesirable film and substrate interaction. To avoid these difficulties, processes have been developed where oxidation is accomplished at lower temperatures by physically activating the oxygen. In one of these techniques, volatilization of resist coats is induced by an rf discharge in flowing oxygen at 5 Torr pressure.[118] Removal is facilitated by the generation of highly reactive atomic oxygen and ion bombardment (sputtering) of the resist. No external heating is required; the wafers subject to the discharge assume temperatures of 100 to 300°C depending on the oxygen pressure and rf energy used. Fully automated ashing machines achieving complete resist removal in 5 to 10 min are commercially available.[118] Resist-removal rates of 1,000 to 2,000 Å min^{-1} at substrate temperatures of 75 to 120°C have been reported by Irving for a similar apparatus.[119]

Another activated oxidation process relies on the chemical reactivity of ozone.[120] The substrates to be stripped are heated to 200 to 250°C in a mixture of oxygen with about 3% ozone. The latter was generated by a commercial ozonator. Complete removal of the resist coat requires 5 min or less depending on the thickness of the coat.

Since volatilization of resist coats is fast and requires much less handling of substrates than solvent stripping, dry oxidation processes are preferable for many applications. While the probability of forming organic residues is very low, inorganic trace impurities from the resist are likely to be left. Their amounts vary depending on the purity of the resist, and they can be removed by a quick dip in dilute hydrofluoric acid.[119] Oxidation of metal film surfaces and radiation damage to sensitive semiconductor devices are possibilities to be considered. It remains to be determined empirically in individual cases whether or not dry oxidation has detrimental effects on the electrical properties of devices.

f. Quality of Thin Film Patterns

The principal factors which determine the quality of etched thin film patterns are the resolution and accuracy of the photoresist mask pattern, the degree of undercutting during the etch, and the occurrence of pinholes or penetration through the polymer mask. These criteria are discussed in the following sections.

(1) Photoresist Pattern Resolution and Precision Questions of prime interest to the thin film worker are: What are the smallest pattern dimensions, and how accurately can they be reproduced with the conventional photoresist-contact mask technology? Assuming for the moment an ideal etch which faithfully reproduces the outline of the resist pattern, the latter determines the answers. Experimental resolution test results from various authors are shown in Table 10.

TABLE 10　Maximum Resolution of Photoresist Patterns as Obtained by Contact Printing with Lippmann Plates

Smallest resolved line, μ	Photoresist coating		Reference
	Thickness, μ	Type	
3	1	KTFR	Martinson[62]
3–4	1	KTFR	Lawson[85]
5	0.4	KTFR	Kodak[83]
3–6	1	KTFR	Kornfeld[79]
5–6	1	KMER	Lawson[85]
6	1	KTFR*	Martinson[55]
5	1.9	KMER	Taylor[87]
12	2.2	KMER	Taylor[87]
20	6	KMER	Lawson[85]

* 25 μ of clear plastic foil between mask and film.

Several conclusions can be drawn from these data. First, the smallest line width obtained with emulsion masks and negative resists is about 3 μ. Most photoresist workers consider the resolution limit of this technique to be about 2.5 μ or 200 lines/ mm.[62,82,121,122] Second, the preferred negative resist for high-resolution patterns is KTFR with KMER not far behind. Third, the resolution is higher the thinner the resist film. This has been demonstrated most convincingly by Tulumello and Harding[123] on somewhat wider (50 μ) lines. These authors investigated 0.5- to 5-μ-thick coatings of 15 different photoresists exposed with different types of light sources and for varying lengths of time. They found that the developed lines of negative resists broadened in proportion to the coating thickness, KPR lines about twice as much as KTFR lines. Lastly, there is some reluctance to use negative-resist coats thinner than 1 μ, although these promise the highest resolution. While it has been inferred that 0.2- to 0.3-μ coatings can resolve 1-μ lines,[62] this has not yet been claimed as an experimental achievement when using emulsion masks. One reason is that polymer masks as thin as 0.5 μ or less are too easily penetrated by chemical etches.

In contrast to the negative resists, the line width of positive resist AZ-1350 is practically independent of thickness.[123] Consequently, one may expect to achieve more reliable (thicker) masks with this system without sacrificing resolution. The manufacturer's claim of 0.5-μ-wide lines in AZ-1350 coats as thick as 1.8 μ confirms this contention, but the etching of 1-μ or smaller lines in general requires exposure techniques other than contact printing through Lippmann-emulsion masks [see Secs. 3a(6) and 4c]. Experimental resolution data for AZ-1350 resist patterns made by emulsion masks and contact printing have not been published as yet, but the feasibility of 2.5-μ-wide lines has been established by Johnson.[124]

The precision with which pattern dimensions can be reproduced is a more complex question because deviations from the specified target values are accumulated throughout the entire mask-fabrication and photoresist process. The quality of the equipment and the experience of the operators represent variables difficult to assess. Assuming perfectly reproducible photoresist application, exposure, development, and etching processes, the final etched pattern will have reproducible dimensions but different from those initially specified. The factors which contribute to this *systematic* deviation are errors committed in artwork cutting, in the reduction ratio, in photomask processing, and the constant but nevertheless present error inherent even in the best photoresist processing. Step-and-repeat positioning and mask-alignment errors are not considered in this context as they affect only the location of objects relative to each other but not their size. An interesting estimate of the error contributions has been published by Schuetze and Hennings.[68] Their diagram shows deviations of 0.5 to 1 μ contributed by any one of the factors listed above if the best available technology is used. Since positive as well as negative deviations may occur in different process steps, the net result is not too catastrophic. However, pattern deviations of ± 2.5 μ are not uncommon for lines wider than about 25 μ, and alignment errors of successive patterns may add about equally large deviations.[23]

A source of dimensional errors which is most difficult to control is line broadening due to light scattering under the dark portions of the mask. With well-adjusted exposures and negligible mask-to-substrate separation, line broadening is dependent on the resist-coat thickness. For coatings of 0.5 μ or less of AZ-1350, KTFR, and KPR Tulumello and Harding[123] found width changes of 1 μ or less. The width of AZ-1350 lines did not change significantly if the coating was several microns thick. The negative resists, however, gave wider lines in proportion to the thickness of the coating, μ producing line broadening of about 3 μ with KTFR and 6 μ with KPR and KPR 2. It has therefore been suggested that this deviation be compensated for by correcting the design dimensions accordingly.

So far, the assumption of perfectly controlled, reproducible photoresist processes has been maintained. In practice, processing conditions are never perfectly constant. Resist-coat thicknesses, exposures, and mask-to-substrate spacings vary within certain limits. Consequently, pattern dimensions vary randomly about their systematic deviation from the target value. The effects of random and systematic deviations from the prescribed values have been studied on small (12 to 75 μ wide) thin film resistors.[93]   The films were about 1,000 Å thick.   Since it was possible to control the sheet resistance to within $\pm 2\%$ and undercutting was avoided by sputter etching, dimensional errors in the resistor geometry were identified as the factor determining the accuracy of the electrical resistance obtained.   With a well-established process and KTFR photoresist, systematic deviations of ± 0.75 to 1.25 μ were found.   The random distributions around the average values were characterized by σ spreads of ± 0.50 to 0.75 μ.   Although worst cases are statistically rare, additive deviations of up to 2.5 and sometimes 3.5 μ were obtained.   In a similar study, maximum deviations of ± 2.5 μ from target line widths of 3 to 25 μ were reported.[125]   Thus, contact printing through emulsion masks permits photoresist line-width control within about ± 2.5 μ for moderately wide lines.   Comparable data for narrow (<10 μ) lines and for positive resists

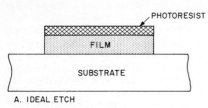

A. IDEAL ETCH

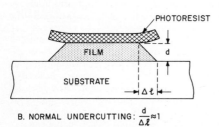

B. NORMAL UNDERCUTTING: $\dfrac{d}{\Delta \ell} \approx 1$

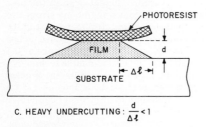

C. HEAVY UNDERCUTTING: $\dfrac{d}{\Delta \ell} < 1$

Fig. 16 Etched thin film line profiles. A = without, B = with norma C = with heavy undercutting.

are not available.   Closer tolerances can be obtained by projection exposure of photoresist [see Sec. $4c(1)$] and with chromium contact masks.

(2) Photoresist Pattern Adhesion and Undercutting   In order to realize the maximum resolution and accuracy of photoresist stencil masks, the underlying film pattern must replicate the polymer pattern exactly, as shown in the cross section of Fig. 16A.   Such ideal behavior is no obtained in practice because chemica etches attack the film material parallel a well as perpendicular to the surface.   In polycrystalline and amorphous materials there is no reason for etch rates to be direction-dependent.   Consequently slightly trapezoidal line profiles as shown in Fig. 16B are the typical result when etching metal films.   The increment of lateral etch (Δl in Fig. 16B and C) is referred to as undercutting, the ratio of film thickness to undercutting as the etch factor.   It follows that normal undercutting (etch factor of about 1) causes etched lines to narrow by increments of about twice the film thickness.

Etch factors considerably greater than 1 may also occur as sketched in Fig. 16C The straightness of the sloping line shoulders is an idealization of the true line profiles which may be curved.   Heavy under cutting of this type is most likely to b encountered when SiO_2 films are etched i hydrofluoric acid solutions.   Because of the resulting pattern distortion, considerable effort has been spent on understanding and preventing this problem.   Since it difficult to rationalize laterally preferential etch rates, the effect is generally attributed to a gradual lifting of the resist coat, thereby exposing increasing areas of the film surface to the etch.   The separation of the polymer coat from the oxide surface must be preceded by a gradual loss of adhesion due to the influence of the etch.   The mechanism of this process has not been clarified as yet, but penetration of small ions like the fluoride anion along the resist-film interface has been suggested as a possible triggering reaction.[87]

The realization that lack of permanent adhesion may be the cause of undercutting when etching SiO_2 films initiated further investigations of factors likely to influence the nature of SiO_2 surfaces.   Taylor[87] reported decreased adhesion of KMER and enhanced undercutting for samples exposed to more than 40% relative humidity. Systematic studies of Bergh,[126] later extended by Lussow,[127] led to the characterization of SiO_2 surfaces in terms of their molecular structure and their wettability by water, and to a correlation with the adhesion of photoresist films. There are three types of SiO_2 surfaces to be distinguished. The first one possesses a firmly adsorbed layer of H_2O molecules and is referred to as hydrated. On the second type, the silicon surface atoms are saturated by hydroxyl groups, thus forming silanol $\left(\diagup \text{Si—OH} \right)$ configurations. The third type of SiO_2 terminates with siloxane groups

$$\left(\begin{array}{ccc} O & O & O \\ \diagdown & \diagup \diagdown & \diagup \\ & Si & Si \\ \diagup & \diagdown \diagup & \diagdown \end{array} \right)$$

at the surface. The wettability of these surfaces as measured by the contact angle of water droplets increases in the order: siloxane < silanol < hydrated, while the adhesion of negative photoresists as revealed by undercutting decreases in the same sequence. According to Lussow's work,[127] the adhesion of KTFR is most sensitive to the surface constitution and only satisfactory on the siloxane type, whereas KPR adheres well on silanol surfaces, too. Hydrated surfaces yield poor adhesion with both resists. KMER coats are relatively unaffected by the surface constitution. Explanations for this behavior usually point to the hydrophobic nature of the resist molecules. It is quite plausible that the binding forces (adsorption energy) of a pure SiO_2 surface should be stronger than those exerted by or through a layer of OH groups or H_2O molecules, if the adsorbate has no strongly polarized groups.

The types of SiO_2 surfaces and hence the degree of resist adhesion are determined by the history of the sample. In general, high temperatures during thermal growth of SiO_2 produce siloxane-type surfaces, while hydrophilic surfaces are obtained from low-temperature deposition processes. Treatment in aqueous solutions, particularly in those containing HF, leaves the surface hydrophilic with relatively poor photoresist adhesion. The presence of P_2O_5 or B_2O_3 at the SiO_2 surface has the same effect. Conversion of hydrophilic surfaces into more or less dehydrated forms is possible by intense desiccation such as annealing in vacuum or dry gases with temperatures of at least 500°C, more effectively 850°C.[126,127]

Other techniques to improve resist adhesion do not require thermal removal of surface water but rely on covering the surface with hydrophobic molecules. Organo-chlorosilanes, for example, are capable of removing adsorbed water under formation of volatile hydrochloric acid:

$$R_2SiCl_2 + 2H_2O(ads) = R_2Si(OH)_2(ads) + 2HCl$$

The resulting silanol compounds can react further, either with each other or with silanol groups on the SiO_2 surface,

$$n \text{ HO—}\underset{\underset{R}{|}}{\overset{\overset{R}{|}}{Si}}\text{—OH} = \left(\text{—}\underset{\underset{R}{|}}{\overset{\overset{R}{|}}{Si}}\text{—O—} \right)_n + n \text{ H}_2\text{O}$$

thereby forming a layer of polymer silicone which renders the oxide surface hydrophobic and greatly improves the adhesion of subsequently applied photoresist.[128] This conversion of the SiO_2 surface is done by dipping the substrate wafer into or spraying it with a dilute solution of the silane in a volatile organic solvent, followed by moderate heating up to 200°C. After the SiO_2 film is etched and the resist coat stripped, the polymer siloxane surface film must be removed. It is chemically quite inert but can be dissolved in HF, strong bases such as KOH and some amines, or in strong acids.[128] An alternate method of improving adhesion is the addition of aluminum stearate to KMER solutions.[87]

Undercutting as the result of poor resist adhesion should be distinguished from overetching. The latter is caused by continued lateral etching after the exposed regions of the film have been dissolved. It may occur when etching metal films without knowing film thickness and etch rate. Manifestations are deeply and irregularly undercut line edges. Overetching is most likely to occur with very rapid etchants which make accurate timing difficult. Continuation of lateral etching due to solution adhering to the sample surface can be limited by quickly transferring the substrate from the etch into a stop bath. The latter must be of such a composition as to inhibit the chemical reaction induced by the etch. An example is the termination of acidic aluminum etches by immersion in dilute alkali baths. Another general remedy is to decrease the etch rate by dilution, possibly with viscous liquids like polyalcohols. The expected benefit from the latter is to slow down the exchange of fresh and spent solution at the line edges. Spray etching is also helpful at times since the exposed film surfaces are more effectively contacted and rinsed than the line edges.

(3) Mask Defects and Etch Penetration   The third criterion to be applied in judging the quality of etched patterns is the integrity of the photoresist mask. In the absence of design errors and easily recognizable gross defects, flaws in the etched pattern may arise from two sources. One is the presence of small defects in the photographic glass mask; the other is insufficient resistance of the polymer stencil mask toward penetration by the etch.

Photographic glass masks are very susceptible to small defects in their fabrication process as well as during their use. These defects may be irregularities in the emulsion such as pinholes and scratches, or dust and lint particles clinging to the surface. When the polymer stencil mask is formed, only those defects are of consequence which lead to a local reversal of image contrast. The possible effects of mask defects on the film pattern are dependent on the type of resist used and on the nature of the thin film being etched. An extensive discussion of this subject has been given by Plough et al.[129] The principal results of defect propagation can be derived from Fig. 13 and are summarized in Table 11.

TABLE 11   Effects of Photographic Mask Defects on Etched Metal and Oxide Film Patterns

	Opaque particle on or under *clear* mask area	Hole or scratch in *dark* part of emulsion
Negative resist:		
Metal etching.............	Pinholes in line pattern	Metal island between lines
Oxide etching.............	Pinholes in insulator film	Protruding oxide island
Positive resist:		
Metal etching.............	Metal island between lines	Pinhole in line pattern
Oxide etching.............	Protruding oxide island	Pinhole in insulator film

The four types of pattern defects which may arise are not all equally harmful. Pinholes in metal line patterns, for example, have negligible effects on the ability of the conductor to carry electric current, provided they are small compared with the line widths. Larger particles like lint fibers are more damaging and may cause discontinuities. However, particles of this size are easily recognized and removed. Metal islands between lines may be harmful, namely, if they are large enough to bridge the gap and provide electrical continuity. The defects in oxide films are principally the same, but their consequences are different. While oxide islands are harmless in most cases, pinholes in an insulator film are a potential source of electrical shorts and therefore constitute the most dreaded type of defect.

The size of defects introduced by the mask is also influenced by the pattern-formation process. Undercutting, for example, tends to reduce islands but to widen pinholes. Light scattering and overexposure decrease the size of pinholes in negative resists but widen holes in positive-resist coats. The same factors produce larger islands in negative resists and smaller ones in positive resists. In practice, neither holes in the dark parts of emulsion masks nor dark spots in the clear areas can be completely avoided. However, the latter are usually the more prevalent type of defect, particularly after the masks have been used several times. Therefore, negative resists are preferable for the etching of dense metal line patterns where bridging is the prime concern, while positive resists are likely to produce fewer pinholes when etching small openings into an oxide film.

In addition to defects generated by the photographic mask and its exposure, photo-resist films may develop pinholes from other causes. Examples are particles of foreign matter embedded in the resist coat, and sharp protrusions on the underlying thin film surface. In such spots, the resist coating is thinner than intended and may be penetrated by the etchant. This condition is most damaging if it arises during an oxide etch because it leads to the formation of pinholes.

Considerable effort has been devoted to identifying pinholes, as they are generally too small to be observed directly. These studies pertain to SiO_2 films on silicon which have been coated with negative photoresists, exposed, developed, subjected to buffered hydrofluoric acid etch, and then stripped of photoresist. Pinhole counts may be obtained by one of several techniques which involve either electrical testing or chemical methods of detection. Electrical testing requires the evaporation of many small metal-film dots to form capacitors with the oxide film constituting the dielectric.[85] Each shorted capacitor indicates the presence of at least one pinhole. Chemical methods include gaseous diffusion of electrically active impurities through the pinhole with subsequent oxide removal and staining of the silicon surface, as well as electrolytic or electroless plating of copper onto the exposed silicon surface in the pinhole. More common are techniques of enlarging the defects by prolonged etching of the silicon surface which leads to undercutting and makes the pinhole microscopically visible. One widely used approach is to expose the sample to chlorine or hydrochloric acid gas at about 900°C.[94,125] This treatment does not affect the SiO_2 film while the silicon reacts to form volatile chlorides. Hence, pinholes are indicated by etch pits in the underlying silicon. Disadvantages are the high temperature to which the substrates must be subjected, and the highly corrosive atmosphere. These problems are avoided in a method described by Sullivan,[116] who used an aqueous etch containing organic reagents* to develop etch pits at 115°C. The required etch time is 6 h. It is claimed that the etch brings forth more pinholes than the chlorine technique on films of comparable quality.

Pinhole counts of several authors are shown in Table 12. Comparison of these data is somewhat tenuous because of variations in SiO_2 film thickness, strength of the etch, cleanliness of the environment, and differences in the methods of identification. How-ever, a reduction in number of pinholes with increasing resist thickness is evident in every case. In general, resist coats of less than 0.5 μ thickness provide inadequate protection against etch penetration and pinholing of SiO_2. Poppert[125] concludes that thicknesses of at least 1 μ are required to keep the pinhole count below 100 cm^{-2}. This is in agreement with capacitor tests performed on 0.4-μ-thick SiO_2 films where Lawson[85] found a significant yield increase for KMER coatings thicker than 1 μ. Taylor[87] made the same observation if the resist thickness was increased from 1.9 to 2.2 μ.

Since not every pinhole in an SiO_2 film causes an electrical short, a quantitative correlation between pinhole density and occurrence of electrical defects is of interest. This problem has been examined by Lawson,[130] who derived a logarithmic relationship between pinhole density and circuit yield of the type

$$\log (\text{yield}) = \text{const} \times (\text{pinhole density})$$

This result is in agreement with the assumption of random pinhole distribution across the surface. The numerical constant in the equation is dependent on the various

---

* Sullivan's pit etch for pinhole identification consists of

        17 ml ethylenediamine (35.6 mol %)
        3 g pyrocatechol (4.1 mol %)
        8 ml water (60.3 mol %)

TABLE 12   Pinhole Counts in SiO₂ Films Which Were Protected by Photoresist Coats and Immersed in Buffered Hydrofluoric Etches

Photoresist coat		Pinholes		Reference
Type	Thickness, μ	Count, cm^{-2}	Method	
KPR	0.1	155	Chlorine etch	Kelly[94]
KPR 2	0.4	0		
KTFR	0.48	1–2		
KMER	1.8	0		
KTFR	0.2	2,000	Chlorine etch	Poppert[125]
KTFR	0.5	250		
KPR	0.15	6,200	Liquid etch	Sullivan[116]
KTFR	0.18	2,000		
KPR 2	0.4	420		

process steps involved and on the relative sizes of metallized and unmetallized areas on top of the oxide film.

4. SPECIAL PATTERN-FORMATION TECHNIQUES

Although conventional photolithography as described in the preceding section is a very universal technique, it has its limitations.   One is the inability of organic resists to withstand strongly oxidizing chemicals which are needed to etch chemically inert films.   Two types of processes, the application of negative relief masks and sputter etching, have been developed for such situations.   The other limitations arise from the use of silver-halide-emulsion masks to transfer the pattern to the substrate.   One possibility of improving pattern resolution and durability of contact masks as well as reducing mask defects is the use of chromium masks [see Sec. 3a(6)].   Other approaches to photoresist pattern formation such as projection techniques and holography do not require a contact mask but merely an enlarged transparency of the pattern to be etched.   High resolution and elimination of mask defects are also obtained with light- and electron-beam writing techniques.   These offer the additional advantage of computer-controlled pattern generation, thereby avoiding the tedious and costly procedure of artwork fabrication.

All these techniques are of relatively recent origin.   Some of them have rapidly become widely accepted as, for example, sputter etching.   Others such as chromium contact masks, projection exposure, and light-beam writing appear to be in the transition phase from the laboratory to large-scale applications.   However, the development of these techniques is still continuing so that changes in procedures and particularly in the equipment used for their implementation are to be expected.

a. Negative Relief Masks

In conventional photolithography, the resist mask is applied on top of the film to be etched and as a positive of the desired pattern.   There are also inverse techniques where a negative relief mask is first applied to the substrate and the film is deposited subsequently.   As a result, the film contacts the substrate directly only in the areas left open by the relief mask.   The sequence of steps is shown in Fig. 17.   The relief mask is finally removed by a solvent which attacks only the mask but not the film material.   Firm adhesion of the latter to the substrate is necessary for this technique to be applicable.

(1) Reverse Photolithography   The simplest way to produce a negative relief mask is to use photoresist.   The techniques required to form such a mask are the same as in conventional photolithography. After film deposition, the resist mask along with the unwanted portions of the film can be removed by swelling in a suitable organic solvent and gentle mechanical action. It is generally necessary to make the relief mask thicker than the film to keep the deposit at the sides (area S in Fig. 17) thin or even discontinuous. This allows the solvent to reach the relief mask under the film and makes it easier to wash off the raised portions of the film.

The main advantage of reverse photolithography is that no strong chemical reagents are required to attack the thin film material. It is therefore primarily of interest for noble-metal films such as platinum which require etchants like aqua regia. However, the adhesion of metal films to the bare substrate areas previously coated with resist is problematical. Substrate heating during deposition to enhance film adhesion must be limited to temperatures below 200°C to avoid outgassing and resist decomposition. A comparison of reverse and conventional photolithography as applied to the fabrication of chromium film masks on glass substrates has been published by Stelter.[64] Although he achieved line widths down to 2.5 to 5 μ, he concluded that reverse photo-

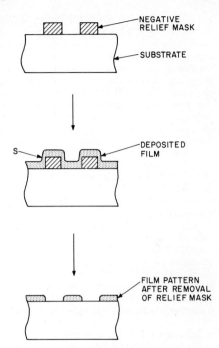

Fig. 17 Film-pattern formation with a negative relief mask.

lithography was more difficult in processing and more likely to yield poor film adhesion than the conventional technique.

(2) Other Reverse Masking Techniques The disadvantages associated with the presence of photoresist in vacuum systems can be avoided by using other easily etchable materials as negative relief masks. These require an additional step, namely, the deposition of the film from which the negative relief mask is formed, followed by conventional photolithography. An example of this type has been reported by Murphy,[131] who utilized 300 to 400 Å or reactively sputtered Bi_2O_3 as the mask film. The oxide adheres well to glasses and can readily be dissolved in 1% hydrochloric acid solutions. During the subsequent deposition of the film to be patterned, the Bi_2O_3 coat may be heated up to 700°C. Because of its fine grain structure, the pattern resolution and accuracy are about the same as obtained with conventional photolithography, namely, 2 to 3 μ.

The principle of negative relief masking has also been employed by Spriggs and Learn[132] but in slightly different form. These authors evaporated a very thin film of indium which consisted of mostly unconnected, large metal islands. After the entire surface was covered with a thin chromium film, the indium was selectively etched, thus leaving behind sparsely connected islands of chromium. The remaining chromium film had a transmittance of 80% and a sheet resistance several hundred times greater than a continuous chromium film of comparable thickness.

b. Sputter Etching

An alternate method for the removal of chemically inert film materials is sputtering. Contrary to the arrangements used in sputter deposition (see Chap. 4), the substrate and film are made part of the target in a glow-discharge circuit. Material from all exposed surfaces is removed by bombardment with positive ions—usually argon—having energies of the order of 1 keV only. To utilize this technique for pattern formation, it is necessary to mask certain portions of the surface against ion bombardment. Furthermore, the positive ions impinging on the surface must have an opportunity to recombine with electrons lest the accumulated charges repel the arriving ions and the removal process stops.   Ways have been found to satisfy these requirements in dc as well as in rf sputter etching.   The latter technique, however, has been found to be more controllable and of more universal applicability since it is not predicated on the presence of a metallic surface to facilitate recombination.

(1) DC Sputter Etching   In dc sputter etching, the recombination of positive ions requires that the substrate surface exposed to the bombarding particle stream is conductive.   This condition is satisfied if both the film to be etched and the mask are metallic.   In practice, there is little incentive to use dc sputter etching in such situations since nearly all metal films can be patterned with conventional photolithography. Of greater practical interest is sputter etching of oxide films, and here one has to rely entirely on the mask to provide for recombination.   Such a technique has been utilized by Valetta.[133]   The mask was made from a thin metal foil and held in close contact with the surface of a silicon wafer bearing an insulator film.   A cathode potential of 1.5 to 2 kV was applied to the mask.   The pattern to be etched into the insulator film consisted of small (0.0025 in. diameter) holes.   Bombardment of these regions was achieved because the argon ions accelerated toward the mask travel in straight lines and through the holes in the metal foil.   The positive charges accumulating on the insulator surface attract some of the secondary electrons ejected by the mask and are therefore neutralized.   This mechanism is effective only over short distances since the electric field across the ion sheath is stronger and tends to draw the secondary electrons away from the surface.   Removal rates of 150 to 300 Å min^{-1} in various glass and SiO_2 films were obtained.   Some scattering of ions under the edges of the hole mask occurred so that the holes had sloping sidewalls with an etch factor $d/\Delta l$ of about 0.4 (see Fig. 16).

To produce etched sidewalls which are steeper, the metal mask must be evaporated onto the film surface so that there is no separation at all.   Spivak et al.[134] employed an evaporated metal grid pattern to etch insulator films but found the removal rates to be nonuniform because of electrostatic effects.   Another problem is metal atoms sputtered off the mask and redepositing in regions between the grid pattern.   Lepselter[13] described a process where the sputter-etching mask was etched from a metal film thicker than the metal film to be etched.   However, all these techniques require an extra metallization and pattern-formation step to produce the mask.   If these extra steps are acceptable, one may as well choose a mask material suitable for subsequent chemical etching which is faster and less expensive [see, for example, Si_3N_4 etching Sec. 3d(13), or Ta etching, Sec. 3d(15)].

(2) RF Sputter Etching   The feasibility of removing thin layers of insulating materials by rf sputtering was demonstrated in 1962 by Anderson, Mayer, and Wehner.[136]   In 1965, Davidse[137] discovered that photoresist stencil masks may be exposed to rf glow discharges for sufficiently long times to etch thin films of any composition.   The simplicity of the rf system, the absence of undercutting due to the straight impingement of ions, the excellent edge definition, the moderately high and well-controlled etch rates obtained, and the possibility to use well-established photoresist techniques to produce the etch mask are significant advantages which led to immediate practical use of this technique.   Within a few years, sputter-etching systems complete with tuning networks to match electrode capacitance and rf power supply became commercially available.[138]

RF sputter-etching systems and their operation have been described by Davidse[13] and Tsui.[139]   They are similar in their construction to rf sputter deposition systems except that the metal plate used as the target also serves as the holder for the sample to be etched.   This is most conveniently done by bringing in the target from the base plate of the system so that the substrates can be placed on top.   It is important to provide water cooling for the target and to use either gallium metal or high-vacuum grease for intimate heat contact between substrate and holder so that the temperature of the photoresist mask can be kept below about 200°C.   The metal surface not occupied by the substrates may be covered with a fused silica plate having suitable openings. Sputtering of metal from the holder and deposition on the sample surface and bell-jar walls is thereby minimized. A counterelectrode is not necessary, as shown by Davidse. Tsui seems to have attained somewhat higher plasma densities and faster removal rates by employing a closely spaced counterelectrode and external magnets, but this arrangement tends to yield somewhat less uniform rates.

The rf glow discharge is established at 2 to 15×10^{-3} Torr of argon pressure and with peak-to-peak voltages of the order of 1 kV at 13.56 MHz. The power applied to the system may vary from 100 to 1,000 W depending on the size of the cathode area and the rates desired. Power densities from 0.5 to 4 W cm^{-2} have been investigated, with values around 1.5 W cm^{-2} being most typical. The mechanisms of charge neutralization which permit sputtering of insulators in an rf glow discharge are discussed in detail in Chaps. 3 and 4.

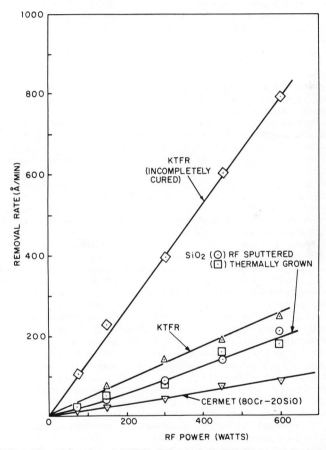

Fig. 18 RF sputter-etch rates of KTFR, SiO$_2$, and Cr-SiO films as functions of power. (*Courtesy of P. D. Davidse.*[137])

The rf sputter-etch rates are proportional to the power input, as shown in Fig. 18. Other film materials, many of them difficult to etch chemically, have been sputter-etched by Tsui[139] with rates shown in Table 13. The rates are constant during prolonged periods of etching. Tsui observed that the rates tend to increase if the surface temperature exceeds 200°C.[139] Complete film removal is recognized by observing the change in reflectivity or color on the surface. In some cases, the sputtered metal also

TABLE 13 RF Sputter-etch Rates of Metals and Insulators after Tsui[139]

Material	Etch rate, $\overset{\circ}{A}\ min^{-1}$
SiO$_2$ (thermally grown)	120
SiO$_2$ (rf sputtered)	120
Fused quartz	120
Si$_3$N$_4$ (pyrolytic)	60
Al$_2$O$_3$ (sapphire)	20–50
KTFR (ultraviolet exposed)	70–300
AZ-340	100–350
Aluminum (evaporated)	120–160
Gold (evaporated)	200–350
Nichrome (sputtered)	70–100
Copper (evaporated)	200–350
Tungsten	50–75

Conditions: 1.6 W cm^{-2} power density, 1.5 kV peak-to-peak rf voltage, 11×10^{-3} Torr argon pressure, 2 cm electrode spacing, quartz substrates of about 190°C surface temperature.

imparts a distinctly different tint to the glow discharge. During the sputtering of Cr-SiO films, for example, the plasma has a bluish hue which changes to purple when the film is completely removed. Termination of the process at exactly the right moment is difficult because of slight local variations in etch rates. However, mild overetching and removal of a thin layer from the material below the film are rarely harmful. Radiation damage from rf sputter etching has not been fully investigated as yet, but the characteristics of bipolar silicon transistors have been found to remain unaffected.[139]

The resolution of rf sputter-etched patterns is solely determined by the photoresist mask since undercutting is virtually absent. Tsui[139] etched 25-μ-diameter holes into 5.5-μ-thick SiO$_2$ films, using a double coating of KTFR, and obtained nearly perpendicular sidewalls with an etch factor $d/\Delta l$ of almost 4 (see Fig. 16). He also etched 3-μ-wide trenches into 6,000-Å-thick SiO$_2$ films. Davidse[137] reported the fabrication of 2-μ-wide lines. RF sputter etching has been used routinely to delineate precision thin film resistors of Cr-SiO with line widths down to 12 μ.[93]

Besides its many virtues, rf sputter etching also has its limitations. As shown in Fig. 18 and Table 13, the photoresist mask is sputtered away at widely variable rates which may exceed those of most other materials. Therefore, the resist mask has to be substantially thicker than the film to be etched. The photoresist sputtering rates are particularly sensitive to increases in temperature and to the presence of residual gases. Tsui[139] found that the deliberate admission of oxygen or hydrogen into the sputtering chamber accelerated the KTFR removal rates about fifteen times while the SiO$_2$ rates were reduced by a factor of about 10. Thus, ions of either species seem to react rather strongly with the polymer molecules, whereas their efficiency in sputtering SiO$_2$ or other film materials is low. Since both hydrogen and oxygen ions can form from water vapor desorbed from the system walls, the residual gas level in the vacuum chamber and the purity of the sputtering gas (argon) have a significant influence on the removal rates and the longevity of the resist mask. The unusually high rates indicated in Fig. 18 for "incompletely cured" photoresist are more likely attributable to temperature and/or gas-purity problems. In the absence of these effects, a KTFR coat of 1 μ thickness is sufficient to etch films up to 2,000 Å thick.[137] Similar considerations apply to positive photoresists. Under extreme conditions (excessive power densities), both types of photoresist deteriorate and char before volatilization occurs.

The removal of photoresist masks after sputtering is extremely difficult. Exposure to the glow discharge probably causes further cross linking of negative photoresists and renders the mask practically insoluble in organic stripper solutions. Swelling in hot solvents and intense swabbing are required to take the polymer off. Although positive resists are not designed to cross-link, they, too, become insoluble. The problem can be mitigated by using post-baking temperatures as low as possible to merely dry and outgas the photoresist coat since no demands are made on mask adhesion. However, this does not completely eliminate the need for some mechanical aid in the stripping process. The deliberate admission of O_2 or H_2 toward the end of the sputter-etching process may be considered as an alternate resist-removal technique [see Sec. $3e(2)$].

c. Pattern Formation without Contact Masks

As the development of integrated circuits progresses toward larger chips and more complex patterns, the yield losses due to pattern defects increase rapidly. Since one of the prime sources of pattern defects is contact printing with emulsion masks, the elimination of this process step promises significant economic advantages. Several alternate techniques such as projection exposure of photoresists and beam writing methods have been investigated.[37] Most of these offer additional advantages such as better line resolution, elimination of the artwork-cutting procedure, and varying degrees of automation to generate the desired pattern.

(1) Image-projection Techniques A photoresist mask for thin film pattern etching can be produced without a contact mask if the reduced image of an intermediate transparency as used in conventional photomask fabrication (see Fig. 10) is projected onto the resist-coated substrate. The optical principles which limit the attainable resolution and image-field sizes in projection printing are the same as discussed in Sec. $3a(2)$. However, important differences arise from three factors. One is the lack of sensitivity of photoresists for the wavelength normally used to expose silver-halide-emulsion plates (5,460 Å). The second factor is interference effects which are encountered when 1-μ or narrower lines are exposed with monochromatic light. The third problem is that step-and-repeat processing as used to fabricate photomasks is too time-consuming and costly to be applied to individual substrates.

The sensitivity of photoresists requires exposure with light of wavelengths shorter than 4,500 Å. To obtain the highest possible line resolution, lenses for projection exposure must be designed to yield minimum aberrations so that their performance is diffraction-limited. This necessitates the use of monochromatic light and of lenses which are corrected for the particular wavelength chosen. Mercury-arc lamps emit at several wavelengths in the ultraviolet spectrum. Two high-intensity lines, at 4,368 and at 4,047 Å, are used in projection systems. For convenience, they are often referred to as the g line (4,368 Å) and the h line (4,047 Å) in allusion to their proximity to corresponding lines in the Fraunhofer spectrum.[35]

The design and performance of lenses intended for use at these two wavelengths have been discussed by several authors.[34,36,38] As stated in Sec. $3a(2)$, the maximum theoretical resolution of diffraction-limited lenses is inversely proportional to the wavelength of the light. Thus, lenses optimized for the g or h lines offer 25 or 35 % better theoretical resolution, respectively, than lenses of equal aperture designed for 5,460 Å. The relative efficiencies of 4,368- and 4,047-Å light in exposing photoresists have been examined by Buzawa et al.[34] and by Lovering.[140] The attainment of uniform field illumination with intensities of at least 100 mW-s cm^{-2} is also important to permit the short exposure times (1 to 10 s) required in practice. Depending on the size of the image field, mercury-arc lamps of 200 to 1,000 W with large-aperture optical systems are commonly utilized.

The various possibilities of projecting images onto photoresist-coated substrates were first discussed systematically by Schuetze and Hennings.[68] The three principal approaches are illustrated in Fig. 19. In every case, an enlarged intermediate transparency—commonly ten times the final size—is required for image projection. The transparency may be produced by reduction of cut artwork or directly by means of an artwork generator [see Sec. $3a(3)$]. The choice of the projection method depends on the resolution and field-size requirements. As previously shown in Fig. 11, maximum resolution and field size of lenses are to some extent mutually exclusive and necessitate practical compromises.

If very high resolution is desired, field size has to be sacrificed. This case is illustrated in Fig. 19A. High-quality microscope objectives yield resolutions of 500 lines/mm and more but only with field sizes of less than 1 mm. A projection system

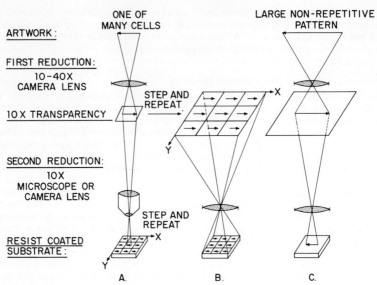

Fig. 19 Image-projection techniques for direct exposure of photoresist.

employing various microscope objectives has been described by Altman and Schmitt.[141]
The line resolution which can be obtained is higher the greater the reduction ratio,
which in turn requires larger numerical apertures. This creates a problem insofar as
the depth of focus of the image is inversely proportional to the square of the numerical
aperture. Consequently, lenses capable of resolving 1-μ-wide or narrower lines have
depths of focus 1 μ or less. This means that very accurate focusing mechanisms are
required, that the image area on the substrate must be flat within 1 μ or better, and
that the resist coat must be very uniform, without graininess, and less than 1 μ thick.
These criteria have been met by several authors who succeeded in etching lines of 1 μ
width or less into various film materials. To illustrate the parameters typical of
submicron line-projection exposure, their results are shown in Table 14.

TABLE 14   Resolution, Field, and Film-thickness Data Reported for Submicron Line-projection Exposure

	Ref. 68			Ref. 66		Ref. 122
Lens:						
N.A..........	0.32	0.63	0.9	0.32	0.65	0.45
Reduction ratio	20:1	25:1	63:1	10:1	25:1	25:1
Etched lines:						
Width, μ......	1.0	0.6	0.4	0.55	0.36	1.0
Useful field di-						
ameter, mm .	2.8	0.7	0.3	Not given		Not given
Etched film:						
Material.......	Tantalum, Nichrome			Chromium		SiO$_2$
Thickness, Å...	300–500			700		1,500, 3,000, 4,500
Photoresist coat:						
Type..........	AZ-1350			KTFR*		AZ-1350
Thickness, Å...	3,000			Not given		600–2,000

* Surface flushed with nitrogen during exposure to avoid oxygen effect.

The most serious problem in the fabrication of micron-size patterns arises from the fact that exposure with monochromatic light causes interference effects. Multiple reflections at the resist-air and resist-substrate interfaces produce variations of light intensity in the photoresist layer perpendicular to the surface. The resulting spatial distributions of light intensity have been discussed by Altman and Schmitt.[142] They are characterized by maxima and minima in different depths which depend on the resist-layer thickness, its refractive index, and the reflectivity of the film material underneath. These one-dimensional intensity distributions, which are comparable with a standing-wave pattern, spread out laterally over small distances because of the diffraction effects discussed in Secs. 3a(2) and 3a(5).

A systematic experimental study of micron-line exposure and the inherent interference effects has been conducted by Middlehoek.[122] His results pertain to photoresist coats on thermally grown SiO_2 films; hence the oxide-silicon interface was one of the reflecting surfaces. When using negative photoresist (KTFR), Middlehoek found it impossible to resolve 1-μ-wide line patterns because either the depth of effectively exposed and cross-linked resist was too shallow to provide protection during the etch (underexposure), or the spaces between lines received enough diffracted light to prevent clear pattern definition in the developer (overexposure). With positive photoresist AZ-1350, underexposure rendered only a partial layer of the resist soluble; hence complete removal of the unwanted portions of the resist was not possible. The effects of overexposure depended on the depth in the resist coat where the light-intensity maximum occurred. If it was near the resist-SiO_2 interface, the bottom layer of the resist became soluble in adjacent regions which were to remain insoluble. Consequently, the entire pattern was washed away in the developer. However, if the intensity maximum was near the resist-air interface, the masking portions of the resist coat became soluble only near the surface, and a usable pattern of somewhat reduced thickness could be obtained.

Projection exposure of photoresist layers on reflecting surfaces has been investigated by Altman and Schmitt.[142] These authors were able to produce micron-line patterns in thin chromium films by exposing KTFR under protective nitrogen gas with light from an unfiltered mercury-arc lamp. However, the top portion of the resist coat separated from the bottom layer near the chromium film and then settled on areas which were to remain bare during the etch. Consequently, unetched patches of chromium film in between the original pattern were obtained. The causes of this phenomenon are, again, standing light waves in the photoresist, producing a laminated structure which is not fully cross-linked and insoluble. The same authors also found considerable line widening as a result of overexposure. Doubling the optimum exposure for 5-μ-wide lines, for example, yielded a 25% increase in width. It was demonstrated that this widening is attributable to diffraction effects rather than to multiple reflections in the resist coat.

These observations and consideration of the oxygen effect associated with KTFR suggest that positive photoresists are to be preferred for projection exposure of micron-line patterns. However, it is necessary to adjust the thin film and resist-coat thicknesses carefully so that constructive interference of the exposing light waves occurs near the resist-air surface. Furthermore, since the remaining resist coat is very thin, the film thickness which can be etched is limited. Middlehoek found thicknesses of 600 to 2,000 Å of AZ-1350 and 1,500, 3,000, and 4,500 Å of SiO_2 to be compatible.[122] For other photoresist and thin film combinations, such data will have to be established in each case.

The extremely small image-field sizes obtained with microscope objectives do not permit the simultaneous exposure of an entire circuit pattern. Typical integrated-circuit chip sizes of current interest are 0.100- to 0.200-in. squares and thus require useful field diameters of 4 to 7 mm diameter. According to the diagram in Fig. 11, such field sizes are compatible with minimum line widths of 1.5 to 2 μ, which are small enough for many applications. Process lenses designed for this performance at ten times reduction have apertures of $F/1.8$ to $F/3.0$, focal lengths of 25 to 30 mm, and depths of focus of 3 to 6 μ (see, for example, Ref. 34). If such a process lens is used instead of the microscope objective shown in Fig. 19A, arrays of integrated-circuit patterns can be projected. While step-and-repeat exposure of individual substrates is time-consuming and not very practical, the method is suitable for the fabrication of chromium masks. The longevity of the latter and their relative freedom from defects are likely to offset the higher cost of this technique.

The use of an enlarged pattern-array mask to process an entire substrate in one exposure is shown in Fig. 19B. This requires that the intermediate transparency of the single-cell artwork is repeatedly photographed on a fairly large emulsion plate. The latter is then projected through a large-field lens. The required field diameter for 1$\frac{1}{4}$-in. silicon wafers is about 30 mm, which permits lines of about 2.5 μ minimum widths. Typically, lenses designed for this application at a 10:1 reduction ratio have apertures of $F/2.0$ to $F/3.0$, focal lengths of 100 to 125 mm, and depths of focus of about 10 μ. Examples have been reported by Schuetze and Hennings[68] (3-μ lines, 28-mm-diameter image field, 10:1 reduction) and by Buzawa et al.[34] (2-μ lines, 18-mm-diameter image field, 4:1 reduction). The same optical system would also be suitable for the projection of large nonrepetitive patterns covering an entire substrate wafer as shown in Fig. 19C.

Since large-field projection printing requires only one exposure per substrate and does not employ contact masks, it is of considerable practical interest. The first commercial projection-exposure systems have already appeared on the market. One model* is capable of projecting transparencies at a 2:1 reduction ratio onto 1$\frac{1}{4}$-in.-diameter substrates with 2-μ line resolution. One of the photorepeaters† is also claimed to permit direct exposure of photoresist.

The most recent technique for direct photoresist exposure without masks is holography. The process has been described by Lu.[143] An enlarged transparency of the mask pattern is projected through the hologram of a pinhole array. Two real images are thereby obtained. These consist of arrays of the reduced mask pattern similar to the arrays formed by multilens systems [see Sec. 3a(4)]. Lu[143] reported resolution of lines wider than 3 μ over fields of 70 mm diameter. At the Bell Telephone Laboratories, 1-μ line resolution over 10-mm-diameter fields are considered to be possible.[144] Thus, holographic image multiplication promises more favorable resolution and field combinations without requiring expensive lenses. At present, the image signal-to-noise ratio is a serious problem. The noise arises from nonlinearity of the recording material, distortion of the hologram surface, and diffraction effects. It appears in the form of graininess and diffraction rings in the reconstructed image. These factors as well as photosensitive materials for preparing holograms have been discussed briefly by Dey and Harrell.[37]

In addition to the more broadly investigated projection techniques involving photoresist masks, methods of forming thin film patterns directly without an etch mask have also been tried. The first example has been reported by Kaplan,[145] who projected a well-focused image of ultraviolet light onto a thin film of lead. Removal of the exposed film portions was accomplished in an atmosphere of nitromethane, which reacts photochemically with the lead and forms volatile tetramethyl lead. The etch rates obtained were below about 15 Å min^{-1}. Sullivan and Kolb[146] attained etch rates of 100 to 1,000 Å min^{-1} by photochemical etching of germanium films in gaseous bromine, which yields volatile $GeBr_4$. They were able to resolve 5-μ-wide lines. The general applicability of this technique is, however, limited to film materials for which a photochemical reaction leading to a volatile compound can be found.

(2) Light-beam Writing The projection-exposure techniques described in the preceding section require enlarged photographic transparencies of the complete circuit layout. The need for intermediate transparencies is eliminated if the patterns to be etched are drawn directly on a resist-coated substrate by means of a programmed light beam. Such a light-beam exposure system has been described by Brennemann et al.[147] and is shown in Fig. 20. The light beam is shaped by a square aperture whose image is reduced ten times and focused on the resist-coated substrate surface. Illumination with yellow light permits viewing of the sample surface and the light spot on a screen

* Kulicke and Soffa Industries, Inc., Fort Washington, Pa., Model 688.
† David W. Mann Company, Burlington, Mass., Model 1595.

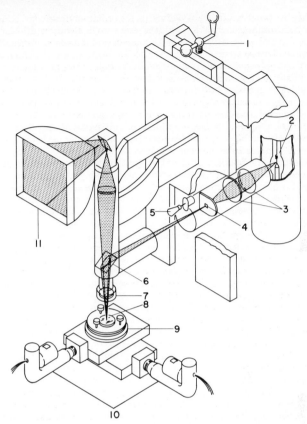

Fig. 20 Light-beam system for the generation of photoresist patterns. 1. Focusing adjustment. 2. Ultraviolet lamp. 3. Condensing-lens system. 4. Square aperture. 5. Tape-controlled shutter. 6. Beam splitter. 7. Lens. 8. Wafer. 9. Adjustable support assembly with vacuum chuck. 10. Tape-controlled X-Y stepping actuators. 11. Viewing screen. (*Courtesy of A. Brennemann and IBM J. Res. Develop.*[147])

at 200× magnification to facilitate registration. The substrate is mounted on an X-Y table with a precision-drive mechanism allowing a positioning accuracy of ±6 μ (250 μin.). Pattern formation is controlled by a programmed magnetic tape. The latter issues instructions to open or close the shutter, determines the duration of the exposure, and selects the direction of travel.

The aperture may be changed manually to vary the size or geometry of the exposed area. In the experiments described,[147] a 50-μ (0.002-in.) square spot was exposed. The table traveled in increments of 25 μ (0.001 in.), thus producing lines by overlapping and double exposure of squares. With this mode of operation, the line ends receive less than the normal amount of exposure while corners receive more. Although it was attempted to compensate for these effects in the program, small pattern irregularities at line ends and corners due to nonoptimized exposure and also to mechanical overshooting were observed. Lines wider than 50 μ (0.002 in.) were made by repeated parallel passes. Subsequent resist development and etching processes were conventional.

The system was used to produce line patterns in 1-μ-thick aluminum films. It is moderately fast, potentially capable of exposing about 5-in. length of line per second. Compared with electron-beam writing, advantages are the lower cost of the apparatus and the ability to process substrate wafers in a normal environment. Satisfactory results have been obtained with line widths down to 5 μ, but widths of 50 μ and greater are considered more amenable to this technique in view of diffraction effects and line-width deviations resulting from variable exposure times. Although an entire substrate may be exposed without image-field limitations, the time required to process wafers individually makes this technique somewhat impractical. The major advantage is the ability to generate line patterns from computer instructions. The variety of beam geometries can be greatly enhanced if the single aperture shown in Fig. 20 is replaced by a series of aperture patterns on a photographic plate. By automatically selecting different aperture shapes, complete circuit patterns can be synthesized.[47] The application of this principle to photographic mask making led to the development of artwork generators as described in Sec. 3a(3).

(3) Electron-beam Techniques Since the limit of line resolution in photolithography is due to the wavelength of ultraviolet light, the growing interest in pattern dimensions of 1 μ has turned the attention of research workers toward resist exposure with electron beams. The latter may readily be focused into 1,000-Å-diameter spots, and even 100 Å diameters have been reported.[148] Image definition is not diffraction-limited because the de Broglie wavelength of 10-kV electrons as commonly employed is of the order of a few tenths of an angstrom. The diameter of image points is limited by spherical aberrations of the electromagnetic lenses, and the latter are negligible as long as the electron-beam angles are very small. For the same reason, the depth of focus for objects of submicron size is greater than 10 μ.[149]

An important prerequisite for electron-beam exposure is a suitable photoresist. The energy of ionizing radiation such as 10-kV electrons is much higher than that of ultraviolet light. Interactions of such electrons with organic polymers are not limited to chromophore groups but may occur anywhere in the molecule. The results are temporarily broken bonds in the macromolecules which may either cross-link to form a three-dimensional polymer network or saturate such that smaller fragments are obtained. In the former case, the solubility of the irradiated product decreases whereas scission of macromolecules causes enhanced solubility. Both mechanisms occur in many organic polymers. Depending on the prevalence of cross linking or degradation, certain compounds act as negative and others as positive resists. The common photoresists as well as other polymer systems have been tested for their suitability as electron-beam resists. A number of such materials and related data are listed in Table 15.

Electron-beam exposure of the common negative photoresists produces cross linking and insolubilization similar to light exposure.[150] However, the reactions induced in the polymer are not well controlled and not confined to the vicinity of the sensitizer molecules. Consequently, the irradiated portions are difficult to strip later on. Furthermore, the electrons are scattered within the resist coat and backscattered from the substrate surface. Therefore, cross linking occurs also in regions adjacent to the irradiated areas. Several authors have reported considerable line broadening[147,149] or slopes as wide as 2 μ along the line edges.[82] Thus, the high resolution of the electron beam cannot easily be realized because of difficulties in controlling the exposure.

The positive resist AZ-1350 seems to suffer mostly scission under electron bombardment because its solubility is enhanced and its working characteristic is maintained. The sensitivity to radiation is somewhat lower than that of the negative photoresists, as demonstrated by the higher minimum dose required. This may be the reason why several authors have succeeded in resolving 1-μ lines with AZ-1350.[149,151]

The difficulties encountered with the common photoresists led several investigators to search for other resist systems which are less likely to show line broadening and which are not sensitive to stray ultraviolet light. Both negative[152] and positive[82] working resists have been found. The principal choice between them depends on whether the area to be etched represents only a small fraction of the total surface, in which case positive resists would be preferable, or whether very narrow openings are required. In the latter case, negative resists offer an advantage because the scattering of electrons tends to reduce the width of openings whereas the opposite effect occurs

TABLE 15 Materials Used as Resists for Electron-beam Exposure

Type	Resist Coat: Material	Thickness [Å]	Developer	Electron Beam Voltage [kV]	Minimum dose [10^{-6} Coul cm^{-2}]	Comments about resist performance as etch mask	Refs.
Negative (crosslinking)	KOR	1,500–2,500	KOR	14	1–3	Pinholes in the resist mask	150)
	KPR	1,500–2,500	KPR	10 and 14	2–5		149, 150)
	KMER	1,500–2,500	3:2:1 toluene/xylene/benzene	14	5		150)
	KTFR	1,500–2,500 / 3,000 / ?	3:2:1 toluene/xylene/benzene / 3:3:1 toluene/xylene/benzene / ?	14 / 14 / 10	3 / ? / 2	Most suitable of all Kodak resists examined.	150) 147) 149)
	Polystyrene, 10% solution in toluene	5,000	xylene	10	100	Resistant to alkali and weak acids. Line edges have < 1 μ wide slopes.	152)
	Poly vinylchloride, 10% solution in cyclohexanone	5,000	cyclohexanone	10	30	Resistant to most acids, alkali, and common solvents.	152)
	Poly acrylamid, 2% solution in water	5,000	water	10	3	Crosslinked polymer swells in aqueous acids and alkali but is resistant to H_3PO_4. Line edges have 2 μ wide slopes.	152)
	AZ-1350	3,600 / ?	AZ-1350	30 / 10	80 / ~30	1 μ wide lines have been attained.	151) 149)
Positive (degrading)	Poly (methyl methacrylate), 10% solution in methyl isobutyl ketone	5,000	3:1 isopropanol/methyl isobutyl ketone. Alternate: 95% ethanol –5% water	10 / 10	50 / 50	< 0.8 μ line resolution. Good resistance to buffered HF and hot conc. HCl etches.	82) 152)
	Poly (α –methylstyrene), 15% solution in trichloro-ethylene	5,000	1:6 benzene/ethanol	10	100	2 μ line resolution. Satisfactory resistance to hot conc. HCl but poor for buffered HF.	82)
	Poly isobutylene, 5% solution in trichloro ethylene	5,000	7:7:6 methylene chloride/benzene/ethanol	10	50	2 μ line resolution, rubbery layer creeps. Satisfactory resistance to buffered HF and hot conc. HCl.	82)
	Cellulose acetate, 4% solution in 1:1 cyclohexanone/methyl ethyl ketone	5,000	1:1:3 methyl ethyl ketone/ethanol/toluene	10	500	< 0.8 μ line resolution. Poor chemical stability, hydrolyses in acids.	82)

with positive resists.[149]   Other factors to be considered are the tendency of the resist material toward line broadening and its resistance to chemical etches.

Negative-working electron-beam resists have been examined by Ku and Scala,[152] and their three systems are listed in Table 15.   For each polymer it was found that the minimum radiation dose required for satisfactory exposure was inversely proportional to the average molecular weight of the polymer in the initial solution.   The physical significance of this observation is that fewer cross-linking bonds are needed to produce an insoluble three-dimensional network the longer the polymer chains are initially. To develop the exposed pattern, a solvent is required which discriminates between uncross-linked chains and cross-linked fractions.   Thus, the developers in this category are the same or very similar to those used to prepare the resist solutions.   Fairly high initial molecular weights are desirable to achieve a high radiation chemical yield. Although monodispersity is neither necessary nor obtainable, a narrow distribution of molecular weights is advantageous to promote line-width control and reduce the chances of pinhole formation.   Ku and Scala[152] found average molecular weights of 60,000 for polystyrene and of 32,000 for polyvinylchloride to yield good results.

Electron bombardment of degrading resists with random scission of bonds leads to an exponential distribution of molecular weights which spreads from very low to high values.[152]   Fractional solution of the irradiated portions in the developer requires significant solubility differences between the degraded and nondegraded polymer. This can be attained only if the molecular-weight distributions of the two products are well separated.   Consequently, the minimum radiation dose does not depend on the initial molecular weight but rather on the radiation chemical yield of the polymer compound; the latter determines how many electrons are needed to attain a distribution of scission products which is sufficiently below the initial molecular-weight distribution.   For good resolution and low pinhole incidence, high initial molecular weights are recommended, i.e., 1 to 2×10^5 for poly(methyl methacrylate) and poly-(α-methyl styrene).[152]

Polymers which are suitable as positive resists have been examined by Haller et al.,[82] and their materials are also shown in Table 15.   Empirically determined mixtures of solvents and nonsolvents are required as developers to separate effectively the degraded from the unexposed fractions.   As indicated in the last column of Table 15, the chemical stability of the polymer styrene masks and their ability to resolve fine lines varied among the materials.   Poly(methyl methacrylate) was found to combine superior line resolution with good stability in the acid etches employed. Pre-baking at 150°C and post-baking at 130°C were found to strengthen the adhesion and minimize undercutting when etching SiO_2.   Two-micron-wide lines with well-defined edges were achieved in 2,600-Å-thick SiO_2 and 1,000-Å-thick aluminum films. Thereafter, the resist mask could be stripped in acetone.

The formation of a resist image by electron-beam exposure may be accomplished in three different ways which have all been investigated.   If an electron beam is utilized, patterns can be drawn by keeping the beam fixed and moving the substrate on an X-Y stage, or the substrate may be fixed and the electron beam deflected.   The third technique employs a large-diameter electron beam to project an image which covers the entire substrate or a major portion of it in one exposure.

Electron-beam exposure with a mechanical stage has been implemented by Brennemann et al.[147]   Their system is shown in Fig. 21.   The electron-microscope gun with two magnetic lenses produced focused spots of 1 μ diameter.   The X-Y stage achieved a positioning accuracy of ± 12.5 μ (0.0005 in.) over a 2- by 2-in. field by translating the revolving motion of the stepping motor through a steel tape into linear travel.   Line patterns were drawn by moving the stage under the electron beam. The stage motion was tape-controlled as described for light-beam writing.   To produce 50-μ (0.002-in.)-wide lines, the electron beam was defocused to this diameter. Broader lines could be obtained by making repeated parallel passes.   Narrower lines of 7.5 μ (0.0003 in.) were also produced, but although they were continuous, their width fluctuated significantly because of variations in speed.

A common problem in electron-beam exposure is the attainment of pattern registration between consecutive process steps.   The technique used by Brennemann et al.[147]

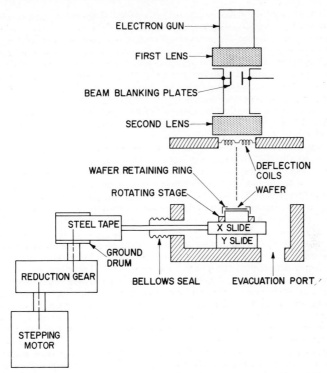

Fig. 21 Schematic of an electron-beam system for the generation of photoresist patterns. (*Courtesy of A. Brennemann and IBM J. Res. Develop.*[147])

required alignment of the substrate on the X-Y stage with a microscope to assure orthogonality of movement. Registration in the exposure system was accomplished by bringing a reference detail on the wafer surface into view. This detail was scanned with the 1-μ electron beam and its secondary electron-emission image observed on a cathode-ray tube (CRT). The starting point of the exposure could thereby be adjusted with a precision of 2.5 μ.

The technique is not particularly fast because of the mechanical stage motion involved. Interconnecting a group of circuits required 30 min on a $1\frac{1}{4}$-in. wafer, but speeds of 500 in. s^{-1} are thought to be possible. The large image field provided by the traveling stage is a definite advantage over other methods of exposure. However, the mechanism also limits the positioning accuracy to a degree which appears incompatible with the fabrication of micron-line patterns.

The second method of resist-mask formation is based on deflecting the electron beam to draw the desired pattern. Several authors have described systems which are modified scanning electron microscopes.[37,149,115] The main beam is focused to a spot of typically 1,000 Å diameter on the substrate surface. To draw a pattern, the beam is directed by scanning coils; additional pulsing coils, which deflect the beam away from the aperture of the electron lens, turn the beam off. The pattern is stored in the form of a photographic image plate[37] or as a chromium mask[149] in front of a CRT. The image is scanned by a microdensitometer[37] or by a flying-spot-scanner pattern generator[151] synchronously with the deflection of the main electron beam. One system permits contrast reversal so that the drawn pattern can be made a positive or a negative of the stored image.[149] The scan generator also drives the scanning coils of a second CRT which displays the composite picture of mask and substrate surface.

This facilitates mask-to-wafer alignment of better than 0.1 μ over a 50-μ-square scanning area.[149]

Registration by viewing the secondary electron-emission image of the reference pattern on the resist-coated wafer is problematical. It involves the risk of exceeding the critical radiation dose while trying to locate the reference detail. Other alignment techniques, which utilize the photovoltage signals from different regions of a field-effect transistor, or the voltage contrast produced by biasing an N-type silicon wafer with diffused P-type regions, have also been tried.[151] Alignment is less critical if transistor structures are made which consist of narrow lines intersecting each other at right angles.[151] Pattern details of 1-μ dimensions have been reported in several instances.[149,151]

Resist exposure with a scanning beam is a relatively slow technique because the beam must sweep the entire image area. However, it is comparatively easy to implement. Some process time may be saved by fabricating the larger parts of a pattern with conventional photolithography and exposing only the fine-line details with the electron beam.[149] An alternate method is to deflect the electron beam such that it traces the desired pattern directly without scanning.[37,149] This method lends itself well for automation because the path of the beam may be controlled by a computer program.

The problem with all beam-deflection techniques including scanning beams is that only small areas of a few thousand square microns can be covered in one exposure. At larger beam deflections, the images are distorted. Therefore, exposure of areas sufficiently large for integrated circuits and of pattern arrays requires some form of step-and-repeat technique. This makes it necessary to align each pattern area prior to exposure, a tedious and rather impractical procedure if performed manually. However, this problem can be overcome by automation. Miyauchi et al.[153] have recently described a scanning-electron-beam exposure system with an X-Y stage that moves in 1-μ steps. Moiré fringes are used to indicate the length of travel and to correct automatically for proper stage position. The smallest reported line width is 0.6 to 0.7 μ and is limited by the resolution of the resist film. The useful image area is 2 mm². No image plate or mask is needed because the system accepts pattern data from a computer.

The third method of electron exposure is analogous to image formation with light optics. It requires a metal mask with the enlarged pattern cut or etched into it and an electron microscope used in reverse to form a reduced image.[149] Disadvantages of this technique are the smaller depth of focus and the limitations imposed by the metal mask. An alternate method which also allows large-area ($\sim$1 in. diameter) single exposures has been described by O'Keeffe et al.[154] It requires a photocathode consisting of a glass plate with a very thin metal film into which the entire pattern array is etched. Illumination of the metal film with ultraviolet light generates secondary electrons. An electron image is formed without reduction by focusing the electrons with coaxial electromagnetic fields. Since part of the ultraviolet light reaches the substrate surface, it is essential that the resist is not photosensitive.[152] The depth of focus ($\sim$25 μ) is greater than that of light optical systems, and lines of 2 μ and wider have been resolved.

In conclusion, electron exposure of resists has definite advantages such as high resolution, linear beam speeds potentially as high as 500 in. s^{-1}.[147] and the possibility of programming the beam path. However, these factors have to be weighed against a number of problems such as the high cost of the equipment, the complexity of the electronic controls, the access delays resulting from working in a vacuum system, the long exposure times inherent in all beam writing techniques, and the difficulties associated with pattern alignment. It remains to be seen if the yield of very small devices fabricated by electron-beam techniques is sufficiently high—because of the vacuum working environment and the absence of a contact mask—to make the method economically attractive.

As in the case of light-projection techniques, there are also thin film materials whose chemical properties are sensitive to exposure with electrons and which can be etched without a resist. One example has been reported by O'Keeffe and Handy,[155] who bombarded thermally grown SiO_2 films of 5,000 to 10,000 Å thickness with 10-kV electrons. This treatment changes the defect structure of the oxide sufficiently to enhance the etch rate [P-etch, see Sec. $3d(12)$] by a factor of 3 to 4. Subsequent etching produces an SiO_2 pattern of somewhat reduced film thickness covering the unexposed regions. Etched windows as small as 0.5 by 5 μ in SiO_2 films have been obtained.[155] An example of electron-beam-induced chemical reaction has been demonstrated with aluminum oxide substrates coated with a thin layer of carbon powder. According to Schwarz,[156] electrons impinging on the surface of this sandwich supply enough energy to reduce the aluminum oxide and form volatile CO_2. Thus, conductive line patterns of aluminum are produced along the path of the electron beam. However, these techniques are not universally applicable and are subject to limitations arising from film and substrate materials.

REFERENCES

1. DeLano, R. B., Jr., U.S. Government Contract NObsr 77508, Final Summary Rept., p. 73, 1961.
2. Stutzman, G. R., *Proc. 3d Ann. Microelectronics Symp.*, 1964, p. 11-D-1, St. Louis IEEE.
3. Jaques, F., and J. Schmidt, *Modern Plastics*, **38**(4), 109 (1960).
4. Lessor, A. E., R. E. Thun, and L. I. Maissel, *IEEE Spectrum*, **1**(4), 72 (1964).
5. Hawkes, P. L., G. W. Bowen, and S. H. Mather-Lees, *Microelectronics Reliability*, **4**, 65 (1965).
6. Fuchs, H., and K. Heine, *Microminiaturization*, **5**, 335 (1966).
7. Nunn, F. A., and D. S. Campbell, *J. Sci. Instr.*, **40**, 337 (1963).
8. Keister, F. Z., R. D. Engquist, and J. H. Holley, *IEEE Trans. Component Pts.*, **CP-11** (1), 33 (1964).
9. Strauss, W. A., Jr., *SCP Solid State Technol.*, **9**(2), 15 (1966).
10. Greenhouse, H. M., R. T. Galla, W. W. Richardson, W. C. Vergara, and T. H. Yaffee, "Thin Film Microcircuit Interconnections," Final Report ECOM-01482-F, The Bendix Corp., Bendix Radio Division, Baltimore, Md., for U.S. Army Electronics Command, Fort Monmouth, N.J., 1967.
11. Thun, R. E., W. N. Carroll, C. J. Kraus, J. Riseman, and E. S. Wajda, in E. Keonjian (ed.), "Microelectronics," p. 200, McGraw-Hill Book Company, New York, 1963.
12. Gregor, L. V., and M. F. Gendron, *IBM Tech. Disclosure Bull.*, **12**, 149 (1968).
13. Randlett, M. R., E. H. Stroberg, and K. L. Chopra, *Rev. Sci. Instr.*, **37**, 1378 (1966).
14. Weimer, P. K., in G. Hass and R. E. Thun (eds.), "Physics of Thin Films," vol. 2, p. 178, Academic Press Inc., New York, 1964.
15. A review of thin film circuit deposition through masks is given in Ref. 11, p. 173.
16. Gregor, L. V., *IBM J. Res. Develop.*, **12**, 149 (1968).
17. Gregor, L. V., and H. L. McGee, *Proc. 5th Intern. Electron Beam Symp.*, 1963, p. 211, Alloyd Electronics Corp., Boston, Mass.
18. Gregor, L. V., H. L. McGee, and E. L. Kenny, Electrochemical Society Meeting, May, 1968, Boston, Mass., Dielectrics and Insulation Division Abstract 26.
19. Caswell, H. L., and Y. Budo, *Solid-State Electron.*, **8**, 479 (1965).
20. Caswell, H. L., L. V. Gregor, and H. L. McGee, U.S. Patent 3,392,051, 1968.
21. Ames, I., L. V. Gregor, A. Leiner, and A. Toxen, U.S. Patent 3,239,374, 1966.
22. Castrucci, P. P., R. H. Collins, and W. R. Marzinsky, *Reprographics*, **4**, 11 (1966).
23. Madland, G. R., H. K. Dicken, R. D. Richardson, R. L. Pritchard, F. H. Bower, and D. B. Kret, "Integrated Circuit Engineering—Basic Technology," 4th ed., Boston Technical Publishers, Inc., 1966.
24. "An Introduction to Photofabrication Using Kodak Photosensitive Resists," Eastman Kodak Company, Rochester, N.Y., 1966.
25. Ostapkovich, P. L., in C. R. Hance (ed.), "Ultra-microminiaturization Precision Photography," p. 11, Society Photographic Scientists and Engineers, Inc., Washington, D.C., 1968; see also *Solid State Technol.*, **12**(7), 53 (1969).
26. Holthaus, D. J., *Proc. 2d Kodak Seminar Microminiaturization*, 1966, p. 12.
27. Maple, T. G., *SCP Solid State Technol.*, **9**(8), 23 (1966).
28. Levine, J. E., *Kodak Photoresist Seminar Proc.*, 1968 ed., vol. 1, p. 5.
29. Koford, J. S., P. R. Strickland, G. A. Sporzynski, and E. M. Hubacher, *Proc. Fall Joint Computer Conf.*, 1966, p. 229, Spartan Books, Washington, D.C; see also Strickland, P. R., and B. J. Crawford, *Solid State Technol.*, **10**(7), 31 (1967).
30. Feller, A., and M. D. Agostino, *Dig. Computer Group Conf.*, June, 1968, p. 23, Institute of Electrical and Electronics Engineers, Inc., New York.

31. Stevens, G. W. W., "Microphotography," Chapman & Hall, Ltd., London, 1957.
32. "Techniques of Microphotography," pamphlet P-52, Eastman Kodak Co., Rochester, N.Y., 1967.
33. Rottmann, H. R., "Optics in Microphotography," presented at the 1968 Conference of the Society of Photographic Scientists and Engineers; see also *Image Technology*, **11**(9), 13 (1969).
33a. Altman, J. H., *Solid State Technol.*, **12**(7), 34 (1969).
34. Buzawa, M. J., G. G. Milne, and A. M. Smith, in C. R. Hance (ed.), "Ultra-microminiaturization Precision Photography," p. 11, Society of Photographic Scientists and Engineers, Inc., Washington, D.C., 1968.
35. Held, S., in C. R. Hance (ed.), "Ultra-microminiaturization Precision Photography," p. 141, Society of Photographic Scientists and Engineers, Inc., Washington, D.C., 1968.
36. Tibbets, R. E., and J. S. Wilczynski, *IBM J. Res. Develop.*, **13**, 192 (1969).
37. Dey, J., and S. Harrell, in C. R. Hance (ed.), "Ultra-microminiaturization Precision Photography," p. 161, Society of Photographic Scientists and Engineers, Inc., Washington, D.C., 1968.
38. Held, S., in C. R. Hance (ed.), "Ultra-microminiaturization Precision Photography," p. 195, Society of Photographic Scientists and Engineers, Inc., Washington, D.C., 1968.
39. Cooperman, H. L., in C. R. Hance (ed.), "Ultra-microminiaturization Precision Photography," p. 1, Society of Photographic Scientists and Engineers, Inc., Washington, D.C., 1968.
40. Held, S., *Solid State Technol.*, **12**(5), 63 (1969).
41. Helmers, T. C., Jr., and J. R. Nall, *Semicond. Prod.*, **4**(1), 37 (1961).
42. Barnes, R. D., *Proc. 1st Kodak Seminar Microminiaturization*, 1965, p. 6.
43. Bell, E., *Proc. 2d Kodak Seminar Microminiaturization*, 1966, p. 15.
44. Beeh, R. C. M., *Electronics*, **41**(3), 78 (1968).
45. *Electronics*, **40**(22), 48 (1967).
46. Fok, S. M., in C. R. Hance (ed.), "Ultra-microminiaturization Precision Photography," p. 115, Society of Photographic Scientists and Engineers, Inc., Washington, D.C., 1968.
47. Cook, P. W., W. E. Donath, G. A. Lemke, and A. E. Brennemann, *IEEE J. Solid State Circuits*, **SC 2**(4), 190 (1967).
48. Rudge, W. E., W. E. Harding, and W. E. Mutter, *IBM J. Res. Develop.*, **7**, 146 (1963).
49. Dill, F. H., "Photographic Mask Making," Research Note NC-434 (1964), IBM Corp., T. J. Watson Research Center, Yorktown Heights, N.Y.
50. Murray, J. J., and R. Maurer, *Semicond. Prod.*, **5**(2), 30 (1962).
51. See Ref. 31, p. 44.
52. De Belder, M., H. Philippaerts, R. Duville, and D. Schultze, in C. R. Hance (ed.), "Ultra-microminiaturization Precision Photography," p. 153, Society of Photographic Scientists and Engineers, Inc., Washington, D.C., 1968.
53. Strauss, W. A., Jr., *Proc. 1st Kodak Seminar Microminiaturization*, 1965, p. 12.
54. McCarny, C. S., *Phot. Sci. Eng.*, **10**, 314 (1966).
55. Martinson, C. E., *Proc. 2d Kodak Seminar Microminiaturization*, 1966, p. 31.
56. Altman, J. H., *Proc. 2d Kodak Seminar Microminiaturization*, 1966, p. 4.
57. Hance, C. R., "Low Density Mask Technique for High Reduction Photography," presented at the Annual Conference of Photographic Scientists and Engineers, San Francisco, Calif., 1966; also available as IBM TR 22.262.
58. Fritz, G. W., and A. Lavoto, *Kodak Photoresist Seminar Proc.*, 1968 ed., vol. 1, p. 39.
59. Levine, J. E., *Kodak Photoresist Seminar Proc.*, 1968 ed., vol. 2, p. 55.
60. Levine, J. E., in C. R. Hance (ed.), "Ultra-microminiaturization Precision Photography," p. 83, Society of Photographic Scientists and Engineers, Inc., Washington, D.C., 1968.
61. Geikas, G. I., and B. D. Ables, *Kodak Photoresist Seminar Proc.*, 1968 ed., vol. 2, p. 47.
62. Martinson, L. E., *Proc. 1st Kodak Seminar Microminiaturization*, 1965, p. 49.
63. Kerwin, R. E., and C. V. Stanionis, *Electrochem. Technol.*, **6**, 463 (1968).
64. Stelter, M. K. J., *SCP Solid State Technol.*, **9**, 60 (March, 1966).
65. Rogel, A., *Rev. Sci. Instr.*, **37**, 1416 (1966).
66. "Incidental Intelligence about Kodak Resists," vol. 6, no. 2, p. 6, Eastman Kodak Co., Rochester, N.Y., 1968.
67. Coates, A. E., *Kodak Photoresist Seminar Proc.*, 1968 ed., vol. 1, p. 34.
68. Schuetze, H. J., and K. E. Hennings, *SCP Solid State Technol.*, **9**(7), 31 (1966).
69. Kosar, J., "Light Sensitive Systems," John Wiley & Sons, Inc., New York, 1965.
70. Bates, T. R., *Proc. 1st Kodak Seminar Microminiaturization*, 1965, p. 58.
71. Oster, G., *Phot. Sci. Eng.*, **4**, 237 (1960).

72. Allen, C. F. H., and J. A. Van Allan, U.S. Patent 2,566,302, 1947.
73. Minsk, L. M., U.S. Patent 2,725,372, 1951.
74. Robertson, E. M., and W. West, U.S. Patent 2,732,301, 1952.
75. Murray, J. J., and G. W. Leubner, U.S. Patent 2,739,892, 1953.
76. Levine, H. A., L. G. Lesoine, and J. A. Offenbach, *Kodak Photoresist Seminar Proc.*, 1968 ed., vol. 1, p. 15.
77. Leubner, G. W., and C. C. Unruh, U.S. Patent 3,257,664, 1961.
78. Harrell, S. A., *Proc. 2d Kodak Seminar Microminiaturization*, 1966, p. 50.
79. Kornfeld, W., *Proc. 2d Kodak Seminar Microminiaturization*, 1966, p. 23.
80. Hepher, M., and H. M. Wagner, U.S. Patent 2,852,379, 1955.
81. See, for example, Neugebauer, W., F. Endermann, and M. K. Reichel, U.S. Patent 3,201,239, 1959.
82. Haller, I., M. Hatzakis, and R. Srinivasan, *IBM J. Res. Develop.*, **12**, 251 (1968).
83. "Applications Data for Kodak Photosensitive Resists," pamphlet P-91, Eastman Kodak Co., Rochester, N.Y., 1966.
84. Bryan, W. S., *Proc. 1st Kodak Seminar Microminiaturization*, 1965, p. 33.
85. Lawson, J. C., *Proc. 1st Kodak Seminar Microminiaturization*, 1965, p. 22.
86. Schwartz, G. C., *Proc. 1st Kodak Seminar Microminiaturization*, 1965, p. 41.
87. Taylor, C. J., *Proc. 1st Kodak Seminar Microminiaturization*, 1965, p. 17.
88. Harrell, S. A., and G. M. Streater, *Kodak Photoresist Seminar Proc.*, 1968 ed., vol. 1, p. 18.
89. Damon, G. F., *Proc. 2d Kodak Seminar Microminiaturization*, 1966, p. 36.
90. Henriksen, G. M., in C. R. Hance (ed.), "Ultra-microminiaturization Precision Photography," p. 205, Society of Photographic Scientists and Engineers, Inc., Washington, D.C., 1968.
91. Damon, G., *Kodak Photoresist Seminar Proc.*, 1968 ed., vol. 2, p. 20.
92. Htoo, M. S., "New Method for Photoresist Exposure Determination," presented at the Annual Meeting of the Society of Photographic Scientists and Engineers, Chicago, Ill., May 15, 1967; also available as IBM TR 22.406; also Htoo, M. S., *Kodak Photoresist Seminar Proc.*, 1968 ed., vol. 1, p. 25.
93. Glang, R., and P. M. Schaible, *Thin Solid Films*, **1**, 309 (1967/68).
94. Kelley, R., *Proc. 1st Kodak Seminar Microminiaturization*, 1965, p. 38.
95. "Process Instructions for AZ-1350 Photoresist," Leaflet PI-1350, Shipley Co., Inc., Massachusetts.
96. Skaggs, C. W., *Electronics*, **37**(18), 94 (1964).
97. Mueller, E., *Z. Phys. Chem.*, **159**, 68 (1932).
98. Woitsch, F., *Solid State Technol.*, **11**(1), 29 (1968).
99. "Incidental Intelligence about Kodak Resists," vol. 7, no. 1, p. 4, Eastman Kodak Co., Rochester, N.Y., 1969.
100. Waits, R. K., *Trans. Met. Soc. AIME*, **242**, 490 (1968).
101. Greer, W. N., *Plating*, **48**, 1095 (1961).
102. Maddocks, F. S., IBM Corp., East Fishkill Facility, Hopewell Junction, N.Y., private communication, 1962.
103. Sayers, J. R., Jr., and J. Smit, *Plating*, **48**, 789 (1961).
104. Zyetz, M. C., and A. M. Despres, *Ext. Abstr. 13th AVS Symp.*, 1966, p. 169, Herbick and Held Printing Co., Pittsburgh, Pa.
105. Huber, F., W. Witt, and I. H. Pratt, *Proc. Electron. Components Conf.*, 1967, p. 66.
106. Dalton, J. V., *Electron. Div. Abstr. 23*, The Electrochemical Society Meeting, Cleveland, Ohio, May, 1966.
107. Wied, O., and M. Tierman, "Planar Microminiaturized Nickel-Chromium Resistors," TP-64-7, Sprague Electric Co., North Adams, Mass., 1964.
108. Chance, D., IBM Corp., East Fishkill Facility, Hopewell Junction, N.Y., private communication.
109. Dey, J., M. Lundgren, and S. Harrell, *Kodak Photoresist Seminar Proc.*, 1968 ed., vol. 2, p. 4.
110. Brown, D. M., W. E. Engeler, M. Garfinkel, and F. K. Heumann, *J. Electrochem. Soc.*, **114**, 730 (1967).
111. Pliskin, W. A., and R. P. Gnall, *J. Electrochem. Soc.*, **111**, 872 (1964).
112. Pliskin, W. A., and H. S. Lehman, *J. Electrochem. Soc.*, **112**, 1013 (1965).
113. Sarace, J. C., R. E. Kerwin, D. L. Klein, and R. Edwards, *Solid-State Electron.*, **11**, 653 (1968).
114. Vromen, B., IBM Corp., East Fishkill Facility, Hopewell Junction, N.Y., private communication.
115. Grossman, J., and D. S. Herman, *J. Electrochem. Soc.*, **116**, 674 (1969).
116. Sullivan, M. V., *Proc. 1st Kodak Seminar Microminiaturization*, 1965, p. 30.
117. Kodak Pamphlet P-194, Eastman Kodak Company, Rochester, N.Y., 1969.

118. Bersin, R. L., *Intern. Plasma Bull.* **1**, nos. 3 and 4, International Plasma Corp., Hayward, Calif., 1968; ashers are also available from Tracerlab, Richmond, Calif.
119. Irving, S. M., *Kodak Photoresist Seminar Proc.*, 1968 ed., vol. 2, p. 26.
120. Burrage, P. M., "Oxidative Removal of Photoresists," recent newspaper presented at the fall, 1967, Meeting of the Electrochemical Society, Chicago, Ill.
121. Schuetze, H. J., and K. E. Hennings, *Electronics*, **38**(25), 237 (1965).
122. Middlehoek, S., "Projection Masking, Thin Photoresist Layers, and Interference Effects," Report RZ-294, IBM Corp., Research Laboratory, Zürich, Switzerland, 1968; to be published.
123. Tulumello, J., and W. B. Harding, *Plating*, **54**, 1234 (1967).
124. Johnson, C., IBM Corp., T. J. Watson Research Laboratory, Yorktown Heights, N.Y., private communication, 1969.
125. Poppert, P. E., *Proc. 5th Ann. Microelectronics Symp.*, 1966, p. 1F-1, St. Louis IEEE.
126. Bergh, A. A., *J. Electrochem. Soc.*, **112**, 457 (1965).
127. Lussow, R. O., *J. Electrochem. Soc.*, **115**, 660 (1968).
128. Bortfield, R. G., *Kodak Photoresist Seminar Proc.*, 1968 ed., vol. 2, p. 30.
129. Plough, C. T., W. F. Crevier, and D. O. Davis, *Proc. 2d Kodak Seminar Microminiaturization*, 1966, p. 44.
130. Lawson, T. R., Jr., *SCP Solid State Technol.*, **9**(7), 22 (1966).
131. Murphy, C. S., *Electron. Reliability Microminiaturization*, **2**, 235 (1963).
132. Spriggs, R. S., and A. J. Learn, *Solid-State Electron.*, **10**, 353 (1967).
133. Valetta, R. M., IBM Corp., Burlington, Vt., private communication.
134. Spivak, G. V., I. N. Prilezhaeva, and O. I. Savochkina, *Doklady Akad. Nauk SSSR*, **88**, 511 (1953); **114**, 1001 (1957).
135. Lepselter, M. P., *Bell System Tech. J.*, **45**, 247 (1966).
136. Anderson, G. S., W. N. Mayer, and G. K. Wehner, *J. Appl. Phys.*, **33**, 2991 (1962).
137. Davidse, P. D., *J. Electrochem. Soc.*, **116**, 100 (1969).
138. Heil, R., S. Hurwitt, and W. Huss, *Solid State Technol.*, **11**(12), 42 (1968).
139. Tsui, R. T. C., *Solid State Technol.*, **10**(12), 33 (1967).
140. Lovering, H. B., in C. R. Hance (ed.), "Ultra-microminiaturization Precision Photography," p. 179, Society of Photographic Scientists and Engineers, Inc., Washington, D.C., 1968.
141. Altman, J. H., and H. C. Schmitt, Jr., in C. R. Hance (ed.), "Ultra-microminiaturization Precision Photography," p. 24, Society of Photographic Scientists and Engineers, Inc., Washington, D.C., 1968.
142. Altman, J. H., and H. C. Schmitt, Jr., *Kodak Photoresist Seminar Proc.*, 1968 ed., vol. 2, p. 12.
143. Lu, S., *Proc. IEEE*, **56**(1), 116 (1968).
144. Field, R. K., *Electron. Design*, **14**(15), 17 (1966).
145. Kaplan, L. N., *J. Phys. Chem.*, **68**, 94 (1964).
146. Sullivan, M. V., and G. A. Kolb, *Electrochem. Technol.*, **6**, 430 (1968).
147. Brennemann, A. E., A. V. Brown, M. Hatzakis, A. J. Speth, and R. F. M. Thornley, *IBM J. Res. Develop.*, **11**, 520 (1967).
148. Pease, R. F. W., and W. E. Nixon, *J. Sci. Instr.*, **42**, 81 (1965).
149. Perkins, K. D., and R. Bennett, *Kodak Photoresist Seminar Proc.*, 1968 ed., vol. 2, p. 39.
150. Thornley, R. F. M., and T. Sun, *J. Electrochem. Soc.*, **112**, 1151 (1965).
151. Larkin, M. W., and R. K. Matta, *Solid-State Electron.*, **10**, 491 (1967).
152. Ku, H. Y., and L. C. Scala, *J. Electrochem. Soc.*, **116**, 980 (1969).
153. Miyauchi, S., K. Tanaka, and J. C. Russ, *Solid State Technol.*, **12**(7), 43 (1969).
154. O'Keeffe, T. W., J. Vine, and R. M. Handy, *Ext. Abstr. Dielectrics Insul. Div.*, The Electrochemical Society, Inc., spring meeting, Boston, Mass., May, 1968, Abstr. 165.
155. O'Keeffe, T. W., and R. M. Handy, *Solid-State Electron.*, **11**, 261 (1968).
156. Schwarz, H. J., U.S. Patent 3,056,881, 1961.

Part Two

The Nature of Thin Films

Chapter **8**

Condensation, Nucleation, and Growth of Thin Films

CONSTANTINE A. NEUGEBAUER

General Electric Research and Development Center, Schenectady, New York

LIST OF SYMBOLS

a_0	adsorbed monomer jump distance
a_1, a_2, a_3	geometric constants describing the shape of an aggregate
α_T	thermal-accommodation coefficient
α_s	sticking coefficient
d	film thickness
D	surface-diffusion coefficient of adsorbed monomers
E	energy of desorbed atom after equilibration with substrate
E_a	activation potential energy for adsorbed monomer desorption (positive)
E_b	bond energy between nearest neighbors in an aggregate
E_d	activation potential energy for surface diffusion for adsorbed monomers (positive)
E_i	dissociation energy of an aggregate containing i monomers into i adsorbed monomers
$E_i{}^*$	dissociation energy of a critical aggregate into adsorbed monomers
E_r	energy of desorbed atom before equilibration with substrate
E_v	incident kinetic energy of vapor atom
F	fraction of substrate covered by aggregates and their associated capture zones
ΔG	total Gibbs free energy of an aggregate with respect to dissociation into the vapor phase
ΔG^*	Gibbs free energy of formation of an aggregate of critical size (positive)
ΔG_{des}	Gibbs free energy of activation for monomer desorption (positive)
ΔG_{sd}	Gibbs free energy of activation for surface diffusion of adsorbed monomers (positive)
ΔG_v	Gibbs free energy of condensation of the film material in the bulk under the conditions of supersaturation of the experiment
γ_i	evaporation rate of aggregates containing i monomers
h	number of monomers in a cluster not in direct contact with the substrate
ΔH_s	heat of sublimation
ΔH_{vap}	heat of vaporization
i	number of atoms or monomers in an aggregate
i^*	number of atoms or monomers in a critically sized aggregate
I	nucleation rate
I^*	rate of formation of critical nuclei, $cm^{-2} sec^{-1}$, nucleation frequency
j	rate at which adsorbed monomers are incorporated into the nucleus
m_a	sites visited by diffusing adsorbed monomer during time τ_a
m_c	sites visited by diffusing adsorbed monomer during time τ_c
$\dot{M}$	mass impingement rate, $g\ cm^{-2}\ sec^{-1}$
$M(t)$	mass of film material condensed at time t
n^*	metastable density of critical nuclei
n_1	steady-state adsorbed monomer concentration
N	density of stable nuclei

N_0	density of adsorption sites on substrate
$N(t)$	density of supercritical clusters at time t
N_∞	saturation density of stable nuclei
ν	frequency of substrate lattice vibrations
ν_1	attempt frequency in small cluster model
ν_0	attempt frequency in capillarity model
r	radius of an aggregate
r^*	radius of a critically sized aggregate
s	capture-zone radius
$\mathcal{R}$	impingement rate
$\mathcal{R}_s$	reevaporation rate of adsorbed monomers from the substrate
$\mathcal{R}_e(b)$	evaporation rate of the film material in the bulk at the substrate temperature
R_g	growth rate of a nucleus within a capture-zone area
ΔS_{vap}	entropy of vaporization
σ	surface free energy
σ_{e-v}	specific free-edge energy
σ_{s-c}	substrate–condensate interfacial free energy
σ_{s-v}	substrate–vapor surface free energy
σ_{v-c}	condensate–vapor surface free energy
t	time
T	substrate temperature
$T_{i-(i+1)}$	substrate temperature at which the critical nucleus changes from an i-atom cluster to an $(i + 1)$-atom cluster
T_0	substrate temperature above which condensation is initially incomplete
T_r	temperature corresponding to energy of desorbed atom before equilibration with substrate
T_v	temperature corresponding to the incident kinetic energy of vapor atom
τ_a	mean free residence time before adsorbed monomer reevaporation
τ_c	mean free time for capture of a monomer by a nucleus
θ	contact angle between aggregates and substrate
V	volume of one molecule of film material
ν_{1i}	number of collisions of monomers with an aggregate
x	radius of the neck which forms when two islands coalesce
Z_i	number of interatomic bonds in an aggregate
π	aggregate periphery

INTRODUCTION

It is well known that the properties of a thin film may be quite different from those of the bulk, particularly if the film thickness is very small. These "anomalous" properties are due to the peculiar structure of the film, and this structure, in turn, is dictated by the processes which occur during film formation. A thin film may be prepared by a process as simple as hammering or rolling down a piece of bulk material. More generally, however, thin films are prepared by depositing the film material, atom by atom, on a substrate. Examples are the condensation of a vapor to give a solid or liquid film, or the plating out of a metal film from solution by electrolysis. Such processes of deposition involve a phase transformation, and the formation of a thin film can be understood by a study of the thermodynamics and kinetics of this phase transformation.

The best-understood process of film formation is that by condensation from the vapor phase. Since the production of thin films by vacuum deposition or vapor reaction in a gaseous-flow system is also the most important practical process film formation by condensation from the vapor phase is the subject of this chapter.

Condensation simply means the transformation of a gas into a liquid or solid. Thermodynamically, the only requirement for condensation to occur is that the partial pressure of the film material in the gas phase be equal to or larger than its vapor pressure in the condensed phase at that temperature. However, this is true only if condensation takes place on film material already condensed or on a substrate made of the same material. In general the substrate will have a chemical nature different from that of the film material. Under these conditions still a third phase must be considered, namely, the adsorbed phase, in which vapor atoms are adsorbed on the substrate but have not yet combined with other adsorbed atoms. Condensation is initiated by the formation of small clusters through combination of several adsorbed atoms. These clusters are called nuclei, and the process of cluster formation is called nucleation. Since small particles display a higher vapor pressure than the bulk material under the same conditions (Gibbs-Thompson equation), a supersaturation ratio larger than unity is required for nucleation to occur. The process of enlargement of the nuclei to finally form a coherent film is termed growth. Frequently both nucleation and growth occur simultaneously during film formation.

Further, it is not sufficient to consider the condensation process to be simply equivalent to a random raining down on the substrate of sticky corkballs which stick where they impinge. Rather, there will be sufficient surface mobility on the substrate to lead to the formation of well-defined islands of film material on the substrate even long after the nucleation step. Eventually these islands coalesce to form a continuous film, but this will generally occur only after the average film thickness has grown to several monolayers.

In this chapter we will examine the nucleation phenomenon from the vapor phase and the subsequent growth of the nuclei to give a continuous film.

2. THEORIES OF THIN FILM NUCLEATION

a. Impingement, Adsorption, and Thermal Accommodation

In all theories of thin film nucleation, the initial step is the impingement of vapor molecules on the substrate. After impingement, the vapor molecules can either adsorb and stick permanently to the substrate, they can adsorb and reevaporate in a finite time, or they can immediately bounce off the substrate like light from a mirror. In general, vapor atoms arrive at the substrate surface with energies appreciably higher than kT, where T is the substrate temperature. The question then arises whether such a vapor atom can equilibrate rapidly enough with the substrate so that it becomes adsorbed, or whether it will rebound without having given up all its incident energy to the substrate. For the latter case, the thermal-accommodation coefficient, which is defined as

$$\alpha_T = \frac{E_v - E_r}{E_v - E} = \frac{T_v - T_r}{T_v - T}$$

will be less than unity. Here

E_v = incident kinetic energy of the vapor atom

E_r = energy of the desorbed atom before equilibration with the substrate

E = energy of the desorbed atom after it has equilibrated with the substrate

T_v, T_r, T = corresponding temperatures

Theoretical investigations by Cabrera,[1] Zwanzig,[2] and McCarroll and Ehrlich[3] indicate that for the case of a "hot" atom impinging on a one-dimensional lattice, the accommodation coefficient is less than unity only if the incident kinetic energy is larger than twenty-five times the energy necessary for desorption after equilibration with the substrate. It was also predicted that the likelihood of complete thermal accommodation, i.e., $\alpha_T = 1$, increases if the ratio of the mass of the impinging atom to that of the mass of the substrate lattice atoms increases.

For the case of a three-dimensional substrate lattice, Goodman[4] has estimated that thermal accommodation is essentially complete if E_v is smaller than the energy necessary for desorption after equilibration with the substrate. Thus, if the activation energy for desorption is 0.5 eV, for example, the incident atom would have to arrive with an energy corresponding to a temperature in excess of 6000°K before thermal accommodation would become incomplete.

It was reported by McFee[5] and Lennard-Jones[6] that the time required for an incident atom to lose its excess kinetic energy and to accommodate thermally with the substrate is of the order of $2/\nu$, where ν is the frequency of the substrate lattice vibrations. Thus, the impinged atoms lose essentially all their excess energy with a few lattice oscillations.

It thus appears that, except for very light impinging atoms or very high incident energies, the incident vapor immediately equilibrates thermally with the substrate. In the subsequent treatment, the assumption of immediate equilibration has therefore been made.

Once the adsorbed vapor molecules have reached a certain population density on the substrate, a steady state is obtained, in the absence of nucleation, in which the flux of reevaporating molecules just equals the impinging flux. The substrate coverage of adsorbed vapor molecules n_1 is therefore a function of the deposition rate R:

$$n_1 = \frac{R}{\nu_0} \exp\left(\frac{\Delta G_{\text{des}}}{kT}\right) \tag{1}$$

where ν_0 = frequency at which the adsorbed molecule attempts to desorb, and is thus equated to its vibrational frequency ($\sim 10^{14} \text{ sec}^{-1}$)

ΔG_{des} = free activation energy for the desorption process

As soon as impingement is stopped and $R \to 0$, the coverage of the substrate by adsorbed atoms approaches zero. The mean residence time τ_a of an adsorbed molecule before reevaporation is

$$\tau_a = \frac{1}{\nu_0} \exp\left(\frac{\Delta G_{\text{des}}}{kT}\right) \tag{2}$$

Condensation of a permanent deposit may thus not be possible even if the substrate temperature is so low that the evaporation rate of the film material in the bulk at that temperature is negligible, i.e., if the supersaturation ratio for film deposition is much larger than unity. This is in marked contrast to ordinary condensation, which will proceed even if the supersaturation ratio only slightly exceeds unity.

Condensation of a permanent deposit on the substrate does occur at sufficiently high impingement rates, because the interaction between adsorbed single atoms cannot be neglected. Adsorbed atoms can migrate over the surface, giving rise to collisions with other atoms, and aggregates of adsorbed atoms can now exist. Aggregates should be more stable toward reevaporation than single adsorbed atoms, since they are bound to each other by the condensation energy. However, the stability of a small aggregate, or cluster, consisting of only a few atoms, is not determined only by the bulk condensation energy. This is because the atoms in such a cluster usually have fewer nearest neighbors than in the bulk, and almost always fewer or no next nearest neighbors. Thus, their surface-to-volume ratio is very high. The resulting high surface energy makes the clusters less stable. Most nucleation theories therefore postulate an equilibrium or steady state to exist between the adsorbed monomers, which diffuse on the substrate surface for a time τ_a, constantly colliding with themselves and other clusters and aggregates of various sizes. Most theories further postulate that once a cluster has reached a certain critical size, it will, on the average, no longer dissociate into monomers but will grow to form a stable condensate. The two principal theories of thin film nucleation, the capillarity model and the atomistic model, differ in their approach to evaluate the energy of formation of these clusters. The important features of both theories are given below, with emphasis on the more recent contributions. Still another model is described, in which the emphasis is placed on the evaporation of the clusters from the substrate. Special attention is given to the case of partial or incomplete condensation. Comparisons among the various theories are made wherever possible, and their limits of validity will be indicated. A survey of the more recent experimental investigations then follows.

The Capillarity Model

Several excellent reviews deal with the heterogeneous nucleation from the vapor phase. Mentioned should be the monograph by Hirth and Pound[7] and the more recent review article by Sigsbee and Pound.[8] The reader is referred to these for a more detailed exposition of the capillarity model than is possible here.

The classical capillarity (droplet) model for homogeneous nucleation from the vapor phase by Volmer and Weber[9] and Becker and Doering[10] has to be modified only slightly to take into account the presence of the substrate. In essence, this model postulates that, in order to form a condensed phase from the supersaturated vapor, positive free-energy fluctuations are required to form stable aggregates of the condensed phase, thus introducing the necessity of overcoming an activation barrier (sometimes called the "nucleation barrier"). It is because of this barrier that a supersaturation greater than unity is required for condensation to take place.

(1) The Critical Nucleus The capillarity model predicts that the free energy of formation of a condensed aggregate goes through a maximum; i.e., the aggregate has a stability minimum with respect to dissociation into the vapor phase as it grows through its "critical" size. This maximum in free energy arises from the very large surface-to-volume ratio of the small aggregates, tending to decrease their stability, and the condensation energy, tending to increase it, as they grow in size.

To calculate the critical radius r^* of such an aggregate, we assume aggregates of surface area $a_1 r^2$ exposed to the vapor phase, a contact area $a_2 r^2$ between the aggregate and substrate, and a volume $a_3 r^3$, where the a's are constants and r is the mean linear dimension of the aggregates. The total free energy of an aggregate with respect to dissociation into the vapor phase as a function of size is given by

$$\Delta G = a_3 r^3 \, \Delta G_v + a_1 r^2 \sigma_{v\text{-}c} + a_2 r^2 \sigma_{s\text{-}c} - a_2 r^2 \sigma_{s\text{-}v} \tag{3}$$

Here ΔG_v (negative) is the free energy of condensation of the film material in the bulk under the same conditions of supersaturation in ergs cm^{-3} and is given by

$$\Delta G_v = \frac{kT}{V} \ln \frac{R}{R_e(b)} \tag{4}$$

$\sigma_{v\text{-}c}$ (positive) and $\sigma_{s\text{-}c}$ (positive or negative) are the surface and interfacial (between deposit and substrate) free energies of the aggregate, respectively, and $\sigma_{s\text{-}v}$ is the surface energy of the substrate, all in ergs cm^{-2}. The $a_2 r^2 \sigma_{s\text{-}v}$ term enters into Eq. (3) because an area of free substrate surface equal to $a_2 r^2$ disappears when the aggregate is created. The volume of one molecule of the film material is V, and $R_e(b)$ is the evaporation rate of monomers from the bulk at the substrate temperature. The ratio $R/R_e(b)$ is the supersaturation ratio.

Differentiating Eq. (3) with respect to aggregate size yields

$$\frac{\partial \Delta G}{\partial r} = 3a_3 r^2 \, \Delta G_v + 2a_1 r \sigma_{v\text{-}c} + 2a_2 r \sigma_{s\text{-}c} - 2a_2 r \sigma_{s\text{-}v} \tag{5}$$

Here it is assumed that the shape of the aggregate does not change as its size changes and also that ΔG_v, $\sigma_{v\text{-}c}$, and $\sigma_{s\text{-}c}$ do not change with size.

The free energy of the aggregate is a maximum for an aggregate of critical size; i.e., for $\partial \, \Delta G / \partial r = 0$:

$$r^* = \frac{-2(a_1 \sigma_{v\text{-}c} + a_2 \sigma_{s\text{-}c} - a_2 \sigma_{s\text{-}v})}{3a_3 \, \Delta G_v} \tag{6}$$

The free energy corresponding to this size is

$$\Delta G^* = \frac{4(a_1 \sigma_{v\text{-}c} + a_2 \sigma_{s\text{-}c} - a_2 \sigma_{s\text{-}v})^3}{27 a_3^2 \, \Delta G_v^2} \tag{7}$$

The dependence of the free energy of an aggregate on its size is shown schematically in Fig. 1. The condition of maximum free energy corresponds to minimum stability of the aggregate, and occurs at the critical size r^*. For values of $r > r^*$, the term in Eq. (3) will predominate, leading to negative free energies and, therefore, stable aggregates for large radii. If an additional atom is added to an aggregate of critical size (called a critical nucleus), it becomes somewhat more stable and will, on the average, not dissociate into single atoms but will grow to form a larger, permanent island. If, on the other hand, an atom is taken away from the critical nucleus, it dissociates again. Therefore, in order to condense a permanent deposit, aggregates of critical size or larger have to be created first.

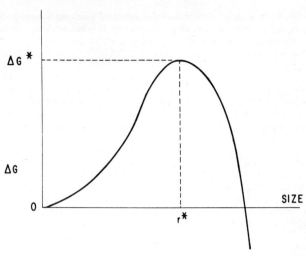

ig. 1   The free energy of formation of an aggregate of film material as a function of size
"he aggregate has minimum stability at the critical radius $r*$.

For a spherical cap nucleus of radius r, making a contact angle θ with the sub-
trate, Eqs. (6) and (7) become, respectively,

$$r^* = \frac{-2\sigma_{v\text{-}c}}{\Delta G_v} \tag{6a}$$

$$\Delta G^* = \frac{4\pi\sigma^3_{v\text{-}c}}{3\,\Delta G_v{}^2}\,(2 + \cos\theta)(1 - \cos\theta) \tag{7a}$$

If the surface free energy of the nucleus is anisotropic, its shape will be a circular
isk of height h.   The number of monomers in a critical cluster is given by[8]

$$i^* = \frac{\pi h\sigma^2_{e\text{-}v}}{V(\Delta G_v + \Sigma\sigma/h)^2} \tag{6b}$$

'here $\sigma_{e\text{-}v}$ = specific free-edge energy of the disk, and

$$\Sigma\sigma = \sigma_{c\text{-}v} + \sigma_{s\text{-}c} - \sigma_{s\text{-}v}$$

quation (7) becomes, for the circular disk,

$$\Delta G^* = \frac{-\pi h\sigma^2_{v\text{-}c}}{\Delta G_v + \Sigma\sigma/h} \tag{7b}$$

(2) The Nucleation Rate   Critical nuclei can grow to supercritical dimensions either
y direct impingement and incorporation of atoms from the gas phase, or by collisions
ith adsorbed monomers diffusing over the surface of the substrate.   If only a rela-
vely small area of the substrate surface is covered by critical nuclei, the latter
techanism is probably the more important one, depending on the diffusion coefficient
f the adsorbed monomers.   For such a case, the rate at which critical nuclei grow
 given by the number of initial nuclei per unit area, and the rate at which adsorbed
tonomers join it.   Assuming a metastable equilibrium between the adsorbed mono-
ters and aggregates of various sizes, one gets for the concentration of nuclei of
ritical size (neglecting small statistical mechanical corrections)

$$n^* = n_1 \exp\left(\frac{-\Delta G^*}{kT}\right) \tag{8}$$

On the other hand, the rate j at which adsorbed monomers are incorporated into the nuclei will depend on the number of adsorbed atoms per unit area, their jump frequency, and their jump distance:

$$j = Cn_1\nu_0 \exp\left(\frac{-\Delta G_{sd}}{kT}\right) \tag{9}$$

where C is a constant containing the size of the critical nucleus and other geometric factors, and ΔG_{sd} is the free energy of activation for the surface diffusion of adsorbed atoms (positive). From Eqs. (1), (8), and (9) one can then express the number of just supercritical aggregates created per unit area and time as

$$I^* = jn^* = CR \exp\left(\frac{\Delta G_{\text{des}} - \Delta G_{sd} - \Delta G^*}{kT}\right) \tag{10}$$

Equation (10) illustrates that I^* (called the nucleation frequency) is a very steep function of the nucleation energetics and thus of the deposition parameters. For example, Fig. 2 illustrates the very steep dependence of the nucleation frequency on the supersaturation ratio.

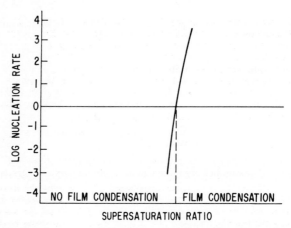

Fig. 2   Schematic of the nucleation frequency vs. the supersaturation ratio, demonstrating a very steep dependence.

Equation (10) illustrates another point, namely, that the probability that some nuclei exist is never zero for any finite film-deposition rate, no matter how low. However, their number may be so small that they escape detection. Because of the very steep dependence of the rate of nucleus formation on the deposition parameters (see below), the rate of one nucleus per cm² per second is often arbitrarily taken as the "onset" of condensation.

For the spherical cap nucleus the constant C is

$$C = \sqrt{\frac{\Delta G^*}{3kTi^{*2}}}\, a_0{}^2\pi r^* \sin\theta\, N_0$$

where i^* = number of monomers in the critical cluster

a_0 = jump distance of the adsorbed monomer (approximately equal to the lattice parameter of the substrate)

N_0 = density of monomer adsorption site ($\sim 10^{15}$ cm^{-2})

The square-root term is a small, nonequilibrium correction term (Zeldovich factor

c. Atomistic Models

(1) The Small-cluster Model of Walton and Rhodin Calculations of the initial cluster size using the capillarity model frequently give sizes in the less than 5 Å radius range, and sometimes even less than one or two atomic radii. Walton[11,12] and Rhodin and Walton[13,14] have pointed out that the application of bulk surface energies to droplets of very small size is not, strictly speaking, correct. This difficulty would be avoided if the nucleation process could be treated by writing the partition functions and potential energies for the reacting species and products. An approximate treatment of this was given by Walton and Rhodin,[11,13,14] and Walton, Rhodin, and Rollins.[15] The principal feature of this approach is to introduce a potential (internal) energy E_i, which is the dissociation energy of a cluster containing i atoms into i *adsorbed monomers*. It thus corresponds to the ΔG term in the capillarity model, except that ΔG is the *free* energy of formation of a cluster with respect to dissociation into the *gas phase*. The dissociation energy of a critical cluster is E_i^*, and the rate of formation of critical clusters becomes (neglecting the Zeldovich factor)

$$I^* = Ra_0 y N_0 \left(\frac{R}{\nu_1 N_0}\right)^{i^*} \exp\left[\frac{(i^* + 1)E_a + E_i^* - E_d}{kT}\right] \quad (11)$$

where y = available cluster periphery for impingement

$\quad \nu_1$ = attempt frequency of the adsorbed monomer for desorption

$\quad i^*$ = number of monomers in a critically sized aggregate

$\quad E_a$ = activation energy for desorption

$\quad E_d$ = activation energy for diffusion (both positive)

The use of potential energies instead of free energies in the small-cluster model implies the hidden presence in Eq. (11) of an entropy term $e^{\Delta S/k}$ (Gibbs-Helmholtz equation). This exponential can be accommodated best by equating the attempt frequencies for the capillarity and small cluster models as

$$\frac{1}{\nu_1} = \frac{e^{\Delta S/k}}{\nu_0}$$

The uncertainty in the surface energy of small clusters in the capillarity model now becomes an uncertainty in i^* and E_i^*. For a detailed comparison between the two models see Sec. 2c(3). The procedure commonly used in comparing Eq. (11) with experiments is to consider only very small values of i^*. Thus, for very high supersaturations, the critical cluster might be one atom only (meaning that a two-atom cluster is the smallest stable aggregate); at somewhat lower supersaturations it will be $i^* = 2$ (a three-atom cluster is now the smallest stable aggregate); then $i^* = 3$, etc. The nucleation rates corresponding to these critical clusters are

$$i^* = 1: \quad I_1^* = Ra_0 y N_0 \frac{R}{\nu_1 N_0} \exp\left(\frac{2E_a - E_d}{kT}\right) \quad (12a)$$

$$i^* = 2: \quad I_2^* = Ra_0 y N_0 \left(\frac{R}{\nu_1 N_0}\right)^2 \exp\left(\frac{3E_a + E_2 - E_d}{kT}\right) \quad (12b)$$

$$i^* = 3: \quad I_3^* = Ra_0 y N_0 \left(\frac{R}{\nu_1 N_0}\right)^3 \exp\left(\frac{4E_a + E_3 - E_d}{kT}\right) \quad (12c)$$

where E_2 and E_3 are the dissociation energies of two-atom and three-atom clusters, respectively. If now the supersaturation is varied over a range of values, such as by varying the substrate temperature, then the transition temperature from one critical cluster to another can be obtained. For example, the temperature at which a two-atom cluster becomes critical is given by [equating Eqs. (12a) and (12b)]

$$T_{1\text{-}2} = \frac{-(E_d + E_2)}{k \ln (R/\nu_1 N_0)} \quad (13)$$

Figure 3 demonstrates the discontinuous change in the slope of the nucleation rate vs. reciprocal temperature which is expected for the case in which the critical cluster goes directly from $i^* = 1$ to $i^* = 3$.[14] The description of nucleation processes involving very small critical aggregates has been the principal success of the small-cluster model.

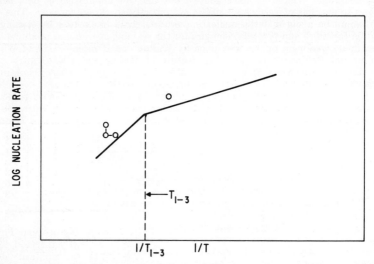

Fig. 3 Schematic dependence of nucleation rate on temperature. (*Rhodin and Walton.*[14]

(2) Bond-energy Formulations for the Small-cluster Model Lewis[16] evaluated the E_i term in the small-cluster model of Walton and Rhodin as the sum of nearest neighbor bonds, each of energy E_b, for the most favorable configuration of atoms. Here E_b is taken to be a constant for a given condensate material and independent of the number of bonds in a cluster.

For a single atom the cluster binding energy E_1 is zero. A pair has one bond; so $E_2 = E_b$. A triplet with the close-packed configuration has three bonds; so $E_2 = 3E_b$. A fourth atom, still on the substrate, can make two more bonds; so $E_4 = 5E_b$. If the fourth atom jumps off the substrate onto the top of a triangular pyramid, the cluster has lost E_a for this atom, and there are six interatomic bonds; so $E_4 = 6E_b - E_a$. Which of these two has the higher energy depends on the relative values of E_b and E_a. In general

$$E_i = X_i E_b - h E_a \qquad (14$$

where X_i is the number of interatomic bonds and h is the number of atoms in the cluster which are off the substrate. The values for X_i and h are chosen so that the value of E_i is maximized.

Lewis then proceeds to define the following reduced relationships:
The reduced nucleation rate $j_i^* = \log (I/\nu_1 N_0)$
The reduced critical cluster binding energy $e_i^* = E_i^*/E_b$
The reduced impingement rate $r = \log (R/\nu_1 N_0)$
The reduced adsorption energy $e_a = E_a/E_b$
The reduced temperature $1/t = \log e/kT$
The reduced surface-diffusion energy $e_d = E_d/E_b$
Substituting into Eq. (11) gives the reduced nucleation rate (neglecting some pre-exponential factors)

$$j_i^* = (1^* + 1)r + \frac{[e_i^* + (i^* + 1)e_a - e_d]E_b}{t} \qquad (1$$

Some values of the reduced bond strength E_b and the E_b/t term are given in Table 1 for various metals at several temperatures. The reduced nucleation rate is given as a function of reciprocal reduced temperature in Fig. 4 for a fixed impingement flux and a range of values for the reduced adsorption energy e_a. The changes in slope shown in Fig. 4 correspond to the conditions at which the size of the critical cluster changes.

TABLE 1 Values of the Reduced Bond Strength E_b, and Temperature Term E_b/t for Various Elements and Temperatures (Lewis[16])

Atom	E_b, eV	E_b/t					
		−196°C 77°K	−148°C 125°K	−73°C 200°K	27°C 300°K	227°C 500°K	527°C 800°K
K...........	0.15	9	5	3	2	1.5	1
Na..........	0.2	13	8	5	3	2	1.5
Pb..........	0.3	20	12	8	5	3	2
Al...........	0.5	32	20	12	8	5	3
Ti...........	0.8	52	32	20	13	8	5
Mo..........	1.2	78	48	30	20	12	8

(3) Comparison of the Small-cluster and Capillarity Models Sigsbee and Pound[8] have compared the small-cluster and capillarity models on a strictly thermodynamic basis. They conclude that Eq. (11) will reduce to Eq. (10) for the capillarity model except for the difference between ΔG^* and E_i^*. Entropy contributions to the free energy of cluster formation were explicitly neglected in the derivation of Eq. (11).

Lewis[16] also has compared the two models. He has shown that the fundamental concepts upon which both are based are identical. Making detailed comparisons for the special case of the cap-shaped nucleus, Lewis finds that the two theories converge if one writes the following equalities:

For the surface energy of the condensate:

$$\sigma_{v\text{-}c}s = \tfrac{1}{2}Z_c E_b \qquad (16a)$$

For the monomer adsorption energy:

$$E_a = (\sigma_{v\text{-}c} + \sigma_{v\text{-}s} - \sigma_{s\text{-}c})s \qquad (16b)$$

For the contact angle:

$$\cos\theta = \frac{2E_a - Z_c E_b}{Z_c E_b} \qquad (16c)$$

For the free energy of formation of a cluster containing i monomers:

$$-\Delta G_{(i)} = kT\ln\frac{N_0}{n_1} + ikT\ln\frac{n_1\nu_0}{N_c\nu_c}$$

$$+ \frac{i}{2}(Z_b E_b - 2E_a) - i^{\frac{2}{3}}2[(Z_c E_b - E_a)^2(Z_c E_b + 2E_a)]^{\frac{1}{3}} \quad (16d)$$

where s = projected effective area per atom in the surface

Z_b = number of nearest atoms surrounding each atom of the condensate in the bulk

Z_c = difference between the number of nearest neighbors in the bulk and at the surface of the condensate

N_c = surface density of atoms in the condensate

ν_c = rate-determining coefficient characteristic of the evaporation process of the bulk material

While Eqs. (16a) to (16d) force agreement between the capillarity model and the small-cluster model, it must be noted that the validity of Eqs. (16a) and (16b), upon which (16c) and (16d) are based, has not been proved.

Lewis correctly concludes that the essential difference between the models is that one (the small-cluster model) uses only discrete arrangements of atoms, whereas the other (the capillarity model) employs simple idealized geometrical shapes for the cluster. The effect is that the capillarity model predicts a continuous variation of

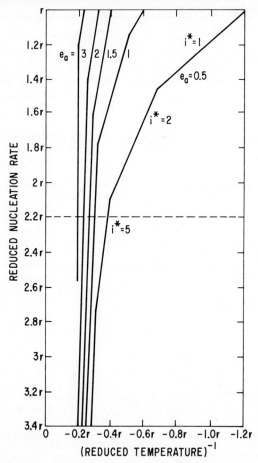

Fig. 4 Nucleation rate as a function of reciprocal temperature for a range of values o adsorption energy and for a fixed incidence rate. (*Lewis.*[16])

the critical cluster size and nucleation rate with supersaturation, while the small cluster model predicts discontinuous changes, which should be particularly striking if the size of the critical cluster is small.

d. Partial vs. Complete Condensation

One of the most striking phenomena in the initial phases of film condensation i that of partial or incomplete condensation. Considerable reevaporation occurs fron the substrate; i.e., the average sticking coefficient, which is the ratio of the amoun of film material condensed on the substrate to the amount which has impinged, i less than unity. In fact, often the sticking coefficient is so small that condensation is not observable by ordinary techniques. On the other hand, the sticking coefficient is found to be strongly dependent on the total time during which the substrate was subject to impingement, and also on the substrate temperature. A nonunity sticking coefficient is usually explained in terms of monomers reevaporating from the areas on the substrate which are outside the capture zones around each stable nucleus. The dependence on time and temperature will be developed now, including the special case of the smallest possible stable nucleus (two monomers). Then follows an additional mechanism to account for partial condensation, namely, the evaporation of aggregates from the substrate.

(1) The Sticking Coefficient and Delayed Condensation In the initial stages of condensation, clusters of various sizes are in metastable equilibrium with adsorbed monomers. As these clusters grow, they deplete the surrounding region of adsorbed monomers, so that further nucleation (or cluster formation) is not possible in this region (called a capture zone). Taking such a capture zone to be of radius r_s, one gets (following Sigsbee and Pound[8])

$$r_s \approx \sqrt{D\tau_a} \qquad (17)$$

where D = surface-diffusion coefficient of adsorbed monomers on the substrate. The fraction of the substrate covered by clusters and their associated capture zones is approximately

$$F(t) = N(t) \, D\tau_a \qquad (18)$$

where $N(t)$ is the number of clusters per unit area above critical size and t is the time elapsed since impingement first began. Further nucleation can occur only on the "empty" substrate area, on which the rate of addition of new nuclei per cm^2-s is

$$\frac{dN(t)}{dt} = I^*[1 - F(t)] \qquad (19)$$

From Eqs. (18) and (19) one gets the substrate area covered at time t:

$$F(t) = 1 - \exp\left(-I^* D\tau_a t\right) \qquad (20)$$

and the number of nuclei at time t:

$$N(t) = \frac{1}{D\tau_a} [1 - \exp\left(-I^* D\tau_a t\right)] \qquad (21)$$

Before coalescence of clusters sets in, $1/D\tau_a$ becomes equal to N_∞, the saturation number of nuclei on the substrate. Thus the number of nuclei will grow toward a saturation value within a certain time, depending on I^* and τ_a. As long as $F(t) \ll 1$, many of the adsorbed monomers are outside the capture zones and can therefore reevaporate. On the other hand, when $F(t) \rightarrow 1$, all impinging vapor will be adsorbed within capture zones and become incorporated into a growing nucleus without having a chance to reevaporate (we here neglect evaporation of the nuclei themselves). The sticking coefficient is therefore approximately equal to $F(t)$.

The instantaneous sticking coefficient can be defined as

$$\alpha_s = \frac{dM/dt}{\dot{M}} \qquad (22)$$

where dM/dt is the instantaneous mass deposit and $\dot{M}$ is the mass impingement rate in g cm^{-2} s^{-1}. Substituting $\alpha_s = F(t)$ and integrating, one gets the mass condensed at time t

$$M(t) = \dot{M} \frac{t + \exp\left(-I^* D\tau_a t\right) - 1}{I^* D\tau_a} \qquad (23)$$

This relationship is illustrated in Fig. 5. It is seen that the mass deposited strongly depends on the total impingement time, the substrate temperature, the diffusion coefficient of monomers on the substrate, the mean residence time of adsorbed mono-mers, and the impingement rate. Experimentally the appearance of some observable mass of condensate on the substrate is taken as the "onset" of condensation, and the corresponding substrate temperature is called the "critical" condensation temperature. Figure 5, however, illustrates that this "critical" temperature depends on the time of observation and sensitivity of the instruments used to sense the condensed mass.

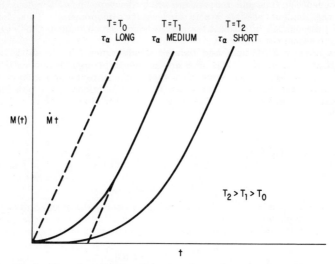

Fig. 5 Mass deposited vs. time at various substrate temperatures.

Below a temperature T_0 in Fig. 5 the mean free residence time is long enough that all impinged monomers are captured by stable nuclei and the condensation is initially complete; i.e., $\alpha_s = 1$, even at $t = 0$. Above T_0 condensation is initially incomplete, i.e., $\alpha_s < 1$ at $t = 0$ or very small values of t. For values of $t > 3/I^*D\tau_a$,

$$M = \dot{M}t\left(1 - \frac{1}{I^*D\tau_a}\right) \tag{24}$$

Substituting $N_\infty = 1/D\tau_a$,

$$I^* = \frac{N_\infty}{1 - M(t)/\dot{M}} \tag{25}$$

From Eq. (25) the nucleation rate can be obtained experimentally by counting the saturation density of nuclei, which can be done long after they have grown to supercritical dimensions, as long as coalescence among nuclei has not taken place. Also it must be pointed out that the above is a high-nucleation-rate treatment applicable only if most nuclei formed do not have time to grow to sizes comparable with the mean free adsorbed monomer diffusion distance before $F(t) \sim 1$. For lower nucleation rates, much of the nucleation and subsequent growth will be on energetically favorable and specific defects on the substrate.

(2) Initially Complete vs. Initially Incomplete Condensation Lewis and Campbell[1] have given a detailed description of the nucleation process near T_0 for the case of the smallest possible critical nucleus, namely, $i^* = 1$; i.e., a two-atom nucleus is already supercritical and stable. In their treatment the saturation density of nuclei is calculated above and below T_0. Above T_0 the fractional area of the substrate covered by nuclei and their associated capture zones is small, and the regime of incomplete condensation presented in Sec. 2d(1) applies, leading to reevaporation of adsorbed monomers. Below T_0, however, mutual capture to form a stable dimer becomes more probable than reevaporation of the monomer. All atoms then condense to form stable pairs, and the nucleation rate becomes the maximum possible, namely, equal to the impingement rate. This method has the advantage that while the theory deals with the initial stages of the condensation process, essentially at $t = 0$, the experiment to test the theory, which consists of counting the saturation density of nuclei, can be carried at a much later stage of film growth when their size has grown much beyond the dimer. The saturation density of nuclei will not change as long as coalescence has not yet taken place.

For the case of initially incomplete condensation, Lewis and Campbell express the area of the capture zone around each nucleus as m_a/N_0, where m_a is the number of adsorption sites which an adsorbed monomer visits during its residence time τ_a on the substrate. For a number N of stable nuclei one has a fractional capture-zone area of Nm_a/N_0, leaving $(1 - Nm_a/N_0)$ of the substrate area for further nucleation and reevaporation of adsorbed monomers.

Growth of nuclei occurs inside the capture-zone area at a rate

$$R_g = R\left(N\frac{m_a}{N_0}\right) \tag{26}$$

while reevaporation outside the capture-zone area occurs at a rate

$$R_e = R\frac{1 - Nm_a}{N_0} \tag{27}$$

The rate of formation of stable nuclei is zero within the capture-zone area, but outside it is

$$\frac{dN}{dt} = R\frac{m_a}{N_0}n^*\left(1 - N\frac{m_a}{N_0}\right) \tag{28}$$

where n^* is the metastable equilibrium density of critical nuclei. The maximum (saturation) number of stable nuclei occurs for $dN/dt = 0$, or

$$N_\infty = N_0 \exp-\left(\frac{E_a - E_d}{kT}\right) \tag{29}$$

which is obtained from Eqs. (11) and (28).

Equation (29) therefore gives the temperature dependence of the saturation density of stable nuclei in the region of initially incomplete condensation. A log N_∞ vs. $1/T$ plot of Eq. (29) has a negative slope. This is physically understandable since, with increasing temperature, as the mean residence time of adsorbed monomers becomes smaller, the size of the capture zone around each nucleus decreases, thereby increasing the maximum number of nuclei possible on the substrate. N_∞ is impingement-rate-independent.

In the temperature range for initially complete condensation, the mean free time for mutual capture of two adsorbed monomers to form a stable dimer τ_c is shorter than the mean free time for reevaporation τ_a. Therefore, the equilibrium density of adsorbed monomers $n_1 = R\tau_a$ gives two monomers at a time within each capture zone m_a/N_0. Adsorbed monomers will now visit only a number of sites $m_c < m_a$ before capture. The maximum density of nuclei is reached if further nucleation no longer takes place. The condition for no further nucleation now is that no other monomer shall be encountered before capture, i.e., that there is only one single monomer for every m_c site on the substrate, or $n_1m_0 = N_0$. Setting $\tau_c = \tau_a(m_c/m_a)$, $n_1 = R\tau_c$, and $N_\infty m_c = N_0$, one gets, using Eq. (11),

$$N_\infty{}^2 = \frac{N_0R}{\nu_1}\exp\left(\frac{E_d}{kT}\right) \tag{30}$$

Thus, under conditions of low temperature or high impingement rate, the impinged monomers immediately form stable pairs of density N_∞, without the possibility of reevaporation. The slope of the log N_∞ vs. $1/T$ plot is now positive, because the diffusion coefficient, and therefore the size of the capture zone, increases with increasing temperature.

The conditions for initially incomplete and complete condensation can be summarized as follows: For initially incomplete condensation, $\tau_a < \tau_c$, $(1 - Nm_a/N_0) < 1$, and the size of the capture zone is proportional to $D\tau_a$. For initially complete condensation, $\tau_a > \tau_c$, $(1 - Nm_a/N_0) = 1$, and the size of the capture zone is proportional to $D\tau_c$.

Figure 6 gives the expected behavior in each region. The size of the capture zone increases with increasing temperature at low temperatures, reaches a maximum at T_0, and then decreases again with increasing temperature.

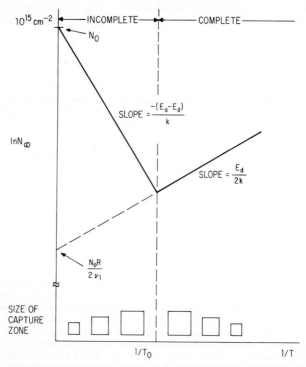

Fig. 6 Saturation density of nuclei vs. reciprocal temperature, illustrating the boundary between initially complete and incomplete condensation. (*Lewis and Campbell.*[17])

A similar figure can be constructed for $i^* = 2, 3, \ldots$, except that there will be another break in the curve at $T_3 < T_2 < T_0$, at which the critical nucleus changes from $i^* = 3$ to $i^* = 2$, and from $i^* = 2$ to $i^* = 1$.

(3) The Evaporation of Aggregates   This theory, proposed by Zinsmeister,[18] attempts to explain incomplete condensation by the evaporation of aggregates from the substrate. The theory is for the case of $i^* = 1$; i.e., a doublet is already stable.

In the theories given above, incomplete condensation is explained on the basis of the evaporation of adsorbed monomers only from the area outside the capture zones [Secs. 2d(1) and (2)]. Even for the case of $i^* = 1$, where dissociation of aggregates into monomers is negligible, the region of incomplete condensation is explained by monomer desorption [see Sec. 2d(2)]. The evaporation of aggregates is taken to be negligible.

Zinsmeister, however, points out that the desorption and evaporation of aggregates are not negligible in spite of the much greater activation energy necessary. The evaporation rate of aggregates containing i monomers is given by

$$\gamma_i = \frac{1}{1 + [\tau_0/w_{1i}R\tau_a^{(i+1)}]} \tag{31}$$

where τ_0 = a constant

w_{1i} = number of collisions of monomers with the aggregate

In order to calculate the sticking coefficient, Zinsmeister solved a system of differential equations expressing the rate of formation of aggregates in terms of the impingement flux, the rate of growth to form bigger aggregates, and its rate of evaporation (see also Courvoisier et al.[19]). The solution yields the instantaneous (at time t) and the dynamic steady-state (for long t) population densities of aggregates of various sizes N_i shown in Fig. 7.

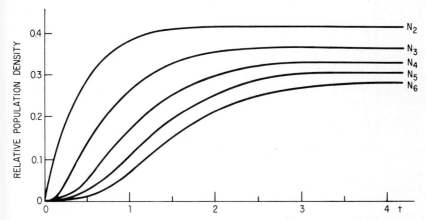

Fig. 7 Instantaneous and steady-state population densities of aggregates of various sizes. (*Zinsmeister.*[18])

This figure indicates that in the beginning of the condensation, monomers and very small aggregates predominate, while later most of the condensate is tied up in larger aggregates. It also shows that the sticking coefficient is low at the beginning of the condensation. The higher the substrate temperature, the longer will the sticking coefficient remain small. This is shown in Fig. 8, where the total number of atoms condensed per cm² is plotted as a function of time for various residence times corresponding to various substrate temperatures. This behavior is identical to that predicted by the theories above on the basis of monomer reevaporation and illustrated in Fig. 5 above. However, in the present theory the point is made that sticking coefficients less than unity and their temperature dependence are to be explained not by monomer desorption but by the evaporation of doublets and higher aggregates.

Perhaps the most important contribution of this theory is to call attention to the fact that the desorption of doublets and small aggregates may not always be negligible, particularly if E_a is small. Also it points out correctly that for most metals the critical nucleus will be very small, quite possibly consisting of two atoms or less.

On the other hand, it is clear that the evaporation of doublets and higher aggregates is much less likely than evaporation of monomers because of the $\exp(-iE_a/kT)$ dependence of τ_a. Small sticking coefficients observed in the condensation of high-boiling-point materials, where the binding energy between atoms is high and the critical nucleus is as small as $i^* = 1$, can be explained using monomer desorption only. Therefore, a theory based solely on aggregate evaporation appears somewhat risky.

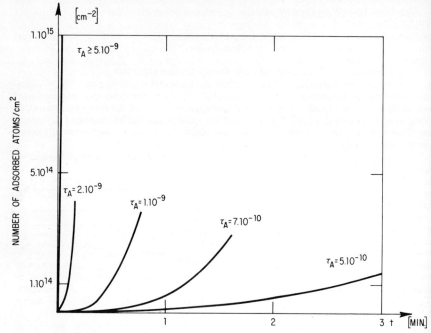

Fig. 8 Total number of condensed atoms vs. time for various residence times (in seconds). (*Zinsmeister*.[18])

e. Conclusions

Conceptually the capillarity model for nucleation is the easiest to understand, and while it may not always give quantitative information about the size of the critical nucleus, it does correctly predict the various dependencies of nucleus size and nucleation rate on substrate temperature, impingement flux, and the nature of the substrate. The atomistic model is almost identical to the capillarity model, except that here emphasis is put on very small critical nuclei. In particular, this theory predicts that if the size of the critical nucleus changes by even one atom with supersaturation, a discontinuity will be found in the nucleation rate vs. supersaturation curve.

The atomistic theory therefore best describes the condensation of those materials which give small critical nuclei, i.e., with a large free energy of condensation in the bulk, or at very high supersaturation. The capillarity model, on the other hand, is quite suitable in describing the condensation of materials of low free energy of formation, or at low supersaturations, where the critical nucleus will be large. In the limit of large critical nuclei, the two models become identical.

The phenomena of delayed condensation or of sticking coefficients less than unity are due to reevaporation of monomers from the substrate area not covered by nuclei and their associated capture zones. It is often thought to be due to the existence of the "nucleation barrier" alone, i.e., the necessity of forming a critical nucleus of some size. However, this is not a prerequisite. Reevaporation of adsorbed monomers will occur as long as they do not collide with other adsorbed monomers, leading to $\alpha_s < 1$, even if the smallest possible unit of condensate, a doublet, is already a stable aggregate.

f. Thin Film Nucleation Experiments

The substrate temperature is perhaps the most important variable in thin film nucleation experiments, since small changes in it result in large changes of the supersaturation. Variation of the nucleation behavior with the substrate temperature is therefore the subject of many experimental investigations. The result frequently is the identification of a "critical" temperature, corresponding to an observed change in the nucleation behavior. The electron-microscope, which allows observation and counting of nuclei when they are still in the 10- to 20-Å range, has recently become an important tool in nucleation experiments, particularly for in situ studies. Most recently, desorption studies have been reported which aim at the direct measurement of the adsorbed monomer population and the reevaporation flux. In this section we will discuss pertinent work in each of these categories.

(1) Critical Temperatures The effect of an increasing substrate temperature is to lower the supersaturation, decrease the mean residence time of an adsorbed monomer, and increase the surface-diffusion coefficient of adsorbed monomers. Consideration of these effects in the theories outlined above allows one to identify the following critical temperatures.

First, there is a "critical condensation temperature" above which the appearance of a condensed deposit connot be observed, because $\alpha_s \ll 1$, τ_a is small, and the area covered by nuclei and their associated capture zones is small. This temperature strongly depends on the method and sensitivity of the detection technique employed, and also on the rate and total deposition time before the observation is made. Above the critical condensation temperature, condensation is "not observable"; below it occurs rapidly. Related to the critical condensation temperature is the temperature T_0 above which the condensation is initially incomplete, and below which it is initially complete.

Second, there are several temperatures at which a sharp break in the log N_∞ vs. $1/T$ curve occurs. These are "transition" temperatures and correspond to a change in the size of the critical nucleus from i to $i + 1$ or $i + 2$, etc. These transition temperatures can be observed only if i is small, so that the addition of one more atom changes its size appreciably. The break in slope in Fig. 3, corresponding to T_{1-3}, is an example of such a transition temperature.

Third, there is an "epitaxial" temperature above which epitaxy of the deposit on a single crystalline substrate is observed and below which it is not. It has been argued[5-8] that a relationship exists between the transition and epitaxial temperatures. Both are deposition-rate-dependent.

(a) *The Critical Condensation Temperature.* A study of the sticking coefficient for mercury vapor impinging on a glass substrate was made by Mayer and Göhre[20] as a function of substrate temperature. Their results, shown in Fig. 9, are very reminiscent of the theoretical behavior predicted in Fig. 5. At temperatures above $-133°C$, the sticking coefficient was initially small and then increased with time, although it did so quite slowly at the highest temperatures. Below $-133°C$, condensation was even initially complete. Thus $T_0 = -133°C$ for these experiments. The mass of material condensed was measured with a sensitive microbalance.

Similar experiments of the condensation as a function of deposition time and substrate temperature of Bi, Pb, Sn, and Sb on glass were made by Palatnik and Komnik.[21] Simultaneous measurements of the optical density and resistivity were made during depositions at an impingement rate of 10^{14} atoms cm^{-2} s^{-1}. It was concluded that the sticking coefficient was initially less than unity above a certain substrate temperature T_0, which for Bi is 50°C, and initially unity below that temperature. This is shown in Fig. 10.

The treatment by Lewis and Campbell[17] predicts a sharp break in the log N_∞ vs. $1/T$ curve at T_0. This has indeed been reported by them for the system gold on rock salt. The island saturation density was counted using electron-microscopic replica techniques. These results are shown in Fig. 11. The impingement rate for the low-temperature leg was 0.15 Å s^{-1}. The two intercepts on the ordinate are V_0 and $N_0 R/2\nu_1$, respectively.

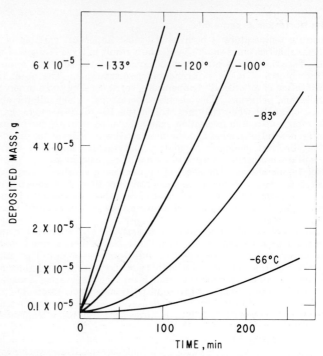

Fig. 9 Mercury condensation on quartz glass as a function of time in vacuo of about 10^{-10} Torr and with an impingement rate of 6.5×10^{13} atoms cm^{-2} s^{-1}. The $-133°C$ curve corresponds to complete condensation ($\alpha_s = 1$). (*Mayer and Göhre.*[20])

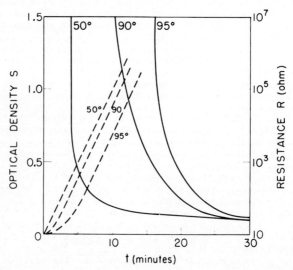

Fig. 10 Optical density and electrical resistivity of Bi vs. deposition time for indicated substrate temperatures, °C. (*Palatnik and Komnik.*[21])

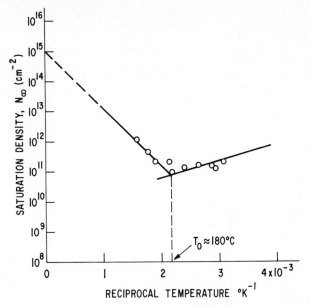

Fig. 11 Experimental saturation island density for gold on vacuum-cleaved rock salt as a function of reciprocal temperature. (*Lewis and Campbell.*[17])

(*b*) *Observation of Transition Temperatures.*   The observation of transition temperatures from one critical nucleus size to the next has been reported by Walton, Rhodin, and Rollins.[15]   In this work silver was deposited at various impingement rates on cleaved rock salt, and the saturation density of nuclei was measured electron microscopically on the areas away from the cleavage steps, as a function of impingement rate and substrate temperature.   A large change in slope in the log I vs. $1/T$ curves was found at a critical temperature $T_{1\text{-}3}$, as shown in Fig. 12.   Above this temperature, the critical cluster is taken to be a triplet, below it a singlet.   Using the interpretation outlined in Sec. 2c(1), the following data were obtained from Fig. 12:

$$T_{1\text{-}3} = 240°C$$
$$E_a = 0.4 \text{ eV}$$
$$E_d = 0.2 \text{ eV}$$
$$E_b = 0.8 \text{ eV}$$

(*c*) *Observation of Epitaxial Temperatures.*   It has been argued[14] that the likelihood of epitaxial orientation of a small cluster on a substrate should depend strongly on its size, and thus the "onset" of epitaxy might well correspond to a temperature such as T_2 above.   Studies of the epitaxy of germanium on single-crystal CaF_2, NaCl, NaF, and MgO, and silver on (100) NaCl have been reported by Sloope and Tiller.[22–24]   Epitaxial temperatures were found to be strongly dependent on deposition rate and residual gas pressure.   Sella and Trillat[25] found a decrease by as much as 200°C in the epitaxial temperature of silver on NaCl when the substrate was cleaved in vacuum as compared with air cleaning.   Experiments by Ino[26] et al. have shown that when rock salt is cleaved in ultrahigh vacuum, silver will not grow on it epitaxially at all, but if cleaved in a gas pressure of 10^{-5} Torr, it will have a lower epitaxial temperature than when cleaved in air.   Bethge and Krohn[27] report that while for epitaxy on rock salt the cleavage steps are of no importance, the adsorption of water vapor and annealing treatments have a very strong influence on the epitaxial temperature.   A detailed study of the dependence of the epitaxial temperature of gold on rock salt on the deposition rate and residual gas pressure was made by Matthews and Gruenbaum.[28]   In contrast to the findings above, the effect of water vapor was found to be small.

Krikorian[29] and Krikorian and Sneed[30] reported the dependence of the epitaxial temperature on deposition rate for sputtered germanium on germanium.   A "triple" point was found in the temperature vs. rate relationship outlining the single-crystalline, polycrystalline, and amorphous regions.   At this point all three phases "coexist." For a detailed exposition of this phenomenon, see Behrndt.[31]

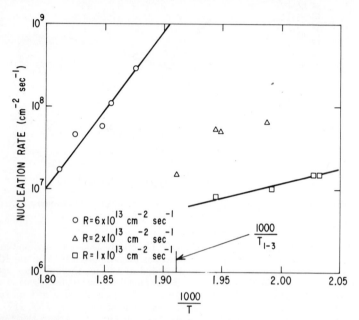

Fig. 12   Nucleation rate vs. reciprocal substrate temperature for silver deposited on rock salt.   (*Walton et al.*[15])

(2) Observation of Nucleation in the Electron Microscope   Perhaps the best technique to study thin film nucleation is to carry out the deposition inside an electron microscope and observe the formation of clusters and measure their sizes directly. Such experiments were first reported by Pashley[32] and Bassett.[33,34]   Experiments recently reported by Poppa[35] will be briefly summarized here.   In these experiments, silver was deposited on thin, amorphous carbon films which were transparent to electrons.   The deposition rate was measured with a quartz-crystal oscillator.   The substrate was kept at 425°C, and the island density and their sizes were measured during deposition.   An increase in both size and density was observed with deposition time until a saturation density was reached.   At saturation the island-size distribution was that shown in Fig. 13.   Poppa used an equation similar to Eq. (19) to analyze his data but made the assumption that the capture zone around each nucleus is negligible.   He then calculated the nucleation rate as a function of time from the observed distribution function dN/dD vs. D, where D is the island diameter.   The result indicates that the nucleation rate is a function of time for both short impingement times and long impingement times, as shown in Fig. 14.   This result, which is quite unexpected, was criticized by Sigsbee and Pound,[8] who pointed out that the depletion zones around islands in calculating the fractional area on which nucleation no longer occurs cannot be neglected but are, in fact, much larger than the area covered by the islands themselves in the early stages of condensation.

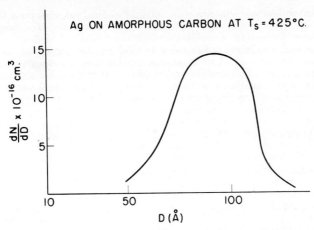

Fig. 13 Distribution function of Ag nuclei diameter 570 s after beginning of deposition. (*Poppa.*[35])

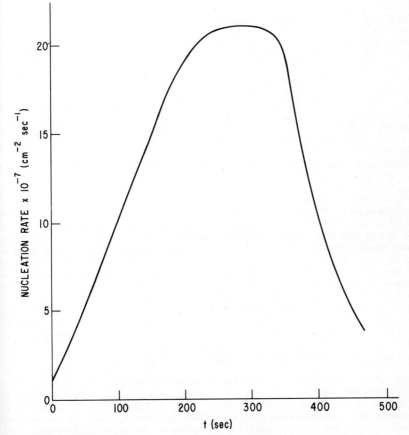

Fig. 14 Derived nucleation rate for Ag on amorphous carbon. (*Poppa.*[35])

(3) Desorption Studies Hudson and Sandejas[36] studied the adsorption and nucleation of Cd vapor on W substrates by using a mass spectrometer to monitor the flux of vapor molecules leaving the substrate while adsorption and nucleation are taking place. The mass spectrometer was tuned to the e/m ratio appropriate to the vapor species being studied. The partial pressure of this species is then proportional to the collector current. Evaporation rates from the substrate as low as the equivalent of 10^{-8} Torr vapor pressure or a desorption rate of 10^{13} molecules $cm^{-2} s^{-1}$ could be measured. The mean residence time of adsorbed Cd on W could be measured by using a shutter to interrupt the beam from the Cd evaporator, and by monitoring the transients in the desorption rate accompanying the opening or closing of the shutter. When the shutter is opened, an exponential rise in signal is observed, with a time constant equal to τ_a. Two distinct regions were found when the adatom coverage and the residence time were investigated as a function of substrate temperature, as shown in Fig. 15a and b. The two exponential regions in τ_a can be characterized by two different heats of desorption, one 41.1 kcal mol^{-1} and the other 21.2 kcal mol^{-1}. This was interpreted in terms of a primary tightly bound adsorbed phase formed on initial exposure of the clean surface to the vapor beam, and which saturates in coverage at a few monolayers. At lower temperatures, however, a secondary, more loosely bound phase forms, which has no apparent limit in coverage.

Hudson and Sandejas conclude, since high coverages (6 to 10 monolayers) are observed in the adatom layer prior to nucleation, that none of the nucleation models discussed above is strictly applicable. Also, since the heat of adsorption of the primary phase is so large compared with the heat of sublimation of bulk cadmium (41.1 kcal mol^{-1} as compared with 26.9 kcal mol^{-1}), it is unlikely that this layer plays any part in the nucleation process. Thus, their conclusions appear to contradict the simple idea that a concentration of adsorbed atoms small compared with one monolayer is sufficient for nucleation.

(4) Conclusions While the general predictions of the nucleation theory are qualitatively obeyed, even for relatively crude experiments, quantitative comparisons require fairly sophisticated experimental techniques. Most important here is the prevention of contamination of the substrate surface. The experiments by Sella and Trillat[25] on the condensation of gold on rock salt, of Caswell[37,38] on tin and indium on glass, and of Melmed[39] on copper on tungsten show the strong influence of the residual gas atmosphere on the nucleation behavior. The observation of a "memory" effect by Sigsbee,[40] in which it was found that Sb nucleated at lower supersaturations on areas of the substrate where Sb had been previously deposited and reevaporated, suggests unexplored areas where investigations might prove fruitful. In the same category fall the striking effect on the growth process of an electric field observed by Chopra[41] and of electron bombardment of the substrate by Lewis and Campbell,[17] Hill,[42] and Stirland.[43] It was found by these authors that metal films tend to be more continuous if the substrate is under electron bombardment than without. This is consistent with an increased nucleation density which has been observed when electron bombardment was used. Chambers and Prutton[44] report that for deposits of nickel on air-cleaved rock-salt substrates at 300 and 380°, electron bombardment improved the epitaxial orientation significantly. In fact, diffraction patterns of electron-bombarded samples frequently showed highly developed Kikuchi patterns, indicative of a high degree of orientation perfection. The quantitative aspects of the nucleation process can be studied best by combining well-controlled conditions of deposition with sophisticated observational techniques, such as ultrahigh vacuum, in situ electron microscopy, mass spectrometry, and in situ electron diffraction.

3. STRUCTURAL CONSEQUENCES OF THIN FILM NUCLEATION

In essence, nucleation theory states that there will be a barrier to the condensation of a permanent deposit. If there is a nucleation barrier, the film will show an island structure in the initial stages of growth. One can further clarify the structural consequences of a nucleation barrier by differentiating between the two extreme cases: the first case involves a very large nucleation barrier, i.e., a large critical nucleus

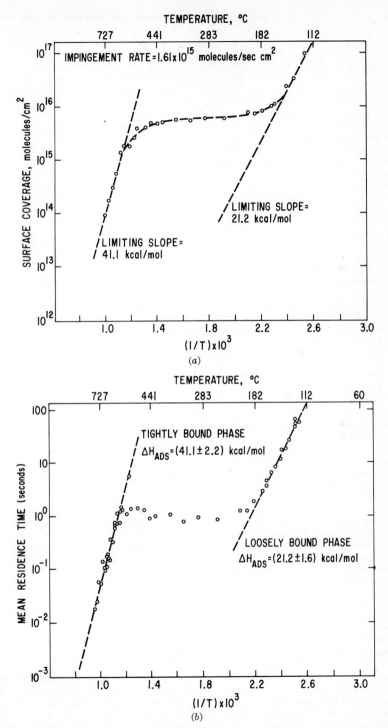

Fig. 15 (a) Surface coverage vs. inverse temperature for cadmium deposited on tungsten. (*Hudson and Sandejas.*[36]) (b) Mean residence times for adsorption of cadmium on tungsten. (*Hudson and Sandejas.*[36])

with a high positive free energy of formation, and the second a very small nucleation barrier. Since the smallest possible condensate consists of two atoms, the latter case would be described for the situation in which a two-atom cluster is already supercritical. Now let us look at the structure of a film after an average thickness of a few monolayers has been deposited. Under the regime of a large nucleation barrier, the film would consist of only a few, but large, aggregates. The aggregates must be large because their minimum possible stable size is large, and there are relatively few of them because the nucleation frequency is relatively small. On the other hand, a film formed under the regime of a small nucleation barrier would consist of many, but small, aggregates, since the minimum stable size is now small, but the nucleation frequency is large.

A film consisting of a dense population of small islands will become continuous at a relatively low average film thickness since these islands touch and grow together quite early in the deposition process. On the other hand, a film consisting of only a few big islands will have an island structure which persists up to relatively high average film thicknesses. (The islands are three-dimensional; i.e., they have an appreciable dimension in the direction perpendicular to the film plane.)

When the islands have joined together to form the grains of a continuous film, the effect of nucleation is often still visible: high-nucleation barrier films will give a coarse-grained film, while low-nucleation barrier films will give a finer-grained film. (It should be noted here, however, that a single grain is generally formed by the coalescence of many nuclei; see Sec. 4b.) Different nucleation barriers can therefore cause drastically different structures in thin films, but even the structure of a thick film can be at least partially traced back to nucleation phenomena. The dependence of the size of the nucleation barrier on the nature of the film and substrate materials, substrate temperature, deposition rate, the activation energies for surface diffusion, and the binding energy to the substrate will now be discussed.[45]

a. Dependence on the Nature of the Film and Substrate

There is a very strong dependence of the size of the critical nucleus on the nature of the metal. Indeed, the principal dependence of i^* is on ΔG_v. For any one deposition rate R, the supersaturation ratio $R/R_e(b)$ (actually the ratio of the equivalent vapor pressures) at the substrate temperature T is given by the bulk heat of sublimation of the film material (Clausius-Clapeyron's equation). Since the heat of vaporization ΔH_{vap} is the major part of the heat of sublimation, and since it is related to the boiling point by Trouton's rule, $\Delta H_{\text{vap}}/T_B \approx 21$ cal mol^{-1} deg^{-1}, it can be stated that high-boiling-point metals have a large ΔH_{vap}, and therefore also a large ΔG_v (Gibbs-Helmholtz equation). Thus, one expects the size of the critical nucleus to decrease as the boiling point of the film material increases. For the high-boiling metals such as W, Mo, Ta, Pt, and Ni, the nuclei are therefore stable even if they are very small, and it follows from Eq. (5) that their stability increases very rapidly with size, making dissociation or reevaporation extremely unlikely. On the other hand, for the low-boiling metals such as Cd, Mg, and Zn, the nuclei must be relatively large before they are stable, and their stability increases only slowly with size, making dissociation or reevaporation much more likely.

From Eq. (6) it is evident that larger critical clusters are required if the surface energy σ_{v-c} of the condensate material is large, and if the surface energy of the substrate material σ_{s-v} is low. Table 2 lists the surface free energies of several common condensate and substrate materials.

b. The Role of Surface Diffusion and the Binding Energy to the Substrate

The size of the critical nucleus is independent of the ability of single metal atoms to diffuse on the substrate surface. However, the rate of formation of critical nuclei clearly must depend on the ability of adsorbed atoms to diffuse and collide with each other. If the activation energy for surface diffusion is very large, the diffusion distance before reevaporation will be small, and nuclei can grow only from material received by direct impingement from the vapor phase.

A comprehensive description of surface processes has been given by Ehrlich.[48]

The activation energy of diffusion has been best studied for the case of metals adsorbed on metals. Table 3 gives some of the measured values. The activation energy for diffusion can often be taken as approximately one-fourth of the activation energy for desorption into the gas phase E_a, at least for metals on metals. Unfortunately, almost nothing is known about the activation energy of atoms adsorbed on non-metallic substrates, except, of course, that it may never be larger than the desorption energy.

TABLE 2 Surface Free Energies of Several Common Condensate and Substrate Materials[46,47]

Material	$T°C$	σ, ergs cm^{-3}	$\delta\sigma/\delta T$, ergs cm^{-3} deg^{-1}
Al	700	900	-0.35
Sb	635	383	
Bi	300	376	-0.06
Cd	370	608	
Cu	1140	1,120	$+0.66$
Ga	30	735	
Au	1120	1,128	-0.10
Fe	1530	1,700	
Pb	350	442	-0.07
Mg	700	542	-0.34
Se	220	105.5	-0.17
Ag	995	923	-0.13
Tl	313	446	
Zn	700	750	-0.25
Sn	700	538	-0.18
Al$_2$O$_3$	2050	580	
B$_2$O$_3$	900	79.5	$+0.04$
FeO	1420	585	
La$_2$O$_3$	2320	560	
PbO	900	132	$+0.03$
SiO$_2$	1400	200–260	
ZnO	1300	455	
Glass	....	250–360	
Polymers containing no polar groups	....	<100	
Polymers containing polar groups	....	<300	

TABLE 3 Some Experimentally Measured Values for the Activation Energy of Diffusion and Reevaporation[18]

	E_d, eV	E_a, eV
Cs on W	0.61	2.8
Ba on W	0.65	3.8
W on W	1.31	5.8
Hg on Hg	0.048	

The binding energy of adsorbed single atoms to the substrate E_a is important in determining both i^* and ΔG^*. The stronger the binding between adsorbed atoms and substrate, the smaller the critical nucleus and the higher the nucleation frequency. If a substrate is not homogeneous but has sites of variable binding energy for adsorbed atoms, smaller critical nuclei are formed at the tight binding sites, but at higher nucleation rates than at the weaker binding sites. A step on a substrate is an often observed strong binding site. Figure 16 shows a relatively high density of small islands in the region of the staircase of cleavage steps at A. The seeding of a substrate with a thin layer of a high-boiling metal increases the binding energy of subsequently adsorbed atoms of a low-boiling metal and thus decreases the nucleation barrier, i.e., decreases i^* and ΔG^*.

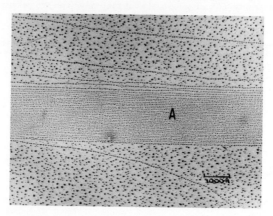

Fig. 16 Evaporated gold on rock salt at very low film thickness, showing a relatively high density of small islands in the region of the "staircase" of cleavage steps at A. (*Bassett et al.*[52])

The binding energy may vary from a weak van der Waals interaction of less than a few tenths electron volt to metallic bonding of several electron volts. Again more information is available for metals on metals than for metals on nonmetals. The simplest case is the binding energy of a metal atom on a lattice plane of a crystal of the same metal. For such a case, the binding energy can be related to the sublimation energy ΔH_s from a knowledge of the crystal structure and the interaction between nearest and next to nearest neighbors (see Ehrlich,[48] Fricke,[49] and Volmer[50]). The situation is more complicated when the metals are dissimilar. The experimental binding energies for several film-substrate pairs are given in Table 4.[48,51]

TABLE 4 Typical Experimental Binding Energies[48,51]

Adsorbed atom	Substrate	E_a, eV
Na	W	2.73
Rb	W	2.60
Cs	W	2.80
Ba	W	3.80
W	W	5.83
Hg	Ag	0.11
Cd	Ag	1.61
Al	NaCl	0.60
Cu	Glass	0.14

Often a very strong chemical bond forms between substrate and film material. An example may be aluminum forming an oxide bond with a glass substrate. For such a case E_a is very large, and the nucleation barrier may be greatly reduced.

c. The Role of Substrate Temperature and Deposition Rate

The dependence of the size of the critical nucleus on temperature can be obtained by differentiating Eq. (6) with respect to temperature. Assuming an inert substrate so that ΔG_{des} and $\sigma_{s\text{-}v}$ are negligible and making the simplifying assumption[45] that $\Delta G_{\mathrm{des}} \approx \sigma_{v\text{-}c} - \sigma_{s\text{-}c} \approx 0$, one obtains, using the capillarity model,

$$-\frac{1}{2}\left(\frac{\partial r^*}{\partial T}\right)_R = \frac{(a_1 + a_2)[\Delta G_v(\partial \sigma_{v\text{-}c}/\partial T) - \sigma_{v\text{-}c}(\partial \Delta G_v/\partial T)]}{3a_3\,\Delta G_v{}^2}$$

With typical values $\sigma_{v\text{-}c} = 1{,}000$ ergs cm^{-2},

$$\frac{\partial \Delta G_v}{\partial T} \approx \Delta S_{\mathrm{vap}} = 8.2 \times 10^7 \text{ ergs cm}^{-3} \text{ deg}^{-1}$$

and

$$\frac{\partial \sigma_{v\text{-}c}}{\partial T} = -0.50 \text{ erg cm}^{-2} \text{ deg}^{-1}$$

this equation yields

$$\left(\frac{\partial r^*}{\partial T}\right)_R > 0 \tag{32}$$

as long as $|\Delta G_v| < 1.64 \times 10^{11}$ ergs cm^{-3}. This will usually be the case if a nucleation barrier exists.

Consequently, increasing the substrate temperature at a constant deposition rate will increase the size of the critical nucleus. Furthermore, an island structure will persist to a higher average coverage than at low temperatures. Finally, a nucleation barrier may be found for metals at a high substrate temperature which may not have existed at lower substrate temperatures.

Again neglecting ΔG_{des} and $\sigma_{s\text{-}v}$, the temperature dependence of the free energy ΔG^* is given by[45]

$$\frac{1}{4}\left(\frac{\partial \Delta G^*}{\partial T}\right)_R = \frac{a_1{}^3\sigma_{v\text{-}c}{}^2[3\,\Delta G_v(\partial \sigma_{v\text{-}c}/\partial T) - 2\sigma_{v\text{-}c}(\partial \Delta G_v/\partial T)]}{27a_3{}^2\,\Delta G_v{}^3}$$

Since again

$$\left|2\sigma_{v\text{-}c}\frac{\partial \Delta G_v}{\partial T}\right| > \left|3\,\Delta G_v\frac{\partial \sigma_{v\text{-}c}}{\partial T}\right|$$

One obtains

$$\left(\frac{\partial \Delta G^*}{\partial T}\right)_R > 0 \tag{33}$$

Because of the exponential dependence of the nucleation frequency on ΔG^*, the rate at which supercritical aggregates are created will decrease rapidly with increasing temperature. This means that at higher substrate temperatures it will take longer before a continuous film is produced.

The dependence of r^* and ΔG^* on the deposition rate is caused by the fact that ΔG_v is a function of this rate. According to Eq. (4), $-\Delta G_v$ increases with increasing rate. Since the surface and interfacial energies are independent of rate, one has[45]

$$\left(\frac{\partial r^*}{\partial R}\right)_T < 0 \tag{34}$$

$$\left(\frac{\partial \Delta G^*}{\partial R}\right)_T < 0 \tag{35}$$

Thus, increasing the deposition rate results in smaller islands and in a higher rate of formation for them. This means that a continuous film is produced at lower average film thicknesses. However, because of the logarithmic dependence of ΔG_v on R, a small increase in R by a factor of 2 or 3 will not substantially alter r^*. A change in R

of several orders of magnitude is necessary before any effects can be observed in the size of the critical nucleus.

In this section we have examined the size and the density of the metal islands which are formed on the substrate in the very first stages of growth, i.e., the first few monolayers of film material. The tendency to form an island structure is increased by (1) a high substrate temperature, (2) a low-boiling-point film material, (3) a low deposition rate, (4) weak binding forces between film and substrate, (5) a high surface energy of the film material, and (6) a low surface energy of the substrate.

Immediately after formation, a critical nucleus will grow by acquiring more atoms which migrate on the substrate surface in search of a tight binding site. If one examines a deposit during the island stage of growth, the average size of the islands will be determined, first, by the total amount of material deposited and, second, by the density of nuclei of or above the critical size. As soon as the islands grow large enough to meet, they can either form a grain boundary or coalesce to form one single grain. This latter process has been observed electron-microscopically. It is aided by a high substrate temperature and closely aligned crystal orientations of the joining islands. The various stages of film growth, from an island structure to a continuous film, are examined in the following section.

4. THE FOUR STAGES OF FILM GROWTH

The general picture of the sequence of the nucleation and growth steps to form a continuous film which emerges from nucleation theory and electron-microscopic observations[52-68] is as follows:

1. Formation of adsorbed monomers.
2. Formation of subcritical embryos of various sizes.
3. Formation of critically sized nuclei (nucleation step).
4. Growth of these nuclei to supercritical dimensions with the resulting depletion of monomers in the capture zone around them.
5. Concurrent with step 4, there will be nucleation of critical clusters in areas not depleted of monomers.
6. Clusters touch and coalesce to form a new island occupying an area smaller than the sum of the original two, thus exposing fresh substrate surface.
7. Monomer adsorbs on these freshly exposed areas, and "secondary" nucleation occurs.
8. Large islands grow together, leaving channels or holes of exposed substrate.
9. The channels or holes fill via secondary nucleation to give a continuous film.

Some of these steps are shown schematically in Fig. 17. Pashley et al.[54] distinguish four stages of the growth process: nucleation and island structure, coalescence of islands, channel formation, and formation of the continuous film. These will be discussed in greater detail below, based principally on the electron-microscopic observations of Pashley and coworkers.[52-63,67,68]

a. The Island Stage

When a substrate under impingement of condensate monomers is observed in the electron microscope, the first evidence of condensation is a sudden burst of nuclei of fairly uniform size. The smallest nuclei detected have a size of 20 to 30 Å. Growth of the nuclei is three-dimensional, but the growth parallel to the substrate is greater than that normal to it. This is probably because growth occurs largely by the surface diffusion of monomers on the substrate, rather than by direct impingement from the vapor phase. For Ag or Au condensed on a MoS_2 substrate kept at 400°C, the density of initial nuclei is 5×10^{10} cm^{-2}, and the minimum diffusion distance is about 500 Å.

b. The Coalescence Stage

Figure 18 illustrates the manner of coalescence of two rounded nuclei. The coalescence occurs in less than 0.1 s for the small nuclei and is characterized by a decrease in total projected area of the nuclei on the substrate (and an increase in their height). In addition, nuclei having well-defined crystallographic shapes before coalescence become rounded during the event.[68] The composite island takes on a crystallographic shape again if left for a sufficiently long time before interacting with its neighbors. The triangular profile of the crystallites is characteristic of the nucleation stage; after coalescence has taken place, the islands assume a more hexagonal profile and are often faulted. A sequence of micrographs illustrating the effects is shown in Fig. 19, where islands A and B have formed a compound island which eventually becomes crystallographically shaped.

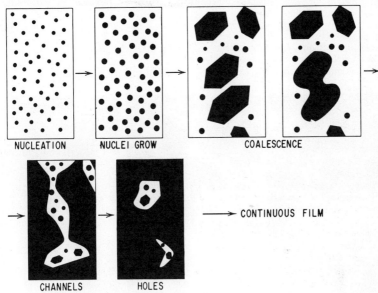

Fig. 17   Schematic of the stages of film growth.

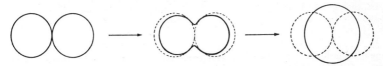

Fig. 18   Schematic of the shape changes during coalescence.   (*Pashley et al.*[54])

The liquidlike character of the coalescence leads to enlargements of the uncovered areas of the substrate, with the result that secondary nuclei form between the islands. This effect becomes noticeable when the primary islands have grown to about 1,000 Å, and continues until the final hole-free film is formed. The small nuclei surrounding island B (Fig. 19a) are examples of these secondary nuclei. A secondary nucleus grows until it touches a neighbor, and if this happens to be a much larger island, the secondary nucleus coalesces very rapidly and becomes completely incorporated in the large island.

Pashley et al.[54] have used the theory of sintering of spherical particles[69-72] to explain the changes in shape during coalescence and the driving force for the process. The driving force for the changes in configuration which occur during sintering is provided by the change in surface energy due to a reentrant region with a high curvature at its base. Transport during sintering is possible by evaporation and condensation, volume diffusion, and surface diffusion. The neck radius X, island radius r, time t, and temperature T can be related by[70]

$$\frac{X^n}{r^m} = A\,(T)t \tag{36}$$

where $n = 3$, $m = 1$ for evaporation-condensation

$\qquad n = 5$, $m = 2$ for volume diffusion

$\qquad n = 7$, $m = 3$ for surface diffusion

$A\,(T)$, a function of temperature, includes the physical constants of the material relevant to the specific transport mechanism. These expressions are valid only for the initial stages of neck growth.

Both volume and surface diffusion are possible mechanisms for mass transfer during coalescence. However, all the evidence suggests that the surface-diffusion effect is the predominant one, the more so the smaller the particles involved. Since appreciable-sized necks have been observed to form within a time as short as 0.06 s, the behavior is most satisfactorily explained in terms of surface diffusion.

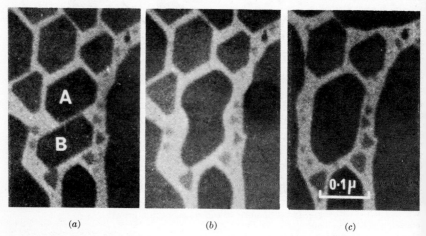

Fig. 19 Successive electron micrographs of gold grown on MoS_2 at 400°C, showing the change in shape of islands during and after coalescence. (*a*) Arbitrary zero. (*b*) 1 to 2 s. (*c*) 60 s. (*Pashley et al.*[54])

The driving force for all the liquidlike behavior is the resultant reduction in surface energy. If the surface energy were independent of crystal orientation, it would act to reduce the surface area to a minimum. Observations show that after the major surface-energy reduction has taken place on coalescence, an additional reduction occurs by the formation of preferred boundary planes, leading to well-developed crystallographically shaped islands. Further, these crystallographic shapes are instantaneously rounded off when such an island takes part in a subsequent coalescence. This can perhaps be understood in terms of a sudden breaking down of a minimum energy configuration as two neighboring islands touch and allow rapid interchange of atoms between each other. Also we might expect the corners of triangular and hexagonal islands to be the most effective sources of fresh mobile atoms, so that they would round off rapidly.

Although the initial stages of coalescence of even very large islands take place in a very short time, an island which has just been formed by coalescence continues to change its shape over a considerably longer period. The area, too, changes during and after coalescence. A large reduction in substrate coverage occurs within a few seconds, after which there is a more gradual increase in area. As the coalescence starts, a reduction in area and an increase in height occur to lower the total surface energy. If the relative values of the substrate and deposit surface energies as well as the interfacial energy are taken into account, an island will have a particular minimum energy shape with a certain ratio of height to diameter. Figure 20 shows the change in the area of a composite island of Au on MoS_2 at 400°C during and after coalescence, as measured in the electron microscope (Pashley et al.[54]).

It has been reported by Adamsky and LeBlanc[73] that a bridge is formed between two islands before coalescence. However, this observation has not yet been confirmed by other workers, and has been claimed by Pashley et al.[54] to be due to contamination effects.

c. The Channel Stage

As the islands grow, there is a decreasing tendency for them to become completely rounded after coalescence. Large shape changes still occur, but these are confined mainly to the regions in the immediate vicinity of the junction of the islands. Consequently, the islands become elongated and join to form a continuous network structure in which the deposit material is separated by long, irregular, and narrow channels of width 50 to 200 Å (Fig. 21a). As deposition continues, secondary nucleation occurs in these channels, and the nuclei are incorporated into the bulk of the film as they grow and touch the sides of the channel. At the same time, channels are bridged at some points and fill in rapidly in a liquidlike manner. A sequence showing such an event is seen in Fig. 21.

Eventually, most of the channels are eliminated and the film is continuous but contains many small irregular holes. Secondary nucleation takes place on the substrate within these holes, and the growing nuclei are incorporated (in a liquidlike manner) into the continuous regions of the deposit. The hole contains many secondary nuclei which coalesce with each other to form secondary islands which then touch the edge of the hole and coalesce with the main film to leave a clean hole. Further secondary nuclei then form, and the process is repeated until the hole finally fills in.

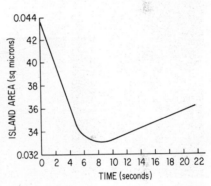

Fig. 20 Change in the area of composite island during and after coalescence. (*Pashley et al.*[54])

The liquidlike behavior of the deposit persists until a complete film is obtained. In the channel and hole stages, secondary nuclei (islands) are pulled into the more massive regions of the film, and these events take place in less than 0.1 s. In addition to this, one can observe a liquidlike channel-filling process, where a channel is bridged and the fronts move along the channel with velocities of the order of 1 to 300 Å s^{-1}. It appears that the channels are not filled completely and only a fairly thin front moves initially, the channel thickening up more slowly over a much longer time. The grooves are usually very irregular, and the bounding deposit is faceted.

It is clear that both the liquidlike behavior of coalescing nuclei and the rapid elimination of channels are manifestations of the same physical effects, namely, the minimization of total surface energy of the overgrowth by the elimination of regions of high surface curvature.

d. The Continuous Film

Considerable changes in the orientation of islands occur during the growth of the film, particularly in the coalescence stage. This is of considerable importance in the growth of epitaxial films and is discussed in Chap. 10. The general mechanism of the growth of polycrystalline layers is similar to that of epitaxial layers, except that the coalescing pairs of islands have a relative orientation which is randomly

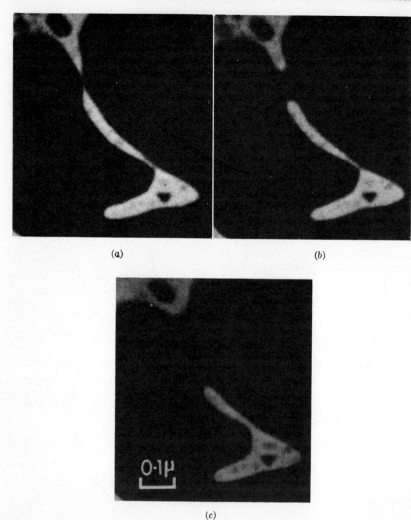

(a)   (b)

(c)

Fig. 21 Successive electron micrographs of gold grown on MoS_2 at 400°C showing channel filling. (a) Arbitrary zero. (b) 0.06 s. (c) 4 s. (Pashley et al.[54])

distributed. However, it is found that some recrystallization occurs during coalescence, so that the grain size of the completed deposit film is large compared with the average separation of the initial nuclei. This is illustrated by the sequence in Fig. 22, which shows a polycrystalline gold film during growth on a carbon substrate.[54]

The deposition was started on all four specimens simultaneously, a moving shutter being used to vary the time of deposition from specimen to specimen. Even with the substrate maintained at room temperature, there is considerable recrystallization, and some 100 or more of the initial nuclei contribute to each grain. Thus, the grain size of the completed film is controlled by the recrystallization which occurs during coalescence of nuclei or islands, not only the initial density of nuclei.

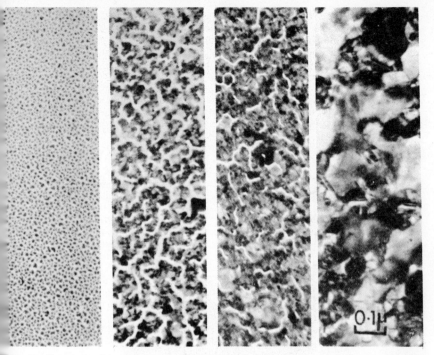

Fig. 22 Sequence of growth of polycrystalline gold film on a carbon substrate at 20°C. (*Pashley et al.*[54])

5. THE INCORPORATION OF DEFECTS DURING GROWTH

When the islands during the initial stages of film growth are still quite small, they are observed to be perfect single crystals.[34,52] However, as soon as the islands become large enough so that they touch, grain boundaries or lattice defects will be incorporated into the film, unless the islands coalesce to form a single grain. This latter phenomenon is indeed frequently observed, even if the two initial nuclei are in completely different orientations. Thus, even in polycrystalline films a recrystallization process occurs continuously at least during the early stages of formation of the film, resulting in a number of grains per unit area which is much less than the density of initial nuclei. Sooner or later, however, a large number of defects are incorporated when these grains grow together, even in single-crystal films grown epitaxially. In this section the process of incorporation of some of these defects during growth is examined.

Observation of defects introduced during the growth of evaporated-metal films became possible with the application of the electron microscope to problems of thin film structure. Much of the work of Hirsch, Whelan, Silcox, Bollmann, Wilsdorf, Matthews, and Phillips,[64,74−80] in addition to the work of Pashley, Bassett, Menter, and coworkers mentioned above,[34,52−63,67,68,81] was directed toward the observation of defects in films, either thinned down from the bulk material or in vapor-deposited form. Usually single-crystal films were studied, since such films display their defects more clearly than polycrystalline deposits. However, the types of defects observed in these films and their mode of formation are probably typical of what might be found in any evaporated film, single-crystal or not, with the possible exception that stacking faults and twin boundaries probably occur much less frequently in fine-grained polycrystalline films than in single-crystal films, and that the grain-boundary area is much larger in polycrystalline films. Emphasis here will be given to polycrystalline films, since defects in single-crystal films (such as stacking faults, and twin and double positioning boundaries) will be discussed in Chap. 10. Such defects may, however, occur within a grain of a polycrystalline film. A complete description of microscopic techniques to study lattice defects has been given by Demny.[82]

a. Dislocations and Minor Defects

The most frequently encountered defects in evaporated films are dislocations. A density of 10^{10} to 10^{11} lines cm^{-2} is frequently encountered. Most information about dislocations in evaporated films comes from electron-microscopic investigations of films of fcc metals. There are five mechanisms by which dislocations can be formed in such films during growth:

1. When two islands coalesce whose lattices are slightly rotated relative to one another, they form a subboundary composed of dislocations (Bassett[34]).

2. Since the substrate and film usually have different lattice parameters, there will be a displacement misfit between islands. Dislocations can result from this misfit when islands grow together.[67]

3. Stresses present in continuous films can generate dislocations at the edges of holes usually present in the earlier stages of film growth (Phillips[78]).

4. Dislocations ending on the substrate surface may continue into the film (Phillips[78]).

Fig. 23 Formation of a dislocation resulting from displacement misfits between three coalescing nuclei. (*Pashley.*[67])

5. When islands containing stacking faults bounded by the surfaces of the island coalesce, partial dislocations must now bound these faults in the continuous film (Matthews[79]).

A dislocation network may be set up at the interface between film and substrate to relieve the strain. Such a network has, for instance, been observed by Amelinckx.[83] The formation of dislocations resulting from displacement misfits between three neighboring islands is shown in Fig. 23.[67]

If the appearance of dislocations during film growth is observed in the electron microscope, it is found that the dislocations all consist of short lines running straight through the metal layers as long as the film is thin. For greater film thicknesses there is evidence of longer dislocation lines running parallel to the film plane. A density of short dislocations of the order of 10^{10} cm^{-2} has been measured in gold[67] and silver[84] films. If the dislocation density is measured during film growth, it is found that most dislocations are incorporated into the film during the channel (or network) and the hole stage. Figure 24 shows the dislocation density as a function of film thickness for a gold film deposited at 300°C on molybdenum disulfide.[8]

The displacement-misfit mechanism is responsible for many of the dislocations at this stage of growth. One of the characteristic features of this mechanism is that dislocations are formed in holes in the growing film. It is commonly found that even after the filling in of the channels in the growing film, a few very small holes (100 to 200 Å diameter) remain in the film. Almost all these holes contain incipient dislocations. It is often found that there are several incipient dislocations in one hole up to six terminating lines (see Fig. 25) have been observed.[67] It is not possible for the pure displacement-misfit mechanism to give rise to more than one dislocation in a hole, and an additional mechanism must be operative.

Pashley[67] has suggested that stresses in the growing film and resulting plastic deformation are responsible for multiple incipient dislocations observed in holes

An incipient dislocation of opposite sign is left in the hole (Fig. 26) for each dislocation which moves into the film, and when the hole eventually fills in this will become
a real dislocation. Since it is possible to nucleate a number of dislocations of the

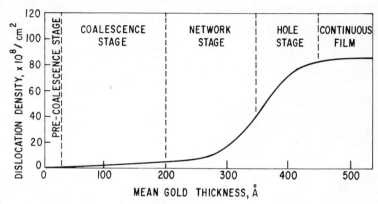

ig. 24 Approximate measured relationship between dislocation density and thickness of
gold films deposited on molybdenum disulfide at 300°C. (*Jacobs et al.*[85])

same sign at the edge of one hole, it is possible for a hole to contain several incipient
dislocations, in the manner observed in Fig. 25. Thus, it is considered that plastic
deformation is the most probable explanation of the multiple incipient dislocations
observed in holes.

The so-called minor defects which
are frequently observed in deposited films
include dislocation loops, stacking-fault
tetrahedra, and small triangular defects.
All are generally attributed to vacancy
collapse.[52,61,78,79] Dislocation loops may
be as large as 100 to 300 Å. Phillips[78]
measured about 2×10^{14} loops cm^{-3} in a
650-Å-thick epitaxial silver film on rock
salt. If each loop was indeed a collapsed
disk of vacancies, one has to postulate
a vacancy concentration of at least

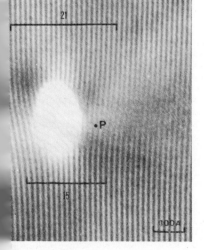

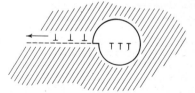

. 25 Incipient dislocations in a hole in
gold film grown on molybdenum disulfide
300°C. (*Jacobs et al.*[85]) Point *P* in the
otograph marks a real dislocation.

Fig. 26 Deformation leaving incipient dislocations in the hole. (*Jacobs et al.*[85])

$\times 10^{-5}$ mole fraction in the film before collapse took place. This is considerably
ther than the concentration which one would expect for bulk silver if it were simply
enched from 300°C, the substrate temperature used by Phillips. A large number
vacancies can be introduced into evaporated films during preparation for two

reasons. First, the effective temperature at which impinging atoms freeze into the lattice is most likely considerably higher than the nominal substrate temperature. Second, metal films are often formed by a rapid condensation in which a deposited layer of atoms is covered by succeeding layers, before reaching thermal equilibrium with the substrate. In this way many vacancies can become trapped in the film.

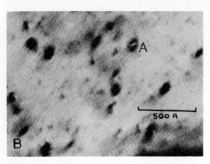

Dotlike features are often observed in electron micrographs of evaporated films.[52] They are believed to be either unresolved dislocation loops, vacancy aggregates, or aggregates of impurity atoms trapped in the film during preparation. An electron micrograph of dot-like and loop features observed in a vapor-deposited gold film is shown in Fig. 27.[5]

Fig. 27 Dot- and loop-like features observed in an evaporated gold film. (*Bassett et al.*[52])

The interpretation of the formation of minor defects by vacancy or interstitial condensation is somewhat in doubt, since it has been shown by Pashley and Presland[86] that such features can be formed in the specimen while it is being observed in the electron microscope, because of bombardment by negative ions originating in the electron gun. It is therefore concluded that there is no conclusive evidence of any defects arising from point-defect aggregation. This is perhaps not a surprising situation, since the mobility of point defects at the temperature of formation of many films will be sufficiently high to allow the vacancies and interstitials to escape to the film surface. However, films formed under conditions where the point-defect mobility is small might be expected to contain imperfections due to point-defect aggregation.

b. Grain Boundaries

In general, a thin film will possess a larger grain-boundary area than bulk material since the average grain size will generally be smaller. In the extreme case of very low surface mobility, the grain size of the film may not be much larger than the size of the critical nucleus. More generally, however, the islands will be much larger when they touch (see Sec. 4 above). The grain size depends on deposition conditions and annealing temperature, as indicated schematically in Fig. 28. Perhaps the common feature in the behavior illustrated in Fig. 28 is that the grain size saturates at some value of the deposition variables. With film thickness as the deposition variable, this must mean that new grains are nucleated on top of the old ones after a certain film thickness has been reached. Nucleation of a new grain may become necessary because of a layer of contamination making coherent growth with the grain below impossible, or if the top surface of the grain below is a nearly perfect closed-packed plane.

Larger grain sizes are naturally expected for increasing substrate or annealing temperature because of an increase in surface mobility, thus allowing the film decrease its total energy by growing large grains and thus decreasing its grain-boundary area. Low temperatures, on the other hand, lead to small grain sizes. The formation, structure, and annealing of films of very small grain size have been described by Buckel and Hilsch and coworkers,[87-92] and Suhrmann and Schnackenberg. It is significant, however, that even at the lowest substrate temperatures explored so far, in the liquid-helium region, metal deposits are still found to be crystalline and therefore possess a small but finite crystal size. An amorphous or liquidlike structure has been obtained only if large amounts of impurities are codeposited to inhibit grain growth.[88,89,94,95] The dependence on deposition rate is less obvious but can be rationalized on the basis that film atoms just impinged on the surface, although they may possess a large surface mobility, become buried under subsequent layers at high deposition rates, before much diffusion can take place.[96] In order for this effect to operate, however, a certain minimum rate must be exceeded. Below this threshold rate, grain size is limited by temperature only; above it the grain size is decreased more and more for higher and higher rates.

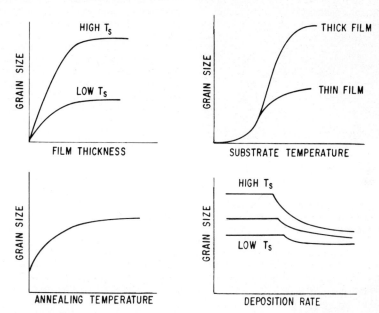

Fig. 28 Dependence of grain size on film thickness, substrate temperature, annealing temperature, and deposition rate. (*Thun.*[96])

c. Surface Area and Roughness

A film can minimize its total energy by keeping its surface area as small as possible, i.e., ideally flat. However, films exhibiting ideally flat surfaces are practically never observed. A certain surface roughness, and therefore an increased surface area, must be expected because of the randomness of the deposition process. If it is assumed that the atoms stick where they impinge on the substrate without subsequent surface diffusion, then statistical fluctuations in the local film thickness will result.[97] The average thickness deviation Δd from the average film thickness d is given by Poisson's probability distribution of a random variable[45] and reduces to $\Delta d = \sqrt{d}$. Thus the surface roughness, and therefore the surface area of such a film, increases with the square root of the film thickness.

In actual practice, however, at least some surface diffusion of the film atoms after impingement is always possible. This makes possible the occupancy of sites in the film lattice left empty above. Surface migration can lead to less surface area by filling in the valleys and leveling the peaks to give a lower surface energy. On the other hand, it may also lead to the development of well-defined crystal faces, particularly those of low energy (low index planes). Also, under a kinetic regime the fastest-growing crystal planes will grow at the expense of the slower ones. This again leads to a larger surface roughness. These surface structures are to be expected at higher substrate temperatures.

It is observed experimentally[98–102] that the largest surface area, as measured by gas adsorption, is obtained for films deposited under conditions where only little surface mobility of the impinged atoms is possible, i.e., relatively low substrate temperatures. Under these conditions the surface area is found to increase linearly with film thickness, and the ratio of the surface area to the geometric area can be larger than 100.[98] This is illustrated in Fig. 29 for nickel films deposited at room temperature. The linear increase in surface area with film thickness suggests a porous film structure with a large internal surface area accessible to the adsorbing gas even in the lowest layers of the film. The simplest explanation for a large internal surface area would be shadowing of film areas from the source by protruding peaks in the deposit.

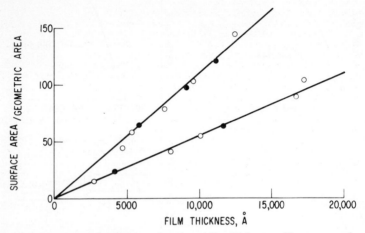

Fig. 29 Surface area of nickel films as a function of film thickness. Upper curve: deposited in 1 Torr of N_2. Lower curve: deposited in high vacuum (*Beeek et al.*[98]). Open circles: measured by adsorption of hydrogen. Full circles: measured by adsorption of carbon monoxide.

If deposition occurs in a system in which the residual gas pressure is high enough for condensation of the vapor atoms into small smoke particles, then a very porous structure with a very large internal surface area must be expected because of the loose packing of the particles on the substrate, leaving much empty space. This is particularly true if the substrate temperature is so low that a substantial rearrangement of the atoms after deposition is unlikely. The surface area of nickel films deposited at room temperature in 1 Torr of nitrogen in Fig. 29 as a function of film thickness is indeed very large (upper curve).

ACKNOWLEDGMENTS

Most stimulating and critical discussions with R. A. Sigsbee are gratefully acknowledged. Thanks are due D. Campbell, J. Hudson, and G. Zinsmeister who, by their conversations with the author, have made the preparation of this chapter much easier.

REFERENCES

1. Cabrera, N., *Discussions Faraday Soc.*, **28,** 16 (1959).
2. Zwanzig, R. W., *J. Chem. Phys.*, **32,** 1173 (1960).
3. McCarroll, B., and G. Ehrlich, *J. Chem. Phys.*, **38,** 523 (1963).
4. Goodman, F. O., *Phys. Chem. Solids*, **23,** 1269 (1962); **24,** 1451 (1963).
5. McFee, J. H., Ph.D. Thesis, Carnegie Institute of Technology, 1960.
6. Lennard-Jones, J. E., *Proc. Roy. Soc. London*, **A163,** 127 (1937).
7. Hirth, J. P., and G. M. Pound, "Condensation and Evaporation," The Macmillan Company, New York, 1963.
8. Sigsbee, R. A., and G. M. Pound, *Advan. Coll. Interf. Sci.*, **1,** 335 (1967).

9. Volmer, M., and A. Weber, *Z. Phys. Chem.*, **119**, 277 (1925).
10. Becker, R., and W. Doering, *Ann. Physik*, **24**, 719 (1935).
11. Walton, D., *J. Chem. Phys.*, **37**, 2182 (1962).
12. Walton, D., *Phil. Mag.*, **7**, 1671 (1962).
13. Rhodin, T., and D. Walton, *Trans. 9th Natl. Vacuum Symp.*, 1962, p. 3, The Macmillan Company, New York.
14. Rhodin, T., and D. Walton, in M. H. Francombe and H. Sato (eds.), "Single Crystal Films," p. 31, Pergamon Press, New York, 1964.
15. Walton, D., T. Rhodin, and R. W. Rollins, *J. Chem. Phys.*, **38**, 2698 (1963).
16. Lewis, B., *Thin Solid Films*, **1**, 85 (1967).
17. Lewis, B., and F. Campbell, *J. Vacuum Sci. Technol.*, **4**, 209 (1967).
18. Zinsmeister, G., *Proc. Intern. Symp. Basic Problems in Thin Film Physics*, 1965, p. 33, Vandenhoeck and Rupprecht, Goettingen.
19. Courvoisier, J. C., L. Jansen, and W. Haidinger, *Trans. 9th Natl. Vacuum Symp.*, 1962, p. 14, The Macmillan Company, New York.
20. Mayer, H., and H. Göhre, *Proc. Northwest German Phys. Soc.*, 1962.
21. Palatnik, L. S., and Y. F. Komnik, *Soviet Phys. Doklady English Transl.*, **4**, 663 (1959).
22. Sloope, B. W., and C. O. Tiller, *J. Appl. Phys.*, **32**, 1331 (1961).
23. Sloope, B. W., and C. O. Tiller, *J. Appl. Phys.*, **33**, 3458 (1962).
24. Sloope, B. W., and C. O. Tiller, *J. Appl. Phys.*, **36**, 3174 (1965).
25. Sella, C., and J. J. Trillat, in M. H. Francombe and H. Sato (eds.), "Single Crystal Films," p. 201, Pergamon Press, New York, 1964.
26. Ino, S., D. Watanabe, and S. Ogawa, *J. Phys. Soc. Japan*, **19**, 881 (1964).
27. Bethge, H., and M. Krohn, *Proc. Intern. Symp. Basic Problems in Thin Film Physics*, 1965, p. 157, Vandenhoeck and Rupprecht, Goettingen.
28. Matthews, J. W., and E. Gruenbaum, *Phil. Mag.*, **11**, 1233 (1965).
29. Krikorian, E., in M. H. Francombe and H. Sato (eds.), "Single Crystal Films," p. 113, Pergamon Press, 1964.
30. Krikorian, E., and R. J. Sneed, *Trans. 10th Natl. Vacuum Symp.*, 1963, p. 368, The Macmillan Company, New York.
31. Behrndt, K. H., *J. Appl. Phys.*, **37**, 3841 (1966).
32. Pashley, D. W., in "Thin Films," p. 59, American Society for Metals, Metals Park, Ohio, 1964.
33. Bassett, G. A., in E. Rutner, et al. (eds.), "Condensation and Evaporation of Solids," p. 521, Gordon and Breach, Science Publishers, Inc., New York, 1964.
34. Bassett, G. A., *Proc. European Conf. Electron Microscopy*, 1960, p. 270, Delft.
35. Poppa, H., *J. Vacuum Sci. Technol.*, **2**, 42 (1965).
36. Hudson, J. B., and J. S. Sandejas, *J. Vacuum Sci. Technol.*, **4**, 230 (1967).
37. Caswell, H. L., *J. Appl. Phys.*, **32**, 105 (1961).
38. Caswell, H. L., *J. Appl. Phys.*, **32**, 2641 (1961).
39. Melmed, A. J., *J. Appl. Phys.*, **37**, 275 (1966).
40. Sigsbee, R. A., Ref. 2, p. 372.
41. Chopra, K. L., *Appl. Phys. Letters*, **7**, 140 (1965).
42. Hill, R. M., *Nature*, **210**, 512 (1966).
43. Stirland, D. J., *Appl. Phys. Letters*, **8**, 326 (1966).
44. Chambers, A., and M. Prutton, *Thin Solid Films*, **1**, 235 (1967).
45. Neugebauer, C. A., in G. Haas and R. Thun (eds.), "Physics of Thin Films," vol. II, p. 1, Academic Press Inc., New York, 1964.
46. Bondi, A., *Chem. Rev.*, **52**, 417 (1953).
47. Schmitt, R. W., *Intern. Sci. Technol.*, 1962, p. 42.
48. Ehrlich, G., in "Structure and Properties of Thin Films," p. 423, John Wiley & Sons, Inc., New York, 1959.
49. Fricke, R., *Kolloid-Z.*, **96**, 211 (1941).
50. Volmer, M., "Kinetik der Phasenbildung," Theodor Steinkopff Verlag, Leipzig, 1939.
51. Wexler, S., *Rev. Mod. Phys.*, **30**, 402 (1958).
52. Bassett, G. A., J. W. Menter, and D. W. Pashley, in C. A. Neugebauer, J. B. Newkirk, and D. A. Vermilyea (eds.), "Structure and Properties of Thin Films," p. 11, John Wiley & Sons, Inc., 1959.
53. Bassett, G. A., and D. W. Pashley, *J. Inst. Metals.*, **87**, 449 (1959).
54. Pashley, D. W., M. J. Stowell, M. H. Jacobs, and T. J. Law, *Phil. Mag.*, **10**, 127 (1964).
55. Pashley, D. W., *Proc. Phys. Soc. London*, **A64**, 1113 (1951).
56. Pashley, D. W., *Proc. Roy. Soc. London*, **A210**, 355 (1952).
57. Pashley, D. W., *Proc. Phys. Soc. London*, **A65**, 33 (1952).
58. Pashley, D. W., *Advan. Phys.*, **5**, 173 (1956).
59. Pashley, D. W., J. W. Menter, and G. A. Bassett, *Nature*, **179**, 752 (1957).

60. Pashley, D. W., *Phil. Mag.*, **4**, 316 (1959).
61. Pashley, D. W., *Phil. Mag.*, **4**, 324 (1959).
62. Bassett, G. A., *Phil. Mag.*, **3**, 72 (1958).
63. Bassett, G. A., J. W. Menter, and D. W. Pashley, *Proc. Roy. Soc. London*, **A246**, 345 (1958).
64. Whelan, M. J., P. B. Hirsch, R. W. Horne, and W. Bollmann, *Proc. Roy. Soc. London*, **A240**, 524 (1957).
65. Whelan, M. J., and P. B. Hirsch, *Phil. Mag.*, **2**, 1121 (1957).
66. Whelan, M. J., and P. B. Hirsch, *Phil. Mag.*, **2**, 1303 (1957).
67. Pashley, D. W., in "Thin Films," p. 59, American Society for Metals, Metals Park, Ohio, 1963.
68. Pashley, D. W., and M. J. Stowell, in S. S. Breese (ed.), *Proc. 5th Intern. Congr. Electron Microscope*, 1962, p. GG1, Academic Press Inc., New York.
69. Kingery, W. D., *J. Appl. Phys.*, **31**, 833 (1961).
70. Kingery, W. D., and M. Berg, *J. Appl. Phys.*, **26**, 1205 (1955).
71. Kuczynski, G. C., *Trans. Met. Soc. AIME*, **185**, 169 (1949).
72. Coble, R. L., *J. Am. Ceram. Soc.*, **41**, 43 (1958).
73. Adamsky, R. F., and R. E. LeBlanc, *J. Appl. Phys.*, **34**, 2518 (1963).
74. Hirsch, P. B., R. W. Horne, and M. J. Whelan, *Phil. Mag.*, **1**(8), 677 (1956); see also Whelan, M. J., and P. B. Hirsch, *Phil. Mag.*, **2**(8), 1121 (1957).
75. Silcox, J., and M. J. Whelan, in C. A. Neugebauer, J. B. Newkirk, and D. A. Vermilyea (eds.), "Structure and Properties of Thin Films," p. 183, John Wiley & Sons Inc., New York, 1959.
76. Bollmann, W., *Phys. Rev.*, **103**, 1588 (1956).
77. Wilsdorf, H. G. F., in "Structure and Properties of Thin Films," p. 151, John Wiley & Sons, Inc., New York, 1959.
78. Phillips, V. A., *Phil. Mag.*, **5**(8), 571 (1960).
79. Matthews, J. W., *Phil. Mag.*, **4**(8), 1017 (1959).
80. Matthews, J. W., *Phil. Mag.*, **7**(8), 915 (1962).
81. Bassett, G. A., and D. W. Pashley, *J. Inst. Metals*, **87**, 449 (1958–1959).
82. Demny, J., *Phys. Bl.*, **18**, 14, 61 (1962).
83. Amelinckx, S., *Phil. Mag.*, **6**(8), 1419 (1961).
84. Kamiya, Y., and R. Uyeda, *Acta Cryst.*, **14**, 70 (1961).
85. Jacobs, M. H., D. W. Pashley, and M. J. Stowell, *Phil. Mag.*, **13**, 129 (1966).
86. Pashley, D. W., and A. E. B. Presland, *Phil. Mag.*, **6**, 1003 (1961).
87. Buckel, W., and R. Hilsch, *Z. Physik*, **138**, 109 (1954).
88. Buckel, W., *Z. Physik*, **138**, 136 (1954).
89. Rühl, W., *Z. Physik*, **138**, 121 (1954).
90. Buckel, W., in "Structure and Properties of Thin Films," p. 53, John Wiley & Sons Inc., New York, 1959.
91. Mönch, W., and W. Sander, *Z. Physik*, **157**, 149 (1959).
92. Mönch, W., *Z. Physik*, **170**, 93 (1962).
93. Suhrmann, R., and H. Schnackenberg, *Z. Physik*, **119**, 287 (1942).
94. Feldtkeller, E., *Z. Physik*, **157**, 64 (1959).
95. Hilsch, R., in V. D. Fréchette (ed.), "Non-crystalline Solids," p. 348, John Wiley & Sons, Inc., New York, 1958.
96. Thun, R., in G. Haas (ed.), "Physics of Thin Films," Academic Press Inc., New York, vol. I, p. 187, 1964.
97. Reinders, W., and L. Hamburger, *Ann. Physik*, **10**(5), 649 (1931).
98. Beeck, O., A. E. Smith, and A. Wheeler, *Proc. Roy. Soc. London*, **A177**, 62 (1940).
99. Porter, A. S., and F. C. Tompkins, *Proc. Roy. Soc. London*, **A217**, 544 (1953).
100. Allen, J. A., and J. W. Mitchell, *Discussions Faraday Soc.*, **8**, 309 (1950).
101. Allen, J. A., C. C. Evans, and J. W. Mitchell, in "Structure and Properties of Thin Films," p. 46, John Wiley & Sons, Inc., New York, 1959.
102. Evans, C. C., and J. W. Mitchell, in "Structure and Properties of Thin Films," p. 263, John Wiley & Sons, Inc., New York, 1959.

Chapter **9**

Determination of
Structures in Films

SIEGFRIED MADER

IBM Watson Research Center, Yorktown Heights, New York

LIST OF SYMBOLS

a, b, c	unit vectors of crystal lattice
a*, b*, c*	unit vectors of reciprocal lattice
a	lattice constant
α	semiangular microscope aperture
b	Burger's vector of a dislocation
β_m	Lorentz deflection
c	velocity of light
C	crystallite size
d	spacing between crystal planes
D	fringe spacing in electron micrographs
δ	microscope resolution
Δ	line broadening
e	electron charge
ϵ	strain
f	atomic scattering factor
F	structure factor
g	vector in reciprocal lattice, diffraction vector
h	Planck's constant
(hkl)	Miller indices of crystal planes
I_0, I_t, I_r	incident, transmitted, and reflected beam intensities
$\mathbf{K^\circ, K'}$	wave vector of incident and diffracted beams
L	specimen-to-plate distance in electron diffraction, out-of-focus distance in microscopy
λ	wavelength
λL	camera constant
m	electron mass
M	(1) magnification; (2) magnetization
μ	absorption coefficient
r	diffraction ring radius
R	displacement vector of stacking faults
s, s_{eff}	vectorial and effective deviation from Bragg position
t	film thickness
T	microscope depth of field
2θ	diffraction angle
$[u, v, w]$	Miller indices of direction normal to film plane
v	electron velocity
V	accelerating potential, also volume of unit cell
z	REL periodicity along axis of fiber texture
Z	atomic number

Other Greek symbols: geometric quantities defined in text.

1. INTRODUCTION

The term "structure" encompasses a variety of concepts which describe, on various scales, the arrangement of the building blocks of a material. On an atomic scale, one deals with the crystal structure, which is defined by the crystallographic data of the unit cell. These data contain the shape and dimensions of the unit cell and the atomic positions within it. They are obtained by diffraction experiments.

On a coarser scale, one deals with microscopic observations of the microstructure which characterizes the sizes, shapes, and mutual arrangements of individual crystal grains. It also includes the morphology of the surface of the material.

An intermediate range is occupied by the defect structure, which is concerned with deviations of the regular arrangement of unit cells within one crystal grain; examples are point defects, dislocations, and stacking faults. In studying the defect structure, one makes use of both direct microscopic (mainly electron microscope) observation and diffraction evidence. In addition, one can utilize measurements of structure-sensitive properties which are related to defects in crystals, e.g., the extra resistivity due to point defects and impurities.

The determination of structures in films proceeds largely along the avenues used in bulk-material science. Numerous textbooks and reference books are available on general crystallographic methods. The reader is referred to Barrett and Massalski[1] and the "International Tables for X-ray Crystallography."[2] The scope of this chapter is more limited and reviews mainly the special aspects encountered in thin film work. A collection of review articles which, in part, deal with the present subject has been edited by Anderson.[3] In thin films, only a small amount of scattering material is available for conventional structure analysis by x-rays. Therefore, electron diffraction is a popular tool, as will be discussed in the next sections.

In the domain of microstructure and defect-structure analysis, the lack of thickness makes films the natural objects for transmission electron microscopy, which will be discussed in the fifth section. Excellent textbooks are also available on electron-beam methods, both diffraction (e.g., Vainshtein[4]) and transmission microscopy (e.g., Heidenreich[5] and Hirsch et al.[6]).

Microstructure and surface-morphology observations of films which are too thick for direct transmission also depend heavily on the high resolving power of electron microscopy. Suitable techniques are surface replication and scanning electron microscopy. The resolving power of optical microscopy is often not sufficient to reveal all the important details because the scale of surface features and of grain sizes in polycrystalline films is generally much smaller than in the same materials solidified from the liquid state.

The classical task of structure analysis, namely, determination of an unknown crystal structure by collecting and evaluating as many diffraction data as possible, is rarely encountered in thin film work. In most cases one can expect only a limited number of possible crystal structures in a given film, and these may be anticipated from the chemical composition and from the preparation conditions. The objective is then to select the proper structure which matches the diffraction data. For this purpose, catalogs of reflections such as the ASTM Powder Diffraction Data File[7] are of great value.

Frequently one has to determine whether a given deposit is a single-crystal film or polycrystalline, either with a random distribution of orientations or with a preferred orientation with respect to the film plane. For a single-crystal film, it is important to know its orientation relationship with respect to the substrate. Other questions which occur often deal with particle size and perfection of crystallites. These can be studied either by broadening of diffraction lines or by microscopy. Several of these topics will be discussed in the fourth section.

At this point it is worthwhile to note that the range of structures of a given material in the form of thin films can be broader than those formed in bulk because the conditions of film formation can vary more widely than the conditions of formation of a bulk solid. As an example, consider the evaporation of Si films onto Si substrates in ultrahigh vacuum.[8] At a substrate temperature of 800°C, the evaporated film is a single crystal of a high degree of perfection; in fact, its structure and defect content do not seem to be different from the perfect bulk crystal used as substrate. At 550°C, the evaporated film is still single-crystalline, but it now contains a high density of stacking faults. At 320°C, the film is polycrystalline, whereas at 25°C, it consists of an amorphous layer.

Besides amorphous structures, which occur more frequently in films than in bulk materials, there are also cases of crystalline structures in films which are not stable modifications of that material in bulk form. For example, Chopra et al.[9] reported face-centered cubic structures of sputtered films of the transition metals Ta, Mo, W, Re, Hf, and Zr which are normally body-centered cubic or hexagonal. Mader[10] described homogeneous metastable structures in alloy films of compositions which have a two-phase equilibrium structure. The films were either amorphous or had the crystal structure of the terminal solid solution. In all these cases, the new structures in films were "simple" ones, i.e., crystallographically not more complex than the structures found in elements or compounds with small unit cells.

2. X-RAY VS. ELECTRON DIFFRACTION FOR THIN FILMS

a. Differences of Diffracted Intensity

X-ray and electron diffraction differ in two important ways, namely, in the wavelength of the diffracted radiation and in the diffracted intensity. The different wavelengths entail a difference of the scale of the diffraction patterns, which will be discussed in Sec. 2c. The differences of intensity are due to differences of the physical scattering mechanisms. The quantity which has to be compared is the atomic scattering factor f, which represents the sum of all elementary scattering acts within one atom. It depends on the atomic number Z of the atom, and for a given atom it is a function of $\sin \theta/\lambda$, where 2θ is the angle between the direction into which the incident beam propagates and the direction under which the scattered radiation is observed.

The elementary scattering act for x-rays can be looked upon as the excitation of a resonating vibration in the electron cloud of an atom which in turn emits electromagnetic radiation of the same frequency. The atomic scattering factor for x-rays f_x is defined as the ratio of the amplitude scattered by the atom to the amplitude scattered by one electron. For small angles, f_x is of the order of Z and decreases with increasing $\sin \theta/\lambda$, reaching about $0.5Z$ at $\sin \theta/\lambda = 0.5$ Å^{-1}. Values of f_x can be found in the "International Tables for X-ray Crystallography."[2]

The absolute value of the intensity I_a scattered by one atom and observed at a distance R is

$$I_a = \left(\frac{e^2}{mc^2}\right)^2 \frac{1 + \cos^2 2\theta}{2} f_x{}^2 \frac{I_0}{R^2} \tag{1}$$

where I_0 is the incident intensity and $e^2/mc^2 = 2.82 \times 10^{-13}$ cm.

Scattering of electrons takes place by refraction of the electron wave as it passes through regions of varying potential within an atom. There are contributions to the potential, the "scattering matter" for electrons, from the positive charge at the nucleus and from the negative charges in the electron cloud. The absolute value of the intensity scattered by one atom observed at a distance R is proportional to the square of the absolute atomic scattering factor for electrons $f_{el\ ab}$,

$$I_e = f^2_{el\ ab} \frac{I_0}{R^2} = \left[\frac{me^2}{2h^2} \left(\frac{\lambda}{\sin \theta}\right)^2 (Z - f_x) \right]^2 \frac{I_0}{R^2} \tag{2}$$

where $me^2/2h^2 = 2.39 \times 10^6$ cm^{-1}.

Relative values of the atomic scattering factor for electrons $f_{el\ rel}$ are also tabulated in "International Tables for X-ray Crystallography."[2] They are referred to the scattering of electrons by a proton (for which $Z = 1$ and $f_x = 0$) with

$$\frac{\sin \theta}{\lambda} = 0.1 \times 10^8 \text{ cm}^{-1} = 0.1 \text{ Å}^{-1}$$

Their magnitudes are of the order of $Z/5$ for small angles and decrease with increasing $\sin \theta/\lambda$, reaching about one-half of their initial value at $\sin \theta/\lambda = 0.25$ Å^{-1}.

The main difference between the scattered intensities of x-rays and of electrons can be seen from Eqs. (1) and (2). First, the scattering power of an atom is much larger for electrons than for x-rays. Taking a typical reflection with $\sin \theta/\lambda \approx 0.2$ Å^{-1}, numerical evaluation shows that the intensity of scattered electrons is 10^8 times larger than the intensity of scattered x-rays. This is the reason for the success of electron diffraction in thin films. A film of 50 Å thickness, for example, diffracts an electron beam sufficiently strong to give a recordable diffraction pattern, whereas the same film usually does not contain enough scattering power for x-rays. On the other hand, a film thicker than several thousand angstroms absorbs most electrons but yields satisfactory x-ray diffraction. The high scattering power for electrons also makes it necessary to take into account multiple scattering and to make use of the dynamic theory of diffraction, as will be shown in Sec. 2b(3).

The second difference between x-ray and electron diffraction is in the angular dependence of the scattered intensities as represented by Eqs. (1) and (2). The absolute electron intensity [Eq. (2)] depends on $1/\sin^4 \theta$ and therefore falls off more rapidly with increasing angle than the x-ray intensity, which depends on $(1 + \cos^2 2\theta)$. The inelastic scattering of electrons follows a similar dependence. As a consequence, the intensities of high-index reflections are relatively weaker for electrons than for x-rays. This angular dependence has to be taken into account if one uses tabulations of x-ray diffraction lines such as the ASTM Powder Diffraction Data File for the evaluation of electron-diffraction patterns. The necessary corrections have been discussed by Sturkey.[11] These corrections are, however, a monotonous function of $\sin \theta/\lambda$ so that relative intensities of neighboring diffraction lines can be inferred directly from the x-ray data files.

b. Elements of Diffraction by Crystals

(1) Bragg Equation and Reciprocal Lattice For a convenient and universally accepted description of directions and planes in a crystal, a coordinate system is introduced whose unit vectors **a**, **b**, **c** are edges or axes of the unit cell. They form an orthogonal system only for cubic, tetragonal, and orthorhombic crystals. A plane in a crystal is given by its Miller indices. These are the triplet (hkl) of reciprocal intercepts of the plane with the a, b, and c axes, reduced to the smallest integers having the same ratio. A direction in a crystal is given as the triplets $[hkl]$ of the smallest-integer components of a vector in this direction, referred to the **a**, **b**, **c** coordinate system. An important quantity is the interplanar spacing d_{hkl} between neighboring (hkl) planes. For orthorhombic crystals, it is

$$\frac{1}{d^2} = \frac{h^2}{a^2} + \frac{k^2}{b^2} + \frac{l^2}{c^2} \tag{3}$$

and for cubic crystals in particular,

$$d = \frac{a}{(h^2 + k^2 + l^2)^{\frac{1}{2}}} \tag{4}$$

Equations for d spacings for other crystal classes are given in textbooks on crystallography and in the "International Tables for X-ray Crystallography."[2]

Diffraction of coherent radiation by the three-dimensional array of atoms in a crystal is governed by Bragg's law, which states that the incident beam is reflected by a set of lattice planes (hkl) if

$$2d \sin \theta = \lambda \tag{5}$$

where θ is the angle between the incident (or reflected) beam and the planes (hkl) and λ the wavelength of the diffracted radiation. It is important to note that the incident beam, the reflected beam, and the normal of the reflecting planes (hkl) are coplanar. This leads to a very convenient description of the multitude of possible Bragg reflections by means of the reciprocal lattice (REL) and construction of Ewald's reflecting sphere.

The REL of a crystal is defined by the vectors normal to all reflecting planes, each with the magnitude equal to the reciprocal d spacing. The REL vector g_{hkl} corresponding to the set of planes (hkl) is

$$g = h\mathbf{a}^* + k\mathbf{b}^* + l\mathbf{c}^* \tag{6}$$

where $\mathbf{a}^*$, $\mathbf{b}^*$, $\mathbf{c}^*$ are the unit vectors of the reciprocal lattice which are related to the unit vectors $\mathbf{a}$, $\mathbf{b}$, $\mathbf{c}$, and the volume V of the unit cell by

$$\mathbf{a}^* = \frac{[\mathbf{b} \times \mathbf{c}]}{V} \qquad \mathbf{b}^* = \frac{[\mathbf{c} \times \mathbf{a}]}{V} \qquad \mathbf{c}^* = \frac{[\mathbf{a} \times \mathbf{b}]}{V} \tag{7}$$

A diffraction event is illustrated in Fig. 1a, where the incident beam is represented by a vector $\mathbf{K}^\circ$ of magnitude $1/\lambda$. This beam is reflected by the lattice planes indicated and deflected by the angle 2θ into the direction of the vector $\mathbf{K}'$.

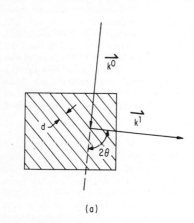

(a)

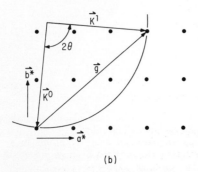

(b)

Fig. 1 (a) Diffraction by lattice planes. (b) The reflecting sphere.

Figure 1b shows the REL construction of this diffraction event. The crystal is replaced by its REL oriented with respect to the crystal according to Eq. (7). The vectors $\mathbf{K}^\circ$ and $\mathbf{K}'$ of magnitude $1/\lambda$ representing incident and reflected beams, respectively, are also shown. The vector $\mathbf{K}^\circ$ is directed at the REL origin. Bragg's equation is satisfied if the vector difference $\mathbf{K}' - \mathbf{K}^\circ$ coincides with a REL vector g of a plane able to reflect. This is the case if a sphere with the radius $\mathbf{K}^\circ$ touches another reflecting REL point besides (000) as shown in Fig. 1b. This construction is commonly referred to as Ewald's sphere.

The REL points which give rise to reflections are determined by the structure factor. This factor takes into account the positions of the individual atoms in the crystal structure. Each atom contributes a scattered wave to the diffracted radiation. Depending on the atomic positions, the scattered waves can be out of phase and interfere destructively for certain directions. These directions in turn are related to reflecting planes or REL points. For example, for a body-centered cubic crystal lattice, the structure factor is zero for all planes (or REL points, respectively) where the sum $(h + k + l)$ is odd. Therefore, the intensity-carrying REL points of the bcc structure are arranged in a face-centered cubic array with the lattice constant $2/a$. Similarly, the REL of a fcc crystal structure is a bcc array with the lattice constant $2/a$ because here the lattice planes with combinations of even and odd indices interfere destructively. Structure factors and absences of reflections are listed in structure-factor tables, e.g., vol. 1 of the "International Tables for X-ray Crystallography."[2]

In a powder specimen or a polycrystalline material consisting of a large number of crystallites with random orientations, a particular set of (hkl) planes can be in reflecting position within many different crystals. The diffracted beams emerge from the specimen on a cone around the primary beam with half apex angle 2θ and intersect the observation plane or surface along "diffraction lines."

A case intermediate between single-crystal and powder specimens which is of importance in thin film work is polycrystalline samples with a fiber texture or preferred orientation. Here all crystallites have one particular set of crystallographic planes parallel to the film plane, but otherwise these crystallites may be randomly oriented. The REL of such a structure can be obtained by rotating the single-crystal REL around the normal vector of the preferred plane. It consists of parallel rings centered on the line of the normal vector. Diffraction from textures will be discussed in more detail in context with Fig. 10 in Sec. 4d.

(2) Reflex Broadening The analysis of diffraction by a finite or small crystal shows that noticeable intensity is reflected even when the reflecting sphere in Fig. 1 passes only through the vicinity of the point g. Hence, there is an intensity region around each REL point, the so-called shape transform of the crystal, whose size is inversely proportional to the crystal size. For a platelike crystal of thickness t, the shape transform is a spike normal to the plate with an intensity distribution

$$I = \text{const} \frac{\sin^2 (\pi t s)}{(\pi s)^2} \qquad (8)$$

In this equation, s denotes the deviation from the exact Bragg position. It is the magnitude of the vector s going from g to the reflecting sphere as illustrated in Fig. 2. The deviation from the Bragg position is particularly important for electron diffraction. Here the reflecting sphere can be approximated by a plane, as will be shown in Sec. 2c. Figure 2 is drawn for this case. It is customary to define s as positive if g falls within the reflecting sphere.

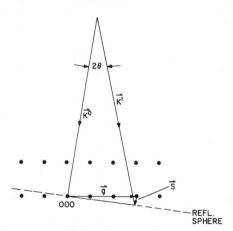

ig. 2 Reflecting-sphere construction for electron diffraction, showing positive deviation from the Bragg position.

The intensity distribution around the point g given by Eq. (8) has a half width of approximately $1/t$ and reaches zero on both sides of g at a distance $s = \pm 1/t$. In addition there are subsidiary maxima at larger values of s. This broadening of the Bragg reflection is the same for all REL points and thus constitutes a measure of crystal size.

In diffraction experiments with polycrystalline films, one deals with powder specimens as discussed in Sec. 2b(1). It can be shown that the angular broadening $\Delta(2\theta)$ of the diffraction lines of powder patterns is related to the crystallite size C by

$$\Delta(2\theta) = \frac{\lambda}{C \cos \theta} \tag{9}$$

Other sources of reflex broadening are nonuniform strains and local variations of the d spacings within the material. This broadening increases with the distance from the REL origin and therefore with diffraction angle. The relative d-spacing variations $\epsilon = \Delta d/d$ are related to the line broadening in powder patterns by

$$\Delta(2\theta) = 2 \tan \theta \frac{\Delta d}{d} \tag{10}$$

Contrary to this behavior, homogeneous strain such as that induced by thermal expansion of a film adhering to a substrate gives rise to a displacement of the diffraction lines since all d spacings are elastically expanded or compressed throughout the specimen.

(3) Dynamic Effects The above concepts of diffraction, known as the kinematic approximation, are applicable to x-ray diffraction and to the diffraction of electrons by small (<100 Å) crystals. For larger crystal sizes, it is not enough to consider the attenuation of the incident electron beam, but the possibility of the diffracted beam itself being scattered by the atoms has to be taken into account.[5,6] This leads to the dynamic theory of diffraction, which is important for the discussion of diffraction contrast in transmission electron microscopy. Its results are summarized in this section, and they will be applied later in Sec. 5b.

An important concept of the dynamic theory is the extinction distance ξ_g. It is given by

$$\xi_g = \frac{\pi V \cos \theta}{\lambda F_g} \tag{11}$$

where V is the volume of the unit cell and F_g the structure factor of the reflecting set of lattice planes which are characterized by their REL vector g. For low-order reflections, ξ_g is of the order of a few hundred angstroms. Tables of extinction distances are given by Hirsch et al.[6] If an electron beam impinges on a crystal exactly in a Bragg condition, all its intensity is in the reflected beam after it has penetrated a distance $\xi_g/2$ into the crystal. Thereafter, the electrons are reflected back into the incident direction, and after penetrating the full extinction distance, all the intensity is in the incident direction. If there is a small deviation s from the exact Bragg position (defined in Fig. 2), this modulation of intensity has a smaller periodicity. It is given by the reciprocal of s_{eff}, the effective deviation parameter

$$s_{eff} = \left(s^2 + \frac{1}{\xi_g{}^2} \right)^{\frac{1}{2}} \tag{12}$$

The intensity I_r of the beam reflected at the exit surface of a crystal of thickness t is

$$I_r = \frac{\pi^2}{\xi_g{}^2} \frac{\sin^2 (\pi t s_{eff})}{(\pi s_{eff})^2} \tag{13}$$

and of the transmitted beam $I_t = 1 - I_r$, where the incident intensity is taken as unity. For large deviations s from the Bragg position, I_r approaches the value of the kinematic approximation [Eq. (8)].

An alternative approach to the dynamical theory is to solve Schrödinger's equation for high-energy electrons in the periodic potential of the crystal with boundary conditions corresponding to incident, transmitted, and reflected waves.[5,6] The solution can be given in terms of Bloch waves representing current flow parallel to the diffracting planes. One wave field has maxima between the atomic planes and is transmitted well, whereas a second wave field has nodes between the atomic planes and is absorbed more strongly. Both wave fields have slightly different components in the transmitted and reflected direction. The beats between these components give rise to the modulation of transmitted or reflected intensity in Eq. (13).

c. Differences of Diffraction Patterns Due to Wavelength

The wavelength of monochromatic x-rays is determined by the characteristic emission of the target material in the x-ray tube. The wavelengths are typically of the order of 1 Å, e.g., 1.54 Å for CuKa and 0.709 Å for MoKa. They cannot in any case be smaller than the short wavelength limit λ_{min} of the continuous spectrum of x-rays,

$$\lambda_{min} = \frac{12,398}{V} \tag{14}$$

where V is the voltage applied to the tube in volts and λ given in angstroms. Equation (14) is derived from the photoeffect quantum relation $eV = h\nu$.

For x-rays, the radii of the reflecting spheres in Fig. 1 are only small multiples of the REL lattice constant. For example, Au with lattice constant $a = 4.08$ Å has a REL lattice constant $g_{200} = 0.49$ Å^{-1}. Therefore, the likelihood of obtaining a reflection with a randomly oriented incident beam is very small except for powder specimens. In order to map out a reasonable number of REL points for a single crystal, various rotating-crystal methods have been devised which sweep the REL points through the reflecting sphere. The reflections are recorded on a cylinder around the rotation axis. From the observed Bragg angles, which can go up to $2\theta = 180°$, the d spacings are calculated according to Eq. (5).

The wavelength of electrons is determined by deBroglie's equation $\lambda = h/mv$, where h is Planck's constant and m and v are the mass and velocity, respectively, of the electrons. Expressed in terms of the accelerating potential V, the rest mass of the electron m, and the velocity of light c, the wavelength is

$$\lambda = \frac{h}{[2meV(1 + eV/2mc^2)]^{\frac{1}{2}}} \tag{15}$$

The second term in the denominator of Eq. (15) is a relativistic correction. An approximation for wavelengths given in Å is $\lambda = (150/V)^{\frac{1}{2}}$, where the potential V has to be expressed in volts.

From Eq. (15) one can derive that electron wavelengths are much shorter than x-ray wavelengths. For $V = 50,000$ V, one obtains $\lambda = 0.054$ Å, and for $V = 100,000$ V, $\lambda = 0.037$ Å. Typically, the radius of the reflecting sphere is now about 50 times larger than the REL lattice constant. Hence, the sphere can be approximated by a REL plane which touches all the REL points normal to the $K°$ direction, as shown in Fig. 2. In addition, only thin crystals are transparent to the electron beam, and their intensity regions are spikes around each REL point extending parallel to $K°$ for normal incidence. This ensures that all reflections in the REL plane normal to $K°$ can be excited, even if the slight curvature of the reflecting sphere is taken into account. Moreover, a noticeable intensity is still diffracted if the REL plane does not coincide exactly with the tangent to the reflecting sphere, as was illustrated in Fig. 2. In practice, this means that Bragg's equation does not have to be satisfied exactly in order to excite a reflection, but that angular deviations of the order of 10^{-2} rad can be tolerated.

The geometry of the reflecting-sphere construction shows further that the diffraction angles for electron beams are very small, of the order of a few degrees. Therefore, the diffracted beams project the REL plane normal to the incident beam practically undistorted onto a screen or photographic plate. If the plate is at a distance L from the specimen and a diffraction spot has the distance r from the spot made by the undiffracted beam, its diffraction angle follows from $\tan 2\theta = r/L$. For the small diffraction angles involved, this can be equated to $2 \sin \theta$ in Eq. (5), and Bragg's equation for electron diffraction reduces to

$$rd = \lambda L \tag{16}$$

The "camera constant" λL is usually obtained from calibration patterns. In order to evaluate an unknown pattern, one generally transforms the measured r values into d spacings with the aid of Eq. (16). In case of x-rays, however, Eq. (5) has to be used.

d. Peculiarities of Electron Diffraction

(1) Inelastic Scattering All the considerations above are valid only for elastically scattered electrons. There are many electrons which suffer energy losses while they traverse the film, and their relative proportion increases with film thickness. They are scattered into the same angular ranges as the elastically scattered electrons; hence they superimpose on the diffraction peaks and also form a strong background intensity. For this reason, intensity data in electron diffraction cannot be used for quantitative analysis with the same confidence as x-ray data. However, the geometric aspects of electron reflections such as angle and direction of diffraction can be precisely translated into d spacings and orientation of the reflecting planes.

The sources of inelastic scattering are discussed in Refs. 5 and 6. The most important ones are (1) the Coulomb interaction between the electrons in the beam and the electrons in the crystal, which leads to the excitation of plasma oscillations and characteristic energy losses of the fast electron; (2) ionization and core excitation of individual atoms; (3) thermal diffuse scattering. An intriguing aspect of inelastically scattered electrons is the fact that some of them are still coherent and that they contribute different amounts to the background between diffraction peaks and to the peaks themselves. Therefore, a simple subtraction of an interpolated background line from observed intensities is not strictly correct.

(2) Double Diffraction and Extra Reflections Electron diffraction can show a multitude of reflections which arise from strongly diffracted beams acting as sources for further diffraction. Consider first a sandwich of two single-crystal films. The diffraction geometry involves only small angles 2θ, and the diffraction vector $\mathbf{g}$ is practically perpendicular to the incident and diffracted wave vectors $\mathbf{K}^{\circ}$ and $\mathbf{K}'$ in Fig. 2. Therefore, each beam diffracted in the first crystal can act as a primary beam for diffraction in the second one. The resultant diffraction pattern consists of the pattern of the second crystal registered around each spot of the first crystal. In terms of the REL construction, the doubly diffracted spots are the vector sums of the $\mathbf{g}$ vectors of the two reflections. If the two crystals have different structures, then the double diffraction spots have no corresponding d spacing in either crystal. The analysis of double diffraction spots from sandwich films is important for the interpretation of moiré patterns in electron-microscope images.

Another type of sample in which double diffraction occurs is single-crystal films containing a large amount of twinned material. In the fcc structure, the REL points of the twins occupy some of the points along $\langle 111 \rangle$ lines through matrix points and divide the distance between matrix points into thirds. If matrix and twin regions overlap in the specimen and double diffraction occurs in the overlapping region, the vector addition of REL points shows that all the third positions along $\langle 111 \rangle$ lines can be reached. Several of these double diffraction spots correspond to d spacings that fit into a hexagonal diffraction pattern. Pashley and Stowell[12] have shown that in gold films these apparent hexagonal spots arise from double diffraction by tracing their origin microscopically into the region of overlapping matrix and twins.

Double diffraction is also responsible for the occurrence of extra reflections which are forbidden by the structure factor. An example is the (200) reflection in diamond cubic structures in films of suitable orientations. Consider a single-crystal film in (011) orientation which gives rise to reflections from (111) and ($\bar{1}$11) planes. Double diffraction between these two reflections within the same crystal adds up to a (200) reflection. This filling in of forbidden reflections is possible only if the REL contains points inside its lattice unit cell. Double diffraction, however, cannot cause the appearance of reflections with mixed indices from fcc structures, or of reflections having an odd sum of indices from bcc structures.

(3) Divergent-beam Effects, Kikuchi Lines Electron-diffraction patterns from single-crystal films often contain Kikuchi lines in addition to diffraction spots. These are pairs of bands with higher and lower intensity than the background intensity. They arise from Bragg diffraction of diffuse inelastically scattered electrons. As illustrated in Fig. 3, a part of the incident beam, after entering the crystal, is divergently spread out by inelastic scattering. Subsequent elastic diffraction gives rise to rays located on cones with the normal of the diffracting plane as axis and $(90 - \theta)$ as the half apex angle. One of the cones contains excess radiation over the background intensity, whereas in the other cone there is a deficiency. The intersection of these cones with the observation plane gives rise to the Kikuchi lines of that particular diffracting plane, with the deficiency line located closer to the primary beam spot.

Kikuchi lines are significant for two reasons. First, their appearance indicates a high degree of perfection of the crystal. The reflecting planes which give rise to a pair of lines must have precisely the same orientation throughout the crystal. If their alignment fluctuates, the Kikuchi lines broaden out and disappear in the background. Second, a pattern of Kikuchi lines can be used to determine the orientation of a single-crystal film with greater precision than is possible from direct Bragg reflections alone. Maps of Kikuchi patterns have been published for the identification of orientations of diamond cubic and fcc[13] as well as for bcc and hcp[14] single-crystal foils. Kikuchi lines are particularly useful to determine the deviation vector s defined in Fig. 2. This quantity and in particular its sign are important for the interpretation of diffraction contrast in transmission electron microscopy.

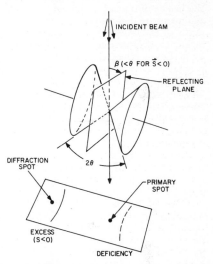

Fig. 3 Cones of rays giving rise to Kikuchi lines in electron-diffraction pattern. (*After Heidenreich.*[5])

For s = 0, the excess and deficiency lines pass through the diffraction spot and primary spot, respectively. For s > 0, both lines are shifted in the direction going from the primary spot to the diffraction spot, whereas for s < 0, they are displaced in the opposite sense, as shown in Fig. 3. The x-ray analogs of Kikuchi lines are Kossel lines which arise when a divergent source of x-rays is placed close to the crystal surface. They seem to have no application in thin film work.

3. IMPLEMENTATION OF DIFFRACTION EXPERIMENTS

a. X-ray Techniques

In x-ray work one can choose between white radiation and monochromatic radiation. With white radiation, a "Laue pattern" is obtained from a stationary crystal, whereby each set of lattice planes selects its proper wavelength for reflection according to the Bragg equation. The symmetry of the arrangement of lattice planes is easily recognizable on a Laue pattern, but the d spacings cannot be determined. The Laue technique is mainly used to determine the orientation of single crystals; it is a convenient tool because a pattern can be recorded on Polaroid film within a few minutes. It has found application in thin film work to determine the orientation relationships of thick epitaxial films on single-crystal substrates where film and substrate have structures of different crystal systems. An example is Cu on sapphire.[15]

X-ray techniques based on monochromatic radiation are generally more important because the d spacings can be calculated from the observed diffraction angles. For thin film work, the powder technique in conjunction with diffractometers is most commonly used. In these instruments, the diffracted radiation is detected by counter tubes which move through the angular range of reflections. The intensities are recorded on synchronously advancing strip charts or by other means suitable for data processing. An important feature of diffractometers is their ability to focus into a sharp diffraction line the radiation which is Bragg-reflected from an extended specimen area. This considerably improves the sensitivity and the signal-to-noise ratio. Focusing is achieved by making the specimen a part of the circumference of a circle, the so-called focusing circle. If x-rays emerge divergently from a point on the circumference of the focusing circle and impinge on the specimen, then all the beams diffracted in different areas by the same family of (hkl) planes cross over again and are detected on this circle. The x-ray source can be placed directly on this circle, or the crossover of a monochromator can be used as the point from which the incident x-rays emerge. With a flat specimen, the focusing condition is approximated by placing it tangential to the circle. The detector, finally, has to face the specimen while it is moved along the circle.

Two different designs of diffractometers are used in thin film work. The first one, the Bragg-Brentano diffractometer, is commercially available and most widely used. It is shown in Fig. 4. Here the specimen is mounted in the center of the

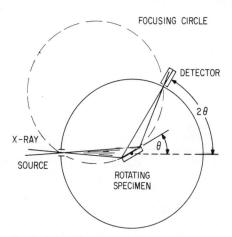

Fig. 4   Bragg-Brentano x-ray diffractometer.

diffractometer and rotated by an angle θ around an axis in the film plane. The counter is attached to an arm rotating around the same axis by angles 2θ twice as large as those of the specimen rotation. It can be seen that the diameter of the focusing circle continuously shrinks with increasing diffraction angle. Only (hkl) planes parallel to the film plane contribute to the diffracted intensity. The effective thickness $t/\sin\theta$ which a film of thickness t exposes to the incident beam also decreases with increasing diffraction angle. Therefore, the effective thickness of a film in the 1,000-Å thickness range is sufficient to excite measurable diffracted radiation at small angles, but the intensity falls off rapidly for higher-index reflections.

The second design of diffractometers utilizes the Seemann-Bohlin ray path, which is shown in Fig. 5. The specimen and the focusing circle remain stationary while the detector tube moves along the circumference of the focusing circle itself. In order always to face the specimen, the tube rotates around an axis going through the circle with half the speed of the rotation of the tube carrier around the circle center. The effective thickness which the film exposes to the beam remains constant and can be made very large by using small angles of incidence. Each family of (*hkl*) planes giving rise to a reflection is inclined with respect to the film plane, and the inclination is greater the higher the index of the reflections.

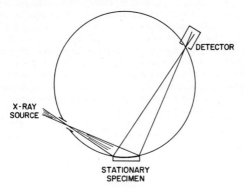

Fig. 5 Seemann-Bohlin x-ray diffractometer.

The Bragg-Brentano diffractometer has been used extensively in thin film work, e.g., by Segmüller,[16] Vook and Witt,[17] and Light and Wagner,[18] whereas the Seemann-Bohlin principle has been employed in only a few studies (e.g., by Keith[19] and Weiner[20]). The main advantage of the latter is the constant angle of incidence, which can be kept as small as 5°, thus giving higher diffracted intensities than the Bragg-Brentano diffractometer in the whole angular range and particularly in the back-reflection region. On the other hand, the Seemann-Bohlin arrangement is limited to polycrystalline specimens of random orientation because the REL remains stationary with respect to the Ewald sphere. The REL of a single crystal or of a fiber texture intersects the Ewald sphere only if the angle of incidence is carefully adjusted. In these cases, the Bragg-Brentano diffractometer is more advantageous because the REL is swept through the Ewald sphere and gives reflections even of single-crystal films as long as the diffractometer plane coincides with a REL plane.

Finally, it is worthwhile to note that for x-ray diffraction a film can remain on its substrate and does not have to be detached as is the case for transmission diffraction of electrons. While this simplifies experimentation and allows "nondestructive testing" of the film-substrate system, the *d* spacings of films attached to their substrates usually contain contributions of homogeneous strains. They are caused either by intrinsic stresses in the film or by differential thermal expansion if the temperatures of fabrication and observation are different. One of the main applications of x-ray diffractometer techniques in films is, in fact, to make use of their good resolution of *d* spacings (of the order of 1 part in 10^4) to investigate strains in films. The relationships between homogeneous strain and line shifts observable in Bragg-Brentano diffractometers have been discussed by Vook and Witt for Cu, Au,[21] and other fcc and bcc metals.[22]

b. Electron-diffraction Techniques

In electron diffraction, only monochromatic radiation and stationary geometric relationships between incident beam and diffracting specimens are employed. However, the possibility of electromagnetically deflecting and focusing electron beams creates a large flexibility in the design of electron-diffraction instruments. A common feature of all electron-diffraction techniques is operation in vacuum and use of a very stable high-voltage supply to ensure a constant wavelength. Beyond this common aspect, one can differentiate between transmission diffraction of thin specimens and reflection diffraction which takes place in a thin surface layer of the specimen. Most of the transmission-diffraction work is carried out as "selected-area diffraction" in an electron microscope; however, some work is also done in special diffraction instruments.   This is particularly the case when structural changes induced by heat treatment or chemical reactions are to be followed in situ, or if a vacuum is required which is better than the 10^{-4} to 10^{-5} Torr used in electron microscopes. In reflection diffraction, high-energy diffraction with voltages in the range of 10 to 100 kV is distinguished from low-energy electron diffraction (LEED) with voltages below 1 kV.   While the former samples a surface layer of the order of 100 Å thickness and projects a diffraction pattern in the angular range close to the primary beam, the latter samples only a few atomic layers at the surface and projects a pattern in the back-reflection region near the specularly reflected incident beam.

A further distinction can be made with regard to the recording of diffracted intensity.   In most cases, and particularly in the diffraction mode of electron microscopes, the patterns are viewed on fluorescent screens and recorded on photographic plates. A modern development is scanning electron diffraction; here a deflection system is used to scan all the diffracted radiation over a stationary detection system, the latter consisting of a scintillation counter-photomultiplier combination with electronic readout.   A velocity filter to reject from the detection system all electrons which were scattered with energy losses may also be incorporated.   This feature greatly improves the clarity of the diffraction pattern by suppressing the incoherent background.   In the following paragraphs, selected aspects of several electron-diffraction schemes are described.

(1) Transmission Electron Diffraction   The most comprehensive and accurate structure studies are possible with transmission electron diffraction.   Basically, a diffraction camera consists of an electron gun, an aperture to define a small beam cross section, a specimen holder which usually accommodates standard electron-microscope grids of $\frac{1}{8}$ in. diameter, and a fluorescent screen and plate camera at the distance L from the specimen.   To improve the sharpness of the diffraction spots, a long-focal-distance lens can be added between electron source and specimen, which focuses the source onto the viewing and recording plane.   The specimen holder should allow angular adjustments of the specimen with respect to the incident beam. Tilts of 30 to 45° between the specimen normal and the incident beam are desirable for the study of single-crystal and fiber-textured films.   Examples of diffraction cameras are given by Vainshtein.[4]   In simple diffraction cameras, the camera length L is defined mechanically by the dimensions of the instrument, and therefore it is very reproducible.

Transmission diffraction in electron microscopes is usually carried out as selected-area diffraction with a ray path as shown in Fig. 6b.   The objective lens generates a diffraction pattern of the specimen in its back focal plane.   In the diffraction mode, the intermediate lens transmits this pattern into its image plane from where it is further magnified by the projector lens.   In the microscopy mode, which is shown for comparison in Fig. 6a, the intermediate lens looks at the image plane of the objective lens, and the specimen image is further magnified.   In the image plane of the intermediate lens is the field-limiting selector aperture which determines the area of the specimen plane that contributes to either the diffraction pattern or the image. One usually has apertures of various sizes allowing diffraction patterns from specimen areas between 1 and 100 μ in diameter.   The great advantage of this arrangement is the possibility of directly viewing the area from which the diffraction pattern arises. One may, for example, locate crystals of unusual morphology, encircle them with the selector aperture, and isolate their diffraction patterns for analytical identification. If a film consists of two phases, the grain sizes of the two types of crystallites are usually different.   In order to distinguish their diffraction patterns, one can decrease the selected area to such an extent that one of the phases still gives a complete powder pattern while the Debye rings of the phase with the larger crystallite size begin to break up into a spot pattern.   This situation is illustrated in Fig. 7.

A disadvantage of selected-area diffraction arises from the optical magnification of the diffraction pattern by lenses which are continually readjusted for use as microscope elements.   This limits the reproducibility of the effective camera length L which contains this magnification, and hence the relative precision of d spacings obtained is only about 5×10^{-3} unless diffraction rings of a known material are present to calibrate each pattern individually.

In many practical situations, transmission diffraction is a destructive analysis because the film has to be detached from its substrate and manipulated onto a supporting grid. The popularity of rock salt and similar water-soluble substrate materials is partly due to the ease of removing a film by floating it onto a water surface. For films that cannot easily be detached or stripped from a substrate, special techniques have been developed. An example is the etching of a hole through the back of a silicon-wafer substrate by means of an acid jet.[23]

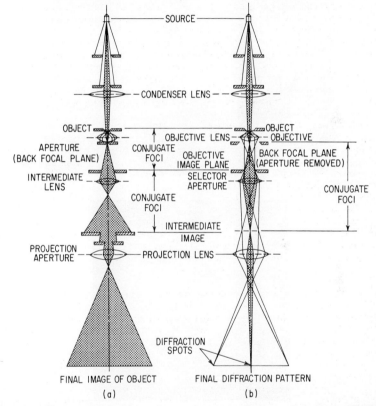

Fig. 6 Ray diagrams for double condenser, three-stage electron microscope. (*a*) Normal image formation. (*b*) Selected-area diffraction operation. (*Courtesy of R. D. Heidenreich*[5] *and Interscience Publishers.*)

(2) Reflection Electron Diffraction Reflection diffraction of high-energy electrons occurs by penetration of a thin surface layer with ensuing diffraction and by transmission through surface protrusions of the film material. Since an electron wave suffers a small refraction when entering and leaving a solid, reflection-diffraction spots generally have a larger spread in their angles of deflection than transmission-diffraction spots. Surface contamination and oxide layers show up markedly on reflection patterns, and the background of inelastically reflected electrons is generally higher than in transmission patterns. All these factors considerably decrease the precision of d spacings determined from reflection patterns. However, two advantages are inherent in this method. The first one is that the film can remain on the substrate and the film-substrate relationship is not disturbed. The second advantage is the possibility of sampling the complete reciprocal lattice of the specimen by rotating it around the normal of its reflecting face, a fact which was pointed out by Holloway.[24] This is in contrast to transmission diffraction, which samples only the REL plane perpendicular to the incident beam and does not allow enough tilting to sweep through the complete REL space.

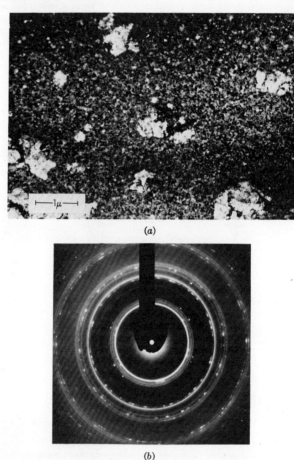

(a)

(b)

Fig. 7 (a) Electron-transmission picture of an evaporated gadolinium film containing large inclusions of GdH_2 (white areas). (b) Diffraction pattern of the same film area showing continuous rings of hexagonal Gd and "spotty" rings of cubic GdH_2.

(3) Low-energy Electron Diffraction (LEED) LEED is a tool strictly for the study of surface structures. In thin film work, it has been used to investigate the initial stages in the formation of epitaxial films.[25] Whereas a clean surface is very desirable for reflection diffraction of high-energy electrons, it is absolutely necessary for LEED. Here diffraction takes place in only the first two or three atom layers at the surface, and the pattern would be seriously masked by an adsorbed gas layer. Therefore, LEED is carried out in bakeable ultrahigh-vacuum systems.

The diffraction geometry of low-energy electrons differs in two ways from that of high-energy electrons which had been discussed with the aid of Fig. 2. First, the magnitude of the wavelength is, as with x-rays, comparable with typical d spacings so that the Ewald sphere cannot be approximated by a plane. Second, as diffraction occurs at a two-dimensional surface grating, the REL consists of rods perpendicular to the surface. These rods always intersect the Ewald sphere without the need for rotating the crystal. The intersections that correspond to back-reflected beams are the ones of interest. The reflected beams are usually projected onto a hemispherical fluorescent screen after passing through a system of filtering and accelerating grids, and the reflection spots are observed through a window in the vacuum system.

The most spectacular result of LEED work has been the discovery of fractional diffraction spots from certain faces of homeopolar crystals, indicating a "reconstructed" surface with a superlattice of the unit cells terminating on the surface. But an unambiguous structure determination of this superlattice has not yet been possible because the atomic scattering factors and their dependence on electron energy are not well enough known in the low-energy region. An interesting aspect of a LEED system is the possibility of identifying atomic species present at the surface by an energy analysis of secondary electrons emitted during LEED observation. Among these secondary electrons are Auger electrons whose energies are characteristic for the emitting element.[26]

(4) Scanning Electron Diffraction Grigson and coworkers developed a very promising detection system for high-energy electrons diffracted in transmission[27,28] and reflection.[29] Figure 8 shows a schematic diagram of such a system for in situ observation of films during deposition. The magnetic lens above the specimen generates a small spot at the observation plane. Magnetic scanning coils below the specimen

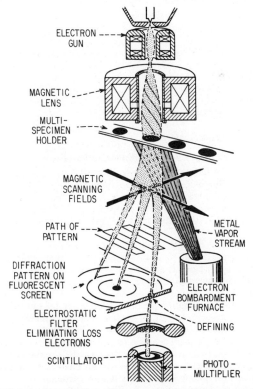

ELECTRON GUN

MAGNETIC LENS

MULTI-SPECIMEN HOLDER

MAGNETIC SCANNING FIELDS

PATH OF PATTERN

METAL VAPOR STREAM

DIFFRACTION PATTERN ON FLUORESCENT SCREEN

ELECTRON BOMBARDMENT FURNACE

ELECTROSTATIC FILTER ELIMINATING LOSS ELECTRONS

DEFINING

SCINTILLATOR

PHOTO-MULTIPLIER

Fig. 8 Diagram of scanning electron diffractometer for diffraction study of thin film growth. (*Courtesy of C. W. B. Grigson and D. B. Dove.*[28])

continuously deflect the diffraction pattern and sweep it over the defining aperture of the detection system. The latter consists of a scintillator and photomultiplier. The detected intensity is displayed on an $X - Y$ recorder as a function of the magnetic deflection or the Bragg angle. Figure 9 shows an example of a two-dimensional scan of a single-crystal film of silver. The figure gives a perspective view of a REL plane with the intensities of the reflections drawn as peaks above their REL points in this plane.

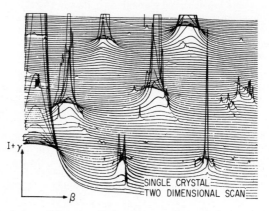

Fig. 9 Scanning electron-diffraction diagram of a single-crystalline silver film. (*Courtesy of C. W. B. Grigson and D. B. Dove*[28] *and J. Vacuum Sci. Technol.*)

An important feature of the scanning diffraction system is the electrostatic filter which eliminates inelastically scattered electrons. Hence the measured intensity yields the proper data for quantitative analysis which can be treated with the same methods developed for structure analysis with x-rays.[1,2] This is particularly important for the Fourier analysis which leads to the radial distribution function of a film with amorphous structure.

4. EVALUATION OF DIFFRACTION DATA

a. Identification of Crystal Structures from Powder Patterns

Frequently, diffraction data are used for qualitative chemical analysis based on the identification of crystalline substances from the d spacings and relative intensities of their reflections. The procedure is described in Ref. 1 and consists of collecting a powder pattern from the material under investigation. This pattern is then matched up with one of the over 10,000 patterns recorded in the ASTM Powder Diffraction Data File.[7]

For thin film analysis, it is convenient to use transmission-electron-diffraction data. Because of the large scattering factors for electrons, their diffraction yields more reflections at high Bragg angles than x-rays. Transmission diffraction is preferable to reflection diffraction because of sharper reflections and higher precision of the d spacings can be achieved. It is desirable to obtain full Debye rings; hence a large selector aperture should be chosen if the diffraction is carried out in an electron microscope. It is also good practice to test for preferred orientation by tilting the specimen with respect to the incident beam to ensure that no ring is missed. This can happen with normal incidence in the presence of a fiber texture (see Sec. 4d).

In order to compare electron-diffraction results with the ASTM file data, all the Debye-ring radii have to be converted into d spacings according to the reduced Bragg equation (16). However, before doing this it is worthwhile to test if all rings or a portion of them form a "cubic sequence." This requires that the ratio of the square of the radius to the sum of the squares of the Miller indices is constant for all reflections according to

$$\frac{r^2{}_{hkl}}{h^2 + k^2 + l^2} = \frac{(\lambda L)^2}{a^2} \tag{17}$$

where a is the lattice constant. This ratio can easily be determined by matching up on a slide rule the r^2 values with the sequence of integers representing the possible or allowed sums $(h^2 + k^2 + l^2)$. For fcc structures, this sequence is 3, 4, 8, 11, 12, 16, 19, 20, 21, . . . , and for bcc structures it is 2, 4, 6, 8, 10, 12, 14, 16, 18, 20, ... If a match is obtained, Eq. (17) immediately gives the lattice constant a of the unknown structure.

b. X-ray Line Broadening and Shifts

Equations (8) and (9) for line broadening are approximations for the effects of particle size and inhomogeneous strains, respectively, if these occur separately. In a real material, there are simultaneous contributions to the line width from both sources as well as from lattice defects, and they superimpose in a complicated fashion. Sophisticated procedures for the separation of the various contributions have been developed[30] and successfully applied to the study of imperfections in cold-worked metals. Evaporated films can also be highly imperfect, and therefore, several authors applied these procedures to thin films of face-centered cubic metals. The methods are worked out for the Bragg-Brentano diffractometer geometry.

To determine particle size and nonuniform strains, Segmüller[16] and Light and Wagner[18] used the Warren-Averbach[30] Fourier analysis of the intensity profile of diffraction lines. This requires at least two orders of reflections from the same set of planes, say (111) and (222). The profiles of these lines are represented as Fourier series. The coefficients A_n of the cosine terms are then plotted logarithmically for a number of values of n, the order of the Fourier coefficient, vs. the values of $\sin^2 \theta$ of the different reflection orders. Extrapolation to $\sin^2 \theta = 0$ gives the particle-size coefficient $A_n{}^s$. From the variation of $A_n{}^s$ with the order n, an effective particle size can be deduced which may subsequently be apportioned to true particle size, stacking-fault, and twin-fault densities. The slopes of the log A_n vs. $\sin^2 \theta$ lines yield the root-mean-square strain averaged over a domain of the size nd, where d is the spacing of the reflecting planes.

Vook and Witt[17] determined effective particle size and strain in a less rigorous approximation by measuring half widths of line profiles directly. The observed broadening $\Delta(2\theta)$ was treated as a sum of Eqs. (8) and (9). A plot of $\Delta(2\theta)$ vs. $\sin \theta$ for several reflections gives the effective particle size from the intercept at $\sin \theta = 0$ and the average nonuniform strain from the slope.

A uniform strain throughout the film and the stacking-fault probability in films of face-centered cubic materials can be deduced from precise measurements of the positions of the diffraction lines.[17,18] Both factors cause shifts of the peak position. A homogeneous tensile stress, for example, causes a contraction of the d spacings perpendicular to the film plane. In crystallites of different orientations, the magnitude of this contraction differs because of the anisotropy of the elastic constants. Consequently, the resulting shifts of the diffraction lines are different for different reflections.[21,22] Stacking faults in fcc structures can be described crystallographically as thin layers of hexagonal material. Their effect on the diffraction pattern is to broaden the reflections and to shift them toward positions of hexagonal reflections. The resulting peak shifts are again different for different reflections and go toward either larger or smaller apparent d spacings.

In the analysis of peak shifts, precise apparent lattice constants are calculated from each reflection and are then plotted against a suitable extrapolation function such as $\cos^2 \theta / \sin \theta$ or $\frac{1}{2}(\cos^2 \theta / \sin \theta - \cos^2 \theta / \theta)$. The peak shifts yield systematic deviations of individual lattice constants from the average or ideal values. The direction and the magnitude of these deviations are evaluated in terms of strain and stacking-fault probabilities. Finally, the stacking-fault broadening can be used to derive true particle sizes from the effective ones which were obtained by the Warren-Averbach Fourier analysis.

The results of such x-ray investigations show a fair amount of scatter between different specimens which is due to the difficulty of reproducing exactly the microstructure in evaporated films. Several experimental data are given in the following to illustrate the orders of magnitude found in thin films. Nickel deposited around 250°C showed particle sizes of 300 Å which increased to 1,500 Å at 300°C. The nonuniform strain was 0.1%.[16] Silver films deposited at 130°C in a vacuum of 10^{-6} Torr or better showed similar values; deposition at 10^{-4} Torr gave, in addition, rise to stacking-fault probabilities of 0.004 and twin-fault probabilities of 0.02.[18] Deposition of copper and gold films at 90°K in 10^{-8} Torr gave about ten times larger values for stacking-fault probability, twin-fault probability, and nonuniform strains, whereas the true particle size after correction of the fault broadening was still in the range of 1,000 Å.[17]

c. Evaluation of Single-crystal Electron-diffraction Patterns

(1) Orientation Determination In transmission electron diffraction, particularly when working with selected-area diffraction in an electron microscope, one frequently encounters spot patterns. They can be reduced to patterns of only one crystal by choosing a sufficiently small selector aperture. Frequently, the chemical nature and crystal structure of the film material are known, and the task is to determine its crystallographic orientation for comparison with features in the microscope image. For this purpose the spot pattern has to be indexed. Since the pattern is equal to a plane in the REL, the spots are arranged in a cross grating. If the cross grating is not clearly recognizable, slight changes of specimen orientation and illumination are likely to reveal the translational symmetry of the pattern. Often one can identify directly the REL plane which the pattern represents by comparing it with a structure model or with catalogs of indexed patterns.[6,31] It is advisable subsequently to measure the r values of the spots and confirm that their d spacings correspond to the assumed indices.

Alternatively, one can determine the d spacings and thereby the crystallographic nature of three spots which together with the origin form a parallelogram. By trial and error one chooses a consistent set of indices such that the indices of the spot opposite to the origin can be obtained as the vector sum of the indices of the two other spots. All other spots can then be indexed by simple vector addition. In order to check the indexing, one can measure the angles which various pairs of reflections form with the origin and compare them with published[1,2,31] or calculated values. The orientation of the crystal can be characterized by the crystallographic direction $[uvw]$, which is normal to the plane of the film. If the incident beam is normal to the plane of the film, the indices $[uvw]$ are obtained as vector products between any two directions in the spot pattern.

(2) Identification of Crystal Structures Sometimes the material which causes a spot pattern is not known. One can then try to obtain a powder pattern by modifying the experimental conditions and identify the material and its structure according to Sec. 4a. In the few cases where this is not possible, one has to determine carefully the d spacings of the reflections which do occur and try to identify them using the Powder Diffraction Data File. However, this is not straightforward because important and intense reflections are likely to be missing in the particular REL plane which is accessible to the diffraction. If the chemical components of the material are known, one can limit the search to the compounds of these elements. Often several structures are found which match the incomplete data of d spacings. At this stage the selection can be narrowed down further by two arguments. First, the observed d spacings must have such indices that the cross-grating pattern can be indexed as described in the preceding section. Second, the angles between pairs of spots and the origin must agree with the ones possible in the structure under consideration.

Frequently, but not always, an unambiguous identification can be carried out in this fashion. It is quite difficult, however, to analyze the structure of a material in thin film form which has not been described in the literature unless this structure is a variation of a known one. The general principles for structure analysis by electron diffraction through Fourier methods are discussed by Vainshtein.[4]

d. Fiber Textures

The crystallites of a polycrystalline film deposited at elevated substrate temperature often have a preferred orientation such that one particular set of crystallographic planes (uvw) is parallel to the film plane in all crystallites. Such a structure is called fiber texture. The normal [uvw] of the preferred planes is the axis of the fiber texture. It is frequently a low-index direction. A well-known example is the (111) texture of fcc films. In x-ray diffractometer work, preferred orientations are indicated if the ratios of integrated intensities of several reflections deviate from the ratios found in randomly oriented powder specimens.[17]

Fiber textures can easily be recognized on electron-diffraction patterns provided the incident beam does not coincide with the axis of the texture. Such patterns

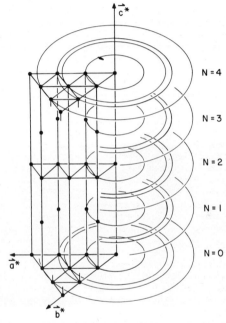

Fig. 10 Section of the reciprocal lattice of a hexagonal close-packed structure and of a fiber texture around the normal of the basal plane ($c/a \approx 0.8$).

consist of arcs of Debye rings, and they are symmetric with respect to a line normal to both, the incident beam, and the texture axis. The centers of the arcs are arranged in a pattern which depends on the crystal structure, the crystallographic direction, and the inclination of the texture axis with respect to the incident beam. Vainshtein[4] has given a complete description of the geometry of these patterns and the location of the arc centers. The analytical procedure is illustrated in Fig. 10, whose left portion shows a REL section of the hcp structure with the c^* axis pointing upward. Densely populated REL rows intersect the basal plane a^*, b^* at right angles. According to the structure factor, the REL points on these rows are separated by a distance $z = 1/d_{(001)}$ except for the rows with $h + k = 3n$, where the distance is $2z$. The REL points, therefore, form layers with the distance z between them.

The REL of a fiber texture of this structure with a [001] fiber axis is indicated by the circles in Fig. 10. It is generated by rotation of the single-crystal REL around the fiber axis whereby the REL points transform into continuous rings. The rings are arranged on cylinders resulting from the vertical REL rows and in layers with the

distance z between each other.   Figure 10 also shows the numbers $N = 0, 1, 2, 3, \ldots$
of the layers.   Finally, the diffraction pattern projected by a beam forming an angle
ϕ with the texture axis is equal to the intersections of a plane tilted by an angle ϕ
with the system of rings.   The intersections are located on ellipses generated by
the cylinders and on layer lines generated by the layers.   The distance between
consecutive layer lines is $z/\sin \phi$.   The top of Fig. 11 shows a diagram of the two sets
of locus lines drawn for the case of $\phi = 30°$, and the bottom shows a diffraction pattern
of hcp Co with a [001] texture taken with a tilt of 30°.   It can be seen that the centers
of the Debye-ring segments of this pattern are arranged in accord with the construction
at the top of the figure.

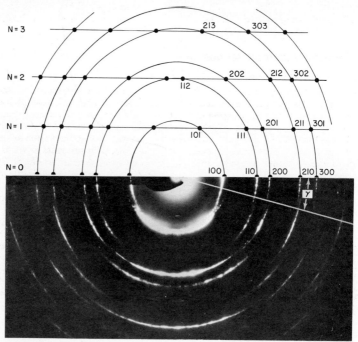

Fig. 11   Diffraction pattern of hcp Co with a [001] fiber-texture axis tilted 30° from the
incident electron beam.

In structures where the densely populated REL rows are inclined with respect
to the fiber axis, the cylinders of Fig. 10 are replaced by cones and hyperboloids,
and in Fig. 11 one would then find other cone sections instead of or in addition to
the ellipses, notably hyperbolas.   These locus lines can best be detected by directly
observing the overall changes of the pattern while the angle ϕ is varied.

For the evaluation of textured patterns, one may first try to establish the periodicity
z along the texture axis and the possible crystallographic identity $[uvw]$ of this axis
from the distance between consecutive layer lines.   Second, the Debye pattern at
$\phi = 0$ has to be consistent with this choice; i.e., all its (hkl) reflections have to obey
the condition

$$hu + kv + lw = N \tag{18}$$

where $N = 0$.   Finally, for $\phi \neq 0$, all reflections on layer lines also have to obey
this condition, with N being the number of the layer line as indicated in Fig. 11.
The assignment of indices to reflections along a particular ellipse can be completed
easily.   If (hkl) is located on layer line N, then $(h + u, k + v, l + w)$ is located on
layer line $N + 1$.

Once the centers of the Debye-ring sections are established, the "sharpness" of texture can be evaluated from the angular length of the arcs, in particular from the half-angular spread γ of reflections on the 0-layer line (see Fig. 11). If the preferred directions of the individual grains are located within a cone of half angle Ω around the ideal-texture axis, then[5]

$$\tan \Omega = \tan \gamma \sin \phi \qquad (19)$$

For the photograph in the bottom part of Fig. 11 ($\gamma = 15°$, $\phi = 30°$), one estimates $\Omega = 7.5°$.

Figure 11 also shows an example of a "forbidden reflection" [see Sec. 2d(2)], the (111) reflection of Co. With $\phi = 30°$, the incident beam is parallel to a [215] direction in several grains and generates a (111) reflection in these grains by double diffraction of the allowed (010) and (101) reflections.

5. TRANSMISSION ELECTRON MICROSCOPY

a. Application and Resolution

For the study of microstructures and defect structures in films, transmission electron microscopy provides the greatest amount of information. The practical limitations which were discussed for transmission diffraction apply here as well; namely, one has to either strip the film from its substrate, or thin the substrate after fabrication of the film, or use very thin substrates which are transparent to electrons such as carbon films or mica flakes. Most microscopes today can use a beam voltage of 100 kV which limits the thickness of specimen films to about 1,000 Å (for Au about 500 Å, for Al about 3,000 Å). Microscopes with higher voltages are also available which permit transmission studies on films close to 1 μ thick.

The ray path of a transmission microscope was shown in Fig. 6a. The majority of electron microscopes employ magnetic lenses. The specimen is illuminated by a double-condenser system which allows one to choose the divergence of the incident beam and the size of the irradiated specimen area. Useful sizes of this area are between 2 and 20 μ. After passing through the specimen, the beam traverses the objective lens which has possibilities of correcting for astigmatic errors. As shown in Fig. 12, the objective aperture which defines the semiangular aperture α of the lens is inserted in the back focal plane of the lens. It intercepts all radiation that forms a larger angle with the optical axis than the angular aperture. This radiation would converge on the image plane with high spherical aberration and reduce contrast and resolution. The specimen image generated by the objective lens is subsequently magnified in one or two more magnification stages by the intermediate and projector lenses and projected onto a fluorescent screen or photographic plate.

It is important to note that magnetic lenses cause the image to rotate around the optical axis by amounts that increase with the strength of the lens. When comparing selected-area diffraction patterns and images of the same area, which, according to Fig. 6, are obtained by changing the strength of the intermediate lens, this rotation has to be taken into account. Moreover, the image is inverted with respect to the diffraction pattern, as can be seen in Fig. 12. For quantitative evaluation of contrast features, one has to rotate the diffraction pattern by 180° in addition to the rotation of the intermediate lens.

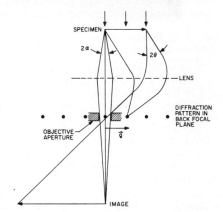

Fig. 12 Electron-ray path in the objective lens showing angular aperture 2α, Bragg angle 2θ, and inversion of image relative to diffraction pattern and specimen.

The resolving power of an electron microscope is of the order of several angstroms, and a resolution of 10 to 20 Å can be expected in routine operation. The smallest resolvable distance δ is approximately equal to the ratio of wavelength λ to the semi-angular aperture α. While α can be made very large (approaching 90°) in light optics so that δ may reach the magnitude of λ, the usable aperture in electron optics is much smaller. The state of the art and possibly intrinsic physical limitations of magnetic lenses and their aberration corrections limit α to 10^{-2} rad in the best cases while 5×10^{-3} rad is a more typical figure. A benefit of the small aperture is the large depth of field. The distance along the optical axis T between two planes which are both still in focus is $T = 2\delta/\alpha$. For $\delta = 10$ Å, the distance T is 0.4 μ, which means that a thin film specimen is simultaneously in focus over its entire thickness.

The angular aperture is defined physically by a round hole of 20 to 50 μ diameter in a thin metal plate. This aperture can be completely retracted so as to allow the complete diffraction pattern to be magnified by subsequent lenses as in Fig. 6b, or it can be inserted and aligned around the primary beam and the optical axis for bright-field microscopy as in Figs. 6a and 12. Furthermore, it can be placed around a diffracted beam in the back focal plane for dark-field microscopy. In this case, image intensity is obtained only from those specimen areas which contributed to the particular diffracted beam or Debye ring. In order to obtain high resolution in dark field, the illuminating beam should be tilted by the deflection angle 2θ so that the diffracted beam travels along the optical axis of the objective lens. Most microscopes have provisions for this manipulation.

The contrast in a transmission picture can be quite complicated as several mechanisms contribute to it. In a bright-field picture of a crystalline film, usually all Bragg diffracted beams are cut out by the aperture; hence the image intensity of a small element of the film is given by the intensity of the transmitted beam at the exit surface of the film element. Similarly, in a dark-field picture the image intensity is given by the intensity of that diffracted beam which is allowed to pass through the aperture. Equation (13) relates the reflected intensity to the thickness t of the crystal and to the parameter s_{eff} containing the local orientation and extinction distance. Strain fields and defects such as stacking faults can change the parameter s_{eff} from one film element to the neighboring one and thus give rise to local variations of the reflected and transmitted intensities. The contrast that arises from such spatial variations of diffraction conditions is known as diffraction contrast and will be discussed further in Sec. 5b.

Periodic structures with a very high d spacing can produce diffracted beams at angular positions within the angular aperture of the objective lens. If an image is formed by allowing two coherent beams from one specimen area to pass through the objective aperture, periodic fringes in the image are observed. They can be considered as projections of the planes of atoms in the crystal. Copper and platinum phthalocyanine crystals with d spacings of 10 to 12 Å are examples for this lattice resolution.[32] The phenomenon is useful for stability and resolution tests of an electron microscope. It is interesting that even the (111) planes of gold films with a d spacing of 2.35 Å have been resolved.[33] A variation of the imaging of small periodic structures is the formation of moiré patterns. They have become important in thin film work, in particular in the early stages of epitaxial growth, and will be discussed in Sec. 5c.

For microstructure work based on the concepts of diffraction contrast or on resolution of high d spacings, it is necessary that the crystallite size is comparable with the field of view at the magnification employed. If the grain size becomes much smaller than the area viewed by the microscope or smaller than the film thickness, the various contrast features overlap to such an extent that they cannot be reasonably separated. In this case it is more advantageous to extract information from the diffraction pattern whose line broadening and line shifts yield parameters averaged over large areas. The grain size, an important parameter in fine-grained films, can still be determined fairly accurately from dark-field pictures with the aperture placed on a strong Debye ring. This causes a few grains to be set off clearly against the dark background, and the technique is applicable down to a size of coherently diffracting regions of the order of 20 Å. In the extreme cases of even smaller particle sizes and of amorphous films, the image contrast is controlled by the product of total scattering cross section and film thickness (mass thickness contrast). In chemically homogeneous specimens, this contrast maps out thickness variations. A thorough discussion of the mass-thickness contrast is given by Heidenreich.[5]

Historically, the oldest metallographic use of electron microscopy is the study of surfaces by replication of their profile in a thin film. If subsequently viewed in electron transmission, this film should ideally give rise only to mass-thickness contrast. Evaporated carbon films[34] are widely used replicas. Their contrast is usually enhanced by evaporating 10 to 30 Å of a heavy metal under a small (10 to 30°) angle of incidence. This causes the well-known shadow effect.[5]

b. Diffraction Contrast

(1) Extinction Contours Diffraction contrast is the physical phenomenon which dominates transmission images of crystalline specimens when only one beam, either the transmitted or a diffracted beam, passes through the objective aperture. The work of Hirsch and coworkers and of Heidenreich reviewed in Refs. 5 and 6 permitted a detailed interpretation of this contrast mechanism and gave rise to the widespread metallurgical applications of transmission electron microscopy. In this section, the elements of diffraction contrast are merely summarized. A concise review has also been given by Stowell.[35]

The contrast effects found in perfect crystals can be discussed by means of Eqs. (12) and (13). Given a crystalline film of constant thickness t which buckles slightly within the field of view, one observes bending contours. In a bright-field picture, these contours are dark bands connecting the regions where one particular set of lattice planes is in exact Bragg position and deflects the intensity into the diffracted beam. As the reflected intensity depends on $\sin^2 (\pi t s_{\text{eff}})$, one can observe subsidiary contours parallel to the main contour at constant deviations from the Bragg condition where $s_{\text{eff}} = n/t$ with $n = 1, 2, 3$, etc. The same periodic dependence gives rise to contours of equal thickness in a specimen with variable thickness t and constant orientation s_{eff}. These contours map out areas where $t = n/s_{\text{eff}}$. In a polycrystalline film with inclined grain boundaries, thickness fringes which connect on the boundary points of equal depth $t' = n/s_{\text{eff}}$ underneath the film surface are observed if one of the adjoining grains is in a reflecting position.

These fringes are particularly prominent at the boundaries of twin lamellae in (001) epitaxial films of noble metals and nickel. With an instrumental magnification M and an inclination γ of the boundary, the observed fringe spacing is

$$D = \frac{M}{s_{\text{eff}}} \tan \gamma \tag{20}$$

The spacing depends through s_{eff} on the extinction distance ξ_g and on the deviation s from the Bragg position [see Eq. (12)]. D is small for low-index reflections and for large values of s. With increasing s, however, the diffracted intensity decreases according to Eqs. (13) and (8), and with it diminishes the contrast of the fringes.

(2) Fault Planes A fault plane in a crystal is a plane along which one part of the crystal has been displaced with respect to the other by an amount $\mathbf{R}$ which is not a translation vector of the crystal structure. Important are stacking faults on close-packed planes of close-packed structures such as on (111) planes in fcc crystals. One distinguishes intrinsic stacking faults which are equivalent to the removal of a close-packed layer, and extrinsic faults equivalent to the insertion of such a layer. They can be described by $\mathbf{R} = +\frac{1}{3}[111]$ or $\mathbf{R} = -\frac{1}{3}[111]$, respectively. Fault planes inclined in the film give rise to a fringe contrast with fringes marking points of equal depth $t' = 1/s_{\text{eff}}$ on the fault planes. However, no contrast is visible if the Bragg reflection $\mathbf{g}$ operating in the field of view is such that

$$\mathbf{g}\mathbf{R} = 0 \tag{21}$$

i.e., if the fault vector is parallel to the reflecting planes. This leads to the determination of the vector **R** by forming dark-field images and finding a reflection with REL vector **g** for which the stacking-fault image disappears. The intrinsic or extrinsic nature of stacking faults can also be established by comparison of bright-field and dark-field images; the methods are described in Ref. 6.

Fringe spacings and contrast of stacking faults are similar to those discussed for inclined grain boundaries. Stacking faults and twin planes parallel to the film plane such as occurring in (111) textures of fcc films give no fringe contrast.

(3) Strain Centers Diffraction contrast also produces images of inhomogeneous strain fields in a crystal. Strain fields may be caused by inclusions of second-phase particles or by dislocations. They are probed on a scale given by the extinction distance and are therefore best visible when low-index reflections are excited in the field of view.

Isotropic compression or dilation centers in the middle of the crystal plate yield images composed of one or more circular isostrain contours which are interrupted in a line of no contrast. Along this line the image intensity is equal to that of the areas unaffected by the strain center. The line of no contrast is perpendicular to the direction of the REL vector **g**.

There are important modifications if the strain center is near the surface of the crystal as was found by Ashby and Brown.[36] The positive print of a dark-field picture shows a black half halo indicating less intensity than in the surroundings and a white one due to excess intensity. Both half haloes are separated by a line of no contrast perpendicular to the direction of **g**. If the strain center *compresses* the matrix, the *deficiency* contrast is on the side of positive **g**, and if the strain center causes *dilation and tensile stress* in the matrix, the *excess* contrast is on the side of positive **g**. The bright-field images are similar if the strain center is near the top of the crystal and complementary if it is near the exit surface. For certain ranges of strain and extinction parameters, exceptions to this Ashby-Brown rule have been discussed.[37] However, the rule seems to be valid for many practical situations, including some cases of x-ray topography (see Sec. 6a).

(4) Dislocations The generation of contrast by dislocations is a complicated process which is discussed in detail in Ref. 6. The elastic strain field of a dislocation is determined by its Burger's vector **b** and the direction of the dislocation line. In a first approximation, the strain field gives no contrast if

$$\mathbf{gb} = 0 \qquad (22)$$

i.e., if the Burger's vector is parallel to the reflecting planes. This may again be used to determine the Burger's vector **b** from those dark-field pictures in which the dislocation disappears.

Equation (22) is an exact condition for no contrast from a screw dislocation whose Burger's vector and elastic displacements are parallel to the dislocation line. In the cases of edge and general dislocation, there are displacement components normal to the slip plane (containing Burger's vector and dislocation line), which can give rise to a contrast (usually weak) unless the normal of the slip plane is also parallel to the reflecting planes. Displacement fields associated with elastic relaxations near the crystal surface give rise to additional contrast effects.

A dislocation image has a width of about $\frac{1}{3}s_{\text{eff.}}$. It is therefore sharper for small extinction distances or for low-order reflections. Also, the image is generally displaced from the location of the projection of the dislocation lines toward the side of negative $\mathbf{g(bs)}$. This displacement can be reversed by reversing the orientation parameter **s**, and thus the sign of **b** can be determined.

For observation and evaluation of diffraction contrast features, it is very important that the orientation of the specimen with respect to the incident beam can be carefully adjusted and changed, and that the same field of view is examined with different diffracting conditions. Otherwise important contrast sources might remain unnoticed.

c. Moiré Patterns

Moiré patterns are sometimes encountered in transmission electron microscopy of thin films. These patterns occur if two crystals are superimposed such as a hetero-epitaxial deposit on a single-crystal-film substrate. They arise from double diffraction, whereby a beam diffracted in the first crystal is reflected back by the second crystal into a position close to the primary beam so that it can pass through the objective aperture. The image consists of fringes perpendicular to the direction of the doubly diffracted reflection spot. If the diffracting planes in the two crystals are parallel and have the spacings d_1 and d_2, a "parallel" moiré pattern is obtained with a fringe spacing of

$$D = \frac{d_1 d_2}{d_1 - d_2} \qquad (23)$$

The fringes project the periodicity with which the two sets of crystal planes go in and out of register, and their spacing may be large enough to be resolved even though the individual d spacings are below the resolution limit. The precise position of the fringes depends on the thicknesses of the two crystals and on the parameters $\mathbf{s}$ of the two diffracting planes. Variation of the thickness of one of the crystals causes a displacement of the fringes. This effect has been used by Jacobs et al.[38] to deduce the profile of small epitaxial islands of Au on MoS_2.

If, in addition, the doubly diffracting lattice with the lattice plane spacing d_2 is rotated by an angle α with respect to the planes of spacing d_1, the moiré fringes have the spacing $D = d_1 M$, where $M = d_2/(d_1{}^2 + d_2{}^2 - 2d_1 d_2 \cos \alpha)^{\frac{1}{2}}$ is the moiré magnification. Moreover, the fringes are rotated with respect to the direction of the parallel moiré pattern. For small angles α, the rotation of the pattern ω is also magnified by the moiré magnification

$$\omega = \alpha M \qquad (24)$$

Therefore, rotation is very sensitive to small misorientations. This effect has been used to study the angular alignment of small epitaxial islands.[39]

It is interesting that dislocations, the lines where incomplete lattice planes terminate, are revealed in moiré patterns even though the lattice planes themselves are not resolved. Optical analogs of this effect are reviewed in Ref. 6. The analysis shows that the number N of extra half fringes produced by a dislocation with Burger's vector $\mathbf{b}$ in one component crystal is $N = \mathbf{gb}$, where $\mathbf{g}$ is the diffraction vector which takes part in the formation of the moiré pattern. It has been possible to observe the formation of dislocations during the coalescence stage of film growth with this technique.[38]

d. Lorentz Microscopy

Any feature of a transmission microscope specimen which deflects the electron beam can be translated into image contrast. In the preceding sections, deflections due to scattering and diffraction have been treated. Electrons are also deflected by internal magnetic and electric fields in the specimen, and the distribution of these fields can be made visible. An important case is the deflection β_m produced by the Lorentz force which acts on an electron while it travels through a ferromagnetic film with magnetization M and thickness t:

$$\beta_m = \frac{e}{mv} 4\pi M t \qquad (25)$$

where e is the charge, m the mass, and v the velocity of the electron. This deflection of the beam can be observed by using the microscope as a small-angle diffraction camera, generating a large camera length with a suitable combination of lenses. β_m can then be evaluated as a measure of the magnetization.[40]

More important is the microscopic determination of the distributions of magnetization directions and of domain walls in films. These features are best visible by the "out-of-focus" method. Figure 13 shows the deflections which occur in a film containing 180° walls. A focused picture which maps out the electron current density at the exit surface of the film would not show an influence of the magnetic deflection because the deflections β_m are so small (of the order of 10^{-4} rad in typical cases) that all deflected electrons can pass through the objective aperture. However, an image of a plane at a distance L below the specimen plane (see Fig. 13) shows a dark band of electron deficiency underneath one domain wall and a light band of excess intensity underneath the neighboring walls. Focusing on a plane above the specimen causes the dark images of a wall to change into light ones and vice versa. As the specimen itself is not in focus, its contrast details appear blurred. Moreover, the specimen has

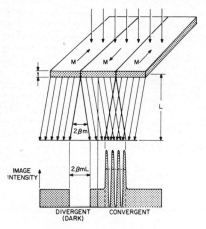

Fig. 13 Out-of-focus Lorentz microscopy showing magnetic deflection β_m and intensity distribution in the image of a plane located at a distance L below the specimen. (*Courtesy of R. D. Heidenreich*[5] *and Interscience Publishers.*)

to be placed outside the magnetic field of the objective lens so that its state of magnetization is not disturbed. These factors limit the resolution to fractions of a micron.

The details of the contrast of the wall images depend on the width of the domain walls and on the gradient of the horizontal magnetization component within the wall. These parameters can be determined approximately by observing the wall images as a function of out-of-focus distance L and by using an evaluation based on geometrical optics.[41-43] If the specimen is illuminated with a coherent electron beam, one may observe interference fringes in the area where the wave fields from two adjoining domains overlap (see Fig. 13). These fringes have been discussed in terms of wave theory, and a more accurate evaluation of the wall parameters on this basis is difficult.[44,45]

Lorentz microscopy allows the observation of another important feature of fine grained films, namely, magnetization ripple which runs normal to the magnetization direction. It is caused by small local deviations of the magnetization direction from the overall easy axis. An example is shown in Fig. 14.

Other contrast effects and observation techniques which are connected with the Lorentz force and its interaction with diffraction contrast are reviewed in Ref. 6. They are less important in thin film work. A recent review of Lorentz microscopy has been given by Grundy and Tebble.[46]

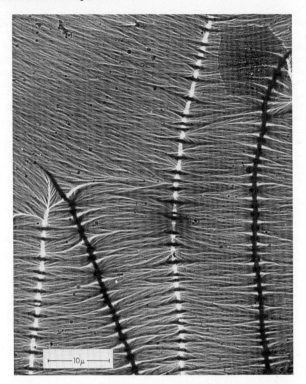

g. 14 Lorentz micrograph of a Co-Au alloy film showing 180° walls and magnetic ripple thin the domains. The magnetization direction is the orthogonal trajectory of the ripple nes.

OTHER MICROSCOPE TECHNIQUES

X-ray Topography

The diffraction of x-rays in crystals is controlled by Bragg's equation, but local riations of diffracting conditions can cause variations of the intensities diffracted om particular regions. These intensities can be recorded as a function of the coordinates of the diffracting regions. The result is an image of the crystal which shows viations from perfect diffraction due to fault planes, strain fields, dislocations, etc. ie contrast of such images, called x-ray topographs, is quite similar to the diffraction contrast in transmission electron microscopy.

X-ray beams cannot be manipulated by lenses and magnification devices, a fact iich has two important consequences for topographic techniques. The first one is imitation to straight-line geometries for incident, transmitted, and diffracted rays. ie wants to use a well-collimated parallel incident beam and probe large areas veral square centimeters) of the specimen, as well as avoid overlap of the images of insmitted and diffracted beams. This led to the development of several geometrilly different methods, which are reviewed in Ref. 1. A very convenient technique large-area topography was described by Schwuttke[47] and is generally known as e scanning oscillator technique. Figure 15 shows the experimental arrangement ed in this technique. A monochromatic and parallel x-ray beam with the cross tion of a narrow rectangle is produced by the x-ray tube and the slit system. The ecimen crystal is oriented such that the x-ray beam impinges under the Bragg

angle of a low-index plane. The specimen and the film for recording the diffracte
intensity are mounted in parallel and moved synchronously through the stationar
beam so that the entire specimen crystal area is scanned. A specimen containin
macroscopic curvatures or slowly varying strain projects an image from the Brag
diffracted beam only in a few areas which happen to fulfill the Bragg equation. I
order to obtain equal diffraction conditions for the entire specimen crystal, the latte
and also the film carrier are made to oscillate around a vertical axis which continuousl
rocks the specimen through the Bragg angle. This is shown in the insert in the lowe
left corner of Fig. 15.

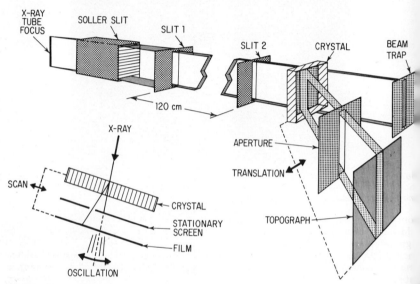

Fig. 15 Ray paths and experimental arrangement for x-ray topography using the scannin
oscillator technique. (*Courtesy of G. H. Schwuttke.*[47])

The second consequence of the lack of lenses for x-rays is the low resolution
x-ray topographs when compared with electron micrographs. Topographs a
recorded without magnification. Subsequent optical magnification is possible
the extent permitted by the resolution of the photographic emulsion, but this
limited to about 100 times. Therefore, the application of x-ray topography
restricted to materials with a low density of defects; it is best used with materi
having a high degree of crystalline perfection.

The materials with the highest degree of crystalline perfection are semiconduct
crystals which are used in solid-state electronics. It is in this area where x-r
topography has found application in thin film technology. The thin films used
microelectronics for either insulation or current conduction must adhere well to t
semiconducting wafers, and they transmit their internal stress to the substrate.
stress gradient arising from areas of poor adhesion or from windows in the film caus
a curvature of the atomic planes in the wafer which can be made visible by x-r
topographs. The details of the contrast depend on the wafer thickness t, the x-r
wavelength, and the absorption coefficient μ of the crystal material. One can choo
to work with kinematic ($\mu t < 1$) or dynamic ($\mu t > 1$) contrast.[48]

Figure 16 is a topograph of a silicon wafer covered with a silver film, taken wi
kinematic contrast. The film contains blisters of poor adhesion which show in t
picture as two semicircles with a line of no contrast normal to the direction of the R
vector **g**. Figure 17 shows a topograph of a silicon wafer with a molybdenum fi

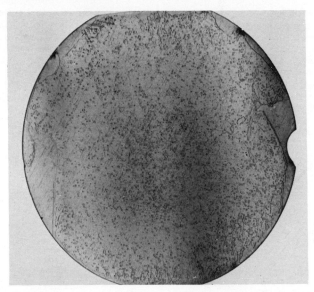

ig. 16 X-ray topograph of a silver film on a silicon wafer. Blisters of poor adhesion shown with kinematic contrast ($\mu t \approx 0.3$). (*Courtesy of G. H. Schwuttke and J. K. Howard,*[48] *and J. Appl. Phys.*)

ontaining a triangular window. This picture was taken with dynamic contrast, which is advantageous for showing the strain gradients at the window edges. The sign of the strain can be determined by the following contrast rule: A topograph of the reflected beam shows *excess* intensity from a region where **g** points toward the *center of curvature* of locally curved Bragg reflecting planes; there is a *deficiency* of intensity where **g** points toward the *concave side* of locally curved planes. Referring to Fig. 17, a window in a film under tensile stress transmits compressive stress into the substrate underneath, and the outline of the window in the topograph has an excess of intensity on the side of positive **g** and vice versa.

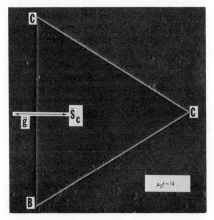

g. 17 X-ray topograph of a molybdenum film with a triangular window on a silicon wafer. The film is under tensile stress ($\mu t \approx 14$). (*Courtesy of G. H. Schwuttke and J. K. Howard,*[48] *and J. Appl. Phys.*)

The above contrast rule is analogous to the Ashby-Brown rule for diffraction contrast in transmission electron microscopy. However, when comparing x-ray topographs with electron micrographs, one has to keep in mind that it is customary to give topographs as photographic negatives and electron micrographs as positive prints. This reverses the meaning of "black-white" contrast in terms of deficiency or excess intensity for the two cases.

Other contrast rules in topography are also similar to those discussed for diffraction contrast in electron microscopy. For example, minimum contrast for stacking faults is obtained if $\mathbf{gR} = 0$, and for dislocations if $\mathbf{gb} = 0$ as in Eqs. (21) and (22).

b. Scanning Electron Microscopy

Scanning electron microscopes are becoming more and more useful and popular for the direct observation of surfaces because they offer better resolution and depth of field than optical microscopes. The resolution approaches that attainable with conventional electron microscopes and surface-replica samples. In addition, in situ observation of surface-morphology changes during heat treatment, etc., of specimens is possible.

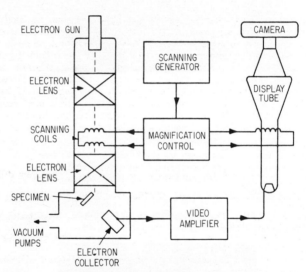

Fig. 18 Schematic of a scanning electron microscope. (*Courtesy of O. C. Wells[49] and John Wiley & Sons.*)

Figure 18 shows a schematic diagram of a scanning electron microscope according to Wells.[49] A beam of very small diameter (of the order of 100 Å) is produced by the electron gun and electron lenses. The scanning coils deflect this beam and sweep it over the specimen surface. A cathode-ray display tube is scanned synchronously with the electron beam. The brightness of the display tube is modulated by the signal which arises from the interaction of the beam with the surface element which is probed. The strength of this signal is thus translated into image contrast. Fig. 18, secondary electrons which the beam probe liberates from the specimen surface are collected and used as the contrast signal. The yield of collected electrons depends on the nature of the specimen surface and on its inclination with respect to the probing beam. Consequently, one obtains pictures with a highly perspective appearance. An example is Fig. 19, which is a scanning-electron-microscope image of an aluminum-film stripe. A corrugated surface and surface growths are visible. These features were formed by passing a high current (3.10^8 A cm^{-2}) through the stripe at an ambient temperature of 100°C.

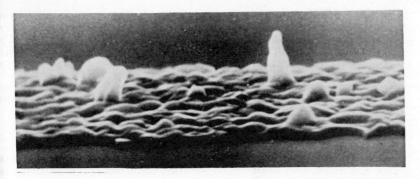

Fig. 19 Scanning electron micrograph of an aluminum film stripe which is 8 μ wide and 3,000 Å thick. (*Courtesy of L. Berenbaum and R. Rosenberg, unpublished material.*)

Contrast signals other than secondary electrons may also be used in scanning microscopy. Examples are backscattered electrons from the incident beam and electron-beam-induced currents in semiconducting specimens. Signals of the latter type can be used to display junctions and other electronic inhomogeneities in integrated circuits.[50] An instrument similar to the scanning electron microscope is the electron microprobe analyzer which is described in Chap. 10. Here, the characteristic x-rays which the beam probe excites in the specimen material are utilized to generate the contrast signal. A review of scanning electron microscopy has been given by Oatley et al.[51] Its main application in the study of thin films has been the direct observation of surface morphologies, as demonstrated by Fig. 19.

REFERENCES

1. Barrett, C. S., and T. B. Massalski, "Structure of Metals," McGraw-Hill Book Company, New York, 1966.
2. "International Tables for X-ray Crystallography," The Kynoch Press, Birmingham, 1952–1962.
3. Anderson, J. C. (ed.), "The Use of Thin Films in Physical Investigations," Academic Press, Inc., London, 1966.
4. Vainshtein, B. K., "Structure Analysis by Electron Diffraction," The Macmillan Company, New York, 1964.
5. Heidenreich, R. D., "Fundamentals of Transmission Electron Microscopy," Interscience Publishers, Inc., New York, 1964.
6. Hirsch, P. B., A. Howie, R. B. Nicholson, D. W. Pashley, and M. J. Whelan, "Electron Microscopy of Thin Crystals," Butterworths, Washington, D.C., 1965.
7. American Society for Testing Materials, X-ray Department, Philadelphia, Pa.
8. Widmer, H., *Appl. Phys. Letters*, **5**, 108 (1964).
9. Chopra, K. L., M. R. Randlett, and R. H. Duff, *Phil. Mag.*, **16**, 261 (1967).
10. Mader, S., Ref. 3, p. 433.
11. Sturkey, L., in "Symposium on Advances in Techniques in Electron Metallography," p. 31, American Society for Testing Materials, 1963.
12. Pashley, D. W., and M. J. Stowell, *Phil. Mag.*, **8**, 1605 (1963).
13. Levine, E., W. L. Bell, and G. Thomas, *J. Appl. Phys.*, **37**, 2141 (1966).
14. Okamoto, P. R., E. Levine, and G. Thomas, *J. Appl. Phys.*, **38**, 289 (1967).
15. Katz, G., *Appl. Phys. Letters*, **12**, 161 (1968).
16. Segmüller, A., *Z. Metallkunde*, **54**, 247 (1963).
17. Vook, R. W., and F. Witt, *J. Vacuum Sci. Technol.*, **2**, 49, 243 (1965).
18. Light, T. B., and C. N. J. Wagner, *J. Vacuum Sci. Technol.*, **3**, 1 (1966).
19. Keith, H. D., *Proc. Roy. Soc. London*, **B69**, 180 (1956).
20. Weiner, K. L., *Z. Krist.*, **123**, 315 (1966).
21. Vook, R. W., and F. Witt, *J. Appl. Phys.*, **36**, 2169 (1965).
22. Witt, F., and R. W. Vook, *J. Appl. Phys.*, **39**, 2773 (1968).
23. Booker, G. R., and R. Stickler, *Brit. J. Appl. Phys.*, **13**, 446 (1962).

24. Holloway, H., Ref. 3, p. 111.
25. MacRae, A. U., Ref. 3, p. 149; see also Jona, F., *Appl. Phys. Letters*, **9**, 235 (1966); Thomas, R. N., and M. H. Francombe, *Appl. Phys. Letters*, **11**, 108 (1967).
26. Weber, R. E., and W. T. Peria, *J. Appl. Phys.*, **38**, 4355 (1967).
27. Grigson, C. W. B., *Rev. Sci. Instr.*, **36**, 1587 (1965).
28. Grigson, C. W. B., and D. B. Dove, *J. Vacuum Sci. Technol.*, **3**, 120 (1966).
29. Tompsett, M. F., and C. W. B. Grigson, *J. Sci. Instr.*, **43**, 430 (1966).
30. Warren, B. E., *Progr. Metal Phys.*, **8**, 147 (1959).
31. Andrews, K. W., D. J. Dyson, and S. R. Keown, "Interpretation of Electron Diffraction Patterns," Hilger and Watts, London, 1967.
32. Menter, J. W., *Advan. Phys.*, **7**, 299 (1958).
33. Komoda, T., *J. Appl. Phys. Japan*, **3**, 122 (1964).
34. Bradley, D. E., *Brit. J. Appl. Phys.*, **5**, 65 (1954); *J. Appl. Phys.*, **27**, 1399 (1956).
35. Stowell, M. J., Ref. 3, p. 131.
36. Ashby, M. F., and L. M. Brown, *Phil. Mag.*, **8**, 1083 (1963).
37. Chik, K. P., M. Wilkens, and M. Rühle, *Phys. Stat. Solidi*, **23**, 113 (1967).
38. Jacobs, M. H., D. W. Pashley, and M. J. Stowell, *Phil. Mag.*, **13**, 129 (1966).
39. Pashley, D. W., M. J. Stowell, M. H. Jacobs, and T. J. Law, *Phil. Mag.*, **10**, 127 (1964).
40. Mader, S., and A. S. Nowick, *Thin Films*, **1**, 45 (1968).
41. Fuller, H. W., and M. E. Hale, *J. Appl. Phys.*, **31**, 238 (1960).
42. Fuchs, E., *Z. Angew. Phys.*, **14**, 203 (1962).
43. Wade, R. H., *Proc. Phys. Soc. London*, **79**, 1237 (1962).
44. Boersch, H., H. Hanisch, K. Grohmann, and D. Wohlleben, *Z. Physik*, **167**, 72 (1962).
45. Wohlleben, D., *J. Appl. Phys.*, **38**, 3341 (1967).
46. Grundy, P. J., and R. S. Tebble, *Advan. Phys.*, **17**, 153 (1968).
47. Schwuttke, G. H., *J. Appl. Phys.*, **36**, 2712 (1965).
48. Schwuttke, G. H., and J. K. Howard, *J. Appl. Phys.*, **39**, 1581 (1968).
49. Wells, O. C., in R. Bakish (ed.), "Introduction to Electron Beam Technology," p. 354, John Wiley & Sons, Inc., New York, London, 1962.
50. Everhart, T. E., O. C. Wells, and R. K. Matta, *J. Electrochem. Soc.*, **111**, 929 (1964).
51. Oatley, C. W., W. C. Nixon, and R. F. W. Pease, *Advan. Electron. Electron Phys.*, **21**, 181 (1965).

Chapter **10**

The Growth and Structure
of Single-crystal Films

IMDAD H. KHAN

National Aeronautics and Space Administration,
Electronics Research Center, Cambridge, Massachusetts

1. INTRODUCTION

Interest in the study of single-crystal films has increased markedly in the past few years. This can be attributed to a number of factors. First, there has been a tremendous surge of interest in the potential applications of thin films in electronic devices. The physical properties of thin films are known to differ widely from those of the corresponding bulk materials. This is evidently connected with the small size of the crystallites forming the film and, in particular, with the large number of defects such as dislocations, vacancies, stacking faults, grain boundaries, and twins. An important role may also be played by the mechanical stress and deformation generated in the film on cooling the substrate, and also by certain peculiarities in the atomic structures of the film, such as the occurrence of polymorphic forms and metastable phases. In view of these facts, it is desirable to obtain films having a structure approaching as closely the ideal single-crystal form as possible, and to exploit the technological potentials of such films. This has led to a considerable volume of work involving the nucleation, growth, and structure of single-crystal films.

The second factor of importance involves the sophistication of experimental techniques for structural investigation. Major advances have been in the development and use of high-resolution electron microscopy, which provides direct observation of lattice imperfections in thin films. In addition, it has been possible to obtain a good deal of information about the early stages of crystal growth by direct observation inside the electron microscope. Other powerful techniques include low- and high-energy electron diffraction for in situ studies of the structure and crystallographic orientation during the initial and successive stages of film growth. Field-emission and field-ion microscopy may also be applied to the problem, but their capabilities are very limited insofar as thin films are concerned.

Studies of single-crystal-film growth have also been augmented by the improvement of vacuum technology. It has now become possible to prepare single-crystal films under cleaner vacuum environments. Information concerning the role of contamination in the growth of single-crystal films is now beginning to emerge, and this will clear up some of the difficulties experienced in understanding the mechanism of single-crystal-film formation.

This chapter will be devoted to the study of the growth and structure of single-crystal films of a wide variety of materials. The materials to be covered will include metals, alloys, semiconductors, and insulators. In principle, it is now possible to grow single-crystal films of any material which has been studied in bulk single-crystal form, by one or more of the preparative techniques available. By far the majority of results apply to vacuum-evaporated films, but mention will be made of films formed by other techniques such as sputtering and vapor deposition. While the different methods lead to films of different degrees of perfection and, sometimes, different structures and orientations, they nevertheless provide a basis for understanding the basic processes involved in the formation of single-crystal films.

2. THE GROWTH OF SINGLE-CRYSTAL FILMS

a. Processes for Single-crystal Growth

There are a variety of methods for the preparation of thin films. They may broadly be divided into two separate groups: (1) physical methods and (2) chemical methods. The physical methods are vacuum evaporation, sublimation, and sputtering, while the chemical methods include vapor deposition, anodization, and electrolytic preparation. The use of these methods for the preparation of single-crystal films of different materials can be found in Sec. 2e.

There are in principle several processes for the growth of single-crystal films on single-crystalline and amorphous substrates. In this section we will briefly describe these processes.

(1) Epitaxy The most commonly used process that leads to the growth of single-crystalline films is based on the phenomenon of epitaxy. This phenomenon has been known for over a century, and since the discovery of electron diffraction, numerous examples of this phenomenon have been observed and extensively reviewed by Pashley.[1] The term epitaxy means the oriented or single-crystalline growth of one material upon another such that there exist crystallographic relations between the overgrowth and the substrate; i.e., certain planes are parallel and certain crystallographic directions are parallel. There can be a multiplicity of crystallographically equivalent orientation relationships, in which case a single-crystalline overgrowth may not be achieved. Epitaxy may be classified as follows:

Isoepitaxy, the oriented or single-crystal growth of a material on a substrate of the same material

Heteroepitaxy, similar growth on a substrate of different material

Rheotaxy, similar growth on a liquid surface of different material

It should be mentioned here that in the present-day semiconductor technology the term epitaxial is commonly used to indicate the growth of single-crystalline overgrowth, e.g., of silicon on silicon by vapor deposition.

The occurrence of epitaxy in thin films has received considerable theoretical treatment which can broadly be classified into two groups: (1) old theories, based on the geometrical fittings between the lattices of the deposit and of the substrate, and (2) current theories, based on the nucleation process. A critical discussion of the old theories can be found in Pashley's review.[1] In the following a brief mention of these theories will be made, followed by discussion of the current attempts to explain the occurrence of epitaxy.

(a) *Old Theories.* Royer[2] has put forward a set of rules for epitaxy and has stated that epitaxy occurs when it involves a small misfit between the deposit film and the substrate. The misfit (in percentage) is expressed as $100(b - a)/a$, where a and b represent the network spacings in the substrate and in the film, respectively. This misfit consideration is not valid, since epitaxy occurs with a wide range of misfits. The general conclusion is that the occurrence of a small misfit is important but not an essential condition for epitaxy. The occurrence of epitaxy with large misfits has been explained by Menzer.[3] His mechanism is based on the concept that an epitaxial layer corresponding to a good fit occurs during the initial growth stage, and subsequent growth gives rise to different orientations. However, numerous examples[4-7] in the literature clearly prove the invalidity of this concept.

The theory put forward by Engel[8,9] shows that there exists a minimum temperature, called the epitaxial temperature, for complete epitaxy, and the occurrence of orientation depends upon the ionization processes and upon the misfit between the film and the substrate. This theory has no validity, since the results on epitaxy (see Sec. 2e) indicate that the epitaxial temperature for a particular film-substrate system varies over a wide range depending upon the conditions of film formation.

According to the Frank and van der Merwe theory,[10,11] the initial stage of growth of an epitaxial film is the formation of an immobile monolayer, homogeneously deformed to fit on the substrate so as to have the same spacing as the substrate. This is called a pseudomorphic monolayer. This occurs provided the misfit is less than a certain critical value, i.e., about 14%. Once this epitaxial monolayer forms, a thick epitaxial film can then grow by repetition of the process. This pseudomorphic film would undergo some transition at a certain growth stage to give a strain-free bulk deposit which would remain epitaxial. The concept of pseudomorphic growth has received considerable experimental confirmation, as will be described later in the chapter (Secs. 2d, 4b, and 5a).

(b) *Current Theories.* The theories discussed above do not provide adequate explanation of the epitaxial growth, since they are all based on unsuitable models for the growth of the initial nuclei of the deposit. We will now describe the general current concepts of epitaxial-film growth in the light of nucleation theory.

In his critical discussion of the theories of epitaxy, Pashley[1] has recognized the importance of nucleation in the epitaxial growth of thin films. He has pointed out that the occurrence of epitaxy depends upon the formation of oriented nuclei, and that a detailed understanding of the epitaxial process depends upon the factors controlling the orientation of these nuclei.

Hirth and Pound [12,13] and Hirth et al.[14] have dealt with the theory of capillary-model heterogeneous nucleation and have considered the relation of nucleation theory to epitaxy. They have predicted that nucleation leading to epitaxy may be coherent, semicoherent, or incoherent. In other words, the orientations of the nuclei are not the critical requirements for epitaxy of thin films. The critical requirement is that a particular orientation has a lower free energy of formation (and hence occurs in greatest numbers) and a much higher nucleation rate than any other orientation. It is suggested that high substrate temperatures and low supersaturations are the requisites for epitaxy. The nucleation at low supersaturations may occur by mutual impingement of large subcritical embryos rather than by addition of a single atom to a critical nucleus.

Moazed[15] has considered the application of the theory of heterogeneous nucleation to epitaxy, and has shown that the main conditions in heterogeneous nucleation which lead to epitaxial nucleus formation are:

1. High substrate temperatures—a decrease in substrate temperature decreases the ratio of epitaxial nucleation rate to random nucleation rate.

2. Low supersaturations—at low supersaturations only preferred sites can act as nucleation sites, while at high supersaturations, random nucleation is more likely. The conditions leading to epitaxy must also take into account other important factors such as growth and recrystallization.

The capillary-model nucleation theory as discussed above is proved to be satisfactory for experimental conditions where the critical nucleus is large. There are many cases, however, involving the condensation of vapors on solid surfaces where the critical nucleus may involve only a few atoms. It is then desirable to use the atomistic model of nucleation for the interpretation of nucleation effects. Rhodin and Walton,[16,17] Walton,[18,19] and Rhodin[20] have extended the nucleation theory into the range of very small nuclei by treating the clusters as simple macromolecules and have considered the role of nucleation in epitaxial-film formation. The nucleation process as described by them is based on the assumption that the critical nucleus contains only a few atoms. In considering the critical size for different arrangements of small clusters of deposit atoms (fcc metals on rock salt), they have pointed out that the nucleus with the fewest atoms has the most probability of being stable. The critical nucleus is defined as containing one atom less than the smallest cluster. It is shown that the critical nucleus for a (111) stable orientation is just a pair of absorbed atoms, for a (100) orientation it is three atoms, and for the rarely observed (110) orientation it is four atoms. This explains why (100) and (111) orientations occur in the fcc metals when grown on rock salt. Rhodin and Walton[17] and Shirai and Fukada[21] have considered the probable mechanism that favors a particular orientation. There are at least two mechanisms by which a particular orientation may be favored:

1. The critical nucleus which leads to the orientation is adsorbed more strongly than any other.

2. The critical nucleus which leads to some other orientation is strongly and energetically favored, but subsequent growth of the cluster requires the addition of atoms in unfavorable positions. Thus a deposit with the orientation can nucleate but may not be able to grow.

It appears that a most likely mechanism for favoring one of the two equally probable orientations of nuclei is by preferential recrystallization during coalescence. Recent experimental results by Matthews[22] lend support to this viewpoint (see Sec. 3b). In situ electron-microscopy studies by Pashley[23,24] have demonstrated the mode of nucleation and the way in which the nuclei develop into a continuous film. The observations have revealed two important processes—"liquid-like" coalescence of nuclei and islands, and recrystallization of islands (see Sec. 3b). It is seen that recrystallization occurs during coalescence of nuclei and islands, and the orientation of the initial nuclei therefore plays a determining role in the final film orientation.

The preceding theoretical treatments indicate that the following factors which contribute to the formation of epitaxial films should be given careful consideration:

Nucleation orientation: initial occurrence of a large number of one type of nuclei with a high growth rate.

Growth orientation: initially there may be many types of nuclei, but a preferred rate of growth of one type may dominate the final film orientation.

Recrystallization: this process occurs after initial nucleation and growth, leading to a film structure with a state of lower free energy. The orientation of the initial nuclei will therefore have a dominant role in providing the final orientation.

It is expected that epitaxy will occur by a combination of two or more of these processes. We should then consider the extent to which these considerations hold good. The contribution of the preferred nucleation and growth of structured nuclei to epitaxial growth as proposed by Rhodin and Walton[16,17] appears to have general applicability. However, the contribution of preferred recrystallization during coalescence to epitaxial-film formation[23,24] may be considered to be a more important factor than that proposed by Rhodin and Walton. Although these treatments are not quite adequate to explain all the experimentally observed features of epitaxial-film growth, they seem to provide a valuable basis for developing a more detailed understanding of the epitaxial growth process.

We will now describe the less common processes that lead to the growth of single-crystal films on single-crystal and amorphous substrates.

(2) Vapor-Liquid-Solid Process In this mechanism, a liquid-alloy droplet of relatively low freezing temperature is first formed on the substrate. The liquid droplet acts as a preferred site for deposition from the vapor, which causes the liquid to become saturated with the arriving atoms, or as a catalyst for the chemical process involved. The crystal grows by precipitation of the evaporant from the droplet. The applicability of this process has been demonstrated in the growth of a silicon crystal on a Si substrate using Au as a catalyst or impurity.[25] This process has the potential of wide application to the growth of single-crystal films. Controlled growth can be obtained through appropriate use of impurities in films on substrate surfaces, and on single-crystal seeds of many materials.

(3) Recrystallization Process In this case, a thin polycrystalline layer is formed on insulating substrates such as sapphire. Recrystallization is then effected by forming a thin liquid-alloy zone in the film and driving this zone across the film using a temperature gradient. Considerable progress has been made in the epitaxial deposition of silicon and germanium on sapphire by the recrystallization process.[26] Recrystallization can also be induced by exposure to electron bombardment and to laser radiation.[27] Silicon and germanium films with large single-crystal areas have been prepared by these techniques.

(4) Ultrathin-alloy-zone Crystallization This is a novel technique, which offers great possibilities in the growth of single-crystal films on amorphous substrates as reported by Filby and Nielson.[28] This technique involves the use of an extremely thin alloy layer on the substrate surface, and this must be differentiated from thin-alloy-zone crystallization (AZC) such as the vapor-liquid-solid process described above. In the AZC process, the depositing material diffuses through the alloy layer to the substrate surface, while in the ultrathin-alloy-zone crystallization process, the alloy layer is too thin for diffusion to be an important factor, and the function of the layer is apparently to change the energetics of the nucleation and growth processes. Filby and Nielson[29] have also used this technique for the epitaxial deposition of silicon on silicon. Their results show that isoepitaxy of silicon can be obtained at 850, 800, 500, 400, and 300°C, using silver, copper, tin, gold, and indium alloys, respectively.

b. Conditions for Epitaxial-film Formation

As indicated above, a small misfit is important but not an essential criterion for the occurrence of epitaxy. In epitaxial growth, a well-defined crystal plane of the deposit grows parallel to the exposed plane of the substrate. If the substrate surface is faceted, the overgrowth will grow parallel to the facets. A match of symmetry at the film-substrate interface is an important factor in determining what orientation is preferred, although sometimes symmetries are not matched; e.g., a threefold (111) plane can grow on a fourfold (100) substrate plane. In many cases the observed orientation is not that which corresponds to the best possible geometrical fitting along one lattice row at the film-substrate interface.

This is the general situation with regard to the fit between the crystal lattices of the film and the substrate. In order to obtain epitaxy, it is necessary to control deposition parameters as well as the substrate parameters. While the deposition parameters depend on the specific deposition technique used, the nature of the substrate remains common. The important substrate and deposition parameters that must be optimized for the growth of epitaxial single-crystal films will be considered in subsequent sections.

(1) Nature of Substrate It is necessary to use single-crystalline substrates for the epitaxial-film growth. However, some progress toward single-crystal growth on amorphous substrates has been reported.[28] The choice of substrates is dictated by a number of factors, such as (1) compatibility of crystal structure, i.e., crystal symmetry, crystal orientation, and lattice parameter; (2) freedom from surface strain and cleavage steps; (3) chemical inertness at deposition temperature; and (4) compatibility of thermal-expansion coefficients between the film material and the substrate (in case of heteroepitaxy). There are, however, exceptions to most of these conditions; e.g., (1) fcc metals can be grown on mica, and (2) cleavage steps enhance nucleation.[20]

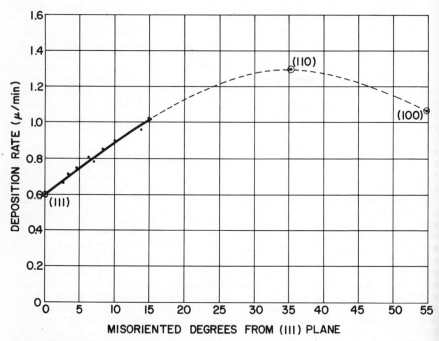

Fig. 1 Deposition rate vs. substrate misorientation for silicon vapor-grown on silicon substrates at 1200°C (Tung[35]). (*Reproduced with permission of Journal of Electrochemical Society.*)

The substrate preparation can have an important influence on epitaxial temperature and film orientation. Important factors to be considered include substrate orientation, surface topography, and surface cleanliness. Surface topography such as surface steps or microfacets can have a direct influence on film orientation. However, it has not been conclusively shown that they have a determining role in epitaxy. Bassett et al.[30] have shown that there is no difference in orientation between gold nuclei formed on steps on rock salt and those formed in between the steps. Epitaxial temperatures vary, depending upon whether the substrate is cleaved in air or in ultrahigh vacuum. This has been demonstrated by Sella and Trillat[31] and Ino et al.[32] They have observed that the epitaxy for common elements such as gold, silver, aluminum, nickel, iron, copper, and germanium on vacuum-cleaved NaCl substrates occurs at temperatures considerably lower than on air-cleaved substrates.   Cleavage in ultrahigh vacuum can produce unexpected deterioration in the orientation of Au, Ag, and Cu films on (100) NaCl surfaces.[33]   These metals produce imperfect (111) orientations on surfaces cleaved at about 10^{-9} Torr.   On the other hand, aluminum produces a perfect (111) epitaxial structure at substrate temperatures as low as 100°C.   It is, however, expected that the epitaxial studies under clean vacuum environment will ensure reproducible film structures and will eventually lead to a better interpretation of the mechanism of epitaxial-film growth.

The influence of substrate orientation on the epitaxial-film orientation can best be revealed by the use of a spherical single-crystal substrate.   Lawless and Mitchell[34] have shown the details of the crystallographic orientations of copper oxide and copper bromide around a single-crystal copper sphere.   It must, however, be recognized that the electropolished surface is not flat on an atomic scale, and that it can have microscopic curvature that may explain the observed orientations.   The influence of substrate orientation on the quality of vapor-grown silicon epitaxial films has been studied in terms of deposition rate and surface topography.   Tung[35] observes that the deposition rate is the lowest on the (111) surface and increases as the degree of misorientation from the ⟨111⟩ direction increases, as illustrated in Fig. 1.   The frequently observed surface pyramidal defects are found to be a function of deposition rate and substrate orientation.   They are not observed for low deposition rates, and when the substrate is misoriented from the ⟨111⟩ direction.

(2) Substrate Temperature   In discussing substrate temperature together with deposition rate, it should be mentioned that an Arrhenius-type relationship is expected on quite general grounds (and most nucleation theories predict this) which connects the logarithm of deposition rate with the

reciprocal substrate temperature (see Fig. 11, for example).   What really has to be explained in many cases is the absence of such a relationship and of a critical epitaxial temperature for a given deposition rate.

Bruck[36] has shown that there is a critical substrate temperature below which perfect epitaxy cannot occur.   He has demonstrated a systematic evidence for this effect for fcc metals on rock-salt substrates.   However, it is now generally recognized that the epitaxial temperature varies widely for different film-substrate combinations, and is influenced by other variables such as substrate surface preparation, deposition rate, and surface contamination.   The substrate temperature together with the deposition rate is sometimes involved in determining whether a film is single-crystalline, polycrystalline, or amorphous.   This has been shown by Krikorian and Sneed[37,38] for germanium films sputtered on single-crystal Ge sub-

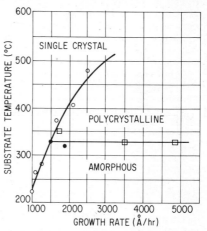

Fig. 2   Effect of growth rate and substrate temperature on the film structure of Ge sputtered on Ge (111) (Krikorian and Sneed[38]).   *(Reproduced with permission of Journal of Applied Physics.)*

strates.   This is illustrated in Fig. 2, which is a diagram relating the film structure to both the substrate temperature and the deposition rate.

Increasing the substrate temperature cleans the substrate surface that results from the desorption of contaminants.   As a result, the epitaxial temperature is lowered, as first demonstrated by Shirai.[39]   Increasing the temperature provides the activation energy required for the depositing atoms to take up positions for alignment with the substrate lattice.   This also increases the surface and volume diffusion for accommodating the misfits which develop when neighboring nuclei grow together.   Interdiffusion between adjacent layers increases with increasing temperature, and this effect may cause alloy films that may have fractionated during evaporation to become homogenized. In general, we see that increasing the substrate temperature improves epitaxy for a variety of film-substrate combinations, and for various methods of film formation. Since the film growth is influenced by many deposition parameters, it is difficult to define a precise value for the epitaxial temperature.

(3) Deposition Rate and Film Thickness The temperature at which epitaxy occurs is dependent on the deposition rate. It decreases with increasing rate. This has been shown to be true for semiconductor films of germanium and silicon formed by vacuum evaporation.[37,40] Some important factors must also be taken into consideration. A higher deposition rate can lead to a rise in substrate temperature, thus favoring epitaxy, and can have effect on the coalescence of nuclei, as shown by Matthews.[22] As agglomeration increases, the film becomes electrically continuous at a higher average thickness called the critical thickness. The critical thickness increases with the deposition rate but approaches a constant value at sufficiently high rates when high supersaturation produces more stable nuclei because of interaction between vapor atoms.[41] At a slower rate, the gas content of the film may be thickness-dependent, and can result in a higher ratio of contamination to the film, which could favor epitaxy.[42,43]

In epitaxial deposition, an orientation is usually maintained up to a certain thickness beyond which further deposition leads to poor orientation and eventually random orientation.[44,45] For example, Spielmann[44] has reported that single-crystal films of β-SiC are limited to 30 μ when deposited on β-SiC substrates at 1800°C. However, little systematic evidence is available. The loss of good orientation is due to the development of twins, stacking faults, and other defects, which are not generally present in the initial stages. The orientation can, however, improve with increasing film thickness, as has been shown in the growth of silver on mica.[46,47] This improvement in orientation is ascribed to reorientation of the initial nuclei.

The type and degree of orientation in a film depend upon the film thickness at which epitaxy is characterized. Changes in the crystal structure and crystallographic orientation occur during film growth, details of which will be found in Sec. 3.

(4) Contamination The presence of impurities in a vacuum system is a serious problem in epitaxial-film studies. The main sources of contamination are contaminants brought into the system on the substrate surface, residual contaminant gases or vapors in the vacuum system, and gases released from the source material or the heater material. An important effect of contamination is the loss of adequate adherence of the deposited film. Contamination in the form of gaseous impurities can be minimized or even eliminated by the use of ultrahigh-vacuum systems. Contamination of a clean surface by exposure to air has been shown to influence the orientation of an epitaxial film. The effect of contamination is to increase the number of initial nuclei, which causes them to coalesce early in film growth, thereby influencing the orientation of a continuous film.[22,43,48,49] Minute amounts of any contamination may influence the nucleation behavior of thin films, but it is difficult to study this aspect because of the difficulty in detecting and identifying minute amounts of contamination.

(5) Electric Field A dc electric field in the plane of the substrate surface can induce coalescence at an early stage of film growth, and the film thus becomes electrically conducting at a small average thickness. Reorientation and recrystallization of the nuclei are also affected, which in turn affects film structure. The work of Chopra and Khan[50] on the epitaxial growth of CdS by vacuum evaporation provides an example of this effect.

(6) Electron Bombardment Electron bombardment of the substrate surface prior to film deposition plays an important role in the growth of single-crystal films.[51] It produces the following effects which enhance single-crystalline film growth:

1. It cleans up a substrate surface by desorption of adsorbed gases.
2. It creates more defects, which in turn creates more nuclei.
3. The electric charge on the surface enhances condensation rate, which create more nuclei.
4. It induces coalescence of nuclei at an early stage of growth in the same way that an electric field does.

The deposition rate as mentioned above is a function of various parameters characteristic of a deposition technique chosen. In sputtering the deposition rate depends upon the sputtering voltage, current, gas pressure, nature of gas used, and cathode-substrate gap. The sputtering voltage, in addition to its effect on the rate, can dramatically affect the structure of a film. In electrodeposition the film properties are affected by the chemistry and physics of the electrolyte and the electrodes and by the current density. The current density controls the deposition rate and has a pronounced effect on the character of the film. The factors that influence the film structure in chemical-vapor deposition involve gas flow rates, pressure, and temperature. An increase in the concentration of the reactive gases or in substrate temperature increases the deposition rate. Impurities affect film growth by modifying the nucleation rate and by causing structural defects. For example, carbon contamination during the isoepitaxial growth of silicon may cause stacking faults and other structural defects.[52]

c. Substrate Preparation

In studies of the epitaxial growth of thin films, the substrate surface is of critical importance, since the surface plays a major role in the nucleation and growth processes. Ideally, the surface should be extremely flat, smooth, and free of crystalline defects. It should also be free of chemical impurities, such as oxide layers. The importance of the detailed structure of the substrate surface has been shown in the recent studies of Sella and Trillat,[31] Ino et al.,[32] and Matthews and Grünbaum.[43] It is therefore important to consider the different types of substrate surfaces that are usually used for the growth of single-crystalline films.

1. Cleavage face of a single crystal: This is the nearest approach to an ideally smooth surface. However, such a surface usually comprises a step structure, the region between steps having areas flat on an atomic scale.

2. Surface of an epitaxially grown film: An epitaxial film may be used as a substrate for the single-crystal-film growth. It is essential to examine the surface structure of the film, since it is usually rough on an atomic scale. A really smooth surface can, however, be prepared by this epitaxial technique. This has been demonstrated by Pashley[46] for silver films grown on mica by vacuum evaporation.

3. Mechanically polished single-crystal surface: This involves the preparation of a single-crystal surface with any desired crystallographic orientation by lapping and mechanical polishing, followed by chemical etching. Procedures for such preparation vary widely with the substrate material used. A surface prepared by this technique usually contains well-developed facets. It should be noted that the film growth occurs on such facets, and this would complicate single-crystalline growth.

In studies of the nucleation and growth of single-crystal thin films, the (100) cleavage surface of an alkali halide, especially rock salt, is by far the most commonly used substrate. The (100) cleavage surface of MgO is also used, when high epitaxial temperature is expected. Orientations other than (100) can be obtained by choosing suitable single-crystal materials. For example, the cleavage faces of CaF_2 and ZnS are (111) and (110), respectively. A wide range of orientations can be obtained by cutting a single crystal parallel to a desired crystallographic plane. After the crystal surface is cut, it is usually lapped, mechanically polished, and then chemically etched with suitable etching solutions. This procedure is usually used in the preparation of single-crystal substrates such as silicon, germanium, sapphire, and spinel.

The substrate cleanliness may be achieved in a number of ways. When a substrate surface is obtained by cleavage in air, it normally carries some debris, which can be removed by thermal treatment in vacuum. This treatment, however, produces etching of the surface. Matthews and Grünbaum[43] have reported that the use of an etched surface as a substrate does not affect crystallographic orientation in the vacuum-deposited films. The process of thermal treatment can be avoided and still a clear, smooth surface can be obtained by cleaving the substrate in ultrahigh vacuum. Another technique that reduces damage and steps on a cleavage surface involves polishing of the crystal surface with moist polishing cloth or paper, followed by a quick rinse in water and methanol. Brockway and Marcus[53] have demonstrated the usefulness of this technique in epitaxial studies of copper on rock salt. Methods of preparing clean surfaces in vacuum have been reviewed by Roberts[54] and Moll.[55] Their methods, in addition to thermal treatment, include ion bombardment and vapor-phase etching.

d. Lattice-parameter Measurements and Pseudomorphism

As mentioned in Sec. 2a, pseudomorphism is a necessary step during epitaxial growth within the groundwork of the Frank and van der Merwe theory.[11] The term pseudomorphism may be defined as either:

1. A nonequilibrium crystal structure in a thin overgrowth whose lattice spacing is similar to that of the substrate
2. A biaxial state of stress in the film

Both of these have been observed experimentally. They will be discussed here and also in Secs. 4 and 5. Evidence for the concept of pseudomorphism has been obtained from lattice-parameter measurements on electron-diffraction patterns. In this section we will first describe the methods used for lattice-parameter measurements and then discuss the experimental confirmation of the concept of pseudomorphism.

(1) Lattice-parameter Measurements Lattice parameters of film materials are usually measured by electron diffraction and x-ray diffraction. The electron-diffraction technique utilizes reflection patterns, and also transmission patterns for thin films, usually less than 1,000 Å. The accuracy of measurements, in most cases, is limited by the sharpness of the diffraction patterns. This is especially serious for reflection measurements. Diffraction patterns may be broadened or elongated such that the reference points may become ill-defined.

The measurements from electron-diffraction patterns require a reference specimen. In situations where the deposited film is very thin, the substrate can be used as a reference, provided its lattice parameters are accurately known. This technique is applicable when diffraction patterns from both the film and the substrate can be obtained for measurements. For accurate measurements, a suitable standard material whose lattice parameter is accurately known is used. For example, thallium chloride has often been used.[56-59]

(2) Concept of Pseudomorphism A critical discussion by Pashley[1] has shown that the concept of pseudomorphism has no valid experimental confirmation, since various investigators[60-69] have reported evidence for and against this concept. Recently Jones[70] has shown by field-emission microscopy that the first few atomic layers of copper deposited on tungsten in ultrahigh vacuum are pseudomorphic. However, this is in contradiction to Schlier and Farnsworth,[71] who find no evidence of such a layer by the low-energy electron-diffraction technique.

Very recently Jesser and Matthews[72-75] have studied the growth of single-crystal films of iron and cobalt on copper (001) surfaces, and of chromium on nickel (001) surfaces at room temperature and at elevated temperatures in ultrahigh vacuum. They have produced conclusive evidence for the pseudomorphic growth of iron, cobalt, and chromium. Details of the pseudomorphic growth can be found in Secs. 4 and 5.

Although reports on the evidence for the concept of pseudomorphism are few, there is ample evidence to show that the film material becomes strained so that its lattice parameters do not exactly match those of the corresponding unstrained bulk material. For example, measurements by Newman and Pashley[76] of the lattice parameters of vacuum-deposited copper and chemically grown silver bromide films on silver (111) substrates show that the spacings of copper and AgBr films are about 0.75 and 0.50% lower than their corresponding bulk values. These results indicate that the strain in the film modified the film lattice parameter toward that of the substrate. Freedman[77] has demonstrated that nickel films deposited on rock salt are in a condition of high compressive strain. The strain is large enough to produce a detectable symmetry modification from cubic to tetragonal. The lattice parameters measured normal and parallel to the substrate surface are respectively 3.546 and 3.500 Å, compared with 3.524 Å for the unstrained bulk material. Similar results have also been observed by Borie et al.[78] on the growth of Cu_2O on Cu (110) surfaces. He has shown from x-ray diffraction measurements that the strain in the Cu_2O film changes its structure from cubic to orthorhombic. The strain is greatest at the oxide-substrate interface and

ecreases with film thickness, becoming almost zero at the oxide surface. The strain
s about 2.5% compared with 18% involved in the pseudomorphic film. Collins and
Ieavens[79] have observed that the FeO lattice is strained to the spacing of Fe in the
lane of the deposit. This gives a contraction of 5 to 6 percent from the normal
pacing in the plane of the film. No evidence of strain in nickel and cobalt films is
btained.

Van der Merwe[80,81] has shown theoretically that the lowest-energy state is one in
hich the deposited film contains a homogeneous lateral strain, so that the difference
i the lattice spacings of the film and the substrate is reduced from what would be
etermined from bulk values. This theoretical treatment is in good agreement with
he experimental observations discussed above.

. Examples of Single-crystal Films

In previous sections we have outlined the theoretical considerations of single-
rystal growth processes and the various parameters that influence film formation.
omprehensive summaries of the epitaxial single-crystalline films have been given by
'ashley.[1,24] Many new observations have since then been reported in the literature.
To attempt to present a comprehensive list of the observed cases of single-crystal
lms has been reported. In this section we will illustrate the structural properties of
ingle-crystal films of materials that are of vital interest for physical investigations.

The film materials of interest may be divided into four broad classes: (1) metal films,
?) alloy films, (3) semiconductor films, and (4) insulating films. Important examples
1 each class will be presented, with special reference to highlighting some of the growth
ictors that critically affect the structural properties of epitaxial single-crystal films.

(1) Metal Films (a) *Gold and Silver.* The formation of single-crystal metal films
n nonmetallic substrates, especially alkali halides, has been extensively studied by
ectron diffraction and electron microscopy. A review of work prior to 1956 has been
ven by Pashley.[1] Numerous investigations have since then been carried out on the
ngle-crystalline growth of metal films by vacuum evaporation and cathode sputter-
g. Among the metals, gold, silver, copper, and nickel have been the subject of
tense investigation. The investigations are of scientific interest because of their
iportance to theories of film nucleation.[16-20]

One of the techniques for growing single-crystal metal films consists of slowly con-
ensing the metal in vacuum onto rock salt and other alkali-halide substrates at
evated temperatures. In the majority of investigations,[1,82-84] the (100) cleavage
arfaces have been prepared by cleavage in air, and the pressure during film growth
as been maintained at 10^{-6} Torr or above. These conditions are such that both the
ibstrate surface and the deposited film must have been badly contaminated.[85] In
rlier studies Pashley[1] has suggested the importance of an epitaxial temperature.
ypical values quoted by him are 250 to 300°C for silver on rock salt or mica and 400°C
r gold on rock salt. Later Bassett and Pashley[86] showed that gold deposited onto
ver substrates at 250 to 300°C is single-crystalline and continuous, even down to a
ickness of 100 Å or less. Transmission electron diffraction from the films shows
at the films grow in (111) orientation on (111) Ag/mica and in (100) orientation on
00) Ag/rock salt. The (100) films give rise to irrational reflections on the diffraction
itterns, normally attributed to twinning.

The sequence of growth of the Au (111) on a Ag(111)/mica substrate at 270°C as
served by Bassett and Pashley is illustrated in Fig. 3a to d. At the earliest observ-
le stage, when the deposit thickness is 5 to 10 Å, a random dispersion of nuclei with
density of $\sim 10^{11}/cm^2$ is observed. Growth proceeds largely by forming together of
:isting nuclei, which take very irregular forms. This process results in a network
'ig. 3c) which gradually fills in, until a uniform film substantially free of holes occurs
a thickness of 50 to 100 Å.

The epitaxial growth of silver on epitaxially grown (001) fcc metal substrates has
een reported by Shirai et al.[87] and Shirai and Fukada.[21] They observe (001) or
11) oriented growth, the occurrence of which is found to depend mainly on the
ercentage misfit. Sloope and Tiller[84] have made a comprehensive study of the effects
deposition conditions on the structural characteristics of single-crystal Ag films

vacuum-deposited onto NaCl substrates. The results show that the Ag films grow in (100) parallel orientation onto rock salt, and that an epitaxial temperature as such does not exist but the minimum temperature of deposition is rate-dependent. Figure is a plot, indicating three regions: continuous nonsingle-crystalline, single-crystalline and highly porous nonsingle-crystalline. In the region of low rate–low temperature up through high rate–high temperature, or at some proper compromise between the two, a single-crystalline film forms. The films formed in the region at or below the lower temperature limit are strongly (100) oriented and all contain the doubly positioned (111) orientation, polycrystallinity, and twins.

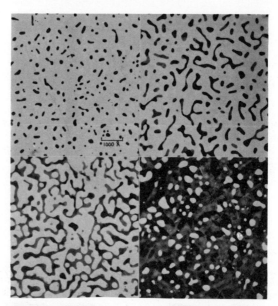

Fig. 3 Sequence illustrating the growth of a (111) gold film on a Ag/mica substrate 270°C (Bassett and Pashley[86]). (*Reproduced with permission of Journal of Institute Metals.*)

The growth of single-crystal films of gold and silver by cathodic sputtering on NaCl and mica substrates has been reported by Chopra et al.,[88] Chopra and Randlett and Campbell and Stirland.[90] Chopra et al.[88] have grown single-crystal Au films air-cleaved mica above 280°C. The resistivities of the films are very close to the bulk value for all thicknesses greater than a limiting thickness at which the film becomes continuous (~250 Å). Epitaxy of both Au and Ag at room temperature on a cleaved NaCl has been observed when grown at low sputtering rates.[90] This result may be due to the enhanced mobility of the sputtered atoms or to the ionic cleaning of the substrate surface by the discharge. Chopra and Randlett[89] have reported the perfect epitaxial growth of low-pressure sputtered Ag films can be obtained on both rock salt and mica at liquid-nitrogen temperatures (77°K). A dependence of epitaxial growth on the sputtering voltage and hence possibly on the kinetic-energy distribution of sputtered atoms is observed.*

The epitaxial temperature can be considerably reduced if deposition is carried o onto vacuum-cleaved rather than air-cleaved substrates. Jaunet and Sella[91] and Sella and Trillat[31] have shown that this so-called epitaxial temperature for Au/NaCl can be reduced to ~200°C, and that for Ag/NaCl to room temperature, when the

* Sputtering voltage influences other things too, e.g., energetic neutrals and electron energy and density at the substrate.

[aCl substrate is vacuum-cleaved at ~10^{-6} Torr.   Ogawa and others[32,43,48,92,93] have
erformed experiments under conditions where the contamination of both the film and
he substrate is small, i.e., when deposition is made onto substrates cleaved in better
acuum (10^{-7} to 10^{-9} Torr).   They have observed that under these very clean condi-
ions, gold, silver, and copper do not grow in single-crystal form on clean NaCl and
CCl substrates, although Matthews and Grünbaum[43,48] have subsequently shown that
ingle-crystal films of Au are obtained on vacuum-cleaved substrates if they are
xposed to air after cleavage, but prior to deposition of gold onto them.   How con-
aminants present in the air aid the single-crystalline growth has been studied in

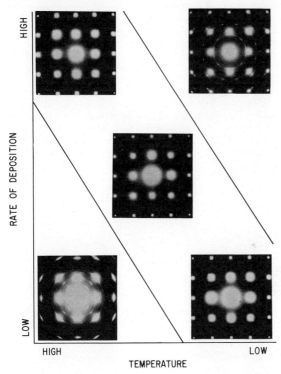

g. 4   Typical electron-diffraction patterns, showing the crystalline structure of Ag
ms on NaCl in three regions: continuous noncrystalline, single-crystalline, and highly
rous nonsingle-crystalline (Sloope and Tiller[84]).   (*Reproduced with permission of Journal
Applied Physics.*)

tail.[22,43,48,49]   The results show that the effect of air is not to improve the alignment
the initial Au nuclei but to increase the number of nuclei generated per unit area.
his increase causes nuclei to coalesce early in film growth.   The stage in film growth
which coalescence occurs influences the orientation changes that accompany coa-
scence, and thereby influences the orientation of the continuous film.   On the basis
these observations, Matthews[94,95] has suggested a technique for growing impurity-
e single-crystal films of fcc metals on clean alkali-halide substrates in ultrahigh
cuum.   The technique he has demonstrated for the growth of Au and Fe on NaCl
nsists of depositing an initial layer of ~20 Å at a very fast rate ($\gtrsim$2,000 Å s^{-1})
lowed by slow growth to a final thickness of several hundred angstroms, i.e., until
continuous film is obtained.   The high deposition rate is used to obtain a high
rticle density and consequently early coalescence, which is considered by Matthews
be essential for the formation of a single-crystal film.   The deposition immediately

after cleaving ensures that the surface does not get contaminated by the residual gas
That impurities are necessary in order to obtain single-crystal film by epitaxy has als
been reported by Harsdorff and others.[42,96-99]

More recently, the above-mentioned results have been disputed. Bauer et al.[1¢
have shown that neither a high density of nuclei nor an air-contaminated surface no
reactive residual gases are necessary for the growth of a single-crystal film. Thei
results show that although "impurity-free" Au single-crystal films (at least on NaC
KCl, and KI, cleaved in high vacuum) may be reasonably free from impurities in th
bulk, there are indications that the surfaces are not clean. Electron-diffractio
studies indicate development of a superstructure, which is associated with an impurit\
gold compound. This indicates that the single-crystal-film growth depends not onl
on the history of the surface and the residual gas pressure but also on the interfac
reactions between film and substrate as demonstrated by low-energy electron di
fraction. More recently these results have been disputed by Palmberg et al.[101]

Electron bombardment of the substrate surface can play an important role in th
formation of single-crystal films.[51,102-104] Stirland[51] has shown that single-cryst;
films of Au in parallel orientation can be produced on vacuum-cleaved NaCl substrate
at 150°C if electron-bombarded prior to evaporation. It is found that a thin bon
barded film contains a larger number of smaller islands than a nonbombarded film
The orientation of the electron-bombarded film is always (100)[110] Au ‖ (100)[11(
NaCl, whereas the nonbombarded films contain, in addition to the parallel orientatio\
several preferred orientations such as (111)[110] Au ‖ (100)[110] NaCl and (111)[11\
Au ‖ (100)[110] NaCl. These orientations have also been discussed by Matthews

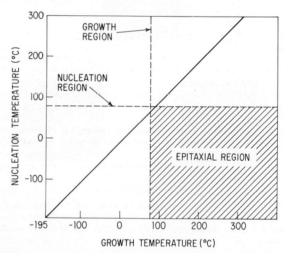

Fig. 5 Nucleation and growth temperatures resulting in epitaxial growth of silver «
electron-bombarded KCl (Palmberg et al.[101,102]). *(Reproduced with permission of Journ\
of Applied Physics and Applied Physics Letters.)*

and Matthews and Grünbaum.[48] More recently, Palmberg et al.[101,102] have show\
that the formation of single-crystal films of Au and Ag depends critically on exposu
of the substrate surface to an electron beam before deposition. Single-crystal fil\
of Au and Ag in parallel orientation are obtained on bombarded KCl surfaces cleav\
in ultrahigh vacuum, i.e., in the absence of contaminating gases. The temperatu
ranges for nucleation and for growth required for epitaxy by deposition are indicat
in Fig. 5. It can be seen that epitaxy by deposition at a single temperature is achiev
for a narrow temperature range corresponding to the section of the 45° line includ
in the epitaxial region. For growth temperatures below 125°C, the orientation of t

film deteriorates. Film of optimum quality is obtained when a total electron-bombardment treatment of $\sim10^{16}$ electrons/cm² is used.

The role of applied electric field in enhancing epitaxial growth has been reported by Chopra.[104] He has shown that an electric field of ~100 V cm⁻¹ applied across a depositing Ag film provides parallel orientation in a freshly cleaved NaCl substrate at 200°C, whereas a similar film without the electric field shows a polycrystalline character.

(b) *Copper.* Studies of the growth and structure of copper films have been less systematic and less numerous than those of gold and silver described above. Bruck[36] has demonstrated that single-crystal copper films can be produced by epitaxy on rock salt. He observes that there is a minimum temperature (300°C) below which epitaxy will not take place. Brockway and Marcus[53] have reported the growth and microstructure of single-crystal copper films formed by vacuum condensation on rock salt at 330°C. They observe that the unannealed film often exhibits the occurrence of (111) twins and stacking faults, which disappear when annealed at high temperatures. The annealed films are more continuous without any small island structure as in the unannealed films. The production of single-crystal films of copper at temperatures as low as −40°C has been described by Yelon and Hoffman.[105] The method consists of depositing a very thin layer of copper on NaCl at 350°C, followed by deposition at low temperatures. Films obtained by this method have single-crystal regions of 1 mm diameter and larger. The epitaxial relationship between the copper and NaCl is completely parallel. This technique appears to be applicable to any of the metals that can be grown by epitaxy on rock salt: Ni, Pd, Al, Ag, Au, Fe, and Co.[79]

The epitaxy of copper on single-crystal substrates other than rock salt has also been studied.[71,106,107] Schlier and Farnsworth[71] have shown that copper deposited by vacuum evaporation onto a clean (0001) surface of titanium forms at nucleation centers in the form of discrete oriented crystallites with the orientation relationship (111)[1$\bar{1}$0] Cu ∥ (0001)[1$\bar{1}$20] Ti. No variation of lattice spacing with thickness is observed by low-energy electron-diffraction measurements. Taylor[106] has studied the growth of very thin single-crystal films of copper on a single-crystal surface of tungsten under ultrahigh-vacuum conditions by low-energy electron diffraction. This investigation shows that evaporation onto clean tungsten at room temperature results in partial alloying and then in the formation of epitaxial (111) copper. Moderate heating results in the diffusion of copper onto the tungsten and in the subsequent alloy formation. A marked improvement in epitaxy is observed when a certain amount of oxygen is chemisorbed or physisorbed onto the tungsten. Krause[107] has studied the isoepitaxy of copper in situ by electron diffraction under high-vacuum conditions, using spherically shaped Cu crystals exhibiting all crystallographic planes on the sample. These results show that on the (001), (110), and (111) Cu planes, the orientation of the epitaxial Cu films formed at room temperature and at 500°C is a twin orientation. For (001) and (110) substrates, all possible (111) twin axes are present, while for the (111) planes, only one twin orientation occurs.

(c) *Aluminum.* Although aluminum is an important metal, not much work on the epitaxial deposition of this material has been reported in the literature. Sella and Trillat[31] have studied the epitaxial growth of Al on (100) NaCl substrates cleaved in air and in vacuum, and have observed that Al grows in (100) orientation on vacuum-cleaved NaCl above 440°C. This temperature is, however, lowered to 300°C when deposition is made onto air-cleaved surfaces. The growth of (111) orientation above 330°C is also observed.

Ogawa et al.[92] have studied the epitaxial deposition of Al on cleaved NaCl and KCl substrates in ultrahigh vacuum. They have observed (111) epitaxial structure on NaCl heated to 80°C, while the KCl substrate favors the (100) orientation. Kunz et al.[103] have also reported the epitaxial deposition of Al on NaCl in ultrahigh vacuum. In very thin films no condensation takes place on the vacuum-cleaved surface. In the thin film deposited at 270°C, the (211) orientation is observed in addition to (100) and (111) orientations. With increasing film thickness, the (211) and (100) orientations become weaker, resulting in the (111) orientation in the final film. On air-cleaved substrates, the (100) orientation predominates.

Epitaxial deposition of Al on Si substrates shows alloy formation in very thin layers. A number of alloy structures have been observed by low-energy electron diffraction (see Sec. 3b).

(d) *Magnetic Metals.* Analysis of magnetic phenomena can be greatly simplified by the use of epitaxial films. Epitaxial single-crystal films provide media for exploring magnetic phenomena, such as thickness dependence of magnetization and Curie temperature,[108] switching behavior and surface pinning effects in magnetic fields, anisotropy of magneto-optical properties,[109] magnetic domain structures,[110-113] and film structure and residual strain on magnetic properties.[77,114] Requirements for the growth of structurally continuous single-crystal magnetic-metal films are essentially similar to those described for metal films. Among the magnetic metals, nickel, cobalt, and iron have been investigated in detail.

Collins and Heavens[79] and Heavens et al.[115] have extensively studied the monocrystalline growth of thin films of nickel, cobalt, and iron onto (100), (110), and (111) crystallographic faces of rock salt. The epitaxial temperatures for Ni, Co, and Fe are reported to be above 330, 480, and 470°C, respectively. The orientation observed is governed mainly by a tendency to continue the substrate structure across the boundary, so that the normal coordination of the atoms at the interface is preserved. The lattice misfit is of less importance than the coordination condition. One or more complete three-dimensional orientations of the deposited film are observed. Twins also form during film growth. Epitaxial growth of these metals on MgO substrates in high vacuum (1×10^{-5} Torr) has been reported by Sato et al.[113,116] The mode of epitaxy for Fe is observed to be (001) Fe ∥ (001) MgO and [100] Fe ∥ [110] MgO. Single-crystalline Ni films in parallel orientation can also be grown on MgO and NaCl substrates at about 450°C. The growth of Co on MgO results in the formation of a fcc form of cobalt instead of the usual hexagonal close-packed structure. This mode of epitaxy is parallel on the (100) face of MgO. Similar results have also been obtained by Honma and Wayman.[117] They observe that the Co films exhibit the best epitaxy when NaCl is vacuum-cleaved at 350°C rather than air-cleaved. The epitaxial films on KCl and NaCl are fcc but contain numerous microtwins and stacking faults. These microtwins can be completely eliminated by annealing at 800°C. Heavens et al.[118] have shown that single-orientation (100) Fc films may be readily grown on KCl and KI at 330°C and at a deposition rate of the order 300 Å min⁻¹. Vacuum cleaving of the substrate results in epitaxial growth at significantly lower temperatures i.e., at 260°C. Deposition during cleaving of the crystal ensures a minimum of interference from residual gas atoms and exposes the deposit material more strongly to the substrate forces.

Freedman[77] has studied the influence of film structure and residual strain on magnetic properties. He has shown that continuous epitaxial films of nickel possess a high compressive strain when grown on the NaCl substrate. The strain is high enough to cause a measurable change in crystalline transformation from cubic to tetragonal. The observed strain can be explained in terms of the difference in thermal expansion coefficient and growth temperature.

(2) Alloy Films (a) *Single-crystalline Growth.* The growth of single-crystal films of alloys including magnetic alloys has been achieved on NaCl and other monocrystalline substrates by vacuum evaporation and cathodic sputtering. Among the alloys studied, thin films of nickel-iron, spinel ferrites, gold-nickel, gold-aluminum, and copper-gold have been of interest, especially for fundamental investigations. Burbank and Heidenreich[119] have reported the growth of thin films of Ni-Fe alloy having a composition near 80:20 (Permalloy) and in the range of 100 to 800 Å thick. The films have been prepared by subliming a Ni-Fe alloy under carefully controlled conditions onto fresh NaCl (100) cleavage surfaces at about 320°C. Selected area diffraction patterns indicate a gross approximation to a single crystal with a cube face in parallel orientation to the NaCl substrate. Thin films in the 100- to 200-Å range consist of discrete particles, three-dimensional in character, with rather irregular shapes. The shapes are suggestive of an agglomeration of hemispheres of various sizes which have coalesced, side by side in one place. Films in the 400- to 800-Å thickness range appear to be made up of particles that tend to be equidimensional

ranging from 300 to 700 Å in diameter.   The films indicate the formation of a micro-twinned layer in between the substrate surface and the single-crystal alloy film.   The microtwinned layer has been interpreted to serve the purpose apparently of accom-modating the cubic alloy to the cubic substrate with minimum strain energy.[119]

Schoening and Baltz[120] have reported similar experiments but in addition have obtained more perfect films by annealing of the partially monocrystalline films in the electron beam.   Annealing results in the formation of (100) orientation containing a high percentage of (111) twins.   The twins give a fringe pattern which could be explained in terms of twin platelets in the thickness range 100 to 400 Å, lying parallel to the {111} planes of the parent crystal.   Alessandrini[121] has identified extra "anom-alous" reflections observed when single-crystal Ni-Fe films grown on cleaved (100) NaCl are annealed at temperatures between 300 and 700°C.   She has shown that all the forbidden reflections observed in the films can be explained by double diffraction and do not appear as a result of a hexagonal-type packing, ordering of the alloy, or an early stage of impurity precipitation in the film matrix.

Single-crystal growth of Ni-Fe alloy films by electron bombardment has been reported by Heavens et al.[115]   This has the advantage that there is no possibility of contamination by the supporting metal.   The Ni-Fe films grow epitaxially on the various crystallographic faces of copper at a deposition temperature of 330°C.   They show considerably less twinning than the films grown directly on rock salt.

Single-crystal Ni-Fe alloy films can also be grown by vacuum evaporation upon (100) and (111) CaF_2 substrates at above 400°C, as has been shown by Verderber and Kostyk.[122]   Differently oriented films can be obtained, depending upon the orienta-tion of the substrate and the temperature of the substrate.   Most of the single-crystal formation occurs in the first 500 Å.   The coercive force in the epitaxial films appears to be dependent on the crystallographic orientation of the deposit-substrate interface.

The growth of large-area transparent single-crystal films of spinel ferrites by chemical-vapor deposition has been reported.[123-125]   Recently Gambino[126] has reported a method for growing ferrite single-crystal films with controlled nucleation on single-crystal MgO substrates by slowly evaporating Na_2CO_3 from a thin layer of Na_2CO_3-ferrite melt.   The best films are formed by using the lowest temperatures that would give complete melting of the Na_2CO_3-ferrite mixture, usually 1100 to 1250°C.   The lithium ferrites grow well at temperatures as low as 1000°C.   Nickel ferrite is the most difficult to grow, requir-ing high temperatures and a low concen-tration in the melt.   The films grow with the same orientation as the MgO sub-strate.   The film surface exhibits trun-cated pyramids with (111) side faces and a (100) top face.   The lattice constants of the as-grown ferrite films are slightly higher than their bulk values.

Alloy films can conveniently be grown on single-crystal substrates by cathodic sputtering.   This technique, under proper deposition conditions, allows formation of the alloy film with the same composition as the starting alloy material.   Khan and Francombe[127,128] have shown that single-phase Au-15 atomic % Ni alloy films grow epitaxially on cleaved surfaces of rock salt and mica at 400°C.   On rock salt the alloy

Fig. 6   Electron-diffraction pattern of Au-15 atomic % Ni alloy film (thickness 700 Å) grown epitaxially on (001) NaCl at 400°C (Khan and Francombe[128]). (*Reproduced with permission of Journal of Applied Physics.*)

is observed to grow with a (100) parallel orientation, while on mica it grows with (111) alloy $\parallel$ (001) mica and [$2\overline{1}\overline{1}$] alloy $\parallel$ [100] mica, and [$1\overline{1}0$] alloy $\parallel$ [100] mica.   The latter orientation is very weak.   Figure 6 shows an electron-diffraction pattern from a (100) Au-Ni alloy film grown on NaCl by sputtering and then stripped from it for observation. The pattern shows the film to have a single phase with the fcc structure. The pattern exhibits evidence of extra reflections consistent with (111) twinning. Annealing at 400°C for 1 h is sufficient to remove most of these twins. The lattice parameter of the film is the same as that for the bulk-alloy cathode. This result is in agreement with Fukano,[129] who prepared Au-Ni alloy films by simultaneous deposition of the constituent metals of the alloy.

The formation of epitaxial alloy films by codeposition of the constituent elements of the alloy onto single-crystal substrates has also been achieved. Sato and Toth[130] have prepared single-crystal thin films of CuAu by successive evaporation of Au and Cu onto NaCl substrate at 400 and 350°C, respectively. The film is annealed at 350°C for 1 h for homogenization. The resulting films exhibit (100) parallel orientation onto the NaCl substrate. Similar studies have also been made by Noreika and Francombe,[131] who prepared epitaxial films with compositions ranging from pure Au to Al on cleaved KCl substrates at 375°C. Phases such as AuAl, Au$_2$Al, Au$_5$Al$_2$, and Au$_4$Al have been examined, and the epitaxial relationships between these phases in the Au-Al system determined. This investigation shows that thin epitaxial films of binary compounds can be easily formed by coevaporation of the elements, and that their crystal symmetry and mutual orientation relationships can be established by electron diffraction.

(b) *Ordering and Precipitation Effects.* Single-crystal thin films provide a suitable medium for investigating the formation of superlattice structures and precipitation effects in alloys. These effects are a result of atomic diffusion in alloy structures. It will be of interest to consider some examples of these effects, demonstrating the influence of growth conditions on the development of atomic ordering in thin alloy films.

Sato and Toth[130] have extensively studied the formation of superlattice structures in the Cu-Au alloy system. Superlattices with the composition of Cu$_3$Au, CuAu, and CuAu$_3$ are known to occur in this alloy system.[132] CuAuII, which is a long-period structure, is a modification of the superlattice formed near the stoichiometric composition CuAu. The alloy at this composition is disordered above 410°C and has a fcc structure. CuAuII is stable in the temperature range between 380 and 410°C, i.e. between the ordered and disordered phases.

Fukano[129] has prepared epitaxial Au-20 atomic % Ni films on rock salt at 430°C and homogenized them by annealing at and then quenching from 700°C. When the films are subsequently annealed at lower temperatures, a nickel-rich phase forms in parallel orientation with the matrix. The precipitation is preceded by the formation of a modulated structure. No superlattice structures have been observed.

In studying ordering and precipitation effects in thin alloy films, Khan and Francombe[128] have observed that these effects are sensitive to the conditions of film formation. Single-crystal thin films (600 to 700 Å thick) of Au-15 atomic % Ni alloy have been deposited by cathodic sputtering onto (100) cleavage faces of rock salt at about 400°C. The films exhibit slight (111) twinning, which can be eliminated by thermal treatment at 450°C. On annealing at and above 500°C, pre-precipitation effects typified by local enrichment of the alloy matrix in Ni, are observed in the electron diffraction patterns. These are associated with the formation, in succession, of the superlattice structures NiAu$_3$ and NiAu, leading to the precipitation of Ni-rich alloy at temperatures above 600°C. The development of these effects is illustrated in Fig. 7. These superlattice structures are isostructural with the cubic CuAu$_3$ and the tetragonal CuAu superlattices.[130] The discrepancy between these results and those reported by Fukano[129] may be ascribed to the conditions of alloy film formation. Since Fukano's films are prepared by vacuum evaporation and homogenized at 700°C subsequent annealing at lower temperatures (~500°C) would be unlikely to promote atom diffusion of the type needed for complete superlattice formation. In Khan and Francombe's experiments, the films have been prepared by sputtering at low substrate temperatures (400°C). Annealing at higher temperatures leads to rapid recrystallization and atom diffusion, causing the Ni atoms to diffuse into lattice sites consistent with the NiAu$_3$ or NiAu structures. The tendency to precipitation would explain the observed transformation: disordered alloy → NiAu₃ → NiAu → Ni-rich precipitate. It should be mentioned here that this is a very striking and very unusual thin film result because, in general, ordering and segregation are mutually exclusive in such binary systems.

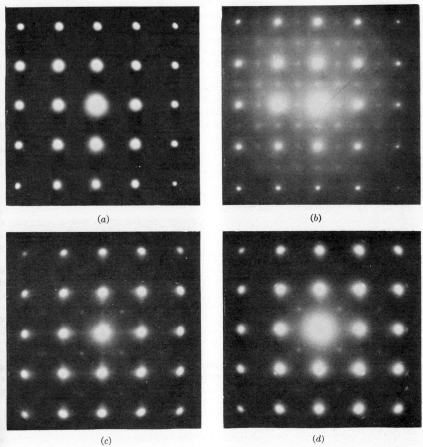

(a)   (b)

(c)   (d)

Fig. 7 Electron-diffraction patterns showing development of superlattice and precipitation effects in Au–15 atomic % Ni alloy film (thickness 700 Å) grown epitaxially by sputtering on cleaved NaCl. (a) Annealed at 400°C in vacuo. (b) Annealed at 500°C. (c) Annealed at 520°C. (d) Annealed at 650°C (outer satellite spots due to Ni-rich precipitate) (Khan and Francombe[128]). (Reproduced with permission of Journal of Applied Physics.)

(3) Semiconductor Films (a) *Silicon.* Most of the early work on the growth of single-crystal films of silicon on silicon has been carried out by vacuum evaporation under high-vacuum conditions, resulting in serious contamination of the deposited films. The contaminations arise from the vacuum ambient or from the source if this is contained in a refractory metal (W,Ta) because of the alloying of silicon with these metals at very high temperatures (1500°C). Reduction in such contaminations has been achieved by the use of electron bombardment.[133–136]

Unvala[133,134] has shown that single-crystal films of silicon up to 50 μ thick can be easily prepared on silicon substrates by the electron-bombardment technique in a vacuum of between 10^{-5} and 10^{-6} Torr. Reflection electron-diffraction patterns from the films give Kikuchi lines showing a high degree of crystalline perfection. The minimum substrate temperature T_s for good single-crystalline films is within a range of 1100 to 1250°C, the exact value depending on the rate of deposition. The relation between T_s and the deposition rate is illustrated in Fig. 8. For comparison, the dotted line shows the temperature dependence of the deposition rate, according to Theuerer,[137] using the method of hydrogen reduction of $SiCl_4$. If T_s is below 1100°C, grey polycrystalline films result. It is claimed that at the pressure used for evaporation, the substrate temperature has to be kept greater than 1100°C to keep the surface free of oxide. If T_s is greater than 1100°C, silicon dioxide formed at the surface reduces to silicon and silicon monoxide, which is volatile at temperatures above 1100°C. This is supported by the observations by Hale[136] of a brown coloration in films prepared at lower temperatures. He observes that if T_s is above 1125°C, single-crystalline growth occurs in a vacuum of 10^{-7} Torr. Best films are grown when fast deposition rates ($\frac{1}{2}$ to 1 μ min^{-1}) are used. Postnikov et al.[138] have also reported the growth of monocrystalline films of silicon on silicon at 1000°C at pressures below 2×10^{-7} Torr and at a high growth rate (20 μ h^{-1}). The isoepitaxy of silicon at temperatures as low as 950°C has also been reported by Petrin and Kurov.[139]

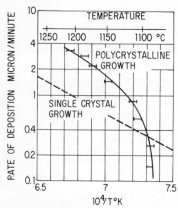

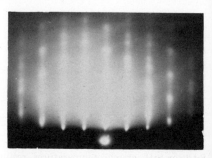

Fig. 8 Dependence of the temperature required for epitaxial growth on the deposition rate (Unvala and Booker[142]). (*Reproduced with permission of Philosophical Magazine.*)

Fig. 9 Electron diffraction pattern showing the formation of an epitaxial β-SiC film on a Si (111) surface as a result of chemical reaction with carbon-bearing organic compounds in the vacuum ambient.

To obtain good epitaxy, it is obvious that the substrate surface should be maintained clean at the high operating temperatures prior to the onset of deposition. Not only must the oxide formation be eliminated but also the formation of SiC, which develops as a result of chemical reactions with organic components of the vacuum ambient,[135,140,141] as illustrated in Fig. 9. The presence of surface contaminants is responsible for the nature of the nucleated growth observed in the initial stages.[136] Unvala and Booker[142,143] have shown that the film growth is markedly dependent on the deposition rate. Beyond a critical value of the deposition rate, no structural defects such as stacking faults occur in the Si film. For $T_s = 1200$°C and slow deposition rates (<0.5 μ min^{-1}), the films contain stacking faults, while for a fast rate, they are free of defects. Newman[144] repeated the experiments of Unvala and Booker and concluded that the detailed composition of the ambient vacuum may be important. For example, if organic vapors are present, SiC crystallites will grow on the Si substrate, and these may give rise to defects in the epitaxial Si films as discussed by Miller et al.[52]

It is now clear from the above observations that better vacuum conditions will be required to improve surface cleanliness for producing single-crystal Si films at lower substrate temperatures. Nannichi[145] has shown that single-crystal Si films up to several microns thick can be deposited by evaporation at T_s as low as 830°C in a vacuum of 10^{-8} to 10^{-9} Torr. Widmer[146] has demonstrated that Si surfaces can be cleaned sufficiently for single-crystalline silicon growth by heating in an ultrahigh vacuum (in the 10^{-10} Torr range during evaporation). He obtains an epitaxial temperature of 550°C, which is considerably lower than has been reported previously. In films formed at $T_s \geq 800$°C, no stacking faults are observed if evaporation is carried out in ultrahigh vacuum at a rate of 6 Å s^{-1}. This shows that in the work performed in high vacuum, structural defects are the results of poor vacuum conditions. At lower temperatures, the formation of single-crystal Si films is associated with structural defects such as stacking faults and dislocations. Extending the work of Widmer, Jona[147] has studied the initial stages of silicon epitaxy by low-energy electron diffraction in ultrahigh vacuum and reported epitaxial growth on unheated Si substrates. This study is confined to deposits only a few monolayers thick. Recently Thomas et al.[148] have shown that isoepitaxy of silicon occurs at substrate temperatures as low as 550°C. They put forward an interesting correlation between the occurrence of stacking faults and the existence of mixed surface structures during deposition. The experiments of Filby and Nielson[29] have shown that epitaxy occurs at temperatures as low as 300°C, using their alloy-zone crystallization technique (see Sec. 2a).

Turning now to the growth of single-crystal Si films by heteroepitaxy, these studies have been rather less systematic and less extensive than those described for the isoepitaxial growth. Unvala[134] has reported the epitaxial growth of silicon on CaF$_2$(111) substrates at 900°C, while Via and Thun[149] have observed epitaxy at a lower temperature (700°C). Other workers have studied the growth of silicon on germanium substrates and have obtained single-crystal films by vacuum sublimation at a rate of about 14 Å min^{-1} at 830°C. Newman[144] has shown that the Si films grown on germanium contain a high density of stacking faults and distortions, which probably arise from the mismatch in the lattice spacings. In addition, cracking of the films is observed because of differences in thermal-expansion coefficients.

The heteroepitaxial growth of silicon on sapphire has been investigated almost entirely by vapor deposition,[150-156] although Salama et al.[157] have reported the epitaxial growth of Si by electron bombardment at 950°C (no Si film being formed above 1100°C). Doo[150] has obtained a large single crystal of silicon on polycrystalline alumina by the application of the melting and regrowth technique. The formation of single-crystal Si films by vapor deposition,[151-154] i.e., by the hydrogen reduction of SiCl$_4$ or SiH$_4$ at high temperatures, results in four modes of epitaxy as confirmed by x-ray diffraction. These epitaxial relationships between Si and sapphire at the interface are listed in Table 1. The modes of epitaxy assume that Si substitutes for Al

TABLE 1 Epitaxial Relationships between Si and Sapphire

1. (100) Si ∥ (1$\bar{1}$02) sapphire and [110] Si ∥ [$\bar{2}$201] sapphire
2. (111) Si ∥ (0001) sapphire and [1$\bar{1}$0] Si ∥ [$\bar{1}$2$\bar{3}$0] sapphire
3. (111) Si ∥ (11$\bar{2}$0) sapphire and [1$\bar{1}$0] Si ∥ [$\bar{2}$201] sapphire
4. (111) Si ∥ (11$\bar{2}$4) sapphire and [1$\bar{1}$0] Si ∥ [$\bar{1}$100] sapphire

and bonds with the oxygen atoms of sapphire to form the first layer for subsequent growth, and this model fits all four cases to varying degrees. The multiplicity of orientations for the (111) depositions suggests that this substrate does not influence the growth of the silicon beyond the plane at the interface. Joyce and Bennett[155] have also performed crystallographic studies of silicon on sapphire by both x-ray and electron diffraction. Their results, which are somewhat at variance with those of Miller and Manasevit,[154] show that the epitaxy of silicon on sapphire consists of discrete nucleation, leading to island formation. This island phenomenon is typical of epitaxial growth in general. These studies indicate that the substrate quality and surface structure play an important role in the epitaxial deposition. Epitaxial growth of Si on (0001) sapphire by the pyrolysis of silane at substrate temperatures of 1120 to 1200°C has also been reported by Chu et al.[156] The observed epitaxial relationships between Si and sapphire are (111) Si ∥ (0001) sapphire and [1$\bar{1}$0] Si ∥ [10$\bar{1}$0] sapphire. These are different from those observed by Miller and Manasevit.

Single-crystal films of silicon have been grown on various crystallographic faces of $MgO \cdot Al_2O_3$ spinel with a cubic structure.[154,158,159] A complete cubic epitaxial relationship is observed. In each case the Si lattice is parallel to the spinel orientation, i.e., (100) Si $\parallel$ (100) spinel, and (110) Si $\parallel$ (110) spinel, and (111) Si $\parallel$ (111) spinel. The epitaxial mechanism appears to be the same as in the Si-sapphire system, i.e., a substitution of Si for the metal ion, filling a Mg^{+2} site and bonding to the oxygen atoms of the spinel. The best matching is obtained when two unit cells of spinel are compared with three of Si. In this case, the mismatch is only 1.9%, while it is 38.2% on a 1:1 basis.

The heteroepitaxy of silicon on other insulating substrates such as BeO, α-SiC, and quartz has also been reported. BeO belongs to a hexagonal system with a wurtzite structure. Single-crystal Si films from silane and $SiCl_4$ are obtained on most of the natural faces of BeO under controlled deposition conditions.[154] On α-SiC substrates, the Si films deposited on the "carbon" face of the substrate in the temperature range 1120 to 1250°C appear to be of better crystalline perfection than those on the "silicon" face.[156] The films on both faces are predominantly of {111} and {110} orientations. Joyce and his coworkers[155,160,161] have shown that single-crystal Si films can be obtained on single-crystal quartz substrates by hydrogen reduction of $SiHCl_3$ at 1120°C. Two orientation dependencies are observed:

$$(100)\ Si\ \parallel\ (11\bar{2}0)\ quartz \tag{1}$$
$$(210)\ Si\ \parallel\ (11\bar{2}0)\ quartz \tag{2}$$

No parallelism of directions has, however, been mentioned. The initial growth process is similar to that observed for evaporated-metal films,[30] i.e., growth and coalescence of nuclei to form islands. Epitaxy can also be obtained on mechanically polished quartz, the observed epitaxial relationship being (001)[010] Si $\parallel$ (0001)[10$\bar{1}$0] quartz.

Very little attempt has so far been made to grow single-crystal Si films on amorphous substrates. Rasmanis[162] has described rheotaxial growth of silicon films by vacuum evaporation or by vapor deposition onto polycrystalline alumina coated with a thin oxide layer which is fluid during film deposition at temperatures between 1000 and 1200°C. This intermediate fluid layer eliminates the effect of the polycrystalline substrate on the Si-film formation. As deposited, the films are of (110) orientation with many defects. Post-deposition annealing at 1300°C results in the (111) orientation with large grains. Filby and Nielson[28] have reported some progress toward epitaxial growth of silicon on fused quartz, using the ultrathin-alloy-zone crystallization process, as discussed in Sec. 2a.

(b) *Germanium.* Let us now discuss the growth of single-crystal films of germanium by iso- and heteroepitaxy. Numerous investigations have been made, involving vacuum-evaporation, cathodic-sputtering, and vapor-deposition techniques. These studies have dealt mainly with the growth of Ge on Ge and on insulating substrates such as CaF_2, quartz, and sapphire. Of special interest has been the influence of deposition parameters on the epitaxial temperature for single-crystal growth.

Semiletov[158] and Kurov et al.[159] have reported the isoepitaxial growth of germanium at 800°C. They have shown that the film structure is amorphous below 370 to 400°C, while it is crystalline above 400 to 500°C. Via and Thun[149] have shown by in situ diffraction studies that germanium films grown on Ge (111) substrates by vacuum evaporation are always single-crystalline above 550°C, while the film structure is textured in the range 350 to 450°C, and amorphous below this temperature range. Weinreich et al.[160] and Davey[161] have also reported the isoepitaxy of germanium with quite perfect structure at temperatures substantially below 800°C. For example, Davey[161] has observed epitaxial growth at 300°C.

The structure of germanium formed by vacuum deposition on a single-crystal substrate has been shown by Kurov[163] to depend substantially on the substrate-surface condition and in particular on traces of oxides. Consideration of the kinetics of oxide formation on germanium shows that the kinetics depend on the reaction with the residual gas in the system, i.e., on the degree of vacuum. The condition that the oxide is removed from the germanium surface, and hence the condition for epitaxial-film formation is expressed by

$$P_{\text{vac}} < \frac{15\,P_1}{\sqrt{T_1}}$$

where P_{vac} denotes residual gas pressure (in torr) and P_1 equilibrium GeO vapor pressure for the substrate at temperature T_1. If P_{vac} is 5×10^{-5} to 1×10^{-4} Torr, the above expression is satisfied for $T_1 > 800°$K, which is the same as the minimum epitaxial temperature T_s as reported by Kurov et al.[159] This indicates that T_s would decrease as the vacuum improves. For $P_{\text{vac}} = 10^{-6}$ to 10^{-7} Torr, T_s lies in the range 300 to 400°C, as experimentally confirmed by Mezentseva et al.[164]

Wolsky et al.[165] have studied the growth of germanium by low-pressure sputtering in an ultrahigh-vacuum system and have shown that isoepitaxy occurs at temperatures as low as 150°C. They have observed the influence of ambient impurity concentration on the epitaxial temperature. This has been supported by Catlin et al.[166] but contradicted by Sloope and Tiller,[40] who have shown that initial growth is not appreciably influenced by the residual gas concentration. The amorphous-polycrystalline transition occurs between 120 and 180°C, while the polycrystalline–single-crystalline transition temperature ranges from 150 to 180°C. Krikorian and Sneed[37,38,167] have reported well-defined diagrams indicating the deposition conditions that yield single-crystal, polycrystal, or amorphous films of germanium formed by vacuum evaporation and cathodic sputtering. Figures 2 and 10 illustrate typical diagrams for sputtered and vacuum-evaporated films, respectively. The results indicate that the transitions are a function of the growth rate and background pressure only. Neither the particular deposition conditions which control the growth rate (voltage and current in sputtering, source temperature and source-substrate distance in evaporation) nor the deposition techniques influence the transitions as long as the background pressure and growth rate are identical. The growth in all states of the film (single-crystal, polycrystal, or amorphous) is a decreasing function of the substrate temperature, and this dependence is consistent with nucleation theory, and yields activation energies of about 10,000 cal mole^{-1} for single-crystal growth, 3,000 cal mole^{-1} for polycrystalline growth, and 900 cal mole^{-1} for amorphous growth.

It is seen from the phase diagram that films formed at the triple point will have heterogeneous structures, consisting of amorphous, polycrystalline, and single-crystalline phases. Similar situations have also been reported by Sloope and Tiller[40] for vaporated films. The results of Wolsky et al.[165] have shown some disagreement over the transition temperatures. The temperatures for amorphous-polycrystalline and polycrystalline–single-crystalline transition are considerably lower than those reported by Krikorian and Sneed. Recently there has been some question about the existence of the so-called "triple point" between regions of crystallinity. Behrndt's[168] analysis of the problem leads to the conclusion that substrate-surface condition will determine to a large extent whether or not a triple point will be found in the crystallographic phase diagram. Recent studies by Adamsky[169] have shown that the amorphous-polycrystalline boundary occurs at lower temperatures than indicated by Krikorian and Sneed, for comparable evaporation rates (Fig. 10) in an ion-pumped ultrahigh-vacuum system. In addition, if the substrates are thermally regenerated by heating to 600°C in ultrahigh vacuum prior to deposition, the epitaxial temperature is significantly lowered to a minimum of about 100°C. Since the thermal regeneration is known to remove GeO$_2$, it appears that surface oxide could exert an appreciable influence on the epitaxial growth process. To test this hypothesis, Adamsky has intentionally leaked pure O$_2$ into the ultrahigh-vacuum system and has grown Ge films at controlled O$_2$ backgrounds from 10^{-10} to 10^{-6} Torr. The results show that O$_2$ ambients as low as 5×10^{-9} Torr raise the epitaxial temperatures by 50 to 75°C, and the amorphous-polycrystalline transition by 125 to 150°C. The experiments indicate that background oxygen can be a significant factor in affecting surface states and subsequent film growth, and can offer an explanation for the appreciable decrease in epitaxial temperature for films deposited in ultrahigh vacuum as compared with the results of Krikorian and Sneed. Recent studies by Layton and Cross[170] have shown that single-crystal Ge films grow on Ge (111) substrates by triode sputtering at temperatures lower than by evaporation at similar rates.   They have considered that the energy of the arrival of sputtered atoms at the surface does play a significant role, possibly because of the cleaning effect produced.

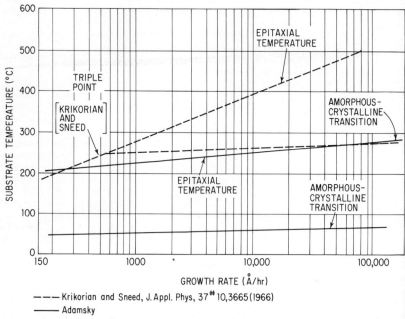

Fig. 10 Effect of growth rate and substrate temperature on film structure of Ge evaporated onto Ge (111) (Krikorian and Sneed[38]; Adamsky[169]). (*Reproduced with permission of Journal of Applied Physics.*)

The heteroepitaxy of germanium on single-crystal substrates such as CaF_2, sapphire and GaAs has been studied in considerable detail.   Most of the work has been don with CaF_2 substrates by vacuum evaporation.[17,149,166,171-175]   The heteroepitaxia growth of germanium has been observed over a wide range of substrate temperature (300 to 915°C).   Of special interest is the work of Pundsack,[174] who has shown tha best-quality Ge films grow on CaF_2 (111) substrates in the temperature range 55 to 575°C.   Below and above this range, the film structure is partially oriented.   Th deterioration of structural orientation at higher temperatures may be ascribed to filn agglomeration.   Sloope and Tiller[175] have observed a rate-dependent epitaxial tem perature and amorphous-crystalline transition temperature.   The relation betwee the logarithm of the deposition rate and the reciprocal of the epitaxial temperature i in good agreement with the predicted linear relationship based on Walton's nucleatio theory of small clusters.[18,19]   This relationship is illustrated in Fig. 11, which show the range of formation conditions over which single-crystal Ge films are produced o CaF_2.   The growth of single-crystal films occurs up to 700°C for the highest rat used (10,000 to 18,000 Å min^{-1}) and to 600°C for all other rates.   Similar result have also been reported by Krikorian and Sneed.[38]   The general behavior with respec to growth rate, activation energies for the epitaxial process, and activation energies growth in different structural phases is quite similar to that for films on germanium discussed previously.

Wallis and Wolsky[176] have studied the growth of Ge sputtered onto sapphire as function of substrate temperature, deposition rate, and film thickness.   The Ge film are generally polycrystalline, although occasional single-crystal areas are observee

Nielson[26] has shown that epitaxial deposition of Ge sapphire can be achieved by the thin-alloy-zone recrystallization process.   The growth of single-crystal Ge films has also been obtained on sapphire by Tramposch[177] using chemical-vapor deposition at substrate temperatures in the range of 750 to 900°C.   The epitaxy of Ge on GaAs (110) substrates above 400°C has been reported by Lever and Huminski.[178]   They conclude that the substrate-surface preparation is a critical factor in the epitaxial process.

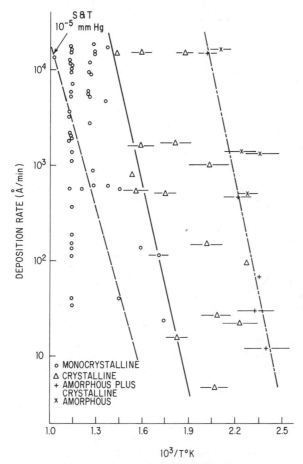

g. 11 Logarithm of deposition rate vs. reciprocal substrate temperature showing the ansition from the amorphous to crystalline to monocrystalline structures for Ge films eposited on polished (111) CaF₂ substrates (Sloope and Tiller[175]). (*Reproduced with ermission of Journal of Applied Physics.*)

Unlike the case of elemental semiconductors, the growth of compound semiconductor ms poses a complicated problem of nonstoichiometry or inhomogeneity in the sultant films.   The problem arises because of several factors including (1) reaction compound evaporant with evaporation source and (2) dissociation during evapora- on or at the substrate surface.   This occurs especially when the components of the mpound have a wide disparity in vapor pressures.   Nonstoichiometry may result a change in film structure.[179-182]   The film inhomogeneity may be eliminated by ermal annealing.[183,184]   Among the deposition techniques used, cathodic sputtering and flash evaporation claim to circumvent the problem of nonstoichiometry or inhomogeneity, provided adequate precautions are taken.   In this section, we will make a brief review of single-crystal growth of compound semiconductor films, which are of potential use in physical investigations.   The film materials to be discussed will include gallium arsenide, silicon carbide, lead sulfide, and cadmium sulfide.

(c) *Gallium Arsenide.*   Single-crystal growth of GaAs in windows on insulating substrates such as SiO_2 has been shown by Tausch et al.[185] by the application of the moving-mask technique.   They observe that the growth initiates from the window and grows out laterally over the SiO_2.   This process is shown to be highly controllable and with proper mask geometry, single-crystal growth of GaAs can be obtained over large areas almost without restraint.   Zanowick[186] has reported the growth of GaAs on sapphire and berrylia, using a thin layer of germanium grown epitaxially on the substrate as a nucleating species for the GaAs growth.   The observed epitaxial relationships are

$$GaAs\ (111)\ \|\ Ge\ (111)\ \|\ sapphire\ (0001)\ and\ (10\bar{1}4)$$

Orientations such as (100), (110), and (111) are observed on berryllia $(10\bar{1}0)$, $(10\bar{1}1)$ and $(10\bar{1}4)$, respectively.   Recently Manasevit[187] has shown that the growth of single-crystal GaAs films can be obtained on a number of single-crystal insulating substrates by chemical-vapor deposition.   The orientation relationships that develop between GaAs and the substrates are given in Table 2.   Although epitaxial GaAs

TABLE 2

Substrate plane	Parallel relationship
(0001) Al_2O_3	(111) GaAs $\|$ (0001) Al_2O_3
(100) $MgAl_2O_4$	(100) GaAs $\|$ (100) $MgAl_2O_4$
(111) $MgAl_2O_4$	(111) GaAs $\|$ (111) $MgAl_2O_4$
$(10\bar{1}0)$ BeO	(100) GaAs $\|$ $(10\bar{1}0)$ BeO
$(10\bar{1}1)$ BeO	(111) GaAs $\|$ $(10\bar{1}1)$ BeO
(0001) BeO	(111) GaAs $\|$ (0001) BeO
(100) ThO_2	(100) GaAs $\|$ (100) ThO_2

film growth has been discussed, there is no mention of deposition conditions, especially the epitaxial temperature.   A measurement of electron mobility in a 15-μm-thick film of 0.014-ohm-cm N-type (111) GaAs on (0001) sapphire gives a value of 3,40 $cm^2\ V^{-1}\ s^{-1}$ at room temperature.   This value is lower than that for the bulk material and is not unexpected for heteroepitaxial films of GaAs on insulating substrate because the film and the substrate possess a considerable difference in lattice parameter and thermal-expansion coefficient.

The heteroepitaxial growth of GaAs by cathodic sputtering onto (100), (110), and (111) surfaces of germanium at 525°C has been reported by Francombe.[188]   At this substrate temperature, however, the surface structure of the sputtered films shows strong evidence of low-angle boundaries and twinning effects.   Muller[189] has shown by the use of the flash-evaporation technique that epitaxy, free of twinning, occurs for above 600°C.   For lower temperatures, structural defects such as stacking faults along the (111) planes of the film occur, leading to a transformation of GaAs from the cubic zinc blende to the hexagonal wurtzite structure.   Chemical-vapor deposition has also been applied to the growth of GaAs on Ge.[190]   Growth temperatures around 700°C usually lead to a grainy surface, while higher temperatures around 750°C result in a smooth GaAs surface.   In either case, the GaAs film is essentially a single-crystal extension of the Ge substrate, as evidenced by Laue back-reflection patterns.   Surface defects have been observed, and these have been found to be correlated with the irregularities at the interface.

The growth of single-crystal GaAs films by isoepitaxy has also been reported by number of investigators.[182,191-193]   Mehal and Cronin[191] have observed epitaxy by chemical-vapor deposition at 750°C, while Ewing and Greene[192] have reported the growth on GaAs (111) surfaces at 780 to 790°C.   Their results show that the epitaxial growth by vapor transport is dependent on the gallium-arsenic ratio in the vapor phase

The growth rates for the surfaces are enhanced by providing excess gallium in the vapor phase during growth on the arsenic surface, and excess arsenic on the gallium surface. This suggests that nucleation on the (111) surfaces is the rate-limiting step in the growth rate of the depositing film, i.e., nucleation of gallium on the arsenic surface and arsenic on the gallium surface. The isoepitaxy of GaAs has also been obtained by Hudock[193] at a substrate temperature of 600°C by the use of a sublimation technique. With this technique, it is possible for the vapor phase to have the same composition as the solid phase because GaAs will sublime congruently below 750°C. If a small temperature difference is maintained between the source and the substrate, sublimation and condensation will occur under near-equilibrium conditions, resulting in a stoichiometric condensed phase. The relative growth rates for four GaAs orientations are $\{311\}$, $> \{100\}$, $> \{110\}$, and $> \{111\}$. This behavior is in agreement with the growth factor, as defined by Sangster[194] for crystallization on these orientations. The sublimation of GaAs under near-equilibrium conditions in ultrahigh vacuum is effective in preparing epitaxial films at comparatively low substrate temperatures and having properties approaching those of the bulk crystals.

Using a three-temperature-zone technique, Davey and Pankey[182] have obtained highly ordered, twin-free GaAs films on the b faces of GaAs and on Ge, in the temperature range 425 to 450°C, while below this range, the epitaxy occurs with twinning. Films formed above 450°C exhibit polymorphic transition (see Sec. 4).

(d) *Silicon Carbide.* Epitaxial films of silicon carbide are of technological interest because of their potential applications in electronic devices capable of operation under conditions of high temperature and radiation. The epitaxial deposition of silicon carbide poses a complicated problem in that it exists in a β form with cubic zinc blende structure and in an α form having a number of polytypes with the hexagonal or rhombohedral structures. The crystal lattice of SiC contains tetrahedrally bonded layers of Si and C atoms; the particular sequence of layers determines whether the symmetry is cubic, rhombohedral, or hexagonal. The cubic SiC is particularly attractive for device applications because of its high carrier mobility.

The growth of β-SiC on silicon and silicon carbide substrates has been accomplished by various techniques. The techniques include (1) hydrogen reduction of $SiCl_4$ in the presence of graphite,[52,195] (2) heating silicon in the presence of graphite in argon,[196] (3) heating silicon in high vacuum,[197,198] (4) simultaneous thermal reduction of silicon-containing and carbon-containing gases,[199-201] and (5) reaction of methane with silicon.[202,203] In all these experiments cubic SiC is formed at high substrate temperatures, usually over 1200°C. Depending upon the deposition technique and deposition conditions used, the film structure ranges from polycrystalline to single-crystalline. Tombs et al.[196] have observed the growth of single-crystal β-SiC films by heating silicon in the presence of graphite. The mechanism of β-SiC formation on single-crystal silicon appears to involve substitutions of C for Si in the Si crystal lattice, leading to the observed parallel epitaxial relationship between β-SiC and Si, i.e., (111) SiC $\parallel$ (111) Si and [110] SiC $\parallel$ [110] Si. Growth proceeds by diffusion of a carbon-bearing species from the gas phase inward to the Si-SiC interface. It is reported that beyond a certain thickness ($\sim$1,000 Å), the growth is polycrystalline, which may be due to the rather large difference between the lattice constants of Si ($a = 5.430$ Å) and β-SiC ($a = 4.358$ Å). Similar results have also been obtained by Nakashima et al.,[203] who have grown β-SiC by reacting with methane at temperatures above 1200°C. From measurements of the growth rate as a function of reaction temperature and partial pressure of methane, they have obtained a parabolic law, which is explained by the diffusion of C atoms through SiC. The β-SiC film is single-crystalline when its thickness is less than 1,000 Å. Jackson et al.[199] have reported β-SiC film growth by vapor-phase decomposition and hydrogen reduction of $SiCl_4$ and C_3H_8, several microns thick, at 1100°C. X-ray diffraction studies have shown no pattern, and do not yield any data as to the structural quality of the film. Only thin films are shown to be single-crystalline by transmission electron diffraction.

Recently Khan and Summergrad[141,204] have reported the growth of single-crystalline β-SiC films on silicon by chemical conversion, i.e., by reaction of unsaturated hydrocarbons such as C_2H_2 and C_2H_4 with silicon at relatively low substrate temperatures.

The β-SiC grows in parallel orientation onto the Si substrate, regardless of orientation. In the temperature range of 800 to 1100°C, the films are epitaxial. The films formed above 900°C are structurally more perfect than those formed at lower temperatures.

A typical example of a single-crystal β-SiC film formed on a Si (110) substrate at a reaction temperature of 950°C is illustrated in Fig. 12. The epitaxial relationship is parallel, i.e.,

(110) SiC ∥ (110) Si and
$$[1\bar{1}1] \text{ SiC} \parallel [1\bar{1}1] \text{ Si}$$

With a Si (111) substrate, the SiC film exhibits the cubic structure with (111) double positioning.

The structural perfection of the epitaxial β-SiC films is influenced by formation conditions such as surface cleanliness, reaction temperature, reaction rate, and film thickness. The use of ultrahigh vacuum ($\sim 10^{-9}$ Torr) prior to the reaction of the hydrocarbon allows one to maintain a clean surface for the epitaxial growth. Films formed

Fig. 12 Electron-diffraction pattern showing a single-crystal β-SiC film formed by chemical conversion of Si (110) at 950°C. Outer streaks due to β-SiC and inner weak ones due to Si (Khan and Summergrad[141]). *(Reproduced with permission of Applied Physics Letters.)*

under such conditions are relatively free from structural imperfections, as is expected. The results of Khan and Summergrad are consistent with thermodynamic considerations which suggest that unsaturated hydrocarbons such as C_2H_2 and C_2H_4 are the most favorable carbon-bearing gases for the formation of SiC by chemical reaction with silicon. Extending this work, Khan[45] has recently demonstrated the applicability of the conversion technique to the epitaxial growth of β-silicon carbide on insulating substrates such as sapphire. In this case, the SiC film is formed by conversion of a single-crystal Si film on sapphire under optimum conditions. It is observed that epitaxy occurs up to a conversion depth of about 6,000 Å, beyond which the structural orientation deteriorates such that the film is polycrystalline at a depth of about 9,000 Å. The SiC film surface exhibits a grainy structure, which shows no

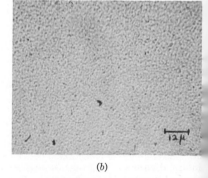

(a) (b)

Fig. 13 (a) Electron-diffraction pattern from an epitaxial β-SiC film formed by conversion of Si (100) on sapphire at 1100°C; SiC [110] azimuth. (b) Optical micrograph of the surface structure of the film in (a) (Khan[45]). *(Reproduced with permission of Materials Research Bulletin.)*

appreciable change with the conversion depth. Figure 13 shows a typical electron-diffraction pattern and a corresponding micrograph obtained from a β-SiC film formed by conversion of Si (100) on sapphire at 1100°C.

(e) *Lead Compounds.* PbS, PbSe, and PbTe belong to the category of materials having a high critical temperature of condensation. They have quite high vapor pressures at temperatures far lower than their respective melting points. When they sublime, their gaseous phase may in principle consist of different types of species, for example, in the case of PbS, atoms of Pb and S, ions of Pb^{++} and S^{--} or PbS^{+} and PbS^{-}, molecules of PbS, Pb_2S_2 etc. Porter[205] has shown that in the case of PbSe and PbTe, the gaseous phase mainly consists of PbSe and PbTe molecules, respectively. Since the dissociation energy of PbS molecules is larger than the heat of sublimation, it can be assumed also that the gaseous phase of PbS comprises PbS molecules.

The lead-compound molecules have fairly high mobility on the substrate. This has been shown by Semiletov,[206] who has deposited a single-crystal film of PbTe on a rock-salt substrate at room temperature. Similar results on PbS have also been obtained by Khan.[207] He has obtained single-crystal growth of PbS films on a cleaved (100) NaCl substrate at room temperature, as illustrated in Fig. 14. The diffraction pattern reveals the characteristics of a single-crystal material.

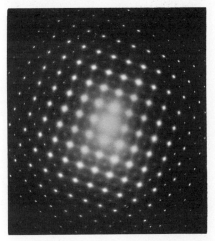

Fig. 14 Transmission electron-diffraction pattern from a single-crystal PbS film grown on a cleaved NaCl substrate at room temperature.

A number of investigations have reported the growth, structure, and physical properties of single-crystal PbS films[1,208–212] grown on rock-salt substrates by vacuum evaporation at temperatures around 300°C in the thickness range 0.2 to 8.0 μ. They have used electron diffraction or x-ray diffraction as a criterion of crystal quality, while others such as Zemel et al.[213,214] and Berlaga[215] have used electrical measurements, in particular the Hall mobility, as the measure of film quality. They have shown that the films behave like single crystals, having band structure, crystal perfection, and dielectric, transport, and mechanical properties comparable with the best available single crystals. Similar observations have also been reported from optical studies of single-crystal films of PbS, PbSe, and PbTe.[216,217]

The nucleation and growth of the compound films can be discussed in terms of the mechanism of heteroepitaxial growth.[218] As the film condenses on the substrate, elementary nuclei form at preferred sites on the substrate surface. These nuclei are strongly oriented with respect to the substrate in the (100) direction. As the nuclei grow, they impinge on neighboring nuclei, resulting in an imperfect matching of the atomic spacings at the boundaries of the crystallites. This mismatch may result in the occurrence of voids when more than two nuclei overlap, as observed by electron microscopy.

(f) *Other Compounds.* CdS is one of the semiconducting compounds whose growth, structure, and physical properties have been extensively investigated. Yet limited information is available on the single-crystalline growth of CdS films. Addis[219] has studied the growth of CdS films by the process of silver-activated recrystallization as reported by Giles and Van Cakenberghe.[220,221] He observes that the vacuum-deposited CdS films attain a high degree of (0001) preferred orientation on a glass substrate, when the film thickness is over 1 μ. On recrystallization with a silver activator, the film consists of large crystals having lateral dimensions up to 1 cm, although the crystals usually contain many small-angle grain boundaries. The activator is distributed uniformly over the surface and is subsequently removed by rinsing the film in a KCN solution. The recrystallization process involves the formation of a nucleus followed by its extended growth across the film. Nucleation usually occurs at the edge of the film or at the edge of the silvered area. Zuleeg and Senkovits[222] have also obtained single-crystal CdS films on glass substrates by the application of a molecular-beam technique, with the substrate temperature varying between 25 and 300°C. X-ray and electron-diffraction studies reveal highly oriented hexagonal CdS films, which approach single-crystal characteristics.

The single-crystal growth of CdS on single-crystal substrates is complicated by the occurrence of the polymorphic forms of the material. Depending upon the deposition conditions and substrate structure, the crystal structure of CdS films is hexagonal wurtzite (α) type or cubic sphalerite (β) type. Details of the structural variations will be found in Sec. 4.

Another semiconducting compound of interest is InSb. Although thin films of this material have been prepared by a number of techniques, very limited work has been carried out on the single-crystal growth of this material. Khan[181] has made a detailed in situ study of the growth of InSb films formed by cathodic sputtering. The details of this investigation will be described in Sec. 3.

(4) Insulating Films Insulating films have been grown in epitaxial form by a number of growth techniques such as thermal oxidation, vacuum evaporation, reactive evaporation, and sputtering. Pashley's review[1] provides information on the epitaxial growth of oxide films by thermal oxidation of metals, i.e., Ni, Cu, Fe, Zn, and Cd. Collins and Heavens[79] have made a detailed study of the epitaxial oxide growth by thermal oxidation of thin epitaxial films of Ni, Co, and Fe on rock salt. It is seen from their experimental results that the lattice mismatch between the substrate and the film is generally less important than the maintenance of correct coordination of atoms in the first layer. In all the oxide growth, the epitaxy occurs up to a few molecular layers, beyond which it degenerates into a polycrystalline form.

Epitaxial oxide films by thermal oxidation of single-crystal Ti films have been obtained by a number of investigators including Conjeaud[223] and Arntz.[224] Conjeaud observes the epitaxial formation of TiO with a NaCl structure, while further oxidation in air results in a fibrous structure of rutile (TiO_2) and an orientation of anatase (another phase of TiO_2) appearing simultaneously. The (001) orientation of TiO_2 is parallel to the (110) of TiO. The TiO_2 films, as observed by Arntz, exhibit various orientational developments, depending primarily upon the structure of the parent metal. Prior to oxidation, the densely packed a axes of Ti are parallel to the ⟨110⟩ directions in this (111) plane of the CaF_2 or BaF_2 substrate. The TiO_2 also shows that the c axes of TiO_2, which are the densely packed directions in the lattice, develop parallel to the same directions, in agreement with the orientation rules described by Gwattmey and Lawless.[225]

The formation of epitaxial oxide films is possible by thermal oxidation of single-crystal binary-alloy films, of which one component is a noble metal and the other an oxidizable transition metal. This has been shown in the work of Khan and Francombe,[127,128] who have obtained single-crystal NiO films by thermal oxidation of single-crystal Au–15 atomic % Ni alloy films on rock salt and mica. In this case, as oxidation proceeds, oxygen is dissolved at the surface of the alloy, and owing to concentration gradient, Ni from the alloy diffuses out and reacts with the adsorbed oxygen to form NiO, while Au remain inert. The oxidation results in the formation of (100) NiO on (100) Au. This parallelism in orientation may be explained on the basis of the diffusion of Ni atoms through the single-crystal Au-Ni alloy structure and also on the basis of the low lattice mismatch between NiO and Au ($\sim +2\%$)

The thermal-oxidation process can also be utilized in the preparation of epitaxial films of spinel-type ferrites. Francombe[226] has shown that a highly oriented ferrite film can be obtained by oxidizing epitaxial sputtered films of composition $NiFe_2$. Somewhat greater success has been obtained in the preparation of epitaxial films of magnesium ferrite ($MgFe_2O_4$) by oxidizing an Fe film grown by sputtering on a single crystal MgO substrate. Here the Fe film grows in the orientation (001) Fe ∥ (001) MgO and [100] Fe ∥ [100] MgO. After the oxidation process and the interdiffusion with the MgO substrate, the epitaxial ferrite film is produced, bearing a parallel orientation relationship to the MgO substrate.

Very limited information has been reported on the epitaxy of insulating material by vacuum evaporation. This is mainly due to the high growth temperatures needed and the dissociation of the film material at high temperatures. However, Müller et al.[227,228] have succeeded in obtaining single-crystal films of certain ferroelectric perovskite materials such as $BaTiO_3$ on LiF substrates by flash evaporation. The flash-evaporation technique, which involves grain-by-grain evaporation, is applicable, in principle, to the preparation of single-crystal films of a great variety of ferroelectric, ferromagnetic, and insulating materials that evaporate incongruently. Single-crystal films of $BaTiO_3$ can also be prepared by chemical thinning of single-crystal plates of this material.[229] The films can be as thin as 1,000 Å and are found to have the same lattice parameters as the bulk crystals.

Reactive evaporation and sputtering have also been used to grow insulating single-crystal oxide films.[230] Krikorian has obtained single-crystal oxide films of aluminum and tantalum on a number of single-crystal substrates (sapphire, MgO, CaF_2, mica, and aluminum). The critical conditions for producing single-crystal films by reactive deposition are fully definable by substrate temperature, growth rate, and oxygen partial pressure. The results appear independent of the growth technique, evaporation or sputtering, once the effect of argon pressure is accounted for in the case of sputtering. There is considerable doubt concerning this. Most people feel that Krikorian's system was subject to more backstreaming (and thus introduction of impurities) at higher pressures and that the presence of argon per se played no clearly identifiable role. Oxides formed at the lower critical partial pressure (below 1×10^{-5} Torr) are single-crystalline, while oxides formed at the higher critical pressures are polycrystalline with some preferred orientation. The plot of substrate temperature against growth rate indicates that at each growth rate, a transition from polycrystalline to single-crystalline structure takes place at a particular temperature and oxygen partial pressure. The transition temperature increases with increasing growth rate of the oxide.

The recrystallization process has been shown to be promising in the growth of single-crystal films of metallic oxides.[231] Francombe et al. have studied the recrystallization of amorphous oxide films of Ta, Nb, Ti, and Zr formed on cleaved NaCl by reactive sputtering. When the amorphous film, removed from the substrate, is annealed at high temperatures, recrystallization occurs, resulting in single-crystal areas many microns in size. The result shows that epitaxial films with single-phase structures can be obtained for the oxides examined, with Ta_2O_5 above 650°C, with Nb_2O_5 above 550°C, and with TiO_2 and ZrO_2 above 500°C. The data for Ta_2O_5 and Nb_2O_5 indicate that pseudo-hexagonal polymorphic forms can be achieved, apparently comprising disordered versions of the intermediate high-temperature polymorphs these oxides.

3. IN SITU OBSERVATIONS OF SINGLE-CRYSTAL-FILM GROWTH

a. Experimental Techniques

In studies of the growth and structure of single-crystal films, it is necessary to have detailed information on the state of the substrate surface on which film nucleation and growth occur, and on the crystal structure and crystallographic orientation in films during the initial and successive stages of film growth. Techniques that are employed for in situ structural investigation include low- and high-energy electron diffraction, electron microscopy, x-ray diffraction, and field-emission microscopy. In this section we will briefly discuss the merits of these techniques insofar as single-crystal-film growth is concerned.

(1) Low-energy Electron Diffraction (LEED) There has been a rapid increase in the applications of LEED to problems related to epitaxial-film growth. The majority of the results reported in the literature are concerned with the structural observation of the initial film nucleation, mainly because of the sensitivity of the technique to the analysis of layers less than one monolayer or so. The technique provides information on the adsorption of gases on single-crystal substrates, substrate surface cleanliness, atomic arrangement of the surface layer, and the orientation of the initial nuclei.

This information is seriously needed for understanding the growth of single-crystal films by any of the growth processes discussed in Sec. 2.

(2) High-energy Electron Diffraction (HEED)　The HEED technique, both transmission and reflection, is a very powerful tool for studying in situ the film structure and the epitaxial relationships that develop between the film and substrate. By the use of this technique, it is possible to obtain a continuous recording of the changes in the diffraction patterns during film growth. The importance of the technique lies in the fact that the growth can be carried on in a controlled way, so that the early stages of epitaxial growth may be studied in detail and all the growth stages may be examined in one experiment, thus avoiding any structural variation from one specimen to another. Any changes in the film structure, i.e., polymorphic transition, may be conveniently recorded and understood. This technique, however, suffers from one disadvantage; i.e., the electron beam may affect the structural orientation of the grown film. Meaningful results are obtained when an ultrahigh-vacuum HEED system is used. Recently Sewell and Cohen,[232] using such a system, have demonstrated that the sensitivity of the HEED technique for detecting thin films, especially the absorbed layer of oxygen onto the (110) Ni surface, is the same as that for the LEED technique. Similar results have also been reported by Mitchell et al.[233]

(3) Electron Microscopy　In situ transmission electron microscopy has been shown to provide detailed information on the growth process of thin films by carrying out the film deposition inside the electron microscope. This reveals the details of the coalescence phenomena that occur when nuclei and islands join each other, orientation changes due to recrystallization and internal crystallographic structure including lattice imperfections such as stacking faults and dislocations. Reflection electron microscopy has found little application to the study of thin film growth, because of its poorer resolution than that of transmission microscopy.

(4) X-ray Diffraction　Although the electron-diffraction and electron microscopy techniques are widely applied for in situ studies of single-crystal-film formation, x-ray diffraction can also be used with some limitations.[234] This technique provides diffraction patterns from the film and substrate, thus resulting in easy determination of crystallographic orientations. Like the HEED, it offers an easy means of detecting structural transformation as a function of film thickness. However, it is not suitable for studying the growth of thin films, especially the initial stage of film formation.

(5) Field-emission Microscopy　The nucleation or the initial stage of growth of metal films on single-crystal metallic substrates can be studied by field-emission microscopy. The growth can be studied continuously, allowing the effect of substrate orientation on the film nucleation to be determined fairly easily. Limited work has been or can be carried out on single-crystal-film growth because of the limitations that the substrate has to be in form of a tip. However, the usefulness of this technique has been demonstrated by Jones,[70] who has studied the growth of copper on tungsten.

b. Structural Changes during Film Growth

In situ observations reveal detailed information on the structural changes that occur during growth by deposition techniques such as vacuum evaporation and sputtering. They also provide detailed information on the mechanism of film formation and on the formation of lattice imperfections in thin films. Typical examples of such observations will be described in some detail.

(1) In Situ Electron-diffraction Observations　(a) *Epitaxy of Ag and Au.* As mentioned in Sec. 2e, Palmberg et al.[102] have studied the epitaxial growth of silver and gold on KCl by low-energy electron diffraction. LEED observations are complicated by contamination from the KCl substrate. The surface dissociates into K and Cl, which migrate into the metal surface, forming a contamination structure. When this contamination is eliminated, a Ag (1×1) pattern is obtained. However, in the case of gold, a (1×5) structure occurs, in agreement with similar observations by others[23] on bulk gold crystals.

The epitaxy of thin vacuum-deposited Ag films on an atomically clean Si (111) surface has been extensively studied by Spiegel[236] by the in situ LEED technique under ultrahigh-vacuum conditions. His observations show that in the initial stages of lm formation, two different film structures develop, depending upon the deposition emperature.   In the temperature range 25 to 200°C, a disordered adsorption layer is rmed, with increasing thickness resulting in three-dimensioned nucleation at film overages greater than monolayer thickness.   Above 200°C, the silver nuclei grow on n ordered two-dimensional silver adsorption structure.   The structure of the silver lm is independent of the structure of the initial adsorption layer.   The epitaxial rowth is always parallel, i.e., Ag (111) ∥ Si (111).

Kunz er al.[103] have studied by high-energy electron diffraction the formation of ingle-crystal films of gold and silver on cleaved alkali-halide substrates in ultrahigh acuum.   Their results show that the different stages of film growth, especially ucleation and coalescence, depend in a complicated manner upon surface condition nd composition.   The growth sequence for wedge films deposited on air-cleaved and hv (ultrahigh-vacuum)-cleaved NaCl at 360°C and KCl at 270°C will be described ere.

Initially only a parallel (100) orientation occurs.   The crystal sizes are considerably rger and better aligned azimuthally on the uhv-cleaved surface than on the air-eaved surface.   The azimuthal alignment increases with film thickness, resulting in ie appearance of a weak (111) orientation.   Upon further increase in film thickness, ie diffraction patterns become different for the various substrates.   On air-cleaved aCl, and on air and uhv-cleaved KCl substrates, streaks along ⟨111⟩ directions ppear, indicating development of (111) twins.   This is followed by a considerable crease in background, the complete disappearance of (111) orientation, and the pearance of 1/5 streaks, resulting from a gold-alkali reaction product.   On an uhv-eaved surface, the diffraction patterns observed with increasing thickness are quite fferent.   In addition to the (100) and (111) orientations, strong approximate (211) ientations develop with ⟨111⟩ twinning.   With increasing film thickness, the back-ound increases strongly, the (100) orientation becomes weaker, the (211) orientations sappear, and strong (111) orientations dominate the diffraction pattern.   These ientation changes are not due to the variation in deposition rate across the wedge it result from the thickness variation.

Au films formed simultaneously onto NaCl and KCl substrates at 270°C exhibit the llowing growth sequence: In the thinnest deposits no condensation takes place on the iv-cleaved surface, while on the air-cleaved surface many crystallites form.   The Traction patterns show that the (100) orientation is dominant and that its azimuthal rfection increases with film thickness.   In films of about 1 Å mean thickness, the imuthal alignment improves sufficiently such that subsidiary orientations are tected.   On the air-cleaved NaCl substrate the (211) orientation is the strongest, lowed by the (110) orientation and some (111) orientations, while on the air-cleaved ıCl substrate, only (211) orientation is observed.   (211), (110), and (111) orienta-ins develop on KCl substrates.

The next important change in crystal orientation occurs when major coalescence gins.   The mean film thickness at which this happens lies between 10 and 100 Å the temperature and deposition rate used.   The diffraction observations show that e subsidiary orientations are weaker than the (100) orientation on air-cleaved NaCl, d on air and uhv-cleaved KCl.   On uhv-cleaved NaCl, however, the (100) orienta-n is weaker and the (111) orientation stronger, but dominant are the (211) orienta-ns.   The film structure during the coalescence stage consists of islands which persist to several hundred angstroms mean thickness.   On the other surfaces, the (100) entation becomes more and more important.   The 1/5 streaks also appear.

The third significant growth stage occurs when the islands begin to interconnect. this stage most of the subsidiary orientations begin to disappear.   When the film :omes continuous, a (100) single-crystal film on uhv-cleaved KCl is formed with ver imperfections than in the film on the air-cleaved surface.   In films on uhv-aved NaCl, the twinned (111) orientation and some of the other orientations remain, ile at 360°C, only the (111) orientation with twinning results.

'n order to check the generality of the in situ reflection electron-diffraction observa-ns made with the Au films, Kunz et al. have studied the growth of silver on KCl ostrates at 340°C.   The following sequence in film growth on air-cleaved surface is revealed: Initially (100), (111), and a number of subsidiary orientations are formed Some random orientation also occurs. With increasing film thickness, the random and subsidiary orientations disappear, and the (100) and (111) orientations become prominent. Random orientation again occurs, followed by strong approximate (21 orientation. Finally for films several hundred angstroms thick, the (111) orientation becomes dominant, with some weak indication of (100) and random orientation. O the uhv-cleaved substrate, the (111) orientation is strong with no indication of an patterns of the (100) orientation and "1/5 streaks."

(b) *Epitaxy of Copper.* Taylor[106] has studied by low-energy electron diffraction the structure of copper films vacuum-deposited on a single-crystal (110) face of tungsten under ultrahigh-vacuum conditions. Figure 15a shows the diffraction

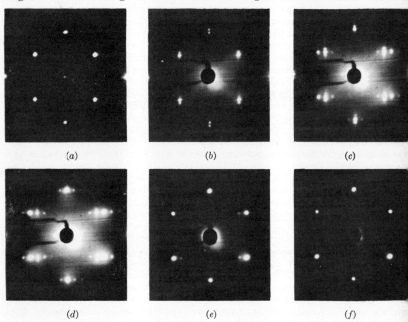

Fig.15 LEED patterns obtained during deposition of Cu onto (110) W. (a) Clean W (10 (b) After deposition of $\frac{1}{2}$ an atomic layer of Cu. (c) One atomic layer of Cu. (d) atomic layers. (e) 3 atomic layers. (f) 10 atomic layers, showing only the Cu patter (Taylor.[106]) (*Reproduced with permission of Surface Science.*)

pattern from a clean W (110) surface. When copper is deposited onto the surface room temperature, there is a marked increase in background intensity and reduct in tungsten spots, until the first appearance of new patterns occurs at about half atomic layer, as evidenced by the diffraction pattern of Fig. 15b. This pattern resu from the formation of a Cu-W alloy structure with a long-period superlattice in W [$\bar{1}$10] direction. Coverage approaching a monolayer results in well-defined fraction spots from both the epitaxial (111) Cu and from multiple scattering betw Cu and W lattices, giving rise to satellite spots as shown in Fig. 15c. Increasing average thickness to $1\frac{1}{2}$ atomic layers results in strong multiple scattering (Fig. 1 The epitaxial relationships between Cu and W are (111) Cu || (110) W and [$\bar{1}$ Cu || [$\bar{1}$10] W. As the copper film thickness increases, the contribution from tungsten lattice gets progressively less, as evidenced first by the disappearance of spots due to multiple scattering (Fig. 15e) and later the complete disappearance spots due to tungsten, leaving just those due to copper (Fig. 15f). This occurs wl the copper is about six atomic layers thick.

No signs of faceting on copper structures are observed, though the presence of surface steps is detected. However, faceting is observed in some experiments, and this is confined to tungsten-oxygen surface. The threefold rather than sixfold symmetry of the pattern of Fig. 15f shows that the double positioning of the Cu (111) on tungsten occurs.

Heating after deposition of about 10 atomic layers of copper results in the appearance of the diffraction pattern of Fig. 15b, but with improved definition. This indicates the diffusion of copper into tungsten, resulting in alloy formation.

The effect of oxygen on the epitaxy of copper can conveniently be illustrated by in situ LEED observations. When half a monolayer of oxygen is first chemisorbed on the tungsten surface (Fig. 16a,b), epitaxy is severely inhibited. Two monolayers of average result in the appearance of rather large poorly oriented Cu diffraction spots (Fig. 16c), indicating the development of islands of copper, probably of average dimensions 50 Å. Further deposition results in a reduction in the size of the Cu spots but improvement in orientation (Fig. 16d). No alloying is observed, showing that the oxygen inhibits its formation. Heating to 300°C results in some improvement in the orientation of the copper and the appearance of weak spots from the tungsten (Fig. 16e). Additional heating at 550°C results in a further orientation improvement and an increase in intensity of tungsten spots (Fig. 16f). New structures involving tungsten and oxygen begin to appear, with full development on heating to 850°C (Fig. 16g). On heating to 1050°C, the Cu spots appear weak because of the copper loss by evaporation, with the result that after 1 min a return to half a monolayer of oxygen on tungsten is reached (Fig. 16h).

Oxygen, when introduced at less than a monolayer coverage of copper, or at any time during alloy formation, displaces the copper from the W-Cu matrix and leaves it in epitaxial form on the clean W surface. When oxygen is introduced after depositing about six monolayers of the copper, no change in the diffraction pattern is observed, indicating that the oxygen is not readily chemisorbed on the copper film at room temperature.

Krause[107] has studied the isoepitaxy of copper by reflection electron diffraction under high-vacuum conditions, using a spherically shaped copper single crystal as a substrate. The important feature of the substrate is that it exhibits all crystallographic planes in the sample, such that film deposition can be made on any desired surface under identical deposition conditions.

In situ observations have been made on three crystal surfaces, i.e., (100), (110), and (111), at room temperature and at 500°C, and at room temperature followed by annealing at 500°C. The reflection patterns observed for epitaxial copper films deposited at 20°C exhibit three different types of single-crystal spots, having three different sets of intensity. These are (1) parallel orientation, (2) double-positioned orientation, and (3) double diffraction. For (100) and (110) planes, all possible ⟨111⟩ twin axes are present, while for the (111) plane only one twin orientation occurs. The double-diffraction spots indicate that the film consists of small crystals. On annealing the film at 500°C, the double-diffraction spots gradually disappear, indicating grain growth, while the twin spots still remain. Annealing at higher temperatures (800 to 1000°C) results in the diffraction pattern consisting of the normal copper single-crystal spots. Films deposited at 500°C exhibit sharp, elongated spots but give no indication of double diffraction. This high-temperature deposition leads to rapid surface diffusion, resulting in large coherent crystal areas immediately upon condensation at 500°C.

In his in situ studies of the early stages of epitaxy of Cu, Cu_2O, and Ag films on copper single-crystal substrates, Krause observes a new kind of reflection electron-diffraction pattern, which is generated from two-dimensional surface structures at the interface and which can be explained by networks of interfacial dislocations. Details of these observations will be described in Sec. 5.

c) *Epitaxy of Silicon.* Jona[147] has studied by low-energy electron diffraction the early stages of the isoepitaxial growth of silicon in ultrahigh vacuum. On a clean (111) substrate, exhibiting the 7×7 surface superstructure structure,[237] the epitaxy occurs above 400°C, as evidenced by the fact that the LEED patterns from the

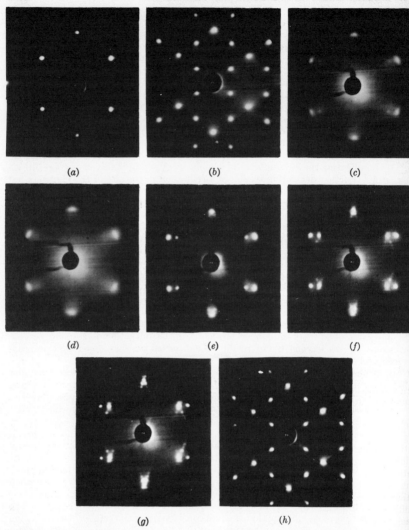

Fig. 16 LEED patterns showing the effect of oxygen on the epitaxy of copper on tungsten (110) surface. (a) Clean tungsten. (b) ½ monolayer of oxygen on the tungsten. (c) monolayers of copper—note the appearance of poorly oriented diffraction spots in copper position (outer). (d) 10 layers of copper. (e) Heating to 300°C for 5 min (result in some orientation improvement and the reappearance of the tungsten beams). Heating at 550°C for 15 min. (g) Heating at 850°C for 1 min (an oxide of tungsten evidenced by the diagonal rows of beams about the tungsten beam positions). (h) 1050°C 1 min and return to ½ monolayer of oxygen on the tungsten. (Taylor.[106]) (*Reproduc with permission of Surface Science.*)

substrate remain unaffected by deposition of several monolayers of silicon. As depositions are made at decreasing substrate temperatures, the diffraction patte gradually broaden, indicating gradual disordering of the film structure. Comp obliteration of the patterns occurs for room-temperature deposition, and thi indicative of an amorphous structure of the deposited film. In situ observations sl that the amorphous film undergoes structural changes when annealed at successiv

▲gher temperatures. Recrystallization of the film begins at about 150°C and takes ▲ace up to about 320°C. The LEED patterns for the first observed phase consist of ▲e normal spots of the bulk structure, flanked by satellites at the one-seventh- and ▲-sevenths-order positions, as illustrated in Fig. 17a. With prolonged heating, the

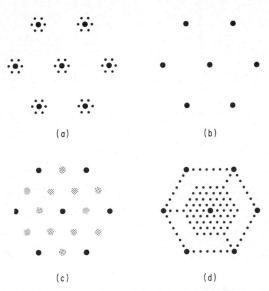

. 17 Schematic LEED patterns observed during annealing of an amorphous Si film ▲ a Si (111) surface. Large circles represent the normal spots and small circles the super-▲tice spots. (a) First phase. (b) Normal phase. (c) Transition stage. (d) 7 × 7 ▲cture. (Jona.[147]) (*Reproduced with permission of Applied Physics Letters.*)

▲ellites gradually disappear, resulting in the normal Si structure at about 350°C ▲g. 17b). Higher-temperature heating leads to the appearance of broad diffuse-▲ttering patches at half-order positions, indicating a transition stage (Fig. 17c), ▲ich then resolve into fractional-order ▲ts of the 7 × 7 structure (Fig. 17d). ▲s sequence of structural changes is ▲ilar to some order-disorder transition ▲nd in alloys.[238] The observed first ▲se may be due to periodic modulation ▲ the lattice parameter, and probably ▲o of the scattering factor. The transi-▲n from the normal to the 7 × 7 struc-▲e involves the formation of antiphase ▲ains, causing broadening of the super-▲ucture spots.

▲Vith a (100) Si substrate, exhibiting ▲ 2 × 2 superstructure, the epitaxy of ▲con occurs at temperatures as low as ▲m temperature. The film growth ▲ults in almost complete obliteration of superlattice spots, leaving the spots of

Fig. 18 Schematic LEED patterns ob-served during annealing of a Si film on a Si (100) surface at room temperature. Large circles represent the normal spots, and the small circles the superlattice spots. (a) Normal Si structure. (b) 2 × 2 struc-ture. (Jona.[147]) (*Reproduced with per-mission of Applied Physics Letters.*)

normal Si (100) structure intense (Fig. 18a). Mild heating of this phase causes ▲ appearance of diffuse half-order spots, which are likely to be caused by antiphase ▲mains of small size, and which then result in sharp superlattice spots upon longer ▲nealing at higher temperatures (Fig. 18b). The clean (100) surface thus appears to be more conducive to epitaxy than the (111) surface, and this is understandable o structural grounds.[239]

The LEED observations by Jona are confined to films only a few monolayers thic. Recently Thomas et al.[148,240] have used LEED to monitor the growth of silicon up several microns thick at various temperatures, and have reported diffraction da differing from those observed by Jona for thin films. Film growth at temperatur between 600 and 1000°C produces essentially no change of the basic Si (111)-7 stru ture at any thickness from about 100 Å to 5 μ. Below 600°C, a mixed Si (111)-7 ar Si (111)-5 structure develops in the initial growth stages and, for a given temperatur is preserved as epitaxy continues. The LEED patterns become progressively mo diffuse with decreasing temperatures until at temperatures below 350°C, the patte: is obliterated completely, indicating the formation of an amorphous structure. Ep taxy on a Si (100) surface occurs at temperatures down to about 300°C. In this cas the original Si (100)-2 surface structure is retained by the film, regardless of thicknes except in the range 650 to 700°C, where a new ordered form of the Si (100)-2 structu is generated.

(d) *Epitaxy of InSb.* In situ observations have been made by Khan[181] on t epitaxial growth of indium antimonide films formed by cathode sputtering on cleaved (001) surfaces of rock salt inside an electron-diffraction system specially bu for such studies. The film is sputtered onto the substrate at a predetermined rate f a certain period of time and then examined by reflection electron diffraction. further deposit is then made and the procedure is repeated until a thick film is forme The diffraction patterns recorded on photographic film are then analyzed to obta information on the structural variation during film growth. An amorphous film formed at room temperature, but it transforms into a crystalline structure at abo 150°C, the degree of orientation increasing with increasing substrate temperatu. Complete èpitaxy occurs at and above 250°C. These changes are illustrated Fig. 19.

The initial and successive stages of epitaxy of InSb sputtered onto a cleaved (0C NaCl substrate at 380°C (Fig. 20a) are described in detail. The first sign of depositi is observed when the average film thickness is about 4 Å. Long streaks appear in t diffraction pattern outside those due to the substrate. As film thickness increas the intensity of these streaks increases and that due to the substrate decreases, illustrated in Fig. 20b and c. Analysis of the pattern shows that the film formed up 20 Å is composed of indium. The tetragonal indium grows epitaxially onto the su strate even at room temperature.

The first appearance of the diffraction pattern due to InSb is observed when t average film thickness is about 32 Å, as shown in Fig. 20d. As the film thickn increases, the intensity of the InSb pattern increases and that due to In decrea (Fig. 20d,e,f). The cubic InSb grows with a parallel orientation onto the NaCl su strate. Analysis of the diffraction data gives evidence of a hexagonal phase of InS the epitaxial relationship between this polymorph and the NaCl substrate be represented by (0001) InSb ∥ (001) NaCl and [10$\bar{1}$0] InSb ∥ [110] NaCl. The hexa onal phase is observed only during the initial stage of film growth. The formati of this phase may be due to the presence of free indium at the surface. Further det: will be found in Sec. 5.

No change in the epitaxy occurs until the film thickness increases to several thousa angstroms. The (111) twinning gradually increases with film thickness, as shown Fig. 20f, g, and h. Figure 20g and h discloses additional twin spots not situated on lines joining the main reflections. They are due to successive twinning around the t different [111] axes in the pattern. Interference lines in the diffraction pattern appe indicating stacking faults of random repetition of the (111) planes.

The above results show that the epitaxial growth of InSb is preceded by the forr tion of a very thin epitaxial layer of indium during the initial stage of film grow The formation of indium at the interface results from the dissociation of InSb at substrate surface, when antimony evaporates off because of wide difference in va pressure. After a certain growth stage, InSb begins to form. Once an epitaxial f is formed, it continues on up to a few microns.

(e) *Epitaxy of Thallium Halides.* Khan[59] has made a detailed in situ high-energy electron-diffraction study of the epitaxial deposition of thallium halides on single-crystal alkali-halide substrates. He observes the occurrence of abnormal fcc structure during the initial stage of film growth, followed by the appearance of the normal bcc structure. Since this study involves epitaxial polymorphism, this will be discussed in the next section.

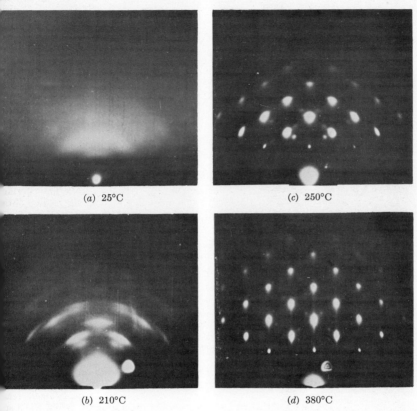

(a) 25°C   (c) 250°C

(b) 210°C   (d) 380°C

Fig. 19 Reflection electron-diffraction patterns showing the effect of substrate temperature on the epitaxy of InSb sputtered onto a clean (001) NaCl substrate; azimuth [110]. (a) 25°C. (b) 210°C. (c) 250°C. (d) 380°C. (Khan.[181]) (*Reproduced with permission of Surface Science.*)

(f) *Surface Reactions.* In many cases of film deposition on metallic substrates, the film material reacts with the substrate, resulting in alloy formation only during the very initial stage of film growth. To obtain a better understanding of the epitaxial process, it is important to have detailed information on the deposition conditions under which alloying occurs between the film and the substrate. The LEED technique is proved to be a very sensitive tool for such studies during the initial stage of film deposition. Typical examples of such studies will be mentioned here. The HEED technique is also useful, and observations by this technique can be found in Sec. 2e(3). In situ LEED observations show that alloying can occur even when the substrate temperature is sufficiently low that the bulk diffusion of each metal with the other is negligible. Taylor[106] observes the formation of a Cu-W alloy structure during the initial stage of growth of copper on a clean (110) surface of tungsten at room temperature. This has previously been described in detail.

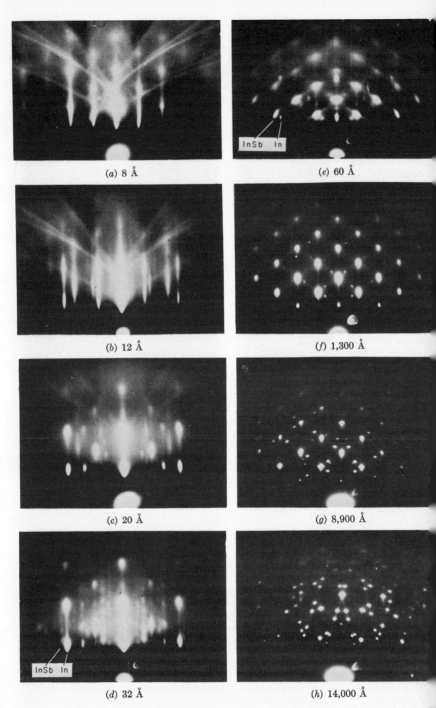

(a) 8 Å   (e) 60 Å

(b) 12 Å   (f) 1,300 Å

(c) 20 Å   (g) 8,900 Å

(d) 32 Å   (h) 14,000 Å

Fig. 20 Reflection electron-diffraction patterns showing the initial and successive stages of epitaxial InSb film growth on a cleaved (001) NaCl surface at 380°C; azimuth [11]. (a) 8 Å. (b) 12 Å. (c) 20 Å. (d) 32 Å. (e) 60 Å. (f) 1,300 Å. (g) 8,900 Å. (h) 14,000 Å. (Khan[181]). (*Reproduced with permission of Surface Science.*)

Lander and Morrison[237,241-243] have made a detailed study by low-energy electron diffraction on the reactions of silicon and germanium with a wide variety of film materials such as aluminum, indium, phosphorus, lead, and tin. Their observations are confined to films only a few monolayers thick, and reaction products are mainly two-dimensional in character. Five alloy structures are observed with aluminum and eight with indium on Si (111) surfaces. Three are common to both systems, and some are very complex. The alloy phases observed for Al on Si are the following: α-Si (111)-7-Al, α-Si (111) $\sqrt{3}$-Al, β-Si (111) $\sqrt{3}$-Al, β-Si (111)-7-Al, and γ-Si (111)-7-Al. These structures depend on composition, ranging from one-third to one monolayer, and on temperature. Some occur by simple addition, and some with substrate reconstruction.

(2) In Situ Electron Microscopy   The mode of nucleation and the general nature of the growth sequence of thin films have been discussed in detail in a previous chapter. In the following we will confine our discussion to in situ electron-microscopy observations of the growth sequence of single-crystal films. Most of the in situ observations reported in the literature[23,244-254] have been carried out with vacuum-deposited metal films, because evaporation inside an electron microscope is easily accomplished. Since there is a large measure of agreement between the various observations, the results reported by Pashley and his coworkers may be considered as typical. Before discussing their results, it may be important to emphasize the limitations of direct observation of the growth sequence in single-crystal films.

The substrates on which film nucleation and growth take place are single crystals in thin film form. Only a limited number of substrates such as molybdenite, mica, and graphite can be prepared in single-crystal form, and surfaces without cleavage steps (which promote nucleation on preferred sites) are difficult to obtain. The most satisfactory method of preparation is that of successive cleavage of a layer structure until it is thin enough for transmission electron microscopy. Other important factors include the stability of the substrate with respect to heating and electron bombardment, and the resistance of the substrate to attack by oxidation or moisture. Consideration must also be given to other limitations such as high background pressure in the electron microscope and formation of carbon layers on the substrate under the influence of the electron beam. Unreliable results concerning the mechanism of film growth could be obtained if adequate precautions are not taken.

Pashley and his coworkers[23,247,253,254] have studied the growth of gold and silver on single-crystal substrates of molybdenite in an electron microscope with a base pressure of 10^{-5} to 10^{-6} Torr. The substrates are heated to above 350°C to avoid the formation of polymer films because of the electron beam. The film growth rate is below 10 Å s^{-1}, and a cinecamera is used to record the rapid changes in structure during growth. An example of the growth sequence of a single-crystal film of gold on molybdenite at 400°C is shown in Fig. 21. In the initial stage of growth, a large number of discrete three-dimensional nuclei are seen. Continued deposition leads to an increase in the size of the nuclei or islands, accompanied by a gradual decrease in their numbers. When the nuclei grow to the point that they touch, coalescence begins, followed by recrystallization or reorientation. The nuclei then join to form a connected network, usually developing into a channeled structure. A continuous film then forms as the film thickness further increases. The larger the number of initial nuclei, the sooner a continuous film is formed.

The in situ observations have revealed two important growth processes—liquid-like coalescence of nuclei and islands, and reorientation or recrystallization of islands. The formation and elimination of lattice imperfections during growth have also been exhibited.

(a) "Liquid-like" Coalescence.   This phenomenon manifests itself during the stage at which nuclei or islands begin to touch each other. Figure 22 illustrates the coalescing sequence of two nuclei, i.e., two gold islands on molybdenite during deposition at 400°C. As the two nuclei coalesce, a neck between them forms, and the nuclei round off, indicating that material is being transported to the neck. Marked changes in the nuclei profiles occur, resulting in the formation of a crystallographic shape, if no further coalescence takes place.

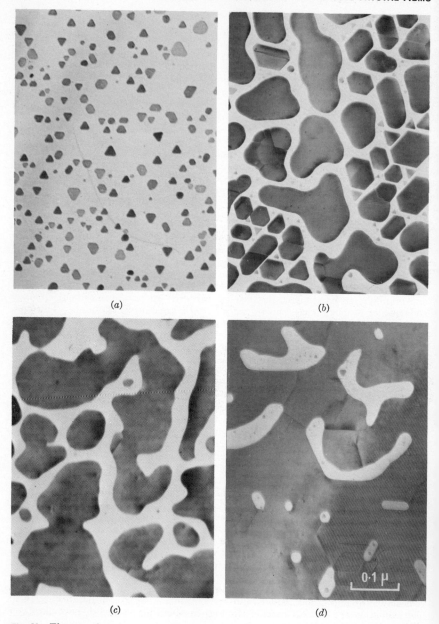

Fig. 21 The growth sequence of gold on molybdenite at 400°C (Pashley et al.[23]).   (*Reproduced with permission of Philosophical Magazine.*)

The general observations are that when the nuclei are small and have a rounded shape, an instantaneous coalescence occurs, resulting in the growth of a single rounded island, as has been shown for gold and silver on molybdenite at 350°C.[23] When a small island touches a larger island, the smaller island tends to run into the larger island, as though a small liquid drop is coalescing with a larger one. For larger islands, a slow coalescence occurs, leading to the formation of a compound island with a greater thickness than the initial islands. If the islands have a crystallographic shape, they assume a rounded shape during coalescence (Fig. 21a and b), and then gradually resume a crystallographic shape if no further coalescence occurs. With increasing thickness, the compound islands develop into a network structure, rapidly exhibiting a channeled structure. The filling in of the channels occurs by the rapid formation of a neck that then broadens. Theoretical considerations indicate that surface self-diffusion of the deposited material is the predominant mass-transfer mechanism.

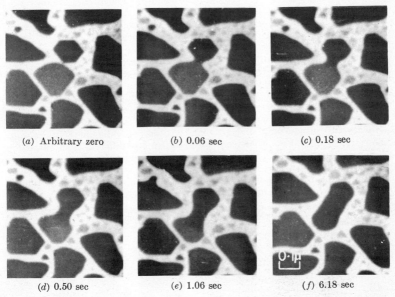

(a) Arbitrary zero　　　(b) 0.06 sec　　　(c) 0.18 sec

(d) 0.50 sec　　　(e) 1.06 sec　　　(f) 6.18 sec

Fig. 22 Coalescing sequence of two gold islands on molybdenite during deposition at 400°C. The micrographs are exposed relative to (a) as the arbitrary zero: (b) 0.06 sec. (c) 0.18 sec. (d) 0.50 sec. (e) 1.06 sec. (f) 6.18 sec. (Pashley and Stowell.[254]) (*Reproduced with permission of Journal of Vacuum Science and Technology.*)

Although the liquid-like coalescence behavior persists throughout the growth of the film, the film is in a solid crystalline form, as evidenced by the normal single-crystal electron-diffraction pattern at all stages. This behavior is attributed entirely to a pronounced surface mobility of the deposited atoms both on the substrate surface and on the deposit islands. The driving force for the mobility is the surface energy of the deposit material. It should be noted that high mobility of gold and silver atoms on the substrate surface is not observed when contamination effects are small.

Bassett[251] has observed small movements of silver islands on graphite, the movements being oscillatory rather than translational. The in situ experiments by Pashley[23] reveal that with gold and silver on molybdenite, the islands themselves are not mobile except for small rotations, but the apparent movement is due to the liquid-like coalescence.

It is expected from the general nature of the coalescence phenomena that the deposition parameters such as deposition rate and substrate temperature would have pronounced effect on the liquid-like coalescence behavior. Faster deposition rate is likely to make the effect of the freshly depositing atoms significant earlier in the coalescence process. This means that as coalescence proceeds, the thickening of the compound islands will be less, resulting in the formation of a continuous film at an earlier stage of deposition. The substrate temperature will affect the rate at which coalescence takes place, and thus will affect the relative contributions of the arriving atoms and the already deposited material to the filling in of the channeled network.

(b) *Reorientation and Recrystallization.* Direct observations have revealed detailed information concerning nuclei rotation and recrystallization during thin film growth. The initial nuclei of gold and silver on molybdenite are aligned with their (111) planes parallel to the (0001) molybdenite surface, but the metal [1$\bar{1}$0] direction is slightly misaligned with respect to the substrate [10$\bar{1}$0] direction. Bassett[245] has observed a slight rotation of the silver nuclei during growth, especially during coalescence when two slightly misorientated nuclei rotate into perfect alignment. Similar observations have been made by Pashley[23] for gold on molybdenite. The mechanism for this effect may be related to the existence of pointed minima in the interfacial energy, as suggested by du Plessis and van der Merwe.[255] This answers the question why the orientation of an epitaxial film improves as growth proceeds.

The (111) nuclei and islands of gold or silver on molybdenite exhibit double positioning; i.e., one orientation is rotated 180° with respect to the other. This is a special form of twinning. The double positioning can easily be identified during the initial stages of film formation, because the triangular nuclei point in opposite directions, with the result that they coalesce to give a compound island. This is the onset of the recrystallization process. Jacobs et al.[253] have made a detailed study of this process by making observations on samples tilted inside the microscope. When tilted, one orientation diffracts more strongly than the other, providing black-white contrast, as shown in Fig. 23. Upon coalescence of two islands of different positions, a double-positioning boundary forms, which then migrates, leaving the island in one position only. Figure 23 illustrates the boundary formation and its movement (as in the compound island *D*). The driving force for such boundary movement is that of the boundary energy, although other driving forces are possible.

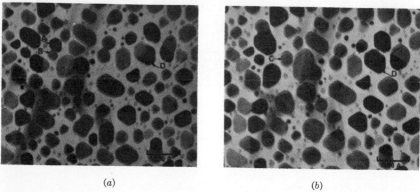

(a)   (b)

Fig. 23 Black-white contrasts showing the double-positioning structure in gold on molybdenite at 400°C. (b) The same as (a) but after further deposition. (Jacobs et al.[253]) (*Reproduced with permission of Philosophical Magazine.*)

The recrystallization process is important in determining the final film structure. The amount of recrystallization, resulting in changes in orientation, depends partly on the relative size of the nuclei involved. Small nuclei are generally consumed by large nuclei. This process can be illustrated by the observations of Matthews[22] on the growth of gold and silver on cleaved NaCl substrates. It is observed that about 10 % of the initial nuclei are in the (111) orientation, while the rest are in the (001) orientation. The (111) nuclei tend to grow more rapidly than the (001) nuclei, with the result that the (111) nuclei are bigger than the (001) nuclei during coalescence. With clean substrates and with a low density of nuclei, the (111) nuclei consume (001) nuclei, resulting in the predominance of the (111) orientation in thicker films. With contaminated air-cleaned surfaces and with a high density of nuclei, coalescence occurs much earlier, before the (111) nuclei have the chance to grow bigger. This results in the consumption of the (111) nuclei by the (001) nuclei and consequently in the (001) orientation rather than the (111). These orientation changes occur by the same recrystallization process as discussed previously for the double-positioning coalescence, although other factors such as the size, shape, and orientation distribution among the nuclei at different stages of growth should also be considered for a better understanding of the process discussed.

(c) *Lattice Imperfections.* The growth of single-crystal films is commonly associated with structural defects such as dislocations, twinning, and stacking faults. Direct observation provides detailed information concerning the nature and origin of such defects. More detailed information is obtained if this technique is coupled with the observation of parallel moiré patterns formed between the film and the substrate. The in situ experiments have revealed the formation and elimination of dislocations and stacking faults, as a result of coalescence of gold islands on molybdenite.[253] These will be discussed in detail in Sec. 5.

The in situ experiments as described above have been performed in poor vacuum, i.e., in the 10^{-5} to 10^{-6} Torr range. Under such vacuum conditions, the adsorption of residual gases on the substrate, or on the growing film, could have a major influence on the detailed mode of nucleation and growth of thin films. Hence the "liquid-like" coalescence process may be influenced by the residual gases, resulting in significant effects on the final film structure. This necessitates in situ observations with an electron microscope with a much better vacuum. Poppa[252] and Valdre et al.[256] have reported modifications of the specimen chamber such that film deposition can be made under ultrahigh-vacuum conditions, but no results of the application of such a system to in situ studies of the nucleation and growth of single-crystal films have yet been reported.

4. POLYMORPHIC TRANSITION IN SINGLE-CRYSTAL FILMS

a. Factors Influencing Polymorphic Transition

In studying the growth of single-crystal films, it is observed that thin films of certain materials exhibit abnormal crystal structures which are different from their normal stable structures, or they undergo polymorphic transition during growth under certain growth conditions. The transition can be induced or influenced by a number of factors. These may be listed as follows:

1. Crystal structure of the substrate
2. Substrate temperature
3. Size of nuclei during initial stages of film growth
4. Strain induced during growth
5. Impurities on the substrate surface
6. Structural imperfections such as dislocations and stacking faults
7. Applied electric field

Polymorphic changes influenced by one or more of the above factors have been observed in single-crystal films of metals, semiconductors, and insulators grown by vacuum evaporation, cathodic sputtering, or chemical-vapor deposition. In the following we will provide some typical examples of polymorphic transition in single-crystal films grown by heteroepitaxy, and discuss the mechanisms by which such transition is effected.

b. Examples of Polymorphic Transition

Materials with several modifications of their equilibrium structures (such as Co, Fe, and several compounds) have a stronger tendency to show polymorphism in thin films than other materials like Ni and noble metals.

(1) Cobalt Films The stable structure of cobalt is hexagonal close-packed (hcp) at temperatures below 420°C, but it deposits with a fcc structure under certain growth conditions. Newman[69] has observed that the cobalt grown on copper substrates exhibits a mixture of the hcp and the fcc structures and that it is not strained to match the copper lattice. Goddard and Wright[257] have grown about 1 μ cobalt films with the fcc structure but have not studied whether the films are strained to match the copper lattice.

Jesser and Matthews[73] have recently reported their observations on the polymorphic transition in single-crystal cobalt films deposited by vacuum evaporation onto (001) copper surfaces in ultrahigh vacuum. Electron-diffraction and electron-microscopy observations show that if the cobalt film is below 20 Å in thickness, it grows as a monolayer with its lattice strained to exactly match the copper substrate. The film shows a fcc structure, similar to that of the substrate. At and above 20 Å, long misfit dislocations are generated to accommodate part of the difference between the cobalt and copper lattices. The intrinsic stacking faults associated with the misfit dislocations convert some of the fcc cobalt into the stable hcp structure. The fraction of cobalt that transforms into the hcp structure depends on the number of dislocations that dissociate and on the average distance moved by the migrating partial dislocations.

The deposition of cobalt at elevated temperatures (350°C) results in the formation of three-dimensional nuclei of the deposit. Nuclei less than 375 Å in radius are fcc and strained to fit the lattice of the copper substrate. Figure 24 shows an electron micrograph of cobalt nuclei formed on a (001) copper surface at 350°C. The arrowed nuclei are small and exhibit the fcc structure, while large nuclei (not shown in the figure) contain moiré fringes formed by reflection from the copper and fcc cobalt lattices. Increased film thickness results in the growth of cobalt in the stable hcp structure.

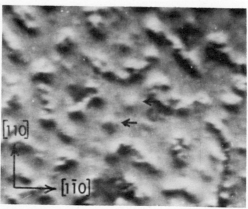

Fig. 24 Micrograph of fcc cobalt nuclei on a copper substrate. Magnification × 80,000 (Jesser and Matthews[73]). (*Reproduced with permission of Philosophical Magazine.*)

Gonzalez and Grünbaum[258] have observed both the hcp and the fcc structures in films deposited on silver substrates at below 400°C, while Sato et al.[113] have obtained the fcc structure, which is retained as long as the film is on the substrate (MgO). This is due to the influence of the crystal structure of the substrate. The observed epitaxial relationship is represented by

$$(001) \; Co \parallel (001) \; MgO \quad \text{and} \quad [100] \; Co \parallel [100] \; MgO$$

Polymorphic changes in epitaxial Co films have also been observed on NaCl and KCl substrates. Collins and Heavens[79] have studied the epitaxial growth of Co films on the (100), (110), and (111) faces of rock salt. They observe the presence of the fcc cobalt, when deposited onto the substrate at room temperature. As the substrate temperature increases, films are formed containing a steadily increasing proportion of the cubic cobalt. Above 480°C, single-crystal films with the cubic structure are formed. The cubic cobalt grows in parallel orientation onto the NaCl substrate, although it is associated with stacking faults and twinning. Honma and Wayman[117] have observed that the Co films on NaCl and KCl substrates at various temperatures in the range 20 to 550°C are substantially fcc, although bulk is stable in only the hcp form below 420°C. The cubic cobalt exhibits mainly the (001) orientation, although sometimes orientations such as (001), (111), and (112) are present. The film contains many microtwins and stacking faults (or hcp phase).

(2) Iron Films Haase[259] has studied the crystal structure of thin films ($\sim$10 Å) of iron deposited in high vacuum onto heated single-crystal copper substrates. He has observed that electron-diffraction patterns from the iron films with the copper substrate could not be distinguished from the undeposited copper surface. This shows that the iron is fcc and its lattice parameter is very close to that of copper.

Jesser and Matthews[72] have repeated the experiments of Haase under cleaner vacuum environment and have studied iron films greater than 10 Å in average thickness. They have observed that thin films ($<$20 Å) possess a fcc structure and are strained to exactly match the copper lattice. In thicker films misfit dislocations are generated to accommodate part of the misfit between the fcc iron and the Cu substrate. The remainder of the misfit is accommodated by elastic strain. Dark-field electron-micrograph observations reveal that the fcc $\rightarrow$ bcc transition begins with the formation of numerous small nuclei. Important orientation relationships between the fcc and bcc iron are

(111) fcc $\parallel$ (110) bcc and [$\bar{1}$10] fcc $\parallel$ [001] and [111] bcc
(001) fcc $\parallel$ ($\bar{1}$10) bcc and [110] fcc $\parallel$ [112] bcc
(001) fcc $\parallel$ (001) bcc and [010] fcc $\parallel$ [$\bar{1}$10] bcc

(3) Nickel Films Freedman[77] has shown that nickel films grown epitaxially on rock salt by vacuum evaporation at 300°C are in a state of high-compression strain. The strain is so high that it causes a structural transition from cubic to tetragonal [see Secs. 2d and 2e(1)].

Wright and Goddard[260] have reported the occurrence of a hcp structure of nickel when deposited on copper and cobalt substrates by electrodeposition. The formation of the fcc structure is associated with the orientation of the substrate as well as the pH of the electrolyte. Nucleation and stabilization of the abnormal phase are effected by impurities present on the substrate surface.

(4) Chromium Films The normal structure of chromium is bcc, but when it is deposited in ultrahigh vacuum onto an epitaxially grown (001) nickel substrate, it exhibits a fcc structure only when the deposit is very thin ($\sim$10 Å).[74] Notably visible are the long straight dislocations to accommodate part of the misfit between the lattices of fcc nickel and fcc chromium. As film thickness increases, the resolution of the misfit dislocations becomes poor, and reflections from the normal bcc chromium appear in the diffraction pattern. Dark-field electron-micrograph observations show that the fcc $\rightarrow$ bcc transformation begins with the formation of very small (70 to 200 Å in diameter) bcc nuclei.

(5) Tantalum Films Marcus[261] has observed phase transition in thin single-crystal films of bcc tantalum grown onto cleaved (001) MgO substrates above 700°C. The transition is due to electron-microscope beam heating. Slow cooling results in a new phase, the transition temperature being within 200 to 500°C. A micrograph of a film containing transformed and untransformed material, together with its diffraction pattern, is illustrated in Fig. 25. Domain boundaries between the two regions run in the $\langle 100 \rangle$ direction. In addition to reflections due to the bcc structure, superlattice spots are observed at points one-fourth the distance to the matrix points, corresponding to a tetragonal structure with $c/a = 1.128$. This phase is reversible, since its growth can be stopped by reducing the electron beam or can be made to disappear by increasing it. The impurities that induce the impurity segregation and the superlattice formation are introduced in the film at some time after film formation (they may originate either in the surface oxide in the film or in the electron-microscope atmosphere).

Denbigh and Marcus[262] have observed a fcc structure in thin ($<$100 Å) films of tantalum, which transforms to the normal bcc structure as the film thickness increases.

Similar observations have been made by Chopra et al.,[263] who have grown Ta films by sputtering onto cleaved NaCl substrates below 450°C. The fcc phase occurs with a parallel orientation onto the substrate. The fcc → bcc transition takes place gradually over a range of temperatures, thus allowing two phases of Ta to be observed simultaneously, as illustrated in Fig. 26. It is interesting to note that the epitaxial fcc structure occurs above 250°C and not below 450°C. As Chopra et al. have

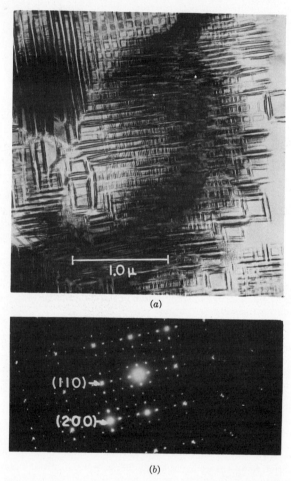

(a)

(b)

Fig. 25 (a) Electron micrograph of a single-crystal Ta film cooled slowly after intense heating in the electron microscope. (b) The corresponding electron-diffraction pattern. (Marcus.[261]) (*Reproduced with permission of Journal of Applied Physics.*)

observed, the crystal structure of the substrate (NaCl and mica) has no effect on the film structure but does influence the growth, indicating that the fcc phase is nucleated. Nucleation and stabilization of the fcc structure may be influenced by the high kinetic energy and electrostatic charges prevalent in the sputtering process.

(6) Metallic Halides Cesium and thallium halides which normally possess a bcc structure undergo polymorphic transformations when grown epitaxially by vacuum evaporation onto alkali-halide and other substrates with the NaCl structure at room temperature. Initially an abnormal NaCl structure of the films is formed. This structure is maintained up to a certain thickness, beyond which the normal bcc structure appears. The abnormal phase grows with a paralled orientation onto the substrate, whereas the bcc structure grows with its (110) plane parallel to the (001) substrate and its [001] direction parallel to the [110] or [1̄10] direction of the substrate. These observations have been made by Khan,[59] Schultz,[264] Ludemann,[265] and Pashley.[266] The abnormal structure appears to be induced by the crystal structure of the substrate. In order to verify such a conclusion, Blackman and Khan[58] and Meyerhoff and Ungelenk[267] have studied the growth of the halides on amorphous substrates such as carbon or Formvar films. The cesium and thallium halides with the exception of CsCl possess their normal structures at room temperature, while CsCl exhibits a mixture of the normal bcc and the abnormal fcc structures if the deposition rate is low. Blackman and Khan have deposited the films at low temperatures (down

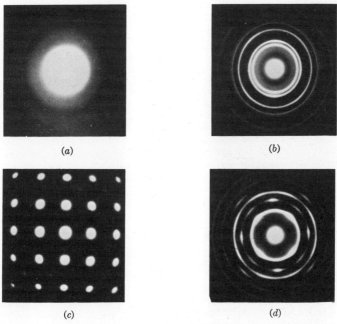

(a)   (b)

(c)   (d)

Fig. 26 Electron-diffraction patterns of ~500-Å-thick Ta films sputtered onto NaCl at (a) 23°C—amorphous, (b) 250°C—fcc, (c) 400°C—fcc, and (d) 450°C—bcc. (Chopra et al.[263]) (*Reproduced with permission of Philosophical Magazine.*)

to −180°C) and studied the structural changes as a result of annealing. The amorphous films, formed at low temperatures, undergo crystalline transformation, exhibiting both the bcc and fcc structures. However, only CsCl shows the abnormal fcc structure persisting up to room temperature. These observations have shown that the formation of the abnormal fcc structure is not solely due to the influence of the substrate structure but is enhanced or stabilized during the epitaxial process.

(7) GaAs Films The stable structure of GaAs is cubic sphalerite, but a hexagonal polymorph has also been observed. Muller[189] has reported the occurrence of the hexagonal wurtzite phase during the epitaxial growth of GaAs on different crystallographic faces of germanium by flash evaporation. Structural defects are observed, which are ascribed to the presence of stacking faults of the (111) planes, the planes with the densest packing of atoms. Stacking faults in regular repetition lead to the appearance of the hexagonal phase. A hexagonal arrangement in GaAs films on NaCl has been observed by Kurdyumova and Semiletov.[180]   This structure is associated with the presence of excess gallium in the films.   Similar results have also been reported by Laverko et al.,[179] who have studied the growth of GaAs on Ge (111) surfaces.   The epitaxial relationship between the hexagonal GaAs and the substrate is

$$(0001) \text{ hex} \parallel (111) \text{ cub and } [10\bar{1}0] \text{ hex} \parallel [110] \text{ cub}$$

Davey and Pankey[182] have recently established that their GaAs films deposited onto Ge (111) substrates by vacuum evaporation contain a hexagonal structure.   They find that the occurrence of this phase is associated with deposition conditions under which excess gallium may exist (above 450°C).   Similar observations have also been made with vacuum-deposited GaP films, when formed above 500°C.[268]

(8) InSb Films   The stable structure of InSb is cubic zinc blende, but a metastable hexagonal phase with the wurtzite structure has been observed when grown by vapor deposition.[269]   Khan[181] has recently observed polymorphic transition in thin InSb films grown epitaxially by cathodic sputtering onto cleaved (100) NaCl substrates (see Sec. 3b).   In the initial stage of film growth, a thin epitaxial film of indium forms, followed by the formation of InSb.   In situ electron-diffraction observations give evidence of a hexagonal phase, mixed with the normal cubic structure.   The formation of this hexagonal structure is observed only during the initial stage of growth of InSb, when free indium is present on the substrate.   It is likely that the nucleation of the hexagonal phase is effected by the presence of surface impurities, i.e., indium.

(9) CdS Films   Studies of the epitaxial growth of CdS films[50,270-275] have shown the formation of both the cubic (β) and the hexagonal (α) modifications.   Semiletov[270] has found the hexagonal phase to increase as a result of annealing.   The epitaxial growth of the cubic as well as the hexagonal CdS films has been reported by Escoffery,[273] Weinstein et al.,[274] and Khan and Chopra.[275]   Recently Chopra and Khan[50] have made a detailed study of the polymorphic transformation in epitaxial CdS films grown by vacuum evaporation on NaCl, mica, and silver substrates under different conditions of film growth.   Epitaxial growth of the hexagonal phase is obtained on cleaved NaCl and mica substrates at room temperature, and on Ag (111) substrates at about 170°C.   The epitaxial relationships are

$$(0001) \text{ CdS} \parallel (100) \text{ NaCl and } [10\bar{1}0] \text{ CdS} \parallel [110] \text{ NaCl}$$
$$[11\bar{2}0] \text{ CdS} \parallel [110] \text{ NaCl (weak)}$$
$$(0001) \text{ CdS} \parallel (0001) \text{ mica and } [10\bar{1}0] \text{ CdS} \parallel [10\bar{1}0] \text{ mica}$$
$$[11\bar{2}0] \text{ CdS} \parallel [10\bar{1}0] \text{ mica (weak)}$$

The α phase transforms gradually into the β phase if depositions are made at higher substrate temperatures.   At temperatures of about 200°C for NaCl and 320°C for mica, the transformation is nearly complete.   Figure 27 illustrates the structural transformation as a function of NaCl substrate temperatures.   The β phase grows in parallel orientation with the (100) NaCl substrate, while with mica the orientation is

$$(111) \text{ CdS} \parallel (0001) \text{ mica and } [11\bar{2}] \text{ CdS} \parallel [10\bar{1}0] \text{ mica}$$

The hexagonal $\rightarrow$ cubic transformation as a result of the increased substrate temperature is observed for film thicknesses up to 4,000 Å, beyond which mixed phases occur with subsequent enhancement of the hexagonal phase at the expense of the cubic phase.   The transformation is dependent on growth rate.   Low growth rates are generally required for the transition, especially with mica when the transition occurs only if the rate is lower than 100 Å min⁻¹.

CdS, when deposited on rock salt at room temperature, has the wurtzite phase, as shown in Fig. 28a.   If an electric field of 100 V cm⁻¹ is applied laterally in the plane of the film during deposition, a partial transformation to the cubic phase occurs, as indicated in Fig. 28b.   However, at an electric field of 300 V cm⁻¹ and a substrate temperature of 50°C, the transformation is complete, as shown in Fig. 28c and d.

Micrographs show that the effect of the applied electric field is to bring about the coalescence of small particles and thus to increase the grain size. The transformation is generally observed at a film thickness of about 50 Å, which is a stage just before film continuity occurs. The application of the field has little effect on thicker films, although the β phase continues to grow, mixed with the α phase. As explained by Chopra and Khan, the early coalescence results in enhanced orientation effects which can be attributed to higher mobility of the islands. This increased mobility plays the

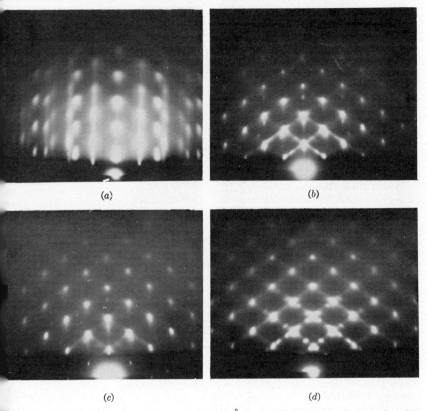

(a)   (b)

(c)   (d)

Fig. 27 Electron-diffraction patterns of ∼1,000-Å-thick CdS/NaCl film deposited at different substrate temperatures. (a) α phase. (b) β phase with weak α phase. (c) β phase transformation almost complete). (d) β phase with twinning. (Chopra and Khan.[50]) *Reproduced with permission of Surface Science.*)

same role as the elevated temperature does in inducing transition. As the film becomes continuous, the electrostatic effects disappear, resulting in little effect on the transition. These observations show that the cubic structure occurs only in a limited range of substrate temperature, film thickness, and applied electric field.

(10) Alloy Films Phase transitions in alloy films have been extensively reported in the literature. These occur depending upon the composition of the constituents of the alloy systems and upon the condition of film formation. Discussion concerning this subject is beyond the scope of this chapter. However, a typical example of phase transitions in epitaxial alloy films can be found in the annealing studies of Au–15 atomic % Ni alloy films grown on rock salt at 400°C [see Sec. 2e(2)]. Similar examples of the development of superlattice structures in the epitaxial Cu-Au alloy system have been made by Sato and Toth.[130]

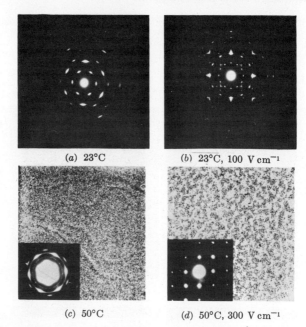

(a) 23°C   (b) 23°C, 100 V cm⁻¹

(c) 50°C   (d) 50°C, 300 V cm⁻¹

Fig. 28   Electron-diffraction patterns and micrographs of a 50-Å-thick CdS film on NaCl (a) Without and (b) with 100 V cm⁻¹ laterally applied electric field during deposition a room temperature.   (c) Without and (d) with 300 V cm⁻¹ laterally applied electric field during deposition at 50°C.   (Chopra and Khan.⁵⁰)   (*Reproduced with permission of Surface Science.*)

5. IMPERFECTIONS IN SINGLE-CRYSTAL FILMS

a. Types of Imperfections Observed

The application of electron diffraction and electron microscopy to the study of th growth of single-crystal films has provided detailed information on the lattice imper fections that are incorporated in the films during growth.   The general types o imperfection structure observed may be listed as follows:

1. Stacking faults
2. Twins
3. Multiple-positioning boundaries
4. Defects due to point-defect aggregation (dislocation loops, stacking-fault tetra hedra, and other configurations)
5. Dislocations

These lattice imperfections have been observed in epitaxial films of a wide variet of materials such as metals, alloys, semiconductors, and insulators.   These have bee considered in the growth studies of single-crystal films (Secs. 2e and 3b).   In th section we will illustrate some typical examples of lattice imperfections in furthe detail.

The growth of planar defects in epitaxial films is very common.   Figure 29 illustrate such defects in a (100) gold film deposited on a rock-salt substrate at 270°C.³⁰   Th main features correspond to stacking faults on {111} planes at 54°45' to the (100) plan of the film.   However, electron-diffraction patterns from such a film exhibit extr reflections characteristic of (111) twinning.   The general interpretation of the micr graph is that it shows stacking faults and narrow twins.   A typical example of (111 twinning is shown in Fig. 13, which is obtained from a β-SiC film formed by chemic: conversion of Si (100) on sapphire at 1100°C.

In many instances, electron-diffraction observations show extra reflections which are not due to the primary diffraction by the twins.   These anomalous reflections appear as a result of double diffraction.   This occurs if a diffracted beam from the matrix passes into the twin and then suffers a second diffraction (as proposed by Thirsk and Whitmore[276]).   That the extra reflections are due to double diffraction can easily be confirmed by dark-field image observations.   The effect of double diffraction, together with twinning and extra reflections due to buckling of the film, is illustrated in Fig.

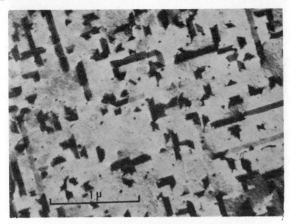

Fig. 29   Electron micrograph of a (100) gold film ($\sim$800 Å thick) grown on NaCl at 270°C, showing stacking faults and twins (Bassett et al.[30]).   (*Reproduced with permission of John Wiley & Sons, Inc.*)

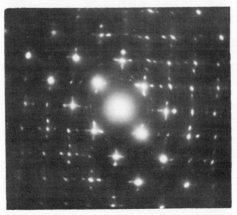

Fig. 30   Electron-diffraction pattern from a Au-15 at. % Ni film formed epitaxially on NaCl at 400°C (*Khan, unpublished*).

30.[207]   Interpretation of the diffraction pattern can be made in the same way as Burbank and Heidenreich[119] have done for their Permalloy films.

Multiple-positioning boundary is a common type of imperfection in single-crystal films, since equivalent orientations occur in many systems.   The commonest example of this defect structure is double positioning, which occurs as a result of two equivalent (111) orientations.   In this case one orientation bears a 180°C rotation with respect to the other.   This double-positioning boundary occurs when two islands with two such orientations join together during growth; this has been discussed in a preceding section.   Figure 31 shows such double positioning in a β-SiC film formed by chemical conversion of Si (111) at 950°C.[45] In addition to the double-positioning structure, the effect of double diffraction is present, as explained in Fig. 32. Additional spots cutting the main reflections by one-third represent the presence of twinning in all the planes of the (111) type. Also present are long interference lines joining the main spots in (111) directions, indicating stacking faults of random repetition of the (111) planes. In

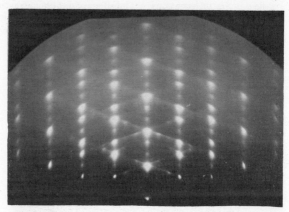

Fig. 31 Electron-diffraction pattern from a β-SiC (111) film grown epitaxially by conversion of Si (111) at 950°C. Azimuth [1Ī0] (Khan[45]). (*Reproduced with permission of Materials Research Bulletin.*)

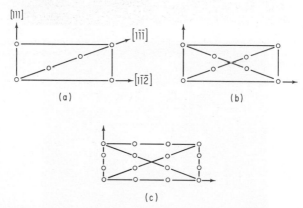

Fig. 32 Generating units of diffraction pattern in azimuths of type [110]. (a) Ideal fcc lattice. (b) Same as (a) but with double positioning (180°C twin pattern). (c) Effect due to double diffraction.

should be mentioned here that in many cases [such as fcc (111) films] double positioning is equivalent to twinning, with the substrate plane being the twin plane. Then the double-positioning boundaries are incoherent twin boundaries.

The commonest structural defects in epitaxial semiconductor films can be illustrated in Fig. 33, which represents the surface structure of an etched silicon film grown epitaxially on a Si (111) surface.[142] The defects consist of three inclined stacking faults sometimes forming a triangular-based tetrahedron. A model showing how faults of this type are formed has been proposed by Booker and Stickler[277] (Fig. 34). The fault originates from the interface and propagates along the three adjacent {111} planes, each making an angle of 70°32' with the substrate surface. These three planes and the grown surface form a regular tetrahedron with one apex at the interface. Chu and

Gavaler[278] have revealed such fault formation by cross-sectioning surfaces of the Si sample. With similar considerations, a square-based tetrahedron occurs on (100) surfaces. The various geometries for stacking-fault defects have been discussed by Pashley.[24]

The major observed imperfections in single-crystal metal films are dislocations, which are as high as 10^{10} dislocations per cm². These consist of lines passing from one end of the film to the other by about the shorted path. They can easily be observed by means of moiré patterns. The initial nuclei do not contain any dislocations, but they appear when the network stage of growth is reached.

Fig. 33 Optical micrograph of an etched Si film formed by evaporation onto a Si (111) surface, showing stacking-fault defects (Unvala and Booker[142]). (*Reproduced with permission of Philosophical Magazine.*)

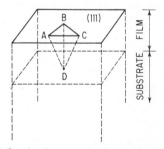

Fig. 34 Model showing how a triangular stacking fault is formed.

Interfacial dislocations have been observed in a number of epitaxial films. The interface between the deposit and the substrate may be considered in terms of a network, consisting of edge dislocations which accommodate the misfit between the two lattices at the interface. Matthews[209] has observed such dislocation networks in superimposed films of PbS and PbSe. Grünbaum and Mitchell[279] have also observed such dislocations for nickel bromide grown on chromic bromide. In this case the misfit is only 0.3%, and the spacing of the network is several hundred angstroms. Similar observations have been reported by Matthews[211] for metal films with a misfit below 5%.

Gradmann[280] has studied the growth of silver on copper (111) surfaces by electron diffraction, and has observed some fine structure in diffraction patterns. This fine structure is explained in terms of strain effects associated with the dislocation network at the interface.   Recently Krause[281] has studied the initial stages of epitaxy of copper and silver films on copper substrates.   He observes a new kind of diffraction pattern, originating from two-dimensional surface structures at the interface, which is interpreted by periodic arrangements of interfacial dislocations.   In the epitaxy of Cu, a hexagonal network of screw dislocations is formed, as can be illustrated by the diffraction pattern in Fig. 35, while for Ag films the diffraction patterns give evidence of networks of interfacial edge dislocations in the very early stages of growth.   The dislocation densities in the film are observed to be in good agreement with values predicted by the theory of interfacial dislocations.[11]

Fig. 35   Electron-diffraction pattern from a hexagonal network of screw dislocations at the interface of a very thin Cu film on a Cu (111) substrate plane, azimuth [1$\bar{1}$0] (Krause[281]) (*Reproduced with permission of Journal of Applied Physics.*)

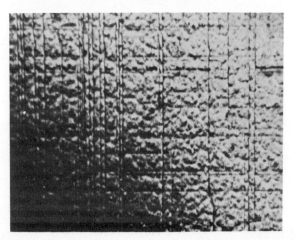

Fig. 36   Electron micrograph showing dislocation lines present to accommodate part of the misfit between copper- and γ-iron.   The plane of the deposit is (001), and the boundaries to the figure are parallel to the [110] and [1$\bar{1}$0] directions.   Magnifications $\times$ 35,000 (Jesser and Matthews[72]).   (*Reproduced with permission of Philosophical Magazine.*)

Jesser and Matthews[72] have recently observed that iron deposited onto copper substrates at room temperature under ultrahigh-vacuum conditions grows as a monolayer and is strained to fit the substrate lattice. As the film thickness increases, long, straight misfit dislocations appear, as illustrated in Fig. 36. Similar observations have also been reported by Jesser and Matthews[73] on the pseudomorphic growth of cobalt on copper and chromium on nickel in ultrahigh vacuum. Long dislocations occur to accommodate part of the misfit between the film and the substrate. They run parallel to the [110] and [1$\bar{1}$0] directions in the (100) plane of the films.

b. Mechanisms of Imperfection Formation

Mechanisms for the formation of lattice imperfections in single-crystal films have been discussed by a number of investigators.[24,30,282,283] These may be listed as follows:
1. Extension of imperfections existing on the substrate surface
2. Accommodation of nuclei misorientations and displacement misfits
3. Point-defect aggregation
4. Plastic deformation during growth
5. Role of surface contamination in defect formation
6. Generation of dislocations to accommodate lattice misfit during pseudomorphic growth

We will now consider these mechanisms in some detail.

(1) Extension of Substrate Imperfections This mechanism does not seem to be very significant, because epitaxially grown films are often found to have dislocation density several orders of magnitude higher than in the substrate surface. As Pashley[24] has pointed out, the extension of substrate dislocations can take place only if the epitaxial film is pseudomorphic with the substrate. Otherwise, the deposit nuclei will be controlled by the crystal structure of the deposited material, rather than that of the substrate. This means that the growing nuclei will ignore the presence of dislocations in the substrate surface. Although pseudomorphic growth of metal films has recently been reported, detailed investigations would be necessary to test the validity of the influence of substrate imperfections.

(2) Accommodation of Nuclei Misorientations and Displacement Misfits It is known that coalescence of small islands during growth plays an important role in the formation of structural imperfections in the growing film. Reorientation of islands occurs during coalescence, and this prevents the formation of dislocations. But when the islands are larger, the misorientations are accommodated by dislocation formation. A low-angle boundary thus results. If a small amount of misorientation exists across a long channel in the film, dislocations will be formed as the channel is gradually filled in. If the channel is terminated at both ends, it will then contain incipient dislocations. It appears that dislocations result from misorientations, and the number of dislocations will therefore depend on the degree of misorientation during coalescence.

Dislocations also result from displacement misfits, as illustrated in Fig. 37, for the case of three coalescing nuclei. The joining occurs in such a way that it minimizes elastic strain at each interface. This will result in the formation of an incipient dislocation in the hole between the three nuclei.

In general, both the nuclei rotation and displacement misfits occur simultaneously such that the two effects may be considered together. In isoepitaxial growth, these defects are normally absent, and hence a low dislocation density is expected. On the other hand, displacement misfits are present, which inevitably introduce dislocations in the film. These effects are applicable to very thin films. For thicker films, it is difficult to know how the dislocation structure changes. It is likely that the dislocation density decreases as a result of dislocation interactions.

Twins and stacking faults occur during coalescence of islands and at the junction plane between the islands. Matthews and Allison[284] have suggested that twins occur particularly when a small poorly oriented nucleus or island coalesces with a large oriented island. In the early stages, large misorientations are present, and if an oriented nucleus close to a twin, it is energetically favorable to rotate into twin orientation as a result coalescence. However, twins occur even if misorientations are small, as has been observed for the growth of Au and Ag on molybdenite.

(3) Point-defect Aggregation The role of point defects in the formation of dislocation loops is not well understood. It is likely that during deposition of a film by evaporation many vacancies should be trapped in the film, particularly when the deposition is done at low temperatures and at fast rates of deposition. In the case where mobility of the point defects is small, the point defects could aggregate slowly, resulting in the formation of dislocation loops and other defects. There is also a good chance that point defects prefer to migrate to the surface of a thin film.

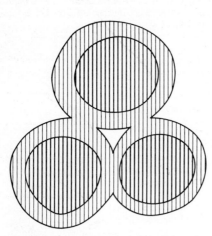 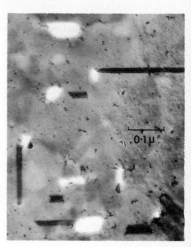

Fig. 37 Formation of incipient dislocation due to displacement misfits between three coalescing nuclei (Pashley[24]). [*Reproduced with permission of Advances in Physics (Philosophical Magazine).*]

Fig. 38 Electron micrograph of a (100) gol film containing holes where partial dislo cations are nucleated because of stres introduced by contamination. Magnifica tion × 70,000 (Pashley[285]). (*Reproduce with permission of American Society fo Metals.*)

(4) Plastic Deformation Epitaxial films are observed to be in a state of high stress which is likely to cause plastic deformation during film growth. The edges of growing islands are the favorable sites for the nucleation of dislocations. This process coul therefore result in dislocations of various configurations.

Localized plastic deformation plays a role in the elimination of movement of faul which follow coalescence. This influences the imperfections in the final film structure Single partial dislocations could be nucleated, producing stacking faults in the film These may occur when the film is in a state of stress because of contamination Figure 38 illustrates stacking faults generated at the edge of the holes of a (100) gol film.[285] In addition to introducing defects, plastic deformation can cause movemer of dislocations produced by other mechanisms. Although the role of plastic deforma tion is not clearly understood, it appears to be a likely possibility for introducing latti imperfections in epitaxially grown films.

(5) Role of Surface Contamination The role of contamination in the production lattice defects can be understood when epitaxial films are formed under extremely clea conditions. The majority of the evidence for defect formation due to substrat surface contamination has been obtained with isoepitaxial films of silicon. In th case, the formation of dislocations is almost absent, because of the zero misfit betwee the lattices of the deposit and the substrate. Tetrahedron stacking faults in epitaxi films of silicon have largely been observed, and these are associated with minute trac of impurities on the substrate surface. This has been evidenced by Unvala ar Booker[143] during their investigations of the different stages of growth of vacuur deposited silicon films. A number of mechanisms for the production of stacking faul have been suggested.[277,285,286] One of them involves the use of a nucleus forming on a (111) substrate surface with a stacking fault at the interface.[277] This nucleus coalesces with its neighbor without any interface stacking faults, resulting in the formation of inclined stacking faults, which in turn form the basis for the development of tetrahedral fault configurations.

(6) Accommodation of Lattice Misfits during Pseudomorphic Growth Recent observations by Jesser and Matthews[72,23] have shown that pseudomorphic growth occurs during the initial stage of film formation under ultrahigh-vacuum conditions; i.e., the lattices of the deposit are strained to match exactly that of the substrate. The misfit is accommodated by dislocation lines which increase with increasing film thickness. The mechanism for the generation of misfit dislocation lines in iron grown on copper is illustrated in Fig. 39.[72] In Fig. 39a, a dislocation line is shown extending from one specimen to the other. The elastic strain of the iron film causes the line in the fcc iron to bow out, the geometry of this bowing portion changing with film thickness until it is as in b. At this point, the glide force on the dislocation in the iron is equal to the line tension of the misfit dislocation. The deposition of further iron will cause the dislocation to glide to the right, as shown in c.

The generation of misfit dislocations is followed by their dissociation into partial dislocations. The migration of partial dislocations as in the case of cobalt on copper[73] creates intrinsic stacking faults and so converts some fcc cobalt into hcp cobalt.

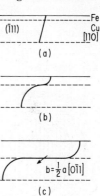

Fig. 39 The generation of a misfit dislocation line in Fe grown on Cu (100). The thick continuous and thick dotted lines represent the intersection of the (111) slip plane with the surfaces of the iron and copper films. The fine dotted line is the intersection of the slip plane and the Cu-Fe interface. The fine continuous line is the dislocation (Jesser and Matthews[72]). (*Reproduced with permission of Philosophical Magazine.*)

6. CONCLUSIONS

In the preceding sections we have discussed the various processes for the growth of single-crystal films and have enumerated the structural characteristics of single-crystal films of a wide variety of thin film materials. Among the various processes discussed, the process based on the phenomenon of epitaxy is most common. Although various theories for the occurrence of epitaxy have been put forward, none of them is quite adequate to explain all the experimentally observed features of single-crystal films. More detailed theoretical considerations should therefore be given to the role of nucleation in epitaxial-film formation, with emphasis placed on the configuration of clusters on surfaces and on the size effect on the most favored orientation. Recent in situ electron-microscopy observations have revealed some basic processes of film formation, which may have a determining influence on the occurrence of epitaxy. Marked orientation changes occur as a result of coalescence and recrystallization during the initial stages of film growth. Since very small amounts of contamination can have a major effect on film growth, this requires careful attention, especially in understanding the basic processes of coalescence and recrystallization. It is thus obvious that more detailed studies of film formation under clean vacuum environment are needed to obtain a clearer picture of the epitaxial process.

The mode of formation of lattice imperfections in single-crystal films is not clearly understood. Although a number of mechanisms have been discussed, the experimental evidence does not rule out any of them. It is established that "liquid-like" coalescence leads to recrystallization and reorientation of the initial nuclei, particularly when they are small. This behavior has to be taken into account when any mechanism for the incorporation of lattice imperfections in growing films is considered. Coalescence of nuclei and islands gives rise to imperfections to accommodate rotational and displacement misfits. Recent studies by Jesser and Matthews illustrate the formation of misfit dislocations and stacking faults as a result of the deposit's being strained to match the lattice of the substrate. These studies have provided strong evidence of the occurrence of pseudomorphism, a long-controversial concept.

On the basis of the above considerations, it is obvious that the general direction in which future studies should be performed will involve in situ electron-diffraction and electron-microscopy investigations on the growth of single-crystalline films under clean vacuum environment. Such investigations will eliminate the role of contamination on film formation, particularly during coalescence and recrystallization processes, and will provide easier means of interpretation. Special emphasis should be given to the mode of formation of lattice imperfections and the occurrence of basal-plane pseudomorphism. Information resulting from such studies would lead to a clearer understanding of the processes involved in the growth of single-crystal films.

REFERENCES

1. Pashley, D. W., *Advan. Phys.*, **5**, 173 (1956).
2. Royer, L., *Bull. Soc. Franc. Mineral.*, **51**, 7 (1928).
3. Menzer, G., *Naturwiss.*, **26**, 385 (1938); *Z. Krist.*, **99**, 378 (1938); **99**, 410 (1938).
4. Kirchner, F., and H. Cramer, *Ann. Physik*, **33**, 138 (1938).
5. Uyeda, R., *Proc. Phys.-Math. Soc. Japan*, **24**, 809 (1942).
6. Raether, H., *Optik*, **1**, 296 (1946).
7. Raether, H., *Ergeb. Exakt. Naturw.*, **24**, 54 (1951).
8. Engel, O. G., *J. Chem. Phys.*, **20**, 1174 (1952).
9. Engel, O. G., *J. Res. NBS*, **50**, 249 (1953).
10. van der Merwe, J. H., *Discussions Faraday Soc.*, no. 5, p. 201, 1949.
11. Frank, F. C., and J. H. van der Merwe, *Proc. Roy. Soc. London*, **A198**, 205 (1949); **A198**, 216 (1949); **A200**, 125 (1949).
12. Hirth, J. P., and G. M. Pound, "Condensation and Evaporation, Nucleation and Growth Kinetics," Pergamon Press, New York, 1963.
13. Pound, G. M., and J. P. Hirth, "Dayton International Symposium on Evaporation and Condensation of Solids," Gordon and Breach, Science Publishers, New York, 1963.
14. Hirth, J. P., S. J. Hruska, and G. M. Pound, in M. H. Francombe and H. Sato (eds.), "Single Crystal Films," Pergamon Press, New York, 1964.
15. Moazed, K. L., in J. C. Anderson (ed.), "The Use of Thin Films in Physical Investigations," p. 203, Academic Press Inc., New York, 1966.
16. Rhodin, T. N., and D. Walton, "Metal Surfaces," p. 259, American Society for Metals, Metals Park, Ohio, 1962.
17. Rhodin, T. N., and D. Walton, in M. H. Francombe and H. Sato (eds.), "Single Crystal Films," p. 31, Pergamon Press, New York, 1964.
18. Walton, D., *J. Chem. Phys.*, **37**, 2182 (1962).
19. Walton, D., *Phil. Mag.*, **7**, 1671 (1962).
20. Rhodin, T. N., in J. C. Anderson (ed.), "The Use of Thin Films in Physical Investigations," p. 187, Academic Press Inc., New York, 1966.
21. Shirai, S., and Y. Fukada, *J. Phys. Soc. Japan*, **17**, 1018 (1962).
22. Matthews, J. W., *Phil. Mag.*, **12**, 1143 (1965).
23. Pashley, D. W., M. J. Stowell, M. H. Jacobs, and T. J. Law, *Phil. Mag.*, **10**, 127 (1964).
24. Pashley, D. W., *Advan. Phys.*, **14**, 327 (1965).
25. Wagner, R. S., and W. C. Ellis, *Appl. Phys. Letters*, **4**, 89 (1964).
26. Nielson, S., *J. Electrochem. Soc.*, **112**, 534 (1965).
27. Zakharov, V. P., Yu. A. Tevirko, and V. N. Chugaev, *Soviet Phys. Doklady English Transl.*, **11**, 899 (1967).
28. Filby, J. D., and S. Nielson, *J. Electrochem. Soc.*, **112**, 957 (1965).
29. Filby, J. D., and S. Nielson, *J. Electrochem. Soc.*, **112**, 535 (1965).
30. Bassett, G. A., J. W. Menter, and D. W. Pashley, in C. A. Neugebauer, J. B. Newkirk, and D. A. Vermilyea (eds.), "Structure and Properties of Thin Films," p. 11, John Wiley & Sons, Inc., New York, 1959.
31. Sella, C., and J. J. Trillat, in M. H. Francombe and H. Sato (eds.), "Single Crystal Films," p. 201, Pergamon Press, New York, 1964.
32. Ino, S., D. Watanabe, and S. Ogawa, *J. Phys. Soc. Japan*, **19**, 881 (1964).
33. Grünbaum, E., and J. W. Matthews, *Phys. Stat. Solidi*, **9**, 731 (1965).

34. Lawless, K. R., and D. F. Mitchell, *Mem. Sci. Rev. Met.*, vol. 62 (special number), pp. 27, 39, May 1965.
35. Tung, S. K., *J. Electrochem. Soc.*, **112**, 436 (1965).
36. Bruck, L., *Ann. Physik*, **26**, 233 (1936).
37. Krikorian, E., and R. J. Sneed, *Trans. 10th AVS Symp.*, 1963, p. 368, The Macmillan Company, New York.
38. Krikorian, E., and R. J. Sneed, *J. Appl. Phys.*, **37**, 3665 (1966).
39. Shirai, S., *Proc. Phys.-Math. Soc. Japan*, **21**, 800 (1939).
40. Sloope, B. W., and C. O. Tiller, *Trans. 10th AVS Symp.*, 1963, p. 339, The Macmillan Company, New York, 1963.
41. Chopra, K. L., and M. R. Randlett, *J. Appl. Phys.*, **39**, 1874 (1968).
42. Shinozaki, S., and H. Sato, *J. Appl. Phys.*, **36**, 2320 (1965).
43. Matthews, J. W., and E. Grünbaum, *Phil. Mag.*, **11**, 1233 (1965).
44. Spielmann, W., *Z. Angew. Phys.*, **19**, 93 (1965).
45. Khan, I. H., *Mater. Res. Bull.*, **4**, S285 (1969).
46. Pashley, D. W., *Phil. Mag.*, **4**, 316 (1959).
47. Matthews, J. W., *Phil. Mag.*, **7**, 415 (1962).
48. Matthews, J. W., and E. Grünbaum, *Appl. Phys. Letters*, **5**, 106 (1964).
49. Matthews, J. W., *J. Vacuum Sci. Technol.*, **3**, 133 (1966).
50. Chopra, K. L., and I. H. Khan, *Surface Sci.*, **6**, 33 (1967).
51. Stirland, D. J., *Appl. Phys. Letters*, **8**, 326 (1966).
52. Miller, D. P., S. B. Watelski, and C. R. Moore, *J. Appl. Phys.*, **34**, 2813 (1963).
53. Brockway, L. O., and R. B. Marcus, *J. Appl. Phys.*, **34**, 921 (1963).
54. Roberts, R. W., *Brit. J. Appl. Phys.*, **14**, 537 (1963).
55. Moll, J., le Vide, no. 105, p. 248, 1963.
56. Boswell, F. W. C., *Phys. Rev.*, **80**, 91 (1950).
57. Boswell, F. W. C., *Proc. Phys. Soc. London*, **A64**, 465 (1951).
58. Blackman, M., and I. H. Khan, *Proc. Phys. Soc. London*, **77**, 471 (1961).
59. Khan, I. H., *Proc. Phys. Soc. London*, **76**, 507 (1960).
60. Finch, G. I., and A. G. Quarrell, *Proc. Roy. Soc. London*, **A141**, 398 (1933).
61. Finch, G. I., and A. G. Quarrell, *Proc. Phys. Soc. London*, **46**, 148 (1934).
62. Cochrane, W., *Proc. Phys. Soc. London*, **48**, 723 (1936).
63. Miyake, S., *Sci. Paper Inst. Phys. Chem. Res. Japan*, **34**, 565 (1938).
64. Raether, H., *J. Phys. Radium*, **11**, 11 (1950).
65. Lucas, L. N. D., *Proc. Phys. Soc. London*, **A64**, 943 (1951).
66. Lucas, L. N. D., *Proc. Phys. Soc. London*, **A215**, 162 (1952).
67. Ehlers, H., and H. Raether, *Naturwiss.*, **39**, 487 (1952).
68. Ehlers, H., *Z. Physik*, **136**, 379 (1953).
69. Newman, R. C., *Proc. Phys. Soc. London*, **B69**, 432 (1956).
70. Jones, J. P., *Proc. Roy. Soc. London*, **A284**, 469 (1965).
71. Schlier, R. E., and H. E. Farnsworth, *J. Phys. Chem. Solids*, **6**, 271 (1958).
72. Jesser, W. A., and J. W. Matthews, *Phil. Mag.*, **15**, 1097 (1967).
73. Jesser, W. A., and J. W. Matthews, *Phil. Mag.*, **17**, 461 (1968).
74. Jesser, W. A., and J. W. Matthews, *Phil. Mag.*, **17**, 475 (1968).
75. Jesser, W. A., and J. W. Matthews, *Phil. Mag.*, **17**, 595 (1968).
76. Newman, R. C., and D. W. Pashley, *Phil. Mag.*, **46**, 927 (1955).
77. Freedman, J. F., *J. Appl. Phys.*, **33** (2), 1148 (1962).
78. Borie, B., C. J. Sparks, and J. V. Cathcart, *Acta Met.*, **10**, 691 (1962).
79. Collins, L. E., and O. S. Heavens, *Proc. Phys. Soc. London*, **B70**, 265 (1957).
80. van der Merwe, J. H., *J. Appl. Phys.*, **34**, 123 (1963).
81. van der Merwe, J. H., in M. H. Francombe and H. Sato (eds.), "Single Crystal Films," p. 139, Pergamon Press, New York, 1964.
82. Gottsche, H., *Z. Naturforsch.*, **11A**, 55 (1956).
83. Kehoe, R. B., *Phil. Mag.*, **2**, 445 (1957).
84. Sloope, B. W., and C. O. Tiller, *J. Appl. Phys.*, **32**, 1331 (1961).
85. Hirth, J. P., and G. M. Pound, *Progr. Mater. Sci.*, **11**, 57 (1963).
86. Bassett, G. A., and D. W. Pashley, *J. Inst. Metals*, **87**, 449 (1958–1959).
87. Shirai, S., Y. Fukada, and M. Nomura, *J. Phys. Soc. Japan*, **16**, 1989 (1961).
88. Chopra, K. L., L. C. Bobb, and M. H. Francombe, *J. Appl. Phys.*, **34**, 1699 (1963).
89. Chopra, K. L., and R. Randlett, *Appl. Phys. Letters*, **8**, 241 (1966).
90. Campbell, D. S., and D. J. Stirland, *Phil. Mag.*, **9**, 703 (1964).
91. Jaunet, J., and C. Sella, *Bull. Soc. Franc. Miner. Crist.*, **87**, 393 (1964).
92. Ogawa, S., D. Watanabe, and S. Ino, *Acta Cryst.*, **16**, A133 (1963).
93. Ino, S., D. Watanabe, and S. Ogawa, *J. Phys. Soc. Japan*, **20**, 242 (1965).
94. Matthews, J. W., *Appl. Phys. Letters*, **7**, 131 (1965).

95. Matthews, J. W., *Appl. Phys. Letters*, **7**, 255 (1965).
96. Harsdorff, M., *Solid State Commun.*, **2**, 133 (1964).
97. Harsdorff, M., and H. Raether, *Z. Naturforsch.*, **19a**, 1497 (1964).
98. Bauer, E., A. K. Green, K. M. Kunz, and H. Poppa, International Symposium, Basic Problems in Thin Film Physics, Clausthal-Göttingen, September, 1965.
99. Bethge, H., and M. Krohn, International Symposium, Basic Problems in Thin Film Physics, Clausthal-Göttingen, September, 1965.
100. Bauer, E., A. K. Green, and K. M. Kunz, *Appl. Phys. Letters*, **8**, 248 (1966).
101. Palmberg, P. W., C. J. Todd, and T. N. Rhodin, *J. Appl. Phys.*, **39**, 4650 (1968).
102. Palmberg, P. W., T. N. Rhodin, and C. J. Todd, *Appl. Phys. Letters*, **10**, 122 (1967).
103. Kunz, K. M., A. K. Green, and E. Bauer, *Phys. Stat. Solidi*, **18**, 441 (1966).
104. Chopra, K. L., *Appl. Phys. Letters*, **7**, 140 (1965).
105. Yelon, A., and R. W. Hoffman, *J. Appl. Phys.*, **31**, 1672 (1960).
106. Taylor, N. J., *Surface Sci.*, **4**, 161 (1966).
107. Krause, G. O., *J. Appl. Phys.*, **37**, 3691 (1966).
108. Glass, S. J., and M. J. Klein, *Phys. Rev.*, **109**, 288 (1958).
109. Heavens, O. S., in M. H. Francombe and H. Sato (eds.), "Single Crystal Films," p. 381, Pergamon Press, New York, 1964.
110. Boersch, H., and M. Lambeck, *Z. Physik*, **159**, 248 (1960).
111. Hale, M. E., H. W. Fuller, and H. Rubinstein, *J. Appl. Phys.*, **30**, 789 (1960).
112. Neugebauer, C. A., in M. H. Francombe and H. Sato (eds.), "Single Crystal Films," p. 361, Pergamon Press, New York, 1964.
113. Sato, H., R. S. Toth, and R. W. Astrue, *J. Appl. Phys.*, **34**, 1062 (1963).
114. Pomeranz, M., J. F. Freedman, and J. C. Suits, *J. Appl. Phys.*, **33(II)**, 1164 (1962).
115. Heavens, O. S., R. F. Miller, G. L. Moss, and J. C. Anderson, *Proc. Phys. Soc. London*, **B78**, 33 (1961).
116. Sato, H., R. S. Toth, and R. W. Astrue, *J. Appl. Phys. Suppl.*, **33**, 1113 (1962).
117. Honma, T., and C. M. Wayman, *J. Appl. Phys.*, **36**, 2791 (1965).
118. Heavens, O. S., R. F. Miller, and M. S. Zafar, *Acta Cryst.*, **20**, 288 (1966).
119. Burbank, R. D., and R. D. Heidenreich, *Phil. Mag.*, **5**, 373 (1960).
120. Schoening, F. R. L., and A. Baltz, *J. Appl. Phys.*, **33**, 1442 (1962).
121. Alessandrini, E. I., *J. Appl. Phys.*, **37**, 4811 (1966).
122. Verderber, R. R., and B. M. Kostyk, *J. Appl. Phys.*, **32**, 696 (1961).
123. Takei, H., and S. Takasu, *J. Appl. Phys. Japan*, **3**, 4 (1964).
124. Pulliam, G. R., et al., National Aerospace Electronics Conference, 1965, p. 241.
125. Archer, J. L., G. R. Pulliam, R. G. Warren, and J. E. Mee, in H. S. Peiser (ed.), *Proc. Intern. Conf. Crystal Growth*, Boston, 1966, p. 337.
126. Gambino, R. J., *J. Appl. Phys.*, **38**, 1129 (1967).
127. Khan, I. H., and M. H. Francombe, *Nature*, **199**, 800 (1963).
128. Khan, I. H., and M. H. Francombe, *J. Appl. Phys.*, **36**, 1699 (1965).
129. Fukano, Y., *J. Phys. Soc. Japan*, **16**, 1195 (1961).
130. Sato, H., and R. S. Toth, *Phys. Rev.*, **124**, 1833 (1961).
131. Noreika, A. J., and M. H. Francombe, *Trans. 13th Natl. Vacuum Symp.*, San Francisco, 1966.
132. Hansen, M., "Constitution of Binary Alloys," p. 198, McGraw-Hill Book Company, New York, 1958.
133. Unvala, B. A., *Nature*, **194**, 966 (1962).
134. Unvala, B. A., *Vide*, **104**, 109 (1963).
135. Nielson, S., D. G. Coates, and J. E. Maines, International Symposium on Condensation and Evaporation of Solids, Dayton, Ohio, Sept. 12–14, 1962.
136. Hale, A. P., *Vacuum*, **13**, 93 (1963).
137. Theuerer, H. C., *J. Electrochem. Soc.*, **108**, 649 (1961).
138. Postnikov, V. V., R. G. Loginova, and M. I. Ovsyannikov, *Soviet Phys.-Cryst. English Transl.*, **10**, 495 (1966).
139. Petrin, A. I., and G. A. Kurov, *Soviet Phys.-Cryst. English Transl.*, **10**, 634 (1966).
140. Thomas, D. J. D., *Phys. Stat. Solidi*, **13**, 359 (1966).
141. Khan, I. H., and R. N. Summergrad, *Appl. Phys. Letters*, **11**, 12 (1967).
142. Unvala, B. A., and G. R. Booker, *Phil. Mag.*, **9**, 691 (1964).
143. Unvala, B. A., and G. R. Booker, *Phil. Mag.*, **11**, 11 (1965).
144. Newman, R. C., *Microelectronics & Reliability*, **3**, 121 (1964).
145. Nannichi, Y., *Nature*, **200**, 1087 (1963).
146. Widmer, H., *Appl. Phys. Letters*, **5**, 108 (1964).
147. Jona, F., *Appl. Phys. Letters*, **9**, 235 (1966).
148. Thomas, R. N., A. J. Noreika, and M. H. Francombe, Electrochemical Society Meeting, Dallas, Tex., 1967.

149. Via, G. G., and R. E. Thun, *Proc. 2d Intern. Congr. Vacuum Technol.*, Washington, D.C., Oct. 16–19, 1961.
150. Doo, V. Y., *J. Electrochem. Soc.*, **111**, 1196 (1964).
151. Manasevit, H. M., and W. I. Simpson, *J. Appl. Phys.*, **35**, 1349 (1964).
152. Nolder, R., and I. Cadoff, *Trans. Met. Soc. AIME*, **233**, 549 (1965).
153. Manasevit, H. M., A. Miller, F. L. Morritz, and R. Nolder, *Trans. Met. Soc. AIME*, **233**, 540 (1965).
154. Miller, A., and H. M. Manasevit, *J. Vacuum Sci. Technol.*, **3**, 68 (1966).
155. Joyce, B. A., R. J. Bennett, R. W. Bicknell, and P. J. Etter, *Trans. Met. Soc. AIME*, **233**, 556 (1965).
156. Chu, T. L., G. A. Gruber, J. J. Oberly, and R. L. Tallman, AFCRL 65-52, 1965.
157. Salama, C. A. T., T. W. Tucker, and L. Young, *Solid-State Electron.*, **10**, 339 (1967).
158. Semiletov, S. A., *Kristallografiya*, **1**, 542 (1956).
159. Kurov, G. A., S. A. Semiletov, and Z. G. Pinsker, *Dokl. Akad. Nauk SSSR*, **110**, 970 (1956).
160. Weinreich, O., J. Dermit, and C. Tufts, *J. Appl. Phys.*, **32**, 1170 (1961).
161. Davey, J. E., *J. Appl. Phys.*, **33**, 1015 (1962).
162. Rasmanis, E., *Proc. Natl. Electron. Conf.*, **20**, 212 (1964).
163. Kurov, G. A., *Soviet Phys.-Solid State English Transl.*, **5**, 2226 (1963).
164. Mezentseva, N. L., A. I. Petrin, and G. A. Kurov, *Soviet Phys.-Solid State English Transl.*, **6**, 1599 (1965).
165. Wolsky, S. P., T. R. Piwkowski, and G. Wallis, *J. Vacuum Sci. Technol.*, **2**, 97 (1965).
166. Catlin, A., A. J. Bellemore, and R. R. Humphris, *J. Appl. Phys.*, **35**, 251 (1964).
167. Krikorian, E., and R. J. Sneed, *J. Vacuum Sci. Technol.*, **1**, 75 (1964).
168. Behrndt, K. H., *J. Appl. Phys.*, **37**, 3841 (1966).
169. Adamsky, R. F., *J. Vac. Sci. Technol.*, **6**, 542 (1969); *J. Appl. Phys.*, **40**, 4301 (1969).
170. Layton, C. K., and K. B. Cross, *Thin Solid Films*, **1**, 169 (1967).
171. Marucchi, J., and N. Nifonstoff, *Compt. Rend.*, **249**, 435 (1959).
172. Schalla, R. L., L. H. Thaller, and A. E. Potter, Jr., *J. Appl. Phys.*, **33**, 2554 (1962).
173. Schalla, R. L., N. W. Tideswell, and F. D. Coffin, in M. H. Francombe and H. Sato (eds.), "Single Crystal Films," p. 301, The Macmillan Company, New York, 1964.
174. Pundsack, A. L., *J. Appl. Phys.*, **34**, 2306 (1963).
175. Sloope, B. W., and C. O. Tiller, *J. Appl. Phys.*, **36**, 3174 (1965).
176. Wallis, G., and S. P. Wolsky, *Trans. 13th AVS Symp.*, San Francisco, Oct. 26–28, 1966.
177. Tramposch, R. F., *Appl. Phys. Letters*, **9**, 83 (1966).
178. Lever, R. F., and E. J. Huminski, *J. Appl. Phys.*, **37**, 3638 (1966).
179. Laverko, E. N., V. M. Merakhonov, and S. M. Polyakov, *Soviet Phys.-Cryst. English Transl.*, **10**, 611 (1966).
180. Kurdyumova, R. N., and S. A. Semiletov, Summary of Contributions to the Second Congress on Electron Diffraction (in Russian), Moscow, 1962.
181. Khan, I. H., *Surface Sci.*, **9**, 306 (1968).
182. Davey, J. E., and T. Pankey, *J. Appl. Phys.*, **39**, 1941 (1968).
183. Semiletov, S. A., *Kristallografiya*, **6**, 314 (1960).
184. Dale, E. B., G. Sevecal, and D. Huebuer, *Trans. 10th AVS Symp.*, 1963, p. 348, The Macmillan Company, New York.
185. Tausch, F. W., Jr., and A. G. Lapierre III, *J. Electrochem. Soc.*, **112**, 150C (1965).
186. Zanowick, R. L., *J. Electrochem. Soc.*, **114**, 146C (1967).
187. Manasevit, H. M., *Appl. Phys. Letters*, **12**, 156 (1968).
188. Francombe, M. H., *Trans. 10th Natl. Vacuum Symp.*, 1963, p. 316, The Macmillan Company, New York.
189. Muller, E. K., *J. Appl. Phys.*, **35**, 580 (1964).
190. Amick, J. A., in M. H. Francombe and H. Sato (eds.), "Single Crystal Films," p. 283, Pergamon Press, New York, 1964.
191. Mehal, E. W., and G. R. Cronin, *Electrochem. Technol.*, **4**, 540 (1966).
192. Ewing, R. E., and P. E. Greene, *J. Electrochem. Soc.*, **111**, 1266 (1964).
193. Hudock, P., Electrochemical Society Spring Meeting, Dallas, Tex., 1967.
194. Sangster, R. C., "Compound Semiconductors," vol. 1, p. 241, Reinhold Publishing Corporation, New York, 1962.
195. Sato, K., *Solid-State Electron.*, **7**, 743 (1964).
196. Tombs, N. C., J. J. Comer, and J. F. Fitzgerald, *Solid-State Electron.*, **8**, 839 (1965).
197. Poser, H., *Z. Phys. Chem.*, **228**, 113 (1965).
198. Biederman, E., and K. Brack, *J. Appl. Phys.*, **37**, 4288 (1966).
199. Jackson, D. M., Jr., and R. W. Howard, *Trans. Met. Soc. AIME*, **233**, 468 (1965).
200. Jennings, V. J., A. Somer, and H. C. Chang, *J. Electrochem. Soc.*, **113**, 728 (1966).
201. Campbell, R. B., and T. L. Chu, *J. Electrochem. Soc.*, **113**, 825 (1966).

202. Spitzer, W. G., D. A. Kleinman, and C. J. Frosch, *Phys. Rev.*, **113**, 133 (1959).
203. Nakashima, H., T. Saguro, and H. Yanai, *Japan. J. Appl. Phys.*, **5**, 874 (1966).
204. Khan, I. H., and R. N. Summergrad, *J. Vaccum Sci. Technol.*, **4**, 327 (1967).
205. Porter, R. F., *J. Chem. Phys.*, **34**, 583 (1961).
206. Semiletov, S. A., *Soviet Phys.-Cryst. English Transl.*, **9**, 65 (1964).
207. Khan, I. H., unpublished.
208. Elleman, A. J., and H. Wilman, *Proc. Phys. Soc. London*, **61**, 164 (1948).
209. Matthews, J. W., *Phil. Mag.*, **6**, 1347 (1961).
210. Matthews, J. W., and K. Isenback, *Phil. Mag.*, **8**, 469 (1963).
211. Matthews, J. W., *Phil. Mag.*, **8**, 711 (1963).
212. Wilson, A. D., R. C. Newman, and R. Bullough, *Phil. Mag.*, **8**, 2035 (1963).
213. Schoolar, R. B., and J. N. Zemel, *J. Appl. Phys.*, **35**, 1848 (1964).
214. Zemel, J. N., J. D. Jensen, and R. B. Schoolar, *Phys. Rev.*, **140**, A330 (1965).
215. Berlaga, R. Ya., I. V. Vinokurov, and P. P. Konorov, *Soviet Phys.-Solid State English Transl.*, **5**, 2523 (1964).
216. Bykova, T. T., *Sov. Phys.-Solid State English Transl.*, **8**, 759 (1966).
217. Semiletov, S. A., I. P. Voronina, and E. I. Kortukova, *Soviet Phys.-Cryst. English Transl.*, **10**, 429 (1966).
218. Pashley, D. W., "Metallurgy of Advanced Electronic Materials," vol. 19, Interscience Publishers, Inc., New York, 1963.
219. Addis, R. R., Jr., *Trans. 10th AVS Symp.*, 1963, p. 354.
220. Giles, J. M., and J. Van Cakenberghe, *Nature*, **182**, 862 (1958).
221. Giles, J. M., and J. Van Cakenberghe, "Solid State Physics in Electronics and Telecommunications," vol. 2, p. 900, part 2, Semiconductors, Academic Press Inc., New York, 1960.
222. Zuleeg, R., and E. J. Senkovits, Electrochemical Society Meeting, Pittsburgh, April, 1963, Abstract 95.
223. Conjeaud, P., *J. Rech. Centre Natl. Rech. Sci.*, **32**, 273 (1953).
224. Arntz, F. O., *Tech. Rept.* AFML-TR-66-122, March, 1966.
225. Gwattmey, A. T., and K. R. Lawless, in H. C. Gatos (ed.), "The Surface Chemistry of Metals and Semiconductors," p. 483, John Wiley & Sons, Inc., New York, 1960.
226. Francombe, M. H., in G. E. Anderson (ed.), "The Use of Thin Films in Physical Investigations," p. 65, Academic Press Inc., New York, 1966.
227. Müller, E. K., B. J. Nicholson, and G. L'E. Turner, *Brit. J. Appl. Phys.*, **13**, 486 (1962)
228. Müller, E. K., B. J. Nicholson, and G. L'E. Turner, *J. Electrochem. Soc.*, **110**, 969 (1963).
229. Tanaka, M., and G. Honjo, *J. Phys. Soc. Japan*, **19**, 954 (1964).
230. Krikorian, E., *Trans. 13th AVS Symp.*, San Francisco, 1966.
231. Francombe, M. H., A. J. Noreika, and S. A. Zeitman, *Trans. 13th AVS Symp.*, San Francisco, 1966.
232. Sewell, P. B., and M. Cohen, *Appl. Phys. Letters*, **7**, 32 (1965).
233. Mitchell, D. F., G. W. Simmons, and K. R. Lawless, *Appl. Phys. Letters*, **7**, 173 (1965)
234. Schossberger, F., and F. Ticulka, *Trans. 8th Vacuum Symp. Intern. Congr.*, 1961 vol. 2, p. 1001, Pergamon Press, New York.
235. Fedak, D. G., and N. A. Gjostein, *Phys. Rev. Letters*, **16**, 171 (1966).
236. Spiegel, K., *Surface Sci.*, **7**, 125 (1967).
237. Lander, J. J., and J. Morrison, *J. Chem. Phys.*, **37**, 729 (1962).
238. Lipson, H., *Progr. Met. Phys.*, **2**, 1 (1950).
239. Jona, F., Proceedings, 13th Segamore Army Materials Conference on the "Physical and Chemical Characteristics of Surfaces and Interfaces," Syracuse University Press, 1967.
240. Thomas, R. N., and M. H. Francombe, *Appl. Phys. Letters*, **11**, 108 (1967).
241. Lander, J. J., and J. Morrison, *Surface Sci.*, **2**, 553 (1964).
242. Lander, J. J., and J. Morrison, *J. Appl. Phys.*, **36**, 1706 (1965).
243. Lander, J. J., and J. Morrison, *J. Appl. Phys.*, **33**, 2089 (1962).
244. McLaughlan, T. A., R. S. Sennett, and G. S. Scott, *Can. J. Res.*, **A28**, 530 (1950)
245. Bassett, G. A., *Proc. European Regional Conf. Electron Microscopy, Delft*, 1960, p. 270 1961.
246. Watt, I. M., *Proc. European Regional Conf. Electron Microscopy, Delft*, 1960, p. 341 1961.
247. Pashley, D. W., and M. J. Stowell, *Proc. Intern. Conf. Electron Microscopy, 5th Conf* Philadelphia, Paper GG-1, 1962.
248. Curzon, A. E., and K. Kimoto, *J. Sci. Instr.*, **40**, 601 (1963).
249. Hanszen, K. J., *Z. Naturforsch.*, **A19**, 820 (1964).
250. Poppa, H., *Z. Naturforsch.*, **A19**, 835 (1964).

251. Bassett, G. A., *Proc. Intern. Symp. Condensation Evaporation of Solids, Dayton,* 1962, p. 599, 1964.
252. Poppa, H., *J. Vacuum Sci. Technol.,* **2,** 42 (1965).
253. Jacobs, M. H., D. W. Pashley, and M. J. Stowell, *Phil. Mag.,* **13,** 129 (1966).
254. Pashley, D. W., and M. J. Stowell, *J. Vacuum Sci. Technol.,* **3,** 156 (1966).
255. du Plessis, J. C., and J. H. van der Merwe, *Phil. Mag.,* **11,** 43 (1965).
256. Valdre, U., D. W. Pashley, E. A. Robinson, M. J. Stowell, K. J. Routledge, and R. Vincent, *Proc. Intern. Conf. Electron Microscopy, 6th Conf.,* Kyoto, 1966, p. 155.
257. Goddard, J., and J. W. Wright, *Brit. J. Appl. Phys.,* **15,** 807 (1964).
258. Gonzalez, C., and E. Grünbaum, *Proc. 5th Intern. Conf. Electron Microscopy,* 1962, vol. 1, p. DD-1, Academic Press Inc., New York.
259. Haase, O., *Z. Naturforsch.,* **A14,** 920 (1959).
260. Wright, J. W., and J. Goddard, *Phil. Mag.,* **11,** 485 (1965).
261. Marcus, R. B., *J. Appl. Phys.,* **37,** 3121 (1966).
262. Denbigh, P. N., and R. B. Marcus, *J. Appl. Phys.,* **37,** 4325 (1966).
263. Chopra, K. L., M. R. Randlett, and R. H. Duff, *Phil. Mag.,* **16,** 261 (1967).
264. Schultz, L. G., *Acta Cryst.,* **4,** 487 (1951).
265. Ludemann, H., *Z. Naturforsch.,* **A12,** 226 (1957).
266. Pashley, D. W., *Proc. Phys. Soc. London,* **A65,** 33 (1952).
267. Meyerhoff, K., and J. Ungelenk, *Acta Cryst.,* **12,** 32 (1959).
268. Davey, J. E., and T. Pankey, *Appl. Phys. Letters,* **12,** 38 (1968).
269. Kurov, G. A., and Z. G. Pinsker, *Soviet Phys.-Tech. Phys. English Transl.,* **3,** 1958 (1958).
270. Semiletov, S. A., *Kristallografiya,* **1,** 306 (1956).
271. Aggarwal, P. S., and A. Goswami, *Indian J. Pure Appl. Phys.,* **1,** 366 (1963).
272. Shalimova, K. V., A. V. Andrushko, and V. A. Dmitriev, *Soviet Phys.-Cryst. English Transl.,* **8,** 618 (1964).
273. Escoffery, C. A., *J. Appl. Phys.,* **35,** 2273 (1964).
274. Weinstein, M., G. A. Wolff, and B. W. Das, *Appl. Phys. Letters,* **6,** 73 (1965).
275. Khan, I. H., and K. L. Chopra, Extended Abstract, Electrochemical Society Meeting, Washington, D.C., Oct. 11–15, 1964.
276. Thirsk, H. R., and E. J. Whitmore, *Trans. Faraday Soc.,* **36,** 565 (1940).
277. Booker, G. R., and R. Stickler, *J. Appl. Phys.,* **33,** 3284 (1962).
278. Chu, T. L., and J. R. Gavaler, *J. Electrochem. Soc.,* **110,** 388 (1963).
279. Grünbaum, E., and J. W. Mitchell, in M. H. Francombe and H. Sato (eds.), "Single Crystal Films," p. 221, Pergamon Press, New York, 1964.
280. Gradmann, U., *Phys. Kondens. Mater.,* **3,** 91 (1964).
281. Krause, G. O., *J. Appl. Phys.,* **37,** 3694 (1966).
282. Matthews, J. W., *Phil. Mag.,* **4,** 1017 (1959).
283. Pashley, D. W., M. J. Stowell, and J. J. Law, *Phys. Stat. Solidi,* **10,** 153 (1965).
284. Matthews, J. W., and D. L. Allison, *Phil. Mag.,* **8,** 1283 (1963).
285. Pashley, D. W., "The Growth and Structure of Thin Films," in "Thin Films," p. 59, American Society for Metals, 1964.
286. Finch, R. H., H. J. Queisser, J. Washburn, and G. Thomas, *J. Appl. Phys.,* **34,** 406 (1963).
287. Mendelson, S., *J. Appl. Phys.,* **35,** 1570 (1964).

Chapter 11

Film Thickness and Composition

WILLIAM A. PLISKIN
and
STELVIO J. ZANIN

IBM Components Division, East Fishkill, New York

LIST OF SYMBOLS

A	absorptivity of a film
A	analyzer setting on an ellipsometer

A	area of a film
d	film thickness
d	channel depth
d'	apparent film thickness
d_G	glass-layer thickness in a multilayer film
d_o	silicon dioxide layer thickness in a multilayer film
d_t	total film thickness in a multilayer film
E_g	band gap, electron volts
F	coefficient of fineness [Eq. (5b)]
I	intensity of transmitted radiation
I_{max}	maximum intensity of transmitted radiation
I_0	intensity of incident radiation
K	dielectric constant
K_i	absorption coefficient of medium i (i = 1, 2, or 3)
k_i	extinction coefficient in medium i (i = 1, 2, or 3)
L	measured distance on a bevel [Eq. (50)]
N	interference order or difference in path length in wavelength units between two (successive) reflected or transmitted beams
N, N', N_2, or ΔN	step height in fringes or number of fringes traversing a step
N_i	a particular order corresponding to incident angle θ_{1i} or angle of refraction θ_{2i} [Eq. (25)]
N_i	a particular order of an extremum occurring at wavelength λ_i
N_i'	approximate order of an extremum occurring at wavelength λ_i or frequency ν_i [Eqs. (31) and (32)]
n	refractive index
n_i	refractive index of medium i (i = 1, 2, or 3)
n_{2i}	refractive index of medium 2 at wavelength λ_i [Eq. (29)]
n_{2G}	refractive index of glass layer G in a multilayered film
n_{2o}	refractive index of silicon dioxide layer in a multilayered film
n_{2Gi}	refractive index at wavelength λ_i of glass layer G in a multilayered film [Eqs. (35) and (36)]
n_{2oi}	refractive index at wavelength λ_i of silicon dioxide layer o in a multilayered film [Eqs. (35) and (36)]
P	polarizer reading on an ellipsometer
$\mathcal{R}$	ratio of reflected to incident energy of radiation
$\mathcal{R}_p$	amplitude of total reflection of the parallel component of radiation [Eq. (38)]
$\mathcal{R}_s$	amplitude of total reflection of the perpendicular component of radiation [Eq. (37)]
r_{12} or r_{12s}	Fresnel-reflection coefficient of the perpendicular component at the 1-2 interface [Eq. (17)]
r_{12p}	Fresnel-reflection coefficient of the parallel component at the 1-2 interface [Eq. (39)]
r_{23p}	Fresnel-reflection coefficient of the parallel component at the 2-3 interface with medium 3 absorbing [Eq. (41)]
r_{23s} and ρ_{23}	Fresnel-reflection coefficient of the perpendicular component at the 2-3 interface with medium 3 absorbing [Eqs. (18) and (40)]
$\mathcal{T}$	ratio of transmitted to incident energy of radiation
	distance between two reflecting surfaces in multiple-beam interferometry
Δt_r	reflectivity correction
Δt_φ	phase-shift thickness correction [Eq. (23)]
V	weight of a film
Z	atomic number
α	lapping angle [Eq. (50)]
δ	phase lag between two successive beams, radians [Eqs. (2) and (4)]
δ'	phase lag [Eq. (22)]
Δ	phase difference $\Delta_p - \Delta_s$

Δ_p	total phase change on reflection of the parallel component
Δ_s	total phase change on reflection of the perpendicular component
δ	phase lag (same as β above)
$\delta\lambda_i$	error in measuring λ_i
θ_i	angle of incidence in medium 1
θ_2	angle of refraction in medium 2
θ_{1i}	angle of incidence corresponding to order N_i [Eq. (25)]
θ_{2Gi}	angle of refraction in glass layer G of multilayered film 2 corresponding to order N_i [Eq. (25)]
θ_{2oi}	angle of refraction in silicon dioxide layer o of multilayered film 2 corresponding to order N_i [Eq. (25)]
θ_{2i}	angle of refraction in medium 2 at wavelength λ_i [Eq. (29)]
θ_{2Gi}	angle of refraction in glass layer G of multilayered film 2 corresponding to wavelength λ_i [Eqs. (35) and (36)]
θ_{2oi}	angle of refraction in silicon dioxide layer o of multilayered film 2 corresponding to wavelength λ_i [Eqs. (35) and (36)]
θ_B	Brewster's angle
λ	wavelength of light or radiation (in vacuum)
λ_i	wavelength of light corresponding to order N_i
ν	reciprocal wavelength or frequency, in wave numbers
ρ	density of a film
ρ_p	amplitude of parallel component of radiation
ρ_s	amplitude of perpendicular component of radiation
ρ_{23}	same as r_{23s}
φ_{23}	phase change on reflection at the 2-3 interface [Eq. (19)]
ψ	arctangent (ρ_p/ρ_s)

1. FILM THICKNESS

a. Introduction

In this chapter, the most useful techniques for determining film thickness and composition will be discussed in sufficient detail for the reader to understand them but references will be given for further details. References will also be given for some film-thickness-measuring techniques which are of limited applicability for general laboratory use. In other cases, we may not go into great detail because excellent reviews, books, and monographs have already been written on these particular techniques. Specifically the reader is referred to the books by Tolansky,[1–] Heavens,[4] Vašíček,[5] Mayer,[6] and Françon,[7] as well as to recent reviews written by Heavens[8] and by the Bennetts.[9] Some of the advantages and disadvantages of many film-thickness-measuring techniques have been listed in a recent article by Gillespie.[ ] The reader may also find of interest a survey by Keinath[11] which covers very briefly thickness-measuring techniques with emphasis on industrial applications up to 1955.

The best technique for a specific application or process depends upon the film type, the thickness of the film, the accuracy desired, and the use of the film. These criteria include such properties as film thickness, film transparency, film hardness, thickness uniformity, substrate smoothness, substrate optical properties, and substrate size. In many cases there is no single best technique, and the particular one chosen will be determined by the personal preferences of the investigator.

Since thin film thicknesses are generally of the order of a wavelength of light, various types of optical interference phenomena have been found to be most useful for the measurement of film thicknesses. We have thus tended to emphasize these. In addition to interference phenomena, there are other optical techniques which can be used to measure thicknesses. Examples are ellipsometry and absorption spectroscopy.

In addition to the optical techniques, there are mechanical, electrical, and magnetic techniques which have been used for film-thickness measurements.[10,11] Among these, the one that has found the widest acceptance is the stylus technique, which is discussed in Sec. 1e(4).

b. Optical Interference Techniques

(1) General Interference Phenomena (Two-beam Interference) The occurrence of interference in a transparent film is shown schematically in Fig. 1. Part of the incident ray B_0 is reflected at the top surface, the interface between media 1 and 2, giving rise to ray B_{12}, and part of it is reflected at the interface between media 2 and 3, giving rise to ray B_{23}. If the reflectivities at the interfaces are not very high, then interference will occur mainly between the first two reflected beams or the first two transmitted beams as shown in Fig. 1. In Fig. 1, the refractive indices of the three media are n_1, n_2, and n_3; the angle of incidence in medium 1 is θ_1; the angle of refraction in medium 2 is θ_2; d is the film thickness; λ is the wavelength of radiation in vacuum; and $k_3 = K_3\lambda/4\pi$ is the extinction coefficient in medium 3, where K_3 is the absorption coefficient of medium 3 at wavelength λ.

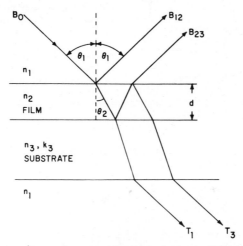

Fig. 1 Schematic diagram for two-beam reflection or transmission through a transparent film (drawn for $n_3 > n_2 > n_1$). In the case of transmission, the extinction coefficient k_3 is zero or sufficiently small to permit detectable radiation through the substrate.

In almost any elementary optics book one can find the simple derivation showing that the optical path-length difference between the two beams is $2n_2d \cos \theta_2$. If this difference is $N\lambda$, where N is an integral number, then the two reflected beams will be in phase resulting in constructive interference; and if N is half integral, the two reflected beams will be 180° out of phase, resulting in destructive interference. It is implied that the phase changes on reflection at the two interfaces are the same. This will be discussed in more detail later. The simple conditions for extrema are thus represented by

$$N\lambda = 2n_2d \cos \theta_2 = 2d(n_2{}^2 - \sin^2 \theta_1)^{\frac{1}{2}} \tag{1}$$

The path-length difference between two successive beams can also be represented in terms of a phase lag in radians by

$$\beta' = \frac{2\pi}{\lambda} 2n_2d \cos \theta_2 = 2N\pi \tag{2}$$

In the case of transmission, the conditions for maxima and minima are reversed.

By referring to Eqs. (1) and (2), it is seen that there are several methods by which the intensity of the reflected light can be made to undergo periodic variations resulting in fringes. The latter can be formed by varying the incident angle θ_1, the film thickness d through the formation of a step, the wavelength λ, or λ/d. These techniques will all be discussed in more detail later.

(2) Multiple-beam Interferometry The sharpness of the fringes increases markedly if interference occurs between many beams. This can be accomplished if the reflectivities at the two interfaces are very high as indicated in Fig. 2, where each of the two parallel glass plates has a thin partially transparent silver film indicated by $CDEF$ and $IJLK$ deposited on it. Another condition for multiple-beam interference is small absorptivity A of the silver film through which light must be transmitted. In the case of multiple-beam reflection, only the silver film $CDEF$ need be fairly transparent (low absorptivity), whereas for multiple-beam transmission both silver films must have low absorptivities. More details on the conditions necessary for good multiple-beam interferometry may be found in the works of Tolansky[3] or Flint.[12]

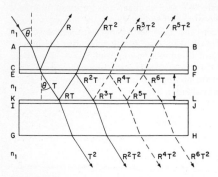

Fig. 2 Schematic representation of multiple-beam interference between two silvered glass plates. In this figure, R and T denote the amplitudes of reflection and transmission.

To understand multiple-beam interferometry better, the equations which describe the fringe positions and relative intensities must be discussed. In Fig. 2, the medium between the two glass plates has a refractive index n_1 and thickness t.

The angle of incidence of the radiation is θ. In the case of transmission the intensity is given by the Airy formula,[1,7,8]

$$I = \frac{I_{\max}}{1 + F \sin^2 (\beta'/2)} \tag{3}$$

where

$$\beta' = \frac{2\pi}{\lambda} 2n_1 t \cos \theta - 2N\pi \tag{4}$$

$$I_{\max} = \left(\frac{T}{T+A}\right)^2 \tag{5a}$$

$$F = \frac{4R}{(1-R)^2} \tag{5b}$$

$$R = \frac{\text{reflected energy}}{\text{incident energy}}$$

and

$$T = \frac{\text{transmitted energy}}{\text{incident energy}}$$

Equation (4) does not rigorously consider the phase changes occurring on reflection;[7] however, in practice this has little effect.[13]

Fabry called F the "co-efficient of finesse." In English texts it is generally called the coefficient of fineness. This coefficient is a measure of how fine or sharp the interference fringes can become. With properly evaporated silver films, a reflectivity of near 0.95 is possible, which gives a value of F of over 1,200 (for details see Tolansky[3]). Thus, as long as $\sin (\beta'/2)$ has any significant value, the intensity of the transmitted light I is very small, as can be seen from Eq. (3). If $\sin (\beta'/2)$ becomes zero, I reaches the maximum of $I_{\max}$. This occurs only if β' is an integral value of 2π, and therefore, very sharp interference fringes with intensity maxima for integral values of N are observed.

In the case of multiple-beam interference by reflection, the interference pattern formed—the so-called interferogram—is just the opposite of that seen in transmission provided the absorptivity is small. In other words, where there are sharp, bright fringes on a dark background in transmission for integral values of N, observation of the reflected light gives sharp, dark fringes on a bright background.

With an increase in the absorption A, the intensity of the entire transmitted pattern is decreased by $[T/(T + A)]^2$. In the reflected-fringe system, however, an increase in the absorption A prevents the fringe minimum from going to zero. Hence, with larger A, the reflected fringes tend to become washed out under the same conditions where they would still be seen on transmission. Consequently, to obtain a sharply delineated reflected-fringe system, the absorption A should be kept to a minimum. In addition to the requirements of high reflectivity and small absorption for good-quality fringes, the interplate distance t should be as small as possible.[3,8]

Referring back to Eq. (4), we can see that there are several factors which contribute to the formation of fringes. For practical applications, fringe systems are identified according to the method of fringe formation, and two cases are distinguished in multiple-beam interferometry. *Fizeau fringes* are generated by monochromatic light and represent contours of equal thickness arising in an area of varying thickness t between two glass plates similar to those shown in Fig. 2. This is accomplished by contacting the two glass plates such that they form a slight wedge at an angle α so that t varies between the two plates. The angle α is generally made very small so that consecutive fringes are spaced as far apart as possible. The angle of incidence θ is typically kept near 0° and the medium is air ($n_1 = 1.0$). Hence, the spacing between fringes corresponds to a thickness difference of $\lambda/2$, where λ is the wavelength of the monochromatic radiation being used.

The second multiple-beam interferometry technique is referred to as *fringes of equal chromatic order*, or *FECO*. In this case, white light is used at an angle of incidence of 0° and the reflected or transmitted white light is dispersed by a spectrograph, thus offering a means of varying λ. According to Eq. (4), fringes will form for certain values of t/λ. Thus, FECO fringes can be obtained with the two silvered surfaces parallel to each other, whereas the plates must be inclined relative to each other to produce Fizeau fringes. The spacing between FECO fringes on the interferogram (or spectrum) is inversely proportional to the thickness.

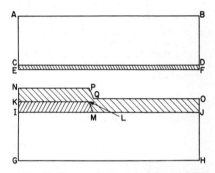

Fig. 3 Semitransparent optical flat ($ABEF$) with film sample ($KLIM$) including step (LM) for multiple-beam interferometric measurement of film thickness.

Multiple-beam interferometry for the measurement of film thickness can be implemented by the method of Donaldson and Khamsavi,[14] which is shown in Fig. 3. The arrangement differs from Fig. 2 in that the substrate $GHIJ$ supports a film $KLMI$ whose thickness is to be measured. A highly reflective opaque metal film $NKJO$ is evaporated on top of the film $KLMI$. Silver films of about 1,000 Å thickness are typically used for this purpose. The step LM in the original film may be formed either by etching the film after deposition or by masking the MJ part of the substrate during deposition. Evaporated silver replicates such steps accurately: the bottom surface of the reference plate $ABCD$ has a thin, highly reflective, semitransparent film as in Fig. 2. To produce Fizeau fringes, the reference plate $ABCD$ is inclined at an angle with respect to the substrate underneath it, as shown in Fig. 4.

In the case of FECO fringes, the film substrate and optical flat are parallel to each other. The distance between the plates is adjusted according to the film thickness since the magnification, i.e., the spacing between the fringes on the interferogram, increases with decreasing separation between the plates.

Note that with either of these techniques the film $KLMI$ may be either opaque or transparent. The requirements for the methods are that a step or channel can be made in the film down to the substrate surface, that the substrate is fairly flat and especially smooth, that the film itself has a smooth surface for fringe formation, and that the film should not be altered by the deposition of the reflective coating. For example, some organic films are altered by the heat generated during the evaporation of the reflective metal and should not be measured by this technique.

(a) *Measurement of 'Fizeau Fringes (Tolansky Technique).* The use of Fizeau fringes for thickness measurements is commonly called the Tolansky technique in recognition of Tolansky's contributions to the field of multiple-beam interferometry. A schematic representation of Fizeau fringes produced by multiple-beam interference is shown in Fig. 4. The sample, channel size, wedge angle, etc., are exaggerated for illustrative purposes and not drawn to scale.

The film thickness is given by $d = \Delta N \lambda / 2$ where ΔN is the number of fringes or fraction thereof traversing the step. In the interferogram shown in Fig. 4, the depth

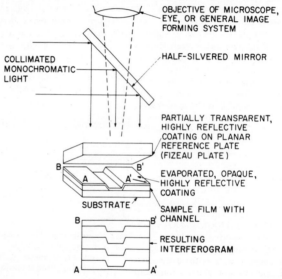

Fig. 4 Schematic view of apparatus for producing multiple-beam Fizeau fringes (Tolansky technique).

of the channel (the film thickness) is exactly one-half of the separation between fringes ($\Delta N = 0.5$), and therefore the film thickness is $\lambda/4$. If the wavelength were that of the green mercury line, the film thickness would be 1,365 Å.

Commercial microscopes which utilize Fizeau multiple-beam interferometry are available. Examples are the Sloan Angstrometer M-100* and the Varian Å-Scope Interferometer.† In addition, conventional metallurgical microscopes can be easily equipped with Fizeau plate attachments for interferometric measurements. These plate attachments are generally equipped with three adjustable screws to determine the tilt of the plate relative to the specimen and thus control the direction and spacing of the interference fringes. These adjustments can be very tedious. An example of such an instrument has been described by Klute and Fajardo,[15] whose stage interferometer facilitates fairly convenient adjustments.

Accurate thickness measurements require careful evaluation of fringe fractions. These may be measured by a calibrated microscope eyepiece, or more accurately and commonly on a photomicrograph of the fringe system. Either way, the evaluation requires a linear measurement, the accuracy of which is strongly dependent on

* Sloan Instruments, Santa Barbara, Calif.
† Varian Vacuum Division, Palo Alto, Calif.

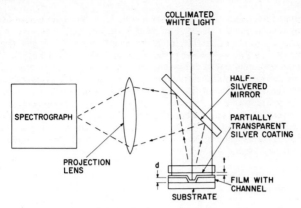

Fig. 5 Schematic of apparatus for producing multiple-beam fringes of equal chromatic order (FECO).

the definition and sharpness of the fringes. As previously mentioned, this would require the optical flat (Fizeau plate) to have high reflectivity and low absorptivity.

Two other prerequisites for an accurate thickness measurement are (1) extremely flat, smooth film surface and (2) very well collimated and narrowband monochromatic light. Thickness measurements from 30 to 20,000 Å can be made routinely to an accuracy of ± 30 Å. With care, film thicknesses can be measured to an accuracy of ± 10 Å.[1]

(b) *Fringes of Equal Chromatic Order* (*FECO*). Fringes of equal chromatic order are more difficult to obtain but yield greater accuracy than Fizeau fringes, especially if the films are very thin. A more detailed discussion of the subject can be found in the works of the Bennetts[9] and Tolansky.[1,3] The principle will be understood from the schematic of the apparatus shown in Fig. 5.

Collimated white light impinges at normal incidence on the two parallel plates. The reflected light is then focused on the entrance slit of a spectrograph. The image of the channel in the film must be perpendicular to the entrance slit of the spectrograph. Assuming an angle of incidence of $\theta = 0°$, then sharp, dark fringes occur for integral values of $N = 2t/\lambda$, as shown schematically in Fig. 6 for two different plate spacings t. The fringes are observed as the wavelength λ is varied by the spectrograph and are recorded on a photographic plate corresponding to $\lambda = 2t/N$. The resulting interferograms are shown schematically in Fig. 6, where the scale is assumed to be linear in wavelength. With

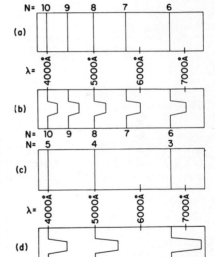

Fig. 6 (a) Interferogram for parallel plate spacing of 2 μ. (b) Interferogram for the same parallel plate spacing of 2 μ but with a channel 1,000 Å deep corresponding to a 1,000-Å film on the lower plate. (c) Interferogram for parallel-plate spacing of 1 μ. (d) Interferogram for the same parallel plate spacing of 1 μ but with a channel 1,000 Å deep corresponding to a 1,000-Å film.

a linear-wavelength scale, the fringes are not equidistant in the interferogram. Note, too, that the order of the fringe increases with decreasing wavelength of the fringe.

It is sometimes preferable to give the order as $N = 2t\nu$, where ν is the reciprocal wavelength or the frequency in wave numbers, as this relation shows N to be linear with wave number. However, this linearity is not found to be exactly true if the phase changes at the two reflecting interfaces vary with wavelength. Hence, there may be a slight dispersion with wavelength, but the effect of phase change on the determined fringes is very small. This has been treated more extensively by Bennett.[16]

In practice, the spacing between plates is not known a priori as assumed in the hypothetical case shown in Fig. 6. It is therefore necessary to deduce t as well as N from the wavelengths of the observed fringes. If N_1 is the order of a fringe corresponding to wavelength λ_1, then $N_1 + 1$ is the order of the next fringe with a shorter wavelength λ_0 on the interferogram. Neglecting the small phase-change dispersion, we have

$$N_1\lambda_1 = (N_1 + 1)\lambda_0 = 2t \tag{6}$$

Solving for N_1 we obtain

$$N_1 = \frac{\lambda_0}{\lambda_1 - \lambda_0} \tag{7}$$

and can now express t solely in terms of measured wavelengths:

$$t = \frac{N_1\lambda_1}{2} = \frac{\lambda_1\lambda_0}{2(\lambda_1 - \lambda_0)} \tag{8}$$

To determine the film thickness, consider that a channel of depth d causes the fringe of the order N_1 to be displaced to a new wavelength λ_1'. Consequently the film thickness d must satisfy the relation

$$t + d = \frac{N_1\lambda_1'}{2} \tag{9}$$

Thus, the film thickness d is given by

$$d = \frac{N_1\lambda_1'}{2} - \frac{N_1\lambda_1}{2} = \frac{\lambda_1' - \lambda_1}{2} \frac{\lambda_0}{\lambda_1 - \lambda_0} \tag{10}$$

In Fig. 6b and d, the fringe displacements due to the channel are clearly related to the order N of the fringes outside the channel. However, if the slope were so steep that the order in the channel could not be related to that outside the channel, the film thickness could still be derived from the equation

$$d = \frac{1}{2} (N_1'\lambda_1' - N_1\lambda_1) \tag{11}$$

whereby the unknown order N_1' must be obtained by observation of a second displaced fringe of order $N_1' + 1$ as in Eq. (8).

The effect of interplate spacing is shown by comparison of Fig. 6a and b with Fig. 6c and d. It is clear from these interferograms that the "magnification" increases with decreasing interplate spacing. The Bennetts[9] have concluded that it is possible to measure film thicknesses with an accuracy of 1 or 2 Å, provided very smooth optical flats are used, the films are carefully evaporated, the plates are carefully aligned, and the fringes are accurately measured.

(3) Michelson Interferometer The Michelson interferometer has been applied in various ways and modifications for the measurement of film thicknesses.[2,7,8] A simple schematic is shown in Fig. 7. In describing the technique we shall consider the central ray. Light from the source L is collimated by the objective O_1. It is split at A by the beam splitter S, which may be a half-silvered mirror, a 50% reflecting dielectric film, or a beam-splitting prism. The two beams obtained at A have equal intensities. The vertical beam is reflected at B on a mirror M_1, thence back to A

and through S toward the objective O_2. The second beam passes through S to be reflected at D on the reference mirror M_2, thence back to A, where it is also reflected down to O_2. The two beams combine under proper conditions to give interference fringes which are seen by the eye E or focused on a photographic plate.

In the system shown, the compensating plate C is made of the same glass and thickness as S so that in white light the paths of the two beams would be identical. The mirror M_2 forms a virtual image at M_2'. The translational position of one of the two mirrors can be accurately adjusted. In addition, each mirror has three tilting screws to adjust the relative positions of mirror M_1 and the virtual image M_2' or mirror M_2. Thus, by adjusting the two mirrors, fringes can be formed. In the case of white light, M_1 and M_2' must intersect or be within one or two wavelengths of each other. If

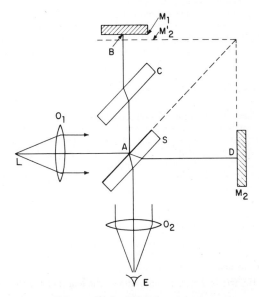

Fig. 7 Schematic of a Michelson interferometer.

mirror M_1 is replaced by a silvered sample with a step representing the film thickness as in Fig. 3, 4, or 5, and if monochromatic light of wavelength λ is used, then the film thickness can be determined by the simple relation $d = \Delta N \lambda / 2$, where ΔN is the step height in fringes. If the step is too steep, the fringe height cannot be determined because the fringe count cannot be made across the step. In such cases it is advisable to check the sample with white light. The zero-order position on both sides of the step can then be determined, and by comparison with monochromatic light, the fringe count ΔN is obtained.

Various models of Michelson interferometers are manufactured by Gaertner Scientific Corporation.* In addition, various microscope manufacturers make instruments which utilize the Michelson interferometer principle for film-thickness measurements. The Zeiss† interference microscope utilizes two prisms in the shape of a cube for the beam splitter. By altering the orientation of one or both of two plane glass plates, one in the reference path and one in the sample path, the spacing and direction of the fringes can be varied.

A simple microscope attachment making use of the Michelson method is manu-

* Gaertner Scientific Corporation, 1201 Wrightwood Ave., Chicago, Ill.
† Carl Zeiss, Oberkochem/Wuertt., West Germany.

factured by W. Watson & Sons Limited* of England. The attachment is made to replace easily the objective of various metallurgical microscopes.[17] Visually, thicknesses can be estimated to about 0.1 fringe and photographically to about 0.05 fringe, provided the sample film is on a reasonably smooth and flat substrate. Similar accuracies can be expected with other interference microscopes based on the Michelson principle. Care must be taken with the attachment to keep vibrations to a minimum.

As with multiple-beam interferometry, the height of a step in a film coated with a reflective metal is given by

$$d = \frac{N\lambda}{2} \tag{12}$$

where N is the step height in fringes. Since in this case N can be determined to about 0.05 fringe (photographically) or to about 0.1 fringe (visually) the thickness is obtained with an accuracy of about 150 to 300 Å.

If an interference microscope or a Michelson interferometer were used with an uncoated transparent film on a reflective substrate, then the film thickness would be given by[18]

$$d = \frac{N'\lambda}{2(n-1)} \tag{13}$$

where N' is the observed fringe count across the step and n is the refractive index of the film. Equation (13) is based on interference between the beams reflected from the reference surface or mirror and the beams reflected from the film-substrate interface. It should not be confused with the situation in Fig. 1, where the interference is due to the beams reflected at the top and bottom of the transparent film. In the latter case, on examination of a step or wedge in the transparent film with a metallurgical microscope, the thickness is given by

$$d = \frac{N_2\lambda}{2n} \tag{14}$$

By making both types of measurements on the same sample and combining the above equations, the refractive index n can be determined [see Sec. 1d(5)].

(4) Polarization Interferometer (Nomarski) Several types of polarization interferometers have been described in the literature,[7,19-21] from which the reader can obtain a detailed understanding of their operation.

In general, interferometers depend on the splitting of light into two beams which, after traversing different optical paths, are recombined with a resulting interference pattern. In the case of the Michelson interferometer, the light is separated into two beams by amplitude splitting. With the Nomarski polarization microscope, a linearly polarized beam, which is incident on and reflected from the sample, is split by a Wollaston prism into two equal-intensity beams with their electric vectors at right angles to each other. After passing through a second polarizer (the analyzer), the electric vectors of the two beams are reoriented into the same direction. The splitting induced by the Wollaston prism is such that the two beams form images which are displaced laterally with respect to each other and also have a phase difference between the two wavefronts. In essence, then, the sample surface produces its own reference. For thickness measurements the sample must have a step representative of the thickness being measured. Fringes indicative of the height of the step occur along the step edges by virtue of the lateral displacement and interference produced. There are two controls to the Nomarski attachment: one is for varying the fringe spacing and the other varies the optical path length and thus sets the position of the zero-order fringe (and other orders) in the field of view. For this operation white light is used, but in general, monochromatic light is used. The fringes form contour lines similar to those observed with a Michelson interferometer

* Available in the United States from William J. Hacker and Co., Inc., West Caldwell, N.J.

or a Watson attachment.   As with instruments based on the Michelson interferometer the film thicknesses are given by Eqs. (12) and (13) and with similar accuracies.

Nomarski polarization interference attachments* are available for many metallographic microscopes equipped with polarizers.   The attachment is very simple to use, even in areas subject to vibration problems.

c. Optical Interference Techniques for Transparent Films

(1) Color Comparisons   The thicknesses of transparent films on reflective substrates can be measured optically by various nondestructive techniques.   The simplest one is a color comparison between the unknown film and a step gauge consisting of films of varying thicknesses but having a refractive index close to that of the unknown film.   The variation of color with film thickness has long been known.   Newton was the first to relate the color of thin films with thickness.   His color chart was reproduced by Rollet[22] many years ago, but it can also be found in more recent publications.[6]   A description of the colors formed by interference on the oxidation of various metals and references to early work in this field are given by Evans.[23]

One of the first precise color thickness gauges was made by Blodgett, who had deposited layers of barium stearate on a lead-glass plate, producing a very colorful and useful thickness gauge.[24,25]   The gauges must be protected by enclosure in a transparent package since barium stearate films are easily damaged.   A much better thickness gauge can be made with silicon dioxide films on silicon.   In this case, a silicon wafer is oxidized to a predetermined thickness, and then various areas are etched away under controlled conditions so as to give films of different thicknesses. This type of gauge is superior to the barium stearate step gauge because thicker films can be grown, and the silicon dioxide is more stable to heat and is immune to attack by most common solvents and chemicals.   In addition, the oxide can be formed on a relatively thin wafer so that direct comparisons with unknown samples can be made more easily.

In making color comparisons, one must be careful not to confuse different film thicknesses which give rise to similar colors but in different orders.   This can be done by comparing the colors of the films while varying the angle of observation.   For films with refractive indices close to that of silicon dioxide, color comparisons are limited to thicknesses between 500 and 15,000 Å with an accuracy of ± 100 Å over most of the region.   A detailed color chart for silicon dioxide on silicon as well as a guide to unusual color differences which help to establish the correct order have been described by Pliskin and Conrad.[26]

The refractive index of the film can influence the color in three ways:

1. The color intensity can be affected by modifying the relative reflectivity at the two interfaces through a change in the refractive index of the transparent film.

2. At normal incidence, Eq. (1) reduces to $d = (N\lambda)/(2n_2)$ and thus similar colors will be obtained with thinner film thicknesses with high-refractive-index films.   A good example of this phenomenon is obtained by comparing the colors of Ta_2O_5 films on tantalum[27] with those of SiO_2 on silicon.[26]   There is also a further complication in that there is a larger phase-shift thickness correction with the films on tantalum than with films on silicon.   Phase-shift thickness corrections will be discussed later.

3. The refractive index of the film also influences the change in color with incident angle θ_1.   From examination of Eq. (1) it follows that low-refractive-index films will undergo more color changes when the viewing angle is varied from normal incidence to higher angles of incidence than high-refractive-index films.

(2) Reflectivity Theory of Transparent Films on Absorbing Reflecting Substrates   Publications about nondestructive techniques for measuring the thickness of transparent films on reflective substrates based on the interference of radiation reflected from the dielectric-air interface with that from the dielectric-substrate interface have increased significantly in recent years.   Many references on this subject can be found in a recent review.[28]   These interferometric techniques can be divided into two general categories.   In the most commonly used technique the radiation is

---

* J. Reichert Optische Werke AG, Vienna, Austria.

reflected from the sample film in a spectrophotometer with fringes being formed as a function of wavelength. This technique has been named CARIS (for constant-angle reflection interference spectroscopy) by Reizman and Van Gelder.[29] In the other technique, interference fringes are formed by varying the angle of observation, and it is therefore called VAMFO (for variable-angle monochromatic fringe observation).[26,30,31] In both techniques, the film thickness can be given approximately by Eq. (1). However, more accurate thicknesses are obtained if fringe minima are used in the relation[32]

$$d = \frac{N\lambda}{2n_2 \cos \theta_2} + \Delta t_\varphi + \Delta t_r \qquad (15)$$

where N = order given by half integers $N = \frac{1}{2}, \frac{3}{2}, \frac{5}{2}, \ldots$
$\quad n_2$ = insulating-film refractive index at wavelength λ
$\quad \theta_2$ = angle of refraction in the insulating film
$\quad \Delta t_\varphi$ = phase-shift thickness correction
$\quad \Delta t_r$ = reflectivity correction

The phase-shift thickness correction for SiO_2 on silicon has been adequately covered for the visible[26,30-32] and ultraviolet[32] regions of the spectrum. Phase-shift thickness corrections for various insulating films on commonly used substrates have been determined, and generalized charts which simplify the calculation of phase-shift thickness corrections have been constructed.[33]

Reflectivity corrections have been determined for the spectrophotometric technique[34] and for VAMFO.[35] With CARIS, approximate compensation for reflectivity corrections is obtained experimentally by reflection off a silicon wafer in the reference beam of the double-beam spectrophotometer.[36] More complete compensation can be obtained by using a silicon wafer with a thick nonuniform SiO_2 film in the reference beam.[34] With both VAMFO and CARIS, the reflectivity corrections are approximately inversely proportional to the order N and thus become insignificant for thicker films.[34,35]

The phase-shift thickness corrections are the same for both VAMFO and CARIS. However, the reflectivity correction for VAMFO differs from that for the spectrophotometric technique. In the former, one must consider the reflectivity variations for a fixed wavelength and varying incident angle, whereas with the latter, the angle of incidence is fixed but the wavelength is varied. Using Vamfo,* reflectivity corrections can be eliminated experimentally by making a 100-Å step in or on the film.[26,3]

The reason for the reflectivity and phase-shift thickness corrections can be seen from examination of the more complete equations describing the reflectivity of a nonabsorbing (transparent) film on a reflective substrate. We will consider only that component of radiation which is perpendicular to the plane of incidence. For experiments, we recommend using a polarizing filter to transmit only the perpendicular (senkrecht) component[26] since this yields more pronounced and sharper minima. Furthermore, at incident angles greater than Brewster's angle for the film, the extrema for the parallel component (parallel to the plane of incidence) occur under conditions opposite to those for the perpendicular component. Therefore, the fringe system tends to be washed out at large angles of incidence if the perpendicular polarizer is not used. The reflectivity for the perpendicular component is given by[5,32,37] [see Fig. 1 and also Eq. (37) of Sec. 1d(1)]

$$R = \frac{r_{12}^2 + \rho_{23}^2 + 2r_{12}\rho_{23} \cos (2\beta - \varphi_{23})}{1 + r_{12}^2\rho_{23}^2 + 2r_{12}\rho_{23} \cos (2\beta - \varphi_{23})} \qquad (16$$

where $\qquad r_{12}^2 = \left(\frac{n_1 \cos \theta_1 - n_2 \cos \theta_2}{n_1 \cos \theta_1 + n_2 \cos \theta_2}\right)^2 \qquad (17$

---

* When we speak of it as a technique we call it VAMFO, but as an instrument it is Vamfo.

is the reflectivity at the air-SiO₂ interface,

$$\rho_{23}{}^2 = \frac{(n_2 \cos \theta_2 - u_3)^2 + v_3{}^2}{(n_2 \cos \theta_2 + u_3)^2 + v_3{}^2} \tag{18}$$

is the reflectivity at the film-substrate interface,

$$\tan \varphi_{23} = \frac{-2n_2v_3 \cos \theta_2}{u_3{}^2 + v_3{}^2 - n_2{}^2 \cos^2 \theta_2} \tag{19}$$

and φ_{23} is the phase change at the film-substrate interface. Furthermore, the following substitutional equations are used:

$$2u_3{}^2 = w + (w^2 + 4n_3{}^2k_3{}^2)^{\frac{1}{2}} \tag{20a}$$
$$2v_3{}^2 = -w + (w^2 + 4n_3{}^2k_3{}^2)^{\frac{1}{2}} \tag{20b}$$
$$w = n_3{}^2 - k_3{}^2 - n_1{}^2 \sin^2 \theta_1 \tag{21}$$

and

$$\beta = 2\pi \, dn_2 \frac{\cos \theta_2}{\lambda} = N\pi \tag{22}$$

Reflectivity minima as given by Eq. (16) do not coincide with those determined by Eq. (1) for two reasons: firstly, the factors r_{12} and ρ_{23} vary with the angle of incidence

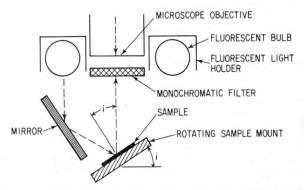

Fig. 8 Schematic sketch of the essential portions of Vamfo. (*Pliskin and Conrad.*[26])

and give rise to reflectivity corrections and secondly, the presence of φ_{23} in the $\cos (2\beta - \varphi_{23})$ term leads to the phase-shift thickness correction. The latter can be expressed in terms of thickness as

$$\Delta t_\varphi = - \left(0.5 - \frac{\varphi_{23}}{2\pi} \right) \frac{\lambda}{2n_2 \cos \theta_2} \tag{23}$$

This will be discussed in more detail later.

(3) Vamfo (a) *Thickness and Refractive-index Measurements.* This nondestructive technique involves the use of a microscope with a stage rotating on an axis normal to the optical axis of the microscope so as to observe the reflected light at various angles. Reflected monochromatic light is produced either by covering the microscope objective with a monochromatic filter or by illuminating the film with monochromatic light. As previously mentioned, improved accuracy is obtained by using a polarizer in conjunction with the monochromatic filter so that only the perpendicular component is transmitted. The light source need not be collimated, and optically flat substrates are not necessary. A schematic drawing of the apparatus is shown in Fig. 8.

The fluorescent bulb can easily be replaced by a mercury bulb,[30] and thus the accuracy of the system is not limited by the lack of monochromaticity of the light source. Even with the fluorescent bulb and filter combination the effective wavelengths of each filter can be determined to an accuracy of 1 to 2 parts in 5,000, which is more than sufficient for most cases.[30]

As the stage and sample are rotated, one observes maxima (bright) and minima (dark) fringes on the film surface. During stage rotation the illuminating mirror is rotated by hand and positioned to reflect light properly onto the sample. With a vertically illuminated microscope, a fixed mirror mounted nearly perpendicular to the rotating stage can be used if it is close to the point on the substrate under examination. Since minima can be determined more accurately than maxima, the angles at which minima occur are noted. The angles of incidence θ_1 are read off a calibrated dial attached to the shaft of the rotating stage or read off a Decitrak unit* which is connected to the shaft.

Various equations and techniques have been used for determining both the film thickness and refractive index from the Vamfo minima positions.[28,30] The thickness is given by

$$d = \frac{N\lambda}{2n_2} \cos \theta_2 \qquad (24)$$

Calculations have been simplified by tabulating the thickness in sets of tables. Each table gives the thickness d as a function of θ_1 for a particular wavelength λ, refractive index n_2, and value of N. In compiling the tables, corrections were made for the stereo angle of the microscope[26] and for phase-change differences at the silicon-film interface. The refractive index and thickness are determined by interpolation between sets of tables to obtain consistent thicknesses for different readings.[28,30] This simplified method has been applied to thin films where different filters are necessary to obtain at least two readings, and to thicker films where two or more minima can be obtained with the same filter. By these techniques refractive indices from 1.30 to 2.10 have been measured. When the refractive index of the film and the order are known, it is only necessary to obtain one reading for a thickness determination from the tables based on Eq. (24).

A recording Vamfo has been constructed by Harvilchuck and Warnecke,[98] thus making the technique more suitable for rapid and simple operations. They utilized a He-Ne laser together with a goniometer for rotating the sample and detector. The angle of incidence and the detector singal are recorded on an X-Y plotter. As with the conventional Vamfo, the fringe amplitude is increased and consequently the accuracy is improved by using perpendicularly polarized light. For greatest accuracy when using the recording Vamfo to measure very thin films ($<3,000$ Å), reflectivity corrections should be made.[35]

(b) *Phase-shift Thickness Corrections.* The phase-shift thickness correction arises because the phase shift of the ray reflected from the dielectric-substrate interface is generally not 180°, whereas the ray reflected from the air-dielectric interface does undergo a 180° phase shift. In the case of SiO_2 on silicon, it has been found that the phase-shift thickness correction is essentially independent of the angle of incidence for the perpendicular component.[26,30,32] Thus in the case of VAMFO, where only the perpendicular component of monochromatic visible radiation is utilized, the small phase-shift thickness corrections were originally given irrespective of angle of incidence.[26,30] In the ultraviolet region of the spectrum, where the phase-shift thickness corrections become appreciable, it was found by Wesson et al.[32,34] that the change with angle was insignificant for the perpendicular component but did vary to some extent for the parallel component.

The general invariance of the phase-shift thickness correction with angle for the perpendicular component has been found to hold with many other combinations of insulators and substrates.[33] On the other hand, in many practical cases, the phase-shift thickness correction for the parallel component increases with an increase in θ_1 as $1/\cos^2 \theta_2$.[33]

* Theta Instrument Corp., 22 Spielman Rd., Fairfield, N.J.

The phase-shift thickness corrections vary with wavelength because of the variation of the refractive index n_2 of the dielectric film and the variation of the refractive index n_3 and extinction coefficient k_3 of the substrate with wavelength.   The optical properties n_3 and k_3 of the reflecting substrate have more influence on the variation of the correction with wavelength than does the refractive index of the transparent film.   Some examples of the variation of the correction with wavelength are given in Fig. 9 for sputtered SiO_2 films on silicon, germanium, copper, and aluminum.[33] Determinations of the phase-shift thickness corrections have been simplified by the use of special phase-shift thickness-correction charts.[33]

(c) *Multiple Films.*   Vamfo can also be used to measure the thickness and refractive index of a second transparent film deposited over an initial transparent film.

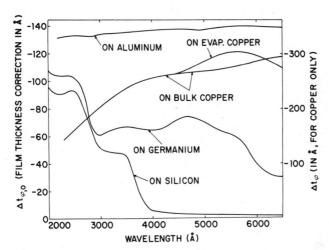

Fig. 9 Phase-shift thickness corrections for sputtered SiO_2 on various substrates. (*Pliskin.*[33])

As with a single transparent film, there are several equations which can be used for measuring the second film thickness.[26,28,31,39] If the order of an observed interference minimum is known, then the second film thickness can be determined by

$$d_G = \frac{N_i\lambda}{2n_{2G}\cos\theta_{2Gi}} - \frac{n_{2o}\cos\theta_{2oi}}{n_{2G}\cos\theta_{2Gi}}d_o \qquad d_o = \frac{N_i\lambda}{2(n_{2G}{}^2 - \sin^2\theta_{1i})^{\frac{1}{2}}} - \frac{(n_{2o}{}^2 - \sin^2\theta_{1i})^{\frac{1}{2}}}{(n_{2G}{}^2 - \sin^2\theta_{1i})^{\frac{1}{2}}}d_o \qquad (25)$$

where the subscripts o or $2o$ refer to the first (lower) transparent film and the subscripts G or $2G$ to the second (upper) transparent film.   In this case we have a glass film G on an oxide film o.

In some cases the thicknesses can be determined more conveniently by the approximate formula[39]

$$d_T = \frac{N\lambda_j}{2n_{2j}\cos\theta_{2j}} = \frac{N\lambda_j}{2(n_{2j}{}^2 - \sin^2\theta_{1j})^{\frac{1}{2}}} \qquad (26)$$

where n_{2j} is the average refractive index at λ_j for the composite film of glass and oxide, and d_T is the total thickness of both films.   If the thickness is sufficiently large for the observation of more than one minimum, then both the average refractive index and d_T can be determined by finding the refractive index for which d_T is the same from sets of tabulated values, as previously discussed in context with Eq. (24).   If the refractive index of the glass is not known beforehand, it can be determined by the use of the relation

$$n_{2G} = \frac{n_{2j}d_T - n_{2o}d_o}{d_G} \tag{27}$$

where

$$d_G = d_T - d_o$$

With thinner films, more than one filter is required and the sets of tables are used in making the interpolation based on a dispersion system corresponding to the type of dispersion expected for the mixed glass-oxide system.

(4) Interference Spectroscopy (a) *Thickness Measurements.* In this technique wavelength is varied rather than the angle of observation. For the case of reflection the technique has been named CARIS, as was previously mentioned. The use of the reflection-spectroscopy technique was given impetus after Corl and Wimpfheimer[·] showed its applicability to the thickness measurement of SiO_2 films on silicon. For homogeneous films, the thickness is often determined by

$$d = \frac{\Delta N \lambda_i \lambda_j}{2n_2(\lambda_j - \lambda_i)\cos\theta_2} = \frac{\Delta N \lambda_i \lambda_j}{2(\lambda_j - \lambda_i)(n_2{}^2 - \sin^2\theta_1)^{\frac{1}{2}}} \tag{28a}$$

or

$$d = \frac{\Delta N}{2n_2(\nu_i - \nu_j)\cos\theta_2} = \frac{\Delta N}{2(\nu_i - \nu_j)(n_2{}^2 - \sin^2\theta_1)^{\frac{1}{2}}} \tag{28b}$$

where ΔN is the number of fringes between wavelengths λ_i and λ_j or frequencies ν_i and ν_j. Equation (28b) is included as it is advantageous to obtain spectra on charts linear with frequency (cm^{-1}) since the fringe system is then more symmetric and the extrema can be measured more accurately. Because of the variation of the refractive index with wavelength, Eqs. (28a) and (28b) are not strictly correct, and for the greatest accuracy the thickness is given by[31]

$$d = \frac{\Delta N \lambda_i \lambda_j}{2(\lambda_j n_{2i} \cos\theta_{2i} - \lambda_i n_{2j} \cos\theta_{2j})} \tag{29}$$

where subscripts i and j refer to wavelengths λ_i and λ_j. The inaccuracy of Eq. (28) depends upon the dispersion of the dielectric film. It can be easily shown that with thick films of SiO_2, using Eq. (28) in the region from about 4,300 to 6,300 Å will give a thickness approximately 2% greater than the more accurate Eq. (29). The thickness obtained from Eq. (28) using the same wavelength region for a higher-dispersion material such as silicon nitride results in a thickness almost 6% greater than that based on Eq. (29). In the ultraviolet region the dispersion of SiO_2 is significantly greater than in the visible, and the resulting thickness error in using Eq. (28) is large. For example, using the range from about 2,100 to 3,900 Å, the film thickness calculated by Eq. (28) is about 9% greater than that calculated by Eq. (29). One could reduce the dispersion error by covering a smaller wavelength range, but then the would increase the relative error in the wavelength measurements. That is $(\delta\lambda_j - \delta\lambda_i)/(\lambda_j - \lambda_i)$ increases with a decrease in $(\lambda_j - \lambda_i)$. The smallest error in thickness measurement due to inaccuracies in measuring λ is obtained by using Eq. (30) below, because $\delta\lambda/\lambda$ is smaller than $(\delta\lambda_i - \delta\lambda_j)/(\lambda_j - \lambda_i)$. $\delta\lambda$ is the error in measuring λ.

Taking into consideration phase-change corrections, dispersion, and wavelength accuracy, it is best to use the relation [see Eq. (15) in Sec. 1c(2)]

$$d = \frac{N\lambda}{2n_2 \cos\theta_2} + \Delta t_\varphi = \frac{N\lambda}{2(n_2{}^2 - \sin^2\theta_1)^{\frac{1}{2}}} + \Delta t_\varphi \tag{30}$$

where for minima the order N is half integral with reflection spectra and integral with transmission spectra for a system where $n_3 > n_2 > n_1$. The opacity of the silicon substrates used in semiconductor-device fabrication necessitates that reflection rather than transmission techniques, be used for these measurements.

Fried and Froot[41] have used the CARIS technique with a microspectrophotometer for the measurement of films as thick as 20,000 Å in areas as small as 0.1 mil in diameter.

(b) Order Determinations. A simple method for determining the order which is most applicable at longer wavelengths is given by the approximate relation[42]

$$N_i' = \frac{\nu_i}{\Delta\nu} \tag{31}$$

where N_i' is approximately the order for the extremum occurring at frequency $\nu_i = 1/\lambda_i$. Equation (31) is only approximately accurate because of variations in the optical properties with wavelength.[32,42] The order N_i' is generally slightly greater than the actual order N_i, and if the film thickness is not too great or if the calculation were made for sufficiently long wavelengths, then the N_i' would be close enough to the

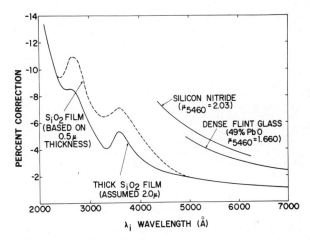

g. 10 Corrections to be applied to the formulas

$$N_i' = \frac{\nu_i}{\Delta\nu} = \frac{\lambda_{i\pm1}}{|\lambda_i - \lambda_{i\pm1}|}$$

obtain the correct order N_i for thickness determinations of various films on silicon. *Pliskin.*[28])

actual N_i as to leave no doubt as to its proper value. With thicker films it is sometimes necessary to make corrections to N_i' to obtain the correct N_i. Equation (31) can be written as

$$N_i' = \frac{\nu_i}{|\nu_j - \nu_i|} = \frac{\lambda_i}{|\lambda_i - \lambda_j|} \tag{32}$$

where the subscripts i and j refer to successive minima. The average N_i' is determined from Eq. (32) using the extrema positions to either side of i. This value is then decreased by the percent given in Fig. 10 to obtain the correct order N_i. The corrections for wavelengths greater than 5,000 Å are practically independent of film thickness, and therefore, there should be no doubt in the determined order N, since it would only be integral or half integral depending on whether it is maximum or a minimum. In the case of insulating films with greater dispersion, the corrections are larger, as shown for dense flint glass and for silicon nitride as compared with that for silicon dioxide.

After establishing the order for one of the longer-wavelength minima (or maxima), each successive minimum (or maximum) toward shorter wavelength is increased by

The film thickness can then be readily calculated for each of the extrema by application of Eq. (30). The phase-change correction Δt_φ can be determined from

Eq. (23) or taken from a plot as shown in Fig. 9. To simplify matters, the factor $2n_2 \cos \theta_2 = 2(n_2{}^2 - \sin^2 \theta_1)^{\frac{1}{2}}$ can also be plotted as a function of wavelength.

Excellent agreement can be obtained between thicknesses determined from each of the minima in the visible and in the ultraviolet. Wesson et al.[32] obtained excellent agreement between thicknesses determined by Vamfo and those determined by CARIS in the ultraviolet region of the spectrum. This agreement would not have been possible without accurate phase-shift thickness corrections.

Film thicknesses can also be determined by use of a nomograph based on Eq. (30).

(c) *Refractive-index Measurements.* The refractive index can be determined as a function of wavelength by spectroscopic techniques provided the exact film thickness at the region of investigation is already known. Solving for n_2 from Eq. (30), we obtain

$$n_2 = \left[\frac{N^2\lambda^2}{4(d - \Delta t_\varphi)^2} + \sin^2 \theta_1 \right]^{\frac{1}{2}} \tag{33}$$

Note that in this formula, the absolute value of the film-thickness correction is added to d since Δt_φ itself is negative by Eq. (23). It should be emphasized that for obtaining accurate film thicknesses and refractive indices by CARIS, the spectrophotometer should be accurately calibrated and the angle of incidence θ_1 accurately determined. If the refractive index and thickness are determined by an external means, such as with Vamfo, then $\sin^2 \theta_1$, or the angle of incidence, can also be determined.

(d) *Multiple Films.* Spectroscopic techniques can also be used for the measurement of a second transparent film deposited over an initial transparent film. Cox and Kosanke[43] obtained the thickness of fused sedimented glass films[44] on oxidized silicon wafers by use of the following equation in the visible and near-infrared spectral regions:

$$d_G = \frac{\Delta N \lambda_i \lambda_j}{(\lambda_j - \lambda_i)(2n_{2G} \cos \theta_{2G})} - \frac{n_{2o} \cos \theta_{2o}}{n_{2G} \cos \theta_{2G}} d_o \tag{34}$$

where the subscripts o or $2o$ refer to the first (lower) transparent film and the subscripts G or $2G$ refer to the second (upper) transparent film. Equation (34) shows that the refractive index and thickness of the first layer must be known as well as the refractive index of the second layer (in this case the glass layer). For most practical purposes Eq. (34) is sufficiently accurate, but since the refractive index varies with wavelength, the film thickness is given more accurately by

$$d_G = \frac{\Delta N \lambda_i \lambda_j}{2(\lambda_j n_{2Gi} \cos \theta_{2Gi} - \lambda_i n_{2Gj} \cos \theta_{2Gj})} - \frac{\lambda_j n_{2oi} \cos \theta_{2oi} - \lambda_i n_{2oj} \cos \theta_{2oj}}{\lambda_j n_{2Gi} \cos \theta_{2Gi} - \lambda_i n_{2Gj} \cos \theta_{2Gj}} d_o \tag{35}$$

Rather than using the involved Eq. (35) for exact determinations, it is better to use a modified form of Eq. (25) which was used for multiple films in Vamfo.

$$d_G = \frac{N_i \lambda_i}{2n_{2Gi} \cos \theta_{2Gi}} - \frac{n_{2oi} \cos \theta_{2oi}}{n_{2Gi} \cos \theta_{2Gi}} d_o = \frac{N_i \lambda_i}{2(n^2{}_{2Gi} - \sin^2 \theta_1)^{\frac{1}{2}}} - \frac{(n^2{}_{2oi} - \sin^2 \theta_1)^{\frac{1}{2}}}{(n^2{}_{2Gi} - \sin^2 \theta_1)^{\frac{1}{2}}} d_o \tag{36}$$

where the subscripts i correspond to the various terms for orders N_i and indicate that these quantities vary with wavelength. Basically the same formula as Eq. (36) was used by Murray et al.[45] in measuring photoresist thicknesses on oxidized silicon wafers.

(e) *Infrared Reflection for Epitaxial-film-thickness Measurements.* The CARIS technique is not limited to dielectric films or to the visible and ultraviolet regions of the spectrum. Although epitaxial semiconductor films are opaque in the visible and ultraviolet regions, nondegenerate (lightly doped) semiconductors are transparent in the infrared region of the spectrum. On the other hand, the difference in the optical constants n and k for the more highly doped material as compared with that for the lightly doped material is sufficient to cause interference if a lightly doped or pure layer of semiconductor material is grown or deposited onto a heavily doped underlayer.[46-49] Spitzer and Tanenbaum[46] were the first investigators to utilize the

CARIS method in the long-wavelength infrared region to determine thickness of epitaxial silicon and germanium films. The interference fringes are more pronounced at longer wavelengths because of greater differences in the semiconductor optical properties with doping at these wavelengths. Grochowski and Pliskin[47] showed the close correlation between film thicknesses determined by the infrared method and by the angle-lap and stain technique [see Sec. 1e(5)]. The phase change at the substrate epitaxial interface is dependent upon the substrate resistivity.[42] Schumann et al.[49] have considered these phase changes in detail.

(5) Reflectivity Intensity In the techniques already discussed, film thicknesses were determined from the positions of interference minima (or maxima) as a function of the angle of incidence θ_1 or the wavelength λ. As mentioned in Sec. 1b(3), an approximate thickness can also be calculated from $N\lambda/2n$, where N is the fringe count on an etched step. With films less than one-half order thick, it is difficult to determine N and thus the film thickness.

Since the intensity of reflected light varies with film thickness [see Eq. (16)] accurate intensity determinations can be utilized for film-thickness determination. In studying the kinetics of film growth, Stebbins and Shreir[50] have made in situ measurements of the intensity of monochromatic light reflected from a uranium surface during anodic oxidation. The light was at near-normal incidence. Lukeš and Schmidt[51] also used near-normal incidence in making intensity measurements of the reflected monochromatic light for thickness determinations of very thin silicon dioxide films on silicon. On a bare silicon surface the reflectivity is near a maximum. It is not exactly at a maximum because the phase change at the silicon-air interface is not exactly 180°. With increasing SiO_2 film thicknesses the reflectivity decreases as given by Eq. (16) until the reflectivity reaches a minimum near a thickness of 900 Å depending upon the wavelength of radiation.

More recently Fried and Froot[41] have extended the reflectivity-intensity technique for the measurement of films 100 to 3,000 Å thick in areas as small as 0.1 mil in diameter utilizing a microspectrophotometer.

A simple technique for very thin films has been devised by Pliskin and Esch.[52] In this technique the observations are made at or near Brewster's angle for silicon with the light polarized in the plane of incidence. Under such conditions there is a minimum of reflectivity for a bare silicon surface, and the intensity of the reflected light increases with film thickness. Thickness measurements are made by direct comparison of the light reflected from a sample with the light reflected from a calibrated SiO_2 on silicon "thickness gauge." Silicon dioxide film thicknesses can be estimated to an accuracy of about ± 30 Å for thicknesses less than 150 Å and ± 50 Å for thickness between 150 and 900 Å. The sensitivity of this technique for very thin films (<300 Å) can be considerably increased by measuring the intensity of the reflected light (from a He-Ne laser) with a photocell as a function of the film thickness.[53]

d. Other Optical Methods for Transparent Films

(1) Ellipsometry Ellipsometry provides another nondestructive method for measuring both thickness and refractive index of transparent films. This technique has also been called polarimetry and polarization spectroscopy. It is very useful for extremely thin films but may also be used for very accurate measurements of thicker films. However, it is more complicated than the relatively simple methods previously described. The body of literature on ellipsometry has become so vast that we cannot give a complete review here. The widespread use of ellipsometry was demonstrated by the symposium sponsored in 1963 by the National Bureau of Standards.[54] For a discussion of elliptically polarized light, the reader may resort to optics texts or more preferably to books on the optics of thin films as, for example, Vašíček,[5] Heavens,[4] or Mayer.[6]

Ellipsometry is based on evaluating the change in the state of polarization of light reflected from a substrate. The state of polarization is determined by the relative amplitude of the parallel (ρ_p) and perpendicular (ρ_s) components of radiation and the phase difference between the two components $\Delta_p - \Delta_s$. On reflection from a surface, bare or film-covered, the ratio of the two amplitudes ρ_p/ρ_s and the phase difference between the two components $\Delta_p - \Delta_s$ undergo changes which are dependent upon the optical constants of the substrate n_3 and k_3, the angle of incidence θ_1, the optical constants of the film n_2 and k_2, and the film thickness d. If the optical constants of the substrate are known, and if the film is nonabsorbing (i.e., $k_2 = 0$), then the only unknowns in the equations describing the state of polarization are the refractive index n_2 and the thickness d of the transparent film. In principle, then, with a complete knowledge of the state of polarization of the incident and reflected light, the refractive index n_2 and thickness d can be determined.

(a) *Theory.* The basic theory of ellipsometry was developed by Drude. Although not shown in Fig. 1, if one sums the amplitudes of all the reflected beams (not just the two shown) from the two interfaces for the perpendicular and for the parallel component, one obtains as the amplitudes of the total reflection[4,5,37]

$$R_s = \rho_s e^{i\Delta_s} = \frac{r_{12s} + r_{23s}e^{-2i\beta}}{1 + r_{12s}r_{23s}e^{-2i\beta}} \tag{37}$$

$$R_p = \rho_p e^{i\Delta_p} = \frac{r_{12p} + r_{23p}e^{-2i\beta}}{1 + r_{12p}r_{23p}e^{-2i\beta}} \tag{38}$$

where the Fresnel reflection coefficients r are given by

$$r_{12s} = \frac{n_1 \cos \theta_1 - n_2 \cos \theta_2}{n_1 \cos \theta_1 + n_2 \cos \theta_2} \tag{17}$$

$$r_{12p} = \frac{n_1 \cos \theta_2 - n_2 \cos \theta_1}{n_1 \cos \theta_2 + n_2 \cos \theta_1} \tag{39}$$

$$r_{23s} = \rho_{23s}e^{i\varphi_{23}} = \frac{n_2 \cos \theta_2 - (u_3 - iv_3)}{n_2 \cos \theta_2 + (u_3 - iv_3)} \tag{40}$$

$$r_{23p} = \frac{(n_3 - ik_3)^2 \cos \theta_2 - n_2(u_3 - iv_3)}{(n_3 - ik_3)^2 \cos \theta_2 + n_2(u_3 - iv_3)} \tag{41}$$

where ρ_{23}, φ_{23}, u_3, v_3, and β are defined in Eqs. (18) through (22). Incidentally, Eq. (16) describing the reflectivity of the perpendicular component was obtained from Eq. (37) by $R = R_s R_s^*$. The state of polarization is given by the amplitude ratio $\tan \psi = \rho_p/\rho_s$ and the phase difference between the two components $\Delta = \Delta_p - \Delta_s$. The basic equation of ellipsometry is thus given by

$$(\tan \psi)e^{i\Delta} = \frac{\rho_p}{\rho_s} e^{i(\Delta_p - \Delta_s)} = \frac{R_p}{R_s} = \frac{r_{12p} + r_{23p}e^{-2i\beta}}{1 + r_{12p}r_{23p}e^{-2i\beta}} \frac{1 + r_{12s}r_{23s}e^{-2i\beta}}{r_{12s} + r_{23s}e^{-2i\beta}} \tag{42}$$

Unfortunately for Drude, he had no electronic computers to help him with the calculations. However, he expanded the equations in terms of d/λ for $d \ll \lambda$ and thus obtained the so-called Drude approximations which are valid for very thin transparent films ($d < 100$ Å) on reflective substrates in air or vacuum.[4]

(b) *Experimental Applications.* With the advent of electronic computers, the application of the exact equation [Eq. (42)] has become possible. Computer programs have been written by Archer[55] and by McCrackin and Colson.[56] For a particular substrate of known optical values of n_3 and k_3 at wavelength λ and for selected film refractive indices n_2, Δ and ψ values can be determined as functions of film thickness d, which is in turn related to β in Eq. (42) by Eq. (43) below. Archer's δ is the same as our β, and if given in degrees, the film thickness is given by

$$d = \frac{(180N + \beta)\lambda}{360(n_2{}^2 - \sin^2 \theta_1)^{\frac{1}{2}}} \tag{43}$$

where the order N takes on integral values depending upon the thickness of the film. Using this technique, Archer made graphical plots for transparent films of various refractive indices on silicon. The example shown in Fig. 11 is applicable for $\lambda = 5,461$ Å, $n_3 = 4.05$, $k_3 = 0.028$, and $\theta_1 = 70.00°$. Each curve represents a particular refractive index for the transparent film, with β values for every 20°

indicated on the curves.   Arrows indicate the direction of increasing β or thickness.
After determining Δ and ψ from ellipsometer readings, the plot gives directly the
refractive index n_2 of the film and β, from which the thickness can be determined
by Eq. (43), provided the order N (or the approximate thickness) is known.   The
difficulty in the technique is that a new plot is needed for any change in λ, θ_1, n_3, or k_3.

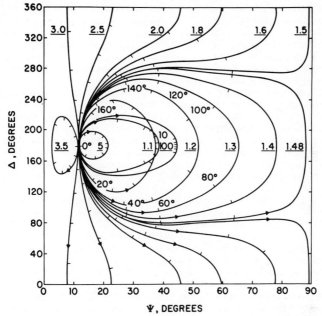

Fig. 11   Variation of Δ and ψ as a function of the index of refraction and thickness (in terms
of δ or β) of transparent films on silicon with $n_3 = 4.05$, $k_3 = 0.028$, $\lambda = 5{,}461$ Å, $\theta_1 = 70.0°$
The underlined numbers are indices of refraction.   (*Archer.*[55])

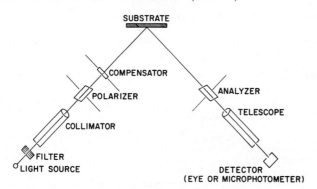

Fig. 12   Schematic representation of the ellipsometer.   (*Archer.*[55])

The schematic of a typical ellipsometer is shown in Fig. 12.   The collimated,
monochromatic light is first linearly polarized by the polarizer.   It is then elliptically
polarized by the compensator, which can be a mica quarter-wave plate or a Babinet-
Soleil compensator.   Although the compensator can be set at any azimuth, for
simplicity it is usually oriented so that its fast axis is at $\pm 45°$ to the plane of incidence.

After reflection from the sample the light is transmitted through a second polarizer which serves as the analyzer. Finally, the light intensity is determined by the eye or more accurately by a photomultiplier detector. The polarizer and analyzer are rotated until extinction is obtained. Under these conditions, the ellipticity caused by the polarizer, compensator, and analyzer combination is the opposite of that produced by reflection from the film and substrate. There are 32 possible sets of polarizer (P) and analyzer (A) readings which will give extinction. However, only four of them are independent. From these extinction readings of P and A, the phase difference Δ and amplitude ratio $\tan \psi$ can be determined. A detailed discussion of the various possible polarizer, compensator, and analyzer settings and the relationship between them and the Δ and ψ values can be found in the work of Winterbottom[57] and McCrackin et al.[58] The latter authors also cover detailed alignment procedures in setting up the apparatus.

The most accurate readings of P and A can be obtained using a technique suggested by Archer.[55] After obtaining a preliminary minimal detector signal by proper adjustment of P and A, the analyzer is kept at the same orientation while readings are taken of the polarizer set on either side of the minimum to give detector signals of equal intensity. The average of the two polarizer settings gives a more accurate value of P. With the polarizer then set at this value of P, the same procedure is repeated with the analyzer, obtaining those settings of A on either side of the minimum which give equal detector signals. The average of these two settings gives an accurate value for A.

In principle the ellipsometer technique is exact. However, near the origin ($\beta = 0°$) where the lines of constant refractive index merge, the refractive indices cannot be determined. The exact techniques of ellipsometry have been used in recent years for the study of various transparent films on reflective substrates.[28,54] Its use has become so widespread that the investigator now has a choice of several good commercially manufactured ellipsometers available to him.

(2) Absorption-band Intensity Film thicknesses can also be determined from characteristic absorption-band intensities. A simple example of this technique is the measurement of silicon dioxide film thicknesses on silicon wafers from the optical density [log (I_0/I)] of the 9.2-μ SiO_2 absorption peak in the infrared transmission spectrum.[59,60] The optical density is not perfectly linear with film thickness because the shape of the absorption band varies with film thickness. Thus, in some cases it is better to use the integrated band intensity.[52] The use of optical density for film-thickness measurements is generally not recommended for deposited films because of variations in the band shape depending on deposition conditions. In reflection spectroscopy, the strong SiO_2 absorption band occurs as a reflection maximum near 9.1 μ. The intensity of this reflection maximum can be used as an approximate measure of the film thickness on wafers which have high carrier concentrations and are therefore opaque to infrared transmission.[61]

(3) Light-section Microscope The light-section microscope* is limited to the measurement of films thicker than 1 μ and therefore will not be discussed in great detail.[62-64] In essence it consists of two microscopes: an illuminating component which projects a slit of light onto the sample at an angle of 45° to the sample and an observing microscope forming an angle of 90° with the illuminating microscope and 45° with the sample. The image of the slit is observed by means of an eyepiece with a reticle and micrometer. In the case of opaque films, the slit is projected across a step in the film. The separation of slit images due to reflection across the step can be measured with an eyepiece micrometer to an accuracy, according to Brown,[64] of 0.1 μ in the 1- to 400-μ range. A step is not necessary in the thickness measurement of transparent films, since two images are obtained because of reflection from both interfaces.[62] The true film thickness d is given by

$$d = d'(2n^2 - 1)^{\frac{1}{2}} \tag{4}$$

where d' is the apparent film thickness (i.e., that which is read off the micrometer

* Carl Zeiss, Inc., 485 5th Ave., New York, or Oberkochem/Wuertt., West Germany.

nd n is the film refractive index.[63] The data of Mansour[63] on the thickness measurement of anodic aluminum coatings indicate an accuracy from a few tenths of a micron o a micron for films from 3 to 25 μ thick, and for thicker films the accuracy approaches $\%$.

(4) Brewster's Angle for Refractive Index In the case of transparent films, some of he film-thickness measuring techniques require a knowledge of the refractive index f the film but cannot in themselves be used to determine the refractive index. We hall therefore describe some additional methods for refractive-index determination.

In vacuum (or for most practical cases in air where actually $n_1 = 1.0003$), we have t Brewster's angle θ_B

$$\tan \theta_B = n_2 \tag{45}$$

vhere n_2 is the refractive index of the film. At this angle the parallel component f the light reflected at the air-film interface vanishes. Therefore, if only the parallel omponent were observed or detected by proper use of polarizing filters, no intererence would be observed. Brewster's angle could then be determined quite easily or films which are nonuniform by the angle at which the interference fringes (due to onuniformity of the film) disappeared.[26] With uniform films the Brewster-angle method of Abelès is most commonly used.[8,9,65] This method requires a step in the lm down to the bare substrate. Using the parallel component of light, Brewster's ngle is then given by that angle at which the intensity of the light reflected from the wo surfaces is equal. The technique is most sensitive for half-integral values of N vhere $n_2d = N\lambda/2 \cos \theta_2$ and least sensitive for integral values of N,[66] provided $n_3 > n_2$. In transparent substrates with refractive indices not much different from that of he film ($n_3 \leq n_2 \pm 0.3$), the point at which the two reflectivities are equal can be etermined easily by visual observation to obtain a refractive-index accuracy of ± 0.002. With more highly reflective substrates, as with dielectric films on metals, he contrast is poorer and a more quantitative measure of reflectivity is necessary o obtain accurate Brewster-angle settings. Kelly and Heavens[67] alternately scanned the coated and uncoated metal surfaces (Ta_2O_5 on Ta) using a photomultiplier for etecting the reflected light intensity. Brewster's angle was determined by the ngle of incidence for which the modulated intensity was a minimum.

The Brewster's-angle technique has been further extended by Hacskaylo.[68] In his case the incident light is polarized in a direction nearly parallel to the plane of cidence and the reflected light is transmitted through another polarizer, the analyzer. he analyzer settings for which the intensities of the reflected light from the two urfaces are equal are plotted as a function of the angle of incidence. Brewster's ngle is accurately given by the incident angle for which the analyzer setting is 0°. or these measurements Hacskaylo used an ellipsometer whose master circle could e read to an accuracy of 20 sec of arc. He claimed accuracies of ± 0.0002 for a m refractive index of 1.2 and ± 0.0006 for a film refractive index of 2.3 independent the refractive index of the transparent substrate.

(5) Wedge or Step Technique for Thickness and Refractive Index The use of a edge or step in a Michelson interferometer for measuring film thickness was discussed in Sec. 1b(3). The thickness of a stepped transparent film can also be meared using monochromatic light with a metallurgical microscope, the film thickness eing given by

$$d = \frac{N\lambda}{2n_2} \tag{14a}$$

ottling and Nicol[69] have discussed a more involved condition which uses a Fizeau ate as in the Tolansky technique but without an overlay reflecting coating on the epped transparent film. By combining either of the equations involving the refracve index [i.e., Eq. (13) or (14)] with an external method for determining the thickss, n can be determined. Booker and Benjamin[18] examined films of silicon dioxide th with and without metallization using a Zeiss interference microscope. From

Eqs. (12) and (13) it is seen that

$$n = 1 + \frac{N'}{N} \tag{46}$$

They estimated the accuracy for a 15,000-Å film to be 1 to 2%, yet they obtained refractive indices of 1.48 to 1.50 for thermally grown SiO_2 films on silicon. This is on the high side of the experimental error when compared with the value of 1.46 obtained by Pliskin and Esch.[30] This difference could be attributed to the effect of large numerical aperture objectives which would require a thickness correction because the effective angle of incidence is not perpendicular to the substrate.[70] At high numerical aperture the values read for N' and N are too low. The relative decrease in N is greater than that for N' and thus the refractive index as given by Eq. (46) would be slightly greater than the true value, as has been observed.

Stebbens and Shreir[50] determined the refractive index of anodically grown uranium oxide in a somewhat similar manner. They formed fringes in the oxide by etching a step, and they formed fringes in air by use of a partially silvered glass flat wedge against the sample. From the ratio of the distances between fringes in the two cases one obtains n. In effect this is the same as solving Eqs. (12) and (14) to obtain

$$n = \frac{N_2}{N} \tag{47}$$

Another comparison technique has been used by Doo et al.[71] in some of the measurements of the refractive indices of pyrolytically deposited silicon nitride. In their technique the sample film (silicon nitride) is deposited on a film (thermally grown silicon dioxide) of known refractive index. A bevel is then formed by lapping the film and part of the substrate at a small angle. The refractive index can be calculated from the interfringe spacing obtained when the beveled sample is illuminated with monochromatic light. It is given by

$$n_s = n_o \frac{d_o}{d_s} \tag{48}$$

where n_o is the refractive index of the silicon dioxide, d_o the interfringe spacing for the oxide, and d_s the interfringe spacing of the sample film. The estimated accuracy for silicon nitride is ± 0.05. The film thickness can also be determined by this technique [see Sec. 1e(5)].

(6) Fluid Matching for Refractive Index If a dielectric film on a reflective substrate is submerged in a fluid whose dielectric constant is the same as that of the film, no reflection will occur at the film-fluid interface and the film would not be detectable by interferometric methods. Two techniques for refractive-index measurement based on this phenomenon will be discussed.

(a) *Immersion Spectrophotometry of Interference Films.* In this method attributable to Ellis[72,73] spectrophotometric peak heights are measured by the CARIS technique [Sec. 1c(4)] with the film and substrate immersed in various fluids as shown in Fig. 1. The peak heights of the interference fringes are plotted against the refractive index of the immersing fluid. The refractive index of the film corresponds to that point on the curve for which the peak height is zero. To minimize the effect of varying dispersion between the fluids and the film, it is best to make measurements with thicker films, which reduces the spacing between fringes. The accuracy could also be improved for low-refractive-index films by proper polarization and increasing the angle of incidence.[74,75]

(b) *Step Disappearance in Cargille Fluids.* A common method for measuring the refractive index of a powdered transparent material is the microscopic examination of the powder immersed in a liquid of known refractive index.[76] If the refractive indices of the powder and liquid do not match, then on raising or lowering the objective of the microscope, a bright border will appear to pass from a powder particle into the liquid or vice versa depending on whether the liquid refractive index is greater ss than that of the powder. If the refractive indices are the same, the particles annot be seen. These liquids (Cargille-index-of-refraction liquids*) have been used the measurement of the refractive indices of thermally grown silicon dioxide lms by Lewis.[77] If a film step is covered with a thick film of a liquid whose refractive dex is the same as that of the film, the step will disappear optically. In this manner, ewis obtained as an average of many thermally grown silicon dioxide films a refrac- ve index (at 5,893 Å) of 1.460 which is 0.001_6 greater than that of fused quartz.[78] sing Vamfo, Pliskin and Esch[30] obtained for medium-thickness thermally grown licon dioxide films a refractive index (at 5,459 Å) of 1.461_8 which is 0.001_7 greater an that of fused quartz, thus showing substantial agreement between the two ethods of measurement. By the Lewis technique the refractive index of an oxide m on silicon can be determined to an accuracy of 0.001 to 0.006 and with Vamfo

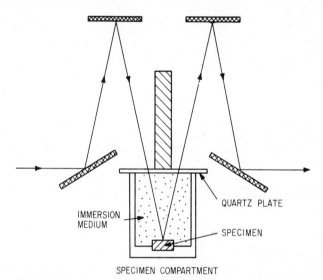

QUARTZ PLATE

IMMERSION
MEDIUM

SPECIMEN

SPECIMEN COMPARTMENT

. 13 Light-beam deflector and specimen-immersion cell for determining refractive index immersion spectrophotometry. (*Ellis.*[72])

ingle calculated refractive index has an accuracy of 0.001 to 0.002 for very thin ns[30] and for films thicker than 2 μ, the accuracy can be better than 0.0005.[31] The wis technique has the advantage of greater simplicity.

Magnetic, Electrical, and Mechanical Methods

There are various mechanical, electrical, and magnetic methods for measuring n thicknesses. Most of these methods will not be considered in detail because of ir inherent limitations and narrow applications. They are described in the reviews Keinath[11] and Kutzelnigg,[79] although the latter is rather limited since it is intended production control of metallic coatings.

(1) Magnetic Magnetic thickness measurements have found little application for n films. Most of them are based on the attractive force between a magnet and nagnetic substrate which is covered by a nonmagnetic film. As the film thickness reases, the force of attraction decreases. The magnetic-force method can also used for nickel coatings on nonmagnetic substrates. Other magnetic methods based on magnetic flux, eddy currents, and magnetic saturation. All the magnetic

These liquids with refractive indices (at 5,893 Å) from 1.400 to 1.700 in intervals of 02 and from 1.700 to 1.800 in intervals of 0.005 can be purchased from R. P. Cargille oratories, 33 Village Park Rd., Cedar Grove, N.J.

techniques are limited to film thicknesses on the order of mils, although in some case film thicknesses of a few microns can be measured.

(2) Electrical The electrical properties which lend themselves for film-thickness measurements are dielectric strength (breakdown voltage), capacitance, and resistance of the film. Because of the unreliability of electrical breakdown voltages as a indicator of film thickness, this technique is not recommended.

Capacitance measurements are sometimes used for determining the thicknesse of dielectric films deposited on conducting substrates. The capacitance is inversel proportional to the film thickness, directly proportional to the area of the electrod which is deposited on the dielectric film (provided the area is large enough to kee edge effects to a minimum), and directly proportional to the dielectric constant the film. Problematical requirements for this measurement are that there must k no pinholes in the film and that the dielectric constant must be accurately know Usually, the dielectric constant of the film is strongly dependent on the depositio conditions (see Chap. 19). Since there are other more accurate thickness-measurin techniques for dielectric films, capacitance measurements are more useful for dete mining the dielectric constant of thin films.

The measurement of film resistance is a simple operation which can be used f thickness determinations of conducting films on nonconducting substrates and f semiconductor epitaxial film layers. The use of resistivity measurements in th evaluation of epitaxial semiconductor films has been reviewed recently by Gar ner et al.[80,81]

The most commonly used instrument for the measurement of film resistance is th four-point probe of Valdes.[82] Like all other thickness measurements based on resi tance, it requires that the film resistivity is accurately known and not variable wi film thickness. This condition is rarely met since the resistivity of the film is strong dependent on deposition conditions and often quite different from that of the bu material. However, it has the advantage of being one of the few nondestructi simple techniques available for thickness measurements of conducting films on no conducting substrates. In some cases it can be used as a process-control techniqu provided the films are not so thin that the drastic resistivity changes associated wi the nucleation and island-growth stages are encountered[83] (see Chap. 18). I combining measurements of conductivity and isothermal Hall coefficient, Leona and Ramey[84] were able to determine film thicknesses of gold, silver, and copp with good accuracy from several thousand angstroms to about 200 Å.[84-86]

(3) Gravimetric One of the oldest techniques for film-thickness determination the weighing of a substrate before and after deposition (or removal) of the fil The average film thickness d is given in angstroms by

$$d = \frac{100W}{A\rho} \qquad (4$$

where W is the weight difference in micrograms, A the area of the sample in cn and ρ the density of the film in g cm^{-3}.

There are several difficulties involved in thickness determinations by gravimet techniques. Even with a substrate which is smooth and of simple geometry, t accuracy of the calculated area is only of the order of $\pm 1\%$. The weight measu ments must be done on the same substrate before and after deposition (or remov of the film. Thus, in a production or process line, adequate provisions for maintaini sample identity must be made. Another requirement is that the substrate must be prone to chipping or other types of material loss, and the deposition process m not result in the accumulation of material other than the desired film. The weighi must be very accurate since relatively small differences in large numbers are be sought. Finally, the film density is dependent on the deposition conditions and n not be known accurately. Assuming that these experimental uncertainties ha been eliminated, a weighing accuracy of ± 1 μg, which is attainable with mod balances, barely permits the detection of about 1 Å of gold and platinum, 8 Å aluminum, and 9 Å of SiO_2 on a typical substrate of 5 cm^2 area.

In actual laboratory practice the gravimetric technique is more useful for det mining the density of a deposited film rather than its thickness. If the film thickness is not constant, extreme precautions must be taken to obtain the proper average thickness for a density determination.

(4) Stylus Instruments Stylus instruments are widely used for the measurement of surface roughness and surface finishes. If a step is made in a deposited film by masking a portion of the substrate during deposition or by removing part of the film from the substrate, then a stylus instrument can also be used for the measurement of film thickness. For the investigation of surface finish, the stylus should have a very small tip to reproduce the surface more accurately, and a very light load to limit possible penetration of the surface. In principle the instrument compares the vertical movement of the stylus traveling across the sample surface with the movement of a "shoe" or "skid" on a smooth and flat reference surface. The latter may be an external flat, or portions of the sample itself may serve the purpose. The difference or vertical displacement is converted to electrical signals by means of a transducer. Various types of transducers can be used.[87] The signal is then amplified and recorded on a strip chart which also amplifies to a lesser extent the horizontal movement of the stylus relative to the sample surface. For the measurement of film thicknesses, the radius of the stylus tip can be increased to reduce the pressure and thus possible penetration of the stylus into the film.

Figure 14a and b shows the traces obtained with the Talystep 1, which is the most recent of the popular Talysurf instruments.* Because of the wide variety of vertical amplifications available with this instrument, it is possible to measure film thicknesses from about 20 Å up to 10 μ with an accuracy of a few percent. In Fig. 14A, quartz was deposited through a mask onto a glass substrate so that the masked portions left grooves in the sample as shown. In Fig. 14b, silver was deposited onto a glass substrate with part of the substrate protected by a mask so that a step was formed on the sample. In this instance, the slopes in the trace on both sides of the step must be considered. The film thickness corresponds to the vertical distance between linear extrapolations of the lower and upper portions of the trace. With sloping traces, the thickness corresponds to the vertical distance rather than the perpendicular distance between the upper and lower traces. This is because the vertical magnification is orders of magnitude greater than the horizontal magnification. Generally, more accurate film thicknesses are obtained with grooves than with steps.

For further details in the operation and application of the Talysurf instruments, the reader is referred to Reason[87] or Schwartz and Brown.[88] Older types of stylus or gauge instruments operating by mechanical displacement of a sensing element have been discussed by Keinath under the headings of dial gauges, pneumatic gauges, and mechanical-electrical methods (see Ref. 11, pages 12–28). The wide acceptance of the Talysurf, Talystep, and similar instruments† is due to the ease and rapidity with which one can obtain a permanent and accurate record of surface finishes or, if a step or channel exists in the film, a film-thickness measurement. Any type of surface can be examined, provided it is not so soft as to be deformed by the stylus or by the reference skid or shoe.

(5) Sectioning Various sectioning methods can be used for film-thickness determinations.[79] They are destructive and not generally recommended. If the film is thick enough (>1 μ), then microscopic measurements can be made on a normal cross section. With thinner films, the sample is beveled at a small angle to "amplify" the thickness of the film. This technique is sometimes used for thickness measurements of epitaxial or diffused semiconductor films.[47,89–92] Since the film boundary cannot be detected directly it is necessary to differentiate the film from the substrate material by chemical-staining techniques. This technique is sometimes called angle lap and stain. The film thickness is given by

$$d = L \sin \alpha \tag{50}$$

* Rank Taylor Hobson, Leicester, England. United States representative: Engis Equipment Company, Mortong Grove, Ill.

† An example of another recent stylus instrument is the Surface Profile Monitor Type SPM 10 Mk2 manufactured by G. V. Planer, Ltd., Middlesex, England.

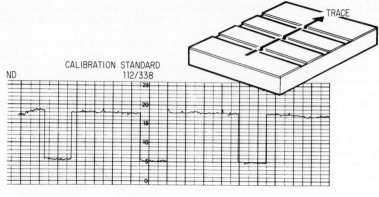

Quartz deposit on glass substrate. Test grooves left by
removal of masking. Graph details:
Vertical magnification   X 1 000 000
   1 small division =   20 angstroms
Horizontal magnification   X 200
   1 small division =   0,025 mm
Thickness of deposit measured between
   mean lines   approximately 260 angstroms
                                (25,9 mm on chart)
                                (a)

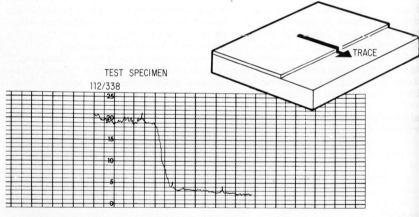

Silver deposit on glass substrate. Step left by removal of
masking. Graph details:
Vertical magnification   X 1 000 000
   1 small division =   20 angstroms
Horizontal magnification   X 200
   1 small division =   0,025 mm
Thickness of deposit measured between
   mean lines   approximately 300 angstroms
                                (30 mm on chart)
                                (b)

Fig. 14   Film-thickness measurements with Talystep 1.   (*Courtesy of Rank Taylor Hobson Leicester, England.*)

where L is the measured distance on the bevel from the film surface to the boundary and α is the lapping angle. Great care must be taken to know the lapping angle precisely. In the case of transparent films on reflective substrates, the tapered film sample is examined with monochromatic illumination and the film thickness is determined by Eq. (14) in the same manner as with a step in the transparent film. If a high-numerical-aperture objective is used without sufficient reduction of the aperture diaphragm, the measured film thickness will be less than the actual film thickness.[70]

Another sectioning technique involves the machining of a cylindrical groove into the film and in some cases even deeper into the substrate.[93,94] From a microscopic measurement of the chord or chords, the film thickness can be determined. This method is seldom used.

f. Radiation Thickness-measurement Techniques

(1) X-ray Methods (a) *Absorption.* The earliest radiation method for measuring thickness was the absorption technique which is based on the measurement of the attenuation by a film of a suitable x-ray beam diffracted by a crystalline substrate.[95] Eisenstein[96] measured the thickness of $SrCO_3$ films on $BaCO_3$ layers on nickel in the 10^{-6} to 10^{-2} cm range. There are serious limitations to this method since the measured intensities are affected by structural properties of the film such as crystallite size, stress, and preferred orientation. It can be used to advantage where the substrate contains measurable amounts of elements present in the film.

An alternate absorption technique is to measure the attenuation by the film of the characteristic radiation of the substrate material. The attenuation for a particular wavelength is an exponential function of the film thickness and dependent on the mass-absorption coefficient of the film material. It is not affected by minor impurities or previous metallurgical history. The range of thickness measurable is dependent on the energy of the substrate radiation and the absorption coefficient of the film. It is generally used for single-component thick films (0.1 to 1,000 μ) with accuracies of $\pm 5\%$. This method is not selective in that any kind of film can be evaluated as long as the substrate produces characteristic radiation which can be measured after attenuation by the film. Examples where this absorption technique has been used successfully are shown in Table 1. Zimmerman[102] and Bertin and Longobucco[103] used a variation of this technique to measure the thickness of coatings on irregularly shaped objects and on fine wires such as nickel and copper on molybdenum wire (0.005 to 0.002 in. diameter) and Au and Pd on nickel-manganese wire (0.002 to .003 in. diameter).

TABLE 1   Examples of X-ray Absorption Techniques

Reference	Film Material	Film Thickness, in.	Substrate
Beeghly[97]	Tin	. . . . .	Steel
Lemany and Liebhafsky[98]	Iron	0.006	Silver, zirconium
Lachey and Serfass[99]	Iron	0.003	Zirconium
Lachey and Serfass[99]	Iron	0.005	Silver
Lambert[100]	Aluminum	0.03	{ Nuclear fuel elements
Lowe, Sierer, and Ogilvie[101]	Stainless steel	0.02	

(b) *Emission.* X-ray emission techniques have been used extensively since their first application in 1952. The film material is excited by a high-energy source such as x-rays, an electron beam, or a radioactive source, and the intensity of a selected wavelength of the characteristic radiation emitted by the film material is measured.

The intensity of the emitted radiation is linearly proportional to thickness for thin films and increases exponentially for thicker films up to a maximum value. The technique is more broadly applicable than the absorption techniques because the only requirement is that the substrate material does not contain any of the elements in the film. Multicomponent films can also be measured. Generally, thickness values are calculated from x-ray data collected for the purpose of film-composition analysis. The technique is selective insofar as elements lighter than $Z = 11$ (Na) cannot be readily detected. Elements from $Z = 11$ to 22 (Na to Ti) require vacuum or helium spectrometers. The method is excellent for films 20 to 10,000 Å thick with an area of 0.2 to 5 cm². Accuracies of $\pm 2\%$ can be obtained.

Typical applications of the x-ray-emission technique are given in Table 2. A comparison of Ni-Fe (81/19) films from 400 to 3,500 Å measured by this x-ray method and by interferometry was made by Silver and Chow,[109] who reported accuracies of ± 20 Å. Cline and Schwartz[110] measured up to 4 μ of Al on Si using the Al characteristic emission and obtained an accuracy of ± 6 Å. Measurements of the absorption of the Si characteristic emission by the same Al films yielded an accuracy of ± 24 Å. Additional examples of thickness measured are given in Sec. 2f. There are also several reviews on x-ray methods.[102,105,111-115]

TABLE 2 Examples of X-ray Emission Techniques

Reference	Film		Substrate
	Material	Thickness	
Koh and Caugherty[104]	Iron, nickel, chromium	Up to 0.002 in.	Aluminum
Liebhafsky and Zemany[105]	Chromium	Up to 10^{-2} cm	Molybdenum
Liebhafsky and Zemany[105]	Tin	Up to 10^{-4} cm	Steel
Birks et al.[106]	Nickel	Up to 10^{-4} in.	Copper plated on steel
Zanin and Szule[107]	Gold	Up to 0.0002 in.	Copper
Keesaer[108]	Chromium	Up to 0.00001 in.	Nickel

X-ray techniques require instruments which are expensive but easy to operate. Furthermore, standards are required which must be calibrated by other thickness measuring techniques. The values obtained represent averages over the entire area measured. Once calibration data for a particular film material are available, the measurement is quickly and easily made and is therefore inexpensive. It is one of the best nondestructive techniques available for opaque films. A variation of the technique using a radioisotope x-ray source instead of the conventional electrical x-ray generator has been described by Cameron and Rhodes.[116] It has the advantages of compactness, stability, and low cost. It has been applied to in-line process control of single-constituent films consisting of several hundred angstroms of zinc and tin on steel, platinum and gold on titanium, and of 10^4-Å films of silver and nickel on copper as well as copper on steel.

(2) Beta Backscattering Film thickness can also be measured from the amount of backscatter from the films of beta particles emitted from a radioactive source.[117-1] The intensity of the backscattered particles depends on the film thickness and increases with the atomic number of the film material, which must be different from the substrate. Standards of the film material in the thickness range of interest are required. The method is inexpensive, very simple to use, fast, and nondestructive. It is suitable for opaque as well as transparent films. With suitable choice of source, thickness measurements from 0.1 to 50 μ on about 1 cm² of film can be made with an accuracy of $\pm 5\%$. This technique is more useful for production control where

TABLE 3 Summary of Film-thickness-measuring Techniques

Method	Range	Accuracy or precision	Remarks
Multiple-beam (Fizeau).....	30–20,000 Å	10–30 Å	Step and evaporated reflective coating needed
Multiple-beam (FECO).....	10–20,000 Å	2 Å	Step, evaporated reflective coating, and spectrograph needed. Very accurate, but time-consuming
Michelson interferometer...	300–20,000 Å	150–300 Å	Step required
Polarization interferometer...............	300–20,000 Å	150–300 Å	Step required
Color comparisons.........	500–15,000 Å	100–200 Å	Values given here are for SiO_2 on Si. Technique limited to transparent films on substrates such that reflectivities at the two interfaces are not too different
VAMFO.................	800–1,300 Å, 2,300 Å–10 μ	0.02–0.05 %	For transparent films on reflective substrates. Nondestructive. Can be used to measure n. The lower limit can be extended to ~ 400 Å and the 1,300–2,300 Å gap can be removed by using a detector system at shorter wavelength ($\sim$2,100 Å)
CARIS..................	400 Å–20 μ	10 Å to 0.1 %	For transparent films. Nondestructive. Must know n if only one angle of incidence is used
Ellipsometry...............	Few Å to few microns	1 Å to 0.1 %	Transparent films. Mathematics complicated, especially with thicker films
Light-section microscope....	1–400 μ	0.2 μ to 2 %	Step required on opaque films. Nondestructive with transparent films. Must know n
Gravimetric...............	Few Å to no limit	<1 Å to 1 %	Average over sample. Must know film density
Stylus...................	20 Å to no limit	Several Å to <3 %	Step required. Film should be sufficiently hard to resist deformation by stylus. Simple and rapid
X-ray absorption...........	0.1–1,000 μ	±5 %	Substrate must produce characteristic radiation
X-ray emission............	20–10,000 Å	±2 %	Substrate must not contain any of the elements in the film. Multicomponent films can be measured
Beta backscattering........	0.1–50 μ	±5 %	Film and substrate must have large differences in atomic number

specific film-substrate combination is measured routinely and lower accuracies are required, than for laboratory work where a wide variety of film and substrate combinations are examined.

g. Summary of Film-thickness-measuring Techniques

To conclude the discussion of film-thickness measurements, some of the more common techniques together with their useful ranges and accuracies are listed in Table 3. In some cases the precision is listed under accuracy because the absolute accuracy is not available. Where the accuracy is given in both a unit of length and a percentage, the former is for thinner films and the latter for thicker films. Quite often the useful ranges and accuracies vary depending on the materials being examined, the operator's ability, the instrumentation used, and other factors. Thus, the values given in Table 3 should be taken only as a guide as to what can be accomplished with each technique. Some techniques are omitted because they are too strongly dependent on particular details to be listed in such a general form. Other tabulated comparisons can be found in the literature.[4,6,120,121]

2. ANALYSIS OF FILM COMPOSITION

a. Introduction

The choice of analytical techniques for the determination of elements in thin films is mainly dependent on the amount of film available since every method has a limit of detection; i.e., there is a minimum amount of the element which can be measured with an acceptable accuracy. Low-level concentrations as represented by impurities are in the order of parts per million and are of importance in single as well as in multicomponent films. Major constituents are of interest only in multicomponent films. The amount of film material available for most thin film analyses is of the order of a few milligrams. Thick films or specially prepared thin films of large areas have larger amounts of material and offer a greater choice among the existing techniques. To illustrate the problem, Table 4 lists the amounts of material available for analysis in several films of about 1-in.-diameter circular area.

TABLE 4 Amounts of Material in 1-μ-thick Films of 5 cm² Area

Elements and concentration, %	Cu 100	Al 100	Ni-Fe 80/20	0.1 % impurity in Cr	0.1 % impurity in SiO$_2$
Density of film, g cm⁻³....	8.9	2.7	8.7	7.2	2.2
Amount, mg..............	4.5	1.3	0.8 Fe, 3.6 Ni	0.004	0.001

The second consideration in choosing an analytical technique is the form of the sample and the type of treatment to which the film can be subjected. A procedure calling for a solution can be used only if the element of interest is not lost in preparing the solution and if the substrate, which may contain the same element, is not attacked. Furthermore, the reagents must not introduce significant amounts of the element of interest. As an example of the first case, gases are lost if the film is dissolved. As an example of the importance of reagent purity, consider a molybdenum film of 10,000 Å by 10 cm² which contains 0.01% Fe (1.0 μg). If this film were dissolved with 1 ml of reagent-grade nitric acid containing 1 ppm Fe impurity, which is not uncommon, then the solution would be contaminated with an equal amount of iron.

A third consideration is that the nature of the element—be it an impurity or a major constituent—may exclude some of the available methods. For example, Ni may be determined by chemical, spectroscopic, neutron-activation, and other methods, whereas Be can be determined only by emission spectroscopy or chemical techniques.

Other considerations in choosing a method are the availability of equipment, cost of the analysis, and time requirements. A single sample submitted infrequently is best analyzed by a chemical method such as colorimetry since the cost of standardization and of the prior investigation needed for instrumental methods may be too high.

Trace analysis usually refers to methods for the determination of minute quantities of less than 100 ppm in bulk material. We shall treat trace analysis in a broader sense as the determination of microgram amounts of material, whereby the latter may be present in parts per million concentrations in a 10-mg sample, or as a 10% major constituent in a 1-mg film, or as an essentially pure element localized in a microvolume but representing an impurity in a large specimen. The methods which are sensitive for microgram amounts may be categorized as follows:

1. Microchemical—These require the element to be in solution. The measurement is made by coulometric, polarographic, colorimetric, atomic absorption, or other techniques.

2. Spectroscopic—Here the sample is generally in a solid form, and the measurement is made by neutron activation, mass spectroscopy, optical, or x-ray emission spectroscopy. In some cases infrared spectroscopy can be used (see Sec. 3).

3. Gaseous elements—The methods of determination rely mostly on gas chromatography, but mass spectroscopy and x-ray emission are also applicable in some cases.

4. Microvolume—These methods deal with the determination of elements in microvolume or with the distribution of an element over an area. Techniques such as x-ray milliprobe, electron microprobe, and mass spectroscopy are utilized.

5. Compounds—These are usually identified from their lattice structure by x-ray or electron-diffraction methods.

Impurities and low concentrations of elements can be determined in films with a minimum of preparation by ultraviolet spectroscopy, neutron activation, or mass spectroscopy, in increasing order of sensitivity. The first method is applicable to about 70 elements at the parts per million (0.1 μg) level in about 10 mg of material. Neutron activation is very sensitive but also very selective because of half-life, interference, and contamination problems. Mass spectroscopy may be used for all the elements with nanogram sensitivity (ppb level) on microgram amounts of material, but only semiquantitative values can be expected.

Major constituents are conveniently determined by x-ray emission on milligram amounts of material with high accuracies for elements heavier than atomic number 11 (Na). Standards are necessary, however, and these must be calibrated by microchemical methods such as colorimetry, atomic absorption, flame spectrometry, polarography, or by x-ray emission on solutions. Ultraviolet emission with spark excitation is accurate but not easily adapted to solid films and of limited value for solutions of films.

The composition analysis of films is generally performed by analytical groups which are staffed by specialists in the various techniques. It is rarely done by personnel working on thin film technology since it is one of the most demanding analytical tasks. The reasons are the diversity of methods required, the small size of the sample, the extremely low concentration levels, the sample-handling problems, the number and kind of elements present, and the possibility of introducing extraneous material from the substrate or process. Because of these complications, the choice of the analytical technique is usually left to the analyst. Accordingly, this section merely introduces the thin film worker to the existing analytical techniques and should guide him in his expectations. It is not intended as a review of analytical procedures. Readers interested in more detail are referred to the analytical literature.[122]

b. Microchemical Techniques

Microchemical techniques require sufficient material to give a solution from 1 to 10 ml with elemental concentrations of 10^{-7} to 10^{-5} molar which is equivalent to a few hundred micrograms of material or less. Manipulative procedures such as dissolving, concentrating, and separating are possible sources of error; part of the element may be lost through sorption on the laboratory-ware surfaces, or contaminants may be introduced from reagents, solution environments, and substrates. Most of these potential sources of error can be corrected by running a blank concurrently. The element of interest may be determined in the initial solution or after chemical manipulation. Examples are separation of the element of interest by concentration, or by removing the matrix and interfering elements. Methods such as ion exchange, extraction, masking, coprecipitation, or electrochemical deposition may be used.

For the quantitative determination of elements in solution, the following techniques are suitable for microchemical work:

1. Volumetry—This is the determination of an element by titrating with a specific reagent using microequipment. The procedure has been largely replaced by spectrophotometric techniques which are more sensitive.

2. Spectrophotometry—The ion of the element is reacted with reagents to produce a complex with a distinct absorption spectrum in the ultraviolet or visible region. The concentration is measured by comparison with standard solutions prepared from the pure element or one of its compounds. Spectrophotometry is used extensively for major-constituent determinations on single samples and for calibrating standards. There are many extensive texts and papers describing the procedures for almost all the elements. The sensitivities are in the order of micrograms.

3. Flame spectrometry—The solution is aspirated into a high-temperature flame and the elements are excited to give off their characteristic visible and ultraviolet emission spectra. The intensity of the selected wavelength of the element is measured by a photodetector. This technique is very good for the alkali- and alkaline-earth elements from very low concentrations to major concentrations. It has been used for about 40 elements with reported sensitivities from 0.001 to 15 μg ml^{-1}. Of particular interest in semiconductor device technology is the work of Barry et al.,[123] who analyzed the Na distribution in 6,000-Å-thick SiO_2 films of 16 cm^2 area. The oxide was removed in layer increments. The initial layer was 50 Å thick and contained 3.4×10^{14} atoms of Na. The next several layers were 1,200 Å thick, and the number of Na atoms ranged from 3.0×10^{13} to 1.3×10^{14}. The final 250 Å of oxide represented the level of sodium contamination at the $Si-SiO_2$ interface and contained about 3×10^{14} Na atoms. The method has a limit of detection of 0.2 ppb.

4. Atomic absorption—The solution is aspirated into a high-temperature flame and the amount of absorption of the radiation from a hollow-cathode discharge tube of the same element as the one to be determined is measured. This procedure is complementary to the flame-emission technique and is more sensitive for some though not all elements. It is more widely used than flame emission, especially for elements other than the alkalies and alkaline earths. Atomic-absorption spectrometry is fast, accurate, and simple to use. The standards are easy to prepare, and the equipment is inexpensive.

In general, microchemical techniques are more time-consuming than the spectroscopic methods and therefore should be used for infrequent analyses or when no standards are available. Their disadvantages such as the problem of contamination, poor sensitivity, destruction of the sample, and high selectivity must be weighed against such advantages as the ease of preparing standards and reducing interelement or matrix effects and the modest cost of the equipment. The most important application of microchemical methods is in conjunction with spectrophotometry (colorimetry) and atomic absorption for the purpose of analyzing a portion of a sample which is then used as a standard for other techniques such as x-ray emission spectrometry.

c. Neutron-activation Analysis

Neutron-activation analysis is based on the identification and measurement of the radioactive species which are formed after nuclear reactions with neutrons, i.e., on the energy spectrum of the radioactive species and their half-life time. The method is extremely sensitive and capable of measuring subnanogram amounts of the elements in milligram-size samples. It has limited application for thin film analysis because surface contamination due to sample handling requires the removal of several hundred angstroms from the surface after activation. There are other disadvantages such as availability of the radiation source and interference due to the activation of the substrate or other impurity elements. Osborne et al.[124] determined the Na and P profiles in a 10,000-Å SiO_2 film. They removed layers of 50 to 100 Å and analyzed the etch solutions. The Na concentration decreased with depth from 500 to 0.01 ppm, the P content from 50,000 to 10 ppm.

d. Emission Spectroscopy

Emission spectroscopy is based on the identification and measurement of the characteristic ultraviolet and visible radiation which is produced when the material is excited by a high-temperature thermal source such as a dc arc, a high-voltage spark, or a flame. About 70 elements can be determined in the microgram range. Assuming a sample quantity of 10 mg, the sensitivities are 1 to 10 ppm for the transition metals, 10 to 100 ppm for alkali metals, and 100 to 1,000 ppm for group V elements. Gases or halogens are rarely determined because their sensitivities are lower than can be obtained by other methods. A minimum of sample preparation is required. The main application is to analyze for low concentrations of elements rapidly and inexpensively with accuracies of about ±25%. An example is the determination of several elements ranging from 0.001 to 0.15% in high-purity gold plate.[125] Similarly, Nohe[126] identified 26 elements at 5 to 1,000 ppm level in 1,000-Å-thick Ta films of 30 cm^2 area (about 5 mg Ta). This technique has been reviewed recently.[127]

Major film constituents can be determined in solution with accuracies around ±5%, but for this application emission spectroscopy has been supplanted by x-ray emission. The advantages of emission spectroscopy are that it is readily available and the analyses are inexpensive, fast, and cover a large number of elements with good sensitivities. The major disadvantages are the requirement of 1 to 10 mg of material and the fact that the sample is consumed.

e. Mass Spectroscopy

Mass spectroscopy is the most sensitive technique available and can be used for determining all the elements. The material to be tested is either vaporized by one of several methods, or it is introduced as a gas into a vacuum chamber. The gaseous sample is ionized, and the ions are accelerated and then analyzed according to their mass-to-charge ratio (see also Chap. 2, Sec. 6c). After separation, the ions are either collected on a photographic plate which records their density and mass position, or the ion currents are collected, amplified, and their relative magnitudes registered.

An excellent review of instrumentation for mass spectroscopy and its application to thin film analysis has been given by Willardson.[128] His comparison of source types and analytical capabilities is reproduced in Table 5. Most widely used is the rf spark source which utilizes a stationary point and a spinning sample. The ion-microprobe instrument bombards the thin film with a beam of 10-keV inert-gas ions, and sputtering is induced at rates of about 80 μ h^{-1} over an area of 0.25 mm^2. This source is very useful to study film composition as a function of depth. Willardson also describes the ion microscope of Castaing and Slodzian and the use of lasers as ion sources. The ion-bombardment sources have a disadvantage compared with the spark sources insofar as the sensitivity to various elements varies with the constitution of the film and also depends on the beam parameters.

The type of thin film analysis which can be performed by mass spectroscopy and the results to be expected are best illustrated by some examples. Of particular interest are applications of the technique to semiconductor surfaces. Thus, Ahearn[129] succeeded in detecting one-hundredth of a monolayer of the indium 113 (4% abundance) in a monolayer of indium on a germanium surface. He used the point-to-stationary plane spark source. The same author also identified Ag, As, Be, Cr, Cu, Pb, and Zn on intentionally contaminated silicon surfaces in quantities of the order of 10^{-9} atom fractions.[130] Hickman and Sweeney[131] used a conventional spark source and a point-to-stationary plane to study volumes of about 25 μ diameter and 3 μ deep. They reported that the detectable amount of 1,000 ppm in the 3-μ-deep crater is equivalent to a 30-Å-thick film over an area of 5 $\times$ 10^{-6} cm^2.

By using a spinning electrode technique, Hickmann and Sandler[132] reduced the depth of the spark penetration to around 2,000 Å with craters of about 4,000 Å diameter. The spectrum of gold from the counter-electrode and a high-purity silicon disk showed Si, Au, C, and O_2, and it was estimated that concentrations of 0.1% are detectable. Malm[133] used a point-to-sample scanning technique with rf spark source and determined 0.02 to 1.6 wt % Mo in 4,000-Å-thick Ta capacitor films deposited on carbon blocks. Sensitivities below 10 ppm (atomic) were predicted as attainable. Malm has recently reviewed the use of rf spark-source mass spectroscopy for the analysis of surface films.[134] Pevera[135] applied the point-to-sample scanning technique to about 1-μ-thick Mo films deposited either on quartz disks or on 10-μ-thick sputtered quartz films on silicon wafers. A conducting film of 0.1 μ of Pt was preevaporated onto the quartz surface to serve as an internal standard. Contamination levels of about 50 ppm of Ar, Cl, Cr, Cu, and Ni in the Mo films could be determined with estimated accuracies of $\pm 25\%$.

TABLE 5 Sources of Ions for Analysis of Thin Films by Mass Spectrometry (Willardson[128])

	Typical energy	Min area analyzed, μ^2	Min depth of analysis, μ	Max penetration rate, μs^{-1}	Comments
Spark sources:					
Conventional......	20–100 keV 1 MHz rf	10^3	3	10^6	Not applicable to most thin film studies
Rotating probe....	20–100 keV 1 MHz rf	0.2	0.2	10^5	Detection limit about 0.1 % unless scanning is used
Ion-bombardment sources:					
Ion microprobe....	1 mA, 10 keV	10^5	10^{-2}	0.01	Sensitivity varies widely for different elements
Ion microscope.....	10 μA, 10 keV	1	10^{-2}	0.01	Qualitative analysis only
Low-voltage source	0.4 keV	Large	10^{-3}	....	Useful only for some organic films
Laser-excited sources:					
Ruby laser.........	4 pulses/min 1 joule (10^{-4} s) pulses	10^3–10^4	10^2–10^3	10^7	Not applicable to most thin film studies
He-Ne laser.......	10^3–10^4 pulses/s 1–100 mJ (2×10^{-7} s) pulses	10^2	10^{-2}	10^5	Scanning required for detection limits below about 0.1 %
CO_2 laser..........	1–100 W continuous duty	10^3	10^{-2}	10^2	Minimum area limited by relatively long wavelength of light used

Mass spectroscopy is the best technique available for the semiquantitative analysis of thin films, including the determination of contaminants on film surfaces and at the substrate-film interface. Sensitivities are of the order of subnanograms of the constituent determined, and the amount of material required is less than 1 mg. All the elements and many compounds such as hydrocarbon gases can be determined in ppb concentrations. Variations of the technique permit examination of micron-size areas, shallow depths, and scanning across surfaces. For very low concentration levels, accuracies of $\pm 25\%$ are obtained. This is generally not sufficient for the determination of major constituents. Other disadvantages are the lack of suitable standards, the considerable time required for an analysis, and the cost of the equipment.

f. X-ray Emission Spectroscopy

X-ray emission spectroscopy is the most useful analytical technique for the composition analysis of films. In this technique (also called x-ray secondary emission) x-ray spectrochemical, or x-ray fluorescence), the specimen is irradiated by an intense x-ray beam from an x-ray tube. The primary x-rays excite the elements in the specimen to emit their characteristic secondary x-ray line spectra. The secondary x-ray beam passes into a spectrometer where it is diffracted by a crystal. In some cases, radiation filters or spectral energy discriminators have been used instead. The intensity of the selected radiation is related to the concentration of the element and is measured. Elements from Na to Ti ($Z = 11$ to 22) require a vacuum or helium-path spectrometer since their radiation is absorbed in air. Elements lighter than $Z = 11$ cannot be readily detected. The x-ray emission technique measures the amount of the elements in mass per unit area, from which the percent composition and the thickness must be calculated. Areas from several square millimeters up to 5 cm² with film thicknesses from 100 Å to 1 μ have been used. Film densities can be determined if the thickness is known.

Generally, the film sample is analyzed nondestructively, provided it is supported on a suitable substrate which does not contain elements present in the film. Trace amounts of such elements in the substrate can cause errors since the primary radiation penetrates films up to several microns thick. The characteristic radiation from these trace elements contributes to the radiation emitted by the film; hence selection of a suitable substrate is critical. The analysis of a solid film specimen requires standards of similar metallurgical history and thickness as well as a composition range close to that of the specimen. Such standards must be calibrated by independent methods such as colorimetry, atomic absorption, or x-ray emission on solutions. This is the major disadvantage of the x-ray emission technique and limits its application for quantitative analysis to samples submitted frequently enough to justify the expense of standards calibration. Substrate-supported films are, however, the most extensively used type of specimen.

An alternate technique is to remove the film from the substrate and prepare a solution, if sufficient material (milligram amounts) is available. The solution can then be analyzed by comparison with standard solutions of the same composition. An advantage of the solution technique is the reduction of interelement effects which are tedious to correct for in solid specimens. A variation of this approach is to absorb and dry the solution on a special filter paper or on polymer films. Standards are prepared in the same fashion. Both these techniques are practical for the analysis of a limited number of samples if no solid standards are available.

The film emission method has been used extensively for film composition and thickness analysis in multicomponent thin films. Koh and Caugherty[104] where the first to apply the x-ray emission technique to determine the composition and thickness of Fe-Ni, Cr-Fe, and Ni-Cr-Fe oxide films. Rhodin[136] studied the compositions and thicknesses of Fe, Ni, Cr, and stainless-steel films up to 300 Å thick. Verderber[137] reported on 80/20 Ni-Fe films of 1,000 to 4,000 Å thickness, using a solution technique and 100 cm² of film. Chow and Cocozza[138] analyzed 500- to 2,500-Å-thick Ni-Fe films with a similar technique. Zanin and Hooser[139] analyzed magnetic tapes from 200 to 3,000 Å thick with a wide variety of compositions: Co-Ni-P (10 to 25% Ni, 1 to 8% P), Ni-Co-P (12 to 50% Co, 3 to 9% P). Zanin et al.[140] described a method for 200- to 2,000-Å films of Ni-Fe on glass. Calibration of the x-ray standards was by colorimetry in both cases. Cocozza and Ferguson[141] analyzed Ni-Fe-Co (79/18/3) films (600 to 1,300 Å thick) and calibrated by analyzing solutions prepared from 10 in.² of selected films by x-ray solution methods. Pluchery[142] described an absolute mathematical method for composition and thickness determinations using pure constituent elements as standards and applied this technique to Ni-Fe films. Bertin[143] determined the composition of Nb_3Sn coatings about 0.0003 in. thick on metallic ribbons and ceramics. Spielberg and Abowitz[144] analyzed 100 to 300 Å of Ni-Cr (75/25) evaporated films and described three techniques for calibration.

The main advantages of the x-ray emission technique are that small amounts of film are sufficient, that high precision ($\pm 2\%$) is attainable, and that is nondestructive. Instruments are commercially available, safe, easy to use, but expensive. The actual analysis time is short, and therefore the costs are low. It can also be easily adapted to in-line process control. Another interesting, although limited, application is the determination of the chemical state of an element by precision measurement of wavelength shifts. These are of the order of 0.01 Å and arise from differences in the energy state of the valence electrons. The main disadvantage of the x-ray emission technique is the cost of standardization. A serious problem with multi-component films over 1,000 Å is interelement effects such as the selective absorption and reemission of radiation of one element by another. Interelement effects can be corrected for mathematically, or experimentally by preparing standards where thickness, composition, and history duplicate the specimen as closely as possible.

g. Microvolume Techniques

(1) Electron-probe Microanalyzer The electron-probe microanalyzer is capable of analyzing micron-size areas up to 1 μ in thickness. An electron beam of about 1 μ diameter impinges upon the area of interest and excites it to emit primary x-rays characteristic of the elements present. Spectrometers and detectors are built into the instrument to select and measure the characteristic radiation. All elements with atomic numbers of 6 or greater can be detected. The microvolume excited by the electron beam represents less than a nanogram of material. Minimum concentrations varying from 0.001 to 0.1% can be measured, depending on the element determined, the matrix, and the beam size. Identification of the elements and semi-quantitative values of their amounts are easily obtained. Quantitative analysis requires standards similar to the specimen. Pure elements may also be used for calibration, but their measured intensity must be corrected for absorption and fluorescence effects caused by the characteristic and continuous radiation.

The electron probe has found wide application in the investigations of thin metal films. Examples are studies of phases, diffusion rates, segregation, inclusions, precipitates, and grain boundaries. The method can be used to scan along a line or over large surface areas of about 200 μ edge length, whereby the output is displayed on a screen to show the elemental distribution. Lublin and Sutkowski[145] reported on the distribution of Al in mixed (Al,Ga)As epitaxial films and evaluated the composition of about 1,000-Å thin films of $BaTiO_3$ and $PbTiO_3$. Ramsey and Weinstein[146] analyzed the distribution of Cr over a 10-μ-square area in Cr-SiO cermet films. They also showed that a multilayer film consisting of the five elements Cu, Cr, Au, Pb, and Sn (each layer 1,000 to 2,000 Å thick) was made up of discrete and unalloyed strata. The application of this technique to thickness determinations of films of small area (500 Å of Au on Si) has been described by Hutchins.[147]

The variation in thickness and composition of phosphosilicate glass on SiO_2 films covering semiconductor devices was measured by Koopman and Gniewek.[148] They also determined the thickness and composition of 1,000-Å-thick Ni-Fe films on glass and of a 400-Å-thick Pt film on Si. The latter was found to consist of 50 Å of unalloyed Pt and 350 Å of Pt-Si alloy. Schreiber[149] identified foreign particles on a Au film plated on a metal substrate. Chodos[150] analyzed Ni-Fe-Co films on a 5-mil wire substrate.

The main applications of the electron probe are composition and thickness analysis of microvolumes. Several elements can be determined simultaneously. Disadvantages are the high cost of the equipment, the time required per sample, and the inadequate supply of standards. Furthermore, the substrate must be free of the elements of interest.

(2) X-ray Milliprobe The x-ray milliprobe uses standard x-ray emission equipment to measure areas of a fraction of a millimeter in diameter. The x-ray emission apparatus must be modified to include an x-ray tube producing a small but intense beam, or a set of collimators with small apertures in the primary and secondary paths, and curved crystals to increase the intensities measured. Scanning can be done by translating the specimen across the x-ray beam. Accessories for sampling, handling, and confining the x-ray beam have been described by Bertin and Longobucco.[151] They applied this technique to the analysis of microgram deposits on electron tube components as well as for point-by-point mapping (1-mm aperture). Sensitivities of the order of 1 μg cm^{-2} were found for Ba, Sr, and Mn. Bertin[152] also reported on the use of a slit probe rather than a pinhole-type aperture to study the diffusion of Mo into ceramics, the behavior of diffusion couples, and to locate the interface of adjacent metals. Sloan[153] used a milliprobe with 100- to 500-μ-diameter apertures to study the metal distribution in 1,500-Å-thick Nichrome films. Glade[154] measured the thickness of plated Au and Rh films (about 0.7 × 25 mm dimensions) on Ni-Fe reed switches.

The x-ray milliprobe is subject to the same limitations as the x-ray-emission technique because of the similarity of equipment and principles. The distinguishing feature is the ability to analyze areas as small as 0.3 mm in diameter. The main advantage of this method is the low cost and simplicity of the apparatus as compared with the electron microprobe. However, the x-ray milliprobe uses lower radiation intensities and is therefore less sensitive than the electron microprobe.

(3) Mass Spectroscopy This technique, which has already been discussed in Sec. 2e, can also be extended for microvolume analysis. The stationary-point-probe spark source is utilized to determine the composition of thin film samples ranging from 1 to 10 μ in diameter. Unlike the electron microprobe, all the elements present can be detected simultaneously. However, the sensitivities are relatively poor, and the material in the analyzed volume is removed so that repeated measurements of the same sample cannot be made.

h. Gases in Films

Gaseous elements are often entrapped during film deposition, particularly in sputtered films. Several standard procedures[155] are available for their determination: nitrogen can be determined by vacuum fusion, inert gas fusion, the Kjeldahl technique, and halogenation; oxygen by vacuum fusion and inert gas fusion; and hydrogen by vacuum fusion. These techniques require fairly large samples. Mass spectroscopy has wider application because it is nonselective, and the gaseous elements can be determined simultaneously along with other elements. It has the advantage of requiring only microgram amounts of sample, and it can be used to analyze very small specimen areas. As pointed out by Guthrie,[156] a complication arises from the fact that the elements most commonly found in metal samples, C, H_2, O_2, and N_2, also make up the residual gases in vacuum systems. These gases contribute to the spectrum if they become involved in the sparking process. Therefore, disagreement between analytical results from different methods such as mass spectroscopy, neutron activation, and microcombustion is not uncommon.

An example of mass spectroscopy as applied to gaseous elements in thin films is the determination of a hundred to several thousand ppm of Ar and Cl in sputtered quartz and molybdenum films by Pevera.[135] The use of x-ray emission for the analysis of gaseous elements has been reported by Hoffmeister and Zuegel.[157] Lloyd[158] measured 0.05 to 7.4 wt % of Ar in sputtered SiO_2 films which were 0.5 to 5 μ thick and had an area of 2.3 cm diameter.

A unique method for the quantitative analysis of gases in thin films was developed by Guldner and Brown.[159] In this technique, which is called flash photolysis, small areas of thin films are vaporized in an evacuated quartz tube by a xenon flash discharge lamp. The gases are then collected through high-vacuum techniques and analyzed in a cryogenic gas chromatograph. Details of the procedure and the required equipment as well as a comparison with other methods may be found in the original publication. Although the method has been applied to a variety of metal films such as Au, Ti, Hf, and V, most of their results pertain to sputtered and evaporated tantalum films. The latter were several thousand angstroms thick, and a typical analysis (in at. %) gave 0.5 H_2, 1.5 Ar, 0.9 O_2, 0.03 N_2, and 0.63 C and required about 4 mg of film material. Similar gas concentrations have also been found in molybdenum films subjected to the same analysis.[160]

A variation of the original flash-photolysis technique was introduced by Winters and Kay[161] to study the Ar gas incorporated into sputtered nickel films. They describe two methods of gas analysis which involve the vaporization of 2- to 6-μ-thick Ni films with the subsequent release of the gases into the vapor phase where they are then analyzed by mass spectroscopy. One method uses a film of known area and weight wrapped in a 0.001-in. Ta foil and placed into a Ta boat which is resistively heated. The second method uses laser-induced flash evaporation of a well-defined volume of film. The area affected by the laser is about 1 mm in diameter. The concentrations reported vary from 10^{-1} to 10^{-4} argon atoms/Ni atom depending on the conditions of film growth.

3. EVALUATION OF GLASS FILMS

a. Introduction

With the increased use of microminiaturization, glass and insulating films become important for use as diffusion masks, crossover insulators, and passivating films for semiconductor devices.[162-164] A combination of optical and spectroscopic techniques has been found to be very useful for evaluation of these films with regard to composition, structure, and water and chemical stability.[42,165-167] No one technique in itself is sufficient for the evaluation of the films, but useful results are obtained by combining several techniques. Complete evaluation should also include various electrical measurements such as resistivity, dielectric constant, dielectric strength, loss factor, and surface-charge densities.[163] Electron microscopy is another useful tool in the evaluation of films.[168,169] Electrical measurements and various techniques such as electron microscopy for investigating the structure of films will not be discussed in detail here since they are covered in other chapters.

The chemical and optical evaluation techniques include the effect on the film properties of various treatments such as thermal annealing and densification, exposure to high humidity and temperature, and etching.

b. Experimental Techniques

(1) Infrared Spectroscopy In some cases compositional and structural evaluation of glass films can be made easily by use of infrared spectroscopy. By comparing the intensities of various infrared absorption bands, one can determine relative amounts of various (though not necessarily all) components in a glass. In the case of transmitted spectra, the absorption-band intensities are roughly proportional to the film thicknesses; i.e., they obey Beer's law fairly well and are much better for quantitative determinations than are reflection spectra.[164] Although the intensity of the absorption is given best by the integrated intensity, in many cases sufficiently accurate composition comparisons can be made by utilizing the optical density (OD) of absorption bands.

In addition to measuring the absorption bands of components which are known to be part of the glass, infrared spectroscopy is especially useful for detecting impurities such as water and silanol groups in the glass from their characteristic absorption bands at about 3,350 cm^{-1} (for water) and at about 3,650 cm^{-1} (for hydrogen-bonded silanol groups). The intensity of these bands is an indication of the quantity of water or silanol groups present and indirectly is an indication of the porosity of the glass film. In the case of SiO_2 showing only a band near 3,650 to 3,700 cm^{-1}, the weight fraction of hydroxyl is equal to the optical density (absorbance) of the band per micron of film.[170,171]

Information can also be obtained from the exact position and half-width of some absorption bands. For example, the exact half-width and position of the Si-O stretching band at ~1,050 to 1,100 cm^{-1} in SiO_2 are strongly influenced by the bond strain, stoichiometry, and porosity of the film[165] as well as being influenced by the presence of other components in the glass.[164,166]

Most published work in this area has been for films deposited on silicon substrates, although other substrates can also be used. Silicon is, however, very useful for a variety of reasons. It is stable at very high temperatures and so is suitable for receiving films formed at high temperatures. It is chemically stable and generally not very reactive. It is excellent for optical studies of deposited films in the visible region using reflection techniques. It has no very strong lattice absorption bands in the useful region of the infrared and thus can be used for transmission studies in this region. To correct for the lattice absorption bands in silicon, a reference silicon sample slightly thinner (~10% for SiO_2 and many silicate glass films on silicon) than the sample silicon substrate containing the deposited glass film is placed in the reference beam of the double-beam spectrophotometer. The reason for using a slightly thinner wafer in the reference beam is that the effective path length through the sample substrate is less than the path length through the bare silicon wafer.[172] The requirement that the silicon substrates be infrared transparent necessitates the use of fairly high resistivity material (≥ 5 ohm-cm). Another requirement for best results is the use of float-zone or Lopex material rather than pulled crystals because the latter contain varying amounts of dissolved oxygen and this can result in a fairly strong absorption band near 1,100 cm^{-1} close to the absorption band for silica and various silicates. The silicon substrates (wafers) should be either chemically polished or mechanically polished (both surfaces) to eliminate radiation losses due to light scattering. Finally, the substrates should be about 30 mils thick (depending on the resolution of the spectrophotometer) so that interference fringes are not formed because of the thickness of the silicon wafer itself.

When checking for the presence of small quantities of water, silanol groups, or other impurities, the absorption bands may be too weak to be detected, especially if the film itself causes interference fringes in the spectrum. This is because the weak absorption band may be at a position corresponding to the position of a steep slope in the interference spectrum where it would cause an imperceptible perturbation of the slope. This problem is avoided by replacing the bare silicon wafer in the reference beam with one containing a pure dry film of approximately the same refractive index and thickness as the sample. The interference fringes thus in effect cancel out and the infrared spectrum can be amplified to allow detection of weak bands. For most glass films having refractive indices in the range of 1.4 to 1.6, the reference wafers and films may be thermally oxidized silicon wafers. Reference wafers which were oxidized in steam should be dried in oxygen or nitrogen at or above 1000°C for at least $\frac{1}{2}$ h, depending on the film thickness. After oxidation the film on one side of the wafer is removed by etching. If the refractive index of the reference film does not coincide with that of the sample film, then the proper reference-film thickness is given by $n_s d_s / n_r$, where d_s is the sample-film thickness and n_s and n_r are the sample- and reference-film refractive indices at the infrared wavelengths being used for examination. A thickness match within about 200 Å of the desired thickness for films thinner than 1 μ is generally sufficient to allow satisfactory amplification of the spectra. With thicker films a better thickness match (within about 100 Å) is desirable.

(2) Optical Techniques The refractive index can be useful for determining the composition and density of glass films.[165,166] For example, the percent composition of lead oxide in the film has a strong influence on the refractive index. Other compounds such as aluminum oxide also influence the refractive index, although to a more limited extent. The refractive index is also influenced by the oxygen stoichiometry in the glass film.[165] For example, glass films which are oxygen-deficient will show higher refractive indices than corresponding films which are completely oxidized. A decrease in density due to porosity will also result in a decreased refractive index. Thus, the refractive index can give an indication as to composition, oxygen deficiency, and density.

Accurate measurements of thicknesses are important for determining any densification of the film that occurs as a result of heating, for determining the stability of the glass film when subjected to various ambients, and for determining its etching characteristics. It is also important for determining the water and chemical stability of glass films, although thickness alone is not usually a good measure for this type of evaluation.[42]

(3) Etching Techniques It has been shown that selective etching is useful for determining the strain and density in silicon dioxide deposited films,[165] and the same can be said for various types of glass films. In addition to strain and porosity, the composition of a glass film also influences its etch rate. For example, glasses which are rich in either lead or boron have faster P-etch* rates than glasses containing no

---

* P-etch consists of 15 parts hydrofluoric acid (49%), 10 parts nitric acid (70%), and 300 parts water.[59]

lead or boron.[164] Similarly, in phosphosilicate glass, the etch rate increases with phosphorus concentration.[59] Snow and Deal[173a] and Eldridge and Balk[173b] found that the P-etch rate of phosphosilicate glass increases logarithmically with its P_2O_5 content. Also, as far as other common compositional components of glasses are concerned, the P-etch rate of a glass can be decreased by increased alumina and silica content as well as by increased oxygen deficiency.

P-etch is an excellent diagnostic tool. However, as a part of present-day semiconductor technology, oxides and glasses are commonly etched in ammonium fluoride buffered hydrofluoric acid solutions whose behavior is somewhat different from that of P-etch. For example, in P-etch, borosilicates etch faster than thermally grown silicon dioxide films; whereas in the buffered hydrofluoric acid solution, the silicon dioxide films etch faster than the borosilicate films. This difference has been utilized by Maissel et al.[174] to show that resputtering from the substrate can occur during the deposition of SiO_2 films by rf sputtering.[175]

TABLE 6   Influence of Some Film Properties and Composition on Refractive Index, Etch Rate, and Infrared Spectra for Silicate Glasses (Pliskin[167])

	Density	Excess silicon	Bond strain	PbO content	Al_2O_3 content
Refractive index	+	+	0	+ +	+
P-etch rate	−	−	+	+	−
~1090 cm^{-1} Si-O absorption band:					
Position	+	−	−	−	−
Half-width	−	+	+	+	+
3650 cm^{-1} SiOH and 3400 cm^{-1} H_2O band intensities	−	0	0	0	0

The influence of various film properties including the composition on the refractive index, etch rate, and infrared spectra for some silicate glasses is summarized in Table 6,[167] which shows that no one technique in itself is sufficient for evaluating the glass films.

c. Application of Techniques

We shall give only a few brief examples of the application of the previously discussed techniques. Many more examples can be found in the cited references.

(1) Water Stability of Borosilicate Glasses by Infrared Spectroscopy   Common glass components which give rise to strong absorption bands in the infrared are B_2O_3 and SiO_2. Therefore, since many of the glasses of interest are borosilicates, infrared spectroscopy is a useful tool for investigating and evaluating borosilicate glass films. One very important advantage of nondestructive techniques such as infrared spectroscopy is that films deposited on silicon wafers can be examined both before and after subjecting them to various chemicals or ambients at elevated temperatures. This is especially true for investigating the water stability of glass films.[42,172]

One of the requirements of an effective semiconductor insulator system is that it be impervious to attack and penetration by the ambient in general and particularly moisture. Since water stability was considered to be important for device passivation, various fused sedimented glass films[44] as well as rf sputtered films[175] were studied for resistance to moisture attack.[42,172] The most rapid and simple accelerated test of this property was found to be exposure of the glassed wafers to boiling water. Some glass films are virtually unaffected by exposure to boiling water (e.g., Pyrex and fused quartz). Those glasses which are affected by boiling water will show a dissolution of the glass which can be detected by weight loss and film-thickness decrease, or they will show significant amounts of moisture in the film as seen by the presence of hydroxyl absorption in the 3-μ region of the spectrum and may show leaching of a soluble component oxide such as B_2O_3. Corning 7050 glass is an example of a glass which is seriously attacked by boiling water without showing any film-thickness decrease. From Fig. 15 it can be seen by the decrease in intensity of the band at 1390 cm^{-1} (7.2 μ) that most of the boron oxide was leached out of the film; and by the increased absorptions at 3650, 3400, and 1630 cm^{-1} silanol groups were formed and water was absorbed by the film. Glasses such as Pyrex (Corning Glass 7740) show no change in the infrared spectrum as a result of a similar moisture exposure. Obviously when looking for a semiconductor encapsulant film, glasses with high moisture stability like Corning glass 7740 are desirable.

Other examples where borosilicate glass films have been examined by use of infrared spectroscopy are (1) in the examination of the water stability of various sedimented, fused glass films;[42] (2) in the investigation by Deal and Sklar[176] of the oxidation of silicon heavily doped with boron to show the incorporation of the boron in the oxide; (3) in a study of the leaching of B_2O_3 from zinc borosilicate glasses by exposure to water saturated with carbon dioxide;[164] and (4) in studies on the structure and composition of rf sputtered borosilicate glass films.[177]

(2) Phosphorus Content in Phosphosilicate by Infrared Spectroscopy A layer of phosphosilicate glass is often included in integrated circuits to help ensure device reliability.[178] Transmission infrared spectra of phosphosilicate films show an absorption band at 1330 cm^{-1}, which could be used for detecting phosphorus in silicon dioxide.[179] However, transmission infrared spectroscopy cannot be used on silicon device wafers because their high doping levels make the silicon opaque to infrared radiation. Corl et al.[180] have made use of reflection spectroscopy for detecting phosphorus in phosphosilicate glasses on actual device wafers as shown in Fig. 16.

(3) Low-temperature Deposited SiO₂ Films There are many different techniques for the deposition of SiO_2 (and other glass films) at relatively low temperatures, i.e., temperatures below those usually used for thermal oxidation or below glass-softening points. These techniques

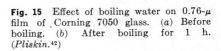

Fig. 15 Effect of boiling water on 0.76-μ film of Corning 7050 glass. (a) Before boiling. (b) After boiling for 1 h. (*Pliskin.*[42])

have been covered in recent reviews.[200] In this section a few examples will be given to illustrate the application of the techniques previously discussed.

(*a*) *Densification of Pyrolytic Oxides.* Pyrolytic films of silicon dioxide can be obtained from the thermal decomposition of alkoxysilanes at deposition temperatures of about 600 to 900°C.[181-183] Depending on the deposition temperature and conditions, films formed by this technique are less dense than those of thermally grown silicon dioxide. In the evaluation of these and other low-temperature deposited silicon dioxide films it is important to examine as many properties as possible such as refractive index, exact infrared Si-O absorption-band position and half-width, close scrutiny of hydroxyl content, and densification effects.[165] As an example, in Fig. 17, spectrum A is that of a pyrolytic oxide formed at 675°C while spectrum B is of the same film after densification by exposure to steam for 15 min at 975°C.

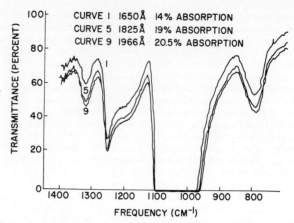

Fig. 16 5×-amplified reflection spectra showing absorptions with small $P{=}O$ differences. (*Corl et al.*[180])

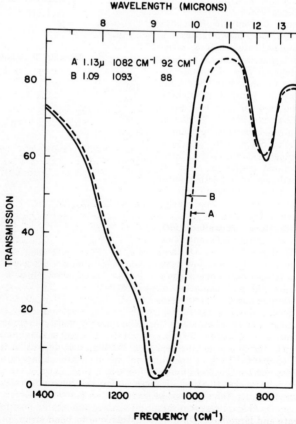

Fig. 17 Steam densification of pyrolytic SiO_2. (*a*) TF pyrolytic SiO_2 formed at 675°C. (*b*) Densified by exposure to steam for 15 min at 975°C. (*Pliskin and Lehman.*[165])

For all practical purposes spectrum B is the same as that obtained for thermal oxides of comparable thicknesses. In the spectrum of the undensified film, the 1100 cm^{-1} band is broader and shifted to lower frequency than in the spectrum of the densified or the thermal oxide, whereas the 800 cm^{-1} band is at higher frequency than in thermal oxide.

A further indication of the porosity of these films was indicated by their lower refractive indices and faster etch rates. The extent of the porosity can also be shown by the intensity of the absorption bands near 3650 cm^{-1} due to silanol groups and near 3400 cm^{-1} due to absorbed water.[165] If the water concentration is sufficiently high, the H_2O deformation band near 1630 cm^{-1} is observable. The 3400 cm^{-1} band also appears to shift to lower frequency with increased amounts of absorbed water. The porosity of the films is shown most clearly by the fact that they can be densified by thermal treatment. Heating in a moist ambient was found to be more effective for densification than heating in a dry ambient.[165]

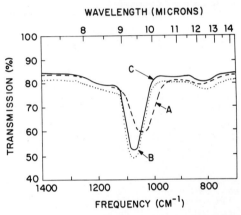

Fig. 18 Spectroscopic comparison. (*a*) Anodized silicon with total oxide on both sides of 2,501 Å. (*b*) Densified by 15 min N_2 at 993°C—total thickness 2,426 Å. (*c*) Thermally grown silicon dioxide with total thickness of 2,310 Å. (*Pliskin and Lehman.*[165])

(*b*) *Anodized SiO₂ Films.* The difference between anodized and thermal oxides is especially noticeable with infrared spectroscopy, as illustrated in Fig. 18.[165] (The anodized films were prepared in a solution of 0.05 M KNO₃ in N methyl acetamide. For some of the films $2\frac{1}{2}$% water was added.) Densification by exposure to nitrogen for 15 min at 993°C produced a film whose infrared spectrum was virtually indistinguishable from that of thermally grown SiO_2, as can be seen in spectra B and C. This densification was sufficient to decrease the film thickness 3%. The band position and half-width of this anodized film indicate either a very porous structure or strained bonding or both. The structure is only slightly porous in view of the 3% densification and the relatively small amount of absorbed water seen in the 3-μ region of the infrared spectrum. The average refractive index of this anodized film was 1.465 as measured by VAMFO. Thus, from a refractive-index measurement alone one would conclude that the film was similar to thermal SiO_2, but the infrared spectrum and the P-etch rate (19 Å s^{-1}) show that it is not. Part of the high refractive index and Si-O stretching band shift may be due to a slight oxygen deficiency. A refractive-index increase due to oxygen deficiency may be compensated by a refractive-index decrease due to the slight porosity such that the overall refractive index is about the same as that of thermal SiO_2. Pliskin and Lehman[165] concluded that the fast etch rate and large band shift were mainly due to bond strain in the oxide. Attempts were made to check for organic impurities by infrared amplification techniques but none could be detected.

Some of the anodized films which were prepared in glycerin solutions containing more water were even poorer in quality. They had P-etch rates as high as 228 Å s^{-1}, low refractive indices (1.30), and showed more water in their spectra, indicating high porosity. At elevated temperatures such films should densify significantly more than the 3% previously observed on heating.

Densification of an anodized SiO$_2$ film was also observed by Schmidt and Rand,[184] who measured a densification of 3.8% on one of their samples which had been heated in argon at 900°C.

(c) *Electron Gun Evaporated SiO$_2$ Films.* As part of a study of electron gun evaporated SiO$_2$ films, the refractive indices, film thicknesses, and etch rates were measured before and after different ambient treatments.[185] These results, summarized in Table 7, show that various humidity or moisture treatments at temperatures less

TABLE 7   Humidity, Room Ambient, and Heating Effects on Electron Gun Evaporated SiO$_2$ Films Deposited at 400°C (Pliskin and Castrucci[185])

Wafer No.	Treatment	Average thickness, Å	Refractive index (at 5,460 Å)	P-etch rate, Å s^{-1}	Thickness change, %
1030-1	Initial	8,127	1.476	28.3	
	1 hr boiling water	8,238	1.482	21.5	+1.4
	30 min steam, 181°C	8,185	1.484	20.5	−0.7
115-3	4 days, 25°C, 25% RH	8,078	1.480		
	24 days, 85°C, 85% RH	8,291	1.483	14.2	+2.6
	10 min N$_2$, 983°C	7,640	1.462	....	−7.9
128-2	Initial	6,475	1.466	32.9	
128-3	2 days, 25°C, 25% RH	6,621	1.465	30.2	
	69 days, 25°C, 25% RH	6,687	1.475	22.6	+1.0
128-4	5 min N$_2$, 985°C	5,989	1.460	3.5(variable)	
128-1	4 days, 85°C, 85% RH	7,081	1.475	16.9	
	66 days, 25°C, 25% RH	7,129	1.469	....	+0.7
	30 min steam, 450°C	6,854	1.441	8 0	−3.8
	36 h, 85°C, 85% RH	6,868	1.457	8.0	+0.2
	10 min N$_2$, 980°C	6,450	1.465	2.0	−6.1

than 100°C both increase the film thickness and refractive index and decrease the P-etch rate. Infrared spectra of the films showed a shift of the Si-O band to higher frequency and a decrease in bandwidth even on exposure to room ambient. As expected, these changes are more pronounced after exposure to 85% RH and 85°C. These results have been attributed to the presence of a high degree of bond strain (in addition to porosity) in the electron gun evaporated films.[165,185] The initial broadness of the Si-O band is due to a great deal of strain and variation in bond energies in addition to the fact that the porosity allows for a greater surface to hold and react with the absorbed water. On reaction of the water with the oxide to form silanol groups, strains in the bonds are relieved and the absorption band due to the Si-O stretching becomes sharper and shifts to higher frequencies. With the relief of the strain, the etch rate of the oxide decreases. Somewhat similar results were obtained with SiO$_2$ films formed by the decomposition of ethyl silicate in oxygen at temperatures of 300 to 500°C.[186]

Care must be taken in interpreting results obtained with strained and porous low-temperature deposited glass films such as electron gun or chemical vapor deposited films. A shift in the Si-O absorption peak to higher frequency combined with a decrease in bandwidth and a decrease in P-etch rate should not in themselves be taken as evidence for densification. The electron gun evaporated films which were exposed to room ambient or to 85°C and 85% RH showed these effects together with an actual increase in film thickness. However, on exposure to moisture at higher temperatures ($\geq 180°C$ in Table 7), these effects were accompanied by densification, as shown by measured decreases in film thicknesses.[185]

The high porosity of the electron gun evaporated films is shown by wafer 115-3, where heating for 10 min in nitrogen at 983°C was sufficient to densify the film by nearly 8%. After this heat treatment, the infrared spectrum and refractive index of the sample indicated that the film was now nearly the same as thermally grown SiO_2 films. An even more surprising effect is shown by sample 128-1 which, according to actual thickness measurements, densified almost 4% by exposure to 30 min of steam at only 450°C. The infrared spectra also indicated significant improvement in film quality due to this treatment which in the past has often been considered a severe stress for testing the passivating quality of glass films. This result emphasizes the need for great care when choosing a stress test for the evaluation of insulating films.

(4) Silicon Nitride Films In recent years many papers have been published on silicon nitride and mixed nitride-oxide films.[71,187-199] The common methods of forming silicon nitride films are reaction at elevated temperatures of silane and a nitrogen-containing compound such as ammonia or hydrazine, reactive sputtering of silicon in a nitrogen atmosphere, and rf sputtering from a silicon nitride target. The interest in silicon nitride films is due to their greater impermeability to the diffusion of various ions.

TABLE 8 Physical Properties of Silicon Oxynitride Films (Brown et al.[197])

Film	Injected % NO	K	n	ρ, g cm^{-3}	E_g, eV	IR, μ	Estimated equivalent % SiO_2
Nitride.....	0.00	7.4 ± 0.2	2.03 ± 0.01	3.11 ± 0.07	4.5	11.7	0
Oxynitride..	0.25	7.3 ± 0.1	1.98 ± 0.02	....	4.7	11.5	7 ± 2
Oxynitride..	1.25	7.0 ± 0.4	1.92 ± 0.01	....	5.1	11.2	17 ± 5
Oxynitride..	2.25	6.7 ± 0.2	1.89 ± 0.01	....	5.3	10.9	25 ± 7
Oxynitride..	5.00	6.5 ± 0.1	1.79 ± 0.03	....	5.6	10.7	34 ± 8
Oxynitride..	10.00	5.8 ± 0.1	1.71 ± 0.03	....	5.8	10.1	50 ± 12
Oxide......	.....	3.9	1.44	2.14 ± 0.07	8.0	9.2	100

With regard to the interrelationship between density, infrared spectra, refractive index, etch rate, excess silicon, etc., these films show the same type of behavior as deposited silicon dioxide films as previously discussed. The main infrared absorption peak in the 11- to 12-μ region is shifted to longer wavelengths and the refractive index decreases with increased porosity or reduced density, which in turn could be due to lower deposition temperatures or faster deposition rates at the same surface temperature. The etch rate also increases with reduced density. On the other hand, excess silicon shifts the infrared absorption peak to longer wavelengths, increases the refractive index, and reduces the etch rate just as excess silicon (or oxygen deficiency) does in silicon dioxide films.

One further complication exists in silicon nitride films in that silicon dioxide forms more easily than silicon nitride and, therefore, in the presence of oxidizing impurities, a mixed nitride-oxide film will form. The presence of oxide in the film will reduce the refractive index, shift the absorption peak to higher frequency, increase the etch rate, reduce the dielectric constant, etc. Mixed oxide-nitride films can be formed purposely by controlled introduction of an oxidizing agent during the formation of the film. Deal et al.[196] recently investigated and compared the electrical properties of such films deposited on silicon. Brown et al.[197] determined the relative oxygen content in mixed films by assuming that various properties exhibited a linear variation in going from pure silicon nitride to pure silicon dioxide. The parameters investigated were the dielectric constant K at low frequency, the refractive index n, the density ρ, the band gap E_g, and the position in microns (μ) of the infrared absorption peak. Their results are shown in Table 8. With regard to the infrared absorption peak, better correlation with the other parameters is obtained if the position of the absorption peak in terms of frequency (or cm^{-1}) is used as a criterion rather than wavelength.

REFERENCES

1. Tolansky, S., "Multiple-beam Interferometry of Surfaces and Films," Oxford University Press, London, 1948.
2. Tolansky, S., "An Introduction to Interferometry," Longmans, Green & Co., Ltd., London, 1955.
3. Tolansky, S., "Surface Microtopography," Interscience Publishers, London, 1960.
4. Heavens, O. S., "Optical Properties of Thin Solid Films," Butterworth Scientific Publications, London, 1955.
5. Vašíček, A., "Optics of Thin Films," North Holland Publishing Company, Amsterdam, 1960.
6. Mayer, H., "Physik dünner Schichten," Wissenschaftliche Verlagsgesellschaft m.b.H., Stuttgart, 1950.
7. Françon, M., "Optical Interferometry," Academic Press Inc., New York, 1966.
8. Heavens, O. S., in G. Hass and R. E. Thun (eds.), "Physics of Thin Films," vol. 2, p. 193, Academic Press, Inc., New York, 1964.
9. Bennett, H. E., and J. M. Bennett, in G. Hass and R. E. Thun (eds.), "Physics of Thin Films," vol. 4, p. 1, Academic Press Inc., New York, 1967.
10. Gillespie, D. J., in B. Schwartz and N. Schwartz (eds.), "Measurement Techniques for Thin Films," p. 102, The Electrochemical Society, New York, 1967.
11. Keinath, G., "The Measurement of Thickness," National Bureau of Standards Circular 585, Washington, D.C., 1958.
12. Flint, P. S., in B. Schwartz and N. Schwartz (eds.), "Measurement Techniques for Thin Films," p. 141, The Electrochemical Society, New York, 1967.
13. Rank, D. H., and H. E. Bennett, *J. Opt. Soc. Am.*, **45**, 69 (1955).
14. Donaldson, W. K., and A. Khamsavi, *Nature*, **159**, 228 (1947).
15. Klute, C. H., and E. R. Fajardo, *Rev. Sci. Inst.*, **35**, 1080 (1964).
16. Bennett, J. M., *J. Opt. Soc. Am.*, **54**, 612 (1964).
17. Terrell, A. C., *Microscope and Crystal Front*, **14**(5), 174 (1964).
18. Booker, G. R., and C. E. Benjamin, *J. Electrochem. Soc.*, **109**, 1206 (1962).
19. Nomarski, G., and A. R. Weill, *Rev. Met. Paris*, **52**(2), 121 (1955).
20. LeMéhauté, C., *IBM J. Res. Develop.*, **6**, 263 (1962).
21. Herzog, F., *Ind.-Anz.*, no. 60, p. 1511, July 27, 1960.
22. Rollet, A. R., *S. B. Akad. Wiss. Wien* III, **77**, 177 (1878).
23. Evans, U. R., "The Corrosion and Oxidation of Metals," p. 787, Edward Arnold (Publishers) Ltd., London, 1960.
24. Blodgett, K., *J. Am. Chem. Soc.*, **57**, 1007 (1935).
25. Blodgett, K., and I. Langmuir, *Phys. Rev.*, **51**, 964 (1937).
26. Pliskin, W. A., and E. E. Conrad, *IBM J. Res. Develop.*, **8**, 43 (1964).
27. Charlesby, A., and J. J. Polling, *Proc. Roy. Soc. London*, **A227**, 434 (1955).
28. Pliskin, W. A., in E. M. Murt and W. G. Guldner (eds.), "Physical Measurement and Analysis of Thin Films," p. 1 (Chap. I), Plenum Publishing Corp., New York, 1969.
29. Reizman, F., and W. Van Gelder, *Solid-State Electron.*, **10**, 625 (1967).
30. Pliskin, W. A., and R. P. Esch, *J. Appl. Phys.*, **36**, 2011 (1965).
31. Pliskin, W. A., and H. S. Lehman, in *Proc. Symp. Manufacturing In-process Control and Measuring Techniques for Semiconductors*, March, 1966, vol. 1, p. 11-1, Phoenix, Ariz. (also available as *IBM Tech. Rept.* TR 22.279).
32. Wesson, R. A., R. P. Phillips, and W. A. Pliskin, *J. Appl. Phys.*, **38**, 2455 (1967).
33. Pliskin, W. A., *Solid-State Electron.*, **11**, 957 (1968).
34. Wesson, R. A., H. W. Young, and W. A. Pliskin, *Appl. Phys. Letters*, **11**, 105 (1967).
35. Pliskin, W. A., and R. A. Wesson, *IBM J. Res. Develop.*, **12**, 192 (1968).
36. Goldsmith, N., and L. A. Murray, *Solid-State Electron.*, **9**, 331 (1966).
37. Born, M., and E. Wolf, "Principles of Optics," 3d rev. ed., Pergamon Press, Oxford, 1965.
38. Harvilchuck, J. M., and A. J. Warnecke, IBM Components Division, Hopewell Junction, N.Y., to be published.
39. Pliskin, W. A., *J. Electrochem. Soc.*, **114**, 620 (1967).
40. Corl, E. A., and H. Wimpfheimer, *Solid-State Electron.*, **7**, 755 (1964).

41. Fried, L. J., and H. A. Froot, *J. Appl. Phys.*, **39**, 5732 (1968).
42. Pliskin, W. A., *Proc. IEEE*, **52**, 1468 (1964).
43. Corl, E. A., and K. Kosanke, *Solid-State Electron.*, **9**, 943 (1966).
44. Pliskin, W. A., and E. E. Conrad, *Electrochem. Tech.*, **2**, 196 (1964).
45. Murray, L. A., N. Goldsmith, and E. L. Jordan, *Electrochem. Tech.*, **4**, 508 (1966).
46. Spitzer, W. G., and M. Tanenbaum, *J. Appl. Phys.*, **32**, 744 (1961).
47. Grochowski, E. G., and W. A. Pliskin, presented at the Electrochemical Society Meeting at Detroit, Mich., October, 1961, *J. Electrochem. Soc.*, **108**, 262C (1961).
48. Albert, M. P., and J. F. Combs, *J. Electrochem. Soc.*, **109**, 709 (1962).
49. Schumann, P. A., Jr., R. P. Phillips, and P. J. Olshefski, *J. Electrochem. Soc.*, **113**, 368 (1966).
50. Stebbins, A. E., and L. L. Shreir, *J. Electrochem. Soc.*, **108**, 30 (1961).
51. Lukeš, F., and E. Schmidt, *Solid-State Electron.*, **10**, 264 (1967).
52. Pliskin, W. A., and R. P. Esch, *Appl. Phys. Letters*, **11**, 257 (1967).
53. Esch, R. P., and W. A. Pliskin, to be published.
54. Passaglia, E., R. R. Stromberg, and J. Kruger (eds.), "Ellipsometry in the Measurement of Surfaces and Thin Films," *NBS Misc. Publ.* 256, Government Printing Office, Washington, D.C., 1964.
55. Archer, R. J., *J. Opt. Soc. Am.*, **52**, 970 (1962).
56. McCrackin, F. L., and J. P. Colson, Ref. 54, p. 61; see also *NBS Tech. Note* 242, 1964.
57. Winterbottom, A. B., "Optical Studies of Metal Surfaces," The Royal Norwegian Scientific Society, *Report* 1.   F. Brun, Trondheim, Norway, 1955.
58. McCrackin, F. L., E. Passaglia, R. R. Stromberg, and H. L. Steinberg, *J. Res. NBS*, **67A**, 363 (1963).
59. Pliskin, W. A., and R. P. Gnall, *J. Electrochem. Soc.*, **111**, 872 (1964).
60. Nakayama, T., and F. C. Collins, *J. Electrochem. Soc.*, **113**, 706 (1966).
61. Murray, L. A., and N. Goldsmith, *J. Electrochem. Soc.*, **113**, 1297 (1966).
62. Illig, W., *Mettalloberflaeche*, **13**, 33 (1959).
63. Mansour, T. M., *Mater. Res. Std.*, **3**, 29 (1963).
64. Brown, R., *Am. Ceram. Soc. Bull.*, **45**, 206 (1966).
65. Abelès, F., in E. Wolf (ed.), "Progress in Optics," vol. II, p. 251, North Holland Publishing Company, Amsterdam, 1963.
66. Kinosita, K., T. Matsumoto, K. Natsume, and M. Yoshida, *J. Appl. Phys. Japan*, **20**, 205, (1960).
67. Kelly, J. C., and O. S. Heavens, *Opt. Acta*, **6**, 339 (1959).
68. Hacskaylo, M., *J. Opt. Soc. Am.*, **54**, 198 (1964).
69. Gottling, J. G., and W. S. Nicol, *J. Opt. Soc. Am.*, **56**, 1227 (1966).
70. Pliskin, W. A., and R. P. Esch, *J. Appl. Phys.*, **39**, 3274 (1968).
71. Doo, V. Y., D. R. Nichols, and G. A. Silvey, *J. Electrochem. Soc.*, **113**, 1279 (1966).
72. Ellis, W. P., *J. Opt. Soc. Am.*, **53**, 613 (1963).
73. Ellis, W. P., *J. Phys. Radium*, **25**, 21 (1964).
74. Muller, R. H., *J. Opt. Soc. Am.*, **54**, 419 (1964).
75. Muller, R. H., *J. Electrochem. Soc.*, **112**, 650 (1965).
76. Morey, G. W., "The Properties of Glass," p. 368, Reinhold Publishing Corporation, New York, 1960.
77. Lewis, A. E., *J. Electrochem. Soc.*, **111**, 1007 (1964).
78. Malitson, I. H., *J. Opt. Soc. Am.*, **55**, 1205 (1965).
79. Kutzelnigg, A., "Testing Metallic Coatings," Robert Draper Ltd., Teddington, England, 1963.
80. Gardner, E. E., in B. Schwartz and N. Schwartz (eds.), p. 240, "Measurement Techniques for Thin Films," The Electrochemical Society, New York, 1967.
81. Gardner, E. E., P. A. Schumann, Jr., and E. F. Gorey, in B. Schwartz and N. Schwartz (eds.), "Measurement Techniques for Thin Films," p. 258, The Electrochemical Society, New York, 1967.
82. Valdes, L. P., *Proc. IRE*, **42**, 420 (1954).
83. Neugebauer, C. A., in B. Schwartz and N. Schwartz (eds.), "Measurement Techniques for Thin Films," p. 191, The Electrochemical Society, New York, 1967.
84. Leonard, W. F., and R. L. Ramey, *J. Appl. Phys.*, **35**, 2963 (1964).
85. Leonard, W. F., and R. L. Ramey, *J. Appl. Phys.*, **37**, 2190 (1966).
86. Marsocci, V. A., and P. Siegel, *J. Appl. Phys.*, **39**, 29 (1968).
87. Reason, R. E., in H. W. Baker (ed.), "Modern Workshop Technology, Part 2," chap. XVII, Cleaver-Hume Press, Ltd., Bristol, England, 1960.
88. Schwartz, N., and R. Brown, *Trans. 8th AVS Symp. and 2d Intern. Congr. Vacuum Sci. Technol.*, 1961, p. 836, Pergamon Press, New York.
89. Joyce, B. A., *Solid-State Electron.*, **5**, 102 (1962).

90. Yeh, T. H., and A. E. Blakeslee, *J. Electrochem. Soc.*, **110**, 1018 (1963).
91. Herzog, A. H., *Semicond. Prod.*, December, 1962, p. 25.
92. Dudley, R. H., and T. H. Briggs, *Rev. Sci. Instr.*, **37**, 1041 (1966).
93. Happ, W. W., and W. Shockley, *Bull. Am. Phys. Soc.*, Ser. II, **1**, 382 (1956).
94. McDonald, B., and A. Goetzberger, *J. Electrochem. Soc.*, **109**, 141 (1962).
95. Friedman, H., and L. S. Birks, *Rev. Sci. Instr.*, **17**, 99 (1946).
96. Eisenstein, A., *J. Appl. Phys.*, **17**, 874 (1946).
97. Beeghly, H. F., *J. Electrochem. Soc.*, **97**, 152 (1950).
98. Zemany, P. D., and H. A. Liebhafsky, *J. Electrochem. Soc.*, **103**, 157 (1956).
99. Achey, F. A., and E. J. Serfass, *J. Electrochem. Soc.*, **105**, 204 (1958).
100. Lambert, M. C., *Advan. X-ray Anal.*, **2**, 193 (1959).
101. Lowe, B. J., P. D. Sierer, and R. B. Ogilvie, *Advan. X-ray Anal.*, **2**, 275 (1959).
102. Zimmerman, R. H., *Metal Finishing*, **59**, 67 (1961).
103. Bertin, E. P., and R. J. Longobucco, *Anal. Chem.*, **34**, 804 (1962).
104. Koh, P. K., and B. Caugherty, *J. Appl. Phys.*, **23**, 427 (1952).
105. Liebhafsky, H. A., and P. D. Zemany, *Anal. Chem.*, **28**, 455 (1956).
106. Birks, L. S., E. J. Brooks, and H. Friedman, *Anal. Chem.*, **25**, 692 (1953).
107. Zanin, S. J., and A. Szule, TR 00.07020.404, IBM Corp., Poughkeepsie, N.Y., 1959.
108. Keesaer, W. C., *Advan. X-ray Anal.*, **3**, 77 (1960).
109. Silver, M. D., and E. T-K. Chow, *J. Vacuum Sci. Technol.*, **2**, 203 (1965).
110. Cline, J. E., and S. Schwartz,·*J. Electrochem. Soc.*, **114**, 605 (1967).
111. Bertin, E. P., in Ref. 28, p. 35 (Chap. II).
112. Liebhafsky, H. A., H. G. Pfeiffer, E. H. Winslow, and P. D. Zemany, chap. 6 in "X-ray Absorption and Emission in Analytical Chemistry," John Wiley & Sons, Inc., New York, 1960.
113. Bertin, E. P., and R. J. Longobucco, *Metal Finishing*, **60**, 42 (1962).
114. Clark, G. L. (ed.), "Encyclopedia of X-rays and Gamma Rays," pp. 149, 373, 412, 772, Reinhold Publishing Corporation, New York, 1963.
115. Cline, J. E., in Ref. 28, p. 83 (Chap. III).
116. Cameron, J. F., and J. R. Rhodes, Ref. 114, p. 150.
117. Clarke, E., J. R. Carlin, and W. E. Barbour, Jr., *Elec. Eng.*, **70**, 35 (1951).
118. Danguy, L., and F. Grard, *Proc. Intern. Conf. Peaceful Uses At. Energy*, vol. **19**, 2d ed., United Nations, 1959.
119. Brown, R., in Ref. 28, p. 93 (Chap. IV).
120. Wright, P., *Electron. Reliability Microminiaturization*, **2**, 227 (1963).
121. Pulker, H., and E. Ritter, *Sonderdruck Vakuum-Tech.*, **4**, 91 (1965).
122. See, for example, Morrison, G. H. (ed.), "Trace Analysis, Physical Methods," Interscience Publishers, Inc., New York, 1965.
123. Barry, J. E., H. M. Donega, and T. E. Burgess, *J. Electrochem. Soc.*, **116**, 257 (1969).
124. Osborne, J. F., G. B. Larrabee, and V. Harrap, *Anal. Chem.*, **39**, 1144 (1967).
125. Shubin, L. D., and J. H. Chaudet, *Appl. Spectry.*, **18**, 137 (1964).
126. Nohe, J. D., *Appl. Spectry.*, **21**, 364 (1967).
127. Nohe, J. D., in Ref. 28, p. 138 (Chap. VI).
128. Willardson, R. K., in B. Schwartz and N. Schwartz (eds.), "Measurement Techniques for Thin Films," p. 58, The Electrochemical Society, New York, 1967.
129. Ahearn, A. J., *Trans. 6th AVS Symp.*, 1960, p. 1, Pergamon Press, New York.
130. Ahearn, A. J., *J. Appl. Phys.*, **32**, 1197 (1961).
131. Hickman, W. M., and G. G. Sweeney, in A. J. Ahearn (ed.), "Mass Spectrometric Analysis of Solids," chap. 5, Elsevier Publishing Company, Amsterdam, 1966.
132. Hickman, W. M., and Y. L. Sandler, "Surface Effects in Detection," p. 194, Spartan Books, Washington, D.C., 1965.
133. Malm, D. L., *Appl. Spectry.*, **22**, 318 (1968).
134. Malm, D. L., in Ref. 28, p. 148 (Chap. VII).
135. Pevera, S. C., IBM Corp. Components Div., Hopewell Junction, N.Y., unpublished work.
136. Rhodin, T. N., *Anal. Chem.*, **27**, 1857 (1955).
137. Verderber, R. R., *Norelco Reptr.*, **10**, 30 (1963).
138. Chow, E. T-K., and E. P. Cocozza, *Appl. Spectry.*, **21**, 290 (1967).
139. Zanin, S. J., and G. E. Hooser, "Analysis of Plated Tape by X-ray Emission Spectrography," TR 22.115 (1964), IBM Corp., Components Div., Hopewell Junction, N.Y.
140. Zanin, S. J., J. C. Lloyd, and G. E. Hooser, 16th Pittsburgh Conference on Analytical Chemistry and Applied Spectroscopy, Mar. 1–5, 1965. Also available as TR 22.186, "Nickel-Iron Thin Film Analysis by X-ray Emission," IBM Corp., Components Div., Hopewell Junction, N.Y.
141. Cocozza, E. P., and A. Ferguson, *Appl. Spectry.*, **21**, 286 (1967).

142. Pluchery, M., *Spectrochem. Acta*, **19**, 533 (1963).
143. Bertin, E. P., *Anal. Chem.*, **36**, 826 (1964).
144. Spielberg, N., and G. Abowitz, *Anal. Chem.*, **38**, 200 (1966).
145. Lublin, P., and W. J. Sutkowski, in T. D. McKinley, K. F. S. Heinrich, and D. B. Wittry (eds.), "The Electron Microprobe," p. 677, John Wiley & Sons, Inc., New York, 1966.
146. Ramsey, J. N., and P. Weinstein, in T. D. McKinley, K. F. S. Heinrich, and D. B. Wittry (eds.), "The Electron Microprobe," p. 715, John Wiley & Sons, Inc., New York, 1966.
147. Hutchins, G. A., in T. D. McKinley, K. F. S. Heinrich, and D. B. Wittry (eds.), "The Electron Microprobe," p. 390, John Wiley & Sons, Inc., New York, 1966.
148. Koopman, N. G., and J. Gniewek, "Electron Beam Microprobe Technique to Measure Phosphosilicate Glass Thickness and Composition," presented at the Third National Conference on Electron Microbeam Analysis, Chicago, Aug. 2, 1968.
149. Schreiber, H., *Advan. X-ray Anal.*, **8**, 363 (1965).
150. Chodos, A. A., *Anal. Chem.*, **40**, 1346 (1968).
151. Bertin, E. P., and R. J. Longobucco, *Advan. X-ray Anal.*, **7**, 566 (1964).
152. Bertin, E. P., *Advan. X-ray Anal.*, **8**, 231 (1965).
153. Sloan, R. D., *Advan. X-ray Anal.*, **5**, 512 (1962).
154. Glade, G. H., *Advan. X-ray Anal.*, **11**, 185 (1968).
155. Guldner, W. G., in F. J. Welcher (ed.), "Standard Methods of Chemical Analysis," chap. 36, D. Van Nostrand Company, Inc., New York, N.Y., 1963.
156. Guthrie, J. W., in A. J. Ahearn (ed.), "Mass Spectrometic Analysis of Solids," chap. IV, Elsevier Publishing Company, New York, 1966.
157. Hoffmeister, W., and M. A. Zuegel, *Thin Solid Films*, **3**, 35 (1969).
158. Lloyd, J. C., "Determination of Argon in RF Sputtered SiO_2 by X-ray Emission," presented at the 17th Annual Conference on Applications of X-ray Analysis, Denver, Colo., Aug. 21–23, 1968.
159. Guldner, W. G., and R. Brown, in B. Schwartz and N. Schwartz (eds.), "Measurement Techniques for Thin Films," p. 82, The Electrochemical Society, New York, 1967.
160. Glang, R., R. A. Holmwood, and P. C. Furois, *Trans. 3d Intern. Vacuum Congr.*, **2**, 643 (1965), Pergamon Press, Oxford.
161. Winters, H. F., and E. Kay, *J. Appl. Phys.*, **38**, 3928 (1967).
162. Perri, J. A., and J. Riseman, *Electronics*, **39**, 108 (1966).
163. Pliskin, W. A., D. R. Kerr, and J. A. Perri, in G. Hass and R. E. Thun (eds.), "Physics of Thin Films," vol. 4, p. 257, Academic Press Inc., New York, 1967.
164. Pliskin, W. A., in B. Schwartz and N. Schwartz (eds.), "Measurement Techniques for Thin Films," The Electrochemical Society, New York, 1967.
165. Pliskin, W. A., and H. S. Lehman, *J. Electrochem. Soc.*, **112**, 1013 (1965).
166. Pliskin, W. A., *Thin Solid Films*, **2**, 1 (1968).
167. Pliskin, W. A., in Ref. 28, p. 168 (Chap. VIII).
168. Stickler, R., and J. W. Faust, Jr., *Electrochem. Technol.*, **4**, 277 (1966).
169. Knopp, A. N., and R. Stickler, *Electrochem. Technol.*, **5**, 37 (1967).
170. Stephenson, G. W., and K. H. Jack, *Trans. Brit. Ceram. Soc.*, **59**, 397 (1960) [in comments on paper by A. J. Moulson and J. P. Roberts, *Trans. Brit. Ceram. Soc.*, **59**, 388 (1960)].
171. Hetherington, G., and K. H. Jack, *Phys. Chem. Glasses*, **3**, 129 (1962).
172. Pliskin, W. A., "The Effect of Moisture on RF Sputtered and Fused Glass Films," American Ceramic Society Meeting, Washington, D.C., May, 1966 (also available as IBM preprint MP 22. 0078).
173a. Snow, E. H., and B. E. Deal, *J. Electrochem. Soc.*, **113**, 263 (1966).
  b. Eldridge, J. M., and P. Balk, *Trans. Met. Soc. AIME*, **242**, 539 (1968).
174. Maissel, L. I., C. L. Standley, and R. E. Jones, to be published.
175. Davidse, P. D., and L. I. Maissel, *J. Appl. Phys.*, **37**, 574 (1966).
176. Deal, B. E., and M. Sklar, *J. Electrochem. Soc.*, **112**, 430 (1965).
177. Pliskin, W. A., P. D. Davidse, H. S. Lehman, and L. I. Maissel, *IBM J. Res. Develop.*, **11**, 461 (1967).
178. Kerr, D. R., J. S. Logan, P. J. Burkhardt, and W. A. Pliskin, *IBM J. Res. Develop.*, **8**, 376 (1964).
179. Pliskin, W. A., *Appl. Phys. Letters*, **7**, 158 (1965).
180. Corl, E. A., S. L. Silverman, and Y. S. Kim, *Solid-State Electron*, **9**, 1009 (1966).
181. Jordan, E. L., *J. Electrochem. Soc.*, **108**, 478 (1961).
182. Klerer, J., *J. Electrochem. Soc.*, **108**, 1070 (1961).
183. Klerer, J., *J. Electrochem. Soc.*, **112**, 503 (1965).
184. Schmidt, P. F., and M. J. Rand, *Solid State Commun.*, **4**, 169 (1966).

185. Pliskin, W. A., and P. P. Castrucci, *Electrochem. Technol.*, **6,** 85 (1968); *J. Electrochem. Soc.*, **112,** 148C (1965).
186. Krongelb, S., *Electrochem. Technol.*, **6,** 251 (1968); Krongelb, S., and T. O. Sedgwick, *J. Electrochem. Soc.*, **113,** 63C (1966).
187. Sterling, H. F., and R. C. G. Swann, *Solid-State Electron.*, **8,** 653 (1965).
188. Hu, S. M., *J. Electrochem. Soc.*, **113,** 693 (1966).
189. Swann, R. C. G., R. R. Mehta, and T. P. Cauge, *J. Electrochem. Soc.*, **114,** 713 (1967).
190. Chu, T. L., C. H. Lee, and G. A. Gruber, *J. Electrochem. Soc.*, **114,** 717 (1967).
191. Bean, K. E., P. S. Gleim, R. L. Yeakley, and W. R. Runyan, *J. Electrochem. Soc.*, **114,** 733 (1967).
192. Hu, S. M., D. R. Kerr, and L. V. Gregor, *Appl. Phys. Letters*, **10,** 97 (1967).
193. Hu, S. M., and L. V. Gregor, *J. Electrochem. Soc.*, **114,** 826 (1967).
194. Yoshioka, S., and S. Takayanagi, *J. Electrochem. Soc.*, **114,** 962 (1967).
195. Doo, V. Y., D. R. Kerr, and D. R. Nichols, *J. Electrochem. Soc.*, **115,** 61 (1968).
196. Deal, B. E., P. J. Fleming, and L. P. Castro, *J. Electrochem. Soc.*, **115,** 300 (1968).
197. Brown, D. M., P. V. Gray, F. K. Heumann, H. R. Philipp, and E. A. Taft, *J. Electrochem. Soc.*, **115,** 311 (1968).
198. Chu, T. L., J. R. Szedon, and C. H. Lee, *J. Electrochem. Soc.*, **115,** 318 (1968).
199. Grieco, M. J., F. L. Worthing, and B. Schwartz, *J. Electrochem. Soc.*, **115,** 525 (1968).
200. Chu, T. L., *S.C.P. and Solid State Technol.*, May, 1967, p. 36, and *J. Vacuum Sci Technol*, **6,** 25 (1969).

Part Three

Properties of Thin Films

Chapter **12**

Mechanical Properties
of Thin Films

DAVID S. CAMPBELL

The Plessey Company Limited, Whiteside Works, Bathgate, West Lothian, Scotland
also Visiting Senior Lecturer in Materials Science, Electrical Engineering Department,
Imperial College, London University, London, England

LIST OF SYMBOLS

A	adhesion
b	length of substrate (uncovered by electrode)
C_0	initial capacity
ΔC	change in capacity
d_f	film thickness
d_s	substrate thickness
E_a	adsorption energy of single atom
E_a'	modified adsorption energy of single atom
E_{ad}	average adsorption energy
$E_b{}^i$	total binding energy of cluster of size i atoms
E_d	diffusion energy of single atom
E_f	Young's modulus of film
E_s	Young's modulus of substrate
G	weight
i	number of atoms in a cluster
i^*	critical size of cluster
J	nucleation rate
K	constant
k	Boltzmann's constant
l	length (of substrate)
m_a	number of sites that an atom visits before reevaporation
N	number of adsorbent atoms per unit volume of substrate
N_s	saturation number of islands
n	rotor speed
n_1	number of sites occupied by single atoms
n_0	number of adsorption sites on surface
n_s	island density
p	pressure

R	rate of arrival of atoms
R_c	critical rate of arrival of atoms
r	rotor radius or distance from center of disk
s	equilibrium distance of condensate atom from surface
T	tensile strength or temperature
$T_{i \to i+1}$	transition temperature from $i^* = i$ to $i^* = i + 1$
V_x	ionization potential of atom of species
v	rate at which atom visits adsorption sites
w	substrate width
α_x	polarizability of atom of species x
Γ_1	probability of evaporation of single atom
Γ_i	probability of atom leaving group of size i
Γ_1^+	probability of atom migrating from one site to another
δ	substrate deflection
σ	stress
σ_0	molecular area of atom
ρ	radius of curvature of beam or film density
ν_y	vibrational frequency of type y
τ	reciprocal of vibrational frequency
ϕ	stripping angle
θ	angle to the vertical

INTRODUCTION

Interest in the mechanical properties of thin films has grown rapidly over the past few years, hand in hand with general interest in all other properties. However, attention had been paid to certain aspects of mechanical behavior, especially the stress present in grown films, as early as the end of the nineteenth century, when Mills,[1] in 1877, made measurements on the stress present in films deposited chemically on glass thermometer bulbs. This work was confirmed by Bouty[2] in 1879, and 30 years later the subject was put on a more quantitative basis for electrodeposited films by Stoney.[3]

Since these early beginnings interest has grown apace, so much so that over the last few years several authoritative reviews have been compiled and published, notably by Hoffman.[4,5,6] In view of the excellence of these reviews, the present author finds a little difficulty in adding new material, and readers will of necessity be constantly referred to these texts.

One difference in the present chapter compared with the reviews mentioned above, however, is that the adhesion of thin films will be considered. This will be examined first, as the degree of adhesion a film has to a substrate governs the observations that can be made of other mechanical properties (e.g., if the adhesion is weak, stress observations cannot be made, as the stress will cause the film to pull away from the substrate). After adhesion has been examined, attention will then be paid to stress and tensile properties.

To understand the mechanical properties of films, it is necessary to correlate the observed behavior with the structure of the films. Many detailed reviews have been given on this subject,[7,8,9,10] which is also considered in Part 2 of this book; so it is not proposed to dwell on it here. One can, however, recognize that the adhesion properties of a film will be determined by the initial stages of growth, and thus in this context, a study of the nucleation behavior is important. Other mechanical properties depend mainly on the subsequent growth and resultant crystallographic type (i.e., amorphous, polycrystalline, fiber-oriented, single-crystal). The crystallographic type and for that matter the initial nucleation will depend on the preparation method, but it is only in this indirect sense that preparation is involved.

It is not necessary in this chapter to examine preparation methods as, again, these are considered in other chapters of the book, and several reviews have also been published.[11,12] Both physical processes (evaporation, sputtering) and chemical processes (vapor-phase deposition, thermal growth, anodization, electroplating, electroless deposition) can give any of the structures possible, though some growth processes are more prone than others to give impurities in the film with a subsequent change in the mechanical behavior.

With this brief introduction, it is now proposed to examine the various aspects of mechanical behavior in detail.

1. ADHESION

The adhesion of deposits has been of interest for a considerable time. The durability of coatings is of prime importance in many fields, and one of the main factors that govern this durability is the adhesion. This is particularly noticeable if the film or substrate is subject to corrosion or to a humid atmosphere, as under these circumstances any tendency for the film to peel from the substrate may well be aggravated. There are, however, circumstances in which poor adhesion is expressly required—for example, in the use of films in replica techniques for electron microscopy.

The variation in the degree of adhesion between different film-substrate pairs has been examined by many authors, and factual information on several pairs was collected by Holland[13] for evaporated deposits. He pointed out that a simple but effective technique to check the different materials is to apply adhesive tape to the surface and subsequently to examine the result of stripping.[14] The different combinations could then be classified into two groups, those which were weakly bonded and came off the substrate with the tape, and those which were strongly bonded and remained on the substrate. Obviously, no degree of adhesion can be examined with this technique. The general picture that emerges is that the oxygen-active metals form chemical bonds to the substrates and are strongly adhesive, whereas those which form only physical bonds (see Sec. 1c) are easily removed.[15] This concept has been examined in some detail by Bateson[16] for the case of metallic films on glass.

As has been remarked, a crude stripping test, although of use as an empirical practical measure of adhesion, needs refinement to make it into a method for examining the physical or chemical nature of the bonding. This has been done, as will be discussed in a later section. However, two other techniques are available for examining adhesion. Studies of the initial growth and nucleation of a deposit enable values of the adhesion energy to be found in certain circumstances, and also the use of micro calorimetry techniques has recently been investigated to determine the adhesion energy at the interface[17] when one film is dissolved from another. It is now proposed to examine the methods and to contrast the results that have been obtained.

a. Mechanical Methods

Mechanical methods of measuring adhesion are more obvious and direct, and various reviews have been given of the methods available.[18,19] It was realized early that if one could measure the work done in mechanically pulling off a film from a substrate, this would give a value for the adhesion energy. If the film is to be pulled off directly, the problem generally is that of attaching some backing to the film to enable it to be detached from the substrate. Furthermore it is difficult to obtain more than comparative values by such a method—either the film comes off with the backing or it does not. Thus other techniques have been employed by various workers, such as scratch tests, abrasion tests, ultrasonic vibration, and the use of an ultracentrifuge.

(1) Tape Methods Most of the early attempts at measuring adhesion used the "scotch tape" method. This was originally suggested by Strong[14] in 1935 and has been used by numerous workers since.[16,20] It consists of pressing a piece of adhesive tape to the film. When the tape is pulled off, the film is either wholly removed, partially removed, or left behind on the substrate. The method is obviously only qualitative and gives no indication of the relative magnitudes of the adhesive forces if the adhesion of the film to the substrate exceeds the adhesion of the tape to the film. If the film is removable, however, it is possible to turn the measurement into a quantitative one. Figure 1 shows a schematic diagram of the apparatus that can be used.[20] To eliminate extraneous effects it is necessary to extrapolate results obtained

at different stripping angles ϕ to zero angle and to zero stripping speeds.   Using these extrapolations, reproducible results are being obtained for films with low adhesion (e.g., gold on glass).   Even without extrapolating so as to obtain absolute values for the adhesion energy, it is possible to use such an apparatus in different ambient atmospheres and determine their effects on apparent adhesion.   Measurements of this type using copper, silver, and gold on soda and borosilicate glass and on quartz show that in every case a higher adhesion value is obtained if the film is stripped off under vacuum conditions (10^{-5} Torr) compared with normal air pressure.[20]   It has been found possible to correlate the pressure at which the adhesion increases with the stripping

speed, showing that the rate at which gas molecules visit the crack formed between film and substrate determines the adhesion-energy value obtained.

Other workers have used other backing techniques.   Williams and Backus[15] studied a whole variety of materials particularly with regard to backing films for electron microscopy.   They used collodion dried in contact with the film and found they could easily remove gold on glass.   They also found that films of gold could be rendered strongly adherent, so that they could not be pulled off, by the preapplication of a thin film of chromium or beryllium.

Flat-headed brass pins have been soldered onto metal films by Belser and Hicklin.[21]   For nickel and certain of the platinum metals this could be done directly, but for other metals an overcoat of sputtered or electroplated nickel was required on top of the film under investigation.   As pointed out by Benjamin and Weaver,[19] it is difficult to forecast the effects of such treatment upon the structure and properties of the

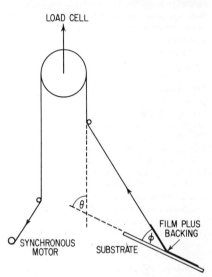

Fig. 1   Diagram of apparatus for stripping experiments.

films; so the results must be accepted with reserve.   However, they did show that the adherence of both evaporated and sputtered films was increased by ion bombardment of the substrate just prior to film deposition, a conclusion which has been verified by many subsequent workers.[22,23,24]

Frederick and Ludema[25] have measured the adhesion of aluminum to glass by pressing 3-mm gold spheres against the film at room temperature, and then pulling to separate it from the glass.   The Au-Al bond, the Al film cohesion, and the Al-glass bond were often strong enough to pull pieces of glass out of the substrate.   Although the stresses present in such an experiment were difficult to calculate, it was possible to conclude that initially clean surfaces gave aluminum films with better adhesion than dirty ones.

(2) Scratch Methods   The first detailed work using a scratch method for measuring adhesion was undertaken by Heavens,[26] who used a smoothly rounded chrome-steel point which was drawn across the film surface.   A vertical load was applied to the point and gradually increased until a critical value of the load was reached at which the film was completely stripped from the substrate, leaving a clear channel behind. The critical load was determined by examining the resultant scratches made in the film under an optical microscope.   The critical load was taken as a measure of the adhesion.

The method was put on a quantitative basis by Benjamin and Weaver.[19]   With tip radii from 0.08 to 0.003 cm, the critical load varied over wide limits (from a few to 100 g or more) dependent on the nature of the film and substrate.   An analysis of the deformation of film and substrate caused by the loaded point showed that the critical vertical load required to remove the films depended primarily on the properties of the interface between film and substrate.   However, a comparison of these results with those obtained by critical-condensation methods[27] showed that the adhesion values obtained from the loaded-probe work were an order of magnitude smaller. This discrepancy has not yet been fully resolved, but the method still has great value as a comparative test.*

A large variety of evaporated film-substrate combinations has been examined using the probe technique.   Double layers of metal on metal on glass have also been examined.   For Al deposited on Cr layers predeposited on glass,[28] it has been found that the adhesion increases with time, in some cases by an order of magnitude over a period of 300 to 400 h, as shown in Fig. 2.   This variation with time has been explained

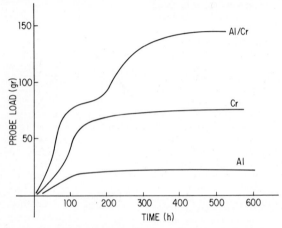

Fig. 2   Adhesion as a function of time for Al, Cr, and Al-Cr on glass at room temperature (Al 700 Å thick, Cr 150 Å thick).   (*After Weaver and Hill.*[28])

as due to alloying at the metal-metal interface.   Similar results have been obtained for Al on underlayers of Ni, Co, and Nichrome (80% Ni 20% Cr)[29] on glass, although in these cases, either a decrease or an increase of the measured adhesion between the two metals may occur according to the compound formed and the actual aging conditions.   Further work on Ag, Cu, Al, and Au on underlayers of Cr on glass has shown that the adhesion is very dependent on the thickness of the Cr layer.[30]   This is illustrated for a fully stabilized film of Ag on Cr in Fig. 3.   The increase in adhesion at ~400 Å of Cr is thought to be due to the chromium layer being discontinuous below this thickness.   Thus, intermediate-layer formation occurs only above Cr thicknesses of 400 Å.

It is of interest to compare these concepts of alloy formation with those used in studies of the adhesion of clean surfaces of like and unlike metals when the metals are pressed together.   In the case of like surfaces, the adhesion appears to depend

---

* Recent work by Butler, Stoddart, and Stuart (private communication, 1968) has shown that the behavior of copper films on glass substrates under a loaded stylus is extremely complex.   Using optical microscopy and scanning electron microscopy, they have shown not only that material is pushed up at the sides of the groove made by the traversing probe but that the center of the groove is thinned long before the pressure is high enough to remove it.   Also the film can become detached from the surface underneath the ridges of material at each side of the groove.   This last effect is more pronounced in the thicker films examined.   These observations mean that further analysis of the mechanical behavior of the film under a moving weighted probe is necessary to determine whether the mechanical strength or the adhesion is the important criterion.

on the degree of contamination of the two surfaces,[31] but with unlike surfaces the position is more complex. Keller[32] has reviewed both situations, and he shows that ideal adhesion between unlike metals depends on the free energy of atomic-bond formation. Certain pairs of metals with a negative free energy will adhere to each other (e.g., Fe-Al, Cu-Ag, Ni-Cu, Ni-Mo), whereas those with a positive free energy are immiscible and will show no adhesion.

The formation of oxide bonds between glass and certain metal films can also be shown by the scratch technique.[33,34] As with the intermetallic formation of two-layer films, the oxide bonding takes time to become established. This is illustrated in Fig. 4. Apparently, if the formation of the interfacial oxide layer is not completed during deposition, diffusion of gas to the metal-glass interface can continue the formation of the oxide layer after deposition.

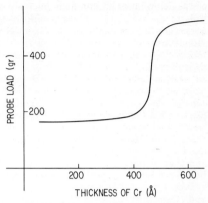

Fig. 3 Variation of adhesion with under-layer thickness for Ag-Cr two-layer structure after aging at 120°C. (*After Weaver and Hill.*[30])

This same type of oxide formation has been shown to be the factor causing good adhesion for bulk metals onto various glasses[35] and onto enamels.[36] Adherent oxides of the metals involved can be deliberately deposited on the glass surface so that a graded transition from oxide to metal can occur at the interface.

Probe measurements of adhesion have also been made on metal films deposited on single-crystal alkali-halide surfaces.[37] No aging effects were found and all the

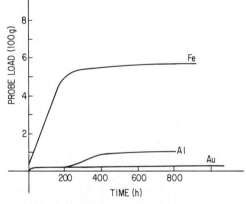

Fig. 4 Variation of adhesion with time for Au, Al, and Fe films evaporated onto glass. (*After Benjamin and Weaver.*[33])

adhesion values were low, indicating that oxide formation played no part in the adhesion and that the bonding was of the van der Waals type. The films used were all polycrystalline, but it was suggested that an oriented overgrowth should show an increased adhesion.

A modified microhardness tester has been used by Karnowsky and Estill[38] for scratch measurements similar to those of Benjamin and Weaver.[33] The results for evaporated metals on quartz and single-crystal alumina showed the importance of surface cleanliness in obtaining good adhesion and confirmed the role of oxide formation at the interface. Using a similar apparatus, Mattox[39] has shown that oxide formation can even play a part in improving the adhesion of gold on fused silica. His deposition equipment allowed rf cleaning of the silica surface and triode sputtering of the gold in an atmosphere containing a variable amount of oxygen. This type of result, indicating the role of oxide formation in films deposited by sputtering as well as by evaporation, is perhaps best illustrated by Maissel and Schaible's work on the bias sputtering of tantalum[40] onto glass. If the anode which holds the substrate is slightly negative with regard to earth, then there is a slight amount of positive ion bombardment which causes oxide and other impurities to be sputtered from the glass surface and the growing films. The adhesion of films deposited under these conditions is poor, but this may be overcome by sputtering initially with the anode positively biased (+20 V) to give a strongly adherent oxide-bonded layer, and then changing to a negative bias (−100 V) to grow pure films. Similar effects have been reported for molybdenum sputtered onto oxidized silicon.[41]

It is of interest to note in passing that measurements have been made by Moore and Thornton[42] on gold pellets melted onto polished fused silica surfaces. With Au[198] as a tracer, they have shown that gold will diffuse into the silica lattice at high oxygen pressure (of the order of 100 Torr), but no diffusion occurs during vacuum (10^{-6} Torr) heating.

(3) Abrasion Methods The abrasion resistance of films has received considerable attention over the years, but these studies have usually been directed at investigating the durability of a deposit. The resistance is determined by rubbing the surface with an emery-loaded rubber.[13,43] However, the mar resistance is found not only to be due to the hardness of the layer but also to depend on the adhesion. The technique has been turned into an adhesion-measuring system[44] by abrading evaporated metal films with a stream of fine silicon carbide particles dropped from a known height. The removal of the film was monitored by measuring the electrical resistance of the film. Differences in evaporation conditions and also annealing caused changes of adhesion that could be detected. The poorest adhesion resulted from evaporation onto a dirty surface and at a slow rate.

(4) Deceleration Methods The adhesion of thin films has been measured by ultra-centrifugal and ultrasonic techniques. In the first method silver films were electro-

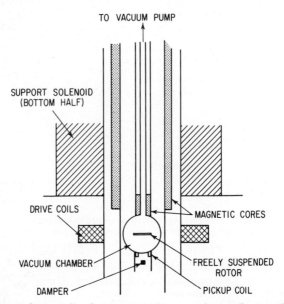

Fig. 5 Diagram of ultracentrifugal arrangement for measuring tensile strength and adhesion.

deposited uniformly onto the cylindrical surfaces of small stainless-steel rotors. The rotors, which were approximately 0.1 in. diameter, were then magnetically suspended in a vacuum and spun at higher and higher speeds until the film was thrown off.[45,46,47] The experimental arrangement is shown in Fig. 5. By this means, accurately known stresses could be applied to the films. It was shown that the tensile strength of the film T and the adhesion A could be related to the rotation speed by the equation

$$4\pi^2 n^2 r^2 \rho = T + \frac{Ar}{d_f} \qquad (1)$$

where n is the rotor speed in Hz, r is the rotor radius, ρ is the density of the film, and d_f is the film thickness.

The point at which the film left the rotor could be detected from the change in speed of the rotor. To measure adhesion alone, it was necessary to make cuts in the film parallel to the axis so as to divide the film into sections. In this case, T equals zero and A is given by

$$A = 4\pi^2 n^2 r d_f \rho \qquad (2)$$

Forces of the order of 10^9 gr wt could be obtained by the method, the only limit being the bursting strength of the rotor. However, it was necessary to contaminate the surface deliberately before deposition to obtain adhesion failure even at the highest possible speeds. Nevertheless, as will be discussed later, the method proved extremely valuable for determining tensile strength.

TABLE 1   Engineering Techniques for Measuring Adhesion

Method	Principle	References
Bending	Substrate bent or twisted until film removed	50, 51
Squashing.	Substrate squashed until film removed	52
Abrasion.	Burnishing or abrasion of surface to remove film	13, 43, 44, 53, 54
Heating and quenching. . .	Heating and sudden quenching will cause film to be removed because of stresses developed by thermal expansion and contraction	55
Scratching.	Film scratches through by probe. Alternatively parallel grooves cut into the film with decreasing separation until with intervening material lifts from substrate	19, 38, 56, 57
Hammering.	Hammering breaks up and removes film	58, 59
Indentation.	Substrate indented from side opposite to film. Coating examined for cracking or flaking off at various stages of indent formation	60, 61
Pulling.	Film pulled off directly if it is thick enough. If not, backing attached using:	62, 63
	Solder	21, 64
	Adhesives	65, 66
	Electroforming	56
Peeling.	Film peeled off using a backing of:	
	Adhesive tape	14, 20
	Electroplated coating	67
Deceleration.	The film and substrate are subject to violent deceleration, which removes the film. Various experimental arrangements are possible:	
	Coated bullet stopped by steel plate	68
	Ultracentrifuge	47, 69, 70
	Ultrasonic vibration	48
Blistering.	Film deposited so that no adhesion exists over a particular area. Air is then introduced into this area and the pressure at which film starts to lift from the edge of the area of no adhesion is measured	71, 72

Ultrasonic vibration as a method of measuring adhesion has been examined by Moses and Witt.[48] At 50 kHz, accelerations of the order of 10^5 gr wt could be obtained, and this can be raised to 10^9 gr wt by using 10 MHz. This force is of the same order as that produced by Beams.[47]

(5) Other Techniques A recent review on engineering practice in the measurement of adhesion, particularly metal on metals, has been given by Davies.[49] Other techniques are discussed, but these are generally more applicable to thick deposits. A list is given in Table 1 of the methods that he examines; it includes several that have already been considered.

b. Nucleation Methods

So far in this chapter, mechanical methods of measuring adhesion have been examined and adhesion has been thought of as a mechanical property of the total film. However, on an atomic scale the removal of a film consists of the breaking of bonds between the individual atoms of the film and of the substrate so that macroscopic adhesion can be considered as the summation of individual atomic forces. In principle, therefore, it should be possible to relate the adsorption energy of a single atom on the substrate E_a to the total adhesion of a film.

The adsorption energy of a single atom is also the term that helps to govern the behavior of condensing atoms on a surface. It controls the lifetime before an arrived atom reevaporates and thus the nucleation of the film on the surface. Electron microscopical observations of the nucleation and initial stages of growth of a film can therefore give measurements from which E_a can be derived.

This section is concerned with the basic theory required to deduce values of E_a from electron microscopy, with the actual observations made, and finally with the comparisons that have been made between mechanical and nucleation methods of measuring total adhesion.

(1) Basic Concepts The theory of the nucleation and initial growth of deposits has recently been discussed in detail, mainly with regard to the physical-growth processes. The chemical processes generally are more involved, as intermediate metastable phases often exist which complicate the energy picture of the system. Sometimes this limitation is not too difficult to avoid, as, for example, in the application of nucleation theory to the vapor-phase growth of silicon on quartz,[73] but nevertheless our concern in this section will be with films grown by evaporation and sputtering.

As has been discussed in Chap. 8, two main theories of nucleation have been evolved, the capillarity theory[74,75] and the atomistic theory.[76,77] As the smallest stable clusters from which the film grows are generally found to contain only one or two atoms, it has been pointed out that the atomistic model is conceptually more appropriate to use,[78] although both theories can be shown to be equivalent if the appropriate terms are introduced.[78] For this reason the subsequent discussion will be entirely in terms of the atomistic model.

The energies needed to set up the atomistic nucleation equations are illustrated in Fig. 6. In this diagram, E_a is the adsorption energy, i.e., the energy required to remove an atom from the substrate to infinity, E_d is the energy for diffusion of atoms from site to site over the surface (it is generally about one-third of E_a), and $E_b^{(2)}$ is the bond energy between two atoms.

The problem is now to set up the type of experiment that enables a measure of E_a to be obtained independent of E_d and E_b. The adsorption energy can be related to the probability of evaporation Γ_1^- of a single atom on the surface by the equation[79]

$$\Gamma_1^- = \nu_0 \exp \frac{-E_a}{kT} \qquad (3)$$

where ν_0 is a vibrational frequency, k is Boltzmann's constant, and T is the substrate, and hence the atom temperature in degrees Kelvin.

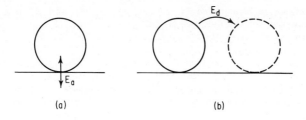

(a)   (b)

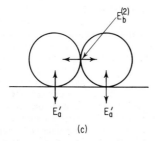

(c)

Fig. 6 Diagram showing the various energies involved in the atomistic theory of nucleation and growth. (*a*) Adsorption of a single atom on a substrate. (*b*) Diffusion of a single atom along the substrate surface. (*c*) Binding of two atoms together on a substrate.

The probability Γ_2^- of an atom leaving a doublet and diffusing away is given by

$$\Gamma_2^- = \nu_2 \exp \frac{-(E_b + E_a)}{kT} \tag{4}$$

where ν_2 is another vibrational frequency.

This equation may be an approximation inasmuch as the value of diffusion energy required for an atom leaving a doublet may not be as much as that required for a single atom migrating from one potential minimum on the surface to the next. When a migrating atom has joined an adsorbed one to form a doublet, the system as a whole has to find a minimum-energy state. Dependent on the potential distribution of the surface it may be that the atoms of the doublet will be adsorbed at a slightly different value E_a' (where $E_a' < E_a$). This will affect the value of E_d for migration away from the doublet. This effect has been ignored in the analysis, although as will be discussed later, it may lead to lower values of adsorption energies being obtained by mechanical techniques.

Finally the diffusion energy can again be introduced by considering the probability Γ_1^+ of a single atom migrating from one site to the next. This will be compounded of the probability of an atom being on a site n_1/n_0 (n_1 is the number of sites occupied by single atoms and n_0 the total number of sites available), and the probability of it having sufficient energy to migrate:

$$\Gamma_1^+ = \frac{n_1}{n_0} \nu_1 \exp \frac{-E_d}{kT} \tag{5}$$

(ν_1 is a vibrational frequency.)

This last equation can be written in a slightly different way by considering the rate v at which an atom visits sites, where v will be given by

$$v = \nu_1 \exp \frac{-E_d}{kT} \tag{6}$$

(2) Nucleation-rate Measurements Using Eqs. (3), (4), and (5), it is possible to set up an equation for the numbers of clusters n_i of size i (i = number of atoms). If R is the rate of arrival of atoms at the substrate, n_i is given by

$$n_i = n_0 \left(\frac{R}{\nu_0 n_0} \right)^i \exp \frac{E_b{}^{(i)} + iE_a}{kT}$$

where $E_b{}^{(i)}$ is the total binding energy of the cluster of size i.

If a critical size i^* is now defined such that there is no decay of groups larger than this (i.e., $\geq i^* + 1$), then the formation rate of stable groups (i.e., the nucleation rate J_{i^*}) can be written [see Chap. 8, Eq. (11)]

$$J_{i^*} = n_0 \left(\frac{R}{\nu_0 n_0} \right)^{i^*} \exp \frac{E_b{}^{i^*} + (i^* + 1)E_a - E_d}{kT} \tag{7}$$

This is a form of the nucleation-rate equation derived by Rhodin and Walton.[77]

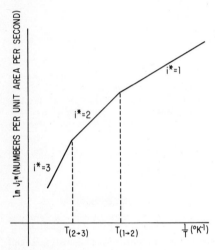

Fig. 7 Expected variation of nucleation rate with temperature for different values of critical nucleus size.

It is possible to plot Eq. (7) for specific values of i^*. Figure 7 shows such a plot for $i^* = 1$, 2, and 3, and it is to be noted that transition temperatures can be defined at which the nucleation rate changes because of the change in the size of the critical cluster. These can be derived by equating the nucleation rates for the different i^* values. The transition temperature from $i^* = 1$ to that of $i^* = 2$, $T_{1 \to 2}$, is given by [see Chap. 8, Eq. (13)]

$$T_{1 \to 2} = \frac{E_a + E_b{}^{(2)}}{k \ln (\nu_0 n_0/R)} \tag{8}$$

Similarly,

$$T_{2 \to 3} = \frac{E_a + E_b{}^{(3)} - E_b{}^{(2)}}{k \ln (\nu_0 n_0/R)} \tag{9}$$

If it is assumed that $E_b{}^{(3)}$ is twice $E_b{}^{(2)}$, then Eq. (9) reduces to

$$T_{2 \to 3} = \frac{E_a + E_b{}^{(3)}/2}{k \ln (\nu_0 n_0/R)} \tag{10}$$

an equation given by Rhodin and Walton.[77]

In a similar manner it is possible to find transitions in the rate of arrival R for changes of i^*.

The circumstances in which the nucleation rate J_{i^*} can be experimentally determined are limited to the early stages when nucleation density is still increasing linearly with time. Under these conditions, i^* can be obtained from the arrival-rate dependence of J, E_a and E_b from the temperature dependence of J and one or more transition relationships.

Walton et al.[80] measured nucleation rates for silver on rock salt. They used an ultrahigh vacuum apparatus, and the metal was deposited on a fresh surface obtained by cleaving the rock salt in vacuum. Before cleavage the evaporation was commenced and kept constant by passing a steady current through the source. The condensation of silver effectively started when the crystal was cleaved, and it was then continued for a specific time and stopped by turning off the source current. The crystal, with deposit, was removed from the chamber, shadowed with platinum, and backed with carbon, and finally the substrate was dissolved away. The specimen was then examined in the electron microscope. The numbers of islands were obtained from counts made on areas containing no cleavage steps.

Walton et al.[80] found a transition from $i^* = 1$ to $i^* = 3$. Values of the various terms could be obtained from their measurements, and were as follows:[78]

$$n_0 = 10^{15} \text{ cm}^{-2}$$
$$E_b{}^{(3)} = 2.1 \text{ eV}$$
$$E_d = 0.2 \text{ eV}$$
$$E_a = 0.4 \text{ eV}$$

The absence of $i^* = 2$ was ascribed to the lack of fit with the substrate for this configuration.

(3) Island-density Measurements It was mentioned in the previous section that nucleation density and hence rate measurements must be taken over the region in which the density is increasing linearly with time. In fact, if a general set of curves for the nucleation and growth of different materials is obtained, they will appear as shown in Fig. 8. In this figure, four different shapes can be distinguished, and nucleation-rate measurements are possible for only one of these (curve A). The four that are drawn are:

Curve A. Initially incomplete condensation with a saturation value being obtained.

Curve B. Initially incomplete condensation but with a high island density so that agglomeration occurs before the saturation value is reached.

Curve C. Initially complete condensation in which the islands only grow in size. It is found that the density value is dependent on rate of arrival.

Curve D. Initially complete condensation but with a high island density so that agglomeration occurs from the very beginning of the growth process.

It is difficult to analyze the complex situations that arise when agglomeration occurs. However, it is possible to

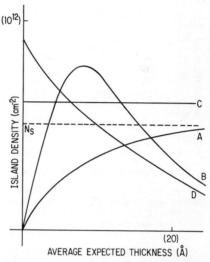

Fig. 8 Diagram of typical changes in island density with average expected thickness. (Arbitrary curves with figures indicated on the axes only as a guide.)

obtain expressions for the saturation island-density values in the case of curves A and C. This has been done, using Eq. (6), by Lewis and Campbell[81] and has been discussed in Chap. 8. For case A, the differential equation representing the curve can be written as

$$\frac{dn_s}{dt} = \frac{R m_a n_i{}^*}{n_0} \left(1 - \frac{m_a n_s}{n_0} \right) \tag{11}$$

where n_s is the island density at time t, and m_a is the number of sites that an atom will visit before reevaporation, i.e., $m_a = v \tau_1$ ($\tau_1 = 1/\nu_1$).

The saturation value N_s is then given from Eq. (11) by putting dn_s/dt equal to zero.

$$N_s = n_0 \exp \frac{-(E_a - E_d)}{kT} \tag{12}$$

This condition is defined by the lifetime on the surface before reevaporation, and the rate R influences only the time taken to reach saturation.

However, at higher rates or lower substrate temperatures, the lifetime before capture becomes the important criterion. Complete saturation is immediately obtained, and N_s is now given by (for the case of $i^* = 1$)

$$N_s = \left(\frac{n_0 R}{\nu_1}\right)^{\frac{1}{2}} \exp \frac{E_d}{2kT} \tag{13}$$

Figure 9 shows the two equations (12) and (13) plotted on one diagram as log N_s vs. $1/T$. If one can obtain a full range of growth parameters, then both the curves of Fig. 9 can be found, and E_d and hence E_a can be found.

Lewis and Campbell[81] have made measurements for the case of gold on vacuum-cleaved rock salt using a high-vacuum system. Rate control was obtained by using a disk ratemeter,[82] and care was taken to back the deposit with carbon prior to letting the system up to air so as to prevent movement of the deposit.[83] Counts of the islands were obtained from the resultant micrographs. Experimental points were obtained which could be fitted to the theoretical lines with the following values:

$$E_a - E_d = 0.4 \text{ eV}$$
$$E_d = 0.2 \text{ eV}$$

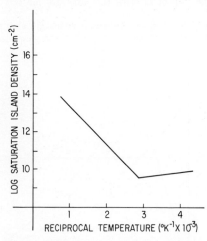

Fig. 9 Diagram of typical saturation island-density values plotted as a function of reciprocal absolute temperature. (*After Lewis and Campbell.*[81])

It was found possible to fit other workers' data to the theoretical lines.*

With some systems (e.g., silicon on silicon grown by vapor deposition[73]) it is possible to obtain the initially complete condensation curve and hence derive diffusion-energy values. However, it has not yet proved possible to obtain points on the other branch of the curve and hence find E_a.

(4) Critical-condensation Measurements The concept behind critical-condensation measurements is that if the rate of arrival of atoms is very low or if the substrate temperature is high, the lifetime of single atoms on the substrate is too short for collisions with other atoms to occur and thus for stable nuclei to be formed. Experimentally this means that the incidence rate must be raised at constant substrate temperature or the substrate temperature lowered at constant incidence rate until condensation is observed.

Using a relatively simple analysis in which a doublet is assumed the smallest stable cluster, Frenkel[79] showed that the critical rate of arrival for no condensation to occur R_c could be related to the substrate temperature by the equation

$$R_c = \frac{\nu_0}{4\sigma_0} \exp \frac{-(E_a + E_b^{(2)})}{kT} \tag{14}$$

where σ_0 is the molecular cross section of the arriving species derived from gas kinetic theory. Thus it should be possible to plot a curve of log R_c against $1/T$ that separates the two regions as shown in Fig. 10; in one of these the conditions of rate of arrival and substrate temperature are such that a film will nucleate and grow, and in the other it will not; the slope of the line will be $-[E_a + E_b^{(2)}]/k$. Cockcroft examined the deposition of cadmium on mica[84] in these terms. Using a Knudsen source so

---

* The latest work by Lewis has shown that the growth of nuclei may play an important part in the nucleatron process in a large number of cases. This modifies the behavior so that no region of positive slope is found in the plot of log N_s vs. $1/T$.

that the flux in any given direction from the source was known, and depositing on a mica strip cemented to a liquid-nitrogen-cooled copper block, he was able to obtain a circular deposit. The time required to obtain the circle varied between 1 or 2 min and several hours dependent on the stream density, and by analyzing the relationship between circle diameter and substrate temperature he was able to calculate the total energy $E_a + E_b^{(2)}$. He found a value of 5 to 6 kcal mol^{-1}, and as similar experiments with glass in place of the mica gave the same result, Cockcroft concluded that the value of E_a was governed by an absorbed gas film on the surface.

To obtain E_a alone, which is the value of interest in the present discussion, it is necessary to establish a value for $E_b^{(2)}$. This is not possible directly in the type of experiment described by Cockcroft, but from data on the heat of sublimation of cadmium crystals, an estimate of 3.2 kcal mol^{-1} was arrived at. Similar types of experiments using zinc sulfide on glass have been performed in the author's laboratories (see Fig. 11a and b), and in this case a total value $E_a + E_b^{(2)}$ of 10 ± 1 kcal mol^{-1} was obtained. One of the experimental difficulties associated with this type of measurement is that of determining exactly where the region of no deposition occurs. Apart from direct visual observations of the sort illustrated by Cockcroft's work, various other techniques have been used. These have included the measurement of resistance,[27,85] the detection of arrival of material by low-energy electron diffraction,[86] and by radioactive methods.[87] The resistance measurement is open to some criticism, as a film will be present in island form long before it is continuous enough to show a finite resistance value— in fact measurements in the laboratory have given a value for the critical-conden-

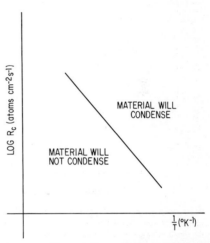

Fig. 10 Plot of log critical-condensation rate against reciprocal absolute temperature.

sation temperature of zinc on glass, which is 80°C lower by the resistance method than by visual observation. Another problem is that with the resistance measurement, the measuring field itself will affect the nucleation behavior. Conflicting results on this effect have been obtained;[88,89] the latest results have shown that an applied field along the surface of the substrate causes the films to become continuous at an earlier stage. Nevertheless, the resistance method of detection has been used by Benjamin and Weaver[27] in a detailed study of the condensation of aluminum, silver, and cadmium on glass, and of the first two of these metals on single-crystal (100) faces of NaCl and KBr. Values of adhesion energies on glass of 4.5 kcal mol^{-1} for Al, 3.4 kcal mol^{-1} for Ag, and 3.6 kcal mol^{-1} for Cd were found, indicating that the adhesion was due to physical adsorption and could be explained in terms of van der Waals forces only. Similarly, low values were obtained for the depositions onto the single-crystal faces.

In certain specific cases it is possible to obtain data from field-emission microscopy. These measurements are usually concerned with gas-adsorption problems,[90] but data have been obtained on various solid combinations (e.g., Cs on Ta and Ta$_2$C,[91] Cs on W[92]). In these cases the presence of the adsorbed material alters the emission behavior of the substrate, which is represented by a very fine wire tip; minute amounts of deposit are sufficient to alter the emission pattern.

Apart from the problem of observing the critical state and also of estimating $E_b^{(2)}$, there are various considerations that make this method less rewarding than others. As has been pointed out by various authors,[76,78,93,94] the approach by Frenkel assumed that the size at which groups become stable was two (i.e., $i^* = 1$), whereas in reality i^* may well take higher values. This in itself makes for difficulty in interpreting critical condensation. Furthermore it must be recognized that both nucleation and growth are involved in critical condensation (see Fuchs[95]). Various criteria have been suggested for determining the critical condition. Thus Yang et al.[96] suggest $J_{i^*} = 1$ nucleus cm^{-2} s^{-1} as such a condition, and the present author has worked with minimum detectable numbers of 10^8 islands cm^{-2} and a minimum detectable radius of 10 Å. This last value is well inside the detection limits of electron microscopy, but in actual fact it turns out that the numbers criterion is the only one

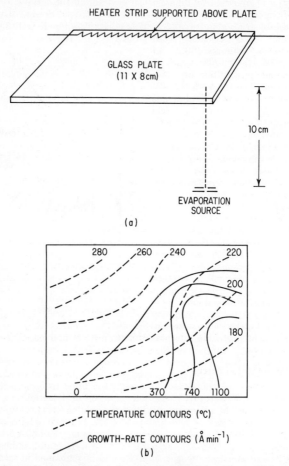

Fig. 11 Experimental arrangements and results for obtaining $(E_a + E_b)$ for zinc sulfide on glass. (a) Experimental arrangement. (b) Experimental results, showing temperature and growth-rate contours on the 11 × 8 cm glass plate.

that generally matters. As an illustration, it can be shown that for gold on rock salt 10^8 islands cm^{-2} will just be visible in 1 h if the rate of arrival is 3.5×10^{11} atoms cm^{-2} s^{-1}, and when this number are present, the islands will be bigger than the size limit suggested. It can be noted that a time criterion has also been brought into the analysis and thus that the strict concept of critical condensation is no longer applicable. The time concept can be introduced from a consideration of Eq. (11), and future measurements of E_a may be expected using this technique.

(5) Residence-time Measurements For certain materials in which the atoms on the substrate are at such a temperature that stable cluster formation is not immediate, it is possible to measure the mean adsorption lifetime directly and so arrive at a figure for the adsorption energy (see Chap. 8, Sec. 6c). Rb^+ on $W^{97,98}$ and Cd on W^{99} have been examined in detail. In the latter case, an adsorption energy of 41 kcal mol^{-1} was found for the atoms in direct contact with the substrate.

(6) Comparison between Mechanical and Nucleation Methods Nucleation methods of measuring the adsorption energy E_a are fairly complicated and generally of limited applicability, especially inasmuch as it is necessary to obtain growth curves of type A or C (Fig. 8). It is necessary to have access to an electron microscope in order to count island densities, and this in its turn means that it must be possible to remove the islands from the substrate without moving them relative to one another. Even if this can be done, it may not always be possible to make enough observations to separate the cohesion-energy term $E_b^{(2)}$ from the adsorption term E_a.

The most consistent results obtained to date have been on gold–rock salt type combinations where the films can be easily dissolved from the substrate. Work with these combinations is at present aimed at determining the effects of parameters such as the arrival energy of the incident atoms or the type of crystallinity of the deposit.

Comparison between adsorption-energy values from nucleation measurements and adhesion as obtained from mechanical methods have been made in only one or two cases.[20,27] This partly arises from the difficulty of making meaningful quantitative measurements by mechanical techniques, but also from the limited applicability of nucleation methods.

Adhesion values per square centimeter may be equated to the average adsorption energy per atom, and it would be expected that the former value would be somewhat lower than the adsorption energy of a single atom as discussed in nucleation atoms. This has already been noted in connection with Eq. (4) and Fig. 6. However, it cannot be lower by more than the diffusion energy of a single atom E_d, so that it should be possible to obtain results which agree within a factor of 2 or 3. Mechanical stripping measurements have been compared with nucleation measurements for gold on glass.[20] It has been found that, with the lowest stripping speeds used to date (500 Å s^{-1}), the average adsorption energy is a factor of 2 greater than that obtained by nucleation methods so that it appears necessary to work at even lower stripping speeds to obtain better agreement.

Other methods of examining adhesion should be useful for obtaining comparative measurements but have not been considered in this article as they are not yet sufficiently proved. These include microcalorimetry[17,100,101] and possibly electron-spin-resonance techniques.[102]

c. Nature of the Adhesion Forces

In the work discussed in the previous sections it has been found that adhesion-energy values between a film and a substrate may vary from tenths of an electron volt up to 10 eV or more. Experimentally it is difficult to measure high adhesion values by nucleation, tape, or deceleration methods, but it has been found possible to make comparative measurements over the whole range by using scratch techniques. To explain adhesion over such a wide range it is necessary to invoke two different mechanisms. For low values of adhesion, the adsorbed atoms keep their electron shells intact and the forces holding them on the surface are of the van der Waals type. This situation holds up to values of approximately 0.4 eV, and the atoms are said to be physisorbed on the substrate. Above 0.4 eV, sharing of electrons can occur, and the atoms are now said to be chemisorbed.

(1) Physisorption The nature of physisorption has been discussed by various authors. Any explanation offered is required to fit into a general picture in which an atom, as it is brought up to a surface, is first attracted and then repelled. There is thus a potential well at a very small distance from the surface into which atoms may be drawn. This situation is similar to that arising when gas is adsorbed on a surface, and the forces attracting the atom are known as van der Waals forces. Of the various attractive forces possible, the orientation effect postulated by Keesom[103] can arise if permanent dipoles are present in the surface. Debye[104] considered that moments could be induced by an arriving dipole[105] in an adjacent molecule of the surface. However, both this effect and the orientation effect do not apply when one is considering the adsorption of nonpolar molecules. This has been explained by London's dispersion effect.[105] In every atom each electron forms an electric dipole with the nucleus whereby the rotating dipoles in adjacent atoms interact causing attraction with respect to each other. London[105] showed that the adsorption energy E_a of an atom of condensate and an atom of the surface could be expressed by

$$E_a = -\frac{3}{2}\frac{\alpha_1\alpha_2}{s^6}\frac{h\nu_1\nu_2}{\nu_1+\nu_2} \tag{15}$$

where α_1 and α_2 are the polarizabilities of the two atoms, ν_1 and ν_2 are their characteristic vibrational frequencies, h is Planck's constant, and s is the equilibrium distance of the condensate atom from the surface. The interaction between the adsorbed atom and the entire substrate may then be calculated from Eq. (15) by integrating over the atoms of the substrate. For an infinite surface the average adsorption energy E_{ad} becomes

$$E_{\mathrm{ad}} = -\frac{N\pi}{4}\frac{\alpha_1\alpha_2}{s^3}\frac{h\nu_1\nu_2}{\nu_1+\nu_2} \tag{16}$$

where N is the number of adsorbent atoms per unit volume in the substrate.

An exact evaluation of E_{ad} from Eq. (18) is seldom possible, as the so-called characteristic energies $h\nu_1$ and $h\nu_2$ of the two materials are not known. However, it has been shown by London[106] that to a first approximation the ionization potentials V_1 and V_2 may be used so that Eq. (16) may be written

$$E_{\mathrm{ad}} = -\frac{N\pi}{4}\frac{\alpha_1\alpha_2}{s^3}\frac{V_1V_2}{V_1+V_2} \tag{17}$$

For an ionic substrate, the attraction between the adsorbed atom and the substrate could be complicated by the fact that ions of the substrate lattice will induce a dipole moment in the nonpolar atoms. However, where calculations have been made,[107,108] this effect has been found to be negligible.

Benjamin and Weaver[27] have applied London's concepts to the physisorption of various metals on glass and alkali halides. The calculation of N for a glass was based on the crystalline silica network described by Zachariasen,[109] with the oxygen ions taken as the adsorbing ones. Table 2 shows some of the results obtained compared with experimental values found using critical-condensation methods with resistive detection.

TABLE 2   Adhesion Energies for Three Metals on Glass[27]

Metal	Theoretical adsorption-energy value E_{ad}, eV	Measured condensation-energy value [see Eq. (14)] $E_a + E_b^{(2)}$, eV
Al........	0.15	0.19
Ag......	0.12	0.15
Ca........	0.09	0.16

The results obtained appear to be of the right order as in all cases the adsorbtion energy obtained is less than the total condensation energy.

Results on alkali halides were more easily calculated, since the lattice-energy characteristics and polarizabilities were known. The value of N can also be more easily determined. Table 3 summarizes some of the results obtained, and again the total condensation energy is greater than the theoretical adsorption-energy value.

The values of adhesion that have been discussed above refer to evaporation onto clean surfaces. The effect of any dirt or adsorbed gas on the surface will be to lower the adhesion. Various techniques have been described for preparing clean surfaces, including special cleaning sequences,[110] ion bombardment,[24] and cleavage in vacuum[111] (see Chap. 6). Even without actual ion bombardment, sputtered deposits adhere better than evaporated ones because the higher kinetic energy of arrival of the atoms in the sputtered case is sufficient to remove loosely adsorbed gas atoms from the surface.[112,113]

TABLE 3 Adhesion Energies for Some Metals on Single-crystal Alkali-halide Faces[27]

Metal	Substrate	Theoretical adsorption-energy value E_{ad}, eV	Measured condensation-energy value [see Eq. (14)] $E_a + E_b^{(2)}$, eV
Ag...............	(100)NaCl	0.12	0.14
Ag...............	(100)KBr	0.11	0.13
Al...............	(100)NaCl	0.17	0.24
Al...............	(100)KBr	0.16	0.19

A further concept has been recently introduced to explain adhesion. If materials with different work functions are initially in close contact, the charges present on the materials will cause an electrostatic attraction to exist between the surfaces. On separation work must be done to overcome these forces. Workers in both Russia[114,115] and the United States[116] have demonstrated that for certain pairs of materials these forces are large. Present work is aimed at determining the degree to which such effects are applicable in the adhesion of thin films.

(2) Chemisorption If the condensation of atoms involves chemical bonding, the high values of E_a and E_d permit little surface migration of condensing atoms. It is thus not possible to obtain nucleated growth from the measurements of which E_a values may be found.

Weyl[117] has suggested that three different growth cases may be distinguished, namely:

1. Epitaxial or oriented growth
2. Intermediate-layer formation that allows continuous transition from one lattice to the other
3. Metal film–ionic substrate pairs where mirror-image forces are of particular importance

The second of these cases has proved the most useful and has been considered as the main cause of adhesion by various workers. Bateson[16] suggested an intermediate oxide formation to explain the high adhesion of various oxygen-active metals to glass (e.g., Cr, Al), while the formation of intermediate alloys has been suggested by Weaver and Hill[28–30] to explain their adhesion results on multilayer films.

2. STRESS

As was mentioned in the introduction to the chapter, interest in the mechanical stresses in thin films started as early as 1877.[1] During the intervening years many stress determinations have been made in thin films produced by various means, and theories have been offered to explain the results obtained. Reviews of this work have already been given with regard to stress in continuous films and also, more specifically, to the stresses that are present in the initial stages of growth when the film consists of completely separate islands.[4–6,118] A further excellent summary of the situation has been given by Wilcock,[119] to whom the author is especially indebted.

Nearly all films, by whatever means they are produced, are found to be in a state of internal stress. The stress may be compressive (i.e., the film would like to expand parallel to the surface) so that in extreme cases it may buckle up on the substrate (cf. the cases reported by Yelon and Voegeli[120]). Alternatively the film may be in tensile stress (i.e., the film would like to contract), and in certain cases the forces may be high enough to exceed the elastic limit of the film so that it breaks up.

For normal deposition temperatures (50 to a few hundred °C), the stress in metal films is typically 10^8 to 10^{10} dyn cm^{-2} and tensile, the refractory metals at the upper, the "soft" ones (Cu, Ag, Au, Al) at the lower end. With dielectric films, stresses are often compressive, and slightly lower values seem to be more common.

Movement of the film relative to the substrate will take place only if the adhesive bond between film and substrate is broken. It is possible to make a calculation of this situation, and it has been shown[121] by considering the simple case of moving atoms along the (111) surface of a fcc structure from one potential minimum to the next by the easiest possible route, that a stress of $\sim 5 \times 10^9$ dyn cm^{-2} will be required to overcome an adsorption energy of 0.2 eV. This adsorption energy, as has been seen in the previous section, is a low figure indicating only physical adsorption onto the substrate, and a value of 5×10^9 dyn cm^{-2} is a relatively high value of stress. Thus, generally speaking, the adhesion between film and substrate is not a limiting factor. If a film does strip or is removed by some external means from a substrate, it is often found that the film will curl up. Such behavior will indicate that the film is stressed in a nonuniform way.

In certain cases the type of stress may depend on the preparation conditions so that films of zero stress may be prepared by slightly altering these. Two technologically important types of film in which the stress may be adjusted are electroplated nickel-iron,[122] where additives can alter the stress, and evaporated silicon monoxide,[123] where residual gas pressure and evaporation rate can affect the type of stress obtained.

If the coefficients of thermal expansion of a film and its substrate are not the same, heating or cooling will produce additional stress which will tend to deform the film-substrate combination. This stress contribution is known as thermal stress, and its type and magnitude depend on the difference between the thermal coefficients of the film and substrate. Even after accounting for the thermal stress, many films are found to have a residual internal stress, and this part is called the intrinsic stress. Thus the total stress σ observed in a film is equal to the sum of any externally applied stress plus thermal plus intrinsic components, i.e.,

$$\sigma = \sigma_{external} + \sigma_{thermal} + \sigma_{intrinsic} \tag{18}$$

In technological applications the total stress must be kept small. The intrinsic stress is the predominant component in many systems and has been the subject of most of the investigations.

a. Stress-measuring Techniques

If a film is deposited in stress on a thin substrate, the substrate will be bent by a measurable degree. A tensile stress will bend it so that the film surface is concave, and a compressive stress so that is convex. The most common methods for measuring the stress in a thin film are based on this principle. The deformation of the substrate due to the stress is measured either by using a thin cantilevered beam as a substrate and calculating the radius of curvature of the beam and hence the stress, from the deflection of the free end, or by observing the displacement of the center of a circular disk.

Stress may also be measured by x-ray or electron-diffraction techniques. The position of the diffracted line gives the interplanar spacing of the set of lattice planes corresponding to the line, and the strain in the crystallites forming the film may be deduced from this. Alternatively, the stress may be deduced from the shape of the diffracted beam by apportioning the line broadening to components resulting from small particle size and from strain.

It should be noted that x-ray and electron-diffraction techniques will give the strain and hence the stress in a crystallite lattice. This is not necessarily the same as that measured by substrate bending since the stress at the grain boundaries may not be the same as that in the crystallites.

Other methods are available for specific materials with particular properties such as semiconductors or magnetics.

All the methods mentioned above will now be considered in more detail.

(1) Disk Methods In the disk method, the stress of a film is measured by observing the deflection of the center of a circular plate when the film is deposited on one side. The interference fringes (Newton's rings) between the disk and an optical flat are used to measure the deflection of the disk, which may be a glass, quartz, or silicon crystal wafer of 1- to 10-mil thickness. Figure 12 shows the apparatus used by Finegan and Hoffman.[124] Other work using the same type of apparatus has been reported by the same authors[125] and by others.[126-130] Because of the limited flatness of available substrates, Finegan and Hoffman dissolved the film from the substrate after use and then used the remaining profile as a reference.

Disks strained by the presence of a film will bend into a paraboloid. The stress σ can then be related to the deflection by the disk at a distance r from the center of the disk, by the equation[130]

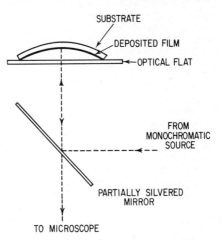

SUBSTRATE

DEPOSITED FILM

OPTICAL FLAT

FROM MONOCHROMATIC SOURCE

PARTIALLY SILVERED MIRROR

TO MICROSCOPE

Fig. 12 Apparatus for observing fringes produced by the variable air gap between a stressed circular substrate and an optical flat. (*After Hoffman.*[6])

$$\sigma = \frac{\sigma}{r^2} \frac{E_s}{3(1 - \nu)} \frac{d_s^2}{d_f} \qquad (19)$$

where E_s is the Young's modulus of the substrate, ν is the Poisson's ratio for the substrate, d_s is the substrate thickness, and d_f is the film thickness.

In certain cases it is not possible to use a substrate that is sufficiently flat to give satisfactory interference patterns. Measurements have been made of the stress in films of molybdenum deposited on silicon wafers by several authors.[41,130,131] Glang et al.[130] have shown that it is then possible to characterize the curvature of a wafer by using a Zeiss light-section microscope. Such an instrument projects a horizontal, slit-shaped beam of light on the sample surface at a 45° angle. The reflected image of the slit is observed in the eyepiece, which has a fine crossline adjustable by a micrometer. Once it is properly focused, a deviation of the reflecting surface from the original focal plane causes a deflection of the beam. If the micrometer is turned until image and crossline coincide, a direct reading in microns is obtained indicating the difference in height between two locations on the sample surface. By moving the sample with a micrometer stage, one obtains a surface profile. Stresses of 10^9 to 10^{11} dyn cm^{-2} can be determined using 1-in.-diameter wafers 0.008 in. thick.*

The disk method is ideal for measuring stress anisotropy, but the sensitivity is limited by the substrate thickness that can be used. The thickness reduction is limited both by the problems of electrostatic attraction between the substrate and the optical flat[124] and by the inapplicability of the usual relationships between stress and beam deflection for very thin substrates (cf. the work of Klokholm[132] on rectangular plates). The method as most commonly used is not an in situ method and hence

---

* A recent paper by P. M. Schaible and R. Glang [in F. Vratny (ed.), "Thin Film Dielectrics," Electrochem. Soc. Inc., New York, 1969] has shown that it is possible to use fiber optics to detect the position of the surface. The amount of light transmitted back through the fibers depends on the displacement of the surface from the ends of the fibers. Sensitivity is ±0.1 μ over a displacement range of 50 μ. The whole process may be automated so that the operator time to obtain stress data is approximately 30 min as compared with 3 h if the profiles are measured with a light-section microscope or by interference techniques and evaluated without a computer.

is not suitable for stress measurements during film deposition—cf. the bending-beam method discussed in the next section.

Two further advances in the use of interference techniques with disk-shaped specimens have been reported. Taloni and Haneman[133] have shown that a laser interferometer may be used for measuring the bending. [The work was concerned with surface stresses in polished and clean (111) surfaces of thin bulk specimens of Ge, InSb, and GaSb.] Also, the use of holography has been demonstrated by Magill and Young[134] using aluminum evaporated onto highly polished silicon.

A method that bears some resemblance to the disk method has been developed by Beams and his coworkers.[47,135] A film is deposited on a dissolvable substrate and then a hole is drilled through the substrate to provide a circular region of film that is unsupported. Care has to be taken in the drilling operation not to rupture the film in the process. The method was originally designed to measure the stress-strain relationships in a thin film by applying pressure on one side of the unsupported region and observing by interference or microscopy techniques how the film bulges. However, initial stress can be detected by the behavior of the unsupported film after removal of

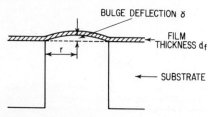

Fig. 13 Diagram of bowing of unsupported film in compressive stress.

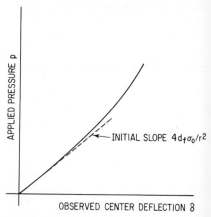

Fig. 14 Applied pressure vs. deflection for bulge determination of initial stress (arbitrary units).

the substrate; if the stress is compressive the film will bow even without pressure applied, as shown in Fig. 13, and the deflection δ will be related to the initial stress by

$$\sigma_0 = \frac{2}{3} \frac{E_f}{1 - \nu} \frac{\delta^2}{r^2}$$

where E_f is the Young's modulus of the film, ν is Poisson's ratio (in this case, of the film), and r is the radius of the hole. For a tensile stress, this must be observed by plotting the applied pressure against bulge height and determining the slope of the first few points. The equation connecting pressure and deflection has been shown to be[47]

$$p = \frac{4d_f\delta}{r^2} \left(\sigma_0 + \frac{2}{3} \frac{E_f}{1 - \nu} \frac{\delta^2}{r^2} \right) \tag{20}$$

where d_f is the film thickness. Figure 14 shows a typical curve obtained for this case, and at small values of δ/r, the second term in Eq. (20) may be ignored.

Disk methods have been applied to electroplated films.[47] Kushner[136] has developed an instrument known as a "Stresometer" based on this principle. Metal disks are used, and the bowing of the disk is observed hydrostatically by having a fluid in contact with one side of the disk and observing the movement of the fluid in a capillary as the disk bows.   Stress in nickel films prepared by chemical reduction from solution[137,138] has been found with this technique.

The method, however, is not suitable for many substrate-film combinations.   It has been suggested by Papirno[139] that the bulge surface is not actually spherical, as had been assumed in the derivation of Eq. (20), so that the stress value would be in error—by a factor of 2 or more (see also the work of Glang et al.[130]).

(2) Bending-beam Methods   In the beam method, the substrate is usually between three and fifty times longer than it is broad.   The beam curvature is found by measuring the deflection either of one end of the beam while the other end is clamped, or of the middle of the beam if it is supported at both ends.   The sensitivities of the various beam methods depend on the detection systems used to observe this movement.   These have been reviewed by Blackburn and Campbell,[140] and their table has been extended by Hoffman.[6]   Some of these methods are illustrated in Fig. 15a to i with references to be found in Table 4.

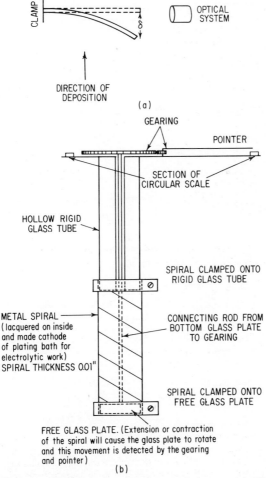

Fig. 15   Methods of measuring stress by bending techniques.   (a) Microscopic observation.   (b) Contractometer.[166]

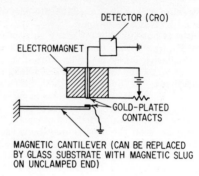

The electromagnet is used to restore the cantilever
to its original position and restoration is detected
by the passing of a current through the gold-
plated contacts.

(c)

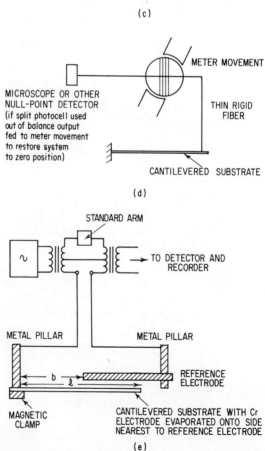

Fig. 15 Methods of measuring stress by bending techniques. (*Continued*) (*c*) Electromechanical. (*After Priest.*[169]) (*d*) Electromechanical. (*After Klokholm.*[132]) (*e*) Capacitive. (*After Wilcock.*[119])

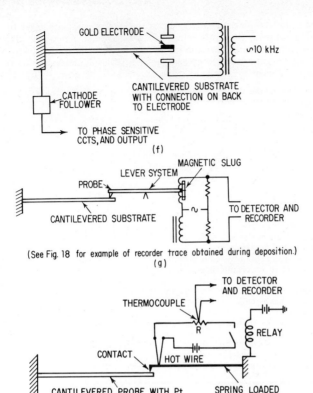

(See Fig. 18 for example of recorder trace obtained during deposition.)
(g)

When contact made, relay open. Hot wire cools and pulls contact open. Overall current in hot-wire circuit measured by thermocouple detection of hot-wire circuit resistor.

(h)

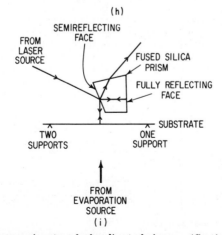

ig. 15 Methods of measuring stress by bending techniques. (*Continued*) (*f*) Capacitive. *After Hill and Hoffman.*[174]) (*g*) Inductive. (*After Blackburn and Campbell.*[140]) (*h*) Iot wire. (*After Shepard et al.*[180]) (*i*) Michelson interferometer. (*After Ennos.*[181])

The most sensitive methods so far developed, which can be used to detect the stresses in films during the initial stages of growth—i.e., when the film consists of completely separate islands—are the inductance-change[141] and capacitance-change[119] methods. Capacitance systems are, in fact, capable of measuring extremely small displacements. Jones[142] has shown that movements of around 3×10^{-10} cm can

TABLE 4 Summary of the Various Methods of Measuring Stress by Bending Beam

Fig. No.	Principle of movement detection	Type of film	References
15a	Microscopic observation of the movement of the free end	Evaporated	145–154
		Electroplated	155, 3
		Anodized	156, 157
		Thermally grown	158–160
		Chemically deposited	161
		Vapor-phase grown	162, 163
15b	Contractometer—expansion or contraction of the spiral detected by gearing and pointer	Electroplated	164–166
15c, d	Electromechanical restoration of beam to its original position	Evaporated	132, 167–170
		Electroplated	171
15e, f	Capacitance change detected by suitable circuit	Evaporated	172–174, 119, 175–177
		Sputtered	178
15g	Inductance change detected by bridge circuit	Evaporated	140, 141, 179
15h	Hot-wire current measured	Evaporated	180
15i	Michelson interferometer detects movement of center of substrates	Evaporated	181

be measured this way. The inductance-change method can use part of a mechanical surface-profile meter[143] of the type used to detect film thickness.[144] It suffers from the disadvantage that measurements have to be made at room temperature. The apparatus can measure very small changes in inductance. At the greatest sensitivity a movement of 100 Å may be detected, and this gives a stress sensitivity of 10^6 dyn cm⁻ in a film 1,000 Å thick on a 3-cm-long, 0.015-cm-thick substrate.

The capacitance system can measure capacitance change to 0.0005 pF, i.e., fractional changes in capacity of 1 part in 30,000, and hence can give a stress sensitivity of 0.2×10^6 dyn cm⁻² /1,000 Å again, on a 3-cm cantilevered length.

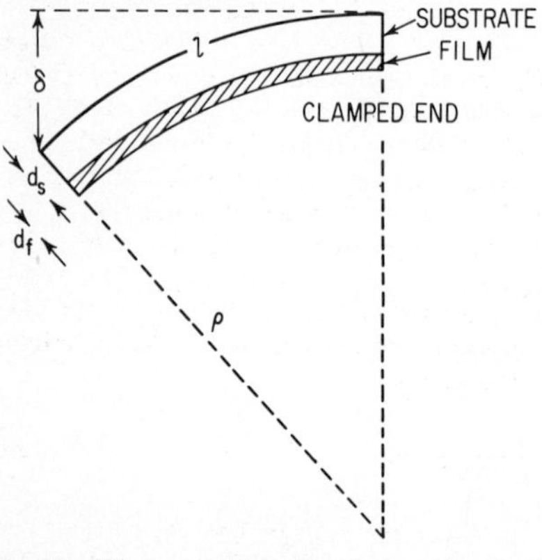

The formula used for calculating stress in terms of the end deflection of the beam has been derived by various workers. The earliest analysis by Stoney[3] assumed that Young's moduli of the film and the substrate were identical and that the stress was isotropic. Under these circumstances he obtained

$$\sigma = \frac{E_s d_s}{6\rho d_f} \qquad (21$$

Fig. 16 Diagram of bending beam, clamped at one end.

where E_s is the Young's modulus of the substrate and film, ρ is the radius of curvature of the beam, d_s is the substrate thickness and d_f is the film thickness (see Fig. 16).

This can be easily turned into an equation relating stress to end deflection by substituting

$$\rho = \frac{l^2}{2\delta} \tag{22}$$

where l is the length of the substrate and δ is the deflection of the free end, viz.,

$$\sigma = \frac{E_s d_s \delta}{3l^2 d_f} \tag{23}$$

The theory used to derive Eq. (23) is limited in its applicability. Klokholm[132] has noted that the substrate length must be at least twice the width. Furthermore, the deflection must not be greater than the substrate thickness. This latter criterion has also been discussed by Hoffman.[6]

Three factors may be noted with regard to the use of this equation. Firstly, if the film is not in the form of a flat continuous slab but is an island structure as during the initial stages of growth, either the stress must be defined from Eq. (21) using an average value for the thickness derived from the known rate of deposition, or the cross-sectional area of the average island must be found. This latter approach has been applied in the case of lithium fluoride[140] where the islands have regular shapes, but generally this is not possible.

Secondly, the stress in a film does not necessarily become zero as the film thickness goes to zero because the film thickness d_f appears in the denominator of Eq. (23). However, the end deflection of the beam does tend to zero so that it is often more convenient to plot the force per unit width σd_f, as this is proportional to the deflection, i.e.,

$$\sigma d_f = \frac{E_s d_s}{3l^2} \delta \tag{24}$$

Thirdly, as has recently been pointed out,[4,182] the derivation for Eq. (24) has neglected the effect of stress in the plane of the substrate and at right angles to the beam length on the beam curvature. Equation (24) therefore needs to be modified by the introduction of Poisson's ratio of the substrate ν.

$$\sigma d_f = \frac{E_s d_s}{3l^2(1 - \nu)} \delta \tag{25}$$

Brenner and Senderoff[183] have reviewed the derivation of Eq. (21) in some detail and obtained formulas to cover all the common experimental arrangements. They considered, among other things, the differences that arose if the beam was completely free during the deposition of the film, if it was constrained from bending but not contracting during the film deposition and then released at one end, and if it was constrained from bending and contracting. They also considered the effect of different elastic moduli in film and substrate. However, as has been pointed out by Hoffman,[6] these corrections are negligible. Davidenkov[182] and Wilcock[119,175] have made even more exact derivations of the relationship for differences in elastic moduli of film and substrate: Wilcock's work[119] has also included the case of anisotropic stress which had previously been considered by Finegan and Hoffman[124,6] (see also Ref. 118).

All the methods for measuring deflection listed in Table 4 can use Eq. (25) directly with the exception of the capacitance method. In this case the curvature of the substrates will give a complex change of capacitance with end deflection. This has been examined by Wilcock,[119,175] who derived the expression

$$d = \frac{3w(l - b)l^2}{4\pi C_0^2(l^2 + bl + b^2)} \Delta C \tag{26}$$

where w is the width of the substrate, $(l - b)$ is the overlap of the reference plate on the cantilever (see Fig. 15e), C_0 is the initial capacitance, and ΔC is the capacity change.

To use Eq. (25) it is necessary to have a knowledge of the elastic constants of the substrate. Poisson's ratio ν has in general been neglected in stress experiments until recently, but where it has been used values between 0.2 and 0.25 have usually been taken. This variation makes a less than 10% difference in stress. The Young's modulus E_s of the substrate, however, is a different matter, and it is necessary to make individual measurements of this for each particular experiment. The simplest way to do this is to hang known weights on the end of the cantilever and observe the deflection of the beam. The equation relating E_s to the deflection is

$$E_s = \frac{4Gl^3}{wd_s^3\delta} \tag{27}$$

where G is the weight placed on the end and w is the substrate width. A plot of G against δ will thus give a value for E_s. More sophisticated techniques have been employed by some workers—for example, Wilcock[119,175] has measured E_s by using an alternating electrostatic field to apply force to the beam and excite the natural resonant frequency of flexural vibration.

(3) X-ray and Electron-diffraction Techniques X-ray methods may be used to measure changes in lattice spacing and hence stresses in films. If, however, the dimensions of the crystallites of the film are less than 1,000 Å, then the diffraction lines will be broadened. To distinguish the broadening effect due to crystallite sizes or film thickness from effects due to stress, careful analysis of the line profiles must be made. Such a technique has been used by Borie et al.[184,185] to measure the stress in single-crystal films of copper oxide. With a reasonable x-ray path length in the material—this can be achieved using glancing-angle techniques—it is possible to measure the lattice spacing directly; this has been done for evaporated films of gold,[129,186,18] copper,[188] nickel,[189-191] and silver.[192] The technique has also been applied to films of silicon on alumina and niobium-tin on nickel alloys.[193]

In the case of single-crystal nickel films deposited on rock-salt substrates, Freedman[190] was able to show that the stress was relieved when the constraint of the substrate was removed. He allowed water vapor to come into contact with the substrate and film, which dissolved the substrate.

Cullity[194] has analyzed the sources of error in x-ray measurements of residual stress. He has shown that two sources of error are possible. The first causes an x-ray line shift whether or not a real macrostress is present and is due to a particular distribution of microstress. It is unavoidable. The second source of error is stacking faults, and their effects may be eliminated by proper x-ray technique.

If the film has been grown on a single-crystal substrate, then it is possible to detect the strain produced in the substrate by the film using x-ray extinction topography. Meieran and Blech[195,196] have studied films of silica on silicon by this technique, and they have also obtained data for platinum films on silicon.

The use of an x-ray interferometer has recently been reported by Hart,[197] who has used such a system to measure the strain in silicon produced by a few thousand angstroms of aluminum or copper. The silicon slices were 0.3 mm thick. The apparatus is capable of measuring strain to 1 part in 10^8.

Electron-diffraction techniques can be applied without the line-broadening effect due to crystal size down to 100 Å crystallite dimensions. It is of interest to note that one of the first determinations of stress in an evaporated-metal film was made by analyzing the line broadening of electron-diffraction patterns.[198]

The most detailed treatment of this type of work, using copper and lithium fluoride films, has been published by Halliday et al.[199] Boswell also made similar measurements.[200] As the two sets of measurements gave conflicting results, the position was discussed by Rymer.[201] He showed that it was not possible to calculate lattice spacings in very small crystallites (<50 Å across) because of the breakdown in electron-diffraction theory when applied to such sizes. Other electron-diffraction methods may also turn out to be of value in this field, e.g., scanning[202] or low-energy[20] techniques.

The Bragg contrast fringes that are obtainable in single-crystal substrates with electron diffraction have themselves been used for stress measurements. Pashley[204] noted diffraction contours in gold films on silver films and suggested that these were due to local regions of strain, the contours representing the loci of areas in the film which diffracted in the same way. Drum[205] used fringe contrast to examine the

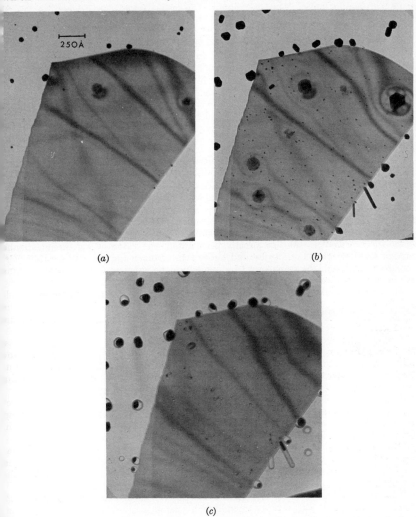

(a)   (b)

(c)

Fig. 17 Electron micrographs showing change in Bragg contrast fringes due to straining of substrate by islands growing on substrate (Ag on Si_3N_4 single-crystal flake with carbon background). (a) Substrate temperature = 400°C. Approximate average thickness of Ag = 1 Å. (b) Substrate temperature = 400°C. Approximate average thickness of Ag = 5 Å. (c) Substrate temperature = 800°C. Ag has evaporated from silicon nitride flake but has not yet completely gone from the surrounding carbon.

trains in aluminum nitride flakes, and recently Stirland[206] has shown that individual islands of silver growing on thin single crystals of silicon nitride stress the substrate. Electron micrographs of this behavior are shown in Fig. 17. To date, although such an observation is of interest in showing the stress effect of islands in the initial stages of growth of a film, the data have not been analyzed in detail to determine the magnitude of the stress.

(4) Other Techniques In some cases the stress in a film may be deduced from its effect on the properties of the film. These measurements usually give order-of-magnitude values only. Measurements of the optical band gap[207] have been used, as have also Hall-effect measurements.[208,209] In superconducting films, changes in the critical temperature can give stress values,[210] and Toxen[211] has shown that the variation of critical temperature with thickness of indium films can be explained by the differential stress induced in cooling. In other materials such as tin,[212] the shift in critical temperature is not only a function of the film thickness but is also a sensitive function of the crystal orientation. A tin film oriented with its c axis normal to the plane of a soda-lime glass substrate is only slightly stressed because the differential thermal contraction of this combination is small. In ferromagnetic films, the ferromagnetic resonance may be measured and the shift of the resonance due to stress observed relative to an unstressed bulk sample.[170,213-216]

Table 5 summarizes the highest sensitivities that have been realized to date by the different methods. Not all the methods discussed are listed, as in many cases the necessary data are not given in the published work. Furthermore, higher sensitivities will be possible in certain cases; for example, bending-beam sensitivities may often be improved beyond those quoted by increasing the length or decreasing the thickness of the cantilever, provided that the limitation with regard to deflection is not ignored. The values of Young's modulus of the substrate E_s and Poisson's ratio ν will also effect the sensitivity. As has already been noted, the value of ν does not change enough to affect the stress value and hence the sensitivity appreciably. However, changes in E_s could be important, and therefore the substrates used together with the values of E_s are included in Table 5. Figures for sensitivities are quoted for the minimum stress (in dyn cm^{-2}) that could be detected in a film 1,000 Å thick.

TABLE 5 Stress Sensitivity for Various Measuring Techniques*

Method	Sensitivity, 10^6 dyn cm^{-2}, for films 1,000 Å thick	Substrate	Young's modulus E_s, 10^{11} dyn cm^{-2}	Reference
Disk:				
Pressure measurement....	∼1,000	Various metals (Ni, Cu)	15–20	136
Interferometric... , ...	15	Soda glass	5	124
Optical.................	∼1,000	Si	20	130
Cantilever:				
Optical.................	250	Cu	12	146
Interferometric..........	0.5	Not quoted		6
Interferometric, supported at both ends..........	60	Silica	7	181
Electromechanical.......	0.5	Soda glass	5	132
Magnetic, restoration.....	250	Ni	20	168
Capacitance, single plate..	0.2	Soda glass	5	119, 175
Capacitance, double plate.................	10	Soda glass	5	174
Inductive...............	1	Soda glass	5	140
Contractometer...........	30	Stainless steel	∼20	166
X-ray....................	500†	....	....	218, 6
Electron diffraction........	0.04 % strain	....	....	193
Ferromagnetic resonance....	1,000	....	....	213

* See Refs. 6, 118, 140, and 217 for other comparative lists.
† Approximate equivalent—a force is not measured.[6]

b. Results

(1) Electrodeposited Films The first observations of the effects of stress in thin films were made in the early days of electrodeposition when it was found that thick electrodeposits tended to crack or to peel from the cathode (Gore, 1858, quoted by Stoney[3]). With the extensive development of the electroplating industry, the effects of internal stresses in the coating assumed great importance.

Three examples of the effects of stress will be considered. Perhaps the most important is the effect of stress on the fatigue range (resistance to repeated vibrational stresses which may be either compressive or tensile) of coated metal parts. When a metal part is built up by plating, the fatigue range is much reduced.[219–221] Although the externally applied stress may be within the safe range, the total stress in the coating, being the sum of the external stress and of the internal stress from the deposition process, can exceed the safe limit. This leads to cracks in the coating which propagate through the metal part, which then fails.

Another example is the stress in nickel films, which can cause appreciable contraction in the film when it is stripped from the cathode. This can be very important in electroforming, particularly in color printing where dimensional changes in the nickel electrotype[222] must be avoided.

Finally, with the advent of plated-wire computer memory systems, the stress in the deposited alloy assumes importance because of its possible effects on the magnetic anisotropy of the films.[223,224] However, this seems to have been overrated because later work[225] has shown that changing the sign of the stress in plated nickel films from tensile to compressive by the use of organic additives does not, within the experimental error, change the magnetic anisotropy.

Mills[1] made the first quantitative measurement of stress in a thin film in 1877. He coated the bulb of a mercury-in-glass thermometer with silver by a chemical technique and used it as a cathode in an electroplating bath. He found that deposited nickel films caused the bulb of the thermometer to contract, thus forcing the mercury higher up the tube than it would have been because of temperature alone. The stress he found in nickel was thus tensile. He also found tensile stress in iron, copper, and silver, but compressive stress in cadmium and zinc films. He calibrated the thermometer by noting the effect of a hydrostatic pressure and thus found the magnitude of the stress. Using the same method, Bouty[2] confirmed Mills' work. Thirty years later, in 1909, Stoney[3] used as a cathode a long, thin, rectangular steel rule which was clamped at one end and had one side lacquered. Using this apparatus, the first use of the bending-beam method to be reported, he found that the stress in nickel films deposited at a plating-bath temperature of 10 to 15°C was of the order of 3×10^9 dyn cm^{-2}.

Much work has been carried out on the stress in electrodeposited films since the time of Stoney, and the work has been reviewed by Soderberg and Graham.[220] More recent reviews have been given by Evens[226] and Popereka.[227] Although there have been many stress measurements, most of these are inadequate, as the many variables which alter the stress have not been recorded. Some of these variables are listed below in order of discovery.

Stoney[3] referred to the fact that increasing the bath temperature reduced the stress in the film, an observation which has been confirmed more recently.[219,228–231] The stress varies with the pH of the plating solution,[232] the current density, and small amounts of inorganic additives. Stager[233] discovered that superimposing an alternating current on the direct plating current reduced the stress. This effect was later confirmed.[234,235] Vuilleumier[236] discovered that the smaller the crystallite size in the deposit, the larger the stress recorded. Kohlschütter et al.[237] also noted this effect but explained it as a fault of their apparatus. MacNaughton and Hothersall[238] discovered just the opposite effect, but they were plating from different solutions.

Marie and Thon[239] found that the stress in copper films could be compressive or tensile depending on whether the cathode was made of bright or dull platinum. Jaquet[240,241] and Fisher[224] found that organic additives had strong and varied effects on the stress. Phillips and Clifton[229] found that the stress also depended on the concentration of the plating solution.

The relationship between stress and preferred orientation has also been considered, both practically and theoretically. Bozorth[242] measured the preferred orientations in deposits using x-ray-diffraction methods, and he decided that these were due to plastic deformation of the metals where the local stresses exceeded the plastic limit. Wyllie[243] cited the work of Hume-Rothery and Wyllie[244,245] in support of this theory. Yang[246] measured the stresses and orientations in electrodeposits of a wide range of metals and showed that the Bozorth-Wyllie theory did not hold for electrodeposits Later the Bozorth-Wyllie theory was also shown to be inapplicable to vacuum-evaporated films by Murbach.[247]

Notwithstanding all these variables, Kohlschütter and Jacober[237] drew up a list of metals in descending order of stress, starting with chromium in tensile stress through palladium, nickel, cobalt, iron, silver (which could be tensile or compressive depending on the deposition conditions), to lead and zinc, which are always compressive.

A theory for the origin of stress was proposed by Wyllie.[243] The stress was considered as due to codeposited hydrogen diffusing out of the film immediately after its deposition, thus allowing it to contract and producing a tensile stress. This was in accord both with the large amounts of hydrogen released at the cathode during chromium deposition and with the effects of a superimposed alternating current which oxidized the hydrogen at the cathode and thus reduced the stress. It was also in accord with annealing experiments[248] in which gas was evolved from the deposit while the tensile stress increased at the same time. However, the formation of compressive stress and the fact that the ratio of metal deposited to hydrogen released at the cathode did not in any case accord with the stress measured were much harder to explain on this theory.

The most recent theory for the origin of stress in electrodeposited films is due to Popereka,[227] who considers the final stress to be the sum of a tensile component arising from effects within the crystallites, and a compressive component arising from effects at grain boundaries. According to this theory, as the crystallites grow, and dislocations which may be formed during the growth process will have like signs repel each other, and be attracted to the crystallite boundary. Thus a contraction in volume of the crystallite will take place, and a tensile stress will result. The compressive component arises from adsorption of impurity particles at grain boundaries. Popereka showed that sufficient energy was available to give the stresses recorded in typical electrodeposits.

(2) Chemically Deposited Films Stress in films produced by chemical reduction from solution has not yet been examined in the detail devoted to stress in electrodeposited films. However, the subject has recently become important in connection with the properties of magnetic films grown by this technique. Hendy, Richards and Simpson[249] found that Ni-Co-P alloys deposited onto glass substrates tended to crack or peel from the substrate if they were allowed to reach thicknesses in excess of 3,000 Å. The value of 1.4×10^9 dyn cm^{-2} measured by the bending-beam method in a Co-P deposit[161] is of the same order as the stress found in electrodeposited nickel by Stoney.[3] The stress may be minimized by choosing suitable preparation conditions.[138]

(3) Thermal Oxide Films It has long been known that oxide films on metals are in a state of stress. Pilling and Bedworth[250] in 1923 divided metals into two groups those in which the oxide film protected the metal from further oxidation, and those in which it allowed the oxidation process. They defined a parameter, the Pilling-Bedworth ratio, which is the ratio of the volume of the oxide to the volume of the metal from which it was formed. This ratio is greater than unity for metals such as silicon (2.04) and aluminum (1.28) in which the oxide is protecting, and it is less than unity for metals such as calcium (0.78) in which the oxide is not protecting. Thus it would be expected that oxide films on the former metals are in compressive stress and on the latter in tensile stress. It is now agreed that the ratio will not apply the metal ions diffuse through the lattice and new oxide is formed at the oxide atmosphere interface. Mechanisms for oxidation using these concepts have been reviewed by Vermilyea,[251] Evans,[226] and Tomashov.[252]

Jaccodine and Schlegel[253] have recently drawn attention to the possible detrimental effects of stress in thermally grown oxide films on silicon on the processing of silicon integrated circuits. These authors used a bending-beam method to measure the stress in films of silicon dioxide produced by the thermal oxidation of chemically and mechanically polished silicon slices at temperatures from 900 to 1200°C. They found a fair agreement between the stresses measured and the stresses to be expected from the difference in expansion coefficients of silicon and silica on cooling from the deposition temperature. The stresses were of the order of 3×10^9 dyn cm^{-2}. Thus, although the type of stress is that predicted by the Pilling-Bedworth ratio, the origin of the stress is thermal. Stresses in such silica films have also been observed by x-ray extinction topography.[191] In order to reduce the thermal stress in silicon oxide films, the temperature at which the films are grown must be reduced. Techniques such as plasma oxidation,[254] which reduces the substrate temperature to room temperature, have been applied to the oxidation of silicon.

However, not all stress in oxide films is thermal in origin. Borie et al.[184,185] have used x-ray line-broadening techniques to measure the stress in single-crystal films of copper oxide grown at 250°C. Their conclusion is that the stress is due to the lattice mismatch at the copper–copper oxide interface.

Recently, Greenbank and Harper[158] have discussed the oxidation of zirconium in detail. They have shown that the stress in the oxide film is compressive at the interface because of the volume expansion on oxidizing the zirconium to the oxide. However, as the film grows the stresses no longer remain compressive. Since the reaction proceeds by the inward diffusion of oxygen, new oxide is formed at the metal-oxide interface and the volume expansion in these layers will generate tensile stresses at the oxide surface. It is probable that this behavior is common to all oxide films in which oxygen diffuses through the oxide to the interface. Even in the initial layers, however, the stress is not as high as the Pilling-Bedworth ratio would predict, and Greenbank and Harper suggest that this is due to the high plasticity which the oxide film has on formation. Such a plasticity has been discussed for zirconium by Douglass.[255]

(4) Anodic Oxide Films The stress in anodic films assumes importance in the field of electronic engineering where such films are grown on materials like tantalum or aluminum and utilized as dielectrics. Stress in anodic films of aluminum has been measured by Bradshaw and Clarke[256] and more recently by Bradhurst and Leach.[156,257] Anodizing from ammonium borate solution, the latter authors found that the stress was not constant with film thickness but passed through a maximum at 200 Å film thickness. The stress was of the order of 5×10^8 dyn cm^{-2} and tensile, thus not agreeing with the Pilling-Bedworth ratio.* However, in the case of zirconium, Nomura et al.[157] have reported compressive stresses. As in the thermal case, this raises again the problem of whether it is the metal ions or the oxygen ions which move during anodization. The problem has been mentioned by Holland[258] and Young,[259] who point out that, in principle, the direction of the stress in the films could be used to determine this. If the metal ions moved the film would be in tensile stress, and vice versa. However, Young noted that any stress from this cause might be masked by the occurrence of stress due to misfit between the oxide and metal lattices.

Holland[258] offered no explanation for the effects commonly observed in anodic films, such as the rupture of the film at the electrolyte surface, which may be due to stress. Croft[260] has ascribed the discontinuities in the anodizing current to the relief of mechanical stresses in the film by cracking. Young[259] suggested that the anomalous behavior of the anodizing voltage, when a rough surface is anodized at constant current,[261,262] arises from the stresses set up in the film because of roughness. Asperities in the substrate are flattened out, and where the asperity has been, the surface of the oxide occupies a smaller area; hence the film is in a state of very high compressive stress. However, it has recently been shown that, at least in the case of aluminum, the oxide film is plastic under electrical fields so that the stress will be reduced[263] [cf. Sec. 2b(3)].

* The tensile stresses found in this work on aluminum were almost certainly due to the method of anodizing. Normally one would expect compressive stresses to develop.

(5) Vapor-phase Films   The stress in films of silicon vapor deposited onto foreign substrates has been found to be strong enough to crack sapphire crystal wafers;[264] it has been measured by bending-beam techniques[162,163] and by x-ray methods.[19] X-ray methods have also been applied to niobium-tin films.   In all cases, the stresses found (9×10^9 dyn cm^{-2} for Si on Al_2O_3) could be entirely explained by the thermal effect of cooling from the deposition temperature.

(6) Sputtered Films   Jackson et al.[265] have found that films of reactively sputtered zirconia wrinkled when the bond between the films and substrate was relieved by etching, suggesting that these films were in compressive stress.

Stress has also been observed in the preparation of single-crystal films by sputtering.[113]   In the case of GaAs films sputtered onto NaCl substrates at a substrate temperature of 300°C, the thermal stress on cooling to room temperature has been high enough to crack the films.[266]

Recent work on the stress of reactively sputtered tantalum films by Stuart[17] has shown that such films can be in either compression or tension.   Low partial pressures of oxygen give films in compressive stress while higher pressures give tensile stress.   It would appear that this behavior can be explained in terms of the oxidation site: for the low oxygen pressures, oxidation is mainly at the substrate with the metal species sputtered, while with high oxygen pressures the oxidation is at the target and the oxide is sputtered.[267]

Stresses in sputtered Mo films on oxidized silicon substrates have been observed by Glang et al.[268] and also by d'Heurle.[41]   Recent work by Maissel[269] has shown that under certain conditions, incorporation of the sputtering gas into the growing film will lead to intrinsic compressive stress.

(7) Evaporated Films   The internal stress in evaporated thin films raises problems in almost every industrial application.   The earliest stress observations were made on optical coatings, where too thick a coating tended to crack, cloud, or even peel from the substrate.   By measuring the stress in films of various materials and by using alternately layers with compressive and tensile stress, Turner[145] was able to increase the total thickness of coatings which could be deposited from 6 to 40 μ. Stresses in a number of optical coatings have recently been measured.[270,181]   Ennos[181] has pointed out that, although in principle the stress in the film is large enough to distort the optical component being coated, in practice this might be important for only the highest-precision optics.

Other measurements of stress in dielectric films have been made because of the application in thin film circuits as capacitor dielectrics.[271]   A small compressive stress is preferred in a dielectric,[217] and the direction of stress in a wide range of materials has been studied.[172]   The stress in evaporated organic dielectrics has also been measured.[172,272]   Hoffman[6] has given an excellent table listing the stress behavior of many dielectrics, and this list has been extended by Ennos.[181]

Stress in evaporated-metal films has been studied principally with regard to the applications of the films to computer memory systems, both magnetic and cryogenic. One of the possible origins of the uniaxial anisotropy parallel to the plane of the substrate in evaporated magnetic films[223] is an anisotropy in the stress, although in the case of single-crystal films, anisotropic stress combined with anisotropic magnetostriction would also give anisotropy of magnetic properties.[189]   The stress anisotropy in Permalloy films is less than 10% of the isotropic stress (10^9 dyn cm^{-2}).   Nevertheless, a stress anisotropy of this magnitude (10^8 dyn cm^{-2}) could account for a large part of the magnetic anisotropy observed in nickel and iron films (cf. the work on electroplated films[223,224,225]).

Stress in cryogenic films is known to affect the critical temperature for superconduction in indium,[211] tin,[212] aluminum,[273] and lead[274] films.   However, the intrinsic stress from the film deposition is relatively unimportant in this case, as cooling to the low temperatures needed for cryogenic operation leads to thermal stress in excess of the elastic limit, and plastic flow in the film takes place.[275]

A large number of stress measurements on metals have been compiled by Hoffman,[4,5,6] and he has included a table listing a range of metals on a variety of substrates in one of his reviews.[6]   Table 6 gives a bibliography of the materials that have been measured.

TABLE 6 Summary of Stress Measurements on Evaporated Films

Material	Substrate	Reference	Material	Substrate	Reference
Ag	Collodion	154	LiF	Glass	172, 278, 145, 179, 140, 217, 141
	Copper	276, 146			
	Glass	172, 119, 177			
	Mica	154, 150, 153		Mica	270
	Silica	154	Mg	Copper	146
AgCl	Glass	145	MgF$_2$	Glass	172, 145, 278, 283
AgF	Glass	145			
AgI	Glass	145		Mica	280, 270, 153
Al	Cellulose	277		Silica	181
	Copper	146	MgPh*	Glass	172
	Silica	181	Mo	Oxidized silicon	131,145
AlPh*	Glass	172			
AlF$_3$	Glass	145	MoO$_3$	Glass	145
Au	Cellulose	277	NaBr	Glass	179
	Copper	146	NaCl	Glass	179
	Glass	119, 177	NaF	Glass	179, 278
	Silica	218		Copper	146
B$_2$O$_3$	Glass	172	Ni	Glass	128, 284, 189, 125, 170, 281
BaF$_2$	Glass	145			
BaO	Glass	172		NaCl	190
Bi	Copper	146		Mica	147, 214, 148
C	Glass	140	Permalloy	Glass	128, 140, 126, 208, 127
Co	Glass	147			
CaF$_2$	Glass	172, 145		Mica	149
	Mica	270	Pd	Copper	146
	Silica	181	Pb	Nickel	278
CdS	Glass	172	PbCl$_2$	Glass	172, 145
CoTe	Silica	181		Silica	181
CeF$_3$	Glass	172	PbF$_2$	Glass	172
	Silica	181	PbTe	Glass	119, 285
Ce$_2$O$_3$	Glass	172		Mica	119, 285
Chiolite	Glass	145	RbI	Glass	179
	Silica	181	Sb	Copper	146, 286
Cryolite	Glass	145, 270, 278		Glass	176
	Silica	181	Sb$_2$O$_3$	Glass	145
Cr	Silica	181	Sb$_2$S$_3$	Glass	145
	Cellulose	199	SrSO$_4$	Glass	145
Cu	Copper	279, 146	Sn	Glass	212
	Mica	167, 147	SnO$_2$	Glass	145
CuI$_2$	Glass	145	SiO	Glass	172, 287, 150, 288, 141, 289, 123, 275, 169, 174, 290
	Copper	146			
Fe	Glass	124, 280, 281			
	Mica	282			
	Silica	280		Nickel	168
Ga	Al foil	152		Silica	181
Ge	Silica	181	Te	Silica	181
In	Silica	211	TlCe	Silica	181
KBr	Glass	179	TlI	Silica	181
KCl	Glass	179	ThOF	Silica	181
KF	Glass	179	Zn	Copper	146
KI	Glass	179	ZnS	Glass	172, 145
LiF	Carbon on glass	141		Mica	270, 153
	Cellulose	135		Silica	181

* Ph = phthalocyanine.

To obtain intrinsic stress values it is of course necessary to eliminate the thermal component. If the stress measurement is made at a temperature other than the deposition temperature, the thermal stress can be calculated and subtracted provided values for the expansion coefficients of film and substrate are available. If these are not known—and generally the expansion coefficient of the film will need to be measured—then one may have to measure the stress at the temperature of deposition. Even with this precaution it is necessary to avoid the thermal transients that are observed during deposition. These are best illustrated by the traces shown in Fig. 18. This trace is the output from the inductance bridge in the inductive method of measuring beam deflection.[140] Each vertical division on the chart corresponds to an end movement of 5,000 Å and each horizontal division is 1 s. An initial transient is observed at the beginning of the deposition on opening the shutter, and a similar one on closing. This can be attributed to the establishment of a temperature gradient across the substrate due to the combined effect of the radiant heat from the source

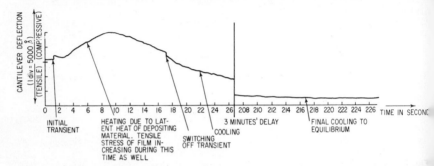

Fig. 18 Output from inductance bridge during growth of LiF on glass showing transients. (Method used is that of Fig. 15g.)

and the latent heat of the evaporant. This explanation has been confirmed by Klokholm.[132] The magnitude of the deflection will depend on the rate of arrival of the deposit. Calculation has shown[140] that for the case of LiF deposited on 0.15-mm-thick glass slides at a rate of 62 Å s^{-1}, the temperature differences across the substrate will be 0.034°C after 1 s.

A second feature in the illustration is that, after the initial transient, the deflection increases with time. This increase is due to the differences in thermal expansion of film and substrate and to nonuniformity in the expansion behavior of the substrate itself. Eventually the deflection due to this cause is masked (at least in the case of an intrinsic tensile stress) by the intrinsic stress developing in the film. Calculation has shown[140] that the temperature of this film-substrate combination (LiF deposited on 0.15-mm-thick glass slides at a rate of 62 Å s^{-1}) will rise by 6°C in 10 s and will reach a final equilibrium temperature of 47°C above the surroundings if evaporation is continued indefinitely. After the shutter is closed, some time is required for such a temperature difference to disappear, and it is only after equilibrium has been obtained that it is possible to measure the final stress value. In the illustration the final stress was tensile after 3 min, a typical delay time that has been employed. In work with the capacitive detection system, times of 10 min were used.

It is difficult to summarize all the stress data that have been collected on evaporated films into a neat picture. However, one or two main guidelines to explain behavior may be recognized:

(a) *Impure Structures.* With films that absorb gas and other impurities into their structures as they grow, compressive stresses may be expected. This behavior is analogous to that shown by thermally grown layers. SiO provides a useful illustration. In Fig. 19, the stress in SiO films is plotted as a function of residual water vapor in the evaporation system;[123] the greater the number of impurities, in this case water molecules, the higher the tendency for the film to be in compressive stress

Similar compressive stresses have been found in Al deposited in poor vacuum conditions[146] and for copper.[199]

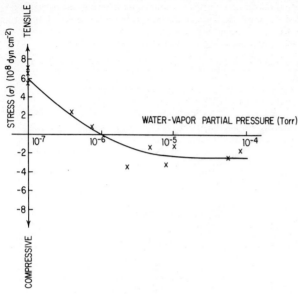

Fig. 19 Stress as a function of residual water-vapor pressure during deposition of SiO on glass. Boat temperature 1183°C. (*After Priest et al.*[123])

(*b*) *Pure Structures.* In this case, the films grow with a tensile stress. The value of the stress depends on the thickness of the film, and Fig. 20 illustrates typical behavior[177] (Ag on glass). The stress reaches a maximum at a stage when the last holes in the island stage of the film are filling up. Similar behavior has been established for films of various alkali halides,[140] for silver on mica,[154] and for gold on glass.[160] In the case of the alkali halides on glass, it is found that the substrate modifies the behavior,[179] and this may often be the case for other materials and substrates as well.

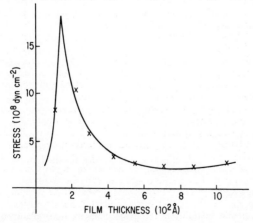

Fig. 20 Stress as a function of film thickness, Ag on glass. (*After Wilcock.*[119])

A large number of explanations have been proposed for the causes of intrinsic stress in pure films and have been reviewed by various authors.[6,118,119,177,179] Table 7 summarizes the situation and gives references to the relevant literature.

Of all these models, the grain-boundary model appears to satisfy much of the experimental data, provided microcrystalline boundaries are found to exist in the islands of the film during the initial stages of growth. The behavior may, however, be complicated in specific cases by interfacial or density effects.

TABLE 7   Summary of the Various Models Used to Explain Stress in Evaporated Films

Model	Features	References
Thermal	Top layers, during deposition, at high temperature. Cooling from recrystallization temperature causes stress[146] Only predicts tensile stress. Main criticism is that atoms lose energy on arrival very quickly (10^{-6} s),[293] hence sufficiently high temperatures are not obtained	146, 147, 291, 292, 167, 293, 140, 294
Imperfection	Imperfections migrate out of film as it grows. Vacancies are the main imperfections.[147] Criticism is that numbers of vacancies required to migrate out are too high to be in equilibrium at normal temperatures[295]	147, 295
Density	Change in density as film goes from liquid to solid. Useful for certain materials (e.g., Ga[152]), but stress not always as predicted.[176] Even if sign correct, magnitude sometimes wrong[179]	296, 297, 152, 176, 179
Grain boundary	Crystalline boundaries move out of film. In island-structure films, these can be the microcrystalline boundaries.[298] Some success in explaining behavior of silver[154,177]	124, 295, 298, 119, 177, 154
Electrostatic	Attraction or repulsion between islands.[299] Applicable to island-structure films on insulating substrates.[177] Magnitude wrong	299, 300, 177, 88, 89
Surface tension	Change in atomic spacing at surface of film equivalent to a surface-tension effect. Many discussions on the reality of surface effects,[201,302–304] but theoretical predictions not in very good agreement with experiment[177]	199, 200, 201, 301, 302, 303, 304, 118, 177, 298, 119
Interfacial	Substrate influence.[306,309] Found to apply in certain cases (e.g., PbS on PbSe[307] and alkali halides on glass[179])	305, 306, 179, 177, 307, 308, 309

(8) Summary   Intrinsic stresses develop in films deposited by all methods. No single model can explain all the results. It appears that in films deposited by electroplating or evaporation, grain-boundary movements during the growth of the film can give the tensile stresses observed. This behavior may be modified by the effect of interfacial misfit or density changes during growth. In "impure" films, in which oxygen and other materials are incorporated into the structure during growth, the resultant stresses are usually sufficiently high to overshadow any of the "pure film" effects. The stresses are then usually compressive, at least in the initial stages of growth and provided that the reacting species migrate through the growing film to the interface. Compressive stresses are observed in thermally grown, anodized, and certain sputtered and evaporated films. If the ion migration takes place from the interface to the film surface, tensile stresses will be observed. In films grown by vapor-phase methods, both effects are generally masked by thermal effects due to the high temperature of preparation.

3. TENSILE PROPERTIES

As mentioned in the introduction to this chapter, several authoritative reviews have already been written on the mechanical properties of films, and these have included data on the tensile properties. The reader is again referred to these,[4,5] especially to the review by Hoffman in "Physics of Thin Films," vol. 3.[6] In this section, we will therefore attempt only to summarize the available information.

The most important feature that is found when one examines the tensile properties is that the films are stronger than bulk material. Several models have been developed to explain this, and these will be discussed in the appropriate contexts.

a. Methods of Measurement

Various types of apparatus have been used for examining the mechanical properties of films, and some of these have already been considered in the section on stress. Hoffman[6] has summarized the various methods in a useful table. A brief review is given here.

(1) Microtensile Apparatus Different machines have been built for measuring tensile properties by observations of the elongations produced by small loads. These have varied from Marsh's apparatus,[286] in which a null-torsion balance is used to apply the forces and the movement is measured optically, to Neugebauer's equipment,[310] in which the strain is applied by means of an electromagnet and the elongation is observed directly in a microscope. In another apparatus due to Blakely,[311] the applied force is again obtained electromagnetically, but the strain is sensed by an interferometric method.

The problem in all these methods is that of separating the film from its substrate without damaging it and then mounting it so as to obtain a uniform load without having the film tear at the edges. Some workers have glued the films on; others have used mechanical grips. Many experimental results have given low-strength figures because of either the rough handling, poor mounting, or large defects in the films themselves. Automatic apparatus for measuring stress-strain curves has been reported.[312]

(2) Bulge Techniques To eliminate the mounting problem, Beams[46] has developed a method for measuring stress-strain characteristics. The method has already been described in the section on stress-measuring techniques. Stress-strain curves may be obtained by applying pneumatic pressure to one side of the film. Equation (20) gives the relationship between pressure, deflection of the center of the bulge, and the stress in the film.

(3) Centrifugal Methods The use of a spinning rotor for measuring adhesion of films deposited on the rotor was devised by Beams[45,46] and has already been discussed. Accurately known stress can be applied to the film by spinning the rotor until the film comes off. The equation relating rotational speed to strength has already been given [see Sec. 2a(4)]. Such an apparatus is useful for measuring the ultimate strength of the film, although the method is applicable only to films with low adhesion so that they are removed by spinning.

(4) X-ray and Electron-diffraction Techniques Changes in lattice spacing can be measured by these techniques, and therefore stress-strain characteristics can be measured (e.g., Refs. 129, 188, 189, and 192). These techniques have already been considered in relationship to intrinsic-stress measurements.

(5) Microhardness Apparatus Microhardness is, in principle, measured by pressing a hard, specially shaped point into the surface and observing the resultant indentation. Read[313] has reviewed the available methods in some detail. Two basic point shapes are used. The Vickers microhardness tester employs a diamond pyramid of four faces with edges of equal length; the Knoop instrument has a pyramid in which the length of two edges is seven times that of the other two. This latter pyramid goes into the work only a third of the distance of the equivalent Vickers pyramid under the same load conditions. The basic trouble with these instruments is that it is not possible to measure very thin films. Angus[314] has claimed that, with a Vickers indenter and a 1 gr wt load, electroplated gold films may be measured provided they are at least 5×10^4 Å thick. This limit is set because the thickness of the sample must not be less than ten times the depth of penetration of the diamond.[314] Palatnik and coworkers[315,316] have used these techniques on copper films.

A similar technique has been used by Raffalovich[317] to measure the hardness of dielectric films. Angus showed[318] that the resistivity of thin films could be affected by loaded probes on the surface, and Raffalovich extended this work to dielectrics. In his apparatus, a hard metal ball (usually of brass, 3.17 mm in diameter) was pressed into the surface of a dielectric film which had previously been deposited on a metal. The force (measured in gr wt) required to reduce the insulation resistance through the film to a very low value—i.e., to cause a short to the base metal—was then taken as a measure of the hardness. Anodic oxide films on tantalum, niobium, and titanium were measured in this way, as well as silicon monoxide on copper and on aluminum. In all cases a weight of around 500 gr wt was required to break through the dielectric.

An ultrasonic technique has been developed by Branton Instrument, Inc.[319] It consists of measuring the frequency of a vibrating rod with one end pointed and applied lightly to the specimen. The other end is clamped. There is a rise in frequency of the natural-length mode vibrations of the rod when the end is applied to the specimen, and the greater the area of the tip of the rod in contact with the specimen, the greater the frequency rise. The area of the rod in contact with the surface is proportional to the hardness of the surface, provided that the force pressing the rod against the surface is constant. The method suffers from the same thickness limitations as the indenter systems inasmuch as very light loads must be applied for thin film measurements. Nevertheless, such a technique may well be developed in the future specifically for thin film use.

Abrasion methods of measuring hardness have already been discussed in the context of adhesion. The complication is usually that adhesion is important to the interpretation of the results.[13,43,44]

b. Results

(1) Elastic Behavior Apart from the general reviews of Hoffman,[4-6] the tensile properties of films have been summarized by Menter and Pashley[320] and Neugebauer.[321] The data presented are not very consistent. A typical stress-strain curve is illustrated in Fig. 21. The initial curve lies above subsequent ones, showing that the initial straining has altered the film in a permanent way. If the film is stretched beyond the elastic limit, plastic deformation will occur, and this is illustrated in Fig. 22. Even below the elastic limit, there is evidence for creep (see Fig. 23) and it is claimed by Neugebauer,[310] for the case of gold films, that the amount depends on load, dimensions, and amount of prestraining. Pashley[204] has identified the slow

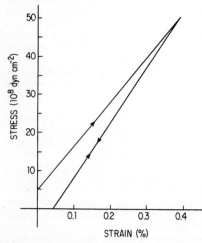

Fig. 21 Stress-strain curve for elastic deformation. Electroplated nickel films. Bulge measurement. (*After Jovanovic and Smith.*[326])

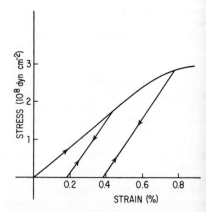

Fig. 22 Stress-strain curve showing plastic deformation at high strain. Gold films. Microtensile measurement. (*After Neugebauer.*[310])

motion of screw dislocations in single-crystal gold films of 200 Å thickness as the responsible mechanism. Without going into detail, it appears that the elastic properties of film are in general very similar to those of bulk materials.

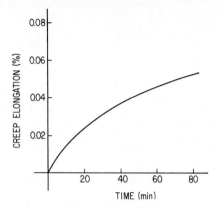

Fig. 23 Creep vs. time for evaporated silver film observed at stresses below the elastic limit. (*After Beams et al.*[46])

(2) Strength Even though a film may undergo creep and plastic deformation, it is found that very high strengths (i.e., yield stress) are obtained in thin films with values up to 200 times as large as those found in the corresponding bulk material. This strength appears to arise partly from the restraint which the film surface exerts on the motion of the large number of grown-in dislocations and also from the prevention of the operation of any dislocation sources.

Machlin[322] has discussed the effect of the surface on strength for bulk materials and has recognized four effects. These have been summarized by Hoffman[6] as follows: (1) pile-up of dislocations at the surface as the surface forms a barrier to their motion; (2) surface drag effects resulting from the formation of a new surface due to the motion of a screw dislocation which intersects the surface; (3) surface anchoring effects due to surface irregularities (etch pits, growth steps, etc.); and (4) lattice-parameter changes at the surface with resultant changes in mechanical properties. Hoffman considers that none of these causes can explain the large increases in strength observed. It is suggested that the strength of films may arise from their relatively perfect surfaces. The energy to nucleate a dislocation at a perfect surface is high,[323] so that the process will not occur. Figure 24 shows a typical strength vs. thickness

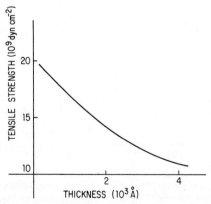

Fig. 24 Fracture strength vs. film thickness. Evaporated nickel films. (*After D'Antonio et al.*[327])

curve obtained for nickel, and it can be seen that the strength increases as the surface becomes more important. Experimentally there is little difference between single-crystal and polycrystalline films, indicating that grain-size effects are not important.

An analogy between yield stress and electrical resistivity has been drawn by some workers.[6] The yield stress σ^*_{film} is considered to be a composite quantity, viz.,

$$\sigma^*_{film} = \sigma_{bulk} + \sigma_{imperfection} + \sigma_{thickness} \tag{28}$$

Following the type of analysis that has been undertaken on resistive films,[324] one can obtain an expression for the yield stress as a function of thickness:

$$\sigma^*_{film} = \sigma_{bulk}\left(1 + \frac{K}{d_f}\right) \tag{29}$$

where K is a constant and d_f is the film thickness. D'Antonio and Tarshis (see Hoffman[6]) have found good agreement with Eq. (29) for 750- to 4,000-Å nickel films. Another analysis[6] has yielded a relationship of the form

$$\sigma^*_{film} = \sigma_{bulk} + \frac{K'}{d_f^{\frac{1}{2}}} \tag{30}$$

where K' is another constant. Grunes et al.[325] have reported that Eq. (30) is more appropriate.

(3) Fracture At very high strains a film will fracture. A very useful review of fracture in single-crystal gold films has been given by Menter and Pashley.[320] It would appear that the mechanism of final fracture in these films is one of localized plastic deformation with a resultant thinning of the film. This gives a rise in the stress level, small cracks appear, and when they join up the film fails. It must be emphasized that the dislocations responsible for this behavior are not those grown into the film but those nucleated in the heavily strained regions of the film.

(4) Hardness Very little is known about the hardness of thin films, mainly because of the thickness limitation on measurement. Palatnik[315,316] has shown that for copper films the hardness is a function of temperature and decreases as the temperature is increased.

ACKNOWLEDGMENTS

I am extremely grateful to a large number of colleagues for their help and criticism in preparing this chapter. It is particularly important to single out for thanks Dr. J. D. Wilcock and Mr. R. Carpenter, who have allowed me to use so much of their material on stress, and B. N. Chapman for the use of his unpublished material on adhesion. I would also like to thank Prof. R. W. Hoffman for the many stimulating discussions I have had with him over the years and to acknowledge, once again, my indebtedness to his reviews on mechanical properties of thin films.

Finally I am grateful to the Plessey Co. for permission to write this chapter.

REFERENCES

1. Mills, E. J., *Proc. Roy. Soc. London*, **26**, 504 (1877).
2. Bouty, M., *Compt. Rend.*, **88**, 714 (1879).
3. Stoney, G. C., *Proc. Roy. Soc. London*, **A32**, 172 (1909).
4. Hoffman, R. W., in "Thin Films," p. 99, American Society for Metals, Metals Park, Ohio, 1963.
5. Hoffman, R. W., in J. C. Anderson (ed.), "The Use of Thin Films in Physical Investigations," p. 261, Academic Press Inc., New York, 1966.
6. Hoffman, R. W., in "Physics of Thin Films," vol. 3, p. 211, Academic Press Inc. New York, 1966.
7. Bassett, G. A., J. W. Menter, and D. W. Pashley, in C. A. Neugebauer, J. B. Newkirk, and D. A. Vermilyea (eds.), "The Structures and Properties of Thin Films," John Wiley & Sons, Inc., New York, 1959.
8. Thun, R. E., in G. Hass (ed.), "Physics of Thin Films," vol. 1, p. 187, Academic Press Inc., New York, 1963.

9. Pashley, D. W., in "Thin Films," p. 59, American Society of Metals, Metals Park, Ohio, 1964.
10. Pashley, D. W., *Advan. Phys.*, **14**, 327 (1965).
11. Behrndt, K. H., in "Thin Films," p. 1, American Society for Metals, Metals Park, Ohio, 1964.
12. Campbell, D. S., in J. Anderson (ed.), "The Use of Thin Films in Physical Investigations," p. 11, Academic Press Inc., New York, 1966.
13. Holland, L., "Vacuum Deposition of Thin Films," pp. 98–103, Chapman & Hall, Ltd., London, 1956.
14. Strong, J., *Rev. Sci. Instr.*, **6**, 97 (1945).
15. Williams, R. C., and R. C. Backus, *J. Appl. Phys.*, **20**, 98 (1949).
16. Bateson, S., *Vacuum*, **2**, 365 (1952).
17. Chapman, B. N., "The Adhesion of Thin Films," Ph.D. Thesis, Electrical Engineering Department, Imperial College, London University, London, 1969.
18. Weaver, C., *Proc. 1st Inst. Conf. Vacuum Tech.*, 1958, p. 734, Pergamon Press, London.
19. Benjamin, P., and C. Weaver, *Proc. Roy. Soc. London*, **254A**, 163 (1960).
20. Chapman, B. N., in: "Aspects of Adhesion," *Proc. 7th Conf. on Adhesion and Adhesives*, 1969, D. J. Alner (ed.) to be published.
21. Belser, R. B., and W. H. Hicklin, *Rev. Sci. Instr.*, **27**, 293 (1956).
22. Holland, L., *Brit. J. Appl. Phys.*, **9**, 410 (1958).
23. Hagstrum, H. D., and C. D'Amico, *J. Appl. Phys.*, **31**, 715 (1960).
24. Maissel, L. I., in G. Hass and R. E. Thun (eds.), "Physics of Thin Films," vol. 3, p. 61, 1966.
25. Frederick J. R., and K. C. Ludema, *J. Appl. Phys.*, **35**, 256 (1964).
26. Heavens, O. S., *J. Phys. Radium*, **11**, 355 (1950).
27. Benjamin, P., and C. Weaver, *Proc. Roy. Soc. London*, **A252**, 418 (1959).
28. Weaver, C., and R. M. Hill, *Phil. Mag.*, **3**, 1402 (1958).
29. Weaver, C., and R. M. Hill, *Phil. Mag.*, **4**, 253 (1959).
30. Weaver, C., and R. M. Hill, *Phil. Mag.*, **4**, 1107 (1959).
31. Ham, J. L., *Am. Soc. Lubrication Trans.*, **6**, 20 (1963).
32. Keller, D. V., *Wear*, **6**, 353 (1963).
33. Benjamin, P., and C. Weaver, *Proc. Roy. Soc. London*, **254A**, 177 (1960).
34. Benjamin, P., and C. Weaver, *Proc. Roy. Soc. London*, **261A**, 516 (1961).
35. Borom, M. P., J. A. Pask, *J. Am. Ceram. Soc.*, **49**, 1 (1966).
36. King, B. W., H. P. Tripp, and W. H. Duckworth, *J. Am. Ceram. Soc.*, **42**, 504 (1959).
37. Benjamin, P., and C. Weaver, *Proc. Roy. Soc. London*, **274A**, 267 (1963).
38. Karnowsky, M. M., and W. B. Estill, *Rev. Sci. Instr.*, **35**, 1324 (1964).
39. Mattox, D. M., *J. Appl. Phys.*, **37**, 3613 (1966).
40. Maissel, L. I., and P. M. Schaible, *J. Appl. Phys.*, **36**, 237 (1965).
41. d'Heurle, F. M., *Trans. Met. Soc. AIME*, **236**, 321 (1966).
42. Moore, D. G., and H. R. Thornton, *J. Res. NBS*, **62**, 127 (1959).
43. Townsley, M. G., *Rev. Sci. Instr.*, **16**, 143 (1945).
44. Schossberger, F., and K. D. Franson, *Vacuum*, **9**, 28 (1959).
45. Beams, J. W., J. L. Young, and J. W. Moore, *J. Appl. Phys.*, **17**, 886 (1946).
46. Beams, J. W., J. B. Freazeale, and W. L. Bart, *Phys. Rev.*, **100**, 1657 (1955).
47. Beams, J. W., in C. A. Neugebauer, J. B. Newkirk, and D. A. Vermilyea (eds.), "Structure and Properties of Thin Films," p. 183, John Wiley & Sons, Inc., New York, 1959.
48. Moses, S. T., and R. K. Witt, *Ind. Eng. Chem.*, **41**, 2334 (1949).
49. Davies, D., *Fulmer Res. Inst. Res. Rept.* P469, London, November, 1964.
50. Brook, G. B., and G. H. Stott, *J. Inst. Metals*, **41**, 73 (1929).
51. Chatfield, H. W., *Paint*, **17**, 295 (1947).
52. Wlodek, S. T., and J. Wulff, *J. Electrochem. Soc.*, **107**, 565 (1960).
53. Benilla, C. F., *Trans. Electrochem. Soc.*, **71**, 263 (1937).
54. Hothersall, A. W., and C. J. Learlveater, *J. Electrodep. Tech. Soc.*, **19**, 49 (1944).
55. Loose, W. S., *Trans. Electrochem. Soc.*, **81**, 215 (1942).
56. Brenner, A., and V. D. Morgan, *Proc. Am. Electroplate Soc.*, **37**, 51 (1950).
57. *Chem. Ind.*, London, **54**, 82 (1944).
58. Piersol, R. J., *Met. Cleaning Finish*, **5**, 516 (1933).
59. Martin, R. C., *Prod. Finishing London*, **2**, 7 (1938).
60. Brown, R. J., *Met. Ind.*, **98**, 167 (1961).
61. Wilson, D. V., and R. D. Butler, *J. Inst. Metals*, **90**, 374 (1961–1962).
62. Ollard, E. A., *Trans. Faraday Soc.*, **21**, 81 (1925).
63. Williams, C., and R. A. F. Hammond, *Trans. Inst. Metal Finishing*, **31**, 124 (1954).
64. McIntyre, J. D., and A. F. McMillan, *Canad. J. Technol.*, **36** (3), 212 (1956).
65. Clark, S. G., *J. Electrodep. Tech. Soc.*, **12**, 26 (1936).

66. Buellet, T. R., and J. L. Prosser, *Trans. Inst. Finishing*, **41**, 112 (1964).
67. Jacquet, P. A., *Trans. Electrochem. Soc.*, **66**, 393 (1934).
68. May, W. D., W. D. P. Smith, and C. O. Snow, *Nature*, **179**, 494 (1957).
69. Malloy, A. M., W. Soller, and A. G. Roberts, *Paint, Oil & Chem. Rev.*, **116**, 14 (1953).
70. Alter, H., and W. Soller, *Ind. Eng. Chem.*, **50**, 922 (1958).
71. Hoffman, E., and O. Georgonssis, *J. Oil & Color Assoc.*, **42**, 267 (1959).
72. Dannenber, H., *J. Appl. Poly. Sci.*, **36**, 125 (1961).
73. Joyce, B. A., R. R. Bradley, and C. R. Booker, *Phil. Mag.*, **15**, 1167 (1967).
74. Hirth, J. P., and G. M. Pound, "Condensation and Evaporation," Pergamon Press, New York, 1963.
75. Hirth, J. P., S. J. Hruska, and G. M. Pound, in M. H. Francombe and H. Sato (eds.), "Single Crystal Films," Pergamon Press, New York, 1964.
76. Walton, D., *J. Chem. Phys.*, **37**, 2182 (1962).
77. Rhodin, T. N., and D. Walton, in M. H. Francombe and H. Sato (eds.), "Single Crystal Films," p. 31, Pergamon Press, New York, 1964.
78. Lewis, B., *Thin Solid Films*, **1**, 85 (1967).
79. Frenkel, J., *Z. Physik*, **26**, 117 (1926) (translated and issued by TPA 3, Tech. Inf. Bureau, Ministry of Supply TB/T3952, 1962).
80. Walton, D., T. N. Rhodin, and R. W. Rollins, *J. Chem. Phys.*, **38**, 2698 (1963).
81. Lewis, B., and D. S. Campbell, *J. Vacuum Sci. Technol.*, **4**, 209 (1967).
82. Beavitt, A. R., *J. Sci. Instr.*, **43**, 182 (1966).
83. Stirland, D. J., and D. S. Campbell, *J. Vacuum Sci. Technol.*, **3**, 258 (1966).
84. Cockcroft, J. D., *Proc. Roy. Soc. London*, **119A**, 293 (1928).
85. Rhodin, T. N., *Discussions Faraday Soc.*, **5**, 215 (1949).
86. MacRae, A. V., in J. C. Anderson (ed.), "The Use of Thin Films in Physical Investigations," Academic Press Inc., New York, 1966.
87. Flanagan, T. P., J. A. Bennett, *Intern. J. Appl. Radiation Isotopes*, **13**, 19 (1962).
88. Chopra, K. L., and L. C. Bobb, in M. H. Francome and H. Sato (eds.), "Single Crystal Films," p. 373, Pergamon Press, New York, 1964.
89. Chopra, K. L., *J. Appl. Phys.*, **37**, 2249 (1966).
90. Ehrlich, G., *Advan. Catalysis*, vol. 14, 1963.
91. Gorbatyi, H. A., and E. M. Ryabchenko, *Soviet Phys. Solid State*, **7**, 921 (1965).
92. Ehrlich, G., in C. A. Neugebauer, J. B. Newkirk, and D. A. Vermilyea (eds.), "Structure and Properties of Thin Films," p. 423, John Wiley & Sons, Inc., New York, 1959.
93. Zinsmeister, G., in R. Niedermayer and H. Mayer (eds.), "Basic Problems in Thin Film Physics," p. 33, Vandenhoeck and Ruprecht, Goettingen, 1966.
94. Zinsmeister, G., *Ext. Abstr. 13th Natl. Vacuum Symp. AVS*, 1966, p. 3.
95. Fuchs, N., *Phys. Z. Sowjetun.*, **4**, 481 (1933).
96. Yang, L., C. E. Birchenall, G. M. Pound, and M. T. Simnad, *Acta Met.*, **2**, 462 (1954).
97. Hughes, F. L., *Phys. Rev.*, **113**, 1036 (1959).
98. Hughes, F. L., and H. Levinstein, *Phys. Rev.*, **113**, 1029 (1959).
99. Hudson, J. B., and J. S. Sandejas, *J. Vacuum Sci. Technol.*, **4**, 230 (1967).
100. Calvet, E., and H. Prat, "Recent Advances in Microcalorimetry," Pergamon Press, New York, 1963.
101. Calvet, E., and H. Prat, "Microcalorimetrie," Masson et Cie, Paris, 1956.
102. Smith, M. J. A., private communication, 1967.
103. Keesom, W. H., *Phys. Z.*, **22**, 129 (1921).
104. Debye, P., *Phys. Z.*, **21**, 178 (1920).
105. London, F., *Z. Phys. Chem.*, **B11**, 222 (1930).
106. London, F., *Trans. Faraday Soc.*, **33**, 8 (1937).
107. Lenel, F. V., *Z. Phys. Chem.*, **B23**, 379 (1933).
108. Orr, W. J. C., *Trans. Faraday Soc.*, **35**, 1247 (1939).
109. Zachariasen, W. H., *J. Am. Chem. Soc.*, **54**, 3841 (1932).
110. Gaffee, D. I., in L. Holland (ed.), "Thin Film Microelectronics," p. 276, Chapman & Hall, Ltd., London, 1965.
111. Stirland, D. J., *Phil. Mag.*, **13**, 1181 (1966).
112. Mattox, D. M., and J. E. McDonald, *J. Appl. Phys.*, **34**, 2493 (1963).
113. Layton, C. K., and D. S. Campbell, *J. Mater. Sci.*, **1**, 367 (1966).
114. Deryagin, B. V., *Res.*, **8**, 70 (1955).
115. Deryagin, B. V., *Proc. 2d Intern. Conf. Surface Activity*, 1957, p. 595, Butterworth & Co. (Publishers), Ltd., London.
116. Skinner, S. K., R. L. Savage, and J. E. Rutzler, *J. Appl. Phys.*, **24**, 438 (1953).
117. Weyl, W. A., *ASTM Proc.*, **46**, 1506 (1946).
118. Campbell, D. S., in R. Niedermayer and H. Mayer (eds.), "Basic Problems in Thin Film Physics," p. 223, Vandenhoeck and Ruprecht, Goettingen, 1966.

119. Wilcock, J. D., "Stress in Thin Films," Ph.D. Thesis, Electrical Engineering Department, Imperial College, London University, London, June, 1967.
120. Yelon, A., and O. Voegeli, in M. H. Francombe and H. Sato (eds.), "Single Crystal Films," p. 321, Pergamon Press, New York, 1964.
121. Carpenter, R., and D. S. Campbell, private communication, 1964.
122. Girard, R., J. Appl. Phys., **38**, 1423 (1967).
123. Priest, J., H. L. Caswell, and Y. Budo, Vacuum, **12**, 301 (1962).
124. Finegan, J. D., and R. W. Hoffman, Trans. 8th Natl. Vacuum Symp., 1961, p. 935, Pergamon Press, New York.
125. Finegan, J. D., and R. W. Hoffman, J. Appl. Phys., **30**, 597 (1959).
126. Cohen, M. S., E. E. Huber, G. P. Weiss, and D. O. Smith, J. Appl. Phys., **31**, 291S (1960).
127. Huber, E. E., and D. O. Smith, J. Appl. Phys., **30**, 267S (1959).
128. Prutton, M., Nature, **193**, 565 (1962).
129. Kinbara, A., and H. Haraki, Japan. J. Appl. Phys., **4**, 243 (1965).
130. Glang, R., R. A. Holmwood, and R. L. Rosenfeld, Rev. Sci. Instr., **36**, 7 (1965).
131. Holmwood, R. A., and R. Glang, J. Electrochem. Soc., **112**, 827 (1965).
132. Klokholm, E., IBM Res. Rept. RC 1352, February, 1965.
133. Taloni, A., and D. Haneman, Surface Sci., **8**, 323 (1967).
134. Magill, P. J., and T. Young, J. Vacuum Sci. Technol., 1967, p. 47.
135. Catlin, A., and W. P. Walker, J. Appl. Phys., **31**, 2135 (1960).
136. Kushner, J. B., Proc. Am. Electroplaters Soc., **41**, 188 (1954).
137. Read, H. J., and T. J. Whalen, Proc. Am. Electroplaters Soc., **46**, 318 (1959).
138. Graham, A. H., R. W. Lindsay, and H. J. Read, J. Electrochem. Soc., **112**, 401 (1965).
139. Papirno, R., J. Appl. Phys., **32**, 1175 (1962).
140. Blackburn, H., and D. S. Campbell, Phil. Mag., **8**, 923 (1963).
141. Campbell, D. S., Trans. 9th Natl. Vacuum Symp., 1962, p. 29, The Macmillan Company, New York.
142. Jones, R. V., Bull. Inst. Phys., **18**, 325 (1967).
143. Reason, R. E., Symposium on the Properties of Metallic Surfaces, Institute of Metals Monograph 13, p. 327, 1953.
144. Campbell, D. S., and H. Blackburn, Trans. 7th Natl. Vacuum Symp., 1960, p. 31, Pergamon Press, New York.
145. Turner, A. F., Thick Thin Films, Bausch & Lomb Tech. Rept. 1951.
146. Murbach, H. P., and H. Wilman, Proc. Phys. Soc. London, **B66**, 905 (1953).
147. Hoffman, R. W., R. D. Daniels, and E. C. Crittenden, Proc. Phys. Soc. London, **B67**, 497 (1954).
148. Hoffman, R. W., F. J. Anders, and E. C. Crittenden, J. Appl. Phys., **24**, 231 (1953).
149. Weiss, G. P., and D. O. Smith, J. Appl. Phys., **33**, 1166S (1962).
150. Novice, M. A., Brit. J. Appl. Phys., **13**, 561 (1962).
151. Kinosita, K., H. Kondo, and I. Sawamura, J. Phys. Soc. Japan, **15**, 942 (1960).
152. Gunther, G., and W. Buckel, in R. Niedermayer and H. Mayer (eds.), "Basic Problems in Thin Film Physics," p. 231, Vandenhoeck and Ruprecht, Goettingen, 1966.
153. Kinosita, K., K. Nakamizo, K. Maki, K. Onuki, and K. Takenchi, Japan. J. Appl. Phys., **4**(suppl. 1), 340 (1965).
154. Kinosita, K., K. Maki, K. Nakamizo, and K. Takenchi, Japan. J. Appl. Phys., **6**, 42 (1967).
155. Hammond, R. A. F., Trans. Inst. Metal Finishing, **30**, 140 (1954).
156. Bradhurst, D. H., and J. S. Llewelyn Leach, J. Electrochem. Soc., **113**, 1245 (1966).
157. Nomura, S., C. Akutsu, and I. Saruyama, Kenki Kagaku, **33**, 725 (1965) (in Japanese).
158. Greenbank, J. C., and S. Harper, Electrochem. Tech., **4**, 88 (1966).
159. Dankov, P. D., and P. V. Churaev, Dokl. Akad. Nauk SSSR, **73**, 1221 (1950).
160. Tylecote, R. F., in "Processus de nucleation dans les reactions des gaz sur les metaux et problemes connexes," p. 241, Centre National de la Recherche Scientifique, Paris, 1965.
161. Judge, J. S., J. R. Morrison, D. E. Spelotis, and G. Bate, J. Electrochem. Soc., **112**, 681 (1965).
162. Ang, C. Y., and H. M. Manasevit, Solid-State Electron., **8**, 994 (1965).
163. Dumin, D. J., J. Appl. Phys., **36**, 2700 (1965).
164. Brenner, A., and J. Senderoff, J. Res. NBS, **42**, 89 (1949).
165. Fry, H., and F. G. Norris, Electroplating Metal Finishing, **12**, 207 (1959).
166. Watkins, H., and A. Kolk, J. Electrochem. Soc., **108**, 1018 (1961).
167. Story, H. S., and R. W. Hoffman, Proc. Phys. Soc. London, **B70**, 950 (1957).
168. Priest, J., and H. L. Caswell, Trans. 8th Natl. Vacuum Symp., 1961, p. 947, Pergamon Press, New York.
169. Priest, J., Rev. Sci. Instr., **32**, 1349 (1961).

170. Klokholm, E., and J. F. Freedman, *J. Appl. Phys.*, **38**, 1354 (1967).
171. Hoar, T. P., and D. J. Arrowsmith, *Trans. Inst. Metal Finishing*, **34**, 354 (1957).
172. Blackburn, H., and D. S. Campbell, *Trans. 8th Natl. Vacuum Symp.*, 1961, p. 943, Pergamon Press, New York.
173. Novice, M. A., *Brit. Sci. Instr. Assoc. Rept.* R. 294, January, 1963.
174. Hill, A. E., and G. R. Hoffman, *Brit. J. Appl. Phys.*, **18**, 13 (1967).
175. Wilcock, J. D., and D. S. Campbell, *Thin Solid Films*, **3**, 3 (1969).
176. Horikoshi, H., and N. Tamura, *Japan. J. Appl. Phys.*, **2**, 328 (1963).
177. Wilcock, J. D., D. S. Campbell, and J. C. Anderson, *Thin Solid Films*, **3**, 13 (1969).
178. Stuart, P. W., private communication, 1967.
179. Carpenter, R., and D. S. Campbell, *J. Mater. Sci.*, **2**, 173 (1967).
180. Shepard, S. C., J. Wales, and M. Yea, *Prod. Design Eng.*, October, 1965, p. 87.
181. Ennos, A. E., *Appl. Opt.*, **5**, 51 (1966).
182. Davidenkov, N. N., *Soviet Phys. Solid State*, **2**, 2595 (1961).
183. Brenner, A., and S. Senderoff, *J. Res. NBS*, **42**, 105 (1949).
184. Borie, B., and C. J. Sparks, in "Thin Films," p. 45, American Society for Metals, Metals Park, Ohio, 1963.
185. Borie, B., C. J. Sparks, and J. V. Cathcart, *Acta. Met.*, **10**, 691 (1962).
186. Vook, R. W., and F. Witt, *J. Appl. Phys.*, **36**, 2169 (1965).
187. Vook, R. W., and F. Witt, *J. Vacuum Sci. Technol.*, **2**, 243 (1965).
188. Vook, R. W., and F. Witt, *J. Vacuum Sci. Technol.*, **2**, 49 (1965).
189. Freedman, J. F., *J. Appl. Phys.*, **30**, 597 (1959).
190. Freedman, J. F., *IBM J. Res. Develop.*, **6**, 449 (1962).
191. Segmuller, A., *IBM J. Res. Develop.*, **6**, 464 (1962).
192. Light, T. B., and C. N. J. Wagner, *J. Vacuum Sci. Technol.*, **3**, 1 (1966).
193. Smith, G. V., private communication, 1967.
194. Cullity, B. D., *J. Appl. Phys.*, **35**, 1915 (1964).
195. Meieran, E. S., and I. A. Blech, *J. Appl. Phys.*, **36**, 3162 (1965).
196. Blech, I. A., and E. S. Meieran, *Appl. Phys. Letters*, **9**, 245 (1966).
197. Hart, M., private communication, 1968.
198. Halteman, E. K., *J. Appl. Phys.*, **23**, 150 (1952).
199. Halliday, J. S., T. B. Rymer, and K. H. R. Wright, *Proc. Roy. Soc. London*, **225**, 548 (1954).
200. Boswell, F. W. C., *Proc. Phys. Soc. London*, **A64**, 465 (1951).
201. Rymer, T. B., *Nuovo Cimento*, **6**, 294 (1957).
202. Denbigh, P. N., and R. B. Marcus, *J. Appl. Phys.*, **37**, 4325 (1966).
203. MacRae, A. U., *Science*, **139**, 379 (1963).
204. Pashley, D. W., *Phil. Mag.*, **4**, 324 (1959).
205. Drum, C. M., *Phil. Mag.*, **13**, 1239 (1966).
206. Stirland, D. J., private communication, 1967.
207. Zemel, J. N., J. D. Jenson, and R. B. Schodar, *Phys. Rev.*, **140**, 330 (1965).
208. Coren, R. L., *J. Appl. Phys.*, **33**, 1168S (1962).
209. Lu, C., and A. A. Milgram, *J. Appl. Phys.*, **38**, 2038 (1967).
210. Ittner, W. B. III, in G. Hass (ed.), "Physics of Thin Films," vol. 1, p. 233, Academic Press Inc., New York, 1963.
211. Toxen, A. M., *Phys. Rev.*, **123**, 442 (1961).
212. Blumberg, R. H., and D. P. Seraphim, *J. Appl. Phys.*, **33**, 163 (1962).
213. MacDonald, J. R., *Proc. Phys. Soc. London*, **A64**, 968 (1951).
214. MacDonald, J. R., *Phys. Rev.*, **106**, 890 (1957).
215. Kuriyama, M., H. Yamanouchi, and S. Hosoya, *J. Phys. Soc. Japan*, **16**, 701 (1961).
216. Pomerantz, M., J. F. Freedman, and J. C. Suits, *J. Appl. Phys.*, **33**, 1164S (1962).
217. Campbell, D. S., *Electron. Reliability Microminiaturization*, **2**, 207 (1963).
218. Kinbara, A., *Oyo Butsuri*, **30**, 496 (1961).
219. Barklie, R. H., and J. J. Davies, *Proc. Inst. Mech. Engrs.*, **1**, 731 (1930).
220. Soderberg, K. G., and A. K. Graham, *Proc. Am. Electroplaters Soc.*, 1947, p. 74.
221. Almen, J. O., *Prod. Eng.*, **22**, 109 (1951).
222. Brenner, A., and C. W. Jennings, *Am. Electroplaters Soc. 34th Tech. Session*, 1947, p. 25.
223. Prutton, M., "Thin Ferromagnetic Films," Butterworth and Co. (Publishers), Ltd., London, 1964.
224. Fisher, R. D., *J. Electrochem. Soc.*, **109**, 479 (1962).
225. Wolff, I. W., *J. Appl. Phys.*, **33**, 1152 (1962).
226. Evans, U. R., "The Corrosion and Oxidation of Metals," Edward Arnold (Publishers) Ltd., London, 1960.
227. Popereka, M. Ya, *Phys. Metals Metallog.*, **20**(5) (1966).

228. Mahla, N. M., *Trans. Am. Electrochem. Soc.*, **77**, 145 (1940).
229. Phillips, W. M., and F. L. Clifton, *Am. Electroplaters Soc. 34th Tech. Session*, 1947, p. 97.
230. Heussner, C. E., A. R. Balden, and L. M. Morse, *Plating*, **35**, 719 (1948).
231. Hothersall, A. W., in Symposium on Internal Stresses in Metals and Alloys, p. 107, Institute of Metals, London, 1948.
232. Kohlschütter, V., and E. M. Vuilleumier, *Z. Elektrochem.*, **24**, 300 (1918).
233. Stager, H., *Helv. Chim. Acta*, **4**, 584 (1920).
234. Kohlschütter, V., *Helv. Chim. Acta*, **5**, 490 (1922).
235. Walker, P. M., N. E. Bentley, and L. E. Hall, *Trans. Instr. Metal. Finishing*, **32**, 349 (1954–1955).
236. Vuilleumier, E. A., *Trans. Am. Electrochem. Soc.*, **42**, 99 (1922).
237. Kohlschütter, V., and F. Jacober, *Z. Elektrochem.*, **33**, 290 (1927).
238. MacNaughton, D. J., and A. W. Hothersall, *Trans. Faraday Soc.*, **24**, 387 (1928).
239. Marie, C., and N. Thon, *Compt. Rend.*, **193**, 31 (1931).
240. Jaquet, P. A., *Compt. Rend.*, **194**, 456 (1932).
241. Jaquet, P. A., *Trans. Am. Electrochem. Soc.*, **66**, 393 (1932).
242. Bozorth, R. M., *Phys. Rev.*, **26**, 390 (1925).
243. Wyllie, M. R. J., *J. Chem. Phys.*, **16**, 52 (1948).
244. Hume-Rothery, W., and M. R. J. Wyllie, *Proc. Roy. Soc. London*, **A181**, 331 (1943).
245. Hume-Rothery, W., and M. R. J. Wyllie, *Proc. Roy. Soc. London*, **A182**, 415 (1943).
246. Yang, L., "Electrodeposition of Metals," D. I. C. Thesis, Imperial College, London University, 1948.
247. Murbach, H. P., "A Study of Stress and Orientation in Crystalline Deposits," D. I. C. Thesis, Imperial College, London University, 1953.
248. MacNaughton, D. J., and A. W. Hothersall, *Trans. Faraday Soc.*, **31**, 1168 (1935).
249. Hendy, J. C., H. D. Richards, and A. W. Simpson, *J. Mater. Sci.*, **1**, 127 (1966).
250. Pilling, N. B., and R. E. Bedworth, *J. Inst. Metals*, **29**, 599 (1923).
251. Vermilyea, D. A., *Acta Met.*, **5**, 492 (1957).
252. Tomashov, N. D., "Theory of Corrosion and Protection of Metals," The Macmillan Company, New York, 1966.
253. Jaccodine, R. J., and W. A. Schlegel, *J. Appl. Phys.*, **37**, 2429 (1966).
254. Ligenza, J. R., *J. Appl. Phys.*, **36**, 2703 (1965).
255. Douglass, D. L., *Corrosion Sci.*, **5**, 255 (1965).
256. Bradshaw, W. N., and S. Clark, *J. Electrodep. Tech. Soc.*, **24**, 147 (1949).
257. Bradhurst, D. H., and J. S. Ll. Leach, *Trans. Brit. Ceram. Soc.*, **62**, 793 (1963).
258. Holland, L., "Vacuum Deposition of Thin Films," p. 348, Chapman & Hall, Ltd., London, 1956.
259. Young, L., "Anodic Oxide Films," Academic Press Inc., New York, 1961.
260. Croft, G. T., *J. Electrochem. Soc.*, **106**, 278 (1959).
261. Vermilyea, D. A., *Acta Met.*, **2**, 476 (1954).
262. Young, L., *Acta Met.*, **5**, 711 (1957).
263. Leach, J. S. Ll., and P. Neufeld, *Proc. Brit. Ceram. Soc.*, **65**, 49 (1966).
264. Miller, A., and H. M. Manasevit, *J. Vacuum Sci. Technol.*, **3**, 68 (1966).
265. Jackson, N. F., E. J. Hollands, and D. S. Campbell, Applications of Thin Films in Electronic Engineering, Paper 13, *IERE Conf. Proc.*, no. 7, 1966.
266. Layton, C. K., private communication, 1967.
267. Hollands, E. J., and D. S. Campbell, *J. Mater. Sci.*, **3**, 544 (1968).
268. Glang, R., R. A. Holmwood, and P. C. Furois, *Trans. 3d Intern. Vacuum Conf.*, **2** (III), 643 (1965).
269. Maissel, L. I., private communication, 1968.
270. Heavens, O. S., and S. D. Smith, *J. Opt. Soc. Am.*, **47**, 469 (1957).
271. Carpenter, R., and D. S. Campbell, *J. Mater. Sci.*, **4**, 526 (1969).
272. White, M., *Vacuum*, **15**, 499 (1965).
273. Notarys, H. A., *Appl. Phys. Letters*, **4**, 79 (1964).
274. Caswell, H. L., J. R. Priest, and Y. Budo, *J. Appl. Phys.*, **34**, 3261 (1963).
275. Joynson, R. E., Applications of Thin Films in Electronic Engineering, Paper 10, *IERE Conf. Proc.*, no. 7, 1966.
276. Kato, H., K. Nagasima, and H. Hasunuma, *Oyo Butsuri*, **30**, 700 (1961).
277. Rymer, T. B., *Proc. Roy. Soc. London*, **A225**, 274 (1956).
278. Doi, Y., *Machine Test. Lab. Rept. (Japan)*, **27**, 44 (1958).
279. Horikoshi, H., Y. Ozawa, and H. Hasunuma, *Japan. J. Appl. Phys.*, **1**, 304 (1962).
280. Riesenfeld, J., and R. W. Hoffman, *AEC Tech. Rept.* 39, Case Institute, Cleveland, Ohio, 1965.
281. Gontarz, R., H. Ratajczak, and P. Suda, *Phys. Stat. Solidi*, **15**, 137 (1966).

282. Finegan, J. D., *AEC Tech. Rept.* 15, Case Institute, Cleveland, Ohio, 1961.
283. Schroder, H., and G. M. Schmidt, *Z. Angew. Phys.*, **18**, 124 (1964).
284. Klokholm, E., *IBM Res. Rept.* RC 1508, 1965.
285. Wilcock, J. D., and D. S. Campbell, private communication.
286. Marsh, D. M., *J. Sci. Inst.*, **38**, 229 (1961).
287. Holland, L., T. Putner, and R. Ball, *Brit. J. Appl. Phys.*, **11**, 167 (1960).
288. Novice, M. A., *Vacuum*, **14**, 385 (1964).
289. Priest, J., H. L. Caswell, and Y. Budo, *Trans. 9th Natl. Vacuum Symp.*, 1962, p. 121.
290. Budo, Y., and J. Priest, *Solid-State Electron.*, **6**, 159 (1963).
291. Wilman, H., *Proc. Phys. Soc. London*, **N:68**, 474 (1955).
292. Hoffman, R. W., and H. S. Story, *J. Appl. Phys.*, **27**, 193 (1956).
293. Gafner, G., *Phil. Mag.*, **5**, 1041 (1960).
294. Buckel, W., *Z. Physik*, **138**, 136 (1954).
295. Finegan, J. D., and R. W. Hoffman, *AEC Tech. Rept.* 18, Case Institute, Cleveland, Ohio, 1961.
296. Buckel, W., in C. A. Neugebauer, J. B. Newkirk, and D. A. Vermilyea (eds.), "Structure and Properties of Thin Films," p. 53, John Wiley & Sons, Inc., New York, 1959.
297. Oswald, W., *Z. Phys. Chem.*, **22**, 289 (1879).
298. Grigson, C. W., and D. B. Dove, *J. Vacuum Sci. Technol.*, **3**, 120 (1966).
299. Dove, D. B., *J. Appl. Phys.*, **35**, 2785 (1964).
300. Hill, R. M., *Nature*, **204**, 35 (1964).
301. Shuttleworth, R., *Proc. Phys. Soc. London*, **A63**, 444 (1950).
302. Benson, G. C., and K. S. Yun, *J. Chem. Phys.*, **42**, 3085 (1965).
303. Piuz, F., and R. Ghez, *Helv. Phys. Acta*, **35**, 507 (1962).
304. McRae, E. G., and C. W. Caldwell, in H. C. Gatos (ed.), "Solid Surfaces," p. 509, North Holland Publishing Company, Amsterdam, 1964.
305. Van der Merwe, J. H., *J. Appl. Phys.*, **34**, 117 (1963).
306. Van der Merwe, J. H., in "Basic Problems in Thin Film Physics," p. 122, R. Niedermayer and H. Mayer (eds.), Vandenhoeck and Ruprecht, Goettingen, 1966.
307. Mathews, J. W., *Phil. Mag.*, **6**, 1347 (1961).
308. Mathews, J. W., *Phil. Mag.*, **13**, 1207 (1966).
309. Cabrera, N., *Mem., Sci. Rev. Metallurg.*, **62**, 205 (1965).
310. Neugebauer, C. A., *J. Appl. Phys.*, **31**, 1096 (1960).
311. Blakely, J. M., *J. Appl. Phys.*, **35**, 1756 (1964).
312. Rudakov, A. P., and N. A. Semenov, *Polymer Mech.*, **1**, 112 (1965).
313. Read, H. J., *Proc. Am. Electroplaters Soc.*, **50**, 37 (1963).
314. Angus, H. C., *Trans. Inst. Metal Finishing*, **39**, 30 (1962).
315. Palatnik, L. S., and A. I. Ll'inskii, *Soviet Phys. Solid State*, **3**, 2053 (1962).
316. Palatnik, L. S., G. V. Fedorov, and A. I. Ll'inskii, *Phys. Met. Metallog.*, **5**, 159 (1961).
317. Raffalovich, A. J., *Rev. Sci. Inst.*, **37**, 368 (1966).
318. Angus, H. G., *Brit. Appl. Phys.*, **13**, 58 (1962).
319. Branson Instrument, Inc., and Dawe Inst., Ltd., *Ultrasonics*, **4**, 88 (1966).
320. Menter, J. W., and D. W. Pashley, in C. A. Neugebauer, J. D. Newkirk, and D. A. Vermilyea (eds.), "Structure and Properties of Thin Films," p. 111, John Wiley & Sons, Inc., New York, 1959.
321. Neugebauer, C. A., in G. Hass and R. E. Thun (eds.), "Physics of Thin Films," vol. 2, p. 1, Academic Press Inc., New York, 1964.
322. Machlin, E. S., in "Strengthening Mechanisms in Solids," p. 375, American Society for Metals, Metals Park, Ohio, 1960.
323. Bilby, B. A., *Nature*, **182**, 296 (1958).
324. Campbell, D. S., in J. C. Anderson (ed.), "The Use of Thin Films in Physical Investigations," p. 299, Academic Press Inc., New York, 1966.
325. Grunes, R. L., C. D. D'Antonio, and F. Kiess, *J. Appl. Phys.*, **36**, 2735 (1965).
326. Jovanovic, S., and C. S. Smith, *J. Appl. Phys.*, **32**, 121 (1961).
327. D'Antonio, C., J. Hirshhorn, and L. Tarshis, *Trans. AIME*, **227**, 1346 (1963).

Chapter **13**

Electrical Properties of Metallic Thin Films

LEON I. MAISSEL

IBM Components Division, East Fishkill, New York

1. SOURCES OF RESISTIVITY IN METALLIC CONDUCTORS

a. Temperature

According to modern quantum electronic theory, electrical conduction in metals is due to electrons, while electrical resistivity results from the scattering of these electrons by the lattice. Because of their wave nature, electrons can pass through a perfect lattice without any attenuation (resistivity being, therefore, a measure of the extent to which a metal lattice departs from perfect regularity). Actually, no lattice is perfect. Electrons always undergo some scattering as they move through a solid, the average distance that they travel between collisions being called the mean free path.[1,2]

Even a lattice which has no structural defects or foreign atoms cannot be completely regular at any temperature, since the atoms will not be fixed (stationary) but will be vibrating about their mean positions. In considering interactions between the electrons and the various vibrational modes of the lattice (phonons), it is convenient to make use of the Debye temperature. In Debye's theory of specific heats, a crystal is assumed to possess a wide spectrum of vibrational modes, with a fixed upper limit. This is so because the minimum phonon wavelength must be of the order of atomic spacing. The Debye temperature θ is then defined in terms of this maximum frequency by

$$\theta = \frac{h\nu_{max}}{k}$$

where k is Boltzmann's constant.

At low temperatures ($T \ll \theta$), resistivity varies as T^n (where n is close to 5), whereas at high temperatures ($T \gg \theta$), resistivity varies linearly with T.

For many metals, the Debye temperature falls close to or below room temperature, so that the temperature variation of resistance above 25°C is approximately linear. This makes it possible to define a temperature coefficient of resistance α which can be measured between two temperatures that are not extremely close together (see Sec. 2c):

$$\alpha = \frac{1}{R} \frac{\Delta R}{\Delta T} \tag{1}$$

The relationship between α and the temperature dependence of the mean free path is shown in Table 1.

TABLE 1 Electronic Mean Free Path of Several Metals

Metal	Calculated mean free path, Å			Temp coeff ppm/°C (0–100)	Room temp resistivity, μohm-cm
	−200°C	0°C	100°C		
Li.........	955	113	79	4,220	8.55
Na........	1,870	335	233	4,400	4.3
K.........	1,330	376	240	5,500	6.1
Cu........	2,965	421	294	4,330	1.69
Ag........	2,425	575	405	4,100	1.47
Au........	1,530	406	290	4,000	2.44
Ni........	.....	133	80	6,750	7.24
Co........	.....	130	79	6,580	9.7
Fe........	2,785	220	156	4,110	8.85
Pt........	720	110	79	3,920	9.83

b. Point Defects

An impurity atom dissolved in a metal will, in general, carry an effective electric charge different from that of the parent metal; it will therefore serve as a source of electron scattering. Even if both the solvent and the solute have the same valence, the screening of the impurity ions by the electron gas will not be quite the same as that given to the parent ions, so that their effective charge will be slightly different. The same general picture holds for a vacancy or an interstitial. As might be expected, resistivity rises with increasing impurity concentration, reaching a maximum for an alloy composition of approximately 50% "impurity." Figure 1 illustrates this for the silver-gold system. The data are taken from Ref. 3.

A striking confirmation of the fact that electrical resistivity is a measure of the irregularity of the lattice is seen in the case of those alloys in which the impurity atoms can assume an ordered arrangement (or superlattice) within the host lattice. An

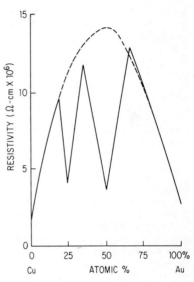

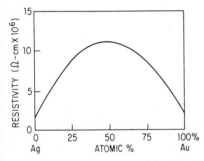

Fig. 1 Resistivity vs. composition for the gold-silver system.

Fig. 2 Resistivity vs. composition for the gold-silver system, both quenched and annealed.

example of such a system is copper-gold. Figure 2 shows the resistivity vs. composition diagram for this system.[3] The solid line is for specimens that were annealed at 200°C to allow ordering to take place, whereas the dotted one is for specimens in which a random impurity distribution has been "frozen in" by quenching at 650°C.

As can be seen from these examples, impurities having the same number of valence electrons as the host increase the resistivity of the latter by a factor of about 0.1 per atomic percent impurity added, giving a maximum increase of a factor of 5 or so.

When the impurity atom has a valence different from that of the host, its effect on the resistivity is, understandably, considerably more pronounced. Contributions have been seen that are as high as a factor of 10 (increase) per atomic percent of impurity added.[4] However, impurities of this sort also cause extensive distortion of the lattice, so that in practice their solid solubility is severely limited. Thus, in bulk materials, resistivity increases due to dissolved impurities will not be much greater than by a factor of 10. In films, however, a considerably greater concentration of impurities than the equilibrium value may be trapped during deposition. As a result, increases in resistivity by factors of several hundred or more are often seen.

In addition, the high resistivities often seen in films prepared under poor deposition conditions may be due to the formation of an insulating phase which is randomly distributed throughout the film. An example of the transition from a dissolved impurity to an insulating phase, and its effect on resistivity, is shown in Fig. 3, taken from Ref. 5.

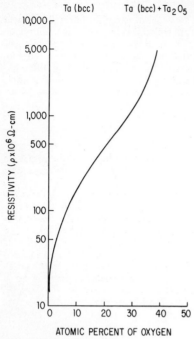

Fig. 3 Resistivity of tantalum as a function of oxygen impurity concentration.

c. Structural Imperfections

The contributions to the resistivity made by vacancies and interstitials have been summarized by Damask and Dienes.[6] Typical values are of the order of 1 μohm-cm per atomic percent of vacancies or interstitials, although values as high as 5.9 μohm-cm per atomic percent of vacancies[7] and 10.5 μohm-cm per atomic percent of interstitials[8] have been recorded.

The scattering power of dislocations in metals is thought to be small.[9] For example, the contribution to the residual resistivity per dislocation per cm[2] has been estimated[10] to be 2.3×10^{-13} μohm-cm. This means that it takes a dislocation density of about 10^{12} per cm[2] to have the same effect as 1% of dissolved chemical impurity.

In general, a solid in equilibrium at absolute temperature T will have a concentration of imperfections in the order of $e^{-w/kT}$, where W is the activation energy for the formation of the imperfections. This corresponds to less than $10^{-4}\%$ at room temperature for a W of about 1 eV. Much larger concentrations can be "frozen in" by quenching from high temperatures; but, even so, the equilibrium temperature for a concentration of as little as 1% is about 3,000°K.

The effect of grain boundaries on the resistivity of a bulk metal is, like that of dislocations, relatively small. This is not surprising since grain boundaries can be treated as a linear array of dislocations. As the size of the grains becomes smaller (and, hence, the role of the grain boundaries becomes more important), the material begins to resemble an amorphous solid—not too far removed in structure from that of a liquid. Thus, an estimate of the maximum possible contribution to the resistivity from grain boundaries may be arrived at by considering the resistivities of liquid metals. In general, the increase in resistivity due to melting is of the order of 5 to 20 μohm-cm. However, this is influenced by factors other than simply the change in grain-boundary density, as is evidenced by the fact that certain metals such as Ga, Sb, and Bi actually have lower resistivities in the liquid state than in the solid. We may also remark that thin films of sufficient thickness prepared under conditions of clean vacuum exhibit resistivities within a few percent of the bulk value despite having grain sizes in the order of only a few hundred angstroms.

It was mentioned above that an insulating phase may be present if sufficient impurities are trapped. It is frequently the case that such phases are concentrated at the grain boundaries. Under these circumstances, the role of the latter is entirely different. Some of these cases will be considered later.

d. Matthiessen's Rule

As indicated above, electrical resistance in metals may arise from a variety of causes, such as temperature, dissolved impurities, and vacancies. Matthiessen's rule states that the resistivity of a given sample will be the simple arithmetic sum of the individual contributions made by all these separate sources of resistance. It is generally convenient to combine all the contributions to the resistivity, other than that of temperature, into one which is referred to as the residual resistivity.

$$\rho = \rho_{\text{temp}} + \rho_{\text{residual}}$$

Experimentally, Matthiessen's rule has been found to hold very well for bulk materials, and although careful measurements[11] have shown some deviation from it, the deviation is small enough to be neglected in most instances.[1]
Since the temperature coefficient of resistance is

$$\alpha = \frac{1}{\rho}\frac{d\rho}{dT} = \frac{1}{\rho}\frac{d\rho_{temp}}{dT}$$

it follows that

$$\alpha\rho = \frac{d\rho_{temp}}{dT} = \text{const} \tag{2}$$

Since both α and ρ are usually known for the bulk, Eq. (2) is a particularly useful form of Matthiessen's rule. It is apparent that, as ρ becomes larger, α will approach zero. Note, however, that (in terms of Matthiessen's rule) α can never become negative.

Matthiessen's rule is useful in situations in which it is easier to measure the temperature coefficient of resistance than the resistivity. This could occur because of problems of accurate thickness measurement—for example, for films on nonplanar substrates. If the temperature coefficient of resistance is positive and exceeds a few hundred parts per million per °C, we can assume that Matthiessen's rule will hold and Eq. (2) can be applied and the resistivity deduced. Experimental confirmation that Matthiessen's rule can hold even for very thin films has been given by Altman[12] for tantalum down to 200 Å and by Young and Lewis[13] for chromium down to 30 Å. Note that in both cases the electron mean free path was substantially less than the thickness of the film. Situations where this may not be the case will be discussed presently.

2. COMMONLY MEASURED QUANTITIES FOR THIN FILMS

a. Sheet Resistance

As can be seen from Fig. 4, the resistance of a rectangularly shaped section of film (measured in a direction parallel to the film surface) is given by

$$R = \frac{\rho}{d}\frac{l}{b} \tag{3}$$

If $l = b$, this then becomes

$$R = \frac{\rho}{d} = R_s$$

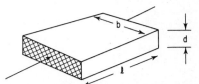

so that the resistance R_s of one square of film is independent of the size of the square—depending only on resistivity and thickness. **Fig. 4** Definition of sheet resistance.
The quantity R_s is called the "sheet resistance" of the film and is expressed in ohms per square. It is a very useful quantity that is widely used for comparing films, particularly those of the same material deposited under similar conditions. If the thickness is known, the resistivity is readily obtained from

$$\rho = dR_s$$

b. Measurement of Sheet Resistance

The most direct method of measuring R_s is to prepare a rectangular sample of film as shown in Fig. 5a, measure its resistance, and divide by the number of squares of film material that lie between the end contacts (four in our example). Where contact resistance between the film and the end terminals may be a problem, a four-terminal method is necessary (see Fig. 5b), the number of squares now being counted between the two voltage terminals (five in this example), and $R = V/I$. It is advisable to make the voltage terminals as narrow as possible at their points of intersection with the film in order to reduce any uncertainty in counting the number of squares lying between them.

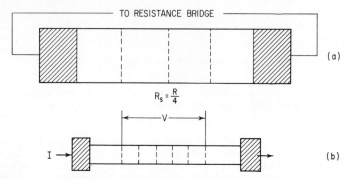

$$R_s = \frac{R}{4}$$

(a)

(b)

Fig. 5   Direct measurement of sheet resistance.

While direct measurement of the number of squares is the simplest method, a technique which does not call for the fabrication of a special sample for each determination of sheet resistance is usually preferred. Furthermore, it may be necessary to measure the variation of sheet resistance in a film from one part of the substrate to another, and the fabrication and measurement of a large number of tiny rectangles become highly impractical (in addition to being a destructive technique). Accordingly, widespread use is made of four-point-probe methods. The most common form of four-point probe is the in-line type illustrated schematically in Fig. 6.

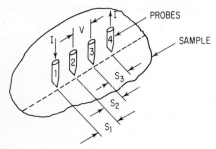

Fig. 6   In-line four-point probe.

This configuration has been treated in detail by Valdes,[14] who showed that when the probes are placed on a material of semi-infinite volume, the resistivity is given by

$$\rho = \frac{V}{I} \frac{2\pi}{1/s_1 + 1/s_3 - 1/(s_1 + s_2) - 1/(s_2 + s_3)}$$

When $s_1 = s_2 = s_3 = s$, this reduces to

$$\rho = \frac{V}{I} 2\pi s \tag{4}$$

If the material on which the probes are placed is an infinitely thin slice resting on an insulating support, it can be shown[14] that Eq. (4) becomes

$$\rho = \frac{V\, d\pi}{I \ln 2}$$

or

$$\frac{\rho}{d} = R_s = 4.532 \frac{V}{I} \tag{5}$$

Although Eq. (5) is independent of the probe spacing, the resolution of any particular probe will, of course, depend on spacing. Thus, a probe of spacing S centrally located inside a square of film measuring $6S \times 6S$ will give a reading that is about 20% too high if the square is surrounded by an insulating film and about 10% too low if it is surrounded by a film of infinite conductivity.   For more exact corrections (to allow for edge effects) the reader is referred to Refs. 14 and 15.   A number of four-point-probe assemblies are commercially available, being extensively used in the semiconductor industry.   Probe spacings of 25 mils are commonly employed for routine use, whereas for high-resolution work, less robust models with spacings down to 10 mils are available.

In cases where very high resolution is needed, a square-probe array such as shown in Fig. 7 rather than a linear one may be used.   To use such a probe, current I is fed in through any two adjacent probes and the voltage V generated across the other two is measured.   The sheet resistance is then computed[17] from

$$R_s = \frac{V}{I} \frac{2\pi}{\ln 2} = 9.06 \frac{V}{I}$$

For very small probes, the requirement that the probe geometry be perfectly square may be difficult to achieve in practice, and the average of two independent measurements using different pairs of current probes may be utilized.   Zrudsky et al.[17] have described a circuit which allows this to be performed automatically.   The constant-current source is alternately switched from one pair of probes to the other, while a voltmeter that responds to the dc average of the two readings is simultaneously connected to the opposite pair of probes.   The authors esti-

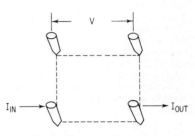

Fig. 7   Square-probe array.

mate that by this means a probe whose geometry deviates from a square by about 7% can be made to read to an accuracy of about 1%.

c. Temperature Coefficient of Resistance (TCR)

α_T, the TCR at temperature T, is experimentally determined from the relationship

$$\alpha_T = \frac{R_1 - R_2}{R_T(T_1 - T_2)}$$

where $T_1 > T > T_2$, and the R's are resistance values (i.e., the geometry of the sample need not be known).   When the TCR is quoted without specifying temperature, T is generally assumed to be 20°C.

A minimum of three resistance measurements must be made in order to obtain a reliable value for α, the third reading being necessary as a check to determine whether the temperature treatment (up or down) has produced any permanent change in the resistance value.   Where the TCR is small, it is usually advisable to plot a series of values of R vs. T (both ascending and descending).   As in all resistance determinations, four terminal measurements should be used whenever possible.

To ensure accurate knowledge of the temperature at which the resistance is measured, immersion of samples in a temperature-controlled bath of a suitable nonconductive liquid (such as mineral oil) is recommended and the usual precautions such as ensuring circulation of the liquid and allowing time to come to equilibrium should be observed.

3. INFLUENCE OF THICKNESS ON THE RESISTIVITY OF A STRUCTURALLY PERFECT FILM

a. Theory

As was discussed above, variations in the resistivity of a metal correspond to changes in the mean free path of the conduction electrons.   Since an electron will suffer a reflection of some sort at the surface (when it happens to reach it), the resistivity increases whenever the specimen becomes thin enough for collisions with the surface to be a significant fraction of the total number of collisions—in other words, whenever one or more dimensions of the specimen become comparable with or less than the mean free path at the temperature concerned.   It should be pointed out here that collisions with the surface will be important from this point of view only if they are nonspecular, i.e., if the direction in which the electron moves after its collision is independent of its direction of motion prior to the collision.   A helpful analogy is to compare the specular and nonspecular cases to the attentuation of light being propagated down light pipes having polished and diffuse surfaces, respectively.

Table 1 shows that surface scattering of electrons at room temperature becomes an important effect for most pure metals if they are less than 2 to 300 Å thick, whereas at temperatures around $-200°C$, the effect is noticeable for thicknesses about one order of magnitude larger.   However, the resistivity of thin films is often much higher than that of the pure bulk material, so that unless special care is taken to ensure chemically pure, imperfection-free films, the electronic mean free path will be a good deal less than the values listed in Table 1.

The possibility that the resistivity might be influenced by a reduction of one of the dimensions of a specimen was first discussed by Thomson[18] in 1901.   A more rigorous treatment was given by Fuchs[19] in 1938, and almost all subsequent work in this area has been based on his approach.   In particular, Sondheimer[20] extended the theory to include other mean-free-path effects in metals.   The subject has also been reviewed, more recently, by Campbell,[21] and the derivation presented below includes material from all these sources.

The starting point for the analysis is the Boltzmann transport equation for electrons (for a simplified description of this, see Ref. 22)

$$-\frac{e}{m}\left(E + \frac{1}{c}\, v \times H\right) \operatorname{grad}_v f + v\, \operatorname{grad}_r f = -\frac{f - f_0}{\tau} \tag{6}$$

where f = nonequilibrium electronic distribution functions (for which we wish to solve this equation)

f_0 = electronic distribution at equilibrium

τ = relaxation time for return to equilibrium, a function only of the absolute value of v, the electron velocity

The other symbols have their usual meaning except that m will be an effective mass rather than the free-electron mass.

The z axis is taken as perpendicular to the plane of a film of thickness d, and current flows through the film in the x direction.

The use of Eq. (6) for the investigation of size effects on conductivity depends on the presence of the second term, $v\, \operatorname{grad}_r f$, which vanishes for bulk material but not in the z direction for a thin film.

It is convenient to write

$$f = f_0 + f_1(v,z) \tag{7}$$

Substitution of (7) in (6) gives

$$\frac{eE}{mv_z}\frac{\partial f_0}{\partial v_x} = \frac{\partial f_1}{\partial z} + \frac{f_1}{\tau v_z} \tag{8}$$

since $H = 0$ and E is in the x direction

The general solution of Eq. (8) is of the form

$$f_1(v,z) = \frac{e\tau E}{m}\frac{\partial f_0}{\partial v_x}\left[1 + F(v)\exp\left(\frac{-z}{\tau v_z}\right)\right] \tag{9}$$

where $F(v)$ is an arbitrary function of v to be determined by the introduction of the appropriate boundary conditions.

To determine $F(v)$, we have to introduce the boundary conditions at the surfaces of the film.   The simplest assumption is to suppose that every free path is terminated by collision at the surface, so that the scattering is entirely diffuse. The distribution function of the electrons leaving each surface must then be independent of direction. Equation (9) shows that this can be satisfied only if we choose $F(v)$ so that $f_1(v,0) = 0$ for all v having $v_z > 0$ (that is, for electrons moving away from the surface $z = 0$), and $f_1(v,d) = 0$ for all v having $v_z < 0$.

There are therefore two distribution functions: f_1^+ for electrons with $v_z > 0$, and f_1^- for electrons with $v_z < 0$. In other words,

$$f_1^+(v,z) = \frac{e\tau E}{m} \frac{\partial f_0}{\partial v_x} \left[1 - \exp\left(\frac{-z}{\tau v_z}\right) \right] \qquad (v_z > 0)$$

$$f_1^-(v,z) = \frac{e\tau E}{m} \frac{\partial f_0}{\partial v_x} \left[1 - \exp\left(\frac{d-z}{\tau v_z}\right) \right] \qquad (v_z < 0)$$

The solution to (8) can then be used to calculate $J(z)$, the current density across the film from

$$J = 2e \left(\frac{m}{h}\right)^3 \int vf \, dv$$

Evaluation of the above integral is facilitated by resorting to polar coordinates. Thus, we substitute $v_z = v \cos \theta$, and the resulting expression for J is then

$$J(z) = \frac{4\pi e^2 m^2 \tau v^{-3}}{h^3} E \int_0^{\frac{\pi}{2}} \sin^3 \theta \left[1 - \exp\left(\frac{-d}{2\lambda \cos \theta}\right) \cosh\left(\frac{d - 2z}{2\lambda \cos \theta}\right) \right] d\theta$$

where $\lambda = \tau \bar{v}$ is the mean free path of the electrons.

By averaging the current density over all values of z from 0 to d, an expression can now be derived for σ, the film conductivity:

$$\sigma = \sigma_0 \left[1 - \frac{3}{2k} \int_1^\infty \left(\frac{1}{t^3} - \frac{1}{t^5}\right) (1 - e^{-kT}) \, dt \right] \tag{10}$$

where $t = \dfrac{1}{\cos \theta}$

$k = \dfrac{d}{\lambda} = \dfrac{\text{film thickness}}{\text{mean free path}}$

$\sigma_0 = $ bulk-conductivity value

Approximations can be made to (10) for large and small k:

$$\frac{\sigma}{\sigma_0} = 1 - \frac{3}{8k} \qquad (k \gg 1) \tag{11}$$

$$\frac{\sigma}{\sigma_0} = \frac{3k}{4} \left(\ln \frac{1}{k} + 0.423 \right) \qquad (k \ll 1) \tag{12}$$

In determining $F(v)$ in (9), it is more general (though somewhat artificial) to allow some fraction p of the electrons to be elastically scattered (v_z is the same before and after a collision) while the remainder are diffusely scattered, as before.

Equations (11) and (12) now become

$$\frac{\sigma}{\sigma_0} = 1 - \frac{3(1 - p)}{8k} \qquad (k \gg 1) \tag{13}$$

$$\frac{\sigma}{\sigma_0} = \frac{3k}{4} (1 + 2p) \left(\ln \frac{1}{k} + 0.423 \right) \qquad (k,p \ll 1) \tag{14}$$

Effect of Thickness on α. Similar reasoning may also be applied to determine the

effect of film thickness on the temperature coefficient of resistance.   The result that
one obtains is

$$\frac{\alpha}{\alpha_0} = 1 - \frac{3(1-p)}{8k} \qquad (k \gg 1)$$

$$\frac{\alpha}{\alpha_0} = \frac{1}{\ln(1/k) + 0.423} \qquad (k \ll 1)$$

In Figs. 8 and 9, curves computed by Campbell[21] for ρ/ρ_0 and α/α_0 as function of k
are shown.

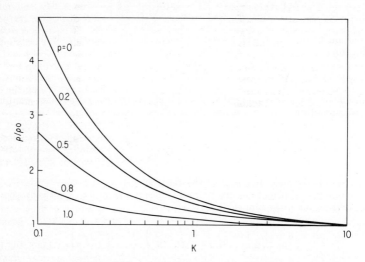

Fig. 8   Effect of film thickness on resistivity.

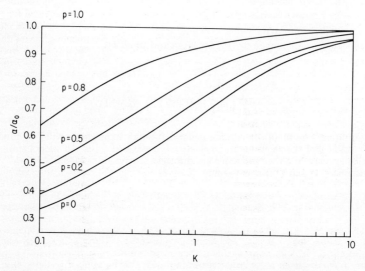

Fig. 9   Effect of film thickness on temperature coefficient of resistance.

b. Experimental Results

The experimental verification of the formulas derived in the preceding section is not as straightforward as might be expected. This is so because of uncertainties in the exact values of the mean free path, the film thickness, and particularly the scattering factor p. In addition, the resistivity for very thin films is not always known. Finally, there are formidable difficulties in preparing sufficiently clean films to make meaningful experiments.

A detailed review of work prior to 1959 has been given by Mayer,[23] in whose own work films were studied under conditions which gave $p = 0$. Continuous measurements of resistivity and thickness were made while the film grew. The work was performed in ultrahigh vacuum with alkali metals which had previously been purified by several distillations. The films were deposited at very low deposition rates onto substrates maintained at temperatures in the order of 90°K. The surface mobility of the depositing atoms was thus extremely low, and the resulting film was therefore rough on an atomic scale, making it possible to realize the condition $p = 0$. At any time during the deposition, the thickness (mass equivalent) was measured using the Langmuir-Taylor ionization method.[24] Films were found to be continuous when their thickness exceeded three to four atomic layers (verified by observing the ohmic behavior of the films). Because of the low temperature of the measurement, k was very much less than 1 for the films investigated. Good agreement between theory and experiment was obtained for a number of alkali-metal films in the range $k = 0.1$ to $k = 0.5$. Similar results using this approach have been reported by other workers.[25] By using atomic beams of known and well-controlled intensity for the deposition of the films, the thickness of the film at any given time could always be deduced without direct measurement.

Attempts to relate theory to the experimental data on the variation of the temperature coefficient of resistivity with thickness have not been as successful as for resistivity. It is possible that these difficulties exist because measurements have been made too close to the Debye temperature. However, a moderately good fit can usually be obtained by assuming a suitable value of p.

Although agreement between theory and experiment has been good for alkali-metal films, there is still considerable interest in films for which $p \neq 0$. For films with finite p, fitting theory to the experimental data introduces several problems. Nossek[26] has discussed some of the difficulties and pitfalls associated with attempts to ascertain the exact values of p and k in such experiments. He points out that, as a practical matter, a surface roughness of at least 10 Å has to be assumed for films. All measurements must therefore be done on films that are at least 10 times thicker than this, i.e., films in the order of 100 Å. This implies that for meaningful comparison between theory and experiment, k has to be greater than 0.1. As a result, it becomes necessary to know the thickness to an accuracy of about 6% in order to distinguish between p values as widely different as $p = 0$ and $p = 0.5$. This is not an easy measurement to make under the conditions for which films of controlled p are being deposited. It might be thought that additional information concerning p could be obtained through supplemental data on temperature coefficient of resistance. However, this is also a problem because it has not been established yet whether or not p itself depends on temperature. Furthermore, even more accurate measurements are needed to obtain the necessary information.

Another difficulty associated with the measurement of p is that it may not be a function solely of the surface roughness but may also depend on the total film thickness. For example, it has been suggested by Parrott[27] that, in analogy with reflection of light, specular reflection takes place only if the angle of incidence of the electron at the film surface is smaller than some critical value. This would imply that specular scattering is a function of k and will be observed only in very thin films. In practice, this has been the case, although largely by default. Experimental evidence for the existence of this effect has been obtained by Chopra[28] by studying magnetoresistance effects in films. It was found that for films in which k was smaller than 0.5, partial specular reflection was observed, whereas for thick films with k greater than 0.5, the data could best be explained by assuming total diffuse scattering.

Abeles and Theye[29] have developed a method where, in addition to measuring resistivity as a function of thickness, the optical reflection and transmittance characteristics of the films in the infrared are measured. The method is applicable to relatively thick films ($k > 1$), which is an advantage, experimentally. By means of this method a value of $p = 0.2$ for gold films 265 Å thick was demonstrated. The mean free path was shown to be 236 Å. Another method which allows specific determination of resistivity p and mean free path is derived from the combined measurement of electrical resistance and Hall coefficient. In this method, also, thicker films ($k > 1$) can be used. Cirkler[30] used this method for potassium films and obtained a p value of 0.5.

It has been found that films of certain metals such as Sn, Pb, Ag, and Au can have values of p very close to unity, provided they have received a suitable annealing treatment. For example, Gillham, Preston, and Williams,[31] working with transparent, highly conducting gold films deposited by sputtering onto bismuth oxide substrates, showed that essentially no variation of resistivity with thickness occurred (down to about 60 Å) if the films had been annealed for a few minutes at 350°C. Typical data are shown in Fig. 10. The apparent increase over bulk resistivity, even for relatively thick films, is not understood and is probably related to the relatively poor conditions under which the films were prepared.

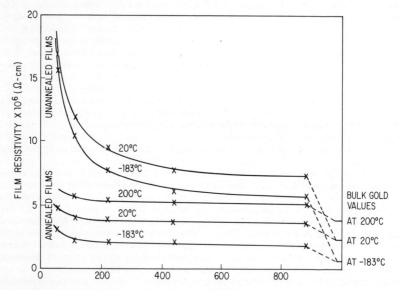

Fig. 10   Resistivity vs. thickness for gold films at several temperatures.

Similar results were obtained by Ennos[32] using evaporated gold films. These films, also, had close-to-bulk resistivity, down to approximately 60 Å. The temperature coefficient was investigated and found to be constant, down to about 60 Å, below which point a discontinuity appeared and the coefficient dropped off very sharply. However, the thickness-independent value was about half that of the bulk's. The reasons for this effect are not clear.

Single-crystal films of gold deposited epitaxially onto mica substrates have also been investigated.[33] Films with $p = 0.8$ were obtained down to about 300 Å. By contrast, films deposited onto cold mica had $p = 0$ even after receiving an annealing treatment of 300°C for 20 min. Annealing had a more pronounced effect on the resistivity of gold films deposited onto niobium oxide; even so, p was still found to be zero.

An interesting demonstration that the resistivity of a very thin film is influenced by the degree of specular scattering of electrons at the surface has been provided by

Lucas.[34] He showed that the resistivity of a "specular" film such as gold on bismuth oxide could be increased by artificially "roughening" the upper surface, i.e., converting it to a nonspecular film. The films were first made specular in the usual way by annealing for a few minutes at 350°C in air; then a small additional amount of gold (less than a monolayer) was deposited onto the specular film by either sputtering or evaporation. An increase in resistivity resulted even though additional conductive material had been added to the film. Subsequent annealing of the "roughened" film then produced the right amount of resistivity decrease, which would be expected from the additional thickness of conductive material. Lucas also found that similar results were obtained when an insulating film such as SiO was deposited over a specular gold film. This was also interpreted as due to a change in p.

Experiments related to those of Lucas have been performed by Chopra and Randlett.[35] The resistance of thin continuous films of gold, silver, copper, and aluminum was studied as a function of superimposed thin layers. It was found that SiO or Permalloy overlayers increased the resistance of gold films, whereas the same materials had little effect on silver. On the other hand, an overlay of germanium increased the resistance of silver films—but had negligible effect on aluminum. These and similar results suggest that an additional "roughening" of the surface does not necessarily lead to an increase in resistance. The explanation which Chopra and Randlett suggest for this is that the additional material behaves in a manner similar to that seen for adsorbed gases on the surface of films,[36,37] where the bonds that form may act as either donors or acceptors. Similarly, the metal overlayers are believed to modify the surface potential for the conduction electrons at the interface, influencing both their number and their mobility near the surface. Depending on the nature of the change at the interface, either an increase or a decrease in resistance may result.

Something of a special case is bismuth. This metal has an unusually large electron mean free path of between 1 and 2 μ, even at room temperature. Consequently, the effect of thickness on electron scattering can readily be observed for relatively thick films at room temperature.[38] An interesting consequence of this phenomenon is that electron scattering at the grain boundaries now becomes an important source of electrical resistance, since some of the grain boundaries behave like free surfaces. As a result, distinct differences occur in the resistivity of bismuth films, depending on their temperature of condensation.

Similar effects were seen by Drumheller,[39] who obtained data on the infrared reflection characteristics of bismuth films and correlated these with their electrical behavior. This represents one of the rare cases where grain-boundary scattering is very significant. According to Drumheller, the grain boundaries in his films fell into two distinct categories, depending on whether they divided crystallites with similar or unrelated orientations. About half the grain boundaries had a mean shunting resistance approximately equal to half the dc resistance, whereas the other grain boundaries had no appreciable effect. The major portion of the electrical resistance parallel to the film surface was contributed by grain boundaries.

Measurements on very thin metal foils rather than on deposited films have also been used to check the Fuchs theory. Such an approach was used by Andrews,[40] who rolled foils of extremely pure tin to a variety of thicknesses (the thinnest being 3 μ) and measured their resistivities at very low temperature (3 to 20°K). Good agreement with theory was found for a value of $p = 0$.

4. HALL EFFECT AND MAGNETORESISTANCE IN THIN FILMS

When a magnetic field is applied to a conductor at right angles to the direction of current flow, a voltage is developed in a plane at right angles to both the current and the magnetic field. The Hall coefficient R_H is then defined by

$$R_H = \frac{\Delta v}{H j \, \Delta x}$$

where Δx is the width of the specimen in the plane of the Hall voltage. The significance of this coefficient stems from the fact that $R_H = 1/Nq$, where N is the number of charge carriers per unit volume. Thus, since $\sigma = Nq\mu$ (where μ is the mobility) we have $\mu = \sigma R_H$, mobility measured in this manner being termed, appropriately, the Hall mobility.

In the previous section it was seen that σ may be reduced when the electron mean free path λ becomes comparable with or greater than the film thickness. The parameter $k = d/\lambda$ was used. When a magnetic field is applied, an additional parameter β is needed; it is defined by $\beta = d/r$ where $r = m\bar{v}c/qH$ is the radius of the circular orbit of an electron in a magnetic field H and $m\bar{v}$ is the mean momentum of an electron. Thus $\beta \propto H$.

The problem of the effect of a magnetic field on the resistance of a thin film was first treated in detail by Sondheimer.[41] His analysis predicts that with increasing magnetic field normal to the plane of the film, the resistivity will oscillate about a mean value that is greater than bulk. As for the zero field case, the increase over the bulk resistivity becomes larger the smaller the value of k. For a given field, the increase over bulk is greatest when the specular reflection coefficient p equals 0. Conductivity equals the bulk value when $p = 1$. Some results are illustrated in Fig. 11.

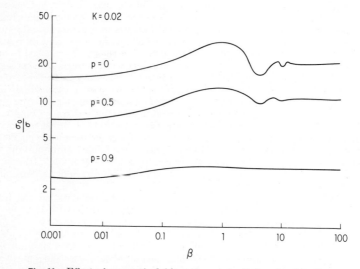

Fig. 11 Effect of magnetic field on the conductivity of a thin film.

For the Hall coefficient, the Sondheimer theory also predicts an increase over the bulk value R_H. For low fields this is given by

$$\frac{R_F}{R_H} = \frac{4}{3k} \frac{1-p}{1+p} \frac{1}{[\ln (1/k)]^2} \qquad (k \ll 1)$$

where R_F is the Hall coefficient of the film. As the field is increased, R_F decreases and returns to the bulk value at $\beta = 1$.

Experimental attempts to verify these predictions have not always met with as much success as for the pure conductivity effects $(H = 0)$. This is not surprising in view of the fact that departures from bulk are greatest for weak fields where accurate measurements are most difficult to make. Jeppesen,[42] for example, measured a Hall coefficient 30% greater than bulk for gold films on glass but found no appreciable variation in its value over the thickness range 30 to 2,000 Å—even though a decrease in Hall mobility was seen. Cirkler[30] obtained the bulk value for potassium films down to 450 Å and saw only a small increase in R_F below this value (all the way down to 50 Å). On the other hand, Forsvoll and Holwech[43,44] found good agreement with the Sondheimer theory for aluminum, for both the magnetoresistance and the Hall coefficient.   Fair agreement with theory for Hall coefficients was also obtained by Chopra and Bahl[45] for gold, silver, and copper films, both epitaxial and poly-crystalline.   For the epitaxial films, the data indicated $p \approx 1$, whereas the poly-crystalline films gave the best fit to Sondheimer's theory for $p = 0$.

In all the above, it was assumed that the magnetic field was applied normal to the plane of the film.   It is, however, possible to apply the field in the plane of the film (but perpendicular to the current) thus generating a Hall voltage across the thin dimension of the film.   This case has been theoretically analyzed by MacDonald and Sarginson,[46] who predicted that the Hall field would be less than the bulk at low magnetic fields, approaching bulk at high fields.   General agreement with this pre-diction was obtained by Holwech[47] for ultrapure aluminum foils measured at tempera-tures of a few °K.

5. NEGATIVE TEMPERATURE COEFFICIENTS OF RESISTANCE IN FILMS

a. Discontinuous Films

In the preceding sections it was assumed that the films were completely continuous and that the film thickness was uniform everywhere.   However, as is discussed in Chap. 8, thin films are anything but continuous during their early stages of growth. Instead, they are made up of small islands which may or may not be physically con-nected to one another.   Even if the individual islands possess resistivities approach-ing the bulk and are multiply interconnected, a high resistivity will still be measured— for purely geometrical reasons—but the temperature coefficient of the film should be close to the bulk value.   In practice, however, the temperature coefficient of very thin but moderately conductive films rarely approaches the bulk value; in fact, it is rarely even positive.   Instead, such films usually exhibit large negative temperature coeffi-cients, and it is with films of this type that we are concerned in this section.   Electrical conduction in films of this type has been reviewed by Neugebauer and Wilson,[48] and the following two sections contain a good deal of material from their article.

Extensive measurements on the electrical properties of thin discontinuous films have been made by Mostovetch and Vodar.[49] Film resistivity, temperature coefficient of resistance, and field dependence of resis-tances were investigated.   The resistance vs. temperature curves obtained by them for discontinuous films of three metals are shown in Fig. 12.   Good fits were obtained using equations of the type

$$R = A_0 T^{-\phi} e^{\theta/kT} \qquad (15)$$

where A_0, ϕ, and θ are constants of the particular film.   The films were generally nonohmic, their resistance decreasing with applied field.   A linear dependence of resis-

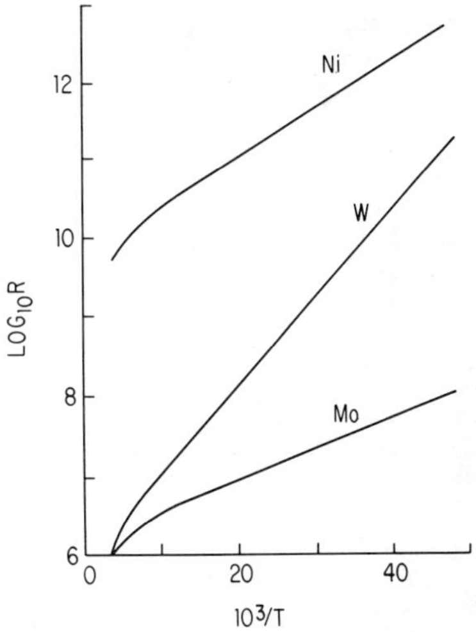

Fig. 12   Resistance vs. temperature for discontinuous films of three different metals.

tance on the square root of the applied field was found for sufficiently high applied fields.

Work published by Minn[50] has clearly indicated (Fig. 13) the rapidly decreasing temperature coefficient of resistance with increasing film resistivity (decreasing film thickness).

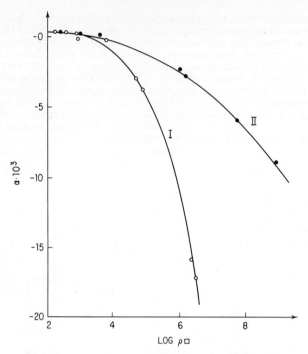

Fig. 13 Effect of sheet resistance on temperature coefficient of resistance.

In the experiments of Neugebauer and Webb,[51] nickel, platinum, or gold was evaporated onto Pyrex glass substrates to give films of average thickness ranging from a few to several tens of angstroms. The resistance of a freshly deposited film always increased with time after the deposition. This was due partly to the decreasing film temperature, which was most likely higher during deposition than the nominal substrate temperature—and partly to some annealing which occurs in films even at room temperature. The film resistance usually stabilized after aging for a few hours, and the measurements were undertaken only with films which were completely stable in the time and temperature interval of the measurements.

Figure 14 is a typical curve showing the dependence of conductance on temperature for a nickel film a few angstroms thick. Good linearity of the log plot of conductance vs. $1/T$ exists. The temperature interval was from 77 to 373°K. Because of the high activation energy for conduction in this particular film it was possible to vary the conductance over six orders of magnitude within a temperature interval of about 200°K.

Many of the films prepared by Neugebauer and Webb obeyed Ohm's law very well, and the conductivity was found to be essentially

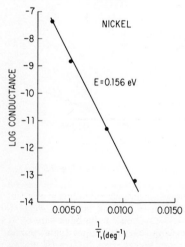

Fig. 14 Typical curve of conductance vs. temperature for a discontinuous nickel film.

independent of field up to several hundred volts per centimeter. However, in some films, particularly the thicker ones, deviations from Ohm's law could be observed at quite low fields, especially at lower temperatures, and a characteristic square-root dependence of applied field was seen. At sufficiently small fields, however, the conductivity always became independent of field, even in these (thicker) films.

Figure 15 shows the log plot of conductance vs. $1/T$ for six platinum films of progressively greater thickness. A film in this series was made thicker by successively evaporating more platinum onto the previous film under otherwise identical conditions. The thickest film was only 10 to 20 Å thick (average thickness). Increasing the average film thickness resulted in an increase in the average particle dimensions, assuming all other conditions remained the same.

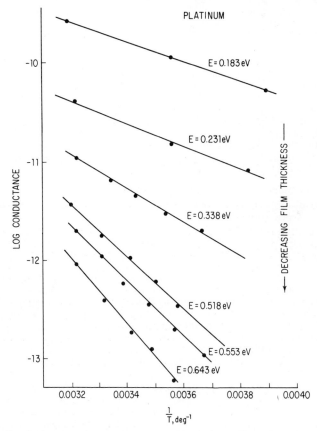

Fig. 15   Conductance vs. temperature curves for platinum films of varying thickness.

Hill[52] has investigated the electrical behavior of discontinuous gold films evaporated onto soda glass or borosilicate substrates. In his configuration, another electrode (to which a dc bias voltage could be applied) was deposited on the back side of the substrate so that the film to be measured was made part of a capacitor structure. The discontinuous metal films measured by Hill were relatively conductive ($R \approx 5.10^3$ ohms) and consisted of rather large islands (260 Å) with large average spacings ($\sim$50 Å). Variations of the resistance of the film with bias voltage were found for both types of glass substrate. The results are shown in Fig. 16.

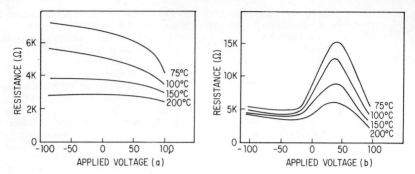

Fig. 16   Resistance of discontinuous films as a function of bias applied to an electrode on the back side of the substrate.   (a) Soda glass.   (b) Borosilicate glass.

Negative temperature coefficients can also occur in films for another reason. Thus, it might be expected that when the substrate has a high thermal expansion and the film a low one, the distance between particles might increase as the substrate expands relative to the films.   An example where a situation of this type prevails is gold on Teflon.   This has been discussed by Neugebauer,[53] who gives data showing, for example, that relatively thin gold films on Teflon substrates have large negative temperature coefficients in the order of $-0.017/°C$.   As the films are increased in thickness, the temperature coefficient becomes progressively more positive, having a value for the thickest films of the order of $0.0012/°C$.

b. Theories of Conduction in Discontinuous Films

(1) Transport by Thermionic Emission   Historically, this was the first mechanism postulated to explain current transport between islands.   Because of the experimental temperature dependence of the thermionic current, functional agreement is obtained with experiment.   However, since the activation energies observed in discontinuous film conduction are very much smaller than the work functions of the corresponding bulk metals, it has been necessary to postulate (1) that this lower activation barrier is due to an overlap of the image-force potentials of two islands very close to each other; (2) that a small particle has a lower work function than bulk material; or (3) that the shape of a small particle brings with it a reduced work function.   However, if electrons are injected into the substrate,[54] it is necessary to consider only the difference between the metal work function and the electron affinity of the insulator.

In the treatment of Minn,[50] for instance, the barrier between particles is taken to be lower than the work function of the bulk material by an amount $\gamma q^2/d$ due to image forces, where γ is a function of the size of the islands and the distance d between them.   The final expression for the conductivity is

$$\sigma = \frac{BqTd}{k} \exp - \left(\frac{\phi - \gamma q^2/d}{kT} \right)$$

where B is a constant (characteristic of each film) and ϕ is the work function for the bulk material.

A remarkably good fit to experimental values of the conductance vs. the square root of the applied potential for thin gold films at various temperatures has been obtained by Skofronick et al.[55] using a thermionic model.   From this fit, the average distance between islands was deduced, and found to be quite large—as much as 59 Å in one example.   On the other hand, the barrier height to be surmounted by the thermal electrons turned out to be very small, ranging from 0.0074 to 0.0258 eV for zero applied voltage in their examples.

(2) Transport by Activated-charge-carrier Creation and Tunneling   This mechanism has been proposed by Neugebauer and Webb[51] for the limit of very small islands of uniform size. For such a case they have postulated that the activation energy is that required to transfer charge from one initially neutral island to another some distance removed. The importance of this electrostatic energy in electrical conduction between small particles has been indicated by Gorter[56] and Darmois.[57] This energy will be of the order of magnitude q^2/r, where q is the electronic charge and r is the average linear dimension of the particles. Only electrons or holes excited to states of at least this energy above the Fermi level will be able to tunnel from one neutral island to another. The process is thus an activated one leading to an equilibrium number of charged islands.

Note that the transfer of charge from a charged island to a neutral one is not activated, because it does not lead to a net increase of the energy of the system, in contrast to the case for two initially neutral particles. Interactions between charges are neglected, leading to the requirement that $n \ll N$, where n is the number of charges and N is the number of particles. This requirement is best met when the islands are very small.

It should also be noted that tunneling from island to island is different from tunneling between two electrodes kept at fixed potentials by a battery, since the Fermi levels of the particles are not held fixed. When an electron tunnels between two electrodes whose Fermi levels are fixed, no electrostatic work has to be done since the positive charge left behind is immediately neutralized and no energy is added to the system. When, however, an electron is removed from one island and is added to another some distance removed, energy is added to the system.

In Neugebauer and Webb's model, a "film" consists of a planar array of many small, discrete metal islands of linear dimension r, separated by an average distance d, which is also small. At equilibrium, and at any temperature above 0°K, a certain number of particles are charged, having lost or gained an electron to or from initially neutral particles some distance away. If the activation energy is supplied by thermal means, the equilibrium number of charge carriers will then be of the order of

$$n = Ne^{-\epsilon/kT}$$

where ϵ is the effective activation energy, which approximately equals q^2/r.

Thus, the model pictures a large number of metal islands N of which a relatively small number n are charged. The equilibrium concentration of these charge carriers is maintained thermally, and there is only random motion of the charge in the absence of a field. Applying a field displaces the relative position of the Fermi levels of neighboring particles, and the tunneling probability increases for a transition of charge from island to island in the field direction, and decreases a corresponding amount against the field direction.

The final expression derived for the conductivity is

$$\sigma = \frac{A\sqrt{2m\phi}}{h^2 D} \exp\left(\frac{-4\pi d}{h}\sqrt{2m\phi}\right) B \exp\left(\frac{-q^2/Kr}{kT}\right) \qquad (16)$$

where A and B are constants, ϕ is the potential barrier between islands, m is the electronic mass, and K is a dielectric constant.

This equation predicts that the conductivity of a film consisting of discrete islands varies exponentially with temperature and is independent of applied field.

(3) Transport by Tunneling between Allowed States In this treatment by Hartman,[58] use is made of the quantization of the energy levels in a small particle. It is postulated that in order for charge transfer to occur between two islands by tunneling, the two energy levels between which the electron tunnels must be "crossed," i.e., their energy widths must overlap. Such an overlap is most likely if the width of the energy levels is great. The width of an energy level in a particle is assumed to be determined by the lifetime of an electron in that level, which, in turn, is assumed to depend on its tunneling probability out of this level. The width of the nth level is, from the uncertainty principle,

$$\delta E_n \approx \frac{\hbar^2}{a}\left(\frac{2E_n}{m}\right)^{\frac{1}{2}} P(E_n)$$

where a = linear dimension of a one-dimensional box, E_n = energy of nth level, and m = electronic mass.

Also, $P(E_n)$ is the probability of tunneling out of level E_n and is given by

$$\exp\left(\frac{-2}{\hbar}\int_{x_1}^{x_2}\{2m[V(x) - E_n]\}^{\frac{1}{2}}\,dx\right)$$

where $V(x)$ is the barrier height at a distance x between islands, and x_1 and x_2 are the values of x where $E_n = V_x$.

Since the tunneling probability is small from levels of energy which are low compared with the barrier height, the width of this level is small, and overlap with a corresponding level in the second island is unlikely. Tunneling could therefore not occur. However, if the electron is activated into a higher level (from which the tunneling probability is larger and, therefore, the level is wide), overlap of the levels of the two islands is much more probable. The energy required to activate an electron from the Fermi level to the first excited level is taken as the activation energy for conduction in this model.

The final equation for the conductivity becomes

$$\sigma = \sigma_0 \exp\left(\frac{-\Delta E_n}{kT}\right) \tag{17}$$

where σ_0 is proportional to $P(E_n + 1)$ and

$$\Delta E_n = E_{n+1} - E_n$$

(4) Transport by Substrate-assisted Tunneling In this model (proposed by Hill[52]), the dielectric substrate is assumed to have a large number of trapping sites between the conduction and valence bands, and the electrons tunnel into these traps rather than make the jump from island to island all at once. The activation energy for conduction in this model is taken to be the energy required to activate an electron from one trap to the trap of the next highest energy in the substrate (Fig. 17). This energy will depend on the distribution of traps in the band gap. Since the process occurs by several steps, the tunneling probability is very much higher than for tunneling in one step, because of the strong e^{-d} dependence.

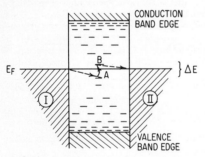

Fig. 17 Model for substrate-assisted tunneling.

In the example shown, if there is no trap at the Fermi level, tunneling must occur from island I into trap A, below the Fermi level. However, since tunneling from A into island II is not possible (because there are no empty states in II), the electron must be thermally activated to trap B (above the Fermi level), from which it can now tunnel into island II. It is clear that through this mechanism tunneling can occur over larger distances.

By making the discontinuous film one electrode of a capacitor, with the substrate as the dielectric, and applying a dc bias (as described in Sec. 5a), the Fermi level in the substrate is lowered toward the valence band. Since the density of the traps and the separation between them depend on their position in the band gap, the activation energy for conduction will change if the position of the Fermi level is changed. In particular, Hill[54] assumed many traps near the band edges, very close to each other, and fewer of them (farther apart) in the middle of the gap. Hill's experimental results on the resistance with applied bias voltage, shown in Fig. 16 for two types of glass substrates, can be explained by taking the Fermi level in the glass at the appropriate position in the gap.

In later papers, Hill has developed a more general version of his theory.[54] In the resulting equations, the current density in the film is expressed as a direct function of the structure of the film, the electrical properties of the substrate, the applied voltage, and the temperature. According to this theory, charge is always transferred between particles by way of the substrate, either through trap hopping (discussed above) or through thermionic emission into the conduction band of the insulator that constitutes the substrate. Experimental evidence is provided to support the theory, and it is also shown that the bias effect (Fig. 16) can be observed only when the substrate contains mobile alkali ions and when the spacing between particles exceeds 30 Å.

(5) Other Models A modified version of Neugebauer and Webb's theory which incorporates some of the ideas discussed by Hill has been proposed by Herman and Rhodin.[59] They examined films of silver, gold, and platinum deposited onto single-crystal sodium chloride substrates in vacua of the order of 10^{-7} to 10^{-8} Torr. Their results suggested that tunneling occurred through the substrate rather than through the vacuum between the particles. The current carriers were believed to be contributed to the insulator surface regions by the metallic microparticles, and in addition, an activation energy was necessary for tunneling to occur. This energy is electrostatic in nature and is dependent upon the average size and separation of the microparticles. The presence of traps in the substrate was suggested, and a distinction was drawn between the filled traps under the metallic particles and those between the particles (traps in the two regions not necessarily being at the same level). Although themselves electrically neutral, the presence of the metallic microparticles could cause energy differences between the traps, because of image forces. The model thus appears to combine some of the features of Neugebauer and Webb and of Hill.

As might be expected, films that are partially continuous have their resistances governed by both an activation mechanism as well as the "normal" lattice-scattering mechanism. Feldman has studied films in this thickness range.[60] Their resistivity can be expressed by an equation of the form

$$\rho = \rho_0[1 + \alpha(T - 273)] + Ce^{\theta/kT}$$

where the first term is the normal temperature dependence and the second represents a thermally activated process of some kind. Figure 18 shows a resistance vs. temperature curve for films of this type. In the first portion of the curve, the activated conductivity contribution dominates—after which the conventional temperature coefficient becomes more important. Note

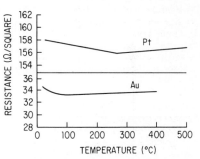

Fig. 18 Sheet resistance vs. temperature for partly continuous films.

that the separate grains that made up the film must have been reasonably pure, since the conventional temperature-coefficient contribution was fairly close to the bulk value.

Another model in which conduction in discontinuous films is the result of two mechanisms operating together has been suggested by Milgram and Lu.[61] It was shown that the I-V characteristics of their films (chromium on glass overcoated with SiO) could be well described by an equation of the form

$$I = AV + BV^n$$

where A and B are constants which are temperature-dependent and n is a number which approaches 2.0 at high temperatures. At low temperatures ($\sim 4°K$), B is the dominant coefficient, being in the order of 10^{-7}, whereas A is less than 10^{-10}. However, at high temperatures ($400°K$), A dominates and is in the order of 10^{-6}, whereas B is approximately 10^{-8}. The authors believe that the linear portion is due to tunneling via thermally activated traps by a mechanism similar to that of Hill, although A could not be fitted by a single activation energy. The B term of their equation is ascribed to space-charge-limited current, which should follow a V^2 law. At high temperature and near zero field the current is carried mainly by thermally excited carriers whose number is a function of the temperature and density of trapping centers. At low temperature the conduction is dominated by the space-charge-limited current.

(6) Range of Applicability of Various Models It appears that the model of Neugebauer and Webb satisfies the qualitative observations on the conductivity of discontinuous films for smaller particle size, whereas Hill's mechanism offers an acceptable explanation for the larger-particle larger-gap films. Quantitative correlation of activation energies with island sizes is necessary to compare these two models further. Field-effect measurements on smaller particles should also be significant.

Figure 19 gives the various regions of island size and separation at which the various conduction mechanisms are most applicable.

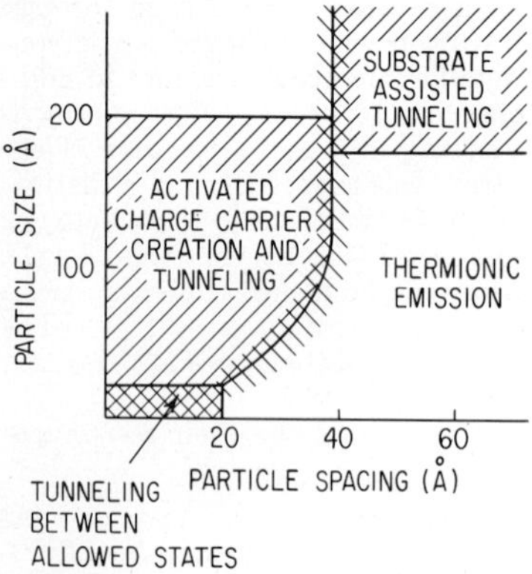

c. Continuous Films

The negative temperature coefficients discussed above apply to films which are still in the island stage of growth and are a consequence of the energy required for electrons to cross the spaces between the islands. However, negative temperature coefficients are commonly seen in metallic films which are well past the island stage and which may be many thousands of angstroms thick.

It appears to be generally true that negative temperature coefficients are not seen in continuous metal films unless impurities are present. Also, the greater the deviation of the film resistivity from the bulk resistivity, the more negative the temperature coefficient. Thus, much work in thin film resistors (see Chap. 19) has been devoted to attempts to find high-resistivity films that do not have large negative temperature coefficients. Since the impurities that are trapped in a film come largely from the background gases present during evaporation, the condensing film will act as a "getter" for these gases. This phenomenon is well known and has been exploited in many areas such as, for example, the titanium sublimation pump, which has a strong affinity for virtually everything except the rare gases.

Fig. 19 Range of particle size and spacing over which various theories of conduction in discontinuous films are applicable.

One consequence of this gettering action by freshly deposited films is that, as deposition proceeds, the residual gas background is reduced and, consequently, a purer film results. The resistivity of a film thus frequently comes closer to the bulk value as the film grows (assuming constant deposition rate) until the point is reached when the rate of gettering just equals the rate at which fresh impurities enter the vacuum chamber. Because of this, the temperature coefficient of the initial layers of the film will often be strongly negative, whereas the coefficient is just as strongly positive for the upper layers. At some point the temperature dependence of the two layers will just cancel, and this has led some workers to report a "magic thickness" at which zero temperature coefficient is obtained.

Evidence that the resistivity and temperature coefficient of tantalum films become independent of their thickness (down to at least 500 Å), if they are sufficiently pure, has been provided by Maissel and Schaible.[62] Using bias sputtering, a process which prevents films from doing any appreciable gettering (see Chap. 4), they showed that resistivity variations due to thickness became smaller and were eventually eliminated as the gettering ability of the films was progressively reduced. This is illustrated in Fig. 20.

As noted above, the contribution of grain boundaries to the resistivity of metal films is small. However, impurities trapped in a film during its deposition can subsequently migrate to the grain boundaries where there is a high probability that precipitation will occur. In addition, it is well known that diffusion along grain boundaries is several orders of magnitude faster than it is in the bulk; so contamination from external sources may also occur on life.

The effects of grain-boundary oxidation have been studied in some detail for tantalum.[63] Films of this material sputtered under reasonably clean conditions and then subsequently heat-treated at 250°C showed only very small increases in resistivity when heat-treated in vacuum. On the other hand, samples heated in air showed increases of over 100%. Similar films depos-

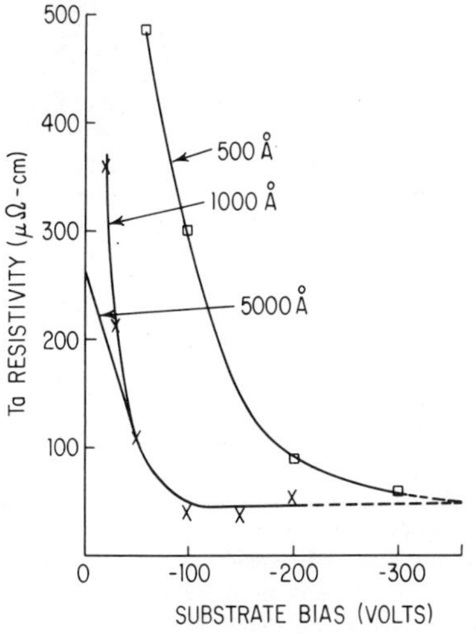

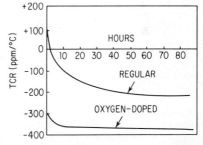

Fig. 20 Lack of dependence of resistivity on film thickness when gettering is suppressed.

Fig. 21 Temperature coefficient of resistance of tantalum films as a function of heat treatment in air at 250°C.

ited in an atmosphere containing 0.1% of oxygen showed increases in resistance when vacuum-heated as well as when air-heated. None of the resistivity increases, even those obtained for the pure films heat-treated in air, could be explained on the basis of simple surface oxidation.[64] It was shown that grain-boundary oxidation of the films was occurring both internally because of migration of trapped oxygen, as well as externally, in from the surface.

Such oxidized grain boundaries give rise to negative temperature coefficients in a manner that is closely analogous to what is observed in discontinuous films. Some results, showing TCR as a function of heat treatment, are shown in Fig. 21. Thus it is clear that films which have undergone grain-boundary oxidation are not electrically continuous, even though they are physically continuous. At this time, no quantitative theory has been developed to deal with negative temperature coefficients in films of this type. This is not surprising, since the problem is complicated by the fact that the effective spacing between grains varies as a function of depth in the film.

6. HIGH-FREQUENCY CHARACTERISTICS OF THIN FILMS

As was discussed above, negative temperature coefficients can occur in thin films because of lack of electrical continuity between the grains. In the case of very thin (agglomerated) films, the grains are separated by air spaces, whereas in thicker films the space between the grains is filled with dielectric material. In either case, the negative temperature coefficient arises because of the activation energy associated with electron transitions from grain to grain.

A structure in which metallic grains are separated by small spaces is electrically equivalent to a series of capacitors, and consequently its ac impedance is less than the dc value. Furthermore, one would expect this difference between ac and dc conductance to become progressively larger as the temperature is reduced, since the dc resistance of discontinuous films increases rapidly with decreasing temperature.

An excellent illustration of this may be seen in Fig. 22, taken from Offret and Vodar,[65] for platinum on Pyrex. As can clearly be seen, under dc conditions the film exhibits a temperature coefficient which is close to zero at room temperature but which becomes increasingly more negative as low temperatures are achieved. However, at a frequency of 1 MHz the spaces between the grains are effectively short-circuited by the alternating current, and a positive temperature coefficient is seen, reflecting the TCR of the metallic grains themselves. Note also that at sufficiently high temperatures the direct current between the grains is large enough for the resistance to be dominated by that of the grains themselves, and no difference is seen between the dc and ac values of resistance.

Fig. 22 Comparison of dc and high-frequency resistance of a discontinuous film as a function of temperature.

In making measurements of the high-frequency falloff in resistance of thin films, allowance must be made for a geometrical effect which has nothing to do with the microstructure of the film. This effect, which has been discussed by Howe,[66,67] arises because of capacitive coupling between one half of the film and the other. This can be visualized by imagining a film initially folded over upon itself as shown schematically in Fig. 23a and then opened up as shown in Fig. 23b. Although the capacitance between the adjacent portions is reduced as the film is "straightened out," it does not go to zero, and finite capacitive coupling between the two halves of the film occurs.

A curve showing the magnitude of this geometrical effect, as calculated by Howe,

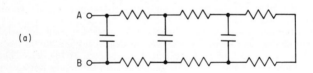

(a)

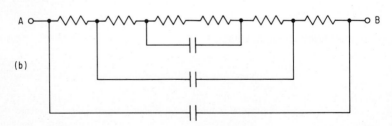

(b)

Fig. 23 Representation of capacitive coupling between the two halves of a thin film.

is shown in Fig. 24. To compute the resistance-frequency behavior of a particular sample of dc resistance R at frequency f, it is necessary to know the capacitance per unit length c. This is a number of the order of 0.1 to 1 pF cm^{-1}, depending on the aspect ratio of the resistor. For example, a resistor of 20 squares has a value of 0.12 pF cm^{-1} while a resistor of 10 squares has a value of 0.17 pF cm^{-1} (dielectric constant of substrate = 1).

Thus, unless the effect of the distributed capacitance is known, no quantitative interpretation can be made of the falloff with high frequencies. For example, Fig. 25[68] shows the frequency-resistance behavior of a lead sulfide film at 25°C where it is seen that at low frequencies the distributed capacity effect dominates, while the intercrys-

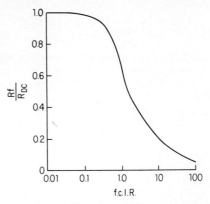

Fig. 24 Falloff in the resistance of a thin film with increasing frequency as a consequence of capacitive coupling.

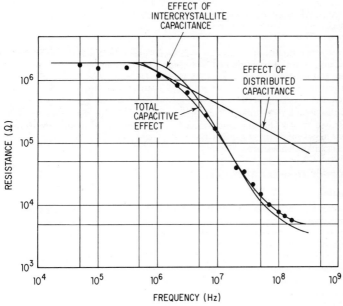

Fig. 25 Resistance vs. frequency behavior of a PbS film, showing both geometrical and structural contributions.

tallite capacitance is the prime factor at high frequencies.

One way to avoid the distributed capacitance effect and thus see only the frequency falloff due to structure is to fabricate the film in the form of an annulus as shown in Fig. 26. This configuration is then made to terminate one end of a transmission line, the strength of the signal reflected off the end being a measure of the ratio in impedance between the sample and the line. The other end of the line is terminated by a short circuit so that capacitance effects are balanced out and only the resistive component is measured.

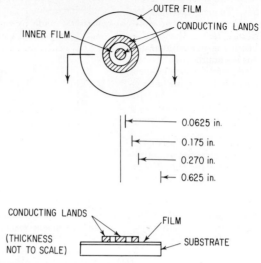

Fig. 26 Annular geometry used for studying resistance vs. frequency characteristics of thin films.

Using measurements of this type, Maissel[63] was able to show that tantalum films of high purity and having a positive temperature coefficient exhibited no resistance falloff to frequencies as high as 10 kMHz. On the other hand, films which had developed large negative temperature coefficients as a result of heating in air showed a pronounced resistance falloff at frequencies in excess of a few hundred MHz. As might be expected, at sufficiently high frequencies the ac resistance approached the dc resistance which the film had exhibited prior to being aged, confirming the fact that heat treatment causes grain-boundary oxidation from the surface of the film down toward the substrate.

7. INFLUENCE OF HEAT TREATMENT

a. Annealing

As we have seen, unless a great deal of special care is taken during their deposition, films will, in general, contain a host of structural defects. In view of the fact that these represent a considerable departure from equilibrium, we would expect some, if not all, of these defects to anneal out of the film on the application of relatively mild heat treatment. This should lead to a corresponding decrease in film resistivity, and in general this is the case. In certain instances, however, heat treatment may lead to an increase in resistivity because of the effects of oxidation and/or agglomeration.

General theories of annealing as well as methods for the analysis of annealing curves are discussed by Damask and Dienes.[6] Of more interest for films are the kinetics of processes distributed in activation energy. These have been reviewed in some detail by Primak.[69] A theory of resistivity changes associated with the annealing of defects specifically aimed for films has been presented by Vand.[70] This will now be discussed in more detail.

Vand begins by postulating that the distortions whose decay is observed during annealing are what he calls the "combined type," that is, vacancies and interstitials in close proximity to one another. These require a characteristic energy E to make them combine with and cancel one another, but the energy required to bring about their migration toward each other is considered to be negligible. E can vary from zero all the way to the activation energy for self-diffusion.

If $r(E)$ is the contribution to the residual resistivity made by one distortion per unit volume, and $N(E) \Delta E$ is the number of distortions per unit volume that have decay energies between E and $E + \Delta E$, then the total contribution to the resistivity made by imperfections is given by

$$\rho_i = \int_{-\infty}^{\infty} r(E)N(E)\, dE$$

The value of ρ_i may change with time as a result of annealing. A distribution function describing the "energy spectrum" of the distortions in the film can then be defined. It will be of the form $F_0(E) = r(E)N(E)$.

As a result of his analysis, Vand shows that

$$F_0(E) = -\frac{1}{kU}\frac{d\rho}{dT} \tag{18}$$

where $d\rho/dT$ is the slope of the resistance vs. temperature curve obtained when the film is allowed to heat up at a uniform rate, and its resistance is measured as function of temperature (and, therefore, also of time). Allowance is made for the portion of $d\rho/dT$ that is due to the normal temperature coefficient of resistance. k is Boltzmann's constant and U is defined by

$$U = \frac{u(u + 2)}{u + 1} \tag{19}$$

where

$$u + \log u = \log 4ntv_{\max} \tag{20}$$

n = number of atoms that can initiate a defect (this number is not known, exactly, but is estimated as being close to 10)

$v_{\max}$ = Debye cutoff frequency for the lattice

t = time that it took to reach the particular temperature at which $d\rho/dT$ was measured

u is readily obtained by successive approximation in Eq. (20) so that $F_0(E)$ may be plotted against E ($= ukT$) to give the "lattice-distortion" spectrum.

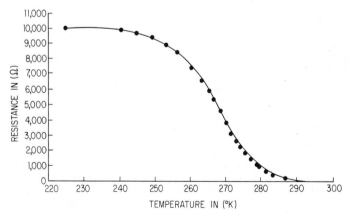

Fig. 27 Annealing curve for an evaporated gold film.

An example of a curve of resistance vs. temperature and time for an evaporated gold film[71] is shown in Fig. 27.

To obtain the distribution function $F_0(E)$, it is now necessary to measure the slope of this curve at successive values of temperature (and, hence, time), correct the resistivities for temperature coefficient, and substitute the results in Eq. (18). In

Eq. (20) the value of u can be calculated for any given values of t [hence U in Eq. (19) is obtained]. The results of this analysis for the curve of Fig. 27 are shown in Fig. 28 for two different deposition rates. The results show that for lower deposition rates, fewer defects are formed, and that these have a smaller decay energy. Separate experiments showed that E_{max} and F_{0max} were independent of film thickness over a range of 100 to 900 Å, for constant deposition rate.

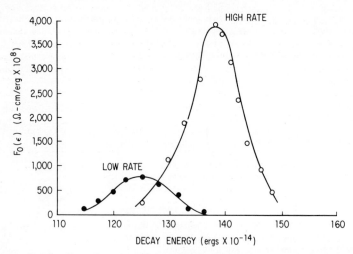

Fig. 28 Lattice-distortion spectrum of gold films deposited at high and low rates.

In a later study, Wilkinson[72] also investigated the effect on the defect spectrum of various background pressures of nitrogen, ranging from 10^{-7} to 0.11 Torr. Three groups of distortions were found at approximately 80, 100, and 120 ($\times 10^{-14}$ erg), the exact values depending on the conditions of preparation. It was shown that the number of distortions decreased as the nitrogen pressure increased, particularly for the two defect types of lower energy. It was argued that the effect of the nitrogen could not be to assist the gold atoms (via collision) to find their correct lattice positions, since the average kinetic energy of nitrogen molecules at 300°K is only 5 to 10% of the distortion decay energy. However, since a distortion is estimated to contain approximately 10 atoms, it is conceivable that during the growth of the film, the nitrogen atoms act separately on the individual atoms that compose the distortion, thus allowing it to be destroyed.

An analysis of defect distribution by Vand's technique has been performed for carbon films by Troyoda and Nagashima.[73] They found the values of the peak energies to be thickness-dependent. Specifically, for films 110 Å thick the peak was at 2.3 eV; for 380-Å-thick films the peak was at 1.6 eV; whereas for 2,000-Å films it was 1.4 eV. Films had to be annealed to about 900°C before a resistivity close to the bulk value could be achieved.

An alternative explanation suggested to account for the decay of resistance on annealing is that vacancies which are present in the film in numbers that far exceed their equilibrium value coalesce into progressively larger groups until they finally organize into sheets and collapse to form dislocation rings.[74] Supporting evidence for this hypothesis was obtained from indirect measurement of the volume change that occurred in evaporated nickel films when they were annealed. This volume change (deduced from observations on strain) was found to correspond fairly closely with the resistivity changes. Initially the film expands because of coalescence of vacancies and then, at higher temperatures, it contracts because of the collapse into dislocation rings.

It was observed earlier that some metals have lower electrical resistivity in the liquid state than in the solid. If the amorphous state can be regarded as closely analogous to a supercooled liquid, we would expect such films to have their lowest resistance while in the amorphous state, and annealing should then lead to an increase rather than a decrease in their resistance. Such behavior has, in fact, been observed in a number of metals. For example, bismuth films[75] deposited at 2°K reached a maximum in resistivity at approximately 80°K as the temperature was slowly raised to 300°K. Thereafter, the resistivity decreased with further heat treatment in the normal fashion. Similar, though more complicated, behavior was seen for gallium films. Additional examples such as antimony are discussed in Ref. 76.

The annealing behavior of metal-alloy films is, in general, similar to that of the elemental films discussed above, provided that the composition of the alloys does not differ from the equilibrium values found in bulk. In many cases, however, it is possible to deposit alloy films in which the constituents are present in proportions considerably different from what is thermodynamically permissible. Mader and coworkers[77] have studied a number of such systems. By performing a two-source deposition onto substrates kept at liquid-nitrogen temperature, nonequilibrium compositions were formed by "vapor quenching."

As an example, we may consider the copper-silver system. Figure 29 shows the annealing behavior of a pure silver film as compared with silver films containing 12 and 50% copper, respectively. After deposition at liquid-nitrogen temperature, the films were heated at a constant rate of 30 to 60°C per hour. Surprisingly, the resistivities of all three films as deposited at 80°K were very close to one another. However, the behavior on annealing is seen to be quite different. The pure-metal films show a steep, as well as continuous, decrease in resistivity from 80°K to room temperature.

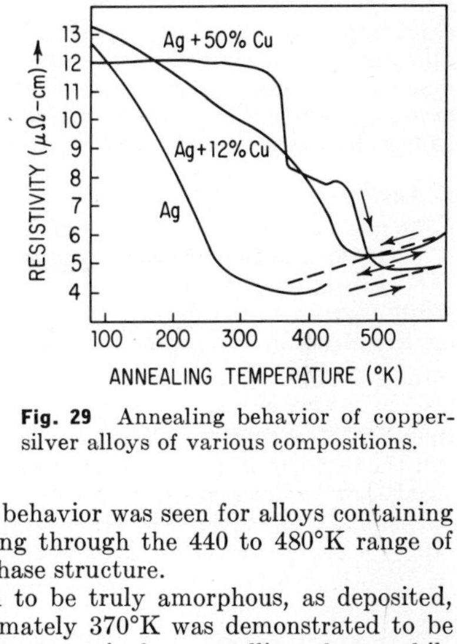

Similar behavior is seen up to 400°K for the films containing 12% copper, although the drop is less steep. These changes appear to be due, as in elemental films, to grain growth and increased perfection. It was concluded that, as deposited, these

Fig. 29 Annealing behavior of copper-silver alloys of various compositions.

films are a single, crystalline phase. Similar behavior was seen for alloys containing up to 35% copper. However, upon annealing through the 440 to 480°K range of temperatures, the films changed into a two-phase structure.

Films containing 50% copper were shown to be truly amorphous, as deposited, and the sharp drop in resistivity at approximately 370°K was demonstrated to be due to a transition from the amorphous state to a single, crystalline phase, while the second resistivity drop corresponded to that which was also seen in the more dilute alloys, namely, reversion to a two-phase system. The as-deposited resistivity of the amorphous phase corresponded quite closely to the resistivity of the liquid alloy (extrapolated back to 80°K).

The fact that the alloy containing 50% copper passed through an intermediate stage as a single-phase structure (rather than going directly to the more stable two-phase structure) was explained in terms of the kinetics of the transformation from the amorphous to the crystalline state. This should be more rapid than the transformation to a two-phase structure. Breaking up of the "log jam" present in an amorphous state requires diffusion over only one or two atomic distances. By contrast, the process of segregation into two phases requires relatively long range diffusion.

Early work on the annealing of metal films condensed at liquid-helium temperatures has been reviewed by Neugebauer.[76] Work similar to that reported by Mader has also been done for thin tin films deposited at 20°K and deliberately contaminated with up to 20% copper.

Studies of the annealing characteristics of these metastable-alloy films have also shown[79] that deposition at a given substrate temperature has considerably more effect on the resistivity of a film than deposition at a lower temperature followed by a subsequent anneal at the higher temperature. For example, films containing 50% copper and 50% silver when deposited at 80°K and then raised to 300°K had a constant resistivity of about 12 μohm-cm. On the other hand, deposition of the same material at a substrate temperature of 300°K produced a film having a resistivity of only 8 μohm-cm. This phenomenon is explained as being due to the fact that greater atomic mobility can be achieved during deposition than in the finished film. This is reasonable, since surface diffusion is much greater than bulk diffusion.

Annealing by a somewhat different mechanism can be seen in systems in which an insulating phase is codeposited with the metal. For example, Feldtkeller[80] has studied the annealing of copper films that have been codeposited with SiO. Films were deposited at liquid-helium temperatures, and their resistance was seen to decrease by many orders of magnitude when they were allowed to warm up to 400°K. These results were interpreted as due to the formation of an insulating interface of SiO around the copper grains. `As heat treatment proceeded, grain growth of the copper occurred by recrystallization, and the number of grain boundaries became less. The SiO was then squeezed out from between the copper grains into small agglomerates whose influence on the resistivity was small.

In systems where the insulating phase and the metallic phase can chemically react to produce an additional metallic phase, the behavior during annealing is rather different. For example, in the system Cr-SiO, the phase Cr_3Si is formed during annealing,[81] and far larger proportions of SiO can be tolerated in such films without excessive resistivities after annealing. This is because the Cr_3Si forms conductive bridges between the chromium crystallites (see also Chap. 18).

b. Agglomeration and Oxidation

As was discussed in Chap. 8, during their early stages of growth, thin films consist of individual islands which have many of the characteristics of liquid droplets. As the substrate temperature increases, the contact angle between droplet and substrate decreases and the droplets grow in size. Films deposited at lower temperatures consist of smaller droplets, since the film particles have lower mobility and remain more or less where they land on the substrate. Thus, a given amount of material deposited on a substrate at a low temperature may be enough to form a continuous film, whereas the same amount of material deposited on a hotter substrate will result in an island structure.

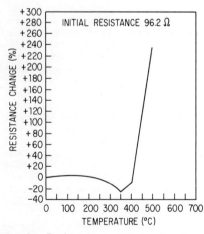

Fig. 30 Resistance vs. temperature for a thin film showing sudden irreversible increase associated with agglomeration.

Because of this phenomenon, it is found that very thin films (in the order of a few hundred angstroms, or less) which possess a continuous structure at their deposition temperature will, when heat-treated, revert to an island structure. This process is referred to as agglomeration and is easily detected by the large increase in resistance that occurs if the film is subjected to too high a temperature. An example is shown in Fig. 30 (taken from Ref. 82). For a given substrate-material combination, there is thus a definite lower limit to the thickness of thin films that are capable of standing up to a given amount of heat treatment. Belser and Hicklin[82] have listed the limits of sheet resistance for various metal films that could survive temperature cycling to 600°C. These vary from a few ohms per square for gold films to more than 4,000 ohms/sq for tungsten films.

Since agglomeration occurs rather suddenly, and since it is associated with a loss of desirable properties, only a few quantitative observations of agglomeration have been reported. In one such study of gold films on zinc sulfide substrates,[83] agglomeration is pictured as occurring in three stages. Initially, small holes are formed in the films, the holes then enlarge, and finally the film separates into islands. The observed resistance changes occur primarily during the second and third stages. It was concluded that the process is a thermally activated one associated with the migration of gold atoms. For this particular system, an activation energy of 1.8 eV was measured.

Skofronick and Phillips[84] have studied the incidence of additional agglomeration in films that, as deposited, are already discontinuous. In films deposited at 90°K and then warmed to room temperature, the resistance first decreased slightly and then increased irreversibly by several orders of magnitude. The model which they invoked to explain the observed behavior was that the effect of heat treatment was to cause motion of the smaller islands, which then coalesced with the larger ones. The mobility of the islands on the substrates and the number of islands with sufficient energy to move are determined by an assumed thermally activated process. As the islands combine, the mean gap between them increases. A typical activation energy was of the order of 0.2 eV. It was also shown that the influence on the results of thermal expansion of the substrate was negligible.

Unless they are suitably protected, many thin films when heated in air undergo increases in resistance because of oxidation. Ultimately they may be converted to insulating films if the treatment is carried on long enough. It has been found that the increase in the resistance of a film is considerably greater than can be accounted for on the basis of only the reduced thickness of the conductive portion of the film resulting from oxidation of the surface. Quantitative evidence for this has been provided by Basseches in a study of the oxidation of tantalum films.[64] He suggested that the oxidation was occurring along the grain boundaries of the films, a hypothesis that had also been advanced to explain the oxidation of iron films.[85] Confirmation of this concept was later obtained by Maissel,[63] who showed that the oxidation of the grain boundaries could occur either as a result of oxygen diffusing in from the surface or from trapped oxygen diffusing internally and then precipitating at the grain boundaries as an insulating phase. The progress of grain-boundary oxidation was correlated with the high-frequency characteristics of the film. It seems likely that these results for tantalum films can be generalized for most other films.

It is interesting to note that even though the change in resistance that occurs during oxidation is much greater than can be accounted for from the equilibrium thickness of oxide at the surface, the actual thickness of the oxide on the surface appears to be less than the corresponding thickness in bulk material, at least for the case of tantalum.[86] The effect is not understood, and it is speculated that it might be due to the known lack of crystallinity of the oxides of thin films deposited on smooth substrates. The fact that the film was even thinner than the usual value furthers the important role of grain-boundary oxidation.

In the case of alloy films, it is possible for preferential oxidation of one of the components to occur. For example, it has been shown[87] that alloy films of the Chromel family (containing nickel, chromium, and iron) exhibit a complicated resistance vs. time curve when heat-treated in air. Initially, the resistance of the film increases with time because of the oxidation of the iron. However, once oxidation of this component is complete, the effect of annealing of defects becomes more important and the resistance begins to decrease with time. Finally, the resistance begins to increase once more as the major fraction of the defects has been annealed out, and the dominant contribution to further change comes from oxidation of the chromium and nickel constituents.

REFERENCES

1. Wilson, A. H., "The Theory of Metals," Cambridge University Press, New York, 1958.

2. Mott, N. F., and H. Jones, "The Theory of the Properties of Metals and Alloys," Dover Publications, Inc., New York, 1958.
3. Bardeen, J., *J. Appl. Phys.*, **11**, 88 (1939).
4. Linde, J. O., *Ann. Physik*, **15**, 239 (1932).
5. Gerstenberg, D., and C. J. Calbick, *J. Appl. Phys.*, **35**, 402 (1964).
6. Damask, A. C., and G. J. Dienes, "Point Defect in Metals," Gordon and Breach, Science Publishers, Inc., New York, 1963.
7. Reale, C., *Phys. Letters*, **2**, 268 (1962).
8. Overhauser, A. W., and R. L. Gorman, *Phys. Rev.*, **102**, 676 (1956).
9. Ziman, J. M., "Electrons and Phonons," p. 237, Oxford University Press, Fair Lawn, N.J., 1960.
10. Blewitt, T. H., R. R. Coltman, and J. K. Redman, *Phys. Rev.*, **93**, 118 (1937).
11. Krantz, E., and H. Schultz, *Z. Naturforsch.*, **9a**, 125 (1954).
12. Altman, C., *Proc. 9th Natl. Vacuum Symp.*, 1962, p. 174.
13. Young, I. G., and C. W. Lewis, *Proc. 10th Natl. Vacuum Symp.*, 1963, p. 428.
14. Valdes, L. B., *Proc. IRE*, **42**, 420 (1954).
15. Logan, M. A., *Bell System Tech. J.*, **40**, 885 (1961).
16. Smits, F. M., *Bell System Tech. J.*, **37**, 699 (1958).
17. Zrudsky, D. R., H. D. Bush, and J. R. Fassett, *Rev. Sci. Instr.*, **37**, 885 (1966).
18. Thomson, J. J., *Proc. Cambridge Phil. Soc.*, **11**, 120 (1901).
19. Fuchs, K., *Proc. Cambridge Phil. Soc.*, **34**, 100 (1938).
20. Sondheimer, E. H., *Advan. Phys.*, **1**, 1 (1952).
21. Campbell, D. S., "The Use of Thin Films in Physical Investigations," p. 299, Academic Press Inc., New York, 1966.
22. Dekker, A. J., "Solid State Physics," Prentice-Hall, Inc., Englewood Cliffs, N.J., 1957.
23. Mayer, H., "Structure and Properties of Thin Films," p. 225, John Wiley & Sons, Inc., New York, 1959.
24. Mayer, H., "Physik duenner Schichten," vol. 2, Wissenschaftliche Verlagsgesellschaft m.b.H., Stuttgart, 1955.
25. Worden, D. G., and G. C. Danielson, *J. Phys. Chem. Solids*, **6**, 89 (1958).
26. Nossek, R., "Basic Problems in Thin Films Physics," p. 550, Vandenhoeck and Ruprecht, Goettingen, 1966.
27. Parrott, J. E., *Proc. Phys. Soc., London*, **85**, 1143 (1965).
28. Chopra, K. L., *Phys. Rev.*, **155**, 660 (1967).
29. Abeles, F., and M. L. Theye, *Phys. Letters*, **4**, 348 (1963).
30. Cirkler, W., *Z. Physik*, **147**, 481 (1957).
31. Gillham, E. J., J. S. Preston, and B. E. Williams, *Phil. Mag.*, **46**, 1051 (1955).
32. Ennos, A., *Brit. J. Appl. Phys.*, **8**, 113 (1957).
33. Chopra, K. L., and L. C. Bobb, *Acta. Met.*, **12**, 807 (1964).
34. Lucas, M. S. P., *Appl. Phys., Letters*, **4**, 73 (1964).
35. Chopra, K. L., and M. R. Randlett, *J. Appl. Phys.*, **38**, 3144 (1967).
36. Suhrmann, R., *Advan. Catalysis*, **7**, 303 (1955).
37. Swietering, R., H. L. T. Koks, and C. VanHeerden, *J. Phys. Chem. Solids*, **11**, 18 (1959).
38. Komnik, Yu F., and L. S. Palatnik, *Soviet Phys. Solid State*, **7**, 429 (1965).
39. Drumheller, C. E., *Proc. 4th Natl. Vacuum Symp.*, 1957, p. 27.
40. Andrews, E. R., *Proc. Phys. Soc. London*, **A62**, 77 (1949).
41. Sondheimer, E. H., *Phys. Rev.*, **80**, 401 (1950).
42. Jeppesen, M. A., *J. Appl. Phys.*, **37**, 1940 (1966).
43. Forsvoll, K., and I. Holwech, *Phil. Mag.*, **9**, 435 (1964).
44. Forsvoll, K., and I. Holwech, *Phil. Mag.*, **10**, 921 (1964).
45. Chopra, K. L., and S. K. Bahl, *J. Appl. Phys.*, **38**, 3607 (1967).
46. MacDonald, D. K. C., and K. Sarginson, *Proc. Roy. Soc. London*, **A203**, 223 (1950).
47. Holwech, I., *Phil. Mag.*, **12**, 117 (1965).
48. Neugebauer, C. A., and R. H. Wilson, "Basic Problems in Thin Film Physics," p. 579, Vandenhoeck and Ruprecht, Goettingen, 1966.
49. Mostovetch, N., and B. Vodar, "Semiconducting Materials," p. 260, Butterworth & Co. (Publishers), Ltd., London, 1951.
50. Minn, S. S., *J. Rech. Centre Natl. Rech. Sci. Lab. Bellevue Paris*, **51**, 131 (1960).
51. Neugebauer, C. A., and M. B. Webb, *J. Appl. Phys.*, **33**, 74 (1962).
52. Hill, R. M., *Nature*, **204**, 35 (1964).
53. Neugebauer, C. A., *Proc. 9th Natl. Vacuum Symp.*, 1962, p. 45.
54. Hill, R. M., *Thin Solid Films*, **1**, 39, 1967.
55. Skofronick, J. G., W. B. Phillips, K. T. McArdle, and R. H. Davis, *Bull. Am. Phys. Soc.*, **10**, 364, 1965.

56. Gorter, C. J., *Physica*, **17**, 778 (1950).
57. Darmois, E., *J. Phys. Radium*, **17**, 210 (1956).
58. Hartman, T. E., *J. Appl. Phys.*, **34**, 943 (1963).
59. Herman, E. S., and T. N. Rhodin, *J. Appl. Phys.*, **27**, 1594 (1966).
60. Feldman, C., *J. Appl. Phys.*, **34**, 1710 (1963).
61. Milgram, A. A., and C. Lu, *J. Appl. Phys.*, **37**, 4773 (1966).
62. Maissel, L. I., and P. M. Schaible, *J. Appl. Phys.*, **36**, 237 (1965).
63. Maissel, L. I., *Proc. 9th Natl. Vacuum Symp.*, 1962, p. 169.
64. Basseches, H., *IRE Trans. Component Pts.*, **CP8**, 51 (1961).
65. Offret, M., and M. D. Vodar, *J. Phys. Radium*, **17**, 237 (1955).
66. Howe, G. W. O., *Wireless Eng.*, **12**, 291 (1935).
67. Howe, G. W. O., *Wireless Eng.*, **17**, 471 (1940).
68. Humphrey, J. N., F. L. Lummis, and W. W. Scanlon, *Phys. Rev.*, **90**, 11 (1953).
69. Primak, W., *Phys. Rev.*, **100**, 1677 (1955).
70. Vand, V., *Proc. Phys. Soc. London*, **55**, 222 (1943).
71. Wilkinson, P. G., and L. S. Birks, *J. Appl. Phys.*, **20**, 1168 (1949).
72. Wilkinson, P. G., *J. Appl. Phys.*, **22**, 419 (1951).
73. Troyoda, H., and M. Nagashima, *J. Phys. Soc. Japan*, **14**, 274 (1959).
74. Hoffman, R. W., F. J. Anders, and E. C. Crittenden, Jr., *J. Appl. Phys.*, **24**, 231 (1953).
75. Buckel, W., and R. Hilsch, *Z. Physik*, **138**, 109 (1954).
76. Neugebauer, C. A., "Physics of Thin Films," vol. 2, p. 1, Academic Press Inc., New York, 1964.
77. Mader, S., A. S. Nowick, and H. Widmer, *Acta. Met.*, **15**, 203 (1967).
78. Ruhl, W., *Z. Physik*, **138**, 121 (1954).
79. Nowick, A. S., and S. Mader, Basic Problems in Thin Film Physics, *Proc. Intern. Symp.*, 1966, p. 212, Vandenhoeck and Ruprecht, Clausthal.
80. Feldtkeller, E., *Z. Physik*, **157**, 65 (1959).
81. Glang, R., R. Holmwood, and S. Herd, *J. Vacuum Sci. Technol.*, **4**, 163 (1967).
82. Belser, R. B., and W. H. Hicklin, *J. Appl. Phys.*, **30**, 313 (1959).
83. Kane, W. M., J. P. Spratt, and L. W. Hershinger, *J. Appl. Phys.*, **36**, 2085 (1966).
84. Skofronick, J. E., and W. B. Phillips, *Proc. Intern. Symp.*, 1966, p. 591, Vandenhoeck and Ruprecht, Clausthal.
85. VanItterbeek, A., L. DeGreve, and M. Francois, *Med. Koninkl. Vlaam.* (Acad. Wetens. Belg. Klasse Wetenschap, 1950), vol. 12, p. 3.
86. Basseches, H., *J. Electrochem. Soc.*, **109**, 475 (1962).
87. Dean, E. R., *J. Appl. Phys.*, **35**, 2930 (1964).

Chapter **14**

Electronic Conduction through Thin Insulating Films

JOHN G. SIMMONS

Department of Electrical Engineering, University of Toronto, Toronto, Canada

1. INTRODUCTION

In this chapter we will discuss conduction mechanisms in thin insulating films *sandwiched* between metal electrodes; we are thus concerned with conduction through, rather than along, the plane of the film. Furthermore, since we are concerned with thin films, generally less than 1 or 2 μ thick, it will also be apparent that we will be concerned primarily with the high-field electrical properties of the films, since applied biases as low as even a few volts will induce fields of the order 10^4 to 10^5 V cm^{-1} in the films. Thin films do in fact manifest much more interesting electrical properties when subjected to high fields: in contrast to the low-field properties which are usually ohmic in nature, that is, current I is linear with voltage V, the high-field I-V characteristics are often rich in structure. Quite often, too, the high-field electrical properties cannot be adequately described by a single conduction process; usually the various field-strength ranges manifest different electrical phenomena.

We will not be concerned with ionic conductivity per se in this chapter. Nor will we be discussing any of the unusual effects that one observes if the electrodes are permitted to go superconducting; this subject is discussed in Chap. 22.

The mks system of units has been used throughout in expressing the equations; this avoids the confusion that often arises when working with the mixed emu and esu cgs system of units. Some of the mks units are, however, unwieldy when applied to thin film calculations. For this reason, then, where appropriate, the equations have also been expressed in the more conventional units used by workers in the field, namely, energies in eV; field in V cm^{-1}; distance, area, and volume, respectively, in cm, cm^2, and cm^3.

2. PRELIMINARY COMMENTS

a. Conductivity of Thin Films

An insulator is a material which contains very few volume-generated carriers, in many instances considerably less than one per cm^3, and thus has virtually no conductivity. It will be clear later to the reader that the conductivity of the thin film materials we will discuss does not necessarily fall into this category; that is, although we will be concerned with materials which have energy gaps greater than about 2 eV or so, the electrical properties may bear no resemblance to what is *intrinsically* expected of such a material. This is because it is becoming increasingly clear that the electrical properties of thin film insulators are determined not by the intrinsic properties of the insulator but by other properties, such as the nature of the electrode-insulator contact. A suitable (ohmic) contact (Sec. 3a) is capable of injecting additional carriers into the insulator, far in excess of the bulk-generated carriers.[1-3] Also the application of a few volts bias is capable of causing inordinately high fields to be generated in a thin film insulator at the cathode-insulator interface. For fields in excess of 10^6 V cm^{-1}, field-emission injection of relatively large currents from the cathode into the conduction band of the insulator is possible [Sec. 5c(1)].

There are also several reasons for believing that the observed conductivity in thin film insulators is due often to extrinsically rather than intrinsically bulk-generated carriers. Consider first the intrinsic current density carried by an insulator.

$$I = e\mu N_c F \exp\left(-\frac{E_g}{2kT}\right) \tag{1}$$

where e is the electronic charge, μ is the mobility, F is the field in the insulator, N_c is the effective density of states in the insulator, E_g is the insulator energy gap, k is Boltzmann's constant, and T is the absolute temperature. At room temperature $N_c = 2.5 \times 10^{19}$ cm^{-3}, and assuming $E_g = 3$ eV and $\mu = 100$ cm^2 V^{-1} s^{-1}, then even for $F = 10^6$ V cm^{-1} the current density is only of the order 10^{-18} A cm^{-2}. This is many orders of magnitude smaller than the current densities we will be discussing for materials which have energy gaps greater than 3 eV. A second point is that the observed thermal-activation energy associated with the conductivity of the films is much smaller than would be expected ($\simeq E_g/2$) if the conductivity were intrinsic in nature.

The source of the extrinsic conductivity is thought to be the inherent defect nature of evaporated chemical compound films. Stoichiometric films of compound insulators, with which we are primarily concerned, are notoriously difficult to prepare by evaporation, because of decomposition and preferential evaporation of the lower-vapor-pressure constituent atom. For example, using the compound as starting material, elemental Cd tends to evaporate more rapidly from CdS; as a result CdS films contain donor centers of free cadmium.[5] SiO yields a film containing a mixture of compounds varying from SiO to SiO$_2$, as well as free Si;[6-9] the free Si per se may act as donor centers in these films[10] or alternatively vacancies existing in the insulator may be the source. A further problem that arises is the contamination of the films by deposits arising from sublimation of the crucible, and by residual gases. Thus it requires dissociation of only one molecule per million, or that the crucible sublimes at one-millionth the rate of the evaporant, to yield an impurity level of the order 10^{17} cm^{-3} within the film. A good example of the existence of donors in evaporated MoO$_3$ films is obtained by examination of capacity measurements obtained from these films.[4] At room temperature the capacity is *independent* of the film thickness, but at liquid-nitrogen temperature it corresponds to the geometrical capacitance (see Fig. 1).

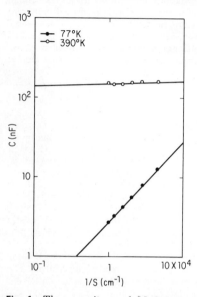

The room-temperature results are explained by assuming that they correspond to the capacitance of Schottky barriers existing at the two electrode-insulator interfaces, which are independent of the insulator thickness [see Secs. 3c and 3d(2) and Fig. 3d], the rest of the bulk being of high conductivity and hence not contributing to the capacity. In order to account for the Schottky barriers and the high bulk conductivity, it is necessary to postulate that the insulator contains a high donor density. The liquid-nitrogen temperature results follow naturally from this model. At the low temperature very few donors are ionized, that is, there are very few carriers in the insulator conduction band. Thus the bulk has a high resistivity and now, together with the Schottky barriers, that is, the total insulator thickness, contributes to the capacity.

Fig. 1 The capacity c of MoO$_3$ as a function of inverse sample thickness $1/S$. Dielectric constant obtained from 77°K results corresponds to bulk dielectric constant.

Another important fact to be considered in thin film insulators is traps. Insulating films deposited onto amorphous (e.g., glass) substrates are usually, at best, polycrystalline, and in many cases are amorphous. For crystallite sizes of 100 Å, trapping levels as high as 10^{18} cm^{-3} are possible because of grain-boundary defects alone; in vacuum-deposited CdS, trapping densities as high as 10^{21} cm^{-3} have been reported.[11] Furthermore, vacuum-deposited films contain large stresses which induce further trapping centers.

It follows then that thin film vacuum-deposited insulators can contain a large density of *both* impurity and trapping centers. A judicious study of electrical conduction in vacuum-deposited thin films cannot be accomplished without consideration of these possibilities.

Throughout most of this chapter, since we are concerned primarily with insulators having large energy gaps ($E_g \gtrsim 3$ eV), we will be concerned mainly with the electron rather than the hole carrier, although the results in general can be applied equally well to either type of carrier. The reason for this is twofold. First, the hole mobility is usually much lower than the electron mobility, and thus the hole contribution to the conductivity can usually, but not always, be neglected. Second, in practical insulating films, where, as previously mentioned, the trapping density is high, the tendency is for a free hole to be trapped quickly and thus become immobilized.

b. Band Structure

It will be noted in the energy diagrams that the forbidden gap separating the conduction and valence bands has been shown with well-defined boundaries. Strictly speaking, a well-defined energy gap is a property of a crystalline solid, and in general we are not dealing with such materials; rather we are concerned with polycrystalline, or even amorphous, insulators. However, it can be shown that the essential features of the band structure of a solid are determined by the *short-range* order within the solid; thus the general properties of the band structure of the crystalline state are carried over to the polycrystalline state. The lack of the long-range order in a noncrystalline solid causes smearing of the conduction and valence-band edges, among other things, so that the energy gap is no longer well defined. What is normally done in the study of thin insulating films is to assume that this smeared-out energy gap, to a first-order approximation, can be represented by a well-defined energy gap which is representative of perhaps an average value of the actual nondiscrete energy gap. With this model we can go a long way toward determining (within the limitation imposed by such a model) features of the band structure of thin film insulators. Indeed, it is found that parameters such as the barrier height at metal-insulator interfaces obtained by independent means (electrical and photoemissive techniques) agree remarkably well, even in what might be considered to be an extreme case of departure from the crystalline state, namely, extremely thin ($\sim$30 Å) layers of amorphous Al_2O_3.

3. METAL-INSULATOR CONTACTS

In order to measure the conductivity of an insulator, it is, of course, necessary to connect electrodes to its surfaces in order to facilitate injection of electrons into, and their withdrawal from, the bulk of the insulator. Clearly, the conductivity of the insulator per se will determine the conductivity of the system, since it is much lower than that of the electrodes. In terms of the energy-band picture, the action of the insulator is to erect between the electrodes a potential barrier, extending from the electrode Fermi level to the bottom of the insulator conduction band (see Figs. 2 and 3). This barrier impedes the flow of electrons from one electrode to the other, which would normally flow virtually unimpeded if the insulator were not there (i.e., metal-metal contact). Clearly, then, the height of the potential barrier is an important parameter in conductivity studies in metal-insulator systems. Furthermore, the height of this barrier is determined by the relative alignment of the electrode and insulator energy bands. How then are these bands aligned? This problem is resolved by making use of the rule that in thermal equilibrium the vacuum and Fermi levels must be continuous throughout the system. (The vacuum level represents the energy of an electron at rest just outside the surface of the material, and the energy difference between the vacuum and Fermi levels is called the work function of the material.) It would appear that the equilibrium conditions can be satisfied only when the work function of the metal ψ_m and that of the insulator ψ_i are equal (see Fig. 2c and d). We shall see a little later, however, that the equilibrium condition is satisfied when $\psi_m \neq \psi_i$, because of charge transfer from the electrode to the insulator or vice versa.

The shape of the potential barrier just within the surface of the insulator depends on whether or not the insulator is intrinsic or extrinsic, and on the relative magnitudes of the work functions of the metal and insulator, among other things.

At reasonable applied fields there will normally be a sufficient supply of carriers available to enter the insulator from the cathode (negatively biased electrode) to replenish the carriers drawn out of the bulk of the insulator. Under these conditions the current-voltage (I-V) characteristics of the sample will be determined by the bulk properties of the insulator; we thus refer to this conduction process as being *bulk-limited*. At high fields, or if the contact is blocking, the current capable of being supplied by the cathode to the insulator will be less than that capable of being carried in the bulk of the insulator. Under these conditions the I-V characteristics of the sample will be controlled primarily by conditions existing at the cathode-insulator interface; this conduction process is referred to as being *emission-limited* or *contact-limited*.

The types of contact that can exist at a metal-insulator interface fall into three categories: (1) ohmic contact, (2) neutral contact, and (3) blocking contact. Each of the contacts will now be discussed in some detail.

a. Ohmic Contact—Mott-Gurney Contact

In order to achieve an ohmic contact at a metal-insulator interface,[1] it is necessary that the electrode work function ψ_m be smaller than the insulator work function ψ_i, as shown in Fig. 2a. Here, the term ohmic contact is used to mean that the electrode can readily supply electrons to the insulator as needed. Under these conditions, in order to satisfy thermal-equilibrium requirements, electrons are injected from the

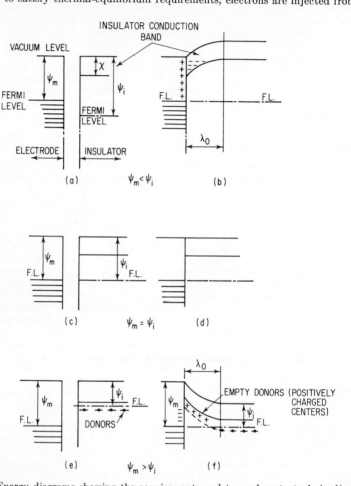

Fig. 2 Energy diagrams showing the requirements and type of contact of: (a, b) ohmic contact; (c, d) neutral contact; (e, f) blocking contact.

electrode into the conduction band of the insulator, thus giving rise to a space-charge region in the insulator. This space-charge region is shown in Fig. 2b to extend a distance λ_0 into the insulator, and is termed the accumulation region. In order to satisfy charge-neutrality requirements an equal amount of positive charge, say, Q_0, accumulates on the electrode surface. The electrostatic interaction between the positive and negative charge induces a local electric field within the surface of the insulator, the strength of which falls off with distance from the interface and is zero at the edge of the space-charge region (i.e., at λ_0 from the interface). This field causes the bottom of the conduction band to rise with distance of penetration into the insulator until it reaches the equilibrium value $\psi_i - \chi$, where χ is the insulator affinity.

The field F within the accumulation region is related to the space-charge density $\rho(x)$ within the accumulation region by Poisson's equation:

$$\frac{dF}{dx} = \frac{\rho(x)}{K\epsilon_0} \tag{2}$$

where K and ϵ_0 are, respectively, the dielectric constant and permittivity of free space. To obtain information regarding conditions in the space-charge region, we substitute $\rho(x) = eN_c \exp(-\psi/kT)$ in (2), where ψ is the potential energy of the insulator conduction-band edge with respect to the electrode Fermi level:

$$\frac{dF}{dx} = \frac{1}{e}\frac{d^2\psi}{dx^2} = -\frac{eN_c}{K\epsilon_0}\exp\left(-\frac{\psi}{kT}\right) \tag{3}$$

On integration, using $d\psi/dx = 0$ at $\psi = \psi_i - \chi$, (3) yields

$$\frac{d\psi}{dx} = \left(\frac{2e^2kTN_c}{K\epsilon_0}\right)^{\frac{1}{2}}\left[\exp\left(-\frac{\psi}{kT}\right) - \exp\left(-\frac{\psi_i - \chi}{kT}\right)\right]^{\frac{1}{2}} \tag{4}$$

Integrating (4) using the boundary conditions $\psi = \psi_m - \chi$ and $(\psi_i - \chi)$ at $x = 0$ and λ_0, respectively, we obtain

$$\lambda_0 = \left(\frac{2kTK\epsilon_0}{e^2N_c}\right)^{\frac{1}{2}}\left\{\frac{\pi}{2} - \sin^{-1}\left[\exp\left(-\frac{\psi_i - \psi_m}{2kT}\right)\right]\right\}\exp\left(\frac{\psi_i - \chi}{2kT}\right) \tag{5}$$

For $\psi_i - \psi_m > 4kT$, which will normally be the case, (5) reduces to

$$\lambda_0 \simeq \frac{\pi}{2}\left(\frac{2kTK\epsilon_0}{e^2N_c}\right)^{\frac{1}{2}}\exp\left(\frac{\psi_i - \chi}{2kT}\right) \tag{6}$$

In Table 1 we have calculated λ_0 for various values of $\psi_i - \chi$, from which it is seen that, at room temperature, a good ohmic contact in the thin insulators we are considering requires that the maximum height of the bottom of the conduction band above the Fermi level be less than about 0.3 eV. This statement is tantamount to the more important condition that, for an effective ohmic contact at room temperature, the interfacial barrier $\psi_m - \chi$ must be less than 0.3 eV.

TABLE 1 Depth of Accumulation Region for Several Values of $\psi_m - \chi$

$\psi_m - \chi$, eV	0.1	0.2	0.3	0.4
λ, cm	1.6×10^{-6}	1.2×10^{-5}	7.2×10^{-5}	7.2×10^{-4}

The total amount of charge injected into the insulator to form the ohmic contact is determined as follows: From (2), using the boundary conditions that $F = F_0$ and 0 at $x = 0$ and λ_0, respectively, we have

$$\int_{F_0}^{0} dF = -\frac{1}{K\epsilon_0}\int_0^{\lambda}\rho(x)\,dx$$

or

$$F_0 = \frac{Q_0}{K\epsilon_0} \tag{6a}$$

where $Q_0 = \int_0^{\lambda}\rho(x)\,dx$ is the total space charge per unit area in the accumulation region. Since $\psi = \psi_m - \chi$ at $x = 0$, we have from (4), assuming $\psi_m < \psi_i - 2kT$, which will normally be the case,

$$F_0 = \left(\frac{2kTN_c}{K\epsilon_0}\right)^{\frac{1}{2}}\exp\left(-\frac{\psi_m - \chi}{kT}\right) \tag{6b}$$

Substituting (6b) into (6a) yields

$$Q_0 = (2K\epsilon_0 kT N_c)^{\frac{1}{2}} \exp\left(-\frac{\psi_m - \chi}{kT}\right) \tag{6c}$$

b. Neutral Contact

When there is no reservoir of charge at the contact ($Q_0 = 0$), this type of contac is known as the neutral contact. The condition $Q_0 = 0$ implies that $\psi_m = \psi_i$, whic means that the conduction band is flat right up to the interface; that is, no ban bending is present, as shown in Fig. 2d.

For initial voltage bias the cathode is capable of supplying sufficient current t balance that flowing in the insulator, so that the conduction process is ohmic. Ther is no theoretical limit to the maximum current an insulator per se may carry, provide a high enough voltage supply is available. (There are, in fact, practical limitation such as Joule heating and dielectric breakdown, which determine the maximum voltag that an insulator can withstand before catastrophic breakdown occurs.) There i however, a limit to the current that the cathode can supply, and this is the saturate thermionic (Richardson) current over the barrier. When this limit is reached, th conduction process is no longer ohmic in nature. The maximum field that may b applied to the insulator before the current supplied by the cathode saturates is obtaine by equating the saturated thermionic current $n_0 ev/4$ to the current flowing in th insulator $n_0 e\mu F$, where n_0 is the density of electrons in the cathode with energie greater than the interfacial potential barrier and v is the thermal velocity of th carriers:

$$\frac{n_0 ev}{4} = n_0 e\mu F$$

or

$$F = \frac{v}{4\mu}$$

In actual fact we will see later that, because of attendant image forces interactin with the electric field at the cathode-insulator interface, the saturation curren increases with applied field [Richardson-Schottky effect; see Sec. 5a(2)].

c. Blocking Contact —Schottky Barrier

A blocking contact occurs when $\psi_m > \psi_i$, and in this case electrons flow from th insulator into the metal to establish thermal-equilibrium conditions. A space-charg region of positive charge, the *depletion* region, is thus created in the insulator and a equal negative charge resides on the metal electrode. As a result of the electrostati interaction between the oppositely charged regions, a local field exists within th surface of the insulator. This causes the bottom of the conduction band to ben downward until the Fermi level within the bulk of the insulator lies ψ_i below th vacuum level. An intrinsic insulator, however, contains such a low density of ele trons that it would have to be inordinately thick to provide the required positiv space-charge region to satisfy the above condition. Thus the conduction band of a intrinsic insulator at a blocking contact does in fact slope only imperceptibly dow ward; that is, we have what is essentially a neutral contact.

Let us assume now that the insulator contains a large density N_d of donors, as show in Fig. 2f. The method of determining quantitative information about this depletio region parallels that applied to the doped semiconductor-metal contact, and w first studied by Schottky.[12,13] We assume that the donors are fully ionized an uniformly distributed (N_d cm^{-3}) in the region extending from the interface to a dept λ_0, the depletion region, into the insulator. Poisson's equation (2) for this system therefore given by

$$\frac{d^2\psi}{dx^2} = \frac{e^2 N_d}{K\epsilon_0}$$

For convenience we measure ψ from the Fermi level of the electrode to the bottom of the conduction band of the insulator. Integrating this expression, and using the boundary condition $d\psi/dx = 0$ at $x = \lambda_0$, we have

$$\frac{d\psi}{dx} = \frac{e^2 N_d}{K\epsilon_0}(x - \lambda_0) \tag{7}$$

Integrating again and using the boundary condition $\psi = \psi_m - \chi$ at $x = 0$, we have

$$\psi = \psi_m - \chi + \frac{e^2 N_d}{K\epsilon_0}\left(\frac{x^2}{2} - \lambda_0 x\right) \tag{8}$$

To determine the depth of the depletion region λ_0, we note that $\psi = \psi_i - \chi$ at $x = \lambda_0$, so that

$$\lambda_0 = \left[\frac{2(\psi_m - \psi_i)K\epsilon_0}{e^2 N_d}\right]^{\frac{1}{2}} \tag{9}$$

or in conventional units

$$\lambda_0 = 1.05 \times 10^3 \left[\frac{(\psi_m - \psi_i)K}{N_d}\right]^{\frac{1}{2}}$$

In Table 2 we calculate a few values for λ_0 using $\psi_m - \psi_s = 2$ eV and $K = 5$. It will be apparent that the donor density is required to be of the order 10^{16} cm^{-3} or higher if the insulator is to be able to accommodate the depletion region adequately.

TABLE 2 Depth of Depletion Region for Several Values of N_d

N_d, cm^{-3}.........	10^{15}	10^{17}	10^{19}	10^{21}
λ, cm............	10^{-4}	10^{-5}	10^{-6}	10^{-7}

To determine the effect of voltage bias on the depletion region, we note that the depletion region has a much lower electron density than the bulk of the insulator; hence its conductivity is much lower. Thus any voltage applied to the system can, to a good approximation, be assumed to be absorbed entirely across the depletion region. Under these conditions, when the metal is negatively biased, the boundary condition leading to (9) is modified to read $\psi = \psi_i - \chi - eV$ at $x = \lambda$. The depth of the depletion region λ now becomes

$$\lambda = \left[\frac{2(\psi_m - \psi_i + eV)K\epsilon_0}{N_d e^2}\right]^{\frac{1}{2}} \tag{10}$$

which increases with increasing voltage bias. The field at the interface $F_0 = e^{-1}\,d\psi/dx$, which we will be needing later, is from (7) and (10)

$$F_0 = -\left[\frac{2N_d(\psi_m - \psi_i + eV)}{K\epsilon_0}\right]^{\frac{1}{2}} \tag{11}$$

Metal-insulator-metal Systems

(1) Two Ohmic Contacts Figure 3a and b illustrates two ohmic contacts on an insulator when the electrodes are at the same potential. In the case of Fig. 3a it is seen that the accumulation regions extend right into the bulk of the insulator. As a result, the bottom of the conduction band is curved throughout its length (that is, an electric field exists at all points within the insulator), the highest point of which is less than the equilibrium position $\psi_i - \chi$ above the Fermi level. This condition is due to either the insulator's being too thin or the interfacial potential barriers' being too high, with the result that insufficient charge is contained within the insulator to screen the interior effectively from the surface. Figure 3b, on the other hand, illustrates the case of good ohmic contacts (small interfacial potential barriers) on an insulator. In this case the accumulation regions effectively screen the interior of the insulator from conditions at the surfaces. The bottom of the conduction band is thus flat within the interior of the insulator (field-free interior) and the height of the conduction band within the interior, reaching its equilibrium level $\psi_i - \chi$ above the Fermi level.

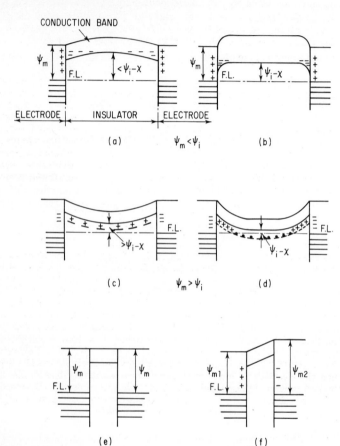

Fig. 3 Energy diagrams of two metal contacts on an insulator. (a) and (b) represent imperfect and good ohmic contacts, respectively; (c) and (d) imperfect and good blocking contacts; (e) and (f) similar and dissimilar neutral contacts.

(2) Two Blocking Contacts Figure 3c and d illustrates two blocking contacts on a insulator. In Fig. 3c the depletion regions extend right into the interior of the insulator, as a result of the insulator's being too thin, or of the doping concentration being too low. As in the case of Fig. 3a, the bottom of the conduction band is curve throughout its length, but in this case concave upward, indicating that an electric field exists at all points throughout the length of the insulator. Because the deple tion regions do not effectively screen the interior from the surfaces, the lowest poin of the bottom of the conduction band is greater than $\psi_i - \chi$ above the Fermi leve In contrast to Fig. 3c the depletion regions in Fig. 3d effectively screen the interic from the surfaces, which means that the insulator is thick or the doping density is high. Under these conditions, as in the case of good ohmic contacts, the interior of the insulator is field-free, and the bottom of the conduction band attains its equilibrium position $\psi_i - \chi$ above the Fermi level, within the interior.

(3) Other Contacts Figure 3e and f illustrates the case of blocking electrodes on an intrinsic or a *very* thin doped insulator. In this case the insulator is incapable of transferring sufficient charge from its interior to the electrodes to give rise to any effective band bending. In the case of similar electrodes (Fig. 3e), the result is that the bottom of the insulator conduction band is flat throughout its entire length.

When the insulator has dissimilar electrodes connected to its surfaces, it is clear that the interfacial potential barriers differ in energy by an amount

$$(\psi_{m2} - \chi) - (\psi_{m1} - \chi) = \psi_{m2} - \psi_{m1}$$

as shown in Fig. 3f. This means that the conduction band must slope upward from the lower barrier with a gradient $(\psi_{m2} - \psi_{m1})/s$, where s is the insulator thickness; this in turn means that a uniform intrinsic field F_{in} of strength $(\psi_{m2} - \psi_{m1})/es$ exists within the insulator. The origin of this zero-bias intrinsic field is a consequence of charge transfer between the *electrodes*. The electrode of lower work function, electrode 1, transfers electrons to electrode 2, so that a positive surface charge appears on electrode 1 and a negative surface charge on electrode 2. The amount of charge Q transferred between the electrodes (the surface charge on the electrodes) depends on the contact potential difference $(\psi_{m2} - \psi_{m1})/e$ existing between the electrodes, and the capacity C of the system as follows:

$$Q = \frac{C(\psi_{m2} - \psi_{m1})}{e} = \frac{(\psi_{m2} - \psi_{m1})AK\epsilon_0}{es} \tag{12}$$

where A is the electrode area.

If the insulator is very thin, the intrinsic field within the insulator can be very large; for example, suppose $s = 20$ Å, as one finds in metal-insulator-metal tunnel junction (see Sec. 4), and $(\psi_{m2} - \psi_{m1})/e = 1$ V, $F_{\text{in}} = 5 \times 10^6$ V cm^{-1}. When electrode 1 is positively biased, the intrinsic field augments the applied field; that is, the field in the insulator F is given by $F = F_{\text{in}} + V/s$. When it is negatively biased, the intrinsic field acts to reduce the effect of the applied field; that is, $F = V/s - F_{\text{in}}$. In the latter case the initial effect of the applied voltage $V < (\psi_{m2} - \psi_{m1})/e$ is to *reduce* the field in the insulator, which is zero when $V = (\psi_{m2} - \psi_{m1})/e$. Further increasing voltage application, $V > (\psi_{m2} - \psi_{m1})/e$, then results in a *reversal* of the field in the insulator which increases with increasing voltage bias. It will be apparent, then, when the electrode of lower work function is negatively biased, the insulator will withstand greater voltage application before it breaks down. By investigating the breakdown voltage of very thin metal-insulator-metal junctions as a function of voltage bias, the effect of the intrinsic field can be detected.[14]

e. Effect of Surface States

(1) Free Surface of an Insulator In the above discussions we have taken the interfacial barrier height to be $\psi_m - \chi$. In actual fact experiment shows, in many instances, that this is an oversimplification of conditions existing at real surfaces; in particular the potential barrier is often found to be independent of the work function of the electrode material. In order to account for similar observations on metal-semiconductor contacts, Bardeen[15] suggested surface states as the cause. These states owe their existence in part to the sudden departure from periodicity of the potential at the surface, and in part to the defect chemical nature of the surface, i.e., absorbed gas, etc.

The effect of surface states on the surface potential of the insulator can be seen by reference to Fig. 4. The surface states, of density N_s (per unit area per unit energy), are assumed to be uniformly distributed throughout the energy gap of the insulator, as shown in Fig. 4a. When filled up to an energy E_0 below the bottom of the conduction band, the surface is assumed to be electrically neutral. However, since the

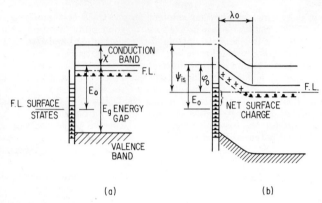

(a)   (b)

Fig. 4 Energy diagram of surface states on a semiconductor. (a) Before surface and bulk are in equilibrium. (b) In equilibrium.

Fermi level associated with the surface states should coincide with that of the bulk, electrons from the conduction band fill up additional surface states. The result is that a sheet of negative charge resides on the surface and a positive space charge (depletion region) of depth λ_0 exists just inside the surface of the insulator, as shown in Fig. 4b. This double-charge layer causes the conduction band to bend upward until the highest filled surface state coincides with the Fermi level of the bulk (Fig. 4b). This band bending is possible only in a *doped* insulator; in an intrinsic insulator too few electrons are available in the bulk to effect this process.

Electrical neutrality of the crystal requires that the number of ionized donors in the depletion region be equal to the excess number of electrons in surface states; thus (from Fig. 4b)

$$N_s(E_0 - \phi_0) = N_d\lambda_0$$

or

$$E_0 - \phi_0 = \frac{N_d\lambda_0}{N_s} \qquad (13)$$

From (13), for an infinite density of surface states we have $\phi_0 = E_0$; *thus the potential barrier at the surface ϕ_0 is determined entirely by surface states*. To see the effect of practical values of surface-charge densities ($N_s \simeq 10^{14}$ cm^{-2} eV^{-1}) let us assume that the bands at the surface are bent upward by an amount 1 eV; therefore, from Table 2 we see that for $N_d = 10^{17}$ cm^{-3}, $\lambda_0 = 10^{-5}$ cm. Using these values and $N_s = 10^{14}$ cm^{-2} eV^{-1} in (13), we have $E_0 - \phi_0 = 10^{-2}$ eV or $\phi_0 \simeq E_0$. Thus for practical values of surface states, ϕ_0 ($\simeq E_0$) is essentially controlled by conditions existing at the surface rather than in the bulk.

(2) Surface States at a Metal-insulator Contact Let us now consider what happens when an electrode is attached to the surface of an insulator on which surface states are prominent. If the work function of the electrode ψ_m is the same as the work function of the insulator surface ψ_{is} the potential barrier ϕ_0 remains undisturbed when contact is made, because the two surfaces are in equilibrium before and after contact. If $\psi_m > \psi_{is}$, electrons are transferred out of the surface states into the surface of the electrode and provided the density of the surface states is sufficiently high, the difference in potential $\psi_m - \psi_{is}$ of the two surfaces can be virtually all dropped in the double-charged surface layers without appreciably altering the potential barrier at the insulator surface. For example, the number of electrons per unit area transferred between the surfaces is $Q/e = (\psi_m - \psi_{is})K\epsilon_0/e^2a$ [cf. Eq. (12)], where a is the distance between the two surfaces, that is, a distance of the order of the interatomic spacing. Assuming $\psi_m - \psi_{is} = 1$ eV and $a = 5 \times 10^{10}$ m, we have $Q/e \simeq 10^{17}$ m$^-$ or 10^{13} cm^{-2}. Assuming $N_s = 10^{14}$ cm^{-2} eV^{-1}, only $\frac{1}{10}$ eV spread of energies at the top of the filled surface states are depleted in order to supply the necessary charge to the electrode to accommodate the difference in work function (see Fig. 5). This means that the Fermi energy of the surface of the insulator is lowered by only $\frac{1}{10}$ eV in releasing the required charge, or in other words, the potential barrier at the surface is increased by only $\frac{1}{10}$ eV as compared with 1 eV if the surface states had not been present. *Thus the interfacial potential barrier is virtually independent of the electrode work function.* Using similar arguments, it can be shown that if ψ_m is 1 eV *less* than ψ_{is}, the surface potential of the insulator *decreases* by only $\frac{1}{10}$ eV.

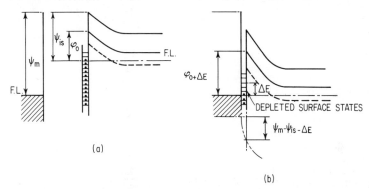

(a)

(b)

Fig. 5 Energy diagram of a metal surface and insulator surface which contains surface states. (*a*) Before contact. (*b*) After contact (the chain-dotted line below the energy diagram shows the change in potential in passing through the interface).

The above arguments have been simplified somewhat in that we have not taken into account changes that occur in the depletion region of the insulator when changes occur at its surface, but this is a matter of detail rather than principle. The important point is that these two examples serve to show why surface states tend to make the surface potential barrier almost independent of the work function of the electrode, and why it is no longer appropriate invariably to assign the value $\psi_m - \chi$ to the height of the interfacial barrier. Henceforth we will define the height of the potential barrier by the parameter ϕ_0, its value being dependent on whether or not surface states play a significant part in determining the barrier height.

4. TUNNEL EFFECT—VERY THIN INSULATORS

a. Tunnel Effect

If the energy of an electron is less than the interfacial potential barrier in a metal-insulator-metal junction upon which it is incident, classical physics predicts certain reflection of the electron at the interface; that is, the electron cannot penetrate the barrier and hence its passage from one electrode to the other is precluded. Quantum theory, however, contradicts this thesis. The quantum-mechanical wave function $\psi(x)$ of the electron has finite values within the barrier (see Fig. 6), and since $\psi(x)\psi^*(x) \cdot dx$ is the probability of finding the electron within the incremental range x to $x + dx$, this means that the electron can penetrate the forbidden region of the barrier. The wave function decays rapidly with depth of penetration of the barrier from the electrode-insulator interface and, for barriers of macroscopic thickness, is essentially zero (Fig. 6*a*) at the opposite interface, indicating zero probability of finding the electron there. However, if the barrier is very thin (<50 Å), the wave function has a nonzero value at the opposite interface. For this case, then, there is a finite probability that the electron can pass from one electrode to the other by penetrating the barrier, as shown in Fig. 6*b*. When the electron passes from one electrode to the other by this process, one speaks of the electron as having tunneled through the barrier.

If two metal surfaces are placed within 50 Å of each other, it follows from the above discussion that the electrons will pass from one electrode to the other by means of the

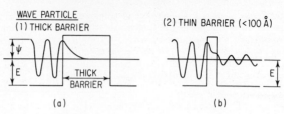

(a)   (b)

Fig. 6 Schematic representation of the tunnel effect. (a) At a thick barrier. (b) At a very thin barrier.

tunnel effect. The close separation of the electrode is achieved in practice by means of a thin insulating film, usually the thermally or anodically grown surface oxide of one of the electrodes, and it is this system we will now consider.

b. Theory of Isothermal Tunneling

The tunnel effect between metal electrodes was first studied, in elementary fashion, by Frenkel.[16] Sommerfeld and Bethe[17] made the first comprehensive studies, in which they included image-force effects but confined their calculations to very low $(V \ll \phi_0/e)$ and very high $(V \gg \phi_0/e)$ voltages. Holm[18] made the next notable investigations and extended the calculation to intermediate voltages, although approximations he used have been found questionable. Stratton[19] and Simmons[20-22] further extended the theory, and the results of these studies are those currently most commonly used in the analysis of experimental data. All the above studies were concerned with a barrier which in its simplest forms is shown in Fig. 3e and f. These models appear to be quite suitable for predicting the salient features of the tunneling I-V characteristics. There have been several other studies of a more detailed nature. These have considered the effect of space charge,[23-25] traps[26] and ions[27] in the insulator, the effect of the shape of forbidden band,[28,29] representation of the insulator by a series of potential wells,[30,31] electric-field penetration of the electrodes,[32] and diffuse reflection.[33] We will be concerned here with tunneling through a barrier of the type shown in Fig. 3e and f.

We will discuss the theory of the tunnel effect using the notation and type of approximation developed by the author.[20,21] This formulation is readily applicable to potential barriers of arbitrary shape and to all practical voltage ranges.

The generalized formula gives the relationship connecting the tunnel current density with the applied voltage for a barrier of arbitrary shape (see Fig. 7) as

$$I = I_0\{ \bar{\phi} \exp(-A\bar{\phi}^{\frac{1}{2}}) - (\bar{\phi} + eV) \exp[-A(\bar{\phi} + eV)^{\frac{1}{2}}]\} \tag{14}$$

where $$I_0 = \frac{e}{2\pi h(\beta \, \Delta s)^2} \quad \text{and} \quad A = \frac{4\pi\beta \, \Delta s}{h}(2m)^{\frac{1}{2}}$$

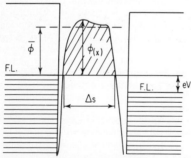

Fig. 7 Energy-diagram arbitrary barrier [see (14)] illustrating the parameters $\bar{\phi}$ and Δs.

Δs = width of the barrier at the Fermi level of the negatively biased electrode

$\bar{\phi}$ = mean barrier height above the Fermi level of the negatively biased electrode

h = Planck's constant

m = mass of the electrons

e = unit of electronic charge

β = a function of barrier shape and is usually approximately equal to unity, a condition we will assume throughout

Expressed in conventional units, except for Δs, which is expressed in angstroms, (14) becomes

$$I = \frac{6.2 \times 10^{10}}{(\Delta s)^2} \left\{ \bar{\phi} \exp\left(-1.025\Delta s\, \bar{\phi}^{\frac{1}{2}}\right) - (\bar{\phi} + V) \exp\left[-1.025\Delta s(\bar{\phi} + V)^{\frac{1}{2}}\right] \right\} \quad (15)$$

This equation was derived using the *WKBJ* approximation and assuming an electronic-energy distribution corresponding to absolute zero of temperature. Since the current density I is only very slightly temperature-dependent, this equation is also suitable for use at higher temperatures. We will, however, discuss the detailed temperature dependence of I later.

We will now use (15) to derive the *I-V* characteristics of an asymmetric junction (see Fig. 3f), as this is the type of junction most often met with in practice. Since the *I-V* characteristic will be shown to be polarity-dependent, in the subsequent discussion the following convention will be adopted: any electrical characteristic is described as the *forward* characteristic when the electrode of lower work function is positively biased, and as the *reverse* characteristic when the electrode of lower work function is negatively biased. J_1 will denote the reverse current density and J_2 the forward current density.

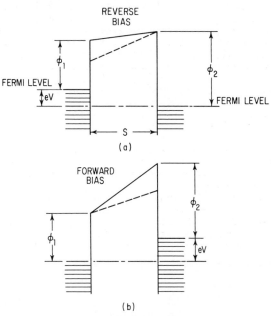

Fig. 8 Energy diagrams of asymmetric potential barrier for $0 \leq V \leq \phi_1/e$. (a) is reverse-bias condition, and (b) is forward-bias condition. The dotted lines illustrate the shape of the barrier (with respect to the Fermi level of the *positively* biased electrode in both cases) under zero-bias condition.

(1) Voltage Range $0 \leq V \leq \phi/e$ Figure 8a and b illustrates the energy diagram for the reverse and forward directions of polarity, respectively. It follows for both these cases that

$$\bar{\phi} = \frac{\phi_1 + \phi_2 - eV}{2} \qquad \Delta s = s \qquad (16)$$

which on substitution in (15) yields

$$J_1 = J_2 = \frac{3.1 \times 10^{10}}{s^2} \left\{ (\phi_1 + \phi_2 - V) \exp\left[-0.725s(\phi_1 + \phi_2 - V)^{\frac{1}{2}}\right] \right.$$
$$\left. - (\phi_1 + \phi_2 + V) \exp\left[-0.725s(\phi_1 + \phi_2 + V)^{\frac{1}{2}}\right] \right\} \quad (17)$$

Since $J_1 = J_2$, the I-V characteristic is symmetric with polarity of bias for the voltage range $0 \le V \le \phi_1/e$.

(2) Voltage Range $V > \phi_2/e$ From Fig. 9a we have for the reverse-biased condition, using $\Delta\phi = \phi_2 - \phi_1$,

$$\bar{\phi} = \frac{\phi_1}{2} \qquad \Delta s = \frac{s\phi_1}{eV - \Delta\phi} \tag{18}$$

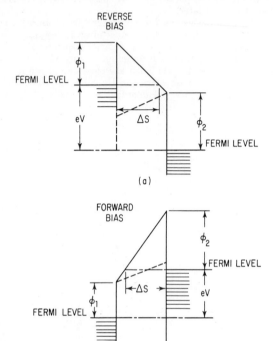

Fig. 9 Energy diagrams of potential barriers for $V > \phi_2/e$. (See Fig. 8 for explanation of dotted lines.)

which on substitution in (15) yields

$$J_1 = \frac{3.38 \times 10^{10}(V - \Delta\phi)^2}{\phi_1 s^2}\left\{\exp\left(-\frac{0.69s\phi_1^{\frac{3}{2}}}{V - \Delta\phi}\right)\right.$$
$$\left. - \left(1 + \frac{2V}{\phi_1}\right)\exp\left[-\frac{0.69s\phi_1^{\frac{3}{2}}(1 + 2V/\phi_1)^{\frac{1}{2}}}{V - \Delta\phi}\right]\right\} \tag{19}$$

From Fig. 9b we have for the forward-biased condition

$$\bar{\phi} = \frac{\phi_2}{2} \qquad \Delta s = \frac{s\phi_2}{eV + \Delta\phi} \tag{20}$$

which on substitution in (15) yields

$$J_2 = \frac{3.38 \times 10^{10}(V + \Delta\phi)^2}{\phi_2 s^2}\left\{\exp\left(-\frac{0.69s\phi_2^{\frac{3}{2}}}{V + \Delta\phi}\right)\right.$$
$$\left. - \left(1 + \frac{2V}{\phi_2}\right)\exp\left[-\frac{0.69s\phi_2^{\frac{3}{2}}(1 + 2V/\phi_2)^{\frac{1}{2}}}{V + \Delta\phi}\right]\right\} \tag{21}$$

In this case, Eqs. (19) and (21) are not equivalent. It follows, then, that the J-V characteristic is asymmetric in this range. In actual fact, not only is the J-V characteristic asymmetric with polarity of bias, but also the direction of easy conductance reverses at some particular voltage, as shown in Fig. 10.

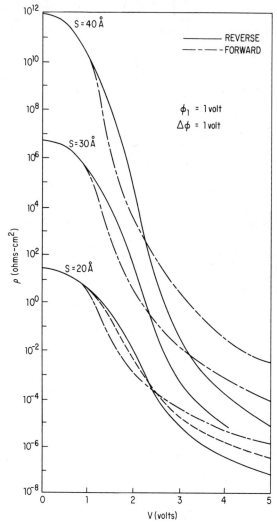

Fig. 10 $\rho(= J/V) - V$ tunnel characteristics for $\phi_1 = 1$ eV, $\phi_2 = 2$ eV, and $s = 30, 30,$ and 40 Å.

At high voltages, i.e., $V \gg \phi_2/e$, both (20) and (21) reduce to the familiar Fowler-Nordheim[34] form:

$$J = \frac{3.38 \times 10^{10} F^2}{\phi} \exp\left(-\frac{0.69\phi^{\frac{3}{2}}}{F}\right) \qquad (22)$$

where F is the field in the insulator.

(3) Numerical Evaluation   Figure 10 illustrates the tunnel resistance $\rho(V/J)$ as a function of V for $s = 20$, 30, and $40 \ \overset{\circ}{\text{A}}$, $\phi_1 = 1$ eV and $\phi_2 = 2$ eV ($\Delta\phi = 1$ eV). The plot of ρ vs. V, rather than I vs. V, is chosen for illustration, because it is a more efficacious way of illustrating the effect of the junction parameters on the electrical characteristics. The reverse and forward directions are depicted, respectively, by the full and dotted curves. Several features of the curves will be immediately apparent. At very low voltages the curves are horizontal, which means that the junction resistance is ohmic. The junction resistance falls off rapidly with increasing voltage bias, and the effect is more pronounced the thicker the film. The junction characteristics are symmetric for $V < \phi_1/e$, as noted above; thereafter they are asymmetric, with the direction of asymmetry *reversing* at a voltage ($V \simeq 2.5$ V) which is practically independent of insulator thickness.

The dotted curve accompanying the characteristic marked $s = 20 \ \overset{\circ}{\text{A}}$ illustrates the ρ-V characteristic for a *symmetric* junction having $\phi = (\phi_1 + \phi_2)/2 = 1.5$ V. This characteristic is, of course, symmetric over the entire voltage range. As would be intuitively expected, this characteristic is approximately equal to the logarithmic mean value of the forward and reverse characteristics.

The above results apply to an ideal barrier inasmuch as image forces have been neglected (see Sec. 5a). The effect of image forces is to reduce the size of the barrier and to lower the impedance of the junction without affecting the general functional form of the curves and the observations noted above. For further information the reader is referred to the literature.[20,21,24]

c. Temperature Dependence of the Tunnel Characteristic

Stratton[19] first derived the I-V characteristic for a tunnel junction for $T \neq 0°$K. The thermal I-V characteristic has also been expressed in terms of the generalized theory, which we will now present.[22]

The generalized formulation connecting the tunnel current $J(V,T)$ at $T°$K to the tunnel current $J(V,0)$ at $0°$K [Eq. (14)] is given by

$$\frac{J(V,T)}{J(V,0)} = \frac{\pi BkT}{\sin \pi BkT}$$

where
$$B = \frac{A}{2\bar{\phi}^{\frac{1}{2}}}$$

In conventional units, but with Δs in angstroms,

$$\frac{J(V,T)}{J(V,0)} \simeq 1 + \frac{3 \times 10^{-9}(\Delta sT)^2}{\bar{\phi}} \tag{23}$$

This result, which states that the thermal component of the tunnel current for a given applied bias V should be proportional to T^2, has been observed experimentally.

A more interesting and important result is apparent, however, when we rewrite (23) in the following form:

$$\hat{J} = \frac{100[I(V,T) - I(V,0)]}{I(V,0)} = \frac{3 \times 10^{-7}(\Delta sT^2)}{\bar{\phi}} \tag{24}$$

$\hat{J}$ will be recognized as the percentage change in current (for a fixed voltage bias) as the temperature is raised from 0 to $T°$K. From (16) and (18) we have, for reverse bias,

$$\hat{J}_1 = \begin{cases} 6 \times 10^{-7} \dfrac{(sT)^2}{\phi_1 + \phi_2 - V} & V \leq \phi_2 \\[2mm] 6 \times 10^{-7}\phi_1 \left(\dfrac{sT}{V - \Delta\phi}\right)^2 & V \geq \phi_2 \end{cases} \tag{25}$$

From (16) and (20), for forward bias,

$$\hat{J}_2 = \begin{cases} 6 \times 10^{-7} \dfrac{(sT)^2}{\phi_1 + \phi_2 - V} & V \leq \phi_1 \\[3mm] 6 \times 10^{-7}\phi_2 \left(\dfrac{sT}{V + \Delta\phi} \right)^2 & V \geq \phi_1 \end{cases} \tag{26}$$

These results are exemplified by the curves in Fig. 11. Both $\hat{J}_1$ and $\hat{J}_2$ increase initially until they reach a maximum value, after which they decrease. However, the maximum of $\hat{J}_1$ and the voltage at which it occurs ($V = \phi_2$) exceed the corresponding maximum and voltage ($V = \phi_1$) for $\hat{J}_2$. The importance of (25) and (26) is now apparent: the $\hat{J}$ maxima occur at voltages equal to the electrode-insulator interfacial barrier heights ϕ_1 and ϕ_2 (or ϕ_0, in the case of a symmetric barrier) and, because of the prominence of the peaks, provide a good method of experimentally determining barrier heights.

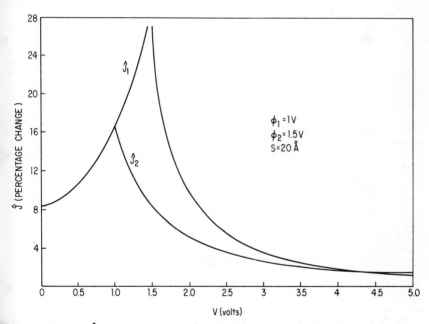

Fig. 11 Graph of $\hat{J}$ vs. V. In this case $\hat{J}$ represents the percentage change in current at a voltage bias V as the temperature of the tunnel junction is raised from 0 to 300°K. Curve abc is the forward-biased ($\hat{J}_2$-V) characteristic and ade the reverse-bias ($\hat{J}_1$-V) characteristic.

d. Experimental

(1) Thermally Grown Oxides Since Fisher and Giaever[35] made the first successful thin film tunnel junctions, there have been many reported observations of tunneling through thin films.[35-48] Here, however, we will confine our attention to the work of Pollack and Morris,[46,47] which offers good experimental examples of such observations. The results of Pollack and Morris[46] for a Al-Al$_2$O$_3$-Al junction are reproduced in Fig. 12. The oxide layer was thermally grown on the surface of the first-deposited Al layer. Their data are compared with the theory, and it is seen that the room-temperature data correlate with theory over *nine* decades of current. Further support for the theory that the conduction phenomena are due to the tunnel effect is seen by the fact that the J_2-V characteristic is only very slightly temperature-

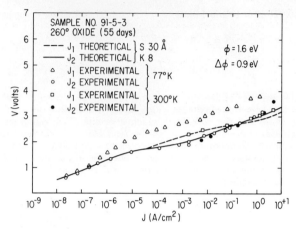

Fig. 12   Tunnel data from Al-Al₂O₃-Al junction.

sensitive.   The fact that the barrier heights (ϕ_1 = 1.6 eV and ϕ_2 = 2.5 eV) are not the same even though the junction is apparently symmetric, and that the J_1-V curve is temperature-sensitive, is attributed to a semiconducting transition region existing between the oxide layer and its parent electrode.

(2) Gaseous Anodized Layers   Pollack and Morris[47] also fabricated Al-Al₂O₃-Al tunnel junctions in which the oxide layer was grown on the Al electrode in an oxygen glow discharge.[49]   Again correlation between theory and experiment was observed over many decades of current density, and the barrier heights for the Al-Al₂O₃-Al system were determined to be ϕ_1 = 1.5 eV and ϕ_2 = 1.85 eV.   These barrier heights were later confirmed remarkably closely by Braunstein et al.,[50] who determined the barriers to be ϕ = 1.49 eV and ϕ = 1.92 eV using photoemissive techniques.

The thermal-tunnel ($\hat{J}$-V characteristic, Sec. 4c) results[47] of Pollack and Morris for Al-Al₂O₃-Al junctions are transcribed in Fig. 13.   These curves are very similar in structure to the theoretical curves of Fig. 11, but the percentage increase in the current is greater than theoretically expected.   The barrier heights determined from these data, that is, the voltage at which the current peaks occur, are ϕ_1 = 2.0 eV and ϕ_2 = 2.2 eV, which are somewhat higher than the isothermal results; these

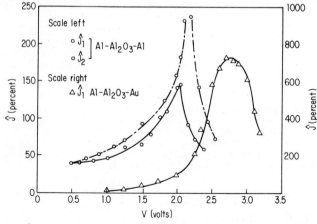

Fig. 13 $\hat{J}$-V tunnel characteristic for Al-Al₂O₃-Al and Al-Al₂O₃-Au junction.

deviations have been attributed to the electric-field penetration of the electrodes.[32,51] The $\hat{J}$-V characteristic for an Al-Al$_2$O$_3$-Au junction is also shown in Fig. 13, but the peak in this characteristic is not as sharp as for the Al-Al$_2$O$_3$-Al junction. The displacement between the peak of this characteristic and that of the corresponding Al electrode of the Al-Al$_2$O$_3$-Al junction is 0.6 V, which is in reasonable agreement with the difference in work junction (0.75 eV) of Al and Au, as required by theory.

e. Other Tunneling Investigations

(1) Al-Al$_2$O$_3$-SnTe Junctions Esaki and Stiles[52,53] have fabricated Al-Al$_2$O$_3$-SnTe junctions by evaporating SnTe onto an oxidized strip of Al. The SnTe layer was highly P-type with about 8×10^{20} carriers cm^{-3}. The I-V characteristics of these junctions (Al electrode negatively biased) taken at $T = 4.2$, 77, and 300°K, are shown in Fig. 14. The interesting feature of these curves is the negative resistance region in the range 0.55 to 0.85 V.

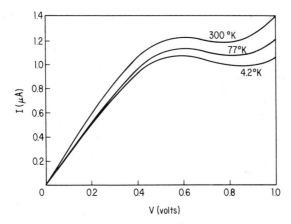

Fig. 14 J-V characteristic for Al-Al$_2$O$_3$-SnTe tunnel junction.

The I-V characteristic can be explained along the following lines, using Fig. 15, which is an idealized energy diagram for the system for various voltage biases (the Al electrode is negatively biased in all cases). For voltages $eV < E_F$ (Fig. 15b) the electrons tunnel from the metal electrode into the empty states lying between the electrode Fermi level and the top of the valence band of the SnTe. The tunnel probability [$\simeq \exp(-\Delta s \, \bar{\phi}^{\frac{1}{2}})$, where Δs and $\bar{\phi}$ are defined in Sec. 4b] for electrons at the Fermi level of the metal increases with increasing bias according to

$$\exp\left[-1.025s \left(E_c + E_g + E_F - \frac{eV}{2} \right)^{\frac{1}{2}} \right] \tag{27}$$

which means that the tunnel current into the SnTe valence band I_v increases with increasing bias.

When $E_F < eV < E_F + E_g$ (Fig. 15c), electrons in energy levels between AB in the electrode only may tunnel into the SnTe. Electrons in levels in the electrode between A and the electrode Fermi level face the energy gap of the SnTe and hence cannot tunnel into the SnTe, because there are no available empty states there. The tunnel probability of electrons entering the SnTe just below the top of the valence band is given by

$$\exp\left[-1.025s \left(E_g + E_c + \frac{eV}{2} \right)^{\frac{1}{2}} \right] \tag{28}$$

The tunnel probability now *decreases* with increasing voltage bias, which means

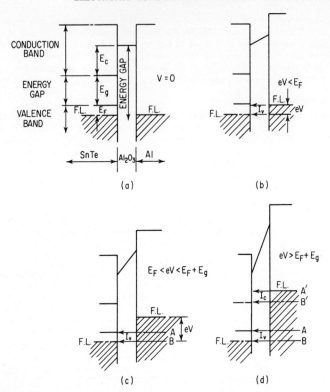

Fig. 15   Energy diagram for Al-Al₂O₃-SnTe tunnel junction for various voltage biases

I_v *decreases* for $E_F < eV < E_F + E_g$, resulting in the negative-resistance region shown in Fig. 14.

For $eV > E_F + E_g$, the tunnel probability for the electrons tunneling into the SnTe valence band is still given by (28). However, we now have an additional component of current I_c which is due to electrons in the electrode in levels between the Fermi level of the electrode and the bottom of the SnTe conduction band (i.e. between A' and B') tunneling into the *conduction* band of the electrode. The tunnel probability of an electron at the Fermi level is

$$\exp\left[-1.025s \left(E_c + E_g + E_F - \frac{eV}{2} \right)^{\frac{1}{2}} \right] \tag{29}$$

which *increases* with increasing voltage bias; thus I_c *increases* with voltage bias. Since the total tunnel current is given by $I_c + I_v$, and since I_c increases much more rapidly than I_v decreases, the tunnel current increases when $eV \gtrsim E_F + E_g$.

In view of the above considerations and by inspection of Fig. 14, it can be seen that $E_F \simeq 0.06$ eV and the energy gap of the SnTe is $E_g \simeq 0.30$ eV.

(2) Si-SiO₂-(Au + Cr)   Dahlke[54] has reported a somewhat different type of tunneling in metal-insulator-semiconductor structures. In this case the semiconductor used was *bulk* degenerate P⁺⁺-type silicon, <111> orientation, 0.0006 ohm-cm resistivity. The oxide was the thermally grown oxide (SiO₂) of the semiconductor; the metal electrodes were evaporated films of Au + Cr.

Figure 16 illustrates the dc tunnel conductance, obtained by differentiation of the measured I-V characteristics, of three samples in which the oxides were grown under different conditions; the polarity of the voltage axis corresponds to that of the metal

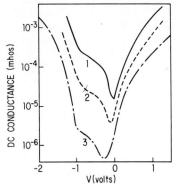
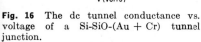

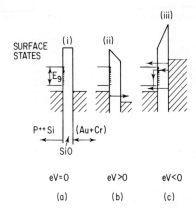

Fig. 16 The dc tunnel conductance vs. voltage of a Si-SiO-(Au + Cr) tunnel junction.

Fig. 17 Energy diagram for Si-SiO(Au + Cr) tunnel junctions for various voltage biases.

electrode. Samples 1 and 2 were grown in dry oxygen and steam, respectively, by brief rf heating under bias at 700°C; sample 3 was treated as sample 2 followed by a 30-min annealing in H_2 at 350°C. The right-hand half of Fig. 16 is the conductance characteristic for electrons tunneling from the valence band into the electrode, as shown in Fig. 17b, and increases monotonically with voltage, as is to be expected. The reason that negative resistance, as observed by Esaki and Stiles in essentially the same type of system, is not observed in these samples when the metal electrode is negatively biased ($V < 0$) is apparently the existence of surface states in the forbidden gap of the semiconductor. Thus electrons in levels in the metal electrode positioned opposite the forbidden band of the semiconductor can, unlike in the case of Esaki and Stiles [Sec. 4e(1)] where the effect of the surface states was apparently negligible, tunnel into the semiconductor by way of the surface states. They then recombine with holes in the valence band, as shown in Fig. 17c. Thus the structure in the characteristics ($V < 0$) is due to the combined effects of the band gap and the distribution of the surface states therein. From these results Dahlke estimates an increase of one to two orders of magnitude in surface-state density in changing from annealed-steam-grown to steam-grown to dry-oxygen-grown oxide layers, which is in qualitative agreement with other workers' results.

(3) Determination of E-K Relationship in Aluminum Nitride Lewicki and Mead[55] have reported a series of measurements carried out on Al-AlN-Mg tunnel junctions of various thicknesses. The junctions were fabricated by exposing freshly deposited Al films in a N_2 glow discharge at a pressure of 200 millitorr for 3 min and subsequently evaporating into the AlN a Mg counterelectrode. From the dependence of the tunneling current upon the insulator thickness they have determined the relationship between the imaginary component of the propagation vector in the forbidden band of the AlN as a function of energy.

The experimental data were analyzed in terms of theory developed by Stratton et al.,[28] which is summarized as follows: For applied voltages higher than the average spread of the tunneling electrons, but less than the barrier energy of the positively biased electrode, the current is related to the average wave vector $\bar{k}(eV)$, corresponding to an incident electron with energy equal to the metal Fermi energy, by

$$I = \frac{B}{s^2} \exp\left[-2\bar{k}(eV)s\right]$$

which is simply the first part of (14) when $\bar{k}(eV) = (2m/h^2)\bar{\phi}^{\frac{1}{2}}$. Thus the ratio of the currents I_1 and I_2 from two samples of thickness s_1 and s_2 for a given voltage bias is related to $\bar{k}(eV)$ by

$$2\bar{k}(eV) = (s_1 - s_2)^{-1} \ln\left(\frac{I_2 s_2{}^2}{I_1 s_1{}^2}\right)$$

Since $\bar{k}$ is an integral of $k(E)$, $k(E)$ is related to $\bar{k}$ as follows:

$$k(\phi_2 - eV) = \frac{1}{e}\frac{\partial}{\partial V}(\phi_1 - \phi_2 + eV)\bar{k}(eV) \tag{30}$$

Lewicki and Mead applied (30) to their results and determined the $k(E)$ vs. E dependence for their junction insulator, which was found to be in agreement with Franz's[55a] empirical relationship for a material with the 4.2-eV energy band gap of AlN, as shown in Fig. 18.

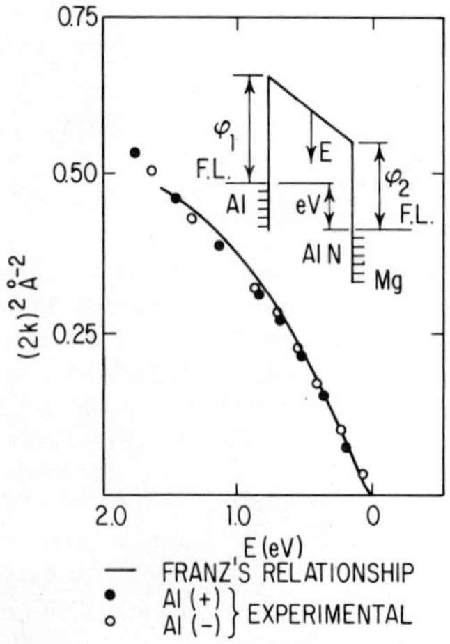

(4) Magnetointernal Field Emission Esaki et al.[56] have prepared tunnel junctions using thin films of Eu chalcogenides, such as EuSe and EuS. We will confine our remarks to observations made on EuSe, which has a magnetic-transition temperature at 4.7°K. The junctions were prepared by the successive evaporation of Al, EuSe, and Au, onto a heated sapphire substrate without breaking vacuum. The thickness of the EuSe ranged from 200 to 600 Å.

The I-V characteristics on a log (I/V^2) vs. V^{-1} plot at 4.2°K are shown in Fig. 19. It is noted that the curve shifts toward lower voltage in a magnetic field, and also that the I-V characteristic obeys a relationship of the form

$$I = \frac{aV^2}{s^2}\exp\left(-\frac{bs}{V}\right) \tag{31}$$

Fig. 18 E-k relationship for AlN. Inset is an energy-band diagram of junction with Al electrode negatively biased.

— FRANZ'S RELATIONSHIP

• Al (+) }
○ Al (−) } EXPERIMENTAL

which is typical of Fowler-Nordheim tunneling [see (22)]. In (31) a and b are constants of the material and s is the EuSe thickness. It was

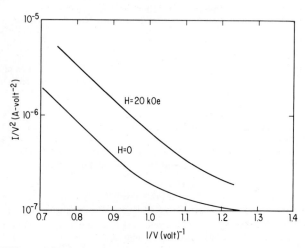

Fig. 19 J/V^2 vs. I/V characteristic for Al-EuSe-Au junction at $H = 0$ and 20 kOe.

verified that application of magnetic fields affects only the constant b, reducing it with increasing field.

The temperature dependence of the EuSe junction is shown in Fig. 20 in terms of the voltage (normalized to its value at 4.2°K) required to maintain a constant current through the junction. The full line illustrates the results with zero magnetic field and the dotted line with a field of 20 kOe.

The interesting features of these curves are: (1) a rapid change near the magnetic-transition temperature of EuSe, and (2) no temperature dependence above and below the transition temperature, except for a gradual decrease above 20°K.

Esaki et al.[56] explain the above observations as follows: They assume that the barrier ϕ of the junction extends from the Fermi level of the electrodes up to the bottom of the conduction band provided by the $5d$ band of the EuSe, and that the conduction process is due to Fowler-Nordheim tunneling through this barrier [see Fig. 9 and Eq. (22)]. They estimate

$$\phi = 0.5 \, \text{eV}$$

at the Al-EuSe interface from the constant b, which is related to ϕ by [see (22)]

$$b = \frac{4 \sqrt{2m}^{\frac{1}{2}} \phi^{\frac{3}{2}}}{3e\hbar} \tag{32}$$

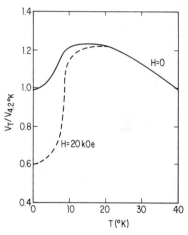

Fig. 20 $V_T/V_{4.2°K}$ vs. temperature for Al-EuSe-Au junction at $H = 0$.

Because b is related to only one of the material constants, namely, ϕ, it is this parameter only that is affected by the magnetic transition at 4.2°K and the applied magnetic field. Esaki et al. conclude, therefore, that the magnetic perturbation of the barrier height is due to spin ordering of the $4f$ electrons of the EuSe, which causes a shift of the $5d$ band with respect to the electrode Fermi level.

5. HIGH-FIELD EFFECTS

a. Electrode-limited Processes

(1) Image Forces The abrupt changes in potential at the metal-insulator interface we have been showing in our energy diagrams are physically unrealistic, since abrupt changes in potential imply infinite electric fields. In actual fact the potential step changes smoothly as a result of the image force.[57] This arises as a result of the metal surface's becoming polarized (positively charged) by an escaping electron, which in turn exerts an attractive force $e^2/16\pi\epsilon_0 K^* x^2$ on the electron. The potential energy of the electron due to the image force is thus

$$\phi_{im} = -\frac{e^2}{16\pi\epsilon_0 K^* x} \tag{33}$$

where x is the distance of the electron from the electrode surface. The dielectric constant K^* used in (33) is the *high-frequency* constant, since in the course of emission from the cathode the escaping electrons spend only an extremely short time in the immediate vicinity of the surface.

Image-force effects play an important role in the conduction process when the current is electrode-limited, as we will see.

(2) Neutral Contact—Theory The potential step (with respect to the Fermi level) at a neutral barrier with attendant image potential as a function of the distance x from the interface is given by

$$\phi(x) = \phi_0 + \phi_{im} = \phi_0 - \frac{e^2}{16\pi\epsilon_0 K^* x} \tag{34}$$

The barrier $\phi(x)$ in the presence of image forces is illustrated by the line AB in Fig. 21. Clearly (34) is not valid at the electrode surface, since $\phi = -\infty$ there. Schottky circumvents this anomaly by assuming that the image force holds only for x greater than some critical distance x. For $x < x_0$, he assumes a constant image force; that is, the potential energy is a linear function of x, and such that it matches the bottom of the electrode conduction band at the surface (see Fig. 21).

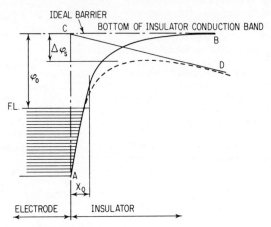

Fig. 21 Schottky effect at a neutral contact.

When an electric field exists at a metal-insulator interface, it interacts with the image force and lowers the potential barrier. This can be seen by reference to Fig. 21. The line CD represents the potential due to a uniform field, which when added to the barrier potential $\phi(x)$ produces the potential step shown by the dotted line, which is seen to be $\Delta\phi_s$ lower than without the electric field. The potential energy of the barrier under the influence of the field with respect to the Fermi level of the electrode is now given by

$$\phi(x) = \phi_0 - \frac{e^2}{16\pi K^* \epsilon_0 x} - eFx$$

This equation has a maximum at $x_m = (e/16\pi K^* \epsilon_0 F)^{\frac{1}{2}}$. The change $\Delta\phi\,[= \phi_0 - \phi(x_m)]$ in the barrier height due to the interaction of the applied field with the image potential is thus given by

$$\Delta\phi_s = \left(\frac{e^3}{4\pi K^* \epsilon_0}\right)^{\frac{1}{2}} F^{\frac{1}{2}} \equiv \beta F^{\frac{1}{2}} \tag{35}$$

In conventional units $\Delta\phi_s = (3.8 \times 10^{-4}/K^{*\frac{1}{2}})F^{\frac{1}{2}}$.

Because of image-force lowering of the barrier, the electrode-limited current does not saturate according to the Richardson law $J = AT^2 \exp(-\phi_0/kT)$ but rather obeys the Richardson-Schottky[57] law:

$$J = AT^2 \exp\left(-\frac{\phi_0 - \Delta\phi}{kT}\right) \tag{36}$$

$$J = AT^2 \exp\left(-\frac{\phi_0}{kT}\right) \exp\frac{\beta_s F^{\frac{1}{2}}}{kT} \tag{37}$$

where $A = 4\pi em(kT)^2/h^2$. When I is expressed in A cm^{-2}, A takes the value 120. Equation (36) was first applied successfully to metal-vacuum interfaces. Its use in insulators has to be qualified: Simmons[58] has pointed out that it holds only if the electron mean free path is of the order of the insulator thickness; otherwise the preexponential factor becomes $2e(2\pi mkT/h^2)^{\frac{1}{2}}\mu V/S$; and Crowell[59] has shown that the effect of the electron's having differing masses in the metal and insulator is to change m to m^*, the effective electron mass, in the constant A.

When high fields exist at the interface, tunneling through the potential barrier can dominate the conduction process, and the relevant I-V characteristic is given by (22). We have seen in Sec. 4 that this mechanism requires that the thickness of the barrier at the electrode Fermi energy be less than about 50 Å. Thus the onset of field emission at a neutral contact occurs at a field given by $F \simeq \phi_0 \times 10^8/50 = 2 \times 10^5 \phi_0$ V cm^{-1}.

A comparison of Richardson-Schottky current and tunnel currents in thin insulators has also been studied; the reader is referred to the literature[60,61] for details.

(3) Neutral Contact—Experimental The Richardson-Schottky effect in insulators appears to have been first observed by Emptage and Tantraporn,[62] who reported a log I vs. $F^{\frac{1}{2}}$ relationship in their samples; since then there have been many other reported similar observations. Mead,[63] however, has pointed out that a log J vs. $F^{\frac{1}{2}}$ relationship does not necessarily imply the Richardson-Schottky effect, but has reported convincing evidence of the effect. His data for Zn-ZnO-Au samples are shown in Fig. 22. The slope of the straight line (β_s/kT) corresponds to a thickness of 61 Å, whereas the capacitively determined thickness was 57 Å. Furthermore, the barrier height determined from thermal-activation measurements gave $\phi = 0.75$ eV, while the intercept of the activation curves yielded an intercept of $\phi \simeq 0.8$ eV; photoemissive measurements also apparently yielded similar barrier heights. These results are good evidence for Schottky emission.

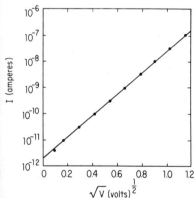

Fig. 22 I-V characteristic of a Zn-ZnO-Au structure.

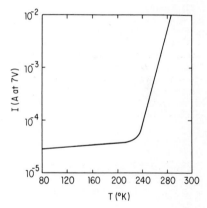

Fig. 23 Temperature variation of current through a Pb-Al$_2$O$_3$-Pb sample. Thickness of Al$_2$O$_3$ = 340 Å and $V = 7$ V.

Because thermal emission falls off rapidly with decreasing temperature, and since field emission is virtually temperature-insensitive (Sec. 4c), it means that low temperatures are amenable to field emission dominating the conduction process. Pollack[64,65] observed the electrode-limited conduction-process change from the Richardson-Schottky to the field-emission mechanism in both relatively thick and very thin insulators as the temperature of the sample was lowered from room to liquid-nitrogen temperatures (see Fig. 23).

(4) Blocking Contact The potential energy of the barrier (Schottky) of a blocking contact (Sec. 3c) with attendant image potential is, from (8) and (33),

$$\phi(x) = \phi_0 - \frac{N_d e^2}{K\epsilon_0}\lambda x - \frac{x^2}{2} - \frac{e^2}{16\pi K^*\epsilon_0 x}$$

This barrier has a maximum at $x_m \simeq 1/[4(\pi N_d \lambda)^{\frac{1}{2}}]$. In this case the image force lowers the potential barrier by an amount

$$\Delta \phi = \phi_0 - \phi(x_m) = \left[\frac{(V + \psi_m - \psi_i)e^7 N_d}{2(8\pi)^2(\epsilon_0 K^*)^3} \right]^{\frac{1}{4}} \tag{38}$$

Substituting (38) into (36) yields, in conventional units,

$$J = 120T^2 \exp\left\{ - \frac{\phi_0 - 8.26 \times 10^{-6}[N_d(V + \psi_m - \psi_i)/K^{*3}]^{\frac{1}{4}}}{kT} \right\} \tag{39}$$

As at the neutral contact, tunneling will occur within the barrier when the field at the interface is sufficiently high to reduce the width of the barrier measured at the electrode Fermi level to about 50 Å or so. For $V > \phi_0$, the barrier at the interface can be considered approximately triangular; so we may use the Fowler-Nordheim expression (22) to estimate the current tunneling through the contact. Hence, substituting (11) into (22) yields, using conventional units,[66]

$$J = \frac{1.22 \times 10^{-11} N_d(V + \psi_m - \psi_i)}{\phi K^{*3}} \exp\left\{ - \frac{3.65 \times 10^{10}(K^* \phi^3)^{\frac{1}{2}}}{[N_d(V + \psi_m - \psi_i)]^{\frac{1}{2}}} \right\} \tag{40}$$

We are not aware of any incontrovertible experimental evidence for these processes in thin film insulators, but the tunneling process is probably the dominant mechanism in the electrode-limited range described in Sec. 5c(3).

b. Poole-Frenkel Effect

(1) Theory The Poole-Frenkel effect[67,68] (field-assisted thermal ionization) is lowering of a Coulombic potential barrier when it interacts with an electric field, as shown in Fig. 24. This process is the bulk analog of the Schottky effect at an interfacial

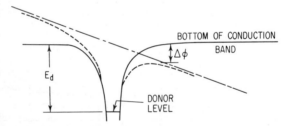

Fig. 24 Poole-Frenkel effect at a donor center.

barrier. Since the potential energy of an electron in a Coulombic field $-e^2/4\pi\epsilon_0 Kx$ is four times that due to image-force effects, the Poole-Frenkel attenuation of a Coulombic barrier $\Delta\phi_{PF}$ in a uniform electric field is twice that due to the Schottky effect at a *neutral* barrier:

$$\Delta\phi_{PF} = \left(\frac{e^3}{\pi\epsilon_0 K^*} \right)^{\frac{1}{2}} F^{\frac{1}{2}} \equiv \beta_{PF}F^{\frac{1}{2}} \tag{41}$$

This result was first applied by Frenkel[67,68] to the host atoms in *bulk* semiconductors and insulators. He argued that the ionization potential E_g of the atoms in a solid are lowered an amount given by (41) in the presence of a uniform field. Thus the conductivity is obtained by substituting $E_g - \Delta\phi_{PF}$ for E_g in (1), yielding a field-dependent conductivity of the form

$$\sigma = \sigma_0 \exp\left(\frac{\beta_{PF}F^{\frac{1}{2}}}{2kT} \right) \tag{42}$$

where σ_0 $[= e\mu N_c \exp(-E_g/2kT)]$ is the low-field conductivity. Equation (42) may be written in the form

$$J = J_0 \exp\left(\frac{\beta_{PF}F^{\frac{1}{2}}}{2kT}\right) \tag{43}$$

where J_0 $(= \sigma_0 F)$ is the low-field current density.

It is interesting to note that although $\Delta\phi_{PF} = 2\Delta\phi_s$, the coefficient of $F^{\frac{1}{2}}$ in the exponential is the same for both the Richardson-Schottky [at a neutral contact, see Eq. (37)] and Poole-Frenkel J-F characteristics (i.e., $\beta_{PF}/2 = \beta_s$). Mead[69] has suggested, however, that since traps abound in an insulator and that a trap having a Coulombic-type barrier would experience the Poole-Frenkel effect at high fields, thereby increasing the probability of escape of an electron immobilized therein, the current density in thin film insulators containing *shallow* traps is given by

$$J = J_0 \exp\left(\frac{\beta_{PF}F^{\frac{1}{2}}}{kT}\right) \tag{44}$$

Note in this case that the coefficient of $F^{\frac{1}{2}}$ is *twice* that in (43). Mead[63] also first reported field-dependent conductivity apparently of the form given by (44), and for this reason (44) is usually the form of Poole-Frenkel equation associated with *thin film* insulators rather than that given by (43).

(2) Experimental From what has been said it follows that it should be possible to differentiate between the Schottky and Poole-Frenkel effects in *thin-film* insulators from their different rates of change of conductivity with field strength;[67] viz., a plot of $\ln J$ vs. $F^{\frac{1}{2}}/kT$ results in a straight line of slope β_s or β_{PF}, depending on whether the conduction process is Richardson-Schottky or Poole-Frenkel. These experimentally determined slopes can be compared with the theoretical β_s and β_{PF}, which can be calculated quite accurately provided the high-frequency dielectric constant K^* for the material is known. Alternatively, the high-frequency dielectric constant may be determined from the slopes assuming the controlling mechanism is known. The resulting K^* should satisfy the equation $K^* = n^2$, where n is the refractive index for the material.

Several investigators have observed a field-dependent conductivity of the form given by (44) in what is apparently *bulk-limited* conduction in Ta_2O_5 and SiO films.[70-72] Johansen[71] has noted, however, that the coefficient of $F^{\frac{1}{2}}/kT$ is compatible with the Schottky effect rather than the Poole-Frenkel effect. On closer examination of the results of Mead[69] and of Hirose and Wada,[70] although they claim Poole-Frenkel emission, it is apparent that their experimental β is actually compatible with Schottky emission.[73] These authors based their conclusions on dielectric-constant values which are typically four times too high for the high-frequency dielectric constants of the materials used. Hartman et al.[72] conclude that neither the Richardson-Schottky equation (37), because conductivity is bulk-limited, nor the Poole-Frenkel equation (44), because of the incompatibility of the experimental and theoretical β, can adequately explain their results.

In actual fact in films of Ta_2O_5 and SiO, because of their wide energy gaps, in order to have any detectable current, even in the absence of traps, it is necessary to have donor or acceptor centers within the insulator to supply the necessary carriers (see Ref. 73 and Sec. 2). Furthermore, if it is assumed that the insulator contains donor centers which lie below the Fermi level—this assumption is supported by the fact that the conductivity of the films continues to increase with increasing temperature above room temperature—and shallow *neutral* traps (see Fig. 25), the bulk J-V characteristic of the film is given by[73]

$$J = J_0 \exp\left(\frac{\beta_{PF}F^{\frac{1}{2}}}{2kT}\right) \tag{45}$$

where

$$J_0 = e\mu N_c \left(\frac{N_d}{N_t}\right)^{\frac{1}{2}} F \exp\left(-\frac{E_d + E_t}{2kT}\right)$$

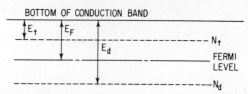

Fig. 25   Energy diagram of insulator film used in deriving (45).

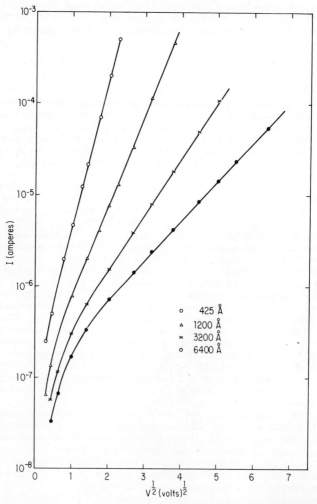

Fig. 26 I-$V^{\frac{1}{2}}$ characteristic of Al-SiO-Al samples illustrating the Poole-Frenkel effect in the higher-voltage range.

Thus in this case the coefficient of $F^{\frac{1}{2}}/kT$ is $\beta_{PF}/2$ ($= \beta_s$) *even though the conductivity is not electrode-limited,* which explains the anomalous experimental results.

Stuart[74,75] has recently shown that his bulk-limited I-V characteristics can be explained in terms of (45). Figure 26 shows a plot of his I-V characteristic for four film thicknesses,[74] which are observed to be linear on a $\ln I$ vs. $V^{\frac{1}{2}}$ plot. In Table 3 we have transcribed his calculation of the ratios of the experimentally determined $\beta(\beta_{\text{exp}})$ and the theoretical $\beta(\beta_{\text{theor}})$ assuming (45) as the relevant J-V characteristic.[74,75] The agreement between β_{exp} and β_{theor} is seen to be excellent over the entire thickness range, which covers almost four decades.

TABLE 3 Stuart's[74,75] Results for $\beta_{\text{exp}}/\beta_{\text{theor}}$ Obtained from SiO Films

s, μ	0.043	0.12	0.32	0.64	1.2	3.3	7.0	13.7
$\beta_{\text{exp}}/\beta_{\text{theor}}$	1.06	1.07	1.09	1.07	1.04	1.05	1.03	1.00

c. Electrode-limited to Bulk-limited Process

(1) Physical Concepts[66] Suppose we have a heavily doped ($N_d \gtrsim 10^{18}$ cm^{-3}) insulator with the band structure shown in Fig. 3d, so that the depletion region is very thin (see Table 2). Although the donor density is high, the conductivity of the bulk will be reasonably low, particularly if traps are present, since the donors lie below the Fermi level, which means a few only will be ionized. Under these conditions it is possible to observe an electrode- to bulk-limited transition in the conduction process, as illustrated by the following qualitative arguments.

For initial voltage bias the conduction process will clearly be electrode-limited, because of the high cathode-insulator barrier. At very low voltages the electrode-limited conduction process is by the thermal excitation of electrons from the cathode over the interfacial barrier. The current is only a slow function of the applied voltage for this process [see (39)] and for barriers in excess of about 0.5 eV will be very low.

At higher voltages one of two processes can occur. If the barrier at the Fermi level becomes thin enough, field emission of electrons from the cathode into the conduction band of the insulator can occur [see Eq. (40)], or when the voltage bias exceeds $3E_g/2$, impact ionization can occur in the depletion region of the insulator. Both these processes are characterized by a rapid increase of current with applied voltage; i.e., the contact resistance decreases extremely rapidly with increasing voltage bias. Since the contact resistance under these conditions at low voltages is much higher than that of the bulk, it follows that while either or both of these processes predominate the I-V characteristic will be virtually *thickness-independent* (because essentially all the applied voltage appears across the contact and very little across the bulk), and very steep (see chain-dotted line in Fig. 27).

The electrode-limited process cannot continue indefinitely, however, because the bulk resistance [see (45)] decreases much more slowly with increasing voltage than does the contact (compare dotted and chain-dotted lines in Fig. 27). Thus, at some voltage V_T, the transition voltage, the contact resistance falls to a value equal to that of the bulk, and when this occurs the applied voltage V_T is shared equally between the contact and the bulk. Thereafter, practically all the voltage in excess of V_T will fall across the bulk and the remaining fraction across the barrier, just sufficient to ensure current continuity throughout the system. Hence, for voltages in excess of V_T, the current will cease to rise as rapidly as for $V < V_T$, since it is controlled by bulk processes, *and will be thickness-dependent.*

(2) Theoretical Results In Fig. 27 the theoretical J-V characteristics for three thicknesses of dielectric ($S = 1, 3$, and 10 μ) have been plotted, using (40) and (45) and the parameters $N_d = 10^{18}$ cm^{-3}, $N_t = 10^{19}$ cm^{-3}, $K^* = 3$, $\mu = 10$ cm^2, (V sec)$^{-1}$, $(E_t + E_d)/2 = 0.55$ eV, and $\phi = 1.5$ eV. For these parameters the current given by (39) is too small to be shown on the graph. The dotted and chain-dotted lines

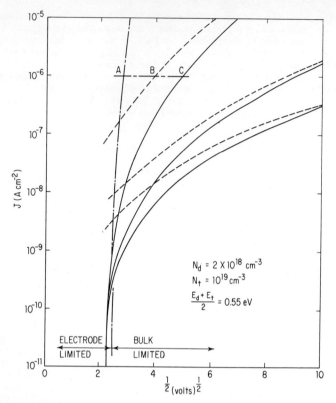

Fig. 27   Theoretical curves showing electrode- to bulk-limited characteristic.

represent the bulk and contact characteristics given by (45) and (40), respectively. The full lines represent the combined effect of the contact and bulk, that is, the actual characteristic of the junction. Since the contact and bulk impedances act in series, the actual curves are arrived at by adding together the contact and bulk voltages corresponding to a given current on the individual curves. Hence, to obtain the point C on the upper curve of Fig. 27 we add the voltages corresponding to the points A and B lying, respectively, on the individual contact and bulk curves.

The curves clearly delineate contact and bulk effects, the transitional stage occurring about 12 V. Below 12 V we have the contact-limited portion, which is characterized by the thickness-independent and rapidly increasing current-voltage tunnel curve. Above 12 V we have the bulk-limited region, which manifests the effect of varying insulator thickness. It will be noted that these curves are displaced to lower current densities and lower slopes than their corresponding individual curves, thus reflecting to a small extent contact effects on the bulk-limited characteristics.

(3) Experimental Results   Stuart's[75] results, obtained from thin film Al-SiO-Al samples, are shown in Fig. 28. The thickness of the SiO films ranges from 1.2 to 13.7 μ. The interesting point about these results is their similarity to the theoretical curves in Fig. 27. Note in particular below $V \simeq 70$ V that the I-V characteristics are essentially independent of the film thickness, even though there is over a decade difference between the two extreme thicknesses. Above $V \simeq 70$ V the J-V characteristics are thickness-dependent and obey a relationship of the form given by (45).

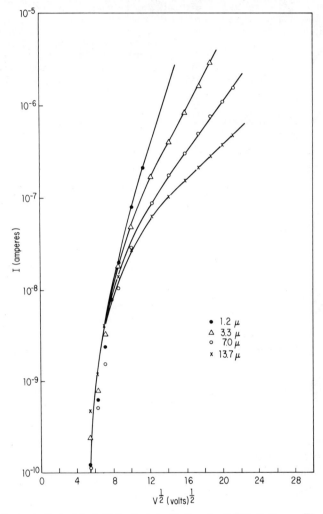

Fig. 28 Electrode-limited to bulk-limited conduction in Al-SiO-Al films. Electrode area 0.1 cm².

6. SPACE-CHARGE-LIMITED (SCL) CURRENTS IN INSULATORS

a. Physical Concepts

We have seen that an insulator which does not contain donors and which is sufficiently thick to inhibit tunneling will not normally conduct significant current. However, if an ohmic contact (Sec. 3a) is made to the insulator, the space charge injected into the conduction band of the insulator is capable of carrying current; this process is termed SCL conduction. In order to gain physical insight into this process, let us consider what happens when a bias is applied to the system shown in Fig. 29, that is, an insulator having two ohmic contacts on its surfaces.

The result of the applied bias is to add *positive* charge to the anode and *negative* charge to the cathode, as would be the case with any such capacitive system. Thus, as the voltage bias increases, the *net* positive charge on the anode *increases* and that on the cathode decreases. Calling the charge on the cathode Q_1, that on the anode Q_2, and the negative space-charge density $\rho(x)$, the condition of electrical neutrality demands that[76]

$$\int_0^s \rho(x)\, dx = Q_1 + Q_2 \qquad (45a)$$

Equation (45a) may be rewritten

$$\int_0^{\lambda_m} \rho(x)\, dx + \int_{\lambda_m}^L \rho(x)\, dx = Q_1 + Q_2$$

where λ_m is chosen such that

$$\int_0^{\lambda_m} \rho(x)\, dx = Q_1 \qquad (46)$$

and

$$\int_{\lambda_m}^s \rho(x)\, dx = Q_2 \qquad (47)$$

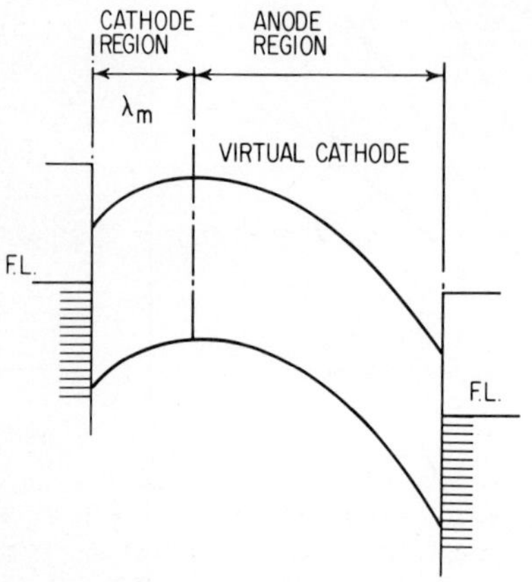

The insulator has thus been divided into two portions (see Fig. 29) with λ_m as the boundary separating the two.

Fig. 29 Energy diagram illustrating virtual cathode, cathode region, and anode region under space-charge-limited condition.

The significance of (46) and (47) is that the *positive* charge on either contact is neutralized by an equal amount of *negative* charge contained between the contact and the plane at $x = \lambda_m$. Thus the field in the insulator due to Q_1 is zero for $x \geq \lambda_m$; that is, the negative charge between $x = 0$ and λ_m screens the insulator beyond $x = \lambda_m$ from conditions at the cathode-insulator interface. Similarly the field due to Q_2 is zero for $x \leq \lambda_m$. Since the field due to Q_1 and Q_2 is zero at $x = \lambda_m$, the net field there must be zero, as shown in Fig. 29, and for this reason the plane at $x = \lambda_m$ is termed the virtual cathode. The region $0 \leq x \leq \lambda_m$ is designated the cathode region, and the region $\lambda_m \geq x \geq s$ the anode region.

From a consideration of (46) and (47) and the fact that Q_1 *decreases* and Q_2 *increases* with increasing voltage, it will be clear that the virtual cathode moves closer to the cathode as the applied voltage increases; that is, the cathode region decreases and the anode region increases. Eventually, when $Q_1 = 0$, the virtual cathode coincides with the physical cathode-insulator interface. Under this condition, then, the anode region extends throughout the whole of the insulator, and an ohmic contact no longer exists at the cathode-insulator interface. Thus, for further increasing voltage bias, the conduction process is no longer space-charge-limited, but rather it is emission-limited.

b. Assumptions and Boundary Conditions

If the cathode makes a good ohmic contact to the insulator (e.g., $\phi \simeq 0$) the cathode region [i.e., the accumulation region at zero bias (see Table 1)] will normally be small compared with thickness of the insulator. We may therefore assume that under voltage bias the anode region occupies the whole of the insulator thickness; or alternatively stated, the virtual cathode coincides with the physical cathode (cathode-insulator interface). This assumption leads us to the following boundary conditions: (1) the field at $x = 0$ is zero, and (2) $V = 0$ and V_a, respectively, at $x = 0$ and s, where V_a is the applied voltage.

c. Trap-free Insulator

Since we are assuming the anode region extends throughout the insulator, we may neglect diffusion effect;[1] thus

$$J = \rho(x)\mu F \tag{48}$$

and from Poisson's equation (2),

$$\frac{dF}{dx} = -\frac{\rho(x)}{K\epsilon_0} \tag{49}$$

Substituting (49) into (48) yields

$$J = -\frac{\mu K\epsilon_0 F\, dF}{dx} = -\frac{(\mu K\epsilon_0/2)\, dF^2}{dx} \tag{50}$$

which on integration, subject to the boundary condition $F = 0$ at $x = 0$, yields

$$F = -\frac{dV}{dx} = \left(\frac{2Jx}{\mu K\epsilon_0}\right)^{\frac{1}{2}} \tag{51}$$

Integrating (51), and using $V = 0$ and V_a at $x = 0$ and $x = s$, we arrive at

$$J = \frac{9\mu K\epsilon_0}{8s^3} V^2 \tag{52}$$

Equation (52) was originally derived by Mott and Gurney.[1] The interesting features of (52) are that it predicts that SCL current is proportional to V^2 and inversely proportional to s^3; both these predictions have since been confirmed by experiment. However, (52) predicts much higher currents than are observed in practice, and also that SCL currents are temperature-insensitive, which is also contrary to observation. These deviations from the simple trap-free theory are readily accounted for when a more realistic insulator, that is, one which contains traps, is considered.

d. Defect Insulator

The theory of SCL currents in defect insulators is due initially to Rose.[2] If the insulator contains traps, a large fraction of the injection space charge will condense therein, which means that the free-carrier density will be much lower than in a perfect insulator. Furthermore, since the occupancy of traps is a function of temperature, the SCL current is temperature-dependent.

Our starting point, as in the trap-free insulator, is (48) and (49), but now it is necessary to separate the space charge into a trapped ρ_t and a free ρ_f component; thus

$$J = \rho_f \mu F \tag{53}$$

and

$$\frac{dF}{dx} = -\frac{\rho_f + \rho_t}{K\epsilon_0} \tag{54}$$

(1) Shallow Traps If the insulator contains N_t shallow traps, positioned an energy E_t below the conduction band (see Fig. 30), then $\rho_f = eN_c \exp(-E_F/kT)$ and $\rho_t = eN_t \exp[-(E_F - E_t)/kT]$; thus

$$\theta \equiv \frac{\rho_f}{\rho_t} = \frac{N_c}{N_t} \exp\left(-\frac{E_t}{kT}\right) \tag{55}$$

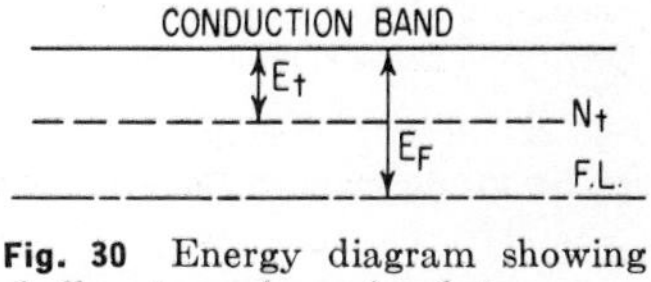

An inspection of (55) shows that ρ_f can normally be neglected in comparison with ρ_t; for example, assuming $N_t = 10^{19}$ cm^{-3} and $E_t = 0.25$ eV, and taking $N_c = 10^{19}$ cm^{-3} at room temperature, we have $\theta < 10^{-5}$. Thus substituting (54) and (55) in (53) yields $J = e\mu K\epsilon_0 \theta F\, dF/dx$. By comparison with (50) and (52), the latter expression integrates to

$$J = \frac{9\mu K\epsilon_0 \theta}{8s^3} V^2 \tag{56}$$

CONDUCTION BAND

Fig. 30 Energy diagram showing shallow traps in an insulator.

It will be clear, then, that the effect of shallow trapping enters into the theory by way of the factor θ, and since it is independent of V, $J \alpha V^2$ as in the trap-free case. We have seen, however, that θ is a very small quantity and temperature-dependent; thus the inclusion of shallow traps in the insulator brings the theory into line with experimental observation (see Sec. 6c).

Lampert[77,78] has noted that only when the injected free-carrier density n_i exceeds the volume-generated free-carrier density n_0 will space-charge effects be observed; when $n_0 > n_i$ the volume (ohmic) conductivity will predominate. Lampert calculates the voltage V_x at which the transition from ohmic to SCL conduction occurs to be

$$V_x = \frac{e n_0 s^2}{\theta K \epsilon}$$

Lampert[77,78] has also pointed out that if sufficient charge is injected into the insulator, the traps will become filled (trap-filled limit TFL). Further injected charge then exists as free charge in the conduction band and contributes in toto to the current. Beyond the TFL, then, the J-V characteristic will be given by (52) rather than (56), thus as V just exceeds V_{TFL}, the current rises rapidly by an amount θ^{-1}. The voltage V_{TFL} at which the TFL occurs is given by

$$V_{\mathrm{TFL}} = \frac{e N_t s}{2 K \epsilon_0}$$

In Fig. 31 we have schematically illustrated a typical I-V characteristic for an insulator having a shallow discrete trapping level. At the lower voltages ($V < V_x$), the characteristic is ohmic, because the bulk-generated current exceeds the SCL current. In the voltage range $V_x < V < V_{\mathrm{TFL}}$ the SCL current predominates and $I \alpha V^2$, (56). When

$$V = V_{\mathrm{TFL}}$$

sufficient charge has been injected into the insulator to fill the traps. Hence, as V just exceeds V_{TFL} the current rises rapidly such that for $V > V_{\mathrm{TFL}}$ the I-V characteristic obeys the trap-free law, (52). Clearly, from the structure exhibited in the characteristic, much information about traps in insulators can be deduced from the experimental data.

(2) Exponential-trap Distribution In amorphous or polycrystalline structures, which are characteristic of thin film insulators, a distribution of trap levels is to be expected, rather than a discrete level of traps. Rose[2] has treated the case of SCL conduction in the presence of a distribution of trap levels that decreases exponentially in density with increasing energy below the conduction band; that is,

$$N_t = A \exp\left(-E / k T_c\right)$$

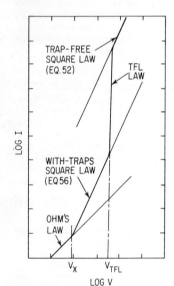

Fig. 31 SCL I-V characteristic for an insulator containing shallow traps.

where E is the energy measured from the bottom of the conduction band, and T_c is a characteristic temperature greater than the temperature at which the currents are measured. In this case

$$I \alpha V^{(T_c/T+1)}$$

which means that, since $T_c > T$, the current increases more rapidly with applied voltage than in the trap-free or discrete trap-level cases.

e. Double Injection

If the cathode is ohmic to electrons and the anode ohmic to holes, SCL currents of both sign of carriers can flow in an insulator. Because the injected electrons and holes are capable of neutralizing each other, it follows that the number of injected carriers at any point in the insulator can exceed the space charge existing there. It follows, then, that double-injection currents will be larger than those due to single-carrier injection under the same condition of voltage bias. Furthermore, in the case of double injection, recombination will take place between the two types of carriers, usually through recombination centers situated in the forbidden gap of the insulator. The effect of charge neutralization and recombination results in the double-injection J-V characteristic's being even richer in structure than that of the single-carrier characteristic.[79,80]

The theory of double injection with attendant recombination processes is clearly much more complex than the single-carrier theory. Furthermore, it is not possible to ignore the effect of diffusion currents, as is done in the single-carrier theory. There is also the problem of making suitable contacts to an insulator, for an electrode that is ohmic to electrons will normally be blocking to holes. The problem of obtaining even one type of ohmic contact to a wideband insulator is difficult; making two different types is formidable. Indeed we are not aware of any unequivocal evidence for double injection in the thin film, wideband insulators with which we are primarily concerned. Thus in view of all the above-mentioned difficulties, it is inappropriate to discuss double injection further here. For further details of the theory of double injection and the experimental evidence for it on bulk crystals, we refer the reader to the review by Lampert.[3]

f. Experimental Results

Very few experimental data on SCLC in solids appear in the literature; there are probably two main reasons for this. First, the problem of making ohmic contacts to

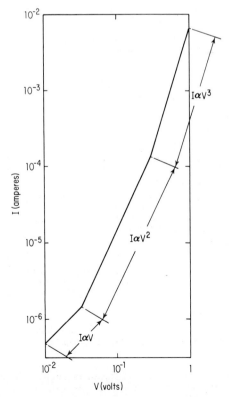

Fig. 32 Space-charge-limited conduction in In-CdS-Au samples.

wideband insulators is formidable. Second, in thin films it is clear that the application of only 1 or 2 V induces fields often sufficient to cause the conduction process to be emission-limited.

Zuleeg[81] has reported SCLC in In-CdS-Au sandwiches, in which the indium electrode made the ohmic contact. His results are shown in Fig. 32. At low voltages the I-V characteristic is ohmic, apparently because of thermally generated carriers. At high voltages $I \alpha V^3$. Over the intermediate-voltage range $I \alpha V^2$, suggesting space-charge-limited conduction. There is no apparent trap-filled limit in this case. Zuleeg and Muller[82] have investigated the effect of the CdS thickness in In-CdS-Au samples in the range where $I \alpha V^2$, by plotting log I (at constant voltage) vs. log s. Their results, which are shown in Fig. 33, were taken from three batches of differing thick-

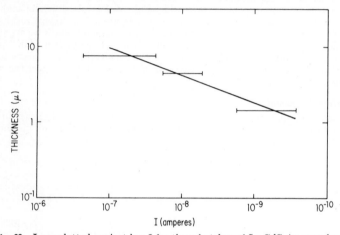

Fig. 33 Logs plotted against log I for three batches of In-CdS-Au samples.

ness; the variation from sample to sample is indicated by the bars. It is seen that a straight line drawn through the average of each batch has a slope of -3, which is in agreement with (56).

7. NEGATIVE RESISTANCE AND MEMORY EFFECTS

It has been observed that when thin insulating films between approximately 200 and 3,000 Å thick have undergone a forming process, the conduction process manifests voltage-controlled negative resistance,[83-96] reversible voltage,[90,94,95] thermal-voltage memory effects,[90,94,95] temperature-independent conductivity,[90,94] and electroluminescence.[91] These effects have been observed in a wide variety of insulators and electrode materials, but we will confine our attention to a series of measurements recently carried out in the Al-SiO-Au system.[94,95]

a. Forming Process and DC Characteristic

The resistance of an unformed sample, when measured at room-ambient pressure, is very high. The V-I characteristic is illustrated in Fig. 34 and is similar to that found in thicker SiO films [see Sec. 5b(2)], that is, log $I \alpha V^{\frac{1}{2}}$ and is very temperature-sensitive. If the sample is now placed in a modest vacuum, a few microns pressure or less, its conductivity is observed to increase sharply when a voltage of about 5 V is applied across the insulator with the gold electrode positively biased. Upon reducing

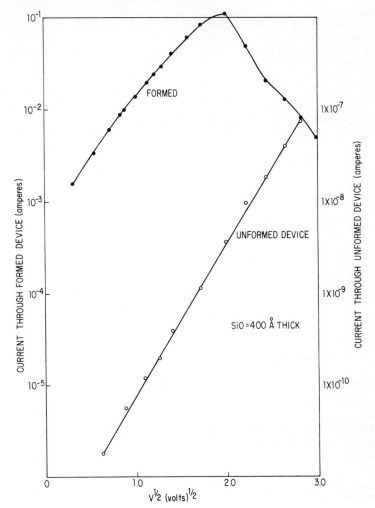

Fig. 34 *I-V* characteristic of unformed and formed sample, on a ln *I* vs. *V*$^{\frac{1}{2}}$ plot, illustrating the large change in conductivity that occurs on forming.

the voltage slowly to zero volts, a *V-I* characteristic with a pronounced voltage-controlled negative-resistance region is observed, as shown in Fig. 35. Any subsequent dc voltage sweep generates an identical characteristic to that shown in Fig. 35; therefore, the change in conductivity is permanent. The sample has clearly undergone some form of forming process.[96]

The interesting feature of the *I-V* curves of the formed sample is the pronounced negative-resistance region which is observed for $V \simeq 4$ V. The peak-to-valley ratios of the curves are typically 100:1, although ratios as high as 1,000:1 have been observed. A second interesting point is that the voltage at which the peak current occurs is not only practically independent of insulator thickness but has also been shown to vary little (3 to 5 V) with insulator and electrode material; the peak current is, however, dependent on the nature and thickness of the insulator.

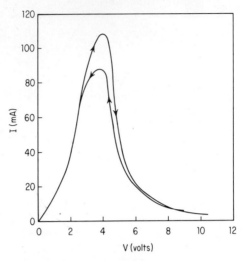

Fig. 35 *I-V* characteristic of formed sample.

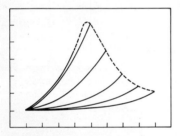

Fig. 36 Oscillogram of *V-I* traces for 1,000-Hz applied voltages for various voltage amplitudes ($y =$ 10 mA cm^{-1}, $x = 1$ V cm^{-1}). The dotted line illustrates the dc characteristic.

b. Dynamic Characteristics

In the voltage range 0 to 2.5 V, the ac ($\gtrsim 1,000$ Hz) *V-I* characteristic follows the dc low-impedance characteristic. However, for applied ac voltages in excess of 2.5 V, which is designated the threshold voltage V_T, a new set of *V-I* characteristics are apparent, as shown by the oscillogram in Fig. 36. In the range $V_T \lesssim V \lesssim 9$ V a distinct ac *V-I* characteristic can be associated with every voltage amplitude, the overall impedance of the characteristic increasing with increasing voltage amplitude; that is, the sample resistance to alternating current is a function of voltage amplitude. These characteristics do not manifest negative resistance, although the locus of their end points, for increasing values of voltage amplitude beyond 4 V, generate the dc negative-resistance characteristic, as shown by the dotted line in Fig. 36.

c. Memory Characteristics

If the voltage corresponding to any operating point on the negative-resistance characteristic shown in Fig. 37 (for example, the points B, C, E), is reduced to zero in about 0.1 ms, it is found on reapplication of voltage that the low-impedance characteristic OA is not generated, but rather a higher-impedance state prevails. For example, a voltage pulse with a trailing edge faster than 0.1 ms, and of amplitude corresponding to the voltage of the operating points E, C, B, generates, respectively, the new impedance states OE', OC', and OB'. These induced, or *memory*, states appear to prevail indefinitely provided that no peak voltage in excess of V_T is applied while the sample is in a memory state. The original low-impedance state AO can be regenerated, that is, the memory erased, by applying a voltage slightly in excess of V_T.

It will be apparent from a consideration of the above observations that the area enclosed by the perimeter $OAB'C'E'O$ represents the electrical stable characteristic of the sample; that is, any operating point therein is obtainable and stable. Operating points lying in the area enclosed by the perimeter $EC'B'ABCEE'$ are unstable except for points on the curve $ABCE$, which are stable.

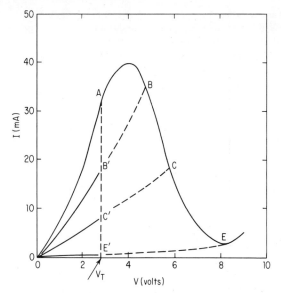

Fig. 37 Typical I-V diagram illustrating several impedance states and the threshold voltage for Al-SiO-Au junction.

Switching from the low- to the high-impedance state, as described above, can be accomplished with a voltage pulse as narrow as 2 or 3 ns wide. Switching from the high-impedance to the low-impedance state can be accomplished by a voltage pulse slightly greater than V_T in height and approximately 100 ns wide. In view of these facts, it would appear that repetitive switching from the high- to low-impedance state and back again should be accomplished in a cycle time of about 100 ns. It transpires, however, that although the sample can be switched from the *high- to the low-imped-ance* state and back again in about 100 ns, any immediate attempt to repeat the process is inhibited for a period of time which varied from milliseconds to fractions of a second dependent, apparently, upon the insulator preparation. This effect is designated the *dead time.*

It has been shown that these type of samples have potential as nonvolatile, non-destructive readout, analog, or binary memory elements.[95]

d. Connection between DC and Dynamic Characteristics and Memory

The difference between the dc characteristic (Fig. 35) and the dynamic characteristic (Fig. 36) is ascribed to the memory phenomena when the rate of fall of the applied sinusoidal voltage waveform is fast enough to set the sample into an induced-memory state. Thus, supposing the sample has impressed on it a voltage of amplitude corresponding to the voltage of the operating point B in Fig. 37, as the voltage falls rapidly from this value to zero, the characteristic $BB'O$ is generated rather than BAO. On the subsequent rising voltage sweep the curve $OB'B$ is again swept out. The reason why the sample does not switch to its low-impedance state as the voltage exceeds V_T is that the time between the voltage being removed and then reapplied to its maximum value is less than the dead time of the sample.

The curve $BB'O$ will continue to be swept out provided the amplitude remains constant. If the amplitude is increased slightly, the sample will partially switch to, say, the point C in Fig. 37, and the operating characteristic is then $OC'C$. The switching thus described is the case of a partial switch as compared with the complete switch described by the transition from A to E shown in Fig. 37.

e. Thermal Characteristics

(1) Thermal-voltage Memory   Figure 38 is an oscillogram of the dc V-I characteristics displayed on a X-Y oscilloscope for an applied sawtooth pulse with a rise time of 10 s and a fall time of 2 s.   Curves a and b are the V-I traces for a sample at

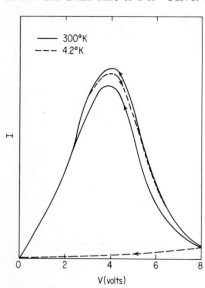

300 and 77°K, respectively.   The slight difference in the two traces for *increasing* voltage indicates that the V-I characteristic is virtually temperature-independent. The return trace to zero volts for the sample at 300°K does not follow the increasing voltage trace because of partial memory storage (cf. Fig. 35).   The return trace at 77°K is generated independent of the rate at which the voltage is removed and never exhibits any partial switching. Any subsequent voltage cycling (fast or slow) of the sample up to 9 V at the low temperatures invariably generates the same high-impedance characteristic. Hence at low temperature the state cannot be erased; the induced state at low temperatures is thus irreversible.   Any similar low-temperature irreversible V-I characteristic lying between the low-impedance and high-impedance characteristic OA and $OE'E$ shown in Fig. 37 can be induced simply by limiting the maximum voltage excursion to lie between $V = V_T$ and $V = 9$ V respectively; for example, limiting the maximum voltage excursion to 5 V results in the V-I characteristic $OB'B$.

Fig. 38   Typical I-V trace for a complete voltage cycle between 0 and 9 V at (a) 300°K and (b) 77°K.

These low-temperature memory states are irreversible only if the sample is held at low temperature.   Raising the sample temperature to 300°K and applying to it a voltage slightly in excess of V_T causes it to switch back into its initial low-impedance state OA.   *In fact, the sample will switch back to the low-impedance state even if the sample temperature is lowered first to 77°K and then the voltage ($V > V_T$) is applied.*   Thus the sample manifests a thermal-voltage memory in addition to a voltage memory.

(2) Voltages below V_T   A somewhat different thermal characteristic is observed if the applied voltage is confined to values less than V_T.   In this case one does not observe a thermal memory; that is, increasing and decreasing voltages generate identical I-V characteristics.   The sample conductivity at liquid-nitrogen temperature is typically 10% less than the room-temperature characteristic and obeys a relationship of the form

$$\sigma = \sigma_0(1 + \alpha T^2) \tag{57}$$

where α for applied voltages above 1 V is approximately 1×10^{-6} per (°K)^2 (see Fig. 39).   Equation (57) suggests that the increase in conductivity with increasing temperature is not due to the thermal creation of free carriers, since no activation-energy term is involved.

f. Theory of Operation

Simmons and Verderber[94,95] have postulated that the forming process introduces a broad band of localized levels within the normally forbidden band of the insulator. The electrons are assumed to move through the insulator by tunneling between adjacent sites within the impurity band; it is also assumed that the electrons can, under certain conditions, be trapped within the localized band.   A model based on these ideas accounts in a self-consistent manner for all the experimental observations.

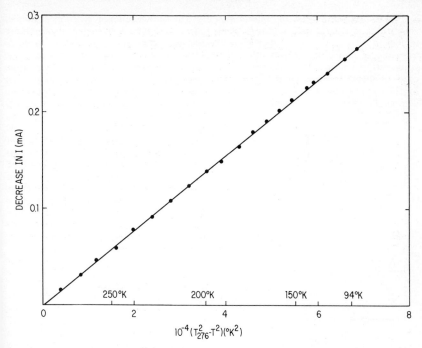

Fig. 39 Plot of decrease in current from 276 to 94°K as a function of the absolute temperature squared. Applied voltage 1.5 V; current at 276°K is 2.25 mA.

A detailed account of this theory is, however, beyond the scope of this chapter, and the reader is referred to the literature for further details.[94]

Hickmott[91] has also presented a theory of operation in order to account for his observations.

8. PRINCIPLES OF HOT-ELECTRON THIN FILM DEVICES

a. Tunnel-cathode Emitter

(1) Principle of Tunnel Cathode The principle of the tunnel cathode, as proposed by Mead,[97] is based on the fact that the electron energy is conserved during the tunnel process. Thus in tunneling from one electrode to the other, such that $eV < \phi_0$, the electrons enter the positively biased electrode with energy eV above the Fermi level of the electrode, as shown in Fig. 40a. The electrons then give up their energy to the lattice and fall into the Fermi sea.

For voltage biases such that $eV > \phi_0$ the electrons first tunnel into the conduction band of the insulator before entering the positively biased electrode. The electron is then assumed to be accelerated by the field within the insulator, without undergoing energy losses, to again enter the positively biased electrode with energy eV above the Fermi level. If eV is less than the positively biased electrode work function ψ, the electron gives up its energy to the electrode lattice as previously described. However, if $eV > \psi$ and the thickness of the positively biased electrode is less than the mean free path of the electrons, Mead[97] suggests that the electrons pass through the electrode to the vacuum interface with little loss of energy, and thus escape into vacuum, as shown by the line abc in Fig. 40b.

In principle, then, with suitable geometry and voltage bias, a large fraction of the tunneling electrons should be able to escape from the tunnel junction into vacuum and be collected by a suitably biased anode. The tunnel junction then, in principle, constitutes a cold cathode. It transpires, however, that such cathodes are extremely inefficient, having *transfer ratios* (ratio of emission current to circulating current) typically of the order 10^{-4} or 10^{-3}.[98-102] These disappointingly low values have apparently been established to result primarily because the majority of the electrons undergo energy losses while traveling in the conduction band of the insulator, as shown by the stepped line bd in Fig. 40b, although losses in the metal film are not negligible. The energy losses in the insulator are such that the electron enters the positively biased electrode with energy less than that of the work function of the electrode and are thus unable to escape. Collins and Davies[104] and Handy[105] estimate the electron mean path in Al_2O_3 to be about 4 or 5 Å.

(2) Attenuation of Hot Electrons in Metal Films The use of a cold-cathode emitter to determine the attenuation length of hot electrons in metal films, the metal film under investigation being the exit electrode of the cathode per se, was first used by Mead.[99] In essence the technique

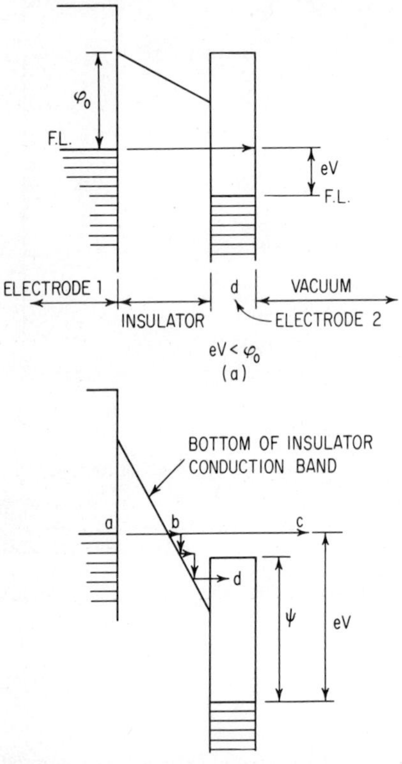

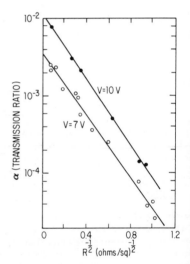

Fig. 40 Energy diagram of cold cathode for (a) $eV < 4$, (b) $eV > \psi$.

Fig. 41 Emission current vs. electrode sheet resistance for two voltage biases: $V = 7$ V, $V = 10$ V.

consists of measuring the relative number of electrons transmitted through the metal film as a function of the film thickness, all other parameters remaining constant, assuming the following relationship holds:

$$n(d,V) = n(0,V) \exp\left(-\frac{d}{\lambda}\right) \tag{58}$$

where $n(d,V)$ is the number of electrons per second escaping through a film of thickness d at an applied voltage V, and λ is the mean free path of the electrons. A more convenient way of expressing (58) is in terms of the transfer ratio α:

$$\alpha = \frac{n(d,V)}{n(0,V)} = \exp\left(-\frac{d}{\lambda}\right) \tag{59}$$

Mead's[97] results for a series of Be-BeO-Au junctions, in which the BeO thickness was held constant and the Au film thickness varied, are shown in Fig. 41 for two voltage biases: $V = 7$ and 10 V. In this figure α is shown plotted against the reciprocal of the square root of the Au sheet resistance ($R^{-\frac{1}{2}}$), which is assumed proportional to the film thickness. (Mead estimates that $R^{-\frac{1}{2}} = 0.95$ corresponds to a Au film thickness of 400 Å.) From the slopes of the two curves, λ_{Au} is seen to be 190 and 180 Å, respectively, for $V = 7$ and 10 V; λ_{Au} is therefore a function of the electron energy, as is to be expected.

It will be noted that the transfer ratios extrapolated to zero thickness ($R^{-\frac{1}{2}} = 0$) are considerably less than unity for both voltage biases, and this is attributed to the electron's undergoing losses in the insulator. This means, of course, that only a fraction of the electrons are reaching the electrode with the energies (7 and 10 eV) that are assumed to correspond to the two plots in Fig. 41.

b. Tunnel-emission Triode

If a second insulator (assumed to be less than the electronic mean free path) and a third electrode are deposited onto the cold cathode described above, we have the basic structure of the thin film, tunnel-emission triode conceived by Mead.[97] The principle of operation of this device can be seen with the aid of the energy diagram of Fig. 42. The energy of electrons tunneling between the emitter and base elec-

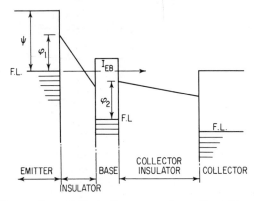

Fig. 42 Energy diagram illustrating proposed operation of tunnel triode.

trodes (the electrodes corresponding to those of the cold cathode described above) is assumed to be conserved; they thus reach the interface existing between the base and the second insulator with energy eV_{EB}, where V_{EB} is the emitter-base bias (base positively biased). For $0 < eV < \phi_2/e$, the electrons do not have sufficient energy to enter the conduction band of the second insulator. If $V > \phi_2/e$ the electrons can enter the conduction band of the second insulator, in which they are then accelerated toward, and collected by, the collector, which is positively biased with respect to the base during operation. Thus the second insulator and collector serve the same function as the vacuum interspace and anode in the cold-cathode emitter.

Unfortunately, the triode suffers from all the problems of the cold cathode, plus additional problems arising from scattering and trapping in the collector insulator, which are not present in the vacuum interspace between the tunnel junction and anode comprising the cold cathode. Thus such a device does not appear viable in light of current state-of-the-art techniques. Nevertheless the principle involved is clearly quite ingenious, and the subsequent research stimulated as a result of the proposal has certainly increased our understanding of the transport properties of thin films.

c. Metal-base Transistor

The operation of the metal-base transistor proposed by Geppert[106] is similar in construction to that of the tunnel transistor, except that the insulators are replaced by semiconducting films. This device suffers from the same kind of problems that bedevil the tunnel triode, and the feasibility of a commercially viable device does not appear very auspicious at present. For further details the reader is referred to the literature.[106]

d. Cold Cathodes with Thick Insulating Films

The tunnel devices described above rely for their operation on direct electrode-electrode tunneling and thus necessitate the use of an extremely thin (<50 Å) layer if significant tunnel currents are to be realized. The use of such a thin oxide is undesirable because even a few volts applied to the structures induces intense electric fields within the oxide during operation—5 V across a 50-Å oxide results in a field of 10^7 V cm^{-1}. Fields of this order are perilously close to dielectric-breakdown fields and must therefore severely limit the life of such cathodes.

Cold cathodes having much thicker insulating films, which thus do not suffer from the above restrictions, have recently been reported;[83,88,90,91,107-110] the transfer ratios obtained from these cathodes are usually a good deal higher than those for the tunnel cathodes described above.[108-110] This cathode is identical in physical structure to the negative-resistance and memory device described in Sec. 7; that is, a typical device would be Al-SiO-Au structure which has undergone forming. It is usual, however, to use somewhat thicker insulating films ($>1,000$ Å) when using the structure as a cold-cathode emitter, if a high α is required.[108-110]

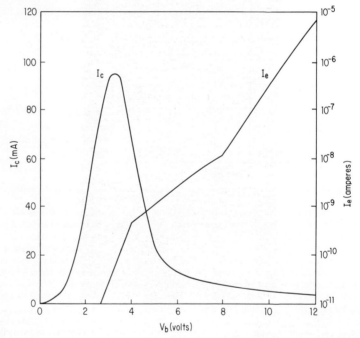

Fig. 43 Emission current I and circulating current I_c vs. applied voltage for a Al-SiO-Au sample.

In Fig. 43 we have plotted the circulating current I_c and the emission current I_E (which was collected by a metal anode biased a few volts positive with respect to the exit electrode) as a function of the applied cathode voltage.[108-110]

The emission process is shown to be due to hot-electron emission rather than thermally stimulated on two counts:[107-110] (1) the emission current in Fig. 43 is observed to increase in the range 3 to 7 V, although the power dissipated in the device is actu ally decreasing; and (2) on reversing the polarity of bias across the cathode no emission is observed, even though the same power is generated in the device independent of polarity of bias.

Examination of the cathodes always revealed pinholes in the upper electrodes,[107–110] and by forming a magnified image of the emitting surface of the cathode at a phosphor screen by means of an electrostatic lens, it was possible to show that the pinhole edges were the source of the emitted electrons.[107–110]

(1) Distribution of Emitted Electrons The spatial distribution of the emitted electrons was observed at a zinc sulfide phosphor screen. The phosphor-coated anode was positioned parallel to the plane of the cathode and biased at 1 to 2 kV positive with respect to the cathode, in order that the emitted electrons would acquire sufficient energy to excite the phosphor.

With the anode positioned a few centimeters from the cathode and the device biased less than 8 V, the emission pattern consists of several small spots within the 0.1 cm² area of the device, which indicate that the electrons are being emitted from small localized regions. If the bias voltage is increased above 7 V, an image much larger than the cathode is observed. The particular interesting aspect of the image is the amount of structure exhibited in it, as shown in Fig. 44. Imposed on a faintly illuminated background are many small brightly illuminated arcs, all of the same radii, none of which subtends an arc greater than 90°.

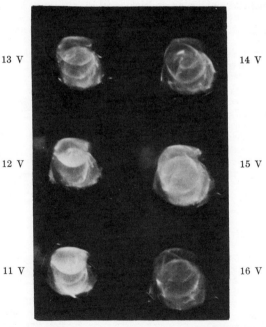

13 V   14 V

12 V   15 V

11 V   16 V

Fig. 44 Full-sized photograph of arcs for various voltage bias. Cathode-screen distance = 4.42 cm; cathode-screen potential 2.0 kV; cathode area $\frac{1}{8}$ by $\frac{1}{8}$ in. (note shape and size of image much different from that of cathode). Exposure times given alongside photographs showing qualitatively that the emission is increasing rapidly with V_b.

All the devices displayed similar images on the screen. It has been shown that these patterns are caused by coherent scattering of hot electrons injected into the gold electrode, followed by their subsequent escape from exposed *edges* of the electrode, such as the circumference of pinholes. Briefly, what this means is that the electrons entering the upper electrode undergo very low energy electron diffraction and are scattered into a hollow cone of diffraction of half angle 2θ (see Fig. 45). The diffracting angle (2θ) is related to the bias voltage V and the material constants

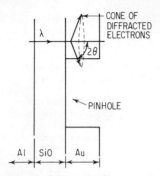

Fig. 45 Schematic diagram of thin film cold cathode, in the vicinity of a pinhole, showing the path of a diffracted electron.

by the Bragg equation

$$\sin \theta = \frac{\lambda}{2d_{hkl}} \qquad (60)$$

where λ is the wavelength of the electron in the top electrode expressed by[107-109]

$$\lambda = \frac{12.27}{(V + \eta)^{\frac{1}{2}}} \quad \text{angstroms} \qquad (61)$$

d_{hkl} is the distance between the diffracting crystal planes, and η is the Fermi energy of the top metal electrode. For the voltage biases shown in Fig. 43, it can be shown that the component of electron momentum parallel to the plane of the film is much greater than the perpendicular component, which means that the electron can escape only at an exposed edge of electrode.

(2) Scattering Processes[109]   The structure in the emission I-V characteristic of Fig. 43 is due to electrons undergoing multiple scattering below 8 V and only a single (coherent) scatter above 8 V. Using $V < 8$ V, $\eta = 5.5$ eV, and d_{111} (the most intense diffracting plane of Au) = 2.35 Å in (61) we find $\sin \theta = 0.71$ and hence $2\theta \simeq 90°$. Thus for $V < 8$ eV the electrons are scattered back toward the insulator and would not normally escape from the electrode unless they underwent further scattering (electron-electron, electron-phonon) to reach an electrode edge. These additional scattering processes lead to substantial energy losses, and the process is obviously an inefficient means of directing electrons toward an edge. For $V > 8$ eV the scattering process is coherent only, and the energy is conserved in this process. In this case, if the scattered electron is within a distance not greater than the electron mean free path from the edge, it will reach the edge without loss of energy and hence will escape into vacuum.

The transmission ratios observed in the cathode are typically 10 to 100 times greater than for tunnel cathodes, yet the insulator and electrode materials are typically ten times thicker. On this basis it is difficult to attribute the low emission currents in tunnel cathodes to small-insulator and electron mean free paths alone. For further details on the transport of the electrons through the exit electrode and insulator the reader is referred to the literature.[109]

(3) Alphanumeric Display[110]   A thin film, cold-cathode, alphanumeric display panel, consisting of a 5 × 5 matrix utilizing the cold cathodes described above, has been shown to be feasible. With this simple system the alphabet and all numerals can be displayed quite legibly, as shown in Fig. 46. These characters are clearly

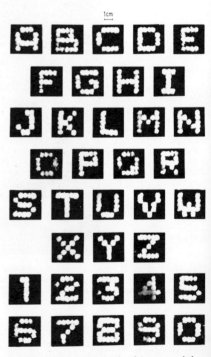

Fig. 46 Characters obtained from an alphanumeric display panel consisting of a 5 × 5 matrix of thin film cold cathodes.

visible under normal room-lighting conditions.   For fabrication details, response time, etc., the reader is referred to Ref. 110.

REFERENCES

1. Mott, N. F., and R. W. Gurney, "Electronic Processes in Ionic Crystals," 2d ed., chap. V, Oxford University Press, Fair Lawn, N.J., 1948.
2. Rose, A., *Phys. Rev.*, **97**, 1538 (1955).
3. Lampert, M. A., *Rept. Progr. Phys.*, **27**, 329 (1964).
4. Simmons, J. G., and G. S. Nadkarni, *J. Vac. Sci. & Tech.*, **6**, (12) 1969.
5. Bujatti, M., and R. S. Muller, *J. Electrochem. Soc.*, **112**, 702 (1965).
6. Hass, G., *J. Am. Ceram. Soc.*, **33**, 353 (1958).
7. Ritter, E., *Opt. Acta*, **9**, 197 (1962).
8. Brady, G. W., *J. Phys. Chem.*, **63**, 1119 (1959).
9. Faessler, Von A., and H. Kramer, *Ann. Physik*, **7**, 263 (1959).
10. Simmons, J. G., *Phys. Rev.*, **155**, 657 (1967).
11. Dresner, J., and F. V. Shallcross, *Solid-State Electron.*, **5**, 205 (1962).
12. Schottky, W., *Z. Physik*, **113**, 367 (1939).
13. Henisch, H. K., "Rectifying Semiconductor Contacts," Oxford University Press, Fair Lawn, N.J., 1957.
14. Simmons, J. G., *Phys. Rev. Letters*, **10**, 10 (1963).
15. Bardeen, J., *Phys. Rev.*, **71**, 717 (1947).
16. Frenkel, J., *Phys. Rev.*, **36**, 1604 (1930).
17. Sommerfeld, A., and H. Bethe, in VonGeiger and Scheel (eds.), "Handbuch der Physik," vol. 24/2, p. 450, Springer-Verlag OHG, Berlin, 1933.
18. Holm, R., *J. Appl. Phys.*, **22**, 569 (1951).
19. Stratton, R., *J. Phys. Chem. Solids*, **23**, 1177 (1962).
20. Simmons, J. G., *J. Appl. Phys.*, **34**, 1793 (1963).
21. Simmons, J. G., *J. Appl. Phys.*, **34**, 2581 (1963).
22. Simmons, J. G., *J. Appl. Phys.*, **35**, 2655 (1964).
23. Geppert, D. V., *J. Appl. Phys.*, **34**, 490 (1963).
24. Geppert, D. V., *J. Appl. Phys.*, **33**, 2993 (1962).
25. Pittelli, E., *Solid-State Electron.*, **6**, 667 (1963).
26. Penley, J. C., *Phys. Rev.*, **128**, 596 (1962).
27. Schmidlin, F. W., *J. Appl. Phys.*, **37**, 2823 (1966).
28. Stratton, R., G. Lewiki, and C. A. Mead, *J. Phys. Chem. Solids*, **27**, 1599 (1966).
29. Grundlock, K. H., and G. Heldmann, *Phys. Solid State*, **21**, 575 (1967).
30. Schnup, P., *Solid-State Electron.*, **10**, 785 (1967).
31. Schnup, P., *Phys. Stat. Solidi*, **21**, 567 (1967).
32. Simmons, J. G., *Brit. J. Appl. Phys.*, **18**, 269 (1967).
33. Stratton, R., *Phys. Rev.*, **90**, 515 (1963).
34. Fowler, R. H., and L. W. Nordheim, *Proc. Roy. Soc. London*, **A119**, 173 (1928).
35. Fisher, J., and I. Giaever, *J. Appl. Phys.*, **32**, 172 (1961).
36. Handy, R. M., *Phys. Rev.*, **126**, 1968 (1962).
37. Penley, J., *J. Appl. Phys.*, **33**, 1901 (1962).
38. Simmons, J. G., and G. J. Unterkofler, *J. Appl. Phys.*, **34**, 1838 (1963).
39. Simmons, J. G., G. J. Unterkofler, and W. W. Allen, *Appl. Phys. Letters*, **2**, 78 (1961).
40. Meyerofer, D., and S. A. Ochs, *J. Appl. Phys.*, **34**, 2535 (1963).
41. Hurst, H. G., and W. Ruppel, *Z. Naturforsch.*, **19a**, 573 (1964).
42. Hartman, T. E., and J. S. Chevian, *Phys. Rev.*, **134**, A1094 (1964).
43. Hartman, T. E., *J. Appl. Phys.*, **35**, 3283 (1964).
44. Pistoulet, B., M. Rouzeyou, and B. Bouat, *L'Onde Elec.*, **44**, 1297 (1964).
45. Nakai, J., and T. Miyazaki, *Japan. J. Appl. Phys.*, **3**, 677 (1964).
46. Pollack, S. R., and C. E. Morris, *J. Appl. Phys.*, **35**, 1503 (1964).
47. Pollack, S. R., and C. E. Morris, *Trans. AIME*, **233**, 497 (1965).
48. McColl, M., and C. A. Mead, *Trans. AIME*, **233**, 502 (1965).
49. Miles, J. L., and P. H. Smith, *J. Electrochem. Soc.*, **110**, 1240 (1963).
50. Braunstein, A., M. Braunstein, G. S. Picus, and C. A. Mead, *Phys. Rev. Letters*, **14**, 219 (1965).
51. Simmons, J. G., *Phys. Letters*, **17**, 104 (1965).
52. Esaki, L., and P. J. Stiles, *Phys. Rev. Letters*, **16**, 1108 (1966).
53. Chang, L. L., P. J. Stiles, and L. Esaki, *J. Appl. Phys.*, **38**, 4440 (1967).
54. Dahlke, W. E., *Appl. Phys. Letters*, **10**, 261 (1967).
55. Lewicki, G., and C. A. Mead, *Phys. Rev. Letters*, **16**, 939 (1966).
55a. Franz, W., in S. Flügge (ed.), "Handbuch der Physik," vol. 17, p. 155, Springer-Verlag OHG, Berlin, 1956.

56. Esaki, L., P. J. Stiles, and S. von Molnar, *Phys. Rev. Letters*, **19**, 852 (1967).
57. Schottky, W., *Physik. Z.*, **15**, 872 (1914).
58. Simmons, J. G., *Phys. Rev. Letters*, **15**, 967 (1965).
59. Crowell, C. R., *Solid-State Electron.*, **8**, 395 (1965).
60. Tantraporn, W., *Solid-State Electron.*, **7**, 81 (1964).
61. Simmons, J. G., *J. Appl. Phys.*, **35**, 2472 (1964).
62. Emptage, P. R., and W. Tantraporn, *Phys. Rev. Letters*, **8**, 267 (1962).
63. Mead, C. A., in R. Niedermayer and H. Mayer (eds.), "Basic Problems in Thin Film Physics," p. 674, Vandenhoeck and Ruprecht, Goettingen, 1966.
64. Pollack, S. R., *J. Appl. Phys.*, **34**, 877 (1963).
65. Flannery, W. E., and S. R. Pollack, *J. Appl. Phys.*, **37**, 4417 (1967).
66. Simmons, J. G., *Phys. Rev.*, **166**, 912 (1968).
67. Frenkel, J., *Tech. Phys.*, **5**, 685 (1938).
68. Frenkel, J., *Phys. Rev.*, **54**, 647 (1938).
69. Mead, C. A., *Phys. Rev.*, **128**, 2088 (1962).
70. Hirose, H., and Y. Wada, *Japan. J. Appl. Phys.*, **3**, 179 (1964).
71. Johansen, I. T., *J. Appl. Phys.*, **37**, 449 (1966).
72. Hartman, T. E., J. C. Blair, and R. Bayer, *J. Appl. Phys.*, **37**, 2468 (1966).
73. Simmons, J. G., *Phys. Rev.*, **155**, 657 (1967).
74. Stuart, M., *Brit. J. Appl. Phys.*, **18**, 1637 (1967).
75. Stuart, M., *Phys. Stat. Solidi*, **23**, 595 (1967).
76. Frank, R., and J. G. Simmons, *J. Appl. Phys.*, **38**, 832 (1967).
77. Lampert, M. A., *Phys. Rev.*, **103**, 1648 (1956).
78. Lampert, M. A., A. Rose, and R. W. Smith, *J. Appl. Chem. Solids*, **8**, 484 (1959).
79. Lampert, M. A., and A. Rose, *Phys. Rev.*, **121**, 26 (1961).
80. Lampert, M. A., *Phys. Rev.*, **125**, 126 (1962).
81. Zuleeg, R., *Solid-State Electron.*, **6**, 645 (1963).
82. Zuleeg, R., and R. S. Muller, *Solid-State Electron.*, **7**, 575 (1964).
83. Kreynina, G. S., L. N. Selivanov, and T. I. Shumskaia, *Radio Eng. Elec. Phys.*, **5**, 8 (1960).
84. Kreynina, G. S., L. N. Selivanov, and T. I. Shumskaia, *Radio Eng. Elec. Phys.*, **5**, 219 (1960).
85. Kreynina, G. S., *Radio Eng. Elec. Phys.*, **7**, 166 (1962).
86. Kreynina, G. S., *Radio Eng. Elec. Phys.*, **7**, 1949 (1962).
87. Hickmott, T. W., *J. Appl. Phys.*, **33**, 2669 (1962).
88. Hickmott, T. W., *J. Appl. Phys.*, **34**, 1569 (1963).
89. Hickmott, T. W., *J. Appl. Phys.*, **35**, 2118 (1964).
90. Cola, R. A., J. G. Simmons, and R. R. Verderber, *Proc. Natl. Aerospace Electron Conf.*, 1964, p. 118.
91. Hickmott, T. W., *J. Appl. Phys.*, **36**, 1885 (1965).
92. Lewowski, T., S. Sendecki, and B. Sujak, *Acta Phys. Polon.*, vol. 28, 1965.
93. Uzan, R., A. Roger, and A. Cachard, *Vide*, **137**, 38 (1967).
94. Simmons, J. G., and R. R. Verderber, *Proc. Roy. Soc. London*, **301**, 77 (1967).
95. Simmons, J. G., and R. R. Verderber, *Radio Elec. Eng.*, **34**, 81 (1966).
96. Verderber, R. R., J. G. Simmons, and B. Eales, *Phil. Mag.*, **16**, 1049 (1967).
97. Mead, C. A., *Proc. IRE*, **48**, 359 (1960).
98. Mead, C. A., *J. Appl. Phys.*, **32**, 646 (1961).
99. Mead, C. A., *Phys. Res. Letters*, **8**, 56 (1962).
100. Kanter, H., and W. A. Feibelman, *J. Appl. Phys.*, **33**, 3580 (1962).
101. Cohen, J., *J. Appl. Phys.*, **33**, 1999 (1962).
102. Cohen, J., *Appl. Phys. Letters*, **1**, 61 (1962).
103. Collins, R. E., and L. W. Davies, *Appl. Phys. Letters*, **2**, 213 (1963).
104. Collins, R. E., and L. W. Davies, *Solid-Stat. Electron.*, **7**, 445 (1964).
105. Handy, R. M., *J. Appl. Phys.*, **37**, 4620 (1967).
106. Geppert, D. V., *Proc. IRE*, **50**, 1527 (1962).
107. Simmons, J. G., R. R. Verderber, J. Lytollis, and R. Lomax, *Phys. Rev. Letters*, **17**, 675 (1965).
108. Simmons, J. G., and R. R. Verderber, *Appl. Phys. Letters*, **10**, 1967.
109. Verderber, R. R., and J. G. Simmons, *Radio Elec. Eng.*, **33**, 347 (1967).
110. Lomax, R., and J. G. Simmons, *Radio Elec. Eng.*, **35**, 265 (1968).

Other Recent Reviews on Conduction in Thin Films

Lamb, D. R., "Electrical Conduction in Thin Insulating Films," Methuen & Co., Ltd., London, 1967.
Ekertova, L., *Phys. Stat. Solidi*, **18**, 3 (1966).
Hill, R. A., *Thin Solid Films*, **1**, 39 (1967).
Jonscher, A. K., *Thin Solid Films*, **1**, 213 (1967).

Chapter **15**

Piezoelectric and Piezoresistive Properties of Films

NORMAN F. FOSTER

Bell Telephone Laboratories, Allentown, Pennsylvania

1. PIEZOELECTRIC FILMS—INTRODUCTION

a. The Necessity Which Mothered the Invention

The original impetus for the work on piezoelectric thin films, which was started in 1963,[1,2] was supplied by the need for extending the range of low-loss wideband electromechanical transducers to much higher frequencies. The piezoelectric transducer structure normally used for the generation or detection of ultrasonic waves in solids is shown in Fig. 1. The mode of vibration of a transducer is determined by

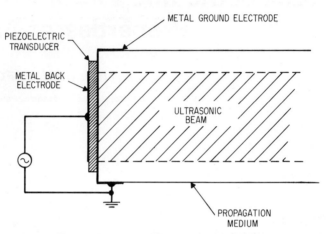

Fig. 1 The piezoelectric transducer structure.

its crystallographic orientation, the geometry, and the direction of the applied field, and the operating frequency is determined by the transducer thickness. Low insertion loss and widest fractional bandwidths are obtained when the thickness is approximately equal to one-half of the wavelength of the mechanical wave in the transducer material. Therefore, the higher the frequency, the thinner the transducer should be. At low frequencies (below 100 MHz) the transducer can be made readily from a suitable insulating piezoelectric material, e.g., single-crystal quartz, and subsequently attached to the propagation medium. For frequencies above 100 MHz (corresponding to transducer thicknesses of less than approximately 20 μ) the practical difficulties of making a discrete transducer and of attaching it to a propagation medium become very severe. Other types of transducer structure have been used for high-frequency operation,[3] but they have not exhibited the high efficiency and large bandwidths required in many ultrasonic applications. It was to overcome these deficiencies that the piezoelectric thin film studies were initiated. Vacuum-deposited films can readily be deposited in micron and fractional-micron thicknesses on a wide variety of substrates. It therefore appeared that if sufficiently well oriented films of a piezoelectric material could be grown with high resistivity, the usefulness of the conventional transducer structure could be extended into the gigahertz frequency range using evaporated films in place of single-crystal plates.

b. Piezoelectric Coupling in Cubic and Hexagonal Crystals

The electromechanical coupling coefficient k is an essential transducer material parameter for calculating the electromechanical characteristics of a piezoelectric transducer. For the transducer geometry used in the thin piezoelectric film work the relevant coupling factors can be defined in terms of the appropriate piezoelectric, dielectric, and elastic constants. The correct choice of these constants depends upon the mode of vibration and the crystallographic structure and orientation of the transducer. All the nonferroelectric materials which have been successfully made into thin film transducers belong to either the cubic crystal structure (zinc blende, or sphalerite) with the point group $\bar{4}3\ m$ or the hexagonal (wurtzite) structure with point group 6 mm. The piezoelectric matrices corresponding to these structures are shown in Fig. 2.

	ELECTRIC FIELD ALONG	LONGITUDINAL STRESS ALONG			SHEAR STRESS PERPENDICULAR TO		
		X_1	X_2	X_3	X_1	X_2	X_3
X_1	a	0	0	0	0	e_{15}	0
X_2		0	0	0	e_{15}	0	0
X_3	c	e_{13}	e_{13}	e_{33}	0	0	0

HEXAGONAL (6 mm) STRUCTURE

X_1	a	0	0	0	e_{14}	0	0
X_2	b	0	0	0	0	e_{14}	0
X_3	c	0	0	0	0	0	e_{14}

CUBIC ($\bar{4}3$ m) STRUCTURE

Fig. 2 Piezoelectric matrices for cubic and hexagonal crystal structures.

(1) Cubic Structure For the cubic structure the positioning of the nonzero coefficients indicates that an electric field applied parallel to any of the cube edges produces only face shear stresses which are not suitable for the generation of a pure mechanical wave. Rotation of the matrix, however, shows that pure longitudinal waves can be generated when the driving electric field is applied in the [111] direction and that pure shear waves can be generated when the field is applied in the [110] direction. The coupling coefficients for the cubic structures are substantially lower than those of the corresponding hexagonal structures.

(2) Hexagonal Structure The lower symmetry of the hexagonal structure results in a slightly more complex piezoelectric matrix (Fig. 2). The nonzero coefficients in the last row of the first three columns show that an electric field applied along the hexagonal or c axis generates longitudinal stresses both along this axis and radially in the plane perpendicular to it. The positions of the two shear constants (e_{15}) show that an electric field applied in any direction perpendicular to the c axis will generate a shear stress in the plane containing the field direction and the c axis.

The application of an oscillating electric field parallel to the c axis produces a longitudinal-thickness mode of vibration. The appropriate coupling coefficient for this mode of vibration in this geometry k_t is given by

$$k_t{}^2 = \frac{e_{33}{}^2}{\bar{c}_{33}\epsilon_{33}{}^s}$$

in which

$$\bar{c}_{33} = c_{33}{}^E + \frac{e_{33}{}^2}{\epsilon_{33}{}^s}$$

where e_{33} is the piezoelectric stress constant relating electric field in the c-axis direction to the mechanical stress in the same direction, $\epsilon_{33}{}^s$ is the dielectric constant in the c-axis direction at constant strain, and $c_{33}{}^E$ is the elastic-stiffness constant in the c-axis direction at constant electric field. The application of an electric field perpendicular to the c axis produces a thickness shear mode of vibration, and the appropriate coupling coefficient k_{15} is given by

$$k_{15}^2 = \frac{e_{15}^2}{\bar{c}\epsilon_{11}^s}$$

in which $$\bar{c} = c_{55}^E + \frac{e_{15}^2}{\epsilon_{11}^s}$$

where e_{15} is the piezoelectric stress constant relating electric field in a direction perpendicular to the c axis to a mechanical shearing stress in the plane containing the c axis and the field direction, ϵ_{11}^s is the dielectric constant perpendicular to the c axis at constant strain, and c_{55}^E is the elastic-stiffness constant at constant electric field for a shearing strain perpendicular to the c axis.

Recent theoretical work has shown that for materials with hexagonal (6 mm) symmetry, the magnitudes of the two coupling coefficients involved change with the angle between the c axis and the electric-field direction in the manner shown in Fig. 3.*

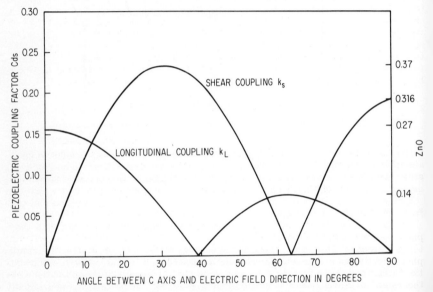

Fig. 3 Piezoelectric coupling in CdS and ZnO as a function of the angle between the electric field and the c axis.

The longitudinal-mode coupling is greatest when the driving field is coincident with the c axis. As the angle between the field and the c axis increases, the coupling falls to zero at about 40°, passes through a maximum at about 63°, and falls to zero again when the field is normal to the c axis. The shear-mode coupling, on the other hand, is zero in the c-axis direction, rises to its highest value at about 30°, falls to zero at 63°, and rises to another maximum value at 90°. A transducer oriented with its thickness direction either parallel to the c axis or at an angle of 63° from the axis will therefore generate a longitudinal wave. On the other hand, a transducer oriented with its thickness direction either 40 or 90° from the c axis will generate a shear wave. On the basis of maximum coupling, the most desirable orientation for longitudinal-

* It should be noted that the modes corresponding to k_S and k_L in Fig. 3 are not exactly pure modes, although the error introduced by making this approximation can be neglected for the materials of practical interest.[4]

wave generation is with the c axis perpendicular to the plate, and for shear-wave generation with the c axis at an angle of 40° to the perpendicular.

c. Piezoelectric Coupling in Thin Films

The above discussion of the piezoelectric coupling was based on the properties of single crystals. Vacuum-deposited piezoelectric films are usually highly polycrystalline with varying degrees of misorientation between individual crystallites. It is therefore necessary to consider what effects this polycrystalline structure will have on the piezoelectric coupling. Each individual crystallite will behave as described above, and provided all the crystallites have the same orientation and are driven in phase, the overall result will be essentially the same as for a single-crystal film. Any misalignment of crystallites will result in a reduced coupling and may also result in the simultaneous generation of both longitudinal and shear waves.

A small angular spread in crystallite orientation, if centered about the optimum, will not give rise to substantial unwanted wave generation because of cancellation effects between oppositely misaligned crystallites. In this case only a small reduction in piezoelectric coupling would be expected. A deviation of the average orientation from the desired angle, however, will result in the simultaneous generation of both shear and longitudinal waves. The presence of a different type of misalignment in which the direction of the polar c axis of some crystallites is reversed with respect to some others will result in out-of-phase mechanical displacements with severe cancellation effects and a substantial decrease in the piezoelectric coupling. For a film containing approximately equal areas of each polarity the apparent coupling would become essentially zero.

2. FILM-DEPOSITION METHODS FOR PIEZOELECTRIC FILMS

A variety of methods have been used for the deposition of piezoelectric films. CdS and ZnS films have been made by vacuum evaporation both of the compounds and of the individual elements. Aluminum nitride and ZnO films have also been grown by reactive evaporation while CdS, ZnO, and lithium niobate films have been deposited by various sputtering techniques.

a. Evaporation Methods

Vacuum-evaporated piezoelectric films, e.g., CdS, ZnS, and ZnO, have been prepared by a number of different methods. As the temperatures required for vaporizing these materials are well in excess of their decomposition temperatures, the vapor phase is in all cases composed of the elemental species. In the CdS system, because of the higher tendency of the sulfur to reevaporate from the substrate as compared with the cadmium, films grown under conditions of equal cadmium and sulfur fluxes (e.g., as from a single CdS source) are normally sulfur-deficient. Such films are semiconducting and therefore too highly electrically conducting for efficient transducer action to be obtained. When using a single CdS source, the stoichiometry is considerably improved by heating the substrate to reduce the residence time of free cadmium on the surface, and by reducing the deposition rate. The stoichiometry can also be improved by including a separate additional sulfur source. An alternative method for increasing the resistivity is to coevaporate a suitable monovalent ion (e.g., Ag or Cu) which acts as an electron acceptor.

Two principal types of evaporation system have been used. In the direct evaporation system[6] there is a direct line of sight between the source and substrate and the mean free path is long compared with the source-substrate distance. In the indirect evaporation system[7] there is no direct path from source to substrate and the mean free path is small compared with the source-to-substrate distance, with a consequent loss of vapor-beam directionality.

There are also two types of thin film transducer structure, one with deposited-metal electrodes on either face of the piezoelectric film, as in Fig. 1, and one without any metal electrodes. The reasons for choosing one type or the other will be discussed in the applications section. From the film-deposition point of view, in the one case the substrate will be a metal film, usually gold, whereas in the other case the substrate will be the material to be insonified, which is sometimes an oriented single crystal. The effects of the nature of the substrate on film orientation will be discussed in Sec. 3.

(1) Direct Evaporation   The first reported work on piezoelectric thin films[1] used a simple direct-evaporation system with a single CdS source. This system required a separate prior evaporation of the underlying metal film (in this case copper) and also required a post-deposition heat treatment to produce the required resistivity by diffusing copper from the underlying metal film into the CdS film. A more recent design of direct-evaporation system[5] is shown diagrammatically in Fig. 4. In this

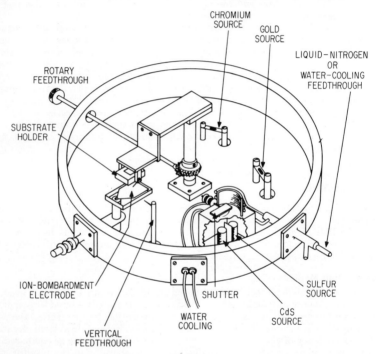

Fig. 4   Direct-evaporation system. (*After Foster.*[5])

system the substrates are held in a heated aluminum holder, which can be rotated over positions for ion bombardment, the vapor deposition of metal electrodes, and the deposition of CdS and sulfur, without breaking vacuum. The holder can also be tilted about its own axis, to adjust the angle of incidence of the vapor stream to achieve orientation control as described in the next section. The individually controlled CdS and sulfur sources are separated and housed in a water-cooled enclosure equipped with a solenoid-operated shutter. Directly above the shutter is a short cooled cylinder to condense out scattered, indirect material. The procedure is as follows: After chemical cleaning, the substrates are heated in the substrate holder in vacuum to approximately 150°C. The surface to be plated is then glow-discharge cleaned in nitrogen and is chromium-gold plated before being swung into place over the CdS deposition source. (It was found necessary to use gold as the substrate metal film because of reaction between the sulfur and Cu or Ag films.) The CdS and sulfur sources are next brought up to temperature and the shutter is opened. The rate of growth and the thickness of the films are continuously monitored by optical or infrared interferometry (see Chap. 2). The rates and temperatures used for the CdS deposition depend upon the desired orientation of the CdS film and will be discussed in the section on film orientation. ZnS and CdSe films have also been deposited using this type of apparatus.

(2) Indirect Evaporation   An example of an indirect-evaporation system[7] is shown in Fig. 5. This arrangement uses individual cadmium and sulfur sources and incorporates a baffle between the sources and the substrate. The whole assembly is enclosed in a heated inner chamber, thereby creating an atmosphere containing substantial vapor pressures of cadmium and sulfur from which deposition takes place on all surfaces including the substrate and quartz-crystal rate monitor. The substrate is heated and cleaned by ion bombardment prior to raising the Cd and S

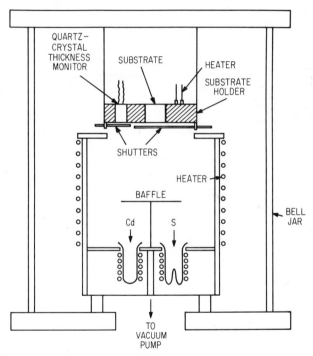

Fig. 5 Indirect-evaporation system. [*After de Klerk.*[7] *Reprinted from Mason, W. P.* (*ed.*), *"Physical Acoustics,"* vol. IVA, p. 202, Academic Press Inc., New York, 1966.]

sources to the temperatures required for the deposition. Typical conditions are substrate temperature 200°C, inner-chamber wall temperature 160°C, sulfur temperature 150°C, cadmium temperature 400°C. ZnS films have also been deposited in this type of system by substituting Zn for Cd.

The evaporation of zinc oxide is considerably more difficult because of a much greater difference in vapor pressure between the components. Successful evaporations have been performed,[6] however, by the evaporation of zinc in an oxygen-containing atmosphere. To obtain these films it was necessary to cool the silver- or gold-plated substrate to approximately −50°C. The oxygen was supplied through a controlled leak to maintain a pressure in the 10^{-3} to 10^{-4} Torr range during the evaporation. The apparatus used is shown diagrammatically in Fig. 6. Film thickness and rates of deposition were monitored by optical interference using a laser beam, and film-deposition rates of up to 900 Å min⁻¹ were achieved. This type of apparatus

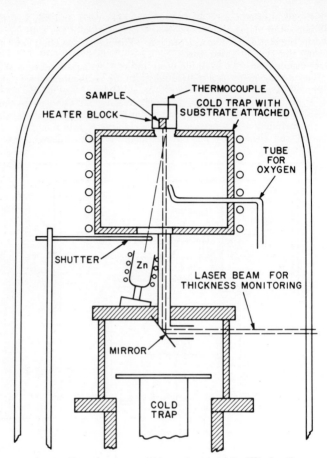

Fig. 6 Reactive evaporation system. (*After Winslow.*[8])

has also been used to deposit AlN films by the evaporation of aluminum in a nitrogen glow discharge.

b. Sputtering Methods

Another approach to the vacuum deposition of thin films is sputtering (see Chap. 4), and several variations of the sputtering process have been used for growing piezoelectric films.[5,9,10] The use of reactive diode sputtering for depositing piezoelectrically active ZnO films was first announced in 1965.[10] The apparatus consisted of a sputtering chamber with facilities for heating the substrate, and a zinc target which was reactively sputtered in an oxygen/argon mixture at 50 μ pressure. In the following year diode sputtering of both ZnO and CdS compounds was reported.[9,11] In this work the diode sputtering was performed in an all-metal bakeable ultrahigh-vacuum system, and rates up to 500 Å min^{-1} were reported on substrates heated to about 200°C.

The pressures required for diode sputtering result in mean free paths of the order of 1 mm, which is many times smaller than the normal source-to-substrate distances. This situation is similar to that present in indirect evaporation in that there can be no directionality to the condensing vapor. Triode sputtering, on the other hand, can be carried out at pressures of about 1 μ where the mean free path is several centimeters, which can be greater than the source-to-substrate distance, permitting some directionality in the condensing vapor. The uniformity of direction is then determined by the relative sizes and separation of the substrate and cathode. CdS, ZnO, and LiNbO₃ films have been deposited by triode sputtering in argon/oxygen mixtures at 1 to 3 μ pressure. The vacuum system used has the same general arrangement as that shown in Fig. 4, but with the CdS evaporation unit removed and the triode-sputtering unit shown in Fig. 7 substituted. Deposition rates were similar to those obtained by diode sputtering, i.e., 1 to 3 μh^{-1}.

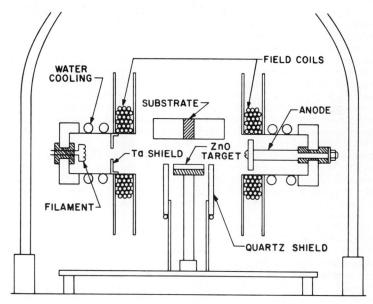

Fig. 7 Triode-sputtering unit. (*After Foster.*[38])

3. ORIENTATION EFFECTS IN PIEZOELECTRIC FILMS

a. Fiber Texture

A crystalline aggregate is said to have a fiber texture if some crystallographic axis is aligned along some preferred direction called the fiber axis. The occurrence of fiber textures in evaporated films has been known for a long time. As early as 1921, Volmer and coworkers[12] observed that zinc films showed a [00.1] fiber axis directed toward the evaporation source. Later studies by Dixit[13] and by Evans and Wilman[14] showed that films of many materials develop fiber textures when evaporated onto amorphous substrates. Dixit postulated that the orientation was determined during the initial stages of film growth when the mobile adsorbed atoms coalesced to form the initial layer. He suggested that the arrangement of the atoms in this initial layer was determined by the substrate temperature, the crystal structure, the atomic radius, and the melting point of the depositing material, and his theory showed remarkably good agreement with experimental results for Ag, Al, and Zn films on quartz surfaces. Evans and Wilman deposited a variety of metallic and nonmetallic materials at oblique incidence onto amorphous substrates and interpreted the results in terms of the formation of two types of orientation depending upon the mobility of the depositing material. One type of orientation, which was similar to that of Dixit, was thought to originate in the initial layers where a densely populated atomic plane formed parallel to the substrate surface. This yielded a fiber axis normal to the surface which, as the film thickness increased, tilted slightly toward the evaporation source. This type of orientation was found with materials that were considered to have high mobilities on the substrate surface. In the other type of orientation the initial layers showed little or no orientation, but a fiber axis, strongly inclined toward the evaporation source, developed at greater film thicknesses. This type of orientation was found with less mobile atomic or molecular species. The inclination of the fiber axis in this case was thought to be due to the preferential growth of crystals oriented with fast-growing natural-crystal faces approximately normal to the incoming vapor stream.

Bauer, in an excellent paper on fiber texture,[15] has reviewed the early models and proposed a more comprehensive general texture model that classifies orientation in terms of (1) the kind of orientation, (2) the location within the film, and (3) the origin of the orientation. The kind of orientation may be random, one-dimensional (i.e., with a single-fiber axis), or two-dimensional (i.e., single-crystal). The orientation may be present at any stage of the film growth, initial, transitional, or final, and may originate either upon nucleation or at some later stage of film growth. As these orientations may depend upon any or all of the many parameters necessary to characterize a deposition (Bauer lists 10 such parameters), the difficulty in formulating a comprehensive model and in obtaining good experimental verification is apparent. The main features of the general texture model proposed by Bauer may be summarized as follows: On substrates in which the interaction between the substrate and the initial nuclei is weak, the nuclei will be bounded by equilibrium crystallographic planes with one such plane preferentially parallel to the surface. This gives rise to a nucleation orientation with a fiber axis normal to the local surface. (On rough surfaces the nuclei will form preferentially on the local surfaces facing the vapor beam, yielding a fiber axis directed toward the evaporation source.) If the interaction with the substrate is strong, the configuration of the nucleus cannot be determined with any certainty. As the film thickness increases, different transitional and final growth orientations may develop from a randomly oriented aggregate formed either directly on the substrate or by renucleation on an oriented initial layer. The growth orientation developed is determined principally by the topography of the surface and the angular distribution of the depositing vapor beam. For a highly directional vapor beam crystals oriented with natural faces approximately normal to the beam will grow preferentially because of the higher density of impinging material, leading to a fiber axis along the beam direction. Further variations caused by substrate temperature and residual ambient gases are discussed in Bauer's paper. More recent studies on the nucleation and growth of thin films have been primarily concerned with growth on single-crystal substrates and have not yielded any important new information on the orientation of films deposited onto amorphous substrates.

b. Cadmium Sulfide Films

(1) Historical Review This review is by no means all-inclusive, but an attempt has been made to include all the major contributions on CdS film orientation. The epitaxial growth on single-crystal substrates such as mica and rock salt is mentioned only briefly, as these materials are not normally used as substrates for piezoelectric film studies.

The crystallographic orientation of evaporated cadmium sulfide films has been examined by many workers in the last 10 years.[5,7,16−27] Much of the earlier work was done in connection with photoresistive and photovoltaic studies. CdS films deposited rapidly onto unheated substrates were characteristically dark and highly conducting whereas slower deposition onto heated substrates yielded orange or yellow films of high electrical resistivity.[22] More recently the interest in thin film transistors and piezoelectric transducers has stimulated a considerably increased effort on CdS films, particularly with respect to crystallographic orientation and structure.

One of the earliest studies on the structure of CdS films was an electron-diffraction study by Semiletov[23] in 1956 on films deposited onto heated mica and rock salt substrates. The films were found to consist of both cubic and hexagonal phases with the hexagonal crystals oriented with their (00.1) planes parallel to the substrate. Two years later, Gilles and van Cakenberghe[16] used x-ray diffractometry to examine films deposited onto heated glass substrates and reported finding hexagonal-phase crystallites about 1,000 Å in size having a highly preferred orientation, again with the (00.1) planes parallel to the substrate. Heating these films in argon at 500 to 600°C did not change the orientation, but heating to much lower temperatures after overplating with silver or copper (or some other metals) caused a massive recrystallization, yielding single-crystal areas several millimeters in size but oriented with the c axis inclined to the substrate normal by from 20 to 30°. Shalimova, Andrusko, and coworkers[18] in a series of papers reported on the structure of CdS films both in the as-deposited state and after recrystallization as described above. Their results essentially confirmed those obtained previously and in addition showed that evaporation at 30° from normal incidence produced a film oriented with the (00.1) planes inclined to the substrate. The results of Dresner and Shallcross[24] also agreed well with those of earlier workers, and a further investigation into the massive recrystallization phenomenon by Addiss[20] showed that under his conditions the initial 1,000 Å of his films were randomly oriented, with the highly preferred normal c-axis orientation appearing at a later stage in the film growth.

By depositing CdS films onto low-index faces of single-crystal rock salt, Aggarwal and Goswami[25] were able to obtain a variety of epitaxial orientations of both the cubic and hexagonal phases, depending upon the substrate temperature. Similar results have been obtained on (200) oriented molybdenum-foil substrates. Further recent work[26,27] has yielded considerable information on the epitaxy of CdS on mica, silver, rock salt, α-Al_2O_3, KCl, GaP, and GaAs. The epitaxial growth of hexagonal CdS films with the (00.1) planes parallel to the c faces of α-alumina (sapphire) is of particular interest as sapphire is a useful material in high-frequency ultrasonics. X-ray diffraction studies of CdS films deposited at oblique incidence have shown these films do not have the usual orientation with the c axis normal to the surface,[17] and more recent work[21] has established a relationship between the orientation obtained and the deposition parameters (see below).

(2) Orientation and Piezoelectric Coupling The strong tendency of CdS films to be oriented with the c axis normal to the film plane leads to the ready deposition of films suitable for the piezoelectric generation of longitudinal waves. Such films have been successfully made and used by several workers using both the direct and indirect evaporation and sputtering techniques.[1,2,5-11] Piezoelectric polarity measurements have been made on several of the highly oriented longitudinal-mode CdS films.[28] These measurements were made by impressing an electrometer probe onto the film and observing the sign of the voltage pulse produced. A negative polarity was usually observed on compression, indicating that the film was orientated with the Cd face at the free-growth surface and the S face next to the substrate. Some films, however, have shown the reverse polarity, and x-ray examination showed that in these films the individual crystallites were oriented with their c axes lying on a cone around the film normal with a half angle of about 20°.[5] This relationship between polarity and growth direction has a close parallel in the vapor-phase growth of single crystals of CdS as reported by Reynolds and Green.[29] This correlation between film and single-crystal growth processes is discussed in more detail elsewhere.[28] As mentioned in Sec. 1b, the orientation required for transducers for the generation of pure shear waves is with the c axis either inclined at an angle of about 40° from the normal, or lying in the film plane. The results of a recent study[21] have shown that CdS films can be deposited with the former type of orientation. In this work a series of films was deposited onto gold film substrates using the direct-evaporation method described earlier. Electron-diffraction analyses of films about 1,000 Å thick showed them to be hexagonal and polycrystalline and normally oriented with the (00.1) planes parallel to the substrate. In cases where the gold film substrates were highly oriented with the (111) planes parallel to the surface, epitaxial growth occurred, giving rise to CdS films which continued to grow with the c axis normal to the surface even under oblique deposition conditions. With randomly oriented gold film substrates, oblique evaporation yielded films with a different type of orientation. The

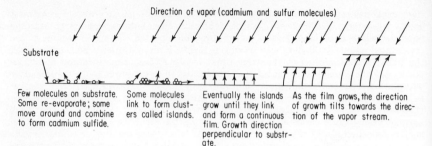

Direction of vapor (cadmium and sulfur molecules)

Substrate

| Few molecules on substrate. Some re-evaporate; some move around and combine to form cadmium sulfide. | Some molecules link to form clusters called islands. | Eventually the islands grow until they link and form a continuous film. Growth direction perpendicular to substrate. | As the film grows, the direction of growth tilts towards the direction of the vapor stream. |

Fig. 8 Diagrammatic representation of the development of an angled orientation under oblique deposition conditions. (*After Beecham.*[30])

development of the crystallographic orientation observed in this case is shown diagrammatically in Fig. 8.[30]   The initial layer of the film ($\sim$1,000 Å) was oriented with the c axis approximately normal to the substrate.   As the film thickness increased, the c axis progressively inclined toward the incoming vapor beam, reaching a final orientation at about a 1 μ film thickness.   The extent to which this inclined orientation developed was influenced by the substrate temperature and rate of deposition as well as the angle of the evaporation.   Table 1 shows the relationship between the final inclination angle of the c axis in the film and the angle of evaporation for the condition used in Ref. 21.   Figure 9 is an x-ray Debye-Scherrer pattern of the 39° film showing the high degree of orientation achieved.   The mechanism by which this oblique one-dimensional final growth orientation develops is presently considered to be by renucleation on the rough surface of the c axis oriented initial layer, followed by the rapid growth of the most favorably oriented nuclei.   Such an explanation could account for the continued growth with the c axis normal to the surface, which occurs when the evaporation rate is low or when the substrate temperature is high, as either of these conditions would result in a supersaturation too low for renucleation.

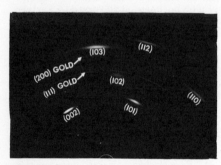

Fig. 9 Debye-Scherrer pattern of an obliquely evaporated CdS film showing the c axis about 40° from the film normal.

TABLE 1 Dependence of c-axis Inclination Angle on Deposition Angle for 4.5-μ-thick CdS Films

Deposition angle (degrees from normal)	Mean c-axis inclination angle (degrees from normal)
35	18
40	33
45	39
50	41
55	47
60	50+

c. Other Piezoelectric Films

Other nonferroelectric materials which have been successfully used for thin film transducers include ZnS, ZnO, and AlN.

(1) Zinc Sulfide The stable phase of bulk ZnS at room temperature is the cubic phase, zinc blende. This has also been the most commonly reported phase for thin films of ZnS when deposited onto either amorphous or single-crystal substrates.[25,31,33] The usual preferred orientation of the nonepitaxial cubic-phase crystals has been with the (111) planes parallel to the substrate. Several workers have reported mixed cubic and hexagonal phases, and predominantly hexagonal-phase films have been obtained by deposition onto suitably heated substrates. In particular, de Klerk[7] has published electron-diffraction patterns showing that either (111) oriented cubic ZnS or (00.1) oriented hexagonal ZnS can be deposited onto the c face of single-crystal sapphire depending upon the substrate temperature. Either orientation will result in a longitudinal-mode transducer film.

(2) Zinc Oxide Zinc oxide is a particularly attractive material for thin film piezoelectric transducers because of its comparatively high electromechanical coupling and the great stability of the hexagonal phase. Information on ZnO thin films, however, has only recently become available, primarily because of disproportionation, which makes its deposition by normal evaporation methods very difficult. Recently successful deposition has been achieved by reactive evaporation (see Chap. 2), and these films were oriented with the (00.1) planes parallel to the substrate surface.[8] Much of the ZnO film deposition has been done by sputtering. The first successful use of this technique for piezoelectric ZnO films was reported in 1965 by Wanuga, Midford, and Dietz.[10] The orientation observed was again with the (00.1) planes parallel to the substrate, and further work by Dietz[34] showed that the angular spread about this orientation was about 10°. A similar orientation has been reported on sapphire and gold film substrates by Rozgonyi and Polito.[9] The polarity along the c axis of these ZnO films was determined by direct piezovoltage probing. In contrast to the results obtained with CdS films, both positive and negative piezovoltage pulses were obtained on films with the c axis oriented normal to the substrate. Polarity inversions were also observed between adjacent areas of a single film. No correlation between polarity, crystallographic orientation, and deposition conditions has been established to date.

In addition to the usual orientation with the c axis approximately normal to the film plane, ZnO sputtered films have been deposited[35] with the c axis inclined to the normal, and with the c axis in the film plane. The inclined c-axis orientation was obtained by angled diode sputtering in a way somewhat similar to that described for CdS evaporation above. The degree of control over the angle of inclination was not sufficient to enable films with the optimum angle for shear-mode transducer performance to be deposited. The orientation with the c axis lying in the film plane does not have any parallel in the reported work on CdS films and is presently not well understood. It has been observed in triode-sputtered films from 0.25 to 4 μ in thickness and appears to be a nucleation orientation rather than a growth orientation. Under perpendicular-incidence deposition conditions the c axes of the crystallites are randomly distributed in the film plane, but under oblique-deposition conditions the axes are aligned in the film plane and directed toward the target. Figure 10 is a Debye-Scherrer x-ray diffraction pattern of such a film taken with the x-ray beam perpendicular to the vapor-beam direction. X-ray diffraction patterns taken with the x-ray beam parallel to the vapor-beam direction show

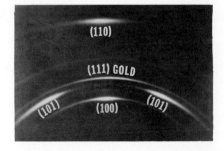

Fig. 10 Debye-Scherrer pattern of a sputtered ZnO film with the c axis in the film plane.

that in this case there was essentially complete freedom of rotation about the c axis so that no crystallographic plane was preferentially oriented parallel to the substrate. Further work is clearly required to elucidate the orientation dependence of these films.

Piezoelectric active films of aluminum nitride have also been deposited by evaporation of aluminum through a nitrogen-glow discharge.[36] These films generated longitudinal waves and were therefore most probably oriented with the c axis normal to the film plane.

(3) Ferroelectric Materials   Many ferroelectric materials have substantially higher electromechanical coupling coefficients than ZnO.[4] Ceramic forms of the materials have been used for many years for lower-frequency transducer applications. The compounds of primary interest have been lead-zirconate-titanate (PZT) and the niobates and tantalates of the alkali metals, with a recent emphasis, for higher-frequency work, on lithium niobate because of its comparatively low dielectric constant.

The deposition of thin films of lead-zirconate-titanate (PZT) showing ferroelectric behavior has been reported.[37] These films were deposited by low-energy reactive sputtering from a ceramic target at rates of approximately 500 Å h^{-1}. The structure and piezoelectric properties of these films have not been reported.

Lithium niobate films have also been deposited both by evaporation[36] and sputtering.[38] Direct vacuum evaporation has not proved a satisfactory method for the deposition of lithium niobate films because of disproportion. Some evaporated films have been deposited, however, and longitudinal-mode piezoelectric activity was reported.[36]

Piezoelectrically active lithium niobate films have been deposited by triode sputtering.[38] At substrate temperature above 300°C the films had a well-defined crystallographic x-ray structure in which the (01.2) planes were frequently oriented about 40° from the film normal plane. The dielectric constant in the films was approximately 65 in good agreement with the single-crystal values of 30 and 80 measured along the c axis and parallel to the c axis, respectively.

4. APPLICATIONS OF PIEZOELECTRIC FILMS

a. Electrical Properties of Piezoelectric Films

Two types of electrical measurements have normally been used to characterize the properties of piezoelectric transducers. These are (1) the frequency dependence of the electrical admittance, usually presented in terms of a parallel combination of a capacitance and a resistance, and (2) the transduction characteristics, normally presented as the frequency dependence of the electrical insertion loss of a structure consisting of two mechanically coupled transducers. The desired behavior of these characteristics depends to some extent upon the application. For low-loss operation the transducer impedance should be close to that of the electrical driving circuit, as any impedance mismatch will result in a reflection of electrical power at the circuit-transducer interface with a consequent increase in loss. The insertion loss is therefore a strong function of the electrical matching between the transducer and the driving circuit.

A typical admittance characteristic for a film transducer is shown in Fig. 11. This characteristic shows the main features predicted from the classical transducer equivalent circuit in that at the fundamental frequency, the conductance passes through a maximum and the capacitance decreases. Significant departures from classical behavior are evident, however, and these have been attributed to resistive losses in the metal electrode films. As the transducer impedance is primarily capacitive, the use of a parallel- or series-tuning inductance can effect a considerable improvement in the electrical matching to a resistive source over some frequency band determined by the Q of the tuning circuit. High-Q elements such as stub stretchers or coaxial cavities can tune and match the transducer impedance to the source impedance at one frequency and can yield very low insertion losses over a narrow band of frequencies. Wideband performance, on the other hand, can be achieved only with low-Q electrical terminations. The minimum insertion loss obtainable for any given structure is a function of the piezoelectric coupling factor of the transducer. It is therefore important to achieve a high coupling factor in transducers for wideband applications where

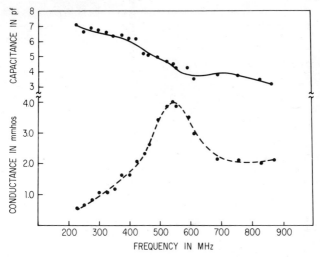

Fig. 11 Admittance characteristic for a ZnO longitudinal-mode thin film transducer on a fused-quartz substrate.

a low insertion loss is required. The values of coupling factors which have been obtained for thin film transducers of various materials are given in Table 2 together with the corresponding single-crystal values. Also included in the table are the orientations and corresponding modes which have been obtained. The coupling factors in Table 2 (except as noted) were calculated from the insertion loss at the fundamental frequency measured between 50-ohm resistive terminations with the transducer area chosen such that the capacitive impedance was 50 ohms at this

TABLE 2

Material	Deposition method	Orientation	Mode	Coupling factor		
				Thin film	Frequency, MHz	Single crystal
CdS........	Evap. (or sputt.)	Hex. c axis $\perp$	L	0.03*	2,500	0.15
				0.12	200	
				0.15†	800	
		Hex. c axis 40°	S	0.16	150	0.21
ZnO........	Sputt.	Hex. c axis $\perp$	L	0.22	500	0.282
		Hex. c axis ‖	S	0.18	500	0.31‡
						0.196§
		Hex. c axis 40°	S			0.35‡
LiNbO₃.....	Sputt.	(01.2) $\sim$ 40°	L	0.08	800	
AlN........	Evap.	Hex. c axis $\perp$	L			$\sim$0.2

* Reeder, T. M., *Proc. IEEE*, **55**, 1099 (1967).
† Ref. 39 (coupling calculated from admittance data).
‡ Ref. 4.
§ Crisler, D. F., J. J. Cupal, and A. R. Moore, *Proc. IEEE*, **56**, 225 (1968)

frequency. Under these conditions the loss per transducer is given by the equation[5]

$$\text{Loss in db} = 10 \log \left(\frac{\pi}{8k^2} \frac{Z_M}{Z_T} \right)$$

where Z_M and Z_T are the acoustic impedances (density times wave velocity) of the propagation material and transducer material, respectively, and k is the appropriate piezoelectric coupling factor. Values of density, velocity, and dielectric constant of the most frequently used materials in this field are given in Table 3.

TABLE 3

Material	Density $\times 10^3$ kgm^{-3}	Orientation	Relative dielectric constant	Velocity, ms^{-1}	
				Long.	Shear
CdS...............	4.82	∥ c axis	10.3	4,500	
		⊥ c axis	9.35	.....	1,800
ZnO...............	5.68	∥ c axis	11.0	6,400	
		⊥ c axis	9.26	.....	2,945
LiNbO₃............	4.7	∥ c axis	∼30	7,200	
		∥ a axis	∼80	6,600	3,780
SiO₂ (fused).......	2.65	....	3.8	5,968	3,770
(Xtal)........	....	∥ x axis	4.34	5,750	5,100
					(31½° to z)*
					3,300
					(58½° to z)*
		∥ z axis	4.27	6,320	
		BC direction	......	.....	5,060 (∥ to x)*
Al₂O₃.............	3.97	∥ c axis	11	11,180	
(sapphire)..........	....	∥ a axis	13	11,180	5,850 (∥ to c)*
					6,670 (⊥ to c)*

* Polarization direction of shear wave.

The coupling factor can also be calculated from the admittance characteristic using the equation[5]

$$k^2 \approx \frac{G_A X_c \pi}{4} \frac{Z_M}{Z_T}$$

where G_A is the conductance above background at antiresonance and X_c is the transducer capacitive reactance also at the antiresonant frequency. The primary difficulty in using the insertion-loss approach lies in obtaining the loss for just one transducer action from the measured data. The main difficulties in the admittance approach lie in obtaining sufficiently accurate data and in estimating the background-conductance contribution to the total conductance.[5] A considerably more sophisticated approach to the calculation of coupling factors from admittance data has been carried out by modifying the equivalent circuit to obtain a fit between calculated and measured conductance curves.[39] This study was carried out using longitudinal-mode CdS film transducers, and values of k equal to single-crystal values were obtained for some films.

Figures 12 and 13 present overall insertion-loss data for longitudinal- and shear-mode CdS and ZnO thin film transducer structures measured between 50-ohm resistive terminations. The loss per transducer is somewhat less than half of these values because of loss incurred in the acoustic transmission path. The minimum losses per transducer under tuned and matched electrical-termination conditions are typically from 2 to 5 db with 10 to 20% bandwidths.

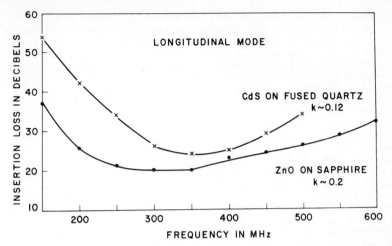

Fig. 12 Insertion-loss characteristics for CdS and ZnO longitudinal-mode thin film transducers driven directly from a 50-ohm source.

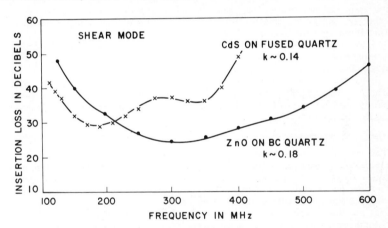

Fig. 13 Insertion-loss characteristics for CdS and ZnO shear-mode thin film transducers driven directly from a 50-ohm source.

b. Ultrasonic Attenuation Studies

Measurements of the attenuation of mechanical waves in solid materials have proved useful in increasing our knowledge of the fundamental properties of the solid state.[40] To make such measurements it is generally necessary to fabricate the material to be studied with flat and parallel end faces and to attach some form of electromechanical transducer to one or more of these faces. An ultrasonic pulse is generated by the transducer and multiply reflected between the end faces. The decay in amplitude of these pulses is monitored by observing the electrical signals generated by the transducer each time the mechanical pulse reflects from it. Transducers for attenuation measurements must therefore be able to generate and detect the desired mode of ultrasonic wave with sufficient efficiency to enable the decaying pulse sequence to be observed readily above the noise, and distinguished from any pulses due to other modes. It is essential, however, that the coupling between the mechanical wave and the electrical measuring circuit be sufficiently loose that the percentage of mechanical energy converted into electrical energy at each reflection from the transducer is negligible compared with the energy loss in the sample itself, since it is not possible to distinguish between the converted energy loss and the material loss. Transducers for attenuation studies should therefore be mismatched (electrically or mechanically) to provide the required loose coupling, consistent with maintaining a sufficient signal-to-noise ratio. Other important requirements for an attenuation-measurement transducer structure, particularly on low-loss materials, are a low internal dissipation and a sufficient uniformity in thickness to avoid significant losses and distortions in the multiply reflecting wave. Thin film transducer structures can avoid the use of either bonds or electrodes, but the internal losses, and possible thickness variations in the transducer itself, may outweigh the advantages in some cases.

It is frequently desirable to obtain attenuation data for longitudinal waves and for shear waves of one or more polarization in the same sample of material. A transducer capable of independently generating and detecting more than one type of wave therefore has some experimental advantage. This advantage has been realized in practice by de Klerk using CdS or ZnS transducer films deposited directly onto insulating single-crystal substrates with the c axis normal to the film plane. With these transducers placed in specially constructed microwave cavities,[41] the driving field can be oriented either normal to the film, thereby yielding longitudinal waves, or parallel to the film, yielding shear waves. Furthermore, as the polarization of the shear wave is always along the electric-field direction, for the parallel-drive case, the polarization can be adjusted in any desired direction with respect to the test sample by rotating it. Another feature of the cavity-drive technique is that it can be used to extremely high frequency (up to 70 GHz).[42]

It is also desirable to be able to make attenuation measurements over a wide range of temperatures. CdS film transducers have been cycled between room temperature and liquid-helium temperatures without degradation and have also been used up to about 200°C. At higher temperatures, the increasing electrical conductivity of the film has caused severe degradation in efficiency. ZnS films have been successfully used up to 600°C,[43] and lithium niobate or aluminum nitride films should be usable to 1000°C.

c. Ultrasonic Devices

The transducer requirements for ultrasonic-device applications are usually somewhat different from those for attenuation measurements outlined in the previous section. In the attenuation-measurement case it is only necessary to achieve a sufficiently low transducer loss to yield a reasonable signal-to-noise ratio, at essentially a single frequency. Ultrasonic devices, on the other hand, usually require the lowest possible insertion loss with fractional bandwidths of from 20 to more than 50% and must therefore utilize transducers with high coupling factors connected to driving and receiving circuits with very low electrical Q. The transducer must also generate essentially a single mode. Although thin film transducers have been used for several years for attenuation-measurement studies, they have appeared in very few device applications primarily because of the more stringent requirements for device work in terms of both mode purity and coupling factor.

The usefulness of the ultrasonic delay line lies in the very much lower velocity of acoustic waves compared with electromagnetic waves which allows delays of tens or hundreds of microseconds to be achieved in a physically small unit. The simplest ultrasonic delay line is physically similar to the attenuation-measurement configuration but usually has transducers on both end faces of the delay medium. An electrical pulse is fed into one transducer where it is converted into an acoustic pulse which traverses the solid delay medium and is reconverted to an electrical pulse by the output transducer. In this type of device a wide bandwidth is required to preserve the pulse shape, and mode purity is required to prevent the simultaneous generation of acoustic pulses with different velocities that would appear as spurious signals at the output with different time delays.

The frequency range of primary interest for delay lines using thin film transducers is from about 100 to 2,000 MHz, with some more limited applications in the 2- to 5-GHz range. The present state of the art for thin film transducers has been summarized in Figs. 12 and 13, and in Table 2. In the 100- to 500-MHz range, longitudinal-mode transducers of CdS and ZnO have shown discriminations against shear modes of greater than 40 db and coupling factors of 0.12 and 0.22, respectively, which represent approximately 75 and 80%, respectively, of the values obtained on single-crystal samples. Shear-mode transducers of controlled polarization have also been made in this frequency range using thin films of CdS and ZnO with coupling factors of 0.16 and 0.18 (75 and 60% of bulk values), respectively, and with discrimination against both longitudinal and perpendicularly polarized shear waves of 40 db.

Much less information is available for frequencies above 1 GHz. A steady apparent degradation in performance is indicated, although to what extent this is due to the increased difficulty in making definitive measurements is not clear. Because of the growth mechanism of the CdS shear-mode films, the discrimination against the longitudinal modes must be expected to be severely degraded in the thinner, higher-frequency films, and such is indeed the case. With the ZnO parallel-oriented shear-mode films, however, the discrimination is not degraded at 3 GHz, although, as with all the shear-mode film transducers, there is a decrease in the effective coupling factor with increasing frequency. To offset some of the difficulties arising from operation at gigahertz frequencies, the multiple-film approach has been investigated to obtain increased conversion efficiency. This approach has provided conversion losses of about 10 db (tuned with high-Q structure) in the 2- to 4-GHz range.[36] More work needs to be done to determine the optimum results obtainable, especially at the high frequencies.

d. Strain-sensitive Transistors

An interesting application which combines the piezoelectric and semiconducting properties of CdS evaporated films has recently been reported based on the field-effect transistor (FET) structure, which is fully discussed in Chap. 20. The operation of the FET is such that the drain-voltage vs. drain-current characteristics are controlled by electrical charges present in the gate region. These charges are normally induced by an externally applied gate voltage, but if the semiconductor material is also piezoelectric, a mechanical strain developed in the gate region can also produce surface charges which will control the device characteristics. This type of strain-sensitive transistor, called a metal-insulator-piezoelectric-semiconductor triode (MIPS), has been studied by Müller and Conragen[44] for static strains and by Fiebiger and Müller[45] for static and dynamic strains. The MIPS was subjected to strains of approximately 6×10^{-4} by bending the glass-microscope-slide substrates, and significant changes in output characteristics were observed. This type of device has several potential advantages as a strain gauge compared with the more usual piezoresistive elements discussed in the next section. Because of the small physical dimensions of the gate region, the spatial resolution, in one direction at least, can be about 10 μ, and the output characteristics are similar to those of a piezoresistive element coupled to its associated amplifier. The sensitivity of the MIPS transducer is at least two orders of magnitude greater than that of the piezoresistive elements, and response times of less than 1 μs have been observed.

5. PIEZORESISTIVE PROPERTIES OF THIN FILMS

a. Introduction—Piezoresistive Properties of Bulk Materials

The phenomenon of piezoresistance, i.e., the change in electrical resistance of a body when mechanically strained, has been known and studied since before the turn of the century. Taking the simplest case of an isotropic cylinder of length l, cross-sectional area A, and resistivity ρ, the fractional change in resistance dR/R produced by tension will be given by

$$\frac{dR}{R} = \frac{d\rho}{\rho} - \frac{dA}{A} + \frac{dl}{l}$$

where the first term accounts for the change in the specific resistivity of the material and the other two terms account for the resistance changes resulting from changes in the physical dimensions. The geometrical terms can be combined using Poisson's ratio ν giving

$$\frac{dR/R}{dl/l} = (1 + 2\nu) + \frac{d\rho/\rho}{dl/l} = \gamma$$

where γ is the electrical-resistance strain coefficient, or strain-gauge factor. As the values of Poisson's ratio for most materials lie between 0.25 and 0.45, the resistance changes due to the alterations in physical dimensions alone will lie in the range of 1.5 to 1.9. Measured strain-gauge factors, however, range from about -20 to $+200$, showing that the piezoresistance effects can be dominated by changes in the resistivity of the materials.

The mechanisms which give rise to electrical resistivity are quite complex and are described fully in Chap. 13. Only a brief mention of the salient factors is made here to provide a basis for understanding the way in which the various resistivity mechanisms change with strain.

When an electric field E is applied across a conductor, the conduction electrons experience a force in a direction opposite to the field direction. In a perfectly periodic lattice these electrons can travel without loss of energy and would therefore accelerate continuously (up to the Brillouin-zone boundary), giving the material a nearly infinite conductivity. In real materials disturbances in the perfect-lattice periodicity occur which scatter the conduction electrons and remove their drift velocity. It is these electron-scattering processes which give rise to the phenomenon of electrical resistivity. As it is only the electrons near the Fermi surface which can participate in the conduction process, the mean time τ and the mean free path λ between scattering interactions are related by $\lambda = v_F\tau$, where v_F is the Fermi velocity. There are many different mechanisms by which the perfect-lattice periodicity can be disturbed, and the relative importance of their contribution to the resistivity depends upon the nature of the conductor.[46]

The alkali metals are the simplest of the real conductors. In these materials the body-centered-cubic crystal structure results in a highly symmetrical first Brillouin zone. As each atom supplies only one conduction electron, this zone is only half filled, and the resulting Fermi surface assumes a nearly spherical shape within the first Brillouin zone and is only slightly distorted by the proximity of the Brillouin-zone boundaries. Under these conditions the conduction electrons can be treated as a nearly free electron gas, yielding the following relationship for the resistivity ρ:

$$\rho = \frac{m^*}{ne^2\tau}$$

where m^* is the effective electron mass, n the number of conduction electrons, and e the electronic charge. In these metals electron scattering can arise from the presence of impurity atoms and from crystallographic imperfections such as the exterior surfaces and grain boundaries. The most important scattering mechanism in relatively pure materials, however, is the scattering from the thermal lattice vibrations. The magnitude of this thermal scattering depends upon the lattice-vibration amplitude. For this reason a compressive strain decreases the electrical resistivity by decreasing the vibration amplitude. This effect gives rise to an increased resistivity in a sample under tension and therefore increases the gauge factor γ above the 1.5 to 1.9 value anticipated from purely geometric factors.

In metals with more complicated crystallographic and electronic structures, other electron-scattering processes can occur in addition to those present in the simple alkali metals. In metals in the first transition series, for example the $3s$ band overlaps the $4d$ band, and the Fermi surface therefore overlaps two Brillouin zones. In this situation an additional scattering process, interband scattering, can occur which greatly increases the resistivity. In the interband-scattering process conduction electrons are scattered into both s and d bands. The increased resistivity which accompanies this process arises from the difference in structure between the two bands. The s band is spatially wide with a low density of states and the s-band electrons can readily take part in the conductive process. The d bands, however, are spatially narrow and have a high density of states. The d-band electrons therefore have a high effective mass and contribute little to the conduction process, and electrons scattered into the d band are no longer available for conduction. When this type of conductor is strained, resistivity changes can occur both from the changes in thermal lattice scattering and also from changes in the relative populations of the s and d bands. This contribution to the strain-gauge factor can be of either sign.

In magnetic materials still another scattering process, magnon scattering, can occur. This scattering arises from the interaction between the conduction electrons and the magnetic-spin disorder. This disorder starts to freeze out below the Curie temperature with a resulting rapid decrease in resistivity. In many magnetic materials magnon scattering is the dominant scattering mechanism and therefore controls the resistivity. The effects of straining magnetic materials are therefore often determined by their magnetic properties. If the strain produces a magnetic field through magnetostriction, the magnetic field in turn can order the magnetic spins and reduce the electrical resistivity. This effect will appear as a negative contribution to the gauge factor, giving rise to the negative values observed for nickel.

Very large positive gauge factors can be obtained with semiconducting materials. As with the transition metals, the large resistivity changes arise from changes in the number of electrons available for conduction. In a semiconductor the electrons are distributed between a conduction band and a lower-energy valence band, the two bands being separated by a narrow forbidden energy gap. The resistivity is determined primarily by the number of electrons in the conduction band. This in turn is a function of energy of the electrons and the width of the forbidden energy gap, both of which can be changed by the application of an external stress.

In materials with anisotropic crystallographic structures the electrical resistivity and piezoresistive coefficients will also be anisotropic. This effect arises because the shape of the Brillouin zone reflects the crystallographic symmetry. The relative energies of the Fermi surface and the Brillouin-zone boundaries are therefore direction-dependent. This in turn makes the number and mobility of the conduction electrons direction-dependent. In an unstrained condition, equivalent crystallographic directions will exhibit equal resistivities, giving rise to the multivalley model proposed by Herring.[47] Straining an anisotropic material, however, distorts the Brillouin zone and destroys the equivalence between the valleys. The resistance even in equivalent crystallographic directions therefore becomes unequal to an extent dependent upon the magnitude and nature of the applied stress. This effect has been utilized to make single-crystal silicon strain gauges which are sensitive only to specific types of strain.[48]

Piezoresistive Properties of Metal Films

As with bulk materials, the piezoresistive properties of thin film conductors are dominated by the changes in resistivity which occur when the material is strained. The structure in thin films, however, is often different from that of the bulk material, and as a result, the relative importance of the various electron-scattering mechanism may be changed. In particular, the close proximity of the major surfaces and the presence of many grain boundaries can make surface-scattering effects the dominant resistivity mechanism.

The structure, and therefore the resistive and piezoresistive properties of thin films, are strongly dependent upon the film thickness and the condition of film deposition. The dependence of film structure and resistivity on thickness and deposition conditions is described in Chap. 13. Films for piezoresistive studies have typically been deposited onto amorphous insulating substrates, e.g., glass or organic resins, held at room temperature. Under these conditions the grain size is usually comparable with the film thickness. The changes in film resistance have usually been measured along the length of the film while subjecting it to a longitudinal strain induced by bending the flexible substrates. With one end of the substrate rigidly clamped, the displacement of the free end gives a measure of the strain induced in the film.

With the exception of some magnetic metals, the piezoresistance characteristics of metal films follow a uniform pattern.[49] The changes in resistance are essentially linear with strain, and the gauge factors are positive and vary with film thickness as shown in Fig. 14, which is for a gold film deposited onto a glass substrate held

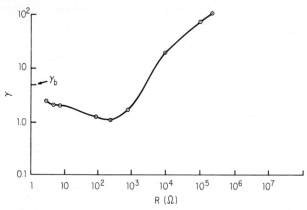

Fig. 14 Gauge factor γ as a function of film resistance (thickness) for gold films deposited onto room-temperature substrates. (*After Parker and Krinsky.*[49])

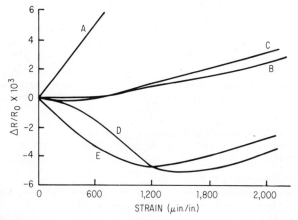

Fig. 15 Relative resistance changes in nickel films vs. strain for various film thicknesses (*a*) Thinnest film $R = 36.3$ kohms. (*b*) $R = 159$ ohms. (*c*) $R = 56$ ohms. (*d*) $R = 4.3$ ohms. (*e*) Wire 0.001 in. diameter. (*After Parker and Krinsky.*[49])

at room temperature. The gauge factor in thick films (low resistance in Fig. 14 approaches that of bulk material but is somewhat lower than the bulk value, probably because of the grain-boundary scattering contribution to the resistivity, which appears to be less strain-sensitive than the lattice scattering contribution. As the film thickness is reduced, the resistance increases and the gauge factor passes through a minimum value in the thickness range of a few hundred angstroms. Calculation of the mean free path λ for the conduction electrons in metals at room temperature yields values of several hundred angstroms. Conduction electrons in films in th thickness range would therefore frequently reach the film surface before interacting with other scattering centers, thus making surface-boundary scattering an important resistivity mechanism. As this scattering mechanism is not strain-sensitive, the overall strain sensitivity is reduced. As the film thickness is reduced still further, the film becomes discontinuous and the overall resistance becomes dominated by the resistance between individual grains. As this intergrain resistance is a strong function of grain separation, and therefore of strain, the gauge factor becomes considerably higher for films below about 50 Å. The structure of these very thin films, however, is extremely sensitive to the film-deposition conditions and cannot be made sufficiently reproducible for the fabrication of practical strain gauges.

The strain sensitivity of the resistance of nickel films does not follow the above pattern except in very thin films where the gauge factor is independent of strain and is large and positive. For thicker films the resistance change is a highly nonlinear function of strain and depends markedly on the method of deposition.[49,50] Figure 15 shows how the longitudinal-resistance change of nickel films varies with strain for films of varying thickness. As with nonmagnetic films, the piezoresistive properties approach those of the bulk materials as the film thickness increases. Calculations of the transverse gauge factor for Ni films, obtained from resistance changes measured in a direction perpendicular to the strain direction, have shown positive gauge factors with a maximum at a film thickness of about 400 Å, somewhat similar to the behavior exhibited by bismuth and antimony films.[50]

c. Piezoresistive Properties of Semimetal and Semiconductor Films

The high values of γ obtainable in bulk semimetals and semiconductors have prompted some investigations into the piezoresistive properties of films of these materials. Antimony and bismuth films in particular have been studied intensively by Thureau, Laniepce, and Jourdain.[50] The films were evaporated onto mica and Plexiglas substrates in a vacuum of 10^{-6} Torr. The films were subjected to tensile strains, both by bending and by stretching the substrates, and the resulting resistance changes were measured both in the direction of strain and perpendicular to the direction of strain, thus yielding both the longitudinal and transverse gauge factors. The magnitude of the room-temperature gauge factors was found to depend somewhat on the substrate used, but the variation of γ with film thickness was generally of the form shown in Fig. 16, exhibiting a pronounced maximum in the region of 1,000 Å.

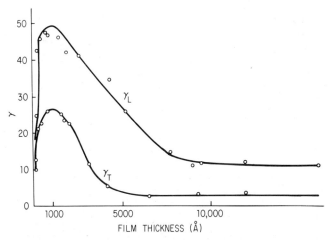

Fig. 16 Variation of longitudinal γ_L and transverse γ_T gauge factors with film thickness for bismuth film deposited onto mica at room temperature. (*After Thureau, Laniepce, and Jourdain.*[50])

Maximum values of 26 and 50 were obtained for the longitudinal and transverse gauge factors, respectively, when deposited on mica substrates. The general shape of this curve was not changed by going to higher or lower temperature, but at 80°K, the film thickness yielding the maximum γ, and the magnitude of γ were approximately twice their room-temperature values. Bismuth films with built-in "impurities" were also examined. These films were made by evaporating several successive layers of equal thickness with exposure to air between each evaporation. These films all showed positive gauge factors of approximately 6 for the longitudinal effect and 20 for the transverse effect, with the highest factor being exhibited by films deposited in 1,000-Å stages. Little variation of γ with total film thickness was observed in these "impure" films.

Piezoresistance studies have also been made on bismuth films deposited at pressures in the range of 5×10^{-8} to 1×10^{-7} Torr.[51] X-ray examination of these films showed a high degree of orientation with the (00.1) planes lying parallel to the substrate surface. The films were strained by bending the glass substrates, and the longitudinal and transverse gauge factors were measured together with the Hall coefficients. The transverse gauge factors were essentially independent of film thickness from 500 Å to 2 μ and were in the range of 25 to 35. The longitudinal gauge factors, however, were close to 10 for films 500 to 2,000 Å thick falling to nearly -10 for films 1 μ and thicker.

Semiconducting indium antimonide films have also been studied.[52] These films were deposited by three different vacuum-evaporation techniques designed to overcome the severe disproportion which occurs with this material. The pressure was below 2×10^{-4} Torr. The methods used were (1) flash evaporation, (2) coevaporation of the elements at individually controlled rates, and (3) the successive slow evaporation to completion of many small charges of indium antimonide followed by a diffusion heat treatment. A suitable choice of conditions enabled indium antimonide films to be formed by each of the above methods, most of which showed a marked preferred orientation with the (111) planes parallel to the substrate surface. Heat treatment at 200°C for 1 h produced considerable grain growth and improved the (111) orientation. Longitudinal gauge factors were measured using the usual temperature-compensating bridge structure deposited onto microscope slides, and gauge factors of -20 were obtained using a 9,000-Å-thick film. When 1,000-Å-thick films were used, gauge factors observed were in the range of $+2$ to $+10$, as would be anticipated from the strain sensitivity of the intergrain resistance dominant in thin discontinuous films.

d. Applications of Piezoresistive Thin Films

The difficulty of obtaining reproducible and stable piezoresistive thin films has resulted in their limited application in practice. For some applications, however, the advantages of small size and easy fabrication could outweigh the disadvantages. One application for which piezoresistive metal films have been examined is for use in the severe environment of a nuclear reactor where the ambient is CO_2 at 600°C.[5] Thin films of chromium or platinum with overlying protective silicon monoxide films, all deposited by vacuum evaporation, were found to be stable in this environment. Other units utilizing silicon films, although giving gauge factors of 50 at 200°C, were much less sensitive at high temperatures and were not sufficiently stable. The platinum or chromium films were deposited onto substrates consisting of a 0.005-in. molybdenum strip coated with 0.001 in. alumina and overcoated with a similar thickness of steatite. This combination achieved the high electrical resistivity required at 600°C and provided a suitably smooth substrate for the metal films. Attempts to use sapphire substrates were not successful, as the sapphire could not be bonded sufficiently strongly to the specimen under test. Four resistive film elements were deposited in the form of a square, and these were incorporated into a bridge circuit to compensate for changes of resistance with temperature. The longitudinal strains were applied parallel to one pair of elements, thus producing an output from the bridge circuit. Gauge factors of from 1 to 2 were measured depending upon the temperature and the metal-film thickness. The gauge factor of the platinum films were in general somewhat greater than those of the chromium films but were also more sensitive to temperature changes. One such device[50] makes use of the low inertia of thin film structures to make a fast-rise-time pressure gauge. In this structure four bismuth films 1,000 Å thick were deposited onto a 20-μ Mylar membrane to form a temperature-compensated bridge circuit. The Mylar membrane was held rigidly at the edges, and displacement of the center portion by differential gas pressures produced opposite transverse strains in the bridge arms, thereby producing an electrical output. Pressure differences of 2 mbar were measured, and the device followed pressure oscillations at 65 kHz. A more conventional type of strain-gauge structure has also been examined[51] consisting of series-connected zigzag bismuth film resistive elements 0.5 to 1 μ thick arranged so that unidirectional strains would produce both longitudinal and transverse strains in the bismuth films. This arrangement was chosen to achieve cancellation of undesirable transient effects. The substrate used was 0.01-mm-thick epoxy resin sheet which was attached to the test specimen using an epoxy resin. Gauge factors of 10 to 20 were obtained and response up to 20 kHz was demonstrated.

A rather different type of application of the piezoresistive effect in metal films is for the fabrication of resistors with a very low temperature coefficient of resistance.[54,55] This effect is achieved by carefully choosing the thin film resistor material and the substrate material such that, as the temperature varies, the resistance change produced by the strain arising from the expansion mismatch between the film and substrate compensates for the resistance change due to the temperature coefficient of resistivity of the film material itself. Units of this type have been made and exhibited an overall temperature coefficient of ± 5.0 ppm/°C, which represents about an order of magnitude over "conventional" thin film resistors.

REFERENCES

1. Foster, N. F., *IEEE Trans. Sonics Ultrasonics*, **SU-11**(2), 63 (November, 1964).
2. de Klerk, J., and E. F. Kelly, *Appl. Phys. Letters*, **5**(1), 2 (July, 1964).
3. Bommel, H. E., and K. Dransfeld, *Phys. Rev.*, **117**(5), 1245–1252 (March, 1960).
4. Berlincourt, D. H., D. R. Curran, and H. Jaffe, in W. P. Mason (ed.), "Physical Acoustics," vol. 1A, chap. 3, Academic Press Inc., New York, 1964.
5. Foster, N. F., G. A. Coquin, S. A. Rozgonyi, and F. A. Vannatta, *IEEE Trans. Sonics Ultrasonics*, **SU-15**, 28–40 (January, 1968).
6. Foster, N. F., *Proc. IEEE*, **53**(10), 1400–1405 (October, 1965).
7. de Klerk, J., and E. F. Kelly, *Rev. Sci. Inst.*, **36**(4), 506 (April, 1965).
8. Malbon, R. M., D. J. Walsh, and D. K. Winslow, *Appl. Phys. Letters*, **10**(1), 9 (January 1967).
9. Rozgonyi, G. A., and W. J. Polito, *Appl. Phys. Letters*, **8**, 220 (1966).
10. Wanuga, S., et al., "Zinc Oxide Film Transducers," paper presented at the IEEE Ultrasonics Symposium, Boston, Dec. 1–4, 1965.
11. Foster, N. F., *Proc. Joint IERE-IEE Conf. Applications of Thin Films in Electronic Engineering*, London, England, July, 1966.
12. Gross, R., and M. Volmer, *Z. Physik*, **2**, 188 (1921).
13. Dixit, K. R., *Phil. Mag.*, **16**, 1049 (1933).
14. Evans, D. M., and H. Wilman, *Acta Cryst.*, **5**, 731 (1952).
15. Bauer, E., *Trans. 9th AVS Symp.*, 1962, p. 35, The Macmillan Company, New York.
16. Gilles, J. M., and J. van Cakenberghe, *Solid State Phys. Electron. Telecommun. Proc. Intern. Conf. Brussels*, **2**, 900 (1960).
17. Okamoto, H., *Japan. J. Appl. Phys.*, **4**, 234 (1967); **4**, 821 (1965); **5**, 251 (1966).
18. Shalimova, K. V., A. F. Andrusko, V. A. Dmitriev, and L. P. Pavlov, *Soviet Phys. Cryst.*, **8**, 618 (1963); **9**, 340 (1964).
19. Behringer, A. J., and L. Corrsin, *J. Electrochem. Soc.*, **110**, 1083 (1963).
20. Addiss, R. R., *Trans. 10th Natl. Vacuum Symp. AVS*, 1963, p. 354, The Macmillan Company, New York.
21. Foster, N. F., *J. Appl. Phys.*, **38**(1), 149 (January, 1967).
22. Wendland, P. H., Evaporated Film of CdS, *J. Opt. Soc. Am.*, **52**, 581 (1962).
23. Semiletov, S. A., *Kristallografiya*, **1**, 304 (1956).
24. Dresner, J., and F. V. Shallcross, *J. Appl. Phys.*, **34**, 2390 (1963).
25. Aggarwal, P. S., and A. Goswami, *Indian J. Pure Appl. Phys.*, **1**, 366 (1963).

26. Zhdan, A. G., R. N. Sheftal, M. E. Chuganova, and M. L. Elinson, *Radio Eng. Electron. Phys.*, **11**, 1339 (August, 1966).
27. K. C. Chopra, and I. H. Khan, *Surface Sci.*, **6**, 33 (1967).
28. Rozgonyi, G. A., and N. F. Foster, *J. Appl. Phys.*, December, 1967.
29. Reynolds, D. C., and L. C. Green, *J. Appl. Phys.*, **29**, 559 (1958).
30. Beecham, D., *Ultrasonics*, **5**, 19 (January, 1967).
31. Kane, W. M., J. P. Spratt, L. W. Hersinger, and I. H. Khan, *J. Electrochem. Soc.*, **113**, 136 (1966).
32. Batailler, S., P. Bugnet, J. Deforges, and S. Durand, *Compt. Rend.*, **264**, 320 (1967).
33. Antcliffe, G. A., *Brit. J. Appl. Phys.*, **16**, 1467 (1965).
34. Dietz, J. P., "Oriented Films of Zinc Oxide on Crystal Quartz and Sapphire," paper presented at the 68th Annual ACS Meeting, Washington, D.C., May, 1966.
35. Foster, N. F., The Deposition and Measurement of Zinc Oxide Shear Mode and Other Thin Film Transducers, *J. Vac. Sci. & Tech.*, **6**, 111, 1969.
36. Winslow, D. K., Microwave Acoustic Transducers—Aluminum Nitride and Lithium Niobate Films, *J. Vacuum Sci. Technol.*, **6**, 111 (1969).
37. Lupfer, D. A., Rutgers University, New Jersey, private communication.
38. Foster, N. F., "Sputtered Lithium Niobate Thin Film Transducers," *J. Appl. Phys.*, **40**, 420, 1969.
39. Bahr, A. J., and I. N. Court, "Determination of the Electromechanical Coupling Coefficient of Thin-film Cadmium Sulphide," paper presented at IEEE Symposium on Sonics and Ultrasonics, Vancouver, B.C., Canada, October, 1967.
40. Mason, W. P., "Physical Acoustics and the Properties of Solids," D. Van Nostrand Company, Inc., Princeton, N.J., 1948.
41. de Klerk, J., *Phys. Rev.*, **139**(5a), A1635 (August, 1965).
42. Thaxter, J. B., and P. E. Tannenwald, *IEEE Trans. Sonics Ultrasonics*, **SU13**, 61 (1966).
43. Lord, A. E., Jr., *J. Appl. Phys.*, December, 1966, p. 4593.
44. Müller, R. S., and J. Conragen, *IEEE Trans. Electron Devices*, **12**, 590 (1965).
45. Fiebiger, J. R., and R. S. Müller, *J. Appl. Phys.*, **38**, 1948 (1967).
46. Kuczynski, G. C., *Phys. Rev.*, **94**, 61 (1954).
47. Herring, C., *Bell System Tech. J.*, **34**, 237 (1955).
48. Pfann, W. G., and R. N. Thurston, *J. Appl. Phys.*, **32**, 2008 (1961).
49. Parker, R. L., and A. Krinsky, *J. Appl. Phys.*, **34**, 2700 (1963).
50. Thureau, P., B. Laniepce, and P. Jourdain, *Rev. Franc. Mecan.*, 1964, pp. 10–11, 81.
51. Koike, R., and H. Kurokawa, *Japan. J. Appl. Phys.*, **5**, 503 (June, 1966).
52. Koike, R., *Elec. Eng. Japan*, **84**, 65 (March, 1964).
53. Maclachlan, D. F. A., *Microelectron. Reliability*, **3**, 227 (1964).
54. Zandman, F., and S. J. Stein, *Proc. Electron. Components Conf.*, 1964, p. 107; 1965, p. S-158.
55. Hall, P. M., *Appl. Phys. Letters*, **12**, 212 (1968).

Chapter **16**

Dielectric Properties
of Thin Films

PETER J. HARROP

The Plessey Company Limited, Whiteside Works, Bathgate, West Lothian, Scotland

and

DAVID S. CAMPBELL

The Plessey Company Limited, Whiteside Works, Bathgate, West Lothian, Scotland
also: Visiting Senior Lecturer in Materials Science, Electrical Engineering
Department, Imperial College, London University, London, England

INTRODUCTION

This chapter represents an attempt to present the basic theory behind the electrical properties of thin insulators as employed in the fabrication of electronic devices. To this end, generalizations about the full spectrum of inorganic and organic solid are preferred to an account of the various detailed analyses of the behavior of individua compounds. Film thicknesses are less than 10,000 Å.

Thin insulating films are used in a wide variety of components, and the films ar usually amorphous or near amorphous, since polycrystalline films are generally foun to be less insulating and single-crystal insulating films are difficult to prepare.

second virtue of amorphous films is their remarkable insensitivity to impurity, a property shared[1] with bulk "glass." For instance, SnO_2 films prepared with up to 10% metallic impurity show unchanged electrical properties.[2]

For the above reasons this survey is primarily concerned with amorphous insulators. Thus ferroelectric materials are not considered, as ferroelectricity is a crystalline property. Behavior within a few hundred degrees of ambient is of primary interest, since it is in this range that most devices are operated.

The theoretical sections begin with a summary of the simple electrical theory assumed in the subsequent sections. DC and ac conduction mechanisms are then considered, followed by permittivity, temperature coefficient of capacitance, and breakdown. Most of these mechanisms are identical with those expected for bulk material. One difference is that conduction is generally lower in the bulk material, if only because "easy paths" are less likely to penetrate from one side of the sample to the other. Also films are often thin enough for quantum processes to be significant. Breakdown fields are lower in the bulk because the geometry admits more mechanisms of breakdown. In general, then, analysis of the properties of dielectrics can use the same approach whether they are bulk single crystals or thin amorphous structures provided the geometry is considered. Indeed, data for bulk material are used, and justified, in the relevant sections.

Following the theoretical sections, a section describes various practical applications of thin dielectrics; the requirements are enumerated, and the general basis for the selection of suitable materials is laid down. Finally, the merits and demerits of a variety of preparative techniques are discussed in this context.

The dielectric properties of thin insulating films have, of course, been studied in the fields of electrochemistry and corrosion research. Much of this is concerned with the added effects of a contacting electrolyte, and these topics are not dealt with here. The reader is referred to recent books on anodic films[3] and corrosion[4,5] for further details.

The balance of the text is such that, where adequate textbooks already exist, the discussion is brief and qualitative. Thus, simple electrical theory as applied to the solid state is well detailed by the books of Kittel[6] and Dekker.[7] Nonohmic dc conduction in films is dealt with by Lamb[8] and by Simmons in this volume.

In the present text mks units are used.

SIMPLE ELECTRICAL THEORY

a. Dielectric Loss

The simple electrical theory necessary for the following sections is now outlined. As mentioned, this takes the form of a qualitative outline only. For a fuller mathematical basis the reader is referred to the textbooks.[6,7,9,10]

Since we are concerned with electrical insulators, these are best considered as capacitor dielectrics. The capacitance C of a parallel-plate capacitor is given by

$$C = \frac{\epsilon\epsilon_0 A}{d}$$

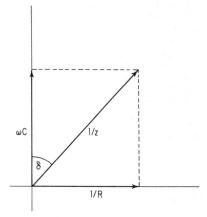

Fig. 1 Vector diagram of practical capacitor.

where ϵ is the relative permittivity, ϵ_0 the permittivity of free space, A the electrode area, and d the electrode separation. Insulators always have a finite parallel resistance R, and the total impedance can be found from a vector diagram as in Fig. 1, where ω is 2π times the frequency. The sine of the angle δ is a measure of the energy absorbed in the insulator; for the present purpose R is always large ($\tan \delta < 10\%$), and one may write

$$\sin \delta \sim \tan \delta = \frac{1}{\omega RC}$$

The quantity $\tan \delta$ is therefore easily calculated from the parallel components of resistance R and capacitance C as measured on an appropriate bridge; $\tan \delta$ is subsequently referred to as the loss of the insulator.

b. Electrode Resistance

If the capacitor is measured in a setup that puts an appreciable resistance (say a few ohms) in series with the parallel R and C of the dielectric, the reading of the bridge will be affected at high frequencies.[11,12] The general behavior is shown in

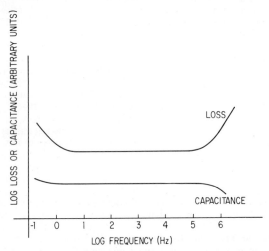

Fig. 2 Typical trends of loss ($\tan \delta$) and capacitance with frequency.

Fig. 2. For the present purpose this is regarded as a measurement error, and it is assumed to be absent in subsequent discussion.

c. Time Constant

A dielectric has a resistance R and capacitance C defined by the equivalent parallel components. If the ideal capacitance C in this equivalent circuit supports a charge Q and a voltage V_c at time t, it will discharge through the resistive part R. However

$$V_c + V_r = 0$$

where V_r is the voltage across the pure (dc) resistance R. Now $V_c = Q/C$ and $V_r = (dQ/dt)R$ by definition, so that

$$\frac{Q}{C} + \frac{dQ}{dt} R = 0$$

Integrating,

$$\log_e Q = -\frac{t}{RC} + K$$

where K is a constant. If $Q = Q_0$ at $t = 0$, $K = \log_e Q_0$ and

$$Q = Q_0 e^{-t/RC}$$

The quantity RC is defined as the time constant τ of the dielectric, being the time for a charge to decay to exp (-1) of its original value with the dielectric on open circuit. In practice τ can be taken as the product of the dc resistance of the dielectric and the capacitance at, say, 1 kHz. Since C is a relatively invariant quantity, τ is usually dominated by the behavior of R.

d. Band Theory

The electrons associated with the atoms in a gas have discrete energy levels. As these atoms are brought together to form a solid, the levels are broadened to bands that may overlap. However, in general there will be some regions of energy that are still forbidden. An insulator constitutes a solid with a forbidden band above a full allowed band, as shown in Fig. 3.

The Fermi level E_F indicated is a quantity defined by the Fermi-Dirac statistics obeyed by electrons[6,7,13] whence

$$n(E) \, dE = Z(E)F(E) \, dE$$

where $n(E) \, dE$ is the number of electrons per unit volume occupying states in the energy range between E and $E + dE$, $Z(E)$ represents the number of possible states per unit volume, and

$$F(E) = \frac{1}{e^{(E-E_F)/kT} + 1}$$

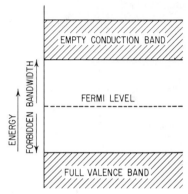

Fig. 3 Energy-band diagram for a simple insulator.

If two solids are put in contact, the Fermi levels equalize at the interface, the other energy levels moving to accommodate this. In a pure insulator, the Fermi level bisects the forbidden band. Impurities may introduce allowed levels into the forbidden band, and this moves the Fermi level up or down. In the extreme, the impurity levels may cause the Fermi level to leave the forbidden band. The conduction then approaches that of a metal and is termed degenerate, meaning that electrons can be in different states at one energy level.[38]

e. Clausius-Mosotti Equation

The permittivity of a solid is usually due to polarization of the constituent electrons and ions. These species approximate to infinitely small mutually interacting dipoles in the lattice. For such dipoles, the local field E_i is related to the applied field E by the Lorentz[9,10] approximation

$$E_i = \frac{\epsilon + 2}{3} E$$

From this approximation the Clausius-Mosotti equation is derived,[9,10] whence

$$\frac{\epsilon - 1}{\epsilon + 2} = \frac{\alpha_m}{3V\epsilon_0} \tag{1}$$

where α_m is the polarizability of a macroscopic volume V.

DC CONDUCTION MECHANISMS

Introduction

Most insulators conduct ohmically below 0.1 MV cm^{-1} and nonohmically above MV cm^{-1}. Since both regions are of interest in practical devices, they will be evaluated separately. It is useful to bear in mind that the ohmic conduction is always extrinsic (i.e., due to impurity, inhomogeneity, etc.) near 20°C in practical materials, whereas the nonohmic components can be intrinsic (i.e., a property of the pure, homogeneous compound). This is true whether the conduction is ionic or electronic, direct or relaxational.

The band theory of conduction was first developed to describe the behavior of metals and semiconductors, and it is not necessarily applicable to insulators, particularly those which are amorphous.[14,15] However, other concepts that have been developed to explain the localized nature of the "conduction" electrons in these solids are still incomplete. Thus, the theory of polarons[15,16] has been widely applied but is now thought to be appropriate to a very few insulators indeed, such as amorphous sulfur.[17] In view of this, the following discussion seeks to evaluate conduction in insulators by suitably modified band theory. Lamb[8] has further justified this approach.

b. High-field Conduction

(1) Electronic Conduction   The basic experimental fact that is observed with thin film dielectrics is that even in the absence of flaws much larger direct electronic currents flow through the films than would be expected from the bulk properties.

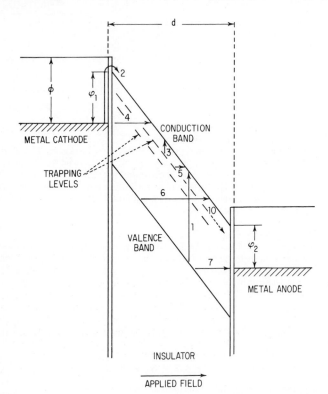

Fig. 4   Diagram of the various conduction processes in a thin film dielectric at high field

Various conduction mechanisms have been advanced to explain this, and sever? reviews have been published.[8,18,19] Many practical examples can be found in thes? texts. The mechanisms can be best classified in four ways and are illustrated ? Figs. 4 and 5. In practice it is often very difficult to distinguish what is actual? happening.

1. Conduction by means of the conduction band of the dielectric. There are various ways in which electrons can be raised into the band: thermally from the valence band if it has a small enough forbidden bandwidth and the temperature is high enough (process 1), by Schottky[8] emission from the metal (process 2), and by thermal excitation into the conduction band from "trapping" levels in the dielectric (process 3).

2. Tunneling processes. The tunneling can be from the metal into the conduction band (process 4), from the trapping levels in the dielectric (process 5), directly between the valence band and the conduction band (process 6), from the valence band directly into the metal electrode (process 7), or directly between the two metal electrodes (process 8).

3. Impurity conduction. This is something of a misnomer, since processes 3 and 5 can be caused by impurity. It is used to refer to electrons hopping from one trapping (e.g., impurity) center to another without going up into the conduction band. This can give an appreciable current flow if there are sufficient centers, and is illustrated as process 9 in Fig. 5.

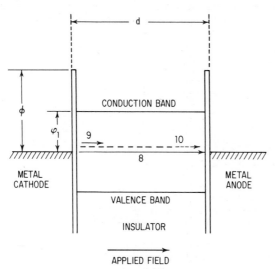

Fig. 5 Diagram of conduction processes in a thin dielectric film at low fields.

4. Space-charge effects. If injection into the conduction band or tunneling or impurity conduction is not the rate-limiting process, it is possible for space-charge buildup to be the major impedance. The charge buildup from the electrons in the conduction band or in traps will in the limit balance out the applied voltage.

(2) Ionic Conduction As well as the processes described above to allow electrons to go through the dielectric, it is also possible for ions, often in the form of impurities or defects, to themselves move through under the influence of an electric field. Under these circumstances, ionic conduction is said to occur. This is shown as process 10 in Fig. 4.

c. Low-field Conduction

At low fields direct tunneling between the electrodes can occur, provided the dielectric is thin enough. Such a method has recently been considered in detail by Schnupp[161,162] for film thicknesses less than 30 Å (10 lattice planes) and is shown in Fig. 5 as process 8. However, such thin continuous films are difficult to realize in practice, and one is generally concerned with processes that are applicable to films thicker than 500 Å.

To appreciate the relative importance of the various possible processes for thicker films at low fields, it is informative to consider a more realistic energy-band diagram. Figures 3, 4, and 5 were oversimplifications in that, in general, the conduction band of the insulator bends near the metal interface. The simplest case of "band bending" is illustrated in Fig. 6a. Here the band gap is small enough for the band bending to take place in a small region of film thickness. The Fermi level still bisects the forbidden band. The metal work function ϕ_m is usually smaller than that of the insulator ϕ_i. Hence the bands usually bend upward as shown.

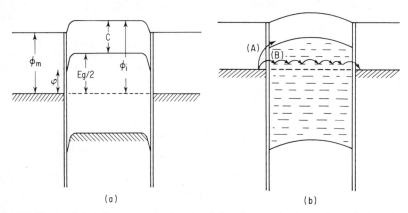

Fig. 6 Energy-band diagrams showing band bending at insulator-electrode interface. (a) Diagram showing symbols used in text. (b) Diagram showing impurity conduction levels.

With thin films with a large value of ϕ, band bending will occur over a distance that will be large compared with the film thickness. Under these conditions, the observed bending will be small in magnitude and the Fermi level will never be in the mid-band-gap position (see Fig. 6b). A detailed discussion of band bending is given by Simmons in Chap. 14.

It is worth noting at this stage that the terminology used to describe the situation at the interface is not consistent. The term "ohmic contact" is quoted in the literature to mean either (1) $\phi_m < \phi_i$, (2) $\phi_m = \phi_i$, or (3) by some means or other the electrode readily supplies electrons into the insulator. The opposite of an ohmic contact is called a blocking contact. Both terms are misleading for insulators as neither requirement 1 nor 2 nor 3 guarantees whether Ohm's law will (or will not) be obeyed. Further, it will shortly be seen that requirement 3 (a ready supply of electrons) can be met for one conduction mechanism but not another, *for a given band picture* (see also Chap. 14).

We can now generalize about low-field conduction mechanisms in the types of thicker film that are of interest in this work. The films are usually amorphous, so that localized "impurity" conduction levels are distributed all over the forbidden band.[27]

The materials of interest, such as SiO_2, Al_2O_3, and MgO, have large forbidden bandwidths, i.e., ≥ 8 eV,[53] and their conduction bandwidths are probably greater than 0.1 eV. By contrast, typical metals have $2 < \phi_m < 6$ eV.[6] Aluminum ($\phi_m = 4.2$ eV[53]) is usually used as a contact for reasons of adhesion and breakdown. Despite the incomplete band bending for such good insulators, we are usually concerned with the case $\phi_m < \phi_i$, and the bands will usually bend up at the interface as shown in Fig. 6b.

The barrier ϕ that an electron must jump (process A) in Fig. 6b to get into the conduction band is usually more than 1 eV (see Chap. 14). Since band-to-band conduction (process 1 in Fig. 4) is negligible, Schottky emission (process 2 in Fig. 4) is required to surmount this. However, Hill[165] has shown that the magnitude of conductivity for Schottky emission over barriers >1 eV is quite unrealistically small to explain low-field conduction.

The various tunneling processes described in Fig. 4 cannot occur at low fields. We have also noted that the one form of tunneling indicated for low field (Fig. 5) occurs only with films less than 30 Å thick. Therefore, only two processes remain; electronic-impurity conduction (process 9 in Fig. 5) and ionic conduction (process 10 in Fig. 5).

Actually, at low fields, the materials in question usually obey Ohm's law and can have a conductivity independent of thickness and the nature of the electrodes (i.e., bulk-limited with the surface having no effect). Thus any electronic-impurity conduction probably starts with a minimal, thermally assisted jump at the electrode interface. Further, since the film is disordered, the localized levels will be relatively random. This impurity conduction is therefore better drawn as process B in Fig. 6b.

We can now return to an earlier point made about so-called "ohmic" contacts, namely, that even if we do define them as contacts that supply a large number of electrons into the insulator, this is insufficient definition. One must define "ohmic" for a given process. Thus, in Fig. 6b, there is not an ohmic contact to process A (Schottky emission), whereas there is to process B (hopping).

The terms "ohmic" and "blocking" will not be further employed. We now consider in more detail the most probable conduction mechanisms in amorphous insulating films at low fields and ambient temperatures, namely, electronic-impurity conduction and ionic conduction.

(1) Impurity Conduction In impurity conduction, electrons will move from one trap to another without going up into the conduction band (process 9). A condition necessary for impurity conduction in crystals is the presence of both donors and acceptors.[24] The acceptors remove the electrons from some of the donors, and hence an electron can move from an occupied donor to an unoccupied one. An amorphous material will have structural traps as well, and there are so many of these that they dominate the conduction processes. Hopping will still occur between these traps, just as in normal impurity conduction. For ease in the future discussion, both types of site are said to provide impurity conduction. The hopping electrons have very low mobility, and thus any effect of them is likely to be masked if there are many electrons in the conduction band. However, since an insulator has a very low density of thermally generated free carriers in the conduction band, impurity conduction is more likely to be observable in an insulator than in a narrow-gap semiconductor.

At low impurity levels, electrons will either tunnel from site to site[8] or jump the potential barrier between sites.[8] These processes have not been distinguished from one another in Fig. 4. For silicon monoxide, impurity tunneling has been considered the most important process.[26] The "impurities" may well be small (10 Å diameter) islands of silicon in silicon dioxide (see the radial-distribution analyses of "SiO" discussed by Coleman and Thomas[25]), a concept which is in agreement with Johansen's model for SiO.[26] Johansen's conduction model is, however, slightly different in that although he considers there to be large numbers of Si islands in a sea of SiO_2, conduction is given by tunneling from the impurity level into the conduction band (process 5).

For high impurity concentrations the distance between sites is decreased and the impurity-band conduction that sets in may be considered in metallic terms (see Mott's work[1,27,32] on this transition).

(2) Ionic Conduction If the actual impurities or defects in the film migrate, ionic conduction results. The ions move by hopping over barriers of energy ϕ and separation l under the applied field E, and three types of behavior occur as E is increased:

1. For $E < 10^5$ V cm^{-1} $Eel \ll kT$ and ohmic conduction occurs of the form

$$J = \frac{C}{kT} \exp - \frac{\phi}{kT}$$

where C is a constant and J the current density.

2. For $Eel \simeq kT$ the barrier is appreciably distorted by the field so that

$$J = \bar{C} \exp - \left(\frac{\phi}{kT} - \frac{Eel}{2kT} \right)$$

3. For $Eel \gg kT$, the number of current carriers is increased by the field. Since multiple hops also occur in this region, the analysis is complex—see Bean,[33] Young,[3] and Vermilyea.[29]

In practice it may not be easy to distinguish between ionic and electronic currents, particularly if the activation energy is large. An activation energy of <0.1 eV and a high charge mobility are usually associated with electronic conduction, whereas a value >0.6 eV with a low mobility can be either ionic conduction or electronic conduction. However, there are several features that may help the distinction.

1. There will be polarization effects,[31,32,34] i.e., the resistivity of the film at constant dc voltage will increase with time. This can be due either to buildup of space charge at the electrodes and the equations will only apply at the *beginning* of measurement or to dielectric relaxation when the equations apply only after some time. In this latter case the initial current is the sum of the true ionic conductivity and an additional transient term.

The dielectric relaxation may occur because of time-dependent dipole formation or rearrangement of permanent dipoles already present in the material (see later) and can obscure ionic polarization near room temperature.[34] Vest[35] has given the theory necessary to determine the proportion of ionic conduction from current-time plots in the absence of dipolar effects.

2. Material transport will occur between the electrodes, and this can be detected chemically if ion transport occurs.

3. Large transit times will be involved so that, as mentioned, loss maxima may occur at low frequencies. These transit times can be observed by using an applied rectangular pulse and observing the delays in the current response. However, electron mobilities can occasionally be just as low as ionic mobilities.

4. To some extent, the magnitude of the ohmic ionic current can be assessed from the prolific data on bulk-diffusion coefficients D in various materials. In this case the Nernst-Einstein[36] equation applies,

$$\frac{\sigma}{D} \simeq \frac{Nz^2e^2}{kT}$$

where σ is the ohmic ionic conductivity, N the number of ions per cm^3, and z their charge in electron charges. Diffusion measurements on oxides up to October, 1967, have been tabulated and their interpretation discussed[37] in this context.

Lidiard[36] has discussed ionic conduction in alkali halides, and Young[3] and Vermilyea[29] have given theory for anodic films. Lamb[8] has also dwelt on the subject for thin films.

d. Temperature Dependence

It is difficult to generalize about the temperature dependence of dc conduction whether it is electronic or ionic since so many processes are possible. However tunneling (high-field) conduction usually exhibits a small temperature dependence. Ohmic (low-field) conduction, whether ionic or electronic, gives an exponential temperature dependence, given by

$$\sigma = \sigma_0 \exp \frac{-Q}{kT}$$

where σ_0 is a constant and Q is the activation energy of the process. Now

$$\sigma = Ne\mu$$

where N is the number of charge carriers, e their charge, and μ their mobility. With extrinsic ionic conduction, it is the mobility that is the activated process, Q being the energy for the ion to hop. With extrinsic electronic conduction, the electrons may move by hopping, i.e., "impurity" electronic conduction (see earlier), in which case $Q = Q_{hop}$. However, if the electronic conduction is by excitation into the conduction band, the production of "free" electrons n, not their mobility μ, is activated. In the latter case, simple band theory, as developed for semiconductors,[6,7,38] can apply.

Whatever the ohmic mechanism, a $\log \sigma$ vs. $1/T$ plot ("Arrhenius plot") will usually exhibit increasing linear slopes (activation energies) as T is raised, as shown in Fig. 7. At the highest temperatures the intrinsic ionic or electronic conduction may finally appear, although the melting point may hide them. In the very simplest case of band-theory intrinsic electronic conduction in semiconductors, the electron is excited directly across the forbidden band and[38] $Q \simeq \bar{E}/2$, where $\bar{E}$ is the forbidden band-width. Intrinsic ionic conduction requires not only the formation but the movement (hopping) of the intrinsic ions. Thus for a binary compound, $Q = Q_i/2 + Q_{hop}$ where Q_i is the energy to form a defect pair.[36]

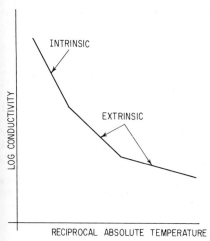

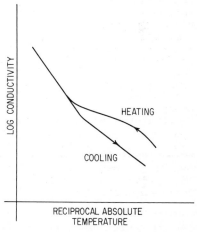

Fig. 7 Arrhenius plot of conductivity of a typical insulator for low fields.

Fig. 8 Diagrammatic illustration of the effects of moisture on the conductivity of ZrO_2 anodic films.

Most oxide films exhibit enhanced conductivity after exposure to moisture. This is easily distinguished by heating and cooling, as illustrated in Fig. 8.[39] With zirconia films, this mechanism is avoided by prior cycling to above 200°C or by measurement in vacuo.[39]

Ionic mobility in any solid is very low near 20°C in the absence of a field. For this reason, films can easily be in a "quenched" or nonequilibrium condition, several days or more being needed for thermodynamic equilibrium to be attained. This can cause hysteresis in conductivity/temperature plots due to transport, aggregation, or precipitation of lattice defects.

3. AC CONDUCTION MECHANISMS

a. Introduction

At any given ac frequency, the dc contributions discussed above will be present. These are seen as a value of R that is constant with frequency and a value of loss tan δ that drops proportionally with frequency, since tan $\delta = 1/\omega RC$ (see Fig. 9). However, ac conduction peaks due to relaxation processes can be superimposed on this; one such peak is shown schematically in Fig. 9.

Since these peaks in conduction generally move to higher frequencies as temperature increases, the dc conduction tends to dominate at high temperatures. This effect is accentuated by the fact that the dc conduction generally increases exponentially with temperature.

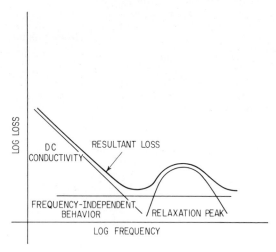

Fig. 9   Loss vs. frequency showing the effect of the various ac conduction processes.

Near room temperature, a second form of ac conduction is exhibited as shown in Fig. 9. This causes the loss to be roughly independent of frequency between about 100 Hz and 10 MHz (i.e., conductivity proportional to frequency); it is common to most dielectrics. These two types of ac conduction will now be analyzed in detail.

b. Relaxation Peaks

At room temperature, peaks in loss have been found with films of SiO[40] and many alkali halides[41,143] near 0.01 Hz. These are ascribed to interfacial polarization effects, i.e., the relaxation of charge buildup at interfaces. Because of their importance, interfacial effects will be dealt with in more detail later. Peaks in loss are rare between 100 Hz and 1 MHz. When found, these have often been ascribed to the relaxation of pairs of point defects, examples being fluorine in ZrO_2 films[42] and Si in SiO_2 films.[43] At higher frequencies, peaks caused by structural vibrations are found with polymer films.[44,45]

In all the above cases, the peaks move to higher frequencies as the temperature is raised, and their behavior is best described by the Debye equations.[46] These express the loss (above the background) and permittivity in terms of the permittivity at frequencies well above ϵ_∞ and well below ϵ_s, the peak. They are based on the assumption that, for a constant impressed voltage, the current decays exponentially, a feature common to most relaxation processes. The Debye equations may be expressed as follows:[46]

$$\epsilon = \epsilon' - j\epsilon'' = \epsilon_\infty + \frac{\epsilon_s - \epsilon_\infty}{1 - j\omega\tau}$$

where ϵ' and ϵ'' are the real and imaginary parts of the permittivity. Hence $\epsilon' = \epsilon_\infty + (\epsilon_s - \epsilon_\infty)/(1 + \omega^2\tau^2)$ and $\tan \delta = \epsilon''/\epsilon' = (\epsilon_s - \epsilon_\infty)\omega\tau/(\epsilon_s + \epsilon_\infty\omega^2\tau^2)$.

The quantity $\epsilon_s - \epsilon_\infty$ can be calculated on various assumptions. For instance, if the relaxing species is large with respect to the ions around it, the internal field approximates to the external field,[47] and for reasonably freely moving dipoles that do not interact such as small concentrations of point-defect pairs, one may write[42,46]

$$\epsilon_s - \epsilon_\infty = \frac{N\mu^2}{3\epsilon_0 kT} \quad \text{and} \quad \epsilon_s + \epsilon_\infty \cong 2\epsilon_\infty$$

where N is the number of dipoles per cubic meter and μ is their moment (length times charge). If the concentration of dipoles is large (say a few percent) then they can interact, and this can complicate the situation, e.g., structural dipoles in plastics[45,48,49] (side branches, etc). Further, if the substance in question is under strain or is inhomogeneous, the individual dipoles have somewhat different environments and relax with a distribution of times. This is often expressed by replacing $j\omega\tau$ by $(j\omega\tau_a)^{1-\beta}$ where β is a constant[50,52] and τ_a is the mean relaxation time, in the equation for permittivity, i.e.,

$$\epsilon = \epsilon' - j\epsilon'' = \epsilon_\infty + \frac{\epsilon_s - \epsilon_\infty}{1 - (j\omega\tau_a)^{1-\beta}}$$

A general observation is that the displacement of a "Debye" peak with frequency at constant temperature gives the activation energy Q of rotation of the "dipole" since[45]

$$\nu = \nu_0 \exp\frac{-Q}{kT}$$

where ν is the frequency of the peak at temperature T and ν_0 is a constant. Further, the peak height above the background conductivity (i.e., when $\omega\tau = 1$) is usually proportional[46,49] to the concentration of dipoles in the solid.

Other types of relaxations occur in bulk solids and could well be found in films. These include charge clouds around defects at high temperatures that may be described by the Debye-Hückel theory,[20] sharp peaks near absolute zero due to electron tunneling dipoles,[21-23] and relaxation around gross defects.[28] Owen[139] has discussed other distributions of relaxation times.

c. Interfacial Polarization

In principle, either ions or electrons can build up at one surface of a film if their rate of arrival exceeds their rate of discharge. If an ac field is applied, a loss peak will occur near that frequency corresponding to the time for the charge carrier to pass from one side of the film to the other. An early mathematical treatment of this phenomenon regarded the film as a two-layer dielectric. This is known as the Maxwell-Wagner theory.[10]

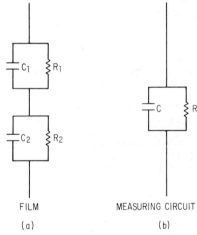

FILM   MEASURING CIRCUIT
(a)   (b)

Fig. 10 Equivalent circuits for Maxwell-Wagner losses.

The film is considered as two layers represented by the circuit of Fig. 10a and measured as the circuit of Fig. 10b. Following Anderson,[10] it emerges that, with the earlier nomenclature,

$$\epsilon' = \epsilon_\infty + \frac{\epsilon_s - \epsilon_\infty}{1 + \omega^2\tau^2}$$

and

$$\tan\delta = \frac{1}{\epsilon'}\left[\frac{1}{\omega C_0(R_1 + R_2)} + \frac{(\epsilon_s - \epsilon_\infty)\omega\tau}{1 + \omega^2\tau^2}\right]$$

where C_0 is a geometrical constant. Thus the Maxwell-Wagner model predicts a

Debye-like loss peak with an additional term in the equation for loss (cf. the appropriate equations in the preceding section).

Although the above theory is a very crude description of interfacial polarization, it is more successful where genuinely two-layer dielectrics exist. Two examples are films supporting adsorbed layers of moisture and porous alumina films.[151]

A more realistic theory of interfacial polarization has been given by Sutton,[84,85] who considers the buildup of charge, and the resultant field distortion, as a function of frequency and thickness. This treatment allows for charge recombination and dissociation and predicts a Debye relaxation between 0.001 and 0.1 Hz (20°C) as observed; i.e., there are no extra terms in the equation for loss.

Having established the validity of the Debye analysis, it is worth considering the common features observed experimentally. Most usually, the activation energy derived in this way is between 0.7 and 1.0 eV for both films[40,41,52] and bulk glasses.[172] For the latter, this is usually equal to the activation energy of dc conduction and is typically considered to be ionic in origin. Unfortunately, flaws usually obscure the uniform low-field dc conduction in films. The second common feature is that the Debye peak is usually much wider than that for a single relaxation time, and a constant β may again be introduced to allow for the distribution of relaxation times (c.f. the preceding section). A value of β for about 0.5 is usual.[40,52,139]

d. Frequency-independent Phenomena

As mentioned earlier, most dielectrics have a loss nearly invariant with frequency at room temperature. The relaxations described above, when present, are superimposed on this loss (see Fig. 9). Since the capacitance of most dielectrics is invariant with frequency, the conductivity usually rises proportionally to frequency.

Frequency-independent loss is found at all loss levels. For example, Morley[52] has found that the loss of SiO films is largely invariant with frequency even when the magnitude is above 1% and the losses are due to weak points in the film ("pinholes"). On the other hand, published data for some of the best insulating films (tan $\delta < 0.1\%$) also have loss relatively invariant with frequency. These materials have a loss that shows some correlation with forbidden bandwidths (see Fig. 11[53]), which shows that in these cases the loss is not just due to pinholes.

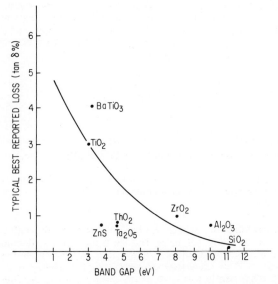

Fig. 11 Typical lowest reported loss vs. band gap for various films.

In this situation, one must closely investigate the factors dominating loss in a given film. However, mathematical models of a broad spread of relaxation frequencies have proved invaluable in calculating the relative invariance of loss with temperature near room temperature[51] and the magnitude of the temperature coefficient of capacitance.[53,54] Gevers[51] and Garton[55] first put forward these models. Two errors have crept into the literature. First, it has often been assumed that the Gevers/Garton models apply to films because they are usually amorphous. However, neither author claimed this restriction, as they were both all too well aware that bulk crystals also commonly exhibit loss invariant with frequency. Gevers and Garton chose to refer to "inhomogeneities" as the general cause of the effect.

Second, a model due to Young[3] by which the loss is due to an exponential variation of conductivity through the thickness of a film has been widely used to explain loss invariant with frequency. The fact that the slightest deviation from exponential caused the model to break down caused Young to retract his suggestion[3] and accept that the process is relaxational. Despite this, the model has been widely and unrealistically used, since it predicts no effects in bulk material and it requires a conductivity gradient through the film that is ridiculously large.

It is now appropriate to summarize the current understanding of this loss invariant with frequency near 20°C. The effect occurs at low fields, and Ohm's law is obeyed. It is present when flaws in films carry the major part of the current, and therefore for typical dielectric films with loss of 1% or more the effect is probably due to a distribution of relaxation processes in these flaws. The effect also occurs with bulk single crystals and with flaw-free amorphous films. Here, the most likely mechanisms at low fields are electronic impurity conduction and ionic conduction. Impurity conduction has been shown to give invariant loss at audio frequencies because of jumps occurring between a distribution of impurity separations[56] although these observations have been made only on silicon near absolute zero. Ionic conduction can also give near-invariant loss at low temperatures, although the theoretical analysis is vague at present.[3,36,139]

4. PERMITTIVITY

a. Introduction

The permittivity of a dielectric is little dependent on structure or loss. It is almost entirely an intrinsic property of the constituent ions and is accordingly more amenable to analysis than loss, which is usually extrinsic near room temperature.

In the absence of electrode effects (see Interfacial Polarization earlier) the permittivity of a dielectric has four components:

$$\epsilon = f(\epsilon_{ex}\epsilon_e\epsilon_n\epsilon_d)$$

where ϵ_{ex} is the extrinsic contribution, ϵ_e the contribution of electronic polarizability, ϵ_n the contribution of that ionic polarizability caused by vibration of the nuclei, and ϵ_d the contribution of the deformation of the ion.

b. Extrinsic Permittivity

The extrinsic permittivity is that part of the permittivity which does not arise from the constituent ions and electrons of the substance. Permittivity is rarely determined to better than 1% absolute. This is usually insufficient to reveal any extrinsic components, which are generally small.

c. Intrinsic Permittivity

(1) Electronic Contribution Where the constituent ions are small and relatively nondeformable, as for many polymers, the electronic term ϵ_e dominates. It also dominates when it is inflated because of a surfeit of conduction electrons, as for the group IV elements. These are plotted as permittivity vs. atomic number in Fig. 12. Atomic number conveniently serves to define position in the periodic table of elements; mean atomic number per molecule is employed for compounds for the same reason.   In the case of polymers and group IV elements,[9]

$$\epsilon = \epsilon_e = n^2$$

where n is the refractive index.   It is a general observation that ϵ_e is little altered from the ultraviolet to all longer wavelengths.

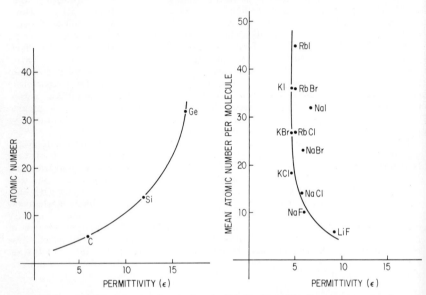

Fig. 12   Permittivity vs. atomic number for elements from group IV.

Fig. 13   Permittivity vs. mean atomic number per molecule for alkali halides.

(2) Nuclear Ionic Contribution   By contrast, the ionic terms ϵ_n and ϵ_d appear only at and beyond the infrared (the "restrahlung" frequencies).   In the simplest case, for nondeformable, fully ionic compounds, ϵ_n dominates.   This is true for the alkali halides as plotted in Fig. 13.   Born has shown[6] that, for this case,

$$\epsilon_n = \frac{e^2}{2\epsilon_0\omega_0{}^2 a^3}\left(\frac{1}{M} + \frac{1}{m}\right)$$

where M and m are the masses of the respective ions, a the lattice constant, and ω_0 the infrared absorption frequency.   More sophisticated analyses are also available.[5]

(3) Deformation Ionic Contribution   Finally, since the oxygen ion is large and very deformable, oxides demonstrate permittivities dominated by ϵ_d.   This has the opposite dependence on mass and atomic number, since these quantities are proportional to the number of electrons in the electron clouds of the ions.   The inertia of the cloud and hence its contribution to ϵ_d, increases with atomic number.   This is shown in Fig. 14.   Ionic volume considerations cause the scatter in this simple plot.   In fact, ϵ_d dominates for nearly all amorphous dielectrics having $\epsilon > 10$.

5. TEMPERATURE COEFFICIENT OF CAPACITANCE AND PERMITTIVITY

a. Introduction

The temperature coefficient of capacitance γ_c is an important practical figure for assessing the expected behavior of thin film circuits. γ_c is defined by the equation

$$\gamma_c = \frac{1}{C}\frac{dC}{dT}$$

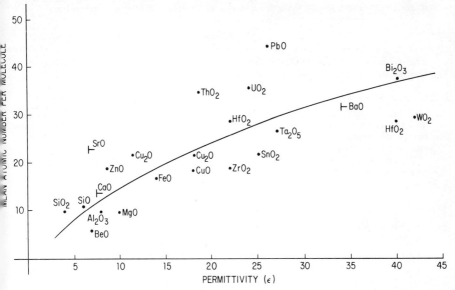

Fig. 14 Permittivity vs. mean atomic number per molecule for oxides.

This is related to the temperature coefficient of permittivity γ_p by the equation

$$\gamma_c = \gamma_p + \alpha \tag{2}$$

where α is the linear-expansion coefficient of the dielectric.

This equation has assumed that γ_p is given by the expression

$$\gamma_p = \frac{1}{\epsilon} \frac{d\epsilon}{dT}$$

However, it should be noted that, strictly speaking, γ_p is defined in partial terms at constant pressure as

$$\gamma_p = \frac{1}{\epsilon} \left(\frac{\partial \epsilon}{\partial T} \right)_P$$

The analysis of γ_p can be split up as with permittivity by examining the extrinsic and intrinsic components. In this case a considerable difference from the permittivity case appears, however, in that the extrinsic component can be all-important.

b. Extrinsic Behavior

We have seen earlier that ac conduction mechanisms rarely exhibit themselves as peaks but that a loss invariant with frequency near room temperature is the norm because of inhomogeneities, etc. Gevers[51] showed that if the invariant loss is described as an exceptionally broad peak, then its (extrinsic) contribution is

$$\gamma_p = A \tan \delta \tag{3}$$

According to Gevers,[51]

$$A = \frac{2}{\pi T} \ln \left(\frac{1}{\omega \tau_0} \right)$$

and Hersping,[60]

$$A = \frac{2}{\pi T} \ln \left(\frac{\tau}{\tau_0} \right)$$

in each case where the relaxation time of a given inhomogeneity is given by

$$\tau = \tau_0 \exp \frac{q}{kT}$$

where τ_0 is a constant and q is the given activation energy of the "dipole." However, Gevers and Hersping agree that for practical purposes $A = 0.05 \pm 0.01°C^{-1}$.

c. Intrinsic Behavior

It has already been noted that three intrinsic contributions to permittivity exist, and obviously all these change with temperature. However, all these contributions can be described by the Clausius-Mosotti equation,[9,10] and it is more convenient to examine them together.

The Clausius-Mosotti equation relates the intrinsic permittivity of a solid ϵ to the polarizability α_m of a macroscopic volume V of that solid,

$$\frac{\epsilon - 1}{\epsilon + 2} = \frac{\alpha_m}{3\epsilon_0 V} \tag{4}$$

By differentiating this equation,

$$\gamma_p = \frac{(\epsilon - 1)(\epsilon + 2)}{\epsilon} \left[\frac{1}{3\alpha_m} \left(\frac{\partial \alpha_m}{\partial T} \right)_P - \frac{1}{3V} \left(\frac{\partial V}{\partial T} \right)_P \right]$$

i.e.,

$$\gamma_p = \frac{(\epsilon - 1)(\epsilon + 2)}{\epsilon} \left[\frac{1}{3\alpha_m} \left(\frac{\partial \alpha_m}{\partial T} \right)_P - \alpha \right] \tag{5}$$

where α is the linear-expansion coefficient.

Bosman and Havinga[58,59] and others[54] have written

$$\frac{1}{3\alpha_m} \left(\frac{\partial \alpha_m}{\partial T} \right)_P = \frac{1}{3\alpha_m} \left(\frac{\partial \alpha_m}{\partial T} \right)_V + \frac{1}{3\alpha_m} \left(\frac{\partial V}{\partial T} \right)_P \left(\frac{\partial \alpha_m}{\partial V} \right)_T$$

thus expressing this quantity in terms of quantities that have been determined for the alkali halides. However, for the present purposes, only Eq. (5) is required.

d. Discussion

One can now write a general expression, including both extrinsic and intrinsic terms, for γ_p and hence, from Eq. (2), for γ_c.

From Eqs. (2), (3), and (5),

$$\gamma_c = \frac{(\epsilon - 1)(\epsilon + 2)}{\epsilon} \left[\frac{1}{3\alpha_m} \left(\frac{\partial \alpha_m}{\partial T} \right)_P - \alpha \right] + A \tan \delta + \alpha$$

i.e.,

$$\gamma_c = \frac{(\epsilon - 1)(\epsilon + 2)}{3\epsilon \alpha_m} \left(\frac{\delta \alpha_m}{\partial T} \right)_P - \frac{\alpha(\epsilon - 1)(\epsilon + 2)}{\epsilon} + A \tan \delta + \alpha \tag{6}$$

Equation (6) may now be simplified for various ranges of loss and permittivity.

(1) Extrinsic Behavior Extrinsic behavior is characterized by high loss (tan $\delta > 0.1\%$) so that

$$A \tan \delta > \frac{(\epsilon - 1)(\epsilon + 2)}{3\epsilon \alpha_m} \left(\frac{\delta \alpha_m}{\partial T} \right)_P$$

Further for most solids, $\epsilon > 2$, and $\alpha \epsilon \gg \alpha$. Therefore, Eq. (6) may be written

$$\gamma_c = A \tan \delta - \alpha \epsilon \tag{7}$$

Equation (7) can be referred to as Gevers' equation since it is a simplified form of the equation he first derived.[51] It gives a series of straight lines for given losses, an example of which is shown in Fig. 15.

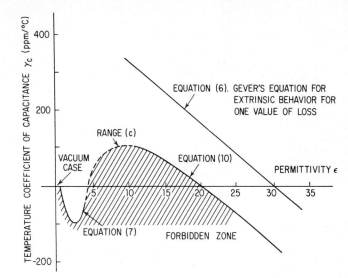

Fig. 15 General form of temperature coefficient of capacitance γ_c vs. permittivity relationship.

(2) Intrinsic Behavior For intrinsic behavior, loss is small, so that the term A tan δ in Eq. (6) can now be ignored. The rest of the equation may be simplified in four ways, dependent on the permittivity value.

1. $\epsilon = 1$. This is a vacuum, where $\gamma_c = 0$.
2. $1.5 < \epsilon < 2.5$. Here the permittivity is so low that the electronic term (see permittivity) dominates and $(\partial\alpha_m/\partial T)_P$ is very small.[59] Hence Eq. (6) becomes

$$\gamma_c = -\alpha \frac{(\epsilon - 1)(\epsilon + 2)}{\epsilon} + \alpha \qquad (8)$$

i.e.,
$$\gamma_c \cong -\alpha \qquad (9)$$

Since these materials are generally plastics with $+50$ ppm/°C $< \alpha < +300$ ppm/°C, γ_c will be negative (see Fig. 15). The more precise Eq. (8) is required if the permittivity is around 2.5.

3. $2.5 < \epsilon < 10$. This region will have a poorly defined value of γ_c where ionic and electronic polarizability are comparable (see Fig. 15, range c).

In the middle of this region, γ_c will pass from negative values through zero to positive values with increasing ϵ. Since the expansion coefficients of materials with permittivities in this region vary by a factor of 10, different materials may possibly have zero γ_c for different permittivities for this reason alone.

4. $\epsilon \geq 10$. In this region the ionic-deformation permittivity dominates. Because ϵ is large, $\alpha\epsilon \gg \epsilon$, and Eq. (6) becomes

$$\gamma_c = \frac{\epsilon}{3\alpha_m}\left(\frac{\partial\alpha_m}{\partial T}\right)_P - \alpha\epsilon \qquad (10)$$

Experimentally, for bulk materials, Hersping[60] and Cockbain and Harrop[54] have shown that the first term is a constant K for some materials.

Therefore,
$$\gamma_c \cong K - \alpha\epsilon \qquad (11)$$

and a straight line can be drawn (see Fig. 15).*

*Note added in proof. It has recently been shown[179] that a positive value of γ_c is found in high-permittivity glasses.

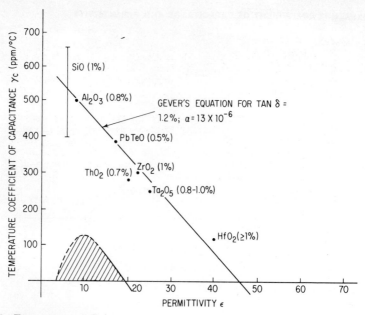

Fig. 16 Temperature coefficient of capacitance γ_c vs. permittivity for films with high losses (tan $\delta > 0.1\%$) (duplicate values for Cu_2O and HfO_2 illustrate the spread in experimental results obtained).

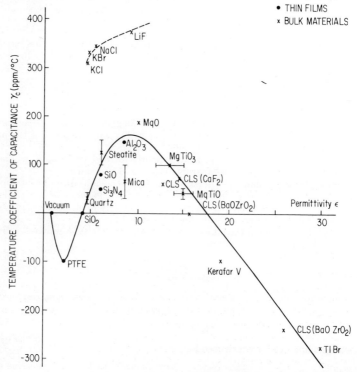

Fig. 17 Temperature coefficient of capacitance γ_c vs. permittivity for materials with low losses (tan $\delta \leq 0.1\%$). (CLS = calcium lanthanum silicate. For further details of bulk compositions, see Ref. 54.)

e. Experimental Data

Cockbain and Harrop[54] first summarized the situation with regard to bulk ceramics, and Harrop[53] has recently shown that thin films behave similarly. Figure 16 gives published data for films that verify Gevers' equation [Eq. (7)]. A reasonable mean for the experimental points is obtained for a value of α of 13 ppm/°C, which is of the order expected.

Figure 17 shows the behavior obtained from both films and bulk materials with a loss less than 0.1%. The general form of the curve is as predicted (Fig. 15) and appears to apply to a wide variety of compounds. The alkali halides are anomalous. This is to be expected in that their permittivities are dominated by the nuclear ionic contribution. Only bulk data are available[54,59] to illustrate this, since work on thin films of alkali halides has to date been confined to a study of extrinsic factors such as water adsorption.[163]

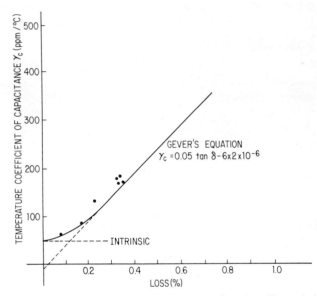

Fig. 18 Temperature coefficient of capacitance γ_c vs. loss for silicon nitride films (rf sputtered).

The minimum achievable value of γ_c for a given material can be obtained by plotting γ_c vs. loss. For high loss a straight line corresponding to Gevers' equation [Eq. (7)] is obtained, and this diverges to a constant (intrinsic) value for zero loss. This is illustrated in Fig. 18, where data are given for rf sputtered silicon nitride films. The value of A has been taken as 0.05, α as 2 ppm/°C,[164] and ϵ as 6.

6. BREAKDOWN FIELD

a. Introduction

Electrical breakdown in thin films is far simpler than in massive materials, because the geometry and structure of films preclude a number of mechanisms. This is reflected in the fact that the best insulating films have breakdown fields near 10 MV cm^{-1}, whereas the best massive insulators, with the exception of mica and a few plastics, break down near 1 MV cm^{-1} at 20°C. Even with the plastics, added mechanisms such as electromechanical deformation[61,64] occur, and the confusion for solids has been well documented by Cooper.[62] Fortunately for the present purposes, we have to confine our attention to films.

Even with films there are many reasons for reduced breakdown field E_B. Electrode material can move through flaws in films to short them out. Silver is notorious in this respect,[39,63] whereas aluminum causes little trouble, probably because it easily forms an insulating oxide. Impurity,[65] especially as a second phase, and inhomogeneity at electrodes[62] can lower E_B, by providing regions of enhanced electrical stress. Thin areas or polycrystalline boundaries can pass high currents, leading to local melting or thermomechanical failure.

As discussed by Klein and Gafni,[67] three types of breakdown are common with carefully prepared films. Firstly, if a low-impedance supply is used, the breakdown points can travel randomly across the electrodes. This "self-propagating" breakdown can give a characteristic "spider" pattern on the electrode and is illustrated in Fig. 19a. It is extrinsic in that it occurs at relatively low fields and is dependent on electrode properties.

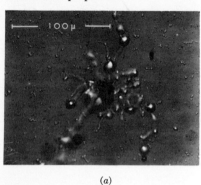
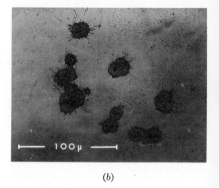

(a)   (b)

Fig. 19 Micrographs of breakdown effects (PbTiO₃ films; Al electrodes). (a) Self-propagating. (b) Local breakdown.

With a high-impedance supply, two other types of breakdown occur. The dielectric can short at "weak" points,[66] determined by dust, etc., on the surface of the substrate (i.e., insufficient cleaning). These regions can also determine the time constant of the final capacitor. Klein and Levanon[77] found that the maximum dc voltage could be determined by thermal effects produced by enhanced current flow through these "weak" points, and Fig. 19b illustrates this form of breakdown. Melting of the electrodes has occurred, and a characteristic splashing appearance is obtained.

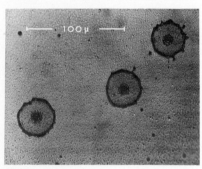

Fig. 20 Micrograph showing isolation of flaws by local melting (PbTiO₃ film; Al electrodes).

If the electrodes are thin enough (e.g., <1,500 Å Al), however, the weak points may be isolated by local melting of the electrodes on application of 1 MV cm⁻¹ or so. This is illustrated in Fig. 20. Even higher fields may then be achieved before an overall "intrinsic" breakdown occurs.

In understanding the "intrinsic" breakdown of films, it is instructive to consider the voltage-current plot leading to such an event. Generally, this divides into three regions, as in Fig. 21.

As discussed earlier, an initial ohmic region A (below 0.1 MV cm⁻¹) is followed by a nonohmic region B (most often log $I \propto V^{\frac{1}{2}}$ ascribed to Schottky emission—see earlier). Finally a "breakaway" region C may precede breakdown.

The maximum achievable breakdown of different materials varies with thickness. Typical trends are shown in Fig. 22, for bulk inorganics[68] and plastics,[68] glasses,[69] and films.[70,71,66]

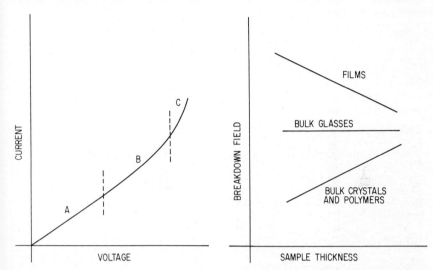

Fig. 21 Typical current-voltage curve for a thin film insulator.

Fig. 22 Typical trends of breakdown field with thickness.

Both "weak-point" and intrinsic breakdown have usually been explained in terms of electron flow. However, ionic flow can be appreciable. This is illustrated in certain cases by comparing the reciprocal of highest reported breakdown for films in Å V^{-1} with the growth-voltage ratio for anodization. Table 1 shows the correlation.

TABLE 1 Comparison of Maximum Breakdown and Growth-Voltage Ratio for Various Thin Films

Material	Reciprocal breakdown, Å V^{-1}	Growth-voltage ratio, Å V^{-1}
SiO_2	5	5
Al_2O_3	16	14
ZrO_2	25	22
Ta_2O_5	17	16

Since the growth-voltage ratio for anodization represents field-assisted ionic transport, it would appear from Table 1 that breakdown itself can be governed by the same process.

One can now list some mechanisms.
1. Electron avalanching by impact ionization of the lattice
2. Collective breakdown by electron-electron interaction
3. Breakdown from ionic transport
4. Joule-heating breakaway

Mechanisms 2 and 3 seem unlikely in good amorphous insulators. The theory of Fröhlich and Paranjape[72] takes collective breakdown as occurring when, by electron-electron collision, the free-electron temperature increases without limit. This requires

$n > 10^{14}$ free electrons per cm^3 in simple inorganic compounds.[62] Since $\sigma = ne\mu$, where mobilities may be 10^{-5} cm^2 V^{-1} s^{-1} or less,[73,17] and conductivity $\leq 10^{-12}$ ohm^{-1} cm^{-1}, n is probably too low in practice. Mechanism 3 requires some destruction of electrodes by the buildup of ions, and this occurs only after long periods.[62] Mechanisms 1 and 4 remain.

b. Impact Ionization

There are many theories to explain how a dielectric may break down by ionization of the lattice ions by itinerant electrons. Tunneling between the valence and conduction bands, as suggested by Zener,[74] is restricted to the very thinnest films (say 20 Å).[62]

Seitz[75] has considered single "streamer" breakdown as in gases, whence

$$E_B = C \ln \left(\frac{d}{E_B \mu \tau} \right) \tag{12}$$

where C is a constant, d the thickness, and μ and τ the mobility and relaxation time of the electron, respectively. This theory may be more appropriate for the single-point breakdown that occurs when electrodes are thick.

Forlani and Minnaja[76] considered the growth of the electron current by collision ionization when the initiating electrons are injected by field emission at the cathode. They find

$$E_B = A d^{-\beta} \tag{13}$$

for a given solid, where $\frac{1}{2} \leq \beta \leq \frac{1}{4}$ and A is a constant.

c. Joule-heating Breakaway

Klein[67,77] and his school have dealt with this case. They do not specify the conduction mechanism prior to breakdown but postulate that the attendant Joule heating increases exponentially with temperature. The heat lost by transport is taken to decrease linearly with T from a temperature T_0, and at breakdown, the heat lost vs. T curve is taken to be tangential to the Joule-heat-supply curve. Thus if

$$\sigma = \sigma_0 \exp\left[a(T - T_0) + bE\right]$$

where σ_0, a, and b are constants

$$E_B = \frac{1}{b} \ln \frac{\Gamma}{ae\, d\, \sigma_0 A\, (E_B)^2} \tag{14}$$

where d is the film thickness, $e = \exp(1)$, and Γ is the thermal conductivity of the film.

d. Experimental

Equations (12), (13), and (14) can each predict the small decrease in E_B with thickness that is found with thin films. Kawamura and Azuma[78] found anodic alumina to obey Eq. (12), whereas Lomer[70] and Merrill and West[71] found that similar films obeyed Eq. (13) with $\beta = 0.3$. Weaver and Macleod[63] found evaporated NaCl, NaF, and cryolite to support Eq. (12), but these films were polycrystalline with crystallite boundaries governing the breakdown. Kawamura and Ryu[79] found $E_B = \text{const } d^{-0.65}$ with thin mica flakes. Budenstein and Hayes[66] found SiO films to obey $E_B = \text{const } d^{-0.5}$ for local breakdown.

The most careful study of "intrinsic" rather than imperfection breakdown has probably been that of Klein.[67,77] He used thin electrodes, and isolated flaws by local melting. He then studied the intrinsic breakdown using a high-impedance voltage source. He found that SiO obeyed Eq. (14) from 4 to 415°K. SiO$_2$ films also behaved as expected from the thermal-breakdown model. These had higher E_B because of their lower conductivity.

(1) AC Effects Very little study has been made of the ac breakdown of thin films. Klein and Levanon[77] found that the ac dielectric strength of SiO films was higher than the dc strength, probably because heating occurred only at the top of a cycle. The general picture with bulk solids is that the shortest pulses give the highest breakdown voltages.[69]

7. PRACTICAL APPLICATIONS

a. Insulators

There are many occasions when a straightforward insulation of one conductor or semiconductor from another is required. In the layout of thin circuits, whether on glass[80] or on silicon, [81] crossovers are needed. The quality of the capacity does not generally have to be of the highest order (tan δ of 1% is quite acceptable), but the design of the junction must ensure that the cross capacity is as small as possible. Other insulation applications occur in storage devices, of both the cryogenic[82] and planar-magnetic-film type,[83] although not in the use of plated wires.[84,85] In the case of the cryogenic stores the problems are different from those normally met with in thin film devices, inasmuch as the films have to stand temperature cycling from room temperature to liquid-helium temperature. This imposes severe mechanical constraint on the films, and the insulator must be one that is thermally matched in expansion coefficient to that of the substrate. It must also be capable of conducting heat from the storage area.

Further examples of the use of dielectric insulators occur in the field of thin film active devices. Both the metal oxide–semiconductor transistor (MOST)[86] (see Chap. 23) and the field-effect transistor[87,88] (see Chap. 20) have an insulating layer separating the metal gate from the semiconductor channel that is carrying the current. Leakage of current through to the gate must be avoided, and the lower the leakage the better the device. In these contexts, therefore, a good-quality dielectric is wanted (tan $\delta \approx 0.1\%$).

b. Capacitors

The second main class of dielectric use is in capacitors. In semiconductor integrated circuits the capacitors may be made from the thermally grown oxide on the semiconductor,[89] in thin film planar circuits generally from evaporated[90] or sputtered layers[91] of various materials. In both cases the need usually is to get a high capacity in a small space, and thus a high permittivity or thin film is required. However, the majority of the requirements do not need the highest-quality films; tan δ of 1% is quite sufficient. The aging and temperature-capacity change is important, however, and methods of deposition that give films with either or both of these characteristics with low value are to be preferred (e.g., a temperature coefficient of capacity of +12 ppm/°C is claimed by Smith and Kennedy for sputtered silica films[92]).

There are some applications for capacitors, however, where the highest-quality films are needed. Active filter networks have now been successfully constructed using thin film planar capacitors (together with integrated-circuit amplifiers),[93] and for this type of application, tan δ figures must be 0.1% or less. Furthermore the aging behavior of both tan δ and capacity must be accurately known, since the performance of filters using such components must not change with time. Figures of total aging of $< 0.25\%$ for capacity and $< \pm 0.002\%$ for tan δ are required.

c. Conductors

It may seem at first sight paradoxical to include the use of dielectric films in such a classification. However, the conduction that can occur through dielectrics has been used in various device designs—the so-called hot-electron diodes and triodes proposed by Mead[94] come in this class (see Chap. 14). Other applications have been suggested in terms of solid-state light sources[95] where conduction of electrons through an insulator into a metal is required. Also heterojunction diodes are improved by using an insulator between the metal and semi-insulator (see Chap. 20).

A further application is that of controlling the conduction behavior of very thin metal films. Such films consist of isolated metal islands which are normally separated by gaps of the order of 100 Å. It has now been shown[96,97] that, provided the gap size is greater than 20 Å, electron flow between the islands will be via the substrate on which the islands rest. Normally this substrate is a glass or ceramic material, but it can equally well be a layer of some thin film dielectric. Variation of the gap size due to applied mechanical strain of the substrate–island structure–film combination will cause large variations in the resistance of the film. Parker and Krinsky[160] have investigated this for the preparation of strain gauges and have been able to obtain gauge factors (change of resistance per unit change in length) as high as 200 using island-structure gold on glass. Conduction will also be through the gap between the islands if the metal islands are dispersed in some dielectric medium, and resistors of this type of structure (cermets of chromium and silicon monoxide, for example[98]) have been studied over the last few years. Degenerate insulators can also be employed in resistors.[99]

In all the applications considered in this section, the essential feature is conduction through the dielectric medium, but obviously this must be a controlled process if any device potential is to be realized.

8. REQUIREMENTS

Now that the physics of dielectric films has been discussed and uses for such films have been outlined, it is possible to consider the specification of the dielectric relative to its uses. Two of the main uses, insulators and capacitors, ask for reasonable-quality dielectrics with loss around 1% (tan δ). However, there is a growing demand for very high quality dielectrics (tan $\delta \leq 0.1\%$, $\tau > 1,000$ s), for low-loss capacitors, gates in field-effect devices, etc. Table 2 has been drawn up to show how the ac and dc conductivity can be reduced, and from this table a list of electrical requirements can be evolved. For nearly all mechanisms of conduction, smooth, uniform, dense, tenacious, and cohesive layers are best. The table points to requirements additional to these. All mechanisms bar tunneling are also reduced by lowering the temperature, but this is rarely practicable.

TABLE 2 Summary of Methods of Reducing Conduction in Dielectric Films
(The mechanisms have the nomenclature used earlier)

Conduction mechanism	Methods of reducing conduction
Thermal excitation of electrons directly into conduction band	Use dielectrics with a large forbidden band
Schottky emission	Use thick films, low voltages, and dielectrics with a large forbidden band
Thermal excitation from donor/acceptor levels	Reduce impurity concentration
Tunneling	Use single-phase films thicker than 500 Å
Electronic "impurity" conduction	Reduce impurity concentration
DC ionic conduction	Eliminate alkali metals and nonstoichiometry and use ion-trapping structures (e.g., Si_3N_4) that are not moisture-sensitive. Allow ionic-charge buildup
AC relaxation peaks	Do not allow ionic charge to buildup. Reduce impurity and nonstoichiometry
Frequency-independent phenomena	Use thick dielectrics with large forbidden bandwidth, low impurity content
Localized breakdown	Use thin electrodes to isolate flaws and thick dielectrics
Intrinsic breakdown	Use thin dielectrics having low dc conduction and large forbidden bandwidth

Other items have to be examined, however, before one has a complete specification. A mechanically stable film is needed. It has been shown that thin films are deposited in a state of mechanical stress[100,101] which may be either compressive (causing the film to buckle) or tensile (causing the film to crack) (see Chap. 12 in this book). Tensile stresses must obviously be avoided, as such films will rapidly develop electrical shorts, but at present it does not appear that one single mechanism can be used to explain the stress behavior,[102,103] and therefore it is not a simple thing to specify the necessary parameters to give a small compressive stress. Carpenter and Campbell[166] have shown that tensile stress markedly increases the loss of SiO capacitors. Figure 23 illustrates the relationship obtained between stress and loss for aluminum–silicon monoxide sandwiches.

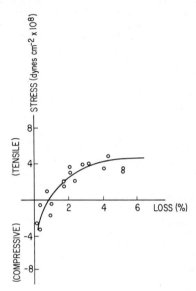

Fig. 23 Stress vs. loss for SiO films (Al electrodes; dielectric thickness between 2,000 and 4,500 Å).

Even if stress is reduced to manageable proportions, the film needs to be stable in other ways. Long-term chemical change of either electrodes or dielectric must be avoided, and the method of preparation used must give long-term electrical stability—i.e., no weak points that can become conducting regions (e.g., pinholes). As already mentioned, pinholes can be avoided by using thin ($<$1,500 Å) electrodes that can be "burnt out" by the application of a high voltage to the film. However, not all structures can allow of this possibility, and in any case pinholes are best avoided in the first place.

With these ideas in mind, a compromise specification for a satisfactory dielectric can be drawn up. The film must have associated with it:

1. Good mechanical stability
2. Good chemical stability
3. A reasonable thickness if possible ($>$1,000 Å)
4. A minimum number of defects and impurities
5. A lack of sensitivity to moisture
6. No pinholes or weak points that have to be "burnt out"
7. Small reaction to physical change—i.e., low aging and temperature behavior
8. Large forbidden bandwidth
9. Amorphous structure

In the case of dielectrics required for conduction purposes the situation is a little different. A small-forbidden-bandwidth crystalline dielectric with deliberately added impurity may be required to give enough conduction. However, reproducibility of the conduction characteristics is required, and the other items in the list will still apply.

In general it would appear that high-permittivity films should be avoided because of the polarizability and hence high loss. Another factor that must be recognized in the use of such films is that appreciable field penetration of the metal electrodes may occur, thereby effectively reducing the apparent permittivity.[104,105]

9. SELECTION OF COMPOUNDS

Harrop[53] has dealt with the selection of dielectrics for thin film capacitors from trends of loss, permittivity, etc., with position in the periodic table of elements. This treatment will now be summarized and extended.

a. Insulators

(1) Crossovers   Crossovers should have as low a capacitance as possible, and since the plastics have the lowest permittivity, these are desirable. SiO_2 can also be used as it is one of the inorganic materials with a low permittivity ($\epsilon = 3.8$). The films should be as thick as possible to reduce the stray capacitance.

(2) Field-effect Devices   Field-effect devices, particularly on silicon, are highly sensitive to the density of charge at the interface between the insulating gate and the semiconductor. Since the best insulators are amorphous and "glasses" are notorious for the ease with which alkali-metal ions pass through them,[106] such impurities are a major embarrassment. Thus, although silica is an obvious choice for gates, it is desirable to coat the device with a substance such as Si_3N_4, that is less easily penetrated by alkali metals. Indeed Si_3N_4, Al_2O_3, or some other light (and hence very insulating) compound may replace SiO_2 as the gate itself if it can be deposited to give as low an interfacial charge density (a formidable technological problem) and can then be shown to resist ingress of alkali metals.

b. Capacitors

(1) Low-loss Capacitors   The important criteria for such capacitors are:
1. Low loss ($<0.1\%$ tan δ).
2. Low γ_c ($< \pm 20$ ppm/°C).
3. As high a capacity per unit area as possible—this implies either a high permittivity or thin film.
4. A working voltage of ≥ 10 V. Taking the criterion that the working voltage is approximately half the breakdown voltage, this implies a high breakdown field, greater than 2 MV cm^{-1} for a film 1,000 Å thick.
5. Time constant $>1,000$ s.
Low loss, high breakdown field, and large time constant imply thick films. On the other hand a large value of capacity per unit area implies a thin film or a high permittivity. It has been shown that a low value of γ_c and a high permittivity are incompatible, so that one is generally considering films that are as thin as possible. For a practical purpose the minimum thickness that can be used is around 2,000 Å. A permittivity of 4 (for zero γ_c) then gives a capacitance of 0.02 μF cm^{-2}.
Materials that have been considered[53] in this context are AlN,[107] Al_2O_3,[108] BN, MgO, Si_3N_4, SiO, SiO_2, and ThO_2.[109,110]   Plastics have also been examined. Only BN and SiO_2 have the correct value of permittivity to give $\gamma_c \sim 0$. The other inorganics will give a positive value of γ_c and the plastics, with low values of permittivity ($\epsilon \sim 2$), a negative value.

(2) High-capacitance Capacitors   The criteria that can be written in this case are:
1. Moderate loss ($<1\%$ tan δ)
2. Moderate γ_c (<500 ppm/°C)
3. As high a capacity per unit area as possible
4. Working voltage > 10 V
5. Time constant > 100 s
Since requirement 3 now has the most emphasis, the highest permittivity consistent with high breakdown strength is required. BaO,[112] Bi_2O_3,[3] HfO_2,[111] Nb_2O_5,[3] Ta_2O_5,[117] and WO_3[3] have been considered[53] in this context. Complex compounds have also

been considered (e.g., $BaTiO_3$,[113,114] $CaTiO_3$,[53] $PbTiO_3$,[116] $SrTiO_3$[115]). In practice, ease of deposition has led to the use of Ta_2O_5 and $PbTiO_3$.

c. Conductors

Resistors in which conduction in an insulator can control the impedance include discrete metal islands on insulating substrates, "cermets" (two-phase metal-insulator films), and degenerate dielectrics. Where the substrate is electrically important, glasses are obviously desirable because of the reproducibility of their electrical properties. Although pure SiO_2 may be ideal, alumino-borosilicates such as Corning 7059 serve well in this respect.

For cermets, the choice is greater. Since most insulators can be coevaporated with a metal, the higher-permittivity materials may be useful because they may exhibit the required tunneling conduction over longer distances. Thus the high-C capacitor dielectrics are worthy of study. To date, a variety of insulators have been employed, such as SiO,[98,119] amorphous Ge,[119] and Al_2O_3.[119]

Resistors and switches employing degenerate conduction in an insulator can be crystalline. Such conduction is easier to achieve with less good insulators, e.g., cations from periods 3 to 7 such as VO_2[118] and $CdSe$.[99]

10. PREPARATION

It is not possible in the space available to go into all the thin film dielectrics that have been prepared by various methods, and in any case the various techniques have been considered in detail in other chapters of this book.

However, a brief survey will be given of the various methods with the intention of highlighting their basic shortcomings and advantages in achieving the specification of an ideal dielectric film.

a. Anodization

The oxide layer produced by anodization is characterized by a mainly amorphous structure and a high breakdown strength. In spite of numerous studies of alternative metals[3,120] only anodized aluminum or tantalum are at present employed commercially.

A semiconducting oxide[121] is often incorporated into the structure in the form of a second layer deposited onto the dielectric, and this improves the frequency and temperature characteristics of the films, compared with those in which a liquid electrolyte is used.

Thin films of sputtered tantalum because of their relatively high purity and surface smoothness compared with bulk material permit the formation of dielectric layers of such improved quality that evaporated metal electrodes may be applied directly without the need of an intervening layer either by electrolyte or semiconducting oxide.[122,123] This is probably the easiest way of obtaining films with high time constants.

b. Thermal Growth

Few satisfactory oxide dielectrics can be prepared by thermal oxidation of the metal in an appropriate atmosphere. Metals like tantalum or aluminum form only very thin amorphous oxides, 50 to 100 Å thick, at temperatures up to about 500°C. At higher temperatures mixed amorphous crystalline oxides are formed possessing inferior dielectric properties. However, satisfactory dielectrics possessing high breakdown strengths have been prepared at temperatures in the region of 900 to 1200°C on silicon, and here a major advantage is the reduction of interfacial states as a consequence of the intimate chemical bonding during growth.

An improved technique has been described that involves a low-pressure oxygen plasma excited by microwave radiation.[126] Quite thick films can be obtained up to 0.01 μF cm^{-2} and with a breakdown of 10 MV cm^{-1}.

c. Vapor-phase Deposition

Catalytic deposition of dielectric layers from the vapor phase onto solid surfaces[169] often calls for external stimulation of the gas phase (e.g., by an rf glow discharge[127]), but the reaction may also occur at hot or cold surfaces in the absence of excitation.[128] A remarkable range of binary inorganic compounds[129] can be made in this way, and polymerization of plastics is also possible.[130,131,169]

Silica may be deposited without excitation from tetraethyl silicate at a pressure of approximately 1 Torr on surfaces heated to at least 650°C. Very good thin films are obtained with high breakdown strengths and low dissipation factors. Similarly, alumina has been prepared by passing aluminum chloride together with hydrogen and carbon dioxide over either molybdenum or nickel strips heated to 1000°C. The deposited films appear to consist of crystals of alpha-alumina deposited on an amorphous phase and because of this do not have very high breakdown strengths. Uniformity and adhesion can be problems with vapor-phase techniques.

d. Evaporation

Silicon monoxide is the most important of the vacuum-evaporated layers at the present time; loss $\sim$0.1% is typical, provided partial pressure, rate of deposition, temperature of substrate, and treatment after the deposition are controlled.[132,133] The actual composition of the dielectric layer is uncertain, and as has been discussed this has led to speculation as to the method of electron conduction through such films (see the work of Johansen,[26] Siddall and Giddens,[134] and Hartman et al.[135]).

Evaporation of silicon dioxide itself is difficult owing to its high melting point. Evaporation can be performed by electron bombardment,[136] and in this case the dielectric layers are close to silicon dioxide in composition. However, the dissipation factor is rather higher than that of evaporated monoxide.

Various attempts have been made to evaporate high-permittivity dielectrics, especially barium titanate. Direct evaporation leads to the partial separation of the oxides, but this may be avoided by flash evaporation of small grains of bulk material.[137] An alternative approach is to coevaporate barium and titanium oxides using electron-beam heating.[138] Both methods give high-permittivity films (500 to 1,300) but high dissipation factors (15% at 1 kHz). Low breakdown strengths are also found (0.2 MV cm^{-1}).

e. Reactive Sputtering

When sputtering is accomplished in an atmosphere of oxygen or an argon-oxygen mixture, metal oxides are deposited. The method has distinct advantages over evaporation, as the high temperatures often associated with evaporation are avoided and materials can therefore be deposited as amorphous films despite the fact that they have extremely high melting points.

Most of the simple oxide dielectrics (e.g., tantalum,[140] zirconium,[141] and silicon[142]) have been prepared by reactive sputtering, but the processes are generally slow. Multicomponent cathodes can be used to give mixed oxide films,[141,116] and such films can be free of pinholes and have breakdown strengths of 2 to 3 MV cm^{-1}.

f. RF Sputtering

It has already been observed that an rf glow discharge can be used to catalyze chemical reactions. A second use of radio frequencies is to maintain a glow discharge at lower pressures than is possible with normal diode sputtering. Reactive sputtering is then possible at 10^{-3} Torr.[168]

Quite distinct from these applications is diode sputtering with the radio frequency directly applied to the electrodes. If one electrode is earthed, insulating targets will be sputtered from the other.[142] This usually gives amorphous films, and much higher rates of deposition are possible than with reactive sputtering. Complex compounds can be sputtered in this way. At the higher rates, however, the film may become anion-deficient. This is avoided to a large extent if oxides are sputtered

TABLE 3 Typical Properties of Thin Film Dielectrics Prepared by a Variety of Techniques

Compound	Method	Capacitance, μF cm^{-2}	Loss, % at 1 kHz	Breakdown, MV cm^{-1}	Permittivity	Reference
Ta_2O_5	Wet anodized	0.1	1	6	25	123
Ta_2O_5	Gaseously anodized	0.0016	1.6	1.6	...	155
Ta_2O_5	Reactively sputtered	0.1	0.9	1	...	117
Ta_2O_5	Reactively evaporated	0.08	0.5	...	16	149
Ta_2O_5	Bulk material	...	0.05	...	25	173
Al_2O_3	Wet anodized	0.02–0.4	0.5	6–8	6	156
Al_2O_3	Gaseously anodized	...	0.5	...	8.5	157
Al_2O_3	RF sputtered	0.01	0.4	1	8	146
Al_2O_3	Bulk material	...	0.02–1.0	0.016–0.064	8.4	170
ZrO_2	Wet anodized	0.1	1	4	22	146
ZrO_2	Reactively sputtered	0.07	0.8	4	20	116
ZrO_2	Bulk material	...	1	...	22	146, 171
TiO_2	Wet anodized	0.3	3	1	30–40	159
TiO_2	Reactively sputtered	2.15	6.5	...	...	117
TiO_2	Bulk material	...	0.02–0.5	0.040–0.084	14–110	170
SiO_2	Thermally grown	0.1	0.1	10	4	89, 124
SiO_2	Vapor phase	...	...	3	4	144
SiO_2	Vapor-glow discharge	0.03	0.01	...	5.9	127
SiO_2	Reactively sputtered	0.005	0.1	1.5	3.8	145
SiO_2	RF sputtered	0.02	0.02	6–10	4	146
SiO_2	Bulk material	...	...	6.5	3.75–4.1	170, 175
$BaTiO_3$	Coevaporated	1.6	18	...	130	138
$BaTiO_3$	Flash evaporated	0.7	15	0.35	420	147
$BaTiO_3$	Bulk material	...	0.01–2.0	0.02–0.12	...	170
$MnTiO_3$	Reactively sputtered	...	3–17	...	20–120	176
$PbTiO_3$	Reactively sputtered	0.1	0.6	2.3	20	116
$PbTiO_3$	Coevaporated	...	~5	...	~100	158
$PbTiO_3$	Bulk material	...	0.01–2.0	0.02–0.12	...	170
$PbTeO_3$	Reactively sputtered	0.1	0.4	...	17	148
ZnS	Evaporated	0.03	1	1	10	150
MgF_2	Evaporated	...	1	2	5.5	144
SiO	Evaporated	0.01	0.1	2	6	52, 92

TABLE 3 Typical Properties of Thin Film Dielectrics Prepared by a Variety of Techniques (Continued)

Compound	Method	Capacitance, $\mu F\ cm^{-2}$	Loss, % at 1 kHz	Breakdown, $MV\ cm^{-1}$	Permittivity	Reference
Si₃N₄	Vapor phase rf	...	0.1	...	6	129
Si₃N₄	RF sputtering	0.02	0.1	1	6.8	146, 167
Si₃N₄	Bulk material	...	0.4	...	7.9	164
Bi₂O₃	RF sputtering	0.32	1.1	...	25	146
Bi₂O₃	Reactive sputtering	0.1	1.0	0.7	38	146
ThO₂	RF sputtering	0.14	0.7	1	30	146
ThO₂	Bulk material	...	<0.1	...	18.6	174
BN	RF sputtering	0.03	0.5	...	...	146
BN	Bulk material	...	0.11	0.32	4.1	172
SnO₂	RF sputtering	0.22	14	...	20	2
Polystyrene	Vapor phase rf	0.002	0.1	...	2.6	152
Polymers	Vapor phase uv	...	1.2	1	2.5–3.5	153, 125
Polymers	Evaporation	0.02	0.1	6	2.1	154
Polymers	Bulk material	...	~0.1–0.002	...	~2	44

Bulk materials included for comparison.

in oxygen-containing environments and nitrides in nitrogen. A major advantage of all types of sputtering is the good adhesion of the films produced.

g. General

It is not possible to specify a single recipe for making very high quality films; all methods are capable of making reasonable dielectric films. The choice of deposition method must mainly rest on the compatibility of the method with the other techniques required to make the device under consideration.

Table 3 gives some guide to the properties that have typically been achieved in the past. The practical quantity C (μF cm^{-2}) is given in addition to permittivity. These are the values quoted in the appropriate reference. If required the thickness used can be calculated from C and ϵ. Work continues rapidly in this field, so that the list cannot give up-to-date information on the best properties obtainable. However, it is hoped that it will act as a guide to the type of material that has been studied. Evaporated aluminum electrodes giving localized breakdown were commonly used. Where permittivities exceed 100, it can be taken that the films are crystalline.

ACKNOWLEDGMENTS

The authors would like to thank numerous colleagues for their help, especially Dr. A. R. Morley for the use of his experimental results obtained at the Electrical Engineering Department, Imperial College, London, as well as for his critical reading of the text. We would also like to thank the Plessey Co. for permission to publish.

REFERENCES

1. Mott, N. F., *Advan. Phys.*, **16**, 49 (1967).
2. Waterworth, P., and P. J. Harrop, unpublished work.
3. Young, L., "Anodic Oxide Films," Academic Press Inc., New York, 1961.
4. Evans, U. R., "An Introduction to Metallic Corrosion," Edward Arnold (Publishers) Ltd., London, 1958; and more advanced, "Corrosion and Oxidation of Metals," Edward Arnold (Publishers) Ltd., London, 1960.
5. Leach, J. S. (ed.), "Experimental Techniques in the Study of Corrosion and Oxidation," D. Van Nostrand Company, Inc., Princeton, N.J., 1970.
6. Kittel, C., "Introduction to Solid State Physics," John Wiley & Sons, Inc., New York, 1966.
7. Dekker, A., "Solid State Physics," The Macmillan Company, New York, 1962.
8. Lamb, D. R., "Electrical Conduction Mechanisms in Thin Insulating Films," Methuen & Co., Ltd., London, 1967.
9. Harnwell, G. P., "Principles of Electricity and Electromagnetism," 2d ed., McGraw-Hill Book Company, New York, 1949.
10. Anderson, J. C., "Dielectrics," Chapman & Hall, Ltd., London, 1964.
11. Maddocks, F. S., and R. E. Thun, *J. Electrochem. Soc.*, **109**, 99 (1962).
12. McLean, D. A., *J. Electrochem. Soc.*, **108**, 48 (1961).
13. Cusack, N., "The Electrical and Magnetic Properties of Solids," Longmans, Green & Co. Ltd., London, 1958.
14. Ioffe, A. F., and A. R. Regel, *Progr. Semicond.*, **4**, 239 (1960).
15. Heikes, R. R., and W. D. Johnson, *J. Chem. Phys.*, **26**, 582 (1957).
16. Sewell, G. L., *Phys. Rev.*, **129**, 597 (1963).
17. Austin, I. G., and A. E. Owen, *Bull. Inst. Phys. and Phys. Soc. London*, **17**, 298 (1966).
18. Seraphim, D. P., *Thin Films ASM 1964*, p. 135.
19. Campbell, D. S., "Advanced Electronic Techniques," *Proc. 2d Oxford Electron Lecture Course*, United Trade Press, London, 1966, p. 35.
20. Kortum, G., and J. O. M. Bockris, "Textbook of Electrochemistry," Elsevier Publishing Company, Amsterdam, 1951.
21. Stevels, J. M., and J. Volger, *Philips Res. Rept.*, **17**, 283 (1962).
22. Stevels, J. M., "Non Crystalline Solids," John Wiley & Sons, Inc., New York, 1960.
23. Volger, J., and J. M. Stevels, *Philips Res. Rept.*, **11**, 79 (1956).
24. Pollak, M., *Phys. Rev.*, **138**, A1822 (1965).
25. Coleman, M. V., and D. J. D. Thomas, *Phys. Stat. Solidi*, **22**, 593 (1967).
26. Johansen, I. T., *J. Appl. Phys.*, **37**, 499 (1966).
27. Mott, N. F., *Endeavour*, **26**, 155 (1967).

28. Van Houten, S., and A. J. Bosman, *Proc. 1st Buhl. Intern. Conf. Transition Metal Compounds*, Pittsburgh, 1963, p. 123.
29. Vermilyea, D. A., in P. Delahay (ed.), "Advances in Electrochemistry and Electrochemical Engineering," p. 211, Interscience Publishers, Inc., New York, 1963.
30. Milward, R. C., and L. J. Neuringer, *Phys. Rev. Letters*, **15**, 664 (1965).
31. Friauf, R. J., *J. Chem. Phys.*, **22**, 1329 (1954).
32. Mott, N. F., and R. W. Gurney, "Electronic Processes in Ionic Crystals," p. 172, Oxford University Press, Fair Lawn, N.J., 1940.
33. Bean, C. P., et al., *Phys. Rev.*, **101**, 551 (1956).
34. Sutter, P. H., and A. S. Nowick, *J. Appl. Phys.*, **34**, 734 (1963).
35. Vest, R. W., and N. M. Tallan, *J. Appl. Phys.*, **36**, 543 (1965).
36. Lidiard, A. B., in S. Flügge (ed.), "Handbuch der Physik," Springer-Verlag OHG, Berlin, 1957.
37. Harrop, P. J., *J. Mater. Sci.*, **3**, 206 (1968).
38. Smith, R. A., "Semiconductors," Cambridge University Press, New York, 1964.
39. Dawson, D. K., and R. H. Creamer, *Brit. J. Appl. Phys.*, **16**, 1643 (1965).
40. Argall, F., Ph.D. Thesis, "Conduction and Dielectric Phenomena in Thin Insulating Films," Physics Department, Chelsea College, London University, 1967.
41. Weaver, C., in J. C. Anderson (ed.), "The Use of Thin Films in Physical Investigations," Academic Press Inc., New York, 1966.
42. Harrop, P. J., and J. N. Wanklyn, *J. Electrochem. Soc.*, **111**, 1133 (1964).
43. Burkhardt, P. J., *IEEE Trans. Electron Devices*, **ED13**, 268 (February, 1966).
44. *Plastics*, February, 1966, p. 1.
45. Mikhailov, G. P., and T. I. Borisova, *Soviet Phys. Usp. English Transl.*, **7**, 375 (1964).
46. Fröhlich, H., "Theory of Dielectrics," Oxford University Press, Fair Lawn, N.J., 1948.
47. Ninomiya, P., *J. Phys. Soc. Japan*, **14**, 30 (1959).
48. Fabian, M. E., *J. Mater. Sci.*, in press (1968).
49. Sastry, P. V., and T. M. Srinivasan, *Phys. Rev.*, **132**, 2445 (1963).
50. Cole, K. S., and R. H. Cole, *J. Chem. Phys.*, **9**, 341 (1941).
51. Gevers, M., *Philips Res. Rept.*, **1**, 279 (1946).
52. Morley, A. R., Ph.D. Thesis, "Temperature Effects in Thin Films of Silicon Oxide,' Electrical Engineering Department, Imperial College, London University, 1968.
53. Harrop, P. J., and D. S. Campbell, *Thin Solid Films*, **2**, 273 (1968).
54. Cockbain, A. G., and P. J. Harrop, *Brit. J. Appl. Phys.*, **18**, 1109 (1968). See also ref. 177.
55. Garton, G. C., *Trans. Faraday Soc.*, **42**, 57 (1946).
56. Pollak, M., and T. H. Geballe, *Phys. Rev.*, **122**, 1742 (1961).
57. Szigeti, B., *Trans. Faraday Soc.*, **45**, 155 (1949); *Proc. Roy. Soc. London*, **A204**, 51 (1950)
58. Havinga, E. E., *J. Phys. Chem. Solids*, **18**, 253 (1961).
59. Bosman, A. J., and E. E. Havinga, *Phys. Rev.*, **129**, 1593 (1963).
60. Hersping, A., *Z. Angew. Phys.*, **5**, 369 (1966).
61. Garton, G. C., and K. H. Stark, *Nature*, **176**, 1225–1226 (1955).
62. Cooper, R., *Brit. J. Appl. Phys.*, **17**, 149 (1966).
63. Weaver, C., and J. E. S. Macleod, *Brit. J. Appl. Phys.*, **16**, 441 (1965).
64. Fava, R. A., *Proc. Inst. Elec. Engrs.*, **112**, 819 (1965).
65. Wood, G. C., and C. Pearson, *Corrosion Sci.*, **7**, 119 (1967).
66. Budenstein, P. P., and P. J. Hayes, *J. Appl. Phys.*, **38**, 2837 (1967).
67. Klein, N., and H. Gafni, *IEEE Trans. Electron Devices*, **ED13**, 281 (1966).
68. Vorob'yev, A. A., *Compt. Rend. Acad. Sci. Moscow*, **86**, 681 (1952); *Soviet Phys. JET English Transl.*, **3**, 225 (1956).
69. Vermeer, J., *Physica*, **22**, 1257 (1956).
70. Lomer, P. D., *Nature*, **166**, 191 (1950).
71. Merrill, R. C., and R. A. West, Electrochemical Society Meeting, Pittsburgh, 1963
72. Fröhlich, H., and B. V. Paranjape, *Proc. Phys. Soc. London*, **B69**, 866 (1956).
73. Davies, D. K., *Proc. Static Electrification Conf. Inst. Phys.*, London, 1967.
74. Zener, C., *Proc. Roy. Soc. London*, **A145**, 523 (1934).
75. Seitz, F., *Phys. Rev.*, **76**, 1328 (1949).
76. Forlani, F., and N. Minnaja, *Phys. Stat. Solidi*, **4**, 311 (1964).
77. Klein, N., and N. Levanon, *J. Appl. Phys.*, **38**, 3721 (1967). See also ref. 178.
78. Kawamura, H., and K. Azuma, *J. Phys. Soc. Japan*, **8**, 797 (1953).
79. Kawamura, H., and T. Ryu, *J. Phys. Soc. Japan*, **9**, 438 (1954).
80. Thun, R. E., et al., in E. Keonjian (ed.), "Microelectronics," p. 173, McGraw-H' Book Company, New York, 1963.
81. Moore, G. E., Ref. 80, p. 309.
82. Newhouse, V. C., and H. H. Edwards, *IERE Conf. Proc. 7*, p. 11/1, 1966.

83. Pohm, A. V., R. J. Zingg, J. H. Hoper, and R. M. Stewart, Jr., *Proc. Intermag. Conf.*, 1964, p. 5-3-1.
84. Sutton, P. M., *J. Am. Ceram. Soc.*, **47**, 188 (1964).
85. Sutton, P. M., *J. Am. Ceram. Soc.*, **47**, 219 (1964).
86. Shepherd, A. A., in L. Holland (ed.), "Thin Film Microelectronics," p. 89, Chapman & Hall, Ltd., London, 1965.
87. Weimer, P. K., G. Sadasiv, L. Meray-Horvath, and W. S. Homa, *Proc. IEEE*, **54**, 354 (1966).
88. Moore, G. E., Ref. 80, p. 320.
89. Yamin, M., and F. C. Worthing, Fall Meeting, *Abstract 75*, p. 182, Electrochemical Society, 1964.
90. Taylor, K., *Trans. 8th AVS Symp.*, 1961, p. 981.
91. Clarke, D. J., and P. Lloyd, *Brit. Commun. Electron.*, **11**, 176 (1964).
92. Smith, E. E., and D. R. Kennedy, *Intern. Conf. Components and Materials Used in Electronic Engineering*, **B22**, 504 (1961).
93. Blackburn, H., D. S. Campbell, and A. J. Muir, *Radio and Elec. Eng.*, **34**, 90 (1967).
94. Mead, C. A., *J. Appl. Phys.*, **34**, 646 (1961).
95. Jaklevic, R. C., D. K. Donald, J. Lambe, and N. C. Vassell, *Appl. Phys. Letters*, **2**, 2 (1963).
96. Hill, R. M., *Nature*, **204**, 35 (1964).
97. Herman, D. S., and T. N. Rhodin, *J. Appl. Phys.*, **37**, 1594 (1966).
98. Neugebauer, C. A., and R. H. Wilson, in R. Niedermayer and H. Mayer (eds.), "Basic Problems in Thin Film Physics," p. 579, Vandenhoeck and Ruprecht, Goettingen, 1966.
99. Satchell, D. W., Private communication from Electrical Research Association, Cleeve Rd., Leatherhead, Surrey, England, 1967.
100. Priest, J., H. L. Caswell, and Y. Budo, *Vacuum*, **12**, 301 (1962).
101. Campbell, D. S., *Electron. Reliability and Microminiaturization*, **2**, 207 (1963).
102. Hoffman, R. W., "Thin Films," p. 128, American Society for Metals, Metals Park, Ohio, 1964.
103. Campbell, D. S., in R. Niedermayer and H. Mayer (eds.), "Basic Problems in Thin Film Physics," Vandenhoeck and Ruprecht, Goettingen, 1966.
104. Mead, C. A., Ref. 103, p. 674.
105. Ku, H. Y., and F. G. Ullman, *J. Appl. Phys.*, **35**, 265 (1964).
106. Stanworth, J. E., "Physical Properties of Glass," Oxford University Press, Fair Lawn, N.J., 1950.
107. Roskovcova, L., *Phys. Stat. Solidi*, **20**, K29 (1967).
108. Loh, E., *J. Chem. Phys.*, **44**, 1940 (1965).
109. Wachtman, J. B., *Phys. Rev.*, **131**, 517 (1963).
110. Burnett, S., Properties of Refractory Materials, *AERE Harwell Rept.*, AERE-R4657, 1964.
111. Huber, F., W. Witt, and W. Y. Pan, *Trans. 3d Vacuum Congr.*, **2**, 359 (1967).
112. Beaver, R. S., and L. R. Sproull, *Phys. Rev.*, **83**, 801 (1951).
113. Brown, V. R., *AVS Trans.* 8-5, 1966.
114. Cardona, M., *Phys. Rev.*, **140**, 651 (1965).
115. Kahn, A. H., et al., *Informal Proc. Buhl Intern. Conf. Mater.*, Pittsburgh, 1963, p. 53, 1964.
116. Jackson, N. F., E. J. Hollands, and D. S. Campbell, *IERE Conf. Proc.* 7, p. 13/1, 1966.
117. Prokharov, Yu. A., et al., *Soviet Phys. Solid State*, **9**, 1091 (1967).
118. Fuls, E. N., D. H. Hensler, and A. R. Ross, *Appl. Phys. Letters*, **10**, 199 (1967).
119. Riddle, G. C., *AVS Symp. Abstr.*, 1966, 5-5.
120. Huber, F., et al., see Ref. 111.
121. McLean, D. A., and F. S. Power, *Proc. IRE*, **44**, 872 (1956).
122. Berry, R. W., and D. J. Sloan, *Proc. IRE*, **47**, 1070 (1959).
123. Vromen, B. H., and J. Klerer, *IEEE Trans. Component Pts.*, **PMP-1**, S1 (1965).
124. Pliskin, W. A., D. R. Kerr, and J. A. Perri, "Physics of Thin Films," vol. 4, Academic Press Inc., New York, 1967.
125. *Ann. Rept. Conf. Elec. Insulation*, *NAS-NRC Publ.* 1080, p. 36, Washington D.C., 1963.
126. Ligenza, J. R., *J. Appl. Phys.*, **36**, 2703 (1965).
127. Ing, S. W., Jr., and W. Davern, *J. Electrochem. Soc.*, **112**, 284 (1965).
128. Pensak, L., *Phys. Rev.*, **75**, 472 (1949).
129. Sterling, H. F., J. H. Alexander, and R. J. Joyce, *Vide*, October, 1966.
130. Goodman, J., *J. Polymer Sci.*, **44**(144), 551 (1960).

131. Connell, R. A., and L. V. Gregor, U.S. Patent 3,297,467, January, 1967.
132. Poat, D. R., *Thin Solid Films*, **4**, 123 (1969).
133. Hirose, H., and Y. Wada, *Japan. J. Appl. Phys.*, **3**, 179 (1964).
134. Siddall, G., and G. S. Giddens, *IERE Conf. Proc.*, **7**, 31 (1966).
135. Hartman, T. E., J. C. Blair, and R. Bauer, *J. Appl. Phys.*, **37**, 2468 (1966).
136. Lewis, B., *Microelectron. Reliability*, **3**, 109 (1964).
137. Mayer, H., in C. A. Neugebauer, J. B. Newkirk, and D. A. Vermilyea (eds.), "Structure and Properties of Thin Films," p. 225, John Wiley & Sons, Inc., New York, 1959.
138. Feuersanger, A. E., A. K. Hagenlocher, and A. L. Solomon, *J. Electrochem. Soc.*, **111**, 1387 (1964).
139. Owen, A. E., *Prog. Ceram. Sci.*, **3**, 110 (1963).
140. Clark, R. S., *IEEE Trans. Component Pts.*, **PMP-1**, S31 (1965).
141. Sinclair, W. R., and F. G. Peters, *Rev. Sci. Instr.*, **33**, 744 (1962).
142. Davidse, P. D., and L. I. Maissel, *J. Appl. Phys.*, **37**, 574 (1966).
143. MacFarlane, J. C., and C. Weaver, *Phil. Mag.*, **13**, 671 (1966).
144. Zinsmeister, G., *Proc. Intern. Meeting Microelectron.*, 1965 Oldenbourg Verlag, Munich.
145. Technical literature, Standard Telephones and Cables, Paignton, Devon, England.
146. Harrop, P. J., unpublished work.
147. Muller, E. K., B. J. Nicholson, and M. H. Francombe, *Electrochem. Tech.*, **5**, 158 (1963).
148. Peters, F. G., *Bull. Am. Ceram. Soc.*, **45**, 1017 (1966).
149. Taylor, K., *Trans. 3d Intern. Vacuum Congr.*, **2**, 329 (1966).
150. Roberts, D. H., and D. S. Campbell, *J. Brit. IRE*, **22**, 281 (1961).
151. Hoar, T. P., and G. C. Wood, *Electrochem. Acta*, **7**, 333 (1962).
152. Stuart, M., *Nature*, **199**, 59 (1963).
153. Gregor, L. V., and H. L. McGee, *Conf. Elec. Insulation*, 1962, p. 36.
154. Oakley, G., and P. J. Harrop, unpublished work.
155. Johnson, M. C., *IEEE Trans. Component Pts.*, June, 1964, p. 1.
156. Jackson, N. F., unpublished work.
157. Tibol, G. J., and R. W. Hull, *J. Electrochem. Soc.*, **111**, 1368 (1964).
158. Feuersanger, A. E., *Bull. Am. Phys. Soc.*, **11**, 35A (1966).
159. Lloyd, P., *Microelectron. Reliability*, **6**, 177 (1967).
160. Parker, R. L., and A. Krinsky, *J. Appl. Phys.*, **34**, 2700 (1963).
161. Schnupp, P., *Phys. Stat. Solidi*, **21**, 567 (1967).
162. Schnupp, P., *Solid-State Electron.*, **10**, 785 (1967).
163. Weaver, C., *Advan. Phys.*, **11**, 83 (1962).
164. West, C. F., private communication, 1967.
165. Hill, R., *Thin Solid Films*, **1**, 39 (1967).
166. Carpenter, R., and D. S. Campbell, *J. Mat. Sci.*, **4**, 526 (1969).
167. Hu, S. M., and L. V. Gregor, *J. Electrochem. Soc.*, **114**, 827 (1967).
168. Gawehn, H., *Z. Angew. Phys.*, **14**, 458 (1962).
169. Gregor, L. V., in G. Hass and R. E. Thun (eds.), "Physics of Thin Films," vol. 3, p. 131, Academic Press Inc., New York, 1967.
170. Hodgman, C. D. (ed.), "Handbook of Chemistry and Physics," Ohio Rubber, 1968.
171. Harrop, P. J., and J. N. Wanklyn, *Brit. J. Appl. Phys.*, **18**, 739 (1967).
172. Technical literature, Carborundum Co., Niagara Falls, N.Y.
173. Pavlovic, A. S., *J. Chem. Phys.*, **40**, 951 (1964).
174. Wachtman, J. B., *Phys. Rev.*, **131**, 517 (1963).
175. Von Hippel, A., and R. Maurer, *Phys. Rev.*, **59**, 820 (1941).
176. Harrop, P. J., G. Oakley, and D. S. Campbell, *Thin Solid Films*, **1**, 475 (1968).
177. Harrop, P. J., *J. Material Science*, **4**, 370 (1969).
178. Klein, N., and N. Levanon, *J. Electrochem. Soc.*, **116**, 963 (1969).
179. Budenstein, P. P., and P. J. Hayes, *J. Vac. Sci. & Tech.*, **6**, 602 (1969).

Chapter **17**

Ferromagnetic Properties
of Films*

MITCHELL S. COHEN

M.I.T. Lincoln Laboratory, Lexington, Massachusetts

* This work was sponsored by the U.S. Air Force.

1. INTRODUCTION

Although it has been known for over 100 years that thin films can exhibit ferromagnetism, the physics of ferromagnetic films has been intensively studied only since 1955. The increasing attention given to this subject in recent years is a direct result of the recognition that such films are extremely useful as magnetic devices, particularly in the technology of information storage and retrieval. The research based on this technological interest not only has yielded a good understanding of ferromagnetism in films but has contributed to the knowledge of ferromagnetism in general. Furthermore, because ferromagnetic properties are very structure-sensitive, insight into the physics of *all* thin films, ferromagnetic or not, has been gained.

It has proved possible to make ferromagnetic films not only of the ferromagnetic metals Fe, Ni, Co, Gd, and their alloys but also of the nonmetallic ferrites, garnets, and chalcogenides. The major portion of the research has been devoted to flat, metallic films, however, and within this category the primary emphasis has been on films of NiFe alloys, sometimes known as Permalloy. The reason for this specialization of interest, which will be reflected in the content of this chapter, is the discovery that NiFe films are eminently suitable as bistable elements in a digital-computer memory. The results of this work are not limited to NiFe alloys, however, but appear to be generally applicable to ferromagnetic films of other metals.

From the standpoint of bulk ferromagnetism, ferromagnetic films can be made with the following unusual and technologically desirable properties: (1) The magnetization $\vec{M}$ tends to remain in the plane of the film because of the film's shape anisotropy. (2) A preferred direction, known as an easy axis (EA), can be induced at an arbitrary azimuth in the plane of the film, so that the lowest-energy state $\vec{M}$ lies along this axis in either sense (uniaxial anisotropy). (3) The sense of $\vec{M}$ can be reversed upon application of small (a few oersted) magnetic fields. Most of this chapter will be concerned with these three interesting properties, their origins, and their influences on other magnetic phenomena. The technological exploitation of these properties is discussed in Chap. 21.

Several review articles and books on ferromagnetic films have been published,[1-9] and a bibliography of the extensive literature of this field is available.[10] For reports on current research, the published proceedings of the Conferences on Magnetism and Magnetic Materials[11] (United States), International Congresses for Magnetism,[11] the All-Union Symposia on the Physics of Ferromagnetic Films[12] (Soviet Union), and the International Colloquia on Magnetic Thin Films[13] are recommended.

2. PHYSICS OF FERROMAGNETISM

This section will be devoted to a review of those basic concepts of (bulk) ferromagnetism which are pertinent to the study of ferromagnetism in films. Only a brief sketch will be presented; the reader desiring a deeper understanding of these topics is referred to the standard treatises on magnetism.[14-18] On the other hand, the reader who is familiar with the physics of ferromagnetism may proceed directly to Sec. 3.

Gaussian cgs units are used throughout this chapter.

a. The Magnetization

Ferromagnetism is a cooperative phenomenon which originates in the alignment of atomic spins (with their associated magnetic moments) within a solid body and is quantitatively expressed by the magnetization $\vec{M}$, the magnetic moment per unit volume. The tendency toward mutual alignment of the spins was assumed by early workers to originate in the action of an ad hoc Weiss *molecular field*, proportional to and parallel to $\vec{M}$, but it is now known that this internal field is caused by a quantum-mechanical exchange effect. As the temperature of the ferromagnet is raised, M decreases because the random thermal agitation of the lattice acts against the aligning molecular field; above a critical temperature T_c, called the *Curie temperature*, M vanishes. It is the task of basic theories of ferromagnetism to predict the temperature dependence of M and the value of T_c; such theories are still undergoing refinement.

It is well established that ferromagnetism is associated with uncompensated spins of the $3d$ electrons in the common ferromagnetic metals Fe, Ni, and Co and their alloys with each other and with nonmagnetic metals. According to *spin-wave* theory, *deviations* from complete spin alignment are represented by "waves" of periodically spaced reversals of these spins. Spin-wave theory has led to the famous Bloch law, which predicts the experimentally confirmed decrease of M proportional to $T^{\frac{3}{2}}$ as the temperature is increased and more spin waves are consequently excited. The original theory is valid only at very low temperatures, where very few spin waves, or *magnons*, are excited. However, later workers improved the theory by using more sophisticated mathematical techniques and extended its range of validity to higher temperatures; the recent Green's-function formulation of the theory covers the entire temperature range to the Curie point, although difficult mathematics are involved.

The mobility of the $3d$ "magnetic" electrons is a matter of some controversy. On the one hand, according to the Heisenberg model, they are considered localized, while alternatively they may be regarded as having a mobile or itinerant character. The latter viewpoint has led to the development of an electron-band theory of ferromagnetism. The true behavior of the magnetic electrons is probably somewhere between the two extreme assumptions.

It should be noted that atomic spins in some materials (usually insulating compounds) often assume ordered arrays which lead to forms of magnetism other than ferromagnetism. Thus, ferrites and garnets, which are said to be *ferrimagnetic*, may be described by two or more interpenetrating sublattices, each of which shows ferromagnetic alignment in a certain direction; the *net* value of M is found from the proper addition of the sublattice moments. If the moments from two sublattices exactly cancel, or if a special helical ordering is present, the material is said to be *antiferromagnetic;* thus, in an antiferromagnet, even though the magnetic spins are ordered, the net value of M vanishes. The temperature, analogous to the Curie temperature in ferromagnets, at which the magnetic order in a ferrimagnet is destroyed is known as the *Néel temperature*.

b. Magnetic Energies

In many of the discussions in this chapter, the quantum-mechanical origins of the various phenomena which are encountered (e.g., the magnetization) need not be considered in detail but can be handled phenomenologically on a scale much coarser than the atomic. In these circumstances, the most general problem is to find the direction of $\vec{M}$ as a function of position everywhere in the material, given the external fields and the characteristics of the material. While this problem, which is known as the *micromagnetic* problem, may be solved exactly only in certain special cases, it is nevertheless illuminating to describe the behavior of a ferromagnet on the micromagnetic level. To do this, it is necessary to consider the various magnetic energies: *field, magnetostatic, anisotropy,* and *exchange*.

The primary technique for determination of the ferromagnetic properties of a sample involves the application of a magnetic field $\vec{H}$ to the sample, with subsequent study of the resulting behavior of $\vec{M}$ in the sample. If a field $\vec{H}$ is applied to a body with the magnetization $\vec{M}$, a torque $\vec{H} \times \vec{M}$ will be exerted on $\vec{M}$, which tends to "pull" $\vec{M}$ parallel to $\vec{H}$. The external *field* energy density E_H may be written as

$$E_H = -\vec{H} \cdot \vec{M} \tag{1}$$

so that the minimum-energy state is reached when $\vec{M}$ and $\vec{H}$ are parallel.

At places in the sample where $\vec{M}$ is discontinuous, i.e., where $\nabla \cdot \vec{M} \neq 0$, *magnetic poles* (charges) are created, and a magnetic field $\vec{H}_{\text{mag}}$ (distinct from the applied field $\vec{H}$) arises from these poles. Such divergences in $\vec{M}$ can occur when strong local perturbations act on $\vec{M}$, e.g., at the boundary of the sample (Fig. 1) or in its interior at local regions which have magnetic properties different from the surrounding material. The *magnetostatic* (or "stray" field) energy density E_{mag} which results from the creation of $\vec{H}_{mag}$ is given by

$$E_{mag} = -\tfrac{1}{2}\vec{H}_{mag} \cdot \vec{M} \tag{2}$$

where the factor $\frac{1}{2}$ arises from the fact that a self-energy is involved. Note that in order to determine $\vec{H}_{mag}$ (and hence E_{mag}) at any arbitrary point, the contributions to $\vec{H}_{mag}$ from $\nabla \cdot \vec{M}$ at *all other* points in the sample must be added.

Because of the properties of the sample, the energy may depend on the orientation of $\vec{M}$ with respect to a set of axes fixed to the sample. If $\vec{M}$ has the direction cosines α_1, α_2, α_3, the *anisotropy* energy density E_K may be expressed as

$$E_K = Kf(\alpha_1, \alpha_2, \alpha_3) \tag{3}$$

where K is the *anisotropy constant* and f is a function of the orientation of $\vec{M}$. Those directions of $\vec{M}$ for which E_K reaches minima and maxima are denoted as *easy* and *hard*, respectively; however, since E_K is usually independent of the sense of $\vec{M}$, they are called easy and hard *axes*.

The quantum-mechanical exchange mechanism which causes the basic alignment of the spins which results in ferromagnetism also tends to prevent $\vec{M}$ from deviating from a straight line. The *exchange* energy density E_{ex} resulting from such a deviation is expressed by

$$E_{ex} = \frac{A}{M^2} |\nabla \vec{M}|^2 \tag{4}$$

where A is the exchange constant, a property of the material ($A \sim 10^{-6}$ erg cm^{-1} for NiFe). This energy represents the "stiffness" of $\vec{M}$, i.e., its tendency to avoid sharp gradients.

The total energy density $E_{tot} = E_H + E_{mag} + E_K + E_{ex}$. Note that each of these energies is in general a function of position in the sample and the orientation of $\vec{M}$ at that position. Thus to find $\vec{M}$ as a function of position, E_{tot} must be minimized at each point in the sample with respect to the angles defining the direction of $\vec{M}$. (Since the derivative of the energy with respect to angle is a torque, this is equivalent to saying that the total torque on $\vec{M}$ must vanish everywhere.) The determination of the energy E_{mag} is usually the hardest part of the micromagnetic calculation, since in order to find E_{mag} at any given point the magnetization configuration at all other points of the sample must first be known; since the unknown $\vec{M}$ configuration is the *goal* of the calculation, a difficult variational problem results.

c. Origins of Magnetic Anisotropy

The magnetic anisotropy expressed by Eq. (3) can originate in any of several mechanisms, which can act simultaneously.

Magnetic anisotropy can occur simply because of the anisotropic geometric shape of the sample. Thus, the distribution of magnetic charge at the surface of the sample is in general dependent on the orientation of $\vec{M}$ (which is here assumed to be uniform everywhere within the sample), thereby causing $\vec{H}_{mag}$ to be a function of the orientation of $\vec{M}$. Now it can be shown that for the special case of an ellipsoidal sample

(Fig. 1), $\vec{H}_{\text{mag}}$ is constant everywhere within the sample, and in fact $\vec{H}_{\text{mag}} = -N\vec{M}$, where the *demagnetizing factor* N is a numerical constant which depends on the

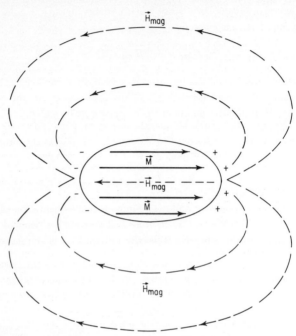

Fig. 1 Magnetostatic field $\vec{H}_{\text{mag}}$ of a ferromagnetic ellipsoid with magnetization $\vec{M}$. The field $\vec{H}_{\text{mag}}$ originates from magnetic poles generated by nonvanishing $\nabla \cdot \vec{M}$ at the surface of the ellipsoid.

orientation of $\vec{M}$. Such ellipsoidal samples are of special interest because E_K may be easily found for them; in fact Eq. (2) shows directly that

$$E_K = \tfrac{1}{2}NM^2 \tag{5}$$

This relation finds wide application because many samples have shapes which closely approximate an ellipsoid. The demagnetizing factors N have been calculated for the principal axes of the generalized ellipsoid and can be expressed approximately in closed form for certain ellipsoids of revolution.

If the sample is a single crystal, a *magnetocrystalline anisotropy* is generated by spin-orbit coupling, so that E_K depends on the direction of $\vec{M}$ relative to the crystal axes. Thus for a cubic crystal, E_K may be described by

$$E_K = K_1 \sum_{i>j} \alpha_i^2\alpha_j^2 + K_2\alpha_1^2\alpha_2^2\alpha_3^2 \tag{6}$$

where K_1 and K_2 are the magnetocrystalline anisotropy constants and α_1, α_2, α_3 are the direction cosines of $\vec{M}$. Similar relations may be written for crystals of other symmetries. If the sample is made of randomly oriented polycrystalline material, E_K averages to zero, but if some preferred crystal orientation is present, E_K may not vanish.

If a ferromagnetic crystal is subjected to an external stress, because of the resulting lattice distortion the magnetocrystalline anisotropy will change by an amount which depends on the direction and magnitude of the strain due to the stress. Therefore, in order to minimize the magnetocrystalline anisotropy energy, an *unconstrained* crystal will exhibit a spontaneous strain which depends on the direction of $\vec{M}$ relative to the crystal axes; in a cubic crystal such strain measured along the [100] or [111] directions, when $\vec{M}$ is parallel to these directions, is defined as the *magnetostriction constant* λ_{100} or λ_{111}, respectively. Now if a uniform stress σ is applied to a randomly oriented polycrystalline sample, it can be shown that the energy of the strained state is

$$E_K = \tfrac{3}{2}\lambda\sigma \sin^2 \varphi_0 \tag{7}$$

where φ_0 is the angle between $\vec{M}$ and the stress direction, and λ is an "average" magnetostriction $\lambda \equiv \tfrac{3}{5}\lambda_{111} + \tfrac{2}{5}\lambda_{100}$. (If there is crystallographic texture in the sample, λ_{100} and λ_{111} will have different coefficients in this equation.) The resulting magnetic anisotropy is called strain-magnetostriction anisotropy.

An anisotropy which is characterized by a $\sin^2 \varphi_0$ dependence of E_K, as is the anisotropy of Eq. (7), is known as a *uniaxial* anisotropy. The general formulation of uniaxial anisotropy energy is given by

$$E_K = K_u \sin^2 \varphi_0 \tag{8}$$

where φ_0 is the angle between $\vec{M}$ and the easy axis (EA) of the material, and K_u is known as the uniaxial anisotropy constant. Two equivalent states of lowest energy are defined ($\varphi_0 = 0$ or π); in these states $\vec{M}$ is directed in either sense along the EA. (For strain-magnetostriction anisotropy, the EA is along the direction of an applied tensile stress if $\lambda > 0$.) As $\vec{M}$ is rotated away from the EA, say by an applied field, E_K increases until a maximum is reached when $\vec{M}$ is normal to the EA ($\varphi_0 = \pi/2$); a *hard* plane is thus defined [in the two-dimensional case, a *hard axis* (HA)]. If the rotation of $\vec{M}$ were continued, E_K would then decrease until another minimum was reached when $\vec{M}$ was again directed along the EA but opposite to its original sense.

Certain characteristics of uniaxial anisotropy can be most easily interpreted in terms of an equivalent field. For this purpose the *anisotropy field* H_K is defined, where $H_K \equiv 2K/M$.

Another induced uniaxial-anisotropy mechanism plays a role in *magnetic annealing*. If certain ferromagnetic materials are subjected to a high-temperature anneal in the presence of a magnetic field, it is found upon quenching to a lower (usually room) temperature that a uniaxial anisotropy has been created, with the EA along the direction of the applied field. An alternative procedure is to subject the sample to bombardment by high-energy nuclear particles, again in the presence of an applied field. In either case the function of the field is only to align $\vec{M}$ everywhere in the sample in a single predetermined direction; i.e., the resulting anisotropy is magnetization-induced, rather than field-induced. The accepted explanation of magnetic annealing is that short-range directional order is induced by atomic rearrangements during the anneal; i.e., because of magnetic interactions a preferred orientation of either individual atoms or like-atom pairs is developed. This directional order is frozen in when the sample is cooled, thus creating uniaxial anisotropy. In NiFe, the anisotropy due to Ni-Ni (or Fe-Fe) pairs is given theoretically by

$$K_u = \frac{Ac^2(1 - c)^2 B^2(T_a)B^2(T_m)}{T_a} \tag{9}$$

where c is the atomic concentration of Ni, T_a is the annealing temperature, T_m is the temperature at which the measurement is taken, $B(T)$ is a mathematical function known as the Brillouin function, and A is a constant. Only nearest-neighbor interactions were considered in the derivation of Eq. (9) and the composition dependence is based on the assumption of ideal solubility. Experiments on bulk binary alloys are in general agreement with the pair-ordering theory, and give an activation energy of about 3 eV for NiFe.

d. Magnetic Domains

Thus far the sample has been considered to be uniformly magnetized throughout, or magnetically *saturated* [although it has been mentioned (Sec. 2*b*) that local inhomogeneities can give rise to local deviations of $\vec{M}$ from the saturation direction]. However, under certain conditions the sample may contain a magnetic *domain* structure, which represents a magnetic state which differs markedly from saturation. A ferromagnetic domain is a region *within* which $\vec{M}$ is saturated in a certain direction, but the domain is bounded by other domains with different directions of $\vec{M}$. The narrow transition regions between the domains are called *domain walls*.

The domain configuration is not fixed in the material but can be changed by the application of a magnetic field $\vec{H}$ which tends to align $\vec{M}$ parallel to $\vec{H}$ [Eq. (1)]. As $\vec{H}$ is increased, $\vec{M}$ will tend to rotate parallel to $\vec{H}$, and favorably oriented domains (domains containing $\vec{M}$ oriented close to $\vec{H}$) will begin to grow at the expense of less favorably oriented domains; i.e., the domain walls will move. For high enough values of H all the domain walls will be swept out of the sample, which will then be *saturated* in the direction of $\vec{H}$. On the other hand, if many antiparallel domains are introduced, e.g., by subjection to a slowly decreasing ac field, the net magnetization in the field direction may vanish, in which case the sample is said to be *demagnetized*. Now the domain configuration of a sample depends not only on $\vec{H}$ but on its physical structure and its previous magnetic history; the latter dependence originates in the fact that a given configuration is commonly not a state of lowest energy, but a metastable state. (The sample's magnetic history therefore usually represents a thermodynamically *irreversible* process.)

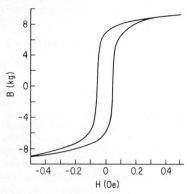

Thus, if the net value of the component of $\vec{M}$ parallel to the applied field $\vec{H}$ is plotted against $\vec{H}$ for a cycle of increasing and decreasing H, the familiar *hysteresis* loop results (Fig. 2). The loop is not single-valued in M, and its shape depends on previously applied $\vec{H}$ cycles. (Note that the value of the net magnetization corresponding to magnetic saturation achieved at high values of H is often denoted by M_s, while the net magnetization found when H is decreased to zero after saturation is called the *remanent* value M_r.)

Fig. 2 BH hysteresis loop of a bulk sample of 78 % Ni, 22 % Fe given the Permalloy heat treatment. (*After R. M. Bozorth, "Ferromagnetism," D. Van Nostrand Company, Inc., Princeton, N.J., 1951.*)

In an ideal material which did not show hysteresis, a unique measurement of the *susceptibility*, which is defined as the ratio M/H, could be made. Although this is not generally possible in a real ferromagnet, the *initial susceptibility* χ is a well-defined quantity, where χ is the ratio M/H for a fully demagnetized sample, measured under the conditions of small H so that reversible, nonhysteretic conditions are maintained.

Many workers have investigated the structure (or shape) of domain walls, i.e., the direction of $\vec{M}$ as a function of the coordinate normal to the plane of the wall.

Domain walls in bulk material are called *Bloch* walls and are characterized by the fact that the component of $\vec{M}$ which is perpendicular to the wall is continuous throughout the wall (see Fig. 28). In this way divergences of $\vec{M}$ are avoided, which permits the neglect of magnetostatic energy in considerations of wall structure, so that the wall assumes a shape which minimizes the sum of the anisotropy and exchange energies. The analysis is simple for a Bloch wall in material having uniaxial anisotropy, particularly for the case of the EA parallel to the wall, on either side of which $\vec{M}$ is antiparallel but also along the EA; such a wall is known as a 180° wall.* In this case an increase in wall width causes an increase in the anisotropy energy [Eq. (8)] but a decrease in the exchange energy [Eq. (4)]; the equilibrium wall width is that width which makes the two energies equal, namely, $\pi(A/K_u)^{\frac{1}{2}}$. The wall energy density γ, i.e., the energy per unit area of the wall, may also be easily calculated.

The *dynamics* of domain-wall motion are of interest. First, it is found that a threshold field known as the wall coercive force H_c (a parameter which is characteristic of the material) must be surpassed in order to produce wall motion. If the sample were perfectly uniform, an infinitesimally small field would be sufficient to sweep a reverse domain completely across it. The fact that a field greater than H_c must be applied for wall motion may be explained by assuming that inhomogeneities cause the energy density $\gamma(x)$ of the domain wall to be a random function of position x. Then the energy provided by the field H_c is just large enough to overcome the potential hills; i.e., for an infinitesimal displacement dx of the wall parallel to itself, the change of field energy [Eq. (1)] may be equated to the change in wall energy, or $2H_cMl\,dx = d(\gamma l d)$ where l is the length of the wall. Thus

$$H_c = (2Md)^{-1}\left[\frac{d(\gamma d)}{dx}\right]_{\max} \tag{10}$$

for a wall of constant length. Now the potential barriers are randomly distributed in position and in height [the subscript max in Eq. (10) refers to the steepest barrier]. An applied field will thus be sufficient to sweep a domain wall through a small region until the wall is stopped by a prohibitively steep barrier; a slight increase in field will then cause a small jump of the wall to a new barrier position. Thus, as the field is gradually increased, the wall moves across the sample in a series of small thermodynamically irreversible *Barkhausen* jumps until saturation in the new direction is achieved.

The wall velocity v has been found to obey the relation

$$v = m(H - H_0) \tag{11}$$

where m is a constant called the wall *mobility* and H_0 may be identified with H_c. The wall mobility will depend on the rate with which the magnetic energy, which is being fed to the sample from the applied field, can be lost to the lattice as heat. For a metallic sample the predominating loss mechanism is the production of eddy currents. For a nonmetallic sample, more fundamental spin-lattice interactions have been postulated, which for a shape-invariant, moving 180° wall, leads to the relation

$$m = 2\gamma\left(\alpha\int_0^\pi \frac{d\phi}{dx}\,d\phi\right)^{-1} \tag{12}$$

where ϕ is the azimuth of $\vec{M}$ within the wall at a position x measured normal to the wall, γ is the gyromagnetic ratio, and α is the damping parameter in the Landau-Lifshitz equation (Sec. 2e).

e. Ferromagnetic Resonance

It was pointed out in Sec. 2a that the basic theory of magnetization is still far from completely understood. Nonetheless, it has been possible to study the motion of the

---

* Note that walls separating $\vec{M}$ at angles other than 180° are possible; e.g., in cubic crystals 90° walls are common.

magnetization which occurs when a ferromagnet is subjected to rf fields in the presence of a dc magnetic field; this motion can be described phenomenologically. For certain rf frequencies and related values of the dc field the system may be said to be in *resonance;* experimental evaluation of the parameters corresponding to resonance leads to important magnetic information.

First consider an atomic system having a total angular momentum $\vec{J}\hbar$ with an associated magnetic moment $\vec{\mu}$. Then $\vec{\mu} = g\mu_B\vec{J}$, or alternatively, $\vec{\mu} = \gamma\hbar\vec{J}$, where g is the Landé g factor, $\mu_B \equiv e\hbar/2mc$ is the Bohr magneton, and $\gamma \equiv g\mu_B/\hbar$ is the gyromagnetic ratio. Speaking in terms of classical physics, if a field $\vec{H}$ is applied to the system, a torque $\vec{T} = \vec{\mu} \times \vec{H}$ will act upon it, or from the previous equation since $\vec{T} = \vec{J}\hbar$,

$$\dot{\vec{\mu}} = \gamma\vec{\mu} \times \vec{H} \tag{13}$$

Equation (13) represents a *precession* of $\vec{\mu}$ about $\vec{H}$ (similar to the precession of a gyroscope) with an angular frequency, called the Larmor frequency, of γH.

By replacing $\vec{\mu}$ by $\vec{M}$, Eq. (13) may be used to describe the behavior of the magnetization in a *macroscopic* ferromagnetic body, if $\vec{M}$ in the body is perfectly uniform. To incorporate a small amount of damping, which is experimentally found, a phenomenological damping term in either the Landau-Lifshitz or (for small damping) equivalent Gilbert formulation is added to the right side of Eq. (13). Thus

$$\dot{\vec{M}} = \gamma\vec{M} \times \vec{H} - M^{-2}\gamma\vec{M} \times (\vec{M} \times \vec{H}) \quad \text{or}$$

$$\dot{\vec{M}} = \gamma\vec{M} \times \vec{H} - M^{-1}\alpha\vec{M} \times \dot{\vec{M}} \tag{14}$$

where λ and α are the Landau-Lifshitz and Gilbert damping parameters, respectively. This equation specifies the general dynamic behavior of the magnetization subject to the field $\vec{H}$. The scalar product of $\vec{M}$ with both sides of Eq. (14) vanishes, thus showing that the magnitude M is conserved since $\dot{\vec{M}}$ is perpendicular to $\vec{M}$; the damping term merely causes $\vec{M}$ to spiral in toward $\vec{H}$. However, if an rf magnetic field at the precession frequency γH is applied perpendicular to the constant field $\vec{H}$, energy will be taken from the rf field to maintain the precession against the damping. Such a system is said to be in ferromagnetic resonance (FMR).

In a practical experiment, the sample is subjected to a microwave field at a fixed frequency ω, while a dc field perpendicular in direction to the microwave field is slowly varied. If the energy absorbed by the sample is suitably measured (Sec. 4c), a peak is found at the resonance field H_r. Both the value of H_r and the width of the resonance peak (related to the damping) may be found from Eq. (14) if in that equation $\vec{H}$ is treated as an effective field, which contains not only $\vec{H}_r$ but also the demagnetizing and anisotropy fields which act upon $\vec{M}$.

3. UNIAXIALLY ANISOTROPIC FILMS—PROPERTIES AND PREPARATION

As stated in Sec. 1, most of the research on ferromagnetic films has been concentrated on uniaxially anisotropic, flat, metallic films having $\vec{M}$ in the plane of the film.* This section is devoted to an introductory survey of the magnetic properties of such films, and to a presentation of the techniques used for their production.

---

* Primarily for technological applications, cylindrical NiFe films have been electroplated on metal wires in a manner such that a circumferential EA is established. The magnetic properties of such films are essentially the same as those of flat NiFe films.

a. Properties

For concreteness, visualize a circular film deposited on a smooth glass substrate (Fig. 3). The film is 1 cm in diameter, 1,000 Å thick, made of 81% Ni, 19% Fe, and contains randomly oriented crystallites 100 Å in diameter on the average. The film is uniaxially anisotropic, with the following magnetic parameters (explained below): $H_K = 3$ Oe, $H_c = 1$ Oe, $\alpha_{90} = 2°$. Such a film may be considered typical for most of the discussion in this chapter.

(1) Single-domain, Coherent Rotation Theory To a good approximation, a thin film of thickness d and diameter L may be considered a very flat oblate ellipsoid of revolution. Then the demagnetizing factors perpendicular to and in the plane of the film are, respectively, $N_\perp \simeq 4\pi$ and

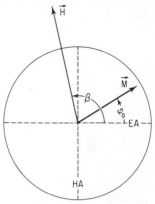

$$N_\| \simeq \frac{4\pi d}{(d + L)}$$

(Sec. 2c). Thus, the magnetic charges created when $\vec{M}$ is normal to the film plane produce a demagnetizing field of $4\pi M$, which for NiFe films is of the order of 10^4 Oe, while for the typical film described above, the parallel demagnetizing field created when $\vec{M}$ is in the plane of the film is only about 0.1 Oe. Clearly then, $\vec{M}$ will tend to lie in the film plane, since if for any reason $\vec{M}$ tilts out of the plane, the resulting large demagnetizing field will pull it back. Alternatively, the film may be said to have a shape anisotropy with a hard axis normal to the film and an easy plane in the film plane. The anisotropy constant $K_\perp$, given by the difference of the energy densities of the two mag-

Fig. 3 Schematic drawing of a typical uniaxially anisotropic, ferromagnetic film. The easy axis (EA) is horizontal, and the hard axis (HA) is vertical. A magnetic field $\vec{H}$ is applied at an angle β to the EA, which causes the magnetization $\vec{M}$ to be inclined at an angle φ_0 to the EA.

netic configurations, is $2\pi M^2$, which is of the order of 10^6 ergs cm^{-3} for NiFe.

From this reasoning, for most purposes the film can be considered magnetically homogeneous throughout its thickness, with $\vec{M}$ everywhere parallel to the film plane. This means that the problem of the magnetization distribution within the film is only two-dimensional, which is a considerable simplification from the case of bulk ferromagnets.

In discussing the magnetic behavior, it is convenient to consider first the case of complete saturation of $\vec{M}$ in a single domain state; i.e., $\vec{M}$ remains *coherent* at all times. Thus $\vec{M}$ is directed uniformly at some angle φ_0 with respect to the EA everywhere in the film (Fig. 3). Because the film is assumed to be uniaxially anisotropic, the states of lowest anisotropy energy are equivalently $\varphi_0 = 0$ and $\varphi_0 = \pi$; i.e., at equilibrium $\vec{M}$ points either left or right along the EA. If $\vec{M}$ is rotated away from the EA, the anisotropy energy is increased [Eq. (8)]; it reaches a maximum at $\varphi_0 = \pm\pi/2$, i.e., when $\vec{M}$ points in either sense along the vertical hard axis (HA).

The external agent which can cause a rotation of $\vec{M}$ out of the EA is, of course, an applied field $\vec{H}$, say at an angle β to the EA. If $\vec{H}$ and K_u are known, the direction taken by $\vec{M}$ can easily be predicted. First note that the total energy E_{tot} is the sum of the anisotropy and field energies E_K and E_H, respectively [Eqs. (8) and (1)],

$$E_{tot} = K_u \sin^2 \varphi_0 - HM \cos (\beta - \varphi_0) \tag{15}$$

Then to find φ_0, the minimum-energy condition $\partial E_{tot}/\partial\varphi_0 = 0$ is applied; i.e., the net torque on $\vec{M}$ vanishes, and φ_0 is found from the resulting equation. Physically speaking, the field $\vec{H}$ exerts a torque on $\vec{M}$ which tends to rotate it out of the EA, while the anisotropy exerts a restoring torque on $\vec{M}$ tending to return it to the EA; the angle φ_0 is determined by the equilibrium between the two torques. [The mechanical analog would be a force $(\vec{H})$ pulling against a spring (K_u).]

It is instructive to consider the simple case where $\vec{H}$ is directed along the HA, i.e., $\beta = 90°$. Then differentiation of Eq. (15) with respect to φ_0 yields

$$\sin \varphi_0 = \frac{H}{H_K} \tag{16}$$

where H_K is called the anisotropy field and is defined as $2K_u/M$. Then as H is increased from zero, φ_0 increases until, when $H = H_K$, $\vec{M}$ is parallel to $\vec{H}$ and remains in that direction for any further increase in H. The anisotropy field H_K (about 3 Oe) is then the minimum HA field necessary to pull $\vec{M}$ from the EA to the HA; it is a very convenient parameter, and in fact is commonly used instead of K_u to characterize the uniaxial anisotropy. (Discussion of the direction of $\vec{M}$ for *arbitrary* values of β will be postponed to Sec. 6a.)

Investigations of the *dynamics* of the change of azimuth of $\vec{M}$ upon application of a pulsed external field show that the sense of $\vec{M}$ along the EA can be reversed in a time of the order of 1 ns for fields of a few oersteds. This behavior, which has great significance for the use of ferromagnetic films as memory elements, may be discussed on the basis of the Landau-Lifshitz equation [Eq. (14)] in the coherent-rotation approximation. Further discussion may be found in Sec. 6b.

(2) Actual Film Behavior The single-domain, coherent rotation model just discussed is insufficient, since marked deviations from magnetic saturation occur. Indeed, a large portion of this chapter will be devoted to discussions of the departures from single-domain behavior. Thus if the film is first saturated along the EA, and then a slowly increasing reverse field is applied antiparallel to the original saturation field, reverse domains are observed to nucleate at the film edge, then grow until the entire film is reversed. For domain movement the reverse EA field must surpass a certain threshold value H_c, the wall coercive force (about 1 Oe). Domain growth can also be observed when the reverse field is applied at an angle to the EA. (A more complete discussion of domain phenomena is given in Sec. 9.)

Perhaps the most useful single instrument for surveying the properties (incoherent as well as coherent) of a magnetic film is a hysteresis-loop tracer, or hysteresigraph (Sec. 4b). The experimental hysteresigram shown in Fig. 4a was taken with the ac field along the EA. The loop indicates remarkable "squareness," showing that the magnetization is completely reversed by the growth of reverse domains soon after H_c is reached; it is seen from the width of the loop that $H_c \simeq 1.8\,\text{Oe}$. The corresponding HA loop is shown in Fig. 4b. The inner loop, taken at small excitation fields, is a straight line indicating coherent rotation, in agreement with Eq. (16), which says that the component of net magnetization along the HA, $M \sin \varphi_0$, is given by HM/H_K, with a saturation value M first attained at $H = H_K$. Thus if the experimental trace is extrapolated to the saturation magnetization value (obtained from Fig. 4a), the corresponding H value should be H_K; from Fig. 4b, $H_K \simeq 3.5\,\text{Oe}$.

The single-domain theory must be modified to cover not only domain phenomena but also the phenomenon of *magnetization ripple*. Even if the film contains only a single domain, $\vec{M}$ is not everywhere uniform in direction, but within the film plane it suffers small, quasi-periodic, local angular deviations (a few degrees) from its average direction. These local perturbations originate in locally random magnetic inhomogeneities which are usually thought to be associated with the randomly oriented crystallites composing the film. Such local directional deviation is called magnetization ripple or *magnetization dispersion*, and is important in many magnetic phenomena. Magnetization ripple can be directly observed by Lorentz microscopy (Sec. 4a) as seen in Fig. 31b, or by high-resolution magneto-optic microscopy (Sec. 4a) as seen in Fig. 7. Note that the "wavefronts" of the ripple are perpendicular to the average $\vec{M}$ direction.

The *dispersion* α_{90} is commonly used as an index of the magnetization dispersion, although there is not a direct relation between this parameter and the average mag-

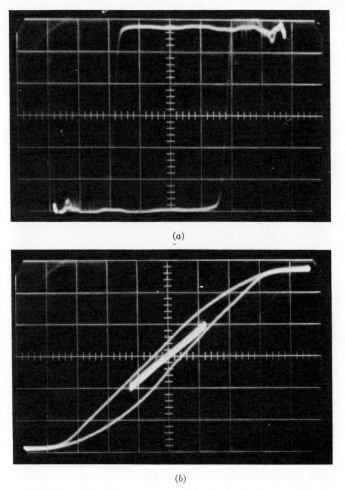

(a)

(b)

Fig. 4 Experimental MH hysteresis loops of a NiFe film. One horizontal division corresponds to 1 Oe. [*After J. I. Raffel, T. S. Crowther, A. H. Anderson, and T. O. Herndon, Proc. IRE, **49**, 155 (1961).*] (a) EA loop. The value of H_c can be found from the intercepts of the loop with the H axis. (b) HA loop. Inner trace, low excitation field. Outer trace, high excitation field. The value of H_K can be found from the slope of the inner trace.

netization-dispersion angle. The determination of α_{90} is a standard film measurement which can easily be carried out with a modified hysteresigraph; typically α_{90} is a few degrees. The measurement is based on the phenomenon of *HA fallback*. If $\vec{M}$ is saturated along the HA and then H is gradually reduced to zero, for sufficiently low values of H, $\vec{M}$ is left in a highly unstable position, since a small rotation of $\vec{M}$ to either the left or the right (toward opposite senses of the EA) results in an energy reduction; the situation is analogous to that of a pencil standing on its tip. As a result of this instability, $\vec{M}$ is sensitive to small perturbations so that many "splitting" domains are joined in a fashion dependent on the ripple. A hysteresigraph study of the fallback for different initial saturation directions near the HA results in a measure of α_{90}. (Further discussion of HA fallback is given in Sec. 11.)

The three quantities H_K, H_c, and α_{90} form a triad of parameters which are often used to describe the magnetic properties of a film. In general, for technical applications it is desired to maintain low values of all three parameters.

b. Preparation

Most of the standard deposition techniques discussed in Part One, "Preparation of Thin Films," can be used to obtain uniaxially anisotropic ferromagnetic films with the typical properties described above. Thus, the techniques of vacuum deposition, electrodeposition, sputtering, chemical reduction or "electroless" deposition, and thermal decomposition of metal-organic vapors have been successfully used. Of these, the first three techniques have found widest application; a discussion of their use in the production of polycrystalline, uniaxial, ferromagnetic films (particularly NiFe films) is given below. Regardless of the deposition technique, however, experience has shown that every effort must be made to maintain constancy of the deposition parameters in order to obtain films with reproducible magnetic parameters.

(1) Vacuum Deposition[6]   Most of the fundamental research on ferromagnetic films has been carried out on films prepared by vacuum deposition because this technique is easy to use and results in films prepared under relatively clean and well-understood conditions. Unless otherwise stated, the films described later in this chapter were all made by vacuum deposition.

The first requirement is a flat, smooth, clean substrate. Fire-polished soft-glass microscope slides or coverslips are commonly used for this purpose after being cut into convenient sizes. The substrates are subjected to a cleaning cycle which includes vapor degreasing, washing in a detergent, and water and organic-solvent rinses (sometimes with ultrasonic agitation). Several substrates are then placed in a holder in the vacuum chamber for simultaneous deposition (Fig. 5). The size of the film is determined by suitable masking of the substrate; for research purposes films approximately 1 cm in diameter are convenient.

In order to provide desirable film properties, it is necessary to heat the substrates during deposition, so that the substrate holder can be a metal block which heats the substrates by conduction and is in turn heated electrically. Alternatively, the substrates may be heated by radiation from electric heaters, although care must be taken to maintain temperature uniformity if this method is used. In either case the substrate temperature is usually monitored with a thermocouple attached to the substrate holder (see Chap. 2). Since the substrate temperature T_d is one of the primary parameters determining the film properties, various values of T_d are used; a value of about 250°C is common.

During deposition and subsequent cooling to room temperature, a magnetic field is applied in the plane of the substrate. The films will then exhibit uniaxial anisotropy with an EA parallel to the direction of the applied field. Since the uniaxial anisotropy is $\vec{M}$- rather than field-induced, the function of the applied field is merely to ensure that the magnetization is saturated during deposition; an applied field of 20 to 50 Oe is more than adequate for this purpose and is conveniently generated by a pair of Helmholtz coils.

The choice of evaporation method is governed by the following considerations: It is desirable to place the source at a large distance from the substrate (perhaps 25 cm) in order to reduce both thickness gradients in the films and effects due to oblique incidence of the vapor beam (Sec. 7a) for films near the edges of the substrate holder. In order to maintain reasonable deposition rates at these large distances, a broad source is preferable. Also, care must be taken to ensure that the source composition does not change with time for alloy depositions. Although depositions have been successfully made by the techniques of flash evaporation, electron bombardment, sublimation of the evaporant from heated wires, evaporation from a small charge surrounded by a hot tungsten wire, and rf induction heating of a charge in a ceramic crucible, the deposition requirements are usually best satisfied by the last technique.* For that technique the charge can be placed in a crucible (often made of alumina) which rests in the rf induction coil (Fig. 5).

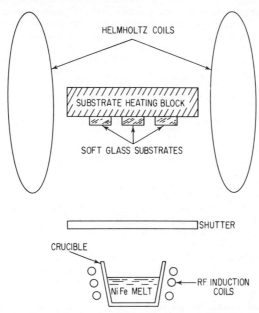

Fig. 5 An apparatus for producing ferromagnetic films of NiFe by vapor deposition in vacuo. The NiFe is evaporated from a melt heated by rf induction. When the shutter is open the vapor from the melt impinges on heated soft-glass substrates. A magnetic field in the plane of the film is provided by Helmholtz coils.

Fortunately, the common ferromagnetic metals Fe, Ni, and Co have approximately the same vapor pressures, so that films of alloys of these metals can be made by placing a mixed charge in the crucible. There is a small fractionation effect which depends on the melt temperature (roughly 1600°C for a deposition rate of several hundred angstroms per minute); thus a melt of 83% Ni, 17% Fe yields films of composition approximately 81% Ni, 19% Fe. However, in order to maintain constancy of composition throughout the film thickness, it is advisable not only to use a large charge but to maintain the charge at the evaporation temperature for several minutes before the shutter is opened to ensure equilibrium conditions. For alloy films containing nonferromagnetic constituents, simultaneous evaporation from separate sources can be used.

The thickness of the film can be monitored during deposition by the standard methods (Chap. 2); the method most commonly used involves measurement of electrical resistance of a film during deposition. In addition, the purely ferromagnetic method of monitoring the net magnetization shown by the EA loop of a hysteresigraph (Fig. 4a) has been employed.

It is usually found that an oil-diffusion-pump system which can maintain a background pressure of 10^{-5} Torr during deposition is adequate for production of films whose physical properties can be profitably studied. Appreciable changes in magnetic parameters are usually found only for depositions at much higher deposition pressures. Thus, of the residual gases in the vacuum during deposition, oxygen seems to be the

* An early report that film properties depend on crucible composition because of contamination from the crucible was not confirmed by later work.

most effective in changing the magnetic parameters, and for detectable effects (usually an increase in H_c and α_{90}) a partial pressure greater than about 10^{-5} Torr is needed. However, certain investigations, e.g., a study of the origins of K_u, may be very sensitive to residual gases and require ultrahigh-vacuum deposition.

(2) Electrodeposition[19,20]   Electrodeposition in the presence of a magnetic field can also be used to produce uniaxially anisotropic films and in fact is a technique which is of considerable technological importance in the production of cylindrical films for plated wire memories (Chap. 21). The critical parameters for this technique include the bath composition (Table 1), pH, temperature, and amount of agitation; the plating current including pulsing program if used; and the substrate preparation. As for vacuum-deposited films, smooth, clean substrates are important. Such substrates can be obtained by coating a basic substrate with a smooth metal film, where the basic substrate can be carefully polished metal or fire-polished glass.

TABLE 1   Bath Compositions for Electroplating NiFe Films*

Reagents	Concentration, g/liter		
	Wolf	Freitag	Girard
$NiSO_4 7H_2O$	227	227	464
$FeSO_4 7H_2O$	5–10	5–10	99
H_3BO_3	25	25	25
NaCl	9	0	0
Sodium lauryl sulfate	42	42	42
NaH_2PO_2	0	Variable	0
$NaAsO_2$	0	Variable	0
Thiourea	0	0	Variable

* After R. Girard, *J. Appl. Phys.*, **38**, 1423 (1967).

Uniaxially anisotropic films have been successfully electroplated under wide variations in the plating parameters.   However, it should be kept in mind that the composition of the film can be different from the composition of the bath, that some baths cause a marked composition gradient throughout the thickness of binary-alloy films, and that large stresses may be present in electroplated films.   The incorporation of thiourea in the Girard bath (Table 1) is said to alleviate some of these problems, since it gives a Ni/Fe ratio in the film which is the same as that in the bath, while the composition gradient is reduced.   In addition, at the proper thiourea concentration (about 0.2 g l^{-1}), the stress in the film can be made to vanish, while the grain size is reduced and the film surface is smoothed or "leveled." On the other hand, additives can become incorporated into the films, thus affecting the magnetic properties.   It is found, for example, that plated films exhibit an aging effect; i.e., H_K and H_c change slowly with time at room temperature; this effect may be attributed to the incorporation of impurities (notably thiourea) during electroplating.   Furthermore, the values of H_K and H_c are known to be affected by the incorporation of P and As into films from the Freitag bath (Table 1).

(3) Sputtering[21]   Sputtering can also yield high-quality ferromagnetic films, if the precautions usually recommended for application of this technique are observed. That is, the cathode (and preferably also the substrate holder) should be cooled, special measures should be taken to maintain low impurity levels in the sputtering gas, and the cathode surface should be precleaned by sputtering while a shutter covers the substrates.   In addition, measures should be taken to ensure that the field applied to obtain uniaxial anisotropy is not distorted by the ferromagnetic cathode.

During sputtering the substrate temperature can rise to several hundred degrees celsius, even with water cooling of the substrate holder, so that the substrate temperature is not always determinable as precisely as in the vacuum-deposition or electrodeposition techniques. On the other hand, for the deposition of alloy films, the fractionation effects encountered in vacuum deposition may not be so serious in sputtering if cathodes of the desired alloy composition are used. Furthermore, in common with electrodeposition, because of the random incidence angles of the incoming material, oblique-incidence effects are absent.

It should be noted that it has been claimed that application of a negative bias to the films during sputtering results in a decrease of impurity-gas concentration and hence better reproducibility, along with lower values of α_{90} and H_c. For more general background, see Chap. 4.

4. APPARATUS AND EXPERIMENTAL TECHNIQUES

Before proceeding to discuss the properties of ferromagnetic films in more detail, it will be useful to describe briefly the experimental methods by which these properties are studied. Many of these techniques are standard in the field of bulk ferromagnetism but have been modified for application to thin films, while other techniques are suitable for thin films alone. In any case, only a brief outline of the principles of each measurement method can be given here; for further details the reader is referred to the original literature.

The measurement techniques may be divided into three categories: techniques for study of the magnetization configuration within the film (e.g., the domain structure), techniques for measurement of the macroscopic quasi-static magnetic properties of the film (e.g., M, K_u, H_K, H_c, or α_{90}), and techniques for investigating the dynamic properties of the film (fast switching and resonance). In general, instruments which are used for measurements of the second category must be designed with more sensitivity than the corresponding instruments used to measure the same parameters in bulk materials. The reason is, of course, that the total mass of ferromagnetic material in the film form is very small.

a. Magnetization Configuration Studies[22]

(1) Bitter Technique In the Bitter technique a colloidal suspension of fine (about 1,000 Å diameter) ferromagnetic particles, e.g., Fe_3O_4, is allowed to flow over the film. A glass coverslip is placed over the colloid, which is then examined by a light microscope, most conveniently under dark field. The particles will be attracted to regions of high fields which originate in areas of high magnetization divergence, e.g., domain walls, thereby inducing easily visible particle aggregation at those areas (Fig. 32).

The resolution of this technique is limited by questions of contrast and particle size; small divergences of $\vec{M}$ such as magnetization ripple cannot be detected (unless the ripple is extraordinarily large). Furthermore, fast magnetization changes cannot be followed since several seconds are required for the colloid to rearrange, and the technique is applicable only within the narrow temperature range within which the colloidal suspension is stable.

(2) Magneto-optic Methods When plane-polarized light is passed through a magnetic film, the plane of polarization is rotated by an angle (about $\frac{1}{3}°$) which is proportional to the component of $\vec{M}$ parallel to the direction of propagation of the light. Thus, to obtain a useful "Faraday" rotation, light may be passed at normal incidence through a film having $\vec{M}$ normal to the film plane, but light must travel obliquely through a film having $\vec{M}$ confined to the film plane (Fig. 6). If the film is then viewed through an analyzing Nicol prism set for extinction of the light traversing a given domain, light traversing an adjacent domain of opposing magnetization sense will not be extinguished owing to its different Faraday rotation. The Faraday (or Kerr) effect thus provides contrast between *domains*, as opposed to the Bitter technique, which distinguishes domain walls.

To utilize the Kerr effect, polarized light is *reflected* from the surface of a film and viewed through an analyzing prism. There are several categories of Kerr effect,

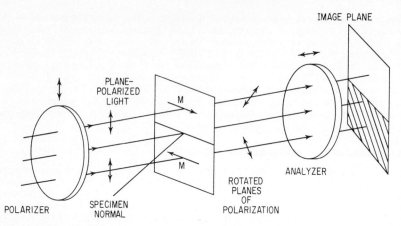

Fig. 6 A Faraday-effect apparatus for observing domain structure. A polarized light beam is passed obliquely through a ferromagnetic film, resulting in rotations of the plane of polarization according to the directions of $\vec{M}$ in the domains of the film. Upon viewing the light beam through an appropriately adjusted analyzer, contrast between domains is achieved. [*After M. Prutton, "Thin Ferromagnetic Films," Butterworth & Co. (Publishers), Ltd., Washington, 1964.*]

depending upon the relative orientations of $\vec{M}$ and the plane of polarization, but only for those in which the electric vector has a component perpendicular to $\vec{M}$ do rotations of the polarization plane occur. Probably the longitudinal meridional configuration (the electric vector of the radiation and $\vec{M}$ are in the plane of incidence of the radiation) is the most commonly used, for which the optimum incidence angle for maximum rotation is about 60°. It is evident that the Kerr and Faraday effects have the advantages of being suited for studying rapid changes in magnetization and of being usable over a wide temperature range.

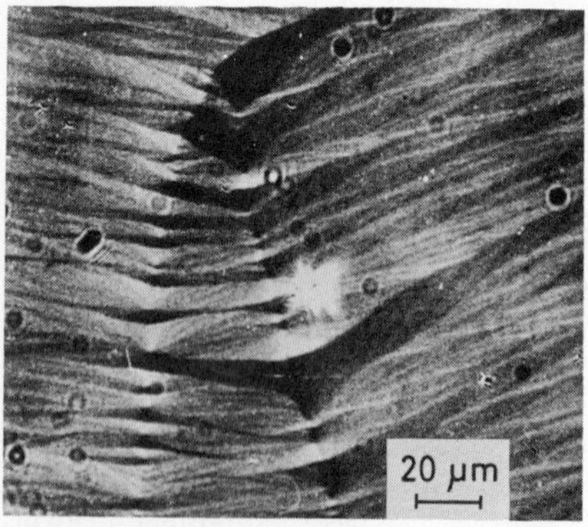

While the Kerr effect is usually used for low-magnification applications, the full resolution of light optics ($\approx 1\ \mu$) is attainable if a high-numerical-aperture objective lens is used (Fig. 7). However, care must be taken to maximize the contrast, which is inherently low. Aside from precautions taken in instrumentation, this goal can be approached by "blooming" the film with a dielectric overlayer which is designed to enhance the Kerr effect by multiple reflections. Optimum thickness and indices of refraction of a dielectric superstructure designed to provide the greatest enhancement of the Kerr and Faraday effects have been calculated.

(3) Lorentz Electron Microscopy With the proper instrumental adjustments, magnetic structure can be viewed by transmission electron microscopy. In the *defocused* mode of operation (Fig. 8), the objective lens is focused at an image plane at some distance (typically a few centi-

Fig. 7 High-resolution Faraday-effect micrograph of Fe film showing ripple structure. (*After H. Boersch and M. Lambeck, "Electric and Magnetic Properties of Thin Metallic Layers," p. 91, Koninkl. Vlaam. Acad. Wetenschap. Letter. Schone Kunsten, Belg. 1961.*)

meters) from the film. Now, because of the Lorentz force $\vec{qv} \times \vec{B}$ acting on them, the electrons are deflected on passage through the film, in senses depending on the $\vec{M}$ direction. Then, as seen from Fig. 8, at the image plane an increased electron intensity will result below some walls and a decreased intensity below others, depending on the orientation of the electron beam relative to the senses of $\vec{M}$ in the adjacent domains. The defocused mode of operation thus detects divergences of $\vec{M}$, e.g., domain walls* (Fig. 31), in analogy with the Bitter technique.

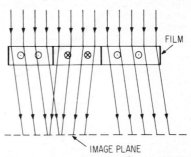

Lorentz microscopy offers the advantage of high resolution, e.g., it can easily detect magnetization ripple. However, the quantitative interpretation of Lorentz micrographs is not straightforward, i.e., the intensity distribution depends upon the defocusing distance (if the microscope is operated *at focus*, the contrast originating in magnetic structure is entirely lost). The situation is further complicated by the fact that the geometric-optical theory of contrast formation presented above is only the first approximation to a more complicated wave-optical theory. Since the geometric-optical theory is only marginally valid for handling the imaging of domain walls, calculations of domain wall shapes based on the geometric-optical interpretation must be treated with caution; a fortiori, magnetization ripple can be quantitatively studied by Lorentz microscopy only if the wave-optical interpretation is employed.

Fig. 8 Illustrating the contrast-formation mechanism of defocused Lorentz electron microscopy. The electron beam passes vertically through the sample and is deflected by the Lorentz force in a sense which depends on the sense of $\vec{M}$ in a given domain. The microscope is focused on an "image" plane at a distance from the sample, so that for the "diverging" or "converging" domain wall, a reduction or increase of electron intensity corresponding to the domain wall is detected, respectively. [*After M. S. Cohen, IEEE Trans. Magnetics*, **MAG-1**, *156* (*1965*).]

b. Magnetic-parameter Determinations[6,23]

The thickness and composition of ferromagnetic films are measured by the usual techniques applicable to all thin films (Chap. 10), but special techniques have been developed for measurement of the parameters pertinent to quasi-static ferromagnetic behavior. Several methods are usually available for measurements of any given parameter. However, in choosing one of these methods, care must always be taken to minimize any spurious results due to the influence of ripple.

(1) Torque Magnetometer Although this instrument is somewhat inconvenient to use because of its mechanical delicacy, it can provide reliable direct measurements of several parameters, notably magnetic-anisotropy constants. A typical instrument is shown in Fig. 9. The sample is suspended from the bottom of the coil of a moving-coil galvanometer so that the plane of the film is horizontal and in the gap of an electromagnet. A rotating dc field can thus be applied to the film, which causes a varying torque $\vec{T}$, originating in the anisotropy of the film, to be exerted on the suspension. However, any resulting incipient rotation of the suspension is detected by the split-photocell arrangement, and current is automatically fed back to the balancing coil so that the suspension is kept in its original position. The feedback current is thus a measure of the torque, and a signal proportional to it is fed to the Y axis of an X-Y recorder. Then by feeding a signal proportional to the azimuth β of the electromagnetic to the X axis, a plot of T vs. β is obtained automatically.

---

* The mean direction of $\vec{M}$ at any point in the film is immediately determinable from a Lorentz micrograph, for $\vec{M}$ is always perpendicular to the ripple lines.

Assuming that $\vec{H}$ is large enough to keep the film saturated in equilibrium, the torque exerted by $\vec{H}$ on $\vec{M}$ is always balanced by the torque $T = \partial E_K / \partial \varphi_0$ exerted by the sample on $\vec{M}$, where E_K is the anisotropy energy (Sec. 2b) and φ_0 is the angle between $\vec{M}$ and the EA (Fig. 3). If H is large enough, φ_0 is approximately equal to the field angle β, so that the magnetometer records a signal proportional to $\partial E_K / \partial \beta$ as a function of β. From Eq. (8) with $\beta = \varphi_0$, it is seen that for a uniaxially anisotropic film this signal is thus proportional to $K_u \sin 2\beta$. Then after calibration of the instrument, K_u can be determined from the full excursion of the $\sin 2\beta$ trace and a

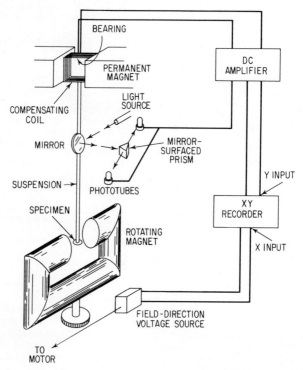

Fig. 9 An automatic torque magnetometer. [*After R. F. Penoyer, Rev. Sci. Instr.*, **30** *711 (1959)*.]

knowledge of the film thickness (since the signal is also proportional to the sample volume). An example of an experimental trace from a film only 35 Å thick is shown in Fig. 10. Because the torque from films is so small (less than 10^{-2} dyn-cm), the instrument must be extremely sensitive.

Since the value of H can be made high enough to suppress the ripple, ripple cannot influence the measurement. On the other hand, if measurements are made at values of H for which ripple is present, rotational hysteresis originating in the ripple can be studied (Sec. 11d).

If the sample is suspended so that the film plane is vertical, application of a small field $\vec{H}$ in the horizontal plane at an angle θ to the film plane induces a torque $MVH \sin \theta$, where V is the film volume, thus permitting the determination of M if the thickness is known. If H is now increased so that $\vec{M}$ is pulled out of the film plane, $K_\perp$ (Sec. 7b) can be measured.

(2) Hysteresigraph Because of its experimental simplicity and its ability to measure most of the pertinent magnetic parameters, the hysteresigraph is the most commonly used instrument for routine examination of magnetic films. In ˜this instrument (Fig. 11) a pickup coil either surrounds or is adjacent to the film, so that the coil

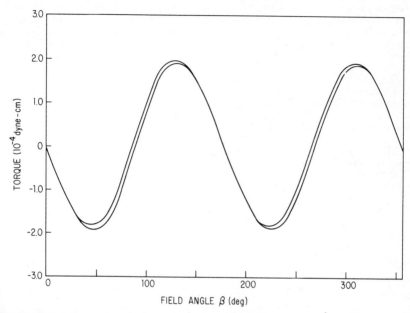

Fig. 10 Torque magnetometer trace; torque vs. field angle β, for a 35 Å, 76% Ni, 24% Fe film. The value of K_u is proportional to the vertical excursion of the curve. [*After F. B. Humphrey and A. R. Johnston, Rev. Sci. Instr.*, **34**, *348 (1963)*.]

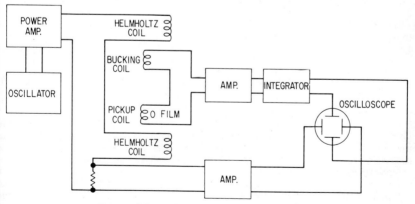

Fig. 11 Schematic diagram of a hysteresigraph.

encloses an appreciable fraction of the magnetic flux originating from magnetic poles at the film edges. The magnetization in the film changes with time in response to an ac excitation field which is produced in the plane of the film by a pair of Helmholtz coils, whose axis is parallel to the axis of the pickup coil. Then by Faraday's law of induction, a voltage is induced in the pickup coil which is proportional to the rate of change of the total flux linking it. This total flux includes a large spurious component due to the mutual inductance of the Helmholtz and pickup coils; this component is customarily canceled by the voltage induced in an auxiliary "bucking" coil, which is within the Helmholtz coil but far from the film. To obtain a hysteresis loop, the compensated pickup voltage, after integration and amplification, is fed to the vertical plates of an oscilloscope, while a voltage proportional to the excitation field is fed to the horizontal plates. In designing the instrument, care must be taken to balance possible phase shifts between various signals, to use a wideband amplifier in order to reproduce the MH loop with good fidelity, and to use a good pickup-coil design in order to obtain the best signal-to-noise ratio.

Since the height of the loop is proportional to both M and d, upon proper calibration either of these quantities may be measured. By applying the excitation field along the EA, H_c may be determined by noting the minimum field for which hysteresis can be detected (Sec. 3a). If a small excitation field is applied along the HA of a film in the single-domain state, H_K may be taken as the field at which the extrapolation of the resulting closed loop cuts the high-drive saturation value of M (Sec. 3a). However, this technique of measuring H_K is valid only for low-dispersion films, since the procedure is ripple-dependent (Sec. 11c).

The hysteresigraph may be used in the determination of the magnetoelastic coupling constant $\eta \equiv dH_K/de$ (Sec. 7a). The film, on a thin ($\sim$0.2 mm) glass substrate, is bent between three or (for better strain uniformity) four knife-edges, while the value of H_K as a function of strain is monitored by the hysteresigraph technique described above. Also, with the provision of a dc field perpendicular as well as parallel to the excitation field, and with appropriate amplification of the pickup signal (preferably using a lock-in amplifier), initial susceptibility studies (Sec. 11c) can be made with the hysteresigraph apparatus.

The HA fallback determination of α_{90} (Sec. 11b) may be made by repositioning the pickup coil so that its axis is perpendicular to the excitation field; the pickup coil now detects the changing flux component normal to the excitation field. A large excitation field (greater than H_K) is then applied along the HA. As the sample is rotated to give a gradually increasing angle α between the HA and the excitation field, or alternatively as a gradually increasing dc EA bias field is applied, a signal is recorded which increases to a saturation value when $\vec{M}$ everywhere in the film rotates in the same sense. The parameter α_{90} is taken as that value of α for which 90% of the saturation signal is recorded.

As discussed above, the ordinary hysteresigraph determination of H_K is subject to spurious ripple influences. However, several flux-pickup techniques are now available which permit measurement of H_K without ripple influence. The techniques employ apparatus very similar to the basic hysteresigraph; the major modifications involve the addition of auxiliary dc field or pickup coils.

It should also be mentioned that flux pickup is not the only detection scheme which can be employed in a hysteresigraph, although it is the most common. Thus, magneto-optic detection can be utilized by replacing the ocular of a Kerr or Faraday apparatus with a photomultiplier tube. Upon proper adjustment of the apparatus, the output of the tube can be made proportional to the component of $\vec{M}$, averaged over the entire film, which is parallel to the direction of light propagation. Similarly, with suitable circuitry, magnetoresistance or Hall-effect detection can also be used for hysteresigraph operation.

c. Dynamic Studies

The measurement techniques described above were designed for quasi-static studies. Special pulse or high-frequency techniques must be used for dynamic studies involving fast magnetization changes.

(1) Switching In order to study fast magnetization reversal, it is necessary to apply short-rise-time (less than 1 ns) step-function field pulses of several oersteds to the film. The commonest arrangement involves application of such pulses along the EA, in the presence of a steady dc HA field, with detection of the reversal process by the flux-pickup method.

The geometry of a stripline is convenient for these measurements. The current pulse is usually generated by charging a length of coaxial cable through a resistor, discharging the cable with a mercury relay, then feeding the resulting current pulse into the stripline. The magnetic film is placed on the stripline so that it experiences an EA magnetic field resulting from the current pulse (Fig. 12). A longitudinal (transverse) pickup loop detects the changes in the component of $\vec{M}$ which is parallel (perpendicular) to the pulse-field direction, i.e., in the EA (HA) direction. The

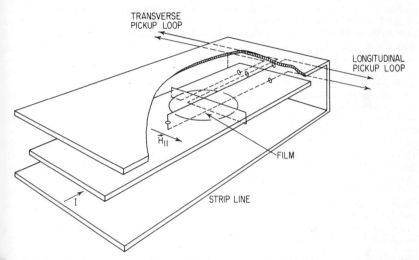

Fig. 12 Stripline arrangement for observing fast magnetization reversal. The stripline is fed by a current pulse I from a charged coaxial cable, which generates a field pulse $\vec{H}_\parallel$ parallel to the EA. [*After K. U. Stein, Z. Angew. Phys.,* **20**, *36 (1965).*]

"air-flux" component of the longitudinal pickup signal corresponding to direct electromagnetic coupling with the stripline must somehow be canceled out, since only the signal originating from the magnetic film is of interest. This cancellation may be accomplished by a series-connected, antiphase dummy loop, or by an arrangement by which the symmetry of the geometry allows automatic compensation of the air-flux signal (Fig. 12). Compensation of the transverse pickup loop is not necessary since it is perpendicular to the pulse field. The signals from both loops are fed to sampling oscilloscopes, with integration (so as to display not dM/dt but M on the oscilloscope) carried out either before or after the sampling process.

Using these techniques, the total time resolution of the system can be brought to a few tenths of a nanosecond.

(2) Ferromagnetic Resonance In order to study ferromagnetic resonance of magnetic films, it is necessary to apply an rf field in the plane of the film and a dc field perpendicular to the rf field. The dc field is usually either in the film plane or normal to it; if it is in the film plane it may have any arbitrary orientation with respect to the EA. In the low-frequency range it is possible to detect the resonance condition (which depends on the properties of the film as well as the magnitude of the dc field and the rf frequency) by the flux detected by pickup coils. However, in the high-frequency (above 1 GHz) range, microwave techniques are necessary.

For microwave-resonance studies, the film is placed inside a resonant cavity which is fed through a waveguide with radiation originating from a klystron. The film is put in a position so that the rf $\vec{H}$ field is maximum and the $\vec{E}$ field is minimum.

Then, since the rf frequency is of course fixed, resonance must be found by varying the dc field. At resonance the energy absorbed by the film (and hence by the cavity) is maximum, so that the resonant condition may be detected by monitoring the electromagnetic energy reflected back from the cavity.

5. THICKNESS DEPENDENCE OF MAGNETIZATION[24,25]

a. Theory

The question of the thickness dependence of the magnetization represents a fundamental aspect of the comparison of bulk and thin film ferromagnetic phenomena. The spin on an atom at the surface of a uniformly magnetized ferromagnetic film is less rigidly constrained to the average direction of $\vec{M}$ than the spin on an interior atom simply because the aligning quantum-mechanical exchange bonds are absent on the exterior side of the surface atom. Then, as the film thickness is gradually decreased and the ratio of surface atoms to interior atoms increases, it is expected that the cooperative aligning forces which cause ferromagnetism will be gradually weakened, so that the saturation magnetization M and the Curie temperature T will decrease. In fact, in the limit of a monatomic layer, calculations show that ferromagnetism may disappear entirely, even at 0°K. The question of the dependence of M and T_c on d, along with the question of the dependence of M on temperature for a given d, has received much experimental and theoretical attention not only because of the implications for thin films, but because this problem provides insight into the fundamental understanding of magnetism.

The theoretical side of the problem has been approached through spin-wave, molecular-field, and electron-band theory. All these theories assume a perfect flat, monocrystalline film, infinite in extent, and with plane-parallel boundaries.

In the original Klein-Smith spin-wave theory, the same periodic boundary conditions were used as were employed for bulk material, while in the summation over all the spin waves to obtain M, the uniform mode term $(\vec{k} = 0)$ was omitted because it led to an awkward divergence which is not found in the bulk case. Later workers pointed out that instead of periodic boundary conditions, special boundary conditions at the two film interfaces should be assumed for spin waves propagating in the direction normal to the film plane. The correct boundary conditions must take account of the fact that the spins at the interfaces may be either entirely free or subject to some constraint in their motion ("pinned") depending on the magnetic anisotropy at the interface layers. Furthermore, the divergence originating in the $\vec{k} = 0$ term disappears if a small external or anisotropy field is assumed to be present, as is always true in practice. An improved theory predicted that M will not decrease with decreasing d until d values less than about 30 Å (for NiFe) are reached; the original theory predicted a much greater threshold d.

While progress was being made on the spin-wave approach, Valenta applied the Heisenberg molecular-field formulation to thin films. The film was divided into sheets of atoms parallel to the substrate, and the magnetic interaction between a given atom in one sheet and nearest-neighbor atoms in the other sheet was found. In agreement with the spin-wave theory, this theory also indicated little change of M with d except for the thinnest films. Valenta's theory was extended to include magnetic anisotropy and was used to calculate not only the *average* magnetization but M as a function of depth in the film. This method also permits calculations of M at high temperatures (and hence calculations of the Curie temperature) where the earlier spin-wave theories were invalid.

Most recently, the sophisticated Green's-function method has been applied; the approach offers the advantage of being valid over the entire temperature range, but the mathematics involved are not easily tractable. Attempts to construct a theory of M vs. thickness based on the electron-band theory of magnetism have also been initiated.

b. Experiment

The first experimental data showed a rapid decrease of M with decreasing film thickness below a threshold thickness of 150 Å and were interpreted to be in agreement with the Klein-Smith theory. However, more recent data, taken from films deposited at pressures $\lesssim 10^{-8}$ Torr and measured directly in the vacuum system at high fields, show that the threshold thickness is much less. Thus (Fig. 13) measurements of M vs. temperature indicate that Ni films with thicknesses greater than about 20 Å exhibit essentially the bulk behavior. Measurements of M at a fixed temperature were made for Ni films of various thicknesses with a torque-magnetometer technique (Fig. 14), and for Fe films with a microbalance technique. In both cases, at room temperature, bulk values of M down to $d = 30$ Å were found.

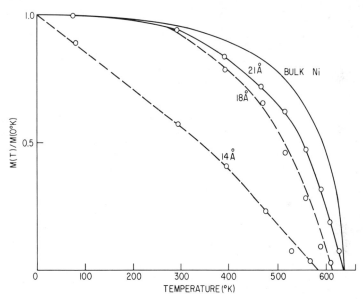

ig. 13 Saturation magnetization relative to value at 0°K vs. temperature for Ni films of arious thicknesses. Films were prepared and measured at pressures less than 10^{-8} Torr. After C. A. Neugebauer, Z. Angew. Phys., **14**, 182 (1962).]

Two reasons have been advanced to explain the disagreement of the early data ith the more recent data: (1) The earlier films were prepared at 10^{-6} Torr and/or easured under conditions such that oxidation occurred. (2) Conditions were such hat the films were not completely magnetically saturated during measurement. lear evidence of the importance of oxidation was provided by experiments which ave the $d = 30$ Å limit for Fe films deposited and measured at 10^{-5} Torr, but the ~ 150 Å limit (similar to the original M vs. d data) after admission of air into the acuum chamber. Similarly, experiments on Ni films showed that oxidation upon dmission of air led to an appreciable decrease in magnetization. On the other hand, ther workers prepared Ni films at 10^{-5} Torr and measured them in air after over-oating with protective SiO layer. The bulk value of M was found for d down to 5 Å as long as fields high enough to saturate the samples were used; sometimes elds in excess of 3×10^3 Oe were required. This experiment emphasizes the requirement of large saturating fields during M measurements rather than good vacuum onditions.

The need for prevention of oxidation and the need for high saturating fields are lated, however. Thinner films have a more granular character, with larger spacing between crystallites; in magnetic terms thinner films assume the character of a collection of superparamagnetic particles, which require a higher saturating field. Thus, the granularity of ultrathin Fe films was clearly demonstrated by measurements of electrical resistance as a function of film thickness, while the superparamagnetic nature of ultrathin Ni films was shown by observations of magnetic-viscosity effects (Sec. 9c). Now besides causing the true magnetic thickness to be less than the assumed thickness, oxidation can promote superparamagnetic behavior because it breaks the intercrystalline exchange bonds. This was proved by an experiment which showed that Fe films deposited at temperatures above 100°C retained bulk values of M for much smaller values of d than did films deposited at room temperature. By monitoring the electrical resistivity as a function of time, the origin of this difference in behavior was demonstrated to lie in a greater susceptibility to oxidation of the low-temperature (and hence presumably fine-grained) films.

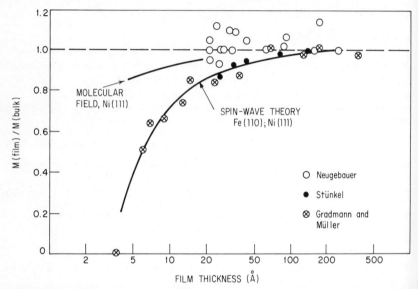

Fig. 14　Dependence of M on film thickness, where M is normalized to the bulk value. Comparison of spin-wave and molecular-field theoretical values with experimental result for polycrystalline Ni (*Neugebauer*), polycrystalline Fe (*Stünkel*), and epitaxial 48% Ni 52% Fe (*Gradmann and Müller*). [*After U. Gradmann and J. Müller, Phys. Stat. Solidi* **27,** *313 (1968)*.]

In summary, the best experimental work indicates, in agreement with the best theoretical analyses, that there is no decrease of M (and T_c) with decreasing d until d is less than 20 or 30 Å. Ultrathin films are granular and superparamagnetic in character, however, and therefore do not fit the flat, planar geometry postulated by the theories. Thus a detailed experimental fit of the theories is not really meaningful for $d \lesssim 50$ Å. For this reason, an attempt has been made to calculate M for a granular film using a sublattice model similar to that of Valenta. This is a formidable theoretical problem, however, especially since the details of the crystallite shapes and the degree of mutual coupling are unknown.

A more promising method of improving the match between theory and experiment may be to fabricate films which are flat, monocrystalline, and planar in the ultrathin range. It has been reported that Ni and NiFe films, when grown epitaxially on Cu films which were in turn epitaxially grown on mica, showed no superparamagnetic behavior even for thicknesses as small as 7 Å, while reflection electron-diffraction spots were narrow and elongated, which indicates flatness. Using a pendulum magnetometer, M of NiFe films was measured as a function of d for films as thin as 5 Å; the results were in better agreement with recent spin-wave theory than with molecular-field theory (Fig. 14). These results are gratifying, although they may be subject to the objection that even if the films are flat, the theoretical conditions are not met because the magnetic material is bounded not by vacuum but by Cu. The importance of this change in boundaries is not clear, however.

6. COHERENT $\vec{M}$ MODEL

As was explained in Sec. 3a, it is well known that the magnetic behavior of uniaxially anisotropic films can depart significantly from the predictions of a model based on the assumption of a single domain with complete magnetization coherence at all times. Nevertheless, the coherent model is extremely useful because it offers a first approximation to the film behavior, and this first approximation provides a convenient basis of comparison for higher-order approximations.

a. Statics—Stoner-Wohlfarth Theory[6]

The Stoner-Wohlfarth (S-W) theory, which was originally developed for single-domain, uniaxially anisotropic, small ferromagnetic particles, may be applied to thin films in which $\vec{M}$ remains in the film plane at all times. In the latter case, the problem is to predict the angle φ_0 of $\vec{M}$ for any arbitrary value of $\vec{H}$ (Fig. 3). As before (Sec. 3a), this problem is solved by finding that value of φ_0 giving the minimum total energy of Eq. (15). It is convenient to define the normalized quantities*

$$\epsilon_{\text{tot}} \equiv E_{\text{tot}}/MH_K, \ \vec{h} \equiv \vec{H}/H_K, \ h_\parallel \equiv H \cos \beta/H_K, \text{ and } h_\perp \equiv H \sin \beta/H_K$$

where $\parallel$ and $\perp$ refer to the "longitudinal" and "transverse" components parallel to and perpendicular to the EA, respectively. Then Eq. (15) becomes

$$\epsilon_{\text{tot}} = \tfrac{1}{2} \sin^2 \varphi_0 - h_\parallel \cos \varphi_0 - h_\perp \sin \varphi_0 \tag{17}$$

The condition for equilibrium is that $\partial \epsilon_{\text{tot}}/\partial \varphi_0 = 0$, or that the net torque Λ_0 vanishes ($\Lambda_0 \equiv \partial \epsilon_{\text{tot}}/\partial \varphi_0$). Application of this condition to Eq. (17) gives

$$\Lambda_0 = \tfrac{1}{2} \sin 2\varphi_0 + h_\parallel \sin \varphi_0 - h_\perp \cos \varphi_0 = 0 \tag{18}$$

Values of φ_0 satisfying Eq. (18) represent stable equilibrium (minimum-energy states) only if $\Lambda_1 > 0$, where

$$\Lambda_1 \equiv \frac{\partial \Lambda_0}{\partial \varphi_0} = \cos 2\varphi_0 + h \cos (\varphi_0 - \beta) \tag{19}$$

Here Λ_1 is called the *uniform effective* or *single-domain* field.

Assume that $\vec{M}$ is pointing to the right along the EA. If now an increasing field $\vec{h}$ (Fig. 15a) is applied at an arbitrary fixed obtuse angle β, $\vec{M}$ will rotate clockwise. If h is less than a certain threshold field h_s, $\vec{M}$ will relax back to its original position upon removal of the field, as in the case of $\vec{h}$ parallel to the HA (Sec. 3a). However, when h exceeds h_s, an irreversible transition process will occur in which $\vec{M}$ discontinuously jumps into a new position in the second quadrant between $\vec{h}$ and the EA (Fig. 15b). The film is now said to have been "switched," for upon setting $\vec{h} = 0$, $\vec{M}$ is found to point *left* along the EA. The switching transition occurs when an energy extremum changes from a minimum to a maximum, i.e., when $\Lambda_1 = 0$. The solution of these conditions is [from Eq. (19)] $h_\parallel = - \cos^3 \varphi_0$, $h_\perp = \sin^3 \varphi_0$, or, suppressing the signs,

$$h_\parallel^{\frac{2}{3}} + h_\perp^{\frac{2}{3}} = 1 \tag{20}$$

* Fields normalized to H_K are denoted by lowercase letters, i.e., $\vec{h} \equiv \vec{H}/H_K$, $h_c \equiv H_c/H_K$.

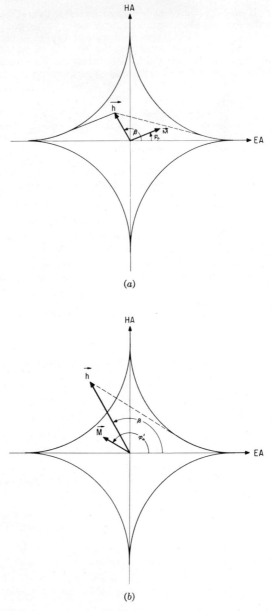

Fig. 15 Illustrating the graphical determination of the azimuth of $\vec{M}$ using the Stoner-Wohlfarth astroid. Field $\vec{h}$ applied at an angle β. (a) $\vec{M}$ at angle φ_0 before switching (b) $\vec{M}$ at angle φ_0' after switching due to growth in magnitude of h.

which represents a mathematical figure known as an astroid (Fig. 15). The uniform effective field Λ_1 thus vanishes when $\overrightarrow{h}$ is on the astroid, or *critical switching curve*. Following this procedure, the angle φ_0 can be calculated [from Eq. (18)] for any given values of h and β. The switching threshold is of particular interest; Table 2 gives the magnitude of the switching field h_s for a given azimuth β, and the $\overrightarrow{M}$ angles φ_0 and φ_0' just before and after switching, respectively.

TABLE 2　Azimuth β and Magnitude h_s of Switching Field, with Angles φ_0 and φ_0' of $\overrightarrow{M}$ before and after Switching*

β	h_s	φ_0	φ_0'
180	1.00000	0.000	180.000
178	0.85929	18.100	179.076
170	0.67381	29.287	175.971
150	0.52402	39.784	169.671
135	0.50000	45.000	165.000
100	0.67381	60.713	144.534
94	0.79237	67.609	133.013
90	1.00000	90.000	90.000

* After E. C. Stoner and E. P. Wohlfarth, *Phil. Trans. Roy. Soc. London*, **A240**, 599 (1948).

However, instead of using Table 2, the orientation of $\overrightarrow{M}$ may be found by a simple graphical construction which is helpful in visualizing switching. First $\overrightarrow{h}$ is plotted in the reduced field space (Fig. 15a) in which the astroid has been drawn. If now a straight line is drawn through the tip of $\overrightarrow{h}$ so that it is tangent to the astroid, then $\overrightarrow{M}$ can be shown to lie parallel to this line. It may be noted, however, that when the tip of $\overrightarrow{h}$ lies within the astroid it is intersected by tangents to *both* upper branches of the astroid (the light lines in Fig. 15a). Which tangent gives the position of $\overrightarrow{M}$ depends on the magnetic history; assuming $\overrightarrow{M}$ pointed right (left) along the EA at $\overrightarrow{h} = 0$, the solid (dotted) tangent would be used. If $\overrightarrow{h}$ is increased in magnitude so that its tip touches the astroid, by the definition of the latter, a switching threshold is achieved ($\Lambda_1 = 0$). A further increase in h gives a state where only *one* tangent representing equilibrium can be drawn, and $\overrightarrow{M}$ jumps to a new position in the second quadrant (Fig. 15b). The film has then been switched.

Since the above considerations give the azimuth of $\overrightarrow{M}$ for any $\overrightarrow{H}$, theoretical quasi-static hysteresis loops (Sec. 3a) may be calculated. Consider, for example, the important special case where $\overrightarrow{H}$ is parallel to the EA, i.e., $\beta = 0$. Then the solutions of Eq. (18) for $h_\perp = 0$ are $h_\parallel = -\cos\varphi_0$, and $\varphi_0 = 0, \pi$. From Eqs. (17) and (19) it is seen that the former and latter solutions are unstable ($\Lambda_1 < 0$) and stable states ($\Lambda_1 > 0$), respectively, and that the transitions between the stable states $\varphi_0 = 0$ and $\varphi_0 = \pi$ occur (at $\Lambda_1 = 0$) at $h_\parallel = \pm 1$. The resulting hysteresis loop obtained by plotting $M\cos\varphi_0$ vs. $h_\parallel$ is shown in Fig. 16a; it is rectangular with a coercivity H_c of H_K. Another important case is the HA loop, given by plotting $M\sin\varphi_0$ vs. $h_\perp$. For this case, solutions of Eq. (18) (with $h_\parallel = 0$) are $h_\perp = \sin\varphi_0$, and $\varphi_0 = \pm\pi/2$; from Eqs. (17) and (19) stable solutions are given by the former expression for $|h_\perp| \leq 1$ and the latter expression for $|h_\perp| \geq 1$ (Fig. 16b). Physically, $\overrightarrow{M}$ rotates with increasing $h_\perp$ until at $h_\perp = 1$ it is parallel to the HA, where it remains for still greater values of $h_\perp$. Hysteresis loops for azimuths other than along the EA and HA may be similarly determined.

The theoretical hysteresis loops of Fig. 16 may be compared with the experimental loops of Fig. 4. It is seen that although the EA loops are roughly similar in appearance, the value of field at which magnetization reversal takes place is very different in the two cases. Thus, experimental EA reversal occurs at a field H_c which is much less than the value H_K predicted by the S-W theory. (As explained in Sec. 3a, the reason for the discrepancy is the nucleation and growth of reverse domains.) On the other hand, the experimental HA loop (Fig. 16b) for low ac excitation fields is in good agreement with the theoretical loop (Fig. 4b). The opening of the HA loop for higher excitation fields indicates the initiation of some incoherent processes.

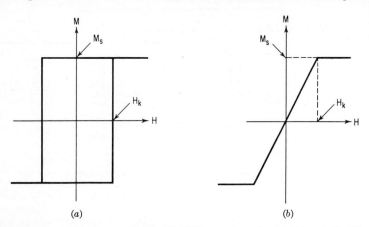

(a)   (b)

Fig. 16 Theoretical Stoner-Wohlfarth MH loops along the (a) EA and (b) HA.

It should be noted that magnetization azimuths, critical switching curves, and hysteresis loops may similarly be calculated, assuming coherent behavior, for films having anisotropies other than uniaxial. Such calculations have been performed for films having biaxial, triaxial, and combinations of biaxial and uniaxial anisotropy.

b. Dynamics[3]

In order to study the switching transition in detail, i.e., the trajectory of $\vec{M}$ between the unswitched (Fig. 15a) and switched (Fig. 15b) positions, the static S-W theory, which is good only for the steady state, must be extended to a dynamic theory. This can be accomplished through the application of the Landau-Lifshitz equation [Eq. (14)].

In experimental investigations of switching, the arrangement of Fig. 17 is usually used. A "transverse" dc field $H_\perp$ is applied along the HA, and a step-function pulse "longitudinal" field $H_\parallel$ is applied along the EA. Pickup coils (shown schematically) are arranged to detect both the EA and HA components of the changing flux from the film; the analysis of these signals gives information about the magnetization during the switching transition. It is therefore useful to express the predictions of the dynamic coherent rotation theory by calculating the theoretical waveforms of the pickup signals.

For this calculation it is first necessary to modify Eq. (17) to take account of the fact that in general the magnetization lifts slightly (a few degrees) out of the film plane during switching. The total normalized energy then becomes

$$\epsilon_{tot} = \tfrac{1}{2} \sin^2 \varphi_0 - h_\parallel \sin \theta \cos \varphi_0 - h_\perp \sin \theta \sin \varphi_0 + 2\pi M^2 \cos^2 \theta \qquad (21$$

Note that Eq. (21) reduces to Eq. (17) when $\vec{M}$ is in the film plane; i.e., $\theta = \pi/2$

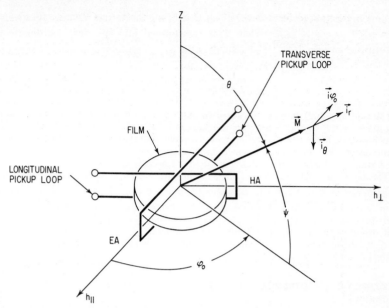

Fig. 17 Schematic diagram of detection arrangement used in switching experiments, along with a coordinate system used in the theoretical analysis. [*After D. O. Smith, in G. T. Rado and H. Suhl (eds.), "Magnetism," vol. 3, p. 465, Academic Press Inc., New York, 1963.*]

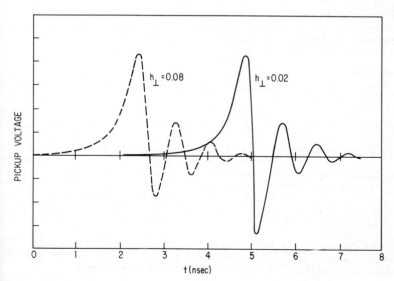

Fig. 18 Theoretical longitudinal switching signal for various values of $h_\perp$. It was assumed that $H_K = 3$ Oe, $\lambda = 10^8$ Hz, and the $h_\parallel$ pulse field was of magnitude 1. [*After D. O. Smith, J. Appl. Phys., 29, 264 (1958).*]

If now a generalized force $\vec{F} \equiv -\nabla \epsilon_{\text{tot}}$ is defined, the torque $\vec{T}$ on $\vec{M}$ is given by $\vec{T} = \vec{r} \times \vec{F} = -\vec{r} \times \nabla \epsilon_{\text{tot}}$. Then $\vec{T}$ is transformed into spherical coordinates and substituted into Eq. (14). Upon equating the $\vec{i}_{\varphi_0}$ and $\vec{i}_\theta$ components, two simultaneous equations for φ_0 and ψ result, which may be combined into

$$\ddot{\varphi}_0 + 4\pi\lambda\dot{\varphi}_0 + \frac{4\pi\gamma^2 \, d\epsilon_{\text{tot}}}{d\varphi_0} = 0 \tag{22}$$

To obtain Eq. (22) it was necessary to use the fact that $\psi \ll 1$. Typical results, obtained with a digital computer, are shown in Fig. 18. It is seen from the figure that the waveform has an oscillatory character and that as the transverse field $h_\perp$ is increased, the duration of the switching process decreases. The magnitude of the oscillations depends on the value assumed for the damping constant λ.

The physical interpretation of the switching process is that the torque exerted on $\vec{M}$ by $h_\parallel$ causes $\vec{M}$ to precess initially about $h_\parallel$ until a ψ value of a few degrees is attained, then $\vec{M}$ precesses in the φ_0 direction about the normal demagnetizing field thus created. This process is completed faster for larger $h_\perp$, since the initial torque is then larger. The origin of the λ used in switching is not understood; it is clear only that for thin ($d \lesssim 5{,}000$ Å) films the eddy-current contribution should be small, whereas in bulk material it would provide the major contribution to λ.

7. MAGNETIC ANISOTROPIES AND THEIR ORIGINS

The magnetic anisotropies which can be found in ferromagnetic films play a central role in determining their magnetic behavior, as was explained for induced uniaxial anisotropy (Sec. 3a). An understanding of the origins of the various magnetic anisotropies is thus a primary goal of research in ferromagnetic films.

Several magnetic anisotropies may be present simultaneously. The resulting magnetic behavior will depend on the net anisotropy energy as a function of the direction of $\vec{M}$ [Eq. (3)]; i.e., the *energy* contribution from each mechanism must be added. For example, two uniaxial anisotropies originating in two different mechanisms, with noncollinear EAs, give a net anisotropy which is still uniaxial, of predictable magnitude, and with an EA in a predictable direction. Thus, if the two values of K_u are equal and the EAs are 90° apart, from Eq. (8) the total energy is

$$(E_K)_{\text{tot}} = K_u \left[\sin^2 \varphi_0 + \sin^2 \left(\frac{\pi}{2} - \varphi_0 \right) \right] = K_u$$

i.e., the film is *isotropic*, and not biaxially anisotropic as may be first thought.

It is assumed that all the anisotropy mechanisms which are operative in bulk material (Sec. 2) may also be present in thin films. Because of the physical structure peculiar to thin films, however, the manifestations of these mechanisms are often different from the bulk case, as explained below.

a. Uniaxial Anisotropy in the Film Plane

(1) Strain-Magnetostriction Anisotropic strain in the plane of the film can create a uniaxial anisotropy, with the anisotropy constant K_u given by Eq. (7). This mechanism acts in films exactly as would be expected from the bulk behavior; i.e., anisotropic strain in the film plane creates an EA along the direction of tension (compression) for positive (negative) λ. The magnetoelastic strain coefficient $\eta \equiv dH_K/de$, which gives the change in H_K for a given increment of strain de, is used to characterize this phenomenon. The measurement of η is conveniently made with a hysteresigraph (Sec. 4b). It is found that η has a strong dependence on the film composition; in fact, η is often used as a convenient index of the composition (Fig. 19).

For many purposes it is preferable to avoid strain-magnetostriction anisotropy. For example, an accidental anisotropic strain in the substrate can easily cause the net EA to deviate the direction of the field applied during deposition. Such a deviation is known as *skew* of the EA and is usually undesirable, particularly in technological applications. However, even if the strain in the film is macroscopically isotropic, there may be local variations in the strain which can contribute to the magnetization dispersion (Sec. 10), which is again undesirable. For these reasons it is often advantageous to fabricate films which are "nonmagnetostrictive," i.e., films of composition such that η is as close to zero as possible. The zero magnetostriction composition (ZMC) is about 81% Ni, 19% Fe (Fig. 19), which corresponds to a melt composition

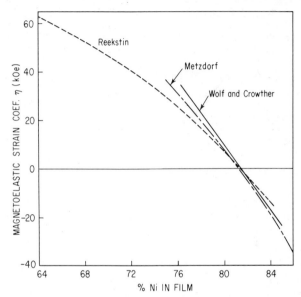

Fig. 19 Magnetoelastic strain coefficient $\eta \equiv dH_K/de$ vs. composition for NiFe films. Metzdorf and Reekstin, vacuum-deposited films; Wolf and Crowther, electroplated films. *After W. Metzdorf, IEEE Trans. Magnetics,* **MAG-2,** *575 (1966); J. P. Reekstin, J. Appl. Phys.,* **38,** *1449 (1967); and I. W. Wolf and T. S. Crowther, J. Appl. Phys.,* **34,** *1205 (1963).]*

of about 83% Ni, 17% Fe for vacuum-deposited films. (The ZMC for NiFeCo alloys is shown in Fig. 21.) It should be noted, however, that for films of a given composition, variations of η equivalent to a composition change of several percent can occur if the substrate temperature or deposition rate is changed.

Because of the strain-magnetostriction anisotropy mechanism, it is frequently desirable to know the (isotropic) planar strain in a film. Results from different authors vary but indicate that the stress, which is tensile at low substrate temperatures T_d, changes with increasing T_d and may even become compressive before a minimum is reached (Fig. 20).

(2) M Induced[26,27,28] The magnetic anisotropy of greatest fundamental and technological interest is the uniaxial anisotropy induced by the application of a field during deposition. This M-induced uniaxial anisotropy, which has been observed in Ni, Fe, Co, Gd, NiFe, FeCo, and many alloys with nonferromagnetic metals, has been extensively investigated. However, in spite of this effort, the basic mechanisms of M-induced anisotropy remain poorly understood.

Nevertheless, several aspects of this phenomenon are immediately clear. First, it is apparent that the anisotropy is indeed induced by the magnetization and not by the deposition field, since films deposited in the absence of a field still show uniaxial anisotropy with local EA's in the directions which $\vec{M}$ happened to take during the deposition; the function of the deposition field is simply to ensure that $\vec{M}$ (and the resulting EA) is uniformly aligned in a desired direction. Furthermore, it is easily demonstrated that the value of K_u is strongly dependent on deposition conditions, particularly the substrate temperature, deposition rate, and any subsequent annealing treatment. The composition of the film is another determining factor, but the inclusion of impurities, e.g., from residual gases during vacuum deposition or from the bath during electroplating, may cause more subtle effects which are perhaps best detected in annealing experiments.

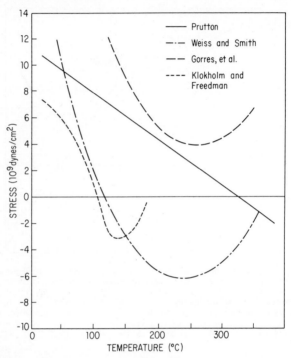

Fig. 20 Isotropic planar stress vs. substrate temperature. Positive values indicate tension; negative values indicate compression. The film compositions were all 81% Ni 19% Fe except for those of Klokholm and Freedman, which were 100% Ni. All stress measurements were made after cooling to room temperature except those of Weiss and Smith, which were made at the deposition temperature. [*After M. Prutton, Nature* **193**, *565 (1962); G. P. Weiss and D. O. Smith, J. Appl. Phys.,* **33**, *1166 (1962); J. M. Gorres M. M. Hanson, and D. S. Lo, J. Appl. Phys.,* **39** *(1968); and E. Klokholm and J. F. Freed man, J. Appl. Phys.,* **38**, *1354 (1967).*]

In Fig. 21, H_K is given for the ternary alloy NiFeCo for a substrate temperature of 250°C. It is seen that a wide range of H_K values is attainable by variation of the composition. In Fig. 22, H_K is plotted as a function of substrate temperature fo NiFe films of various compositions, while in Fig. 23, K_u is shown as a function o composition for a substrate temperature of about 250°C. As was noted in Sec. 3b the true substrate temperature may be different from that of the substrate support in the preceding two figures, Wilts was the only author who took care to give the true temperature. It is seen from Fig. 22 that H_K decreases with increasing T_d, and i fact H_K extrapolates to zero at temperatures within about 20°C of the Curie tem perature. From Fig. 23 it is clear that although the results of various authors are different, K_u passes through a minimum in the vicinity of 85% Ni.

Two mechanisms which can cause uniaxial anisotropy in bulk material have been invoked to explain these results: pair ordering (Sec. 2c) and strain-magnetostriction (Secs. 2c and 7a). The pair-ordering mechanism is expected to act in films in the same way that it acts in bulk material, except that the annealing times and temperatures may be reduced from the bulk case because of the high atomic mobility pre-

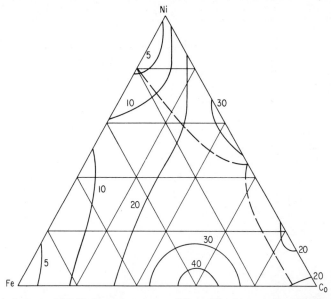

Fig. 21 Anisotropy field H_K for various compositions of the ternary alloy NiFeCo deposited at 250°C (solid curves). Some of the data in the center of the diagram are conjectural. Dashed curves give the zero magnetostriction composition. [*After C. H. Wilts and F. B. Humphrey, J. Appl. Phys., 39 (1968).*]

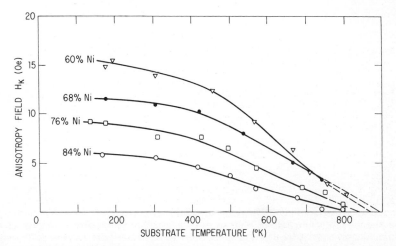

Fig. 22 Anisotropy field H_K vs. substrate temperature for NiFe films of various compositions. [*After C. H. Wilts, in R. Niedermayer and H. Mayer (eds.), "Basic Problems in Thin Film Physics," p. 422, Vandenhoeck and Ruprecht, Goettingen, 1966.*]

vailing during deposition, and the high defect concentration found in thin films. On the other hand, even if pair ordering is operative in ferromagnetic films, it is clear from the simple fact that uniaxial anisotropy can be induced in pure Ni, Fe, Co, and Gd films that at least one additional anisotropy mechanism must also be present. Several authors have suggested that a strain-magnetostriction phenomenon may provide this additional mechanism.

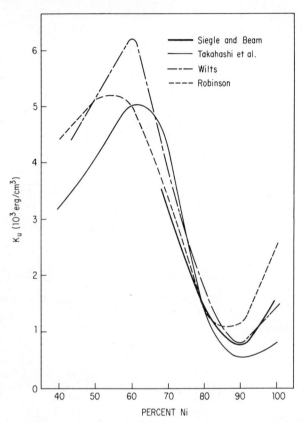

Fig. 23 Uniaxial anisotropy constant K_u vs. NiFe composition. Substrate temperatures: Robinson, 240°C; Siegle and Beam, 250°C; Takahashi et al., 300°C; Wilts, 250°C. [*After G. Robinson, J. Phys. Soc. Japan,* **17**, *Suppl. B-I, 558 (1962); W. T. Siegle and W. R. Beam, J. Appl. Phys.,* **36**, *1721 (1965); M. Takahashi, D. Watanabe, T. Kōno, and S. Ogawa, J. Phys. Soc. Japan,* **15**, *1351 (1960); and C. H. Wilts in R. Niedermayer and H. Mayer (eds.), "Basic Problems in Thin Film Physics," p. 422, Vandenhoeck and Ruprecht, Goettingen, 1966.*]

This strain-magnetostriction mechanism is thought to act in the following manner: At a temperature T, a film which is free from its substrate will sustain a temperature-dependent anisotropic strain $\lambda(T)$, the saturation magnetostriction suitably averaged for a polycrystalline film. The direction of this strain is along $\vec{M}$ and is tensile or compressive according as the sign of $\lambda(T)$ is positive or negative. Assume that as the film is cooled after deposition, a critical temperature T_0 is reached below which the atoms lose their mobility, so that the film is constrained by the substrate to maintain the strain $\lambda(T_0)$ at temperatures less than T_0. Thus a stress $\sigma = Y\lambda(T_0)$, where Y is Young's modulus, is exerted by the substrate on the film, and is tensile (compressive) if $\lambda(T_0)$ is positive (negative). At the measuring temperature T_m (usually room temperature) the saturation magnetostriction is $\lambda(T_m)$; since the film is under the stress σ, a uniaxial anisotropy is created with [Eq. (7)]

$$K_u = \tfrac{3}{2}\lambda(T_m)\sigma = \tfrac{3}{2}Y\lambda(T_0)\lambda(T_m) \tag{23}$$

It is difficult to estimate T_0; presumably it is within the range between T_m and the deposition temperature T_d.

West has pointed out that K_u calculated by Eq. (23) is fundamentally incorrect, for rather than the magnetostriction, the magnetoelastic *energy* should be averaged over all the crystallographic directions. Using the correct formulation, he obtained values of K_u of 4.2, 1.8, and 42×10^3 ergs cm^{-3} for Ni, Fe, and Co, respectively; these are maximum values since it was assumed that $T_0 = T_m = $ room temperature. Experimental results are roughly consistent with these estimates, but it must be emphasized that the strength of the bond between the film and the substrate determines the constraint which the substrate can exert on the film, and hence the effective value of T_0. This bond can change with substrate cleanness, deposition temperature, and background pressure, thereby explaining the large scatter in K_u values found for films having high values of λ.

Robinson attempted to fit his K_u data for NiFe (Fig. 23) by assuming the simultaneous presence of *both* pair-ordering and strain-magnetostriction anisotropies; i.e., he used both Eqs. (9) and (23). The constant A was fixed by assuming that the strain contribution vanished at the zero magnetostriction composition (81% Ni, 19% Fe), while T_a and T_0 were set equal to T_d. This procedure gave a fair fit to the data, but the theoretical peak value of K_u was less than the experimental value and occurred at a Ni concentration above the experimental value. However, West showed that his reformulation of Eq. (23) led, for Ni concentrations less than 65%, to a significant increase in the strain contribution to K_u, while the theoretical peak occurred at a Ni concentration (about 50%) near that found experimentally. Siegle and Beam also successfully fitted their K_u vs. composition data (Fig. 23) with Eqs. (9) and (23) both before and after magnetic annealing treatments changed the values of K_u; the annealing or deposition temperatures were used for T_a and T_0 in these equations.

However, these fits of the pair-strain model to the K_u data are only semiquantitative because of the doubt about the values of T_0 and T_a, and ignorance of the correct concentration dependence of Eq. (9) and the temperature dependence of the parameters of Eq. (23). Also, the experimental data in Fig. 23 may not be pertinent for a theoretical fit for Ni concentrations less than 40% because of a γ-to-α phase change below this Ni concentration; upon proper extrapolation it is found that if the γ phase were retained no peak in the K_u vs. Ni concentration curve would be found at all. Furthermore, determinations of K_u in nonmagnetostrictive films as a function of temperature or as a function of magnetization are often not in accord with Eq. (9).

Thus a better insight into the origins of K_u is needed; this insight is provided by magnetic-annealing studies. In such studies H_K is monitored as a function of time during a high-temperature anneal of a film in the presence of a dc field. Upon application of a HA field during the anneal, H_K decreases monotonically (Fig. 24a); for a sufficiently high annealing temperature H_K can even assume negative values, i.e., an interchange of the EA and HA occurs. A subsequent EA anneal can restore the anisotropy to nearly its original value, and the procedure can be repeated.

Experiments of this nature have been made in which H_K was measured in vacuo either continuously or periodically by a magnetoresistive method. The resulting H_K vs. time t data were then analyzed by assuming that for n annealing processes

$$H_K(t) = \sum_{j=1}^{n} H_{Kj}\left[2\exp\left(-\frac{t}{\tau_j}\right) - 1\right] \tag{24}$$

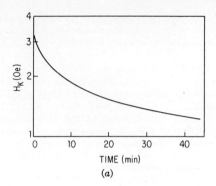

(a)

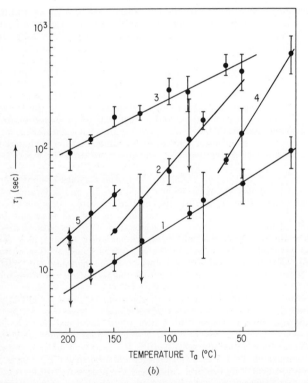

(b)

Fig. 24 (a) Semilog plot of H_K vs. annealing time of an 81 % Ni, 19 % Fe film.   Annealing temperature was 100°C.   A HA field was applied during the anneal.   [*After D. O. Smith, G. P. Weiss, and K. J. Harte, J. Appl. Phys.*, **37**, *1464* (*1966*).]   (b) Semilog plot of the relaxation time τ_j as a function of the annealing temperature T_a.   The data were obtained from analyses of annealing curves like that of Fig. 24a.   Each numbered line refers to a different atomic process with an activation energy E_j which can be found from the graph; the values of E_j are given in Table 3.   [*After D. O. Smith, G. P. Weiss, and K. J. Harte, J. Appl. Phys.*, **37**, *1464* (*1966*).]

where τ_j is the jth relaxation time with activation energy E_j,

$$\tau_j = \tau_{0j} \exp\left(\frac{E_j}{kT_a}\right) \tag{25}$$

Using Eq. (24), several relaxation times τ_j were obtained for each value of T_a (e.g., $T_a = 100°C$ in Fig. 24a); the experiment was then repeated, after returning the EA to its original position, for different values of T_a. In this way curves of τ_j vs. T_a were obtained (Fig. 24b) which in turn gave values of E_j and τ_{0j} upon analysis in terms of Eq. (25). Results are shown in Table 3; note that the results of the two groups of investigators agree for processes I and 3.

TABLE 3 Anisotropy Processes*

Process	E_j, eV	τ_{0j}, s	Fraction of total H_K
Smith et al.:			
3	0.15	2	0.10
1	0.18	7×10^{-2}	0.08
5	0.26	3×10^{-2}	0.05
2	0.34	1.5×10^{-3}	0.07
4	0.48	6×10^{-6}	0.08
Kneer and Zinn:			
I	0.15	1.5	0.20
II	0.4	3×10^{-2}	0.08
III	0.4	3×10^{-1}	0.16
IV	1.5	2×10^{-8}	0.56

* The processes of Smith et al. correspond to Fig. 24b. Considerable experimental uncertainty is associated with processes 4 and 5. Note agreement between processes 3 and I. After D. O. Smith, G. P. Weiss, and K. J. Harte, *J. Appl. Phys.*, **37**, 1464 (1966); and G. Kneer and W. Zinn, in R. Niedermayer and H. Mayer (eds.), "Basic Problems in Thin Film Physics," p. 437, Vandenhoeck and Ruprecht. Goettingen, 1966.

It is difficult to match the activation energies and relaxation times in Table 3 with processes for which these parameters are known. Thus, all the activation energies of Table 3 are considerably lower than the 3-eV activation energy of Ni-Ni pairs, as measured in bulk material. On the other hand, it was found that after a heat treatment which recrystallized the film, the activation energy of the predominant process (process IV, Table 3) rose to 2.1 eV, thus approaching the bulk value. Process IV may therefore be associated with pairs after all; it is possible that the high defect concentration in films somehow makes pair orientation easier than in bulk material. The remaining anisotropy mechanisms might be sought in the directional order of some of the structural defects like interstitials or vacancies which are found in high concentration in thin films, but here again, the agreement of known data for rearrangement of such defects with data of Table 3 is poor. Perhaps some defect structures which are not common in bulk material are involved. On the other hand, some investigators have taken the alternative view that pair ordering is the *only* operative anisotropy mechanism and that the various annealing relaxation processes should be identified with annealing of specific defects which affect the ease of reorientation of the pairs.

Annealing experiments can also be performed on films having nonvanishing magnetostriction, so that the strain-magnetostriction mechanism is operative. Thus, magnetostrictive NiFe, Ni, or Co films vacuum-deposited at room temperature exhibit room-temperature annealing effects with relaxation times of several days, while films deposited at several hundred degrees celsius show no room-temperature annealing. This behavior is explainable on the strain-magnetostriction model if it is assumed that the film-substrate bond is weak for a room-temperature deposition so that the constraint of the substrate on the film may be changed by a magnetic anneal. Room-temperature annealing or "aging" effects are also common in electroplated films (Sec. 3b), even if their composition is such that they are nonmagnetostrictive. Here, however, the effect is attributed to impurities incorporated into the film from the electroplating bath. The properties of such films can be stabilized by subjecting them to a high-temperature anneal.

(3) Oblique Incidence[29] Microscopic geometric inhomogeneities in a magnetic film can easily cause magnetic anisotropy effects due to the magnetic-shape anisotropy (Sec. 2c) of the inhomogeneities. Thus, unidirectional grinding of the substrate can induce uniaxial ferromagnetic anisotropy in the magnetic film deposited on it, with the EA parallel to the direction of the scratches. This effect becomes understandable if it is noted that the demagnetizing factor $N_\parallel$ associated with the major axis of an elongated ellipsoid of revolution is much less than the minor-axis factor $N_\perp$, so that $\vec{M}$ tends to lie along the major axis, thus making it the EA (Sec. 2c); the effect of the grinding is to build many such elongated ellipsoids into the film. On the other hand, if the substrate has small-scale randomly shaped geometric inhomogeneities scattered throughout its surface, the overlying magnetic film will have corresponding regions of local ferromagnetic shape anisotropy, with randomly directed EAs. The magnetization dispersion will accordingly be greater than that of a film on a flat substrate. It is for these reasons that it is important to use smooth, clean substrates in order to obtain films with well-defined M-induced uniaxial anisotropy with low magnetization dispersion (Sec. 3b).

However, under certain circumstances, microscopic-shape anisotropy can still play a decisive role even if the substrate is perfectly smooth and clean. If a ferromagnetic film is vacuum-deposited with the vapor beam not normal to the substrate, as tacitly assumed thus far, but at oblique incidence, uniaxial anisotropy will be induced. This will be true even in the absence of an applied field during deposition, or in the presence of a field in the film plane which rapidly rotates during deposition so that the M-induced anisotropy is averaged out.

The magnitude of H_K and the EA direction depend upon the incidence angle, the film composition, and the substrate temperature. Thus, films of 81% Ni, 19% Fe deposited at $T_d = 200°C$ have H_K values which increase monotonically with incidence angle θ (measured from the film normal) for $\theta < 55°$, while the EA is in the plane of the film and is perpendicular to the plane of incidence (Fig. 25). Such anisotropy is known as perpendicular or positive oblique-incidence anisotropy. If the incidence angle is increased beyond 55°, however, H_K begins to decrease; for $\theta > 70°$, "negative" values are found, meaning that the EA is now *parallel* to the plane of incidence. The values of H_K can be very large compared with the M-induced anisotropy, about 300 Oe at $\theta \simeq 50°$ for perpendicular anisotropy and about $-1,000$ Oe at $\theta \simeq 85°$ for parallel anisotropy. Oblique-incidence anisotropy has also been observed in films of Fe, Co, and Ni, but not in electrodeposited or sputtered ferromagnetic films because the incidence directions of the incoming atoms are randomized in these processes (Sec. 3b).

Oblique-incidence anisotropy of either sign can be explained on the basis of shape anisotropy. For low values of θ ($H_K > 0$), a self-shadowing mechanism is invoked according to which regions behind growing crystallites do not receive incoming vapor because they are in the "shadows" of the crystallites; as the crystallites grow in size they will have a statistical tendency to agglomerate into parallel crystallite "chains" elongated normal to the plane of incidence and separated from each other by voids. The shape anisotropy of these chains provides the magnetic anisotropy. This anisotropy can be very high; the value of H_K of an infinitely long NiFe ellipsoid is about 5,000 Oe. Evidence for this interpretation comes not only directly from electron micrographs which show the crystallite chains but from observations of anisotropic electrical resistance and anisotropic absorption of polarized light (dichroism).

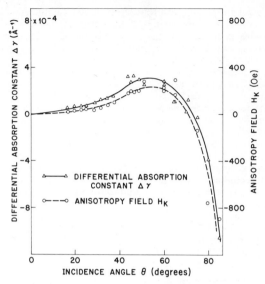

Fig. 25 Anisotropy field H_K, and differential absorption coefficient $\Delta\gamma$ for polarized light vs. incidence angle for oblique incidence 81 % Ni, 19 % Fe films deposited at 200°C substrate temperature. [*After M. S. Cohen, J. Appl. Phys.*, **32**, 87S (1961).]

The dichroism was investigated by measuring the differential absorption constant $\Delta\gamma \equiv \gamma_\| - \gamma_\perp$ where $\gamma_\|$ and $\gamma_\perp$ are the optical-absorption constants when the plane of polarization is respectively parallel and perpendicular to the axis of greatest absorption. In Fig. 25 $\Delta\gamma$ is plotted against θ; it is seen that, as anticipated, the optical-anisotropy and magnetic-anisotropy dependences on θ are in agreement.

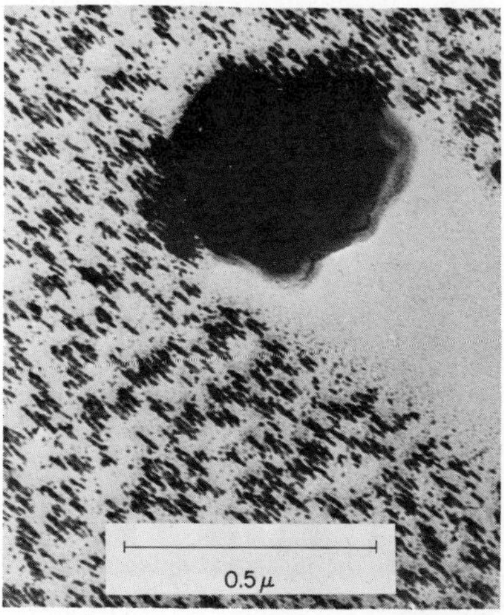

The production of crystallite chain structure by oblique incidence of the vapor is not confined to magnetic materials; its presence in nonmagnetic materials has been detected by electron microscopy and observations of dichroism in oblique incidence films. If NiFe films are deposited at normal incidence or by electrodeposition on underlayers of nonmagnetic metals which were previously deposited at oblique incidence, uniaxial anisotropy with the EA normal to the incidence plane is produced; i.e., the geometric anisotropy of the underlayer is "replicated" by the NiFe.

For high values of θ ($H_K < 0$) a different geometric anisotropy mechanism is involved; the crystallites tend to elongate in the direction of the incoming vapor

Fig. 26 Electron micrograph of an 81 % Ni, 19 % Fe film deposited at 85° incidence angle and 200°C substrate temperature. The shadow cast by the large dirt particle reveals the vapor-beam direction. [*After M. S. Cohen, J. Appl. Phys.*, **32**, 87S (1961).]

beam to form crystallite "columns," as is seen from the micrograph of Fig. 26. Thus, there is a competition between the shape anisotropy of the crystallite columns, contributing to parallel anisotropy, and that of crystallite chains (which can also be seen in Fig. 26), contributing to perpendicular anisotropy. The columnar growth overbalances the chains at higher angles. That this geometric interpretation is correct is shown by the agreement of the $\Delta\gamma$ and the H_K dependences on θ over the *entire* θ range.

The work just described was carried out at relatively low values of T_d ($\lesssim 250°C$). Other work showed that in the range $300°C < T_d < 400°C$ the magnitude of oblique-incidence anisotropy decreased, presumably because enhanced atomic mobility at higher substrate temperatures allowed filling of the voids, thus tending to destroy the geometric inhomogeneities. Also, at higher substrate temperatures the negative portion of the H_K vs. θ curve (Fig. 25) shifted toward lower values of θ (at $T_s \simeq 350°C$ negative H_K values could be observed at θ values of 20 to 30°); i.e., at higher values of T_d the chain-column balance shifts in favor of columns. This interpretation was again verified by good correlation with dichroism measurements. Annealing treatments of positive oblique-incidence anisotropy film emphasized the delicate nature of the chain-column balance, for it was found that under certain circumstances annealing treatments converted perpendicular into parallel anisotropy; i.e., the annealing suppressed the chains, but not the columns.

It should be mentioned that NiFe films with a Ni content $>90\%$ show parallel anisotropy under conditions where 81% Ni 19% Fe films exhibit perpendicular anisotropy. It was postulated that a surface-tension mechanism puts the chains in tension, thus creating parallel anisotropy for these negative-magnetostriction films.

b. Perpendicular Anisotropy[30]

If $\vec{M}$ is inclined at an angle θ to the film normal, and if any anisotropy within the plane is ignored, the resultant anisotropy energy may be written as $E_\perp = -K_\perp \sin^2 \theta$, where $K_\perp$ is the perpendicular anisotropy constant. For $K_\perp > 0$, this equation implies that the film normal is a hard axis while the film plane is an easy plane. Thus, if the only contribution to $K_\perp$ is given by the shape-demagnetizing factor of the film (Sec. 3a), then $K_\perp = 2\pi M^2$. However, several measurements have indicated that there can be an additional structural contribution $K_\perp$ so that $K_\perp = 2\pi M^2 + K_\perp{}^S$. Often $K_\perp{}^S$ is negative and thus tends to maintain $\vec{M}$ normal to the film plane. [It is useful to define a perpendicular anisotropy field $H_K{}^\perp \equiv 2K^\perp/M = 4\pi M + (H_K{}^\perp)_S$, where $(H_K{}^\perp)_S \equiv 2K_\perp{}^S/M$.]

Measurements of $K_\perp{}^S$ can be made by measuring $K_\perp$ and subtracting off the $2\pi M^2$ contribution; such measurements have been made with torque-magnetometer and magnetic-resonance techniques. In addition, indirect measurements of $K_\perp{}^S$ can be made by studying a certain nonplanar magnetization distribution (Sec. 15b) known as stripe domains, for which $K_\perp{}^S < 0$. There is still some controversy over the details of the origins of the $K_\perp{}^S$ contributions, but two mechanisms have been clearly identified: strain-magnetostriction and microscopic-shape anisotropy.

The strain-magnetostriction contribution of $3\lambda\sigma/2$ to $K_\perp{}^S$ arises from the (isotropic) planar stress σ exerted on the film by the substrate. Since σ is strongly dependent on T_d, and λ is strongly dependent on composition (Sec. 7a) the strain-magnetostriction contribution is strongly dependent on both these quantities. A clear demonstration of the strain-magnetostriction contribution to $K_\perp$ is afforded by observations that the measured value of $K_\perp$ of epitaxially grown Ni films reverted to $2\pi M^2$ upon stripping from the substrate, while the strain, as measured by x-rays, disappeared. Furthermore, the value of $K_\perp{}^S$ calculated from λ and σ was consistent with the measured value of $K_\perp{}^S$. By measuring $K_\perp$, σ, and M, it was also shown that the only contribution to $K_\perp{}^S$ for polycrystalline Ni films deposited at various substrate temperatures came from strain-magnetostriction.

On the other hand, other observations indicated that for high deposition pressures the value of $K_\perp{}^S$ could not be explained *only* by the strain-magnetostriction mechanism. For example, a $K_\perp{}^S$ contribution is found for 81% Ni 19% Fe films for which $\lambda \simeq 0$. The necessary additional contribution was found to increase as the pressure during deposition increased, while the measured value of M simultaneously decreased. (In Fig. 27, $H_K \equiv 2K_\perp / M_{\text{bulk}}$ and $4\pi M$ are plotted against pressure.) A mechanism involving growth of crystallite columns parallel to the film normal was invoked to provide a microscopic-shape anisotropy contribution to $K_\perp{}^S$. The specific model envisions the film as broken into blocks which are separated by thin oxide layers; the oxide thus accounts both for the shape anisotropy and the measured decrease in M.

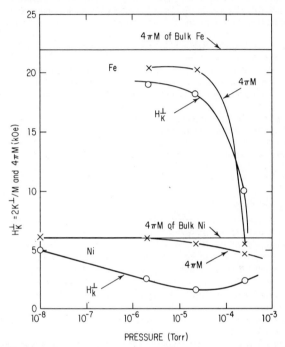

Fig. 27 Perpendicular anisotropy field $H_K{}^\perp \equiv 2K_\perp / M_{\text{bulk}}$, and $4\pi M$ (measured in film) vs. pressure during deposition for Fe and Ni films. [*After H. Fujiwara, Y. Sugita, and N. Saito, J. Phys. Soc. Japan,* **20**, *2088 (1965).*]

The postulated columnar growth has been directly observed in Ni and NiFe films by low-angle electron diffraction, and by electron microscopy in Al films. Moreover, this phenomenon is clearly related to the oblique-incidence parallel-anisotropy effect (Sec. 7a), where the columns are *inclined* to the film normal. Thus, anisotropy with an EA not normal to the plane but inclined nearly parallel to the direction of the columnar growth has been observed in oblique-incidence films. The oblique-incidence anisotropy discussed in Sec. 7a is only that part of the microscopic-shape anisotropy which is effective in the plane of the film.

It should be noted, however, that after stripping films from the substrate to eliminate the strain-magnetostriction contribution, in some cases the remaining value of $K_\perp{}^S$ was too large to be explained by the columnar-growth model. Other anisotropy mechanisms may also be acting. One such mechanism is magnetocrystalline anisotropy, which can be important in monocrystalline films or films with crystallographic texture of the proper orientation.

c. Nonuniaxial Anisotropy in the Film Plane

The sole macroscopic anisotropy within the film plane which is exhibited by the polycrystalline films which are of most interest in this chapter (Sec. 3a) is uniaxial anisotropy. However, special films can be prepared which have other types of anisotropy.

(1) Magnetocrystalline The only true anisotropy of order higher than uniaxial originates in magnetocrystalline anisotropy. It is true that for small applied fields, the magnetic behavior may be interpreted in terms of pseudo-biaxial, triaxial, or n-axial anisotropy if two or more uniaxial regions (or films) are coupled magnetostatically or otherwise. However, under the rigorous test condition of large applied fields, only a net uniaxial anisotropy is admissible of consideration since, as discussed above, the correct addition of uniaxial anisotropies always yields another uniaxial anisotropy.

Single-crystal films of Fe, NiFe, Co, and MnBi have been made by epitaxy. The conditions for obtaining good single crystals are the same as those for nonmagnetic films (Chap. 9). Measurements of the anisotropy constants [usually only K_1 in Eq. (6)] are generally in good agreement with bulk data, while room-temperature measurements of anisotropy constants of cubic Co and hexagonal Ni phases, both unstable in bulk form, have been made. Departures from bulk results and the presence of uniaxial anisotropy components have been attributed to a strain-magnetostriction mechanism, as has been verified by experiments in which the strain was released by stripping from the substrate.

(2) Unidirectional It is possible to prepare special films whose lowest-energy magnetic state is achieved when $\vec{M}$ is parallel (but not antiparallel) to a specified easy *direction* (not axis) in the film plane. Such unidirectional anisotropy stands in contrast with uniaxial anisotropy, where neither of the two senses of the EA is energetically preferred. Unidirectional anisotropy, which is also found in bulk material, depends on the exchange interaction between a ferromagnet and an antiferromagnet which are in intimate contact; it is therefore sometimes called exchange anisotropy. The antiferromagnetic component is assumed to have a high uniaxial anisotropy and a Néel temperature lower than the Curie temperature of the ferromagnet. The EA direction in the antiferromagnet is then fixed, through ferromagnetic-antiferromagnetic exchange, by the position of $\vec{M}$ in the ferromagnet as the system is cooled through the Néel temperature. The unidirectional anisotropy is thereafter maintained by the exchange interaction; it cannot be changed below the Néel temperature since an external field cannot act directly on the antiferromagnet.

Unidirectionally anisotropic films in the systems Co-CoO, Fe-FeS, Ni-NiO, NiFe-NiFeMn, NiFe-Cr, and NiFe-Cr_2O_3 have been prepared. These systems show one or more of the three properties characteristic of ferromagnetic-antiferromagnetic coupling: a unidirectional ($\sin \theta$) component in the torque curve at high fields, a BH loop displaced from the origin along the H axis, and rotational hysteresis at high fields. A good quantitative theory explaining the measurements does not exist; in particular it is not even understood how these three characteristics can be simultaneously present at the same value of the applied field. Furthermore, some unexplained time-dependent effects have been noted.

8. DOMAIN WALLS[31,32]

Because the formation of domain structure is a serious departure from the coherent magnetization model presented in Sec. 6, the physics of domains and domain walls has received much attention. Although the general concepts of domain-wall behavior learned from the investigation of bulk materials can be transferred to film studies, considerable differences are observed in wall phenomena in these two states of matter.

a. One-dimensional Structures

In bulk material all domain walls are Bloch walls (Sec. 2d), but in thin films walls known as Néel and intermediate can appear. Such walls are usually described in terms of a "one-dimensional" magnetization distribution; i.e., $\vec{M}$ inside the wall is assumed to be constant throughout the film thickness and along the length of the wall, and to vary only with the coordinate normal to the wall. Even with this simplifying assumption, however, calculations of wall energy density and shape are involved because of the difficulty of calculating the magnetostatic energy contribution (Sec. 2b).

(1) Bloch Wall The determination of the thickness and energy density of a Bloch wall in bulk material does not involve magnetostatic energy because the rotation of $\vec{M}$ through the wall is such that div $\vec{M} = 0$ (Sec. 2d). However, in a thin film there *is* a magnetostatic contribution due to magnetic poles at the film interfaces, because $\vec{M}$ has a component normal to the plane of the film in the transition region (Fig. 28). As the film thickness decreases, this magnetostatic contribution becomes comparable with the anisotropy and exchange-energy contributions which are present in the thin film as well as the bulk case.

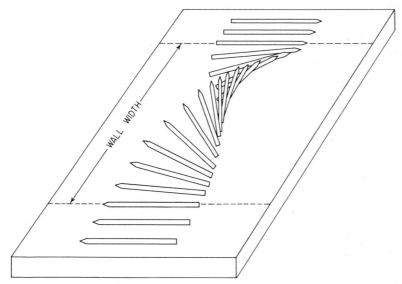

Fig. 28 Illustrating the magnetization distribution in a 180° Bloch wall in a ferromagnetic film. The magnetization rotates in a plane normal to the film plane.

A good theoretical approximation for a 180° Bloch wall (separating antiparallel domains) can be obtained by assuming the simple distribution

$$
\begin{aligned}
\phi &= \frac{-\pi}{2} & x &\le -\frac{a}{2} \\
&= \frac{\pi x}{a} & \frac{-a}{2} &\le x \le \frac{a}{2} \\
&= \frac{\pi}{2} & x &\ge \frac{a}{2}
\end{aligned}
\tag{26}
$$

where x is the coordinate normal to the wall, ϕ is the angle measured in the plane of the wall between $\vec{M}$ and the normal to the film, and a is the wall width. Using this distribution it is found that the average anisotropy energy density in the wall is [Eq. (8)] $E_K = K_u/2$, while the exchange energy density [Eq. (4)] is $E_{\text{ex}} = A(\pi/a)^2$. To estimate the magnetostatic energy density the wall is approximated by a cylinder of elliptical cross section, where d is the major axis of the ellipse and a is the minor axis. Then since the demagnetizing factor $N = 4\pi a/(a + d)$ when $\vec{M}$ is parallel to the major axis, from Eq. (5) it can be shown that $E_{\text{mag}} = \pi a M^2/(a + d)$. The

FERROMAGNETIC PROPERTIES OF FILMS

energy per unit area of the wall γ_B is found by multiplying each of the above expressions by a, then adding them:

$$\gamma_B = \frac{aK_u}{2} + \frac{\pi^2 A}{a} + \frac{\pi a^2 M^2}{a + d} \tag{27}$$

It is seen from Eq. (27), as expected on physical grounds, that an increase in a causes an increase in the anisotropy and magnetostatic energies but a decrease in the exchange energy. The value of a giving the *minimum* energy is determined by setting $d\gamma_B/da = 0$, while the value of γ_B corresponding to this equilibrium value of a is

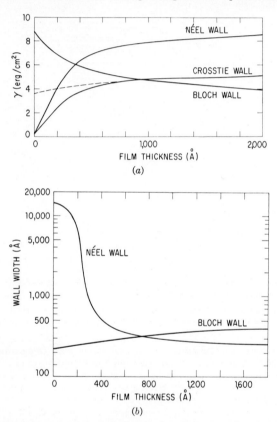

Fig. 29 (a) Theoretical surface energy density γ of a domain wall as a function of thickness. Assumed $A = 10^{-6}$ ergs cm^{-1}, $M = 800$ G, $K_u = 10^3$ ergs cm^{-3}. The dashed curve shows the estimate of crosstie-wall energy density when the energy of Bloch lines is included. (b) Theoretical widths of Bloch and Néel walls as a function of film thickness. Assumed $A = 10^{-6}$ ergs cm^{-1}, $M = 800$ G, $K_u = 10^3$ ergs cm^{-3}. [*After S. Middelhoek, in J. C. Anderson (ed.), "The Use of Thin Films in Physical Investigations," p. 385, Academic Press Inc., New York, 1966; and J. Appl. Phys.*, **34**, *1054 (1963).*]

found by substituting back into Eq. (27). In Fig. 29a, γ_B is plotted as a function of d assuming $K_u = 10^3$ ergs cm^{-3}, $M = 800$ G, and $A = 10^{-6}$ ergs cm^{-1}, while in Fig. 29b, a is plotted vs. d. For these typical parameters γ_B consists primarily of exchange and magnetostatic energies only. The rapid increase of γ_B with decreasing d is thus caused by an increase in the magnetostatic energy.

(2) Néel Wall The magnetostatic energy can be decreased by a different magnetization distribution in the wall, in which $\vec{M}$ rotates not out of the film plane, but in the plane of the film (Fig. 30). In such a domain wall, known as a *Néel wall*, there are no poles at the surfaces of the film, but instead volume poles are created within the wall since now the component of $\vec{M}$ normal to the wall is not continuous. The dependence of γ_N and a upon d may be calculated in the same way as in the Bloch-wall case (Fig. 29). Note the rapid increase in a as d decreases because of the decrease in the magnetostatic energy, which predominates in γ_N for $d < 400$ Å.

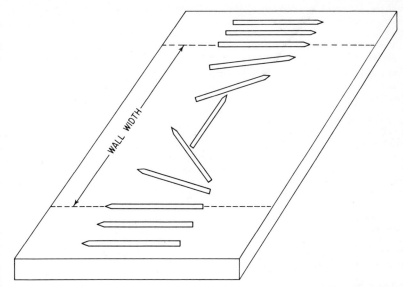

Fig. 30 Illustrating the magnetization distribution in a 180° Néel wall in a ferromagnetic film. The magnetization rotates in the film plane.

(3) Intermediate Wall Thus far only 180° walls have been considered. Walls separating domains whose $\vec{M}$ vectors lie not at 180° but at the general angle θ are found in monocrystalline films and can be created in uniaxially anisotropic polycrystalline films by the application of fields in the plane of the film and normal to a wall, i.e., a HA field. In the latter case, considerations similar to the above indicate that when the wall angle is decreased below a certain critical value (by applying a sufficiently large HA field), the Bloch wall transforms into a Néel wall.

However, closer examination of the problem shows that under certain circumstances it is possible to have an *intermediate* (or mixed) wall of a mixed Néel *and* Bloch character. Thus, in a Bloch wall $\vec{M}$ may be thought of as rotating about an axis in the film plane and normal to the wall, while in a Néel wall the rotation occurs about an axis normal to the film plane; in an intermediate wall the rotation occurs about an inclined axis. An intermediate wall will then have magnetostatic fields both in the film plane and normal to it. It is found that intermediate walls occur for $\theta_c < \theta < 180°$, where θ_c is a critical angle, with the ratio of Bloch to Néel character increasing as θ increases; for $\theta < \theta_c$ pure Néel walls occur, while only for $\theta = 180°$ can pure Bloch walls exist. Intermediate 180° walls cannot occur.

(4) Further Calculations The simple calculations for the shape and energy of Bloch and Néel walls presented above have been improved considerably by several approaches. The ideal method would consist of solving the full variational problem (Sec. 2b), but this has proved possible only if certain simplifying assumptions for the magnetostatic energy are made. An easier approach involves the use of the Ritz method, where a parametric form of the magnetization distribution is assumed, and the total energy, *including* the magnetostatic contribution, is evaluated in terms of those parameters. The values of the parameters yielding the minimum value of the energy are then found. Unfortunately, however, the Ritz technique gives no indication of which class of functional forms to choose for the computation. A technique not suffering from this disadvantage involves the use of a digital computer to obtain numerical solutions. A general magnetization distribution is assumed which is not a continuous but a discrete function of the distance x measured normal to the wall, i.e., the film is divided into cells parallel to the x axis, and $\vec{M}$ is assumed continuous within each cell. The magnetostatic energy can then be computed explicitly, and the computer can be used to find the minimum-energy state.

These computational methods confirm the general ideas of Bloch and Néel walls as shown in Figs. 28 to 30, but more accurate magnetization distributions are obtained.

Thus, it has been shown that at the outer edges of a Bloch wall, $\vec{M}$ is reverse-rotated to give a magnetization component normal to the plane of the film opposite to that obtaining at the wall center. This unexpected configuration serves to lower the wall energy. The Néel-wall calculations show that a central region in the wall has a high value of $d\phi/dx$ and is surrounded on either side by a region of low $d\phi/dx$; the behavior may be described by the equation $\sin \phi = \tanh (x/a)$. The ratio of the width of the "tails" of the Néel wall to that of the central region has been estimated to be about 30. Thus there can still be appreciable values of ϕ several microns from the center of a Néel wall, while the entire magnetization distribution of a Bloch wall should be confined to a region a few hundred angstroms wide. These results have also been confirmed by high-resolution Lorentz electron-microscopy investigations (Sec. 4a).

The various Bloch-wall calculations give values of γ and wall shapes which are in fairly good agreement with each other; furthermore they all obey a necessary self-consistency condition developed by Aharoni. However, the agreement among Néel-wall calculations is not so satisfactory, and the Aharoni criterion is far from satisfied for the extant Néel calculations.

b. Two-dimensional Structures

There are several theoretical indications that the calculated values of the wall energy density would be lowered if a two-dimensional $\vec{M}$ distribution in the wall were permitted, i.e., if $\vec{M}$ were allowed to vary not only in the x direction normal to the wall but in the direction normal to the film plane as well. Furthermore, there are direct experimental indications that magnetization variations *along* the wall can lower the wall energy.

Thus, it is seen from Fig. 29a that a 180° Néel wall is energetically more favorable than a Bloch wall for values of $d < 350$ Å. It is found experimentally, however, that as the film thickness decreases, a Bloch wall does not suddenly change into a Néel wall at a certain value of d but instead a transition structure, known as a crosstie wall, is observed (Fig. 31).

A crosstie wall may be regarded as a Néel wall having alternating intervals of oppositely directed senses of rotation of $\vec{M}$ within the wall. This arrangement provides some flux closure in the wall and thus decreases the magnetostatic energy. At the boundaries of the intervals of opposite-rotation sense are transition regions known as Bloch lines, in which $\vec{M}$ is normal to the film plane. The Bloch lines themselves are small in extent (about 100 Å diameter), while the magnetization around them may follow either a circular or a convergent pattern. These patterns alternate in a crosstie wall (Fig. 31); the convergent pattern is associated with a *crosstie*, which is a piece of Néel wall. The calculation of the energy of a crosstie wall is not as straightforward as the Bloch and Néel cases; the estimate given in Fig. 29a shows a Bloch-to-crosstie transition at 850 Å, and a crosstie-to-Néel transition at 200 Å.

The analog of a crosstie wall is found in thick films which support Bloch walls. Bloch walls can have alternating intervals of oppositely directed rotation senses separated by transition regions known as Néel lines. The magnetization in the small region of the Néel line is normal to the wall and in the plane of the film. Just as for a crosstie wall, this arrangement serves to decrease the demagnetizing energy by providing flux closure.

The change of wall type with film thickness can be clearly seen in Fig. 32, which shows a Bitter pattern of a 180° wall in a tapered film. Crossties are easily visible,

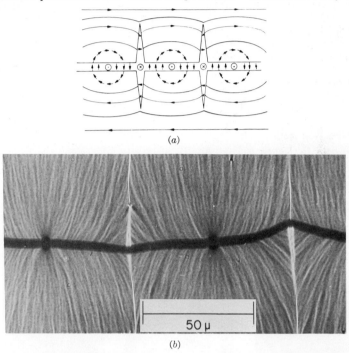

(a)

(b)

Fig. 31 (a) Illustrating the magnetization distribution of a crosstie wall. The magnetization is normal to the film plane at Bloch lines. [*After E. E. Huber, Jr., D. O. Smith, and J. B. Goodenough, J. Appl. Phys.*, **29**, *294 (1958)*.] (b) Lorentz micrograph of a crosstie wall.

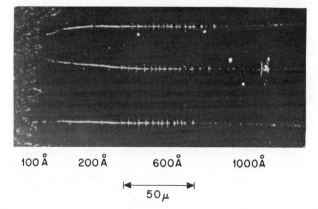

100 Å   200 Å   600 Å   1000 Å

|← 50 μ →|

Fig. 32 Bitter pattern of a 180° wall in a tapered film showing transition from a Néel wall, through a crosstie wall, to a Bloch wall. The film thickness increases from left to right. (*After S. Middelhoek, Thesis, University of Amsterdam, 1961.*)

while low contrast at the thick end of the film indicates a Bloch wall and high contrast at the thin end indicates a Néel wall.   In contradiction to the behavior expected from examination of Fig. 29a, the transitions between the various wall types are experimentally found to occur gradually, rather than abruptly.   For values of $d > 1,000$ Å, Bloch walls are found, with alternating sections of opposite polarity separated by Néel lines.   At $d < 900$ Å Néel walls begin to be of comparable energy to Bloch walls, so that various intervals of the wall are made of Néel-wall segments of alternating polarity.   As d decreases, the proportion of Néel to Bloch wall length increases, until around 600 Å the wall has become a crosstie wall.   The spacing between Bloch lines increases as d continues to decrease, until a pure single-polarity (Bloch-line-free) Néel wall is produced for $d < 300$ Å.

9. DOMAIN-WALL MOTION[33]

Depending on the properties of the film and upon the nature of the applied fields, various kinds of wall motion can be observed.

a. Ordinary Wall Motion

(1) Domain Nucleation and Growth   If, after saturation along the EA, a slowly increasing reverse field (antiparallel to the original saturating field) is applied, the magnetic state will change not by the uniform rotational process predicted by the S-W theory, but noncoherently by the nucleation and growth of reverse domains.

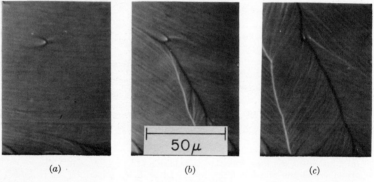

(a)   (b)   (c)

Fig. 33   Growth of reverse domains as revealed by Lorentz microscopy.   Magnetization first saturated along EA (vertical), then reverse fields applied of (a) 0 Oe, (b) 1.3 Oe, (c) 1.9 Oe.

Two threshold fields are involved: that for nucleation of a reverse domain H_n and that for growth of the nucleated domain (movement of domain walls) H_0.   It is found that nucleation preferentially occurs at inhomogeneities like scratches or holes, but especially at the film edges;* subsequently (Fig. 33) as the reverse field continues to

---

* A uniformly magnetized film would support a uniform demagnetizing field if it were a perfect ellipsoid (Sec. 3a).   Since the film is not a *perfect* ellipsoid, the demagnetizing field originating from poles at an edge is greater at that edge than in the interior.   A lower-energy state than one of uniform magnetization may therefore be achieved if $\vec{M}$ at the edge is parallel to the edge; the transition region between the interior of the film (where $\vec{M}$ is perpendicular to the edge) and the edge depends on both d and the applied field.   As $\vec{H}$ is decreased in magnitude from a value giving saturation, a 90° "curling" of $\vec{M}$ will appear in the transition region; reverse domains can nucleate in the interface between two peripheral regions of opposing curl senses at reverse fields $\vec{H}_n$ whose magnitude decreases with an increase in d.   However, the detailed nucleation process, which must involve magnetization rotation, is not well understood theoretically.

increase, the nucleated domains grow in length and width by discontinuous movements of the domain walls in a series of "Barkhausen" jumps (Sec. 2d). The walls always move in a direction such that domains in which $\vec{M}$ is oriented parallel to the applied field $\vec{H}$ grow at the expense of domains containing $\vec{M}$ oriented antiparallel to $\vec{H}$. As the reverse field increases, the growth continues until the film is saturated in the reverse direction. The threshold value of the reverse field necessary for the full reversal process is called the wall coercive force H_c, although slightly different definitions of H_c are possible within the narrow range of field necessary for complete reversal. Thus, H_c should equal H_n or H_0, whichever is larger. For films of about 1 cm diameter and 1,000 Å thick, $H_c = H_0$; the coercive force for wall movement will therefore be called H_c in the following.

Domain-wall motion can also occur when the reversing field is not along the EA but at an angle β to it. In describing the resulting reversal processes it is customary to assume that the film is first saturated along the *negative* EA direction (Fig. 15) and that a reversing field $\vec{h}$ is applied in the first quadrant in a direction at an angle β to the positive EA direction. When β is less than a critical angle β_c, reversal occurs when* $h > h_w(\beta)$ by the growth of domains nucleated at the film edges, as just described, but if $\beta > \beta_c$ and $h > h_q(\beta)$ a parallel array of long, slender, inclined reverse domains appears (Fig. 34). If h is increased beyond h_q, for moderate angles ($\beta \lesssim 60°$), the reversal is completed at $h \geq h_w(\beta)$ by ordinary edge domains sweeping through, while for high angles ($\beta \gtrsim 60°$) the slender domains grow in width before the edge domains come in. The angular dependence of the threshold fields $\vec{h}_q$ and $\vec{h}_w$ may be displayed by plotting these quantities in reduced field space, where they may be compared with the astroid (Fig. 35). Thus $(h_q)_{\parallel} = h_q \cos \beta$ and $(h_q)_{\perp} = h_q \sin \beta$, (while the angle β may be measured directly from the $h_{\parallel}$ axis). It is seen from such plots that for $\beta < \beta_c$, $h_w \cos \beta \sim h_c$, i.e., application of h at an angle β is essentially equivalent to EA reversal. If films of increasing h_c are examined, it is found that β_c decreases and the curves move away from the origin. Since the

Fig. 34 Lorentz micrograph of labyrinth domains generated by application of a 4.0 Oe reversing field at $\beta = 80°$ from the EA. Labyrinthine paths of $\vec{M}$ within the spike-shaped reverse domains are indicated by the ripple configurations. [*After M. S. Cohen, J. Appl. Phys.*, **34**, *1841 (1963)*.]

value of h_c is correlated with the magnetization dispersion, it may be inferred that these effects originate in a dispersion-dependent phenomenon.

The character of the high-β reversal mechanism has been controversial. Some workers observed *propagation* of the slender inclined domains from the film edge, and called them labyrinth domains because of the labyrinthine path of the magnetization within the domains. Labyrinths were postulated to grow from their tips by sequential switching of adjacent regions by magnetostatic fields from the tips, although because of the ripple the orientation of $\vec{M}$ was so unfavorable in certain regions during labyrinth propagation that those regions did not switch. Upon removal of the field the labyrinthine appearance is enhanced by magnetostatic fields created when the magnetization relaxes toward the EA.

---

* The threshold field for reversal by motion of domains nucleated at the film edges is defined as H_w for a reverse field applied at an angle β to the EA. At $\beta \simeq 0°$, $H_w = H_c$.

On the other hand, other investigators postulated a process called "partial rotation," in which the slender domains are created by *simultaneous* reversal without propagation features. In this process the magnetization throughout the film rotates uniformly as h is increased, until at h_q switching occurs in bands which happen to have $\vec{M}$ favorably oriented because of the ripple. The resulting magnetostatic fields prevent the unswitched bands from also switching as h is further increased. This controversy between labyrinth and partial-rotation generation was resolved when it was shown

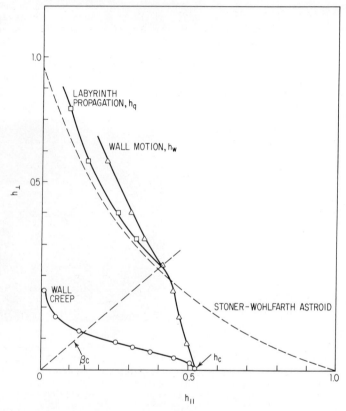

Fig. 35 Critical curves for wall motion, labyrinth propagation, and wall creep as compared with astroid. [*After S. Middelhoek and D. Wild, IBM J. Res. Develop.*, **11**, *93 (1967)*.]

that labyrinths occurred in a film with low dispersion, but upon increasing the dispersion in the film the reversal proceeded by partial rotation. The aid of the magnetostatic field from a labyrinth tip is apparently not needed if the magnetization dispersion is large enough.

(2) Coercive Force The coercive force H_c for reversal by domain-wall motion, which is usually a few oersteds, cannot yet be quantitatively predicted. It is clear only that inhomogeneities in the film must cause a moving wall to experience a variation of wall energy with position, and thus cause a nonzero value of H_c [Eq. (10)]. For example, Lorentz microscopy shows directly that a large inhomogeneity such as a hole in the film can impede the motion of a wall. In general, however, the inhomogeneities which are effective and the precise wall-imperfection interaction mechanisms are as yet obscure.

Inhomogeneities which could contribute to H_c include spatial variations in film thickness, magnetic anisotropy, magnetization, and the exchange constant. In an early theoretical work, the effect of thickness variations due to the surface roughness of a film was considered, and it was predicted that $H_c \propto d^{-\frac{4}{3}}$. Several investigators have measured the thickness dependence of H_c in an effort to check this theoretical prediction. In an attempt to reduce the considerable experimental scatter found in experimental plots of H_c vs. d, some workers have measured H_c of a single film while it was still growing in thickness in the vacuum chamber. Nevertheless, the results reported by the authors who used this method (Fig. 36) differ substantially, while each individual author had difficulty in obtaining reproducible results. It is thus clear that the value of H_c is very sensitive to deposition conditions, and that the $\frac{4}{3}$ power law is not generally satisfied. On the other hand, it is noted that peaks in the H_c vs. d curves can be found (Ahn's curves in Fig. 36) at values of d which correspond to transitions between the various wall types. This is a gratifying result, since it is consistent with the predictions of Eq. (10), because different types of walls have different values of $[d(\gamma d)/dx]_{max}$.

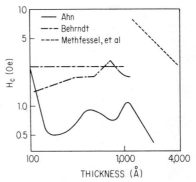

Fig. 36 Wall coercive force H_c as a function of film thickness for films deposited and measured in vacuum. Substrate temperature was 200°C. Ahn, 81% Ni, 19% Fe; Behrndt, 81% Ni, 19% Fe films at two different deposition rates and background pressures; Methfessel et al., 76% Ni, 24% Fe. [*After K. Y. Ahn, J. Appl. Phys.*, **37**, *1481 (1966); K. H. Behrndt, J. Appl. Phys.*, **33**, *193 (1962); and S. Methfessel, A. Segmüller, and R. Sommerhalder, J. Phys. Soc. Japan*, **17**, *Suppl. B-I, 575 (1962).*]

There are many indications that a successful theory of the coercive force must include the effect of the ripple. Thus, it is well known that conditions which lead to low values of magnetization dispersion (e.g., smooth, clean substrates) also favor low values of H_c. Furthermore, the value of h_c increases monotonically with α_{90} (Fig. 37); in fact, h_c is a very useful index of magnetization dispersion. A fruitful approach to the theoretical problem may thus be to regard the domain wall as embedded in the magnetization ripple; the periodically distributed magnetic poles thus produced at the wall will cause the necessary $d(\gamma d)/dx$.

In some films with very high magnetization dispersion, values of $h_c > 1$ have been found. The occurrence of such films is unexpected, because on the S-W theory, magnetization reversal by coherent rotation would be expected at $H = H_K$ before H_c was reached. Observations of inverted films (Fig. 38) show, however, that before ordinary reverse domains can be introduced, an EA "locking" process occurs: the ripple gradually intensifies as the reversing field grows, until domain walls are produced, precisely like the HA fallback process in low-dispersion films (Secs. 3a and 11b). Because of the manner in which the Néel walls were formed during the locking process, $\vec{M}$ in the wall center is antiparallel to the reversing field. Thus, in order to propagate reverse domains, the wall sense must be reversed. This is accomplished by the nucleation and subsequent separation of two Bloch lines on each wall, which then allows the reverse domains to propagate through the wall (Fig. 38).

(3) Wall Mobility Using special techniques, the velocity of domain walls can be measured for various values of applied fields, thus giving the mobility m defined by Eq. (11). A small region in the film having good uniformity must be chosen so that only a single Barkhausen jump is studied, thereby ensuring that the motion of a single wall can be reproducibly followed. With care, the linear relation of Eq. (11) can be verified, as shown in Fig. 39, although it is seen that some films show an unexplained decrease in m for high fields.

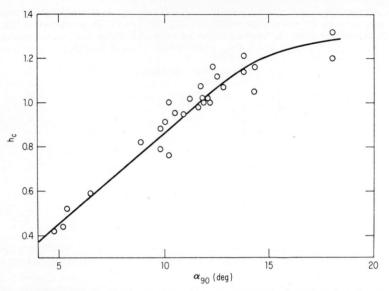

Fig. 37　Normalized wall coercive force h_c vs. α_{90} for 81 % Ni, 19 % Fe rectangular films 0.25 × 1.50 mm with the EA parallel to the long dimension.　The films were about 1,000 Å thick.　The substrate temperatures varied from 300 to 400°C, thus causing the indicated range of α_{90} values.　The hysteresigraph measurements of H_K and hence α_{90} were affected by the shape anisotropy of the films.　(*Courtesy of T. S. Crowther.*)

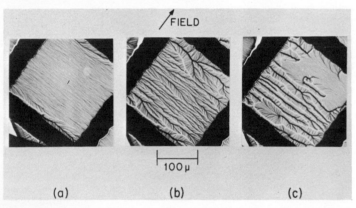

Fig. 38　Lorentz micrographs illustrating EA locking in an 81 % Ni, 19 % Fe film.　(*a*) Saturated single-domain film, no applied field.　(*b*) EA reversing field of 4.0 Oe applied. (*c*) EA reversing field increased to 5.4 Oe.

Measurements of m as a function of film thickness have been made (Fig. 40). The variation among the results of different authors is large, but there seems to be a minimum in the curves which may be associated with the transition from crosstie to Bloch walls.　This may be expected from Eq. (12), which shows that walls of different shapes have different mobilities.

Just as for wall motion in bulk material, the mobility is determined by the rate at which energy can be lost to the lattice.　There is general agreement that the basic mechanism for this energy loss cannot be eddy-current damping, since m

calculated from that mechanism is proportional to $1/d$, and calculated values of m result which are greater than the experimental values. It is suspected that the intrinsic energy-loss mechanism which is operating here may be the same mechanism which is effective in fast switching (Sec. 12); this loss mechanism is poorly under-

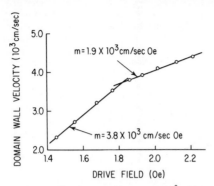

Fig. 39 Domain wall velocity vs. EA drive field for a 500 Å thick 77 % Ni, 23 % Fe film. $H_K = 3.3$ Oe and $H_c = 1.5$ Oe. [*After C. E. Patton and F. B. Humphrey, J. Appl. Phys.,* **37**, *4269 (1966)*.]

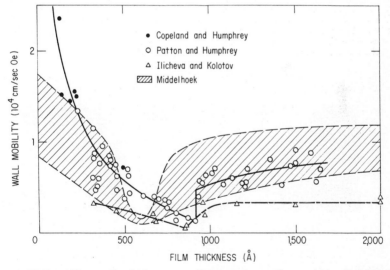

Fig. 40 Wall mobility vs. film thickness according to various authors. (*After E. Feldtkeller, Internat. Konf. Magnetismus, Deutscher Verlag für Grundstoffindustrie, Leipzig, 1967.*)

stood. Thus, although some attempts have been made to fit Eq. (12) to the mobility data, too little is known about wall shapes and the damping parameter α to obtain conclusive results. There are indications that ripple plays a role in the damping process, since it has been observed that m decreases with increasing h_c, while α shows a corresponding trend.

b. Creep[34]

If a dc EA and a dc HA field are applied to a film containing a domain wall, that wall will move and reverse the magnetization of the film only if the net field exceeds the threshold field h_w plotted in Fig. 35. However, if a dc EA field and a *pulse* (or ac) HA field are applied, it is found that even if the net field lies well within the h_w threshold, the wall can still move. The wall will be translated a few microns in the direction favored by the EA field upon the application of each HA pulse in a process known as *creep*. Creep is a statistical process; a HA pulse will *sometimes* help certain parts of the wall to overcome potential barriers, so that repeated HA pulses cause uneven, jerky wall motion. Upon application of several thousand HA pulses, the creep process can cause appreciable demagnetization. Creep is thus of great technological concern because information stored in a film bit in a digital-computer memory can be slowly lost under the repeated application of HA pulse fields whose dc values would be too small to cause wall motion.

Creep can be investigated by examining the demagnetization of a film resulting from the application of a specified number of pulses of HA field in the presence of a given EA field, or by studying the creep velocity of individual walls. In either case threshold curves for creep can be drawn in field space in the same manner as wall-motion thresholds for dc fields; such threshold curves are far inside the astroid and the dc field curves (Fig. 35).

Creep does not occur for a dc HA and a pulsed EA field but only for a dc EA and changing HA field. The shifts of a creeping wall occur during the rising or falling edges of the HA field pulses, so that only the HA field-pulse height and total number of pulses are important, but not the repetition rate and pulse length. The creep threshold is also critically dependent upon the structure of the wall; films thin enough to contain only Néel walls do not exhibit creep under unipolar HA pulses, although some creep is seen with bipolar HA pulses. This fact is demonstrated by Fig. 41.

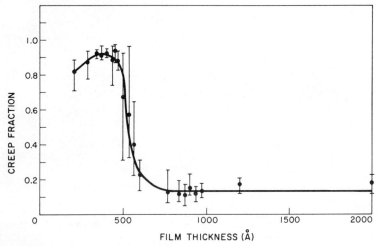

Fig. 41   Creep fraction vs. film thickness. A 1-Oe dc EA field was applied along with 15,000 HA pulses of magnitude such that an additional 10% of demagnetization occurred by creep from a state of the film initially demagnetized 10% from saturation (to ensure the presence of some creepable walls). The creep fraction is defined as the ratio of the magnitude of the multipulse HA field to that of a single-pulse HA field (representing ordinary wall motion) capable of causing the same 10% additional demagnetization. [*After G. P. Gagnon and T. S. Crowther, J. Appl. Phys.*, **36**, *1112 (1965)*.]

The creep fraction, which is plotted in that figure against film thickness, is so defined that a value near 1 corresponds to a high-threshold creep field, while a low value means that creep occurs at relatively small fields. It is seen that for the unipolar pulses applied, films thinner than about 400 Å, i.e., films supporting Néel walls, indeed show little creep.

The mechanism causing creep is not well understood. It is probable that each HA pulse causes a local change in the magnetization distribution of a wall, so that a consequent change in the local potential energy distribution at the wall is also made. This change may be favorable, so that the wall can move locally until another large-potential hill is encountered. Specific reasons for such changes in the wall mobility [Eq. (10)] may lie in changes in wall mass or wall stiffness with wall shape. Thus, it is well documented that Bloch lines move laterally along crosstie walls in Bark-hausen jumps during creep, just as domain walls move in Barkhausen jumps upon application of a field parallel to the wall. The Bloch-line motion causes a local change in the sense of rotation of the wall, which may be adequate to cause the necessary wall-mobility change, and hence initiate creep. Several other explanations have been proposed, but there are fundamental objections to many of these mechanisms and none is on a firm theoretical or experimental footing.

c. Thermally Enhanced Wall Motion[22]

In certain specially prepared films, it is found that in the presence of a *constant* dc field, a domain wall will move slowly across the film without large Barkhausen jumps. For example, Fig. 42 contains successive Lorentz micrographs of a wall

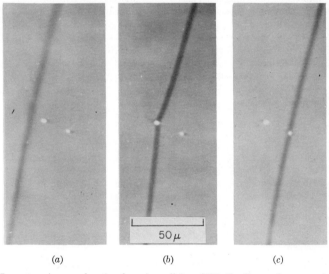

      (a)   (b)   (c)

Fig. 42 Lorentz micrographs of a domain wall in a NiFeCu film under a constant 2.1-Oe dc field applied parallel to the wall. (a) Time $t = 0$. (b) $t = 30$ s. (c) $t = 60$ s. The white circles are holes in the film. [*After M. S. Cohen, J. Appl. Phys.*, **38**, *860 (1967)*.]

moving in the presence of a constant field in a special NiFeCu film. For a fixed field, the wall velocity in such films increases with the temperature of the film. The explanation of the phenomenon is that thermal energy aids the wall in overcoming the potential barriers which it cannot surmount with the applied field energy alone. Such thermally enhanced wall motion is classified under the category of magnetic-viscosity effects and is not to be confused with the slow domain-wall motion observed in creep, where the activation comes not from thermal energy but from the presence of an additional varying HA field.

A simple statistical theory can explain these results. The probability per unit time of a volume V of the film being reversed in a field $\vec{H}$ with the aid of the thermal activation is $p = C \exp\left[-(E_0 - 2HMV)/kT\right]$, where E_0 is the barrier height and C is a constant. If all barriers are assumed to be of equal height E_0 and spaced a distance l apart, then the wall velocity may be written as

$$v = Pl = Cl \exp \left[\frac{2MV(H - H_v)}{kT} \right]$$

where $H_v \equiv E_0/(2MV)$, and $H \le H_v$. Since H_v is the field necessary for wall motion at $T = 0°K$ (no thermal activation), H_v may be identified with H_c. Furthermore, the fact that V appears in the exponent means that for large V, thermal-activation effects will be unobservable since values of H impractically close to H_v would be required, while for small V wall motion should occur at $H \sim 0$. Thus films having only a small range of particle size will exhibit magnetic-viscosity effects.

Several workers have observed magnetic viscosity in magnetic films, but only in selected films because difficulty was encountered in finding films having sufficient uniformity for meaningful measurements (spatial variations of H_v can cause Barkhausen jumps between intervals of slow wall motion). Thus it was reported that magnetic-viscosity effects became more pronounced in Ni films as d decreased ($d < 300$ Å). Such behavior is not surprising since the increasing crystallite separation found in very thin films means increasing magnetic independence of the crystallites, so that the sample may be viewed as a planar aggregate of crystallites with relatively weak mutual magnetostatic interactions instead of a continuous film (a superparamagnetic aggregate is approached). Thermal activation may thus reverse $\vec{M}$ at the domain wall in some crystallites which correspond to the volumes V of the above theory. In the same way, the Cu in the NiFeCu films of Fig. 42 may be preferentially concentrated at the grain boundaries, thus breaking exchange bonds and making the crystallites magnetically more independent.

d. Wall Streaming[35]

In a film which is thick enough to support Bloch walls, a peculiar wall motion known as *wall streaming* can be observed if a train of short-rise-time field pulses is applied along the HA. In the presence of a dc EA field such HA pulses would cause creeping. However, wall streaming is not the same as creeping because it occurs when all EA fields are eliminated, including the demagnetizing field from the film edge (a multidomain, demagnetized film is used). Furthermore, a streaming wall does not move uniformly but instead moves back and forth in an oscillating, fluttering fashion (Fig. 43). It is found that the wall displacement per HA pulse increases with the

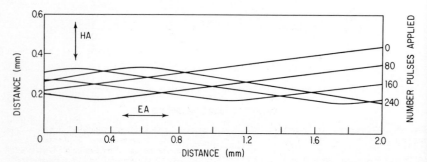

Fig. 43 Trace of a Bloch wall in a 1,500-Å-thick 81 % Ni, 19 % Fe film after the application of the noted number of HA pulses of magnitude 0.3 H_K, illustrating wall streaming. No EA field was applied. [*After K. U. Stein and E. Feldtkeller, J. Appl. Phys.*, **38**, *4401 (1967)*.]

magnitude of the pulse but decreases as the pulse rise time increases. Also, the thinner the film, the shorter the rise time which is required for such wall streaming, but films thin enough for crosstie or Néel walls do not exhibit streaming no matter how short the rise time.

The streaming motion is explained by the precession of the spins in a Bloch wall about the HA. This precession causes an "unwinding" of the Bloch wall (Fig. 28) in such a way that the wall moves; if the Bloch wall were of the opposite sense, the precession would cause wall motion in the opposite direction. Now Néel lines separate Bloch-wall segments of opposite sense so that the walls on opposite sides of Néel lines move in opposite directions. The characteristic fluttering motion arises because the HA pulses can cause translation of the positions of the Néel lines along the wall.

As in the case of creep, wall streaming is of technological interest because it can lead to loss of information stored in a memory bit.

10. MAGNETIZATION RIPPLE[36,37]

Even if a film remains in a single-domain state without domain walls, its behavior will still depart from the S-W model. The reason for this departure is the presence of magnetization dispersion, or magnetic ripple in the film. Ripple is detected not only directly, as seen in Lorentz or high-resolution Faraday micrographs (e.g., Figs. 31b and 7), but its effects are found in macroscopic measurements of film properties (Sec. 11). Examples of such macroscopic measurements are α_{90} and h_c. From Fig.

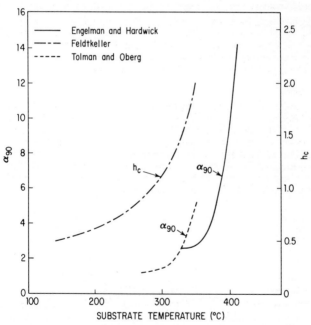

Fig. 44 Normalized wall coercive force h_c and α_{90} vs. substrate temperature for films of approximate composition 81% Ni, 19% Fe. [*After J. H. Engelman and A. J. Hardwick, in Trans. 9th AVS Symp., 1962, p. 100, The Macmillan Company, New York; E. Feldtkeller, J. Appl. Phys., **34**, 2646 (1963); and C. H. Tolman and P. E. Oberg, Proc. Intermag Conf. (IEEE, New York, 1963).*]

44 it is clear that as the substrate temperature increases, the ripple, as expressed by α_{90} and h_c, becomes more pronounced. This effect is evidently caused by the increase in the strength of the ripple-inducing perturbations as the sizes of the crystallites in a film are increased by higher deposition temperatures.

The first attempts to explain the effects of magnetization dispersion on the macroscopic properties were based on an *anisotropy-dispersion* model. In this model the film was thought of as divided into a large number of regions, each one of which had its own EA direction and magnitude of H_K. The regions were considered to be *noninteracting*, by either exchange or magnetostatic forces, so that the macroscopic behavior of the film was determined simply by the composite of the S-W predictions for all the independent regions. To describe such an ensemble of regions, the statistical distributions of the EA directions and the values of H_K in the film were denoted by quantities called the "angular dispersion of easy axes" and the "magnitude dispersion," respectively. The "angular dispersion" was identified with the quantity α_{90} which was obtained from the hard-axis-fallback experiment (Sec. 11b), while the magnitude dispersion was identified with the width of the $\chi''_{t,\pi/2}$ imaginary initial susceptibility peak, called Δ_{90}, obtained with a dc field along the HA (Sec. 11c).

The anisotropy-dispersion model was successful, at least qualitatively, in explaining many general properties of magnetic films. However, to explain particular features of film behavior in detail, many ad hoc assumptions had to be made. Thus, it was sometimes necessary to postulate that the independent regions had not uniaxial anisotropy but uniaxial anisotropy mixed with certain proportions of biaxial or even higher-order anisotropy. An even more important objection to this model is that it is known on general grounds that the regions of the film *cannot* act independently but are in fact coupled by exchange and magnetostatic forces. For these reasons, the problem has more recently been attacked by the micromagnetic method in which the magnetic properties are related to the measurable physical structure of the film without the use of ad hoc assumptions. Although the micromagnetic theory cannot yet quantitatively predict all the aspects of magnetic-film behavior, this theory is more acceptable than the anisotropy-dispersion theory because its basic assumptions correspond far better with physical reality.

The initial postulates of the micromagnetic theory are that $\vec{M}$ is independent of the coordinate normal to the film plane, so that the desired distribution of $\vec{M}$ is two-dimensional and that (in the static case) $\vec{M}$ is everywhere parallel to the film plane because of the shape-demagnetizing field (Sec. 3a). It is also assumed that uniaxial anisotropy, uniform in magnitude and direction, has been induced throughout the film. The film is visualized as being composed of randomly oriented, contiguous, generalized crystallites (which may or may not correspond to the actual crystallites composing the film); within each crystallite, in addition to the uniform uniaxial anisotropy, there is a crystallite anisotropy whose magnitude and orientation vary from crystallite to crystallite. It is this spatially randomly varying local anisotropy which causes the perturbations of the direction of $\vec{M}$ which are known as magnetization ripple. However, $\vec{M}$ does not simply take the direction within each crystallite which would be dictated by the joint action of the uniform uniaxial and crystallite anistropy because exchange and magnetostatic forces prevent abrupt changes in $\vec{M}$. Instead, the magnetization distribution in the film must be determined by a micromagnetic calculation which takes into consideration the many components of the local total energy (Sec. 2b) E_{tot}. It is here that the micromagnetic theory departs from the anisotropy-dispersion theory.

This section will be devoted to an exposition of the two most prominent micromagnetic theories of ripple, those by Hoffmann and Harte. However, before entering a detailed theoretical treatment it is possible to present a few well-confirmed facts about ripple: The ripple is primarily *longitudinal* rather than *transverse* (Fig. 45); i.e., the ripple wavefronts are perpendicular to the average $\vec{M}$ direction, as is indicated by Lorentz micrographs (Fig. 31b). The reason for the preference of longitudinal over transverse ripple is clear from a comparison of Fig. 45a and b, which indicates that although the exchange and anisotropy energies are the same in both cases, the magnetostatic energy is much higher for transverse ripple. The average ripple wavelength and ripple angle have been found by Lorentz microscopy to be about 1 μ and 1°, respectively. Lorentz microscopy does not easily provide more detailed information on the *ripple spectrum*, i.e., on the magnitude of the Fourier components

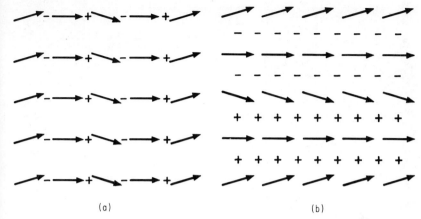

(a)   (b)

Fig. 45 Schematic diagram of magnetization ripple. Mean magnetization direction is horizontal. (a) Longitudinal ripple. (b) Transverse ripple. [*After K. J. Harte, J. Appl. Phys.*, **39**, *1503 (1968)*.]

as a function of ripple wavelength, because it is necessary to use wave optics and not geometric optics for interpretation of the micrographs (Sec. 4a).

It should be pointed out that the micromagnetic theory has been worked out in terms of generalized perturbing anisotropies. For convenience, each perturbing anisotropy is usually assumed to be uniaxial and associated with a randomly oriented crystallite of diameter D, but the theory does not demand the specification of the physical origin of this anisotropy. Higher-order perturbing anisotropy, e.g., biaxial, as well as preferred crystallite orientation could be introduced into the theory.

a. Origin of the Perturbations

Many attempts have been made to determine the physical origin of the random perturbing anisotropies, which are usually ascribed to either strain-magnetostriction, or magnetocrystalline anisotropy.* Thus, several workers have measured α_{90} vs. Ni content in NiFe films. A broad, flat minimum around the composition 81% Ni, 19% Fe, the composition for zero average magnetostriction, is revealed (Fig. 46), but this curve does not permit a choice between the two mechanisms because the magnetocrystalline constant K vanishes at 74% Ni, 26% Fe. Furthermore, strain-magnetostriction could be important even if the *average* magnetostriction vanished, since the *local* magnetostriction and the *local* strain, which may be spatially dependent, are important (Sec. 2c).

Lorentz-microscopy investigations have given conflicting conclusions regarding the origin of ripple. On the other hand, Faraday-effect observations of HA domain splitting (Sec. 11b) support the strain-magnetostriction model. However, perhaps the most reliable results have been obtained with the Faraday effect by illuminating only a small (a few hundred microns diameter) area of a film in order to obtain local measurements of α_{90}. Both long-range (many millimeters) and local contributions to α_{90} were found; the long-range contribution originated in strain-magnetostriction, while the local portion originated in both strain-magnetostriction and magnetocrystalline anisotropy. The latter conclusion was based on studies of local α_{90} vs. composition at various values of T_d (Fig. 47), which showed a minimum at the composition for vanishing magnetocrystalline anisotropy only when T_d was high enough to minimize local strains; otherwise the minimum was near the composition for vanishing magnetostriction.

* The contribution from the spatial variations in the $\vec{M}$-induced anisotropy caused by the random orientations of the crystallites is small, while an early report that the film composition is spatially dependent has not been confirmed.

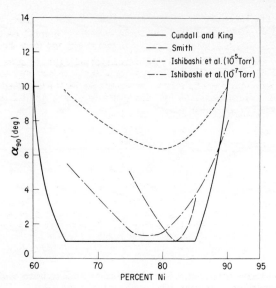

Fig. 46 Dispersion angle α_{90} vs. composition of NiFe films. Substrate temperature 250°C. [*After D. O. Smith, J. Appl. Phys.,* **32**, *70S (1961); J. A. Cundall and A. P. King, Phys. Stat. Solidi,* **16**, *613 (1966); and S. Ishibashi, K. Hida, M. Masuda, and S. Uchiyama, Elec. Eng. Japan,* **86**, *23 (1966).*]

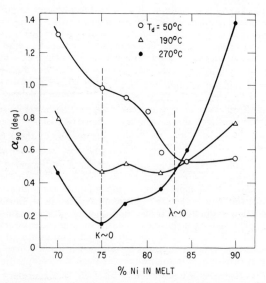

Fig. 47 Dispersion angle α_{90} vs. composition of NiFe films for various substrate temperatures. Measurements were made with Faraday effect over 0.1 mm² area so as to detect local variations only. [*After S. Uchiyama, T. Fujii, M. Masuda, and Y. Sakaki, Japan. J. Appl. Phys.,* **6**, *512 (1967).*]

b. Micromagnetic Theories

The micromagnetic calculations are made in the same spirit as the simple S-W calculations (Sec. 6a); the total energy E_{tot} is calculated and then the derivative with respect to the angle of $\vec{M}$, i.e., the total torque, is set equal to zero. However, the total energy is now far more complicated, for it includes not only the external field and uniaxial anisotropy energies but also exchange, local anisotropy, and magnetostatic energies. Thus, using the coordinate system of Fig. 48, if ϕ is the local angle of $\vec{M}$ and β is the angle of the applied field, then

(a) External field $\qquad E_H = -HM \cos (\beta - \phi)$
(b) Uniaxial anisotropy $\quad E_{K_u} = K_u \sin^2 \phi$
(c) Exchange $\qquad\qquad E_{\text{ex}} = A (\nabla \phi)^2$ $\qquad\qquad\qquad$ (28)
(d) Local anisotropy $\qquad E_K = K f(\phi, x, y)$
(e) Magnetostatic $\qquad\quad E_{\text{mag}} = -\frac{1}{2} \vec{H}_{\text{mag}} \cdot \vec{M}$

It is the presence of E_K, which is a function of position in the plane (x,y) of the sample, which generates the energy contributions of the last three terms.

Minimizing the total energy with respect to ϕ leads to the equation

$$\delta \int_V (E_H + E_{K_u} + E_{\text{ex}} + E_K + E_{\text{mag}})\, dV = 0 \qquad (29)$$

where V is the volume of the entire film. Because of the E_{mag} term, Eq. (29) represents a difficult variational problem (Sec. 2b). It has been attacked most successfully with the different approaches developed by Harte and by Hoffmann. Each of these two approaches has its own advantages and disadvantages. Harte's treatment of the magnetostatic energy is more precise than that of Hoffmann's and is especially suitable for handling approximations to the ripple of higher order than the linear. On the other hand, Hoffmann's approach is more appealing to the physical intuition, and more important, it permits an examination of the spatial distribution of the magnetostatic field, i.e., a study of local ripple-instability conditions. Harte's theory is not as suitable for this purpose, which is important for discussing macroscopic manifestations of ripple (Sec. 11). On the other hand, for those phenomena covered by both theories, excellent agreement is found, particularly within the linear approximation.

The calculation of the average ripple angle will now be sketched for both theories.

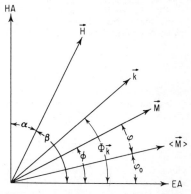

Fig. 48 Coordinate system for micromagnetic ripple theories. In Hoffmann's theory the x axis is along $\langle \vec{M} \rangle$, and the y axis is perpendicular to $\langle \vec{M} \rangle$; in Harte's theory, the x and y axes are parallel to the EA and HA, respectively.

(1) Hoffmann's Theory Hoffmann expressed the direction of $\vec{M}$ at any point $\phi(\vec{r})$ as the sum of the mean magnetization $\langle \vec{M} \rangle$ direction φ_0 (independent of position) and the angular deviation from it $\varphi(\vec{r})$:

$$\phi(\vec{r}) = \varphi_0 + \varphi(\vec{r}) \qquad (30)$$

where $\vec{r} = \vec{i}x + \vec{j}y$ is in the film plane (Fig. 48). By retaining only the initial terms in a Taylor series expansion in φ of Eqs. (28), he showed that only magnetostatic interactions originating within a small *coupling region* surrounding the point r are important in determining $\varphi(\vec{r})$. Thus Eq. (29) can be applied to the coupling region alone as if it were magnetically isolated, resulting in an enormous simplification in the micromagnetic problem. To first order, Hoffmann found that* $\varphi(\vec{r})$ depends on the uniform effective field Λ_1 [which is a function of the applied field $\vec{h}$ and the uniaxial anisotropy K_u (Sec. 6a)].

The coupling region may be shown to be elliptical in shape with the minor axis parallel to the average magnetization $\langle \vec{M} \rangle$. *The length of the minor axis is determined by exchange forces, while the length of the major axis is determined by magnetostatic fields.* The length of the semiminor axis is given by $[A/K\Lambda_1]^{\frac{1}{2}}$ while the semimajor axis is much longer (by a factor of about 30 for $\Lambda_1 \sim 1$ for a typical film). This interpretation is consistent with Lorentz micrographs of ripple (Fig. 31b) which indeed show structure elongated in a direction perpendicular to $\langle \vec{M} \rangle$. Again in correspondence with ripple the coupling region is not anchored in the film; its size and orientation and the resulting $\varphi(\vec{r})$ depend upon the applied field through Λ_1.

For comparison with macroscopic measurements, it is useful to calculate the magnetization dispersion δ, which is defined as the value of φ^2 averaged over the entire film. (A non-squared average vanishes.) It can be shown that, assuming the perturbing anisotropies are randomly-oriented uniaxial anisotropies,

$$\delta = \frac{0.17S}{(AK_u\Lambda_1)^{\frac{3}{8}}d^{\frac{1}{4}}M^{\frac{1}{2}}} \tag{31}$$

Here the structure constant $S \equiv KD/\sqrt{2n}$, where n is the number of crystallite layers in the film, characterizes those properties of the film which determine the ripple magnitude.

The linear theory just presented is valid only for large values of Λ_1 (corresponding to field points far from the astroid), because only then is the ripple state *reversible* and the higher order terms in φ in Eq. (29) negligible. That is, the linear theory is valid only for the case of small ripple amplitude (δ less than a few degrees). To discuss the limits of the linear theory, an effective field $h_{\text{eff}}(\beta)$ is defined in terms of the second variational derivative of the total energy [Eq. (28)] integrated over the film. This definition is made in analogy with the S-W theory [Eq. (19)] where the uniform effective field $\Lambda_1 \equiv \partial^2\epsilon_{\text{tot}}/\partial\varphi^2$. Then it can be shown that

$$h_{\text{eff}}(\beta) = \Lambda_1 + \frac{Kf_2}{2K_u} - \frac{M^2C_x}{K_u}\left[\varphi\frac{\partial^2\varphi}{\partial x^2} + \frac{1}{2}\left(\frac{\partial\varphi}{\partial x}\right)^2\right] \tag{32}$$

where C_x is a "magnetostatic" constant and f_2 originates in the second order term in the expansion of E_K. Note that $h_{\text{eff}}(\beta)$ is a *function of position* in the film. The second term in Eq. (32) gives the magnitude of an effective field arising from the local anisotropy, while the last two terms represent an effective longitudinal magnetostatic field originating in magnetic poles on the ripple. The second term is usually negligible, and the last two terms are the predominating high order terms. Thus, when the ripple is small, this longitudinal magnetostatic field is small, so that the value of $h_{\text{eff}}(\beta)$ averaged over the film is nearly equal to Λ_1, the uniform effective field, and the linear theory is valid. However, when the field point approaches the astroid and the ripple grows, the longitudinal magnetostatic field also grows; the resulting ripple state becomes *blocked* and is no longer described by the linear theory. The condition for validity of the linear theory is that $h_{\text{eff}}(\beta) > 0$ everywhere in the film, as will be discussed further in Sec. 11a.

(2) Harte's Theory   Harte's theory begins by expanding $\varphi(\vec{r})$ in the Fourier series

$$\varphi(\vec{r}) = \sum_{\vec{k}\neq 0} \varphi_{\vec{k}}e^{i\vec{k}\cdot\vec{r}} \tag{33}$$

---

* Thus a film could have the same ripple state for two entirely different applied fields, as long as Λ_1 corresponding to each field was the same. Consider, for example, fields along the EA and HA. If the HA field is greater than the EA field by the quantity $2H_K$, the ripple states will be identical, since Λ_1 will then be identical for the two cases [Eq. (19)].

where the wavevector $\vec{k}$ is at an angle Φ_k to the x axis (Fig. 48). Then $\varphi_{\vec{k}}$ is found for all $\vec{k}$ by first expressing the various interactions [Eqs. (28)] in terms of the torques $T_i(\vec{r}) \equiv \delta E_i/\delta\phi$, then Fourier analyzing each torque, and finally equating the sum of the kth components of the torques to zero.

By this method Harte found not only a first-order solution, but a higher-order solution as well. These solutions depended on the scale of the perturbing inhomogeneities compared with the exchange and magnetostatic lengths R_e and R_m, defined (in the linear approximation) by

$$R_e \equiv \left(\frac{A}{K_u\Lambda_1}\right)^{\frac{1}{2}} \quad \text{and} \quad R_m \equiv \frac{\pi M^2 d}{K_u\Lambda_1} \tag{34}$$

respectively. The results are shown in Table 4. Thus for fine inhomogeneities the ripple is attenuated by magnetostatic and exchange fields, but for coarse inhomogeneities exchange is not important. In examining Table 4 it must be remembered

TABLE 4   Magnetization Dispersion δ for Thin Films* (Harte's Theory)

Condition	Linear	Nonlinear	Longitudinal coherence length
$d, D \ll R_e$ Fine-scale inhomogeneity	$\dfrac{0.17KD(AK_u\Lambda_1)^{-\frac{3}{8}}}{d^{\frac{1}{4}}M^{\frac{1}{2}}} \underset{3°}{\to} 0.29\left(\dfrac{KD}{dM^2}\right)^{\frac{2}{8}}$		R_e
$d, R_e \ll D \ll R_m$ Coarse-scale inhomogeneity	$\dfrac{0.41KD^{\frac{1}{4}}(K_u\Lambda_1)^{-\frac{3}{4}}}{d^{\frac{1}{4}}M^{\frac{1}{2}}} \underset{6°}{\to} 0.29\left(\dfrac{KD}{dM^2}\right)^{\frac{2}{8}}$		D

* This table gives δ and the longitudinal coherence length for various conditions on the scale of inhomogeneity D, exchange length R_e, and magnetostatic length R_m. Arrows indicate transition from linear to nonlinear theory as ripple amplitude increases, while the transition value of δ is given below the arrow for a 500-Å-thick film with $\Lambda_1 = 6$ Oe and $M = 800$ g. After K. J. Harte, *J. Appl. Phys.*, **39**, 1503 (1968).

that R_e, R_m, and Λ_1 are dependent upon the external field; in particular, as the astroid is approached ($\Lambda_1 \to 0$), δ increases and the nonlinear $(KD/dM^2)^{\frac{2}{8}}$ dependence is valid rather than the linear dependences on the left. For a 500-Å film with $\Lambda_1 = 6$ Oe and $M = 800$ G, the transition δ is $3°$ for fine, but $6°$ for coarse inhomogeneities, where $D = 1\ \mu$ was assumed in the latter case.

Variations of $\vec{M}$ in distances smaller than the critical lengths R_e and R_m [Eq. (34)] are suppressed because of the action of exchange and magnetostatic fields, respectively, i.e., components $\varphi_{\vec{k}}$ associated with large k values in the longitudinal and transverse directions are suppressed, thus determining *short-wavelength* cutoffs. Then since the autocorrelation function of $\varphi(\vec{r})$ is simply the Fourier transform of the ripple spectrum, there will be corresponding maximum correlation distances, or *coherence lengths*, in these directions (Table 4). For the thin film, fine-inhomogeneity, linear case the longitudinal and transverse coherence lengths are found to be close to R_e and $(R_eR_m)^{\frac{1}{2}}$, respectively, which are approximately the dimensions of Hoffmann's coupling region (see above). Note also that the corresponding calculation of the magnetization dispersion δ (first entry of Table 4) is in agreement with Hoffmann's calculation [Eq. (31)]. Thus for this case Harte's results agree in all respects with those of Hoffmann's.

11. RIPPLE-DETERMINED BEHAVIOR

Magnetization ripple has some influence in many magnetic-film phenomena; in the phenomena discussed below, however, its role is decisive. These phenomena are all quasi-static; i.e., the applied field changes in magnitude in times long compared with the relaxation time of the magnetization configuration. Since the relaxation time is of the order of 1 ns, the category of quasi-static phenomena includes magnetic behavior under all conditions except those which involve special apparatus capable of applying fields having very high frequency Fourier components (Secs. 12 and 13).

Hoffmann's theory, which is the most suitable theory for the interpretation of these quasi-static phenomena, does not yet provide quantitative results for all their aspects, but at least a qualitative explanation can always be given. The difficulty in dealing with these problems originates in the necessity of using the less well-developed nonlinear theory when the uniform effective field Λ_1 is small, i.e., when the ripple is pronounced. The theory is far better when Λ_1 is large and the ripple is small, but unfortunately the phenomena in question are then of much less interest.

In view of its importance in the interpretation of quasi-static phenomena, the physical interpretation of this loss of validity of the linear theory as $\Lambda_1 \to 0$, i.e., the theory of ripple blocking, will first be explored.

a. Ripple Blocking and Locking (Theory)

According to the S-W theory, the condition for stable equilibrium is that $\Lambda_1 > 0$; when $\Lambda_1 = 0$, i.e., for the field point on the astroid, an unstable condition is achieved, and the film switches. In analogy with the S-W theory, the condition that $h_{\mathrm{eff}}(\beta) > 0$ ensures that the *ripple* is stable and "free" to rearrange itself into the state of minimum energy dictated by the magnitude and direction of the applied field. However, when $h_{\mathrm{eff}}(\beta)$ is reduced to zero or assumes negative values, the nonlinear magnetostatic fields *block* the ripple, so that it is no longer free to rearrange. The penultimate and final terms of Eq. (32) represent two magnetostatic fields, say $h_1(\beta)$ and $h_2(\beta)$, originating from the ripple, which are respectively parallel and antiparallel to $\langle \vec{M} \rangle$. The field $h_1(\beta)$ tends to attenuate the ripple and ensure stability, while the antiparallel field $h_2(\beta)$ tends to promote ripple instability.

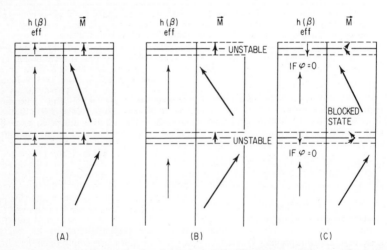

(A)   (B)   (C)

Fig. 49 Schematic diagram showing initiation of the blocking condition, as the uniform effective field Λ_1 is decreased below the blocking field $\Lambda_1{}^b$. Right side shows $\vec{M}$ configuration; left side shows effective field, $h_{\mathrm{eff}}(\beta)$. (A) $\Lambda_1 > \Lambda_1{}^b$, ripple is free. (B) $\Lambda_1 = \Lambda_1{}^b$, blocking initiates at regions of instability, where $\varphi = 0$. (C) $\Lambda_1 < \Lambda_1{}^b$, ripple is no longer free, but blocked. [*After H. Hoffmann, IEEE Trans. Magnetics,* **MAG-2,** *566 (1966).*]

Thus for large values of Λ_1 the ripple is free and the linear theory is valid, so that as Λ_1 is gradually reduced in magnitude (say by reduction of H), the ripple angle (i.e., δ) will increase (Fig. 49A), as will the longitudinal ripple period R_e [Eq. (34)]. Finally, a value of Λ_1 will be reached for which ripple blocking occurs at certain places in the film; the ripple will become blocked in those regions where $h_{\text{eff}}(\beta)$ first vanishes, then goes negative (Fig. 49B and C). Since $h_1(\beta)$ is always positive, the initial blocking regions will be those regions where $\varphi = 0$ [$h_1(\beta)$ vanishes]. The condition for the initiation of ripple instability at these regions thus becomes $\Lambda_1 = h_2(\beta)$, so that ripple is free only if $\Lambda_1 > h_2(\beta)$. [Since $|h_1(\beta)| \simeq |h_2(\beta)|$, the latter condition also justifies the neglect of terms of higher order in ripple theory.]

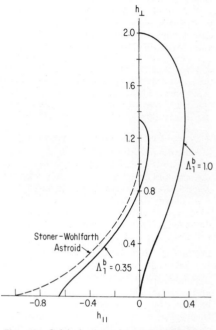

Now because of the random distribution of the crystallites, blocking does not initiate simultaneously everywhere in the film where $\varphi = 0$ but occurs in a random fashion throughout the film corresponding to a statistical distribution of $h_2(\beta)$. The average value of $h_2(\beta)$, or the average *blocking field* $\Lambda_1{}^b$, is

$$\Lambda_1{}^b = \frac{1}{AK_u}\left(\frac{Md^{\frac{1}{2}}S^2}{4\pi 2^{\frac{1}{2}}}\right)^{\frac{2}{3}} \qquad (35)$$

The condition for free ripple and validity of the linear theory is that $\Lambda_1 > \Lambda_1{}^b$.

Once blocking has initiated, as Λ_1 is reduced still further even the ripple which is outside the blocked region will no longer be able to rearrange as it did before blocking, because of the immobility of $\vec{M}$ in the blocked region. The ripple periodicity will thus remain fixed at approximately $R_e(\Lambda_1{}^b)$. At the start of blocking, small-angle walls will develop at the blocked region, but large-angle walls will develop if Λ_1 is reduced even further; such large-angle walls correspond to the much more stable condition of *locking*. Blocking is thus the first stage of locking.

The field point at which blocking initiates can be plotted directly on the astroid with the aid of Eqs. (19), (32), and (35). Examples of such plots are given in Fig. 50 for the cases $\Lambda_1{}^b = 0.35$ and $\Lambda_1{}^b = 1.0$. As h is decreased from saturation (from a free-ripple condition), blocking occurs when $\vec{h}$ first intersects the blocking curves. These curves show the S-W astroid can never be reached; hence coherent switching can never be achieved in a real film because blocking will occur first.

Fig. 50 Critical curve for blocking fields showing fields at which the transition from the free ripple to the blocked state occurs. Curves are plotted for the two values of $\Lambda_1{}^b(\beta)$ of 0.35 and 1.0. These values correspond to films with the parameters $A = 10^{-6}$ ergs cm^{-1}, $K_u = 1.5 \times 10^3$ ergs cm^{-3}, $D = 200$ Å, $d = 1,000$ Å; and $K = 3 \times 10^4$ ergs cm^{-3} and 6×10^4 ergs cm^{-3}, respectively. [*After H. Hoffmann, IEEE Trans. Magnetics,* **MAG-2**, *566 (1966).*]

b. Ripple Blocking and Locking (Experiment)

A very important experimental manifestation of the blocking and locking phenomena is provided by the HA fallback, or HA domain-splitting experiment (Sec. 3a). In this experiment the film is saturated by a high field parallel to the HA; then the magnitude of the field is slowly reduced. As expected, the ripple magnitude and average wavelength grow during this process, until at a certain value of H the phenomenon of HA fallback begins (Fig. 51): In regions where, because of the ripple, M is locally deviated to the "right" ("left") of the HA, $\vec{M}$ rotates clockwise (counterclockwise) toward the EA. Finally an array of approximately equal-spaced domain walls appears parallel to the EA. If the experiment is repeated with $\vec{H}$ applied at an angle α to the HA (Fig. 48), a smaller density of "splitting domains" is observed, and domains containing $\vec{M}$ directed along the EA sense closest to $\vec{H}$ are wider than oppositely directed domains. Thus the EA remanence (at $H = 0$), which is 0 at $\alpha = 0$, increases monotonically with α. As pointed out in Sec. 3a, measurements of the remanence as a function of α are useful in characterizing the magnetization dispersion of magnetic films; the "dispersion" α_r, i.e., the field angle α giving $r\%$ remanence, is often quoted; typically α_{90} is a few degrees.

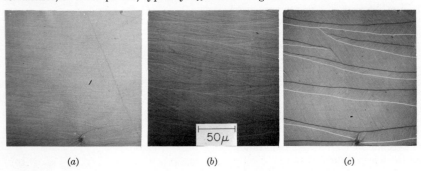

(a)   (b)   (c)

Fig. 51 HA fallback as revealed by Lorentz microscopy. Sample was saturated along the HA (vertical), after which the applied field was slowly reduced to the following values: (a) 3.8 Oe. (b) 2.2 Oe. (c) Zero.

It is clear from Sec. 11 that HA fallback is simply a special case of ripple blocking and locking, and that all aspects of this phenomenon should be interpretable on the basis of the micromagnetic theory. However, the quantitative predictions of the theory are not yet in satisfactory agreement with measurements of the most accessible experimental parameters, namely, the dependences on film parameters of α_{90} and the widths of the splitting domains. The major difficulty in the theory is the fact that different regions of the film have different values of $\Lambda_1{}^b$. [Thus experiments show a *distribution* of the widths of the splitting domains, which are expected from the theory to have the width of half the ripple wavelength at the initiation of blocking, say $\frac{1}{2}R_e(\Lambda_1{}^b)$.] On the other hand, if it is assumed that $\Lambda_1{}^b$ is constant throughout the film, it can be shown* that $\alpha_{90} \propto S^2$. In any case it is clear that α_{90} is *not* to be identified simply as δ but is related to it in a complicated fashion.

As expected from the considerations of Sec. 11a, the phenomena of blocking and locking are observable for field directions other than merely those near to the HA. Thus in high-dispersion films, locking is often observed for reverse fields along the EA (Sec. 9a) or at small angles to the EA. The reason why this phenomenon is not observed in low-dispersion films is simply that in such films the magnetization configuration is destroyed by the propagation of reverse domains before the locking process is completed. In contrast, HA splitting can be observed in *all* uniaxial films.

It should be noted that historically, HA fallback was first interpreted with the anisotropy-dispersion model (Sec. 10). On the basis of that model, when $\vec{H}$ is applied at an angle α to the right of the HA, only those regions with EAs which deviate from

---

* Experimental agreement with this relation was obtained, and from the experimental curves it was possible to deduce that K was too large to be explained by magnetocrystalline anisotropy and must originate in strain-magnetostriction anisotropy (see Sec. 10a).

the mean EA direction by an angle greater than α can support local counterclockwise $\vec{M}$ rotation; in all other regions $\vec{M}$ rotates clockwise, thus giving an EA remanence to the right. As α increases, the splitting-domain density decreases and the EA remanence increases, until when α is greater than the maximum local EA dispersion angle, $\vec{M}$ rotates uniformly clockwise toward the EA. From this reasoning α_{90} was incorrectly considered to give a direct measure of the angular dispersion of the EA direction.

c. Initial Susceptibility

Measurements of the initial susceptibility χ (Sec. 2d) are extremely sensitive to the presence of magnetization ripple. These measurements are made by applying, in the plane of the film, a dc bias field $\vec{H}$ at an angle β to the EA together with a small ac excitation field, while detecting magnetic-flux changes in the ac field direction with a pickup coil. If the ac field is parallel (perpendicular) to $\vec{H}$, then $\chi_{l,\beta}$, the longitudinal ($\chi_{t,\beta}$, the transverse) susceptibility is studied. For either $\chi_{l,\beta}$ or $\chi_{t,\beta}$, either the in-phase component χ' or the $\pi/2$ out-of-phase component χ'' (corresponding to an energy loss) can be measured.

For transverse, in-phase susceptibility it is easy to show that the S-W model predicts

$$\chi'_{t,\beta} = \frac{M}{H_K \Lambda_1} = \frac{M}{H_K[h \cos(\beta - \varphi_0) + \cos 2\varphi_0]} \tag{36}$$

Measurements of $\chi'_{t,\beta}$ are most commonly made with the dc field along the EA or HA for which $\beta = \varphi_0 = 0$ or $\pi/2$, respectively, so that

$$\chi'_{t,0 \atop \pi/2} = \frac{M}{H_K(h \pm 1)} \tag{37}$$

It is seen from Eq. (37) that in the absence of a dc field, $H_K = M/\chi'_{t,0}$; this equation is the justification for the commonly used hysteresigraph determination of H_K (Sec. 3a). It is also noted that as $h \to \infty$, $\chi'_{t,\beta} \to 0$, while $\chi'_{t,0}(\chi'_{t,\pi/2})$ becomes infinite at $h = -1(h = +1)$. Furthermore, Eq. (37) states that the $\chi'_{t,0}$ and $\chi'_{t,\pi/2}$ vs. h curves are congruent but displaced by 2, in H_K normalized units.

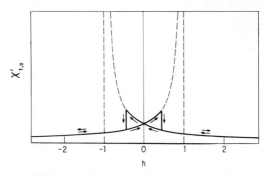

Fig. 52 Initial susceptibility $\chi'_{t,0}$ (dc field along EA) vs. applied field $h \equiv H/H_K$ for film with $h_c = 0.45$. Dotted line is theoretical curve according to Stoner-Wohlfarth theory. Arrows indicate history of dc field magnitude. [*After E. Feldtkeller, Z. Physik,* **176**, *510 (1963)*.]

Experimental curves of $\chi'_{t,0}$ and $\chi'_{t,\pi/2}$ are shown in Figs. 52 and 53, respectively. At high fields good agreement was found with the S-W predictions (shown dotted), but the congruency conditions were not satisfied over the complete range of h because of the introduction of reverse domains at fields $h < -h_c$ in the $\chi'_{t,0}$ measurements. Moreover, the agreement of $\chi'_{t,\pi/2}$ with the S-W theory was poor for $h \sim 1$; e.g., the field h_p for the peak value of $\chi'_{t,\pi/2}$ was not 1, and $\chi'_{t,\pi/2}$ at the peak was of course not infinite. This disagreement with the S-W theory is again caused by magnetization dispersion; as the magnetization dispersion increases, the departure from the S-W theory (e.g., $h_p - 1$) increases. It is also noted from Fig. 53 that nonzero values of $\chi''_{t,\pi/2}$ are detected. On the S-W theory, such loss should occur only over that miniscule range of h near 1 which permits the net field vector $\vec{h}$ to cross the astroid near its tip and thus cause repeated switching of $\vec{M}$.

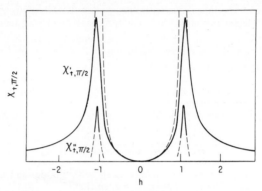

Fig. 53 Initial susceptibility $\chi'_{t,\pi/2}$ and $\chi''_{t,\pi/2}$ (dc field along HA) vs. applied field $h \equiv H/H_K$ for film with $h_c = 0.45$. Dotted line is theoretical curve according to Stoner-Wohlfarth theory. After saturation, the magnitude of the dc field was slowly decreased. [*After E. Feldtkeller, Z. Physik*, **176**, *510 (1963)*.]

The first attempts to explain these results, particularly the χ'' results, were made on the anisotropy-dispersion model. However, the micromagnetic theory explains the experimental facts, at least qualitatively, in a much more satisfactory way. According to that theory, the torque on the average magnetization comes not only from the applied field and the uniaxial anisotropy but also from the perturbing anisotropies and the nonlinear magnetostatic fields. To find $\chi'_{t,\beta}$, Λ_1 in Eq. (36) is simply replaced by $h_{\text{eff}}(\beta)$; i.e., the nonlinear theory must be used to evaluate $h_{\text{eff}}(\beta)$. However, it is necessary to average over the entire film, while taking account of the fact that certain portions of the film are blocked and other portions are free (Sec. 11a); this difficult calculation has not yet been performed. In any case it is clear that $\chi'_{t,\beta}$ is reduced by the ripple to a value below the S-W value; in particular, since it can be proved that $h_{\text{eff}}(\beta)$ averaged over the entire film is always positive, $\chi'_{t,0}(\chi'_{t,\pi/2})$ cannot diverge at $h = -1(h = 1)$.

This result shows that exaggerated values of H_K may result from measurements of $\chi'_{t,0}$ by the common hysteresigraph technique [Eq. (37)], which is a well-known experimental fact. As the ripple magnitude increases and $h_{\text{eff}}(\beta)$ departs further from Λ_1, this technique becomes less and less reliable. However, it is interesting to note that the congruency condition for $\chi'_{t,0}$ and $\chi'_{t,\pi/2}$ predicted on the basis of the S-W theory holds even in the presence of ripple. This statement is true because the ripple state depends on the applied field only through the magnitude of Λ_1, even in the nonlinear theory (Sec. 10b). Then if EA and HA dc field values are found which correspond to identical values of $\chi'_{t,0}$ and $\chi'_{t,\pi/2}$, respectively, the fact that these field values differ by $2H_K$ offers a ripple-insensitive measurement of H_K.

Measurements of the longitudinal susceptibility $\chi'_{l,\beta}$ are also of interest. For $\chi'_{l,0}$ or $\chi'_{l,\pi/2}$, i.e., $\vec{H}$ along the EA or HA, respectively, $\vec{M}$ is always parallel to $\vec{H}$ for high values of H. In that case it is clear from the S-W theory that $\chi'_{l,0} = \chi'_{l,\pi/2} = 0$, so that any measurable susceptibility must originate in the ripple. Now $\chi'_{l,0}$ is proportional to δ^2/Λ_1 so that from Eq. (31), $\chi'_{l,0} \propto S^2/\Lambda_1^{\frac{7}{4}}$. This field dependence has been experimentally verified for very uniform films with little skew. Such experiments provide strong support for the micromagnetic theory and may be useful for providing measurements of δ as well as S.

The origin of the imaginary part of the susceptibility χ'' lies in ripple hysteresis. Unfortunately, ripple hysteresis has been treated so far only on the basis of the linear theory, which gives unreliable quantitative predictions. Nevertheless, the validity of the ripple-hysteresis explanation has been demonstrated by observations of isolated local ripple rearrangements in susceptibility measurements by Lorentz microscopy. Also, measurements of $\chi''_{t,0}$, $\chi''_{t,\pi/2}$, and $\chi''_{t,\pi/4}$ have been made with special dc field configurations, each of which involved a fixed-bias field $\vec{h}_b$ and a field $\vec{h}$ which was variable in magnitude. The field configurations were chosen so that $h = \Lambda_1$ for each susceptibility measurement. Since the ripple-state field dependence involves only Λ_1, the $\chi''_{t,\beta}$ curves should be, and were, congruent as a function of $h = \Lambda_1$ for all three configurations (Fig. 54). The congruency condition was destroyed only

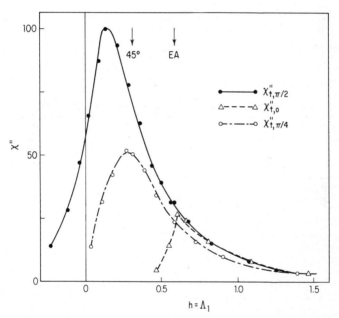

Fig. 54 Imaginary part of the transverse initial susceptibility $\chi''_{t,\beta}$ vs. $h = \Lambda_1$ for three different field configurations: $\beta = 0$, $\pi/4$, and $\pi/2$. For each configuration fields were applied so that Λ_1 was given by the magnitude of an external field $\vec{h}$. The three curves show good congruence except after reverse domains enter. Independent measurements of the values of h at which these reverse domains enter are indicated by the arrows. [*After* K. J. Harte, M. S. Cohen, G. P. Weiss, and D. O. Smith, Phys. Stat. Solidi, **15**, 225 (1966).]

when reverse domains swept through the film. (Domain propagation also prevents detection of appreciable values of $\chi''_{t,0}$ for a low-dispersion film; reverse domains sweep through before appreciable values of $\chi''_{t,0}$ are attained.)

d. Rotational Hysteresis

The rotational hysteresis W_r is defined as the work done when a field $\vec{H}$ applied in the film plane is rotated 360°. Thus, since the torque $\vec{T}$ exerted by $\vec{H}$ upon $\vec{M}$

is given by $HM \sin (\beta - \varphi_0)$,

$$W_r = \int_0^{2\pi} T \, d\varphi_0 = HM \int_0^{2\pi} \sin (\beta - \varphi_0) \, d\varphi_0 \tag{38}$$

where the angles β and φ_0 are defined in Fig. 48. This work W_r is a consequence of irreversible processes occurring in the film during the rotation of $\vec{H}$; W_r is lost to the lattice as heat. Rotational hysteresis is most conveniently measured with a torque magnetometer by integrating the measured torque over a 360° rotation [Eq. (38)].

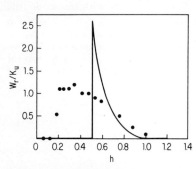

Fig. 55 Rotational hysteresis vs. applied field h for a low-dispersion film ($h_c = 0.2$). Solid curve calculated from Stoner-Wohlfarth theory. [*After W. D. Doyle, J. E. Rudisill, and S. Shtrikman, J. Phys. Soc. Japan,* **17**, *Suppl B-1, 567 (1962).*]

According to the S-W theory, rotational hysteresis occurs only for applied fields in the range $0.5 < h < 1$, since irreversible processes should occur only within that range. Experimental results show, however, that even in films with small ripple, i.e., small values of h_c (e.g., $h_c < 0.2$), non-zero values of W_r appear for values of h less than 0.5 and greater than 1 (Fig. 55). Furthermore, films which have large magnetization dispersion can exhibit a tail of the W_r vs. h curve which extends to very high values (several hundred) of h.

These experimental results were again initially interpreted on the basis of the anisotropy-dispersion model, but it is now clear that the correct explanation must be based on ripple hysteresis. Rotational hysteresis is thus simply another method, in addition to observations of the imaginary part of the susceptibility, of studying ripple hysteresis. The quantitative theory of rotational hysteresis is not yet satisfactory because the nonlinear ripple-hysteresis theory has not yet been worked out for that problem.

e. Effects of High S/K_u

As the structure constant S is increased, the uniaxial character becomes increasingly masked by the random perturbing anisotropies, and the magnetic behavior is increasingly determined by the latter alone. This fact can be demonstrated by studying the effects of a series of annealing treatments given to a single film. It is found that as the annealing treatment proceeds, the crystal size D increases, as does the ripple intensity h_c and α_{90} (see, e.g., Fig. 44). Splitting domains become more pronounced upon EA reversal (EA locking), and finally "rotatable initial susceptibility" (RIS) films are produced. These RIS films [not to be confused with the RIS properties of stripe-domain films (Sec. 15b)] have such high magnetization dispersion that $\vec{M}$ remains in any arbitrary direction in which the film has been previously saturated; the large local anisotropies prevent $\vec{M}$ from relaxing back to the nearest EA. The initial susceptibility in the absence of a dc field is then largest in the direction perpendicular to $\vec{M}$, so that the initial susceptibility can be "rotated" at will. While it is clear that the high magnetization dispersion is the determining factor in establishing the RIS property, the exact mechanisms of RIS in ac fields are still under discussion.

12. FAST SWITCHING[38]

The theory of coherent switching, based on the dynamics described by the Landau-Lifshitz equation, has been presented in Sec. 6b. However, using the apparatus described in Sec. 4c, it is found that this theory is often in appreciable disagreement with experimental switching results. It is well established that the reasons for the departure from coherent behavior lie in ripple effects, but the mechanisms of the ripple action are not yet clear and are in fact the goals of much experimental and theoretical research.

a. Experimental Observations

In switching experiments, a dc HA field $h_\perp$ is applied, together with a pulse EA field $h_\|$, and the resulting magnetization behavior is deduced from the longitudinal and

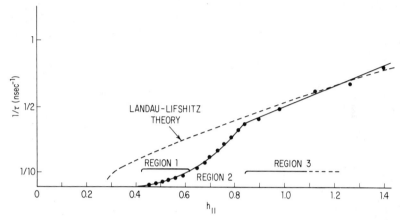

Fig. 56 Reciprocal switching time $1/\tau$ vs. normalized EA switching field $h_\|$ for a normalized dc HA $h_\perp$ field of 0.45 for $H_K = 3.0$ Oe, $h_c \sim 0.5$. The switching regions 1, 2, and 3 are shown. [*After K. U. Stein, Z. Angew. Phys.*, **20**, *36 (1965).*]

transverse voltage waveforms. A convenient way to present the results of these experiments is to plot the reciprocal of the time τ needed to complete the switching transition vs. $h_\|$. There are various ways to define τ, however. Perhaps the most common definition is the time necessary for the longitudinal signal to decay to say 10% of its peak value, but an alternative definition is the time interval necessary for the magnetization component along the HA to reassume the value it had before the application of the EA field pulse. An experimental plot based on the latter definition is shown in Fig. 56; here $1/\tau$ is plotted against $h_\|$ for a constant value of $h_\perp$.

It is seen from Fig. 56 that for high values of $h_\|$, the experimental values of τ fit the predictions of the coherent notation theory (for an assumed value* of the damping constant α of 0.03). In general such agreement is found as long as τ is less than a few nanoseconds, which is achieved only for low magnetization dispersion, and large values of $h_\|$ and $h_\perp$, i.e., only for field points well outside the S-W astroid. A typical longitudinal waveform from a film switching under these conditions is shown in Fig. 57; comparison with Fig. 18 shows the same type of oscillations as are predicted theoretically.

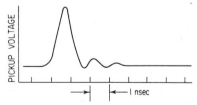

Fig. 57 Switching voltage waveform from a longitudinal pickup loop for $h_\perp = 0.5$, $h_\| = 1$, $h_c = 0.47$. The entire waveform is attributed to coherent rotation. [*After J. H. Hoper, IEEE Trans. Magnetics*, **MAG-3**, *166 (1967).*]

* The value of α that fits the data is often several times that found from resonance or free-oscillation experiments (Sec. 13a).

Switching curves taken for higher-dispersion films at low values of $h_\perp$ are shown in Fig. 58. (The first definition of τ is used here.) A direct comparison of the results of Figs. 56 and 58 is difficult because of the different conditions in the two cases, but it is clear that in both cases the switching curve can be divided into three regions. Region 3 in both figures has been attributed to coherent rotation and region 1 is interpreted as representing complete incoherence (e.g., domain-wall motion), while region 2 suggests a transition between 1 and 2. However, if region 3 of Fig. 58 really does represent coherent rotation, it is probably different from region 3 of Fig. 56, since only the latter fits the Landau-Lifshitz predictions.

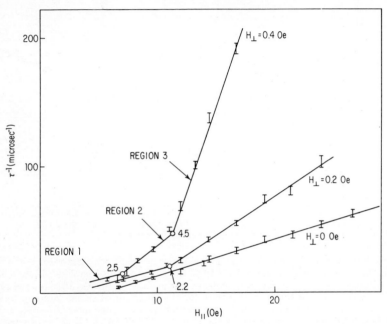

Fig. 58 Reciprocal switching time τ^{-1} vs. $H_\parallel$ for various values of $H_\perp$ for $H_K = 3.0$ Oe, $h_c = 1.2$. Values of the products $H_\parallel \cdot H_\perp$ are given at the inflection points. The switching regions 1, 2, and 3 are indicated on the $H_\perp = 0.4$ Oe curve. [*After R. V. Telesnin, E. N. Ilicheva, O. S. Kolotov, T. N. Nikitina, and V. A. Pogozhev, Phys. Stat. Solidi*, **14**, *371 (1966)*.]

The process of breakdown of coherent into incoherent rotation is well portrayed by plotting the trajectory of $\vec{M}$ as a function of time (Fig. 59). Here $\vec{M}(t)$ is obtained from the combination of both the integrated longitudinal and transverse signals. The initial angles of $\vec{M}$ are indicated for two different values of $h_\perp$, while the corresponding trajectories, which result after the application of an $h_\parallel$ pulse, are plotted at 0.5-ns intervals. For coherent rotation the trajectory should be the arc of a circle; this condition is well fulfilled for the initial portions of both the $h_\perp = 0.15$ and 0.60 curves. After a certain time interval, however, the angular velocity $\dot\varphi_0$ and the value of $|\vec{M}|$ decrease rapidly, the former by a factor of 10 to 100, indicating the initiation of incoherent behavior. The shape of the $\vec{M}$ trajectory depends not only upon $h_\perp$ and $h_\parallel$ but upon the magnetization dispersion; e.g., high-dispersion films can show conservation of $|\vec{M}|$ even after the initiation of incoherence as revealed by a marked decrease in $\dot\varphi_0$.

Further insight into the switching process is furnished by "interrupted switching" experiments; i.e., $h_\parallel$ is turned off before the switching process is completed. Analysis of the transverse and longitudinal waveforms showed that even in the "coherent" region 3, coherent switching was never completed if $h_\parallel$ was interrupted before the full switching time; instabilities occurred, and $\vec{M}$ collapsed into an incoherent configuration. Presumably the same incoherent processes begin soon after an (uninterrupted) initial coherent rotation for switching at low fields (region 2). Several investigators have also tried to use this interrupted-switching technique, in conjunction with Kerr or Bitter observations made *after* switching, to infer the detailed magnetization con-

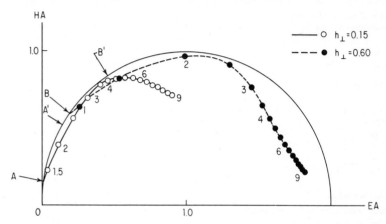

Fig. 59 Vector locus of net magnetization during switching for $h_\parallel = 1.92$ and two different values of $h_\perp$ for $H_K = 2.4$ Oe, $h_c = 1.0$. Time in nanoseconds after application of $h^\parallel$ pulse is marked on each locus. The initial directions of $\vec{M}$ are indicated at A and B, while the directions corresponding to switching according to the Stoner-Wohlfarth theory are indicated at A' and B', for $h_\perp = 0.15$ and $h_\perp = 0.60$, respectively. [*After Y. Sakurai, T. Kusuda, S. Konishi, and S. Sugatani, IEEE Trans. Magnetics,* **MAG-2**, *570 (1966).*]

figuration existing *during* switching. However, more recent investigations based on ultrashort-exposure Kerr micrographs taken *during* switching have cast doubt on this technique by demonstrating large changes in the magnetization distribution after interruption of the switching field. The newer investigations show rotational processes during high-speed switching but reversal by propagation of "diffuse" boundaries at low speeds.

b. Interpretation

It is clear that ripple plays a major role in determining the departure from coherent rotation. Even in the coherent region 3 of Fig. 56, the effective damping constant α which fits the experimental results apparently depends upon the magnetization dispersion, while it is found that large values of $h_\perp$ and $h_\parallel$ are needed to switch higher-dispersion films coherently. Thus, the *switching coefficient* S_w, i.e., the reciprocal of the slope of the curves in region 3, is directly proportional to the magnetization dispersion, and the positions of the inflection points in those curves depend on the dispersion.

Several attempts have been made to explain such results by ripple effects. Thus, for example, a switching theory based on Harte's magnetization-ripple theory has been presented (Sec. 10b). The later stages of the incoherent state were ignored, and an attempt was made to predict only the conditions under which the coherent state became unstable at some stage during switching (see, e.g., Fig. 59). It was noted that if, at any instant during switching, the ripple pattern for some reason suddenly became immobile while the average $\vec{M}$ continued to rotate, the ripple state would change from longitudinal to transverse. The transverse ripple would then exert on the average $\vec{M}$ a magnetostatic "reaction torque," which could become large enough to overbalance the torque exerted by the applied field and thus "lock" the rotation. The ripple will in fact become immobilized when it grows large enough during switching so that the associated magnetostatic fields cause violation of the Hoffmann stability criterion (Sec. 10b). From these considerations the conditions for the beginning of incoherence can be predicted quantitatively.

This theory assumes that the relaxation time of the ripple is short so that the ripple, when not immobilized by becoming unstable, can reach equilibrium with the average $\vec{M}$ in a time small compared with the switching time. That this is a good assumption was shown by an investigation which indicated that the ripple relaxation time was about 1 ns; this result is consistent with values of the free-oscillation damping constant (Sec. 13a).

13. RESONANCE

Two types of resonance experiments are performed with ferromagnetic films: ferromagnetic resonance (FMR) in which the magnetization moves coherently throughout the film, and spin-wave resonance (SWR) in which inhomogeneous magnetization distributions, i.e., spin waves, are excited.

a. Ferromagnetic Resonance[8]

The general theory of FMR was presented in Sec. 2e, while the experimental apparatus was discussed in Sec. 4c. In order to find the conditions for resonance, Eq. (14) is used without the damping term. However, $\vec{H}$ in that equation must be treated as an effective field $\vec{H}_{eff}$ which contains not only the applied field $\vec{H}_r$ but demagnetizing fields and anisotropy fields which act upon $\vec{M}$. Now $\vec{H}_r$ is usually applied normal to the film plane ("perpendicular" FMR) or in the film plane ("parallel" FMR), while the rf field is always perpendicular to $\vec{H}_r$ and in the film plane. Let H_K be the ordinary in-plane uniaxial anisotropy field and $H_K^{\perp}$ be the perpendicular anisotropy field (Sec. 7b). Then from Eq. (14) the resonance condition is given by

$$\omega_{\perp} = \gamma[(H_r - H_K^{\perp} - H_K)(H_r - H_K^{\perp})]^{\frac{1}{2}} \tag{39}$$

For parallel FMR, the resonance condition depends on the direction of $\vec{M}$ with respect to the EA, which is given by the S-W theory. Thus, for example,

$$\omega_{\parallel} = \gamma \left\{ \left[H_r + H_K^{\perp} + \frac{H_K}{2}(1 \pm 1) \right] (H_r \pm H_K) \right\}^{\frac{1}{2}} \tag{40}$$

where the $\pm$ signs refer to the special cases of $\vec{H}_r$ perpendicular or parallel to the in-plane EA, respectively.

Ferromagnetic resonance has been experimentally observed in films in the frequency range from 100 kHz up to 70 GHz; most of the work reported has employed microwaves in the centimeter range. For films less than about 5,000 Å thick, the skin depth is larger than the film thickness (for frequencies less than 10 GHz), so that, in contrast with FMR in bulk materials, the rf field is constant throughout the depth of the film and eddy-current damping can be neglected.

For parallel FMR in the range from about 100 kHz to 1 GHz, H_r and H_K are small enough to be neglected in the first bracket of Eq. (40) so that for a given frequency $\omega_{\parallel}$, the resonance peaks corresponding to H_r parallel to the EA and HA are separated by $2H_K$. This fact may be used as the basis of a measurement of H_K; if $\omega_{\parallel}^2$ is plotted against H_r for $\vec{H}_r$ in the two directions, two straight lines separated at a given value of $\omega_\parallel$ by $2H_K$ result (Fig. 60). From Eq. (40) the line representing the average of these two lines should pass through the origin. Experiments show that instead it intercepts the field axis at a distance H_i to the right of the origin, indicating the presence of an internal field $\vec{H}_i$ parallel to $\vec{M}$. This internal field, which may attain values of several oersteds, increases monotonically with magnetization dispersion, but is nevertheless probably not directly related to the magnetostatic fields predicted by nonlinear ripple theory.

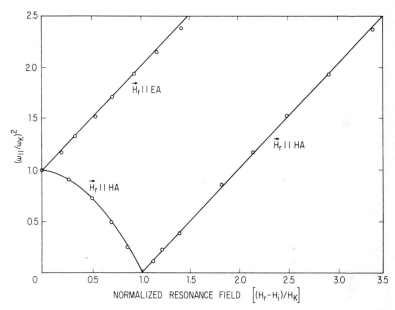

Fig. 60 Normalized resonance frequency vs. normalized resonance field for parallel resonance. Here $\omega_K \equiv \gamma(4\pi M H_K)^{\frac{1}{2}}$. Internal field H_i is about 1 Oe, $H_K = 5.19$ Oe. [*After D. T. Ngo, J. Appl. Phys.*, **36**, *1125 (1965)*.]

In the higher-frequency range, large enough values of H_r are required so that H_K can be ignored in Eqs. (39) and (40). If parallel *and* perpendicular FMR experiments are performed, simultaneous solution of these equations will yield γ and hence g; for NiFe, g values of about 2.1 are usually found. In addition, $H_K^\perp$ can be determined, thus yielding a measurement of the magnetization if $K_\perp{}^S$ is small enough so that $H_K^\perp = 4\pi M$.

If the damping term in Eq. (14) is included in the calculations leading to the above equations, the width of the resonance peak is predictable. For parallel FMR it is found that, if the frequency $\omega_\parallel$ is fixed while H is varied,

$$\Delta H_r = \frac{2\lambda\omega_\parallel}{\gamma^2 M} = \frac{2\alpha\omega_\parallel}{\gamma} \tag{41}$$

where ΔH_r is the field interval between half-power points. It is found experimentally that ΔH_r can range from a few to several hundred oersteds, depending on the film. However, the experimental data do not fit Eq. (41) but rather the equation $\Delta H_r = (\Delta H_r)_0 + 2\alpha\omega_\parallel/\gamma$; i.e., there is a residual broadening even at zero frequency. This residual broadening is attributable to magnetization dispersion, which causes different parts of the film to come into resonance at slightly different fields. The residual broadening has even been used as an index of magnetization dispersion;

$(\Delta H_r)_0$ increases monotonically with increasing magnetization dispersion of films as measured by susceptibility techniques.

The FMR experiments just described are studies in *forced* oscillations, where the forcing rf field is equal to the natural precession field at resonance. It is also possible to study *free* oscillations if a step-function pulse field is applied to the film. The transverse-loop signal resulting from such a step-function field applied along the HA is shown in Fig. 61, which demonstrates that $\vec{M}$ oscillates about its final equilibrium position with an amplitude which decays with time. The values of the free-oscillation frequencies and damping constants (deduced from the envelope of Fig. 61) are close to the values of the resonance frequencies and damping constants determined from ordinary forced-oscillation FMR experiments. However, the damping constants deduced from switching experiments are somewhat higher than these values, which are typically about 10^8 Hz.

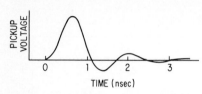

Fig. 61 Experimental voltage waveform from transverse loop showing free oscillations of a film ($H_K = 4.2$ Oe) subjected to a step-function HA pulse of 3.3 Oe. No EA field. [*After G. Matsumoto, T. Satoh, and S. Iida, J. Phys. Soc. Japan,* **21**, *231 (1966)*.]

The phenomena causing damping in FMR are poorly understood. Various mechanisms have been proposed and several experiments have been performed to test these mechanisms, but an unambiguous conclusion has not yet been reached.

b. Spin-wave Resonance[39,40]

As discussed in Sec. 2a, spin waves were first developed to explain the temperature dependence of the magnetization. However, it was later found that under proper conditions spin waves could be separately excited to resonance by rf fields, thus providing dramatic evidence for the validity of the spin-wave concept. The resonance conditions are similar to those which govern FMR, so that in SWR experiments the same apparatus and techniques are employed as in FMR. However, attention is now focused on spin-wave excitations rather than on the dynamics of the *uniform mode* (the coherent component of the magnetization). Usually *perpendicular* SWR is studied; i.e., the dc field $\vec{H_r}$ is applied normal to the plane of the film.

The dynamics of SWR are still described by Eq. (14), but a term must now be added which corresponds to the exchange energy which is associated with spin-wave excitation (Sec. 2b). This term takes the form of an equivalent field of magnitude $2M^{-2}A\nabla^2\vec{M}(r)$. Thus Eq. (14) becomes (without damping)

$$\vec{M}(r) = \gamma\vec{M}(r) \times \left[\vec{H}_{\text{eff}} + \left(\frac{D}{\gamma M}\right) \nabla^2\vec{M}(r) \right] \tag{42}$$

where $\vec{H}_{\text{eff}}$ includes applied, demagnetizing, and anisotropy fields, as before, and where $D \equiv 2\gamma A/M$. (If $K_\perp{}^S = 0$, then $H_{\text{eff}} = H_r - 4\pi M$.) It can be shown that this equation leads to the dispersion relation between the frequency ω and the propagation vector $\vec{k}$ of a traveling spin wave

$$\omega = \gamma H_{\text{eff}} + Dk^2 + \cdots \tag{43}$$

If rf energy of frequency ω satisfying Eq. (43) could be coupled to the spin wave, the latter would be maintained in resonance. However, this condition can be realized only if the spin waves are in a coherent state, such as represented by *standing* spin waves. Such standing spin waves are set up by reflections at the film interfaces. If the magnetic anisotropy at the interfaces is strong enough to immobilize or "pin" the spins at the interfaces, the problem is analogous to that of a vibrating string clamped at both ends. The boundary conditions then demand a standing-wave solution of the form sin kz where $k = p\pi/d$ and p, the *mode number*, is an integer (Fig. 62). Since the rf field is uniform within the film, in order to excite the spin wave p must be an *odd* integer, for when p is even the transverse component of the magnetization averages to zero over the film thickness, so that the rf field cannot couple to the magnetization.

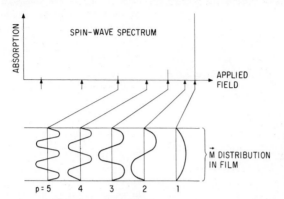

Fig. 62 Schematic diagram of theoretical spin-wave spectrum under assumption of perfect pinning in a uniform film. Because of the symmetry of the $\vec{M}$ distribution corresponding to the even-numbered modes, the rf field cannot couple with them, so that only odd-numbered modes are excited. [*After P. Wolf, in R. Neidermayer and H. Mayer (eds.), "Basic Problems in Thin Film Physics," p. 392, Vandenhoeck and Ruprecht, Goettingen, 1966.*]

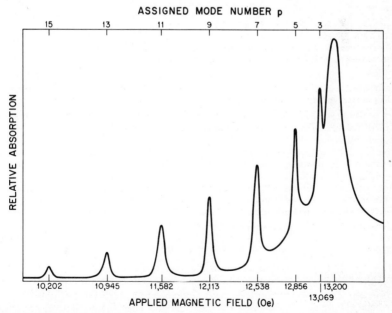

Fig. 63 Experimental spin-wave spectrum of a carefully prepared 4,100-Å-thick 85 % Ni, 15 % Fe film taken at 9.18 GHz. Peaks follow quadratic-dispersion law over the entire range. (*Courtesy of R. Weber.*)

As in FMR it is convenient to keep ω fixed and vary H_r, which leads to the prediction [Eq. (43)] that H_r depends linearly on p^2. An experimental SWR spectrum is shown in Fig. 63, where it is seen that, as expected, the even modes are not excited. The predicted p^2 dependence is experimentally confirmed* by loglog plots of the field separation from the uniform mode vs. mode number (Fig. 64). These plots were made for films of various thicknesses in order to ensure the accuracy of the mode-number assignments by covering a wide range of mode numbers. [The low-order modes are hidden by the breadth of the uniform-mode peak for the thicker films (Fig. 63).] From such plots it is possible to find the exchange constant A if d, γ, and M are known. (The latter two quantities can be found from FMR, Sec. 13a.) Thus from Fig. 64, $A = 0.92 \times 10^{-6}$ ergs cm^{-1}. Determinations of A by SWR constitute a major application of this technique; values of A thus found compare favorably with values found by other means.

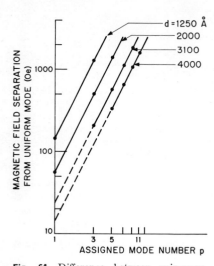

Fig. 64 Difference between spin-wave resonance-field and uniform-mode resonance field as a function of assigned mode number p for 80% Ni, 20% Fe films of various thicknesses. The curves indicate a p^2 dependence. [*After R. Weber, IEEE Trans. Magnetics,* **MAG-4**, *28 (1968).*]

Large deviations from the quadratic p dependence have been observed in many films, particularly at low p values where linear p dependences have sometimes been found. To explain the SWR spectrum of such films, an effective variation of M through the thickness of the film, in either parabolic or other (even linear) fashion, has been postulated instead of pinning at the interfaces as discussed above. Alternatively, variation of magnetic anisotropy normal to the film, caused, for example, by variation of strain through the film thickness, can produce the same effect. In any case these deviations from quadratic dependence have been shown to originate in poorly prepared samples; when deposition conditions are carefully controlled so that very homogeneous films result (e.g., the films of Fig. 64), a quadratic dependence results.

It is also possible to excite spin waves in *parallel* SWR, i.e., with $\vec{H}$ parallel to the plane of the film. However, parallel SWR intensities are usually much weaker than the perpendicular SWR intensities just described; parallel SWR is observed only when film inhomogeneities are such that favorable boundary conditions result.

Further refinements of SWR experiments permit fundamental investigations into the physics of ferromagnetic exchange. Thus, experiments at 70 GHz have yielded SWR spectra containing over 20 peaks, thereby permitting the determination of the coefficient F of an additional Fk^4 term in the dispersion relation [Eq. (43)]. The temperature dependences of D and F have also been studied. The results of these experiments can be compared with various theories of basic mechanisms of ferromagnetism. The results indicate that in ferromagnetic metals the exchange interaction may extend beyond nearest-neighbour atoms.

14. MULTILAYER FILMS[41,42]

The films discussed thus far have been composed of a single layer of ferromagnetic material. However, considerable research has also been carried out on multilayer

---

* It should be noted that although such data strongly support the interfacial-pinning model, the physical origin of the pinning action is poorly understood.

films consisting of two or more layers of ferromagnetic material. Various kinds of magnetic interactions between these layers are possible, depending on the magnetic properties of the ferromagnetic layers (which may differ from layer to layer) and the properties of nonmagnetic intermediate layers, if any, which separate the ferromagnetic layers. The systems discussed here will consist of only two ferromagnetic layers, i.e., a *couple*, but similar behavior is expected for laminated films with more layers.

a. Interactions between Layers

If there is no intermediate layer, a *composite* film is formed for which there is an exchange interaction which tends to maintain $\vec{M}$ in the two films *parallel*. If the properties of the constituent layers of a composite film are different, e.g., because of compositional differences, such a film may not be equivalent to a single film whose thickness is the sum of the thicknesses of the constituent layers. Another interaction which is present even with an intermediate layer originates in the effect of the demagnetizing field from one film on the other film, which tends to maintain $\vec{M}$ in the two films *antiparallel*. This interaction becomes less important for films of large diameter, which have small demagnetizing fields.

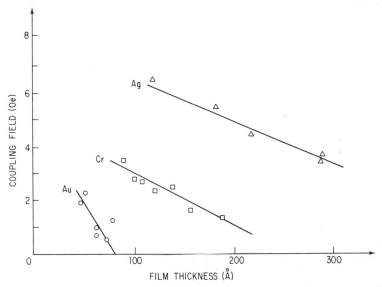

Fig. 65 Coupling field vs. intermediate-layer thickness for sandwiches of 81 % Ni, 19 % Fe and 45 % Co, 45 % Ni, 10 % Fe films with various intermediate layers. [*After J. C. Bruyère, O. Massenet, R. Montmory, and L. Néel, IEEE Trans. Magnetics, MAG-1, 10 (1965).*]

In these considerations idealized film structures with perfectly planar interfaces are assumed. However, a real vacuum-deposited sandwich made of two magnetic films separated by a nonmagnetic intermediate film, e.g., SiO, contains small-scale interface roughnesses because of the nonzero crystallite size. Magnetic charges are thus created at both intermediate-layer interfaces so that the resulting magnetostatic energy is minimized when the magnetizations in both layers are parallel.

A more subtle exchange-coupling mechanism is associated with two superposed vacuum-deposited films separated by a thin (up to a few hundred angstroms) nonmagnetic, metallic intermediate layer. The magnetic films are usually of different compositions, e.g., NiFe and NiFeCo; so they have markedly different values of H_c. The strength of the interaction can be measured by several means; perhaps the most striking is a shift along the field axis of the MH loop associated with the magnetically softer (lower H_c) film.* Such measurements demonstrate the presence of an equivalent coupling field of several oersteds which tends to align $\vec{M}$ in both films. This coupling field, which can be ten times the field expected from the surface-roughness model just discussed, decreases with increasing thickness of the intermediate layer (Fig. 65). This effect can be explained for intermediate layers of Ag, Au, and Cr by postulating the production of tiny bridges of ferromagnetic material which penetrate through small pores in the intermediate layer and thus transmit the exchange interaction. On the other hand, it has been shown that Pd intermediate layers form low Curie-temperature ferromagnetic alloys by interdiffusion with the ferromagnetic layers.

Assuming the absence of domain walls, calculations of the $\vec{M}$ positions in both films, along with switching thresholds, can be made for the interactions discussed above or indeed for a generalized interaction, in analogy with the S-W calculations for single films (Sec. 6a). In these calculations, the azimuth of $\vec{M}$ within a layer is usually allowed to vary as a function of distance in the normal direction (thus requiring consideration of exchange), while d, M, H_K, and the EA directions of the films are regarded as known quantities. The values of any or all of these parameters may differ between the films, which can necessitate involved calculations for $\vec{M}$ positions and switching thresholds. When such coupled-film systems have been experimentally investigated, good correlation with the theory has usually been found.

b. Interactions between Domain Walls

If the thickness of the intermediate nonmagnetic layer is less than about a domain-wall width, interesting interaction effects between juxtaposed walls in the two ferromagnetic layers can be observed. While these effects are always caused by the magnetostatic fields originating from the domain *walls*, the resulting wall structure depends on whether the *domain* structures of both ferromagnetic films are identical or different. Here identical domain structure means that, except within the domain walls, at any given position in the plane of the couple, $\vec{M}$ in one film is parallel to $\vec{M}$ in the other film. If the exchange or interface-roughness coupling between the two ferromagnetic films is strong and/or the films have nearly the same ferromagnetic properties, their domain structures may be identical; otherwise, domains may grow in one film but not in the other.

In either case interacting walls assume shapes which result in lower total magneto-static energy compared with juxtaposed walls unmodified from the single-film configuration. Possible wall configurations for couples with identical domain structures are shown in Fig. 66a and b, where the $\vec{M}$ direction in one layer is shown as a solid arrow and in the other layer as a dotted arrow. Good flux closure is achieved in the configuration of Fig. 66a by Néel walls of opposing rotation senses and in Fig. 66b by walls which contain *quasi-Néel* portions. (These quasi-Néel walls are not real walls because $\vec{M}$ on both sides of the wall are parallel.) Even in the case of non-identical domain structures, a domain wall in one film can induce a quasi-Néel wall in the other film (Fig. 66c) to give a resultant net lowering of the total energy.

Using micromagnetic techniques similar to those discussed in Sec. 8 for single films, the energy and shape of interacting walls of a couple can be calculated. The results show that the wall energies are less while the wall widths will be considerably greater than in corresponding single films. Furthermore, juxtaposed Néel walls of opposite senses or Néel, quasi-Néel combinations are energetically favored over crosstie or Bloch walls for far greater film thicknesses than in single-layer films (Fig. 67). These theoretical predictions are in good agreement with Lorentz-microscopy observations.

---

* The hysteresigraph excitation field exceeds H_c of the soft, but not that of the hard film, which remains saturated during the measurement.

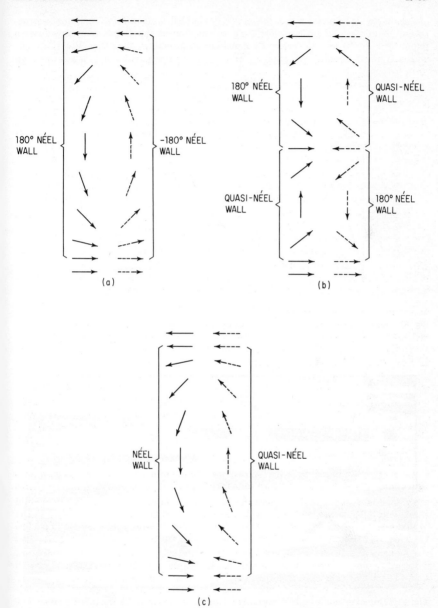

Fig. 66 Possible $\vec{M}$ directions in juxtaposed walls of a ferromagnetic couple. $\vec{M}$ in one layer is indicated by solid arrows, while $\vec{M}$ in the other layer is indicated by dotted arrows; the two patterns are thought of as being superposed by a horizontal translation. [*After E. Feldtkeller, J. Appl. Phys.*, **39**, *1181* (*1968*).] (*a*) Néel walls of opposite senses. (*b*) Walls of a Néel, quasi-Néel combination. (*c*) Néel wall in one film, quasi-Néel wall in other film.

Peculiar wall behavior can be observed in coupled films of differing domain structures. A domain which is propagating in one film of the couple may encounter a region of high H_c and subsequently continue its growth in the other film (Fig. 68). This results in a 180° wall normal to $\vec{M}$ (circle a in Fig. 68), which is a state of high

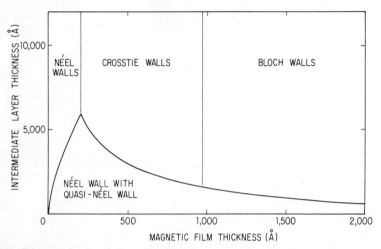

Fig. 67 Diagram showing the type of walls to be expected in a thin film couple as a function of the thickness of the ferromagnetic layers (assumed equal) and the intermediate non-magnetic layer. Here K_u was assumed to be 2×10^3 ergs cm^{-3}. [*After S. Middelhoek, J. Appl. Phys.*, **37**, *1276 (1966)*.]

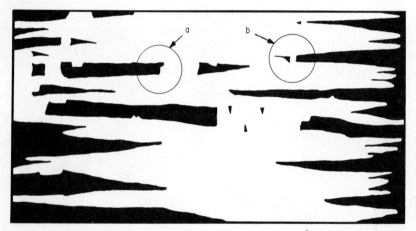

Fig. 68 Domains in one layer of a coupled film consisting of 350-Å-thick NiFe layers and a 500-Å-thick SiO intermediate film; $H_c = 0.3$ Oe and 0.55 Oe for the ferromagnetic layers. A domain propagating in one ferromagnetic film may continue its growth in the other film. [*After S. Middelhoek, J. Appl. Phys.*, **37**, *1276 (1966)*.]

energy in single but not in multilayer films. The wall may even retransfer to the original film (circle b in Fig. 68).

The wall coercive force is determined by the spatial variation of the wall energy (Sec. 2d). Thus, since walls in multilayer films have lower energies and hence lower energy variations (especially when averaged over their greater widths), it is not surprising to find that H_c in these films is lower than in corresponding single films.

Since the flux closure is more complete the thinner the intermediate layer, H_c decreases as the intermediate-layer thickness is decreased. On the other hand, the creep threshold (Sec. 9b) is raised over that for a single film of the same thickness because of the retention of the less creep-sensitive Néel wall for larger film thicknesses.

It might also be anticipated that the wall mobility would be higher in coupled than in single films [see Eq. (12)], but experimental results are contradictory on this point. The high-speed switching behavior is also controversial.

15. $\vec{M}$ OUT OF FILM PLANE

The shape anisotropy of ferromagnetic films is so high that, as pointed out in Sec. 3a for the films discussed thus far, $\vec{M}$ nearly always lies in the film plane. For certain films, however, additional perpendicular anisotropy contributions $K_\perp{}^S$ can originate in various structural mechanisms (Sec. 7b). If the easy axes associated with these anisotropies are along the film normal, reductions in the net value of $K_\perp$ can be caused which are so large that magnetization configurations are possible in which $\vec{M}$ has an appreciable component out of the plane. The nature of the resulting magnetization configurations depends on whether $|K_\perp{}^S|$ is large enough so that the net value of $K_\perp$ becomes negative, i.e., the resultant EA is normal to the film plane, or whether $K_\perp$ remains positive but is appreciably reduced from the value of $2\pi M^2$.

a. $K_\perp < 0$ (Manganese-Bismuth[43])

One of the few materials having a structural anisotropy which is high enough to overcome $2\pi M^2$ is MnBi. The crystals of MnBi have hexagonal symmetry with an EA along the c axis and an associated magnetocrystalline anisotropy of about 10^7 ergs cm^{-3}. Thus if films of MnBi are prepared with the c axes of the constituent crystallites normal to the film plane, the magnetocrystalline anisotropy field $(H_K{}^\perp)_S$ will be high enough ($\sim 3 \times 10^4$ Oe) to overcome the $4\pi M$ demagnetizing field ($\sim 7 \times 10^3$ Oe) so that $K_\perp < 0$; these MnBi films therefore have the unique property that $\vec{M}$ in the remanent state is normal to the film plane. Such MnBi films have been made by interdiffusion of Mn and Bi layers successively deposited on glass. However, the Mn-Bi chemical reaction which yields ferromagnetic MnBi is peritectic and hence slow and unreliable; the reaction seems to propagate from nucleation sites of previously reacted magnetic material. Production of single-crystal MnBi films by epitaxy on mica appears to be a more reproducible procedure.

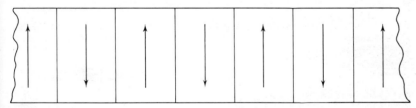

Fig. 69 Domain structure in a material with a high enough value of $K_\perp{}^S$ to create an EA normal to the film plane.

While $\vec{M}$ is everywhere normal to the MnBi film plane, the *sense* of $\vec{M}$ may vary across the film. That is, contiguous domains of oppositely directed $\vec{M}$ may be present (Fig. 69) because such a domain structure provides a state of low magnetostatic energy. Calculations based on such a model show that in the minimum-energy state the domain width depends on the film thickness. However, owing to the presence of local energy barriers, the magnetization does not actually assume such an equilibrium configuration. Nevertheless, it is possible to induce reverse domains in a uniformly magnetized film by the application of reversing fields normal to the film plane. The details of the reversal process can be followed by the Faraday effect, which shows that a reversing field causes the formation not of slab-shaped reverse domains but of a mosaic pattern of hexagonal areas of reversal. Within each small area a ring of unreversed material is retained until swept out by the applied field.

b. $K_\perp > 0$ (Stripe Domains[44])

Theory and experiment show that in some cases $\vec{M}$ can have components directed out of the film plane if $K_\perp$ is positive but is reduced to values less than $2\pi M^2$. Thus, very thick films approach the bulk state, so that the shape-anisotropy contribution decreases below the value $2\pi M^2$, and $K_S^\perp$ can predominate. The equilibrium configuration then consists of an array of domains alternately magnetized normal to the film plane as indicated in Fig. 69, often with additional small "closure" domains at the film interfaces, which serve to decrease the magnetostatic energy. However, for small film thicknesses, $\vec{M}$ will lie *in the film plane* because then the shape-anisotropy contribution will approach $2\pi M^2$ and overbalance $K_\perp^S$ so that the considerations of Sec. 3a will hold. On the other hand, a transition configuration for intermediate film thicknesses has been predicted for a thickness greater than a critical thickness d_c. In this configuration $\vec{M}$ is not alternately magnetized normal to the film plane (Fig. 69) but is *tilted* out of the film plane in alternate senses in adjacent domains (Fig. 70), with perhaps some curling of $\vec{M}$ at the film interfaces.

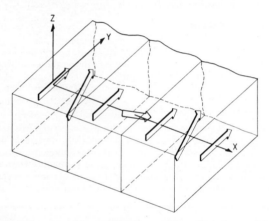

Fig. 70 Schematic diagram of magnetization configuration in a stripe-domain film. Film is in the xy plane, and the mean magnetization is in y direction. [*After H. Fujiwara, Y. Sugita, and N. Saito, Appl. Phys. Letters, 4, 199 (1964).*]

This *stripe-domain* configuration has actually been observed in vacuum-deposited films having a large value of $|K_\perp^S|$ (Fig. 71). Because of the tilting of $\vec{M}$ out of the film plane, the hysteresis loop of such films has a characteristic shape, sometimes called a "transcritical" shape (Fig. 72), indicating low remanence at zero field. Thus, for film thicknesses greater than d_c, the remanence normalized to the saturation magnetization M_r/M_s monotonically decreases with increasing d, while the saturating field H_s monotonically increases (Fig. 73). For d less than d_c, which is usually several thousand angstroms, the stripe domains and transcritical loop vanish, and the film assumes the usual behavior, with $\vec{M}$ in the film plane. Calculations of d_c, the stripe-domain width, and the shape of the transcritical loop have been made on the basis of the model of Fig. 70, and fair agreement with experiment was obtained. On the other hand, it has been reported that in the thickness range 2 to 10 μ while the transcritical loop is maintained, the magnetization configuration involves domains alternately magnetized *normal* to the film (Fig. 69), but with closure domains at the film surfaces which contribute to the remanence. Films thicker than 10 or 12 μ revert to bulk magnetization configurations without an associated transcritical loop.

Stripe-domain films can also display RIS properties as shown by hysteresigraph and torque-magnetometer investigations. The explanation offered for this phenomenon (differing from that of Sec. 11e) is that while $\vec{M}$ between the stripe walls can rotate under the application of small excitation fields, the walls themselves remain immobile until larger fields are applied; "resetting" the wall direction with a high field thus changes the initial susceptibility characteristics. A relaxation phenomenon is also observed in stripe-domain films; when a small ac field is applied at a large angle to a previously applied high ac field, the MH loop shows hysteresis which increases with time in a time interval which depends on the magnitude of the applied filed. It was demonstrated that this effect is caused by a creeplike (Sec. 9b) growth of regions of stripe domains oriented along the new field direction.

It is found that d_c and the domain width depend upon film composition and substrate temperature, because of the dependence of $K_\perp^S$ on these quantities. Originally $K_\perp^S$ and hence stripe domains were attributed to a strain-magnetostriction mechanism, i.e., either tension in negative-magnetostriction films or compression in positive-magnetostriction films. It is now realized that other mechanisms can contribute to $K_\perp^S$ and

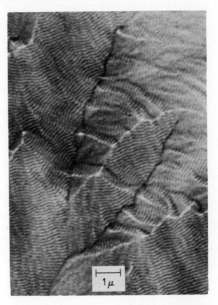

Fig. 71 Lorentz micrograph of stripe-domain film. [*Courtesy of S. Chikazumi. See T. Koikeda, K. Suzuki, and S. Chikazumi, Appl. Phys. Letters, 4, 160 (1964).*]

thus aid in the formation of stripe domains (Sec. 7b). For example, stripe domains have been observed in single-crystal platelets and thinned-down foils, for which $K_\perp^S$ is provided by magnetocrystalline anisotropy.

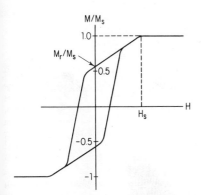

Fig. 72 Hysteresis loop of a stripe-domain film, illustrating the transcritical shape. [*After E. E. Huber, Jr., and D. O. Smith, J. Appl. Phys., 30, 267S (1959).*]

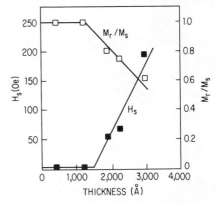

Fig. 73 Normalized remanence M_r/M_s and saturating field H_s (see Fig. 72) vs. film thickness for 86% Ni, 14% Fe stripe-domain films. Here $d_c = 1,200$ Å. [*After E. E. Huber, Jr., and D. O. Smith, J. Appl. Phys., 30, 267S (1959).*]

REFERENCES

1. Goodenough, J. B., and D. O. Smith, in "Magnetic Properties of Metals and Alloys," p. 112, American Society for Metals, Metals Park, Ohio, 1959.
2. Jaggi, R., S. Methfessel, and R. Sommerhalder, in Landolt-Börnstein, "Zahlenwerte und Funktionen," vol. 2, pt. 9, pp. 1–141, Springer-Verlag OHG, Berlin, 1962.
3. Smith, D. O., in G. T. Rado and H. Suhl (eds.), "Magnetism," vol. 3, p. 465, Academic Press Inc., New York, 1963.
4. Andrä, W., Z. Frait, V. Kamberský, Z. Málek, U. Rösler, W. Schüppel, P. Šuda, L. Valenta, and G. Volger, Phys. Stat. Solidi, 2, 99, 112, 136, 345, 941, 1227, 1241, 1417 (1962); 3, 3 (1963).
5. Pugh, E. W., in G. Haas (ed.), "Physics of Thin Films," vol. 1, p. 277, Academic Press Inc., New York, 1963.
6. Prutton, M., "Thin Ferromagnetic Films," Butterworth & Co. (Publishers), Ltd., London, 1964.
7. Feldtkeller, E., Z. Angew. Phys., 17, 121 (1964).
8. Soohoo, R. F., "Magnetic Thin Films," Harper & Row, Publishers, Incorporated, New York, 1965.
9. Cohen, M. S., in K. L. Chopra, "Thin Film Phenomena," chap. 10, McGraw-Hill Book Company, New York, 1969.
10. Chang, H., and G. C. Feth, IEEE Trans Commun. Electron., 83, 706 (1964); Chang, H., and Y. S. Lin, IEEE Trans. Magnetics, MAG-3, 653 (1967) (a bibliography).
11. For example, J. Appl. Phys., 39 (3) (1968).
12. For example, IEEE Trans. Magnetics, MAG-4 (1) (1968).
13. For example, Izv. Akad. Nauk SSSR, Ser. Fiz., 31 (3) (1967). [English translation in Bull. Acad. Sci. USSR, Phys. Ser., 31 (3) (1967).]
14. Kneller, E., "Ferromagnetismus," Springer-Verlag OHG, Berlin, 1962.
15. Rado, G. T., and H. Suhl (eds.), "Magnetism," vols. 1–4, Academic Press Inc., New York, 1963–66.
16. Chikazumi, S., "Physics of Magnetism," John Wiley & Sons, Inc., New York, 1964.
17. Mathis, D. C., "The Theory of Magnetism," Harper & Row, Publishers, Incorporated, New York, 1965.
18. Martin, D. H., "Magnetism in Solids," The M.I.T. Press, Cambridge, Mass., 1967.
19. Wolf, I., J. Appl. Phys., 33, 1152 (1962).
20. Girard, R., J. Appl. Phys., 38, 1423 (1967).
21. Mayer, W. M., IEEE Trans. Magnetics, MAG-2, 166 (1966).
22. Craik, D. J., and R. S. Tebble, "Ferromagnetism and Ferromagnetic Domains," North Holland Publishing Company, Amsterdam, 1965.
23. Humphrey, F. B., J. Appl. Phys., 38, 1520 (1967).
24. Abbel, R., in R. Niedermayer and H. Mayer (eds.), "Basic Problems in Thin Film Physics," p. 375, Vandenhoeck and Ruprecht, Goettingen, 1966.
25. Corciovei, A., IEEE Trans. Magnetics, MAG-4 (1968).
26. West, F. G., J. Appl. Phys., 35, 1827 (1964).
27. Kneer, G., and W. Zinn, in R. Niedermayer and H. Mayer (eds.), "Basic Problems in Thin Film Physics," p. 437, Vandenhoeck and Ruprecht, Goettingen, 1966.
28. Slonczewski, J. C., IEEE Trans. Magnetics, MAG-4, 15 (1968).
29. Smith, D. O., M. S. Cohen, and G. P. Weiss, J. Appl. Phys., 31, 1755 (1960).
30. Fujiwara, H., and Y. Sugita, IEEE Trans. Magnetics, MAG-4, 22 (1968).
31. Middelhoek, S., in J. C. Anderson (ed.), "The Use of Thin Films in Physical Investigations," p. 385, Academic Press Inc., New York, 1966; J. Appl. Phys., 34, 1054 (1963)
32. Feldtkeller, E., in R. Niedermayer and H. Mayer (eds.), "Basic Problems in Thin Film Physics," p. 451, Vandenhoeck and Ruprecht, Goettingen, 1966.
33. Feldtkeller, E., in "Magnetismus," p. 215, Deutscher Verlag für Grundstoffindustrie Leipzig, 1967.
34. Middelhoek, S., and D. Wild, IBM J. Res. Develop., 11, 93 (1967).
35. Stein, K. U., and E. Feldtkeller, J. Appl. Phys., 38, 4401 (1967).
36. Hoffmann, H., IEEE Trans. Magnetics, MAG-4, 32 (1968).
37. Harte, K. J., J. Appl. Phys., 39, 1503 (1968).
38. Hagedorn, F. B., IEEE Trans. Magnetics, MAG-4, 41 (1968).
39. Wolf, P., in R. Niedermayer and H. Mayer (eds.), "Basic Problems in Thin Film Physics," p. 392, Vandenhoeck and Ruprecht, Goettingen, 1966.
40. Weber, R., IEEE Trans. Magnetics, MAG-4, 28 (1968).
41. Middelhoek, S., J. Appl. Phys., 37, 1276 (1966).
42. Feldtkeller, E., J. Appl. Phys., 39, 1181 (1968).
43. Chen, D., J. Appl. Phys., 37, 1486 (1966); 38, 1309 (1967).
44. Saito, N., H. Fujiwara, and Y. Sugita, J. Phys. Soc. Japan, 19, 1116 (1964).

Part Four

Applications of Thin Films

Chapter **18**

Thin Film Resistors

LEON MAISSEL

IBM Components Division, East Fishkill, New York

1. INTRODUCTION

The use of thin films for the construction of resistors goes back at least 50 years.[1] When used for the fabrication of discrete resistors, thin films offer improved performance and reliability as compared with resistors of the composition type—and lower cost for comparable performance when compared with precision wirewound resistors. It is in the area of integrated circuitry, however, that thin film resistors have really come into their own. For resistors having minimum dimensions of 5 to 10 mils, fired glazes can compete very well with thin films; but where precision resistors are needed (with dimensions of 5 mils or less), the use of thin films becomes mandatory. The application of thin films to discrete resistors has been reviewed in a number of places.[2,3] More recent work has been done on thin film resistors in integrated-circuit applications.[4-7]

2. CHOICE OF MATERIALS

a. Film-resistor Requirements

Most film-resistor requirements can be met with films having R_s (sheet resistance) in the range 10 to 1,000 ohms/sq. Resistors below 10 ohms are rarely needed, whereas resistors with values in the megohm range can be realized through use of very long path lengths. There remains, however, a limited, but urgent, need for films with R_s greater than 1,000 ohms/sq, and much of current research on thin film resistors is devoted to finding a solution to this problem.

Besides a suitable sheet resistance, films must possess a low temperature coefficient of resistance (generally less than 100 ppm/°C). They must also be sufficiently stable so that any changes in resistance value that occur during their operating life may reliably be expected to fall below some prespecified value. Finally, the process that is used to prepare the resistors must be such that the final product can be made to meet its specifications at a reasonable cost.

b. Sources of Resistivity in Films

It can be inferred from the foregoing remarks that materials used for resistive thin films should have resistivities in the range 100 to 2,000 μohm-cm. It will be recalled, however, that metals in bulk cannot have resistivities much in excess of the *lower* limit of this range (as is summarized in Table 1). Bulk semiconductors can readily satisfy these resistivity requirements, but this is invariably at the price of a large negative temperature coefficient. Semimetals such as bismuth and antimony (and their alloys)[8] show about an order of magnitude increase in resistivity over the metals but their low melting points and relatively large temperature coefficients make these materials unattractive for resistor applications.

TABLE 1 Approximate Maximum Contribution to the Residual Resistivity by Various Types of Defect

Type of defect	Contribution, $\mu ohm\text{-}cm$
Dislocations	0.1
Vacancies	0.5
Interstitials	1
Grain boundaries	40
Impurities in equilibrium	180

Fortunately, many materials, when deposited in film form, achieve resistivities that are significantly higher than their bulk counterparts, without necessarily acquiring large temperature coefficients. Some of the ways in which this comes about include:

1. There may be a significant amount of scattering of the conduction electrons at the film surface (Fuchs-Sondheimer effect), leading to high resistivity as well as low temperature coefficient. However, because of the very small thickness normally required to produce the effect, this increased resistivity is extremely sensitive to any changes in the thickness. In addition, such films are liable to agglomerate rather easily and therefore have very limited mechanical integrity. Practical thin film resistors rarely rely directly on this phenomenon as a source of resistivity.

2. The material may contain impurities or imperfections in concentrations greatly in excess of thermodynamic equilibrium. This, too (by Mathiessen's rule), will lead to a low temperature coefficient. However, drastic departures from equilibrium are liable to lead to precipitation later (during the operating life of the component). Even if excessive defect concentrations are not present, any change in the defect concentration (for whatever reason) will be reflected as a resistance change during the life of the film. In practice, this problem is overcome either by incorporating a stabilizing heat treatment into the resistor fabrication process or by employing only very refractory materials, or both.

3. Two-phase systems (metal-insulator or cermet films). This type of system "dilutes" a conductive film by dispersing it in an insulator matrix so that the physical thickness of the film is considerably greater than its electrical thickness. The resistivity of such a film may, consequently, include a significant contribution from the surface scattering of electrons. The film itself will be much more robust than a film in which surface scattering results from a straightforward reduction in thickness. A significant problem with such films is the control of composition which, if lost, may lead to large negative temperature coefficients as well as to poor stability.

4. Low-density or porous films. These are similar to those of type 3 above, in that they have a physical thickness considerably greater than their electrical thickness. An example of such a film is low-density tantalum. One problem with this type of film is that it has a very large surface area and is therefore very susceptible to oxidation effects. If suitably protected, however, such a film can have high resistance with low temperature coefficient and adequate stability.

5. Semicontinuous films. These are films that are still in the "island" stage of growth. The spacing between islands is such that the positive temperature coefficient of the metal islands just balances the negative temperature coefficient associated with electron transfer between islands (see Fig. 18, Chap. 13). In such films, there is always a danger of agglomeration. The films are also susceptible to oxidation effects, as well as presenting a control problem during deposition. Successful resistors of this type have, however, been reported for the case of rhenium (see below).

6. Stratified films. A thin layer with a positive temperature coefficient and low resistivity may overlay a thicker layer having negative temperature coefficient and high resistivity, giving a combination that has high resistivity and low temperature coefficient. Such films are obtained as a natural result of gettering during deposition (see Chap. 13). Many chromium and Nichrome films are in this category. Their

principal problem is control since the exact amount of contaminant taken up by such films varies with the deposition conditions.

7. New crystal structures. Certain materials may assume, when in thin film form, a crystalline structure which does not exist in bulk. These structures often exhibit relatively high resistivity and low TCR, probably as a result of having a low density of conduction electrons.[9]   The best-known example is β-tantalum. The phenomena are summarized in Table 2.

TABLE 2   Mechanisms Causing Metal Films to Have Resistivities Greater than the Bulk

Description	Mechanism for resistivity increase	Effect on TCR	Example
Ultrathin..................	Fuchs-Sondheimer effect	→ 0	
Trapped gas...............	Impurity scattering	→ 0	Ta nitride
Insulating phase...........	Intergrain barriers	→ − ∞	Cermets
Netlike...................	Construction resistance	→ 0	Low-density Ta
Discontinuous.............	Particle separation	→ − ∞	Rhenium
Double layer..............	TCRs cancel	→ 0	Cr
New structure.............	Fewer carriers	→ 0	β Ta

c. Deposition Methods

Usually, the choice of deposition method is made after the material has been selected. In a limited number of cases, however, a particular deposition technique may be preferred if it fits more easily into a larger process. At any rate, before the final choice can be made, three questions must be asked: Will the method work for this type of material? What degree of control will it allow? How much will it cost?

(1) Vacuum Evaporation   The most widely used method for resistor film deposition is vacuum evaporation, since most materials lend themselves to deposition through this technique. The commonest exceptions are the refractory metals and materials such as tin oxide which may decompose on evaporation. Major problems associated with vacuum evaporation are the great sensitivity of the amount of contamination to the deposition conditions, and the difficulty of obtaining uniform film thickness over large areas. This, in turn, is intimately related to cost, since if large numbers can be processed at one time, the process will obviously be cheaper. The problems of uniformity of the deposit in vacuum-evaporation systems are discussed in Chap. 1.

Resistance monitoring during vacuum evaporation is straightforward and easily implemented, provided the rate of deposition is not too great. Considerable engineering work has already been done in industry on large vacuum-evaporation systems so that much tooling and fixturing are already commercially available. To date, most evaporation systems have been of the batch rather than of the continuous-feed type because it is difficult to replenish the evaporant source constantly without breaking the vacuum. In cases where the tolerances involved allow the use of masks to define the resistor patterns, evaporation is the preferred method, since manipulation of mask changers in vacuum does not present any serious problem.

(2) Cathode Sputtering   This is a preferred method for very refractory metals (such as tantalum) and for alloy systems (such as Nichrome) when a very high degree of control is required. During conventional sputtering, there is a greater likelihood of contamination than during evaporation. However, the introduction of techniques such a s bias sputtering and getter sputtering (see Chap. 4) has considerably improved this situation. Resistance monitoring during sputtering is difficult because of interference from the discharge plasma. However, control of thickness through deposition time alone is usually easier during sputtering than during evaporation.[10]

One of the major limitations to sputtering is that the material to be deposited may not always be available as a sheet large enough to form a cathode. Relatively little work has been done on the use of very large cathodes in batch systems. However, sputtering is well suited for use in a continuous-feed system, since there is no source-replenishment problem.[11,12] Masking, however, is difficult during sputtering, unless in-contact masks are used. Substrate temperatures are comparable with those required for evaporated films, but their control is much more difficult during sputtering (see Chap. 2).

(3) Pyrolytic Decomposition This method has not been widely used for resistors, except for carbon films. One of the principal limitations to the method is the relatively high substrate temperature needed. In addition, control of film thickness is a problem, partly because of the problems of in situ monitoring, and particularly because it is difficult to obtain good uniformity over large areas as a result of the great sensitivity of deposition rate to substrate temperatures. A related problem is the lack of uniformity resulting from variations in the composition of the gases entering and leaving the reaction chamber. The method is, however, readily adaptable to mass production and is widely used in industry for the epitaxial deposition of semiconductor films. The films tend to be relatively pure compared with most evaporated or sputtered films. In situ masking is not feasible, because of the high substrate temperature as well as the nonshadowing nature of the deposit.

(4) Hydrolysis The use of this method for resistor films is limited to tin oxide films and requires substrate temperatures of 500°C or more. It tends to produce films which have high surface roughness, and control to precise value is difficult. In situ masking is not possible in this case. The method lends itself well to mass production, and the films are noted for their unusually good adhesion to the substrate.

(5) Electroless Deposition Thin film resistors using electrolessly deposited films are still in the exploratory stage. This method may be amenable to mass production but is limited to a relatively small group of metals. Control presumably would be by time alone, and it seems likely that uniformity problems comparable with those seen in the pyrolytic technique could arise.

One aspect of resistor fabrication common to all systems is the need for a suitable conductor metallurgy. For many resistor materials it is important that the conductive layer be deposited in the same system as the resistor film. This is particularly true when material having low sheet resistance is involved. In general, the deposition of films of two different materials in rapid succession presents no problems for evaporation or pyrolytic decomposition. For sputtering, however, special tooling is needed, and several multicathode systems which allow rapid subsequent deposition of the second metal have been described in the literature.[13]

The various deposition methods and their pros and cons are summarized in Table 3.

TABLE 3 Methods for the Deposition of Resistive Films

Method	Advantages	Disadvantages
Evaporation	In situ masking; easy monitoring; almost universal	Refractories a problem; gas contamination
Sputtering	Refractories easy; long source life; low-density forms	Cathode required; monitor problem; gas contamination
Pyrolysis	High deposition rate; high purity; well annealed	High substrate temperature; thickness variations
Hydrolysis	Good film-to-substrate bond	High substrate temperature, thickness variations, and film roughness
Electroless plating	Cheap? Flexible substrate	Control problem

3. AVAILABLE MATERIAL SYSTEMS

a. Metal-alloy Films

(1) Resistivity of Alloys Even when they are alloyed with one another, metals in bulk rarely achieve resistivities in excess of 20 to 30 μohm-cm (see Chap. 13). Some

notable exceptions among the elements are hafnium (30.6 μohm-cm), zirconium (42.4 μohm-cm), titanium (43.1 μohm-cm), and manganese (139 μohm-cm), the values quoted being at 22°C. A limited number of alloys have resistivities as high as 160 μohm-cm, and these are used for the manufacture of discrete resistors. One of their characteristics is a very low TCR over a limited, but useful, temperature range. The state of the art has been summarized by Jackson et al.,[14] and in Table 4 we have

TABLE 4   Electrical Properties of Some Metal Alloys

	Resistivity, μohm-cm*	TCR, ppm/°C	TCR range, °C
Pd-Ag...................	38	±50	0 to +100
Cu(83) Mn(13) Ni(4).......	48	±10	+15 to +35
Ni(80) Cr(20)..............	110	+85	−55 to +100
Ni(75) Cr(20) Al(3) Cu(2)..	130	±20	−55 to +100
Ni(76) Cr(20) Al(2) Fe(2)...	133	±5	−65 to +250
Pd-Au-Fe................	158	<10	0 to +100

* 20°C.

summarized the properties of some of the alloys which they discuss. It can be seen that every composition includes at least one transition metal—suggesting that it is the presence of the latter that is responsible for the special properties of these alloys. This is in fact so because in these metals a nearly filled d band overlaps the s band. At the Fermi level, the d band has a large density of states relative to the s band, and there is a high probability that conduction electrons will be scattered from the s band into the d band, where they no longer contribute significantly to conduction. As a result, in alloys of this type the number of free carriers is less than is usual for a metal.

The low TCR arises because, as the temperature increases, the density of states in the d band (at the Fermi level) decreases (fewer electrons are scattered out the s band). Accordingly, the number of conduction electrons increases with increasing temperature. Over a narrow range of temperature this increase in the number of free carriers just compensates for the normal increase in phonon scattering and a very low TCR results. In practice, discrete resistors made of alloys of this type require very carefully controlled heat treatments if reproducible results are to be obtained.

(2) Nichrome Films   Once the advantages of metal films as resistor elements were recognized, most early workers thought that it would be necessary to use the same materials that had been proved to be optimum in bulk. This was so because the additional sources of resistivity available to materials in thin film form (as outlined in Table 2) were not fully appreciated. One of the most successful of the resistor alloys had been Nichrome, either with composition in the vicinity of 80% nickel and 20% chromium or modified as indicated in Table 4. Consequently, most early work in the thin film resistor field was done with deposited Nichrome films.[15,16] It was soon found that most Nichrome films had higher resistivities than the bulk alloy, so that interest in their application was sustained and Nichrome films continue to be widely used in industry even today.

The most popular method for the deposition of Nichrome films has been vacuum evaporation. The major problems associated with this (other than contamination by background gases) are the very considerable difference in vapor pressure between the nickel and the chromium and the high reactivity of Nichrome with many crucible materials. In attempting to overcome the latter problem, some workers have sublimed the alloy from a wire source,[17] rather than evaporating it from the melt. Unfortunately, the lower the temperature, the larger the difference in vapor pressure between the two constituents. For example, at 1000°C, chromium evaporates 300 times faster than nickel, whereas at 1300°C, the difference is only a factor of 8.

Good control through sublimation has, however, been reported by some workers. For example, Monnier[18] used a broad strip as a Nichrome source running it at a temperature of 1170°C. The deposition rates were as low as 13 Å min⁻¹ in.⁻² of source area, but the level of uniformity was such that resistance monitoring was considered unnecessary, and sheet resistance could be controlled through evaporation time alone. As expected, the chromium content of the film was about 40% higher than that of the source.

Because of the higher vapor pressure of the chromium component, the composition (and hence the resistivity) of films evaporated from a molten source of finite mass will vary as a function of time. For example, when Degenhart and Pratt[19] evaporated about 12% of a 1.2-g charge by heating it at about 1450°C, they found that the composition of the films varied systematically with sheet resistance (illustrated in Fig. 1).

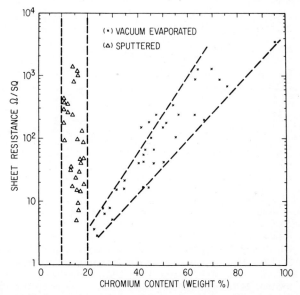

Fig. 1 Influence of chromium content of Nichrome films on sheet resistance, for both evaporated and sputtered films.

An alternative approach to the problem is to accept the fact that the film composition will be different from that of the starting material but, by using a source of sufficient mass, cause the variations in film composition during a given run to be negligible. This approach has been taken by, for example, Wiedd and Tierman,[20] who found that a melt containing 14% chromium had to be used to obtain a film having the approximate composition 50% chromium 50% nickel. The source itself consisted of a 200-g conical charge supported inside a 3- by 2½-in. thin-wall conical ceramic crucible. It was brought to operating temperature by rf induction heating and controlled through a thermocouple.

Another problem in the deposition of Nichrome films is the partial oxidation of the chromium constituent during deposition (the exact extent depending, obviously, on deposition rate, background gas concentration, and substrate temperature). In addition, since films normally receive a subsequent stabilization treatment, changes in resistance due to oxidation will be a function of the amount of chromium at the surface of the film. It is, thus, particularly important to have a reproducible composition at the surface.

Wiedd and Berner[21] found a critical film thickness above which the film characteristics were predictable, but below which the films were erratic in their electrical behavior. This critical thickness was in the order of 100 Å and was independent of substrate temperature, substrate material, and deposition rate. It varied only slightly with film composition. Since part of their experimental procedure included a post-deposition heat treatment in air, it was concluded that the critical depth represented a diffusion depth for oxygen during the heat treatment. Related phenomena have been seen in tantalum films (see next section), and are associated with grain-boundary oxidation.

An interesting relation between TCR and composition for Nichrome films has been reported by Campbell and Hendry.[22] The TCR was found to be more negative for films with high chromium content, and for one particular composition a condition was

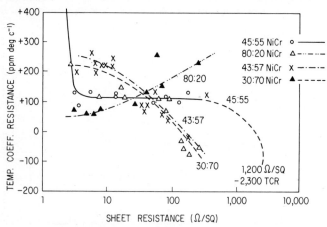

Fig. 2 TCR of Nichrome films as a function of sheet resistance for several different compositions.

found where the TCR was independent of sheet resistance (Fig. 2). The increasingly more negative TCR at higher ohms per square for compositions other than 80-20 is the normal behavior seen in all metal films. The explanation suggested for the increasingly positive TCR of the 80-20 composition at high ohms per square is that the films have an islandlike structure of nickel grains in a matrix of chromium oxide. Usually, such a structure would have a negative temperature coefficient of resistance. However, near a composition of 80-20, the difference in coefficient of expansion between the film and the glass substrate is believed to lead to an increase in the gap between grains—so that a positive TCR results.

The problem of control of metal composition in Nichrome films can be solved by the flash-evaporation technique (see Chap. 1). By using powders of the desired composition and dropping these onto a hot filament, Campbell and Hendry[22] showed that the composition of the films was the same as that of the starting material (to within 1%). A related method that closely resembles flash evaporation and lends itself well to production techniques has been used by Siddall and Probyn.[23] A Nichrome wire is fed down a guide tube onto a stage where the tip of the wire is evaporated through electron-bombardment heating. Since the entire tip evaporates before the wire moves on, the composition of the wire is transferred to the film.

The problem of composition control of Nichrome films can also be solved by using sputtering, rather than evaporation, as the deposition method. In a comparative study of evaporated and sputtered Nichrome films, Pratt[24] found that, for the sputtered films, the composition remained very nearly constant over an extended range of sheet resistance. The TCR of the sputtered films was also found to be much less variable than for the evaporated ones. Thus, the TCR of evaporated films varied from +350 ppm/°C for films around 3 ohms/sq to −300 ppm/°C for films around

3,000 ohms/sq, whereas sputtered films had a TCR of $+150$ ppm/°C in the range 5 to 1,200 ohms/sq. The percent resistance change due to heat treatment as a function of sheet resistance for both vacuum-evaporated and sputtered Nichrome is shown in Fig. 3. A slight decrease in resistance is seen for the vacuum-evaporated films below 10 ohms/sq. Above this, there is a slight increase, with a more rapid rise as the sheet resistance increases. The sputtered films show a similar pattern, but the increase in resistance is larger and there is considerable scatter for the higher sheet resistances. This suggests that the sputtered films were either more susceptible to oxidation or that decreases due to annealing effects are less in sputtered films. Both explanations seem possible.

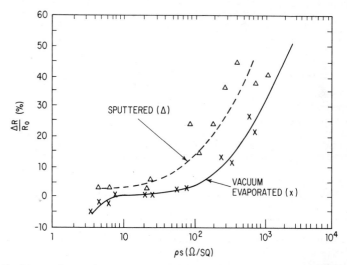

Fig. 3 Resistance change due to heat treatment of sputtered and evaporated Nichrome films.

Stern[25] has applied the technique of bias sputtering to the deposition of Nichrome films. He has shown that it is possible in this way to produce films whose properties resemble those of the bulk alloy very closely, and thus have a maximum useful sheet resistance of about 40 ohms/sq. The films were very stable and, when heat-treated in air without protection, underwent little or no change. Because the method is extremely reproducible, sheet-resistance control to $\pm 2\%$ was possible. The primary problem that had to be solved in order to achieve this level of control was the variation in deposition rate associated with varying degrees of hydrogen contamination (see Chap. 4).

Bicknell et al.[26] have investigated the structure of Nichrome films as deposited through evaporation at 10^{-4} Torr. The film composition was 40 Ni 60 Cr, and it was observed that only about one-third of the chromium was alloyed with the nickel, the remaining chromium being in a poorly crystallized chromium-oxygen "skin" phase. After annealing at temperatures in excess of 500°C, the film was seen to consist of monocrystalline islands, up to 100 Å in size, of Ni Cr in a matrix of αCr_2O_3.

Siddall and Probyn[23] have summarized a number of the practical requirements associated with the production of Nichrome films through evaporation: (1) substrates should be kept at 2 to 300°C during deposition in order to relieve internal stresses: (2) the oxidation of the film during deposition must be controlled by adjustment of residual gas pressure and rate of deposition; and (3) the deposited film must be annealed. This can be done in air at 250 to 350°C, but the final film must be protected to improve its stability under widely varying atmospheric conditions.

(3) Other Alloy Films Attempts to obtain alloy films of higher resistivity than Nichrome have been made by replacing the nickel in the Nichrome with silicon. Studies of such films have been reported by Layer et al.[27] Relatively stable films appear to have been obtained with sheet resistances ranging up to several thousand ohms per square. However, the temperature coefficients were in the order of ±500 ppm/°C and, for the highest-resistivity films, as low as −2,000 ppm/°C. Insufficient work has been done with this system to determine what the ultimate degree of control might be, but reports published to date have not been encouraging.[28] The material has, however, been found to be unusually rugged and could be operated in air for several hours at temperatures as high as 750°C. Exposure to air at 250°C for several thousand hours produced changes of only 1 to 2%.

Thin film resistors in which nickel has been alloyed with phosphorus have been reported by Foley.[29] These nickel phosphide resistors were deposited by electroless deposition. A wide range of sheet resistance, up to 2,000 ohms/sq, was reported, with TCRs less than 100 ppm/°C. The high resistance characteristics of the films were due to the phosphorus, whose percentage varied between 0.5 and 13%. Control of sheet resistance was, however, rather difficult, and to obtain resistors having useful tolerances, adjustment of individual resistors by an abrasion method was necessary.

b. Single-metal Systems

With the growing realization that "pure" metal films could have resistivities substantially higher than the bulk, the attractiveness of alloys for the achievement of high resistances steadily diminished. On the other hand, the attractiveness of single-component systems was obvious since composition control, precipitation, etc., could no longer be a problem. A number of such single-metal systems have been investigated.

(1) Tantalum Interest in tantalum came about initially as a by-product of work on thin film capacitors, for which tantalum provides many advantages,[30] and this interest has increased steadily since 1959.[31] Although the initial attractiveness of a unified system consisting entirely of one metal has largely disappeared, tantalum still remains an attractive candidate as a thin film resistor. In addition to its refractory nature (which implies that any imperfections frozen in during deposition will not anneal out on life) tantalum belongs to a class of metals known as valve metals which form tough, self-protective oxides, either through heat treatment in oxygen or through anodic oxidation. The latter process allows for precise control of thickness and may be used as a method of trimming (see below).

Because of the refractory nature of tantalum, the preferred method for its deposition has been sputtering, rather than evaporation—although the latter method is quite feasible.[32] Since tantalum is such a reactive material, the sputtered films have a tendency to be contaminated, unless special precautions are taken. However, as was discussed above, some degree of contamination is desirable in order to achieve useful properties. Attempts to improve the uniformity and to control the purity of tantalum films have led to several advances in the state of the art of sputtering in general.

The technology of tantalum resistors is complicated by the fact that tantalum films can exist in at least three forms whose properties are summarized[33] in Table 5. The α structure is the conventional bcc tantalum structure corresponding to the bulk. The β structure was first reported by Read and Altman.[34] The exact conditions which determine whether α- or β-tantalum are formed have not been fully defined as yet.[33] However, it is established that the β form will not appear in systems in which there is an appreciable degree of gaseous contamination or if the substrate temperature exceeds about 600°C.[35] Frequently, films which contain a mixture of both the α and the β structures are obtained. Problems of this type make precisely reproducible films difficult to achieve, and sheet resistances are best controlled through resistance monitoring rather than by time alone.[36] The low-density form is entirely different in origin, and will be discussed presently.

Since some form of contamination is necessary if usable resistivities are to be obtained, much effort has been devoted to investigations in which the impurities were added in controlled fashion, as opposed to relying on accidental contamination from

TABLE 5 Comparison of Forms of Tantalum Films

Property	Bulk material	Thin film I, α structure*	Thin film II, β structure*	Thin film III, low density†
Crystal structure............	bcc	bcc	Tetragonal	Tetragonal or bcc
Lattice constant.............	$a_0 = 3.303$	$a_0 = 3.31\text{–}3.33$	$a = 5.34, c = 9.94$	
Density, g/cm⁻³.............	16.6	15.6	15.9	12.1
Resistivity, μohm-cm........	13	25–50	180–220	5,000
Temp coeff, ppm/°C........	+3,800	+500 to +1,800	−100 to +100	−100 to +100
Superconductive transition, °K.....................	4.4	3.3	0.5	

* Sputtered at 4 to 5 kV.
† Sputtered at 1.5 kV. These data obtained with low density β-tantalum, but similar properties are found for low-density bcc tantalum.

the background gas. While oxygen has a profound effect on both the resistivity and the TCR of tantalum films, the change in resistivity with percent dissolved oxygen is extremely rapid, and control is difficult. After suitable heat treatment, oxygen-doped tantalum films have been found to be extremely stable.[37,38] However, even small amounts of oxygen when present in the films cause the TCR to increase very rapidly to a negative value of the order of −400 ppm/°C (or more).

A more attractive contaminant for deliberate introduction during the sputtering of tantalum is nitrogen. The resistivity vs. %N₂ curve,[39] as shown in Fig. 4, includes a plateau, and levels out at a resistivity of about 250 μohm-cm and at a TCR in the order

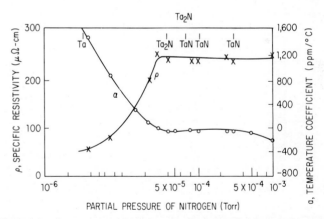

Fig. 4 Effect of varying amounts of nitrogen in the sputtering gas on the resistivity and TCR of Ta films.

of −75 ppm/°C. An important feature of these nitrogen-doped films is that they can be anodized just as well as pure tantalum.[33] In practice, the film composition is chosen so as to correspond as closely as possible to that of Ta₂N, since it has been found that resistors having this composition display the greatest stability during load-life tests.

Even though tantalum films do not readily anneal, their susceptibility to oxidation makes it necessary to include a stabilizing treatment as part of the fabrication procedure. A common practice is first to lightly anodize the film (25 to 75 V), then heat-treat it for about 5 h at 250 to 400°C, completing the procedure with a second anodization at higher voltage (generally combined with trimming the resistor to value). In cases where anodization is not desirable or not practical, it is possible to monitor the heat treatment of groups of resistors, stopping the heat treatment when the correct sheet resistance has been reached.[40] Extremely stable tantalum films, which undergo only very small changes when heat-treated in air, can be obtained by doping with gold.[41] The gold is applied as a separate layer before or after tantalum deposition and is subsequently diffused into the film, where it is believed to settle at the grain boundaries.

The existence of low-density tantalum films was first observed by Schuetze et al.[42] In the course of studying the deposition of tantalum, they observed that the resistivity of their films depended on only one parameter—the sputtering voltage. Their results are shown in Fig. 5. At a given voltage, deposition was performed at several

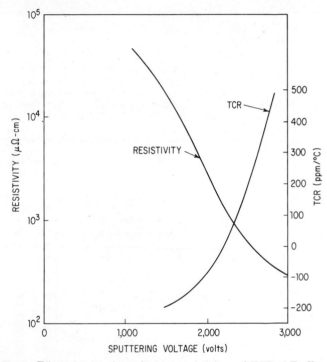

Fig. 5 Effect of sputtering voltage on resistivity and TCR of Ta films.

different currents which were obtained by varying the pressure. The only parameter that affected resistivity significantly was found to be the sputtering voltage. Films sputtered at about 2,000 V exhibited temperature coefficients very close to zero, even though they had resistivities in the order of several thousand microhm-centimeters. Electron micrographs showed that the high-resistivity films were very porous and had the appearance of an open network. This is illustrated in Fig. 6, in which the upper picture shows an electron micrograph of a "normal" tantalum film surface, while a low-density tantalum film is seen in the lower picture. These films thus have an electrical thickness which is less than their physical thickness. In addition, the porous structure makes for many narrow regions at which current is constricted. Such regions have higher-than-normal resistivity because of surface scattering of the conduction electrons.

Unfortunately, in the as-deposited condition low-density tantalum films are unstable. For example, when a stabilizing heat treatment of 1 to 2 h at 200°C is applied, relatively little change in sheet resistance occurs—but the TCR drops very

Fig. 6 Electron micrographs comparing "regular" with low-density Ta films. The low-density films are in the lower half of the picture.

rapidly to values in the order of -300 ppm/°C, as illustrated in Fig. 7.[40] Presumably, such films are particularly sensitive to oxidation because of the very small local thickness at the constrictions in the network structure. Work is in progress at various locations to develop methods for protecting these films against the effect of oxidation— for example, by a suitable combination of anodization and thermal oxidation. However, control of sheet resistance is still difficult at present, and films that have not received adequate heat treatment cannot be considered to be stable.

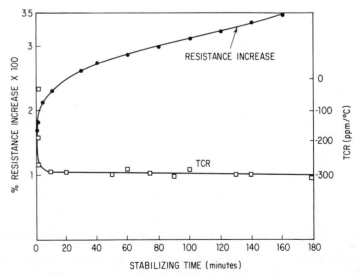

Fig. 7 Change in resistance and TCR of low-density Ta films when heat-treated.

Recent reviews of tantalum technology include those of Schwartz and Berry,[4] McClean,[33] McClean et al.,[43] and Berry et al.[44]

(2) Other Refractory Metals   Several other valve metals have also been investigated for possible application as thin film resistors.   Amongst these are titanium,[45,46] hafnium,[47] molybdenum,[48] and tungsten.[49]   It seems likely that any of these materials could be engineered into a reasonably good resistor system having many of the characteristics of tantalum.   None, however, has shown any great advantage over tantalum, and this, coupled with the fact that these materials (with the possible exception of hafnium) produce capacitors inferior to tantalum, has considerably reduced the incentive for undertaking the kind of extensive development that the tantalum system has received.

A refractory metal that has been applied as a thin film resistor in a totally different manner from tantalum is rhenium.[50]   The principle employed here is to use extremely thin films that are just on the edge of agglomeration, so that additional resistivity comes about because of the finite spacing between film particles.   Use of such films for resistors has generally not been attempted because of their highly unstable nature; but the fact that rhenium is so refractory does allow it to be used, even though films having a sheet resistance of 1,000 ohm/sq are only about 40 Å thick.   Good stability is claimed to temperatures up to 500°C (in vacuum).   Suitable protective overcoats are essential for stability in air, since very small changes in the size of the film particles can cause large variations in film properties.

Zimmerman[51] has investigated a number of single-metal systems, including aluminum, chromium, tungsten, and rhenium.   His results suggested that rhenium was, by a considerable margin, the best choice for a thin film resistor.   The survey includes data on the as-deposited sheet-resistance tolerances he was able to achieve for various materials as a function of their sheet resistance.   These are shown in Fig. 8.

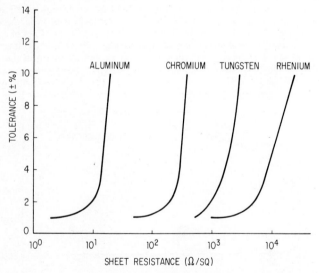

Fig. 8   Sheet-resistance control achievable with various single-metal systems as a function of sheet resistance.

To have a usable TCR, the rhenium had to be deposited at the correct substrate temperature and then given a 500°C annealing treatment.   For films deposited at 450°C to 5,000 ohms/sq, a TCR in the order of −200 ppm/°C could be achieved. Films deposited at other temperatures had more negative TCRs and showed a resistance decrease of 15 to 20% during the heat treatment.

(3) Chromium As was previously discussed, the percentage of chromium in Nichrome films is often much higher than the 20% present in the starting material. Because of the limited mutual solid solubility of nickel and chromium, evaporated Nichrome films frequently contain more chromium in solution than is thermodynamically permissible. This introduces a source of instability since the excess chromium will eventually precipitate out. Furthermore, the gas gettered by the chromium during its deposition has a more pronounced effect on the resistivity of chromium films than does the addition of nickel. As a result, "pure" chromium films have considerably higher resistivities than Nichrome films of optimum composition. Because of these factors, as well as the greater simplicity of a single-component system, considerable interest has been generated in chromium as a thin film resistor material.[52,53]

Although the use of pure chromium eliminates the problem of metal compositional control as well as the danger of precipitation, the sensitivity of film properties to deposition conditions is greater for chromium than for Nichrome because of the absence of any "built-in" contamination. On the other hand, the well-known adhesive strength of chromium to glass substrates is attractive for a resistor component since it is associated with a low tendency to agglomerate. Also, chromium is a good "glue" for any conductive metallurgy associated with it. An additional attraction possessed by chromium for film deposition is the ease with which it can be sublimed. This is commonly done from a tungsten filament coated with chromium by electroplating. Heat treatment of the coating in hydrogen prior to use is advisable since the electroplated layer generally contains large quantities of oxide.

Since chromium is not a particularly refractory metal, there is a limit to the temperatures at which chromium films can be continuously operated. Heat treatment of chromium films in vacuum causes decreases in resistance because of annealing effects which are absent in more refractory films such as tantalum.[54,55] At present, chromium appears to be most widely used in discrete resistors where units can be trimmed to value by spiraling or by a related abrasion technique. These are not suited to microelectronic applications.[56]

Studies of the structure of chromium films[57] confirm that (like most resistive films) they consist of relatively pure islands of metal in a matrix of insulating chromium oxide. The influence of deposition conditions on the resistivity of chromium films has been studied in some detail by Scow and Thun.[58] One interesting finding was that

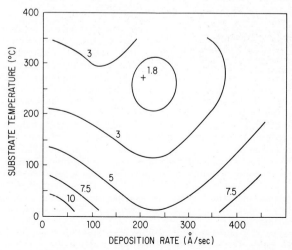

Fig. 9 Influence of substrate temperature and deposition rate on the resistivity ratio of films to bulk for chromium.

films of minimum resistivity could be obtained for only one combination of substrate temperature and deposition rate (Fig. 9).

c. Cermets

Once it became apparent that most thin film resistors owe their desirable electronic properties to the incorporation of impurities, it was a logical step to include the impurities deliberately instead of accidentally. In addition, it was realized that there is no need to be limited to gaseous contaminants. Many solid impurities are available, whereas contaminants of gaseous origin are restricted, in effect, to nitrogen, oxygen, or carbon. Finally, a sticking coefficient very close to unity could reasonably be expected for most solid contaminants so that a greater degree of compositional control, as compared with gaseous contaminants, would seem likely.

(1) Cr-SiO Although a number of metal-insulator combinations have been studied in film form, by far the most successful to date has been the chromium–silicon monoxide system.[59] An important feature of such films, and one of the original motivations for their development, is their high resistivity and stability without large negative temperature coefficients.

There is some confusion in the literature as to the resistivity of Cr-SiO films as a function of their composition—primarily because of uncertainties in measuring the composition of the actual films but also because the resistivity is strongly dependent on the thermal history of the film. Through careful use of x-ray fluorescence to provide compositional data to an accuracy of $\pm 1\%$, Glang et al.[60] obtained the curves shown in Fig. 10. Resistivity vs. composition is given for films as deposited at 200°C as well as after annealing treatments at 400, 500, and 600°C (in argon for 1 h each).

Figure 11 shows TCR data for Cr-SiO films, both as deposited at 200°C and after annealing for 1 h at 400°C. In practice, 400°C for 1 h has been found to be an adequate stabilizing treatment for most applications. It is interesting to note that after their stabilization treatment, films containing as much as 50 atomic % SiO have temperature coefficients which are not only close to zero but are actually

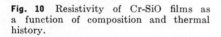

Fig. 10 Resistivity of Cr-SiO films as a function of composition and thermal history.

slightly positive. It was shown by Glang et al.[60] that the films, as deposited, were amorphous but that with heat treatment, crystalline phases (including Cr_3Si) appeared.

On the basis of Hall-effect measurements, Lood[61] has suggested that for films containing up to 10% SiO, the role of the latter is the creation of impurity levels in the band structure of the chromium. Pure Cr films had a positive Hall coefficient, but with the addition of SiO, the coefficient became more negative, passing through zero at 5% SiO and reaching a minimum at 10% SiO. As a result of x-ray diffraction observations on films containing about 25% SiO, Scott[62] has suggested that chromium is uniformly dispersed in the SiO for the as-deposited films but that on annealing, small (~ 20 Å) particles of Cr appear and form short chains along which increased conduction can take place.

The appearance of a positive TCR in films with higher SiO content, despite their high resistivity, suggests that the deposited film consists of chromium grains (containing some dissolved silicon) dispersed in a matrix of silicon monoxide. The physical separation of the particles leads to a high resistivity as well as to a large negative TCR, since electrons must be thermally activated in order to cross the gaps between particles. During annealing, some of the SiO disproportionates, giving free silicon which reacts at the surface of each grain to form a skin of Cr_3Si. The oxide between grains is "squeezed out," and the grains now touch one another so that insulation resistance is replaced by constriction resistance.

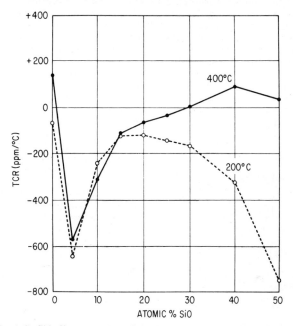

Fig. 11 TCR of Cr-SiO films as a function of composition, both as deposited at 200°C and after heat treatment at 400°C.

When heated in air, unprotected Cr-SiO films are subject to oxidation and undergo an increase in resistance whose magnitude depends on film composition as well as on film thickness. Although excellent stability can be obtained in unprotected cermet films if they are adequately stabilized, some degree of sheet-resistance control is lost, and most workers prefer to include a protective overcoat along with the films. For example, Himes, Stout, and Thun[63] used silicon monoxide, Wagner and Ourlau[64] used a silicone and epoxy layer, and Glang et al.[60] used a film of rf sputtered SiO_2. The latter is particularly suitable for integrated-circuit applications, being readily etched and able to withstand high processing temperatures. This is important, since Cr-SiO films are finding increasing application in integrated circuits because of their high resistivity and stability characteristics.[65]

Some of the early work on the deposition of Cr-SiO films was performed using two sources,[59,66] the prime objective being to be able to prepare films with a wide range of composition. For routine work, most investigators have used flash evaporation.[67] This method allows the coating of relatively large numbers of substrates with reasonable compositional uniformity. In most designs, a premixed powder of chromium and silicon monoxide particles is fed down a chute onto a hot filament. As pointed out by Wagner and Mertz,[68] however, the vapor stream from the filament is liable to preferentially deflect SiO particles out of its way so that the resulting films are low in SiO content, often in a nonreproducible fashion.

The problem of composition control was further analyzed by Glang et al.[69] and effectively solved by substituting small, presintered pellets of the desired composition for powder. Each pellet fractionates and is evaporated to completion; however, several pellets in various stages of evaporation are always present on the filament so that intimate mixing of the components occurs and the composition of the film is the same as that of the pellets. By using this technique, run-to-run control of the sheet resistance of Cr-SiO films (including stabilizing treatments) was found to be in the order of $\pm 2\%$ for compositions having about 20% SiO. For films having other compositions, the control gradually deteriorated, particularly with increasing concentration of SiO.

When protected against oxidation, Cr-SiO films exhibit excellent thermal stability and do not drift in value unless they are raised to a temperature equal to or greater than the maximum temperature to which they had previously been exposed (Fig. 12). Note that the temperature of anneal plays a significantly more important

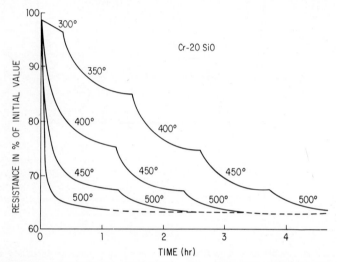

Fig. 12   Resistance of Cr-SiO films (20 % SiO) as a function of time, when heated at several different temperatures.

role than the time of anneal. The maximum temperature to which Cr-SiO films can be subjected for any extended period of time appears to be about 500°C. Above this, the sheet-resistance changes become less predictable, although the films retain their integrity to temperatures at least as high as 900°C. It has been suggested that this loss of control is associated with the instability above 500°C of the compound Cr_3Si.[68]

Films with a chemical composition similar to Cr-SiO (but presumably with a different structure) have been obtained by flash evaporation of mixtures of Si and Cr_2O_3.[70] The films obtained had resistivities in the order of 10^4 μohm-cm with a TCR close to zero and appeared to be usable to temperatures of at least 500°C. It was shown that the important phase formed in these films is $CrSi_2$.

(2) Other Cermets   As part of their original investigation of the Cr-SiO system, Beckerman and Thun[59] examined the systems $Cr-MgF_2$ and Au-SiO. However, none of these systems had either the resistivity range or the stability of the Cr-SiO system.

Farrell and Lane[71] have conducted a search for films having sheet resistances considerably larger than those obtainable with Cr-SiO. They used a two-source sputtering system in which the substrates were placed on a rotating holder beneath two cathodes from which the components were simultaneously sputtered.

One cathode was a nonreactive metal such as platinum or gold, whereas the other cathode was a reactive metal such as tungsten or tantalum. By including some oxygen in the sputtering gas, a mixture of the noble metal and the reactive metal oxide was produced.

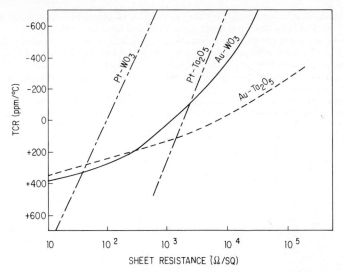

Fig. 13 TCR vs. sheet resistance for several types of cermet film.

The combinations Pt-WO$_3$, Pt-Ta$_2$O$_5$, Au-WO$_3$, and Au-Ta$_2$O$_5$ were investigated. Promising results were obtained with the gold-tantalum oxide system, as illustrated in Fig. 13. Although the gold-tantalum oxide films increased in resistance upon heating, after suitable time periods (typically 200 h at 100°C) the resistance value leveled out and appeared to have stabilized. However, considerable control was lost. It was speculated that some of this may have been due to problems with the contacts. In view of its apparent promise, this system will doubtless be further investigated.

Equipment somewhat similar to that of Farrell and Lane was used by Miller and Shirn[72] for investigating the Au-SiO$_2$ system, except that the SiO$_2$ component was produced through direct rf sputtering rather than through reactive sputtering. The films were found to have resistivities about one order of magnitude larger than evaporated Au-SiO films of analogous composition. These authors have also investigated the Nichrome-SiO$_2$ system.[73] Their data on resistivity vs. TCR for both systems are shown in Fig. 14. The compositions (by weight) at zero TCR were approximately 90% gold and 70% NiCr.

Wilson and Terry[74] have studied films consisting of mixtures of Cr$_3$Si, TaSi$_2$, and Al$_2$O$_3$. The composition range investigated covered mixtures containing equal amounts of Cr$_3$Si and TaSi$_2$, with increasing additions of Al$_2$O$_3$. The results are shown in Fig. 15. Baking in air at 450 to 490°C was used as a means for adjusting the sheet

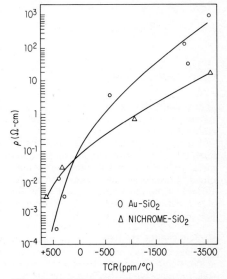

Fig. 14 Resistivity vs. TCR for the cermet systems Au-SiO$_2$ and NiCr-SiO$_2$.

resistance so that a precision of about $\pm 10\%$ was obtained. This also ensured stability, so that drift during storage tests was "small."

An investigation of several metal-insulator combinations has been made by Riddle.[75] The choice of metal-insulator combinations studied was dictated by hopes of finding a combination of materials having similar vapor pressures so that steady, rather than flash, evaporation could be used. Among the compounds investigated were germanium–chromium, aluminum–silicon monoxide, silicon–silicon dioxide, and aluminum–aluminum oxide. The most promising system was germanium–chromium, which, for a 50-50 composition, produced films having a resistivity of 1,000 μohm-cm

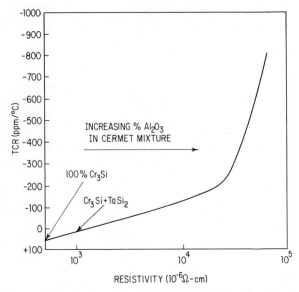

Fig. 15 TCR vs. resistivity for the cermet $Cr_3Si\text{-}TaSi_2\text{-}Al_2O_3$ as a function of composition.

and a TCR very close to zero. The film was found to be "quite stable." Higher resistivities could be obtained by using less chromium, but at the price of reduced stability.

Among the earliest cermets to be investigated was the system chromium–titanium nitride.[76] The process that was used for their preparation was first to deposit the metal constituents of the film (chromium–titanium) and then to convert this to the nitride through heating in an ammonia atmosphere at from 900 to 1200°C for about 10 min. The films thus formed were hard and were observed to be stable at quite elevated temperatures. Films with zero TCR up to about 500 ohms/sq could be prepared.

d. Semiconductor Films

Wherever relatively large temperature coefficients can be tolerated in order to achieve high sheet resistance, semiconductors hold out considerable promise as materials for resistive films. Over the years, a number of semiconductors such as silicon and germanium have been investigated for possible thin film resistor applications. However, the two materials that have had the most success have been carbon and tin oxide.

(1) Carbon Films Because of high processing temperatures and difficulties in sheet-resistance control, carbon films have not gained acceptance for integrated-circuit applications. Nevertheless, they have enjoyed considerable success as discrete resistors.   The most recent review of their technology and properties appears to have been in 1960.[77]

Carbon resistor films are usually deposited onto ceramic substrates.   The latter are needed because of the high temperatures (in the order of 1000°C) involved in the deposition process, which is the pyrolysis of a carbon-bearing gas such as methane. Usually, the hydrocarbon gas is diluted with a neutral gas such as nitrogen to allow better control.   By varying temperature, gas concentration, etc., films of various thicknesses can be obtained.   Devices made in this fashion are commonly referred to as "deposited-carbon" resistors.

Because it is not possible to achieve adequate sheet-resistance control in present-day pyrolysis methods, the resistors must be individually trimmed to value by binning and spiraling (see Sec. 4c).   The TCRs of pure carbon films are relatively large, varying from about −250 ppm/°C at 10 ohms/sq up to −400 ppm/°C at 1,000 ohms/sq. To allow for small changes in resistance associated with the attachment of end contacts, resistors are deliberately trimmed to a value about 1% too low and a final adjustment is made to the device through gentle abrasion of the film just prior to application of a protective coating.

Much harder and more stable films can be produced by combining other elements such as silicon and oxygen with the carbon to give so-called alloy films.[78]   Unlike the conventional types, which must be carefully protected, the alloy films are immune to oxidation, even without protective coatings.   However, the TCR of these films is no smaller than that of conventional carbon films.

A drastic decrease in the TCR of carbon films can be obtained by including a boron-bearing gas along with the methane.[79]   Films of this sort have TCRs as low as −20 ppm/°C at 10 ohms/sq (for 4% boron) and −250 ppm/°C at 1,000 ohms/sq. Boron hydride mixed with methane and benzene has also been used to prepare boron-doped films, as well as one-compound systems such as tripropylborane.   However, the commonest additive has been boron trichloride.

(2) Tin Oxide Films   All the resistor systems discussed thus far are susceptible in varying degrees to the effects of oxidation.   However, a material which is already fully oxidized would be expected to be free of this defect.   Tin oxide is such a material.   In addition, because of its refractory nature its chances of annealing or agglomeration are low.

The commonest method for the preparation of tin oxide films is the surface hydrolysis of stannous chloride ($SnCl_4$), according to the reaction

$$SnCl_4 + 2\ H_2O \rightleftharpoons SnO_2 + 4\ HCl$$

Since pure stannous chloride is likely to hydrolyze too rapidly, it is common practice to moderate the reaction by the addition of an alcohol, such as ethyl alcohol, an organic acid, such as acetic acid or, frequently, HCl.   Typically, a mixture containing equal parts by volume of the various ingredients is sprayed onto a heated glass or ceramic substrate at whose surface the reaction occurs.   The rate of reaction is slow at 500°C and becomes excessively rapid at about 800°C.   Because of the high temperature at which it is produced, the tin oxide forms an extremely adherent film.   Improved control has been claimed for systems in which only a narrow spray jet is used while the substrate rotates.[80]

Tin oxide is a semiconductor with a large band gap, and when its constituents are present in their correct stoichiometry, it has a high resistivity.   However, films formed by hydrolysis (as described above) may be oxygen-deficient or they may contain some chloride ions in solution.   This causes them to be N-type semiconductors.   It is common practice to further modify the conductivity of the tin oxide films through the addition of suitable dopants.   For example, antimony acts as a donor, increasing the conductivity still further and reducing the temperature coefficient.   Indium, on the other hand, acts as an acceptor and compensates for the oxygen vacancies, causing both resistivity and TCR to increase.[81]

SnO_2 films can have high resistivity.   Thus, films of 1,000 ohms/sq may be as thick as 1 μ.[82]   The films are very rugged and will operate in oxidizing atmospheres without

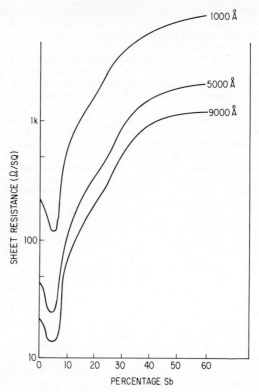

Fig. 16 Sheet resistance of tin oxide films vs. percent antimony in the films, for several film thicknesses.

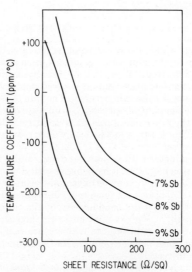

Fig. 17 TCR vs. sheet resistance of tin oxide films, for several antimony concentrations.

deterioration at temperatures as high as 450°C.[82] This feature of high-temperature stability considerably lessens the danger of runaway deterioration due to hot spots. In particular, antimony-doped films are the most stable in oxidizing atmospheres since undoped films may undergo some changes in conductivity due to refilling of some of the oxygen vacancies. The relationship between sheet resistance for a given thickness and percent antimony in the film[83] is shown in Fig. 16, whereas Fig. 17 shows the relationship between TCR and sheet resistance for several antimony concentrations.[83]

One interesting feature of tin oxide films is that they possess a high degree of transparency. They have therefore found wide application in areas such as the manufacture of so-called "conducting glass" and as heating elements. However, the method of preparation, the high temperatures involved, and the fact that films produced through surface hydrolysis are very coarse-grained and rough preclude the widespread use of films of this type for integrated-circuit applications. Smooth films deposited at lower temperatures might be very attractive, but the little work that has been done to evaporate or sputter tin oxide films has indicated that post-deposition treatments to about 800°C are needed if films with useful properties are to be obtained.[84] For further details on the optical and electrical properties of tin oxide films the reader is referred to Refs. 85 and 86.

4. TRIMMING OF THIN FILM RESISTORS

a. Factors Determining Precision of Thin Film Resistors

(1) Sheet-resistance Control Several factors determine the degree of sheet-resistance control that can be achieved through any given process. The ultimate limitation is the degree to which films of uniform thickness can be deposited over a significant area. In evaporation, excellent thickness uniformity can be achieved by using large source-to-substrate distances and collecting evaporant from only a small solid angle. Similarly in sputtering, a large cathode can be used and only the material deposited under the central portion may be collected.

Analogous solutions can be visualized for other deposition methods. For commercial processes, however, cost effectiveness demands that the number of substrates coated per deposition run be maximized; so "tricks" of this type, though frequently used in exploratory studies, are of little value when price is a consideration.

In addition to thickness, the resistivity of a film may vary from one portion of the deposition area to another. This may be due, for example, to local differences in substrate temperature during deposition[13] which influence the degree of oxidation as well as the subsequent annealability of the film. It is also possible for leaks or sources of outgassing to lead to differences in sheet resistance in their vicinity because of heavier local contamination. If perfectly smooth substrates are not used, the degree of roughness also plays a role in determining the level of sheet-resistance control. Generally, a substrate finish of 2 to 5 μin. (or better) is necessary for substrate roughness not to influence control of sheet resistance.

Finally, the design of the monitor used to control the point at which deposition is stopped may be critical. The monitor should be designed so that it "sees" the same temperature, composition, and deposition rate which the substrates do. Also, as is now well known, the nucleation properties of a film can be quite different if they are deposited in the presence of an electric field.[87] For example, the dependence of resistance on thickness for silver films[88] as deposited in the presence or absence of a small electric field is compared in Fig. 18. It is thus important to delay application of the monitoring field until a relatively substantial film thickness has been deposited. Also, quite aside from the nucleation phenomenon, the premature application of a measuring field can lead to error since extremely thin films may overheat at the contacts and burn out.

The ability to stop precisely at a particular sheet-resistance value obviously varies from one process to another. For example, at high deposition rates, the automatic monitoring equipment may not be able to shut off the deposition equipment quickly enough, and the stop value chosen must allow for some overshoot. Present-day deposition techniques appear to have a lower limit of sheet-resistance control of $\pm 2\%$. This degree of control has been reported for bias sputtering, as applied to Nichrome by Stern,[25] and for pellet flash evaporation as applied to Cr-SiO by Glang et al.[69]

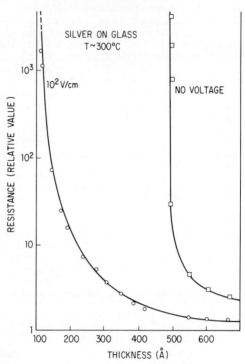

Fig. 18 Resistance vs. thickness of silver films on glass, deposited with no applied voltage between the end contacts or with a field of 100 V cm^{-1} applied across the substrate.

(2) Resistor Dimensions Unless relatively large area resistors are to be used, delineation of shapes by evaporation through masks is not sufficiently accurate. Evaporation masks commonly display dimensional errors in the order of several tenths of a mil. In addition, variations in the amount of material spreading under the mask as a function of the mask's distance from the substrate, angle of incidence of the incoming evaporant, etc., all make for errors in excess of $\pm 10\%$ in resistors having dimensions of 5 mils or less. Finally, masks cannot be used for other deposition processes such as sputtering or pyrolytic deposition. It is thus becoming increasingly common practice to generate the patterns of thin film resistors by means of photolithography.

Even though it provides much closer tolerances than does masking, at the present stage of the art, photolithography also has definite limitations. A detailed discussion of this has been presented by Glang and Schaible,[89] who show that one of the major sources of error is the limits in precision of the photomasks themselves. For example, the developed silver grains in lines finer than 1 mil are not completely continuous. Also, the transition from exposed to nonexposed regions is gradual rather than abrupt.

Phenomena of this type lead to random errors in the dimensions of the final resistor. In addition, a systematic error will frequently be superimposed over this, depending on how carefully the original master was cut. The latter class of error can be reduced significantly by going through several iterations. However, the cost and inconvenience of such a procedure are considerable. In Table 6 are summarized experimental data showing the random errors found in the dimensions of a number of resistors of varying size plus the systematic errors that were superimposed on these (for this particular case).

TABLE 6   Random and Systematic Errors as a Function of Resistor Dimensions

Resistor dimensions, mils	Random variations (95% limits, ± %)	Superimposed systematic error, %
2 × 10	2.5	+0.3
3 × 3	3.5	−0.5
2 × 2	4.5	−2.8
0.5 × 5	8.5	+17.8
0.5 × 2.5	9.5	+12.2

Similar random errors were obtained in all cases, whereas the sign and magnitude of the systematic error varied from mask to mask; these results do not include a random error of ±2% due to variations in sheet resistance.

Even if "perfect" masks could be obtained, additional errors would still be introduced by the photoresist and etching steps. For example, the exact width of the photoresist image developed after exposure depends very critically on the distance between the mask and the substrate, as illustrated in Table 7.

TABLE 7   Photoresist Image Widening as Obtained by Separating Mask and Substrate Surfaces during Exposure

Mask-to-substrate separation, mils	Reproduction of 1-mil mask line, mils
"0"	1.04
0.5	1.2
1	1.3
2	1.5

b. Substrate Trimming

When the major source of resistor error lies in poor sheet-resistance control and when all resistors on a given substrate have uniform composition, it may be economically worthwhile to adjust the sheet resistance of entire substrates. A number of obvious methods for this exist, and have been used. Most commonly, substrate trimming is performed after resistors have been fabricated, and one or more resistors on the substrate are monitored during the adjustment procedure. This is frequently some sort of heat treatment which causes the sheet resistance to change, either through annealing or through oxidation.

Very good control of sheet resistance can also be achieved by reducing the thickness through rf sputter etching. This method allows adjustment to an accuracy of at least ±3% of the amount removed, and can be applied to several substrates simultaneously. Similarly, anodic oxidation (where applicable) may be used as a method for adjusting the sheet resistance of all resistors on the substrate.

c. Individual Resistor Trimming

Substrate trimming as described above is often not feasible, either because the resistor tolerances are limited by dimensional errors or because the local sheet resistance varies considerably from one resistor to another, even though they are on the same substrate.

(1) Discrete Devices   The most widely used method for the adjustment of discrete resistors is spiraling.   This method takes advantage of the cylindrical shape of these resistors by cutting a spiral groove in the film, thus effectively increasing the number of squares of material between the end contacts.   The resistance of the film is monitored during spiraling, the latter being automatically terminated when the desired resistance value is reached.   The magnification in resistance achieved in this manner may be as high as 500, so that very wide variations in sheet resistance can be tolerated.

In practice, it does not pay to vary (to any great extent) the amount of spiraling given to any one group of resistors, since the automatic cutoff works best over only a limited range.   It is thus necessary for all resistors that go into a given series for adjustment to be fairly close to one another in value.   This is accomplished by means of binning.   After a given film-deposition run, all resistors are measured in the unspiraled condition and are separated, or binned, into batches according to their resistance values.   Generally, all resistors in a given bin are within 5 to 10% of one another.   The resistors from each bin are then spiraled, the equipment being adjusted to cut the best groove for that particular group.   An important feature of any binning procedure is that spot checks of the TCR are made for each bin, it being assumed that the TCR is reasonably constant within a given bin.

Another method of trimming that is applicable only to discrete devices is abrasion as described, for example, by Owens and Lesh.[90]   In this method, rough substrates are deliberately used and a raised pattern may also be embossed onto the substrate to define a shape to the resistor.   After coating, each unit is individually abraded. Material is removed from the high spots, and the process is continued until the desired resistance value is obtained.

The remaining methods to be discussed are also suitable for the adjustment of individual resistors even though they may form part of a resistor network or an integrated circuit.

(2) Change of Geometry   In this approach, the resistor is deliberately given too low a value initially and is then adjusted upward by one of the general techniques schematically illustrated in Fig. 19: (*a*) shows a resistor having several parallel legs; (*b*) shows

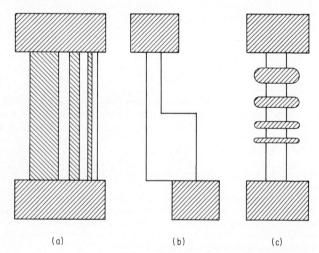

(a)   (b)   (c)

Fig. 19 Schematic representation of three different methods for mechanical trimming of resistors.

a resistor that is too wide, at least for a portion of its width; (*c*) shows a resistor which has portions of its length shorted out by conductive overlays.   To trim resistors of this type, either one or more of the parallel legs are open-circuited, the wide region of the resistor is partly cut away, or one or more of the shorting paths are removed.   Methods for effecting these changes include grinding, sandblasting, chemical etching, and local evaporation through heating with hot gas jets, with radiation, with electron beams, etc.   Many variations of these three basic approaches have been developed over the years.   For example, 15 different techniques are outlined in a review of resistor adjustment by Langford.[91]   Trimming through these methods is invariably slow, usually of limited precision, often requires a skilled operator, and gives low spatial resolution.   For these reasons, the methods are rapidly falling into disfavor for all but discrete resistors.

(3) Thickness Adjustment through Anodic Oxidation   In this method, which obviously is limited to valve-metal films, the effective resistor thickness is reduced by growing an anodic oxide on its surface and monitoring the resistance of the remaining portion.   The method has the advantage that an additional protective overcoating is generated as a by-product of the process.

In implementing the method, all resistors on a given substrate are necessarily at the same anode potential.   However, to allow adjustment of individual resistors, a separate cathode is supplied for each resistor.   In one approach, the electrolyte used for anodizing each resistor is electrically isolated by use of a grease mask which is applied through silk screening.   Thus, each resistor can be anodized by a different amount.   A typical fixture used for anodizing to a preset value[92] is shown in Fig. 20.

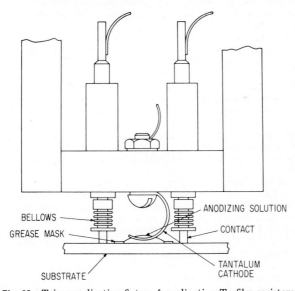

Fig. 20   Trim anodization fixture for adjusting Ta film resistors.

If a low enough anodizing current is used, the oxide builds up relatively slowly and the process can readily be terminated to achieve resistor tolerances to within $\pm 0.01\%$.

Resistor trimming through anodization has been used in production.[44]   The anodizing current is obtained through half-wave rectification of 60-Hz alternating current.   The maximum anodizing current is limited by a resistor in series with each cathode, and the maximum voltage is limited by a zener diode.   The anodizing voltage therefore appears as half a sine wave until the limiting voltage is reached, at which time the peak of the wave begins to be clipped.   During the interval when the anodizing current is zero (once every cycle), the resistor is measured and compared with some preset value.

In order to trim several individual resistors, devices are scanned sequentially (at an adjustable rate) for a maximum of 24 resistors in a given group. The measurement time is in the order of 50 ms, so that each resistor is monitored once every 1.2 s, anodization occurring during the interval between measurements. To avoid overshooting, the circuitry is arranged so that each resistor is initially trimmed at a fast rate, in the order of 1,000 V min^{-1}; but when the unit being trimmed reaches some given percentage of the preset value, the trim rate is automatically reduced.

The grease-mask approach for isolation of the electrolyte has been found to be of limited utility for production applications and has been replaced by a fixture made up of a number of slots. The substrate is pressed against the open end of the fixture and electrolyte is drawn up into the slots by capillary action. A serious disadvantage of this technique is that it is limited to substrates that have good planarity. On substrates that are not perfectly flat, electrolyte creep causes leakage paths to the contact areas.

(4) Joule Heating Since contact has, in any case, to be made to precision thin film resistors in order to monitor their values during trimming, it is an obvious notion to use this as a means for passing sufficient current through the films to cause Joule heating. The local heat thus generated can cause changes, either through oxidation (if the films are not protected) or through annealing (if they are). The latter method causes a resistance decrease during trimming and is to be preferred, in general, since any hot spots that may be present in the resistor prior to heat treatment will be diminished as a result of the Joule heating. On the other hand, in resistors trimmed through oxidation, hot spots are likely to increase in magnitude and, in extreme cases, will lead to open circuits.

If a steady current is used to effect the Joule heating, the change in resistance may be followed by measuring the voltage drop across the resistor. For adequate precision, allowance must be made for the TCR of the film. An additional problem with the dc method is that a good deal of the applied power is wasted through thermal conduction into the substrate, into the conducting lands, and into the measuring probes.

The dc approach has been used by Monnier,[93] for example, to trim Nichrome film resistors, which could be adjusted either upward or downward, depending on whether or not they were protected. This is illustrated in Fig. 21. The process is very sensitive to the temperature at which trimming takes place, and burnout of the devices occurred quite readily when too high a power was used. To control the latter, the resistor temperature was monitored during trimming through infrared-emission measurements.

DC trimming of resistors formed from the alloy platinum-rhodium has been performed by Drukaroff and Fabula.[94] They encountered a tendency to overtrim because of continued annealing after the desired resistance value had been reached. In order to obtain increased precision, resistors were first deliberately trimmed to values slightly short of the final value, followed by a brief trim under carefully controlled conditions. By "creeping up" on the final value in this fashion, precision in the order of $\pm 0.0001\%$ could be achieved.

Using much higher power levels, Bullard et al.[95] were able to achieve trimming of Cr-SiO cermet films in much shorter times—in the order of 1 to 2 s. Because of the high temperature of trimming, even the small TCR of the cermet films could not be neglected, and since a relatively large fraction of the total time was spent in cooling down from the trim temperature back to room temperature, there was a tendency to overshoot. With practice, this could be allowed for by stopping slightly short of the desired value. The resistors in this case were protected with overcoats of SiO and, consequently, trimming occurred downward, so that resistor uniformity was improved.

Many of the problems associated with the approach discussed above are eliminated if short-duration high-current pulses are used in place of direct current. Employment of a sufficiently brief pulse ensures that all the energy is dissipated within the resistor itself. Substantial temperatures can be achieved which, because of the short period of time involved, can be tolerated by the resistor without difficulty. In addition, because of the small amount of energy per pulse, rapid cooling occurs between pulses. If the duty cycle employed is not too heavy, it is also possible to measure resistance in between pulses (generally, just prior to the application of each pulse) and to terminate pulsing automatically when the desired resistance values have been reached.

Such an approach has been used successfully to trim tantalum film resistors.[96] It was found that, even when protected against external oxidation, tantalum films

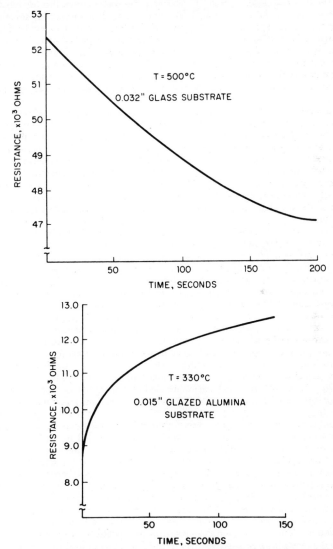

Fig. 21 Trimming (both upward and downward) of NiCr resistors through Joule heating.

increased in resistance during pulse trimming. This was ascribed to the migration of trapped oxygen in the film to the grain boundaries where it precipitated as an insulating phase.

Pulse trimming has enjoyed what is perhaps its greatest success when applied to cermet resistors.[65,97] In a detailed study of this system, Glang et al.[97] have concluded that the optimum pulse duration is in the order of 2 to 5 μs at a repetition rate of 100 pulses per second. Under these conditions, resistors could be trimmed to a precision of at least 0.03 %. Pulse-repetition rates as high as 1,000 pulses per second could be used, but this reduced precision to the order of ±0.3 %. It was shown that each successive pulse reduces the resistance by a progressively smaller amount. This is illustrated in Fig. 22 for several different power levels. It is evident from the figure that there is little to be gained by using more than about 100 trimming pulses—but that this number of pulses is sufficient to bring about a resistance change of up to 30%. The process therefore requires less than 1 s. It was estimated that the instantaneous temperature reached by resistors during trimming was in the order of 1100°C. Life tests indicated no significant change in resistor stability as a result of trimming.

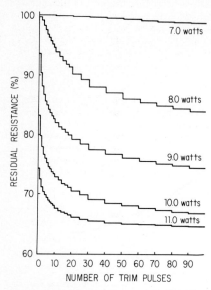

In cases where the conducting land area is too small to allow placement of probes for trimming purposes, it is possible to etch a small hole through the protective overcoat and then evaporate a large land that sits on the top surface and reaches down into the hole to make contact to the resistor. This temporary land is used to perform the trimming and is then removed at the end of the process.[98] Such an arrangement also allows for resistors to be isolated from the rest of the circuit during trimming, connection into the circuit being made on the surface of the overcoat rather than at the resistor level.

Fig. 22 Relative resistance of Cr-SiO films vs. number of trim pulses, for several different power levels.

(5) Contactless Methods In many circuits, optimum performance is best secured by adjusting the resistors to values that depend on the particular active and passive components that make up the remainder of the circuit. In such cases, a very precise value for any particular resistor has no meaning, and resistors can be adjusted to their optimum value only by operating the circuit and observing one or more circuit parameters while trimming is taking place. This type of trimming is referred to as functional trimming.

Since additional contacts to the resistors may be impractical or even undesirable during functional trimming, various methods have been used in the past to achieve contactless trimming. Approaches that have been taken include heating with hot gas streams, electron-beam heating, etc. Recently, the most successful methods have centered around the use of laser beams. A pulsed laser beam is repeatedly directed to various parts of the resistor, each spot that is struck being thus raised to a very high local temperature. The method works best in systems which decrease in resistance on heating, such as Cr-SiO.[65,99] The method has an additional advantage in that, if necessary, trimming may also be made to go in an upward direction, either by increasing the pulse power or by allowing the laser beam repeatedly to strike the same spot rather than moving it over the resistor. This leads to burnout at that spot, and a local increase in resistance results. The latter procedure, can, however, be dangerous, since a "hot spot" is created.

5. RELIABILITY OF THIN FILM RESISTORS

Any electronic component is judged on the basis of three criteria—performance, cost, and reliability. Furthermore, these three criteria interact with one another at all stages of the manufacturing process and can be properly assessed only by recognizing this fact. Thin film resistors are no exception, and it is thus impossible to assign precise reliability ratings to a materials system as such. Once minimum reliability specifications have been met, the user must always check carefully to ascertain what properties, if any, are being traded off in return for any additional gains in reliability.[100]

Since, by definition, a reliable component will take many years to fail under normal operating conditions, various accelerated tests have been devised in attempts to provide an indication as to the future expected reliability of a given component. These tests fall into three broad categories, (1) temperature, (2) humidity, and (3) load, and may be used alone or in combination.

a. Temperature

This test is the easiest to implement and is therefore the most widely used. Thermal aging gives a measure of the amount of oxidation as well as of internal changes such as annealing and precipitation that may occur.

A good review of some of the general aging mechanisms that may occur in metal film resistors has been given by Bohrer and Lewis,[101] while some specific effects of heat treatment have been discussed by Levinson and Stewart,[102] whose conclusions are summarized in Table 8. The data relate to a two-component system (in this case, the alloy "evanohm").

TABLE 8 Summary of Metallurgical Changes Which May Influence Stability

Process	Effect on resistivity	Effect on temp coeff	Effect on resistance of component	Magnitude of effect
Stress relief.........	Decrease	Increase	Decrease	A few parts per hundred maximum
Vacancy condensation	Usually decrease	Usually increase	Negligible	A few parts per million maximum
Long-range ordering of solute	Decrease	Increase	Decrease	Large but limited to a few cases
Precipitation of solute	Decrease	Increase	Decrease	May be very large
Clustering of solute*..	Usually increase	Usually decrease	Increase	Normally less than 1% but may be larger
Oxidation†..........	None	None	Increase	No limit
Evaporation.........	None	None	Increase	No limit
Solution of impurities.	Increase	Decrease	Increase	May be very large
Disordering of solid solution..........	Increase	Decrease	Increase	May be very large
Mechanical stressing..	Increase	Decrease	Increase	A few percent maximum
Homogenization‡.....	May go either way	May go either way	May go either way	Can be large
Agglomeration........	None	None	Increase	No limit

* Such, for example, as the formation of Guinier-Preston zones in aluminum-copper alloys. In special cases the process is specific and amounts to short-range ordering.

† Certain metals such as Nb, Ta, Ti, Zr, and V can dissolve appreciable oxygen. This portion of the oxidation process is then identical to contamination by solution of impurities.

‡ As, for example, of an evaporated film with a composition gradient through the thickness.

By plotting the logarithm of some criterion (such as mean time to failure) against the reciprocal of temperature, it is possible (in principle) to obtain an activation energy for the process which is causing the resistor to drift and, by extrapolating back to room temperature, to predict the operational life of the resistor. Such a procedure can, however, be misleading, since the processes causing change in the resistor are quite likely to possess more than one activation energy, and it is only the process that dominates at or near room temperature that is significant for extrapolation purposes. It is essential, therefore, to determine this lower activation energy. Lewis and Bohrer[103] consider the hypothetical case depicted in Fig. 23. Mechanism 1 is rate-controlling up to approximately 125°C, whereas above 175°C, mechanism 2 dominates. As a result, any accelerated testing program above 125°C will lead to false conclusions concerning the behavior of the resistor at room temperature. The only way out of such a dilemma (assuming that the existence of these two activation energies has been established) is to develop test conditions which can suppress any changes resulting from mechanism 2, while allowing, at high temperatures, only changes that result from mechanism 1. In practice, this is easier said than done.

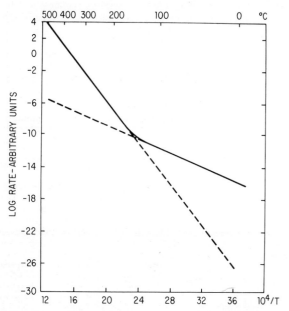

Fig. 23 Schematic illustration of log failure rate vs. $1/T$ for a degradation process having more than one activation energy.

Although thermal-aging data may therefore not guarantee anything in terms of long-term reliability, certain accelerated tests have come to be generally accepted in the industry. For most resistors, a change of 0.5% or less after storage for 1,000 h at 150°C is generally considered acceptable, although many high-stability resistors will exhibit changes of less than 0.25% after 1,000 h at 200°C. It is not uncommon to terminate thermal-heating tests at 1,000 h; since heat treatment is itself a source of stabilization for the resistor, changes during the second thousand hours of test are almost invariably substantially less than during the first thousand hours. In fact, since most resistor systems can achieve a high degree of stability if given sufficient heat treatment, a thermal-aging test is useful only for setting minimum specifications. Any resistor system that does unusually well on this test must also be assessed in terms of its performance and cost.

In certain cases, it is possible to overstabilize a resistor film. This has been pointed out by, for example, Walker et al.,[104] whose studies of the drift of tantalum films when heated at 420°C indicate a plateau between approximately 1 h and 3 h where the rate of resistance change with time is a minimum—so that 1 h represents the optimum stabilization time at 420°C.

b. Humidity

In practice, exposure to humidity is really a test of whatever protective coating the resistor may have. Commonly used test conditions are 85 to 95°C at 85 to 95%

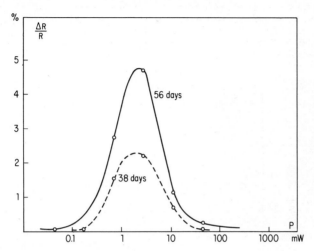

Fig. 24 Resistance change as a function of applied power for a resistor kept at very high humidity.

relative humidity. The test is most meaningful when combined with electrical load, since the principal problem anticipated is electrolytic corrosion of the resistors in the presence of a film of condensed moisture.

The most drastic changes that occur under the combined stress of humidity and load take place when the maximum load is applied that does not prevent the condensation of moisture on the surface of the resistor. The importance of this has been discussed in some detail by Futschik.[105] For example, Fig. 24 shows the percentage resistance change in a metal film after storage at 100% relative humidity for 63 days as a function of various applied loads. As can be seen, the failure rate is noticeably *less* at high loads, and significant changes occur over only a narrow range of applied power.

The extent to which a resistor drifts under conditions that give rise to electrolytic corrosion is also dependent on geometry. For example, the maximum electric field possible between two points on the substrate surface will be greater for a meandering resistor than for a straight-line resistor (Fig. 25). Similarly, in a spiraled resistor substantial electric fields may be generated across the grooves.

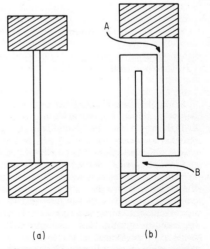

Fig. 25 Schematic representation of two resistors (both containing 20 squares), one straight and one meandering, illustrating existence of regions of high local field in the latter.

c. Load

Resistor drift under this test should be no different from what would result from storage at the temperature generated by the load[106] unless:

1. *Electrolytic corrosion* occurs because of condensed moisture, as discussed above.

2. *Metal migration* occurs. This effect is normally not seen in discrete resistors but only in the very small resistors used in integrated circuits. At very high current densities (in the order of 10^6 A cm^{-2} through the cross section of the resistor) a redistribution of material may occur such that, under conditions of direct current, metal migrates away from the cathode. Thinning of the material in the vicinity of the cathode occurs if the process is continued long enough. Certain metals, such as aluminum, are particularly susceptible to this effect, which is discussed more fully in Chap. 23.

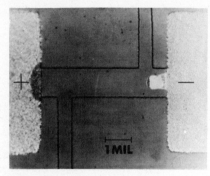

Even if the resistor material is itself not subject to electromigration at the current densities used for the test, the contact metals may still be. For example, in Fig. 26 (due to Schaible et al.[107]) the migration of aluminum can be clearly seen. It is of interest to note that in this particular case the migration of the aluminum was not the cause of resistor failure. This was due, instead, to the dark region at the anode, which can also be seen in the picture. This dark spot was identified as

Fig. 26 Micrograph of cermet resistor with Al end contacts after prolonged stress by high current densities.

a high-resistance area immediately under the aluminum contact and was thought to result from the formation of a chromium-aluminum intermetallic compound which, in turn, caused a depletion of chromium from the Cr-SiO film. This resulted in a higher local concentration of SiO and, therefore, in an increase in resistance in that region.

The appearance, under conditions of load life, of contact resistance when none was present initially has also been observed in other resistor-conductor systems—for example, in tantalum films in cases where the conductor and resistor films were not deposited during the same pumpdown.[36] It is thus important, when performing load-life tests, always to use a configuration such as in Fig. 26 which allows both the central (true) resistor value as well as the cathode and anode contributions to be measured during the life test.

3. *The resistor contains hot spots.* This is a difficult test to perform, as the incidence of hot spots is normally rather random and a large number of resistors need to be tested before sufficient statistical data can be gathered. If hot spots are a major cause of drift, however, the test should lead to catastrophic failure fairly quickly. A system in which resistance decreases with heat treatment is obviously an advantage in such cases, since hot spots will be eliminated on load life.

4. *Electrolysis of the substrate* may occur because of the presence of mobile ions. As in the electrolysis of condensed moisture, this effect will depend on the geometry of the resistor, but unlike the former, it will increase rapidly with temperature, which causes the mobility of the ions (almost always sodium) to increase. Although the migration of sodium ions usually takes place through the substrate, it is also possible for sodium-ion migration to occur within the overcoat provided as protection to the resistor. An example of this is in SiO films that have been deposited under conditions which allow contamination to occur.[108]

Because the various mechanisms which cause serious drift under load are quite different from one another, it is important to identify which one is responsible in a particular case, and to tailor the accelerated life test accordingly. Thus, for electrolytic corrosion, the test condition is maximum applied voltage compatible with reten-

on of some moisture on the resistor surface. For metal migration, the effect depends
n the current density through the thickness of the film, whereas the drift due to hot
ots is proportional to the applied power unit area of resistive material.

DESIGN OF THIN FILM RESISTORS

Choice of Resistor Shape

The choice of what sheet-resistance value to use for a particular group of resistors
ill be dictated by the lowest-value resistor in the circuit.   Experience has shown that
he number of squares for any resistor should always exceed 0.5.   If this is not the
ase, there is a danger of inaccuracies due to poor control of the spacing between the
onducting lands, as well as a rapidly increasing sensitivity to contact-resistance
ffects.

If possible, all resistors should be shaped as simple straight lines, and patterns that
se meandering lines should be employed only if absolutely necessary. Simple
traight lines are to be preferred because of their superior high-frequency performance,
educed sensitivity to sodium-ion migration, and lesser likelihood of failure under
umidity and load.   In addition, it is more difficult to predict the exact value of the
nal resistance for anything but a straight line.

The resistance associated with a square corner has been computed by Dow[109] and
 illustrated in Fig. 27a.   The resistance associated with a curved path is as illustrated

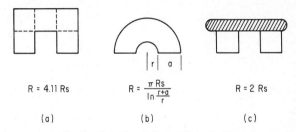

$$R = 4.11\ Rs \qquad\qquad R = \frac{\pi\ Rs}{\ln \frac{r+a}{r}} \qquad\qquad R = 2\ Rs$$

(a)   (b)   (c)

ig. 27   Resistance contribution introduced by various designs of meandering resistors.

 Fig. 27b.   In some cases, the problem of how much to allow for going around a
orner is bypassed by using the configuration shown in Fig. 27c, where the corner is
horted out by conducting material.

General formulas for the resistance of various shapes are given by Hall.[113]

Choice of Resistor Area

Unless there is good reason to do otherwise, it is desirable for thin film resistors to
ccupy all the area available to them.   This is so as to decrease errors due to dimen-
ional inaccuracies as well as to increase power-dissipative ability.   In practice, space
 usually at a premium, and the major portion of the available area must be given to
he resistors that dissipate the most power.   It then becomes necessary to form an
stimate of the power-dissipative capability of the system.

It must be emphasized that power dissipation is not a characteristic of a given
naterial system.   All that can be specified is the maximum temperature and current
lensity at which the resistor can be guaranteed to operate.   The amount of power
.eeded for a given resistor to reach that temperature will be a complex function of the
•articular substrate material, the mode of mounting of the substrate, etc.

As resistors are made smaller and smaller, the amount of power per unit area that a
esistor is able to dissipate grows progressively larger since the resistor begins to look
ncreasingly like a point source of heat.   For example, Fig. 28 shows the increase in
emperature above room temperature reached by a cermet resistor on an oxidized
ilicon substrate as a function of its area for a power density of 50 mW mil^{-2}.   In this
•articular example, the silicon substrate was mounted on a suitable heat sink so that

the temperature which each resistor reached was mainly determined by the rate at which it conducted heat through the thermal oxide to the silicon below. Thus, the temperature that each resistor reached was only weakly dependent on the size of the substrate or on how many other resistors were simultaneously under load.

The temperature that a resistor reaches depends both on its ability to conduct heat into the substrate as well as on the ability of the substrate to dissipate this heat. The

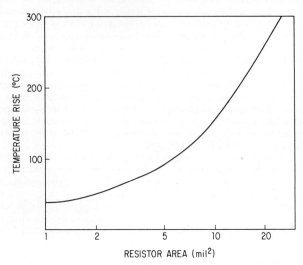

Fig. 28 Temperature rise of cermet resistors when subjected to a load of 50 mW mil⁻² as a function of resistor area.

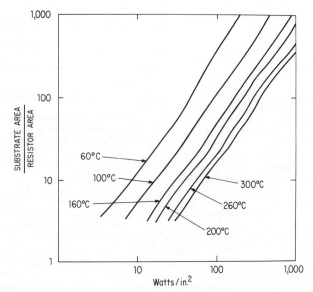

Fig. 29 Isothermals for Ta film resistors on glass substrates as a function of power density and ratio of substrate area to resistor area.

latter, in turn, will be a function of the substrate temperature. The temperature rise of the resistor will then depend, among other things, on its own size as well as on the ratio of its area to that of the substrate.[110] In Fig. 29, for example, are shown isothermals for tantalum films on ceramic substrates at different power levels for various substrate-to-resistor area ratios.[111]

In cases where the substrate temperature may rise appreciably, it is sometimes possible to calculate the maximum power dissipation by regarding all the resistors on the substrate as one large resistor.[110] Also, where relatively massive electrical leads to the substrate are used, their effect on the power dissipation of the circuit cannot be neglected, since they represent an important path for the removal of heat.[110]

c. Other Considerations

Among other considerations that determine resistor layout, we may include minimizing the incidence of crossovers,[112] choosing the correct conductor metallurgy and protective overcoat, and ensuring compatibility between them. It is also recommended that the resistive layer always be laid down prior to the conductive layer (contacts). It is not good practice to reverse the order (for example, in attempting to avoid contact-resistance problems), since undue thinning of the resistive film invariably occurs at the step junction between the conductive film and the substrate.

Also to be considered is the temperature of processing. The maximum temperature to which the system must be subjected can be *minimized* by using as high a deposition temperature as possible and then dispensing with the post-deposition-annealing step. However, depositing the resistor film at this relatively high temperature is liable to result in poor sheet-resistance control because of local differences in substrate temperature during deposition. With this in mind, one should deposit at the lowest temperature consistent with good adhesion and perform the stabilizing heat treatment later.

REFERENCES

1. Kruger, F., British Patent 157,909, 1919.
2. Marsten, L., *Proc. IRE*, **50**, 902 (1962).
3. Dummer, G. W. A., "Fixed Resistors," Pitman Publishing Corporation, New York, 1967.
4. Schwartz, N., and R. W. Berry, in G. Hass and R. Thun (eds.), "Physics of Thin Films," vol. 2, p. 363, Academic Press Inc., New York, 1964.
5. Siddall, G., in L. Holland (ed.), "Thin Film Microelectronics," John Wiley & Sons, Inc., New York, 1965.
6. Glang, R., *J. Vacuum Sci. Technol.*, **3**, 37 (1966).
7. Coombe, R. A. (ed.), "The Electrical Properties and Applications of Thin Films," Pitman Publishing Corporation, New York, 1967.
8. Edwards, A. J., *Extended Abstract, 14th Natl. Vacuum Symp.*, 1967, p. 115.
9. Sosniak, J., *Extended Abstract, 13th Natl. Vacuum Symp.*, 1966, p. 135.
10. Maissel, L. I., in G. Hass and R. Thun (eds.), "Physics of Thin Films," vol. 3, p. 61, Academic Press Inc., New York, 1966.
11. Charsan, S. S., R. W. Glenn, and H. Westgaard, *Western Elec. Engr.*, **7**, 9 (1963).
12. Hanfmann, A. M., *Western Elec. Engr.*, **10**, 41 (1966).
13. Maissel, L. I., and J. H. Vaughn, *Vacuum*, **13**, 421 (1963).
14. Jackson, C. M., J. G. Dunleavy, and A. M. Hall, *Proc. Electron. Components Conf.*, 1961, p. 36-1.
15. Polytechnic Institute of Brooklyn, *Repts.* R-16-45, R-103-45, 1945.
16. Alderson, R. H., and F. Ashworth, *Brit. J. Appl. Phys.*, **8**, 205 (1957).
17. Lakshmanan, T. K., *Trans. 8th Natl. Vacuum Symp.*, 1961, p. 868.
18. Monnier, M. H., *Proc. IEEE*, **53**, 539 (1965).
19. Degenhart, H. J., and I. H. Pratt, *Trans. 10th Natl. Vacuum Symp.*, 1963, p. 408.
20. Wiedd, O., and M. Tierman, *Trans. 3d Ann. Microelectronic Symp.*, St. Louis, Mo., 1964; also Sprague Tech. Paper TP-64-7.
21. Wiedd, O. J., and W. E. Berner, *Trans. 3d Intern. Vacuum Congr.*, Stuttgart, 1965, p. 59.
22. Campbell, D. S., and B. Hendry, *Brit. J. Appl. Phys.*, **16**, 1719 (1965).

23. Siddall, G., and B. A. Probyn, *Brit. J. Appl. Phys.*, **12**, 668 (1961).
24. Pratt, I. H., *Proc. Natl. Electron. Components Conf.*, **20**, 215 (1964).
25. Stern, E., *Proc. Natl. Electron. Components Conf.*, 1966, p. 233.
26. Bicknell, R. W., H. Blackburn, D. S. Campbell, and D. J. Stirland, *Microelectron. Reliability*, **3**, 61 (1964).
27. Layer, E. H., C. M. Chapman, and E. R. Olson, *Proc. 2d Natl. Conf. Military Electron.*, 1958, p. 96.
28. Calcatelli, A., and G. Cerutti, *Vacuum*, **16**, 373 (1966).
29. Foley, M. A., *Proc. Electron. Components Conf.*, 1967, p. 422.
30. Berry, R. W., and D. J. Sloane, *Proc. IRE*, **47**, 1070 (1959).
31. McClean, D. A., *WESCON Conv. Record*, 1959, p. 87.
32. Gerstenberg, D., and P. M. Hall, *J. Electrochem. Soc.*, **111**, 936 (1964).
33. McClean, D. A., *J. Electrochem. Soc. Japan*, **34**, 1 (1966).
34. Read, M. H., and C. Altman, *Appl. Phys. Letters*, **7**, 51 (1965).
35. Sosniak, J., W. J. Polito, and G. A. Rozgonyi, *J. Appl. Phys.*, **38**, 3041 (1967).
36. Maissel, L. I., R. J. Hecht, and N. W. Silcox, *Proc. Electron. Components Conf.*, 1963, p. 190.
37. Maissel, L. I., *Trans. 9th Natl. Vacuum Symp.*, 1962, p. 169.
38. Kuo, C. Y., J. S. Fischer, and J. C. King, *Proc. Electron. Components Conf.*, 1965, p. 123.
39. Gerstenberg, D., and C. J. Calbick, *J. Appl. Phys.*, **35**, 402 (1964).
40. Walker, M. J., *IEEE Trans. Electron. Devices*, **ED13**, 472 (1966).
41. Schaible, P. M., and L. I. Maissel, *Trans. 9th Natl. Vacuum Symp.*, 1962, p. 190.
42. Schuetze, H. J., H. W. Ehlbeck, and G. G. Doerbeck, *Trans. 10th Natl. Vacuum Symp.*, 1963, p. 434.
43. McClean, D. A., N. Schwartz, and E. D. Tidd, *Proc. IEEE*, **52**, 1450 (1964).
44. Berry, R. W., P. M. Hall, and M. T. Harris (eds.), "Thin Film Technology," D. Van Nostrand Company, Inc., Princeton, N.J., 1968.
45. Huber, F., *IEEE Trans. Component Pts.*, **CP11**, 38 (1964).
46. Lakshmanan, T. K., C. A. Wysocki, and W. Slegesky, *IEEE Trans. Component Pts.*, **CP11**, 14 (1964).
47. Huber, F., W. Witt, W. Y. Pan, and I. H. Pratt, *Proc. Electron. Components Conf.*, 1966, p. 324.
48. Degenhart, H. J., *Proc. Electron. Components Conf.*, 1967, p. 84.
49. Bowerman, E. R., J. W. Culp, E. F. Hudock, F. Leder, and H. J. Degenhart, *Proc. Electron. Components Conf.*, 1963, p. 158.
50. Hemmer, F. J., C. Feldman, and W. T. Layton, *Proc. Natl. Electron. Conf.*, **20**, 201 (1964).
51. Zimmerman, D. D., *Proc. Microelectronics Symp.*, 1965, p. 2C-7, St. Louis, Mo.
52. Hoffman, D. M., and J. Riseman, *Trans. 4th Natl. Vacuum Symp.*, 1957, p. 42.
53. Hoffman, D. M., and J. Riseman, *Trans. 7th Natl. Vacuum Symp.*, 1960, p. 218.
54. Chapman, R. M., *Vacuum*, **13**, 213 (1963).
55. Layton, W. T., and H. E. Culver, *Proc. Electron. Components Conf.*, 1966, p. 225.
56. Fowler, A. O., *Proc. Electron. Components Conf.*, 1963, p. 12.
57. Gould, P. A., *Brit. J. Appl. Phys.*, **16**, 1481 (1965).
58. Scow, K. B., and R. E. Thun, *Trans. 9th Natl. Vacuum Symp.*, 1962, p. 151.
59. Beckerman, M., and R. E. Thun, *Trans. 8th Natl. Vacuum Symp.*, 1961, p. 905.
60. Glang, R., R. A. Holmwood, and S. R. Herd, *J. Vacuum Sci. Technol.*, **4**, 163 (1967).
61. Lood, D. E., *J. Appl. Phys.*, **38**, 5087 (1967).
62. Scott, R. E., *J. Appl. Phys.*, **38**, 2652 (1967).
63. Himes, W., C. M. Stout, and R. E. Thun, *Trans. 9th Natl. Vacuum Symp.*, 1962, p. 144.
64. Wagner, R., and R. R. Ourlau, *Proc. Electron. Components Conf.*, 1967, p. 412.
65. Braun, L., and D. E. Lood, *Proc. IEEE*, **54**, 1521 (1966).
66. Ostrander, W. J., and C. W. Lewis, *Trans. 8th Natl. Vacuum Symp.*, 1961, p. 881.
67. Beckerman, M., and R. L. Bullard, *Proc. Electron. Components Conf.*, 1962, p. 53.
68. Wagner, R., and K. M. Mertz, *Proc. Electron. Components Conf.*, 1964, p. 97.
69. Glang, R., R. A. Holmwood, and L. I. Maissel, *Thin Solid Films*, **1**, 151 (1967).
70. MacLachlan, F. A., G. Planar, and R. Arnold, *Ceramics*, **17**, 30 (1966).
71. Farrell, J. P., and C. H. Lane, *Proc. Electron. Components Conf.*, 1966, p. 213.
72. Miller, N. C., and G. H. Shirn, *Appl. Phys. Letters*, **10**, 86 (1967).
73. Miller, N. C., and G. H. Shirn, *SCP Solid State Technol.*, 1967, p. 28.
74. Wilson, R. W., and L. E. Terry, *Proc. Electron. Components Conf.*, 1967, p. 397.
75. Riddle, G. C., *Trans. 6th Ann. Microelectronic Symp.* (IEEE), 1967, p. B4-1.
76. Olson, E. R., E. H. Layer, and A. E. Mittleton, *J. Electrochem. Soc.*, **102**, 73 (1955).

77. Wellard, C. L., in "Resistance and Resistors," Chap. 16, McGraw-Hill Book Company, New York, 1960.
78. Wellard, C. L., and K. Gentner, *Proc. Electron. Components Conf.*, 1956, p. 179.
79. Grisdale, R. O., A. C. Pfister, and G. K. Teal, *Bell System Tech. J.*, **30**, 271 (1951).
80. Sasaki, H., Y. Nishimura, and T. Yamamoto, *Proc. Electron. Components Conf.*, 1966, p. 79.
81. Aitchison, R. E., *Australian J. Appl. Phys. Sci.*, **5**, 11 (1965).
82. Burkett, R. H. W., *J. Brit. IRE*, **21**, 301 (1961).
83. Phillips, L. S., *Res. Develop.*, **19**, 46 (1963).
84. Sinclair, R. W., F. G. Peters, and S. E. Koonce, *J. Electrochem. Soc.*, **110**, 560 (1963).
85. Ishiguro, K., T. Sasaki, T. Arai, and I. Imai, *J. Phys. Soc. Japan*, **13**, 296 (1958).
86. Loch, L. D., *J. Electrochem. Soc.*, **110**, 1081 (1963).
87. Chopra, K. L., *Appl. Phys. Letters*, **7**, 140 (1965).
88. Kennedy, D. I., R. E. Hayes, and R. W. Alsford, *J. Appl. Phys.*, **38**, 1986 (1967).
89. Glang, R., and P. M. Schaible, *Thin Solid Films*, **1**, 309 (1967/68).
90. Owens, J. L., and N. G. Lesh, *Proc. Electron. Components Conf.*, 1967, p. 405.
91. Langford, R. C., *Proc. Electron. Components Conf.*, 1960, p. 66.
92. Jackson, W. H., and R. J. Moore, *Proc. Electron. Components Conf.*, 1965, p. 45.
93. Monnier, M. H., *IEEE Trans. Parts, Mater. Packag.*, June, 1966, p. 44.
94. Drukaroff, I., and J. Fabula, *Trans. 5th Ann. Microelectronic Symp.*, St. Louis, Mo., July, 1966, p. 4S-1.
95. Bullard, R. L., R. A. Hasbrouck, P. S. Schlemmer, and R. E. Thun, U.S. Patent 3,308,528, March, 1967.
96. Maissel, L. I., and D. R. Young, U.S. Patent 3,261,082, 1966.
97. Glang, R., K. Jaeckel, M. Perkins, and L. I. Maissel, *IEEE Spectrum*, **6**, 71, 1969.
98. Glang, R., and P. M. Schaible, *IBM Tech. Discl. Bull.*, **9**, 128 (1966).
99. Lins, S. J., and R. D. Morrison, paper presented at WESCON, August, 1966.
100. Berry, R. W., W. H. Jackson, G. I. Parisi, and A. H. Schafer, *Proc. Electron. Components Conf.*, 1964, p. 86.
101. Bohrer, J. J., and C. W. Lewis, in M. E. Goldberg and J. Vaccaro (eds.), "Physics of Failure in Electronics," vol. 2, p. 338, 1964.
102. Levinson, D. W., and R. G. Stewart, in "Physics of Failure in Electronics," M. E. Goldberg and J. Vaccaro (eds.), vol. 3, p. 281, Spartan Books Inc., 1965.
103. Lewis, C. W., and J. J. Bohrer, in M. E. Goldberg and J. Vaccaro (eds.), "Physics of Failure in Electronics," Spartan Books Inc., 1963.
104. Walker, M., A. McKelvey, G. Schnable, and M. Sharp, in M. E. Goldberg and J. Vaccaro (eds.), "Physics of Failure in Electronics," vol. 4, p. 179, 1966.
105. Futschik, F., *Proc. Electron. Components Conf.*, 1965, p. 143.
106. McLean, W. E., J. A. Thornton, and H. R. Aschan, *Proc. Electron. Components Conf.*, 1963, p. 4.
107. Schaible, P. M., J. C. Overmeyer, and R. Glang, in T. S. Shilliday and J. Vaccaro (eds.), "Physics of Failure in Electronics," vol. 5, p. 143, 1967.
108. Smith, P. C., and M. Genser, in M. E. Goldberg and J. Vaccaro (eds.), "Physics of Failure in Electronics," vol. 3, p. 306, 1965.
109. Dow, R. J., *Proc. Electron. Components Conf.*, 1962, p. 7.
110. Peek, J. R., *Electron. Design*, **24**, 64 (1966).
111. Peek, J. R., *Proc. Electron. Components Conf.*, 1966, p. 68.
112. Hawkes, P. L., Ref. 7, Chap. 7.
113. Hall, P. M., *Thin Solid Films*, **1**, 277 (1967/68).

Chapter **19**

Thin Film Capacitors

DIETER GERSTENBERG

Bell Telephone Laboratories, Inc., Allentown, Pennsylvania

INTRODUCTION

The increasing quantity and complexity of electronic systems have stimulated the development of thin film microelectronics. This approach to integrated circuits involves the deposition of many resistors, capacitors, and their connections on stable, insulating substrates. From the discussion of the dielectric properties of films in Chap. 16, it is evident that a large number of dielectric materials are available for the fabrication of thin film capacitors. But only those dielectric films are of interest for application in thin film capacitors which are thermally and chemically stable. In addition the processes used for their deposition have to be compatible within other steps of the integrated-circuit fabrication. The thin film dielectrics which are extensively used in exploratory RC circuit applications are mainly inorganic substances, namely, oxides and halides of metals and semiconductors. Table 1 lists some of the more important materials, their dielectric constants, and the methods for their formation. This list, which also includes an organic polymer, is by no means complete, and it is expected that as time goes on new materials and structures with different or improved properties will become available.

TABLE 1   Most Commonly Used Thin Film Capacitor Dielectrics

Dielectric material	Dielectric constant	Method of formation
Ta_2O_5.	25	Anodic oxidation, reactive sputtering
Al_2O_3.	9	Anodic oxidation, reactive sputtering, rf sputtering
SiO.	6	Evaporation
SiO_2. . , ...	4	RF sputtering, thermal oxidation
Parylene*.	2.65	Vacuum polymerization

* Union Carbide trade name.

The two most common dielectric materials that have been thoroughly evaluated in thin film capacitors are evaporated SiO and anodized Ta_2O_5. A large part of this chapter, therefore, will be devoted to reviewing the properties of these two materials and the processing and design parameters which affect their dielectric constant, loss, dielectric strength, and long-term stability under voltage and temperature stress. Only those deposition or formation techniques will be discussed in this chapter which combine good control of thickness, structure, and chemical composition of the dielectric layers, because these variables determine to a large extent the electrical characteristics of thin film capacitors.

A number of other materials which have shown promise as thin film capacitor dielectrics will be described only briefly. These materials include TiO_2 and the titanates, which are of interest because of their high dielectric constant. Preliminary results indicate that the oxides of hafnium,[1] yttrium,[2,3] and other rare-earth metals[2] are extremely stable thermodynamically when formed by anodic oxidation or reactive sputtering. The low temperature dependence of capacitance and dissipation factor of dysprosium borosilicate capacitors has resulted in the development of high-temperature capacitors which can be used up to 450°C.[4] Organic-polymer materials have shown very attractive dielectric and mechanical properties. They are able to undergo plastic deformation when stresses are applied, thus minimizing residual stresses in the film structure which may present a problem with some inorganic dielectric materials. The drawbacks of polymer films are the lack of precise control of film composition and the difficulty of producing high-resolution film patterns.[5] Although the material properties which are of primary importance for film capacitors are the properties of the dielectric film, in many instances, the metal electrodes and the substrate can affect the properties of the dielectric or the properties of the capacitor, or both. Thus the materials which form a thin film capacitor will be considered together. The requirements a thin film dielectric material must meet for thin film electronic applications, and the properties of major types of thin film capacitors have been reviewed by a number of authors.[6,7,8]

1. CAPACITOR PROPERTIES

Before specific materials are discussed, some important parameters used to describe the properties of thin film capacitors will be introduced.

a. Capacitance

In electronic circuits a capacitor is the basic component for storing electric charge. Thin film capacitors usually have a parallel-plate configuration, as shown in Fig. 1, where two conducting layers separated by a layer of a suitable dielectric are supported by an insulating substrate 100 to 1,000 times thicker. The magnitude of the capacitance is a function of the overlapping electrode area $(A = LW)$, the thickness of the dielectric d, and the dielectric constant ϵ of the dielectric layer:

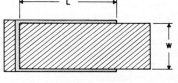

$$C = \epsilon_0 \epsilon \frac{A}{d} \qquad (1)$$

where $\epsilon_0 = 8.85 \times 10^{-12}$ F/m is the permittivity of free space.

Fringing effects at the edges of the thin film capacitor are negligible since the thickness of the dielectric is usually small compared with its lateral dimensions. The geometric and the measured capacitance can be different if the electric field at the

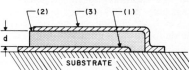

Fig. 1 Schematic thin film capacitor showing the length L and width W of top electrode covering the capacitor dielectric of thickness d. (1) Bottom electrode. (2) Dielectric. (3) Top electrode.

insulator-metal interface varies in the insulator over a region which is not insignificant compared with the total insulator thickness.[9] For dielectrics several thousand angstroms thick this problem, however, does not exist. In comparing the properties of different dielectric materials it is often convenient to use the capacitance density, which is defined as the capacitance per unit area

$$\frac{C}{A} = \frac{\epsilon_0 \epsilon}{d} \qquad (2)$$

This equation relates dielectric constant and the dielectric thickness of a material with capacitance density. For a given material the film thickness alone establishes the capacitance density, which in turn can be used to determine the area needed for a particular capacitance value. There are limits for the capacitance densities which can be realized for a particular dielectric material, because below 1,000 Å the dielectric strength might be reduced rapidly owing to pinholes and discrete defects in the dielectric film. Thus for SiO and the alkalide halides the upper limit for capacitance density is in the order of 0.01 μF cm^{-2}, while for Ta$_2$O$_5$, TiO$_2$, and the titanates it is in the order of 0.1 to 0.5 μF cm^{-2}.

b. Dielectric Strength, Breakdown, and Voltage Rating

For most dielectric film materials the dielectric strength is between 10^5 and 10^7 V cm^{-1}. If it is assumed that the dielectric strength is independent of film thickness, then

$$V_D = E_D d \qquad (3)$$

where V_D is the breakdown voltage of the capacitor, E_D the dielectric breakdown strength of the material, and d the dielectric thickness. Knowledge of the dielectric strength is very important for the design of thin film capacitors because it determines the voltage at which destructive breakdown of the dielectric occurs. Theoretical treatments of intrinsic dielectric breakdown, usually developed for bulk dielectric materials, have been summarized by Whitehead,[10] O'Dwyer,[11] and Franz.[12] These theories of electrical breakdown are based either on the generation of carriers or on the transfer of excess energy from existing carriers to phonons. A number of criteria have been developed for calculating a critical field at which electrons start to be accelerated to ionizing velocities, and its dependence on temperature and dielectric thickness. But these criteria will not be discussed here because for thin film dielectrics breakdown due to flaws in the dielectric layer is much more typical[13,14] than intrinsic breakdown, while the maximum dielectric strength of the dielectric in a thin film capacitor structure is often characterized by breakdown due to thermal instability.[13] These two forms of dielectric breakdown will be discussed next.

Since the SiO layers in SiO-film capacitors are deposited by evaporation while the Ta$_2$O$_5$ layers in tantalum-film capacitors are formed in an electrolyte in the presence of an electric field in the order of 6 $\times$ 10^6 V cm^{-1}, it is often assumed that tantalum-film capacitors contain fewer flaws and defects than SiO films.[15] But there is no solid evidence that this is the case. From all the available experimental evidence it appears that both types of dielectric are not completely free of defective areas and flaws, and that a certain range of dielectric-breakdown strengths can be expected. For SiO-film capacitors breakdown has been observed when fields between 4 $\times$ 10^5 and 4 $\times$ 10^6 V cm^{-1} are applied, no matter whether the SiO layer is 1,000 or 10,000 Å thick. Siddall showed that the dielectric strength is higher when the SiO film is deposited slowly.[16] But he also found, and his results have been confirmed by numerous investigators since, that breakdown frequently originates at pinholes, voids, and defects in the films because of dust on the substrate, spitting of the evaporant, or some localized inhomogeneity in the SiO film. Optical and electron micrographs of SiO-film capacitors which were stripped off the substrate and subjected to electrical stress in the electron microscope showed that in vacuum breakdown originates at inhomogeneities in the SiO film and not at pinholes, dust spots, or fissures.[17] The inhomogeneities appear as irregularly shaped dark spots about 0.5 μm in diameter, but their structure has not been determined. The presence of defects in the SiO therefore results in a certain variability of the breakdown field among capacitors simultaneously deposited on the same substrate. The values for the breakdown field were also found to depend on the substrate material. Chaiken and St. John[18] found that they decrease according to the substrate in the following sequence: Vycor, optical polish; Pyrex, optical polish, microscope slide. Breakdowns were correlated to microscopic blemishes about 1 to 2 μm in size which were visible in dark-field illumination. The breakdown field of the capacitors could be raised significantly by precoating the substrates with a 1-μm layer of SiO.

Very often the metal selected for the capacitor electrodes affects the breakdown characteristics of the structure. Breakdown measurements in evaporated NaCl, LiF, and cryolite (Na$_3$AlF$_6$) layers indicated that, out of Ag, Al, and Au which were used as electrodes, Ag behaved anomalously.[19] For capacitors with Ag electrodes lower breakdown fields were observed than for the other two types of electrodes. This was attributed to the high mobility of Ag in a strong electric field, which might have resulted in migration of Ag atoms into the dielectric layer. Tantalum-film capacitors also show a dependence of the breakdown properties upon the top electrode material.[7,20] This is illustrated in Table 2 for 0.035-μF tantalum-film capacitors. Au, Pd,

TABLE 2 Effect of Counterelectrode on Anodic-breakdown Field of Tantalum-film Capacitors with 2,200 Å of Ta_2O_5

Counterelectrode metal	Deposition	Breakdown field,* 10^6 V cm^{-1}
Gold.............	Evaporation	4.12
Palladium........	Evaporation	4.09
Copper...........	Evaporation	3.64
Antimony........	Evaporation	3.40
Cadmium.........	Evaporation	3.22
Iron.............	Evaporation	2.22
Indium..........	Evaporation	1.86
Aluminum........	Evaporation	1.09
Tantalum.........	Sputter	0.68

* Value when substantial current increase occurs. Catastrophic-breakdown fields may be 5 to 20 % higher.

and Cu form loosely adhering layers whereas most of the other ones adhere well to the anodic Ta_2O_5 layer and may introduce strain into the dielectric layer. Au and Cu also tend to agglomerate during deposition while iron, aluminum, and tantalum are likely to enter microfissures and pores in the oxide film resulting in capacitors with lower breakdown fields.[20] But further studies are required to elucidate the effect of the top electrode material on the breakdown characteristics of thin film capacitors in more detail.

If the metal counterelectrode of a thin film capacitor is thin enough (between 500 and 2,500 Å) breakdown at weak spots does not necessarily destroy the whole capacitor as shown in Fig. 2 for a tantalum-film capacitor. The metal electrode is vaporized

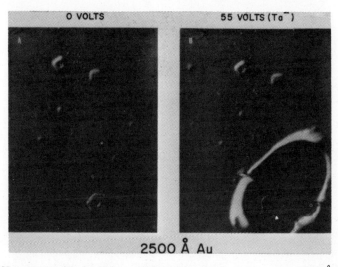

Fig. 2 Noncatastrophic breakdown of tantalum-film capacitor with 2,500-Å gold top electrode. Before (0 V) and after voltage of 50 V between top and bottom electrode was applied; tantalum electrode negatively biased (N. N. Axelrod[14]).

over an area of about 50 μm in diameter, while the oxide is destroyed over a much smaller area. Such breakdowns are nonshorting, and the capacitor can still be used. Klein and Gafni have investigated nonshorting breakdowns in thin film silicon oxide capacitors.[13] The purpose of their study was to obtain information about the maximum dielectric strength of evaporated silicon oxide films between 3,000 and 50,000 Å thick. Three forms of breakdown can be observed: single-hole, self-propagating, and maximum-voltage breakdowns. Single-hole and self-propagating breakdowns occur at flaws, but self-propagating breakdowns, which usually destroy the whole capacitor, can be prevented by putting a sufficiently large resistor (>10 kilohms) in series with the capacitor during testing. After weak spots are eliminated in the silicon oxide by utilizing self-healing breakdowns, a maximum breakdown voltage can be obtained which destroys the whole capacitor. On applying a dc voltage to a SiO-film capacitor and increasing it slowly the current increases quasi-exponentially as shown in Fig. 3.

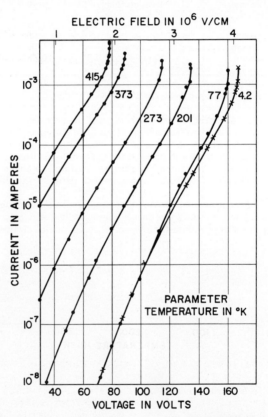

Fig. 3 Current-voltage characteristics of silicon oxide film.[21]

After reaching a maximum voltage, the current increases more rapidly and the voltage shows a slight decrease. This behavior has been attributed to thermal instability. Values for the maximum dielectric strength calculated on the basis of the electrical and thermal conductance of the SiO-film capacitor structure agree quite well with those observed experimentally in the temperature range between −145 and 65°C[21] (Fig. 4). At room temperature the maximum dielectric strength for SiO was found to be 2.95×10^6 V cm^{-1} for a 4,500-Å film and 1.97×10^6 V cm^{-1} for a film about

22,000 Å thick. These values are somewhat smaller than those observed for SiO_2[21] and Si_3N_4,[22] which are at least 6.00×10^6 V cm^{-1} and 9.30×10^6 V cm^{-1}, respectively, for films a few thousand angstroms thick.

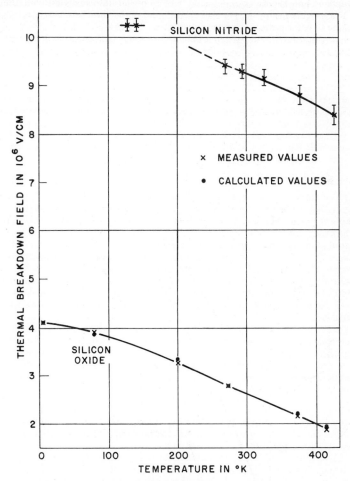

Fig. 4 Maximum breakdown field for SiO^{21} and $Si_3N_4^{14}$ films as a function of temperature.

It is obvious from the previous discussion that the rated voltage, or the working voltage, of a capacitor must be less than the breakdown voltage. How much less depends on the operating temperature, the capacitor area, dielectric film thickness, the type of substrate and electrode materials, and the maximum allowable failure rates. But the rated voltage is rarely more than half the breakdown voltage, and in some cases it may be as low as 10% of the breakdown voltage. The rated voltage is then

$$V_R = kV_D = kE_D d \qquad (4)$$

where V_R is the rated voltage and k a number between 0.1 and 0.5. For Eq. (3) it was assumed that the dielectric strength is independent of film thickness. But for very thin films the density of defects is higher, so that the dielectric strength is lower. Safe lower limits are 200 to 300 Å for anodic films and 1,000 to 3,000 Å for evaporated films.[7]

There is also an upper limit on film thickness for most dielectrics. For example, evaporated silicon monoxide films more than 50,000 Å thick frequently show rupturing and peeling due to mechanical stresses. The upper limit for anodically formed films usually depends on the voltage beyond which scintillation and breakdown during formation of the oxide occur. For anodic Ta_2O_5 the upper limit is about 6,500 Å, which corresponds to a formation voltage of 400 V.

The charge-storage factor, which is the rated voltage times the capacitance density, can be written as follows:

$$V_R \frac{C}{A} = k\epsilon_0\epsilon E_D \tag{5}$$

The charge-storage factor can be used as a figure of merit for the evaluation of different dielectric materials because it provides information about the substrate area needed for a given capacitance at a certain voltage. Tantalum oxide film capacitors have a rating of about 5.5 μF V^{-1} cm^{-2}. The corresponding value for evaporated silicon monoxide capacitors is only about 0.5 μF V^{-1} cm^{-2} because of the lower dielectric constant and a somewhat lower dielectric strength of SiO.

c. Capacitor Losses

The dielectric materials used in thin film capacitors, ionic solids such as the alkali halides, anodic oxides of the valve metals, or silicon oxide have a net dipole moment in the absence of an electric field which is zero. They may have dipolar groups of atoms, but they are frozen in and are not free to change their equilibrium positions in the range of audio and microwave frequencies.[23] The natural frequencies of the molecular dipoles of these materials are mostly in the infrared and optical range, and they are therefore characterized by infrared absorption. Their polarizability in the 10^2 to 10^8 kHz range arises from the displacement of electrons, and since the polarizable constituents have a wide distribution of relaxation times the dielectric constant and dissipation factor are to a first approximation independent of frequency over a wide range of frequencies.

There are two other sources of loss in thin film capacitors besides the dielectric loss. Both electrode and lead resistance can also contribute substantially to the total loss in the planar thin film capacitor structure. Since the electrode resistance and the lead resistance are in series with the dielectric, it is convenient to represent the capacitor with a series-equivalent circuit.[24,25] Thus, the total dissipation factor of the thin film capacitor is

$$\tan \delta = \tan \delta' + \omega C R_e \tag{6}$$

where tan δ is the measured dissipation factor, tan δ' is the frequency-independent contribution from the dielectric, C is the capacitance, ω is the angular frequency, and R_e is the electrode resistance. The contribution from the lead resistance has been neglected in Eq. (6), and it has been assumed that it can be kept small compared with the electrode resistance. The resistance of one electrode in the overlapping area of the capacitor (Fig. 1) is $R = \omega\rho L/Wd$, where ρ is the resistivity of the electrode material and d the electrode thickness. If we assume that ρ and d are equal for both bottom and top electrode and with R_e equal to $R/3$,[25] Eq. (6) can be rewritten as follows:

$$\tan \delta = \tan \delta' + \frac{2\omega C\rho L}{3Wd} \tag{7}$$

It is obvious from Eq. (7) that ρ, d, and the aspect ratio are strongly affecting the high-frequency performance of thin film capacitors. For example, in tantalum-film capacitors the major contribution to R_e originates from the tantalum anode whose value for ρ is about 200 μohm-cm. Since tan δ' for anodic tantalum oxide is about 0.005, an electrode resistance of only 1 ohm will double the value for tan δ of a 500-pF capacitor at 1 MHz. Within limits, the high-frequency performance of thin film capacitors can be improved by using thicker and highly conductive metal films for the capacitor electrodes. In the case of tantalum, an underlay of a more conductive metal, like aluminum, has been suggested. Aluminum has the advantage that it is anodizable itself and permits oxidation at pinholes in the tantalum layer during anodic oxidation.[25,26]

2. TYPES OF FILM DIELECTRICS

a. Inorganic Materials

(1) Silicon Monoxide Evaporated silicon monoxide is probably the most widely investigated dielectric that can be used for thin film capacitors.[7,8] Previously it had proved useful as a protective layer on mirrors and for preparing replica as well as support films for electron-microscope and electron-diffraction studies.[27] The effect of deposition parameters on the properties of silicon oxide films has also been extensively studied to provide insulating layers between different levels of integrated circuits, and for protective coatings.[28]

One of the attractive features of SiO is its high vapor pressure, which increases from 0.1 Torr at 1150°C to over 20 Torr at 1600°C.[29] It evaporates at much lower temperatures than either silicon or silicon dioxide, and it condenses on cool substrates as a smooth, uniform film. The compound SiO is readily available in the form of powder or pellets. A baffled, optically dense source is used for evaporation to avoid spitting of discrete particles onto the substrates.[29] SiO films are amorphous and the x-ray diffraction pattern shows only one diffuse ring corresponding to an average atomic distance of 3.60 Å when the films are deposited at pressures less than 1×10^{-5} Torr and at rates between 6 and 12 Å sec^{-1}.[30] These films have an amber color and the density of bulk SiO, which is 2.2 g cm^{-3}. Films deposited below 5 Å sec^{-1} oxidize rapidly in air because of their porous nature and show a change in color; while films deposited at rates in the 20 to 30 Å sec^{-1} range have larger densities than bulk SiO, and electron-diffraction patterns indicate a solid solution of silicon and silicon dioxide.[27] These results suggest that the compound SiO is stable in the gaseous state but metastable in the solid state where it might show deviations from the stoichiometric compositions. Further proof for this has been presented by White, who found that silicon monoxide films can be prepared by evaporation from a boat containing silicon dioxide at high temperature and low oxygen pressure.[31] Apparently gas-phase decomposition takes place during the evaporation of silicon dioxide, indicating that the composition of the final evaporated layer is independent of the composition of the source material.

Successful fabrication of thin film capacitors based on silicon monoxide depends to a large extent on the formation of stable films. The stability of silicon monoxide films is closely related to the mechanical stresses which are present in the films. Large compressive stresses can cause crinkling or peeling of the silicon monoxide, whereas large tensile stresses result in cracking of the films.[16,32] A number of studies revealed that source temperature, deposition rate, residual gas pressure, and type of the residual gas have a significant effect on the mechanical stresses of silicon monoxide films,[7] and that on exposure to air the stress becomes more compressive.[33,34] Figure 5 shows how the stresses depend on N, where N is defined as the number of silicon monoxide molecules impinging on unit area of the substrate in unit time divided by the corresponding number of residual gas molecules: $N = K (R/P)$. R is the deposition rate in angstroms per second, P is the residual gas pressure, and K is a constant which depends on the type of reactive gas (for oxygen $K = 3.5 \times 10^{-7}$ Torr sec Å^{-1}). In the range $10^{-2} < N < 10^{-1}$ the films consist of SiO$_2$, whereas films deposited at $N > 1$ have the stoichiometric composition of SiO. The observed compressive stress occurring when $N < 1$ has been attributed to the absorption of oxygen in the SiO structure during deposition, while the tensile stress for $N > 1$ is that expected for evaporated SiO films. The results plotted in Fig. 5 also indicate that the stresses in silicon oxide films do not depend directly on the source temperature in the range between 1200 and 1500°C. But it should be emphasized that care was taken that the rate of deposition was kept constant by varying the aperture of a vane between source and substrate. For a certain value of N the film stress remains constant during the evaporation and for as long as the sample remains under vacuum. But when the films are exposed to air, oxygen, or water vapor, the stress becomes more compressive (Fig. 6). This

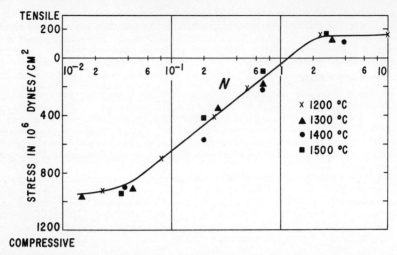

Fig. 5 Stress of silicon oxide films as a function of log N. The films were evaporated in residual oxygen atmospheres and the source temperature was varied as shown.[34] N is defined as the number of silicon monoxide molecules divided by the number of residual gas molecules impinging on the substrate per unit area and time.

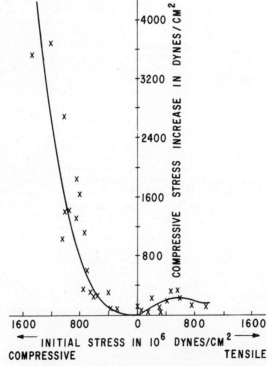

Fig. 6 Increase in compressive stress observed on evaporated SiO films after exposure to air.[34]

change is caused by further oxidation of the film. The observed changes in refractive index and loss of amber color have been attributed to a gradual conversion to silicon dioxide. If the stress is initially tensile, the change is usually much smaller, as shown in Fig. 6. Films evaporated at increasing angles of incidence between the vapor stream and the substrate normal become increasingly susceptible to oxidation with an associated increase in compressive stress. This stress was found to be a maximum for films evaporated at incident angles of about 55°, after which it rapidly decreases.[34]

According to Hill et al.[34] stable silicon monoxide films can be obtained by evaporating the films at low partial pressures of oxygen and water vapor, and at higher rates; e.g., the value for N should always be above 1. For a partial pressure of 2.5×10^{-6} Torr of water vapor or 3.5×10^{-6} Torr of oxygen the deposition rate should be above 10 Å sec^{-1}. The value of the source temperature is not too critical as long as there is no dissociation of the silicon monoxide source material. This usually does not occur below 1400°C.

Since Si and SiO are strong absorbers in the ultraviolet and visible region, the optical properties have been used as a measure for the composition of evaporated silicon oxide films. Films evaporated at a residual gas pressure of 5×10^{-6} Torr show an increase in transmission at 3,600 Å from 42 to 87 % as the deposition rate is lowered from 100 to 5 Å sec^{-1}, respectively.[35] One hundred percent transmission is obtained at 3,600 Å for the composition range from Si_2O_3 and SiO_2,[36] but measurements at 3,200 Å indicate that at lower wavelengths the absorption characteristics for films in this composition range do depend on the degree of oxidation.[37,38] The effect of source temperature, rate of evaporation, and residual gas pressure has also been studied by determining the infrared absorption characteristics.[39,40] Films evaporated at a fast rate (375 Å sec^{-1}) and a pressure of less than 10^{-4} Torr show one strong absorption band at 10 μm and the spectrum of SiO films. The spectrum of SiO_2 films, obtained at slow evaporating conditions, shows a strong band at 9.2 μm, a moderate band at 12.5 μm, and a distinct shoulder in the 8.9-μm region. Films evaporated at intermediate rates indicate weak absorption at 12.5 μm due to SiO$_2$, a moderate-intensity band near 11.5 μm, and a strong absorption at 9.6 μm, which have been attributed to the presence of Si_2O_3.[37,39,40] Deposition on substrates heated to 400°C during deposition can cause decomposition of SiO to form Si, Si_2O_3, and SiO_2. The presence of Si in the film then results in a refractive index which is greater than that of pure SiO, which is 1.996.

The work of Hass[27] on the optical properties of evaporated silicon oxide films led Siddall[16] to study the influence of deposition rate on the electrical properties, and to use SiO in thin film capacitors. The dielectric constant and the dissipation factor have been shown to decrease with decreasing rate and increasing oxygen pressure.[16,42] More recent studies of the effect of the deposition parameters on the dielectric properties of silicon oxide films have revealed that these properties are a unique function of the ratio of incidence rates of O_2 and SiO at the substrate independent of the absolute values of pressure or rate from 2×10^{-5} to 2×10^{-4} Torr of oxygen and for deposition rates between 10 and 110 Å sec^{-1}.[41] Figure 7 shows how the dielectric constant of evaporated silicon oxide films depends on N, the number of silicon monoxide molecules reaching the substrate, divided by the corresponding number of oxygen gas molecules multiplied by K (see also Fig. 5). For $N > 1$ the value of the dielectric constant of the film is about 5, somewhat lower than the value for bulk silicon monoxide; whereas for $N < 0.1$ the value for the film is about 3.8, which corresponds to that of silicon dioxide. In the region between these two values the dielectric constant depends linearly on log N. The loss factor of the samples was also found to increase with N according to log tan δ $\propto$ log N within $10^{-2} < N < 1$.[34] Another property of the silicon oxide film which depends on the degree of oxidation is the conductivity. Using the optical properties as a criterion for the composition of SiO_x films, Johansen[38] found that the conductivity varies over five decades, from 10^{-11} ohm-cm^{-1} for $SiO_{1.5}$ to less than 10^{-16} ohm-cm^{-1} for SiO_2 films.

The effect of the substrate material and the electrode metals on yield of short-free capacitors has been discussed in detail.[7,18,43] It is well established that the best yields are obtained on smooth glass or glazed ceramic substrates with aluminum electrodes.

Aluminum can be evaporated at relatively low temperatures, and its atoms have a low surface mobility. Silicon oxide film capacitors with aluminum electrodes, however, need protection against humidity because of possible corrosion of aluminum under high-humidity conditions.[44]   The number of shorted capacitors can also be reduced by evaporating a 2-μm layer of SiO before the lower metal electrode is deposited, because such a layer covers rough areas which might be present even on glass substrates.[43]

In general, capacitors with dielectric films of a composition near SiO have high leakage currents, high dissipation factors between 0.05 and 0.1, and large values for the temperature coefficient of capacitance (TCC) in the order of 1000 ppm/°C. Lower values for the leakage current, dissipation factor, and TCC can be obtained by either evaporating the silicon oxide films at lower rates or heat-treating the dielectric films in air at temperatures between 250 and 450°C for periods of one to several hours. The resulting devices, which have silicon oxide films with a composition closer to $SiO_{1.5}$

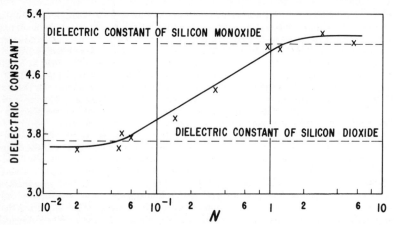

Fig. 7 Dielectric constant of evaporated silicon oxide films as a function of log N, in residual oxygen atmosphere.[34] N is defined as the number of silicon monoxide molecules divided by the number of residual gas molecules impinging on the substrate per unit area and time.

than to SiO, appear to be best suited for capacitor applications. Their dissipation factors are between 0.0005 and 0.005 at 1 kHz, and they have TCC values which are well below 100 ppm/°C between 25 and 125°C. Silicon oxide film capacitors of this kind have been tested under voltage and temperature stress for extended periods of time by St. John and Chaiken.[44]   Capacitors with a dielectric layer of 10,000 Å were tested at 47 V and 85°C. They did not show any failure in 1.2 $\times$ 10^6 component hours, whereas specimens with only 1,000 Å of silicon oxide, when tested at 10 V and 85°C, began to display capacitor failures after 0.75 $\times$ 10^6 component hours. The changes in capacitance during the 1.5-year test were in the order of 1.5%. The values for the capacitance density, 0.03 μF cm^{-2} for the 1,000-Å film and 0.003 μF cm^{-2} for the thicker silicon oxide film, are relatively small but silicon oxide film capacitors in the 10- to 1,000-pF range requiring an area well below 0.05 cm^2 are quite useful in microwave circuits. Attempts have been made to increase the capacitance density by depositing multilayer capacitors with more than one dielectric layer where the dielectric layers are separated by metal electrodes.[45,46]   But such structures have not been used very often because of high fabrication cost and low yields. Multiple dielectrics consisting of 5,000 Å of silicon oxide and 1,000 Å of silicon dioxide, magnesium fluoride, or aluminum oxide have also been made in order to reduce the leakage current of single-layer silicon oxide capacitors to values below 1 nA.[47]

(2) Silicon Dioxide   Silicon dioxide films are of interest as diffusion masks and for the surface passivation of silicon devices.   But the films are also gaining in importance as insulating layers for conductor crossovers and as thin film capacitor dielectric. The techniques for the formation of SiO_2 films and the required substrate temperatures have been reviewed by Glang.[28] The substrate temperatures range from 1100°C used for thermal oxidation to 25°C for the anodic oxidation of silicon. For the silicon dioxide formation of films to be used in RC integrated-circuit applications only low-temperature methods are of interest since temperatures in excess of 500°C can easily damage finished devices or components which are already present on a substrate. Films produced by evaporating SiO at high oxygen pressures $(2 \times 10^{-3}$ Torr) and low rates are porous and are still not sufficiently oxidized to show the ultraviolet absorption spectrum of used silica. They require heat treatment at 600°C in air for 2 h to approach the composition of SiO_2.[40] Before heat treatment these evaporated films have unstable capacitor characteristics and low yields even with aluminum electrodes. Vacuum evaporation of SiO_2 from a heated crucible or with an electron gun is unsatisfactory because of the tendency of the oxide to dissociate. The most successful methods for depositing stress-free dense silicon dioxide films with properties which approach those of fused quartz are reactive sputtering of silicon in the presence of oxygen[48,49] or rf sputtering from a fused-quartz cathode.[50]

The refractive index of SiO_2 films deposited by reactive sputtering in argon with 0.1% oxygen at a rate of 100 Å min^{-1} was found to be 1.463. This value agrees well with that obtained for thermally grown SiO_2. A study of the properties of reactively sputtered SiO_2 films indicated that their physical properties as determined by etch rates and infrared spectra are dependent on oxygen pressure, substrate temperature, and deposition rate. Deposition rates above 250 Å min^{-1} result in porous SiO_2 films which contain large amounts of water. Increased porosity can also be observed for substrate temperatures below 400°C and high partial oxygen pressures. It has been speculated that the porosity is caused by incomplete polymerization of the SiO_4 tetrahedra.[49] SiO_2 films used in capacitor structures were deposited by sputtering in pure oxygen atmospheres at a rate of 40 Å min^{-1}. The dielectric constant of these films was 3.90 and the dissipation factor at 1 kHz was 0.0003. Both values agree quite well with those reported for bulk fused silica.[48] The temperature dependence of capacitance and dissipation factor was measured in the temperature range between -200 and $+300$°C. In the vicinity of 25°C the TCC was found to be less than 10 ppm/°C. But values for TCC increase considerably above 100°C depending on film thickness and substrate temperature. Breakdown fields for SiO_2-film capacitors in the order of 6×10^6 to 10×10^6 V cm^{-1} have been reported and capacitors stressed at about one-third of the breakdown field and 125°C have shown good reliability.[46]

The method of rf sputtering from a fused-quartz cathode has also been used successfully for the deposition of SiO_2 films. This technique, which is described in some detail in Chap. 4, consists of applying a high-frequency potential to a metal electrode behind the insulator.[50,51] Deposition rates varying from 100 Å min^{-1} at an electrode potential of 1 kV to over 1,000 Å min^{-1} at 3.5 kV have been reported,[51] but they are also strongly dependent upon substrate temperature. The rate for SiO_2 drops by 50% as the substrate temperature is raised from 100 to 500°C. The rf-sputtered SiO_2 films have excellent mechanical properties. The films are smooth and adhere well to the surface onto which they are deposited. The films are generally under compressive stress, which is a function of the substrate temperature and can be explained by the thermal mismatch between substrate and film. The composition of the rf-sputtered silica films is $SiO_{1.95}$, indicating that the oxygen deficiency is insignificant.[52] Good capacitor yields were reported for 2-μm-thick SiO_2 films sputtered on silicon wafers. Out of 100 capacitors with aluminum electrodes on a surface area of 0.114 cm^2 per capacitor only one was shorted and this has been attributed to the presence of dust particles prior to sputtering.[50] Other dielectric properties of rf-sputtered SiO_2 films like dielectric constant and dissipation factor are equal to the values reported for reactively sputtered SiO_2 films.

(3) Silicon Nitride (Si_3N_4) The recent interest in silicon nitride films is primarily due to their possible use as diffusion masks and for the passivation of planar semiconductor devices.[53] However, since the dielectric constant of bulk Si_3N_4 of 9.4 is higher than that of SiO ($\epsilon = 6$) and SiO_2 ($\epsilon = 3.8$), silicon nitride films are also of interest for thin film capacitor applications. The dielectric strength of Si_3N_4 films deposited at 1000°C by reacting $SiCl_4$ and NH_3 was found to be higher than that of either SiO or

SiO_2 films (see also Fig. 4), but the reaction temperature of 1000°C is certainly too high for the formation of films to be used in thin film integrated circuits. Preliminary results indicate that stoichiometric Si_3N_4 films can be deposited at low substrate temperatures using reactive dc sputtering[54,55] or rf sputtering[55] from polycrystalline Si cathodes in a nitrogen atmosphere. Of the two methods rf sputtering appears to be preferable. Supported dc glow-discharge sputtering has also been used for the preparation of Si_3N_4 films,[54] but the film density is about 15 to 20% lower than the value reported for bulk silicon nitride, which is 3.44 g cm^{-3}, and the dielectric constant is only 6.4. The density of the rf-sputtered films is also lower than that of bulk silicon nitride, but it can be improved by raising the rf power density, and values as high as 8.5 have been reported for the dielectric constant of rf-sputtered Si_3N_4 films. For reactively sputtered films the dissipation factor measured at 1 kHz is below 1%. But certainly more work is required to establish the usefulness of Si_3N_4 films in thin film capacitor applications.

(4) Alkali Halides The values of the dielectric constant of bulk alkali halides range from 4.7 (RbBr) to 9.2 (LiF). They have low dissipation factors which are in the order of 0.1% and breakdown strengths between 0.5×10^6 and 3×10^6 V cm^{-1} depending on the size of the halogen atoms.[56] In contrast to silicon oxide and silicon nitride films which are amorphous, evaporated alkali-halide films are generally polycrystalline and have a tendency to recrystallize at low temperatures. The electrical properties of a large number of evaporated alkali-halide films have been studied by Weaver.[57] Observations of the electrical breakdown[19] were made in a vacuum of 1×10^{-5} Torr immediately after film deposition. The results indicated that the breakdown fields of NaCl and LiF films between 1,000 and 10,000 Å thick were lower than the corresponding bulk values of single crystals by an order of magnitude.[18] The lower values for the electrical-breakdown strength were attributed to the presence of voids and lattice defects in the polycrystalline structure of the films.

The losses of the evaporated alkali-halide films are generally an order of magnitude or more higher than those of bulk samples because of the same reasons. Aging of the films in vacuum at room temperature produces decreases in dissipation factor and capacitance due to recrystallization and structural changes as shown in Fig. 8 for MgF_2 films which were aged for about 1 week. The rise of the capacitance and tan δ curves toward low frequencies suggests a loss peak at about 1 kHz. From similar results obtained for other halide films, Weaver concluded that the magnitude of the losses depended upon the size of the respective ions. Larger losses were observed for

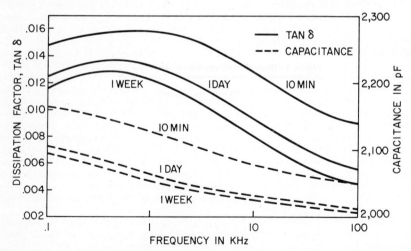

Fig. 8 Effect of vacuum aging on frequency dependence of capacitance and dissipation factor of evaporated MgF_2 film capacitors.[57]

the films with increasing anion size and/or decreasing cation size when they were measured immediately after deposition. The rate of aging as determined from dielectric loss and capacitance measurements was also found to be related to the ion size. The changes in electrical properties during aging can be explained qualitatively by the diffusion of cation and anion vacancies which migrate to grain boundaries. For example, blocking of cation vacancies at intercrystalline boundaries and at the electrodes has been responsible for two maxima in the tan δ vs. frequencies curves below 1 kHz for NaCl and NaBr films.[58]

Exposure of alkali-halide films to moisture after film deposition results in large increases of the values for dissipation factor and capacitance during the first 5 min. This increase in tan δ due to adsorbed moisture is often larger than that due to either normal aging or the basic loss mechanism. In the case of insoluble salts such as LiF the adsorption and desorption of water do not affect the aging process while in other cases such as NaCl films the presence of moisture accelerates the long-term recrystallization process, which results in a reduction of the losses at frequencies below 1 kHz. The decrease of the dissipation factor and the capacitance of NaCl films during the aging process due to adsorption of moisture is illustrated in Fig. 9. Table 3 sum-

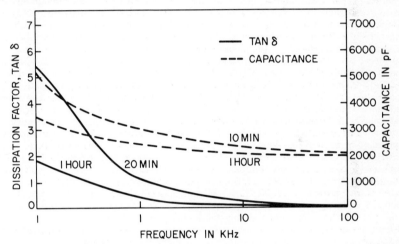

Fig. 9 Effect of humidity on frequency dependence of capacitance and dissipation factor of evaporated NaCl film capacitors.[57]

TABLE 3   Dielectric Properties of Metal Halides[57]

Type	Dielectric constant		tan δ*
	Bulk	Film*	
MgF₂.........	8.4	4.9	0.007
CaF₂.........	8.4	3.2	0.015
LiF...........	9.2	5.2	0.031
NaF..........	4.96	4.0	0.016
NaCl.........	5.77	5.1	0.18
NaBr.........	5.99	5.9	0.36
KBr..........	4.78	4.96	0.15
RbBr.........	4.7	4.65	0.30

* Measured at 100 Hz after aging in vacuum.

marizes the values for the dielectric constant and dissipation factor of the various metal halides found by Weaver,[57] but it is quite obvious that at present alkali-halide films do not have the required chemical and structural stability to be useful as capacitor dielectrics.

(5) Barium and Lead Titanate   The methods which have been used for the preparation of barium titanate films ($BaTiO_3$) include direct evaporation of the powder, flash evaporation, and simultaneous evaporation of barium and titanium in an oxygen atmosphere.   All three methods have been only moderately successful in obtaining titanate films with high dielectric constants possessing the perovskite structure.   For example, the evaporation of barium titanate powder from a tungsten filament onto platinum substrates at oxygen pressures below 5×10^{-5} Torr resulted in decomposition of the source material into BaO and TiO_2 during evaporation.   In addition, a reduction of TiO_2 to conducting titanium oxides occurred.   Firing of the deposit for extended periods at temperatures above 1100°C in air was necessary to form barium titanate layers after evaporation of the barium titanate paste.   After firing, the films have a dielectric constant of 240 and a dissipation factor of 5% at 1 kHz.[59]   X-ray diffraction analysis showed that these films had predominantly the bulk tetragonal structure with a small amount of the hexagonal phase present.   Addition of TiO_2 to the source material in order to compensate for the preferential volatilization of the BaO during evaporation as well as high-temperature firing after deposition results in films which have the stoichiometric Ba/Ti ratio.   But besides tetragonal $BaTiO_3$ several other titanate and oxide phases were found to be present as well.[60]

Another technique which permits better control of the stoichiometry of the barium titanate films is that of flash evaporation,[61] where small particles of $BaTiO_3$ are dropped onto a tungsten strip heated to about 2200°C.   Moll[61] reported for flash-evaporated $BaSrTiO_3$ films between 1,000 and 6,000 Å thick a dielectric constant of about 1,500 at 1 GHz.   In an attempt to promote single-crystalline formation an electric field of 8 kV cm^{-1} and a substrate temperature of 200°C were applied.   Other investigators[62,63] found that substrate temperatures of at least 400°C were needed for the deposition of crystalline films.   The dielectric constant of films evaporated onto substrates heated to 400°C was 440 and for substrates heated to 500°C it was 50. Films evaporated from an iridium strip showed the cubic structure.   Cubic films were also obtained by Sekine and Toyoda,[64] and a film about 2.7 μm thick, deposited at 700°C and heated at 1200°C in air after deposition, had a dielectric constant of 1,200.   For flash-evaporated films, contamination of the films from the source was found to be high, about 10% for tungsten and 2% for iridium.[63]

Simultaneous evaporation by electron-beam-heating titanium and barium in an oxygen atmosphere has also been explored as a technique for the formation of $BaTiO_3$ films.[65]   Films between 1,000 and 13,000 Å thick had the cubic structure and dielectric constants above 1,000.   They were deposited onto Pt-Rh substrates which had to be heated to over 600°C to form crystalline $BaTiO_3$ films.   Films formed by simultaneous evaporation on unheated substrates were amorphous and had dielectric constants of less than 100.

Only films which are well crystallized, with crystallites several thousand angstroms in diameter, show the maxima for capacitance and dissipation factor at 120°C[59,65] and the ferroelectric behavior to be expected for tetragonal $BaTiO_3$.[59]

Thick films in the order of several micrometers have been produced by flame spraying,[66] by pulling foil from the melt,[67] and by a powder-melting technique.[68] Single-crystal films on platinum substrates between 1 and 100 μm thick which were grown from the melt possessed a domain structure below 120°C.   Their dielectric constant measured as a function of the electric field depended strongly upon the thickness of the film, and the field at which nonlinear polarization is observed increases sharply as the film thickness is reduced.[68]

The formation of $PbTiO_3$ films by reactive sputtering from a lead-titanium cathode in an argon-oxygen atmosphere has been reported by Bickley and Campbell.[69]   Control of the area of the titanium cathode which was coated with lead, and rotating the cathode during sputtering resulted in films with a Pb/Ti ratio of unity which after annealing at 500°C in air showed the perovskite structure, but the dielectric constant of these films was only about 33.

Thin films of $PbTiO_3$ with dielectric constants between 200 and 400 have been prepared by simultaneous evaporation of lead oxide and titanium oxide onto Pt-Rh foil and passivated silicon substrates. These films have a tetragonal structure with crystallites about 3,000 Å in diameter, but their electrical properties depended on the composition.[70]

b. Metal Oxide Dielectrics

Oxide coatings can be grown on a large number of metals by various processes. Such coatings are not only useful for protecting the underlying metal from atmospheric oxidation or chemical corrosion but they are also of importance for their electrical properties. For example, the high dielectric constant and high dielectric strength of anodic oxide films on aluminum and tantalum have been utilized for decades for the fabrication of wet electrolytic capacitors.[71] Solid electrolytic tantalum capacitors[72] and more recently tantalum-film capacitors[73] based on anodic tantalum oxide have proved their usefulness as electronic components. Only tantalum and aluminum are at present used for capacitor anodes, but other metals like Ti, Zr, Hf, W, Nb, Y, and La have also been investigated for this application.

The oxide layer can be grown by forming metal ions and oxygen ions which combine either at the metal surface, inside the oxide, or at the oxide-oxygen interface, depending on which ion migrates through the oxide. The energy barriers which impede the migration of the anions and/or cations through the oxide can be lowered by application of an electric field in the order of 10^7 V cm^{-1} while the probability for crossing the barrier can be enhanced by supplying sufficient thermal energy. Thus anodic oxidation in a wet electrolyte, plasma oxidation, and thermal oxidation have been successfully used for the growth of coherent and uniform oxide layers between 50 and several thousand angstroms thick. It is also possible to form metal oxide films by reactive sputtering, evaporation of the oxide, or reactive evaporation of the metal.

(1) **Methods of Oxide Formation** (a) *Anodic Oxidation.* For a limited number of metals the application of a potential between two metal electrodes in an electrolytic cell results in oxide growth on the metal anode. The metals which most readily form a coherent, uniform film of oxide[71] are listed in Table 4. Tantalum is an example of a

TABLE 4 Oxide-forming Metals

Metal	Anodic oxide	Dielectric constant of anodic oxide
Aluminum	Al_2O_3	10
Antimony	Sb_2O_3 or Sb_2O_4	~20
Bismuth	Bi_2O_3	18
Hafnium	HfO_2	40
Niobium	Nb_2O_5	41
Tantalum	Ta_2O_5	27
Tungsten	WO_3	42
Yttrium	Y_2O_3	12
Zirconium	ZrO_2	24.7

metal which forms uniform oxide films in aqueous solutions of almost any electrolyte, except in hydrofluoric acid, which dissolves the oxide. But most of the metals can be anodized in only a few electrolytes, or within a limited range of pH values. This is the case for aluminum, which forms soluble products in electrolytes with a pH above 8, and anodization of aluminum in sulfuric acid results in a composite film consisting of a dense oxide layer on the metal surface and a porous layer between the dense layer and the electrolyte. Such composite films are widely used for decorative and protective purposes, but their insulation resistance is rather low. Dense, nonporous oxide layers on aluminum which are useful in capacitors can be formed in buffered solutions of tartrates, citrates, phosphates, or carbonates.

The oxide film which forms on tantalum during anodic oxidation is a product of the following reaction, which takes place at the anode of the electrolytic cell:

$$2Ta^{5+} + 10OH^- \rightarrow Ta_2O_5 + 5H_2O \tag{8}$$

This equation describes the formation of one molecule of oxide. The rate-controlling process for anodization of tantalum and other metals is the migration of ions within the oxide films. Empirically anodization current and field strength were found to be related by the following expression:[71]

$$J = Ae^{BE}$$

where A and B are constants, E is the field strength, and J is the current density. A number of theoretical expressions which have been derived to relate the ionic current and field strength are based on the assumption that only one type of ion is mobile and that the number of charge carriers is independent of field strength. The results of more recent experimental studies, however, strongly suggest that both cations and anions move during the anodization of tantalum, niobium, aluminum, and tungsten.[74] Detailed reviews of the models for ionic conduction through anodic oxides have been published by Young,[71] Vermilyea,[75] and more recently by Goruk et al.[76]

At high fields the current density during the anodization of tantalum depends exponentially on the field strength, and the rate of oxide growth is proportional to the density of the current. Since the current efficiency for the oxide growth is about 100%, Faraday's law can be used for determining the amount of oxide formed from the quantity of electricity passed through the electrolytic cell.[71] For example, at a current density of 1×10^{-3} A cm^{-2} the growth rate is about 5.7 Å s^{-1}. At this current density the measured voltage rise is 0.33 V s^{-1}, resulting in a growth constant of 17.3 Å V^{-1} for the formation of tantalum oxide.

The growth of the anodic oxide on tantalum and other metals is limited by scintillation and local breakdown. The high temperatures at points in the oxide during arcing lead to the formation of small crystallites of β-Ta$_2$O$_5$ at the metal-oxide interface. Evidence for this behavior is a change in appearance from the bright uniform interference color of the amorphous oxides to a dull gray coloration. The potential at which sparking and breakdown in the oxide occur depends on the structure and purity of the tantalum, and on the nature and concentration of the electrolyte. From the results shown in Table 5 it appears that the highest forming voltages can be

TABLE 5   Scintillation Voltages of Ta$_2$O$_5$ Formed in Various Electrolytes[77]

Electrolyte	Scintillation voltages of sputtered tantalum
Conc. HNO$_3$	145
10% H$_2$SO$_4$	130
Conc. H$_2$SO$_4$	50
10% H$_3$PO$_4$	250
Conc. H$_3$PO$_4$	65
10% NaH$_2$PO$_4$	100–140
10% KNO$_3$	150
10% citric acid	250–300
0.01% citric acid	350–400
10% HAc	>300

expected for electrolytes with a low conductivity such as dilute citric and acetic acid solutions.[77]

Thus the wrong choice for the electrolyte solution can lead to low breakdown voltages or it may affect the formation of anodic films by dissolving some of the reaction products. Solutions containing fluoride ions affect the usually very strong adhesion of the oxide layer on tantalum and, to a somewhat lesser degree, on niobium.[78] But the quality of the anodically formed oxide films depends mainly on the cleanliness, purity, and character of the metal surface to be oxidized. For example, areas of well-crystallized oxide are formed under certain conditions on tantalum foil.[79] Defects in the oxide may consist of cracks, local nonstoichiometry, or impurities which probably affect the electronic conduction and the breakdown characteristics of the oxide. In fact, decoration of the oxide surface at flaws can be seen if the oxide-coated metal is made cathodic in a solution containing metal ions, indicating that there are weak spots in the oxide which have sufficiently high electronic conduction to permit neutralization of the metal ions and the deposition of metal atoms from the solution.[71] The origin of the imperfections in the oxide and their relationship to the metal surface are still largely unexplained. In spite of the fact that the anodic oxides occupy a larger volume than the metal from which they are formed, the oxide films are not under high compressive stress, as might be expected. Studies of the stresses in anodic films grown on tantalum foil indicate that they are under tensile stress.[80] It was also found that anodic oxides formed on tantalum, zirconium, and aluminum wire have a certain degree of ductility and that they do not crack immediately during elongation of the wire. Anodic tantalum oxide shows the highest ductility, and some of the films could be stretched by as much as 50% before fracture of the oxide was observed.[81]

The formation of tantalum oxide as a capacitor dielectric usually consists of anodization at constant current density until a preselected voltage is reached. During this period the oxide increases at a constant rate according to the increase in potential necessary to maintain the current. The anodization is continued at constant voltage until the ion current has decayed to a sufficiently low level of about 1% of its original value. The addition of the second anodization step at constant voltage yields better-quality dielectrics, and it permits better control of the dielectric thickness because the rate of growth decreases rapidly. For example, during the first stage at a constant current density of 10^{-3} A cm^{-2} the growth rate is about 340 Å min^{-1}; while after reaching the formation voltage the rate decreases with time and has dropped to less than 2 Å min^{-1} during the first 30 min of the second stage of anodization. The change of the anodization current and voltage during the formation of tantalum oxide is shown in Fig. 10. Sputtered hafnium films show a somewhat different anodization behavior.

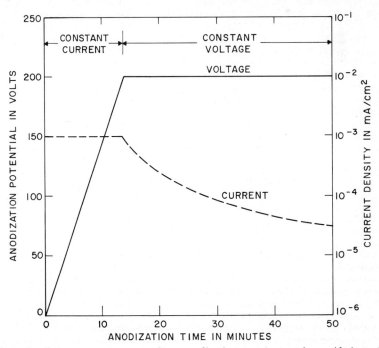

Fig. 10 Constant current–constant voltage anodization steps for tantalum oxide formation.

After reaching the anodization voltage, the current decreases at first but it begins to increase again after 30 min to 1 h.[1] It is not clear whether this current increase is electronic or ionic in nature, but it appears well established that for the fabrication of a high-quality capacitor dielectric the anodization has to be discontinued when the anodization current reaches its lowest value. During anodization of tantalum such an increase in current has been observed only when tantalum foil is anodized in certain electrolytes (for example, in nitric acid[79]), leading to the formation of crystalline oxide which replaces the amorphous oxide during the later stages of anodization.

(b) *Plasma Oxidation.* Oxide films on tantalum, aluminum, and other metals have also been prepared in an oxygen plasma. In this technique the positively biased sample which is to be oxidized is immersed in a low-pressure oxygen plasma. The plasma is generated by a low-voltage discharge between two metal electrodes. The potential between the two electrodes is typically 0.5 to 1 kV, resulting in a plasma current of about 20 mA at an oxygen pressure of 50 mTorr.[82,83] Even if no direct electrical connection to the film is made, a thin layer of oxide up to a few hundred angstroms is formed. After this initial period of rapid oxide growth the thickness of the oxide is proportional to the positive bias voltage, and the current flowing through the sample shows the same decay as a function of time as for films formed in a wet electrolyte after reaching a preset voltage. Jennings et al.[84] used an electrode arrangement in a magnetic field which improved the efficiency of the generation of negatively charged oxygen ions, and heating effects due to electron and positive-ion bombardment of the sample to be oxidized were reduced by placing it outside of the plasma.

Tantalum oxide films grown at 1 and 2 mA cm^{-2} and at bias voltages ranging from 75 to 250 V revealed a linear dependence of film thickness on bias voltage. This dependence was found to hold over a wide range of oxide film thicknesses from a few hundred to over 3,000 Å. The physical properties, density, and dielectric constant of plasma-oxidized films are comparable with those of oxide films formed in a wet electrolyte, but their dielectric strength is somewhat lower and their dissipation factors are about an order of magnitude higher.[84] The growth rate for the oxide formation is much lower for plasma-oxidized films. This has been attributed to a lower current efficiency. For example, the current efficiency for plasma oxidation of tantalum is only between 10 and 20% of the value commonly found for wet anodization.[84] Plasma-oxidized films are generally purer than films formed by wet anodization which frequently contain residue from the electrolyte.[85,86] The method of plasma oxidation has also been found useful for the formation of oxide layers on metals like Ge and Be which are difficult to anodize in liquid electrolytes because their oxides tend to dissolve in the electrolyte.[82]

(c) *Reactive Evaporation.* This technique is described in more detail in Chap. 1, and some of the difficulties encountered in the reactive evaporation of metal oxide films have been discussed already in this chapter in connection with the formation of titanate films. A proper balance between deposition rate and oxygen pressure in the vacuum system has to be maintained for the deposition of oxide films which have the desired stoichiometric composition. For example, insulating aluminum oxide films prepared by evaporating aluminum at a rate of 1 Å s^{-1} in an oxygen atmosphere of 10^{-3} Torr had a dielectric constant of only 6 when measured in vacuum immediately after film deposition. This low value has been attributed to the presence of suboxides which have a lower dielectric constant.[87] After exposure to air the dielectric constant increased to 11, and it was suggested by DaSilva and White that oxidation from AlO to Al_2O_3 was responsible for this increase. Aluminum oxide films with a dielectric constant of 8.6, which is only 3% less than the value reported for bulk alumina, were evaporated by electron bombardment of a sapphire rod at a deposition rate of 40 Å s^{-1} in an oxygen atmosphere of about 10^{-4} Torr.[88] After exposure to air these films showed no change of the dielectric constant, indicating that the dissociation products which form during the evaporation process recombine on the substrate to form Al_2O_3. But in general, the technique of reactive evaporation is limited by the fact that the oxygen pressure cannot be rasied much above 1 mTorr, which is below the dissociation pressure of a large number of oxides at the temperature of deposition.

(d) *Reactive Sputtering.* The method of reactive sputtering in a dc glow discharge, which has been described in Chap. 4, is more appropriate for the deposition of metal oxides because sufficiently high oxygen pressures, up to 100 mTorr, can be used.[89] By adding controlled amounts of oxygen to an argon sputtering atmosphere the degree of oxidation of the sputtered films can be controlled within wide limits. This technique has been applied by Holland et al. for the preparation of semiconducting cadmium oxide.[90] In sputtering the more reactive metals, among them tantalum, niobium, and titanium, insulating oxide films can be formed by adding only between 1 and 5 % of oxygen to the argon flow.[2,91]

The sputtering conditions and some of the dielectric properties of reactively sputtered oxide films are listed in Table 6. The oxide films between 1,000 and 2,000 Å thick were sputtered onto aluminum and in some cases onto gold electrodes.[2] Except for hafnium and yttrium oxide no difference in the dielectric strength of the test samples was reported whether aluminum or gold films were used for the electrodes. In the case of the hafnium oxide, gold electrodes resulted in a higher dielectric strength, and for reactively sputtered yttrium oxide, aluminum films were found to give better results.

TABLE 6 Sputtering Conditions and Properties of Reactively Sputtered Metal Oxide Dielectrics[2]

Metal oxide	Sputter voltage, kV	Oxygen conc., % of argon Pr.	Deposition rate, Å min^{-1}	Dielectric constant	Dissipation factor at 1 kHz	Dielectric strength, V cm^{-1} × 10^6
HfO$_2$...........	2.2	2.0	200	24.5	0.0067	2.63
La$_2$O$_3$...........	2.0	1.1	175	30	0.0055	1.87
Nb$_2$O$_5$..........	2.0	0.9	30	38	0.018	
Ta$_2$O$_5$..........	3.0	1.2	240	26	0.010	1.36
TiO$_2$...........	3.0	1.0	20	60	0.010	0.24
Y$_2$O$_3$...........	1.5	2.1	175	11.5	0.0045	3.56
ZrO$_2$...........	2.4	1.9	80	5.7	0.0051	1.46

(2) Tantalum Oxide Among the metal oxides anodically formed tantalum oxide is the most thoroughly studied dielectric which is available for thin film capacitor applications.[71] Anodic tantalum oxide has gained such an importance because it is an integral part of the tantalum-film-circuit technology which permits the use of the same basic material for capacitors, resistors, and interconnections.[7,92] Tantalum oxide also has a high dielectric constant (Table 4), and when formed by anodic oxidation the films are amorphous, pore-free, and chemically resistant. It was first reported by Berry and Sloan[73] that anodic films formed on sputtered tantalum films have a high dielectric strength corresponding more closely to their growth under high fields than those formed on bulk tantalum. The reason for this is that sputtered and also evaporated films are cleaner and have a smoother surface, and the impurities are more uniformly distributed throughout the film provided that the tantalum films are deposited on smooth glass or glazed ceramic substrates.

The preparation of tantalum-film capacitors and their properties have been discussed in detail by a number of authors.[7,92,93] A 4,000-Å tantalum film is usually sputtered in an argon atmosphere. The capacitor area is defined either by sputtering the tantalum through a mechanical mask or by photoetching. Part of the tantalum film is then anodically oxidized in a dilute aqueous electrolyte in the manner described before. Frequently the films are also electrochemically etched at 70 % of the anodization voltage in a LiCl or AlCl$_3$ solution. This etch removes metal underneath defects in the anodic film.[7] The oxide is then re-formed for 30 min in the anodizing bath, resulting in capacitors with lower anodic leakage currents and fewer shorts. Another technique for removing weak spots in the oxide consists of reversing the anodizing voltage for a few seconds during the final part of the anodization.[94] Finally, the capacitor structure is completed by evaporating the top electrode onto the anodized portion of the tantalum film, normally a gold layer about 2,000 Å thick. Figure 11 illustrates how the capacitance per unit area of the anodic film formed on various types of sputtered tantalum films depends on the formation voltage and film thickness, assuming a growth constant of about 17.3 Å V^{-1}.[95] The dissipation factor of tantalum-film capacitors is between 0.002 and 0.01 for frequencies between 0.1 and 10 kHz, and the capacitance changes by about 1% in this frequency range. The temperature coefficient of capacitance between 0 and 100°C is about 200 ppm/°C.

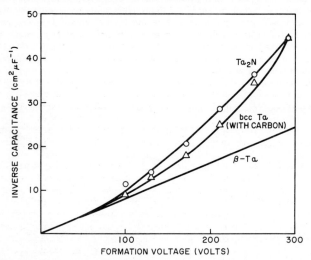

Fig. 11 Inverse capacitance density of anodic films formed on various types of sputtered tantalum as a function of anodization voltage.[97]

The electronic leakage current has been used as a criterion for the quality of the tantalum-film capacitors. Units are acceptable if their leakage current is below 1 μA μF^{-1} at about 60% of the anodization voltage.[7,92,93] Based on this test, yields of 90% and higher have been obtained for 0.01-μF capacitors (anodized at 130 V), and their performance under load has been evaluated.[93,96,97] The results indicated that the reliability of tantalum-film capacitors depends on the structure, film stresses, and composition of the sputtered tantalum films. Vratny et al. reported that capacitors prepared at 130 V from β-tantalum,[96,98] a high-resistivity material which has been discussed in Chaps. 4 and 18, showed fewer failures during a long-term life test at 50 V and 85°C than capacitors formed on pure bcc tantalum films. This has been attributed to the difference in the mechanical properties and the size of the crystallites of the two types of tantalum films which are used for the formation of the anodic layers. β-tantalum films generally show a lower incidence of mechanical faults and have smoother surface characteristics than pure bcc tantalum films. Extended life testing of 0.01-μF tantalum-film capacitors anodized at 130 V revealed that there were no failures at 30 V and 85°C during 2.5 $\times$ 10^5 component hours. Gerstenberg et al.[97] found that reactively sputtered Ta$_2$N films and bcc tantalum films with about 10 to 15 atomic % of carbon interstitially dissolved,[99] when used as capacitor anodes, are mechanically more stable than either pure bcc tantalum or β-tantalum, permitting the formation of anodic films with a low density of defects or flaws. The use of these reactively sputtered tantalum films resulted in an improvement in the yield of non-shorted units, and it reduced the number of failures during accelerated life testing.[97] The capacitance densities of the anodic films formed on reactively sputtered tantalum, however, are somewhat lower than those of Ta_2O_5 layers formed at the corresponding voltage on pure tantalum anodes as shown in Fig. 11.   In the case of Ta_2N film this decrease in capacitance density is due to a lower dielectric constant for the oxynitride, which is formed anodically.

The conduction characteristics of capacitors based on anodic oxide films generally depend on the polarity of the applied voltage.   Capacitors formed on pure tantalum films, for example, have breakdown voltages which are about 80 to 90% of the anodization voltage when the tantalum electrode is positive (anodic polarization), while upon cathodic polarization this value is only 5 to 10% of the anodization voltage. There is evidence that the low cathodic breakdown voltage is related to flaws in the oxide which have different electrical characteristics from the bulk of the oxide.[100] Tantalum-film capacitors with almost symmetrical conduction properties have been prepared from tantalum films to which nitrogen, oxygen, or carbon had been added during the sputtering process.   Addition of about 1 atomic % molybdenum to the tantalum films by cosputtering techniques[101] also results in capacitors with nonpolar conduction characteristics.

Capacitors based on tantalum oxide formed by plasma oxidation or reactive sputtering have not been evaluated as extensively as units using anodic oxide.   Preliminary results suggested that the nonshort yield of tantalum-film capacitors prepared by all three techniques is comparable, but differences were found for the dielectric strength, the dissipation factor, and a number of other properties, indicating that the films produced by aqueous anodization are superior to those formed by the other two methods.[102]

(3) Other Metal Oxides   A large number of other metal oxides, especially those of refractory metals, have also been investigated.   The methods used for their formation were anodic oxidation and reactive sputtering, and some of the properties of the oxides are summarized in Tables 4 and 6.   Tungsten, titanium, and hafnium have been considered for thin film component applications.   Bowermann[103] has used sputtered tungsten films for the fabrication of resistors and capacitors, but the results found for the anodized tungsten oxide layers were somewhat variable and the dissipation factor of about 0.6 was quite high.   Chemically deposited and anodized titanium films have been prepared by Fuller[104] on ceramic substrates.   This technique was used for the formation of capacitors and resistors on the same substrate.   Huber[105] has deposited titanium films by evaporation on glass substrates.   Wet anodization at room temperature was used for the formation of oxide layers for capacitors and diodes and for adjusting resistors to value.   The anodization constant of 22 Å V^{-1} agreed quite well with results reported earlier by Hass,[106] but the dielectric constant of the anodically formed TiO_2 layers of 40[105,107] was considerably lower than the values reported for bulk TiO_2 with the exception of anatase, a crystalline modification of TiO_2.   The various modifications of bulk TiO_2 have dielectric constants varying from 117 for polycrystalline rutile and 78 for brookite to 31 for anatase.   The dissipation factor of anodized TiO_2 layers increases from 2 to about 5% between 0.1 and 100 kHz while the capacitance decreases by about 10%.[105]   The temperature coefficient of capacitance is about 300 ppm/°C between ±200°C.   The conduction of titanium-film capacitors shows a large asymmetry, and no information is available about the reliability of titanium-film capacitors during accelerated life tests.

More recent results reported by Huber[1] indicated that HfO_2 formed by anodizing sputtered hafnium films with about 2 atomic % zirconium is also of interest for thin film capacitor applications.   The hafnium-film capacitors have almost nonpolar conduction characteristics, which have been attributed to the small amount of ZrO_2 present in the anodic film, and breakdown voltages which are between 80 and 90% of the formation voltage.   Capacitance densities from 0.54 to 0.082 $\mu F\ cm^{-2}$ have been reported for anodization voltages between 10 and 150 V.   The change in capacitance and dissipation factor as a function of frequency between 0.1 and 100 kHz was found to be quite low.   The temperature coefficient of capacitance is 125 ppm/°C in the temperature range from -196 to 350°C.   Only a few samples formed at 200 V were subjected to an accelerated life test at 40 V and 125°C.   None of the units failed but capacitance changes in the order of 1% were observed.

The preparation of anodized aluminum-film capacitors has been described by Martin,[108] who studied the influence of the deposition conditions during the evaporation of aluminum films on polarity effects, capacitance, dissipation factor, and capacitor yield. Careful control of substrate temperature, residual gas pressure, source-to-substrate distance, and thickness of the base layer of aluminum resulted in good capacitor yields for units ranging in size from 100 pf to 1 μF. The dissipation factor of the aluminum-film capacitors was between 0.4 and 1%, and the temperature dependence of capacitance was found to be linear between -60 and $+180°C$ with a value of 390 ppm/°C for the TCC. Storage tests at elevated temperatures revealed that these capacitors remain unaffected by treatment at temperatures as high as 350°C. However, it appears that aluminum-oxide-film capacitors have to be encapsulated in order to withstand operation under high-humidity conditions.

c. Organic Dielectrics

Although most of the development effort on thin film dielectrics has centered around inorganic materials, organic polymers offer highly desirable electrical and mechanical properties which make them suitable as capacitor dielectrics. The desirable properties of organic polymers which are hard to obtain in inorganic materials and ceramics include mechanical flexibility, which makes winding possible, and the availability of such techniques as spraying, dipping, or casting for making films in the range of 0.1 to 2 mils. In addition, they show high dielectric strength and low dissipation factors, and among the organic materials a wide variety of electrical characteristics is available.[109] Table 7 lists the electrical and mechanical properties

TABLE 7 Properties of Discrete Polymer-film Capacitors[110]

Property	Polyethylene terephthalate (Mylar*)	Polystyrene	Polytetra-fluoroethylene (Teflon*)	Poly-carbonates
ϵ at 1 kHz	3.23	2.57	2.0	3.10
tan δ at 1 kHz	0.004	0.0001	0.0002	0.001
Dielectric strength, V cm^{-1}	4×10^6	2×10^6	$>4 \times 10^5$	1.2×10^6
DC insulation resistance, ohm-F	1×10^5	1×10^6	2×10^6	1×10^5
Tensile strength, psi	25×10^3	10×10^3	2.5×10^3	2×10^4
Modulus of elasticity, psi	5×10^5	5×10^5	5×10^4	3×10^5
Density, g cm^{-3}	1.38	1.05	2.1–2.3	1.20
Approx dielectric film thickness, μm	4–5	7.5–12.5	50	3–4

* Du Pont trade name.

of a number of polymers which are commercially used as capacitor dielectrics in discrete capacitors.[110]

During the last decade a number of other film-forming techniques have been investigated which permit the deposition of much thinner organic films in the order of a few hundred to about 20,000 Å. Some of these techniques have shown promise for the deposition of thin organic films as a thin film capacitor dielectric and in other microelectronics applications.[111]

(1) Methods of Deposition In contrast to inorganic materials, most organic materials cannot be thermally evaporated in vacuum and condensed on the substrate surface because they decompose below their vaporization temperature. Thus polymer insulating films are predominantly prepared by the interaction of a monomeric or low-molecular-weight polymeric species in gaseous form with a solid surface in the presence of some form of energy. These processes can be classified in the following ways:[111]

1. Gaseous discharge
2. Electron bombardment
3. Photolysis
4. Pyrolysis

Polymer-film deposition by glow discharge appears to be the most widely used process, and in its simplest form it employs a pair of parallel electrodes. A discharge in the monomer vapor of about 10^{-1} Torr leads to polymer-film formation on substrates which are placed on one of the electrodes. The electrical properties of a large number of thin polymer films made by this technique have been described by Bradley and Hammes.[112] Stuart[113] deposited polymerized styrene films on aluminum foil in a glow discharge generated by a high-frequency (180 kHz) voltage between two electrodes at a styrene vapor pressure of about 1 Torr. The film thickness used was about 10,000 Å, and uniformity of film thickness within 10 % was achieved by moving the aluminum foil through the discharge at a constant rate. The completed capacitors had dissipation factors between 0.001 and 0.002 in the frequency range 0.1 to 100 kHz, and the dielectric constant of the polystyrene films of 2.6 to 2.7 was in good agreement with the value of the bulk material. The use of masks for depositing butadiene and vinyl chloride films in selected areas on a substrate surface has been explored by DaSilva et al.,[114] who injected electrons for sustaining a discharge in the monomer vapor at low pressures, but substrate heating due to the radiation from the cathode and shadowing still proved to be a problem. Both these problems were minimized in a technique described by Connell and Gregor[115] which employed a magnetic field for confining the glow discharge in a tube using rf electrodeless excitation. The authors showed that insulating films based on styrene, divinylbenzene, and other organic molecules could be obtained which adhered well to a large variety of substrates. In addition, the shadowing was reduced to less than 0.002 cm.

The conversion of organic molecules adsorbed on a surface to solid polymer films can also be accomplished by low-energy electron bombardment. It has been shown by Christy[116] that adsorbed pump-oil vapor of DC704 can be polymerized by electron bombardment and form films with good insulating properties. The rate of growth of the films was studied[116,117] as a function of the oil (siloxane) vapor pressure, the accelerating voltage for the electrons and current density, substrate potential, and temperature. The highest values for dielectric strength and resistivity, 10^6 V cm^{-1} and 10^{15} ohm-cm, respectively, were obtained for an electron-beam energy range of 50 to 300 eV and a substrate potential of 120 V. Under these conditions, the films were deposited at a rate of 100 Å min^{-1} and their dielectric constant was between 2.5 and 3.0; but after exposure to atmosphere the polymerized siloxane films still show increases in capacitance due to saturation of phenyl radicals formed during the polymerization process.[117] The presence of unpolymerized material and free radicals was greatly reduced by Brenneman and Gregor,[118] who prepared polymer films by electron bombardment of an evaporated epoxy resin. Films ranging from 200 to 10,000 Å were deposited on glass, aluminum, and cast epoxy substrates at rates between 100 and 900 Å min^{-1}. All the samples were free of shorts, and measurements of the dielectric properties gave the results listed in Table 8.

TABLE 8 Some Dielectric Properties of Polymerized Epoxy-resin Films at 23°C and 1 kHz

Dielectric film thickness, Å	Dielectric constant ϵ	Dissipation factor tan δ	Dielectric strength, V cm^{-1}	Breakdown voltage V
2,680	6.2	0.0070		
2,400	5.7	0.0063		
2,100	...	...	...	45
1,980	5.3	0.004	$\geq 1.9 \times 10^6$	
1,900	...	...	...	40
1,700	5.4	0.0062	$\geq 1.2 \times 10^6$	
1,630	...	...	...	25

The properties of polymer films formed by exposing substrates to ultraviolet light in the presence of vaporized butadiene, styrene, acrylonitrile, and methyl isopropenyl ketone have been studied by White.[119]   Butadiene polymer films had a dielectric constant of 2.65, dielectric strength of 5.10^6 V cm^{-1}, and dissipation factors of 0.2 to 1 % in the frequency range 100 Hz to 100 kHz.   As in the case of the electron-bombardment technique, selective coating of the substrate is possible by irradiating only that part of the substrate where film deposition is required.   Epoxy-resin films also reveal interesting properties when exposed to either ultraviolet light or electron bombardment.   Caswell and Budo[120] observed that evaporation of tin and indium resulted in coatings on exposed areas of the polymer film only.   This phenomenon has been explained by the creation of active surface sites which favor reaction and bonding between these sites and arriving atoms or molecules.   It would be of interest to explore whether this effect can be utilized for the generation of metallic thin film patterns.

In a more recent development[121,122] poly-p-xylene polymer films have been grown by a vacuum process without any external activation during the growth step.   These films possess a combination of very useful electrical and mechanical properties which will be discussed in more detail.

(2) Parylene   Poly-p-xylylene films are formed by evaporating di-p-xylylene at 150 to 250°C, converting the dimer to the monomer by pyrolysis between 500 and 750°C, and condensing the p-xylylene vapor on the substrate.[122]   The films consist of linear chains of about 5,000 molecules, and they can be grown to any desired thickness from less than 100 Å up to several hundred micrometers on a variety of substrate materials including glass, metals, and ceramics.   The films can be deposited in a conventional oil-diffusion pumped-vacuum system at pressures of less than 10^{-3} Torr.   Pattern generation of the dielectric films is possible by either contact masking or photolithography because dipolymerization occurs when the polymer film is irradiated with ultraviolet light in the presence of oxygen.   During irradiation a peroxide is formed which can be decomposed and removed from the irradiated area.   For contact masking line definitions of ± 25 μm have been achieved, but no information is available about the pattern definition which can be obtained for the photolithographic process.

The film-growth rate depends mainly upon the p-xylylene vapor pressure and the substrate temperature (which should be well below 100°C).   As shown in Fig. 12,

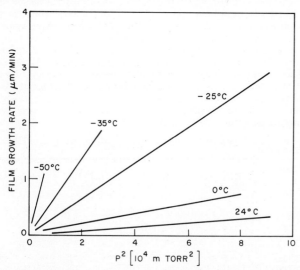

Fig. 12   Parylene polymer-film growth rate as a function of the square of p-xylylene pressure for different substrate temperatures.[122]

rates ranging from 0.01 to over 3 μm min^{-1} have been reported. Examination of the electrical properties of the Parylene films revealed that they are independent of film thickness from 0.1 to over 10 μm. The dielectric constant of Parylene is 2.65, and by varying the film thickness capacitance densities from 0.00025 to 0.025 μF cm^{-2} have been obtained. The dissipation factor of the dielectric is in the order of 0.01 to 0.1%, and dissipation factor and capacitance are essentially independent of frequency between 0.1 kHz and 1 MHz. Only the dielectric strength shows a change with film thickness which is plotted in Fig. 13. During the evaluation of the electrical

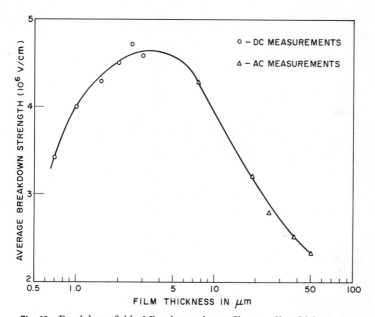

Fig. 13 Breakdown field of Parylene polymer films vs. film thickness.[122]

properties of arrays of test capacitors it was found that contamination from adsorbed gases or moisture resulted in leakage paths across the surface of the Parylene film. Increased electrode separation as well as an overcoating of Parylene helped to eliminate this problem.

3. DESIGN CONSIDERATIONS

The usefulness of the various types of insulating films for thin film capacitor applications is determined, to a large extent, by how well the properties and the geometrical dimensions of the films can be controlled during the formation process. In contrast to thin film resistors, which can be trimmed to value by a variety of techniques (see Chap. 18), the values of thin film capacitors cannot be changed conveniently after the capacitor structure is completed (Fig. 1). Thus one of the most important requirements for an insulating material to be utilized in thin film capacitors is good control of the capacitance density and of the lateral dimensions of the active area of the capacitor during the formation process. Other properties which affect capacitor design and performance are device yield and capacitor stability as a function of voltage, temperature, and frequency under various humidity conditions. But before some of these requirements are discussed in more detail a few special capacitor structures which have shown great promise will be described.

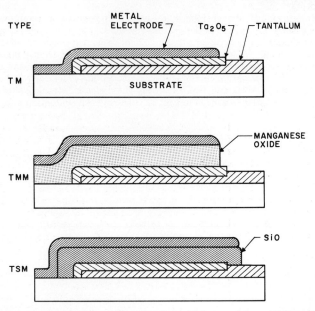

Fig. 14   Various types of tantalum-film capacitors: TM parent type, TMM with MnO_{2-x} layer between dielectric and top electrode,[123] and TSM duplex capacitor structure with silicon oxide–tantalum oxide dielectric.[129]

a. Special Capacitor Structures

(1) MnO_{2-x}-Ta_2O_5 Film Capacitors   The area of tantalum oxide and other thin film insulators which can be utilized in thin film capacitor structures is usually limited to about 1 cm^2 because of the presence of defects in the insulating layer.   McLean and Rosztoczy[123] found that the formation of a semiconducting layer of manganese oxide deposited directly over the tantalum oxide improves the quality of tantalum-film capacitors and permits the use of thinner tantalum oxide films.   A cross section of the four-layered structure with a manganese oxide layer in between the Ta_2O_5 dielectric and the metal top electrode is shown in Fig. 14b.   The methods for the deposition of the manganese oxide film include surface pyrolysis of a manganous nitrate solution at about 200°C,[123] reactive sputtering,[124,125] and reactive evaporation.[126]   The resulting manganese oxide layer is usually a mixture of Mn_2O_3 and β-MnO_2 with specific resistivities between 5 and 1×10^4 ohm-cm depending on the concentration of the respective oxide phases.   The manganese oxide layers are between 2,000 and 5,000 Å thick, and the protective mechanisms of this film are probably due to several processes occurring during localized breakdown of the Ta_2O_5 dielectric.   It has been shown that the manganese oxide can act as a solid-state electrolyte for the anodization of defective areas in the tantalum oxide.[127]   Heat generated during local breakdown in the tantalum oxide can also cause reduction of the conducting β-MnO_2 to insulating Mn_2O_3, which protects the defective areas in the dielectric.[128]   The self-healing mechanism of tantalum-film capacitors with manganese oxide was found to increase the yield of short-free capacitors and working voltage.   It also permits an increase of the capacitance density, and it improves the ability of the tantalum-film capacitors to withstand elevated temperatures.   One advantage of this capacitor structure, therefore, is that it permits the fabrication of large-value capacitors, up to several microfarads, with good yields.   Another advantage is that it can also be used with unglazed ceramic substrates, thus providing a wider choice of substrate materials than is generally available for thin film capacitors.[123]   In comparison with tantalum-film capacitors, units with a manganese oxide layer have somewhat higher dissipation factors and higher and more variable temperature coefficients of capacitance. It is expected that this situation can be improved by achieving better control of the composition of the manganese oxide.

(2) SiO-Ta₂O₅ Film Capacitors This structure, which has been described recently,[129] consists of a duplex dielectric of anodic tantalum oxide and evaporated silicon monoxide where the bottom electrode is tantalum and the top electrode is either gold or a double layer of Nichrome and gold (Fig. 14c). Since it behaves like a series-capacitor structure with the capacitance determined essentially by the low-valued SiO layer, this duplex dielectric has proved useful in tantalum-circuit technology where low-valued capacitors in the order of 10 to several hundred picofarads are required. A typical capacitance density for a Ta₂O₅ layer, anodized to 130 V, is about 0.1 μF cm^{-2}, whereas the use of SiO films with thicknesses between 2,500 and 25,000 Å results in capacitance densities from 0.02 to less than 0.002 μF cm^{-2} so that an error in defining the capacitor area results in a relatively small error in capacitance.

A considerable advantage of such a duplex dielectric is a dielectric strength which is higher than could be realized for each dielectric layer alone because it is unlikely that defects in the tantalum oxide and silicon oxide films are at the same sites. Capacitors of 0.09 μF with an area of about 5 cm² have been made successfully, but the duplex capacitor structure is of primary interest for low-value capacitor applications.

b. Pattern Layout and Capacitor Trimming

The properties which determine the characteristics of a thin film capacitor are primarily those of the dielectric film. In a number of applications, however, the pattern chosen for the thin film capacitor structure will also affect the precision and the frequency characteristics as well as the polarization behavior of the capacitor. Thus a number of capacitor patterns which have been discussed by Berry et al.[130] will be treated briefly in this section.

To obtain a desired value for a thin film capacitor, it is important to achieve good control over the capacitance density of the dielectric layer. For anodic oxides, especially tantalum oxide, the capacitance density is determined by the anodization voltage, and its value can be controlled to better than 1%. By monitoring the deposition conditions during reactive sputtering or evaporation of the dielectric layer, the capacitance density can usually be controlled to about 5%. The value of the capacitor is then determined by the capacitance density and the area where bottom electrode, dielectric, and top electrode overlap [see Eq. (1)]. Since the capacitor structure normally consists of three layers, the area accuracy depends on the precision of each pattern and pattern registration. With both mechanical masks and photomasks pattern definition of ±0.0008 cm can be achieved, but the positional tolerances due to mask registration are usually in the order of ±0.003 cm. A capacitor pattern which employs crossed electrodes[130] provides a way of obtaining good area definition without precise mask alignment. It can be shown that the increase in area due to a rotational misalignment of the top and bottom electrode patterns by 1° is less than 0.02%. This error is independent of the capacitor area, but for small-area capacitors pattern definition itself might become the determining factor for defining the accuracy of the capacitor area.

The next two patterns are of special interest for tantalum-film capacitors because the one shown in Fig. 15b can be used to reduce the effective resistance of the tantalum base electrode and the pattern shown in Fig. 15c results in a nonpolar capacitor. It was shown earlier that the ohmic resistance of the bottom electrode contributes significantly to the series resistance of the capacitor electrodes [see Eq. (7)]. In the pattern shown in Fig. 15b the base electrode has been enlarged on three sides of the capacitor. Such a pattern significantly lowers the contribution of the bottom electrode to the series resistance of the capacitor electrodes so that it is less than 20% of the resistance of the bottom electrode underneath the dielectric. The nonpolar pattern (Fig. 15c) consists of two tantalum-film capacitors in series. In this layout the capacitors are back to back so that one of the capacitors is always positively biased during operation and protects the negatively biased unit from destructive breakdown by limiting the current flow.

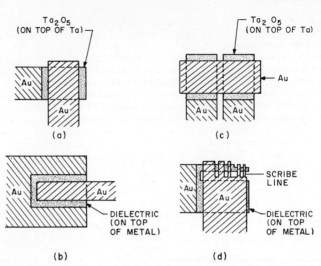

Fig. 15 Typical thin film capacitor patterns.[130] (a) Crossed electrodes. (b) Low resistance. (c) Nonpolar pattern. (d) Adjustable capacitor pattern.

Various techniques have been suggested for adjusting the value of a thin film capacitor after the capacitor has been completely fabricated. Removal of part of the top metal electrode by spark erosion,[131] laser-beam heating, or chemical etching has been considered, but it is not certain whether there is a detrimental effect on the dielectric layer. The pattern shown in Fig. 15d offers a safe way of lowering the capacitance by as much as 10%. It consists of a larger capacitance area and a number of smaller areas where the capacitance can be reduced by disconnecting the top electrode which leads up to the small trimming capacitor. The value for the trimming capacitor can be chosen in such a way that it is possible to come within 0.2% of the desired value.

c. Capacitor Parasitics

As mentioned earlier, the two major contributions to thin film capacitor parasitics are the dielectric loss and the series resistance of the capacitor electrodes. Other parasitics are the series inductance and the direct leakage current. An estimate of the series conductance of a thin film capacitor has been given by Berry et al.[130] By neglecting the dielectric, the induction of the electrodes can be calculated if it is assumed that they form a single conducting path which is much longer than it is wide. The contribution of the series inductance from the capacitor electrodes is usually in the nanohenry range. The frequency at which a capacitor and an inductance in series resonate is given by

$$f = \tfrac{1}{2}\pi \sqrt{LC}$$

where f is the resonance frequency, L is the series inductance, and C is the capacitance. Thus the resonance frequency not only depends upon the series inductance but is also a function of the capacitance value.

The direct leakage current is usually more a measure of the quality of the dielectric than a capacitor parasitic because for most thin film dielectrics the leakage current is so low that it does not affect the ac properties of the capacitor. In fact, the direct leakage current can be used to determine the insulation resistance of a capacitor which is the product of the dc resistance and the capacitance value. For example, the insulation resistance of discrete polymer-film capacitors is in the order of 10^5 to 10^6 ohm-F (Table 7), and for tantalum-film capacitors typical values are between 10 and 10^3 ohm-F.

d. Capacitor Yield and Stability

(1) Capacitor Yield A major advantage of the integrated-circuit approach is in the simultaneous fabrication of large numbers of components on a single substrate, thus eliminating interconnections and handling operations. The success of this approach depends largely on the component yield which can be achieved. Assuming that faulty components are randomly distributed over all substrates produced and that a single faulty component on a substrate results in rejection of the whole substrate, the following equation describes the relationship between component yield and substrate yield:

$$Y_s = Y_c{}^n \qquad (9)$$

where Y_s is the substrate yield, Y_c is the component yield, and n is the number of components per substrate. From Eq. (9) one finds that the component yield has to be improved from 71% for a substrate with two components to over 93% for a substrate with 10 components if a substrate yield of 50% is to be maintained. In the case of thin film capacitors the yield is mainly affected by the quality of the insulating film. As pointed out earlier, the dielectric strength of most insulating films, and consequently the breakdown voltage of the thin film capacitor structure, depends on the nature and the density of the defective areas in the dielectric. Since breakdown during electrical testing usually occurs at point defects in the dielectric film which are characterized by high leakage currents, the capacitor yield can be based on the results of leakage currents measured at a test voltage which is well above the working voltage.

Not much is known about the nature of the defects which affect capacitor yield. But it is well established that the choice of the materials for the fabrication of thin film capacitors and the amount of care taken during the processing have a significant influence on yield. The highest yields are obtained for thin film capacitors formed on glass or glazed-ceramic substrates with a smoothness of better than 250 Å.[7] The sintered structure of alumina results in irregular surfaces with peak-to-valley distances of several micrometers, and the yield of tantalum-film capacitors formed on such substrates is quite low.[7] Similar results have been found for capacitors based on anodized Al_2O_3[108] and evaporated SiO.[132] At present even fine-grained polished-alumina substrates cannot be used for thin film capacitor applications because of low yields. Other factors which affect capacitor yield are the choice of the metal electrodes[19,20] (see also Table 2) and the degree of cleanliness during capacitor fabrication. It is obvious from the small dimensions of the metal electrodes and the dielectric films, which are only several thousand angstroms thick, that contamination of the substrate or the various layers which make up the capacitor has to be avoided. Chemical contamination by organic films or inorganic salts is usually minimized by including special cleaning steps during capacitor preparation, while contamination by particulate matter can be significantly reduced by operating under clean-room conditions where the particle density is as low as 30 to 150 particles/m^3 [25] for particles greater than 0.3 μm.

Various ways have been described for removing defective areas after forming the insulating layer either before or after deposition of the top metal electrode. The beneficial effects of an electrochemical etch[7] or the reversal of the formation potential[94] on the quality of anodic films has been mentioned earlier. Another technique for clearing defective areas in SiO film capacitors consists of repeatedly discharging a large-value capacitor in the order of 1 μF charged to twice or three times the working voltage of the thin film capacitor.[132] The yield of thin film capacitors has also been found to depend on the area of the capacitors on a substrate. Martin reported that the yield for aluminum oxide film capacitors decreased from 95 to about 80% as the area was increased from 3×10^{-4} to 3 cm^2.[108] At present the highest yields for large-value capacitors which cover more than 1 cm^2 of substrate area can be achieved by adding a manganese oxide film between the capacitor dielectric and the top metal electrode.[123]

(2) Capacitor Stability The last step in thin film capacitor development, after a suitable materials system has been selected and once the processing steps are well enough under control to permit fabrication at high yields, is the evaluation of the stability of the capacitor under voltage, temperature, and humidity conditions. In order to obtain this information, groups of capacitors are subjected to a wide range of stress conditions and the distributions of the failures at each stress level are plotted as a function of time. The term failure refers to capacitors which fail electrically by destructive breakdown as well as to units showing changes in capacitance of more than a few percent. This type of accelerated life testing allows extrapolation of the results to a stress condition at which the number of failures is at an acceptable level. The data also permit determination of a maximum working voltage for a particular capacitor structure and of the change in capacitance and dissipation factor which can be expected during its lifetime under operating conditions. The characteristics of thin film capacitor structures for which extensive accelerated-life-test experiments have been carried out are summarized in Table 9. All these structures with the exception of the tantalum-film capacitors with adherent Nichrome-gold top electrodes have been described earlier in some detail. The upper limit of the capacitance range refers to the total capacitance per substrate, and it is based on satisfactory yields. The values of the working voltages given in Table 9 are an estimate based on available life test results rather than on the average dielectric strength of the particular thin film dielectric. Only in the case of the Parylene capacitors has the dielectric strength of a 1-μm film (Fig. 13) been used for determining the working voltage assuming that $k = 0.1$ [see Eq. (4)].

The characteristics of tantalum-film capacitors with NiCr-Au top electrodes have been included in Table 9 because of the importance of this capacitor for precision RC networks which require stability under varying humidity conditions of the order of 0.1 %.[133] Tantalum-film capacitors with loosely bound top electrodes such as gold adsorb and desorb moisture at the interface region between top electrode and dielectric during humidity cycling at room temperature. During cycling from 0 to 87 % relative humidity capacitance changes between 1 and 3 % are not uncommon. Protective coatings delay but do not reduce the magnitude of the capacitance change. Klerer et al.[134] found that the absorption of moisture at the dielectric-metal interface can be greatly reduced by applying a counterelectrode of Nichrome and gold which adheres well to the tantalum oxide. Capacitors of this type change less than 0.3 % during humidity cycling.

None of the film capacitors listed in Table 9, with the possible exception of tantalum-film capacitors with manganese oxide and aluminum-film capacitors,[108] are capable of operating at temperatures much above 100°C because the dielectric strength of Ta_2O_5, SiO, and parylene decreases rapidly with increasing temperature. In addition, the temperature coefficient of capacitance for these dielectric increases significantly between 100 and 200°C, and above 200°C irreversible changes in capacitance and dissipation factor take place. A preliminary survey of evaporated thin film dielectrics which were believed to possess potential for high-temperature operation was conducted by Feldman and Hacskaylo.[135] Reaction between substrate, metal electrodes, and the insulating layer during high-temperature testing was reduced by using fused-silica substrates and aluminum or gold films as capacitor electrodes. The capacitors, based on a wide variety of inorganic insulating materials, showed changes in capacitance of about 10 % at the following temperature: 180°C (SiO), 225°C (ZnS), 350°C (Yb_2O_3), 375°C (SiO_2), 410°C (Al_2O_3), 470°C (Si_3N_4), and 490°C (CeO_2), but it was not stated whether these capacitance changes were due to the temperature dependence of capacitance alone or whether irreversible capacitance changes also played a role. A more complete study of the high-temperature properties of thin film capacitors based on evaporated rare-earth borosilicates by Hacskaylo and Smith[4] indicated that dysprosium borosilicate films deposited by evaporating a mixture of 80 wt % Dy_2O_3, 10 wt % B_2O_3, and 10 wt % SiO_2 from a tungsten boat were mechanically and electrically more stable than the materials mentioned above. The dysprosium atoms might inhibit cationic conductivity in the film at elevated temperature, since in the range between room temperature and 470°C the temperature coefficient of capacitance of samples with a dielectric layer about 1 μ thick is only 120 ppm/°C. As expected the dielectric strength shows a decrease, from 4×10^6 V cm^{-1} at room temperature to

TABLE 9 Characteristics of Thin Film Capacitors

Thin film dielectric	Substrate	Bottom electrode	Top electrode	Capacitance range, pF	Dissipation factor $\tan \delta$ at 1 kHz	Temp coeff of capacitance TCC, ppm/°C	Max working voltage V	Reference
Ta_2O_5	7059 glass, glazed ceramic	Ta	Au	$5 \times 10^2 - 5 \times 10^4$	0.003	200	50	93,97
Ta_2O_5	7059 glass, glazed ceramic	Ta	NiCr-Au	$5 \times 10^2 - 5 \times 10^4$	0.003	200	30	133
Ta_2O_5	7059 glass, glazed ceramic, unglazed ceramic	Ta	MnO_{2-x}-Au	$5 \times 10^2 - 5 \times 10^6$	0.01	400–700	75	123
Ta_2O_5 + SiO	7059 glass, glazed ceramic	Ta	NiCr-Au	$10 - 10^3$	0.01	$\ldots$	50	129
Al_2O_3	7059 glass, unglazed ceramic	Al	Al	$10 - 10^5$	0.004–0.001	300–500	approx. 50	108
SiO	7059 glass, glazed ceramic	Al	Al	$10 - 10^3$	0.005–0.01	100–500	50	43,44,132
Parylene	7059 glass, glazed ceramic, unglazed ceramic	Al	Al	$5 - 10^3$	0.0001–0.0002	−100	approx. 60	122

about 1.7 V cm⁻¹ at 500°C. Small irreversible changes in capacitance were found during prolonged treatment at elevated temperatures; but these changes were only in the order of a few percent, indicating that the rare-earth borosilicate films should be useful for high-temperature capacitor applications.

REFERENCES

1. Huber, F., W. Witt, and I. H. Pratt, *Proc. Electron. Components Conf.*, 1967, p. 66, Washington, D.C.
2. Goldstein, R. M., and F. W. Leonhard, *Proc. Electron. Components Conf.*, 1967, p. 312, Washington, D.C.
3. Rairden, J. R., *J. Electrochem. Soc.*, **114**, 75 (1967).
4. Hacskaylo, M., and R. C. Smith, *J. Appl. Phys.*, **37**, 1967 (1966).
5. Gregor, L. V., in G. Hass and R. Thun (eds.), "Physics of Thin Films," vol. 3, Academic Press Inc., New York, 1966.
6. Thun, R. E., W. N. Carroll, C. J. Kraus, J. Riseman, and E. S. Wajda, in E. Keonjian (ed.), "Microelectronics," p. 191, McGraw-Hill Book Company, New York, 1963.
7. Schwartz, N., and R. W. Berry, in G. Hass and R. Thun (eds.), "Physics of Thin Films," vol. 2, p. 398, Academic Press Inc., New York, 1964.
8. Siddall, L., in L. Holland (ed.), "Thin Film Microelectronics," p. 21, John Wiley & Sons, Inc., New York, 1965.
9. Ku, H. Y., and F. G. Ullman, *J. Appl. Phys.*, **35**, 265 (1964).
10. Whitehead, S., "Dielectric Breakdown in Solids," Oxford University Press, Fair Lawn, N.J., 1951.
11. O'Dwyer, J. J., "The Theory of Dielectric Breakdown in Solids," Oxford University Press, Fair Lawn, N.J., 1964.
12. Franz, W., "Handbuch der Physik," vol. 17, p. 155, Springer-Verlag OHG, Berlin, 1956.
13. Klein, N., and H. Gafni, *IEEE Trans. Electron Devices*, **ED-13**, 281 (1966).
14. Axelrod, N. N., *J. Electrochem. Soc.*, **116**, 460 (1969).
15. Halaby, S. A., L. V. Gregor, and S. M. Rubens, *Electro-Technol.*, **57**, 95 (1963).
16. Siddall, G., *Vacuum*, **9**, 274 (1959).
17. Budenstein, P. P., and P. J. Hayes, *J. Appl. Phys.*, **38**, 2837 (1967).
18. Chaiken, S. W., and G. A. St. John, *Electrochem. Technol.*, **1**, 291 (1963).
19. Weaver, C., and J. E. S. McLeod, *Brit. J. Appl. Phys.*, **16**, 441 (1965).
20. Silcox, N. W., and L. I. Maissel, *J. Electrochem. Soc.*, **109**, 1151 (1962).
21. Klein, N., and Z. Lisak, *Proc. IEEE*, **54**, 979 (1966).
22. Sze, S. M., *J. Appl. Phys.*, **38**, 2951 (1967).
23. Anderson, J. C., "Dielectrics," p. 53, Reinhold Publishing Corporation, New York, 1964.
24. McLean, D. A., *J. Electrochem. Soc.*, **108**, 1 (1961).
25. McLean, D. A., *IEEE Intern. Conv. Record*, pt. 7, Electron Devices, Materials and Microwave Components, p. 108, 1967.
26. Maissel, L. I., R. J. Hecht, and N. W. Wilcox, *Proc. Electron. Components Conf.*, 1963, p. 190, Washington, D.C.
27. Hass, G., *J. Am. Ceram. Soc.*, **33**, 353 (1950).
28. Glang, R., *J. Vacuum Sci. Technol.*, **3**, 41 (1966).
29. Drumheller, C. E., *Trans. Vacuum Symp.*, 1960, p. 306, Pergamon Press, New York.
30. Hass, G., and C. D. Salzberg, *J. Opt. Soc. Am.*, **44**, 181 (1954).
31. White, P., *Vacuum*, **12**, 15 (1962).
32. Priest, J., H. L. Caswell, and Y. Budo, *Vacuum*, **12**, 301 (1962).
33. Holland, L., and T. Putner, *Brit. J. Appl. Phys.*, **12**, 581 (1961).
34. Hill, A. E., and G. R. Hoffman, *Brit. J. Appl. Phys.*, **18**, 13 (1967).
35. York, D. B., *J. Electrochem. Soc.*, **110**, 271 (1963).
36. Cremer, E., T. Kraus, and E. Ritter, *Z. Elektrochem.*, **62**, 939 (1958).
37. Ritter, E., *Opt. Acta*, **9**, 197 (1962).
38. Johansen, I. T., *J. Appl. Phys.*, **37**, 499 (1966).
39. Pliskin, W. A., and H. S. Lehman, *J. Electrochem. Soc.*, **112**, 1014 (1965).
40. Nishimura, Y., T. Inagaki, and H. Sasaki, *Fujitsu Sci. Technol. J.*, **2**, 87 (1966).
41. Anastasio, T., *J. Appl. Phys.*, **38**, 2606 (1967).
42. Hirose, H., and Y. Wadda, *Japan. J. Appl. Phys.*, **3**, 179 (1964).
43. Schenkel, F. W., *Proc. Electron. Components Conf.*, 1964, p. 194, Washington, D.C.
44. St. John, G. A., and S. W. Chaikin, *IEEE Trans. Pts., Mater. Packag.*, **PMP-2**, 29 (1966).

45. Degenhart, H. J., and I. H. Pratt, *Trans. 8th Natl. Vacuum Symp.*, 1961, p. 859, Pergamon Press, New York.
46. Clark, R. S., *Trans. Met. Soc.*, *AIME*, **233**, 592 (1965).
47. Rossmeisl, R. A., and F. M. Uno, *Trans. 10th Natl. Vacuum Symp.* 1963, p. 449, The Macmillan Company, New York.
48. Peters, F. G., *Am. Ceram. Soc. Bull.*, **45**, 1017 (1966).
49. Valletta, R. M., J. A. Perri, and J. Riseman, *Electrochem. Technol.*, **4**, 403 (1966).
50. Davidse, P. D., and L. I. Maissel, *J. Appl. Phys.*, **37**, 574 (1966).
51. Anderson, G. S., W. N. Mayer, and G. K. Wehner, *J. Appl. Phys.*, **33**, 2991 (1962).
52. Pliskin, W. A., P. D. Davidse, H. S. Lehman, and L. I. Maissel, *IBM J.*, **11**, 461 (1967).
53. Symposium on Silicon Nitride, Electrochemical Society Meeting (Philadelphia, October, 1966), *J. Electrochem. Soc.*, **113**, 212C (1966).
54. Janus, A. R., and G. A. Shirn, *J. Vacuum Sci. Technol.*, **4**, 37 (1967).
55. Hu, S. M., and L. V. Gregor, *J. Electrochem. Soc.*, **114**, 826 (1967).
56. Anderson, J. C., in "Dielectrics," p. 109, Reinhold Publishing Corporation, New York, 1964.
57. Weaver, C., *Advan. Phys.*, **11**, 83 (1962).
58. Weaver, C., *Vacuum*, **15**, 171 (1965).
59. Feldman, C., *Rev. Sci. Instr.*, **26**, 463 (1955).
60. Becker, W. M., *Bull. Am. Phys. Soc.*, **4**, 184 (1959).
61. Moll, A., *Z. Angew. Phys.*, **10**, 410 (1958).
62. Roder, O., *Z. Angew. Phys.*, **12**, 323 (1960).
63. Muller, E. K., B. J. Nicholson, and M. H. Francombe, *Electrochem. Tech.*, **1**, 158 (1963).
64. Sekine, E., and H. Toyoda, *Rev. Commun. Lab. Japan*, **10**, 457 (1962).
65. Feuersanger, A. E., A. K. Hagenlocher, and A. L. Solomon, *J. Electrochem. Soc.*, **111**, 1387 (1964).
66. Bliton, J. L., and R. Havell, *Am. Ceram. Soc. Bull.*, **41**, 762 (1952).
67. de Vries, R. C., *J. Am. Ceram. Soc.*, **45**, 225 (1962).
68. Bursian, E. V., and N. P. Smirnova, *Fiz. Tverd. Tela*, **6**, 1818 (1964); *Soviet Phys.-Solid State English Transl.*, **6**, 1429 (1964).
69. Bickley, W. P., and D. S. Campbell, *Vide*, **99**, 214 (1962).
70. Feuersanger, A. E., *Bull. Am. Phys. Soc.*, **11**, 35 (1966).
71. Young, L., in "Anodic Oxide Films," p. 12, Academic Press Inc., New York, 1961.
72. McLean, D. A., and F. S. Power, *Proc. IRE*, **44**, 872 (1956).
73. Berry, R. W., and D. J. Sloan, *Proc. IRE*, **47**, 1070 (1959).
74. Davis, J. A., B. Dormeij, J. P. S. Pringle, and F. Brown, *J. Electrochem. Soc.*, **112**, 675 (1965).
75. Vermilyea, D. A., in P. Delahay (ed.), "Advances in Electrochemistry," vol. 3, p. 211, Interscience Publishers, Inc., New York, 1963.
76. Goruk, W. S., L. Young, and F. G. R. Zobel, in J. O'M. Bockris (ed.), "Modern Aspects of Electrochemistry," p. 176, Plenum Press, New York, 1966.
77. Klerer, J., unpublished results.
78. Young, L., *Trans. Faraday Soc.*, **53**, 841 (1957).
79. Vermilyea, D. A., *J. Electrochem. Soc.*, **104**, 52 (1957).
80. Vermilyea, D. A., *J. Electrochem. Soc.*, **110**, 345 (1963).
81. Vermilyea, D. A., *J. Electrochem. Soc.*, **114**, 882 (1967).
82. Miles, J. L., and P. H. Smith, *J. Electrochem. Soc.*, **110**, 1240 (1965).
83. Tibol, G. J., and R. W. Hull, *J. Electrochem. Soc.*, **111**, 1368 (1964).
84. Jennings, T. A., W. McNeill, and R. E. Salomon, *J. Electrochem. Soc.*, **114**, 1134 (1967).
85. Randall, J. J., W. J. Bernard, and R. R. Wilkinson, *Electrochem. Acta*, **10**, 183 (1965).
86. Vratny, F., *J. Electrochem. Soc.*, **112**, 289 (1965).
87. DaSilva, E. M., and P. White, *J. Electrochem. Soc.*, **109**, 12 (1962).
88. Lewis, B., *Microelectron. Reliability*, **3**, 109 (1964).
89. Holland, L., "Vacuum Deposition of Thin Films," John Wiley & Sons, Inc., New York, 1956.
90. Holland, L., and G. Siddall, *Vacuum*, **3**, 375 (1953).
91. Vratny, F., *J. Electrochem. Soc.*, **114**, 505 (1967).
92. McLean, D. A., N. Schwartz, and E. D. Tidd, *Proc. IEEE*, **52**, 1450 (1964).
93. Vromen, B. H., and J. Klerer, *Proc. Electron. Components Conf.*, 1965, p. 194, Washington, D.C.
94. Standley, C. L., and L. I. Maissel, *J. Appl. Phys.*, **35**, 1530 (1964).
95. Gerstenberg, D., *J. Electrochem. Soc.*, **113**, 542 (1966).

96. Vratny, F., B. H. Vromen, and A. J. Harendza-Harinxma, *Electrochem. Technol.*, **5**, 283 (1967).
97. Gerstenberg, D., and J. Klerer, *Proc. Electron. Components Conf.*, 1967, p. 77, Washington, D.C.
98. Altman, C., and M. H. Read, *Appl. Phys. Letters*, **7**, 51 (1965).
99. These films were formed by adding less than 1 % of methane to the argon sputtering atmosphere; see also D. Gerstenberg and C. J. Calbick, *J. Appl. Phys.*, **35**, 402 (1964).
100. Schwartz, N., and M. Gresh, *J. Electrochem. Soc.*, **112**, 295 (1965).
101. Axelrod, N., B. H. Vromen, H. D. Guberman, D. J. Harrington, and N. Schwartz, *J. Electrochem. Soc.*, **113**, 51C (1966).
102. Vratny, F., *J. Am. Ceram. Soc.*, **50**, 283 (1967).
103. Bowermann, E. R., *IEEE Trans. Component Pts.*, **CP10**, 86 (1963).
104. Fuller, W. D., *Proc. Natl. Electron. Conf.*, **17**, 32 (1961).
105. Huber, F., *Microelectron. Reliability*, **4**, 283 (1965).
106. Hass, G., *Vacuum*, **2**, 331 (1952).
107. Rudenberg, H. G., J. R. Johnson, and L. C. White, *Proc. Electron. Components Conf.*, 1962, p. 90, Washington, D.C.
108. Martin, J. H., *Proc. Electron. Components Conf.*, 1965, p. 267, Washington, D.C.
109. McLean, D. A., *Proc. IEE*, **109**, (B), Suppl. 22, 457 (1961).
110. McMahon, W., in P. Bruins (ed.), "Plastics in Insulation," p. 141, John Wiley & Sons, Inc., New York, 1968.
111. Gregor, L. V., in G. Hass and R. Thun (eds.), "Physics of Thin Films," vol. 3, p. 131, Academic Press Inc., New York, 1966.
112. Bradley, A., and J. P. Hammes, *J. Electrochem. Soc.*, **110**, 15 (1963).
113. Stuart, M., *Nature*, **199**, 59 (1963).
114. DaSilva, E. H., and R. E. Miller, *Electrochem. Technol.*, **2**, 147 (1964).
115. Connell, R. A., and L. V. Gregor, *J. Electrochem. Soc.*, **112**, 1198 (1964).
116. Christy, R. W., *J. Appl. Phys.*, **31**, 1980 (1960).
117. Hill, G. W., *Microelectron. Reliability*, **4**, 109 (1965).
118. Brenneman, A. E., and L. V. Gregor, *J. Electrochem. Soc.*, **112**, 1194 (1965).
119. White, P., Paper 4, Joint Meeting IEE and British IRE, Symposium on Microminiaturization, Edinburgh, 1964.
120. Caswell, H. L., and Y. Budo, *Solid-State Electron.*, **8**, 479 (1965).
121. Valley, D. J., and J. S. Wagener, *IEEE Trans. Component Pts.*, **CP11**, 205 (1964).
122. Cariou, F. E., D. J. Valley, and W. E. Loeb, *Proc. Electron. Components Conf.*, 1965, p. 54, Washington, D.C.
123. McLean, D. A., and F. E. Rosztoczy, *Electrochem. Technol.*, **4**, 523 (1966).
124. Cash, J. H., Jr., and R. Scot Clark, *J. Electrochem. Soc.*, **113**, 58C (1966).
125. Valletta, R. M., and W. A. Pliskin, *J. Electrochem. Soc.*, **114**, 944 (1966).
126. Slack, L. H., unpublished results.
127. Smyth, D. M., *J. Electrochem. Soc.*, **113**, 19 (1966).
128. Sharp, D., paper presented at the fall meeting of the Electrochemical Society, Buffalo, N.Y., 1965.
129. Keller, H. N., C. T. Kemmerer, and C. L. Naegele, *IEEE Trans.*, **PMP3**, 97 (1967).
130. Berry, R. W., P. M. Hall, and M. T. Harris (eds.), "Thin Film Technology," p. 510, D. Van Nostrand Company, Inc., Princeton, N.J., 1968.
131. Sandbank, C. P., T. M. Jackson, and A. W. Horsley, *Electron. Commun.*, **41**, 422 (1966).
132. Nowak, W. B., and J. J. O'Connor, *Proc. Electron. Components Conf.*, 1965, p. 186, Washington, D.C.
133. Orr, W. H., Solid State Circuits Conf., *Dig. Tech. Papers*, **7**, 56 (1964).
134. Klerer, J., W. H. Orr, and D. Farrell, *Proc. Electron. Components Conf.*, 1966, p. 348, Washington, D.C.
135. Feldman, C., and M. Hacskaylo, *Rev. Sci. Instr.*, **33**, 1459 (1962).

Chapter 20

Thin Film Active Devices

PAUL K. WEIMER

**RCA Laboratories, David Sarnoff Research Center
Princeton, New Jersey**

1. ACTIVE COMPONENTS DEPOSITED BY EVAPORATION

The deposition of thin film transistors upon the same substrate with passive components permits the fabrication of complex integrated circuits entirely by evaporation. Experimental thin film circuits having thousands of active and passive components have been produced in the laboratory with excellent yields.[1,2]   An entire circuit can be deposited upon an insulating substrate in one pump-down of the vacuum system using movable masks for pattern delineation.   A noncrystalline substrate such as glass is particularly attractive for large-area circuits such as television image sensors[3] or displays.   Substrate capacitance is eliminated and greater freedom in circuit layouts is possible.

Two types of thin film transistors which can be produced by evaporation have been investigated.   These are the "hot-electron" devices[4,5] (see Chap. 14) and the insulated-gate field-effect transistors[6,7] known at TFTs.   Of these, only the field-effect transistors have shown sufficiently good electrical characteristics to permit useful circuits to be built.   Numerous semiconductor materials which can be evaporated have been studied, but the best results to date have been obtained with cadmium sulfide,[6] cadmium selenide,[2,8,9] and tellurium.[10,11]   The TFTs so produced are electrically similar to the well-known silicon MOS transistors, particularly the silicon-on-sapphire version of the MOS.[12]   Although presently inferior to silicon transistors in stability and frequency response, TFTs are potentially important for special applications such as imaging devices.   Research now in progress on TFTs promises improvements in stability and gain-bandwidth capabilities.   A brief survey of the TFT is given in the following section.

Other thin film components to be discussed in this chapter are diodes and photoconductive elements.   The use of these components in complex evaporated circuits will be illustrated in Sec. 2.

a. Thin Film Field-effect Transistors (TFTs)

(1) Structures and Processing   The TFT structure comprises two laterally spaced electrodes called a "source" and "drain" joined by a thin semiconductor film whose conductivity can be modulated by a third electrode called the "gate."   The source and drain make an ohmic contact to the semiconductor, but the gate is separated from the semiconductor by a thin layer of insulator.   In the commonest form of TFT, all materials are deposited by evaporation upon a glass substrate, using metal masks[2] or photoresist techniques[9] for defining the patterns.   Other deposition methods[48] which have been tested include chemical spraying,[13] silk-screen printing,[14,15] reactive sputtering,[16] chemical transport,[17] thermal oxidation,[16] and anodization[18,19] of metals.

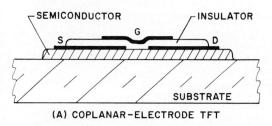

(A) COPLANAR–ELECTRODE TFT

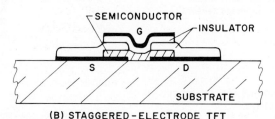

(B) STAGGERED–ELECTRODE TFT

Fig. 1   Cross-sectional views of two thin film transistor structures.

Figure 1 shows in cross section two forms of thin film transistors which have been used in integrated circuits.   An advantage of the coplanar-electrode structure[20] shown in Fig. 1a is that the semiconductor can be deposited and processed at elevated temperatures without requiring the electrodes to undergo such treatment.   This structure was used successfully for many of the early cadmium sulfide transistors having aluminum source and drain electrodes.   Figure 1b shows a "staggered-electrode" TFT structure which has been used in integrated thin film circuits having thousands of transistors on one substrate. Rows of such transistors have been deposited side by side on 2-mil centers using a fine-wire grill for defining all strip patterns. Since all parts of each transistor are deposited successively through the same gap between wires, all parts have the same width unless the substrate is moved during the evaporation to broaden strips or to connect adjacent devices. Although more evaporation steps are required with the wire-grill technique (since source and drain must be evaporated in sequence) complex circuits of a repetitive nature can be deposited in one pump-down of the vacuum system, provided the jigs permit the substrate to be moved with respect to the grill (see Sec. 2).

The utility of the wire-grill masking technique was greatly enhanced by the development of a TFT fabrication procedure[2] in which all components could be deposited upon an unheated substrate. A typical evaporation sequence for producing an array of CdSe transistors (staggered-electrode structure) is as follows:

1. Gold source electrode
2. Gold drain electrode
3. Indium contact over the drain
4. Indium contact over the source
5. Cadmium selenide film joining source and drain
6. Thin silicon monoxide TFT insulator over semiconductor
7. Thick silicon monoxide insulator over source
8. Thick silicon monoxide insulator over drain
9. Aluminum gate electrode over TFT insulator

Such transistors normally require a final air bake at about 150°C to give the desired characteristics. Figure 2 shows the source-drain current at various stages in fabrication. Electrical monitoring of a sample transistor during fabrication has been used as a guide in depositing the semiconductor, although this is not actually necessary. Film-thickness measurements are normally monitored during evaporation using a quartz-crystal oscillator or by measuring optical transmission. Semiconductor thickness is held to about 1,500 Å and the TFT insulator to about 100 Å. The source and drain insulators which serve to reduce unnecessary gate capacitance at the overlap of source and drain are thicker than the TFT insulator and less critical. The most critical step is the semiconductor evaporation since the electrical properties of the film are strongly dependent on the evaporation parameters. In practice, the evaporation procedure can be controlled sufficiently well in the laboratory that circuits containing hundreds of transistors can be made with good yield (see Sec. 2).

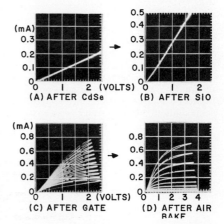

Fig. 2 Source-drain current measured at various stages of fabrication of a CdSe TFT.

Figure 3 shows the steps in fabricating a CdSe TFT using photoresist techniques[9] for pattern delineation instead of mechanical masking. In this case all electrodes and the gate insulator were deposited before the semiconductor. A single photoresist exposure serves to define the gate electrode and locate the edges of the source and drain close to the gate without overlap. The aluminum gate was anodically oxidized to produce the insulator layer prior to the semiconductor deposition. Individual TFT structures of this type have been made with a high degree of precision, but their use in integrated circuits has not been reported. The need to connect a voltage supply to each transistor gate for anodization may be inconvenient in complex circuits.

Cadmium sulfide and cadmium selenide yield N-type transistors, while evaporated

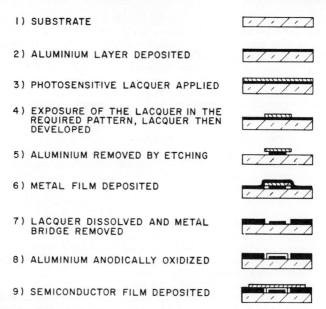

1) SUBSTRATE

2) ALUMINIUM LAYER DEPOSITED

3) PHOTOSENSITIVE LACQUER APPLIED

4) EXPOSURE OF THE LACQUER IN THE REQUIRED PATTERN, LACQUER THEN DEVELOPED

5) ALUMINIUM REMOVED BY ETCHING

6) METAL FILM DEPOSITED

7) LACQUER DISSOLVED AND METAL BRIDGE REMOVED

8) ALUMINIUM ANODICALLY OXIDIZED

9) SEMICONDUCTOR FILM DEPOSITED

Fig. 3   Steps in fabrication of a CdSe TFT using photoresist techniques (Philips process[9]).

tellurium[10,11] has been found to give good P-type units.   Tellurium units can be made entirely without baking, although when they are used in circuits with CdSe transistors they can be designed to require the same post-evaporation bake that would be given to cadmium selenide.   Figure 4 shows the current-voltage characteristics for 264 complementary pairs[36] of cadmium selenide and tellurium TFTs, measured in parallel.

(2) Electrical Characteristics of TFTs   Modulation of the semiconductor conductivity in the TFT by variation of gate potential results from the electrostatic attraction or repulsion of free carriers into source-drain region.   In a "depletion-type" unit the semiconductor conductivity is normally high but can be reduced by the action of the

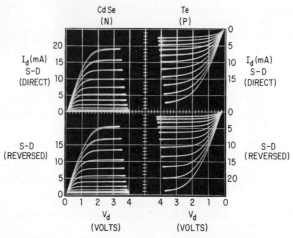

Fig. 4   Current-voltage characteristics for 264 complementary pairs of CdSe and Te TFTs, measured in parallel.

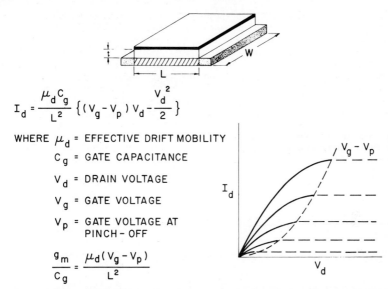

Fig. 5 Theoretical drain current–voltage curves for an insulated-gate transistor.

gate. In an "enhancement-type" unit the initial conductivity is small but can be greatly increased by gate action. Idealized drain current-voltage characteristics for an N-type insulated-gate field-effect transistor are shown in Fig. 5. Increasing voltage steps applied to the gate yields a family of drain-current curves which saturate at higher values of drain voltage. For drain voltages up to the knee of the curve the drain current is given by the approximate expression[21]

$$I_d = \frac{\mu_d C_g}{L^2} \left[(V_g - V_p)V_d - \frac{V_d^2}{2} \right] \tag{1}$$

where μ_d = field-effect mobility of the carriers
L = source-drain spacing
C_g = capacitance across gate insulator
V_g = gate voltage (source is grounded)
V_p = threshold gate voltage required for onset or pinch-off of drain current
For $V_d > V_g - V_p$ the drain current saturates at the value

$$I_{d\,max} = \frac{\mu_d C_g}{2L^2} (V_g - V_p)^2 \tag{2}$$

Although these expressions were derived assuming the mobility invariant with gate voltage and a semiconductor thin compared with normal space-charge layers, they agree sufficiently well with experiment to be useful.

The threshold voltage V_p is an important parameter which must be carefully controlled in the fabrication of TFTs for integrated circuits. Its value can range from several volts positive for an enhancement-type unit to several volts negative for a depletion-type unit. V_p is determined by the total number of charges N_0 initially present in the semiconductor. In the derivation of Eq. (1),

$$V_p \equiv -\frac{N_0 q}{C_g} \tag{3}$$

where q is the electronic charge. With a depletion-type unit N_0 represents the total number of free carriers present at zero gate bias. For an enhancement-type unit N_0

is negative and represents the total number of traps which must be filled before the free-carrier density will start to increase with increasing gate bias.

In most applications it is preferable to have the saturated drain current at zero gate bias I_{d0} be small (i.e., V_p should be near zero). Both the N-type CdSe TFTs and P-type tellurium TFTs can be made with a low I_{d0}, even though they are depletion-type units with appreciable carrier density at zero gate bias. From Eq. (3) it is clear that by making the semiconductor thin (N_0 small) and the gate insulator thin (C_g large) V_p can be small. CdS TFTs can be made with characteristics ranging from strongly enhancement-type (excess acceptor states) to strongly depletion-type (excess donor states). Excess cadmium provides shallow donor levels in CdS. Oxygen centers produced by baking in air act as acceptors.

(3) Performance Limitations The gain-bandwidth (GB) product for the TFTs can be calculated from the ratio of transconductance g_m to gate capacitance:

$$\text{GB} \approx \frac{1}{2\pi} \frac{g_m}{C_g} \tag{4}$$

where

$$\frac{g_m}{C_g} = \frac{\mu_d V_d}{L^2} \quad \text{for} \quad V_d \leq V_{d\,\text{knee}} \tag{5}$$

or

$$\frac{g_m}{C_g} = \frac{\mu_d (V_g - V_p)}{L^2} \quad \text{for} \quad V_d \geq V_{d\,\text{knee}} \tag{6}$$

Evaporated TFTs normally have a lower GB product than silicon MOS transistors because the mobility is lower in most evaporated films. GB products up to about 30 MHz have been obtained with CdS and CdSe. GB products up to 170 MHz have been reported for Te TFTs.[11] InAs has been proposed as an attractive material for TFTs because field-effect mobilities up to about 1,500 cm² V⁻¹ s⁻¹ have been measured in evaporated films[22] (compared with approximately 100 cm² V⁻¹ s⁻¹ for CdS and CdSe). Although the same authors have made InAs TFTs with GB $\sim$ 8 MHz, experimental units have not been reported with the high GB products which might be expected from the mobility measurements. The experimental InAs TFTs did not have as close a source-drain spacing as the 0.25 mil normally used for the CdSe and Te TFTs. As noted from Eqs. (5) and (6), the gain-bandwidth product is a very sensitive function of the source-drain spacing.

Other evaporated semiconductors have been used to make experimental TFTs. These include Si,[23] PbS,[24] InSb,[25] PbTe,[26] HgSe,[27] and SnO₂.[18] Unfortunately, none of these materials has shown particular promise of being superior to CdS, CdSe, or Te. Evaporated silicon transistors[23] reported to date do not compare in performance with MOS transistors formed in monolithic silicon or in silicon on sapphire. PbS and InSb transistors are interesting because both N- and P-type films can be produced, but processing may be difficult. Conclusions about the relative value of new semiconducting materials for TFTs must be regarded as tentative since they have not been investigated as intensively as CdS, CdSe, and Te. The choice of insulator, source-drain contacts, gate material, and a myriad of processing details will have great influence on performance of the final device.

For some applications the gain-bandwidth product of existing TFTs is adequate. A more serious limitation is found in the tendency for V_p to shift during operation. This instability is analogous to the drift problems experienced with silicon MOS transistors.[28] Two types of instability have been observed. Consider an N-type TFT in which a positive voltage is applied to the gate. Electrons are attracted to the semiconductor surface, causing an increase of drain current. Some of these electrons, however, may fall into traps in the insulator near the semiconductor interface. A slow decrease of drain current results which is equivalent to an increase of the threshold voltage. The slumping of drain current with time is the commonest instability in TFTs. A second type of drift can arise from the motion of positively charged ions in the insulator. With a positive gate the ions move toward the semiconductor, attracting more electrons and causing the drain current to increase. Drift in this direction corresponds to a reduction in threshold voltage. Both effects are temperature- and voltage-dependent and are influenced by ambient water vapor.

Work now in progress shows promise of significant improvement of stability of the TFT. Experimental CdSe units made with aluminum oxide insulator formed by gaseous anodization have been found to be more stable[19,19a] than the usual transistors using silicon monoxide. Greater resistance to the effects of ambient moisture has also been noted. In these units the aluminum gates are evaporated first onto the insulating substrate. The Al_2O_3 is formed anodically on top of the gate by applying voltage to the electrode in the presence of an oxygen glow discharge. The oxide builds up at the rate of 23 Å V^{-1} applied to the gate. The CdSe semiconductor is then deposited by evaporation, followed by source-drain electrodes of gold-indium alloy. Such units can be fabricated entirely in one vacuum system, but the need for electrical connections to each gate during fabrication is inconvenient for complex circuits. In some digital applications such as the shift-register scan generator[2] described in Sec. 2, the all-evaporated CdSe TFT appears to yield adequate stability.

b. Thin Film Diodes

Two types of diodes which have been used in thin film integrated circuits are (1) field-effect diodes and (2) Schottky-barrier diodes. The field-effect diodes are the equivalent of a TFT with the gate tied to the drain. Figure 6 shows how the diode characteristics depend upon the pinch-off voltage of the transistor. For most appli-

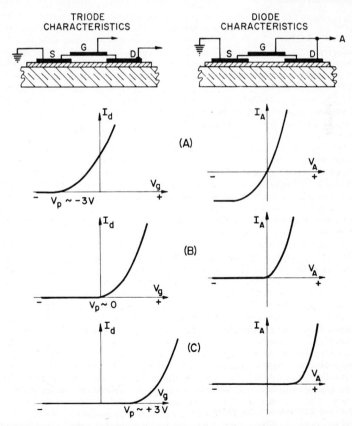

Fig. 6 Field-effect diode characteristics obtained by connecting the source to the drain of a TFT.

cations V_p should be near zero or slightly positive. In practice, the gate electrode may be omitted in the field-effect diode if the edge of the drain electrode nearest the source is separated from the semiconductor by a thin layer of insulator. The 180-stage integrated shift register (Fig. 19) contains 180 CdSe field-effect diodes.

Thin film Schottky-barrier diodes are made by contacting a semiconductor with a metal or other material of a type which forms a potential barrier at the contact. Such diodes are usually layer-type structures which include a second metal which makes an ohmic contact to the semiconductor. Typical materials forming blocking contacts to cadmium sulfide are tellurium,[29] platinum,[30] or gold,[31] which yield barrier heights of 1.1, 0.88, and 0.77 eV, respectively. Aluminum, chromium, or indium are used for the ohmic contacts. CdS diodes are usually made with the blocking contact on top of the semiconductor. Dissociation of the CdS during evaporation ordinarily causes a cadmium-rich layer to be formed initially which gives an ohmic contact to the underlying metal even if the metal is gold. A barrier contact can be formed to an underlying metal only if special precautions are taken in evaporating the cadmium sulfide.[32]

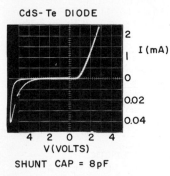

CdS- Te DIODE

I (mA)

2
0
0.02
0.04

4 2 0 2 4
V (VOLTS)

SHUNT CAP = 8 pF

Fig. 7 Current-voltage characteristics for a cadmium sulfide–tellurium Schottky-barrier diode.

A Schottky diode which has been used frequently in experimental thin film circuits consists of layers of gold, indium, cadmium sulfide, tellurium, and gold, deposited in that order on an unheated substrate. The completed device is subsequently baked in air at 150°C. The gold, in this case, merely provides low-resistance connecting electrodes. Such diodes yield rectification ratios of greater than 10^4 at $1\frac{1}{2}$ volts with a forward current density of tens of amperes per square centimeter. They will withstand reverse voltages of 5 to 10 V, but the depletion region is relatively thin, giving higher shunt capacitance than would exist in a comparable silicon diode. Figure 7 shows typical current-voltage characteristics for a diode of this type.

Although the reverse leakage and high shunt capacitance of Schottky diodes arise from the large trap density in the CdS, it has been possible to effect some improvement in performance by interposing a thin layer of high-resistivity material between the CdS and the blocking contact.[33] The insulator layer must be made thin enough that it will transmit useful currents in the forward direction by field emission. Heterojunction diodes consisting of Al-CdS-CdTe-Au, Al-CdS-SiOx-Au, and Au-CdSe-ZnSe-Au[34] have been made. A Au-CdS-ZnS-Au diode[35] has been reported to be useful at temperatures up to 500°C.

Thin film diodes comprising Ti-TiO$_2$-Pd have also been made.[36] The TiO$_2$ was produced by anodic oxidation of the underlying titanium-metal film. Current-rectification ratios up to 10^6:1 at 1.5 V and reverse-breakdown voltages of 10 to 15 V were obtained.

Organic diodes consisting of gold–copper thallocyanine–chromium have also been produced by evaporation.[37] Although capable of good rectification ratios, such diodes have proved to be unstable when exposed to air.

In spite of the lower performance of the thin film diodes compared with silicon diodes both field-effect and barrier-type diodes have been used successfully in thin film circuits operating in the 10-MHz range of frequencies.[3] An interesting variation of the barrier diodes is illustrated in the photoconductive image-sensor arrays described in Sec. 2. By using laterally spaced electrodes and a thin semiconductor the shunt capacitance at reverse bias can be reduced and the breakdown voltage increased.

c. Evaporated Photoconductor Films

Thin film photoconductors commonly used in devices fall into two categories:
1. High-gain materials such as CdS, CdSe, and PbS in which many charge carriers flow between electrodes for each photon absorbed.

2. Low-gain materials, such as antimony sulfide, lead oxide, zinc oxide, or amorphous selenium, in which no more than one carrier flows for each photon.

High-gain photoconductors[38] are used in photoconductive cells and in the image-sensor arrays described below.   In these devices the current flow is parallel to the plane of the film between laterally spaced coplanar electrodes.   At least one of the electrodes must make an ohmic contact with the photoconductor in order to supply the large number of carriers released for each photon absorbed in the photoconductor. Photoconductive gain[39] is equal to the ratio of carrier lifetime τ to transit time τ_r between electrodes:

$$G = \frac{\tau}{\tau_r}$$

The lifetime may range from 10^{-4} to 10^{-2} s for a sensitive photoconductor with gains exceeding 10,000 electrons/photon.

Low-gain photoconductors are used in imaging devices such as television camera tubes and electrostatic copying machines.[40]   In these devices the photoconductor must have high resistivity in the dark ($>10^{12}$ ohm-cm) because their operation requires charge storage for a period of time.   The photoconductor is deposited by evaporation upon a conducting electrode and the current flow is perpendicular to the plane of the film.   In the vidicon camera tube,[41] whose cross section is shown in Fig. 8,

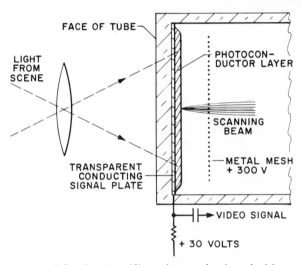

Fig. 8 Cross-sectional drawing of a vidicon photoconductive television-camera tube.

the underlying electrode is a transparent conductor[42] such as tin oxide or thin gold. The opposite electrode is provided by the electron-scanning beam.   In the electrostatic copier the opposite electrode is provided by a corona discharge.   In both devices, the electrodes act as blocking contacts, thereby limiting the photoconductive gain to near unity.   Charge stored in the internal capacitance across the photoconductive film is discharged in a variable amount in different areas depending upon the light and shade in the picture.   The antimony sulfide layer[43] in the vidicon is normally evaporated in an inert atmosphere of about 10^{-3} Torr instead of in high vacuum in order to produce a porous layer having higher resistivity and lower effective dielectric constant.

2. EVAPORATED CIRCUITS INCORPORATING THIN FILM ACTIVE DEVICES

Inasmuch as evaporated thin film transistors are themselves experimental in nature, integrated circuits incorporating TFTs have been produced only in the laboratory.

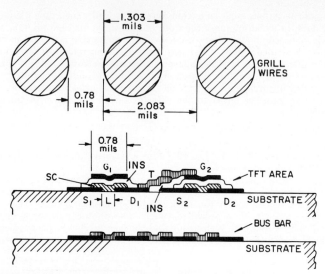

Fig. 9 Details of interconnection of TFTs using wire-grill masking techniques. Cross section of transistors taken along a line perpendicular to the grill wires.

Nevertheless, relatively complex circuits having thousands of active and passive components have been deposited upon a glass substrate. These were made in a single pump-down of the vacuum system using wire-grill masking.[44] Figure 9 shows how successive TFTs can be interconnected and continuous busbars produced by movement of the substrate relative to the grill during evaporation. The grills are made of 256 stretched Nichrome wires approximately 1.4 mils in diameter spaced upon 2.08-mil centers. The substrate is mounted within a fraction of a mil of the plane of the wires. Although a single evaporation through a fixed grill produces an array of lines of width equal to the spacing between wires, the line widths may be broadened to the extent of overlapping by moving the substrate relative to the grill during evaporation. A movable photoetched mask below the grill determines the length of each set of strips deposited upon the substrate.

The grill technique is particularly useful because of its versatility and accuracy. Critical dimensions such as the TFT source-drain spacings will be accurately duplicated over large areas provided the substrate motion is rectilinear and the wire diameters are constant. A single set of masks and grills can be made to serve for many circuit variations by varying relative positions of masks and sequences of evaporation. The use of wire-grill masking for pattern delineation is described in greater detail in Chap. 7.

A review of some of the experimental circuits which have been fabricated is given in the following section.

a. Complementary Inverters

An inverter stage consists of a transistor in series with a load impedance. The load may consist of a resistor, a similar type of transistor, or a complementary transistor. The complementary-pair inverter is advantageous because it is faster and draws less standby power. A flip-flop element formed from two complementary inverters[45] will draw very little current regardless of which state the flip-flop is in. This feature is particularly useful for multistage shift registers and memory-storage arrays.

Thin film inverters have been made in large arrays using pairs of CdSe and Te TFTs. Figure 10 shows the structure for a row of 264 complementary inverters used in the 264-stage shift register described below. Busbars connecting all sources in parallel

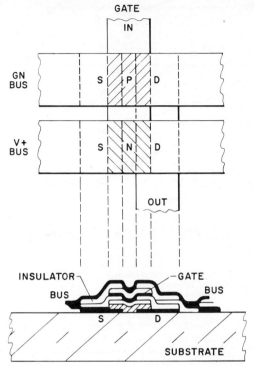

Fig. 10 Cross section and plan view of a row of complementary inverter stages used in a 264-stage shift register.

were deposited over the top of the transistors to save space.   The characteristics of 264 P-type and N-type transistors measured in parallel are shown in Fig. 4.

b. A 264-stage Complementary Shift Register

A 264-stage parallel-output shift register[46] which was designed for scanning a photosensitive array is illustrated in Fig. 11.   A voltage pulse of either polarity is trans-

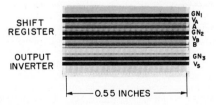

Fig. 11 A photomicrograph of a 264-stage complementary shift register using CdSe and Te TFTs.   Output strips are spaced on 2.08-mil centers.

mitted from one stage to the next at a rate determined by the clock frequency applied to terminals A and B.   The 264 output terminals are spaced upon 2.08-mil centers. Each stage of the register comprises two complementary inverters and two or three additional transistors, as shown in Fig. 12.   An additional output inverter was included in each stage for driving an array, making a total of 1,320 CdSe TFTs and 792 Te TFTs deposited upon a single glass substrate.   The feedback transistor was

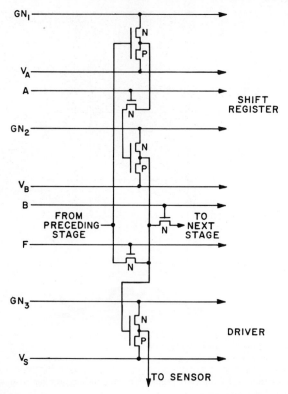

Fig. 12 Circuit for one stage of a complementary shift register.

omitted from the experimental samples because it was found that excellent flat-topped pulses could be obtained at the relatively slow clock rate (15.75 kHz) required for vertical scanning of the array. Slow-speed operation in a dynamic register requires the coupling transistors to have a very low conductance at zero bias. For very much slower rates of transfer the feedback transistor driven by F would be useful in stabilizing operation during the hold period by internally connecting each stage as a flip-flop.

c. Photoconductor-diode Arrays

Photoconductor-diode arrays[3,47] having address strips coupled to multiple-output pulse generators can serve the same function as a television camera tube for the transmission of images. Each element of the array consists of a photoconductive cell in series with a diode connected to mutually perpendicular address strips. The coincidence of pulses at an element causes its diode to conduct and allows the photocurrent of that particular element to be measured uniquely. Thin film PC-diode arrays have been produced in the laboratory with up to 360×360 elements spaced upon 1-mil centers. Work now in progress is aimed at producing arrays with even more elements.

Figure 13 is a photomicrograph of a portion of a 180×180 element PC-diode array spaced upon 2.08-mil centers. The CdS-CdSe photoconductor squares were deposited upon the glass substrate prior to the metallic address strips. As shown by the cross-sectional drawings in Fig. 14, the current flow in the photoconductor is from A to K parallel to the plane of the film. The horizontal cathode strips make ohmic contact to the photoconductor (indium contact) while the vertical strips form a blocking contact (tellurium contact). The gap between A and K is determined by the width of the

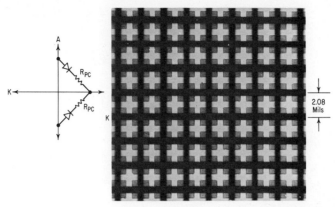

Fig. 13 A photomicrograph of a portion of a 180 × 180 element photoconductor-diode array.   Elements are spaced upon 2.08-mil centers.

insulating crossover which separates the two sets of strips.   The photoconductor film is processed for high sensitivity by baking in the presence of copper-doped CdS powder prior to the deposition of the electrodes.[48]

All elements of the photoconductive array could be tested in parallel by applying voltage to all anode strips and cathode strips.   As shown by Fig. 15, each element of the panel is equivalent to a high-gain photoconductor in series with a diode.   A sensitivity of several amperes per lumen was obtained with a few volts applied in the forward direction.   The rectification ratio of the photocurrent at ±2 V was 10⁴ to 10⁵.

Figure 16 shows a photoconductor-diode array having 360 × 360 elements spaced upon 1.04-mil centers.   This array was made with the same 480 wires per in. grill used in making the 180 × 180 array.   Two sets of interleaved strips were deposited in sequence with the substrate displaced 1.04 mils between successive evaporations.   The photoconductor in this case was CdS deposited in strips parallel to the cathode strips.

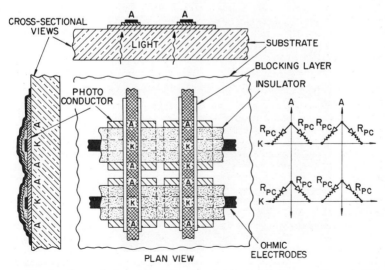

Fig. 14   Cross section and plan view of the photoconductor-diode array shown in Fig. 13.

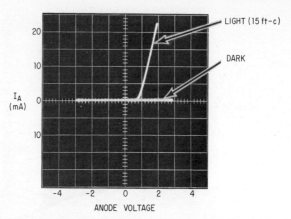

Fig. 15 Forward and reverse currents for the 180×180 element CdS-CdSe array having all anodes and all cathodes connected in parallel.

d. A Self-scanned Image Sensor

A 180×180 element image-sensor array with integrated scan generators has been constructed for use in an experimental television camera.[3] Figure 17 is a photograph of the sensor array with two 180-stage shift registers and a column of 180 TFTs for separating out the video signal. Each of the four subcircuits was deposited upon a separate 1-in.-square glass substrate. These were subsequently joined with epoxy and interconnected by means of 180 metallic strips deposited across the epoxy joint. An equivalent circuit of the integrated sensor is shown in Fig. 18.

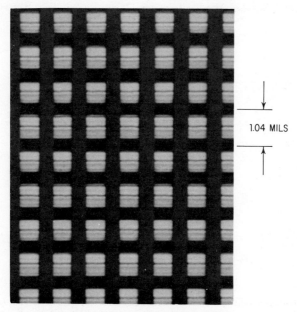

1.04 MILS

Fig. 16 Photomicrograph of a 360×360 element photoconductor-diode array having the elements spaced upon 1.04-mil centers.

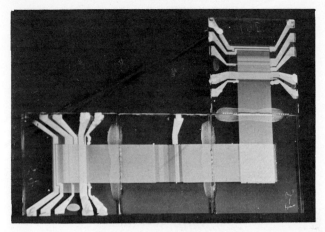

Fig. 17 Photograph of completely integrated 180 × 180 image sensor comprising the photosensitive array, the horizontal-scan generator, the vertical-scan generator, and the video-coupling circuit.

The 180-stage parallel-output shift register[2] used for scanning is shown in greater detail in Figs. 19 and 20. The thin film circuit contains 540 CdSe TFTs, 360 Nichrome resistors, and 180 capacitors. The circuit is shown in Fig. 21. The diodes were of the field-effect type, made by connecting the gates and drains of 180 of the transistors. The deposition of the circuit required about 25 successive evaporations during one pump-down of the vacuum system. Yield was relatively good, and many experimental units were made in which pulses could be transferred through the entire register. Registers were life-tested successfully for up to 10,000 hours at temperatures up to 85°C. The unloaded register was operated at clock frequencies ranging from 5 kHz to more

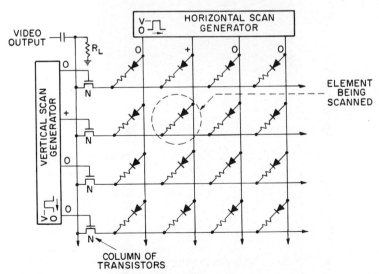

Fig. 18 Equivalent circuit for the completely integrated 180 × 180 element array showing method of attaching scan generators and coupling out the video signal. Storage of excited carriers in the photoconductor provides light integration.

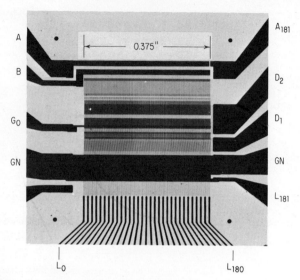

Fig. 19 A photomicrograph of the 180-stage thin film scan generator. This pattern contains a total of 1,080 active and passive elements deposited upon a glass substrate. The output stages are spaced on 0.00208-in. centers.

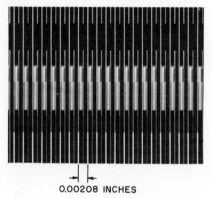

0.00208 INCHES

Fig. 20 An enlarged view of an interconnection region of the 180-stage thin film scan generator. The distance between adjacent strips is 0.26 mil.

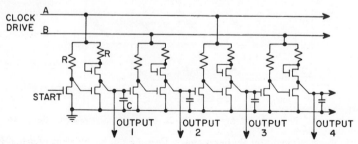

Fig. 21 Circuit for the 180-stage thin film shift register that was used for scanning the 180 × 180 element integrated sensor.

Fig. 22 A photograph of a television picture transmitted by the thin film image-sensor camera. In this test 125 × 140 elements were being scanned in $\frac{1}{60}$ s.

than 2 MHz. When driving the array, some difficulty was experienced in getting the register to operate at frequencies above 1.4 MHz.

The integrated sensor was tested at standard television scan rates in an experimental battery-powered camera. Figure 22 shows a transmitted picture reproduced upon a conventional television receiver. The sensor area being scanned was approximately 125 by 140 elements. Although the transmitted pictures were relatively poor in quality, the results were considered very encouraging as a demonstration of the capabilities of solid-state sensors. Work now in progress is aimed at producing integrated sensors suitable for pictures of much higher quality.

SUMMARY

The ability to deposit thin film transistors, diodes, and photoconductors as well as passive components on an inexpensive, insulating substrate offers a potentially important alternative to silicon for large-scale integrated circuits. Although presently inferior to silicon circuits in performance, evaporated circuits can be made with fewer defects, thus permitting larger, more complex circuits to be made without selective connections. With continued improvement in performance and in methods of fabrication, completely integrated thin film circuits may find use in specialized applications such as solid-state imaging and display devices.

REFERENCES

1. Weimer, P. K., H. Borkan, G. Sadasiv, L. Meray-Horvath, and F. V. Shallcross, *Proc. IEEE*, **52**, 1479–1486 (December, 1964).
2. Weimer, P. K., G. Sadasiv, L. Meray-Horvath, and W. S. Homa, *Proc. IEEE*, **54**, 354–360 (March, 1966).
3. Weimer, P. K., G. Sadasiv, J. E. Meyer, Jr., L. Meray-Horvath, and W. S. Pike, *Proc. IEEE*, **67**, 1591–1602 (September, 1967).
4. Mead, C. A., *Proc. IRE*, **48**, 359–361 (1960).
5. Gutierrez, W. A., and H. L. Wilson, *Proc. IEEE*, **54**, 1114 (August, 1966).
6. Weimer, P. K., *Proc. IRE*, **50**, 1462–1469 (June, 1962).
7. For a review of early work, see P. K. Weimer, in R. Thun and G. Hass (eds.), "Physics of Thin Films," vol. 2, p. 147, Academic Press Inc., New York, 1964.
8. Shallcross, F. V., *Proc. IEEE*, **51**, 851 (1963).
9. de Graaff, H. C., and H. Koelmans, *Philips Tech. Rev.*, **27**, 200–206 (Nov. 7, 1966).
10. Weimer, P. K., *Proc. IEEE*, **52**, 608–609 (May, 1964).
11. Wilson, H. L., and W. Gutierrez, *Proc. IEEE*, **55**, 415–416 (March, 1967).
12. Mueller, C. W., and P. H. Robinson, *Proc. IEEE*, **52**, 1487–1490 (December, 1964).
13. Heime, K., *Solid-State Electron.*, **10**, 732 (1967).
14. Sihvonen, Y. T., S. G. Parker, and D. R. Boyd, *J. Electrochem. Soc.*, **114**, 96–102 (January, 1967).
15. Witt, W., F. Huber, and W. Lazanowsky, *Proc. IEEE*, **54**, 897–898 (June, 1966).
16. Feuersanger, A. E., *Proc. Electron. Components Conf.*, 1967, p. 104, Washington, D.C.
17. Hegyi, I. J., *Extended Abstracts, Electron. Div., Electrochem. Soc. Meeting*, Washington,

D.C., October, 1964, vol. 13, no. 2, "Semiconductors," p. 72, Electrochemical Society, New York, 1963.

18. Klasen, H. A., and H. Koelmans, *Solid-State Electron.*, **7**, 701–702 (1964).
19. Waxman, A., "Interface Properties of Thin-film Al-Al$_2$O$_3$-CdSe Transistors," paper presented at the IEEE Solid-State Device Research Conference, Santa Barbara, Calif., June, 1967.
19a. Waxman, A., *Electronics*, **41**, 88–93 (Mar. 18, 1968).
20. Weimer, P. K., F. V. Shallcross, and H. Borkan, *RCA Rev.*, **24**, 661–675 (December, 1963).
21. Borkan, H., and P. K. Weimer, *RCA Rev.*, **24**, 153–165 (June, 1963).
22. Brody, T. P., and H. E. Kunig, *Appl. Phys. Letters*, **9**, 259–260 (October, 1966).
23. Salama, C. A. T., and L. Young, *Proc. IEEE*, **53**, 2156–2157 (December, 1965).
24. Pennebaker, W. B., *Solid-State Electron.*, **8**, 509–515 (1965).
25. Frantz, V. L., *Proc. IEEE*, **53**, 760 (July, 1965).
26. Skalski, J. F., *Proc. IEEE*, **53**, 1792 (November, 1965).
27. Malachowski, M. J., *Phys. Stat. Solidi*, **14**, K35–K37 (1966).
28. Hofstein, S. R., *Solid-State Electron.*, **10**, 657–670 (July, 1967).
29. Dresner, J., and F. V. Shallcross, *Solid-State Electron.*, **5**, 205–210 (July/August, 1962).
30. Learn, A. J., J. A. Scott-Monck, and R. Spencer Sprigg, *Appl. Phys. Letters*, **8**, 144–146 (Mar. 15, 1966).
31. Zuleeg, R., and R. S. Muller, *Solid-State Electron.*, **7**, 575–582 (1964).
32. Scott-Monck, J. A., and A. J. Learn, *Proc. IEEE*, **56**, 68 (January, 1968).
33. Muller, R. S., and R. Zuleeg, *J. Appl. Phys.*, **35**, 1550–1556 (May, 1964).
34. Gutierrez, W. A., and H. L. Wilson, *Proc. IEEE*, **53**, 749 (July, 1965).
35. Spence, W., W. A. Gutierrez, and H. L. Wilson, *Proc. IEEE*, **54**, 694 (February, 1966).
36. Huber, F., *Solid-State Electron.*, **5**, 410–411 (1962).
37. Sussman, A., paper presented at the Thin-film Symposium, Electrochemical Society, Pittsburgh, Pa., April, 1963.
38. Bube, R. H., "Photoconductivity of Solids," John Wiley & Sons, Inc., New York, 1960.
39. Rose, A., "Concepts in Photoconductivity and Allied Problems," Interscience Tracts on Physics and Astronomy, no. 19, John Wiley & Sons, Inc., New York, 1963.
40. Schaffert, R. M., and C. D. Oughton, *J. Opt. Soc. Am.*, **38**, 991–998 (December, 1948).
41. Weimer, P. K., S. V. Forgue, and R. R. Goodrich, *Electronics*, **23**, 70–73 (May, 1950).
42. Holland, L., "Vacuum Deposition of Thin Films," pp. 491–509, John Wiley & Sons, Inc., New York, 1956.
43. Forgue, S. V., R. R. Goodrich, and A. D. Cope, *RCA Rev.*, **12**, 335–349 (September, 1951).
44. Gray, S., and P. K. Weimer, *RCA Rev.*, **20**, 413–425 (September, 1959).
45. Weimer, P. K., U.S. Patent 3,191,061.
46. Sadasiv, G., P. K. Weimer, and W. S. Pike, *IEEE Trans. Electron Devices*, **ED-15**, 215, (1968).
47. Weimer, P. K., G. Sadasiv, H. Borkan, L. Meray-Horvath, J. E. Meyer, Jr., and F. V. Shallcross, "A Thin-film Solid State Image Sensor," International Solid State Circuits Conference Digest of Technical Papers, February, 1966, pp. 122–123.
48. Shallcross, F. V., *RCA Rev.*, December, 1967, pp. 569–584.

Chapter **21**

Magnetic Devices

JACK I. RAFFEL

M.I.T. Lincoln Laboratories,* Lexington, Massachusetts

* Operated with support from the U.S. Air Force.

1. INTRODUCTION

a. Film-device Characteristics

The most extensive engineering application of magnetic films has been to random-access static storage. This use has been stimulated by a number of potential advantages of films over ferrite cores, the standard existing component for such applications. The most important of these are (1) relative ease of batch fabrication, (2) switching in nanoseconds at fields of a few oersteds, (3) elimination of threading of access wires through holes, (4) possible use of extremely small bits, and (5) utilization of selection modes which allow very high tolerances on magnetic parameters. Against these must be weighed a considerable reduction in signal level.

b. Historical Development

The earliest efforts in magnetic-memory design were concerned with exploiting the speed advantage of magnetic films rather than the potential for dense, batch-fabricated arrays. Recently the effort has concentrated on minimizing cost as well as cycle time. The first attempt[1] to utilize films in a static store depended on a conventional three-coordinate selection system in which all conductor sets applied both easy- and hard-axis field to a selected bit. The required control of magnetic properties for such an arrangement becomes unduly severe, and the system is of little practical interest.

2. RANDOM-ACCESS MEMORIES

a. Conventional Selection System

Almost all memories which utilize rotational switching have the ONE and ZERO storage states along the easy axis and employ word selection with a field that rotates

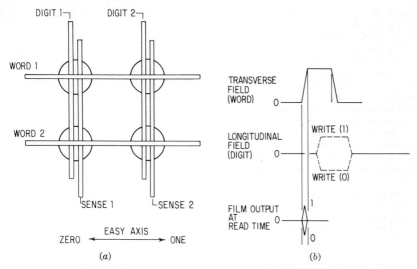

Fig. 1a Magnetic-film array with word and digit lines.
Fig. 1b Pulse sequence for reading and writing in magnetic-film array.

the magnetization to the hard axis, as shown in Figs. 1a and b.[2] The storage states are read out by the leading edge of the word-field pulse and induce opposite polarities of voltage on the sense line, which is linked by the easy-axis flux. Writing is accomplished by applying an easy-axis field in the ONE or ZERO direction while the word pulse is still on. Terminating the word pulse in the presence of the easy-axis (or digit) field then causes the flux to relax to the required storage state. (It is sometimes desirable to provide separate word pulses for reading and writing to minimize driver dissipation.)

The readout process destroys the information so that it is necessary to rewrite after reading. However, the wide tolerances for this system result from this very fact (i.e., the transverse field may be larger than H_k, the hard-axis saturation field), and variations in H_k are therefore not critical. (The limitation on transverse field is imposed only by interaction with neighboring word lines, which will be described below and is known as "creep disturbing.")

Additional advantages of this system are:

1. Only a unidirectional pulse is required on the word line.

2. The difference in polarity between ONEs and ZEROs means that the effect of variations in film output is reduced.

3. The word line is orthogonal to the sense line, resulting in low capacitive and inductive coupling between them at signal time.

4. All drive conductors may be straight lines, thus providing simple fabrication and good transmission characteristics.

b. Magnetic Properties

A detailed discussion of the properties of magnetic films is given in Chap. 17. For engineering purposes the parameters of primary significance are:

1. *The anisotropy field H_k*, defined as the value of the hard-axis field required to rotate the magnetization to the hard direction. In general, this will be composed of two components, H_{k_i}, the intrinsic anisotropy field where

$$H_{k_i} = \frac{2K_u}{M}$$

and the shape component due to hard-axis demagnetizing field H_{k_s}.

2. *The wall coercive force H_c*. The easy-axis field which will cause existing domains to propagate is measured by the field-axis intercept for a saturated hysteresis loop.

In general, the easy-axis loop will not be perfectly square. The shear of this loop may be due to local variations in coercive force. Even in films with an abrupt threshold, however, the demagnetizing field will cause shear.

3. *Skew angle,* the angle between the median value of the easy-axis angular distribution and the expected easy axis (direction of the deposition field).

4. *Angular dispersion* α, the angular distribution of local skew of some percentage, usually 50 or 90% (α_{50} or α_{90}) of the total flux. Measurement standards have been proposed by the IEEE for using a conventional low-frequency hysteresigraph to determine these parameters.[3]

(1) Operating Margins In an ideal uniaxial film negligible digit field (easy axis) is required to cause the magnetization to relax to the ONE or the ZERO direction on release of the word field. In real films the effective dispersion of the easy-axis direction (which may be thought of as equivalent to skewed areas on the microscopic scale) as well as the overall average skew[4] provides a lower limitation of write field. This may be thought of as due to the worst-case equivalent skew and requires a field as shown in Fig. 2. The upper limit on digit field is set by the lowest threshold for wall motion over the area of the film. Figure 2 shows the operating region for a word-organized memory using orthogonal fields.

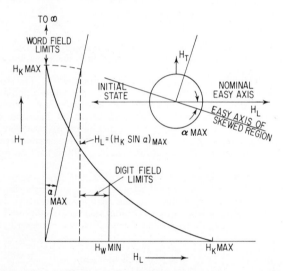

Fig. 2 Operating region for word and digit currents in magnetic-film array.

Ideally, it is desirable to have as high a wall coercive force as possible, and consequently a high digit threshold and as low an anisotropy field as possible to reduce word-current requirements. In practice, an H_c/H_k ratio near 1 is desirable since severely inverted films ($H_c/H_k > 1$) usually have large dispersion.

A variation of the conventional system[5] rotates the film easy axis with respect to the drive conductors by an angle large enough to overcome maximum skew, thus allowing the word field to guarantee storage in one direction in the absence of the digit field. In order to write bits in the opposite direction, the digit field is applied which must be large enough to overcome the easy-axis field due to the skewed word line as well as the inherent skew and dispersion of the film. This system has the advantage of requiring only a single polarity of digit drive but requires twice the amplitude of digit field and thus increases minimum digit field by a factor of 2.

(2) Demagnetizing Effects One of the principal differences between the basic planar film structure and conventional bulk magnetic devices is that in the former the magnetic material does not provide a closed flux path. The demagnetizing field of a particular storage bit will be determined by the thickness, shape, and magnetization of the sample; and, in particular, for successful storage it is necessary that the wall coercive force of the bit be larger than its demagnetizing fields. Except for the case of a perfect ellipsoid, the demagnetizing field may, in general, be extremely nonuniform so that except in special cases reverse edge domains exist under remanent storage conditions. However, such bits are often treated as single domains for simplicity because experimentally it has been found that an ellipsoid approximation is adequate for engineering purposes even for rectangular bits.[6]

The demagnetizing factors N_e along the length and N_w across the width of a very thin ellipsoid of thickness t, length l, and width w are shown in Table 1, where $l \geq w \gg t$. For example, a circular spot ($w/l = 1$) 1,000 Å thick and with a diameter of 1 cm will have a demagnetizing factor of

$$N_e = N_w = \frac{0.785 \times 4\pi \times t}{l} 10^{-4}$$

For a magnetic material with an M of 800 G (typical of Permalloy), this would result in a demagnetizing field of approximately 0.08 Oe.

TABLE 1 Demagnetizing Factors*
$l \geq w \gg t$ **(Very Flat Ellipsoid)**

$\dfrac{W}{l}$	$L = \dfrac{N_e l}{4\pi t}$	$W = \dfrac{N_w l}{4\pi t}$
1.0	0.785	0.785
0.9	0.764	0.895
0.8	0.744	1.026
0.7	0.710	1.212
0.6	0.674	1.453
0.5	0.630	1.792
0.4	0.576	2.300
0.3	0.505	3.150
0.2	0.410	4.843
0.1	0.262	9.899
0.05	0.170	17.951
0.0	0.0	∞

* From R. L. Coren, "Shape Effects in Permalloy Films," 1965 Proceedings of the Intermag Conference, Institute of Electrical and Electronics Engineers, Inc., New York, 1965.

The main effect of self-demagnetizing, aside from providing a lower limit on wall coercive force, is to add to the inherent easy-axis dispersion an added component of digit field required to overcome the demagnetizing field. In other words, the apparent dispersion of a continuous film when it is etched into discrete bits is enhanced by an amount equivalent to the application of an easy-axis field equal to the demagnetizing field.

A second effect of the demagnetizing field is to provide bit-to-bit coupling fields which reduce the overall digit margins. The worst-case pattern consists of all bits in the same digit magnetized in the same direction as the bit under test and all other bits magnetized oppositely.[7]

(3) Creep The term "creep switching" is used to describe the easy-axis magnetization reversal that may occur upon the application of a large number of small transverse pulses (only a few percent of H_k) in the presence of digit-writing fields.[8,9] This occurs in films greater than 400 Å thick which can support Bloch or crosstie walls. Such a situation arises in practical memories because of sneak currents in word-selection switches (diodes or cores) which allow small word currents on unselected word lines to occur in the presence of normal digit fields or where transverse fields can occur at words adjacent to the selected word because of fringing of the word field or because of the hard-axis film-to-film coupling.

Creep effects may be minimized by (1) using films which have high creep thresholds and (2) choosing a geometry which minimizes the field fringing. The effect of creep in some designs may be small until many thousands of pulses have been applied so that production testing of films may be extremely time-consuming if worst-case creep testing is required. Typical creep-threshold curves are shown in Fig. 3. The principal effect of creep on memory design is to limit the density of word lines unless other provision is made for reducing creep effects.

Another technique which may be used[10] to reduce the effect of creep is to limit single-layer film thickness to less than 400 Å and then to build up the total flux required by using a number of films. It is necessary in such a multilayer structure to separate the layers to minimize magnetostatic interactions which can lower coercive force.

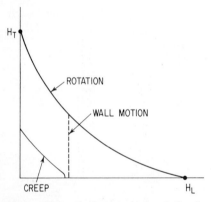

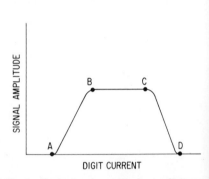

Fig. 3 "Creep," rotation, and wall thresholds in magnetic film.

Fig. 4 Peak signal amplitude vs. digit current for writing.

(4) Pulse Testing The hysteresigraph is generally used to display film properties on large samples for experimental purposes or in order to measure gross properties of an entire array. A more detailed presentation of individual bit properties is often provided by subjecting a film to the standard word and digit pulse sequences for reading, writing, and digit disturbing and plotting the signal output as a function of drive current. Figure 4 shows a plot of peak signal output as a function of digit field (the same for writing and disturbing) for a typical bit under test, with word current held constant. Point A indicates roughly the digit current necessary to overcome skew. The rising portion from A to B corresponds to effective dispersion including the effects of demagnetizing fields. Point C indicates the onset of digit disturbing and point D the point of total disturbing. The sloping section from C to D represents the effect of demagnetizing fields in reducing the threshold for wall switching.

c. The Sensing Problem

(1) Digit Noise The term "noise" is generally used to denote both random noise and spurious coupling of word and digit pulses to the sense winding.

When writing, a large voltage is imposed on the sense amplifier from the coupling to the digit field; this coupling varies greatly for different sense-digit configurations which fall into two broad categories, those using separate digit and sense lines and those employing a single line for both. If separate lines are used, cancellation techniques can be employed to provide low coupling, such as the examples shown in Figs. 5a and 5b.[11] In Fig. 5a the sense line is wound with two halves in opposition, providing no net coupling to the digit line. In Fig. 5b, a balanced planar-sense line is shown

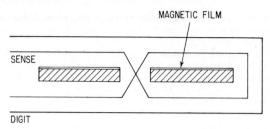

Fig. 5a Sense winding with crossover for low coupling to digit winding.

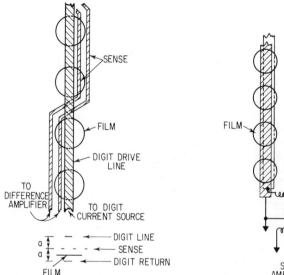

Fig. 5b Planar sense winding with low coupling to digit winding.

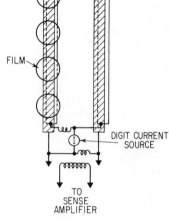

Fig. 5c Common digit-sense line balanced bridge.

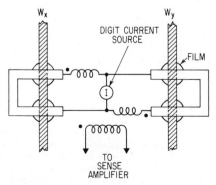

Fig. 5d Planar digit-sense line bridge.

Another approach is to use the directional coupler familiar to microwave engineers.[12] When a common line is used for sense and digit, bridge arrangements such as shown in Fig. 5c and d may be employed.[2] In Fig. 5c, the sense return passes under the substrate, and a dummy line which has no film coupling provides a bucking voltage when the two are hooked in a bridge arrangement. In Fig. 5d, a planar sense is used which links two elements per bit and is balanced against another active planar line in a bridge arrangement. In any case, depending on the number of bits on the digit line, the cancellation technique must operate on digit voltages of volts and produce a null such that signals of the order of a millivolt or less can be sensed. This puts severe overload restrictions on sense-amplifier design, and the recovery from digit transients may be a significant fraction of overall cycle time.

(2) Word Noise The principal source of word noise is the capacitive coupling between the word and sense lines when the word line is excited with a high current pulse and may cause an excursion of tens of volts. Depending on the particular configuration of word and digit lines, this capacitive transient can cause: (1) difference-mode coupling to the sense line due to capacitive imbalance between word-line and sense-line pairs, (2) a large common-mode voltage on the sense line which is not properly rejected at the amplifier, and (3) ground currents in neighboring digit lines which are asymmetrically coupled to the sense line, producing a net coupled flux. In addition, noise may arise from inductive coupling of the word line to the sense line due to asymmetries in either sense or word lines. This situation is further complicated by the fact that mutual coupling between the selected word line and near neighbors causes an induced voltage on these lines which falls off monotonically with distance. If common-mode rejection is adequate and care is taken to eliminate line-edge defects and other asymmetries, the primary source of word noise will be due to capacitive imbalance between halves of the sense line and word lines (e.g., Fig. 5d). Care must be taken to ensure that separation and line geometries are well controlled. A given capacitive imbalance will result in a noise component which increases with word-line distance from the shorted end of the sense line, so that this problem becomes increasingly severe for longer sense lines.

(3) Random Noise The ultimate limit on bit size is set by the random-noise level in the sensing circuit. An acceptable signal-to-noise ratio for a specified mean time between failures is determined from the formula[13]

$$\text{MTBF} = \frac{2T}{(1 - \text{erf}\, A/\sqrt{2}\,\sigma)M}$$

where M is the number of digits, T is the cycle time, and A/σ is the peak signal-to-rms-noise ratio. Since signal is directly proportional to film-switching speed, and the noise is proportional to the square root of amplifier bandwidth, signal-to-noise ratio improves as the square root of switching speed or bandwidth. For a given bit size the signal may, of course, be increased by using a thicker film; but this will require higher coercive force and hence higher drive currents.

The smallest bits employed in a memory to date have had a total sensed flux of 2.3 mV-ns (130 μV peak signal, 35 ns wide triangular pulse) which after step-up through a transformer provides a peak signal-to-rms-noise ratio of 27:1 for the particular amplifier used.[14] For this particular memory the signal-to-noise is approaching the tolerable limit of 8:1 required for a MTBF of approximately 1 year for 400 digits operating at 1 μs. For low-density thicker-plated wire films, signals of the order of 10 mV are common.

d. Peripheral Electronics

(1) Digit Circuitry The principal requirements of sense-amplifier circuitry are high common-mode rejection of word and digit voltages, high gain (1,000 to 10,000), and low noise input. For small signals ($<$1 mV) and long digit lines ($>$1 ft) it is common practice to use a differential-amplifier input stage in combination with common-mode chokes and/or an input transformer. In some cases the entire amplifier is made of difference stages to simplify the problem of recovery from the digit transient. Digit

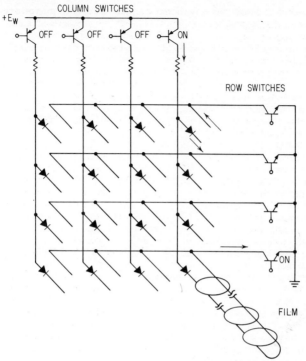

Fig. 6a Diode matrix for word selection.

current requirements for writing vary from tens of milliamperes for closed structures to hundreds of milliamperes for open-flux structures.

(2) Word Circuitry The principal word-selection techniques use diode and transistor matrices. The two commonest examples are shown in Figs. 6a and b. For a diode matrix the word lines are arranged in groups shorted together at one end and connected individually to diodes on the other. The other ends of corresponding diodes in each group are shorted together to make a conventional two-coordinate diode matrix. The row and column busses of the matrix are connected to a transistor switch such that the selection of one row switch and one column switch provides a single series path through the selected diode and word line. In such a system it is usual practice to preselect the row switch such that the noise that results from the voltage excursion on all lines in a group has subsided by the time the column switch is selected and the film signal occurs. The transient voltage due to the column switch is isolated by diodes from all but the selected line.

The transistor matrix has the advantage that two-coordinate selection occurs at relatively low (base-to-emitter) currents and that these excitations are isolated from the sense lines, the selected line being the only one which receives significant voltage. For either system current rise times are generally of the order of 10 to 30 ns, and depending on geometry, currents of the order of 500 mA are required.

e. Storage-cell Configurations

(1) Continuous Sheet A variety of cell configurations have been devised in an attempt to improve on the simple planar cell. Continuous sheets have been used in order to eliminate the necessity for discrete bit definition.[5] This has generally not been justified because it introduces extraneous disturb interactions and increased demagnetizing fields due to interbit materials.

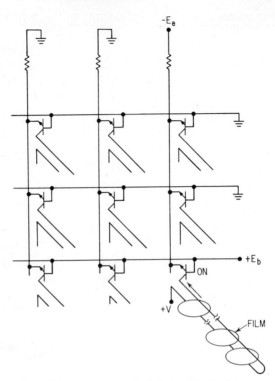

Fig. 6b   Transistor matrix for word selection.

(2) Multilayer Films with Flux Closure   Three basic multilayer-film configurations have been investigated which provide flux closure through a second film in the hard, the easy, and both directions, respectively.[15]   These are shown in Fig. 7 schematically.   These structures are formed by evaporating both magnetic and conducting layers and then etching both together, thus eliminating registration problems as well as providing partial flux closure.   The relative advantages and disadvantages of these structures are also shown in Fig. 7.   A fourth structure uses integral word lines as in 2 of Fig. 7 but without a second magnetic layer.   Arrays of 100,000 bits using this technique have been reported.[14]

The hard- or easy-axis closure may be made essentially perfect by eliminating the air gaps found in evaporated flat film by electroplating a magnetic layer on round wire.

(3) Plated Wire   Although both word wires (longitudinal easy axis) and digit wires (circumferential easy axis) can be made, the latter is the most commonly employed.[16,17,18]   By closing the storage flux path and thus eliminating easy-axis demagnetizing effects as shown in Fig. 8, it is possible to use thick films (typically 10,000 Å) with relatively low coercive force.   This results in much larger signals than are obtained with planar films and lower digit currents.   Thick films, however, require large transverse fields because of transverse demagnetizing effects and provide low word densities because of the large stray fields and possible creep due to the restrictions on the way in which word straps are wrapped around the bit wire.   In addition, plated wire is more susceptible to magnetostrictive effects due to mechanical stress and aging effects which require special annealing treatment.   Wire must, of course, be attached to terminals and "fixed" into a matrix, whereas planar films may be fabricated in place with integral attachment flags.   In addition, the beryllium copper usually used as the substrate is three to four times as lossy as copper.   Bit wires up to

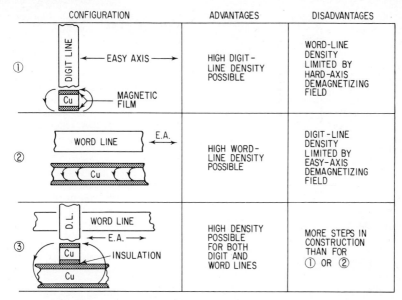

CONFIGURATION	ADVANTAGES	DISADVANTAGES
① DIGIT LINE ←— EASY AXIS —→ Cu MAGNETIC FILM	HIGH DIGIT-LINE DENSITY POSSIBLE	WORD-LINE DENSITY LIMITED BY HARD-AXIS DEMAGNETIZING FIELD
② WORD LINE E.A. ←— Cu	HIGH WORD-LINE DENSITY POSSIBLE	DIGIT-LINE DENSITY LIMITED BY EASY-AXIS DEMAGNETIZING FIELD
③ D.L. WORD LINE ←— E.A. —→ Cu INSULATION Cu	HIGH DENSITY POSSIBLE FOR BOTH DIGIT AND WORD LINES	MORE STEPS IN CONSTRUCTION THAN FOR ① OR ②

Fig. 7 Coupled film configuration with (1) easy-axis closure, (2) hard-axis closure, (3) closure in both directions.

3 ft long with up to 1,000 bits have been produced. This "linear" batch fabrication is midway between the single-bit core technology and the "area" fabrication of planar arrays.

(4) Keeper Another approach to minimizing demagnetizing fields uses a bulk keeper of either magnetic rubber or a flat ferrite plate to provide closure[19] paths in order to reduce creep effects due to word-line fringing and easy-axis demagnetizing which cuts into writing-current margins on both the high and low ends. It is necessary when employing keepers to be sure that the material is magnetically homogeneous so that asymmetries do not lead to net flux coupling and noise when current pulses are excited.

(5) Ground Planes Deposition of films directly on a ground plane (or an insulating layer over a ground plane) provides low-impedance drive and sense lines and signal doubling but leads to ground-current spreading, which is frequency-dependent. Another effect of a keeper is to eliminate such effects by reducing the flux penetration into the ground plane.

(6) Sharp Films So-called "sharp" films are designed to reduce demagnetizing fields in discrete bits by shaping the bit in the form of an elongated diamond.[20] While there is evidence that sharp films are free of edge domains and thus have less tendency to creep and improve nondestructive rotation threshold, they must be relatively thin (500 Å), do not eliminate bit-to-bit demagnetizing, and are relatively difficult to fabricate. It is not clear how much reduction of signal results from the fact that edge poles are distributed under the sense line which covers the bit.

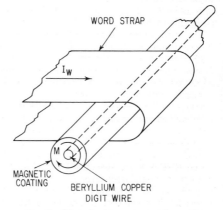

Fig. 8 Plated-wire memory element.

(7) Rods Cylindrical rods on which magnetic films are coated with longitudinal easy axis have received limited but persistent attention over a number of years.[21] This structure seems to combine all the undesirable features of planar films and bit wire; i.e., it does not possess flux closure, is relatively difficult to handle, and is not particularly amenable to batch fabrication.

(8) Waffle Iron The waffle-iron memory is a thin film or ferrite device, depending on one's point of view.[22] Basically, it employs a block of ferrite with channels milled out for orthogonal current-carrying hairpins. The pillars thus created are bridged by a planar anisotropic film providing an essentially closed-flux structure thus allowing low currents and reasonable signal levels (e.g., $I_{\text{word}} = 90$ mA, $I_{\text{digit}} = 10$ mA, $V_{\text{signal}} = 1$ mV for 5-mil slots on 15-mil centers). In addition, propagation of the flux pattern into the ferrite allows sensing in an area remote from the effects of digit transient. Unfortunately, the basic structure is relatively complicated to manufacture, requires precise machining, and promises only a modest bit density.

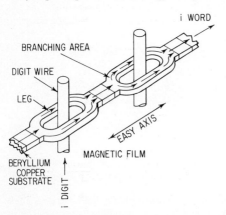

Fig. 9 Magnetic chain element.

(9) The Chain The chain,[23] a device which provides flux closure in both hard- and easy-axis directions, may be produced by plating a copper strip of interconnected flat toroids with Ni-Fe film as shown in Fig. 9. The resultant structure is excited with a transverse field by running word current through the copper core; digit current is provided by wires which thread the toroids and cause the magnetization to run circumferentially. The magnetic material is oriented with the easy axis along the length of the chain.

Word current flows in the same direction in both legs of each toroid, providing no net coupling to the copper toroid core. Demagnetizing effects are essentially eliminated so long as branch areas are small compared with the toroidal legs. The advantage must be balanced against the need to thread digit holes and limitations on the size of the basic element.

(10) Saturable Shielding It has been proposed that one method of providing for word selection without the need for semiconductor switches would be to use one film to shield another and thus provide a coincidence threshold for transverse fields.[24] The shield is a very low H_k film while the storage film has relatively high intrinsic H_k. If both films are essentially at the same points in space and are subjected to the same transverse field, film 1 will essentially shield film 2 until its shape anisotropy has been overcome and the field at film 2 will be zero until this threshold is reached. Unfortunately, the demagnetizing field is nonuniform and the threshold is not very sharp in practice. Furthermore, the currents must still flow in shielded word lines, so that it is a relatively inefficient means of accomplishing word selection. It may be useful, however, in shielding against fringing fields which produce creep.

(11) Biaxial Films It has been proposed[25] that films having biaxial anisotropy be used as storage elements. The main feature of such a device is that word and digit pulses need not be applied in time coincidence. No particular system advantage seems to accrue from this property and no practical application has been realized.

f. NDRO

(1) Applications The motivation for providing nondestructive readout in certain applications is threefold.

1. For those conventional applications where high speed is essential, NDRO eliminates rewrite time on those machine cycles during which new information is not written. This can represent typically 80% of machine cycles, depending on the

particular program. Furthermore, the write time may be considerably longer than read time because of digit-transient recovery, so that the saving on read cycles could be considerable.

2. In an application such as an aerospace mission, writing may be very infrequent; but an error in writing (which cannot be readily corrected) may be catastrophic.

3. For the case of very small bits, it is possible to recover the signal from the random noise by repeated nondestructive interrogation and averaging at the expense of cycle time (i.e., the effective time integral of the signal may be increased indefinitely).

(2) Partial Rotation   A variety of methods have been proposed for providing NDRO of magnetic films. The most straightforward approach is to apply a transverse field during readout which is below the threshold for irreversible rotation. Ideally, this threshold should provide a rotation of up to 90° providing full signal, but in practice variations in the dispersion and creep disturbing generally restrict the effective signal to approximately one-third of full signal. For planar bits without flux closure, the presence of self-demagnetizing as well as the demagnetizing fields from adjacent bits provides a limit due to creep disturbing which is not present in bit wires in which easy-axis demagnetizing and bit-to-bit coupling are absent. However, disturb effects are introduced by interbit areas along the continuous films on the digit wire. In order to reduce these effects, an extra wire may be inserted between bits which applies a dc bias opposite to the direction of the usual word current.[26] This has the effect of increasing readout signals and disturb thresholds.

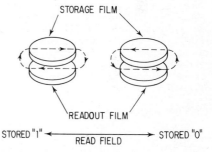

(3) Bicore   One attempt to increase the effective signal for NDRO utilizes magnetostatic coupling between a high-anisotropy film and a low-anisotropy film. This "bicore"[27] structure is shown in Fig. 10.

Fig. 10   Bicore nondestructive-readout element.

The magnetization of the low-anisotropy film may be completely reversed at low easy-axis fields to provide full readout signal with full restoration to the storage state because of the demagnetizing field of the high-coercive-force film, which remains essentially unchanged by the low-readout field.

(4) Thick-coupled Films with Hard-axis Interrogation   Another variation[28] on the bicore approach uses a high-coercive-force storage film which is coincidentally written in a direction orthogonal to the easy axis of a low-anisotropy read film. This read film is biased to hard-axis saturation if the storage film is demagnetized. The application of an interrogate pulse in the hard direction thus encounters a high- or low-permeability read film and provides a large or negligible output depending on the storage state. This type of memory is suitable for a fast NDRO store with low-frequency write capability.

(5) Eddy-current Restore   Another approach is to provide the necessary restore field for the readout film by inducing a current in a surrounding conductor during the rapid change of field during readout.[29,30] It is essential for proper NDRO that the time constant of the easy-axis field due to the induced current be long enough that sufficient write field is present when the transverse current pulse terminates. The control of the time constant for current decay is determined by the copper thickness, separation between the two conductors, and the film dimensions, and may be typically from 10 to 100 ns.

It is necessary to use longer word pulses in order to write new information into such a memory.

(6) Ferromagnetic Resonance   Some experimental work has been attempted to apply the well-known ferromagnetic resonance in films as a form of nondestructive readout.[31] This technique makes use of the fact that in the presence of a small easy-axis bias field (below the wall coercive force to the film) the resonance frequency will shift according to

$$f_0 = f_k \left(1 \pm \frac{H}{H_k} \right)^{\frac{1}{2}}$$

where f_k is the frequency in the absence of bias and the $+$ and $-$ signs correspond to the antiparallel storage states. Unfortunately, this technique requires distribution of interrogate power in the 500-MHz range, which is impractical in a complex matrix of word and digit lines. Furthermore, signal levels are still quite low (typically a few millivolts for a 100-mil bit) in spite of the high frequencies because magnetization excursion is extremely small.

(7) Fluted Films Films with alternate strips approximately 10 μ wide of thin and thick Permalloy are called fluted films and have been tested extensively.[32] The thin strips are restricted to less than 200 Å, while the thick strips are in the range of 500 Å. Storage states in these films are reached by a positive hard-axis pulse followed by a positive digit pulse for the ONE and a negative hard-axis pulse followed by a negative digit pulse for the ZERO. This can only be achieved in a word-organized memory by angling the digit line to provide both hard- and easy-axis components. This introduces noise voltage at read time due to inductive coupling which must be compensated by a dummy line. In addition, nondestructive-readout signals for the ONE have both positive and negative components which must be resolved by the amplifier. Integration of this signal due to inadequate bandwidth causes loss of signal-to-noise ratio. Limitations in bit-size reduction due to the complications of evaporating the strips through a wire grid must be balanced against the wide margins for writing and NDRO.

g. Film-preparation Techniques

The most important parameters controlling the magnetic characteristics of evaporated magnetic films for engineering use are composition and substrate surface. The first influences magnetostriction H_k and H_c, and the second affects H_c and dispersion. Secondly, but still important, are substrate temperature and angle of incidence; the former affects H_k and H_c and, in combination with the latter, skew. The choice of optimum preparation conditions is determined by the specific design and especially the requirements that the wall coercive force be large enough to overcome demagnetizing effects for a specific bit size. It is possible to make films which are slightly "inverted" (i.e., $H_c > H_k$), but generally an H_c/H_k ratio between 0.8 and 1 is acceptable and avoids the problem of high dispersion.

Angle-of-incidence effects are usually minimized either by having large-area (filament) sources or by placing the source at a large distance from the substrate. The significance of magnetostriction depends largely on the type of substrate used; thin substrates or wires a few mils thick have to be relatively free of strain sensitivity and close to the zero magnetostriction composition. Films deposited on massive substrates, on the other hand, are essentially insensitive to stress normally encountered.

The compositions yielding the lowest H_c and H_k are nonmagnetostrictive Permalloy evaporated at relatively high temperatures. For high-density applications which require higher coercive force, ternary alloys using Ni-Co-Fe can provide a wide range of coercive forces up to 30 Oe. In some cases two films have been used in direct contact with one another (and hence exchange-coupled) to provide a convenient way of averaging the H_c and H_k of each individually.[14] Finally, different annealing procedures may be used to manipulate H_c and H_k after deposition. In one such arrangement, H_k may be reduced by annealing in a field at 90°[33] or by deposition in a field which is pulsed in easy and hard directions.[34] In other cases the controlled diffusion of an overlay of copper has been effective in raising coercive force and producing slightly inverted films which do not have excessive dispersion.[35]

Deposition of films having acceptable H_c and dispersion requires a surface flatness comparable with that of fire-polished glass. Attempts to evaporate directly onto metal have been largely unsuccessful, although the deposition of a "smoothing" layer of SiO or SiO_2 of a thickness of 5,000 Å, for instance, can alleviate this problem. Mechanical polishing of solid-metal substrates is usually not adequate to provide a satisfactory surface. This limitation applies also to plated-wire substrates where rather elaborate preparation of the wire substrate is required. Even here, with few exceptions, it has not been possible to prepare films with satisfactory characteristics much less than 10,000 Å thick.

3. SEQUENTIAL MEMORIES

a. Domain Propagation Parallel to Easy Axis

The earliest successful cell-to-cell transfer of information in magnetic films utilized the well-known fact that there may be a large difference between the magnetic field required to nucleate a reverse domain and that required to propagate such a domain.[36] A shift-register configuration based on this principle is shown in Fig. 11a. Consider two sets of conductors each of which winds back and forth in a series of interconnected alternating hairpins. If these conductors are placed over a magnetic film whose easy axis is perpendicular to the hairpins, then current in either conductor produces regions of alternating easy-axis field. By pulsing the two conductors (which are

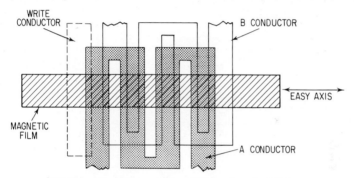

Fig. 11a   Conductor configuration for domain shifting.

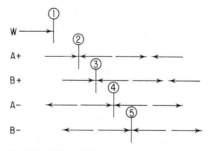

Fig. 11b   Film shift-register storage states.

displaced from one another as shown in Fig. 11a) in a sequence $A+$, $B+$, $A-$, $B-$, where A and B are the two conductors and $+$ and $-$ refer to positive and negative pulses, the field pattern shown in Fig. 11b results. If the current levels are adjusted such that the field under the conductors is at all times less than the nucleation field H_N but greater than the wall coercive force H_W, then a wall introduced by the write conductor at say point 1 of Fig. 11b will be seen to propagate successively to positions 2, 3, 4, etc. If no wall were present, no new walls will be nucleated. The operating margins of such a system clearly depend on the ratio of H_N/H_W which can be made as large as 6:1 by using films of 74% Ni, 15% Fe, 11% Co.[37] A writing conductor is used to introduce the initial domain at the input terminal of a register by applying a reversing field which exceeds H_N. Readout is accomplished at the exit point of the register either by sensing the inductively coupled flux or by using the magnetoresistive effect.

Three significant characteristics of this device should be emphasized.

1. It requires a basic four-pulse pattern to advance a bit of information a single stage.

2. The register is bidirectional.

3. Switching is by wall motion, and speeds are governed essentially by the speed of domain propagation and bit size.

Speeds up to 1 MHz have been reported.

One limitation of this system is the fact that domain propagation is parallel to the easy axis of the films, resulting in jagged bit boundaries which ultimately limit bit density.

b. Domain Propagation Perpendicular to Easy Axis

A system which shifts domains perpendicular to the easy axis, using sidewise wall motion, can use a single wall to form a boundary of a bit. One such system[38] is shown in Fig. 12, where two sets of conductors are excited alternately to supply the currents

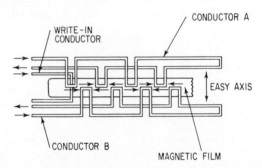

Fig. 12   Conductor configuration for sidewise domain shifting.

shown at times t_1 and t_2. Each conductor produces a field pattern which alternates in polarity spatially along the length of the film. On the t_1 pulse the domain is moved along by causing both the left- and right-hand walls to advance to the right such that they are then placed under the influence of the second field pattern applied at t_2, which again advances the domain so that on the t_3 and t_4 cycles when the current is reversed the walls will be advanced again by the field patterns applied at those times. It can be seen that this is perfectly analogous to the system described above except that domain motion is perpendicular rather than parallel to the easy-axis and wall direction. With simple conductors it is possible only to approximate roughly the "square-wave" spatial distribution of field due to the field contributions from lead-on wires. Nevertheless, a shift register has been operated at 14 bits/in. and a clock rate of 125 KHz. Difficulty in providing conductor configurations with the necessary field distribution may well impose the very limitation which the system is meant to eliminate, namely, low bit density.

In order to eliminate the problem of supplying the complicated field patterns in the system described above, a technique has been proposed[38] which utilizes spatially uniform fields but provides spatial variation in the wall coercive force of films to produce the same net effect. The basic scheme is illustrated by the coercive-force profile of Fig. 13a. In order to provide directionality and still allow only uniform fields to be applied to the entire strip, it is necessary to provide asymmetry in the coercive-force profile which favors propagation in one direction. The profile itself is not adequate because any absolute barrier would block propagation from either side; it must be combined with a pulse-duration control that allows interruption of the domain

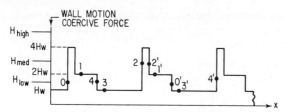

Fig. 13a Storage states and graded coercive-force profile.

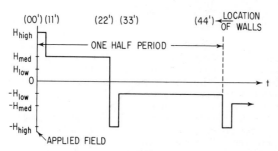

Fig. 13b Current waveform for graded coercive-force shift register.

propagation before it has gone to equilibrium and thus provides for varying the location of the wall with respect to the barrier.   Figure 13b shows the required pulse sequence.   Consider the two walls located initially at points 0 and 0′.   Any uniform applied field drives the wall of a domain in opposite directions either toward one another or apart.   If a high-coercive-force barrier is placed immediately to the right of the left-hand wall and at some distance to the left of the right-hand wall, a pulse may be applied which is greater than the high coercive force and long enough in duration to sweep the wall at 0 past the left-hand barrier but short enough in duration so that it does not sweep the wall at 0′ past the right-hand barrier.   The resulting wall positions will be 1 and 1′.   (The medium-level barrier will be seen later to ensure the initial asymmetric position of the walls with respect to the barriers.)   If a field lying between the high and medium barriers is applied in the same direction, the walls will be driven closer together to positions 2–2′ located at the right-hand barrier.   A high field pulse of reverse polarity now drives the walls apart to positions 3–3′; and when this field is reduced to a low level, the walls reach equilibrium at points 4–4′ with one wall close to the barrier and one far away as in the initial state 0–0′.   If the entire pulse sequence is now repeated, with reverse polarity, by a similar process the pattern will be seen to have advanced by 1 bit at the end of this entire sequence.

Operating margins for the two devices just described depend on maintaining a high nucleation factor.   This means that edge domains along the length of the film must be completely eliminated throughout the film in its normal condition of saturation across the film strip.   To achieve this condition, the film is tapered by evaporation through a mask placed $\frac{1}{8}$ to $\frac{1}{4}$ in. from the deposition surface.   To obtain the high- and medium-coercive-force regions in the second system, a 250-Å layer of aluminum is evaporated selectively on the substrates prior to the magnetic-film deposition.   By proper choice of substrate deposition temperature, coercive force may be varied over the necessary 4:1 range.   No working system using this scheme has been reported.

c. Domain-tip Propagation

A shift register has been proposed[39] which utilizes channels of low-coercive-force material surrounded by high-coercive-force material in which reverse domains once inserted may be propagated by fields below the nucleation level.   The basic configura-

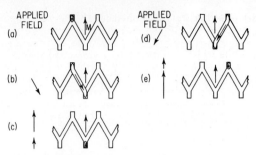

Fig. 14 Domain-tip-propagation shift register showing domain locations and applied fields.

tion is shown in Fig. 14. The principle of operation makes use of the fact that under the influence of a nearly transverse applied field the domain tip may be distorted such that at a junction of two channels it can be made to propagate down one channel in preference to the other. In the example shown, the magnetization of the surrounding high-coercive-force material is always up.

A small domain inserted as in (a) may be expanded down and to the right by the application of a field in that direction (b). A reverse easy-axis field (c) will cause the domain to be restored to original saturation everywhere except in the lower stub where this is prevented by the application of an inhibiting field in the down direction. We have thus succeeded in transferring the information from the upper left-hand to the lower right-hand stub. The transfer to the next upper right-hand stub will be accomplished by a similar sequence of pulses, which will employ a transverse field component in the opposite direction (d). The major advantage of this system over the system of Ref. 36 is the ability to make very long, continuous registers by using the zigzag patterns to turn corners. In addition, higher domain-propagation velocities are claimed, but detailed quantitative data have not been published.

d. Sonic-strain-wave Interrogation and Writing

A serial store has been proposed which uses a propagating strain wave to modulate sequentially the switching asteroid of a register of bits for both coincidence writing and nondestructive read.[40,41] A piezoelectric transducer attached to a quartz substrate holding the bits is excited and produces a strain wave along the strip of magnetostrictive magnetic film and at an angle φ to the easy axis of the storage bits. The simplest conception of this device uses an easy axis perpendicular to the strip length such that the applied stress is parallel to the easy axis, and the stress and magnetostriction are adjusted to provide a strain anisotropy component somewhat less than the inherent anisotropy causing the net H_k to be substantially reduced. A hard-axis bias and a pulsed easy-axis field, which is below the wall threshold, are applied to the entire strip, the latter in time coincidence with the strain wave at some desired bit position. The bit will then assume the remanent storage state corresponding to the pulsed field.

While the easy-axis strain wave provides good margin for writing, the resultant H_k reduction cannot be used directly for nondestructive interrogation. Since no rotation occurs, it must be combined with an easy-axis bias field to produce readout signals. This field is then severely limited by the need to stay below the threshold for irreversible switching.

If the stress wave is applied at an intermediate angle to the direction of propagation, it is possible to obtain signal output due to rotation of the easy axis without hard-axis bias but at the expense of writing margins. The angle of rotation is given by

$$\tan 2\theta_r = K \frac{\sin 2\varphi}{1 + K \cos 2\varphi}$$

where φ is the angle of applied strain and K is the normalized anisotropy change $\Delta H_k / H_k$.

The magnitudes of strain waves required for the best magnetostrictive materials tested so far are on the order of 10^{-4}.   Packing densities are essentially limited by available transducers which must supply pulses with rise times on the order of 5 ns in order to resolve bits which correspond to perhaps 25 to 50 ns of acoustical-delay length.   For quartz, with an acoustic velocity of 0.14 mil ns^{-1}, this corresponds to a bit density of between 150 to 300/in.   This, however, puts stringent requirements on the transducer since voltage pulses of up to 1,000 V may be required to produce the strain waves.   In addition, readout levels are extremely low, <1 mV, for films of relatively large dimensions, 100 to 200 mils.   As bit densities increase, shape anisotropy severely limits the signal output unless a magnetic structure closed in both hard and easy directions is used.

4. LOGIC

a. Wire-coupled Film Logic

(1) Open Structures   Attempts have been made to apply the techniques of wire-coupled core logic to films,[42] at the same time exploiting their high-speed rotation and orthogonal field properties.   These devices have an effective flux gain because steering currents have to be only large enough to overcome effective dispersion using the guided-fallback technique of standard memory practice.   Energy is thus supplied from the transverse field, which is a non-information-bearing pump.   The principal problem relates to the difficulty of providing an efficient coupling loop in which leakage inductance and resistance do not dominate the effective film inductance.   The problem arises from the need to limit film thickness due to demagnetizing fields in open-flux structures.   The performance of such structures depends fundamentally then on conductor thickness and spacing.

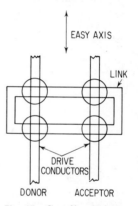

Fig. 15   Coupling link for information transfer between films.

The most practical method for providing low-impedance coupling loops is to evaporate films above a conducting ground plane such that the spacing between coupling loop and ground is determined by a relatively thin insulator layer.   The loop geometry is shown in Fig. 15, where two elements per cell provide balanced capacitive coupling to the drive conductors.   In order to estimate the current in the loop when the transmitting film is switched, we assume that sufficient transverse field is applied to the transmitting film that its magnetization rotates to the hard direction essentially instantaneously (i.e., in a time short compared with the L/R time constant of the coupling loop).   Under these conditions an impulse of flux is applied, providing an initial current given by

$$I_0 = \frac{\Phi}{L}$$

where L is the total initial inductance of the loop.   An exact expression for the total time response of current is difficult to obtain because of the time-varying nature of both R and L due to skin effect.

An approximate engineering solution may be obtained by assuming that the skin depth is time-dependent and given by

$$\delta = \sqrt{\frac{t}{4\pi\sigma}}$$

This assumes that the skin depth is independent of the link current, which is a reasonable approximation for determining the initial time constant.   For typical bit dimensions, i.e., intersections of conductors 40 mils wide, a coupling loop long enough to accommodate one donor and two acceptor cells is found to have a peak current of approximately 470 mA for a film thickness of 1,700 Å. This current is just sufficient to write one acceptor film fully but decays to one-half this value in less than 2 ns. Such a system would require pulses shorter than 2 ns and would still be only marginal and would have limited fan-out capability.

(2) Plated-wire Shift Register Many of the problems arising from the use of an open-flux structure may be eliminated by using a closed structure so that thicker films and lower digit currents may be used. A shift register using bit wire and resistive coupling loops which are connected to the wire is shown in Fig. 16a.[43] This

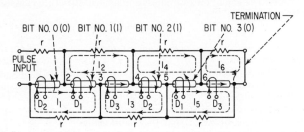

Fig. 16a Shift register using resistively coupled plated wires.

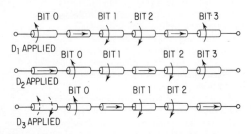

Fig. 16b Three-pulse sequence for information transfer in resistively coupled bit wires.

system requires an extra element for every two active bits. The shift register is designed such that each wire element is coupled to its left- and right-hand neighbors through a resistive loop. The system uses three drive phases of transverse field excitation so that at any given time every two adjacent elements storing information are separated from the next two by a bit which is held in transverse saturation. This blank bit is a receiver from the transmitter on its left, which in turn becomes a receiver on the next pulse from its left-hand neighbor, and so on, so that every third bit is transmitting to its right-hand neighbor while its left-hand neighbor is undisturbed by the backward bit wire current which flows during transfer.

The three-phase shift sequence is illustrated by the magnetization pattern of Fig. 16b. In Fig. 16a, current and magnetization direction are shown for the D_3 phase when element 3–4, for instance, is saturated in the hard direction, having just transmitted to element 4–5 and ready to receive 2–3 on the next D pulse. In addition to the current l_3 due to 3–4 switching, there is also a current l_2 which has no effect on element 2–3. Typical bit currents are from 2 to 6 mA with clock rates of 3 MHz for bit wires 0.2 in. long with 10,000 Å of Permalloy.

(3) Bit Steering for Content-address Memories A structure which combines both closed flux path and low impedance, tightly coupled ground plane is the coaxial configuration proposed for an associative memory.[44] Here the word current flows in the plated wire, and digit current is supplied by the transverse field strap. In such a memory 1 to 4 bits of all words are searched simultaneously to obtain a match condition which is then stored in a separate flag bit on the same wire. The transfer of energy required is essentially that which occurs in ordinary wired film logic except that many more inactive bits are in series in the transfer loop. That is, a mismatch signal causes a voltage to be generated which sends a current down the plated wire. This provides the easy-axis field necessary to condition the storage state of the flag bit during the fallback on the release of the transverse field. The coaxial structure is accomplished by essentially burying the plate wire in a copper ground plane. Impedances on the order of from 2 to 5 ohms allow digit currents of 20 to 30 mA to flow, located on a line which may be 30 to 40 bits long, sufficient to write information into a receiving bit.

b. Domain-tip-propagation Logic

The high fields provided by the free poles of a domain tip of the kind described in the shift register may be used as the basis for controlling the propagation of other domain tips, thus providing the means for performing all the basic logic functions.[45,46] Clearly, the operability of any such system will depend on the ratio between the stray field from the tip H_i, the field for nucleation H_N, and the field for tip propagation H_t. In general, the larger the ratio of H_i to H_t the greater the controlling effect and the higher the margins, while the higher the ratio of H_N to H_t, the greater the range of fields which may be applied to provide faster domain propagation.

Thus far two classes of structures have been investigated: those involving interaction in the plane,[45] as shown in Fig. 17, and those utilizing multilayer films having free-pole interactions at crossovers,[46] as shown in Fig. 18a. Planar interactions have

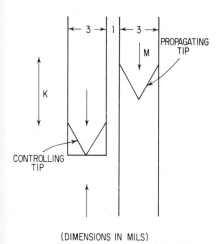

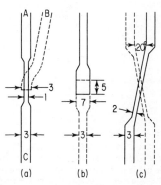

Fig. 17 Inhibition of domain propagation by stray field from a domain tip.

Fig. 18 Two-layer DTPL structures. (a) Inhibit gate. (b) Film-to-film information-transfer element. (c) Noninteracting channel crossovers. Dimensions are given in thousandths of an inch. Film-film separation is 0.0002 in. The easy-axis and applied-drive fields are vertical.

been demonstrated, but the devices are of marginal performance, typically yielding an interaction field of 1 to 3 Oe for propagation thresholds of from 3 to 6 Oe. Furthermore, the planar version is of only academic interest because any general practical system must provide for crossovers. Figure 18 shows experimental structures of the multilayer type illustrating (a) an inhibit gate, (b) film-to-film transfer element, and (c) a noninteracting crossover.

For a simple gate, as in Fig. 18a, in which one domain tip is used to inhibit the propagation of another, the limits on applied field are given by

$$H_{min} = H_t$$
$$H_{max} - H_i = H_t$$
$$H_{max} + H_i = H_N$$

That is, the minimum applied field just equals the threshold necessary simply to propagate a tip in the absence of any free poles. The maximum field must be small enough so that the free-pole interaction can prevent the main-channel domain from propagating. A further restriction is that the combined field of the free poles and applied field does not exceed the nucleation field.

An experimental DTPL structure having $H_N = 12$ Oe, $H_t = 4$ Oe has been made. This structure achieved the maximum predicted tolerances since (from the previous equations)

$$H_{max} = \tfrac{1}{2}(H_N + H_t) = 8 \text{ Oe}$$

resulting in threshold tolerances of

$$\text{Tol.} = \pm\tfrac{1}{2}\frac{H_{max} - H_{min}}{(H_{max} + H_{min})/2} = \pm 33\%$$

Note that this is the upper limit on performance if one assumes that H_i cannot exceed H_t, since $H_i = H_{max} - H_t$, $H_i = H_t$, and $H_N = 3H_t$. To first order this is a reasonable assumption since the sending film cannot support a demagnetizing field in excess of its propagation threshold.

c. Domain-wall Logic

While most film-logic and storage systems rely on the magnetization direction in domains as the information-bearing mechanism, it has been proposed[47] that the direction of magnetization within the narrow walls between domains may be used for this purpose. There are four distinct combinations of Néel wall and domain-magnetization direction, as shown in Fig. 19. For both upper patterns the magnetization is

NÉEL WALLS

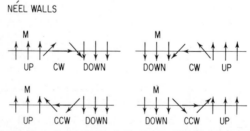

Fig. 19 Clockwise and counterclockwise magnetization across a Néel wall.

seen to rotate clockwise as we proceed from left to right across the wall separating the domains. Similarly for both lower patterns the magnetization rotates counterclockwise as we cross the wall. Walls of the first type (clockwise) are known as P (positive) walls, while counterclockwise walls are N (negative). It is important to note that neither the wall nor domain-magnetization direction by itself, but rather the combination of the two, determines whether a wall is of P or N type.

The proposed logic system utilizes the fact that two walls of unlike type when brought together will extinguish one another easily, while two walls of like type tend to remain intact for relatively high applied fields. In the only experimental work done thus far,[48] films less than 100 Å thick were necessary in order to prevent "unraveling" of the walls at their extremities. For films in the 50 to 100-Å region, a ratio of better than 10:1 was obtained for the field required to erase walls of like sense compared with walls of opposite sense. Two conductors, one for writing ONEs and one for ZEROs, angled at approximately $+60°$ and $-60°$ to the film strip ($\pm 30°$ to the easy axis) provide the means of injecting a wall at the input to the register. These two conductors supply easy-axis fields which are equal and hard-axis fields which are opposite in direction, thus producing P- and N-type walls. For a shift register the walls may be propagated by two continuous hairpins also at 60° which essentially have the domains along preserving the clockwise and counterclockwise orientations within the walls. Conceptually, the ability to extinguish or preserve walls by virtue of the direction of the magnetization change across the wall provides a mechanism for forming AND and OR gates and ultimately realizing general-purpose logic, although as yet little has been done to test the feasibility of such a system.

d. Parametrons

The basic operating principles of the parametron[49] require a time-varying reactance which is pumped at a constant frequency 2ω. If such a reactance is resonated at a frequency which is one-half the pump frequency, then the phase of the resonant signal is indeterminate, and by the injection of a small amount of energy at the resonant frequency it is possible to condition this phase and use this as the basic information-storage mechanism.

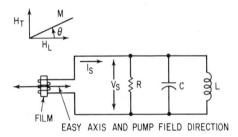

Fig. 20   Magnetic-film parametron circuit.

The use of magnetic films as the time-varying inductive has been studied by a number of workers.[50,51] The basic device is shown in Fig. 20, where a bias field of a constant value H_0 and a varying pump component $H_p(t) = H_{pm} \sin 2\omega t$ are applied along the easy axis. A signal winding linking the hard-axis flux is resonated with an external capacitor. The equilibrium conditions for the easy-axis pump and bias fields and the hard-axis field due to the signal winding current are given by

$$(H_0 + H_p) \sin \theta + i_s K_1 \cos \theta = 0$$
$$\lambda = \lambda_m \sin \theta$$

where the intrinsic H_k of the film is assumed small compared with H_0 and where λ = hard-axis component of flux. The easy-axis field may be thought of as varying the effective H_k of the film and hence the effective inductance of the tank circuit such that

$$L = \frac{L_0 \sqrt{1 - \lambda^2/\lambda_m{}^2}}{1 + H_p/H_0}$$

where $L_0 = K_0 \lambda_m/H_0$ is the small-signal unmodulated inductance. The circuit equation then becomes

$$C\ddot{\lambda} + \frac{1}{R_e} \dot{\lambda} + \frac{1}{L_0} \lambda \frac{1 + Hp/H_0}{\sqrt{1 - \lambda^2/\lambda_m{}^2}}$$

where the effect of the pump is indicated by substituting the equivalent modulated inductance in the resonant-circuit equation. Taking into account the fundamental damping which appears in the modified Landau-Lifshitz equation of Gilbert, the effective resistance R_e includes a component $(\alpha/\nu L_0 H_0)$. Expanding $(1 - \lambda^2/\lambda_m{}^2)^{-\frac{1}{2}}$, we get

$$C\ddot{\lambda} + \frac{1}{R_e}\dot{\lambda} + \frac{1}{L_0}\left(1 + \frac{H_p}{H_0}\right)\left(\lambda + \frac{\lambda^3}{2\lambda_m{}^2}\right) = 0$$

This equation assumes that the capacitor C has been chosen so that the resonance is considerably below the natural resonance condition given by the Landau-Lifshitz equation, thus considerably simplifying the equations.

If the signal flux $\lambda(t)$ is assumed to build exponentially, it will be of the form

$$\lambda(t) = A \exp\frac{\alpha\omega t}{2\pi \sin(\omega t + \theta)}$$

where $\alpha = \pi[H_p/2(H_0) - 1/Qe]$ for small θ.

The minimum threshold for stable oscillation will then occur when α goes positive or

$$\frac{H_{pm}}{2(H_0)} = \frac{1}{Qe}$$

The maximum value of α will occur when $H_{pm}/H_0 = \frac{1}{2}$. For $Q_0 = \gg 1$ this gives a maximum gain per cycle of $e^{\pi/2}$. This derivation neglects the presence of leakage inductance in the signal winding. For an open-flux structure, demagnetizing effects will limit film thickness as dimensions decrease, which means that air inductance becomes increasingly significant relative to effective film inductance. This adds a constant to the effective film inductance and effectively lowers the negative resistance, the threshold condition being modified to

$$\frac{H_{pm}}{H_0}\frac{L_0}{L_a + L_0} = \frac{1}{Qe}$$

While individual film parametrons have been demonstrated in the laboratory, no significant system has been reported. The complexities of timing and supplying pump power as well as the complicated wiring necessary to provide useful fan-out made this technique impractical, particularly at the low signal levels and densities possible with simple open-flux structures.

Typical characteristics of an operating film parametron are[52] $2\omega = 14.4$ MHz, $Q_0 = 12$, and $\alpha = 1.7$.

5. OPTICAL DEVICES

a. Magneto-optic Readout

Increasing attention is being given to the problem of using optical techniques for interrogating information stored magnetically.

(1) Sequential Access In the case of sequential rotating stores (e.g., tapes, disks, drums) present limitations on track and bit density due to mechanical constraints on head design and speed limitations imposed by mechanical rotation suggest the potential advantages of magneto-optic readout using a highly focused, scanning photon beam. This approach was originally investigated and found to be unfeasible[53] until the advent of the laser provided a photon beam of essentially unlimited energy density.[54] The basic system for interrogating stored information in a sequential store is to focus a light source to illuminate a single bit with polarized light and to detect the effective rotation of the plane of polarization, due to the Kerr effect, of the reflected wave. There are three basic sources of noise in such a system: shot noise, due to the discrete nature of light; surface noise, due to surface imperfections; and fluctuations in light level, due to temporal variations in the light beam and spatial variations in the reflectivity of the surface.

For a beam of polarized light viewed through an analyzer after reflection from the surface, the average light power will be given by

$$W = W_0 \sin^2 \theta$$

where W_0 is the incident light and θ is the angle of rotation of the analyzer from minimum transmission. The change in light power transmitted by the analyzer due to a small rotation of polarization $\Delta\theta$ is

$$\Delta W = W_0(2 \sin\theta \cos\theta)\Delta\theta = W_0\varphi \sin 2\theta$$

where φ is the Kerr angle, typically of the order of a few minutes. This results in a signal current from a photocathode with a conversion coefficient P given by

$$I_s = PW_0\varphi \sin 2\theta$$

with a direct current due to the average light of

$$I_B = PW_0 \sin^2\theta$$

which results in a shot-noise current given by

$$I_n = (2efPW_0)^{\frac{1}{2}} \sin\theta$$

where f is the bandwidth and e is the charge or the electron. This produces a signal-to-shot-noise ratio of

$$\frac{I_s}{I_n} = \left(\frac{2PW_0}{ef}\right)^{\frac{1}{2}} \varphi \cos\theta$$

which is essentially independent of angle θ for all angles up to $10°$.

The dependence of signal-to-shot-noise ratio on illumination intensity in the above equation implies that the essential limit is determined by Δt, the allowable temperature rise of the film, for a coherent source which can be focused down to the diffraction limit. The temperature rise assuming no diffusion (e.g., for a thin film on an insulating substrate) is given by

$$\Delta t = \frac{W_a}{fGC}$$

where W is the absorbed light power. For a sequential store where the time a bit is illuminated is inversely proportional to the bandwidth or rotational frequency of the system, the signal-to-noise ratio will therefore be independent of frequency and is given by

$$\frac{I_s}{I_n} = \left[\frac{2P \, \Delta t \, GCR}{e(1-R)}\right]^{\frac{1}{2}} \varphi \cos\theta$$

which yields a signal-to-noise ratio of 30 db for $A = 10^{-6}$ cm^2 $\Delta t = 100°$C, $P = 3 \times 10^{-2}A/W$, the reflection coefficient $R = 0.5$, and $\varphi = 10^{-3}$ rad, $C = 0.5J/g$ deg, and a specific gravity of 8 g cm^{-2}. This is a worst-case solution in which no heat diffusion is assumed. If a conducting substrate or other provision is made for removing heat, the allowable illumination level may be raised accordingly so that shot noise is no longer a limiting factor. A similar analysis using a conventional incoherent light source shows that it is possible to attain signal-to-noise ratios of only about one for bits approximately 1 mil in diameter, scanned at a 1-MHz rate.

Noise which is generated by diffraction from small pinholes and other such imperfections is essentially depolarized and can therefore be discriminated against by simply setting the analyzer angle to large angles to increase the signal level, since this source of noise is essentially insensitive to angle. This, however, reduces contrast and makes the system more sensitive to fluctuations in incident light intensity. The analyzer setting must therefore be chosen as a compromise between opposing conditions.

If the spot size is larger than most pinholes and other imperfections, one may use the large difference in the cone angles in the far-field region to insert a stop which blocks the diffracted light from the imperfections without losing a significant amount of signal energy.

Finally, in order to eliminate noise due to light fluctuations, the reflected beam may be divided by a beam splitter and fed into two polarizers, one set at $+\theta$ and the other at $-\theta$ from "extinction," and each fed into a photocell whose outputs are in turn fed to opposite sides of a difference amplifier, thus providing a null. A change in magnetization will cause the outputs to rotate to $\theta + \varphi$ and $-\theta + \varphi$, respectively, thus producing magneto-optic signals of opposite signs whose difference will appear at the output of the amplifier.

(2) Random Access The problem of accessing static stores optically is considerably more complex.[55] The motivation for this approach lies in the possibility of achieving higher bit densities than might otherwise be feasible since signal energy no longer depends in any simple way on magnetic flux. Thus, the usual interdependence of signal and demagnetizing energy is no longer controlling. The principal difficulty with developing such an approach is that no simple method is available for precise deflection of light beams so that, in general, it is necessary to accomplish at least some part of the addressing process by other means. This requires that many bits be illuminated, simultaneously, while using a selective array of light detectors and/or modulating the selected bit magnetically to produce a distinguishable signal. Because the magneto-optic interactions are fairly weak to begin with, this implies being able to supply fairly large amounts of optical power but more important the signal-to-shot-noise ratio is reduced by $\sqrt{N}$, where N is the total number of background bits illuminated simultaneously along with the selected bit.

It should be pointed out that any attempt to cancel out this background (e.g., by interference effects, etc.) usually has the effect of reducing the signal at the same time. This results from the fact that the output signal is given by

$$\Delta W = W_0{}^2 \sin \theta \cos \theta$$

Cancellation of background is equivalent to making $\theta = 0$ as in the case for crossed polarizer and analyzer and causes the signal to decrease as well as the background.

At the present time no system has been provided which overcomes the shot-noise limitation. Even if magneto-optical detection proves feasible, it still requires an entirely separate apparatus for writing in addition to that used for magnetic modulation. This must be provided by conductors in the conventional manner or by some other means (e.g., Curie-point writing with electron beams).[55] If the former is used, there are serious mechanical problems related to mechanical interference between conductors and light paths. If the latter are used, the usual problems attendant to operation in a vacuum are encountered as well as the added complications of achieving a suitable modulation mechanism.

b. Thermostrictive Recording

One method for recording on films which might be compatible with magneto-optic readout utilizes high local stress induced by heating with either an electron or photon beam.[56] For an externally applied stress σ at an angle β with respect to the easy axis it has been shown that the anisotropy will undergo a rotation through an angle θ_0 given by

$$\sigma = \frac{M_s H_k}{3\lambda_s} \frac{\sin 2\theta_0}{\sin 2(\beta - \theta_0)}$$

where M_s is the saturation magnetization and λ_s is the magnetostriction, which is assumed to be isotropic. If a temperature distribution of the form

$$\Delta T(r) = \Delta T_0 \left(1 - \frac{r^2}{a^2} \right)$$

is assumed where ΔT_0 is the maximum temperature rise within the beam spot of diameter a, the stress on a thin circular disk resulting from such a temperature is given by

$$\sigma_r = -\frac{1}{4} \alpha Y \Delta T_0 \left(1 - \frac{r^2}{a^2} \right) \qquad \text{in the radial direction}$$

$$\sigma_\theta = -\frac{1}{4} \alpha Y \Delta T_0 \left(1 - \frac{3r^2}{a^2} \right) \qquad \text{in the tangential direction}$$

where α is the coefficient of linear expansion and Y is Young's modulus. The total energy required to satisfy this temperature distribution is given by

$$E = 2\pi tC\rho \int^a r\,\Delta T(r)\,dr$$

$$= \frac{\pi t}{2} C\rho a^2\,\Delta T_0$$

where t is film thickness, ρ the mass density, and C the specific heat.

For $\beta = 90°$, we find that the amount of the hard-axis stress necessary to just flip the easy axis is

$$\sigma = \frac{M_s H_k}{3\lambda_s}$$

which will occur at the center of the spot for a temperature rise

$$\Delta T_0 = \frac{M_s H_k/3\lambda_s}{\alpha Y/4}$$

Thus, the energy required for recording by thermostriction is

$$E = \frac{2}{3}\frac{M_s H_k}{\lambda_s}\frac{V\rho C}{\alpha Y}$$

here V is the film volume. If in coincidence with this thermal stress a longitudinal field of either polarity below the wall threshold is applied, the magnetization will relax into the corresponding two-easy-axis remanent state. Writing may also be accomplished by a hard-axis field below the normal anisotropy field due to the effective rotation of the switching asteroid due to the applied stress. Only a limited demonstration of this technique has been reported. It should be noted that stress-control requirements should be considerably less severe than for sonic-delay lines which have stringent threshold limits on both stress and field excitations. The ultimate resolution and speed of this technique will depend on thermal diffusion, demagnetizing considerations, and deflection technology, as well as the magneto-optic or other methods necessary for readout.

c. Variable-diffraction Display

In certain films having negative magnetostriction, it has been observed that so-called "rotatable anisotropy" is present. These films tend to break up into "stripe domains" with a periodic variation in surface pole magnetization. When a Bitter solution is applied to the surface of such a film, it is possible to form a diffraction grating, which is electrically alterable by switching the domain pattern between its two stable states, which are $90°$ apart,[57,58] as in Fig. 21.

This configuration can be used to write in an arbitrary combination of bits, which thus provides a display with built-in storage. While the switching time of the indi-

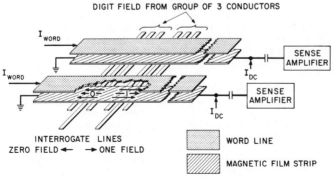

Fig. 21 Associative memory cell.

vidual bits is less than 100 ns, the time for the colloid to redistribute after reorientation of the magnetization is approximately 10 ms, which is still adequate to provide a flicker-free display. The crucial display parameters depend on the efficiency of the grating and the resultant contrast and absolute light levels obtainable.

A typical grating[57] illuminated by a gas laser at 6,328 Å at an angle of 45° produces a diffracted beam which peaks at $-29.5°$ and a reflected ray at 45° with a background intensity over all angles due to scattering which contains more energy than the diffracted beam. Thus, while the contrast is about 100 within the 5° beam width of the diffracted beam, any attempt to widen the angle of view results in serious degradation of the contrast.

A detailed understanding of the interaction between the incident light and the Bitter particles depends upon the structure which is assumed to collect at the free poles of the stripe domains.

Analysis for a sinusoidal magnetization distribution in the film shows that there should be negligible grating structure, and indeed normal viewing of the colloid solution fails to show any significant structure unless special pains are taken to adjust the microscope illumination. If either a square-wave or trapezoidal distribution is assumed, significant structure can be demonstrated theoretically; but much stronger effects can be shown to occur for the cases where a large external field is applied either normal to the film plane or perpendicular to the stripe domains in the plane of the film. Under these conditions the grating period increases by a factor of 2, and the structure is readily visible under normal conditions. In the experiment described previously, the original wavelength of 5,250 Å was doubled, and the ratio of diffracted to scattered energy was improved by over a factor of 4 by applying a normal field of 49 Oe.

Further evidence for the trapezoidal distribution is the experimental verification of the predicted variation of grating spacing as a function of film thickness t

$$d \alpha t^{\frac{1}{2}}$$

The technological problems associated with obtaining a display with sufficient contrast and viewing angle are further complicated by the need for a colloid solution and package which will eliminate the need for regular refreshing of the solution. Progress has been made in all these areas, but ultimate practicality of a working system has yet to be established. The overall ratio of diffracted power to incident power of the best films to date has been about 12% using the normal field enhancement.

6. MISCELLANEOUS

a. Associative Memory

An associative or content-addressed memory requires that a bit or bits of an interrogate word be compared simultaneously with a corresponding bit or bits of every word in the memory and all match locations indicated. Such a memory necessarily uses nondestructive readout. For fully parallel access, the simplest magnetic realization requires that

1. A perfect match produces no output.
2. Mismatch of from 1 to n bits produces an output which ranges from δ to $n\delta$, where δ is the output from a single mismatched bit. The detection problem, then, for the worst case reduces to discriminating between a signal δ and no output.

In general most magnetic-storage devices cannot satisfy either condition above directly because

1. Nondestructive readout usually produces an output for both polarities of input, and/or
2. The outputs for ZEROs and ONEs are opposite in polarity.

The usual technique for achieving the required outputs is to use a pair of elements or even two pairs to provide unipolar mismatch and nulled-match outputs. Aside from requiring extra elements, these techniques inevitably depend upon the ability to cancel outputs from many pairs and therefore require very strict control of film-output uniformity. It is possible, however, to realize the required functions in a single film cell.

Consider an interior region of a saturated magnetized film line (removed from boundaries or walls) to which is applied an easy-axis reversing field greater than H_c but below the field for wall nucleation H_N.[59]   In 1,000-Å films, in the absence of transverse fields, "birotation" occurs, that is, incoherent rotation which results in regions with positive and regions with negative transverse components (often with walls in between).   When the field is removed, the region returns to its original state.   A field of opposite polarity has no significant effect on the domain already saturated in that direction.

Figure 22 shows a number of cells of a proposed memory.[60]   The structure consists of a normal word line over a strip of magnetic film and a digit line split into three parallel conductors.   Writing is accomplished using the standard method of coincidence of transverse and digit fields (all three digit conductors are excited equally). The result is a region at the intersection which is magnetized either to the right for a ONE or to the left for a ZERO.   To interrogate the memory in all bits simultaneously, the middle-digit conductors alone are excited with a field to the left if searching for a ZERO, and with a field to the right if searching for a ONE.   For either match condition, there will be essentially no change in the magnetization of the interrogated cell. For a mismatch, the magnetization will rotate noncoherently to the multidomain pattern which occurs just below nucleation, as described above.   It has been proposed

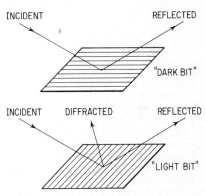

Fig. 22   "Rotatable" diffraction grating for display.

that the magnetoresistive effect be used to sense the mismatch condition.   Experimentally, it is found that a resistance change corresponding to approximately 10% of that for full transverse switching is obtained.   The advantages of this technique are:

1. The sense line (the film in this case) is perpendicular to the interrogate line.
2. No extra conductors are required.
3. The use of the magnetoresistive effect allows frequency discrimination between the rise-time transients and the signal, which is proportional to flux rather than to the rate of change of flux.
4. The output for both mismatch conditions is of the same polarity.

While a system of the type proposed has not been built, experiments indicate that signal levels of the order of 1 mV can be obtained for cells 10 mils wide by 100 mils long, 1,000 Å thick.[61]

REFERENCES

1. Pohm, A. V., and S. M. Rubens, *Eastern Joint Computer Conf. Proc.*, December, 1956, pp. 120–123.
2. Raffel, J. I., T. S. Crowther, A. H. Anderson, and T. O. Herndon, *Proc. IRE*, **49**(1), 155 (January, 1961).
3. Katz, H. W., A. H. Anderson, W. Kayser, J. F. Gianola, H. P. Louis, J. H. Glaser, C. D. Olson, J. Hart, and I. W. Wolf, *IEEE Trans. Mag.*, **MAG-1**(3), 218 (September, 1965).
4. Crowther, T. S., *J. Appl. Phys.*, **34**(3), 580 (March, 1963).
5. Bradley, E. M., *Brit. IRE*, **20**, 765 (1960).
6. Coren, R. L., *Intermag Conf. Proc.*, 1965, p. 5.5-1.
7. Anderson, A. H., T. S. Crowther, and J. I. Raffel, *J. Appl. Phys.*, **34**(4), pt. 2, 1165 (April, 1963).
8. Olson, A. L., and E. J. Torok, *J. Appl. Phys.*, **37**(3), 1297 (March, 1966).
9. Gagnon, G. P., and T. S. Crowther, *J. Appl. Phys.*, **36**(3), pt. 2, 1112 (March, 1965).
10. Siegle, W. T., and B. L. Flur, *IEEE Trans. Mag.*, **MAG-3**(3), 484 (September, 1967).
11. Raffel, J. I., *J. Appl. Phys.*, **30**(4), 60 (April, 1959).
12. Bland, G. F., *IBM J. Res. Develop.*, **7**(3), 252 (July, 1963).
13. Blatt, H., *M.I.T. Lincoln Lab. Tech. Note* 1964–6, August, 1964.

14. Raffel, J. I., A. H. Anderson, T. S. Crowther, T. O. Herndon, and C. Woodward, *Spring Joint Computer Conf. Proc.*, Atlantic City, N.J., April, 1968.
15. Raffel, J. I., *J. Appl. Phys.*, **35**(3), pt. 2, 748 (March, 1964).
16. Danylchuk, I., A. J. Perneski, and M. W. Sagal, *Intermag Conf. Proc.*, 1964, p. 5.4-1.
17. Fedde, G. A., *J. Appl. Phys.*, **37**(3), 1373 (March, 1966).
18. Finch, T. R., and S. Waaben, *IEEE Trans. Mag.*, **MAG-2**(3), 529 (September, 1966).
19. Agajanian, A. H., and C. G. Ravi, *IEEE Trans. Mag.*, **MAG-3**(3), 500 (September, 1967).
20. Barker, P. D., and E. J. Torok, *J. Appl. Phys.*, **37**(3), 1363 (March, 1966).
21. Meier, D. A., and A. J. Kolk, in M. C. Yovits (ed.), "The Magnetic Rod—A Cylindrical, Thin Film Memory Element, Large Capacity Memory Techniques for Computing Systems," p. 195, The Macmillan Company, New York, 1961.
22. Smith, J. L., *J. Appl. Phys.*, **34**(4), pt. 2, 1167 (April, 1963).
23. Leilich, H. O., *J. Appl. Phys.*, **37**(3), 1361 (March, 1966).
24. Berger, R., T. S. Crowther, and J. I. Raffel, *J. Appl. Phys.*, **37**(3), 1359 (March, 1966).
25. Pugh, E. W., *Intermag Conf. Proc.*, 1963, p. 15.1-1.
26. Kashiwagi, E., and H. Murakami, *IEEE Trans. Mag.*, **MAG-2**(3), 524 (September, 1966).
27. Oakland, L. J., *J. Appl. Phys.*, **30**(4), 54 (April, 1959).
28. Sie, C. H., *J. Appl. Phys.*, **37**(3), 1375 (March, 1966).
29. Maltz, M. S., *J. Appl. Phys.*, **36**(3), 1121 (March, 1965).
30. Kohn, G., W. Jutzi, T. Mohr, and D. Seitzer, *Intermag Conf. Proc.*, 1965, p. 8.3-1.
31. Toombs, H. D., and T. E. Hasty, *Proc. IRE*, **50**(6), 1526 (June, 1962).
32. Billing, H., A. Rudiger, and R. Schilling, *IEEE Trans. Mag.*, **MAG-2**(3), 520 (September, 1966).
33. Chu, W. W. L., J. E. Wolfe, and B. C. Wagner, *J. Appl. Phys.*, **30**(4), 272 (April, 1959).
34. Beam, W. R., and W. T. Siegle, *IEEE Trans. Mag.*, **1**(1), 66.
35. Crowther, T. S., *Intermag Conf. Proc.*, 1965, p. 2.8-1.
36. Broadbent, K. D., *IEEE Trans. Electron Components*, **EC-9**(3), 321 (September, 1960).
37. Spain, R. J., and H. Jauvtis, *J. Appl. Phys.*, **36**(3), pt. 2, 1101 (March, 1965).
38. Spain, R. J., H. Jauvtis, and H. Fuller, *J. Appl. Phys.*, **36**(3), pt. 2, 1103 (March, 1965).
39. Spain, R. J., *J. Appl. Phys.*, **37**(7), 2572 (June, 1966).
40. Weinstein, H., L. Onyshkevych, K. Karstad, and R. Shahbender, *Proc. Joint Computer Conf.*, **29**, 333 (November, 1966).
41. Cohler, E. U., and R. Rubinstein, *IEEE Trans. Mag.*, **MAG-2**(3), 528 (September, 1966).
42. Edwards, J. G., *Proc. IEE*, **112**(1), 40 (January, 1965).
43. Dick, G. W., and W. D. Farmer, *IEEE Trans. Mag.*, **MAG-2**(3), 343 (September, 1966).
44. Chow, W. F., and L. M. Spandorfer, *IEEE Trans. Mag.*, **MAG-3**(3), 504 (September, 1967).
45. Spain, R. J., and H. Jauvtis, *J. Appl. Phys.*, **37**(7), 2584 (June, 1966).
46. Spain, R. J., and H. Jauvtis, *J. Appl. Phys.*, **38**(3), 1201 (March, 1967).
47. Smith, D. O., *IRE Trans. Electron. Components*, **EC-10**(4), 708 (December, 1961).
48. Ballantyne, J. M., *J. Appl. Phys.*, **33**(suppl.)(3), 1067 (March, 1962).
49. Goto, E., *Proc. IRE*, **47**, 1304 (August, 1959).
50. Pohm, A. V., A. A. Read, R. M. Steward, Jr., and R. F. Schauer, *J. Appl. Phys.*, **31**(suppl.)(5), 119 (May, 1960).
51. Johnson, A. R., *IEEE Trans. Mag.*, **MAG-1**(1), 26 (March, 1965).
52. Pohm, A. V., A. A. Read, R. M. Stewart, Jr., and R. F. Schauer, *Proc. Natl. Electron. Conf.*, 1959, p. 202.
53. Fan, G., E. Donath, E. S. Barrekette, and A. Wirgin, *IEEE Trans. Electron. Computers*, **EC-12**(1), 3 (February, 1963).
54. Treves, D., *J. Appl. Phys.*, **38**(3), 1192 (March, 1967).
55. Smith, D. O., *IEEE Trans. Mag.*, **MAG-3**(4), (December, 1967).
56. Kump, H. J., and P. T. Chang, *IBM J. Res. Develop.*, **10**(3), 255 (May, 1966).
57. Cohler, E. U., H. Rubinstein, and C. Jones, *IEEE Trans. Mag.*, **MAG-2**(3), 304 (September, 1966).
58. Spain, R. J., H. W. Fuller, and R. J. Webber, *IEEE Trans. Mag.*, **MAG-2**(3), 288 (September, 1966).
59. Methfessel, S., S. Middlehoek, and H. Thomas, *J. Appl. Phys.*, **32**(suppl.)(4), 294 (March, 1961).
60. Raffel, J. I., and T. S. Crowther, *IEEE Trans. Electron. Computers*, **EC-13**(5), 611 (October, 1964).
61. Naiman, Mark, *Intermag Conf. Proc.*, 1965, p. 11-2.

Chapter **22**

Superconductive
Thin Films and Devices

R. EDWIN JOYNSON

General Electric Research, Schenectady, New York

1. SUPERCONDUCTIVE FILMS

a. Review of Superconductivity in Bulk Material

When cooled below 21°K, a number of elements, alloys, and compounds display superconductivity.[1] In this state a pure, unstrained sample will have two electromagnetic properties: (1) in the absence of transport current or applied magnetic field, the electrical resistance abruptly drops to zero at the temperature T_c, a characteristic of the material; and (2) magnetic flux is excluded.

At temperatures below T_c, superconductivity can be destroyed by the application of a magnetic field sufficient to produce the critical field H_c at the surface of the sample. For samples with zero demagnetization factor,[2] the field reaches the critical value at the same time over the entire surface, and the transition into the normal state proceeds simultaneously and completely throughout the body. For other factors the transitions take place over a range of field values. The temperature dependence of the bulk critical field H_c is very nearly parabolic:

$$H_c = H_0(1 - t^2)$$

where H_0 is the critical field at absolute zero and t is the reduced temperature $t = T/T_c$. The critical temperatures and fields of several superconductive elements are listed in Table 1. Superconductivity can be destroyed also by an impressed current. In bulk samples of type I and defect-free, homogeneous type II superconductors (the types to be discussed later), the critical current is that which produces the critical magnetic field at the surface of the sample (Silsbee's hypothesis[3]).

TABLE 1 Superconductive Parameters of Some Elements

Element	Critical temp T_c, °K	Critical field H_0, Oe	Penetration depth $\lambda(0)$, Å	Ginzburg-Landau parameter κ	Correlation distance ξ_0, Å
Al..........	1.20	99	515	0.03	16,000
In..........	3.40	293	515	0.1	2,600
Sn..........	3.72	309	510	0.2	2,300
Ta.........	4.48	830	(500)	0.5	
Pb.........	7.19	803	390	0.4	830
Nb.........	9.46	1,980	(500)	1.1	

The exclusion of flux is a fundamental property of superconductivity. It cannot be deduced from the property of zero resistance alone, which implies that the flux through a superconductor would remain unchanged after the transition, rather than being expelled from the interior. Both properties are contained in the London equations,[4] which adequately describe many macroscopic superconducting phenomena:

$$\mathbf{H} = -\frac{4\pi\lambda_L{}^2}{c} \operatorname{curl} \mathbf{J} \tag{1}$$

$$\mathbf{E} = \frac{4\pi\lambda_L{}^2}{c} \dot{\mathbf{j}} \tag{2}$$

where
$$\lambda_L = \left(\frac{mc^2}{4\pi ne^2}\right)^{-\frac{1}{2}} \tag{3}$$

$\mathbf{H}$ and $\mathbf{E}$ are the magnetic and electric fields, $\mathbf{J}$ is the supercurrent density, and n is the density of superconducting electrons. From Eq. (2) it follows that $\mathbf{E} = 0$ for stationary conditions. Current can still flow, however, as it is linked to the magnetic field.

When Eq. (1) is combined with Maxwell's equations, one obtains

$$\lambda_L{}^2 \nabla^2 \mathbf{H} = \mathbf{H} \tag{4}$$

and
$$\lambda_L{}^2 \nabla^2 \mathbf{J} = \mathbf{J} \tag{5}$$

Upon applying Eq. (4) to a bulk superconductor in a magnetic field, it can be shown that, deep inside the superconductor, the field is vanishingly small (the Meissner effect).

The magnetic field is not discontinuous at the surface, however, but in the London theory decays exponentially with distance x into the surface according to H exp $(-x/\lambda_L)$. The quantity λ_L is seen to be a penetration depth, of magnitude 10^2 to 10^3 Å, for the magnetic field. A surface current having the same exponential factor exists in the penetrated region. Since λ_L depends upon the density n of superconducting electrons, it is a function of temperature. From the two-fluid model of superconductivity proposed by Gorter and Casimir,[5] the fraction of superconducting electrons at a reduced temperature t is given by

$$\frac{n_s}{n_0} = 1 - t^4$$

If this is used in Eq. (3), there results

$$\lambda_L(T) = \lambda_L(0)(1 - t^4)^{-\frac{1}{2}}$$

The penetration depth at zero temperature $\lambda_L(0)$ is another material characteristic (Table 1).

Equation (1) can be written in terms of the vector potential $\mathbf{A}$ ($\mathbf{H} = \operatorname{curl} \mathbf{A}$),

$$\mathbf{J} = \frac{-c}{4\pi\lambda_L{}^2} \mathbf{A} \tag{6}$$

The London theory is seen to be local in character in that it relates the current density at a point in space to the vector potential [or to the magnetic field, Eq. (1)] at that point. It does not consider the long-range correlations now known to exist between electrons. Further, λ_L in Eq. (3) contains only the material properties n and m (the effective mass of an electron), which are insufficient to describe the effects of impurities or sample size on the measured penetration depth. To account for these effects, Pippard[6] proposed that the supercurrent at a given point in space be obtained by averaging the vector potential over the range of correlation between electrons. He suggested replacing Eq. (6) by a "nonlocal" expression

$$\mathbf{J} = \frac{-3c}{16\pi^2\xi_0\lambda_L{}^2} \int \frac{(\mathbf{A} \cdot \mathbf{r})\mathbf{r}}{r^4} e^{-r/\xi} \, dv \tag{7}$$

where ξ_0 is the correlation distance for a pure superconductor and ξ, the effective correlation distance, is a function of the electron mean free path l,

$$\frac{1}{\xi} = \frac{1}{\xi_0} + \frac{1}{l}$$

The penetration depth λ in the nonlocal theory cannot be expressed explicitly except in certain simplifying circumstances. It can be seen, however, that the material properties are contained in λ_L and ξ_0. The size of the specimen enters in the limits of integration, and in the mean free path if a dimension is less than the mean free path of the bulk material. The effect of impurities is to lower the mean free path, making the effective correlation distance smaller and hence reducing the volume over which the vector potential is averaged.

For pure bulk superconductors at temperature not too close to T_c, $\lambda \ll \xi$. The nonlocal penetration depth for these conditions has been shown[7] to be

$$\lambda = \left(\frac{\sqrt{3}}{2\pi} \xi_0 \lambda_L{}^2 \right)^{\frac{1}{3}}$$

The opposite limit, $\lambda \gg \xi$, may occur for a pure bulk superconductor at a temperature close to T_c where λ is large because of its temperature dependence, or for an alloy or thin film where ξ is limited by a short mean free path. The vector potential $\mathbf{A}$ is then constant over the correlation distance ξ, and Eq. (7) reduces to a local expression. The penetration depth becomes

$$\lambda = \lambda_L \left(\frac{\xi_0}{\xi} \right)^{\frac{1}{2}}$$

A result of the nonlocal relation [Eq. (7)] is that the decays of current density and magnetic field into a superconductor are no longer exponential. At some distance from the surface they reverse direction. Drangeid et al.[8] have obtained experimental verification for the reversal.

Not only is the London theory limited to the condition $\lambda \gg \xi$, but it is also a weak-field theory, valid only for magnetic fields much less than the critical field. A theory invented explicitly to handle strong magnetic fields is that of Ginzburg and Landau.[9] They included in the free-energy function a term to account for the spatial variation with field of the superconducting electron density,

$$F_s(H) = F_s(0) + \frac{H^2}{8\pi} + \frac{\hbar^2}{2m} \left(-i\nabla - \frac{e}{c}\mathbf{A} \right) |\psi|^2$$

$|\psi|^2$ may be considered as the density n_s of superelectrons. To apply this equation, $F_s(0)$, the free energy in the absence of a magnetic field, must be specified as a function of the order parameter ψ. Finding an exact form good for all temperatures has been difficult. Ginzburg and Landau expanded $F_s(0)$ in a power-series in $|\psi|^2$,

$$F_s(0) = F_n(0) + \alpha(T)\psi^2 + \tfrac{1}{2}\beta(T)\psi^4$$

expected to be good only near T_c. The coefficients α and β are then related to parameters such as the thermodynamic critical field,

$$\frac{1}{2} \frac{\alpha^2}{\beta} = \frac{H_c{}^2}{8\pi} = F_n(0) - F_s(0)$$

and the penetration depth,

$$\frac{\alpha}{\beta} = -\frac{m}{4\pi} \left(\frac{c}{2e\lambda_L} \right)^2$$

The Ginzburg-Landau theory distinguishes between two types of superconductors: (1) type I, in which the field H_{c1} at which flux penetrates the sample and the field H_{c2} at which full resistance is restored are both equal to the thermodynamic critical field H_c; and (2) type II, in which $H_{c1} < H_c < H_{c2}$. This is not the kind of field penetration (which can occur even with type I superconductors) due to specimen geometry. A type II superconductor can be penetrated by flux and still display zero resistance in the field range from H_{c1} to H_{c2}. The criteria for distinguishing the two types of superconductors are given by the dimensionless parameter

$$\kappa = \frac{\sqrt{8}\, eH_c\lambda_L{}^2}{\hbar c}$$

a measure of the interphase boundary energy. For the "soft" superconductors of type I, $\kappa < 1/\sqrt{2}$. These include indium, tin, and lead. Type II superconductors, for which $\kappa > 1/\sqrt{2}$, include niobium, many superconductive compounds, and most alloys. Any treatment of a type I superconductor that sufficiently reduces the electron mean path turns it into a type II superconductor.

The Ginzburg-Landau theory reduces to the London theory in the weak-field limit. It has also been shown[10] that the Ginzburg-Landau equations can be derived from the microscopic theory of superconductivity for temperatures near T_c and with the condition that the penetration depth is large compared with the correlation distance (the London limit).

In the microscopic theory of Bardeen, Cooper, and Schrieffer (BCS),[11] superconductivity is attributed to an electronic interaction in which an attractive force exists between pairs of conduction electrons having equal and opposite momenta. The mechanism by which the electrons are coupled to form "Cooper pairs" is a mutual interaction with the vibrations of the crystal lattice. Materials in which the attraction due to interaction with the phonon spectra can become greater than the Coulombic repulsion exhibit the phenomenon of superconductivity at sufficiently low temperatures. The range of coherence, or the correlation distance ξ, over which the attractive force between the pairs extends is of the same order as the phonon mean free path (10^{-4} cm). An energy gap 2ϵ exists between the condensed phase (the states of coupled pairs) and the normal phase. At absolute zero the BCS theory predicts

$$2\epsilon = 3.52kT_c$$

The energy gap decreases with increasing temperature and vanishes at T_c, the critical temperature. A minimum energy of ϵ is required to decouple the pairs of electrons.

b. The Critical Temperature of Films

The critical temperatures of well-annealed thin films of superconductive materials are usually close to that of the bulk material. There are a number of mechanisms that may operate to cause the critical temperature of a film to differ, however. Among these are (1) film structure different from bulk, (2) impurities, (3) proximity to another conductor or superconductor, and (4) stress.

(1) Structure The structures of films deposited at ordinary temperatures are in general similar to those of bulk. At lower deposition temperatures, where the atom mobility of the condensate is low and crystalline growth is inhibited, the structure may be amorphous-like.

Films of a number of elements, obtained by vapor quenching (depositing onto a substrate maintained at temperatures below 10°K), have critical temperatures quite different from the crystalline material.[12] That of aluminum is increased from 1.2 to 2.5°K, while that of gallium is increased from 1.1 to 8.4°K. Bismuth and beryllium, both nonsuperconductive in their ordinary crystalline states, exhibit critical temperatures of 6 and 8°K, respectively.[13] Chopra et al.[14] have shown that enhancement of the critical temperature of tungsten films may be attributed to a metastable fcc phase, stabilized by small grain size, with a critical temperature of about 4.6°K. They have obtained amorphous-like films on low-temperature substrates with an ion-beam sputtering arrangement. As the films anneal, the grain size increases, with the amorphous structure converting to the normal structure via the fcc phase, and the critical temperature of the film decreases.

Enhancement of critical temperature has also been observed when other deposition techniques were used. Abeles et al.,[15] using reactive deposition in oxygen onto room-temperature substrates, have obtained enhancement in aluminum, gallium, tin, and indium. Strongin et al.[16] have achieved enhancements greater than that obtained in vapor quenching by depositing structures consisting of alternate 60-Å layers of superconductor S and barrier material B onto low-temperature substrates. The effect was observed for the sequence SBS for the metals Al, Zn, In, Sn, and Pb with barrier materials of SiO, LiF, anthracene, and Al deposited in partial pressures of oxygen and argon. Cohen and Douglass[17] have suggested that the results are due to superconductive pairing of electrons across the barrier. They argue that such pairing may be possible for all metals including the ferromagnets.

(2) Impurities The preparation of superconductive thin films usually introduces some degree of contamination into the film, the amount and kind depending upon the fabrication process. For sputtered or vacuum-deposited films, gaseous impurities predominate. The addition of small amounts of impurities, or alloying elements, to a pure superconductor generally decreases the critical temperature linearly with decreasing electronic mean free path. Assuming that Matthiessen's rule holds and that $\rho l = $ constant, the decrease in critical temperature can be expressed as

$$\Delta T_c = -k\rho_{\mathrm{res}}$$

where k depends upon the solute-solvent pair and ρ_{res} is the temperature-independent part of the resistivity. ΔT_c is often reported in the literature in terms of the atomic percent of solute,

$$\Delta T_c = -k_a \qquad \text{atomic \%}$$

or in terms of the reciprocal resistance ratio

$$\Delta T_c = -\frac{k_r}{\Gamma}$$

with

$$\Gamma = \frac{\rho - \rho_{\mathrm{res}}}{\rho_{\mathrm{res}}} = \frac{\rho_{\mathrm{temp}}}{\rho_{\mathrm{res}}} = \frac{R - R_{\mathrm{res}}}{R_{\mathrm{res}}}$$

where R is the sample resistance at some convenient high temperature, say $300°K$, and R_{res} is taken at a temperature low enough that the temperature-dependent part has vanished.

The critical temperature of vacuum-deposited films of materials with relatively low gettering capabilities, such as tin and indium, is less affected than that of strong getterers vanadium, tantalum, and niobium. Only with strong doping of indium with oxygen is its critical temperature changed, the decrease being about $0.05°K$ μohm-cm^{-1}.[18] This is to be contrasted with $0.2°K$ μohm-cm^{-1} for light alloying of bulk superconductors.[19,20] The effect of oxygen on the critical temperature of tin is severely masked by stress effects,[21] but it appears to be small.

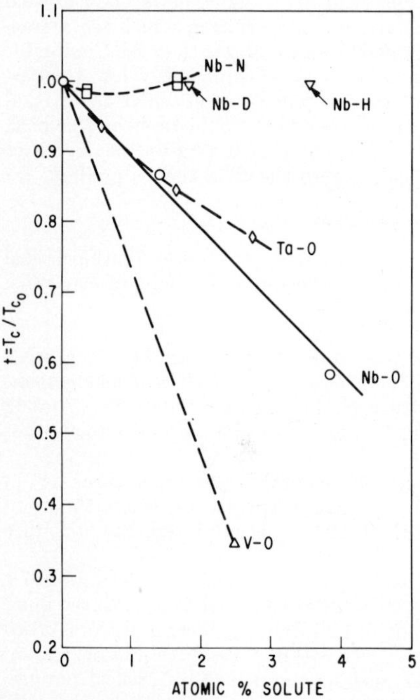

Fig. 1 The effect of concentration of some interstitial solute atoms on the critical temperature of niobium, vanadium, and tantalum, where T_c is the critical temperature of the solution and T_{c0} is the critical temperature of the pure metal.

The solubility of oxygen in vanadium, tantalum, and niobium is of the order of several atomic percent, and the critical temperatures are quite sensitive to small amounts of dissolved oxygen. Measurements by Seraphim et al.[22] yield about $0.5°K$/atomic %. DeSorbo[23] has found values $0.93°K$/atomic % and $1.3°K$/atomic %, respectively, for niobium and vanadium (Fig. 1). Accordingly, special techniques are required to produce films of these materials with critical temperatures approaching that of bulk. These include ultrahigh-vacuum deposition,[24] very high deposition rates (up to 1,200 Å sec^{-1}),[25] asymmetric sputtering,[26] and getter sputtering.[27]

The gettering of oxygen and oxygen-containing compounds can be used to advantage. Neugebauer and Ekvall[28] were able to produce films of niobium, tantalum, and vanadium with critical temperatures near bulk using low rates of deposition in

10^{-5} Torr vacuum by depositing the material from electron-bombarded hanging drops onto large areas of the evaporation chamber. The constantly renewed active surface gettered in competition with the film, reducing the oxygen partial pressure. Incorporation of oxygen was further reduced by increasing the temperature of the film substrate, thus decreasing the sticking coefficient of oxygen. Rairden and Neugebauer[29] have investigated the influence of substrate temperature upon the resistance ratios, lattice parameters, and critical temperatures of tantalum and niobium films. The relation between critical temperature and resistance ratio is compared with the predicted results in Fig. 2.

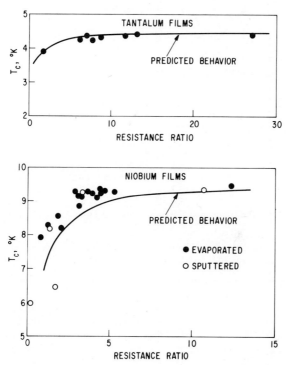

Fig. 2 Critical temperatures of niobium and tantalum films as a function of their resistance ratio.

(3) Proximity Effects The critical temperature of a superconductor can be depressed by superimposing a film of normal metal, or a film of another superconductor with a lower critical temperature. Smith et al.,[30] studying the Pb-Ag system, found that not only was the critical temperature of a thin film of lead depressed with increasing thickness of silver but also that films of silver or gold, sandwiched between films of lead and tin, showed evidence of being superconducting. Simmons and Douglass[31] investigated the effect of relative film thicknesses on the critical temperature of the Sn-Ag system when the thicknesses were less than either the penetration depth or the coherence length. They found that the data could be represented by

$$T_c = T_{cSn} \left(\frac{1 - \alpha d_{Ag}/d_{Sn}}{1 + \alpha d_{Ag}/d_{Sn}} \right)^{\frac{1}{2}}$$

where d_{Sn} and d_{Ag} are the film thicknesses. α is a function of electron densities and critical fields of the metals:

$$\alpha = \frac{(-H^2_{cAg})/N_{Ag}}{H^2_{cSn}(0)/N(0)_{Sn}}$$

The concept of a critical field (imaginary) for the normal metal arises from considering the normal metal as one for which the free energy of the superconducting state is higher than that of the normal state. The positive free-energy difference is then described by $-H_{cn}^2/8\pi$. The comparison of experimental critical temperatures with those predicted by a phenomenological theory of Douglass[32] is given in Fig. 3.

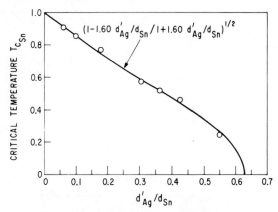

Fig. 3 Critical temperature of superimposed films of Ag and Sn vs. ratio of Ag thickness to Sn thickness.

The nature and magnitude of electron-electron interactions in normal metals can be estimated from theories and experiments on superimposed superconductive-normal films. Hauser and Theuerer[33] have used sandwiches in which both metals were superconductive in order to check the validity of a theory. They obtained the critical temperatures of superimposed films of lead and aluminum to compare with a theory of De Gennes'.[34] The closeness of the agreement, shown in Fig. 4, is taken to establish the validity of the theory and the fact that they observed true proximity effects unobscured by spurious diffusion.

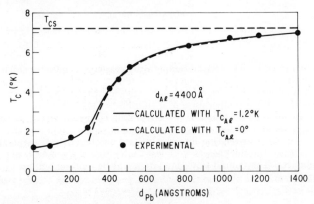

Fig. 4 Critical temperature of Pb-Al sandwiches for a constant Al-film thickness of 4,400 Å as a function of the Pb-film thickness.

(4) Stress The effect of stress on the critical temperature of superconductive thin films was first suggested by Lock.[35] He found that the critical temperature of tin deposited upon various substrates ranged from 3.53 to 4.0°K, depending upon the coefficient of expansion of the substrate. He concluded that the shifts were due to strains caused by differential contraction of film and substrate when cooled from room temperature to that of liquid helium. Films removed from their substrates had bulk critical temperatures.

The critical temperatures of tin films, 3,000 Å thick and deposited simultaneously upon various insulated (5,000 Å SiO) substrates, are plotted in Fig. 5 against the room-temperature coefficient of expansion of the substrate.

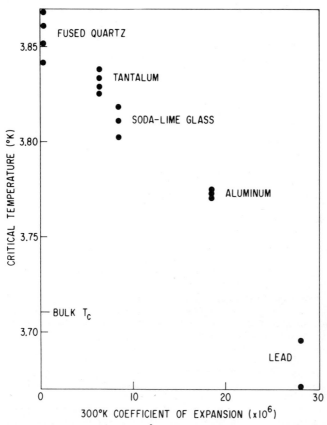

Fig. 5 The critical temperature of 3,000-Å-thick tin films on SiO (5,000 Å) insulated substrates vs. the room-temperature coefficient of expansion of the substrate.

The variation of critical temperature with film thickness has been studied by Toxen.[36] Using indium on fused quartz, he found the systematic variation shown in Fig. 6. The critical temperatures of thick films (>10,000 Å) are within a few millidegrees of the bulk value. The sharp increase with decreasing thickness is explained as a result of the variation of critical yield stress with thickness. From the resultant biaxial tensile stress in the film and the hydraulic pressure shift in critical temperature, the variation in film critical temperature was calculated to be

$$\Delta T_c = \frac{52}{d} - \frac{750}{d^2}$$

where d is the film thickness in angstroms. This expression (the solid line in Fig. 6) is in good agreement with the data for films thicker than 600 Å.

Vogel and Garland[37] have found that strain is insufficient to produce the critical temperature shifts of up to 0.5°K observed in indium films less than 600 Å thick.

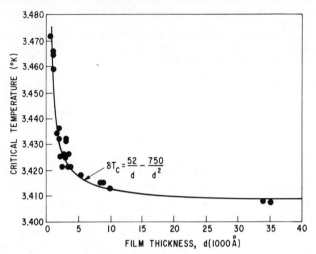

Fig. 6 Thickness dependence of superconducting critical temperature of indium films on fused quartz.

They consider the effect of the mean free path on the electron interaction due to phonons, and arrive at the expression

$$\Delta T_c = \frac{51}{2.2d - 5 \times 10^{-3}d^2}$$

for indium films in the thickness range 100 to 1,000 Å.

Blumberg and Seraphim[38] found that, in addition to variations of critical temperature with thickness, the orientation of crystallites in a thin film had a pronounced influence on the critical temperature. Tin is tetragonal and highly anisotropic. The tetrad axis (c), with the larger coefficient of expansion, has a high change of critical temperature with uniaxial stress. Shifts of critical temperature can be calculated ranging from almost zero for films with the crystallite tetrad axes normal to the plane of the film, up to $+0.23°K$ for films with the tetrad axes in the plane and aligned. For the more usual case of the diad axes normal to the plane and the tetrad axes randomly oriented in the plane, a shift of about 0.1°K is observed for tin on soda-lime glass.

Hall[39] has added stress to tin films on glass by bending the substrates at liquid-helium temperature. He finds that, although the stresses due to thermal differential contraction may be as high as 4,000 atm, no irreversible change takes place in the films even if an additional 4,000 atm is applied.

Stress effects in thin superconductive films can be summarized as follows: Figure 7 shows schematically the behavior of stress with decreasing temperature for films of the same thickness on two different substrates each having a coefficient of expansion less than that of the superconductor. Initially, as the temperature is lowered, the film is stressed until the elastic limit is reached and the film plastically deforms. The elastic limit is determined by the film thickness, being greater for the thinner films. To account for the variation of critical temperature of films of the same thickness, it is necessary to postulate that the yield stress increases with decreasing temperature until a temperature is reached where it is greater than the stress produced by differential contraction. Plastic deformation stops, and decreasing temperature only increases stress. The critical temperature is determined by the final stress in the film.

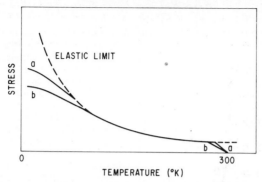

Fig. 7 Schematic representation of stress in superconductive films on substrates with different coefficients of expansion, both lower than that of the superconductor.

c. The Critical Field of Films

In a bulk superconductor an applied magnetic field and the resultant shielding supercurrents are confined to a thin layer, the penetration depth, at the surface of the material. In a film whose thickness is comparable with the penetration depth, the flux is not totally excluded from any portion of the film. It may be expected that the critical field of such a film is different from that of bulk material.

The field variation within a thin film may be obtained by applying the London theory to a film having thickness d and infinite width immersed in a uniform magnetic field H_0, parallel to the surface of the film. With this geometry, Eq. (4) reduces to

$$\frac{d^2 H_z}{dx^2} = \frac{H_z}{\lambda_L^2}$$

Choosing an origin at the center of the film, the solution for the field internal to the film is

$$H(x) = H_0 \frac{\cosh x/\lambda_L}{\cosh d/2\lambda_L} \tag{8}$$

The thermodynamic critical field is defined in terms of the free-energy difference between the normal and the superconducting state, in zero field:

$$F_n - F_s = \frac{H_c^2}{8\pi} \tag{9}$$

In the normal state, $F_n = H_c^2/8\pi - H^2/8\pi$ in an applied field H. F_s is zero in a bulk specimen, except in the penetrated region, whose volume is so small compared with the total volume that the contribution from this region can be neglected. When the film thickness is comparable with, or smaller than, the penetration depth, magnetic contributions to the free energy of the superconducting state cannot be neglected. In this case the critical field for the transition will be given by

$$\frac{H_c^2}{8\pi} = \frac{1}{2} H_c{}^f \Delta M \tag{10}$$

where $H_c{}^f$ is the film critical field. The difference in magnetization ΔM of the film in its normal and superconducting states is given by

$$\Delta M = \frac{H_c{}^f}{4\pi} - \frac{1}{4\pi d} \int_{-d/2}^{+d/2} H(x)\, dx \tag{11}$$

Combining Eqs. (8), (10), and (11), the critical field of the film is found to be

$$H_c{}^f = H_c \left(1 - \frac{2\lambda_L}{d} \tanh \frac{d}{2} \lambda_L\right)^{-\frac{1}{2}} \tag{12}$$

The London theory predicts that the critical field of a film, shown in Fig. 8 as the ratio $H_c{}^f/H_c$ vs. d/λ_L, is greater than the critical field of bulk material.

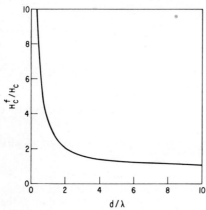

Experimentally it is found that the magnitudes of the applied longitudinal field necessary to destroy superconductivity in films are greater than predicted by the London theory. Although that theory serves a tutorial purpose in that it is readily tractable and gives qualitatively correct results, other theories must be used when conditions are outside the weak field and local limits inherent in the London theory.

A number of approaches have been employed to describe the critical magnetic fields of thin superconducting films of type I materials. Ittner[40] found it convenient to replace λ_L in Eq. (12) with an effective penetration depth λ_e. Experimental penetration depths are determined by solving Eq. (12) for λ_e using measured values of bulk and film critical fields and

Fig. 8 The ratio of film critical field $H_c{}^f$ to bulk critical field H_c vs. the ratio of film thickness d to penetration depth λ.

film thickness. The theoretical calculation of λ_e depends upon being able to express the spatial variation of magnetic field within the film in a form similar to Eq. (8). One approach is to use a method outlined by Schrieffer,[41] based on the BCS theory. The effective penetration depth, calculated in this manner, includes not only the material properties and temperature, as does λ_L, but also the long-range correlation, specimen purity, and sample dimensions.

Blumberg[42] has obtained the effective penetration depth, extrapolated to zero temperature, for a number of tin films of various thickness using the above approach. His results, compared with calculated penetration depths as a function of film thickness for bulk mean free paths of 1,000 and 10,000 Å, are shown in Fig. 9.

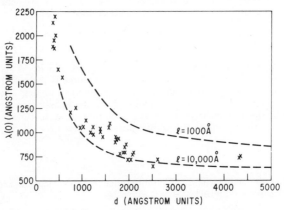

Fig. 9 The penetration depth extrapolated to absolute zero $\lambda_e(0)$ vs. tin-film thickness. The dotted curves are calculated for bulk mean free paths of 1,000 and 10,000 Å.

Somewhat analogous studies of the mean free path and thickness effects in indium films were made by Toxen[43] and Toxen and Burns.[44] They chose the Ginzburg-Landau theory as being more appropriate to describe phenomena occurring at the critical field. The nonlocal parameters were introduced into the BCS calculation of the magnetic susceptibility.

Precise critical-field measurements of tin films in the London limit were made by Douglass and Blumberg.[45] They ensured that the London limit was satisfied by working very near to the critical temperature and by working with thin evaporated specimens where the coherence length would be limited by size. They used an expression for the film critical field,

$$H_c^f = \sqrt{24}\, H_c(T)\, \frac{\lambda(T,d)}{d} \tag{13}$$

derived from the Ginzburg-Landau theory, which is valid for $d \leq \sqrt{5}\, \lambda(T,d)$. The temperature dependence of the bulk critical field H_c and the penetration depth were derived from data to give

$$H_c^f = 1{,}510\, \frac{\lambda(0,d)}{d}\, (\Delta t)^{\frac{1}{2}} (1 + \epsilon\, \Delta t)$$

where $\epsilon = 0.14 \pm 0.10$ and $\Delta t = 1 - t$. They found also that the variation of penetration with film thickness could be well represented by

$$\lambda = \lambda_{\text{bulk}} \left(1 + \frac{\xi_0}{d} \right)^{\frac{1}{2}}$$

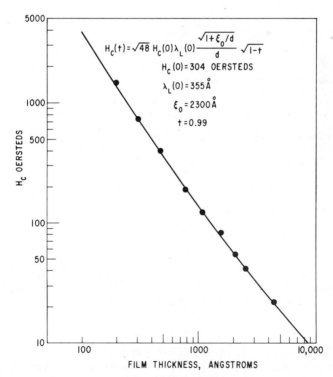

$$H_c(t) = \sqrt{48}\, H_c(0)\lambda_L(0)\, \frac{\sqrt{1 + \xi_0/d}}{d}\, \sqrt{1-t}$$

$$H_c(0) = 304 \text{ OERSTEDS}$$

$$\lambda_L(0) = 355\,\text{Å}$$

$$\xi_0 = 2300\,\text{Å}$$

$$t = 0.99$$

Fig. 10   Critical magnetic field H_c for thin tin films vs. film thickness at $t = 0.99$.

for films ranging in thickness from 196 to 4,330 Å. Glover[46] has shown that the Douglass-Blumberg data fit the relatively simple semiempirical formula

$$H_c{}^f = \frac{\sqrt{48}\, H_c(0)\lambda_L(0)\, \sqrt{1 + \xi_0/d}\, \sqrt{1 - t}}{d}$$

remarkably well for $d < \sqrt{5}\, \lambda(T,d)$ and temperatures down to $t = 0.5$, as is shown in Fig. 10. For thicker films, $d > \lambda(T,d)$, the expression

$$H_c{}^f = H_c(0) \left[2(1 - t) + \frac{\sqrt{2}\, \lambda_L(0)\, \sqrt{1 + \xi_0/1}\, \sqrt{1 - t}}{d} \right]$$

gives qualitatively correct results for temperature down to $t = 0.5$.

The Ginzburg-Landau theory predicts a magnetic hysteresis associated with the first-order phase transition in films with $d > \sqrt{5}\, \lambda$. Both superheating and super-cooling have been observed by Caswell.[47] Hysteresis in smaller-grained films could be attributed solely to supercooling, but superheating also occurred in larger-grained films. The phase transition in thin films, $d \leq \sqrt{5}\, \lambda$, is second-order, and magnetic hysteresis is absent.

d. The Critical Current of Films

Current flowing in a bulk superconductor is confined to a surface layer having a thickness on the order of the penetration depth. According to Silsbee,[3] the critical current is that which produces the critical field at the sample surface. Thus, for a cylindrical wire of radius $a \gg \lambda_L$, the surface critical-current density, with no applied field, can be obtained:

$$2\pi a \lambda_L \frac{4\pi}{c} J_c = 2\pi a H_c$$

or
$$J_c = \frac{cH_c}{4\pi\lambda_L} \tag{14}$$

where the current density is given by

$$J(x) = J_c \exp\left(-\frac{x}{\lambda_L} \right)$$

and the limits of integration over depth are from the surface, $x = 0$, to very deep inside the superconductor.

If the wire is replaced by a thin film of thickness $d \ll \lambda_L$, deposited on the surface of an insulating cylinder of radius $a \gg \lambda_L$, the current distribution will be approximately uniform throughout the film thickness. If the current is uniform around the cylindrical film, the field on the inner surface of the film will be zero. The relation between the current density and the field at the outer surface is

$$J = \frac{cH}{4\pi d}$$

If Silsbee's rule were correct for thin films, the critical-current density would be much larger than is experimentally observed. In fact, the measured critical-current densities are almost constant with changes in thickness. The field at the outer surface of the cylindrical film carrying a critical-current density becomes

$$H_J = \frac{4\pi}{c} J_c d = \frac{d}{\lambda_L} H_c$$

where H_c is the critical field for bulk material. Thus, it is the critical-current density that is fundamental in both current and field transitions. The critical field is that which produces a screening current of critical density.

The supercurrent density $J_s = n_s e v_s$ becomes critical when the electron-drift velocity v_s approaches a critical value comparable with the velocity of the phonons responsible for the long-range correlation of electron pairs. The kinetic energy of the superconducting electron is

$$E_s = \frac{1}{2} n_s m v_s^2$$

or, in terms of the current,

$$E_s = \frac{1}{2} \frac{m}{n_s e^2} J_s^2$$

For a film much thinner than the penetration depth, the magnetic-field energy difference between the normal and the superconducting states is much smaller than the supercurrent energy. Thus, the free-energy difference, evaluated in terms of the bulk parameter H_c, is

$$\frac{H_c^2}{8\pi} = \frac{1}{2} \frac{m}{n_s e^2} J_c^2$$

This result, first obtained by London,[48] leads immediately to Eq. (14).

The temperature dependence of the critical-current density, for temperatures near the critical temperature, is obtained from the temperature dependences of the critical field $(1 - t^2)$ and the penetration depth $[2(1 - t)]^{-\frac{1}{2}}$ from the BCS theory. These give

$$J_c = \frac{cH_0}{\sqrt{2}\,\pi\lambda_L(0)} (1 - t)^{\frac{3}{2}}$$

where H_0 and $\lambda_L(0)$ are the critical field and penetration depth, respectively, for bulk material at $0°K$, and only the leading-temperature factor has been retained.

The Ginzburg-Landau theory gives the result for $d \ll \lambda$, T near T_c,

$$J_c = \frac{2cH_0(1 - t)^{\frac{3}{2}}}{3\sqrt{3}\,\pi\lambda_L(0)(1 + \xi_0/l)^{\frac{1}{2}}}$$

where the effect of scattering has been introduced. It is only through the mean free path that the film thickness affects the critical current, which decreases slowly with decreasing film thickness.

Bardeen[49] has discussed modifications of Eq. (14) valid for various temperature ranges and scattering conditions. The modifications reduce the theoretical estimates of critical current by not more than a factor of 2.

The total critical current per unit width of film I_c/w is proportional to the thickness in the range $d \ll \lambda$, where the current density is uniform through the thickness. In the range $d \gg \lambda$, where the current is limited to near the surface, I_c/w is constant. The London theory, applied to a flat, unshielded film, predicts

$$\frac{I_c}{w} = \frac{c}{2\pi} H_c \tanh \frac{d}{2\lambda_L} \tag{15}$$

If the flat film is shielded on one side by a superconducting ground plane,

$$\frac{I_c}{w} = \frac{c}{4\pi} H_c \tanh \frac{d}{\lambda_L} \tag{16}$$

These equations were derived with the assumption of uniform current across the width of the film. They predict that the critical current for very thin films is independent of shielding since the current density through the film is nearly uniform. For thick films, they predict that an unshielded film should carry twice the current of a shielded film of the same width. Current can flow on both sides of an unshielded film, but image currents in the ground plane restrict the current to one side of the shielded film.

But, by analogy with the distribution of electric charge over the surface of a conductor of the same shape, it can be seen that the current in a flat, unshielded superconductor is far from uniform, being carried primarily by the film edges. Measured critical-current densities, determined by dividing the critical current by the film cross section, lead to values 5 to 35 times lower than calculated on the basis of the uniform current approximation.[50,51]

A typical current-induced transition for an unshielded tin film is shown in Fig. 11.[52] As the current is increased from zero, the current density at the edges becomes critical, but then stays constant as the edge material goes into the intermediate state.[53] The intermediate state spreads inward as the current increases, resulting in a more uniform distribution of current and the generation of normal components of magnetic field. When the entire width of the film is in the intermediate state, resistance appears, but at a value considerably less than predicted because of the normal-field components. Further increase of current increases the resistance until Joule heating causes thermal propagation of resistive areas, and the entire film becomes resistive. The current value at which this occurs is strongly dependent upon the thermal conductivity of the substrate and the film resistivity. To return to the superconducting state, the current has to be reduced well below the above value because of Joule heating. The effect of Joule heating can be suppressed by the use of short current pulses. The isothermal transitions, shown in Fig. 11 by the dashed line, give current densities at the half-resistance point that are in fair agreement with the uniform-current assumption.

Two film configurations, in which the currents are inherently more uniform, are the cylindrical film and the shielded plane film. Critical-current density measurements by Ginzburg and Shalnikov[54] gave a temperature dependence $(1 - t)^{\frac{3}{2}}$ as pre-

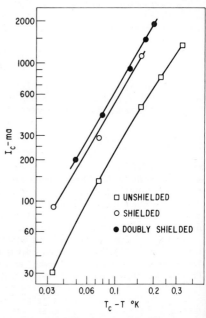

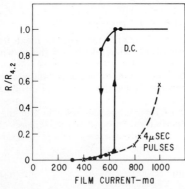

Fig. 11 Current induced transition for 0.3-μ-thick, 4.05-mm-wide tin film on sapphire substrate at $\Delta T = 0.09°$K for direct current and for 4-μs pulses.

Fig. 12 Critical currents (at onset of resistance) for 3,000 Å $\times$ 2 mm unshielded, singly shielded, and doubly shielded tin films on glass.

dicted, but magnitudes 22% lower than theoretical. Their results may have been due to heating effects from use of direct currents. Hagedorn,[55] using current pulses less than 10^{-9} s in duration, has found that the critical current of cylindrical tin films is in good agreement, over a wide temperature range, with the Silsbee hypothesis, if latent-heat effects are taken into account.

By analogy to a capacitor plate over a ground plane, current in a flat, shielded film can be seen to be uniform except for a region comparable with the insulation thickness near the edges. The average surface density at which resistance appears should be higher than for an unshielded film of the same dimensions. The results of Edwards and Newhouse,[56] comparing unshielded, single and double (shield plane on both sides) shielded films, are shown in Fig. 12. The currents corresponding to onset of resistance are larger for the more shielded films as predicted, but their magnitudes are still too small.

Working with shielded, highly agglomerated indium films of various degrees of connectedness, Learn and Spriggs[57] observed a peak in the critical current at constant temperature as the film thickness was varied. The peak occurred at thicknesses between those in which the film first became conductive and those in which the room-temperature resistivity became independent of thickness. The location of the peak, shown in Fig. 13, was dependent upon the deposition conditions. The peak values of

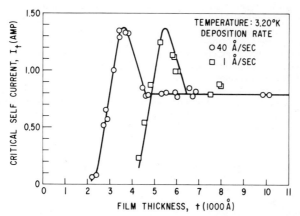

Fig. 13 Critical current (onset of resistance) of 480-μ-wide indium films at $T = 3.20°K$ vs. film thickness for two different deposition rates on room-temperature substrates.

the critical current were in substantial agreement with the critical-current-density switching hypothesis. It was postulated that structure and thickness inhomogeneities, providing a pinning effect on current fibers to inhibit displacement by the Lorentz force of adjacent fibers, served to maintain a uniform current distribution over the surface of the film.

Hagedorn[58] has considered that current flow in the edge of a shielded film may be responsible for triggering the transition into the normal state prematurely. By the use of a special structure providing two current paths through the same film, only one of which includes the edges, he was able to show that critical currents near the Silsbee limit are observed for tin films when the edges are excluded.

The question of criterion for phase transition in superconductive films has been discussed by Cheng,[59] who considers maxima of magnetic field (Silsbee), current density (London), and local energy density (the sum of field and current energy densities). He demonstrates that a maximum average energy density, where the specimen volume is the domain of averaging, is in far better agreement with observed behavior than any of the others.

2. CRYOTRONS

a. Thin Film Cryotrons

The electrical resistance of a superconductor can be changed by variations in a magnetic field generated by current in an adjacent conductor. The term "cryotron" was coined for such a device by Buck,[60] who proposed the use of a tantalum-wire gate surrounded by a niobium-wire solenoid as a computer element operating at liquid-helium temperature. Current passing through the permanently superconducting niobium control wire produced a magnetic field strong enough to quench superconductivity in the tantalum gate. The cryotron can be designed so that the minimum required control current is smaller than the critical current of the gate. Under these conditions the gain of the device is greater than 1, and the gate current from one cryotron can control the current through other cryotrons without additional amplification.

The basic time constant of a cryotron is given by the ratio of the control inductance L to the resistance R of that portion of the gate driven normal by the control. The inductance of the control solenoid of a wirewound cryotron is large, and the resistance of the solid wire gate is small, resulting in time constants of several hundred microseconds for even the smallest devices.

Since the current in a superconductor is carried in a thin layer, the penetration depth, only a film of gate material is required, and the resistance of a normal gate is greatly increased. Flat, thin film cryotrons were constructed by Newhouse and Bremer.[61] They used a narrow film of lead, as a control, crossing a wider gate film of tin. The control was insulated from the gate by a film of SiO. The inductances of the control and of other circuit elements were drastically reduced by depositing them over a superconducting ground plane of lead, also insulated by SiO.[50,52] This has the additional advantages of increasing the critical gate current I_{gc} by distributing the gate current uniformly across the gate width, and of sharpening the resistance transition of the gate with control-current change by confining the control field to the region of gate directly under the control. The structure of a shielded crossed-film cryotron is shown in Fig. 14.

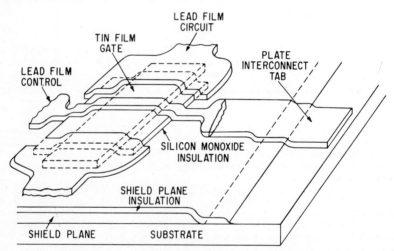

Fig. 14   The structure of a shielded crossed-film cryotron.

The static gain of a cryotron is defined as

$$G = \frac{I_{gc}}{I_{cc}}$$

where I_{gc} is the critical gate current in the absence of control current, and I_{cc} is the current in the control that produces a critical field at the surface of the gate in the absence of gate current. The gain may also be expressed in terms of the fields produced by the respective currents. Assuming that the shielding causes uniform current distributions in the gate and the control, the fields (not necessarily equal as explained below) are

$$H_{gc} = 0.4\pi \frac{I_{gc}}{W} \quad \text{Oe}$$

and

$$H_{cc} = 0.4\pi \frac{I_{cc}}{w} \quad \text{Oe}$$

where H_{gc} is the field at the gate surface produced by the critical gate current, H_{cc} is the critical field of the gate, and W and w are the gate width and the control width, respectively. The gain becomes

$$G = \frac{H_{gc}W}{H_{cc}w}$$

The fields H_{gc} and H_{cc} are independent of the widths of the gate and control, but both depend upon the superconducting properties and the thickness of the gate film. For films thick compared with the penetration depth λ of the gate superconductor, the ratio H_{gc}/H_{cc} approaches unity, and the gain of the crossed-film cryotron is just the crossing ratio W/w. If the films are thin compared with the penetration depth, as would be the case for operation near T_c, the ratio of the fields decreases. Ginzburg and Landau[9] have given expressions for the fields of thin films,

$$H_{gc} = \frac{2\sqrt{2}\,d}{3\sqrt{3}\,\lambda} H_c$$

and

$$H_{cc} = \sqrt{24}\,\frac{\lambda}{d} H_c$$

where H_c is the critical field of the bulk superconductor and d is the film thickness. Making these substitutions, one obtains for the gain

$$G = \frac{1}{9}\left(\frac{d}{\lambda}\right)^2 \frac{W}{w}$$

At first sight it might be supposed that the speed of a cryotron might be increased by increasing the resistance of the gate. If the thickness is decreased, the gain is decreased. If impurities are added, the gain is also decreased, since the penetration depth increases.

The I_g-I_c characteristic of a tin-gate cryotron vacuum-deposited through a stencil is shown by the dashed line of Fig. 15. The solid curves are linear recordings of the gate resistance as a function of control current with the zero of resistance set level with the value of the gate current shown on the log scale. The characteristic is drawn through the points of onset of resistance for increasing control current. The crossing ratio W/w for this device is 10. This is to be compared with the realized gain of about 3.

The tail of the I_g-I_c curve for low gate current is characteristic of cryotrons that are deposited through a stencil onto clean, room-temperature substrates. Shadowing by the edges of the stencil produces a penumbra, a region of diminishing thickness extending away from the geometric edge of the gate. The critical field for the thinner material is higher than for the central portion of the film. The thin edges can still carry a small gate current for quite large control currents. If the gate current is large, it is pushed to the edges of the film by the normal central material, causing the edges, which have a small critical current, to become normal at lesser control currents.

A number of methods have been used to eliminate or reduce the edge effects caused by the penumbra. The simplest, mechanically trimming,[62] is frequently used for critical-field studies. Other methods must be employed for high-density arrays. Behrndt et al.[63] have deposited tin onto silver- or copper-seeded substrates held at

100°C.　The tin forms a fine-grained continuous film over the seeded areas with a sharp boundary at the edge of the seeding.　Elsewhere the tin agglomerates into isolated large-grain particles.　Considerable improvement of the edge structure may be obtained even without seeding by depositing at temperatures (60°C) just below the agglomeration temperature.[64]　Caswell[18] was able to stencil deposit tin with sharp edges onto room-temperature substrates using deposition rates of 50 Å s^{-1} at system pressures of 10^{-8} Torr.

The proximity effect has been used by a number of workers to suppress the edge effect.　Tyler and Walker[65] have used silver as the normal metal adjacent to tin, while

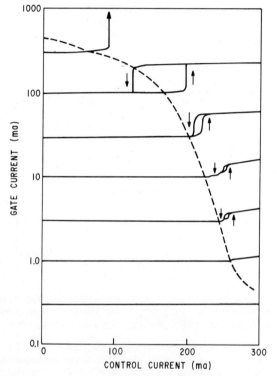

Fig. 15　I_g-I_c characteristic of a tin-gate CFC, vacuum-deposited through a stencil onto a substrate at $T = 20°C$.　$T_c = 3.85°K$, $W = 0.5$ mm, $d = 5,000$ Å, $w = 50$ μ.　The operating temperature was 3.7°K.　The ground plane is anodized niobium film.

Ames and Seki[66] report that either copper or gold will serve as a suppressant for both tin and indium.　The very thin (about 100 Å) normal films depress the critical temperature, and field, of the thin penumbral material while leaving the thicker gate regions unaffected.

In the photomask-photoresist technique for fabrication of cryotron arrays,[67] gates are formed from a continuous film of superconductor by chemical etching, and a penumbra does not occur.　An I_g-I_c characteristic of an etched-gate cryotron is given in Fig. 16.[68]　The gate film was deposited under conditions similar to that of Fig. 15. The control-current cutoff is sharp down to 0.3 mA gate current, the limit of measurement.　Also, the resistance transitions are abrupt, reaching full value at currents a few milliamperes beyond the onset of resistance.

The hysteresis in the resistance transition for the higher gate currents is due to

Joule heating in the normal film. It increases with gate current up to the point where thermal propagation occurs and the entire film becomes normal.[69] The hysteresis for low-gate-current transitions, suggested in Fig. 15 and fully developed in Fig. 16, is due to supercooling and superheating effects arising from the first-order phase transitions which occur in films with $d > \sqrt{5}\,\lambda$. Supercooling, in which the normal phase can persist in a metastable state at fields below the thermodynamic critical field as the field is decreased, is predominant, so that superheating is usually ignored.[70] Superheating, generally a much smaller effect, contributes to the hysteresis of large-grained films obtained at higher substrate temperatures.[47]

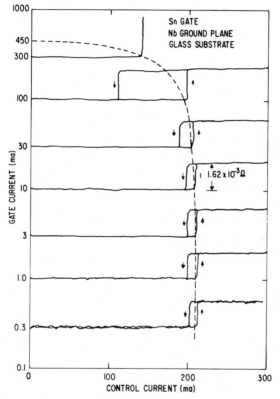

Fig. 16 I_g-I_c characteristic of an etched tin-gate CFC. $T_c = 3.86°K$, $W = 0.5$ mm, $d = 5,000$ Å, $w = 50\ \mu$, $T_{op} = 3.7°K$, $T_s = 20°C$.

The basic time constant $\tau = L/R$ neglects the inductance of all leads and connecting circuitry. Although this somewhat arbitrary τ is not realistic in terms of a working circuit, it does display the basic limitations of the cryotron. L is the inductance of the control of width w, length W (equal to the width of the gate), and separated a distance t above the ground plane. The gate film is considered to be normal, so that its presence is ignored, and the penetration depths into the control and ground-plane superconductors are included in t. The field under the control is related to the control current by

$$H = 0.4\pi \frac{I}{w} \qquad \text{Oe}$$

The magnetic-field energy under the control in the volume wWt is

$$E = \frac{1}{8\pi} H^2 wWt \times 10^{-7}\,\text{J}$$

or

$$E = 2\pi I^2 \frac{Wt}{w} \times 10^{-9}\,\text{J}$$

Since $E = \frac{1}{2}LI^2$,

$$L = 4\pi \frac{Wt}{w} \times 10^{-9}\,\text{H}$$

The resistance is that of the normal gate film directly under the control, enclosed in the volume wWd,

$$R = \rho \frac{w}{Wd} \quad \text{ohms}$$

where ρ is the resistivity of the normal gate material at the operating temperature, and d is the gate thickness. The time constant is then

$$\tau = 4\pi \frac{td}{\rho} \left(\frac{W}{w}\right)^2 \times 10^{-9}\,\text{s}$$

It is instructive to express the time constant in terms of the gain through the quantity $(W/w)^2$ for the cases $d \gg \lambda$ and $d \ll \lambda$,

$$\frac{\tau}{G^2} = 4\pi \frac{td}{\rho} \times 10^{-9}\,\text{s} \qquad\qquad d \gg \lambda$$

$$\frac{\tau}{G^2} = 4\pi \frac{td}{\rho} 9^2 \left(\frac{\lambda}{d}\right)^4 \times 10^{-9}\,\text{s} \qquad d \ll \lambda$$

For $d \gg \lambda$, τ/G^2 increases linearly with d, but for the very thin films it rises rapidly as d^{-3}. Ittner[71] has calculated that the minimum of τ/G^2 is about 10^{-8} s for film thicknesses between 3,000 to 5,000 Å, depending on the resistivity. The resistivity of a 5,000-Å tin gate at 4.2°C is about 0.2 μohm-cm. The resistance of a cryotron with a 10-mil-wide gate and a 2-mil-wide control would be 0.8 mohm. With a control-to-ground-plane distance t of 10,000 Å, the inductance would be 2π pH, leading to a time constant of just less than 10 ns. The speed of a circuit will always be greater, however, because of the inductance of the leads.

The in-line cryotron was developed to overcome some of the speed-limiting factors inherent in the crossed-film cryotron.[72] By placing the gate and the control superimposed and parallel, or in line, the resistance of the normal gate can be increased by increasing its length. For this arrangement, with a gate length l,

$$L = 4\pi \frac{lt}{W} \times 10^{-9}\,\text{H}$$

$$R = \rho \frac{l}{Wd} \quad \text{ohms}$$

and

$$\tau = 4\pi \frac{td}{\rho} \times 10^{-9}\,\text{s}$$

independent of the length or width of the gate. Using the same values as before for t, d, and ρ as typical for thin film cryogenic devices, one calculates $\tau = \pi \times 10^{-10}$ s, or about 0.3 ns. The ratio l/W of the in-line cryotron is available to increase device density, or to match impedances of interconnections.

The static gain of an in-line cryotron must be less than unity, since the widths of the gate and the control are equal. But, whereas the I_g-I_c characteristic of a crossed-film cryotron is symmetric, that of an in-line cryotron is asymmetric with respect to the polarities of the gate and control currents as in Fig. 17.   By applying a bias field, either from an external source or from current in an additional in-line control film, the origin of I_g can be shifted to the right in the figure so that the gain of the device is greater than unity.

Although current can be switched rapidly in the in-line cryotron once it is in the resistive state, the minimum time required to develop the full resistance ranges from 10 to 40 ns.   An early model of phase transition held that the propagation of a normal phase into a superconducting metal was delayed by eddy currents in the normal metal.[73]   Applied to thin films, this model predicted delay times of 1 ns for propagation normal to the plane of the film.   The additional delays were attributed to two effects.[74]

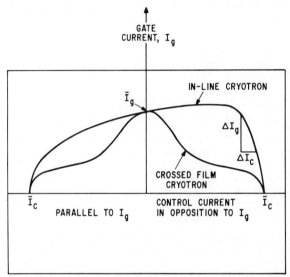

Fig. 17   Comparison of characteristic curves of crossed-film and in-line cryotrons with unity aspect ratios, $w/W = 1$, and with the same λ/d.

In fields with small overdrives (control currents 10 to 15% greater than I_c), the latent heat of the superconducting-to-normal phase transition lowers the temperature of the superconductor and increases the critical field to a value comparable with that of the applied field.   The time of complete transition will then depend upon the rate of heat flow into the superconductor as its temperature returns to that of the liquid-helium bath.   These times range from 100 ns to 1 μs.

For overdrive fields greater than the isentropic critical magnetic field, in which the latent-heat delay should be negligible, delays of 10 to 40 ns are associated with the time required for the normal phase boundaries to propagate laterally between nucleation sites.   Subsequent studies of the dynamic operation of the in-line cryotron in bistable circuits have indicated that 25 ns is close to the minimum transfer (transition plus switching) times attainable with either crossed-film or in-line cryotron structures.[75]

Currents in a superconducting circuit can be switched by varying the inductances of the branches.   The inductance of a current-carrying element can be changed by causing an adjacent shield to undergo phase transitions, either by using an external field[76] or by passing current through the shield.[77]   The same limitations to switching speed will apply to these devices.

b. Amplifiers

If a cryotron gate is maintained at a point midway between the superconducting and normal states, small changes in the control current can produce large changes in the voltage across the gate, or in the current passing through the gate, depending upon the circuit.

A crossed-film cryotron amplifier is shown in Fig. 18.[78] A balanced arrangement is used to minimize the effect of temperature fluctuations in the liquid-helium bath. The cryotrons are held in the resistive transition region by the control and gate-biasing currents I_c and I_g, respectively. With no signal current applied, the gate-bias current will divide equally between the balanced input cryotron gates. An input-current signal, as shown in the figure, will increase the control field of the upper input cryotron, and decrease it on the lower. Thus, the resistance of the upper (lower)

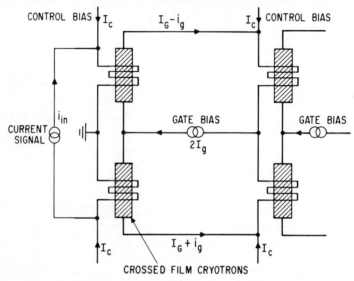

Fig. 18   A crossed-film cryotron amplifier.

gate is increased (decreased). The gate-bias current will then redistribute, with less current flowing in the upper output cryotron control, decreasing its resistance. With constant current flowing in the output cryotrons, the voltage across the upper cryotron will decrease; that across the lower will increase.

If the output cryotrons are replaced with another balanced loop, the amplifier may be cascaded. The dc gain per stage $\partial I_g/\partial I_c$ is found to depend strongly on the edge structure (penumbra) of the gate as well as the bath temperature and biases, i.e., the shape of the resistive transition. The electrical time constant is L/r_g, where L is the control inductance and r_g is the differential gate resistance $\partial V_g/\partial I_g$. The upper frequency limit is usually determined by heat transfer rather than by the electrical time constant, however, since the gates are carrying current in a partially resistive state. With the use of single-crystal quartz, or aluminum, substrates, frequencies up to 2 MHz have been obtained. Low-frequency noise caused by bubble formation in the liquid helium at the surface of the gate can be severe. Attempts to reduce this source of noise include isolation of the amplifier from the liquid and cooling by vapor,[79] providing 100% negative feedback at frequencies below 1 kHz,[78] and using pulsed gate biases to reduce the average gate heating.[80]

An eight-stage amplifier with a current gain of 22,000 and upper frequency limit of 100 kHz has been used to measure amplifier self-noise. Referred to the input, noise-current levels of 0.068 μA rms at 45 kHz were found, compared with calculated Johnson noise of 0.017 μA rms.

A crossed-film cryotron with many controls crossing the gate can be used as a single-stage high-gain amplifier.[80] Pulsed operation of the gate bias allows amplification of dc signals, as well as reduction of bubble noise. An amplifier with 512 crossovers exhibited a bandwidth of 1 MHz and transimpedances up to $10\,\mathrm{mV\,mA^{-1}}$. An amplifier with 8,064 crossovers has been used by Mundy[81] as the sense amplifier for an experimental memory array.

c. Logic and Memory Circuits

Certain properties of a cryotron make it an almost ideal gating device. It is a four-terminal component with no coupling from the gate back to the control. Currents can flow in either direction in both elements. Once current is switched into a branch of a circuit, it persists without the need of special holding circuits. Complete systems of logic and memory may be made entirely of cryotrons, without auxiliary components such as resistors or capacitors in the liquid-helium environment.

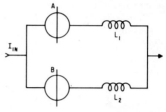

Fig. 19 A cryotron circuit.

Consider the circuit of Fig. 19. The cryotron is represented by the circle (gate) with a line (control) drawn through it. The inductances L_1 and L_2 represent the controls of other cryotrons (their electrical equivalent) in series with the gates A and B. With no current flowing in either of the controls A or B, an applied current I_{in} will divide between the branches inversely as the inductances

$$I_A = \frac{L_2}{L_1 + L_2} I_{in}$$

$$I_B = \frac{L_1}{L_1 + L_2} I_{in}$$

If now a current is applied to control A, driving gate A normal, the current flowing in that branch will be diverted entirely into the other branch by the voltage developed across the normal gate. The time constant for current switching is

$$\tau = \frac{L_1 + L_2}{R}$$

where R is the gate resistance.

The control current A may now be removed. As there is no driving force (all elements are superconducting), the current will continue to flow in the lower branch. The behavior of the cryotron is seen to be not that of an inverter; when the control current was removed, current did not reappear in the gate. It does behave as a latching relay or as a flip-flop, and the circuit will "remember" the state left by the last "on" control currents. It should be noted that as a flip-flop, this circuit does not require the cross-coupled holding gates needed in flip-flops made from more conventional devices, although they are sometimes added to speed up and to complete the flip-flop action.

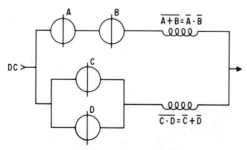

Fig. 20 Cryotron logic functions.

The circuit in Fig. 19, called the J cell,[82] has been used in shift registers, parallel-to-serial converters, and ring oscillators[83] with frequencies up to 6 MHz.

Logical functions may be performed by adding series and parallel cryotrons to the branches as in Fig. 20. In the upper branch, current in either control will divert current out of the branch to give the NOR operation. The NAND function is performed by the parallel cryotrons in the lower branch. At least one superconducting path between the dc terminals must be maintained to prevent driving current through resistive gates. Also, only one path containing controls should be left open to prevent divided or circulating current conditions that would interfere with the proper logical operation of the circuit.

Figure 21 shows a multivariable switch, or decoder tree, which can be expanded to any number of inputs. It provides selection of a single line corresponding to the binary inputs. For example, line 5 may be connected to the common terminal by

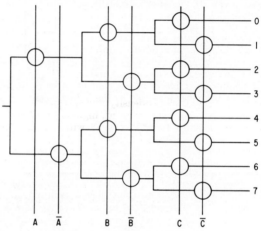

Fig. 21 A cryotron decoding tree.

applying current to the lines $A\bar{B}C$. The tree, or modifications such as the rectangular switch, is used in location selection in a memory. Numerous examples of cryotron circuits for logic applications have appeared in the literature. A rather complete review is given by Bremer.[82]

Let us refer again to Fig. 19 and resume with the conditions of both control currents off and the input current I_{in} flowing in the lower branch. If now the input current is interrupted, the energy stored in the inductance L_2 will cause a current of magnitude

$$I = \frac{L_2}{L_1 + L_2} I_{in}$$

to circulate in the loop formed by the two branches. The magnitude of the current can be calculated by dividing the stored energy between the inductances with the condition of equal currents, or more simply from the following consideration of current superposition. With I_{in} still on, add a reverse current $-I_{in}$ to the input. This current will divide between the inductances as before, leaving no net current from the dc terminals, but a current

$$I_B = I_{in} - \frac{L_1}{L_1 + L_2} I_{in} = \frac{L_2}{L_1 + L_2} I_{in}$$

in the lower branch, and a current

$$I_A = -\frac{L_2}{L_1 + L_2} I_{in}$$

in the upper branch. The signs of the currents indicate that the current is circulating counterclockwise in the figure. The stored current can be quenched by activating either of the controls A or B.

The sequence of shunting current into one leg of a branch, then removing first the control current and second the input current to store circulating current in a loop, forms the basis of most cryotron memory cells. The circuit of a crossed-film cryotron random-access memory cell is shown in Fig. 22. Writing into the cell is accomplished by putting currents into both the "write" and "write enable" lines at the same time, but turning off "write enable" prior to turning off "write." Current will then be stored in loop A. Current on the "read enable" line will divert "read" current through the central cryotron if no current is present in the storage loop, or through an external detection circuit if that path is normal because of the presence of stored current.

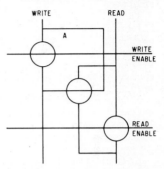

Fig. 22 A cryotron random-access memory cell.

Many other arrangements of cryotrons are possible to form memory cells. They have been designed with as few as one cryotron per cell in a linear-select, or word-organized type of memory.[84] It is generally true, however, that as the complexity of the cell is reduced to achieve higher density, the operating tolerances of the memory system decrease also. A particularly tolerant, high-density cell with three cryotrons has been described by Mundy and Newhouse[81] with an operating range of drive-current variations of 40% and with a bit density of 2,500 bits/in.[2].

A type of organization in which the memory is addressed by stored content has received considerable interest. In contrast to the location-addressed memory in which a specified word or bit location is interrogated and the content divulged, all words in an associative memory are interrogated simultaneously. The bits within the interrogation word may be masked so that only selected portions of the word are compared with the corresponding bit locations of the words in memory. This allows various fields of inquiry in a word. If a match, or a "hit," of the selected bits is found in one or more words in the memory, the words are "flagged" for further processing. Control circuitry then allows the flagged words to be read from the memory in a sequential manner. The criteria for a hit may be identity (catalog or data-addressed memory) or any logical operation (associative memory) such as "greater than" or "less than."

The associative memory requires a large number of active elements to give the features of nondestructive readout, logic functions, and control operations. The cryotron is one of the few electronic devices suitable in terms of cost of fabrication, component density, and power dissipation for making large associative memory arrays. Fruin et al.[85] and Adams et al.[64] have described the circuits, devices, and fabrication of a multiplate cryotron associative memory. Active-device densities greater than $100/\text{cm}^2$ with up to 10,000 devices per substrate were achieved.

A cryogenic location-addressed memory that held great promise was the continuous-film memory. A CFM storage cell is depicted in Fig. 23. It consists of two orthogonal lead drive lines over a superconductive tin or indium continuous-sheet storage plane. The drive currents are selected such that, along either drive line, the field is below the critical field of the storage-film material and magnetic flux will not penetrate, but at the intersection of the current-carrying X and Y lines, components of the field add to produce a greater than critical field and a portion of the

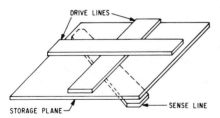

Fig. 23 A continuous-film memory cell.

plane below the intersection becomes normal. Storage occurs by the trapping of magnetic flux in the plane. Readout is accomplished by pulsing the lines with currents of opposite polarity. The change of trapped flux may be detected by the voltage induced either on a sense line crossing the intersection on the opposite side of the storage plane, or on one of the drive lines using the current-stretching-mode amplifier.[81]

The possibility of packing densities of at least 10,000 cells/in.[2] and the simplicity of fabrication, including the incorporation of cryotron selection trees, made this type of memory extremely attractive. Unfortunately, the theoretical drive-current tolerances are only 14%, and in practice can be much less. Aside from the obvious variations that can occur in fabrication such as line widths, etc., the properties of the storage plane itself are difficult to control within the necessary tolerances over sufficiently large areas.[86,87]

A list, composed early in the development of the cryotron, contained numerous advantages of the cryotron over other active devices for logic and memory applications. Experience with the cryotron and advances and new developments in competing technologies have taken their toll from the list. It is instructive to examine the old list and to update it.

1. There is no interaction from the gate back to the control circuit. A similar characteristic is found in light-activated semiconductors.

2. Currents can flow in either direction in both elements of a crossed-film cryotron. The voltage-activated device, the metal-oxide-semiconductor field-effect transistor (MOSFET), can be made symmetric with respect to source-drain current.

3. Current switched into one leg of a multiple-branched cryotron circuit will persist without an additional holding mechanism. Even this most unique built-in memory feature of the cryotron has analogies in the bistable resistance states of some metal-oxide-metal junctions and in tunnel-diode circuits.

4. Complete computer systems can be built solely with cryotrons; no auxiliary components such as capacitors or resistors are required, except for system input/output circuitry at room temperature. With the continuing simplification of MOSFET processing and higher yields, large-scale integrated MOSFET systems are being built in which the MOSFET is used as resistors and capacitors as well as an active device.

5. The off-on impedance ratio of a cryotron is infinite. This is, perhaps, the one unique feature of the cryotron, since it depends upon the zero-resistance state of the superconductor.

6. The cryotron is extremely compact, giving high-density arrays. MOSFET circuits with densities an order of magnitude higher are being made. Both technologies have the same basic density limitation—photolithography capabilities.

7. The cryotron switches rapidly. The initial promise of subnanosecond speeds for cryotrons was not fulfilled. Bipolar transistors are considerably faster. The MOSFET is now competitive in speed with the cryotron and may become even faster.

8. Power consumption of cryotron circuits is very small. No power at all is dissipated by a persistent current. The switching of 10^6 cryotrons at a rate of 10^7 times per second dissipates about 1 W of power at the operating temperature. But this requires 1 kW of power at room temperature for cooling the cryostat. The complementary MOSFET is almost competitive with the cryotron in its low power consumption—and at room temperature.

9. The cryotron should lend itself to low-cost, simple, batch-fabricated large-scale arrays. Again, the MOSFET is emerging as a device that will have these capabilities of fabrication.

The major disadvantage of the cryotron is that it requires a low-temperature environment. Any practical system would use a closed-cycle liquid-helium refrigerator, costing somewhere between $25,000 and $50,000. This is a rather large albatross about the neck of the cryogenic-system designer. It has forced him to increase the size of an "economical" system continually to rather astronomical proportions. Whereas early memory targets were about 10^5 bits, they are now approaching 10^8 bits and higher.

From these considerations the future of cryogenic computing devices looks bleak.

Whether a newer device, as the Josephson tunnel junction with its faster speed, but still requiring a refrigerator, can compete with large-scale integration of bipolars and MOSFET remains to be seen.

3. RADIATION AND PARTICLE DETECTORS

The use of thin film superconductive devices designed to respond to temperature changes produced by radiation or energetic particles has not been widespread.   There are two examples that deserve mention, however.

Several advantages may be obtained from superconductive bolometers.   First, the thermal fluctuations in the detector are low because of its cryogenic environment, and second, the sensitivity of a bolometer is inversely proportional to its specific heat, which decreases with decreasing temperature.   These, together with the very strong dependence of resistance on temperature, lead to extremely sensitive devices.

Martin and Bloor,[88] using thin tin films deposited on 3-μ-thick mica, have developed a bolometer capable of detecting 10^{-12} W at a time constant of about 1 s.   The tin film, loosely coupled to a liquid-helium reservoir, forms part of an ac bridge circuit. The reservoir provides a coarse temperature control to $10^{-3}°$K.   Fine temperature control is provided by a heater in a negative-feedback arrangement from the bridge circuit.   The film can be held to within $10^{-5}°$K at the midpoint of its transition.   The limitation in sensitivity is due to noise.   The noise sources include Johnson noise, photon noise, and possibly noise due to the superconducting transition.   Calculations by Martin and Bloor for the first two of these sources yield a noise power of 3×10^{-14} W, compared with the 10^{-12} W observed.   An analysis by Newhouse[2] indicates that bolometers with a minimum detectable power of 10^{-16} W might be possible.

The passage of an energetic particle through a superconductor can serve as a source of thermal energy.   Spiel and Boom[89] have shown that alpha particles can be detected in a current-carrying tin film, provided that the film is sufficiently narrow.   Their films, 1,000 Å thick and 75 μ wide, were held at a temperature just below T_c and carried currents just less than critical.   The passage of the alpha particle produces a narrow normal region across the film (hence the requirement in width).   A voltage is developed while the region remains normal.   By control of temperature and current, the duration of the voltage pulse can be kept less than 100 ns.

4. FABRICATION OF THIN FILM CRYOGENIC DEVICES

a. General Considerations

The fabrication problems involved in the making of thin film cryogenic devices can be considered in two categories: material and electrical.   Unlike film devices that are to be used at room temperature, cryogenic film devices must meet the more severe environmental requirements imposed by the cycling between room temperature and the liquid-helium operating temperature, as well as surviving the various fabrication steps.   The material problems include adhesion of films to substrates and to previously deposited films, strain-induced recrystallization, and corrosion of metal films by atmospheric moisture.   The electrical problems include those of obtaining the desired characteristics for the labile superconductors, maintaining their uniformity and reproducibility, and preventing flux trapping in the stabile superconductive ground plane and interconnections due to heat generation in the active devices.

In general the choice of material for a particular application is quite limited.   The material must have both the proper superconductive properties and characteristics compatible with a fabrication procedure.   There are several examples.   For controls and device intraconnections of cryotrons, only lead meets all the requirements of having a sufficiently high transition temperature, a high current-carrying capability, and the ability to be readily stencil-deposited, or etched, without damage to previously deposited circuit layers.   Niobium films, which must be deposited on hot substrates to preserve their superconductive properties, can be used for ground planes where damage to underlying layers is not of concern. Among the elements and alloys considered, only tin and indium possess the characteristics suitable for gates or storage media in practical cryotron arrays or continuous-film memory systems.

The earliest method of fabricating thin film cryotrons consisted of vacuum deposition of circuit layers through stencils, or masks. An evaporation system for depositing multilayer films has been described by Caswell and Priest.[90] All layers, both metallic and dielectric, are deposited through stencils in one pump-down cycle of the evaporator. Because of the complexity of the circuits, a large number of stencils are required.

An entirely different approach to multilayer-film-device fabrication has been described by Pritchard et al.[67] In this technique a metal film is deposited all over the surface of the substrate or previously processed films. The circuit configuration of this layer then is delineated by the exposure of an applied photoresist through a photographic mask. After development to remove the photoresist from the areas in which no metal is desired, the excess metal is etched away. The remaining photoresist is removed, an insulator may be applied, and the process is repeated until the circuit is complete. They have found that some photoresists are suitable for use as insulators.[91]

A hybrid approach, having the advantages of both the above methods, has been described by Fruin et al.[85] and Adams et al.[64] Anodized niobium film on glass was used as the insulated ground plane. Tin gates, some lead runs, and silicon monoxide interelement insulation were deposited through their respective stencils. The lead circuit layer, containing the device intraconnections and the dimensionally critical controls, was formed by photoetching an all-over-deposited film of lead. The underlying metal layers were protected from the etchant by the silicon monoxide layer.

b. Substrate, Ground Plane, and Insulation

An ideal substrate should be smooth, rigid, and nonbrittle. In consideration of its role as a support for superconductive gates, it should have an expansion coefficient near to that of the gate material to minimize the effect of stress. It should also have a high heat conductivity at the operating temperature to carry off the Joule heat generated by current in a gate that has been driven normal. Of the economically feasible materials, soda-lime glass (lantern or microscope slides) and polished aluminum plates are reasonable choices, although each is a compromise of the desired properties. Glass has a low thermal conductivity and aluminum is difficult to fabricate with the desired surface smoothness.

The requirements for a ground plane are: (1) it must be superconducting at the operating temperature of the cryotron circuit and must remain superconducting under all magnetic-field conditions encountered in operation; (2) it must exclude magnetic fields and not trap flux; and (3) it must be capable of being insulated with a very thin layer of dielectric material.

Lead and niobium films have been used as ground planes. They both meet requirements 1 and 2. Lead, although relatively easy to vacuum-deposit without degradation of its superconductive properties, is difficult to insulate. The deposition of niobium, on the other hand, requires elaborate techniques but can be simply, and reliably, insulated.

While good devices have been made with lead ground planes, the yields were low. Vacuum-deposited films of dielectric materials, such as SiO, characteristically have pinholes unless a thick layer is used. Also, they may have strains as a result of the conditions of deposition[92] and deteriorate by cracking when thermally cycled. Attempts to deposit insulation on relatively low temperature substrates, to avoid melting or recrystallization of the ground plane, led to the use of organic materials. These have included materials condensed on the surface from the gas phase and polymerized by exposure to an electron beam,[93,94] a glow discharge, or ultraviolet light.[95,96] Photoresist materials, properly applied and cured, have shown some promise as a lead-film insulator.[91] The role of organic polymers in thin film electronics has been reviewed by Allam and Stoddart.[97]

However, the behavior of lead films undergoing temperature cycling makes it suspect that a good insulation can be obtained. Caswell et al.[98] have shown that large stresses occur at low temperature in lead films that were deposited on room-temperature substrates with low thermal-expansion coefficients. The films are unable to withstand the stresses and plastic deformation results, causing localized crystallite growth. With repeated cycling, the crystallites increase in size and density, removing material from their surroundings and growing in thickness. Such a mechanism could easily explain the difficulties of trying to insulate lead.

Polished niobium sheet has been used as a combination ground plane and substrate. Both baked SiO films and an anodized surface layer have been used for insulation. Although it was demonstrated that niobium serves as a ground plane, the insulation systems were marginal because of impurity intrusions in the bulk niobium surface which caused shorts.

The superconducting properties of niobium are seriously degraded by impurities, particularly oxygen, as has been described by DeSorbo.[23] Thin niobium films with high critical temperatures are difficult to prepare. Neugebauer and Ekvall[28] and Rairden and Neugebauer[29] have found that a high substrate temperature during deposition leads to high-quality niobium films with critical temperatures near that of the bulk material. Joynson et al.[68] have made ground planes on soda-lime glass and polished aluminum with electron-beam-deposited niobium. The temperature of deposition was about 550°C, near the deformation temperature of the glass.

Niobium film can be anodized to form a pinhole-free, extremely tough, insulating layer of oxide, which maintains its integrity through repeated thermal cycles to liquid-helium temperature.

c. Cryotron Gate Material

The requirements for the active elements, or gates, in thin film cryogenic amplifiers and switches are similar. A gate is operated only a few tenths of a degree below its transition temperature in order to be easily switched from superconducting to normal by a relatively small current in an overlying control film. It is expected to carry high currents in the superconducting state and to have a high resistance in the normal state. In a cryotron array, the thousands of gates on a plate of large area must be uniform to within certain tolerances and the properties must be reproducible for the many plates that make up a system. Further, practical considerations dictate the use of vacuum evaporators with fast cycle times, which for the present state of the art implies working pressures in the 10^{-6} Torr range of pressure.

Among the superconductive elements, tin and indium have convenient transition temperatures for gate operation and are relatively easy to deposit. There are differences between them, however, that may provide criteria for choice in a particular application. Tin requires a higher evaporation temperature than does indium. The temperatures for 1-μ vapor pressure are 1092°C and 837°C, respectively, for tin and indium.[99] This may permit higher deposition rates of indium with less source radiation than with tin. Caswell[18] has found that indium is less reactive with water vapor, the major constituent of conventional vacuum systems.

The effect of stress on the superconductive properties of indium is less than for tin because of the low yield stress of indium even at low temperatures. As a result the width of the temperature transition of indium films is sharp and reproducible, while that of tin films is broader and may occur over a wide range depending upon the coefficient of expansion of the substrate. Tin has a higher resistivity than does indium, the values being 0.2 and 0.12 μohm-cm, respectively. The switching time of a tin gate would be faster than for an indium gate of the same dimensions. Tin films are considerably more stable than indium films, which tend to break up into electrically isolated islands when stored at room temperature for even a few days. Brennemann[72] reported that lead from Pb-In junctions diffuses into indium films at room temperature, changing the superconductive properties of the film. The diffusion rates are 100 times higher than for bulk materials. The diffusion was not observed for Pb-Sn junctions or for Pb-In junctions stored at 77°K. It is, perhaps, these instabilities of indium that make tin a more favored gate material.

Studies of the critical currents and fields of superconductive tin have been reported extensively in the literature. In particular, the decrease of critical current, the increases of critical field and resistivity, and the production of a penumbra have been noted for depositions in high partial pressures of oxygen and, to some extent, water vapor. One of the most effective ways to reduce the partial pressure of oxygen is by gettering with either titanium or silicon monoxide. If SiO is used as an insulating layer over lead ground planes, or even as an additional layer over anodized niobium, the amount of oxygen in the system is reduced as a matter of course prior to the tin deposition. The reduction of water vapor pressure presents more of a problem, but as the effect of water is less, reasonable results can be obtained in a well-trapped (liquid-nitrogen) system. Other techniques such as breaking vacuum with dry nitrogen gas, maintaining a 300-μ argon bleed through the deposition chamber and the (cold) diffusion pump overnight, and general good housekeeping can keep the water vapor at a minimum.

The all-stencil and the hybrid fabrication schemes provide for depositions of the gate, the interelement insulation, and the top lead layer(s) during the same evaporator pump-down. This protects the tin-gate surfaces from oxidation. When the gates are photoetched, the surface oxide, formed in air during the process, must somehow be removed prior to the lead deposition. An effective means of oxide removal is the use of atomic hydrogen as a reducing agent.[100] Hydrogen is admitted to the deposition chamber at about 1 μ pressure. It impinges against a tungsten filament heated to 1800°C and dissociates. After 5 min exposure of the gate surface to the atomic hydrogen, the system is pumped down and the lead film is deposited. Full superconducting current can be carried by the resulting Pb-Sn junctions. Contamination of the junction by tungsten oxide is not a problem.

d. Intra- and Interconnections

We have seen that lead is unique in its superconductive and mechanical properties for use as the control element and for the connections among gates and controls on a cryotron circuit plate. Some of the techniques evolved to handle this material are discussed.

The sticking coefficient of lead is between 0.8 and 0.9. Thus, when evaporated lead strikes a substrate, a substantial amount reevaporates. If the deposition is through a stencil, the reevaporating lead may strike the backside of the stencil and reevaporate again to the substrate. The result is that a stencil-deposited lead run may be broadened hundreds of microns for stencil-substrate separation of a few mils. This problem does not occur in the photoetch technique. Lines 1.3 mils wide with width tolerances of ± 0.1 mil can be achieved for films 1 μ thick. The width tolerance is apparently determined by the grain size of the lead and its etching characteristics.

The adhesion of vacuum-deposited lead films to the useful cryogenic insulators— glass, SiO, Nb_2O_5, epoxy, Mylar—is minimal at most. Where the lead is not to be in contact with a tin gate, i.e., a lead ground plane or soldering tabs, various thin film "glues" such as Nichrome have been used. A general-purpose glue, even useful in the Pb-Sn junction, is a flash of tin itself. It adheres well to the insulators and lead sticks to it. It does not degrade the junction. If its thickness is less than 200 Å, it is etched away along with the lead in the photoetch process.

A similar flash of tin on the top surface of the lead has been a solution to the following problems:

1. The adhesion of photoresists, either negative or positive types of resist, to lead is very poor.

2. A lead film becomes completely disrupted after a few thermal cycles because of recrystallization.

3. Lead films corrode very quickly in humid air or in the condensed water that accumulates when a sample is removed from a cold cryostat.

The photoresists adhere well to the thin layer of tin. The tin also prevents corrosion of the lead film. Samples of Sn/Pb/Sn films in cryogenic structures have been cycled in excess of 36 times to liquid-helium temperature, allowed to warm to room temperature in humid air and to dry unforced, and stored in undesiccated boxes for months between cycles with no visible or electrical change in the films. Apparently recrystallization of the lead films is inhibited by the addition of the tin layers. The superconducting properties of the Sn/Pb/Sn films appear to be the same as those of pure lead.

The interconnection of cryotrons on separate substrates has been a major problem. It is necessary to connect power leads to the plates with superconductors so that Joule heating does not occur at the plate. Signal leads between cryotron circuits have the much more stringent requirement that their inductance be low, preferably the same as the conductors on a plate.

The simplest approach, if not the most satisfactory, is to solder lead wire to lead bonding pads on the plate with a superconductive solder. Not only is the inductance high, but the solder joints fail in thermal cycling. In another approach, lead-foil strips, bonded to $\frac{1}{4}$-mil Mylar, are used in a pressure contact against the film pad. After a few thermal cycles, the surface of the foil becomes corroded, and superconducting contacts can no longer be made. An advantage to this method is that lead foil can be bonded to the opposite surface of the Myler to serve as a ground plane and lower the inductance.

An improvement is to replace the lead foil with vacuum-deposited Sn/Pb/Sn on both sides of the Mylar and photoetch the conductors. A light pressure contact between the conductors and Sn/Pb/Sn pads on the plate will carry adequate supercurrent.

Low-inductance interconnections have been made by depositing ground plane Sn/Pb/Sn on Mylar, insulating with 8,000 Å of photoresist, and depositing Sn/Pb/Sn conductors.[81] Such interconnections, fragile as they may be, can be pressure-contacted without shorting and cycled several times.

5. SUPERCONDUCTIVE TUNNELING DEVICES

While cryotrons and detectors can be understood in terms of the macroscopic and phenomenological theories of superconductivity, tunneling devices require the details of electronic motion given by the microscopic theories for their comprehension. In the following sections the results of the theories will be introduced as required. More detailed discussions can be found in the referenced literature.

a. Giaever Tunneling

Electrons can pass between two metals separated by a thin insulating film via the quantum-mechanical process of tunneling. The current that flows between the metals decreases exponentially with the thickness of the insulating barrier. This restricts the insulator thickness to less than about 100 Å for observable currents. For normal metals in which the density of states in the conduction band is constant over the applied voltage range, the current through the insulator will vary linearly (ohmic behavior) with the applied voltage.[101] In a superconductor the density of states is not constant, however, but changes rapidly in a narrow energy range centered at the Fermi level. The current-voltage characteristic for tunneling is nonlinear when either or both of the metals is in the superconducting state.

It was Giaever[102] who first showed that low-voltage tunneling experiments could reveal the electronic structure of superconductors. He prepared junctions of aluminum-aluminum oxide-lead by vacuum depositing an aluminum strip onto glass, oxidizing the aluminum in air, and then depositing a lead-film strip at right angles to the aluminum strip. The intersection of the strips formed the tunneling junction. Current-voltage characteristics of a typical junction are shown in Fig. 24. The oxide is thought to be about 15 to 20 Å thick. Ohmic behavior was obtained when both films were maintained in the normal state with a magnetic field. When the lead was allowed to become superconducting, a low, nonlinear current was observed in the low-voltage range.

Under the assumption that tunneling current is proportional to differences in the density of states near the Fermi level, the relative change in the density of states can be obtained directly by plotting the conductance when one of the metals is superconducting $(dI/dV)_{ns}$, divided by the conductance when both metals are normal $(dI/dV)_{nn}$, against energy. The relative conductance of curve 3 of Fig. 24, plotted vs. potential difference (= energy/e) in Fig. 25, expresses the density of states in superconducting lead relative to the density of states in normal lead, as a function of energy measured from the Fermi level.   Since conduction through the junction is symmetric, the change in density of states is symmetric about the Fermi level.   The

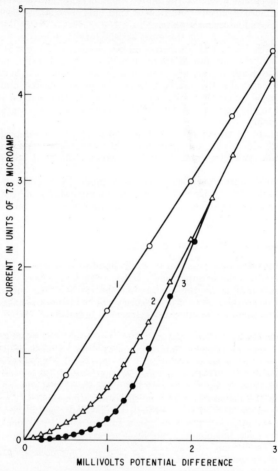

Fig. 24   Tunnel current between Al and Pb through Al$_2$O$_3$ film as a function of voltage. (1) $T = 4.2$ and $1.6°$K, $H = 2.7$ kOe (Pb normal). (2) $T = 4.2°$K, $H = 0$. (3) $T = 1.6°$K, $H = 0$.   The Pb is superconducting in (2) and (3).

point in Fig. 25 at which the relative conductance is unity is an approximate measure of half the energy gap.   The curve is similar to the Bardeen-Cooper-Schrieffer[11] density of states for quasi-particle excitations (the normal electrons) in a superconductor,

$$\rho(E) = N(0) \left| \frac{E}{[E^2 - \epsilon(T)^2]^{\frac{1}{2}}} \right| \qquad |E| \geq \epsilon(T)$$
$$\rho(E) = 0 \qquad\qquad\qquad\qquad |E| < \epsilon(T)$$

where $N(0)$ is the density of electrons and $\epsilon(T)$ is the half-energy gap.   This is shown in Fig. 25 by the dashed curve.

When both metals are superconducting, the current-voltage characteristic exhibits a negative-resistance region for $T > 0°K$ if the energy gaps of the two superconductors are different.[103] If the energy gaps are the same, the negative-resistance region is suppressed, and there is a sharp rise in current at the full band-gap energy. Examples of characteristic curves for tunneling between two superconductors are given in Fig. 26.[104] The current present for energies below the band gap for the Al/Al oxide/Al sandwich is ascribed to thermal broadening which occurs at finite temperatures.

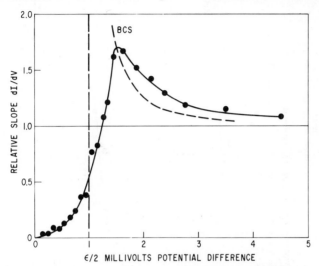

Fig. 25   The relative conductance of curve 3 of Fig. 24, $(dI/dV)_{NS}/(dI/dV)_{NN}$ as a function of voltage. The dashed curve corresponds to the BCS density of states.

An expression for the tunneling current may be obtained by considering two metals, 1 and 2, separated by an insulating barrier.   Assuming that the tunneling probability is constant over the range of energies of interest,[105,106] the tunneling current from metal 2 to metal 1 is proportional to an integral over all energies of the product of the number of electrons in metal 1 and the number of unoccupied states, or holes, in metal 2.   As tunneling occurs in both directions, the net current will be the difference of the opposed one-way currents.   Taking the zero of energy at the Fermi energy of metal 2, and applying a positive potential to metal 2, the current can be written

$$I = C\int\{\rho_1(E - V)f(E - V)\rho_2(E)[1 - f(E)] \\ - \rho_1(E - V)[1 - f(E - V)]\rho_2(E)f(E)\}\,dE \quad (17)$$

where C is a constant of proportionality (including the exponential dependence on insulator thickness), $\rho(E)$ is the density-of-states function for the subscripted metal at the energy E, $f(E)$ is the Fermi function

$$f(E) = \frac{1}{1 + e^{\beta E}}$$

and V is the energy equivalent to the applied potential.   Equation (17) reduces to

$$I = C\int\rho_1(E - V)\rho_2(E)[f(E - V) - f(E)]\,dE$$

For the case of both metals normal, and with the density-of-states functions constant over the applied voltage range, it is easy to show that the current is proportional to the voltage.   This situation is pictured in Fig. 27a,[107] where the occupancy of states is denoted by hatching in the energy diagrams.   The application of a potential difference

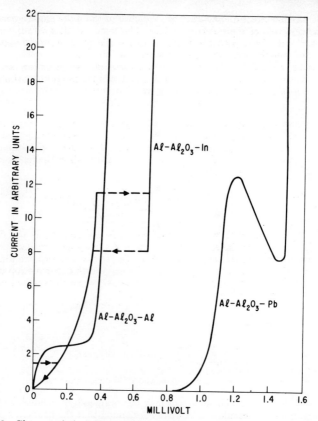

Fig. 26   Characteristic I-V curves for tunneling between two superconductors.

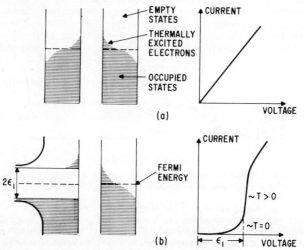

Fig. 27   Energy diagrams and current-voltage characteristics for the cases (*a*) both metals normal and (*b*) one metal normal and the other superconducting.

to the metals has the effect of elevating the diagram of the more negative metal with respect to the diagram of the positive metal. The result of integrating over energy the electron transfers from occupied states of one metal to unoccupied states of the other yields the linear current-voltage relationship.

The current-voltage relation for the case of one metal superconducting can be found readily only for the limit $T = 0°K$ for which the Fermi functions take on simple forms. Then,

$$I_{ns}(0) \approx [V^2 - \epsilon^2(0)]^{\frac{1}{2}} \qquad V \geq \epsilon(0)$$

The current is zero for voltages less than the half-energy gap of the superconductor. It is the thermal broadening that produces current at lesser voltages at nonzero temperatures, as can be seen by consideration of Fig. 27b. An expression for I_{ns} at temperatures greater than zero, and for applied voltages less than the half-energy gap, has been given by Giaever and Megerle:[107]

$$I_{ns} = 2C_{nn} \frac{\epsilon}{e} \sum_{m=1}^{\infty} (-1)^{m+1} K_1(m\beta\epsilon) \sinh (m\beta eV)$$

where C_{nn} is the conductance when both metals are normal, K_1 is the first-order modified Bessel function of the second kind, and V is the applied voltage.

Extensive machine computations have been made for the two-superconductor case by Shapiro et al.[108] A comparison of calculated results with experimental data for a tin-lead sandwich at 3.60°K is given in Fig. 28. Qualitative understanding of tun-

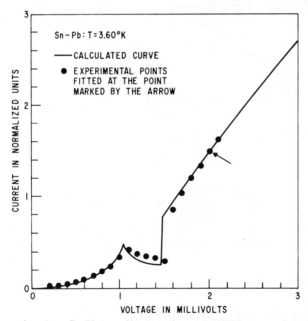

Fig. 28 Comparison for a Sn-Pb tunneling junction at 3.60°K between the calculated I-V characteristics (solid curve) and experimental points. Current was normalized at the point marked by the arrow.

neling between two superconductors can be obtained from the energy diagrams of Fig. 29.[104] Thermally excited electrons and holes are shown for the smaller-gap material. There will be relatively few thermally excited electrons for the larger gap. As the voltage is raised from zero in Fig. 29a, no current will flow until the upper band edge of metal 1 is within kT of the upper band edge of metal 2. Current will increase with voltage, then, as more of the thermally excited electrons in metal 1 are raised above the forbidden gap in metal 2. This will continue until the upper bands are coincident at a voltage equivalent to $\epsilon_2 - \epsilon_1$ as in Fig. 29b. When the voltage is raised above this, the electrons in metal 1 tunnel into the decreasing density of states in the upper band of metal 2, and the current decreases, producing the negative-resistance region. This situation persists until the potential difference is raised to $\epsilon_1 + \epsilon_2$, where the lower band edge of metal 1 is coincident with the upper band edge

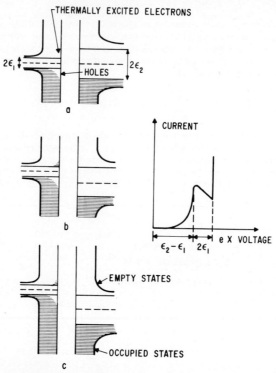

Fig. 29 Analysis of the current-voltage characteristic of dissimilar superconductors. (a) The two superconductors with no voltage applied. (b) A voltage $(\epsilon_2 - \epsilon_1)/e$ is applied. The current has increased with increasing voltage. Further increase of voltage will cause a decreasing current since the tunneling electrons face a less favorable density of states (the negative-resistance region). (c) A voltage $(\epsilon_2 + \epsilon_1)/e$ is applied. The current is at a minimum. Further voltage increase will cause a rapid increase of current.

of metal 2 (Fig. 29c). Note that for $T = 0°K$, the thermally excited electrons in metal 1 would be absent and no current would flow for potentials between zero and $\epsilon_1 + \epsilon_2$. Finally, when a voltage greater than $\epsilon_1 + \epsilon_2$ is applied, the current increases rapidly because the electrons below the gap in metal 1 can tunnel.

The current-voltage characteristic for dissimilar superconductors thus allows a direct measurement of both band gaps. The voltage difference between the maximum and minimum currents at the ends of the negative-resistance region is a direct measure of the full band gap of the lower-gap material. The voltage at the minimum is the sum of the half-energy gaps from which the wider-gap value can be obtained.

Giaever tunneling junctions have been proposed as active-circuit devices for switches and oscillators[109] and as detectors for microwave and submillimeter-wave radiation.[110] Their major uses, however, have been as tools in the study of superconductivity and the tunneling process.

The results of measurements of the band gaps of a number of elements and the compound Nb_3Sn are given in Table 2. Unique in the list is that of bismuth, which is not normally superconducting. Samples were prepared by vapor quenching onto substrates at liquid-helium temperature.[114] Such films are superconducting with $T_c = 6°K$[12] and are reported to show a well-defined energy gap.

TABLE 2   The Band Gap 2ϵ of Superconductors Measured at Temperature T by Giaever Tunneling.   Values for $T = (0)$ Are Extrapolated

Material	$2\epsilon/kT_c$	T	T_c	Reference
Al	$3.20 \neq 0.3$	1.0	1.2	107
	$4.2 \neq 0.6$	(0)		107
	$2.5 \neq 0.3$	(0)	1.3	108
Tl	3.2–3.3	(0)	2.43	111
W(β)	2.5–3.7	(0)	3.1	112
In	$3.63 \neq 0.1$	1.0	3.40	107
	~ 3	(0)	3.40	108
Sn	$3.46 \neq 0.1$	1.0	3.72	107
	$3.10 \neq 0.05$	(0)	3.80	108
La	3.2	(0)	5.0	113
Bi(quenched)	3.9		6	114
Pb	$4.33 \neq 0.1$	1.0	7.2	107
	$4.04 \neq 0.1$	(0)	7.2	108
Nb	3.59	1.9	9.2	115
Nb_3Sn	1.3	(0)	17.5	116

Energy-gap measurements by tunneling techniques have been made on super-conductors containing small amounts of paramagnetic impurities.[114] It is found that the gap disappears long before superconductivity (zero electrical resistance) is destroyed by increasing concentrations of the impurity. Gapless superconductivity in dirty superconductors (alloys of Sn-In and Pb-Bi in which the electronic mean free path of a film is much less than its thickness) in near-critical magnetic fields has been found in tunneling experiments by Guyon et al.[117] This is in agreement with a theory of De Gennes[118] which predicts an absence of a gap in the density of states of a superconductor for the experimental conditions employed.

The temperature dependence of the energy gap of a pure, type I superconductor is quite consistent with the BCS theory.[107,108] A typical comparison is given in Fig. 30,[111] where the reduced energy gap of thallium is plotted against the reduced temperature.

The dependence of the energy gap of aluminum on a longitudinal applied field has been measured by Douglass,[119] using lead as the reference superconductor. For values of the ratio of film thickness to penetration depth d/λ less than $\sqrt{5}$, the energy gap decreases smoothly to zero with increasing magnetic field according to

$$\left[\frac{\epsilon(H)}{\epsilon(0)}\right]^2 = 1 - \left(\frac{H}{H_c}\right)^2$$

indicating that the phase transition is second-order. For $d/\lambda > \sqrt{5}$, the gap remains finite near the critical field H_c. At the critical field the gap drops abruptly from the critical-gap value to zero, indicating a first-order transition.

Various characteristics of the tunneling-current fine structure have been noted. Townsend and Sutton[120] have found evidence in thick lead films for two energy gaps arising from two different zones in the Fermi surface of lead. Taylor and Burstein[121] have attributed to double-particle tunneling the excess current between two super-conductors at voltages $V = \epsilon_1/e$ and $V = \epsilon_2/e$ for which the band edge of one super-conductor is coincident with the Fermi level of the other. In this mechanism[122] tunneling can take place either by the recombination of two electrons from singly occupied states of one superconductor into an electron pair at the Fermi level of the other, or by the dissociation of a pair in one superconductor to two unpaired electrons in the other.

Since the attraction between superconducting pairs of electrons is due to electron-phonon interactions, it might be expected that effects should arise from the detailed shape of the phonon spectra. Giaever et al.[123] first pointed out structure in the

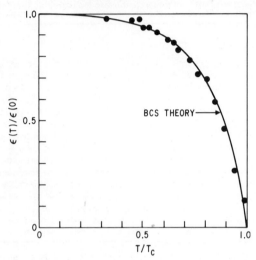

Fig. 30   Reduced energy gap of thallium vs. reduced temperature.

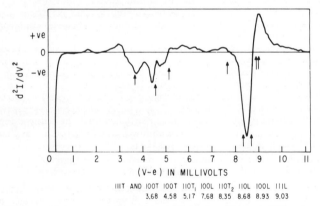

Fig. 31 d^2I/dV^2 vs. V (measured from ϵ) for a Pb-Pb junction at 1.3°K.   Arrows indicate the bias at which singularities in the phonon distribution appear.

relative conductance vs. energy plot of normal metal-lead tunneling junctions that could be related to the phonon spectrum of lead.   Rowell and Chynoweth[124] found periodic structure in the density-of-states plot of lead related to harmonics of the transverse acoustic phonons.   Extending the technique, Rowell et al.[125] resolved the structure in detail and were able to assign much of it to specific singularities arising from the phonon distribution as shown in Fig. 31.   Phonon structures were found also for aluminum and tin.   Phonon impurity bands caused by light, substitutional impurity atoms in a cubic crystal have been observed for dilute solutions of indium in lead.[126]

In the voltage ranges where effects due to the superconducting nature of the metal are observed, no structure has been identified as arising from the insulator. These ranges include those involving the band gap of the superconductor (0 to 3 mV), the phonon spectrum (3 to 10 mV), and the multiphonon spectrum (5 to 25 mV). However, Jaklevic and Lambe[127] have found structure in the 50- to 500-mV region corresponding to the infrared region from 25 to 2.5 μ. These are associated with the bending and stretching modes of OH groups that are present either chemically in the oxide or as absorbed water vapor and are independent of the superconductive properties of the metal electrodes.

b. Josephson Tunneling

The Giaever tunneling process between two superconductors involves the transfer of single particles (the normal electrons), or the transfer of pairs of electrons which form condensed pairs only in one of the superconductors. That superconducting pairs of electrons could tunnel from one superconductor to another through a sufficiently thin insulator was predicted by Josephson.[128] He showed that two forms of tunneling could occur. In the first, the dc Josephson effect, a direct current can flow without any voltage across the junction. If the current exceeds the critical current J_1 for the junction, a finite voltage V appears. In addition to a direct current, there is now an alternating current of frequency $\nu = 2eV/h$. This is the ac Josephson effect.

In a single superconductor the centers of mass of all the pairs of electrons move with the same momentum. This correlation of pair motion, or long-range order, meets the requirement of the exclusion principle. The result is that the quantum-mechanical wavelengths of the electron pairs with center-of-mass momentum p must be the same, namely, h/p. In addition, energy considerations demand that the phases of all pair waves be the same. It is these equalities of pair momenta and of phases that lead to the zero resistance of the superconducting state. The energy difference between electron pairs at opposite ends of a superconductor, caused by a voltage difference, would develop a time-varying phase difference between the pairs. The resulting breakdown of pair phase locking would raise the superconductor to a higher-energy state. Instead, as long as the critical current of the superconductor is not exceeded, it will carry the current in the lower-energy state without allowing a voltage difference to appear.

The pair phases between two separated superconductors are randomly related. If the superconductors are brought sufficiently close together, on the order of 10 Å, so that pair tunneling can occur, then a unique, time-independent relation between the phases develops by virtue of the coupling energy of the tunneling pairs. Because the phase relation is time-independent, direct current can flow across the junction without a voltage appearing. The junction current density is obtained as the difference of opposed tunneling currents where the probability of tunneling is related to the degree of phase mismatch encountered by the tunneling pairs. The result of the microscopic analysis gives the current density as

$$J = J_1 \sin \varphi$$

where φ is the phase difference between the quantum-mechanical wave functions of the two superconductors. The maximum direct Josephson current density for a symmetric junction (identical superconductors) is[129]

$$J_1 = \tfrac{1}{2}\pi R_n^{-1}\epsilon(T) \tanh \tfrac{1}{2}\beta\epsilon(T) \qquad (18a)$$

where $\epsilon(T)$ is the half-energy gap of the superconductor, R_n is the normal resistance of the junction, and $\beta = 1/kT$. For an asymmetric junction,

$$J_1 = \pi R_n^{-1} \frac{\epsilon_1(T)\epsilon_2(T)S}{\beta} \qquad (18b)$$

where
$$S = \sum_{l=-\infty}^{+\infty} \{[\omega_l^2 + \epsilon_1^2(T)][\omega_l^2 + \epsilon_2^2(T)]\}^{-\frac{1}{2}}$$

with $\omega_l = \pi(2l + 1)/\beta$. This can be represented as

$$J_1 = \pi R_n^{-1} \frac{\epsilon_1 \epsilon_2}{\epsilon_1 + \epsilon_2}$$

for $T/T_c < 0.5$.[130]

Undoubtedly the zero voltage current had been observed on many occasions both by those working with cryotrons, where inadvertently a tunnel junction would be formed in a lead-tin contact causing "critical gate currents" of a few milliamperes, and by those working with Giaever tunnel junctions. This effect was generally ascribed to metallic bridges, of low critical current, shorting the two superconductors. The first reported observation of the direct Josephson current was made by Anderson and Rowell,[131] who ruled out bridges in their films from the following considerations. The currents were quite sensitive to magnetic fields as predicted by Josephson, whereas the critical field of a filament should be very high. The critical temperatures of the junctions were the same as that of the lower T_c superconductor, which would be unlikely for a filament. If bridges were present, the critical currents would indicate normal conductivity several orders of magnitude higher than observed. And the "bridges" could not be burned out. Figure 32 shows the I-V characteristics for a

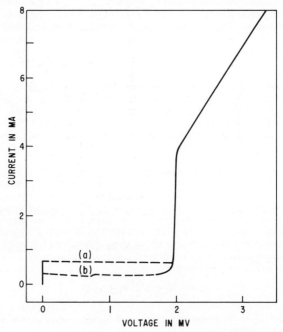

Fig. 32 Current-voltage characteristic for a Sn-Sn oxide-Pb tunnel junction at 1.5°K. (a) $H = 6 \times 10^{-3}$ G. (b) $H = 0.4$ G.

tin-tin oxide-lead junction at 1.5°K for two values of magnetic field. For currents greater than the junction critical current, a voltage develops and the Giaever tunneling characteristic for two superconductors is obtained.

The magnetic-field dependence of the direct Josephson tunnel current was investigated by Rowell.[132] His results are shown in Fig. 33 for a magnetic field parallel to the plane of the junction. The periodic dependence of the maximum current on magnetic field arises from an additional term that must be added to the phase difference between the superconductors to account for the magnetic field. This addition takes the form

$$\Delta\varphi = -2\pi \frac{2e}{hc} \int_1^2 \mathbf{A} \cdot d\mathbf{s} \tag{19}$$

where $\mathbf{A}$ is the vector potential related to the magnetic field by $\mathbf{H} = \nabla \times \mathbf{A}$, and the integration is carried out along a curve joining the two superconductors. This results in a spatial variation of the difference between the pair phases in the two superconductors. For a uniform, constant field in the junction barrier, the phase

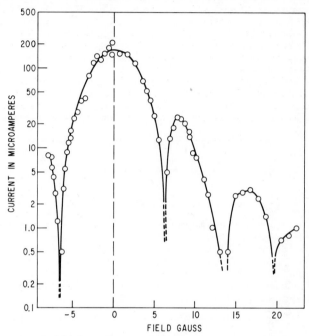

Fig. 33 The magnetic-field dependence of the dc Josephson current in a Pb-I-Pb junction at 1.3°K.

difference varies in the plane of the junction in a direction at right angles to the direction of the field and at a rate proportional to the field:

$$\operatorname{grad} \varphi = 2\pi \frac{2e}{hc} t(\mathbf{H} \times \mathbf{n}) \tag{20}$$

where $\mathbf{H}$ is the field in the junction barrier, $\mathbf{n}$ is a unit vector normal to the plane of the barrier, and

$$t = \lambda_1 + \lambda_2 + d$$

is the effective thickness of the barrier occupied by the field, i.e., the barrier thickness d and a penetration depth into either superconductor. Since the Josephson current density is periodic in phase, the direct current in the junction will be spatially periodic as shown in Fig. 34. From Eq. (20) it may be seen that a complete period of 2π occurs over a distance w (Fig. 34) such that the flux $\Phi = Hwt$ is one unit of the flux quantum

$$\Phi_0 = \frac{hc}{2e} = 2.1 \times 10^{-7} \text{ G cm}^2$$

The total current is obtained by integrating the current density over the junction area.   It takes the form

$$I = I_0 \left| \frac{\sin \pi H/H_J}{\pi H/H_J} \right|$$

where H_J is the field for which the junction contains one flux quantum.

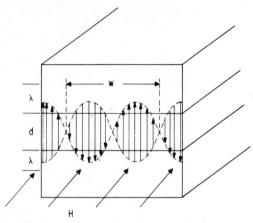

Fig. 34   The spatial variation of Josephson tunneling current in a magnetic field.

The observed junction critical currents are generally less than that predicted by the current-density equations (18a) and (18b) even in zero magnetic field.   Fiske,[130] using parallel strip junctions 0.025 cm wide and with junction resistances in the range 0.01 to 1 ohm, has observed currents as high as 0.8 of the theoretical maximum.   He found none that was higher.   The temperature dependence of both symmetric and asymmetric junctions was in good agreement with the theory of Ambegaokar and Baratoff[129] [Eqs. (18a) and (18b)], as shown in Fig. 35.

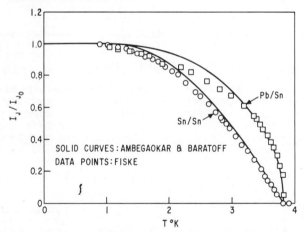

Fig. 35   Temperature dependence of the direct Josephson current for symmetric Sn-Sn and asymmetric Pb-Sn junctions in comparison with theory.

The predominant causes for the reduction of the direct Josephson critical current are thermal fluctuations and self-fields.   The energy of coupling between the superconductors,

$$E_J = -\frac{h}{2\pi e} J_1 \cos \varphi$$

is small in magnitude, ranging from 10^{-2} eV for thick barrier junctions to a few electron volts for thin barriers. Thermal fluctuations, particularly from portions of the measuring circuit not at the cryogenic temperature, will make the thicker barrier junctions unstable.[133] Thin barriers will become unstable to thermal fluctuations as the current increases because of the cosine dependence of coupling energy on the phase difference.

Ferrell and Prange[134] have pointed out that the current through a Josephson junction is self-limited by the magnetic field set up by the current. The current is then restricted to flow only in a region near the periphery of the junction. The width of the region is twice the "Josephson penetration depth"

$$\lambda_J = \left(32\pi^2 \frac{\lambda e^2 J_1}{hc^2}\right)^{-\frac{1}{2}}$$

where λ is the penetration depth of the superconductor. Values of λ_J are of the order 0.1 to 1.0 mm for Pb/oxide/Pb junctions operating at 4.2°K. The reduction of the current by the self-field limitation is least in junctions formed from parallel strips. The maximum current in crossed-film junctions is only 0.3 of theoretical, even for small low-resistance junctions.[135] In addition, the magnetic-field dependence is modified such that the maximum current is nonsymmetric with field.

The concept of the periodic dependence of the direct Josephson current upon the magnetic flux contained within the junction was extended by Jaklevic et al.[136] to include multiple junctions in parallel connected by superconducting links, as in Fig. 36.

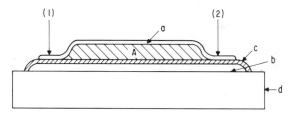

Fig. 36 Cross section of a Josephson junction pair, vacuum-deposited on a quartz substrate (*d*). A thin oxide layer (*c*) separates thin tin films (*a*) and (*b*) in the junction areas (1) and (*a*). The area (*A*), enclosed by the superconducting links, is 3.5 mm long by 3 to 0.3 μ thick. Current flow is measured between films (*a*) and (*b*).

This configuration will lead to two periodicities of the current with flux. The first is associated with the flux within the junctions (assumed to be of equal size). The second period involves a quantum-mechanical interference between the currents flowing through the separate junctions. It is associated with the flux enclosed in the area A (Fig. 36) between the junctions. Applying the phase-difference addition [Eq. (19)] to the structure of Fig. 36, one obtains the doubly periodic relation

$$I = \left|\frac{\sin \pi H/H_J}{\pi H/H_J}\right| \left|\sin \left(\varphi - \pi \frac{H}{H_A}\right)\right|$$

where, again, H_J is the field for which one flux quantum is contained in a junction, and H_A is the field for which one flux quantum is contained in the larger area A. Observations of the second periodicity for junctions separated by 3.5 mm confirms the superconducting state as a truly long-range single-quantum state.

The interference of coupled Josephson junctions is shown in Fig. 37. The junction pairs were made by Edward Kaplan in the author's laboratory using thin film niobium and lead. The insulator, corresponding to the material in area A of Fig. 36, was niobium oxide formed by anodization to a thickness of 1,000 Å.   The junction barriers were formed by anodization to a few tenths of a volt to give thicknesses of about 10 Å.[137]   The junctions were separated by 2 mm.   The area enclosed by the link was 4×10^{-6} cm², where the penetration depths of the superconductors have been included in the thickness.   This gave a periodicity of 0.05 Oe/flux quantum, or 1.6 mA of field current/flux quantum in a superposed "control" film.

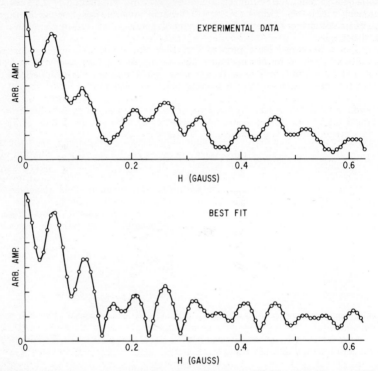

Fig. 37   Interference of coupled Josephson junctions, Nb-Nb oxide-Pb, at 4.2°K.

The "best-fit" curve of Fig. 37 is an attempt to fit the experimental data to an expression derived for junctions of unequal areas,

$$\frac{I}{I_{\max}} = \left[(i_2 - i_1)^2 + 4 i_1 i_2 \cos^2 \left(\frac{\pi H}{H_A} \right) \right]^{\frac{1}{2}}$$

where

$$i_1 = \frac{a_1}{a_1 + a_2} \frac{\sin \pi H/H_1}{\pi H/H_1}$$

$$i_2 = \frac{a_2}{a_1 + a_2} \frac{\sin \pi H/H_2}{\pi H/H_2}$$

a_1 and a_2 are the junction areas, and H_1, H_2, and H_A are the values of the applied field which produce one flux quantum in junction 1, in junction 2, and in the loop, respectively.   The major features of the interference pattern have been reproduced by the expression.   The deviations in the fine detail may be due to nonuniformity of the individual junctions.

Two developments have led to the consideration of using dc Josephson junctions as active devices for digital memory and logic circuits.   Smith[138] observed that persistent current would flow in a superconducting lead loop containing a junction.

In a zero magnetic field, he was able to trap a current of 7.2 mA whose magnitude was equal to the critical current of the junction. The persistent current remained constant for 7 h, implying an upper bound of 4×10^{-16} V for the potential across the junction. An upper limit for the switching time, or the time of transition from pair tunneling to single-particle tunneling, has been determined by Matisoo[139] to be less than 0.8 ns. This limit was imposed by the measuring circuit. The actual transition time may be substantially smaller. He reports there is no delay in the transition after the threshold current is exceeded, nor does the rise time of the voltage across the junction change with changes in driving-current amplitude or rise time.

Matisoo[140] has constructed Josephson-junction switches with current gains of 4. Geometrically the switches are analogous to the in-line cryotron. Unlike the cryotron gate, however, which requires 10 to 40 ns for transition from superconducting to normal and becomes a milliohm resistance in an inductive circuit, the Josephson junction becomes a low-impedance voltage source in less than 0.8 ns. In a flip-flop constructed of Josephson junctions used in a current-steering mode, and having reasonable dimensions (large enough to fabricate but small enough to be interesting for high density), the current-transfer times are calculated to be 160 ps. The total time to change the flip-flop state would be less than 960 ps.

Clark and Baldwin[141] have proposed a superconducting memory device using Josephson junctions in which persistent current is stored in a superconducting loop containing a junction (Fig. 38). The structure, together with a control strip, is

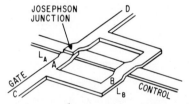

Fig. 38   A superconducting memory device using Josephson junctions.

deposited on an insulated ground plane. Information is written into the structure by passing current J_a, just less than the maximum Josephson current J_1 for the junction, from C to D (or from D to C for the complement) while applying a field to the junction by means of current through the control strip. The majority of the current will be shunted into the path CBD. Upon removing first the field and then the current, a circulating current will be left in the loop. Readout is performed by turning on the field and observing the sign of the transient voltage appearing across the junction. They estimate switching speeds of a cell to be less than 10 ns.

When the current through a junction exceeds the maximum direct Josephson current, a voltage V appears across the junction. The phase difference across the barrier is no longer constant in time but varies with a frequency $\nu = 2eV/h$. Because of the sinusoidal dependence of the Josephson current on the phase, the current oscillates between the superconductors at the frequency ν. This is the ac Josephson effect. The quantity $2e/h$ has the value 483.6 MHz $\mu\mathrm{V}^{-1}$. Since typical junction voltages range from a few microvolts to several millivolts, frequencies as high as hundreds of gigahertz may be expected.

Josephson predicted[128] that rf fields would modulate the ac supercurrent. The current would then have Fourier components at frequencies $2eV/h \pm n\nu$, where ν is the frequency of the applied rf voltage and n is an integer. A dc component of the supercurrent would develop for some n such that $2eV/h = n\nu$. The I-V characteristic would display a step structure, or abrupt increase in junction current at the voltage V. This effect was found by Shapiro,[142] using microwaves of frequencies 9.3 and 24.85 GHz. His results constituted the first indirect confirmation of the ac Josephson effect.

Further indirect evidence of the ac Josephson effect was obtained by Eck et al.[143] in I-V measurements on Pb-Pb oxide-Pb junctions. At low temperature and in fields large enough to quench the direct current, they observed temperature-independent resonance-shaped dc peaks at voltages less than the energy gap of the superconductor. The voltage at which the peak occurred was a linear function of the applied magnetic field. The results were ascribed to coupling between the electromagnetic modes of the junction and the alternating Josephson current, which has a spatial variation due to the field and a time variation due to the voltage.

Coon and Fiske[144] have shown that the resonance peak is a special case of the interaction of Josephson current-density waves with the electromagnetic fields in the junction acting as an open-ended cavity resonator. In applied fields of a few gauss, the current-voltage characteristics become steplike, as in Fig. 39,[145] with appreciable

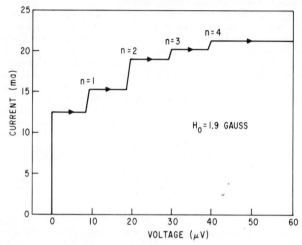

Fig. 39 The steplike structure for a Sn-Sn oxide-Sn junction in an external magnetic field $H_0 = 1.9$ G. The voltage separation of the modes (n) corresponds to a frequency separation of 4.6 GHz.

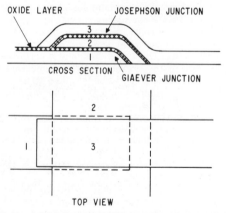

Fig. 40 An arrangement of Josephson and Giaever junctions to detect the ac Josephson effect.

currents being carried in constant voltage steps at regularly spaced voltages. The steps correspond to the resonance modes of the cavity. They report modes ranging from 5 GHz, set by the junction dimensions, up to 250 GHz for tin. The upper frequency is limited by the half-energy gap of the superconductor where the frequency becomes high enough to raise single electrons above the energy gap.

Direct observation of the alternating Josephson current by detection of the electromagnetic fields radiated from the junction is complicated by the fact that the amount of energy radiated is very small because of the bad impedance match between the junction and free space.   Giaever,[146] however, was able to build a detector tightly coupled to the Josephson junction and to detect 10^{-7} W of power.   The detector consisted of a Giaever tunnel junction over which the Josephson junction was constructed as in Fig. 40.   Radiation generated in the Josephson junction was coupled into the Giaever junction, where it produced current steps in the single-particle tunneling I-V characteristic, analogous to the microwave-stimulated junctions of Dayem and Martin.[147]

The direct detection of the microwave power from a biased Josephson junction in a magnetic field (and the direct confirmation of the ac Josephson effect) was first reported by Yanson et al.,[148] who detected 10^{-13} W outside the cryostat containing the junction.   Shortly thereafter, Langenberg et al.[145] observed a power of 10^{-12} W at 9.2 GHz with a spectral purity better than 1 part in 10^4.

Numerous applications of the Josephson effects are being used, or considered, including the generation of electromagnetic radiation between the microwaves and infrared and the precise determination of physical constants.

REFERENCES

1. Roberts, B. W., *Progr. Cryog.*, **4**, 159 (1964).
2. Newhouse, V. L., "Applied Superconductivity," p. 248, John Wiley & Sons, Inc., New York, 1964.
3. Silsbee, F. B., *J. Wash. Acad. Sci.*, **6**, 597 (1916).
4. London, F., "Superfluids," vol. 1, Dover Publications, Inc., New York, 1961.
5. Gorter, C. J., and H. B. G. Casimir, *Physica*, **1**, 305 (1934).
6. Pippard, A. B., *Proc. Roy. Soc. London*, **A216**, 547 (1953).
7. Faber, T. E., and A. B. Pippard, *Proc. Roy. Soc. London*, **A231**, 336 (1955).
8. Drangeid, K. E., R. Sommerhalder, H. Muller, and H. Seitz, *IBM J. Res. Develop.*, **10**, 13 (1964).
9. Ginzburg, V. L., and L. D. Landau, *J. Exptl. Theoret. Phys. USSR*, **20**, 1064 (1950).
10. Gorkov, L. P., *Soviet Phys. JETP*, **10**, 998 (1960).
11. Bardeen, J., L. N. Cooper, and J. R. Schrieffer, *Phys. Rev.*, **108**, 1175 (1957).
12. Buckel, W., and R. Hilsch, *Z. Physik*, **138**, 109 (1954).
13. Lazarev, B. G., A. I. Sudovtsev, and A. P. Smirnov, *Soviet Phys. JETP*, **6**, 816 (1957).
14. Chopra, K. L., M. R. Randlett, and R. H. Duff, *Appl. Phys. Letters*, **9**, 402 (1966).
15. Abeles, B., R. W. Cohen, and G. W. Cullen, *Phys. Rev. Letters*, **17**, 632 (1966).
16. Strongin, M., O. F. Kammerer, D. H. Douglass, Jr., and M. H. Cohen, *Phys. Rev. Letters*, **19**, 121 (1967).
17. Cohen, M. H., and D. H. Douglass, Jr., *Phys. Rev. Letters*, **19**, 118 (1967).
18. Caswell, H. L., *J. Appl. Phys.*, **32**, 2461 (1961).
19. Chanin, G., E. A. Lynton, and B. Serin, *Phys. Rev.*, **114**, 719 (1959).
20. Seraphim, D. P., C. Chiou, and D. J. Quinn, *Acta Met.*, **9**, 861 (1961).
21. Caswell, H. L., *J. Appl. Phys.*, **32**, 105 (1961).
22. Seraphim, D. P., D. T. Novick, and J. I. Budnick, *Acta Met.*, **9**, 446 (1961).
23. DeSorbo, W., *Phys. Rev.*, **132**, 107 (1963).
24. Gerstenberg, D., and P. M. Hall, *J. Electrochem. Soc.*, **111**, 936 (1964).
25. Fowler, P., *J. Appl. Phys.*, **34**, 3538 (1963).
26. Frerichs, R., and C. J. Kircher, *J. Appl. Phys.*, **34**, 3541 (1963).
27. Theuerer, H. C., and J. J. Hauser, *J. Appl. Phys.*, **35**, 554 (1964).
28. Neugebauer, C. A., and R. A. Ekvall, *J. Appl. Phys.*, **35**, 547 (1964).
29. Rairden, J. R., and C. A. Neugebauer, *Proc. IEEE*, **52**, 1234 (1964).
30. Smith, P. H., S. Shapiro, J. L. Miles, and J. Nicol, *Phys. Rev. Letters*, **6**, 686 (1961).
31. Simmons, W. A., and D. H. Douglass, Jr., *Phys. Rev. Letters*, **9**, 153 (1962).
32. Douglass, D. H., Jr., *Phys. Rev. Letters*, **9**, 155 (1962).
33. Hauser, J. J., and H. C. Theuerer, *Phys. Rev. Letters*, **14**, 270 (1965).
34. De Gennes, P. G., *Rev. Mod. Phys.*, **36**, 225 (1964).
35. Lock, J. M., *Proc. Roy. Soc. London*, **A208**, 391 (1951).
36. Toxen, A. M., *Phys. Rev.*, **123**, 442 (1961).
37. Vogel, H. E., and M. M. Garland, *J. Appl. Phys.*, **38**, 5116 (1967).
38. Blumberg, R. H., and D. P. Seraphim, *J. Appl. Phys.*, **33**, 163 (1962).
39. Hall, P. M., *J. Appl. Phys.*, **36**, 2471 (1965).

40. Ittner, W. B. III, *Phys. Rev.*, **119**, 1591 (1960).
41. Schrieffer, J. R., *Phys. Rev.*, **106**, 47 (1957).
42. Blumberg, R. H., *J. Appl. Phys.*, **33**, 1822 (1962).
43. Toxen, A. M., *Phys. Rev.*, **127**, 382 (1962).
44. Toxen, A. M., and M. J. Burns, *Phys. Rev.*, **130**, 1808 (1963).
45. Douglass, D. H., Jr., and R. H. Blumberg, *Phys. Rev.*, **127**, 2038 (1962).
46. Glover, R. E., III, *Thin Films*, ASM, 1964, p. 173.
47. Caswell, H. L., *J. Appl. Phys.*, **36**, 80 (1965).
48. London, F., *Proc. Roy. Soc. London*, **A152**, 24 (1935).
49. Bardeen, J., *Rev. Mod. Phys.*, **34**, 667 (1962).
50. Smallman, C. R., A. E. Slade, and M. L. Cohen, *Proc. IRE*, **48**, 1562 (1960).
51. Schmidlin, F. W., A. J. Learn, E. C. Crittenden, and J. N. Cooper, *Solid-State Electron.*, **1**, 323 (1960).
52. Newhouse, V. L., J. W. Bremer, and H. H. Edwards, *Proc. IRE*, **48**, 1395 (1960).
53. Shoenberg, D., "Superconductivity," 2d ed., Cambridge University Press, New York, 1960.
54. Ginzburg, N. I., and A. I. Shalnikov, *Soviet Phys. JETP*, **10**, 285 (1960).
55. Hagedorn, F. B., *Phys. Rev. Letters*, **12**, 322 (1964).
56. Edwards, H. H., and V. L. Newhouse, *J. Appl. Phys.*, **33**, 868 (1962).
57. Learn, A. J., and R. S. Spriggs, *J. Appl. Phys.*, **35**, 3507 (1964).
58. Hagedorn, F. B., *J. Appl. Phys.*, **36**, 352 (1965).
59. Cheng, K. L., *J. Appl. Phys.*, **35**, 1302 (1964).
60. Buck, D. A., *Proc. IRE*, **44**, 482 (1956).
61. Newhouse, V. L., and J. W. Bremer, *J. Appl. Phys.*, **30**, 1458 (1959).
62. Delano, R. B., Jr., *Solid-State Electron.*, **1**, 381 (1960).
63. Behrndt, M. E., R. H. Blumberg, and G. R. Giedd, *IBM J. Res. Develop.*, **4**, 184 (1960).
64. Adams, C. N., J. W. Bremer, A. K. Oka, and M. L. Ummel, *IEEE Trans. Mag.*, **MAG-2**, 385 (1966).
65. Tyler, A. R., and P. A. Walker, *Phys. Letters*, **9**, 226 (1964).
66. Ames, I., and H. Seki, *J. Appl. Phys.*, **35**, 2066 (1964).
67. Pritchard, J. P., J. T. Pierce, and B. G. Slay, *Proc. IEEE*, **52**, 1207 (1964).
68. Joynson, R. E., C. A. Neugebauer, and J. R. Rairden, *J. Vacuum Sci. Technol.*, **4**, 171 (1967).
69. Bremer, J. W., and V. L. Newhouse, *Phys. Rev. Letters*, **1**, 282 (1958).
70. Baldwin, J. P., *Phys. Letters*, **3**, 223 (1963).
71. Ittner, W. B. III, *Solid-State Electron.*, **1**, 239 (1960).
72. Brennemann, A. E., *Proc. IEEE*, **51**, 442 (1963).
73. Pippard, A. B., *Phil. Mag.*, **41**, 243 (1950).
74. Brennemann, A. E., J. J. McNichol, and D. P. Seraphim, *Proc. IEEE*, **51**, 1009 (1963).
75. Brennemann, A. E., H. Seki, and D. P. Seraphim, *Proc. IEEE*, **52**, 228 (1964).
76. Meyerhoff, A. J., C. R. Cassidy, C. G. Hebeler, and C. C. Huang, "Advances in Cryogenic Engineering," vol. 9, Plenum Press, New York, 1965.
77. Gange, R. A., *Proc. IEEE*, **52**, 1216 (1964).
78. Newhouse, V. L., and H. H. Edwards, *Proc. IEEE*, **52**, 1191 (1964).
79. Johnson, A. K., and P. M. Chirlian, *IEEE Trans. Mag.*, **MAG-2**, 390 (1966).
80. Newhouse, V. L., J. L. Mundy, R. E. Joynson, and W. H. Meiklejohn, *Rev. Sci. Instr.*, **38**, 798 (1967).
81. Mundy, J. L., and V. L. Newhouse, *GE Res. Develop. Ctr. Rept.* 67-C-451, December, 1967.
82. Bremer, J. W., "Superconductive Devices," McGraw-Hill Book Company, New York, 1962.
83. Cohen, M. L., *Proc. IRE*, **49**, 371 (1961).
84. Newhouse, V. L., and H. H. Edwards, *Proc. IERE-IEE Conf. on Applications of Thin Films in Electronic Engineering*, Institute of Electronic and Radio Engineers, London, 1966.
85. Fruin, R. E., A. K. Oka, and J. W. Bremer, *IEEE Trans. Mag.*, **MAG-2**, 381 (1966).
86. Burns, L. L., *Proc. IEEE*, **52**, 1164 (1964).
87. Barnard, D. D., R. H. Blumberg, and H. L. Caswell, *Proc. IEEE*, **52**, 1177 (1964).
88. Martin, D. H., and D. Bloor, *Cryogenics*, **1**, 159 (1961).
89. Spiel, D. E., and R. W. Boom, *U.S. ANDL Rept.*, ASTIA Doc. 612677, 1965.
90. Caswell, H. L., and J. R. Priest, *Trans. 9th Natl. Vacuum Symp.*, 1962, p. 138, The Macmillan Company, New York.
91. Pierce, J. T., and J. P. Pritchard, Jr., *IEEE Trans. Component Pts.*, March, 1965, p. 8.
92. Priest, J., and H. L. Caswell, *Trans. 8th Natl. Vacuum Symp.*, 1962, p. 947, Pergamon Press, New York.

93. Christy, R. W., *J. Appl. Phys.*, **31**, 1680 (1960).
94. Hill, G. W., *Microelectron. Reliability*, **4**, 109 (1965).
95. White, P., *Electrochem. Technol.*, **4**, 468 (1966).
96. Wright, A. N., *Nature*, **215**, 953 (1967).
97. Allam, D. S., and C. T. H. Stoddart, *Chem. Brit.*, September, 1965, p. 410.
98. Caswell, H. L., J. R. Priest, and Y. Budo, *J. Appl. Phys.*, **34**, 3261 (1963).
99. Dushman, S., "Scientific Foundations of Vacuum Technology," 2d ed., John Wiley & Sons, Inc., New York, 1962.
100. Bergh, A. A., *Bell System Tech. J.*, **44**, 261 (1965).
101. Holm, R., *J. Appl. Phys.*, **22**, 569 (1951).
102. Giaever, I., *Phys. Rev. Letters*, **5**, 147 (1960).
103. Nicol, J., S. Shapiro, and P. H. Smith, *Phys. Rev. Letters*, **5**, 461 (1960).
104. Giaever, I., *Phys. Rev. Letters*, **5**, 464 (1960).
105. Bardeen, J., *Phys. Rev. Letters*, **6**, 57 (1961).
106. Cohen, M. H., L. M. Falicov, and J. C. Phillips, *Phys. Rev. Letters*, **8**, 316 (1962).
107. Giaever, I., and K. Megerle, *Phys. Rev.*, **122**, 1101 (1961).
108. Shapiro, S., P. H. Smith, J. Nicol, J. L. Miles, and P. F. Strong, *IBM J. Res. Develop.*, **6**, 34 (1962).
109. Giaever, I., and K. Megerle, *IRE Trans. Electron Devices*, **ED-9**, 459 (1962).
110. Burstein, E., D. N. Langenberg, and B. N. Taylor, *Phys. Rev. Letters*, **6**, 92 (1961).
111. Neel, J. C., A. Ballonoff, and G. J. Santos, Jr., *Appl. Phys. Letters*, **6**, 2 (1965).
112. Pollack, S. R., and S. Basavaiah, *Bull. Am. Phys. Soc.*, **13**(II), 476 (1968).
113. Hauser, J. J., *Phys. Rev. Letters*, **17**, 921 (1966).
114. Reif, F., and M. A. Woolf, *Phys. Rev. Letters*, **9**, 315 (1962).
115. Sherrill, M. D., and H. H. Edwards, *Phys. Rev. Letters*, **6**, 460 (1961).
116. Goldstein, Y., *Rev. Mod. Phys.*, **36**, 213 (1964).
117. Guyon, E., A. Martinet, J. Matricon, and P. Pincus, *Phys. Rev.*, **138A**, 746 (1965).
118. De Gennes, P. G., *Phys. Condensed Matter*, **3**, 79 (1964).
119. Douglass, D. H., Jr., *Phys. Rev. Letters*, **7**, 14 (1961).
120. Townsend, P., and J. Sutton, *Phys. Rev. Letters*, **11**, 154 (1963).
121. Taylor, B. N., and E. Burstein, *Phys. Rev. Letters*, **10**, 14 (1963).
122. Schrieffer, J. R., and J. W. Silkins, *Phys. Rev. Letters*, **10**, 17 (1963).
123. Giaever, I., H. R. Hart, Jr., and K. Megerle, *Phys. Rev.*, **126**, 941 (1962).
124. Rowell, J. M., A. G. Chynoweth, and J. C. Phillips, *Phys. Rev. Letters*, **9**, 59 (1962).
125. Rowell, J. M., P. W. Anderson, and D. E. Thomas, *Phys. Rev. Letters*, **10**, 334 (1963).
126. Rowell, J. M., W. L. McMillan, and P. W. Anderson, *Phys. Rev. Letters*, **14**, 633 (1965).
127. Jaklevic, R. C., and J. Lambe, *Phys. Rev. Letters*, **17**, 1139 (1966).
128. Josephson, B. D., *Phys. Letters*, **1**, 251 (1962).
129. Ambegaokar, V., and A. Baratoff, *Phys. Rev. Letters*, **11**, 104 (1963).
130. Fiske, M. D., *Rev. Mod. Phys.*, **36**, 221 (1964).
131. Anderson, P. W., and J. M. Rowell, *Phys. Rev. Letters*, **10**, 230 (1963).
132. Rowell, J. M., *Phys. Rev. Letters*, **11**, 200 (1963).
133. Josephson, B. D., *Rev. Mod. Phys.*, **36**, 216 (1964).
134. Ferrell, R. A., and R. E. Prange, *Phys. Rev. Letters*, **10**, 479 (1963).
135. Yamashita, T., and Y. Onodera, *J. Appl. Phys.*, **38**, 3523 (1967).
136. Jaklevic, R. C., J. Lambe, A. H. Silver, and J. E. Mercereau, *Phys. Rev.*, **140**, 1628 (1965).
137. Young, L., "Anodic Oxide Films," p. 148, Academic Press Inc., New York, 1961.
138. Smith, T. I., *Phys. Rev. Letters*, **15**, 460 (1965).
139. Matisoo, J., *Appl. Phys. Letters*, **9**, 167 (1966).
140. Matisoo, J., *Proc. IEEE*, **55**, 172 (1967).
141. Clark, T. D., and J. P. Baldwin, *Electron. Letters*, **3**, 178 (1967).
142. Shapiro, S., *Phys. Rev. Letters*, **11**, 80 (1963).
143. Eck, R. E., D. J. Scalapino, and B. N. Taylor, *Phys. Rev. Letters*, **13**, 15 (1964).
144. Coon, D. D., and M. D. Fiske, *Phys. Rev.*, **138A**, 744 (1965).
145. Langenberg, D. N., D. J. Scalapino, B. N. Taylor, and R. E. Eck, *Phys. Rev. Letters*, **15**, 294 (1965).
146. Giaever, I., *Phys. Rev. Letters*, **14**, 904 (1965).
147. Dayem, A. H., and R. J. Martin, *Phys. Rev. Letters*, **8**, 246 (1962).
148. Yanson, I. K., V. M. Svistunov, and I. M. Dmitrenko, *Zh. Eksperim. i Teor. Fiz.*, **48**, 976 (1965).

Chapter **23**

Thin Films in
Integrated Circuits

ILAN BLECH

HARRY SELLO

Fairchild Semiconductor Research and Development, Palo Alto, California

and

LAWRENCE V. GREGOR

IBM Components Division, East Fishkill, New York

1. INTRODUCTION

An integrated circuit can very simply be described as a tiny block of semiconductor, usually silicon, which contains precisely doped regions and on the surface of which have been placed precisely formed, geometrically patterned, thin films of metals and insulators.

To help in visualizing the thin film arrangement, a schematic cutaway cross section of an integrated circuit is shown in Fig. 1. The integrated circuit consists of a silicon

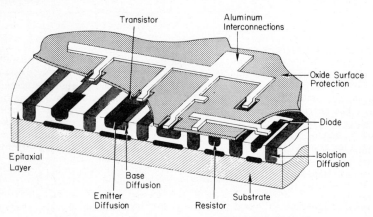

Fig. 1 Schematic cross section of an integrated circuit.

single-crystal substrate upon which has been grown an epitaxial layer of additional silicon. In the epitaxial layer, doped regions have been formed by diffusion. These doped regions are the PN-junction regions and constitute the "active" electrical part of the circuit. The individual active regions are diodes and transistors which form the integrated circuit when they are interconnected. The "active" regions are electrically separated from each other by isolation-diffusion regions. As a by-product of the diffusion processes, a thin ($\sim$1.0 μ) film of SiO_2 is grown on the surface of the silicon. This film of SiO_2 performs the vital basic function of protecting the P-N junctions from the outside world. The SiO_2 also serves as an insulating layer upon which is placed a thin (0.5 to 1 μ) film of metal, often aluminum. The Al, when properly geometrically patterned, makes electrical contact to the underlying Si through openings, or "windows," in the oxide and interconnects the active areas in the Si. Metal films may also be used from time to time as resistors or capacitors. To gain perspective of size, the dimensions of the metal interconnects are typically 0.5 mil wide separated by 0.5-mil spaces, while at the contact area they can be smaller. In advanced devices the contact dimensions have been pushed down to 0.1 mil or even less. The overall Si block (or chip) size can vary but is usually in the range of 40 $\times$ 40 to about 100 $\times$ 100 mils.

The thin films with which we will be concerned are those which appear on the surface of the silicon, i.e., metals and insulators. We will not be concerned directly with the Si itself. Not shown in Fig. 1, but to be discussed later, are the terminations of the metal interconnects at the periphery of the chip. These terminations are the contact pads to which are made the connections to the outside world.

2. THIN FILM CONDUCTORS

a. Metal-to-Semiconductor (M-S) Contacts. Theory of M-S Contacts

It is desired in semiconductor devices to obtain M-S contacts that will exhibit low-resistance, linear (nonrectifying) I-V characteristics over a wide range of current. Such contacts are commonly termed ohmic contacts. Despite the great importance of M-S contacts to the semiconductor industry, only scant quantitative data about such contacts exist today. In this section, we briefly outline the simple M-S theory, followed by a discussion of the practical contact practices. As will be seen, the simple theory does not explain the behavior of actual ohmic contacts but is used for purposes of illustration.

The basic characteristics of M-S contacts can be better understood by using a simple band-diagram approach.[1] In contacting a metal with Si, the Fermi levels of the metal and Si must coincide. If the work function of the metal differs from that of the Si, significant accumulation or depletion of carriers can occur in the semiconductor because of carrier diffusion at the boundary. An electric field will build up in such a space-charge region and will serve to balance the diffusion current. The Si energy bands are then forced to bend near the contact and will accommodate the difference in the work functions between the metal and the semiconductor.

For example, for N-type Si, if the Si work function is larger than the metal work function, electrons will be permitted to diffuse into the Si, leading to an accumulation layer. The charge buildup establishes an electric field pointing from the metal to the semiconductor, causing electrons to drift toward the metal and thus balance the diffusion current. The electric field results in a downward bending of the Si energy bands (Fig. 2) near the M-S contact. An accumulation layer at the boundary has little effect on the contact resistance since the junction will be essentially nonrectifying (for majority carriers). Thus, the carriers will flow easily under both forward and reverse bias. However, if the Si work function is lower than that of the metal, an upward bending of the energy bands is expected and a depletion region will form at the metal-Si junction. If the depletion region is wide enough, the M-S contact will exhibit a nonlinear I-V characteristic.[1]

For P-type Si, these considerations are reversed, and a nonlinear I-V characteristic is obtained when the metal work function is smaller than the Si work function.

The presence of a depletion layer does not necessarily lead to a rectifying contact. If the depletion layer is very thin, tunneling can take place[2] and a linear I-V characteristic may be expected. Tunneling may play an important role in the case of Al ohmic contacts to silicon. An increase in the carrier concentration of Si reduces the

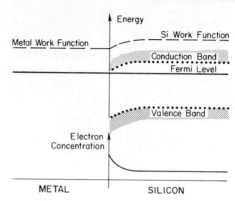

Fig. 2 A simple energy-band diagram and electron concentration near a metal-silicon contact at equilibrium, illustrating an accumulation layer in N-type silicon.

depletion-layer thickness, resulting in nonrectifying contacts.[3] In practice, highly doped Si is generally used to form an ohmic contact.

Despite the great importance of ohmic contacts to the semiconductor industry, relatively few quantitative studies have been made.[4,5,6] A number of inherent difficulties prevent carrying out such studies: (1) Little is known about the work functions of thin metal films. Work functions depend on the condition of the metal surface and the crystal structure. The metal surface is often very difficult to define and prohibits any serious quantitative calculations of M-S contacts from being made. (2) The presence of a native SiO_2 skin, or contamination on the Si contact surface, further complicates the situation. The interface under these conditions is no longer a simple metal-Si interface. In order to penetrate such layers successfully and contact the Si, it is necessary to heat-treat (or "alloy") the metal at high temperatures. Under special conditions such heat treatments can result in a nonohmic barrier because of the formation of metal-semiconductor compounds or phases. (3) Relatively little is known about the effect of surface states. These can cause further energy-band bending, independent of the metal or semiconductor work function, and modify the depletion layers.

In summary, the simple band model predicts the general behavior of M-S contacts. However, this model does not serve to describe the contact characteristics quantitatively.

(1) Specific Contact Materials Al is the most commonly used contact metal in integrated circuits today. Al thin films are vacuum-evaporated onto Si wafers containing the integrated circuits. The desired interconnection configuration is then etched into the metal by photolithographic techniques. To ensure a low-resistance ohmic M-S contact, the circuits are heat-treated, typically 10 min at 550°C. During the heat treatment, the Al reduces some of the native SiO_2, Si-Al interdiffusion occurs, and physical contact to the semiconductor surface is made. At the alloying temperatures, Si diffuses more rapidly into Al than Al into Si.[7] The diffusion of Si in Al can proceed until the Al is saturated with Si (about 1.5 atomic % at 550°C). At this temperature and for 10 min, the diffusion length of Si in Al bulk is about 10^{-3} cm. The dissolution of Si in Al is reported to be anisotropic. The (111) Si atom planes dissolve slowly[8] so that for a wafer cut in the (111) orientation, shallow "dissolution pits" result. Rapid lateral dissolution of Si can proceed under the SiO_2 adjacent to the contact. Figure 3a is a photomicrograph of an Al-Si contact area. Faceted dissolution pits are clearly seen around the contact areas, extending laterally under the oxide. On cooling the structure after alloying, the Al becomes supersaturated with Si, and Si precipitation takes place. The precipitated Si is seen on the SiO_2 (Fig. 3b) around the contact cuts once the aluminum is etched away. Figure 4 shows

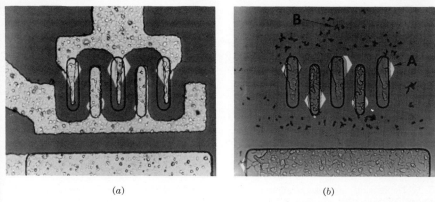

(a)   (b)

Fig. 3 Photomicrograph of dissolution pits in silicon wafer after 10-min heat treatment at 550°C. P-type silicon 5×10^{17} atoms/cm² boron concentration. Wafer surface is the (111) plane (1,000×).

an electron photomicrograph (replicate) of a single precipitate. Transmission electron micrographs identified the precipitate as Si, and showed the Si to be primarily located on grain boundaries. The lateral dissolution of Si under the SiO_2 can, in extreme cases, cause shorts in narrow-spaced emitters. Very shallow emitters are also susceptible to Si dissolution. In this case, complete penetration normal to the emitter surface can occur. If the deposited film is Al saturated with Si, the detrimental Si dissolution will be minimized when the contact is heat-treated.

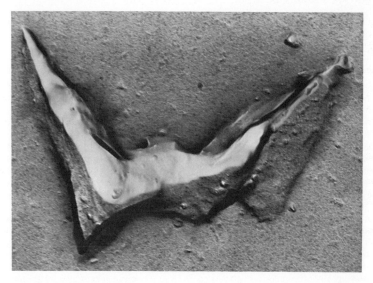

Fig. 4 Electron photomicrograph of a single Si precipitate remaining on the SiO_2 surface after etching of the Al (37,500×).

Another contacting scheme that has received wide attention in recent years was developed by Lepselter[9] at the Bell Telephone Laboratories. According to this method, a thin layer of Pt, 0.05 μ, is sputtered on the Si wafer and sintered at 600 to 700°C. Gray-colored platinum silicides are formed in the contact areas upon sintering. The Pt is etched from the field, leaving the silicides behind. The interconnection metallization Ti-Pt-Au is then applied. No significant lateral Si diffusion is encountered in this system. Examples of other contacting metals that have been used are Ti deposition followed by Al in place of Al alone[10] and Mo followed by a deposition of Au.[11]

b. Interconnections

The interconnection metallization on an integrated circuit provides the means by which the undivided functions (active areas) are connected to form the complete circuit function. Interconnections also provide the contact to the outside world.

In this section, we will briefly describe the use of thin film interconnections in integrated circuits in terms of their requirements.

(1) Requirements The requirements for a thin film metal interconnection can be summarized as follows:

1. Deposition processes should be compatible with modern (high-resolution) photolithographic techniques. A practical thickness for deposition of interconnects is about 1 μ, and practical etched widths may be as small as 10 μ.

2. In many integrated circuits a sheet resistance of 0.04 ohms/sq or lower is required. This means that the metal should have a resistivity of 4×10^{-6} ohm-cm or less for a 1-μ-thick film.

3. The metal should form good ohmic contacts with highly doped N-type and P-type Si.

4. The film should handle a reasonable current density. Most present-day circuits require 1 to 3×10^5 A/cm^{-2}.

5. The metal should not oxidize or corrode.

6. The film should adhere well to the Si and SiO$_2$ and be metallurgically compatible with external bonds.

7. Scratch resistance is highly desirable.

8. The metal should withstand high-temperature processing, e.g., packaging or operation.

9. Deposition and fabrication should be economical.

c. Specific Materials

Unfortunately, no single metal will fulfill all the above requirements. Therefore, trade-offs are necessary. The following discussion is limited to the three common interconnection schemes used today in integrated circuitry.

(1) Aluminum Aluminum is by far the commonest metal used today for interconnections on integrated circuits. It fulfills most of the requirements rather well: it is easy to deposit and fabricate; the resistivity, about 3×10^{-6} ohm-cm, is adequate; it has good adhesion to Si and SiO$_2$, external leads are easily bonded; it forms good ohmic contacts to Si; it can handle current densities of more than 10^5 A cm^{-2} at 150°C. Furthermore, Al is relatively resistant to further oxidation and is stable up to the Si-Al eutectic temperature (577°C) for short periods of time. Al has some disadvantages, however: corrodibility and low scratch resistance.

Al is evaporated on the wafer surface and then photolithographically etched into the desired pattern. The Al metallization is heat-treated or "alloyed" to form ohmic contacts to the Si. Al or Au wires are attached to the Al film for external leads.

(2) Mo-Au The molybdenum-gold system has recently been proposed as an interconnection scheme.[11] According to this scheme, a thin (0.1 to 0.15 μ) layer of sputtered Mo is used as the ohmic contact to Si and at the same time serves as an adherent layer to SiO$_2$. A second, thicker (~ 1 μ) Au layer, which is the main conductor, is deposited. This metal composite is relatively stable up to the Au-Si eutectic (370°C) temperature. Prolonged heating above this temperature can be catastrophic; Au can penetrate through the Mo into the Si. Au wires serve as external leads to the Mo-Au composite.

Variations of the Mo-Au system have been described. For example, Pt or Al is frequently deposited on the wafers prior to Mo to ensure good ohmic contacts to Si. Some advantages claimed[12] are higher scratch resistance (than Al) and reduced effects due to electrotransport (see below). However, the Mo-Au system is more costly and does not offer corrosion resistance.

(3) Ti-Pt-Au This system has been developed at the Bell Telephone Laboratories.[9] According to their procedure, a thin layer of Pt $(0.07\ \mu)$ is sputtered onto Si and then sintered to form contacts. This is followed by a layer of Ti $(0.05\ \mu)$, Pt $(0.15\ \mu)$ and finally by Au. Ti and Pt are deposited by diode sputtering and Au, either by evaporation or by electroplating. In this system Ti serves as an adherent layer to the dielectric, while the Pt is an inert barrier under the Au. The Au, again, serves as the conductor. The main advantage of this system is its high resistance to corrosion and storage stability at moderately high temperatures (e.g., 350°C).

3. MULTILAYER STRUCTURES

a. Introduction

The evolution of multilayer metallization on a silicon integrated-circuit chip is a logical result of the capability achieved for the simpler "single" level circuits. This evolution is the trade-off between increasing functional complexity, which demands ever-larger area chips, and maintaining high yield, which in turn pushes toward reducing chip area. For a given size (area) of silicon chip, the complexity of metallized patterns can be increased by employing a sandwich structure of more than one layer of metal. The first metal layer of a typical simple integrated circuit is covered by a dielectric insulating layer, and a second metal layer is then deposited on top of this dielectric material. Such a structure is illustrated in Fig. 5, which is a photomicrograph of a metal-oxide-semiconductor (MOS) complex array chip which has two levels of metallization.[13,14] This is an array of about 100-logic-gate (or 300- to 500-com-

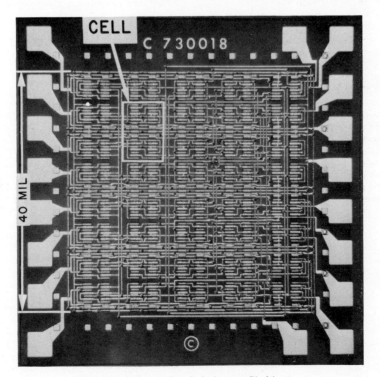

Fig. 5 MOS complex digital-array Si chip.[13]

ponent) complexity on a 60×60 mil chip with an "active" area of about 1,600 mil². Each of the "cells," shown in detail in Fig. 6, is a matrix of MOS resistors and MOS transistors (see Chap. 20) connected with aluminum metallization to form a fundamental logic building block. By making use of the insulating dielectric layer and a second Al metal layer, the cells are interconnected to form a subsystem logic function. The communication between metal levels is achieved by the small circular holes known as vias, in the dielectric (Fig. 6). A similar multilayer approach has been described for bipolar complex arrays[15] and is illustrated (Fig. 7) by an 80×110 mil chip with an "active" area of 6,000 mil² and which has a complexity of 32 logic gates.

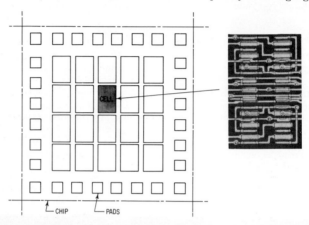

Fig. 6 Individual logic cell of array in Fig. 5.[13]

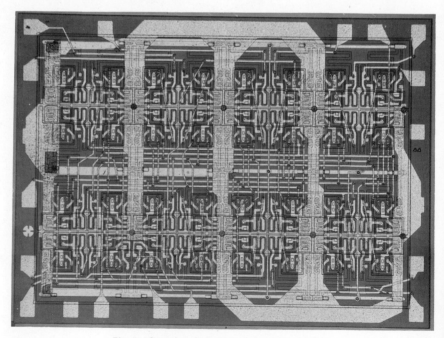

Fig. 7 Complex digital bipolar array $(40\times)$.[16]

The particular geometrical layout of components and circuits on a silicon chip to perform the desired logic functions can be achieved by a variety of multilayer structures. This varies according to the specific design needs. Such design considerations include complexity of function, the choice of bipolar vs. MOS logic, the type of metal employed, or the number of metal layers. These design choices have been amply described elsewhere[16,17,18] and are not the subject of the present discussion. We shall confine our attention to the technology involved in achieving multilayer structures. We shall consider this technology from the aspects of materials, communication between levels, crossovers, surface effects on underlying semiconductor, and external contacts.

b. Materials

The materials employed as dielectric insulation for multilayer integrated circuits are characterized not only by their composition but also by the process used for deposition. This has led to a wide variety of choices.

A selection of the more prominent metal-dielectric combinations reported is shown in Table 1. The complexity of the multilayer structure varies from the simplest,

TABLE 1 Selected Materials Used for Multilayer Integrated Circuits (Film Thickness in Microns)

Metal	Dielectric		No. of layers		Reference
	Material	Process of deposition	Metal	Dielectrics	
Al (0.5–1.5)	SiO$_2$ (0.5–1.7)	RF-sputtered	3	2	17
			2	1	19
			2	1	21
	SiO$_2$ (0.5–1.0)	Vapor-deposited by oxidation of silane	2	1	19
			2	1	24
	Lead borosilicate glass (1.5)	Fired frit	2	2	20
	SiO (0.5)	Evaporated	2	1	23
	SiO (0.5)	Reactively sputtered in oxygen	2	1	24
	Alumina borosilicate (0.6)	Vapor-deposited by oxidation of doped silane	4	3	16 (p. 167)
			2	1	26
Mo/Au/Mo (0.1/0.7/0.1)	SiO$_2$ (2.0–3.0)	RF-sputtered	3	2	26
			2	1	16 (p. 154)

which consists of two layers of metal plus one of dielectric, to four metal layers separated by three layers of dielectric. Two dielectric materials are favored, rf-sputtered silicon dioxide and vapor-deposited SiO$_2$ from the oxidation of silane. RF-sputtered SiO$_2$ is used up to 3.0 μ thick while none of the others exceeds 1.5 μ. The lowest thickness is 0.5 μ. Thickness control is important in order to ensure adequate isolation and to avoid pinholes.

For vapor-deposited SiO$_2$ films, essentially two serious problems arise in considering film thickness,[19] namely, nonuniformity due to gas-flow control in the reactor ($+20\%$ across a slice) and stress cracking for films greater than 2 μ thick. On the other hand, less stress is seen for the rf-sputtered films while thickness control across a slice is $+5\%$.

Glass films of low enough firing temperature applied by a fired-frit process are limited to a maximum of 1.5 μ in thickness. Above this, cracking occurs because of the thermal-expansion mismatch.[20] At this thickness, control is claimed to be $\pm 2\%$ on a wafer and $\pm 4\%$ wafer to wafer.

Other dielectric materials have been tested for multilayer insulation.[21] These include oxides of metals (Al, Mo, Ni), silicon dioxide by the decomposition of tetraethyl orthosilicate, and organic insulation. All were found to be undesirable. An interesting possibility is silicon nitride because it is such an excellent diffusion barrier for contaminants likely to affect devices. However, tests on simple devices show that more work is needed as silicon nitride is inferior to silicon dioxide as a dielectric.[22]

While a variety of dielectrics have been reported, the predominant metal used is Al. This wide use is, of course, an extension of conventional integrated-circuit technology. The Al depositions are not critical and can be filament-, crucible-, or electron-beam-evaporated. However, Al metallization requires that dielectric deposition processes be carried out at maximum temperatures consistent with Al-Si metallurgy, i.e., 577°C eutectic temperature. Thus, the dielectrics shown in Table 1, used with Al, involve "low" temperature depositions, e.g., room temperature for sputtering (see Chap. 4) and about 360°C for the oxidized silane films.[19] The thicknesses required for Al do not seem to be critical (above a minimum). It is necessary that the step at the oxide or dielectric "via" hole be completely covered for purposes of electrical continuity. As current levels increase in integrated circuits to the region of 10^6 A cm^{-2} limitations due to Al electrotransport must be considered. This is discussed in Sec. 5.

Only one metallization scheme other than Al has been seriously considered, i.e., rf-sputtered Mo-Au. Lathrop et al.[21] report that the primary reason for investigating this system is the apparent tendency for Al to react with rf-sputtered SiO_2 and form insoluble compounds, not removed by etching. Mo serves in this scheme as an ohmic contact metal and as a seal to the SiO_2 layers. The buried layers are composed of Mo-Au-Mo (0.1 μ-0.7 μ-0.1 μ), while the top layer is just Mo-Au.[21]

c. Communications between Levels

The communication between levels of metallization is provided by the contact made between first, second, and subsequent metal layers by means of openings in the dielectric layers. Poor contacts, with the resulting series resistance, are undesirable for effective communication between levels. Several factors play an important role, among them the size of the via hole and residues on the metal surfaces.

(1) Size of the Via Hole The higher the complexity desired, the smaller will be the dimension of the via hole. In turn, dielectric etching problems are increased with smaller dimensions. It becomes difficult to maintain control of the cutout size as well as of etching depth. The danger is the tendency to overetch the bottom metal layer in an effort to remove all the dielectric. Good interlevel Al connections were reported for about 0.5 $\times$ 0.5 mil vias in 0.3-μ-thick rf-sputtered SiO_2.[23] Lathrop et al.[21] employed vias of similar size for the Mo-Au system. On the other hand, this via size was judged too small to achieve low resistances for Al/vapor-deposited SiO_2 structures.[24]

(2) Residues on the Metal Surfaces Residues deposited upon the metal as a result of etching and/or cleaning steps can create serious interlevel-resistance problems. Such process difficulties led Condon et al.[23] to abandon the use of SiO (evaporated or sputtered) as a dielectric in favor of SiO_2. Even the use of a glow discharge to clean the first-layer metal prior to depositing the second metal was inadequate to overcome the difficulties.

As mentioned earlier, rf-sputtered SiO_2 can apparently react with Al to form insoluble residues which require damaging overetching of the oxide in order to ensure removal.

In the Mo-Au metal system,[21] other problems leading to poor communications between levels can occur as a consequence of the additional steps required to process the more complex metal films peculiar to this system.

.5 $\times 10^{-3}$ inch $\times$ 25.4 $\frac{mm}{in}$ = 12.7 $\times 10^{-6}$ m

d. Crossovers

A crossover in a multilayer circuit is shown schematically in Fig. 8. The number of crossovers in a complex array can vary widely, depending upon the width of the metal runs (area of crossover) and the complexity (number of logic gates) in the circuit. As a rough approximation, present designs seem to fall in the range of 2 to 20 crossovers per logic gate. Thus, a 100-gate complex array may have from 200 to 2,000 crossovers. From Fig. 8, it can be seen that a pinhole, crack, or similar defect in the area of the crossover can cause a short between the two metal layers and result in a loss of circuit function.

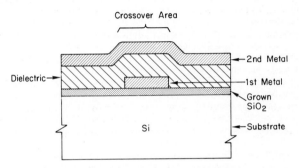

Fig. 8 Schematic diagram of a crossover.

Defects which destroy crossovers can occur for many reasons. Most frequently, these include the presence of contamination or residue of the metal surface or, during dielectric deposition, cracking of the dielectric, pinholes in the dielectric, overetching of the dielectric, or excessive penetration by the top-layer metal during heat treating or "alloying." Obviously, the thinner the dielectric layer the more prone the crossovers will be to exhibit defects such as pinholes. It is largely for this reason that a minimum thickness of dielectric layer is required. A common method for evaluating the merit of a given process for the preparation of multilayer structures is to construct test patterns containing varying numbers of crossovers and test the structures for electrical shorts. Lathrop et al.[21] have used test patterns containing up to 1,000 crossovers of 1 mil^2 area. They report figures of merit in terms of the number of defects per 100,000 such crossovers. On this basis, the following table compares the defect levels reported by two investigators.

Dielectric	Thickness, μ	Defects per 100,000 crossovers	Reference
RF-sputtered SiO$_2$	0.3	1.25	19
	1.0	6	21
	2.0	1	21
Vapor-deposited SiO$_2$	0.1	500	19
	1.0	0	19

The results depend at least in part on the different techniques used for the preparation and testing of the test structures. The lack of correlation of thickness with defect level may be due to differences in cleanliness of processing.

e. External Contacts

The high functional complexity of multilayer integrated circuits requires that there be a large number of metal-contact bonding pads on the silicon chip. For example, the chip shown in Fig. 5 has 18 bonding pads. The number can go as high as 50 or more.[16] Such bonding pads are part of the top-layer metallization and are the areas to which the connections to the outside world are bonded. Two general schemes of connections have been described, wire bonding and face-down (or bump) bonding. The two schemes differ basically in that the wire bonds are made one at a time while the face-down bonding involves inverting the chip into a face-down (or flip-chip) configuration and bonding all contacts immediately. The beam-lead technique[9] also employs face-down bonding. Au beams are formed on the chip by electroplating, and the chip is bonded directly to a substrate by thermocompression.

(1) Wire Bonding This is the "classical" method of bonding integrated circuits and has been pushed to a very high degree of complexity in systems where as many as 128 such bonds are used.[21] Disadvantages of wire bonding, at this complexity, are the high cost and low reliability of the large number of such bonds.

The metals used are generally Al wires to Al pads (ultrasonic wedge bonding) or Au wires to Au pads (thermocompression and ultrasonic bonding).

(2) Face-down (Bump) Bonding This is an approach in which the bonding pads on an integrated-circuit chip are built up with additional metal to form raised areas in the form of either bumps or beams. Bonding to suitable substrates is accomplished by ultrasonic means, thermocompression, or combination of both. The metal systems used vary according to the chip metallization and the bonding process. Examples of the types of chip contacts are Al bumps,[25] Au beams,[9,25] and solder bumps.[20,25]

Face-down bonding is particularly advantageous to multilayer structures because of the high complexity of the chip. The process is economical, is reliable from a thermal and mechanical stress point of view, and makes possible the assembly of many chips to a common substrate in order to achieve a very high degree of complexity of functions. Finally, the advantage of "repairability" becomes practical. A given silicon chip can easily be removed if defective, and another one can be bonded in its place. The ease of repairability is influenced greatly by the metal-contact interfaces of the chip top-layer metal, the bump metal, and the metallization of the substrate to which the chip is bonded.

4. FILMS AS ENCAPSULANTS

In the preceding section, we discussed the use of dielectric films as insulating layers in a multilayer complex integrated circuit. Many of the same properties which are desirable in such an application can be extended to the use of dielectric films as encapsulants. Some of these uses are discussed below. The properties and preparation of thin glass films will not be discussed here, as they have been described in an excellent review by Pliskin, Kerr, and Perri.[26]

a. Semiconductor Surface Passivation

Unless it is suitably protected, the surface of a semiconductor is usually quite unstable in its electrical behavior. It is therefore standard practice to coat the active surfaces of any semiconductor device with a suitable passivating layer.

The electrical requirements which must be met by an ideal layer of this type have been defined by Thomas and Young.[27]

1. The semiconductor surface potential must not change significantly with time under the stress conditions that are encountered by the device.

2. The semiconductor surface potential should be optimum for the particular device under consideration.

3. The surface-state density and the surface recombination velocity must remain at a level that is low enough for the proper operation of the particular device.

Amorphous SiO_2 grown through high-temperature oxidation is the most widely used passivation layer for silicon transistors and integrated circuits. Although this insulator has been found to be more suitable than other materials, it by no means fulfills all the above requirements. This is partly because, as passivation techniques improve, devices are operated over a wider and more severe range of stress conditions.

In the bipolar transistor the device is strongly stressed by electric fields and local temperatures in the junction region, whereas in the insulated-gate field-effect transistor (IGFET) the device is strongly stressed both in the junction region as well as on surfaces near the junction (see Chap. 20). The latter is through normal electric fields which approach 10^6 V cm^{-1}. In the bipolar transistor fringing fields may reach well outside the junction even to the outside surface of the passivating layer, while in the IGFET this potential is generated by means of a control electrode located on the outside of the passivating layer. Thus, IGFETs comprise a number of phases which are in intimate contact one with another and which influence each other via a set of electrical and chemical potentials which are not necessarily constant with time.

The complexity of the interactions between the components and phases in the system gives rise to a large variety of experimental phenomena which have yet to be adequately described by a suitable theory. The phases, even when taken individually, are extremely complex in their behavior, each one being fundamentally sensitive in its electrical properties to contamination by an extremely low density of electrically active defects. Effects that occur in the semiconductor are probably the easiest to analyze because of the wide variety of experimental techniques and experience available. Similarly, the behavior of metals and the relationship of this behavior to a known variety of defects are also fairly well understood (see Chap. 13).

Theories and relationships for the behavior of insulators are, however, not well established in spite of the large volume of experimental and theoretical work that has been done. In particular, quantitative relationships for amorphous insulators such as SiO_2 are often orders of magnitude in disagreement with experiment, and even the experimental data are often not reproducible. In the case of interfacial properties, it was well appreciated by Tamm[28] that in a real crystal the external surfaces represent regions of major discontinuity in some of the properties of the solid. Using a one-dimensional periodic model, he showed that a solution of the Schrödinger equation using Bloch functions gave a new set of stationary solutions when the one-dimensional model was assumed to have finite size. These represented states in which the electrons are localized at the boundary. These states were called "surface states" and were subsequently shown by other authors to be present in more refined three-dimensional models; they arise solely from the termination of the periodic array and hence would be present even for the case of an ideal free surface.

On real surfaces, the density and distribution of surface states are profoundly altered by adsorbed molecules and ions, structural defects, impurities in the host crystal, etc. Of particular interest is the nature of a surface which is in intimate contact with another solid phase, such as an oxide layer formed by chemical reaction of the crystalline surface with an oxidizing environment. The formation of the oxide layer would be expected to use up most of the states that were present on the clean surface through bond formation. Since the first observation of surface states on germanium by Bardeen,[29] it has always been observed that the number of surface states actually detected in any measurement is two or more orders of magnitude smaller than that predicted by Tamm. The latter is, in principle, of the order of the number of surface atoms.

The presence of an oxide layer on the surface of a crystal can give rise to a non-uniform potential distribution across the interface. This is called the "surface potential" ϕ_s. In a metal, such a static surface potential causes a perturbation of the electron density for a distance of 10^{-8} cm into the metal. For a semiconductor, the effects of surface potential are observed to much greater depths. If the carrier density is low, the fractional change in electronic properties caused by the surface effect may be very large.

A number of models have been proposed to account for the origin of the surface potential as well as the number and energy distribution of the surface states. These vary considerably in response to different fabrication treatments as well as during the application of various stresses. It has been established that the oxidation of silicon results in an effective potential gradient at the interface which causes an accumulation of electrons at the silicon surface. The excess carriers near the surface cause the surface conductivity to vary considerably from bulk behavior, particularly where the bulk of the silicon conductivity is due to holes; in this case, the space-charge region is termed an inversion layer since the majority carrier in the surface region is the minority carrier in the bulk. The phenomenon is of great technological importance because thermal growth of oxide layers is universally used in the fabrication of silicon devices for electronic use.

In addition to serving as a passivating layer, the thermally grown silicon dioxide is used for the masking of appropriate areas during the formation of planar junctions by diffusion as well as for straightforward electrical insulation.

b. Unencapsulated Chips

A suitable dielectric film placed directly upon the integrated-circuit chip can hermetically seal the active areas "in situ." This will permit the chip to be used in an unpackaged form or to be packaged within a low-cost nonhermetic package. Beam-lead integrated circuits are an excellent example.[30] The combination of a noncorrodible metal system plus silicon nitride passivation on the chip results in a device which is very insensitive to the ambient. Bilous et al.[17] describe an analogous approach in which the chip is covered with rf-sputtered SiO_2 and "refractory-metal" pads are placed on top of the dielectric.

c. Protection of Metallization

The relatively soft metallization patterns of an integrated circuit are easily abraded and scratched during assembly and testing. A dielectric-film "cover" will greatly reduce the incidence of damage and will enhance the chip reliability. Fused frit glass[20,31] has been used for this purpose.

In this process a glass powder is suspended in a nonaqueous solvent and the powder is deposited, by centrifugation, onto an integrated-circuit wafer. The wafer is then fired at a temperature near the softening point of the glass, which fuses the glass into a continuous film. By varying the composition of the glass powder, a wide range of glass films can be so fired, varying from temperatures of 500 to 1200°C.

Besides being amenable to mass production, it is possible by this technique to coat wafers with metallized interconnect patterns already on them, and also to control the degree of penetration into the silicon oxide.

d. Improvement of Solder Contacts

An interesting specific application of dielectric-film encapsulants has been as an aid in achieving bonding by the solder-bump technique. Solder balls are fused into holes in a protective glass film,[20] which also permits automated assembly of this chip to its substrate (face-down bonding). In an alternative approach, Wagner and Doelp[32] describe the formation of solder pads by dipping a metallized wafer in a molten-solder bath. The wafer was previously covered with vapor-deposited SiO_2 and the contact pads exposed by etching. While the authors demonstrated feasibility, the yields were too low to be economically practical.

5. RELIABILITY CONSIDERATIONS

a. Introduction

In the earlier sections, we have discussed the use of thin films in planar integrated-circuit devices. It was seen that the thin metallic films which make the contacts to the various regions of the semiconductor serve as electrical interconnections for the integrated circuit. External leads are bonded to these films.

An operating device, dissipating power, can reach a temperature as high as 125°C. The metal interconnections operating at such temperatures often carry electric-current densities greater than 10^5 A cm^{-2}. Temperature gradients of from 10^2 to 10^3°C cm^{-1} are possible along the metal pathway.

During long-term device operation, the metal films are required to maintain their desired properties; i.e., the films should retain:

1. Low contact resistance between the metallic film and the semiconductor
2. Low electrical resistance along the film
3. Low contact resistance between the film and external connections (such as wire leads or "bumps")
4. Good mechanical properties such as high film adhesion to the substrate or good bond strength with the external connection

It has been found that all the above properties are affected during device operation by a variety of failure mechanisms. Devices are designed so that under normal operating conditions for a properly manufactured device, failure mechanisms will proceed at negligible rates. However, defects resulting from poor control in manufacturing can greatly accelerate the degradation of the metal films.

A comprehensive review on metallization and bond-failure mechanisms has already been given.[33] Here we will only briefly discuss a few of the commoner failure mechanisms.

Thin film Al interconnections are used on the great majority of the integrated circuits manufactured today. It is not surprising, therefore, that most of our knowledge of wear-out or failure mechanisms pertains to Al. Much less information is available on the other metal interconnection systems, such as Mo-Au and Ti-Pt-Au. Most of the failures which have been reported in these systems can be attributed to poor control in manufacturing. Some are "inherent" failures, e.g., Au diffusion into emitters.

b. Contact Resistance

When pure Al films are connected to the semiconductor, there is a potential danger that Si will dissolve in the Al upon temperature aging (Fig. 3). This solution may cause an increase in the electrical resistance when Al contacts to highly doped N-type Si are annealed at 500 to 600°C. In extreme cases, where contacts are made to very shallow emitters (1,000 Å), the solution can actually cause the metal to penetrate through the semiconductor junction and effectively destroy it. This metal-semiconductor contact degradation is not an important reliability consideration, particularly since measures can be taken to prevent the solution of Si.

c. Film Resistance

Many of the reliability problems in integrated circuits are related to the loss of high conductance along the thin film path which occurs after extended device operation or high-temperature storage.

(1) Metal-SiO$_2$ Reaction The reduction of SiO$_2$ films by overlying Al films is thermodynamically favored. Fortunately, the reaction rate is very small. The sheet resistance of Al on SiO$_2$ was reported to degrade only when very thin films were aged at high temperatures for very long periods of time. Aging a 500-Å-thick film at 500°C for several hours increased its sheet resistance by an order of magnitude.[34] It was not clear how much of this large change was caused by Al-SiO$_2$ interaction and how much was due to oxidation of the film. However, Al films (1,000 Å thick) are very stable when aged in various ambients, at 350°C, even for 500 h.[34,35]

From this stability, it is clear that the loss of the conductivity of Al films on SiO$_2$ is of little practical importance in integrated circuits. Similarly, no metal-SiO$_2$ degradation was reported when other metal systems such as Mo-Au or Ti-Pt-Au were used.

(2) Interdiffusion of Metals If two or more metal films are used for interconnections, interdiffusion can occur on aging. This leads to an increase in electrical resistivity. Extensive studies of such resistivity changes, due to solute scattering and compound formation, have shown the limitations of various metal combinations.[34,35]

Mo-Au combinations are stable at elevated temperatures because of the lack of compound formation and the low solubility of Mo in Au. Ti-Pt-Au sandwiches are relatively stable because of the very slow interdiffusion of Pt and Au. Such sandwiches were aged in excess of 1,000 h at 300°C in dry N$_2$ without appreciable increase in their resistivity.

(3) Oxidation and Hydration Oxidation and corrosion of interconnections have been occasionally found in integrated circuits.[33,34,36,37] In general, these types of failures are not found in hermetically sealed packages. Deterioration of Al thin films due to oxidation in air is slow: 2,000-Å Al on SiO_2 aged at 350°C in air for 100 h showed only 11% increase in sheet resistance.[34]

(4) Electrotransport Increased resistance or discontinuities in metal interconnections can also be brought about by electrotransport (electromigration). The high current densities used today in advanced integrated circuits (more than 10^5 A cm^{-2}) induce a significant metal mass transport.[38,39] This electrotransport can eventually cause an interconnection to open.

Mass transport due to an electric current (dc) in bulk specimens was observed by various investigators.[40-42] Penney[41] found that Al atoms move in the direction of electron flow, i.e., toward the anode. Heumann and Meiners[43] confirmed this observation and also indicated that the amount of mass transported in Al depends on the Al grain size. More transport was observed for finer-grained samples.

Huntington[40] estimated this atomic flux J_a due to a current of density j to be

$$J_a = \frac{ND}{kT} Z^* e \rho j \qquad (1)$$

where N is the density of metallic atoms in the strip, D is the diffusion coefficient of vacancies, k is the Boltzmann constant, T is the absolute temperature, Z^*e is an "effective charge" of the metal atoms, and ρ is the strip resistivity. J_a depends, then, through D, ρ, and Z^*, on the temperature, composition, and structure of the metal film.

Discontinuities in interconnections will occur if the divergence of the mass flow is negative (i.e., net loss of material). It follows that gradients in the interconnect temperature or composition, or changes in its structure, are responsible for void formation.[44] For Al, where the atoms are transported in the direction of the electron flow and the mass flow is greater at higher temperatures, we find voids forming whenever the electron flow is in the direction of increasing temperature.[44]

Figure 9 is an electron photomicrograph of an Al emitter (where electrons flow in the direction of increasing temperature) showing such an open due to electrotransport. If the temperature gradients are small, the composition or structure changes become

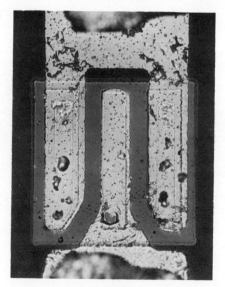

Fig. 9 Discontinuous emitter fingers on an NPN transistor that carried about 5×10^5 A cm^{-2} at 250°C for 500 h. Electron flow in the two emitter outside stripes is in the direction of increasing temperature. Note the voids and hillock formation (750×).

relatively important. For example, voids will form in Al when electrons flow from a low-diffusivity area into a high-diffusivity area, "blowing" the fast diffusant toward the anode. A random distribution of structural variations does not have significant temperature gradients when carrying current.

Whisker growth is occasionally observed. These whiskers are Al and are single-crystal in large segments (Fig. 10). Such whiskers can grow in excess of 100 μ in length and are typically 1 μ wide. If the whiskers touch adjacent metal strips, they can cause additional device failures. If Al strips are protected with a dielectric, hillock formations cause occasional cracking of such films.

Fig. 10 Scanning electron micrograph of a whisker formed during passage of current through an Al strip (15 μ wide $\times$ 0.6 μ thick). Conditions: 50 h at 180°C, current density, 1.9×10^6 A cm^{-2}.

The Al mass transport was directly observed by transmission electron microscopy.[44] Al strips 1,800 Å thick and 100 to 200 μ long were suspended over 1,000-Å SiO$_2$. At current densities of 1 to 2 $\times 10^6$ A cm^{-2}, large temperature gradients were created because of the limited heat removal through the strips. These gradients led to the formation of a clearly visible cluster of voids (Fig. 11) near the cathode (electron flow is in the direction of increasing temperature). A cluster of hillocks was observed close to the anode.

Fig. 11 A composite transmission electron micrograph of a 1,500-Å-thick Al strip (over 1,000 Å SiO$_2$) after passage of about 1.6×10^6 A cm^{-2} for 75 min. Electron flow is from left to right (1,500$\times$).

The use of alternating current—60 Hz—did not cause any observable mass transport. Bulk diffusion seems to be much too slow to account for the observed large mass transport in thin films. Grain-boundary diffusion offers a more plausible explanation. At the same time, as seen by transmission electron microscopy, various voids change their shape rapidly during the experiment, indicating large amounts of surface diffusion. This possibility is supported by the fact that in some areas, mass is clearly lost at the film surface (Fig. 12). It is reasonable then that both grain-bound-

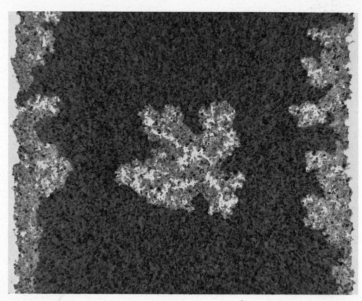

Fig. 12 Transmission electron photomicrograph of 1,500-Å Al film after passage of current. Note the large thinned areas (light gray patches) (3,000×).

ary and surface transport are taking active part in the void formation. Void formation is accelerated in many occasions on grain boundaries. Lifetime data were taken for Al conductors as a function of current and temperature. These conductors, which saw only very small temperature gradients, showed a strong dependence of the lifetime on the metal structure. Various heat treatments were found to change the lifetime by an order of magnitude.

The time to 50% failure t has been found to fit reasonably well the relation

$$t \alpha \frac{(T/j)^n}{D} \tag{2}$$

The index n is close to 2 for current densities around 10^6 A cm^{-2}. However, it is not clear whether this value stays constant for low current densities and low temperature. A typical time to 50% failure for Al strips is 1,000 h at 150°C and 10^6 A cm^{-2}.

In Al, electrotransport is generally not appreciable at 10^5 A cm^{-2}, but it becomes a serious failure mode for integrated circuits operating close to 10^6 A cm^{-2}.

Electrotransport is strongly dependent on the self-transport or self-diffusion coefficients of the interconnect metals. It is expected that metals with smaller self-diffusion coefficients will tend to exhibit less electrotransport. Mutter has confirmed this for metals such as Mo, Cu, and Ag. In Au mass transport was also found to be in the direction of electron flow. Figure 13 is an electron photomicrograph of the cathode and anode of a gold strip before and after passage of 2×10^6 A cm^{-2} for 2 h at 350°C. As before, it is clearly seen that the voids form near the cathode while hillocks grow near the anode. Whisker growth and changes in strip width are also seen in these films, after such prolonged stressing. When enough of the voids form and coalesce, a catastrophic failure (or open) will result.

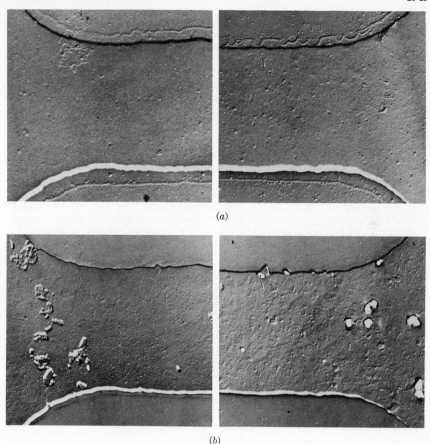

(a)

(b)

Fig. 13 Electron photomicrograph of a cathode and anode of a Au film 6,000 Å thick, before and after passage of 2×10^6 A cm^{-2} at 350°C for 2 h. Note voids at cathode and hillocks at anode (4,000×).

d. Contact Resistance between the Film and External Leads

When a contact is made by attaching an external lead of a different metal to the thin film interconnect, it is possible to observe changes on aging in the resistance of such a contact. The classical example of such changes is the bond between a gold wire and an Al film. High-temperature storage tends to deplete the Al from the area around the Au wire, resulting in increased electrical resistance.[35,38,45] At elevated temperatures, for example, 300°C, the contact resistance between the gold-ball bonds and Al interconnects increases considerably on aging[38] and also shows intermittent characteristics.[46] The use of dissimilar metals for external connections can also increase the possibility of electrochemical attack of the thin films.

e. Film Adhesion and Bond Strength

Loss of thin film adhesion to the underlying dielectric is occasionally observed in integrated circuits. Schnable and Keen[35] have reported "flaking" of Al on very long time operation at moderate ambient temperatures (125°C). In many cases this flaking can be attributed to electrotransport.

Occasional loss of Al film adhesion to SiO_2 under gold-ball bonds is seen after long-time aging.[38] In this case, Au-Al intermetallics form whose adhesion to the SiO_2 is less than that of Al.

f. Surface Failure

Almost all silicon integrated circuits employ the planar process, which includes oxide passivation of all active surfaces. This technique of forming junctions under silicon oxide largely reduces, but has not entirely eliminated, the surface problem. Integrated circuits are hermetically sealed more to protect the surface than for any other reason. Protection of that surface throughout processing and during its field life is vital to high reliability. There are two types of surface effect that can cause great difficulty:

1. Ionic contamination occurring in or on top of the glass layer, but not inside the silicon itself
2. Effects induced at the silicon surface itself

Contamination on top of the glass causes shunt paths which give rise to leakage currents. Sources of this difficulty might be moisture, weld gases, gas desorption from other parts, and mobile contamination migrating to the junction area from within the sealed package (or from the outside if the seal has been broken). This may be detected in the device characteristics as high leakage or soft breakdown.

Surface-induced effects can take any of three forms:

1. Excessive accumulation of N or P impurities on similarly doped material (P+ on P or N+ on N)
2. Depletion of normal impurities (the reverse of 1)
3. Inversion or channeling where the surface has an excess of the opposite impurity type (N on P or P on N)

These effects could be induced temporarily, as a result of externally applied fields, or more or less permanently by a field resulting from electrolysis, ionic contamination, or ionizing radiation on the surface. Figure 14 shows the modification of the charge-distribution pattern resulting from ionic contamination. Designers sometimes use a "guard-ring" diffusion to limit the possible extent of channeling (Fig. 15). This has the effect of minimizing the channel area and hence the shunting effects.

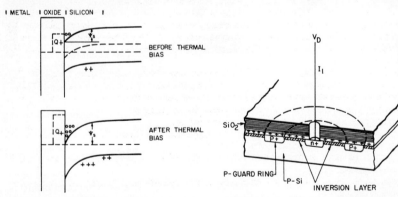

Fig. 14 The positive space charge Q_+ in the oxide is increased by ion drift during thermal bias, causing an increase in band bending and inversion of the P silicon.

Fig. 15 The diffused P+ guard ring is not inverted by the positive charge in the oxide; hence the leakage current I_1 at V_D is minimized.

Depletion manifests itself in low gain, while accumulation results in abnormally low breakdown voltage. Very thin inversion layers may be detected by their "roll-out" (Fig. 16). The initial sweep of the breakdown characteristic on a curve tracer will be

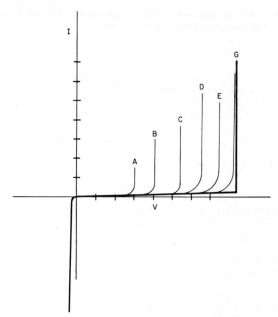

Fig. 16 Schematic representation of current-voltage characteristic of a PN junction. The light traces A to E occur immediately and disappear as "roll-out" occurs. Final characteristic is heavy line G.

very low, the next somewhat higher, and after two or three sweeps, the apparent breakdown voltage will have "rolled" up to the level expected from the relative doping levels on either side of the junction.

Recently, several theories which partly explain some of the observed phenomena have been advanced. One theory suggests an oxygen-ion vacancy in the SiO_2 layer as the principal cause of shifts in the surface potential.[47] Another suggests impurity ions (e.g., sodium) as contaminants within the oxide.[48] Any other ions that remain on the surface of the SiO_2 following the chemical processing of wafers will also affect the stability of devices.

The effect of ions on NPN and PNP transistors has been explained in terms of ionic motion on the surface of the oxide. The oxygen-vacancy model predicts additional features but does not completely explain the effects of hydrogen. The alkali-ion-migration model seems to be the most complete at present.

Some specific effects that have been observed are:

1. Hydrogen can apparently diffuse through SiO_2 and cause silicon to appear more N-type, even inverting surfaces from P to N.

2. Oxygen at high temperatures reverses the effect of hydrogen and apparently stabilizes surfaces.

3. The presence of P_2O_5 in the oxide has a marked effect in improving the stability of silicon surfaces.

4. Metal placed over PN junctions with SiO_2 as an insulating layer can also stabilize devices.

The IGFET is far more sensitive to ionic contamination than other types of devices. The use of lightly doped material for the gate and the fact that the zero bias current depends entirely on surface states have made the production of stable devices a considerable technological challenge. As with most other electrical systems, IGFETs are subject to two distinct classes of failure. These are catastrophic failure, which includes any sudden alteration in one or more operational parameters resulting in device uselessness, and degradation failure in which drift of the characteristics gradually reaches the point where specification limits are exceeded.

Generally, catastrophic failure in IGFET circuits is restricted to gate failure, which comes about by a breakdown of the insulation between the gate and the source, channel, or drain region. Other modes of catastrophic failure involve loss of lead wires, etc., but these are common to all silicon devices and are considered above. Prevention of excessive voltage between the gate electrode and the rest of the device is achieved by placing Zener diodes in the gate circuit[49] or by using a reversed-biased IGFET.

The more insidious modes of failure of IGFETs are the degradative failures which result in the continuous drift of the gate voltage, the source-to-drain current, and other parameters. These drifts are generally caused by a buildup of positive space charge in the oxide at the oxide-silicon interface, usually as a result of the presence of impurity ions (such as $Na+$) which move toward the silicon when the appropriate voltage is applied across the oxide. Drift may also result from exposure of the oxide to certain solvents before metallization[50] or from electrochemical reactions between the oxide and the metallic gate electrode. Finally, space charges can be created by radiation damage in the oxide. Although radiation damage to the silicon is not significant in the IGFET since it is a majority-carrier device, the effects of radiation on the oxide can be quite important and can affect the operation of the device.[51]

These slow changes due to positive space charge in the oxide are much more pronounced when the gate electrode is positive. This means that a P-channel device is inherently more stable than an N-channel device since in the former the gate is normally negative. Another mode of degradation failure, however, is common to both types of IGFET. This is the increase in source-to-drain current due to removal of surface traps or scattering sites by heating in a reducing atmosphere (H_2, forming gas, etc.). This type of degradation, however, is less common since either fairly high concentrations of H_2 or elevated temperatures ($\sim 200°C$) are required for it to be effective.

Pinholes in the oxide can cause many types of failure, ranging from an immediate aluminum-to-silicon short to a slow degradation of the junction because of contamination. This is more of a problem in integrated circuits than in the fabrication of single devices since larger areas of the wafer are involved in the fabrication of integrated-circuit chips.

Degradation failures are minimized through several different approaches. The drift of ionic space charge from the metal-oxide interface can be prevented by leaving a $P_2O_5 + SiO_2$ glass on the oxide surface. This glass acts as an effective trap for $Na+$, $H+$, etc., and also prevents metal-oxide electrochemical reactions from occurring. Another approach to minimizing ionic contamination is to employ ultraclean methods. There is still some question as to the efficiency of the latter in preventing space-charge formation when it is due to species other than $Na+$. Finally, claims have been made that IGFET stability problems can be solved by doing away with the oxide altogether and using silicon nitride as the gate insulation. This is because silicon nitride is much less permeable than SiO_2 to all the known diffusants.

REFERENCES

1. Spenke, E., "Electronic Semiconductors," McGraw-Hill Book Company, New York, 1958.
2. Kroger, F. A., G. Diemer, and H. A. Klasens, *Phys. Rev.*, **103**, 279, 1956.
3. Lewis, A., Fairchild Semiconductor, *Tech. Rept.* 316, Dec. 11, 1967.
4. Archer, R. J., and M. M. Atalla, Bell Telephone System, Technical Publication Monograph 4527, 1963.
5. Crowell, C. R., and S. M. Sze, *Solid-State Electron.*, **9**, 1035 (1966).
6. Geppert, D. V., A. M. Cowley, and B. V. Dore, *J. Appl. Phys.*, **37**, 2458 (1966).
7. Smithells, C. J., "Metals Reference Book," vol. II, Butterworths, Washington, D.C., 1962.
8. Schnable, G. L., R. S. Keen, and L. R. Loewenstern, Rome Air Development Center, *Tech. Rept.* RADC-TR-67-311, September, 1967.
9. Lepselter, M. P., *Bell System Tech. J.*, **45**, 233 (1966).

10. Bergh, A., "Reliability Physics," 6th Annual Symposium, Los Angeles, November, 1967.
11. Cunningham, J. A., *Solid-State Electron.*, **8**, 735 (1965).
12. Cunningham, J. A., and J. G. Harper, *Electron. Engr.*, January, 1967.
13. Vadasz, L., R. Nevala, W. Sander, and R. Seeds, "A Systematic Engineering Approach to Complex Arrays," presented at the International Solid State Circuit Conf., Feb. 10–12, 1966; *Proc. ISSCC*, **9**, 120–121 (1966).
14. Vadasz, L., and W. Sander, *Proc. INEA Conf.*, Munich, Germany, October, 1966.
15. Nevala, R. D., *NEREM Record*, 1966, p. 212.
16. *Electronics*, Feb. 20, 1967, pp. 123–182.
17. Bilous, O., I. Feinberg, and J. L. Langdon, *IBM J. Res. Develop.*, **10**, 370 (1966).
18. Martin, J. W., *Electron. Engr.*, October, 1966, p. 118.
19. Condon, D., et al., Philco-Ford Microelectronics Division, WPAFB Contract AF(615)-3620, *Interim Tech. Rept.* 6, May 15–Aug. 15, 1967, pp. 52–60.
20. Perri, J. A., and J. Riseman, *Electronics*, Oct. 3, 1966, pp. 108–116.
21. Lathrop, J. W., et al., Texas Instruments, Inc., WPAFB Contract AF33(615)-3546, *Interim Tech. Rept.* 3, November, 1966, pp. 82–85, 90, 95.
22. Gregor, L. V., IBM Components Division, East Fishkill Facility, WPAFB AF33(615)-5386, *Final Rept.*, Sept. 30, 1967, p. 53.
23. Condon, D., et al., IBM Components Division, East Fishkill Facility, WPAFB AF33(615)-5386, *Interim Tech. Rept.* 4, Nov. 15, 1966–Feb. 15, 1967, pp. 63, 66; *Rept.* 3, Aug. 15–Nov. 15, 1966, p. 13.
24. Herzog, G. B., et al., RCA, WPAFB Contract AF33(615)-3491, *Interim Tech. Rept.* 7, November, 1967, pp. 33, 37.
25. *Electron. Design*, **4**, Feb. 15, 1967, p. 17.
26. Pliskin, W. A., D. R. Kerr, and J. A. Perri, Thin Glass Films, in "Physics of Thin Films," vol. 4, Academic Press Inc., New York, 1967.
27. Thomas, J. E., Jr. and D. R. Young, *IBM J. Res. Develop.*, **8**, 368 (1964).
28. Tamm, I., *Phys. Zeit. Sowjet.*, **1**, 733 (1932).
29. Bardeen, J., *Phys. Rev.*, **71**, 717 (1947).
30. Forster, J. H., and J. B. Singleton, *Bell Lab. Record*, October–November, 1966, p. 313.
31. Pliskin, W. A., and E. E. Conrad, *Electrochem. Technol.*, **2**, 196 (1964).
32. Wagner, S., and W. Doelp, Philco-Ford Corp. Lansdale, Pa., Contract DA28-043AMC-01424(E) U.S. Army Electronics Command, Fort Monmouth, N.J., *Final Rept.*, May, 1967.
33. Schnable, G. L., and R. S. Keen, "Reliability Physics," Sixth Annual Symposium, Los Angeles, November, 1967.
34. Schnable, G. L., R. S. Keen, and L. R. Loewenstern, Rome Air Development Center, *Tech. Rept.* RADC-TR-67-311, September, 1967.
35. Schnable, G. L., and R. S. Keen, Rome Air Development Center, *Tech. Rept.* RADC-TR-66-165, April, 1966.
36. Patridge, J., L. D. Hanley, and E. C. Hall, "Progress Report on Attainable Reliability of Integrated Circuits for Systems Application," E-1679, MIT Instrumentation Laboratory, Cambridge, Mass.
37. Brandewie, G. V., P. H. Eisenberg, and R. A. Meyer, "Physics of Failure in Electronics," Fourth Annual Symposium, Chicago, November, 1965.
38. Sello, H., I. A. Blech, et al., Rome Air Development Center, *Tech. Rept.* RADC-TR-67-13, April, 1967.
39. Ghate, P. B., *Appl. Phys. Letters*, **11**, 14 (1967).
40. Huntington, H. B., and A. R. Grone, *J. Phys. Chem. Solids*, **20**, 76 (1961).
41. Penney, R. V., *J. Phys. Chem. Solids*, **25**, 335 (1964).
42. Ho, P. S., and H. B. Huntington, *J. Phys. Chem. Solids*, **27**, 1319 (1966).
43. Heumann, T., and H. Meiners, *Z. Metallkunde*, **57**, 571 (1966).
44. Blech, I. A., and E. S. Meieran, *Appl. Phys. Letters*, **11**, 263 (1967).
45. Blech, I. A., and H. Sello, *J. Electrochem. Soc.*, **113**, 1052 (1966).
46. Browning, G. V., L. E. Colteryahn, and D. G. Cummings, "Physics of Failure in Electronics," Fourth Annual Symposium, Chicago, November, 1965.
47. Seraphim, D. P., A. E. Brennemann, F. M. D'Heurle, and H. L. Friedman, *IBM J. Res. Develop.*, **8**, 400 (1964).
48. Snow, E. H., A. S. Grove, B. E. Deal, and C. T. Sah, *J. Appl. Phys.*, **36**, 1664 (1965).
49. Richman, P., "Characteristics and Operation of MOS Field-effect Devices," chap. 5, McGraw-Hill Book Company, New York, 1967.
50. Gregor, L. V., paper presented at Electrochemical Society Meeting, May 3, 1966, Cleveland, Ohio, Abstract 12.
51. Mitchell, J. P., and D. K. Wilson, *Bell System Tech. J.*, **46**, 1 (1967).

Index

Aberrations, optical, **7**-14, **7**-17
Abnormal glow, **4**-4
Abrasion measurements to determine film adhesion, **12**-10
Absorption coefficient, **11**-5
Accommodation coefficient, **1**-28, **1**-82, **8**-6
Accumulation layer at metal-insulator interface, **14**-6 to **14**-7
Adhesion of photoresist film patterns, **7**-44 to **7**-46
Adhesion of thin films, **12**-5 to **12**-21
  determination by nucleation studies, **12**-12 to **12**-19
  forces between deposit atoms and substrate, **12**-12 to **12**-21
  in integrated circuits, **23**-19
  mechanical test methods, **12**-6 to **12**-12
  to determine surface cleanliness, **6**-37 to **6**-38
  relationship to nucleation and growth, **12**-5, **12**-11 to **12**-13
Adhesion energy, **12**-19 to **12**-21
Adsorbed atoms:
  binding energy of, **1**-30, **2**-40 to **2**-43, **2**-46, **8**-28 to **8**-30, **12**-12 to **12**-21
  diameters of, **2**-42
  monolayer of, **2**-43
  residence time of, **2**-41
Adsorbent materials, **2**-22
Adsorption of gases on solids, **2**-40 to **2**-43
Adsorption isotherms:
  for atmospheric gases on molecular sieve 5A, **2**-24
  Langmuir's, **2**-41 to **2**-43
  at low-gas pressures (Henry's law), **2**-41
  role in cryosorption pumping, **2**-21 to **2**-22, **2**-25 to **2**-28
Adsorption traps, **2**-7, **2**-13 to **2**-14, **2**-24 to **2**-28
Aerial image, optical quality of, **7**-13 to **7**-15
Agglomeration, **13**-30 to **13**-31
Aggregates, evaporation of, **8**-18
Aging mechanisms in films, **18**-33 to **18**-34
  in SiO₂ films, **11**-48
Air, partial pressures of gases in, **2**-58
Airy disk, **7**-14, **7**-21
Airy formula, **7**-14, **11**-6
Alkali halide films, dielectric properties of, **19**-14 to **19**-16

Alloys:
  activity of constituents, **1**-73 to **1**-74
  annealing behavior of thin films, **12**-29
  compositional changes during evaporation, **1**-75 to **1**-77
  electroplating of films, **5**-7 to **5**-8
  evaporation of, **1**-73 to **1**-80
  fractionation of, **1**-75
  growth of epitaxial films of, **10**-16 to **10**-19
  phase transitions in, **10**-51 to **10**-52
  sputtering of, **3**-29, **4**-39 to **4**-41
  sublimation of, **1**-76 to **1**-78
  vapor pressure of, **1**-73 to **1**-74
Alphanumeric display based on cold cathodes, **14**-48
Altered region at target surface during compound sputtering, **4**-39
Aluminum:
  vapor pressure of, **1**-13 to **1**-14
  vapor sources for, **1**-48, **1**-50
  wire gaskets of, **2**-83, **2**-104
Aluminum films:
  epitaxial, deposition of, **10**-15
  etching of, **7**-35 to **7**-36
  as interconnections for integrated circuits, **23**-6
  rf sputter-etch rates for, **7**-52
Aluminum nitride, tunneling in thin films of, **14**-23 to **14**-24
Aluminum oxide:
  relative cost of substrates of, **6**-43
  single-crystalline: properties of, **6**-12
  rf sputter-etch rates of, **7**-52
  sintered bodies and substrates: mechanical strength of, **2**-76
  pore volume of, **6**-22 to **6**-23
  properties of, **1**-45, **6**-10
  surface texture of, **6**-15 to **6**-17
Aluminum oxide films:
  anodized, **5**-17 to **5**-18, **5**-20
  electron-gun evaporated, **1**-71
  as gate insulation for thin film transistors, **20**-7
  thermally grown, **5**-21
  tunneling in, **14**-21 to **14**-22
  vapor-phase deposition of, **5**-14
Amplifiers, thin film cryotron, **22**-23 to **22**-25
Analysis of thin films, **11**-34 to **11**-36

1

Selected Units of Measurement, Their Symbols and Conversion Factors,* Which Are Used in This Book

Physical quantity		Unit of measurement		Unit system	One unit is equal to
Type	Symbol	Name	Symbol		
Length...........	l	centimeter	cm	CGS	0.393701 in.
		meter	m	MKSA	100 cm
		micron	μ	....	10^{-4} cm
		angstrom	Å	....	10^{-8} cm
		inch	in.	British	2.540 cm
		mil	mil	British	10^{-3} in.
Volume..........	V	cubic centimeter	cm³	CGS	0.061024 in.³
		cubic meter	m³	MKSA	10^6 cm³
		liter	l	....	10^3 cm³
		cubic inch	in.³	British	16.387 cm³
Time.............	t	second	s	CGS, MKSA	
		minute	min	....	60 s
		hour	h	....	3,600 s
Frequency.........	ν	hertz	Hz	CGS, MKSA	1 s^{-1}
Mass.............	m	gram	g	CGS	
		kilogram	kg	MKSA	10^3 g
		pound	lb	British	453.592 g
Molar mass........	M	gram	g	....	$M = 12$ for C^{12}
Amount of substance	n	gram mole	mol	....	N$_A$ molecules
Force............	F	dyne	dyn	CGS	1 g cm s^{-2}; 1.0197 · 10^{-3} g. wt.
		newton	N	MKSA	1 kg m s^{-2}; 10^5 dyn
Weight...........	G	gram weight	g. wt.		980.665 dyn
		pound-force	lbf	British	453.592 g. wt.
Pressure..........	p	microbar	μbar	CGS	1 dyn cm^{-2}
		newton per square meter	N m^{-2}	MKSA	10 μbar
		torr (= mmHg)	Torr	....	1,333.22 μbar
		standard atmosphere	atm	....	{1,013,250 μbar {760 Torr
		pound-force per square inch	lbf in.$^{-2}$	British	6.895 μbar
Temperature:					
Thermodynamic..	T	degree Kelvin	°K		
Practical........	t	degree Celsius	°C	....	$t = T - 273.15$
		also used:	deg		
Energy, work, heat	E, U, W	erg	erg	CGS	1 dyn cm; 2.39 · 10^{-8} cal
		joule	J	MKSA	1 N m^{-1}; 10^7 erg
		calorie (thermochemical)	cal	....	4.184 · 10^7 erg
		kilocalorie	kcal	....	10^3 cal
		British thermal unit	Btu	British	252 cal; 1,055 J
Entropy...........	S	Clausius	Cl	....	1 cal deg^{-1}
Power............	P	erg per second	erg s^{-1}	CGS	10^{-7} W
		watt	W	MKSA	1 J s^{-1}
Electric current.....	I	ampere	A	MKSA	0.1 emu; $c \times 0.1$ esu†
Electromotive force, potential	E, V	volt	V	MKSA	{10^8 emu; {$1/c \times 10^{-8}$ esu†
Resistance..........	R	ohm	Ω	MKSA	{10^9 emu; {$1/c^2 \times 10^9$ esu†
Electric charge......	Q	coulomb	C	MKSA	{1 A s; 0.1 emu; {$c \times 0.1$ esu†
Capacitance........	C	farad	F	MKSA	{1 C V^{-1}; 10^{-9} emu; {$c^2 \times 10^{-9}$ esu†
Magnetic field strength	H	ampere-turns per meter	A-turns m^{-1}	MKSA	$4\pi \times 10^{-3}$ Oe
		oersted	Oe	CGS	$10^3/4\pi$ A-turns m^{-1}
Magnetic flux.......	Φ	weber	Wb	MKSA	1 V s; 10^8 Mx
		maxwell	Mx	CGS	10^{-8} Wb
Magnetic flux density	B	tesla	T	MKSA	1 Wb m^{-2}; 10^4 G
		gauss	G	CGS	10^{-4} Wb m^{-2}
Inductance.........	L	henry	H	MKSA	1 Wb A^{-1}; 10^9 emu

* After Kaye, G. W. C., and T. H. Laby, "Tables of Physical and Chemical Constants," John Wiley and Sons, Inc., 13th ed., New York, 1966.
† emu = electromagnetic units of the CGS system
  esu = electrostatic units of the CGS system
  $c = 2.997925 \times 10^{10}$ cm s^{-1} = velocity of light in free space

Fundamental Physical Constants, Their Symbols and Numerical Values*

Physical constant	Symbol	Numerical value
Velocity of light	c	2.997925×10^{10} cm s^{-1}
Permittivity of free space	ϵ_0	8.85416×10^{-14} F cm^{-1}
Permeability of free space	μ_0	1.256637×10^{-8} H cm^{-1}
Standard value of the acceleration of gravity	g	980.665 cm s^{-2}
Avogadro's number	N_A	6.02252×10^{23} mol^{-1}
Unit atomic mass ($= \frac{1}{12}$ of the mass of a C^{12} atom)	m_u	1.66043×10^{-24} g
Faraday's constant	F	9.64870×10^{4} C mol^{-1}
Charge of electron	e	1.60210×10^{-19} C; 4.80298×10^{-10} esu
Mass of electron	m_e	9.1091×10^{-28} g
Charge-to-mass ratio of electron	e/m_e	1.758796×10^{8} C g^{-1}
Planck's constant	h	6.6256×10^{-27} erg s
Planck's constant, $h/2\pi$	$\hbar$	1.0545×10^{-27} erg s
Boltzmann's constant	k	1.38054×10^{-16} erg deg^{-1}; 8.6170×10^{-5} eV deg^{-1}
Gas constant, $k \cdot N_A$	R	8.3143×10^{7} erg deg^{-1} mol^{-1} 1.9872 cal deg^{-1} mol^{-1} 82.057 atm cm^3 deg^{-1} mol^{-1} 62.364 Torr 1 deg^{-1} mol^{-1}
Loschmidt's constant (number of gas particles per unit volume at STP)	N_L	2.6870×10^{19} cm^{-3} atm^{-1}
Volume of 1 mole of an ideal gas at STP	V_0	2.24136×10^{4} cm^3 mol^{-1}
Thermal energy at 273.15°K	kT	2.3537×10^{-2} eV
	kT	3.7709×10^{-14} erg
	RT	0.5428 kcal mol^{-1}
Electron energy after acceleration through 1 volt	eV	1.60210×10^{-12} erg
for 1 mole of electrons	....	23.061 kcal mol^{-1} 9.6487×10^{4} J mol^{-1}

* After Kaye, G. W. C., and T. H. Laby, "Tables of Physical and Chemical Constants," John Wiley and Sons, Inc., 13th ed., New York, 1966.

1-MONTH